Solutions Manual

to accompany

Technical Mathematics

and

Technical Mathematics
With Calculus

Second Edition

John C. Peterson
with
Alan Herweyer

Chattanooga State Technical College

Delmar Publishers

I(T)P˙ An International Thomson Publishing Company

Albany • Bonn • Boston • Cincinnati • Detroit • London • Madrid • Melbourne
Mexico City • New York • Pacific Grove • Paris • San Francisco • Singapore • Tokyo
Toronto • Washington

NOTICE TO THE READER

COPYRIGHT © 1997
By Delmar Publishers
a division of International Thomson Publishing Inc.

The ITP logo is a trademark under license.

Printed in the United States of America

For more information, contact:

Delmar Publishers
3 Columbia Circle, Box 15015
Albany, New York 12212-5015

International Thomson Publishing Europe
Berkshire House 168-173
High Holborn
London, WC1V 7AA
England

Thomas Nelson Australia
102 Dodds Street
South Melbourne, 3205
Victoria, Australia

Nelson Canada
1120 Birchmont Road
Scarborough, Ontario
Canada, M1K 5G4

International Thomson Editores
Campos Eliseos 385, Piso 7
Col Polanco
11560 Mexico D F Mexico

International Thomson Publishing GmbH
Konigswinterer Strasse 418
53227 Bonn
Germany

International Thomson Publishing Asia
221 Henderson Road
#05-10 Henderson Building
Singapore 0315

International Thomson Publishing--Japan
Hirakawacho Kyowa Building, 3F
2-2-1 Hirakawacho
Chiyoda-ku, Tokyo 102
Japan

2 3 4 5 6 7 8 9 10 XXX 01 00 99 98

ISBN: 0-8273-8431-9

Contents

Online Services

Delmar Online
To access a wide variety of Delmar products and services on the World Wide Web,
point your browser to:
   http://www.delmar.com
   or email: info@delmar.com

thomson.com
To access International Thomson Publishing's
home site for information on more than 34 publishers
and 20,000 products, point your browser to:
   http://www.thomson.com
   or email: findit@kiosk.thomson.com

A service of **I(T)P**®

1

The Real Number System

≡ 1.1 SOME SETS OF NUMBERS

1. 15 is a natural number, whole number, integer, rational number, and a real number

2. $-\frac{2}{3}$ is a rational number and a real number

3. $\frac{-\sqrt{7}}{8}$ is an irrational number and a real number

4. 0 is a whole number, integer, rational, and real number

5. $|15| = 15$

6. $\left|\frac{-2}{3}\right| = \frac{2}{3}$

7. $\left|\frac{-\sqrt{7}}{8}\right| = \frac{\sqrt{7}}{8}$

8. $|0| = 0$

9. $|4 - 7| = |-3| = 3$

10. $\left|8 - \frac{7}{2}\right| = \left|\frac{9}{2}\right| = \frac{9}{2}$

11.

12.

13.

14.

15. $2 < 3; 2 - 3 = -1$ negative

16. $5 > 3; 5 - 3 = 2$ positive

17. $-4 < 7; -4 - 7 = -11$ negative

18. $9 > -7; 9 - (-7) = 16$ positive

19. $-3 > -8; -3 - (-8) = 5$ positive

20. $-15 < -7; -15 - (-7) = -8$ negative

21. $-\frac{2}{3} < -\frac{1}{2}; -\frac{2}{3} - \left(-\frac{1}{2}\right) = -\frac{4}{6} - \left(-\frac{3}{6}\right) = -\frac{1}{6}$ negative

22. $-\frac{7}{2} < -\frac{8}{3}$ because $-\frac{7}{2} - \left(-\frac{8}{3}\right) = -\frac{5}{6}$ which is negative

23. $0.7 > 0.5; 0.7 - 0.5 = 0.2$ positive

24. $-2.1 < -2.0$ because $-2.1 - (-2.0) = -0.1$ which is negative

25. $|2.3| < |-4.1|$ because $|2.3| - |-4.1| = 2.3 - 4.1 = -1.8$ which is negative

26. $|-5.1| < |3.7|$ because $|-5.1| - |3.7| = 5.1 - 3.7 = 1.4$ which is positive

27. The reciprocal of -5 is $\frac{1}{-5} = -\frac{1}{5}$

28. The reciprocal of $\frac{1}{2}$ is $\frac{1}{1/2} = 1 \times \frac{2}{1} = 2$

29. The reciprocal of $\frac{17}{3}$ is $\frac{1}{17/3} = 1 \times \frac{3}{17} = \frac{3}{17}$

30. The reciprocal of $\frac{-2}{\pi}$ is $\frac{1}{-2/\pi} = 1 \times \left(-\frac{\pi}{2}\right) = -\frac{\pi}{2}$

31. $-5, -|4|, -\frac{2}{3}, \frac{-1}{3}, \frac{16}{3}, |-8|$

32. $-\sqrt{7}, -\left|\frac{7}{4}\right|, -\sqrt{2}, \sqrt{7} - \sqrt{2}, \frac{7}{5}, \pi, \frac{22}{7}$

33. (a) $\frac{1}{32}$ in., (b) $\frac{1}{8}$ in., (c) $\frac{1}{16} = 0.0625$ in.

34. $\frac{3}{4}$ h = 0.75 h, $1\frac{1}{2}$ h = 1.5 h, $2\frac{1}{4}$ h = 2.25 h

35. $|-19.4 - (-16.8)| = |-2.6| = 2.6$ so 2.6 V

36. $|-6.5 - 14.7| = |-21.2| = 21.2$ V

37. $\frac{5}{7}$ (Sheila), $\frac{2}{3}$ (Hazel), $\frac{3}{5}$ (José), $\frac{9}{16}$ (Robert), and $\frac{4}{9}$ (Lamar)

38. $|23.7 - (-5.2)| = |23.7 + 5.2| = |28.9| = 28.9$ V

39. Helen began at mile marker 157. If the mile markers were increasing, then after one hour she was at mile marker $157 + 64 = 221$. If the mile markers were decreasing, then after one hour she was at mile marker $157 - 64 = 93$.

40. $|14 - 19| = |-5| = 5 \ \Omega$

☰ 1.2 BASIC LAWS OF REAL NUMBERS

1. Commutative law for addition (CA)

2. Associative law for multiplication (AM)

3. Commutative law for multiplication (CM)

4. Distributive law (DL)

5. Associative law for addition (AA)

6. Distributive law

7. Identity element for multiplication

8. Multiplicative inverses

9. Additive inverses

10. Commutative law for addition

11. Associative law for multiplication

12. Identity element for addition

13. Additive inverse of 91 is −91

14. 8

15. $-\sqrt{2}$

16. $-\frac{1}{3}$

17. Multiplicative inverse of $\frac{1}{2}$ is 2

18. $-\frac{1}{5}$

19. $\frac{2}{\sqrt{2}} = \frac{2}{\sqrt{2}}\frac{\sqrt{2}}{\sqrt{2}} = \frac{2\sqrt{2}}{2} = \sqrt{2}$

20. $-\frac{7}{3}$

21. $16 - 8 \div 4 = 16 - 2 = 14$

22. $16 \div 8 + 2 = 2 + 2 = 4$

23. $24 + 3 - 10 \div 5 \times 8 + 2 = 24 + 3 - 2 \times 8 + 2 = 24 + 3 - 16 + 2 = 27 - 16 + 2 = 11 + 2 = 13$

24. $13 \times 7 - 26 \div 5 + 5 = 91 - 5.2 + 5 = 85.8 + 5 = 90.8$

25. $(7 - 2) - (3 + 8 - 7) = 5 - (11 - 7) = 5 - 4 = 1$

26. $7 - 2 - 3 + 8 - 7 = 5 - 3 + 8 - 7 = 2 + 8 - 7 = 10 - 7 = 3$

27. $7 \times 3 + 5 \times 2 = 21 + 10 = 31$

28. $6 \times 4 - 3 \times 5 = 24 - 15 = 9$

29. $\{-[5 - (8 - 4) + (3 - 7)] - (4 - 2)\} = -[5 - 4 + (-4)] - 2 = -[1 + (-4)] - 2 = -[-3] - 2 = 3 - 2 = 1$

30. $(14 + 3(8 - 6) + 4(9 - 5)) = 14 + 3 \times 2 + 4 \times 4 = 14 + 6 + 16 = 20 + 16 = 36$

31. Press $\boxed{(-)}$ 39 $\boxed{-}$ 72. The result is −111.

32. Press $\boxed{(-)}$ 257 $\boxed{-}$ 63. The result is −320.

33. Press 192 $\boxed{-}$ $\boxed{(-)}$ 78. The result is 270.

34. Press $\boxed{(-)}$ 56 $\boxed{-}$ $\boxed{(-)}$ 344. The result is 288.

35. (a) Press 243 $\boxed{+}$ $\boxed{(}$ $\boxed{(-)}$ 691 $\boxed{+}$ 89 $\boxed{)}$. The result is −359.
    (b) Press $\boxed{(}$ 243 $\boxed{+}$ $\boxed{(-)}$ 691 $\boxed{)}$ $\boxed{+}$ 89. The result is −359.

36. (a) Press $\boxed{(-)}$ 41 $\boxed{(}$ 63 $\boxed{+}$ $\boxed{(-)}$ 177 $\boxed{)}$. The result is 4674.
    (b) Press $\boxed{(-)}$ 41 $\boxed{\times}$ 63 $\boxed{+}$ $\boxed{(-)}$ 41 $\boxed{\times}$ $\boxed{(-)}$ 177. The result is 4674.

37. (a) Press $\boxed{(}$ $\boxed{(-)}$ 342 $\boxed{+}$ $\boxed{(-)}$ 18 $\boxed{)}$ $\boxed{\times}$ 91. The result is −32760
    (b) Press $\boxed{(-)}$ 342 $\boxed{\times}$ 91 $\boxed{+}$ $\boxed{(-)}$ 18 $\boxed{\times}$ 91. The result is −32760.

38. (a) Press 3 $\boxed{\div}$ 7 $\boxed{\times}$ 7 $\boxed{\div}$ 3. The result is 1.
    (b) Press $\boxed{(-)}$ 3 $\boxed{\div}$ 11 $\boxed{+}$ 3 $\boxed{\div}$ 11. The result is 0.

≡ 1.3 BASIC OPERATIONS WITH REAL NUMBERS

1. $27+(+23)$. Same sign $27+23 = 50$; both positive so, $+50$

2. $8+(-19)$. Different signs $|-19| - |8| = 19 - 8 = 11$; larger negative so, -11

3. $27 + (-13)$. Different signs $|27| - |-13| = 27 - 13 = 14$; larger positive so, 14

4. $-9+(-8)$. Same sign $|-9| + |-8| = 9 + 8 = 17$; both negative so, -17

5. $7 - 16 = 7 + (-16)$. Different signs $|-16| - |7| = 16 - 7 = 9$; larger negative so, -9

6. $29 - (-8) = 29 + (+8) = 37$

7. $-8 - 16 = -8 + (-16)$. Same sign $|-8| + |-16| = 8 + 16 = 24$; both negative so, -24

8. $-25 - (-13) = -25 + (+13)$. Different signs $|-25| - |13| = 12$; larger negative so, -12

9. $-37 - (-49) = -37 + (+49)$. Different signs $|49| - |37| = 49 - 37 = 12$; larger positive so, $+12$

10. $(-2)(6)$; $|-2| \cdot |6| = 2 \cdot 6 = 12$. Different signs so, -12

11. $(-3)(-5)$; $|-3| \cdot |-5| = 3 \cdot 5 = 15$. Same sign so $+15$

12. $(7)(-8)$; $|7| \cdot |-8| = 7 \cdot 8 = 56$. Different signs so, -56

13. $-38 \div 4$; $|-38| \div |4| = 38 \div 4 = \frac{38}{4} = \frac{19}{2}$. Different signs so $-\frac{19}{2}$

14. $-45 \div (-9)$; $\frac{|-45|}{|-9|} = \frac{45}{9} = 5$. Same sign so $+5$

15. $\frac{3}{4} + \frac{-5}{8}$; $\frac{3}{4} = \frac{6}{8}$; $\frac{6}{8} + \frac{-5}{8} = \frac{6+(-5)}{8} = \frac{1}{8}$

16. $-1\frac{3}{4} + \frac{-2}{3} = -\frac{7}{4} + \frac{-2}{3}$. Same sign $\frac{7}{4} + \frac{2}{3} = \frac{21}{12} + \frac{8}{12} = \frac{29}{12}$. Both negative so $-\frac{29}{12} = -2\frac{5}{12}$

17. $\frac{-9}{5} + \frac{7}{3}$ common denominator is 15;
$\frac{-9}{5} + \frac{7}{3} = \frac{-27}{15} + \frac{35}{15} = \frac{-27+35}{15} = \frac{8}{15}$

18. $\frac{-2}{3} + \frac{5}{6}$ common denominator is 6;
$\frac{-2}{3} + \frac{5}{6} = \frac{-4}{6} + \frac{5}{6} = \frac{-4+5}{6} = \frac{1}{6}$

19. $\frac{2}{5} + \frac{-1}{4} = \frac{8}{20} + \frac{-5}{20} = \frac{8+(-5)}{20} = \frac{3}{20}$

20. $\frac{-4}{5} + \frac{-5}{6} = \frac{-24}{30} + \frac{-25}{30} = \frac{-24+(-25)}{30} = \frac{-49}{30} = -1\frac{19}{30}$

21. $\frac{3}{8} - \frac{-1}{4} = \frac{3}{8} - \frac{-2}{8} = \frac{3-(-2)}{8} = \frac{5}{8}$

22. $1\frac{1}{3} - \frac{-5}{6} = \frac{4}{3} - \frac{-5}{6} = \frac{8}{6} - \frac{-5}{6} = \frac{8-(-5)}{6} = \frac{13}{6} = 2\frac{1}{6}$

23. $\frac{-9}{10} - \frac{2}{3} = \frac{-27}{30} - \frac{20}{30} = \frac{-27-20}{30} = \frac{-47}{30} = -1\frac{17}{30}$

24. $\frac{-5}{16} - \frac{-3}{8} = \frac{-5}{16} - \frac{-6}{16} = \frac{-5-(-6)}{16} = \frac{1}{16}$

25. $\frac{5}{32} - \frac{1}{8} = \frac{5}{32} - \frac{4}{32} = \frac{5-4}{32} = \frac{1}{32}$

26. $-\frac{7}{3} - \frac{-6}{7} = \frac{-49}{21} - \frac{-18}{21} = \frac{-49-(-18)}{21} = \frac{-31}{21} = -1\frac{10}{21}$

27. $\frac{-2}{3} \times \frac{4}{5} = \frac{-2 \times 4}{3 \times 5} = \frac{-8}{15} = -\frac{8}{15}$

28. $\frac{3}{4} \times \frac{-5}{8} = \frac{3 \times (-5)}{4 \times 8} = -\frac{15}{32}$

29. $\frac{-1}{8} \times \frac{-3}{4} = \frac{-1 \times (-3)}{8 \times 4} = \frac{3}{32}$

30. $\frac{9}{16} \times \frac{1}{2} = \frac{9 \times 1}{16 \times 2} = \frac{9}{32}$

31. $-\frac{4}{3} \times \frac{5}{2} = \frac{-4 \times 5}{3 \times 2} = \frac{-20}{6} = -\frac{10}{3}$
or $\frac{-2 \times 2 \times 5}{3 \times 2} = -3\frac{1}{3}$

32. $\frac{-9}{5} \times \frac{-3}{8} = \frac{-9 \times (-3)}{5 \times 8} = \frac{27}{40}$

33. $-\frac{3}{4} \div \frac{-5}{8} = -\frac{3}{4} \times \frac{8}{-5} = \frac{-3 \times 4 \times 2}{4 \times (-5)} = \frac{-6}{-5} = \frac{6}{5} = 1\frac{1}{5}$

34. $1\frac{3}{4} \div \frac{-2}{3} = \frac{7}{4} \div \frac{-2}{3} = \frac{7}{4} \times \frac{3}{-2} = \frac{7 \times 3}{4 \times (-2)} = \frac{21}{-8} = -\frac{21}{8} = -2\frac{5}{8}$

35. $-\frac{3}{5} \div 4 = \frac{-3}{5} \times \frac{1}{4} = \frac{-3 \times 1}{5 \times 4} = \frac{-3}{20} = -\frac{3}{20}$

36. $-\frac{3}{8} \div \frac{1}{4} = \frac{-3}{8} \times \frac{4}{1} = \frac{-3 \times 4}{2 \times 4} = -\frac{3}{2} = 1\frac{1}{2}$

37. $-\frac{7}{5} \div \left(-\frac{5}{7}\right) = -\frac{7}{5} \times \frac{-7}{5} = \frac{-7 \times (-7)}{5 \times 5} = \frac{49}{25} = 1\frac{24}{25}$

38. $\frac{2}{3} \div \left(-\frac{7}{3}\right) = \frac{2}{3} \times \frac{-3}{7} = \frac{2 \times (-3)}{3 \times 7} = \frac{-2}{7} = -\frac{2}{7}$

39. $\left(-2 + \frac{2}{3}\right) \times \frac{-1}{2} - \left[\frac{3}{2} \div (-3)\right] + \frac{7}{3} = \frac{-4}{3} \times \frac{-1}{2} - \left(\frac{-1}{2}\right) + \frac{7}{3} = \frac{-4 \times (-1)}{3 \times 2} - \left(\frac{-1}{2}\right) + \frac{7}{3} = \frac{2}{3} - \left(\frac{-1}{2}\right) + \frac{7}{3} = \frac{4}{6} - \frac{-3}{6} + \frac{7}{3} = \frac{7}{6} + \frac{7}{3} = \frac{7}{6} + \frac{14}{6} = \frac{21}{6} = \frac{7}{2} = 3\frac{1}{2}$

40. $\frac{4}{3} \times \frac{-7}{8} \times \frac{3}{-5} \div \frac{4}{5} + \frac{-5}{8} - \frac{7}{8} = \frac{4 \times (-7)}{3 \times 8} \times \frac{3}{-5} \div \frac{4}{5} + \frac{-5}{8} - \frac{7}{8} = \frac{-7 \times 3}{6 \times (-5)} \div \frac{4}{5} + \frac{-5}{8} - \frac{7}{8} = \frac{7}{10} \div \frac{4}{5} + \frac{-5}{8} - \frac{7}{8} =$

$\frac{7}{10} \times \frac{5}{4} + \frac{-5}{8} - \frac{7}{8} = \frac{7 \times 5}{2 \times 5 \times 4} + \frac{-5}{8} - \frac{7}{8} = \frac{7}{8} + \frac{-5}{8} - \frac{7}{8} = \frac{2}{8} - \frac{7}{8} = -\frac{5}{8}$

41. Add: $17\frac{1}{2} + 15\frac{7}{8} + 29\frac{3}{4} + 15\frac{3}{8} = 17\frac{4}{8} + 15\frac{7}{8} + 29\frac{6}{8} + 15\frac{3}{8} = 76\frac{20}{8} = 78\frac{4}{8} = 78\frac{1}{2} = 6'6\frac{1}{2}''$

42. Add: $\frac{5}{32} + \frac{3}{16} = \frac{5}{32} + \frac{6}{32} = \frac{11}{32}$

43. We add the voltages of the batteries: $30 + 15 + (-12) + 24 = 57$. So, the total voltage is 57 V.

44. Subtract: $1\frac{5}{8} - \frac{1}{5} = \frac{13}{8} - \frac{1}{5} = \frac{65}{40} - \frac{8}{40} = \frac{57}{40} = 1\frac{17}{40}''$

45. **(a)** $4\frac{5}{8} + 3\frac{1}{5} = \frac{37}{8} + \frac{16}{5} = \frac{185}{40} + \frac{128}{40} = \frac{313}{40} = 7\frac{33}{40}$ miles;

(b)   We need to add the distance between the second and third checkpoints to the answer for (a).
$7\frac{33}{40} + 3\frac{3}{8} = \frac{313}{40} + \frac{27}{8} = \frac{313}{40} + \frac{135}{40} = \frac{448}{40} = 11\frac{8}{40} = 11\frac{1}{5}$ miles;

(c)   Using the answer from (b), we subtract that answer from the total length of the race.
$14\frac{1}{2} - 11\frac{1}{5} = \frac{29}{2} - \frac{56}{5} = \frac{145}{10} - \frac{112}{10} = \frac{33}{10} = 3\frac{3}{10}$ miles

46. We divide $427\frac{1}{5}$ by $13\frac{3}{4}$.
$427\frac{1}{5} \div 13\frac{3}{4} = \frac{2136}{5} \div \frac{55}{4} = \frac{2136}{5} \times \frac{4}{55} = \frac{8544}{275} = 31\frac{19}{275} \approx 31.07$ mi/gal

47. We want $37 \times 2'3\frac{1}{2}''$. There are $12''$ in 1 foot, so there are $2 \times 12'' = 24''$ in 2 feet. Thus, $2'3\frac{1}{2}'' = 24'' + 3\frac{1}{2}'' = 27\frac{1}{2}''$. Now, $37 \times 27\frac{1}{2} = \frac{37}{1} \times \frac{55}{2} = \frac{2035}{2} = 1017\frac{1}{2}'' = 84'9\frac{1}{2}''$

48. Each strip requires $1\frac{5}{16} + \frac{1}{8} = 1\frac{7}{17}$ in. of wood for its width. To determine the total number of strips we can get from the sheet of plywood, we will divide 8 ft by $1\frac{7}{16}$ in. Converting 8 feet to $8 \times 12 = 96$ in., we want the answer to $96 \div 1\frac{7}{16} = 96 \div \frac{23}{16} = 96 \times \frac{16}{23} = \frac{1536}{23} \approx 66.78$. You can get a total of 66 strips. (Note that the scrap is a strip $1\frac{1}{8}''$ wide.)

49. We subtract the two smaller lengths from the total length. $7\frac{5}{8} - 3\frac{3}{8} - 3\frac{3}{8} = \frac{61}{8} - \frac{27}{8} - \frac{27}{8} = \frac{34}{8} - \frac{27}{8} = \frac{7}{8}$
The missing length is $\frac{7}{8}''$.

50. Adding the voltages, we get
$-4 + 22 + (-9) + (-7) + 13 + (-6) + (-14) + 5 = 0$
The sum of the voltages is 0 V.

51. **(a)** $-6.5 - 14.7 = -21.2$ V
(b) The negative sign indicates that the voltage dropped 21.2 V.

52. **(a)** $23.7 - (-5.2) = 23.7 + 5.2 = 28.9$ V
(b) The positive sign indicates that the voltage increased 28.9 V.

≡ 1.4 LAWS OF EXPONENTS

1. $5^3 = 5 \times 5 \times 5 = 125$

2. $3.8^0 = 1$ by Rule 6

3. $\left(\frac{2}{3}\right)^{-1} = \frac{2^{-1}}{3^{-1}} = \frac{1/2}{1/3} = \frac{1}{2} \times \frac{3}{1} = \frac{3}{2}$

4. $\left(\frac{3}{5}\right)^{-2} = \frac{3^{-2}}{5^{-2}} = \frac{5^2}{3^2} = \frac{25}{9}$

5. $(-4)^2 = (-4)(-4) = 16$

6. $(-5)^4 = (-5)(-5)(-5)(-5) = 625$

7. $\frac{7}{7^3} = \frac{7^1}{7^3} = 7^{1-3} = 7^{-2} = \frac{1}{7^2} = \frac{1}{49}$

8. $-3^2 = -(3^2) = -(3 \cdot 3) = -9$

9. $3^2 \cdot 3^4 = 3^{2+4} = 3^6$

10. $d^8 d^5 = d^{8+5} = d^{13}$

11. $2^4 \cdot 2^3 \cdot 2^5 = 2^{4+3+5} = 2^{12}$

12. $f^3 f^4 f^1 = f^{3+4+1} = f^8$

13. $\dfrac{2^5}{2^3} = 2^{5-3} = 2^2$

14. $\dfrac{5^{14}}{5^3} = 5^{14-3} = 5^{11}$

15. $(2^3)^2 = 2^{3\times 2} = 2^6$

16. $(5^7)^3 = 5^{7\times 3} = 5^{21}$

17. $(x^4)^5 = x^{4\times 5} = x^{20}$

18. $(xy^3)^4 = (x^1 y^3)^4 = x^{1\times 4} y^{3\times 4} = x^4 y^{12}$

19. $(a^{-2}b)^{-3} = a^{-2\times(-3)} b^{-3} = a^6 b^{-3} = \dfrac{a^6}{b^3}$

20. $\left(\dfrac{2}{3}\right)^4 = \dfrac{2^4}{3^4}$; Rule 4a

21. $\left(\dfrac{x}{4}\right)^3 = \dfrac{x^3}{4^3}$

22. $\left(\dfrac{a}{b^3}\right)^5 = \dfrac{a^5}{b^{3\times 5}} = \dfrac{a^5}{b^{15}}$

23. $\left(\dfrac{a^2 b}{c^3}\right)^4 = \dfrac{a^{2\times 4} b^4}{c^{3\times 4}} = \dfrac{a^8 b^4}{c^{12}}$

24. $4^{-3} = \dfrac{1}{4^3}$; Rule 7

25. $x^{-7} = \dfrac{1}{x^7}$

26. $\dfrac{1}{p^{-5}} = p^5$

27. $\left(\dfrac{1}{5}\right)^3 = \dfrac{1}{5^3}$

28. $\dfrac{x^4}{x^2} = x^{4-2} = x^2$

29. $\dfrac{7^3}{7^8} = 7^{3-8} = 7^{-5} = \dfrac{1}{7^5}$

30. $\dfrac{5^2}{5^{10}} = 5^{2-10} = 5^{-8} = \dfrac{1}{5^8}$

31. $\dfrac{a^2 y^3}{a^5 y^7} = a^{2-5} y^{3-7} = a^{-3} y^{-4} = \dfrac{1}{a^3}\dfrac{1}{y^4} = \dfrac{1}{a^3 y^4}$

32. $\dfrac{x^4 y b^2}{x y^3 b^5} = x^{4-1} y^{1-3} b^{2-5} = x^3 y^{-2} b^{-3} = \dfrac{x^3}{y^2 b^3}$

33. $\dfrac{a^2 p^5 y^3}{a^6 p^5 y} = a^{2-6} p^{5-5} y^{3-1} = a^{-4} p^0 y^2 = \dfrac{y^2}{a^4}$

34. $\dfrac{p^3 q^4 r^2}{p^4 r^2} = p^{3-4} q^4 r^{2-2} = p^{-1} q^4 r^0 = \dfrac{q^4}{p}$

35. $(pr^2)^{-1} = \dfrac{1}{pr^2}$

36. $\left(\dfrac{2x^2}{y}\right)^{-1} = \dfrac{y}{2x^2}$

37. $\left(\dfrac{4y^3}{5^2}\right)^{-1} = \dfrac{5^2}{4y^3}$

38. $\left(\dfrac{4x^3}{y^2}\right)^{-2} = \dfrac{4^{-2} x^{-6}}{y^{-4}} = \dfrac{y^4}{4^2 x^6}$ or $\left(\dfrac{4x^3}{y^2}\right)^{-2} = \left(\dfrac{y^2}{4x^3}\right)^2 = \dfrac{y^4}{16x^6}$

39. $\left(\dfrac{2b^2}{y^5}\right)^{-3} = \dfrac{2^{-3} b^{-6}}{y^{-15}} = \dfrac{y^{15}}{2^3 b^6}$ or $\left(\dfrac{2b^2}{y^5}\right)^{-3} = \left(\dfrac{y^5}{2b^2}\right)^3 = \dfrac{y^{15}}{8b^6}$

40. $(-8pr^2)^{-3} = (-8)^{-3} p^{-3} r^{-6} = \dfrac{1}{-8^3 p^3 r^6} = \dfrac{-1}{8^3 p^3 r^6}$

41. $(-b^4)^6 = (-b^4)(-b^4)(-b^4)(-b^4)(-b^4)(-b^4) = b^{24}$

42. $ap^2(-a^2 p^3)^2 = ap^2(a^4 p^6) = a^{1+4} p^{2+6} = a^5 p^8$

43. First we simplify $\left((2\pi f C)^{-1}\right)^2$ as $(2\pi f C)^{-2}$. Next, we compute $(2\pi f C)^{-2}$. Substituting the given values of f and C and using the $\boxed{\pi}$ key on the calculator, we obtain $(2\pi f C)^{-2} = 0.030159289^{-2} = 1099.405205$. To this, add 40^2, with the result of 2699.405205. Then $\sqrt{2699.405205} = 51.95580049$ gives the answer of $51.96\ \Omega$ when rounded off to 2 decimal places.

44. Substituting the given values, we obtain
$$R = \left(\dfrac{1}{0.8} + \dfrac{1}{0.45}\right)^{-1} + 0.76 = \left(\dfrac{0.45 + 0.8}{(0.8)(0.45)}\right)^{-1} + 0.76 = \dfrac{(0.8)(0.45)}{0.8 + 0.45} + 0.76 = \dfrac{0.36}{1.25} + 0.76 = 0.288 + 0.76 = 1.048$$
The total resistance is $1.048\ \Omega$.

Note: On a calculator such as a TI-8x[1], you could have keyed this as

$\boxed{(}$ 1 $\boxed{\div}$ $.8$ $\boxed{+}$ 1 $\boxed{\div}$ $.45$ $\boxed{)}$ $\boxed{x^{-1}}$ $+.76$ $\boxed{\text{ENTER}}$

[1] The notation *TI-8x* refers to something that is true for any of the TI-81, TI-82, or TI-85 calculators.

45. Substituting the given values, we obtain

$$R = 6.5 + \left(\frac{1}{5} + \frac{1}{6} + \frac{1}{7.5}\right)^{-1} = 6.5 + \left(\frac{6+5}{(5)(6)} + \frac{1}{7.5}\right)^{-1} = 6.5 + \left(\frac{11}{30} + \frac{1}{7.5}\right)^{-1} = 6.5 + \left(\frac{11}{30} + \frac{4}{30}\right)^{-1} =$$

$$6.5 + \left(\frac{11+4}{30}\right)^{-1} = 6.5 + \left(\frac{15}{30}\right)^{-1} = 6.5 + \left(\frac{1}{2}\right)^{-1} = 6.5 + 2 = 8.5$$

The total resistance is 8.5 Ω.

Note: On a calculator such as a TI-8x, you could have keyed this as

6.5 $\boxed{+}$ $\boxed{(}$ 1 $\boxed{\div}$ 5 $\boxed{+}$ 1 $\boxed{\div}$ 6 $\boxed{+}$ 1 $\boxed{\div}$ 7.5 $\boxed{)}$ $\boxed{x^{-1}}$ $\boxed{\text{ENTER}}$

46. Substituting the given values, we obtain

$$A = 1{,}200\left(1 + \frac{0.068}{12}\right)^{12 \cdot 5} = 1{,}200\left(1 + \frac{0.068}{12}\right)^{60} \approx 1684.32$$

The total amount of money accumulated after 5 years is $1684.32.

47. **(a)** For -7^2, press $\boxed{(-)}$ 7 $\boxed{\wedge}$ 2 $\boxed{\text{ENTER}}$. The result should be -49. For $(-7)^2$, press $\boxed{(}$ $\boxed{(-)}$ 7 $\boxed{)}$ $\boxed{\wedge}$ 2 $\boxed{\text{ENTER}}$. The result should be 49. Note that you could have used the $\boxed{x^2}$ key rather than the $\boxed{\wedge}$ 2 keystroke combination.
(b) Different
(c) The notation -7^2 is a short way of writing $(-1)7^2 = (-1)49 = -49$
The notation $(-7)^2$ means $(-7)(-7) = 49$.

48. **(a)** For $(2^3)^2$, press $\boxed{(}$ 2 $\boxed{\wedge}$ 3 $\boxed{)}$ $\boxed{\wedge}$ 2 $\boxed{\text{ENTER}}$. The result should be 64. For 2^{3^2}, the keystrokes you should use will depend on the type of calculator you are using. On a TI-8x, press 2 $\boxed{\wedge}$ $\boxed{(}$ 3 $\boxed{\wedge}$ 2 $\boxed{)}$ $\boxed{\text{ENTER}}$; on a Casio calculator, press 2 $\boxed{x^y}$ $\boxed{(}$ 3 $\boxed{x^y}$ 2 $\boxed{)}$ $\boxed{\text{ENTER}}$; on a Sharp calculator, press 2 $\boxed{a^b}$ 3 $\boxed{a^b}$ 2 $\boxed{\text{ENTER}}$. The result should be 512.
(b) Different
(c) The notation $(2^3)^2 = (8)^2 = 64$. The notation 2^{3^2} means $2^{(3^2)} = 2^9 = 512$.

≡ 1.5 SIGNIFICANT DIGITS AND ROUNDING OFF

1. Exact		**15.**	**(a)** 6.05	**24.**	**(a)** 4	**30.**	**(a)** 0.04		
			(b) 6.05		**(b)** 4.4		**(b)** 0.037		
2. Approximate					**(c)** 4.36		**(c)** 0.0373		
		16.	**(a)** 6.324						
3. Approximate			**(b)** both	**25.**	**(a)** 10	**31.**	**(a)** 300		
					(b) 14		**(b)** 310		
4. Exact; exact		**17.**	**(a)** 5.01		**(c)** 14.4		**(c)** 305		
			(b) 0.027						
5. Approximate				**26.**	**(a)** 4	**32.**	**(a)** 4		
		18.	**(a)** 19,020		**(b)** 4.1		**(b)** 4.4		
6. Approximate			**(b)** 19,020		**(c)** 4.07		**(c)** 4.36		
7. 3		**19.**	**(a)** 27,0$\tilde{0}$0	**27.**	**(a)** 7	**33.**	**(a)** 10		
			(b) 27,0$\tilde{0}$0		**(b)** 7.0		**(b)** 14		
8. 3					**(c)** 7.04		**(c)** 14.4		
		20.	**(a)** both						
9. 3			**(b)** 0.003	**28.**	**(a)** 0.006	**34.**	**(a)** 4		
					(b) 0.0062		**(b)** 4.1		
10. 3		**21.**	**(a)** 86		**(c)** 0.00616		**(c)** 4.06		
			(b) 0.2						
11. 1				**29.**	**(a)** 400	**35.**	**(a)** 7		
		22.	**(a)** 305.00		**(b)** 4$\tilde{0}$0		**(b)** 7.0		
12. 3			**(b)** both		**(c)** 403		**(c)** 7.04		
13. 4		**23.**	**(a)** 140.070						
			(b) 140.070						
14. 6									

36.	(a) 0.006	39.	(a) 300	42.	(a) 130	45.	(a) 440
	(b) 0.0062		(b) 3$\tilde{0}$0		(b) 125.4		(b) 438.0
	(c) 0.00616		(c) 305		(c) 125.376		(c) 437.998
37.	(a) 400	40.	(a) 30	43.	(a) 240	46.	(a) 10
	(b) 4$\tilde{0}$0		(b) 25.3		(b) 237.3		(b) 12.3
	(c) 403		(c) 25.335		(c) 237.302		(c) 12.341
38.	(a) 0.04	41.	(a) 90	44.	(a) 100	47.	(a) 80
	(b) 0.037		(b) 89.9		(b) 97.0		(b) 78.7
	(c) 0.0372		(c) 89.899		(c) 96.999		(c) 78.671

48. Absolute error = 0.005 m, relative error = 0.0004; percent error = 0.04%

49. Length: Absolute error is measured value minus true value, so $23.72 - 24 = -0.28$ mm. Relative error is absolute error divided by true value, so $\frac{-0.28}{24} = -0.0117$. Percent error is relative error $\times 100$, so $-0.0117 \times 100 = -1.17\%$

Width: Absolute error; $8.35 - 8 = 0.35$ mm. Relative error; $\frac{0.35}{8} = 0.0438$. Percent error; $0.0438 \times 100 = 4.38\%$

Thickness: Absolute error; $2.98 - 3 = -0.02$ mm. Relative error; $\frac{-0.02}{3} = -0.0067$. Percent error; $-0.0067 \times 100 = -0.67\%$

50.	24.68	52.	410.8	54.	82	56.	0.0085
51.	99.37	53.	1618	55.	1020	57.	17

58. There are 2 front lamps, so they draw a total of $2 \times 0.417 = 0.834$ A. The 2 tail lamps draw a total current of $2 \times 0.457 = 0.914$ A. Thus the total for the 5 lamps is $0.834 + 0.914 + 0.736 = 2.484$. Rounded off to 3 significant digits, this is 2.48 A.

59. To find the total length, we need to multiply $37 \times 132 \times 0.072$. The product is 351.648, which is 351.65 when rounded to 5 significant digits.

60. **(a)** We want the product of 379 and 11 feet $10\frac{1}{8}$ inches. Convert 11 ft $10\frac{1}{8}$ in. to $11 \times 12 +$ $10\frac{1}{8}$ in. $= 142\frac{1}{8}$ in. $= 142.125$ in. Now when we multiply this by 379, we obtain 53865.375 in. This is correct to 3 decimal places, so the answer to **(a)** is 53,865.375 in. $= 4,488$ ft $9\frac{3}{8}$ in.

(b) Rounded off to 3 significant digits, the answer is 53,900 in. $= 4,491$ ft 8 in.

61. The cutter removes 18×0.086 mm $= 1.548$ mm with each revolution. The total of 597.3 revolutions will remove $597.3 \times 1.548 = 924.6204$ mm. Rounded off to 4 significant digits, this is 924.6 mm.

1.6 SCIENTIFIC AND ENGINEERING NOTATION

1.	4.2×10^4	10.	2.07×10^0	19.	5.6×10^0	28.	0.96
2.	3.7×10^8	11.	74×10^3	20.	3.08×10^0	29.	75 000
3.	3.8×10^{-4}	12.	910×10^6	21.	4500	30.	19 000 000
4.	7.5×10^{-6}	13.	470×10^{-6}	22.	370000	31.	47 500 000 000
5.	9.80700×10^9	14.	7.5×10^{-6}	23.	40500000	32.	317 $\tilde{0}$00 000 000
6.	8.700×10^7	15.	9.80700×10^9	24.	305000000	33.	0.000 0392 0
7.	9.70×10^{-5}	16.	57.00×10^6	25.	0.000063	34.	0.001 75
8.	4.00×10^{-1}	17.	53.10×10^{-6}	26.	0.0000000187	35.	83.15
9.	4.3×10^0	18.	700×10^{-3}	27.	72	36.	0.000 000 391 25

37. $(7.6 \times 10^5) \times (2.04 \times 10^{10}) = 15.504 \times 10^{15} = 1.5504 \times 10^{16}$

38. $(4.32 \times 10^7) \times (8.5 \times 10^8) = 36.72 \times 10^{15} = 3.672 \times 10^{16}$

39. $(3.5 \times 10^{-5}) \times (7.6 \times 10^{-7}) = 26.6 \times 10^{-12} = 2.66 \times 10^{-11}$

40. $(4.2 \times 10^{-4}) \times (7.5 \times 10^{-5}) = 31.5 \times 10^{-9} = 3.15 \times 10^{-8}$

41. $(8.4 \times 10^8) \times (3.5 \times 10^{-4}) = 29.4 \times 10^4 = 2.94 \times 10^5$

42. $(4.2 \times 10^{-6}) \times (2.3 \times 10^4) = 9.66 \times 10^{-2}$

43. $(7.04 \times 10^4) \times (3.2 \times 10^{-6}) = 22.528 \times 10^{-2} = 2.2528 \times 10^{-1}$

44. $(3.02 \times 10^{-4}) \times (4.37 \times 10^9) = 13.1974 \times 10^5 = 1.31974 \times 10^6$

45. $(2.88 \times 10^{10}) \div (2.4 \times 10^5) = 1.2 \times 10^5$

46. $(5.55 \times 10^{10}) \div (3.7 \times 10^5) = 1.5 \times 10^5$

47. $(3.75 \times 10^5) \div (1.5 \times 10^8) = 2.5 \times 10^{-3}$

48. $(7.98 \times 10^4) \div (8.4 \times 10^8) = 0.95 \times 10^{-4} = 9.5 \times 10^{-5}$

49. $(3.2 \times 10^{-3}) \div (1.6 \times 10^{-7}) = 2 \times 10^4$

50. $(4.8 \times 10^{-4}) \div (3 \times 10^{-7}) = 1.6 \times 10^3$

51. $(3.6 \times 10^{-7}) \div (2 \times 10^{-4}) = 1.8 \times 10^{-3}$

52. $(9.8 \times 10^{-9}) \div (1.4 \times 10^{-5}) = 7 \times 10^{-4}$

53. $(8.76 \times 10^6)(2.46 \times 10^8)(6.4 \times 10^9) = 137.91744 \times 10^{6+8+9} = 137.91744 \times 10^{23} = 1.3791744 \times 10^{25} \approx 1.38 \times 10^{25}$

54. $(4.36 \times 10^6)(6.25 \times 10^8)(3.87 \times 10^{10}) = 105.4575 \times 10^{24} = 1.054575 \times 10^{26} \approx 1.05 \times 10^{26}$

55. $(2.5 \times 10^5)(6.3 \times 10^8) \div (6.3 \times 10^{-9}) = 2.5 \times 10^{5+8-(-9)} = 2.5 \times 10^{22}$

56. $(2.52 \times 10^4)(8 \times 10^6) \div (3.97 \times 10^9) = 5.0780856 \times 10^{4+6-9} = 5.0780856 \times 10^1 \approx 5.08 \times 10^1$

57. $(5.2 \times 10^{-6})(4.8 \times 10^8) \div (6.4 \times 10^{-12}) = 3.9 \times 10^{-6+8-(-12)} = 3.9 \times 10^{14}$

58. $(9.6 \times 10^7)(8.1 \times 10^5) \div (2.43 \times 10^{14}) = 32 \times 10^{7+5-14} = 3.2 \times 10^{-1}$

59. 675 (answers may vary)

60. 2 200 (answers may vary)

61. $(3 \times 10'')(3.7 \times 10^{-6}) = 11.1 \times 10^5 = 1.11 \times 10^6$ mm

62. $(1.5 \times 10^8) \div (3 \times 10^5) = 0.5 \times 10^3 = 5 \times 10^2 = 500$ seconds; $500 \div 60 = 8.33$ minutes

63. $(1.6606 \times 10^{-27})(1.2 \times 10^1)(1.4 \times 10^7) = 2.789808 \times 10^{-27+1+7} = 2.789808 \times 10^{-19}$ kg

64. $(1.6606 \times 10^{-27})(5.5 \times 10^1)(2.3 \times 10^8 = 21.00659 \times 10^{-18} = 2.100659 \times 10^{-17}$ kg

65. One neutron; approximately $(1.6750 \times 10^{-27}) \div (9.1095 \times 10^{-31}) = 1.84 \times 10^3$ times heavier

66. $X_L = 2\pi f L = 2\pi(1 \times 10^7)(1.5 \times 10^{-2}) \approx 9.425 \times 10^5 = 942\,500$ Ω

67. $X_L = 2\pi f L = 2\pi(18 \times 10^6)(2.5 \times 10^{-3}) \approx 282.74 \times 10^3$. The inductive reactance is about 282.74×10^3 Ω.

68. Since the density is the mass divided by the volume, we have

$$\text{density} = \frac{5.975 \times 10^{24}}{1.083 \times 10^{21}}$$
$$= \frac{5.975}{1.083} \times \frac{10^{24}}{10^{21}}$$
$$= \frac{5.975}{1.083} \times 10^{24-21}$$
$$\approx 5.517 \times 10^3$$

The density of the earth is about 5.517×10^3 kg/m^3.

69. Easier to express very large and very small numbers. Historically, to use slide rules, numbers were first written in scientific notation.

70. Makes conversion from one unit to another in the metric system more convenient.

$\equiv$ 1.7 ROOTS

1. 5   **6.** -4 **11.** $\frac{2}{3}$ **16.** 0.5

2. 6   **7.** -3 **12.** $\frac{-3}{10}$ **17.** 3

3. 12   **8.** 3   **13.** 0.2   **18.** 5

4. 11   **9.** 2   **14.** 0.5   **19.** 8.32

5. 2   **10.** -3 **15.** -0.1 **20.** 7.91

21. $\sqrt[3]{5}\sqrt[3]{25} = \sqrt[3]{5 \cdot 25} = \sqrt[3]{5^3} = 5$

22. $\sqrt[5]{-3}\sqrt[3]{81} = \sqrt[5]{-3}\sqrt[5]{(-3)^4} = \sqrt[5]{(-3)^5} = -3$ **23.** $\sqrt[4]{8}\sqrt[4]{9} = \sqrt[4]{8 \cdot 9} = \sqrt[4]{72}$

24. $\sqrt[6]{12}\sqrt[6]{48} = \sqrt[6]{12 \cdot 48} = \sqrt[6]{576} = \sqrt[6]{64}\sqrt[6]{9} = 2\sqrt[6]{3^2} = 2\sqrt[3]{3}$

25. $\dfrac{\sqrt{75}}{\sqrt{3}} = \sqrt{\dfrac{75}{3}} = \sqrt{25} = 5$ **31.** $\sqrt[3]{\left(\dfrac{2}{3}\right)^3} - \sqrt[4]{\left(\dfrac{1}{3}\right)^4} = \dfrac{2}{3} - \dfrac{1}{3} = \dfrac{1}{3}$

26. $\dfrac{\sqrt{112}}{\sqrt{7}} = \sqrt{\dfrac{112}{7}} = \sqrt{16} = 4$ **32.** $\sqrt[5]{\left(\dfrac{3}{4}\right)^5} + \sqrt[3]{\left(\dfrac{5}{4}\right)^3} = \dfrac{3}{4} + \dfrac{5}{4} = \dfrac{8}{4} = 2$

27. $\dfrac{\sqrt[3]{5}}{\sqrt[3]{40}} = \sqrt[3]{\dfrac{5}{40}} = \sqrt[3]{\dfrac{1}{8}} = \dfrac{1}{2}$ **33.** $\sqrt{5^{2/3}} = \left(\sqrt{5^{1/3}}\right)^2 = 5^{1/3}$

34. $\sqrt{7^{2/5}} = \left(\sqrt{7^{1/5}}\right)^2 = 7^{1/5}$

28. $\dfrac{\sqrt[3]{11}}{\sqrt[3]{297}} = \sqrt[3]{\dfrac{11}{297}} = \sqrt[3]{\dfrac{1}{27}} = \dfrac{1}{3}$ **35.** $\sqrt[3]{16^{3/4}} = \left(\sqrt[3]{16^{1/4}}\right)^3 = 16^{1/4} = \sqrt[4]{16} = 2$

29. $\sqrt[3]{2^3} + \sqrt[4]{5^4} = 2 + 5 = 7$ **36.** $\sqrt[4]{27^{4/3}} = 27^{1/3} = 3$

30. $\sqrt{3^2} - \sqrt[3]{2^3} = 3 - 2 = 1$

37. $16^{3/4} = \sqrt[4]{16^3} = 2^3 = 8$ **38.** $(-27)^{2/3} = (-3)^2 = 9$ **39.** $(-8)^{2/3} = (-2)^2 = 4$

40. $25^{3/2} = \sqrt{25}^3 = 5^3 = 125$

45. $\sqrt{\dfrac{81}{(8)(0.01)}} = \dfrac{\sqrt{81}}{\sqrt{8}\sqrt{0.01}} = \dfrac{9}{2\sqrt{2}(0.1)} = \dfrac{9}{0.2\sqrt{2}}$

41. $8^{-2/3} = \left(\sqrt[3]{8}\right)^{-2} = 2^{-2} = \dfrac{1}{2^2} = \dfrac{1}{4}$

42. $9^{-3/2} = \sqrt{9}^{-3} = 3^{-3} = \dfrac{1}{3^3} = \dfrac{1}{27}$ **46.** $\sqrt[3]{\dfrac{(27)(0.008)^2}{0.027}} = \dfrac{\sqrt[3]{27}(\sqrt[3]{0.008})^2}{\sqrt[3]{0.027}} = \dfrac{3(0.2)^2}{0.3} =$

43. $(-27)^{4/3} = (\sqrt[3]{-27})^4 = (-3)^4 = 81$ $\dfrac{3(0.04)}{0.3} = 0.4$

44. $(-32)^{3/5} = \sqrt[5]{-32}^3 = -2^3 = -8$

47. $\sqrt[3]{\dfrac{(0.125)^3\sqrt{144}}{3/2}} = \dfrac{\sqrt[3]{(0.125)^3}\sqrt[3]{\sqrt{144}}}{\sqrt[3]{3/2}} = \dfrac{\sqrt[3]{(0.125)^3}\sqrt[3]{12}}{\frac{\sqrt[3]{3}}{\sqrt[3]{2}}} = 0.125\sqrt[3]{\dfrac{12 \cdot 2}{3}} = 0.125\sqrt[3]{8} = 0.125 \times 2 = 0.25$

48. $\sqrt{\dfrac{64}{(0.25)(0.16)}} = \dfrac{8}{(0.5)(0.4)} = \dfrac{8}{0.2} = 40$ **49.** $\lambda = \dfrac{3 \times 10^8}{1.435 \times 10^8} = 2.091$ m

50. about 4.601 days

51. $v_L = \sqrt{\dfrac{K}{\rho}} = \sqrt{\dfrac{2.1 \times 10^9}{1000}} = \sqrt{\dfrac{2.1 \times 10^9}{10^3}} = \sqrt{2.1 \times 10^{9-3}} = \sqrt{2.1 \times 10^6} \approx 1449.1377$ m/s

52. 2.473 m/s   **53.** (a) $\sqrt{81} = 9$ and $-\sqrt{81} = -9$
    (b) 9

(c) $\sqrt[3]{-27} = -3$ and $-\sqrt{3}27 = 3$

(d) -3

54. No solutions found

55. (a) 4
    (b) 4
    (c) yes
    (d) Both represent the cube root of 64.

CHAPTER 1 REVIEW

1. (a) Integers, rational numbers, real numbers
    (b) Rational numbers, real numbers
    (c) Irrational numbers, real numbers

4. (a) 5
    (b) -17
    (c) $-4\sqrt{2}$

2. (a) $\sqrt{42}$
    (b) 16
    (c) $\frac{5}{8}$

5. (a) Commutative law for addition
    (b) Distributive law
    (c) Additive identity
    (d) Multiplicative inverse

3. (a) $\frac{3}{2}$
    (b) $\frac{-1}{8}$
    (c) -5

6. 64

7. -44

8. 53

9. $37 - (-61) = 37 + 61 = 98$

10. -32

11. $\frac{2}{3} + \frac{-5}{6} = \frac{4}{6} + \frac{-5}{6} = \frac{4-5}{6} = -\frac{1}{6}$

12. $\frac{4}{5} - \frac{5}{6} = \frac{24}{30} - \frac{25}{30} = -\frac{1}{30}$

13. -3

14. -3

15. $-\frac{2}{5}$

16. $3\frac{3}{4} \times -4\frac{1}{3} = \frac{15}{4} \times \frac{-13}{3} = \frac{5}{4} \times \frac{-13}{1} = -\frac{65}{4}$

17. $\frac{2}{3} \div \frac{1}{4} = \frac{2}{3} \times \frac{4}{1} = \frac{8}{3}$

18. $\frac{1}{5} \div \frac{-2}{15} = \frac{1}{5} \times \frac{15}{-2} = \frac{1}{1} \times \frac{3}{-2} = -\frac{3}{2}$

19. $2\frac{1}{2} \div 3\frac{1}{4} = \frac{5}{2} \div \frac{13}{4} = \frac{5}{2} \times \frac{4}{13} = \frac{5}{1} \times \frac{2}{13} = \frac{10}{13}$

20. $-4\frac{1}{3} \div -3\frac{1}{6} = -\frac{13}{3} \div -\frac{19}{6} = -\frac{13}{3} \times -\frac{6}{19} = \frac{26}{19}$

21. $2^5 = 2 \times 2 \times 2 \times 2 \times 2 = 32$

22. $(-3)^4 = (-3)(-3)(-3)(-3) = 81$

23. $(-4)^3 = (-4)(-4)(-4) = -64$

24. $8^{1/3} = \sqrt[3]{8} = 2$

25. $4^{1/2} = \sqrt{4} = 2$

26. $(-64)^{1/3} = \sqrt[3]{-64} = -4$

27. $2^5 \cdot 2^3 = 2^{5+3} = 2^8$

28. $3^5 \cdot 3^4 = 3^{5+4} = 3^9$

29. $2^{-3} \cdot 2^5 = 2^{-3+5} = 2^2$

30. $16^{-4} \cdot 16^{-3} = 16^{-7} = \frac{1}{16^7}$

31. $(4^3)^5 = 4^{3 \cdot 5} = 4^{15}$

32. $(2^{-3})^4 = 2^{(-3)4} = 2^{-12} = \frac{1}{2^{12}}$

33. $(4^{1/3})^3 = 4^1 = 4$

34. $(5^{1/4})^{2/3} = 5^{1/4 \cdot 2/3} = 5^{1/6}$

35. $\frac{a^2 b^3}{ab^4} = \frac{a^{2-1}}{b^{4-3}} = \frac{a}{b}$

36. $\frac{x^2 y^3 z}{x^3 yz} = \frac{y^{3-1}}{x^{3-2}} = \frac{y^2}{x}$

37. $(ax^2)^{-2} = a^{-2}x^{-4} = \frac{1}{a^2 x^4}$ or $(ax^2)^{-2} = \frac{1}{(ax^2)^2} = \frac{1}{a^2 x^4}$

38. $\frac{(by^{-2})^2}{(cy^{-4})^{1/4}} = \frac{b^2 y^{-4}}{c^{1/4} y^{-1}} = \frac{b^2}{c^{1/4} y^3}$

39. (a) 2.37
    (b) 2.37

40. (a) 2.02
    (b) 0.002

41. (a) both
    (b) 0.7

42. (a) both
    (b) 0.0021

43. (a) 7.4
    (b) 7.35
    (c) 7.4
    (d) 7.35

44. (a) 18
    (b) 18.3
    (c) 18.3
    (d) 18.29

45. (a) 2.1
(b) 2.05
(c) 2.1
(d) 2.05

46. (a) 4.0
(b) 4.03
(c) 4.0
(d) 4.03

47. 3.71×10^{11}

48. 2.54×10^{15}

49. 2.4×10^{-11}

50. 4.91×10^{-20}

51. 29.08×10^{15}

52. 753×10^{21}

53. 75×10^{-24}

54. 193×10^{-18}

55. 12

56. -4

57. 5

58. $\frac{6}{11}$

59. (a) An order of 418 items is $\frac{418}{12} \approx 34.83$ dozen. The discount is given only for a full dozen; thus the manufacturer gets a discount on 34 dozen items. The total amount of the discount is $34 \times 12 \times 0.14 = 57.12$, or $57.12.
(b) The total cost before the discount of the 418 items is $418 \times 13.79 = 5764.22$. Subtracting the discount of $57.12, you get $5764.22 - 57.12 = 5707.10$. Thus, the total cost of the order is $5,707.10.

60. (a) Substituting the given data in the above formula, we have

$$E = \frac{3}{2}kT = \frac{3}{2}(8.617\,33 \times 10^{-5})(295)$$
$$\approx 0.0381316853$$

The average kinetic energy of particles at room temperature ($T = 295$ K) is about 0.038 eV.
(b) Substituting the given data in the above formula, we have

$$E = \frac{3}{2}kT = \frac{3}{2}(8.617\,33 \times 10^{-5})(5.714 \times 10^{3})$$
$$\approx 0.7385913543$$

The average kinetic energy of particles at the surface of the sun ($T = 5.714 \times 10^{3}$ K) is about 0.739 eV.

CHAPTER 1 TEST

1. (a) $\left|\frac{4}{3}\right| = \frac{4}{3}$

   (b) $\left|\frac{1}{2} - \frac{5}{8}\right| = \left|\frac{4}{8} - \frac{5}{8}\right| = \left|-\frac{1}{8}\right| = \frac{1}{8}$

   (c) $|-6| = 6$

2. $-\frac{7}{3}$

3. $\frac{5}{8}$

4. 111

5. -112

6. -70

7. $\frac{5}{3} + 5\frac{2}{3} = \frac{5}{3} + \frac{17}{3} = \frac{22}{3}$ or $7\frac{1}{3}$

8. $\frac{7}{4} - \left(-\frac{3}{5}\right) = \frac{35}{20} + \frac{12}{20} = \frac{47}{20}$

9. $-4\frac{1}{2} \times 2\frac{1}{3} = -\frac{9}{2} \times \frac{7}{3} = -\frac{21}{2}$

10. $-\frac{5}{7} \div \frac{15}{28} = -\frac{5}{7} \times \frac{28}{15} = -\frac{4}{3}$

11. $-2\frac{1}{3} \div -4\frac{5}{6} = -\frac{7}{3} \div -\frac{29}{6} = -\frac{7}{3} \times -\frac{6}{29} = \frac{14}{29}$

12. $(-5)^3 = (-5)(-5)(-5) = -125$

13. $4^{3/2} = \left(\sqrt{4}\right)^3 = 2^3 = 8$

14. $2^6 \cdot 2^{-4} = 2^{6-4} = 2^2 = 4$

15. $\left(5^{1/4}\right)^8 = 5^{(1/4)8} = 5^2 = 25$

16. $3^{5/2} \div 3^{-3/2} = 3^{5/2-(-3/2)} = 3^{5/2+3/2} = 3^4 = 81$

17. $\dfrac{a^3 b^5}{a^4 b^2} = \dfrac{b^{5-2}}{a^{4-3}} = \dfrac{b^3}{a}$

18. 4.516

19. 0.000 51

20. 4.75×10^{13}

21. $\sqrt{\dfrac{49}{25}} = \dfrac{\sqrt{49}}{\sqrt{25}} = \dfrac{7}{5}$

22. 92.5×10^{12}

CHAPTER
2
Algebraic Concepts and Operations

≡ 2.1 ADDITION AND SUBTRACTION

1. 4 is a constant; x and y are variables.

3. 8 and π are constants; r is a variable.

2. 5 and 3 are constants; x, t, and w are variables.

4. $\sqrt{7}$ is a constant; a and b are usually constants; x and y are variables.

5. $3x^3$ and $4x$

10. 3, 5, $x - y$, and $\frac{1}{x^2+3y}$

15. $3x^2y$ and $17x^2y$

6. $47x^4$ and $-9y^5$

11. 47

16. $\sqrt{5}xy$ and $\frac{2}{3}bxy$, if b is a constant

7. $(2x^3)(5y)$, $\sqrt{3}ab$, and $-\frac{7a}{b}$

12. $2\sqrt{3}$

8. $-3x^5n$, $-(4a)\left(\frac{b}{5}\right)$, and $9\sqrt{y}$

13. $\frac{\pi a}{4}$, if a is a constant

17. $(x+y)^2$ and $5(x+y)^2$

18. $-9xy^2$ and $5(xy^2)$

9. -1, 5, x, and $\frac{1}{y}$

14. $\frac{\sqrt{17b}}{\sqrt{5a}}$, if a and b are constants

19. $a+b-c$ and $-2(a+b-c)$

20. There are two pairs of like terms. One pair is ax^2 and $5x^2$, and the other pair is a^2x and $5a^5x$, if a is a constant.

33. $(a+6b)-(a-6b) = a+6b-a+6b = a-a+6b+6b = (1-1)a+(6+6)b = 0a+12b = 12b$

21. $4x+7x = (4+7)x = 11x$

34. $(x-7y)-(7y-x) = x-7y-7y+x = x+x-7y-7y = 2x-14y$

22. $5y-2y = (5-2)y = 3y$

23. $3z-z = 3z-1z = (3-1)z = 2z$

35. $(2a^2+3b)+(2b+4a) = 2a^2+3b+2b+4a = 2a^2+(3+2)b+4a = 2a^2+5b+4a$

24. $7w+4w-w = (7+4-1)w = 10w$

36. $(7c^2-8d)+(6d-8c = 7c^2-8d+6d-8c) = 7c^2+(-8+6)d-8c = 7c^2-2d-8c$

25. $8x+9x^2-2x = 8x-2x+9x^2 = (8-2)x+9x^2 = 6x+9x^2$ or $9x^2+6x$

37. $(4x^2+3x)-(2x^2-3x) = 4x^2+3x-2x^2+3x = (4-2)x^2+(3+3)x = 2x^2+6x$

26. $11y-7y+6y^2 = (11-7)y+6y^2 = 4y+6y^2$ or $6y^2+4y$

38. $(3y^2-4x)-(4y+2x) = 3y^2-4x-4y-2x = 3y^2+(-4-2)x-4y = 3y^2-6x-4y$ or $3y^2-4y-6x$

27. $10w+w^2-8w^2 = 10w+(1-8)w^2 = 10w-7w^2$ or $-7w^2+10w$

39. $2(6y^2+7x) = 2\cdot 6y^2+2\cdot 7x = 12y^2+14x$

28. $y^2-6y^2+4y = (1-6)y^2+4y = -5y^2+4y$

40. $5(3a+4b) = 5\cdot 3a+5\cdot 4b = 15a+20b$

41. $-3(4b-2c) = -3\cdot 4b+(-3)(-2c) = -12b+6c$

29. $ax^2+a^2x+ax^2 = ax^2+ax^2+a^2x = 2ax^2+a^2x$

42. $-2(-6b+3a) = (-2)(-6b)+(-2)(3a) = 12b-6a$

30. $by-by^2+by = by+by-by^2 = 2by-by^2$

43. $4(a+b)+3(b+a) = 4a+4b+3b+3a = 7a+7b$

31. $7xy^2-5x^2y+4xy^2 = 7xy^2+4xy^2-5x^2y = 11xy^2-5x^2y$

44. $2(c+d)+8(d+c) = 2c+2d+8d+8c = 10c+10d$

32. $12wz-8w^2z+6w^2z = 12wz+(-8+6)w^2z = 12wz-2w^2z$

45. $3(x+y)-2(x+y) = 3x+3y-2x-2y = (3-2)x+(3-2)y = x+y$

46. $3(x^2 - y) - 2(y + x^2) = 3x^2 - 3y - 2y - 2x^2 = x^2 - 5y$

47. $2(a + b + c) + 3(a + b - c) = 2a + 2b + 2c + 3a + 3b - 3c = 5a + 5b - c$

48. $4(x - y + z) + 2(x + y - z) = 4x - 4y + 4z + 2x + 2y - 2z = 6x - 2y + 2z$

49. $3[2(x + y)] = 3[2x + 2y] = 6x + 6y$

50. $4[3(x - y)] = 4[3x - 3y] = 12x - 12y$

51. $3(a + b) + 4(a + b) - 2(a + b) = 3a + 3b + 4a + 4b - 2a - 2b = 5a + 5b$ or $(3 + 4 - 2)(a + b) = 5(a + b) = 5a + 5b$

52. $2(x+y) - 3(x+y) - 4(x+y) = (2-3-4)(x+y) = -5(x + y) = -5x - 5y$ or $2x + 2y - 3x - 3y - 4x - 4y = -5x - 5y$

53. $2(a + b + c) + 3(a - b + c) + (a - b - c) = 2a + 2b + 2c + 3a - 3b + 3c + a - b - c = 6a - 2b + 4c$

54. $3(x - y + z) - 2(x + y - z) + 4(-x - y + z) = 3x - 3y + 3z - 2x - 2y + 2z - 4x - 4y + 4z = -3x - 9y + 9z$

55. $2(x + 3y) - 3(x - 2y) + 5(2x - y) = 2x + 6y - 3x + 6y + 10x - 5y = 9x + 7y$

56. $3(y - 2a) - (a + 3y) + 4(a - 3y) = 3y - 6a - a - 3y + 4a - 12y = -12y - 3a$

57. $3(x + y - z) - 2(3x + 2y - z) - 3(x - y + 4z) = 3x + 3y - 3z - 6x - 4y + 2z - 3x + 3y - 12z = -6x + 2y - 13z$

58. $5(x - y + z) - (y - x + z) + 2(x - 2y + z) = 5x - 5y + 5z - y + x - z + 2x - 4y + 2z = 8x - 10y + 6z$

59. $(x + y) - 3(x - z) + 4(y + 4z) - 2(x + y - 3z) = x + y - 3x + 3z + 4y + 16z - 2x - 2y + 6z = -4x + 3y + 25z$

60. $(a+b) - 2(b-c) + 4(c+2d) - 5(d+2c-3b-a) = a + b - 2b + 2c + 4c + 8d - 5d - 10c + 15b + 5a = 6a + 14b - 4c + 3d$

61. $x + [3x + 2(x + y)] = x + [3x + 2x + 2y] = x + [5x + 2y] = x + 5x + 2y = 6x + 2y$

62. $x + [5y + 3(y - x)] = x + [5y + 3y - 3x] = x + [8y - 3x] = -2x + 8y$

63. $y - [2z - 3(y+z) + y] = y - [2z - 3y - 3z + y] = y - [-z - 2y] = y + z + 2y = 3y + z$

64. $2w - [4z - 5(z+w) + 2w] = 2w - [4z - 5z - 5w + 2w] = 2w - [-z - 3w] = 2w + z + 3w = 5w + z$

65. $[2x + 3(x+y) - 2(x - y) + y] - 2x = [2x + 3x + 3y - 2x + 2y + y] - 2x = [3x + 6y] - 2x = x + 6y$

66. $[4a - 8(a+b) + 2(a - b) + 3a] - 3b = [4a - 8a - 8b + 2a - 2b + 3a] - 3b = [a - 10b] - 3b = a - 13b$

67. $-\{-[2a - (3b + a)]\} = -\{-[2a - 3b - a]\} = -\{-[a - 3b]\} = -\{-a + 3b\} = a - 3b$

68. $-\{-3[4x - (5x - 4y)]\} = -\{-3[4x - 5x + 4y]\} = -\{-3[-x + 4y]\} = -\{3x - 12y\} = -3x + 12y$

69. $5a - 2\{4[a + 2(4a+b) - b] + a\} - a = 5a - 2\{4[a + 8a + 2b - b] + a\} - a = 5a - 2\{4[9a + b] + a\} - a = 5a - 2\{36a + 4b + a\} - a = 5a - 2\{37a + 4b\} - a = 5a - 74a - 8b - a = -70a - 8b$

70. $7x - 3\{-[x + 2(x - y) - y] + 2x\} - 3y = 7x - 3\{-[x + 2x - 2y - y] + 2x\} - 3y = 7x - 3\{-[3x - 3y] + 2x\} - 3y = 7x - 3\{-3x + 3y + 2x\} - 3y = 7x - 3\{-x + 3y\} - 3y = 7x + 3x - 9y - 3y = 10x - 12y$

71. $p + \frac{1}{2}p + \frac{2}{3}p = \frac{6}{6}p + \frac{3}{6}p + \frac{4}{6}p = \frac{6+3+4}{6}p = \frac{13}{6}p$

72. $x + 2x + 3x + 3x = (1 + 2 + 3 + 3)x = 9x$

73. Follow the process in Example 2.7. The common denominator is C_1C_2. This produces the following result:

$\frac{1}{C_T} = \frac{C_2}{C_1C_2} + \frac{C_1}{C_1C_2} = \frac{C_2 + C_1}{C_1C_2}$

74. Follow the process in Example 2.7. The common denominator is $L_1L_2L_3$. This produces the following result:

$\frac{1}{L_T} = \frac{L_2L_3}{L_1L_2L_3} + \frac{L_1L_3}{L_1L_2L_3} + \frac{L_1L_2}{L_1L_2L_3} = \frac{L_2L_3 + L_1L_3 + L_1L_2}{L_1L_2L_3}$

2.2 MULTIPLICATION

1. $\left(a^2x\right)\left(ax^2\right) = a^{2+1} \cdot x^{1+2} = a^3x^3$

2. $\left(by^2\right)\left(b^2y\right) = b^{1+2}y^{2+1} = b^3y^3$

3. $(3ax)\left(2ax^2\right) = 3 \cdot 2 \cdot a^{1+1}x^{1+2} = 6a^2x^3$

4. $(5by)\left(3b^2y\right) = 5 \cdot 3b^{1+2}y^{1+1} = 15b^3y^2$

5. $\left(2xw^2z\right)\left(-3x^2w\right) = 2(-3)x^{1+2}w^{2+1}z = -6x^3w^3z$

6. $\left(-4ya^2b\right)\left(6y^2b\right) = (-4)6 \cdot y^{1+2}a^2b^{1+1} = -24y^3a^2b^2$

7. $(3x)(4ax)\left(-2x^2b\right) = 3 \cdot 4(-2)x^{1+1+2}ab = -24x^4ab$

8. $(4y)\left(3y^2b\right)\left(-5by^2\right) = 4 \cdot 3(-5)y^{1+2+2}b^{1+1} = -60b^2y^5$

9. $2(5y - 6) = 2 \cdot 5y + 2(-6) = 10y - 12$

10. $4(3x - 5) = 4(3x) + 4(-5) = 12x - 20$

11. $-5(4w - 7) = -5(4w) + (-5)(-7) = -20w + 35$ or $35 - 20w$

12. $-3(8 + 5p) = -24 - 15p$

13. $3x(7y + 4) = 3x(7y) + 3x(4) = 21xy + 12x$

14. $6x(8y - 7) = 6x(8y) + 6x(-7) = 48xy - 42x$

15. $-5t(-3 + t) = (-5t)(-3) - 5t(t) = 15t - 5t^2$

16. $-3n(2n - 5) = -3n(2n) - 3n(-5) = -6n^2 + 15n$ or $15n - 6n^2$

17. $\frac{1}{2}a(4a - 2) = \frac{1}{2}a(4a) + \frac{1}{2}a(-2) = 2a^2 - a$

18. $\frac{1}{3}x(-21x - 15) = \frac{1}{3}x(-21x) + \frac{1}{3}x(-15) = -7x^2 - 5x$

19. $2x(3x^2 - x + 4) = 2x(3x^2) + 2x(-x) + 2x(4) = 6x^3 - 2x^2 + 8x$

20. $3y(4y^2 - 5y - 7) = 3y(4y^2) + 3y(-5y) + 3y(-7) = 12y^3 - 15y^2 - 21y$

21. $4y^2(-5y^2 + 2y - 5 + 3y^{-1} - 6y^2) = 4y^2\left(-5y^{-2}\right) + 4y^2(2y) + 4y^2(-5) + 4y^2\left(3y^{-1}\right) + 4y^2\left(-6y^{-2}\right) - 20y^4 + 8y^3 - 20y^2 + 12y - 24$

22. $5p^2\left(-4p^3 - 3p + 2 + p^{-1} - 7p^{-2}\right) 5p^2\left(-4p^3\right) + 5p^2(-3p) + 5p^2(2) + 5p^2\left(p^{-1}\right) + 5p^2\left(-7p^{-2}\right) = -20p^5 - 15p^3 + 10p^2 + 5p - 35$

23. $(a + b)(a + c) = (a + b)a + (a + b)c = a^2 + ab + ac + bc$ or by FOIL $a^2 + ac + ab + bc$

24. $(s + t)(s + 2t) = s^2 + 2st + st + 2t^2 = s^2 + 3st + 2t^2$

25. $(x + 5)(x^2 + 6) = x^3 + 6x + 5x^2 + 30$ or $x^3 + 5x^2 + 6x + 30$

26. $(y + 3)(y^2 + 7) = y^3 + 7y + 3y^2 + 21$ or $y^3 + 3y^2 + 7y + 21$

27. $(2x + y)(3x - y) = 6x^2 - 2xy + 3xy - y^2 = 6x^2 + xy - y^2$

28. $(4a + b)(8a - b) = 32a^2 - 4ab + 8ab - b^2 = 32a^2 + 4ab - b^2$

29. $(2a - b)(3a - 2b) = 6a^2 - 4ab - 3ab + 2b^2 = 6a^2 - 7ab + 2b^2$

30. $(4p + q)(3p - 2q) = 12p^2 - 8pq + 3pq - 2q^2 = 12p^2 - 5pq - 2q^2$

31. $(b - 1)(2b + 5) = 2b^2 + 5b - 2b - 5 = 2b^2 + 3b - 5$

32. $(4x - 1)(3x - 2) = 12x^2 - 8x - 3x + 2 = 12x^2 - 11x + 2$

33. $\left(7a^2b + 3c\right)\left(8a^2b - 3c\right) = 56a^4b^2 - 21a^2bc + 24a^2bc - 9c^2 = 56a^4b^2 + 3a^2bc - 9c^2$

34. $\left(6p^2r + 2t\right)\left(5p^2r + 4t\right) = 30p^4r^2 + 24p^2rt + 10p^2rt + 8t^2 = 30p^4r^2 + 34p^2rt + 8t^2$

35. $(x + 4)(x - 4) = x^2 - 16$ (Difference of Squares)

36. $(a + 8)(a - 8) = a^2 - 64$ (Difference of Squares)

37. $(p - 6)(p + 6) = p^2 - 36$

38. $(b - 10)(b + 10) = b^2 - 100$

39. $(ax + 2)(ax - 2) = a^2x^2 - 4$

40. $(xy - 3)(xy + 3) = x^2y^2 - 9$

41. $\left(2r^2 + 3x\right)\left(2r^2 - 3x\right) = 4r^4 - 9x^2$

42. $\left(4p^3 - 7d\right)\left(4p^3 + 7d\right) = 16p^6 - 49d^2$

43. $\left(5a^2x^3 - 4d\right)\left(5a^2x^3 + 4d\right) = 25a^4x^6 - 16d^2$

44. $\left(3p^2st - \frac{11}{3}w^3\right)\left(3p^2st + \frac{11}{3}w^3\right) = 9p^4s^2t^2 - \frac{121}{9}w^6$

45. $\left(\frac{2}{3}pa^2f + \frac{3}{4}tb^3\right)\left(-\frac{2}{3}pa^2f + \frac{3}{4}tb^3\right) = \left(\frac{3}{4}tb^3 + \frac{2}{3}pa^2f\right)\left(\frac{3}{4}tb^3 - \frac{2}{3}pa^2f\right) = \frac{9}{16}t^2b^6 - \frac{4}{9}p^2a^4f^2$

46. $\left(\frac{\sqrt{3}}{2} + \frac{7}{5}t^2u\right)\left(-\frac{\sqrt{3}}{2} + \frac{7}{5}t^2u\right) = \left(\frac{7}{5}t^2u + \frac{\sqrt{3}}{2}\right)\left(\frac{7}{5}t^2u - \frac{\sqrt{3}}{2}\right) = \frac{49}{25}t^4u^2 - \frac{3}{4}$

47. $(x + y)^2 = x^2 + 2xy + y^2$ (Square of a Binomial)

48. $(p + r)^2 = p^2 + 2pr + r^2$ (Square of a Binomial)

49. $(x - 5)^2 = x^2 - 10x + 25$

50. $(b - 7)^2 = b^2 - 14b + 49$

51. $(a + 3)^2 = a^2 + 6a + 9$

52. $(w + 5)^2 = w^2 + 10w + 25$

53. $(2a + b)^2 = 4a^2 + 4ab + b^2$

54. $(3c + d)^2 = 9c^2 + 6cd + d^2$

55. $(3x - 2y)^2 = 9x^2 - 12xy + 4y^2$

56. $(5a - 6f)^2 = 25a^2 - 60af + 36f^2$

57. $4x(x+4)(3x-2) = 4x\left(3x^2 - 2x + 12x - 8\right) = 4x\left(3x^2 + 10x - 8\right) = 12x^3 + 40x^2 - 32x$

58. $5y(y-6)(2y+3) = 5y\left(2y^2 - 9y - 18\right) = 10y^3 - 45y^2 - 90y$

59. $(x+y-z)(x-y+z) = (x+y-z)(x)+(x+y-z)(-y)+(x+y-z)(z) = x^2+xy-xz-xy-y^2+yz+xz+yz-z^2 = x^2-y^2-z^2+2yz$ or $[x+(y-z)][x-(y-z)] = x^2-(y-z)^2 = x^2-y^2-z^2+2yz$

60. $(a+b+c)(a-b-c) = [a+(b+c)][a-(b+c)] = a^2-(b+c)^2 = a^2-\left(b^2 + 2bc + c^2\right) = a^2-b^2-c^2-2bc$

61. Think of $(a+2)^3$ as $(a+2)^2(a+2)$.
$(a+2)^3 = (a+2)^2(a+2) = \left(a^2 + 4a + 4\right)(a+2) = \left(a^2 + 4a + 4\right)(a) + \left(a^2 + 4a + 4\right)(2) = \left(a^3 + 4a^2 + 4a\right) + \left(2a^2 + 8a + 8\right) = a^3 + 6a^2 + 12a + 8$

62. Think of $(m-n)^3$ as $(m-n)^2(m-n)$.
$(m-n)^3 = (m-n)^2(m-n) = \left(m^2 - 2mn + n^2\right)(m-n) = \left(m^2 - 2mn + n^2\right)(m) + \left(m^2 - 2mn + n^2\right)(-n) = \left(m^3 - 2m^2n + mn^2\right) + \left(-m^2n + 2mn^2 - n^3\right) = m^3 - 3m^2n + 3mn^2 - n^3$

63. $2[n(n+2)+n] = 2\left[n^2 + 2n + n\right] = 2\left[n^2 + 3n\right] = 2n^2 + 6n$

64. $(2R-x)^2 - x^2 - R^2 = \left(4R^2 - 4Rx + x^2\right) - x^2 - R^2 = 3R^2 - 4Rx$

65. $\frac{1}{2}(y_2 - y_1)(y_2 + y_1) = \frac{1}{2}\left(y_2^2 - y_1^2\right) = \frac{1}{2}y_2^2 - \frac{1}{2}y_1^2$

66. $500(1+r)^3 = 500(1+r)^2(1+r) = 500\left(1 + 2r + r^2\right)(1+r) = 500\left(1 + 3r + 3r^2 + r^3\right) = 500 + 1500r + 1500r^2 + 500r^3$

67. $x = 3(t+4)(t+1) = 3\left(t^2 + 5t + 4\right) = 3t^2 + 15t + 12$

68. $Q = \bar{e}A\left\{\left(\frac{T_1}{100} - \frac{T_2}{100}\right)\left(\frac{T_1}{100} + \frac{T_2}{100}\right)\left[\left(\frac{T_1}{100}\right)^2 + \left(\frac{T_2}{100}\right)^2\right]\right\}10^8 t = \bar{e}A\left\{\left[\left(\frac{T_1}{100}\right)^2 - \left(\frac{T_2}{100}\right)^2\right]\left[\left(\frac{T_1}{100}\right)^2 + \left(\frac{T_2}{100}\right)^2\right]\right\}10^8 t = \bar{e}A\left[\left(\frac{T_1}{100}\right)^4 - \left(\frac{T_2}{100}\right)^4\right]10^8 t$

≡ 2.3 DIVISION

1. x^7 by $x^3 = \frac{x^7}{x^3} = x^{7-3} = x^4$

2. y^8 by $y^6 = \frac{y^8}{y^6} = y^{8-6} = y^2$

3. $2x^6$ by $x^4 = \frac{2x^6}{x^4} = 2x^2$

4. $3w^4$ by $w^2 = \frac{3w^4}{w^2} = 3w^2$

5. $12y^5$ by $4y^3 = \frac{12y^5}{4y^3} = \frac{12}{4} \cdot \frac{y^5}{y^3} = 3y^2$

6. $15a^7$ by $3a^4 = \frac{15a^7}{3a^4} = 5a^3$

7. $-45ab^2$ by $15ab = \frac{-45ab^2}{15ab} = -3b$

8. $-55xy^3$ by $-11xy = \frac{-55xy^3}{-11xy} = 5y^2$

9. $33xy^2z$ by $3xyz = \frac{33xy^2z}{3xyz} = 11y$

10. $65x^2yz$ by $5xyz = \frac{65x^2yz}{5xyz} = 13x$

11. $96a^2xy^3$ by $-16axy^2 = \frac{96a^2xy^3}{-16axy^2} = -6ay$

12. $105b^3yw^2$ by $-15b^2yw = \frac{105b^3yw^2}{-15b^2yw} = -7bw$

13. $144c^3d^2f$ by $8cf = \frac{144c^3d^2f}{8cf} = 18c^2d^2$

14. $162x^2yz^3$ by $9x^2z^2 = \frac{162x^2yz^3}{9x^2z^2} = 18yz$

15. $9np^3$ by $-15n^3p^2 = \frac{9np^3}{-15n^3p^2} = \frac{3p}{-5n^2}$ or $\frac{-3p}{5n^2} = -\frac{3}{5}\frac{p}{n^2}$

16. $15rs^2t$ by $-27r^2st^3 = \frac{15rs^2t}{-27r^2st^3} = \frac{-5s}{9rt^2}$

17. $8abcdx^2y$ by $14adxy^2 = \frac{8abcdx^2y}{14adxy^2} = \frac{4bcx}{7y}$

18. $9efg^2hr$ by $24e^2fh^3r = \frac{9efg^2hr}{24e^2fh^3r} = \frac{3g^2}{8eh^2}$

19. $2a^3 + a^2$ by $a = \frac{2a^3+a^2}{a} = \frac{2a^3}{a} + \frac{a^2}{a} = 2a^2 + a$

20. $4x^4 - x^3$ by $x^2 = \frac{4x^4-x^3}{x^2} = \frac{4x^4}{x^2} - \frac{x^3}{x^2} = 4x^2 - x$

21. $36b^4 - 18b^2$ by $9b = \frac{36b^4-18b^2}{9b} = \frac{36b^4}{9b} - \frac{18b^2}{9b} = 4b^3 - 2b$

22. $49y^5 + 35y^3$ by $7y^2 = \frac{49y^5+35y^3}{7y^2} = \frac{49y^5}{7y^2} + \frac{35y^3}{7y^2} = 7y^3 + 5y$

23. $42x^2 + 28x$ by $7 = \frac{42x^2+28x}{7} = \frac{42x^2}{7} + \frac{28x}{7} = 6x^2 + 4x$

24. $56z^6 - 48z^3$ by $8 = \frac{56z^6 - 48z^3}{8} = \frac{56z^6}{8} - \frac{48z^3}{8} = 7z^6 - 6z^3$

25. $34x^5 - 51x^2$ by $17x^2 = \frac{34x^5 - 51x^2}{17x^2} = \frac{34x^5}{17x^2} - \frac{51x^2}{17x^2} = 2x^3 - 3$

26. $105w^6 + 63w^4$ by $21w^2 = \frac{105w^6 + 63w^4}{21w^2} = 5w^4 + 3w^2$

27. $24x^6 - 8x^4$ by $-4x^3 = \frac{24x^6 - 8x^4}{-4x^3} = \frac{24x^6}{-4x^3} - \frac{8x^4}{-4x^3} = -6x^3 + 2x$

28. $42y^7 - 24y^5$ by $6y^4 = \frac{42y^7 - 24y^5}{6y^4} = \frac{42y^7}{6y^4} - \frac{24y^5}{6y^4} = 7y^3 - 4y$

29. $5x^2y + 5xy^2$ by $xy = \frac{5x^2y + 5xy^2}{xy} = \frac{5x^2y}{xy} + \frac{5xy^2}{xy} = 5x + 5y$

30. $7a^2b - 7ab^2$ by $ab = \frac{7a^2b - 7ab^2}{ab} = \frac{7a^2b}{ab} - \frac{7ab^2}{ab} = 7a - 7b$

31. $10x^2y + 15xy^2$ by $5xy = \frac{10x^2y + 15xy^2}{5xy} = \frac{10x^2y}{5xy} + \frac{15xy^2}{5xy} = 2x + 3y$

32. $25p^2q - 15pq^2$ by $5pq = \frac{25p^2q - 15pq^2}{5pq} = \frac{25p^2q}{5pq} - \frac{15pq^2}{5pq} = 5p - 3q$

33. $ap^2q - 2pq$ by $pq = \frac{ap^2q - 2pq}{pq} = \frac{ap^2q}{pq} - \frac{2pq}{pq} = ap - 2$

34. $bx^2w + 3xw$ by $xw = \frac{bx^2w + 3xw}{xw} = \frac{bx^2w}{xw} + \frac{3xw}{xw} = bx + 3$

35. $a^2bc + abc$ by $abc = \frac{a^2bc + abc}{abc} = \frac{a^2bc}{abc} + \frac{abc}{abc} = a + 1$

36. $x^3yz - xyz$ by $xyz = \frac{x^3yz}{xyz} - \frac{xyz}{xyz} = x^2 - 1$

37. $9x^2y^2z - 3xyz^2$ by $-3xyz = \frac{9x^2y^2z}{-3xyz} - \frac{3xyz^2}{-3xyz} = -3xy + z$

38. $12a^2b^2c + 4abc^2$ by $-4abc = \frac{12a^2b^2c}{-4abc} + \frac{4abc^2}{-4abc} = -3ab - c$

39. $b^3x^2 + b^3$ by $-b = \frac{b^3x^2}{-b} + \frac{b^3}{-b} = -b^2x^2 - b^2$

40. $c^5y^3 - cy^2$ by $-c = \frac{c^5y^3}{-c} - \frac{cy^2}{-c} = -c^4y^3 + y^2$

41. $x^2y + xy - xy^2$ by $xy = \frac{x^2y}{xy} + \frac{xy}{xy} - \frac{xy^2}{xy} = x + 1 - y$

42. $ab^2 - ab + a^2b$ by $ab = \frac{ab^2}{ab} - \frac{ab}{ab} + \frac{a^2b}{ab} = b - 1 + a$

43. $18x^3y^2z - 24x^2y^3z$ by $-12x^2yz = \frac{18x^3y^2z}{-12x^2yz} - \frac{24x^2y^3z}{-12x^2yz} = -\frac{3}{2}xy + 2y^2$

44. $36a^4b^2c - 27a^2b^4c^2$ by $-27a^2bc = \frac{36a^4b^2c}{-27a^2bc} - \frac{27a^2b^4c^2}{-27a^2b^4c} = -\frac{4}{3}a^2b + b^3c$

45.
$$
\begin{array}{r}
x + 4 \\
x + 3 \overline{\smash{)}\, x^2 + 7x + 12} \\
\underline{x^2 + 3x} \\
4x + 12 \\
\underline{4x + 12} \\
0
\end{array}
$$

46.
$$
\begin{array}{r}
x - 3 \\
x + 4 \overline{\smash{)}\, x^2 + x - 12} \\
\underline{x^2 + 4x} \\
-3x - 12 \\
\underline{-3x - 12} \\
0
\end{array}
$$

47.
$$
\begin{array}{r}
x - 1 \\
x - 2 \overline{\smash{)}\, x^2 - 3x + 2} \\
\underline{x^2 - 2x} \\
-x + 2 \\
\underline{-x + 2} \\
0
\end{array}
$$

48.
$$
\begin{array}{r}
x + 3 \\
x - 5 \overline{\smash{)}\, x^2 - 2x - 15} \\
\underline{x^2 - 5x} \\
3x - 15 \\
\underline{3x - 15} \\
0
\end{array}
$$

49.
$$
\begin{array}{r}
x - 1 \\
x + 2 \overline{\smash{)}\, x^2 + x - 2} \\
\underline{x^2 + 2x} \\
-x - 2 \\
\underline{-x - 2} \\
0
\end{array}
$$

50.
$$
\begin{array}{r}
x - 2 \\
x + 3 \overline{\smash{)}\, x^2 + x - 6} \\
\underline{x^2 + 3x} \\
-2x - 6 \\
\underline{-2x - 6} \\
0
\end{array}
$$

51.
$$
\begin{array}{r}
2a + 1 \\
3a + 7 \overline{\smash{)}\, 6a^2 + 17a + 7} \\
\underline{6a^2 + 14a} \\
3a + 7 \\
\underline{3a + 7} \\
0
\end{array}
$$

52.
$$
\begin{array}{r}
b + 3 \\
4b - 2 \overline{\smash{)}\, 4b^2 + 10b - 6} \\
\underline{4b^2 - 2b} \\
12b - 6 \\
\underline{12b - 6} \\
0
\end{array}
$$

53.
$$
\begin{array}{r}
4y + 2 \\
2y - 3 \overline{\smash{)}\, 8y^2 - 8y - 6} \\
\underline{8y^2 - 12y} \\
4y - 6 \\
\underline{4y - 6} \\
0
\end{array}
$$

54.
$$
\begin{array}{r}
3t - 2 \\
4t + 3 \overline{)\, 12t^2 + \ t - 6\,} \\
\underline{12t^2 + 9t} \\
-8t - 6 \\
\underline{-8t - 6} \\
0
\end{array}
$$

55.
$$
\begin{array}{r}
x - 2 \\
x^2 + 2x - 1 \overline{)\, x^3 + 0x^2 - 5x + 2\,} \\
\underline{x^3 + 2x^2 - x} \\
-2x^2 - 4x + 2 \\
\underline{-2x^2 - 4x + 2} \\
0
\end{array}
$$

56.
$$
\begin{array}{r}
d \\
d^2 + d + 2 \overline{)\, d^3 + d^2 - 3d + 2\,} \\
\underline{d^3 + d^2 + 2d} \\
-5d + 2
\end{array}
$$

57.
$$
\begin{array}{r}
2a - 1 \\
3a^2 - 2a + 4 \overline{)\, 6a^3 - 7a^2 + 10a - 4\,} \\
\underline{6a^3 - 4a^2 + \ 8a} \\
-3a^2 + \ 2a - 4 \\
\underline{-3a^2 + \ 2a - 4} \\
0
\end{array}
$$

58.
$$
\begin{array}{r}
3y \ - 3 \\
3y^2 + 3y - 4 \overline{)\, 9y^3 + 0y^2 - 16y + \ 8\,} \\
\underline{9y^3 + 9y^2 - 12y} \\
-9y^2 - \ 4y + \ 8 \\
\underline{-9y^2 - \ 9y + 12} \\
5y - \ 4
\end{array}
$$

59.
$$
\begin{array}{r}
2x^2 \ + x - 1 \\
2x - 1 \overline{)\, 4x^3 + 0x^2 - 3x + 4\,} \\
\underline{4x^3 - 2x^2} \\
2x^2 - 3x + 4 \\
\underline{2x^2 - \ x} \\
-2x + 4 \\
\underline{-2x + 1} \\
3
\end{array}
$$

60.
$$
\begin{array}{r}
\frac{7}{3}p^2 - \frac{14}{9}p + \frac{46}{27} \\
3p + 2 \overline{)\, 7p^3 + \ 0p^2 + \ 2p - \ 5\,} \\
\underline{7p^3 + \frac{14}{3}p^2} \\
-\frac{14}{3}p^2 + \ 2p - \ 5 \\
\underline{-\frac{14}{3}p^2 - \frac{28}{9}p} \\
\frac{46}{9}p - \ 5 \\
\underline{\frac{46}{9}p + \frac{92}{27}} \\
-\frac{227}{27}
\end{array}
$$

61.
$$
\begin{array}{r}
r^2 - 3r \\
r + 2 \overline{)\, r^3 - \ r^2 - 6r + 5\,} \\
\underline{r^3 + 2r^2} \\
-3r^2 - 6r + 5 \\
\underline{-3r^2 - 6r} \\
5
\end{array}
$$

62.
$$
\begin{array}{r}
2c^2 \ + c + 3 \\
c - 2 \overline{)\, 2c^3 - 3c^2 + \ c - 4\,} \\
\underline{2c^3 - 4c^2} \\
c^2 + \ c - 4 \\
\underline{c^2 - 2c} \\
3c - 4 \\
\underline{3c - 6} \\
2
\end{array}
$$

63.
$$
\begin{array}{r}
x^3 \ + 3x^2 \ + 9x + 27 \\
x - 3 \overline{)\, x^4 \qquad\qquad\quad - 81\,} \\
\underline{x^4 - 3x^3} \\
3x^3 \qquad\qquad - 81 \\
\underline{3x^3 - 9x^2} \\
9x^2 \qquad - 81 \\
\underline{9x^2 - 27x} \\
27x - 81 \\
\underline{27x - 81} \\
0
\end{array}
$$

64.
$$
\begin{array}{r}
y^3 \ - 3y^2 \ + 9y - 27 \\
y + 3 \overline{)\, y^4 \qquad\qquad\quad - 81\,} \\
\underline{y^4 + 3y^3} \\
-3y^3 \qquad\qquad - 81 \\
\underline{-3y^3 - 9y^2} \\
9y^2 \qquad - 81 \\
\underline{9y^2 - 27y} \\
-27y - 81 \\
\underline{-27y - 81} \\
0
\end{array}
$$

65.

$$3x^2 + x - 2 \overline{)\begin{array}{r} 4x^3 - 3x^2 - 2x + 6 \\ 12x^5 - 5x^4 - 14x^3 + 23x^2 + 8x - 12 \end{array}}$$

$$\begin{array}{r} \underline{12x^5 + 4x^4 - 8x^3} \\ -9x^4 - 6x^3 + 23x^2 \\ \underline{-9x^4 - 3x^3 + 6x^2} \\ -3x^3 + 17x^2 + 8x \\ \underline{-3x^3 - x^2 + 2x} \\ 18x^2 + 6x - 12 \\ \underline{18x^2 + 6x - 12} \\ 0 \end{array}$$

66.

$$7a^3 + 2a - 7 \overline{)\begin{array}{r} 3a^4 - 2a^3 + 1 \\ 21a^7 - 14a^6 + 6a^5 - 25a^4 + 21a^3 + 0a^2 + 10a - 35 \end{array}}$$

$$\begin{array}{r} \underline{21a^7 + 6a^5 - 21a^4} \\ -14a^6 + 0a^5 - 4a^4 + 21a^3 + 0a^2 + 10a - 35 \\ \underline{-14a^6 - 4a^4 + 14a^3} \\ 7a^3 + 0a^2 + 10a - 35 \\ \underline{7a^3 + 2a - 7} \\ 8a - 28 \end{array}$$

67.

$$x - y \overline{)\begin{array}{r} x + y \\ x^2 - y^2 \end{array}}$$

$$\begin{array}{r} \underline{x^2 - xy} \\ xy - y^2 \\ \underline{} \\ 0 \end{array}$$

68.

$$a + b \overline{)\begin{array}{r} a - b \\ a^2 - b^2 \end{array}}$$

$$\begin{array}{r} \underline{a^2 + ab} \\ -ab - b^2 \\ \underline{-ab - b^2} \\ 0 \end{array}$$

69.

$$w - z \overline{)\begin{array}{r} w^2 + wz + z^2 \\ w^3 - z^3 \end{array}}$$

$$\begin{array}{r} \underline{w^3 - w^2z} \\ w^2z - z^3 \\ \underline{w^2z - wz^2} \\ wz^2 - z^3 \\ \underline{wz^2 - z^3} \\ 0 \end{array}$$

70.

$$x + y \overline{)\begin{array}{r} x^2 - xy + y^2 \\ x^3 + y^3 \end{array}}$$

$$\begin{array}{r} \underline{x^3 - x^2y} \\ -x^2y - y^3 \\ \underline{-x^2y - xy^2} \\ xy^2 + y^3 \\ \underline{xy^2 + y^3} \\ 0 \end{array}$$

71.

$$x + y \overline{)\begin{array}{r} x^2 + y^2 \\ x^3 + x^2y + xy^2 + y^3 \end{array}}$$

$$\begin{array}{r} \underline{x^3 + x^2y} \\ 0 + xy^2 + y^3 \\ \underline{xy^2 + y^3} \\ 0 \end{array}$$

72.

$$a - b \overline{)\begin{array}{r} a^2 + b^2 \\ a^3 - a^2b + ab^2 - b^3 \end{array}}$$

$$\begin{array}{r} \underline{a^3 - a^2b} \\ 0 + ab^2 - b^3 \\ \underline{ab^2 - b^3} \\ 0 \end{array}$$

73.

$$cd - 2 \overline{)\begin{array}{r} c^2d^2 + 2cd + 4 \\ c^3d^3 - 8 \end{array}}$$

$$\begin{array}{r} \underline{c^3d^3 - 2c^2d^2} \\ 2c^2d^2 - 8 \\ \underline{2c^2d^2 - 4cd} \\ 4cd - 8 \\ \underline{4cd - 8} \\ 0 \end{array}$$

74.

$$ef + 3 \overline{)\begin{array}{r} e^2f^2 - 3ef + 9 \\ e^3f^3 + 27 \end{array}}$$

$$\begin{array}{r} \underline{e^3f^3 + 3e^2f^2} \\ -3e^2f^2 + 27 \\ \underline{-3e^2f^2 - 9ef} \\ 9ef + 27 \\ \underline{9ef + 27} \\ 0 \end{array}$$

75.
$$\begin{array}{r}
x \;-\; y \\
x-y \overline{\smash{\big)}\ x^2 - 2xy + y^2} \\
\underline{x^2 -\; xy} \\
-\,xy + y^2 \\
\underline{-\,xy + y^2} \\
0
\end{array}$$

76.
$$\begin{array}{r}
a \;+\; 3b \\
a+3b \overline{\smash{\big)}\ a^2 + 6ab + 9b^2} \\
\underline{a^2 + 3ab} \\
3ab + 9b^2 \\
\underline{3ab + 9b^2} \\
0
\end{array}$$

77.
$$\begin{array}{r}
p^2 r - 2p \;+\; 3r^2 \\
5p-r \overline{\smash{\big)}\ 5p^3 r -\; p^2 r^2 - 10p^2 + 2pr + 15pr^2 - 3r^3} \\
\underline{5p^3 r -\; p^2 r^2} \\
0 - 10p^2 + 2pr + 15pr^2 - 3r^3 \\
\underline{-10p^2 + 2pr} \\
0 + 15pr^2 - 3r^3 \\
\underline{0 + 15pr^2 - 3r^3} \\
0
\end{array}$$

78.
$$\begin{array}{r}
4x^2 - 4xy + y^2 \\
2x-y \overline{\smash{\big)}\ 8x^3 - 12x^2 y + 6xy^2 - y^3} \\
\underline{8x^3 -\; 4x^2 y} \\
-8x^2 y + 6xy^2 - y^3 \\
\underline{-8x^2 y + 4xy^2} \\
2xy^2 - y^3 \\
\underline{2xy^2 - y^3} \\
0
\end{array}$$

79.
$$\begin{array}{r}
a \;+\; d \;+\; 4 \\
a-3d-1 \overline{\smash{\big)}\ a^2 - 2ad - 3d^2 + 3a - 13d - 8} \\
\underline{a^2 - 3ad \qquad\quad -\, a} \\
ad - 3d^2 + 4a - 13d - 8 \\
\underline{ad - 3d^2 \qquad\quad -\, d} \\
4a - 12d - 8 \\
\underline{4a - 12d - 4} \\
-4
\end{array}$$

80.
$$\begin{array}{r}
x \;-\; 2y \;+\; 3 \\
2x+3y-5 \overline{\smash{\big)}\ 2x^2 -\; xy - 6y^2 +\; x + 19y - 15} \\
\underline{2x^2 + 3xy \qquad\qquad -\, 5x} \\
-4xy - 6y^2 + 6x + 19y - 15 \\
\underline{-4xy - 6y^2 \qquad\quad +\, 10y} \\
6x + 9y - 15 \\
\underline{6x + 9y - 15} \\
0
\end{array}$$

81.
$$\begin{array}{r}
af \\
a-f \overline{\smash{\big)}\ a^2 f - af^2} \\
\underline{a^2 f - af^2} \\
0
\end{array}$$

82.
$$\begin{array}{r}
d^2 - 2m^2 \\
d-m \overline{\smash{\big)}\ d^3 - 2dm^2 + 2m^3 - d^2 m} \\
\underline{d^3 \qquad\qquad\qquad -\, d^2 m} \\
-2dm^2 + 2m^3 \\
\underline{-2dm^2 + 2m^3} \\
0
\end{array}$$

83.
$$\begin{array}{r}
a \;+\; b \;-\; c \\
a-b+c \overline{\smash{\big)}\ a^2 \qquad\qquad - b^2 + 2bc - c^2} \\
\underline{a^2 - ab + ac} \\
ab - ac - b^2 + 2bc - c^2 \\
\underline{ab \qquad\;\; - b^2 +\; bc} \\
-ac \qquad\;\; + bc - c^2 \\
\underline{-ac \qquad\;\; + bc - c^2} \\
0
\end{array}$$

84.
$$\begin{array}{r}
e \;-\; f \;+\; h \\
e+f+h \overline{\smash{\big)}\ e^2 \qquad\quad + 2eh \qquad\quad - f^2 + h^2} \\
\underline{e^2 \;+ ef + eh} \\
-ef + \; eh \qquad\quad - f^2 + h^2 \\
\underline{-ef \qquad\quad - fh - f^2} \\
eh + fh \qquad\quad + h^2 \\
\underline{eh + fh \qquad\quad + h^2} \\
0
\end{array}$$

85.
$$\begin{array}{r}
a^2 \;-\; a + 1 \\
a^2+a+2 \overline{\smash{\big)}\ a^4 + 0a^3 + 2a^2 -\; a + 2} \\
\underline{a^4 +\; a^3 + 2a^2} \\
-\, a^3 + 0a^2 -\; a + 2 \\
\underline{-\, a^3 -\; a^2 - 2a} \\
a^2 +\; a + 2 \\
\underline{a^2 +\; a + 2} \\
0
\end{array}$$

86.
$$\begin{array}{r}
x^3 \qquad\qquad -\, 1 \\
x^3-x+1 \overline{\smash{\big)}\ x^6 - x^4 \qquad + 2x^2 \qquad - 1} \\
\underline{x^6 - x^4 + x^3} \\
-x^3 + 2x^2 \qquad - 1 \\
\underline{-x^3 \qquad\quad + x - 1} \\
2x^2 - x
\end{array}$$

87. Reciprocal is $\dfrac{R_2 R_3 + R_1 R_3 + R_1 R_2}{R_1 R_2 R_3} = \dfrac{R_2 R_3}{R_1 R_2 R_3} +$

$\dfrac{R_1 R_3}{R_1 R_2 R_3} + \dfrac{R_1 R_2}{R_1 R_2 R_3} = \dfrac{1}{R_1} + \dfrac{1}{R_2} + \dfrac{1}{R_3}$

88. $\bar{e} = \dfrac{\sigma}{\frac{e_1+e_2+e_3}{e_1e_2}} = \dfrac{\sigma}{\frac{e_1}{e_1e_2} + \frac{e_2}{e_1e_2} - \frac{e_1e_2}{e_1e_2}} =$

$\dfrac{\sigma}{\frac{1}{e_2} + \frac{1}{e_1} - 1} = \dfrac{\sigma}{\frac{1}{e_1} + \frac{1}{e_2} - 1}$

89. $V_2 = V_1\left(1 + \frac{T_2 - T_1}{T_1}\right) = V_1\left(1 + \frac{T_2}{T_1} - 1\right) =$

$V_1\left(\frac{T_2}{T_1}\right) = \frac{V_1 T_2}{T_1}$

90. **(a)** Divide $x^3 + 6x^2 - 7x$ by $x + 7$. The quotient is $x^2 - x$, and the length of the rectangle is $x^2 - x$, ft.
(b) The area is $4^3 + 6 \cdot 4^2 - 7 \cdot 4 = 132$ ft^2, the length is $4 + 7 = 11$ ft, and the width is $4^2 - 4 = 12$ ft.

≡ 2.4 SOLVING EQUATIONS

1. $x - 7 = 32;\ (x-7)+7 = 32+7;\ x = 39$

2. $y - 8 = 41;\ (y-8)+8 = 41+8;\ y = 49$

3. $a + 13 = 25;\ (a+13) - 13 = 25 - 13;\ a = 12$

4. $b + 21 = 34;\ b + 21 - 21 = 34 - 21;\ b = 13$

5. $25 + c = 10;\ 25 + c - 25 = 10 - 25;\ c = -15$

6. $28 + d = 12;\ 28 + d - 28 = 12 - 28;\ d = -16$

7. $4.3 + w = 8.7;\ 4.3 + w - 4.3 = 8.7 - 4.3;\ w = 4.4$

8. $5.1 + z = 9.1;\ 5.1 + z - 5.1 = 9.1 - 5.1;\ z = 4$

9. $4x = 18;\ \frac{4x}{4};\ x = \frac{9}{2}$ or $4\frac{1}{2}$

10. $5y = 12;\ \frac{5y}{5} = \frac{12}{5};\ y = \frac{12}{5}$ or $2\frac{2}{5}$ or 2.4

11. $-3w = 24;\ \frac{-3w}{-3} = \frac{24}{-3};\ w = -8$

12. $6z = -42 = \frac{6z}{6} = \frac{-42}{6};\ w = -7$

13. $12a = 18;\ \frac{12a}{12} = \frac{18}{12};\ a = \frac{3}{2}$

14. $15b = -25;\ \frac{15b}{15} = \frac{-25}{15};\ b = -\frac{5}{3}$

15. $21c = -14;\ \frac{21c}{21} = \frac{-14}{21};\ c = -\frac{2}{3}$

16. $24d = 16;\ \frac{24d}{24} = \frac{16}{24};\ d = \frac{2}{3}$

17. $\frac{p}{3} = 5;\ 3 \cdot \frac{p}{3} = 5 \cdot 3;\ p = 15$

18. $\frac{r}{5} = 4;\ 5 \cdot \frac{r}{5} = 5 \cdot 4;\ r = 10$

19. $\frac{t}{4} = -6;\ 4 \cdot \frac{t}{4} = -6 \cdot 4;\ t = -24$

20. $\frac{s}{-3} = -5;\ -3\frac{s}{-3} = -5(-3);\ s = 15$

21. $4a + 3 = 11;\ 4a + 3 - 3 = 11 - 3;\ 4a = 8;\ \frac{4a}{4} = \frac{8}{4};\ a = 2$

22. $3b + 4 = 16;\ 3b + 4 - 4 = 16 - 4;\ 3b = 12;\ \frac{3b}{3} = \frac{12}{3};\ b = 4$

23. $7 - 8d = 39;\ 7 - 8d - 7 = 39 - 7;\ -8d = 32;\ \frac{-8d}{-8} = \frac{32}{-8};\ d = -4$

24. $9 - 7c = 44;\ 9 - 7c - 9 = 44 - 9;\ -7c = 35;\ \frac{-7c}{-7} = \frac{35}{-7};\ c = -5$

25. $4x - 3 = -37;\ 4x - 3 + 3 = -37 + 3;\ 4x = -34;\ \frac{4x}{4} = \frac{-34}{4};\ x = -\frac{17}{2}$

26. $5y - 4 = -41;\ 5y - 4 + 4 = -41 + 4;\ 5y = -37;\ \frac{5y}{5} = \frac{-37}{5};\ y = \frac{-37}{5}$

27. $2.3w + 4.1 = 13.3;\ 2.3w + 4.1 - 4.1 = 13.3 - 4.1;\ 2.3w = 9.2;\ \frac{2.3w}{2.3} = \frac{9.2}{2.3};\ w = 4$

28. $3.5z + 5.2 = 22.7;\ 3.5z + 5.2 - 5.2 = 22.7 - 5.2;\ 3.5z = 17.5;\ \frac{3.5z}{3.5} = \frac{17.5}{3.5};\ z = 5$

29. $2x + 5x = 28;\ 7x = 28;\ \frac{7x}{7} = \frac{28}{7};\ x = 4$

30. $3y + 8y = 121;\ 11y = 121;\ \frac{11y}{11} = \frac{121}{11};\ y = 11$

31. $3x = 7 - 10;\ 3x = -3;\ \frac{3x}{3} = \frac{-3}{3};\ x = -1$

32. $4x = 16 - 24;\ 4x = -8;\ \frac{4x}{4} = \frac{-8}{4};\ x = -2$

33. $3a + 2(a+5) = 45;\ 3a + 2a + 10 = 45;\ 5a + 10 = 45;\ 5a = 35;\ a = 7$

34. $4b + 3(7+b) = 56;\ 4b + 21 + 3b = 56;\ 7b + 21 = 56;\ 7b = 35;\ b = 5$

35. $4(6+c) - 5 = 21;\ 24 + 4c - 5 = 21;\ 4c + 19 = 21;\ 4c = 2;\ c = \frac{1}{2}$

36. $5(7+d) + 4 = 31;\ 35 + 5d + 4 = 31;\ 5d + 39 = 31;\ 5d = -8;\ d = \frac{-8}{5}$

37. $2(p-4)+3p=16$; $2p-8+3p=16$; $5p-8=16$; $5p=24$; $p=\frac{24}{5}$

38. $7(n-5)+4n=16$; $7n-35+4n=16$; $11n-35=16$; $11n=51$; $n=\frac{51}{11}$

39. $3x=2x+5$; $3x-2x=5$; $x=5$

40. $4y=3y+7$; $4y-3y=7$; $y=7$

41. $4w=6w+12$; $4w-6w=12$; $-2w=12$; $w=-6$

42. $7z=10z+42$; $7z-10z=42$; $-3z=42$; $z=-14$

43. $9a=54+3a$; $9a-3a=54$; $6a=54$; $a=9$

44. $8b=55+3b$; $8b-3b=55$; $5b=55$; $b=11$

45. $\frac{5x}{3}=\frac{4x}{2}-7$; $6\left(\frac{5x}{2}\right)=\left(\frac{4x}{3}-7\right)$; $3\cdot 5x=\frac{6\cdot 4x}{3}-6\cdot 7$; $15x=8x-42$; $15x-8x=-42$; $7x=-42$; $x=-6$

46. $\frac{3y}{7}=\frac{2y}{3}+4$; $21\left(\frac{3y}{7}\right)=21\left(\frac{2y}{3}+4\right)$; $3\cdot 3y=7\cdot 2y+21\cdot 4$; $9y=14y+84$; $-5y=84$; $y=\frac{-84}{5}$

47. $\frac{6p}{5}=\frac{3p}{2}+4$; $10\frac{6p}{5}=10\left(\frac{3p}{2}+4\right)$; $2\cdot 6p=5\cdot 3p+40$; $12p=15p+40$; $-3p=40$; $p=-\frac{40}{3}$

48. $\frac{5z}{3}=\frac{4z}{5}-3$; $15\cdot\frac{5z}{3}=15\left(\frac{4z}{5}-3\right)$; $5\cdot 5z=3\cdot 4z-15\cdot 3$; $25z=12z-45$; $13z=-45$; $z=-\frac{45}{13}$

49. $8n-4=5n+14$; $8n-5n=14+4$; $3n=18$; $n=6$

50. $9p-5=6p+37$; $9p-6p=37+5$; $3p=42$; $p=14$

51. $7r+3=11r-21$; $7r-11r=-21-3$; $-4r=-24$; $r=6$

52. $8s+7=15s-56$; $8s-15s=-56-7$; $-7s=-63$; $s=9$

53. $9t+6=3t-5$; $9t-3t=-5-6$; $6t=-11$; $t=-\frac{11}{6}$

54. $11u-4=6u+5$; $11u-6u=5+4$; $5u=9$; $u=\frac{9}{5}$

55. $\frac{6x-3z}{2}=\frac{7x+2}{3}$; $6\left(\frac{6x-3}{2}\right)$; $6\left(\frac{7x+2}{3}\right)$; $3(6x-3)=2(7x+2)$; $18x-9=14x+4$; $18x-14x=4+9$; $4x=13$; $x=\frac{13}{4}$

56. $\frac{4r-3}{3}=\frac{5r+2}{2}$; $6\left(\frac{4r-3}{3}\right)=6\left(\frac{5r+2}{2}\right)$; $2(4r-3)=3(5r+2)$; $8r-6=15r+6$; $8r-15r=6+6$; $-7r=12$; $r=-\frac{12}{7}$

57. $\frac{3t+4}{4}=\frac{2t-5}{2}$; $8\left(\frac{3t+4}{4}\right)=8\left(\frac{2t-5}{2}\right)$; $2(3t+4)=4(2t-5)$; $6t+8=8t-20$; $6t-8t=-20-8$; $-2t=-28$; $t=14$

58. $\frac{6a-5}{3}=\frac{7a+5}{6}$; $6\left(\frac{6a-5}{3}\right)=6\left(\frac{7a+5}{6}\right)$; $2(6a-5)=1(7a+5)$; $12a-10=7a+5$; $12a-7a=5+10$; $5a=15$; $a=3$

59. $3(x+5)=2x-3$; $3x+15=2x-3$; $3x-2x=-3-15$; $x=-18$

60. $2(y-3)=4+3y$; $2y-6=4+3y$; $2y-3y=4+6$; $-y=10$; $(-1)(-y)=(-1)10$; $y=-10$

61. $5(w-7)=2w+4$; $5w-35=2w+4$; $5w-2w=4+35$; $3w=39$; $w=13$

62. $6(z+5)=2z-9$; $6z+30=2z-9$; $6z-2z=-9-30$; $4z=-39$; $z=-\frac{39}{4}$

63. $\frac{x}{2}+\frac{x}{3}-\frac{x}{4}=2$; $12\left(\frac{x}{2}+\frac{x}{3}-\frac{x}{4}\right)=12\cdot 2$; $12\frac{x}{2}+\frac{12x}{3}-\frac{12x}{4}=24$; $6x+4x-3x=24$; $7x=24$; $x=\frac{24}{7}$

64. $\frac{9}{2}-\frac{p}{3}-\frac{p}{4}=3$; $12\frac{9}{2}-12\frac{p}{3}-12\frac{p}{4}=12\cdot 3$; $6p-4p-3p=36$; $-p=36$; $p=-36$

65. $\frac{4(a-3)}{5}=\frac{3(a+2)}{4}$; $20\left(\frac{4(a-3)}{5}\right)=20\left(\frac{3(a+2)}{4}\right)$; $4(4(a-3))=5(3(a+2))$; $16a-48=15a+30$; $16a-15a=30+48$; $a=78$

66. $\frac{5(b+4)}{3}=\frac{4(b-5)}{5}$; $15\left(\frac{5(b+4)}{3}\right)=15\left(\frac{4(b-5)}{5}\right)$; $5(5(b+4))=3(4(b-5))$; $25b+100=12b-60$; $25b-12b=-60-100$; $13b=-160$; $b=-\frac{160}{13}$

67. Solve $ax+b=3ax$ for x; $ax-3ax=-b$; $x(a-3a)=-b$; $x(-2a)=-b$; $x=\frac{-b}{-2a}=\frac{b}{2a}$

68. Solve $2by=6+4by$ for y; $2by-4by=6$; $-2by=6$; $y=\frac{6}{-2b}=-\frac{3}{b}$

69. Solve $ax-3a+x=5a$ for a; $ax+x=8a$; $ax-8a=-x$; $a(x-8)=-x$; $a=\frac{-x}{x-8}=\frac{x}{x-8}$ or $\frac{x}{8-x}$

70. Solve $2(by-c)=3\left(\frac{y}{2}-c\right)$ for y; $2by-2c=\frac{3y}{2}-3c$; $4by-4c=3y-6c$; $4by-3y=-2c$; $y(4b-3)=-2c$; $y=\frac{-2c}{4b-3}$

71. $\frac{3}{x} + \frac{4}{x} = 3$; $x\left(\frac{3}{x} + \frac{4}{x}\right) = x \cdot 3$; $3 + 4 = 3x$; $7 = 3x$; $3x = 7$; $x = \frac{7}{3}$

72. $\frac{5}{y} - \frac{3}{y} = 6$; $y\left(\frac{5}{y} - \frac{3}{y}\right) = 6y$; $5 - 3 = 6y$; $2 = 6y$; $6y = 2$; $y = \frac{2}{6} = \frac{1}{3}$

73. $\frac{3}{4p} + \frac{1}{p} = \frac{7}{4}$; $4p\left(\frac{3}{4p} + \frac{1}{p}\right) = 4p \cdot = \frac{7}{4}$; $3 + 4 = 7p$; $7 = 7p$; $p = 1$

74. $\frac{6}{5q} - \frac{2}{q} = \frac{6}{5}$; $5q\left(\frac{6}{5q} - \frac{2}{q}\right) = 5q\left(\frac{6}{5}\right)$; $6 - 10 = 6q$; $-4 = 6q$; $q = -\frac{4}{6} = -\frac{2}{3}$

75. $\frac{1}{x+1} - \frac{2}{x-1} = 0$; $(x+1)(x-1)\left[\frac{1}{x+1} - \frac{2}{x-1}\right] = 0(x+1)(x-1)$; $(x-1) - 2(x+1) = 0$; $x - 1 - 2x - 2 = 0$; $-x - 3 = 0$; $-x = 3$; $x = -3$

76. $\frac{3}{x+2} - \frac{4}{x-2} = 0$; $(x+2)(x-2)\left[\frac{3}{x+2} - \frac{4}{x-2}\right] = 0$; $3(x-2) - 4(x+2) = 0$; $3x - 6 - 4x - 8 = 0$; $-x - 14 = 0$; $-x = 14$; $x = -14$

77. $\frac{3}{2x} = \frac{1}{x+5}$; $(x+5)2x \cdot \frac{3}{2x} = (x+5)2x\left(\frac{1}{x+5}\right)$; $3(x+5) = 2x$; $3x + 15 = 2x$; $3x - 2x = -15$; $x = -15$

78. $\frac{4}{3x} = \frac{2}{x+1}$; $4(x+1) = 2 \cdot 3x$; $4x + 4 = 6x$; $4 = 2x$; $2x = 4$; $x = 2$

79. $\frac{2x+1}{2x-1} = \frac{x-1}{x-3}$; $(x-3)(2x+1) = (x-1)(2x-1)$; $2x^2 - 5x - 3 = 2x^2 - 3x + 1$; $2x^2 - 2x^2 - 5x + 3x = 1 + 3$; $-2x = 4$; $x = -2$

80. $\frac{2x+3}{2x+5} = \frac{5x+4}{5x+2}$; $(2x+3)(5x+2) = (5x+4)(2x+5)$; $10x^2 + 19x + 6 = 10x^2 + 33x + 20$; $19x - 33x = 20 - 6$; $-14x = 14$; $x = -1$

81. A common denominator of these expressions is 12. Multiplying by 12 and simplifying produces the following:
$(12)\left(\frac{6b-5a}{3}\right) + (12)\left(\frac{5a+b}{2}\right) = (12)\left(\frac{5(a+2b)}{4}\right) + (12)5$; $4(6b - 5a) + 6(5a + b) = 15(a + 2b) + 60$; $24b - 20a + 30a + 6b = 15a + 30b + 60$; $30b + 10a = 15a + 30b + 60$; $-5a = 60$; $a = -12$

82. A common denominator of these expressions is $15t$. Multiplying by $15t$ and simplifying produces the following:
$(15t)\left(\frac{3r+t}{5t}\right) - (15t)\left(\frac{r+5t}{3t}\right) = (15t)\left(\frac{2(r-4t)}{15t}\right) + (15t)$; $23(3r+t) - 5(r+5t) = 2(r-4t) + 30t$; $9r + 3t - 5r - 25t = 2r - 8t + 30t$; $4r - 22t = 2r + 22t$; $2r = 44t$

83. Multiplying by a common denominator of $6x$ produces the following:
$(6x)\left(\frac{2z+a}{3x}\right) - (6x)\left(\frac{9z+a}{6x}\right) = (6x)\left(\frac{z-a}{2x}\right) + (6x)\left(\frac{4a}{3x}\right)$; $2(2z+a) - (9z+a) = 3(z-a) + 2(4a)$; $4z + 2a - 9z - a = 3z - 3a + 8a$; $-5z + a = 3z + 5a$; $a - 5a = 3z + 5z$; $-4a = 8z$; $a = -2z$

84. Multiplying by a common denominator of $6y$ produces the following:
$(6y)\left(\frac{3p+2x}{2y}\right) - (6y)\left(\frac{5p-3x}{3y}\right) = (6y)\left(\frac{x-2p}{y}\right) + (6y)\left(\frac{3x+p}{6y}\right)$; $3(3p+2x) - 2(5p-3x) = 6(x-2p) + (3x+p)$; $9p + 6x - 10p + 6x = 6x - 12p + 3x + p$; $-p + 12x = 9x - 11p$; $12x - 9x = -11p + p$; $3x = -10p$; $x = -\frac{10}{3}p$

85. $F - 32 = \frac{9}{5}C$; $\frac{9}{5}C = F - 32$; $\frac{5}{9} \cdot \frac{9}{5}C = \frac{5}{9}(F - 32)$; $C = \frac{5}{9}(F - 32)$

86. **(a)** $v - v_0 = at$; $at = v - v_0$; $t = \frac{v - v_0}{a}$

   (b) $t = \frac{97-12}{9.8} = \frac{85}{9.8} \approx 8.7$ seconds

2.5 APPLICATIONS OF EQUATIONS

1. $\frac{79+85+74+x}{4} = 80$; $\frac{238+x}{4} = 80$; $238 + x = 320$; $x = 82$

2. $\frac{65+72+x}{3} = 75$; $\frac{137+x}{3} = 75$; $137 + x = 225$; $x = 88$

3. $\frac{85+82+x}{3} = 75$; $167 + x = 225$; $x = 58$; minimum of 60

4. $\frac{69+73+68+t}{4} = 72$; $210 + t = 288$; $t = 78°$

5. $.80w = 920$; $w = \frac{920}{.8} = \$1150$; $1150 - 920 = \$230$

6. $c + .15c = 920$; $1.15c = 920$; $c = \frac{920}{1.15} = \$800$

7. $0.30c = 1839$; $c = \frac{1839}{0.30}$; $c = \$6130$

8. $p + .03p + .04(p + .03p) = 227.63$; $1.03p + .04(1.03p) = 227.63$; $103p + .0412p = 227.63$; $1.0712p = 227.63$; $p = \frac{227.63}{1.0712}$; $p = \$212.50$

9. a = amount at 7.5%; $(4500 - a)$ = amount at 6%;
$0.075 \times a + .06(4500 - a) = 303$; $0.075a + 270 -$
$.06a = 303$; $0.015a = 33$; $a = \frac{33}{.015} = \$2200$
at 7.5%; $4500 - 2200 = \$2300$ at 6%

10.
$$a = \text{amount at } 8.2\%$$
$$6200 - a = \text{amount at } 7.25\%$$
$$0.082 \times a + .0725(6200 - a) = 482.75$$
$$.082a + 449.5 - .0725a = 482.75 0.0095a$$
$$0.0095a = 33.25$$
$$a\frac{33.25}{0.0095} = 3500 \text{ at } 8.2\%$$
$$6200 - 3500 = 2700 \text{ at } 7.25\%$$

11.
$$40 \times 8.50 + x(1.5)8.50 = 429.25$$
$$340 + 12.75x = 429.25$$
$$12.75x = 89.25$$
$$x = 80.25 \div 12.75 = 6.29 \text{ h}$$

12.
$$w = \text{weekend}$$
$$14 - w = \text{regular}$$
$$\text{overtime}$$
$$40 \times 8.20 + (14 - w)(1.5)8.20 + 2(w)8.20 = 524.80$$
$$328 + 172.220 - 12.30w + 16.40w = 524.80$$
$$500.2 + 4.1w = 524.80$$
$$4.1w = 24.60w = 6 \text{ h}$$

13. $d = rt$; $d = 38 \text{ mph} \times 7 \text{ h} = 266 \text{ mi}$

14. $d = rt$; $475 = 38t$; $\frac{475}{38} = t$; $t = 12.5 \text{ h}$

15.
$$t = \text{days of ships}$$
$$380 - 12 \times (t + 1) = 80t$$
$$380 - 12t - 12 = 80t$$
$$368 = 80t + 12t$$
$$368 = 92t \frac{368}{92} = t$$
$$t = 4 \text{ days}$$
$$80 \times 4 = 320 \text{ km}$$

16. $60(t) = 80\left(t - \frac{1}{2}\right)$; $60t = 80t - 40$; $-20t = -40$; $t = 2 \text{ h}$; $1:00 + 2 = 3:00 \text{ p.m.}$; $60 \times 2 = 120 \text{ mi}$

17. h = h together $\frac{1}{6} + \frac{1}{4} = \frac{1}{h}$; $12h = LCD 12h \cdot$
$\frac{1}{6} + 12h \cdot \frac{1}{4} = 12h \cdot \frac{1}{h} 2h + 3h = 125h = 12h =$
$\frac{12}{5} = 2\frac{2}{5}$; 2 hours 24 minutes

18. $\frac{1}{45} + \frac{1}{e} = \frac{1}{30}$; $90e\frac{1}{45} + 90e\frac{1}{e} = 90e \cdot \frac{1}{30}$; $2e + 90 = 3e$; $90 = e$; 90 days

19. $\frac{1}{4} + \frac{1}{2} = \frac{1}{h}$; $4h \cdot \frac{1}{4} + 4h \cdot \frac{1}{2} = 4h \cdot \frac{1}{h}$; $h + 2h = 4$; $3h = 4$; $h = \frac{4}{3} = 1 \text{ h } 20 \text{ min}$

20. $\frac{70}{12} + \frac{70}{4} = \frac{70}{h}$; $12h \cdot \frac{70}{12} + 12h \cdot \frac{70}{4} = 12h \cdot \frac{70}{h}$; $70h + 210h = 840$; $280h = 840$; $h = 3 \text{ min}$

21. $50 \times 0.86 = (50 + w) \times 0.40$; $43 = 20 + 0.40$; $23 = .4w$; $\frac{23}{.4} = w$; $57.5 \text{ mL} = w$

22. x = amount of 8% alcohol; $1\,000\,000 - x$ = amount of 14% alcohol; $0.08x + 0.14(1\,000\,000 - x) = 0.09 \times 1\,000\,000$; $140\,000 - 0.06x = 90\,000$; $-0.06x = -50,000$; $x = \frac{50\,000}{0.06} = 833\,333$ of 8%; $1\,000\,000 - 833\,333 = 166\,667$; the tank should be filled with $166\,667$ L of 14% alcohol and $833\,333$ L of 8% alcohol.

23. x = amount of 35%; $750 - x$ = amount of 75%; $0.35x + 0.75(750 - x) = 0.60 \times 750$; $0.35x + 562.5 - 0.75x = 450$; $-0.4x = -112.5$; $x = 281.25$ of 35% copper; $750 - 281.25 = 468.75$ of 75% copper.

24.
$$0.20x + 0.30 \times 50 = 0.27(50 + x)$$
$$0.2x + 15 = 13.5 + 0.27x$$
$$15 - 13.5 = .27x - 0.2x$$
$$1.5 = 0.07x$$
$$x = \frac{1.5}{0.07} = 21.428571 \text{ lb}$$

25. $850 - 500 = 350 \text{ lb}$; $500x = 350(20 - x)$; $500x = 7000 - 350x$; $(500 + 350)x = 7000$; $850x = 7000$; $x = 8.24 \text{ ft}$ from 500 lb end

26.
$$140\,000 - 60\,000 = 80\,000 \text{ N on rear wheels}$$
$$x = \text{distance from front}$$
$$60\,000x = 80\,000(4.2 - x)$$
$$60\,000x = 336\,000 - 80\,000x$$
$$60\,000x + 80\,000x = 336\,000$$
$$140\,000x = 336\,000x = 2.4 \text{ m}$$
$$\text{from front axle}$$

27.
$$x = \text{distance from left}$$
$$25x = 15(12 - x)$$
$$25x = 180 - 15x$$
$$40x = 180$$
$$x = 180 \div 40$$
$$x = 4.5 \text{ ft from the left end}$$

28. Let d be the location of the center of gravity in cm from the right end. The torque to the left of the center of gravity is $48 + 8(18 - d)$. To the right of the center of gravity, it is $8d$. Thus, $48 + 8(18 - d) = 8d$ or $192 = 16d$ and $d = 12$. The center of gravity is 12 cm from the right end.

29. Let d be the location of the center of gravity in cm from the right end. The torque to the left of the center of gravity is $4 + 2(8 - d)$. To the right of the center of gravity, it is $2d$. Thus, $4 + 2(8 - d) = 2d$ or $20 = 4d$ and $d = 5$. The center of gravity is 5 in. from the right end.

30. Let x be the thickness in cm of the right side. The torque to the left of the center of gravity is

$72 + 48x$. To the right of the center of gravity, it is $84x$. Thus, $72 + 48x = 84x$ or $72 = 36x$ and $x = 2$. The thickness of the right side is 2 cm.

31. Let n represent the number of ccs (cm^3) of pure alcohol that the nurse must add.
$$100\%(n) + 60\%(10) = 90\%(n + 10)$$
$$n + 6 = 0.9n + 9$$
$$0.1n = 3$$
$$n = 30$$

The nurse must add 30 cc of pure alcohol.

32. Let n represent the number of grams of the 80% solution of hydrochloric acid that must be added.
$$20\%(300 - n) + 80\%(n) = 25\%(300)$$
$$60 - 0.2n + 0.8n = 75$$
$$0.6n = 15$$
$$n = 25$$

The nurse must add 25 g of the 80% solution of hydrochloric acid.

CHAPTER 2 REVIEW

1. $8y - 5y = (8 - 5)y = 3y$

2. $4z + 15z = (4 + 15)z = 19z$

3. $7x - 4x + 2x - 8 = (7 - 4 + 2)x - 8 = 5x - 8$

4. $-9a + 4a - 3a + 2 = (-9 + 4 - 3)a + 2 = -8a + 2$

5. $(2x^2 + 3x + 4) + (5x^2 - 3x + 7) = (2 + 5)x^2 + (3 - 3)x + (4 + 7) = 7x^2 + 11$

6. $(3y^2 - 4y - 3) + (5y - 3y^2 + 6) = (3 - 3)y^2 + (-4 + 5)y + (-3 + 6) = y + 3$

7. $2(8x + 4) = 2 \cdot 8x + 2 \cdot 4 = 16x + 8$

8. $-3(4a - 2) = -3 \cdot 4a - 3(-2) = -12a + 6$

9. $-3(x - 1) = -3x + 1$ or $1 - 3x$

10. $-(2z + 5) = -2z - 5$

11. $(4x^2 + 3x) - (2x - 5x^2 + 2) = 4x^2 + 3x - 2x + 5x^2 - 2 = 9x^2 + x - 2$

12. $(7y^2 + 6y) - (6y - 7y^2) + 2y - 5 = 7y^2 + 6y - 6y + 7y^2 + 2y - 5 = 14y^2 + 2y - 5$

13. $2(a + b) - 3(a - b) + 4(a + b) = 2a + 2b - 3a + 3b + 4a + 4b = 3a + 9b$

14. $6(c - d) - 4(d - c) + 2(c + d) = 6c - 6d - 4d + 4c + 2c + 2d = 12c - 8d$

15. $(ax^2)(a^2x) = (aa^2)(x^2x) = a^{1+2}x^{2+1} = a^3x^3$

16. $(cy^3)(dy) = cdy^{3+1} = cdy^4$

17. $(9ax^2)(3x) = (9 \cdot 3)(ax^{2+1}) = 27ax^3$

18. $(6cy^2z)(2cz^3) = 6 \cdot 2c^{1+1}y^2z^{1+3} = 12c^2y^2z^4$

19. $4(5x - 6) = 4 \cdot 5x - 4 \cdot 6 = 20x - 24$

20. $3(12y - 5) = 3 \cdot 12y - 3 \cdot 5 = 36y - 15$

21. $2x(4x - 5) = 2x \cdot 4x - 2x \cdot 5 = 8x^2 - 10x$

22. $3a(6a + a^2) = 3a \cdot 6a + 3a \cdot a^2 = 18a^2 + 3a^3$

23. $(a + 4)(a - 4) = a^2 - 16$ (Difference of Squares)

24. $(x - 9)(x + 9) = x^2 - 81$ (Difference of Squares)

25. $(2a - b)(3a - b) = 6a^2 - 2ab - 3ab + b^2 = 6a^2 - 5ab + b^2$ FOIL

26. $(4x + 1)(3x - 7) = 12x^2 - 28x + 3x - 7 = 12x^2 - 25x - 7$ FOIL

27. $(3x^2 + 2)(2x^2 - 3) = 6x^4 - 9x^2 + 4x^2 - 6 = 6x^4 - 5x^2 - 6$ FOIL

28. $(4a^3 + 2)(6a - 3) = 24a^4 - 12a^3 + 12a - 6$ FOIL

29. $(x + 2)^2 = x^2 + 2 \cdot x \cdot 2 + 2^2 = x^2 + 4x + 4$
Binomial Squared

30. $(3-y)^2 = 3^2 - 2 \cdot 3 \cdot y + y^2 = 9 - 6y + y^2$
Binomial Squared

31. $5x(3x-4)(2x+1) = 5x[6x^2 + 3x - 8x - 4] = 5x[6x^2 - 5x - 4] = 30x^3 - 25x^2 - 20x$

32. $6a(4a+3)(a-2) = 6a[4a^2 - 8a + 3a - 6] = 6a[4a^2 - 5a - 6] = 24a^3 - 30a^2 - 36a$

33. $a^5 \div a^2 = \frac{a^5}{a^2} = a^{5-2} = a^3$

34. $x^7 \div x^3 = \frac{x^7}{x^3} = x^{7-3} = x^4$

35. $8a^2 \div 2a = \frac{8a^2}{2a} = \frac{8}{2} \cdot a^{2-1} = 4a$

36. $27b^3 \div 3b = \frac{27b^3}{3b} = \frac{27}{3} \cdot b^{3-1} = 9b^2$

37. $45a^2x^3 \div -5ax = \frac{45a^2x^3}{-5ax} = -9ax^2$

38. $52b^4c^2 \div -4b^2 = \frac{52b^4c^2}{-4b^2} = -13b^2c^2$

39. $(36x^2 - 16x) \div 2x = \frac{36x^2 - 16x}{2x} = \frac{36x^2}{2x} - \frac{16x}{2x} = 18x - 8$

40. $(39b^3 + 52b^5) \div 13b^2 = \frac{39b^3 + 52b^5}{13b^2} = \frac{39b^3}{13b^2} + \frac{52b^5}{13b^2} = 3b + 4b^3$

41.
$$\begin{array}{r} x - 4 \\ x+3 \overline{)\,x^2 - x - 12} \\ \underline{x^2 + 3x} \\ -4x - 12 \\ \underline{-4x - 12} \\ 0 \end{array}$$

42.
$$\begin{array}{r} x + 5 \\ x-6 \overline{)\,x^2 - x - 30} \\ \underline{x^2 - 6x} \\ 5x - 30 \\ \underline{5x - 30} \\ 0 \end{array}$$

43.
$$\begin{array}{r} x^2 + 3x + 9 \\ x-3 \overline{)\,x^3 \qquad\quad - 27} \\ \underline{x^3 - 3x^2} \\ 3x^2 \qquad - 27 \\ \underline{3x^2 - 9x} \\ 9x - 27 \\ \underline{9x - 27} \\ 0 \end{array}$$

44.
$$\begin{array}{r} 4a^2 - 8a + 16 \\ 2a+4 \overline{)\,8a^3 \qquad\qquad\qquad + 64} \\ \underline{8a^3 + 16a^2} \\ -16a^2 \qquad + 64 \\ \underline{-16a^2 - 32a} \\ 32a + 64 \\ \underline{32a + 64} \\ 0 \end{array}$$

45.
$$\begin{array}{r} x - y \\ x+y \overline{)\,x^2 \qquad - y^2} \\ \underline{x^2 + xy} \\ -xy - y^2 \\ \underline{-xy - y^2} \\ 0 \end{array}$$

46.
$$\begin{array}{r} y - a \\ y+a \overline{)\,y^2 \qquad - a^2} \\ \underline{y^2 + ya} \\ -ya - a^2 \\ \underline{-ya - a^2} \\ 0 \end{array}$$

47.
$$\begin{array}{r} x - 2y \\ x^2-y^2 \overline{)\,x^3 - 2x^2y - xy^2 + 2y^3} \\ \underline{x^3 \qquad\qquad - xy^2} \\ -2x^2y \qquad + 2y^3 \\ \underline{-2x^2y \qquad + 2y^3} \\ 0 \end{array}$$

48.
$$\begin{array}{r} a^2 + 4ab + b^2 \\ a-b \overline{)\,a^3 + 3a^2b + 2ba - 3ab^2 - b^3} \\ \underline{a^3 - a^2b} \\ 4a^2b + 2ba - 3ab^2 - b^3 \\ \underline{4a^2b \qquad - 4ab^2} \\ 2ba + ab^2 - b^3 \\ \underline{ab^2 - b^3} \\ 2ba \end{array}$$

49. $x + 9 - 9 = 47 - 9; \ x = 38$

50. $y - 19 + 19 = -32 + 19; \ y = -13$

51. $\frac{2x}{2} = \frac{15}{2}; \ x = \frac{15}{2}$

52. $\frac{-3y}{-3} = \frac{14}{-3}; \ y = -\frac{14}{3}$

53. $4 \cdot \frac{x}{4} = 4 \cdot 9; \ x = 36$

54. $\frac{y}{3} = -7; \ 3\frac{y}{3} = 3(-7); \ y = -21$

55. $4x - 3 = 17; \ 4x - 3 + 3 = 17 + 3; \ 4x = 20; \ \frac{4x}{4} = \frac{20}{4}; \ x = 5$

56. $7 + 8y = 23$; $7 + 8y - 7 = 23 - 7$; $8y = 16$; $\frac{8y}{8} = \frac{16}{8}$; $y = 2$

57. $3.4a - 7.1 = 8.2$; $3.4a - 7.1 + 7.1 = 8.2 + 7.1$; $3.4a = 15.3$; $a = \frac{15.3}{3.4} = 4.5$

58. $6.2b + 19.1 = 59.4$; $6.2b = 59.4 - 19.1$; $6.2b = 40.3$; $b = \frac{40.3}{6.2} = 6.5$

59. $4x + 3 = 2x$; $4x - 2x = -3$; $2x = -3$; $x = -\frac{3}{2}$

60. $7a - 2 = 2a$; $7a - 2a = 2$; $5a = 2$; $a = \frac{2}{5}$

61. $4b + 2 = 3b - 5$; $4b - 3b = -5 - 2$; $b = -7$

62. $7c + 9 = 12c - 4$; $7c - 12c = -4 - 9$; $-5c = -13$; $c\frac{13}{5}$

63. $\frac{4(x-3)}{3} = \frac{5(x+4)}{2}$; $6\left(\frac{4(x-3)}{3}\right) = 6\left(\frac{5(x+4)}{2}\right)$; $2 \cdot 4(x-3) = 3 \cdot 5(x+4)$; $8x - 24 = 15x + 60$; $-7x = 84$; $x = -12$

64. $\frac{3(y-7)}{5} = \frac{5(y+4)}{2}$; $10\left(\frac{3(y-7)}{5}\right) = 10\left(\frac{5(y+4)}{2}\right)$; $2 \cdot 3(y-7) = 5 \cdot 5(y+4)$; $6y - 42 = 25y + 100$; $-19y = 142$; $y = -\frac{142}{19}$

65. $\frac{2}{a} - \frac{3}{a} = 5$; $a\left(\frac{2}{a} - \frac{3}{a}\right) = a \cdot 5$; $2 - 3 = 5a$; $-1 = 5a$; $a = -\frac{1}{5}$

66. $\frac{4}{x} + \frac{5}{x} = \frac{1}{8}$; $8x\left(\frac{4}{x} + \frac{5}{x}\right) = 8x \cdot \frac{1}{8}$; $32 + 40 = x$; $72 = x$

67. $\frac{3}{2a} = \frac{3}{a+2}$; $2a(a+2)\frac{3}{2a} = 2a(a+2) \cdot \frac{3}{a+2}$; $(a+2)3 = 2a \cdot 3$; $3a + 6 = 6a$; $6 = 3a$; $a = 2$

68. $\frac{9}{4b} = \frac{12}{b+4}$; $(b+4)9 = 4b(12)$; $9b + 36 = 48b$; $36 = 39b$; $b = \frac{36}{39} = \frac{12}{13}$

69. $\frac{68 + 70 + 74 + x}{4} = 72$; $\frac{212 + x}{4} = 72$; $212 + x = 72 \times 4$; $212 + x = 288$; $x = 288 - 212 = 76$

70. $p + 0.065p = \$342.93$; $1.065p = 342.93$; $p = 342.93 \div 1.065 = \$322$

71. $d = rt$; $\frac{2718}{755} = \frac{755t}{755}$; $t = 3.6$ hours or 3 hrs 36 min

72. $t =$ satellite time; $t - 1\frac{3}{4} =$ shuttle time; $330t = 430(t - 1.75)$; $330t = 430t - 752.5$; $-100t = -752.5$; $t = 7.525$ hours or at $7:31\frac{1}{2}$ min $= 7:31:30$

73. $40 \text{ kg} \times 50\% + 15 = 20 + 15 = 35$ kg lead; $\frac{35}{40+15} = \frac{35}{55} = 63.64\%$ lead

74. $460 - 320 = 140N$; $320x = 140(8 - x)$; $320x = 1120 - 140x$; $(320 + 140)x = 1120$; $460x = 1120x = 2.43$ m from 320 force

75. Use the formula $F = ma$. $F_1 = 1538.6$ N; $F_2 = 1107.4$ N; $m_1 = \frac{F_1}{a} = \frac{1538.6}{9.8} = 157$ kg; $m_2 = \frac{F_2}{a} = \frac{1107.4}{9.8} = 113$ kg

76. Let n represent the number of grams of the 40% solution of the medicine that the pharmacist must use. Then $20 - n$ is the number of grams of the 70% solution that must be used.
$$40\%(n) + 70\%(20 - n) = 52\%(20)$$
$$0.4n + 14 - 0.7n = 10.4$$
$$-0.3n = -3.6$$
$$n = 12$$

The pharmacist should use 12 g of the 40% solution and $20 - 12 = 8$ g of the 70% solution.

≡ CHAPTER 2 TEST

1. $5x^2 + (2 - 5)x = 5x^2 - 3x$

2. $(4a^3 - 2b) - (3b + a^3) = 4a^3 - 2b - 3b - a^3 = (4a^3 - a^3) - (2b + 3b) = 3a^3 - 5b$

3. $4x + [3(x + y - 2) - 5(x - y)] = 4x + [3x + 3y - 6 - 5x + 5y] = (4x + 3x - 5x) + (3y + 5y) - 6 = 2x + 8y - 6$

4. $(4xy^3z)\left(\frac{1}{2}xy^{-2}z^2\right) = 4\left(\frac{1}{2}\right)x^{1+1}y^{3-2}z^{1+2} = 2x^2yz^3$

5. $(2b - 3)(2b + 3) = (2b)^2 - 3^2 = 4b^2 - 9$

6. $(x^3 + 3x) \div x$; $\frac{x^3 + 3x}{x} = \frac{x^3}{x} + \frac{3x}{x} = x^{3-1} + 3x^{1-1} = x^2 + 3x^0 = x^2 + 3$

7. $(6x^5 + 4x^3 - 1) \div 2x^2$; $\frac{6x^5 + 4x^3 - 1}{2x^2} = \frac{6x^5}{2x^2} + \frac{4x^3}{2x^2} - \frac{1}{2x^2}$; $= 3x^3 + 2x - \frac{1}{2x^2}$

8.

$$
\begin{array}{r}
3x^2 - 8x + 17 \\
x + 2 \overline{\smash{\big)}\,3x^3 - 2x^2 + x - 3} \\
\underline{3x^3 + 6x^2} \\
-8x^2 + x - 3 \\
\underline{-8x^2 - 16x} \\
17x - 3 \\
\underline{17x + 34} \\
-37
\end{array}
$$

9. $\dfrac{y}{3y+2} - \dfrac{4}{y-1} = \dfrac{y(y-1)}{(3y+2)(y-1)} - \dfrac{4(3y+2)}{(3y+2)(y-1)} =$

$\dfrac{y(y-1)-4(3y+2)}{(3y+2)(y-1)} = \dfrac{y^2-y-12y-8}{(3y+2)(y-1)} = \dfrac{y^2-13y-8}{(3y+2)(y-1)}$

10. $5x-8 = 3x$; $5x-3x = 8$; $2x = 8$; $x = \dfrac{8}{2}$; $x = 4$

11. $\dfrac{7x+3}{2} - \dfrac{9x-12}{4} = 8$; $4\left(\dfrac{7x+3}{2}\right) - 4\left(\dfrac{9x-12}{4}\right) = 4 \cdot 8$; $14x + 6 - 9x + 12 = 32$; $5x + 18 = 32$; $5x = 14$; $x = \dfrac{14}{5}$

12. $p + 0.7p = 74.85$; $1.07p = 74.85$; $p = 74.85 \div 1.07$; $p = \$69.95$

13. Let n be the amount of original solution to be removed. Then $9 - n$ is the amount left. You want to end with 60% of 9 qt or 5.4 qt of antifreeze. The antifreeze left from the original solution is 50% of $9 - n$, so $50\%(9-n)+n = 5.4$; $0.5(9-n)+n = 5.4$; $4.5 - 0.5n + n = 5.4$; $0.5n = 0.9$; $n = 1.8$ qt

CHAPTER
3
Geometry

≡ 3.1 LINES AND ANGLES

1. Using the proportion $\frac{D}{180°} = \frac{R}{\pi}$, we substitute $15°$ for D and solve for R. Thus, $\frac{15°}{180°} = \frac{R}{\pi}$ or $R = \frac{15°}{180°}\pi = \frac{1}{12}\pi = \frac{\pi}{12} \approx 0.2617994$.

2. Substituting $75°$ for D in the proportion $\frac{D}{180°} = \frac{R}{\pi}$, we obtain $\frac{75°}{180°} = \frac{R}{\pi}$ or $R = \frac{75°}{180°}\pi = \frac{5}{12}\pi = \frac{5\pi}{12} \approx 1.3089969$.

3. Multiplying $210°$ by $\frac{\pi}{180°}$, we obtain $\frac{210°}{180°}\pi = \frac{7}{6}\pi = \frac{7\pi}{6} \approx 3.6651914$.

4. We first convert $10°45'$ to decimal degrees. Now $45' = \frac{45°}{60} = 0.75°$, so $10°45' = 10.75°$. Multiplying $10.75°$ by $\frac{\pi}{180°}$, we obtain $\frac{10.75°}{180°}\pi = \frac{215}{3600} = \frac{43}{720}\pi = \frac{43\pi}{720} \approx 0.1876229$. You could also have multiplied $10.75°$ by 0.01745 with the result $10.75° \times 0.01745 = 0.1875875$. Notice that the two methods give different decimal approximations because 0.01745 is rounded off before it is multiplied.

5. Multiplying $85.4°$ by $\frac{\pi}{180°}$ produces $\frac{85.4°}{180°}\pi = \frac{854}{1800}\pi = \frac{427}{900}\pi \approx 1.4905112$. Multiplying $85.4 \times 0.01745 = 1.49023$.

6. As in Exercises 1–5, multiplying $45.6°$ by $\frac{\pi}{180°}$ produces $\frac{48.6°}{180°}\pi = \frac{27}{100}\pi \approx 0.84823$. Multiplying $48.6°$ by 0.01745 gives 0.84807.

7. Multiplying $163.5°$ by $\frac{\pi}{180°}$ we obtain $\frac{163.5°}{180°}\pi = \frac{109}{120}\pi \approx 2.8536133$. Multiplying $163.5 \times 0.01745 = 2.853075$.

8. Changing $242°35'$ to decimal degrees results in $242°35' = 242\frac{35}{60}° \approx 242.583°$. Multiplying this by $\frac{\pi}{180°}$ produces $\frac{2911}{2160}\pi \approx 4.2338779$ and multiplying 242.583 by 0.01745 yields 4.23307345.

9. Multiplying by $\frac{180°}{\pi}$ produces $\frac{4\pi}{3} \cdot \frac{180°}{\pi} = 240°$.

10. Multiplying by $\frac{180°}{\pi}$ results in $\frac{\pi}{6} \cdot \frac{180°}{\pi} = 30°$.

11. Multiplying by $\frac{180°}{\pi}$ yields $(1.3\pi)\left(\frac{180°}{\pi}\right) = 1.3 \times 180° = 234°$.

12. $2.15 \times \frac{180°}{\pi} = \frac{387°}{\pi} \approx 123.18593$ or multiplying by $57.296°$ results in $123.1864°$.

13. Multiplying by $57.296°$ results in $14.324°$.

14. Multiplying 1.1 by $57.296°$ yields $63.0256°$.

15. **(a)** To find the supplement of a $35°$ angle in degrees, subtract $35°$ from $180°$ with the result $180° - 35° = 145°$.
(b) Converting $145°$ to radians, multiply by $\frac{\pi}{180°}$ and obtain $\frac{145}{180}\pi = \frac{29}{36}\pi \approx 2.5307274$.

16. **(a)** Two angles are complementary if the sum of their measures is $90°$. The complement of a $65°$ angle is $90° - 65° = 25°$.
(b) Converting $25°$ to radians, we obtain $25° \times \frac{\pi}{180°} = \frac{25\pi}{180} = \frac{5\pi}{36} \approx 0.4363323$.

17. Angle A and the given angle are supplementary, so $\angle A = 180° - 33° = 147°$. Angle B and the given angle are vertical angles and so they are congruent. Thus, $\angle B = 33°$.

18. Angles A and the given angle are complementary, so $\angle A = 90° - 42° = 48°$. Angle B is a vertical angle with the indicated right angle, so $\angle B = 90°$.

19. Angle A and the given angle are alternate exterior angles and so congruent. Then, $\angle A = \frac{3\pi}{8}$. Angle

B and $\angle A$ are supplementary, so $\angle B = \pi - \frac{3\pi}{8} = \frac{5\pi}{8}$.

20. Angle A and the given angle are corresponding angles, so $\angle A = \frac{3\pi}{4}$. Angle B and $\angle A$ are supplementary, so $\angle B = \pi - \angle A = \pi - \frac{3\pi}{4} = \frac{\pi}{4}$.

21. Angle B and the given angle are opposite interior angles and so they are congruent. Thus, $\angle B = 70°30'$. Angle A and $\angle B$ are supplementary, so $\angle A = 180° - \angle B = 180° - 70°30' = 109°30'$.

22. Angle A and the given angle are complementary so $\angle A = \frac{\pi}{2} - \frac{5\pi}{16} = \frac{3\pi}{16}$. Angle B and the given angle are alternate interior angles and so they are congruent. Thus, $\angle B = \frac{5\pi}{16}$.

23. Since corresponding segments of transversals that intersect parallel lines are proportional, we have $\frac{AB}{BC} = \frac{DE}{EF}$. Substituting the given information in this proportion we obtain $\frac{AB}{3} = \frac{8}{5}$. Solving this proportion we obtain $AB = \frac{3\cdot 8}{5} = \frac{24}{5} = 4.8$.

24. As in Exercise 23, we have the proportion $\frac{AB}{BC} = \frac{DE}{EF}$. Substituting the given data, we obtain $\frac{AB}{22.65} = \frac{8.64}{15.1}$ and so, $AB = \frac{(22.65)(8.64)}{15.1} = 12.96$.

25. The three angles form a straight angle. We have $\angle A + 85° + \angle B = 180°$ and so $\angle A + \angle B = 95°$. Since $\angle A \cong \angle B$, we have $\angle A = \frac{95°}{2} = 47.5°$.

26. Here we have three parallel lines (the cables of the bridge) and two transversals (the girder and the deck), so corresponding segments are proportional. Thus, $\frac{AB}{4.5} = \frac{3.6}{4.8}$ and $AB = \frac{(4.5)(3.6)}{4.8} = 3.375$m.

27. Here $\omega = 2\pi f$ and, since $f = 60$, $\omega = 2\pi(60) = 120\pi$ rad/s.

28. **(a)** Since $\theta = 285°$ and $t = 0.6$ s, we have $\omega = \frac{285°}{0.6} = 475°$/s.

(b) Since $\theta = \frac{11\pi}{16}$ and $t = 0.9$ s, we have $\omega = \frac{11\pi/16}{0.9} = \frac{110\pi}{144} = \frac{55\pi}{72}$ rad/s.

29. Angle x is the supplement of $28°$. Hence, $x = 180° - 28° = 152°$.

30. The following program will work on a TI-82 graphing calculator.

☰ 3.2 TRIANGLES

1. Since $x + 35° + 75° = 180°$, we have $x = 180° - (35° + 75°) = 180° = 110° = 70°$.

2. Since $x + 122°20' + 15°30' = 180°$, we know that $x = 180° - (122°20' + 15°30') = 180° - 137°50' = 179°60' - 137°50' = 42°10'$.

3. This is an isosceles triangle. Base angles (the angles opposite the congruent sides) of an isosceles triangle are congruent. So, x and the $55°$ angle are congruent and $x = 55°$. To find y, we add to obtain $55° + y + x = 180°$. Solving for y produces $y = 180° - 2(55°) = 180° - 110° = 70°$.

4. This is a right triangle, so $x + 2x = 90°$ and so $x = 30°; 2x = 60°$.

5. This is an equilateral triangle, and so it is also an equiangular triangle and each angle is $\frac{180°}{3} = 60°$. Thus, $x = y = 60°$.

6. As in Exercise 3, base angles of isosceles triangles are congruent. Labeling the triangle as shown, we have $\angle A$ and $\angle ABD$ the base angles of $\triangle ABD$ and $\angle C$ and $\angle CBD$ the base angles of $\triangle CBD$. Thus, from $\triangle ABC$, we have $x + x + y + y = 180°$ or $2x + 2y = 180°$. Now, from $\triangle BCD$, we have $2x + 116° = 180°$, so $x = 32°$. Since $2x + 2y = 180°$ then $x + y = 90°$ and $y = 90° - x = 90° - 32° = 58°$.

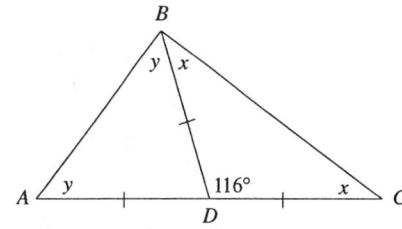

7.   Labeling the triangle as shown, we see that $\angle ABC$ is a vertical angle of 53.13°. Since $\triangle ABC$ is a right triangle, $x + 53.13° = 90°$ and we have $x = 90° - 53.13° = 36.87°$.

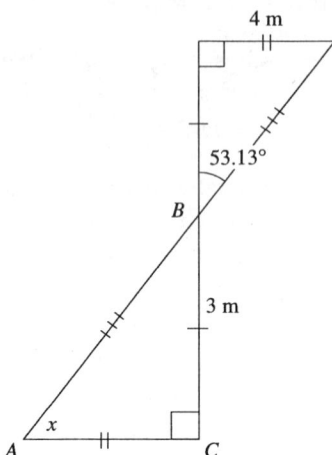

8.   Label the figure as shown. From the given information, we know that $\angle ABE \cong \angle CBD$ because they are vertical angles, and $\angle D \cong \angle E$ because they are right angles. That means that $\triangle ABE$ and $\triangle CBD$ are similar triangles and so corresponding sides are proportional, or $\frac{AB}{CB} = \frac{AE}{CD} = \frac{BE}{BD}$. Substituting the given data, we have $\frac{AB}{y} = \frac{x}{8} = \frac{2}{5}$. Solving the right-hand proportion, $\frac{x}{8} = \frac{2}{5}$, we obtain $x = \frac{16}{5} = 3.2$. Using the Pythagorean Theorem on $\triangle BCD$, we see that $5^2 + 8^2 = y^2$ and so $89 = y^2$ and $y = \sqrt{89}$.

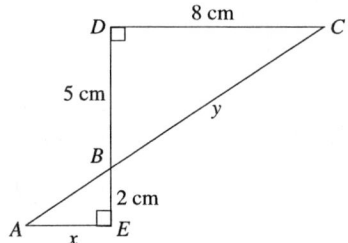

9.   The perimeter $P = 6.5 + 7.2 + 9.7 = 23.4$ units. The area, $A = \frac{1}{2}bh$. Here $b = 7.2$ and $h = 6.5$, so $A = \frac{1}{2}bh = \frac{1}{2}(7.2)(6.5) = 23.4$ sq. units (or you may write this as 23.4 units^2).

10. $P = 15 + 41 + 52 = 108$ units. $A = \frac{1}{2}bh$ with $b = 52$ and $h = 9; A = \frac{1}{2}(52)(9) = 234$ units^2.

11. $A = \frac{1}{2}bh = \frac{1}{2}(28)(9) = 126$ units^2. Notice that the dotted line $\overline{BD}$ is not part of the indicated

triangle. To determine the perimeter, we need to find the length of $\overline{AC}$. We will do this using the Pythagorean Theorem and $\triangle ADC$. That means we need the length of $\overline{DC} = DB + 28$. Using the Pythagorean Theorem, we determine $BD = \sqrt{15^2 - 9^2} = \sqrt{144} = 12$, so $DC = 40$ and $AC = \sqrt{9^2 + 40^2} = \sqrt{1681} = 41$. Using this information, $P = AB + BC + CA = 15 + 28 + 41 = 84$ units.

12.   Using the Pythagorean Theorem on $\triangle ABD$ we see that $BD = \sqrt{25^2 - 20^2} = \sqrt{225} = 15$ and from $\triangle CAD$ we find $DC = \sqrt{101^2 - 20^2} = \sqrt{9801} = 99$. Thus, $BC = BD + DC = 15 + 99 = 114$ units. $P = 25 + 101 + 114 = 240$ units and $A = \frac{1}{2}(20)(114) = 1140$ units^2.

13.   **(a)** The amount of fencing is the same as the perimeter, or $23.2 + 47.6 + 62.5 = 133.3$ m.
    (b) The area will provide the amount of sod that is needed. Using Hero's formula, we first determine the semiperimeter, $s = \frac{23.2 + 47.6 + 62.5}{2} = 66.65$. So, by Hero's formula, we find the area is $A = \sqrt{66.65(66.65 - 23.2)(66.65 - 47.6)(66.65 - 62.5)} = \sqrt{66.65(43.45)(19.05)(4.15)} = \sqrt{228,945.9742} \approx 478.5$ m^2.

14.   Using proportions we have

$$\frac{\text{height of building}}{\text{shadow of building}} = \frac{\text{height of stick}}{\text{shadow of stick}}$$

or $\frac{h}{123'6''} = \frac{3'}{2'3''}$. Changing $123'6''$ to $12.5'$ and changing $2'3''$ to $2.25'$ and substituting in the proportion, we obtain $\frac{h}{123.5'} = \frac{3'}{2.25'}$. Solving for h, produces $h = \frac{(123.5)(3)}{2.25} = \frac{370.5}{2.25} = 164\frac{2}{3}' = 164'8''$.

15.   Assuming that the building is perpendicular to the ground, we have a right triangle with the ladder forming the hypotenuse. So, $5^2 + s^2 = 15^2$ or $s^2 = 15^2 - 5^2 = 200$ and $s = \sqrt{200}m = 10\sqrt{2}m \approx 14.121$ m.

16.   The area of the triangle is $\frac{1}{2}$ the area of the rectangle, so $A = \frac{1}{2}(23)(16) = 184$ cm^2.

17.   The diagonal, d, is the hypotenuse of the right triangle formed by the two streets. So, $d = \sqrt{45^2 + 62^2} = \sqrt{5869} \approx 76.6$ ft.

18.   The length of the rafter, indicated by r on the drawing, is $x + 0.5$ m. The roof forms an isosceles triangle, so a perpendicular from the ridge to the base bisects the base. Thus, we have a right triangle with sides 3 and 4. Using the Pythagorean Theorem, we determine that $x = 5$ m and so, $r = 5.5$ m.

19. **(a)** Let s represent the length of the top of the second beam. The figure below shows two similar triangles $\triangle ABC$ and $\triangle ADE$. Similar triangles have proportional sides and altitudes, so $\frac{s}{2400} = \frac{5700}{2600}$ and we obtain $s = 5261.54$ mm (5262 mm to 4 significant digits). Let t represent the length of the top of the third beam. Reasoning as above, we have $\frac{t}{2400} = \frac{8800}{2600}$ or $t = 8123$ mm.
(b) Reasoning as in (a), we have the proportion $\frac{b}{2400} = \frac{10\,000}{2600}$ or $b = 9231$ mm.

20. What we have are three right triangles each with one side 175 m and the other sides the length of the cable from the antenna base. We need to determine the length of the hypotenuse of each triangle. For the first, the length is $\sqrt{175^2 + 50^2} = \sqrt{33125} \approx 182$. For the second, the length is $\sqrt{175^2 + 75^2} = \sqrt{36250} \approx 190.04$. The length of the third cable is $\sqrt{175^2 + 100^2} = \sqrt{40625} \approx 201.6$.

21. Since the total length of the conduit is 30 m, the length of the "bent" section is $30 - 8 - 10 = 12$ m. This bent section is the hypotenuse of a right triangle with one leg 5 m. Using the Pythagorean Theorem, we see that $x^2 + 5^2 = 12^2$ and so $x = \sqrt{119}$ $m \approx 10.9$ m.

22. The amount needed for the straight conduit is the length of the hypotenuse of a right triangle with one leg of length $x = \sqrt{119}$ and the other leg $8 + 5 + 10 = 23$. Using the Pythagorean Theorem, we obtain $\sqrt{119 + 23^2} = \sqrt{648} \approx 25.5$ m. Subtracting this from the length of the conduit that was actually used, produces $30 - 25.5 = 4.5$ m.

23. **(a)** The amount of pipe used is $97 + 62 + 53 = 212$ m.
(b) We need to find the length of the hypotenuse AB of a right triangle. One side of the triangle is $65 - 28 = 37$ m. The length of the other side is made up of three horizontal distances, two for the pipe sections on the sides of the ravine and the 62 m section across the bottom. For the left-hand section, the horizontal distance is $\sqrt{97^2 + 65^2} = \sqrt{5184} = 72$ m. For the right-hand section, the horizontal distance is $\sqrt{53^2 - 28^2} = \sqrt{2025} = 45$ m. So, the length of the base of the right triangle is $72 + 62 + 45 = 179$ m and the length of the hypotenuse is $\sqrt{179^2 + 37^2} = \sqrt{33410} \approx 182.8$ m.

24. The three triangles are similar, so $\frac{AB}{BF} = \frac{BC}{EB}$. Substituting the given information in this proportion, we have $\frac{24}{0.5} = \frac{BC}{0.5}$. Solving the porportion for BC produces $BC = 24$ ft. Similarly, we have the proportion $\frac{EB}{DG} = \frac{BC}{CD}$ or $\frac{0.5}{0.2} = \frac{24}{CD}$. Solving for CD results in $CD = \frac{24(0.2)}{0.5} = 9.6$ ft.

25. The triangle formed by the building and its shadow and the triangle formed by the meter stick and its shadow are similar. This leads to the proprotion $\frac{\text{height of building}}{\text{shadow of building}} = \frac{\text{height of meter stick}}{\text{shadow of meter stick}}$ or $\frac{\text{height of building}}{128.45 \text{ m}} = \frac{1 \text{ m}}{1.75 \text{ m}}$. Solving the proportion for the height of the building produces $\frac{128.45 \text{ m})(1 \text{ m})}{1.75 \text{ m}} = 73.4$ m.

26. The inside length is obtained using the Pythagorean theorem. So, the inside length is $\sqrt{12.7^2 + 9.6^2} = \sqrt{161.29 + 92.16} = \sqrt{253.45} \approx 15.9$. The inside length plus the lengths of the two ends add up to 17.9 cm.

▬ 3.3 OTHER POLYGONS

1. The figure is a square with $s = 15$ cm. $P = 4s = 4(15) = 60$ cm. $A = s^2 = 15^2 = 225$ cm^2.

2. The figure is a rectangle. We will let $a = 25$ and $b = 9$. $P = 2(a + b) = 2(25 + 9) = 2(34) = 68$ in. $A = ab = (25)(9) = 225$ in^2.

3. This figure is a parallelogram with $a = 15\frac{1}{4}, b = 33$, and $h = 11\frac{1}{2}$. $P = 2(a + b) = 2(15\frac{1}{4} + 33) = 2(48\frac{1}{4}) = 96.5$ in. $A = bh = 33(11\frac{1}{2}) = 379\frac{1}{2} = 379.5$ in^2.

4. Here the figure is a trapezoid. Since the left side is perpendicular to the base, it will act as the altitude. For this trapezoid, we have $b_1 = 23.4, b_2 = 24.9$, and $a = h = 11.2$. We must determine the length of c. If we drop a perpendicular from the top right vertex, we obtain a right triangle with legs 11.2 and $24.9 - 23.4 = 1.5$. Then $c = \sqrt{11.2^2 + 1.5^2} = 11.3$. Using these values, produces $P = a + b_1 + c + b_2 = 11.2 + 23.4 + 11.3 + 24.9 = 70.8$ mm. $A = \frac{1}{2}(b_1 + b_2)h = \frac{1}{2}(23.4 + 24.9)11.2 = \frac{1}{2}(48.3)11.2 = 270.48$ mm^2.

5. The figure for this exercise is a regular hexagon with $s = 11\frac{1}{2}$. Thus, $P = 6s = 6\left(11\frac{1}{2}\right) = 69$ in. Using the second formula for the area of regular hexagon that is given in Figure 3.27, we have $A = \frac{3\sqrt{3}}{2}s^2 = \frac{3\sqrt{3}}{2}\left(11\frac{1}{2}\right)^2 = \frac{3\sqrt{3}}{2}132.25 = 198.375\sqrt{3} \approx 343.6$ in^2.

6. This figure is a rhombus with $s = 12$ in and $h = 6\frac{1}{2}$ in. $P = 4s = 4(12) = 48$ in. $A = sh = 12\left(6\frac{1}{2}\right) = 78$ in^2.

7. Here we have a kite with $a = 10$ cm, $b = 12$ cm, $d_1 = 16$ cm, and $d_2 = 21$ cm. Thus, $P = 2(a + b) = 2(10 + 12) = 44$ cm. $A = \frac{1}{2}d_1 d_2 = \frac{1}{2}(16)(21) = 168$ cm^2.

8. This is a trapezoid with the bases being the two vertical lines. Thus, $a = h = 20.8$ mm, $b_1 = 213$ mm, and $b_2 = 318$ mm. As in Exercise 4, we use the Pythagorean Theorem to determine that $c = \sqrt{20.8^2 + 105^2} = \sqrt{11457.64} \approx 107$ mm. Using all this, we determine that $P = a + b_1 + c + b_2 = 20.8 + 213 + 107 + 318 = 658.8$ mm. $A = \frac{1}{2}(b_1 + b_2)h = \frac{1}{2}(213 + 318)(20.8) = 5522.4$ mm^2.

9. The figure is a parallelogram with $a = 15.7$ mm, $b = 21.2$ mm, and $h = 13.2$ mm. $P = 2(a + b) = 2(15.7 + 21.2) = 73.8$ mm. $A = bh = (21.2)(13.2) = 279.84$ mm^2.

10. Here the figure is a trapezoid with $a = 12\frac{1}{2}$ in, $b_1 = 14\frac{3}{4}$ in, $c = 13$ in, $b_2 = 23\frac{1}{4}$ in, and $h = 12$ in. Using the given formulas we determine that $P = a + b_1 + c + b_2 = 12\frac{1}{2} + 14\frac{3}{4} + 13 + 23\frac{1}{4} = 63.5$ in. $A = \frac{1}{2}(b_1 + b_2)h = \frac{1}{2}\left(14\frac{3}{4} + 23\frac{1}{4}\right)(12) = \frac{1}{2}(38)(12) = 228$ in^2.

11. Draw a line across one side of the "L" to make this into two rectangles. If the line is drawn as shown in the figure, one rectangle will be $26' \times 14'$ with an area of 364 ft^2 and the other rectangle will be $8' \times 10'$ with an area of 80 ft^2. Adding the two areas together, we obtain a total area of 444 ft^2.

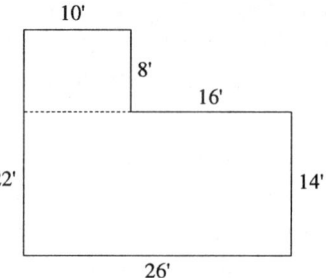

12. If there is no mortar and no space between the bricks, then this is almost the same as asking how many $8''$ bricks will divide into the perimeter of the patio. The patio has a perimeter of $22' + 10' + 8' + 16' + 14' + 26' = 96'$. Changing $96'$ to inches and dividing by 8, we obtain the number of bricks that will be needed $96 \times 12 \div 8 = 144$. But, if we use 144 bricks, there will be four corners with cut-outs. Each cut-out takes a half-brick, so the total is $144 + 4\left(\frac{1}{2}\right) = 146$ bricks.

13. Following the hint, we think of this as a rectangle that measures 100 mm by 300 mm for an area of 30 000 mm^2. Each trapezoid has bases $b_1 = 240$ mm and $b_2 = 260$ mm with $h = 40$ mm. Thus, each trapezoid has an area of $A = \frac{1}{2}(b_1 + b_2)h = \frac{1}{2}(240 + 260)(40) = 10\,000$ mm^2. Thus, the area of the I-beam is $A = 30\,000 - 2(10\,000) = 30\,000 - 20\,000 = 10\,000$ mm^2.

14. As in Exercise 13, we will think of this as a rectangle with two trapezoids removed. The area of the rectangle is $50' \times 45' = 2250$ ft^2. Each trapezoid has bases $30'$ and $35'$ and height $12'$ for an area $\frac{1}{2}(65)(12) = 390$ ft^2. Thus, the area of the highway support is $2250 - 2(390) = 1470$ ft^2.

15. Consulting Figure 3.27, we can see that the given dimension is $2h$ and so $h = \frac{7}{16}''$ and the area is $A = 2\sqrt{3}h^2 = 2\sqrt{3}\left(\frac{7}{16}\right)^2 = \frac{98}{256}\sqrt{3} \approx 0.663$ in^2. There are several ways we could determine s, but perhaps the easiest is to use the area formula $A = 3sh$. Since we know that $A = \frac{98}{256}\sqrt{3}$ and $h = \frac{7}{16}$, we find $s = A/3h = \left(\frac{98}{256}\sqrt{3}\right)/3\left(\frac{7}{16}\right) =$

$\frac{7}{8\sqrt{3}} \approx 0.505$. Since the perimeter is $P = 6s = 6\frac{7}{8\sqrt{3}} = \frac{21}{4\sqrt{3}} \approx 3.031$ in.

16. As in Exercise 15, $h = 7.5$ mm and so $A = 2\sqrt{3}(7.5)^2 = 112.5\sqrt{3} \approx 194.856$ mm^2. Again, $s = A/3h = \left(112.5\sqrt{3}\right)/3(7.5) = 5\sqrt{3}$ and so $P = 6s = 6\left(5\sqrt{3}\right) = 30\sqrt{3} \approx 51.96$ mm.

17. (a) The cross-sectional area of this trapezoid is $A = \frac{1}{2}(85 + 40)12 = 750$ ft^2.
(b) For the number of linear feet of concrete, we need the length along the bottom and the two sides. One side is $13'$ and the bottom is $40'$. We need to determine the length of the right-hand side. That side is the hypotenuse of a right triangle with sides 12 and 40 and so is $\sqrt{1744} \approx 41.76$ ft. Thus, the linear feet is $13 + 40 + 41.76 = 94.76$ ft.

18. The area of this floor is $20'6'' \times 15'3'' = 20.5' \times 15.25' = 312.625'$. Dividing by 9, we determine that there are 34.736 yd^2. Thus, the cost is $\$6.75 \times 34.736 = \$234.47$.

19. As in Exercise 11, we can draw a line to divide this into two rectangles. The area of one rectangle will be $5'' \times \frac{1}{4}'' = \frac{5}{4}$ in^2 and the area of the other rectangle is $3\frac{3}{4}'' \times \frac{1}{4}'' = \frac{15}{16}$ in^2. Adding these two areas together, we get the total area of $\frac{5}{4} + \frac{15}{16} = \frac{35}{16} = 2\frac{3}{16}$ in$^2 = 2.1875$ in^2 or 2.19 in^2.

20. You can put $12' \div 9'' = 144'' \div 9'' = 16$ tiles along one wall and $17'3'' \div 9'' = 207'' \div 9'' = 23$ tiles along the other side. This produces a total of $16 \times 23 = 368$ tiles.

21. Drawing a line across the top of the base of the "T" we divide the figure into two rectangles. The vertical stem of the "T" measures 18 mm by 391 mm and has an area of 7038 mm^2. The top of the "T" is a rectangle that measures 29 mm by 400 mm with an area of 11 600 mm^2. The total area of the "T" is $7038 + 11\,600 = 18638$ mm^2.

22. Draw a line across the house at the top of the vertical walls. The side is now divided into a rectangle which measures $38'6'' \times 12'$ with a trapezoid on top. The area of the rectangle is 462 ft^2. The area of the trapezoid is $A = \frac{1}{2}(38'6'' + 24'0'')(6'6'') = \frac{1}{2}(62.5')(6.5') = 203.125$ ft^2. Thus, the total area is $462 + 203.125 = 665.125$ ft^2.

≡ 3.4 CIRCLES

1. $A = \pi r^2 = \pi(4)^2 = 16\pi$ cm^2; $C = 2\pi r = 2\pi(4) = 8\pi$ cm

2. Here $d = 16$ in, so $r = 8$ in. $A = \pi r^2 = \pi(8)^2 = 64\pi$ in^2; $C = 2\pi r = 2\pi(8) = 16\pi$ in.

3. $A = \pi r^2 = \pi(5)^2 = 25\pi$ in^2; $C = 2\pi r = 2\pi(5) = 10\pi$ in.

4. Since $d = 23$ mm, $r = 11.5$ mm. Thus, $A = \pi(11.5)^2 = 132.25\pi$ mm^2 and $C = 2\pi(11.5) = 23\pi$ mm

5. $A = \pi(14.2)^2 = 201.64\pi$ mm^2; $C = 2\pi(14.2) = 28.4\pi$ mm.

6. $A = \pi\left(13\frac{1}{4}\right)^2 = \pi\left(\frac{53}{4}\right)^2 = \frac{2809}{16}\pi = 175.5625$ in^2; $C = 2\pi\left(13\frac{1}{4}\right) = 26.5\pi$ in.

7. Here $d = 24.20$ mm, so $r = 12.10$ mm and $A = \pi(12.10)^2 = 146.41\pi$ mm^2; $C = 2\pi(12.10) = 24.20\pi$ mm

8. Here $d = 23\frac{1}{2}$ in, so $r = 11\frac{3}{4}$ in. $A = \pi\left(11\frac{3}{4}\right)^2 = (11.75)^2\pi = 138.0625\pi$ in^2; $C = 2\pi\left(11\frac{3}{4}\right) = 23.5\pi$ in.

9. (a) This table top has a diameter, d, of 48 in and so, $r = 24$ in. The area, A, is $A = \pi(24)^2 = 576\pi$ in^2.
(b) The amount of metal edging is the same as the circumference. $C = \pi d = \pi(48) = 48\pi$ in ≈ 151 in.

10. Each turn of the wire is the same as the cirumference of the coil. So, the total circumference of 42 turns is $42C = 42(\pi d) = 42\pi(0.5) = 21\pi$ m ≈ 66 m.

11. The diameter of this drum is 764 mm and so its circumference is $764\pi \approx 2400$ mm. If the rivets are 75 mm apart, then there are $2400 \div 75 = 32$ rivets.

12. Because the pulleys have the same radii, the belt can be considered as the length of two opposite sides of a rectangle and the circumference of two semicircles. Thus, the length, as shown in the figure, is $2a + 2\pi r$. The length a is the distance between the centers of the pulleys, so $a = 11'9\frac{1}{4}''$ and $r = 3'2''$. The length is $2\left(11'9\frac{1}{4}''\right) + 2\pi(3'2'') = 23'6\frac{1}{2}'' + 6'4''\pi \approx 23.542 + 19.897 = 43.439$ ft.

13. Since $\theta = 110°$, we need to change it to radians. $\theta = 110° = \frac{11}{18}\pi$.

(a) The arc length $s = r\theta = 85\left(\frac{11}{18}\pi\right) = 51\frac{17}{18}\pi \approx 163.2$ mm.

(b) For the area, $A = \frac{1}{2}r^2\theta = \frac{1}{2}\left(85^2\right)\left(\frac{11}{18}\pi\right) \approx 6935.5$ mm^2.

14. This is a semicircle on top of a rectangle.
(a) The molding will be placed around the two vertical sides, the bottom of the rectangle, and the semicircle. This is $4'11'' + 4'2'' + 4'11'' + 2'1''\pi = 14' + 2'1''\pi \approx 20.545' \approx 20'6.5''$.
(b) The area of the rectangle is $4'2'' \times 4'11'' = 20\frac{35}{72} \approx 20.486$ ft^2. The area of the semicircle is $\frac{1}{2}\pi(2'1'')^2 \approx 6.818$ ft^2. So, the total area is 27.304 ft^2.

15. **(a)** The amount of molding is the length of $\overset{\frown}{AB}$ + the length of $\overset{\frown}{BC}$ + the length of segment $\overline{AC}$. Since $60° = \frac{\pi}{3}$, $m\overset{\frown}{AB} = 28''\left(\frac{\pi}{3}\right) = m\overset{\frown}{BC}$. So, the length of molding needed is $28''\left(\frac{\pi}{3}\right) + 28''\left(\frac{\pi}{3}\right) + 28'' \approx 86.643$ in.
(b) The area of this window is the area of the sector bounded by $\overset{\frown}{BC}$ and the sides $\overline{AB}$ and $\overline{AC}$ plus the area bounded by $\overline{AB}$ and $\overset{\frown}{AB}$. To find the area of this last figure, we will find the area of the sector bordered by $\overset{\frown}{AB}, \overline{BC}$ and $\overline{AC}$ and subtract the area of $\triangle ABC$. The area of each sector is $\frac{1}{2}(28'')^2\left(\frac{\pi}{3}\right) \approx 410.50$ in^2. Using Hero's formula, the area of $\triangle ABC$ is $\sqrt{42 \times 14^3} = 196\sqrt{3} \approx 339.48$. Subtracting this from the area of a sector produces $410.50 - 339.48 = 71.02$ and adding this to the area of a sector gives the desired answer, $410.50 + 71.02 = 481.52$ in^2.

16. **(a)** As in Exercise 15a, we need to find the length of the two arcs and add that to the length of three sides of the rectangle. The length of each arc is 2400 mm $\left(\frac{\pi}{3}\right)$. So, the amount of molding needed is $3000 + 2400 + 3000 + 2(2400)\left(\frac{\pi}{3}\right)$ mm $= 8400 + 1600\pi \approx 13,426.5$ mm.
(b) The area of the glass needed is the area of the rectangle plus the area of the curved shape above the rectangle. As in Exercise 15b, the area of the curved shape is approximately $(2400$ mm$)^2\left(\frac{\pi}{3}\right) - 2\,494\,153.16$ mm$^2 \approx 6\,031\,857.90 - 2\,494\,153.16 = 3\,537\,704.74$ mm^2. The area of the rectangular base is $(2400)(3000) = 7\,200\,000$ mm^2, so the total area is $3\,537\,704.74$ mm$^2 + 7\,200\,000$ mm$^2 = 10\,737\,704.74 \approx 10\,737\,705$ mm^2. Since 1 m$^2 = 1000$ mm$^2 = 1\,000\,000$ mm^2, then $10\,737\,705$ mm$^2 = 10\,737\,705 \div 1\,000\,000$ m$^2 \approx 10.738$ m^2.

17. **(a)** This is the area of a rectangle measuring $42'' \times 36''$ and two semicircles of radius $18''$. $A = 42 \times 36 + \pi(18)^2 = 1512 + 324\pi \approx 2529.876$ in^2.
(b) The edging needed is $2(42'') + 2\pi(18'') = 84'' + 36\pi'' \approx 197.1$ in.

18. The area is the difference between the area of the circles. So, $A = 40^2\pi - 30^2\pi = 1600\pi - 900\pi = 700\pi \approx 2199$ mm^2.

19. The area of a wire with a diameter of 8.42 mm is $4.21^2\pi = 17.7241\pi$. A wire with twice this area has an area of 35.4482π. If $\pi r^2 = 35.4482\pi$, then $r^2 = 35.4482$ and $r \approx 5.954$, so the diameter of the new wire is 11.91 mm.

20. The dotted line from the satellite to the Earth is tangent to the circle representing Earth, so it is perpendicular to the radii shown in the Figure. A line from the satellite to the Earth's center is the hypotenuse of two congruent right triangles and bisects the $120°$ angle. Thus, each triangle is a $30° - 60°$ right triangle. The length of the hypotenuse of a $30° - 60°$ right triangle is double the length of the shortest side, so the hypotenuse is 7920 mi. Subtracting the radius of the Earth, we see that the satellite is 3960 mi above the Earth.

21. **(a)** $3\frac{1}{7} - \pi \approx 0.00126$
(b) $\frac{3\frac{1}{7} - \pi}{\pi}\% \approx 0.040\%$

22. **(a)** $3.14 - \pi \approx -0.00159$
(b) $\frac{3.14 - \pi}{\pi}\% \approx -0.051\%$

3.5 GEOMETRIC SOLIDS

1. This is a cylinder with $r = 6''$ and $h = 18''$. $L = 2\pi rh = 2\pi(6'')(18'') = 216\pi$ in^2 ≈ 678.6 in^2. $T = 2\pi r(r+h) = 2\pi(6'')(6''+18'') = 2\pi(6'')(24'') = 288\pi$ in^2 ≈ 904.8 in^2. $V = \pi r^2 h = \pi(6'')^2(18'') = 648\pi$ in^3 $\approx 2,035.8$ in^3.

2. This figure is a rectangular prism with $h = 6$ cm, $l = 14$ cm, and $w = 8$ cm. $L = 2h(l + w) = 2(6$ cm$)(14$ cm $+ 8$ cm$) = 2(6$ cm$)(22$ cm$) = 264$ cm^2. $T = 2(lw + hw) = 2[(14$ cm$)(8$ cm$) + (14$ cm$)(6$ cm$) + (6$ cm$)(8$ cm$)] = 2(112$ cm$^2 + 84$ cm$^2 + 48$ cm$^2) = 488$ cm^2. $V = lwh = (14$ cm$)(8$ cm$)(6$ cm$) = 672$ cm^3.

3. This is a right circular cone with $r = 8$ mm, $h = 15$ mm, and $s = 17$ mm. $L = \pi rs = \pi(8$ mm$)(17$ mm$) = 136\pi$ mm^2 ≈ 427.3 mm^2. $T = \pi r(r + s) = \pi(8$ mm$)(8$ mm $+ 17$ mm$) = 200\pi$ mm^2 ≈ 628.3 mm^2. $V = \frac{1}{3}\pi r^2 h = \frac{1}{3}\pi(8$ mm$)^2(15$ mm$) = 320\pi$ mm^3 ≈ 1005.3 mm^3.

4. This is a square pyramid with $s = 17$ in and $h = 15$ in. Because it is a square, $p = 4(16$ in$) = 64$ in and $B = (16$ in$)^2 = 256$ in^2. $L = \frac{1}{2}ps = \frac{1}{2}(64$ in$)(17$ in$) = 544$ in^2. $T = \frac{1}{2}ps + B = L + B = 544$ in$^2 + 256$ in$^2 = 800$ in^2. $V = \frac{1}{3}Bh = \frac{1}{3}(256$ in$^2)(15$ in$) = 1,280$ in^3.

5. This triangular prism has $h = 21$ in and $p = 10$ in $+ 9$ in $+ 16$ in $= 35$ in. Using Hero's formula, we see that $B \approx 40.91$ in^2. $L = ph = (35$ in$)(21$ in$) = 735$ in^2; $T = L + 2B = 735$ in$^2 + 2(40.91$ in$^2) \approx 816.8$ in^2; $V = Bh = (40.91$ in$^2)(21$ in$) \approx 859.1$ in^3.

6. Here the figure is a frustum of a cone with $s = 10$ cm, $r_1 = 8$ cm, $r_2 = 16$ cm, and $h = 6$ cm. $L = \pi s(r_1 + r_2) = \pi(10$ cm$)(8$ cm $+ 16$ cm$) = 240\pi$ cm^2 ≈ 754 cm^2; $T = L + \pi r_1^2 + \pi r_2^2 = 240\pi$ cm$^2 + \pi(8$ cm$)^2 + \pi(16$ cm$)^2 = 240\pi + 64\pi + 256\pi$ cm$^2 = 560$ cm$^2\pi \approx 1759.3$ cm^2. $V = \frac{1}{3}\pi h \left(r_1^2 + r_2^2 + r_1 r_2\right) = \frac{1}{3}\pi(6)\left(8^2 + 16^2 + (8)(16)\right) = 2\pi(64 + 256 + 128) = (640 + 256)\pi = 896\pi \approx 2814.9$ cm^3.

7. This is a sphere with radius 8 cm. There is no lateral surface area. $T = 4\pi r^2 = 4\pi(8$ cm$)^2 = 256\pi$ cm$^2 \approx 804.25$ cm^2. $V = \frac{4}{3}\pi r^3 = \frac{4}{3}\pi(8$ cm$)^3 = \frac{2048}{3}\pi$ cm$^3 = 682.7\pi$ cm$^3 \approx 2144.7$ cm^3.

8. Here the figure is the frustum of a pyramid with square bases. We have $s = 50$ cm and $h = 45.83$ cm. We can determine that the perimeter of the top base is $p_1 = 80$ cm and its area is $B_1 = 400$ cm^2. Similarly, the perimeter of the bottom base is $p_2 = 240$ cm and its area is $B_2 = 3600$ cm^2. $L = \frac{s}{2}(p_1 + p_2) = \frac{50}{2}(80 + 240) = 25(320) = 8000$ cm^2. $T = L + B_1 + B_2 = 8000 + 400 + 3600 = 12\,000$ cm^2. $V = \frac{h}{3}\left(B_1 + B_2 + \sqrt{B_1 B_2}\right) = \frac{45.83}{3}\left(400 + 3600 + \sqrt{(400)(3600)}\right) \approx (15.28)(4000 + 1200) = 79\,456$ cm^3.

9. The cross-section is a triangle on top of a rectangle. The area of the rectangle is 24 ft^2 (Note: $24' \times 12'' = 24' \times 1' = 24$ ft^2) and the area of the triangle is $\frac{1}{2}(24')(6'') = \frac{1}{2}(24')(\frac{1}{2}'') = 6$ ft^2. Thus, the total area of the cross-section is $24 + 6 = 30$ ft^2. Multiplying this area by the length of the road gives a volume of $30 \times 5280 = 158,400$ ft^3. Thus, there are $158,400 \div 27 \approx 5866.7$ yd^3.

10. We can think of the I-beam as a prism with the "I" the base.
(a) The volume is the product of the area of the base and the altitude. The area of the base is 92 cm^2, so the volume is 92 cm$^2 \times 1000$ cm $= 92\,000$ cm^3.
(b) The painted portion amounts to the lateral surface area of the beam. The perimeter of the "I" is 96 cm, so the lateral surface area is 96 cm $\times 10\ m = 96\,000$ cm^2.
(c) The mass is the product of the volume ($92\,000$ cm^3) and the density (0.008 kg/cm^3) or 736 kg.

11. **(a)** The volume is $30(10)(12) = 3,600$ ft^3.
(b) $T = 2[(30)(10) + (30)(12) + (10)(12)] = 2(300 + 360 + 120) = 1,560$ ft^2

12. $V = \pi r^2 h = \pi(48)^2 140 = 322,560\pi$ ft$^3 \approx 1,013,352.1$ ft^3; $T = 2\pi r(r + h) = 2\pi(48)(48 + 140) = 96\pi(188) = 18,048\pi \approx 56,699.5$ ft^2.

13. Here $r = 33$ mm.
(a) $V = \pi(33)^2(95) = 103,455\pi \approx 325,013.5$ mm^3.
(b) Because the label overlaps 5 mm, we need to add 5 mm to the circumference in order to determine the area of the paper needed for the label. $L = (2\pi r + 5)h = [2\pi(33) + 5]95 = (66\pi + 5)95 = 6270\pi + 475 \approx 20\,172.79$ mm^2.

14. The volume is the area of the cross-section multiplied by the length of the pipe. The cross-section has an area of $\pi(57^2 - 42^2) = 1485\pi$ mm^2, so the volume is $(1485\pi \text{ mm}^2)(2000 \text{ mm}) = 2\,970\,000\pi$ $\text{mm}^3 \approx 9\,330\,530.2 \text{ mm}^3$.

15. $V = \frac{4}{3}\pi r^3 = \frac{4000}{3}\pi \text{ m}^3 \approx 1333.3\pi \text{ m}^3 \approx 4188.8 \text{ m}^3$.

16. $L = \frac{s}{2}(p_1 + p_2)$. Here the perimeter of the top base is $p_1 = 4(380 \text{ mm}) = 1520 \text{ mm}$ and the perimeter of the bottom base is $p_2 = 4(900 \text{ mm}) = 3600 \text{ mm}$. Since the slant height is 730 mm, the lateral surface area is $L = \frac{730}{2}(1520 + 3600) = (365)(5120) = 1\,868\,800 \text{ mm}^2 \approx 1.8688 \text{ m}^2$.

17. In Exercise Set 3.3, Exercise #14, we determined that the cross-sectional area of this highway support is 1470 ft². Since it is $2'6'' = 2.5$ ft thick, the total volume of the support is $(1470 \text{ ft}^2)(2.5 \text{ ft}) = 3675 \text{ ft}^3 = \frac{3675}{27} \text{ yd}^3 \approx 136.1 \text{ yd}^3$.

18. In Exercise Set 3.3, Exercise #17(a), we determined that the cross-sectional area of this river bed is 750 ft². Thus, the volume of a $\frac{1}{4}$-mile section is the product $(750 \text{ ft}^2)(1320 \text{ ft}) = 990,000 \text{ ft}^3$.

19. When the piece of copper is joined along the straight edges, a cone is formed as shown in this figure below. To find the volume of this cone, we need the radius of the base and the height of the cone. We can determine the radius of the base from the corcumference of the base. That circumference is the same as the arc length of the original sheet of copper. So, $C = 2\pi r = 85(110°)\left(\frac{\pi}{180°}\right)$ and $r = \frac{935}{36} \approx 25.97$. Using the Pythagorean Theorem, we get $h = \sqrt{85^2 - \left(\frac{935}{36}\right)^2} \approx 80.94$. So, $V = \frac{1}{3}\pi r^2 h \approx 18\,198.4\pi \approx 57\,172 \text{ mm}^3$.

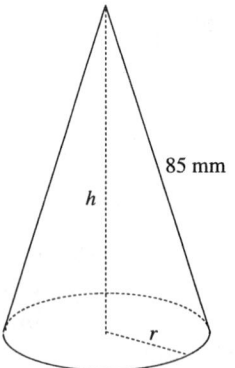

20. We need to find the lateral area of one frustum. By erecting a perpendicular to the bottom from its outer rim, we get a right triangle like the one shown. Its hypotenuse is the slant height of the frustum. So, $s = \sqrt{260^2 + 55^2} = \sqrt{70\,625} \approx$

265.8 mm. The lateral area is $L = \pi s (r_1 + r_2) \approx 265.8\pi(245 + 190) = 115\,623\pi \text{ mm}$. Thus, 500 tubs would require $500L \approx 57\,811\,500\pi \text{ mm}^2 \approx 181\,620\,180 \text{ mm}^2$.

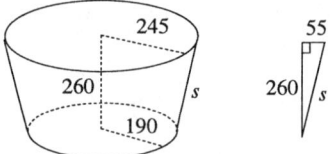

21. $L = \pi 12(1.5 + 4) = 66\pi \text{ in}^2 \approx 207.3 \text{ in}^2$.

22. This funnel is formed from two frustums of cones. The top frustum has a lateral area of $7.5\pi(5 + 1.5) = 48.75\pi \text{ cm}^2$. For the bottom frustum, the slant height, $s = \sqrt{13^2 + 1^2} = \sqrt{170}$. So, the lateral area is $\sqrt{170}\pi(1.5 + 0.5) = 2\sqrt{170}\pi \approx 26.08\pi \text{ cm}^2$. So, the total metal needed is $48.75\pi + 26.08\pi = 74.83\pi \approx 235.1 \text{ cm}^2$.

23. Draw the square formed by the centers of the silos. Each side of the square is 6 m, so the square has an area of 36 m². What remains outside the square are 4 sectors, each of which is $\frac{3}{4}$ of the original circle (silo). The area of each sector is $\frac{1}{2}r^2\theta = \frac{1}{2}(3^2)\left(\frac{3\pi}{2}\right) = \frac{27\pi}{4}$. So, the shaded area in Figure 3.34 is $4\left(\frac{27\pi}{4}\right) + 36 = 27\pi + 36 \text{ m}^2$. The volume of grain is this area multiplied by the height of the silo, 10 m, or $270\pi + 360 \text{ m}^3 \approx 1208.23 \text{ m}^3$ of grain.

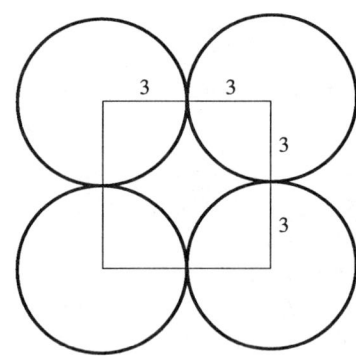

24. (a) The bore is 7 cm, so the radius of the cylinder is 3.5 cm and its height is 8 cm. Thus, the volume is $V = \pi r^2 h = \pi(3.5^2)8 = 98\pi \approx 307.9 \text{ cm}^3$.
(b) A 6-cylinder engine would have a total displacement of $6 \times 98\pi = 588\pi \approx 1847 \text{ cm}^3$.

25. Here $h = 4$ ft and $r = 6$ ft, so $V = \frac{1}{3}\pi r^2 h = \frac{1}{3}(6^2)4 = 48\pi \approx 150.8 \text{ ft}^3$.

26. The surface area of the cylinder is $2\pi rh$ or $2\pi \cdot 8 \cdot 40 \approx 2010.62$ ft^2. The top of the silo is a hemisphere, or half a sphere, and has a surface area of $\frac{1}{2} \cdot 4\pi r^2 = 2\pi \cdot 8^2 \approx 402.12$ ft^2. The bottom of the silo is a circle with a radius of 8 ft, and its area is $\pi r^2 = \pi \cdot 8^2 \approx 201.06$ ft^2. The total surface area of the silo is the sum of the three surface areas we just found, or $2010.62 + 402.12 + 201.06 = 2613.80$ ft^2.

≡ CHAPTER 3 REVIEW

1. Using the proportion $\frac{D}{180°} = \frac{R}{\pi}$, we substitute 27° for D and solve for R. Thus, $\frac{27°}{180°} = \frac{R}{\pi}$ or $R = \frac{27°}{180°}\pi = \frac{3}{20}\pi \approx 0.4712$.

2. Using the proportion $\frac{D}{180°} = \frac{R}{\pi}$, we substitute 212° for D and solve for R. Thus, $\frac{212°}{180°} = \frac{R}{\pi}$ or $R = \frac{212°}{180°}\pi = \frac{53}{45}\pi \approx 3.700$.

3. Multiplying by $\frac{180°}{\pi}$ produces 198°.

4. Multiplying by $\frac{180°}{\pi}$ produces $\frac{135°}{\pi} \approx 42.97°$.

5. The supplement of a 137° angle is $180° - 137° = 43°$.

6. The complement of a $\frac{\pi}{6}$ angle is $\frac{\pi}{2} - \frac{\pi}{6} = \frac{2\pi}{6} = \frac{\pi}{3}$.

7. Because $\angle A$ and the given angle are supplementary, $\angle A = 180° - 28° = 152°$. Angle A and $\angle B$ are corresponding angles formed by parallel lines cut (or intersected) by a transversal and so they are congruent. Thus, $\angle B = 152°$.

8. Angle B and the given angle are alternate exterior angles formed by parallel lines cut by a transversal and so they are congruent. Thus, $\angle B = \frac{7\pi}{16}$. Angle A and $\angle B$ are supplementary, so $\angle A = \pi - \angle B = \pi - \frac{7\pi}{16} = \frac{9\pi}{16}$.

9. Because corresponding segments are proportional, we have $\frac{2.7}{6.3} = \frac{AB}{7.7}$. Solving for AB produces $AB = \frac{(2.7)(7.7)}{6.3} = \frac{20.79}{6.3} = 3.3$.

10. Because the sum of the angles of a triangle is 180°, we have $x = 180° - (32° + 21°) = 180° - 53° = 127°$.

11. This is an isosceles triangle, so the base angles are congruent and $x = 180° - 2(75°) = 180° - 150° = 30°$.

12. Using the Pythagorean Theorem, the missing side is the hypotenuse, so its length is $\sqrt{120^2 + 209^2} = \sqrt{58081} = 241$.

13. In this right triangle, the hypotenuse is 37 so, using the Pythagorean Theorem, the length of the missing side is $\sqrt{37^2 - 16^2} = \sqrt{1113} \approx 33.36$.

14. The perimeter of this triangle is $P = 18 + 5 + 16 = 39$ units. Using Hero's formula, we determine that the semiperimeter is $s = \frac{P}{2} = \frac{39}{2} = 19.5$ units and so the area is
$A = \sqrt{19.5(19.5) - 18)(19.5 - 5)(19.5 - 16)} = \sqrt{19.5(1.5)(14.5)(3.5)} = \sqrt{1484.4375} \approx 38.53$ units^2.

15. This rectangle has an area of $A = (24.5)(12) = 294$ units^2. The perimeter is $P = 2(l + w) = 2(24.5 + 12) = 2(36.5) = 73$ units.

16. This parallelogram has a perimeter of $P = 2(l + w) = 2(23 + 14) = 2(37) = 74$ units. The area is $A = bh = 23(10) = 230$ units^2.

17. The circumference of a circle is $C = 2\pi r = 2\pi(9) = 18\pi \approx 56.55$ units. The area is $A = \pi r^2 = \pi 9^2 = 81\pi \approx 254.5$ units^2.

18. For this circle, we are given the diameter of 21 units and so the radius is $r = 10.5$ units. $C = \pi d = 21\pi \approx 65.97$ units. The area is $A = \pi r^2 = \pi(10.5)^2 = 110.25\pi \approx 346.36$ units^2.

19. Using the Pythagorean Theorem, we find that the length of the hypotenuse is $\sqrt{56^2 + 33^2} = \sqrt{4225} = 65$. Thus, $P = 33 + 56 + 65 = 154$ units and, since this is a right triangle, $A = \frac{1}{2}(33)(56) = 924$ units^2.

20. Converting 110° to radians, we get $110° = \frac{11}{18}\pi$, so the arc length is $r\theta = 7\left(\frac{11}{18}\pi\right) = \frac{77}{18}\pi \approx 4.277\pi \approx 13.44$ units. Adding the lengths of the two straight sides (radii) we get a perimeter of $13.44 + 7 + 7 = 27.44$ units. The area is $A = \frac{1}{2}r^2\theta = \frac{1}{2}7^2\left(\frac{11}{18}\pi\right) = \frac{539}{36}\pi \approx 14.972\pi \approx 47.04$ units^2.

21. The perimeter of this trapezoid is $P = 15 + 12 + 10.6 + 29.5 = 67.1$ units. The area is

$\frac{1}{2}(b_1 + b_2)h = \frac{1}{2}(12 + 29.5)(9) = \frac{1}{2}(41.5)(9) = 186.75$ units^2.

22. This is a rectangular prism. We will let $l = 15.5$, $w = 8.4$, and $h = 35$. $L = 2h(l + w) = 2(35)(15.5 + 8.4) = 2(35)(23.9) = 1673$ units^2. $T = 2(lw + lh + hw) = 2[(15.5)(8.4) + (15.5)(35) + (35)(8.4)] = 2[130.2 + 542.5 + 294] = 2(966.7) = 1933.4$ units^2. $V = lwh = (15.5)(8.4)(35) = 4557$ units^3.

23. This is a right circular cylinder with $r = 12.1$ and $h = 9.6$. $L = 2\pi rh = 2\pi(12.1)(9.6) = 232.32\pi \approx 729.85$ units^2. $T = 2\pi r(r + h) = 2\pi(12.1)(12.1 + 9.6) = 2\pi(12.1)(21.7) = 525.14\pi \approx 1649.78$ units^2. $V = \pi r^2 h = \pi(12.1)^2(9.6) = \pi(146.41)(9.6) = 1405.54\pi \approx 4415.62$ units^3.

24. This figure is a right triangular prism with $p = 19 + 12 + 21 = 52$ units, $s = \frac{p}{2} = 26$ units, and $h = 17$ units. Using Hero's formula, we determine $B = \sqrt{26(26 - 21)(26 - 19)(26 - 12)} = \sqrt{26(5)(7)(14)} + \sqrt{12740} \approx 112.87$ units^2. $L = ph = (52)(17) = 884$ units^2. $T = ph + 2B = 884 + 2(112.87) = 1109.74$ units^2. $V = Bh \approx (112.87)(17) = 1918.79$ units^3.

25. This is a square pyramid with $p = 4(40) = 160$ units, $B = 40^2 = 1600$ units^2, $s = 29$ units, and $h = 21$ units. $L = \frac{1}{2}ps = \frac{1}{2}(160)(29) = 2320$ units^2. $T = \frac{1}{2}ps + B = L + B = 2320 + 1600$ units$^2 = 3920$ units^2. $V = \frac{1}{3}Bh = \frac{1}{3}(1600)(21) = 11,200$ units^3.

26. This figure is a right circular cone with $r = 8$ units and $h = 15$ units. Using the Pythagorean Theorem, we determine that $s = \sqrt{15^2 + 8^2} = \sqrt{225 + 64} = \sqrt{289} = 17$ units. $L = \pi rs = \pi(8)(17) = 136\pi \approx 427.26$ units^2. $T = \pi r(r + s) = \pi(8)(8 + 17) = 200\pi \approx 628.32$ units^2. $V = \frac{1}{3}\pi r^2 h = \frac{1}{3}\pi 8^2(15) = 320\pi \approx 1005.31$ units^3.

27. This is a frustum of a cone with $r_1 = 7$ units, $r_2 = 19$ units, $h = 16$ units, and $s = 20$ units.

$L = \pi s(r_1 + r_2) = \pi(20)(7 + 19) = 520\pi \approx 1633.6$ units^2. $T = \pi[r_1(r_1 + s) + r_2(r_2 + s)] = \pi[(7)(7 + 20) + (19)(19 + 20)] = \pi(189 + 741) = 930\pi \approx 2921.68$ units^2. $V = \frac{1}{3}\pi h \left(r_1^2 + r_2^2 + r_1 r_2\right) = \frac{1}{3}\pi(16)\left(7^2 + 19^2 + (7)(19)\right) = \frac{1}{3}\pi(16)(49 + 361 + 133) = \frac{1}{3}\pi(16)(543) = 2,896\pi \approx 9098.05$ units^3.

28. This is a frustum of a square pyramid with $p_1 = 4(24) = 96$ units, $p_2 = 4(14) = 56$ units, $s = 13$ units, and $h = 12$ units. Since the bases are squares we can determine that $B_1 = 24^2 = 576$ units^2 and $B_2 = 14^2 = 196$ units^2. $L = \frac{1}{2}s(p_1 + p_2) = \frac{1}{2}(13)(96 + 56) = 988$ units^2. $T = \frac{1}{2}s(p_1 + p_2) + B_1 + B_2 = L + B_1 + B_2 = 988 + 576 + 196 = 1760$ units^2. $V = \frac{1}{3}h(B_1 + B_2 + \sqrt{B_1 B_2}) = \frac{1}{3}(12)\left(576 + 196 + \sqrt{(576)(196)}\right) = \frac{1}{3}(12)(576 + 196 + 336) = 4432$ units^3.

29. This sphere has a radius of 9 units. Because it is a sphere, it has no lateral area. $T = 4\pi r^2 = 4\pi 9^2 = 324\pi \approx 1017.88$ units^2. $V = \frac{4}{3}\pi r^3 = \frac{4}{3}\pi 9^3 = 972\pi \approx 3053.63$ units^3.

30. Using the Pythagorean Theorem, we can find the length of each of the four cables. The length of c_1, the cable that is attached 50 m above the ground, is $c_1 = \sqrt{75^2 + 50^2} \approx 90.14$ m. The length of c_2, the cable that is attached 100 m above the ground, is $c_2 = \sqrt{75^2 + 100^2} = 125.00$ m. c_3, the cable that is attached 150 m above the ground, has a length of $c_3 = \sqrt{75^2 + 150^2} \approx 167.71$ m. The length of c_4, the cable that is attached to the top of the antenna, is $c_4 = \sqrt{75^2 + 175^2} \approx 190.39$ m. One set of four cables has a total length of about $90.14 + 125.00 + 167.71 + 190.39 = 573.24$ m. The three sets have a total length of $3 \times 573.24 = 1,719.72$ m.

31. Assuming $\overline{AF}, \overline{BE},$ and $\overline{CD}$ are all parallel, we obtain the proportion $\frac{d}{100} = \frac{352.6}{82}$. Solving for d, we get $d = \frac{100 \times 352.6}{82} = 430$ ft.

32. **(a)** Only 95 cm of the strip is used to make washers. Each strip can make $\frac{95}{3.20} \approx 29.7$ washers. Do not round this up to 30 washers because the last 5 cm of each strip cannot be used. So, each strip will only produce 29 washers. Hence, to make 100,000 washers, a total of $\frac{100,000}{29} \approx 3,448.3$ strips are needed. Since 3,448 strips will give a total of 99,992 washers, we must need 3,449 strips.
(b) Each washer has a volume of $\pi(R^2 - r^2)h = \pi(1.6^2 - 0.525^2)(0.240) = 1.722378$ cm^3. The total metal in the 100,000 washers is 172,237.8 cm^3. (Actually, since the 3,449 strips will produce 100,021 washers, a total of 172,238 cm^3 will probably be used to produce washers.)

(c) Each strip contains $3.20 \times 100 \times 0.240 = 76.8$ cm^3 of metal. The total volume of metal in the 3,449 strips is $3,449 \times 76.8 = 264{,}883.2$ cm^3. In (b), we found that 172,237.8 cm^3 of metal is actually used for the washers. Since 172,238.8 cm^3 is used to produce 100,000 washers, the rest must be scrap. Thus, there is a total of $264{,}883.2 - 172{,}237.8 = 92{,}645.4$ cm^3 in scrap metal.

CHAPTER 3 TEST

1. $35\left(\frac{\pi}{180}\right) = \frac{35\pi}{180} = \frac{7\pi}{36} \approx 0.611$

2. $\left(\frac{7\pi}{15}\right)\left(\frac{180}{\pi}\right) = \frac{7\pi180}{15\pi} = 84°$

3. $180° - 76° = 104°$

4. $133°$

5. $\frac{5}{7} = \frac{9}{AB}$. Solving for $\overline{AB}$, we get $\overline{AB} = \frac{63}{5} = 12.6$ units.

6. Using the Pythagorean Theorem, with $c = 3.9$ and $b = 3.6$, we have $a^2 + 3.6^2 = 3.9^2$ or $a^2 = 3.9^2 = 3.6^2 = 15.21 - 12.96 = 2.25$. Since $\sqrt{2.25} = 1.5$, $a = 1.5$ cm.

7. This rectangle has $l = 36.4$ cm and $w = 14.3$ cm, so $P = 2(l+w) = 2(36.4 + 14.3) = 2(50.7)101.4$ cm.

8. This trapezoid has been divided into a rectangle and two triangles. The length of the bottom is $4 + 12 + 4 = 20$ in and so the perimeter is $P = 12 + 5 + 20 + 5 = 42$ in.

9. This is a right trapezoid, so $h = 10.4$ cm and the area is $A = \frac{1}{2}h(b_1 + b_2) = \frac{1}{2}(10.4)(21.6 + 32.4) = 280.8$ cm^2

10. This circle has a radius of $r = 9$ in and so its area is $A = \pi r^2 = \pi 9^2 = 81\pi \approx 254.47$ in^2

11. $\frac{14}{11.6} = \frac{x}{120}$. Solving for x produces $\frac{(14)(120)}{11.6} \approx 144.8$.

12. The circumference of the tire is $C = 2\pi r = \pi d = \pi(62)$ cm. To travel 25 m, the tire will have to make $\frac{25 \text{ m}}{\pi(62) \text{ cm}}$ revolutions. Converting 25 m to cm, we obtain 2500 cm. Substituting this in the previous statement, we get that the number of revolutions is $\frac{2500 \text{ cm}}{\pi(62) \text{ cm}} \approx 12.84$.

13. $V = lwh = (162)(52)(231) = 1\,945\,944$ mm$^2 = 1\,945\,944$ mm$^3 = 1945.944$ cm^3

14. $V = \frac{4}{3}\pi r^3$. We are given the diameter as $35'$, so $r = \frac{35}{2} = 17.5$ ft. Thus $V = \frac{4}{3}\pi(17.5)^3 \approx 7145.83333\pi \approx 22{,}449.29750$ feet.

15. The label covers the lateral surface area of the cylinder plus an overlap. The lateral surface area is $L = 2\pi rh = \pi dh$ and the amount needed for the overlap is $0.8h$. So, the area is $\pi(6.5)(9.5) + (0.8)(9.5) = 61.75\pi + 7.6 \approx 201.59$ cm^2.

CHAPTER

4

Functions and Graphs

≣ 4.1 FUNCTIONS

1. This is a function because each value of x produces just one value for y.

2. This is a function because each value of x produces just one value for y.

3. This is not a function of x because the same value of x can give two different values for y. For example, the ordered pairs (3, 2) and (3, −2) both satisfy the relation.

4. This is a function because each value of x produces just one value for y.

5. Notice that $\sqrt{x}$ is defined only when $x \geq 0$. Since $\sqrt{x}$ produces a unique number for each x, this is a function for $x \geq 0$.

6. This is a function because each value of x produces just one value for y.

7. Since the square of any real number is nonnegative, this equation will be defined only for those values of x where $2x - 7 \geq 0$ or when $x \geq \frac{7}{2}$. However, this is not a function since (4, 1) and (4, −1) both satisfy the equation.

8. This is not a function of x because the same value of x can give two different values for y. For example, the ordered pairs (10, 5) and (10, −15) both satisfy the relation.

9. Yes, because for each value of x there is just one value of y.

10. Yes, because for each value of x there is just one value of y.

11. No, because when $x = 4$ there are two values of y: − 2 and 2, or when $x = 1, y$ is 1 or −1.

x	−40	−30	−20	−10	0	10	20	30	40
y	−40	−22	−4	14	32	50	68	86	104

12. Yes, because for each value of x there is just one value of y.

13. Domain: $\{-3, -2, -1, 0, 1, 2\}$;
Range: $\{-7, -5, -3, -1, 1, 3\}$

14. Domain: $\{-3, -2, -1, 0, 1, 2\}$;
Range: $\{-6, -5, -2, 3\}$

x	−3	−2	−1	0	1	2	3
y	−25	−6	1	2	3	10	25

15. Domain: $\{-3, -2, -1, 0, 1, 2, 3\}$;
Range: $\{-25, -6, 1, 2, 3, 10, 25\}$

16. Domain: $\left\{-2, -\frac{3}{2}, -1, -\frac{1}{2}, 0, \frac{1}{2}, 1, \frac{3}{2}, 2\right\}$;
Range: $\left\{-5, -\frac{7}{3}, -2, -1, 5, 20, 25\right\}$
In Exercises 17–28, determine the domain of each function. In Exercises 17–22, also determine the range of each function.

17. Domain and range are both all real numbers.

18. Domain and range are both all real numbers.

19. Domain is all real numbers except 5; that is, the domain is $\{x : x \neq 5\}$. Rewrite $y = \frac{x}{x-5}$ as $y = 1 + \frac{5}{x-5}$. Since $\frac{x}{x-5}$ can never be 0, then $y = 1 + \frac{x}{x-5}$ can never be 1. Thus, the range is all real numbers except 1; that is, the range is $\{y : y \neq 1\}$.

20. Domain is all real numbers except −13; that is, the domain is $\{x : x \neq -13\}$. Rewrite $y = \frac{x+1}{x+13}$ as $y = 1 + \frac{-12}{x+13}$. Since $\frac{-12}{x+13}$ can never be 0, then $y = 1 + \frac{-12}{x+13}$ can never be 1. Thus, the range is all real numbers except 1; that is, the range is $\{y : y \neq 1\}$.

21. Domain is all nonnegative real numbers, $x \geq 0$. Range is all real numbers greater than or equal to −2, that is the range is $\{y : y \geq -2\}$.

22. Domain is all real numbers. Range is all nonnegative real numbers; that is, the range is $\{y : y \geq 0\}$.

23. Domain is all real numbers except 2 and -3; that is, the domain is $\{x : x \neq 2, x \neq -3\}$.

24. Domain is all real numbers except -1 and 5; that is, the domain is $\{x : x \neq -1, x \neq 5\}$.

25. Domain is all real numbers except -1 and 5; that is, the domain is $\{x : x \neq -1, x \neq 5\}$.

26. Domain is all real numbers except 2 and -3; that is, the domain is $\{x : x \neq 2, x \neq -3\}$.

27. Domain is all real numbers except -4 and 2; that is, the domain is $\{x : x \neq -4, x \neq 2\}$.

28. Domain is all real numbers except 2 and -5; that is, the domain is $\{x : x \neq 2, x \neq -5\}$.

29. Any function with a denominator of $x - 5$ will work as long as the numerator is defined for all real numbers. For example, $y = \frac{1}{x-5}$ is one answer. Other possible answers include $y = \frac{x}{x-5}$, $y = \frac{x-5}{x-5}$, and $y = -10(x-5)^2$.

30. Any function with a denominator of $x + 7$ will work as long as the numerator is defined for all real numbers. For example, $y = \frac{1}{x+7}$ is one answer. Other possible answers include $y = \frac{x}{x+7}, y = \frac{x+7}{x+7}, y = \frac{-10}{(x+7)^2}$, and $y = \frac{-10(x+5)}{(x+7)^3}$.

31. Any function with a denominator of $(x+1)(x-2)$ will work as long as the numerator is defined for all real numbers. For example, $y = \frac{1}{(x+1)(x-2)}$ is one answer. Other possible answers include $y = \frac{x}{(x+1)(x-2)}, y = \frac{(x+1)(x-2)}{(x+1)(x-2)}, y = \frac{-10}{(x+1)(x-2)^2}$, and $y = \frac{3x(x+1)}{(x+1)^2(x-2)^5}$.

32. Any function with a denominator of $(x+3)(x-1)(x-4)$ will work as long as the numerator is defined for all real numbers. For example, $y = \frac{1}{(x+3)(x-1)(x-4)}$ is one answer. Other possible answers include $y = \frac{x}{(x+3)(x-1)(x-4)}, y = \frac{(x+3)(x-1)(x-4)}{(x+3)(x-1)(x-4)}, y = \frac{-10}{(x+3)(x-1)(x-4)^2}$, and $y = \frac{3(x+3)^7(x-1)}{(x+3)^2(x-1)(x-4)^5}$.

33. Any function with an even root will work if the quantity under the radical sign is nonnegative when $x \geq 5$. Since $x \geq 5$ is the same as $x - 5 \geq 0$, then $y = \sqrt{x-5}$ is one answer. Other possible answers include $y = \sqrt[4]{x-5}$, $y = -7\sqrt[4]{x-5}$, and $y = \sqrt[6]{x-5}$.

34. Any function with an even root where the quantity under the radical sign is nonnegative when $x \leq 7$. Since $x \leq 7$ is the same as $7 - x \geq 0$, then $y = \sqrt{7-x}$ is one answer. Other possible answers include $y = \sqrt[4]{7-x}$, $y = -5\sqrt[4]{7-x}$, and $y = \sqrt[6]{7-x}$.

35. Any function with $x - 4$ in the denominator and with an even root where the quantity under the radical sign is nonnegative when $x \geq -1$. For example, $y = \frac{\sqrt{x+1}}{x-4}$ is one answer and $y = \frac{\sqrt{x+1}}{\sqrt[3]{x-4}}$ is another.

36. Any function in which the denominator include *both* $x - 3$ and an even root where the quantity under the radical sign is positive when $x > -5$. For example, $y = \frac{1}{(x-3)\sqrt{x+5}}$ is one answer and $y = \frac{12x}{\sqrt[3]{x-3}\sqrt{x+5}}$ is another.

37. $f(x) = 3x - 2$, so $f(0) = 3(0) - 2 = 0 - 2 = -2$.

38. $f(2) = 3(2) - 2 = 6 - 2 = 4$.

39. $f(-3) = 3(-3) - 2 = -9 - 2 = -11$.

40. $f\left(\frac{4}{3}\right) = 3\left(\frac{4}{3}\right) - 2 = 4 - 2 = 2$

41. Since $f(0) = -2$ and $f(-3) = -11$, then $f(0) + f(-3) = -2 + -11 = -13$.

42. $f(b) = 3(b) - 2 = 3b - 2$.

43. $f(b+3) = 3(b+3) - 2 = 3b + 9 - 2 = 3b + 7$.

44. Since $f(x+h) = 3(x+h) - 2 = 3x + 3h - 2$ and $f(x) = 3x - 2$, then $f(x+h) - f(x) = (3x + 3h - 2) - (3x - 2) = (3x + 3h - 2 - 3x + 2 = 3h$.

45. Since $g(x) = x^2 - 5x$, then $g(0) = 0^2 - 5(0) = 0 - 0 = 0$.

46. Since $g(x) = x^2 - 5x$, then $g(-3) = (-3)^2 - 5(-3) = 9 - (-15) = 9 + 15 = 24$.

47. $g(2) = 2^2 - 5(2) = 4 - 10 = -6$.

48. $g(0.4) = (0.4)^2 - 5(0.4) = 0.16 - 2 = -1.84$

49. $g\left(\frac{2}{5}\right) = \left(\frac{2}{5}\right)^2 - 5\left(\frac{2}{5}\right) = \left(\frac{4}{25}\right) - 2 = \frac{4}{25} - \frac{50}{25} - \frac{46}{25}$

50. $g(-x) = (-x)^2 - 5(-x) = x^2 5(-x) = x^2 + 5x$.

51. $g(x-5) = (x-5)^2 - 5(x-5) = (x^2 - 10x + 25) - 5x + 25 = x^2 - 15x + 50$.

52. $g(3-t) = (3-t)^2 - 5(3-t) = (9 - 6t + t^2) - (15 - 5t) = 9 - 6t + t^2 - 15 + 5t = -6 - t + t^2 = t^2 - t - 6$.

53. $g(4m^2) = (4m^2)^2 - 5(4m^2) = (4^2 m^4) - (20m^2) = 16m^4 - 20m^2$.

54. Since $g(x + h) = (x + h)^2 - 5(x + h) = x^2 + 2xh + h^2 - 5x - 5h$ and $g(h) = x^2 - 5x$, then $g(x + h) - g(x) = (x^2 + 2xh + h^2 - 5x - 5h) - (x^2 - 5x) = x^2 + 2xh + h^2 - 5x - 5h - x^2 + 5x = 2xh + h^2 - 5h$.

55. We have $F(x) = \frac{2-x}{x^2+2}$, and so $F(0) = \frac{2-0}{0^2+2} = \frac{2}{0+2} = \frac{2}{2} = 1$.

56. $F(2) = \frac{2-2}{2^2+2} = \frac{0}{4+2} = \frac{0}{6} = 0$.

57. $F(1.5) = \frac{2-1.5}{(1.5)^2+2} = \frac{0.5}{2.25+2} = \frac{0.5}{4.25} = \frac{2}{17}$.

58. $F\left(\frac{1}{3}\right) = \frac{2-\frac{1}{3}}{\left(\frac{1}{3}\right)^2+2} = \frac{\frac{5}{3}}{\frac{1}{9}+2} = \frac{\frac{5}{3}}{\frac{19}{9}} = \frac{5}{3} \times \frac{9}{19} = \frac{15}{19}$.

59. $F(2x) = \frac{2-2x}{(2x)^2+2} = \frac{2-2x}{4x^2+2} = \frac{2(1-x)}{2(2x^2+1)} = \frac{1-x}{2x^2+1}$.

60. $\frac{F(2x)}{F(x)} = \frac{(1-x)/(2x^2+1)}{(2-x)/x^2+2} = \frac{1-x}{2x^2+1} \cdot \frac{x^2+2}{2-x} = \frac{(1-x)(x^2+2)}{(2x^2+1)(2-x)}$.

61. $C(32) = \frac{5}{9}(32 - 32) = \frac{5}{9}(0) = 0$.

62. $C(98.6) = \frac{5}{9}(98.6 - 32) = \frac{5}{9}(66.6) = 37$.

63. $C(-40) = \frac{5}{9}(-40 - 32) = \frac{5}{9}(-72) = -40$.

64. $C(72 - 32) = \frac{5}{9}(40) = \frac{200}{9} \approx 22.222222$.

65. Explicit.

66. Implicit. Solving for y we get an explicit form $y = -2x + 7$.

67. Implicit. Solving for y we get

$$y = x^2y + x$$
$$y - x^2y = x$$
$$y(1 - x^2) = x$$
$$y = \frac{x}{1 - x^2}$$

Thus, an explicit form is $y = \frac{x}{1-x^2}$.

68. Explicit.

69. Since $f(x, y) = 3x - 2y + 4$, then $f(1, 0) = 3(1) - 2(0) + 4 = 3 - 0 + 4 = 7$.

70. $f(0, 1) = 3(0) - 2(1) + 4 = 0 - 2 + 4 = 0$.

71. $f(-1, 2) = 3(-1) - 2(2) + 4 = -3 - 4 + 4 = -3$.

72. $f(2, -3) = 3(2) - 2(-3) + 4 = 6 + 6 + 4 = 16$.

73. Here $g(x, y) = x^2 - y^2 + 2xy$, and so $g(-2, 0) = (-2)^2 - 0^2 + 2(-2)(0) = 4 - 0 + 0 = 4$.

74. $f(0, -2) = 0^2 - (-2)^2 + 2(0)(-2) = 0 - 4 + 0 = -4$.

75. $g(5, 4) = 5^2 - 4^2 + 2(5)(4) = 25 - 16 + 40 = 49$.

76. $g(-3, 5) = (-3)^2 - 5^2 + 2(-3)(5) = 9 - 25 - 30 = -46$.

77. Since $h(x, y, z) = 3x - 4y - 2z + xyz$, then $h(-1, 2, 3) = 3(-1) - 4(2) - 2(3) + (-1)(2)(3) = -3 - 8 - 6 - 6 = -23$.

78. $h(2, -1, -3) = 3(2) - 4(-1) - 2(-3) + (2)(-1)(-3) = 6 + 4 + 6 + 6 = 22$.

79. $h(3, 2, -1) = 3(3) - 4(2) - 2(-1) + (3)(2)(-1) = 9 - 8 + 2 - 6 = -3$.

80. $h(a, a^2, a^3) = 3a - 4a^2 - 2a^3 + a(a^2)(a^3) = 3a - 4a^2 - 2a^3 + a^6$.

81. If $r = 20, V = 16,000\pi$ ft^3, $S = 2400\pi$ ft^2; if $r = 30, V = 36,000\pi$ ft^3, $S = 4200\pi$ ft^2

82. (a) $s(1) = 4.91(1^2) = 4.91 \times 1 = 4.91$ m;
(b) $s(2) = 4.91(2^2) = 4.91 \times 4 = 19.64$ m;
(c) $s(5) = 4.91(5^2) = 4.91 \times 25 = 122.75$ m.

83. Here $R(7) = 20 + 5(7 - 4) = 20 + 5(3) = 20 + 15 = 35$, so the cost of renting this chain saw is \$35.

84. (a) The total perimeter is $2(l + w) = 450$. So, $l + w = 225$ and $l = 225 - w$. The area is $lw = (225 - w)w$. Thus, the area as a function of the width is $A(w) = w(225 - w) = 225w - w^2$, where w is the width of the rectangular yard,
(b) $8,750$ ft^2,
(c) $11,250$ ft^2,
(d) $-6,250$ ft^2,
(e) $0 < w < 225$.

85. Since, $P(d) = 62.4d$ and $d = 80$, then $P(80) = 62.4 \cdot 80 = 4,992$ lb/ ft^2.

86. (a) {birth, 2 days, 14 day, 3 mo, 6 mo, 1 yr, 2 yr, 4 yr, 8 - 21 yr}
(b) $\{4.3, 4.6, 4.7, 4.8, 5.0, 5.1, 5.3\}$

▤ 4.2 OPERATIONS ON FUNCTIONS; COMPOSITE FUNCTIONS

1. Solving $y = x - 5$ for x, we obtain $x = y + 5$. Exchanging the x- and y-variables produces the inverse, $y = x + 5$.

2. Solving $y = x + 7$ for x, we obtain $x = y - 7$. Exchanging the $x-$ and y-variables produces the inverse, $y = x - 7$.

3. Solving $y = 2x - 8$ for x, we obtain $x = \frac{y+8}{2}$. Exchanging the x- and y-variables produces the inverse, $y = \frac{x+8}{2}$. This can also be written as $y = \frac{1}{2}(x + 8)$ or $y = \frac{1}{2}x + 4$.

4. Solving $y = 5x + 12$ for x, we obtain $x = \frac{y-12}{5}$. Exchanging the x- and y-variables produces the inverse, $y = \frac{x-12}{5}$. This can also be written as $y = \frac{1}{5}(x - 12)$.

5. Solving $6y = 2x - 4$ for x, we obtain $2x = 6y + 4$, which simplifies to $x = 3y + 2$. Exchanging the x- and y-variables produces the inverse, $y = 3x + 2$.

6. Solving $9y = 3x + 5$ for x, we obtain $3x = 9y = -5$, and then $x = \frac{9y-5}{3}$. Exchanging the x- and y-variables produces the inverse, $y = \frac{9x-5}{3}$. This can be simplified as $y = 3x - \frac{5}{3}$.

7. Solving $y^2 = x - 5$ for x, we obtain $x = y^2 + 5$. Exchanging the x- and y-variables produces the inverse $y = x^2 + 5$.

8. Solving $x^3 + 7$ for x, we first obtain $x^3 = y - 7$ and then $x = \sqrt[3]{y - 7}$. Exchanging the x- and y-variables produces the inverse, $y = \sqrt[3]{x - 7}$.

9.

x	-3	-2	-1	0	1	2	3	4
$f(x)$	0	1	2	3	4	5	6	7
$g(x)$	-10	5	0	-1	5	1	-10	7
$(f+g)(x)$	-10	6	2	2	9	6	-4	14

10.

x	-3	-2	-1	0	1	2	3	4
$f(x)$	0	1	2	3	4	5	6	7
$g(x)$	-10	5	0	-1	5	1	-10	7
$(f-g)(x)$	10	-4	2	4	-1	4	16	0

11.

x	-3	-2	-1	0	1	2	3	4
$f(x)$	0	1	2	3	4	5	6	7
$g(x)$	-10	5	0	-1	5	1	-10	7
$(g-f)(x)$	-10	4	-2	-4	1	-4	-16	0

12.

x	-3	-2	-1	0	1	2	3	4
$f(x)$	0	1	2	3	4	5	6	7
$g(x)$	-10	5	0	-1	5	1	-10	7
$(f \cdot g)(x)$	0	5	0	-3	20	5	-60	49

13.

x	-3	-2	-1	0	1	2	3	4
$f(x)$	0	1	2	3	4	5	6	7
$g(x)$	-10	5	0	-1	5	1	-10	7
$(f/g)(x)$	0	$\frac{1}{5}$	undefined	-3	$\frac{4}{5}$	5	$-\frac{3}{5}$	1

14.

x	-3	-2	-1	0	1	2	3	4
$f(x)$	0	1	2	3	4	5	6	7
$g(x)$	-10	5	0	-1	5	1	-10	7
$(g/f)(x)$	undefined	5	0	$-\frac{1}{3}$	$\frac{5}{4}$	$\frac{1}{5}$	$-\frac{5}{3}$	1

15.

x	-3	-2	-1	0	1	2	3	4
$g(x)$	-10	5	0	-1	5	1	-10	7
$(f \circ g)(x)$	-7	8	3	2	8	4	-7	10

16.

x	-6	-5	-4	-3	-2	-1	0	1	2	3	4
$f(x)$	-3	-2	-1	0	1	2	3	4	5	6	7
$(g \circ f)(x)$	-10	5	0	-1	5	1	-10	7			

17. The domain and range of f are all real numbers. The domain of g is $\{-3, -2, -1, 0, 1, 2, 3, 4\}$. The range of g is $\{-10, -1, 0, 1, 5, 7\}$.

18. (a) The domain of $f+g$ is $\{-3, -2, -1, 0, 1, 2, 3, 4\}$. By reading the bottom line of the table in the solution to Exercise 9, we can see that the range of $f + g$ is $\{-10, -4, 2, 6, 9, 14\}$.

(b) The domain of $f-g$ is $\{-3, -2, -1, 0, 1, 2, 3, 4\}$. By reading the bottom line of the table in the solution to Exercise 10, we can see that the range of $f - g$ is $\{-1, -4, 0, 2, 4, 10, 16\}$.

(c) The domain of $f \cdot g$ is $\{-3, -2, -1, 0, 1, 2, 3, 4\}$. By reading the bottom line of the table in the solution to Exercise 1, we can see that the range of $f \cdot g$ is $\{-60, -3, 0, 5, 20, 49\}$.

19. (a) The domain of f/g is $\{-3, -2, 0, 1, 2, 3, 4\}$. Notice that -1 is not in the domain of f/g because the denominator of f/g is 0 when $x = -1$. By reading the bottom line of the table in the solution to Exercise 13, we can see that the range of f/g is $\{-3, -\frac{3}{5}, 0, \frac{1}{5}, 1, 5\}$.

(b) The domain of g/f is $\{-2, -1, 0, 1, 2, 3, 4\}$. Notice that -3 is not in the domain of g/f because the denominator g/f is 0 when $x = -3$. By reading the bottom line of the table in the solution to Exercise 14, we can see that the range of g/f is $\{-\frac{5}{3}, -\frac{1}{3}, 0, \frac{1}{5}, 1, \frac{5}{4}, 5\}$.

20. **(a)** The domain of $f \circ g$ is $\{-3, -2, -1, 0, 1, 2, 3, 4\}$. By reading the bottom line of the table in the solution to Exercise 15, we can see that the range of $f \circ g$ is $\{-7, 2, 3, 4, 8, 10\}$.

(b) The domain of $g \circ f$ is $\{-6, -5, -4, -3, -2, -1, 0, 1\}$. Notice that 2, 3, and 4 are not in the domain of $g \circ f$ because for these values of x, the corresponding values of $f(x)$ are not in the domain of g. By reading the bottom line of the table in the solution to Exercise 16, we can see that the range of $g \circ f$ is $\{-10, -1, 0, 1, 5, 7\}$.

21. $f(+g)(x) = (x^2 - 1) + (3x + 5) = x^2 - 1 + 3x + 5 = x^2 + 3x + 4$.

22. $(f + g)(4) = 4^2 + 3(4) + 4 = 16 + 12 + 4 = 32$.

23. $(f - g)(x) = (x^2 - 1) - (3x + 5) = x^2 - 1 - 3x - 5 = x^2 - 3x - 6$.

24. $(f - g)(4) = 4^2 - 3(4) - 6 = 16 - 12 - 6 = -2$.

25. $(g - f)(x) = (3x + 5) - (x^2 - 1) = 3x + 5 - x^2 + 1 = 3x - x^2 + 6$.

26. $(g - f)(5) = 3(5) - 5^2 + 6 = 15 - 25 - 6 = -4$.

27. $(f \cdot g)(x) = (x^2 = 1)(3x + 5) = 3x^3 + 5x^2 - 3x - 5$.

28. $(f \cdot g)(5) = 3(5^3) + 5(5^2) - 3(5) - 5 = 3(125) + 5(25) - 15 - 5 = 375 + 125 - 15 - 5 = 480$.

29. $(f/g)(x) = \left(\frac{f}{g}\right)(x) = \frac{f(x)}{g(x)} = \frac{x^2 - 1}{3x + 5}$.

30. $(f/g)(-2) = \left(\frac{f}{g}\right)(-2) = \frac{f(-2)}{g(-2)} = \frac{(-2)^2 - 1}{3(-2) + 5} = \frac{4 - 1}{-6 + 5} = \frac{3}{-1} = -3$.

31. $(g/f)(x) = \left(\frac{g}{f}\right)(x) = \frac{g(x)}{f(x)} = \frac{3x + 5}{x^2 - 1}$.

32. $(g/f)(-2) = \left(\frac{g}{f}\right)(-2) = \frac{g(-2)}{f(-2)} = \frac{3(-2) + 5}{(-2)^2 - 1} = \frac{-6 + 5}{4 - 1} = \frac{-1}{3} = -\frac{1}{3}$.

33. The domains of both f and g are all real numbers.

34. The domain of f/g is all real numbers except $-\frac{5}{3}$, because when $x = -\frac{5}{3}$ the denominator of f/g is 0. The domain of g/f is all real numbers except -1 and $+1$, because the denominator of g/f, $x^2 - 1$, is 0 when $x = -1$ or $x = 1$.

35. $(f \circ g)(x) = f(g(x)) = f(3x + 5) = (3x + 5)^2 - 1 = 9x^2 + 30x + 24$.

36. One way to work this is: $(f \circ g)(-3) = f(g(-3)) = f(3(-3) + 5) = f(-4) = (-4)^2 - 1 = 16 - 1 = 15$. Another way is to use the value of $(f \circ g)(x) = 9x^2 + 30x + 24$ from the previous exercise. Then $(f \circ g)(-3) = 9(-3)^2 + 30(-3) + 24 = 9(9) - 90 + 24 = 81 - 90 + 24 = 15$.

37. $(g \circ f)(x) = g(f(x)) = g(x^2 - 1) = 3(x^2 - 1) + 5 = 3x^2 - 3 + 5 = 3x^2 + 2$.

38. Using the result of the previous exercise, $(g \circ f)(-3) = 3(-3)^2 + 2 = 3(9) + 2 = 27 + 2 = 29$.

39. $(f + g)(x) = f(x) + g(x) = (3x - 1) + (3x^2 + x) = 3x^2 + 4x - 1$.

40. $f(+g)(4) = 3(4)^2 + 4(4) - 1 = 3 \cdot 16 + 16 - 1 = 48 + 16 - 1 = 63$.

41. $f(-g)(x) = f(x) - g(x) = 3x - 1) - (3x^2 + x) = (3x - 1 - 3x^2 - x = -3x^2 + 2x - 1$.

42. $(f - g)(-2) = -3(-2)^2 + 2(-2) - 1 = -3(4) + (-4) - 1 = -12 - 4 - 1 = -17$.

43. $(g - f)(x) = g(x) - f(x) = (3x^2 + x) - (3x - 1) = 3x^2 + x - 3x_1 = 3x^2 - 2x + 1$.

44. $(g - f)(-2) = 3(-2)^2 - 2(-2) + 1 = 3(4) + 4 + 1 = 12 + 4 + 1 = 17$. Note: $(f - g)(-2) = -[(g - f)(-2)]$.

45. $(f \cdot g)(x) = (3x - 1)(3x^2 + x) = 9x^3 + 3x^2 - x = 9x^3 - x$.

46. $(f \cdot g)(-1) = 9(-1)^3 - (-1) = -9 + 1 = -8$.

47. $(f/g)(x) = \frac{f(x)}{g(x)} = \frac{3x - 1}{3x^2 + x}$.

48. $(f/g)(-1) = \frac{3(-1) - 1}{3(-1)^2 + (-1)} = \frac{-3 - 1}{3 - 1} = \frac{-4}{2} = -2$.

49. $(g/f)(x) = \frac{g(x)}{f(x)} = \frac{3x^2 + x}{3x - 1}$.

50. $(g/f)(-1) = \frac{3(-1)^2 + (-1)}{3(-1) - 1} = \frac{3 - 1}{-3 - 1} = \frac{2}{-1} = -\frac{1}{2}$. Note: $(g/f)(-1) = \frac{1}{(f/g)(-1)}$.

51. $(f \circ g)(x) = f(x(x)) = f(3x^2 + x) = 3(3x^2 + x) - 1 = 9x^2 + 3x - 1$.

52. $(f \circ g)(5) = f(g(5)) = f(3 \cdot 5^2 + 5) = f(75 + 5) = f(80) = 3 \cdot 80 - 1 = 239$ or using #31$(f \circ g)(5) = 9(5)^2 + 3(5) - 1 = 9 \cdot 25 + 15 - 1 = 225 + 15 - 1 = 239$.

53. $(g \circ g)(x) = g(f(x)) = g(3x - 1) = 3(3x - 1)^2 + (3x - 1) = 3(9x^2 - 6x + 1) + (3x - 1 = 27x^2 - 18x + 3 + 3x - 1 = 27x^2 - 15 + 2$.

54. $(g \circ f)(5) = 27(5)^2 - 15(5) + 2 = 27 \cdot 25 - 5 + 2 = 675 - 75 + 2 = 602$.

55. The domains of both f and g are all real numbers.

56. The domain of f/g is all real numbers except 0 and $-\frac{1}{3}$. The domain of g/f is all real numbers except $\frac{1}{3}$.

57. **(a)** $C(n) = F(n) + V(n) = 7500 + 15n$ or $15n + 7500$;
(b) $C(100) = 15(100) + 7500 = 1500 + 7500 = \$9,000$;
(c) $C(1000) = 15(1000) + 7500 = 15000 + 7500 = \$22,500$.

58. **(a)** $C(n) = S(n) + F(n) + V(n) = \frac{2,750,000}{n} + 12,500 + 8n$ dollars;
(b) $C(100) = \frac{2,750,000}{100} + 12,500 + 8(100) = 27,500 + 12,500 + 800 = \$40,800$;

(c) $C(1000) = \frac{2,750,000}{1000} + 12,500 + 8(1000) = 2750 + 12,500 + 8000 = \$23,250$

59. **(a)** $P(n) = R(n) - C(n) = 90n - (30n + 275) = 60n - 275$;
(b) $P(50) = 60(50) - 275 = 3000 - 275 = \$2,725$.

60. **(a)** $R(p) = f(p) \times p = (750 - 4.2p)p = 750p - 4.2p^2$;
(b) $f(100) = 750 - 4.2(100) = 750 - 420 = 330$;
(c) $R(100) = 750(100) - 4.2(100)^2 = 75000 - 42000 = 33000$ or $330 \times 100 = \$33,000$;
(d) $f(80) = 750 - 4.20(80) = 750 - 336 = 414$;
(e) $R(80) = 750(80) - 4.2(80)^2 = 60,000 - 4.2(6400) = 60,000 - 26880 = \$33,120$ or $414 \times 80 = \$33,120$

61. $(P \circ n)(a) = P(7a + 4) = 300(7a + 4)^2 - 50(7a + 4) = 300(49a^2 + 56a + 16) - 50(7a + 4) = 14700a^2 + 16800a + 4800 - 350a - 200 = 14700a^2 + 16450a + 4600$.

62. **(a)** The volume V of a sphere of radius r is $V(r) = \frac{4}{3}\pi r^3$. We are given $r(t) = 3t$, where t represents the number of seconds since the weather balloon was begun to be inflated. Then $V(t) = (V \circ r)(t) = \frac{4}{3}\pi(3t)^3 = 36\pi t^3$ cm^3.
(b) $V(10) = 36\pi(10^3) = 36,000\pi$ cm^3.

≡ 4.3 RECTANGULAR COORDINATES

1.

2.

3.

4.

5.

6.

7.

8.

9.

10.

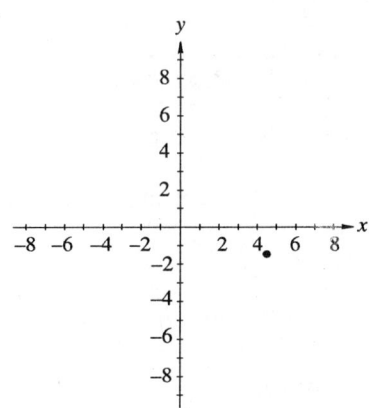

11. The fourth vertex will have the same x-coordinate as $B, -1$, and the same y-coordinate as $C, -4$. So its coordinates are $(-1, -4)$.

12.

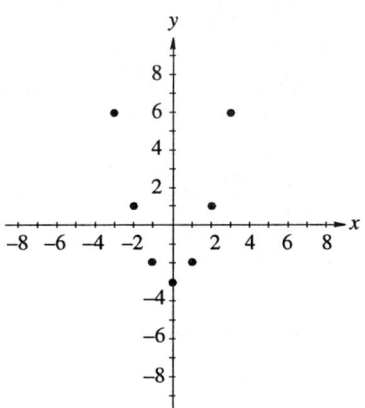

13. They are on a horizontal line and have the same y-coordinate.

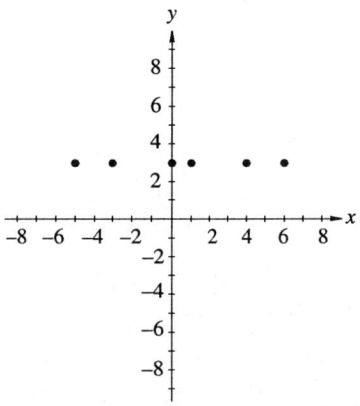

14. They are all on the same vertical line and have the same x-coordinate.

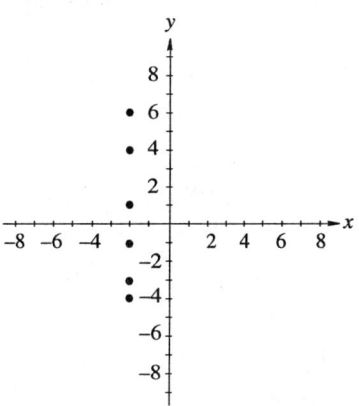

15.   They all lie on the same straight line.

16.   As in problem #14, these points will all lie on a vertical line through $x = 0$. This is the y-axis.

17.   As in problem #13, these points all lie in a horizontal line through $y = -2$.

18.   On the horizontal line $y = 7$

19.   On the vertical line $x = -5$

20.   On the vertical line $y = 0$. This is the x-axis.

21.   These points must all lie to the right of the vertical line $x = -3$.

22.   These points must be to the right of the y-axis (see problems 16 and 18) and below the x-axis. This is the fourth quadrant.

23.   To the right of the vertical line through the point $(1,0)$ and below the horizontal line through the point $(0, -2)$.

24.   **(a)**

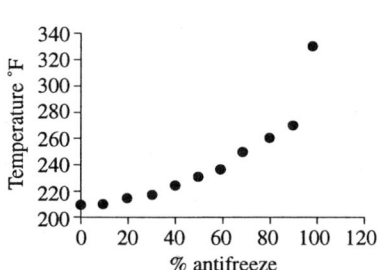

(b) 228°F;
(c) 236°F.

25.   **(a)**

(b) 41.9 psi;
(c) 99.6 psi.

26.   **(a)**

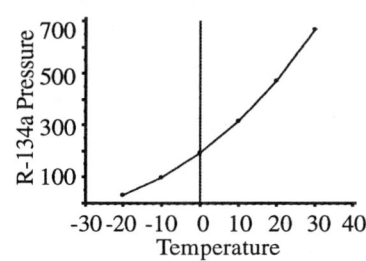

(b) Around 250 kPa (actual answer is closer to 248.1 kPa)
(c) Around 780 kPa (actual answer is closer to 782.6 kPa)
(d) Around 21.7°C.

27.   **(a)**

(b) Around 317 min.;
(c) Around 23.5 min.

28.   **(a)**

(b) Around 153 board feet;
(c) Around 37.5 ft.

≣ 4.4 GRAPHS

1.   (a)

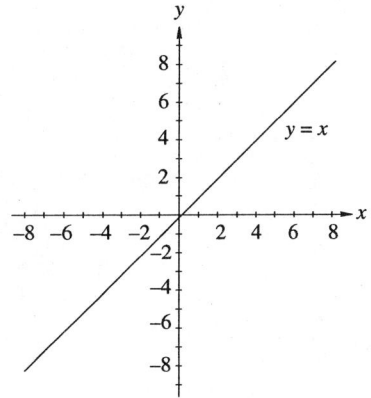

(b) x-intercept is 0; y-intercept is 0;
(c) Let $(x_1, y_1) = (0, 0)$ and $(x_2, y_2) = (2, 2)$; thus
$m = \frac{2-0}{2-0} = \frac{2}{2} = 1$.

2.   (a)

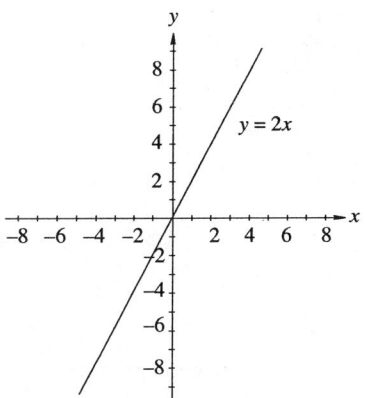

(b) x-intercept is 0; y-intercept is 0;
(c) Let $(x_1, y_1) = (0, 0)$ and $(x_2, y_2) = (1, 2)$; then
$m = \frac{2-0}{1-0} = \frac{2}{1} = 2$.

3.   (a)

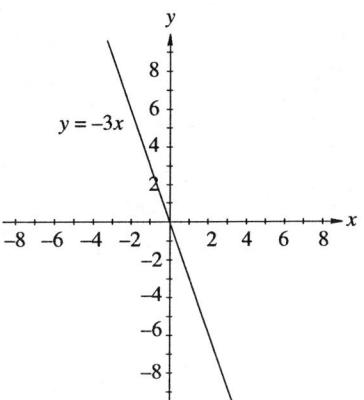

(b) x-intercept is 0; y-intercept is 0;

(c) Let $(x_1, y_1) = (0, 0)$ and $(x_2, y_2) = (1, -3)$; then $m = \frac{-3-0}{1-0} = \frac{-3}{1} = -3$.

4.   (a)

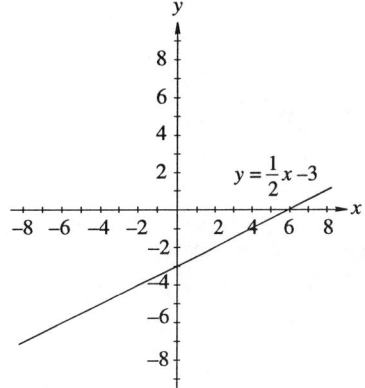

(b) x-intercept is 6; y-intercept is -3;
(c) Let $(x_1, y_1) = (0, 0)$ and $(x_2, y_2) = (6, 0)$; then
$m = \frac{0-(-3)}{6-0} = \frac{3}{6} = \frac{1}{2}$.

5.   (a)

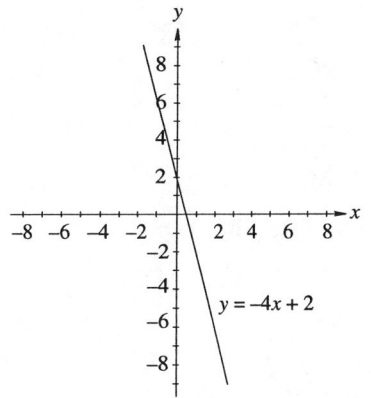

(b) x-intercept is $\frac{1}{2}$; y-intercept is 2;
(c) $m = \frac{0-2}{\frac{1}{2}-0} = \frac{-2}{\frac{1}{2}} = -4$.

6.   (a)

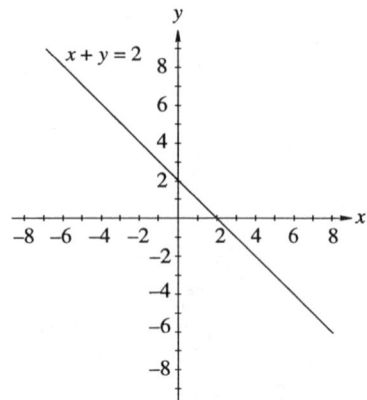

(b) x-intercept is 2; y-intercept is 2;

(c) $m = \frac{0-2}{2-0} = -1$.

7.   (a)

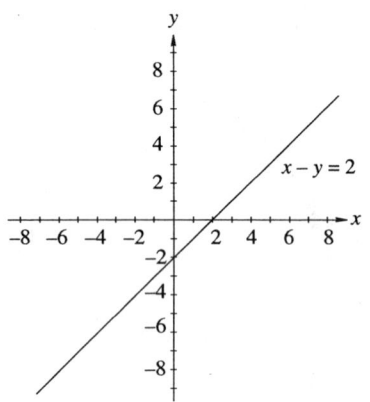

(b) x-intercept is 2; y-intercept is -2;

(c) $m = \frac{0-(-2)}{2-0} = \frac{2}{2} = 1$.

8.   (a)

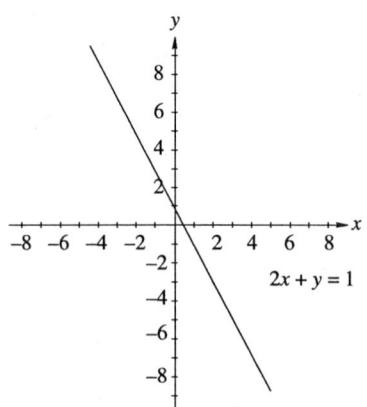

(b) x-intercept is $\frac{1}{2}$; y-intercept is 1;

(c) $m = \frac{0-1}{\frac{1}{2}-0} = \frac{-1}{\frac{1}{2}} = -2$.

9.   (a)

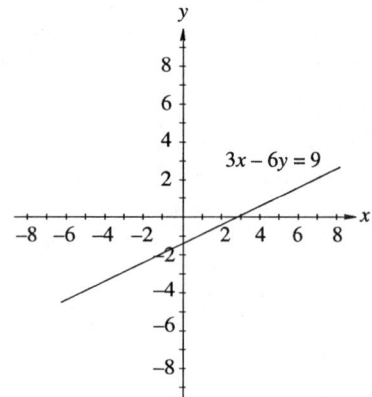

(b) x-intercept is 3; y-intercept is $-\frac{3}{2}$;

(c) $m = \dfrac{0-(\frac{3}{2})}{3-0} = \dfrac{\frac{3}{2}}{\frac{2}{3}} = \frac{1}{2}$.

10.   (a)

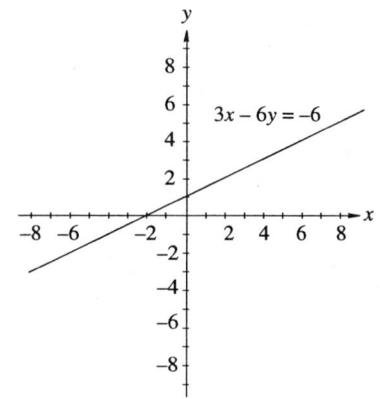

(b) x-intercept is -2; y-intercept is 1;

(c) $m = \frac{0-1}{-2-0} = \frac{-1}{-2} = \frac{1}{2}$.

11.   (a)

x	-3	-2	-1	0	1	2	3
y	9	4	1	0	1	4	9

(b)

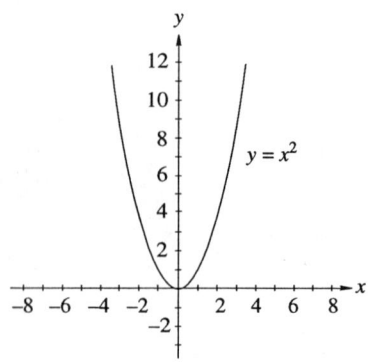

(c) Domain: all real numbers; range: $\{y : y \geq 0\}$.

12. **(a)**

x	-3	-2	-1	0	1	2	3
y	12	7	4	3	4	7	12

(b)

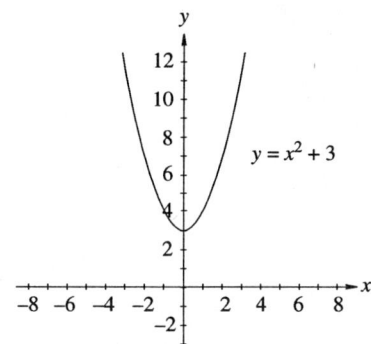

$y = x^2 + 3$

(c) Domain: all real numbers; range: $\{y : y \geq 3\}$.

13. **(a)**

x	-3	-2	-1	0	1	2	3
y	7	2	-1	-2	-1	2	7

(b)

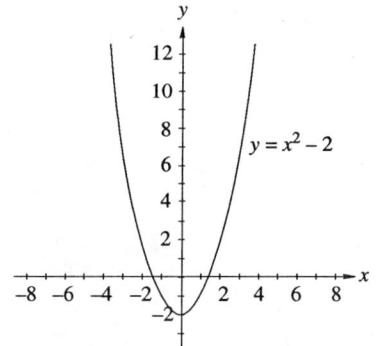

$y = x^2 - 2$

(c) Domain: all real numbers; range: $\{y : y \geq -2\}$.

14. **(a)**

x	-3	-2	-1	0	1	2	3
y	4	1	0	1	4	9	16

(b)

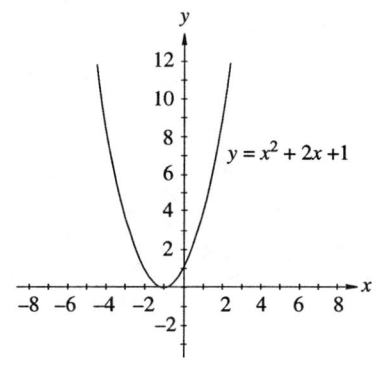

$y = x^2 + 2x + 1$

(c) Domain: all real numbers; range: $\{y : y \geq 0\}$.

15. **(a)**

x	-3	-2	-1	0	1	2	3	4	5	6
y	36	25	16	9	4	1	0	1	4	9

(b)

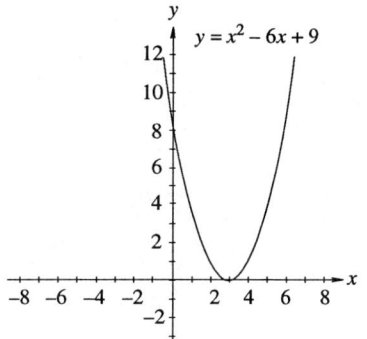

$y = x^2 - 6x + 9$

(c) Domain: all real numbers; range: $\{y : y \geq 0\}$.

16. **(a)**

x	-3	-2	-1	0	1	2	3
y	-7	-2	1	2	1	-2	-7

(b)

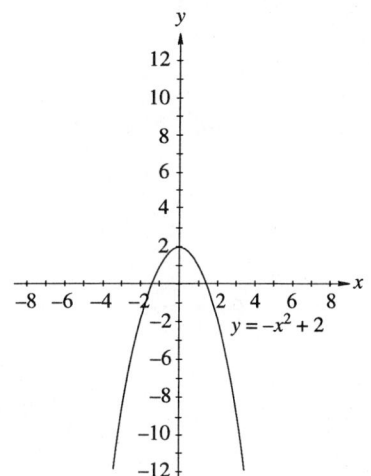

$y = -x^2 + 2$

(c) Domain: all real numbers; range: $\{y : y \leq 2\}$.

17.

x	-3	-2	-1	$-\frac{1}{2}$	$-\frac{1}{4}$	0	$\frac{1}{4}$	$\frac{1}{2}$	1	2	3
y	$-\frac{1}{3}$	$-\frac{1}{2}$	-1	-2	-4	undefined	4	2	1	$\frac{1}{2}$	$\frac{1}{3}$

(b)

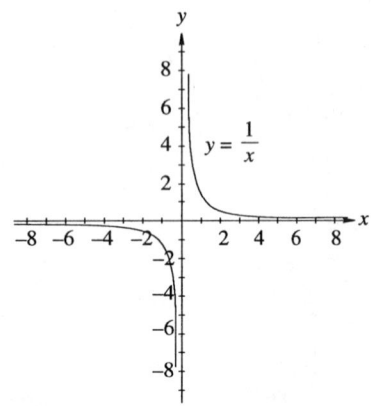

$y = \dfrac{1}{x}$

(c) Domain: $\{x : x \neq 0\}$; range; $\{y : y \neq 0\}$.

18. (a)

x	-6	-5	-4	$-3\frac{1}{2}$	$-3\frac{1}{4}$	-3	$-2\frac{3}{4}$	$-2\frac{1}{2}$	-2	-1	0	1
y	$-\frac{1}{3}$	$-\frac{1}{2}$	-1	-2	-4	undefined	4	2	1	$\frac{1}{2}$	$\frac{1}{3}$	$\frac{1}{4}$

(b)

$y = \dfrac{1}{x+3}$

(c) Domain: $\{x : x \neq -3\}$; range; $\{y : y \neq 0\}$.

19. (a)

x	-3	-2	-1	0	1	2	3
y	-27	-8	-1	0	1	8	27

(b)

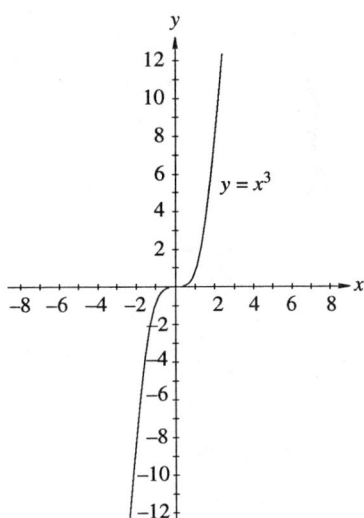

(c) Domain: all real numbers; range: all real numbers.

20. **(a)**

x	-27	-8	-1	0	1	8	27
y	-3	-2	-1	0	1	2	3

(b)

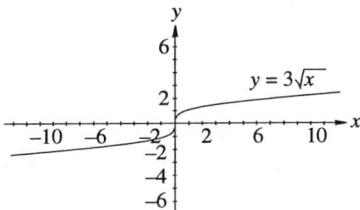

(c) Domain: all real numbers; range: all real numbers.

21.

x	0	0	5	-5	3	-3	3	-3	4	4	-4	-4
y	5	-5	0	0	4	4	-4	-4	3	-3	3	-3

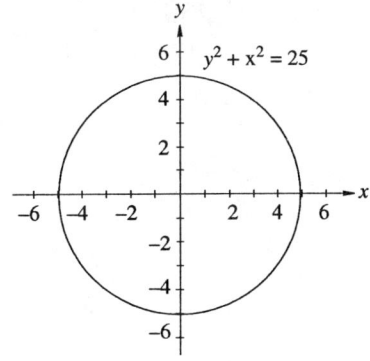

22.

x	0	3	-3	4	-4	5	-5
y	5	4	4	3	3	0	0

23.

x	-5	-4	-3	-2	-1	0	1	2
y	3	2	1	0	1	2	3	4

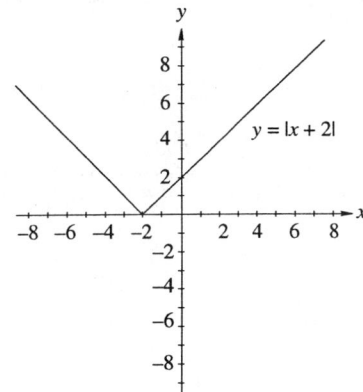

24.

x	0	1	-1	2	-2
y	± 3	± 2.6	± 2.6	0	0

25.

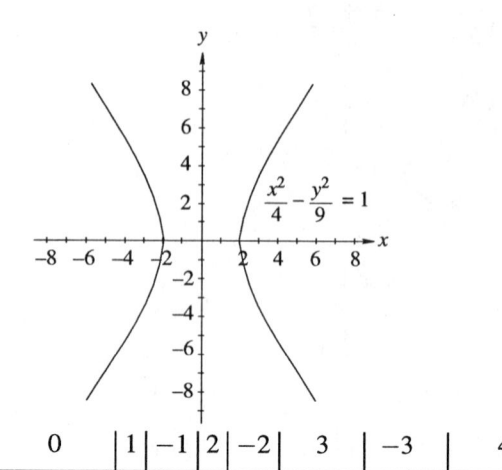

x	0	1	−1	2	−2	3	−3	4	−4
y	undefined			0	0	±3.35	±3.35	±5.2	±5.2

26.

x	−3	−2	−1	0	$\frac{1}{4}$	$\frac{1}{2}$	$\frac{3}{4}$	1	2	3
y	−36	−12	−2	0	−0.05	$-\frac{1}{8}$	−0.14	0	4	18

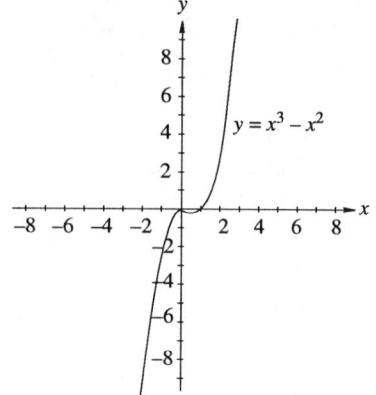

27. Function.   **28.** Not a function.   **29.** Function.   **30.** Not a function.

31. **(a)**

k	0	1	2	3	4	5	6	7	8	9
$p(k)$	25.0	8.33	5.00	3.57	2.78	2.27	1.92	1.67	1.47	1.32

$$p(k) = \frac{25}{2k+1}$$

(b) Yes.

32.

k	0	1	2	3	4	5	6	7	8	9
$p(k)$	25.00	17.68	14.43	12.50	11.18	10.21	9.45	8.84	8.33	7.91

$$p(k) = \frac{25}{\sqrt{k}+1}$$

33. **(a)** $0 < p < \$4.65$;

(b)

p	0.25	0.50	0.75	1.00	1.25	1.50	1.75	2.00
$d(p)$	10.56	11	11.31	11.50	11.56	11.50	11.31	11

(c)

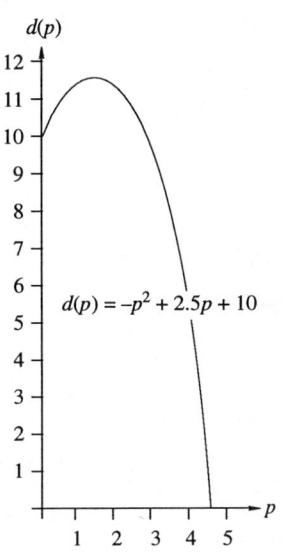

$d(p) = -p^2 + 2.5p + 10$

(d) $1.25.

34. **(a)**

t	1	2	3	4	5	6	7	8
m	2.4	3.0	3.8	5.0	6.7	9.0	12.3	17.0

t	9	10	11	12	13	14	15	
m	23.6	33.0	46.3	65.0	91.5	129.0	182.0	

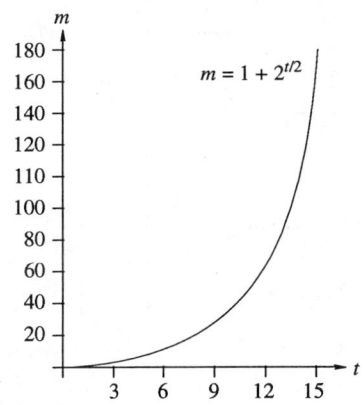

$m = 1 + 2^{t/2}$

(b) About 13.3 from graph. More exactly it's 13.2587.

$\equiv 4.5$ GRAPHING CALCULATORS AND COMPUTER-AIDED GRAPHING (OPTIONAL)

Note: All graphs were generated on a TI-81 calculator. In each figure, the tick marks are 1 unit apart.

1.   Graph $y = x$.

2. $y = 2x$.

3. $y = -3x$.

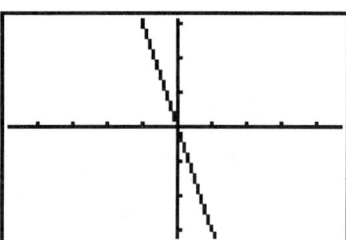

4. $y = \frac{1}{2}x - 3$.

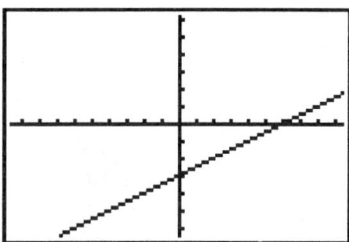

5. $y = -4x + 2$.

6. $x + y = 2$; graph either $y = 2 - x$ or $y = -x + 2$.

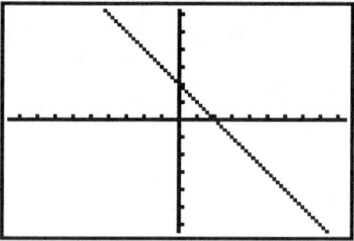

7. $x - y = 2$; graph as $y = x - 2$.

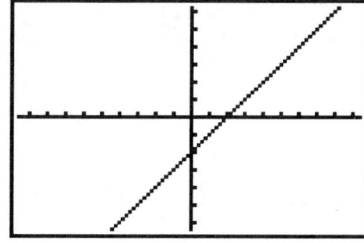

8. $2x + y = 1$; graph as $y = -2 + 1$.

9. $3x - 6y = 9$; graph as $y = \frac{9-3x}{-6}$ or $y = \frac{1}{2}x - \frac{3}{2}$.

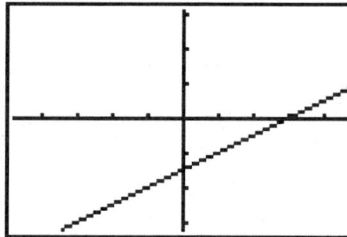

10. $3x - 6y = -6$; graph as $y = \frac{-6-3x}{-6}$ or $y = \frac{1}{2}x + 1$.

11. $y = x^2.$

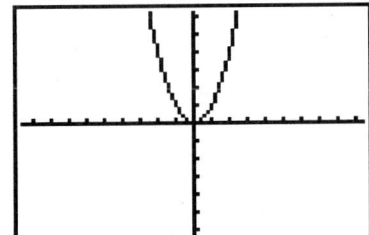

12. $y = x^2 + 3.$

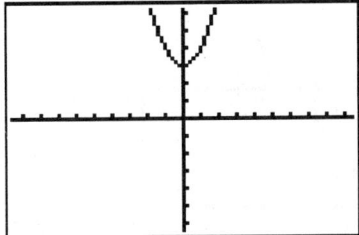

13. $y = x - 2.$

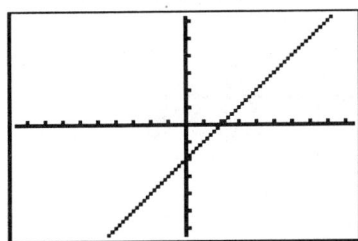

14. $y = x^2 + 2x + 1.$

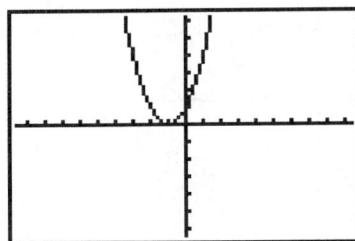

15. $y = x^2 - 6x + 9.$

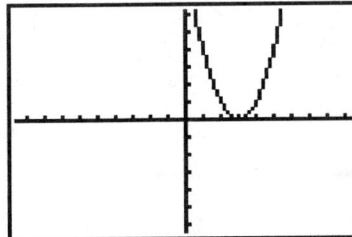

16. $y = -x^2 + 2.$

17. $y = \frac{1}{x}.$

18. $y = \frac{1}{x+3}.$

19. $y = x^3.$

20. $y = \sqrt[3]{x}.$

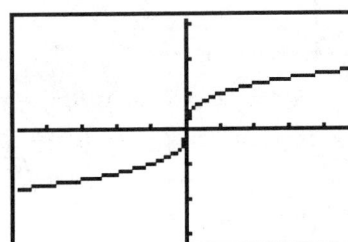

21. $y^2 + x^2 = 25$ or $y^2 = 25 - x^2$; graph both $y = \sqrt{25 - x^2}$ and $y = -\sqrt{25 - x^2}$.

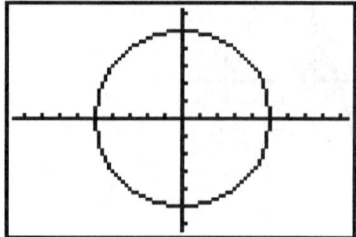

22. $y = \sqrt{25 - x^2}$.

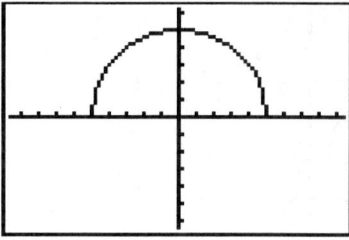

23. $y = |x + 2|$.

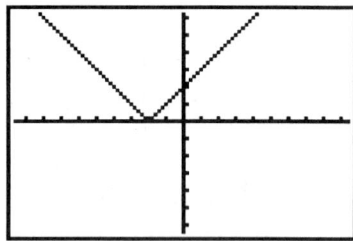

24. $\frac{x^2}{4} + \frac{y^2}{9} = 1$ or $9x^2 + 4y^2 = 36$, which becomes $4y^2 = 36 - 9x^2$ or $2y = \pm\sqrt{36 - 9x^2}$; graph as $y = \frac{\pm\sqrt{36-9x^2}}{2}$; make sure you graph both.

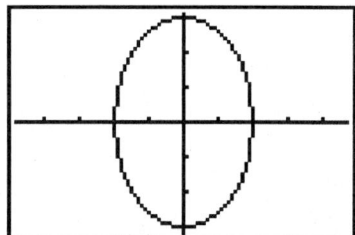

25. Here $\frac{x^2}{4} - \frac{y^x}{9} = 1$ can be written as $9x^2 - 4y^2 = 36$ or $-4y^2 = 36 - 9x^2$ or $4y^2 = 9x^2 - 36$ and so $2y = \pm\sqrt{9x^2 - 36}$; graph as $y = \frac{\pm\sqrt{9x^2-36}}{2}$; make sure you graph both.

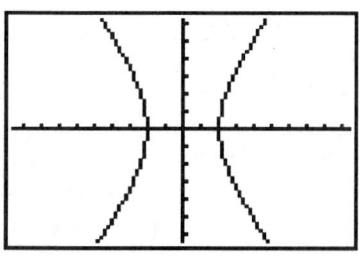

26. $y = x^3 - x^2$

27.

28.

29.

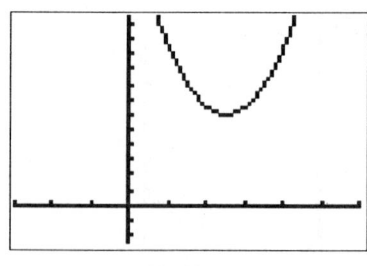

[–6, 12, 2] x [–5, 25, 2]

30.

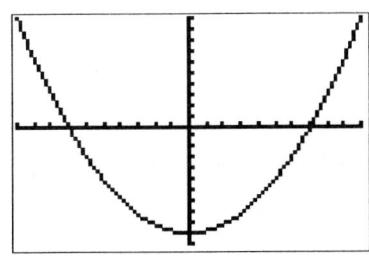

[–20, 20, 2] x [–22, 20, 2]

31.

32.

33.

34.

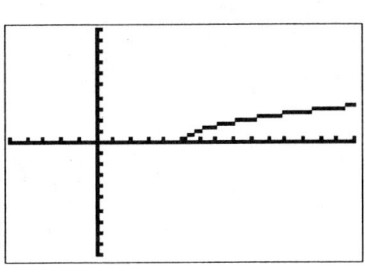

[–5, 15, 1] x [–10, 10, 1]

35.

36.

37.

38.

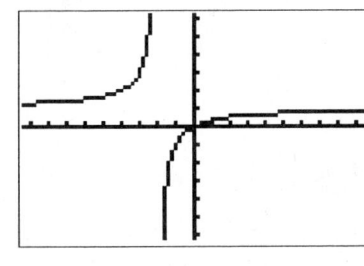

[–9.4, 9.4] x [–6.2, 6.2]

39.

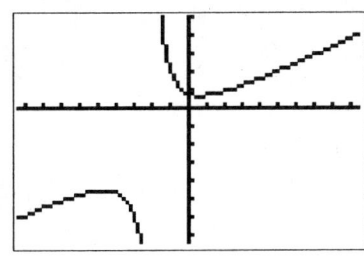

[–9.4, 9.4, 1] x [–10, 32, 2]

40.

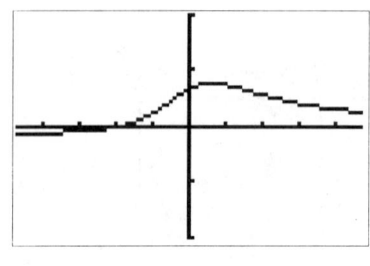

[–4.7, 4.9] x [–2, 2]

42.

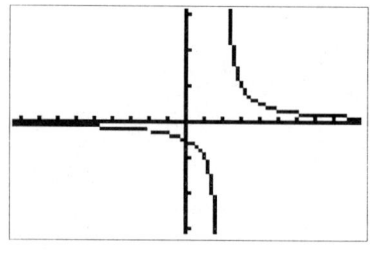

[–9.4, 9.4] x [–3.1, 3.1]

41.

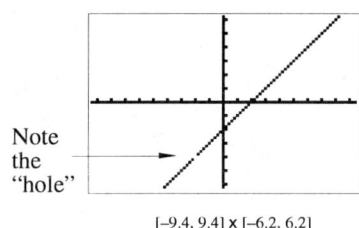

Note the "hole"

[–9.4, 9.4] x [–6.2, 6.2]

≡ 4.6 USING GRAPHS TO SOLVE EQUATIONS

1.   The root is $x = -\frac{5}{2}$. .

2.   The root is $x = -\frac{9}{5}$.

$y = 2x + 5$

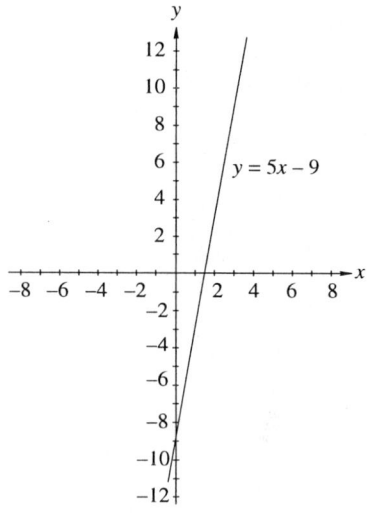

$y = 5x - 9$

3.   The root is $x = -3$ and $x = 3$.

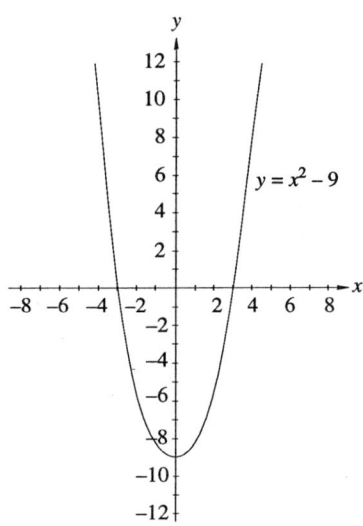

$y = x^2 - 9$

4.   The roots are $x \approx -1.5811$ and $x \approx 1.5811$.

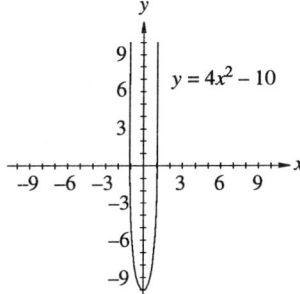

$y = 4x^2 - 10$

5.   The roots are $x = 0$ and $x = 5$.

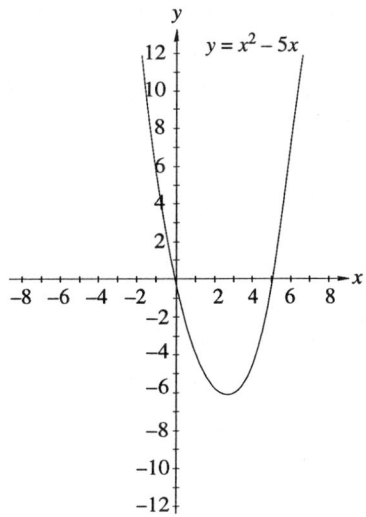

$y = x^2 - 5x$

6.   The roots are $x = 1$ and $x = 3$.

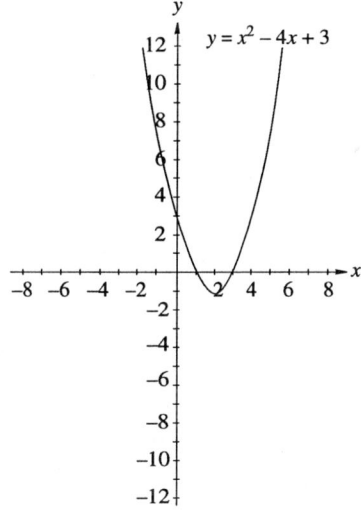

$y = x^2 - 4x + 3$

7.   The roots are $x \approx -5.5414$ and $x \approx 0.5414$.

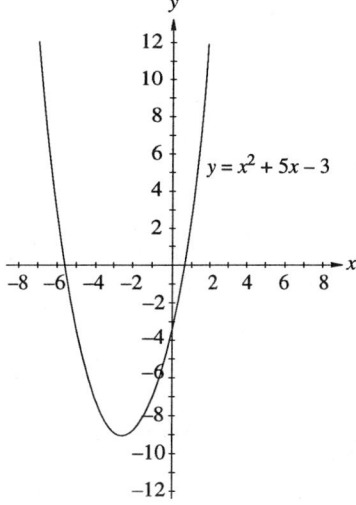

$y = x^2 + 5x - 3$

8.   The roots are $x \approx -2.1350$ and $x \approx 0.4684$.

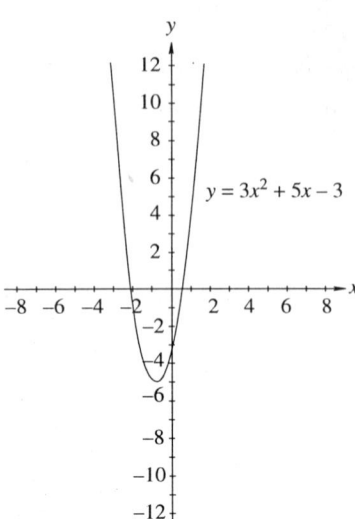

9.   The roots are $x = -\frac{9}{4}$ and $x = \frac{6}{5}$.

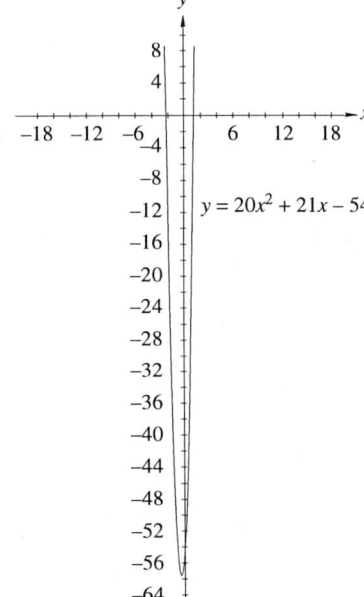

10.   The roots are $x \approx -3.3635, x \approx -0.4916, x \approx 0.4929$, and $x \approx 7.3622$.

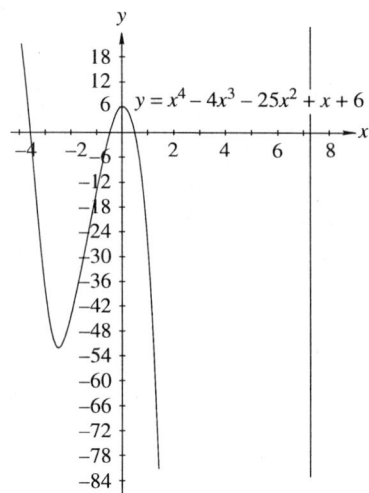

11.   The root is $x = -1$.

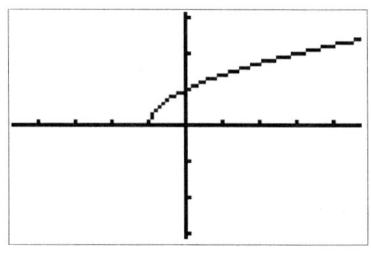

$[-4.7, 4.7] \times [-3.1, 3.1]$

12.   The root is $x = 2.5$.

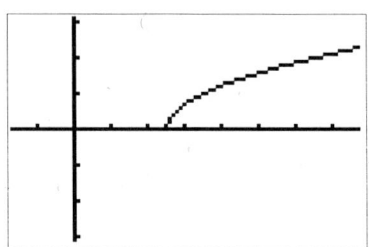

$[-1.7, 7.7] \times [-3.1, 3.1]$

13. Graph $y = \sqrt{x+1} - 3$. The root is where the graph crosses the x-axis. The root is $x = 8$.

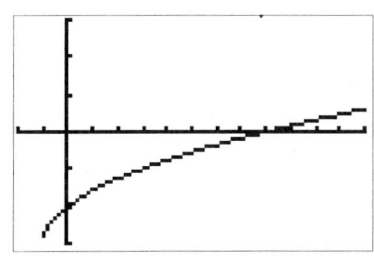

$[-2, 12] \times [-3, 3]$

14. Graph $y = \sqrt{x-3} - 2$. The root is $x = 7$.

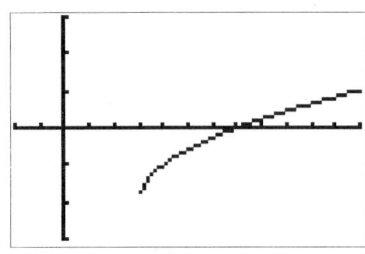

$[-2, 12] \times [-3, 3]$

15. Graph $y = \frac{x}{x+1} + 3$. The toot is $x = -0.75$.

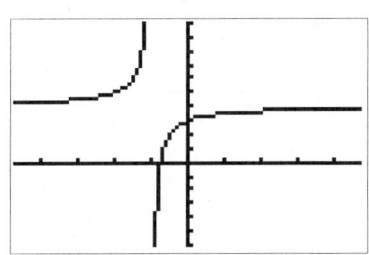

$[-4.7, 4.7] \times [-6, 10]$

16. Graph $y = \frac{x}{x-2} - 5$. The root is $x = 2.5$.

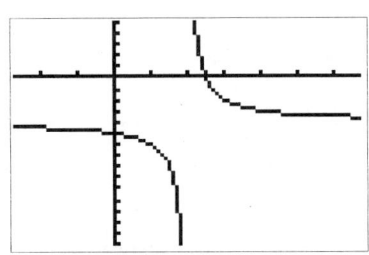

$[-2.7, 6.7] \times [-15, 5]$

17. Graph $y = \frac{x}{x-2} + x^2 - 2$. The root is $x \approx -1.2695$.

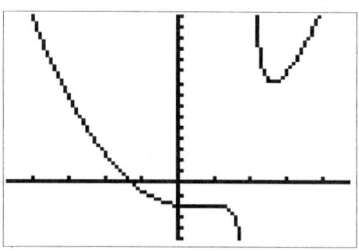

$[-4.7, 4.7] \times [-1, 15]$

18. Graph $y = \frac{x^2+5}{x-3} - x^2 + 2$. The roots are $x \approx -0.7, x \approx 0.3$, and $x \approx 4.4$.

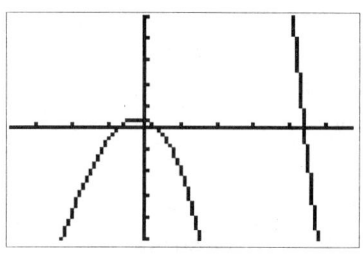

$[-3.7, 5.7] \times [-5, 5]$

19. (a)

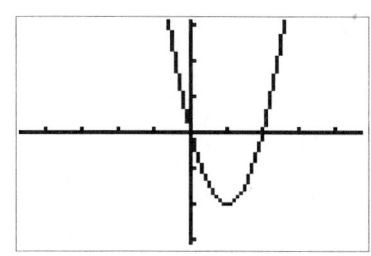

$[-9.4, 9.2, 2] \times [-6.2, 6.2, 2]$

(b) Domain: all real numbers; range $\{y:y \geq -4\}$.

20. (a)

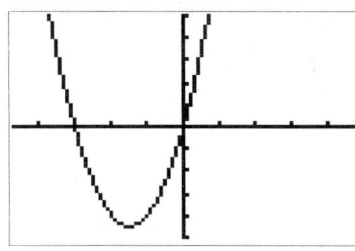

$[-9.4, 9.2, 2] \times [-10, 10, 2]$

(b) Domain: all real numbers; range $\{y:y \geq -9\}$.

21. **(a)**

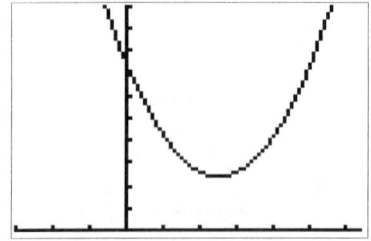

$$[-6, 12.8, 2] \times [0, 50, 5]$$

(b) Domain: all real numbers; range $\{y{:}y \geq 12\}$.

22. **(a)**

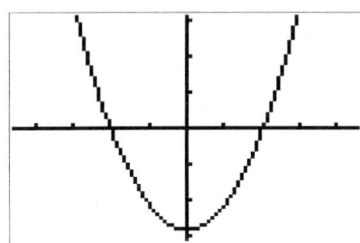

$$[-32, 32, 7] \times [-22, 22, 7]$$

(b) Domain: all real numbers; range $\{y{:}y \geq -20\}$.

23. **(a)**

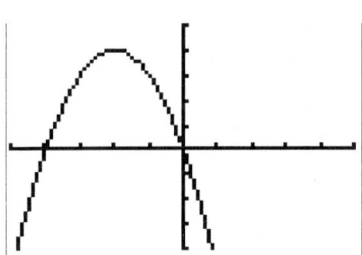

$$[-5, 5, 1] \times [-4, 5, 1]$$

(b) Domain: all real numbers; range $\{y{:}y \leq 4\}$.

24. **(a)**

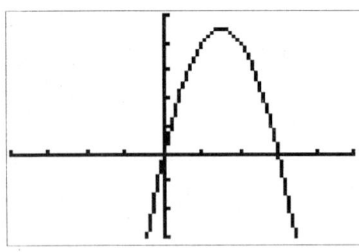

$$[-8, 10, 2] \times [-6, 10, 2]$$

(b) Domain: all real numbers; range $\{y{:}y \leq 9\}$.

25. **(a)**

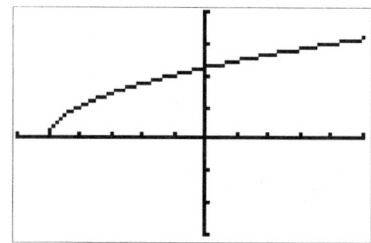

$$[-6, 5, 1] \times [-3, 4, 1]$$

(b) Domain: $\{x{:}x \geq -5\}$; range $\{y{:}y \geq 0\}$.

26. **(a)**

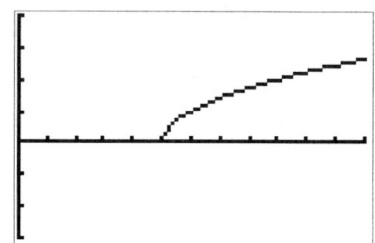

$$[0, 12, 1] \times [-3, 4, 1]$$

(b) Domain: $\{x{:}x \geq 5\}$; range $\{y{:}y \geq 0\}$.

27. **(a)**

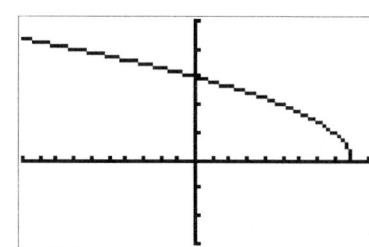

$$[-10, 10, 1] \times [-3, 5, 1]$$

(b) Domain: $\{x{:}x \leq 9\}$; range $\{y{:}y \geq 0\}$.

28. **(a)**

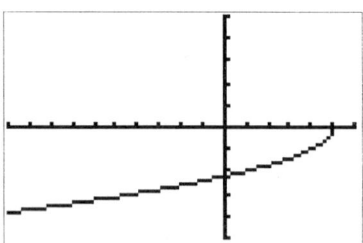

$$[-10, 6, 1] \times [-5, 5, 1]$$

(b) Domain: $\{x{:}x \leq 5\}$; range $\{y{:}y \leq 0\}$.

29. (a)

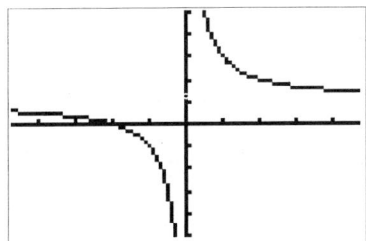

[–4.7, 4.7, 1] x [–5, 5, 1]

(b) Domain: $\{x : x \neq 0\}$; range $\{y : y \neq 1\}$

30. (a)

[–9.4, 9.4, 1] x [–8, 8, 1]

(b) Domain: $\{x : x \neq -2\}$; range $\{y : y \neq 1\}$

31. (a)

[–9.4, 9.4, 1] x [–15, 8, 1]

(b) Domain: $\{x : x \neq -2\}$; range $\{y : y < -7.46\}$ (decimal values are approximate).

32. (a)

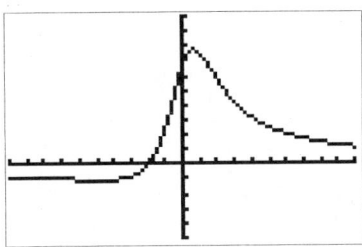

[–10, 10, 1] x [–0.5, 1, 0.1]

(b) Domain: all real numbers; range $\{y : -0.11 < y \leq 0.78\}$ (decimal values are approximate).

33. (a)

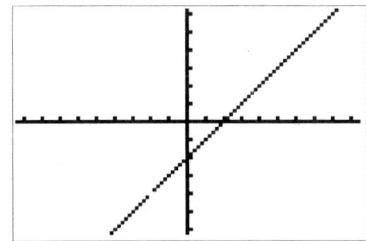

[–9.4, 9.4, 1] x [–6.2, 6.2, 1]

(b) Domain: $\{x : x \neq -2\}$; range $\{y : y \neq -4\}$.

34. (a)

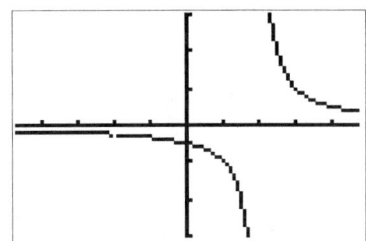

[–4.7, 4.7, 1] x [–3, 3, 1]

(b) Domain: $\{x : x \neq -2, x \neq 2\}$; range $\{y : y \neq 0, y \neq -\frac{1}{4}\}$.

35. Has an inverse function.

36. Has an inverse function.

37. Does not have an inverse function.

38. Does not have an inverse function.

39.

40.

41.

42.

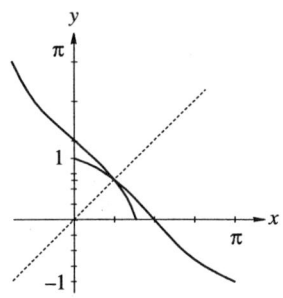

43. (a) 30%;
(b) 60%;
(c) 73%.

44. (a) $P = 5\sqrt{n} - 10; P + 10 = 5\sqrt{n}; \frac{P+10}{5} = \sqrt{n}; n = \left(\frac{P+10}{5}\right)^2$ or $P^{-1}(n) = \left(\frac{n+10}{5}\right)^2$;

(b) $n = \left(\frac{5+10}{5}\right)^2 = 3^2 = 9$;

(c)

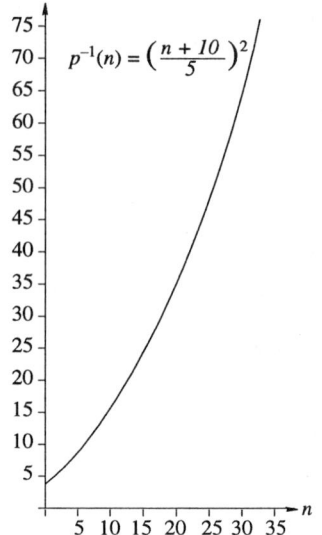

45. No.

46. (a) $C(30) = \frac{6.4(30)}{100-30} = \frac{192}{70} \approx 2.74$ million or 2,700,000;

(b) $C(60) = \frac{6.4(60)}{100-60} = \frac{384}{40} = 9.6$ million or 9,600,000;

(c) $C = \frac{6.4x}{100-x}; C(100-x) = 6.4x; 100C - Cx = 6.4x; 100C = 6.4x + Cx; 100C = x(6.4+C); x = \frac{100C}{6.4+C}$; or $P(x) = \frac{100x}{6.4+x}$ where x is the amount of money (in millions) that can be spent and $P(x)$ is the percent of pollutant that can be removed;

(d) $P(12) = \frac{100(12)}{6.4+12} = \frac{1200}{18.4} = 65.217 \approx 65.2\%$.

1. **(a)**

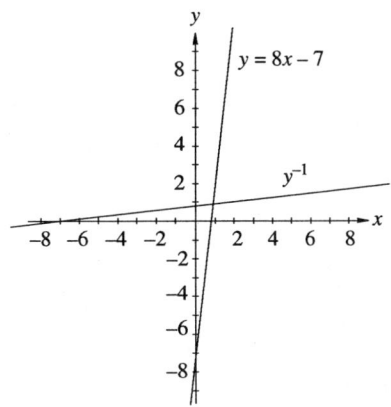

(b) Domain: all real numbers; Range: all real numbers; set $y = 0$ so $0 = 8x - 7$; $7 = 8x$; or $x = \frac{7}{8}$; x-intercept $\frac{7}{8}$; set $x = 0$; $y = 8(0) - 7 = -7$; y-intercept -7;
(c) Function;
(d) Has an inverse function.

2. **(a)**

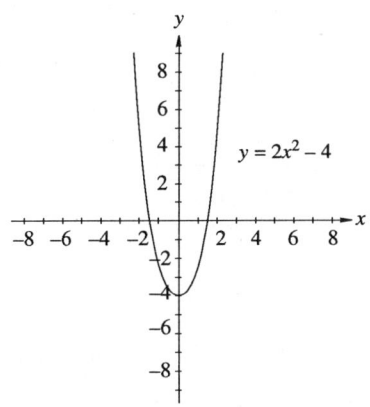

(b) Domain: all real numbers; Range: all real numbers greater than or equal to -4; set $y = 0$; $0 = 2x^2 - 4$; $2x^2 = 4$; $x = \pm\sqrt{2}$ x-intercept $\pm\sqrt{2}$; set $x = 0$; $y = 2(0)^2 - 4 = 0 - 4 = -4$; y-intercept -4;
(c) Function;
(d) Does not have an inverse function.

3. **(a)**

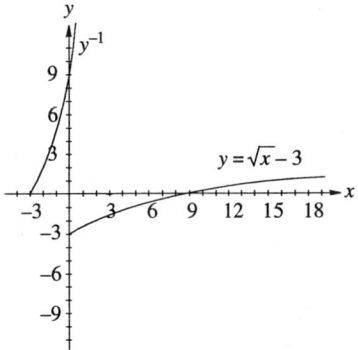

(b) Domain: all non-negative real numbers; $x \geq 0$; Range: all real numbers greater than or equal to -3. To find the x-intercept, set $y = 0$; Then, $0 = \sqrt{x} - 3$ or $\sqrt{x} = 3$ and so; $x = 3^2 = 9$ and the x-intercept $= 9$. To find the y-intercept, set $x = 0$. Then, $y - \sqrt{0} - 3 = -3$. and hence, the y-intercept $= -3$;
(c) Function;
(d) Has an inverse function.

4. **(a)**

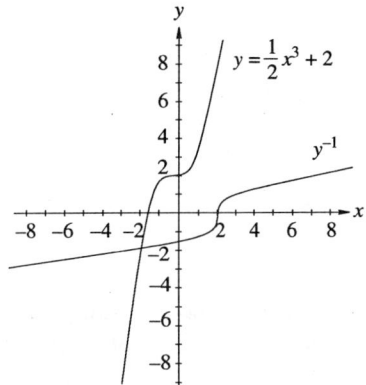

(b) Domain: all real numbers; Range: all real numbers; $0 = \frac{1}{2}x^3 + 2$; $\frac{1}{2}x^3 = -2$; $x^3 = -4$; $x = \sqrt[3]{-4}$; x-intercept $\sqrt[3]{-4}$; $y = \frac{1}{2}(0)^3 + 2 = 2$; y-intercept 2;
(c) Function;
(d) Has an inverse function.

5.   (a)

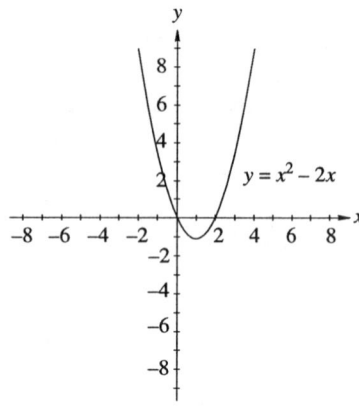

(b) Domain: all real numbers; Range: all real numbers greater than or equal to -1; $0 = x^2 - 2x$; $x(x - 2) = 0$, so $x = 0$ or $x = 2$; x-intercepts $0, 2$; $y = 0^2 - 2(0) = 0$; y-intercept 0;
(c) Function;
(d) Does not have an inverse function.

6.   (a)

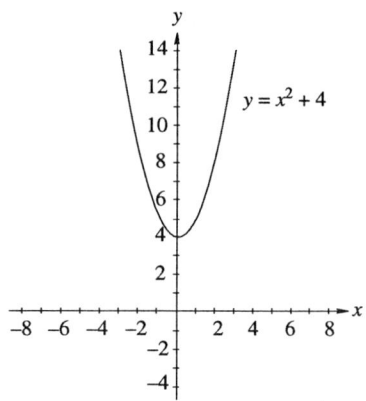

(b) Domain: all real numbers; Range: all real numbers 4 or larger; no x-intercept; y-intercept 4;
(c) Function;
(d) Does not have an inverse function.

7. $f(0) = 4(0) - 12 = 0 - 12 = -12$

8. $f(-2) = 4(-2) - 12 = -8 - 12 = -20$

9. $f(3) = 4(3) - 12 = 12 - 12 = 0$

10. $f(a) = 4a - 12$

11. $f(a - 2) = 4(a - 2) - 12 = 4a - 8 - 12 = 4a - 20$

12. $f(x + h) = 4(x + h) - 12 = 4x + 4h - 12$

13. $g(0) = \frac{0^2 - 9}{0^2 + 9} = \frac{-9}{9} = -1$

14. $f(3) = \frac{3^2 - 9}{3^2 + 9} = \frac{9 - 9}{9 + 9} = \frac{0}{18} = 0$

15. $g(-3) = \frac{(-3)^2 - 9}{(-3)^2 + 9} = \frac{9 - 9}{9 + 9} = \frac{0}{18} = 0$

16. $g(-2) = \frac{(-2)^2 - 9}{(-2)^2 + 9} = \frac{4 - 9}{4 + 9} = \frac{-5}{13} \approx -0.3846$

17. $g(4) = \frac{4^2 - 9}{4^2 + 9} = \frac{16 - 9}{16 + 6} = \frac{7}{25} = 0.28$

18. $g(-5) = \frac{(-5)^2 - 9}{(-5)^2 + 9} = \frac{25 - 9}{25 + 9} = \frac{16}{34} = \frac{8}{17} \approx 0.4706$

19.

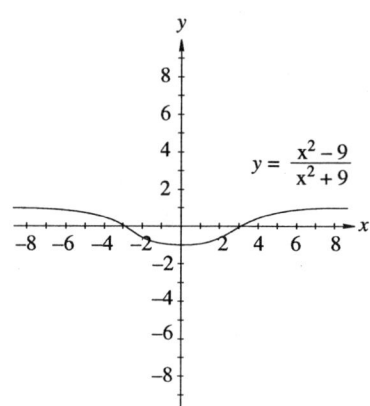

The zeros are -3 and 3.

20. $(f + g)(x) = f(x) + g(x) = 4x - 12 + \frac{x^2 - 9}{x^2 + 9}$

21. $(f + g)(3) = f(3) + g(3) = (4 \cdot 3 - 12) + \frac{3^2 - 9}{3^2 + 9} = 0 + 0 = 0$

22. $(f - g)(x) = f(x) - g(x) = 4x - 12 - \frac{x^2 - 9}{x^2 + 9}$

23. $(f - g)(-2) = f(-2) - g(-2) = 4(-2) - 12 - \frac{-5}{13} = -20 + \frac{5}{13} = \frac{-260}{13} + \frac{5}{13} = \frac{-255}{13}$

24. $(f \cdot g)(x) = f(x) \cdot g(x) = (4x - 12)\left(\frac{x^2 - 9}{x^2 + 9}\right)$

25. $(f \cdot g)(0) = f(0) \cdot g(0) = (-12)(-1) = 12$

26. $(f/g)(x) = \frac{4x - 12}{\frac{x^2 - 9}{x^2 + 9}} = (4x - 12)\left(\frac{x^2 + 9}{x^2 - 9}\right) = \frac{4(x - 3)(x^2 + 9)}{(x - 3)(x + 3)} = \frac{4(x^2 + 9)}{x + 3}$

27. $(f/g)(5) = \frac{4(5^2 + 9)}{5 + 3} = \frac{4(25 + 9)}{8} = \frac{1(34)}{2} = 17$

28. $(g/f)(x) = \frac{1}{(f/g)(x)} = \frac{x + 3}{4(x^2 + 9)}$

29. $(g/f)(2) = \frac{2 + 3}{4(2^2 + 9)} = \frac{5}{4(4 + 9)} = \frac{5}{4(13)} = \frac{5}{52}$

30. $(f \circ g)(x) = f(g(x)) = f\left(\frac{x^2 - 9}{x^2 + 9}\right) = 4\left(\frac{x^2 - 9}{x^2 + 9}\right) - 12$

31. $(f \circ g)(4) = 4\left(\frac{4^2-9}{4^2+9}\right) - 12 = 4\left(\frac{16-9}{16+9}\right) - 12 =$

$4\left(\frac{7}{25}\right) - 12 = \frac{28}{25} - \frac{300}{25} = \frac{-272}{25}$

32. $(g \circ f)(x) = g(f(x)) = g(4x-12) = \frac{(4x-12)^2-9}{(4x-12)^2+9} =$

$\frac{(16x^2-96x+144)-9}{(16x^2-96+144)+9} = \frac{16x^2-96x+135}{16x^2-96x+153}$

33. $(g \circ f)(3) = \frac{16(3)^2-96(3)+135}{16(3)^2-96(3)+153} = \frac{144-288+135}{144-288+153} =$

$\frac{-9}{9} = -1$

34.

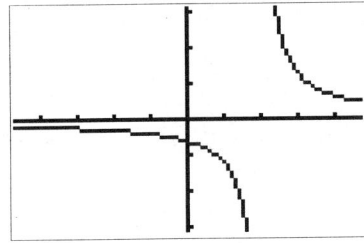

[–9.4, 9.4] x [–6.2, 6.2]

35.

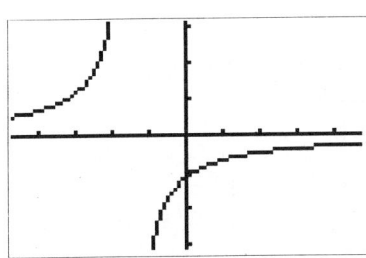

[–9.4, 9.4] x [–6.2, 6.2]

36.

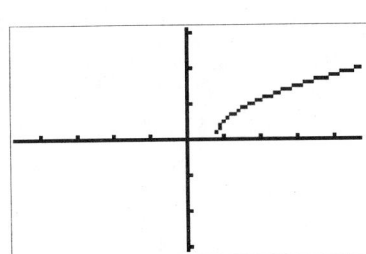

[–9.4, 9.4] x [–6.2, 6.2]

37.

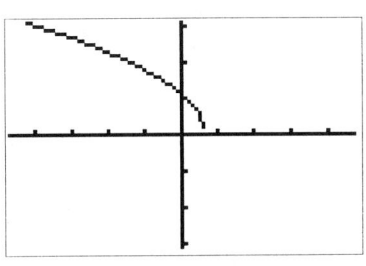

[–9.4, 9.4] x [–6.2, 6.2]

38.

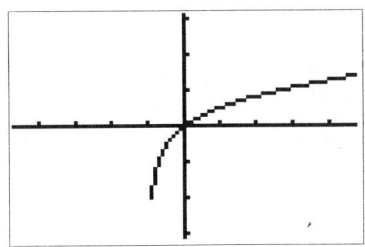

[–9.4, 9.4] x [–6.2, 6.2]

39. Solve for y to get two functions: $y = \sqrt{9 - x^2}$ and $y = -\sqrt{9 - x^2}$. When both of these are graphed on the same coordinate system you get the following figure.

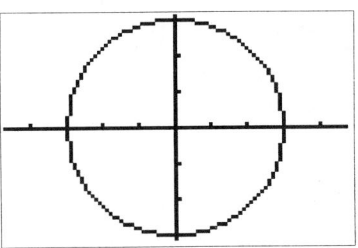

[–4.7, 4.7] x [–3.1, 3.1]

40. Solve for y to get two functions: $y = \sqrt{16 - 0.5x^2}$ and $y = -\sqrt{16 - 0.5x^2}$. When both of these are graphed on the same coordinate system, you get the following figure.

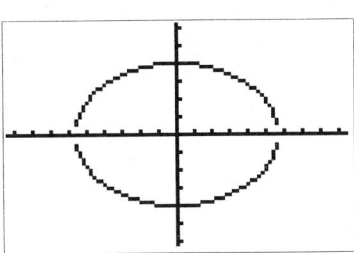

[–9.4, 9.4] x [–6.2, 6.2]

41. $4x + 7y = 0; 7y = -4x; y = \frac{-4}{7}x$; roots are $0 = -\frac{4}{7}x$ or $x = 0$.

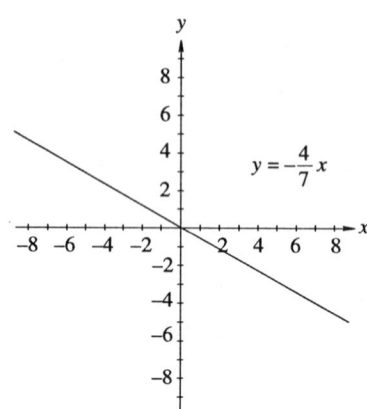

42. $x^2 - 20 = y$ or $y = x^2 - 20$; roots are $0 = x^2 - 20$; $x^2 = 20; x = \pm\sqrt{20} = \pm\sqrt{5} \approx \pm 4.4721$.

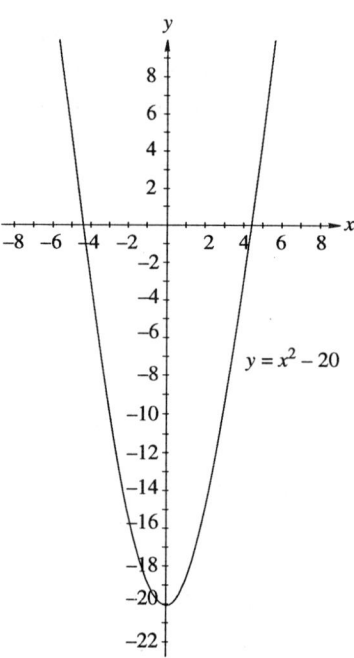

43. $2x^2 + 10x + 4 = 0$ or $2(x^2 + 5x + 2) = 0$ roots are about -0.44 and -4.50.

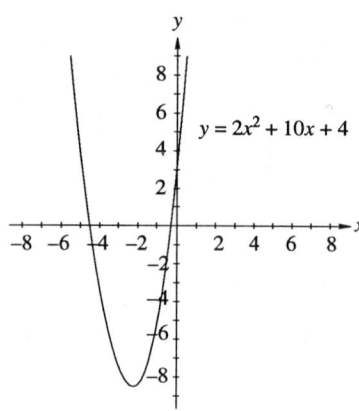

44. $y = 8x^3 - 20x^2 - 34x + 21$; roots: $x = -1.5, x = 0.5$, and $x = 3.5$.

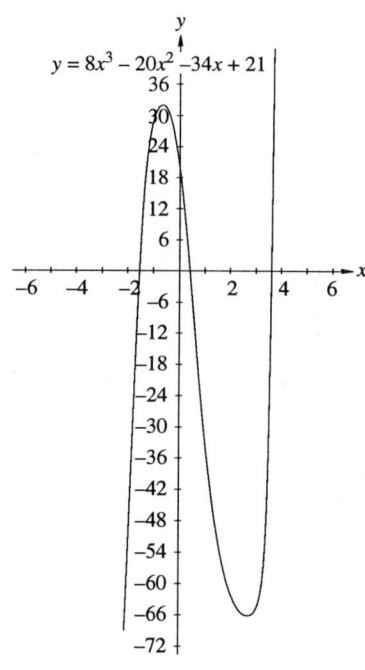

45. **(a)** $P(101) = 35 - \frac{101}{20} = 35 - 5.05 = \$29.95$;

$P(350) = 35 - \frac{350}{20} = 35 - 17.50 = \$17.50$;

$P(400) = 35 - \frac{400}{20} = 35 - 20 = \$15.00$;

(b) $R = D \times P = n\left(35 - \frac{n}{20}\right) = 35n - \frac{n^2}{20}$;

(c) $R = 101(29.95) = 3024.95$ or $R(n) = 35(101) - \frac{(101)^2}{20} = 3535 - 510.05 = 3024.95$;

$R(350) = 35(350) - \frac{350^2}{20} = 6125$;

$R(400) = 35(400) - \frac{400^2}{20} = 6000$

46. (a) $P(n) = \left(30n - \frac{n^2}{20}\right) - (550 + 10n) = 20n - \frac{n^2}{20} - 550$;

(b) $P(30) = 20(30) - \frac{30^2}{20} - 550 = \$5$;

$P(100) = 20(100) - \frac{100^2}{20} - 550 = \$950$; $P(150) = \$1325$; $P(300) = \$950$; $P(400) = -\$550$.

47.

t	0	1	2	3	4	5
$h(t)$	0	3.08	6.48	10.68	15.68	21.00

t	6	7	8	9	10
$h(t)$	25.68	28.28	26.88	19.08	2.00

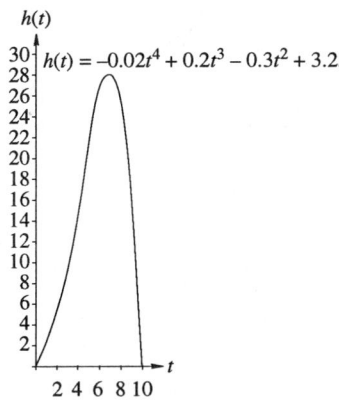

$h(t) = -0.02t^4 + 0.2t^3 - 0.3t^2 + 3.2t$

48. (a)

v	0	10	20	30	40	50	60	70
$s(v)$	0	14	36	66	104	150	204	266

(b) about 69.8 mph;

(c)

$s(v) = 0.04v^2 + v$

$s^{-1}(v)$

CHAPTER 4 TEST

1. $f(-2) = 7(-2) - 5 = -14 - 5 = -19$

2. $f(3 - a) = 7(3 - a) - 5 = 7 \cdot 3 - 7a - 5 = 21 - 7a - 5 = 16 - 7a$

3. $g(0) = \frac{0^2 - 2 \cdot 0 - 15}{0 + 3} = \frac{-15}{3} = -5$

4. $g(5) = \frac{5^2 - 2 \cdot 5 - 15}{5 + 3} = \frac{25 - 10 - 15}{8} = \frac{0}{8} = 0$

5. (a)

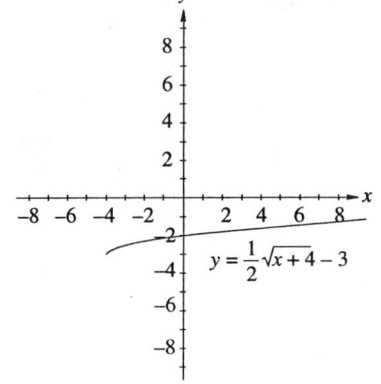

$y = \frac{1}{2}\sqrt{x + 4} - 3$

(b) This function is defined whenever $\sqrt{x + 4} \geq 0$ or when $x \geq -4$;

(c) The low point of the graph is when $x = -4$. Here the y-value is -3. So, the graph can assume all values of $y \geq -3$;

(d) The x-intercept is when $y = 0$. Solving $0 = \frac{1}{2}\sqrt{x + 4} - 3$, we get $3 = \frac{1}{2}\sqrt{x + 4}$ or $6 = \sqrt{x + 4}$. Squaring both sides we get $36 = x + 4$ or $x = 32$;

(e) The y-intercept is when $x = 0$. Solving $y = \frac{1}{2}\sqrt{0 + 4} - 3 = \frac{1}{2}\sqrt{4} - 3 = \frac{1}{2}2 - 3 = 1 - 3 = -2$.

6. $3x - 15 + \frac{x - 5}{x + 5} = \frac{(3x - 15)(x + 5)}{x + 5} + \frac{x - 5}{x + 5} = \frac{3x^2 - 75}{x + 5} + \frac{x - 5}{x + 5} = \frac{3x^2 + x - 80}{x + 5}$

7. $3x - 15 - \frac{x - 5}{x + 5} = \frac{(3x - 15)(x + 5)}{x + 5} - \frac{x - 5}{x + 5} = \frac{3x^2 - 75}{x + 5} - \frac{x - 5}{x + 5} = \frac{3x^2 + x - 70}{x + 5}$

8. $(3x - 15)\left(\frac{x - 5}{x + 5}\right) = \frac{(3x - 15)(x - 5)}{x + 5} = \frac{3(x - 5)(x - 5)}{x + 5} = \frac{3(x - 5)^2}{x + 5}$

9. $(3x-15) \div \frac{x-5}{x+5} = (3x-15)\left(\frac{x+5}{x-5}\right) = \frac{3(x-5)(x+5)}{x-5} =$
$3(x+5) = 3x+15$

10. $3\left(\frac{x-5}{x+5}\right) - 15 = \frac{3x-15}{x+5} - \frac{15x+75}{x+5} = \frac{3x-15-15x-75}{x+5} =$
$\frac{-12x-90}{x+5} = -6\left(\frac{2x+15}{x+5}\right)$

11. $\frac{(3x-15)-5}{(3x-15)+5} = \frac{3x-20}{3x-10}$

12.

13.

14.

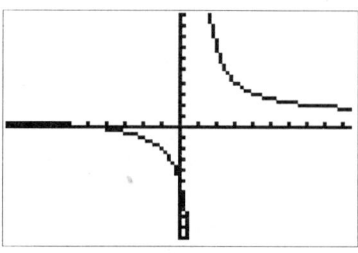

[−9.4, 9.4] x [−10, 10]

15. Graph $y = \frac{x^2}{x+1} - 2x + 1$.

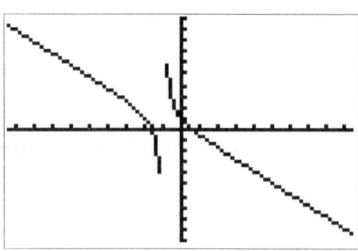

[−9.4, 9.4] x [−10, 10]

16. Graph $y = \sqrt{x^2 - 1} - 3 + x$.

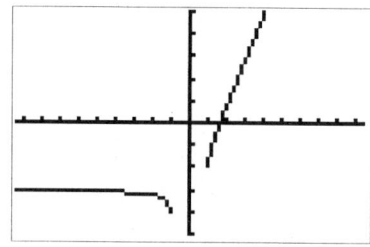

[−9.4, 9.4] x [−5, 5]

17. **(a)**

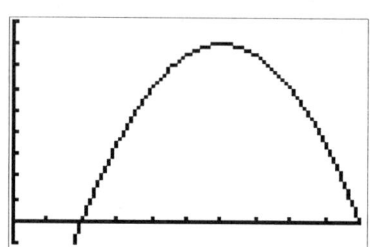

[0, 10, 1] x [−2, 18, 2]

(b) $r(3) = -3^2 + 12(3) - 20 = -9 + 36 - 20 = 7$.
So, seven responses can be expected after 3 ms.
(c) Here $-s^2 + 12s - 20 = 16$ or $s^2 - 12s + 36 = 0$.
This factors as $(s-6)^2 = 0$, so $s = 6$. It has been
6 ms since the nerve was stimulated.
(d) We need to solve $-s^2 + 12s - 20 = 12$ or
$s^2 - 12s + 32 = 0$. This factors as $(s-4)(s-8) = 0$,
so $s = 4$ or $s = 8$. It has been either 4 ms or 8
ms since the nerve was stimulated.

18.

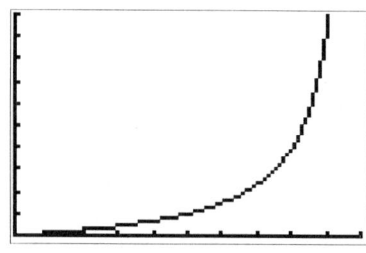

[0, 100, 10] x [0, 50, 5]

(b) \$5,000,
(c) \$45,000,
(d) $y = \frac{5x}{100-x}$. Solving for x, we have $y(100 - x) = 5x$ or $100y - xy = 5x$. Grouping x's
we obtain $5x + xy = 100y$, and factoring yields
$x(y + 5) = 100y$ and so, $x = \frac{100y}{y+5}$. Interchanging
x and y, and using function notation, we obtain
$C^{-1}(x) = \frac{100x}{x+5}$,
(e) 75%.

CHAPTER

5

Systems of Linear Equations and Determinants

≡ 5.1 LINEAR EQUATIONS

1. $(2,5)$, $(3,8)$, $m = \frac{y_2-y_1}{x_2-x_1} = \frac{8-5}{3-2} = \frac{3}{1} = 3$

2. $(4,7)$, $(-2,1)$, $m = \frac{1-7}{-2-4} = \frac{-6}{-6} = 1$

3. $(1,8)$, $(5,3)$, $m = \frac{3-8}{5-1} = \frac{-5}{4} = -\frac{5}{4}$

4. $(0,4)$, $(5,0)$, $m = \frac{0-4}{5-0} = \frac{-4}{5} = -\frac{4}{5}$

5. $(9,3)$, $(2,-7)$, $m = \frac{-7-3}{2-9} = \frac{-10}{-7} = \frac{10}{7}$

6. $m = \frac{-5-1}{2-(-6)} = \frac{-6}{8} = -\frac{3}{4}$

7. $\frac{5-5}{4-0} = \frac{0}{4} = 0$, so the slope is 0, or $m = 0$.

8. $\frac{7-5}{-3-(-3)} = \frac{2}{0}$. You cannot divide by 0, and so the slope is undefined.

9. $\frac{\frac{3}{4} - 17}{2 - 2} = \frac{\frac{-65}{4}}{0}$. You cannot divide by 0, and so the slope is undefined.

10. $\frac{-7-(-7)}{-7-2} = \frac{0}{-9} = 0$. Thus the slope is 0, or $m = 0$.

11. $m = 4$, point: $(5,3)$. Using $y - y_1 = m(x - x_1)$ we get $y - 3 = 4(x - 5)$

12. $m = -3$, point: $(-6,1)$, $y - 1 = -3(x - (-6))$ or $y - 1 = -3(x + 6)$

13. $m = \frac{2}{3}$, point: $(1,-5)$, $y - (-5) = \frac{2}{3}(x - 1)$ or $y + 5 = \frac{2}{3}(x - 1)$

14. $m = \frac{3}{2}$, point: $(0,5)$, $y - 5 = \frac{3}{2}(x - 0)$ or $y - 5 = \frac{3}{2}x$

15. Since $m = 0$, this is a horizontal line. All horizontal lines have the same y-coordinate, so the equation of this line is $y = -5$.

16. Since the slope m is undefined, this is a vertical line. All vertical lines have the same x-coordinate, so the equation of this line is $x = -3$.

17. $m = -\frac{5}{3}$, point: $(2,0)$, $y - 0 = -\frac{5}{3}(x - 2)$ or $y = -\frac{5}{3}(x - 2)$

18. $m = -\frac{3}{4}$, point: $(-3,-1)$, $y - (-1) = -\frac{3}{4}(x - (-3))$ or $y + 1 = -\frac{3}{4}(x + 3)$

19. Since the slope m is undefined, this is a vertical line. All vertical lines have the same x-coordinate, so the equation of this line is $x = 7$.

20. Since $m = 0$, this is a horizontal line. All horizontal lines have the same y-coordinate, so the equation of this line is $y = 5$.

21. points: $(1,5)$ and $(-3,2)$. First $m = \frac{2-5}{-3-1} = \frac{-3}{-4} = \frac{3}{4}$; then using $(-3,2)$ we get $y - 2 = \frac{3}{4}(x + 3)$; using $(1,5)$ we get $y - 5 = \frac{3}{4}(x - 1)$

22. points: $(-5,6)$ and $(1,-6)$, $m = \frac{-6-6}{1+5} = \frac{-12}{6} = -2$, $y + 6 = -2(x - 1)$ or $y - 6 = -2(x + 5)$

23. $m = 0$, so this is a horizontal line. All horizontal lines have the same y-coordinate, so the equation of this line is $y = 3$.

24. $m = \frac{\frac{25}{8} - \frac{3}{4}}{0 - \frac{2}{3}} = -\frac{57}{16}$. We are given the y-intercept: $(0, \frac{25}{8})$, so the equation is $y = -\frac{57}{16}x + \frac{25}{8}$.

25. $m = 2$, $b = 4$. Using $y = mx + b$ we get $y = 2x + 4$

26. $m = -3, b = 5, y = -3x + 5$

27. $m = 5, b = -3, y = 5x - 3$

28. $m = -4, b = -2, y = -4x - 2$

29. $y - 3x = 6$; adding $3x$ to both sides we get $y = 3x + 6$ so $m = 3, b = 6, x\text{-intercept} = -2$

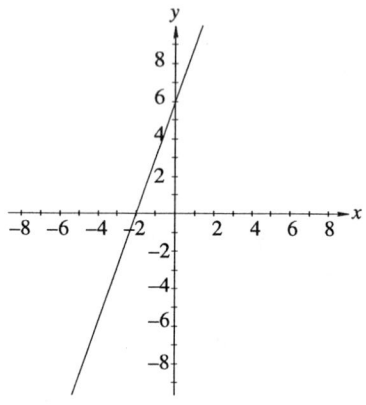

30. $2x - y = 5$. Subtracting $2x$ yields $-y = -2x + 5$, multiplying by -1 yields $y = 2x - 5$ so $m = 2, b = -5; x\text{-intercept} = \frac{5}{2}$

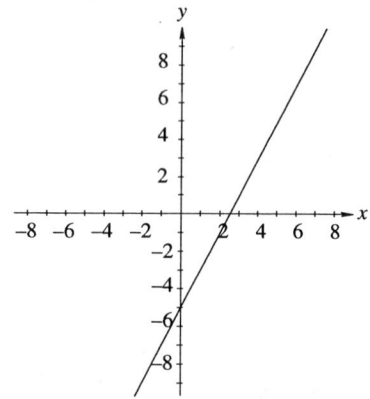

31. $2y - 5x = 8 \Rightarrow 2y = 5x + 8 \Rightarrow y = \frac{5}{2}x + 4; m = \frac{5}{2}; b = 4; x\text{-intercept} = -\frac{8}{5}$

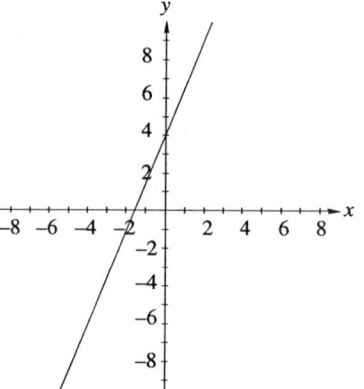

32. $2x - 3y = 9 \Rightarrow -3y = -2x + 9 \Rightarrow y = \frac{2}{3}x - 3; m = \frac{2}{3}; b = -3; x\text{-intercept} = \frac{9}{2}$

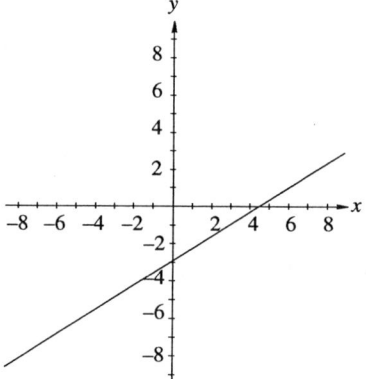

33. $5x - 2y - 10 = 0 \Rightarrow -2y = -5x + 10 \Rightarrow y = \frac{5}{2}x - 5; m = \frac{5}{2}; b = -5; x\text{-intercept} = 2$

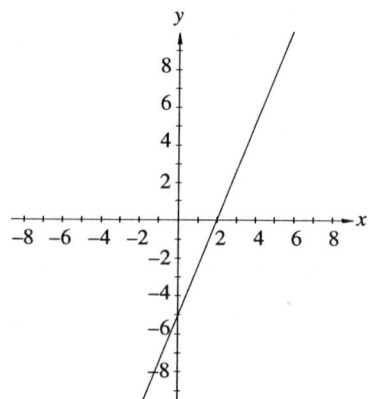

34. $4y - 3x - 4 = 0; 4y = 3x + 4; y = \frac{3}{4}x + 1; m = \frac{3}{4}; b = 1; x\text{-intercept} = -\frac{4}{3}$

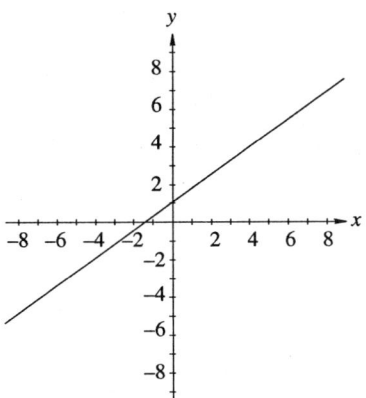

35. $x = -3y + 7; 3y = -x + 7; y = -\frac{1}{3}x + \frac{7}{3}; m = -\frac{1}{3}; b = \frac{7}{3}; x\text{-intercept} = 7$

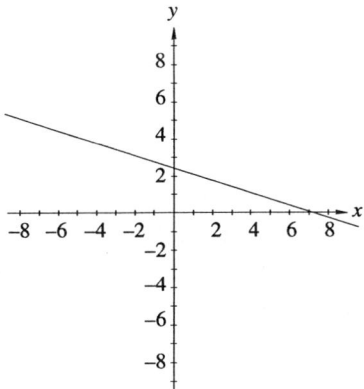

36. $3x = 5y - 6, 5y = 3x + 6; y = \frac{3}{5}x + \frac{6}{5}; m = \frac{3}{5}, b = \frac{6}{5}, x\text{-intercept} = -2$

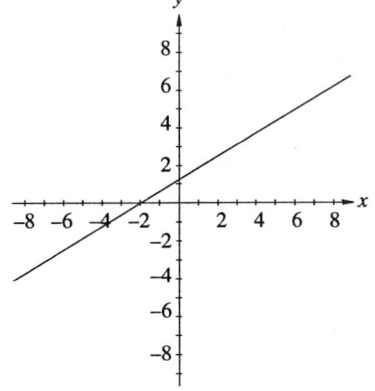

37. **(a)** This line is horizontal, so $m = 0$.
(b)

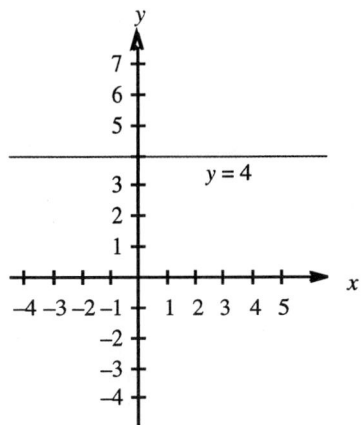

38. **(a)** This is a vertical line, so the slope is undefined.
(b)

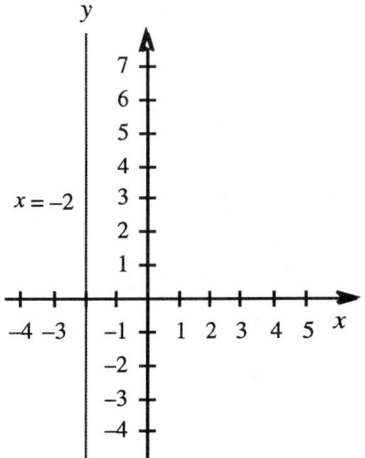

39. **(a)** This is a vertical line, so the slope is undefined.
(b)

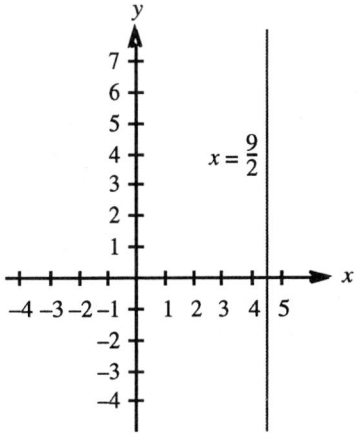

40. **(a)** This is a horizontal line, so $m = 0$.
(b)

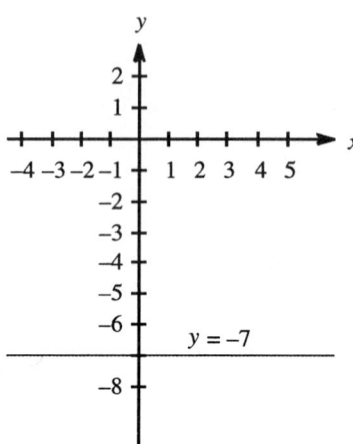

41. $s = md = \left(\frac{1780\pi}{60}\right)d$ so $m = \frac{1780\pi}{60} = \frac{178}{6}\pi = \frac{89\pi}{3} \approx 93.2$

42. **(a)** $m = \frac{0-(-40)}{32-(-40)} = \frac{40}{72} = \frac{5}{9}$

(b) Equation is $y - 0 = \frac{5}{9}(x - 32) \Rightarrow y = \frac{5}{9}x - \frac{160}{9}$.
Hence the y-intercept is $-\frac{160}{9}$.

(c) $y = \frac{5}{9}(x - 32)$ or $y + 40 = \frac{5}{9}(x + 40)$ which has the slope-intercept form $y = \frac{5}{9}x - \frac{160}{9}$.

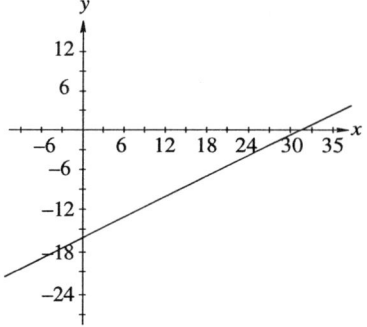

43. **(a)** $F = kx = k(L - L_0)$ or $F = kL - kL_0$;
(b) $F = 4.5(L - 6)$ or $F = 4.5L - 27$
(c)

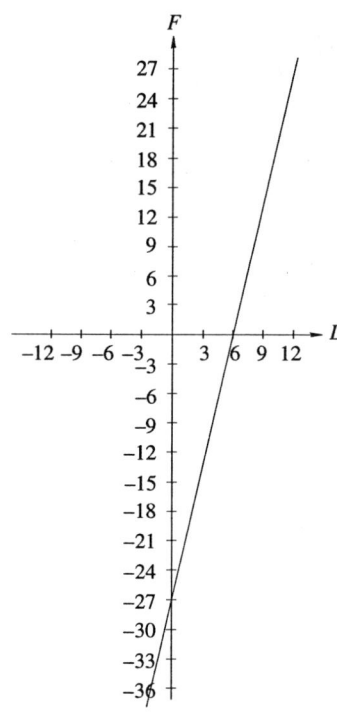

44. Using $(4, 250)$ and $(9, 562.5)$ as two points the slope $= \frac{562.5-250}{9-4} = \frac{312.5}{5} = 62.5$; hence $P - 250 = 62.5(h - 4)$; $P - 250 = 62.5h - 250$ or $P = 62.5h$.

☰ 5.2 GRAPHICAL AND ALGEBRAIC METHODS FOR SOLVING TWO LINEAR EQUATIONS IN TWO VARIABLES

1.

$(4, 2)$

2.

$(3, 5)$

3.

$(4.5, 2)$

4.

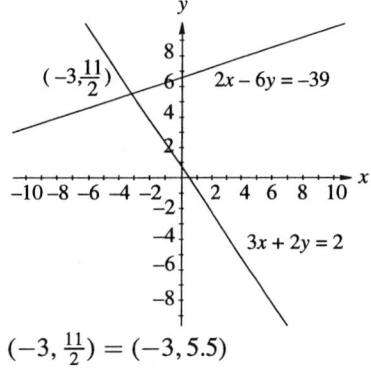

$(-3, \frac{11}{2}) = (-3, 5.5)$

5.

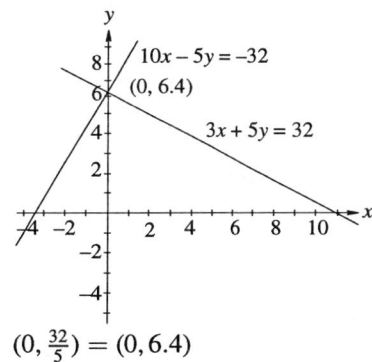

$(0, \frac{32}{5}) = (0, 6.4)$

6.

$(-2.2, 0)$

7.

$(-2.2, -1.1)$

8.

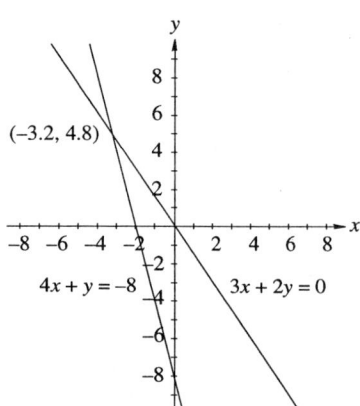

$(-3.2, 4.8)$

9. $\begin{cases} y = 3x - 4 & (1) \\ x + y = 8 & (2) \end{cases}$

Substituting $(3x - 4)$ for y in equation (2) yields $x + (3x - 4) = 8$. Solving $4x - 4 = 8; 4x = 12$, $x = 3$. Substituting $x = 3$ in (1) $\Rightarrow y = 3(3) - 4 = 9 - 4 = 5; (3, 5)$

10. $\begin{cases} x = -2y + 12 & (1) \\ x + y = 5 & (2) \end{cases}$

Substituting $-2y + 12$ for x in (2) $\Rightarrow -2y + 12 + y = 5$. Solving $-y + 12 = 5; -y = -7; y = 7$. Substituting 7 for y in (1) $\Rightarrow x = -2(7) + 12 = -14 + 12 = -2; (-2, 7)$

11. $\begin{cases} y = -2x - 2 & (1) \\ 3x + 2y = 0 & (2) \end{cases}$

Substituting $-2x - 2$ for y in (2) $\Rightarrow 3x + 2(-2x - 2) = 0; 3x - 4x - 4 = 0; -x = 4$ or $x = -4$. Back substituting in (1) $y = -2(-4) - 2 = +8 - 2 = 6; (-4, 6)$

12. $\begin{cases} x = 7 + 2y & (1) \\ 3x + 4y = 1 & (2) \end{cases}$

Substituting $7 + 2y$ in (2) $\Rightarrow 3(7 + 2y) + 4y =$

$1; 21 + 6y + 4y = 1; 21 + 10y = 1; 10y = -20; y = -2; x = 7 + 2(-2); x = 7 - 4 = 3$. Solution $(3, -2)$

13. $\begin{cases} 2x + 5y = 6 & (1) \\ x - y = 10 & (2) \end{cases}$

Solving (2) for $x \Rightarrow x = 10 + y$. Substituting in (1) $\Rightarrow 2(10 + y) + 5y = 6; 20 + 2y + 5y = 6; 7y = -14; y = -2. x = 10 + (-2) = 8$. Solution $(8, -2)$

14. $\begin{cases} 3x - 2y = 5 & (1) \\ -7x + 4y = -7 & (2) \end{cases}$

Solving (1) for y yields $-2y = 5 - 3x$ or $y = \frac{3x-5}{2}$ (3). Substituting in (2) yields $-7x + 4(\frac{3x-5}{2}) = -7; -7x + 6x - 10 = -7; -x = 3$ or $x = -3$. Substituting in (3) yields $y = \frac{3(-3)-5}{2} = \frac{-14}{2} = -7; (-3, -7)$

15. $\begin{cases} 2x + 3y = 3 & (1) \\ 6x + 4y = 15 & (2) \end{cases}$

Solving for x in (1) $\Rightarrow 2x = 3 - 3y$ or $x = \frac{3-3y}{2}$ (3). Substituting this in (2) yields $6(\frac{3-3y}{2}) + 4y = 15; 3(3 - 3y) + 4y = 15; 9 - 9y + 4y = 15; -5y = 6$ or $y = -\frac{6}{5}$ or -1.2. Substituting this for y in (3) yields $x = \frac{3-3(-1.2)}{2} = \frac{3+3.6}{2} = \frac{6.6}{2} = 3.3$. Solution $(3.3, -1.2)$

16. $\begin{cases} 2x + 2y = -3 & (1) \\ 4x + 9y = 5 & (2) \end{cases}$

Solving (1) for x we get $2x = -2y - 3$ or $x = \frac{-2y-3}{2}$. Substituting in (2) yields $4(\frac{-2y-3}{2}) + 9y = 5; -4y - 6 + 9y = 5; 5y = 11; y = \frac{11}{5} = 2.2; x = \frac{-2(2.2)-3}{2} = \frac{-4.4-3}{2} = \frac{-7.4}{2} = -3.7; (-3.7, 2.2)$

17. $\begin{cases} 4.8x - 1.3y = 16.9 & (1) \\ -7.2x - 2.8y = -9.2 & (2) \end{cases}$

Solving equation (1) for x, we obtain $4.8x = 1.3y + 16.9$ or $x = 0.27083y + 3.52083$ (3)

Now, substituting for x in equation (2), we have

$$-7.2(0.27083y + 3.52083) - 2.8y = -9.2$$
$$-1.95y - 25.35 - 2.8y = -9.2$$
$$-4.75y = 16.15$$
$$y = -3.4$$

We still need to find the x-coordinate of the solution. Back-substituting into equation (3) yields $x = 0.27083(-3.4) + 3.52083 = 2.6$. The solution is $(2.6, -3.4)$.

18. $\begin{cases} 2.3x + 1.7y = 8.5 & (1) \\ -6.7x + 3.7y = 38.4 & (2) \end{cases}$
Solving equation (1) for y we get $1.7y = -2.3x + 8.5$, which simplifies to

$$y = -1.352941176x + 5 \ (3)$$

Now, by substituting for y in equation (2), we obtain

$$\begin{aligned} -6.7x + 3.7(-1.352941176x + 5) &= 38.4 \\ -6.7x - 5.005882353x + 18.5 &= 38.4 \\ \text{or} \qquad\qquad -11.70588235x &= 19.9 \\ x &= -1.7 \end{aligned}$$

Back-substituting this result into equation (3), we obtain $y = -1.352941176(-1.7) + 5$ or $y = 7.3$. The answer is $(-1.7, 7.3)$.

19. $\begin{cases} 4.2x + 3.7y = 10.79 & (1) \\ 6.5x - 0.3y = -15.24 & (2) \end{cases}$
Solving equation (1) for x, we obtain
$4.2x = -3.7y + 10.79$ or
$x = -0.88095231y + 2.569047619 \ (3)$
Substituting this result for x in equation (2) gives

$$\begin{aligned} 6.5(-0.88095231y + 2.569047619) - 0.3y &= -15.24 \\ -5.726190476y + 16.69880952 - 0.3y &= -15.24 \\ -6.026190476y &= -31.93880952 \\ y &= 5.3 \end{aligned}$$

Back-substituting into equation (3) yields $x = -0.880952381(5.3) + 2.569047619$ or $x = -2.1$. The solution is $(-2.1, 5.3)$.

20. $\begin{cases} 0.75x + 1.5y = -0.225 & (1) \\ 3.13x + 1.74y = 13.073 & (2) \end{cases}$
Solving equation (1) for y yields
$1.5y = -0.75x - 0.225$ or
$y = -0.5x - 0.15 (3)$
Substituting this result into equation (2), we obtain

$$\begin{aligned} 3.13x + 1.74(-0.5x - 0.15) &= 13.073 \\ 3.13x - 0.87x - 0.261 &= 13.073 \\ 2.26x &= 13.334 \\ x &= 5.9 \end{aligned}$$

Back-substituting into equation (3), we obtain $y = -0.5(5.9) - 0.15 = -3.1$. The solution is $(5.9, -3.1)$.

21. $\begin{cases} x + y = 9 & (1) \\ x - y = 5 & (2) \end{cases}$
Adding $(1) + (2) \Rightarrow 2x = 14; x = 7$; Substituting in (1) yields $7 + y = 9; y = 2; (7, 2)$

22. $\begin{cases} 2x + 3y = 5 & (1) \\ -2x + 5y = 3 & (2) \end{cases}$
Adding $(1) + (2)$ yields $8y = 8$ or $y = 1$. Substituting 1 for y yields $2x + 3 = 5$ or $2x = 2$ and so $x = 1$. The solution is $(1, 1)$.

23. $\begin{cases} -x + 3y = 5 & (1) \\ 2x + 7y = 3 & (2) \end{cases}$
Multiply (1) by $2 \Rightarrow -2x + 6y = 10\,(3)$. Adding

$(3) + (2)$ yields $13y = 13$ or $y = 1$. Back substituting in (1) yields $-x + 3 = 5$ or $-x = 2$ and so we see that $x = -2$. The solution is $(-2, 1)$.

24. $\begin{cases} 3x - 2y = 8 & (1) \\ 5x + y = 9 & (2) \end{cases}$
Multiply (2) by $2 \Rightarrow 10x + 2y = 18\,(3)$. Adding $(1) + (3)$ gives $13x = 26$ or $x = 2$. Substituting 2 for x in (1) yields $3(2) - 2y = 8$ or $6 - 2y = 8$ and so $-2y = 2; y = -1$. The solution is $(2, -1)$.

25. $\begin{cases} 3x - 2y = -15 & (1) \\ 5x + 6y = 3 & (2) \end{cases}$
Multiply (1) by $3 \Rightarrow 9x - 6y = -45(3)$. Add $(3) + (2): 14x = -42 \Rightarrow x = -3$. Substituting $x = -3$ in (1) we get $3(-3) - 2y = -15 \Rightarrow -9 - 2y = -15; -2y = -6$, so $y = 3$ and the solution is $(-3, 3)$.

26. $\begin{cases} 2x - 3y = 11 & (1) \\ 6x - 5y = 13 & (2) \end{cases}$
Multiply (1) by $3 \Rightarrow 6x - 9y = 33(3)$; Subtract $(3) - (2) \Rightarrow -4y = 20 \Rightarrow y = -5$. Thus, $2x - 3(-5) = 11$ or $2x + 15 = 11; 2x = -4$ and so $x = -2$. The solution is $(-2, -5)$.

27. $\begin{cases} 3x - 5y = 37 & (1) \\ 5x - 3y = 27 & (2) \end{cases}$

Multiply (1) by 5 $\Rightarrow$ $15x - 25y = 185\,(3)$. Multiply (2) by 3 $\Rightarrow$ $15x - 9y = 81\,(4)$. Subtract $(3)-(4) \Rightarrow -16y = 104$ and so $y = -6.5$. Substituting in (1) $\Rightarrow 3x - 5(-6.5) = 37$ or $3x + 32.5 = 37 \Rightarrow x = 1.5$. The solution is $(1.5, -6.5)$.

28. $\begin{cases} x + \frac{1}{2}y = 7 & (1) \\ 4x - 2y = 5 & (2) \end{cases}$

Multiply (1) by 4 $\Rightarrow$ $4x + 2y = 28$ (3). Adding $(3) + (2) \Rightarrow 8x = 33$ and so $x = \frac{33}{8} = 4\frac{1}{8} = 4.125$. Substituting $x = 4.125$ in (1), we get $4.125 + 0.5y = 7$ or $0.5y = 2.875$ and so we see that $y = 5.75$. The solution is $(4.125, 5.75)$.

29. $\begin{cases} 6.4x - 1.7y = 66.7 & (1) \\ -4.2x + 5.1y = -62.1 & (2) \end{cases}$

Multiply equation (1) by 3 to obtain

$$19.2x - 5.1y = 200.1 \quad (3)$$

Adding equations (2) and (3), we get $15x = 138$ or $x = 9.2$. Now, substitute this value for x in equation (1) to get $6.4(9.2) - 1.7y = 66.7$ or $-1.7y = 7.82$, which means that $y = -4.6$. The solution is $(9.2, -4.6)$.

30. $\begin{cases} 1.6x - 2.9y = -2.645 & (1) \\ 2.4x + 1.4y = 11.27 & (2) \end{cases}$

Multiply equation (1) by 3 to get

$$4.8x - 8.7y = -7.935 \quad (3)$$

Multiply equation (2) by -2 to get

$$-4.8x - 2.8y = -22.54 \quad (4)$$

Add equations (3) and (4) to obtain $-11.5y = -30.475$, or $y = 2.65$. Back-substituting into equation (1) produces $1.6x - 2.9(2.65) = -2.645$ or $1.6x = 5.04$ and so, $x = 3.15$. Thus, we have the solution $(3.15, 2.65)$.

31. $\begin{cases} 2x + 3y = 5 & (1) \\ x - 2y = 6 & (2) \end{cases}$

Solving (2) for x, we get $x = 2y + 6$. Substituting in (1) produces $2(2y + 6) + 3y = 5$ or $4y + 12 + 3y = 5; 7y = -7; y = -1; x = 2(-1) + 6 = -2 + 6 = 4; (4, -1)$

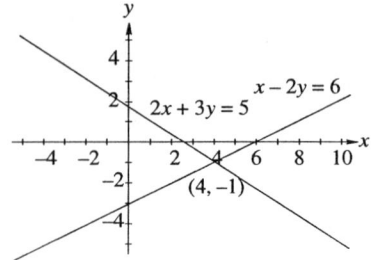

32. $\begin{cases} 2x - 3y = -14 & (1) \\ 3x + 2y = 44 & (2) \end{cases}$

Multiplying (1) by 2 we get $4x - 6y = -28\,(3)$. Multiplying (2) by 3 produces $9x + 6y = 132\,(4)$. Adding (3) and (4) we get $13x = 104$ and so $x = 8$. Substituting in (1) gives $2(8) - 3y = -14$ or $16 - 3y = -14 \Rightarrow -3y = -30$, and so we see that $y = 10$. The solution is $(8, 10)$.

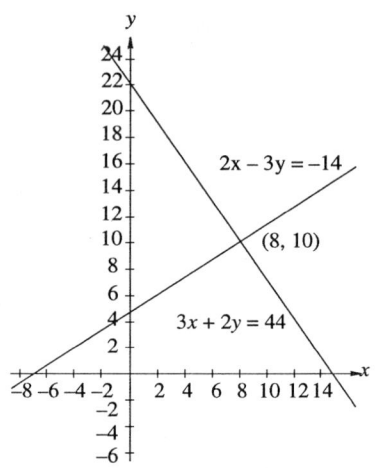

33. $\begin{cases} 8x + 3y = 13 & (1) \\ 3x + 2y = 11 & (2) \end{cases}$

$(1) \times 2 \Rightarrow 16x + 6y = 26\,(3)$. $(2) \times 3 \Rightarrow 9x + 6y = 33\,(4)$. $(3) - (4) \Rightarrow 7x = -7$ or $x = -1$. Substituting in (1) we get $8(-1) + 3y = 13 \Rightarrow -8 + 3y = 13$ and so $3y = 21$ or $y = 7$. The solution is $(-1, 7)$.

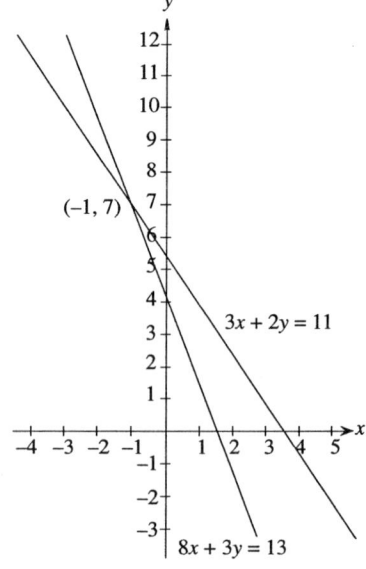

34. $\begin{cases} 6x + 12y = 7 & (1) \\ 8x - 15y = -1 & (2) \end{cases}$

$(1) \times 4 \Rightarrow 24x + 48y = 28\,(3).\,(2) \times 3 \Rightarrow 24x - 45y = -3\,(4)$. Subtracting: $(3) - (4) \Rightarrow 93y = 31$ and so $y = \frac{1}{3}$; Substituting in (1) yields $6x + 12\left(\frac{1}{3}\right) = 7$ or $6x + 4 = 7$ which simplifies to $6x = 3$, and so $x = \frac{1}{2}$. The solution is $\left(\frac{1}{2}, \frac{1}{3}\right)$.

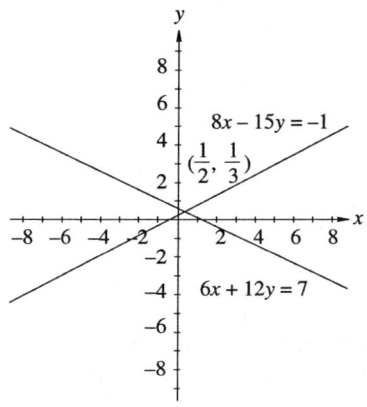

35. $\begin{cases} 10x - 9y = 18 & (1) \\ 6x + 2y = 1 & (2) \end{cases}$

$(1) \times 2 \Rightarrow 20x - 18y = 36\,(3);\,(2) \times 9 \Rightarrow 54x + 18y = 9\,(4)$. Adding $(3) + (4) \Rightarrow 74x = 45$ and so $x = \frac{45}{74} \approx 0.608$. Substituting in (1), we get $10\left(\frac{45}{74}\right) - 9y = 18$ or $\frac{450}{74} - 9(y = 18 \Rightarrow -9y = 18 - \frac{450}{74} = \frac{441}{37}$, and so $y = -\frac{1}{9}\left(\frac{441}{37}\right) = -\frac{49}{37} \approx -1.324$. The solution is $\left(\frac{45}{74}, -\frac{49}{37}\right) \approx (0.608, -1.324)$.

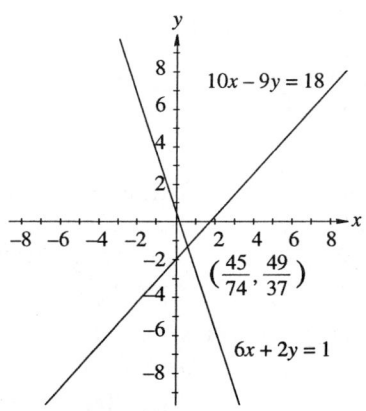

36. $\begin{cases} 4x - 5y = 7 & (1) \\ -8x + 10y = -30 & (2) \end{cases}$

$(1) \times 2 \Rightarrow 8x - 10y = 14\,(3);\,(3) + (2) \Rightarrow 0 = -16$. Since $0 \neq -16$ there is no solution. The system is inconsistent.

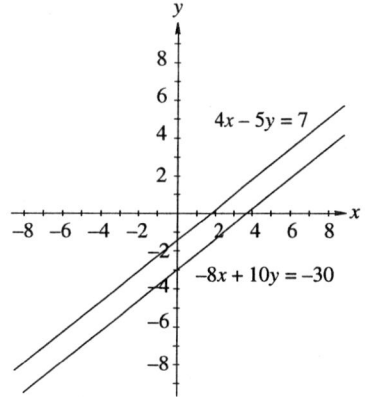

37. $\begin{cases} x - 9y = 0 & (1) \\ \frac{x}{3} = 2y + \frac{1}{3} & (2) \end{cases}$

$(2) \times (3) \Rightarrow x = 6y + 1$ or $x - 6y = 1\,(3)$. Subtracting: $(1) - (3) \Rightarrow -3y = -1$ or $y = \frac{1}{3}$. Substituting in (1), we obtain $x - 9\left(\frac{1}{3}\right) = 0$ or $x - 3 = 0$ and so $x = 3$. The solution is $\left(3, \frac{1}{3}\right)$.

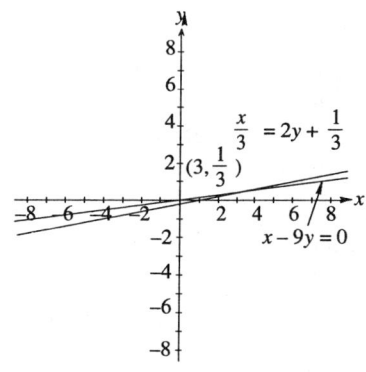

38. $\begin{cases} 5x + 3y = 7 & (1) \\ \frac{3}{2}x - \frac{3}{4}y = 9\frac{1}{4} & (2) \end{cases}$

$(2) \times 4 \Rightarrow 6x - 3y = 37\,(3)$. Adding: $(3) + (1) \Rightarrow 11x = 44$ or $x = 4$. Substituting for x in (1), we get $5(4) + 3y = 7$ or $20 + 3y = 7 \Rightarrow 3y = -13$, and so $y = -\frac{13}{3}$ or $-4\frac{1}{3}$. The solution is $\left(4, -4\frac{1}{3}\right)$.

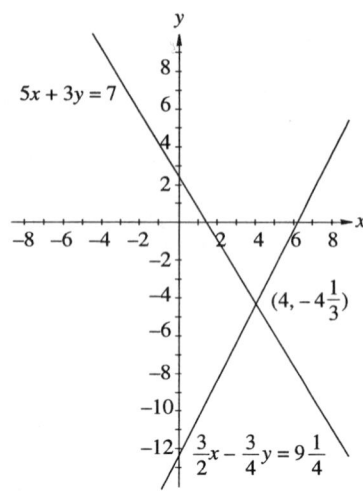

39. $\begin{cases} 4.9x + 1.7y = 10.6 & (1) \\ 3.6x - 1.2y = 14.4 & (2) \end{cases}$

Multiply equation (1) by 1.2 to obtain

$$5.88x + 2.04y = 12.72 \,(3)$$

Multiply equation (2) by 1.7 with the result

$$6.12x - 2.04y = 24.48 \quad (4)$$

Now, add equations (3) and (4) to get $12x = 37.2$ or $x = 3.1$. Back-substituting into equation (1) yields $4.9(3.1) + 1.7y = 10.6$ or $1.7y = -4.59$. Hence, $y = -2.7$, and the solution is $(3.1, -2.7)$.

40. $\begin{cases} 3.14x + 4.57y = -7.9 & (1) \\ 2.48x + 11.84y = 15.16 & (2) \end{cases}$

Multiply equation (1) by 2.48 to get

$$7.7872x + 11.3336y = -19.592 \quad (3)$$

Multiply equation (2) by -3.14 to get

$$-7.7872x - 37.1776y = -47.6024 \quad (4)$$

Add equations (3) and (4) to get $-25.844y = -67.1944$. Hence, $y = 2.6$. Back-substituting into equation (1) gives $3.14x + 4.57(2.6) = -7.9$ or $3.14x = -19.782$ or $x = -6.3$. The solution is $(-6.3, 2.6)$.

41. Substituting $w + 8$ for L in the first equation yields $2(w + 8) + 2w = 36; 2w + 16 + 2w = 36 \Rightarrow 4w + 16 = 36$ or $4w = 20$ and so $w = 5$. $L = w + 8$ or $5 + 8 = 13$. Length is 13 km and the width is 5 km.

42. Since the field is a rectangle, the perimeter is $2L + 2w = 72$. We are given $L = w + 9$. By substitution $2(w + 9) = 2w = 72$ and so $4w = 54$ or $w = 13.5$. Now, $L = w + 9 = 13.5 + 9 = 22.5$. Thus, $L = 22.5$ mi and $w = 13.5$ mi.

43. $2L + 2w = 45$ and $L = 3w$ are two equations that fit this situation. By substitution $2(3w) + 2w = 45 \Rightarrow 6w + 2w = 45$ or $8w = 45$ and so $w = 5.625$. Substituting, we get $L = 3(5.625) = 16.875$. Thus, the length is 16.875 km and the width is 5.625 km.

44. Let $x = $ length of third side and $y = $ length of the two equal sides. Equations are $y = 1\frac{1}{2}x$ and $x + y + y = 96$. By substitution, $x + 1\frac{1}{2}x + 1\frac{1}{2}x = 96 \Rightarrow 4x = 96$ and so $x = 24$. By substitution, $y = 1\frac{1}{2}(24) = 36$. Thus, the three sides measure 36 yd, 36 yd, and 24 yd.

45. Multiplying (2) by 100 $\Rightarrow 5x + 13y = 80,000$ and multiplying (1) by 5 $\Rightarrow 5x + 5y = 50,000$. Subtracting yields $8y = 30,000$ and so $y = 3750$. Substituting in (1), we see that $x + 3750 = 10,000$ and so $x = 6,250$. As a result of these calculations, we have determined that we need 6,250 L of 5% gasohol and 3,750 L of 13% gasohol.

46. As in Exercise 37, $x + y = 20,000; 0.04x + 0.12y = 0.09(20,000)$. Multiplying equation (2) $\times 100 \Rightarrow 4x + 12y = 180,000$ and multiplying equation (1) $\times 4 \Rightarrow 4x + 4y = 80,000$. Subtracting gives $8y = 100,000$ or $y = 12,500$. Since $x + y = 20,000$ we have $x + 12,500 = 20,000$ and so $x = 7,500$. Thus, we will need 7,500 L of 4% gasohol and 12,500 L of 12% gasohol.

47. Equations that fit this situation are $8x = 5y$ (1) and $12x = 5(y + 3.2)$ (2). Solving (1) for x produces $x = \frac{5}{8}y$. Substituting into (2) yields $12\left(\frac{5}{8}y\right) = 5(y + 3.2); \frac{15}{2}y = 5y + 16; 2.5y = 16$ or $y = 6.4$. $x = \frac{5}{8} \times 6.4$ or $x = 4$. Thus, the unknown mass is 6.4 kg and the 8-kg force is at a distance of 4 m.

48. Let x be the amount replaced. Then $(12 - x)(0.25) + x(1.00) = 12(.45); 3 - 0.25x + x = 5.4; 0.75x + 3 = 5.4; 0.75x = 2.4; x = 2.4 \div 0.75$ or 3.2. You will need to replace 3.2 L of antifreeze.

49. $8 = I_1(3) + I_2(5); 5 = 5(I_2) - 6(I_1 - I_2)$. The second equation equation simplifies to $5 = 5I_2 - 6I_1 + 6I_2$ or $5 = 11I_2 - 6I_1$. Taking the first equation times 2 yields $16 = 6I_1 + 10I_2$. Adding these two gives $21 = 21I_2$ or $I_2 = 1$. Replacing I_2 with 1 in the first equation you get $8 = 3I_1 + 5$ or $3 = 3I_1$ or $I_1 = 1$. Since $I_1 = I_2 + I_3; I_3 = I_1 - I_2$ or $I_3 = 1 - 1 = 0$.

50. Using the set up of 49 you get $10 = 2I_1 + 4I_2$ and $15 = 4I_2 - 8(I_1 - I_2)$ or $15 = 12I_2 - 8I_1$.

Multiplying the first equation by 4 gives $40 = 8I_1 + 16I_2$. Adding the last two equations gives $55 = 28I_2$ or $I_2 = \frac{55}{28} \approx 1.96$. Substituting in the first $10 = 2I_1 + 4\left(\frac{55}{28}\right)$ or $10 = 2I_1 + \frac{55}{7}$ and so $2I_1 = \frac{15}{7}$ and $I_1 = \frac{15}{14} \approx 1.07$. $I_1 = I_2 + I_3$ so $\frac{15}{14} = \frac{55}{28} + I_3$ and so $I_3 = \frac{15}{14} - \frac{55}{28} = -\frac{25}{28} \approx -0.89$; $I_1 = \frac{15}{14} \approx 1.07$ A, $I_2 = \frac{55}{28} \approx 1.96$ A; $I_3 = -\frac{25}{28} \approx -0.89$ A. Note that I_3 is in the reverse direction of the diagram.

51. Using $I_2 = I_1 + I_3$ is equivalent to $I_2 - I_1 = I_3$. Hence $E_2 = R_3 I_3 + R_2 I_2$ is equivalent to $E_2 = R_3(I_2 - I_1)$. Substituting the given values yields $6 = 8I_1 + 4I_2$ and $10 = 7(I_2 - I_1) + 4I_2$ or $10 = 11I_2 - 7I_1$. Taking the first equation times $11 \Rightarrow 66 = 88I_1 + 44I_2$. The second times 4 gives $40 = 44I_2 - 28I_1$. Subtracting gives $26 = 116I_1$ or $I_1 = \frac{26}{116} = 0.224$ A. Thus, $6 = 8(0.224) + 4I_2 \Rightarrow 6 = 1.792 + 4I_2 \Rightarrow 4.208 = 4I_2$. Thus, $I_2 = 1.052$ A; $I_3 = 1.052 - 0.224 = .828$ A.

52. See *Computer Programs* in main text.

▤ 5.3 ALGEBRAIC METHODS FOR SOLVING THREE LINEAR EQUATIONS IN THREE VARIABLES

1.
$$\begin{cases} 2x + y + z = 7 & (1) \\ x - y + 2z = 11 & (2) \\ 5x + y - 2z = 1 & (3) \end{cases}$$
Solving (2) for x yields $x = 11 + y - 2z$ (4). Substituting in (3) yields $5(11 + y - 2z) + y - 2z = 1$; $55 + 5y - 10z + y - 2z = 1$; or $6y - 12z = -54$ (5). Substituting (4) into (1) we get $2(11 + y - 2z) + z = 7$ or $22 + 2y - 4z + y + z = 7$ or $3y - 3z = -15$(7). Substituting using (6) into (7) we get $3(2z - 9) - 3z = -15$; $6z - 27 - 3z = -15$; $3z = 12$; $z = 4$. Using (6) we get $y = 2 \cdot 4 - 9 = 8 - 9 = -1$. Using (4) we get $x = 11 + (-1) - 2(4) = 11 - 1 - 8 = 11 - 9 = 2$. The solution is $x = 2$; $y = -1$; $z = 4$.

2.
$$\begin{cases} x + y + 2z = 0 & (1) \\ 2x - y + z = 6 & (2) \\ 4x + 2y + 2z = 0 & (3) \end{cases}$$
Solving equation (1) for y yields $y = -x - 2z$(4). Substituting this into equation (2), we get $2x - (-x - 2z) + z = 6$, which simplifies to $3x + 3z = 6$ or $x + z = 2$(5). We again use (4) to substitute into equation (3) to get $4x + 2(-x - 2z) + 2z = 0$ which simplifies to $2x - 2z = 0$ or $x - z = 0$ (6). Solving equation (6) for x we obtain $x = z$ which we substitute into (5) to obtain $z + z = 2$ or $2z = 2$ and so $z = 1$. Back substitution into equation (6) yields $x = 1$ and further back substitution into equation (4) yields $y = -1 - 2 = -3$. The solution is $(1, -3, 1)$.

3.
$$\begin{cases} 2x - y - z = -8 & (1) \\ x + y - z = -9 & (2) \\ x - y + 2z = 7 & (3) \end{cases}$$
Solving equation (1) for y we obtain $y = 8 + 2x - z$ (4). Substituting this into (2) yields $x + (8 + 2x - z) - z = -9$ which simplifies to $3x - 2z = -17$(5). Again, using equation (4), we substitute into equation (3), which yields $x - (8 + 2x - z) + 2z = 7$.

This simplifies to $-x + 3z = 15$(6). Solving equation (6) for x, we have $x = 3z - 15$(7). Using equation (7) and substituting into equation (5) yields $3(3z - 15) - 2z = -17$ or $9z - 45 - 2z = -17$ or $7z = 28$. Hence, $z = 4$. Back substituting into equation (7) we get $x = 3(4) - 15 = 12 - 15 = -3$. Now, back substitution into equation (4) yields $y = 8 + 2(-3) - 4 = 8 - 6 - 4 = -2$. The solution is $(-3, -2, 4)$.

4.
$$\begin{cases} x + y + 5z = -10 & (1) \\ x - y - 5z = 11 & (2) \\ -x + y - 5z = 13 & (3) \end{cases}$$
Since all three equations have $5z$ in them, let's solve equation (1) for $5z$. This gives $5z = -10 - x - y$(4). Now, substituting for $5z$ in equation (2), we get $x - y - (-10 - x - y) = 11$ or $x - y + 10 + x + y = 11$, which simplifies to $2x = 1$ or $x = \frac{1}{2}$. Using equation (4) again, we substitute for $5z$ in equation (3) with the result $-x + y + 10 + x + y = 13$. This simplifies to $2y = 3$ or $y = \frac{3}{2}$. Back substituting into equation (1) we get $\frac{1}{2} + \frac{3}{2} + 5z = -10$ or $2 + 5z = -10$ or $5z = -12$ and so, $z = -\frac{12}{5} = -2.4$. The solution is $(\frac{1}{2}, \frac{3}{2}, -\frac{12}{5}) = (0.5, 1.5, -2.4)$.

5.
$$\begin{cases} 2x + y + z = 7 & (1) \\ x - y + 2z = 11 & (2) \\ 5x + y - 2z = 1 & (3) \end{cases}$$
Adding $(1) + (2)$ yields $3x + 3z = 18$ (4). Adding $(2) + (3)$ yields $6x = 12$ or $x = 2$ (5). Substituting 2 for x in (4) yields $3 \cdot 2 + 3z = 18$; $6 + 3z = 18$; $3z = 12$, $z = 4$. Substituting $z = 4$, $x = 2$ in (1) yields $2(2) + y + 4 = 7$ or $4 + y + 4 = 7$; $y = -1$. The solution is $x = 2, y = -1, z = 4$.

6. $\begin{cases} x+y+2z=0 & (1) \\ 2x-y+z=6 & (2) \\ 4x+2y+2z=0 & (3) \end{cases}$

$(1)+(2) \Rightarrow 3x+3z=6$ or $x+z=2\,(4). 2\times(2) \Rightarrow 4x-2y+2z=12\,(5). (5)+(3) \Rightarrow 8x+4z=12\,(6). 4\times(4) \Rightarrow 4x+4z=8\,(7). (6)-(7) \Rightarrow 4x=4$ or $x=1$. Substituting into (4) yields $1+z=2$ or $z=1$. Substituting into (1) yields $1+y+2(1)=0$ or $y=-3$. The solution is $x=1, y=-3, z=1$.

7. $\begin{cases} 2x-y-z=-8 & (1) \\ x+y-z=-9 & (2) \\ x-y+2z=7 & (3) \end{cases}$

$(1)+(2) \Rightarrow 3x-2z=-17\,(4). (2)+(3) \Rightarrow 2x+z=-2\,(5). 2\times(5) \Rightarrow 4x+2z=-4\,(6). (4)+(6) \Rightarrow 7x=-21$ or $x=-3$. Substituting into (5) yields $2(-3)+z=-2$ or $-6+z=-2$ or $z=4$. Substituting into (2) yields $(-3)+y-4=-9$ or $-7+y=-9$ or $y=-2$. The solution is $x=-3, y=-2, z=4$.

8. $\begin{cases} x+y+5z=-10 & (1) \\ x-y-5z=11 & (2) \\ -x+y-5z=13 & (3) \end{cases}$

$(1)+(2) \Rightarrow 2x=1$ or $x=0.5. (2)+(3) \Rightarrow -10z=24$ or $z=-2.4$. Substituting into (1) yields $0.5+y+5(-2.4)=-10$ or $0.5+y-12=-10$ or $y-11.5=-10$ so $y=1.5$. The solution is $x=0.5, y=1.5, z=-2.4$.

9. $\begin{cases} x+y+z=2 & (1) \\ 8x-2y+4z=-3 & (2) \\ 6x-4y-3z=3 & (3) \end{cases}$

$2\times(1) \Rightarrow 2x+2y+2z=4\,(4). (4)+(2) \Rightarrow 10x+6z=1\,(5). 4\times(1) \Rightarrow 4x+4y+4z=8\,(6). (6)+(3) \Rightarrow 10x+z=11\,(7). (5)-(7) \Rightarrow 5z=-10$ or $z=-2$. Substituting in (7) yields $10x-2=11$ or $10x=13$ or $x=1.3$. Substituting in (1) yields $1.3+y-2=2; y-0.7=2$ so $y=2.7$. The solution is $x=1.3, y=2.7, z=-2$.

10. $\begin{cases} x+y-z=7 & (1) \\ 8x+4y+2z=21 & (2) \\ 4x+3y+6z=2 & (3) \end{cases}$

$2\times(1) \Rightarrow 2x+2y-2z=14\,(4).$ Adding: $(4)+(2) \Rightarrow 10x+6y=35\,(5). 6\times(1) \Rightarrow 6x+6y-6z=42\,(6).$ Adding: $(6)+(3) \Rightarrow 10x+9y=44\,(7).$ Subtracting: $(7)-(5) \Rightarrow 3y=9$ or $y=3$. Substituting in (5) yields $10x+6(3)=35; 10x+18=35; 10x=17$ and so $x=1.7$. Substituting in (1) we get $1.7+3-z=7$ or $4.7-z=7$ and so $z=-2.3$. The solution is $x=1.7, y=3, z=-2.3$.

11. $\begin{cases} 3x-y-2z=11 & (1) \\ -x+3y+2z=-1 & (2) \\ 2x-2y-4z=17 & (3) \end{cases}$

$2\times(2) \Rightarrow -2x+6y+4z=-2\,(4).$ Adding: $(4)+(3) \Rightarrow 4y=15$ or $y=3.75\,(5)$, and adding

$(1)+(2) \Rightarrow 2x+2y=10$ or $x+y=5\,(6)$. Substituting 3.75 for y into (6) yields $x+3.75=5; x=1.25$. Substituting into (1) yields $3(1.25)-(3.75)-2z=11; 3.75-3.75-2z=11$ or $-2z=11$ and so $z=-5.5$. The solution is $x=1.25, y=3.75, z=-5.5$.

12. $\begin{cases} x-2y+z=-4 & (1) \\ 2x+y+3z=5 & (2) \\ 6x+3y+12z=6 & (3) \end{cases}$

$3\times(2) \Rightarrow 6x+3y+9z=15\,(4)$. Subtracting: $(4)-(3) \Rightarrow -3z=9$ or $z=-3\,(5). 2\times(2) \Rightarrow 4x+2y+6z=10\,(6). (1)+(6) \Rightarrow 5x+7z=6\,(7)$. Substituting -3 for z in (7) yields $5x+7(-3)=6$ or $5x-21=6$ or $5x=27$ and we get $x=5.4$. Substituting in (1) yields $5.4-2y+(-3)=-4; 2.4-2y=-4; 2y=-6.4$ and so $y=3.2$. The solution is $x=5.4, y=3.2, z=-3$.

13. $\begin{cases} 2x+3y+3z=9 & (1) \\ 5x-2y+8z=6 & (2) \\ 4x-y+5z=-1 & (3) \end{cases}$

To eliminate ys: $(3)\times2 \Rightarrow 8x-2y+10z=-2\,(4). (4)-(2) \Rightarrow 3x+2z=-8\,(5). (3)\times3 \Rightarrow 12x-3y+15z=-3\,(6). (1)+(6) \Rightarrow 14x+18z=6\,(7). 9\times(5) \Rightarrow 27x+18z=-72\,(8). (8)-(7) \Rightarrow 13x=-78$ or $x=-6$. Substituting in (5) yields $3(-6)+2z=-8; -18+2z=-8; 2z=10; z=5$. Substituting in (1) yields $2(-6)+3y+3(5)=9; -12+3y+15=9; 3y+3=9; 3y=6; y=2$. The solution is $x=-6, y=2, z=5$.

14. $\begin{cases} x+2y+3z=4 & (1) \\ 2x-3y-4z=-1 & (2) \\ 3x-4y+5z=6 & (3) \end{cases}$

Eliminating xs: $2\times(1) \Rightarrow 2x+4y+6z=8\,(4). (4)-(2) \Rightarrow 7y+10z=9\,(5). 3\times(1) \Rightarrow 3x+6y+9z=12\,(6). (6)-(3) \Rightarrow 10y+4z=6$ or $5y+2z=3\,(7)$. Next, eliminate the ys: $5\times(5)-7\times(7) \Rightarrow 36z=24; z=\frac{24}{36}=\frac{2}{3}$. Substituting in (7) $\Rightarrow 5y+2(\frac{2}{3})=3$ or $5y+\frac{4}{3}=3$ and $y=\frac{1}{3}$. Substituting in $x+2(\frac{1}{3})+3(\frac{2}{3})=4$, we get $x=\frac{4}{3}$. Thus, we get the solution $x=1\frac{1}{3}, y=\frac{1}{3}, z=\frac{2}{3}$.

15. $\begin{cases} I_1-I_2+I_3=0 & (1) \\ 6I_1+6I_2=18 \Rightarrow I_1+I_2=3 & (2) \\ 6I_2+I_3=14 & (3) \end{cases}$

$(3)-(1) \Rightarrow -I_1+7I_2=14\,(4)$. Adding $(2)+(4) \Rightarrow 8I_2=17; I_2=\frac{17}{8}=2.125$. Substituting in (2) yields $I_1+2.125=3$ or $I_1=0.875$. Substituting in (3) yields $6(2.125)+I_3=14; 12.75+I_3=14$ or $I_3=1.25$. The values of the currents are $I_1=0.875$ A, $I_2=2.125$ A, $I_3=1.25$ A.

16. $\begin{cases} I_1 + I_2 - I_3 = 0 & (1) \\ 3I_1 - 5I_2 - 10 = 0 & (2) \\ 5I_2 + 6I_3 - 5 = 0 & (3) \end{cases}$

$3 \times (1) \Rightarrow 3I_1 + 3I_2 - 3I_3 = 0\,(4)$. Subtracting: $(4) - (2) \Rightarrow 8I_2 - 3I_3 + 10 = 0\,(5)$. $2 \times (5) \Rightarrow 16I_2 - 6I_3 + 20 = 0\,(6)$. Adding: $(6) + (3) \Rightarrow 21I_2 + 15 = 0; 21I_2 = -15; I_2 = -\frac{5}{7}$. Substituting in (2) yields $3I_1 - 5(-\frac{5}{7}) - 10 = 0$ or $3I_1 + \frac{25}{7} - \frac{70}{7} = 0; 3I_1 = \frac{45}{7}; I_1 = \frac{15}{7} = 2\frac{1}{7}$. Substituting in (1), we get $2\frac{1}{7} - \frac{5}{7} = I_3$ and so $I_3 = 1\frac{3}{7}$. The solutions are $I_1 = 2\frac{1}{7}$ A, $I_2 = -\frac{5}{7}$ A, $I_3 = 1\frac{3}{7}$A.

17. Substituting the coordinates of $P(5, 1)$ in the standard equation for a circle, we get $5^2 + 1^2 + 5a + b + c = 0 \Rightarrow 5a + b + c = -26\,(1)$. Next we substitute $x = -2$ and $y = -6$, the coordinates of Q, in the standard equation of a circle, with the result: $(-2)^2 + (-6)^2 - 2a - 6b + c = 0 \Rightarrow -2a - 6b + c = -40$ or $2a + 6b - c = 40\,(2)$. Finally, substituting the coordinates of $R(-1, -7)$ in the standard equation of a circle yields $(-1)^2 + (-7)^2 - a - 7b + c = 0$ or $-a - 7b + c = -50$ or $a + 7b - c = 50\,(3)$. We now proceed to solve this system of three equations. We begin by adding $(1) + (2) \Rightarrow 7a + 7b = 14$ or $a + b = 2\,(4)$ and adding $(1) + (3) \Rightarrow 6a + 8b = 24$ or $3a + 4b = 12\,(5)$. Then, $3 \times (4) \Rightarrow 4a + 4b = 8\,(6)$ and subtracting $(6) - (5)$, we find that $a = -4$. Substituting -4 for a in $(4) \Rightarrow -4 + b = 2; b = 6$. Substituting into (1) yields $5(-4) + 6 + c = -26$ or $-20 + 6 + c = -26; c - 14 = -26$ and so $c = -12$ and so the equation of the circle through points P, Q, and R is $x^2 + y^2 - 4x + 6y - 12 = 0$.

18. As in Exercise 17, we will substitute the coordinates of S, T and U into the standard equation for a circle. Substituting the coordinates of S, we get $4^2 + 16^2 + 4a + 16b + c = 0$ or $16 + 256 + 4a + 16b + c = 0 \Rightarrow 4a + 16b + c = -272\,(1)$. Next, substituting the coordinates of T produces $(-6)^2 + (-8)^2 - 6a - 8b + c = 0$ or $36 + 64 - 6a - 8b + c = 0 \Rightarrow 6a + 8b - c = 100\,(2)$. Finally, the coordinates of U are substituted in the standard equation for a circle, with the result $11^2 + (-1)^2 + 11a - b + c = 0 \Rightarrow 11a - b + c = -122\,(3)$. We now solve this system of three equations. Adding $(1) + (2) \Rightarrow 10a + 24b = -172$ or $5a + 12b = -86\,(4)$ and adding $(2) + (3) \Rightarrow 17a + 7b = -22\,(5)$. $7 \times (4) \Rightarrow 35a + 84b = -602\,(6)$. $12 \times (5) \Rightarrow 204a + 84b = -264\,(7)$. $(7) - (6) \Rightarrow 169a = 338$ or $a = 2$. Substituting in (4) yields $5(2) + 12b = -86; 12b = -96; b = -8$. Substituting in (3) produces $11(2) + 8 + c = -122; 30 + c = -122; c = -152; a = 2, b = -8, c = -152$ and the equation of the circle is $x^2 + y^2 + 2x - 8y - 152 = 0$.

19. Organizing the information as in Example 5.18, we get $\begin{cases} 7L + 6M + 8S = 64 & (1) \\ 6L + 3M + 1S = 33 & (2) \\ 4L + 2M + 2S = 26 & (3) \end{cases}$

$(2) \times 3 \Rightarrow 12L + 6M + 2S = 66\,(4)$. Subtracting $(4) - (3) \Rightarrow 8L + 4M = 40\,(5)$. Multiplying $4 \times (3) \Rightarrow 16L + 8M + 8S = 104\,(6)$ and then subtracting $(6) - (1) \Rightarrow 9L + 2M = 40\,(7)$. $2 \times (7) \Rightarrow 18L + 4M = 80\,(8)$. $(8) - (5) \Rightarrow 10L = 40$ or $L = 4$. Substituting into (5) yields $8(4) + 4M = 40; 32 + 4M = 40; 4M = 8$ and $M = 2$. Substituting into (2) yields $6(4) + 3(2) + S = 33$ and $S = 3$. So, the company needs 4 large, 2 medium and 3 small trucks.

20.

	Light	Medium	Heavy
Crude oil A	10%	20%	70%
Crude oil B	30%	40%	30%
Crude oil C	43%	44%	13%

Organizing the information as in Example 5.17.

$\begin{cases} 0.10A + 0.30B + 0.43C = 0.24(450,000) \\ \qquad\qquad\qquad\qquad = 108,000 \quad (1) \\ 0.20A + 0.40B + 0.44C = 0.32(450,000) \\ \qquad\qquad\qquad\qquad = 144,000 \quad (2) \\ 0.70A + 0.30B + 0.13C = 0.44(450,000) \\ \qquad\qquad\qquad\qquad = 198,000 \quad (3) \end{cases}$

$2 \times (1) \Rightarrow 0.20A + 0.60B + 0.86C = 216,000\,(4)$. $(4) - (2) \Rightarrow 0.2B + 0.42C = 72,000\,(5)$. $7 \times (1) \Rightarrow 0.70A + 2.10B + 3.01C = 756,000\,(6)$. $(6) - (3) \Rightarrow 1.80B + 2.88C = 558,000\,(7)$. $9 \times (5) \Rightarrow 1.80B + 3.78C = 648,000\,(8)$. $(8) - (7) \Rightarrow 0.90C = 90,000 \Rightarrow C = 100,000$. Substituting in $(5) \Rightarrow 0.2B + 42,000 = 72,000; 0.2B = 30,000; B = 150,000$. Substituting in $(1) \Rightarrow 0.1A + 0.30 \times (150,000) + 43,000 = 108,000; 0.1A + 45,000 + 43,000 = 108,000; 0.1A = 20,000 \Rightarrow A = 200,000$. Thus, we see that 200 000 t of A, 150 000 t of B, and 100 000 t of C should be mixed. Note $A + B + C = 450\,000$.

21. The final 400 lb of fertilizer contains $0.16 \times 400 = 64$ lb of nitrogen, $0.3 \times 400 = 12$ lb of phosphorus, and $0.9 \times 400 = 36$ lb of potash. Let x be the amount, in pounds, of 10-12-15 fertilizer; y the amount of 10-0-6; and z the amount of the 30-6-15 fertilizer. This leads to the following system of equations:

$$\begin{cases} 0.10x + 0.10y + 0.30z = & 64 \quad (1) \\ 0.12x + 0y + 0.06z = & 12 \quad (2) \\ 0.15x + 0.06y + 0.15z = & 36 \quad (3) \end{cases}$$

$$\begin{cases} (1) + -5(2) \Rightarrow -0.5x + 0.10y = & 4 \quad (4) \\ (1) + -2(3) \Rightarrow -0.2x = & -8 \quad (5) \end{cases}$$

Equation (5) yields $x = \frac{-8}{-0.2} = 40$. Back-substitution into equation (4) produces $y = 240$. Finally, back-substitution into equation (1) gives $0.30z =$

$64 - 4 - 24 = 36$ or $z = \frac{36}{0.30} = 120$. The solution is 40 lb of the 10-12-15 fertilizer; 240 lb of the 10-0-5 fertilizer, and 120 lb of the 30-6-15 fertilizer.

22. Let c, t, and r be the number of chairs, tables, and recliners, respectively, that will be made. Then we obtain the following system of equations.

$$\begin{cases} c + 2t + r = & 500 \quad (1) \\ c + 4t + 2r = & 900 \quad (2) \\ 2c + 5t + 3r = & 1{,}300 \quad (3) \end{cases}$$

$$\begin{cases} -1(1) + (2) \Rightarrow 2t + 5 = & 400 \quad (4) \\ -2(1) + (3) \Rightarrow t + r = & 300 \quad (5) \end{cases}$$

Then, adding equations $(4) + -(5) \to t = 100$. Back substitution yields $r = 200$ and $c = 100$. The solution is 100 chairs, 100 tables, and 200 recliners.

≡ 5.4 DETERMINANTS

1. $\begin{vmatrix} 2 & 3 \\ 4 & -1 \end{vmatrix} = 2(-1) - 4(3) = -2 - 12 = -14$

2. $\begin{vmatrix} 4 & -5 \\ 6 & 2 \end{vmatrix} = 4(2) - (6)(-5) = 8 + 30 = 38$

3. $\begin{vmatrix} 5 & -1 \\ 8 & 1 \end{vmatrix} = 5(1) - 8(-1) = 5 + 8 = 13$

4. $\begin{vmatrix} 9 & -1 \\ -2 & 3 \end{vmatrix} = 9(3) - (-2)(-1) = 27 - 2 = 25$

5. $\begin{vmatrix} 4 & 7 \\ -3 & 1 \end{vmatrix} = 4(1) - (-3)7 = 4 + 21 = 25$

6. $\begin{vmatrix} -1 & -4 \\ 0 & 1 \end{vmatrix} = (-1)(1) - 0(-4) = -1 - 0 = -1$

7. $\begin{vmatrix} -9 & \frac{1}{2} \\ 2 & 1 \end{vmatrix} = (-9)(1) - (2)\frac{1}{2} = -9 - 1 = -10$

8. $\begin{vmatrix} 6 & -3 \\ \frac{1}{3} & -\frac{2}{3} \end{vmatrix} = 6\left(-\frac{2}{3}\right) - \frac{1}{3}(-3) = -4 + 1 = -3$

9. (a) 2,
(b) $\begin{vmatrix} 3 & -4 \\ -2 & 1 \end{vmatrix} = -5$,
(c) Since $1 + 2 = 3$ is odd, the cofactor is the negative of the minor or $-(-5) = 5$.

10. (a) 3,
(b) $\begin{vmatrix} 2 & -1 \\ 1 & 1 \end{vmatrix} = 3$,
(c) Since $2 + 1 = 3$ is odd, the cofactor is the negative of the minor, or -3.

11. (a) 1,

(b) minor $= \begin{vmatrix} 4 & -1 \\ 3 & -4 \end{vmatrix} = -16 + 3 = -13$,
(c) $3 + 2$ is odd so the cofactor is $-(-13) = 13$.

12. (a) -4,
(b) minor $= \begin{vmatrix} 4 & 2 \\ -2 & 1 \end{vmatrix} = 4 + 4 = 8$,
(c) $2 + 3$ is odd so the cofactor is $-(-8) = 8$.

13. (a) -2,
(b) minor $= \begin{vmatrix} 2 & -1 \\ 7 & -4 \end{vmatrix} = -8 + 7 = -1$,
(c) $3 + 1$ is even so cofactor = minor = -1.

14. (a) -1,
(b) minor $= \begin{vmatrix} 3 & 7 \\ -2 & 1 \end{vmatrix} = 3 + 14 = 17$,
(c) $1 + 3$ is even so cofactor = 17 also.

15. $-3\begin{vmatrix} 2 & -1 \\ 1 & 1 \end{vmatrix} + 7\begin{vmatrix} 4 & -1 \\ -2 & 1 \end{vmatrix} - (-4)\begin{vmatrix} 4 & 2 \\ -2 & 1 \end{vmatrix} =$

$-3(3) + 7(2) + 4(8) = -9 + 14 + 32 = 37$

16. $-1\begin{vmatrix} 3 & 7 \\ -2 & 1 \end{vmatrix} - (-4)\begin{vmatrix} 4 & 2 \\ -2 & 1 \end{vmatrix} + 1\begin{vmatrix} 4 & 2 \\ 3 & 7 \end{vmatrix} =$

$-1(17) + 4(8) + 1(22) = -17 + 32 + 22 = 37$

17. 0, rows 1 and 3 are identical.

18. 0, row 2 is a multiple of row 1.

19. 0, because row 3 is a multiple of row 1 (or because column 2 is a multiple of column 1).

20. $\begin{vmatrix} 2 & 4 & 3 \\ 0 & 1 & 19 \\ 0 & 0 & -3 \end{vmatrix} = 2(1)(-3) = -6$

21. Using the first row to evaluate, we obtain

$$\begin{vmatrix} 5 & 2 & 3 \\ 4 & -5 & -6 \\ -2 & 5 & -9 \end{vmatrix} = 5\begin{vmatrix} -5 & -6 \\ 5 & -9 \end{vmatrix} - 2\begin{vmatrix} 4 & -6 \\ -2 & -9 \end{vmatrix} + 3\begin{vmatrix} 4 & -5 \\ -2 & 5 \end{vmatrix}$$

$$= 5[9-5)(-9) - 5(-6)] - 2[4(-9) - (-2)(-6)]$$
$$\quad + 3[4(5) - (-2)(-5)]$$
$$= 5[45 + 30] - 2[-36 - 12] + 3[20 - 10]$$
$$= 5(75) - 2(-48) + 3(10)$$
$$= 375 + 96 + 30 = 501$$

22. First add $-1(R3)$ to $R1$ to get $\begin{vmatrix} 3 & 0 & 0 \\ -2 & 0 & 6 \\ 6 & 3 & 3 \end{vmatrix}$ and

then, interchange $C2 + C3$ to get $\begin{vmatrix} 3 & 0 & 0 \\ -2 & 6 & 0 \\ 6 & 3 & 3 \end{vmatrix} =$

$-(3)(6)(3) = -54$

23. $\begin{vmatrix} 4 & 3 & 9 \\ -4 & -6 & 16 \\ 2 & 3 & 2 \end{vmatrix} \overset{R1+R2}{=} \begin{vmatrix} 4 & 3 & 9 \\ 0 & -3 & 25 \\ 2 & 3 & 2 \end{vmatrix} \overset{-2R3+R1}{=}$

$\begin{vmatrix} 0 & -3 & 5 \\ 0 & -3 & 25 \\ 2 & 3 & 2 \end{vmatrix} \overset{-R2+R1}{=} \begin{vmatrix} 0 & 0 & -20 \\ 0 & -3 & 25 \\ 2 & 3 & 2 \end{vmatrix} =$

$-\begin{vmatrix} 2 & 3 & 2 \\ 0 & -3 & 25 \\ 0 & 0 & -20 \end{vmatrix} = -2(-3)(-20) = -120$

24. $\begin{vmatrix} 9 & 18 & -1 \\ 0 & -7 & 5 \\ 0 & 0 & 2 \end{vmatrix} = 9(-7)(2) = -126$

25. 252.86583

26. $\begin{vmatrix} \sqrt{3} & \frac{1}{2} & -5 \\ \frac{1}{4} & \sqrt{2} & 4 \\ \frac{2}{5} & 7 & -1 \end{vmatrix} = -55.94248.$ Use $\sqrt{3} = 1.732$

and $\sqrt{2} = 1.414.$

27. -0.057525

28. $\begin{vmatrix} \sqrt{7} & 3 & \sqrt{5} \\ -1.4 & \sqrt{6} & 2 \\ \sqrt{8} & 3 & -6 \end{vmatrix} = -87.873415.$ Use $\sqrt{6} =$

$2.4495, \sqrt{7} = 2.6458, \sqrt{5} = 2.2361,$ and $\sqrt{8} = 2.8284$

29. $\begin{vmatrix} r & s & t \\ a & b & c \\ x & y & z \end{vmatrix} = -(-5) = 5$

30. $\begin{vmatrix} a & b & c \\ r & s & t \\ x & y & z \end{vmatrix} = -\begin{vmatrix} r & s & t \\ a & b & c \\ x & y & z \end{vmatrix} = \begin{vmatrix} s & r & t \\ b & a & c \\ y & x & z \end{vmatrix} =$

$-\begin{vmatrix} s & t & r \\ b & c & a \\ y & z & x \end{vmatrix} = -(-5) = 5$

31. $\begin{vmatrix} a & b & c \\ 3r & 3s & 3t \\ -2x & -2y & -2z \end{vmatrix} = 3(-2)(-5) = 30$

32. $\begin{vmatrix} a & b & c \\ r & s & t \\ x & y & z \end{vmatrix} = \begin{vmatrix} a & b & c \\ r & s & t \\ x-r & y-s & z-t \end{vmatrix} =$

$\begin{vmatrix} a & b & c \\ r-a & s-b & c-t \\ x-r & y-s & z-t \end{vmatrix} = -5$

33. $\begin{vmatrix} 1 & 2 & 3 & 4 \\ 5 & 0 & 7 & 0 \\ 9 & 10 & 11 & 12 \\ 13 & 14 & 15 & 16 \end{vmatrix} \overset{-3(R1)+R3}{\underset{-4(R1)+R4}{=}}$

$\begin{vmatrix} 1 & 2 & 3 & 4 \\ 5 & 0 & 7 & 0 \\ 6 & 4 & 2 & 0 \\ 9 & 6 & 3 & 0 \end{vmatrix} =$

0, since $R4 = 1.5 \times R3$.

34. $\begin{vmatrix} 2 & 0 & 3 & -1 \\ 4 & 0 & -2 & 5 \\ 9 & -5 & 0 & -2 \\ 0 & 3 & 1 & -6 \end{vmatrix} = -(-5)\begin{vmatrix} 2 & 3 & -1 \\ 4 & -2 & 5 \\ 0 & 1 & -6 \end{vmatrix} +$

$3\begin{vmatrix} 2 & 3 & -1 \\ 4 & -2 & 5 \\ 9 & 0 & -2 \end{vmatrix} = 5(82) + 3(149) = 410 +$

$447 = 857$ since $\begin{vmatrix} 2 & 3 & -1 \\ 4 & -2 & 5 \\ 0 & 1 & -6 \end{vmatrix} = 2\begin{vmatrix} -2 & 5 \\ 1 & -6 \end{vmatrix} -$

$4\begin{vmatrix} 3 & -1 \\ 1 & -6 \end{vmatrix} = 2(12 - 5) - 4(-18 + 1) = 2(7) -$

$4(-17) = 14 + 68 = 82.$ $\begin{vmatrix} 2 & 3 & -1 \\ 4 & -2 & 5 \\ 9 & 0 & -2 \end{vmatrix} =$

$-3\begin{vmatrix} 4 & 5 \\ 9 & -2 \end{vmatrix} - 2\begin{vmatrix} 2 & -1 \\ 9 & -2 \end{vmatrix} = -3(-8-45) - 2(-4 +$

$9) = -3(-53) - 2(5) = 159 - 10 = 149.$

35. Using a calculator, we obtain approximately -321.545

36. Using a calculator, we obtain approximately $1433.260.$

37. Using a calculator, we obtain 0.

38. Using a calculator, we obtain approximately 59.603.

39. See *Computer Programs* in main text.

40.
$$\begin{vmatrix} 5 & 2 & 3 \\ 4 & -5 & -6 \\ -2 & 5 & -9 \end{vmatrix}\begin{matrix} 5 & 2 \\ 4 & -5 \\ -2 & 5 \end{matrix} = 5(-5)(-9)+$$

$2(-6)(-2) + 3(4)(5) - (-2)(-5)(3) - 5(-6)(5) - (-9)(4)(2) = 225+24+60-30+150+72 = 501$

$$\begin{vmatrix} 9 & 3 & 3 \\ -2 & 0 & 6 \\ 6 & 3 & 3 \end{vmatrix}\begin{matrix} 9 & 3 \\ -2 & 0 \\ 6 & 3 \end{matrix} = 9(0)(3) + 3(6)(6)+$$

$3(-2)(3) - 6(0)(3) - (3)(6)(9) - 3(-2)(3) = 0 + 108 - 18 - 0 - 162 + 18 = -54$

$$\begin{vmatrix} 4 & 3 & 9 \\ -4 & -6 & 16 \\ 2 & 3 & 2 \end{vmatrix}\begin{matrix} 4 & 3 \\ -4 & -6 \\ 2 & 3 \end{matrix} = 4(-6)(2)+3(16)(2)+$$

$9(-4)(3)-2(-6)(9)-3(16)(4)-2(-4)(3) = -48+$

$96 - 108 + 108 - 192 + 24 = -120$

$$\begin{vmatrix} 9 & 18 & -1 \\ 0 & -7 & 5 \\ 0 & 0 & 2 \end{vmatrix}\begin{matrix} 9 & 18 \\ 0 & -7 \\ 2 & 3 \end{matrix} = 9(-7)(2) + 18(5)(0)+$$

$(-1)(0)(0) - 0(-7)(-1) - 0(5)(9) - 2(0)(18) = -126$

41. See *Computer Programs* in main text.

42. **(a)** $1\begin{vmatrix} 5 & 6 \\ 8 & 9 \end{vmatrix} - 2\begin{vmatrix} 4 & 6 \\ 7 & 9 \end{vmatrix} + 3\begin{vmatrix} 4 & 5 \\ 7 & 8 \end{vmatrix} = 1(45-48)-$

$2(36 - 42) + 3(32 - 35) = -3 + 12 - 9 = 0$
(b) 0 on a TI-82
(c) $-2.4\mathrm{E}-12$ on a TI-85, 0 on a Sharp EL 9300
(Remember: $-2.4\mathrm{E}-12$ means -2.4×10^{-12}.)

≡ 5.5 USING CRAMER'S RULE TO SOLVE SYSTEMS OF LINEAR EQUATIONS

1. $D = \begin{vmatrix} 2 & 1 \\ 3 & -2 \end{vmatrix} = -4-3 = -7, D_x = \begin{vmatrix} 7 & 1 \\ -7 & -2 \end{vmatrix} = -14 + 7 = -7, D_y = \begin{vmatrix} 2 & 7 \\ 3 & -7 \end{vmatrix} = -14 - 21 = -35, x = \frac{-7}{-7} = 1, y = \frac{-35}{-7} = 5$

2. $D = \begin{vmatrix} 3 & 1 \\ 2 & -3 \end{vmatrix} = -9-2 = -11, D_x = \begin{vmatrix} 3 & 1 \\ 13 & -3 \end{vmatrix} = -9 - 13 = -22, D_y = \begin{vmatrix} 3 & 3 \\ 2 & 13 \end{vmatrix} = 39 - 6 = 33,$

$x = \frac{-22}{-11} = 2, y = \frac{33}{-11} = -3$

3. $D = \begin{vmatrix} 4 & 3 \\ 3 & -2 \end{vmatrix} = -8 - 9 = -17, D_x = \begin{vmatrix} 4 & 3 \\ -14 & -2 \end{vmatrix} = -8+42 = 34, D_y = \begin{vmatrix} 4 & 4 \\ 3 & -14 \end{vmatrix} = -56 - 12 = -68, x = \frac{34}{-17} = -2, y = \frac{-68}{-17} = 4$

4. $D = \begin{vmatrix} \frac{1}{2} & -\frac{2}{3} \\ \frac{1}{3} & 2 \end{vmatrix} = 1 + \frac{2}{9} = \frac{11}{9}, D_x = \begin{vmatrix} \frac{3}{4} & -\frac{2}{3} \\ \frac{5}{6} & 2 \end{vmatrix} = \frac{3}{2} + \frac{5}{9} = \frac{37}{18}, D_y = \begin{vmatrix} \frac{1}{2} & -\frac{3}{4} \\ \frac{1}{3} & \frac{5}{6} \end{vmatrix} = \frac{5}{12} - \frac{3}{12} = \frac{2}{12} = \frac{1}{6}, x = \frac{37/18}{11/9} = \frac{37}{18} \times \frac{9}{11} = \frac{37}{22} \approx 1.6818, y = \frac{1/6}{11/9} = \frac{1}{6} \cdot \frac{9}{11} = \frac{3}{22} \approx 0.13636$

5. $D = \begin{vmatrix} 1.2 & 3.7 \\ 4.3 & -5.2 \end{vmatrix} = -22.15, D_x = \begin{vmatrix} 9.1 & 3.7 \\ 8.3 & -5.2 \end{vmatrix} = -78.03, D_y = \begin{vmatrix} 1.2 & 9.1 \\ 4.3 & 8.3 \end{vmatrix} = -29.17, x = \frac{-78.03}{-22.15} \approx 3.5228, y = \frac{-29.17}{-22.15} = 1.3169$

6. $D = \begin{vmatrix} 0.02 & -1.22 \\ -0.14 & 0.32 \end{vmatrix} = -0.1644, D_x = \begin{vmatrix} 3.74 & -1.22 \\ -1.32 & 0.32 \end{vmatrix} = -0.4136, x = \frac{-4.136}{-.1644} \approx +2.5158, D_y = \begin{vmatrix} 0.02 & 3.74 \\ -.14 & -1.32 \end{vmatrix} = 0.4972, y = \frac{.4972}{-.1644} = -3.0243$

7. $D = \begin{vmatrix} 2.5 & 3.8 \\ .5 & .76 \end{vmatrix} = 0$, inconsistent system, no solution.

8. $D = \begin{vmatrix} 4.3 & -2.7 \\ 3.5 & 4.2 \end{vmatrix} = 27.51, D_x = \begin{vmatrix} -4 & -2.7 \\ -1 & 4.2 \end{vmatrix} = -19.5, D_y = \begin{vmatrix} 4.3 & -4 \\ 3.5 & -1 \end{vmatrix} = 9.7, x = \frac{-19.5}{27.51} \approx -7.088, y = \frac{9.7}{27.51} \approx 0.3526$

9. $D = \begin{vmatrix} -5.3 & 2.1 \\ 6.2 & -3.1 \end{vmatrix} = 3.41, D_x = \begin{vmatrix} 4.6 & 2.1 \\ 6 & -3.1 \end{vmatrix} = -26.86, D_y = \begin{vmatrix} -5.3 & 4.6 \\ 6.2 & 6 \end{vmatrix} = -60.32, x = \frac{-26.86}{3.41} \approx -7.8768, y = \frac{-60.32}{3.41} \approx -17.689$

10. $D = \begin{vmatrix} 1.3 & -0.8 \\ 1.7 & -0.7 \end{vmatrix} = .45, D_x = \begin{vmatrix} 2.9 & -0.8 \\ 0.4 & -0.7 \end{vmatrix} =$

$-1.71, D_y = \begin{vmatrix} 1.3 & 2.9 \\ 1.7 & 0.4 \end{vmatrix} = -4.41, x = \frac{-1.71}{.45} =$

$-3.8, y = \frac{-4.41}{.45} = -9.8$

11. $D = \begin{vmatrix} 6 & 2.5 \\ 13.8 & 5.75 \end{vmatrix} = 0$, inconsistent system, no
solution.

12. $D = \begin{vmatrix} 3.5 & -6.5 \\ 5.5 & 3.3 \end{vmatrix} = 47.3, D_x = \begin{vmatrix} 22.45 & -6.5 \\ -1.21 & 3.3 \end{vmatrix} =$

$66.22, D_y = \begin{vmatrix} 3.5 & 22.45 \\ 5.5 & -1.21 \end{vmatrix} = -127.71, x = \frac{66.22}{47.3} =$

$1.4, y = \frac{-127.71}{47.3} = -2.7$

13. $D = \begin{vmatrix} 4 & 3 \\ 3 & -2 \end{vmatrix} = -8 - 9 = -17, D_x =$

$\begin{vmatrix} 2 & 3 \\ -24 & -2 \end{vmatrix} = -4 + 72 = 68, D_y = \begin{vmatrix} 4 & 2 \\ 3 & -24 \end{vmatrix} =$

$-96 - 6 = -102, x = \frac{68}{-17} = -4, y = \frac{-102}{-17} = 6$

14. $D = \begin{vmatrix} 1 & 3 & -1 \\ 5 & -7 & 1 \\ 2 & -1 & -2 \end{vmatrix} = 42, D_x = \begin{vmatrix} 7 & 3 & -1 \\ 3 & -7 & 1 \\ 0 & -1 & -2 \end{vmatrix} =$

$126, D_y = \begin{vmatrix} 1 & 7 & -1 \\ 5 & 3 & 1 \\ 2 & 0 & -2 \end{vmatrix} = 84, D_z = \begin{vmatrix} 1 & 3 & 7 \\ 5 & -7 & 3 \\ 2 & -1 & 0 \end{vmatrix} =$

$84, x = \frac{126}{42} = 3, y = \frac{84}{42} = 2, z = \frac{84}{42} = 2, (3, 2, 2)$

15. $D = \begin{vmatrix} 1 & -.5 & 1.5 \\ 1 & .5 & -3.75 \\ -3 & -2.5 & 4 \end{vmatrix} = -12.5,$

$D_x = \begin{vmatrix} 2 & -.5 & 1.5 \\ .25 & .5 & -3.75 \\ -.75 & -2.5 & 4 \end{vmatrix} = -16.03125,$

$D_y = \begin{vmatrix} 1 & 2 & 1.5 \\ 1 & .25 & -3.75 \\ -3 & -.75 & 4 \end{vmatrix} = 12.6875,$

$D_z = \begin{vmatrix} 1 & -.5 & 2 \\ 1 & .5 & .25 \\ -3 & -2.5 & -.75 \end{vmatrix} = -1.75,$

$x = \frac{-16.03125}{-12.5} = 1.2825, y = \frac{12.6875}{-12.5} = -1.015, z =$

$\frac{-1.75}{-12.5} = 0.14$

16. $D = \begin{vmatrix} 1 & 1 & -1 \\ .5 & -3.5 & 2.4 \\ 1.5 & 4.5 & -3.6 \end{vmatrix} = -0.3,$

$D_x = \begin{vmatrix} -4.7 & 1 & -1 \\ 17.4 & -3.5 & 2.4 \\ -21.6 & 4.5 & -3.6 \end{vmatrix} = -.36,$

$x = \frac{-.36}{-.3} = 1.2,$

$D_y = \begin{vmatrix} 1 & -4.7 & -1 \\ .5 & 17.4 & 2.4 \\ 1.5 & -21.6 & -3.6 \end{vmatrix} = .72,$

$y = \frac{.72}{-.3} = -2.4,$

$D_z = \begin{vmatrix} 1 & 1 & -4.7 \\ .5 & -3.5 & 17.4 \\ 1.5 & 4.5 & -21.6 \end{vmatrix} = -1.05, z = \frac{-1.05}{-.3} =$

$3.5.$ The solution is $x = 1.2, y = -2.4,$ and
$z = 3.5.$

17. First, we evaluate $D = -150, D_x = 315, D_y = -25.5, D_z = 652.5,$ and $D_w = -1437.$ Then, by Cramer's rule, we have $x = -2.1, y = 0.17, z = -4.35,$ and $w = 9.58.$

18. First, we evaluate $D = 60, D_a = -120, D_b = 60, D_c = 180,$ and $D_d = -60.$ Then, by Cramer's rule, we have $a = -2, b = 1, c = 3,$ and $d = -1.$

19. See *Computer Programs* in main text.

20. $D = \begin{vmatrix} 1 & 1 & 1 \\ 8 & 0 & -10 \\ 6 & -3 & 0 \end{vmatrix} = -114,$

$D_{I_1} = \begin{vmatrix} 0 & 1 & 1 \\ 8 & 0 & -10 \\ 12 & -3 & 0 \end{vmatrix} = -144,$

$D_{I_2} = \begin{vmatrix} 1 & 0 & 1 \\ 8 & 8 & -10 \\ 6 & 12 & 0 \end{vmatrix} = 168,$

$D_{I_3} = \begin{vmatrix} 1 & 1 & 0 \\ 8 & 0 & 8 \\ 6 & -3 & 12 \end{vmatrix} = -24,$

$I_1 = \frac{-144}{-114} = \frac{24}{19}, I_2 = \frac{168}{-114} = -\frac{28}{19},$
$I_3 = \frac{-24}{-144} = \frac{4}{19}.$

21. First, we make the substitutions for t and s to obtain the following system of three equations.

$$\begin{cases} s_0 + 2v_0 + 2a = 212 \\ crs_0 + 5v_0 + \frac{25}{2}a = 156.5 \\ crs_0 + 7v_0 + \frac{49}{2}a = 70.5 \end{cases}$$

$$D = \begin{vmatrix} 1 & 2 & 2 \\ 1 & 5 & 12.5 \\ 1 & 7 & 24.5 \end{vmatrix} = 15,$$

$$D_{s_0} = \begin{vmatrix} 212 & 2 & 2 \\ 156.5 & 5 & 12.5 \\ 70.5 & 7 & 24.5 \end{vmatrix} = 3000,$$

$$D_{v_0} = \begin{vmatrix} 1 & 212 & 2 \\ 1 & 156.5 & 12.5 \\ 1 & 70.5 & 24.5 \end{vmatrix} = 237,$$

$$D_a = \begin{vmatrix} 1 & 2 & 212 \\ 1 & 5 & 156.5 \\ 1 & 7 & 70.5 \end{vmatrix} = -147,$$

$$s_0 = \frac{3000}{15} = 200,$$

$v_0 = \frac{237}{15} = 15.8, a = \frac{147}{15} = -9.8.$ Thus,
$s = 200 + 15.8t - 4.9t^2.$

22. $D = \begin{vmatrix} 1 & 1 & 1 \\ 1 & -2 & 0 \\ -4 & 0 & 1 \end{vmatrix} = -11,$

$$D_A = \begin{vmatrix} 100 & 1 & 1 \\ 0 & -2 & 0 \\ 0 & 0 & 1 \end{vmatrix} = -200,$$

$$D_B = \begin{vmatrix} 1 & 100 & 1 \\ 1 & 0 & 0 \\ -4 & 0 & 1 \end{vmatrix} = -100 \begin{vmatrix} 1 & -2 \\ -4 & 1 \end{vmatrix} = -100,$$

$$D_C = \begin{vmatrix} 1 & 1 & 100 \\ 1 & -2 & 0 \\ -4 & 0 & 0 \end{vmatrix} = 100 \begin{vmatrix} 1 & -2 \\ -4 & 0 \end{vmatrix} = -800,$$

$A = \frac{-200}{-11} \approx 18.18, B = \frac{100}{11} \approx 9.09,$

$C = \frac{800}{11} \approx 72.73.$ Thus, the alloy consists of 18.18% metal A, 9.09% metal B, and 72.73% metal C.

23. Let A, B, and C represent the three distances. Then we get the equations $B = 90+A$ and $C = 130+B$, which leads to the system of three equations:

$$\begin{cases} -A + B = 90 \\ -B + C = 130 \\ A + B + C = 1900 \end{cases}$$

$$D = \begin{vmatrix} -1 & 1 & 0 \\ 0 & -1 & 1 \\ 1 & 1 & 1 \end{vmatrix} = 3,$$

$$D_A = \begin{vmatrix} 90 & 1 & 0 \\ 130 & -1 & 1 \\ 1900 & 1 & 1 \end{vmatrix} = 1590,$$

$$D_B = \begin{vmatrix} -1 & 90 & 0 \\ 0 & 130 & 1 \\ 1 & 1900 & 1 \end{vmatrix} = 1860,$$

$$D_C = \begin{vmatrix} -1 & 1 & 90 \\ 0 & -1 & 130 \\ 1 & 1 & 1900 \end{vmatrix} = 2250.$$

$A = \frac{1590}{3} = 530,$

$B = \frac{1860}{3} = 620, C = \frac{2250}{3} = 750.$ Hence, the car traveled 530 km on fuel A, 620 km on fuel B, and 750 km on fuel C.

24. $D = \begin{vmatrix} 0 & \frac{6}{7} & -\frac{2}{3} \\ -1 & -\frac{3}{7} & \frac{1}{3} \\ 0 & \frac{2}{7} & \frac{2}{3} \end{vmatrix} = \frac{1}{21} \begin{vmatrix} 0 & 6 & -2 \\ -1 & -3 & 1 \\ 0 & 2 & 2 \end{vmatrix} = \frac{1}{21}(16) = \frac{16}{21}, D_{FA} = \begin{vmatrix} 2{,}000 & \frac{6}{7} & -\frac{2}{3} \\ 0 & -\frac{3}{7} & \frac{1}{3} \\ 1{,}200 & \frac{2}{7} & \frac{2}{3} \end{vmatrix} =$

$\frac{1}{21} \begin{vmatrix} 2{,}000 & 6 & -2 \\ 0 & -3 & 1 \\ 1{,}200 & 2 & 2 \end{vmatrix} = \frac{1}{21}(-16{,}000) = -\frac{16{,}000}{21}, D_{FB} = \begin{vmatrix} 0 & 2{,}000 & -\frac{2}{3} \\ -1 & - & \frac{1}{3} \\ 0 & 1{,}200 & \frac{2}{3} \end{vmatrix} =$

$\frac{1}{3} \begin{vmatrix} 0 & 2{,}000 & -2 \\ -1 & 0 & 1 \\ 0 & 1{,}200 & 2 \end{vmatrix} = \frac{1}{3}(6{,}400) = \frac{6{,}400}{3}, D_{FC} = \begin{vmatrix} 0 & \frac{6}{7} & 2{,}000 \\ -1 & -\frac{3}{7} & 0 \\ 0 & \frac{2}{7} & 1{,}200 \end{vmatrix} = \frac{1}{7} \begin{vmatrix} 0 & 6 & 2{,}000 \\ -1 & -3 & 0 \\ 0 & 2 & 1{,}200 \end{vmatrix} = \frac{1}{7}(3{,}200) =$

$\frac{3{,}200}{7}.$ Thus, $F_A = \frac{-16{,}000/21}{16/21} = \frac{-16{,}000}{16} = -1{,}000, F_B = \frac{6{,}400/3}{16/21} = 2{,}800, F_C = \frac{3{,}200/7}{16/21} = 600.$ The three forces are $F_A = -1{,}000$ N, $F_B = 2{,}800$ N, $F_C = 600$ N.

25. **(a)** Let $A, B,$ and C represent the three companies. The 5 units of salt will satisfy $2A + 2B + C$. Similarly, for the sand we get the equation $2A + 2B + 5C = 6$, and for the inhibiting agent, the equation $A + 2B + C = 4$. Thus, we have the system
$$\begin{cases} 2A + 2B + C = 5 \\ A + 2B + 5C = 6 \\ A + 2B + C = 4 \end{cases}$$
From this sytem, we get the determinants

$$D = \begin{vmatrix} 2 & 2 & 1 \\ 1 & 2 & 5 \\ 1 & 2 & 1 \end{vmatrix} = -8,$$

$$D_A = \begin{vmatrix} 5 & 2 & 1 \\ 6 & 2 & 5 \\ 4 & 2 & 1 \end{vmatrix} = -8,$$

$$D_B = \begin{vmatrix} 2 & 5 & 1 \\ 1 & 6 & 5 \\ 1 & 4 & 1 \end{vmatrix} = -10, \text{ and}$$

$$D_C = \begin{vmatrix} 2 & 2 & 5 \\ 1 & 2 & 6 \\ 1 & 2 & 4 \end{vmatrix} = -4. \text{ Thus, } A = \frac{-8}{-8} = 1, B = \frac{-10}{-8} = \frac{5}{4}, \text{ and } C = \frac{-4}{-8} = \frac{1}{2}. \text{ Thus, we have the proportion}$$

$1 : \frac{5}{4} : \frac{1}{2} = 4 : 5 : 2$.

(b) If we let x represent the basic number of units that must be purchased, then the ratio of $4 : 5 : 2$ can be thought of as $4x + 5x + 2x = 3,297,283$, or $11x = 3,297,283$. Thus, $x = \frac{3,297,283}{11} = 299,753$. The city must therefore purchase $4 \times 299,753 = 1,199,095$ units from company A, it must purchase $5 \times 299,753 = 1,498,765$ units from company B, and $2 \times 299,753 = 599,506$ units from company C.

26. Substituting the three sets of values for v and d, we obtain the following system of equations:
$$\begin{array}{lll} a \cdot 20^2 + b \cdot 20 + c = 40 & \text{or} & 400a + 20b + c = 40 \\ a \cdot 55^2 + b \cdot 55 + c = 206.25 & \text{or} & 3025a + 65b + c = 206.25 \\ a \cdot 65^2 + b \cdot 65 + c = 276.25 & \text{or} & 4225a + 65b + c = 276.25 \end{array}$$
The required determinants are $D = -15750, D_a = -787.5, D_b = -15750,$ and $D_c = 0$. These lead to $a = 0.05, b = 1,$ and $c = 0$.

CHAPTER 5 REVIEW

1. $\begin{vmatrix} 2 & 5 \\ -3 & 6 \end{vmatrix} = 12 + 15 = 27$

2. $\begin{vmatrix} 4 & -7 \\ 3 & -6 \end{vmatrix} = -24 + 21 = -3$

3. $\begin{vmatrix} 5 & 9 \\ 8 & 1 \end{vmatrix} = 5 - 72 = -67$

4. $\begin{vmatrix} 3 & -12 \\ 0 & 5 \end{vmatrix} = 15 - 0 = 15$

5. $\begin{vmatrix} 6 & -2 \\ -4 & 3 \end{vmatrix} = 6 \times 3 - (-4)(-2) = 18 - 8 = 10$

6. $\begin{vmatrix} 8 & 9 \\ -1 & -2 \end{vmatrix} = -16 + 9 = -7$

7. We first get this determinant in triangular form.

$\begin{vmatrix} 9 & 2 & 1 \\ 3 & -4 & 6 \\ 7 & 2 & 1 \end{vmatrix} \overset{-R3+R1}{\longrightarrow} \begin{vmatrix} 2 & 0 & 0 \\ 3 & -4 & 6 \\ 7 & 2 & 1 \end{vmatrix}$

$\overset{-6R3+R2}{\longrightarrow} \begin{vmatrix} 2 & 0 & 0 \\ -39 & -16 & 0 \\ 7 & 2 & 1 \end{vmatrix}$

The determinant is now triangular, so $\begin{vmatrix} 9 & 2 & 1 \\ 3 & -4 & 6 \\ 7 & 2 & 1 \end{vmatrix} =$

$\begin{vmatrix} 2 & 0 & 0 \\ -39 & -16 & 0 \\ 7 & 2 & 1 \end{vmatrix} = 2(-16)(1) = -32.$

8. $\begin{vmatrix} 9 & -2 & 3 \\ 4 & -5 & 2 \\ 3 & 2 & 1 \end{vmatrix} = 9(-5)1 + (-2)2(3) + 3(4)2-$

$3(-5)3 - 2(2)(9) - 1(4)(-2) = -45 - 12 + 12 + 24 + 45 - 36 + 8 = -16$

9. (a) $m = \frac{1-3}{5-(-2)} = \frac{-2}{7}$,

(b) $y - 1 = \frac{-2}{7}(x - 5)$ or $y - 3 = \frac{-2}{7}(x + 2)$

10. (a) $m = \frac{-2-(-4)}{3-8} = \frac{2}{-5} = \frac{2}{-5}$.

(b) $y + 2 = \frac{-2}{5}(x - 3)$ or $y + 4 = \frac{-2}{5}(x - 8)$

11. (a) $m = \frac{4-(-1)}{-2-9} = \frac{5}{-11} = \frac{-5}{11}$,

(b) $y - 4 = \frac{5}{11}(x + 2)$ or $y + 1 = \frac{-5}{11}(x - 9)$

12. (a) $m = \frac{-1-(-3)}{2-(-4)} = \frac{2}{6} = \frac{1}{3}$,

(b) $y + 1 = \frac{1}{3}(x - 2)$ or $y + 3 = \frac{1}{3}(x + 4)$

13. $4x - 2y = 6, -2y = -4x + 6, y = 2x - 3$

14. $9x + 3 = 5, 3y = -9x, y = -3x + \frac{5}{3}$

15. $4x = 5y - 8, 5y = 4x + 8, y = \frac{4}{5}x + \frac{8}{5}$

16. $7x + 3y - 8 = 0, 3y = -7x + 8, y = \frac{-7}{3}x + \frac{8}{3}$

17.

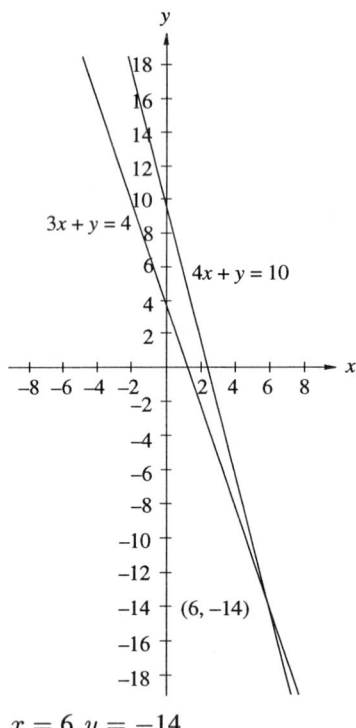

$3x + y = 4$

$4x + y = 10$

$(6, -14)$

$x = 6, y = -14$

18.

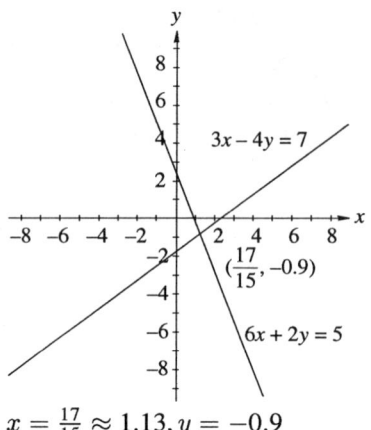

$3x - 4y = 7$

$(\frac{17}{15}, -0.9)$

$6x + 2y = 5$

$x = \frac{17}{15} \approx 1.13, y = -0.9$

19. $\begin{cases} 3x + y = 5 & (1) \\ 4x - y = 16 & (2) \end{cases}$
$y = 5 - 3x; 4x - (5 - 3x) = 16;$
$7x - 5 = 16; 7x = 21; x = 3, y = 5 - 3(3) = 5 - 9 = -4; (3, -4)$

20. $\begin{cases} 3x - 2y = 4 & (1) \\ x + 8y = 3 & (2) \end{cases}$
Solving (2) for x, we get $x = 3 - 8y$ (3). Substituting in (1) yields $3(3 - 8y) - 2y = 4$ or $9 - 24y - 2y = 4 \Rightarrow -26y = -5$ and so $y = \frac{5}{26} = 0.1923$. Substituting this value of y in (3) produces $x = 3 - 8(0.1923) = 1.4615$. The solution is $(1.4615, 0.1923)$.

21. $\begin{cases} 4x - y = 7 & (1) \\ 3x + 2y = 5 & (2) \end{cases}$
Solving (1) for y, we get $y = 4x - 7$ (3). Substituting this in (2) yields $3x + 2(4x - 7) = 5 \Rightarrow 3x + 8x - 14 = 5$ or $11x = 19$ and so, $x = \frac{19}{11} = 1.7273$. Substituting this value of x in (3), we see that $y = 4\left(\frac{19}{11}\right) - 7 = \frac{76}{11} - \frac{77}{11} = \frac{-1}{11} = -0.0909$. Thus, the solution is $\left(\frac{19}{11}, \frac{-1}{11}\right) = (1.7273, -0.0909)$.

22. $\begin{cases} 6x + y - 5 = 0 & (1) \\ 4y - 3x = -7 & (2) \end{cases}$
$y = 5 - 6x; 4(5 - 6x) - 3x = -7; 20 - 24x - 3x = -7; -27x = -27, x = 1; y = 5 - 6(1) = 5 - 6 = -1$

23. $\begin{cases} x - y = -2 & (1) \\ x + y = 8 & (2) \end{cases}$
Adding gives $2x = 6; x = 3; 3 - y = -2; -y = -5; y = 5$

24. $\begin{cases} 6x + 5y = 7 & (1) \\ 3x - 7y = 13 & (2) \end{cases}$
$(2) \times 2 \Rightarrow 6x - 14y = 26$ (3); $(1) - (3) \Rightarrow 19y = -19; y = -1; 6x + 5(-1) = 7; 6x = 12; x = 2$

25. $\begin{cases} x + \frac{1}{2}y = 2 & (1) \\ 3x - y = 1 & (2) \end{cases}$

$(1) \times 2 \Rightarrow 2x + y = 4\ (3); (3) + (2) \Rightarrow 5x = 5; x = 1\ 3(1) - y = 1; 3 - y = 1; -y = -2; y = 2$

26. $\begin{cases} \frac{1}{6}x + \frac{1}{4}y = \frac{1}{3} & (1) \\ \frac{1}{4}x - \frac{1}{2}y = 1 & (2) \end{cases}$

$(1) \times 12 \Rightarrow 2x + 3y = 4\ (3); (2) \times 8 \Rightarrow 2x - 4y = 8\ (4); (3) - (4) \Rightarrow 7y = -4; y = -\frac{4}{7}; 2x + 3(-\frac{4}{7}) = 4; 2x - \frac{12}{7} = 4; 2x = \frac{40}{7}; x = \frac{20}{7}$

27. $\begin{cases} 3x + 2y + 3z = -7 & (1) \\ 5x - 3y + 2z = -4 & (2) \\ 7x + 4y + 5z = 2 & (3) \end{cases}$

Begin by multiplying $(1) \times 3 \Rightarrow 9x + 6y + 9z = -21$ (4). Next, multiply $(2) \times 2 \Rightarrow 10x - 6y + 4z = -8$ (5). Add: $(4) + (5) \Rightarrow 19x + 13z = -29$ (6). Now, multiply $(1) \times 2 \Rightarrow 6x + 4y + 6z = -14$ (7), and subtract $(7) - (3) \Rightarrow -x + z = -16$ (8). If you now multiply $19 \times (8) \Rightarrow -19x + 19z = -304$ (9), and add $(9) + (6) \Rightarrow 32z = -333$. As a result, we obtain $z = -\frac{333}{32} = -10.40625$. From (8), we obtain $x = z + 16$, and substituting for z yields $x = 16 - 10.40625 = 5.59375$. Substituting these values for x and z in (1) produces $3(5.59375) + 2y + 3(-10.40625) = -7 \Rightarrow 2y = -7 - 3(5.59375) + 3(10.40625)$ and so $2y = 7.4375$ or $y = 3.71875$. Thus, the solution is $x = 5.59375, y = 3.71875, z = -10.40625$.

28. $\begin{cases} x + y + z = 2.7 & (1) \\ 6x + 7y - 5z = -8.8 & (2) \\ 10x + 16y - 3z = -6.4 & (3) \end{cases}$

First, multiply $(1) \times 6 \Rightarrow 6x + 6y + 6z = 16.2$ (4), and subtract: $(4) - (2) \Rightarrow -y + 11z = 25$ (5). Next, multiply $(1) \times 10 \Rightarrow 10x + 10y + 10z = 27$ (6), and subtract: $(3) - (6) \Rightarrow 6y - 13z = -33.4$ (7). Finally, multiply $(5) \times 6 \Rightarrow -6y + 66z = 150$ (8), and add $(7) + (8) \Rightarrow 53z = 116.6$, and so $z = 2.2$. From (5), $y = 11z - 25 = 11(2.2) - 25 = -0.8$. From (1), $x = 2.7 - y - z = 2.7 - (-.8) - 2.2 = 1.3$, and the solution is $(1.3, -0.8, 2.2)$.

29. $D = \begin{vmatrix} 3 & -2 \\ 5 & 2 \end{vmatrix} = 6 + 10 = 16;$

$D_x = \begin{vmatrix} 4 & -2 \\ 12 & 2 \end{vmatrix} = 8 + 24 = 32;$

$D_y = \begin{vmatrix} 3 & 4 \\ 5 & 12 \end{vmatrix} = 36 - 20 = 16; x = \frac{32}{16} = 2;$

$y = \frac{16}{16} = 1$

30. $D = \begin{vmatrix} 4 & 3 \\ 2 & -5 \end{vmatrix} = -20 - 6 = -26;$

$D_x = \begin{vmatrix} 27 & 3 \\ -19 & -5 \end{vmatrix} = -78;$

$D_y = \begin{vmatrix} 4 & 27 \\ 2 & -19 \end{vmatrix} = -130;$

$x = \frac{-78}{-26} = 3; y = \frac{-130}{-26} = 5$

31. $D = \begin{vmatrix} 4 & -3 \\ 11 & 4 \end{vmatrix} = 16 + 33 = 49;$

$D_x = \begin{vmatrix} 9 & -3 \\ 7 & 4 \end{vmatrix} = 57; x = \frac{57}{49};$

$D_y = \begin{vmatrix} 4 & 9 \\ 11 & 7 \end{vmatrix} = 28 - 99 = -71; y = \frac{-71}{49}$

32. $D = \begin{vmatrix} 2 & -5 \\ 4 & 6 \end{vmatrix} = 12 + 20 = 32;$

$D_x = \begin{vmatrix} -8 & -5 \\ 17 & 6 \end{vmatrix} = 37; x = \frac{37}{32};$

$D_y = \begin{vmatrix} 2 & -8 \\ 4 & 17 \end{vmatrix} = 34 + 32 = 66; y = \frac{66}{32} = \frac{33}{16}$

33. $D = \begin{vmatrix} 5 & 3 & -2 \\ 3 & -4 & 3 \\ 1 & 6 & -4 \end{vmatrix} - 9;$

$D_x = \begin{vmatrix} 5 & 3 & -2 \\ 13 & -4 & 3 \\ -8 & 6 & -4 \end{vmatrix} = -18;$

$D_y = \begin{vmatrix} 5 & 5 & -2 \\ 3 & 13 & 3 \\ 1 & -8 & -4 \end{vmatrix} = 9;$

$D_z = \begin{vmatrix} 5 & 3 & 5 \\ 3 & -4 & 13 \\ 1 & 6 & -8 \end{vmatrix} = -9;$

$x = \frac{-18}{-9} = 2;$

$y = \frac{9}{-9} = -1; z = \frac{-9}{-9} = 1. (2, -1, 1)$

34. $D = \begin{vmatrix} 5 & -2 & 3 \\ 6 & -3 & 4 \\ -4 & 4 & -9 \end{vmatrix} = 15;$

$D_x = \begin{vmatrix} 6 & -2 & 3 \\ 10 & -3 & 4 \\ 4 & 4 & -9 \end{vmatrix} = 10;$

$D_y = \begin{vmatrix} 5 & 6 & 3 \\ 6 & 10 & 4 \\ -4 & 4 & -9 \end{vmatrix} = -110;$

$D_z = \begin{vmatrix} 5 & -2 & 6 \\ 6 & -3 & 10 \\ -4 & 4 & 4 \end{vmatrix} = -60;$

$x = \frac{10}{15} = \frac{2}{3};$

$y = \frac{-110}{15} = -7.3; z = \frac{-60}{15} = -4. (\frac{2}{3}, -7\frac{1}{3}, -4)$

35. Let x be the amount of Colombian Supreme and y the amount Mocha Java coffee. Then we obtain two equations (1) $x + y = 50$ and (2) $4.99x + 5.99y = 50 \times 5.39$ or $499x + 599y = 26950$. Multiplying (1) by 499 gives $499x + 499y = 24950$. Subtracting yields $100y = 2000$ or $y = 20$. This leaves 30 for x. Solution: 30 pounds of Colombian Supreme and 20 pounds of Mocha Java coffee.

36. Organizing the information in chart form we have

computer type	Chip A	Chip A
PCx	4	11
BCy	9	6

$4x + 9y = 670; 11x + 6y = 1055$. Using Cramer's

Rule $D = \begin{vmatrix} 4 & 9 \\ 11 & 6 \end{vmatrix} = 24 - 99 = -75,$

$D_x = \begin{vmatrix} 670 & 9 \\ 1055 & 6 \end{vmatrix} = -5475,$

$D_y = \begin{vmatrix} 4 & 670 \\ 11 & 1055 \end{vmatrix} = -3150,$

$x = \frac{-5475}{-75} = 73, y = \frac{-3150}{-75} = 42$. They should produce 73 PC and 42 BC computers.

37. Let x be the number of 1 room offices and y be the number of 2 room offices and z be the number of 3 room offices; then
$$\begin{cases} x + y + z = 66 \\ x + 2y + 3z = 146 \\ 300x + 520y + 730z = 37,160 \end{cases}$$
Using Cramer's Rule $D = \begin{vmatrix} 1 & 1 & 1 \\ 1 & 2 & 3 \\ 300 & 520 & 730 \end{vmatrix} = -10,$

$D_x = \begin{vmatrix} 66 & 1 & 1 \\ 146 & 2 & 3 \\ 37,160 & 520 & 730 \end{vmatrix} = -100,$

$D_y = \begin{vmatrix} 1 & 66 & 1 \\ 1 & 146 & 3 \\ 300 & 37,160 & 730 \end{vmatrix} = -320,$

$D_z = \begin{vmatrix} 1 & 1 & 66 \\ 1 & 2 & 146 \\ 300 & 520 & 37,160 \end{vmatrix} = -240, x = \frac{-100}{-10} = 10,$

$y = \frac{-320}{-10} = 32, z = \frac{-240}{-10} = 24$. There are 10 small, 32 medium, and 24 large offices.

CHAPTER 5 TEST

1. **(a)** $m = \frac{6-2}{5-(-4)} = \frac{4}{9}.$

(b) $y - 2 = \frac{4}{9}(x + 4)$ or $y - 6 = \frac{4}{9}(x - 5); y - 2 = \frac{4}{9}x + \frac{16}{9}$ and so $y = \frac{4}{9}x + \frac{34}{9}$

2. **(a)** $-5y = -6x + 12; y = \frac{6}{5}x - \frac{12}{5},$

(b) slope: $m = \frac{6}{5},$

(c) y-intercept is $-\frac{12}{5},$

(d) $6x - 5y = 12$

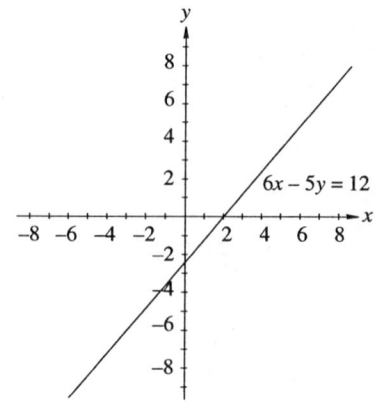

3. $\begin{cases} 2x + 3y = 1 & (1) \\ -3x + 6y = 16 & (2) \end{cases}$

(1) × 2 ⇒ $4x + 6y = 2$ (3), (3) − (2) ⇒ $7x = -14$, and we see that $x = -2$. Substitution in (1) yields $2(-2) + 3y = 1$ or $-4 + 3y = 1 \Rightarrow 3y = 5$ and so, $y = \frac{5}{3}$. Thus, the solution is $(-2, \frac{5}{3})$.

4. $\begin{vmatrix} 4 & -2 \\ 5 & 6 \end{vmatrix} = 4(6) - 5(-2) = 24 + 10 = 34$

5. $\begin{vmatrix} 4 & 2 & -1 \\ 3 & 0 & 4 \\ -3 & 1 & 1 \end{vmatrix}$. Using the 2nd column and minors,

$-2\begin{vmatrix} 3 & 4 \\ -3 & 1 \end{vmatrix} - 1\begin{vmatrix} 4 & -1 \\ 3 & 4 \end{vmatrix} = 2(15) - 1(19) = -30 - 19 = -49$

6. $\begin{cases} 2x + 3y = 5 \\ x - 4y = -14 \end{cases}$

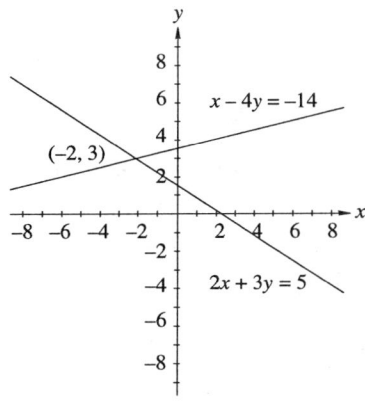

7. $\begin{cases} 4x + 3y = 9 & (1) \\ 2x + y = 2 & (2) \end{cases}$

From (2), we get $y = 2 - 2x$ (3). Substituting in (1) yields $4x + 3(2 - 2x) = 9 \Rightarrow 4x + 6 - 6x = 9$, or $-2x + 6 = 9$ and so, $x = -\frac{3}{2}$. Substituting in (3) produces $y = 2 - 2(-\frac{3}{2}) = 2 + 3 = 5$. Hence, the solution is $(-\frac{3}{2}, 5)$.

8. We have $D = \begin{vmatrix} 2 & -4 \\ -6 & 8 \end{vmatrix} = 16 - 24 = -8$,

$D_x = \begin{vmatrix} 2 & -4 \\ -9 & 8 \end{vmatrix} = 16 - 36 = -20$, and

$D_y = \begin{vmatrix} 2 & 2 \\ -6 & -9 \end{vmatrix} = -18 + 12 = -6$.

Using Cramer's rule, we see that $x = \frac{-20}{-8} = \frac{5}{2}$ and $y = \frac{-6}{-8} = \frac{3}{4}$. The solution is $(\frac{5}{2}, \frac{3}{4})$.

9. Using the addition method on the system
$\begin{cases} -x + 3y - 2z = 7 & (1) \\ 3x + 4y - 7z = -8 & (2) \\ x + 2y + z = 2 & (3) \end{cases}$
we obtain

(1) + (2) ⇒ $5y - z = 9$ (4)

$3(1) + (2) \Rightarrow 13y - 13z = 13$ or $y - z = 1$ (5)

(4) − (5) ⇒ $4y = 8$ or $y = 2$

Back-substitution into equation (5) yields $z = 1$ and finally $x = -3$. So, the solution is $(-3, 2, 1)$.

10. Using Cramer's rule, $D = 2, D_{x_1} = 220, D_{x_2} = -246, D_{x_3} = -130$, and $D_{x_4} = -102$. This produces the solution $x_1 = 110, x_2 = -123, x_3 = -65$, and $x_4 = -51$.

11. $x =$ cost of 1st, $y =$ cost of 2nd, $z =$ cost of 3rd
$\begin{cases} x + y + z = 60 & (1) \\ x = y + z & (2) \\ y = 2z + 3 & (3) \end{cases}$
or (2) can be written as $x - y - z = 0$ (4). Adding (1) + (4) we get $2x = 60$ or $x = 30$. Substituting this value for x in (2), we get $y + z = 30$ (5). Substituting $2z + 3$ for y in (5) yields $2z + 3 + z = 30$ or $3z = 27$, and so $z = 9$. Finally, substituting for z in (3), we see that $y = 2(9) + 3 = 18 + 3 = 21$. So, the cost of the first part is $30, the cost of the second part $21, and the cost of the third part is $9.

12. By Kirchhoff's laws, we have the following equations:

$$I_1 - I_2 - I_3 = 0 \quad (1)$$
$$6I_1 + 4I_2 = 20 \quad (2)$$
$$4I_2 - 2I_3 = 10 \quad (3)$$

Now

$$6 \times (1) - (2) \Rightarrow -10I_2 - 6I_3 = -20$$

or

$$5I_2 + 3I_3 = 10 \quad (4).$$

Then,

$$2 \times (4) + 3 \times (3) \Rightarrow 22I_2 = 50$$

and so $I_2 = \frac{25}{11}$. Back-substituting into equation (4) produces $I_3 = -\frac{5}{11}$ and, from equation (1), we find that $I_1 = I_2 + I_3 = \frac{20}{11}$. The solutions are $I_1 = \frac{20}{11}$ A, $I_2 = \frac{25}{11}$ A, and $I_3 = -\frac{5}{11}$ A.

CHAPTER

6

Ratio, Proportion, and Variation

6.1 RATIO AND PROPORTION

1. $1.38 for 16 bolts is 1.38 : 16

2. 725 revolutions per minute is 725 : 1

3. 86 L per km is 86 : 1

4. 236 miles in 4 hours is 236 : 4

5. $\frac{9}{2} = \frac{4.5}{1}$ or 4.5 : 1

6. $\frac{7}{5} = \frac{1.4}{1}$ or 1.4 : 1

7. $\frac{23}{7} \approx \frac{3.2857}{1}$ or 3.2857 : 1

8. $\frac{37}{4} = \frac{9.25}{1}$ or 9.25 : 1

9. $2 \times 15 = 30; 3 \times 10 = 30$ so equal

10. $1 \times 12 = 12; 3 \times 4 = 12$ so equal

11. $145 \times 5 = 725; 25 \times 29 = 725$ so equal

12. $3 \times 33 = 99; 8 \times 12 = 96$ so not equal

13. $8c = 7 \times 32; c = \frac{7 \times 32}{8}; c = 7 \times 4 = 28$

14. $9b = 3 \times 24; b = \frac{3 \times 24}{9} = 8$

15. $124d = 62 \times 158; d = \frac{62 \times 158}{124} = 79$

16. $20a = 8 \times 3.5; a = \frac{8 \times 3.5}{20} = 1.4$

17. $0.15a = 4 \times 0.16; a = \frac{4 \times 0.16}{0.15} \approx 4.267$

18. $10.5x = 7.5 \times 6.3; x = \frac{7.5 \times 6.3}{10.5} = 4.5$

19. $8b = 20 \times 5.6; b = \frac{20 \times 5.6}{8} = 14$

20. $2.4d = 10.8 \times 1.6; d = \frac{10.8 \times 1.6}{2.4} = 7.2$

21. $9x = 4 \times 4; x = \frac{4 \times 4}{9} \approx 1.78$

22. $xx = 3 \times 12; x^2 = 36; x = \pm\sqrt{36} = \pm 6$

23. $\frac{2}{x} = \frac{5}{14} = \frac{9}{z}; 5x = 2 \times 14; x = 5.6; 5z = 9 \times 14; z = \frac{9 \times 14}{5} = 25.2$

24. $\frac{a}{20} = \frac{6}{y} = \frac{9}{45}; \frac{a}{20} = \frac{9}{45}; 45a = 9 \times 20; a = \frac{9 \times 20}{45} = 4; 9y = 6 \times 45; y = \frac{6 \times 45}{9} = 30$

25. 54 : 13 or $\approx$ 4.1538 : 1

26. $\frac{15}{12} = \frac{5}{4}$ or 5 : 4

27. A steering wheel that makes $4\frac{2}{3}$ complete turns turns a total of $4\frac{2}{3} \times 360° = \frac{14}{3} \times 360° = 1680°$. Thus, the steering ratio is $\frac{1680}{60} = \frac{28}{1}$ or 28 : 1

28. 19 : 1 or just 19

29. $\frac{5 \times 10^{-4}}{300} \approx 1.67 \times 10^{-6}$ F $= 1.67 \mu$F

30. $\frac{20 \text{ mm}}{80 \text{ mm}} = \frac{1}{4}$ or 1 : 4

31. $\frac{8.7}{1} = \frac{96}{x}; x = \frac{96}{8.7} \approx 11.03$ or 11.03 cm^3

32. $\frac{90 \text{ mL}}{20 \text{ L}} = \frac{x \text{mL}}{54 \text{L}}; x = \frac{90 \times 54}{20}$ mL $= 243$ mL

33. $\frac{80 \text{ cal}}{1 \text{ g}} = \frac{x \text{ cal}}{785 \text{ kg}} = \frac{x \text{ cal}}{785,000 \text{ g}}; x = 80 \times 785000; x = 62,800,000$ calories

34. $\frac{120}{V} = \frac{100}{750}; V = \frac{120 \times 750}{100} = 900$ V

35. $\frac{25.4 \text{ mm}}{1 \text{ in.}} = \frac{88.9 \text{ mm}}{x \text{ in.}}; x = \frac{88.9}{25.4} = 3.5; 3.5$ in.

36. $\frac{9.78039}{32.0878} = \frac{9.83217}{x}; x = \frac{9.83217 \times 32.0878}{9.78039}; x = 32.257681$ ft/s^2

37. We have the proportion $\frac{x \text{ L}}{14.2 \text{ gal}} = \frac{1 \text{L}}{0.2642 \text{ gal}}$. Multiplying by 14.2 gal produces 53.7472 L.

38. We have the proportion $\frac{1.25 \text{ oz}}{140 \text{ cal}} = \frac{14.1 \text{ oz}}{x \text{ cal}}$ or x cal $=$ 14.1 oz $\cdot \frac{140 \text{ cal}}{1.25 \text{ oz}} = 1579.2$ cal. Similarly, $\frac{1.25 \text{ oz}}{3 \text{ g fiber}}$ or x g fiber $= 14.1$ oz. $\cdot \frac{3 \text{ g}}{1.25 \text{ oz}} = 33.84$ g of dietary fiber.

6.2 SIMILAR GEOMETRIC SHAPES

1. $\frac{a}{2} = \frac{3}{4} = \frac{b}{5}$; $4a = 3 \times 2$; $a = 6 \div 4 = 1.5$; $4b = 3 \times 5$; $b = 15 \div 4 = 3.75$

2. $\frac{a}{13} = \frac{2}{b} = \frac{9}{15}$; $15a = 9 \times 13$; $15a = 117$; $a = 7.8$; $9b = 2 \times 15$; $9b = 30$; $b = 3.33$ or $3\frac{1}{3}$

3. $\frac{a}{33} = \frac{24.3}{c} = \frac{22}{b}$; $66a = 33 \times 34$; $a = 17$; $34b = 22 \times 66$; $b \approx 42.71$; $34c = 24.3 \times 66$; $c \approx 47.17$

4. $\frac{a}{6} = \frac{b}{11} = \frac{c}{5} = \frac{d}{8} = \frac{4}{7}$; $7a = 4 \times 6$; $a = \frac{24}{7} \approx 3.43$; $7b = 4 \times 11$; $b = \frac{44}{7} \approx 6.29$; $7c = 5 \times 4$; $c = \frac{20}{7} \approx 2.86$; $7d = 4 \times 8$; $d = \frac{32}{7} \approx 4.57$

5. $\frac{a}{3} = \frac{b}{4} = \frac{c}{5} = \frac{d}{5} = \frac{1.7}{2}$; $2a = 3 \times 1.7$; $a = 2.55$; $2b = 4 \times 1.7$; $b = 3.4$; $2c = 5 \times 1.7$; $c = 4.25$; $d = c = 4.25$

6. $\frac{a}{5} = \frac{b}{3} = \frac{c}{4} = \frac{16}{9}$; $9a = 5 \times 16$; $a = \frac{80}{9} = 8\frac{8}{9} \approx 8.89$; $9b = 3 \times 16$; $b = \frac{48}{9} = 5\frac{3}{9} = 5\frac{1}{3}$; $9c = 4 \times 16$; $c = \frac{64}{9} = 7\frac{1}{9}$

7. $\frac{A}{20} = \frac{1^2}{(\frac{1}{8})^2}$, $\frac{1}{64}A = 20$, $A = 1280$ ft^2; 1280 ft$^2 = 144 \times 1280$ in.$^2 = 184,320$ in.^{2}

8. Discharge is proportional to cross sectional area; $\frac{2000}{3000} = \frac{75^2}{x^2}$, $x^2 = \frac{75^2 \times 3000}{2000} = 8437.5$; $x = \sqrt{8437.5} \approx 91.86$

9. For a fixed length, the mass is proportional to the cross sectional area; $\frac{19^2}{38^2} = \frac{x}{22}$, $x = \frac{19^2 \times 22}{38^2} = 5.5$ kg

10. Cost is proportional to length or perimeter; $\frac{652}{3 \times 652} = \frac{982^2}{x^2} = \frac{1}{3}$, $x^2 = 3 \times 982^2$; $x^2 = 2,892,972$, $x = \sqrt{2,892,972}$, $x = \$1,700.87$

11. Amount of paint is proportional to area; $\frac{700}{x} = \frac{20^2}{35^2}$; $x = \frac{35^2 \times 700}{20^2} = 2\,143.75$ L

12. $\frac{20}{x} = \frac{12^3}{30^3}$; $x = \frac{20 \times 30^3}{12^3} = 312.5$ kL

13. $\frac{3930}{x} = \frac{15^3}{5^3} = \frac{3375}{125}$, $x = \frac{125 \times 3930}{3375} \approx 145.56$ mm^3

14. $\frac{4}{5} = \frac{2.5}{h}$; $4h = 5 \times 2.5$; $h = 3.125''$, $V_1 = \pi r^2 h = \pi \cdot 2^2 \times 5 \approx 62.83$ in.^{3}; $V_2 = \pi(1.25)^2 \cdot 3.125 \approx 15.34$ in.^{3}

15. $\frac{10^2}{2^2} = \frac{1256.64}{A}$; $A = \frac{4 \times 1256.64}{100} \approx 50.27$ cm^2, $\frac{10^3}{2^3} = \frac{4188.79}{V}$; $V = \frac{8 \times 4188.79}{1000} \approx 33.51$ cm^3

16. For similar cans, volume is proportional to the cube of the diameter or the height. For the diameter we have $\frac{66^3}{d^3} = \frac{1}{2}$ or $d^3 = 2 \cdot 66^3 = 574\,992$. Taking the cube root, we get $d \approx 83.2$ mm. Similarly, for the height, $\frac{95^3}{h^3} = \frac{1}{2}$, so $h^3 = 2 \cdot 95^3 = 1\,714\,750$. Thus, $h \approx 119.7$ mm.

6.3 DIRECT AND INVERSE VARIATION

1. $R = kl$ or $\frac{R}{l} = k$

2. $P = kh$ or $\frac{P}{h} = k$

3. $A = kd^2$ or $\frac{A}{d^2} = k$

4. $E = kf$ or $\frac{E}{f} = k$

5. $IN = k$ or $I = \frac{k}{N}$

6. $Ed^2 = k$ or $E = \frac{k}{d^2}$

7. $k = \frac{r}{t} = \frac{4}{6} = \frac{2}{3}$

8. $ab = k$; $k = 8 \times 3 = 24$

9. $d = kr^2$; $k = \frac{d}{r^2} = \frac{8}{4^2} = \frac{8}{16} = \frac{1}{2}$

10. $ps^3 = k$; $k = 8 \times 2^3 = 8 \times 8 = 64$

11. $\frac{a_1}{d_1} = \frac{a_2}{d_2} \Rightarrow \frac{4}{8} = \frac{a}{18}$; $a = 4 \times 18 \div 8 = 9$

12. $r_1 t_1 = r_2 t_2 \Rightarrow r \times 9 = 6 \times 3$, $r = \frac{18}{9} = 2$

13. Note 1 min equals 60 s; $\frac{v_1}{t_1} = \frac{v_2}{t_2} \Rightarrow \frac{v}{60} = \frac{60}{20} \Rightarrow$

$v = \frac{60 \times 60}{20} = 180$ m/s

14. $p_1 q_1 = p_2 q_2 \Rightarrow p \times 40 = 30 \times 10 \Rightarrow p = \frac{30 \times 10}{40} =$

$\frac{30}{4} = 7.5$ mm^3

15. $\frac{n_1}{d_1^2} = \frac{n_2}{d_2^2} \Rightarrow \frac{n}{2^2} = \frac{10890}{10^2} \Rightarrow n = \frac{4 \times 10890}{100} = 435.6$

neutrons

16. $m_1 r_1^3 = m_2 r_2^3 \Rightarrow m5^3 = 25 \times 30^3$ or $125m =$

25×30^3, and so $m = \frac{25 \times 27000}{125} = 5\,400$ kPA

17. **(a)** $V = kT \Rightarrow = V/T = \frac{25}{293.18} \approx 0.08527 =$
constant;
(b) $V = 0.08527 \times (85 + 273.18) \approx 30.542$ m^3; ΔV
$= 30.542 - 25 = 5.542$ m$^3 = $ change in volume

18. $d = kt; k = \frac{d}{t} = \frac{475}{1} = 475$ mph; $1235 = 475t \Rightarrow$

$t = \frac{1235}{475} = 2.6$ hr; or $\frac{d_1}{t_1} = \frac{d_2}{t_2} \Rightarrow \frac{475}{1} = \frac{1235}{t} \Rightarrow$

$475t = 1235; \Rightarrow t = \frac{1235}{475} = 2.6$ hr or 2 hr 36 min

19. $R = k\ell$ or $\frac{R_1}{l_1} = \frac{R_2}{l_2}; \frac{32}{1500} = \frac{R}{5000}, 1500R = 32 \times$

$5000; R\frac{32 \times 5000}{1500} = 106.67\,\Omega$

20. $d = kt^2$ or $\frac{d_1}{t_1^2} = \frac{d_2}{t_2^2};$

(a) $k = \frac{d}{t^2} = \frac{29.4}{3^2} = \frac{29.4}{9} \approx 3.267; d = 3.267t^2;$
(b) $d = 3.267 \times 8^2 = 3.267 \times 64 \approx 209.09$ m;
(c) $150 = 3.267t^2 \Rightarrow t^2 = \frac{150}{3.267} \approx 45.91$ which
means that $t = \sqrt{45.91} \approx 6.78$ s.

21. $P_1 V_1 = P_2 V_2; 40 \times 200 = P \times 50 \Rightarrow P = \frac{40 \times 200}{50} =$
$40 \times 4 = 160$ psi

22. $Ed^2 = k$ or $E_1 d_1^2 = E_2 d_2^2; 600 \times 2^2 = E \times 6^2; E =$
$\frac{600 \times 4}{36} \approx 66.67$ lux

23. $In = k,\ I = $ current, $n = $ number of turns or
$I_1 n_1 = I_2 n_2; I \times 200 = 0.3 \times 50; I = \frac{0.3 \times 50}{200} -$
0.075 A

24. $V = kn$ or $\frac{v_1}{n_1} = \frac{v_2}{n_2}; \frac{120}{300} = \frac{V}{40}; V = \frac{120 \times 40}{300} = 16$ V

25. $wr^2 = k$ or $w_1 r_1^2 = w_2 r_2^2; 180 \times 4000^2 = w \times$
$(4000 + 500)^2; w = \frac{180 \times 4000^2}{4500^2} \approx 142.22$ lb

26. From the first sentence we have the general expression $td^2 = k$. Because we have two different diameter pipes and two different times, we have the equation $t_1 d_1^2 = t_2 d_2^2$. Letting $d_1 = 3, t_2 = 16,$ and $d_2 = 2$, we have $t_1 \cdot 3^2 = 16 \cdot 2^2$ or $9t_1 = 16 \cdot 4 = 64$. Solving for t_1 we get $t_1 = \frac{64}{9} = 7\frac{1}{9}$ h or 7 h 6 min 40 s.

≣ 6.4 JOINT AND COMBINED VARIATION

1. $K = kmv^2$

2. $V = klwh$

3. $f = \frac{kv}{l}$

4. $r = \frac{ks}{d}$

5. $E = \frac{klI}{d^2}$

6. $L = \frac{kN^2 A}{l}$

7. $a = kbc \Rightarrow 20 = k4(2); k = \frac{20}{4 \cdot 2} = \frac{20}{8} = 2.5$

8. $p = krst;\ 15 = k \cdot 3 \cdot 10 \cdot 25; k = \frac{15}{3 \cdot 10 \cdot 25} = \frac{1}{50} = 0.02$

9. $u = \frac{kv}{w}; 20 = \frac{k \cdot 5}{2}; k = \frac{20 \times 2}{5} = 8$

10. $x = \frac{kyw^3}{z^2}; 175 = \frac{k \cdot 3 \cdot 5^3}{4^2}; k = \frac{175 \times 16}{3 \times 125} \approx 7.47$

11. $r = kst; 10 = k \cdot 6 \cdot 5; k = \frac{10}{30} = \frac{1}{3}; r = \frac{1}{3} \cdot 3 \cdot 4 = 4$

12. $f = \frac{kr}{\ell}; 125 = \frac{k \cdot 10}{5}; k = \frac{125 \times 5}{10} = \frac{125}{2} = 62.5; f = \frac{125}{2} \cdot \frac{20}{10} = 125$

13. $a = kxy^2; 9 = k \cdot 3 \cdot 5^2; k = \frac{9}{3 \cdot 5^2} = \frac{3}{25} = 0.12;$
$a = 0.12 \cdot 6 \times 15^2 = 162$

14. $10 = 0.12 \cdot 9 \cdot y^2; y^2 = \frac{10}{0.12 \times 9} \approx 9.2593; y \approx \pm\sqrt{9.2593} \approx \pm 3.04$

15. $p = \frac{kr\sqrt{t}}{w}; 9 = \frac{k18\sqrt{9}}{6}; k = \frac{9 \times 6}{18 \times 3} = 1; p = \frac{36\sqrt{4}}{2} = 36$

16. $18 = \frac{9\sqrt{t}}{4}; \sqrt{t} = \frac{18 \cdot 4}{9} = 8; t = 8^2 = 64$

17. **(a)** $A = kbh; 12 = k \cdot 3 \cdot 8; k = \frac{12}{24} = \frac{1}{2};$
(b) $A = \frac{1}{2} \cdot 10 \cdot 5 = 25$

18. $W = kIVt; 2.4 = k \cdot 15 \cdot 240 \cdot \frac{45}{60}; k = \frac{2.4 \times 60}{15 \times 240 \times 45} \approx$
$8.89 \times 10^{-4}; W = 8.89 \times 1.25 \times 120 \times 8 \times 10^{-4}; W \approx$
1.07 kWh

19. $w = \frac{kV^2}{R}; 2 = \frac{k6^2}{18}; k = \frac{2 \times 18}{36} = 1; V^2 = wR = 50 \times 10 = 500; V = \sqrt{500} \approx 22.36$ V

20. $C = \frac{kA}{d}; A = s^2 = 4^2; 0.3 = \frac{k4^2}{0.5}; k = \frac{0.3 \times 0.5}{16}; 0.009375; C = \frac{.009375 \times 16}{0.1} = 1.5$ μF

21. $R = \frac{k\ell}{A}; 0.56 = \frac{k80}{2.5}; k = \frac{0.56 \times 2.5}{80} = 0.0175; 3 = \frac{0.0175\ell}{0.1}; \ell = \frac{3 \times 0.1}{0.0175} \approx 17.14$ m

22. $\Delta R = k\Delta C; 2.25 - 2.09 = k(25 - 10); 0.16 = k15; k = \frac{0.16}{15} \approx 0.01067; R - 2.25 = 0.01067(50 - 25); R - 2.25 \approx 0.267; R \approx 0.267 + 2.25; R \approx 2.517\,\Omega$

23. $H = \frac{kA(\Delta T)}{d}; 50 = \frac{k12(72-32)}{3}; k = 0.3125; H = \frac{0.3125(72-10) \times 12}{3} = 77.5$ Btu/hr

24. $F = \frac{k\sqrt{T}}{\ell}; 256 = \frac{k\sqrt{25}}{500}; k = 25\,600; F = \frac{25\,600\sqrt{30}}{450} \approx 311.59$ Hz

25. $P = \frac{kT}{V}; 15 = \frac{k(70+460)}{5}; k = \frac{15 \times 5}{530} \approx 0.1415; 75 = \frac{0.1415(200+460)}{V}; V = \frac{0.1415 \times 660}{75} = 1.2452$ or 1.25 ft^3

26. **(a)** $S = \frac{kwd^2}{l}$

(b) $\frac{k \cdot 2w(\frac{1}{2}d)^2}{l} = \frac{1}{2}\left(\frac{kwd^2}{l}\right)$, so the strength is halved

(c) Since the strength varies directly as d^2, an increase of 75% means that the new (increased) strength will be 100% of the old strength + 75% of the old strength = 175% = 1.75 of the old strength. Thus, the new diameter is $\sqrt{1.75} \approx 1.323$ times the old. This is about a 32% increase.

CHAPTER 6 REVIEW

1. $3 : x = 4 : 6; 4x = 3 \times 6; 4x = 18; x = \frac{18}{4} = \frac{9}{2} = 4.5$

2. $x : 5 = 3 : 15; 15x = 5 \times 3; x = 1$

3. $\frac{7}{9} = \frac{21}{d}; 7d = 21 \cdot 9; d = \frac{21 \cdot 9}{7} = 27$

4. $\frac{14}{6} = \frac{c}{27}; 6c = 14 \times 27; c = \frac{14 \times 27}{6} = 63$

5. $4 : 8 = 19 : x; 4x = 8 \times 19; x = \frac{8 \times 19}{4} = 38$

6. $x : 12 = 15 : 32; 32x = 12 \times 15; x = \frac{12 \times 15}{32} = 5.625$

7. $7 : 24.5 : x = 8 : y : 42; \frac{7}{8} = \frac{24.5}{y} = \frac{x}{42};$ $8x = 7 \times 42; x = 36.75; 7y = 8 \times 24.5; y = 28$

8. $\frac{12.5}{x} = \frac{y}{47} = \frac{8}{5}; 8x = 5 \times 12.5; x = 7.8125; 5y = 8 \times 47; y = 75.2$

9. $TR = \frac{N_p}{N_s}; 25 = \frac{4000}{N_s}; N_s = \frac{4000}{25} = 160$

10. $\Delta S = kI; 12.50 = k6.5; k \approx 1.923; \Delta S = 1.923 \times 9.6 \approx \$18.46$

11. $\frac{s_1}{4} = \frac{s_2}{5} = \frac{180}{8}; S_1 = \frac{180 \times 4}{8} = 90$ mm; $S_2 = \frac{180 \times 5}{8} = 112.5$ mm

12. TMA $= \frac{90}{5} = 18$

13. $E = \frac{\text{AMA}}{\text{TMA}} = \frac{16}{18} = \frac{8}{9}$ or ≈ 0.889 or 88.9%

14. If rate of flow is constant, then $V_1 A_1 = V_2 A_2; A \propto d^2 \Rightarrow V_1 d_1^2 = V_2 d_2^2;$
(a) $3 \cdot 1^2 = (.5)^2 \cdot V_2; V_2 = \frac{3}{.25} = 12$ m/s;
(b) $30d^2 = 3; d^2 = \frac{1}{10}; d = \sqrt{.1} \approx .316$ cm $= 3.16$ mm diameter

15. $5 \times .5^2 = V \times 2.5^2; V = \frac{5 \times 0.25}{6.25} = 0.2$ m/s

16. $W = kT^4; k = \frac{W}{T^4} = \frac{6.5 \times 10^7}{(5800)^4} = 5.7438 \times 10^{-8}$

17. $F = \frac{kC_1 C_2}{d^2}; k = \frac{Fd^2}{C_1 C_2} = \frac{7.192 \times 10^{-4}(.05)^2}{4 \times 10^{-9} \times 5 \times 10^{-8}} = 8.99 \times 10^9$

18. $M = \frac{kI_1}{d}; 4 \times 10^{-6} = \frac{k100}{5}; k = \frac{4 \times 10^{-6} \times 5}{100}; k = 2 \times 10^{-7}; M = \frac{2 \times 10^{-7} \times 150}{15} = 2 \times 10^{-6}T$

19. $2.42 = \frac{3 \times 10^8}{s}; s = \frac{3 \times 10^8}{2.42} \approx 1.24 \times 10^8$ m/s

CHAPTER 6 TEST

1. $4 : 9 = x : 51; 9x = 4 \times 51; x = \frac{4 \times 51}{9} = \frac{68}{3} = 22\frac{2}{3} \approx 22.67$

2. $8 : x = 68 : 93.5; 68x = 8 \times 93.5; x = 11$

3. $\frac{a}{28} = \frac{12}{20}; 20a = 12 \times 28; a = 16.8$

4. $\frac{24}{42} = \frac{78}{d}$; $24d = 78 \times 42$; $d = 136.5$

5. $6 : 8 : x = y : 14 : 24.5$; $\frac{6}{y} = \frac{8}{14} = \frac{x}{24.5}$; $8y = 6 \times 14$; $y = 10.5$; $14x = 8 \times 24.5$; $x = 14$

6. $\frac{a}{9} = \frac{b}{30} = \frac{75}{45}$; $45a = 75 \times 9$; $a = 15$; $45b = 75 \times 30$; $b = 50$

7. $E = kd^2$

8. $h = \frac{kr}{t^3}$

9. $\frac{a}{5} = \frac{b}{12} = \frac{71.5}{13}$; $13a = 5 \times 71.5$; $a = 27.5$; $13b = 12 \times 71.5$; $b = 66$

10. $3x + 2x = \frac{540}{2}$; $5x = 270$; $x = 54$; $\ell = 3 \cdot 54 = 162$ cm; $w = 2 \cdot 54 = 108$ cm

11. $A = ks^2$; $72 = k(12)^2$ and so $k = 0.5$. Thus, $A = (0.5)(18^2) = 162$ m^2 or $A_1 S_2^2 = A_2 S_1^2$; $A \cdot 12^2 = 72 \cdot 18^2$; $A = 162$ m^2

12. $d = \frac{k}{ph}$ and so $45 = \frac{k}{(80)(8)}$. Thus, $k = 28,800$ and $d = \frac{28,800}{(60)(10)} = 48$

CHAPTER
7
Factoring and Algebraic Fractions

≡ 7.1 SPECIAL PRODUCTS

1. $3(p + q) = 3p + 3q$

2. $7(2 + x) = 14 + 7x$

3. $3x(5 - y) = 15x - 3xy$

4. $2a\left(4 + a^2\right) = 8a + 2a^3$

5. $(p + q)(p - q) = p^2 - q^2$

6. $(3a + b)(3a - b) = (3a)^2 - b^2 = 9a^2 - b^2$

7. $(2x - 6p)(2x + 6p) = (2x)^2 - (6p)^2 = 4x^2 - 36p^2$

8. $\left(\frac{y}{2} + \frac{2p}{3}\right)\left(\frac{y}{2} - \frac{2p}{3}\right) = \left(\frac{y}{2}\right)^2 - \left(\frac{2p}{3}\right)^2 = \frac{y^2}{4} - \frac{4p^2}{9}$

9. $(r + w)^2 = r^2 + 2rw + w^2$

10. $(q - f)^2 = q^2 - 2af + f^2$

11. $(2x + y)^2 = (2x)^2 + 2 \cdot 2x \cdot y + y^2 = 4x^2 + 4xy + y^2$

12. $\left(\frac{1}{2}a + b\right)^2 = \left(\frac{1}{2}a\right)^2 + 2 \cdot \frac{1}{2}ab + b^2 = \frac{1}{4}a^2 + ab + b^2$

13. $\left(\frac{2}{3}x + 4bx\right)^2 = \left(\frac{2}{3}x\right)^2 + 2 \cdot \frac{2}{3}x(4b) + (4b)^2 = \frac{4}{9}x^2 = \frac{16}{3}xb + 16b^2$

14. $\left(a - \frac{1}{2}y\right)^2 = a^2 - 2 \cdot a\frac{1}{2}u + \left(\frac{1}{2}y\right)^2 = a^2 - ay + \frac{1}{4}y^2$

15. $\left(2p - \frac{3}{4}r\right)^2 = (2p)^2 - 2 \cdot 2p \cdot \frac{3}{4}r + \left(\frac{3}{4}r\right)^2 = 4p^2 - 3pr + \frac{9}{16}r^2$

16. $\left(\frac{2}{3}r - 5t\right)^2 = \left(\frac{2}{3}r\right)^2 - 2 \cdot \frac{2}{3}r \cdot 5t + (5t)^2 = \frac{4}{9}r^2 - \frac{23}{r}t + 25t^2$

17. $(a + 2)(a + 3) = a^2 + (2 + 3)a + 2 \cdot 3 = a^2 + 5a + 6$

18. $(x + 5)(x + 7) = x^2 + (5 + 7)x + 5 \cdot 7 = x^2 + 12x + 35$

19. $(x - 5)(x + 2) = x^2 + (-5 + 2)x + (-5)(2) = x^2 - 3x - 10$

20. $(a + 9)(x - 12) = ax - 12a + 9x - 108$ or $ax + 9x - 12a - 108$

21. $(2a + b)(3a + b) = 2 \cdot 3a^2 + (2 + 3)ab + b^2 = 6a^2 + 5ab + b^2$

22. $(4x + 2)(3x + 5) = 4 \cdot 3x^2 + [4 \cdot 5 + 2 \cdot 3]x + 2 \cdot 5 = 12x^2 + (20 + 6)x + 10 = 12x^2 + 26x + 10$

23. $(3x + 4)(2x - 5) = 3 \cdot 2x^2 + [3(-5) + 4 \cdot 2]x + 4(-5) = 6x^2 + [-15 + 8]x - 20 = 6x^2 - 7x - 20$

24. $(3m - n)(4m + 2n) = 3 \cdot 4m^2 + [3 \cdot 2 + (-1)4]mn + (-1)2n^2 - 12m^2 + 2mn - 2n^2$

25. $(a + b)^3 = a^3 + 3a^2b + 3ab^2 + b^3$

26. $(r - s)^3 = r^3 - 3r^2s + 3rs^2 - s^3$

27. $(x + 4)^3 = x^3 + 3 \cdot x^2 \cdot 4 + 3x \cdot 4^2 + 4^3 = x^3 + 12x^2 + 48x + 64$

28. $(2a - b)^3 = (2a)^3 - 3(2a)^2b + 3(2a)b^2 - b^3 = 8a^3 - 12a^2b + 6ab^2 - b^3$

29. $(2a - b)^3 = (2a)^3 - 3(2a)^2b + 3(2a)b^2 - b^3 = 8a^3 - 12a^2b + 6ab^2 - b^3$

30. $(3x + 2y)^3 = (3x)^3 + 3(3x)^2(2y) + 3(3x)(2y)^2 + (2y)^3 = 27x^3 + 54x^2y + 36xy^2 + 8y^3$

31. $(3x - 2y)^3 = (3x)^3 - 3(3x)^2(2y) + 3(3x)(2y)^2 - (2y)^3 = 27x^3 - 54x^2y + 36xy^2 - 8y^3$

32. $(4r - s)^3 = (4r)^3 - 3(4r)^2s + 3(4r)s^2 - s^3 = 64r^3 - 48r^2s + 12rs^2 - s^3$

33. $(m + n)(m^2 - mn + n^2) = m^3 + n^3$

34. $(a + 2)(a^2 - 2a + 4) = a^3 + 8$

35. $(r - t)\left(r^2 + rt + t^2\right) = r^3 - t^3$

36. $(h - 3)\left(h^2 + 3h + 9\right) = h^3 - 27$

37. $(2x + b)\left(4x^2 - 2xb + b^2\right) = (2x + b)\left[(2x)^2 - 2x \cdot b + b^2\right] = (2x)^3 + b^3 = 8x^3 + b^3$

38. $\left(3a + \frac{2}{3}c\right)\left(9a^2 - 2ac + \frac{4}{9}c^2\right) = \left(3a + \frac{2}{3}c\right)$
$\left[(3a)^2 - 3a \cdot \frac{2}{3}c + \left(\frac{2}{3}c\right)^2\right] = (3a)^3 + \left(\frac{2}{3}c\right)^3 =$
$27a^3 + \frac{8}{27}c^3$

39. $(3a - d)(9a^2 + 3da + d^2) = (3a - d)\left[(3a)^2 + 3ad + d^2\right]; = (3a)^3 - d^3 = 27a^3 - d^3$

40. $\left(\frac{2e}{5} - \frac{5r}{4}\right)\left(\frac{4e^2}{25} + \frac{er}{2} + \frac{25r^2}{16}\right) = \left(\frac{2e}{5} - \frac{5r}{4}\right)$
$\left[\left(\frac{2}{3}t\right)^2 + \frac{2e}{5} \cdot \frac{5r}{4} + \left(\frac{5r}{4}\right)^2\right] = \left(\frac{2e}{5}\right)^3 - \left(\frac{5r}{4}\right)^3 =$
$\frac{8e^3}{125} - \frac{125r^3}{64}$

41. $3(a + 2)^2 = 3(a^2 + 4a + 4) = 3a^2 + 12a + 12$

42. $5(2a - 4)^2 = 5(4a^2 - 16a + 16) = 20a^2 - 80a + 80$

43. $\frac{5r}{t}\left(t + \frac{r}{5}\right)^2 = \frac{5r}{t}\left(t^2 + \frac{2rt}{5} + \frac{r^2}{25}\right) = \frac{5r}{t} \cdot t^2 + \frac{5r}{t} \cdot$
$\frac{2rt}{5} + \frac{5r}{t} \cdot \frac{r^2}{25} = 5rt + 2r^2 + \frac{r^3}{5t}$

44. $(x^2 + 4)(x + 2)(x - 2) = (x^2 + 4)(x^2 - 4) = x^4 - 4^2 = x^4 - 16$

45. $(x^2 - 6)(x^2 + 6) = x^4 - 6^2 = x^4 - 36$

46. $\left(\frac{2a}{b} + \frac{b}{2}\right)^2 = \left(\frac{2a}{b}\right)^2 + 2\frac{2a}{b}\frac{b}{2} + \left(\frac{b}{2}\right)^2 = \frac{4a^2}{b^2} + 2a + \frac{b^2}{4}$

47. $\left(\frac{3x}{y} - \frac{y}{x}\right)\left(\frac{3x}{y} + \frac{y}{x}\right)x^2y^2 = \left(\left(\frac{3x}{y}\right)^2 - \left(\frac{y}{x}\right)^2\right)$
$x^2y^2 = \left(\frac{9x^2}{y^2} - \frac{y^2}{x^2}\right)x^2y^2 = 9x^4 - y^4$

48. $(x + 1)(x + 2)(x + 3) = \left[x^2 + 3x + 2\right](x + 3) = x^3 + 3x^2 + 3x^2 + 9x + 2x + 6 = x^3 + 6x^2 + 11x + 6$

49. $(x - y)^2 - (y - x)^2 = (x^2 - 2xy + y^2) - (y^2 - 2yx + x^2) = x^2 - 2xy + y^2 - y^2 + 2xy - x^2 = 0$

50. $(x - y)^2 - (y - x)^2 = (x^2 - 2xy + y^2) - (y^2 - 2yx + x^2) = 4xy$

51. $(x + 3)(x - 3)^2 = (x + 3)(x^2 - 6x + 9) = x^3 - 6x^2 + 9x + 3x^2 - 18x + 27 = x^3 - 3x^2 - 9x + 27$

52. $(y - 4)(y + 4)^2 = (y - 4)(y^2 + 8y + 16) = y^3 + 8y^2 + 16y - 4y^2 - 32y - 64 = y^3 + 4y^2 - 16y - 64$

53. $r(r - t)^2 - t(t - r)^2 = r\left(r^2 - 2rt + t^2\right) - t(t^2 - 2rt + r^2) = r^3 - 2r^2 + rt^2 - t^3 + 2rt^2 - r^2t = r^3 - 3r^2t + 3rt^2 - t^3$

54. $5(x + 4)(x - 4)\left(x^2 + 16\right) = 5\left(x^2 - 16\right)\left(x^2 + 16\right) = 5\left(x^4 - 256\right) = 5x^4 - 1280$

55. $(5 + 3x)(25 - 15x + 9x^2) = (5 + 3x)\left[5^2 - 5 \cdot 3x + (3x)^2\right] = 5^3 + (3x)^3 = 125 + 27x^3$

56. $(2 - \sqrt{x})(2 + \sqrt{x})(4 + x) = \left(2^2 - \sqrt{x^2}\right)(4 + x) = (4 - x)(4 + x) = 16 - x^2$

57. $[(x + y) - (w + z)]^2 = (x + y)^2 - 2(x + y)(w + z) + (x + z)^2$

58. $[(r + s) + (t + u)]^2 = (r + s)^2 + 2(r + s)(t + u)^2 + (t + u)^2$

59. $(x + y - z)(x + y + z) = [(x + y) - z][(x + y) + z] = (x + y)^2 - z^2 = x^2 + 2xy + y^2 - z^2$

60. $(a + b + 2)(a - b + 2) = [(a + b) + b][(a + 2) - b] = (a + 2)^2 - b^2 = a^2 + 4a + 4 - b^2$

61. $z^2 = R^2 + \left[x_L^2 - 2x_Lx_C + x_C^2\right] = R^2 + x_L^2 - 2x_Lx_C + x_C^2$

62. $KE = \frac{1}{2}m(3t + 1)^2 = \frac{1}{2}m\left(9t^2 + 6t + 1\right) = \frac{9}{2}mt^2 + 3mt + \frac{1}{2}m$

63. $a_c = \frac{(2r^2 - t)^2}{r} = \frac{4t^4 - 4t^3 + r^2}{r}$

64. $W = \frac{1}{2}m\left(v_1^2 - v_1^2\right)$

65. $d = \left(\frac{P}{3EI}\right)\left(l_1^3 - l_2^3\right)$

66. $d = \left(\frac{w}{8EI}\right)(l_1 - l_2)(l_1 + l_2)\left(l_1^2 + l_2^2\right)$
$= \left(\frac{w}{8EI}\right)\left(l_1^2 - l_2^2\right)\left(l_1^2 + l_2^2\right)$
$= \left(\frac{w}{P8EI}\right)\left(l_1^4 - l_2^4\right)$

≡ 7.2 FACTORING

1. $6x + 6 = 6(x + 1)$

2. $12x + 12 = 12(x + 1)$

3. $12a - 6 = 6(2a - 1)$

4. $15d - 5 = 5(3d - 1)$

5. $4x - 2y + 8 = 2(2x - y + 4)$

6. $6a + 9b - 3c = 3(2a + 3b - c)$

7. $5x^2 + 10x + 15 = 5\left(x^2 + 2x + 3\right)$

8. $16a^2 + 8b - 24 = 8(2a^2 + b - 3)$

9. $10x^2 - 15 = 5\left(2x^2 - 3\right)$

10. $14x^2 + 21 = 7(2x^4 + 3)$

11. $4x^2 + 6x = 2x(2x + 3)$

12. $8a - 4a^2 = 4a(2 - a)$

13. $7b^2y + 28b = 7b(by + 4)$

14. $9ax^2 + 27bx = 9x(ax + 3b)$

15. $3ax + 6ax^2 - 2ax = ax(3 + 6x - 2) = ax(1 + 6x)$

16. $6by - 12b^2y + 7by^2 = by(6 - 12b + 7y)$

17. $4ap^2 + 6a^2pq + 8apq^2 = 2ap(2p + 3aq + 4q^2)$

18. $12p^2r^2 - 8p^3r + 24pr^2 = 4pr(3pr - 2p^2 + 6r)$

19. $a^2 - b^2 = (a + b)(a - b)$

20. $p^2 - r^2 = (p + r)(p - r)$

21. $x^2 - 4 = x^2 - 2^2 = (x + 2)(x - 2)$

22. $a^2 - 16 = a^2 - 4^2 = (a + 4)(a - 4)$

23. $y^2 - 81 = y^2 - 9^2 = (y + 9)(y - 9)$

24. $m^2 - 49 = (m + 7)(m - 7)$

25. $4x^2 - 9 = (2x)^2 - 3^2 = (2x + 3)(2x - 3)$

26. $49y^2 - 64 = (7y)^2 - 8^2 = (7y + 8)(7y - 8)$

27. $9a^4 - b^2 = (3a^2)^2 - b^2 = (3a^2 + b)(3a^2 - b)$

28. $16t^6 - a^2 = (4t^3)^2 - a^2 = (4t^3 + a)(4t^3 - a)$

29. $25a^2 - 49b^2 = (5a)^2 - (7b)^2 = (5a + 7b)(5a - 7b)$

30. $121r^2 - 81t^2 = (11r)^2 - (9t)^2 = (11r + 9t)(11r - 9t)$

31. $144 - 25b^4 = (12)^2 - (5b^2)^2 = (12 + 5b^2)(12 - 5b^2)$

32. $81 - 49r^4 = 9^2 - (7r^2)^2 = (9 + 7r^2)(9 - 7r^2)$

33. $5a^2 - 125 = 5(a^2 - 25) = 5(a + 5)(a - 5)$

34. $7x^2 - 63 = 7(x^2 - 9) = 7(x + 3)(x - 3)$

35. $28a^2 - 63b^4 = 7(4a^2 - 9b^4) = 7[(2a)^2 - (3b^2)^2] = 7(2a + 3b^2)(2a - 3b^2)$

36. $81x^2 - 36t^6 = 9(9x^2 - 4t^6) = [(3x)^2 - (2t^3)^2] = 9(3x + 2t^3)(3x - 2t^3)$

37. $a^4 - 81 = (a^2 + 9)(a^2 - 9) = (a^2 + 9)(a + 3)(a - 3)$

38. $b^4 - 256 = (b^2 + 16)(b^2 - 16) = (b^2 + 16)(b + 4)(b - 4)$

39. $16x^4 - 256y^4 = 16(x^4 - 16y^4) = 16(x^2 + 4y^2)(x^2 - 4y^2) = 16(x^2 + 4y^2)(x + 2y)(x - 2y)$

40. $25a^2 - 400ab^8 = 25a(a^4 - 16b^8) = 25a(a^2 + 4b^4)(a^2 - 4b^4) = 25a(a^2 + 4b^4)(a + 2b^2)(a - 2b^2)$

41. $\pi r \left(r + \sqrt{h^2 + r^2} \right)$

42. $m \left(c\Delta t + L_f \right)$

43. $\frac{1}{2}d \left(v_2^2 - v_1^2 \right) = \frac{1}{2}d \left(v_2 + v_1 \right) \left(v_2 - v_1 \right)$

44. $\pi R^2 - \pi r^2 = \pi \left(R^2 - r^2 \right) = \pi(R + r)(R - r)$

45. $\frac{(\omega_f - \omega_0)(\omega_f + \omega_0)}{2\theta}$

46. $e\sigma T^4 - e\sigma T_0^4 = e\sigma \left(T^4 - T_0^4 \right) = e\sigma \left(T^2 + T_0^2 \right) \times (T + T_0)(T - T_0) = e\sigma (T - T_0)(T + T_0) \left(T^2 + T_0^2 \right)$

47. $W = \frac{1}{2}I\omega_2^2 - \frac{1}{2}I\omega_1^2$

$\qquad = \frac{1}{2}I \left(\omega_2^2 - \omega_1^2 \right)$

$\qquad = \frac{1}{2}I \left(\omega_2 - \omega_1 \right) \left(\omega_2 + \omega_1 \right)$

48. **(a)** Factoring by grouping, we obtain
$f^{-1} = n \left(r_1^{-1} - r_2^{-1} \right) - 1 \left(r_1^{-1} r_2^{-1} \right)$
$\quad = (n - 1) \left(r_1^{-1} - r_2^{-1} \right)$
(b) $\frac{1}{f} = (n - 1) \left(\frac{1}{r_1} - \frac{1}{r_2} \right)$

☰ 7.3 FACTORING TRINOMIALS

1. $x^2 + 9x - 8$; $b^2 - 4ac = 113$; $\sqrt{113} \approx 10.6$; factors, but does not factor using rational numbers.

2. $x^2 + 7x - 8$; $b^2 - 4ac = 81$; $\sqrt{81} = 9$; factors.

3. $3x^2 - 10x - 8$; $b^2 - 4ac = 196$; $\sqrt{196} = 14$; factors.

4. $2x^2 + 16x + 14$; $b^2 - 4ac = 144$; $\sqrt{144} = 12$; factors.

5. $5x^2 + 23x + 18$; $b^2 - 4ac = 169$; $\sqrt{169} = 13$; factors.

6. $7x^2 - 5x + 16$; $b^2 - 4ac = -423$; $\sqrt{-423}$ is not a real number; so, $7x^2 - 5x + 16$ does not factor using real numbers

7. $x^2 + 7x + 10 = (x + 2)(x + 5)$

8. $x^2 + 8x + 15 = (x + 3)(x + 5)$

9. $x^2 - 12x + 27 = (x - 3)(x - 9)$

10. $x^2 - 14x + 33 = (x - 3)(x - 11)$

11. $x^2 - 27x + 50 - (x - 2)(x - 25)$

12. $x^2 + 19x + 48 = (x + 3)(x + 16)$

13. $x^2 - x - 2 = (x-2)(x+1)$ **14.** $x^2 - 4x - 5 = (x-5)(x+1)$ **15.** $x^2 - 3x - 10 = (x-5)(x+2)$

16. $p^2 - 16p + 64 = p^2 - 2 \cdot 8p + 8^2 = (p-8)^2$

17. $r^2 + 10r + 25 = r^2 + 2 \cdot 5r + 5^2 = (r+5)^2$

18. $v^2 - 14v + 49 = v^2 - 2 \cdot 7v + 7^2 = (v-7)^2$

19. $a^2 + 22a + 121 = a^2 + 1 \cdot 11a + 11^2 = (a+11)^2$

20. $e^2 + 26e + 169 = e^2 + 2 \cdot 13e + 13^2 = (e+13)^2$

21. $f^2 = 30f + 225 = f^2 - 2 \cdot 15f + 15^2 = (f-15)^2$

22. $3x^2 + 4x + 1 = (3x+1)(x+1)$ **23.** $6y^2 - 7y + 1 = (6y-1)(y-1)$ **24.** $3p^2 + 5p + 2 = (3p+2)(p+1)$

25. $7 \times 2 = 14; 7 + 2 = 9; 7t^2 + 9t + 2 = 7t^2 + 7t + 2t + 2; = 7t(t+1) + 2(t+1) = (7t+2)(t+1)$

26. $5(-3) = -15; 15 \times (-1) = -15; 15 + (-1) = 14; 5a^2 + 14a - 3 = 5a^2 + 15a - 1a - 3 = 5a(1 + 3) - 1(a+3) = (5a-1)(a+3)$

27. $7(-5) = -35; -35 \times 1 = -35; -35 + 1 = -34; 7b^2 - 34b - 5 = 7b^2 - 35b + b - 5 = 7b(b-5) + 1(b-5)(7b+1)(b-5)$

28. $2(-6) = 12; 4 \times (-3) = -12; 4 + (-3) = 1; 2y^2 + y - 6 = 2y^2 + 4y - 3y - 6 = 2y(y+2) - 3(y+2) = (2y-3)(y+2)$

29. $4(-5) = -20; 20 \times (-1) = -20; 20 + (-1) = 19; 4e^2 + 19e - 5 = 4e^2 + 20e - 3 - 5 = 4e(e+5) - 1(e+5) = (4e-1)(e+5)$

30. $6 \times 3 = 18; (-18)(-1) = 18; (-18) + (-1) = -19; 6m^2 - 19m + 3 = 6m^2 - 18m - m + 3 = 6m(m-3) - 1(m-3) = (6m-1)(m-3)$

31. $3 \times 8 = 24; 4 \times 6 = 24; 4 + 6 = 10; 3u^2 + 10u + 8 = 3u^2 + 6u + 4u + 8 = 3u(u+2) + 4(u+2) = (3u+4)(u+2)$

32. $7(-2) - 14; 14(-1) = -14; 14 + (-1) = 13; 7r^2 + 13r - 2 = 7r^2 + 14r - r - 2 = 7r(r+2) - 1(r+2) = (7r-1)(r+2)$

33. $9(-6) = -54; (-27)(2) = -54; (-27) + 2 = -25; 9t^2 - 25t - 6 = 9t^2 - 27t + 2t - 6 = 9t(t-3) + 2(t-3) = (9t+2)(t-3)$

34. $4 \times 3 = 12; 2 \times 6 = 12; 2 + 6 = 8; 4x^2 + 8x + 3 = 4x^2 + 6x + 2x + 3 = 2x(2x+3) + 1(2x+3) = (2x+1)(2x+3)$

35. $6(-5) = -30; 15(-2) = -30; 15 + (-2) = 13; 6x^2 + 13x - 5 = 6x^2 + 15x - 2x - 5 = 3x(2x+5) - 1(2x+5) = (3x-1)(2x+5)$

36. First factor out a 2, with the result $8y^2 - 8y - 6 = 2(4y^2 - 4y - 3)$. Then, $4(-3) = -12; -6 \times 2 = -12; -6 + 2 = -4; 8y^2 - 8y - 6 = 2(4y^2 - 4y - 3) = [4y^2 - 6y + 2y - 3] = 2[2y(2y-3) + 1(2y-3)] = 2(2y+1)(2y-3)$

37. $15(-15) = -225; -25 \times 9 = -225; -25 + 9 = -16; 15a^2 - 16a - 15 = 15a^2 - 25a + 9a - 15 = 5a(3a-5) + 3(3a-5) = (5a+3)(3a-5)$

38. $15(-15) - 225; 25(-9) = -225; 25 - 9 = 16; 15d^2 + 16d - 15 = 15d^2 + 25d - 9d - 15 = 5d(3d+5) - 3(3d+5) = (5d-3)(3d+5)$

39. $15 \times 15 = 225; 25 \times 9 = 225; 25 + 9 = 34; 15e^2 + 34e + 15 = 15e^2 + 25e + 9e + 15 = 5e(3e+5) + 3(3e+5) = (5e+3)(3e+5)$

40. $14 \times 10 = 140; (-35)(-4) = 140; -35 - 4 = -39; 14a^2 - 39a + 10 = 14a^2 - 35a - 4a + 10 = 7a(2a-5) - 2(2a-5) = (7a-2)(2a-5)$

41. $10 \times 6 = 60; (-15) \times (-4) = 60; -15 + (-4) = -19; 10x^2 - 19x + 6 = 10x^2 - 15x - 4x + 6 = 5x(2x-3) - 2(2x-3) = (5x-2)(2x-3)$

42. $3x^2 + 18x + 27 = 3(x^2 + 6x + 9) = 3(x+3)^2$

43. $3r^2 - 18r - 21 = 3(r^2 - 6r - 7) = 3(4-7)(r+1)$

44. First factor out a 5, with the result $15x^2 + 50x + 35 = 5(3x^2 + 10x + 7)$. Then, $3 \times 7 = 21; 3 + 7 = 10; 15x^2 + 50x + 35 = 5(3x^2 + 10x + 7) = 5[3x^2 + 3x + 7x + 7] = 5[3x(x+1) + 7(x+1)] = 5(3x+7)(x+1)$

45. $49t^4 - 105t^3 + 14t^2 = 7t^2(7t^2 - 15t + 2) = 7t^2(7t-1)(t-2)$

46. $-2 \times -7 = 14; = 2(-7) = -9; 2y^4 - 9y^2 + 7 = 2y^4 - 2y^2 + 7 = 2y^2(y^2 - 1) - 7(y^2 - 1) = (2y^2 - 7)(y^2 - 1) = (2y^2 - 7)(y-1)(y+1)$

47. $6(-10) = -60; (-15)4 = -60; -15 + 4 = -11; 6x^2 - 11xy - 10y^2 = 6x^2 - 15xy + 4xy - 10y^2 = 3x(2x-5y) + 2y(2x-5y) = (3x+2y)(2x-5y)$

48. $4p^2 + 20pq + 25q^2 = (2p)^2 + 2(2p)(5q) + (5q)^2 = (2p+5q)^2$

49. $8 \times (-9) = -72; -18(4) = -72; -18 + 4 = -14; 8a^2 - 14ab - 9b^2 = 8a^2 - 18ab + 4ab - 9b^2 = 2a(4a-9b) + b(4a-9b) = (2a+b)(4a-9b)$

50. $6d^9 + 15d^5e^2 + 6de^4 = 3d(2d^8 + 5d^4e^2 + 2e^4) = 3d(2d^4 + e^2)(d^4 + 2e^2)$

51. $a^3 - b^3 = (a - b)\left(a^2 + ab + b^2\right)$

52. $y^3 - 8 = y^3 - 2^3 = (y - 2)\left(y^2 + 2y + 4\right)$

53. $8x^3 - 27 = (2x)^3 - 3^3 = (2x - 3) \times \left[(2x)^2 + 2x \cdot 3 + 3^2\right] = (2x - 3)\left(4x^2 + 6x + 9\right)$

54. $64p^3 + 125t^6 = (4p)^3 + \left(5t^2\right)^3 = \left(4p + 5t^2\right)\left[(4p)^2 - 4p \cdot 5t^2 + \left(5t^2\right)^2\right] = \left(4p + 5t^2\right)\left(16p^2 - 20pt^2 + 25t^4\right)$

55. $i = 0.7\left(t^2 - 3t - 4\right) = 0.7(t - 4)(t + 1)$

56. $V = 180x - 58x^2 + 4x^3 = 2x\left(90 - 29x + 2x^2\right) = 2x(10 - x)(9 - 2x)$

57. $0.0001n^2(n - 2000) - 3(n - 2000) = \left(0.0001n^2 - 3\right)(n - 2000) = 0.0001\left(n^2 - 30{,}000\right)(n - 2000)$

58. $-16t^2 + 48t + 448 = -16\left(t^2 - 3t - 28\right) = -16(t - 7)(t + 4)$

59. **(a)** $(6 + 2x)(10 + 2x) - 60 = 36$; $60 + 32x + 4x^2 - 60 - 36 = 0$; $4x^2 + 32x - 36 = 0$.
(b) $4\left(x^2 + 8x - 9\right) = 0$; $4(x + 9)(x - 2) = 0$

60. Note: $2(-21) = -42$ has factors of -6 and 7 which add up to 1, the coefficient of Q. We proceed as follows using the grouping method: $2Q^2 + Q - 21 = 2Q^2 - 6Q + 7Q - 21 = 2Q(Q - 3) + 7(Q - 3) = (2Q + 7)(Q - 3)$.

≡ 7.4 FRACTIONS

1. $\frac{7}{8}$ (by 5); $\frac{7}{8} \cdot \frac{5}{5} - \left(\frac{3}{5}\right)$ 40; $7 \times 40 = 280$; $8 \times 35 = 280$

2. $\frac{-5}{9}$ (by -4); $\frac{-5}{9} \cdot \frac{-4}{-4} = \frac{20}{-36}$; $(-5)(-36) = 180$; $9(20) = 180$

3. $\frac{x}{y}$ (by a); $\frac{x}{y} \cdot \frac{a}{a} = \frac{ax}{ay}$. The cross product is $axy = axy$.

4. $\frac{r}{t}$ (by z); $\frac{rz}{tz}$. The cross-product is $rtz = rtz$.

5. $\frac{x^2 y}{a}$ (by $3ax$); $\frac{3ax^3 y}{3c \cdot x}$. The cross-product is $3a^2 x^3 y = 3a^2 x^3 y$.

6. $\frac{a^3 b}{ca}$ (by $3ab$); $\frac{3a^4 b^2}{3ca^2 b}$. The cross-product is $3a^5 b^2 c = 3a^5 b^2 c$.

7. $\frac{4}{x-y}$ (by $x + y$); $\frac{4}{(x-y)} \cdot \frac{(x+y)}{(x+y)} = \frac{4(x+y)}{x^2 - y^2} = \frac{4x + 4y}{x^2 - y^2}$. The cross-product is $4\left(x^2 - y^2\right) = (4x + 4y)(x - y) = 4x^2 - 4y^2$.

8. $\frac{a+b}{4}$ (by $a - 4$); $\frac{(a+b)}{4} \frac{(a-4)}{(a-4)} = \frac{a^2 - 4a + ab - 4b}{4a - 16}$. The cross-product is $(a + b)(4a - 16) = 4a^2 - 16a + 4ab - 16b$.

9. $\frac{a+b}{a-b}$ (by $a + b$); $\frac{(a+b)}{(a-b)} \cdot \frac{(a+b)}{(a+b)} = \frac{a^2 + 2ab + b^2}{a^2 - b^2}$. The cross-product is $(a + b)\left(a^2 - b^2\right) = a^3 + a^2 b - ab^2 - b^3 = (a - b)\left(a^2 + 2ab + b^2\right)$.

10. $\frac{x-2}{x+3}$ (by $x - 3$); $\frac{(x-2)}{(x+3)} \cdot \frac{(x-3)}{(x-3)} = \frac{x^2 - 5x + 6}{x^2 - 9}$. The cross-product is $(x - 2)\left(x^2 - 9\right) = x^3 - 2x^2 - 9x + 18 = (x + 3)(x^2 - 5x + 6)$.

11. $\frac{38}{24}$ (by 2)$= \frac{19}{12}$; $12 \times 38 = 456$; $24 \times 19 = 456$

12. $\frac{51}{119}$ (by 17)$= \frac{3}{7}$; $7 \times 51 = 357$; $119 \times 3 = 357$

13. $\frac{3x^2}{12x}$ (by $3x$) $= \frac{x}{4}$ $4 \cdot 3x^2 = 12x^2$ $12x \cdot x = 12x^2$

14. $\frac{15a^3 x^2}{3a^4 x}$ (by $3a^3 x$) $= \frac{5}{xa}$; $a \cdot 15a^3 x^2 = 15a^4 x^2$; $3a^4 x \cdot 5x = 15a^4 x^2$

15. $\frac{4(x+2)}{(x+2)(x-3)}$ (by $x + 2$) $= \frac{4}{x-3}$; $4(x + 2)(x - 3) = 4(x + 2)(x - 3)$.

16. $\frac{7(x-3)(x+5)}{14(x+5)(x-1)}$ $\left[\text{by } 7(x + 5)\right] = \frac{x-3}{2(x-1)}$; $2(x-1)(7)(x-3)(x + 5) = 14(x - 1)(x - 3)(x + 5)$

17. $\frac{x^2 - 16}{x^2 + 8x + 16}$ (by $x + 4$) $= \frac{x^2 - 16}{x^2 + 8x + 10} = \frac{(x+4)(x-4)}{(x+4)^2} = \frac{x-4}{x+4}$; $(x + 4)\left(x^2 - 16\right) = (x + 4)^2 (x - 4)$

18. $\frac{(x-a)(x-b)(x-c)}{(x-c)(x-b)(x-d)}$ $\left[\text{by } (x - c)(x - b)\right] = \frac{x-a}{x-d}$; $(x - a)(x - c)(x - b)(x - d) = (x - d)(x - a)(x - b)(x - c)$

19. $\frac{4x^2}{12x} = \frac{4x \cdot x}{4x \cdot 3} = \frac{x}{3}$

20. $\frac{9y}{3y^2} = \frac{3y \cdot 3}{3y \cdot y} = \frac{3}{y}$

21. $\frac{x^2 + 3x}{x^3 + 5x} = \frac{x(x+3)}{x\left(x^2 + 5\right)} = \frac{x+3}{x^2 + 5}$

22. $\frac{y^2 - 4y}{2y + y^3} = \frac{y(y-4)}{y(2 + y^2)} = \frac{y-4}{2 + y^2}$

23. $\frac{6m^2 - 3m^3}{9m + 18m^3} = \frac{3m^2(2 - m)}{9m(1 + 2m^2)} = \frac{m(2-m)}{3(1 + 2m^2)}$ or $\frac{2m - m^2}{3 + 6m^2}$

24. $\frac{4r^2 + 12r^3}{8r + 12r^2} = \frac{4r^2(1 + 3r)}{4r(2 + 3r)} = \frac{4(1 + 3r)}{2 + 3r} = \frac{4 + 3r^2}{2 + 3r}$

25. $\frac{x^2 + 3x}{x^2 - 9} = \frac{x(x+3)}{(x+3)(x-3)} = \frac{x}{x-3}$

26. $\frac{a^2 - 9a}{a^2 - 81} = \frac{a(a-9)}{(a+9)(a-9)} = \frac{a}{a+9}$

27. $\frac{2b^2 - 10b}{3b^2 - 75} = \frac{2b(b-5)}{3(b+5)(b-5)} = \frac{2b}{3(b+5)}$ or $\frac{2b}{3b+15}$

28. $\dfrac{4e^2-196}{14e-2e^2} = \dfrac{4\left(e^2-49\right)}{2e(7-3)} = \dfrac{4(e+7)(e-7)}{2e(7-e)} = \dfrac{-2\cdot2(e+7)(7-e)}{2e(7-e)} = \dfrac{-2(e+7)}{e}$

29. $\dfrac{z^2-9}{z^2-6z+9} = \dfrac{(z+3)(z-3)}{(z-3)^2} = \dfrac{z+3}{z-3}$

30. $\dfrac{x^2-16}{x^2+8x+16} = \dfrac{(x+4)(x-4)}{(x+4)^2} = \dfrac{x-4}{x+4}$

31. $\dfrac{x^2+4x+3}{x^2+7x+12} = \dfrac{(x+1)(x+3)}{(x+4)(x+3)} = \dfrac{x+1}{x+4}$

32. $\dfrac{a^2-5a+6}{a^2+5a-14} = \dfrac{(a-2)(a-3)}{(a+7)(a-2)} = \dfrac{a-3}{a+7}$

33. $\dfrac{2x^2+9x+4}{x^2+9x+20} = \dfrac{(2x+1)(x+4)}{(x+5)(x+4)} = \dfrac{2x+1}{x+5}$

34. $\dfrac{15m^2-22m-5}{3m^2+4m-15} = \dfrac{(5m+1)(3m-5)}{(3m-5)(m+3)} = \dfrac{5m+1}{m+3}$

35. $\dfrac{12y^3+12y^2+3y}{6y^2-3y-3} = \dfrac{3y(4y^2+4y+1)}{3(2y^2-y-1)} = \dfrac{y(2y+1)^2}{(2y+1)(y-1)} = \dfrac{y(2y+1)}{y-1}$

36. $\dfrac{45x^2-60x+20}{6x^2+5x-6} = \dfrac{5(9x^2-12x+4)}{(2x+3)(3x-2)} = \dfrac{5(3x-2)^2}{(2x+3)(3x-2)} = \dfrac{5(3x-2)}{2x+3}$

37. $\dfrac{x^3y^6-y^3x^6}{2x^3y^4-2x^4y^3} = \dfrac{x^3y^3\left(y^3-x^3\right)}{2x^3y^3(y-x)} = \dfrac{(y-x)\left(y^2+yx+x^2\right)}{2(y-x)} = \dfrac{y^2+xy+x^2}{2}$

38. $\dfrac{x^3-y^3}{y^2-x^2} = \dfrac{(x-y)\left(x^2+xy+y^2\right)}{(y-x)(y+x)} = -\dfrac{x^2+xy+y^2}{(x+y)}$

39. $\dfrac{x^2-y^2}{x+y} = \dfrac{(x+y)(x-y)}{(x+y)} = \dfrac{x-y}{1} = x-y$

40. $\dfrac{y-x}{x^2-y^2} = \dfrac{y-x}{(x+y)(x-y)} = \dfrac{-1}{x+y}$

41. $p = \dfrac{6whs+6wh}{3s^2+6s+3} = \dfrac{6wh(s+1)}{3(s+1)^2}$

$= \dfrac{6wh}{3(s+1)}$

$= \dfrac{2wh}{s+1}$

42. $p = \dfrac{2whs+2whk+2amhs+2amhk-2bhs-2bhk}{s^2+2sk+k^2}$

$= \dfrac{2h[2ws+2wk+ams+amk-bs-bk]}{(s+k)^2}$

Regrouping, we get

$p = \dfrac{2h[(2ws+ams-bs)+(wk+amk-bk)]}{(s+k)^2}$

$= \dfrac{2h[s(w+am-b)+k(w+am-b)]}{(s+k)^2}$

$= \dfrac{2h(s+k)(w+am-b)}{(s+k)^2}$

$= \dfrac{2h(w+am-b)}{(s+k)}$

≡ 7.5 MULTIPLICATION AND DIVISION OF FRACTIONS

1. $\dfrac{2}{x} \cdot \dfrac{5}{6} = \dfrac{2\times5}{x\times y} = \dfrac{10}{xy}$

2. $\dfrac{4}{y} \cdot \dfrac{x}{3} = \dfrac{4\cdot x}{y\cdot3} = \dfrac{4x}{3y}$

3. $\dfrac{4x^2}{5} \cdot \dfrac{3}{y^3} = \dfrac{4x^2\cdot3}{5y^3} = \dfrac{12x^2}{5y^3}$

4. $\dfrac{7x}{6} \cdot \dfrac{5y}{2t} = \dfrac{7x\cdot5y}{6\cdot2t} = \dfrac{35xy}{12t}$

5. $\dfrac{3}{x} \div \dfrac{7}{y} = \dfrac{3}{x} \times \dfrac{y}{7} = \dfrac{3y}{7x}$

6. $\dfrac{a}{3} \div \dfrac{b}{4} = \dfrac{a}{3} \times \dfrac{4}{b} = \dfrac{4a}{3b}$

7. $\dfrac{2x^2}{3} \div \dfrac{7y}{4x} = \dfrac{2x^2}{3} \times \dfrac{4x}{7y} = \dfrac{8x^3}{21y}$

8. $\dfrac{9x}{2y} \div \dfrac{4y}{3a} = \dfrac{9x}{2y} \times \dfrac{3a}{4y} = \dfrac{27ax}{8y^2}$

9. $\dfrac{2x}{3y} \cdot \dfrac{5}{4x^2} = \dfrac{2\cdot x\cdot5}{3\cdot y\cdot2\cdot2\cdot x\cdot x} = \dfrac{5}{3y2x} = \dfrac{5}{6xy}$

10. $\dfrac{3xy}{7} \cdot \dfrac{14x}{5y^2} = \dfrac{3\cdot x\cdot y\cdot7\cdot2\cdot x}{7\cdot5\cdot y\cdot y} = \dfrac{3\cdot x\cdot2\cdot x}{5\cdot y} = \dfrac{6x^2}{5y}$

11. $\dfrac{3a^2b}{5d} \cdot \dfrac{25ad^2}{6b^2} = \dfrac{3a^2b\cdot5\cdot5\cdot add}{5d\cdot3\cdot2\cdot bb} = \dfrac{5a^3d}{2b}$

12. $\dfrac{x^2y^2t}{abc} \cdot \dfrac{b^2c}{y^3t} = \dfrac{x^2y^2tbbc}{abcy^2yt} = \dfrac{x^2b}{ay}$

13. $\dfrac{3y}{5x} \div \dfrac{15x^2}{8xy} = \dfrac{3y}{5x} \times \dfrac{8xy}{15x^2} = \dfrac{3y8xy}{5x\cdot3\cdot5xx} = \dfrac{8y^2}{25x^2}$

14. $\dfrac{4y^2}{7x} \div \dfrac{8y^3}{21x} = \dfrac{4y^2}{7x} \times \dfrac{21x}{8y^3} = \dfrac{4y^2\cdot3\cdot7x}{7x\cdot2\cdot4y^2\cdot y} = \dfrac{3}{2y}$

15. $\dfrac{3x^2y}{7p} \div \dfrac{15x^2p}{7y^2} = \dfrac{3x^2y}{7p} \times \dfrac{7y^2}{15x^2p} = \dfrac{3x^2y\cdot7y^2}{7p\cdot3\cdot5x^2p} = \dfrac{y^3}{5p^2}$

16. $\dfrac{9xyz}{7a} \div \dfrac{3ayx}{14z} = \dfrac{9xyz}{7a} \times \dfrac{14z}{3ayz} = \dfrac{3\cdot3xyz\cdot2\cdot7z}{7a\cdot3ayz} = \dfrac{6xz}{a^2}$

17. $\dfrac{4y+16}{5} \cdot \dfrac{15y}{3y+12} = \dfrac{4(y+4)\cdot3\cdot5y}{5\cdot3(y+4)} = \dfrac{4y}{1} = 4y$

18. $\dfrac{x^2+3x}{6a} \cdot \dfrac{a^2}{x^2-9} = \dfrac{x(x+3)a^2}{6a(x+3)(x-3)} = \dfrac{ax}{6(x-3)}$ or $\dfrac{ax}{6x-18}$

19. $\dfrac{a^2-b^2}{a+3b} \cdot \dfrac{5a+15b}{a+b} = \dfrac{(a+b)(a-b)5(a+3b)}{(a+3b)(a+b)} = 5(a-b)$ or $5a-5b$

20. $(x+y)\dfrac{x^2+2x}{x^2-y^2} = \dfrac{(x+y)(x)(x+2)}{(x+y)(x-y)} = \dfrac{x(x+2)}{x-y}$ or $\dfrac{x^2+2x}{x-y}$

21. $\dfrac{x^2-100}{10} \div \dfrac{2x+10}{15} = \dfrac{x^2-100}{10} \times \dfrac{15}{2x+10} = \dfrac{(x+10)(x-10)\cdot3\cdot5}{2\cdot5\cdot2(x+5)} = \dfrac{3(x^2-100)}{4(x+5)}$

22. $\dfrac{5a^2}{x^2-49} \div \dfrac{25ax-25a}{x^2+7x} = \dfrac{5a^2}{x^2-49} \times \dfrac{x^2+7x}{25ax-25a} = \dfrac{5\cdot aax(x+7)}{(x+7)(x-7)25a(x-1)} = \dfrac{ax}{5(x-7)(x-1)}$

23. $\dfrac{4x^2-1}{9x-3x^2} \div \dfrac{2x+1}{x^2-9} = \dfrac{4x^2-1}{9x-3x^2} \times \dfrac{x^2-9}{2x+1} = \dfrac{(2x+1)(2x-1)(x+3)(x-3)}{3x(3-x)(2x+1)} = -\dfrac{(2x-1)(x+3)}{3x}$

24. $\frac{x+y}{3x-3y} \div \frac{(x+y)^2}{x^2-y^2} = \frac{x+y}{3(x-y)} \times \frac{x^2-y^2}{(x+y)^2} =$

$\frac{(x+y)(x+y)(x-y)}{3(x-y)(x+y)^2} = \frac{1}{3}$

25. $\frac{a^2-8a}{a-8} \cdot \frac{a+2}{a} = \frac{a(a-8)(a+2)}{(a-8)a} = a+2$

26. $\frac{49-x^2}{x+y} \cdot \frac{x}{7-x} = \frac{(7+x)(7-x)\cdot x}{(x+y)(7-x)} = \frac{x(7+x)}{x+y}$

27. $\frac{2a-b}{4a} \cdot \frac{2a-b}{4a^2-4ab+b^2} = \frac{(2a-b)^2}{4a(2a-b)^2} = \frac{1}{4a}$

28. $\frac{x^4-81}{(x-3)^2} \cdot \frac{x-3}{4-x^2} = \frac{(x^2+9)(x+3)(x-3)}{(x-3)^2} \cdot$

$\frac{(x-3)}{(2+x)(2-x)} = \frac{(x^2+9)(x+3)}{4-x^2}$

29. $\frac{y^2}{x^2-1} \div \frac{y^2}{x-1} = \frac{y^2}{(x+1)(x-1)} \times \frac{x-1}{y^2} = \frac{1}{x+1}$

30. $\frac{m^2-49}{m^2-5m-14} \div \frac{m+7}{2m^2-13m-7} = \frac{(m+7)(m-7)}{(m-7)(m+2)} \times$

$\frac{(2m+1)(m-7)}{(m+7)} = \frac{(2m+1)(m-7)}{m+2}$

31. $\frac{2y^2-y}{4y^2-4y+1} \div \frac{y^2}{8y-4} = \frac{y(2y-1)}{(2y-1)^2} \cdot \frac{(4(2y-1)}{y^2} = \frac{4}{y}$

32. $\frac{a-1}{a^2-1} \div \frac{(a-1)^2}{a^2-1} = \frac{a-1}{(a^2-1)} \cdot \frac{(a^2-1)}{(a-1)^2} = \frac{1}{a-1}$

33. $\frac{x^2-3x+2}{x^2+5x+6} \cdot \frac{x+3}{3x-6} = \frac{(x-2)(x-1)}{(x+2)(x+3)} \cdot \frac{x+3}{3(x-2)} = \frac{x-1}{3(x+2)}$

34. $\frac{2x+2}{x^2+2x-8} \cdot \frac{x^2-4}{x^2+4x+4} = \frac{2(x+1)}{(x+4)(x-2)} \cdot$

$\frac{(x+2)(x-2)}{(x+2)^2} = \frac{2(x+1)}{(x+4)(x+2)}$

35. $\frac{x^2+xy-6y^2}{x^2+6xy+8y^2} \cdot \frac{x^2-9xy+20y^2}{x^2-4xy-21y^2} = \frac{(x+3y)(x-2y)}{(x+4y)(x+2y)} \cdot$

$\frac{(x-4y)(x-5y)}{(x-7y)(x+3y)} = \frac{(x-2y)(x-4y)(x-5y)}{(x+4y)(x+2y)(x-7y)}$

36. $\frac{y^2+14xy+49x^2}{y^2-7xy-30x^2} \cdot \frac{y^2-100x^2}{y^3+7xy^2} = \frac{(y+7x)^2}{(y-10x)(y+3x)} \cdot$

$\frac{(y+10x)(y-10x)}{y^2(y+7x)} = \frac{(y+7x)(y+10x)}{y^2(y+3x)}$

37. $\frac{9x^2-25}{x^2+6x+9} \div \frac{3x+5}{x+3} = \frac{(3x+5)(3x-5)}{(x+3)^2} \times \frac{x+3}{3x+5} = \frac{3x-5}{x+3}$

38. $\frac{x^2-16}{x^2-6x+8} \div \frac{x^3+4x^2}{x^2-9x+14} = \frac{(x+4)(x-4)}{(x-4)(x-2)} \cdot \frac{(x-7)(x-2)}{x^2(x+4)} =$

$\frac{x-7}{x^2}$

39. $\frac{x^2+4xy+4y^2}{x^2-4y^2} \div \frac{x^2+xy-2y^2}{x^2-xy-2y^2} = \frac{(x+2y)^2}{(x+2y)(x-2y)} \times$

$\frac{(x-2y)(x+y)}{(x+2y)(x-y)} = \frac{x+y}{x-y}$

40. $\frac{p^3-27q^3}{3p^2+9pq+27q^2} \div \frac{9q^2-p^2}{6p+18q} = \frac{(p-3q)\left(p^2+3pq+9q^2\right)}{3\left(\left(p^2+3pq+9q^2\right)\right)} \cdot \frac{6(p+3q)}{(3q+p)(3q-p)} = -2$

41. $\frac{x+y}{4x-4y} \div \left[\frac{(x+y)^2}{x^2-y^2} \cdot \frac{x^3-y^3}{x^3+y^3}\right] = \frac{x+y}{4(x-y)} \cdot \left[\frac{\left(x^2-y^2\right)\left(x^3+y^3\right)}{(x+y)^2\left(x^3-y^3\right)}\right]$

$= \frac{x+y}{4(x-y)} \cdot \left[\frac{(x-y)(x+y)(x+y)\left(x^2-xy+y^2\right)}{(x+y)^2(x-y)\left(x^2+xy+y^2\right)}\right]$

$= \frac{x+y}{4(x-y)} \cdot \frac{\left(x^2-xy+y^2\right)}{\left(x^2+xy+y^2\right)}$

$= \frac{x^3+y^3}{4\left(x^3-y^3\right)}$

42. $\frac{2.4m^2n}{0.8mn^2} \div \left[\frac{0.6m}{0.3n} \cdot \frac{3.6m}{2.4n}\right] = \frac{2.4m^2n}{0.8mn^2} \cdot \frac{0.3n}{0.6m} \cdot \frac{2.4n}{3.6m}$

$= \frac{1.728m^2n^3}{1.728m^3n^2}$

$= \frac{n}{m}$

43. $\frac{x^2-25}{5x^2-24x-5} \cdot \frac{2x^2+12x+2}{6x^2-12x} \div \frac{x^2+6x+1}{5x^2-9x-2}$

$= \frac{x^2-25}{5x^2-24x-5} \cdot \frac{2x^2+12x+2}{6x^2-12x} \cdot \frac{5x^2-9x-2}{x^2+6x+1}$

$= \frac{(x+5)(x-5)}{(5x+1)(x-5)} \cdot \frac{2\left(x^2+6x+1\right)}{6x(x-2)} \cdot \frac{(5x+1)(x-2)}{x^2+6x+1}$

$= \frac{x+5}{3x}$

44. $\dfrac{a-b}{a^3-b^3} \cdot \dfrac{a^3+ab^2}{4a+4b} \div \dfrac{a^4-b^4}{8a^2-8b^2} = \dfrac{a-b}{a^3-b^3} \cdot \dfrac{a^3+ab^2}{4a+4b} \cdot \dfrac{8a^2-8b^2}{a^4-b^4}$

$$= \dfrac{a-b}{(a-b)\left(a^2+ab+b^2\right)} \cdot \dfrac{a\left(a^2+b^2\right)}{4(a+b)} \cdot \dfrac{(8(a-b)(a+b)}{(a-b)(a+b)\left(a^2+b^2\right)}$$

$$= \dfrac{2a}{\left(a^2+ab+b^2\right)(a+b)}$$

45. $\left(\dfrac{2x^2-5x-3}{x^2-x-12} \div \dfrac{2x^2+5x+2}{3x+9}\right) \div \dfrac{x^2-9}{x^2-2x-8}$

$$= \dfrac{(2x+1)(x-3)}{(x-4)(x+3)} \cdot \dfrac{3(x+3)}{(2x+1)(x+2)} \cdot \dfrac{(x-4)(x+2)}{(x-3)(x+3)}$$

$$= \dfrac{3}{x+3}$$

46. $\dfrac{a^2+4a-5}{a^2-3a-4} \cdot \dfrac{a^2+3a-28}{(a-3)^2} \div \dfrac{a^2+12a+35}{a^2-2a-3}$

$$= \dfrac{(a+5)(a-1)}{(a-4)(a+1)} \cdot \dfrac{(a+7)(a-4)}{(a-3)^2} \cdot \dfrac{(a-3)(a+1)}{(a+7)(a+5)}$$

$$= \dfrac{a-1}{a-3}$$

47. $\dfrac{(a^3-a'^3)/a^3}{(a^2-a'a)/a^2} = \dfrac{a^3-a'^3}{a^3} \cdot \dfrac{a^2}{a^2-a'a}$

$$= \dfrac{(a-a')(a^2+aa'-a'^2)}{a(a-a')(a+a')}$$

$$= \dfrac{a^2+aa'-a'^2}{a(a+a')}$$

48. The ratio of charge to energy is $\dfrac{Q}{E} = \dfrac{CV}{\frac{1}{2}\frac{Q^2}{V}} =$

$\dfrac{CV}{1} \cdot \dfrac{2V}{Q^2} = \dfrac{2CV^2}{(CV)^2} = \dfrac{2}{C}.$

☰ 7.6 ADDITION AND SUBTRACTION OF FRACTIONS

1. $\frac{2}{7}+\frac{5}{7}=\frac{2+5}{7}=\frac{7}{7}=1$

2. $\frac{4}{5}+\frac{-11}{5}=\frac{4-11}{5}=\frac{-7}{5}$

3. $\frac{7}{3}-\frac{5}{3}=\frac{7-5}{3}=\frac{2}{3}$

4. $\frac{-2}{9}=\frac{8}{9}=\frac{-2-8}{9}=\frac{-10}{9}$

7. $\frac{4}{5}-\frac{2}{3}=\frac{12}{15}-\frac{10}{15}=\frac{12-10}{15}=\frac{2}{15}$

5. $\frac{1}{2}+\frac{1}{3}=\frac{3}{6}+\frac{2}{6}=\frac{3+2}{6}=\frac{5}{6}$

8. $-\frac{5}{7}-\frac{3}{5}=\frac{-25}{35}-\frac{21}{35}=\frac{-25-21}{35}=\frac{-46}{35}$

6. $\frac{3}{4}+\frac{-2}{3}=\frac{9}{12}+\frac{-8}{12}=\frac{9-8}{12}=\frac{1}{12}$

9. $\frac{1}{x}+\frac{5}{x}=\frac{1+5}{x}=\frac{6}{x}$

10. $\frac{2}{y}+\frac{-5}{y}=\frac{2-5}{y}=\frac{-3}{y}$

11. $\frac{4}{a}-\frac{3}{a}=\frac{4-3}{a}=\frac{1}{a}$

12. $\frac{-5}{p}-\frac{-7}{p}=\frac{-5+7}{p}=\frac{2}{p}$

13. $\frac{2x}{y}+\frac{3x}{y}=\frac{2x+3x}{y}=\frac{5x}{y}$

15. $\frac{3r}{2t}+\frac{-r}{2t}-\frac{5r}{2t}=\frac{3r-r-5r}{2t}=\frac{-3r}{2t}$

14. $\frac{4p}{q}-\frac{6p}{q}=\frac{4p-6p}{q}=\frac{-2p}{q}$

16. $\frac{3x}{2y}-\frac{5x}{2y}+\frac{x}{2y}=\frac{3x-5x+x}{2y}=\frac{-x}{2y}$

17. $\frac{3}{x+2}+\frac{x}{x+2}=\frac{3+x}{x+2}$

18. $\frac{5}{y-3}+\frac{y}{y-3}=\frac{5+y}{y-3}$

19. $\frac{t}{t+1}-\frac{2}{t+1}=\frac{t-2}{t+1}$

20. $\frac{a}{b-3}-\frac{4}{3-b}=\frac{a}{b-3}-\frac{4}{3-b}\frac{(-1)}{(-1)}=\frac{a}{b-3}-\frac{-4}{b-3}=\frac{a+4}{b-3}$

23. $\frac{x+2}{a+b}-\frac{x-5}{a+b}=\frac{(x+2)-(x-5)}{a+b}=\frac{7}{a+b}$

21. $\frac{y-3}{x+2}+\frac{3+y}{x+2}=\frac{(y-3)+(3+y)}{x+2}=\frac{2y}{x+2}$

24. $\frac{x+4}{y-5}-\frac{2-x}{y-5}=\frac{(x+4)-(2-x)}{y-5}=\frac{2x+2}{y-5}$

22. $\frac{x+4}{x-2}+\frac{x-5}{x-2}=\frac{(x+4)+(x-5)}{x-2}=\frac{2x-1}{x-2}$

25. $\frac{2}{x}+\frac{3}{y}=\frac{2}{x}\cdot\frac{y}{y}+\frac{x}{x}\cdot\frac{3}{y}=\frac{2y}{xy}+\frac{3x}{xy}=\frac{2y+3x}{xy}$

26. $\frac{x}{y} + \frac{5}{x} = \frac{x}{y} \cdot \frac{x}{x} + \frac{5}{x} \cdot \frac{y}{y} = \frac{x^2}{xy} + \frac{5y}{xy} = \frac{x^2+5y}{xy}$

27. $\frac{a}{b} - \frac{4}{d} = \frac{a}{b} \cdot \frac{d}{d} - \frac{4}{d} \cdot \frac{b}{b} = \frac{ad}{bd} - \frac{4b}{bd} = \frac{ad-4b}{bd}$

28. $\frac{2x}{y} - \frac{3y}{x} = \frac{2x}{y} \cdot \frac{x}{x} - \frac{3y}{x} \cdot \frac{y}{y} = \frac{2x^2}{xy} - \frac{3y^2}{xy} = \frac{2x^2-3y^2}{xy}$

29. $\frac{3}{x(x+1)} + \frac{4}{x^2-1} = \frac{3}{x(x+1)} + \frac{4}{(x+1)(x-1)} = \frac{3(x-1)}{x(x+1)(x-1)} +$
$\frac{4x}{x(x+1)(x-1)} = \frac{3x-3+4x}{x(x+1)(x-1)} = \frac{7x-3}{x(x+1)(x-1)}$

30. $\frac{5}{y(x+1)} + \frac{x}{y(x+2)} = \frac{5(x+2)}{y(x+1)(x+2)} + \frac{x(x+1)}{y(x+2)(x+1)} =$
$\frac{5x+10+x^2+x}{x(x+1)(x+2)} = \frac{x^2+6x+10}{x(x+1)(x+2)}$

31. $\frac{2}{x^2-1} - \frac{4}{(x+1)^2} = \frac{2}{(x+1)(x-1)} - \frac{4}{(x+1)^2} = \frac{2(x+1)}{(x+1)^2(x-1)} -$
$\frac{4(x-1)}{(x+1)^2(x-1)} = \frac{(2x+2)-(4x-4)}{(x+1)^2(x-1)} = \frac{-2x+6}{(x+1)^2(x-1)}$

32. $\frac{6}{y-2} - \frac{3}{y+2} = \frac{6}{y-2} \cdot \frac{(y+2)}{(y+2)} - \frac{3}{(y+2)} \cdot \frac{(y-2)}{(y-2)} =$
$\frac{(6y+12)-(3y-6)}{(y+2)(y-2)} = \frac{3y+18}{(y+2)(y-2)}$

33. $\frac{x}{x^2-11x+30} + \frac{2}{x^2-36} = \frac{x}{(x-6)(x-5)} + \frac{2}{(x-6)(x+6)} =$
$\frac{x(x+6)+2(x-5)}{(x-6)(x+6)(x-5)} = \frac{x^2+6x+2x-10}{(x-6)(x+6)(x-5)} = \frac{x^2+8x-10}{(x-6)(x+6)(x-5)}$

34. $\frac{a}{(a-3)(a-6)} + \frac{a}{(a+3)(a-3)} = \frac{a(a+3)}{(a+3)(a-3)(a-6)} +$
$\frac{a(a-6)}{(a+3)(a-3)(a-6)} = \frac{(a^2+3a)+(a^2-6a)}{(a+3)(a-3)(a-6)} =$
$\frac{2a^2-3a}{(a+3)(a-3)(a-6)}$

35. $\frac{2}{x^2-x-6} - \frac{5}{x^2-4} = \frac{2}{(x-3)(x+2)} - \frac{5}{(x+2)(x-2)} =$
$\frac{2(x-2)}{(x-3)(x+2)(x-2)} - \frac{5(x-3)}{(x-3)(x-2)(x+2)} = \frac{(2x-4)-(5x-15)}{(x-3)(x+2)(x-2)} =$
$\frac{3x+11}{(x-3)(x+2)(x-2)}$

36. $\frac{b}{b^2-10b+21} - \frac{b}{b^2-9} = \frac{b}{(b-7)(b-3)} - \frac{b}{(b+3)(b-3)} =$
$\frac{b(b+3)}{(b-7)(b-3)(b+3)} - \frac{b(b-7)}{(b+3)(b-3)(b-7)} = \frac{(b^2+3b)-(b^2-7b)}{(b+3)(b-3)(b-7)} =$
$\frac{10b}{(b+3)(b-3)(b-7)}$

37. $\frac{x-1}{3x^2-13x+4} + \frac{3x+1}{4x-x^2} = \frac{x-1}{(3x-1)(x-4)} + \frac{3x+1}{x(4-x)} +$
$\frac{x(x-1)}{x(3x-1)(x-4)} + \frac{(-1)(3x+1)(3x-1)}{x(x-4)(3x-1)} = \frac{(x^2-x)-(9x^2-1)}{x(x-4)(3x-1)} =$
$\frac{-8x^2-x+1}{x(x-4)(3x-1)}$

38. $\frac{x-3}{x^2+3x+2} + \frac{2x-5}{x^2+x-2} = \frac{x-3}{(x+2)(x+1)} + \frac{2x-5}{(x+2)(x-1)} =$
$\frac{(x-3)(x-1)}{(x+2)(x+1)(x-1)} + \frac{(2x-5)(x+1)}{(x+2)(x+1)(x-1)} =$
$\frac{(x^2-4x+3)+(2x^2-3x-5)}{(x+2)(x+1)(x-1)} = \frac{3x^2-7x-2}{(x+2)(x+1)(x-1)}$

39. $\frac{x-3}{x^2-1} + \frac{2x-7}{x^2+5x+4} = \frac{x-3}{(x+1)(x-1)} + \frac{2x-7}{(x+1)(x+4)} =$
$\frac{(x-3)(x+4)}{(x+1)(x-1)(x+4)} + \frac{(2x-7)(x-1)}{(x+1)(x-1)(x+4)} =$
$\frac{(x^2+x-12)+(2x^2-9x+7)}{(x+1)(x-1)(x+4)} = \frac{3x^2-8x-5}{(x+1)(x-1)(x+4)}$

40. $\frac{x+4}{x^2-9} - \frac{x-3}{x^2+6x+9} = \frac{x+4}{(x+3)(x-3)} - \frac{x-3}{(x+3)^2} = \frac{(x+4)(x+3)}{(x+3)^2(x-3)} -$
$\frac{(x-3)(x-3)}{(x+3)^2(x-3)} = \frac{(x^2+7x+12)-(x^2-6x+9)}{(x+3)^2(x-3)} = \frac{13x+3}{(x+3)^2(x-3)}$

41. $\frac{y+3}{y^2-y-2} - \frac{2y-1}{y^2+2y-8} = \frac{y+3}{(y-2)(y+1)} - \frac{2y-1}{(y+4)(y-2)} =$
$\frac{(y+3)(y+4)}{(y-2)(y+1)(y+4)} - \frac{(2y-1)(y+1)}{(y-2)(y+1)(y+4)} =$
$\frac{(y^2+7y+12)-(2y^2+y-1)}{(y-2)(y+1)(y+4)} = \frac{-y^2+6y+13}{(y-2)(y+1)(y+4)}$

42. $\frac{1}{(a-b)(a-c)} + \frac{1}{(b-a)(b-c)} - \frac{1}{(b-c)(a-c)} =$
$\frac{(b-c)-1(a-c)-(a-b)}{(a-b)(a-c)(b-c)} = \frac{-2a+2b}{(a-b)(a-c)(b-c)} =$
$\frac{-2(a-b)}{(a-b)(a-c)(b-c)} =$
$\frac{-2}{(a-c)(b-c)}$

43. $\frac{x}{(x^2+3)(x-1)} + \frac{3x^2}{(x-1)^2(x+2)} - \frac{x+2}{x^2+3} =$
$\frac{x(x-1)(x+2)+3x^2(x^2+3)-(x+2)(x+2)(x-1)^2}{(x^2+3)(x-1)^2(x+2)} =$
$\frac{(x^3+x^2-2x)+(3x^4+9x^2)-(x^4+2x^3-3x^2-4x+4)}{(x^2+3)(x-1)^2(x+2)} =$
$\frac{2x^4-x^3+13x^2+2x-4}{(x^2+3)(x-1)^2(x+2)}$

44. $\frac{2x-1}{(x+2)(x+3)} - \frac{x-2}{(x+3)(x+1)} + \frac{x-4}{(x+2)(x+1)} =$
$\frac{(2x-1)(x+1)-(x-2)(x+2)+(x-4)(x+3)}{(x+2)(x+3)(x+1)} =$
$\frac{(2x^2+x-1)-(x^2-4)+(x^2-x-12)}{(x+2)(x+3)(x+1)} =$
$\frac{2x^2+x-1-x^2+4+x^2-x-12}{(x+2)(x+3)(x+1)} = \frac{2x^2-9}{(x+2)(x+3)(x+1)}$

45. $\frac{1+\frac{2}{x}}{1-\frac{3}{x}} = \frac{\left(1+\frac{2}{x}\right)x}{\left(1-\frac{3}{x}\right)x} = \frac{x+2}{x-3}$

46. $\frac{x+\frac{1}{x}}{2-\frac{1}{x}} = \frac{\left(x+\frac{1}{x}\right)x}{\left(2-\frac{1}{x}\right)x} = \frac{x^2+1}{2x-1}$

47. $\frac{x-1}{1+\frac{1}{x}} = \frac{(x-1)x}{\left(1+\frac{1}{x}\right)x} = \frac{x^2-x}{x+1} = \frac{x(x-1)}{x+1}$

48. $\frac{x^2-25}{\frac{1}{x}-\frac{1}{5}} = \frac{(x^2-25)5x}{\left(\frac{1}{x}-\frac{1}{5}\right)5x} = \frac{\left(x^2-25\right)5x}{5-x} =$
$\frac{(x+5)(x-5)5x}{5-x} = -5x(x+5)$

49.
$$\frac{\frac{x}{x+y} - \frac{y}{x-y}}{\frac{x}{x+y} + \frac{y}{x-y}} = \frac{\left(\frac{x}{x+y} - \frac{y}{x-y}\right)}{\frac{x}{x+y} + \frac{y}{x-y}} \cdot$$
$$\frac{(x+y)(x-y)}{(x+y)(x-y)} = \frac{x(x-y) - y(x+y)}{x(x-y) + y(x+y)} =$$
$$\frac{x^2 - xy - xy - y^2}{x^2 - xy + xy + y^2} = \frac{x^2 - 2xy - y^2}{x^2 + y^2}$$

50.
$$\frac{\left(x+3 - \frac{16}{x+3}\right)}{\left(x-6 + \frac{20}{x+6}\right)} \cdot \frac{(x+3)(x+6)}{(x+3)(x+6)} =$$
$$\frac{\left(x^2 + 6x + 9 - 16\right)(x+6)}{\left(x^2 - 36 + 20\right)(x+3)} =$$
$$\frac{\left(x^2 + 6x - 7\right)(x+6)}{\left(x^2 - 16\right)(x+3)}$$

51.
$$\frac{1 + \frac{3}{x}}{1 + \frac{2}{x}} = \frac{\frac{x+3}{x}}{\frac{x+2}{x}} = \frac{x+3}{x} \times \frac{x}{x+2} = \frac{x+3}{x+2}$$

52.
$$\frac{y + \frac{1}{y}}{3 + \frac{2}{y}} = \frac{\frac{y^2+1}{y}}{\frac{3y+2}{y}} = \frac{y^2+1}{y} \times \frac{y}{3y+2} = \frac{y^2+1}{3y+2}$$

53.
$$\frac{t-1}{t + \frac{1}{t}} = \frac{t-1}{\frac{t^2+1}{t}} = \frac{t-1}{1} \times \frac{t}{t^2+1} = \frac{t^2-t}{t^2+1}$$

54.
$$\frac{x^2 - 36}{\frac{1}{6} - \frac{1}{x}} = \frac{x^2 - 36}{\frac{x-6}{6x}} = \frac{x^2 - 36}{1} \times \frac{6x}{x-6} = (x+6)(6x)$$

55.
$$\frac{\frac{x}{x-y} - \frac{y}{x+y}}{\frac{1}{x-y} + \frac{1}{x+y}} = \frac{\frac{x^2+xy}{x^2-y^2} - \frac{xy-y^2}{x^2-y^2}}{\frac{x+y+x-y}{x^2-y^2}} = \frac{\frac{x^2+y^2}{x^2-y^2}}{\frac{2x}{x^2-y^2}} =$$
$$\frac{x^2+y^2}{x^2-y^2} \times \frac{x^2-y^2}{2x} = \frac{x^2+y^2}{2x}$$

56.
$$\frac{t - 5 + \frac{25}{t-5}}{t + 3 + \frac{10}{t-3}} = \frac{\frac{t^2-10t+25+25}{t-5}}{\frac{t^2-9+10}{t-3}} = \frac{t^2 - 10t + 50}{t-5} \times$$
$$\frac{t-3}{t^2+1} = \frac{\left(t^2 - 10t + 50\right)(t-3)}{(t-5)\left(t^2+1\right)}$$

57. $\dfrac{1}{R_1} + \dfrac{1}{R_2} = \dfrac{R_2}{R_1 R_2} + \dfrac{R_1}{R_1 R_2} = \dfrac{R_1+R_2}{R_1 R_2}$

58. $\dfrac{1}{p} + \dfrac{1}{q} = \dfrac{q}{pq} + \dfrac{p}{pq} = \dfrac{q+p}{pq} = \dfrac{p+q}{pq}$

59. $\dfrac{1}{C_1} + \dfrac{1}{C_2} + \dfrac{1}{C_3} = \dfrac{C_2 C_3}{C_1 C_2 C_3} + \dfrac{C_1 C_3}{C_1 C_2 C_3} + \dfrac{C_1 C_2}{C_1 C_2 C_3} =$
$\dfrac{C_1 C_2 + C_1 C_3 + C_2 C_3}{C_1 C_2 C_3}$

60. $\dfrac{d_1 + d_2}{\frac{d_1}{v_1} + \frac{d_2}{v_2}} \cdot \dfrac{v_1 v_2}{v_1 v_2} = \dfrac{v_1 v_2 (d_1 + d_2)}{d_1 v_1 + d_2 v_1}$

61.
$$\left(\frac{m_1}{m_1 + m_2} - \frac{m_2}{m_1 + m_2}\right)v + 2\left(\frac{m_2}{m_1 + m_2}\right)v$$
$$= \frac{vm_1}{m_1 + m_2} - \frac{vm_2}{m_1 + m_2} + \frac{2vm_2}{m_1 + m_2}$$
$$= \frac{vm_1 + vm_2}{m_1 + m_2}$$
$$= \frac{v(m_1 + m_2)}{m_1 + m_2} = v$$

62.
$$\frac{1}{\left(r - \frac{d}{2}\right)^2} - \frac{1}{\left(r - \frac{d}{2}\right)^2} = \frac{\left(r + \frac{d}{2}\right)^2 - \left(r - \frac{d}{2}\right)^2}{\left(r - \frac{d}{2}\right)^2 \left(r + \frac{d}{2}\right)^2} = \frac{\left(r^2 + rd + \frac{d}{4}\right) - \left(r^2 - rd + \frac{d}{4}\right)}{\left(r - \frac{d}{2}\right)^2 \left(r + \frac{d}{2}\right)^2}$$
$$= \frac{2rd}{\left[\left(r - \frac{d}{2}\right)\left(r + \frac{d}{2}\right)\right]^2}$$
$$= \frac{2rd}{\left[\left(r - \left(\frac{d}{2}\right)^2\right]^2}$$
$$= \frac{2rd}{\left(\frac{4r^2 - d^2}{4}\right)^2} = \frac{32rd}{(4r^2 - d^2)^2}$$

63. $\dfrac{\frac{V_1}{R_1} + \frac{V_2}{R_2} + \frac{V_3}{R_3}}{\frac{1}{R_1} + \frac{1}{R_2} + \frac{1}{R_3}}$. Multiply by $\dfrac{R_1 R_2 R_3}{R_1 R_2 R_3}$ to obtain $\dfrac{V_1 R_2 R_3 + V_2 R_1 R_3 + V_3 R_1 R_2}{R_2 R_3 + R_1 R_3 + R_1 R_2}$.

64. $\dfrac{d_1+d_2}{\frac{d_1}{v_1}+\frac{d_2}{v_2}} = \dfrac{d_1+d_2}{\frac{d_1v_2+d_2v_1}{v_1v_2}} = \dfrac{d_1+d_2}{1}\cdot\dfrac{v_1v_2}{d_1v_2+d_2v_1} = \qquad \dfrac{(d_1+d_2)\,v_1v_2}{d_1v_2+d_2v_1}$

CHAPTER 7 REVIEW

1. $5x(x-y)=5x^2-5xy$

2. $(3+x)^2=9+6x+x^2$

3. $(x-2y)^3 = x^3-3x^2(2y)+3x(2y)^2-(2y)^3 = x^3-6x^2y+12xy^2-8y^3$

4. $(x+y)(x-6)=x^2-6x+xy-6y$ using FOIL

5. $(2x+3)(x-6)=2x^2-12x+3x-18=2x^2-9x-18$

6. $(x+7)(x-7)=x^2-7^2=x^2-49$

7. $\left(x^2-5\right)\left(x^2+5\right)=\left(x^2\right)^2-5^2=x^4-25$

8. $(7x-1)(x+5)=7x^2+35x-x-5=7x^2+34x-5$

9. $(2+x)^3=2^3+3\cdot2^2x+3\cdot2x^2+x^3=8+12x+6x^2+x^3$

10. $(x-7)^2=x^2-2\cdot x\cdot7+7^2=x^2-14x+49$

11. $9+9y=9(1+y)$

12. $x^2-4=(x+2)(x-2)$

13. $7x^2-63=7\left(x^2-9\right)=7(x+3)(x-3)$

14. $x^2-12x+36=x^2-2\cdot x\cdot6+6^2=(x-6)^2$

15. $x^2-11x+30=(x-5)(x-6)$ or $(x-6)(x-5)$

16. $x^2+15x+36=(x+12)(x+3)$

17. $x^2+6x-16=(x+8)(x-2)$

18. $x^2-4x-45=(x+5)(x-9)$

19. $2(-9)=-18;\ -6\cdot3=-18;\ -6+3=-3;$
$2x^2-3x-9=2x^2-6x+3x-9=2x(x-3)+3(x-3)=(2x+3)(x-3)$

20. Begin by factoring out $2x$, getting $8x^3+6x^2-20x=2x\left(4x^2+3x-10\right)$. Then, for the $4x^2+3x-10$ expression, we have $4(-10)=-40$; $8(-5)=-40$; $8+(-5)=3$; $2x\left(4x^2+3x-10\right)=2x\left[4x^2+8x-5x-10\right]=2x\left[4x(x+2)-5(x+2)\right]=2x(4x-5)(x+2)$

21. $\dfrac{2x}{6y}=\dfrac{x}{3y}$

22. $\dfrac{7x^2y}{9xy^2}=\dfrac{7x}{9y}$

23. $\dfrac{x^2-9}{(x+3)^2}=\dfrac{(x+3)(x-3)}{(x+3)^2}=\dfrac{x-3}{x+3}$

24. $\dfrac{x^2-4x-45}{x^2-81}=\dfrac{(x-9)(x+5)}{(x-9)(x+9)}=\dfrac{x+5}{x+9}$

25. $\dfrac{x^3+y^3}{x^2+2xy+y^2}=\dfrac{(x+y)\left(x^2-xy+y^2\right)}{(x+y)^2}=\dfrac{x^2-xy+y^2}{x+y}$

26. $\dfrac{x^3-16x}{x^2+2x-8}=\dfrac{x(x+4)(x-4)}{(x+4)(x-2)}=\dfrac{x(x-4)}{x-2}$

27. $\dfrac{x^2}{y}\cdot\dfrac{3y^2}{7x}=\dfrac{3x^2y^2}{7xy}=\dfrac{3xy}{7}$

28. $\dfrac{x^2-9}{x+4}\cdot\dfrac{x^3-16x}{x-3}=\dfrac{(x+3)(x-3)(x)(x+4)(x-4)}{(x+4)(x-3)}=x(x+3)(x-4)$

29. $\dfrac{4x}{3y}\div\dfrac{2x^2}{6y}=\dfrac{4x}{3y}\cdot\dfrac{6y}{2x^2}\dfrac{4x\cdot3\cdot2y}{3y\cdot2x^2}=\dfrac{4}{x}$

30. $\dfrac{x^2-25}{x^2-4x}\div\dfrac{2x^2+2x-40}{x^3-x}=\dfrac{x^2-25}{x^2-4x}\cdot\dfrac{x^3-x}{2x^2+2x-40}=\dfrac{(x+5)(x-5)x(x+1)(x-1)}{x(x-4)2(x+5)(x-4)}=\dfrac{(x-5)(x+1)(x-1)}{2(x-4)^2}$

31. $\dfrac{4x}{y}+\dfrac{3x}{y}=\dfrac{4x+3x}{y}=\dfrac{7x}{y}$

32. $\dfrac{4}{x-y}+\dfrac{6}{x+y}=\dfrac{4(x+y)}{x^2-y^2}+\dfrac{6(x-y)}{x^2-y^2}=\dfrac{4x+4y+6x-6y}{x^2-y^2}=\dfrac{10x-2y}{x^2-y^2}$

33. $\dfrac{3(x-3)}{(x+2)(x-5)^2}+\dfrac{4(x-1)}{(x+2)^2(x-5)}=\dfrac{3(x-3)(x+2)+4(x-1)(x-5)}{(x+2)^2(x-5)^2}$
$=\dfrac{3x^2-3x-18+4x^2-24x+20}{(x+2)^2(x-5)^2}=\dfrac{7x^2-27x+2}{(x+2)^2(x-5)^2}$

34. $\dfrac{8a}{b}-\dfrac{3}{b}=\dfrac{8a-3}{b}$

35. $\dfrac{x}{y+x}-\dfrac{x}{y-x}=\dfrac{x(y-x)-x(y+x)}{(y+x)(y-x)}=\dfrac{xy-x^2-xy-x^2}{y^2-x^2}=\dfrac{-2x^2}{y^2-x^2}$ or $\dfrac{2x^2}{x^2-y^2}$ or $\dfrac{2x^2}{(x+y)(x-y)}$

36. $\dfrac{2(x+3)}{(x+1)^2(x+2)}-\dfrac{3(x-1)}{(x+1)(x+2)^2}=\dfrac{2(x+3)(x+2)-3(x-1)(x+1)}{(x+1)^2(x+2)^2}$
$=\dfrac{2x^2+10x+12-3x^2+3}{(x+1)^2(x+2)^2}=\dfrac{-x^2+10x+15}{(x+1)^2(x+2)^2}$

37. $\dfrac{x^2-5x-6}{(x+6)(x+2)}+\dfrac{x^2+7x+6}{(x+2)(x-6)}=\dfrac{\left(x^2-5x-6\right)(x-6)}{(x+6)(x+2)(x-6)}+\dfrac{\left(x^2+7x+6\right)(x+6)}{(x+2)(x-6)(x+6)}=\dfrac{x^3-11x^2+24x+36+x^3+13x^2+48x+36}{(x+2)(x-6)(x+6)}$
$=\dfrac{2x^3+2x^2+72x+72}{(x+2)(x-6)(x+6)}=\dfrac{2\left(x^3+x^2+36x+36\right)}{(x+2)(x-6)(x+6)}=\dfrac{2\left[x^2(x+1)+36(x+1)\right]}{(x+2)(x-6)(x+6)}=\dfrac{2\left(x^2+36\right)(x+1)}{(x+2)(x-6)(x+6)}$

38. $\dfrac{2x-1}{4x^2-12x+5}-\dfrac{x+1}{4x^2-4x-15}=\dfrac{2x-1}{(2x-5)(2x-1)}-\dfrac{x+1}{(2x-5)(2x+3)}=\dfrac{(2x-1)(2x+3)}{(2x-5)(2x-1)(2x+3)}-\dfrac{(x+1)(2x-1)}{(2x-5)(2x+3)(2x-1)}=\dfrac{\left(4x^2+4x-3\right)-\left(2x^2+x-1\right)}{(2x-5)(2x+3)(2x-1)}$

$$= \frac{2x^2+3x-2}{(2x-5)(2x+3)(2x-1)} = \frac{(2x-1)(x+2)}{(2x-5)(2x+3)(2x-1)}$$

$$= \frac{x+2}{(2x-5)(2x+3)}$$

39. $\dfrac{x^2-5x-6}{x^2+8x+12} \div \dfrac{x^2+7x+6}{x^2-4x-12} = \dfrac{(x+1)(x-6)}{(x+6)(x+2)} \cdot \dfrac{(x-6)(x+1)}{(x+6)(x+1)} =$

$\dfrac{(x-6)^2}{(x+6)^2}$

40. $\dfrac{x^2+x-2}{7a^2x^2-14a^2x+7a^2} \cdot \dfrac{14ax-28a}{1-2x+x^2} = \dfrac{(x+2)(x-1)}{7a^2(x-1)^2} \cdot \dfrac{14a(x-2)}{(x-1)^2} =$

$\dfrac{2(x+2)(x-2)}{a(x-1)^3} = \dfrac{2\left(x^2-4\right)}{a(x-1)^3}$

41. $\dfrac{\frac{1}{x}-\frac{1}{y}}{\frac{1}{x}+\frac{1}{y}} = \dfrac{\frac{y-x}{xy}}{\frac{y+x}{xy}} \cdot \dfrac{xy}{xy} = \dfrac{y-x}{y+x}$

42. $\dfrac{\frac{1}{x}+\frac{1}{y}}{x+y} = \dfrac{\frac{y+x}{xy}}{x+y} \cdot \dfrac{xy}{xy} = \dfrac{y+x}{(x+y)(xy)} = \dfrac{1}{xy}$

43. $\dfrac{\frac{1}{x}-\frac{1}{y}}{\frac{x-y}{xy}} = \dfrac{\frac{y-x}{xy}}{\frac{x-y}{xy}} \cdot \dfrac{xy}{xy} = \dfrac{y-x}{x-y} = -1$

44. $\dfrac{1-\frac{1}{x}}{x-2+\frac{1}{x}} = \dfrac{\frac{x-1}{x}}{\frac{x^2-2x+1}{x}} \cdot \dfrac{x}{x} = \dfrac{x-1}{x^2-2x+1}$

$= \dfrac{x-1}{(x-1)^2} = \dfrac{1}{x-1}$

45. $\dfrac{\frac{x}{1+x}-\frac{1-x}{x}}{\frac{x}{1+x}+\frac{1-x}{x}} = \dfrac{\frac{x^2-(1-x)(1+x)}{x(1+x)}}{\frac{x^2+(1+x)(1-x)}{x(1+x)}} \cdot \dfrac{x(1+x)}{x(1+x)} =$

$\dfrac{x^2-\left(1-x^2\right)}{x^2+1-x^2} = \dfrac{2x^2-1}{1} = 2x^2-1$

46. $\dfrac{x-\frac{xy}{x-y}}{\frac{x^2}{x^2-y^2}-1} = \dfrac{\frac{x(x-y)-xy}{x-y}}{\frac{x^2-(x^2-y^2)}{x^2-y^2}} \cdot \dfrac{(x-y)(x+y)}{(x-y)(x+y)} =$

$\dfrac{x(x+y)(x-y)-xy(x+y)}{x^2-\left(x^2-y^2\right)} =$

$\dfrac{\frac{x^3-xy^2-x^2y-xy^2}{y^2}}{} = \dfrac{x^3-x^2y-2xy^2}{y^2} =$

$\dfrac{x\left(x^2-xy-2y^2\right)}{y^2} = \dfrac{x(x+y)(x-2y)}{y^2}$

▤ CHAPTER 7 TEST

1. $(x+5)(x-3) = x^2 - 3x + 5x - 15 = x^2 + 2x - 15$

2. $(2x-3)(2x+3) = (2x)^2 - 3^2 = 4x^2 - 9$

3. $(3x^2-4)(2-5x) = 6x^2 - 15x^3 - 8 + 20x =$
$-15x^3 + 6x^2 + 20x - 8$

4. $(x-4)^3 = (x-4)(x-4)^2 = (x-4)(x^2 - 8x + 16)$
$= x^3 - 8x^2 + 16x - 4x^2 + 32x - 64$
$= x^3 - 12x^2 + 48x - 64$

5. $2x^2 - 128 = 2\left(x^2 - 64\right) = 2(x+8)(x-8)$

6. $x^2 - 12x + 32 = (x-8)(x-4)$

7. $10 \times (-21) = -210$ which has factors 15 and -14.
Hence, $10x^2 + x - 21 = 10x^2 + 15x - 14x - 21$
$= 5x(2x+3) - 7(x+3)$
$= (5x-7)(2x+3)$

8. $x^3 - 125 = x^3 - 5^3$. This is a difference of cubes and it factors as $(x-5)(x^2 + 5x + 25)$.

9. $\dfrac{x^2-25}{x^2+6x+5} = \dfrac{(x+5)(x-5)}{(x+5)(x+1)} = \dfrac{x-5}{x+1}$

10. $\dfrac{3(a+b)^3-x(a+b)}{a^2-b^2} = \dfrac{(a+b)[3(a+b)^2-x]}{(a+b)(a-b)} = \dfrac{3(a+b)^2-x}{a-b}$

11. $\dfrac{3x}{x+2} \cdot \dfrac{x-1}{x+2} = \dfrac{3x(x-1)}{(x+2)(x+2)} = \dfrac{3x(x-1)}{(x+2)^2} = \dfrac{3x^2-3x}{x^2+4x+4}$

12. $\dfrac{2x+6}{x-2} \div \dfrac{3x+9}{x^2-4} = \dfrac{2x+6}{x-2} \cdot \dfrac{x^2-4}{3x+9} = \dfrac{2(x+3)(x+2)(x-2)}{(x-2)3(x+3)} =$
$\dfrac{2(x+2)}{3}$

13. $\dfrac{6}{x-5} + \dfrac{x^2-2x}{x-5} = \dfrac{x^2-2x+6}{x-5}$

14. $\dfrac{2x}{x+3} - \dfrac{x+4}{x-2} = \dfrac{2x(x-2)}{(x+3)(x-2)} - \dfrac{(x+4)(x+3)}{(x+3)(x-2)} =$
$\dfrac{\left(2x^2-4x\right)-\left(x^2+7x+12\right)}{(x+3)(x-2)} = \dfrac{x^2-11x-12}{(x+3)(x-2)}$

15. $\dfrac{x-\frac{1}{x}}{x-\frac{2}{x+1}} = \dfrac{\frac{x^2-1}{x}}{\frac{x^2+x-2}{x+1}} = \dfrac{(x^2-1)}{x} \cdot \dfrac{(x+1)}{x^2+x-2} =$
$\dfrac{(x+1)(x-1)(x+1)}{x(x+2)(x-1)} = \dfrac{(x+1)^2}{x(x+2)}$

16. The LCD is $(2x+1)^2$. Hence, $\dfrac{1}{2x+1} - \dfrac{2}{4x^2+4x+1} =$
$\dfrac{2x+1-2}{(2x+1)^2} = \dfrac{2x-1}{(2x+1)^2}$.

17. $\dfrac{2d}{\frac{d}{r_1}+\frac{d}{r_2}} = \dfrac{2d}{\frac{d(r_1+r_2)}{r_1r_2}} = \dfrac{2d}{1} \cdot \dfrac{r_1r_2}{d(r_1+r_2)} = \dfrac{2r_1r_2}{r_1+r_2}$

18. $\dfrac{1}{R_1} + \dfrac{1}{R_2} + \dfrac{1}{R_3} = \dfrac{R_2R_3}{R_1R_2R_3} + \dfrac{R_1R_3}{R-1R-2R_3} + \dfrac{R_1R_2}{R_1R_2R_3} =$
$\dfrac{R_1R_2+R_1R_3+R_2R_3}{R_1R_2R_3}$

CHAPTER

8

Fractional and Quadratic Equations

≡ 8.1 FRACTIONAL EQUATIONS

1. $\frac{x}{2} + \frac{x}{3} = \frac{1}{4}$; LCD = 12; $12\left(\frac{x}{2} + \frac{x}{3}\right) = 12 \cdot \frac{1}{4}$; $12 \cdot \frac{x}{2} + 12 \cdot \frac{x}{3} = 12 \cdot \frac{1}{4}$; $6x + 4x = 3$; $10x = 3$; $x = \frac{3}{10}$ or 0.3

2. $\frac{x}{3} - \frac{x}{4} = \frac{1}{2}$; LCD = 12; $12\left(\frac{x}{3} - \frac{x}{4}\right) = 12 \cdot \frac{1}{2}$; $12 \cdot \frac{x}{3} - 12 \cdot \frac{x}{4} = 12 \cdot \frac{1}{2}$; $4x - 3x = 6$; $x = 6$

3. $\frac{y}{2} + 3 = \frac{4y}{5}$; LCD = 10; $10\left(\frac{y}{2} + 3\right) = 10\frac{4y}{5}$; $10 \cdot \frac{y}{2} + 10 \cdot 3 = 10 \cdot \frac{4y}{5}$; $5y + 30 = 2 \cdot 4y$; $5y + 30 = 8y$; $30 = 3y$; $10 = y$

4. $\frac{y}{5} - 5\frac{1}{2} = \frac{3y}{4}$; LCD = 20; $20\left(\frac{y}{5} - \frac{11}{2}\right) = 20\frac{3y}{4}$; $20\frac{y}{5} - 20 \cdot \frac{11}{2} = 20\frac{3y}{4}$; $4y - 10 \cdot 11 = 5 \cdot 3y$; $4y - 110 = 15y$; $-110 = 11y$; $y = -10$

5. $\frac{x-1}{2} + \frac{x+1}{3} = \frac{x-1}{4}$; LCD = 12; $12 \cdot \frac{x-1}{2} + 12 \cdot \frac{x+1}{3} = 12 \cdot \frac{x-1}{4}$; $6(x-1) + 4(x+1) = 3(x-1)$; $6x - 6 + 4x + 4 = 3x - 3$; $10x - 2 = 3x - 3$; $7x = -1$; $x = -\frac{1}{7}$

6. $\frac{x+2}{3} - \frac{x-4}{2} = \frac{x-1}{6}$; LCD = 6; $6 \cdot \frac{x+2}{3} - 6 \cdot \frac{x+4}{2} = 6 \cdot \frac{x-1}{6}$; $2(x+2) - 3(x+4) = x - 1$; $2x + 4 - 3x - 12 = x - 1$; $-x - 8 = x - 1$; $-7 = 2x$; $x = -\frac{7}{2}$

7. $\frac{1}{x} + \frac{2}{x} = \frac{1}{3}$; LCD = $3x$; $x \neq 0$; $3x \cdot \frac{1}{x} + 3x \cdot \frac{2}{x} = 3x \cdot \frac{1}{3}$; $3 + 6 = x$; $9 = x$

8. $\frac{3}{x} - \frac{4}{x} = \frac{2}{5}$; LCD = $5x$; $x \neq 0$; $5x \cdot \frac{3}{x} - 5x \cdot \frac{4}{x} = 5x \cdot \frac{2}{5}$; $15 - 20 = 20x$; $-5 = 2x$; $x = -\frac{5}{2}$

9. $\frac{7}{w-4} = \frac{1}{2w+5}$; LCD = $(-4)(2w+5)$; $\neq 4$; $w \neq -\frac{5}{2}$; $(w-4)(2w+5) \cdot \frac{7}{w-4} = (w-4)(2w+5) \cdot \frac{1}{2w+5}$; $(2w+5)7 = w-4$; $14w + 35 = w - 4$; $13w = -39$; $w = -3$

10. $\frac{5}{y+1} = \frac{3}{y-3}$; LCD = $(y+1)(y-3)$; $y \neq -1$; $y \neq 3$; $(y+1)(y-3) \cdot \frac{5}{y+1} = (y+1)(y-3) \cdot \frac{3}{y-3}$; $(y-3)5 = (y+1)3$; $5y - 15 = 3y + 3$; $2y = +18$; $y = 9$

11. $\frac{2}{2x-1} = \frac{5}{x+5}$; LCD = $(2x-1)(x+5)$; $x \neq \frac{1}{2}$; $x \neq -5$; $(2x-1)(x+5) \cdot \frac{2}{2x-1} = (2x-1)(x+5) \cdot \frac{5}{x+5}$; $(x+5)2 = (2x-1)5$; $2x + 10 = 10x - 5$; $15 = 8x$; $x = \frac{15}{8}$

12. $\frac{3}{4x+2} = \frac{1}{x+2}$; LCD = $(4x+2)(x+2)$; $x \neq -\frac{1}{2}$; $x \neq -2$; $(4x+2)(x+2) \cdot \frac{3}{4x+2} = (4x+2)(x+2) \cdot \frac{1}{x+2}$; $(x+2)3 = 4x + 2$; $3x + 6 = 4x + 2$; $4 = x$

13. $\frac{4x}{x-3} - 1 = \frac{3x}{x+2}$; LCD = $(x-3)(x+2)$; $x \neq 3$; $x \neq -2$; $4x(x+2) - (x-3)(x+2) = 3x(x-3)$; $(4x^2 + 8x) - (x^2 - x - 6) = 3x^2 - 9x$; $3x^2 + 9x + 6 = 3x^2 - 9x$; $18x = -6$; $x = -\frac{1}{3}$

14. $7 - \frac{3x}{x+2} = \frac{4x}{x+2}$; LCD = $x + 2$; $\neq -2$; $7(x+2) - (x+2) \cdot \frac{3x}{x+2} = (x+2)\frac{4x}{x+2}$; $7x + 14 - 3x = 4x$; $4x + 14 = 4x$; $0 = 14$; no solution

15. $\frac{4}{x+2} - \frac{3}{x-1} = \frac{5}{(x-1)(x+2)}$; LCD = $(x+2)(x-1)$; $x \neq -2$; $x \neq 1$; $(x+2)(x-1) \cdot \frac{4}{x+2} - (x+2)(x-1) \cdot \frac{3}{x-1} = (x+2)(x-1) \cdot \frac{5}{(x+2)(x-1)}$; $4(x-1) - 3(x+2) = 5$; $4x - 4 - 3x - 6 = 5$; $x - 10 = 5$; $x = 15$

16. $\frac{3}{x-3} + \frac{2}{2-x} = \frac{5}{(x-3)(x-2)}$; LCD = $(x-3)(x-2)$; $x \neq 3$; $x \neq 2$; $(x-3)(x-2) \cdot \frac{3}{x-3} + (x-3)(x-2) \cdot \frac{2}{2-x} = (x-3)(x-2)\frac{5}{(x-3)(x-2)}$; $3(x-2) - 2(x-3) = 5$; $3x - 6 - 2x + 6 = 5$; $x = 5$

17. $\frac{x+1}{x+2} + \frac{x+3}{x-2} = \frac{2x^2+3x-5}{x^2-4}$; LCD = $(x+2)(x-2)$ or $x^2 - 4$; $x \neq 2$; $x \neq -2$; $(x+2)(x-2) \cdot \frac{x+1}{x+2} +$

$(x+2)(x-2) \cdot \frac{x+3}{x-2} = (x^2-4) \cdot \frac{2x^2+3x-5}{x^2-4}$; $(x-2)(x+1) + (x+2)(x+3) = 2x^2 + 3x - 5$; $x^2 - x - 2 + x^2 + 5x + 6 = 2x^2 + 3x - 5$; $2x^2 + 4x + 4 = 2x^2 + 3x - 5$; $x = -9$

18. $\frac{x+2}{x+3} - \frac{x+5}{x-3} = \frac{2x-1}{x^2-9}$; LCD $= (x+3)(x-3)$ or $x^2 - 9$; $x \neq 3$; $x \neq -3$; $(x-3)(x+2) - (x+3)(x+5) = 2x - 1$; $x^2 - x - 6 - x^2 - 8x - 15 = 2x - 1$; $-9x - 21 = 2x - 1$; $-11x = 20$; $x = -\frac{20}{11}$

19. $\frac{3}{a+1} + \frac{a+1}{a-1} = \frac{a^2}{a^2-1}$; LCD $= (a+1)(a-1)$ or $a^2 - 1$; $a \neq 1$; $a \neq -1$; $3(a-1) + (a+1)(a+1) = a^2$; $3a - 3 + a^2 + 2a + 1 = a^2$; $5a - 2 = 0$; $a = \frac{2}{5}$

20. $\frac{5}{x-4} - \frac{x+2}{x+4} = \frac{x^2}{16-x^2}$; LCD $= (x-4)(x+4)$ or $x^2 - 16$; $x \neq 4$; $x \neq -4$; $5(x+4) - (x+2)(x-4) = (x^2-16)\frac{x^2}{16-x^2}$; $5x + 20 - x^2 + 2x + 8 = -x^2$; $7x + 28 = 0$; $x = -4$; no solution, since -4 makes the denominator 0.

21. $\frac{2}{x-1} + \frac{5}{x+1} = \frac{4}{x^2-1}$; LCD $= (x-1)(x+1)$ or $x^2 - 1$; $x \neq 1$; $x \neq -1$; $2(x+1) + 5(x-1) = 4$; $2x + 2 + 5x - 5 = 4$; $7x - 3 = 4$; $x = 1$; no solution since x cannot equal 1.

22. $\frac{3x+4}{x+2} - \frac{3x-5}{x-4} = \frac{12}{x^2-2x-8}$; LCD $= (x+2)(x-4)$ or $x^2 - 2x - 8$; $x \neq -2$; $x \neq 4$; $(3x+4)(x-4) - (3x-5)(x+2) = 12$; $(3x^2 - 8x - 16) - (3x^2 + x - 10) = 12$; $-9x - 6 = 12$; $-9x = 18$; $x = -2$; no solution, since x cannot equal -2.

23. $\frac{5x-2}{x-3} + \frac{4-5x}{x+4} = \frac{10}{x^2+x-12}$; LCD $= (x-3)(x+4)$ or $x^2 + x - 12$; $x \neq 3$; $x \neq -4$; $(5x-2)(x+4) + (4-5x)(x-3) = 10$; $(5x^2 + 18x - 8) + (-5x^2 + 19x - 12) = 10$; $37x - 20 = 10$; $37x = 30$; $x = \frac{30}{37}$

24. $\frac{2x}{x-1} - \frac{3}{x+2} = \frac{4x}{x^2+x-2} + 2$; LCD $= (x-1)(x+2) = x^2 + x - 2$; $x \neq 1$; $x \neq -2$; $2x(x+2) - 3(x-1) = 4x + 2(x^2+x-2)$; $2x^2 + 4x - 3x + 3 = 4x + 2x^2 + 2x - 4$; $-5x = -7$; $x = \frac{7}{5}$

25. $\frac{5}{x} + \frac{3}{x+1} = \frac{x}{x+1} - \frac{x+1}{x}$; LCD $= x(x+1)$; $x \neq 0$; $x \neq -1$; $5(x+1) + 3x = x^2 - (x+1)(x+1)$; $5x + 5 + 3x = x^2 - (x^2 + 2x + 1)$; $8x + 5 = -2x - 1$; $10x - 6$; $x = -\frac{6}{10} = -\frac{3}{5}$

26. $\frac{y}{y+2} + \frac{5}{y-1} = \frac{3}{y+2} + \frac{y}{y-1}$; LCD $= (y+2)(y-1)$; $y \neq -2$; $y \neq 1$; $y(y-1) + 5(y+2) = 3(y-1) + y(y+2)$; $y^2 - y + 5y + 10 = 3y - 3 + y^2 + 2y$; $4y + 10 = 5y - 3$; $13 = y$

27. $\frac{2t-4}{2t+4} = \frac{t+2}{t+4}$; LCD $= (2t+4)(t+4)$; $t \neq -2$; $t \neq -4$; $(2t-4)(t+4) = (t+2)(2t+4)$; $2t^2 + 4t - 16 = 2t^2 + 8t + 8$; $4t - 16 = 8t + 8$; $-4t = 24$; $t = -6$

28. $\frac{3x+5}{x-5} = \frac{3x-1}{x+3}$; LCD $= (x-5)(x+3)$; $x \neq 5$; $x \neq -3$; $(3x+5)(x+3) = (3x-1)(x-5)$; $3x^2 + 14x + 15 = 3x^2 - 16x + 5$; $30x = -10$; $x = -\frac{1}{3}$

29. $\frac{3x+1}{x-1} - \frac{x-2}{x+3} = \frac{2x-3}{x+3} + \frac{4}{x-1}$; LCD $= (x-1)(x+3)$; $x \neq 1$; $x \neq -3$; $(3x+1)(x+3) - (x-2)(x-1) = (2x-3)(x-1) + 4(x+3)$; $3x^2 + 10x + 3 - (x^2 - 3x + 2) = 2x^2 - 5x + 3 + 4x + 12$; $2x^2 + 13x + 1 = 2x^2 - x + 15$; $14x = 14$; $x = 1$; no solution

30. $\frac{7x+2}{x+2} + \frac{3x-1}{x+3} = \frac{6x+1}{x+3} + \frac{4x-3}{x+2}$; LCD $= (x+2)(x+3)$; $x \neq -2$; $x \neq -3$; $(7x+2)(x+3) + (3x-1)(x+2) = (6x+1)(x+2) + (4x-3)(x+3)$; $7x^2 + 23x + 6 + 3x^2 + 5x - 2 = 6x^2 + 13x + 2 + 4x^2 + 9x - 9$; $10x^2 + 28x + 4 = 10x^2 + 22x - 7$; $6x = -11$; $x = -\frac{11}{6}$

31. $\frac{1}{r} + \frac{1}{s} = \frac{1}{t}$; LCD $= rst$; $rst \cdot \frac{1}{r} + rst \cdot \frac{1}{s} = rst \cdot \frac{1}{t}$; $st + rt = rs$; $rs - st = rt$; $s(r-t) = rt$; $s = \frac{rt}{r-t}$

32. $\frac{P_1 V_1}{T_1} = \frac{P_2 V_2}{T_2}$; LCD $= T_1 T_2$; $P_1 V_1 T_2 = P_2 V_2 T_1$; $T_1 = \frac{P_1 V_1 T_2}{P_2 V_2}$

33. $\frac{1}{R} = \frac{1}{R_1} + \frac{1}{R_2} + \frac{1}{R_3}$; LCD $= RR_1 R_2 R_3$; $R_1 R_2 R_3 = RR_2 R_3 + RR_1 R_3 + RR_1 R_2$; $R(R_2 R_3 + R_1 R_3 + R_1 R_2) = R_1 R_2 R_3$; $R = \frac{R_1 R_2 R_3}{R_2 R_3 + R_1 R_3 + R_1 R_2}$

34. $P = \frac{E^2}{R+r} - \frac{E^2}{(R+r)^2}$; LCD $= (R+r)^2$; $P(R+r)^2 = E^2(R+r) - E^2$; $E^2(R+r-1) = P(R+r)^2$; $E^2 = \frac{P(R+r)^2}{R+r-1}$

35. $V = 2\pi rh + 2\pi r^2$; $2\pi rh = V - 2\pi r^2$; $h = \frac{V - 2\pi r^2}{2\pi r}$

36. $\frac{5}{9}(F-32) = C$; $(F-32) = \frac{9}{5}C$; $F = \frac{9}{5}C + 32$

37. $\frac{1}{f} = (n-1)\left(\frac{1}{R_1} + \frac{1}{R_2}\right)$; LCD $= fR_1 R_2$; $R_1 R_2 = (n-1)(fR_2 + fR_1)$; $R_1 R_2 - (n-1)(fR_2) = (n-1)fR_1$; $R_2(R_1 - fn + f) = (n-1)fR_1$; $R_2 = \frac{(n-1)fR_1}{R_1 - nf + f}$

38. $\frac{P_1}{g} + \frac{V_1^2}{2g} + h_1 = \frac{P_2}{dg} + \frac{V_2^2}{2g} + h_2$; LCD $= 2gd$; $2gd \cdot \frac{P_1}{g} + 2gd \cdot \frac{V_1^2}{2g} + 2gdh_1 = 2dg \cdot \frac{P_2}{dg} + 2dg \cdot \frac{V_2^2}{2g} + 2dgh_2$; $2dP_1 + dV_1^2 + 2gdh_1 = 2P_2 + dV_2^2 + 2dgh_2$; $2gdh_1 -$

$2dgh_2 = 2P_2 + dV_2^2 - 2dP_1 - dV_1^2$; $g \cdot 2d \cdot (h_1 - h_2) = 2P_2 + dV_2^2 - 2dP_1 - dV_1^2$; $g = \frac{2P_2 + dV_2^2 - 2dP_1 - dV_1^2}{2d(h_1 - h_2)}$ or

$g = \frac{2dP_1 + dV_1^2 - 2P_2 - dV_2^2}{2d(h_2 - h_1)}$

39. First simplify $\frac{1}{R_2} - \frac{1}{R_1} = \frac{R_1 - R_2}{R_1 R_2}$. Now, solving for C, we get $C\left(\frac{R_1 - R_2}{R_1 R_2}\right) = \frac{d}{9 \times 10^9}$ and so $C = \frac{d}{9 \times 10^9} \cdot \frac{R_1 R_2}{R_1 - R_2} = \frac{dR_1 R_2}{(9 \times 10^9)(R_1 - R_2)}$.

40. $\frac{1}{C} = \frac{1}{C_1} + \frac{1}{C_2} + \frac{1}{C_3} = \frac{C_2 + C_3 + C_1 C_3 + C_1 C_2}{C_1 C_2 C_3}$. Hence, $C = \frac{C_1 C_2 C_3}{C_2 C_3 + C_1 C_3 + C_1 C_2}$.

41. $(n-1)\left(\frac{1}{r_1} - \frac{1}{r_2}\right) = \frac{(n-1)(r_2 - r_1)}{r_1 r_2}$. Hence,

$f = \frac{r_1 r_2}{(n-1)(r_2 - r_1)}$.

42. First we rewrite this as $\frac{1}{p} = \frac{1}{f} - \frac{1}{q}$. The right-hand side simplifies to $\frac{q-f}{fq}$. Hence, $p = \frac{qf}{q-f}$.

43. $\frac{h}{D-d} = 2$ is equivalent to $h = 2(D-d) = 2D - 2d$. So, $2D = 2d + h$ or $D = \frac{2d+h}{2} = d + \frac{h}{2}$.

44. Eff $= 1 - \frac{T_4 - T_1}{\alpha(T_3 - T_2)}$ can be rewritten as $1 -$ Eff $= \frac{T_4 - T_1}{\alpha(T_3 - T_2)}$. Multiplying both sides of this equation by $\alpha(T_3 - T_2)$ produces $\alpha(T_3 - T_2)(1 - \text{Eff}) = T_4 - T_1$, and so $T_1 = T_4 - \alpha(T_3 - T_2)(1 - \text{Eff})$.

45. $\frac{1}{x} = \frac{1}{10} + \frac{1}{6} = \frac{6}{60} + \frac{10}{60} = \frac{16}{60}$. Thus, $x = \frac{60}{16} = 3\frac{3}{4}$. It will take $3\frac{3}{4}$h or 3 h 45 min to complete the payroll.

46. $\frac{1}{t} = \frac{1}{12} + \frac{1}{15} = \frac{5}{60} + \frac{4}{60} = \frac{5+4}{60} = \frac{9}{60} = \frac{3}{20}$. This means that $t = \frac{20}{3} = 6\frac{2}{3}$. It will take $6\frac{2}{3}$ h or 6 h 40 min to fill the tank with both pipes.

47. $h =$ hours B is open; $\frac{h+1}{6} + \frac{h}{4} = 1$; LCD $= 12$; $12 \cdot \frac{h+1}{6} + 12 \cdot \frac{h}{4} = 12$; $2h + 2 + 3h = 12$; $5h = 10$; $h = 2$; $h + 1 = 3$. It takes a total of 3 hours.

48. $t =$ time together; $\frac{t}{5} + \frac{t}{8}$; $8t + 5t = 40$; $13t = 40$; $t = \frac{40}{13} \approx 3.08 \ \mu s$.

49. $\frac{h+4}{18} + \frac{h}{12} = 1$; LCD $= 36$; $2(h+4) + 3h = 36$; $5h + 8 = 36$; $5h = 28$; $h = \frac{28}{5} = 5.6$ hr together; Total time is $5.6 + 4 = 9.6$ hr

50. $r =$ speed of airplane in still air; $r + 20 =$ speed of airplane with tail wind; $r - 20 =$ speed of airplane with head wind; distance $=$ rate $\times$ time so time $= \frac{\text{distance}}{\text{rate}}$; $\frac{500}{r-20} = \frac{650}{r+20}$; $500(r+20) = 650(r-20)$; $500r + 10{,}000 = 650r - 13000$; $23{,}000 = 150r$; $r = \frac{23000}{150} \approx 153.3$ km/h

51. $\frac{1}{t} = \frac{1}{3} - \frac{1}{4} = \frac{4-3}{12} = \frac{1}{12}$. Hence, $t = 12$, so it will take 12 h to fill the tank.

52. $\frac{45{,}000}{6} = \frac{12{,}000}{8} + \frac{45{,}000}{t}$. Multiplying by $24t$ yields

$$180{,}000t = 36{,}000t + 1{,}080{,}000$$
$$144{,}000t = 1{,}080{,}000$$
$$t = 7.5$$

It will take Collector B 7.5 h working alone to collect 45,000 Btu.

8.2 QUADRATIC EQUATIONS AND FACTORING

1. Factoring $x^2 - 9 = 0$ we get $(x+3)(x-3) = 0$. By the zero product rule we have $x + 3 = 0$ and $x = -3$ or $x - 3 = 0$ and so $x = 3$.

2. Factoring $x^2 - 100 = 0$ produces $(x+10)(x-10) = 0$. By the zero product rule we have $x + 10 = 0$ and so $x = -10$ or $x - 10 = 0$ and we get $x = 10$.

3. $x^2 + x - 6 = 0$; $(x+3)(x-2) = 0$; $x + 3 = 0 \Rightarrow x = -3$ or $x - 2 = 0 \Rightarrow x = 2$

4. $x^2 - 6x - 7 = 0$; $(x-7)(x+1) = 0$; $x - 7 = 0 \Rightarrow x = 7$ or $x + 1 = 0 \Rightarrow x = -1$

5. $x^2 - 11x - 12 = 0$; $(x-12)(x+1) = 0 \Rightarrow x = 12$ or $x + 1 = 0 \Rightarrow x = -1$

6. $x^2 - 5x + 4 = 0$; $(x-4)(x-1) = 0$; $x - 4 = 0 \Rightarrow x = 4$ or $x - 1 = 0 \Rightarrow x = 1$

7. $x^2 + 2x - 8 = 0$; $(x+4)(x-2) = 0$; $x + 4 = 0 \Rightarrow x = -4$ or $x - 2 = 0 \Rightarrow x = 2$

8. $x^2 + 2x - 15 = 0$; $(x+5)(x-3) = 0$; $x + 5 = 0 \Rightarrow x = -5$ or $x - 3 = 0 \Rightarrow x = 3$

9. $x^2 - 5x = 0$; $x(x-5) = 0$; $x = 0$ or $x - 5 = 0 \Rightarrow x = 5$

10. $x^2 + 10x = 0$; $x(x+10) = 0$; $x = 0$ or $x + 10 = 0 \Rightarrow x = -10$

11. $x^2 + 12 = 7x$; $x^2 - 7x + 12 = 0$; $(x-3)(x-4) = 0$; $x - 3 = 0 \Rightarrow x = 3$; $x - 4 = 0 \Rightarrow x = 4$

12. $x^2 = 7x - 10$; $x^2 - 7x + 10 = 0$; $(x-5)(x-2) = 0$; $x - 5 = 0 \Rightarrow x = 5$; $x - 2 = 0 \Rightarrow x = 2$

13. $2x^2 - 3x - 14 = 0$; $(2x - 7)(x + 2) = 0$; $2x - 7 = 0 \Rightarrow x = \frac{7}{2}$; $x + 2 = 0 \Rightarrow x = -2$

14. $2x^2 + x - 15 = 0$; $(2x - 5)(x + 3) = 0$; $2x - 5 = 0 \Rightarrow x = \frac{5}{2}$; $x + 3 = 0 \Rightarrow x = -3$

15. $2x^2 + 12 = 11x$; $2x^2 - 11x + 12 = 0$; $(2x - 3)(x - 4) = 0$; $2x - 3 = 0 \Rightarrow x = \frac{3}{2}$; $x - 4 = 0 \Rightarrow x = 4$

16. $2x^2 + 18 = 15x$; $2x^2 - 15x + 18 = 0$; $(2x - 3)(x - 6) = 0$; $2x - 3 = 0 \Rightarrow x = \frac{3}{2}$; $x - 6 = 0 \Rightarrow x = 6$

17. $3x^2 - 8x - 3 = 0$; $(3x + 1)(x - 3) = 0$; $3x + 1 = 0 \Rightarrow x = -\frac{1}{3}$; $x - 3 = 0 \Rightarrow x = 3$

18. $3x^2 - 4x - 4 = 0$; $(3x + 2)(x - 2) = 0$; $3x + 2 = 0 \Rightarrow x = -\frac{2}{3}$; $x - 2 = 0 \Rightarrow x = 2$

19. $4x^2 - 24x + 35 = 0$; $(2x - 5)(2x - 7) = 0$; $2x - 5 = 0 \Rightarrow x = \frac{5}{2}$; $2x - 7 = 0 \Rightarrow x = \frac{7}{2}$

20. $6x^2 - 13x + 6 = 0$; $(3x - 2)(2x - 3) = 0$; $3x - 2 = 0 \Rightarrow x = \frac{2}{3}$; $2x - 3 = 0 \Rightarrow x = \frac{3}{2}$

21. $6x^2 + 11x - 35 = 0$; $(3x - 5)(2x + 7) = 0$; $3x - 5 = 0 \Rightarrow x = \frac{5}{2}$; $2x + 7 = 0 \Rightarrow x = -\frac{7}{2}$

22. $10x^2 + 9x - 9 = 0$; $(5x - 3)(2x + 3) = 0$; $5x - 3 = 0 \Rightarrow x = \frac{3}{5}$; $2x + 3 = 0 \Rightarrow x = -\frac{3}{2}$

23. $10x^2 - 17x + 3 = 0$; $(5x - 1)(2x - 3) = 0$; $5x - 1 = 0 \Rightarrow x = \frac{1}{5}$; $2x - 3 = 0 \Rightarrow x = \frac{3}{2}$

24. $14x^2 - 29x - 15 = 0$; $(7x + 3)(2x - 5) = 0$; $7x + 3 = 0 \Rightarrow x = -\frac{3}{7}$; $2x - 5 = 0 \Rightarrow x = \frac{5}{2}$

25. $6x^2 = 31x + 60$; $6x^2 - 31x - 60 = 0$; $(3x - 20)(2x + 3) = 0$; $3x - 20 = 0 \Rightarrow x = \frac{20}{3}$; $2x + 3 = 0 \Rightarrow x = -\frac{3}{2}$

26. $15x^2 - 23x + 4 = 0$; $(5x - 1)(3x - 4) = 0$; $5x - 1 = 0 \Rightarrow x = \frac{1}{5}$; $3x - 4 = 0 \Rightarrow x = \frac{4}{3}$

27. $(x - 1)^2 = 4$; $x - 1 = \pm\sqrt{4}$; $x - 1 = \pm 2$; $x - 1 = 2 \Rightarrow x = 3$; $x - 1 = -2 \Rightarrow x = -1$

28. $(x + 2)^2 = 9$; $x + 2 = \pm\sqrt{9}$; $x + 2 = \pm 3$; $x + 2 = 3 \Rightarrow x = 1$; $x + 2 = -3 \Rightarrow x = -5$

29. $(5x - 2)^2 = 16$; $5x - 2 = \pm\sqrt{16}$; $5x - 2 = \pm 4$; $5x - 2 = 4 \Rightarrow x = \frac{6}{5}$; $5x - 2 = -4 \Rightarrow x = -\frac{2}{5}$

30. $(3x + 2)^2 = 64$; $3x + 2 = \pm 8$; $3x + 2 = 8 \Rightarrow x = 2$; $3x + 2 = -8 \Rightarrow x = -\frac{10}{3}$

31. $\frac{x}{x+1} = \frac{x+2}{3x}$; $3x \cdot x = (x + 2)(x + 1)$; $3x^2 = x^2 + 3x + 2$; $2x^2 - 3x - 2 = 0$; $(2x + 1)(x - 2) = 0$; $x = -\frac{1}{2}$ or 2

32. Note that $x = 1$ and $x = 0$ are *not* possible solutions. $\frac{4x}{x-1} = \frac{7x+2}{x}$. Cross-multiplying produces $4x^2 = 7x^2 - 5x - 2$ or $3x^2 - 5x - 2 = 0$. This factors as $(3x + 1)(x - 2) = 0$. Thus, $x = -\frac{1}{3}$ or $x = 2$.

33. $(x + 5)^3 = x^3 + 1385$. Expanding the left-hand side, we can rewrite the equation as $x^3 + 15x^2 + 75x + 125 = x^3 + 1385$. Combining like terms, this becomes $15x^2 + 75x - 1260 = 0$ or $x^2 + 5x - 84 = 0$. This factors as $(x + 12)(x - 7) = 0$ and so $x = -12$ or $x = 7$.

34. $(x - 3)^3 = x^3 - 63$. Expanding the left-hand side, we can rewrite the equation as $x^3 - 9x^2 + 27x - 27 = x^3 - 63$. Combining like terms, this becomes $-9x^2 + 27x + 36 = 0$ or $x^2 - 3x - 4 = 0$. This factors as $(x - 4)(x + 1) = 0$, and so $x = 4$ or $x = -1$.

35. $(x - 4)^3 - x^3 = -316$. Expanding the left-hand side, we can rewrite the equation as $x^3 - 12x^2 + 48x - 64 - x^3 = -316$. Combining like terms, we get $-12x^2 + 48x + 252 = 0$ or $12x^2 - 48x - 252 = 0$. Dividing by 12, this becomes $x^2 - 4x - 21 = 0$ or $(x - 7)(x + 3) = 0$. Thus, $x = 7$ or $x = -3$.

36. $(x + 2)^3 - x^3 = 56$; $x^3 + 3x^2 \cdot 2 + 3x2^2 + 2^3 = 56$; $6x^2 + 12x + 8 = 56$; $6x^2 + 12x - 48 = 0$; $x^2 + 2x - 8 = 0$; $(x - 2)(x + 4) = 0$ $x = 2$ or $x = -4$

37. $\frac{1}{x-3} + \frac{1}{x+4} = \frac{1}{12}$; $12(x + 4) + 12(x - 3) = (x - 3)(x + 4)$; $12x + 48 + 12x - 36 = x^2 + x - 12$; $24x + 12 = x^2 + x - 12$; $0 = x^2 - 23x - 24$; $(x - 24)(x + 1) = 0$; $x = 24$ or $x = -1$.

38. $\frac{1}{x-5} + \frac{1}{x+3} = \frac{1}{3}$; $3(x+3) + 3(x-5) = (x-5)(x+3)$; $3x + 9 + 3x - 15 = x^2 - 2x - 15$; $6x + 9 = x^2 - 2x$; $x^2 - 8x - 9 = 0$; $(x - 9)(x + 1) = 0$; $x = 9$ or $x = -1$

39. Multiplying $\frac{1}{x-1} + \frac{1}{x-2} = \frac{7}{12}$ by the LCD, $12(x-1)(x-2)$, we obtain $12(x-2) + 12(x-1) = 7(x-1)(x-2)$. Distributing produces $12x - 24 + 12x - 12 = 7\left(x^2 - 3x + 2\right)$ which simplifies to:

$$24x - 36 = 7x^2 - 21x + 14$$
$$0 = 7x^2 - 45x + 50$$
$$0 = (7x - 10)(x - 5)$$

Thus, $x = \frac{10}{7}$, 5.

40. Multiplying $\frac{1}{x+1} - \frac{1}{x+2} = \frac{1}{20}$ by the LCD of $20(x+1)(x+2)$, we obtain $20(x+2) - 20(x+1) = (x+1)(x+2)$ or $20x + 40 - 20x - 20 = x^2 + 3x + 2$. Combining terms on the left-hand side produces $20 = x^2 + 3x + 2$. This simplifies to $x^2 + 3x - 18 = 0$ and factors as $(x+6)(x-3) = 0$. So, $x = -6$, 3.

41. Multiplying $\frac{2}{x-24} + \frac{1}{x-9} = \frac{1}{6}$ by the LCD of $6(x-4)(x-9)$, we obtain $12(x-9) + 6(x-4) = (x-4)(x-9)$ or $18x - 132 = x^2 - 13x + 36$. This simplifies to $x^2 - 31x + 168 = 0$ and factors as $(x-24)(x-7) = 0$. So, $x = 24$, 7.

42. Multiplying $\frac{x}{x+1} + \frac{1}{x} = \frac{13}{12}$ by the LCD of $12(x+1)$, we obtain $12x + 12(x+1) = 13x(x+1)$ which expands to $12x^2 + 12x + 12 = 13x^2 + 13x$. This simplifies to $x^2 + x - 12 = 0$ and factors as $(x-3)(x+4) = 0$. So, $x = 3, -4$.

43. Multiplying $\frac{x}{x+1} - \frac{2x}{x+3} = -\frac{1}{15}$ by the LCD of $15(x+1)(x+3)$, we obtain $15x(x+3) - 30x(x+1) = -(x+1)(x+3)$ or $15x^2 + 45x - 30x^2 - 30x = -x^2 - 4x - 3$. This simplifies to $14x^2 - 19x - 3 = 0$ and factors as $(7x+1)(2x-3) = 0$. Thus, $x = -\frac{1}{7}, \frac{3}{2}$.

44. Multiplying $\frac{3x}{x-1} - \frac{9x}{x+2} = 8.4$ by the LCD of $(x-1)(x+2)$, we obtain $3x(x+2) - 9x(x-1) = 8.4(x-1)(x+2)$ or $3x^2 + 6x - 9x^2 + 9x = 8.4(x^2 + x - 2)$ or $-6x^2 + 15x = 8.4x^2 + 8.4x - 16.8$. Collecting like terms, we have

$$0 = 14.4x^2 - 6.6x - 16.8$$
$$0 = 144x^2 - 66x - 168$$
$$0 = 24x^2 - 11x - 28$$
$$0 = (8x + 7)(3x - 4)$$

Thus, $x = -\frac{7}{8}, \frac{4}{3}$.

45. $-16t^2 + 64t + 192 = 192$; $-16t^2 + 64t = 0$; $-16t(t-4) = 0$; $t = 0$ or $t = 4$; we want $t = 4$ s.

46. $-16t^2 + 64t + 192 = 0$; $-16(t^2 - 4t - 12) = 0$; $-16(t-6)(t+2) = 0$; $t = 6$ or $t = -2$; we want $t = 6.0$ s.

47. If we let ℓ represent the length in cm and w its width in cm, then we are given $\ell = w + 5$. The area of a rectangle is $\ell w = (w+5)w = 104$. Multiplying, we get $w^2 + 5w - 104 = 0$, which factors as $(w+13)(w-8) = 0$. Thus, by the zero product rule, we have $w = -13$ or $w = 8$. Choose $w = 8$ cm and then $\ell = w + 5 = 8 + 5 = 13$ cm.

48. $4.9t^2 = 411$; $t^2 = \frac{411}{4.9}$; $= \sqrt{\frac{411}{4.9}} \approx 9.158$ s

49. Pythagorean Theorem yields $(x-7)^2 + x^2 = 13^2$; $x^2 - 14x + 49 + x^2 = 169$; $2x^2 - 14x - 120 = 0$; $x^2 - 7x - 60 = 0$; $(x-12)(x+5) = 0$; $x = 12$ or $x = -5$; cannot be negative so $x = 12$; $x - 7 = 5$. The rafters are 5 m and 12 m.

50. $x = $ time for B alone; $x - 6 = $ time for A alone; $\frac{4}{x} + \frac{4}{x-6} = 1 \Rightarrow 4(x-6) + 4x = x(x-6) \Rightarrow 8x - 24 = x^2 - 6x$, which can be written as $x^2 - 14x + 24 = 0$. This last equation factors as $(x-12)(x-2) = 0$. From this we see that B is 12 hr and A is $12 - 6 = 6$ hr

51. Since total area $-$ inside area $=$ area of concrete, we have the equation $(2x+16)(2x+10) - 160 = 192$ or $4x^2 + 52x + 160 - 160 = 192$. Collecting like terms, this becomes $4x^2 + 52x - 192 = 4\left(x^2 + 13x - 48\right) = 0$. This factors as $4(x+16)(x-3)$ and so, $x = -16$ or $x = 3$. The only answer that fits the situation is $x = 3.00$ in.

52. Let x be the amount added to both the length and width. We then get the equation $(50 + x)(80 + x) = 18{,}000$ or $x^2 + 130x + 4{,}000 = 18{,}000$ or $x^2 + 130x - 14{,}000 = 0$. This factors as $(x+200)(x-70) = 0$. Since $x = -200$ does not make physical sense, we must add 70 m to the width and length to obtain a parking lot that measures 120 m by 150 m.

53. The radius of the pool is 12 m. Its area is 144π m^2. The area of the deck plus the pool is $(12 + w)^2\pi$. This leads to the equation $(12 + w)^2\pi - 144\pi = 432\pi$ or $(12 + w)^2 - 144 = 432$. Expanding, we get $144 + 24w + w^2 - 144 = 432$. Collecting terms, we can rewrite this as $w^2 + 24w - 432 = 0$. This factors as $(x+36)(x-12) = 0$. Thus, $x = -36$ or $x = 12$. The only answer that makes physical sense is $x = 12$. The width of the deck should be 12.0 m.

54. We use the fact that the area of pool + deck − area of pool = area of deck. If x is the desired width of the deck, then the area of pool + deck is $(30 + 2x)(22 + 2x)$. The area of the pool is $30 \cdot 22$ and the area of the deck is 480. Thus, we get the equation $(30 + 2x)(22 + 2x) - 30 \cdot 22 = 480$ or $4x^2 + 104x - 480 = 0$. Dividing by 4, we get $x^2 + 26x - 120 = 0$. This factors as $(x + 30)(x - 4) = 0$. The width of the deck is 4.0 ft.

55. $\frac{1}{B} + \frac{1}{B+6} = \frac{1}{4}$ or $4(B + 6) + 4B = B(B + 6)$ or $8B + 24 = B^2 + 6B$. Collecting like terms, this simplifies to $B^2 - 2B - 24 = 0$ or $(B-6)(B+4) = 0$. Thus, $B = 6$ or $B = -4$. Since $B > 0$, $B = 6$ h and $A = 12$ h.

56. Since $R = 10$ and $E = 510$, then $RI^2 + EI = 18\,000$ can be written as $10I^2 + 510I = 18\,000$ or $10I^2 + 510I - 18\,000 = 0$. Dividing by 10, the equation is $I^2 - 51I = 1\,800 = 0$. This factors as $(I + 75)(I - 24) = 0$. Since the current, I, must be positive, we reject $I = -75$. Hence, $I = 24$ A.

57. Substituting the given values, we obtain $47.75 = 0.002837D^2(75) - 0.127248$. This simplifies to $0.212775D^2 = 47.877248$ or $D^2 = 225.01$. Thus, $D = \sqrt{225.01} \approx 15.00$ in.

58. Substituting the given values, we obtain $61.35 = 0.002835D^2(85) - 0.337655$ or $0.240975D^2 = 61.687655$. Thus, $D^2 = 255.99$, so $D = \sqrt{255.99} \approx 16.00$ in.

≡ 8.3 COMPLETING THE SQUARE

1. $x^2 + 6x + 8 = 0$; $x^2 + 6x + 9 = -8 + 9$; $(x+3)^2 = 1$; $x + 3 = \pm 1$; $x = -3 + 1 = -2$ or $x = -3 - 1 = -4$

2. $x^2 - 7x - 8 = 0$; $x^2 - 7x + \frac{49}{4} = 8 + \frac{49}{4}$; $\left(x - \frac{7}{2}\right)^2 = \frac{81}{4}$; $x - \frac{7}{2} = \pm\sqrt{\frac{81}{4}}$; $x - \frac{7}{2} = \pm\frac{9}{2}$; $x = \frac{7}{2} + \frac{9}{2} = 8$ or $x = \frac{7}{2} - \frac{9}{2} = -1$

3. $x^2 - 10x = 11$; $x^2 - 10x + 25 = 11 + 25$; $(x-5)^2 = 36$, $x - 5 = \pm 6$ and so, $x = 5 + 6 = 11$ or $x = 5 - 6 = -1$

4. $2x^2 - 3 = 14$; $x^2 - \frac{3}{2}x = 7$; $x^2 - \frac{3}{2}x + \frac{9}{16} = 7 + \frac{9}{16}$; $\left(x - \frac{3}{4}\right)^2 = \frac{121}{16}$; $x - \frac{3}{4} = \pm\frac{11}{4}$; $x = \frac{3}{4} + \frac{11}{4} = \frac{7}{2}$; $x = \frac{3}{4} - \frac{11}{4} = -2$

5. $x^2 + 6x + 3 = 0$; $x^2 + 6x + 9 = -3 + 9$; $(x+3)^2 = 6$; $x + 3 = \pm\sqrt{6}$ and so $x = -3 \pm \sqrt{6}$

6. $x^2 + 8x + 10 = 0$; $x^2 + 8x + 16 = -10 + 16$; $(x+4)^2 = 6$; $x + 4 = \pm\sqrt{6}$; $x = -4 \pm \sqrt{6}$

7. $x^2 - 5x + 5 = 0$; $x^2 - 5x + \frac{25}{4} = -5 + \frac{25}{4}$; $\left(x - \frac{5}{2}\right)^2 = \frac{5}{4}$; $x - \frac{5}{2} = \pm\frac{\sqrt{5}}{2}$; $x = \frac{5}{2} \pm \frac{\sqrt{5}}{2}$ or $x = \frac{5 \pm \sqrt{5}}{2}$

8. $x^2 - 7x + 11 = 0$; $x^2 - 7x + \frac{49}{4} = -11 + \frac{49}{4}$; $\left(x - \frac{7}{2}\right)^2 = \frac{5}{4}$; $x - \frac{7}{2} = \pm\frac{\sqrt{5}}{2} = \pm\frac{7 \pm \sqrt{5}}{2}$

9. $2x^2 - 6x - 10 = 0$; $x^2 - 3x - 5 = 0$; $x^2 - 3x + \frac{9}{4} = 5 + \frac{9}{4}$; $\left(x - \frac{3}{2}\right)^2 = \frac{29}{4}$; $x - \frac{3}{2} = \pm\frac{\sqrt{29}}{2}$; $x = \frac{3 \pm \sqrt{29}}{2}$

10. $3x^2 + 12x - 18 = 0$; $x^2 + 4x - 6 = 0$; $x^2 + 4x + 4 = 6 + 4$; $(x+2)^2 = 10$; $x + 2 = \pm\sqrt{10}$; $x = -2 \pm \sqrt{10}$

11. $4x^2 - 12x - 18 = 0$; $x^2 - 3x - \frac{9}{2} = 0$; $x^2 - 3x + \frac{9}{4} = \frac{9}{2} + \frac{9}{4}$; $\left(x - \frac{3}{2}\right)^2 = \frac{27}{4}$; $x - \frac{3}{2} = \frac{\pm\sqrt{27}}{2}$; $x = \frac{3 \pm \sqrt{27}}{2} = \frac{3 \pm 3\sqrt{3}}{2}$

12. $3x^2 - 9x = 33$; $x^2 - 3x = 11$; $x^2 - 3x + \frac{9}{4} = 11 + \frac{9}{4}$; $\left(x - \frac{3}{2}\right)^2 = \frac{53}{4}$; $x - \frac{3}{2} = \frac{\pm\sqrt{53}}{2}$; $x = \frac{3 \pm \sqrt{53}}{2}$

13. $x^2 + 2kx + c = 0$; $x^2 + 2kx = -c$; $x^2 + 2kx + k^2 = -c + k^2$; $(x+k)^2 = -c + k^2$; $x + k = \pm\sqrt{k^2 - c}$; $x = -k \pm \sqrt{k^2 - c}$

14. $px^2 + 2qx + r = 0$; $x^2 + \frac{2q}{p}x = -\frac{r}{p}$; $x^2 + \frac{2q}{p}x + \left(\frac{q}{p}\right)^2 = \frac{-r}{p} + \left(\frac{q}{p}\right)^2 = \frac{q^2 - rp}{p^2}$; $\left(x + \frac{q}{p}\right)^2 = \frac{q^2 - rp}{p^2}$; $x + \frac{q}{p} = \frac{\pm\sqrt{q^2 - rp}}{p}$; $x = \frac{-q \pm \sqrt{q^2 - rp}}{p}$

15. $196 - 16t^2 = 0$; $16t^2 = 196$; $t^2 = 12.25$; $t = \pm\sqrt{12.25}$ cannot be negative so $\sqrt{12.25} = 3.5$ s

16. $q(2520 - 3q) = 140{,}400$; $-3q^2 + 2520q = 140{,}400$; $q^2 - 840q = -46{,}800$; $q^2 - 840q + 176{,}400 = -46{,}800 + 176{,}400 = 129{,}600$; $(q - 420)^2 = 360^2$; $q = 420 \pm 360$; $q = 60$ or 780 objects

17. $q(1560 - 4q) = 29,600$; $-4q^2 + 1560q = 29,600$; $q^2 - 390q = -7400$; $q^2 - 390q + (195)^2 = -7400 + 195^2$; $(q - 195)^2 = 30625$; $q - 195 = \pm 175$; $q = 195 + 175 = 370$ objects; or $q = 195 - 175 = 20$ objects

18. $p = 2(\ell + w) = 1500$ or $\ell + w = 750$ and so $\ell = 750 - w$. $A = \ell \cdot w = (750 - w)w = 137,600$; $-w^2 + 750w = 137,600$; $w^2 - 750w = -137,600$; $w^2 - 750w + 375^2 = -137,600 + 375^2$; $(w - 375)^2 = 3025$; $w - 375 = \pm 55$; $w = 375 + 55 = 430 \Rightarrow \ell = 750 - 430 = 320$; or $w = 375 - 55 = 320$ and $\ell = 430$. Thus, we see that the dimensions are 430 ft by 320 ft

19. $P = 4w + 3\ell = 2700$; $3\ell = 2700 - 4w$; $\ell = 900 - \frac{4}{3}w$; $A = 2\ell w = 270,000$; $\ell w = 135,000$; $\left(900 - \frac{4}{3}w\right)w = 135,000$; $900w - \frac{4}{3}w^2 = 135,000$; $-\frac{4}{3}w^2 + 900w = 135,000$; $w^2 - 675w = -101,250$;

$w^2 - 675w + \left(\frac{675}{2}\right)^2 = -101,250 + \left(\frac{675}{2}\right)^2$; $\left(w - \frac{675}{2}\right)^2 = 12656.25$; $w - 337.5 = \pm \sqrt{12656.25}$; $w - 337.5 = \pm 112.5$; $w = 337.5 + 112.5 = 450 \Rightarrow \ell = 300$; or $w = 337.5 - 112.5 = 225 \Rightarrow \ell = 600$; 450 ft by 300 ft or 225 ft by 600 ft

20. Let $R_1 = x$ and $R_2 = x + 5$. Then we obtain $\frac{1}{6} = \frac{1}{x} + \frac{1}{x+5}$ which, when multiplied by the LCD of $6x(x + 5)$ yields $x(x + 5) = 6(x + 5) + 6x$ or $x^2 + 5x = 12x + 30$. Hence, $x^2 - 7x - 30 = 0$. Completing the square, we get $x^2 - 7x + \frac{49}{4} = 30 + \frac{49}{4}$ or $\left(x - \frac{7}{2}\right)^2 = \frac{169}{4}$. So, $x - \frac{7}{2} = \pm\frac{13}{2}$. Since x must be positive, we get $x = \frac{7}{2} + \frac{13}{2} = \frac{20}{2} = 10$. Thus, $R_1 = 10\,\Omega$ and $R_2 = 15\,\Omega$.

≡ 8.4 THE QUADRATIC FORMULA

1. $x^2 + 3x - 4 = 0$; $x = \frac{-3 \pm \sqrt{(3)^2 - 4(1)(-4)}}{2 \cdot 1} = \frac{-3 \pm \sqrt{9 + 16}}{2} = \frac{-3 \pm \sqrt{25}}{2} = \frac{-3 \pm 5}{2}$; $\frac{-3 + 5}{2} = 1$; $\frac{-3 - 5}{2} = -4$; $\{1, -4\}$

2. $x^2 - 8x - 33 = 0$; $x = \frac{-(-8) \pm \sqrt{(-8)^2 - 4(-33)}}{2 \cdot 1} = \frac{8 \pm \sqrt{64 + 132}}{2} = \frac{8 \pm \sqrt{196}}{2} = \frac{8 \pm 14}{2}$; $\frac{8 + 14}{2} = \frac{22}{2} = 11$; $\frac{8 - 14}{2} = -\frac{6}{2} = -3$; $\{-3, 11\}$

3. $3x^2 - 5x - 2 = 0$; $x = \frac{-(-5) \pm \sqrt{(-5)^2 - 4(3)(-2)}}{2 \cdot 3} = \frac{5 \pm \sqrt{25 + 24}}{6} = \frac{5 \pm \sqrt{49}}{6} = \frac{5 \pm 7}{6}$; $\frac{5 + 7}{6} = \frac{12}{6} = 2$; $\frac{5 - 7}{6} = \frac{-2}{6} = -\frac{1}{3}$; $\left\{-\frac{1}{3}, 2\right\}$

4. $7x^2 + 5 - 2 = 0$; $x = \frac{-5 \pm \sqrt{5^2 - 4 \cdot 7(-2)}}{2 \cdot 7} = \frac{-5 \pm \sqrt{25 + 56}}{14} = \frac{-5 \pm \sqrt{81}}{14} = \frac{-5 \pm 9}{14}$; $\frac{-5 + 9}{14} = \frac{4}{14} = \frac{2}{7}$; $\frac{-5 - 9}{14} = \frac{-14}{14} = -1$; $\left\{-1, \frac{2}{7}\right\}$

5. $7x^2 + 6x - 1 = 0$; $x = \frac{-6 \pm \sqrt{6^2 - 4 \cdot 7(-1)}}{14} = \frac{-6 \pm \sqrt{36 + 28}}{14} = \frac{-6 \pm \sqrt{64}}{14} = \frac{-6 \pm 8}{14}$; $\frac{-6 + 8}{14} = \frac{2}{14} = \frac{1}{7}$; $\frac{-6 - 8}{14} = \frac{-14}{14} = -1$; $\left\{-1, \frac{1}{7}\right\}$

6. $2x^2 - 3x - 20 = 0$; $x = \frac{-(-3) \pm \sqrt{(-3)^2 - 4 \cdot 2(-20)}}{2 \cdot 2} = \frac{3 \pm \sqrt{9 + 160}}{4} = \frac{3 \pm \sqrt{169}}{4} = \frac{3 \pm 13}{4}$; $\frac{3 + 13}{4} = \frac{16}{4} = 4$; $\frac{3 - 13}{4} = \frac{-10}{4} = \frac{-5}{2}$; $\left\{4, -\frac{5}{2}\right\}$

7. $2x^2 - 5x - 7 = 0$; $x = \frac{-(-5) \pm \sqrt{(-5)^2 - 4 \cdot 2(-7)}}{2 \cdot 2} = \frac{5 \pm \sqrt{25 + 56}}{4} = \frac{5 \pm \sqrt{81}}{4} = \frac{5 \pm 9}{4}$; $\frac{5 + 9}{4} = \frac{14}{4} = \frac{7}{2}$; $\frac{5 - 9}{4} = \frac{-4}{4} = -1$; $\left\{-1, \frac{7}{2}\right\}$

8. $3x^2 + 4x - 7 = 0$; $x = \frac{-4 \pm \sqrt{4^2 - 4(3)(-7)}}{2 \cdot 3} = \frac{-4 \pm \sqrt{16 + 84}}{6} = \frac{-4 \pm \sqrt{100}}{6} = \frac{-4 \pm 10}{6}$; $\frac{-4 + 10}{6} = \frac{6}{6} = 1$; $\frac{-4 - 10}{6} = \frac{-14}{6} = -\frac{7}{3}$; $\left\{1, -\frac{7}{3}\right\}$

9. $3x^2 + 2x - 8 = 0$; $x = \frac{-2 \pm \sqrt{2^2 - 4(3)(-8)}}{2 \cdot 3} = \frac{-2 \pm \sqrt{4 + 96}}{6} = \frac{-2 \pm \sqrt{100}}{6} = \frac{-2 \pm 10}{6}$; $\frac{-2 + 10}{6} = \frac{8}{6} = \frac{4}{3}$; $\frac{-2 - 10}{6} = \frac{-12}{6} = -2$; $\left\{-2, \frac{4}{3}\right\}$

10. $9x^2 - 6x + 1 = 0$; $x = \frac{-(-6) \pm \sqrt{-(-6)^2 - 4 \cdot 9(1)}}{2 \cdot 9} = \frac{6 \pm \sqrt{36 - 36}}{18} = \frac{6}{18} = \frac{1}{3}$ double root

11. $9x^2 + 12x + 4 = 0$; $x = \frac{-12 \pm \sqrt{12^2 - 4 \cdot 9 \cdot 4}}{2 \cdot 9} = \frac{-12 \pm \sqrt{144 - 144}}{18} = \frac{-12}{18} = \frac{-2}{3}$ double root

12. $3x^2 + 3x - 7 = 0$; $x = \frac{-3 \pm \sqrt{3^2 - 4(3)(-7)}}{2 \cdot 3} = \frac{-3 \pm \sqrt{9 + 84}}{6} = \frac{-3 \pm \sqrt{93}}{6}$

13. $2x^2 - 3x - 1 = 0$; $x = \frac{-(-3) \pm \sqrt{(-3)^2 - 4 \cdot 2 \cdot (-1)}}{2 \cdot 2} = \frac{3 \pm \sqrt{9 + 8}}{4} = \frac{3 \pm \sqrt{17}}{4}$

14. $2x^2 - 5x + 1 = 0$; $x = \frac{-(-5)\pm\sqrt{(-5)^2-4\cdot2\cdot1}}{2\cdot2} =$
$\frac{5\pm\sqrt{25-8}}{4} = \frac{5\pm\sqrt{17}}{4}$

15. $x^2 + 5x + 2 = 0$; $x = \frac{-5\pm\sqrt{5^2-4\cdot2\cdot1}}{2\cdot1} = \frac{-5\pm\sqrt{25-8}}{2} =$
$\frac{-5\pm\sqrt{17}}{2}$

16. $3x^2 - 6x - 2 = 0$; $x = \frac{-(-6)\pm\sqrt{(-6)^2-4\cdot3\cdot(-2)}}{2\cdot3} =$
$\frac{6\pm\sqrt{36+24}}{6} = \frac{6\pm\sqrt{60}}{6} = \frac{6\pm2\sqrt{15}}{6} = \frac{2\left(3\pm\sqrt{15}\right)}{2\cdot3} =$
$\frac{3\pm\sqrt{15}}{3}$

17. $2x^2 + 6x - 3 = 0$; $x = \frac{-6\pm\sqrt{6^2-4(2)(-3)}}{2\cdot2} = \frac{-6\pm\sqrt{36+24}}{4}$
$= \frac{-6\pm\sqrt{60}}{4} = \frac{-6\pm2\sqrt{15}}{4} = \frac{2\left(-3\pm\sqrt{15}\right)}{2\cdot2} = \frac{-3\pm\sqrt{15}}{2}$

18. $5x^2 + 2x - 1 = 0$; $x = -\frac{-2\pm\sqrt{2^2-4(5)(-1)}}{2\cdot5} =$
$\frac{-2\pm\sqrt{4+20}}{10}$
$= \frac{-2\pm\sqrt{24}}{10} = \frac{-2\pm2\sqrt{6}}{10} = \frac{2(-1\pm\sqrt{6})}{2\cdot5} = \frac{-1\pm\sqrt{6}}{5}$

19. $x^2 - 2x - 7 = 0$; $x = \frac{2\pm\sqrt{(-2)^2-4\cdot1(-7)}}{2} = \frac{2\pm\sqrt{4+28}}{2}$
$= \frac{2\pm\sqrt{32}}{2} = \frac{2\pm4\sqrt{2}}{2} = 1\pm2\sqrt{2}$

20. $x^2 + 3 = 0$; $x = \frac{0\pm\sqrt{0-4(3)}}{2} = \frac{\pm\sqrt{-12}}{2}$ no real roots
since the discriminant is -12

21. $2x^2 - 3 = 0$; $x = \frac{0\pm\sqrt{0-4(2)(-3)}}{2\cdot2} = \frac{\pm\sqrt{24}}{4} =$
$\frac{\pm2\sqrt{6}}{4} = \frac{\pm\sqrt{6}}{2}$

22. $2x^2 = 5$; $2x^2 - 5 = 0$; $x = \frac{0\pm\sqrt{0^2-4\cdot2(-5)}}{2\cdot2} =$
$\frac{\pm\sqrt{40}}{4} = \frac{\pm2\sqrt{10}}{4} = \frac{\pm\sqrt{10}}{2}$

23. $3x^2 + 4 = 0$; $x = \frac{0\pm\sqrt{0^2-4(3)(4)}}{2\cdot3} = \frac{\pm\sqrt{-48}}{6}$ no real
roots because the discriminant is -48

24. $3x^2 + 1 = 5x$; $3x^2 - 5x + 1 = 0$; $x = \frac{5\pm\sqrt{(-5)^2-4\cdot3\cdot1}}{2\cdot3}$
$= \frac{5\pm\sqrt{25-12}}{6} = \frac{5\pm\sqrt{13}}{6}$

25. $\frac{2}{3}x^2 - \frac{1}{9}x + 3 = 0$; $x = \frac{\frac{1}{9}\pm\sqrt{\left(-\frac{1}{9}\right)^2-4\cdot\frac{2}{3}\cdot3}}{2\cdot\frac{2}{3}} =$
$\frac{\frac{1}{9}\pm\sqrt{\frac{1}{81}-8}}{\frac{4}{3}} = \frac{\frac{1}{9}\pm\sqrt{\frac{-647}{81}}}{\frac{4}{3}}$ no real roots because the discriminant is $\frac{-647}{81}$

26. $\frac{1}{2}x^2 - 2x + \frac{1}{3} = 0$; $x = \frac{2\pm\sqrt{(-2)^2-4\cdot\left(\frac{1}{2}\right)\left(\frac{1}{3}\right)}}{2\cdot\frac{1}{2}}$
$= \frac{2\pm\sqrt{4-\frac{2}{3}}}{1} = 2\pm\sqrt{\frac{10}{3}}$; or $2\pm\frac{\sqrt{30}}{3}$ or $\frac{6\pm\sqrt{30}}{3}$

27. $0.01x^2 + 0.2x = 0.6$; $0.01x^2 + 0.2x - 0.6 = 0$;
$x = \frac{-0.2\pm\sqrt{.2^2-4\cdot(0.01)(-0.6)}}{2\cdot(0.01)} = \frac{-0.2\pm\sqrt{0.064}}{0.02}$ or see
the alternate solution.
Alternate Solution:
$0.01x^2 + 0.2x - 0.6 = 0$; $x^2 + 20x - 60 = 0$; $x =$
$\frac{-20\pm\sqrt{20^2-4(-60)}}{2} = \frac{-20\pm\sqrt{400+240}}{2} = \frac{-20\pm\sqrt{640}}{2} =$
$\frac{-20\pm2\sqrt{160}}{2} = -10\pm\sqrt{160} = -10\pm4\sqrt{10}$

28. $0.16x^2 = 0.8x - 1$; $0.16x^2 - 0.8x + 1 = 0$; $x =$
$\frac{0.8\pm\sqrt{(-0.8)^2-4(0.16)(1)}}{2(0.16)} = \frac{0.8\pm\sqrt{0}}{0.32} = 2.5$ double root

29. $\frac{1}{4}x^2 + 3 = \frac{5}{2}x$; $\frac{1}{4}x^2 - \frac{5}{2}x + 3 = 0$ or $x^2 - 10x + 12 =$
0; $x = \frac{10\pm\sqrt{(-10)^2-4\cdot12}}{2} = \frac{10\pm\sqrt{52}}{2} = \frac{10\pm2\sqrt{13}}{2} =$
$5\pm\sqrt{13}$

30. $\frac{3}{2}x^2 + 2x = \frac{7}{2}$; $\frac{3}{2}x^2 + 2x - \frac{7}{2} = 0$; $3x^2 + 4x - 7 = 0$;
$x = \frac{-4\pm\sqrt{4^2-4(3)(-7)}}{2\cdot3} = \frac{-4\pm\sqrt{16+84}}{6} = \frac{-4\pm\sqrt{100}}{6} =$
$\frac{-4\pm10}{6}$; $\frac{-4+10}{6} = 1$; $\frac{-4-10}{6} = \frac{-7}{3}$

31. $1.2x^2 = 2x - 0.5$; $1.2x^2 - 2x + 0.5 = 0$; $12x^2 -$
$20x + 5 = 0$; $x = \frac{20\pm\sqrt{(-20)^2-4(12)5}}{2\cdot12} = \frac{20\pm\sqrt{400-240}}{24}$
$= \frac{20\pm\sqrt{160}}{24} = \frac{20\pm4\sqrt{10}}{24} = \frac{5\pm\sqrt{10}}{6}$ or $x =$
$\frac{2\pm\sqrt{(-2)^2-4(1.2)(0.5)}}{2\cdot1.2} = \frac{2\pm\sqrt{4-2.4}}{2.4} = \frac{2\pm\sqrt{1.6}}{2.4}$

32. $1.4x^2 + 0.2x = 2.3$ or $1.4x^2 + 0.2x - 2.3 = 0$;
Multiplying by 10 produces $14x^2 + 2x - 23 = 0$;
and so $x = \frac{-2\pm\sqrt{2^2-4(14)(-23)}}{2(14)} = \frac{-2\pm\sqrt{4+1288}}{28} =$
$\frac{-2\pm\sqrt{1292}}{28} = \frac{-2\pm\sqrt{323}}{28} = \frac{-1\pm\sqrt{323}}{14}$; or $x =$
$\frac{-0.2\pm\sqrt{(0.2)^2-4\cdot1.4(-2.3)}}{2(1.4)} = \frac{-0.2\pm\sqrt{.04+12.88}}{28} =$
$\frac{-0.2\pm\sqrt{12.92}}{2.8}$

33. $3x^2 + \sqrt{3}x - 7 = 0$; $x = \frac{-\sqrt{3}\pm\sqrt{\sqrt{3}^2-4(3)(-7)}}{2\cdot3} =$
$\frac{-\sqrt{3}\pm\sqrt{3+84}}{6} = \frac{-\sqrt{3}\pm\sqrt{87}}{6}$

34. $2x^2 - \sqrt{89}x + 5 = 0$. Here $a = 2$, $b = -\sqrt{89}$, and
$c = 5$. $x = \frac{\sqrt{89}\pm\sqrt{(\sqrt{89})^2-4(2)(5)}}{2\cdot2} = \frac{\sqrt{89}\pm\sqrt{89-40}}{4} =$
$\frac{\sqrt{89}\pm7}{4}$

35. $\frac{x-3}{7} = 2x^2$; $x - 3 = 14x^2$; $14x^2 - x + 3 = 0$; $x = \frac{1\pm\sqrt{(-1)^2-4\cdot14\cdot3}}{2} = \frac{1\pm\sqrt{-167}}{2}$. Since the discriminant is negative, there are no real roots.

36. $\frac{x-5}{3} = 5x^2$; Multiplying by 3 produces $x - 5 = 15x^2$ and so, we get $15x^2 - x + 5 = 0$. $x = \frac{1\pm\sqrt{1-4\cdot15\cdot5}}{2(15)} = \frac{1\pm\sqrt{-299}}{30}$. Since the discriminant -299 is negative, there are no real roots.

37. $\frac{2}{x-1} + 3 = \frac{-2}{x+1}$; LCD $= (x-1)(x+1)$. Multiplying by the LCD produces $2(x + 1) + 3(x - 1)(x + 1) = -2(x - 1)$; $2x + 2 + 3x^2 - 3 = -2x + 2$; $3x^2 + 4x - 3 = 0$; $\frac{-4\pm\sqrt{4^2-4(3)(-3)}}{2(3)}$; $\frac{-4\pm\sqrt{16+36}}{6}$; $\frac{-4\pm\sqrt{52}}{6} = \frac{-4\pm2\sqrt{13}}{6} = \frac{-2\pm\sqrt{13}}{3}$

38. $\frac{3x}{x+2} + 2x = \frac{2x^2-1}{x+1}$; LCD$(x+2)(x+1) = x^2+3x+2$. Multiplying by the LCD produces $3x(x + 1) + 2x\left(x^2 + 3x + 2\right) = \left(2x^2 - 1\right)(x+2)$; $3x^2 + 3x + 2x^3 + 6x^2 + 4x = 2x^3 + 4x^2 - x - 2$; $9x^2 + 7x = 4x^2 - x - 2$; $5x^2 + 8x + 2 = 0$; $x = \frac{-8\pm\sqrt{8^2-4\cdot5\cdot2}}{2\cdot5} = \frac{-8\pm\sqrt{64-40}}{10} = \frac{-8\pm\sqrt{24}}{10} = \frac{-8\pm2\sqrt{6}}{10} = \frac{-4\pm\sqrt{6}}{5}$

39. $-4.9t^2 + 411 = 0$; $\frac{0\pm\sqrt{0-4(-4.9)(411)}}{2(-4.9)} = \frac{\pm\sqrt{8055.6}}{-9.8} \approx 9.16$ s; only positive answer works

40. $\frac{20\pm\sqrt{(-20)^2-4(-4.9)(411)}}{2(-4.9)} = \frac{20\pm\sqrt{400+8055.6}}{-9.8} = \frac{20\pm\sqrt{8455.6}}{-9.8} \approx \frac{20\pm91.95}{-9.8}$. So, $x \approx \frac{20-91.95}{-9.8} \approx 7.34$ or $x \approx \frac{20+91.95}{-9.8} = -11.42$. Only the positive answer makes sense, and so the ball hits the ground about 7.34 s after it is thrown.

41. $\frac{-20\pm\sqrt{20^2-4(-4.9)(411)}}{2(-4.9)} = \frac{-20\pm91.95}{-9.8} = \frac{-20-91.95}{-9.8} \approx +11.42$; $\frac{-20+91.95}{-9.8} \approx -7.34$; only positive answer fits the problem situation 11.42 s

42. $4\cdot(x - 8)^2 = 100$; $(x - 8)^2 = 25$; $x - 8 = \pm5$; $x = 8 + 5 = 13$; only 13 cm fits problem since $x = 8 - 5 = 3$ gives negative box dimensions, so the desired dimensions are 13 cm $\times$ 13 cm.

43. $V = \ell\cdot w\cdot h = (1.5w-6)(w-6)\cdot3 = 578$; $\left(1.5w^2 - 15w + 36\right)3 = 578$; $4.5w^2 - 45w + 108 = 578$; $4.5w^2 - 45w - 470 = 0$; $x = \frac{45\pm\sqrt{(-45)^2-4(4.5)(-470)}}{2\cdot4.5}$ $= \frac{45\pm\sqrt{10,485}}{9} \approx \frac{45+102.4}{9} = \frac{147.4}{9} \approx 16.38$; $w = 16.38$ cm; $\ell = 1.5(16.38) = 24.57$ cm.

44. $20x + 15x - x^2 = 81.25$; $-x^2 + 35x - 81.25 = 0$; $x^2 - 35x + 81.25 = 0$; $x = \frac{35\pm\sqrt{(-35)^2-4(81.25)}}{2} =$

$\frac{35\pm\sqrt{900}}{2} = \frac{35\pm30}{2}$; $\frac{35-30}{2} = 2.5$; $\frac{35+30}{2} = 32.5$; x must be less than 15 so answer is 2.5 cm

45. $x(12-2x) = 16.875$; $12x - 2x^2 = 16.875$; $-2x^2 + 12x - 16.875 = 0$; $2x^2 - 12x + 16.875 = 0$; $x = \frac{12\pm\sqrt{(-12)^2-4\cdot(2)(16.875)}}{2\cdot2} = \frac{12\pm3}{4}$; $\frac{12+3}{4} = \frac{15}{4} = 3\frac{3}{4}$; $\frac{12-3}{4} = \frac{9}{4} = 2\frac{1}{4}$; $12 - 2\cdot3\frac{3}{4} = 4.5$; $12 - 2\cdot2\frac{1}{4} = 7.5$; 2.25 in. by 7.5 in.; or 3.75 in. by 4.5 in.

46. $A = 2\pi r^2 + 2\pi rh$; $2\pi r^2 + 2\pi\cdot10 = 245$; $r^2 + 10r = \frac{245}{2\pi} \approx 39$; $r^2 + 10r - 39 = 0$; $r = \frac{-10\pm\sqrt{10^2-4\cdot1\cdot(-39)}}{2} = \frac{-10\pm\sqrt{100+156}}{2}$; $\frac{-10+16}{2} = \frac{6}{2} = 3$ cm. Only positive answers fit the problem.

47. $x = $ amount of increase; new price is $30+x$; number of customers is $1000 - 5x$; revenue $= (1000 - 5x)(30+x)$; $(1000-5x)(30+x) = 45,000$; $30000 + 850x - 5x^2 = 45,000$; $-5x^2 + 850x - 15,000 = 0$; $x^2 - 170x + 3,000 = 0$; $x = \frac{170\pm\sqrt{(-170)^2-4(3,000)}}{2} = \frac{170\pm130}{2}$. Thus, one value produces $x = \frac{170-130}{2} = \frac{40}{2} = 20$. The other value yields $x = \frac{170+130}{2} = 150$; a \$20 increase will lose the fewest customers.

48. $V^2 = V_R^2 + (V_L - V_c)^2$; $5.8^2 = 5^2 + (V_L - 10)^2$; $33.64 = 25 + (V_L - 10)^2$; $8.64 = (V_L - 10)^2$; $V_L - 10 = \pm\sqrt{8.64} \approx \pm2.94$; $V_L \approx 10 + 2.94 = 12.94$ V or $10 - 294 = 7.06$ V

49. $610^2 = 300^2 + (X_L - 531)^2$; $372100 = 90000 + (X_L - 531)^2$; $282100 = (X_L - 531)^2$; $x_L - 531 = \pm\sqrt{282100} \approx \pm531.1$; $x_L \approx 531 \pm 531.1$. Since x_L cannot be negative, we have $x_L = 1062.13$ Ω

50. $\frac{1}{2}w\ell x - \frac{1}{2}wx^2 = 0$; $\ell x - x^2 = 0$; $x(\ell - x) = 0$; $x = 0$ or $x = \ell$

51. Since the total area $-$ inside area $=$ area of concrete, we have the equation $(2x + 16)(2x + 10) - 160 = 245$ which simplifies to $4x^2 + 52x - 245 = 0$. Using the quadratic formula we have $x = \frac{-52\pm\sqrt{52^2-4(4)(-245)}}{2(4)} = \frac{-52\pm\sqrt{6624}}{8} = \frac{-52\pm12\sqrt{46}}{8} = \frac{-13\pm3\sqrt{46}}{2}$. Since $x > 0$, we get $\frac{-13+3\sqrt{46}}{2} \approx 3.67$ in.

52. Let x represent the amount that should be added to the length and width of the parking lot. The area of the new lot will be $(x+50)(x+80) = 2\cdot50\cdot80$ or $x^2 + 130x + 4000 = 8000$. This simplifies to $x^2 + 130x - 4000 = 0$. By the quadratic formula, we get $x = \frac{-130\pm\sqrt{130^2+16,000}}{2} = \frac{-130\pm\sqrt{32,900}}{2} =$

$\frac{-130\pm10\sqrt{329}}{2} = -65 \pm 5\sqrt{329}$. Since x must be positive, we take $-65 + 5\sqrt{329} \approx 29.7$ m. So, the new width is $50 + 29.7 = 79.7$ m and the new length is $80 + 29.7 = 109.7$ m.

53. Substituting, we get $926 = 0.79D^2 - 2D - 4$. Collecting like terms produces $0.79D^2 - 2D - 930 = 0$. By the quadratic formula,
$D = \frac{2\pm\sqrt{(-2)^2+4(0.79)(930)}}{2(0.79)} \approx 35.6$ in. or -33.1 in. Since, $D > 0$, we ignore the negative answer.

54. Substituting, we get $1078 = 0.79D^2 - 2D - 4$ or $0.79D^2 - 2D - 1082 = 0$. By the quadratic formula we obtain $D = \frac{2\pm\sqrt{(-2)^2+4(0.79)(1082)}}{2(0.79)} \approx -35.76$ or 38.29. Since D must be positive, the diameter is about 38.3 in.

55. See *Computer Programs* in main text.

56. The following program will work on a *TI-82* graphing calculator.
Prompt A
Prompt B
Prompt C

$B^2 - 4AC \to D$
If $D < 0$
Goto 1
$(^-B - \sqrt{D})/2A \to S$
$(^-B + \sqrt{D})/2A \to T$
Disp S
Disp T
Stop
Lbl 1
Disp "COMPLEX ROOTS"
Disp "THE REAL PART IS"
$^-B/2A \to R$
Disp R
$-\sqrt{D}/2A \to I$
Disp "THE IMAGINARY"
Disp I

57. **(a)** $3x^2 + 5x - 2 = (3x-1)(x+2)$ and so, $x = -2$ or $x = \frac{1}{3}$.

(b) $\frac{-5\pm\sqrt{25+24}}{6} = \frac{-5\pm7}{6}$. $\frac{-5+7}{6} = \frac{2}{6} = \frac{1}{3}$ and $\frac{-5-7}{6} = \frac{-12}{6} = -2$

CHAPTER 8 REVIEW

1. $\frac{x}{3} + \frac{x}{2} = 5$; LCD $= 6$; $6 \cdot \frac{x}{3} + 6 = 6 \cdot \frac{x}{2} = 6 \cdot 5 \Rightarrow 2x + 3x = 30 \Rightarrow 5x = 30$; $x = 6$

2. $\frac{2}{x} - \frac{3}{x} = \frac{1}{5}$; LCD $= 5x$, $x \neq 0$; $5 \cdot 2 - 5 \cdot 3 = x$; $10 - 15 = x$; $-5 = x$

3. $\frac{x-2}{4} - \frac{x+2}{5} = \frac{x}{2}$; LCD 20; $20\left(\frac{x-2}{4}\right) - 20\left(\frac{x+2}{5}\right) = 20 \cdot \frac{x}{2}$; $(5x-10)-(4x+8) = 10x$; $5x-10-4x-8 = 10x$; $x - 18 = 10x$; $-18 = 9x$; $x = -2$

4. $\frac{2}{x-1} + \frac{3}{x-2} = \frac{4}{x^2-3x+2}$; LCD $(x-1)(x-2) = x^2-3x+2$; $x \neq 1$; $x \neq 2$; $2(x-2)+3(x-1) = 4$; $2x-4+3x-3 = 4$; $5x = 11$; $x = \frac{11}{5}$

5. $\frac{3x}{x-1} - \frac{3x+2}{x} = 3$; LCD $(x-1)(x) = x^2-x$; $3x(x) - (3x+2)(x-1) = 3\left(x^2-x\right)$; $3x^2-3x^2+x+2 = 3x^2-3x$; $0 = 3x^2-4x-2$; $\frac{4\pm\sqrt{(-4)^2-4\cdot(3)(-2)}}{2\cdot3} = \frac{4\pm\sqrt{40}}{6} = \frac{4\pm2\sqrt{10}}{2\cdot3} = \frac{2\pm\sqrt{10}}{3}$

6. $\frac{4x}{2x-3} + \frac{1}{x-1} = \frac{2x+1}{x-1}$; We see that the LCD $= (2x-3)(x-1)$ and that $x \neq \frac{3}{2}$ and $x \neq 1$. Multiplying by the LCD, we get $4x(x-1)+(2x-3) = (2x+1)(2x-3)$ or $4x^2-4x+2x-3 = 4x^2-4x-3$. Collecting terms, we get $2x = 0$ and so, $x = 0$.

7. $A = 2lw + 2(l+w)h$; $Aa - 2lw = 2(l+w)h$; $\frac{A-2lw}{2(l+w)} = h$

8. $A = 2lw + 2(l+w)h$; $A = 2lw + 2lh + 2wh$; $2lw + 2lh = A - 2wh$; $l(2w+2h) = A - 2wh$; $l = \frac{A-2wh}{2w+2h}$

9. $\frac{1}{f} = \frac{1}{p} + \frac{1}{q}$; LCD $= fpq$; $fpq \cdot \frac{1}{f} = fpq \cdot \frac{1}{p} + fpq \cdot \frac{1}{q}$; $pq = fq + fp$; $pq = f(q+p)$; $f = \frac{pq}{p+q}$

10. $X = wL - \frac{1}{wc}$; $Xwc = wLwC - 1$; $Xwc = w^2Lc - 1$; $w^2Lc - Xwc = 1$; $c\left(w^2L - Xw\right) = 1$; $c = \frac{1}{w^2L-Xw}$ or $c = \frac{1}{w(wL-X)}$

11. $x^2 - 8x + 7 = 0$; $(x-7)(x-1) = 0$; $x - 7 = 0 \Rightarrow x = 7$; $x - 1 = 0 \Rightarrow x = 1$

12. $x^2 + 4x + 3 = 0$; $(x+3)(x+1) = 0$; $x+3 = 0 \Rightarrow x = -3$; $x+1 = 0 \Rightarrow x = -1$

13. $x^2 - 11x + 10 = 0$; $(x-10)(x-1) = 0$; $x - 10 = 0 \Rightarrow x = 10$; $x - 1 = 0 \Rightarrow x = 1$

14. $x^2 + 15x + 14 = 0$; $(x+14)(x+1) = 0$; $x + 14 = 0 \Rightarrow x = -14$; $x + 1 = 0 \Rightarrow x = -1$

15. $x^2 + 15x + 56 = 0$; $(x + 8)(x + 7) = 0$; $x + 7 = 0 \Rightarrow x = -7$; $x + 8 = 0 \Rightarrow x = -8$

16. $x^2 - 8x + 12 = 0$; $(x - 6)(x - 2) = 0$; $x - 6 = 0 \Rightarrow x = 6$; $x - 2 = 0 \Rightarrow x = 2$

17. $2x^2 - 5x + 3 = 0$; $(2x - 3)(x - 1) = 0$; $2x - 3 = 0 \Rightarrow x = \frac{3}{2}$; $x - 1 = 0 \Rightarrow x = 1$

18. $3x^2 + 10x + 7 = 0$; $(3x + 7)(x + 1) = 0$; $3x + 7 = 0 \Rightarrow x = -\frac{7}{3}$; $x + 1 = 0 \Rightarrow x = -1$

19. $6x^2 + 7x - 10 = 0$; $(6x - 5)(x + 2) = 0$; $6x - 5 = 0 \Rightarrow x = \frac{5}{6}$; $x + 2 = 0 \Rightarrow x = -2$

20. $8x^2 + 22x + 9 = 0$; $8 \times 9 = 72$; $18 \times 4 = 72$; $18 + 4 = 22$; $8x^2 + 18x + 4x + 9 = 0$; $2x(4x + 9) + 1(4x + 9)$; $(2x + 1)(4x + 9) = 0$; $2x + 1 = 0 \Rightarrow x = -\frac{1}{2}$; $4x + 9 = 0 \Rightarrow x = -\frac{9}{4}$

21. $4x^2 - 9 = 0$; $(2x + 3)(2x - 3) = 0$; $2x + 3 = 0 \Rightarrow x = -\frac{3}{2}$; $2x - 3 = 0 \Rightarrow x = \frac{3}{2}$

22. $9x^2 - 16 = 0$; $(3x + 4)(3x - 4) = 0$; $3x + 4 = 0 \Rightarrow x = -\frac{4}{3}$; $3x - 4 = 0 \Rightarrow x = \frac{4}{3}$

23. $x^2 + 4x - 5 = 0$; $x^2 + 4x + 4 = 5 + 4$; $(x + 2)^2 = 9$; $x + 2 = \pm 3$; $x = -2 + 3 = 1$; $x = -2 - 3 = -5$

24. $x^2 - 7x + 6 = 0$; $x^2 - 7x + \frac{49}{4} = -6 + \frac{49}{4}$; $\left(x - \frac{7}{2}\right)^2 = \frac{25}{4}$; $x - \frac{7}{2} = \pm \frac{5}{2}$; $x = \frac{7}{2} + \frac{5}{2} = \frac{12}{2} = 6$; $x = \frac{7}{2} - \frac{5}{2} = \frac{2}{2} = 1$

25. $x^2 + 19x = 11$; $x^2 + 19x + \left(\frac{19}{2}\right)^2 = 11 + \left(\frac{19}{2}\right)^2$; $\left(x = \frac{19}{2}\right)^2 + \frac{44}{4} + \frac{361}{4} = \frac{405}{4}$; $x + \frac{19}{2} = \frac{\pm\sqrt{405}}{2} = \frac{\pm 9\sqrt{5}}{2}$; $x = \frac{-19 \pm 9\sqrt{5}}{2}$

26. $2x^2 + 7x = 15$, upon dividing by 2 becomes $x^2 + \frac{7}{2}x = \frac{15}{2}$ and after completing the square yields $x^2 + \frac{7}{2}x + \left(\frac{7}{4}\right)^2 = \frac{15}{2} + \left(\frac{7}{4}\right)^2 = \frac{120}{16} + \frac{49}{16}$ or $\left(x + \frac{7}{4}\right)^2 = \frac{169}{16}$, and after taking the square roots of both sides produces $x + \frac{7}{4} = \pm\frac{13}{4}$ or $x = -\frac{7}{4} + \frac{13}{4} = \frac{6}{4} = \frac{3}{2}$ and $x = \frac{-7}{4} - \frac{13}{4} = \frac{-20}{4} = -5$.

27. $3x^2 + 5x - 14 = 0$ or $3x^2 + 5x = 14$, and, after dividing by 3 yields $x^2 + \frac{5}{3}x = \frac{14}{3}$. Completing the square produces $x^2 + \frac{5}{3}x + \left(\frac{5}{6}\right)^2 = \frac{14}{3} + \left(\frac{5}{6}\right)^2$ or $\left(x + \frac{5}{6}\right)^2 = \frac{168}{36} + \frac{25}{36} = \frac{193}{36}$. Taking the square

roots of both sides produces $x + \frac{5}{6} = \pm\frac{\sqrt{193}}{6}$ and so $x = \frac{-5\pm\sqrt{193}}{6}$.

28. $4x^2 + 2x - 5 = 0$ or $4x^2 + 2x = 5$, and, after dividing by 4 yields $x^2 + \frac{1}{2}x = \frac{5}{4}$. Completing the square produces $x^2 + \frac{1}{2}x + \frac{1}{16} = \frac{5}{4} + \frac{1}{16}$ or $\left(x + \frac{1}{4}\right)^2 = \frac{21}{16}$. Taking the square roots of both sides produces $x = -\frac{1}{4} \pm \frac{\sqrt{21}}{4}$ and so $x = \frac{-1\pm\sqrt{21}}{4}$.

29. $x^2 - 8x + 5 = 0$; $x = \frac{8\pm\sqrt{(-8)^2 - 4(5)}}{2} = \frac{8\pm\sqrt{64-20}}{2} = \frac{8\pm\sqrt{44}}{2} = \frac{8\pm2\sqrt{11}}{2} = 4 \pm \sqrt{11}$

30. $x^2 + 7x - 6 = 0$; $x = \frac{-7\pm\sqrt{7^2 - 4(-6)}}{2} = \frac{-7\pm\sqrt{49+24}}{2} = \frac{-7\pm\sqrt{73}}{2}$

31. $2x^3 + 3x - 5 = 0$; $x = \frac{-3\pm\sqrt{3^2 - 4(2)(-5)}}{2\cdot2} = \frac{-3\pm\sqrt{9+40}}{4} = \frac{-3\pm7}{4}$; $\frac{-3+7}{4} = 1$; $\frac{-3-7}{4} = \frac{-5}{2}$

32. $2x^2 - 7x + 4 = 0$; $x = \frac{7\pm\sqrt{(-7)^2 - 4(2)(4)}}{2\cdot2} = \frac{7\pm\sqrt{49-32}}{4} = \frac{7\pm\sqrt{17}}{4}$

33. $3x^2 + 2x - 4 = 0$; $x = \frac{-2\pm\sqrt{2^2 - 4(3)(-4)}}{2\cdot3} = \frac{-2\pm\sqrt{4+48}}{6} = \frac{-2\pm\sqrt{52}}{6} = \frac{-2\pm2\sqrt{13}}{2\cdot3} = \frac{-1\pm\sqrt{13}}{3}$

34. $4x^2 - 3x = 1$; $4x^2 - 3x - 1 = 0$; $x = \frac{3\pm\sqrt{(-3)^2 - 4(4)(-1)}}{2\cdot4} = \frac{3\pm\sqrt{9+16}}{8} = \frac{3\pm5}{8}$; so $x = \frac{3+5}{8} = 1$ or, $x = \frac{3-5}{8} = -\frac{1}{4}$

35. $5x^2 + 2 = 8x$; $5x^2 - 8x + 2 = 0$; $x = \frac{8\pm\sqrt{(-8)^2 - 4(5)(2)}}{2\cdot5} = \frac{8\pm\sqrt{64-40}}{2\cdot5} = \frac{8\pm\sqrt{24}}{2\cdot5} = \frac{8\pm2\sqrt{6}}{2\cdot5} = \frac{2(4\pm\sqrt{6})}{2\cdot5} = \frac{4\pm6}{5}$

36. $6x^2 + 2x = 3$; $6x^2 + 2x - 3 = 0$; $x = \frac{-2\pm\sqrt{2^2 - 4(6)(-3)}}{2\cdot6} = \frac{-2\pm\sqrt{4+72}}{12} = \frac{-2\pm\sqrt{76}}{12} = \frac{-2\pm2\sqrt{19}}{2\cdot6} = \frac{-1\pm\sqrt{19}}{6}$

37. $3x^2 - 8x + 10 = 0$; $x = \frac{8\pm\sqrt{(-8)^2 - 4(3)(10)}}{2\cdot3} = \frac{8\pm\sqrt{64-120}}{6} = \frac{8\pm\sqrt{-56}}{6}$; no answer because the discriminant is negative

38. $8x^2 = 4x + 3$; $8x^2 - 4x - 3 = 0$; $x = \frac{4\pm\sqrt{(-4)^2 - 4(8)(-3)}}{2\cdot8} = \frac{4\pm\sqrt{16+96}}{16} = \frac{4\pm\sqrt{112}}{16} = \frac{4\pm4\sqrt{7}}{16} = \frac{1\pm\sqrt{7}}{4}$

39. $\frac{x}{x-1} + \frac{2}{x+1} = 3$; $(x+1)x + 2(x-1) = 3(x+1)(x-1)$; $x^2 + x + 2x - 2 = 3x^2 - 3$; $0 = 2x^2 - 3x - 1$; $x = \frac{3 \pm \sqrt{(-3)^2 - 4(2)(-1)}}{2 \cdot 2} = \frac{3 \pm \sqrt{9+8}}{4} = \frac{3 \pm \sqrt{17}}{4}$

40. $\frac{2}{x} - \frac{3}{x+2} = 4$; $2(x+2) - 3(x) = 4(x)(x+2)$; $2x + 4 - 3x = 4x^2 + 8x$; $0 = 4x^2 + 9x - 4$; $x = \frac{-9 \pm \sqrt{9^2 - 4(4)(-4)}}{2 \cdot 4} = \frac{-9 \pm \sqrt{81+64}}{8} = \frac{-9 \pm \sqrt{145}}{8}$

41. $12 = (i_1 + 0.4)^2 \cdot 50$; $(i_1 + 0.4)^2 = \frac{12}{50} = .24$; $(i_1 + 0.4) = \pm\sqrt{.24} \approx \pm.4899$; $i_1 \approx -0.4 + .4899$; $i_1 \approx 0.0899$ A; Only positive answers fit the problem

42. $\frac{h}{7} + \frac{h}{8} = 1$; $8h + 7h = 56$; $15h = 56$; $h = \frac{56}{15} \approx 3.73$ hr

43. $144 - 16t^2 = 0$; $9 - t^2 = 0$; $t^2 = 9$; $t = \pm\sqrt{9} = \pm 3$; only 3 fits the problem, so it takes 3 s for the stone to reach the bottom

44. $a^2 + (a-6)^2 = 27^2$; $a^2 + a^2 - 12a + 36 = 729$; $2a^2 - 12a - 693 = 0$; $a = \frac{12 \pm \sqrt{(-12)^2 - 4(2)(-693)}}{2 \cdot 2} = \frac{12 \pm \sqrt{5688}}{4} \approx \frac{12 + 75.42}{4} \approx 21.86$. Then $a - 6 = 21.86 - 6 = 15.86$ and the lengths are 15.86 mm and 21.86 mm

▤ CHAPTER 8 TEST

1. $\frac{x}{8} + \frac{3x}{4} = \frac{7}{2}$; LCD $= 8$; $8 \cdot \frac{x}{8} + 8 \cdot \frac{3x}{4} = 8 \cdot \frac{7}{2}$; $x + 6x = 28$; $7x = 28$; $x = 4$

2. $\frac{12}{x^2 - 9} = \frac{x}{x-3} - \frac{x}{x+3}$; LCD $= (x-3)(x+3) = x^2 - 9$; $x \neq 3$; $x \neq -3$; $12 = x(x+3) - x(x-3)$; $12 = x^2 + 3x - x^2 + 3x$; $12 = 6x$; $x = 2$

3. $x^2 - 4x - 5 = 0 \Rightarrow (x-5)(x+1) = 0$; $x - 5 = 0 \Rightarrow x = 5$; $x + 1 = 0 \Rightarrow x = -1$

4. $3x^2 + x - 2 = 0$; Using the quadratic formula produces $x = \frac{-1 \pm \sqrt{1^2 - 4(3)(-2)}}{2 \cdot 3} = \frac{-1 \pm \sqrt{1 + 24}}{6} = \frac{-1 \pm 5}{6}$. Thus, we see that $x = \frac{-1 - 5}{6} = -1$ or $x = \frac{-1 + 5}{6} = \frac{2}{3}$. However, the given equation factors, and so this could have been solved by factoring where $(x+1)(3x-2) = 0$. By the zero product rule, $x + 1 = 0 \Rightarrow x = -1$ or $3x - 2 = 0 \Rightarrow x = \frac{2}{3}$.

5. $16x^2 + 9 = 24x$; $16x^2 - 24x + 9 = 0$; $(4x-3)^2 = 0$; $4x - 3 = 0 \Rightarrow x = \frac{3}{4}$ double root

6. $2x^2 + 3x + 3$; $x = \frac{-3 \pm \sqrt{3^2 - 4(2)(3)}}{2 \cdot 2} = \frac{-3 \pm \sqrt{-15}}{4}$; Discriminant is -15 so no solution

7. $x^2 + 3x + 1 = 0$; $\frac{-3 \pm \sqrt{3^2 - 4}}{2} = \frac{-3 \pm \sqrt{5}}{2}$

8. $x^2 - 8x + 16 = -5 + 16$; $(x-4)^2 = 11$; $x - 4 = \pm\sqrt{11}$; $x = 4 \pm \sqrt{11}$

9. Since $x^2 + 7x - 30 = (x+10)(x-3)$, the solutions are $x = -10$ and $x = 3$.

10. Since $3x^2 + 5x - 28 = (x+4)(3x-7)$, the solutions are $x = -4$ and $x = \frac{7}{3}$.

11. $V = 3D^2H - 5$ so, $V + 5 = 3D^2H$ and $H = \frac{V+5}{3D^2}$.

12. $V = 3D^2H - 5$ so $3D^2H = V + 5$ and $D^2 = \frac{V+5}{3H}$. Thus, $D = -\sqrt{\frac{V+5}{3H}}$ and $D = \sqrt{\frac{V+5}{3H}}$.

13. $l = w + 3$; $(w+3)w = 46.75$; $w^2 + 3w - 46.75 = 0$; $w = \frac{-3 \pm \sqrt{9 + 4(46.75)}}{2} = \frac{-3 \pm \sqrt{196}}{2}$. Since a negative value does not make sense in this situation, we have $w = \frac{-3 + 14}{2} = \frac{11}{2} = 5.5$. Thus, $w = 5.5$ m and $l = 8.5$ m.

14. $s = vt + \frac{at^2}{2}$; $s - \frac{at^2}{2} = vt$ or $vt = \frac{2s - at^2}{2}$ and so $v = \frac{2s - at^2}{2t}$

9

Trigonometric Functions

☰ 9.1 ANGLES, ANGLE MEASURE, AND TRIGONOMETRIC FUNCTIONS

1. $90° = 90° \times \frac{\pi}{180°} = \frac{\pi}{2}$ or 1.5708 or $90° \times 0.01745 = 1.5705$

2. $45° = 45° \times \frac{\pi}{180°} = \frac{\pi}{4}$ or 0.7854 or $45° \times 0.01745 = 0.78525$

3. $80° = 80° \times \frac{\pi}{180°} = \frac{4\pi}{9}$ or 1.3963 or $80° \times 0.01745 = 1.396$

4. $-15° = -15° \times \frac{\pi}{180°} = \frac{-\pi}{12} = -0.2618$ or $-15° \times 0.01745 = -0.26175$

5. $155° = 155° \times \frac{\pi}{180°} = \frac{31\pi}{36} = 2.7053$ or $155° \times 0.01745 = 2.70475$

6. $-235° = -235° \times \frac{\pi}{180°} = -4.1015$ or $-235° \times 0.01745 = -4.10075$

7. $215° = -215° \times \frac{\pi}{180°} = 3.7525$ or $215° \times 0.01745 = 3.75175$

8. $180° = \frac{180°\pi}{180°} = \pi \approx 3.1416$ or $180 \times 0.01745 = 3.141$

9. $2 \text{ rad} = 2 \cdot \frac{180°}{\pi} = \frac{360°}{\pi} \approx 114.592°$ or $2 \times 57.296° = 114.592°$

10. $3 = 3 \cdot \frac{180°}{\pi} \approx 171.887°$ or $3 \times 57.296° = 171.888°$

11. $1.5 = 1.5 \times \frac{180°}{\pi} \approx 85.944°$ or $1.5 \times 57.296 = 85.944°$

12. $\pi = \pi \cdot \frac{180°}{\pi} = 180°$

13. $\frac{\pi}{3} = \frac{\pi}{3} \cdot \frac{180°}{\pi} = 60°$

14. $\frac{5\pi}{6} = \frac{5\pi}{6} \times \frac{180°}{\pi} = 150°$

15. $-\frac{\pi}{4} = -\frac{\pi}{4} \cdot \frac{180°}{\pi} = -45°$

16. $-1.3 = -1.3 \times \frac{180°}{\pi} \approx -74.485°$ or $-1.3 \times 57.296° = -74.4848°$

17. (c) $150° = 150° + 360° = 510°$; $150° - 360° = -210°$

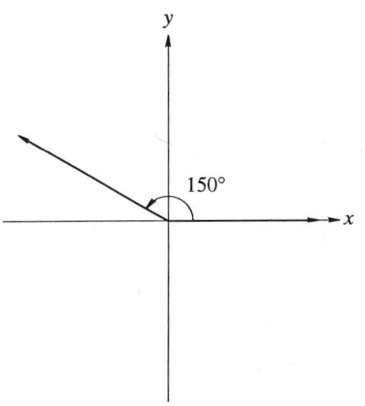

18. (c) $315° = 315° + 360° = 675°$; $315° - 360° = -45°$

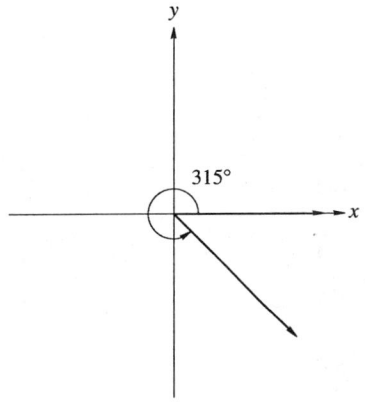

19. (c) $-135° = -135° + 360° = 225°$; $-135° - 360° = -495°$

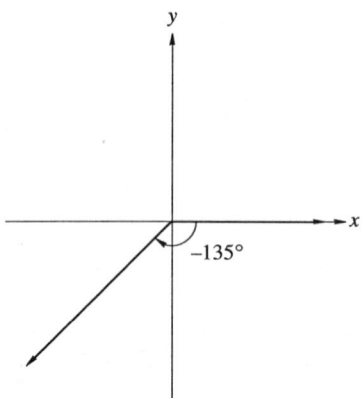

20. (c) $-30° = -30° + 360° = 330°$; $-30° - 360° = -390°$

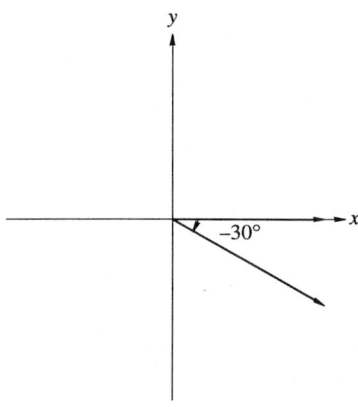

21. $r = \sqrt{4^2 + 3^2} = \sqrt{25} = 5$; $\sin\theta = \frac{3}{5}$; $\cos\theta = \frac{4}{5}$; $\tan\theta = \frac{3}{4}$; $\csc\theta = \frac{5}{3}$; $\sec\theta = \frac{5}{4}$; $\cot\theta = \frac{4}{3}$

22. $r = \sqrt{(-6)^2 + (-8)^2} = \sqrt{36 + 64} = \sqrt{100} = 10$; $\sin\theta = -\frac{4}{5}$; $\cos\theta = \frac{-3}{5}$; $\tan\theta = \frac{4}{3}$; $\csc\theta = \frac{-5}{4}$; $\sec\theta = \frac{-5}{3}$; $\cot\theta = \frac{3}{4}$

23. $r = \sqrt{8^2 + (-15)^2} = 17$; $\sin\theta = \frac{-15}{17}$; $\cos\theta = \frac{8}{17}$; $\tan\theta = \frac{-15}{8}$; $\csc\theta = \frac{-17}{15}$; $\sec\theta = \frac{17}{8}$; $\cot\theta = \frac{-8}{15}$

24. $r = \sqrt{(-20)^2 + (21)^2} = 29$; $\sin\theta = \frac{21}{29}$; $\cos\theta = \frac{-20}{29}$; $\tan\theta = -\frac{21}{20}$; $\csc\theta = \frac{29}{21}$; $\sec\theta = \frac{-29}{20}$; $\cot\theta = \frac{-20}{21}$

25. $r = \sqrt{1^2 + 2^2} = \sqrt{5}$; $\sin\theta = \frac{2}{\sqrt{5}}$; $\cos\theta = \frac{1}{\sqrt{5}}$; $\tan\theta = 2$; $\csc\theta = \frac{\sqrt{5}}{2}$; $\sec\theta = \sqrt{5}$; $\cot\theta = \frac{1}{2}$

26. $r = \sqrt{(-2)^2 + 4^2} = \sqrt{20} = 2\sqrt{5}$; $\sin\theta = \frac{2}{\sqrt{5}}$; $\cos\theta = \frac{-1}{\sqrt{5}}$; $\tan\theta = -2$; $\csc\theta = \frac{\sqrt{5}}{2}$; $\sec\theta = -\sqrt{5}$; $\cot\theta = -\frac{1}{2}$

27. $r = \sqrt{10^2 + (-8)^2} = \sqrt{164} = 2\sqrt{41}$; $\sin\theta = -\frac{4}{\sqrt{41}}$; $\cos\theta = \frac{5}{\sqrt{41}}$; $\tan\theta = \frac{-4}{5}$; $\csc\theta = -\frac{\sqrt{41}}{4}$; $\sec\theta = \frac{\sqrt{41}}{5}$; $\cot\theta = \frac{-5}{4}$

28. $r = \sqrt{(-5)^2 + (-2)^2} = \sqrt{25 + 4} = \sqrt{29}$; $\sin\theta = \frac{-2}{\sqrt{29}}$; $\cos\theta = -\frac{5}{\sqrt{29}}$; $\tan\theta = \frac{2}{5}$; $\csc\theta = -\frac{\sqrt{29}}{2}$; $\sec\theta = -\frac{\sqrt{29}}{5}$; $\cot\theta = \frac{5}{2}$

29. $r = \sqrt{\left(\sqrt{11}\right)^2 + 5^2} = \sqrt{11 + 25} = \sqrt{36} = 6$, $\sin\theta = \frac{5}{6}$; $\cos\theta = \frac{\sqrt{11}}{6}$; $\tan\theta = \frac{5}{\sqrt{11}}$; $\csc\theta = \frac{6}{5}$; $\sec\theta = \frac{6}{\sqrt{11}}$; $\cot\theta = \frac{\sqrt{11}}{5}$

30. $r = \sqrt{(-6)^2 + (\sqrt{13})^2} = \sqrt{36 + 13} = \sqrt{49} = 7$, $\sin\theta = \frac{\sqrt{13}}{7}$; $\cos\theta = \frac{-6}{7}$; $\tan\theta = -\frac{\sqrt{13}}{6}$; $\csc\theta = \frac{7}{\sqrt{13}}$; $\sec\theta = \frac{-7}{6}$; $\cot\theta = \frac{-6}{\sqrt{13}}$

31. $r = 3$, $\sin\theta = 0$; $\cos\theta = 1$; $\tan\theta = 0$; $\csc\theta =$ Does not exist; $\sec\theta = 1$; $\cot\theta =$ Does not exist.

32. $r = \sqrt{0^2 + (-4)^2} = \sqrt{16} = 4$; $\sin\theta = -1$; $\cos\theta = 0$; $\tan\theta$: Does not exist; $\csc\theta = -1$; $\sec\theta$: Does not exist; $\cot\theta = 0$

33. $r = 5$; $\sin\theta = 0$; $\cos\theta = -1$; $\tan\theta = 0$; $\csc\theta =$ Does not exist; $\sec\theta = -1$; $\cot\theta =$ Does not exist.

34. $r = 6$; $\sin\theta = 1$; $\cos\theta = 0$; $\tan\theta$: Does not exist; $\csc\theta = 1$; $\sec\theta$: Does not exist; $\cot\theta = 0$

35. $x = 6$, $r = 10$, $y > 0$, $y^2 = r^2 - x^2 = 10^2 - 6^2 = 100 - 36 = 64$, $y = \pm\sqrt{64}$ so $y = 8$; $\sin\theta = \frac{8}{10} = \frac{4}{5}$; $\cos\theta = \frac{x}{r} = \frac{6}{10} = \frac{3}{5}$; $\tan\theta = \frac{4}{3}$; $\csc\theta = \frac{5}{4}$; $\sec\theta = \frac{5}{3}$; $\cot\theta = \frac{3}{4}$

36. $y = -9$, $r = 15$, $x > 0$, $x^2 = r^2 - y^2$, $x^2 = 15^2 - (-9)^2 = 225 - 81 = 144$, $x = \pm\sqrt{144} = \pm12$ so $x = 12$; $\sin\theta = \frac{-3}{5}$; $\cos\theta = \frac{4}{5}$; $\tan\theta = \frac{-3}{4}$; $\csc\theta = \frac{-5}{3}$; $\sec\theta = \frac{5}{4}$; $\cot\theta = \frac{-4}{3}$

37. $x = -20$, $r = 29$, $y > 0$, $y^2 = r^2 - x^2 = 29^2 - (-20)^2 = 841 - 400 = 441$, $y = \pm\sqrt{441} = \pm21$ so $y = 21$, $\sin\theta = \frac{21}{29}$; $\cos\theta = \frac{-20}{29}$; $\tan\theta = \frac{-21}{20}$; $\csc\theta = \frac{29}{21}$; $\sec\theta = \frac{-29}{20}$; $\cot\theta = \frac{-20}{21}$

38. $y = 5$, $r = 13$, $x < 0$, $x^2 = r^2 - y^2 = 13^2 - 5^2 = 169 - 25 = 144$, $x = \pm\sqrt{144} = \pm 12$ so $x = -12$, $\sin\theta = \frac{5}{13}$; $\cos\theta = \frac{-12}{13}$; $\tan\theta = \frac{-5}{12}$; $\csc\theta = \frac{13}{5}$; $\sec\theta = \frac{-13}{12}$; $\cot\theta = \frac{-12}{5}$

39. $x = -7$, $r = 8$, $y < 0$, $y^2 = 8^2 - (-7)^2 = 64 - 49 = 15$, $y = \pm\sqrt{15}$ so $y = -\sqrt{15}$; $\sin\theta = -\frac{\sqrt{15}}{8}$; $\cos\theta = \frac{-7}{8}$; $\tan\theta = \frac{\sqrt{15}}{7}$; $\csc\theta = -\frac{8}{\sqrt{15}}$; $\sec\theta = \frac{-8}{7}$; $\cot\theta = \frac{7}{\sqrt{15}}$

40. $y = 5$, $r = 30$, $x < 0$, $x^2 = 30^2 - 5^2 = 900 - 25 = 875$, $x = \pm\sqrt{875} = \pm 5\sqrt{35}$ so $x = -5\sqrt{35}$, $\sin\theta = \frac{y}{r} = \frac{5}{30} = \frac{1}{6}$; $\cos\theta = \frac{x}{r} = \frac{-5\sqrt{35}}{30} = \frac{-\sqrt{35}}{6}$; $\tan\theta = \frac{y}{x} = \frac{5}{-5\sqrt{35}} = -\frac{1}{\sqrt{35}}$, $\csc\theta = \frac{r}{y} = \frac{6}{1} = 6$; $\sec\theta = \frac{r}{x} = \frac{30}{-5\sqrt{35}} = \frac{-6}{\sqrt{35}}$; $\cot\theta = \frac{x}{y} = \frac{-5\sqrt{35}}{5} = -\sqrt{35}$

9.2 VALUES OF THE TRIGONOMETRIC FUNCTIONS

1. $a = 5$, $c = 13$, $b = \sqrt{c^2 - a^2} = \sqrt{13^2 - 5^2} = \sqrt{169 - 25} = \sqrt{144} = 12$; $\sin\theta = \frac{5}{13}$; $\cos\theta = \frac{12}{13}$; $\tan\theta = \frac{5}{12}$; $\csc\theta = \frac{13}{5}$; $\sec\theta = \frac{13}{12}$; $\cot\theta = \frac{12}{5}$

2. $a = 7$, $b = 8$, $c = \sqrt{7^2 + 8^2} = \sqrt{49 + 64} = \sqrt{113}$; $\sin\theta = \frac{7}{\sqrt{113}}$; $\cos\theta = \frac{8}{\sqrt{113}}$; $\tan\theta = \frac{7}{3}$; $\csc\theta = \frac{\sqrt{113}}{7}$; $\sec\theta = \frac{\sqrt{113}}{8}$; $\cot\theta = \frac{8}{7}$

3. $a = 1.2$, $c = 2$, $b = \sqrt{2^2 - 1.2^2} = \sqrt{4 - 1.44} = \sqrt{2.56} = 1.6$, $\sin\theta = \frac{1.2}{2} = \frac{12}{20} = \frac{3}{5}$, $\cos\theta = \frac{1.6}{2} = \frac{16}{20} = \frac{4}{5}$, $\tan\theta = \frac{3}{4}$; $\csc\theta = \frac{5}{3}$; $\sec\theta = \frac{5}{4}$; $\cot\theta = \frac{4}{3}$

4. $a = 2.1$, $b = 2.8$, $c = \sqrt{2.1^2 + 2.8^2} = \sqrt{4.41 + 7.84} = \sqrt{12.25} = 3.5$; $\sin\theta = \frac{3}{5}$; $\cos\theta = \frac{4}{5}$; $\tan\theta = \frac{3}{4}$; $\csc\theta = \frac{5}{3}$; $\sec\theta = \frac{5}{4}$; $\cot\theta = \frac{4}{3}$

5. $a = 1.4$, $b = 2.3$, $c = \sqrt{1.4^2 + 2.3^2} = \sqrt{1.96 + 5.29} = \sqrt{7.25}$; $\sin\theta = \frac{1.4}{\sqrt{7.25}}$; $\cos\theta = \frac{2.3}{\sqrt{7.25}}$; $\tan\theta = \frac{14}{23}$; $\csc\theta = \frac{\sqrt{7.25}}{1.4}$; $\sec\theta = \frac{\sqrt{7.25}}{2.3}$; $\cot\theta = \frac{23}{14}$

6. $c = 3.5$, $b = 2.1$; $a = \sqrt{3.5^2 - 2.1^2} = \sqrt{12.25 - 4.41} = \sqrt{7.84} = 2.8$; $\sin\theta = \frac{4}{5}$; $\cos\theta = \frac{3}{5}$; $\tan\theta = \frac{4}{3}$; $\csc\theta = \frac{5}{4}$; $\sec\theta = \frac{5}{3}$; $\cot\theta = \frac{3}{4}$

7. Since $\cos\theta = \frac{8}{17}$, the side adjacent is 8 and the hypotenuse is 17, and so the side opposite is 15. $\sin\theta = \frac{\text{opp}}{\text{hyp}} = \frac{15}{17}$; $\tan\theta = \frac{\text{opp}}{\text{adj}} = \frac{15}{8}$, $\csc\theta = \frac{\text{hyp}}{\text{opp}} = \frac{17}{15}$, $\sec\theta = \frac{\text{hyp}}{\text{adj}} = \frac{17}{8}$, $\cot\theta = \frac{\text{adj}}{\text{opp}} = \frac{8}{15}$

8. Since $\tan\theta = \frac{21}{20}$, the side opposite is 21 and the side adjacent is 20, so the hypotenuse is 29. $\sin\theta = \frac{21}{29}$; $\cos\theta = \frac{20}{29}$; $\tan\theta = \frac{21}{20}$; $\csc\theta = \frac{29}{21}$; $\sec\theta = \frac{29}{20}$; $\cot\theta = \frac{20}{21}$

9. Since $\sin\theta = \frac{3}{5}$, the side opposite is 3 and the hypotenuse is 5, so side adjacent is 4. $\sin\theta = \frac{3}{5}$; $\cos\theta = \frac{4}{5}$; $\tan\theta = \frac{3}{4}$; $\csc\theta = \frac{5}{3}$; $\sec\theta = \frac{5}{4}$; $\cot\theta = \frac{4}{3}$

10. Since $\sec\theta = \frac{10}{6}$, the hypotenuse is 10 and the side adjacent is 6, so the side opposite is 8. $\sin\theta = \frac{8}{10} = \frac{4}{5}$; $\cos\theta = \frac{6}{10} = \frac{3}{5}$; $\tan\theta = \frac{8}{6} = \frac{4}{3}$; $\csc\theta = \frac{10}{8} = \frac{5}{4}$; $\sec\theta = \frac{10}{6} = \frac{5}{3}$; $\cot\theta = \frac{6}{8} = \frac{3}{4}$

11. Here $\csc\theta = \frac{29}{20}$, so the hypotenuse is 29 and the side opposite is 20. Thus, the side adjacent is 21. $\sin\theta = \frac{20}{29}$; $\cos\theta = \frac{21}{29}$; $\tan\theta = \frac{20}{21}$; $\csc\theta = \frac{29}{20}$; $\sec\theta = \frac{29}{21}$; $\cot\theta = \frac{21}{20}$

12. Since $\csc\theta = \frac{15}{12}$, the hypotenuse is 15 and the side opposite is 12, and so the side adjacent is 9. $\sin\theta = \frac{12}{15} = \frac{4}{5}$; $\cos\theta = \frac{9}{15} = \frac{3}{5}$; $\tan\theta = \frac{12}{9} = \frac{4}{3}$; $\csc\theta = \frac{15}{12} = \frac{5}{4}$; $\sec\theta = \frac{15}{9} = \frac{5}{3}$; $\cot\theta = \frac{9}{12} = \frac{3}{4}$

13. $\sin\theta = 0.866$, $\cos\theta = 0.5$, $\tan\theta = \frac{0.866}{0.5} = 1.732$, $\csc\theta = \frac{1}{0.866} \approx 1.155$, $\sec\theta = 2$, $\cot\theta = \frac{0.5}{0.866} \approx 0.577$

14. $\sin\theta = 0.975$, $\cos\theta = 0.222$, $\tan\theta = \frac{0.975}{0.222} \approx 4.392$, $\csc\theta = \frac{1}{0.975} \approx 1.026$, $\sec\theta = \frac{1}{0.222} \approx 4.505$, $\cot\theta = \frac{0.222}{0.975} \approx 0.228$

15. $\sin\theta = 0.085$, $\sec\theta = 1.004$, $\cos\theta = \frac{1}{1.004} \approx$
0.996, $\tan\theta \approx \frac{0.085}{0.996} \approx 0.085$, $\csc\theta = \frac{1}{0.085} \approx$
11.765, $\cot \approx \frac{0.996}{0.085} \approx 11.718$

16. $\csc\theta = 4.872$, $\cos\theta = 0.979$, $\sin\theta = \frac{1}{\csc\theta} =$
$\frac{1}{4.872} \approx 0.205$, $\tan\theta \approx \frac{0.205}{0.979} \approx 0.209$; $\cot\theta \approx \frac{0.979}{0.205}$
≈ 4.776; $\sec\theta = \frac{1}{\cos\theta} = \frac{1}{0.979} \approx 1.021$

17. $\sin 18.6° = 0.3189593$

18. $\cos 38.4° = 0.7836935$

19. $\tan 18.3° = 0.3307184$

20. $\sin 20°15' = \sin 20.25° = 0.3461171$

21. $\tan 76°32' = \tan 76\frac{32}{60}° = 4.1760011$

22. $\sec 24°14' = \dfrac{1}{\cos\left(24 + \frac{14}{60}\right)} = 1.0966337$

23. $\cot 82.6° = \frac{1}{\tan 82.6°} = 0.1298773$

24. $\csc 19°50' = \dfrac{1}{\sin\left(19 + \frac{50}{60}\right)} = 2.9473725$

25. $\sin 0.25\text{ rad} = 0.247404$

26. $\cos 0.4\text{ rad} = 0.921061$

27. $\tan 0.63\text{ rad} = 0.7291147$

28. $\sec 1.35\text{ rad} = \frac{1}{\cos 1.35} = 4.5660706$

29. $\cot 1.43\text{ rad} = \frac{1}{\tan 1.43} = 0.1417341$

30. $\sin 1.21\text{ rad} = 0.935616$

31. $\csc 0.21\text{ rad} = \frac{1}{\sin 0.21} = 4.7970857$

32. $\tan 1.555\text{ rad} \approx 63.300591$

33. $R = \frac{15^2}{9.8}\sin(2 \cdot 40) \approx 22.61$ m

34. If the angle is 45°, then $R = \frac{15^2}{9.8}\sin(2 \times 45°) \approx$
22.96 and the football would travel $22.96 - 22.61 =$
0.35 m farther than if it were thrown at an angle
of 40°. If the angle is 48°, then $R = \frac{15^2}{9.8}\sin(2 \cdot$
$48°) \approx 22.83$ and the football would travel $22.83 -$
$22.61 = 0.22$ m farther than at an angle of 40°.

35. $V_R = 5.8\cos 32° \approx 4.92$ V

36. $V_R = 11.25\cos 71° \approx 3.66$ V

37. See *Computer Programs* in main text.

38. See *Computer Programs* in main text.

≡ 9.3 THE RIGHT TRIANGLE

1. $\sin A = 0.732$, $A = 47.05°$

2. $\cos B = 0.285$, $B = 73.44°$

3. $\tan C = 4.671$, $C = 77.92°$

4. $\sin D = 0.049$, $D = 2.81°$

5. $\cos E = 0.839$, $E = 32.97°$

6. $\tan F = 0.539$, $F = 28.32°$

7. $\sec G = 3.421$, $G = 73.00°$

8. $\csc H = 1.924$, $H = 31.32°$

9. $\cot I = 0.539$, $I = 61.68°$

10. $\sec J = 1.734$, $J = 54.78°$

11. $\csc K = 4.761$, $K = 12.12°$

12. $\cot L = 4.021$, $L = 13.97°$

13. $B = 90 - 16.5 = 73.5$; $\sin A = \frac{a}{c}$, $c = \frac{a}{\sin A} =$
$\frac{7.3}{\sin 16.5} \approx 25.7$, $\tan A = \frac{a}{b}$, $b = \frac{a}{\tan A} = \frac{7.3}{\tan 16.5} \approx$
24.6. In summary, we have

$a = 7.3 \quad A = 16.5°$
$b \approx 24.6 \quad B = 73.5°$
$c \approx 25.7 \quad C = 90°.$

14. $A = 90° - B = 90° - 53° = 37°$; $\tan B = \frac{b}{a}$ and
so $a = \frac{b}{\tan B} = \frac{9.1}{\tan 53°} \approx 6.9$; $\sin B = \frac{b}{c}$ and so,
$c = \frac{b}{\sin B} = \frac{9.1}{\sin 53°} \approx 11.4$. In summary, we have

$a \approx 6.9 \quad A = 37°$
$b = 9.1 \quad B = 53°$
$c \approx 11.4 \quad C = 90°.$

15. $B = 90 - 72.6 = 17.4$, $\sin A = \frac{a}{c}$, $a = c\sin A =$
$20\sin 72.6 \approx 19.1$, $\cos A = \frac{b}{c}$, $b = 20\cos 72.6 \approx$
6.0. In summary, we have

$a \approx 19.1 \quad A = 72.6°$
$b \approx 6.0 \quad B = 17.4°$
$c = 20.0 \quad C = 90°.$

16. $\cos B = \frac{a}{c}$ and so $c = \frac{19.4}{\cos 12.7} \approx 19.9$. $A =$
$90° - 12.7° = 77.3°$; $\tan B = \frac{b}{a}$ which means that
$b = 19.4\tan 12.7 \approx 4.4$. In summary, we have

$a = 19.4 \quad A = 77.3°$
$b \approx 4.4 \quad B = 12.7°$
$c \approx 19.9 \quad C = 90°.$

17. $B = 90° - 43° = 47°$ and $\cos A = \frac{b}{c}$ which means that $c = \frac{34.6}{\cos 43} \approx 47.3$. Finally, $\tan A = \frac{a}{b}$, and so $a = b \tan A = 34.6 \tan 43 = 32.3$. Summarizing, we have

$a \approx 32.3 \quad A = 43°$
$b = 34.6 \quad B = 47°$
$c \approx 47.3 \quad C = 90°.$

18. We first find that $A = 90° - 67° = 23°$. Next, we see that $\sin B = \frac{b}{c}$ and so $b = 32.4 \sin 67° \approx 29.8$. Finally, we have $\cos B = \frac{a}{c}$, which means that $a = 32.4 \cos 67 \approx 12.7$. Summarizing, we have the following results:
$a \approx 12.7 \quad A = 23°$
$b \approx 29.8 \quad B = 67°$
$c = 32.4 \quad C = 90°.$

19. We begin by determining that $B = \frac{\pi}{2} - 0.92 \approx 0.65$ rad. Next, we determine that $\sin A = \frac{a}{c}$ and so $c = \frac{a}{\sin A} = \frac{6.5}{\sin 0.92} \approx 8.2$. Finally, we find that $\tan A = \frac{a}{b}$ and so $b = \frac{6.5}{\tan 0.92} \approx 4.9$. In summary, we have the following results:

$a = 6.5 \quad A = 0.92$ rad
$b \approx 4.9 \quad B \approx 0.65$ rad

$c \approx 8.2 \quad C = \frac{\pi}{2}$ rad

20. Using the given information, we see that $A = C - B = \frac{\pi}{2} - 1.13 \approx 0.44$ rad. Then $\tan B = \frac{b}{a}$ and so $a = \frac{24}{\tan 1.13} \approx 11.3$. Finally, $\sin B = \frac{b}{c}$ leads to $c = \frac{24}{\sin 1.13} \approx 26.5$. In summary, we have the following results:
$a \approx 11.3 \quad A \approx 0.44$ rad
$b = 24.0 \quad B = 1.13$ rad

$c \approx 26.5 \quad C = \frac{\pi}{2}$ rad

21. We first determine that $B = \frac{\pi}{2} - 0.15 \approx 1.42$ rad. Next, we find that $\sin A = \frac{a}{c}$ and so $a = c \sin A = 18 \sin 0.15 \approx 2.7$. Finally, we have $\cos A = \frac{b}{c}$ which produces $b = 18 \cos 0.15 \approx 17.8$. Summarizing, we have:

$a \approx 2.7 \quad A = 0.15$ rad
$b \approx 17.8 \quad B \approx 1.42$ rad

$c = 18.0 \quad C = \frac{\pi}{2}$ rad

22. We begin by finding that $A = \frac{\pi}{2} - 0.23 \approx 1.34$ rad. Then, $\cos B = \frac{a}{c}$ and so $c = \frac{a}{\cos B} = \frac{9.7}{\cos .23} \approx 10.0$. Third, we find that $\tan B = \frac{b}{a}$ and as a result, $b = a \tan B = 9.7 \tan 0.23 \approx 2.3$. In summary, we have the following results:

$a = 9.7 \quad A \approx 1.34$ rad
$b \approx 2.3 \quad B = 0.23$ rad

$c \approx 10.0 \quad C = \frac{\pi}{2}$ rad

23. Using the given information, we find that $B \approx 1.57 - 1.41 \approx 0.16$ rad. Next, we see that $\tan A = \frac{a}{b}$ produces $a = 40 \tan 1.41 \approx 246.6$. Finally, we have $\cos A = \frac{b}{c}$, and so $c = \frac{40}{\cos 1.41} \approx 249.8$. In summary, we have the following results:

$a \approx 246.6 \quad A = 1.41$ rad
$b = 40.0 \quad B \approx 0.16$ rad

$c \approx 249.8 \quad C = \frac{\pi}{2}$ rad

24. We begin by determing that $B \approx 1.57 - 1.15 \approx 0.42$ rad. Using the given information, we can use $\sin A = \frac{a}{c}$ to yield $a = 18 \sin 1.15 \approx 16.4$. Finally, we use $\cos A = \frac{b}{c}$ to find $b = 18 \cos 1.15 \approx 7.4$. Summarizing, we see that:

$a \approx 16.4 \quad A = 1.15$ rad
$b \approx 7.4 \quad B \approx 0.42$ rad

$c = 18.0 \quad C = \frac{\pi}{2}$ rad

25. $a = 9$, $b = 15$, $c = \sqrt{9^2 + 15^2} \approx 17.5$, $\tan A = \frac{a}{b} = \frac{9}{15}$, $A = \tan^{-1} \frac{9}{15} = 31.0°$ or 0.54 rad, $B \approx 90° - 31° = 59°$ or $B \approx 1.57 - 0.54 = 1.03$ rad

26. $a = 19.3$, $c = 24.4$; $b = \sqrt{24.4^2 - 19.3^2} \approx 14.9$, $\sin A = \frac{a}{c} = \frac{19.3}{24.4}$, $A \approx 52.3°$ or 0.91 rad; $B \approx 90° - 52.3° = 37.7°$ or $B = 1.57 - 0.91 = 0.66$ rad

27. $b = 9.3$, $c = 18$, $a = \sqrt{18^2 - 9.3^2} \approx 15.4$, $\cos A = \frac{b}{c} = \frac{9.3}{18}$, $A \approx 58.9°$ or 1.03 rad, $B \approx 90° - 58.9° = 31.1°$ or $1.57 - 1.03 = 0.54$ rad

28. $a = 14$, $b = 9.3$, $c = \sqrt{14^2 + 9.3^2} \approx 16.8$, $\tan A = \frac{14}{9.3}$, $A \approx 56.4°$ or 0.98 rad, $B \approx 90° - 56.4° = 33.6°$ or $B = 1.57 - 0.98 = 0.59$ rad

29. $a = 20$, $c = 30$, $b = \sqrt{30^2 - 20^2} \approx 22.4$, $\sin A = \frac{a}{c} = \frac{20}{30}$, $A \approx 41.8°$ or 0.73 rad, $B \approx 90° - 41.8° = 48.2°$ or $B \approx 1.57 - 0.73 = 0.84$ rad

30. $b = 15$, $c = 25$, $a = \sqrt{25^2 - 15^2} = 20$, $\cos A = \frac{b}{c} = \frac{15}{25}$, $A \approx 53.1°$ or 0.93 rad, $B = 90° - 53.1° = 36.9°$ or $B \approx 1.57 - 0.93 = 0.64$ rad

31. $\tan \phi = \frac{12.3}{19.7}$, $\phi = \tan^{-1} \frac{12.3}{19.7} \approx 31.9792$ or about $32°$

32. $\tan 23.2° = \frac{222}{x}$, $x = \frac{222}{\tan 23.2°} \approx 518$ ft

33. $\sin 37.6° = \frac{700}{d}$, $d = \frac{700}{\sin 37.6°} \approx 1147$ m

34. $\tan 57.62 = \frac{d}{95}$, $d = 95 \tan 57.62 \approx 149.8$ m

35. $F_x = 10 \cos 30° \approx 8.7$ N, $F_y = 10 \sin 30° = 5$ N

36. $c = \sqrt{20^2 + 12^2} \approx 23.3$ lb, $\tan \phi = \frac{20}{12}$, $\phi = \tan^{-1} \frac{20}{12} \approx 59.0°$

37. $\tan \phi = \frac{(80.67)^2}{32 \cdot 1200} \approx 0.1695$, $\phi = \tan^{-1}(0.1695) \approx 9.62°$

38. $\tan \phi = \frac{24.44^2}{9.8 \times 500} \approx 0.1219$, $\phi = \tan^{-1} 0.1219 \approx 6.95°$

39. $\cos 67° = \frac{x}{300}$, $x = 3.00 \cos 67° \approx 117.2$ m, $\sin 67° = \frac{d}{300}$ and so $d = 300 \sin 67° \approx 276.2$ m. So 117.2 m from intersection, 276.2 m from service station.

40. $\tan 58.6° = \frac{h}{950}$, so $h = 950 \tan 58.6° \approx 1,556.3$ ft

41. $\tan 67.3° = \frac{h}{25.75}$, so $h = 25.75 \tan 67.3° \approx 61.56$ ft

42. **(a)** Let d be the distance from the base of the ladder to the base of the building. Then $\cos 65.0° = \frac{d}{32}$ and $d = 32 \cos 65.0° \approx 13.5$ m
(b) No, the firefighter can reach only $2.7 + 32 \sin 65.0 + 1.92 \approx 33.62$ m. If we assume that the firefighter can reach another 1 m, then he or she will still be $35.85 - (33.62 + 1) = 1.23$ m too short.

≡ 9.4 TRIGONOMETRIC FUNCTIONS OF ANY ANGLE

1. $87°$ is in Quadrant I, so $\theta_{ref} = 87°$

2. $200°$ is in Quadrant III, so $\theta_{ref} = 200° - 180° = 20°$

3. $137°$ is in Quadrant II, so $\theta_{ref} = 180° - 137° = 43°$

4. $298°$ is in Quadrant IV, so $\theta_{ref} = 360° - 298° = 62°$

5. $\frac{9\pi}{8}$ rad is in Quadrant III, so $\theta_{ref} = \frac{9\pi}{8} - \pi = \frac{9\pi}{8} - \frac{8\pi}{8} = \frac{\pi}{8}$

6. 2.1 rad is in Quadrant II, so $\theta_{ref} = \pi - 2.1 \approx 3.14 - 2.1 = 1.04$

7. 4.5 rad is in Quadrant III, so $\theta_{ref} = 4.5 - \pi \approx 4.5 - 3.14 = 1.36$

8. 5.85 rad is in Quadrant IV, so $\theta_{ref} = 2\pi - 5.85 \approx 6.28 - 5.85 = 0.43$

9. $518°$; $518° - 360° = 158°$

10. $1673°$; $1673° - 4(360°) = 1673° - 1440° = 233°$

11. $-871°$; $3 \times 360° - 871° = 1080° - 871° = 209°$

12. $-137°$; $360° - 137° = 223°$

13. 7.3 rad; $7.3 - 2\pi \approx 7.3 - 6.28 = 1.02$

14. $\frac{16\pi}{4}$ rad; $\frac{16\pi}{3} - 4\pi = \frac{16\pi}{3} - \frac{12\pi}{3} = \frac{4\pi}{3}$

15. -2.17 rad; $-2.17 + 2\pi \approx -2.17 + 6.28 = 4.11$

16. -8.43 rad; $-8.43 + 4\pi \approx -8.43 + 12.56 = 4.13$

17. $165°$; Quadrant II

18. $285°$; Quadrant IV

19. $-47°$; Quadrant IV

20. $312°$; Quadrant IV

21. $250°$; Quadrant III

22. $197°$; Quadrant III

23. $98°$; Quadrant II

24. $\sin \theta$ is positive in Quadrants I and II.

25. $\tan \theta$ is negative in Quadrants II and IV.

26. $\cos \theta$ is positive in Quadrants I and IV.

27. $\sec \theta$ is negative in Quadrants II and III.

28. $\cot \theta$ is negative in Quadrants II and IV.

29. $\csc \theta$ is negative and $\cos \theta$ is positive in Quadrant IV.

30. $\tan \theta$ and $\sec \theta$ are negative in Quadrant II.

31. $\sin 105°$ is in Quadrant II, so the expression is positive.

32. $\sec 237°$ is in Quadrant III, so the expression is negative.

33. $\tan 372°$ is in Quadrant I, so the expression is positive.

34. $\cos(-53°)$ is in Quadrant IV, so the expression is positive.

35. $\cos 1.93$ rad is in Quadrant II, so the expression is negative.

36. $\cot 4.63$ rad is in Quadrant III, so the expression is positive.

37. $\sin 215°$ is in Quadrant III, so the expression is negative.

38. $\csc 5.42$ rad IVQ $=$ negative

39. $\sin 137° = 0.6820$

40. $\cos 263° = -0.1219$

41. $\tan 293° = -2.3559$

42. $\sin 312° = -0.7431$

43. $\tan 164.2° = -0.2830$

44. $\csc 197.3° = -3.3628$

45. $\sin 2.4$ rad $= 0.6755$

46. $\cos 1.93$ rad $= -0.3515$

47. $\tan 6.1$ rad $= -0.1853$

48. $\sec 4.32$ rad $= -2.6151$

49. $\tan 1.37$ rad $= 4.9131$

50. $\sin 3.2$ rad $= -0.0584$

51. $\sin 415.5° = 0.8241$

52. $\tan 512.1° = -0.5295$

53. $\cot -87.4° = -0.0454$

54. $\cos 372.1° = 0.9778$

55. $\csc 432.4° = -1.0491$

56. $\cos 357.3° = 0.9989$

57. $\sin 6.5$ rad $= 0.2151$

58. $\sec -4.3$ rad $= -2.4950$

59. $\tan 8.35$ rad $= -1.8479$

60. $\sin 9.42$ rad $= 0.0048$

61. $\cos -9.43$ rad $= 0.9090$

62. $\sec 9.34$ rad $= -1.0036$

63. $I = 32.65 \sin 132° \approx 24.26$ mA

64. $V = 115.2 \sin 113.4° \approx 105.73$ V

65. The unlabeled angle at B is $180° - 123° - 20° = 37°$. Thus, $AB = \frac{96}{\sin 37°} \approx 159.52$; $BC = \frac{264}{\cos 20°} \approx 280.94$. The total length is $159.52 + 280.94 = 440.46$ mm.

66. Note that the drawing is not to scale. First, we compute the extra length of the base to be $308.2 - 166.0 - 142.2$ cm. Half of this is 71.1 cm. Then, $\theta = 90° + \tan^{-1}\left(\frac{71.1}{25.5}\right) = 90° + 70.3° = 160.3°$.

67. Note that the drawing is not to scale. The ground that needs to be sodded is the shape of a trapezoid.

To find h, the height of the trapezoid, we divide the trapezoid into a rectangle and a right triangle. The leg of the right triangle that joins the base of the trapezoid is h. The other leg is $223 - 156 = 67$ ft. The angle opposite h is $19.7°$, so $h = 67 \tan 19.7° \approx 23.99$ ft. Thus, the area of the trapezoid is $A = \frac{1}{2} h (b_1 + b_2) = \frac{1}{2}(23.99)(223 + 156) = 4{,}546$ ft^2.

68. From B, draw perpendiculars to the two opposite sides. The horizontal segment has length $2.17 \sin 75° = 2.10$ in. The bottom of the right triangle with hypotenuse AB has length $2.49 - 2.10 = 0.39$. The length of $AB = \frac{0.39}{\cos 70°} \approx 1.14$ in.

9.5 INVERSE TRIGNOMETRIC FUNCTIONS

1. $\sin \theta = \frac{1}{2}$ in Quadrants I and II.

2. $\tan \theta = \frac{3}{4}$ in Quadrants I and III.

3. $\cot \theta = -2$ in Quadrants II and IV.

4. $\cos \theta = -0.25$ in Quadrants II and III.

5. $\sec \theta = 4.3$ in Quadrants I and IV.

6. $\cos \theta = 0.8$ in Quadrants I and IV.

7. $\csc \theta = -6.1$ in Quadrants III and IV.

8. $\sin \theta = -\frac{\sqrt{3}}{2}$ in Quadrants III and IV.

9. $\sin \theta = \frac{1}{2}$; $\theta_r = \sin^{-1} 0.5 = 30.0°$; $\theta_I = 30.0°$, $\theta_{II} = 180° - 30° = 150°$

10. $\tan \theta = \frac{3}{4}$; $\theta_r = \tan^{-1} 0.75 = 36.9°$; $\theta_I = 36.9°$; $\theta_{III} = 180° + \theta_r = 180° + 36.9° = 216.9°$

11. $\cot \theta = -2$; $\theta_r = \cot^{-1} 2 = \tan^{-1} \frac{1}{2} = 26.6°$; $\theta_{II} = 180° - \theta_r = 153.4°$; $\theta_{IV} = 360° - \theta_r = 333.4°$

12. $\cos \theta = -0.25$; $\theta_r = \cos^{-1} 0.25 = 75.5°$; $\theta_{II} = 180° - \theta_r = 104.5°$; $\theta_{III} = 180° + \theta_r = 255.5°$

13. $\sec\theta = 4.3$; $\theta_r = \sec^{-1} 4.3 = \cos^{-1}\frac{1}{4.3} = 76.6°$; $\theta_{\mathrm{I}} = \theta_r = 76.6°$; $\theta_{\mathrm{IV}} = 360 - \theta_r = 283.4°$

14. $\cos\theta = 0.8$; $\theta_r = \cos^{-1} 0.8 = 36.9°$; $\theta_{\mathrm{I}} = \theta_r = 36.9°$; $\theta_{\mathrm{IV}} = 360 - \theta_r = 323.1°$

15. $\csc\theta = -6.1$; $\theta_r = \csc^{-1} 6.1 = \sin^{-1}\frac{1}{6.1} = 9.4°$; $\theta_{\mathrm{III}} = 180° + \theta_r = 189.4°$; $\theta_{\mathrm{IV}} = 360° - \theta_r = 350.6°$

16. $\sin\theta = -\frac{\sqrt{3}}{2}$; $\theta_r = \sin^{-1}\frac{\sqrt{3}}{2} = 60°$ exactly; $\theta_{\mathrm{III}} = 180° + \theta_r = 240°$; $\theta_{\mathrm{IV}} = 360° - \theta_r = 300°$

17. $\sin\theta = 0.75$; $\theta_r = \sin^{-1} 0.75 = 0.85$; $\theta_{\mathrm{I}} = \theta_r = 0.85$; $\theta_{\mathrm{II}} = \pi - \theta_r = 2.29$

18. $\tan\theta = 1.6$; $\theta_r = \tan^{-1} 1.6 = 1.01$; $\theta_{\mathrm{I}} = \theta_r = 1.01$; $\theta_{\mathrm{III}} = \pi + \theta_r = 4.15$

19. $\cot\theta = -0.4$; $\theta_r = \cot^{-1} 0.4 = \tan^{-1}\frac{1}{0.4} = 1.19$; $\theta_{\mathrm{II}} = \pi - \theta_r = 1.95$; $\theta_{\mathrm{IV}} = 2\pi - \theta_r = 5.09$

20. $\cos\theta = 0.25$; $\theta_r = \cos^{-1} 0.25 = 1.32$; $\theta_{\mathrm{I}} = \theta_r = 1.32$; $\theta_{\mathrm{IV}} = 2\pi - \theta_r = 4.96$

21. $\csc\theta = 4.3$; $\theta_r = \csc^{-1} 4.3 = \sin^{-1}\frac{1}{4.3} = 0.23$; $\theta_{\mathrm{I}} = \theta_r = 0.23$; $\theta_{\mathrm{II}} = \pi - \theta_r = 2.91$

22. $\sin\theta = -0.08$; $\theta_r = \sin^{-1} 0.08 = 0.08$; $\theta_{\mathrm{III}} = \pi + \theta_r = 3.22$; $\theta_{\mathrm{IV}} = 2\pi - \theta_r = 6.20$

23. $\sec\theta = 2.7$; $\theta_r = \sec^{-1} 2.7 = \cos^{-1}\frac{1}{2.7} = 1.19$; $\theta_{\mathrm{I}} = \theta_r = 1.19$; $\theta_{\mathrm{IV}} = 2\pi - \theta_r = 5.09$

24. $\cos\theta = -0.95$; $\theta_r = \cos^{-1} 0.95 = 0.32$; $\theta_{\mathrm{II}} = \pi - \theta_r = 2.82$; $\theta_{\mathrm{III}} = \pi + \theta_r = 3.46$

25. $\arcsin 0.84 = 57.1°$

26. $\arccos(-0.21) = 102.1°$

27. $\arctan 4.21 = 76.6°$

28. $\arccos(-0.25) = \arctan\left(\frac{1}{-0.25}\right) = -76.0°$ but range of arccot is $0°$ to $180°$ so $180° - 76° = 104°$ is the correct answer.

29. $\sin^{-1} 0.32 = 18.7°$

30. $\cos^{-1} 0.47 = 62.0°$

31. $\tan^{-1}(-0.64) = -32.6°$

32. $\csc^{-1}(-3.61) = -16.1°$

33. $\arccos(-0.33) = 1.91$

34. $\arctan 1.55 = 1.00$

35. $\arccos 0.29 = 1.28$

36. $\operatorname{arcsec}(-3.15) = \arccos\left(\frac{1}{-3.15}\right) = 1.89$

37. $\sin^{-1} 0.95 = 1.25$

38. $\cos^{-1}(-0.67) = 2.31$

39. $\tan^{-1} 0.25 = 0.24$

40. $\cot^{-1}(-0.75) = \tan^{-1}\frac{1}{-0.75} = -0.93$ but range of $\cot^{-1}$ is 0 to π so correct answer is $\pi - 0.93 = 2.21$

41. $\theta = \tan^{-1}\left(-\frac{33}{40}\right) \approx -0.6898$ rad $\approx -39.52°$

42. $\theta = \cos^{-1}\left(\frac{30}{50}\right) \approx 53.1° \approx 0.927$ rad

43. $\theta = \sin^{-1}\left(\frac{4.635}{6.725}\right) \approx 43.57°$

44. (a) Since $h = \frac{v_0^2 \sin^2\theta}{2g}$, then $\sin^2\theta = \frac{2gh}{v_0^2}$, so $\sin\theta = \frac{\sqrt{2gh}}{v_0}$ and $\theta = \sin^{-1}\frac{\sqrt{2gh}}{v_0}$.

(b) Substituting the given values, we obtain $\theta = \sin^{-1}\frac{\sqrt{2\times 9.8\times 431.1}}{172.5} \approx 32.2°$

45. See *Computer Programs* in main text.

46. See *Computer Programs* in main text.

47. See *Computer Programs* in main text.

48. See *Computer Programs* in main text.

▤ 9.6 APPLICATIONS OF TRIGONOMETRY

1. $47° = 47 \times \frac{\pi}{180}$ rad ≈ 0.8203 rad; $s = r\theta = 1050.250 \times 0.8203 \approx 861.520$ m

2. $v = r\omega = 8 \cdot 5 = 40$ in./s

3. 30 rpm $= 30 \cdot 2\pi$ rad/min ≈ 188.5 rad/min. $v = r\omega = 188.5 \times 12 = 2262$ in./min $= 2262$ in/min $\left(\frac{1\,\min}{60\,\text{s}}\right) = 37.7$ in/s

4. $v = \frac{4}{2} = 2$m/s; $v = r\omega$; $\omega = \frac{v}{r} = \frac{2\text{m/s}}{0.18\text{m}} \approx 11.11$ rad/s

5. $s = r\theta = 15 \times \frac{5\pi}{8} \approx 29.45$ cm

6. $85° = \frac{85\pi}{180} \approx 1.48353$ rad; $s = r\theta = 235 \times 1.48353 \approx 348.63$ mm

7. $A = \frac{1}{2}r^2\theta = \frac{1}{2}(15)^2\left(\frac{5\pi}{8}\right) \approx 220.89$ cm^2

8. $A = \frac{1}{2}r^2\theta = \frac{1}{2}(235)^2 \times 1.483\,53 \approx 40\,963.97$ mm^2

9. $\theta = 2 \times 0.07 = 0.14$; $s = r\theta = 1.2 \times 0.14 = 0.168$ m

10. $\omega = 1\,\text{rev}/24\,\text{h} = 0.2617994$ rad/h $= 15°/\text{h}$; $v = r\omega = (22{,}300 + 3963) \times (0.2617994) \approx 6876$ mph

11. (a) $\omega = 45\frac{\text{rev}}{\text{min}} \times \frac{2\pi}{\text{rev}} \approx 282.74\frac{\text{rad}}{\text{min}}$ or 282.74 $\frac{\text{rad}}{\text{min}}\frac{1\,\text{min}}{60\,\text{s}} \approx 4.712$ rad/s; $r = 175.26/2 = 87.63$ mm; $v = r\omega = 87.63 \times 282.74 \approx 24\,776.5$ mm/min or $v = 87.63 \times 4.712 \approx 412.9$ mm/s;
(b) 24 776.5 mm

12. $\alpha = \frac{\omega_f - \omega_0}{t} = \frac{0 - 4.2}{11.92} \approx -0.352$ rad/s^2

13. (a) We use the given information to determine that $\cos D = (\sin 35°30')(\sin 40°40') + (\cos 35°30')$ $(\cos 40°40')\big(\cos[-10°10']\big)$ and so $\cos D \approx 0.98624$ and so the angular distance between the cities is $D \approx 9.515\,478°$.
(b) To find the linear distance s between the cities, we use $s = r\theta$ with $\theta = 9.515\,478°\big(\frac{\pi}{180°}\big) \approx 0.1660764$ rad. Using $r = 3960$ mi, we get $s = 3960(0.1660764) \approx 657.66$ mi.

14. $r = \frac{d}{2} = \frac{1}{2}$ ft; $v = r\omega = \frac{1}{2}$ ft $\cdot 1200\frac{\text{rev}}{\text{min}} \cdot \frac{2\pi}{\text{rev}} = 3769.9$ ft/min

15. $\frac{8}{1} \times \frac{\pi}{3} \times 2.5 = 83.77$. Since you cannot have a fraction of a byte, the answer is 83.

16. $C = 2\pi r = 2\pi \cdot 2.5 = 15.708$; $l = 15.708 \div 16 = 0.9818$ in.

17. $850\frac{\text{rev}}{\text{min}} \cdot \frac{360°}{1\,\text{rev}} \cdot \frac{1\,\text{min}}{60\,\text{s}} = 5100°/\text{s}$

18. $r = 7.5$ in.; $v = r\omega = 7.5$ in. $\times 850\frac{\text{rev}}{\text{min}} \cdot \frac{2\pi}{\text{rev}} \cdot \frac{1\,\text{min}}{60\,\text{s}} \approx 667.59$ in./s

19. $v = r\omega = 4.2$ m $\cdot \frac{20\,\text{rev}}{\text{min}} \cdot \frac{2\pi}{\text{rev}} \cdot \frac{1\,\text{min}}{60\,\text{s}} \approx 8.80$ m/s

20. $\sin 23° = \frac{F_x}{F}$; $F_x = 2500 \sin 23° \approx 977$ N; $F_y = F \cos 23° \approx 2301$ N

21. $\theta = 66.5° \times \frac{\pi}{180°} = 1.160\,644$ rad, $s = r\theta = 6370 \times 1.1606 = 7393.3$ km

22. $\theta = \tan^{-1}\big(\frac{n_b}{n_a}\big) = \tan^{-1}\big(\frac{1.516}{1.00}\big) \approx 56.6°$.

23. $\cos^2 \theta = \frac{I}{I_m} = \frac{3}{4}$, so $\cos \theta = \sqrt{\frac{3}{4}}$, and so $\theta = \cos^{-1}\sqrt{\frac{3}{4}} = 30°$.

24. The length of $AB = CD = 9.2 \cos 23° = 8.47$ m. The covered area is $8.47 \times 14.5 = 122.8$ m^2.

25. (a) 12 hr $= 12 \cdot 60$ min $= 12 \cdot 60 \cdot 60$ s $= 43200$ s. In 12 hr, a satellite completes 1 revolution, amounting to 2π radians. Its angular speed is thus $\omega = \frac{\theta}{t} = \frac{2\pi}{43200}$, or $\frac{\pi}{21600}$ rad/sec $\approx 0.000\,145\,444$ rad/s.
(b) The radius of the orbit of a satellite is $3960 + 11{,}000 = 14{,}960$ mi. Thus, the linear speed of a satellite is given by $v = r\omega = (14{,}960)\big(\frac{\pi}{21600}\big)$, or about 2.175843801 mi/s.

26. (a) First, convert $1'$ to radians. $1' = \big(\frac{1}{60}\big)° = \frac{1}{60} \cdot \frac{\pi}{180} = \frac{\pi}{10{,}800}$ radians. Using $s = r\theta$ produces $s = 3960\big(\frac{\pi}{10{,}800}\big) \approx 1.1519$ mi, and 1.1519 mi $= (1.1519)(5280) \approx 6082$ ft.
(b) $6{,}082 - 5{,}280 = 802$ ft

CHAPTER 9 REVIEW

1. $60° \cdot \frac{\pi}{180} = \frac{\pi}{3}$ or $60° \times 0.01745 = 1.047$

2. $198° \times 0.01745 = 3.4551$ or $198 \times \frac{\pi}{180} \approx 3.456$

3. $325 \times 0.01745 = 5.67125$ or $325 \times \frac{\pi}{180} \approx 5.672$

4. $180° \times \frac{\pi}{180} = \pi \approx 3.1416$

5. $-115° \times 0.01745 = -2.00675$; $-115° \times \frac{\pi}{180} \approx -2.007$

6. $435° \times 0.01745 = 7.59075$; $435 \cdot \frac{\pi}{180} \approx 7.592$

7. $\frac{3\pi}{4}$ rad $= \frac{3\pi}{4} \cdot \frac{180°}{\pi} = 135°$

8. 1.10 rad $= 1.10 \times \frac{180°}{\pi} \approx 63.025$ or $1.10 \times 57.296 \approx 63.026$

9. 2.15 rad $= 2.15 \times 57.296 \approx 123.19°$ or $2.15 \times \frac{180°}{\pi} \approx 123.19°$

10. $\frac{7\pi}{3}$ rad $= \frac{7\pi}{3} \cdot \frac{180°}{\pi} = 420°$

11. -4.31 rad $= -4.31 \times 57.296 \approx -246.95°$ or $-4.31 \times \frac{180°}{\pi} \approx -246.94°$

12. $5.92 \text{ rad} = 5.92 \times 57.296 \approx 339.19°$ or $5.92 \times \frac{180°}{\pi} \approx 339.19°$

13. (a) I,
(b) $60°$,
(c) $60° + 360 = 420°$; $60 - 360 = -300°$

14. (a) III,
(b) $\theta_r = 198 - 180 = 18°$,
(c) $198 + 360 = 558°$; $198 - 360 = -162°$

15. (a) IV,
(b) $\theta_r = 360 - 325 = 35°$,
(c) $325 + 360 = 685°$, $325 - 360 = 35°$

16. (a) Quadrantal angle between II and III,
(b) $0°$,
(c) $180 + 360 = 540°$; $180 - 360° = -180°$

17. (a) III,
(b) $\theta_r = 245 - 180 = 65°$,
(c) $-115 + 360 = 245°$; $-115 - 360° = -475°$

18. (a) I,
(b) $75°$,
(c) $435 + 360 = 795°$; $435 - 360 = 75°$; $75° - 360° = -285°$

19. (a) II,
(b) $\pi - \frac{3\pi}{4} = \frac{\pi}{4}$,
(c) $\frac{3\pi}{4} + 2\pi = \frac{11\pi}{4}$, $\frac{3\pi}{4} - 2\pi = \frac{-5\pi}{4}$

20. (a) I,
(b) 1.10,
(c) $1.10 + 2\pi = 1.10 + 6.28 = 7.38$; $1.10 - 6.28 = -5.18$

21. (a) II,
(b) $\pi - 2.15 = 3.14 - 2.15 = 0.99$,
(c) $2.15 + 2\pi = 2.15 + 6.28 = 8.43$; $2.15 - 2\pi = 2.15 - 6.28 = -4.13$

22. (a) I,
(b) $\frac{7\pi}{3} - 2\pi = \frac{\pi}{3}$,
(c) $\frac{7\pi}{3} + 2\pi = \frac{13\pi}{3}$, $\frac{7\pi}{3} - 2\pi = \frac{\pi}{3}$, or $\frac{7\pi}{3} - 4\pi = -\frac{5\pi}{3}$

23. (a) II,
(b) $4.31 - \pi = 1.17$,
(c) $-4.31 + 6.28 = 1.97$; $-4.31 - 6.28 = -10.59$

24. (a) IV,
(b) $2\pi - 5.92 = 6.28 - 5.92 = 0.36$,
(c) $5.92 - 2\pi = -0.36$; $5.92 + 6.28 = 12.20$

25. $r = \sqrt{3^2 + (-4)^2} = 5$; $\sin\theta = \frac{y}{r} = \frac{-4}{5}$; $\cos\theta = \frac{x}{r} = \frac{3}{5}$; $\tan\theta = \frac{y}{x} = \frac{-4}{3}$; $\csc\theta = \frac{r}{y} = \frac{-5}{4}$; $\sec\theta = \frac{r}{x} = \frac{5}{3}$; $\cot\theta = \frac{x}{y} = \frac{-3}{4}$

26. $r = \sqrt{5^2 + 12^2} = 13$; $\sin\theta = \frac{y}{r} = \frac{12}{13}$; $\cos\theta = \frac{x}{r} = \frac{5}{13}$; $\tan\theta = \frac{y}{x} = \frac{12}{5}$; $\csc\theta = \frac{r}{y} = \frac{13}{12}$; $\sec\theta = \frac{r}{x} = \frac{13}{5}$; $\cot\theta = \frac{x}{y} = \frac{5}{12}$

27. $r = \sqrt{(-20)^2 + 21^2} = \sqrt{841} = 29$; $\sin\theta = 21/29$, $\cos\theta = -20/29$, $\tan\theta = -21/20$, $\csc\theta = 29/21$, $\sec\theta = -29/20$, $\cot\theta = -20/21$

28. $r = \sqrt{(-4)^2 + (-7)^2} = \sqrt{16 + 49} = \sqrt{65}$; $\sin\theta = -7/\sqrt{65}$, $\cos\theta = -4/\sqrt{65}$, $\tan\theta = 7/4$, $\csc\theta = -\sqrt{65}/7$, $\sec\theta = -\sqrt{65}/4$, $\cot\theta = 4/7$

29. $r = \sqrt{7^2 + 1^2} = \sqrt{50} = 5\sqrt{2}$; $\sin\theta = \frac{1}{5\sqrt{2}}$; $\cos\theta = \frac{7}{5\sqrt{2}}$; $\tan\theta = \frac{1}{7}$; $\csc\theta = 5\sqrt{2}$; $\sec\theta = \frac{5\sqrt{2}}{7}$; $\cot\theta = 7$

30. $r = \sqrt{144 + 64} = \sqrt{208} = 4\sqrt{13}$; $\sin\theta = \frac{8}{4\sqrt{13}} = \frac{2}{\sqrt{13}}$; $\cos\theta = \frac{-12}{4\sqrt{13}} = \frac{-3}{\sqrt{13}}$; $\tan\theta = \frac{-8}{12} = \frac{-2}{3}$; $\csc\theta = \frac{\sqrt{13}}{2}$; $\sec\theta = -\frac{\sqrt{13}}{3}$; $\cot\theta = -\frac{3}{2}$

31. $b = \sqrt{17^2 - 8^2} = 15$; $\sin\theta = \frac{a}{c} = \frac{8}{17}$; $\cos\theta = \frac{b}{c} = \frac{15}{17}$; $\tan\theta = \frac{a}{b} = \frac{8}{15}$; $\csc\theta = \frac{c}{a} = \frac{17}{8}$; $\sec\theta = \frac{c}{b} = \frac{17}{15}$; $\cot\theta = \frac{b}{a} = \frac{15}{8}$

32. $c = \sqrt{5^2 + 12^2} = 13$; $\sin\theta = \frac{a}{c} = \frac{5}{13}$; $\cos\theta = \frac{b}{c} = \frac{12}{13}$; $\tan\theta = \frac{a}{b} = \frac{5}{12}$; $\csc\theta = \frac{c}{a} = \frac{13}{5}$; $\sec = \frac{c}{b} = \frac{13}{12}$; $\cot\theta = \frac{b}{a} = \frac{12}{5}$

33. $a = \sqrt{\left(\frac{40}{3}\right)^2 - 8^2} = \sqrt{\frac{1600}{9} - \frac{576}{9}} = \sqrt{\frac{1024}{9}} = \frac{32}{3}$; $\sin\theta = \frac{32}{40} = \frac{4}{5}$; $\cos\theta = \frac{8}{\frac{40}{3}} = \frac{24}{40} = \frac{3}{5}$; $\tan\theta = \frac{\frac{32}{3}}{8} = \frac{4}{3}$; $\csc\theta = \frac{5}{4}$; $\sec\theta = \frac{5}{3}$; $\cot = \frac{3}{4}$

34. $b = \sqrt{7.5^2 - 6^2} = 4.5$; $\sin\theta = \frac{6}{7.5} = .8$; $\cos\theta = \frac{4.5}{7.5} = 0.6$; $\tan\theta = \frac{6}{4.5} = 1.33$; $\csc\theta = \frac{7.5}{6} = 1.25$; $\sec\theta = \frac{7.5}{4.5} = 1.67$; $\cot\theta = \frac{4.5}{6} = 0.75$

35. $a = \sqrt{18.2^2 - 7^2} = 16.8$; $\sin\theta = \frac{16.8}{18.2} = \frac{168}{182} = \frac{84}{91}$; $\cos\theta = \frac{7}{18.2} = \frac{70}{182} = \frac{35}{91}$; $\tan\theta = \frac{16.8}{7} = \frac{168}{70} = \frac{84}{35}$; $\csc\theta = \frac{91}{84}$; $\sec\theta = \frac{91}{35}$; $\cot\theta = \frac{35}{84}$

36. $c = \sqrt{42^2 + 44.1^2} = 60.9$; $\sin\theta = \frac{42}{60.9} = \frac{420}{609}$;

$\cos\theta = \frac{44.1}{60.9} = \frac{441}{609}$; $\tan\theta = \frac{42}{44.1} = \frac{420}{441} = \frac{20}{21}$;

$\csc\theta = \frac{609}{420}$; $\sec\theta = \frac{609}{441}$; $\cot\theta = \frac{21}{20}$

37. $a = 12.8$; $c = 27.2$; $b = \sqrt{27.2^2 - 12.8^2} = 24$;

$\cos\theta = \frac{24}{27.2}$; $\tan\theta = \frac{12.8}{24}$; $\csc\theta = \frac{27.2}{12.8}$; $\sec\theta = \frac{27.2}{24}$; $\cot\theta = \frac{24}{12.8}$

38. $a = 16$; $b = 16.8$; $c = \sqrt{16^2 + 16.8^2} = 23.2$;

$\sin\theta = \frac{16}{23.2}$; $\cos\theta = \frac{16.8}{23.2}$; $\tan\theta = \frac{16}{16.8}$; $\csc\theta = \frac{23.2}{16}$; $\sec\theta = \frac{23.2}{16.8}$; $\cot\theta = \frac{16.8}{16}$

39. $c = 4$; $b = 2.5$; $a = \sqrt{4^2 - 2.5^2} \approx 3.12$; $\sin\theta = \frac{3.12}{4}$; $\cos\theta = \frac{2.5}{4}$; $\tan\theta = \frac{3.12}{2.5}$; $\csc\theta = \frac{4}{3.12}$; $\sec\theta = \frac{4}{2.5}$; $\cot\theta = \frac{2.5}{3.12}$

40. $\cos\theta = \frac{\sin\theta}{\tan\theta} = \frac{0.532}{0.628} \approx 0.847$; $\csc\theta = \frac{1}{0.532} \approx 1.880$; $\sec\theta = \frac{1}{0.847} \approx 1.181$; $\cot\theta = \frac{1}{0.628} \approx 1.592$

41. $\tan\theta = \frac{0.5}{0.866} \approx 0.577$; $\csc\theta = \frac{1}{0.5} = 2$; $\sec\theta = \frac{1}{0.866} \approx 1.155$; $\cot\theta = \frac{0.866}{0.5} = 1.732$

42. $\sin\theta = \frac{1}{1.364} \approx 0.733$; $\tan\theta \approx \frac{0.733}{0.680} \approx 1.078$; $\sec\theta = \frac{1}{0.680} \approx 1.471$; $\cot\theta \approx \frac{0.680}{0.733} \approx 0.928$

43. $\sin 45° = 0.7071$

44. $\cos 82.5° = 0.1305$

45. $\tan 213.5° = 0.6619$

46. $\sec(-81°) = 6.3925$

47. $\cos 2.3 \text{ rad} = -0.6663$

48. $\sin 4.75 \text{ rad} = -0.9993$

49. $\tan(-3.2) \text{ rad} = -0.0585$

50. $\csc 0.21 \text{ rad} = 4.7971$

51. $\sin\theta = 0.5$; $\theta_r = \sin^{-1} 0.5 = 30°$; $\theta_I = \theta_r = 30°$; $\theta_{II} = 180 - 30° = 150°$

52. $\tan\theta = 2.5$; $\theta_r = \tan^{-1} 2.5 = 68.2$; $\theta_I = 68.2$; $\theta_{III} = 180 + 68.2 = 248.2$

53. $\cos\theta = -0.75$; $\theta_r = \cos^{-1} 0.75 = 41.4$; $\theta_{II} = 180 - 41.4 = 138.6$; $\theta_{III} = 180 41.4 = 221.4$

54. $\csc\theta = 3.0$; $\theta_r = \sin^{-1}\frac{1}{3} = 19.5$; $\theta_I = 19.5$; $\theta_{II} = 180 - 19.5 = 160.5$

55. $\cos\theta = -0.5$; $\theta_r = \cos^{-1} 0.5 = 1.047 \approx 1.05$; $\theta_{II} = \pi - 1.05 = 3.14 - 1.05 = 2.09$; $\theta_{III} = \pi + 1.05 = 3.14 + 1.05 = 4.19$

56. $\sin\theta = 0.717$; $\theta_r = \sin^{-1} 0.717 = 0.80$; $\theta_I = 0.80$; $\theta_{II} = \pi - 0.80 = 3.14 - 0.80 = 2.34$

57. $\theta_r = \tan^{-1} 0.95 = 0.76$; $\theta_{II} = \pi - 0.76 = 3.14 - 0.76 = 2.38$; $\theta_{IV} = 2\pi - 0.76 = 6.28 - 0.76 = 5.52$

58. $\theta_r = \cos^{-1}\frac{1}{2.25} = 1.11$; $\theta_I = 1.11$; $\theta_{IV} = 2\pi - 1.11 = 6.28 - 1.11 = 5.17$

59. $\arcsin 0.866 = 60.0°$

60. $\arccos 0.5 = 60.0°$

61. $\arctan(-1) = -45.0°$

62. $\cos^{-1}(-0.707) = 135.0°$

63. $\sin^{-1} 0.385 = 22.6°$

64. $\cot^{-1}(3.5) = \tan^{-1}\frac{1}{3.5} = 15.9°$

65. $1.33 = \frac{\sin 30°}{\sin r}$, $\sin r = \frac{\sin 30°}{1.33} \approx 0.3759$; $r \approx \sin^{-1} 0.3759 \approx 22.1°$

66. $n = \frac{\sin 0.90}{\sin 0.55} \approx 1.50$

67. $I = 12.6 \sin\frac{\pi}{5} \approx 7.41 \text{ A}$

68. $\sin\theta = \frac{4.732}{7.3} \approx 0.6482$; $\theta \approx \sin^{-1} 0.6482 \approx 40.4°$

69. $\tan\theta = \frac{44}{50}$; $\theta = \tan^{-1}\frac{44}{50} \approx 41.3°$; $V = \sqrt{V_L^2 + V_R^2} = \sqrt{44^2 + 50^2} \approx 66.6 \text{ V}$

70. $\frac{1}{2}(57 - 46) = 5.5$; $\tan 3° = \frac{5.5}{h}$; $h = \frac{5.5}{\tan 3°} \approx 104.9$ mm

71. We use the equation $s = r\theta$ with $\theta = 37° \times \frac{\pi}{180} \approx 0.6458$. Thus, we obtain $s \approx 900 \times 0.6458 \approx 581.2$ ft.

72. $300 \frac{\text{rev}}{\text{min}} \cdot \frac{2\pi \text{ rad}}{\text{rev}} \cdot \frac{1 \text{ min}}{60 \text{ sec}} = 10\pi \text{ rad/s} \approx 31.416 \text{ rad/s}$; $v = r\omega = \frac{5.25 \text{ in.}}{2} \times 31.416 = 82.467 \text{ in./s}$

73. $\omega = \frac{1 \text{ rev}}{365 \text{ days}} \cdot \frac{2\pi}{1 \text{ rev}} \cdot \frac{1 \text{ day}}{24 \text{ hr}} \approx 7.17 \times 10^{-4} \frac{\text{rad}}{\text{hr}}$; $v = r\omega = 9.3 \times 10^7 \times 7.17 \times 10^{-4} \approx 66{,}700$ mph

74. $\tan 68° = \frac{h}{15}$; $h = 15 \tan 68° \approx 37.1$ ft

75. Let $y = $ length of common side, then $\cos 36° = \frac{y}{110.5}$; $y = 110.5 \cos 36° = 89.4$; $\cos x = \frac{82.3}{89.4}$; $x = \cos^{-1}\frac{82.3}{89.4}$; $x = 23.0°$

76. $\tan \theta = \frac{56}{43}$; $\theta = \tan^{-1} \frac{56}{43} \approx 52.5°$

77. $\tan 53.7° = \frac{h}{150}$; $h = 150 \tan 53.7° \approx 204.2$ ft

78. Draw a segment parallel to the axis from the top of the smaller circle. This makes a right triangle with legs of 128.3 m and $(50.2 - 26.1) = 24.1$ m. Then $\tan \phi = \frac{24.1}{128.3}$; $\phi \approx 10.64$; $\theta = 2\phi \approx 21.3°$

79. $F_x = 3500 \sin 34.3° \approx 1972.3$ lb; $F_y = 3500 \cos 34.3$ 2891.3 lb

80. Let h be height of the tower and x be the distance from the base of the tower to the closer anchor. Then $\tan 23.7° = \frac{h}{x+35}$ and $\tan 36.4° = \frac{h}{x}$; $\tan 23.7° \approx 0.439$; $\tan 36.4 \approx 0.737$, so $0.439(x+35) = h$ and $0.737x = h$; by substitution $0.439(x + 35) = 0.737x$; $0.439x + 15.365 = 0.737x$; $15.365 = 0.298x$; $x = \frac{15.365}{0.298} \approx 51.56$; $h = x \tan 36.4$; $h = 51.56 \tan 36.4 \approx 38.0$ m

CHAPTER 9 TEST

1. $50° \times \frac{\pi}{180°} = \frac{5\pi}{18}$ or $50 \times 0.01745 = 0.8725$

2. $\frac{4\pi}{3} = \frac{4\pi}{3} \cdot \frac{180°}{\pi} = \frac{4}{3} \cdot 180° = 4(60°) = 240°$

3. **(a)** III;
(b) $237° - 180° = 57°$

4. $r = \sqrt{(-5)^2 + 12^2} = 13$; $\sin \theta = \frac{12}{13}$; $\cos \theta = \frac{-5}{13}$; $\tan \theta = \frac{-12}{5}$; $\csc \theta = \frac{13}{12}$; $\sec \theta = \frac{-13}{5}$; $\cot \theta = \frac{-5}{12}$

5. $b = \sqrt{c^2 - a^2} = \sqrt{23.4^2 - 9^2} = \sqrt{466.56} = 21.6$;
(a) $\sin \theta = \frac{9}{23.4} \approx 0.38462$;
(b) $\tan \theta = \frac{9}{21.6} \approx 0.41667$

6. Here the side opposite $\theta = 30$ and the side adjacent to θ is 6.75. So, the hypotenuse $r = \sqrt{30^2 + 6.75^2} = 30.75$;
(a) $\tan \theta = \frac{1}{\cot \theta} = \frac{30}{6.75} \approx 4.44444$;

(b) $\sin \theta = \frac{30}{30.75} \approx 0.97561$

7. Using a calculator, we get
(a) 0.79864;
(b) 2.47509;
(c) -1.66164; **(d)** -0.49026 [In (d) make sure that your calculator is in radian mode.]

8. **(a)** $\theta_r = \tan^{-1} 1.2 \approx 50.2°$; $\theta_{\text{III}} = 180° + \theta_r = 230.2°$;
(b) $\theta_r = \sin^{-1} 0.72 = 46.1$; $\theta_{\text{III}} = 180° + \theta_r = 226.1°$; $\theta_{\text{IV}} = 360° - \theta_r = 313.9°$

9. $\tan 53° = \frac{h}{135}$; $h = 135 \tan 53° \approx 179.15$ ft

10. One revolution is π radians, so it rotates $3600 \times 2\pi = 7200\pi$ radians per minute. There are 60 seconds in one minute, so it rotates $7200\pi \div 60 = 120\pi$ radians per second.

10

Vectors and Trigonometric Functions

☰ 10.1 INTRODUCTION TO VECTORS

1.

2.

3.

5.

6.

7.

8.

4.

9.

10.

11.

12.

13.

14.

15.

16.

17.

18.

19.

20.

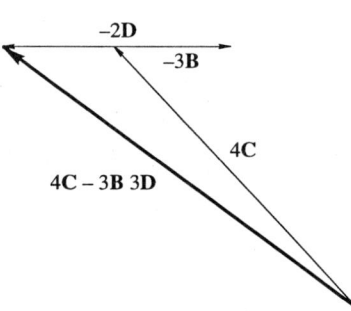

21. $\mathbf{P_x} = 20\cos 75° = 5.176$; $\mathbf{P_y} = 20\sin 75° = 19.319$

22. $\mathbf{P_x} = 16\cos 212° = -13.569$; $\mathbf{P_y} = 16\sin 212° = -8.479$

23. $\mathbf{P_x} = 18.4\cos 4.97 \text{ rad} = 4.688$; $\mathbf{P_y} = 18.4\sin 4.97 = -17.793$

24. $\mathbf{P_x} = 23.7\cos 2.22 = -14.328$; $\mathbf{P_y} = 23.7\sin 2.22 = 18.879$

25. $\mathbf{V_x} = 9.75\cos 16° = 9.372$; $\mathbf{V_y} = 9.75\sin 16° = 2.687$

26. $\mathbf{P_x} = 24.6\cos 317° = 17.991$; $\mathbf{P_y} = 24.6\sin 317° = -16.777$

27. $A = \sqrt{(-9)^2 + 12^2} = 15$; θ is in Quadrant II; $\theta_{Ref} = \tan^{-1}\frac{12}{9} = 52.13$; $\theta = 180° - 53.13° = 126.87°$

28. $B = \sqrt{10^2 + (-24)^2} = 26$; θ is in Quadrant IV; $\theta_{Ref} = \tan^{-1}\frac{24}{10} = 67.38°$; $\theta = 360 - 67.38 = 292.62°$

29. $C = \sqrt{8^2 + 15^2} = 17$; θ is in Quadrant I; $\theta_{Ref} = \tan^{-1}\frac{15}{8} = 61.93°$; $\theta = 61.93°$

30. $D = \sqrt{(-14)^2 + (-20)^2} = 24.413$; θ is in Quadrant III; $\theta_{Ref} = \tan^{-1}\frac{20}{14} = 55.01°$; $\theta = 180° + 55.01° = 235.01°$

31. $v = \sqrt{5^2 + 12^2} = 13$ km/hr

32. $\mathbf{V_x} = 765.0$, $\mathbf{V_y} = 73$, $V = \sqrt{756^2 + 73^2} = 759.5$ km/hr; $\theta = \tan^{-1}\frac{73}{756} = 5.52°$ north of east

33. $\mathbf{V_x} = 976\cos 72.4° = 295$ N; $\mathbf{V_y} = 976\sin 72.4° = 930$ N

34. $F = \sqrt{125.0^2 + 26.50^2} = 127.8$ lb; $\theta = \tan^{-1}\frac{26.5}{125} = 11.97°$ with the large force or 11.97° from vertical.

35. Parallel to ramp $153 \times \sin 12° = 31.8$ lb. Perpendicular to ramp $153\cos 12° = 149.7$ lb.

36. $V = \sqrt{V_R^2 + (V_L - V_C)^2} = \sqrt{12^2 + (5 - 10)^2} = 13$ V; $\phi = \tan^{-1}\frac{-5}{12} = -22.6°$ or 22.6° from $\mathbf{V}_R$ toward $(\mathbf{V}_L - \mathbf{V}_C)$.

37. $V = \sqrt{15^2 + (8 - 17)^2} = 17.5$ V; $\phi = \tan^{-1}\frac{-9}{5} \approx -30.96°$ or 30.96° from $\mathbf{V}_R$ towards $(\mathbf{V}_L - \mathbf{V}_C)$.

38. Since $(\mathbf{V}_L - \mathbf{V}_C) = 28.3 - 15.2 = 13.1$, we have $V = \sqrt{28.3^2 + 13.1^2} \approx 31.2$ V, $\phi = \tan^{-1}\left(\frac{13.1}{28.3}\right) \approx 24.84°$ or $-24.84°$ from $\mathbf{V}_R$ toward $(\mathbf{V}_L - \mathbf{V}_C)$.

10.2 ADDING AND SUBTRACTING VECTORS

1. $R = |20.0 - 32.5| = 12.5$; $\theta_R = $ angle of larger so 180°

2. $R = |14.3 + 7.2| = 21.5$; $\theta_R = 90°$

3. $R = |121.7 - 86.9| = 34.8$; $\theta_R = 270°$ (angle of the larger magnitude)

4. $R = |63.1 + 43.5| = 106.6$; $\theta_R = 180°$

5. $R = \sqrt{55^2 + 48^2} = 73.$; $\theta = \tan^{-1}\frac{55}{48} = 48.9$; since it's in Quadrant II $\theta_R = 180° - 48.9° = 131.1°$. (Note $\mathbf{C}$ is a vertical vector, $\mathbf{D}$ is a horizontal vector. $\theta_R = \tan^{-1}\frac{\text{vertical}}{\text{horizontal}}$)

6. $R = \sqrt{65^2 + 72^2} = 97$; $\theta = \tan^{-1}\frac{65}{72} = 42.08°$; Quadrant III so $180° + 42.08 = 222.08° = \theta_R$

7. $R = \sqrt{81.4^2 + 37.6^2} = 89.7$; $\theta_R = \tan^{-1}\frac{37.6}{81.4} = 24.79°$

8. $R = \sqrt{63.4^2 + 9.4^2} = 64.1$; $\theta = \tan^{-1}\frac{63.4}{9.4} = 81.57$; Quadrant IV so $\theta_R = 360° - 81.57 = 278.43$

9. $A = \sqrt{33^2 + 56^2} = 65$; $\theta_A = \tan^{-1}\frac{33}{56} = 59.49°$

10. $B = \sqrt{231^2 + 520^2} = 569$; $\theta_B = \tan^{-1}\frac{231}{520} = 66.05°$

11. $C = \sqrt{11.7^2 + 4.4^2} = 12.5$; $\theta_C = \tan^{-1}\frac{11.7}{4.4} = 20.61°$

12. $D = \sqrt{31.9^2 + 36.0^2} = 48.1$; $\theta_D = \tan^{-1}\frac{31.9}{36.0} = 48.46°$

13. $E = \sqrt{6.3^2 + 1.6^2} = 6.5$; $\theta_E = \tan^{-1}\frac{6.3}{1.6} = 14.25°$

14. $F = \sqrt{5.1^2 + 14.0^2} = 14.9$; $\theta_F = \tan^{-1}\frac{5.1}{14.0} = 69.98°$

15. $G = \sqrt{8.4^2 + 12.6^2} = 15.1$; $\theta_G = \tan^{-1}\frac{8.4}{12.6} = 56.31°$

16. $H = \sqrt{15.3^2 + 9.2^2} = 17.9$; $\theta_H = \tan^{-1}\frac{15.3}{9.2} = 31.02°$

17. From these values for R_x and R_y we can determine $R = \sqrt{R_x^2 + R_y^2} = 12.33$ and $\theta_R = \tan^{-1}\frac{R_y}{R_x} = 32.03°$

vector	horizontal component	vertical component
A	$\mathbf{A_x} = 4\cos 60° = 2.000$	$\mathbf{A_y} = 4\sin 60° = 3.464$
B	$\mathbf{B_x} = 9\cos 20° = 8.457$	$\mathbf{B_y} = 9\sin 20° = 3.078$
R	$\mathbf{R_x}$ 10.457	$\mathbf{R_y}$ 6.542

18. $R = \sqrt{R_x^2 + R_y^2} = 25.55$; $\theta = \tan^{-1}\frac{R_y}{R_x} = 53.81°$

vector	horizontal component	vertical component
C	$\mathbf{C_x} = 12\cos 75° = 3.106$	$\mathbf{C_y} = 12\sin 75° = 11.591$
D	$\mathbf{D_x} = 15\cos 37° = 11.980$	$\mathbf{D_y} = 15\sin 37° = 9.027$
R	$\mathbf{R_x}$ 15.086	$\mathbf{R_y}$ 20.618

19. $R = \sqrt{R_x^2 + R_y^2} = 53.00$; $\theta = \tan^{-1}\frac{R_y}{R_x} = 178.11°$

vector	horizontal component	vertical component
C	$\mathbf{C_x} = 28\cos 120° = -14.000$	$\mathbf{C_y} = 28\sin 120° = 24.249$
D	$\mathbf{D_x} = 45\cos 210° = -38.971$	$\mathbf{D_y} = 45\sin 210° = -22.500$
R	$\mathbf{R_x}$ -52.971	$\mathbf{R_y}$ 1.749

20. $R = \sqrt{R_x^2 + R_y^2} = 97.00;$

$\theta = \tan^{-1} \frac{R_y}{R_x} = -30.92°;$ since
this is Quadrant IV, we have
$\theta = 360 - 30.92 = 329.08°$

vector	horizontal component	vertical component
E	$E_x = 72 \cos 287° = 21.051$	$E_y = 72 \sin 287° = -68.854$
F	$F_x = 65 \cos 17° = 62.160$	$F_y = 65 \sin 17° = 19.004$
R	R_x \hfill 83.211	R_y \hfill -49.850

21. $R = \sqrt{R_x^2 + R_y^2} = 88.50;$

$\theta = \tan^{-1} \frac{R_y}{R_x} = 223.83°$

vector	horizontal component	vertical component
A	$A_x = 31.2 \cos 197.5° = -29.756$	$A_y = 31.2 \sin 197.5° = -9.382$
B	$B_x = 62.1 \cos 236.7° = -34.094$	$B_y = 62.1 \sin 236.7° = -51.904$
R	R_x \hfill -63.850	R_y \hfill -61.286

22. $R = \sqrt{R_x^2 + R_y^2} = 57.42;$

$\theta = \tan^{-1} \frac{R_y}{R_x} = 247.28°$

vector	horizontal component	vertical component
C	$C_x = 53.1 \cos 324.3° = 43.122$	$C_y = 53.1 \sin 324.3° = -30.986$
D	$D_x = 68.9 \cos 198.6° = -65.301$	$D_y = 68.9 \sin 198.6° = -21.976$
R	R_x \hfill -22.179	R_y \hfill -52.962

23. $R = \sqrt{R_x^2 + R_y^2} = 22.44;$

$\theta = \tan^{-1} \frac{R_y}{R_x} = 348.90°$

vector	horizontal component	vertical component
E	$E_x = 12.52 \cos 46.4° = 8.634$	$E_y = 12.52 \sin 46.4° = 9.067$
F	$F_x = 18.93 \cos 315° = 13.386$	$F_y = 18.93 \sin 315° = -13.386$
R	R_x \hfill 22.020	R_y \hfill -4.319

24. $R = \sqrt{R_x^2 + R_y^2} = 116.86;$

$\theta = \tan^{-1} \frac{R_y}{R_x} = 65.61°$

vector	horizontal component	vertical component
G	$G_x = 76.2 \cos 15.7° = 73.357$	$G_y = 76.2 \sin 15.7° = 20.620$
H	$H_x = 89.4 \cos 106.3° = -25.092$	$H_y = 89.4 \sin 106.3° = 85.807$
R	R_x \hfill 48.265	R_y \hfill 106.427

25. $R = \sqrt{R_x^2 + R_y^2} = 13.04;$ $\theta = \tan^{-1} \frac{R_y}{R_x} = 150.30°$

vector	horizontal component	vertical component
A	$A_x = 9.84 \cos 215.5° = -8.011$	$A_y = 9.84 \sin 215.5° = -5.714$
B	$B_x = 12.62 \cos 105.25° = -3.319$	$B_y = 12.62 \sin 105.25° = 12.176$
R	R_x \hfill -11.330	R_y \hfill 6.462

26. $R = \sqrt{R_x^2 + R_y^2} = 104.99;$ $\theta = \tan^{-1} \frac{R_y}{R_x} = 240.29°$

vector	horizontal component	vertical component
C	$C_x = 79.63 \cos 262.75° = -10.049$	$C_y = 79.63 \sin 262.75° = -78.993$
D	$D_x = 43.72 \cos 196.2° = -41.984$	$D_y = 43.72 \sin 196.2° = -12.197$
R	R_x \hfill -52.033	R_y \hfill -91.190

27. $R = \sqrt{R_x^2 + R_y^2} = 63.58;$

$\theta = \tan^{-1} \frac{R_y}{R_x} = 4.62°$

vector	horizontal component	vertical component
E	$E_x = 42.0 \cos 3.4 = -40.606$	$E_y = 42.0 \sin 3.4 = -10.733$
F	$F_x = 63.2 \cos 5.3 = 35.036$	$F_y = 63.2 \sin 5.3 = -52.599$
R	R_x \hfill -5.570	R_y \hfill -63.332

28. $R = \sqrt{R_x^2 + R_y^2} = 58.91;$

$\theta = \tan^{-1} \frac{R_y}{R_x} = 1.23°$

vector	horizontal component	vertical component
G	$G_x = 37.5 \cos 0.25 = 36.334$	$G_y = 37.5 \sin 0.25 = 9.278$
H	$H_x = 49.3 \cos 1.92 = -16.868$	$H_y = 49.3 \sin 1.92 = 46.325$
R	R_x \hfill 19.466	R_y \hfill 55.603

29. $R = \sqrt{R_x^2 + R_y^2} = 13.13;$

$\theta = \tan^{-1} \frac{R_y}{R_x} = 64.41°$

vector	horizontal component	vertical component
A	$A_x = 15 \cos 25° = 13.59$	$A_y = 15 \sin 25° = 6.34$
B	$B_x = 20 \cos 120° = -10.00$	$B_y = 20 \sin 120° = 17.32$
C	$C_x = 12 \cos 280° = 2.08$	$C_y = 12 \sin 280° = -11.82$
R	R_x \hfill 5.67	R_y \hfill 11.84

30. $R = \sqrt{R_x^2 + R_y^2} = 236.56;$

$\theta = \tan^{-1} \frac{R_y}{R_x} = 187.51°$

vector	horizontal component	vertical component
A	$A_x = 320 \cos 105° = -82.82$	$A_y = 320 \sin 105° = 309.10$
B	$B_x = 480 \cos 215° = -393.19$	$B_y = 480 \sin 215° = -275.32$
C	$C_x = 250 \cos 345° = 241.48$	$C_y = 250 \sin 345° = -64.70$
R	R_x \hfill -234.53	R_y \hfill -30.92

31. Draw the coordinate axes and label the angles as shown in the figure below. The origin is placed at the ring where the three cables meet, because the tension on all three cables acts on this ring.

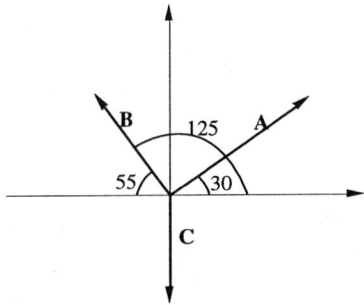

We use the following table listing the horizontal and vertical components of each vector.

vector	horizontal component	vertical component
A	$\mathbf{A_x} = A\cos 30°$	$\mathbf{A_y} = A\sin 30°$
B	$\mathbf{B_x} = B\cos 125°$	$\mathbf{B_y} = B\sin 125°$
C	$\mathbf{C_x} = C\cos 270° = 0$	$\mathbf{C_y} = C\sin 270° = -C = -215$
R	$\mathbf{R_x}$	$\mathbf{R_y}$

Since, $R_x = A\cos 30° + B\cos 125° = 0$ and $R_y = A\sin 30° + B\sin 125° - 215 = 0$, we are led to the system of two equations in two variables.

$$\begin{cases} 0.86603A - 0.57358B = 0 \\ 0.5A + 0.81915B = 215 \end{cases}$$

Using Cramer's rule, we determine that $A \approx 123.8$ N and $B = 186.9$ N. The tension in C is 215 kg $\times 9.8$ m/s$^2 \approx 2107$ N.

32. A diagram of the vectors for the forces is shown in the figure below. Notice that, since the brace is pushing on the sign we have drawn the vector as if it were leaving the center of the sign.

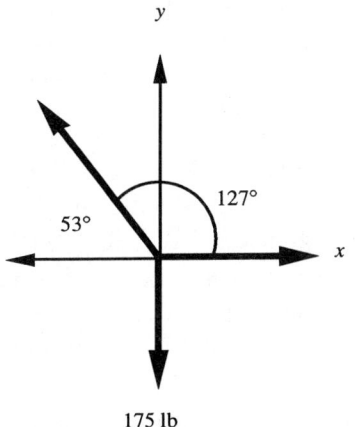

We use the following table listing the horizontal and vertical components of each vector.

vector	horizontal component	vertical component
A	$\mathbf{A_x} = A\cos 0° = A$	$\mathbf{A_y} = A\sin 0° = 0$
B	$\mathbf{B_x} = B\cos 127°$	$\mathbf{B_y} = B\sin 127°$
S	$\mathbf{S_x} = 175\cos 270° = 0$	$\mathbf{S_y} = 175\sin 270° = -175$
R	$\mathbf{R_x}$	$\mathbf{R_y}$

Here $R_x = A + B\cos 127° + 0 = 0$ and $R_y = 0 + B\sin 127° - 175 = 0$. From the second equation we obtain $0.7986B - 175 = 0$ and so, $B = \frac{175}{0.7986} \approx 219.1$ Substituting this value in the first equation produces $A + (219.1)\cos 127° = 0$ and this leads to $A \approx 131.9$. Thus, the brace exerts a force of 131.9 lb and the cable a force of 219.1 lb.

33. Use a diagram similar to the one used to solve the previous exercise. This should lead to the following table listing the horizontal and vertical components of each vector.

vector	horizontal component	vertical component
A	$A_x = A\cos 0° = A$	$A_y = A\sin 0° = 0$
B	$B_x = B\cos 143°$	$B_y = B\sin 143°$
S	$S_x = 235\cos 270° = 0$	$S_y = 235\sin 270° = -235$
R	R_x	R_y

Here $R_x = A + B\cos 143° + 0 = 0$ and $R_y = 0 + B\sin 143° - 235 = 0$. From the second equation we obtain $0.6018B - 235 = 0$ and so, $B = \frac{235}{0.6018} \approx 390.5$. Substituting this value in the first equation produces $A + (390.5)\cos 143° = 0$ and this leads to $A \approx 311.9$. Thus, the brace exerts a force of 311.9 lb and the cable a force of 390.5 lb.

34. The vertical cable has a downward tension of $F = 85 \text{ kg} \cdot 9.8 \text{ m/s}^2 = 833$ N. Let F_L be the tension on the left cable and F_R the tension on the right cable. We use the following table listing the horizontal and vertical components of each vector.

vector	horizontal component	vertical component
F	$F_x = 833\cos 270° = 0$	$F_y = 833\sin 270° = -833$
F_L	$F_{L_x} = F_L\cos 122°$	$F_{L_y} = F_L\sin 122°$
F_R	$F_{R_x} = F_R\cos 38°$	$F_{R_y} = F_R\sin 38°$
R	R_x	R_y

Hence, $F_L\cos 122° + F_R\cos 38° = 0$. In the same way, the vertical components add up to 0 or $-833 + F_L\sin 122° + F_R\sin 38° = 0$. Solving this system, we get $F_L = 660$ N and $F_R = 448$ N.

35. The components for the airplane are $A_x = 480\cos 27° = 427.68$, $A_y = 480\sin 27° = 217.92$; while the components for the wind are $W_x = 45\cos -55° = 25.81$, $W_y = 45\sin -55° = -36.86$. Using these, we can determine that the components for the resultant are $R_x = 427.68 + 25.81 = 453.49$, $R_y = 217.92 - 36.86 = 181.06$ and the length of the component produces a ground speed of $R = \sqrt{R_x^2 + R_y^2} = 488.3$ mph. To determine the course and drift angle, we first find that $\theta = \tan^{-1}\frac{R_y}{R_x} = 21.8°$. Hence the course is $90 - 21.8 = 68.2°$ and the drift angle is $68.2 - 63 = 5.2°$.

36. The components for the airplane are $A_x = 320\cos -82° = 44.54$ and $A_y = 320\sin -82° = -316.89$; while the components for the wind are $W_x = 72\cos 133° = -49.10$ and $W_y = 72\sin 133° = 52.66$. Using these, we can determine that the components for the resultant are $R_x = 44.54 - 49.10 = -4.56$, $R_y = -316.89 + 52.66 = -264.23$ and so the length of the component produces a ground speed of $R = \sqrt{R_x^2 + R_y^2} = 264.3$ mph. To determine the course and drift angle, we first find that $\theta = \tan^{-1}\frac{R_y}{R_x} = -91°$. Hence, the course is $\approx 181°$ and the drift angle is $\approx 9°$.

37. We will assume that one voltage, vector **A**, is horizontal and that the other voltage, **B**, makes an angle of 68° with the horizontal vector. We can form the following component table.

vector	horizontal component		vertical component	
A	$A_x = 196\cos 0° =$	196.0	$A_y = 196\sin 0° =$	0
B	$B_x = 196\cos 68° \approx$	73.4	$B_y = 196\sin 68° \approx$	181.7
R	R_x	269.4	R_y	181.7

Using the value of R_x and R_y, we see that

$$R = \sqrt{R_x^2 + R_y^2}$$
$$= \sqrt{269.4^2 + 181.7^2} \approx 324.9$$

The total voltage is about 324.9 V.

38. We will assume that the first voltage, vector **A**, is horizontal and that the second voltage, **B**, makes an angle of 120° with the horizontal vector. We can form the following component table.

vector	horizontal component		vertical component	
A	$A_x = 86 \cos 0° =$	86.0	$A_y = 86 \sin 0° =$	0
B	$B_x = 110 \cos 120° =$	−55.0	$B_y = 116 \sin 120° \approx$	95.3
R	R_x	31.0	R_y	95.3

Using the value of R_x and R_y, we see that

$$R = \sqrt{R_x^2 + R_y^2}$$
$$= \sqrt{31.0^2 + 95.3^2} \approx 100.2$$

The total voltage is about 100.2 V.

39. See *Computer Programs* in main text.   40. See *Computer Programs* in main text.

10.3 APPLICATIONS OF VECTORS

1. Let the 70 lb force be at 0° and the 50 lb force at 35°. Then
$$R = \sqrt{R_x^2 + R_y^2} = 114.60 \text{ and}$$
$$\theta_R = \tan^{-1} \frac{R_y}{R_x} = 14.49° \text{ from the 70 lb force.}$$

vector	horizontal component		vertical component	
A	$A_x = 70 \cos 0° =$	70.000	$A_y = 70 \sin 0° =$	0.000
B	$B_x = 50 \cos 35° =$	40.958	$B_y = 50 \sin 35° =$	28.679
R	R_x	110.958	R_y	28.679

2. Horizontal component is $35 \cos 25° = 31.72$ lb. Vertical component is $35 \sin 25° = 14.79$ lb.

3. (a) Downward force is $25 \sin 55° = 20.48$ lb.
   (b) Horizontal force is $25 \cos 55° = 14.34$ lb.
   (c) for a 40° angle, Downward force $= 25 \sin 40° = 16.07$ lb, Horizontal force is $25 \cos 40° = 19.15$ lb.

4. The component forces perpendicular to the axis of the ship must be equal but opposite. Hence $1500 \sin 37° = F \sin 36°$ or $F = \frac{1500 \sin 37°}{\sin 36°} = 1535.80$ N.

5. The forward force is $F_f = 1700 \cos 18° = 1616.80$ N; and the sideward force is $F_s = 1700 \sin 18° = 525.33$ N.

6. The ship's components are $S_x = 15 \cos 135° = -10.61$ and $S_y = 15 \sin 135° = 10.61$ and the components for the flow of the river are $R_x = 7$ and $R_y = 0$. Adding these, we get the components for the ship's velocity relative to the Earth's surface as $V_x = -3.61$ and $V_y = 10.61$. The resultant of the ship's components gives the ship's velocity as $V = \sqrt{V_x^2 + V_y^2} = 11.21$ km/h. Next

we determine that $\theta = \tan^{-1} \frac{V_y}{V_x} = 108.79°$ and so, the angle is 18.79° west of north, which is a compass direction of 341.21°.

7. Since the heading and wind are perpendicular, the ground speed is $\sqrt{370^2 + 40^2} = 372.16$ mi/h. The angle east of south is $\theta = \tan^{-1} \frac{40}{370} = 6.17°$. This is a compass direction of $180° - 6.17° = 173.83°$.

8. $A_v = 420 \cos 115° = -177.50$, $A_h = 420 \sin 115° = 380.65$, $W_v = 62 \cos 32° = 52.58$, $W_h = 62 \sin 32° = 32.85$, $G_v = -124.92$, $G_h = 413.50$, $G = \sqrt{G_v^2 + G_h^2} = 431.96$ mi/h; $\theta = \tan^{-1} \frac{G_h}{G_v} = 106.81°$.

9. Parallel component is $1200 \sin 22° = 449.53$ kg. Perpendicular component is $1200 \cos 22° = 1112.62$ kg.

10. $F = 30 \sin 15° = 7.76$ lb

11. The angle the guy wire makes with the ground is $\theta = \tan^{-1} \frac{35}{27} = 52.352°$. Horizontal tension $T_x = 195 \cos 52.352° = 119.11$ lb. Vertical-downward $T_y = 195 \sin 52.352° = 154.40$ lb.

12. Since I_C and I_R are perpendicular the magnitude of $I_T = \sqrt{I_C^2 + I_R^2} = \sqrt{2.4^2 + 1.6^2} = 2.88$ A. The phase angle $\theta = \tan^{-1} \frac{2.4}{1.6} = 56.31°$.

13. $I_C + I_R = I$; $I_C = I - I_R$; $I_C^2 = 9.6^2 - 7.5^2$, $I_C = 5.99$ or 6 A; $\phi = \tan^{-1} \frac{I_C}{I_R} = 38.66°$.

14. $I = \sqrt{I_L^2 + I_R^2} = \sqrt{6.2^2 + 8.4^2} = 10.44$ A; $\theta = \tan^{-1} \frac{I_C}{I_R} = \tan^{-1} \frac{6.2}{8.4} = 36.43°$; Since θ is clockwise, $\theta = -36.43°$

15. $I_L = \sqrt{I^2 - I_R^2} = \sqrt{12.4^2 - 6.3^2} = 10.68$; $\theta = -\tan^{-1} \frac{I_L}{I_R} = -\tan^{-1} \frac{10.68}{6.3} = -59.46°$

16.

vector	horizontal component	vertical component
X_C	$X_{Cx} = 0$	$X_{Cy} = -50$
X_L	$X_{Lx} = 0$	$X_{Ly} = 90$
R	$R_x = 12$	$R_y = 0$
Z	$Z_x = 12$	$Z_y = 40$

$Z = \sqrt{Z_x^2 + Z_y^2} = \sqrt{12^2 + 40^2} = 41.76$; $\theta = \tan^{-1} \frac{40}{12} = 73.30°$

17.

vector	horizontal component	vertical component
X_C	$X_{Cx} = 0$	$X_{Cy} = -60$
X_L	$X_{Lx} = 0$	$X_{Ly} = 40$
R	$R_x = 24$	$R_y = 0$
Z	$Z_x = 24$	$Z_y = -20$

$Z = \sqrt{Z_x^2 + Z_y^2} = \sqrt{24^2 + -20^2} = 31.24$; $\theta = \tan^{-1} \frac{-20}{24} = -39.81°$

18.

vector	horizontal component	vertical component
X_L	$X_{Lx} = 0$	$X_{Ly} = 38$
X_C	$X_{Cx} = 0$	$X_{Cy} = -256$
R	$R_x = 75$	$R_y = 0$
Z	$Z_x = 75$	$Z_y = -227$

$Z = \sqrt{75^2 + -227^2} = 239.07$; $\theta = \tan^{-1} \frac{-227}{75} = -71.72$

19. $I = \sqrt{I_x^2 + I_y^2} = 55.84$; $\theta = \tan^{-1} \frac{I_y}{I_x} = \tan^{-1} \frac{6.342}{55.478} = 6.52°$

vector	horizontal component		vertical component	
I_A	$I_{Ax} = 20 \cos 35°$	$= 16.383$	$I_{Ay} = 20 \sin 35°$	$= 11.472$
I_B	$I_{Bx} = 15 \cos -20°$	$= 14.095$	$I_{By} = 15 \sin -20°$	$= -5.130$
I_C	I_{Cx}	$= 25.000$	I_{Cy}	$= 0.000$
I	I_x	$= 55.478$	I_y	$= 6.342$

20. $V = \sqrt{V_N^2 + V_E^2} = \sqrt{7.222^2 + 3.941^2} = 8.23$ km/hr; $\theta = \tan^{-1} \frac{V_E}{V_N} = \tan^{-1} \frac{3.941}{7.222} = 28.62°$

vector	horizontal component		vertical component	
S	$S_N = 12 \cos 10°$	$= 11.818$	$S_E = 12 \sin 10°$	$= 2.084$
W	$W_N = 2 \cos 270°$	$= 0.000$	$W_E = 2 \sin 270°$	$= -2.000$
C	$C_N = 6 \cos 140°$	$= -4.596$	$C_E = 6 \sin 140°$	$= 3.857$
V	V_N	$= 7.222$	V_N	$= 3.941$

21. $V_x = 120$ m/s; $V_y = -9.8 \cdot 4$ m/s $= -39.2$ m/s; $V = \sqrt{V_x^2 + V_y^2} = 126.24$ m/s; $\theta = \tan^{-1} \frac{V_y}{V_x} = \tan^{-1} \frac{-39.2}{120} = -18.09°$

22. $4.9t^2 = 5000$; $t^2 = \frac{5000}{4.9}$; $t = \sqrt{\frac{5000}{4.9}}$; $t = 31.94$ sec. $V_y = -9.8(31.94) = -313.012$ m/s; $V = \sqrt{V_x^2 + V_y^2} = \sqrt{120^2 + 313.012^2} = 335.23$ m/s; $\theta = \tan^{-1} \frac{V_y}{V_x} = \tan^{-1} \frac{-313.012}{120} = -69.02°$

23. We will let the longer side be vector **A** and assume that it is horizontal. The other (shorter) side will be **B** and it makes an angle of 40° with the horizontal vector. Before we make the component table, we will convert 1′9″ to $12 + 9 = 21″$ and 2′3″ to $2 \cdot 12 + 3 = 24 + 3 = 27″$. We now form the following component table.

vector	horizontal component		vertical component	
A	$A_x = 27 \cos 0°$ =	27.0	$A_y = 27 \sin 0°$ =	0
B	$B_x = 21 \cos 40°$ ≈	16.1	$B_y = 21 \sin 40°$ ≈	13.5
R	R_x	43.1	R_y	13.5

Thus, the smallest rectangle that can be used is one that measures 43.1 in. × 13.5 in. or $3'7\frac{1}{10}'' \times 1'1\frac{1}{2}''$.

24. We will let the longer side be vector **A** and assume that it is horizontal. The other (shorter) side will be **B** and it makes an angle of 35° with the horizontal vector. We can form the following component table.

vector	horizontal component		vertical component	
A	$A_x = 32 \cos 0° =$	32.0	$A_y = 32 \sin 0° =$	0
B	$B_x = 15 \cos 35° \approx$	12.3	$B_y = 15 \sin 35° \approx$	8.6
R	R_x	44.3	R_y	8.6

The smallest rectangle will be one that measures 44.3 cm × 8.6 cm.

≡ 10.4 OBLIQUE TRIANGLES: LAW OF SINES

1. $A = 19.4°$; $B = 85.3°$; $c = 22.1$; $C = 180° - A - B = 180° - 19.4° - 85.3° = 75.3°$; $\frac{c}{\sin A} = \frac{c}{\sin C}$; $\frac{a}{\sin 19.4°} = \frac{(22.1)}{\sin 75.3°}$; $a = \frac{(22.1) \sin 19.4°}{\sin 75.3°} = 7.59$; $\frac{b}{\sin B} = \frac{c}{\sin C}$; $\frac{b}{\sin 85.3°} = \frac{22.1}{\sin 75.3°}$; $b = \frac{(22.1) \sin 85.3°}{\sin 75.3°} = 22.77$. Summary: $C = 75.3°$, $a = 7.59$, $b = 22.77$

2. $A = 180° - (62.4° + 43.9°) = 73.7°$, $\frac{a}{\sin A} = \frac{b}{\sin B}$; $\frac{12.4}{\sin 73.7°} = \frac{b}{\sin 62.4°}$; $b = \frac{(12.4) \sin 62.4°}{\sin 73.7°} \approx 11.45$; $\frac{12.4}{\sin 73.7°} = \frac{c}{\sin 43.9°}$; $c = \frac{(12.4) \sin 43.9°}{\sin 73.7°} \approx 8.96$. Summary: $A = 73.7°, b = 11.45, c = 8.96$

3. $a = 14.2$; $b = 15.3$; $B = 97°$; $\frac{a}{\sin A} = \frac{b}{\sin B}$; $\sin A = \frac{a \sin B}{b} = \frac{14.2 \sin 97°}{15.3}$; $\sin A = .9211866$; $A = 67.1°$; Since B is obtuse, A must be acute. $C = 180° - A - B = 180° - 67.1° - 97° = 15.9°$; $\frac{c}{\sin C} = \frac{b}{\sin B}$; $c = \frac{15.3 \sin 15.9°}{\sin 97°} = 4.22$; $A = 67.1°$; $C = 15.9°$; $c = 4.22$

4. $A = 27.42°$, $a = 27.3$, $b = 35.49$; $\frac{b}{\sin B} = \frac{a}{\sin A}$; $\sin B = \frac{b \sin A}{a} = \frac{35.49 \sin 27.42°}{27.3}$; $\sin B = 0.59866$; $B_1 = 36.77°$; $B_2 = 180° - B_1 = 143.23°$; $C_1 = 180° - A - B_1 = 180° - 27.42° - 36.77° = 115.81°$; $C_2 = 180° - A - B_2 = 180° - 27.42° - 143.23° = 9.35°$; $\frac{c_1}{\sin C_1} = \frac{a}{\sin A}$; $c_1 = \frac{27.3 \sin 115.81°}{\sin 27.42°} = 53.37$; $\frac{c_2}{\sin C_2} = \frac{a}{\sin A}$; $c_2 = \frac{27.3 \sin 9.35°}{\sin 27.42°} = 9.63$; $B_1 = 36.77°$; $C_1 = 115.81°$; $c_1 = 53.37$; $B_2 = 143.23°$; $C_2 = 9.35°$; $c_2 = 9.63$

5. $A = 86.32°$, $a = 19.19$, $c = 18.42$; $\frac{c}{\sin C} = \frac{a}{\sin A}$; $\sin C = \frac{18.42 \sin 86.32}{19.19}$; $\sin C = 0.9578958$, $C_1 = 73.31°$; $C_2 = 180° - C_1 = 106.69°$; but $C_2 + A > 180°$ is not possible. $B = 180° - A - C = 180° - 86.32° - 73.31° = 20.37°$; $\frac{b}{\sin B} = \frac{a}{\sin A}$; $b = \frac{19.19 \sin 20.37}{\sin 86.32} = 6.69$; $C = 73.31°$; $B = 20.37°$; $b = 6.69$

6. $B = 75.46°$, $b = 19.4$, $C = 44.95°$; $\frac{c}{\sin C} = \frac{b}{\sin B}$; $c = \frac{b \sin C}{\sin B} = \frac{19.4 \sin 44.95°}{\sin 75.46°} \approx 14.16$; $A = 180° - (75.46° + 44.95°) = 59.59°$; $\frac{a}{\sin A} = \frac{b}{\sin B}$; $a = \frac{19.4 \sin 59.59°}{\sin 75.46°} \approx 17.28$. Summary: $c = 14.16$, $A = 59.59°$, $a = 17.28$

7. $B = 39.4°$, $b = 19.4$, $c = 35.2$; $\frac{c}{\sin C} = \frac{b}{\sin B}$; $\sin C = \frac{c \sin B}{b} = \frac{35.2 \sin 39.4}{19.4}$; $\sin C = 1.15$. Since the range of the sine function is $[-1, 1]$, there is no solution, and so this is not a triangle.

8. $A = 84.3°$, $b = 9.7$, $C = 12.7°$; $B = 180° - A - C = 180° - 84.3° - 12.7° = 83°$; $\frac{a}{\sin A} = \frac{b}{\sin B}$; $a = \frac{b \sin A}{\sin B} = \frac{9.7 \sin 84.3}{\sin 83} = 9.72$; $\frac{c}{\sin C} = \frac{b}{\sin B}$; $c = \frac{b \sin C}{\sin B} = \frac{9.7 \sin 12.7}{\sin 83} = 2.15$. Summary: $B = 83°$, $a = 9.72$, $c = 2.15$

9. $A = 45°$, $a = 16.3$, $b = 19.4$; $\frac{b}{\sin B} = \frac{a}{\sin A}$; $\sin B = \frac{b \sin A}{a} = \frac{19.4 \sin 45°}{16.3}$; $\sin B = 0.841587$; $B_1 = 57.31°$; $B_2 = 180° - B_1 = 122.69°$. $C_1 = 180° - A - B_1 = 180° - 45° - 57.31° = 77.69°$. $C_2 = 180° - A - B_2 = 180° - 45° - 122.69° = 12.31°$. $\frac{c_1}{\sin C_1} = \frac{a}{\sin A}$; $c_1 = \frac{16.3 \sin 77.69°}{\sin 45°} = 22.52$. $\frac{c_2}{\sin C_2} = \frac{a}{\sin A}$; $c_2 = \frac{16.3 \sin 12.31}{\sin 45} = 4.91$; $B_1 = 57.31°$; $C_1 = 77.69°$; $c_1 = 22.52$; $B_2 = 122.69°$; $C_2 = 12.31°$; $c_2 = 4.91$.

10. $a = 10.4$, $c = 5.2$, $C = 30°$; $\frac{a}{\sin A} = \frac{c}{\sin C}$; $\sin A = \frac{a \sin C}{c} = \frac{10.4 \sin 30}{5.2} = 1$; $A = 90°$, $B = 180° - A - C = 180° - 90° - 30° = 60°$; $\frac{b}{\sin B} = \frac{c}{\sin C}$; $b = \frac{5.2 \sin 60°}{\sin 30°} = 9.01$. Summary: $A = 90°$, $B = 60°$, $b = 9.01$.

11. $a = 42.3$, $B = 14.3°$, $C = 16.9°$; $A = 180° - B - C = 180° - 14.3° - 16.9° = 148.8°$; $\frac{b}{\sin B} = \frac{a}{\sin A}$; $b = \frac{42.3 \sin 14.3}{\sin 148.8} = 20.17$; $\frac{c}{\sin C} = \frac{a}{\sin A}$; $c = \frac{42.3 \sin 16.9}{\sin 148.8} = 23.74$. Summary: $A = 148.8°$, $b = 20.17$, $c = 23.74$

12. $A = 105.4°$, $B = 68.2°$, $c = 4.91$; $C = 180° - A - B = 180° - 105.4° - 68.2° = 6.4°$; Using $\frac{a}{\sin A} = \frac{c}{\sin C}$, we get $a = \frac{4.91 \sin 105.4°}{\sin 6.4°} = 42.47$. Next we use $\frac{b}{\sin B} = \frac{c}{\sin C}$, to get $b = \frac{4.91 \sin 68.2}{\sin 6.4} = 40.90$. Summary: $c = 6.4°$, $a = 42.47$, $b = 40.90$.

13. $b = 19.4$, $c = 12.5$, $C = 35.6°$; $\frac{b}{\sin B} = \frac{c}{\sin C}$; $\sin B = \frac{b \sin C}{c} = \frac{19.4 \sin 35.6}{12.5}$; $\sin B = 0.9034549$; $B_1 = 64.62°$; $B_2 = 180° - B_1 = 115.38°$; $A_1 = 180° - B_1 - C = 180° - 64.62° - 35.6° = 79.78°$; $A_2 =$

$180° - B_2 - C = 180° - 115.38° - 35.6° = 29.02°$; $a_1 = \frac{c \sin A_1}{\sin C} = \frac{12.5 \sin 79.78}{\sin 35.6°} = 21.13$; $a_2 = \frac{c \sin A_2}{\sin C} = \frac{12.5 \sin 29.02}{\sin 35.6°} = 10.42$. Summary: $B_1 = 64.62, A_1 = 79.78, a_1 = 21.13$, or $B_2 = 115.38$, $A_2 = 29.02, a_2 = 10.42$.

14. $a = 121.4, A = 19.7°, c = 63.4$; $\sin C = \frac{c \sin A}{a} = \frac{63.4 \sin 19.7°}{121.4} = 0.176045$; $C = 10.1°$. The other answer will not work. $B = 180° - A - C = 180° - 19.7° - 10.1° = 150.02°$; $b = \frac{a \sin B}{\sin A} = \frac{121.4 \sin 150.2°}{\sin 19.7°} = 179.0$.

15. $a = 19.7, b = 8.5, B = 78.4°$; $\sin A = \frac{a \sin B}{b} = \frac{19.7 \sin 78.4}{8.5} = 2.27$. Since 2.27 is not in the range of the sine function, there is no solution.

16. $b = 9.12, B = 1.3$ rad, $C = 0.67$ rad; $A = \pi - B - C = 3.14 - 1.3 - .67 = 1.17$ rad. $a = \frac{b \sin A}{\sin B} = \frac{9.12 \sin 1.17}{\sin 1.3} = 8.71$; $c = \frac{b \sin C}{\sin B} = \frac{9.12 \sin 0.67}{\sin 1.3} = 5.88$: $A = 1.17$ rad, $a = 8.71, c = 5.88$

17. $b = 8.5, c = 19.7, C = 1.37$ rad; $\sin B = \frac{b \sin C}{c} = \frac{8.5 \sin 1.37}{19.7} = .4228$; $B_1 = 0.44$ rad; $B_2 = 3.14 - 0.44 = 2.7$; $A_1 = \pi - B_1 - C = 3.14 - .44 - 1.37 = 1.33$; $A_2 = \pi - B_2 - C = 3.14 - 2.7 - 1.37 = -0.93$ cannot be so only $B_1 + A_1$ are answers. $a = \frac{c \sin A}{\sin C} = \frac{19.7 \sin 1.33}{\sin 1.37} = 19.52$: $B = 0.44, A = 1.33, a = 19.52$.

18. $b = 19.7, c = 36.4, C = 0.45$ rad; $\sin B = \frac{b \sin C}{c} = \frac{19.7 \sin 0.45}{36.4} = 0.2354$. $B = 0.24$ rad only answer $A = \pi - B - C = 3.14 - 0.24 - 0.45 = 2.45$ rad. $a = \frac{c \sin A}{\sin C} = \frac{36.4 \sin 2.45}{\sin .45} = 53.37$: $B = 0.24, A = 2.45, a = 53.37$.

19. $A = 0.47$ rad, $b = 195.4, C = 1.32$ rad; $B = \pi - A - C = 3.14 - 0.47 - 1.32 = 1.35$. $a = \frac{b \sin A}{\sin B} = \frac{195.4 \sin 0.47}{\sin 1.35} = 90.7$; $c = \frac{b \sin C}{\sin B} = \frac{195.4 \sin 1.32}{\sin 1.35} = 194.0$: $B = 1.35, a = 90.7$, $c = 194.0$.

20. $a = 29.34, A = 1.23$ rad, $C = 1.67$ rad; $B = \pi - A - C = 3.14 - 1.23 - 1.67 = 0.24$ rad. $b = \frac{a \sin B}{\sin A} = \frac{29.34 \sin 0.24}{\sin 1.23} = 7.40$; $c = \frac{a \sin C}{\sin A} = \frac{29.34 \sin 1.67}{\sin 1.23} = 30.98$: $B = 0.24, b = 7.40, c = 30.98$.

21. The length between towers A and B is $AB = c = 360$ m. The angle at tower $B = 67.4°$ and at tower $A = 49.3°$. Hence, the angle at tower $C = 180° - A - B = 63.3°$. The distance between towers A and C is $AC = b = \frac{c \sin B}{\sin C} = \frac{360 \sin 67.4}{\sin 63.3} = 372.0$ m and the distance between towers B and C is $BC = a = \frac{c \sin A}{\sin C} + \frac{360 \sin 49.3}{\sin 63.3} = 305.5$ m.

22. $AC = b = 250$ m, $BC = a = 275$ m, $A = 43.62°$: $\sin B = \frac{b \sin A}{a} = \frac{250 \sin 43.62}{275} = 0.627157$; $B = 38.84°$; $C = 180° - A - B = 97.54°$. $AB = c = \frac{a \sin C}{\sin A} = \frac{275 \sin 97.54}{\sin 43.62} = 395.2$ m.

23. Let C designate the present location of the airplane. Since the plane has a heading of 313°, then $\angle A$ of $\triangle ABC$ is 43°. We are given that the heading from B is 27° and so $B = 90° - 27° = 63°$. Using these two angles, we find that $C = 180° - 43° - 63° = 74°$. Now, using the Sine law, if the distance the plane has flown is b, then we have $\frac{b}{\sin B} = \frac{c}{\sin C}$. We were given that $c = 37$ mi, and so we have $b = \frac{c \sin B}{\sin C} = \frac{37 \sin 63°}{\sin 74°} \approx 34.3$ mi.

24. A football field has two 40-yard lines. The angle given is to the farther of the two lines from the end of the stadium whose height we want to determine. Let A be the top of the given end of the stadium, B a point on the 50-yard line, and C a point on the given 40-yard line, so that A, B, and C are colinear and the line $\overleftrightarrow{BC}$ is perpendicular to the 40- (and 50-) yard lines. The distance, d, from the top of the stadium to the 40-yard line is AC. We can determine the following angle measures: $\angle ACB = 10.45°$, $\angle CAB = 0.91°$, and $\angle CBA = 168.64°$. From the Law of Sines, we see that

$$d = \frac{10 \sin 168.64°}{\sin 0.91°} = 124.02.$$

If h is the height of the stadium, then $\sin 10.45° = \frac{h}{d}$ and so $h = d \sin 10.45° = 22.49$ yards. Thus, the height of this one end of the stadium is about 22.5 yards.

25. $c = 180° - 47° - 67° = 66°$; $CB = a = \frac{400 \sin 47°}{\sin 66°} = 320.22$; $\sin 67° = \frac{h}{a}$; $h = 320.22 \sin 67° = 294.77$ m.

26. Since the size of $\angle A = 47°$ and $\angle B = 102°$, we find that the size of $\angle C = 180° - (47° + 102°) = 31°$. We use the Sine Law to determine the length of AC.

$$\frac{AC}{\sin B} = \frac{AB}{\sin C}$$
$$\frac{AC}{\sin 102°} = \frac{15.6}{\sin 31°}$$
$$AC = \frac{15.6 \sin 102°}{\sin 31°}$$
$$\approx 29.6$$

So, AC is 29.6 cm long. Next, we use the Sine Law to find the length of BC.

$$\frac{BC}{\sin A} = \frac{AB}{\sin C}$$
$$\frac{BC}{\sin 47°} = \frac{15.6}{\sin 31°}$$
$$BC = \frac{15.6 \sin 47°}{\sin 31°}$$
$$\approx 22.2$$

So, BC is 22.2 cm long.

27. Let h be the height of the building above the eye of the observer and d the distance of the observer from the building. Then, $\tan 57.8° = \frac{h}{d}$ and $\tan 67.4° = \frac{h+225}{d}$. Hence $d = \frac{h}{\tan 57.8°} = \frac{h+225}{\tan 67.4°}$. Solving $\frac{h}{\tan 57.8°} = \frac{h}{\tan 67.4°} + \frac{225}{\tan 67.4°}$ or $h\left(\frac{1}{\tan 57.8°} - \frac{1}{\tan 67.4°}\right) = \frac{225}{\tan 67.4°}$ or $h = 438.74$ ft or $438'9''$. Adding the height of the observer's eyes, we get $438'9'' + 5'7'' = 444'4''$.

28. Call the center of the circle C and the two adjacent holes A and B. Since the holes are equally spaced, $\angle ACB = \frac{360°}{5} = 72°$. A and B are on a circle

centered at C, so $\overline{CA} = \overline{CB} = 6.25$ in. and $\triangle ABC$ is an isosceles triangle. Thus, $\angle CAB = \angle CBA = \frac{180° - 72°}{2} = 54°$. We can use the Sine Law.

$$\frac{AB}{\sin C} = \frac{AC}{\sin B}$$
$$AB = \frac{AC \sin C}{\sin B}$$
$$= \frac{6.25 \sin 72°}{\sin 54°} \approx 7.35$$

The holes are about 7.35 in. apart.

29. Call the center of the circle C and the two adjacent holes A and B. Since the holes are equally spaced, $\angle ACB = \frac{360°}{12} = 30°$. A and B are on a circle centered at C, so $\overline{CA} = \overline{CB} = 16.40$ cm and $\triangle ABC$ is an isosceles triangle. Thus, $\angle CAB = \angle CBA = \frac{180° - 30°}{2} = 75°$. We can use the Sine Law.

$$\frac{AB}{\sin C} = \frac{AC}{\sin B}$$
$$AB = \frac{AC \sin C}{\sin B}$$
$$= \frac{16.40 \sin 30°}{\sin 75°} \approx 6.42$$

The holes are about 8.49 cm apart.

30. The angle at the top (between the guy wire and the line of elevated sight) is 21.25°. Thus, you can use the Law of Sines to determine that the length of the guy wire, w, is $\frac{w}{\sin 42.5°} = \frac{8.2}{\sin 21.25°}$ and $w = \frac{8.2 \sin 42.5°}{\sin 21.25°} \approx 15.28$ m.

≡ 10.5 OBLIQUE TRIANGLES: LAW OF COSINES

1. $a = 9.3, b = 16.3, C = 42.3°$ SAS so Cosine Law. First we find c by using the cosine law: $c^2 = a^2 + b^2 - 2ab \cos C = 9.3^2 + 16.3^2 - (9.3)(16.3) \cos 42.3°$, $c^2 = 127.9386, c = 11.3$. Next find the angle opposite the short side, namely A using Law of Sines. $\frac{a}{\sin A} = \frac{c}{\sin C}$; $\sin A = \frac{a \sin C}{c} = \frac{9.3 \sin 42.3}{11.3}$; $\sin A = 0.553895, A = 33.6°, B = 180° - A - C = 180° - 42.3° - 33.6° = 104.1°$. (Note: If you use the Law of Sines to find B, you may not recognize that it is obtuse. That's why you should find the angle opposite the shorter side since it must be acute.) $c = 11.31, A = 33.6°, B = 104.1°$.

2. $A = 16.25°, b = 29.43, c = 36.52$; $a^2 = b^2 + c^2 - 2bc \cos A = 29.43^2 + 36.52^2 - 2(29.43)(36.52) \cos 16.25° = 136.1436; a = 11.67$.

$$\sin B = \frac{b \sin A}{a} = \frac{29.43 \sin 16.25}{11.67} = .705687;$$

$B = 44.89; C = 180° - A - B = 180° - 16.25° - 44.89° = 118.86°$. (Note: If you used Law of Sines to find C, you may not recognize that it is obtuse. That is why you should find the angle opposite the shorter side first since it must be acute.) $a = 11.67, B = 44.80°, C = 118.86°$.

3. $a = 47.85, B = 113.7°, c = 32.79; b^2 = a^2 + c^2 - 2ac \cos B = 47.85^2 + 32.79^2 - 2(47.85)(32.79) \cos 113.7° = 4626.1199$; hence, $b = 68.02$ and so, $\sin A = \frac{a \sin B}{b} = \frac{47.85 \sin 113.7°}{68.02} = 0.64414$. Thus, $A = 40.1°$. $C = 180° - A - B = 180° - 40.1° - 113.7° = 26.2°$. $b = 68.02, A = 40.1°, C = 26.2°$: (Note: Since B is obtuse, A and C must be acute).

4. $a = 19.52, b = 63.42, c = 56.53$: Use the alternate form of the Law of Cosines to find one of the angles. Choose B since it is the largest and if it's obtuse, the Law of Cosines will yield this answer directly.
$\cos B = \frac{a^2 + c^2 - b^2}{2ac} = \frac{19.52^2 + 56.53^2 - 63.42^2}{2(19.52)(56.53)}$; $\cos B \approx -0.2018$; $B = 101.6°$. $\sin A = \frac{a \sin B}{b} = \frac{19.52 \sin 101.6°}{63.42} = 0.3015$; $A = 17.5°$; $C = 180° - A - C = 180° - 17.5° - 101.6° = 60.9°$: $A = 17.5°, B = 101.6°, C = 60.9°$

5. $a = 29.43, b = 16.37, c = 38.62$: Use the alternate form of Law of Cosines to find one of the angles. Choose C since it's the largest and if it's obtuse, the Law of Cosines will yield this answer directly. $\cos C = \frac{a^2 + b^2 - c^2}{2ab} = \frac{29.43^2 + 16.37^2 - 38.62^2}{2(29.43)(16.37)}$; $\cos C \approx -0.37093$; $C \approx 111.8°$; $\sin A = \frac{a \sin C}{c} = \frac{29.43 \sin 111.8°}{38.62} \approx 0.7075$; $A = 45.0°$; $B = 180° - A - C = 180° - 45.0° - 111.8° = 23.2°$: $A = 45.0°, B = 23.2°, C = 111.8°$.

6. $A = 121.37°, b = 112.37, c = 93.42; a^2 = b^2 + c^2 - 2bc \cos A = 112.37^2 + 93.42^2 - 2(112.37)(93.42) \cos 121.37° = 32283.636$; $a = 179.68$. $\sin C = \frac{c \sin A}{a} = \frac{93.42 \sin 121.37}{179.68} = 0.443924$; $C = 26.35°$. $B = 180° - A - C = 180° - 121.37° - 26.35° = 32.28°$. $a = 179.68, B = 32.28°, C = 26.35°$.

7. $a = 63.92, B = 92.44°, c = 78.41; b^2 = a^2 + c^2 - 2ac \cos B = 63.92^2 + 78.41^2 - 2(63.92)(78.41) \cos 92.44° = 10660.645$; $b = 103.25$. $\sin A = \frac{a \sin B}{b} = \frac{63.92 \sin 92.44}{103.25} = .6185186$; $A = 38.21$. $C = 180° - A - B = 180° - 38.21° - 92.44° = 49.35°$: $b = 103.25, A = 38.21°, C = 49.35°$

8. $a = 19.53, b = 7.66, C = 32.56°; c^2 = a^2 + b^2 - 2ab \cos C = 19.53^2 + 7.66^2 - 2(19.53)(7.66) \cos 32.56° = 187.9226$; $c = 13.71$. $\sin B = \frac{b \sin C}{c} = \frac{7.66 \sin 32.56}{13.71} = .300691$; $B = 17.50°$. $A = 180° - B - C = 180° - 17.50° - 32.56° = 129.94°$: $c = 13.71, A = 129.94°, B = 17.50°$

9. $a = 4.527, b = 6.239, c = 8.635$: Find C first, $\cos C = \frac{a^2 + b^2 - c^2}{2ab} = \frac{4.527^2 + 6.239^2 - 8.635^2}{2(4.527)(6.239)}$; $\cos C = -0.268099$; $C = 105.55°$ $\sin A = \frac{a \sin C}{c} = \frac{4.527 \sin 105.55°}{8.635} = 0.505072$; $A = 30.34$. $B = 180° - A - C = 180° - 30.34° - 105.55° = 44.11°$: $A = 30.34°, B = 44.11°, C = 105.55°$.

10. $A = 7.53°, b = 37.645, c = 42.635$; SAS, $a^2 = b^2 + c^2 - 2bc \cos A = 52.58182$; $a = 7.251$. $\sin B = \frac{b \sin A}{a} = 0.6803476$; $B = 42.87°$. $c = 180° - A - B = 129.6°, a = 7.251, B = 42.87°, C = 129.6°$.

11. $a = 8.5, b = 15.8, C = 0.82$ rad; SAS, $c^2 = a^2 + b^2 - 2ab \cos C = 138.6454$; $c = 11.77$; $\sin A = \frac{a \sin C}{c} = .5280$; $A = 0.56$ rad. $B = \pi - A - C = 3.14 - 0.82 - 0.56 = 1.76$ rad: $c = 11.77, A = 0.56$ rad, $B = 1.76$ rad.

12. $A = 0.31, b = 15.8, c = 38.47$; SAS, $a^2 = b^2 + c^2 - 2bc \cos A = 571.8747$; $a = 23.91$; $\sin B = \frac{b \sin A}{a} = 0.20159$; $B = 0.203$ rad. $C = \pi - A - B = 3.14 - 0.31 - 0.20 = 2.63$ rad: $a = 23.91, B = 0.20$ rad, $C = 2.63$ rad.

13. $a = 52.65, B = 1.98, c = 35.8; b^2 = a^2 + c^2 - 2ac \cos B = 5553.5624$; $b = 74.52$. $\sin C = \frac{c \sin B}{b} = .440744, C = 0.46$; $A = \pi - B - C = 3.14 - 1.98 - 0.46 = 0.70$: $b = 74.52, A = 0.70, C = 0.46$.

14. $a = 43.5, b = 63.4, c = 37.3$; SSS, $\cos B = \frac{a^2 + c^2 - b^2}{2ac} = -.2268, B = 1.80$ rad. $\sin A = \frac{a \sin B}{b} = 0.668176$; $A = 0.73$ rad. $C = \pi - A - B = 3.14 - 0.73 - 1.80 = 0.61$. $A = 0.73, B = 1.80, C = 0.61$

15. $a = 36.27, b = 24.55, c = 44.26$; $\cos C = \frac{a^2 + b^2 - c^2}{2ab} = -0.02287$; $C = 1.59$ rad. $\sin A = \frac{a \sin C}{c} = 0.8193$; $A = 0.96$. $B = \pi - A - C = 0.59$: $A = 0.96, B = 0.59, C = 1.59$

16. $a = 2.41, b = 153.21, c = 87.49$; SAS; $a^2 = b^2 + c^2 - 2bc \cos A = 51076.452$; $a = 226.00$. $\sin B = \frac{b \sin A}{b} = 0.7981$; $A = 0.924$. $C = \pi - A - B = 3.142 - 0.924 - 1.625 = 0.593$: $b = 68.56, A = 0.924, C = 0.593$

17. $a = 54.8, B = 1.625, c = 38.33$; SAS; $b^2 = a^2 + c^2 - 2ac \cos B = 4699.8253$; $b = 68.56$. $\sin A = \frac{a \sin B}{b} = 0.7981$; $A = 0.924$. $C = \pi - A - B = 3.142 - 0.924 - 1.625 = 0.593$: $b = 68.56, A = 0.924, C = 0.593$

18. $a = 7.621, b = 3.429, C = 0.183$; SAS, $c^2 = a^2 + b^2 - 2ab \cos C = 18.4456$; $c = 4.295$. $\sin B = \frac{b \sin C}{c} = 0.1452876$; $B = 0.146$. $A = \pi - B - C = 3.142 - 0.146 - 0.183 = 2.813$. $c = 4.295, A = 2.813, B = 0.146$

19. $a = 2.317, b = 1.713, c = 1.525$; SSS, $\cos A = (b^2 + c^2 - a^2)/2bc = -0.020766$, $A = 1.59$; $\sin B = \frac{b \sin A}{a} = .7392$, $B = 0.83$; $C = \pi - A - B = 3.14 - 1.59 - 0.83 = 0.72$: $A = 1.59, B = 0.83, C = 0.72$

20. $A = 0.09, b = 40.75, c = 50.25$; SAS, $a^2 = b^2 + c^2 - 2bc \cos A = 106.825$; $a = 10.34$. $\sin B = \frac{b \sin A}{a} = .35421188$; $B = 0.36$. $C = \pi - A - B = 3.14 - 0.09 - 0.36 = 2.69$: $a = 10.34, B = 0.36, C = 2.69$

21. From the diagram we know that $a = 90 + 34 = 124.7°$, $\overline{AC} = b = 6335$, $\overline{AB} = c = 12325$, $\overline{BC} = a$; $a^2 = b^2 + c^2 - 2bc \cos A = 6335^2 + 12525^2 - 2(6335)(12325) \cos 124.7° = 280\ 935\ 259.5$ and so $a = 16\ 761$. Distance above the Earth's surface is $16\ 761 - 6\ 335 = 10\ 426$ km.

22. $a = 21 \times 3 = 63$ km. $b = 21 \times 4 = 84$ km. $C = 180° - 37 = 143°$. $c^2 = a^2 + b^2 - 2ab \cos C = 63^2 + 84^2 - 2 \cdot 63 \cdot 84 \cos 143$; $c^2 = 19477.76$; $c = 139.6$ km. Bearing $= B = \sin^{-1}\left(\frac{84 \sin 143}{139.6}\right) = 21.2°$.

Distance from port is 139.6 km; bearing is 21.2° East of north.

23. The angle between the antenna and the hill is $90 + 12.37° = 102.37°$ If the length of the wire is ℓ then $\ell^2 = 40^2 + 75^2 - 2 \cdot 40 \cdot 75 \cos 102.37°$; $\ell^2 = 8510.34$; $\ell = 92.25$ ft.

24. $R^2 = 50^2 + 35^2 - 2 \cdot 50 \cdot 35 \cos 147.85° = 6688.3$, $R = 81.78$ lb

25. $R^2 = 12^2 + 23^2 - 2 \cdot 12 \cdot 23 \cos 121.27° = 959.53$; $R = \sqrt{959.53} = 30.98$ or 31 N. $\theta = \sin^{-1}\left(\frac{23 \sin 121.27}{31}\right) = 39.36°$

26. $\cos \phi = (36.4^2 + 15.5^2 - 30.1^2)/(2 \times 36.4 \times 15.5)$; $\cos \phi = 0.58419$; $\phi = 54.25°$. $\theta = 180° - \phi = 180° - 54.25° = 125.75°$

27. The other angle of the parallelogram is $180° - 74° = 106°$. By the Law of Cosines, $Z^2 = 56^2 + 38^2 - 2 \cdot 56 \cdot 38 \cos 106° = 5753.11$. Taking the square root, we find that $Z = \sqrt{5753.11} \approx 75.8$ or about 76 kΩ.

28. By the Law of Cosines $x^2 = 3.38^2 + 2.14^2 - (3.38)(2.14) \cos 42° \approx 10.6287$. Hence, $x = \sqrt{10.6287} \approx 3.26$ m.

CHAPTER 10 REVIEW

1.

2.

3.

4.

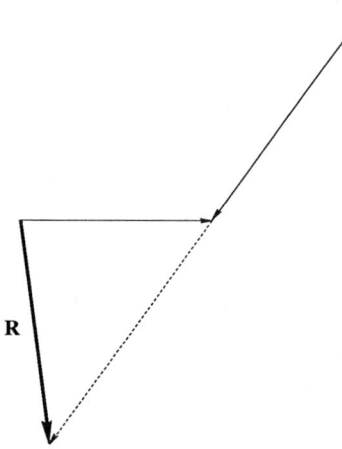

5. $P_x = 35\cos 67° = 13.68; P_y = 35\sin 67° = 32.22$

6. $P_x = 19.7\cos 237° = -10.73; P_y = 19.7\sin 237° = -16.52$

7. $P_x = 23.4\cos 172.4° = -23.19; P_y = 23.4\sin 172.4° = 3.09$

8. $P_x = 14.5\cos 338° = 13.44; P_y = 14.5\sin 338° = -5.43$

9. $A = \sqrt{A_x^2 + A_y^2} = \sqrt{16^2 + (-8)^2} = 17.89. \theta = \tan^{-1}\left(\frac{A_y}{A_x}\right) = \tan^{-1}\frac{-8}{16} = \tan^{-1}-\frac{1}{2} = -26.57°$ or $333.43°$

10. $B = \sqrt{B_x^2 + B_y^2} = \sqrt{(-27)^2 + 32^2} = 41.87;$
$\theta_R = \tan^{-1}\frac{32}{27} = 49.84$ Second Quadrant so $180° - 49.84° = 130.16°$

11. $A_x = 38\cos 15° = 36.71; A_y = 38\sin 15° = 9.84$

12. $B_x = 43.5\cos 127° = -26.18; A_y = 43.5\sin 127° = 34.74$

13. $C_x = 19.4\cos 1.25 = 6.12;$
$C_y = 19.4\sin 1.25 = 18.41$

14. $D_x = 62.7\cos 5.37 = 38.32; D_y = 62.7\sin 5.37 = -49.62$

15. $R = \sqrt{R_x^2 + R_y^2} = 50.41;$
$\theta = \tan^{-1}\frac{R_y}{R_x} = 20.69°$

vector	horizontal component	vertical component
A	$A_x = 19\cos 32° = 16.11$	$A_y = 19\sin 32° = 10.07$
B	$B_x = 32\cos 14° = 31.05$	$B_y = 32\sin 14° = 7.74$
R	R_x 47.16	R_y 17.81

16. $R = \sqrt{R_x^2 + R_y^2} = 36.96;$
$\theta_R = \tan^{-1}\frac{5.88}{36.49} = 9.15°.$ Since θ is in 4th Quadrant $\theta = 360 - 9.15 = 350.85°$

vector	horizontal component	vertical component
C	$C_x = 24\cos 57° = 13.07$	$C_y = 24\sin 57° = 20.13$
D	$D_x = 35\cos 312° = 23.42$	$D_y = 35\sin 312° = -26.01$
R	R_x 36.49	R_y -5.88

17. $R = \sqrt{R_x^2 + R_y^2} = 72.77;$
$\theta_R = \tan^{-1}\frac{.92}{72.76} = .01264.$ θ in 2nd Quadrant so $\theta = \pi - \theta_R = 3.13$ rad

vector	horizontal component	vertical component
E	$E_x = 52.6\cos 2.53 = -43.07$	$E_y = 52.6\sin 2.53 = 30.20$
F	$F_x = 41.7\cos 3.92 = -29.69$	$F_y = 41.7\sin 3.92 = -29.28$
R	R_x -72.76	R_y 0.92

18. $R = \sqrt{R_x^2 + R_y^2} = 57.67;$
$\theta_R = \tan^{-1}\frac{57.57}{3.41} = 1.51.$ θ is in 3rd Quadrant so $\theta = \pi + 1.51 = 4.65$

vector	horizontal component	vertical component
G	$G_x = 43.7\cos 4.73 = 0.77$	$G_y = 43.7\sin 4.73 = -43.69$
H	$H_x = 14.5\cos 4.42 = -4.18$	$H_y = 14.5\sin 4.42 = -13.88$
R	R_x -3.41	R_y -57.57

19. $a = 14, b = 32, c = 27$: SSS so use Law of Cosines to find the largest angle, namely $B \cos B = \frac{a^2+c^2-b^2}{2ac} = -0.1310; B = 97.5. \sin A = \frac{a\sin B}{b} =$ $.4338; A = 25.7. C = 180° - A - B = 56.8°: A = 25.7°, B = 97.5°, C = 56.8°$

20. $a = 43, b = 52, B = 86.4°$: SSA so Law of Sines. $\sin A = \frac{a \sin B}{b} = .8253; A = 55.6. C = 180° - A - B = 38.0°. c = \frac{b \sin C}{\sin B} = 32.08.$

21. $b = 87.4, B = 19.57°, c = 65.3$: SSA, Sine Law. $\sin C = \frac{c \sin B}{b} = 0.25026; C = 14.49°. A = 180° - B - C = 145.94°. a = \frac{b \sin A}{\sin B} = 146.14. A = 145.94°, a = 146.14, C = 14.49°$

22. $A = 121.3°, b = 42.5, c = 63.7$: SAS, Cosine Law. $a^2 = b^2 + c^2 - 2bc \cos A = 8676.8762; a = 93.15. \sin B = \frac{b \sin A}{a} = 0.3898497; B = 22.9°; C = 180° - A - B = 35.8°: a = 93.15, B = 22.9°, C = 35.8°$

23. $a = 127.35, A = 0.12, b = 132.6$: SSA, Sine Law. $\sin B = \frac{b \sin A}{a} = 0.1246; B = 0.12, C = \pi -$

$A - B = 3.14 - .12 - .12 = 2.90. c = \frac{a \sin C}{\sin A} = 254.51. B = 0.12, C = 2.90, c = 254.51$

24. $b = 84.3, c = 95.4, C = 0.85$: SSA so Sine Law. $\sin B = \frac{b \sin C}{c} = 0.6639; B = 0.73. A = \pi - B - C = 1.56. a = \frac{c \sin A}{\sin C} = 127.0: B = 0.73, A = 1.56, a = 127.0.$

25. $a = 67.9, b = 54.2, C = 2.21$: SAS so Cosine Law. $c^2 = a^2 + b^2 - 2ab \cos C = 11938.92; c = 109.3. \sin B = \frac{b \sin C}{c} = 0.39798; B = 0.41: A = \pi - B - C = 0.52: c = 109.3, A = 0.52, B = 0.41.$

26. $a = 53.1, b = 63.2, c = 74.3$: SSS so use the Cosine Law. $\cos C = (a^2 + b^2 - c^2)/2ab = 0.1927; C = 1.38. \sin A = \frac{a \sin C}{c} = 0.7017; A = 0.78. B = \pi - A - C = 0.98: A = 0.78, B = 0.98, C = 1.38$

27. $R = \sqrt{R_x^2 + R_y^2} = 2831.17;$ $\theta = \tan^{-1} \frac{R_y}{R_x} = 83.97°.$ Resultant force is 2831 kg directed at an angle of 83.97° to the axis of the vehicle.

vector	horizontal component		vertical component	
A	$A_x = 1650 \cos 68° =$	618.10	$A_y = 1650 \sin 68° = 1529.85$	
B	$B_x = 1325 \cos 76° =$	-320.55	$B_y = 1325 \sin 76° = 1285.64$	
R	R_x	297.55	R_y	2815.49

28. Call the alternate route a. Then by the Cosine Law we see that $a^2 = 3621^2 + 2342^2 - 2(3621)(2342) \cos 72.4°; a^2 = 13468181; a = 3669.9$ ft. Over the hill $3621 + 2342 = 5963; 2.3 \times 3669.9 = 8440.77$. Over the hill is less expensive.

29. Parallel component is $126.5 \sin 31.7° = 66.47$ lb. Perpendicular component $126.5 \cos 31.7° = 107.63$ lb.

30. $X_C + X_L = -72 + 52 = -20\Omega \; \theta = \tan^{-1}\left(\frac{-20}{35}\right) = -29.74°$

31. $AB = c, AC = b = 73, A = 123.4, C = 42.1. B = 180° - A - C = 14.5. c = \frac{b \sin C}{\sin B} = 195.47$ m. The distance is 195.47 m.

32. By Cosine Law $d^2 = 1235^2 + 962^2 - 2(1235)(962) \cos 52.57°; d^2 = 1006471; d = 1003.23$ m

CHAPTER 10 TEST

1. $V_x = 47 \cos 117° = -21.34;$ $V_y = 47 \sin 117° = 41.88$

2. $A = \sqrt{A_x^2 + A_y} = \sqrt{12.91^2 + (-14.36)^2} = 19.31;$ $\theta = \tan^{-1} \frac{-14.36}{12.91} = -48.04°$ or $311.96°$

3.

vector	horizontal component		vertical component	
A	$A_x = 25 \cos 64° =$	10.96	$A_y = 25 \sin 64° = 22.47$	
B	$B_x = 40 \cos 112° =$	-14.98	$B_y = 40 \sin 112° = 37.09$	
R	R_x	-4.02	R_y	59.56

$R = \sqrt{R_x^2 + R_y^2} = 59.70; \theta = \tan^{-1} \frac{59.56}{-4.02} = 93.86°$

4. $b = \frac{a \sin B}{\sin A} = \frac{9.42 \sin 67.5}{\sin 35.6} = 14.95.$

5. $a^2 = b^2 + c^2 - 2bc \cos A = 4.95^2 + 6.24^2 - 2(4.95)(6.24) \cos 13.4; a^2 = 87.9743; a = 9.38.$

6. $A = 24°, b = 36.5,$ and $C = 97°: B = 180° - A - C = 59°. a = \frac{b \sin A}{\sin B} = \frac{36.5 \sin 24°}{\sin 59°} = 17.32. c =$

$\frac{b \sin C}{\sin B} = \frac{36.5 \sin 97°}{\sin 59°} = 42.26. B = 59°, a = 17.32, c = 42.26.$

7. $c^2 = 30^2 + 36^2 - 2 \cdot 30 \cdot 36 \cos 97° = 2459.237782; c = 49.59$

CHAPTER
11
Graphs of Trigonometric Functions

≡ 11.1 SINE AND COSINE CURVES: AMPLITUDE AND PERIOD

1. Period $= \frac{2\pi}{2} = \pi$,
   Amplitude $= |3| = 3$,
   Frequency $= \frac{1}{\pi}$

2. Period $= \frac{2\pi}{6} = \frac{\pi}{3}$,
   Amplitude $= |5| = 5$,
   Frequency $= \dfrac{1}{\frac{\pi}{3}} = \frac{3}{\pi}$

3. Period $= 2\pi$,
   Amplitude $= |2| = 2$,
   Frequency $\frac{1}{2\pi}$

4. Period $= \frac{2\pi}{3}$,
   Amplitude $= 7$,
   Frequency $\dfrac{1}{\frac{2\pi}{3}} = \frac{3}{2\pi}$

13. Period $= 2\pi$,
    Amplitude $= |-\frac{1}{2}| = \frac{1}{2}$,
    Frequency $\frac{1}{2\pi}$

14. Period $= \frac{2\pi}{3}$,
    Amplitude $= |-\frac{1}{2}| = \frac{1}{2}$,
    Frequency $\frac{3}{2\pi}$

5. Period $= \frac{2\pi}{2\pi} = 1$,
   Amplitude $= 8$,
   Frequency 1

6. Period $= \frac{2\pi}{\pi} = 2$,
   Amplitude $= 7$,
   Frequency $\frac{1}{2}$

7. Period $= \frac{2\pi}{4} = \frac{\pi}{2}$,
   Amplitude $= \frac{1}{2}$,
   Frequency $\frac{2}{\pi}$

8. Period $= \dfrac{2\pi}{\frac{1}{3}} = 6\pi$,
   Amplitude $= 5$,
   Frequency $\frac{1}{6\pi}$

9. Period $= \dfrac{2\pi}{\frac{1}{2}} = 4\pi$,
   Amplitude $= \frac{1}{3}$,
   Frequency $\frac{1}{4\pi}$

10. Period $= \frac{2\pi}{3}$,
    Amplitude $= \frac{1}{5}$,
    Frequency $\frac{3}{2\pi}$

11. Period $= \dfrac{2\pi}{\frac{1}{4}} = 8\pi$,
    Amplitude $= |-3| = 3$,
    Frequency $\frac{1}{8\pi}$

12. Period $= \dfrac{2\pi}{\frac{1}{3}} = 6\pi$,
    Amplitude $= |-5| = 5$,
    Frequency $\frac{1}{6\pi}$

15.

16.

17.

18.

19.

20.

21.

22.

23.

24.

25.

26.

27.

28.

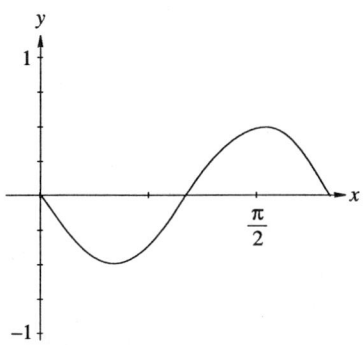

29. Period $= \frac{2\pi}{120\pi} = \frac{1}{60}s$, Amplitude $= 6.5$ A, Frequency 60 Hz

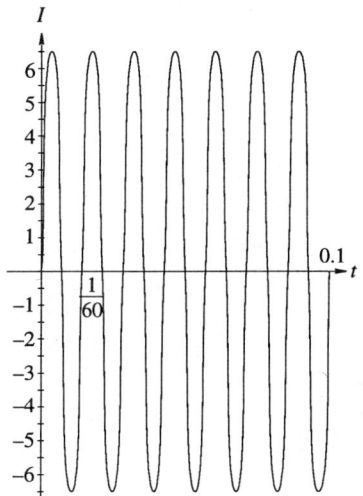

30. The given equation, $E = 12\sin 120\pi t$, is of the form $E = A\sin Bt$, where $A = 12$ and $B = 120\pi$.

(a) Amplitude: $|A| = |12| = 12$, period: $\frac{2\pi}{|B|} = \frac{2\pi}{|120\pi|} = \frac{1}{60}$, frequency: $\frac{1}{\text{period}} = \frac{1}{\frac{1}{60}} = 60$

(b)

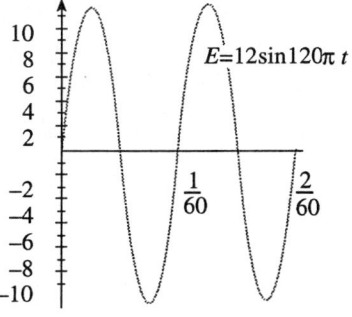

31. The given equation, $G_i = 1094\cos i$, is of the form $E = A\cos Bi$, where $A = 1094$ and $B = 1$.

(a) Amplitude: $|A| = |1094| = 1094$, period: $\frac{2\pi}{|B|} = \frac{2\pi}{|1|} = 2\pi$, frequency: $\frac{1}{\text{period}} = \frac{1}{2\pi}$

(b)

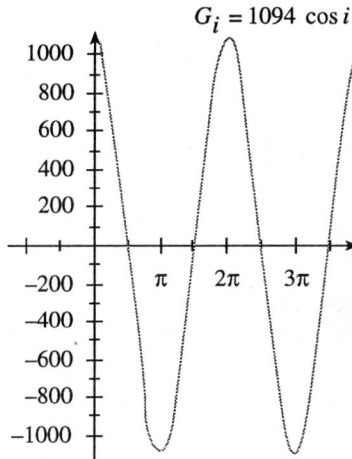

32. The given equation, $y = 2.1\cos 0.45t$, is of the form $y = A\cos Bx$, where $A = 2.1$ and $B = 0.45$.

(a) Amplitude: $|A| = |2.1| = 2.1$, period: $\frac{2\pi}{|B|} = \frac{2\pi}{|0.45|} = \frac{40}{9}\pi \approx 13.96$, frequency: $\frac{1}{\text{period}} = \frac{1}{\frac{2x}{|0.45|}} = \frac{0.45}{2\pi} = \frac{9}{40\pi} \approx 0.0716$

(b)

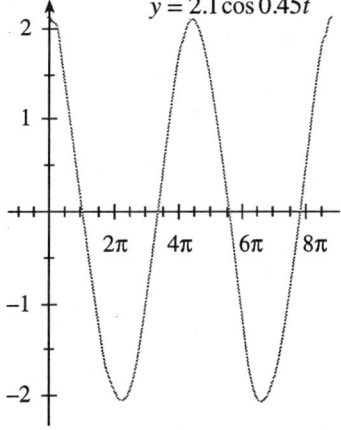

11.2 SINE AND COSINE CURVES: DISPLACEMENT OR PHASE SHIFT

1. Amplitude $= 2$,
Period $= 2\pi$,
Phase Shift $= -\frac{\pi}{4}$ or $\frac{\pi}{4}$ left,
Vertical displacement $= 0$

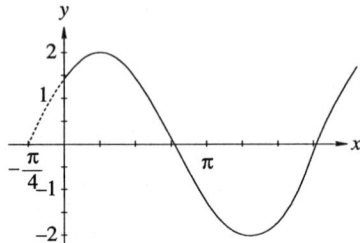

2. Amplitude $= 1.5$,
Period $= 2\pi$,
Phase Shift $= \frac{\pi}{3}$ or $\frac{\pi}{3}$ to the right,
Vertical displacement $= 0$

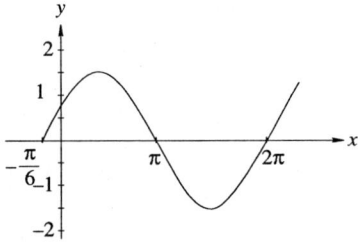

3. Amplitude $= 2.5$,
Period $= \frac{2\pi}{3}$,
Phase Shift $= \frac{\pi}{3}$ or $\frac{\pi}{3}$ to the right,
Vertical displacement $= 0$

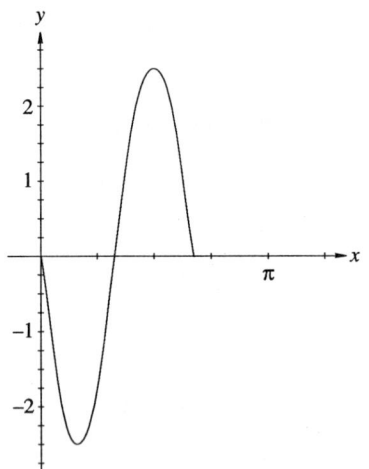

4. Amplitude $= 3$,
Period $= \pi$,
Phase Shift $= -\frac{\pi}{4}$ or $\frac{\pi}{4}$ to the left,
Vertical displacement $= 0$

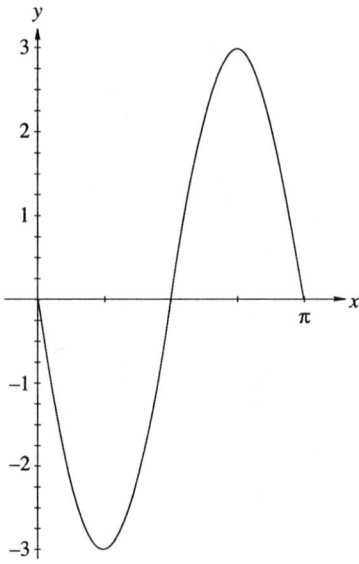

5. Amplitude $= 6$,
Period $= \frac{360°}{1.5} = 240° = \frac{4\pi}{3}$,
Phase Shift $= -\frac{180°}{1.5} = -120° = -\frac{2\pi}{3}$ to the left,
Vertical displacement $= 0$

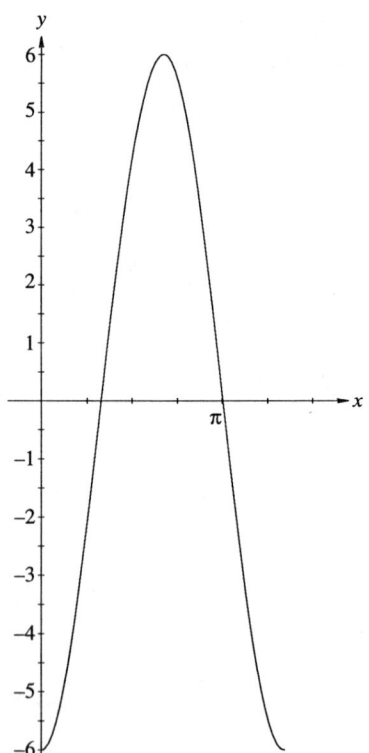

6. Amplitude $= 8$,
Period $= \frac{360°}{2.5} = 102.86° = 4\pi/7$,
Phase Shift $= \frac{90°}{3.5} = 25.7° = \frac{\pi}{7}$,
Vertical displacement $= 0$

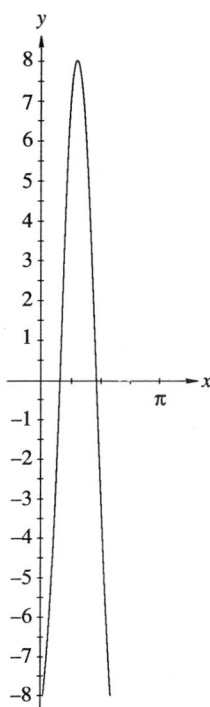

8. Amplitude $= 4$,
Period $= \frac{360°}{5} = 72° = 2\pi/5$,
Phase Shift $= -\frac{600°}{5} = -120° = -\frac{2\pi}{3}$,
Vertical displacement $= 0$

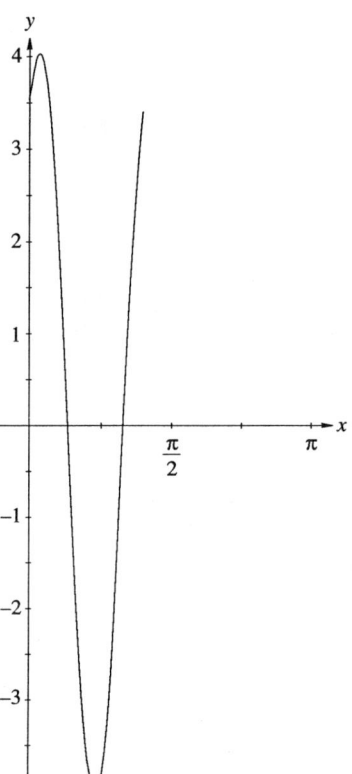

7. Amplitude $= 2$,
Period $= \frac{360°}{3} = 120° = \frac{2\pi}{3}$,
Phase Shift $= \frac{90°}{3} = 30° = \frac{\pi}{6}$,
Vertical displacement $= 0$

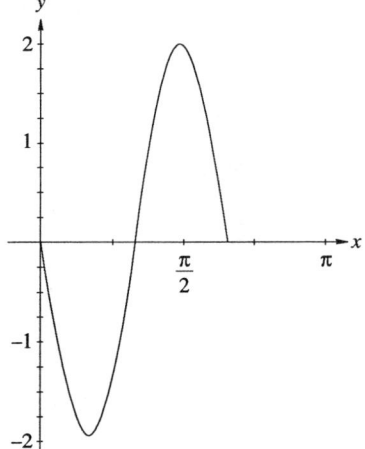

9.   Amplitude $= 4.5$

Period $= \frac{2\pi}{6} = \frac{\pi}{3}$,

Phase Shift $= \frac{4}{3}$,

Vertical displacement $= 0$

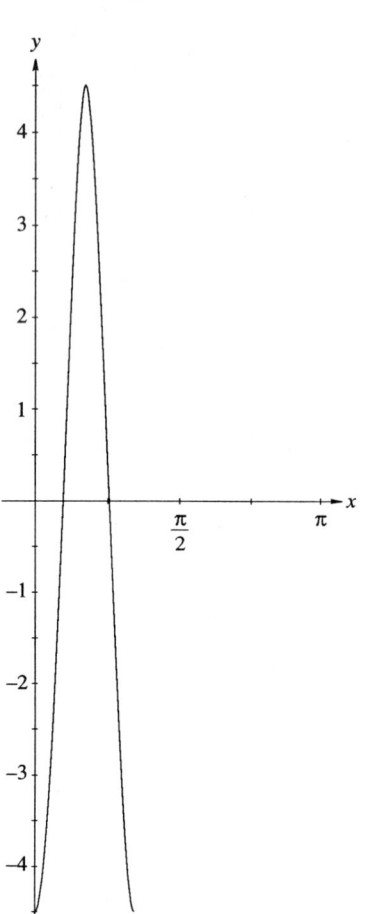

10.   Amplitude $= 9$,

Period $= 2\pi/5$,

Phase Shift $= -\frac{3}{5}$,

Vertical displacement $= 0$

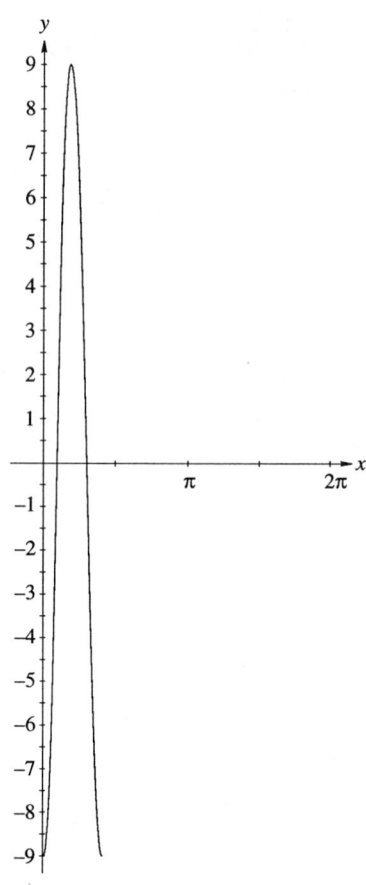

11.   Amplitude $= 0.5$,

Period $= \frac{2\pi}{\pi} = 2$,

Phase Shift $= -\frac{\pi}{8} \div \pi = -\frac{\pi}{8} \cdot \frac{1}{\pi} = -\frac{1}{8}$,

Vertical displacement $= 0$

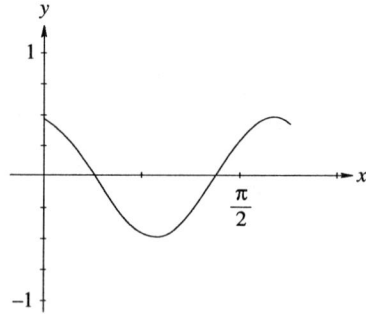

12. Amplitude $= 0.25$,
     Period $= 2\pi^2$,
     Phase Shift $= \frac{\pi}{4}$,
     Vertical displacement $= 0$

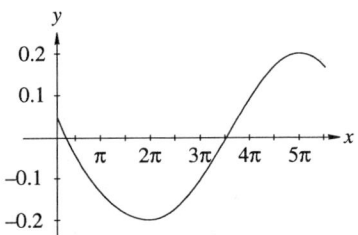

13. Amplitude $= 0.75$,
     Period $= \frac{2\pi}{\pi} = 2$,
     Phase Shift $= \frac{\pi^2}{3} \div \pi = \frac{\pi^2}{3} \cdot \frac{1}{\pi} = -\frac{\pi}{3}$,
     Vertical displacement $= 0$

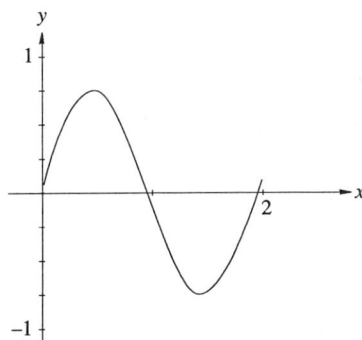

14. Amplitude $= \pi$,
     Period $= 2\pi^2$,
     Phase Shift $= -\frac{2}{3} \div \frac{1}{\pi} = -\frac{2\pi}{3}$,
     Vertical displacement $= 0$

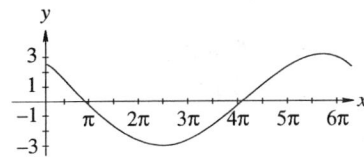

15. Amplitude $= 2$,
     Period $= 2$,
     Phase Shift $= -\frac{1}{3\pi}$,
     Vertical displacement $= 4$

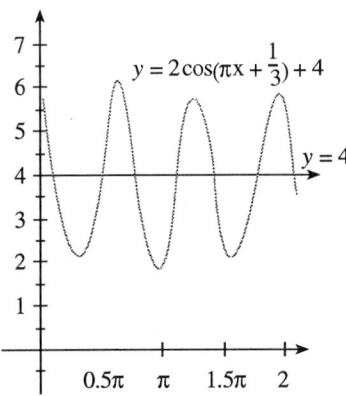

$$y = 2\cos\left(\pi x + \frac{1}{3}\right) + 4$$

$y = 4$

16. Amplitude $= 3.25$,
     Period $= \frac{2\pi}{3}$,
     Phase Shift $= \frac{4}{3}$,
     Vertical displacement $= -\frac{4}{2}$

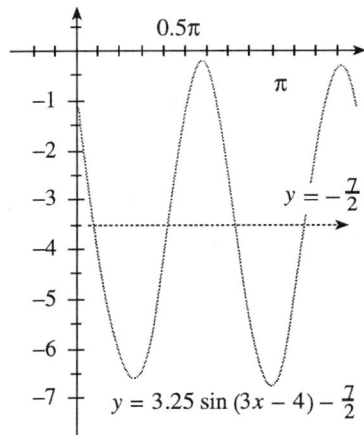

$y = -\frac{7}{2}$

$$y = 3.25 \sin(3x - 4) - \frac{7}{2}$$

17. Amplitude $= \frac{1}{2}$,
     Period $= 2$,
     Phase Shift $= -\frac{1}{\pi}$,
     Vertical displacement $= \frac{3}{2}$

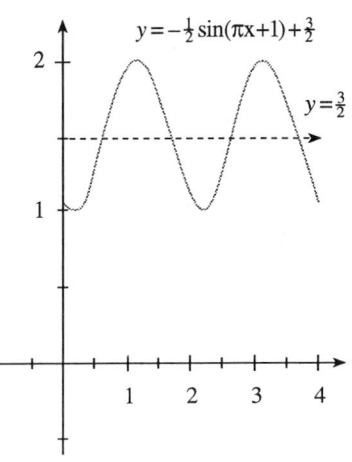

$$y = -\frac{1}{2}\sin(\pi x + 1) + \frac{3}{2}$$

$y = \frac{3}{2}$

18. Amplitude $= 1.75$,
Period $= \pi$,
Phase Shift $= -\frac{\pi}{2}$,
Vertical displacement $= 2.25$

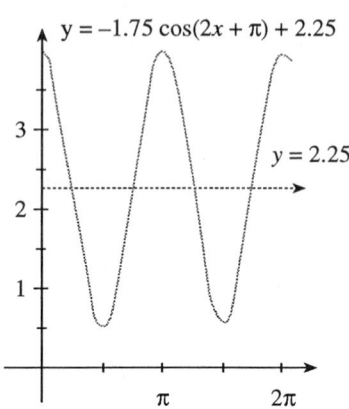

$y = -1.75 \cos(2x + \pi) + 2.25$

$y = 2.25$

19. Amplitude $= 2$,
Period $= 180° = \pi$,
Phase Shift $= 22.5°$,
Vertical displacement $= 3.5$

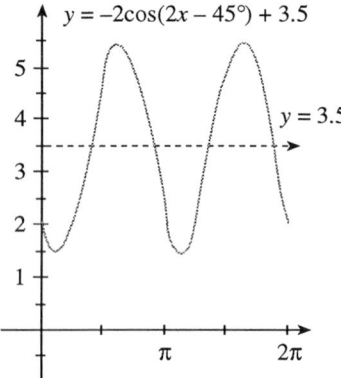

$y = -2\cos(2x - 45°) + 3.5$

$y = 3.5$

20. Amplitude $= 3$,
Period $= 60° = \frac{\pi}{3}$,
Phase Shift $= -85°$,
Vertical displacement $= -2.5$

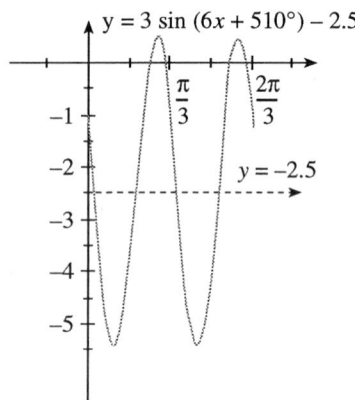

$y = 3 \sin(6x + 510°) - 2.5$

$y = -2.5$

21. Amplitude $= 65$ A,
Period $= \frac{2\pi}{120\pi} = \frac{1}{60}$ s,

Phase Shift $= \dfrac{-\frac{\pi}{2}}{120\pi} = -\frac{1}{240}$ s

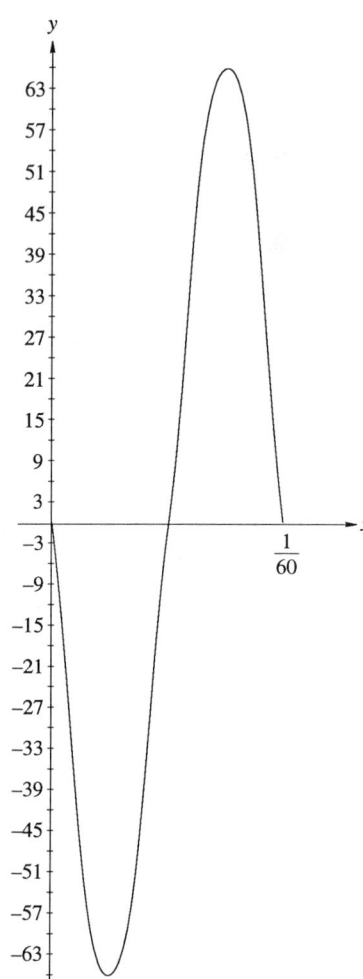

22. Rewrite the given equation as $y = 3 \sin\left(\frac{\pi}{6}t - \frac{5\pi}{6}\right)$.
Then, we get the following:
Amplitude $= 3$ m,
Period $= \dfrac{2\pi}{\frac{\pi}{6}} = 12$ s,

Phase Shift $= \dfrac{\frac{5\pi}{6}}{\frac{\pi}{6}} = 5$ s.

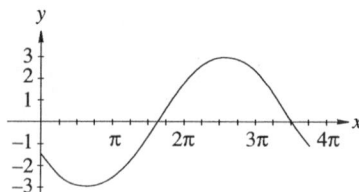

23. **(a)** The amplitude $A = \frac{1}{2}(11.8 - 6.2) = 2.8$. Since the period is 12.4 h, we have $12.4 = \frac{2\pi}{B}$ yields $B = \frac{\pi}{6.2} = \frac{5\pi}{31}$. Since high tide is at 4:00 A.M., no tide is $\frac{1}{4}$ period or $\frac{12.4}{4} = 3.1$ hr before or at $4 - 3.1 = 0.9$ hr. $\frac{C}{B} = 0.9$ and so, $C = 0.9B = \frac{0.9\pi}{6.2} = \frac{9\pi}{62}$. The average water height is $\frac{1}{2}(6.2+11.8) = 9$ m, so $D = 9$. Putting these all together, we get $d = 2.8\sin\left(\frac{\pi}{6.2}t - \frac{0.9\pi}{6.2}\right) + 9 = 2.8\sin\left(\frac{5\pi}{31}t - \frac{9\pi}{62}\right) + 9$.

(b)

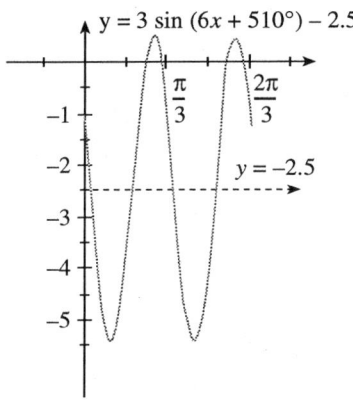

$y = 3\sin(6x + 510°) - 2.5$

$y = -2.5$

24. **(a)** Amplitude: 12,

Period: $\frac{2\pi}{120\pi} = \frac{1}{60}$,

Phase Shift: $-\frac{0.5}{120\pi} = -\frac{1}{240}$,

Vertical displacement: 0

(b)

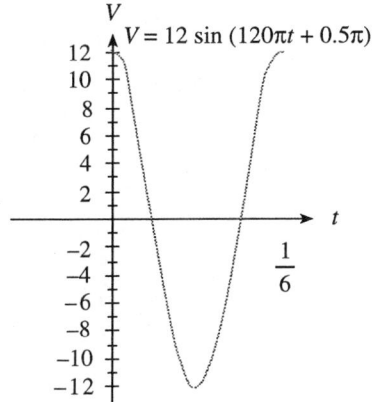

$V = 12\sin(120\pi t + 0.5\pi)$

25. $A = \frac{1}{2}(72 - 11) = 30.5$ $D = \frac{1}{2}(72 + 11) = 41.5$. The period is 52 weeks so $52 = \frac{2\pi}{B}$ which leads to $B = \frac{2\pi}{52} = \frac{\pi}{26}$. The phase shift is 12 weeks, so $\frac{C}{B} = 12$ and so $C = 12\left(\frac{\pi}{26}\right) = \frac{6\pi}{13}$. Putting this all together, we get $T = 30.5\sin\left(\frac{\pi}{26}t - \frac{6\pi}{13}\right) + 41.5$.

(b)

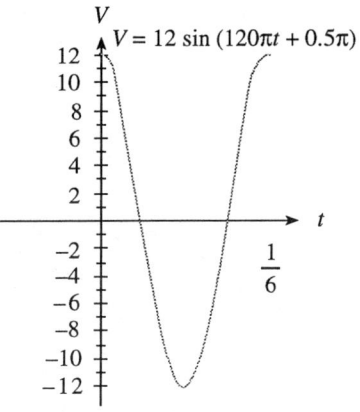

$V = 12\sin(120\pi t + 0.5\pi)$

26. **(a)** $A = \frac{1}{2}(89.3 - 29.2) = 30.05$ $D = \frac{1}{2}(89.3 + 29.2) = 59.25$. The period is 52 weeks so $52 = \frac{2\pi}{B}$ which leads to $B = \frac{2\pi}{52} = \frac{\pi}{26}$. The phase shift is 12 weeks, so $\frac{C}{B} = 12$ and so $C = 12\left(\frac{\pi}{26}\right) = \frac{6\pi}{13}$. Putting this all together, we get $T = 30.05\sin\left(\frac{\pi}{26}t - \frac{6\pi}{13}\right) + 59.25$.

(b)

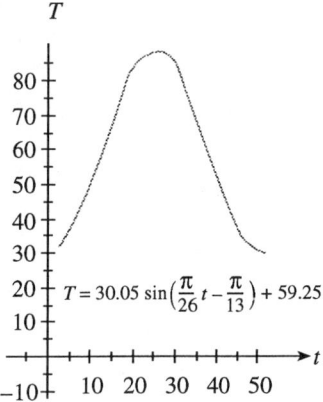

$T = 30.05\sin\left(\frac{\pi}{26}t - \frac{\pi}{13}\right) + 59.25$

☰ 11.3 COMPOSITE SINE AND COSINE CURVES

1.

2.

3.

4.

5.

6.

7.

8.

9.

10.

11.

12.

13.

14.

15.

16.

17.

18.

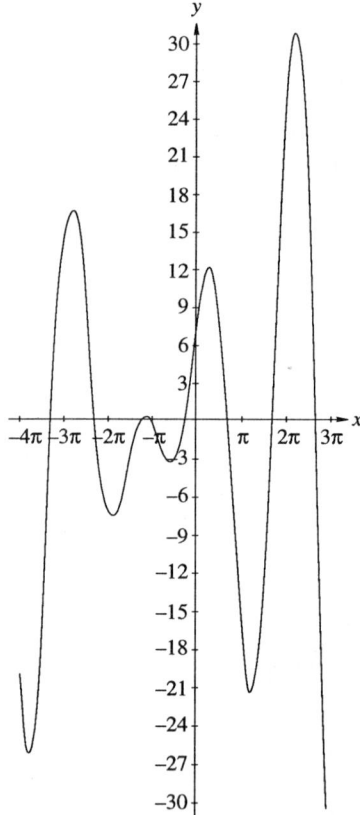

19.   (a) Amplitude (maximum): 5, Period: π,

(b) Viewing window: one cycle: $\left[0, \pi, \frac{\pi}{4}\right] \times \left[-3, 7, 1\right]$, two cycles: $\left[0, 2\pi, \frac{\pi}{4}\right] \times \left[-3, 7, 1\right]$,

(c)

$$\left[0, 2\pi, \frac{\pi}{4}\right] \cdot [-3, 7, 1]$$

20.   (a) Amplitude (maximum): 7, Period: $\frac{2\pi}{3}$,

(b) Viewing window: one cycle: $\left[0, \frac{2\pi}{3}, \frac{\pi}{6}\right] \times \left[-11, 3, 1\right]$, two cycles: $\left[0, \frac{4\pi}{3}, \frac{\pi}{6}\right] \times \left[-11, 3, 1\right]$,

(c)

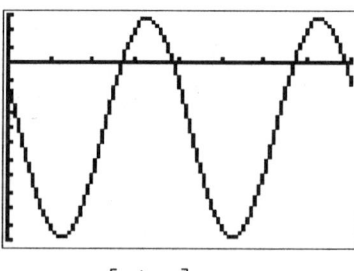

$$\left[0, \frac{4\pi}{3}, \frac{\pi}{6}\right] \cdot [-11, 3, 1]$$

21. (a) Amplitude (maximum): 6, Period: The least common multiple of $\frac{2\pi}{3}$ and $\frac{2\pi}{5}$ is 2π,

(b) Viewing window: one cycle: $\left[0, 2\pi, \frac{\pi}{4}\right] \times \left[-9, 3, 1\right]$, two cycles $\left[0, 4\pi, \frac{\pi}{4}\right] \times [-9, 3, 1]$, and

(c)

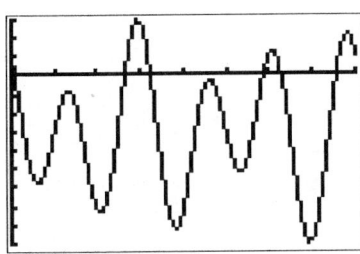

$$\left[0, 2\pi, \tfrac{\pi}{4}\right] \cdot [-9, 3, 1]$$

22. (a) Amplitude (maximum): $\frac{7}{2}$, Period: The least common multiple of $\frac{\pi}{2}$ and $\frac{2\pi}{5}$ is 2π,

(b) Viewing window: one cycle: $\left[0, 2\pi, \frac{\pi}{4}\right] \times [0, 10, 1]$, two cycles: $\left[0, 4\pi, \frac{\pi}{4}\right] \times [0, 10, 1]$,

(c)

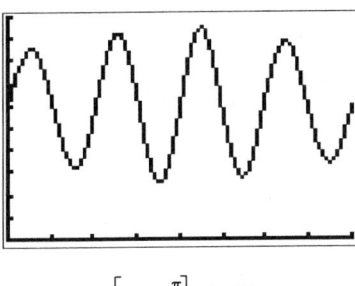

$$\left[0, 2\pi, \tfrac{\pi}{4}\right] \cdot [0, 10, 1]$$

23. (a) Amplitude (maximum): 5.5, Period: Since the individual periods are 1 and 2π, and one is rational and the other irrational, they have no least common multiple. Hence, this function is not periodic.

(b) Viewing window: $\left[0, 4\pi, \frac{\pi}{2}\right] \times [-9, 2, 1]$,

(c)

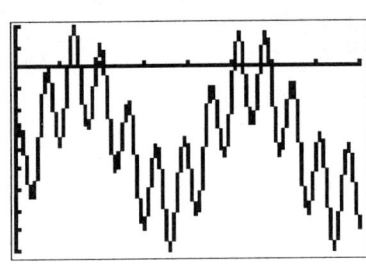

$$\left[0, 4\pi, \tfrac{\pi}{2}\right] \cdot [-9, 2, 1]$$

24. (a) Amplitude (maximum): $\frac{5}{6}$, Period: 2π,

(b) Viewing window: one cycle: $\left[0, 2\pi, \frac{\pi}{4}\right] \times [0, 3, 1]$, two cycles: $\left[0, 4\pi, \frac{\pi}{4}\right] \times [0, 3, 1]$,

(c)

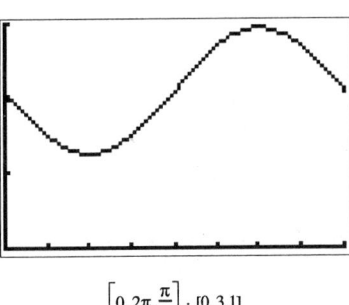

$$\left[0, 2\pi, \tfrac{\pi}{4}\right] \cdot [0, 3, 1]$$

25.

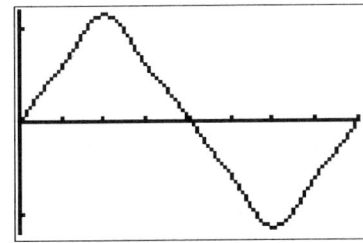

$$\left[0, 2\pi, \tfrac{\pi}{4}\right] \cdot [0, 3, 1]$$

26.

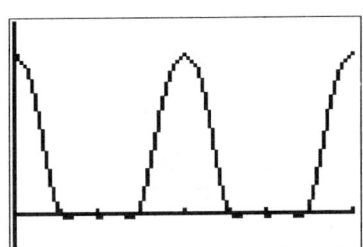

$$\left[0, 4\pi, \tfrac{\pi}{2}\right] \cdot [-0.2, 1.2, 1]$$

11.4 GRAPHS OF THE OTHER TRIGONOMETRIC FUNCTIONS

1.

2.

3.

4.

5.

6.

7.

8.

9.

10.

11.

12.

15.

13.

16.

14.

17.

18.

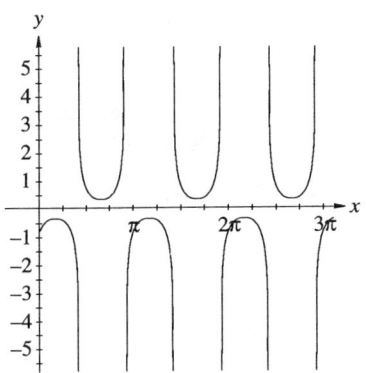

19. (a) $\left[0, 4\pi, \frac{\pi}{2}\right] \times [-8, 6, 2]$,

(b)

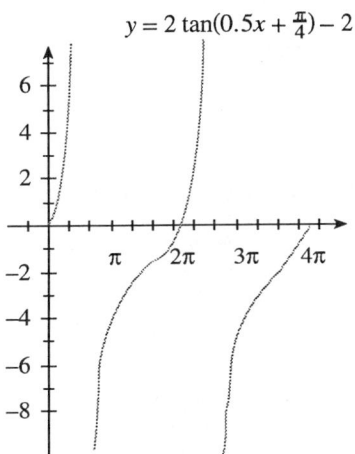

$$y = 2\tan(0.5x + \tfrac{\pi}{4}) - 2$$

20. (a) $[0.4, 1] \times [-6, 8, 2]$,

(b)

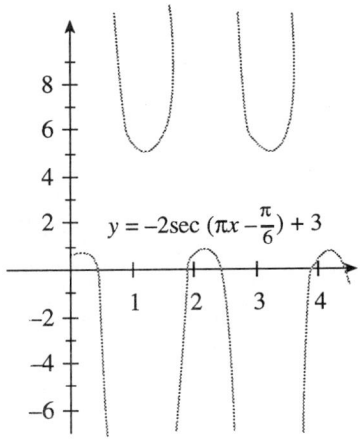

$$y = -2\sec\left(\pi x - \tfrac{\pi}{6}\right) + 3$$

21. For the given tapered hole,
$$\begin{aligned}
d &= -2h\tan\theta + D \\
&= -2(8)\tan\theta + 4.5 \\
&= -16\tan\theta + 4.5
\end{aligned}$$

(a) If $\theta = 10°$, then $d = -16\tan 10° + 4.5 \approx 1.6788$. Thus, the small-end diameter is about 1.68 cm.

(b) Here, we let $d = 0$ and solve the equation for θ.
$$\begin{aligned}
0 &= -16\tan\theta + 4.5 \\
16\tan\theta &= 4.5 \\
\tan\theta &= \frac{4.5}{16} \\
\theta &= \tan^{-1}\left(\frac{4.5}{16}\right) \approx 15.71°
\end{aligned}$$
Thus, an angle of about 15.71° will produce a small-end diameter of 0 cm.

(c) The figure shows the graph of $d = -16\tan\theta + 4.5$ for $0° \leq \theta < 16°$, as drawn by a TI-82. Note that you need to put your calculator in degree mode to obtain this graph.

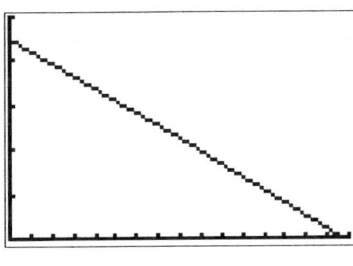

[0, 16] x [0, 5]

22. (a) The maximum distance the camera can televise to the left will be determined by rotating the camera 40° to the left. If this distance from P is x_1, and we are given $CP = 75$, then $\tan 40° = \frac{x_1}{75}$ and so $x_1 = 75\tan 40° = 62.9$ ft

(b) Similarly, if x_2 is the farthest distance from P to the right that the camera can televise, then $x_2 = 75\tan 55° \approx 107.1$ ft

(c) If d is the distance from the camera to the parade, and θ represents the angles the camera can see, then $\cos\theta = \frac{75}{d}$ and so we want the graph of $d = \frac{75}{\cos\theta} = 75\sec\theta$ where $-40° \leq \theta \leq 55°$. Notice that angles to the left of $\overline{CP}$ are considered as negative. The resulting graph, as drawn on a TI-82 is shown below.

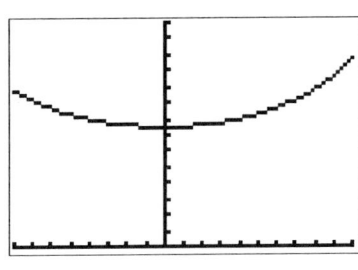

[−40, 50, 5] × [0, 140, 10]

11.5 APPLICATIONS OF TRIGONOMETRIC GRAPHS

1. The amplitude is 10 so $A = 10$. Since $f = 4$, $\omega = 2\pi f = 8\pi$. $y = A \sin \omega t = 10 \sin 8\pi t$.

2. Using the answer for Exercise 1 and inserting a phase shift ϕ we have $y = 10 \sin(8\pi t - \phi)$. Substituting 8 for y and 0.1 for t solve for ϕ. $8 = 10 \sin(8\pi(0.1) - \phi)$; $\sin(0.8\pi - \phi) = 0.8$; $(0.8\pi - \phi = \sin^{-1} 0.8 = 0.927295 \approx 0.927$; $-\phi = 0.927 - 0.8\pi = -1.586$ or $\phi = 1.586$. This gives the equation $y = 10 \sin(8\pi t - 1.586)$.

3. Amplitude $A = 0.8$ m, $\omega = \frac{\pi}{3}$ rad/s. Since $y = 0$ when $t = 0$ there is no phase shift. Substituting into the form $y = A \sin \omega t$ we get $y = 0.8 \sin \frac{\pi}{3} t$.

4. $\omega = 2\pi f$ or $f = \frac{\omega}{2\pi} = \frac{\frac{\pi}{3}}{2\pi} = \frac{1}{6}$ Hz.

5. $a = -g \sin \theta = -32 \sin 5° = -2.79$ ft/s^2.

6. $a = -g \sin \theta = -9.8 \sin(0.05) = -0.49$ m/s^2

7. $y = 8.5 \cos 2.8t$
   (a) Amplitude is 8.5 cm,
   (b) Period $= \frac{2\pi}{2.8} = \frac{\pi}{1.4} \approx 2.24$ s,
   (c) $f = \frac{2.8}{2\pi} = \frac{1.4}{\pi} \approx 0.45$ Hz.

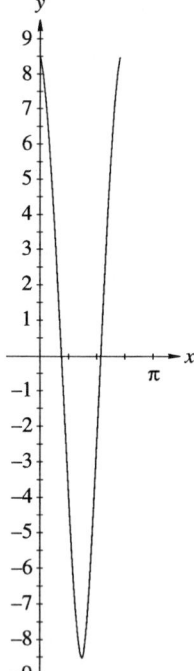

8. (a) $f = 3000$ rpm $= 50$ rps
   (b) Using $y = A \cos 2\pi f t$ we get $y = \frac{10.5}{2} \cos 2\pi 50t$ or $y = 5.25 \cos 100\pi t$.
   (c) $y = 5.25 \cos \frac{4500}{60} \cdot 2\pi t = 5.25 \cos 150\pi t$

9. $I = 10 \sin 120\pi t$: Amplitude 10 A, Period $= \frac{2\pi}{120\pi} = \frac{1}{60}$ s; $f = \frac{120\pi}{2\pi} = 60$ Hz, $\omega = 2\pi f = 120\pi$ rad/s

10. Given that $I_{max} = 6.8$ and $f = 80$ Hz, then $I = I_{max} \sin 2\pi f t$, and so $I = 6.8 \sin 160\pi t$

11. If $\phi = \frac{\pi}{3}$ then $I = 6.8 \sin \left(160\pi t + \frac{\pi}{3}\right)$

12. $V = I_{max} \sin(2\pi f t) = 220 \sin(80\pi t)$

13. $V = 220 \sin \left(80\pi t - \frac{\pi}{3}\right)$

14. Since cosine and sine have the same shape and cosine is out of phase from the sine by $\frac{\pi}{2}$ we get
    $$V = 220 \cos \left(80\pi t + \frac{\pi}{2}\right)$$

15. Since cosine is the same as sine except for a phase shift of $\frac{\pi}{2}$ we have $V = 220 \cos \left(80\pi t - \frac{\pi}{3} + \frac{\pi}{2}\right) = 220 \cos \left(80\pi t + \frac{\pi}{6}\right)$

16. $V = V_{max} \sin 2\pi f t = 250,000 \sin \left(2\pi 10^{19} t\right)$

17. $y = 10^{-12} \sin 2\pi 10^{23} t$, frequency is 10^{23} Hz, amplitude is 10^{-12}

18. $I = I_{max} \sin(2\pi f t + \phi) = 3.6 \sin \left(2\pi \cdot 60t + \frac{\pi}{2}\right) = 3.6 \sin \left(120\pi t + \frac{\pi}{2}\right)$

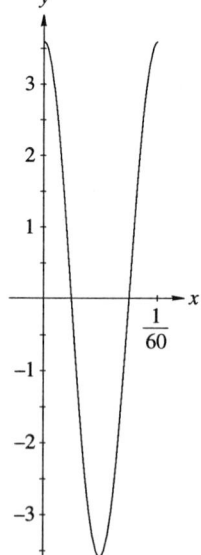

19. $I = I_{\max} \sin(2\pi f t + \phi) = 1.2 \sin(800\pi t + 37°)$ or
$I = 1.2 \sin(800\pi t + 0.6458)$

20.
$\phi = \tan^{-1}\left(\dfrac{V_L - V_C}{V_R}\right) = \tan^{-1}\left(\dfrac{80-130}{120}\right) \approx$
$\tan^{-1}(-0.4167); \theta = -22.62° = -0.3948$ rad

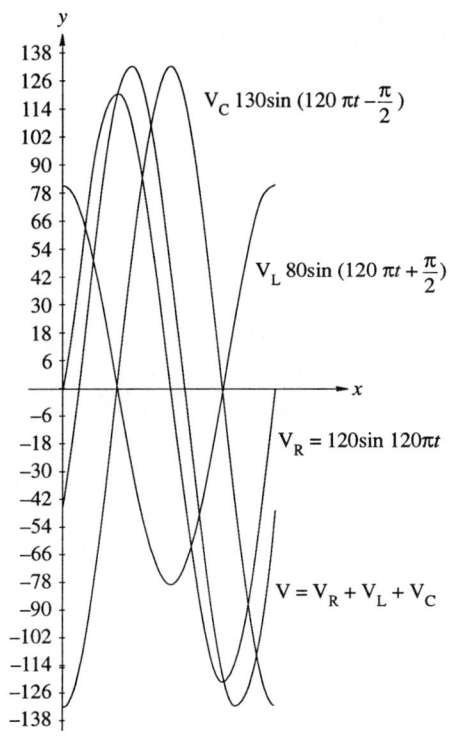

21. **(a)** Period: $\dfrac{2\pi}{2\pi \times 10^{18}} = 10^{-18}$ Hz, Amplitude: 10^{-10} m,
(b)

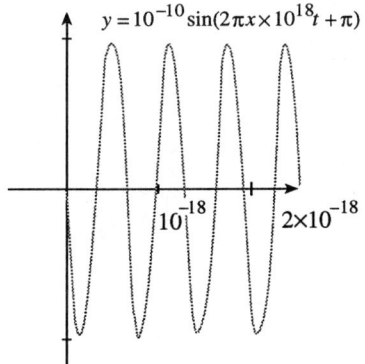

22. **(a)** maximum: $25,000 + 7,200 = 32,250$; minimum: $25,000 - 7,250 = 17,750$,
(b) Here $t = 12$, so $P = 25,000 + 7,250\cos(\pi) = 25,000 - 7,250 = 17,750$
(c) $[0, 36, 3] \times [15000, 35000, 2000]$,

(d)

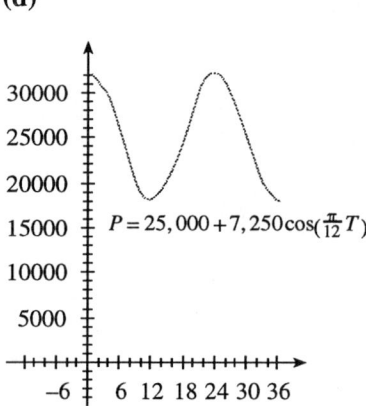

$$P = 25,000 + 7,250\cos(\tfrac{\pi}{12}T)$$

≡ 11.6 PARAMETRIC EQUATIONS

1.

t	-3	-2	-1	0	1	2	3
x	-3	-2	-1	0	1	2	3
y	-9	-6	-3	0	3	6	9

Given $x = t$ and $y = 3t$, substituting x for t in the second equation yields $y = 3x$.

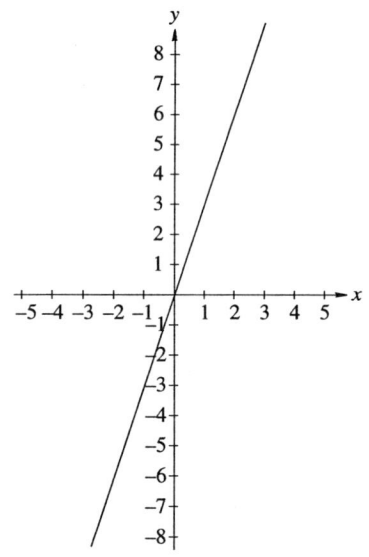

2.

t	-3	-2	-1	0	1	2	3
x	-6	-4	-2	0	2	4	6
y	-11	-7	-3	1	5	9	13

Here $x = 2t$ and $y = 4t + 1$. The first equation gives $t = \frac{x}{2}$. Substituting into the second equation yields $y = 4\left(\frac{x}{2}\right) + 1 = 2x + 1$ which simplifies to $y = 2x + 1$.

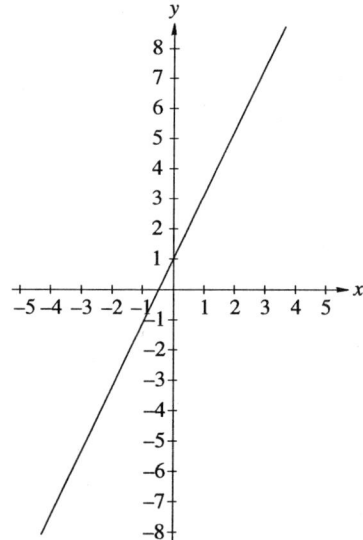

3.

t	-3	-2	-1	0	1	2	3
x	-3	-2	-1	0	1	2	3
y	$-\frac{1}{3}$	$-\frac{1}{2}$	-1	$*$	1	$\frac{1}{2}$	$\frac{1}{3}$

* Not defined

Here $x = t$ and $y = \frac{1}{t}$. Direct substitution yields $y = \frac{1}{x}$.

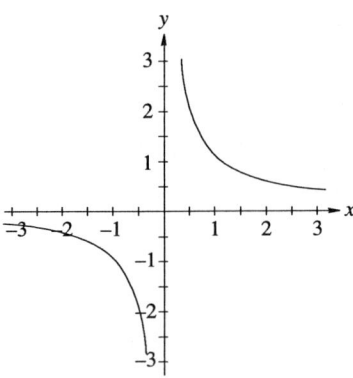

4.

t	-1	0	1	2	3	4	5
x	4	5	6	7	8	9	10
y	-5	-2	1	4	7	10	13

Here $x = t + 5$ and $y = 3t - 2$. The first equation gives $t = x - 5$. Substitution into the second equation yields $y = 3(x - 5) - 2 = 3x - 15 - 2 = 3x - 17$ which simplifies to $y = 3x - 17$.

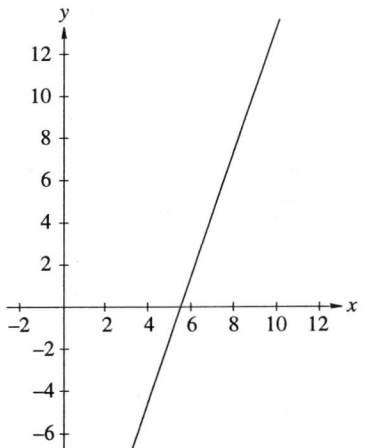

5.

t	-5	-4	-3	-2	-1	0	1	2	3	4	5
x	8	7	6	5	4	3	2	1	0	-1	-2
y	16	7	0	-5	-8	-9	-8	-5	0	7	16

Here $x = 3 - t$ and $y = t^2 - 9$. The first equation gives $t = 3 - x$. Substituting into the second equation yields $y = (3-x)^2 - 9 = 9 - 6x + x^2 - 9 = x^2 - 6x$ or $y = x^2 - 6x$.

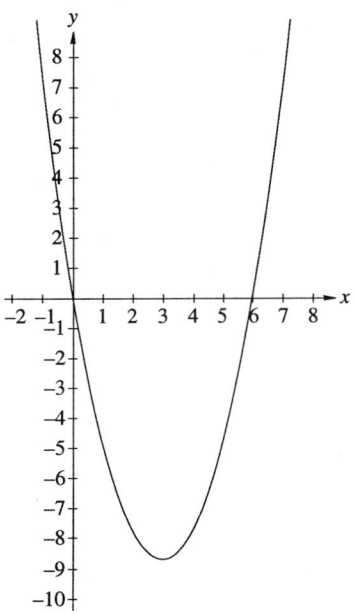

6.

t	-5	-4	-3	-2	-1	0	1	2	3	4	5
x	-3	-2	-1	0	1	2	3	4	5	6	7
y	30	20	12	6	2	0	0	2	6	12	20

Here $x = t + 2$ and $y = t^2 - t$. The first equation gives $t = x - 2$. Substitution into the second equation yields $y = (x-2)^2 - (x-2) = x^2 - 4x + 4 - x + 2 = x^2 - 5x + 6$ and so $y = x^2 - 5x + 6$.

7.

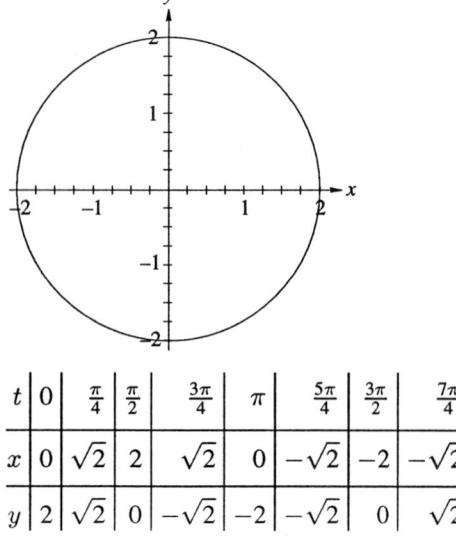

t	0	$\frac{\pi}{4}$	$\frac{\pi}{2}$	$\frac{3\pi}{4}$	π	$\frac{5\pi}{4}$	$\frac{3\pi}{2}$	$\frac{7\pi}{4}$
x	0	$\sqrt{2}$	2	$\sqrt{2}$	0	$-\sqrt{2}$	-2	$-\sqrt{2}$
y	2	$\sqrt{2}$	0	$-\sqrt{2}$	-2	$-\sqrt{2}$	0	$\sqrt{2}$

8.

t	0	$\frac{\pi}{4}$	$\frac{\pi}{2}$	$\frac{3\pi}{4}$	π	$\frac{5\pi}{4}$	$\frac{3\pi}{2}$	$\frac{7\pi}{4}$
x	0	$\sqrt{2}$	5	$\sqrt{2}$	0	$-\sqrt{2}$	-5	$-\sqrt{2}$
y	2	$\frac{5\sqrt{2}}{2}$	0	$-\frac{5\sqrt{2}}{2}$	-2	$-\frac{5\sqrt{2}}{2}$	0	$\frac{5\sqrt{2}}{2}$

9.

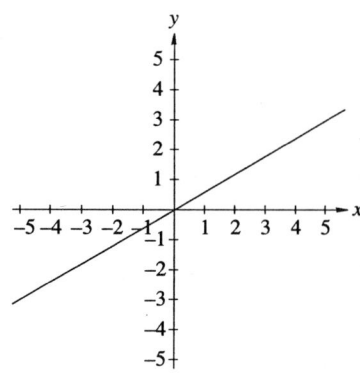

t	0	$\frac{\pi}{4}$	$\frac{\pi}{2}$	$\frac{3\pi}{4}$	π	$\frac{5\pi}{4}$	$\frac{3\pi}{2}$	$\frac{7\pi}{4}$	2π
x	2	$\frac{5\sqrt{2}}{2}$	5	$\frac{5\sqrt{2}}{2}$	0	$-\frac{5\sqrt{2}}{2}$	-5	$-\frac{5\sqrt{2}}{2}$	0
y	0	$\frac{3\sqrt{2}}{2}$	3	$-\frac{3\sqrt{2}}{2}$	0	$-\frac{3\sqrt{2}}{2}$	-3	$-\frac{3\sqrt{2}}{2}$	0

10.

t	0	$\frac{\pi}{4}$	$\frac{\pi}{2}$	$\frac{3\pi}{4}$	π	$\frac{5\pi}{4}$	$\frac{3\pi}{2}$	$\frac{7\pi}{4}$	2π
x	2	$\sqrt{2}$	0	$-\sqrt{2}$	-2	$-\sqrt{2}$	0	$-\sqrt{2}$	2
y	6	$3\sqrt{2}$	0	$-3\sqrt{2}$	-6	$-3\sqrt{2}$	0	$-3\sqrt{2}$	6

11.

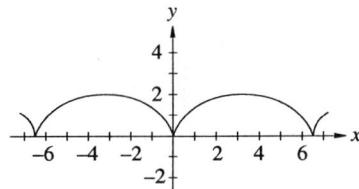

t	-5	-4	-3	-2	-1	0	1	2	3	4	5
x	-5.96	-4.76	-2.86	-1.09	-0.16	0	0.16	1.09	2.86	4.76	5.96
y	0.72	1.65	1.99	1.41	0.46	0	0.46	1.41	1.99	1.65	0.72

12.

t	0	$\frac{\pi}{6}$	$\frac{\pi}{4}$	$\frac{\pi}{3}$	$\frac{\pi}{2}$	$\frac{2\pi}{3}$	$\frac{3\pi}{4}$	$\frac{5\pi}{6}$	π	$\frac{7\pi}{6}$	$\frac{5\pi}{4}$	$\frac{4\pi}{3}$
x	0	$\frac{\sqrt{3}}{3}$	1	$\sqrt{3}$	$*$	$-\sqrt{3}$	-1	$-\frac{\sqrt{3}}{3}$	0	$\frac{\sqrt{3}}{3}$	1	$\sqrt{3}$
y	$*$	$6\sqrt{3}$	6	$\frac{6}{\sqrt{3}}$	0	$-\frac{6}{\sqrt{3}}$	-6	$-6\sqrt{3}$	$*$	$6\sqrt{3}$	6	$\frac{6}{\sqrt{3}}$

$*$ Not defined

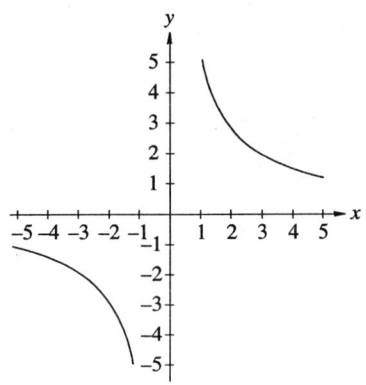

13.

t	0	$\frac{\pi}{6}$	$\frac{\pi}{4}$	$\frac{\pi}{3}$	$\frac{\pi}{2}$	$\frac{2\pi}{3}$	$\frac{3\pi}{4}$	$\frac{5\pi}{6}$	π	$\frac{7\pi}{6}$	$\frac{5\pi}{4}$	$\frac{4\pi}{3}$	$\frac{3\pi}{2}$	$\frac{5\pi}{3}$	$\frac{7\pi}{4}$	$\frac{11\pi}{6}$	2π
x	1	$\frac{2}{\sqrt{3}}$	$\sqrt{2}$	2	$*$	-2	$-\sqrt{2}$	$-\frac{2}{\sqrt{3}}$	-1	$-\frac{2}{\sqrt{3}}$	$-\sqrt{2}$	-2	$*$	2	$\sqrt{2}$	$\frac{2}{\sqrt{3}}$	1
y	$*$	4	$2\sqrt{2}$	$\frac{4}{\sqrt{3}}$	2	$\frac{4}{\sqrt{3}}$	$2\sqrt{2}$	4	$*$	-4	$-2\sqrt{2}$	$-\frac{4}{\sqrt{3}}$	-2	$-\frac{4}{\sqrt{3}}$	$-2\sqrt{2}$	-4	$*$

$*$ Not defined

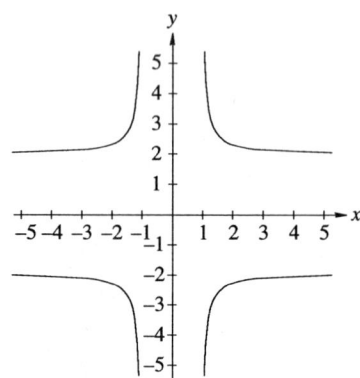

14.

t	0	$\frac{\pi}{6}$	$\frac{\pi}{4}$	$\frac{\pi}{3}$	$\frac{\pi}{2}$	$\frac{2\pi}{3}$	$\frac{3\pi}{4}$	$\frac{5\pi}{6}$	π	$\frac{7\pi}{6}$	$\frac{5\pi}{4}$	$\frac{4\pi}{3}$	$\frac{3\pi}{2}$	$\frac{5\pi}{3}$	$\frac{7\pi}{4}$	$\frac{11\pi}{6}$	2π
x	3	$2\sqrt{3}$	$3\sqrt{2}$	6	$*$	-6	$-3\sqrt{2}$	$-2\sqrt{3}$	-3	$-2\sqrt{3}$	$-3\sqrt{2}$	-6	$*$	6	$3\sqrt{2}$	$2\sqrt{3}$	3
y	0	$\frac{\sqrt{3}}{3}$	1	$\sqrt{3}$	$*$	$-\sqrt{3}$	-1	$-\frac{\sqrt{3}}{3}$	0	$\frac{\sqrt{3}}{3}$	1	$\sqrt{3}$	$*$	$-\sqrt{3}$	-1	$-\frac{\sqrt{3}}{3}$	0

$*$ Not defined

15. **(a)** 1:1;

16. **(a)** 1:2;

17. **(a)** 2:1;

18. **(a)** 1:3;

19. **(a)** 1:4;

20. **(a)** 1:2;

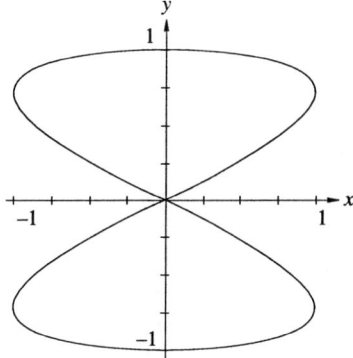

21. **(a)** 2:3 or 1:1.5;

22. **(a)** 3:4;

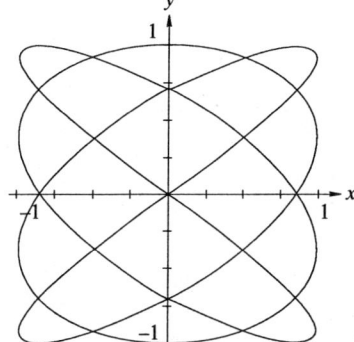

23.   **(a)** 2:5 or 1:2.5;

24.   **(a)** 3:5;

25.

26.

27.

28.

29.

30.

31. (a)

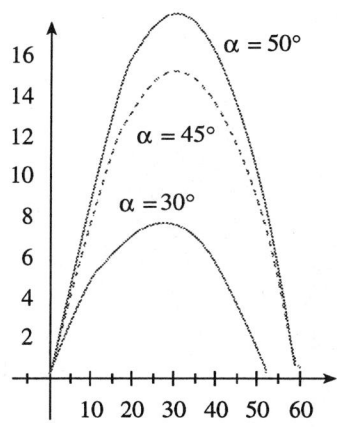

(b) Solving $x = (V \cos \alpha)t$ for t, we get $t = \frac{x}{V \cos \alpha}$. Substituting into the second equation, we get

$$y = (V \sin \alpha)\left(\frac{x}{V \cos \alpha}\right) - \frac{1}{2}g\left(\frac{x}{V \cos \alpha}\right)^2$$

$$= x \tan \alpha - \frac{gx^2}{2V^2} \sec^2 \alpha$$

(c) When the gravel hits the road $y = 0$ which means that $x \tan \alpha = \frac{gx^2}{2V^2} \sec^2 \alpha$. Solving for x, we get

$$x = \frac{2V^2 \tan \alpha}{g \sec^2 \alpha}$$

$$= \frac{2V^2 \left(\frac{\sin \alpha}{\cos \alpha}\right)}{g \left(\frac{1}{\cos^2 \alpha}\right)}$$

$$= \frac{2V^2 \sin \alpha \cos \alpha}{g}$$

$$= \frac{V^2 \sin 2\alpha}{g}$$

Note that the last step used the identity $2 \sin \cos \alpha = \sin 2\alpha$. $x = \frac{V^2 \sin 2\alpha}{g}$

32. (a)

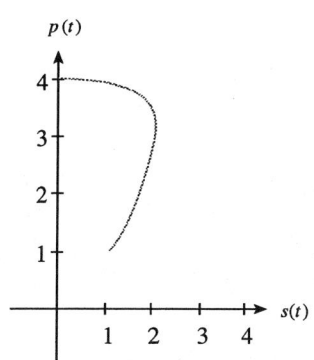

(b) $400, $350, $150
(c) 200,000 units, 153,846 units

11.7 POLAR COORDINATES

1.

2.

3.

4.

5.

6.

7.

8.

9.

10.

11.

12.

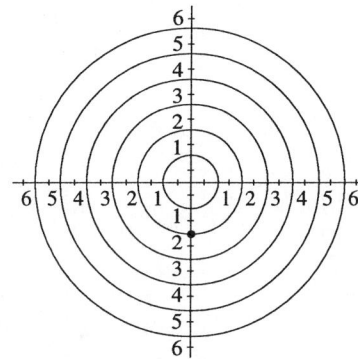

13. Using $x = r\cos\theta$, we find $x = 4\cos\frac{\pi}{3} = 4\cdot\frac{1}{2} = 2$. Using $y = r\sin\theta$ yields $y = 4\sin\frac{\pi}{3} = 4\cdot\frac{\sqrt{3}}{2} = 2\sqrt{3}$. Thus, the polar coordinates $\left(4, \frac{\pi}{3}\right)$ have the rectangular coordinates $\left(2, 2\sqrt{3}\right)$ or $(2, 3.46)$.

14. Using $x = r\cos\theta$, we find $x = 5\cos 75° = 1.294$, and using $y = r\sin\theta$ yields $y = 5\sin 75° = 4.830$. Thus, the polar coordinates $(5, 75°)$ have the rectangular coordinates $(1.29, 4.83)$.

15. Here $x = r\cos\theta$ yields $x = 2\cos 135° = 2 \cdot \frac{-\sqrt{2}}{2} = -\sqrt{2}$, and $y = r\sin\theta$ produces $y = 2\sin 135° = 2 \cdot \frac{\sqrt{2}}{2} = \sqrt{2}$. Thus, the polar coordinates $(2, 135°)$ have the rectangular coordinates $(-\sqrt{2}, \sqrt{2})$ or $(-1.414, 1.414)$.

16. Here $x = r\cos\theta$ yields $x = 3\cos\frac{3\pi}{2} \cdot 0 = 0$, and $y = r\sin\theta$ gives $y = 3\sin\frac{3\pi}{2} = 3 \cdot -1 = -3$. Thus, the polar coordinates $\left(3, \frac{3\pi}{2}\right)$ have the rectangular coordinates $(0, -3)$.

17. Here $x = -6\cos 20° = -5.64$ and $y = -6\sin 20° = -2.05$ with the result $(-5.64, -2.05)$.

18. Here $x = -3\cos\frac{5\pi}{3} = -3 \cdot \frac{1}{2} = -\frac{3}{2}$ or -1.5 and $y = -3\sin\frac{5\pi}{3} = -3 \cdot \frac{-\sqrt{3}}{2} = \frac{3\sqrt{3}}{2}$ or 2.60 and we get $(-1.5, 2.60)$.

19. Here $x = -2\cos 4.3 = 0.80$ and $y = -2\sin 4.3 = 1.83$ with the result $(0.80, 1.83)$.

20. Here $x = -5\cos 255° = 1.29$ and $y = -5\sin 255° = 4.83$, hence $(1.29, 4.83)$.

21. $x = 3\cos -170° = -2.95$, $y = 3\sin -170° = -0.52$; $(-2.95, -0.52)$

22. $x = 4\cos\frac{-\pi}{8} = 3.70$, $y = 4\sin\frac{-\pi}{8} = -1.53$; $(3.70, -1.53)$

23. $x = -6\cos(-2.5) = 4.81$, $y = -6\sin(-2.5) = 3.59$; $(4.81, 3.59)$

24. $x = -3\cos -195° = 2.90$, $y = -3\sin -195° = -0.78$; $(2.90, -0.78)$

25. Here $r = \sqrt{x^2 + y^2} = \sqrt{4^2 + 4^2} = 4\sqrt{2} \approx 5.66$ and $\theta = \tan^{-1}\frac{4}{4} = \tan^{-1} 1 = 45°$ or $\frac{\pi}{4}$, so the rectangular coordinates $(4,4)$ have a polar coordinate $(4\sqrt{2}, 45°)$ or $(4\sqrt{2}, \frac{\pi}{4})$ or $(5.66, 45°)$.

26. Here $r = \sqrt{x^2 + y^2} = \sqrt{3^2 + 6^2} = \sqrt{45} = 3\sqrt{5}$ or 6.71 and $\theta = \tan^{-1}\frac{6}{3} = \tan^{-1} 2 = 63.43°$ or 1.11, so the rectangular coordinates $(3,6)$ have a polar coordinate $(6.71, 1.11)$ or $(6.71, 63.43°)$.

27. Here $r = \sqrt{4^2 + 3^2} = 5$ and $\theta = \tan^{-1}\frac{3}{4} = 36.87° = 0.64$ rad, so the rectangular coordinates $(4, 3)$ have a polar coordinate $(5, 36.87°)$ or $(5, 0.64°)$.

28. Here $r = \sqrt{5^2 + 12^2} = 13$ and $\theta = \tan^{-1}\frac{12}{5} = 67.38°$ or 1.18 rad, so the rectangular coordinates $(5,12)$ have a polar coordinate $(13, 67.38°)$ or $(13, 1.18)$.

29. Here $r = \sqrt{(-20)^2 + 21^2} = 29$, and $\theta_R = \tan^{-1}\frac{21}{20} = 46.40°$ or 0.81 rad. Since θ is in Quadrant II, $\theta = 180 - \theta_r = 133.60°$ or, $\theta = \pi - 0.81 = 2.33$, so the rectangular coordinates $(-20, 21)$ have a polar coordinate $(29, 133.60°)$ or $(29, 2.33)$.

30. Here $r = \sqrt{(-12)^2 + 5^2} = 13$, and $\theta_R = \tan^{-1}\left(\frac{5}{12}\right) = 22.62°$ or 0.39 rad. Since θ is in Quadrant II, we have $180 - 22.62 = 157.38°$ and so $\theta = \pi - 0.39 = 2.75$. Hence, the rectangular coordinates $(-12, 5)$ have a polar coordinate $(13, 157.38°)$ or $(13, 2.75)$.

31. $r = \sqrt{(-3)^2 + 4^2} = 5, \theta_R = \tan^{-1}\frac{4}{3} = 53.13° = 0.73$ rad, θ in Quadrant II so $180 - 53.13 = 126.87°$ or, $\theta = \pi - 0.93 = 2.21$; $(5, 126.87°)$ or $(5, 2.21)$

32. $r = \sqrt{9^2 + (-5)^2} = 10.30, \theta_R = \tan^{-1}\frac{5}{9} = 29.05° = 0.507, \theta$ in Quadrant IV so $360 - 29.05 = 330.95°$ or, $\theta = 2\pi - 0.507 = 5.776$; $(10.30, 330.95°)$ or $(10.30, 5.776)$

33. $r = \sqrt{(-7)^2 + (-10)^2} = 12.21, \theta_R = \tan^{-1}\frac{10}{7} = 55.01°$ or 0.96 rad. θ in Quadrant III so $180 + 55.01 = 235.01°$ or, $\theta = \pi + .96 = 4.102$; $(12.21, 235.01°)$ or $(12.21, 4.102)$

34. $r = \sqrt{(-8)^2 + (-3)^2} = 8.54, \theta_R = \tan^{-1}\frac{3}{8} = 20.56°$ or 0.359. θ in Quadrant III so $180 + 20.56 = 200.56°$ or, $\theta = \pi + 0.359 = 3.501$; $(8.54, 200.56°)$ or $(8.54, 3.501)$

35 $r = \sqrt{2^2 + 9^2} = 9.22, \theta_R = \tan^{-1}\left(\frac{9}{2}\right) = 77.47°$ or 1.352; $(9.22, 77.47°)$ or $(9.22, 1.352)$

36. $r = \sqrt{(-6)^2 + 1^2} = \sqrt{37} = 6.08, \theta_R = \tan^{-1}\left(\frac{1}{6}\right) = 9.46°$ or 0.165 rad. θ is in Quadrant II so $\theta = 180 - 9.46 = 170.54°$ or, $0° = \pi - 0.165 = 2.977$; $(6.08, 170.54°)$ or $(6.08, 2.977)$

37.

38.

39.

40.

41.

42.

43.

44.

45.

46.

47.

48.

49.

50.

51.

52.

53.

54.

55.

56.

57.

58.

59.

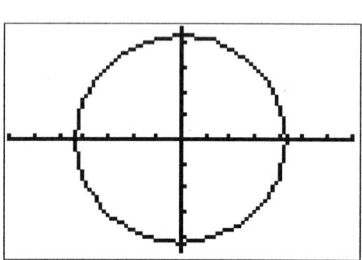

$[-7050, 7050, 1000] \times [-4650, 4650, 1000]$

60. **(a)** Multiply both sides by $E^2 \cos^2 \theta + \sin^2 \theta$ to obtain $r^2 E^2 \cos^2 \theta + r^2 \sin^2 \theta = E^2$

Thus, $E^2 x^2 + y^2 = E^2$ or $x^2 + \frac{y^2}{E^2} = 1$

(b)

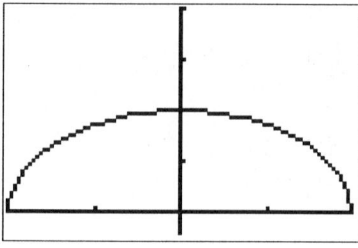

$[-1, 1, 0.5] \times [-0.1, 1.0, 0.25]$

61.

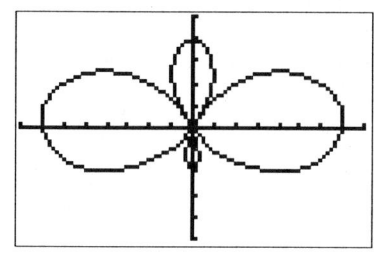

$[-8, 8, 1] \times [-5, 5, 1]$

62. **(a)** Multiply $r = 1 - 3 \sin \theta$ by r to get $r^2 = r - 3r \sin \theta$ or $x^2 + y^2 = r - 3y$. Thus, $r = x^2 + y^2 + 3y$. Squaring both sides, we get $r^2 = x^4 + 2x^2 y^2 + 6x^2 y + y^4 + 6y^3 + 9y^2$. Since $r^2 = x^2 + y^2$, this becomes $x^4 + 2x^2 y^2 + 6x^2 y + y^4 + 6y^3 + 8y^2 - x^2 = 0$

(b)

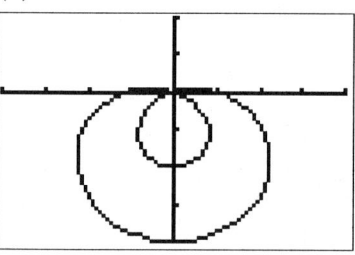

$[-4, 4, 1] \times [-4, 2, 1]$

CHAPTER 11 REVIEW

1. Period $= \frac{2\pi}{4} = \frac{\pi}{2}$,
   Amplitude $= |8| = 8$,
   Frequency $= \frac{2}{\pi}$, displacement or
   Phase Shift $= 0$

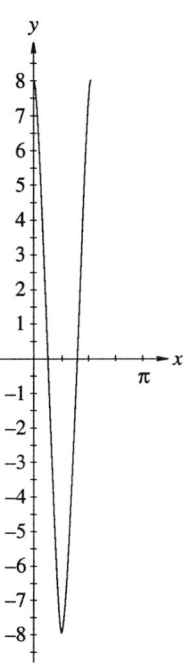

3. Period $= \frac{\pi}{3}$,
   Amplitude $= \infty$,
   Frequency $= \frac{3}{\pi}$, displacement or
   Phase Shift $= 0$

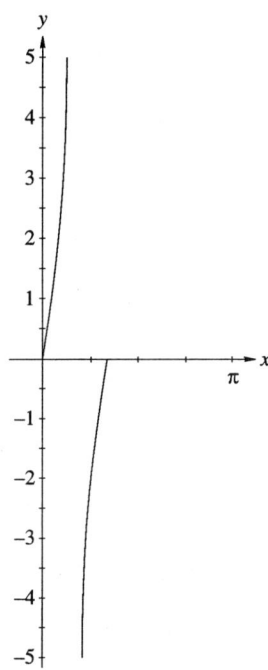

2. Period $= \frac{2\pi}{2} = \pi$,
   Amplitude $= 3$,
   Frequency $= \frac{1}{\pi}$, displacement or
   Phase Shift $= 0$

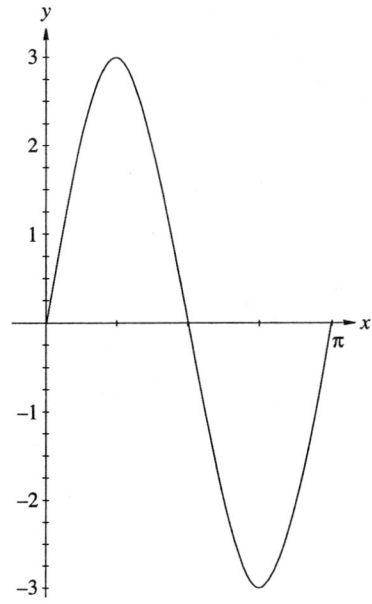

4. Period $= \frac{2\pi}{2} = \pi$,
   Amplitude $= 3$,
   Frequency $= \pi$, displacement $= -\dfrac{\frac{\pi}{2}}{2} = -\frac{\pi}{4}$

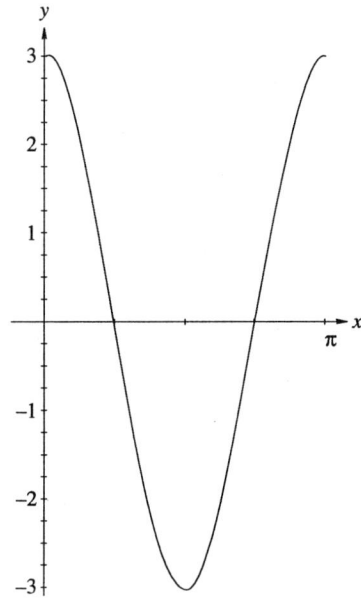

5.   Period = $\frac{2\pi}{3}$,

Amplitude = $\frac{1}{2}$,

Frequency = $\frac{3}{2\pi}$, displacement = $\dfrac{\frac{\pi}{3}}{3} = \frac{\pi}{9}$

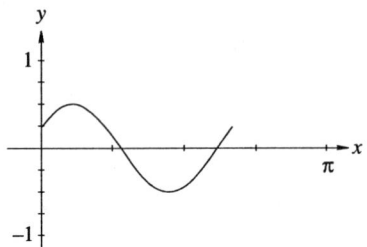

6.   Period = $\frac{2\pi}{2} = \pi$,
Amplitude = ∞,

Frequency = $\frac{1}{\pi}$, displacement = $\dfrac{\frac{\pi}{6}}{2} = \frac{\pi}{12}$

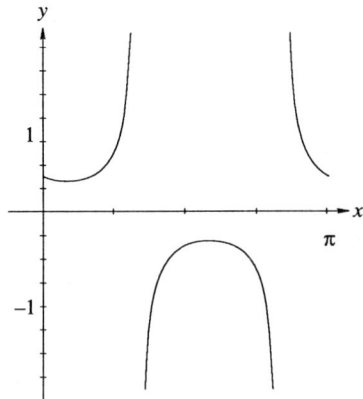

7.   Period = π,
Amplitude = ∞,
Frequency = $\frac{1}{\pi}$, displacement = $-\frac{\pi}{4}$

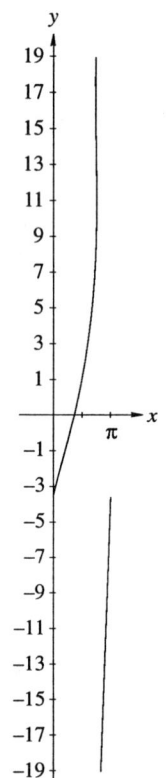

8.   Period = $\dfrac{2\pi}{\frac{1}{3}} = 6\pi$,

Amplitude = ∞,

Frequency = $\frac{1}{6\pi}$, displacement = $\dfrac{\frac{\pi}{5}}{\frac{1}{3}} = \frac{3\pi}{5}$

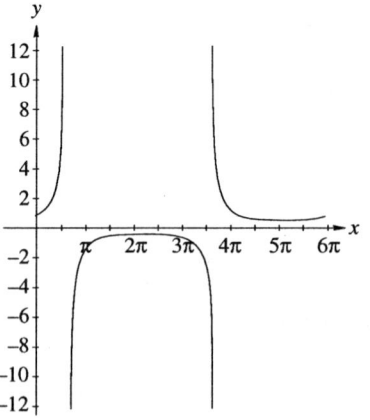

9. Period $= \dfrac{2\pi}{\frac{2}{3}} = 3\pi$,

Amplitude $= 2$,

Frequency $= \frac{1}{3\pi}$, displacement $= -\dfrac{\frac{\pi}{6}}{\frac{2}{3}} = -\frac{\pi}{4}$

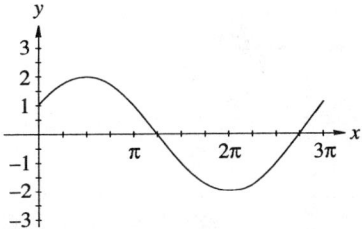

10. Period $= \dfrac{2\pi}{\frac{1}{2}} = 4\pi$,

Amplitude $= \frac{1}{4}$,

Frequency $= \frac{1}{4\pi}$, displacement $= \dfrac{\frac{2\pi}{3}}{\frac{1}{2}} = \frac{4\pi}{3}$

11.

12.

13.

14.

15.

16.

17.

18.

19.

20.

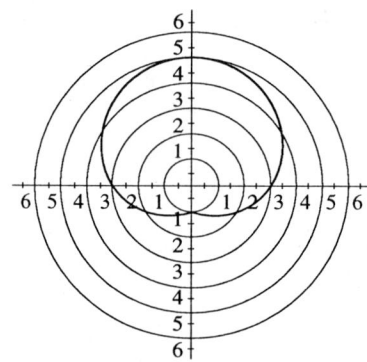

21. $y = A \sin 2\pi f t = 85 \sin 2\pi 5 t, y = 85 \sin 10\pi t$

22.   (a) $A = 1.7$ m,
    (b) Period $= \frac{2\pi}{3.4} = 1.85$ s,
    (c) $f = \frac{3.4}{2\pi} = 0.54$ Hz

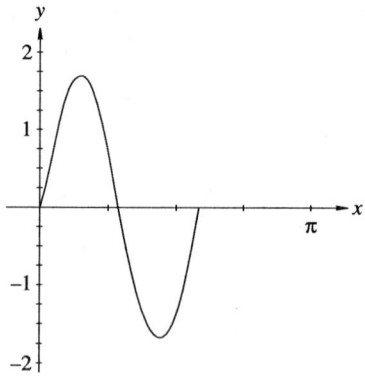

23. $V = V_{\text{max}} \sin 2\pi ft, \quad I = I_{\text{max}} \sin(2\pi ft + \phi)$
$= 4.8 \sin(2\pi 60t + \frac{\pi}{2}), I = 4.8 \sin\left(120\pi t + \frac{\pi}{2}\right)$

24.

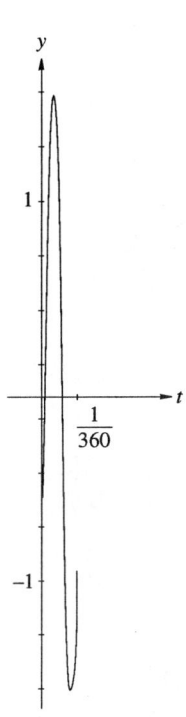

CHAPTER 11 TEST

1. (a) Period $= \frac{2\pi}{5}$, amplitude $= |-3| = 3$, frequency, $= \frac{5}{2\pi}$, displacement $= 0$

(b) Period $= \frac{2\pi}{3}$, Amplitude $= 2.4$, frequency $= \frac{3}{2\pi}$, displacement $= \frac{\pi}{12}$

(c) Period $= \frac{\pi}{2}$, Amplitude $= \infty$, frequency $= \frac{2}{\pi}$, displacement $= -\frac{\pi}{10}$

2.

3.

4.

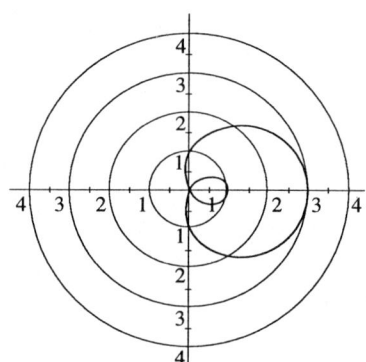

5.   Using $x = r\cos\theta$, we find $x = 2\cos 55° = 1.147$, and using $y = r\sin\theta$ yields $y = 2\sin 55° = 1.638$. Thus, the polar coordinates $(2, 55°)$ have the rectangular coordinates $(1.147, 1.638)$.

6.   Here $r = \sqrt{5^2 + (-12)^2} = 13$ and $\theta_R = \tan^{-1}\frac{\pm 12}{5} = 67.38°$ or 1.176 rad. Since θ is Quadrant IV, we see that $\theta = 360 - 67.38 = 292.52°$, or $\theta = 2\pi - 1.175 = 5.107$. Thus, the rectangular coordinates $(5, -12)$ have polar coordinates $(13, 292.62°)$ or $(13, 5.107)$.

7. $y = I_{\max}\sin(2\pi ft) = 5.7\sin(2\pi \cdot 60 \cdot t), y = 5.7\sin 120\pi t$

12
Exponents and Radicals

☰ 12.1 FRACTIONAL EXPONENTS

1. $25^{1/2} = \sqrt{25} = 5$

2. $27^{1/3} = \sqrt[3]{27} = 3$

3. $64^{1/3} = \sqrt[3]{64} = 4$

4. $125^{1/3} = \sqrt[3]{125} = 5$

5. $25^{-1/2} = \frac{1}{25^{1/2}} = \frac{1}{\sqrt{25}} = \frac{1}{5}$

6. $81^{-1/4} = \frac{1}{81^{1/4}} = \frac{1}{\sqrt[4]{81}} = \frac{1}{3}$

7. $32^{-1/5} = \frac{1}{\sqrt[5]{32}} = \frac{1}{2}$

8. $64^{-1/3} = \frac{1}{\sqrt[3]{64}} = \frac{1}{4}$

9. $27^{2/3} = \sqrt[3]{27}^2 = 3^2 = 9$

10. $81^{3/4} = \sqrt[4]{81}^3 = 3^3 = 27$

11. $125^{2/3} = \sqrt[3]{125}^2 = 5^2 = 25$

12. $32^{3/5} = \sqrt[5]{32}^3 = 2^3 = 8$

13. $16^{-3/4} = \frac{1}{\sqrt[4]{16}^3} = \frac{1}{2^3} = \frac{1}{8}$

14. $(-8)^{2/3} = \sqrt[3]{-8}^2 = (-2)^2 = 4$

15. $(-8)^{-1/3} = \frac{1}{\sqrt[3]{-8}} = -\frac{1}{2}$

16. $(-27)^{-2/3} = \frac{1}{\sqrt[3]{-27}^2} = \frac{1}{(-3)^2} = \frac{1}{9}$

17. $\left(\frac{1}{8}\right)^{1/3} = \frac{1}{\sqrt[3]{8}} = \frac{1}{2}$

18. $\left(\frac{1}{25}\right)^{3/2} = \sqrt{\frac{1}{25}}^3 = \left(\frac{1}{5}\right)^3 = \frac{1}{125}$

19. $\left(\frac{1}{16}\right)^{-5/4} = \sqrt[4]{16}^5 = 2^5 = 32$

20. $\left(\frac{-1}{27}\right)^{4/3} = \sqrt[3]{\left(\frac{-1}{27}\right)}^4 = \left(-\frac{1}{3}\right)^4 = \frac{1}{81}$

21. $3^2 \cdot 3^5 = 3^{2+5} = 3^7$

22. $5^9 5^8 = 5^{9+8} = 5^{17}$

23. $7^6 7^{-2} = 7^{6-2} = 7^4$

24. $11^9 11^{-6} = 11^{9-6} = 11^3$

25. $x^4 x^6 = x^{4+6} = x^{10}$

26. $y^7 y^9 = y^{7+9} = y^{16}$

27. $y^6 y^{-4} = y^{6-4} = y^2$

28. $x^8 x^{-2} = x^{8-2} = x^6$

29. $\left(9^5\right)^2 = 9^{5\times 2} = 9^{10}$

30. $\left(11^8\right)^{-5} = 11^{8(-5)} = 11^{-40} = \frac{1}{11^{40}}$

31. $\left(x^7\right)^3 = x^{7\times 3} = x^{21}$

32. $\left(P^9\right)^{-5} = P^{9\times(-5)} = P^{-45} = \frac{1}{P^{45}}$

33. $(xy)^5 = x^5 y^5$

34. $(yt)^3 = y^3 t^3$

35. $(ab)^{-5} = \frac{1}{(ab)^5} = \frac{1}{a^5 b^5}$

36. $(xyz)^{-9} = \frac{1}{x^9 y^9 z^9}$

37. $\frac{x^{10}}{x^2} = x^{10-2} = x^8$

38. $\frac{P^9}{P^3} = P^{9-3} = P^6$

39. $\frac{x^2}{x^8} = \frac{1}{x^{8-2}} = \frac{1}{x^6}$

40. $\frac{a^3}{a^{12}} = \frac{1}{a^{12-3}} = \frac{1}{a^9}$

41. $x^{1/2} x^{3/2} = x^{1/2+3/2} = x^2$

42. $a^{1/3} a^{4/3} = a^{1/3+4/3} = a^{1/3+4/3} = a^{5/3}$

43. $r^{3/4} r = r^{3/4+1} = r^{7/4}$

44. $a^2 a^{2/3} = a^{2+2/3} = a^{8/3}$

45. $a^{1/2} a^{1/3} = a^{1/2+1/3} = a^{3/6+2/6} = a^{5/6}$

46. $b^{2/3} b^{1/4} = b^{2/3+1/4} = b^{8/12+3/12} = b^{11/12}$

47. $d^{2/3} d^{-1/4} = d^{2/3-1/4} = d^{8/12-3/12} = d^{5/12}$

48. $x^{3/5} y^{-2/3} = \frac{x^{3/5}}{y^{2/3}}$

49. $\frac{a^2 b^5}{a^5 b^2} = \frac{b^{5-2}}{a^{5-2}} = \frac{b^3}{a^3}$

50. $\frac{x^3 y^2}{x^7 y} = \frac{y^{2-1}}{x^{7-3}} = \frac{y}{x^4}$

51. $\frac{r^5 s^2 t}{t r^3 s^5} = r^{5-3} s^{2-5} t^{1-1} = r^2 s^{-3} t^0 = \frac{r^2}{s^3}$

52. $\dfrac{a^2bc^3}{(abc)^3} = \dfrac{a^2bc^3}{a^3b^3c^3} = \dfrac{1}{ab^2}$

53. $\dfrac{(xy^2z)^4}{x^4(yz^2)^2} = \dfrac{x^4y^8z^4}{x^4y^2z^4} = y^6$

54. $\dfrac{m^6n^7}{(m^2n)^3} = \dfrac{m^6n^7}{m^6n^3} = n^4$

55. $\left(\dfrac{a}{b^2}\right)^3\left(\dfrac{a}{b^3}\right)^2 = \dfrac{a^3}{b^6}\cdot\dfrac{a^2}{b^6} = \dfrac{a^5}{b^{12}}$

56. $\left(\dfrac{x^2}{y}\right)^4\left(\dfrac{x}{y^2}\right)^2 = \dfrac{x^8}{y^4}\cdot\dfrac{x^2}{y^4} = \dfrac{x^{10}}{y^8}$

57. $\dfrac{(xy^2b^3)^{1/2}}{(x^{1/4}b^4y)^2} = \dfrac{x^{1/2}yb^{3/2}}{x^{1/2}b^8y^2} = \dfrac{1}{b^{8-3/2}y} = \dfrac{1}{b^{13/2}y}$

58. $\dfrac{(x^{1/3}y^3)^3}{(y^{10}x^5)^{1/5}} = \dfrac{xy^9}{y^2x} = y^7$

59. $\left(\dfrac{2x}{p^2}\right)^{-2}\left(\dfrac{p}{4}\right)^{-1} = \dfrac{p^4}{4x^2}\cdot\dfrac{4}{p} = \dfrac{p^3}{x^2}$

60. $\left(\dfrac{5a^2}{6b}\right)^{-2}\left(\dfrac{6}{a}\right)^{-4} = \dfrac{6^2b^2}{5^2a^4}\cdot\dfrac{a^4}{6^4} = \dfrac{b^2}{5^26^2} = \dfrac{b^2}{25\cdot36} = \dfrac{b^2}{900}$

61. $\dfrac{(x^2y^{-1}z)^{-2}}{(xy)^{-4}} = \dfrac{(xy)^4}{(x^2y^{-1}z)^2} = \dfrac{x^4y^4}{x^4y^{-2}z^2} = \dfrac{y^6}{z^2}$

62. $\left\{\left[(2x^2)^3\right]^{-4}\right\}^{-1} = \left\{\left[8x^6\right]^{-4}\right\}^{-1} = [8x^6]^4 = 8^4x^{24} = 4096x^{24}$

63. $\dfrac{(27x^6y^3)^{-1/3}}{(16x^4y^{12})^{-1/4}} = \dfrac{(16x^4y^{12})^{1/4}}{(27x^6y^3)^{1/3}} = \dfrac{2xy^3}{3x^2y} = \dfrac{2y^2}{3x}$

64. $\dfrac{(64x^6y^{12})^{-5/6}}{(9x^4y^2)^{-3/2}} = \dfrac{(9x^4y^2)^{3/2}}{(64x^6y^{12})^{5/6}} = \dfrac{\left(\sqrt{9x^4y^2}\right)^3}{\left(\sqrt[6]{64x^6y^{12}}\right)^5} = \dfrac{(3x^2y)^3}{(2xy^2)^5} =$

$\dfrac{27x^6y^3}{32x^5y^{10}} = \dfrac{27x}{32y^7}$

65. $\left(\dfrac{x^2}{y^3}\right)^{-1/5}\left(\dfrac{y^2}{x^5}\right)^{1/3} = \left(\dfrac{y^3}{x^2}\right)^{1/5}\left(\dfrac{y^2}{x^5}\right)^{1/3} = \left(\dfrac{y^3}{x^2}\right)^{3/15}$

$\left(\dfrac{y^2}{x^5}\right)^{5/15} = \left(\dfrac{y^9}{x^6}\right)^{1/15}\left(\dfrac{y^{10}}{x^{25}}\right)^{1/15} = \left(\dfrac{y^{19}}{x^{31}}\right)^{1/15} =$

$\dfrac{y^{19/15}}{x^{31/15}}$

66. $\left(\dfrac{a^2x^3}{b^5y}\right)^{-4/3}\left(\dfrac{axy^{-2}}{b}\right)^{5/3} = \left(\dfrac{b^5y}{a^2x^3}\right)^{4/3}\left(\dfrac{axy^{-2}}{b}\right)^{5/3}$

$= \dfrac{b^{20/3}y^{4/3}}{a^{8/3}x^{12/3}}\cdot\dfrac{a^{5/3}x^{5/3}y^{-10/3}}{b^{5/3}} = \dfrac{b^{20/3}y^{4/3}}{a^{8/3}x^{12/3}}\cdot\dfrac{a^{5/3}x^{5/3}}{b^{5/3}y^{10/3}} =$

$\dfrac{b^{15/3}}{a^{3/3}x^{7/3}y^{6/3}} = \dfrac{b^5}{ax^{7/3}y^2}$

67. $\left(\dfrac{9x^3}{t^5}\right)^{-1/2}\left(\dfrac{8t^2}{x^5}\right)^{-1/3} = \left(\dfrac{t^5}{9x^3}\right)^{1/2}\left(\dfrac{x^5}{8t^2}\right)^{1/3} =$

$\dfrac{t^{5/2}}{3x^{3/2}}\cdot\dfrac{x^{5/3}}{2t^{2/3}} = \dfrac{t^{15/6}x^{10/6}}{6x^{9/6}t^{4/6}} = \dfrac{t^{11/6}x^{1/6}}{6}$

68. $\left(\dfrac{125a^6}{8b^3}\right)^{-2/3}\left(\dfrac{8b}{5a^2}\right)^{-1/2} = \left(\dfrac{8b^3}{125a^6}\right)^{2/3}\left(\dfrac{5a^2}{8b}\right)^{1/2} =$

$\dfrac{4b^2}{25a^4}\cdot\dfrac{\sqrt{5a}}{\sqrt{8b^{1/2}}} = \dfrac{4b^2}{25a^4}\cdot\dfrac{\sqrt{5a}}{2\sqrt{2b^{1/2}}} = \dfrac{2\sqrt{5}b^{3/2}}{25\sqrt{2}a^3} = \dfrac{\sqrt{10}b^{3/2}}{25a^3}$

69. $\dfrac{(9x^4y^{-6})^{-3/2}}{(8x^{-6}y^3)^{-2/3}} = \dfrac{(8x^{-6}y^3)^{2/3}}{(9x^4y^{-6})^{3/2}} = \dfrac{2^2x^{-4}y^2}{3^3x^6y^{-9}} = \dfrac{4y^{11}}{27x^{10}}$

70. $\dfrac{(81x^5y^{-8}z)^{-3/4}}{(64xy^{-3}z^2)^{-5/3}} = \dfrac{(64xy^{-3}z^2)^{5/3}}{(81x^5y^{-8}z)^{3/4}} = \dfrac{4^5x^{5/3}y^{-5}z^{10/3}}{3^3x^{15/4}y^{-6}z^{3/4}} =$

$\dfrac{4^5x^{20/12}y^6z^{40/12}}{3^3x^{45/12}y^5z^{9/12}} = \dfrac{4^5yz^{31/12}}{3^3x^{25/12}} = \dfrac{1024yz^{31/12}}{27x^{25/12}}$

71. $8.3^{2/3} \approx 4.0993852$

72. $7.3^{2/5} \approx 2.21477261$

73. $92.47^{5/7} \approx 25.368006$

74. $81.94^{3/5} \approx 14.06363465$

75. $(-81.52)^{2/7} = \left((-81.42)^2\right)^{1/7} \approx 3.5162154$

76. $(-78.64)^{1/3} \approx -4.284312769$

77. $(432.61)^{1/4} \approx 4.5606226$

78. $(-537.15)^{2/3} = \left((-537.15)^2\right)^{1/3} \approx 66.07903926$

79. Using a calculator: $\left((-32.35)\wedge7\boxed{x^{-1}}\right)\wedge\boxed{(-)}$
$3 = -0.2253777406$

80. Using a calculator: $\left((-0.1439074804)\wedge5\boxed{x^{-1}}\right)\wedge$
$\boxed{(-)}\ 3 = -3.2$ on a TI-82, or -3.20000000021
on a TI-85.

81. $9.8(4.75)^2 \approx 221.1125$ or 221 m

82. $\frac{4}{3}\pi r^3 = \frac{4}{3}\pi(19.25)^3 = 9511.1042\pi = 29880.01498$
or $29,880$ in.^{3}

83. $P = 5^5\left(\dfrac{15}{0.5}\right)^{5/3} = 905\ 146.3$ N/m^2

$T = 315\left(\dfrac{15}{0.5}\right)^{5/3} = 91\ 238.747$ K

84. $q(2000) = 75\cdot2^{-2000/1600} = 31.53361557$ or about
32 mg

85. Substituting for α, we see that the exponent becomes $\dfrac{\alpha}{1-\alpha} = \dfrac{1.4}{1-1.4} = \dfrac{1.4}{-0.4} = -\dfrac{7}{2}$. Hence, we get
$E = 1 - \left(\dfrac{p_1}{p_2}\right)^{-7/2} = 1 - \left(\dfrac{p_2}{p_1}\right)^{7/2}$.

86. Here, $A = 15,000$, $r = 0.09$, $n = 12$, and $t = 5$.
Substituting these values in the given equation, we obtain

$$R = A\left[\dfrac{\dfrac{r}{n}}{1 - \left(1+\dfrac{r}{n}\right)^{-nt}}\right]$$

$$= 15,000\left[\dfrac{\dfrac{0.09}{12}}{1 - \left(1+\dfrac{0.09}{12}\right)^{-12\cdot5}}\right] \approx 311.38$$

Each monthly payment would be about $311.38.

12.2 LAWS OF RADICALS

1. $\sqrt[3]{16} = \sqrt[3]{2^4} = \sqrt[3]{2^3}\sqrt[3]{2^1} = 2\sqrt[3]{2}$

2. $\sqrt[3]{81} = \sqrt[3]{3^3 \cdot 3^1} = 3\sqrt[3]{3}$

3. $\sqrt{45} = \sqrt{9}\sqrt{5} = 3\sqrt{5}$

5. $\sqrt[3]{y^{12}} = y^{12/3} = y^4$

7. $\sqrt[5]{a^7} = \sqrt[5]{a^5 a^2} = a\sqrt[5]{a^2}$

4. $\sqrt[3]{40} = \sqrt[3]{8 \cdot 5} = 2\sqrt[3]{5}$

6. $\sqrt[4]{p^8} = p^{8/4} = p^2$

8. $\sqrt[7]{b^{10}} = \sqrt[7]{b^7 b^3} = b\sqrt[7]{b^3}$

9. $\sqrt{x^2 y^7} = \sqrt{x^2}\sqrt{y^6}\sqrt{y} = xy^3\sqrt{y}$

13. $\sqrt[3]{8x^4} = \sqrt[3]{2^3 x^3 x} = 2x\sqrt[3]{x}$

10. $\sqrt[3]{x^5 y^3} = \sqrt[3]{x^3}\sqrt[3]{x^2}\sqrt[3]{y^3} = xy\sqrt[3]{x^2}$

14. $\sqrt[4]{81y^9} = \sqrt[4]{3^4 y^8 y} = 3y^2\sqrt[4]{y}$

11. $\sqrt[4]{a^5 b^3} = \sqrt[4]{a^4}\sqrt[4]{ab^3} = a\sqrt[4]{ab^3}$

15. $\sqrt{27x^3 y} = \sqrt{3^2 x^2 3xy} = 3x\sqrt{3xy}$

12. $\sqrt[5]{p^{12} y^8} = \sqrt[5]{p^{10}}\sqrt[5]{y^5}\sqrt[5]{p^2 y^3} = p^2 y\sqrt[5]{p^2 y^3}$

16. $\sqrt[3]{32a^5 b^2} = \sqrt[3]{2^3 a^3 2^2 a^2 b^2} = 2a\sqrt[3]{4a^2 b^2}$

17. $\sqrt[3]{-8} = \sqrt[3]{(-2)^3} = -2$

18. $\sqrt[5]{-243} = \sqrt[5]{(-3)^5} = -3$

19. $\sqrt[3]{a^2 b^4}\sqrt[3]{ab^5} = \sqrt[3]{a^3 b^9} = ab^3$

20. $\sqrt[5]{x^3 y^2 z^4}\sqrt[5]{x^2 y^8 z} = \sqrt[5]{x^5 y^{10} z^5} = xy^2 z$

22. $\sqrt[6]{m^3 n^2 e^7}\sqrt[6]{m^2 n^4 e^5} = \sqrt[6]{m^5 n^6 e^{12}} = ne^2\sqrt[6]{m^5}$

21. $\sqrt[4]{p^3 q^2 r^6}\sqrt[4]{pq^6 r} = \sqrt[4]{p^4 q^8 r^7} = pq^2 r\sqrt[4]{r^3}$

23. $\sqrt[3]{\dfrac{8x^3}{27}} = \dfrac{\sqrt[3]{2^3 x^3}}{\sqrt[3]{3^3}} = \dfrac{2x}{3}$

24. $\sqrt[4]{\dfrac{81y^8}{16}} = \dfrac{\sqrt[4]{3^4 y^8}}{\sqrt[4]{2^4}} = \dfrac{3y^2}{2}$

25. $\sqrt[5]{\dfrac{x^5 y^{10}}{z^5}} = \dfrac{\sqrt[5]{x^5 y^{10}}}{\sqrt[5]{z^5}} = \dfrac{xy^2}{z}$

26. $\sqrt[3]{\dfrac{a^3 b^9}{c^6}} = \dfrac{\sqrt[3]{a^3 b^9}}{\sqrt[3]{c^6}} = \dfrac{ab^2}{c^2}$

28. $\sqrt{\dfrac{125a^5 b^2}{c^4}} = \dfrac{\sqrt{5^2 \cdot 5a^4 \cdot a \cdot b^2}}{\sqrt{c^4}} = \dfrac{5a^2 b\sqrt{5a}}{c^2}$

27. $\sqrt[3]{\dfrac{16x^3 y^2}{z^6}} = \dfrac{\sqrt[3]{2^3 \cdot 2x^3 y^2}}{\sqrt[3]{z^6}} = \dfrac{2x\sqrt[3]{2y^2}}{z^2}$

29. $\sqrt{\dfrac{64x^3 y^4}{9z^4 p^2}} = \dfrac{\sqrt{8^2 x^2 xy^4}}{\sqrt{3^2 z^4 p^2}} = \dfrac{8xy^2\sqrt{x}}{3z^2 p}$

31. $\sqrt{\dfrac{16}{3}} = \dfrac{\sqrt{16}}{\sqrt{3}} \cdot \dfrac{\sqrt{3}}{\sqrt{3}} = \dfrac{4\sqrt{3}}{3}$

33. $\sqrt[3]{\dfrac{27}{4}} = \dfrac{\sqrt[3]{3^3}}{\sqrt[3]{2^2}} \cdot \dfrac{\sqrt[3]{2}}{\sqrt[3]{2}} = \dfrac{3\sqrt[3]{2}}{2}$

30. $\sqrt[3]{\dfrac{8a^5 b^3}{27r^6 s^9}} = \dfrac{\sqrt[3]{2^3 a^3 a^2 b^3}}{\sqrt[3]{3^3 r^6 s^9}} = \dfrac{2ab\sqrt[3]{a^2}}{3r^2 s^3}$

32. $\sqrt{\dfrac{4}{5}} = \dfrac{\sqrt{4}}{\sqrt{5}} \dfrac{\sqrt{5}}{\sqrt{5}} = \dfrac{2\sqrt{5}}{5}$

34. $\sqrt[3]{\dfrac{16}{25}} = \dfrac{\sqrt[3]{2^4}}{\sqrt[3]{5^2}} \cdot \dfrac{\sqrt[3]{5}}{\sqrt[3]{5}} = \dfrac{2\sqrt[3]{10}}{5}$

35. $\sqrt{\dfrac{25}{2x}} = \dfrac{\sqrt{25}}{\sqrt{2x}} \cdot \dfrac{\sqrt{2x}}{\sqrt{2x}} = \dfrac{5\sqrt{2x}}{2x}$

43. $\sqrt{4 \times 10^4} = \sqrt{4} \times \sqrt{10^4} = 2 \times 10^2 = 200$

44. $\sqrt{9 \times 10^6} = \sqrt{9} \times \sqrt{10^6} = 3 \times 10^3 = 3000$

45. $\sqrt{25 \times 10^3} = \sqrt{25}\sqrt{10^1 10^2} = 5\sqrt{10} \times 10 = 50\sqrt{10}$

36. $\sqrt[3]{\dfrac{8}{5y^2}} = \dfrac{\sqrt[3]{2^3}}{\sqrt[3]{5y^2}} = \dfrac{\sqrt[3]{5^2 y}}{\sqrt[3]{5^2 y}} = \dfrac{2\sqrt[3]{25y}}{5y}$

46. $\sqrt{16 \times 10^7} = \sqrt{16}\sqrt{10^1(10^6)} = 4\sqrt{10} \times 10^3 = 4000\sqrt{10}$

37. $\sqrt[4]{\dfrac{81}{32z^2}} = \dfrac{\sqrt[4]{3^4}}{\sqrt[4]{2^5 z^2}} \cdot \dfrac{\sqrt[4]{2^3 z^2}}{\sqrt[4]{2^3 z^2}} = \dfrac{3\sqrt[4]{8z^2}}{4z}$

47. $\sqrt{4 \times 10^7} = \sqrt{4}\sqrt{10(10^6)} = 2\sqrt{10} \times 10^3 = 2000\sqrt{10}$

38. $\sqrt[3]{\dfrac{-2}{25r^2}} = \dfrac{\sqrt[3]{-2}}{\sqrt[3]{5^2 r^2}} \cdot \dfrac{\sqrt[3]{5r}}{\sqrt[3]{5r}} = \dfrac{3\sqrt[3]{-10r}}{5r}$

48. $\sqrt{9 \times 10^9} = \sqrt{9}\sqrt{10}\sqrt{10^8} = 3\sqrt{10} \times 10^4 = 30{,}000\sqrt{10}$

39. $\sqrt[3]{\dfrac{16x^2 y}{x^5}} = \sqrt[3]{\dfrac{16y}{x^3}} = \dfrac{\sqrt[3]{(2^3)2y}}{\sqrt[3]{x^3}} = \dfrac{2\sqrt[3]{2y}}{x}$

49. $\sqrt[3]{1.25 \times 10^{10}} = \sqrt[3]{125 \times 10^8} = \sqrt[3]{125}\sqrt[3]{10^2}\sqrt[3]{10^6}$
$= 5\sqrt[3]{100} \times 10^2 = 500\sqrt[3]{100}$ or $\sqrt[3]{1.25 \times 10^{10}} =$
$\sqrt[3]{1.25 \times 10 \times 10^9} = \sqrt[3]{12.5} \times 10^3$

40. $\sqrt[4]{\dfrac{25a^3 b^5}{a^7 b}} = \sqrt[4]{\dfrac{5^2 b^4}{a^4}} = \dfrac{b}{a}\sqrt[4]{5^2} = \dfrac{b}{a}5^{2/4} = \dfrac{b5^{1/2}}{a} = \dfrac{b\sqrt{5}}{a}$

41. $\sqrt[3]{\dfrac{-8x^3 yz}{27b^2 z^4}} = \sqrt[3]{\dfrac{-8x^3 y}{27b^2 z^3}} = \dfrac{\sqrt[3]{(-2)^3 x^3 y}}{\sqrt[3]{3^3 b^2 z^3}} \cdot \dfrac{\sqrt[3]{b}}{\sqrt[3]{b}} = \dfrac{-2x\sqrt[3]{by}}{3bz}$

50. $\sqrt[5]{3.2 \times 10^{14}} = \sqrt[5]{32 \times 10^{13}} = \sqrt[5]{2^5 \cdot 10^{10} \times 10^3} =$
$2 \times 100 \times \sqrt[5]{1000} = 200 \times \sqrt[5]{1000}$ or $\sqrt[5]{3.2 \times 10^{14}} =$
$\sqrt[5]{3.2 \times 10^4 \times 10^{10}} = \sqrt[5]{3.2 \times 10^4 \times 10^{10}} =$
$\sqrt[5]{3.2 \times 10^4} \times 10^2$

42. $\sqrt[4]{\dfrac{25a^2 b^3}{16c^3 b^6}} = \sqrt[4]{\dfrac{5^2 a^2}{2^4 c^3 b^3}} = \dfrac{\sqrt[4]{5^2 a^2}}{\sqrt[4]{2^4 c^3 b^3}} \cdot \dfrac{\sqrt[4]{cb}}{\sqrt[4]{cb}} = \dfrac{\sqrt[4]{25a^2 bc}}{2bc}$

51. $\sqrt{\frac{x}{y}+\frac{y}{x}} = \sqrt{\frac{x^2}{xy}+\frac{y^2}{xy}} = \sqrt{\frac{x^2+y^2}{xy}} = \frac{\sqrt{x^2+y^2}}{\sqrt{xy}}\cdot$

$\frac{\sqrt{xy}}{\sqrt{xy}} = \frac{\sqrt{xy(x^2+y^2)}}{xy} = \frac{\sqrt{x^3y+xy^3}}{xy}$

52. $\sqrt{\frac{a}{b}-\frac{b}{a}} = \sqrt{\frac{a^2}{ab}-\frac{b^2}{ab}} = \sqrt{\frac{a^2-b^2}{ab}}\cdot\frac{\sqrt{ab}}{\sqrt{ab}} =$

$\frac{\sqrt{ab(a^2-b^2)}}{ab} = \frac{\sqrt{a^3b-ab^3}}{ab}$

53. $\sqrt{a^2+2ab+b^2} = \sqrt{(a+b)^2} = |a+b|$

54. $\sqrt{x^2-2xy+y^2} = \sqrt{(x-y)^2} = |x-y|$

55. $\sqrt{\frac{1}{a^2}+\frac{1}{b}} = \sqrt{\frac{b}{a^2b}+\frac{a^2}{a^2b}} = \sqrt{\frac{b+a^2}{a^2b}} =$

$\frac{\sqrt{b+a^2}}{\sqrt{a^2b}}\cdot\frac{\sqrt{b}}{\sqrt{b}} = \frac{\sqrt{b^2+a^2b}}{ab}$

56. $\sqrt{\frac{x}{y^2}+\frac{y}{x^2}} = \sqrt{\frac{x^3}{x^2y^2}+\frac{y^3}{x^2y^2}} = \sqrt{\frac{x^3+y^3}{x^2y^2}} =$

$\frac{\sqrt{x^3+y^3}}{\sqrt{x^2y^2}} = \frac{\sqrt{x^3+y^3}}{xy}$

57. $\sqrt[6]{\sqrt[3]{27x^2}}$. Multiply the indexes to get $\sqrt[18]{27x^2}$.

58. $\sqrt[4]{\sqrt[6]{125x^3y^5}}$. Multiply the indexes to get $\sqrt[24]{125x^3y^5}$.

59. Multiply the indexes to get $\sqrt[21]{-624.2x^{15}y^{10}z}$.

60. Multiply the indexes to get $\sqrt[8]{9x^8y^3} = |x|\sqrt[8]{9y^3}$.

61. $f = \frac{1}{2L}\sqrt{\frac{T}{\mu}} = \frac{1}{2L}\cdot\frac{\sqrt{T}}{\sqrt{\mu}}\cdot\frac{\sqrt{\mu}}{\sqrt{\mu}} = \frac{1}{2L}\cdot\frac{\sqrt{T\mu}}{\mu} = \frac{1}{2L\mu}\cdot\sqrt{T\mu}$

62. When T is quadrupled the frequency is doubled.

63. $z = \dfrac{1}{\sqrt{\frac{1}{x^2}+\frac{1}{R^2}}} = \dfrac{1}{\sqrt{\frac{R^2+x^2}{x^2R^2}}} = \sqrt{\frac{x^2R^2}{R^2+x^2}} =$

$\frac{xR\sqrt{R^2+x^2}}{R^2+x^2}$

64. $\sqrt[3]{\frac{M}{2\rho N}}\cdot\frac{\sqrt[3]{2^2\rho^2N^2}}{\sqrt[3]{2^2\rho^2N^2}} = \frac{\sqrt[3]{4M\rho^2N^2}}{2\rho N}$

65. First, we simplify the expression under the radical to obtain $1+\sqrt{\frac{m+2h}{m}}$ which equals $1+\frac{m^2+2mh}{m^2}$ or $1+\frac{\sqrt{m^2+2mh}}{m} = \frac{m+\sqrt{m^2+2h}}{m}$

66. $f = \frac{1}{2\pi}\sqrt{\frac{3EI}{ml^3}} = \frac{1}{2\pi}\frac{\sqrt{3EI}}{\sqrt{ml^3}}\frac{\sqrt{ml}}{\sqrt{ml}} = \frac{\sqrt{3EIml}}{2\pi ml^2}$

≡ 12.3 BASIC OPERATIONS WITH RADICALS

1. $2\sqrt{3}+5\sqrt{3} = 7\sqrt{3}$

2. $5\sqrt{6}-3\sqrt{6} = 2\sqrt{6}$

3. $\sqrt[3]{9}+4\sqrt[3]{9} = 5\sqrt[3]{9}$

4. $\sqrt[4]{8}+3\sqrt[4]{8} = 4\sqrt[4]{8}$

5. $2\sqrt{3}+4\sqrt{2}+6\sqrt{3} = 8\sqrt{3}+4\sqrt{2}$

6. $5\sqrt{3}-6\sqrt{5}-9\sqrt{3} = -4\sqrt{3}-6\sqrt{5}$

7. $\sqrt{5}+\sqrt{20} = \sqrt{5}+2\sqrt{5} = 3\sqrt{5}$

8. $\sqrt{8}+\sqrt{2} = 2\sqrt{2}+\sqrt{2} = 3\sqrt{2}$

9. $\sqrt{7}-\sqrt{28} = \sqrt{7}-2\sqrt{7} = -\sqrt{7}$

10. $\sqrt{8}-\sqrt{32} = 2\sqrt{2}-4\sqrt{2} = -2\sqrt{2}$

11. $\sqrt{60}-\sqrt{\frac{5}{3}} = 2\sqrt{15}-\sqrt{\frac{5}{3}}\sqrt{\frac{3}{3}} = 2\sqrt{15}-\frac{\sqrt{15}}{3} = \frac{6\sqrt{15}}{3}-\frac{\sqrt{15}}{3} = \frac{5\sqrt{15}}{3}$

12. $\sqrt{84}-\sqrt{\frac{3}{7}} = 2\sqrt{21}+\frac{\sqrt{21}}{7} = \left(\frac{14}{7}+\frac{1}{7}\right)\sqrt{21} = \frac{15}{7}\sqrt{21}$

13. $\sqrt{\frac{1}{2}}-\sqrt{\frac{9}{2}} = \frac{\sqrt{2}}{2}-\frac{3\sqrt{2}}{2} = \frac{-2\sqrt{2}}{2} = -\sqrt{2}$

14. $\sqrt{\frac{4}{3}}-\sqrt{\frac{25}{3}} = \frac{2\sqrt{3}}{3}-\frac{5\sqrt{3}}{3} = \frac{-3\sqrt{3}}{3} = -\sqrt{3}$

15. $\sqrt{x^3y}+2x\sqrt{xy} = x\sqrt{xy}+2x\sqrt{xy} = 3x\sqrt{xy}$

16. $\sqrt{a^5b^3}-3ab\sqrt{a^3b} = a^2b\sqrt{ab}-3a^2b\sqrt{ab} = -2a^2b\sqrt{ab}$

17. $\sqrt[3]{24p^2q^4}+\sqrt[3]{3p^8q} = \sqrt[3]{8\cdot 3p^2q^3q}+\sqrt[3]{3p^6p^2q} = 2q\sqrt[3]{3p^2q}+p^2\sqrt[3]{3p^2q} = (2q+p^2)\sqrt[3]{3p^2q}$

18. $\sqrt[4]{16a^2b}-\sqrt[4]{81a^6b} = 2\sqrt[4]{a^2b}-3a\sqrt[4]{a^2b} = (2-3a)\sqrt[4]{a^2b}$

19. $\sqrt{\frac{x}{y^3}}-\sqrt{\frac{y}{x^3}} = \sqrt{\frac{xy}{y^4}}-\sqrt{\frac{xy}{x^4}} = \left(\frac{1}{y^2}-\frac{1}{x^2}\right)\sqrt{xy}$ or $\frac{x^2-y^2}{x^2y^2}\sqrt{xy}$

20. $\sqrt{\frac{a^3}{b^3}}+\sqrt{\frac{b}{a^5}} = \frac{a}{b}\sqrt{\frac{a}{b}}+\frac{1}{a^2}\sqrt{\frac{b}{a}} = \frac{a}{b^2}\sqrt{ab}+\frac{1}{a^3}\sqrt{ab} = \left(\frac{a}{b^2}+\frac{1}{a^3}\right)\sqrt{ab}$ or $\frac{a^4+b^2}{a^3b^2}\sqrt{ab}$

21. $a\sqrt{\dfrac{x}{3a}} + b\sqrt{\dfrac{a}{3b}} = \dfrac{a\sqrt{3ab3a}}{3a} + \dfrac{b\sqrt{3ab}}{3b} = \dfrac{\sqrt{3ab}}{3} + \dfrac{\sqrt{3ab}}{3} = \dfrac{2\sqrt{3ab}}{3}$

22. $x\sqrt{\dfrac{y}{5x}} - y\sqrt{\dfrac{x}{5y}} = \dfrac{x\sqrt{5xy}}{5x} - \dfrac{y\sqrt{5xy}}{5y} = \dfrac{\sqrt{5xy}}{5} - \dfrac{\sqrt{5xy}}{5} = 0$

23. $\sqrt{5}\sqrt{8} = \sqrt{40} = 2\sqrt{10}$

24. $\sqrt[5]{-7}\sqrt[5]{11} = \sqrt[5]{-77}$

25. $\sqrt{3x}\sqrt{5x} = \sqrt{15x^2} = x\sqrt{15}$

26. $\sqrt[3]{7x^2}\sqrt[3]{3x} = \sqrt[3]{21x^3} = x\sqrt[3]{21}$

27. $\left(\sqrt{4x}\right)^3 = \left(2\sqrt{x}\right)^3 = 2^3\sqrt{x^3} = 8\sqrt{x^3} = 8x\sqrt{x}$

28. $\left(\sqrt[3]{2x^2y}\right)^4 = \sqrt[3]{2^4x^8y^4} = 2x^2y\sqrt[3]{2x^2y}$

29. $\sqrt{\dfrac{5}{8}} \cdot \sqrt{\dfrac{9}{10}} = \sqrt{\dfrac{5\cdot9}{8\cdot10}} = \sqrt{\dfrac{9}{16}} = \dfrac{3}{4}$

30. $\sqrt{\dfrac{7}{6}}\sqrt{\dfrac{12}{3}} = \sqrt{\dfrac{7\cdot12}{6\cdot3}} = \dfrac{\sqrt{7\cdot2}}{\sqrt{3}} \cdot \dfrac{\sqrt{3}}{\sqrt{3}} = \dfrac{\sqrt{42}}{3}$

31. $\sqrt{2}(\sqrt{x} + \sqrt{2}) = \sqrt{2x} + 2$

32. $\sqrt{3}(\sqrt{12} - \sqrt{y}) = \sqrt{36} - \sqrt{3y} = 6 - \sqrt{3y}$

33. $(\sqrt{x}+\sqrt{y})^2 = \sqrt{x}^2 + 2\sqrt{xy} + \sqrt{y}^2 = x + 2\sqrt{xy} + y$

34. $\left(\sqrt{a} + 3\sqrt{b}\right)^2 = \sqrt{a}^2 + 6\sqrt{ab} + (3\sqrt{b})^2 = a + 6\sqrt{ab} + 9b$

35. $\sqrt[3]{\dfrac{5}{2}} \cdot \sqrt[3]{\dfrac{2}{7}} = \sqrt[3]{\dfrac{5\cdot2}{2\cdot7}} = \sqrt[3]{\dfrac{5}{7}} \cdot \sqrt[3]{\dfrac{7^2}{7^2}} = \dfrac{\sqrt[3]{245}}{7}$

36. $\sqrt{\dfrac{ab}{5c}}\sqrt{\dfrac{abc}{5}} = \sqrt{\dfrac{a^2b^2c}{5^2c}} = \sqrt{\dfrac{a^2b^2}{5^2}} = \dfrac{ab}{5}$

37. $\left(\sqrt{a} + \sqrt{b}\right)\left(\sqrt{a} - \sqrt{b}\right) = \sqrt{a}^2 - \sqrt{b}^2 = |a| - |b|$

38. $\left(\sqrt{x} - 2\sqrt{y}\right)\left(\sqrt{x} + 5\sqrt{y}\right) = x + 3\sqrt{xy} - 10y$

39. $\sqrt[3]{x}\sqrt{x} = x^{1/3}x^{1/2} = x^{2/6}x^{3/6} = x^{5/6} = \sqrt[6]{x^5}$

40. $\sqrt{5x}\sqrt[3]{2x^2} = (5x)^{1/2}(2x^2)^{1/3} = (5x)^{3/6}(2x^2)^{2/6} = (5^3x^3)^{1/6}(2^2x^4)^{1/6} = \sqrt[6]{5^3x^3}\sqrt[6]{4x^4} = x\sqrt[6]{5^3x4} = x\sqrt[6]{500x}$

41. To obtain a common index, rewrite $\sqrt[3]{49x^2}\sqrt[4]{3x}$ as $(49x^2)^{1/3}(3x)^{1/4} = (49x^2)^{4/12}(3x)^{3/12} = (49^4x^8)^{1/12}(3^3x^3)^{1/12} = (49^43^3x^{11})^{1/12} = \sqrt[12]{49^43^3x^{11}} = \sqrt[12]{3^349^4x^{11}} = \sqrt[12]{155{,}649{,}627x^{11}}$

42. To obtain a common index, rewrite $\sqrt[3]{9y}$ as $\sqrt[15]{(9y)^5}$ and $\sqrt[5]{-8x^2}$ as $\sqrt[15]{(-8x^2)^3}$. Multiplying produces $\sqrt[15]{(9y)^5}\,\sqrt[15]{(-8x^2)^3} = \sqrt[15]{9^5y^5(-8)^3x^6} = \sqrt[15]{-9^58^3y^5x^6} = \sqrt[15]{-3^{10}2^9y^5x^6} = \sqrt[15]{-30{,}233{,}088y^5x^6}$

43. $\dfrac{\sqrt{32}}{\sqrt{2}} = \sqrt{\dfrac{32}{2}} = \sqrt{16} = 4$

44. $\dfrac{\sqrt[3]{a^2}}{\sqrt[3]{4a}} = \sqrt[3]{\dfrac{a^2}{4a}} = \sqrt[3]{\dfrac{a}{4}} \cdot \sqrt[3]{\dfrac{2}{2}} = \dfrac{\sqrt[3]{2a}}{\sqrt[3]{8}} = \dfrac{\sqrt[3]{2a}}{2}$

45. $\dfrac{\sqrt[3]{4b^2}}{\sqrt[3]{16b}} = \sqrt[3]{\dfrac{4b^2}{16b}} = \sqrt[3]{\dfrac{b}{4}}\sqrt[3]{\dfrac{2}{2}} = \dfrac{\sqrt[3]{2b}}{2}$

46. $\dfrac{\sqrt{5a^3b}}{\sqrt{15ab^3}} = \sqrt{\dfrac{5a^3b}{15ab^3}} = \sqrt{\dfrac{a^2}{3b^2}}\sqrt{\dfrac{3}{3}} = \dfrac{a\sqrt{3}}{3b}$

47. $\dfrac{\sqrt{3x^2}}{\sqrt[3]{9x}} = \dfrac{\sqrt[6]{(3x^2)^3}}{\sqrt[6]{(9x)^2}} = \dfrac{\sqrt[6]{27x^6}}{\sqrt[6]{81x^2}} = \sqrt[6]{\dfrac{27x^6}{81x^2}} = \sqrt[6]{\dfrac{x^4}{3}} \cdot \dfrac{\sqrt[6]{3^5}}{\sqrt[6]{3^5}} = \dfrac{\sqrt[6]{243x^4}}{3}$

48. $\dfrac{\sqrt[3]{15x}}{\sqrt{5x}} = \sqrt[6]{\dfrac{(15x)^2}{(5x)^3}} = \sqrt[6]{\dfrac{225x^2}{125x^3}} = \sqrt[6]{\dfrac{9}{5x}} = \sqrt[6]{\dfrac{9}{5x}}\dfrac{\sqrt[6]{(5x)^5}}{\sqrt[6]{(5x)^5}} = \dfrac{\sqrt[6]{28{,}125x^5}}{5x}$

49. $\dfrac{\sqrt[3]{16a^2b}}{\sqrt[4]{2ab^3}} = \dfrac{\sqrt[12]{(16a^2b)^4}}{\sqrt[12]{(2ab^3)^3}} = \sqrt[12]{\dfrac{(16a^2b)^4}{(2ab^3)^3}} = \sqrt[12]{\dfrac{(2^4a^2b)^4}{(2ab^3)^3}} = \sqrt[12]{\dfrac{2^{16}a^8b^4}{2^3a^3b^9}} = \sqrt[12]{\dfrac{2^{16}a^8b^4}{2^3a^3b^9}} = \sqrt[12]{\dfrac{2^{13}a^5}{b^5}} = 2\sqrt[12]{\dfrac{2a^5}{b^5}} = 2\sqrt[12]{\dfrac{2a^5}{b^5}} \cdot \dfrac{\sqrt[12]{b^7}}{\sqrt[12]{b^7}} = 2\dfrac{\sqrt[12]{2a^5b^7}}{b}$

50. $\dfrac{\sqrt[5]{75a^3x^2}}{\sqrt[3]{-15ax^2}} = \dfrac{\sqrt[5]{75a^3x^2}}{\sqrt[3]{-15ax^2}} \cdot \dfrac{\sqrt[3]{15^2a^2x}}{\sqrt[3]{15^2a^2x}} = \dfrac{\sqrt[5]{75a^3x^2}\sqrt[3]{15^2a^2x}}{15ax} = \dfrac{\sqrt[15]{75^3a^9x^6}\,\sqrt[15]{15^{10}a^{10}x^5}}{-15ax} = \dfrac{\sqrt[15]{3^35^6a^9x^63^{10}5^{10}a^{10}x^5}}{-15ax} = \dfrac{\sqrt[15]{3^{13}5^{16}a^{19}x^{11}}}{-15ax} = \dfrac{\sqrt[15]{3^{13}5^{16}a^4x^{11}}}{-15x}$

51. $\dfrac{\sqrt[3]{4}}{\sqrt{2}} = \dfrac{\sqrt[3]{4}}{\sqrt{2}} \cdot \dfrac{\sqrt{2}}{\sqrt{2}} = \dfrac{\sqrt[3]{2}\sqrt{2}}{2} = \dfrac{\sqrt[6]{2^4\cdot2^3}}{2} = \dfrac{\sqrt[6]{2^7}}{2} = \sqrt[6]{2}$

52. $\dfrac{\sqrt[5]{-8}}{\sqrt[3]{4}} = \dfrac{\sqrt[5]{-8}}{\sqrt[3]{2^2}} \cdot \dfrac{\sqrt[3]{2}}{\sqrt[3]{2}} = \dfrac{\sqrt[5]{-1(2^3)}\sqrt[3]{2}}{\sqrt[3]{2^3}} = \dfrac{\sqrt[15]{(-1)2^9}\,\sqrt[15]{2^5}}{2} = \dfrac{\sqrt[15]{(-1)2^{14}}}{2}$

53. $\dfrac{1}{x+\sqrt{5}} \cdot \dfrac{x-\sqrt{5}}{x-\sqrt{5}} = \dfrac{x-\sqrt{5}}{x^2-5}$

54. $\dfrac{1}{x+\sqrt{3}} \cdot \dfrac{x-\sqrt{3}}{x-\sqrt{3}} = \dfrac{x-\sqrt{3}}{x^2-3}$

55. $\dfrac{\sqrt{5}-\sqrt{3}}{\sqrt{5}+\sqrt{3}} \cdot \dfrac{\sqrt{5}-\sqrt{3}}{\sqrt{5}-\sqrt{3}} = \dfrac{5-2\sqrt{15}+3}{5-3} = \dfrac{8-2\sqrt{15}}{2} = \dfrac{2(4-\sqrt{15})}{2} = 4 - \sqrt{15}$

56. $\dfrac{\sqrt{5}+\sqrt{7}}{\sqrt{7}-\sqrt{5}} \cdot \dfrac{\sqrt{7}+\sqrt{5}}{\sqrt{7}+\sqrt{5}} = \dfrac{5+2\sqrt{35}+7)}{7-5} = \dfrac{12+2\sqrt{35}}{2} = \dfrac{2(6+\sqrt{35})}{2}$
$= 6 + \sqrt{35}$

57. $\dfrac{\sqrt{x+1}}{\sqrt{x-1}} + \dfrac{\sqrt{x-1}}{\sqrt{x+1}} = \dfrac{\sqrt{x+1}\sqrt{x-1}}{\sqrt{x-1}\sqrt{x-1}} + \dfrac{\sqrt{x-1}\sqrt{x+1}}{\sqrt{x+1}\sqrt{x+1}} = \dfrac{\sqrt{x^2-1}}{x-1} +$
$\dfrac{\sqrt{x^2-1}}{x+1} = \dfrac{(x+1)\sqrt{x^2-1}}{x^2-1} + \dfrac{(x-1)\sqrt{x^2-1}}{x^2-1} = \dfrac{2x\sqrt{x^2-1}}{x^2-1}, x \neq$
$1, -1$

58. $\dfrac{\sqrt{x+3}}{\sqrt{x-3}} - \dfrac{\sqrt{x-3}}{\sqrt{x+3}} = \dfrac{\sqrt{x+3}\sqrt{x-3}}{\sqrt{x-3}\sqrt{x-3}} - \dfrac{\sqrt{x-3}\sqrt{x+3}}{\sqrt{x+3}\sqrt{x+3}} = \dfrac{\sqrt{x^2-9}}{x-3} -$
$\dfrac{\sqrt{x^2-9}}{x+3} = \dfrac{(x+3)\sqrt{x^2-9}}{x^2-9} - \dfrac{(x-3)\sqrt{x^2-9}}{x^2-9} = \dfrac{6\sqrt{x^2-9}}{x^2-9}, x \neq$
$3, -3$

59. $\dfrac{\sqrt{x+y}}{\sqrt{x-y}-\sqrt{x}} \cdot \dfrac{\sqrt{x-y}+\sqrt{x}}{\sqrt{x-y}+\sqrt{x}} = \dfrac{\sqrt{x^2-y^2}+\sqrt{x^2+xy}}{x-y-x} =$
$-\dfrac{\sqrt{x^2-y^2}+\sqrt{x^2+xy}}{y}, y \neq 0$

60. $\dfrac{\sqrt{1+y}}{\sqrt{1-y}+\sqrt{y}} \cdot \dfrac{\sqrt{1-y}-\sqrt{y}}{\sqrt{1-y}-\sqrt{y}} = \dfrac{\sqrt{1-y^2}-\sqrt{y+y^2}}{(1-y)-y} =$
$\dfrac{\sqrt{1-y^2}-\sqrt{y+y^2}}{1-2y}, y \neq \frac{1}{2}$

61. $\dfrac{-b+\sqrt{b^2-4ac}}{2a} + \dfrac{-b-\sqrt{b^2-4ac}}{2a} = \dfrac{-2b}{2a} = \dfrac{-b}{a}$

62. $v = \sqrt{\dfrac{\pi}{4\lambda d}} + \sqrt{\dfrac{4\pi}{\lambda d}} = \dfrac{1}{2}\sqrt{\dfrac{\pi}{\lambda d}} + \dfrac{2}{1}\sqrt{\dfrac{\pi}{\lambda d}} = \dfrac{5}{2}\sqrt{\dfrac{\pi}{\lambda d}} =$
$\dfrac{5}{2\lambda d}\sqrt{\pi\lambda d}$

63. **(a)** $R = \dfrac{R_1 R_2}{R_1+R_2} = \dfrac{x^{3/2}\sqrt{x}}{x^{3/2}+\sqrt{x}} = \dfrac{x^{3/2}x^{1/2}}{x^{3/2}+x^{1/2}} = \dfrac{x^2}{\sqrt{x^3}+\sqrt{x}} \cdot$
$\dfrac{\sqrt{x^3}-\sqrt{x}}{\sqrt{x^3}-\sqrt{x}} = \dfrac{x^2 x^{3/2}-x^2 x^{1/2}}{x^3-x} = \dfrac{x^{7/2}-x^{5/2}}{x^3-x} = \dfrac{x^{5/2}(x-1)}{x(x^2-1)} =$
$\dfrac{x \cdot x^{3/2}(x-1)}{x(x+1)(x-1)} = \dfrac{x^{3/2}}{x+1}$

(b) if $x = 20$; $\dfrac{20^{3/2}}{20+1} = \dfrac{89.44279191}{21} = 4.2591771 \approx$
$4.259 \ \Omega$

64. $\dfrac{\sqrt{d_1}-\sqrt{d_2}}{\sqrt{d_1}+\sqrt{d_2}} \cdot \dfrac{\sqrt{d_1}-\sqrt{d_2}}{\sqrt{d_1}-\sqrt{d_2}} = \dfrac{d_1-2\sqrt{d_1 d_2}+d_2}{d_1-d_2}$

65. $\left(\dfrac{-b+\sqrt{b^2-4ac}}{2a}\right)\left(\dfrac{-b-\sqrt{b^2-4ac}}{2a}\right) = \dfrac{b^2-\left(\sqrt{b^2-4ac}\right)^2}{4a^2}$
$= \dfrac{b^2-(b^2-4ac)}{4a^2} = \dfrac{4ac}{4a^2} = \dfrac{c}{a}$

66. $\dfrac{3}{8\sqrt[3]{2}} \cdot \dfrac{KA}{\sqrt[3]{L}} = \dfrac{3}{8\sqrt[3]{2}} \cdot \dfrac{KA}{\sqrt[3]{L}} \cdot \dfrac{\sqrt[3]{2^2 L^2}}{\sqrt[3]{2^2 L^2}} = \dfrac{3KA\sqrt[3]{4L^2}}{8 \cdot 2L} =$
$\dfrac{3KA\sqrt[3]{4L^2}}{16L}$

≡ 12.4 EQUATIONS WITH RADICALS

1. $\sqrt{x+3}=5; \left(\sqrt{x+3}\right)^2 = 5^2; x+3 = 25; x = 22.$
Check: $\sqrt{22+3} = \sqrt{25} = 5$. Solution is 22.

2. $\sqrt{x-16}=5; \left(\sqrt{x-16}\right)^2 = 5^2; x-16 = 25; x = 41$. Check: $\sqrt{41-16} = \sqrt{25} = 5$. Solution is 41.

3. $\sqrt{2x+4}-7=0; \sqrt{2x+4}=7; \left(\sqrt{2x+4}\right)^2 = 7^2; 2x+4 = 49; 2x = 45; x = \frac{45}{2}$. Check:
$\sqrt{2 \cdot \frac{45}{2}+4}-7 = \sqrt{45+4}-7 = \sqrt{49}-7 = 0.$
Solution is $\frac{45}{2}$ or 22.5.

4. $\sqrt{3y-9}-9-18=0; \sqrt{3y-9}=18; \left(\sqrt{3y-9}\right)^2 = 18^2; 3y-9 = 324; 3y = 333; y = 111$. Check: $\sqrt{3 \cdot 111 -9}-18 = \sqrt{333-9}-18 = \sqrt{324}-18 = 18-18 = 0$. Solution is 111.

5. $\sqrt{x^2+24}=x-4; \left(\sqrt{x^2+24}\right)^2 = (x-4)^2; x^2 + 24 = x^2-8x+16; 24 = -8x+16; 8 = -8x; x = -1$. Check: $\sqrt{(-1)^2+24} = \sqrt{1+24} = \sqrt{25} = 5; -1-4 = -5$ but $5 \neq -5$ so no solution.

6. $\sqrt{x^2-75}=x+5; \left(\sqrt{x^2-75}\right)^2 = (x+5)^2; x^2-75 = x^2 + 10x + 25; -100 = 10x; x = -10.$

Check: $\sqrt{(-10)^2-75} = \sqrt{100-75} = \sqrt{25} = 5; -10+5 = -5$ but $5 \neq -5$ so no solution.

7. $(y+12)^{1/3}=3; \left((y+12)^{1/3}\right)^3 = 3^3; y+12 = 27; y = 15$. Check: $(15+12)^{1/3} = 27^{1/3} = 3$. Solution is 15.

8. $(r+9)^{1/4}=7; \left[(r+9)^{1/4}\right]^4 = 7^4; r+9 = 2401; r = 2392$. Check: $(2392+9)^{1/4} = (2401)^{1/4} = 7$. Solution is 2392.

9. $\left(\frac{x}{2}+1\right)^{1/2}=3; \left[\left(\frac{x}{2}+1\right)^{1/2}\right]^2 = 3^2; \frac{x}{2}+1 = 9;$
$\frac{x}{2} = 8; x = 16$. Check: $\left(\frac{16}{2}+1\right)^{1/2} = (8+1)^{1/2} = 9^{1/2} = 3$. Solution is 16.

10. $\left(\frac{y}{3}+4\right)^{1/3} = \frac{5}{2}; \left[\left(\frac{y}{3}+4\right)^{1/3}\right]^3 = \left(\frac{5}{2}\right)^3; \frac{y}{3}+$
$4 = \frac{125}{8}; \frac{y}{3} = \frac{125}{8} = \frac{32}{8} = \frac{93}{8}; y = \frac{279}{8}$. Check:
$\left(\dfrac{\frac{279}{8}}{3}+4\right)^{1/3} = \left(\frac{93}{8}+4\right)^{1/3} = \left(\frac{93}{8}+\frac{32}{8}\right)^{1/3} =$
$\left(\frac{125}{8}\right)^{1/3} = \frac{5}{2}$. Solution is $\frac{279}{8}$ or $34\frac{7}{8}$.

11. $(x-1)^{3/2} = 27; \left((x-1)^{3/2}\right)^2 = 27^2; (x-1)^3 = 729; (x-1) = \sqrt[3]{729} = 9; x = 10.$ Check: $(10-1)^{3/2} = 9^{3/2} = 3^3 = 27.$ Solution is 10.

12. $(x+1)^{3/4} = 8; \left((x+1)^{3/4}\right)^{4/3} = 8^{4/3}; x+1 = 16$ so $x = 15.$ Check: $(15+1)^{3/4} = 16^{3/4} = \sqrt[4]{16^3} = 2^3 = 8.$ Solution is 15.

13. $\sqrt{x+1} = \sqrt{2x-1}; \sqrt{x+1}^2 = \sqrt{2x-1}^2; x+1 = 2x-1; -x = -2; x = 2.$ Check: $\sqrt{2+1} = \sqrt{3};$
$\sqrt{2 \cdot 2 - 1} = \sqrt{4-1} = \sqrt{3}.$ Solution is 2.

14. $\sqrt{x^2-4} = \sqrt{2x-1}; \sqrt{x^2-4}^2 = \sqrt{2x-1}^2; x^2 - 4 = 2x-1$ or $x^2 - 2x - 3 = 0; (x-3)(x-1) = 0; x = 3$ or $x = -1.$ Check: $\sqrt{3^2-4} = \sqrt{2 \cdot 3 - 1};$ $\sqrt{9-4} = \sqrt{6-1} = \sqrt{5}; \sqrt{(-1)^2-4} = \sqrt{2(-1)-1}; \sqrt{-3} = \sqrt{-3}$ not real. Solution is 3 only.

15. $\sqrt{x+1} + \sqrt{x-5} = 7; \sqrt{x+1} = 7 - \sqrt{x-5};$ $\sqrt{x+1}^2 = (7 - \sqrt{x-5})^2;$ $x + 1 = 49 - 14\sqrt{x-5} + (x-5); x + 1 = x - 14\sqrt{x-5} + 44; -43 = -14\sqrt{x-5}; \frac{43}{14} = \sqrt{x-5}; \left(\frac{43}{14}\right)^2 = \sqrt{x-5}^2; \frac{1849}{196} = x - 5; x = \frac{2829}{196}.$ Check: $\sqrt{\frac{2829}{196} + 1} + \sqrt{\frac{2829}{196} - 5} = \sqrt{\frac{3025}{196}} + \sqrt{\frac{1849}{196}} = \frac{55}{14} + \frac{43}{14} = \frac{98}{14} = 7.$ Solution is $\frac{2829}{196} \approx 14.43.$

16. $\sqrt{x-3} - \sqrt{x+2} = 19; \sqrt{x-3}^2 = \sqrt{x+2} + 19; \sqrt{x-3} = (\sqrt{x+2} + 19)^2; x - 3 = (x+2) + 38\sqrt{x+2} + 361; -366 = 38\sqrt{x+2}.$ No solution because the right-hand side must be positive.

17. $\sqrt{x-1} + \sqrt{x+5} = 2; \sqrt{x-1} = 2 - \sqrt{x+5}; \sqrt{x-1}^2 = (2 - \sqrt{x+5})^2; x - 1 = 4 - 4\sqrt{x+5} + (x+5); -10 = -4\sqrt{x+5}; \frac{5}{2} = \sqrt{x+5}; \left(\frac{5}{2}\right)^2 = \sqrt{x+5}^2; \frac{25}{4} = x + 5; x = \frac{25}{4} - \frac{20}{4} = \frac{5}{4}.$ Check: $\sqrt{\frac{5}{4} - 1} + \sqrt{\frac{5}{4} + 5} = \sqrt{\frac{1}{4}} + \sqrt{\frac{25}{4}} = \frac{1}{2} + \frac{5}{2} = \frac{6}{2} = 3.$ Does not check so there is no solution.

18. $\sqrt{x+3} - \sqrt{2x-4} = 3; \sqrt{x+3} = \sqrt{2x-4} + 3; \left(\sqrt{x+3}\right)^2 = \left(\sqrt{2x-4} + 3\right)^2; x+3 = (2x-4) + 6\sqrt{2x-4} + 9; -x - 2 = 6\sqrt{2x-4}; (-x-2)^2 = (6\sqrt{2x-4})^2; x^2 + 4x + 4 = 36(2x-4); x^2 + 4x + 4 = 72x - 144; x^2 - 68x + 148 = 0;$ by the quadratic formula $x \approx 2.25$ or $x \approx 65.75.$ Neither checks so there is no solution. (Note: you can conclude there is no solution when $-x - 2 = 6\sqrt{2x-4}$ since x must be > 2 for the radicand

$2x - 4$ to be positive. This yields a negative expression on the left and a positive on the right which cannot be equal.)

19. $\sqrt{\frac{1}{x}} = \sqrt{\frac{4}{3x-1}};$ square both sides to get $\frac{1}{x} = \frac{4}{3x-1}; 3x - 1 = 4x; -1 = x.$ This does not check as it gives negative radicands so no solution.

20. $\sqrt{\frac{2}{x-1}} = \sqrt{\frac{5}{x+1}};$ square both sides to get $\frac{2}{x-1} = \frac{5}{x+1}; 2(x+1) = 5(x-1); 2x + 2 = 5x - 5; 7 = 3x$ or $x = \frac{7}{3}.$ Check: $\sqrt{\frac{2}{\frac{7}{3} - 1}} = \sqrt{\frac{2}{\frac{4}{3}}} = \sqrt{\frac{6}{4}} = \frac{\sqrt{6}}{2}; \sqrt{\frac{5}{\frac{7}{3} + 1}} = \sqrt{\frac{5}{\frac{10}{3}}} = \sqrt{\frac{3}{2}} = \frac{\sqrt{6}}{2}$ so the solution is $\frac{7}{3}$

21. Here $v = \sqrt{\mu g_s R}.$ Squaring both sides give $v^2 = \mu_s g R$ and so $R = \frac{v^2}{\mu_s g}.$

22. Here $v = \sqrt{v_0^2 - 2gh}.$ Squaring both sides yields $v^2 = v_0^2 - 2gh$ or $2gh = v_0^2 - v^2,$ and so $h = \frac{v_0^2 - v^2}{2g}.$

23. We are given $\frac{d_A}{d_B} = \frac{\sqrt{I_A}}{\sqrt{I_B}}.$ Substituting the given data produces $\frac{11}{20} = \frac{3.5 I_B - 15}{\sqrt{I_B}}.$ Now, square both sides to get $\frac{121}{400} = \frac{3.5 I_B - 15}{I_B}$ and cross-multiply, with the result $121 I_B = 1400 I_B - 6000$ or $-1279 I_B = -6000$ and so, $I_B = \frac{6000}{1279} \approx 4.691.$ Since $I_A = 3.5 I_B - 15,$ we have $I_A = 3.5 \times 4.691 - 15 = 1.419.$ The final result is $I_A = 1.419, I_B = 4.691.$

24. **(a)** $\frac{\sqrt{I_x}}{r_x} = \frac{\sqrt{I_y}}{r_y}.$ Square both sides to get $\frac{I_x}{r_x^2} = \frac{I_y}{r_y^2}$ or $I_x = \frac{I_y r_x^2}{r_y^2} = \left(\frac{r_x}{r_y}\right)^2 I_y.$
(b) Substituting the given data yields
$I_x = \left(\frac{2r_y + 3}{r_y}\right)^2 \cdot 30 = \left(2 + \frac{3}{r_y}\right)^2 \cdot 30 = \left(4 + \frac{12}{r_y} + \frac{9}{r_y^2}\right) 30 = 120 + \frac{360}{r_y} + \frac{270}{r_y^2},$ if $r_y \neq 0.$

25. Substituting for $t,$ we have $8.6 = \frac{1}{4}\sqrt{s}.$ Multiply by 4 to get $\sqrt{s} = 34.4.$ Squaring both sides, we get $s = 34.4^2 = 1,183.36$ ft.

26. Substituting, we have $43 = \sqrt{25^2 + X^2}.$ Squaring both sides, we get $1,849 = 625 + X^2$ or $X^2 = 1,224.$ Thus, taking the principal square root, we obtain $X = 34.99\,\Omega.$

CHAPTER 12 REVIEW

1. $49^{1/2} = \sqrt{49} = 7$

2. $81^{1/4} = \sqrt[4]{81} = 3$

3. $25^{3/2} = \sqrt{25}^3 = 5^3 = 125$

4. $64^{2/3} = \sqrt[3]{64}^2 = 4^2 = 16$

7. $5^4 5^9 = 5^{13}$

5. $9^{-3/2} = \frac{1}{9^{3/2}} = \frac{1}{\sqrt{9}^3} = \frac{1}{3^3} = \frac{1}{27}$

8. $13^6 13^{-9} = 13^{6-9} = 13^{-3} = \frac{1}{13^3}$

9. $(2^4)^8 = 2^{32}$

6. $125^{-4/3} = \frac{1}{125^{4/3}} = \frac{1}{\sqrt[3]{125}^4} = \frac{1}{5^4} = \frac{1}{625}$

10. $(5^3)^6 = 5^{18}$

11. $(x^2 y^3)^4 = x^8 y^{12}$

12. $(y^6 x)^4 = y^{24} x^4$

13. $\left(\frac{a^3}{ya^4}\right)^{-5} = \left(\frac{1}{ya}\right)^{-5} = (ya)^5 = y^5 a^5$

15. $x^{1/2} x^{1/3} = x^{3/6} x^{2/6} = x^{5/6}$

16. $y^{2/3} y^{3/4} = y^{8/12} y^{9/12} = y^{17/12}$

14. $\left(\frac{a^5}{ab^4}\right)^3 = \left(\frac{a^4}{b^4}\right)^3 = \frac{a^{12}}{b^{12}}$ or $\left(\frac{a}{b}\right)^{12}$

17. $\frac{(xy^2 a)^5}{x^4 (ya^3)^2} = \frac{x^5 y^{10} a^5}{x^4 y^2 a^6} = \frac{xy^8}{a}$

18. $\frac{(ab^3 c^2)^4}{(a^2 bc)^3} = \frac{a^4 b^{12} c^8}{a^6 b^3 c^3} = \frac{b^9 c^5}{a^2}$

19. $\sqrt[3]{-27} = -3$

20. $\sqrt[4]{81^3} = \sqrt[4]{81}^3 = 3^3 = 27$

28. $3\sqrt{7} - 6\sqrt{7} = -3\sqrt{7}$

21. $\sqrt{a^2 b^5} = \sqrt{a^2} \sqrt{b^4} \sqrt{b} = ab^2 \sqrt{b}$

29. $3\sqrt{6} + 5\sqrt{6} - 2\sqrt{3} = 8\sqrt{6} - 2\sqrt{3}$

22. $\sqrt[4]{a^3 b^7} = \sqrt[4]{b^4} \sqrt[4]{a^3 b^3} = b\sqrt[4]{a^3 b^3}$

30. $\sqrt{6} + \sqrt{24} = \sqrt{6} + 2\sqrt{6} = 3\sqrt{6}$

23. $\sqrt[3]{\frac{-8x^3 y^6}{z}} = \frac{-2xy^2}{\sqrt[3]{z}} = \frac{-2xy^2 \sqrt[3]{z^2}}{\sqrt[3]{z}\sqrt[3]{z^2}} = \frac{-2xy^2 \sqrt[3]{z^2}}{z}$ or $\frac{-2xy^2}{z}\sqrt[3]{z^2}$

31. $\sqrt[4]{16a^3 b} + \sqrt[4]{81a^7 b} = 2\sqrt[4]{a^3 b} + 3a\sqrt[4]{a^3 b}; (2+3a)\sqrt[4]{a^3 b}$

32. $\sqrt[4]{32x^5} - x^2 \sqrt[4]{2x} = 2x\sqrt[4]{2x} - x^2 \sqrt[4]{2x}; = (2x - x^2)\sqrt[4]{2x}$

24. $\sqrt[4]{\frac{32a^6 b^8}{cd^2}} = \sqrt[4]{\frac{2^5 a^6 b^8}{cd^2}} \cdot \sqrt[4]{\frac{c^3 d^2}{c^3 d^2}} = \frac{2ab^2 \sqrt[4]{2a^2 c^3 d^2}}{cd} = \frac{2ab^2}{cd}\sqrt[4]{2a^2 c^3 d^2}$

33. $a\sqrt{\frac{a}{7b}} - b\sqrt{\frac{b}{7a}} = \frac{a}{7b}\sqrt{7ab} - \frac{b}{7a}\sqrt{7ab} = \frac{a^2}{7ab}\sqrt{7ab} - \frac{b^2}{7ab}\sqrt{7ab} = \frac{(a^2 - b^2)\sqrt{7ab}}{7ab}$

25. $\sqrt{\frac{a}{b^2} - \frac{b}{a^2}} = \sqrt{\frac{a^3}{a^2 b^2} - \frac{b^3}{a^2 b^2}} = \frac{\sqrt{a^3 - b^3}}{a^2 b^2} = \sqrt{\frac{a^3 - b^3}{ab}}$

34. $\sqrt{\frac{x-2}{x+2}} + \sqrt{\frac{x+2}{x-2}} = \frac{\sqrt{x-2}\sqrt{x+2}}{x+2} + \frac{\sqrt{x+2}\sqrt{x-2}}{x-2} = \frac{\sqrt{x^2-4}}{x+2} + \frac{\sqrt{x^2-4}}{x-2} = \frac{(x-2)\sqrt{x^2-4}}{(x-2)(x+2)} + \frac{(x+2)\sqrt{x^2-4}}{(x+2)(x-2)} = \frac{x\sqrt{x^2-4} - 2\sqrt{x^2-4}}{x^2-4} + \frac{x\sqrt{x^2-4} + 2\sqrt{x^2-4}}{x^2-4} = \frac{2x\sqrt{x^2-4}}{x^2-4}$

26. $\sqrt{a^2 + 8a + 16} = \sqrt{(a+4)^2} = a + 4$

27. $2\sqrt{5} + 6\sqrt{5} = 8\sqrt{5}$

35. $\sqrt{3}\sqrt{5} = \sqrt{15}$

36. $\sqrt{7}\sqrt{6} = \sqrt{42}$

37. $\sqrt{2x}\sqrt{7x} = \sqrt{14x^2} = x\sqrt{14}$

38. $\sqrt[3]{5y}\sqrt[3]{25y} = \sqrt[3]{125y^2} = 5\sqrt[3]{y^2}$

43. $\sqrt{x-1} = 9$ so $x - 1 = 81$ and $x = 82$. Check: $\sqrt{82-1} = \sqrt{81} = 9$. Solution is 82.

39. $\left(\sqrt{a} + \sqrt{2b}\right)\left(\sqrt{a} - \sqrt{2b}\right) = \sqrt{a}^2 - \sqrt{2b}^2 = a - 2b$

44. $\sqrt{2x-7} - 5 = 0$ or $\sqrt{2x-7} = 5$. Squaring both sides gives $2x - 7 = 25$ or $2x = 32$, and so $x = 16$. Check: $\sqrt{2 \cdot 16 - 7} - 5; = \sqrt{32 - 7} - 5 = \sqrt{25} - 5 = 5 - 5 = 0$. Solution is 16.

40. $\left(\sqrt{x} - 3\sqrt{y}\right)\left(\sqrt{x} + 3\sqrt{y}\right) = \sqrt{x}^2 - \left(3\sqrt{y}\right)^2 = x - 9y$

41. $\frac{\sqrt[3]{2a^2}}{\sqrt[3]{16a}} = \sqrt[3]{\frac{2a^2}{16a}} = \sqrt[3]{\frac{a}{8}} = \frac{\sqrt[3]{a}}{2}$

45. Here $\sqrt{x^2 - 7} = 2x - 4$. Squaring yields $\sqrt{x^2 - 7}^2 = (2x - 4)^2$ or $x^2 - 7 = 4x^2 - 16x + 16$ and so $0 = 3x^2 - 16x + 25$. The discriminant, $b^2 - 4ac = -44$, is negative, so there is no solution.

42. $\frac{\sqrt[3]{5a^2 b}}{\sqrt[3]{25ab^2}} = \sqrt[3]{\frac{5a^2 b}{25ab^2}} = \sqrt[3]{\frac{a}{5b}} \cdot \sqrt[3]{\frac{5^2 b^2}{5^2 b^2}} = \frac{\sqrt[3]{25ab^2}}{5b}$

46. Multiplying $\frac{3x}{4} - 1 = 25x^2 - 30x + 9$ by 4 gives $3x - 4 = 100x^2 - 120x + 36$, or $100x^2 - 123x + 40 = 0$. The discriminant, $b^2 - 4ac = -871$, so there is no solution.

47. Here $\sqrt{x+1} + \sqrt{x-5} = 8$ can be written as $\sqrt{x+1} = 8 - \sqrt{x-5}$. Squaring both sides produces $\sqrt{x+1}^2 = (8 - \sqrt{x-5})^2$ or $x + 1 = 64 - 16\sqrt{x-5} + (x-5)$ which simplifies to $-58 = -16\sqrt{x-5}$ or $\frac{58}{16} = \sqrt{x-5}$ or $3.625 = \sqrt{x-5}$. Squaring again yields $3.625^2 = x - 5$ or $13.140625 = x - 5$ and so $x = 18.140625$. This answer does check so $x \approx 18.14$.

48. Here $\sqrt{x+3} + \sqrt{x-2} = 20$ can be written as $\sqrt{x+3} = 20 - \sqrt{x-2}$. Squaring both sides gives $\sqrt{x+3}^2 = (20 - \sqrt{x-2})^2$ or $x + 3 = 400 - 40\sqrt{x-2} + (x-2)$ and so $-395 = -40\sqrt{x-2}$ or $\sqrt{x-2} = 9.875$. Squaring again gives $x - 2 = 9.875^2$ or $x - 2 = 97.51562$ and so, $x = 99.51562$. This answer does check so the solution is about 99.52.

49. Square both sides to get $\frac{1}{x} = \frac{3x}{4} - 1 = \frac{3x-4}{4}$. Multiplying by $4x$, we get $4 = 3x^2 - 4x$ or $3x^2 - 4x - 4 = 0$. This factors as $(3x+2)(x-2) = 0$ and so $x = -\frac{2}{3}$ or $x = 2$. We see that $-\frac{2}{3}$ yields a negative radicand so does not check. Thus, the only solution is 2.

50. Here $\sqrt{\frac{3}{x-1}} = \sqrt{\frac{4}{x+1}}$. Square both sides to get $\frac{3}{x-1} = \frac{4}{x+1}$ and so $3(x+1) = 4(x-1)$ or $3x+3 = 4x-4$ which simplifies to $x = 7$. Check: $\sqrt{\frac{3}{7-1}} = \sqrt{\frac{3}{6}} = \sqrt{\frac{1}{2}}; \sqrt{\frac{4}{7+1}} = \sqrt{\frac{4}{8}} = \sqrt{\frac{1}{2}}$. Solution is 7.

51. $q(t) = q_0 2^{-t/12.5}$
(a) $q(10) = 50 \cdot 2^{-10/12.5} = 28.71745887$; or about 28.72 mg;
(b) $q(20) = 50 \cdot 2^{-20/12.5} = 16.49384888$ or 16.49 mg.

52. (a) $v = \sqrt{\frac{2KE}{m}} = \frac{\sqrt{2KEm}}{m}$;

(b) $v = \sqrt{\frac{2KE}{m}}; v^2 = \frac{2KE}{m}; m = \frac{2KE}{v^2}$

53. By the Pythagorean theorem $v^2 = v_x^2 + v_y^2$ or $v^2 = (3t+2)^2 + (4t-1)^2 = 9t^2 + 12t + 4 + 16t^2 - 8t + 1; v^2 = 25t^2 + 4t + 5; v = \sqrt{25t^2 + 4t + 5}$

54. From the diagram and the given information we have that $AB = 12, AC = x$ and let $BC = y$. Then $y - x = 5$ and $x^2 + 12^2 = y^2; y = 5 + x$ by substitution $x^2 + 144 = (5+x)^2; x^2 + 144 = 25 + 10x + x^2; 10x = 119; x = 11.9$ m; $y = 5 + 11.16 = 16.9$ m. $AC = 11.9$ m, $BC = 16.9$ m

CHAPTER 12 TEST

1. $8^5 8^{-7} = 8^{-2} = \frac{1}{8^2}$

2. $(3^5)^4 = 3^{20}$

3. $\left(\frac{x^2}{xy^3}\right)^5 = \left(\frac{x}{y^3}\right)^5 = \frac{x^5}{y^{15}}$

4. $x^{1/4} x^{4/3} = x^{3/12} x^{16/12} = x^{19/12}$

5. $\sqrt[3]{-64} = \sqrt[3]{(-4)^3} = -4$

6. $\sqrt[4]{81} = \sqrt[4]{3^4} = 3$

7. $\sqrt{x^4 y^3} = \sqrt{x^4}\sqrt{y^2}\sqrt{y} = x^2 y\sqrt{y}$

8. $\sqrt[3]{\frac{-27x^6 y}{z^2}} = \frac{\sqrt[3]{-27x^6 yz}}{\sqrt[3]{z^3}} = \frac{-3x^2}{z}\sqrt[3]{yz}$

9. $\sqrt{x^2 + 6x + 9} = \sqrt{(x+3)^2} = x + 3$

10. $3\sqrt{6} + 5\sqrt{24} = 3\sqrt{6} + 10\sqrt{6} = 13\sqrt{6}$

11. $\sqrt{2x}\sqrt{6x} = \sqrt{12x^2} = 2\sqrt{3}x$

12. $(3\sqrt{x} + \sqrt{5y})(3\sqrt{x} - \sqrt{5y}) = (3\sqrt{x})^2 - \sqrt{5y}^2 = 9x - 5y$

13. $\frac{\sqrt[3]{4x^2 y}}{\sqrt[3]{2x^2 y^2}} = \sqrt[3]{\frac{4x^2 y}{2x^2 y^2}} = \sqrt[3]{\frac{2}{y}}\sqrt[3]{\frac{y^2}{y^2}} = \frac{\sqrt[3]{2y^2}}{y}$

14. $\sqrt{x+3} = 6$. Square both sides to get the result $x + 3 = 36$ or $x = 33$. Check: $\sqrt{33+3} = \sqrt{36} = 6$. Solution is 33.

15. $\sqrt{2x+1} = 5$. Squaring both sides produces the result $2x + 1 = 25$ or $2x = 24$ and so $x = 12$. Check: $\sqrt{2 \cdot 12 + 1} = \sqrt{24+1} = \sqrt{25} = 5$. Solution is 12.

16. $\sqrt{2x^2 - 7} = x + 3$. Squaring both sides gives the result $\sqrt{2x^2 - 7}^2 = (x+3)^2$ or $2x^2 - 7 = x^2 + 6x + 9$ and so $x^2 - 6x - 16 = 0$. This factors as $(x-8)(x+2) = 0$, so $x = 8$ or $x = -2$. Check: $\sqrt{2 \cdot 8^2 - 7} = \sqrt{2 \cdot 64 - 7} = \sqrt{128 - 7} = \sqrt{121} = 11; 8 + 3 = 11; 8$ works; $\sqrt{2(-2)^2 - 7} = \sqrt{2(4) - 7} = \sqrt{8 - 7} = 1; -2 + 3 = 1$ also, so -2 works. Solutions are 8 and -2.

17. $\sqrt{\frac{5}{x+3}} = \sqrt{\frac{4}{x-3}}$. Square both sides to get $\frac{5}{x+3} = \frac{4}{x-3}$ and so $5(x-3) = 4(x+3)$ or $5x-15 = 4x+12$ which simplifies to $x = 27$. Check: $\sqrt{\frac{5}{27+3}} = \sqrt{\frac{5}{30}} = \sqrt{\frac{1}{6}}$; $\sqrt{\frac{4}{27-3}} = \sqrt{\frac{4}{24}} = \sqrt{\frac{1}{6}}$. Solution is 27.

18. **(a)** $d = (1.12 \times 10^{-2})\sqrt[3]{F\ell}$ or $\sqrt[3]{F\ell} = \frac{d}{1.12 \times 10^{-2}} = 8.92857 \times 10^{1}d$. Cubing both sides produces $F\ell = (8.92857 \times 10^{1}d)^3 = 7.12 \times 10^{5}d^3$ and so $F = \frac{7.12 \times 10^{5}d^3}{\ell}$,

(b) $F = \frac{7.12 \times 10^{5}(1.2)^3}{5} = 246,067.2$ kg

CHAPTER

13

Exponential and Logarithmic Functions

≡ 13.1 EXPONENTIAL FUNCTIONS

1. $3^{\sqrt{2}} = 4.7288$

2. $4^{\pi} = 77.8802$

3. $\pi^3 = 31.0063$

4. $\left(\sqrt{3}\right)^{\pi} = 5.6164$

5. $\left(\sqrt{3}\right)^{\sqrt{5}} = 3.4154$

6. $\left(\sqrt{4}\right)^{4/3} = 2^{4/3} = 2.5198$

7.

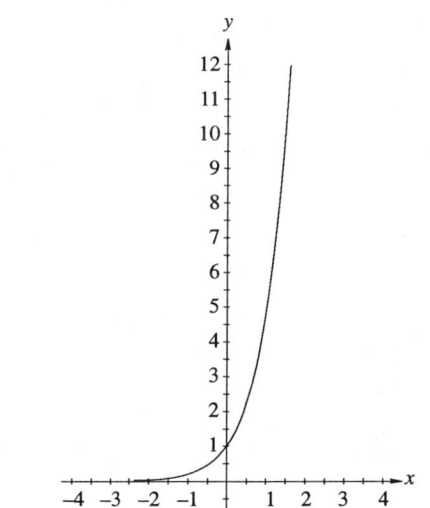

x	-3	-2	-1	0	1	2	3	4	5
$f(x)$	$\dfrac{1}{64}$	$\dfrac{1}{16}$	$\dfrac{1}{4}$	1	4	16	64	256	1024

8.

x	-3	-2	-1	0	1	2	3	4	5
$g(x)$	$\dfrac{1}{27}$	$\dfrac{1}{9}$	$\dfrac{1}{3}$	1	3	9	27	81	243

9.

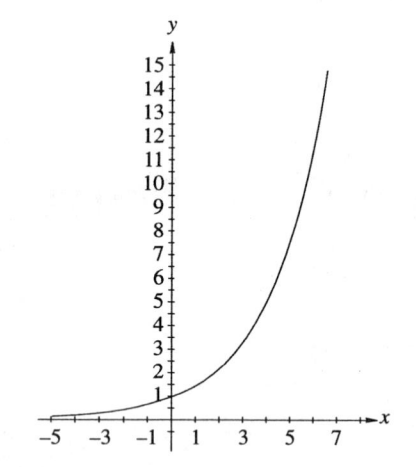

x	-3	-2	-1	0	1	2	3	4	5
$h(x)$	$\dfrac{8}{27}$	$\dfrac{4}{9}$	$\dfrac{2}{3}$	1	1.5	2.25	3.375	5.0625	7.59375

10.

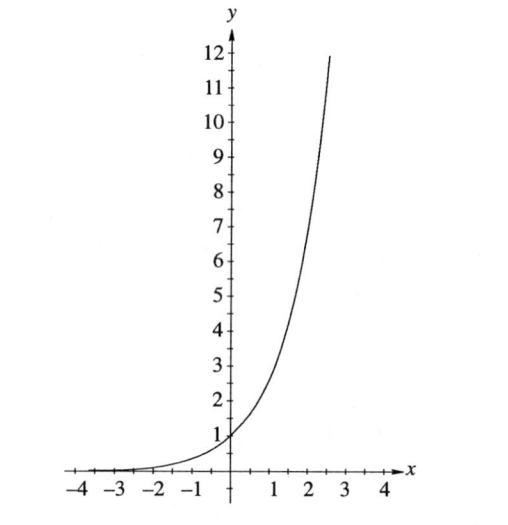

x	-3	-2	-1	0	1	2	3	4	5
$k(x)$	0.064	0.16	0.4	1	2.5	6.25	15.625	39.0625	97.65625

11.

x	-3	-2	-1	0	1	2	3	4	5
$f(x)$	27	9	3	1	$\dfrac{1}{3}$	$\dfrac{1}{9}$	$\dfrac{1}{27}$	$\dfrac{1}{81}$	$\dfrac{1}{243}$

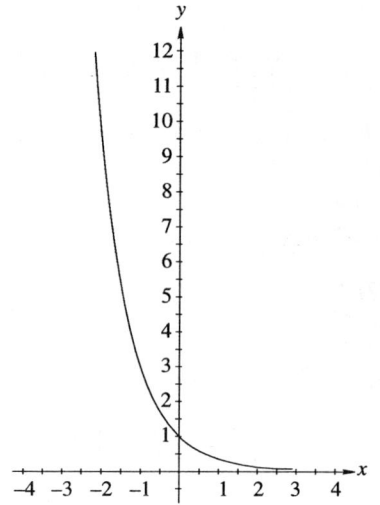

12.

x	-3	-2	-1	0	1	2	3	4	5
$g(x)$	125	25	5	1	$\dfrac{1}{5}$	$\dfrac{1}{25}$	$\dfrac{1}{125}$	$\dfrac{1}{625}$	$\dfrac{1}{3125}$

13.

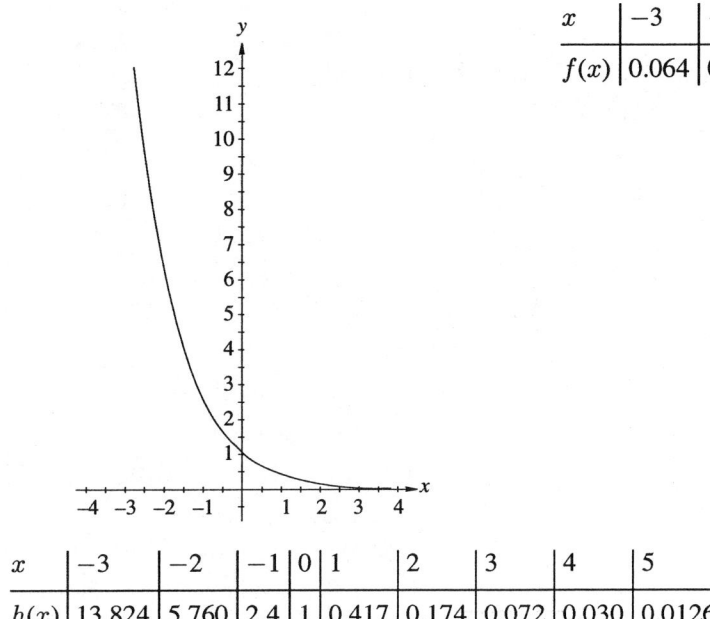

x	-3	-2	-1	0	1	2	3	4	5
$h(x)$	13.824	5.760	2.4	1	0.417	0.174	0.072	0.030	0.0126

14.

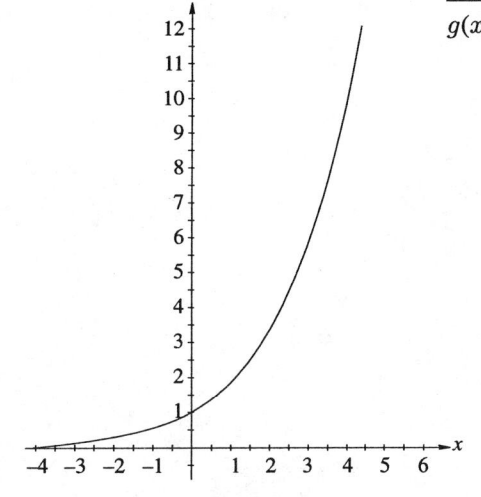

x	-3	-2	-1	0	1	2	3	4	5
$k(x)$	0.192	$\frac{1}{3}$	0.577	1	$\sqrt{3}$	3	$3\sqrt{3}$	9	$9\sqrt{3}$

15.

x	-3	-2	-1	0	1	2	3	4
$f(x)$	0.064	0.192	0.577	1.732	5.196	15.588	46.765	140.296

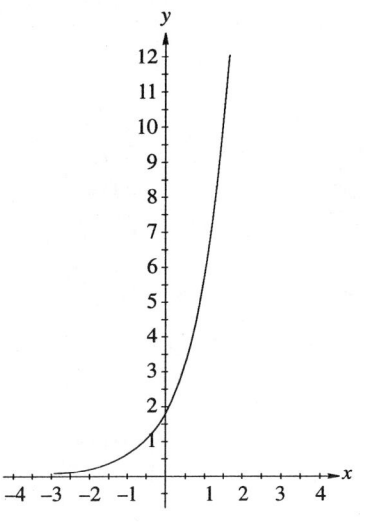

16.

x	-3	-2	-1	0	1	2	3	4	5
$g(x)$	0.105	0.210	0.420	0.841	1.682	3.364	6.727	13.454	26.908

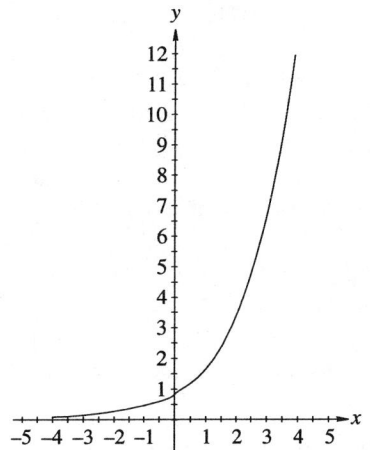

17. **(a)** $S = 1000(1 + 0.06)^5 = \$1338.23$,

(b) $S = 1000\left(1 + \frac{0.06}{2}\right)^{2 \times 5} = 1000(1.03)^{10} = \$1343.92$,

(c) $S = 1000\left(1 + \frac{0.06}{4}\right)^{4 \times 5} = 1000(1.015)^{20} = \$1346.86$,

(d) $S = 1000\left(1 + \frac{0.06}{12}\right)^{12 \times 5} = 1000(1.005)^{60} = \$1348.85$.

18. $S = P\left(1 + \frac{r}{k}\right)^{kt}$,

(a) $S = 2000(1 + 0.08)^{10} = \$4317.85$,

(b) $S = 2000\left(1 + \frac{0.08}{2}\right)^{2 \times 10} = \$4382.25$,

(c) $S = 2000\left(1 + \frac{0.08}{4}\right)^{4 \times 10} = \$4416.08$,

(d) $S = 2000\left(1 + \frac{0.08}{12}\right)^{12 \times 10} = \$4439.28$

19. First Bank: $S = 1000\left(1 + \frac{0.06}{2}\right)^{2 \times 5} = 1343.92$.

Second Bank: $S = 1000\left(1 + \frac{0.06}{12}\right)^{12 \times 5} = 1348.85$.

Since $1348.85 - 1343.92 = \$4.93$, you will earn $\$4.93$ more at the second bank.

20. First Bank: $S = 2000\left(1 + \frac{.08}{1}\right)^{1 \times 10} = 4317.85$.

Second Bank: $S = 2000\left(1 + \frac{.08}{4}\right)^{4 \times 10} = 4416.08$,

$4416.08 - 4317.85 = \$98.23$

21. **(a)** $200{,}000 = Q_0 2^2$ or $200{,}000 = 4Q_0$ and so $Q_0 = \frac{200{,}000}{4} = 50{,}000$.

(b) $Q = Q_0 2^4 = 50{,}000 \cdot 2^4 = 50{,}000 \cdot 16 = 800{,}000$

(c) We want $100{,}000 = 50{,}000 \cdot 2^t$ or $2 = 2^t$. Thus, $t = 1$, and the doubling time is 1 hr.

(d) We want $8 \cdot 50{,}000 = 50{,}000 \cdot 2^t$ or $8 = 2^t$. Since $8 = 2^3 = 2^t$, we see that $t = 3$ and it takes 3 hr for Q to become eight times as large as Q_0.

22. **(a)** $2{,}760 = Q_0 3^1$ and so $Q_0 = \frac{2{,}760}{3} = 970$.

(b) $Q = 920 \cdot 3^5 = 920 \cdot 243 = 223{,}560$

(c) $3 \cdot 970 = 970 \cdot 3^t$ and so $3 = 3^t$ and $t = 1$. It will take 1 hr for Q to become 3 times as large as Q_0.

23. **(a)** When $t = 0$, you get $400 = Q_0 3^0$ and so, $Q_0 = 400$.

(b) Here, $t = 2.5$ gives $1{,}200 = 400 \cdot 3^{2.5k}$ or $3 = 3^{2.5k}$. Hence, $2.5k = 1$ and $k = \frac{1}{2.5} = \frac{2}{5}$.

(c) $Q = 400 \cdot 3^{(2/5)5} = 400 \cdot 3^2 = 400 \cdot 9 = 3{,}600$.

24. **(a)** Here, $t = 2000 - 1995 = 5$, and so $P = P_0(1.018)^t = 5.7 \times 10^9 (1.018)^5 = 6.23 \times 10^9$.

(b) This time, $t = 2050 - 1995 = 55$, and so $P = P_o(1.018)^t = 5.7 \times 10^9 (1.018)^{55} \approx 1.52 \times 10^{10}$ or over 15 billion people.

(c)

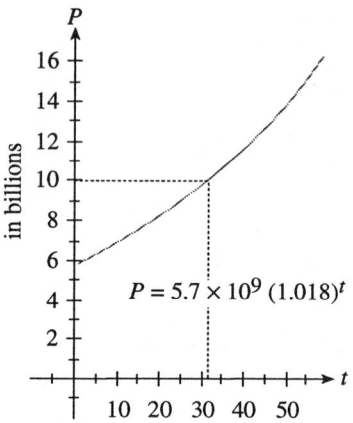

(d) Locate 10^{10} on the P-axis. Draw a horizontal line from there until it intersects the graph of $P = P_0(1.018)^t = 5.7 \times 10^9 (1.018)^t$. From there draw a vertical line toward the t-axis. It intersects the t-axis near 2027, and so we conclude that the world's population of $10{,}000{,}000{,}000$ will occur at about the year 2027.

25.

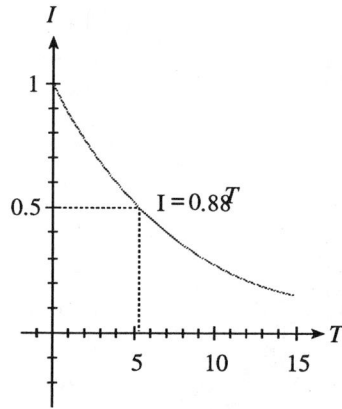

26. $I_p + I_g = 0.0005\left(8 + \frac{350}{14}\right)^{3/2} = 0.0948$ A

≡ 13.2 THE EXPONENTIAL FUNCTION e^x

1. $e^3 = 20.085\ 537$

3. $e^{4.65} = 104.584\ 986$

5. $e^{-4} = 0.018\ 316$

2. $e^7 = 1096.633\ 158$

4. $e^{5.375} = 215.939\ 872$

6. $e^{-9} = 1.234\ 098 \times 10^{-4}$ or $0.000\ 123\ 4$

7. $e^{-2.75} = 0.063\ 928$

8. $e^{-0.25} = 0.778\ 801$

9.

x	-2	-1	0	1	2	3	4
$f(x)$	0.5413	1.4715	4	10.8731	29.5562	80.3421	218.3926

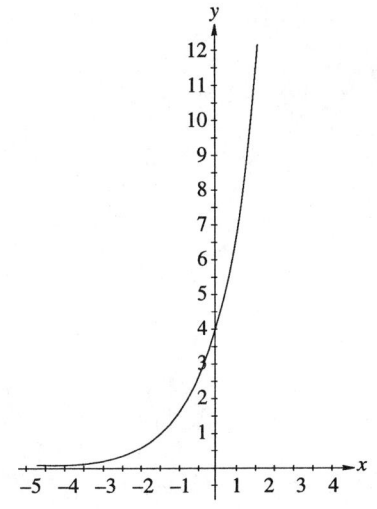

10.

x	-1	0	1	2	3	4
$g(x)$	0.0236	3.5	519.446	77 092.630	11 441 560.8	1.6981×10^9

11.

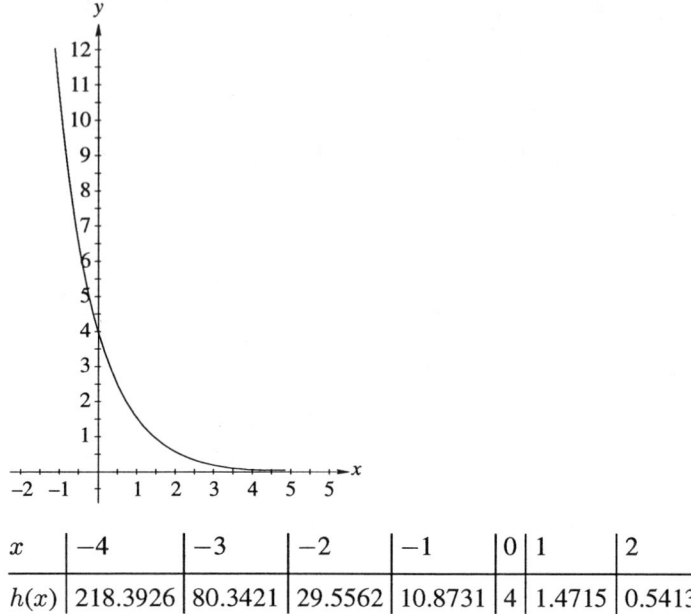

x	-4	-3	-2	-1	0	1	2
$h(x)$	218.3926	80.3421	29.5562	10.8731	4	1.4715	0.5413

12.

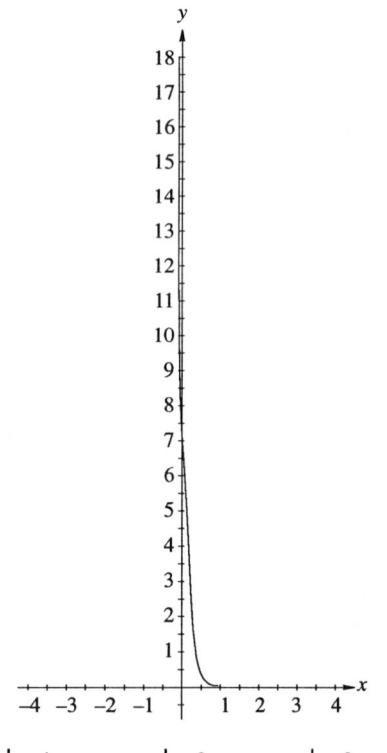

13. $Q = 125e^{-0.375t}$
(a) when $t = 0$, we get $e^0 = 1$ so Q is 125 mg
(b) when $t = 1$, we have $Q = 125e^{-0.375} = 85.91$ mg
(c) when $t = 16$, we obtain $Q = 125e^{-0.375 \times 16} = 0.3098$ mg or 0.31 mg

14.   A half life of 1600 years yields $\frac{1}{2} = e^{1600K}$. The initial amount is 100. $2000 \div 1600 = 1.25$. $y = 100e^{2000K} = 100\left(e^{1600K}\right)^{1.25} = 100\left(\frac{1}{2}\right)^{1.25} = 42.04$.

15.   Initial amount is 5000. $y = ce^{kt}$. $15,000 = 5000e^{20kt}; 3 = e^{20kt}$.
(a) when $t = 10$, $y = 5000e^{k10} = 5000\left(e^{20k}\right)^{1/2} = 5000 \cdot 3^{1/2} = 8,660$.
(b) when $t = 30$, $y = 5000e^{30k} = 5000\left(e^{20k}\right)^{1.5} = 5000 \times 3^{1.5} = 25,981$.
(c) when $t = 3$ days $= 3 \times 24$ hr $= 72$ hr. $72 \div 20 = 3.6$, $y = 5000e^{72k} = 5000\left(e^{20k}\right)^{3.6} = 5000 \cdot 3^{3.6} = 260,980$.

16. $S = Pe^{rt} = 5000e^{.05 \times 10} = \$8243.61$.

x	-4	-3	-2	-1	0	1	2
$k(x)$	2.252×10^{11}	5.581×10^8	1.383×10^6	3429.14	8.5	0.0211	5.223×10^{-5}

17. Method 1. Use the compound interest formula
$S = P(1 + r)^t$
   (a) $S = 200,000(1.07^5 = 280,510.$
   (b) $S = 2000,000(1.07)^{10} = 393,430.$

Method 2. Use the growth formula $y = ce^{kt}$ with $c = 200,000$ and $e^k = 1.07$.
   (a) $y = 200,000e^{5k} = 200,000(e^k)^5 = 200,000(1.07)^5 = 280,510.$
   (b) $y = 200,000e^{10k} = 200,000(e^k)^{10} = 200,000(1.07)^{10} = 393,430.$

18. Use $S = Pe^{rt}$, with $S = 1000$, $r = 0.045$, and $t = 7 \times 24$. This gives $1000 = Pe^{rt} = Pe^{0.045 \times 7 \times 24} = Pe^{7.56}$, and so $P = \frac{1000}{e^{7.56}} = 0.521$ units.

19. Use $T = T_m + (T_0 - T_m)e^{-kt}$, with $T = 150°$, $T_0 = 180°$, $T_m = 60°$, and $t = 20$ min. This gives $150 = 60 + (180 - 60)e^{-k \cdot 20}$, or $90 = (120)e^{-k20}$, and so $e^{-k20} = 0.75$. When $t = 1$ hr $= 60$ min, we have $60 = 20 \times 3$. $T = 60 + (180 - 60)e^{-k \cdot 60} = 60 + 120(e^{-k \cdot 20})^3$ or $T = 60 + 120(0.75)^3 = 60 + 50.625 \approx 110.6°$F.

20. Use $T = T_m + (T_0 - T_m)e^{-kt}$, with $T = 90°$, $T_m = 30°$, $T_0 = 150°$, and $t = 15$ min. This gives $90 = 30 + (150 - 30)e^{-k15}$, or $60 = (120)e^{-k \cdot 15}$, which leads to $\frac{1}{2} = e^{-k \cdot 15}$. When $t = 30$, you get $T = 30 + (150 - 30)e^{-k30} = 30 + 120(e^{-k \cdot 15})^2 = 30 + 120(\frac{1}{2})^2 = 30 + 30 = 60°$C.

21. Using the formula for Newton's law of cooling $T = T_m + (T_0 - T_m)e^{-kt}$, with $T = 70$ when $t = 25, T_0 = 87°$, and $T_m = 37°$. This gives $70 = 37 + (87 - 37)e^{-k25}$, or $33 = 50e^{-k25}$ or $e^{-25k} = 0.66$ which is $(e^{-k})^{25} = 0.66$, and so $e^{-k} = 0.66^{1/25} = 0.9835$. Using this value of $e^{-k} = 0.9835$ with $T = 45$ you get $45 = 37 + (87 - 37)e^{-kt}$ or $8 = 50e^{-kt}$ or $e^{-kt} = 0.16$, and so $0.9835^t = 0.16$. Thus, $t = \frac{\log 0.16}{\log 0.9835} = 110.26$, and we have $t = 110.26$ or about 1 hr and 50 min.

22. $I = I_0(1 - e^{-t/T}); I_0 = \frac{V}{R}, T = \frac{L}{R}; V = 120, R = 40, L = 30; I_0 = \frac{120}{40} = 3, T = \frac{3.0}{40} = \frac{3}{40}$
   (a) When $t = 0.01$ s; $-t/T = \frac{-0.01}{3/40} = -0.1333$; $I = 3(1 - e^{-.1333}) = 0.374$ A;
   (b) When $t = 0.1$ s; $-t/T = \frac{-0.1}{3/40} = -1.333$; $I = 3(1 - e^{-1.333}) = 2.209$ A
   (c) When $t = 1.0$ s; $-t/T = -13.33; I = 3(1 - e^{-13.33}) = 3.000$ A

23. Use $Q = Q_o e^{-t/T}$, where $Q_0 = CV$, and $T = RC$. We have $C = 130$ μF $= 130 \times 10^{-6}$F, $V = 120$ V, and $R = 4500 \Omega$. So, $Q_0 = (130 \times 10^{-6})120 = 1.560 \times 10^{-2}$, and $T = 4500 \times 130 \times 10^{-6} = 0.585$. Hence $Q = 1.56 \times 10^{-2} \cdot e^{\frac{-0.5}{0.585}} = 0.0066$ C.

24. A half-life of 12.5 years means $e^{12.5k} = \frac{1}{2}$ so $y = 100e^{40k} = 100(e^{12.5k})^{3.2} = 100(\frac{1}{2})^{3.2} = 10.88$ g. (Note $40 \div 12.5 = 3.2$.)

25. A half-life of 5.4 yields $\frac{1}{2}y_0 = y_0 e^{k(5.4)}$ or $\frac{1}{2} = e^{k(5.4)}$. Hence, $5.4k = \ln\frac{1}{2}$ or $k = \frac{\ln\frac{1}{2}}{5.4} = -0.12836$. Thus, the amount of material still radioactive after 30 days is $y = 1200e^{-0.12836(30)} \approx 25.5$ μCi.

26. Since the doubling time is 100 days, we solve for k in $y = y_0 e^{kt}$ as follows:
$$2y_0 = y_0 e^{100k}$$
$$2 = e^{100k}$$
$$\ln 2 = 100k$$
$$k = \frac{\ln 2}{100} \approx 0.0069314718.$$

Hence, $y = y_0 e^{0.0069314718t}$.
(a) Using 365.25 days $= 1$ year we see that 9 yr $= 365.25 \times 9 = 3287.25$ days. The initial number of cancer cells is $y_0 = 1$. Substituting, we obtain $y = y_0 e^{(0.0069314718)(3287.25)} \approx 7,863,366,762 = 7.863 \times 10^9$ cells.
(b) Ten years is 3652.5 days. Substituting, we get $y = y_0 e^{(0.0069314718)(3652.5)} \approx 9.88828 \times 10^{10}$ cells.
(c) Assuming that a lump is spherical, its volume is $V = \frac{4}{3}\pi r^3 = \frac{4}{3}\pi(0.5)^3 = \frac{\pi}{6} \approx 0.523599$ cm^3. Each of the cells has a volume of $\frac{0.523599}{9.88828 \times 10^{10}} \approx 5.295 \times 10^{-12}$ cm^3.

27. Here $C = 500 \times 10^{-6} = 5 \times 10^{-4}$ F, $R = 1$ k$\Omega = 10^3 \Omega$. Using the other given information, we have $i = \frac{V}{R}e^{-t/RC} = \frac{50}{10^3}e^{-t/(10^3 \cdot 5 \times 10^{-4})} = 0.05e^{-2t}$
(a) When $t = 0$, we get $i = 0.05e^{-2 \cdot 0} = 0.05e^0 = 0.05$ A
(b) Here $t = 0.02$, so $i = 0.05e^{-2 \cdot 0.02} = 0.05e^{-0.04} \approx 0.048$ A
(c) Here $t = 0.04$, so $i = 0.05e^{-2 \cdot 0.04} = 0.05e^{-0.08} \approx 0.046$ A

28. $2 + \frac{1}{2} + \frac{1}{6} + \frac{1}{24} + \frac{1}{120} + \frac{1}{720} = 2.7181$

☰ 13.3 LOGARITHMIC FUNCTIONS

1. $\log_6 216 = 3$; $6^3 = 216$

2. $\log_9 6591 = 4$; $9^4 = 6561$

3. $\log_4 16 = 2$; $4^2 = 16$

4. $\log_7 16{,}807 = 5$; $7^5 = 16{,}807$

5. $\log_{1/7} \frac{1}{49} = 2$; $\left(\frac{1}{7}\right)^2 = \frac{1}{49}$

6. $\log_{1/2} \frac{1}{64} = 6$; $\left(\frac{1}{2}\right)^6 = \frac{1}{64}$

7. $\log_2 \frac{1}{32} = -5$; $2^{-5} = \frac{1}{32}$

8. $\log_9 \frac{1}{243} = -5$; $3^{-5} = \frac{1}{243}$

9. $\log_9 2187 = \frac{7}{2}$; $9^{\frac{7}{2}} = 2187$

10. $\log_8 2048 = \frac{11}{3}$; $8^{\frac{11}{3}} = 2048$

11. $5^4 = 625$; $\log_5 625 = 4$

12. $3^5 = 243$; $\log_3 243 = 5$

13. $2^7 = 128$; $\log_2 128 = 7$

14. $4^3 = 64$; $\log_4 64 = 3$

15. $7^3 = 343$; $\log_7 343 = 3$

16. $11^2 = 121$; $\log_{11} 121 = 2$

17. $5^{-3} = \frac{1}{125}$; $\log_5 \frac{1}{125} = -3$

18. $3^{-5} = \frac{1}{243}$; $\log_3 \frac{1}{243} = -5$

19. $4^{7/2} = 128$; $\log_4 128 = \frac{7}{2}$

20. $125^{5/3} = 3125$; $\log_{125} 3125 = \frac{5}{3}$

21. $\ln 5 = 1.609\ 437\ 912$

22. $\ln 19 = 2.944\ 438\ 979$

23. $\ln 4.751 = 1.558\ 355\ 122$

24. $\ln 35.62 = 3.572\ 907\ 278$

25. $\log 4 = 0.602\ 059\ 991\ 3$

26. $\log 23 = 1.361\ 727\ 836$

27. $\log 12.67 = 1.102\ 776\ 615$

28. $\log 78.143 = 1.892\ 890\ 08$

29. $\log_5 8 = \frac{\ln 8}{\ln 5} = 1.292\ 029\ 674$

30. $\log_3 20 = \frac{\ln 20}{\ln 5} = 2.726\ 833\ 028$

31. $\log_{12} 16.4 = \frac{\ln 16.4}{\ln 12} = 1.125\ 708\ 821$

32. $\log_8 691.45 = \frac{\ln 691.45}{\ln 8} = 3.144\ 493\ 707$

33.

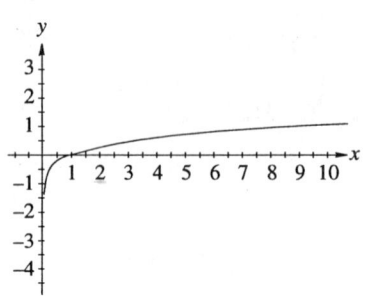

x	0.1	0.5	1	2	4	6	8	10
$\ln x$	−2.30	−0.69	0	0.69	1.39	1.79	2.08	2.30

34.

x	0.1	0.5	1	2	4	6	8	10
$\log x$	−1	−0.30	0	0.30	0.60	0.78	0.90	1

35.

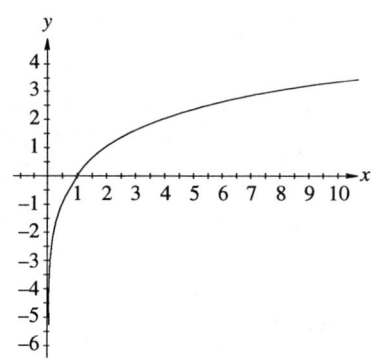

x	0.1	0.5	1	2	4	6	8	10
$\log_2 x$	−3.32	−1	0	1	2	2.58	3	3.32

36.

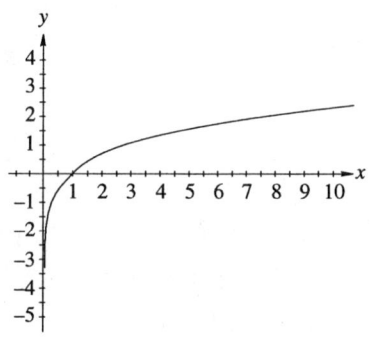

x	0.1	0.5	1	2	4	6	8	10
$\log_3 x$	−2.10	−0.63	0	0.63	1.26	1.63	1.89	2.10

37.

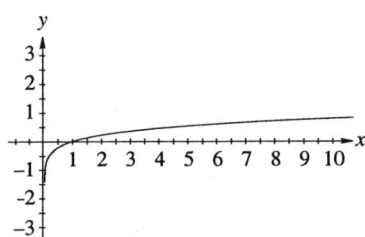

x	0.1	0.5	1	2	4	6	8	10
$\log_{12} x$	-0.93	-0.28	0	0.28	0.56	0.72	0.84	0.93

38.

x	0.1	0.5	1	2	4	6	8	10
$\log_{1/2} x$	3.32	1	0	-1	-2	-2.58	-3	-3.32

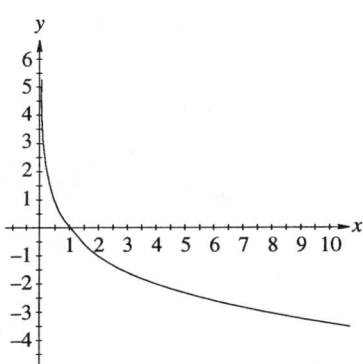

39.

x	0.1	0.5	1	2	4	6	8	10
$\log_{1/4} x$	1.66	0.5	0	-0.5	-1	-1.29	-1.5	-1.66

40. $R = \log I = \log_{10} I; \ 10^R = I$

41. (a) $\beta = 10 \log \frac{I}{I_0}; \ \frac{\beta}{10} = \log \frac{I}{I_0}, \ 10^{\beta/10} = \frac{I}{I_0}; I = I_0 10^{\beta/10}$

(b) $\beta = 10 \log \left(\frac{10^{-7}}{10^{-12}} \right) = 10 \log 10^5 = 10 \cdot 5 = 50$ dB

(c) $\beta = 10 \log \left(\frac{10^{-3}}{10^{-12}} \right) = 10 \log 10^9 = 10 \cdot 9 = 90$ dB

42. (a) The current population will be when $t = 0$. So,
$$P(0) = 47{,}000 + 9{,}000 \ln(0.7(0) + 1)$$
$$= 47{,}000 + 9{,}000 \ln(1)$$
$$= 47{,}000 + 9{,}000(0)$$
$$= 47{,}000$$
The city's current population is 47,000.

(b) Here, $t = 10$, so we want $P(10)$.
$$P(10) = 47{,}000 + 9{,}000 \ln(0.7(10) + 1)$$
$$= 47{,}000 + 9{,}000 \ln(8)$$
$$= 66{,}715$$
The city's population in 10 years is expected to be 66,715.

(c)
$$75{,}000 = 47{,}000 + 9{,}000 \ln(0.7t + 1)$$
$$28{,}000 = 9{,}000 \ln(0.7t + 1)$$
$$3.111 = \ln(0.7t + 1)$$
$$e^{3.111} = e^{\ln(0.7t+1)}$$
$$e^{3.111} = 0.7t + 1$$
$$22.4 = 0.7t + 1$$
$$21.4 = 0.7t$$
$$30.6 = \frac{21.4}{0.7} = t$$
It will take about 30.6 years.

43. Substituting the given values of A, S, and BA, we get
$$\ln Y = 5.36437 - 101.16296 S^{-1} - 22.00048 A^{-1}$$
$$+ 0.97116 \ln \text{BA}$$
$$= 5.36437 - 101.16296(110)^{-1} -$$
$$22.00048(40)^{-1} + 0.97116 \ln 90$$
$$= 8.26473$$
$$Y = e^{8.26473} \approx 3884$$
The predicted yield is about 3,884 ft^3/acre.

44.
$$Z_0 = \frac{87}{\sqrt{\epsilon' + 1.4}} \ln \left(\frac{6H}{0.8W + t} \right)$$
$$\frac{Z_0 \sqrt{\epsilon' + 1.4}}{87} = \ln \left(\frac{6H}{0.8W + t} \right)$$
$$e^{Z_0 \sqrt{\epsilon' + 1.4}/87} = \frac{6H}{0.8W + t}$$
$$0.8W + t = \frac{6H}{e^{Z_0 \sqrt{\epsilon' + 1.4}/87}}$$
$$t = \frac{6H}{e^{Z_0 \sqrt{\epsilon' + 1.4}/87}} - 0.8W$$

13.4 PROPERTIES OF LOGARITHMS

1. $\log \frac{2}{3} = \log 2 - \log 3$

2. $\log \frac{5}{4} = \log 5 - \log 4$

3. $\log 14 = \log(2 \cdot 7) = \log 2 + \log 7$

4. $\log 55 = \log(5 \cdot 11) = \log 5 + \log 11$

5. $\log 12 = \log(2 \cdot 2 \cdot 3) = \log 2 + \log 2 + \log 3$

6. $\log 28 = \log(2 \cdot 2 \cdot 7) = \log 2 + \log 2 + \log 7$

7. $\log \frac{150}{7} = \log \left(\frac{2 \cdot 3 \cdot 5 \cdot 5}{7} \right) = \log 2 + \log 3 + \log 5 + \log 5 - \log 7$

8. $\log \left(\frac{588}{5x} \right) = \log 588 - \log 5x = \log(2 \cdot 2 \cdot 3 \cdot 7 \cdot 7) - \log(5 \cdot x) = \log 2 + \log 2 + \log 3 + \log 7 + \log 7 - \log 5 - \log x$

9. $\log 2x = \log 2 + \log x$

10. $\log \frac{1}{5x} = \log 1 - \log(5x) = 0 - \log 5 - \log x; = - \log 5 - \log x$

11. $\log \frac{2ax}{3y} = \log 2ax - \log 3y = \log 2 + \log a + \log x - \log 3 - \log y = \log 2 + \log a + \log x - (\log 3 + \log y) = \log 2ax - \log 3y$

12. $\log \frac{4bc}{3xy} = \log 4bc - \log 3xy = \log 4 + \log b + \log c - \log 3 - \log x - \log y = \log 4 + \log b + \log c - (\log 3 + \log x + \log y)$

13. $\log 2 + \log 11 = \log(2 \cdot 11) = \log 22$

14. $\log 3 + \log 13 = \log(3 \cdot 13) = \log 39$

15. $\log 11 - \log 3 = \log \frac{11}{3}$

16. $\log 17 - \log 23 = \log \frac{17}{23}$

17. $\log 2 + \log 2 + \log 3 = \log(2 \cdot 2 \cdot 3) = \log 12$

18. $\log 3 + \log 3 + \log 5 = \log(3 \cdot 3 \cdot 5) = \log 45$

19. $\log 4 + \log x - \log y = \log \frac{4x}{y}$

20. $\log 5 + \log 7 - \log 11 = \log \frac{5 \cdot 7}{11} = \log \frac{35}{11}$

21. $5 \log 2 + 3 \log 5 = \log 2^5 + \log 5^3 = \log \left(2^5 \cdot 5^3 \right) = \log 4{,}000$

22. $7 \log 3 - 2 \log 8 = \log 3^7 - \log 8^2 = \log \frac{3^7}{8^2} = \log \frac{2187}{64}$

23. $\log \frac{2}{3} + \log \frac{6}{7} = \log \left(\frac{2}{3} \cdot \frac{6}{7} \right) = \log \frac{4}{7}$

24. $\log \frac{3}{24} + \log \frac{4}{7} - \log \frac{1}{3} = \log \left(\frac{3}{24} \cdot \frac{4}{7} \cdot 3 \right) = \log \frac{3}{14}$

25. $\log 8 = \log 2^3 = 3 \log 2 = 3(0.3010) = 0.9030$

26. $\log 9 = \log 3^2 = 2 \log 3 = 2(0.4771) = 0.9542$

27. $\log 12 = \log 2^2 \cdot 3 = 2 \log 2 + \log 3 = 2(0.3010) + (0.4771) = 0.6020 + 0.4771 = 1.0791$

28. $\log 36 = \log 2^2 \cdot 3^2 = 2 \log 2 + 2 \log 3 = 2(0.3010) + 2(0.4771) = 0.6020 + .9542 = 1.5562$

29. $\log 15 = \log 3 \cdot 5 = \log 3 + \log 5 = 0.4771 + 0.6990 = 1.1761$

30. $\log 30 = \log(2 \cdot 3 \cdot 5) = \log 2 + \log 3 + \log 5 = 0.3010 + 0.4771 + 0.6990; = 1.4771$

31. $\log \frac{75}{2} = \log(3 \cdot 5 \cdot 5/2) = \log 3 + 2 \log 5 - \log 2 = 0.4771 + 2(0.6990) - (0.3010) = 1.5741$

32. $\log \frac{45}{8} = \log(5 \cdot 3^2/2^3) = \log 5 + 2 \log 3 - 3 \log 2; = 0.6990 + 2(0.4771) - 3(0.3010) = 0.7502$

33. $\log 200 = \log 2 \cdot 10^2 = \log 2 + 2 \log 10 = 0.3010 + 2 \cdot 1 = 2.3010$ (Note: $\log 10 = 1$)

34. $\log 150 = \log 10 \cdot 3 \cdot 5 = \log 10 + \log 3 + \log 5 = 1 + 0.4771 + 0.6990 = 2.1761$

35. $\log 5000 = \log 10^3 \cdot 5 = 3 \log 10 + \log 5 = 3 \cdot 1 + 0.6990 = 36990$

36. $\log 4500 = \log 10^2 \cdot 5 \cdot 3^2 = 2 \log 10 + \log 5 + 2 \log 3 = 2 \cdot 1 + 0.6990 + 2(0.4771) = 3.6532$

37. $\log_b 16 = \log_b 2^4 = 4 \log_b 2 = 4(0.3869) = 1.5476$

38. $\log_b 15 = \log_b(3 \cdot 5) = \log_b 3 + \log_b 5 = 0.6131 + 0.8982 = 1.5113$

39. $\log_b \frac{5}{3} = \log_b 5 - \log_b 3 = 0.8082 - 0.6131 = 0.2851$

40. $\log_b \frac{24}{5} = \log_b(2^3 \cdot 3/5) = 3 \log_b 2 + \log_b 3 - \log_b 5 = 3(0.3869) + 0.6131 - 0.8082 = 0.8756$

41. $\log 2 = \frac{\ln 2}{\ln b} = 0.3869; \ln b = \ln 2/0.3869 = 1.7915;$ Since e^x and $\ln x$ are inverse functions; $b = e^{1.7915} = 5.99$ or $b = 6;$ (Note: $\log_b 6 = \log_b 2 + \log_b 3 = 0.3869 + 0.6131 = 1.0000$)

42. $N = 20(\log I_1 - \log I_2) + 10(\log R_1 - \log R_2)$

$= 20\log\left(\dfrac{I_1}{I_2}\right) + 10\log\left(\dfrac{R_1}{R_2}\right)$

$= \log\left(\dfrac{I_1}{I_2}\right)^{20} + \log\left(\dfrac{R_1}{R_2}\right)^{10}$

$= \log\left[\left(\dfrac{I_1}{I_2}\right)^{20}\left(\dfrac{R_1}{R_2}\right)^{10}\right]$

43. $\Delta dB = 20(\log I_2 + \dfrac{1}{2}\log R_2 - \log I_1 - \dfrac{1}{2}\log R_1)$

$= 20(\log I_2 + \log\sqrt{R_2} - \log I_1 - \log\sqrt{R_1})$

$= 20\log\left(\dfrac{I_2\sqrt{R_2}}{I_1\sqrt{R_1}}\right)$

$= \log\left(\dfrac{I_2\sqrt{R_2}}{I_1\sqrt{R_1}}\right)$

$= \log\left(\dfrac{I_2^{20}R_2^{10}}{I_1^{20}R_1^{10}}\right)$

44. The percent of the fish population alive after any given time, when $t = 0$, is given by $\frac{P}{P_0} = \frac{P}{100}$.

(a) If there is only 50% of the fish population remaining, then $\frac{P}{P_0} = \frac{50}{100} = \frac{1}{2}$ and we have $\ln\frac{1}{2} = -0.0435t$ and $t = \dfrac{\ln\frac{1}{2}}{-0.0435} \approx 15.93$. It will take about 15.93 yr for the fish population to be reduced to 50%.

(b) If the fish population is reduced by 90%, then 10% of the population remains. Thus, $\frac{P}{P_0} = \frac{10}{100} = 0.1$ and $\ln 0.1 = -0.0435t$, so $t = \frac{\ln 0.1}{-0.0435} \approx 52.93$. It will take about 52.93 yr for the fish population to be reduced by 90%.

(c) In the first year, we have $t = 1$. The unknown percent is $\frac{P}{100}$ and so, $\ln\frac{P}{P_0} = \ln\frac{P}{100} = -0.0435(1) = -0.0435$. Thus, $\frac{P}{100} = e^{-0.0435}$. Solving for P, we get $P = (100)e^{-0.0435} \approx (100)(0.9574) = 95.74$. After one year, 95.74% of the population is alive.

13.5 EXPONENTIAL AND LOGARITHMIC EQUATIONS

1. $5^x = 29; \log 5^x = \log 29; x\log 5 = \log 29; x = \frac{\log 29}{\log 5} = 2.0922$

2. $6^y = 32; \log 6^y = \log 32; y\log 6 = \log 32; y = \frac{\log 32}{\log 6} = 1.9343$

3. $4^{6x} = 119; \log 4^{6x} = \log 119; 6x\log 4 = \log 119; x = \frac{\log 119}{6\log 4} = 0.5746$

4. $3^{-7x} = 4; \log 3^{-7x} = \log 4; -7x\log 3 = \log 4; x = \frac{\log 4}{-7\log 3} = -0.1803$

5. $3^{y+5} = 16; \ln 3^{y+5} = \ln 16; (y+5)\ln 3 = \ln 16; y+5 = \frac{\ln 16}{\ln 3}; y = \frac{\ln 16}{\ln 3}; -5 = -2.4763$

6. $2^{y-7} = 67; \ln 2^{y-7} = \ln 67; (y-7)\ln 2 = \ln 67; y-7 = \frac{\ln 67}{\ln 2}; y = \frac{\ln 67}{\ln 2} + 7 = 13.066$

7. $e^{2x-3} = 10; \ln e^{2x-3} = \ln 10; 2x-3 = \ln 10; 2x = \ln 10 + 3; x = \frac{\ln 10 + 3}{2} = 2.6513$

8. $4^{3y+5} = 30; \ln 4^{3y+5} = \ln 30; (3y+5)\ln 4 = \ln 30; 3y+5 = \frac{\ln 30}{\ln 4}; 3y = \frac{\ln 30}{\ln 4} - 5; y = \frac{(\ln 30/\ln 4)-5}{3} = -0.8489$

9. $3^{4x+1} = 9^{3x-5}; 3^{4x+1} = (3^2)^{3x-5}; 3^{4x+1} = 3^{6x-10}; \log_3 3^{4x+1} = \log_3 3^{6x-10}; 4x+1 = 6x-10; -2x = -11; x = \frac{11}{2} = 5.5$

10. $2^{3-2x} = 8^{1+2x}; 2^{3-2x} = (2^3)^{1+2x}; 2^{3-2x} = 2^{3+6x};$ Taking $\log_2$ on both sides gives $3 - 2x = 3 + 6x; -2x = 6x; 8x = 0$ or $x = 0$.

11. $e^{3x-1} = 5e^{2x}; \ln e^{3x-1} = \ln(5e^{2x}); 3x-1 = \ln 5 + 2x; x = \ln 5 + 1 = 2.6094$

12. $3^{2-5x} = 8(9^{x-1}); 3^{2-5x} = 8 \cdot 3^{2x-2}; \log_3 3^{2-5x} = \log_3(8 \cdot 3^{2x-2}); 2 - 5x = \log_3 8 + 2x - 2; -7x = \log_3 8 - 4; x = \frac{4-\log_3 8}{7} = \frac{4-(\ln 8/\ln 3)}{7} = 0.3010$

13. $\log x = 2.3$; rewrite as $x = 10^{2.3} \approx 199.53$

14. $\ln x = 5.4$; rewrite as $x = e^{5.4} = 221.406$

15. $\log(x - 5) = 17; x - 5 = 10^{17}; x = 10^{17} + 5$

16. $\ln(y + 3) = 19; y + 3 = e^{19}; y = e^{19} - 3$

17. $\ln 2x + \ln x = 9; \ln 2x^2 = 9; 2x^2 = e^9; x^2 = \frac{e^9}{2}; x = \sqrt{\frac{1}{2}e^9} \approx 63.6517$

18. $2\ln x = \ln 4; \ln x^2 = \ln 4; x^2 = 4; x = \pm 2$. Positive 2 is ok but -2 is not in the domain of ln so the only answer is 2.

19. $2\ln 3x + \ln 2 = \ln x$; $\ln 9x^2 + \ln 2 = \ln x$; $\ln 18x^2 = \ln x$; $18x^2 = x$; $18x = e^0$; $18x = 1$; $x = \frac{1}{18} \approx 0.056$

20. $2\ln(x+3) = \ln x + 4$; $\ln(x+3)^2 - \ln x = 4$; $\ln\left(\frac{(x+3)^2}{x}\right) = 4$; $\frac{(x+3)^2}{x} = e^4 = 54.59815$; $(x+3)^2 = 54.59815x$; $x^2 + 6x + 9 = 54.59815x$; $x^2 + (6 - 54.59815)x + 9 = 0$. Using the quadratic formula gives $x = 48.4122$ or $x = 0.1859$

21. $\log(x - 1) = 2$; $x - 1 = 10^2 = 100$; $x = 101$

22. $\log(x - 4) = \log x - 4$; $\log(x - 4) - \log x = -4$; $\log\frac{x-4}{x} = -4$; $\frac{x-4}{x} = 10^{-4} = 0.0001$; $x - 4 = .0001x$; $0.9999x = 4$; $x = \frac{4}{0.9999} = 4.0004$

23. $2\log(x - 4) = 2\log x - 4$; $\log(x - 4)^2 - \log x^2 = -4$; $\log\left(\frac{x-4}{x}\right)^2 = -4$; $\left(\frac{x-4}{x}\right)^2 = 10^{-4} = 0.0001$; $\frac{x-4}{x} = \sqrt{0.0001} = 0.1$; $x - 4 = .01x$; $0.99x = 4$; $x = \frac{4}{0.99} = 4.040404$

24. $\ln x = 1 + 3\ln x$; $\ln x - 3\ln x = 1$; $\ln x - \ln x^3 = 1$; $\ln\left(\frac{x}{x^3}\right) = 1$; $\ln\left(\frac{1}{x^2}\right) = 1$; $\frac{1}{x^2} = e^1$; $x^2 = \frac{1}{e}$; $x = \sqrt{\frac{1}{e}} = e^{-0.5} = 0.6065$

25. $t = \frac{\ln 2}{r} = \frac{\ln 2}{0.06} = 11.55$ years

26. $t = \frac{\ln 2}{0.08} = 8.66$ years

27. $t = \frac{\ln 2}{0.005} = 138.63$ years

28. $t = \frac{\ln 2}{0.03} = 23.10$ years

29. $12.5 = \frac{\ln 2}{r}$; $r = \frac{\ln 2}{12.5} = 0.055$ or 5.5%

30. $15 = \frac{\ln 2}{r}$; $r = \frac{\ln 2}{15} = 0.046 = 4.6\%$ per hour

31. $8.5 = \frac{\ln 2}{r}$; $r = \frac{\ln 2}{8.5} = 0.0815$ or 8.15%

32. $m = 200 - 2^{t/2}$; $100 = 200 - 2^{t/2}$; $-100 = -2^{t/2}$; $2^{t/2} = 100$; $\ln 2^{t/2} = \ln 100$; $t/2 \ln 2 = \ln 100$; $t/2 = \frac{\ln 100}{\ln 2}$; $t = \frac{2\ln 100}{\ln 2} = 13.29$ min

33. $-\log(2.5 \times 10^{-6}) = 5.6$

34. $5.3 = -\log\left[H^+\right]$; $\log H^+ = -5.3$; $H^+ = 10^{-5.3} = 5.01 \times 10^{-6}$ moles per liter.

35. $H = (30T + 8000)\ln\frac{P_0}{P} = (30(-5) + 8000)\ln\frac{76}{43.29}$
$= (7850)\ln\left(\frac{76}{43.29}\right) = 4418.07$ m

36. $\beta = 10\log\frac{I}{I_0}$; $105 = 10\log\frac{I}{10^{-12}}$;
$10.5 = \log\frac{I}{10^{-12}}$; $10.5 = \log I - \log 10^{-12}$; $10.5 = \log I + 12$; $-1.5 = \log I$; $I = 10^{-1.5}$ W/m$^2 \approx 0.0316$ W/m^2

37. $Q = Q_0 e^{-t/T}$; $T = RC = 500 \times 100 \times 10^{-6} = 0.05$; $0.05 = 0.40e^{-t/0.05}$; $0.125 = e^{-t/0.05}$; $\ln 0.125 = -t/0.05$; $t = -0.05\ln 0.125 = 0.104$ s

38. $T = T_0 + ce^{-kt}$; $c = T_1 - T_0 = 1200 - 20 = 1180$; $40 = 20 + (1180)e^{-.08t}$; $20 = 1180e^{-.08t}$; $0.016949 = e^{-0.08t}$; $\ln 0.016949 = -0.08t$; $t = \frac{\ln 0.016949}{-0.08} = 50.97$ hours

39. $I = I_0(1 - e^{-Rt/L})$; $L = 0.2$, $R = 4$; $I_0 = 1.5$; $I = 1.4$; $1.4 = 1.5(1 - e^{-4t/0.2})$; $0.9\bar{3} = 1 - e^{-20t}$; $-0.06\bar{6} = -e^{-20t}$; $e^{-20t} = 0.0\bar{6}$; $-20t = \ln 0.0\bar{6}$; $t = \frac{\ln 0.0\bar{6}}{-20} = 0.1354$ seconds

40. Substituting $Y = 4{,}720$, $A = 60$, and $S = 110$, we have
$$\ln Y = 5.36437 - 101.16296S^{-1} - 22.00048A^{-1} + 0.97116\ln BA$$
$$\ln 4{,}720 = 5.36437 - 101.16296(110)^{-1} - 22.00048(60)^{-1} + 0.97116\ln BA$$
$$0.97116\ln BA = 4.381532$$
$$\ln BA = 4.511648$$
$$BA = e^{4.511648} \approx 91.07$$
The basal area of this particular is about 91.07 ft^2/acre.

41.
$$E = \frac{U^2C}{2}\left(1 - e^{-2t/RC}\right)$$

$$\frac{2E}{U^2C} = 1 - e^{-2t/RC}$$

$$e^{-2t/RC} = 1 - \frac{2E}{U^2C}$$

$$\frac{-2t}{RC} = \ln\left(1 - \frac{2E}{U^2C}\right)$$

$$t = -\frac{RC}{2}\ln\left(1 - \frac{2E}{U^2C}\right)$$

42.
$$E = \frac{U^2C}{2}\left(1 - e^{-2t/RC}\right)$$

$$\frac{2E}{U^2C} = 1 - e^{-2t/RC}$$

$$e^{-2t/RC} = 1 - \frac{2E}{U^2C}$$

$$\frac{-2t}{RC} = \ln\left(1 - \frac{2E}{U^2C}\right)$$

$$\frac{RC}{-2t} = \frac{1}{\ln\left(1 - \frac{2E}{U^2C}\right)}$$

$$R = \frac{-2t}{C\ln\left(1 - \frac{2E}{U^2C}\right)}$$

$$= \frac{2t}{C}\ln\left(1 - \frac{2E}{U^2C}\right)$$

13.6 GRAPHS USING SEMILOGARITHMIC AND LOGARITHMIC PAPER

1.

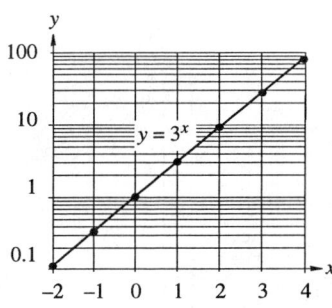

x	-3	-2	-1	0	1	2	3	4
y	0.037	0.11	0.33	1	3	9	27	81

2.

x	-3	-2	-1	0	1	2	3	4
y	$-.008$	$-.04$	$-.2$	1	5	25	125	625

3.

x	-3	-2	-1	0	1	2	3	4
y	0.014	0.083	0.5	3	18	108	648	3888

4.

x	-2	-1	0	1	2	3	4
y	0.102	0.714	5	35	245	1715	12005

5.

x	-5	-4	-3	-2	-1	0	1	2	3
y	32	16	8	4	2	1	1/2	1/4	1/8

6.

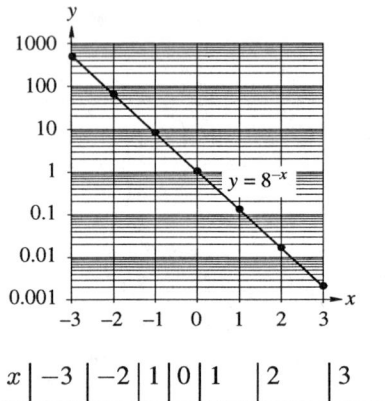

x	-3	-2	1	0	1	2	3
y	512	64	8	1	1/8	1/64	1/512

7.

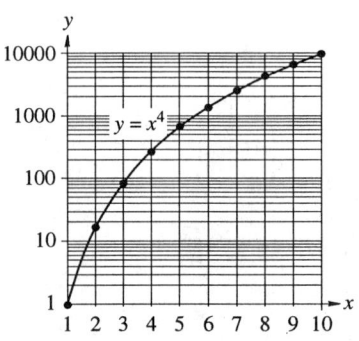

x	1	2	3	4	5	6	7	8	9	10
y	1	16	81	256	625	1296	2401	4096	6561	10000

8.

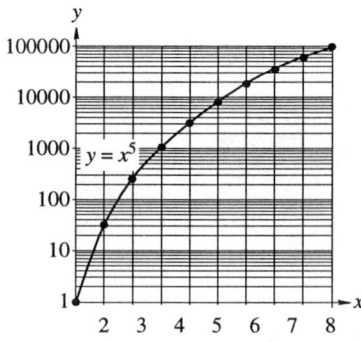

x	1	2	3	4	5	6	7	8	9
y	1	32	243	1024	3125	7776	16807	32768	59049

9.

x	1	2	3	4	5	6	7	8	9	10
y	8	64	216	512	1000	1728	2744	4096	5842	8000

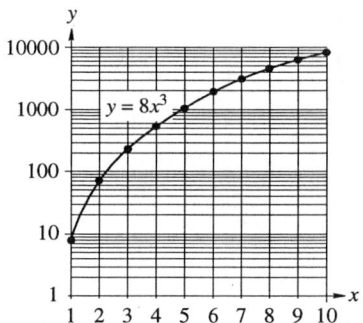

10.

x	1	2	3	4	5	6	7	8	9
y	5	80	405	1280	3125	6480	12005	20480	32805

11.

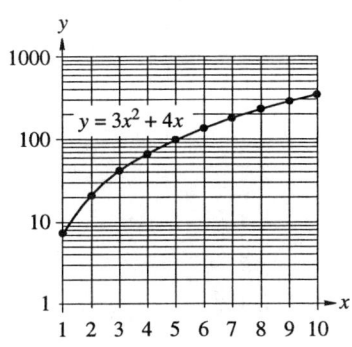

x	1	2	3	4	5	6	7	8
y	7	20	39	64	95	132	175	224

12.

x	1	2	3	4	5	6	7	8	9	10
y	11	74	237	548	1055	1806	2849	4232	6003	8210

13.

x	.1	.25	1	4	9	16	25	36	49	64	81
y	.316	.5	1	2	3	4	5	6	7	8	9

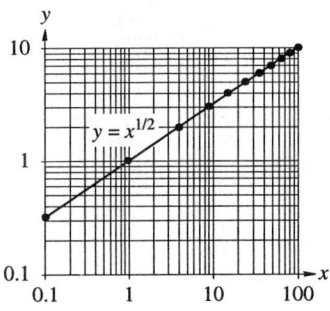

14.

x	.1	1	8	27	64	125	100
y	.464	2	2	3	4	5	4.64

15.

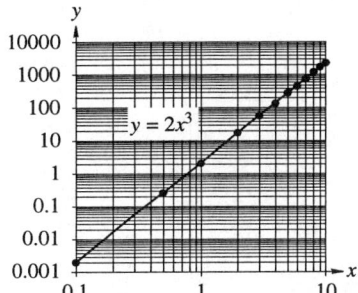

x	.1	.5	1	2	3	4	5	6	7	8
y	.002	.25	2	16	54	128	250	432	686	1024

16.

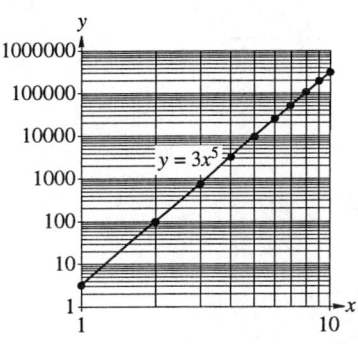

x	1	2	3	4	5	6	7	8	9
y	3	96	729	3072	9357	23328	50421	98304	177147

17.

x	1	2	3	4	5	6	7	9
y	2	5.41	10.7	18	27.24	38.45	51.65	84

18.

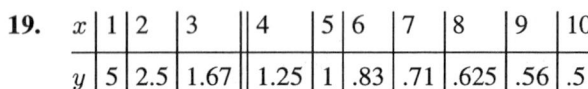

x	.1	1	2	3	4	5	6	8	10
y	.101	2	10	30	68	130	222	520	1010

19.

x	1	2	3	4	5	6	7	8	9	10
y	5	2.5	1.67	1.25	1	.83	.71	.625	.56	.5

20.

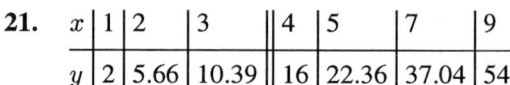

x	1	2	3	4	5	6	7	8	9	10
y	8	1	.3	.125	.064	.037	.023	.016	.011	.008

21.

x	1	2	3	4	5	7	9
y	2	5.66	10.39	16	22.36	37.04	54

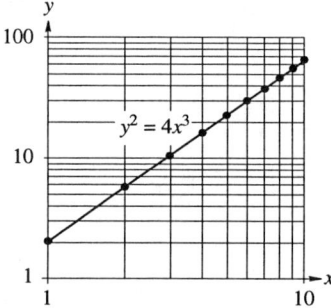

22. Rewrite $y^4 = 6x^5$ as $y = (6x^5)^{1/4}$ and use this second equation to complete the following table:

x	1	2	3	4	6	8	10
y	1.57	3.72	6.18	8.85	14.70	21.06	27.83

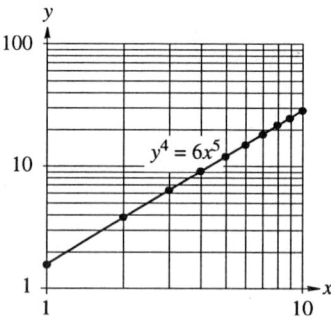

23. Rewrite $x^2 y^3 = 8$, first as $y^3 = 8x^{-2}$, and then, solve for y with the result that $y = (8x^{-2})^{1/3}$ and use this third equation to complete the following table:

x	1	2	3	4	6	8	10
y	1	1.26	0.96	0.79	0.61	0.5	0.43

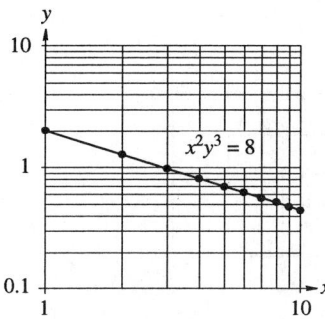

24. Rewrite $x^3 y = 100$ as $y = 100x^{-3}$ and use this second equation to complete the following table:

x	1	2	3	4	6	8	10
y	100	12.5	3.7	1.56	0.46	0.195	0.1

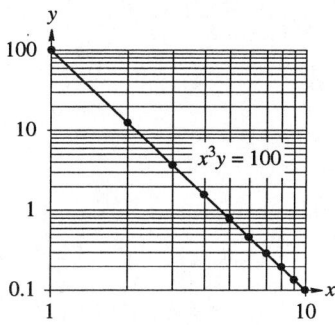

25. $P = e^{(-1.25 \times 10^{-4})h}$:

h	0	10	100	1000	5000	10000
P	1	0.999	0.987	0.882	0.535	0.287

26. $P = 100e^{-0.3h}$:

h	0	2	4	6	8	10
P	100	54.9	30.1	16.5	9.07	4.98

27. Rewrite $RD^2 = 0.9$ as $D^2 = 0.9R^{-1}$ and then, solve for D with the result $D = (0.9R^{-1})^{1/2}$. Now, use this last equation to complete the following table:

R	0.1	0.9	10	100	90
D	3	1	0.3	0.095	0.1

28. Rewrite $PV = 5$ as $P = 5V^{-1}$ and complete the following table:

P	0.1	0.5	1	5	10
V	50	10	5	1	0.5

29.

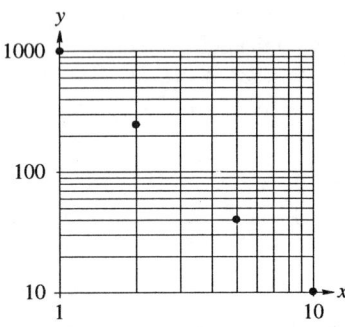

From the graph on the log-log paper we can conclude that the equation is of the form $y = ax^m$ when $x = 1\, y = 1000$, so we can conclude that $a = 1000$. To find m take

$$m = \frac{\log 250 - \log 1000}{\log 2 - \log 1} = \frac{2.39794 - 3}{0.30103 - 0} - 2$$

hence $y = 1000x^{-2}$

30.

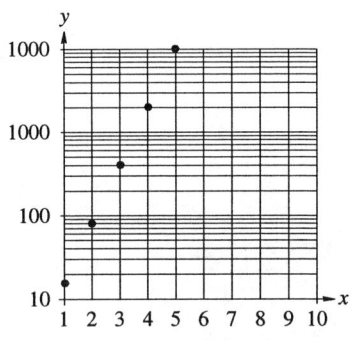

Since the graph is a straight line on semi-log paper, the equation is of the form $y = ab^x$ when $x = 1$, $y = 16$ so $16 = ab^1$ or $ab = 16$ when $x = 2, y = 80$ so $80 = ab^2$ or $ab^2 = 80$; $\frac{ab^2}{ab} = \frac{80}{16}$; $b = 5$.

Since $ab = 16, a = \frac{16}{5} = 3.2$. Hence the equation is $y = 3.2(5^x)$.

31.

32.

CHAPTER 13 REVIEW

1. $e^5 = 148.41316$

2. $e^{-7} = 9.118819 \times 10^{-4} = 0.00091188$

3. $e^{4.67} = 106.69774$

4. $e^{-3.91} = 0.0200405$

5. $\log 8 = 0.90309$

6. $\log 196.5 = 2.2933626$

7. $\ln 81.3 = 4.398146$

8. $\ln 325.6 = 5.7856696$

9.

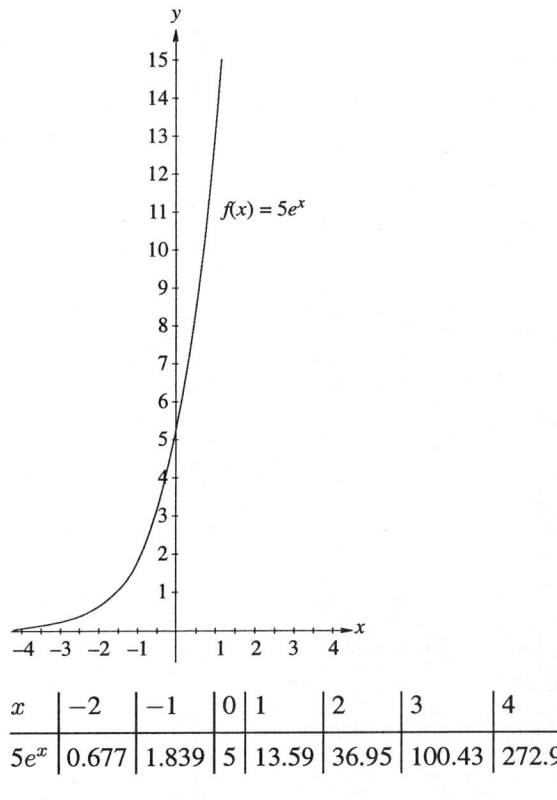

x	-2	-1	0	1	2	3	4
$5e^x$	0.677	1.839	5	13.59	36.95	100.43	272.99

10.

x	-4	-3	-2	-1	0	1	2
$9e^{-0.5x}$	66.50	40.34	24.46	14.84	9	5.46	3.31

11.

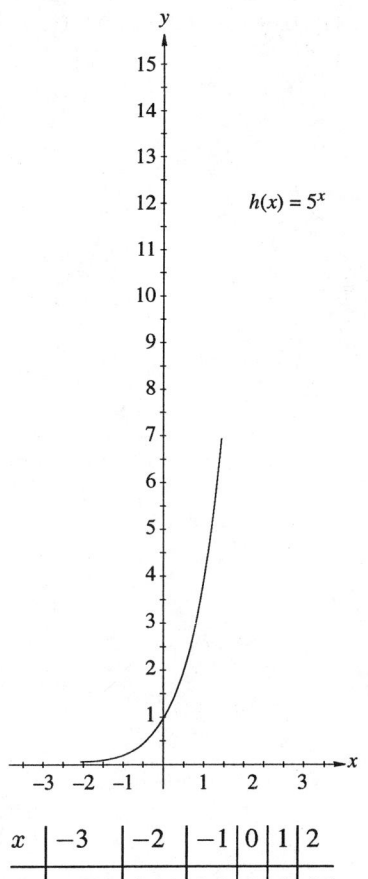

x	-3	-2	-1	0	1	2
5^x	0.008	0.04	0.2	1	5	25

12.

x	-4	-3	-2	-1	0	1	2
2.1^{-x}	19.45	9.26	4.41	2.1	1	0.48	0.23

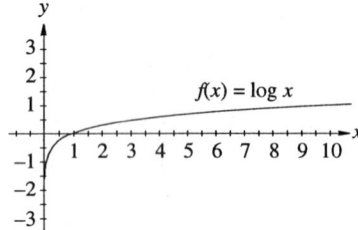

13.

x	0	1	2	3	4	5	6	7	8	9	10
$\log x$	not defined	0	0.301	0.477	0.602	0.699	0.778	0.845	0.903	0.954	1

14.

x	0	1	2	3	4	5	6	7	8	9	10
$3\log 2x$	not defined	0.9031	1.806	2.334	2.709	3	3.238	3.438	3.612	3.766	3.903

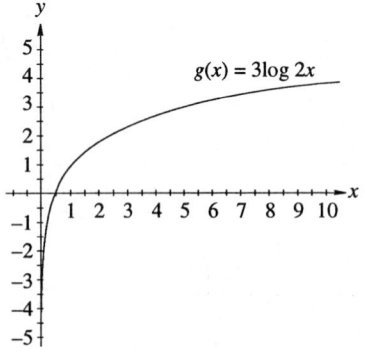

15.

x	$-\frac{1}{2}$	0	1	2	3	4	5	6	7	8	9	10
$\ln(2x+1)$	not defined	0	1.099	1.609	1.946	2.197	2.398	2.565	2.708	2.833	2.944	3.045

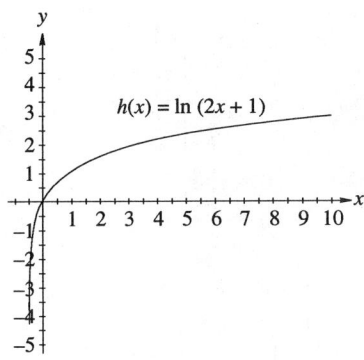

16.

x	0	1	2	3	4	5	6	7	8	9	10
$2+3\ln x$	not defined	2	4.079	5.296	6.159	6.828	7.375	7.838	8.238	8.592	8.908

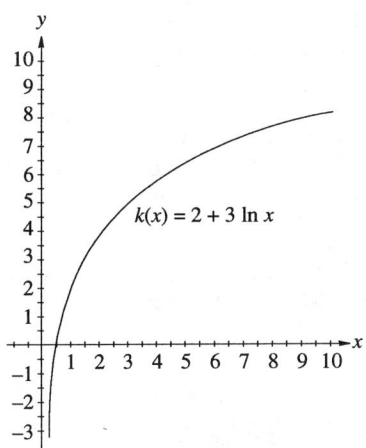

17. $\log \frac{3}{4} = \log 3 - \log 4$

18. $\log 77 = \log(7 \cdot 11) = \log 7 + \log 11$

19. $\log 5x = \log 5 + \log x$

20. $\ln \frac{7x}{3} = \ln 7 + \ln x - \ln 3$

21. $\ln(4x)^3 = 3\ln 4x = 3[\ln 4 + \ln x]$

22. $\ln 4x^3 = \ln 4 + 3\ln x$

23. $\log \sqrt{48} = \log 48^{1/2} = \frac{1}{2}\log 48 =$

$\frac{1}{2}[\log 6 + \log 8]$. Answers may vary.

24. $\log \frac{\sqrt[3]{2x^5}}{4x} = \log(2x^5)^{1/3} - \log 4x =$

$\frac{1}{3}[\log 2 + 5\log x] - \log 4 - \log x$. Answers may vary.

25. $\log 5 + \log 9 = \log(5 \cdot 9) = \log 45$

26. $\log 19 - \log 11 = \log \frac{19}{11}$

27. $4\log x + \log x = \log x^4 + \log x = \log x^5$

28. $\log 3a + \log b - \log a = \log\left(\frac{3ab}{a}\right) = \log 3b$

29. $7\log x + 2\log x - \log x = 8\log x = \log x^8$

30. $\ln \frac{2}{3} + \ln \frac{15}{8} = \ln\left(\frac{2}{3} \cdot \frac{15}{8}\right) = \ln\left(\frac{5}{4}\right)$

31. $4^x = 28; \log 4^x = \log 28; x \log 4 = \log 28; x = \frac{\log 28}{\log 4} = 2.404$

32. $5^{3x} = 20; \ln 5^{3x} = \ln 250; 3x \ln 5 = \ln 250; x = \frac{\ln 250}{3 \ln 5} = 1.1436$

33. $e^{5x} = 11; \ln e^{5x} = \ln 11; 5x = \ln 11; x = \frac{\ln 11}{5} = 0.4796$

34. $e^{5x-1} = 2e^x; \ln^{5x-1} = \ln(2e^x) = \ln 2 + \ln e^x; 5x - 1 = \ln 2 + x; 4x = \ln 2 + 1; x = \frac{\ln 2 + 1}{4} = 0.4233$

35. $\ln x = 9.1; e^{\ln x} = e^{9.1}; x = e^{9.1} = 8955.29$

36. $4 \log x = 49; \log x = \frac{49}{4}; x = 10^{49/4} = 1.778 \times 10^{12}$

37. $\log(2x + 1) = 50; 2x + 1 = 10^{50}; x = \frac{10^{50}-1}{2}$ or $0.5 \times 10^{50} - 0.5$ (Note: to 50 significant digits this is $0.5 \times 10^{50} = 5 \times 10^{49}$.)

38. $2 \ln 4x = 100 + \ln 2x; \ln 16x^2 - \ln 2x = 100;$ $\ln \left(\frac{16x^2}{2x} \right) = 100; \ln 8x = 100; 8x = e^{100}; x = \frac{e^{100}}{8} = 3.36 \times 10^4 2$

39.

x	1	2	3	4	5	6	7	8	9
y	8	16	32	64	128	256	512	1024	2048

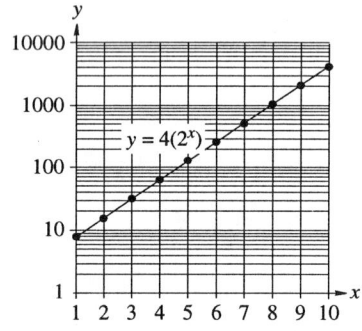

40.

x	1	2	4	6	8	10
y	8	128	2048	10,368	32,768	80,000

41.

x	1	2	3	4	5	6
y	32	128	512	2048	8192	32,768

42.

x	1	2	3	4	6	8
y	3	12	48	192	3072	49,152

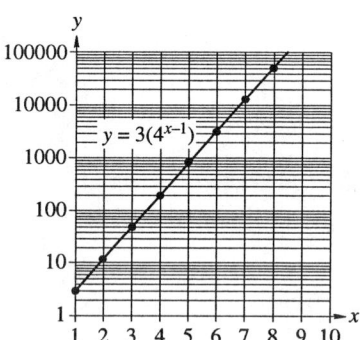

43.

x	1	10	16	81	256	625	2401	10000
y	1	1.78	2	3	4	5	7	10

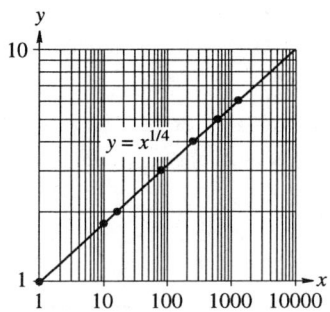

44.

x	1	2	3	4	6	8	10
y	5	40	135	320	1080	2560	5000

45.

x	0.001	0.01	0.1	1	10
y	211	37.6	6.69	1.18	0.211

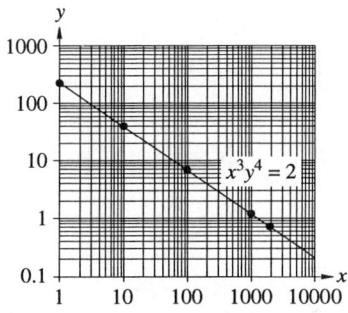

46. $y^3 = 10x^2$ or $y = (10x^2)^{1/3}$.

x	0.1	.1	10	100	1000
y	0.46	2.15	10	46.4	215.4

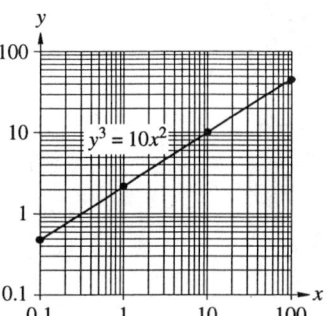

47. First Bank $S = 1000\left(1 + \frac{.075}{2}\right)^{2\times10} 2088.15$, Second Bank $S = 1000\left(1 + \frac{0.75}{12}\right)^{2\times10} = 2112.06$, Third Bank $S = 1000 \times e^{.075\times10} = 2117.00$, Double time $= \frac{\ln 2}{r} = \frac{\ln 2}{.075} = 9.24$ years

48. $e^{1600k} = \frac{1}{2}$; After 100 years $y = 100e^{100k} = 100\left(e^{1600k}\right)^{1/16} = 100(0.5)^{1/16} = 95.76$ mg. After 1000 yr, $y = 100e^{1000k} = 100\left(e^{1600k}\right)^{\frac{10}{16}} = 100(.5)^{5/8} = 64.84$ mg.

49. $y = ce^{kt}$; $25,000 = 4000e^{24k}$; $e^{24k} = \frac{25}{4}$; $24k = \ln\frac{25}{4}$; $k = 0.0763571$.
(a) $y = 4000e^{.07635761\times12} = 10000$.
(b) $y = 4000e^{0.7635761\times48} = 156,250$.

50. $R = \log I$ or San Francisco of 1906 $I = 10^{8.3}$;
(a) San Francisco of 1979 $I = 10^6$; $\frac{10^{8.3}}{10^6} = 10^{2.3} = 199.53$ times
(b) San Francisco of 1989 $I = 10^{7.1}$; $\frac{10^{8.3}}{10^{7.1}} = 10^{1.2} = 15.85$ times

51. Since the graph is a straight line on semi log paper the equation must be of the form $y = ab^x$. Hence $15.76 = ab^2$ and $620.84 = ab^4$. This gives $\frac{ab^4}{ab^2} = \frac{620.84}{15.76}$ or $b^2 = 39.3934$ or $b = 6.2764$. Back substituting to find a, $15.76 = a(6.2764)^2$ $a = \frac{15.76}{6.2764^2} = 0.4$. Answer: $y = 0.4(6.2764)^x$

52. Since the graph is a straight line on log-log paper, the equation must be of the form $y = ax^m$. $m = \frac{\log 2.1875 - \log 8.75}{\log 4 - \log 2} = -2$. Back substituting to find a, $8.75 = a2^{-2}$ or $2^2 \cdot 8.75 = 35$. Answer $y = 35x^{-2}$

53. $M = 6 - 2.5 \log \frac{I}{I_0}$

(a) $M - 6 = -2.5 \log \frac{I}{I_0}$; $\log \frac{I}{I_0} = \frac{6-M}{2.5}$; $\frac{I}{I_0} = 10^{\frac{6-M}{2.5}}$; $I = I_0 \cdot 10^{\frac{6-M}{2.5}}$

(b) $\dfrac{I_0 10^{\frac{6+12.5}{2.5}}}{I_0 10^{\frac{6-2}{2.5}}} = \dfrac{10^{18.5/2.5}}{10^{4/2.5}} = 10^{5.8} = 630957.34$

54. Using $Q = \frac{MQ_0}{Q_0+(m-Q_0)e^{-kmt}}$ we have $800 = \frac{2000 \cdot 400}{400+(2000-400)e^{-k \cdot 2000 \cdot 1/2}}$ or $800 = \frac{800000}{400+(1600)e^{-1000k}}$ and so $400 + 1600e^{-1000k} = \frac{800000}{800}$ or $400 + 1600e^{-1000k} = 1000$. This yields $1600e^{-1000k} = 600$ or $e^{1000k} = \frac{600}{1600} = \frac{3}{8} = 0.375$. Taking the natural logarithm of both sides produces $-1000k = \ln 0.375 = -0.9808$ and so $k = \frac{0.9808}{1000} = 0.0009808 = 9.808 \times 10^{-4}$.

CHAPTER 13 TEST

1. $e^{5.3} = 200.33681$

2. $\log 715.3 = 2.854488225$

3. $\ln 72.35 = 4.281515453$

4. $\log \frac{9}{15} = \log 9 - \log 15$ or $\log \frac{9}{15} = \log \frac{3}{5} = \log 3 - \log 5$

5. $\log 65x = \log 65 + \log x = \log 5 + \log 13 + \log x$

6. $\ln(7x)^{-2} = -2 \ln 7x = -2[\ln 7 + \ln x] = -2 \ln 7 - 2 \ln x$

7. $\log \frac{\sqrt[5]{5x^3}}{9x} = \log \sqrt[5]{5x^3} - \log 9x$; $\frac{1}{5} \log 5x^3 - \log 9x = \frac{1}{5}[\log 5 + 3 \log x] - \log 9 - \log x$

8. $\log 11 + \log 3 = \log(11 \cdot 3) = \log 33$

9. $\log 17 - 2 \log x = \log 17 - \log x^2 = \log \frac{17}{x^2}$

10. $\ln 4 - \ln 5 + \ln 15 - \ln 8 = \ln \frac{4 \cdot 15}{5 \cdot 8} = \ln \frac{3}{2}$

11. $\ln x = 17.2$; $x = e^{17.2} = 29502925.92$

12. $3 \log x = 55$; $\log x = \frac{55}{3}$, $x = 10^{55/3} = 2.154434 \times 10^{18}$

13. $3^x = 35$; $\ln 3^x = \ln 35$; $x \ln 3 = \ln 35$; $x = \frac{\ln 35}{\ln 3} = 3.23621727$

14. $e^{4x+1} = 2e^x$. Taking the natural logarithm of both sides gives $\ln e^{4x+1} = \ln 2e^x = \ln 2 + \ln e^x$ or $4x + 1 = \ln 2 + x$ and so $3x = \ln 2 - 1$, which means that $x = \frac{\ln 2 - 1}{3} \approx -0.10228$.

15. $\log(2x^2+4x) = 5 + \log 2x$; $\log(2x^2+4x) - \log 2x = 5$; $\log \left(\frac{2x^2+4x}{2x} \right) = 5$; $\log(x + 2) = 5$; $x + 2 = 10^5$; $x = 10^5 - 2 = 99,998$

16.

x	-3	-2	-1	0	1	2	3	4
$f(x)$	0.037	0.111	0.333	1	3	9	27	81

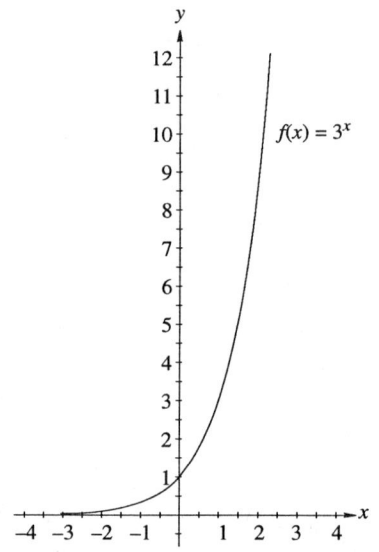

$f(x) = 3^x$

17.

x	1	2	3	4	5	6	7
y	1.2	3.6	10.8	32.4	97.2	291.6	874.8

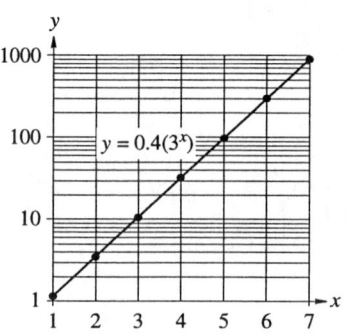

$y = 0.4(3^x)$

18.

x	0.1	1	2	10	32	243	1024
y	1.26	2	2.30	3.17	4	6	8

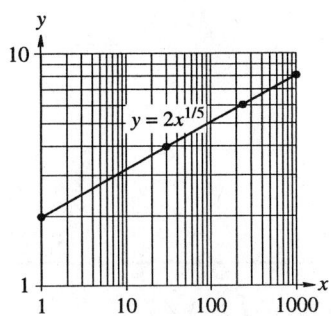

19. $\ln Y = 5.36437 - 101.16296 S^{-1} - 22.00048 A^{-1} + 0.97116 \ln \mathrm{BA}$

$\qquad = 5.36437 - 101.16296(120)^{-1} - 22.00048(60)^{-1} + 0.97116 \ln 130$

$\qquad = 8.8818$

$\quad Y = e^{8.8818} \approx 7199.9$

The predicted yield is about 7,200 ft^3/acre.

20. $S = P \left(1 + \dfrac{r}{k} \right)^{kt}$

$\quad = 5{,}000 \left(1 + \dfrac{0.055}{12} \right)^{12 \cdot 8}$

$\quad = 7{,}753.74$

21. We use the formula $y = c e^{kt}$. If 7.5% decays in one year, then $\frac{y}{c} = 1 - 0.75\% = 1 - 0.0075 = 0.9925$. Hence, $0.9925 = e^k$ and so $\ln 0.9925 = k$ or $k \approx -0.007528$. To find the half-life, we need to solve $\frac{y}{c} = \frac{1}{2}$ or $\frac{1}{2} = e^{-0.007528 t}$ and so, $\ln(1/2) = \ln 0.5 = -0.007528 t$ and $t = \frac{\ln 0.5}{-0.007528} \approx 92.07$. The half-life is about 92.07 yr.

CHAPTER
14
Complex Numbers

1. $\sqrt{-25} = \sqrt{25}\sqrt{-1} = 5j$

2. $\sqrt{-81} = \sqrt{81}\sqrt{-1} = 9j$

3. $\sqrt{-0.04} = \sqrt{0.04}\sqrt{-1} = 0.2j$

4. $\sqrt{-1.44} = \sqrt{1.44}\sqrt{-1} = 1.2j$

5. $\sqrt{-75} = \sqrt{25}\sqrt{3}\sqrt{-1} = 5\sqrt{3}j$ or $5j\sqrt{3}$

6. $-\sqrt{-72} = -\sqrt{36}\sqrt{2}\sqrt{-1} = -6j\sqrt{2}$

7. $-3\sqrt{-20} = -3\sqrt{4}\sqrt{5}\sqrt{-1} = -3 \cdot 2 \cdot \sqrt{5}j = -6j\sqrt{5}$

8. $5\sqrt{-30} = 5\sqrt{30}\sqrt{-1} = 5j\sqrt{30}$

9. $\sqrt{\frac{-9}{16}} = \sqrt{\frac{9}{16}} = \sqrt{-1} = \frac{3}{4}j$

10. $\sqrt{\frac{-25}{36}} = \sqrt{\frac{25}{36}}\sqrt{-1} = \frac{5}{6}j$

11. $-4\sqrt{\frac{-9}{16}} = -4\sqrt{\frac{9}{16}}\sqrt{-1} = -4 \cdot \frac{3}{4}j = -3j$

12. $-3\sqrt{-\frac{10}{81}} = \frac{-3\sqrt{10}\sqrt{-1}}{\sqrt{81}} = \frac{-3\sqrt{10}j}{9} = \frac{-\sqrt{10}}{3}j = \frac{-j\sqrt{10}}{3}$

13. $\left(\sqrt{-11}\right)^2 = \left(\sqrt{11}j\right)^2 = \sqrt{11}^2 j^2 = 11(-1) = -11$

14. $\left(\sqrt{-7}\right)^2 = \left(\sqrt{7}j\right)^2 = \sqrt{7}^2 j^2 = 7(-1) = -7$

15. $\left(3\sqrt{-2}\right)^2 = 3^2\sqrt{2}^2 j^2 = 9 \cdot 2(-1) = -18$

16. $\left(-2\sqrt{-3}\right)^2 = (-2)^2\sqrt{3}^2 j^2 = 4 \cdot 3(-1) = -12$

17. $\sqrt{-4}\sqrt{-25} = 2j \cdot 5j = 10j^2 = -10$

18. $\sqrt{-9}\sqrt{-16} = 3j \cdot 4j = 12j^2 = -12$

19. $\left(\sqrt{-49}\right)\left(2\sqrt{-9}\right) = 7j \cdot 2 \cdot 3j = 42j^2 = -42$

20. $\left(3\sqrt{-16}\right)\left(\sqrt{-36}\right) = 3 \cdot 4j \cdot 6j = 72j^2 = -72$

21. $\left(\sqrt{-5}\right)\left(-\sqrt{5}\right) = \sqrt{5}j\left(-\sqrt{5}\right) = -5j$

22. $\left(-\sqrt{-7}\right)\left(\sqrt{7}\right) = -\sqrt{7}j \cdot \sqrt{7} = -7j$

23. $j^7 = j^4 j^3 = 1(-j) = -j$

24. $j^{26} = j^{24+2} = j^{24}j^2 = \left(j^4\right)^6 j^2 = 1^6 j^2 = 1(-1) = -1$

25. $\sqrt{-\frac{1}{4}}\sqrt{\frac{16}{9}} = \left(\sqrt{\frac{1}{4}}j\right)\sqrt{\frac{16}{9}} = \left(\frac{1}{2}j\right)\left(\frac{4}{3}\right) = \left(\frac{1}{2}j\right)\left(\frac{4}{3}\right) = \frac{2}{3}j$

26. $\sqrt{\frac{4}{25}}\sqrt{-\frac{49}{9}} = \left(\sqrt{\frac{4}{25}}\right)\left(\sqrt{\frac{49}{9}}j\right) = \left(\frac{2}{5}\right)\left(\frac{7}{3}j\right) = \frac{14}{15}j$

27. $\sqrt{-\frac{3}{25}}\sqrt{\frac{3}{16}} = \left(\sqrt{\frac{3}{25}}j\right)\left(\sqrt{\frac{3}{16}}\right) = \left(\frac{\sqrt{3}}{5}\right)j\left(\frac{\sqrt{3}}{4}\right) = \frac{3}{20}j$

28. $\sqrt{\frac{75}{36}}\sqrt{-\frac{49}{5}} = \left(\sqrt{\frac{75}{36}}\right)\left(\sqrt{\frac{49}{5}}j\right) = \left(\frac{5\sqrt{5}}{6}\right)\left(\frac{7}{\sqrt{5}}j\right) = \frac{35}{6}j$

29. $\sqrt{-\frac{25}{36}}\sqrt{-\frac{9}{16}} = \left(\sqrt{\frac{25}{36}}j\right)\left(\sqrt{\frac{9}{16}}j\right) = \left(\frac{5}{6}j\right)\left(\frac{3}{4}j\right) = \left(\frac{5}{6}\right)\left(\frac{3}{4}\right)j^2 = \frac{5}{8}j^2 = -\frac{5}{8}$

30. $\sqrt{-\frac{16}{81}}\sqrt{-\frac{9}{64}} = \left(\sqrt{\frac{16}{81}}j\right)\left(\sqrt{\frac{9}{64}}j\right) = \left(\frac{4}{9}j\right)\left(\frac{3}{8}j\right) = \left(\frac{4}{9}\right)\left(\frac{3}{8}\right)j^2 = \frac{1}{6}j^2 = -\frac{1}{6}$

31. $\left(\sqrt{-0.5}\right)\left(\sqrt{0.5}\right) = \sqrt{0.5}j \cdot \sqrt{0.5} = 0.5j$

32. $\left(\sqrt{0.8}\right)\left(\sqrt{-0.2}\right) = \sqrt{0.8}\sqrt{0.2}j = \sqrt{0.16}j = 0.4j$

33. $\left(-2\sqrt{2.7}\right)\left(\sqrt{-3}\right) = \left(-2\sqrt{2.7}\right)\left(\sqrt{3}j\right) = -2\sqrt{8.1} = -2\sqrt{9}\sqrt{0.9}j = -2\cdot3\sqrt{0.9}j = -6j\sqrt{0.9}$

34. $\left(-4\sqrt{1.6}\right)\left(2\sqrt{-0.4}\right) = \left(-4\sqrt{1.6}\right)\left(2j\sqrt{0.4}\right) = -8\sqrt{0.64}j = -8(0.8)j = -6.4j$

35. $\left(\sqrt{-5}\right)\left(\sqrt{-6}\right)\left(\sqrt{-2}\right) = \sqrt{5}j\sqrt{6}j\sqrt{2}j = \sqrt{60}j^3 = 2\sqrt{15}(-j) = -2\sqrt{15}j = -2j\sqrt{15}$ (Note: $j^3 = -j$.)

36. $\left(\sqrt{-3}\right)\left(\sqrt{-9}\right)\left(\sqrt{-15}\right) = \sqrt{3}\sqrt{9}\sqrt{15}j^3 = 9\sqrt{5}j^3 = -9j\sqrt{5}$ (Note: $j^3 = -j$.)

37. $\left(-\sqrt{-7}\right)^2\left(\sqrt{-2}\right)^2 j^3 = \left(-\sqrt{7}j\right)^2\left(\sqrt{2}j\right)^2 j^3 = 7j^2\cdot2j^2j^3 = 14j^3 = -14j$ (Note: $j^3 = -j$.)

38. $\left(\sqrt{-3}\right)^2\left(\sqrt{-5}\right)^2 j^2\left(\sqrt{-2}\right) = \left(\sqrt{3}j\right)^2\left(\sqrt{5}j\right)^2 j^2\sqrt{2}j = 3j^25j^2j^2\sqrt{2}j = 15\sqrt{2}j^7 = -15j\sqrt{2}$ (Note: $j^7 = -j$.)

39. $x + yj = 7 - 2j$ so $x = 7, y = -2$

40. $x + yj = -9 + 2j$ so $x = -9, y = 2$

41. $x + 5 + yj = 15 - 3j$ so $x + 5 = 15$ or $x = 10$ and $y = -3$

42. $6j - x + yj = 4 + 2j$ so $-x = 4$ or $x = -4$ and $6 + y = 2$ or $y = -4$

43. $x - 5j + 2 = 4 - 3j + yj$ so $x + 2 = 4$ or $x = 2$ and $-5 = -3 + y$ or $y = -2$

44. $2x - 4j = 6j + 4 - yj$ so $2x = 4$ or $x = 2$ and $-4 = 6 - y$ or $y = 10$

45. The real part yields $\frac{1}{2}x = -\frac{1}{4}$, so $x = -\frac{1}{2}$ and the imaginary part gives $\frac{3}{4} = 2+y$, so $y = \frac{3}{4}-2 = -\frac{5}{4}$.

46. Comparing the real parts, we have $\frac{2}{5}x + 1 = 5$ or $\frac{2}{5}x = 4$ and $x = 4 \cdot \frac{5}{2} = 10$. Comparing the imaginary parts, we see that $-\frac{2}{3}j = -\frac{4}{3}j - yj$, so $y = -\frac{4}{3} + \frac{2}{3} = -\frac{2}{3}$

47. Comparing the real parts, we see that $1.2x + 3 = 7.2$ or $1.2x = 4.2$ and so, $x = 4.2 \div 1.2 = 3.5$. Comparing the imaginary parts, we obtain $y = -4.3$

48. The real parts are $-1.5x = -4.8$ and thus $x = -4.8 \div -1.5 = 3.2$. The imaginary parts are $3.7 = 2.4 + 0.5y$ or $1.3 = 0.5y$, and so $y = 2.6$.

49. $2(4 + 5j) = 8 + 10j$

50. $3(2 - 4j) = 6 - 12j$

51. $-5(2 + j) = -10 - 5j$

52. $-3(4 + 7j) = -12 - 21j$

53. $j(3 - 2j) = 3j - 2j^2 = 3j + 2 = 2 + 3j$

54. $j(5 + 4j) = 5j + 4j^2 = 5j - 4 = -4 + 5j$

55. $2j(4 + 3j) = 8j + 6j^2 = 8j - 6 = -6 + 8j$

56. $-3j(2 - 5j) = -6j + 15j^2 = -15 - 6j$

57. $\frac{1}{2}(6 - 8j) = 3 - 4j$

58. $\frac{2}{3}(6 + 9j) = 4 + 6j$

59. $\frac{5-10j}{5} = \frac{5}{5} - \frac{10}{5}j = 1 - 2j$

60. $\frac{6+12j}{3} = \frac{6}{3} + \frac{12}{3}j = 2 + 4j$

61. $\frac{6+\sqrt{-18}}{3} = \frac{6+3j\sqrt{2}}{3} = \frac{6}{3} + \frac{3j\sqrt{2}}{3} = 2 + j\sqrt{2}$

62. $\frac{7-\sqrt{-98}}{7} = \frac{7-\sqrt{98}j}{7} = \frac{7-7\sqrt{2}j}{7} = \frac{7}{7} - \frac{7j\sqrt{2}}{7} = 1 - j\sqrt{2}$

63. $\frac{8-\sqrt{-24}}{4} = \frac{8-\sqrt{24}j}{4} = \frac{8-2\sqrt{6}j}{4} = \frac{8}{4} - \frac{2j\sqrt{6}}{4} = 2 - \frac{1}{2}j\sqrt{6}$

64. $\frac{9+\sqrt{-27}}{6} = \frac{9+3j\sqrt{3}}{6} = \frac{9}{6} + \frac{3j\sqrt{3}}{6} = \frac{3}{2} + \frac{1}{2}j\sqrt{3}$

65. $\frac{2}{3}\left(\frac{3}{4} - \frac{1}{3}j\right) = \frac{2}{3}\left(\frac{3}{4}\right) - \frac{2}{3}\left(\frac{1}{3}\right)j = \frac{6}{12} - \frac{2}{9}j = \frac{1}{2} - \frac{2}{9}j$

66. $\frac{1}{2}\left(\frac{8}{5} + \frac{9}{4}j\right) = \frac{1}{2}\left(\frac{8}{5}\right) + \frac{1}{2}\left(\frac{9}{4}\right)j = \frac{8}{10} + \frac{9}{8}j = \frac{4}{5} + \frac{9}{8}j$

67. $-\frac{5}{3}\left(-\frac{3}{8} + \frac{6}{15}j\right) = -\frac{5}{3}\left(-\frac{3}{8}\right) + \left(-\frac{5}{3}\right)\frac{6}{15}j = \frac{15}{24} - \frac{30}{45}j = \frac{5}{8} - \frac{2}{3}j$

68. $-\frac{7}{4}\left(\frac{8}{21} - \frac{16}{35}j\right) = -\frac{7}{4}\left(\frac{8}{21}\right) - \left(-\frac{7}{4}\right)\frac{16}{35}j = -\frac{2}{3} + \frac{4}{5}j$

69. $1.5(2.4 - 3j) = 1.5(2.4) - 1.5(3j) = 3.6 - 4.5j$

70. $-7.2(-0.25+1.75j) = -7.2(0.25)+(-7.2)(1.75j) = 1.8 - 12.6j$

71. The conjugate of $7 + 2j$ is $7 - 2j$

72. The conjugate of $9 + \frac{1}{2}j$ is $9 - \frac{1}{2}j$

73. The conjugate of $6 - 5j$ is $6 + 5j$

74. The conjugate of $\frac{1}{2} - 9j$ is $\frac{1}{2} + 9j$

75. The conjugate of 19 is 19

76. The conjugate of $7j$ is $-7j$

77. The conjugate of $-8j$ is $8j$

78. The conjugate of -11 is -11

79. $\sqrt{2} - 7.3j$

80. $-4.5 + \sqrt{19}j$

81. $x^2 + x + 2.5 = 0$. Here $a = 1, b = 1$, and $c = 2.5$. Thus $x = \frac{-1 \pm \sqrt{1^2 - 4 \cdot 1 \cdot 2.5}}{2 \cdot 1} = \frac{-1 \pm \sqrt{1-10}}{2} = \frac{-1 \pm \sqrt{-9}}{2} = \frac{-1 \pm 3j}{2}$. Hence $x = -\frac{1}{2} + \frac{3}{2}j$ and $x = -\frac{1}{2} - \frac{3}{2}j$.

82. $x^2 + 2x + 5 = 0$. Here $a = 1, b = 2$, and $c = 5$. Thus $x = \frac{-2 \pm \sqrt{2^2 - 4 \cdot 1 \cdot 5}}{2 \cdot 1} = \frac{-2 \pm \sqrt{4-20}}{2} = \frac{-2 \pm \sqrt{-16}}{2} = \frac{-2 \pm 4j}{2}$. Hence $x = -1 + 2j$ and $x = -1 - 2j$.

83. $x^2 + 9 = 0$. Here $a = 1, b = 0$, and $c = 9$. Thus $x = \frac{-0 \pm \sqrt{0^2 - 4 \cdot 9}}{2} = \frac{\pm \sqrt{-36}}{2} = \frac{\pm 6j}{2}$. Hence $x = 3j$ and $x = -3j$.

84. $x^2 + 25 = 0$. Here $a = 1, b = 0$, and $x = 25$. Thus $x = \frac{-0 \pm \sqrt{0^2 - 4 \cdot 25}}{2} = \frac{\pm \sqrt{-100}}{2} = \frac{\pm 10j}{2}$. Hence $x = 5j$ and $x = -5j$.

85. $2x^2 + 3x + 7 = 0$. Here $a = 2, b = 3$, and $c = 7$. Thus $x = \frac{-3 \pm \sqrt{3^2 - 4 \cdot 2 \cdot 7}}{2 \cdot 2} = \frac{-3 \pm \sqrt{9-56}}{4} = \frac{-3 \pm \sqrt{-47}}{4} = \frac{-3 \pm j\sqrt{47}}{4}$. Hence $x = \frac{-3}{4} + \frac{\sqrt{47}}{4}j$ and $x = \frac{-3}{4} - \frac{\sqrt{47}}{4}j$.

86. $2x^2 + 7x + 9 = 0$. Here $a = 2, b = 7$, and $c = 9$. Thus $x = \frac{-7 \pm \sqrt{7^2 - 4 \cdot 2 \cdot 9}}{2 \cdot 2} = \frac{-7 \pm \sqrt{49-72}}{4} = \frac{-7 \pm \sqrt{-23}}{4} = \frac{-7 \pm j\sqrt{23}}{4}$. Hence $x = \frac{-7}{4} + \frac{\sqrt{23}}{4}j$ and $x = \frac{-7}{4} - \frac{\sqrt{23}}{4}j$.

87. $5x^2 + 2x + 5 = 0$. Here $a = 5, b = 2$, and $c = 5$. Thus $x = \frac{-2 \pm \sqrt{2^2 - 4 \cdot 5 \cdot 5}}{2 \cdot 5} = \frac{-2 \pm \sqrt{4-100}}{10} = \frac{-2 \pm \sqrt{-96}}{10} = \frac{-2 \pm 4j\sqrt{6}}{10}$. Hence $x = \frac{-1}{5} + \frac{2\sqrt{6}}{5}j$ and $x = \frac{-1}{5} - \frac{2\sqrt{6}}{5}j$.

88. $3x^2 + 2x + 10 = 0$. Here $a = 3, b = 2$, and $c = 10$. Thus $x = \frac{-2 \pm \sqrt{2^2 - 4 \cdot 3 \cdot 10}}{2 \cdot 3} = \frac{-2 \pm \sqrt{4-120}}{6} = \frac{-2 \pm \sqrt{-116}}{6} = \frac{-2 \pm \sqrt{116}j}{6} = \frac{-2}{6} \pm \frac{2j\sqrt{29}}{6}$. Hence $x = \frac{-1}{3} - \frac{\sqrt{29}}{3}j$ and $x = -\frac{1}{3} - \frac{\sqrt{29}}{3}j$.

89. Substituting the given data, we obtain $0.08 = \frac{X}{50^2 + X^2}$ and so $0.08(50^2 + X^2) = X$ or $0.08X^2 - X + 200 = 0$. Using the quadratic formula with $a = 0.08, b = -1$, and $c = 200$, we get $6.25 + 49.61j\,\Omega$ or $6.25 - 49.61j\,\Omega$.

90. Substituting the given data, we obtain $0.05 = \frac{X}{12^2 + X^2}$ or $0.05(12^2 + X^2) = X$ which simplifies to $0.05X^2 - X + 7.2 = 0$. Using the quadratic formula with $a = 0.05, b = -1$, and $c = 7.2$, we get $10 + 6.63j\,\Omega$ or $10 - 6.63j\,\Omega$.

91.
$$R^2 + X^2 = \frac{X}{B}$$
$$R^2 = \frac{X}{B} - X^2$$
$$R = \pm\sqrt{\frac{X}{B} - X^2}$$

92.
$$R^2 + X^2 = \frac{X}{B}$$
$$B\left(R^2 + X^2\right) = X$$
$$BR^2 + BX^2 = X$$
$$BX^2 - X + BR^2 = 0$$
$$X = \frac{1 \pm \sqrt{1 - 4B^2 R^2}}{2B}$$

≡ 14.2 OPERATIONS WITH COMPLEX NUMBERS

1. $(5 + 2j) + (-6 + 5j) = -1 + 7j$

2. $(9 - 7j) + (6 - 8j) = 15 - 15j$

3. $(11 - 4j) + (-6 + 2j) = 5 - 2j$

4. $(21 + 3j) + (-7 - 6j) = 14 - 3j$

5. $(4 + \sqrt{-9}) + (3 - \sqrt{-16}) = (4 + 3j) + (3 - 4j) = 7 - j$

6. $\left(-11 + \sqrt{-4}\right) + \left(9 + \sqrt{-36}\right) = (-11 + 2j) + (9 + 6j) = -2 + 8j$

7. $\left(2 + \sqrt{-9}\right) + \left(8j - \sqrt{5}\right) = (2 + 3j) + \left(8j - \sqrt{5}\right) = 2 - \sqrt{5} + 11j$

8. $(3+\sqrt{-8})+(3-\sqrt{8}) = (3+2j\sqrt{2})+(3-2\sqrt{2}) = 6 - 2\sqrt{2} + 2j\sqrt{2}$

9. $(14+3j) - (6+j) = 8+2j$

10. $(-8+3j) - (4-3j) = -12+6j$

11. $(9-\sqrt{-4}) - (\sqrt{-16}+6) = (9-2j)-(4j+6) = 3-6j$

12. $(\sqrt{-25}-3) - (3-\sqrt{-25}) = (5j-3)-(3-5j) = -6+10j$

13. $(4+2j) + j + (3-5j) = 7-2j$

14. $(2-3j) - j - (6+\sqrt{-81}) = (2-4j)-(6+9j) = -4-13j$

15. $(2+j)3j = 6j+3j^2 = -3+6j$

16. $(5-3j)2j = 10j-6j^2 = 6+10j$

17. $(9+2j)(-5j) = -45j-10j^2 = 10-45j$

18. $(11-4j)(-3j) = -33j+12j^2 = -12-33j$

19. $(2+j)(5+3j) = 10+6j+5j+3j^2 = 7+11j$

20. $(3-2j)(4+5j) = 12+15j-8j-10j^2 = 22+7j$

21. $(6-2j)(5+3j) = 30+18j-10j-6j^2 = 36+8j$

22. $(4-2j)(7-3j) = 28-12j-14j+6j^2 = 22-26j$

23. $(2\sqrt{-9}+3)(5\sqrt{-16}-2) = (6j+3)(20j-2) = 120j^2 - 12j + 60j - 6 = -126+48j$

24. $(6\sqrt{-25}-4)(-3-2\sqrt{-49}) = (30j-4)(-3-14j) = -90j - 420j^2 + 12 + 56j = 432 - 34j$

25. $(\sqrt{-3})^4 = (\sqrt{3}j)^4 = \sqrt{3}^4 j^4 = 9(1) = 9$; (Note: $j^4 = 1$)

26. $(\sqrt{-9})^3 = (3j)^3 = 27j^3 = -27j$; (Note: $j^3 = 1$)

27. $(1+2j)^2 = 1+4j+4j^2 = -3+4j$

28. $(3+4j)^2 = 9+24j+16j^2 = -7+24j$

29. $(7-j)^2 = 49-14j+j^2 = 48-14j$

30. $(4-3j)^2 = 16-24j+9j^2 = 7-24j$

31. $(5+2j)(5-2j) = 5^2+2^2 = 25+4 = 29$ (Note: $(a+bj)(a-bj) = a^2+b^2$)

32. $(7+3j)(7-3j) = 7^2+3^2 = 49+9 = 58$ (Note: $(a+bj)(a-bj) = a^2+b^2$)

33. $\frac{6-4j}{1+j} \cdot \frac{1-j}{1-j} = \frac{6-6j-4j+4j^2}{1+1} = \frac{2-10j}{2} = 1-5j$

34. $\frac{4-8j}{2-2j} \cdot \frac{2+2j}{2+2j} = \frac{8+8j-16j-16j^2}{4+4} = \frac{24-8j}{8} = 3-j$

35. $\frac{6-3j}{1+2j} \cdot \frac{1-2j}{1-2j} = \frac{6-12j-3j+6j^2}{1+4} = \frac{-15j}{5} = -3j$

36. $\frac{5-10j}{1-2j} \cdot \frac{1+2j}{1+2j} = \frac{5+10j-10j-20j^2}{1+4} = \frac{25}{5} = 5$

37. $\frac{4+2j}{1-2j} \cdot \frac{1+2j}{1+2j} = \frac{4+8j+2j+4j^2}{1+4} = \frac{10j}{5} = 2j$

38. $\frac{9+5j}{3+j} \cdot \frac{3-j}{3-j} = \frac{27-9j+15j-5j^2}{9+1} = \frac{32+6j}{10} = 3.2+0.6j$

39. $\frac{2j}{5+j} \cdot \frac{5-j}{5-j} = \frac{10j-2j^2}{25+1} = \frac{2+10j}{26} = \frac{1}{13} + \frac{5}{13}j$

40. $\frac{5j}{6-j} \cdot \frac{6+j}{6+j} = \frac{30j+5j^2}{36+1} = \frac{-5+30j}{37} = -\frac{5}{37} + \frac{30}{37}j$

41. $\frac{\sqrt{3}-\sqrt{-6}}{\sqrt{-3}} = \frac{\sqrt{3}-\sqrt{6}j}{\sqrt{3}j} \cdot \frac{-\sqrt{3}j}{-\sqrt{3}j} = \frac{-3j+3\sqrt{2}j^2}{3} = \frac{-3\sqrt{2}-3j}{3} = -\sqrt{2}-j$

42. $\frac{\sqrt{5}+\sqrt{-10}}{\sqrt{-5}} = \frac{\sqrt{5}+\sqrt{10}j}{\sqrt{5}j} \cdot \frac{-\sqrt{5}j}{-\sqrt{5}j} = \frac{-5j-5\sqrt{2}j^2}{5} = \frac{5\sqrt{2}-5j}{5} = \sqrt{2}-j$

43. $\frac{(5+2j)(3-j)}{4+j} = \frac{15-5j+6j-2j^2}{4+j} = \frac{17+j}{4+j} \cdot \frac{4-j}{4-j} = \frac{68-17j+4j-j^2}{16+1} = \frac{69-13j}{17} = \frac{69}{17} - \frac{13}{17}j \approx 4.059 - 0.765j$

44. $\frac{(6-j)(2+3j)}{-1+3j} = \frac{12+18j-2j-3j^2}{-1+3j} = \frac{15+16j}{-1+3j} \cdot \frac{-1-3j}{-1-3j} = \frac{-15-45j-16j-48j^2}{1+9} = \frac{33-61j}{10} = 3.3 - 6.1j$

45. $(1+j)^4 = [(1+j)(1+j)]^2 = (1+2j+j^2)^2 = (2j)^2 = 4j^2 = -4$

46. $(1-j)^4 = [(1-j)^2]^2 = (1-2j+j^2)^2 = (-2j)^2 = 4j^2 = -4$

47. $\left(\frac{1}{2}+\frac{\sqrt{3}}{2}j\right)^2 = \frac{1}{4}+\frac{\sqrt{3}}{2}j+\frac{3}{4}j^2 = -\frac{1}{2}+\frac{\sqrt{3}}{2}j = -0.5 + \frac{j\sqrt{3}}{2}$

48. $\frac{4+j}{(3-2j)+(4-3j)} = \frac{4+j}{7-5j} \cdot \frac{7+5j}{7+5j} = \frac{28+20j+7j+5j^2}{49+25} = \frac{23+27j}{74} = \frac{23}{74} + \frac{27}{74}j \approx 0.3108 + 0.3649j$

49. $(a+bj)+(a-bj) = (a+a)+(b-b)j = 2a+0j = 2a$ which is a real number

50. $(a+bj) - (a-bj) = (a-a) + (b+b)j = 2bj$ which is a pure imaginary number

51. $(a+bj)(a-bj) = a^2 - (bj)^2 = a^2 - b^2j^2 = a^2+b^2$ which is a real number

52. $(9.32-6.12j)+(7.24+4.31j) = 16.56 - 1.81j$ V

53. $(6.21-1.37j)+(4.32-2.84j) = 10.53 - 4.21j$ V

54. $(19.2-3.5j) - (12.4+1.3j) = 6.8 - 4.8j$ V

55. $(7.42 + 1.15j) - (2.34 - 1.73j) = 5.08 + 2.88j$ V

56. $(0.25 + 0.20j) + (0.15 - 0.25j) = (0.40 - 0.05j)$ Ω

57. $(9.13 - 4.27j) - (3.29 - 5.43j) = 5.84 + 1.16j$ Ω

58. $Z = \frac{6(3j)}{6+3j} = \frac{18j}{6+3j} \cdot \frac{6-3j}{6-3j} = \frac{54+108j}{36+9} = \frac{54+108j}{45} =$
$\frac{9(6+12j)}{9 \cdot 5} = \frac{6}{5} + \frac{12}{5}j$ or $1.2 + 2.4j$ Ω

59. $Z = \frac{(20+10j)(10-20j)}{(20+10j)+(10-20j)} = \frac{200-400j+100j-200j^2}{30-10j} =$
$\frac{15000-5000j}{1000} = 15 - 5j$ Ω

60. $V = IZ; V = (12.3+4.6j)(16.4-9.0j) = 201.72 - 110.7j + 75.44j - 41.4j^2 = 243.12 - 35.26j$ V

61. $I = \frac{V}{Z} = \frac{5.2+3j}{4-2j} \cdot \frac{4+2j}{4+2j} = \frac{20.8+10.4j+12j+6j^2}{16+4} =$
$\frac{14.8+22.4j}{20} = 0.74 + 1.12j$ A

62. $Z = \frac{V}{I} = \frac{10.6-6.0j}{4+j} \cdot \frac{4-j}{4-j} = \frac{42.4-10.6j-24j+6j^2}{16+1} =$
$\frac{36.4-34.6j}{17} = 2.1412 - 2.0353j$ Ω

63. See *Computer Programs* in main text.

64. See *Computer Programs* in main text.

65. See *Computer Programs* in main text.

66. The following program will work on a TI-82 graphing calculator.

```
Disp "COMPLEX"
:ClrHome
:Disp "ENTER THE FIRST"
:Disp "NUMBER AS A+BJ"
:Prompt A,B
```

```
:ClrHome
:Disp "ENTER THE SECOND"
:Disp "NUMBER AS C+DJ"
:Prompt C,D
:ClrHome
:Disp "THE SUM IS"
:Disp "     REAL"
:Disp A+C
:Disp "     IMAGINARY"
:Disp B+D
:Disp ""
:Disp "PRESS <ENTER>"
:Pause
:ClrHome
:Disp "THE DIFFERENCE"
:Disp "IS      REAL"
:Disp A-C
:Disp "     IMAGINARY"
:Disp B-D
:Disp ""
:Disp "PRESS <ENTER>"
:Pause
:ClrHome
:Disp "THE PRODUCT IS"
:Disp "     REAL"
:Disp AC-BD
:Disp "     IMAGINARY"
:Disp BC+AD
:Disp ""
:Disp "PRESS <ENTER>"
:Pause
:ClrHome
:Disp "THE QUOTIENT IS"
:Disp "     REAL"
:Disp (AC + BD)/(C² + D²)▶Frac
:Disp "     IMAGINARY"
:Disp (BC − AD)/(C² + D²)▶Frac
```

☰ 14.3 GRAPHING COMPLEX NUMBERS; POLAR FORM OF A COMPLEX NUMBER

1. $4(\cos 30° + j \sin 30°) = 2\sqrt{3} + 2j; a = 4\cos 30° = 2\sqrt{3} \approx 3.4641; b = 4\sin 30° = 2$

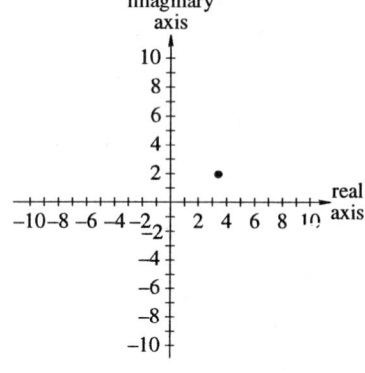

2. $10(\cos 45° + j \sin 45°) = 5\sqrt{2} + 5\sqrt{2}j \approx 7.0711 + 7.0711j$

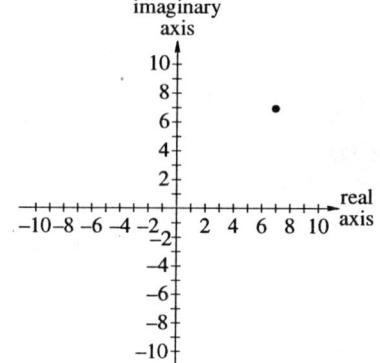

3. $5(\cos 135° + j\sin 135°) = \frac{-5\sqrt{2}}{2} + \frac{5j\sqrt{2}}{2} \approx$
$-3.5355 + 3.5355j$

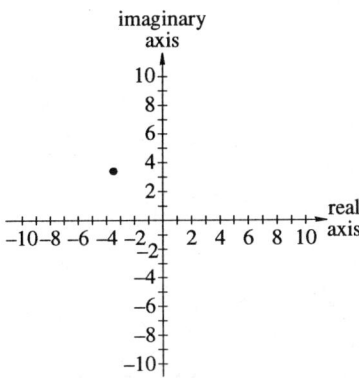

4. $8(\cos 305° + j\sin 305°) \approx 4.5886 - 6.5532j$

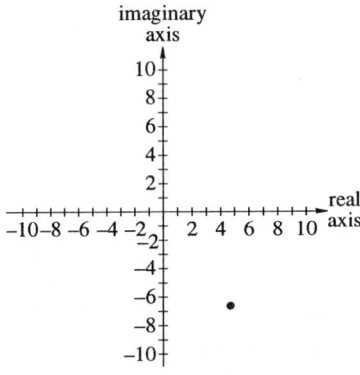

5. $7 \operatorname{cis} 260° = 7\cos 260° + 7j\sin 260° = -1.2155 -$
$6.8937j$

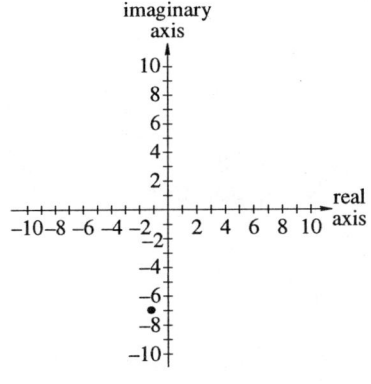

6. $3 \operatorname{cis} 340° = 3\cos 340° + 3j\sin 340° = 2.8191 -$
$1.0261j$

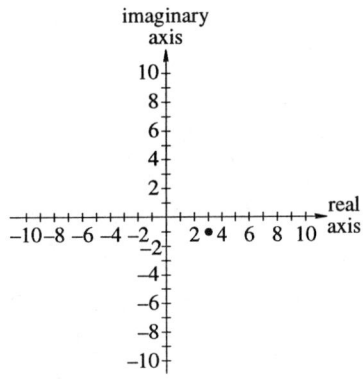

7. $2 \operatorname{cis} 115° = 2\cos 115° + 2j\sin 115° = -0.8452 +$
$1.8126j$

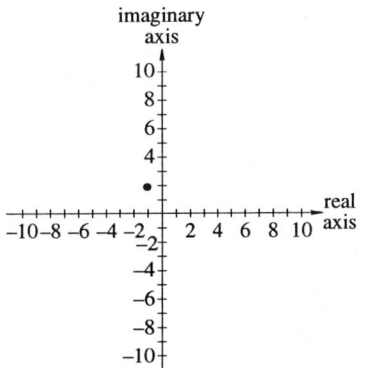

8. $5 \operatorname{cis} 285° = 5\cos 285° + 5j\sin 285° = 1.2941 -$
$4.8296j$

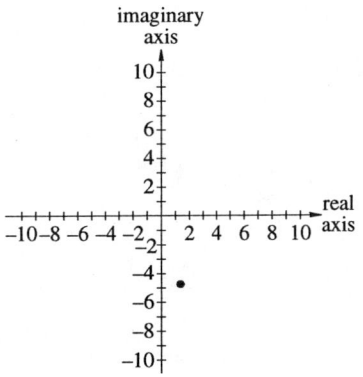

9. $3\underline{/25°} = 3\cos 25° + 3j\sin 25° = 2.7189 + 1.2679j$

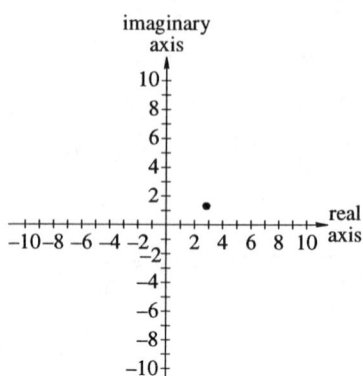

10. $4\underline{/240°} = 4\cos 240° + 4j\sin 240° = -2 - 3.4641j$

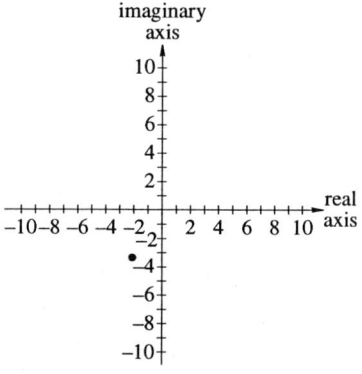

11. $5\underline{/340°} = 5\cos 340° + 5j\sin 340° = 4.6985 - 1.7101j$

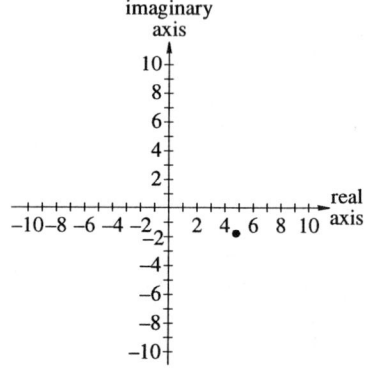

12. $6\underline{/90°} = 6\cos 90° + 6j\sin 90° = 6j$

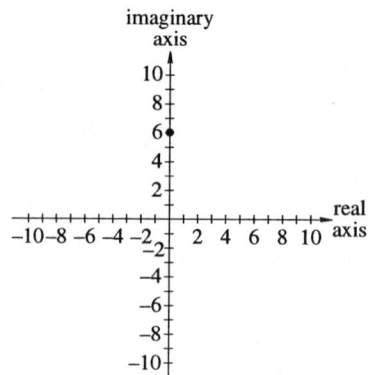

13. $4.5\underline{/245°} = 4.5\cos 245° + 4.5j\sin 245° = -1.9018 - 4.0784j$

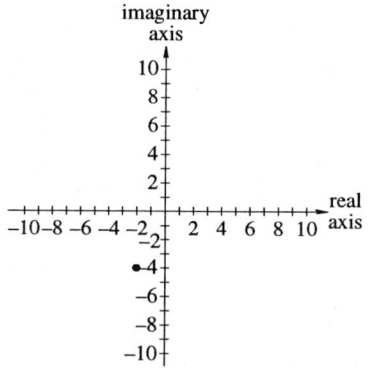

14. $6.8\underline{/10°} = 6.8\cos 10° + 6.8j\sin 10° = 6.6967 + 1.1808j$

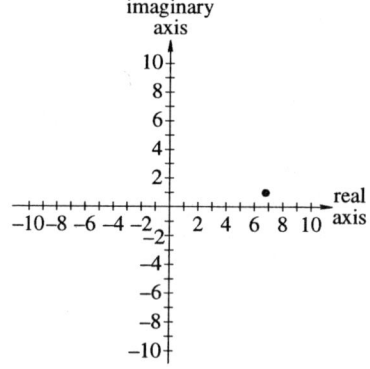

15. $2.5/\underline{180°} = 2.5\cos 180° + 2.5j\sin 180° = -2.5$

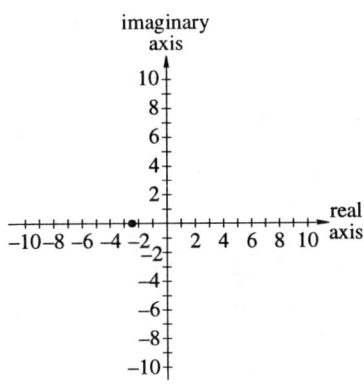

16. $5.9/\underline{270°} = 5.9\cos 270° + 5.9j\sin 270° = -5.9j$

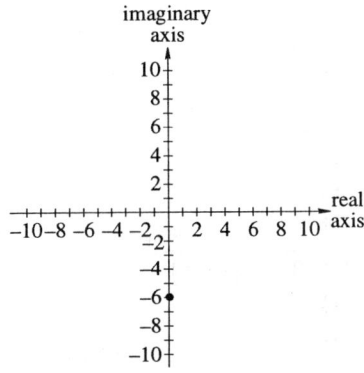

17. $5+5j; r = \sqrt{5^2 + 5^2} = \sqrt{50} = 5\sqrt{2} \approx 7.071; \theta = \tan^{-1}\frac{5}{5} = 45°; 7.071/\underline{45°}$ or $7.071\operatorname{cis}45°$

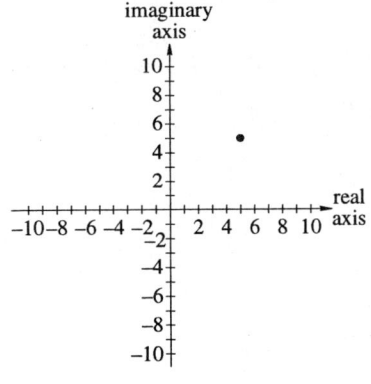

18. $6 + 3j; r = \sqrt{6^2 + 3^2} = 6.708; \theta = \tan^{-1}\frac{3}{6} = 26.6;$
$6.708/\underline{26.6°}$ or $6.708\operatorname{cis}26.6°$

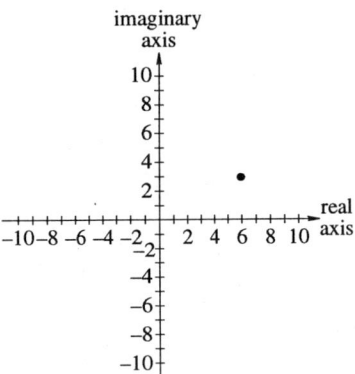

19. $4 - 8j; r = \sqrt{4^2 + (-8)^2} = 8.944; \theta = \tan^{-1}\left(-\frac{8}{4}\right) = -63.4°$ or $296.6°; 8.944/\underline{296.6°}$ or $8.944\operatorname{cis}296.6°$

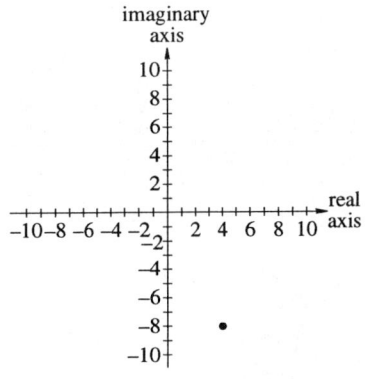

20. $8 - 2j; r = \sqrt{8^2 + (-2)^2} = \sqrt{68} = 8.246; \theta = \tan^{-1}\frac{-2}{8} = -14.0°$ or $346°; 8.246/\underline{346°}$ or $8.246\operatorname{cis}346°$

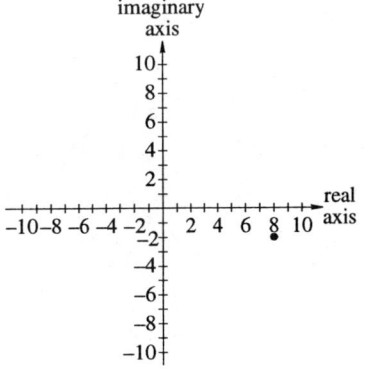

21. $-4 + 7j; r = \sqrt{(-4)^2 + 7^2} = 8.0623; \theta =$
$\tan^{-1} \frac{-7}{4} = 119.7°; 8.0623\underline{/119.7°}$

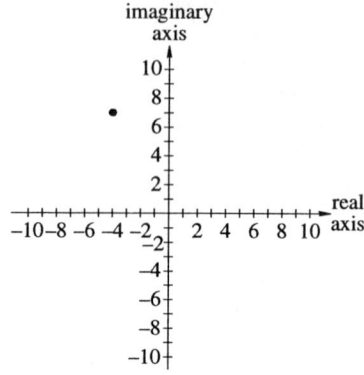

24. $-10 - 2j; r = \sqrt{(-10)^2 + (-2)^2} = 10.1980; \theta =$
$\tan^{-1} \frac{-2}{-10} = 191.3; 10.1980\underline{/191.3°}$

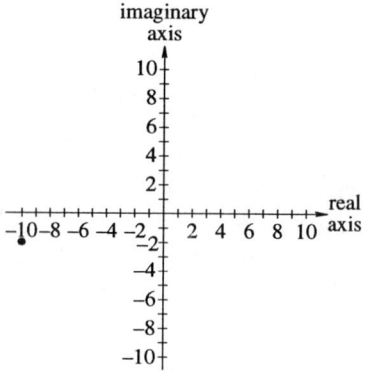

22. $-9 + 3j; r = \sqrt{(-9)^2 + 3^2} = 9.4868; \theta =$
$\tan^{-1} \frac{3}{-9} = 161.6; 9.4868\underline{/161.6°}$

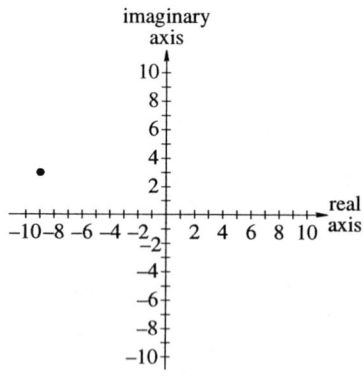

25. $6; r = 6; \theta = 0; 6\underline{/0°}$

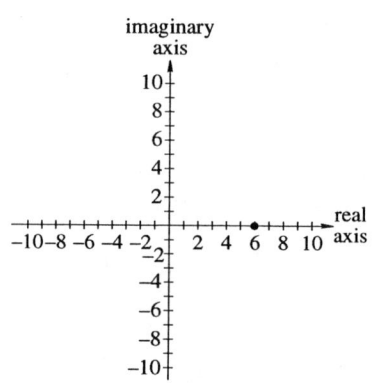

23. $-6 - 2j; r = \sqrt{(-6)^2 + (-2)^2} = 6.3246;$
$\theta = \tan^{-1} \frac{-2}{-6} = 198.4°; 6.3246\underline{/198.4°}$

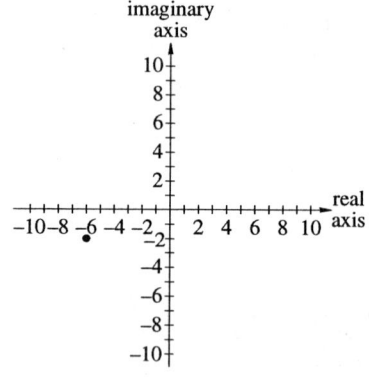

26. $1.2 + 7.3j; r = \sqrt{1.2^2 + 7.3^2} = 7.3980; \theta =$
$\tan^{-1} \frac{7.3}{1.2} = 80.7; 7.3980\underline{/80.7°}$

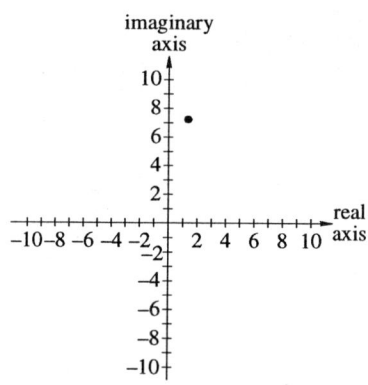

27. $4.2 - 6.3j; r = \sqrt{4.2^2 + (-6.3)^2} = 7.5717; \theta =$
$\tan^{-1} \frac{-6.3}{4.2} = -56.3$ or $303.7; 7.5717 \underline{/303.7°}$

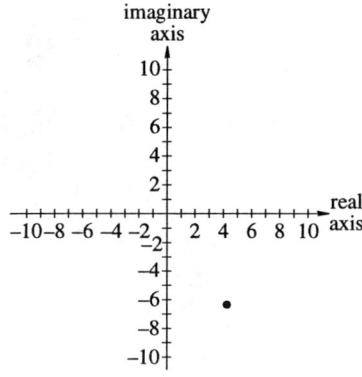

28. $9j; r = 9, \theta = 90°; \underline{/90°}$

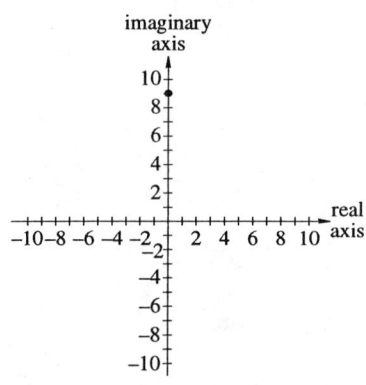

29. $-5.8 + 0.2j; r = \sqrt{(-5.8)^2 + 0.2^2} = 5.8034; \theta =$
$\tan^{-1} \frac{0.2}{-5.8} = 178.0°; 5.8034 \underline{/178.0°}$

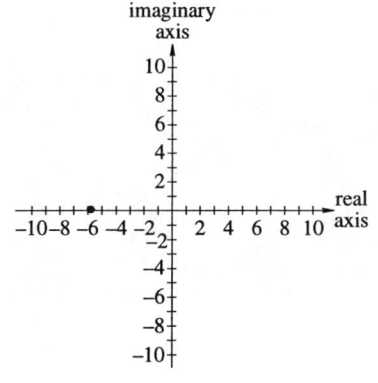

30. $-7j; r = 7; \theta = 270°; 7 \underline{/270°}$

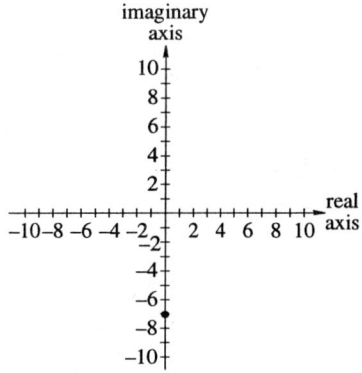

31. $-2.7; r = 2.7; \theta = 180°; 2.7 \underline{/180°}$

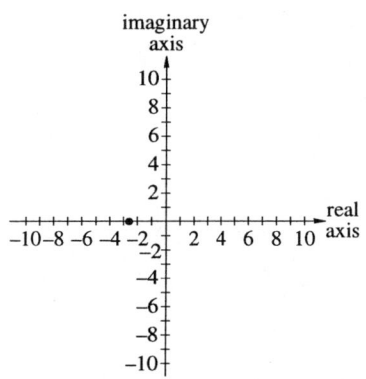

32. $-4.7 - 1.1j; r = \sqrt{(-4.7)^2 + 1.1^2} = 4.8270;$
$\theta = \tan^{-1} \frac{-1.1}{-4.7} = 193.2°; 4.8270 \underline{/193.2°}$

33.

34.

35.

36.

37.

38.

39.

40.

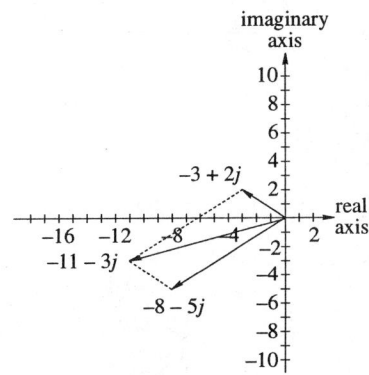

41. First, $r = \sqrt{4.7^2 + (-6.5)^2} = \sqrt{64.34} \approx 8.0$. Then we have $\tan\theta = \frac{-6.5}{4.7}$. Since θ is in the

fourth quadrant, $\theta = \tan^{-1}\frac{-6.5}{4.7} \approx -54.1°$ or $-54.1° + 360° = 305.9°$. Thus, $4.7 - 6.5j$ written in polar form is $8.0\underline{/-54.1°} = 8.0\underline{/305.9°}$A.

42. First, $r = \sqrt{8.2^2 + 5.4^2} = \sqrt{96.4} \approx 9.8$. Then, $\tan\theta = \frac{5.4}{8.2}$. Since θ is in the first quadrant, $\theta = \tan^{-1}\frac{5.4}{8.2} \approx 33.4°$. Thus, $8.2 + 5.4j$ written in polar form is $9.8\underline{/33.4°}$ A.

43. $r = \sqrt{(-110.4)^2 + 46.1^2} = \sqrt{14{,}313.37} \approx 119.6$. $\tan\theta = \frac{46.1}{-110.4}$. Since θ is in the second quadrant, $\theta = \tan^{-1}\frac{46.1}{-110.4} + 180° \approx 157.3°$. Thus, $-110.4 + 46.1j$ V is written in polar form as $119.6\underline{/157.3°}$V.

44. $r = \sqrt{108.5^2 + (-57.6)^2} \approx 122.8$. $\tan\theta = \frac{-57.6}{108.5}$. Since θ is in the fourth quadrant, $\theta = \tan^{-1}\frac{-57.6}{108.5} \approx -28.0° + 360° = 332.0°$. Thus, $108.5 - 57.6j$ V is written in polar form as $122.8\underline{/-28.0°} = 122.8\underline{/332.0°}$ V.

45. $a = r\cos\theta = 2.5\cos(-50°) \approx 1.61$. $b = r\sin\theta = 2.5\sin(-50°) \approx -1.92$. Thus, $2.5\underline{/-50°}$ A is written in rectangular form as $1.61 - 1.92j$ A.

46. $a = r\cos\theta = 8.5\cos\left(\frac{2\pi}{3}\right) = 8.5(-0.5) = -4.25$. $b = r\sin\theta = 2.5\sin\left(\frac{2\pi}{3}\right) \approx 7.30$. Thus, $8.5\underline{/2\pi/3}$ Ω is written in rectangular form as $-4.25 + 7.36j$ Ω.

47. See *Computer Programs* in main text.

☰ 14.4 EXPONENTIAL FORM OF A COMPLEX NUMBER

1. $3(\cos\frac{3\pi}{2} + j\sin\frac{3\pi}{2}); r = 3; \theta = \frac{3\pi}{2}$; Answer: $3e^{3\pi/2 j} \approx 3e^{4.7j}$

2. $7(\cos 1.4 + j\sin 1.4); r = 7, \theta = 1.4$; Answer: $7e^{1.4j} 2e^{1.05j}$

3. $2(\cos 60° + j\sin 60°); r = 2, \theta = 60° = \frac{\pi}{3}$; Answer: $2e^{\frac{\pi}{3}j} \approx 2e^{1.05j}$

4. $11(\cos 320° + j\sin 320°); r = 11, \theta = 320° = 5.59$; Answer: $11e^{5.59j}$

5. $1.3(\cos 5.7 + \sin 5.7) = 1.3e^{5.7j}$

6. $9.5(\cos 2.1 + j\sin 2.1) = 9.5e^{2.1j}$

7. $25° = 0.44; 3.1(\cos 25° + j\sin 25°) = 3.1e^{0.44j}$

8. $195° = 3.40; 10.5(\cos 195° + j\sin 195°) = 10.5e^{3.40j}$

9. $8 + 6j = 10\operatorname{cis} 0.64 = 10e^{0.64j}$

10. $12 - 5j = 13\operatorname{cis} 5.89 = 13e^{5.89j}$

11. $-9 + 12j = 15\operatorname{cis} 2.21 = 15e^{2.21j}$

12. $-8 - 2j = 8.25\operatorname{cis} 3.39 = 8.25e^{3.39j}$

13. $5e^{0.5j} = 5\operatorname{cis} 0.5 = 4.3879 + 2.3971j$

14. $8e^{1.9j} = 8\operatorname{cis} 1.9 = -2.5863 + 7.5704j$

15. $2.3e^{4.2j} = 2.3\operatorname{cis} 4.2 = -1.1276 - 2.0046j$

16. $4.5e^{\frac{7\pi}{6}j} = 4.5\operatorname{cis}\frac{7\pi}{6} = -3.8971 - 2.25j$

17. $2e^{3j} \cdot 6e^{2j} = (2 \cdot 6)e^{3j+2j} = 12e^{5j}$

18. $3e^{j} \cdot 4e^{2j} = (3 \cdot 4)e^{j+2j} = 12e^{3j}$

19. $e^{1.3j} \cdot 2.4e^{4.6j} = (1 \cdot 2.4)e^{(1.3+4.6)j} = 2.4e^{5.9j}$

20. $1.5e^{4.1j} \cdot 0.2e^{1.7j} = (1.5 \cdot 0.2)e^{(4.1+1.7)j} = 0.3e^{5.8j}$

21. $7e^{4.3j} \cdot 4e^{5.7j} = (7\cdot4)e^{(4.3+5.7)j} = 28e^{10j} = 28e^{3.72j}$ (Note: $10 - 2\pi \approx 3.72$)

22. $3.6e^{5.4j} \cdot 2.5e^{6.1j} = 9e^{11.5j} = 9e^{5.22j}$ (Note: $11.5 - 2\pi \approx 5.22$)

23. $8e^{3j} \div 2e^{j} = (8/2)e^{3j-j} = 4e^{2j}$

24. $28e^{5j} \div 7e^{2j} = (28/7)e^{5j-2j} = 4e^{3j}$

25. $17e^{4.3j} \div 4e^{2.8j} = 17/4e^{(4.3-2.8)j} = 4.25e^{1.5j}$

26. $8.5e^{3.4j} \div 2e^{5.3j} = (8.5/2)e^{(3.4-5.3)j} = 4.25e^{-1.9j} = 4.25e^{4.38j}$ (Note: $-1.9 + 2\pi \approx 4.38$)

27. $(3e^{2j})^4 = 3^4 e^{2j\cdot4} = 81e^{8j} \approx 81e^{1.72j}$ (Note: $8 - 2\pi \approx 1.72$.)

28. $(4e^{3j})^5 = 4^5 e^{15j} = 1024e^{15j} \approx 1024e^{2.43j}$ (Note: $15 - 4\pi \approx 2.43$.)

29. $(2.5e^{1.5j})^4 = 2.5^4 e^{1.5j\cdot4} = 39.0625e^{6j}$

30. $(7.2e^{2.3j})^5 = 7.2^5 e^{5\times2.3j} = 19349e^{11.5j}$ $= 19349e^{5.22j}$ or $7.2^5 e^{5.22j}$

31. $(4e^{6j})^{1/2} = 4^{1/2} e^{\frac{6j}{2}} = 2e^{3j}$

32. $(16e^{6j})^{1/4} = 16^{1/4} e^{\frac{6j}{4}} = 2e^{1.5j}$

33. $(6.25e^{4.2j})^{1/2} = 6.25^{1/2} e^{\frac{4.2j}{2}} = 2.5e^{2.1j}$

34. $(1.728e^{2.1j})^{1/3} = 1.728^{1/3} e^{\frac{2.1j}{3}} = 1.2e^{0.7j}$

35. $V = 56.5 + 24.1j = 61.41\operatorname{cis}0.4031 = 61.41e^{0.4031j}$ V

36. $I = 4.90 - 4.11j = 6.40\operatorname{cis}5.59 = 6.40e^{5.59j}$ A; more exactly, $6.40e^{5.5852j}$ A

37. $Z = 135\underline{/-52.5°}\ \Omega = 135e^{5.37j}\ \Omega = 82.2 - 107.1j\ \Omega$

38. $X_L = 40.5\underline{/-\pi/2}\ \Omega = 40.5e^{-\pi/2j}\ \Omega = 40.5e^{-1.5708j}\ \Omega = 40.5e^{4.7124j}\ \Omega = -40.5j\ \Omega$

39. $V = IZ = (12.5e^{-0.7256j})(6.4e^{1.4285j}) = (12.5 \times 6.4)e^{(-0.7256+1.4285)j} = 80e^{0.7029j}$ V

40. $V_L = IX_L = \left(4.24e^{0.5627j}\right)\left(28.5e^{-1.5708j}\right) = (4.24 \times 28.5)e^{(0.5627-1.5708)j} = 120.84e^{-1.0081j}$ V $\approx 120.84e^{5.2751j}$ V

41. $V = IZ \Rightarrow I = \frac{V}{Z} = \frac{115e^{-0.2125j}}{2.5e^{0.5792j}} = \frac{115}{2.5}e^{(-0.2145-0.5792)j} = 46e^{-0.7937j}$ A $\approx 46e^{5.49j}$ A

42. $R = \frac{V_R}{I} = \frac{35.1e^{1.3826j}}{0.78e^{1.3826j}} = 45e^{0j} = 45\ \Omega$

43. $Z = \frac{V}{I} = \frac{122.4e^{0.2551j}}{36e^{-0.8189j}} = \frac{122.4}{36}e^{0.2551j-(-0.8189j)} = \frac{122.4}{36}e^{(0.2551+0.8189)j} = 3.4e^{1.074j}\ \Omega$

44. $X_L = \frac{V_L}{I} = \frac{119.7e^{0.7254j}}{4.2e^{-0.1246j}} = \frac{119.7}{4.2}e^{0.7254j-(-0.1246j)} = \frac{119.7}{4.2}e^{(0.7254+0.1246)j} = 28.4e^{0.8500j}\ \Omega$

≡ 14.5 OPERATIONS IN POLAR FORM; DEMOIVRE'S FORMULA

1. $(3\operatorname{cis}46°) \cdot (5\operatorname{cis}23°) = 3 \cdot 5\operatorname{cis}(46° + 23°)$ $= 15\operatorname{cis}69° = 15(\cos69° + j\sin69°)$

2. $(4\operatorname{cis}135°) \cdot (5\operatorname{cis}63°) = 4 \cdot 5\operatorname{cis}(135° + 63°) = 20\operatorname{cis}198° = 20(\cos198° + j\sin198°)$

3. $(2.5\operatorname{cis}1.4°) \cdot (4\operatorname{cis}2.67) = 2.5 \cdot \operatorname{cis}(1.43 + 2.67) = 10\operatorname{cis}4.1 = 10(\cos4.1° + j\sin4.1°)$

4. $(6.4\operatorname{cis}0.25°) \cdot (3.5\operatorname{cis}1.1) = 6.4 \times 3.5\operatorname{cis}(0.25 + 1.1) = 22.4\operatorname{cis}1.35 = 22.4(\cos1.35° + j\sin1.35°)$

5. $\frac{8\operatorname{cis}85°}{2\operatorname{cis}25°} = \frac{8}{2}\operatorname{cis}(85° - 25°) = 4\operatorname{cis}60° = 4(\cos60° + j\sin60°)$

6. $\frac{6\operatorname{cis}273°}{3\operatorname{cis}114°} = \frac{6}{3}\operatorname{cis}(273° - 114°) = 2\operatorname{cis}159° = 2(\cos159° + j\sin159°)$

7. $\frac{9\underline{/137°}}{2\underline{/26°}} = \frac{9}{2}\underline{/137° - 26°} = 4.5\underline{/111°}$

8. $\frac{8\underline{/3.52°}}{5\underline{/2.14°}} = \frac{18}{5}\underline{/3.52 - 2.14} = 3.6\underline{/1.38°}$

9. $(3\underline{/2.7})(4\underline{/5.3}) = 3\cdot4\underline{/2.7 + 5.3} = 12\underline{/8} = 12\underline{/1.72}$ (Note: $8 - 2\pi \approx 1.72$.)

10. $[3\operatorname{cis}20°]^4 = 3^4\operatorname{cis}(4 \cdot 20) = 81\operatorname{cis}80°$

11. $[5\operatorname{cis}84°]^6 = 5^6\operatorname{cis}(6 \cdot 84) = 5^6\operatorname{cis}(504) = 15,625\operatorname{cis}144°$

12. $[2.5\operatorname{cis}118°]^3 = 2.5^3\operatorname{cis}(3\cdot118) = 15.625\operatorname{cis}354°$

13. $[10.4\operatorname{cis}3.42]^3 = 10.4^3\operatorname{cis}(3 \cdot 3.42)$ $= 1124.864\operatorname{cis}10.26 = 1124.864\operatorname{cis}3.98$

14. $(2\underline{/1.38})^5 = 2^5\underline{/5 \cdot 1.38} = 32\underline{/6.9} = 32\underline{/0.62}$

15. $(3.4\underline{/5.3})^4 = 3.4^4\underline{/4 \cdot 5.3} = 133.6\underline{/21.2} \approx 133.6\underline{/2.35}$ (Note: $21.2 \approx 2.35 + 6\pi$.)

16. $(4.41\underline{/124°})^{1/2} = 4.41^{1/2}\underline{/1/2 \cdot 124°} = 2.1\underline{/62°}$

17. $\left(4e^{2.1j}\right)\left(3e^{1.7j}\right) = 12e^{(2.1+1.7)j} = 12e^{3.8j}$

18. $6e^{1.5} \div 4e^{0.9j} = \frac{6}{4}e^{(1.5-0.9)j} = 1.5e^{0.6j}$

19. $(0.5e^{0.3j})^3 = 0.5^3 e^{3\times 0.3j} = 0.125 e^{0.9j}$

20. $(0.0625 e^{4.2j})^{1/2} = 0.0625^{1/2} e^{\frac{4.2j}{2}} = 0.25 e^{2.1j}$

21. Cube roots of 1. First the polar form of 1 is $1\operatorname{cis}0° = \operatorname{cis}0°$. $w_0 = \sqrt[3]{1}\operatorname{cis}\left(\frac{0}{3}+\frac{360\cdot 0}{3}\right)° = 1\operatorname{cis}0° = 1; w_1 = \sqrt[3]{1}\operatorname{cis}\left(\frac{0}{3}+\frac{360\cdot 1}{3}\right)° = 1\operatorname{cis}120° = -0.5 + 0.866j; w_2 = \sqrt[3]{1}\operatorname{cis}\left(\frac{0}{3}+\frac{360\cdot 2}{3}\right)° = 1\operatorname{cis}240° = -0.5 - 0.866j°$

22. Cube roots of j. First $j = 1\operatorname{cis}90°$. $w_0 = \sqrt[3]{1}\operatorname{cis}\left(\frac{90}{3}+\frac{360\cdot 0}{3}\right)° = 1\operatorname{cis}30° = 0.866 + 0.5j; w_1 = \sqrt[3]{1}\operatorname{cis}\left(\frac{90}{3}+\frac{360\cdot 1}{3}\right)° = 1\operatorname{cis}150° = -0.866 + 0.5j; w_2 = \sqrt[3]{1}\operatorname{cis}\left(30+\frac{360\cdot 2}{3}\right)° = 1\operatorname{cis}270° = -j$

23. We want the cube roots of $-8j$. Now, $-8j = 8\operatorname{cis}270°$. $w_0 = \sqrt[3]{8}\operatorname{cis}\left(\frac{270}{3}+\frac{360\cdot 0}{3}\right)° = 2\operatorname{cis}90° = 2j; w_1 = \sqrt[3]{8}\operatorname{cis}\left(90+\frac{360\cdot 1}{3}\right)° = 2\operatorname{cis}210° = -1.732 - j; w_2 = 2\operatorname{cis}\left(90+\frac{360\cdot 2}{3}\right)° = 2\operatorname{cis}330° = 1.732 - j$

24. Fourth roots of -16; $-16 = 16\operatorname{cis}180°$. $w_0 = \sqrt[4]{16}\operatorname{cis}\left(\frac{180}{4}+\frac{360\cdot 0}{4}\right)° = 2\operatorname{cis}45° = \sqrt{2}+\sqrt{2}j; = 1.414 + 1.414j; w_1 = 2\operatorname{cis}\left(45+\frac{360\cdot 1}{4}\right)° = 2\operatorname{cis}135° = -1.414 + 1.414j; w_2 = 2\operatorname{cis}\left(45+\frac{360\cdot 2}{4}\right)° = 2\operatorname{cis}(255°) = -1.414-1.414j; w_3 = 2\operatorname{cis}\left(45+\frac{360\cdot 3}{4}\right)° = 2\operatorname{cis}(315°) = 1.414 - 1.414j$

25. Fourth roots of $-16j$. $-16j = 16\operatorname{cis}270° w_0 = \sqrt[4]{16}\operatorname{cis}\left(\frac{270}{4}+\frac{360\cdot 0}{4}\right)° = 2\operatorname{cis}67.5° = 0.7654 + 1.8478j; w_1 = 2\operatorname{cis}\left(67.5+\frac{360\cdot 1}{4}\right)° = 2\operatorname{cis}157.5° = -1.8478 + 0.7654j; w_2 = 2\operatorname{cis}\left(67.5+\frac{360\cdot 2}{4}\right)° = 2\operatorname{cis}247.5° = -0.7654 - 1.8478j; w_3 = 2\operatorname{cis}\left(67.5+\frac{360\cdot 3}{4}\right)° = 2\operatorname{cis}337.5° = 1.8478 - 0.7654j$

26. Fourth roots of $1 - j$. $1 - j = \sqrt{2}\operatorname{cis}315° w_0 = \sqrt[4]{\sqrt{2}}\operatorname{cis}\left(\frac{315}{4}+\frac{360\cdot 0}{4}\right)° = 1.0905\operatorname{cis}78.75° = 0.2127 + 1.0696j; w_1 = \sqrt[8]{2}\operatorname{cis}\left(\frac{315}{4}+\frac{360\cdot 1}{4}\right)° = 1.0905\operatorname{cis}168.75; = -1.0696 + 0.2127j; w_2 = \sqrt[8]{2}\operatorname{cis}\left(\frac{315}{4}+\frac{360\cdot 2}{4}\right)° = 1.0905\operatorname{cis}258.75° = -0.2127 - 1.0696j; w_3 = \sqrt[8]{2}\operatorname{cis}\left(\frac{315}{4}+\frac{360\cdot 3}{4}\right)° = 1.0905\operatorname{cis}348.75° = 1.0696 - 0.2127j$

27. Fifth roots of $1 + j$. $1 + j = \sqrt{2}\operatorname{cis}45°$. $w_0 = \sqrt[5]{\sqrt{2}}\operatorname{cis}\left(\frac{45}{5}+\frac{360\cdot 0}{5}\right)° = 1.0718\operatorname{cis}9° = 1.0586 + 0.1677j; w_1 = \sqrt[10]{2}\operatorname{cis}\left(9+\frac{360\cdot 1}{5}\right)° = 1.0718\operatorname{cis}$

$81° = 0.1677 + 1.0586j; w_2 = \sqrt[10]{2}\operatorname{cis}\left(9+\frac{360\cdot 2}{5}\right)° = 1.0718\operatorname{cis}153° = -0.9550 + 0.4866j; w_3 = \sqrt[10]{2}\operatorname{cis}\left(9+\frac{360\cdot 3}{5}\right)° = 1.0718\operatorname{cis}225° = -0.7579 - 0.7579j; w_4 = \sqrt[10]{2}\operatorname{cis}\left(9+\frac{360\cdot 4}{5}\right) = 1.0718\operatorname{cis}297° = 0.4866 - 0.9550j$

28. Sixth roots of $-1-j = \sqrt{2}\operatorname{cis}225°; w_0 = \sqrt[6]{\sqrt{2}}\operatorname{cis}\left(\frac{225}{6}+\frac{360\cdot 0}{6}\right)° = 1.0595\operatorname{cis}37.5° = 0.8405 + 0.6450j; w_1 = \sqrt[12]{2}\operatorname{cis}\left(37.5+\frac{360}{6}\right)° = 1.0595\operatorname{cis}97.5° = -0.1383 + 1.0504j; w_2 = \sqrt[12]{2}\operatorname{cis}\left(37.5=\frac{360\times 2}{6}\right)° = 1.0595\operatorname{cis}157.5° = -0.9788 + 0.4054j; w_3 = \sqrt[12]{2}\operatorname{cis}\left(3.75+\frac{360\times 3}{6}\right)° = 1.0595\operatorname{cis}217.5° = -0.8405 - 0.6450j; w_4 = \sqrt[12]{2}\operatorname{cis}\left(37.5+\frac{360\times 4}{6}\right)° = 1.0595\operatorname{cis}277.5° = 0.1383 - 1.0504j; w_5 = \sqrt[12]{2}\operatorname{cis}(37.5+60\cdot 5)° = 1.0595\operatorname{cis}337.5° = 0.9788 - 0.4054j$

29. $x^3 = -j$: the solutions are the three cube roots of $-j = 1\operatorname{cis}270°$. $w_0 = \sqrt[3]{1}\operatorname{cis}(270/3)° = 1\operatorname{cis}90° = j; w_1 = 1\operatorname{cis}(90+360/3)° = 1\operatorname{cis}210° = -0.8660 - 0.5j; w_2 = 1\operatorname{cis}(90+120\cdot 2)° = 1\operatorname{cis}330° = 0.8660 - 0.5j$

30. $x^3 = 125j$: the solutions are the three cube roots of $125j = 125\operatorname{cis}90°$. $w_0 = \sqrt[3]{125}\operatorname{cis}(90/3)° = 5\operatorname{cis}30° = 4.3301 + 2.5j; w_1 = 5\operatorname{cis}(30+360/3)° = 5\operatorname{cis}150° = -4.3301 + 2.5j; w_2 = 5\operatorname{cis}(30+120\times 2)° = 5\operatorname{cis}270° = -5j$

31. The solution of $x^6 - 64j = 0$ are the six sixth roots of $64j = 64\operatorname{cis}90°$. $w_0 = \sqrt[6]{64}\operatorname{cis}(90/6)° = 2\operatorname{cis}15° = 1.9319 + 0.5176j; w_1 = 2\operatorname{cis}(15+\frac{360}{6})° = 2\operatorname{cis}75° = 0.5176 + 1.9319j; w_2 = 2\operatorname{cis}(15+60\times 2)° = 2\operatorname{cis}135° = -1.4142 + 1.4142j; w_3 = 2\operatorname{cis}(15+60\times 3)° = 2\operatorname{cis}195° = -1.9319 - 0.5176j; w_4 = 2\operatorname{cis}(15+60\times 4)° = 2\operatorname{cis}255° = -0.5176 - 1.9319j; w_5 = 2\operatorname{cis}(15+60\times 5)° 2\operatorname{cis}315° = 1.412 - 1.4142j$

32. The solutions of $x^6 - 1 = j$ are the six sixth roots of $1 + j = \sqrt{2}\operatorname{cis}45°$. $w_0 = \sqrt[6]{\sqrt{2}}\operatorname{cis}(45/6)° = 1.0595\operatorname{cis}7.5° = 1.0504 + 0.1383j; w_1 = 1.0595\operatorname{cis}(7.5+\frac{360}{6})° = 1.0595\operatorname{cis}67.5° = 0.4054 + 0.9788j; w_2 = 1.0595\operatorname{cis}(7.5+60\times 2)° = 1.0595\operatorname{cis}127.5° = -0.6450 + 0.8405j; w_3 = 1.0595\operatorname{cis}(7.5+60\times 3)° = 1.0595\operatorname{cis}187.5° = -1.0504 - 0.1383j; w_4 = \sqrt[12]{2}\operatorname{cis}(7.5+60\times 4)° = 1.0595\operatorname{cis}247.5° = -0.4054 - 0.9788j; w_5 = \sqrt[12]{2}\operatorname{cis}(7.5+60\times 5)° = 1.0595\operatorname{cis}307.5° = 0.6450 - 0.8405j$

33. $V = IZ; I = 12\underline{/-23°}, Z = 9\underline{/42°}; V = (12\underline{/-23°})(9\underline{/42°}) = (12 \times 9)\underline{/-23° + 42°} = 108\underline{/19°}$

34. $V = IZ; Z = \frac{V}{I} = \frac{20\underline{/30°}}{5\underline{/40°}} = \frac{20}{5}\underline{/30° - 40°} = 4\underline{/-10°}$ or $4\underline{/350°}$

35. We first convert Z_T to polar form. $Z_T = 4\underline{/-90°} + 3\underline{/0°} = -4j + 3 = 5\underline{/-53.13°}$ Ω. Then,

$$V_x = \frac{Z_x E}{Z_T} = \frac{(4\underline{/-90°})(100\underline{/0°})}{5\underline{/-53.13°}}$$
$$= \frac{4 \cdot 100}{5}\underline{/-90° + 0° - (-53.13°)}$$
$$= 80\underline{/-36.87°} \text{ V}$$

36. We first convert Z_T to polar form. $Z_T = 6\underline{/0°} + 9\underline{/90°} + 17\underline{/-90°} = 6 + 9j - 17j = 6 - 8j = 10\underline{/-53.13°}$ Ω.

$$V_x = \frac{Z_x E}{Z_T} = \frac{(6\underline{/0°})(50\underline{/30°})}{10\underline{/-53.13°}}$$
$$= \frac{6 \cdot 50}{10}\underline{/0° + 30° - (-53.13°)}$$
$$= 30\underline{/-83.13°} \text{ V}$$

37. First, convert each of the expressions to polar form: $1 + j = \sqrt{2}\underline{/45°}; 1 - j = \sqrt{2}\underline{/-45°};$ and $3 - 4j = 5\underline{/-53.13°}$. Substituting these polar expressions in the given equation, we obtain

$$Z = \frac{(1 + j)^2(1 - j)^2}{(3 - 4j)^2}$$
$$= \frac{(\sqrt{2}\underline{/45°})^2(\sqrt{2}\underline{/-45°})^2}{(5\underline{/-53.13°})^2}$$
$$= \frac{(2\underline{/90°})(2\underline{/-90°})}{25\underline{/-106.26°}}$$
$$= \frac{4}{25}\underline{/106.26°} = 0.16\underline{/106.26°} \text{ Ω}$$

If we change this to rectangular form, we see that

$Z = 0.16\underline{/106.26°} = 0.16\cos 106.26° + 0.16j\sin 106.26° = -0.0448 + 0.1536j$ Ω.

38. **(a)** $Z = 12 - 5j = 13\underline{/-22.62°}$. Thus, $Y = Z^{-1} = \frac{1}{13\underline{/-22.62°}} = \frac{1}{13}\underline{/22.62°} \approx 0.0769\underline{/22.62°}$ S

(b) $Y = Z^{-1} = \frac{1}{12-5j} = \frac{1}{12-5j}\frac{12+5j}{12+5j} = \frac{12+5j}{169} = \frac{12}{169} + \frac{5}{169}j \approx 0.0710 + 0.0296j$ S

39. **(a)** $Y = Z^{-1} = \frac{1}{Z} = \frac{1}{4.68\underline{/20.56°}} = \frac{1}{4.68}\underline{/-20.56°} \approx 0.2137\underline{/-20.56°}$ S

(b) Converting $Y = Z^{-1} \approx 0.2137\underline{/-20.56°}$ S to rectangular form, we get $a = 0.2137\cos -20.56° \approx 0.200$ and $b = 0.2137\sin -20.56° \approx -0.075$. Thus, $Y = 0.200 - 0.075j$ S.

40. **(a)** Since $Y = Z^{-1}$, then $Z = Y^{-1}$. Converting Y to polar form, we get $Y = 0.087 - 0.034j \approx 0.0934\underline{/-21.35°}$ S. Thus, $Z = \frac{1}{Y} = \frac{1}{0.0934\underline{/-21.35°}}$

$= \frac{1}{0.0934}\underline{/21.35°} \approx 10.71\underline{/21.35°}$ Ω.

(b) Here we have $a = 10.71\cos 21.35° \approx 9.975$ and $b = 10.71\sin 21.35° \approx 3.899$. Thus, $Z = 10.71\underline{/21.35°} = 9.975 + 3.899j$ Ω.

41. See *Computer Programs* in main text.

42. See *Computer Programs* in main text.

43. See *Computer Programs* in main text.

≡ 14.6 COMPLEX NUMBERS IN AC CIRCUITS

1. **(a)** in Series
$Z = Z_1 + Z_2 = (2 + 3j) + (1 - 5j) = 3 - 2j;$
(b) in Parallel $Z = \frac{Z_1 Z_2}{Z_1 + Z_2} = \frac{(2+3j)(1-5j)}{3-2j} =$
$\frac{2-7j-15j^2}{3-2j} = \frac{17-7j}{3-2j} \cdot \frac{3+2j}{3+2j} = \frac{51+34j-21j-14j^2}{9+4} =$
$\frac{65+13j}{13} = 5 + j$

2. **(a)** in Series $Z = Z_1 + Z_2 = (4-7j)+(-3+4j) = 1 - 3j;$
(b) in Parallel $Z = \frac{Z_1 Z_2}{Z_1 + Z_2} =$

$\frac{(4-7j)(-3+4j)}{1-3j} = \frac{-12+16j+21j-28j^2}{1-3j} = \frac{16+37j}{1-3j} \cdot \frac{1+3j}{1+3j} =$
$\frac{16+48j+37j+111j^2}{1+9} = \frac{-95+85j}{10} = -9.5 + 8.5j$

3. **(a)** in Series $Z = Z_1 + Z_2 = (1-j)+3j = 1+2j;$
(b) in Parallel
$Z = \frac{Z_1 Z_2}{Z_1 + Z_2} = \frac{(1-j)(3j)}{1+2j} = \frac{3+3j}{1+2j} \cdot \frac{1-2j}{1-2j} =$
$\frac{3-6j+3j-6j^2}{1+4} = \frac{9-3j}{5} = 1.8 - 0.6j$

4. **(a)** in Series
$Z = Z_1 + Z_2 = (2 + j) + (4 - 3j) = 6 - 2j;$

(b) in Parallel

$Z = \frac{Z_1 Z_2}{Z_1 + Z_2} = \frac{(2+j)(4-3j)}{6-2j} = \frac{8-6j+4j-3j^2}{6-2j} = \frac{11-2j}{6-2j} \cdot$

$\frac{6+2j}{6+2j} = \frac{66+22j-12j-4j^2}{36+4} = \frac{70+10j}{40} = \frac{7}{4} + \frac{1}{4}j = $

$1.75 + 0.25j$

5. $Z_1 = 2.19\underline{/18.4°} = 2.0780 + 0.6913j;$
$Z_2 = 516\underline{/67.3°} = 1.9913 + 4.7603j;$
(a) $Z = Z_1 + Z_2 = 4.0693 + 5.4516j = $
$6.803\underline{/53.26°};$

(b) $Z = \frac{Z_1 Z_2}{Z_1 + Z_2} = \frac{(2.19\underline{/18.4°})(5.16\underline{/67.3°})}{6.803\underline{/53.26°}} = $

$\frac{2.19 \cdot 5.16}{6.803}\underline{/18.4 + 67.3 - 53.26} = 1.6611\underline{/32.44°}$

6. $Z_1 = 2\sqrt{3}\underline{/30°} = 3 + \sqrt{3}j;$
$Z_2 = 2\underline{/120°} = -1 + \sqrt{3}j;$
(a) $Z = Z_1 + Z_2 = 2 + 2\sqrt{3}j = 4\underline{/60°};$

(b) $Z = \frac{Z_1 Z_2}{Z_1 + Z_2} = \frac{(2\sqrt{3}\underline{/30°})(2\underline{/120°})}{4\underline{/60°}} = $

$\frac{2 \cdot 2\sqrt{3}}{4}\underline{/30° + 120° - 60°} = \sqrt{3}\underline{/90°} = \sqrt{3}j$ or
$1.732j$

7. $Z_1 = 3\sqrt{5}\underline{/\pi/7} = 6.0439 + 2.9106j;$
$Z_2 = 1.5\underline{/0.45} = 1.3507 + 0.6524j;$
(a) $Z = Z_1 + Z_2 = 7.3946 + 3.5630j = $
$8.2082\underline{/0.449};$

(b) $Z = \frac{Z_1 Z_2}{Z_1 + Z_2} = \frac{(3\sqrt{5}\underline{/\pi/7})(1.5\underline{/0.45})}{8.2082\underline{/0.449}} = $

$\left(\frac{3\sqrt{5} \cdot 1.5}{8.2082}\right)\underline{/\pi/7 + 0.45 - 0.449} = 1.226\underline{/0.45}$

8. $Z_1 = 2.57\underline{/0.25} = 2.4901 + 0.6358j;$
$Z_2 = 1.63\underline{/1.38} = 0.3091 + 1.6004j;$
(a) $Z = Z_1 + Z_2 = 2.7992 + 2.2362j = 3.583\underline{/0.67};$

(b) $Z = \frac{Z_1 Z_2}{Z_1 + Z_2} = \frac{(2.57\underline{/0.25})(1.63\underline{/1.38})}{3.583\underline{/0.674}} = $

$\frac{2.57 \times 1.63}{3.583}\underline{/0.25 + 1.38 - 0.67} = 1.17\underline{/0.96}$

9. **(a)** $Z = Z_1 + Z_2 + Z_3 = $
$(4 + 3j) + (3 - 2j) + (5 + 4j) = 12 + 5j;$
(b) $Z = \frac{Z_1 Z_2 Z_3}{Z_1 Z_2 + Z_1 Z_3 + Z_2 Z_3} = $

$\frac{(4+3j)(3-2j)(5+4j)}{(4+3j)(3-2j)+(4+3j)(5+4j)+(3-2j)(5+4j)} = $

$\frac{(12-8j+9j-6j^2)(5+4j)}{(12-8j+9j-6j^2)+(20+16j+15j+12j^2)+(15+12j-10j-8j^2)} = $

$\frac{(18+j)(5+4j)}{(18+j)+(8+31j)+(23+2j)} = \frac{86+77j}{49+34j} \cdot \frac{49-34j}{49-34j} = $

$\frac{6832+849j}{2401+1156} = 1.9207 + 0.2387j$

10. **(a)** $Z = Z_1 + Z_2 + Z_3 = $
$(3 - 4j) + (1 + 5j) + (-2j) = 4 - j;$
(b) $Z = \frac{Z_1 Z_2 Z_3}{Z_1 Z_2 + Z_1 Z_3 + Z_2 Z_3} = $

$\frac{(3-4j)(1+5j)(-2j)}{(3-4j)(1+5j)+(3-4j)(-2j)+(1+5j)(-2j)} = $

$= \frac{(23+11j)(-2j)}{(23+11j)+(-8-6j)+(10-2j)} = $

$\frac{22-46j}{25+3j} \cdot \frac{25-3j}{25-3j} = \frac{412-1216j}{625+9} = 0.6498 - 1.9180j$

11. **(a)** $Z_1 = 1.64\underline{/38.2°} = 1.2888 + 1.0142j;$
$Z_2 = 2.35\underline{/43.7°} = 1.6990 + 1.6236j;$
$Z_3 = 4.67\underline{/-39.6°} = 3.5983 - 2.9768j; Z = $
$Z_1 + Z_2 + Z_3 = 6.5861 - 0.3390j = 6.595\underline{/-2.95°};$
(b) $Z_1 Z_2 = (1.64\underline{/38.2})(2.35\underline{/43.7°}) = $
$3.854\underline{/81.9°} = 0.5430 + 3.8156j;$
$Z_1 Z_3 = (1.64\underline{/38.2°})(4.67\underline{/-39.6°}) = $
$7.6588\underline{/-1.4°} = 7.6565 - 0.1871j;$
$Z_2 Z_3 = (2.35\underline{/43.7°})(4.67\underline{/-39.6°}) = $
$10.9745\underline{/4.1} = 10.9464 + 0.7846j;$
$Z_1 Z_2 + Z_1 Z_3 + Z_2 Z_3 = 19.1459 + 4.4131j = $
$19.6479\underline{/12.98°}; Z = \frac{Z_1 Z_2 Z_3}{Z_1 Z_2 + Z_1 Z_3 + Z_2 Z_3} = $
$\frac{(1.64\underline{/38.2°})(2.35\underline{/43.7°})(4.67\underline{/-39.6°})}{19.648\underline{/12.98°}} = 0.916\underline{/29.32°}$

12. **(a)** $Z_1 = 0.15\underline{/0.95} = 0.0873 + 0.1220j;$
$Z_2 = 2.17\underline{/1.39} = 0.3902 + 2.1346j;$
$Z_3 = 1.10\underline{/0.40} = 1.0132 + 0.4284j;$
$Z = Z_1 + Z_2 + Z_3 = 1.4907 + 2.6850j = $
$3.071\underline{/1.064};$
(b) $Z_1 Z_2 = (0.15\underline{/0.95})(2.17\underline{/1.39}) = $
$0.3255\underline{/2.34} = -0.2264 + 0.2339j; Z_1 Z_3 = $
$(0.15\underline{/0.95})(1.10\underline{/0.40}) = 0.165\underline{/1.35} = 0.0361 + $
$0.1610j; Z_2 Z_3 = (2.17\underline{/1.39})(1.10\underline{/0.40}) = $
$2.387\underline{/1.79} = -0.5191 + 2.3299; Z_1 Z_2 + Z_1 Z_3 + $
$Z_2 Z_3 = -0.7094 + 2.7248j = 2.8156\underline{/1.825}; Z = $
$\frac{Z_1 Z_2 Z_3}{Z_1 Z_2 + Z_1 Z_3 + Z_2 Z_3} = \frac{(0.15\underline{/0.95})(2.17\underline{/1.39})(1.10\underline{/0.40})}{2.8156\underline{/1.825}} = $
$0.127\underline{/0.915}$

13. $V = IZ = (4 - 3j)(8 - 15j) = 32 - 60j - 24j$
$+ 45j^2 = -13 - 84j$ V

14. $Z = \frac{V}{I} = \frac{5+5j}{4+3j} \cdot \frac{4-3j}{4-3j} = \frac{20-15j+20j-15j^2}{16+9} = \frac{35+5j}{25}$
$= \frac{7}{5} + \frac{1}{5}j \, \Omega = 1.4 + 0.2j \, \Omega$

15. $V = IZ = (1 - j)(1 + j) = 2$ V

16. $I = \frac{V}{Z} = \frac{3+4j}{5-12j} \cdot \frac{5+12j}{5+12j} = \frac{15+36j+20j+48j^2}{25+144} = $
$\frac{-33+56j}{169} = 0.1953 + 0.331j$ A

17. $Z = \frac{V}{I} = \frac{7\underline{/36.3°}}{2.5\underline{/12.6°}} = 2.8\underline{/23.7°} \, \Omega$

18. $I = \frac{V}{Z} = \frac{3\underline{/1.37}}{4\underline{/0.16}} = 0.75\underline{/1.21}$ A

19. $X_L = 2\pi f L = 2\pi 60 \cdot 0.2 = 75.40 \, \Omega;$
$X_C = \frac{1}{2\pi f C} = \frac{1}{2\pi 60 \cdot 40 \times 10^{-6}} = 66.31 \, \Omega;$
$X = X_L - X_C = 9.09 \, \Omega;$
$R + Xj = 38 + 9.09j = 39.07\underline{/13.45°}; Z = $
$39.07 \, \Omega, \phi = 13.45°$

20. $X_L = 2\pi f L = 2\pi 60 \cdot 0.15 = 56.55 \ \Omega;$
$X_C = \frac{1}{2\pi f C} = \frac{1}{2\pi 60 \cdot 80 \times 10^{-6}} = 33.16 \ \Omega;$
$X = X_L - X_C = 23.39 \ \Omega;$
$R + Xj = 35 + 23.39j = 42.10\underline{/33.75°};$
$Z = 42.10 \ \Omega, \phi = 33.75°$

21. $X_L = 2\pi f L = 2\pi \times 60 \times 0.4 = 150.80 \ \Omega;$
$X_C = \frac{1}{2\pi f C} = \frac{1}{2\pi 60 \cdot 60 \times 10^{-6}} = 44.21 \ \Omega;$
$X = X_L - X_C = 106.59 \ \Omega;$
$R + Xj = 20 + 106.59j = 108.45\underline{/79.37°};$
$Z = 108.45 \ \Omega, \phi = 79.37°$

22. $X_L = \omega L = 80 \times 0.3 = 24 \ \Omega;$
$X_C = \frac{1}{\omega C} = \frac{1}{80 \times 250 \times 10^{-6}} = 50 \ \Omega;$
$X = X_L - X_C = -26 \ \Omega;$
$R + Xj = 12 - 26j = 28.64\underline{/-65.22°};$
$Z = 28.64 \ \Omega, \phi = -65.22°$

23. $X_L = \omega L = 50 \times 0.25 = 12.5 \ \Omega; \ X_C = \frac{1}{\omega C} =$
$\frac{1}{50 \times 200 \times 10^{-6}} = 100 \ \Omega; \ X = X_L - X_C = -87.5 \ \Omega;$
$R + Xj = 28 - 87.5j = 91.87\underline{/-72.26°}; \ Z =$
$91.87 \ \Omega, \phi = -72.26°$

24. $X_L = \omega L = 1000 \times 3 = 3000 \ \Omega; \ X_C = \frac{1}{\omega C} =$
$\frac{1}{1000 \times 0.5 \times 10^{-6}} = 2000 \ \Omega; \ X = X_L - X_C = 1000 \ \Omega;$
$R + Xj = 2000 + 1000j = 2236.07\underline{/26.57°};$
$Z = 2236.07 \ \Omega, \phi = -26.57°$

25. $X = X_L - X_C = 60 - 40 = 20 \ \Omega;$
$R + Xj = 75 + 20j = 77.62\underline{/14.93°}; \ Z = 77.62 \ \Omega,$
$\phi = 14.93°; V = IZ = 77.62 \times 3.50 = 271.67 \text{ V}$

26. $X = X_L - X_C = 30 - 60 = -30 \ \Omega;$

27. $X = X_L - X_C = 6.0 - 5.0 = 1.0 \ \Omega;$
$R + Xj = 3.0 + 1.0j = 3.162\underline{/18.43°}; Z = 3.162 \ \Omega,$
$\phi = 18.43°; V = IZ = 2.85 \times 3.162 = 9.01 \text{ V}$

28. $X = X_L - X_C = 11.4 - 2.4 = 9 \ \Omega; R + X_j =$
$12.0 + 9.0j = 15\underline{/36.87°}; Z = 15 \ \Omega, \phi = 36.87°;$
$V = IZ = 0.60 \times 15 = 9V$

29. $I_2 = I_1 + I_3 \Rightarrow I_3 = I_2 - I_1 = (9 - 7j) - (7 + 2j) = 2 - 9j$

30. $\omega L = \frac{1}{\omega C}; 100 \times 0.5 = \frac{1}{100 \times C};$
$C = \frac{1}{100 \times 100 \times .5} = 2 \times 10^{-4} = 200\mu F$

31. $\frac{1}{2\pi f C} = 2\pi f L \Rightarrow 1 = 4\pi^2 f^2 LC; f^2 = \frac{1}{4\pi^2 LC}; f = $
$\frac{1}{2\pi\sqrt{LC}}; f = \frac{1}{2\pi\sqrt{2.5 \times 20 \times 10^{-6}}} = 22.51 \text{ Hz}$

32. $Y = \frac{1}{Z} = \frac{1}{3-2j} = \frac{1}{3-2j} \cdot \frac{3+2j}{3+2j} = \frac{3+2j}{9+4} = \frac{3}{13} + \frac{2}{13}j = 0.2308 + 0.1538j$

33. $Z = \frac{1}{Y} = \frac{1}{4+3j} = \frac{1}{4+3j} \cdot \frac{4-3j}{4-3j} = \frac{4-3j}{25} = 0.16 - 0.12j$

34. $B = \frac{X}{Z^2}j = \frac{4}{(8+7j)^2}j = \frac{4}{15+112j}j;$
$\frac{4j}{15+112j} \cdot \frac{15-112j}{15-112j} = \frac{448+60j}{225+12544} = \frac{448+60j}{12769} = 0.0351 + 0.0047j$

35. See *Computer Programs* in main text.

36. See *Computer Programs* in main text.

≡ CHAPTER 14 REVIEW

1. $\sqrt{-49} = \sqrt{49}\sqrt{-1} = 7j$

2. $\sqrt{-36} = \sqrt{36}\sqrt{-1} = 6j$

3. $\sqrt{-54} = \sqrt{9}\sqrt{6}\sqrt{-1} = 3\sqrt{6}j \text{ or } 3j\sqrt{6}$

4. $(2j^3)^3 = 2^3 j^9 = 8j \text{ (Note: } j^8 = 1)$

5. $\sqrt{-2}\sqrt{-18} = (\sqrt{2}j)(3j\sqrt{2}) = 6j^2 = -6$

6. $\sqrt{-9}\sqrt{-27} = (3j)(3j\sqrt{3}) = 9j^2\sqrt{3} = -9\sqrt{3}$

7. $(2 - j) + (7 - 2j) = 9 - 3j$

8. $(9 + j)(4 + 7j) = 36 + 63j + 4j + 7j^2 = 29 + 67j$

9. $(5 + j) - (6 - 3j) = -1 + 4j$

10. $\frac{1}{11-j} = \frac{1}{11-j} \cdot \frac{11+j}{11+j} = \frac{11+j}{11^2+1^2} = \frac{11}{122} + \frac{j}{122} = 0.0902 + 0.0082j$

11. $(6 + 2j)(-5 + 3j) = -30 + 18j - 10j + 6j^2 = -36 + 8j$

12. $\frac{2-5j}{6+3j} \cdot \frac{6-3j}{6-3j} = \frac{12-6j-30j+15j^2}{36+9} = \frac{-3-36j}{45} = -0.0667 - 0.8j$

13. $(4 - 3j)(4 + 3j) = 4^2 + 3^2 = 16 + 9 = 25$

14. $\frac{-4}{\sqrt{3}+2j} \cdot \frac{\sqrt{3}-2j}{\sqrt{3}-2j} = \frac{-4\sqrt{3}+8j}{3+4} = \frac{-4\sqrt{3}}{7} + \frac{8}{7}j = -0.9897 + 1.1429j$

15. $9 - 6j; r = \sqrt{9^2 + 6^2} = 10.82; \tan\theta = \frac{-6}{9}; \theta = -33.69$ or 326.31; Answer: $10.82\underline{/326.31°}$

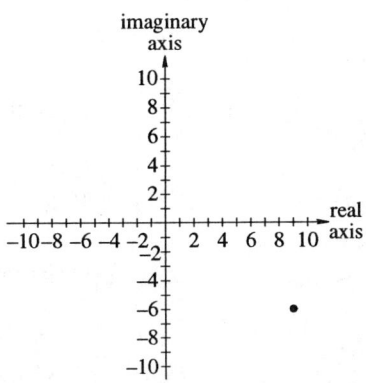

16. $-8 + 2j; r = \sqrt{8^2 + 2^2} = \sqrt{68} = 8.246; \tan\theta = \frac{2}{-8} = 165.96°$; Answer: $8.246\underline{/165.96°}$

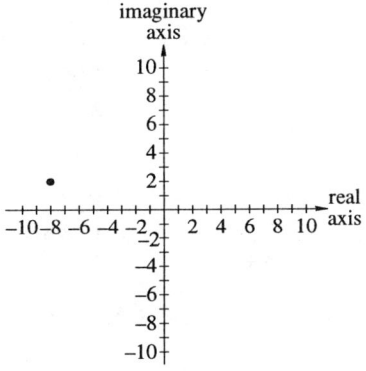

17. $4 - 4j = 4\sqrt{2}\underline{/315°}$ or $5.657\underline{/315°}$

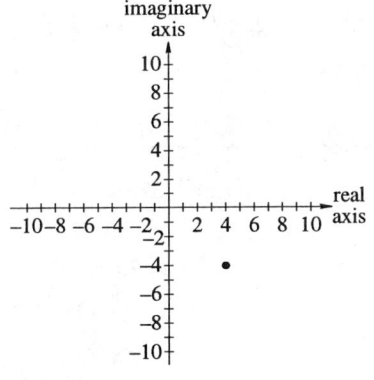

18. $4 \operatorname{cis} 60° = 4\cos 60° + 4j\sin 60° = 2 + 3.464j$

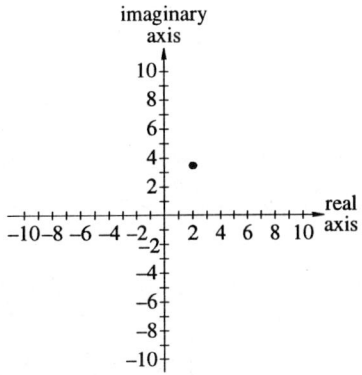

19. $6.5\underline{/2.3} = 6.5\cos 2.3 + 6.5j\sin 2.3 = -4.331 + 4.847j$

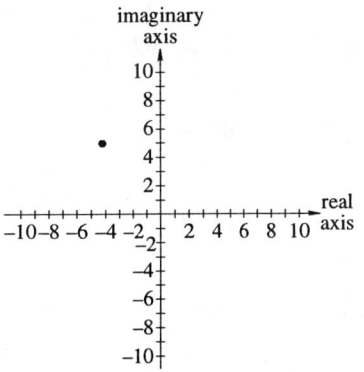

20. $10\underline{/20°} = 10\cos 20° + 10j\sin 20° = 9.397 + 3.420j$

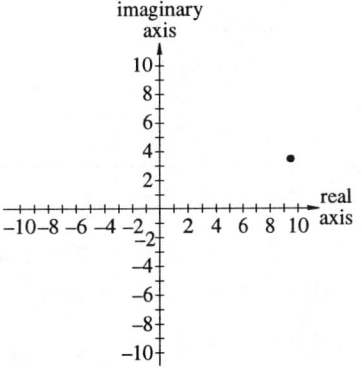

21. $(2 \operatorname{cis} 30°)(5 \operatorname{cis} 150°) = 10 \operatorname{cis} 180° = -10$

22. $\frac{3 \operatorname{cis} 20°}{6 \operatorname{cis} 80°} = 0.5 \operatorname{cis}(-60°) = 0.25 - 0.433j$

23. $(3 \operatorname{cis} \frac{5\pi}{4})^{14} = 3^{14} \operatorname{cis} \frac{14 \cdot 5\pi}{4} = 4,782,969 \operatorname{cis} \frac{35\pi}{2} = -4,782,969j$

24. $(324 \operatorname{cis} 225°)^{1/5} = \sqrt[5]{324} \operatorname{cis} \frac{225}{5} = 3.178 \operatorname{cis} 45° = 2.247 + 2.247j$

25. $(3\underline{/\pi/4})(9\underline{/2\pi/3}) = 3 \cdot 9 \underline{/\pi/4 + 2\pi 3} = 27\underline{/11\pi/12} = -26.080 + 6.988j$

26. $(44\underline{/125°}) \div (4\underline{/97°}) = \frac{44}{4}\underline{/125 - 97} = 11\underline{/28} = 9.7124 + 5.1642j$

27. $(2\underline{/\pi/6})^{12} = 2^{12}\underline{/\frac{12\pi}{6}} = 4096\underline{/2\pi} = 4096$

28. $(2048\underline{/330°})^{1/11} = \sqrt[11]{2048}\underline{/300/11} = 2\underline{/30} = 1.732 + j$

29. Cube roots of $-j = 1 \operatorname{cis} 270°$; $w_0 = \sqrt[3]{1} \operatorname{cis} \frac{270}{3} = 1 \operatorname{cis} 90° = j$; $w_1 = 1 \operatorname{cis}(90 + \frac{360}{3}) = 1 \operatorname{cis} 210° = -0.866 - 0.5j$; $w_2 = 1 \operatorname{cis}(90 + 120 \times 2) = 1 \operatorname{cis} 330° = 0.866 - 0.5j$

30. Fourth roots of $16 = 16 \operatorname{cis} 0°$; $w_0 = \sqrt[4]{16} \operatorname{cis} \frac{0}{4} = 2$; $w_1 = 2 \operatorname{cis}(0 + \frac{360}{4}) = 2 \operatorname{cis} 90° = 2j$; $w_2 = 2 \operatorname{cis} 180° = -2$; $w_3 = 2 \operatorname{cis} 270° = -2j$

31. Square roots of $16 \operatorname{cis} 120°$; $w_0 = \sqrt[4]{16} \operatorname{cis} \frac{120}{2} = 4 \operatorname{cis} 60° = 2 + 3.464j$; $w_1 = 4 \operatorname{cis}(60 + 180) - 4 \operatorname{cis} 240° = -2 - 3.464j$

32. Cube roots of $27j = 27 \operatorname{cis} 90°$; $w_0 = \sqrt[3]{27} \operatorname{cis} \frac{90}{3} = 3 \operatorname{cis} 30° = 2.598 + 1.5j$; $w_1 = 3 \operatorname{cis}(30 + \frac{360}{3}) = 3 \operatorname{cis} 150° = -2.598 + 1.5j$; $w_2 = 3 \operatorname{cis}(30 + 120 \times 2) = 3 \operatorname{cis} 270° = -3j$

33. $(3 + 2j)(5 - j) = (3.60555e^{0.588j})(5.099e^{-0.197j}) = 18.385e^{0.391j}$

34. $(4 - 7j)/(3 + j) = (8.062e^{-1.052j})/(3.162e^{0.322j}) = 2.55e^{-1.374j}$

35. $(5 + 3j)^5 = (5.831e^{0.5404j})^5 = 5.831^5 e^{5 \times 0.5404j} = 6740.6e^{2.702j}$

36. $(-7 - 2j)^{1/3} = (7.280e^{-2.863j})^{1/3} = \sqrt[3]{7.280}e^{\frac{-2.863}{3}j} = 1.938e^{-0.954j}$

37. $7.3 - 1.4j = 7.433 \operatorname{cis}(-10.86°)$; magnitude is 7.433, direction is $-10.86°$

38. $X = X_L - X_C = 7.0 - 4.5 = 2.5 \; \Omega$; $R + Xj = 3 + 2.5j = 3.905\underline{/39.81°}$; $Z = 3.905 \; \Omega$, $\phi = 39.81°$; $V = IZ = 1.5 \times 3.905 = 5.858$ V

39. $X_L = 2\pi fL = 2\pi(0.3)60 = 113.10 \; \Omega$; $X_C = \frac{1}{2\pi fC} = \frac{1}{2\pi 60 \times 5 \times 10^{-6}} = 53.05 \; \Omega$; $X = X_L - X_C = 60.05 \; \Omega$; $R + Xj = 55 + 60.05j = 81.43\underline{/47.51°}$; $Z = 81.43 \; \Omega$, $\phi = 47.51°$

40. **(a)** in series $Z = Z_1 + Z_2 = (3 + 5j) + (6 - 3j) = 9 + 2j$;

(b) in parallel $Z = \frac{Z_1 Z_2}{Z_1 + Z_2} = \frac{(3+5j)(6-3j)}{9+2j} = \frac{18 - 9j + 30j - 15j^2}{9+2j} = \frac{33+21j}{9+2j} \cdot \frac{9-2j}{9-2j} = \frac{297 - 66j + 189j - 42j^2}{81+4} = \frac{339 + 123j}{85} = 3.988 + 1.447j$

CHAPTER 14 TEST

1. $\sqrt{-80} = \sqrt{80}\sqrt{-1} = \sqrt{16}\sqrt{5}j = 4j\sqrt{5}$

2. $7 - 2j$; $r = \sqrt{7^2 + 2^2} = \sqrt{53} \approx 7.28$; $\tan\theta = \frac{-2}{7} = -15.9$ or $344.1°$; Answer: $7.28\underline{/344.1°}$

3. $8 \operatorname{cis} 150° = 8\cos 150° + 8j\sin 150° = -6.9282 + 4j$ or $-4\sqrt{3} + 4j$

4. $(5 + 2j) + (8 - 6j) = 13 - 4j$

5. $(-5 + 2j) - (8 - 6j) = -13 + 8j$

6. $(2 + 3j)(4 - 5j) = 8 - 10j + 12j - 15j^2 = 23 + 2j$

7. $\frac{6+5j}{-3-4j} \cdot \frac{-3+4j}{-3+4j} = \frac{-18 + 24j - 15j + 20j^2}{9+16} = \frac{-38 + 9j}{25} = -1.52 + 0.36j$

8. $(7 \operatorname{cis} 75°)(2 \operatorname{cis} 105°) = 14 \operatorname{cis} 180° = -14$

9. $\frac{4 \operatorname{cis} 115°}{3 \operatorname{cis} 25°} = \frac{4}{3} \operatorname{cis} 90° = \frac{4}{3}j$ or $1.333j$

10. $\left(9 \operatorname{cis} \frac{2\pi}{3}\right)^{5/2} = (9)^{5/2} \operatorname{cis} \frac{5\pi}{3} = 243 \operatorname{cis} \frac{5\pi}{3} = \frac{243}{2} - \frac{243}{2}j\sqrt{3}$

11. $(27\underline{/129°})^{1/3} = \sqrt[3]{27} \operatorname{cis} \frac{129°}{3} = 3 \operatorname{cis} 43° \approx 2.194 + 2.046j$

12. Since $j = 1 \operatorname{cis} 90°$, then the four fourth roots of j are: $w_0 = \sqrt[4]{1} \operatorname{cis} \frac{90°}{4} = 1 \operatorname{cis} 22.5° = 0.924 + 0.383j$; $w_1 = 1 \operatorname{cis}(22.5 + 90)° = 1 \operatorname{cis} 112.5° = -0.383 + 0.924j$; $w_2 = 1 \operatorname{cis}(22.5 + 90 \times 2)° = 1 \operatorname{cis} 202.5° = -0.924 - 0.383j$; $w_3 = 1 \operatorname{cis}(22.5 + 90 \times 3) = 1 \operatorname{cis} 292.5° = 0.383 - 0.924j$

13. **(a)** in series $Z = Z_1 + Z_2 = (4 + 2j) + (5 - 3j) = 9 - j$;

(b) in parallel $Z = \frac{Z_1 Z_2}{Z_1 + Z_2} = \frac{(4+2j)(5-3j)}{9-j} = \frac{20 - 12j + 10j - 6j^2}{9-j} = \frac{26-2j}{9-j} \cdot \frac{9+j}{9+j} = \frac{234 + 26j - 18j - 2j^2}{81+1} = \frac{236+8j}{82} = \frac{118}{41} + \frac{4}{41}j = 2.878 + 0.098j$

14. $Z^{-1} = \frac{1}{9-3j} \frac{9+3j}{9+3j} = \frac{9+3j}{90} = \frac{1}{10} + \frac{1}{30}j \approx 0.1 + 0.033j$ S.

15. $V = 5 + 2j \approx 5.385\underline{/21.80°}$ and $I = 3 - 4j \approx 5\underline{/-53.13°}$. Since $Z = \frac{V}{I} = \frac{5.385\underline{/21.80°}}{5\underline{/-53.13°}} = 1.077\underline{/74.93°} = 0.28 + 1.04j\ \Omega$.

16. **(a)** $Z = Z_1 + Z_2 = (9 + 3j) + (7 - 2j) = (9 + 7) + (3 - 2)j = 16 + j\ \Omega$

(b) $\frac{1}{Z_1} = \frac{1}{9+3j} = \frac{1}{9-3j} \cdot = \frac{9-3j}{9-3j} = \frac{9-3j}{90} \approx 0.1 -$

$0.0333j$ and $\frac{1}{Z_2} = \frac{1}{7-2j} = \frac{1}{7-2j} \cdot \frac{7+2j}{7+2j} = \frac{7-2j}{53} \approx 0.1321 + 0.0377j$.

Then,

$$
\begin{aligned}
Z &= \frac{1}{\frac{1}{Z_1} + \frac{1}{Z_2}} \\
&= \frac{1}{(0.1 - 0.0333j) + (0.1321 + 0.0377j)} \\
&= \frac{1}{0.2321 + 0.0044j} \\
&= \frac{1}{0.2321 + 0.0044j} \cdot \frac{0.2321 - 0.0044j}{0.2321 - 0.0044j} \\
&= \frac{0.2321 - 0.0044j}{0.2321^2 + 0.0044^2} \\
&= \frac{0.2321 - 0.0044j}{0.0539} \\
&= \frac{0.2321}{0.0539} - \frac{0.0044}{0.0539}j \\
&\approx 4.3061 - 0.0816\ \Omega
\end{aligned}
$$

CHAPTER
15

An Introduction to Plane Analytic Geometry

1. $d = \sqrt{(2-7)^2 + (4+9)^2} = \sqrt{(-5)^2 + 13^2} = \sqrt{25 + 169} = \sqrt{194} \approx 13.928$

2. $d = \sqrt{(-3-0)^2 + (5-1)^2} = \sqrt{(-3)^2 + 4^2} = \sqrt{9 + 16} = \sqrt{25} = 5$

3. $d = \sqrt{(5+3)^2 + (-6+5)^2} = \sqrt{8^2 + (-1)^2} = \sqrt{64 + 1} = \sqrt{65} \approx 8.062$

4. $d = \sqrt{(-8-6)^2 + (-4-10)^2} = \sqrt{(-14)^2 + (-14)^2} = \sqrt{392} = 14\sqrt{2} \approx 19.799$

5. $d = \sqrt{(12-3)^3 + (1+13)^2} = \sqrt{9^2 + 14^2} = \sqrt{81 + 196} = \sqrt{277} \approx 16.643$

6. $d = \sqrt{(11-4)^2 + (-2+8)^2} = \sqrt{7^2 + 6^2} = \sqrt{49 + 36} = \sqrt{85} \approx 9.220$

7. $d = \sqrt{(-2-5)^2 + (5-5)^2} = \sqrt{(-7)^2} = 7$

8. $d = \sqrt{(-4+4)^2 + (-4-7)^2} = \sqrt{(-11)^2} = 11$

9. midpoint $= \left(\frac{2+7}{2}, \frac{4-9}{2}\right) = \left(\frac{9}{2}, \frac{-5}{2}\right)$

10. midpoint $= \left(\frac{-3+0}{2}, \frac{5+1}{2}\right) = \left(-3/2, 3\right)$

11. midpoint $= \left(\frac{5-3}{2}, \frac{-6-5}{2}\right) = \left(1, \frac{-11}{2}\right)$

12. midpoint $= \left(\frac{-8+6}{2}, \frac{-4+10}{2}\right) = (-1, 3)$

13. midpoint $= \left(\frac{12+3}{2}, \frac{1-13}{2}\right) = \left(\frac{15}{2}, -6\right)$

14. midpoint $= \left(\frac{11+4}{2}, \frac{-2-8}{2}\right) = \left(\frac{15}{2}, -15\right)$

15. midpoint $= \left(\frac{-2+5}{2}, \frac{5+5}{2}\right) = \left(\frac{3}{2}, -5\right)$

16. midpoint $= \left(\frac{-4-4}{2}, \frac{-4+7}{2}\right) = (-4, 3/2)$

17. slope $= \frac{-9-4}{7-2} = \frac{-13}{5} = -13/5 = -2.6$

18. slope $= \frac{1-5}{0+3} = \frac{-4}{3} = -4/3$

19. slope $= \frac{-5+6}{-3-5} = \frac{1}{-8} = -1/8$

20. slope $= \frac{10+4}{6+8} = \frac{14}{14} = 1$

21. slope $= \frac{-13-1}{3-12} = \frac{-14}{-9} = \frac{14}{9}$

22. slope $= \frac{-8+2}{4-11} = \frac{-6}{-7} = \frac{6}{7}$

23. slope $= \frac{5-5}{5+2} = \frac{0}{7} = 0$

24. slope $= \frac{7+4}{-4+4} = \frac{11}{0}$ is undefined

25. $y - 4 = -2.6(x - 2)$ or $y = -2.6x + 9.2$

26. $y - 5 = -\frac{4}{3}(x + 3)$ or $y - 5 = -\frac{4}{3}x - 4$ or $y = -\frac{4}{3}x + 1$

27. $y + 6 = -\frac{1}{8}(x - 5)$ or $y + 6 = -\frac{1}{8}x + \frac{5}{8}$ or $y = -\frac{1}{8}x - 5\frac{3}{8}$

28. $y + 4 = 1(x + 8)$ or $y = x + 4$

29. $y - 1 = \frac{14}{9}(x - 12)$ or $y - 1 = \frac{14}{9}x - \frac{56}{3}$ or $y = \frac{14}{9}x - 17\frac{2}{3}$

30. $y + 2 = \frac{6}{7}(x - 11)$ or $y + 2 = \frac{6}{7}x - \frac{66}{7}$ or $y = \frac{6}{7}x - \frac{80}{7}$ or $y = \frac{6}{7}x - 11\frac{3}{7}$

31. $y - 5 = 0(x + 2)$ or $y - 5 = 0$ or $y = 5$

32. Special case vertical line so $x = x$-intercept or $x = -4$

33. $m = \tan 45° = 1$

34. $m = \tan 175° \approx -0.0874887$

35. $m = \tan 0.15 \approx 0.1511352$

36. $m = \tan 1.65 \approx -12.599265$

37. $\alpha = \tan^{-1} 2.5 = 68.19859°$

38. $\alpha = \tan^{-1} 0.3 = 16.699244°$

39. $\alpha = 180 + \tan^{-1}(-0.50) = 153.43495°$

40. $\alpha = 180 + \tan^{-1}(-1.475) = 124.13594°$

41. $m = -\frac{1}{3} = -1/3$

42. $m = -\frac{1}{2/5} = -5/2$

43. $m = -\frac{1}{-1/2} = 2$

44. $m = -\frac{1}{-5} = 1/5$

45. $y + 5 = 6(x - 2)$ or $y + 5 = 6x - 12$ or $y = 6x - 17$

46. $y + 2 = -\frac{1}{2}(x - 6)$ or $y + 2 = -\frac{1}{2}x + 3$ or $y = -\frac{1}{2}x + 1$

47. $m = \frac{-2-3}{4-2} = \frac{-5}{2}; y + 2 = -\frac{5}{2}(x - 4)$ or $y + 2 = -\frac{5}{2}x + 10$ or $y = \frac{-5}{2}x + 8$

48. $m = \frac{-4-2}{-3+5} = \frac{-6}{2} = -3; y - 2 = -3(x + 5)$ or $y = -2 = -3x - 15$ ro $y = -3x$

49. $m = \tan 60° = 1.732; y + 4 = 1.732(x + 2)$ or $y = 1.732x - 0.536$

50. $m = \tan 0.75 = 0.932; y - 1 = 0.932(x - 5)$ or $y = 0.932x - 3.66$

51. $m = \tan 20° = 0.364; y = 0.364x + 3$

52. $m = \tan 2.5 = -0.747; y = -0.747x - 2$

53. $2x - 3y + 4 = 0 \Rightarrow -3y = -2x + 4 \Rightarrow y = \frac{2}{3}x - \frac{4}{3}$; so the slope is $\frac{2}{3}; y - 5 = \frac{2}{3}(x - 2)$ or $3y - 15 = 2x - 4$. Combining terms produces $-2x + 3y = 11$ or $-2x + 3y - 11 = 0$ or $2x - 3y + 11 = 0$

54. $3x + 4y = 12 \Rightarrow 4y = -3x + 12 \Rightarrow y = -\frac{3}{4}x + 3$; so the slope is $-\frac{3}{4}$. Thus, we obtain $y - 2 = -\frac{3}{4}(x + 3)$ or $4y - 8 = -3x - 9$ or $3x + 4y = -1$ or $3x + 4y + 1 = 0$

55. $2x + 5y = 20 \Rightarrow 5y = -2x + 20 \Rightarrow y = -\frac{2}{5}x + 4$; the slope of this line is $-\frac{2}{5}$ and the slope of the perpendicular is $\frac{5}{2}$. Using the point-slope form, we get $y + 1 = \frac{5}{2}(x - 4)$ or $2y + 2 = 5x - 20$ or $-5x + 2y = -22$ or $5x - 2y = 22$

56. $8x - 3y = 24 \Rightarrow -3y = -8x + 24 \Rightarrow y = \frac{8}{3}x - 8$; $m = \frac{8}{6}$ so the slope of perpendicular is $-\frac{3}{8}$. Using the point-slope form, we get $y + 6 = -\frac{3}{8}(x + 1)$ or $8y + 48 = -3x - 3$ or $3x + 8y = -51$

57. $3x + 2y = 12 \Rightarrow 2y = -3x + 12 \Rightarrow y = -\frac{3}{2}x + 6; m = -\frac{3}{2}$, y-intercept $= 6$; when $y = 0, x = 4$ so the x-intercept $= 4$.

58. $5x - 3y = 15 \Rightarrow -3y = -5x + 15 \Rightarrow y = \frac{5}{3}x - 5; m = \frac{5}{3}$, y-intercept $= -5$; when $y = 0, x = 3$ and so, the x-intercept $= 3$

59. $x - 3y = 9 \Rightarrow -3y = -x + 9 \Rightarrow y = \frac{1}{3}x - 3; m = \frac{1}{3}$, y- intercept $= -3$, when $y = 0, x = 9$ so x-intercept$= 9$

60. $6x + y = 9 \Rightarrow y = -6x + 9; m = -6, y$-intercept $= 9$. When $y = 0$, we see that $x = \frac{3}{2}$ and so, the x-intercept $= \frac{3}{2}$.

61. $v = v_0 + at, v_0 = 2.6, v = 5.8$ when $t = 8; 5.8 = 2.6 + a \cdot 8; 3.2 = 8a; a = 0.4; v = 2.6 + 0.4t$

62. **(a)** $\alpha = \frac{\text{change in length}}{\text{change in temp.}} = \frac{1.000084 - 1.000000}{15 - 10} = \frac{0.000084}{5} = 1.68 \times 10^{-5}$;

(b) $\alpha = \frac{.000336}{5} = 6.72 \times 10^{-5}$;

(c) The amount of expansion is proportional to the original length of the solid. You need to divide by the original length of the solid; in part (b) you need to divide by 4;

(d) $\dfrac{\frac{72.005208 - 72.000024}{18 - 10}}{72.000024} = 9.000 \times 10^{-6}$

63. $E = IR + Ir; 12 = 2.5 \times 4 + 2.5 \times r; 12 = 10 + 2.5r; 2 = 2.5r; r = \frac{2}{2.5} = 0.8\,\Omega$

64. **(a)** $R = 7 + 0.006T$ or $R = 0.006T + 7.000$, **(b)** $R = 0.006 \times 17 + 7.000 = 7.102\,\Omega$

65. **(a)** $C = 1,225 + 1.25n$, **(b)** Substituting 20,000 for n, we get $26,225.

66.

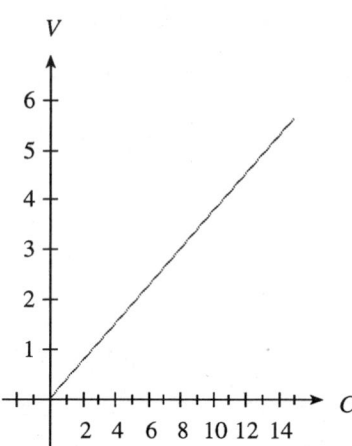

67. **(a)** Here, $t = 1995 - 1940 = 55$. Thus, $C = 1.44(55) + 282.88 = 362.08$ ppm.
(b) Here, $t = 2050 - 1940 = 110$. Thus, $C = 1.44(110) + 282.88 = 441.28$ ppm.

(c) In this exercise, we have t = current year -1940. Answer varies with year.

68. **(a)** For Chattanooga, Tennessee, $L = 31°2' \approx 31.03°$. Thus, $B = 31.03° - 23.5° = 7.53°$. Since, $H = 8'$, we have $W = H \tan B = 8 \tan 7.53° \approx 1.06$. The overhang should have a width of about 1.06 ft $\approx 1'\frac{3}{4}''$.
(b) The latitude of Vancouver, British Columbia is $49°16'$ N, so $L \approx 49.27°$ and $B = 49.27° - 23.5° = 25.77°$. Thus, $W = 2.7 \tan 25.77° \approx 1.303$. The overhang should be about 1.303 m.
(c) The latitude of Anchorage, Alaska is $61°13'$ N, so $L = 61°13' \approx 61.22°$ and $B = 61.22° - 23.5° = 37.72°$. Converting $8'6''$ to 8.5 ft and substituting, we see that $W = 8.5 \tan 37.72° \approx 6.57$. Thus, this overhang will have to be about $6.57' = 6'6\frac{11}{16}''$ in order to prevent summer sunshine.

≡ 15.2 THE CIRCLE

1. $C = (2, 5), r = 3$:   The standard equation for the circle is $(x-2)^2 + (y-5)^2 = 3^2$ or $(x-2)^2 + (y-5)^2 = 9$. This gives $x^2 - 4x + 4 + y^2 - 10y + 25 = 9$ or $x^2 + y^2 - 4x - 10y + 20 = 0$ for the general equation.

2. $C = (4, 1), r = 8$: $(x-4)^2 + (y-1)^2 = 64$ is the standard form. $x^2 - 8x + 16 + y^2 - 2y + 1 = 64$ or $x^2 + y^2 - 8x - 2y - 47$ is the general form.

3. $C = (-2, 0), r = 4$: $(x+2)^2 + y^2 = 16$ is the standard form. $x^2 + y^2 + 4x - 12 = 0$ is the general form.

4. $C = (0, -7), r = 2$: $x^2 + (y+7)^2 = 4$ is the standard form. $x^2 + y^2 + 14y + 45 = 0$ is the general form.

5. $C = (-5, -1), r = \frac{5}{2}$: $(x+5)^2 + (y+1)^2 = \frac{25}{4}$ is the standard form. $x^2 + 10x + 25 + y^2 + 2y + 1 = \frac{25}{4}$ or $x^2 + y^2 + 10x + 2y + 19.75$ is the general form.

6. $C = (-4, -5), r = \frac{5}{4}$: $(x+4)^2 + (y+5)^2 = \frac{25}{16}$ is the standard form. Expanding the standard form and collecting terms, we get $x^2 + 8x + 16 + y^2 + 10y + 25 = \frac{25}{16}$; $x^2 + y^2 + 8x + 10y + \frac{631}{16} = 0$ which is the general form.

7. $C = (2, -4), r = 1$: $(x-2)^2 + (y+4)^2 = 1$ is the standard form. $x^2 - 4x + 4 + y^2 + 8y + 16 = 1$ or $x^2 + y^2 - 4x + 8y + 19 = 0$ is the general form.

8. $C = (5, -2), r = \sqrt{3}$: $(x-5)^2 + (y+2)^2 = 3$ is the standard form. $x^2 - 10x + 25 + y^2 + 4y + 4 = 3$ or $x^2 + y^2 - 10x + 4y + 26 = 0$ is the general form.

9. $C = (3, 4), r = 3$

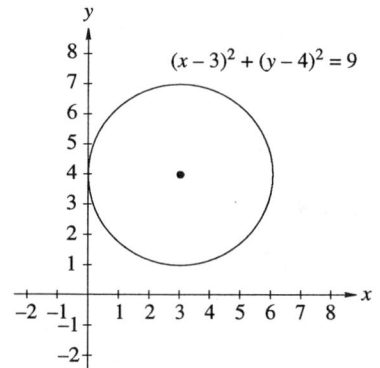

10. $C = (7, -5), r = 5$

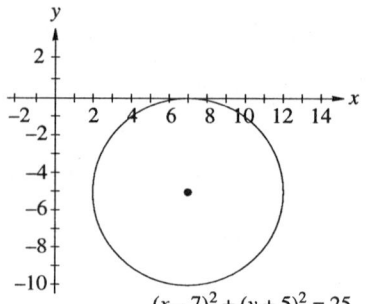

11. We first write this in the standard form for the equation of a circle, $(x-h)^2 + (y-k)^2 = r^2$ or $\left(x - (-\frac{1}{2})\right)^2 + \left(y - (-\frac{13}{4})\right)^2 = \left(\sqrt{7}\right)^2$. Thus, the center is $C = (h,k) = (-\frac{1}{2}, -\frac{13}{4})$, and the radius is $r = \sqrt{7} \approx 2.646$.

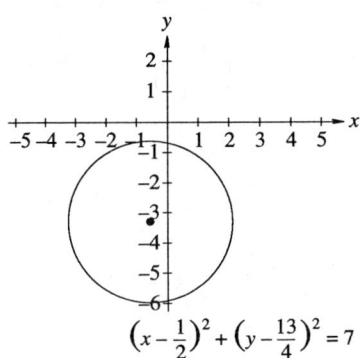

$$\left(x - \tfrac{1}{2}\right)^2 + \left(y - \tfrac{13}{4}\right)^2 = 7$$

12. $C = (-\frac{7}{2}, \frac{7}{3}), r = \sqrt{\frac{11}{6}} \approx 1.354$

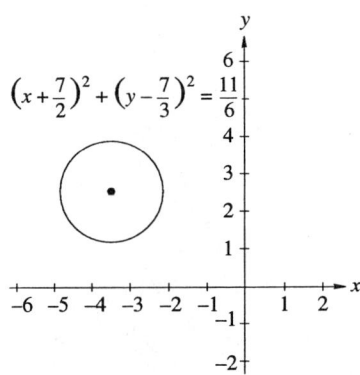

$$\left(x + \tfrac{7}{2}\right)^2 + \left(y - \tfrac{7}{3}\right)^2 = \tfrac{11}{6}$$

13. $C = (0, \frac{7}{3}), r = \sqrt{6} \approx 2.449$

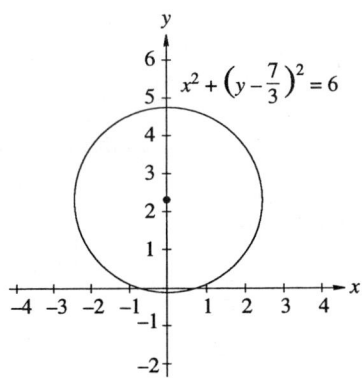

$$x^2 + \left(y - \tfrac{7}{3}\right)^2 = 6$$

14. $C = (-3, 0), r = \sqrt{1.21} = 1.1$

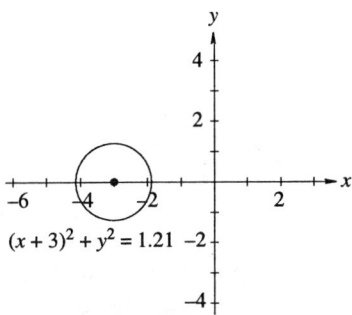

$$(x+3)^2 + y^2 = 1.21$$

15. $x^2 + y^2 + 4x - 6y + 4 = 0$. Completing the square leads to $x^2 + 4x + 4 + y^2 - 6y + 9 = -4 + 4 + 9$ or $(x+2)^2 + (y-3)^2 = 9$. This is the standard equation for a circle with center $(-2, 3)$ and radius $r = 3$.

16. $x^2 + y^2 - 10x + 2y + 22$. Completing the square leads to $x^2 - 10x + 25 + y^2 + 2y + 1 = -22 + 25 + 1$ or $(x-5)^2 + (y+1)^2 = 4$. This is the standard equation for a circle with center $C = (5, -1)$ and radius $r = 2$.

17. $x^2 + y^2 + 10x - 6y - 47 = 0$; $x^2 + 10x + 25 + y^2 - 6y + 9 = 47 + 25 + 9$; $(x+5)^2 + (y-3)^3 = 81$; Circle: $C = (-5, 3), r = 9$

18. $x^2 + y^2 + 2x - 12y - 27 = 0$; $x^2 + 2x + 1 + y^2 - 12y + 36 = 27 + 1 + 36$; $(x+1)^2 + (y-6)^2 = 64$; Circle: $C = (-1, 6), r = 8$

19. $x^2 + y^2 - 2x + 2y + 3 = 0$; $x^2 - 2x + 1 + y^2 + 2y + 1 = -3 + 1 + 1$; $(x-1)^2 + (y+1)^2 = -1$; $r^2 = -1$ so it is not a circle.

20. $x^2 + y^2 + 2x + 2y - 3 = 0$; $x^2 + 2x + 1 + y^2 + 2y + 1 = 3 + 1 + 1$; $(x+1)^2 + (y+1)^2 = 5$; Circle: $C = (-1, -1), r = \sqrt{5}$

21. $x^2 + y^2 + 6x - 16 = 0$; $x^2 + 6x + 9 + y^2 = 16 + 9$; $(x+3)^2 + y^2 = 25$; Circle: $C = (-3, 0), r = 5$

22. $x^2 + y^2 - 8y - 9 = 0$; $x^2 + y^2 - 8y + 16 = 9 + 16$; $(x^2) + (y-4)^2 = 25$; Circle: $C = (0, 4), r = 5$

23. $x^2 + y^2 + 5x - 9y = 9.5$; $x^2 + 5x + \frac{25}{4} + y^2 - 9y + \frac{81}{4} = \frac{19}{2} + \frac{25}{4} + \frac{81}{4}$; $(x + \frac{5}{2})^2 + (y - \frac{9}{2})^2 = \frac{144}{4} = 36$; Circle: $C = (-\frac{5}{2}, \frac{9}{2}), r = 6$

24. $x^2 + y^2 - 7x + 3y + 2.5 = 0$; $x^2 - 7x + \frac{49}{4} + y^2 + 3y + \frac{9}{4} = -\frac{5}{2} + \frac{49}{4} + \frac{9}{4}$; $(x - \frac{7}{2})^2 + (y + \frac{3}{2})^2 = 12$; Circle: $C = (\frac{7}{2}, -\frac{3}{2})$; $r = \sqrt{12} = 2\sqrt{3} \approx 3.464$

25. $9x^2 + 9y^2 + 18x - 15y + 27 = 0; 9(x^2 + 2x) + 9(y^2 - \frac{5}{3}y) = -27; 9(x^2 + 2x + 1) + 9(y^2 - \frac{5}{3}y + \frac{25}{36}) = -27 + 9 \cdot 1 + 9 \cdot \frac{25}{36}; 9(x + 1)^2 + 9(y - \frac{5}{6})^2 = -11.75; (x+1)^2 + (y - \frac{5}{6})^2 = -1.306$; Not a circle since r^2 is negative.

26. $25x^2 + 25y^2 - 10x + 30y + 1 = 0; 25(x^2 - \frac{2}{5}x) + 25(y^2 + \frac{6}{5}y) = -1; 25(x^2 - \frac{2}{5}x + \frac{1}{25}) + 25(y^2 + \frac{6}{5}y + \frac{9}{25}) = -1 + 25 \cdot \frac{1}{25} + 25 \cdot \frac{9}{25}; 25(x - \frac{1}{5})^2 + 25(y + \frac{3}{5})^2 = 9; (x - \frac{1}{5})^2 + (y + \frac{3}{5})^2 = \frac{9}{25}$; Circle: $C = (\frac{1}{5}, -\frac{3}{5}), r = \frac{3}{5}$

27. $\sigma = \frac{mv}{\beta q} = \frac{(1.673 \times 10^{-27})(1.186 \times 10^7)}{(4 \times 10^{-4})(1.5 \times 10^{-19})} = 330.70$. Choosing the center of the circle to be at the origin of the coordinate system we get $x^2 + y^2 = (330.70)^2$ or $x^2 + y^2 \approx 1.094 \times 10^5$.

28. $x^2 + y^2 = (1.495 \times 10^8)^2$ or $x^2 + y^2 = 2.235 \times 10^{16}$

29. $(x - 1.495 \times 10^8)^2 + y^2 = (3.844 \times 10^5)^2$ or $(x - 1.495 \times 10^8)^2 + y^2 = 1.478 \times 10^{11}$

30. First circle is centered at (0, 0) with radius 6 so $x^2 + y^2 = 36$. Second circle center with x coordinate $6 + 5 - 1 = 10$ has center $C = (10, 0)$ and radius 5 so the equation is $(x - 10)^2 + y^2 = 25$.

31. If the center is 0.9 directly above the origin, the center is (0, 0.9), the radius is 1.1/2 so $x^2 + (y - 0.9)^2 = 0.55^2$ or $x^2 + (y - 0.9)^2 = 0.3025$.

32. The x-coordinate of the center is $-(b + c)$. The y-coordinate is $-(a - c)$, the radius is c
$(x + (b + c))^2 + (y + (a - c))^2 = c^2; (x + 1.25)^2 + (y + 3)^2 = 0.25^2$

33. Center is at (0, 5) and $r = \frac{26}{2} = 13$
(a) $x^2 + (y - 5)^2 = 13^2 = 169$
(b) Drawing a radius to where the floor would intersect the flywheel and the y-axis, we get a right triangle with hypotenuse 13 and one leg of 5. Using the Pythagorean theorem we find the other leg is 12, hence the chord is 2×12 or 24. Allowing for a 2 cm clearance on both sides, the opening must be $24 + 4 \times 2 = 28$ cm.

34. **(a)** $Z = \sqrt{R^2 + (X_L - X_C)^2} = \sqrt{6^2 + (10 - 12)^2} \approx 6.32\,\Omega$,
(b) $Z = \sqrt{R^2 + (X_L - X_C)^2} = \sqrt{8^2 + (X_L - 12.5)^2}\,\Omega$,
(c) Solving for X_L, we first square both sides to obtain $Z^2 = R^2 + (X_L - X_C)^2$. This leads to $(X_L - X_C)^2 = Z^2 - R^2$ or $X_L - X_C = \pm\sqrt{Z^2 - R^2}$. Hence $X_L = X_C \pm \sqrt{Z^2 - R^2}$. Substituting the given data, we obtain $X_L = 50 \pm \sqrt{40^2 - R^2} = 50 \pm \sqrt{1,600 - R^2}\,\Omega$.
(d)

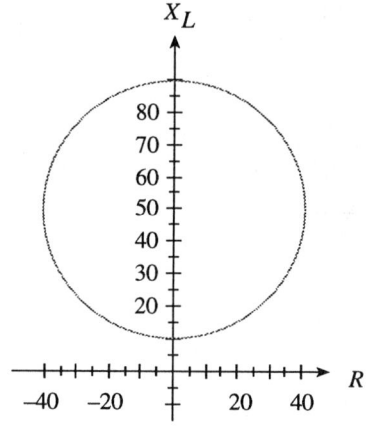

☰ 15.3 THE PARABOLA

1. $x^2 = 4y, p = 1$, vertical so $F = (0, 1)$, directrix: $y = -1$, opens upward

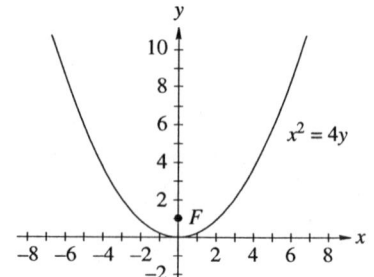

2. $x^2 = 12y, p = 3$, vertical so $F = (0, 3)$, directrix: $y = -3$, opens upward.

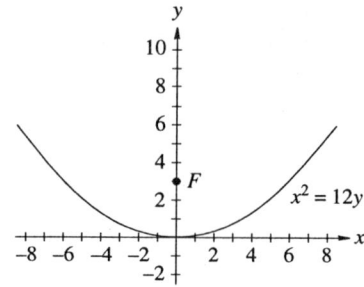

3. $y^2 = -4x, p = -1$, horizontal so $F = (-1, 0)$, directrix: $x = 1$, opens left.

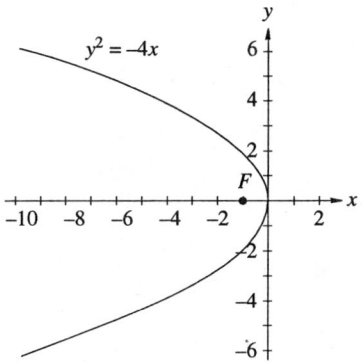

4. $y^2 = -20x, p = -5$, horizontal so $F = (-5, 0)$, directrix: $x = 5$, opens left.

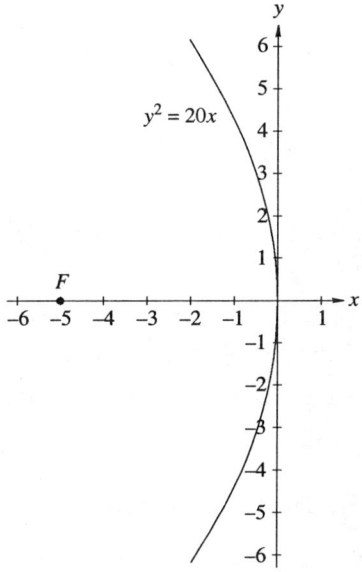

5. $x^2 = -8y, p = -2$, vertical so $F = (0, -2)$, directrix: $y = 2$, opens downward.

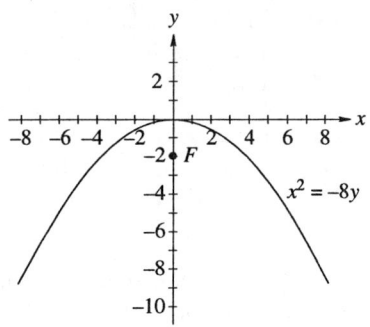

6. $x^2 = -16y, p = -4$, vertical so $F = (0, -4)$, directrix: $y = 4$, opens downward.

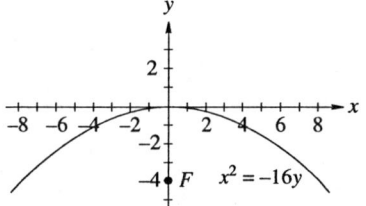

7. $y^2 = 10x, p = \frac{10}{4} = 2.5$, horizontal so $F = (2.5, 0)$, directrix: $x = -2.5$, opens right.

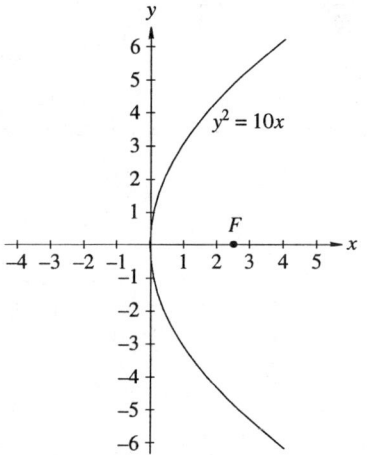

8. $y^2 = -14x, p = \frac{-14}{4} = -3.5$, horizontal so $F = (-3.5, 0)$, directrix: $x = 3.5$, opens left.

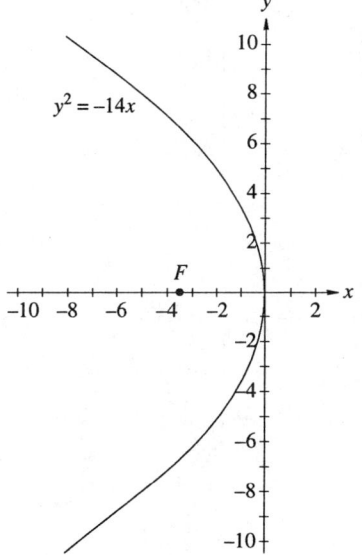

9. $x^2 = 2y, p = \frac{2}{4} = 0.5$, vertical so $F = (0, 0.5)$, directrix: $y = -0.5$, opens upward.

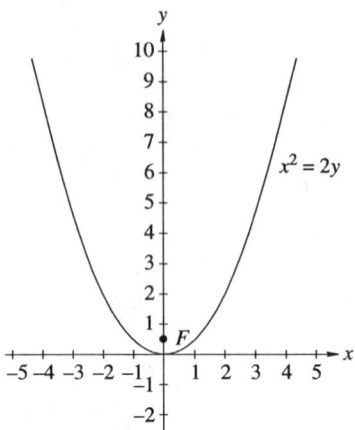

10. $x^2 = -17y, p = \frac{-17}{4} = -4.25$, vertical so $F = (0, -4.25)$, directrix: $y = 4.25$, opens downward.

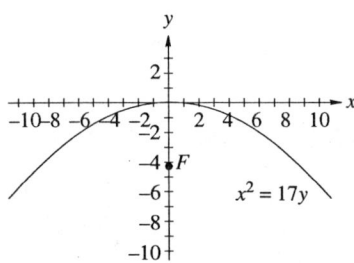

11. $y^2 = -21x, p = \frac{-21}{4} = -5.25$, horizontal so $F = (-\frac{21}{4}, 0)$, directrix: $x = \frac{21}{4}$, opens left.

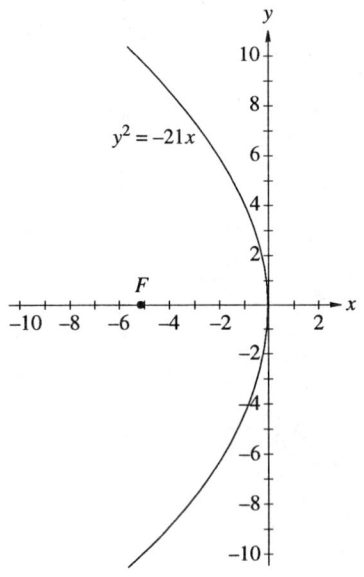

12. $y^2 = 5x, p = \frac{5}{4} = 1.25$, horizontal so $F = (1.25, 0)$, directrix: $x = -1.25$, opens right.

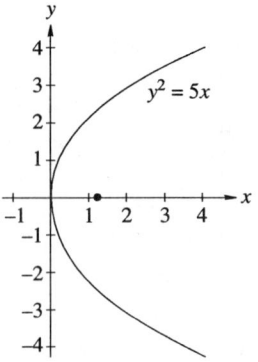

13. Focus $(0, 4)$, directrix: $y = -4, p = 4$ opens upward so $x^2 = 4 \cdot 4y$ or $x^2 = 16y$.

14. Focus $(0, -5)$, directrix: $y = 5, p = -5$ opens downward so $x^2 = -5 \cdot 4y$ or $x^2 = -20y$.

15. Focus $(-6, 0)$, directrix: $x = 6, p = -6$ opens left so $y^2 = 4(-6)x$ or $y^2 = -24x$.

16. Focus $(3, 0)$, directrix: $x = -3, p = 3$, opens right so $y^2 = 4 \cdot 3x$ or $y^2 = 12x$.

17. Focus $(2, 0)$, vertex $(0, 0)$, $p = 2$, opens right so $y^2 = 4 \cdot 2x$ or $y^2 = 8x$

18. Focus $(0, -7)$, vertex $(0, 0), p = -7$, opens downward so $x^2 = 4(-7)y$ or $x^2 = -28y$.

19. Vertex $(0, 0)$, directrix: $x = -\frac{3}{2}, p = \frac{3}{2}$, opens right so $y^2 = 4 \cdot \frac{3}{2}x$ or $y^2 = 6x$.

20. Vertex $(0, 0)$, directrix $y = \frac{3}{4}, p = -\frac{3}{4}$, opens downward so $x^2 = 4(-\frac{3}{4})y$ or $x^2 = -3y$.

21. Choosing the vertex to be at the center of the bridge and the origin, we get $x^2 = 4py$ where $(180, 90)$ is a point on the parabola. Solving for p we get $(180)^2 = 4 \cdot 90p$ or $p = 90$. $x^2 = 360y$. Now setting $x = 100$ and solving for y, $100^2 = 360y$ or $y = \frac{10000}{360} = 27.78$ m.

22. If the diameter is 2 the radius is 1. Using $x^2 = 4py$ and the point $(1, 0.4)$ we can solve for p. $1 = 4(0.4)p$ or $p = \frac{1}{1.6} = 0.625$. The location is 0.625 m from the bottom or vertex.

23. $d = 5 \Rightarrow r = 2.5$. Using $x^2 = 4py$ and the point $(2.5, 1.5)$ we solve for p. $2.5^2 = 4(1.5)p$ or $p = \frac{6.25}{6}$ or 1.042 m from vertex.

24. $P = 10I^2$

25.

26.

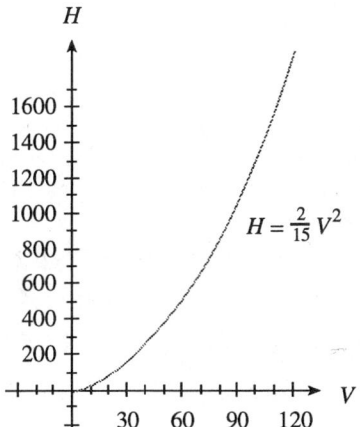

27. Position the "dish" so that the vertex is at $(0, 0)$ and the focus at $(0, f)$. The directrix is $y = -f$. One point on the dish is at $(5, 2)$ and, since the dish is a parabola, we know that distance from the focus to $(5, 2)$ is equal to the distance from the focus to the directrix. This leads to the equation

$$(f + 2)^2 = 5^2 + (f - 2)^2$$
$$f^2 + 4f + 4 = 25 + f^2 - 4f + 4$$
$$8f = 25$$
$$f = \frac{25}{8} = 3.125$$

The focus is 3.125 ft from the vertex.

28. If the focal length of a mirror is f, then $f = kp^2$. A mirror spinning at 45 rpm has a rotational period of $p = \frac{1}{45 \text{ rpm}} = \frac{1 \text{ min}}{45 \text{ rev}}$. Thus, if a mirror spinning at 45 rpm has a focal length of 22.1 cm, and if $f = kp^2$, then $22.1 = k\left(\frac{1 \text{ min}}{45 \text{ rev}}\right)^2$ and $k = 44752.5$ cm · rev/ min and we have established that $f = 44752.5p^2$

(a) For a mirror spinning at $33\frac{1}{3} = \frac{100}{3}$ rpm, the focal length is $f = 44752.5 \left(\dfrac{1}{\frac{3}{100}}\right)^2 = 44752.5\left(\frac{3}{100}\right)^2 = 40.27725 \approx 40.3$ cm.

(b) For a mirror spinning at 15 rpm, the focal length is $f = 44752.5\left(\frac{1}{15}\right)^2 = 198.9$ cm.

15.4 THE ELLIPSE

1. $F(4, 0), V(6, 0)$, so $c = 4, a = 6, b^2 = a^2 - c^2$ or $b^2 = 6^2 - 4^2 = 20$. Major axis is horizontal so the equation is $\frac{x^2}{36} + \frac{y^2}{20} = 1$.

2. $F(9, 0), V = (12, 0), c = 9, a = 12, b^2 = a^2 - c^2 = 12^2 - 9^2 = 63$. Major axis is horizontal so the equation is $\frac{x^2}{144} + \frac{y^2}{63} = 1$.

3. $F(0,2), V = (0,-4)$. Major axis is vertical so $b = 4, c = 2, a^2 = b^2 - c^2 = 4^2 - 2^2 = 12$. $\frac{x^2}{12} + \frac{y^2}{16} = 1$.

4. $F(0,-3), V(0,5)$. Major axis is vertical so $b = 5, c = 3, b^2 = a^2 - c^2 = 25 - 9 = 16$. $\frac{x^2}{2}5 + \frac{y^2}{16} = 1$.

5. $F(3,0), V(5,0)$. Major axis is horizontal so $a = 5, c = 3, b^2 = a^2 - c^2 = 25 - 9 = 16; \frac{x^2}{25} + \frac{y^2}{16} = 1$.

6. $F(0,-2), V(0,-3)$. Major axis vertical so $b = 3, c = 2, a^2 = b^2 - c^2 = 9 - 4 = 5$. $\frac{x^2}{5} + \frac{y^2}{9} = 1$.

7. Minor axis length 6, vertex $(4, 0)$. Major axis is horizontal so $a = 4, b = \frac{6}{2} = 3$;

$$\frac{x^2}{16} + \frac{y^2}{9} = 1$$

8. Minor axis length 10, $V(0,-6)$. Major axis is vertical so $b = 6, a = \frac{10}{2} = 5, \frac{x^2}{25} + \frac{y^2}{36} = 1$.

9. $\frac{x^2}{4} + \frac{y^2}{9} = 1$. Major axis vertical $b = 3, a = 2, c^2 = b^2 - a^2 = 9 - 4 = 5, c = \sqrt{5}, V = (0,3), V' = (0,-3), F = (0, \sqrt{5}), F' = (0, -\sqrt{5}), M = (2,0), M' = (-2,0)$

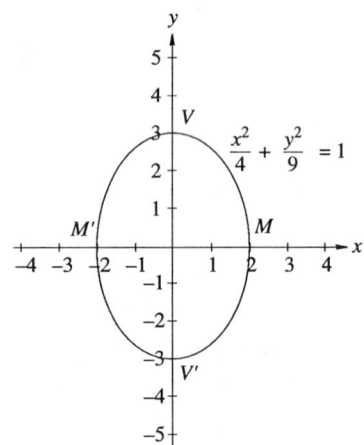

10. $\frac{x^2}{9} + \frac{y^2}{4} = 1$. Major axis is horizontal, $a = 3, b = 2, c^2 = 9 - 4 = 5, c = \sqrt{5}$. $V = (3,0), V' = (-3,0), F = (\sqrt{5},0), F' = (-\sqrt{5},0), M = (0,2), M' = (0,-2)$

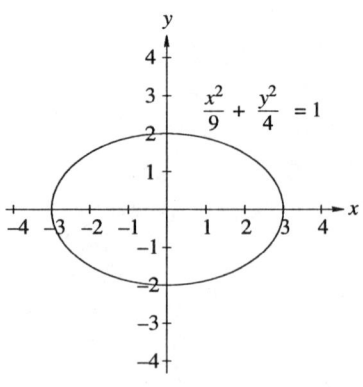

11. $\frac{x^2}{4} + \frac{y^2}{1} = 1, a = 2, b = 1, c^2 = a^2 - b^2 = 3, c = \sqrt{3}, V = (2,0), V' = (-2,0), F = (\sqrt{3},0), F' = (-\sqrt{3},0), M = (0,1), M' = (0,-1)$.

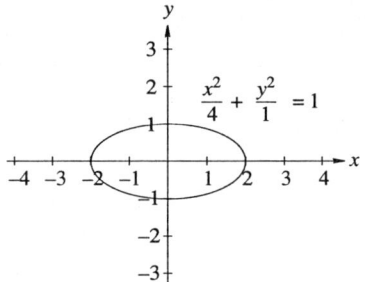

12. $\frac{x^2}{16} + \frac{y^2}{36} = 1, b = 6, a = 4, c^2 = 36 - 16 = 20, c = \sqrt{20} = 2\sqrt{5}, V = (0,6), V' = (0,-6), F = (0, 2\sqrt{5}), F' = (0,-2\sqrt{5}), M = (4,0), M' = (-4,0)$

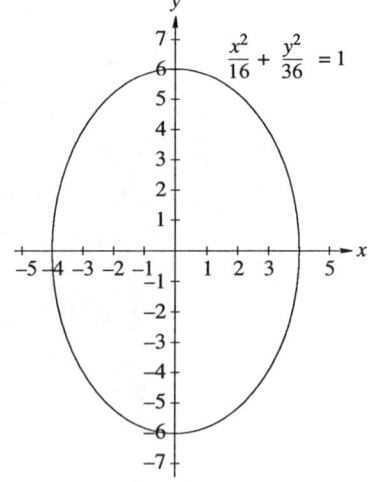

13. $4x^2 + y^2 = 4 \Rightarrow \frac{x^2}{1} + \frac{y^2}{4} = 1, a = 1, b = 2, c = \sqrt{3}, V = (0, \pm 2), F = (-0, \pm \sqrt{3}), M = (\pm 1, 0)$

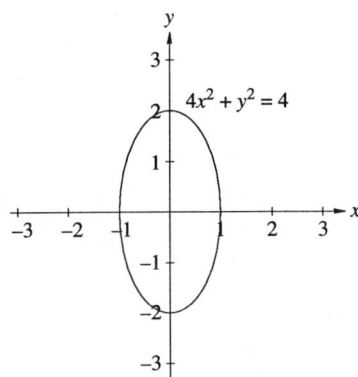

14. $9x^2 + 4y^2 = 36 \Rightarrow \frac{x^2}{4} + \frac{y^2}{9} = 1, a = 2, b = 3, c = \sqrt{5}, V = (0, \pm 3), F = (0, \pm \sqrt{5}), M = (\pm 2, 0)$

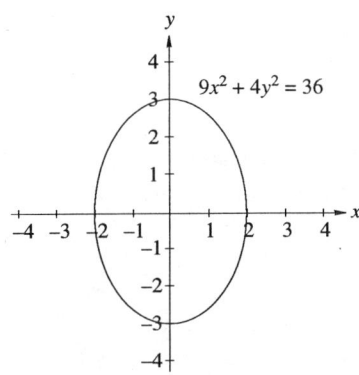

15. $25x^2 + 36y^2 = 900 \Rightarrow \frac{x^2}{36} + \frac{y^2}{25} = 1, a = 6, b = 5, c = \sqrt{11}, V = (\pm 6, 0), F = (\pm \sqrt{11}, 0), M = (0, \pm 5)$

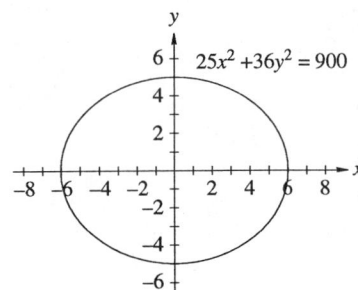

16. $9x^2 + y^2 = 18 \Rightarrow \frac{x^2}{2} + \frac{y^2}{18} = 1, a = \sqrt{2}, b = 3\sqrt{2}, c = 4, V = (0, \pm \sqrt{2}), F = (0, \pm 4), M = (\pm \sqrt{2}, 0)$

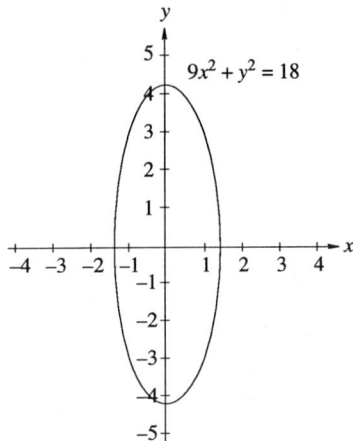

17. $a = 2.992 \times 10^8 / 2 = 1.496 \times 10^8, c = a/60 = 2.493 \times 10^6, b^2 = a^2 - c^2$ or $b = \sqrt{a^2 - c^2} = 1.49579 \times 10^8$ Minor axis $= 2b = 2.9916 \times 10^8$ km.

18. Major axis is 80 so $a = \frac{80}{2} = 40$ and $b = 24$. Substituting in the standard equation for an ellipse, we obtain $\frac{x^2}{40^2} + \frac{y^2}{24^2} = 1$ or $\frac{x^2}{1600} + \frac{y^2}{576} = 1$.

19. Minor axis is the diameter of the pipe is 10 in. Major axis is $10\sqrt{2}$ or 14.14 in.

20. Distance between foci used in the two ellipses is $2a$ so $a = 3$. $c = a - 1 = 2, b^2 = 3^2 - 2^2 = 9 - 4 = 5$ so $b = \sqrt{5}$. Minor axis $= 2b = 2\sqrt{5}$ cm.

21. $a = 3600/2 = 1800, \ b = 44/2 = 22,$
$c = \sqrt{a^2 - b^2} = \sqrt{1800^2 - 22^2} = 1799.865551,$
eccentricity $= \frac{c}{a} = \frac{1,799.865551}{1800} = 0.9999253.$

22. Adding the Earth's radius to both altitudes, we obtain 7 137.4 and 9 390.5. Thus, $2a = 7\ 137.4 + 9\ 390.5 = 16\ 527.9$, and so $a = 8\ 263.95$. Now, $c = 8\ 263.95 - 7\ 137.4 = 1\ 126.55$. From the relationship $b^2 = a^2 - c^2 = (8\ 263.95)^2 - (1\ 126.55)^2 = 67\ 023\ 754.7$, we get $b \approx 8\ 186.80$. Substituting these values into the equation for an ellipse, we obtain $\frac{x^2}{8263.95^2} + \frac{y^2}{8186.80^2} = 1.$

23. Looking at the cross-section we see that the length of the major axis is the same as the length of the diagonal of a square with sides 8.0 in. Hence, $2a = 8\sqrt{2} \approx 11.3$ in.

24. Using $a = \frac{52}{2} = 26$ and $b = 12.5$, we get the equation $\frac{x^2}{26^2} + \frac{y^2}{12.5^2} = 1$. Setting $x = 14$ and solving, we have $\frac{14^2}{26^2} + \frac{y^2}{12.5^2} = 1$ or $y^2 = 12.5^2\left(1 - \frac{14^2}{26^2}\right) = 110.9$. Thus, $y = \sqrt{110.9} \approx 10.5$ ft.

≡ 15.5 THE HYPERBOLA

1. $F(6,0), V(4,0)$: Transverse axis is on the x-axis so $a = 4, c = 6$, and $b^2 = c^2 - a^2 = 36 - 16 = 20$. The equation is $\frac{x^2}{16} - \frac{y^2}{20} = 1$.

2. $F(12,0), V = (9,0)$: Transverse axis is on the x-axis so $c = 12, a = 9$, and $b^2 = c^2 - a^2 = 12^2 - 9^2 = 63$. The equation is $\frac{x^2}{81} - \frac{y^2}{63} = 1$.

3. $F(0,5), V(0,-3)$: Transverse axis is on the y-axis so $c = 5, a = 3$, and $b^2 = 5^2 - 3^2 = 16$. The equation is $\frac{y^2}{9} - \frac{x^2}{16} = 1$.

4. $F(0,-4), V(0,2)$: Transverse axis is on the y-axis and $c = 4, a = 2$, and $b^2 = 16 - 4 = 12$. The equation is $\frac{y^2}{4} - \frac{x^2}{12} = 1$.

5. $F(5,0), V = (3,0)$: $c = 5, a = 3$, and $b^2 = 5^2 - 3^2 = 16$. Transverse axis is on the x-axis so the equation is $\frac{x^2}{9} - \frac{y^2}{16} = 1$.

6. $F(0,-3), V(0,-2), c = 3, a = 2$, and so $b^2 = 9 - 4 = 5$. The transverse axis is on the y-axis so the equation is $\frac{y^2}{4} - \frac{x^2}{5} = 1$.

7. $F(4,0)$, length of conjugate axis is 6 so $b = \frac{6}{2} = 3$, $c = 4, a^2 = c^2 - b^2 = 16 - 9 = 7$. The Transverse axis is on the x-axis so the equation is $\frac{x^2}{7} - \frac{y^2}{9} = 1$.

8. $F(0,-6)$, conjugate axis length is 10 so $b = \frac{10}{2} = 5, c = 6, a^2 = 6^2 - 5^2 = 11$. The transverse axis is on the y-axis so the equation is $\frac{y^2}{11} - \frac{x^2}{25} = 1$.

9. $\frac{x^2}{4} - \frac{y^2}{9} = 1, a = 2, b = 3, c = \sqrt{4+9} = \sqrt{13}$. $V = (\pm 2, 0), F(\pm\sqrt{13}, 0), M = (0, \pm 3)$.

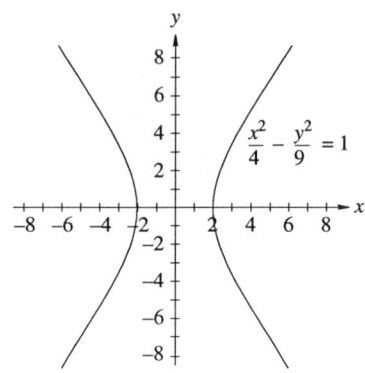

10. $\frac{x^2}{9} - \frac{y^2}{4} = 1, a = 3, b = 2, c = \sqrt{9+4} = \sqrt{13}$. The transverse axis is on the x-axis, so $V = (\pm 3, 0)$, $F = (\pm\sqrt{13}, 0), M = (0, \pm 2)$.

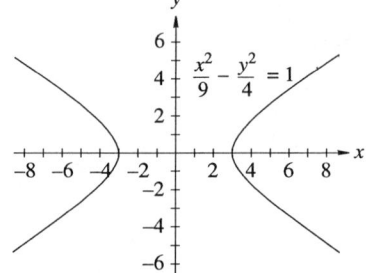

11. $\frac{y^2}{4} - \frac{x^2}{1} = 1, a = 2, b = 1, c = \sqrt{4+1} = \sqrt{5}$. The transverse axis is on the y-axis, so $V = (0, \pm 2), F = (0, \pm\sqrt{5}), M = (\pm 1, 0)$.

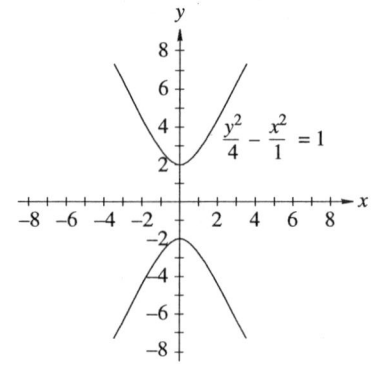

12. $\frac{y^2}{36} - \frac{x^2}{16} = 1, a = 6, b = 4, c = \sqrt{36 + 16} = \sqrt{52} = 2\sqrt{13}$. The transverse axis is on the y-axis, so $V = (0, \pm 6), F = (0, \pm 2\sqrt{13}), M = (\pm 4, 0)$.

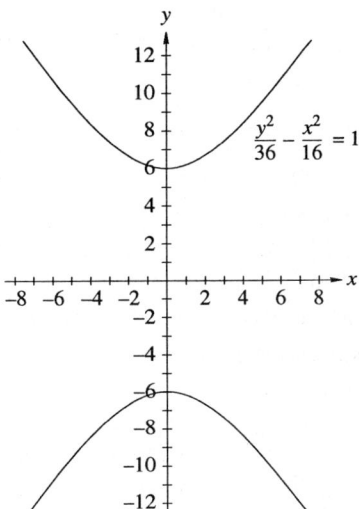

13. $4x^2 - y^2 = 4 \Rightarrow \frac{x^2}{1} - \frac{y^2}{4} = 1, a = 1, b = 2, c = \sqrt{1 + 4} = \sqrt{5}$. The transverse axis is on the x-axis, so $V = (\pm 1, 0), F = (\pm \sqrt{5}, 0), M = (0, \pm 2)$.

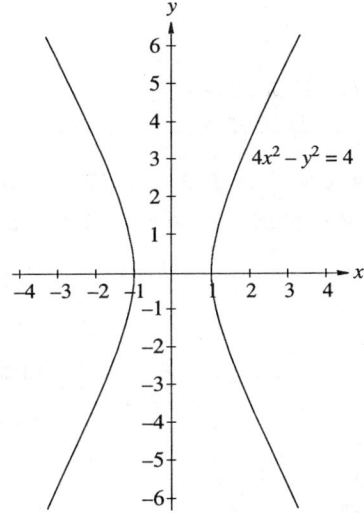

14. $9x^2 - 4y^2 = 36 \Rightarrow \frac{x^2}{4} - \frac{y^2}{9} = 1, a = 2, b = 3, c = \sqrt{4 + 9} = \sqrt{13}$. The transverse axis is on the x-axis, so $V = (\pm 2, 0), F = (\pm \sqrt{13}, 0), M = (0, \pm 3)$.

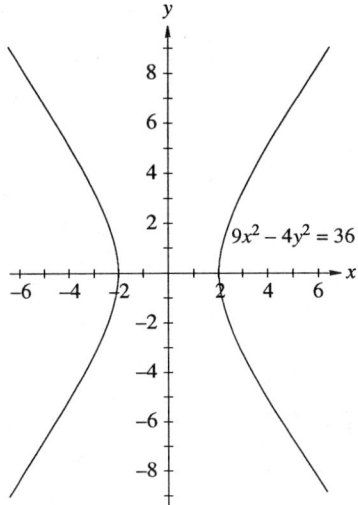

15. $36y^2 - 25x^2 = 900 \Rightarrow \frac{y^2}{25} - \frac{x^2}{36} = 1, a = 5, b = 6, c = \sqrt{25 + 36} = \sqrt{61}$. The transverse axis is on the y-axis so $V = (0, \pm 5), F = (0, \pm \sqrt{61}), M = (\pm 6, 0)$.

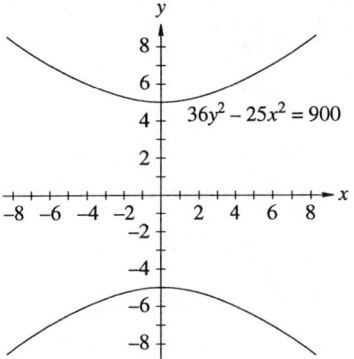

16. $y^2 - 9x^2 = 18 \Rightarrow \frac{y^2}{18} - \frac{x^2}{2} = 1, a = \sqrt{18} = 3\sqrt{2}, b = \sqrt{2}, c = \sqrt{18+2} = \sqrt{20} = 2\sqrt{5}$. The transverse axis is on the y-axis so $V = (0, \pm 3\sqrt{2})$, $F = (0, \pm 2\sqrt{5}), M = (\pm\sqrt{2}, 0)$.

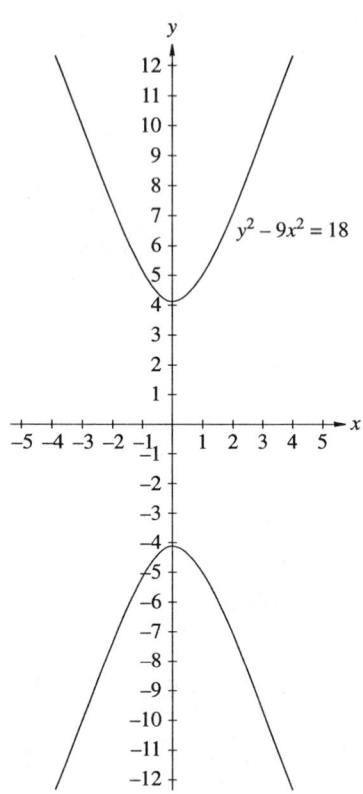

17. If the vertex is $(7, 0)$ then $a = 7, \frac{c}{a} = 1.5$ so $c = 1.5 \times 7 = 10.5, b^2 = c^2 - a^2 = 61.25$. The transverse axis is on the x-axis so the equation is $\frac{x^2}{49} - \frac{y^2}{61.25} = 1$.

18. **(a)**

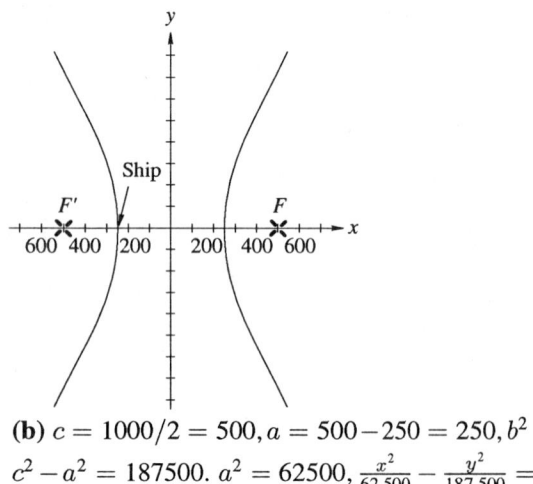

(b) $c = 1000/2 = 500, a = 500 - 250 = 250, b^2 = c^2 - a^2 = 187500. a^2 = 62500, \frac{x^2}{62,500} - \frac{y^2}{187,500} = 1$

19. If $a = b$ then $c = \sqrt{a^2 + b^2} = \sqrt{2a^2} = a\sqrt{2}, e = \frac{c}{a} = \frac{a\sqrt{2}}{a} = \sqrt{2}$

20. $a = 153, \frac{x^2}{153^2} - \frac{y^2}{b^2} = 1, \frac{238^2}{153^2} - \frac{(-616)^2}{b^2} = 1, \frac{616^2}{b^2} = \frac{238^2}{153^2} - 1, b^2 = \dfrac{616^2}{\frac{238^2}{153^2} - 1} = 267{,}269; b = 516.98$.

The equation is $\frac{x^2}{153^2} - \frac{y^2}{516.98^2} = 1$

21. The product of the current I and resistance R in a simple circuit as a constant is represented by $IR = k$. Since $i = 3.2$ and $R = 18$, we see that $IR = 3.2 \times 18 = 57.6$. Thus, $k = 57.6$ is the constant. Solving for I, we obtain $I = \frac{57.6}{R}$. Graphing this function produces the following figure.

22. From the figure we see that $a = 82$. Since the slope of the asymptotes is $\pm\frac{b}{a} = \pm 1.25$, we have $b = 1.25a = 102.5$. Hence, we get $\frac{x^2}{82^2} - \frac{y^2}{102.5^2} = 1$ or $25x^2 - 16y^2 = 168{,}100$.

23. Letting $a = 42$, we get $\frac{x^2}{42^2} - \frac{y^2}{b^2} = 1$. Substituting for x and y, we obtain $\frac{(-126)^2}{42^2} - \frac{(30\sqrt{3})^2}{b^2} = 1$ or

$$126^2 b^2 - (30\sqrt{3})^2 42^2 = 42^2 b^2$$
$$b^2(126^2 - 42^2) = 42^2(30\sqrt{3})^2$$
$$14{,}112 b^2 = 4{,}762{,}800$$
$$b^2 = 337.5$$
$$b \approx 18.37$$

The desired equation is $\frac{x^2}{1764} - \frac{y^2}{337.5} = 1$.

24. The graph of $I = 14/R$ is shown in the figure below.

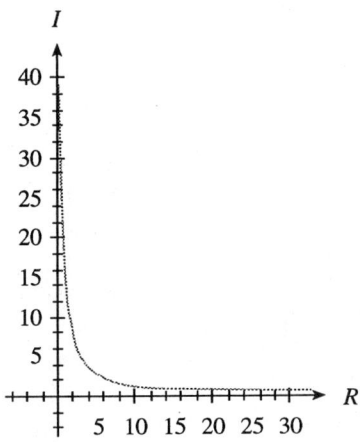

25. Substituting, we obtain $2250 = 8.34FC$. Solving for F, we get $F = \frac{2250}{8.34C} = \frac{270}{C}$. The graph is shown below.

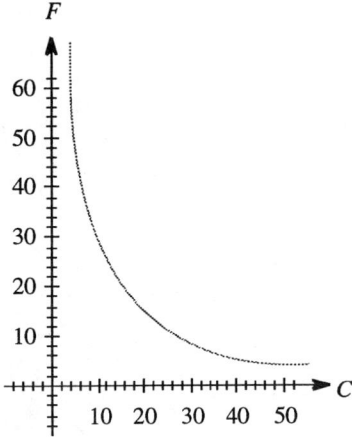

26. Since $750 = 8.34FC$, we graph $F = \frac{750}{8.34C} = \frac{90}{C}$ as shown below.

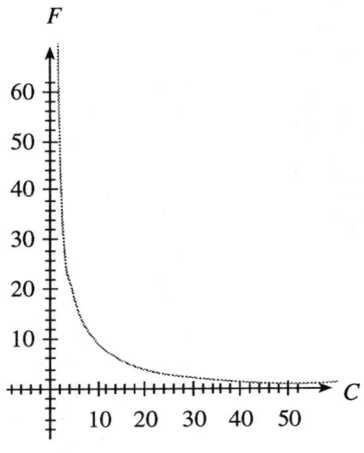

15.6 TRANSLATION OF AXES

1. $\frac{(x-4)^2}{9} + \frac{(y+3)^2}{4} = 1$, $x' = x - 4, y' = y + 3$ so $(h, k) = (4, -3)$; $\frac{x'^2}{9} + \frac{y'^2}{4} = 1$. This is an ellipse with $a = 3, b = 2$, and $c = \sqrt{9 - 4} = \sqrt{5}$. The major axis is horizontal.

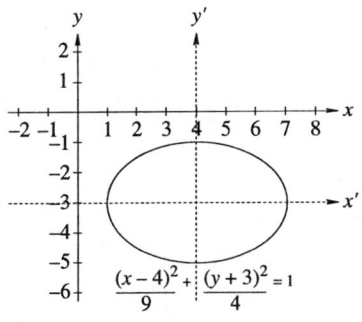

	$x'y'$-system	xy-system
center	$(0', 0')$	$(4, -3)$
vertices	$(\pm 3', 0')$	$(7, -3), (1, -3)$
foci	$(\pm\sqrt{5}', 0')$	$(\sqrt{5} + 4, -3),$ $(-\sqrt{5} + 4, -3)$
endpoints of minor axis	$(0', \pm 2')$	$(4, -1), (4, -5)$

2. $\frac{(x+5)^2}{16} - \frac{(y+3)^2}{25} = 1$. This is a hyperbola with $x' = x + 5$ and $y' = y + 3$ so $(h, k) = (-5, -3)$; $a = 4, b = 5$ so $c = \sqrt{16 + 25} = \sqrt{41}$. The transverse axis is horizontal.

	$x'y'$-system	xy-system
center	$(0', 0')$	$(-5, -3)$
vertices	$(\pm 4', 0')$	$(-9, -3),\ (-1, -3)$
foci	$(\pm\sqrt{41}, 0')$	$(\sqrt{41} - 5, -3),$ $(-\sqrt{41} - 5, -3)$
endpoints of conjugate axis	$(0', \pm 5')$	$(-5, 2),\ (-5, -8)$

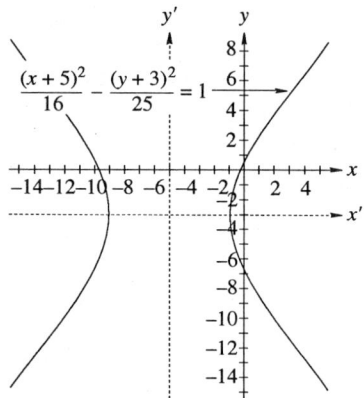

3. $\frac{(y-3)^2}{36} - (x+4)^2 = 1$, is a hyperbola with $y' = y - 3$ and $x' = x + 4$, so $(h, k) = (3, -4), a = 6, b = 1, c = \sqrt{37}$. The transverse axis is horizontal.

	$x'y'$-system	xy-system
center	$(0', 0')$	$(-4, 3)$
vertices	$(0', \pm\sqrt{37'})$	$(-4, 9),\ (-4, -3)$
foci	$(0', \pm 6)$	$(-4, \sqrt{37} + 3)$ $(-4, -\sqrt{37} + 3)$
endpoints of conjugate axis	$(\pm 1', 0')$	$(-3, 3),\ (-5, 3)$

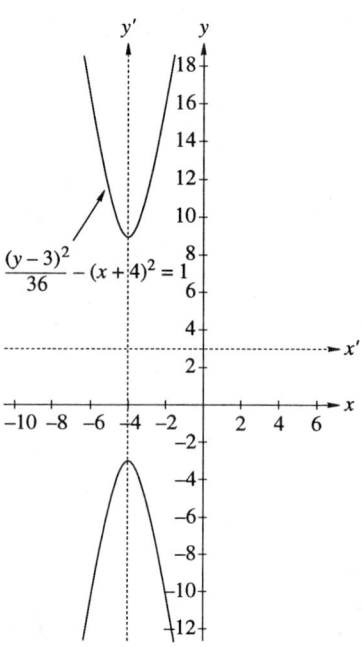

4. $(y + 5)^2 = 12(x + 1), x' = x + 1, y' = y + 5$ so $(h, k) = (-1, -5); p = 12/4 = 3$ parabola opens right

	$x'y'$-system	xy-system
vertex	$(0', 0')$	$(-1, -5)$
focus	$(3', 0')$	$(2, -5)$

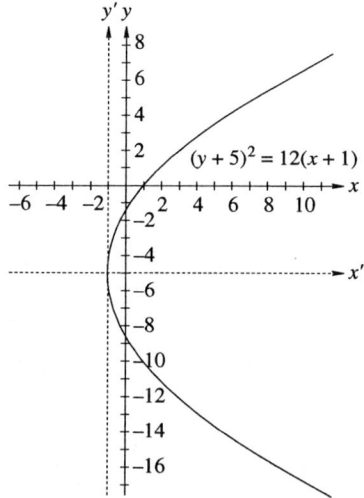

5. $100(x+5)^2 - 4y^2 = 400 \Rightarrow \frac{(x+5)^2}{4} - \frac{y^2}{100} = 1$;
This is a hyperbola with $x' = x+5$ and $y' = y$ so $(h,k) = (-5,0); a = 2, b = 10, c = \sqrt{104} = 2\sqrt{26}$. The transverse axis is vertical.

	$x'y'$-system	xy-system
center	$(0',0')$	$(-5,0)$
vertices	$(\pm 2', 0')$	$(-3,0), \ (-7,0)$
foci	$(\pm 2\sqrt{26}', 0'$	$(2\sqrt{26}-5, 0),$ $(-2\sqrt{26}-5, 0)$
endpoints of conjugate axis	$(0', \pm 10')$	$(-5,10), \ (-5,-10)$

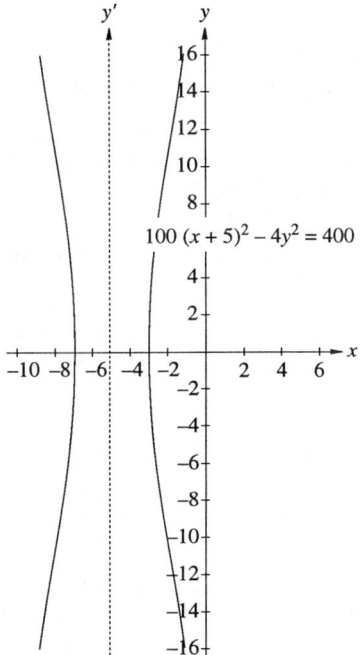

$100\,(x+5)^2 - 4y^2 = 400$

6. $(y+3)^2 = 8(x-2)$ is a parabola with $x' = x-2$ and $y' = y+3$. The vertex is $(h,k) = (2,-3)$ and $p = \frac{8}{4} = 2$, so the parabola opens right.

	$x'y'$-system	xy-system
vertex	$(0',0')$	$(2,-3)$
focus	$(2',0')$	$(4,-3)$

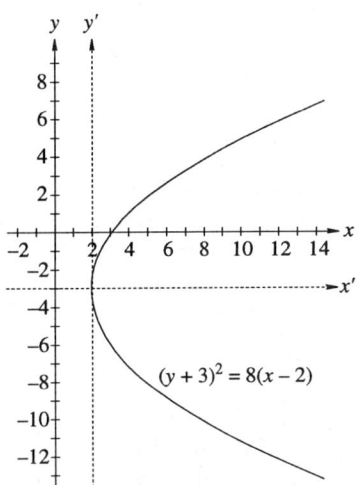

$(y+3)^2 = 8(x-2)$

7. $16x^2 + 4y^2 + 64x - 12y + 57 = 0$ or $16(x^2 + 4x) + 4(y^2 - 3y) = -57$. Completing the square produces $16(x^2 + 4x + 4) + 4(y^2 - 3y + \frac{9}{4}) = -57 + 16 \cdot 4 + 4 \cdot \frac{9}{4}$ or $16(x+2)^2 + 4(y - \frac{3}{2})^2 = 16$. This has the standard form equation $\frac{(x+2)^2}{1} + \frac{(y - \frac{3}{2})}{4} = 1$ and so it is an ellipse with $x' = x+2$ and $y' = y - \frac{3}{2}$. Thus, $(h,k) = (-2, \frac{3}{2}), a = 1, b = 2$, and $c = \sqrt{4-1} = \sqrt{3}$. The major axis is vertical.

	$x'y'$-system	xy-system
center	$(0',0')$	$(-2, \frac{3}{2})$
vertices	$(0', \pm 2')$	$(-2, \frac{7}{2}), \ (-2, -\frac{1}{2})$
foci	$(0', \pm \sqrt{3}')$	$(-2, \sqrt{3} + \frac{3}{2}),$ $(-2, -\sqrt{3} + \frac{3}{2})$
endpoints of minor axis	$(\pm 1', 0')$	$(-1, \frac{3}{2}), \ (-3, \frac{3}{2})$

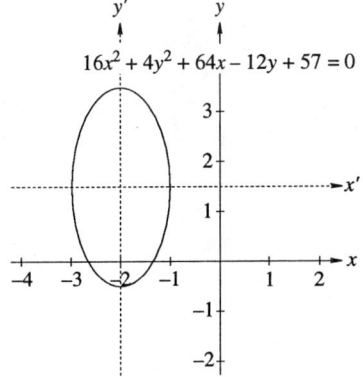

$16x^2 + 4y^2 + 64x - 12y + 57 = 0$

8. $x^2 + y^2 - 2x + 2y - 2 = 0; (x^2 - 2x + 1) + (y^2 + 2y + 1) = 2 + 1 + 1; (x-1)^2 + (y+1)^2 = 4$; circle; center: $(1, -1)$, radius 2

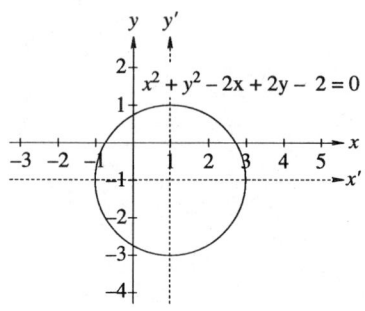

9. $25x^2 + 4y^2 - 250x - 16y + 541 = 0$ or $25(x^2 - 10x) + 4(y^2 - 4y) = -541$. Completing the square produces $25(x^2 - 10x + 25) + 4(y^2 - 4y + 4) = -541 + 25 \cdot 25 + 4 \cdot 4$ or $25(x-5)^2 + 4(y-2)^2 = 100$. This has the standard form $\frac{(x-5)^2}{4} + \frac{(y-2)^2}{25} = 1$ and so it is an ellipse with $x' = x - 5, y' = y - 2$. Here $(h, k) = (5, 2), a = 2, b = 5, c = \sqrt{21}$. The major axis is vertical.

	$x'y'$-system	xy-system
center	$(0', 0')$	$(5, 2)$
vertices	$(0', \pm 5')$	$(5, 7),\ (5, -3)$
foci	$(0', \pm \sqrt{21}')$	$(5, 2 + \sqrt{21}),$ $(5, 2 - \sqrt{21})$
endpoints of minor axis	$(\pm 2', 0')$	$(7, 2),\ (3, 2)$

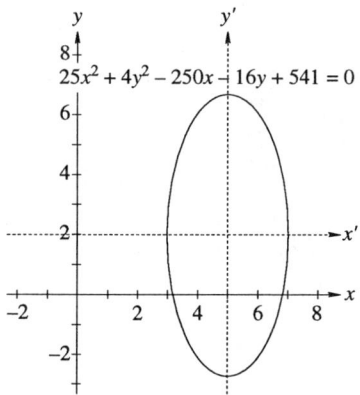

10. $100x^2 - 180x - 100y + 81 = 0$ or $100x^2 - 180x = 100y - 81$. Factoring, we get $100(x^2 - 1.8x) = 100y - 81$ and then completing the square yields $100(x^2 - 1.8x + 0.81) = 100y - 81 + 81$ or $100(x - 0.9)^2 = 100y$ which is equivalent to $(x - 0.9)^2 = y$. This is the standard form equation for a parabola with $(h, k) = (0.9, 0)$ and $p = \frac{1}{4} = 0.25$. The parabola opens upward.

	$x'y'$-system	xy-system
vertex	$(0', 0')$	$(0.9, 0)$
focus	$(0', 0.25')$	$(0.9, 0.25)$

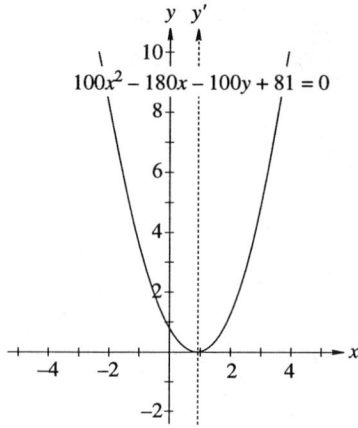

11. $2x^2 - y^2 - 16x + 4y + 24 = 0$ or $2(x^2 - 8x) - (y^2 - 4y) = -24$. Completing the square yields $2(x^2 - 8x + 16) - (y^2 - 4y + 4) = -24 + 2 \cdot 16 - 1 \cdot 4$ or $2(x - 4)^2 - (y - 2)^2 = 4$ which has the standard form equation $\frac{(x-4)^2}{2} - \frac{(y-2)^2}{4} = 1$. This is an hyperbola with $a = \sqrt{2}, b = 2; c = \sqrt{6}, x' = x - 4, y' = y - 2, (h, k) = (4, 2)$. The transverse axis is horizontal.

	$x'y'$-system	xy-system
center	$(0', 0')$	$(4, 2)$
vertices	$(\pm \sqrt{2}', 0')$	$(4 + \sqrt{2}, 2),$ $(4 - \sqrt{2}, 2)$
foci	$(\pm \sqrt{6}', 0')$	$(4 + \sqrt{6}, 2)$ $(4 - \sqrt{6}, 2)$
endpoints of conjugate axis	$(0', \pm 2')$	$(4, 0),\ (4, 4)$

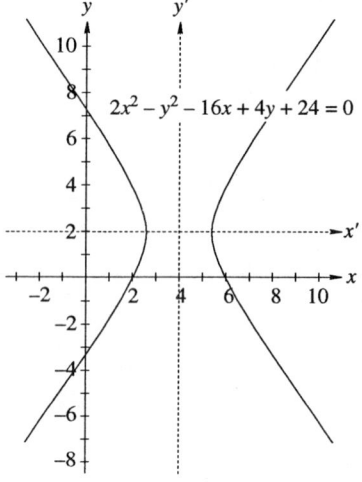

12. $9x^2 - 36x + 16y^2 - 32y - 524 = 0$ or $9(x^2 - 4x) + 16(y^2 - 2y) = 524$. Completing the square, we get $9(x^2 - 4x + 4) + 16(y^2 - 2y + 1) = 524 + 9 \cdot 4 + 16 \cdot 1$ or $9(x - 2)^2 + 16(y - 1)^2 = 576$. This has the standard form equation $\frac{(x-2)^2}{64} + \frac{(y-1)^2}{36} = 1$ and so we see that it is an ellipse, with $a = 8, b = 6; c = \sqrt{64 - 36} = \sqrt{28} = 2\sqrt{7}, x' = x - 2, y' = y - 1; (h, k) = (2, 1)$. The major axis is horizontal.

	$x'y'$−system	xy−system
center	$(0', 0')$	$(2, 1)$
vertices	$(\pm 8', 0')$	$(10, 1), (-6, 1)$
foci	$(\pm 2\sqrt{7}, 0')$	$(2 + 2\sqrt{7}, 1),$ $(2 - 2\sqrt{7}, 1)$
endpoints of minor axis	$(0', \pm 6')$	$(2, 7), (2, -5)$

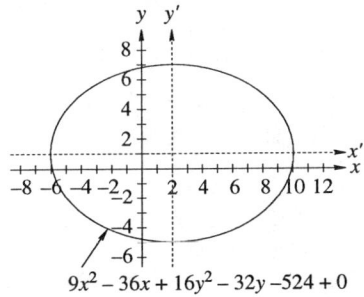

$9x^2 - 36x + 16y^2 - 32y - 524 + 0$

13. Parabola, Vertex $= (2, -3), p = 8$, axis vertical $x' = x - 2, y' = y + 3, x'^2 = 4py'$ or $(x - 2)^2 = 32(y + 3)$

14. Hyperbola, vertex at $(2, 1)$, foci at $(-6, 1)$ and $(8, 1)$. Foci are horizontally displaced so the transverse axis is horizontal. The center is the midpoint between the foci or at $(1, 1), c = |8 - 1| = 7, a = |2 - 1| = 1, b = \sqrt{c^2 - a^2} = \sqrt{49 - 1} = \sqrt{48}, x' = x - 1, y' = y - 1, \frac{x'^2}{1} - \frac{y'^2}{48} = 1$ or $\frac{(x-1)^2}{1} - \frac{(y-1)^2}{48} = 1$.

15. Ellipse, center at $(4, -3)$, focus at $(8, -3)$, vertex at $(10, -3), a = |10 - 4| = 6, c = |8 - 4| = 4, b^2 = $

$a^2 - c^2 = 36 - 16 = 20$. Major axis is horizontal. $x' = x - 4, y' = y + 3$, and so $\frac{x'^2}{36} + \frac{y'^2}{20} = 1$ or $\frac{(x-4)^2}{36} + \frac{(y+3)^2}{20} = 1$.

16. Ellipse, center at $(-2, 0)$, focus at $(6, 0)$ vertex at $(9, 0)$. Major axis is horizontal, $a = |9 - -2| = 11, c = |6 - -2| = 8, b^2 = a^2 - c^2 = 11^2 - 8^2 = 121 - 64 = 57, x' = x + 2, y' = y$ and so $\frac{x'^2}{121} + \frac{y'^2}{57} = 1$ or $\frac{(x+2)^2}{121} + \frac{y^2}{57} = 1$.

17. Hyperbola, center at $(-3, 2)$ focus at $(-3, 7)$, transverse axis 6 units. Center and focus vertically displaces so transverse axis is vertical. $c = |7 - 2| = 5, a = \frac{6}{2} = 3, b^2 = c^2 - a^2 = 16, x' = x + 3, y' = y - 2$ and so $\frac{y'^2}{9} - \frac{x'^2}{16} = 1$ or $\frac{(y-2)^2}{9} - \frac{(x+3)^2}{16} = 1$.

18. Ellipse, center at $(3, 5)$, focus at $(3, 8)$ minor axis 2 units. Major axis is vertical $c = |8 - 5| = 3, a = \frac{2}{2} = 1, b^2 = c^2 + a^2b^2 = 9 + 1 = 10; x' = x - 3, y' = y - 5$ and so $\frac{x'^2}{1} + \frac{y'^2}{10} = 1$ or $\frac{(x-3)^2}{1} + \frac{(y-5)^2}{10} = 1$.

19. Parabola, vertex at $(-5, 1)$ so $x' = x + 5, y' = y - 1, p = -4$, axis parallel to x axis $y'^2 = 4px'$ or $(y - 1)^2 = -16(x + 5)$.

20. Parabola, vertex at $(-3, -6)$ so $x' = x + 3, y' = y + 6, p = -12$, axis parallel to y-axis $x'^2 = 4py'$ or $(x + 3)^2 = -48(y + 6)$.

21. Ellipse, center at $(-4, 1)$ so $x' = x + 4, y' = y - 1$, and the focus is at $(-4, 9)$, so major axis is vertical and $c = |9 - 1| = 8$, minor axis is 12. Hence, $a = \frac{12}{2} = 6, b^2 = a^2 + c^2 = 64 + 36 = 100. \frac{x'^2}{36} + \frac{y'^2}{100} = 1$ or $\frac{(x+4)^2}{36} + \frac{(y-1)^2}{100} = 1$.

22. Hyperbola, center at $(2, 8)$ so $x' = x - 2, y' = y - 8$; vertex at $(6, 8)$ so $a = |6 - 2| = 4$, transverse axis is horizontal, conjugate axis is 10 so $b = \frac{10}{2} = 5. \frac{x'^2}{16} - \frac{y'^2}{25} = 1$ or $\frac{(x-2)^2}{16} - \frac{(y-5)^2}{25} = 1$.

23. $s = 29.4t - 4.9t^2$ or $4.9t^2 - 29.4t = -s, 4.9(t^2 - 6t) = -s, 4.9(t^2 - 6t + 9) = -s + 4.9 \times 9, 4.9(t - 3)^2 = -s + 44.1$ or $(t - 3)^2 = -\frac{1}{4.9}(s - 44.1)$. This is a parabola with vertex at $(3, 44.1)$ it opens downward so the maximum height is 44.1 m.

24. $2a = 90 + 140 + 2 \cdot 4000 = 8230$ so $a = 4115$ and $c = \frac{140-90}{2} = \frac{50}{2} = 25$. Hence $b^2 = a^2 - c^2 = 4115^2 - 25^2 = 16932600$ and so $\frac{x^2}{4115^2} + \frac{y^2}{16\,932\,600} = 1$ or $\frac{x^2}{4115^2} + \frac{y^2}{4114.924^2} = 1$.

25. Since the signal from A arrives $600\mu s$ after B the common difference on the hyperbola is $300 \times 600 = 180,000$ m $= 180$ km so $2a = 180$ hence $a = 90, b^2 = c^2 - a^2 = 250000 - 8100 = 241,900$. The equation is $\frac{x^2}{8100} - \frac{y^2}{241,900} = 1$. If the transverse axis passes through A and B and the conjugate axis passes through the midpoint of $\overline{AB}$, then A is at $(-500, 0)$ and B at $(500, 0)$. $AP + BP = 300 \times 8,000 = 2,400,000$ m $= 2,400$ km; $AP + BP = 2,400; AP - BP = 180; AB = 1000; 2AP = 2580; AP = 1290; BP = 1110; \alpha = \cos^{-1}\left(\frac{1000^2 + 1290^2 - 1110^2}{2 \cdot 1000 \cdot 1290}\right) = 56.2866°; y = AP \sin\alpha = 1073.054$ km; $x = AP \cos\alpha - 500 = 216$ km. The plane is at $(216, 1073.054)$.

26. Here $a = 18.09$ A.U.; $b = 4.56$ A.U. Combining these produces $c^2 = a^2 - b^2$, or $c = \sqrt{18.09^2 - 4.56^2} = 17.51$ A.U. The distance from a focus to the center is $c = 17.51$ A.U., so if the sun is at one focus and has coordinates $(0, 0)$, then the center is at $(17.51, 0)$ and we have the ellipse: $\frac{(x-17.51)^2}{18.09^2} + \frac{y^2}{4.56^2} = 1$. The maximum distance is $a + c = 18.09 + 17.51 = 35.60$ A.U. The minimum distance is $a - c = 0.58$ A.U.

27. **(a)** $x^2 + 200y - 4.00 = 0$ or $x^2 = -200y + 4.00$ and so $x^2 = -200(y - 0.02)$
(b) In the no load position $y - 0.02 = 0$ or $y = 0.02$ m $= 2$ cm.

28. **(a)** $x^2 + 100y - 4.00 = 0$
$$x^2 = -100y + 4.00$$
$$= -100(y - 0.04)$$
Thus, the standard form for this equation is $x^2 = -100(y - 0.04)$.
(b) The vertical displacement is the y-shift, or 0.04 ft.

29. We begin by using the distance formula. The distance from the airplane to the transmitter at $(0, 0)$ is $\sqrt{x^2 + y^2}$. Similarly, the distance from the airplane to the transmitter at $(0, 250)$ is $\sqrt{x^2 + (y - 250)^2}$. If the place is 50 mi further from $(0, 0)$ than from $(0, 250)$, we have $\sqrt{x^2 + y^2} = \sqrt{x^2 + (y - 250)^2} + 50$. We solve this equation as follows

$$\sqrt{x^2 + y^2} = \sqrt{x^2 + (y - 250)^2} + 50$$
$$\sqrt{x^2 + y^2} - 50 = \sqrt{x^2 + y^2 - 500y + 250^2}$$
$$x^2 - 100\sqrt{x^2 + y^2} + y^2 + 50^2 = x^2 + y^2 - 500y + 250^2$$
$$-100\sqrt{x^2 + y^2} = -500y + 250^2 - 50^2$$
$$= -500y + 60,000$$
$$100^2(x^2 + y^2) = (-500y + 60,000)^2$$
$$10,000x^2 + 10,000y^2 = 250,000y^2 - 60,000,000y + 3,600,000,000$$
$$24y^2 - 6,000y - x^2 = -360,000$$
$$24(y^2 - 250y) - x^2 = -360,000$$
$$24(y^2 - 250y + 125^2) - x^2 = -360,000 + 24 \cdot 125^2$$
$$24(y - 125)^2 - x^2 = 15,000$$

Thus, the equation of the hyperbola is $24(y - 125)^2 - x^2 = 15,000$. Since it is closer to $(0, 250)$ than to $(0, 0)$, it is on the upper half of the hyperbola.

(b) Since the plane is on the line $y = x$ and the hyperbola $24(y - 125)^2 - x^2 = 15,000$, we can substitute x for y and solve the resulting equation.

$$24(x - 125)^2 - x^2 = 15,000$$
$$24(x^2 - 250x + 15,625) - x^2 = 15,000$$
$$23x^2 - 6,000x + 360,000 = 0$$

Using the quadratic formula, we find that $x = \frac{3000 \pm 600\sqrt{2}}{23}$. Because the plane is on the upper-half of the hyperbola, we have $x = \frac{3000 + 600\sqrt{2}}{23}$ and the plane is located at $\left(\frac{3,000+600\sqrt{2}}{23}, \frac{3,000+600\sqrt{2}}{23} \right) \approx (167.3, 167.3)$.

30. Using the same hyperbola and substituting $2x$ for y, we obtain

$$24(2x - 125)^2 - x^2 = 15,000$$
$$24(4x^2 - 500x + 15,625) - x^2 = 15,000$$
$$95x^2 - 12,000x + 360,000 = 0$$
$$19x^2 - 2,400x + 72,000 = 0$$

The quadratic formula yields $x = \frac{1200 \pm 120\sqrt{5}}{19}$. As we need the larger value on the upper branch, we find that the plane is at $\left(\frac{1,200+120\sqrt{5}}{19}, \frac{2,400+240\sqrt{5}}{19} \right) \approx$ $(77.3, 154, 6)$.

(b)

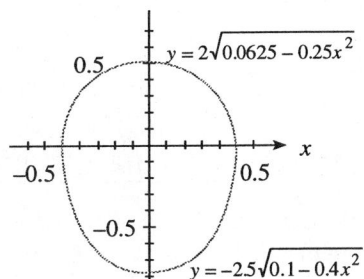

31. (a) Solving the equation for the circle produces the equation $y = \pm 2\sqrt{0.0625 + 0.25x^2}$. We will select the positive square root, $y = 2\sqrt{0.0625 - 0.25x^2}$, to get the graph of the upper semicircle as the top part of the cam. Solving the equation for the ellipse we get $y = \pm 2.5\sqrt{0.1 - 0.4x^2}$. Here we take the negative square root, $y = -2.5\sqrt{0.1 - 0.4x^2}$, to get its graph as the bottom part of the cam.

32.

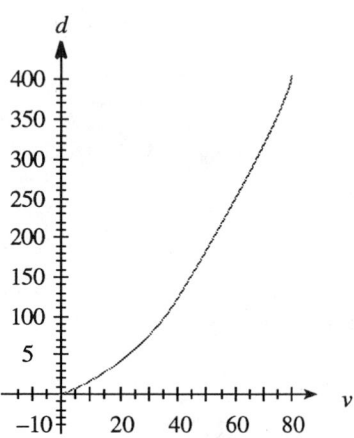

33. **(a)** The cost is 0.9 times the number of items, less 0.01 times the number of items over 24. This yields the function $C = 0.9x - 0.01x(x - 24) = 0.9x - 0.01x^2 + 0.24x = 1.14x - 0.01x^2$.

(b)

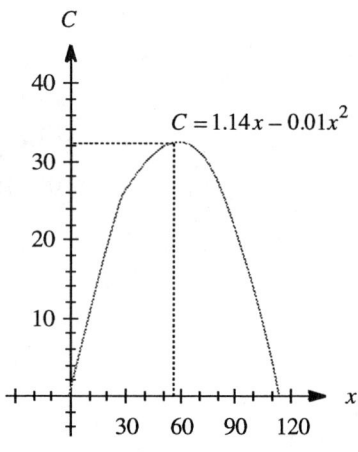

(c) From the graph, we estimate the maximum cost when 57 items are packaged.

(d) The maximum cost is \$32.49.

34. **(a)**

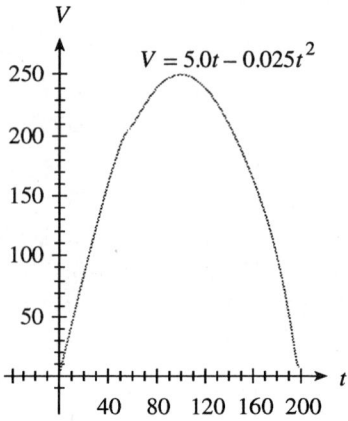

(b) When $V = 0$, we have $5.0T - 0.025T^2 = (5.0 - 0.025T)T = 0$. Thus, $T = 0$ or $5.0 - 0.025T = 0$. The latter occurs when $T = \frac{5}{0.025} = 200$. We conclude the $V = 0$ when $T = 0°C$ or when $T = 200°C$.

≡ 15.7 ROTATION OF AXES; THE GENERAL SECOND-DEGREE EQUATION

1. **(a)** $xy = -9$. Here, $A = 0, B = 1$, and $C = 0$, and the discriminant is $B^2 - 4AC = 1 > 0$; so the curve is a hyperbola,

(b) $\cot 2\theta = \frac{A-C}{B} = 0$ or $2\theta = 90°$ so the angle needed to rotate the coordinate axes to remove the xy-term is $\theta = 45°$,

(c) The rotation formulas are $x = x'' \cos\theta - y'' \sin\theta$ and $y = y'' \cos\theta + x'' \sin\theta$ we get $x = 0.7071x'' - 0.7071y''$ and $y = 0.7071y'' + 0.7071x''$,

(d) Substituting the equations from (c) into the original equation $xy = -9$ we get $(0.7071x'' - 0.7071y'')(0.7071y'' + 0.7071x'') = -9$ or $\frac{1}{2}x''^2 - \frac{1}{2}y''^2 = -9$ which is equivalent of $\frac{y''^2}{18} - \frac{x''^2}{18} = 1$.

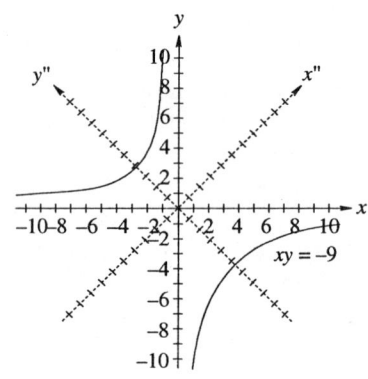

2.   **(a)** $xy = 9$. Here $A = 0, B = 1$, and $C = 0$, so the discriminant is $B^2 = 4AC = 1 > 0$, so the curve is a hyperbola,

(b) $\cot 2\theta = \frac{A-C}{B} = \frac{0}{1} = 0 \Rightarrow 2\theta = 90°$ so the angle needed to rotate the coordinate axes to remove the xy-term is $\theta = 45°$,

(c) The rotation formulas are $x = x'' \cos 45° - y'' \sin 45°$ and $y = y'' \cos 45° + x'' \sin 45°$,

(d) Substituting the equations from (c) into the original equation $xy = 9$ we get
$\left(\frac{\sqrt{2}}{2} x'' - \frac{\sqrt{2}}{2} y'' \right) \left(\frac{\sqrt{2}}{2} y'' + \frac{\sqrt{2}}{2} x'' \right) = 9 \Rightarrow \frac{1}{2} x''^2 - \frac{1}{2} y''^2 = 9$ which is equivalent to $\frac{x''^2}{18} - \frac{y''^2}{18} = 1$.

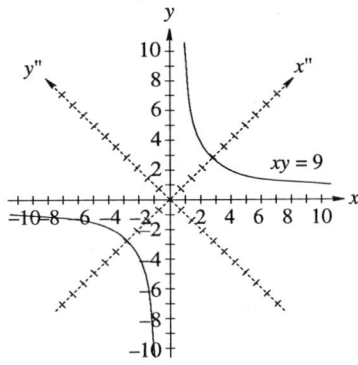

3.   **(a)** $x^2 - 6xy + y^2 - 8 = 0$. Here, $A = 1, B = -6$, and $C = 1$. The discriminant is $B^2 - 4AC = 36 - 4 = 32 > 0$, so the curve is a hyperbola

(b) $\cot 2\theta = \frac{A-C}{B} = \frac{0}{-6}$ so $2\theta = 90°$ and so the angle needed to rotate the coordinate axes to remove the xy-term is $\theta = 45°$,

(c) The rotation formulas are $x = x'' \cos 45° - y'' \sin 45°, y = y'' \cos 45° + x'' \sin 45°$,

(d) Substituting the equations from (c) into the original equation $x^2 - 6xy + y^2 - 8 = 0$ we get
$\left(\frac{\sqrt{2}}{2} x'' - \frac{\sqrt{2}}{2} y'' \right)^2 - 6\left(\frac{\sqrt{2}}{2} x'' - \frac{\sqrt{2}}{2} y'' \right) \left(\frac{\sqrt{2}}{2} y'' + \frac{\sqrt{2}}{2} x'' \right) + \left(\frac{\sqrt{2}}{2} y'' + \frac{\sqrt{2}}{2} x'' \right)^2 - 8 = 0$; $\frac{1}{2} x''^2 - x'' y'' + \frac{1}{2} y''^2 - 6\left(\frac{1}{2} x''^2 - \frac{1}{2} y''^2 \right) + \left(\frac{1}{2} y''^2 + x'' y'' + \frac{1}{2} x''^2 \right) - 8 = 0$ or $-2x''^2 + 4y''^2 - 8 = 0$.

This is equivalent to $\frac{y''^2}{2} - \frac{x''^2}{4} = 1$.

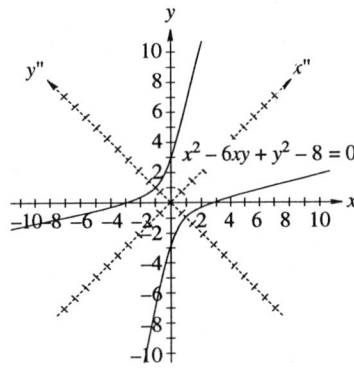

$x^2 - 6xy + y^2 - 8 = 0$

4.   **(a)** $x^2 + 4xy - 2y^2 - 6 = 0$. Here $A = 1, B = 4$, and $C = -2$, so the discriminant is $B^2 - 4AC = 16 + 8 = 24 > 0$ and we see that this curve is a hyperbola,

(b) $\cot 2\theta = \frac{A-C}{B} = \frac{1+2}{4} = \frac{3}{4}, 2\theta = 53.13$ so the angle needed to rotate the coordinate axes to remove the xy-term is $\theta = 26.565°$,

(c) The rotation formulas are $x = x'' \cos 26.565° - y'' \sin 26.565°$ or $x = 0.8944x'' - 0.4472y''$, and $y = y'' \cos 26.565° + x'' \sin 26.565°$ or $y = 0.8944y'' + 0.4472x''$

(d) Substituting the equations from (c) into the original equation $x^2 + 4xy - 2y^2 - 6 = 0$ we get
$(0.8944x'' - 0.4472y'')^2 + 4(0.8944x'' - 0.4472y'')(0.8944y'' + 0.4472x'') - 2(0.8944y'' + 0.4472x'')^2 - 6 = 0$. Multiplying, we get $(0.8x''^2 - 0.8x''y'' + 0.2y''^2) + 4(0.4x''^2 + 0.6x''y'' - 0.4y''^2) - 2(0.8y''^2 + 0.8x''y'' + 0.2x''^2) - 6 = 0$. Collecting like terms leads to $2x''^2 - 3y''^2 - 6 = 0$ which is equivalent to $\frac{x''^2}{3} - \frac{y''^2}{2} = 1$.

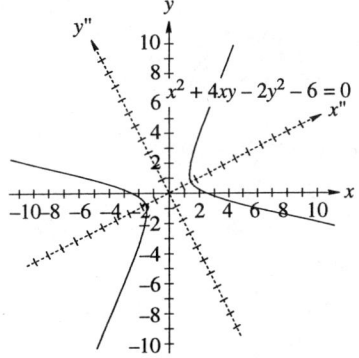

$x^2 + 4xy - 2y^2 - 6 = 0$

5. **(a)** $52x^2 - 72xy + 73y^2 - 100 = 0$. Here $A = 52, B = -72$, and $C = 73$, so the discriminant is $B^2 - 4AC = -10000 < 0$ so the curve is an ellipse,

(b) $\cot 2\theta = \frac{A-C}{B} = \frac{52-73}{-72} = 0.29167, 2\theta = 73.7398$ so the angle needed to rotate the coordinate axes to remove the xy-term is $\theta = 36.870$ with $\cos\theta = 0.8$ and $\sin\theta = 0.6$,

(c) The rotation formulas are $x = 0.8x'' - 0.6y''$ and $y = 0.8y + 0.6x''$,

(d) $52(0.8x'' - 0.6y'')^2 - 72(0.8x'' - 0.6y'')(0.8y'' + 0.6x'') + 73(0.8y'' + 0.6x'')^2 - 100 = 0$ which expands to $52(0.64x''^2 - 0.96x''y'' + 0.36y''^2) - 72(0.48x''^2 + 0.28x''y'' - 0.48y''^2) + 73(0.64y''^2 + 0.96x''y'' + 0.36x''^2) - 100 = 0$ or $25x''^2 + 100y''^2 = 100$ which is equivalent to $\frac{x''^2}{4} + y''^2 = 1$.

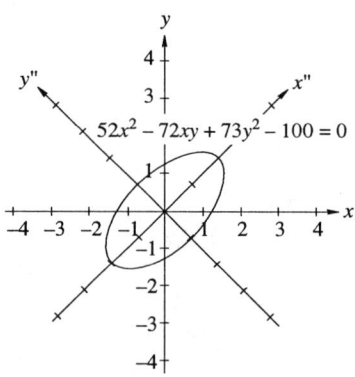

6. **(a)** $11x^2 + 10\sqrt{3}xy + y^2 - 4 = 0$. Here $A = 11, B = 10\sqrt{3}$, and $C = 1$, so the discriminant is $B^2 - 4AC = 300 - 44 = 256 > 0$ so the curve is a hyperbola,

(b) $\cos 2\theta = \frac{A-C}{B} = \frac{11-1}{10\sqrt{3}} = \frac{1}{\sqrt{3}}, 2\theta = 60°$ so the angle needed to rotate the coordinate axes to remove the xy-term is $\theta = 30°$; so, $\cos\theta = \frac{\sqrt{3}}{2} \approx 0.866$ and $\sin\theta = 0.5$.

(c) The rotation formulas are $x = \frac{\sqrt{3}}{2}x'' - \frac{1}{2}y''$ and $y = \frac{\sqrt{3}}{2}y'' + \frac{1}{2}x''$,

(d) Substituting the equations from (c) into the original equation $11x^2 + 10\sqrt{3}xy + y^2 - 4 = 0$ we get $11\left(\frac{\sqrt{3}}{2}x'' - \frac{1}{2}y''\right)^2 + 10\sqrt{3}\left(\frac{\sqrt{3}}{2}x'' - \frac{1}{2}y''\right)\left(\frac{\sqrt{3}}{2}y'' + \frac{1}{2}x''\right) + \left(\frac{\sqrt{3}}{2}y'' + \frac{1}{2}x''\right)^2 - 4 = 0; 11\left(\frac{3}{4}x''^2 - \frac{\sqrt{3}}{2}x''y'' + \frac{1}{4}y''^2\right) + 10\sqrt{3}\left(\frac{\sqrt{3}}{4}x''^2 + \frac{1}{2}x''y'' - \frac{\sqrt{3}}{4}y''\right) + \left(\frac{3}{4}y'' + \frac{\sqrt{3}}{2}x''y'' + \frac{1}{4}x''^2\right) - 4 = 0$. Collecting terms results in $\frac{33}{4}x''^2 - \frac{11\sqrt{3}}{2}x''y'' + \frac{11}{4}y''^2 + \frac{30}{4}x''^2 + \frac{10\sqrt{3}}{2}x''y'' - \frac{30}{4}y''^2 + \frac{3}{4}y''^2 + \frac{\sqrt{3}}{2}x''y'' + \frac{1}{4}x''^2 - 4 = 0; 16x''^2 - 4y''^2 = 4$

which is equivalent to the standard form equation $4x''^2 - y''^2 = 1$ or $\frac{x''^2}{1/4} - \frac{y''^2}{1} = 1$.

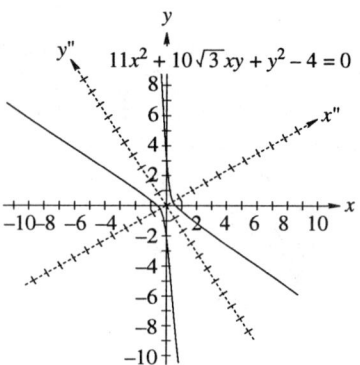

7. **(a)** $x^2 - 2xy + y^2 + x + y = 0$. The discriminant is $B^2 - 4AC = (-2)^2 - 4 \cdot 1 \cdot 1 = 4 - 4 = 0$ so the curve is a parabola,

(b) $\cot 2\theta = \frac{A-C}{B} = \frac{1-1}{-2} = \frac{0}{-2} = 0$ so $2\theta = 90°$ and so $\theta = 45°$,

(c) The rotation formulas are $x = \frac{\sqrt{2}}{2}x'' - \frac{\sqrt{2}}{2}y''$ and $y = \frac{\sqrt{2}}{2}y'' + \frac{\sqrt{2}}{x}x''$,

(d) Substituting the equations from (c) into the original equation $x^2 - 2xy + y^2 + x + y = 0$ we get $\left(\frac{\sqrt{2}}{2}x'' - \frac{\sqrt{2}}{2}y''\right)^2 - 2\left(\frac{\sqrt{2}}{2}x'' - \frac{\sqrt{2}}{2}y''\right)\left(\frac{\sqrt{2}}{2}y'' + \frac{\sqrt{2}}{2}x''\right) + \left(\frac{\sqrt{2}}{2}y'' + \frac{\sqrt{2}}{2}x''\right)^2 + \left(\frac{\sqrt{2}}{2}x'' - \frac{\sqrt{2}}{2}y''\right) + \left(\frac{\sqrt{2}}{2}y'' + \frac{\sqrt{2}}{2}x''\right) = 0; \left(\frac{1}{2}x''^2 - x''y'' + \frac{1}{2}y''^2\right) - 2\left(\frac{1}{2}x''^2 - \frac{1}{2}y''^2\right) + \left(\frac{1}{2}y''^2 + x''y'' + \frac{1}{2}x''^2\right) + \sqrt{2}x'' = 0; 2y''^2 + \sqrt{2}x'' = 0$ or $2y''^2 = -\sqrt{2}x''$ which is equivalent to $y''^2 = \frac{-\sqrt{2}}{2}x''$.

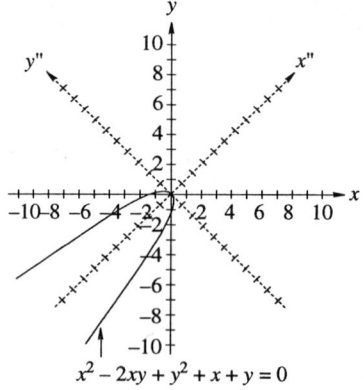

8. **(a)** $3x^2 + 2\sqrt{3}xy + y^2 - 2x + 2\sqrt{3}y = 0$. The discriminant is $B^2 - 4AC = 12 - 12 = 0$ so the curve is a parabola,

(b) $\cot 2\theta = \frac{A-C}{B} = \frac{3-1}{2\sqrt{3}} = \frac{2}{2\sqrt{3}} = \frac{1}{\sqrt{3}}$ so $2\theta = 60°$ and so $\theta = 30°$,

(c) The rotation formulas are $x = \frac{\sqrt{3}}{2}x'' - \frac{1}{2}y''$ and $y = \frac{\sqrt{3}}{2}y'' + \frac{1}{2}x''$,

(d) Substituting the equations from (c) into the original equation $3x^2 + 2\sqrt{3}xy + y^2 - 2x + 2\sqrt{3}y = 0$ we get

$3\left(\frac{\sqrt{3}}{2}x'' - \frac{1}{2}y''\right)^2 + 2\sqrt{3}\left(\frac{\sqrt{3}}{2}x'' - \frac{1}{2}y''\right)\left(\frac{\sqrt{3}}{2}y'' + \frac{1}{2}x''\right) + \left(\frac{\sqrt{3}}{2}y'' + \frac{1}{2}x''\right)^2 - 2\left(\frac{\sqrt{3}}{2}x'' - \frac{1}{2}y''\right) + 2\sqrt{3}\left(\frac{\sqrt{3}}{2}y'' + \frac{1}{2}x''\right) = 0; 3\left(\frac{3}{4}x''^2 - \frac{\sqrt{3}}{2}x''y'' + \frac{1}{4}y''^2\right) + 2\sqrt{3}\left(\frac{\sqrt{3}}{4}x''^2 + \frac{1}{2}x''y'' - \frac{\sqrt{3}}{4}y''^2\right) + \left(\frac{3}{4}y''^2 + \frac{\sqrt{3}}{2}x''y'' + \frac{1}{4}x''^2\right) - 2\left(\frac{\sqrt{3}}{2}x'' - \frac{1}{2}y''\right) + 2\sqrt{3}\left(\frac{\sqrt{3}}{2}y'' + \frac{1}{2}x''\right) = 0; \frac{9}{4}x''^2 - \frac{3\sqrt{3}}{2}x''y'' + \frac{3}{4}y''^2 + \frac{6}{4}x''^2 + \frac{2\sqrt{3}}{2}x''y'' - \frac{6}{4}y''^2 + \frac{3}{4}y''^2 + \frac{\sqrt{3}}{2}x''y'' + \frac{1}{4}x''^2 - \sqrt{3}x'' + y'' + 3y'' + \sqrt{3}x'' = 0; 4x''^2 + 4y'' = 0$ or $x''^2 = -y''$.

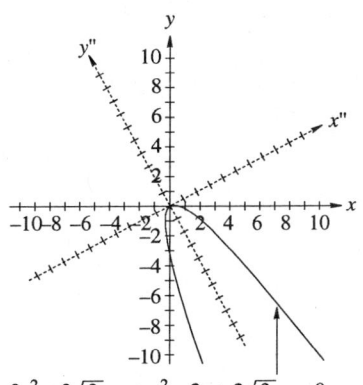

$3x^2 + 2\sqrt{3}\ xy + y^2 - 2x + 2\sqrt{3}\ y = 0$

9. **(a)** $2x^2 + 12xy - 3y^2 - 42 = 0$. The discriminant is $B^2 - 4AC = 144 + 24 = 168 > 0$ so the curve is a hyperbola,

(b) $\cot 2\theta = \frac{A-C}{B} = \frac{2+3}{12} = \frac{5}{12}, 2\theta = 67.380$ so the angle needed to rotate the coordinate axes to remove the xy-term is $\theta = 33.690, \cos\theta = 0.8321, \sin\theta = 0.5547$,

(c) The rotation formulas are $x = 0.8321x'' - 0.5547y''$ and $y = 0.8321y'' + 0.5547x''$,

(d) Substituting the equations from (c) into the original equation $2x^2 + 12xy - 3y^2 - 42 = 0$ we get $2(0.8321x'' - 0.5547y'')^2 + 12(0.8321x'' - 0.5547y'')(0.8321y'' + 0.5547x'') - 3(0.8321y'' + 0.5547x'')^2 - 42 = 0$ or $2(0.6923x''^2 - 0.9231x''y'' + 0.3077y''^2) + 12(0.4615x''^2 + 0.3846x''y'' - 0.4615y''^2) - 3(0.6923y''^2 + 0.9231x''y'' +$

$0.3077x''^2) - 42 = 0; 1.3846x''^2 - 1.8462x''y'' + 0.6154y''^2 + 5.538x''^2 + 4.6152x''y'' - 5.538y''^2 - 2.0769y''^2 - 2.7693x''y'' - .9231x''^2 - 42 = 0; 6x''^2 - 7y''^2 = 42$ which is equivalent to $\frac{x''^2}{7} - \frac{y''^2}{6} = 1$.

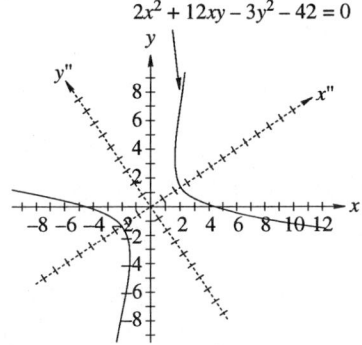

$2x^2 + 12xy - 3y^2 - 42 = 0$

10. **(a)** $7x^2 - 20xy - 8y^2 + 52 = 0$. The discriminant is $B^2 - 4AC = 400 + 224 = 624 > 0$ so the curve is a hyperbola,

(b) $\cot 2\theta = \frac{A-C}{B} = \frac{7+8}{-20} = \frac{15}{-20} = \frac{-3}{4}, 2\theta = 126.8699$ so the angle needed to rotate the coordinate axes to remove the xy-term is $\theta = 63.435°; \cos\theta = 0.4472, \sin\theta = 0.8944$,

(c) The rotation formulas are $x = 0.4472x'' - 0.8944y''$ and $y = 0.4472y'' + 0.8944x''$,

(d) Substituting the equations from (c) into the original equation $7x^2 - 20xy - 8y^2 + 52 = 0$ produces $7(0.4472x'' - 0.8944y'')^2 - 20(0.4472x'' - 0.8944y'')(0.4472y'' + 0.8944x'') - 8(0.4472y'' + 0.8944x'')^2 + 52 = 0; 7(0.2x''^2 - 0.8x''y'' + 0.8y''^2) - 20(0.4x''^2 - 0.6x''y'' - 0.4y''^2) - 8(0.2y''^2 + 0.8x''y'' + 0.8x''^2) + 52 = 0$ or $1.4x''^2 - 5.6x''y'' + 5.6y''^2 - 8x''^2 + 12x''y'' + 8y''^2 - 1.6y''^2 - 6.4x''y'' - 6.4x''^2 + 52 = 0; -13x''^2 + 12y''^2 + 52 = 0$ which is equivalent to $\frac{x''^2}{4} - \frac{y''^2}{4.33} = 1$.

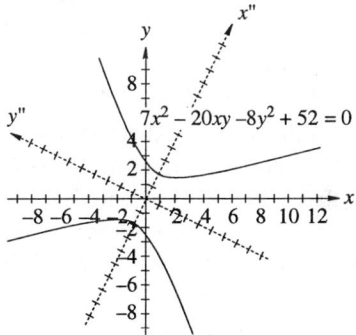

$7x^2 - 20xy - 8y^2 + 52 = 0$

11. **(a)** $6x^2 - 5xy + 6y^2 - 26 = 0$. The discriminant is $B^2 - 4AC = 25 - 144 = -119$ so the curve is an ellipse,

(b) $\cot 2\theta = \frac{A-C}{B} = \frac{6-6}{-5} = 0$ so $2\theta = 90°$ and we see that $\theta = 45°$,

(c) The rotation formulas are $x = \frac{\sqrt{2}}{2}x'' - \frac{\sqrt{2}}{2}y''$ and $y = \frac{\sqrt{2}}{2}y'' + \frac{\sqrt{2}}{2}x''$,

(d) Substituting the equations from (c) into the original equation $6x^2 - 5xy + 6y^2 - 26 = 0$ yields $6(\frac{\sqrt{2}}{2}x'' - \frac{\sqrt{2}}{2}y'')^2 - 5(\frac{\sqrt{2}}{2}x'' - \frac{\sqrt{2}}{2}y'')(\frac{\sqrt{2}}{2}y'' + \frac{\sqrt{2}}{2}x'') + 6(\frac{\sqrt{2}}{2}y'' + \frac{\sqrt{2}}{2}x'')^2 - 26 = 0$ or $6(\frac{1}{2}x''^2 - x''y'' + \frac{1}{2}y''^2) - 5(\frac{1}{2}x''^2 - \frac{1}{2}y''^2) + 6(\frac{1}{2}y''^2 + x''y'' + \frac{1}{2}x''^2) - 26 = 0$; $3x''^2 - 6x''y'' + 3y''^2 - \frac{5}{2}x''^2 + \frac{5}{2}y''^2 + 3y''^2 + 6x''y'' + 3x''^2 - 26 = 0$ which simplifies to $\frac{7}{2}x''^2 + \frac{17}{2}y''^2 = 26$ or $7x''^2 + 17y''^2 = 52$ or the standard form equation $\frac{x^2}{\frac{52}{7}} + \frac{y''^2}{\frac{52}{17}} = 1$.

Note: $a = \sqrt{\frac{52}{7}} \approx 2.726, b = \sqrt{\frac{52}{17}} \approx 1.749$.

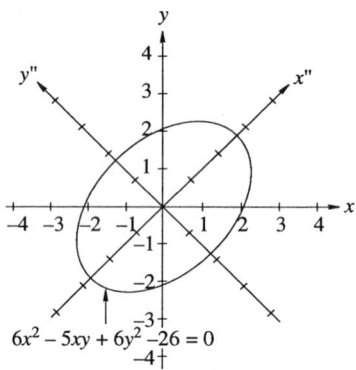

$6x^2 - 5xy + 6y^2 - 26 = 0$

12. **(a)** $9x^2 - 6xy + y^2 - 12\sqrt{10}x - 36\sqrt{10}y = 0$. The discriminant is $B^2 - 4AC = 36 - 36 = 0$ so the curve is a parabola,

(b) $\cot 2\theta = \frac{A-C}{B} = \frac{9-1}{-6} = \frac{8}{-6} = \frac{-4}{3}, 2\theta = 143.1301$ and so the angle needed to rotate the co-ordinate axes to remove the xy-term is $\theta = 71.565$. Thus, we see that $\cos\theta = 0.3162, \cos^2\theta = 0.1, \sin\theta = 0.9487, \sin^2\theta = 0.9$, and $\sin\theta\cos\theta = 0.3$,

(c) The rotation formulas are $x = 0.3162x'' - 0.9487y''$ and $y = 0.3162y'' + 0.9487x''$,

(d) Substituting the equations from (c) into the original equation $9x^2 - 6xy + y^2 - 12\sqrt{10}x - 36\sqrt{10}y = 0$ yields $9(0.3162x'' - 0.9487y'')^2 - 6(0.3162x'' - 0.9487x'') + (0.3162y'' + 0.9487y'') + (0.3162y'' + 0.9487x'')^2 - 12\sqrt{10}(0.3162x'' - 0.9487y'') - 36\sqrt{10}(0.3162y'' + 0.9487x'') = 0$; $9(0.1x''^2 - 0.6x''y'' + 0.9y''^2) - 6(0.3x''^2 - 0.8x''y'' - 0.3y''^2) + (0.1y''^2 + 0.6x''y'' + 0.9x''^2) -$

$12\sqrt{10}(0.3162x'' - 0.9487y'') - 36\sqrt{10}(0.3162y'' + 0.9487x'') = 0$ which simplifies to $10y''^2 - 120x'' = 0$ which is equivalent to $y''^2 = 12x''$.

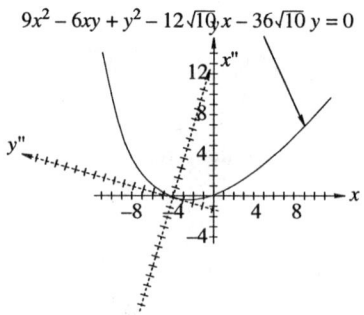

$9x^2 - 6xy + y^2 - 12\sqrt{10}x - 36\sqrt{10}y = 0$

13. $k = PV = 15 \times 400 = 6000, PV = 6000$

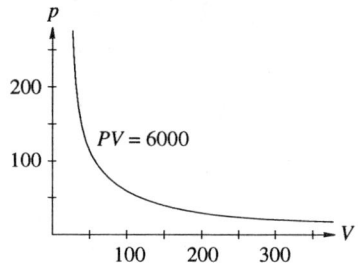

$PV = 6000$

14. If $w = $ constant 280, $X_c = \frac{1}{280C}$ or $X_cC = \frac{1}{280}$. This is an equilateral hyperbola.

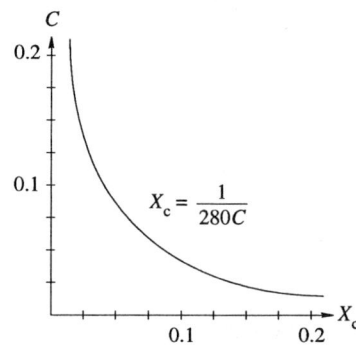

$X_c = \frac{1}{280C}$

15. $(\overline{v_1} - \overline{v_2})^2 = 19.62h_L$ is a parabola; If $h_L < 5$ then $(\overline{v_1} - \overline{v_2}) < 9.9$

16. $(3x)^2 + y^2 = (x + 10)^2$ or $9x^2 + y^2 = x^2 + 20x + 100$; $8x^2 - 20x + y^2 = 100$; $8(x^2 - 2.5x + 1.5625) + y^2 = 100 + 12.5$; $8(x - 1.25)^2 + y^2 = 112.5$; $\frac{(x-1.25)^2}{14.0625} + \frac{y^2}{112.5} = 1$; $a = 3.75, b = 10.61$. This is an ellipse with center at (1.25, 0).

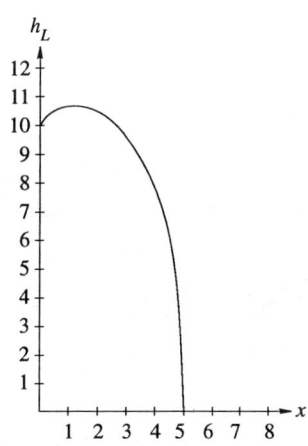

≡ 15.8 CONIC SECTIONS IN POLAR COORDINATES

1. $r = \frac{6}{1+3\cos\theta}, e = 3 > 1$, so this is a hyperbola.

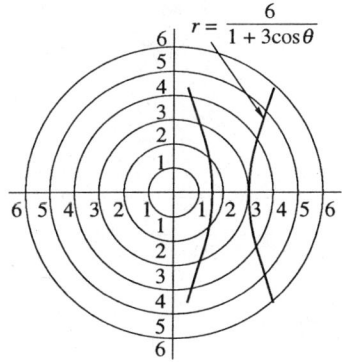

2. $r = \frac{8}{1-2\sin\theta}, e = 2$, so this is a hyperbola.

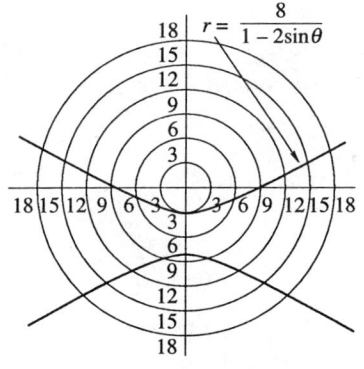

3. $r = \frac{12}{3-\cos\theta} = \frac{4}{1 - \frac{1}{3}\cos\theta}, e = \frac{1}{3}$, so this is an ellipse.

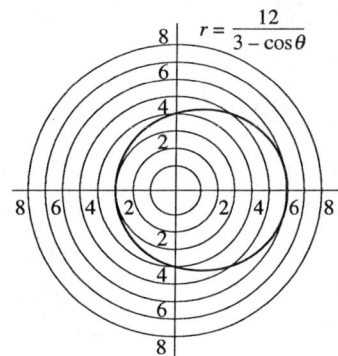

4. $r = \frac{15}{3+9\sin\theta} = \frac{5}{1+3\sin\theta}, e = 3$, so this is a hyperbola.

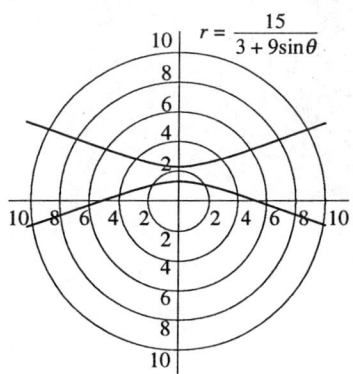

5. $r = \frac{12}{4-4\cos\theta} = \frac{3}{1-1\cos\theta}$, $e = 1$, so this is a parabola.

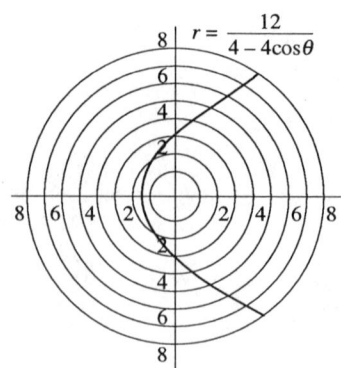

6. $r = \frac{8}{6+6\sin\theta} = \frac{\frac{4}{3}}{1+\sin\theta}$, $e = 1$, so this is a parabola.

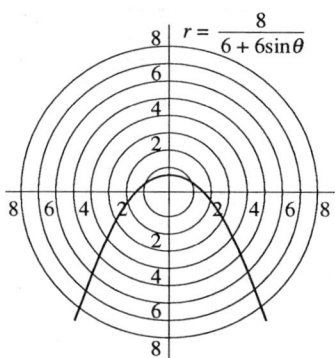

7. $r = \frac{12}{3+2\sin\theta} = \frac{4}{1+\frac{2}{3}\cos\theta}$, $e = \frac{2}{3}$, so this is an ellipse.

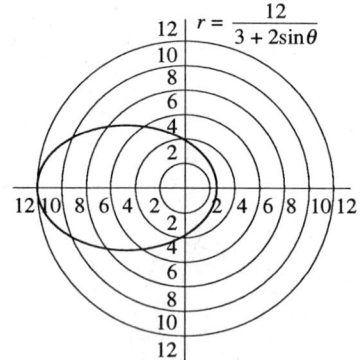

8. $r = \frac{12}{2-3\cos\theta} = \frac{6}{1-\frac{3}{2}\cos\theta}$, $e = \frac{3}{2}$, so this is a hyperbola.

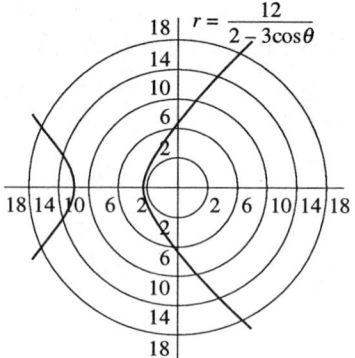

9. $r = \frac{12}{2-4\cos(\theta+\frac{\pi}{3})} = \frac{6}{1-2\cos(\theta+\frac{\pi}{3})}$, $e = 2$, so this is a hyperbola rotated $\frac{\pi}{3}$ counterclockwise.

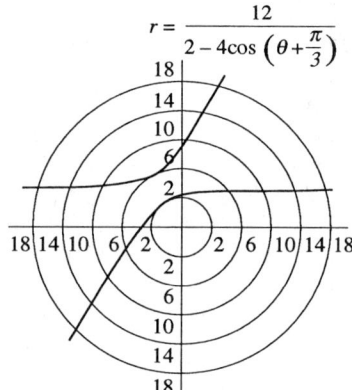

10. $r = \frac{6}{2+\sin(\theta+\frac{\pi}{6})} = \frac{3}{1+\frac{1}{2}\sin(\pi+\frac{\pi}{6})}$, $e = \frac{1}{2}$, so this is an ellipse rotated $\frac{\pi}{6}$ counterclockwise.

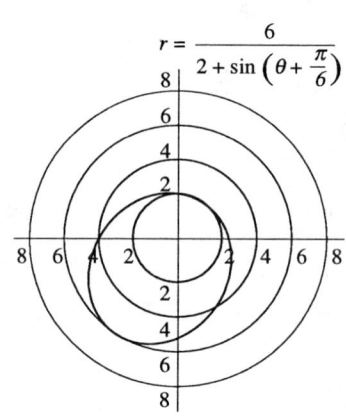

11. $r = \frac{pe}{1 - e\cos\theta} = \frac{4 \cdot \frac{3}{2}}{1 - \frac{3}{2}\cos\theta} = \frac{12}{2 - 3\cos\theta}$, so this is a hyperbola.

12. $r = \frac{2 \cdot \frac{3}{4}}{1 + \frac{3}{4}\cos\theta} = \frac{6}{4 + 3\cos\theta}$, so this is an ellipse.

13. $r = \frac{pe}{1 - e\sin\theta} = \frac{5}{1 - \sin\theta}$, so this is a parabola.

14. $r = \frac{3 \cdot 2}{1 + 2\sin\theta} = \frac{6}{1 + 2\sin\theta}$, so this is a hyperbola.

15. $r = \frac{1 \cdot \frac{2}{3}}{1 - \frac{2}{3}\cos\theta} = \frac{2}{3 - 2\cos\theta}$, so this is an ellipse.

16. $r = \frac{1 \cdot 5}{1 - \cos\theta} = \frac{5}{1 - \cos\theta}$, so this is a parabola.

17. Greatest distance is when $\theta = 0$, so $r = 4.335 \times 10^7$. The shortest distance is when $\theta = \pi$ so $r = 2.854 \times 10^7$.

18.

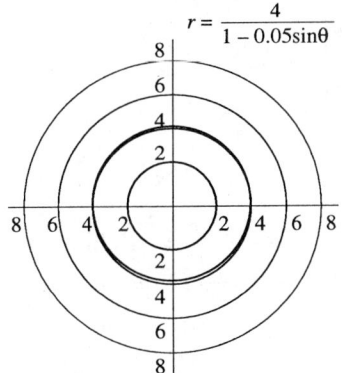

$r = \frac{4}{1 - 0.05\sin\theta}$

19.

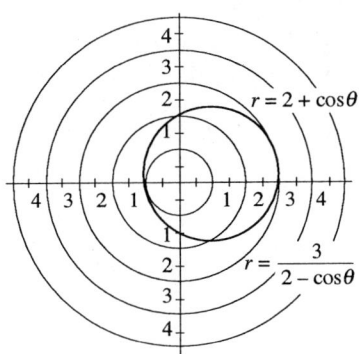

$r = 2 + \cos\theta$

$r = \frac{3}{2 - \cos\theta}$

20. **(a)** Using the polar axis as the axis of the parabola, this equation is of the form $r = \frac{p}{1 - \cos\theta}$. When $\theta = 45°, r = 50{,}000{,}000$ and solving for p, we have

$p = r(1 - \cos\theta)$
$= 50{,}000{,}000(1 - \cos 45°)$
$= 50{,}000{,}000\left(1 - \frac{\sqrt{2}}{2}\right)$
$= 25{,}000{,}000 - 25{,}000{,}000\left(2 - \sqrt{2}\right)$

Hence, the equation is $r = \frac{25\ 000\ 000(2 - \sqrt{2})}{1 - \cos\theta}$ or $r = \frac{14\ 644\ 661}{1 - \cos\theta}$

(b)

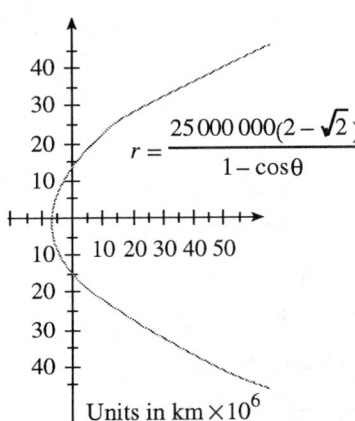

$r = \frac{25\,000\,000(2 - \sqrt{2})}{1 - \cos\theta}$

Units in km$\times 10^6$

(c) The comet is closest to the Sun when $\theta = 180°$. Then, $r = \frac{14\ 644\ 661}{1 - \cos 180°} = \frac{14\ 644\ 661}{2} \approx 7\ 322\ 330$ km.

CHAPTER 15 REVIEW

1. (a) $d = \sqrt{(2+1)^2 + (5-9)^2} = \sqrt{3^2 + (-4)^2} = 5$

(b) midpoint $= \left(\frac{2-1}{2}, \frac{5+9}{2}\right) = \left(\frac{1}{2}, 7\right)$

(c) $m = \frac{9-5}{-1-2} = \frac{4}{-3} = -\frac{4}{3}$

(d) $y - 5 = -\frac{4}{3}(x - 2)$ or $3y - 15 = -4x + 8$ or $4x + 3y = 23$

2. (a) $d = \sqrt{12^2 + 5^2} = 13$

(b) midpoint $= \left(\frac{8}{2}, \frac{-15}{2}\right) = \left(4, \frac{-15}{2}\right)$

(c) $m = \frac{-10+5}{10+2} = \frac{-5}{12}$

(d) $y + 5 = -\frac{5}{12}(x + 2)$ or $12y + 60 = -5 - 10$ or $5x + 12y = -70$

3. (a) $d = \sqrt{2^2 + 10^2} = \sqrt{104} = 2\sqrt{26}$

(b) midpoint $= \left(\frac{4}{2}, \frac{2}{2}\right) = (2, 1)$

(c) $m = \frac{6+4}{3-1} = \frac{10}{2} = 5$

(d) $y + 4 = 5(x - 1)$ or $y + 4 = 5x - 5$ or $5x - y = 9$ or $5x - 9$ or $y - 5x = -9$

4. (a) $d = \sqrt{8^2 + 8^2} = 8\sqrt{2}$

(b) midpoint $= \left(\frac{2-6}{2}, \frac{-5+3}{2}\right) =$

$\left(\frac{-4}{2}, \frac{-2}{2}\right) = (-2, -1)$

(c) $m = \frac{8}{-8} = -1$

(d) $y + 5 = -1(x - 2)$ or $y + 5 = -x + 2$ or $x + y = -3$

5. #1: $\frac{3}{4}$; #2: $\frac{12}{5}$; #3: $-\frac{1}{5}$; #4: 1

6. #1: $y - 7 = \frac{3}{4}\left(x - \frac{1}{2}\right)$ or $4y - 28 = 3x - \frac{3}{2}$ or $8y - 56 = 6x - 3$ or $8y - 6x = 53$;

#2: $y + \frac{15}{2} = \frac{12}{5}(x - 4)$ or $5y + \frac{75}{2} = 12x - 48$ or $10y + 75 = 24x - 96$ or $10y - 24x = -171$;

#3: $y - 1 = -\frac{1}{5}(x - 2)$ or $5y - 5 = -x + 2$ or $x + 5y = 7$;

#4: $y + 1 = 1(x + 2)$ or $y + 1 = x + 2$ or $y = x + 1$ or $y - x = 1$ or $x - y = -1$

7. The slope of $2y + 4x = 9$ is obtained by solving for y: This gives $y = -2x + \frac{9}{2}$ so the slope is -2. $y - 5 = -2(x + 3)$ or $y - 5 = -2x - 6$ or $y + 2x = -1$

8. $y + 7 = 4(x - 2)$ or $y + 7 = 4x - 8$ or $y - 4x = -15$ or $4x - y = 15$ or $y = 4x - 15$

9. $x^2 + y^2 = 16$: circle with center at $(0, 0)$ and $r = 4$

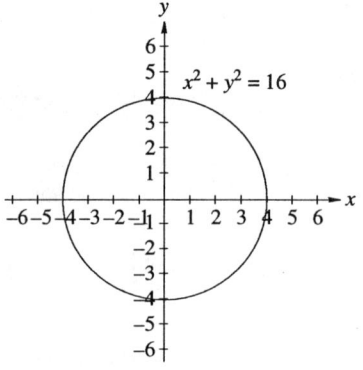

10. $x^2 - y^2 = 16 \Rightarrow \frac{x^2}{16} - \frac{y^2}{16} = 1$, hyperbola, center at $(0, 0)$, $a = 4, b = 4, c = 4\sqrt{2}$

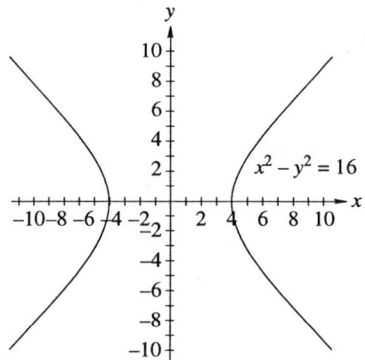

11. $x^2 + 4y^2 = 16 \Rightarrow \frac{x^2}{16} + \frac{y^2}{4} = 1$, ellipse, center at $(0, 0)$, $a = 4, b = 2, c = \sqrt{12} = 2\sqrt{3}$

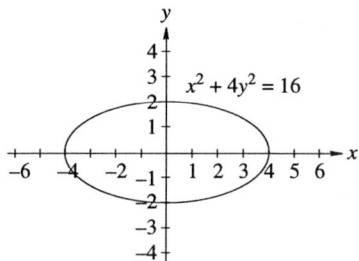

12. $y^2 = 16x$, parabola, vertex at $(0, 0)$ focus at $(4, 0)$, directrix $x = -4$

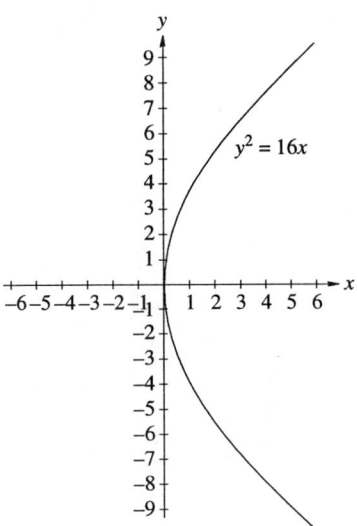

13. $(x - 2)^2 + (y + 4)^2 = 16$, circle center $(2, -4)$, radius 4

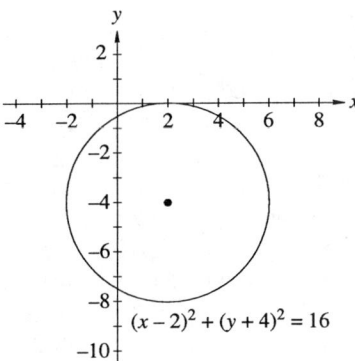

14. $(x - 2)^2 - (y + 4)^2 = 16 \Rightarrow \frac{(x-2)^2}{16} - \frac{(y+4)^2}{16} = 1$ hyperbola, center at $(2, -4), a = 4, b = 4, c = 4\sqrt{2}$

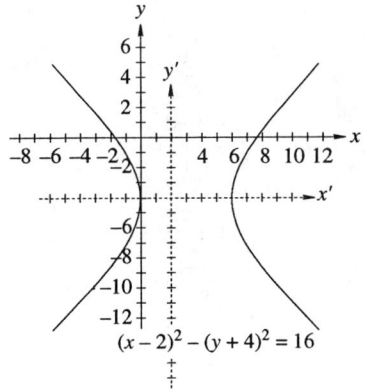

15. $(x - 2)^2 + 4(y + 4)^2 = 16 \Rightarrow \frac{(x-2)^2}{16} + \frac{(y+4)^2}{4} = 1$ ellipse, center $(2, -4), a = 4, b = 2, c = \sqrt{12} = 2\sqrt{3}$

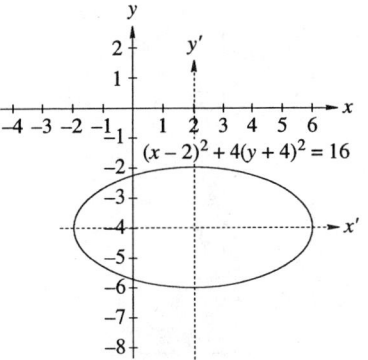

16. $(y+4)^2 = 16(x-2)$, parabola, vertex $(2, -4)$ opens right, focus at $(6, -4)$

17. $x^2 + y^2 + 6x - 10y + 18 = 0, x^2 + 6x + 9 + y^2 - 10y + 25 = -18 + 9 + 25, (x+3)^3 + (y-5)^2 = 16$ circle, center $(-3, 5)$ radius 4

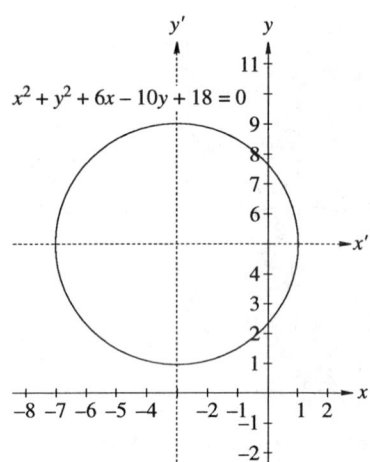

18. $x^2 - 4y^2 + 6x + 40y - 107 = 0, x^2 + 6x - 4(y^2 - 10y) = 107, (x^2 + 6x + 9) - 4(y^2 - 10y + 25) = 107 + 9 - 4 \cdot 25, (x+3)^2 - 4(y-5)^2 = 16, \frac{(x+3)^2}{16} - \frac{(y-5)^2}{4} = 1$ Hyperbola, center $(-3, 5), a = 4, b = 2, c = \sqrt{20} = 2\sqrt{5}$

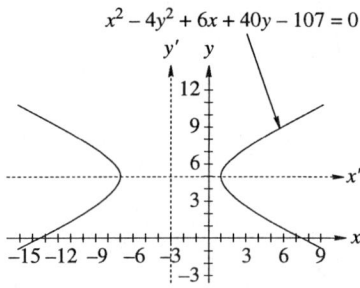

19. $x^2 + 4y^2 + 6x - 40y + 93 = 0, (x^2 + 6x) + 4(y^2 - 10y) = -93, (x^2 + 6x + 9) + 4(y^2 - 10y + 25) = -93 + 9 + 4 \cdot 25, (x+3)^2 + 4(y-5)^2 = 16, \frac{(x+3)^2}{16} + \frac{(y-5)^2}{4} = 1$, ellipse, center $(-3, 5), a = 4, b = 2, c = \sqrt{12} = 2\sqrt{3}$

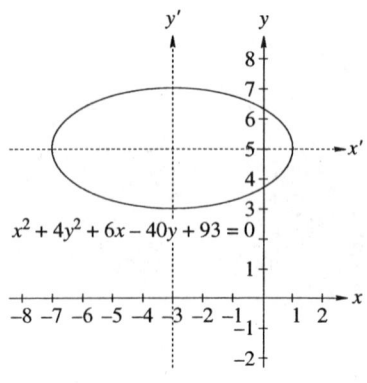

20. $x^2 + 6x - 4y + 29 = 0, x^2 + 6x = 4y - 29, (x^2 + 6x + 9) = 4y - 29 + 9, (x+3)^2 = 4y - 20, (x+3)^2 = 4(y-5)$ parabola, vertex $(-3, 5), p = 1$, opens upward, Focus $(-3, 6)$, directrix, $y = 4$

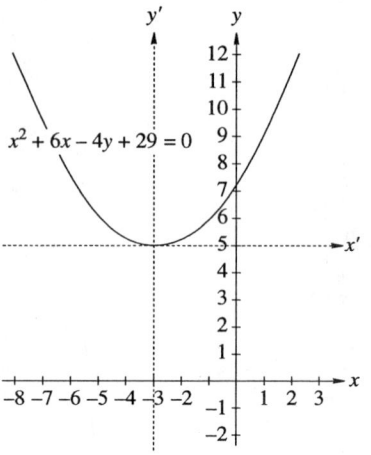

21. $2x^2 + 12xy - 3y^2 - 42 = 0$. The discriminant is $B^2 - 4AC = 144 + 24 = 168$ so the curve is a hyperbola. Thus, $\cot 2\theta = \frac{A-C}{B} = \frac{2+3}{12} = \frac{5}{12}$, and so $2\theta = 67.3801°$. Hence, the angle of rotation is $\theta = 33.69°$ and we obtain $\cos\theta = 0.83205, \cos^2\theta = 0.6923, \sin\theta = 0.5547, \sin^2\theta = 0.3077$, and $\sin\theta\cos\theta = 0.4615$. The rotation equations are $x = 0.83205x'' - 0.5547y''$ and $y = 0.83205y'' + 0.5547x''$. When these are substituted into the given equation, we obtain
$2(0.83205x'' - 0.5547y'')^2 + 12(0.83205x'' - 0.5547y'') \times (0.83205y'' + 0.5547x'') - 3(0.83205y'' + 0.5547x'')^2 - 42 = 0$ or $2(0.6923x''^2 - 0.9230x''y'' + 0.3077y''^2) + 12(0.4615x''^2 + 0.3846x''y'' - 0.4615y''^2) - 3(0.6923y''^2 + 0.9230x''y'' + 0.3077x''^2) - 42 = 0$. This simplifies to $6x''^2 - 7y''^2 = 42$ or the standard form equation $\frac{x''^2}{7} - \frac{y''^2}{6} = 1$ with $a = \sqrt{7}, b = \sqrt{6}$, and $c = \sqrt{13}$.

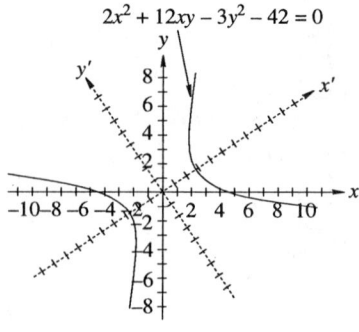

22. $5x^2-4xy+8y^2-36=0$. The discriminant is $B^2-4AC=16-160=-144$ so the curve is an ellipse. Thus, $\cot 2\theta = \frac{A-C}{B} = \frac{5-8}{-4} = \frac{-3}{-4} = \frac{3}{4}$, so $2\theta = 53.1301$, and the angle of rotation is $\theta = 26.565$. As a result, we have $\cos\theta = 0.8944, \cos^2\theta = 0.8, \sin\theta = 0.4472, \sin^2\theta = 0.2$, and $\cos\theta\sin\theta = 0.4$. The rotation equations are $x = 0.8944x'' - 0.4472y''$ and $y = 0.8944y'' + 0.4472x''$. When these are substituted into the given equation, we obtain $5(0.8944x'' - 0.4472y'')^2 - 4(0.8944x'' - 0.4472y'')(0.8944y'' + 0.4472x'') + 8(0.8944y'' + 0.4472x'')^2 - 36 = 0$ or $5(0.8x''^2 - 0.8x''y'' + 0.2y''^2) - 4(0.4x''^2 + 0.6x''y'' - 0.4y''^2) + 8(0.8y''^2 + 0.8x''y'' + 0.2x''^2) - 36 = 0$. This simplifies to $4x''^2 + 9y''^2 = 36$ which has the standard form equation $\frac{x''^2}{9} + \frac{y''^2}{4} = 1$ with $a = 3, b = 2$, and $c = \sqrt{5}$.

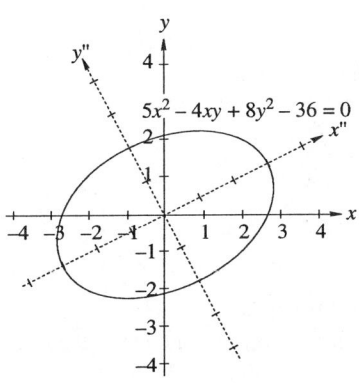

23. $3x^2 + 2\sqrt{3}xy + y^2 + 8x - 8\sqrt{3}y = 32$. The discriminant is $B^2 - 4AC = 12 - 12 = 0$ so the curve is a parabola. Thus, $\cot 2\theta = \frac{3-1}{2\sqrt{3}} = \frac{2}{2\sqrt{3}} = \frac{1}{\sqrt{3}}$ and so $2\theta = 60°$, and the angle of rotation is $\theta = 30°$. As a result, we have $\cos 30° = \frac{\sqrt{3}}{2}, \cos^2 30° = \frac{3}{4}, \sin 30° = \frac{1}{2}$, and $\sin^2 30° = \frac{1}{4}$. The rotation equations are $x = \frac{\sqrt{3}}{2}x'' - \frac{1}{2}y''$ and $y = \frac{\sqrt{3}}{2}y'' + \frac{1}{2}x''$. When these are substituted into the given equation, we obtain $3(\frac{\sqrt{3}}{2}x'' - \frac{1}{2}y'')^2 + 2\sqrt{3}(\frac{\sqrt{3}}{2}x'' - \frac{1}{2}y'')(\frac{\sqrt{3}}{2}y'' + \frac{1}{2}x'') + (\frac{\sqrt{3}}{2}y'' + \frac{1}{2}x'')^2 + 8(\frac{\sqrt{3}}{2}x'' - \frac{1}{2}y'') - 8\sqrt{3}(\frac{\sqrt{3}}{2}y'' + \frac{1}{2}x'') = 32$ or $3(\frac{3}{4}x''^2 - \frac{\sqrt{3}}{2}x''y'' + \frac{1}{4}y''^2) + 2\sqrt{3}(\frac{\sqrt{3}}{2}x''^2 - \frac{\sqrt{3}}{4}y''^2 + \frac{1}{2}x''y'') + (\frac{3}{4}y''^2 + \frac{\sqrt{3}}{2}x''y'' + \frac{1}{4}x''^2) + 4\sqrt{3}x'' - 4y'' - 12y'' - 4\sqrt{3}x'' = 32$. This simplifies to $4x''^2 = 16y'' + 32$ which has the standard form equation for a parabola $x''^2 = 4(y'' + 2)$ with $p = 1$. The parabola opens upward and has its vertex at $(0'', -2'')$.

$3x^2 + 2\sqrt{3}\,xy + y^2 + 8x - 8\sqrt{3}\,y = 32$

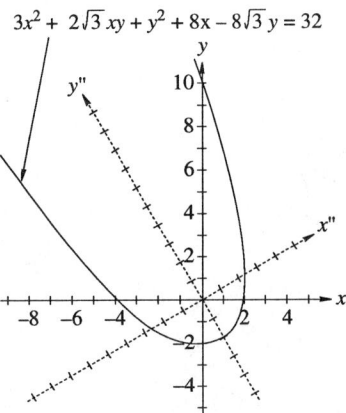

24. $2x^2 - 4xy - y^2 = 6$. The discriminant is $B^2 - 4AC = 16 + 8 = 24 > 0$ so the curve is a hyperbola. Thus, $\cot 2\theta = \frac{A-C}{B} = \frac{2+1}{-4} = -\frac{3}{4}$ and so $2\theta = 126.8699$ and the angle of rotation is $\theta = 63.4349$. As a result, we have $\cos\theta = 0.447, \cos^2\theta = 0.2, \sin\theta = 0.8944, \sin^2\theta = 0.8, \cos\theta\sin\theta = 0.4$. The rotation equations are $x = x''\cos\theta - y''\sin\theta = 0.447x'' - 0.8944y''$ and $y = y''\cos\theta + x''\sin\theta = 0.8944y'' + 0.447x''$. When these are substituted into the given equation, we obtain $2(x''\cos\theta - y''\sin\theta)^2 - 4(x''\cos\theta - y''\sin\theta)(y''\cos\theta + x''\sin\theta) - (y''\cos\theta + x''\sin\theta)^2 = 6$ or $2(0.2x''^2 - 0.8x''y'' + 0.8y''^2) - 4(0.4x''^2 - 0.6x''y'' - 0.4y''^2) - (0.2y''^2 + 0.8x''y'' + 0.8x''^2) = 6$. This simplifies to $-2x''^2 + 3y''^2 = 6$ or the standard form equation $\frac{y''^2}{2} - \frac{x''^2}{3} = 1$ with $a = \sqrt{2}, b = \sqrt{3}$, and $x = \sqrt{5}$.

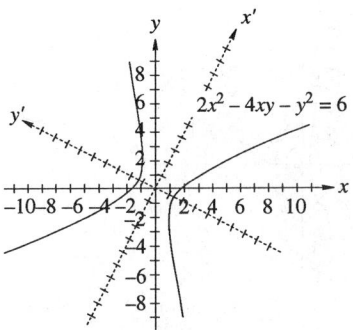

25. $r = \frac{16}{5-3\cos\theta} = \frac{\frac{16}{5}}{1 - \frac{3}{5}\cos\theta}$; $e = \frac{3}{5}$ so ellipse

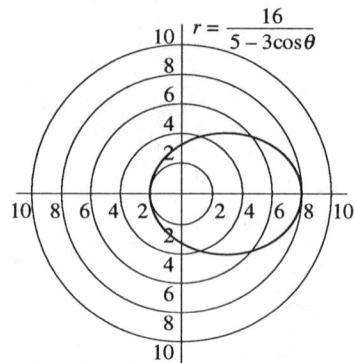

26. $r = \frac{9}{3-5\cos\theta} = \frac{3}{1 - \frac{5}{3}\cos\theta}$; $e = \frac{5}{3}$ so hyperbola

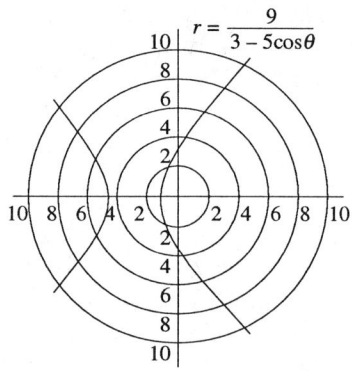

27. $r = \frac{9}{3+3\sin\theta} = \frac{3}{1+\sin\theta}$; $e = 1$ so parabola

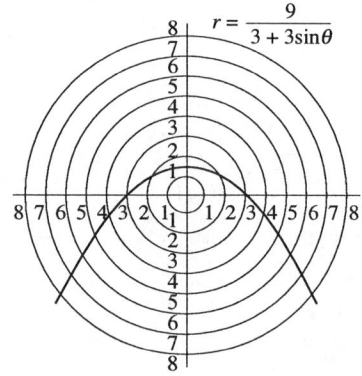

28. $r = \frac{2}{1+\cos\theta}$; $e = 1$, so parabola

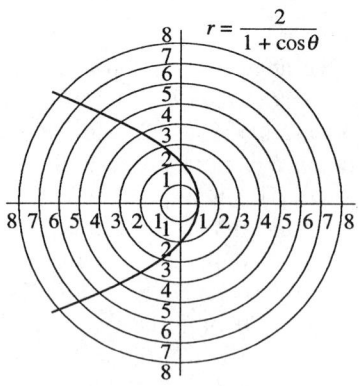

29. **(a)** Completing the square, we obtain $(y+3)^2 = 16x + 4$. We graph $y_1 = \sqrt{16x+4} - 3$ and $y_2 = -\sqrt{16x+4} - 3$. Their graphs produce the figure shown below.

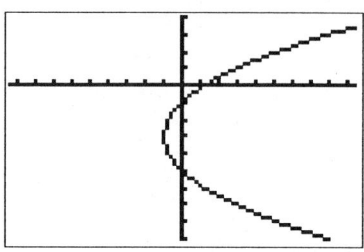

[–9.4, 9.4] x [–9, 4]

(b) This is a parabola, which has the standard form equation $(y+3)^2 = 4(x+1)$.

(c) The parabola has a horizontal axis, opens to the right with vertex $(-1, -3)$, directrix $x = -2$, and focus $(0, -3)$.

30. **(a)**

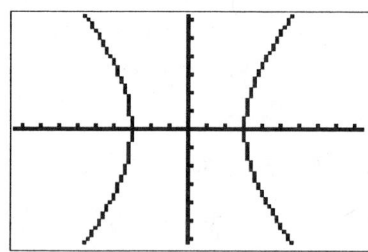

[–9.4, 9.4] x [–6.2, 6.2]

(b) This is a hyperbola with standard form equation $\frac{x^2}{9} - \frac{y^2}{16} = 1$.

(c) From the standard form equation, we see that $a = 3, b = 4$, and so, $c = 5$. Thus, the "important" points are, Center $(0, 0)$, Vertices: $(-3, 0)$ and $(3, 0)$, Foci: $(-5, 0)$ and $(5, 0)$; endpoints of conjugate axis: $(0, 4)$ and $(0, -4)$

31. Since $2a = 100, a = 50$, this gives $\frac{x^2}{50^2} + \frac{y^2}{b^2} = 1$. The clearance constraint gives the point $(45, 6.1)$. Substituting and solving for b^2 results in $\frac{45^2}{50^2} + \frac{6.1^2}{b^2} = 1$ or $b^2 \cdot 45^2 + 6.1^2 \cdot 50^2 = 50^2 b^2$ or $50^2 b^2 - 45^2 b^2 = 50^2 \cdot 6.1^2$. As a result, $b^2 = \frac{50^2 \cdot 6.1^2}{50^2 - 45^2} = 195.84$ and we get the standard form equation $\frac{x^2}{2500} + \frac{y^2}{195.84} = 1$.

32. Taking the road to be the x-axis and the center of the road as the origin, the vertex of the parabola is at $(0, 10)$. This gives an equation of the form $x^2 = 4p(y - 10)$. The top of the tower gives the point $(200, 120)$. Substituting and solving for $4p$ we obtain $4p = \frac{x^2}{y - 10} = \frac{200^2}{120 - 10} = \frac{40000}{110} = 363.64$. The equation is $x^2 = 363.64(y - 10)$.

33. The equation is a hyperbola. The common difference is $AP - BP = 320 \times 500 = 160{,}000\,\text{m} = 160$ km. Thus, $2a = 160, a = 80, 2c = 400$, and so $c = 200$ and $b^2 = c^2 - a^2 = 33600$. The equation is $\frac{x^2}{6400} - \frac{y^2}{33600} = 1$. The y-coordinate for p is -50. Solving for x we get $\frac{x^2}{6400} = 1 + \frac{50^2}{33600} = 1.0744$ or $x^2 = 6876.19$ and so $x = 82.92$. The plane is at $(82.92, -50)$.

34. $x^2 + y^2 - 2x + 4y = 6361, x^2 - 2x + 1 + y^2 + 4y + 4 = 6361 + 1 + 4 = 6366, (x - 1)^2 + (y + 2)^2 = (79.7872)^2$. The radius is Earth's radius plus the number of units the satellite is orbiting above Earth. This is $79.7872 + 0.8 = 80.587$ km and so $r^2 = 6494.3$. Thus, we get the equation for a circle: $(x - 1)^2 + (y + 2)^2 = 6494.3, x^2 - 2x + 1 + y^2 + 4y + 4 = 6494.3, x^2 + y^2 - 2x + 4y - 6489.3 = 0$.

≡ CHAPTER 15 TEST
_____

1. Comparing the given equation with the basic equation $y^2 = 4px$, we see that $4p = -18$ and so $p = -\frac{18}{4} = -\frac{9}{2} = -4.5$. The focus of the parabola is at $(p, 0) = (-4.5, 0)$ and the directrix is the line $x = -p = 4.5$.

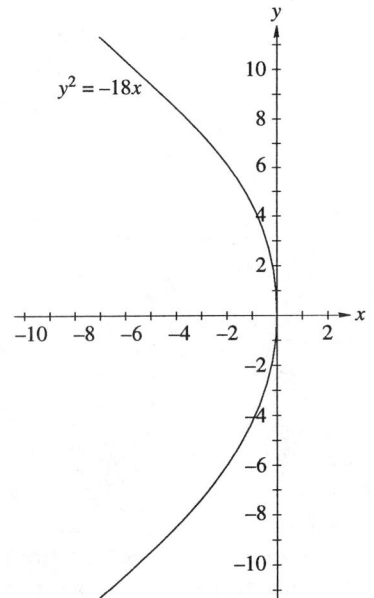

2. To write this in standard form, divide both sides by 36, obtaining $\frac{9x^2}{36} - \frac{4y^2}{36} = 1$ which simplifies to $\frac{x^2}{4} - \frac{y^2}{9} = 1$. From this we see that $a^2 = 4$ and $b^2 = 9$. Since $c^2 = a^2 + b^2 = 4 + 9 = 13$, we have $c = \sqrt{13}$. The foci are at $F' = (-c, 0) =$

$(-\sqrt{13}, 0)$ and $F = (c, 0) = (\sqrt{13}, 0)$; the vertices are $V' = (-a, 0) = (-2, 0)$ and $V = (a, 0) = (2, 0)$; and the endpoints of the conjugate axis are $M' = (0, -b) = (0, -3)$ and $M = (0, b) = (0, 3)$.

3. **(a)**

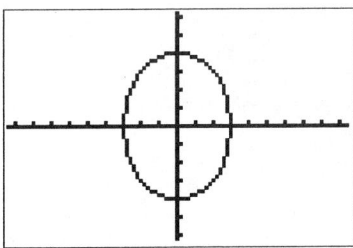

$[-9.4, 9.4] \times [-6.2, 6.2]$

(b) This is an ellipse with standard form equation $\frac{x^2}{9} + \frac{y^2}{16} = 1$.

(c) From the standard form equation, we see that $a = 3, b = 4$, and so, $c = \sqrt{4^2 - 3^2} = \sqrt{7}$. Thus, the "important" points are, Center: $(0, 0)$, Vertices $(-3, 0)$ and $(3, 0)$, Foci: $(-\sqrt{7}, 0)$ and $(\sqrt{7}, 0)$; endpoints of minor axis: $(0, 4)$ and $(0, -4)$

4. Using the distance formula for the distance between two points P_1 and P_2, we obtain

$d(P_1, P_2) = \sqrt{(x_2 - x_1)^2 + (y_2 - y_1)^2} =$
$\sqrt{(5 - (-2))^2 + (-6 - 4)^2} = \sqrt{7^2 + (-10)^2} =$
$\sqrt{49 + 100} = \sqrt{149}$

5. The midpoint between these points is $\left(\frac{x_1 + x_2}{2}, \frac{y_1 + y_2}{2}\right)$
$= \left(\frac{4+2}{2}, \frac{-5 + (-8)}{2}\right) = \left(3, \frac{-13}{2}\right)$. The slope of the line through the two given points is $m = \frac{y_2 - y_1}{x_2 - x_1} =$
$\frac{-8 - (-5)}{2 - 4} = \frac{-3}{-2} = \frac{3}{2}$ and so the slope of a line perpendicular to this line is $-\frac{2}{3}$. Hence the desired line is $y - \left(\frac{-13}{2}\right) = -\frac{2}{3}(x - 3)$ or $y + \frac{13}{2} = -\frac{2}{3}(x - 3)$ or $6y + 4x + 27 = 0$.

6. A line with an angle of inclination 30° has a slope $m = \tan 30° = \frac{\sqrt{3}}{3}$. An x-intercept of 3 means it passes through $(3, 0)$. The equation has the point-slope form $y - 0 = \frac{\sqrt{3}}{3}(x - 3)$. Changing this to the slope-intercept form produces $y = \frac{\sqrt{3}}{3}x - \sqrt{3}$.

7.   (a)

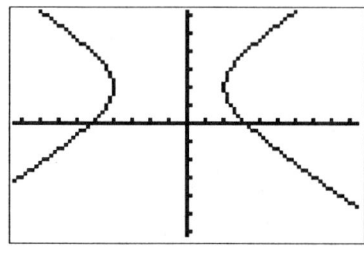

[−9.4, 9.4] x [−6.2, 6.2]

(b) This is a hyperbola with standard form equation $\frac{(x+1)^2}{9} - \frac{(y-2)^2}{4} = 1$.

(c) From the standard form equation, we see that $a = 3, b = 2$, and so, $c = \sqrt{3^2 + 2^2} = \sqrt{13}$. Thus, the "important" points are, Center: $(-1, 2)$, Vertices: $(-4, 2)$ and $(2, 2)$, Foci: $(-1 - \sqrt{13}, 2)$ and $(-1 + \sqrt{13}, 2)$; endpoints of minor axis: $(-1, 4)$ and $(-1, 0)$

8.   (a) Completing the square we obtain $(3x^2 - 6x) + (4y^2 + 16y) = -7$ or $3(x^2 - 2x + 1) + 4(y^2 +$

$4y + 4) = -7 + 3 + 16$, which can be written as $3(x - 1)^2 + 4(y + 2)^2 = 12$. Dividing both sides by 12, we obtain a standard form equation for a horizontal ellipse $\frac{(x-1)^2}{4} + \frac{(y+2)^2}{3} = 1$;

(b) Ellipse;

(c) center: $(1, -2)$, foci: $(0, -2), (2, -2)$, vertices: $(-1, -2), (3, -2)$, endpoints of minor axis: $(1, -2 - \sqrt{3}), (1, -2 + \sqrt{3})$

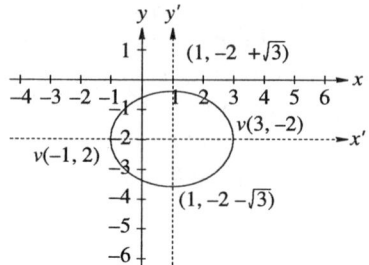

9.   (a) The angle is determined by the equation $\cot 2\theta = \frac{A - C}{B}$. Here $A = 1, B = 1$, and $C = 1$, and so $\cot 2\theta = \frac{0}{1}$, which means that $\theta = \frac{\pi}{4}$.

(b) Using $B^2 - 4AC = 1^2 - 4(1)(1) = -3$. Since this is negative, the conic is an ellipse.

10. The denominator has a constant term of 1 so $e = 4$. Since $e > 1$, the conic is a hyperbola

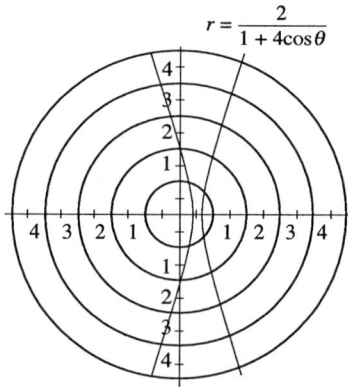

$$r = \frac{2}{1 + 4\cos\theta}$$

11. Consider this parabola to have its lowest point (vertex) at the origin and its axis the y-axis. Then the equation is of the form $x^2 = 4py$. Then the supports can be represented by the lines $x = -60$ and $x = 60$. The distance of 48 ft in from each support and 3 ft above the lowest point are the points $(-12, 3)$ and $(12, 3)$. Substituting the point $(12, 3)$ into the equation for the parabola we obtain $12^2 = 4p(3)$ and so $p = 12$. The equation is now $x^2 = 48y$. Substituting $x = 60$ in this equation and solving for y, we get $y = 75$.

CHAPTER

16

Systems of Equations and Inequalities

≡ 16.1 SOLUTIONS OF NONLINEAR SYSTEMS OF EQUATIONS

1. $\begin{cases} x - 2y = 5 & (1) \\ x^2 - 4y^2 = 45 & (2) \end{cases}$. Solve (1) for x : $x = 2y + 5$. Substitute this result into (2) to get $(2y + 5)^2 - 4y^2 = 45$ or $4y^2 + 20y + 25 - 4y^2 = 45$ which simplifies to $20y = 20$ and so $y = 1$. Back substitution produces $x = 2 \cdot 1 + 5 = 7$. The only solution is $(7, 1)$

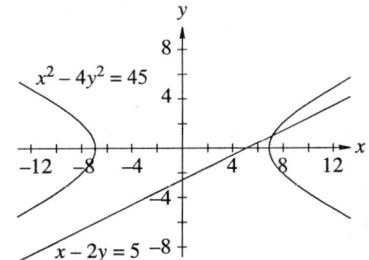

2. $\begin{cases} x^2 + y^2 = 13 & (1) \\ 2x - y = 4 & (2) \end{cases}$. Solve (2) for y : $y = 2x - 4$. Substitute into (1) to obtain $x^2 + (2x - 4)^2 = 13$ or $x^2 + 4x^2 - 16x + 16 = 13$. This simplifies to $5x^2 - 16x + 3 = 0$, which factors as $(5x - 1)(x - 3) = 0$ so $x = \frac{1}{5}$ or $x = 3$. If $x = \frac{1}{5}$, then $y = 2 \cdot \frac{1}{5} - 4 = -\frac{18}{5}$. If $x = 3$, then $y = 2 \cdot 3 - 4 = 2$. The solutions are $\left(\frac{1}{5}, -\frac{18}{5}\right)$ and $(3, 2)$

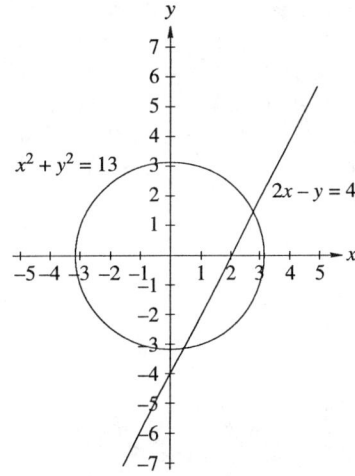

3. $\begin{cases} x^2 + 4y^2 = 32 & (1) \\ x + 2y = 0 & (2) \end{cases}$. Solving (2) for x : $x = -2y$. Substituting into (1) we get $(-2y)^2 + 4y^2 = 32, 4y^2 + 4y^2 = 32, 8y^2 = 32, y^2 = 4; y = \pm 2$. If $y = 2 \Rightarrow x = -2(2) = -4$; if $y = -2 \Rightarrow x = -2(-2) = 4$. Solutions $(-4, 2)$ and $(4, -2)$

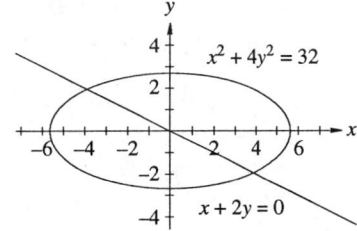

4. $\begin{cases} x - y = 4 & (1) \\ x^2 - y^2 = 32 & (2) \end{cases}$. Solving (1) for x we get $x = y + 4$. Substituting into (2) we get $(y + 4)^2 - y^2 = 32$, or $y^2 + 8y + 16 - y^2 = 32$, or $8y = 16$, and so $y = 2$. Back substituting, we have $x = 2 + 4 = 6$, and the solution is $(6, 2)$.

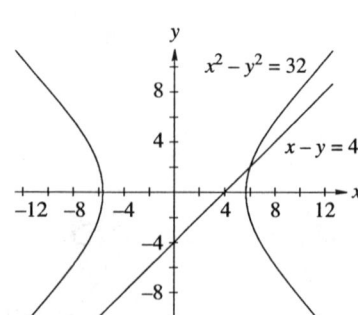

5. $\begin{cases} x^2 + y^2 = 4 & (1) \\ x^2 - 2y = 1 & (2) \end{cases}$. Subtracting (2) from (1) yields $y^2 + 2y = 3$ or $y^2 + 2y - 3 = 0$. This factors as $(y+3)(y-1) = 0$ and so $y = -3$ or $y = 1$. If $y = 1$, then, from (2), we obtain $x^2 - 2 = 1$, or $x^2 = 3$ and so $x = \pm\sqrt{3}$. If $y = -3$, then, from (2), we get $x^2 - 2 \cdot (-3) = 1$, or $x^2 + 6 = 1, x^2 = -5, x = \pm j\sqrt{5}$, which is not real. Solution: $(\sqrt{3}, 1), (-\sqrt{3}, 1), (j\sqrt{5}, -3), (-j\sqrt{5}, -3)$

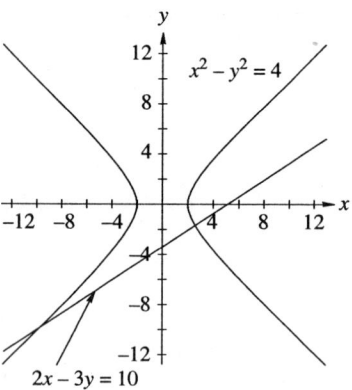

7. $\begin{cases} x^2 + y^2 = 7 & (1) \\ y^2 = 6x & (2) \end{cases}$. Substituting $6x$ for y^2 in (1) yields: $x^2 + 6x = 7, x^2 + 6x - 7 = 0, (x + 7)(x - 1) = 0, x = -7$ or $x = 1, x = -7$ yields $y^2 = 6(-7) = -42, y = \pm j\sqrt{42}$. $x = 1$ yields $y^2 = 6 \cdot 1 = 6, y = \pm\sqrt{6}$. Solutions $(1, \sqrt{6}), (1, -\sqrt{6}), (-7, j\sqrt{42}), (-7, -j\sqrt{42})$. (Note: the last two solutions are not real numbers.)

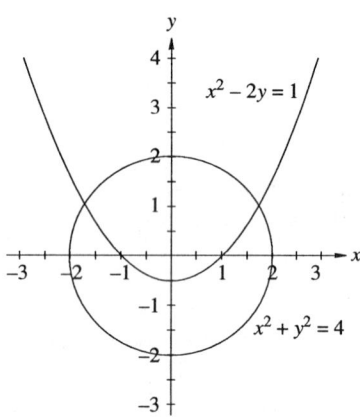

6. $\begin{cases} x^2 - y^2 = 4 & (1) \\ 2x - 3y = 10 & (2) \end{cases}$. Solving (2) for x : $x = \frac{3y+10}{2}$. Substituting into (1) we get $\left(\frac{3y+10}{2}\right)^2 - y^2 = 4$, or $\frac{9y^2 + 60y + 100}{4} - y^2 = 4$. Multiplying by 4 produces $9y^2 + 60y + 100 - 4y^2 = 16$, which simplifies as $5y^2 + 60y + 84 = 0$. By the quadratic formula $y = -1.618$ or $y = -10.382$. If $y = -1.618 \Rightarrow x = 2.573$. If $y = -10.382 \Rightarrow x = -10.573$. Thus, the solutions are $(2.573, -1.618)$ and $(-10.573, -10.382)$

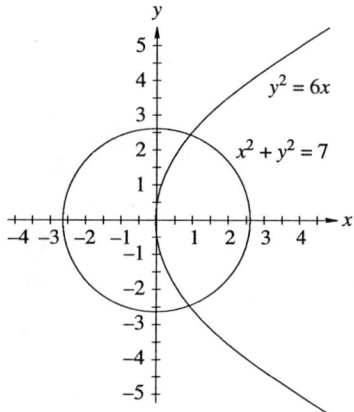

8. $\begin{cases} x^2 - y^2 + 6 = 0 & (1) \\ y^2 = 5x & (2) \end{cases}$. Substituting $5x$ for y^2 into (1) yields: $x^2 - 5x + 6 = 0, (x - 2)(x - 3) = 0, x = 2$ or $x = 3, x = 2$ gives $y^2 = 10$ or $y = \pm\sqrt{10}, x = 3$ gives $y^2 = 15$ or $y = \pm\sqrt{15}$. Solutions $(2, \sqrt{10}), (2, -\sqrt{10}), (3, \sqrt{15}), (3, -\sqrt{15})$

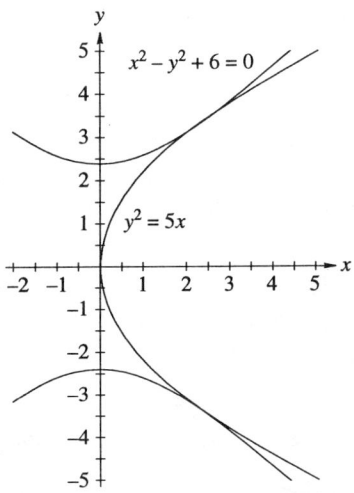

9. $\begin{cases} xy = 3 & (1) \\ 2x^2 - 3y^2 = 15 & (2) \end{cases}$. Solving (1) for x : $x =$ $\frac{3}{y}$. Substituting into (2) $2 \cdot \left(\frac{3}{y}\right)^2 - 3y^2 = 15$, or $\frac{18}{y^2} - 3y^2 = 15$, or $18 - 3y^4 = 15y^2$, or $3y^4 + 15y^2 - 18 = 0$, which simplifies to $y^4 + 5y^2 - 6 = 0$, which factors as $(y^2 + 6)(y^2 - 1) = 0$. Thus, $y^2 = -6$ and $y = \pm j\sqrt{6} = \pm 3j\sqrt{\frac{2}{3}}$ or $y^2 = 1$ and $y = \pm 1$. Now, $y = 1 \Rightarrow x = 3$ which is the solution $(3, 1)$ and $y = -1 \Rightarrow x = -3$ and the solution $(-3, -1)$. Also, $y = j\sqrt{6} \Rightarrow x = \frac{3}{j\sqrt{6}} = -\frac{\sqrt{6}}{2}j = -j\sqrt{\frac{3}{2}}$ which produces the solution $\left(-\frac{\sqrt{6}}{2}j, j\sqrt{6}\right)$ or $\left(-j\sqrt{\frac{3}{2}}, 3j\sqrt{\frac{2}{3}}\right)$. Similarly, $y = -j\sqrt{6} \Rightarrow x = \frac{\sqrt{6}}{2}j = j\sqrt{\frac{3}{2}}$ and we have the solution $\left(\frac{\sqrt{6}}{2}j, -j\sqrt{6}\right)$ or $\left(j\sqrt{\frac{3}{2}}, -3j\sqrt{\frac{2}{3}}\right)$.

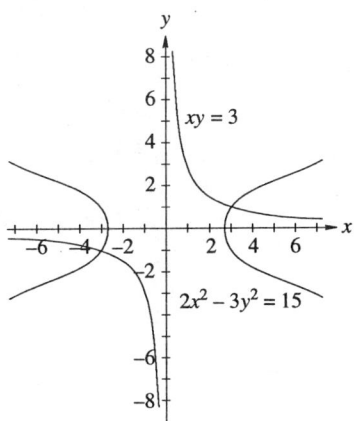

10. $\begin{cases} y = x^2 - 4 & (1) \\ x^2 + 3y^2 + 4y - 6 = 0 & (2) \end{cases}$. Substituting $x^2 - 4$ for y into (2) yields: $x^2 + 3(x^2 - 4)^2 + 4(x^2 - 4) - 6 = 0$, or $x^2 + 3(x^4 - 8x^2 + 16) + 4x^2 - 16 - 6 = 0$,

or $x^2 + 3x^4 - 24x^2 + 48 + 4x^2 - 22 = 0$, which simplifies to $3x^4 - 19x^2 + 26 = 0$. Thus, either $x^2 = \frac{13}{3}$ and $x = \frac{\pm\sqrt{39}}{3}$ or $x^2 - 2 = 0$ which means that $x^2 = 2$, and so $x = \pm\sqrt{2}$. Now, $x = \frac{\pm\sqrt{39}}{3} \Rightarrow y = \frac{39}{9} - 4 = \frac{13}{3} - 4 = \frac{1}{3}$, and $x = \pm\sqrt{2} \Rightarrow y = 2 - 4 = -2$. So the solutions are $\left(\frac{\sqrt{39}}{3}, \frac{1}{3}\right), \left(-\frac{\sqrt{39}}{3}, \frac{1}{3}\right), \left(\sqrt{2}, -2\right), \left(-\sqrt{2}, -2\right)$.

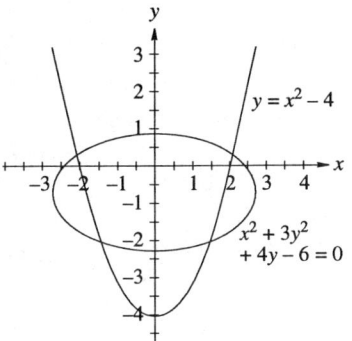

11. $t = \frac{d}{r}, t_1 = \frac{100}{r}, t_2 = \frac{20}{r-10}, t_1 + t_2 = \frac{100}{r} + \frac{20}{r-10} = 3, (r - 10)100 + 20r = 3r(r - 10), 100r - 1000 + 20r = 3r^2 - 30r, 3r^2 - 150r + 1000 = 0$. Using the quadratric formula, $r = 42.08$ mph for the first 100 miles, 32.08 mph for the last 20 miles.

12. The perimeter is $2\ell + 2w = 160$ and the area is $\ell \cdot w = 1200$. From the equation for the perimeter, we get $\ell = 80 - w$ and substituting this in the equation for the area gives $(80 - w)w = 1200$, or $80w - w^2 = 1200$, or $w^2 - 80w + 1200 = 0$ which factors as $(w - 60)(w - 20) = 0$. Thus, $w = 60$ or $w = 20$ and we get that the length is 60 m and the width is 20 m.

13. When the resistors are connected in series, we get $R_1 + R_2 = 100$ and when they are connected in parallel $\frac{R_1 R_2}{R_1 + R_2} = 24$. Since $R_1 + R_2 = 100$, the second equation is equivalent to $R_1 R_2 = 2400$. Now, $R_1 = 100 - R_2$ and so $(100 - R_2)R_2 = 2400$, or $100R_2 - R_2^2 = 2400$, or $R_2^2 - 100R_2 + 2400 = 0$. This factors as $(R_2 - 60)(R_2 - 40) = 0$ which means that $R_2 = 60$ or 40. The resistances are $40\,\Omega$ and $60\,\Omega$.

14. $R_1 + R_2 = 10, \frac{R_1 R_2}{R_1 + R_2} = 2.1, \frac{R_1 R_2}{10} = 2.1, R_1 R_2 = 21, R_1 = 10 - R_2, (10 - R_2)R_2 = 21, 10R_2 - R_2^2 = 21, R_2^2 - 10R_2 + 21 = 0, (R_2 - 3)(R_2 - 7) = 0, R_2 = 3$ or 7. One is $3\,\Omega$, the other is $7\,\Omega$.

15. $40v_1 + 60v_2 = 450 \Rightarrow 4v_1 + 6v_2 = 45, 20v_1^2 + 30v_1^2 = 1087.5 \Rightarrow 2v_1^2 + 3v_2^2 = 108.75$. Solving the first equation for v_1 yields: $v_1 = \frac{45 - 6v_2}{4}$. Substituting into the second equation yields $2\left(\frac{45 - 6v_2}{4}\right)^2$

$+ 3v_2^2 = 108.75$ or $2\left(\frac{2025 - 540v_2 + 36v_2^2}{16}\right) + 3v_2^2 = 108.75$, or $2025 - 540v_2 + 36v_2^2 + 24v_2^2 = 870$, which is equivalent to $60v_2^2 - 540v_2 + 1155 = 0$ or $4v_2^2 - 36v_2 + 77 = 0$. Factoring the last equation produces $(2v_2 - 7)(2v_2 - 11) = 0$, and so $v_2 = \frac{7}{2}$ or $\frac{11}{2}$. If $v_2 = \frac{7}{2} = 3.5$, then $v_1 = 6$. If $v_2 = 5.5$, then $v_1 = 3$. Either $v_1 = 6$ m/s and $v_2 = 3.5$ m/s or $v_1 = 3$ m/s and $v_2 = 5.5$ m/s.

16. $20x^2 + 40y^2 = 8000 \Rightarrow x^2 + 2y^2 = 400$ (1), and $30x^2 + 20y^2 = 6000 \Rightarrow 3x^2 + 2y^2 = 600$ (2). Subtracting (1) from (2) yields $2x^2 = 200$ or $x^2 = 100$ and so $x = \pm 10$. Since $2y^2 = 400 - x^2$, then if $x = \pm 10$, we get $2y^2 = 400 - 100 = 300$, or $y^2 = 150$, and so $y = \pm\sqrt{150} \approx \pm 12.25$. The points of intersection are approximately $(10, 12.25), (-10, 12.25), (10, -12.25)$ and $(-10, -12.25)$.

17. Since $F\ell = 200(0.8) = 160 \Rightarrow F = \frac{160}{\ell}$ and so $(F + 50)(\ell - 1) = 160 \Rightarrow \left(\frac{160}{\ell} + 50\right)(\ell - 1) = 160$, or $(160 + 50\ell)(\ell - 1) = 160\ell$, or $50\ell^2 + 110\ell - 160 = 160\ell$, which is equivalent to $50\ell^2 - 50\ell - 160 = 0$ and simplifies as $5\ell^2 - 5\ell - 16 = 0$. Using the quadratic formula $\ell = 2.36$ or -1.36. But, ℓ cannot be negative so 2.36 is the only answer. $F = \frac{160}{2.36} = 67.80$ gives $F = 67.80$ N, $\ell = 2.36$ m.

18. $y = 15t - 4.9t^2, y = 10t$ so $10t = 15t - 4.9t^2, 4.9t^2 - 5t = 0, t(4.9t - 5) = 0$, so $t = 0$ or $t = 1.02$. If $t = 0, y = 0$; If $t = 1.02, y = 1.02$, so the solutions are $(0, 0)$ and $(1.02, 10.2)$

19. The first equation can be rewritten as $x = y + 5.59$. Subtracting the two areas we obtain the equation $\pi x^2 - y^2 = 303.22$. Substituting 3.14 for π and $y + 5.59$ for x we obtain

$$3.14(y + 5.59)^2 - y^2 = 303.22$$

Expanding and collecting like terms produces the equation

$$2.14y^2 + 35.11y - 205.10 = 0$$

The quadratic formula yields the answers $y = -20.98$ and $y = 4.57$. Since y must be positive, the answers are $x = 10.16$ mm and $y = 4.57$ mm.

20. The second equation can be rewritten as $T_2 = T_1 + 250$. Substituting this into the first equation yields

$$0.02\left((T_1 + 250)^2 - T_1^2\right) = 4280$$
or $T_1^2 + 500T_1 + 62,500 - T_1^2 = 214,000$

Hence, $500T_1 = 151,500$ and $T_1 = 303$. Back substitution yields $T_2 = 303 + 250 = 553$. The answers are $T_1 = 303$ K and $T_2 = 553$ K.

21. Substituting $0.334v^2$ for r into the first equation and simplifying we have $L = (0.334v^2)x = 1.336v^2x$.

22. Solving the second equation for v_2 yields $v_2 = -\frac{2}{3}v_1$. Substituting this into the first equation and clearing the fractions we get $52v_1^2 + 78\left(-\frac{2}{3}v_1\right)^2 = 46,800$ or $52v_1^2 + \frac{104}{3}v_1^2 = 46,800$. Hence, $260v_1^2 = 140,400$ and $v_1^2 = 540$, so $v_1 \approx 23.24$ m/s. Back substitution into the equation $v_2 = -\frac{2}{3}v_1$ yields $v_2 \approx -15.49$ m/s. The velocities are in opposite directions.

23. Solving the first equation for n produces $15n = 4291 - 4g^2$ or $n = 286.07 - 0.267g^2$. Substituting into the second equation we get

$$8g^2 + 15g - 5(286.07 - 0.267g^2) = 8607$$
or $9.33g^2 + 15g - 10,037.34 = 0$

The quadratic formula yields $g = 32$ and back substitution gives $n = 13$.

24. Solving the second equation for x yields $x = -1.25y + 3,158$. Now, substituting this result for x into the first equation, we obtain

$$(-1.25y + 3,158)^2 + 1.5y^2 - (-1.25y + 3,158)y = 2,891,189$$

Expanding and collecting like terms, we get

$$4.3125y^2 - 11,053y + 7,081,775 = 0$$

The quadratic formula yields $y = 1271$ or 1292. The corresponding x values are 1569 and 1543. Hence, $x = 1,543$ and $y = 1,292$ or $x = 1,569$ and $y = 1,271$.

16.2 PROPERTIES OF INEQUALITIES; LINEAR INEQUALITIES

1. $x < 10 \Rightarrow x + 5 < 15$

3. $x < 10 \Rightarrow 3x < 30$

5. $x < 10 \Rightarrow -\frac{x}{2} > -5$

2. $x < 10 \Rightarrow x - 7 < 3$

4. $x < 10 \Rightarrow -4x > -40$

6. $x < 10 \Rightarrow \frac{x}{5} < 2$

7. $x < 10 \Rightarrow x^2 < 100$ (Note $x > 0$)

8. $x < 10 \Rightarrow \sqrt{x} < \sqrt{10}$

9. Dividing $3x > 6$ by 3 leads to $x > 2$.

10. Dividing $4x < -8$ by 4 leads to $x < -2$.

11. Subtracting 5 from both sides of the inequality $2x + 5 > -7$ produces $2x > -12$, which is equivalent to $x > -6$.

12. Adding 4 to both sides of $3x - 4 \leq 8$ produces $3x \leq 12$, is equivalent to $x \leq 4$.

13. Adding 5 to both sides of the inequality $2x - 5 < x$ produces $2x < x + 5$ and subtracting x from both sides yields $x < 5$

14. Adding $-x - 4$ to both sides of the inequality $3x + 4 \geq x$, we get $2x \geq -4$, and dividing by 2 produces $x \geq -2$.

15. Adding $7 - 2x$ to both sides of the inequality $4x - 7 \geq 2x + 5$ gives $2x \geq 12$, and dividing by 2, produces $x \geq 6$.

16. Adding $-7 - 5x$ to both sides of $7 - 3x \leq 3 + 5x$ gives $-8x \leq -4$. Then, dividing both sides by -8, we get $x \geq \frac{1}{2}$.

17. First we multiply the inequality $\frac{2}{3}x - 4 < \frac{1}{3}x + 2$ by 3, with the result $2x - 12 < x + 6$. Adding $12 - x$ to both sides, yields the result $x < 18$.

18. Multiplying $\frac{1}{5}x - \frac{2}{5} > \frac{3}{5}x + \frac{4}{5}$ by 5, produces $x - 2 > 3x + 4$. Adding $2 - 3x$ to both sides gives $-2x > 6$; and dividing by -2 leads to $x < -3$.

19. Multiplying $\frac{x-2}{4} \leq \frac{3}{8}$ by 8 produces $2(x - 2) \leq 3$ or $2x - 4 \leq 3$. Adding 4 to both sides, gives $2x \leq 7$, and then dividing by 2, we get $x \leq \frac{7}{2}$ or $x \leq 3\frac{1}{2}$.

20. Multiplying $\frac{x+2}{3} \geq \frac{5}{6}$ by 3 gives $x + 2 \geq \frac{5}{2}$ and subtracting 2 produces $x \geq \frac{1}{2}$.

21. Multiplying $\frac{x-3}{4} \leq \frac{2x}{3}$ by 12 gives $3(x-3) \leq 4 \cdot 2x$ which simplifies as $3x - 9 \leq 8x$, or $-5x \leq 9$, and so, after dividing by -5, we get $x \geq -\frac{9}{5}$.

22. Multiplying the inequality $\frac{2x+5}{3} > \frac{3x-1}{2}$ by 6 produces $2(2x+5) > 3(3x-1)$ or $4x + 10 > 9x - 3$. This simplifies to $-5x > -13$ and, after dividing by -5, to $x < \frac{13}{5}$.

23. The absolute value inequality $|x + 1| < 5$ is equivalent to $-5 < x + 1 < 5$ which simplifies to $-6 < x < 4$.

24. The absolute value inequality $|2x-1| < 7$ is equivalent to $-7 < 2x - 1 < 7$, which simplifies to $-6 < 2x < 8$ or $-3 < x < 4$.

25. The inequality $|x + 4| > 6$ is equivalent to $x + 4 < -6$ or $x + 4 > 6$. Simplifying each of these inequalities, we get $x < -10$ or $x > 2$.

26. The inequality $|3x + 9| \geq 6$ is equivalent to $3x + 9 \leq -6$ or $3x + 9 \geq 6$. Adding -9 to each of these inequalities, produces $3x \leq -15$ or $3x \geq -3$ which simplify to $x \leq -5$ or $x \geq -1$.

27. $|x+5| < -3$. But, an absolute value can never be negative, so there are no real numbers that satisfy this inequality. This is a contradictory inequality and so there is no solution.

28. The inequality $|3x + 2| < 12$ is equivalent to $-12 < 3x + 2 < 12$, or $-14 < 3x < 10$, and so $-\frac{14}{3} < x < \frac{10}{3}$.

29. The inequality $-7 \leq 3x + 5 < 26$ is equivalent to $-12 \leq 3x < 21$, or $-4 \leq x < 7$

30. The inequality $-6 < 5 - 3x \leq 17$ is equivalent to $-11 < -3x \leq 12$, or, after dividing by -3, to $\frac{11}{3} > x \geq -4$ or $-4 \leq x < \frac{11}{3}$.

31. The inequality $3x + 1 < 5 < 2x - 3$ is equivalent to the two inequalities $3x + 1 < 5$ and $5 < 2x - 3$. Subtracting 1 from the first inequality, gives $3x < 4$ or $x < \frac{4}{3}$. Adding 3 to the second inequality, gives $8 < 2x$ which can be simplified as $4 < x$ or $x > 4$. Thus, the solution is $x < \frac{4}{3}$ and $x > 4$. This is a contradictory statement, there are no real numbers that satisfy the given inequality.

32. The inequality $4x - 6 \leq 11 < 9x + 1$ is equivalent to $4x - 6 \leq 11$ and $11 < 9x + 1, 4x \leq 17$ and $10 < 9x, x \leq \frac{17}{4}$ and $\frac{10}{9} < x$. Thus, we get $\frac{10}{9} < x \leq \frac{17}{4}$.

33. $6357 \text{ km} \leq r \leq 6378 \text{ km}$

34. $357\,000 \text{ km} \leq d \leq 407\,000 \text{ km}$

35. $0.60 - 0.05 \leq c \leq 0.60 + 0.05, 0.10° \leq c \leq 1.10°$

36. $2 \leq R - 12 \leq 4$ or $32 \text{ oz} \leq R - 12 \leq 64 \text{ oz}$

37. $200 < \frac{800R}{800+R} < 500, 200(800 + R) < 800R < 500(800+R), 160000+200R < 800R < 400000+ 500R$, so $160000 + 200R < 800R$ and $800R < 400000 + 500R, 160000 < 600R$ and $300R < 400000, 266.67 < R$ and $R < 1333.33, 266.67 \, \Omega < R < 1333.33 \, \Omega$

38. $C = \frac{5}{9}(F - 32) \Rightarrow \frac{9}{5}C = F - 32, F = \frac{9}{5}C + 32, 1800°C \Rightarrow \frac{9}{5} \cdot 1800 + 32 = 3272°F, 2200°C \Rightarrow \frac{9}{5} \cdot 2200 + 32 = 3992°$ F, so the temperature range is between $3272°$ F and $3992°$ F.

39. $R - C > 0, 60x - (1,500+25x) > 0, 35x - 1,500 > 0, 35x > 1,500, x > 42.86$ or, since we cannot have a fraction, $x \geq 43$

40. $\ell \cdot h \cdot 600 =$ or $\ell \cdot 1.5 \cdot 600 = 900\ell, 2400 < 900\ell < 4000, 2.67 \text{ m} < \ell < 4.444 \text{ m}$.

41. (a) $|\epsilon| < 0.6$,
(b) $|P - 9.0| < 0.6$,
(c) Since $|P - 9.0| < 0.6$, then $-0.6 < P - 9.0 < 0.6$. Adding 9.0 to all three terms, we get $8.4 < P < 9.6$ W.

42. Since $c = \frac{Ad}{150}$ we know that $\frac{c}{d} = \frac{A}{150}$. Since $25\% = \frac{25}{100}$ and $50\% = \frac{50}{100}$, we want to know when $\frac{25}{100} < \frac{c}{d} < \frac{50}{100}$ or $\frac{25}{100} < \frac{A}{150} < \frac{50}{100}$. Multiplying this last inequality by 150, we obtain $37.5 < A < 75$. Thus, the child's dosage is between 25% and 50% of an adult's dose when the child is between the ages of 37.5 mo and 75 mo (or 3 yr 1.5 mo and 6 yr 3 mo).

43. Let x represent the number of copies sold. Since profit P is the difference between the revenue R and the cost C, we have $P = R - C = 0.35x + (0.10)(0.35)(x - 10,000) - 0.38x$. Solving this equation, we obtain

$P = 0.35x + (0.10)(0.35)(x - 10,000) - 0.38x$
$\quad = 0.35x - 350 + 0.35x - 0.38x$
$\quad = 0.005x - 350$

The break-even point is when $P = 0$ or $0.005x - 350 = 0$ or $x = 70,000$. A profit is made on the 70,001st copy. So, the company must sell at least 70,001 copies in order for the company to make a profit on this magazine.

44. $|x - 29.4°C| \leq 3.7°C$ is equivalent to $-3.7°C \leq x - 29.4°C \leq 3.7°C$. Adding 29.4 to all three parts, we obtain $(-3.7 + 29.4)°C \leq x \leq (3.7 + 29.4)°C$ or $25.7°C \leq x \leq 33.1°C$.

≡ 16.3 NONLINEAR INEQUALITIES

1. $\{x | x < -1 \text{ or } x > 3\}$

2. $\{x | x \leq -2 \text{ or } x \geq 4\}$

3. $\{x | -4 \leq x \leq 1\}$

4. $\{x | -5 < x < 2\}$

5. $x^2 - 1 < 0$ is equivalent to $(x + 1)(x - 1) < 0$. The zero values are -1 and 1, so the solution set is $\{x | -1 < x < 1\}$.

6. $x^2 \leq 16$ is equivalent to $x^2 - 16 \leq 0$ or $(x + 4)(x - 4) \leq 0$. The zero values are -4 and 4, so the solution set is $\{x | -4 \leq x \leq 4\}$.

7. $x^2 - 2x - 15 \leq 0$ is equivalent to $(x + 3)(x - 5) \leq 0$. The zero values are -3 and 5, so the solution set is $\{x | -3 \leq x \leq 5\}$.

8. $x^2 + x - 2 > 0$ is equivalent to $(x + 2)(x - 1) > 0$. The zero values are -2 and 1, so the solution set is $\{x | x < -2 \text{ or } x > 1\}$.

9. $x^2 - x - 2 \geq 0$ is equivalent to $(x + 1)(x - 2) \geq 0$. The zero values are -1 and 2, so the solution set is $\{x | x \leq -1 \text{ or } x \geq 2\}$.

10. $x^2 - 3x - 10 < 0$ is equivalent to $(x+2)(x-5) < 0$. The zero values are -2 and 5, so the solution set is $\{x | -2 < x < 5\}$.

11. $x^2 - 5x > -6$ is equivalent to $x^2 - 5x + 6 > 0$ which factors as $(x - 2)(x - 3) > 0$. The zero values are 2 and 3, so the solution set is $\{x | x < 2 \text{ or } x > 3\}$.

12. $x^2 + 7x \leq -12$ is equivalent to $x^2 + 7x + 12 \leq 0$ which factors as $(x + 3)(x + 4) \leq 0$. The zero values are -4 and -3, so the solution set is $\{x | -4 \leq x \leq -3\}$.

13. $2x^2 + 7x + 3 < 0$ is equivalent to $(2x + 1)(x + 3) < 0$. The zero values are -3 and $-\frac{1}{2}$, so the solution set is $\{x | -3 < x < -\frac{1}{2}\}$

14. $2x^2 - 5x + 2 \geq 0$ is equivalent to $(2x - 1)(x - 2) \geq 0$. The zero values are $\frac{1}{2}$ and 2, so the solution set is $\{x | x \leq \frac{1}{2} \text{ or } x \geq 2\}$.

15. $2x^2 - x < 1$ is equivalent to $2x^2 - x - 1 < 0$, or $(2x + 1)(x - 1) < 0$. The zero values are $-\frac{1}{2}$ and 1, so the solution set is $\{x | -\frac{1}{2} < x < 1\}$.

16. $2x^2 - x \leq 3$ is equivalent to $2x^2 - x - 3 \leq 0$ or $(2x - 3)(x + 1) \leq 0$. The zero values are -1 and $\frac{3}{2}$, so the solution set is $\{x | -1 \leq x \leq \frac{3}{2}\}$

17. $4x^2 + 2x > x^2 - 1$ is equivalent to $3x^2 + 2x + 1 > 0$. The discriminant of the equation $3x^2 + 2x + 1 = 0$ is $b^2 - 4ac = 4 - 4 \cdot 3 \cdot 1 = -8 < 0$, so this equation has no real solutions. When $x = 0$, we get $1 > 0$ which is true. Thus, the given inequality is an absolute inequality, all real numbers satisfy this inequality.

18. $2x^2 + 5x < 2 - x^2$ is equivalent to $3x^2 + 5x - 2 < 0$ or $(3x - 1)(x + 2) < 0$. The zero values are $\frac{1}{3}$ and -2, so the solution set is $\{x | -2 < x < \frac{1}{3}\}$.

19. The zero values are $-1, 2$, and -3, so the solution is $x < -3$ or $-1 < x < 2$.

20. $x \leq -2$ or $1 \leq x \leq 3$

21. The zero values are $-3, \frac{5}{2}$, and -4, so the solution set is $\{x| -4 < x < -3 \text{ or } x > \frac{5}{2}\}$.

22. The zero values are $-4, \frac{1}{3}$, and 5, so the solution set is $\{x| -4 \leq x \leq \frac{1}{3} \text{ or } x \geq 5\}$.

$$
\begin{array}{cccc}
- & + & - & + & \text{product} \\
- & + & + & + & x+4 \\
- & - & - & + & x-5 \\
- & - & + & + & 3x-1 \\
\end{array}
$$

23. $(x-2)^2(x+4) < 0$. The zero values are 2 and -4, so the solution set is $\{x|x < -4\}$.

24. $(x+3)^2(x-5) \geq 0$. The zero values are -3 and 5, so the solution set is $\{x|x = -3 \text{ or } x > 5\}$.

25. $x^3 - 4x > 0$ is equivalent to $x(x+2)(x-2) > 0$. The zero values are $0, -2$, and 2, so the solution set is $\{x| -2 < x < 0 \text{ or } x > 2\}$.

26. $x^4 - 9x^2 \leq 0$ is equivalent to $x^2(x^2 - 9) \leq 0$ which factors as $x^2(x+3)(x-3) \leq 0$. The zero values are $0, -3$, and 3, so the solution set is $\{x| -3 \leq x \leq 3\}$.

$$
\begin{array}{ccc}
+ & - & + & \text{product} \\
- & - & + & x-3 \\
- & + & + & x+3 \\
+ & + & + & x^2 \\
\end{array}
$$

27. $x^2 + 2x + 3 \leq 0$. The discriminant of the equation $x^2 + 2x + 3 = 0$ is $2^2 - 4 \cdot 3 = -8 < 0$, so

there are no real solutions to this equation. When $x = 0, 3 \leq 0$ is false so the given inequality is a contradictory inequality and there are no real solutions.

28. $x^2 + x + 1 > 0$. The discriminant is $b^2 - 4ac = 1 - 4 = -3 < 0$, so the quadratic equation $x^2 + x + 1 = 0$ has no real solutions. When $x = 0$, the given inequality gives $1 > 0$, which is always true. This is an absolute inequality, so all real numbers satisfy this inequality.

29. The zero values are 2, 5, and -1, so the solution set is $\{x|x < -1 \text{ or } 2 < x < 5\}$

$$
\begin{array}{cccc}
- & + & - & + & \text{product} \\
- & - & - & + & x-5 \\
- & - & + & + & x-2 \\
- & + & + & + & x+1 \\
\end{array}
$$

30. The zero values are -2, 4 and 2. Since the denominator is 0 when $x = 2$, we know that 2 cannot be in the solution set. We see that the solution set is $\{x| -2 \leq x < 2 \text{ or } x \geq 4\}$.

$$
\begin{array}{cccc}
- & + & - & + & \text{product} \\
- & - & - & + & x-4 \\
- & - & + & + & x-2 \\
- & + & + & + & x+2 \\
\end{array}
$$

31. $\frac{x}{(x+1)(x-2)} > 0$. The zero values are $0, -1$, and 2, so the solution set is $\{x| -1 < x < 0 \text{ or } x > 2\}$.

32. $\frac{x(x-3)}{(x+1)(x-4)} \leq 0$. The zero values are 0, 3, -1, and 4, so the solution set is $\{x| -1 < x \leq 0 \text{ or } 3 \leq x < 4\}$.

33. $\frac{(3x+1)(x-3)}{2x-1} \leq 0$. The zero values are $-\frac{1}{3}, \frac{1}{2}$, and 3, so the solution set is $\{x|x \leq -\frac{1}{3} \text{ or } \frac{1}{2} < x \leq 3\}$.

34. $\frac{(x-1)(x+6)}{(x+1)(x-3)} > 0$. The zero values are $-6, -1, 1$, and 3, so the solution set is $\{x|x < -6 \text{ or } -1 < x < 1 \text{ or } x > 3\}$.

35. The inequality $\frac{4}{x-1} < \frac{5}{x+1}$ is equivalent to $\frac{4}{x-1} - \frac{5}{x+1} < 0$ or $\frac{4(x+1)-5(x-1)}{(x-1)(x+1)} < 0$. This simplifies to $\frac{-x+9}{(x-1)(x+1)} < 0$. The zero values are $9, -1$, and 1, so the solution set is $\{x| -1 < x < 1 \text{ or } x > 9\}$.

36. The inequality $\frac{-3}{x+2} \geq \frac{6}{x-3}$ is equivalent to $\frac{-3}{x+2} - \frac{6}{x-3} \geq 0$ or $\frac{-3(x-3)-6(x+2)}{(x+2)(x-3)} \geq 0$, which simplifies to $\frac{-9x-3}{(x+2)(x-3)} \geq 0$ which is equivalent to $\frac{3x+1}{(x+2)(x-3)}$

≤ 0. The zero values are $-\frac{1}{3}, -2$, and 3, so the solution set is $\{x \mid x < -2 \text{ or } -\frac{1}{3} \leq x < 3\}$.

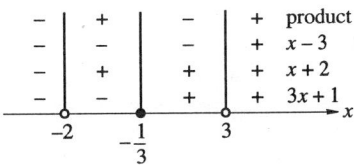

37. $\frac{x^2+2x+3}{x-1} \geq 0$. The numerator x^2+2x+3 is always > 0, so the original inequality is true when $x > 1$.

38. $\frac{x^2+x+1}{x+2} < 0$; $x^2 + x + 1$ is always > 0 so the original inequality is true when $x < -2$

39. $|x^2 + 3x + 2| \leq 4$ is equivalent to $-4 \leq x^2 + 3x + 2 \leq 4$. We first examine the right-hand inequality $x^2 + 3x + 2 \leq 4$, which simplifies to $x^2 + 3x - 2 \leq 0$. Using the quadratic formula, we see that the equation $x^2 + 3x - 2 = 0$ is zero when $x = \frac{-3\pm\sqrt{9+8}}{2} = \frac{-3\pm\sqrt{17}}{2}$. Next, we examine the left-hand inequality and see that $-4 \leq x^2 + 3x + 2$ is always true. So the solution set is $\left\{x \mid \frac{-3-\sqrt{17}}{2} \leq x \leq \frac{-3+\sqrt{17}}{2}\right\}$.

40. $|x^2-3x+2| > 5$ is equivalent to $x^2-3x+2 < -5$ or $x^2 - 3x + 2 > 5$. $x^2 - 3x + 2 < -5$ is never true, so the only answers are those that satisfy $x^2 - 3x + 2 > 5$ or, equivalently, $x^2 - 3x - 3 > 0$. By the quadratic formula, we see that $x^2-3x-3 = 0$ when $x = \frac{3\pm\sqrt{9+12}}{2}$ or $x = \frac{3\pm\sqrt{21}}{2}$. Thus, the solution set is $\left\{x \mid x < \frac{3-\sqrt{21}}{2} \text{ or } x > \frac{3+\sqrt{21}}{2}\right\}$

41. Since $I = 75d^2$, the solution to $75 < I < 450$ is equivalent to $75 < 75d^2 < 450$ or $1 < d^2 < 6$. Since the context of the problem means that d is positive, and so we see that $1 \text{ m} < d < \sqrt{6} \text{ m}$.

42. We have the inequality $34.3t - 4.9t^2 > 49$ which is equivalent to $-4.9t^2 + 34.3t - 49 > 0$ which simplifies to $t^2 - 7t + 10 < 0$. This factors as $(t - 2)(t - 5) < 0$ and so $2 < t < 5$. Thus, the solution is $2 \text{ s} < t < 5 \text{ s}$.

43. $x^2 - 1.1x + 0.2 > 0.08$ is equivalent to $x^2 - 1.1x + 0.12 > 0$. The equation $x^2 - 1.1x + 0.12 = 0$ has solutions $x = \frac{1.1\pm\sqrt{1.21-.48}}{2} = \frac{1.1\pm\sqrt{0.73}}{3}$. So, the solution set to the given inequality is $x < 0.123$ or $x > 0.977$.

44. $L = \frac{90 \cdot 6d^2}{18} = 30d^2$. Since $L = 3000$, we need to solve the inequality $30d^2 > 3000$, which is equivalent to $d^2 > 100$ and so $d > \pm10$. Since d is positive and < 12, we see that $10'' < d < 12''$.

45. **(a)** The height will be greater than 500 ft when $h(t) = -16t^2 + 64t + 452 > 500$. Solve the inequality as follows
$$-16t^2 + 64t + 452 > 500$$
$$-16t^2 + 64t + 452 - 500 > 0$$
$$-16t^2 + 64t - 48 > 0$$
$$t^2 - 4t + 3 < 0$$
$$(t - 3)(t - 1) < 0$$
A sign chart, or your graphing calculator, will show that $t^2 - 4t + 3 < 0$ between $1 < t < 3$ s and so the ball is more than 500 ft above the ground when $1 \text{ s} < t < 3 \text{ s}$.

(b) The height will be less than 395 ft when $h(t) = -16t^2 + 64t + 452 < 395$ or $-16t^2 + 64t + 452 - 395 < 0$. This is equivalent to $-16t^2 + 64t + 57 < 0$. The quadratic formula shows that $-16t^2 + 64t + 57 = 0$ when $t = -0.75$ or $t = 4.75$. The first result is not possible since that is before the ball is thrown. Examining a sign chart or a graph we see that the solution will be when $t > 4.75$ s.

46. Substituting and solving the inequality yields
$$0.002837D^2(75) - 0.127248 > 47.75$$
$$0.212775D^2 - 0.127248 > 47.75$$
$$0.212775D^2 > 0.127248 + 47.75$$
$$0.212775D^2 > 47.877248$$
$$D^2 > 225.0135$$
$$D > 15.00$$
$$\text{or} \quad D < -15.00$$
It does not make physical sense for a tree to have a negative diameter. Hence, the only solution is when $D > 15.00$ in.

47. $$0.79D^2 - 2D - 4 > 926$$
$$0.79D^2 - 2D - 9308 > 0$$
By the quadratic formula, the roots of $0.79D^2 - 2D - 9308 = 0$ are $D = -33.1$ and $D = 35.6$. Hence, by a sign chart or graphing calculator, you can see that $D \geq 35.6$ in. is the only realistic solution.

48. The absolute value inequality is equivalent to $-10,000 < -2.8T^2 + 820T - 44,000 < 10,000$. There are two equations associated with this inequality, namely $-10,000 = -2.8T^2 + 820T - 44,000$ and $-2.8T^2 + 820T - 44,000 = 10,000$. Solving for the roots of the first equation we have
$$-10,000 = -2.8T^2 + 820T - 44,000$$
$$0 = -2.8T^2 + 820T - 34,000$$
By the quadratic formula we see that $T = 500$ or $T = 242.9$. Solving the second equation, we get
$$-2.8T^2 + 820T - 44,000 = 10,000$$
$$-2.8T^2 + 820T - 54,000 = 0$$
Its roots are $T = 100.0$ and $T = 192.9$. Using a sign chart with the original inequality yields

$50.0 < T < 100.0$ or $192.9 < T < 242.9$. Check your answer by looking at the graphs of $y_1 = |-2.8T^2 + 820T - 44,000|$ and $y_2 = 10,000$.

≡ 16.4 INEQUALITIES IN TWO VARIABLES

1.

2.

3.

4.

5.

6.

7.

8.

9.

10.

11.

12.

13.

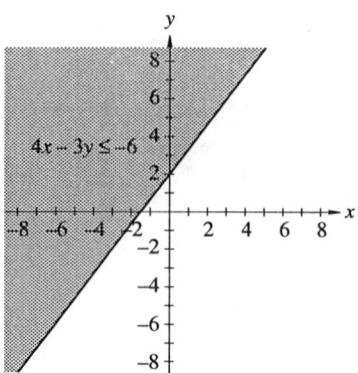

$4x - 3y \leq -6$

14.

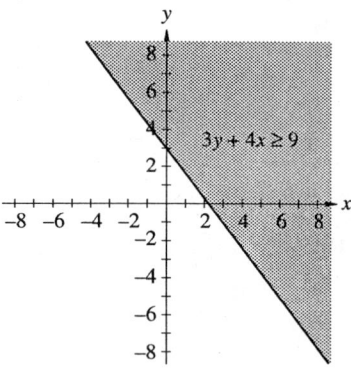

$3y + 4x \geq 9$

15.

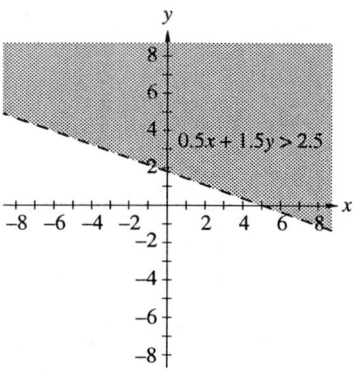

$0.5x + 1.5y > 2.5$

16.

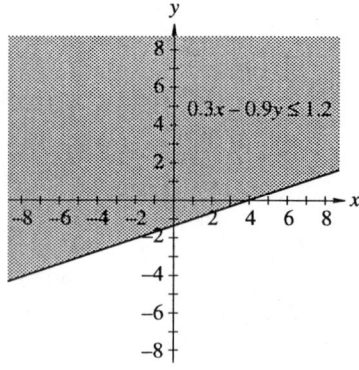

$0.3x - 0.9y \leq 1.2$

17.

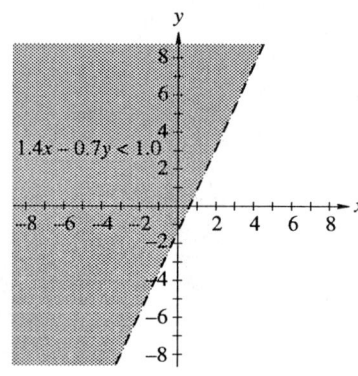

$1.4x - 0.7y < 1.0$

18.

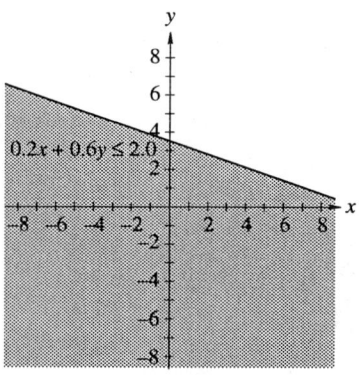

$0.2x + 0.6y \leq 2.0$

19.

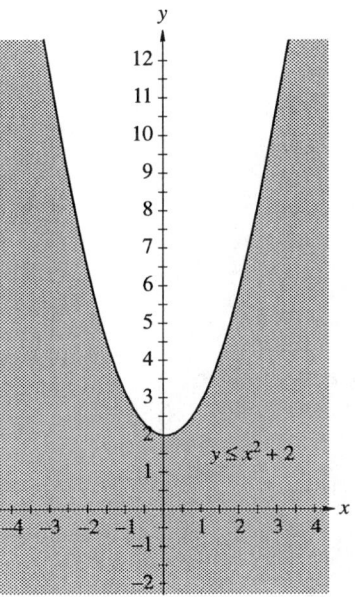

$y \leq x^2 + 2$

20.

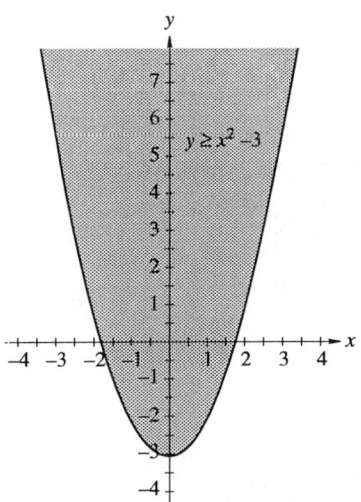

$y \geq x^2 - 3$

21.

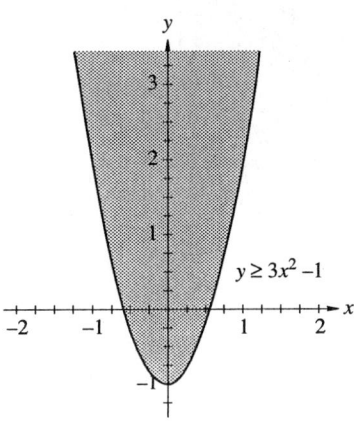

$y \geq 3x^2 - 1$

22.

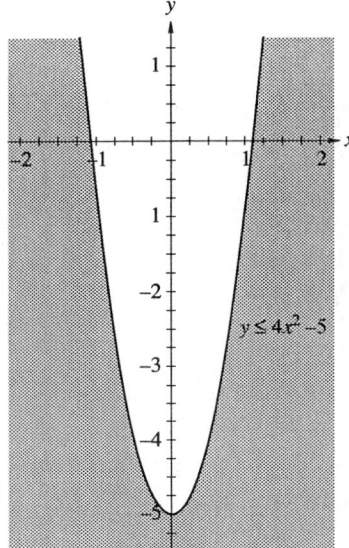

$y \leq 4x^2 - 5$

23.

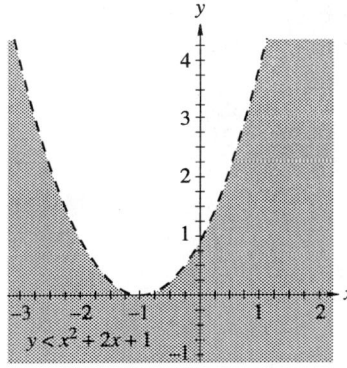

$y < x^2 + 2x + 1$

24.

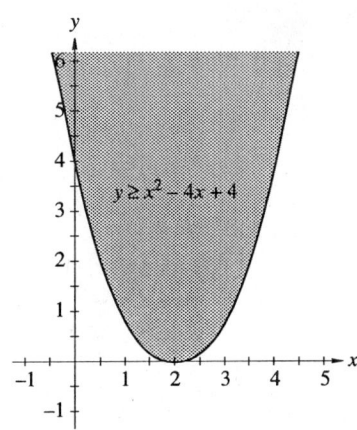

$y \geq x^2 - 4x + 4$

25.

26.

27.

28.

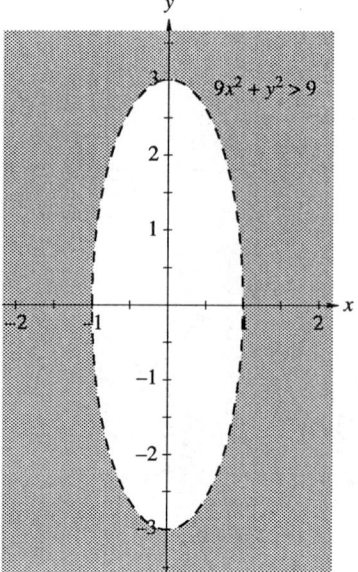

29. $|x + 1| < 4, -4 < x + 1 < 4, -5 < x < 3$

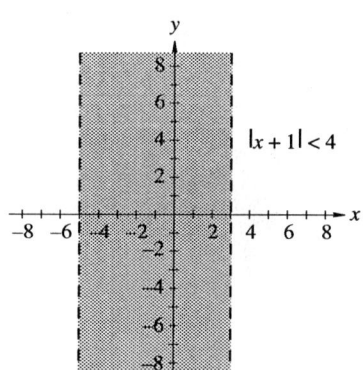

30. $|x - 5| \le 3, -3 \le x - 5 \le 3, 2 \le x \le 8$

31.

$1.8P + T < 270$
if $T > 0$ and $P \geq 0$

32.

$C < n^2 + 10n + 35$
$n \geq 0$

33. The inequality is $0.085x + 0.0675y \geq 3.6$ or $y \geq 53.5 - 1.25925x$.

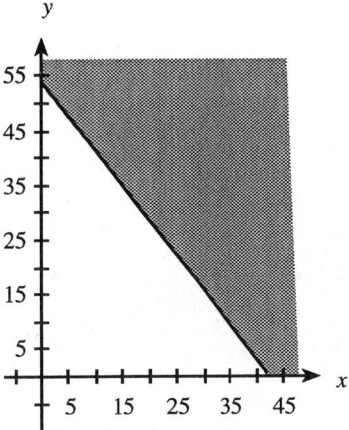

34. The inequality is $15x + 9y \geq 90$ or $y \geq 10 - \frac{5}{3}x$.

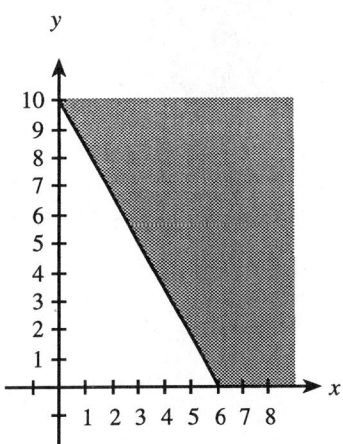

16.5 SYSTEMS OF INEQUALITIES: LINEAR PROGRAMMING

1.

4.

2.

5.

3.

6.

7.

8.

9.

10.

11.

12.

13.

14.

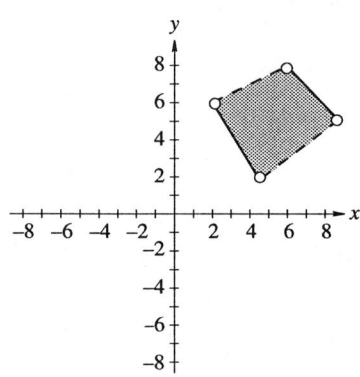

15. First graph the constraints. The vertex points of the feasible region are (2, 3), (2, 4.5) and (4, 3). (Fig. 16.79s). Evaluating $P = 2x + y$ at each of these points (2, 3): $2 \cdot 2 + 3 = 7, (2, 4.5): 2 \cdot 2 + 4.5 = 8.5, (4, 3) : 2 \cdot 4 + 3 = 11$. Maximum of 11 at (4, 3).

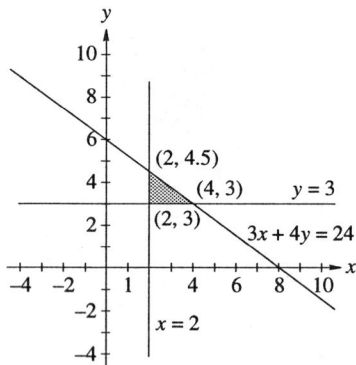

16. After graphing the constraints, the vertex points of the feasible region are at $(0, 6.25), (15, 0)$ and $(3, 4)$. $C(0, 6.25) = 0 + 5 \times 6.25 = 31.25, C(15, 0) = 9 \cdot 15 + 4 \cdot 0 = 135, C(4, 3) = 9 \cdot 4 + 5 \cdot 3 = 51$. Minimum is 31.25 at (0, 6.25).

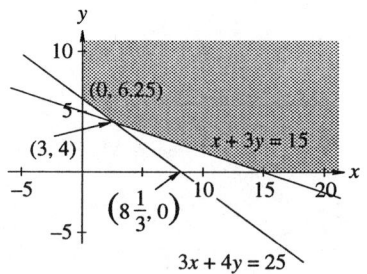

17. After graphing the constraints, the vertex points of the feasible region are $(0, 0), (0, 23\frac{1}{3}), (26\frac{2}{3}, 0)$, and $(20, 10)$. $F = 15x + 10y$, so $F(0, 0) = 15 \cdot 0 + 10 \cdot 0 = 0, F(0, 23\frac{1}{3}) = 10 \times 23\frac{1}{3} = 233.3, F(26\frac{2}{3}, 0) = 15 \times 26\frac{2}{3} = 400, F(20, 10) = 15 \cdot 20 + 10 \cdot 10 = 400$ also. The maximum of 400 occurs at any point on the segment $3x + 2y = 80$ and $20 \leq x \leq 26\frac{2}{3}$.

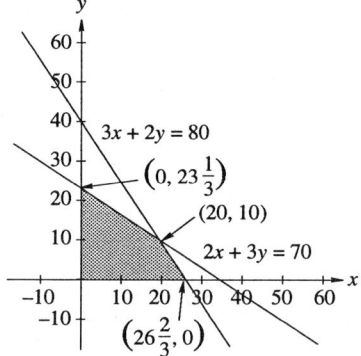

18. After graphing the constraints, the vertex points of the feasible region are at $(0, 2), (0, 10), (5, 7)$, and $(3\frac{1}{3}, 8\frac{2}{3})$. $F(0, 2) = 6 \cdot 0 + 9 \cdot 2 = 18, F(0, 10) = 6 \cdot 0 + 9 \cdot 10 = 90, F(5, 7) = 6 \cdot 5 + 9 \cdot 7 = 93, F(3\frac{1}{3}, 8\frac{2}{3}) = 6 \cdot 3\frac{1}{3} + 9 \cdot 8\frac{2}{3} = 20 + 78 = 98$. Minimum of 18 at (0, 2)

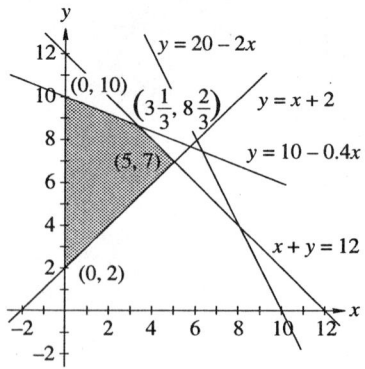

19. After graphing the constraints, the vertex points of the feasible region are at (0,3), (0,18), (12,12), and (16.5,3). Evaluating $P = 8x + 10y$ at each of these vertex points, we obtain $P(0,3) = 30, P(0,18) = 180, P(12,12) = 216$, and $P(16.5,3) = 162$. Maximum of 216 at (12, 12).

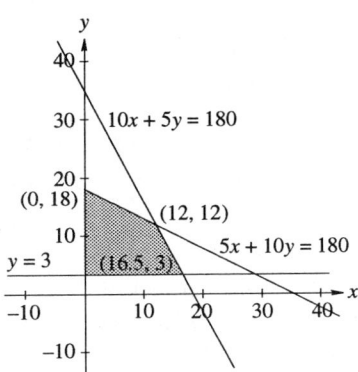

20. After graphing the constraints, the vertex points of the feasible region are at $(-2,-3), (-5,0), (2,7)$, and $(2,-3)$. Evaluating $M = 10x - 8y + 15$ at each of these vertex points, we obtain $M(-2,-3) = -20 + 24 + 15 = 19, M(-5,0) = -50 + 15 = -35, M(2,7) = 20 - 56 + 15 = -21$, and $M(2,-3) = 20 + 24 + 15 = 59$. Minimum of -35 at $(-5,0)$.

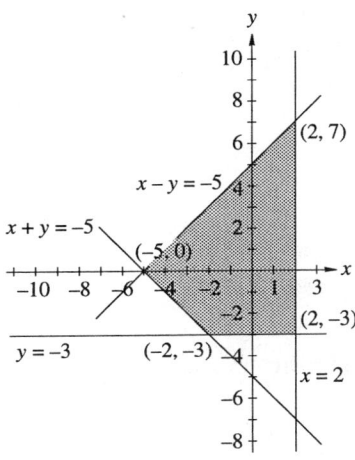

21. The constraints are as follows: $X \geq 0, Y \geq 0, \frac{1}{2}X + \frac{1}{4}Y \leq 100$, or $Y \leq 400 - 2X, \frac{3}{4}X + \frac{1}{6}Y \leq 80$ or $Y \leq 480 - \frac{9}{2}X$. Graphing the constraints gives vertex points at (0, 0), (0, 400), $(106\frac{2}{3}, 0)$, and (32, 336). Evaluating $P(x,y) = 10x + 8y$ at each of these vertex points, we get $P(0,0) = 0, P(0,400) = 3200, P(106\frac{2}{3},0) = 1067$, and $P(32, 336) = 3008$. Maximum profit of \$3200 when you produce 0 of X and 400 of Y. Adding people at A will increase profits.

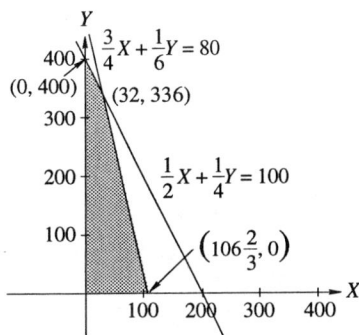

22. Labor constraint $A + B \leq 120$. Parts constraint $4A + 3B \leq 390$ along with the constraints $A \geq 0$ and $B \geq 0$. Graphing the constraints gives vertex points of the feasible region at (0,0), (0, 120), (30, 90), and (97.5,0). $P(A,B) = 7A + 5.5B, P(0,0) = 0, P(0,120) = 660, P(30,90) = 705, P(97.5,0) = 682.5$. Thus, we see that 30 of A and 90 of B will produce a maximum profit of \$705.00.

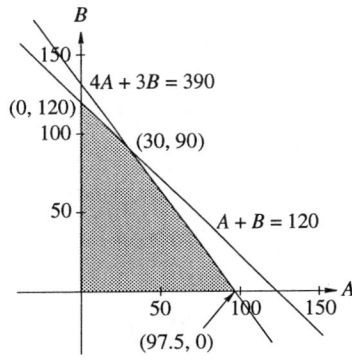

23. $P(A,B) = 8A + 9.50B, P(0,0) = 0, P(0,120) = 1140, P(30,90) = 1095, P(97.5,0) = 780$. The maximum profit will be when the company makes 120 of B and none of A for a profit of \$1,140.00.

24. Let P represent the number of PCs that the company makes and B the number of BCs that the company makes. The constraint on the AB chip means that $2P + 3B \leq 200$, and the constraint on the EP chip means that $3P + 8B \leq 450$. The nature of the problem means that $P \geq 0$ and $B \geq 0$. The feasible region is shown in the figure. Vertex points of the feasible region are at (0, 0), (100, 0), (0, 56.25), and (35.71, 42.86). The profit function is $P(P,B) = 145P + 230B$. Evaluating the profit function at each of the vertex points, we see that $P(0,0) = 0, P(100,0) = 14,500, P(35.71,42.86) = 15,035.75$, and $P(0, 56.25) = 12,937.5$. So, making 35.71 of PC and 42.86 of BC will produce a maximum profit of \$15,035.75.

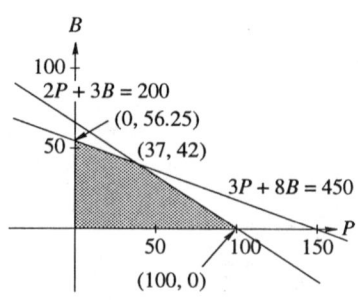

25. Let x be the amount in millions of dollars for home loans and y the amount for commercial loans. Then the constant equations are $x \geq 0, y > 0, x + y \leq 50, x \geq 5y$, and $0.085x + 0.0675y \geq 3.6$. The vertices are at $\left(\frac{7200}{197}, \frac{1440}{197}\right) \approx (36.5, 7.3)$, $\left(\frac{125}{3}, \frac{25}{3}\right) \approx (41.7, 8.3), (50, 0)$, and $\left(\frac{720}{17}, 0\right) \approx (42.4, 0)$. The bank can make as few as 37 home loans and 7 commercial loans in order to reach its goal of at least \$3.6 million dollars.

26. The constraints are $15X + 9Y \geq 90, 10X + 12Y \geq 90, X \geq 0$, and $Y \geq 0$. The vertices of the feasible region are $(0, 10)$, $(3, 5)$, and $(19, 0)$. The cost equation is $C = 5X + 3.5Y$. Substitution yields costs of 35, 32.5, and 45. The patient should take 3 of brand X and 5 of brand Y to minimize cost.

27. Let x represent the number of small tubes and y the number of large ones. The constraints are $x \geq 8,000, y \geq 5,000, x + y \geq 15,000$, and $x \geq 2y$. The only vertex of the feasible region is $(10000, 5000)$. Buy 10,000 small and 5,000 large for a total cost of \$2,100.

28. The constraints are $3A + 2B \leq 120, A + 2B \leq 80, A \geq 0$, and $B \geq 0$. The vertices of the feasible region are at $(0, 40), (20, 30)$, and $(40, 0)$. The profit equation is $P = 32A + 45B$. Substituting these values gives respective profits of 1800, 1990, and 1280. The company should manufacture 20 As and 30 Bs for a profit of \$1,990.

 CHAPTER 16 REVIEW

1. $3x < -12$ or $x < -4$

2. $4x + 5 < 6$ or $4x < 1$ and so $x < \frac{1}{4}$

3. $2x - 7 \geq 15$ or $2x \geq 22$ which gives $x \geq 11$

4. $7 - 5x < 2 + 3x$ or $-8 < -5$, and so $x > \frac{5}{8}$

5. $\frac{2x+5}{4} < \frac{4x-1}{3}$ or $6x + 15 < 16x - 4$ which simplifies to $-10x < -19$, and so $x > \frac{19}{10}$.

6. $|2x - 1| \leq 5$ is equivalent to $-5 \leq 2x - 1 \leq 5$ or $-4 \leq 2x \leq 6$, and so $-2 \leq x \leq 3$.

7. $|3x + 2| > 7$ is equivalent to

$3x + 2 < -7$ or $3x + 2 > 7$
$3x < -9$ or $3x > 5$
$x < -3$ or $x > \frac{5}{3}$

8. $2x + 3 < 13 \leq 3x - 9$ is equivalent to
$2x + 3 < 13$ and $13 \leq 3x - 9$
$2x < 10$ and $22 \leq 3x$
$x < 5$ and $x \geq \frac{22}{3}$
This is a contradictory inequality and so there is no solution.

9. $-2x + y < 5$ or $y < 2x + 5$

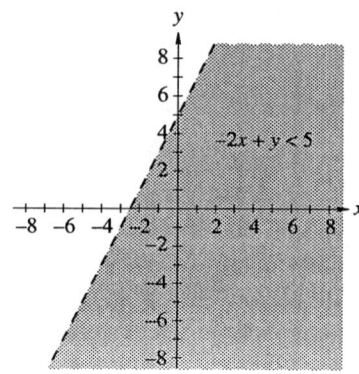

10. $2x + 3y \geq 6$ or $y \geq \frac{-2x}{3} + 2$

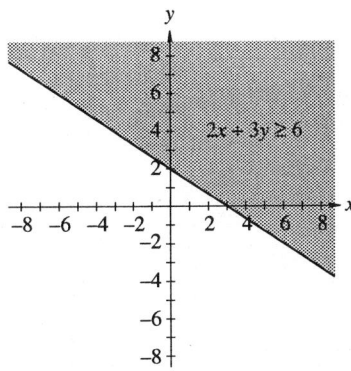

11. $4x - 3y < 3$ or $y > \frac{4}{3}x - 1$

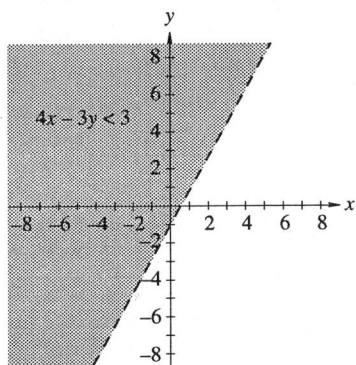

12. $y > 2x^2 - 5$

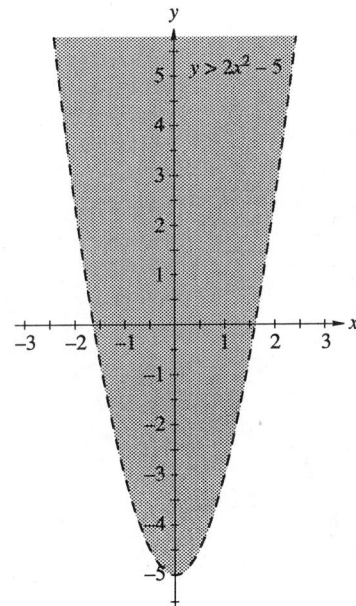

13. $y \leq x^2 - 6x + 9$

14.

15.

16.

17.

18.

19.

20.

21.

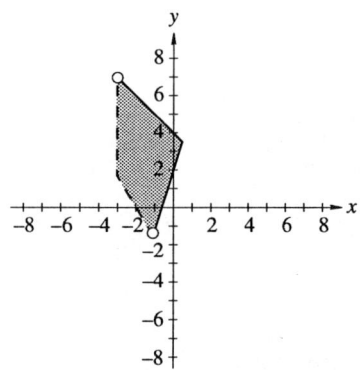

22. Graphing the constraints as shown, we see that the vertices of the feasible region are $(-1, 6), (-1, -3)$, $(3, 1)$ and $(3, 2)$. We have $P(x, y) = 3x + 5y$, $P(-1, 6) = -3 + 30 = 27$, $P(-1, -3) = -3 - 15 = -18$, $P(3, 1) = 9 + 5 = 14$, and $P(3, 2) = 9 + 10 = 19$. We can see that there is a maximum of 27 at $(-1, 6)$

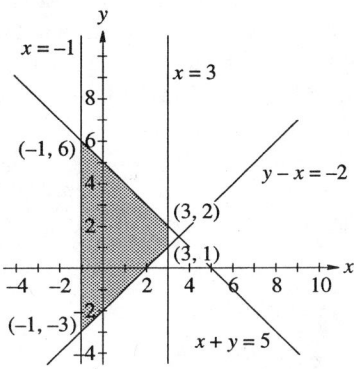

23. After graphing, we see that the vertices of the feasible region are $\left(\frac{-4}{7}, \frac{-23}{7}\right), \left(\frac{18}{5}, -\frac{6}{5}\right)$, and $\left(\frac{7}{6}, \frac{11}{3}\right)$. Checking $F(x, y) = 4x - 3y$ at these points, we see that $F\left(\frac{4}{7}, \frac{23}{7}\right) = 7\frac{4}{7}, F\left(\frac{7}{6}, \frac{11}{3}\right) = -6\frac{1}{3}$, and $F\left(\frac{18}{5}, -\frac{6}{5}\right) = 18$. So, the minimum of $-6\frac{1}{3}$ is at $\left(\frac{7}{6}, \frac{11}{3}\right)$ or $\left(1\frac{1}{6}, 3\frac{2}{3}\right)$.

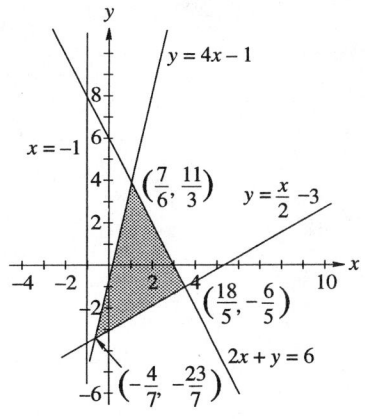

24. $A = lw., 7.0 \times 1.4 < A < 7.5 \times 1.6, 9.8 \text{ mm}^2$ $A < 12 \text{ mm}^2$

25. Since $R_1 R_2 = 20$, we see that $\frac{R_1 R_2}{R_1 + R_2} \geq 4$ is equivalent to $\frac{20}{R_1 + R_2} \geq 4$ or $20 \geq 4(R_1 + R_2)$ and so $R_1 + R_2 \leq 5\,\Omega$.

26.

	r	t	d
normal	r	8	$8r$
increased	$r + 4$	7.5	$7.5(r + 4)$

Thus, we have the equation $8r = 7.5r + 30$, or $0.5r = 30$, and so $r = 60$. Usual speed is 60 mph, distance is $8 \cdot 60 = 480$. mi.

27. We have the system: $12C + 18S \leq 220, 15C + 10S \leq 180, C \geq 0, S \geq 0, C$ and S must be integers. We graph C on the horizontal axis, and S on the vertical axis. Solving the equation derived from the second inequality, we get $S = 18 - 1.5C$. Substituting this value of S into the equation from the first inequality yields $12C + 18(18 - 1.5C) = 220$ or $12C + 324 - 27C = 220$ which simplifies to $-15C = -104$. Thus, we find that these two lines intersect when $C = 6.9\overline{3}$. Possible vertex points are $(0, 0), (0, 12), (6, 8), (7, 7)$ and $(12, 0)$. Evaluating $P(C, S) = 1250C + 1620S$ at each of these vertex points, we find $P(0, 0) = 0, P(0, 12) = 19,400, P(6, 8) = 20,460, P(7, 7) = 20,090$, and $P(12, 0) = 15,000$. Maximum profit of \$20,460 at $(6, 8)$ or 6 of C and 8 of S.

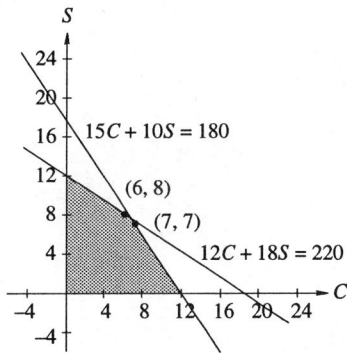

CHAPTER 16 TEST

1. $5x > 30, x > 6$

2. $|4x + 5| < 17, -17 < 4x + 5 < 17, -22 < 4x < 12, -\frac{11}{2} < x < 3$

3. $\frac{4x-5}{2} \leq \frac{7x-2}{3}, 12x - 15 \leq 14x - 4, -2x \leq 11, x \geq -\frac{11}{2}$

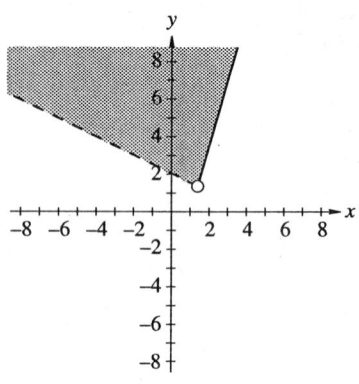

4. $(x - 2)(x - 5) > 0$ has two cases. In Case I, we have $x - 2 > 0$ or $x > 2$ and that $x - 5 > 0$ or $x > 5$. These both occur when $x > 5$. In Case II, we have $x - 2 < 0$ or $x < 2$ and $x - 5 < 0$ or $x < 5$. These both occur when $x < 2$. Thus the solution set is $\{x | x < 2 \text{ or } x > 5\}$

5. $5x^2 + 3x \leq 2x^2 - x - 1, 3x^2 + 4x + 1 \leq 0, (3x + 1)(x + 1) \leq 0$. The solution set is $\{x | -1 \leq x \leq -\frac{1}{3}\}$

6.

7. Iron constraint gives the inequality $12x + 5y \geq 80$, the zinc constraint gives the inequality $10x + 8y \geq 90$, while $x \geq 0$, and $y \geq 0$. The vertex points on the axes are (0, 16) and (9, 0). The two equations $10x + 8y = 90$ and $12x + 5y = 80$ intersect at $\left(\frac{95}{23}, \frac{140}{23}\right) \approx (4.13, 6.09)$. The points (5, 5) and (4, 7) are the closest to this vertex. The cost is given by the equation $C(x, y) = 4x + 6y$. Evaluating the cost function at each of these four points we get $C(0, 15) = 90¢, C(4, 7) = 58¢, C(5, 5) = 50¢$, and $C(9, 0) = 36¢$.

(a) The minimum cost is obtained by taking 9 of brand X and none of Y.

(b) The minimum cost is 36¢.

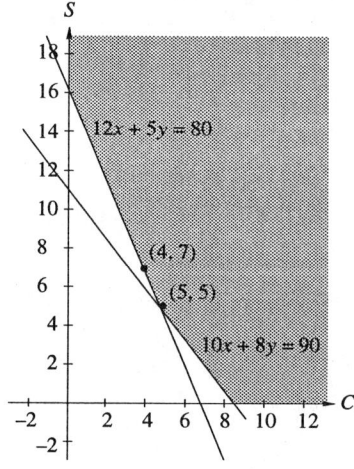

17

Matrices

≡ 17.1 MATRICES

1. 2×3

2. 3×3

3. 3×6

4. 4×1

5. 4×2

6. 1×7

7. $a_{11} = 1; a_{24} = 13; a_{21} = 8; a_{32} = 6$

8. $b_{12} = 0; b_{21} = 4; b_{23} = 9; b_{32} = -8$

9. $x = 12, y = -6, z = 2, w = 5$

10. $x - 2 = 7$, so $x = 9$; $y + 5 = 3$, so $y = -2$; $3z = 3$, so $z = 1$; $w + 6 = 7$, so $w = 1$; and $2p = 10$, so $p = 5$.

11. $\begin{bmatrix} -3 & 6 & 6 & -8 \\ 12 & 9 & 17 & 0 \\ 9 & 0 & 5 & 10 \end{bmatrix}$

12. $\begin{bmatrix} 7 & -2 \\ -3 & 10 \\ 11 & 6 \end{bmatrix}$

13. $\begin{bmatrix} 2 & -1 & 4 \\ 5 & 9 & -2 \end{bmatrix}$

14. $\begin{bmatrix} 2 & 0 & 3 & 3 \\ 3 & 0 & 9 & 1 \\ 14 & 8 & 0 & -14 \\ 10 & 10 & 2 & 10 \\ 0 & 6 & 23 & 4 \end{bmatrix}$

15. $A + B = \begin{bmatrix} -1 & 7 & 4 \\ 8 & 1 & 15 \end{bmatrix}$

16. $B - C = \begin{bmatrix} -10 & -4 & -2 \\ -3 & 3 & -1 \end{bmatrix}$

17. $3A = \begin{bmatrix} 12 & 9 & 6 \\ 15 & 0 & 21 \end{bmatrix}$

18. $4B = \begin{bmatrix} -20 & 16 & 8 \\ 12 & 4 & 32 \end{bmatrix}$

19. $2B + C = \begin{bmatrix} -10 & 8 & 4 \\ 6 & 2 & 16 \end{bmatrix} + \begin{bmatrix} 5 & 8 & 4 \\ 6 & -2 & 9 \end{bmatrix} = \begin{bmatrix} -5 & 16 & 8 \\ 12 & 0 & 25 \end{bmatrix}$

20. $4A - 3C = \begin{bmatrix} 16 & 12 & 8 \\ 20 & 0 & 28 \end{bmatrix} - \begin{bmatrix} 15 & 24 & 12 \\ 18 & -6 & 27 \end{bmatrix} = \begin{bmatrix} 1 & -12 & -4 \\ 2 & 6 & 1 \end{bmatrix}$

21. $2B - 3A = \begin{bmatrix} -10 & 8 & 4 \\ 6 & 2 & 16 \end{bmatrix} - \begin{bmatrix} 12 & 9 & 6 \\ 15 & 0 & 21 \end{bmatrix} = \begin{bmatrix} -22 & -1 & -2 \\ -9 & 2 & -5 \end{bmatrix}$

22. $4C + 2B = \begin{bmatrix} 20 & 32 & 16 \\ 24 & -8 & 36 \end{bmatrix} + \begin{bmatrix} -10 & 8 & 4 \\ 6 & 2 & 16 \end{bmatrix} = \begin{bmatrix} 10 & 40 & 20 \\ 30 & -6 & 52 \end{bmatrix}$

23. $D - E = \begin{bmatrix} -3 \\ -7 \\ 3 \\ -5 \end{bmatrix}$

24. $E + F = \begin{bmatrix} 21 \\ 27 \\ -12 \\ 15 \end{bmatrix}$

25. $3E = \begin{bmatrix} 21 \\ 27 \\ -12 \\ 30 \end{bmatrix}$

26. $F - 2E = \begin{bmatrix} 14 \\ 18 \\ -8 \\ 5 \end{bmatrix} - \begin{bmatrix} 14 \\ 18 \\ -8 \\ 20 \end{bmatrix} = \begin{bmatrix} 0 \\ 0 \\ 0 \\ -15 \end{bmatrix}$

27. $3D + 2E = \begin{bmatrix} 12 \\ 6 \\ -3 \\ 15 \end{bmatrix} + \begin{bmatrix} 14 \\ 18 \\ -8 \\ 20 \end{bmatrix} = \begin{bmatrix} 26 \\ 24 \\ -11 \\ 35 \end{bmatrix}$

28. $2F - 7D = \begin{bmatrix} 28 \\ 36 \\ -16 \\ 10 \end{bmatrix} - \begin{bmatrix} 28 \\ 14 \\ -7 \\ 35 \end{bmatrix} = \begin{bmatrix} 0 \\ 22 \\ -9 \\ -25 \end{bmatrix}$

29. $7D - 2F = \begin{bmatrix} 28 \\ 14 \\ -7 \\ 35 \end{bmatrix} - \begin{bmatrix} 28 \\ 36 \\ -16 \\ 10 \end{bmatrix} = \begin{bmatrix} 0 \\ -22 \\ 9 \\ 25 \end{bmatrix}$

30. $5E - 2F = \begin{bmatrix} 35 \\ 45 \\ -20 \\ 50 \end{bmatrix} - \begin{bmatrix} 28 \\ 36 \\ -16 \\ 10 \end{bmatrix} = \begin{bmatrix} 7 \\ 9 \\ -4 \\ 40 \end{bmatrix}$

31. $\begin{bmatrix} 18 & 7 & 9 \\ 7 & 11 & 4 \end{bmatrix} + \begin{bmatrix} 9 & 5 & 6 \\ 11 & 4 & 12 \end{bmatrix} = \begin{bmatrix} 27 & 12 & 15 \\ 18 & 15 & 16 \end{bmatrix}$

27 chip A, 12 chip B, 15 EPROMS, 18 keyboards, 15 motherboards and 16 disk drives

32.   The answer is the second matrix minus the first

Store	195/70SR-14	205/70SR-14	185/70SR-15	205/70SR-15	215/70SR-15
A	8	33	32	17	28
B	11	45	13	0	21
C	29	43	27	43	23

33. $\begin{bmatrix} 70 & 40 \\ 30 & 60 \end{bmatrix} + \begin{bmatrix} 65 & 42 \\ 35 & 58 \end{bmatrix} = \begin{bmatrix} 135 & 82 \\ 65 & 118 \end{bmatrix}$

34.   The answer is the inventory matrix minus the sales matrix.

Chip type :	286	386	486	586
Warehouse A	930	1070	1590	635
Warehouse B	530	350	4076	625
Warehouse C	800	1974	3457	188

35.   The answer is the last sales matrix times 1.1. Note: Answers are rounded to the nearest integer.

Chip Type :	286	386	486	586
Warehouse A	297	2343	3531	292
Warehouse B	1232	4653	3436	83
Warehouse C	352	3439	3017	1113

36.   Model

	S	M	L	XL
Jogging	20	16	12	16
Sweating	28	48	60	28
Walking	16	32	48	56

37.   See *Computer Programs* in main text.

38.   See *Computer Programs* in main text.

39.   See *Computer Programs* in main text.

≡ 17.2 MULTIPLICATION OF MATRICES

1. $\begin{bmatrix} 2 & 3 \\ 1 & -4 \end{bmatrix} \times \begin{bmatrix} 4 & -3 \\ 6 & 5 \end{bmatrix} = \begin{bmatrix} 8+18 & -6+15 \\ 4-24 & -3-20 \end{bmatrix} =$

$\begin{bmatrix} 26 & 9 \\ -20 & -23 \end{bmatrix}$

2. $\begin{bmatrix} 1 & -2 & 4 \end{bmatrix} \times \begin{bmatrix} 3 \\ 9 \\ -7 \end{bmatrix} = [3 - 18 - 28] = [-43]$

3. $\begin{bmatrix} 5 & 1 & 6 & 2 \end{bmatrix} \times \begin{bmatrix} 2 \\ 9 \\ -8 \\ 3 \end{bmatrix} = [10 + 9 - 48 + 6] =$

$[-23]$

4. $\begin{bmatrix} 2 & 3 & 1 \\ -4 & 2 & 0 \end{bmatrix} \times \begin{bmatrix} 4 & 1 \\ -5 & 2 \\ 3 & 0 \end{bmatrix} =$

$\begin{bmatrix} 8-15+3 & 2+6+0 \\ -16-10+0 & -4+4+0 \end{bmatrix} = \begin{bmatrix} -4 & 8 \\ -26 & 0 \end{bmatrix}$

5. $\begin{bmatrix} 1 & 2 & 4 \\ 2 & 3 & 6 \end{bmatrix} \times \begin{bmatrix} 5 & -7 & 4 \\ 4 & 6 & 0 \\ -4 & 2 & 1 \end{bmatrix} =$

$\begin{bmatrix} 5+8-16 & -7+12+8 & 4+0+4 \\ 10+12-24 & -14+18+12 & 8+0+6 \end{bmatrix} =$

$\begin{bmatrix} -3 & 13 & 8 \\ -2 & 16 & 14 \end{bmatrix}$

6.
$$\begin{bmatrix} 4 & 7 \\ 2 & 3 \\ 6 & 2 \\ 9 & 1 \end{bmatrix} \times \begin{bmatrix} 3 & 7 & 6 & 1 & -5 \\ 4 & -3 & 2 & 0 & 10 \end{bmatrix} =$$

$$\begin{bmatrix} 12+28 & 28-21 & 24+14 & 4+0 & -20+70 \\ 6+12 & 14-9 & 12+6 & 2+0 & -10+30 \\ 18+8 & 42-6 & 36+4 & 6+0 & -30+20 \\ 27+4 & 63-3 & 54+2 & 9+0 & -45+10 \end{bmatrix}$$

$$= \begin{bmatrix} 40 & 7 & 38 & 4 & 50 \\ 18 & 5 & 18 & 2 & -20 \\ 26 & 36 & 40 & 6 & -10 \\ 31 & 60 & 56 & 9 & -35 \end{bmatrix}$$

7.
$$\begin{bmatrix} 1 & 2 & 3 \\ 6 & 5 & 4 \\ -1 & 0 & 2 \end{bmatrix} \times \begin{bmatrix} 9 & 8 & 7 \\ -2 & 3 & -1 \\ 0 & 1 & 0 \end{bmatrix} =$$

$$\begin{bmatrix} 9-4+0 & 8+6+3 & 7-2+0 \\ 54-10+0 & 48+15+4 & 42-5+0 \\ -9+0+0 & -8+0+2 & -7+0+0 \end{bmatrix} =$$

$$\begin{bmatrix} 5 & 17 & 5 \\ 44 & 67 & 37 \\ -9 & -6 & -7 \end{bmatrix}$$

8.
$$\begin{bmatrix} 9 & 8 & 7 \\ -2 & 3 & -1 \\ 0 & 1 & 0 \end{bmatrix} \times \begin{bmatrix} 1 & 2 & 3 \\ 6 & 5 & 4 \\ -1 & 0 & 2 \end{bmatrix} =$$

$$\begin{bmatrix} 9+48-7 & 18+40+0 & 27+32+14 \\ -2+18+1 & -4+15+0 & -6+12-2 \\ 0+6+0 & 0+5+0 & 0+4+0 \end{bmatrix} =$$

$$\begin{bmatrix} 50 & 58 & 73 \\ 17 & 11 & 4 \\ 6 & 5 & 4 \end{bmatrix}$$

9.
$$\begin{bmatrix} 3 & 4 & -2 \\ 7 & 8 & 10 \end{bmatrix} \times \begin{bmatrix} 0 & 4 & 1 \\ 1 & 0 & 2 \\ 8 & -1 & 0 \end{bmatrix} =$$

$$\begin{bmatrix} 0+4-16 & 12+0+2 & 3+8+0 \\ 0+8+80 & 28+0-10 & 7+16+0 \end{bmatrix} =$$

$$\begin{bmatrix} -12 & 14 & 11 \\ 88 & 18 & 23 \end{bmatrix}$$

10.
$$\begin{bmatrix} 1 & 2 & -1 \\ 2 & -1 & 3 \end{bmatrix} \times \begin{bmatrix} 0 & 1 & 0 & 2 \\ 1 & 0 & 2 & -1 \\ 3 & 4 & -1 & 0 \end{bmatrix} =$$

$$\begin{bmatrix} 2-3 & 1-4 & 4+1 & 2-2 \\ -1+9 & 2+12 & -2-3 & 4+1 \end{bmatrix} =$$

$$\begin{bmatrix} -1 & -3 & 5 & 0 \\ 8 & 14 & -5 & 5 \end{bmatrix}$$

11. $AB = \begin{bmatrix} 1 & 2 \\ 3 & 4 \end{bmatrix} \times \begin{bmatrix} 3 & 6 \\ 2 & 5 \end{bmatrix} = \begin{bmatrix} 3+4 & 6+10 \\ 9+8 & 18+20 \end{bmatrix} =$
$\begin{bmatrix} 7 & 16 \\ 17 & 38 \end{bmatrix}$

12. $BA = \begin{bmatrix} 3 & 6 \\ 2 & 5 \end{bmatrix} \times \begin{bmatrix} 1 & 2 \\ 3 & 4 \end{bmatrix} = \begin{bmatrix} 3+18 & 6+24 \\ 2+15 & 4+20 \end{bmatrix} =$
$\begin{bmatrix} 21 & 30 \\ 17 & 24 \end{bmatrix}$

13. $AC = \begin{bmatrix} 1 & 2 \\ 3 & 4 \end{bmatrix} \begin{bmatrix} 4 & -2 \\ -6 & 2 \end{bmatrix} =$
$\begin{bmatrix} 4-12 & -2+4 \\ 12-24 & -6+8 \end{bmatrix} = \begin{bmatrix} -8 & 2 \\ -12 & 2 \end{bmatrix}$

14. $BC = \begin{bmatrix} 3 & 6 \\ 2 & 5 \end{bmatrix} \begin{bmatrix} 4 & -2 \\ -6 & 2 \end{bmatrix} =$
$\begin{bmatrix} 12-36 & -6+12 \\ 8-30 & -4+10 \end{bmatrix} = \begin{bmatrix} -24 & 6 \\ -22 & 6 \end{bmatrix}$

15. $A(BC) = A\begin{bmatrix} -24 & 6 \\ -22 & 6 \end{bmatrix} = \begin{bmatrix} 1 & 2 \\ 3 & 4 \end{bmatrix}\begin{bmatrix} -24 & 6 \\ -22 & 6 \end{bmatrix} =$
$\begin{bmatrix} -24-44 & 6+12 \\ -72-88 & 18+24 \end{bmatrix} = \begin{bmatrix} -68 & 18 \\ -160 & 42 \end{bmatrix}$

16. $(AB)C = \begin{bmatrix} 7 & 16 \\ 17 & 38 \end{bmatrix}\begin{bmatrix} 4 & -2 \\ -6 & 2 \end{bmatrix} =$
$\begin{bmatrix} 28-96 & -14+32 \\ 68-228 & -34+76 \end{bmatrix} = \begin{bmatrix} -68 & 18 \\ -160 & 42 \end{bmatrix}$

17. $A(B+C) = \begin{bmatrix} 1 & 2 \\ 3 & 4 \end{bmatrix}\left(\begin{bmatrix} 3 & 6 \\ 2 & 5 \end{bmatrix} + \begin{bmatrix} 4 & -2 \\ -6 & 2 \end{bmatrix}\right) =$
$\begin{bmatrix} 1 & 2 \\ 3 & 4 \end{bmatrix}\begin{bmatrix} 7 & 4 \\ -4 & 7 \end{bmatrix} = \begin{bmatrix} 7-8 & 4+14 \\ 21-16 & 12+28 \end{bmatrix} =$
$\begin{bmatrix} -1 & 18 \\ 5 & 40 \end{bmatrix}$

18. $(B+C)A = \begin{bmatrix} 7 & 4 \\ -4 & 7 \end{bmatrix}\begin{bmatrix} 1 & 2 \\ 3 & 4 \end{bmatrix} =$
$\begin{bmatrix} 7+12 & 14+16 \\ -4+21 & -8+28 \end{bmatrix} = \begin{bmatrix} 19 & 30 \\ 17 & 20 \end{bmatrix}$

19. $(2A)(3B)$: Scaler multiplication is commutative
so
$(2A)(3B) = 6(AB) = 6 \cdot \begin{bmatrix} 7 & 16 \\ 17 & 38 \end{bmatrix} = \begin{bmatrix} 42 & 96 \\ 102 & 228 \end{bmatrix}$

20. $(2A + \frac{1}{2}C)B = \left(2\begin{bmatrix} 1 & 2 \\ 3 & 4 \end{bmatrix} + \frac{1}{2}\begin{bmatrix} 4 & -2 \\ -6 & 2 \end{bmatrix}\right) \cdot$
$\begin{bmatrix} 3 & 6 \\ 2 & 5 \end{bmatrix} = \left(\begin{bmatrix} 2 & 4 \\ 6 & 8 \end{bmatrix} + \begin{bmatrix} 2 & -1 \\ -3 & 1 \end{bmatrix}\right)\begin{bmatrix} 3 & 6 \\ 2 & 5 \end{bmatrix} =$
$\begin{bmatrix} 4 & 3 \\ 3 & 9 \end{bmatrix} \cdot \begin{bmatrix} 3 & 6 \\ 2 & 5 \end{bmatrix} = \begin{bmatrix} 12+6 & 24+15 \\ 9+18 & 18+45 \end{bmatrix} =$
$\begin{bmatrix} 18 & 39 \\ 27 & 63 \end{bmatrix}$

21. $A = \begin{bmatrix} 2 & 10 \\ 3 & 15 \end{bmatrix}$, $B = \begin{bmatrix} -10 & 25 \\ 2 & -5 \end{bmatrix}$, $AB =$

$\begin{bmatrix} -20+20 & 50-50 \\ -30+30 & 75-75 \end{bmatrix} = \begin{bmatrix} 0 & 0 \\ 0 & 0 \end{bmatrix}$. We cannot *conclude* that, if $AB = 0$, then either $A = 0$ or $B = 0$. Hence, matrix multiplication does *not* have a zero product principal.

22. $\begin{bmatrix} 2 & x \\ y & 5 \end{bmatrix}\begin{bmatrix} 4 & 6 \\ 9 & -1 \end{bmatrix} = \begin{bmatrix} 8+9x & 12-x \\ 4y+45 & 6y-5 \end{bmatrix} = \begin{bmatrix} 35 & 9 \\ 49 & 1 \end{bmatrix}$. So, $8 + 9x = 35$ and $12 - x = 9$. From the first, we obtain $9x = 27$ or $-x = -3$ and so $x = 3$. Also, $4y + 45 = 49$ and $6y - 5 = 1$. From the first, we get $4y = 4$ and so $y = 1$. The solutions are $x = 3$ and $y = 1$.

23. Here we have the product

$$\begin{bmatrix} 270 & 2130 & 3210 & 265 \\ 1120 & 4230 & 3124 & 75 \\ 320 & 3126 & 2743 & 1012 \end{bmatrix} \times \begin{bmatrix} 25 \\ 35 \\ 55 \\ 95 \end{bmatrix} =$$

$$\begin{bmatrix} 283,025 \\ 354,995 \\ 364,415 \end{bmatrix}$$

From this we see that Warehouse A generated sales of \$283,025, Warehouse B sales of \$354,995, and Warehouse C generated sales of \$364,415.

24. **(a)** The amount of material needed for each model house is determined by finding the product

$$PA = \begin{bmatrix} 30 & 20 & 15 \\ 10 & 10 & 15 \end{bmatrix} \cdot \begin{bmatrix} 10 & 2 & 2 & 3 \\ 15 & 3 & 4 & 4 \\ 25 & 5 & 6 & 3 \end{bmatrix} =$$

$$\begin{bmatrix} 975 & 195 & 230 & 215 \\ 625 & 125 & 150 & 115 \end{bmatrix};$$

(b) The cost to build each size house is found from the product $AC = \begin{bmatrix} 10 & 2 & 2 & 3 \\ 15 & 3 & 4 & 4 \\ 25 & 5 & 6 & 3 \end{bmatrix} \cdot \begin{bmatrix} 25 \\ 210 \\ 75 \\ 40 \end{bmatrix} =$

$$\begin{bmatrix} 940 \\ 1465 \\ 2245 \end{bmatrix};$$

(c) The cost for each house model is the product of all three of the given matrices,

$$PAC = \begin{bmatrix} 30 & 20 & 15 \\ 10 & 10 & 15 \end{bmatrix} \cdot \begin{bmatrix} 10 & 2 & 2 & 3 \\ 15 & 3 & 4 & 4 \\ 25 & 5 & 6 & 3 \end{bmatrix} \cdot \begin{bmatrix} 25 \\ 210 \\ 75 \\ 40 \end{bmatrix}$$

$$= \begin{bmatrix} 91,175 \\ 57,725 \end{bmatrix};$$

(d) To find the total cost for the subdivision you need to add the columns in the answer to (c) with the result that the total is \$148,900.

25. $A = \begin{bmatrix} 0 & 1 \\ 1 & 0 \end{bmatrix}$:

$A^2 = \begin{bmatrix} 0 & 1 \\ 1 & 0 \end{bmatrix} \times \begin{bmatrix} 0 & 1 \\ 1 & 0 \end{bmatrix} = \begin{bmatrix} 1 & 0 \\ 0 & 1 \end{bmatrix} = I$,

$B = \begin{bmatrix} 0 & -i \\ i & 0 \end{bmatrix}$: $B^2 = \begin{bmatrix} 0 & -i \\ i & 0 \end{bmatrix}\begin{bmatrix} 0 & -i \\ i & 0 \end{bmatrix} =$

$\begin{bmatrix} -i^2 & 0 \\ 0 & -i^2 \end{bmatrix} = \begin{bmatrix} 1 & 0 \\ 0 & 1 \end{bmatrix} = I; C = \begin{bmatrix} 1 & 0 \\ 0 & -1 \end{bmatrix}$:

$C^2 = \begin{bmatrix} 1 & 0 \\ 0 & -1 \end{bmatrix} \cdot \begin{bmatrix} 1 & 0 \\ 0 & -1 \end{bmatrix} = \begin{bmatrix} 1 & 0 \\ 0 & 1 \end{bmatrix} = I$

26. $AB = \begin{bmatrix} 0 & 1 \\ 1 & 0 \end{bmatrix}\begin{bmatrix} 0 & -i \\ i & 0 \end{bmatrix} = \begin{bmatrix} i & 0 \\ 0 & -i \end{bmatrix}$

$BA = \begin{bmatrix} 0 & -i \\ i & 0 \end{bmatrix}\begin{bmatrix} 0 & 1 \\ 1 & 0 \end{bmatrix} = \begin{bmatrix} -i & 0 \\ 0 & i \end{bmatrix} = -AB$.

$AC = \begin{bmatrix} 0 & 1 \\ 1 & 0 \end{bmatrix}\begin{bmatrix} 1 & 0 \\ 0 & -1 \end{bmatrix} = \begin{bmatrix} 0 & -1 \\ 1 & 0 \end{bmatrix}$

$CA = \begin{bmatrix} 1 & 0 \\ 0 & -1 \end{bmatrix}\begin{bmatrix} 0 & 1 \\ 1 & 0 \end{bmatrix} = \begin{bmatrix} 0 & 1 \\ -1 & 0 \end{bmatrix} = -AC$,

$BC = \begin{bmatrix} 0 & -i \\ i & 0 \end{bmatrix}\begin{bmatrix} 1 & 0 \\ 0 & -1 \end{bmatrix} = \begin{bmatrix} 0 & i \\ i & 0 \end{bmatrix}$,

$CB = \begin{bmatrix} 1 & 0 \\ 0 & -1 \end{bmatrix} \times \begin{bmatrix} 0 & -i \\ i & 0 \end{bmatrix} =$

$\begin{bmatrix} 0 & -i \\ -i & 0 \end{bmatrix} = -BC$

27. $AB - BA = \begin{bmatrix} i & 0 \\ 0 & -i \end{bmatrix} - \begin{bmatrix} -i & 0 \\ 0 & i \end{bmatrix} =$

$\begin{bmatrix} 2i & 0 \\ 0 & -2i \end{bmatrix} = 2i\begin{bmatrix} 1 & 0 \\ 0 & -1 \end{bmatrix} = 2iC. AC - CA =$

$\begin{bmatrix} 0 & -1 \\ 1 & 0 \end{bmatrix} - \begin{bmatrix} 0 & 1 \\ -1 & 0 \end{bmatrix} = \begin{bmatrix} 0 & -2 \\ 2 & 0 \end{bmatrix} = -2iB$.

$-2i\begin{bmatrix} 0 & -i \\ i & 0 \end{bmatrix} = \begin{bmatrix} 0 & 2i^2 \\ -2i^2 & 0 \end{bmatrix} = \begin{bmatrix} 0 & -2 \\ 2 & 0 \end{bmatrix}$.

$BC - CB = \begin{bmatrix} 0 & i \\ i & 0 \end{bmatrix} - \begin{bmatrix} 0 & -i \\ -i & 0 \end{bmatrix} = \begin{bmatrix} 0 & 2i \\ 2i & 0 \end{bmatrix} =$

$2i\begin{bmatrix} 0 & 1 \\ 1 & 0 \end{bmatrix} = 2iA$

28. $[x \ y]\begin{bmatrix} 5 & -7 \\ 7 & 3 \end{bmatrix}\begin{bmatrix} x \\ y \end{bmatrix} =$

$[5x + 7y \ -7x + 3y]\begin{bmatrix} x \\ y \end{bmatrix} =$

$5x^2 + 7xy - 7xy + 3y^2 = 30$. Collecting terms, we get $5x^2 + 3y^2 = 30$ or $\frac{x^2}{6} + \frac{y^2}{10} = 1$, which is the equation for an ellipse.

29. See *Computer Programs* in main text.

30. **(a)** If $\theta = \pi$, then $\begin{bmatrix} \cos\theta & \sin\theta \\ -\sin\theta & \cos\theta \end{bmatrix} = \begin{bmatrix} -1 & 0 \\ 0 & -1 \end{bmatrix}$.
From the given values for P_1, P_2, P_3, and P_4, we

have $\begin{bmatrix} x_1 & y_1 \\ x_2 & y_2 \\ x_3 & y_3 \\ x_4 & y_4 \end{bmatrix} \begin{bmatrix} \cos\theta & \sin\theta \\ -\sin\theta & \cos\theta \end{bmatrix} =$

$\begin{bmatrix} 1 & 2 \\ 5 & 4 \\ -2 & 7 \\ -1 & -5 \end{bmatrix} \begin{bmatrix} -1 & 0 \\ 0 & -1 \end{bmatrix} = \begin{bmatrix} -1 & -2 \\ -5 & -4 \\ 2 & -7 \\ 1 & 5 \end{bmatrix}$.

(b) If $\theta = \frac{\pi}{2}$, then $\begin{bmatrix} \cos\theta & \sin\theta \\ -\sin\theta & \cos\theta \end{bmatrix} = \begin{bmatrix} 0 & 1 \\ -1 & 0 \end{bmatrix}$.
From the given values for P_1, P_2, P_3, and P_4, we

have $\begin{bmatrix} 1 & 2 \\ 5 & 4 \\ -2 & 7 \\ -1 & -5 \end{bmatrix} \begin{bmatrix} 0 & 1 \\ -1 & 0 \end{bmatrix} = \begin{bmatrix} -2 & 1 \\ -4 & 5 \\ -7 & 2 \\ 5 & -1 \end{bmatrix}$.

(c) $\begin{bmatrix} x_1 & y_1 \\ x_2 & y_2 \\ x_3 & y_3 \\ x_4 & y_4 \\ x_5 & y_5 \end{bmatrix} \begin{bmatrix} \cos\theta & \sin\theta \\ -\sin\theta & \cos\theta \end{bmatrix}$

≡ 17.3 INVERSES OF MATRICES

1. $\begin{vmatrix} 6 & 1 \\ 5 & 1 \end{vmatrix} = 6 - 5 = 1$. Inverse is $\begin{bmatrix} 1 & -1 \\ -5 & 6 \end{bmatrix}$

2. $\begin{vmatrix} 5 & 1 \\ 9 & 2 \end{vmatrix} = 10 - 9 = 1$. Inverse is $\begin{bmatrix} 2 & -1 \\ -9 & 5 \end{bmatrix}$

3. $\begin{vmatrix} 10 & 4 \\ 8 & 2 \end{vmatrix} = 30 - 32 = -2$. Inverse is

$\frac{1}{-2} \begin{bmatrix} 3 & -4 \\ -8 & 10 \end{bmatrix} = \begin{bmatrix} \frac{-3}{2} & 2 \\ 4 & -5 \end{bmatrix} = \begin{bmatrix} -1.5 & 2 \\ 4 & -5 \end{bmatrix}$

4. $\begin{vmatrix} 9 & 6 \\ 4 & 3 \end{vmatrix} = 27 - 24 = 3$. Inverse is

$\frac{1}{3} \begin{bmatrix} 3 & -6 \\ -4 & 9 \end{bmatrix} = \begin{bmatrix} 1 & -2 \\ -\frac{4}{3} & 3 \end{bmatrix}$

5. $\begin{vmatrix} 8 & -6 \\ -6 & 4 \end{vmatrix} = 32 - 36 = -4$. Inverse is

$-\frac{1}{4} \begin{bmatrix} 4 & 6 \\ 6 & 8 \end{bmatrix} = \begin{bmatrix} -1 & -1.5 \\ -1.5 & -2 \end{bmatrix}$

6. $\begin{vmatrix} 12 & 9 \\ -9 & -7 \end{vmatrix} = -84 + 81 = -3$. Inverse is

$-\frac{1}{3} \begin{bmatrix} -7 & -9 \\ 9 & 12 \end{bmatrix} = \begin{bmatrix} \frac{7}{3} & 3 \\ -3 & -4 \end{bmatrix}$

7. $\begin{vmatrix} 15 & 10 \\ 4 & 3 \end{vmatrix} = 45 - 40 = 5$. Inverse is

$\frac{1}{5} \begin{bmatrix} 3 & -10 \\ -4 & 15 \end{bmatrix} = \begin{bmatrix} .6 & -2 \\ -0.8 & 3 \end{bmatrix}$

8. $\begin{bmatrix} 15 & 10 \\ 3 & 4 \end{bmatrix} = 60 - 30 = 30$. Inverse is

$\frac{1}{30} \begin{bmatrix} 4 & -10 \\ -3 & 15 \end{bmatrix} = \begin{bmatrix} \frac{2}{15} & -\frac{1}{3} \\ -\frac{1}{10} & \frac{1}{2} \end{bmatrix} =$

$\begin{bmatrix} 0.133 & -0.333 \\ -0.1 & 0.5 \end{bmatrix}$

9. $\begin{bmatrix} 0 & -3 & 0 & \vdots & 1 & 0 & 0 \\ 1 & 0 & 0 & \vdots & 0 & 1 & 0 \\ 0 & 0 & 4 & \vdots & 0 & 0 & 1 \end{bmatrix} \begin{matrix} -\frac{1}{3}R_1 \\ \Rightarrow \\ \frac{1}{4}R_3 \end{matrix}$

$\begin{bmatrix} 0 & 1 & 0 & \vdots & -\frac{1}{3} & 0 & 0 \\ 1 & 0 & 0 & \vdots & 0 & 1 & 0 \\ 0 & 0 & 1 & \vdots & 0 & 0 & \frac{1}{4} \end{bmatrix}$. Interchange R_1 and

$R_2 \begin{bmatrix} 1 & 0 & 0 & \vdots & 0 & 1 & 0 \\ 0 & 1 & 0 & \vdots & -\frac{1}{3} & 0 & 0 \\ 0 & 0 & 1 & \vdots & 0 & 0 & \frac{1}{4} \end{bmatrix}$. Inverse is

$\begin{bmatrix} 0 & 1 & 0 \\ -\frac{1}{3} & 0 & 0 \\ 0 & 0 & \frac{1}{4} \end{bmatrix}$

10. $\begin{bmatrix} 1 & 0 & 0 & \vdots & 1 & 0 & 0 \\ 0 & 4 & 0 & \vdots & 0 & 1 & 0 \\ 0 & 0 & 2 & \vdots & 0 & 0 & 1 \end{bmatrix} \begin{matrix} \frac{1}{4}R_3 \Rightarrow \\ \frac{1}{2}R_2 \end{matrix}$

$\begin{bmatrix} 1 & 0 & 0 & \vdots & 1 & 0 & 0 \\ 0 & 1 & 0 & \vdots & 0 & \frac{1}{4} & 0 \\ 0 & 0 & 1 & \vdots & 0 & 0 & \frac{1}{2} \end{bmatrix}$, so the inverse is

$\begin{bmatrix} 1 & 0 & 0 \\ 0 & \frac{1}{4} & 0 \\ 0 & 0 & \frac{1}{2} \end{bmatrix}$

11. $\begin{bmatrix} 1 & 2 & 6 \\ 0 & 0 & 2 \\ -3 & -6 & -9 \end{bmatrix}$ Since column 2 is twice column
1, this is a singular matrix and it has no inverse.

12. $\begin{bmatrix} 4 & 5 & 1 \\ 1 & 0 & 1 \\ 4 & 5 & 1 \end{bmatrix}$. Since $R_1 = R_3$ this is a singular
matrix and it has no inverse.

13.
$$\begin{bmatrix} 8 & 7 & -1 & \vdots & 1 & 0 & 0 \\ -5 & -5 & 1 & \vdots & 0 & 1 & 0 \\ -4 & -4 & 1 & \vdots & 0 & 0 & 1 \end{bmatrix} 2R_3 + R_1$$

$$\begin{bmatrix} 0 & -1 & 1 & \vdots & 1 & 0 & 2 \\ -5 & -5 & 1 & \vdots & 0 & 1 & 0 \\ -4 & -4 & 1 & \vdots & 0 & 0 & 1 \end{bmatrix} -1R_2$$

$$\begin{bmatrix} 0 & -1 & 1 & \vdots & 1 & 0 & 2 \\ 5 & 5 & -1 & \vdots & 0 & -1 & 0 \\ -4 & -4 & 1 & \vdots & 0 & 0 & 1 \end{bmatrix} R_3 + R_2$$

$$\begin{bmatrix} 0 & -1 & 1 & \vdots & 1 & 0 & 2 \\ 1 & 1 & 0 & \vdots & 0 & -1 & 1 \\ -4 & -4 & 1 & \vdots & 0 & 0 & 1 \end{bmatrix} 4R_2 + R_3$$

$$\begin{bmatrix} 0 & -1 & 1 & \vdots & 1 & 0 & 2 \\ 1 & 1 & 0 & \vdots & 0 & -1 & 1 \\ 0 & 0 & 1 & \vdots & 0 & -4 & 5 \end{bmatrix} \text{Swap } R_1 + R_2$$

$$\begin{bmatrix} 1 & 1 & 0 & \vdots & 0 & -1 & 1 \\ 0 & -1 & 1 & \vdots & 1 & 0 & 2 \\ 0 & 0 & 1 & \vdots & 0 & -4 & 5 \end{bmatrix} R_2 + R_1$$

$$\begin{bmatrix} 1 & 0 & 1 & \vdots & 1 & -1 & 3 \\ 0 & -1 & 1 & \vdots & 1 & 0 & 2 \\ 0 & 0 & 1 & \vdots & 0 & -4 & 5 \end{bmatrix} \begin{matrix} -R_3 + R_1 \\ -R_3 + R_2 \end{matrix}$$

$$\begin{bmatrix} 1 & 0 & 0 & \vdots & 1 & 3 & -2 \\ 0 & -1 & 0 & \vdots & 1 & 4 & -3 \\ 0 & 0 & 1 & \vdots & 0 & -4 & 5 \end{bmatrix} -1R_2$$

$$\Rightarrow \begin{bmatrix} 1 & 0 & 0 & \vdots & 1 & 3 & -2 \\ 0 & 1 & 0 & \vdots & -1 & -4 & 3 \\ 0 & 0 & 1 & \vdots & 0 & -4 & 5 \end{bmatrix} . \text{ Inverse is}$$

$$\begin{bmatrix} 1 & 3 & -2 \\ -1 & -4 & 3 \\ 0 & -4 & 5 \end{bmatrix}$$

14.
$$\begin{bmatrix} 1 & 2 & 3 & \vdots & 1 & 0 & 0 \\ 2 & 5 & 7 & \vdots & 0 & 1 & 0 \\ 1 & 1 & 1 & \vdots & 0 & 0 & 1 \end{bmatrix} \begin{matrix} -2R_1 + R_2 \\ -R_1 + R_3 \end{matrix}$$

$$\begin{bmatrix} 1 & 2 & 3 & \vdots & 1 & 0 & 0 \\ 0 & 1 & 1 & \vdots & -2 & 1 & 0 \\ 0 & -1 & -2 & \vdots & -1 & 0 & 1 \end{bmatrix} \begin{matrix} 2R_3 + R_1 \\ R_2 + R_3 \end{matrix}$$

$$\begin{bmatrix} 1 & 0 & -1 & \vdots & -1 & 0 & 2 \\ 0 & 1 & 1 & \vdots & -2 & 1 & 0 \\ 0 & 0 & -1 & \vdots & -3 & 1 & 1 \end{bmatrix} \begin{matrix} -R_3 + R_1 \\ R_3 + R_2 \\ -R_3 \end{matrix}$$

$$\begin{bmatrix} 1 & 0 & 0 & \vdots & 2 & -1 & 1 \\ 0 & 1 & 0 & \vdots & -5 & 2 & 1 \\ 0 & 0 & 1 & \vdots & 3 & -1 & -1 \end{bmatrix} .$$

The inverse is $\begin{bmatrix} 2 & -1 & 1 \\ -5 & 2 & 1 \\ 3 & -1 & -1 \end{bmatrix}$

15. $\begin{bmatrix} 3 & -1 & 0 \\ -6 & 2 & 0 \\ 1 & 0 & 5 \end{bmatrix}$. Since $R_2 = -2R_1$ the matrix is singular, so there is no inverse.

16.
$$\begin{bmatrix} 1 & -1 & 1 & \vdots & 1 & 0 & 0 \\ 7 & -8 & 5 & \vdots & 0 & 1 & 0 \\ -4 & 5 & -3 & \vdots & 0 & 0 & 1 \end{bmatrix} \begin{matrix} -7R_1 + R_2 \\ 4R_1 + R_3 \end{matrix}$$

$$\begin{bmatrix} 1 & -1 & 1 & \vdots & 1 & 0 & 0 \\ 0 & -1 & -2 & \vdots & -7 & 1 & 0 \\ 0 & 1 & 1 & \vdots & 4 & 0 & 1 \end{bmatrix} \begin{matrix} R_3 + R_1 \\ R_2 + R_3 \end{matrix}$$

$$\begin{bmatrix} 1 & 0 & 2 & \vdots & 5 & 0 & 1 \\ 0 & -1 & -2 & \vdots & -7 & 1 & 0 \\ 0 & 0 & -1 & \vdots & -3 & 1 & 1 \end{bmatrix} \begin{matrix} 2R_3 + R_1 \\ 2R_3 - R_2 \\ -R_3 \end{matrix}$$

$$\begin{bmatrix} 1 & 0 & 0 & \vdots & -1 & 2 & 3 \\ 0 & 1 & 0 & \vdots & 1 & 1 & 2 \\ 0 & 0 & 1 & \vdots & 3 & -1 & -1 \end{bmatrix} . \text{ The inverse is}$$

$$\begin{bmatrix} -1 & 2 & 3 \\ 1 & 1 & 2 \\ 3 & -1 & -1 \end{bmatrix} .$$

17. $\begin{bmatrix} 2 & 0 & 0 & \vdots & 1 & 0 & 0 \\ 2 & 2 & 0 & \vdots & 0 & 1 & 0 \\ 2 & 2 & 2 & \vdots & 0 & 0 & 1 \end{bmatrix} .$

First multiply the entire matrix by $\frac{1}{2}$ to get

$$\begin{bmatrix} 1 & 0 & 0 & \vdots & \frac{1}{2} & 0 & 0 \\ 1 & 1 & 0 & \vdots & 0 & \frac{1}{2} & 0 \\ 1 & 1 & 1 & \vdots & 0 & 0 & \frac{1}{2} \end{bmatrix} \begin{matrix} -R_1 + R_2 \\ -R_1 + R_3 \end{matrix}$$

$$\begin{bmatrix} 1 & 0 & 0 & \vdots & \frac{1}{2} & 0 & 0 \\ 0 & 1 & 0 & \vdots & -\frac{1}{2} & \frac{1}{2} & 0 \\ 0 & 1 & 1 & \vdots & -\frac{1}{2} & 0 & \frac{1}{2} \end{bmatrix} -R_2 + R_3$$

$$\begin{bmatrix} 1 & 0 & 0 & \vdots & \frac{1}{2} & 0 & 0 \\ 0 & 1 & 0 & \vdots & -\frac{1}{2} & \frac{1}{2} & 0 \\ 0 & 0 & 1 & \vdots & 0 & -\frac{1}{2} & \frac{1}{2} \end{bmatrix} .$$

The inverse is $\begin{bmatrix} \frac{1}{2} & 0 & 0 \\ -\frac{1}{2} & \frac{1}{2} & 0 \\ 0 & -\frac{1}{2} & \frac{1}{2} \end{bmatrix}$ or

$$\begin{bmatrix} 0.5 & 0 & 0 \\ -0.5 & 0.5 & 0 \\ 0 & -0.5 & 0.5 \end{bmatrix} .$$

18. $\begin{bmatrix} 3 & 1 & -1 & \vdots & 1 & 0 & 0 \\ 1 & -2 & 0 & \vdots & 0 & 1 & 0 \\ 0 & 3 & 1 & \vdots & 0 & 0 & 1 \end{bmatrix} \begin{matrix} -3R_2 + R_1 \\ \\ \end{matrix}$

$\begin{bmatrix} 0 & 7 & -1 & \vdots & 1 & -3 & 0 \\ 1 & -2 & 0 & \vdots & 0 & 1 & 0 \\ 0 & 3 & 1 & \vdots & 0 & 0 & 1 \end{bmatrix}$. Swap R_1 and R_2

to get $\begin{bmatrix} 1 & -2 & 0 & \vdots & 0 & 1 & 0 \\ 0 & 7 & -1 & \vdots & 1 & -3 & 0 \\ 0 & 3 & 1 & \vdots & 0 & 0 & 1 \end{bmatrix} R_3 + R_2$

$\begin{bmatrix} 1 & -2 & 0 & \vdots & 0 & 1 & 0 \\ 0 & 10 & 0 & \vdots & 1 & -3 & 1 \\ 0 & 3 & 1 & \vdots & 0 & 0 & 1 \end{bmatrix} \begin{matrix} \frac{2}{10}R_2 + R_1 \\ \frac{1}{10}R_2 \rightarrow \\ -\frac{3}{10}R_2 + R_3 \end{matrix}$

$\begin{bmatrix} 1 & 0 & 0 & \vdots & 0.2 & 0.4 & 0.2 \\ 0 & 1 & 0 & \vdots & 0.1 & -0.3 & 0.1 \\ 0 & 0 & 1 & \vdots & -0.3 & 0.9 & 0.7 \end{bmatrix}$.

The inverse is $\begin{bmatrix} 0.2 & 0.4 & 0.2 \\ 0.1 & -0.3 & 0.1 \\ -0.3 & 0.9 & 0.7 \end{bmatrix}$.

19. $\begin{bmatrix} 1 & -1 & 1 & \vdots & 1 & 0 & 0 \\ 0 & 2 & -1 & \vdots & 0 & 1 & 0 \\ 2 & 3 & 0 & \vdots & 0 & 0 & 1 \end{bmatrix} \begin{matrix} \\ \\ -2R_1 + R_3 \end{matrix}$

$\begin{bmatrix} 1 & -1 & 1 & \vdots & 1 & 0 & 0 \\ 0 & 2 & -1 & \vdots & 0 & 1 & 0 \\ 0 & 5 & -2 & \vdots & -2 & 0 & 1 \end{bmatrix} \begin{matrix} \frac{1}{2}R_2 + R_1 \\ \\ -\frac{5}{2}R_2 + R_3 \end{matrix}$

$\begin{bmatrix} 1 & 0 & \frac{1}{2} & \vdots & 1 & \frac{1}{2} & 0 \\ 0 & 2 & -1 & \vdots & 0 & 1 & 0 \\ 0 & 0 & \frac{1}{2} & \vdots & -2 & -\frac{5}{2} & 1 \end{bmatrix} \begin{matrix} -R_3 + R_1 \\ 2R_3 + R_2 \\ \end{matrix}$

$\begin{bmatrix} 1 & 0 & 0 & \vdots & 3 & 3 & -1 \\ 0 & 2 & 0 & \vdots & -4 & -4 & 2 \\ 0 & 0 & \frac{1}{2} & \vdots & -2 & -\frac{5}{2} & 1 \end{bmatrix} \begin{matrix} \\ \frac{1}{2}R_2 \\ 2R_3 \end{matrix}$

$\begin{bmatrix} 1 & 0 & 0 & \vdots & 3 & 3 & -1 \\ 0 & 1 & 0 & \vdots & -2 & -2 & 1 \\ 0 & 0 & 1 & \vdots & -4 & -5 & 2 \end{bmatrix}$.

The inverse is $\begin{bmatrix} 3 & 3 & -1 \\ -2 & -2 & 1 \\ -4 & -5 & 2 \end{bmatrix}$.

20. $\begin{bmatrix} 1 & 2 & -1 & \vdots & 1 & 0 & 0 \\ 2 & -2 & 1 & \vdots & 0 & 1 & 0 \\ 6 & 4 & 3 & \vdots & 0 & 0 & 1 \end{bmatrix} \begin{matrix} \\ -2R_1 + R_2 \\ -6R_1 + R_3 \end{matrix}$

$\begin{bmatrix} 1 & 2 & -1 & \vdots & 1 & 0 & 0 \\ 0 & -6 & 3 & \vdots & -2 & 1 & 0 \\ 0 & -8 & 9 & \vdots & -6 & 0 & 1 \end{bmatrix} \begin{matrix} \frac{1}{3}R_2 + R_1 \\ \\ \end{matrix}$

$\begin{bmatrix} 1 & 0 & 0 & \vdots & \frac{1}{3} & \frac{1}{3} & 0 \\ 0 & -6 & 3 & \vdots & -2 & 1 & 0 \\ 0 & -8 & 9 & \vdots & -6 & 0 & 1 \end{bmatrix} \begin{matrix} \\ \\ -4R_2 + 3R_3 \end{matrix}$

$\begin{bmatrix} 1 & 0 & 0 & \vdots & \frac{1}{3} & \frac{1}{3} & 0 \\ 0 & -6 & 3 & \vdots & -2 & 1 & 0 \\ 0 & 0 & 15 & \vdots & -10 & -4 & 3 \end{bmatrix} -\frac{1}{5}R_3 + R_2$

$\begin{bmatrix} 1 & 0 & 0 & \vdots & \frac{1}{3} & \frac{1}{3} & 0 \\ 0 & -6 & 0 & \vdots & 0 & \frac{9}{5} & -\frac{3}{5} \\ 0 & 0 & 15 & \vdots & -10 & -4 & 3 \end{bmatrix} \begin{matrix} \\ -\frac{1}{6}R_2 \\ \frac{1}{15}R_3 \end{matrix}$

$\begin{bmatrix} 1 & 0 & 0 & \vdots & \frac{1}{3} & \frac{1}{3} & 0 \\ 0 & 1 & 0 & \vdots & 0 & -0.3 & 0.1 \\ 0 & 0 & 1 & \vdots & -\frac{2}{3} & -\frac{4}{15} & 0.2 \end{bmatrix}$.

The inverse is $\begin{bmatrix} \frac{1}{3} & \frac{1}{3} & 0 \\ 0 & -0.3 & 0.1 \\ -\frac{2}{3} & -\frac{4}{15} & 0.2 \end{bmatrix}$.

21. All problems should check.

22. See *Computer Programs* in main text.

23. (a) Not a square matrix
(b) Every element in a row (or column) is 0
(c) Every element in a row (or column) is a constant multiple of every corresponding element in another row (or column)
(d) The determinant is 0.

≡ 17.4 MATRICES AND LINEAR EQUATIONS

1. The coefficient matrix of $A = \begin{bmatrix} 6 & 1 \\ 5 & 1 \end{bmatrix}$ has an inverse $A^{-1} = \begin{bmatrix} 1 & -1 \\ -5 & 6 \end{bmatrix}$. Thus, we see that $\begin{bmatrix} x \\ y \end{bmatrix} = \begin{bmatrix} 1 & -1 \\ -5 & 6 \end{bmatrix} \begin{bmatrix} -4 \\ -3 \end{bmatrix} = \begin{bmatrix} -1 \\ 2 \end{bmatrix}$, and so $x = -1$ and $y = 2$

2. The coefficient matrix $A = \begin{bmatrix} 10 & 4 \\ 8 & 3 \end{bmatrix}$ has the inverse $A^{-1} = \begin{bmatrix} -1.5 & 2 \\ 4 & -5 \end{bmatrix}$. Thus, $\begin{bmatrix} x \\ y \end{bmatrix} = \begin{bmatrix} -1.5 & 2 \\ 4 & -5 \end{bmatrix} \cdot \begin{bmatrix} 8 \\ 7 \end{bmatrix} = \begin{bmatrix} 2 \\ -3 \end{bmatrix}$, and so we see that $x = 2$ and $y = -3$

3. Here the coefficient matrix is $A = \begin{bmatrix} 8 & -6 \\ -6 & 4 \end{bmatrix}$, which has an inverse matrix

$A^{-1} = \begin{bmatrix} -1 & -1.5 \\ -1.5 & -2 \end{bmatrix}$. Thus, we see that the solution is

$\begin{bmatrix} x \\ y \end{bmatrix} = \begin{bmatrix} -1 & -1.5 \\ -1.5 & -2 \end{bmatrix} \begin{bmatrix} -27 \\ 19 \end{bmatrix} = \begin{bmatrix} -1.5 \\ 2.5 \end{bmatrix}$, and so $x = -1.5$ and $y = 2.5$

4. This coefficient matrix is $A = \begin{bmatrix} 15 & 10 \\ 4 & 3 \end{bmatrix}$ which has an inverse of $A^{-1} = \begin{bmatrix} 0.6 & -2 \\ -0.8 & 3 \end{bmatrix}$. The solution then is found from

$\begin{bmatrix} x \\ y \end{bmatrix} = \begin{bmatrix} 0.6 & -2 \\ -0.8 & 3 \end{bmatrix} \begin{bmatrix} -5 \\ 0 \end{bmatrix} = \begin{bmatrix} -3 \\ 4 \end{bmatrix}$, and we find that $x = -3$ and $y = 4$

5. This coefficent matrix $A = \begin{bmatrix} 1 & 2 & 6 \\ 0 & 0 & 2 \\ -3 & -6 & -9 \end{bmatrix}$, is a singular matrix, and so there is no solution.

6. The coefficient matrix is $A = \begin{bmatrix} 8 & 7 & -1 \\ -5 & -5 & 1 \\ -4 & -4 & 1 \end{bmatrix}$ which has an inverse $A^{-1} = \begin{bmatrix} 1 & 3 & -2 \\ -1 & -4 & 3 \\ 0 & -4 & 5 \end{bmatrix}$

and constant matrix $K = \begin{bmatrix} 9 \\ -1 \\ 0 \end{bmatrix}$. Thus, we have

$\begin{bmatrix} x \\ y \\ z \end{bmatrix} = A^{-1}K = \begin{bmatrix} 6 \\ -5 \\ 4 \end{bmatrix}$; which gives the result that $x = 6, y = -5$, and $z = 4$.

7. The coefficient matrix is $A = \begin{bmatrix} 1 & 2 & 3 \\ 2 & 5 & 7 \\ 1 & 1 & 1 \end{bmatrix}$; which has the inverse $A^{-1} = \begin{bmatrix} 2 & -1 & 1 \\ -5 & 2 & 1 \\ 3 & -1 & -1 \end{bmatrix}$; and the

constant matrix is $K = \begin{bmatrix} 4 \\ 7.5 \\ 1 \end{bmatrix}$. Thus we have

$\begin{bmatrix} x \\ y \\ z \end{bmatrix} = A^{-1}K = \begin{bmatrix} 1.5 \\ -4 \\ 3.5 \end{bmatrix}$; with the result that $x = 1.5, y = -4$, and $z = 3.5$.

8. We have $A = \begin{bmatrix} 1 & -1 & 1 \\ 7 & -8 & 5 \\ -4 & 5 & -3 \end{bmatrix}$; and so $A^{-1} = \begin{bmatrix} -1 & 2 & 3 \\ 1 & 1 & 2 \\ 3 & -1 & -1 \end{bmatrix}$; and we have a constant matrix

of $K = \begin{bmatrix} -6.6 \\ -43 \\ 26.9 \end{bmatrix}$. As a result, we have $\begin{bmatrix} x \\ y \\ z \end{bmatrix} =$

$A^{-1}K = \begin{bmatrix} 1.3 \\ 4.2 \\ -3.7 \end{bmatrix}$, which gives the result that $x = 1.3, y = 4.2$, and $z = -3.7$.

9. The coefficient matrix $A = \begin{bmatrix} 2 & 0 & 0 \\ 2 & 2 & 0 \\ 2 & 2 & 2 \end{bmatrix}$ has inverse $A^{-1} = \begin{bmatrix} .5 & 0 & 0 \\ -.5 & .5 & 0 \\ 0 & -.5 & .5 \end{bmatrix}$ and the constant

matrix is $K = \begin{bmatrix} 11 \\ 4 \\ -9 \end{bmatrix}$. Thus, we have

$\begin{bmatrix} x \\ y \\ z \end{bmatrix} = A^{-1}K = \begin{bmatrix} 5.5 \\ -3.5 \\ -6.5 \end{bmatrix}$, which produces the solution $x = 5.5, y = -3.5$, and $z = -6.5$.

10. The coefficient matrix $A = \begin{bmatrix} 1 & -1 & 1 \\ 0 & 2 & -1 \\ 2 & 3 & 0 \end{bmatrix}$ has inverse $A^{-1} = \begin{bmatrix} 3 & 3 & -1 \\ -2 & -2 & 1 \\ -4 & -5 & 2 \end{bmatrix}$. The constant

matrix is $K = \begin{bmatrix} 22 \\ -23 \\ -11.2 \end{bmatrix}$ and so we use $\begin{bmatrix} x \\ y \\ z \end{bmatrix} =$

$A^{-1}K = \begin{bmatrix} 8.2 \\ -9.2 \\ 4.6 \end{bmatrix}$ which leads to the solution that $x = 8.2, y = -9.2$, and $z = 4.6$.

11. $A = \begin{bmatrix} 2 & 1 \\ -3 & 2 \end{bmatrix}, K = \begin{bmatrix} 1 \\ 16 \end{bmatrix}$,

$A^{-1} = \frac{1}{7}\begin{bmatrix} 2 & -1 \\ 3 & 2 \end{bmatrix}, \begin{bmatrix} x \\ y \end{bmatrix} = A^{-1}K = \begin{bmatrix} -2 \\ 5 \end{bmatrix}, x = -2, y = 5$.

12. $A = \begin{bmatrix} 4 & 5 \\ 3 & -2 \end{bmatrix}, A^{-1} = \frac{1}{-23}\begin{bmatrix} -2 & -5 \\ -3 & 4 \end{bmatrix}, K = \begin{bmatrix} 2 \\ 13 \end{bmatrix}, \begin{bmatrix} x \\ y \end{bmatrix} = A^{-1}K = \begin{bmatrix} 3 \\ -2 \end{bmatrix}, x = 3, y = -2$.

13. $A = \begin{bmatrix} 2 & 2 \\ 4 & 3 \end{bmatrix}, A^{-1} = -\frac{1}{2}\begin{bmatrix} 3 & -2 \\ -4 & 2 \end{bmatrix},$

$K = \begin{bmatrix} 4 \\ 1 \end{bmatrix}, \begin{bmatrix} x \\ y \end{bmatrix} = A^{-1}K = \begin{bmatrix} -5 \\ 7 \end{bmatrix}, x = -5, y = 7.$

14. $A = \begin{bmatrix} 3 & 1 \\ 4 & 3 \end{bmatrix}, A^{-1} = \frac{1}{5}\begin{bmatrix} 3 & -1 \\ -4 & 3 \end{bmatrix},$

$K = \begin{bmatrix} -5 \\ 2 \end{bmatrix}, \begin{bmatrix} x \\ y \end{bmatrix} = A^{-1}K = \begin{bmatrix} -3.4 \\ 5.2 \end{bmatrix},$
$x = -3.4, y = 5.2.$

15. $A = \begin{bmatrix} 1.5 & 2.5 \\ 3.2 & 2.6 \end{bmatrix}, A^{-1} = \frac{1}{-4.1}\begin{bmatrix} 2.6 & -2.5 \\ -3.2 & 1.5 \end{bmatrix},$

$K = \begin{bmatrix} 0.3 \\ 7.2 \end{bmatrix}, \begin{bmatrix} x \\ y \end{bmatrix} = A^{-1}K = \begin{bmatrix} 4.2 \\ -2.4 \end{bmatrix}, x = 4.2,$
$y = -2.4.$

16. $A = \begin{bmatrix} 7 & 2 & 1 \\ 3 & -2 & 4 \\ 4 & 5 & -1 \end{bmatrix}, K = \begin{bmatrix} 2 \\ 13 \\ 1 \end{bmatrix}, \begin{bmatrix} x \\ y \\ z \end{bmatrix} =$

$A^{-1}K = \begin{bmatrix} -1 \\ 2 \\ 5 \end{bmatrix}, x = -1, y = 2, z = 3.$

17. $A = \begin{bmatrix} 5 & 2 & 3 \\ -3 & 2 & -8 \\ 4 & -2 & 9 \end{bmatrix}, K = \begin{bmatrix} 1 \\ 6 \\ -7 \end{bmatrix},$

$\begin{bmatrix} x \\ y \\ z \end{bmatrix} = A^{-1}K = \begin{bmatrix} -2 \\ 4 \\ 1 \end{bmatrix}, x = -2, y = 4, z = 1.$

18. $A = \begin{bmatrix} 2 & 4 & 1 \\ 4 & 2 & 1 \\ 6 & 4 & 7 \end{bmatrix}, K = \begin{bmatrix} 10 \\ 8 \\ -2 \end{bmatrix},$

$\begin{bmatrix} x \\ y \\ z \end{bmatrix} = A^{-1}K = \begin{bmatrix} 1.5 \\ 2.5 \\ -3 \end{bmatrix}, x = 1.5, y = 2.5,$
$z = -3.$

19. $A = \begin{bmatrix} 1 & 2 & 4 \\ 3 & 1 & 4 \\ 2 & 9 & -2 \end{bmatrix}, K = \begin{bmatrix} 7 \\ -2 \\ 10 \end{bmatrix},$

$\begin{bmatrix} x \\ y \\ z \end{bmatrix} = A^{-1}K = \begin{bmatrix} -3.4 \\ 2.2 \\ 1.5 \end{bmatrix}, x = -3.4, y = 2.2,$
$z = 1.5.$

20. Det $\begin{bmatrix} 2 & -1 & 1 \\ 4 & -2 & 1 \\ 6 & -3 & 5 \end{bmatrix}, = 0$, so no solution.

21. $A = \begin{bmatrix} 1 & -1 & -1 \\ 16 & 0 & 4 \\ 0 & 12 & -4 \end{bmatrix}, K = \begin{bmatrix} 0 \\ 100 \\ 60 \end{bmatrix},$

$\begin{bmatrix} x \\ y \\ z \end{bmatrix} = A^{-1}K = \begin{bmatrix} 6.053 \\ 5.263 \\ 0.789 \end{bmatrix}, I_A = 6.05, I_B = $
$5.26, I_C = 0.79$ all in Amperes

22. $A = \begin{bmatrix} -1 & -1 & 1 \\ 20.5 & -15.4 & 0 \\ 0 & 15.4 & 10.6 \end{bmatrix}, K = \begin{bmatrix} 0 \\ 40 \\ 140 \end{bmatrix}$ and so

$\begin{bmatrix} I_1 \\ I_2 \\ I_3 \end{bmatrix} = \begin{bmatrix} 4.59 \\ 3.51 \\ 8.10 \end{bmatrix} = A^{-1}K$, and we obtain $I_1 = 4.59, I_2 = 3.51, I_3 = 8.10$ all in amperes.

23. $A = \begin{bmatrix} 7.18 & -1 & 2.2 \\ -1 & 5.8 & 1.5 \\ 2.2 & 1.5 & 8.4 \end{bmatrix}, K = \begin{bmatrix} 10 \\ 15 \\ 20 \end{bmatrix},$

$\begin{bmatrix} I_1 \\ I_2 \\ I_3 \end{bmatrix} = A^{-1}K = \begin{bmatrix} 1.22 \\ 2.37 \\ 1.64 \end{bmatrix}, I_1 = 1.22, I_2 = 2.37$
and $I_3 = 1.64$ all in amperes.

24. Let x represent the number of RC-1 units that can be made and y the number of RC-2 units that can be made. The conditions lead to the system
$\begin{cases} 8x + 9y = 1596 \\ 4x + 5y = 860 \end{cases}$, with $A = \begin{bmatrix} 8 & 9 \\ 4 & 5 \end{bmatrix}$ and $K = \begin{bmatrix} 1596 \\ 860 \end{bmatrix}$, so $A^{-1}K = \begin{bmatrix} 60 \\ 124 \end{bmatrix}$. They should make 60 of the RC-1 model and 124 units of the RC-2 model.

25. See *Computer Programs* in main text.

26. The coefficient matrix is $\begin{bmatrix} 1 & 1 & 1 \\ 1 & -2 & 0 \\ -4 & 0 & 1 \end{bmatrix}$ and the constant matrix is $\begin{bmatrix} 100 \\ 0 \\ 0 \end{bmatrix}$. Hence, $\begin{bmatrix} A \\ B \\ C \end{bmatrix} =$

$\begin{bmatrix} 1 & 1 & 1 \\ 1 & -2 & 0 \\ -4 & 0 & 1 \end{bmatrix}^{-1}\begin{bmatrix} 100 \\ 0 \\ 0 \end{bmatrix} = \begin{bmatrix} 200/11 \\ 100/11 \\ 800/11 \end{bmatrix}$, and we

see that $A = \frac{200}{11} \approx 18.18, B = \frac{100}{11} \approx 9.09, C = \frac{800}{11} \approx 72.73.$

27. The problem yields the system of equations
$\begin{cases} A + B + C = 1{,}900 \\ -A + B = 90 \\ -B + C = 130 \end{cases}$. Using matrices we get

$\begin{bmatrix} 1 & 1 & 1 \\ -1 & 1 & 0 \\ 0 & -1 & 1 \end{bmatrix}^{-1}\begin{bmatrix} 1900 \\ 90 \\ 130 \end{bmatrix} = \begin{bmatrix} 530 \\ 620 \\ 750 \end{bmatrix}$. Hence,
$A = 530$ km, $B = 620$ km, and $C = 750$ km.

28. $\begin{bmatrix} 0 & \frac{6}{7} & -\frac{2}{3} \\ -1 & -\frac{3}{7} & \frac{1}{3} \\ 0 & \frac{2}{7} & \frac{2}{3} \end{bmatrix}^{-1}\begin{bmatrix} 2000 \\ 0 \\ 1200 \end{bmatrix} = \begin{bmatrix} -1000 \\ 2800 \\ 600 \end{bmatrix}.$
Hence, $F_A = -1{,}000$ N, $F_B = 2{,}800$ N, and $F_C = 600$ N.

29. **(a)** The given information leads to the system of equations
$$\begin{cases} 2A+2B+C=5 \\ A+2B+5C=6. \\ A+2B+C=4 \end{cases}$$ Using matrices we get

$$\begin{bmatrix} 2 & 2 & 1 \\ 1 & 2 & 5 \\ 1 & 2 & 1 \end{bmatrix}^{-1} \begin{bmatrix} 5 \\ 6 \\ 4 \end{bmatrix} = \begin{bmatrix} 1 \\ 1.25 \\ 0.5 \end{bmatrix}.$$ Multiplying this

last matrix by the common denominator of 4, we get the ratio 4 : 5 : 2. Hence, the city should buy from Companies A, B, and C in the proportion $A : B : C = 4 : 5 : 2$ units.

(b) The proportion of 4 : 5 : 2 units is a total of 11 units. Dividing the number of units the city expects to need for this winter by 11, we get $3,630,000 \div 11 = 330,000$. Multiplying each number in the continued proportion by 330,000, we determine that the city should purchase the following number of units from the three companies. Company A: 1,320,000 units, Company B: 1,650,000 units, Company C: 660,000 units.

30. The given information leads to the system of equations
$$\begin{cases} 40 = a \cdot 20^2 + b \cdot 20 + c \\ 206.25 = a \cdot 55^2 + b \cdot 55 + c \\ 276.25 = a \cdot 65^2 + b \cdot 65 + c \end{cases}$$
or
$$\begin{cases} 400a + 20b + c = 40 \\ 3025a + 55b + c = 206.25 \\ 4225a + 65b + c = 276.25 \end{cases}$$
Using matrices we get
$$\begin{bmatrix} 400 & 20 & 1 \\ 3025 & 55 & 1 \\ 4225 & 65 & 1 \end{bmatrix}^{-1} \begin{bmatrix} 40 \\ 206.25 \\ 276.25 \end{bmatrix} = \begin{bmatrix} 0.05 \\ 1 \\ 0 \end{bmatrix}.$$
Thus, we see that $a = 0.05, b = 1$, and $c = 0$.

31. Using matrices, we have $\begin{bmatrix} 1 & 1 & 1 & 1 \\ 1 & 1 & -1 & 1 \\ -1.64 & 0 & 0 & 1 \\ 3 & 0 & -2 & 2 \end{bmatrix}^{-1}$

$= \begin{bmatrix} 100 \\ 0 \\ 0 \\ -1 \end{bmatrix} = \begin{bmatrix} 12.5 \\ 27.25 \\ 39.75 \\ 20.5 \end{bmatrix}.$ Hence, $A = 12.5, B = 27.25, C = 39.75$, and $D = 20.5$.

CHAPTER 17 REVIEW

1. $A + C = \begin{bmatrix} 4 & 3 & 2 & 5 \\ 6 & 7 & -1 & 4 \\ 9 & 10 & -8 & 3 \end{bmatrix} + \begin{bmatrix} 0 & 1 & 0 & 2 \\ 3 & 0 & 4 & 0 \\ 0 & -5 & 0 & -6 \end{bmatrix} = \begin{bmatrix} 4 & 4 & 2 & 7 \\ 9 & 7 & 3 & 4 \\ 9 & 5 & -8 & -3 \end{bmatrix}$

2. $B + C = \begin{bmatrix} 3 & -2 & 1 & 0 \\ 5 & -1 & 2 & 0 \\ 4 & 3 & -2 & 0 \end{bmatrix} + \begin{bmatrix} 0 & 1 & 0 & 2 \\ 3 & 0 & 4 & 0 \\ 0 & -5 & 0 & -6 \end{bmatrix} = \begin{bmatrix} 3 & -1 & 1 & 2 \\ 8 & -1 & 6 & 0 \\ 4 & -2 & -2 & -6 \end{bmatrix}$

3. $A - B = \begin{bmatrix} 4 & 3 & 2 & 5 \\ 6 & 7 & -1 & 4 \\ 9 & 10 & -8 & 3 \end{bmatrix} - \begin{bmatrix} 3 & -2 & 1 & 0 \\ 5 & -1 & 2 & 0 \\ 4 & 3 & -2 & 0 \end{bmatrix} = \begin{bmatrix} 1 & 5 & 1 & 5 \\ 1 & 8 & -3 & 4 \\ 5 & 7 & -6 & 3 \end{bmatrix}$

4. $C - B = \begin{bmatrix} 0 & 1 & 0 & 2 \\ 3 & 0 & 4 & 0 \\ 0 & -5 & 0 & -6 \end{bmatrix} - \begin{bmatrix} 3 & -2 & 1 & 0 \\ 5 & -1 & 2 & 0 \\ 4 & 3 & -2 & 0 \end{bmatrix} = \begin{bmatrix} -3 & 3 & -1 & 2 \\ -2 & 1 & 2 & 0 \\ -4 & -8 & 2 & -6 \end{bmatrix}$

5. $2A - 3C = \begin{bmatrix} 8 & 6 & 4 & 10 \\ 12 & 14 & -2 & 8 \\ 18 & 20 & -16 & 6 \end{bmatrix} - \begin{bmatrix} 0 & 3 & 0 & 6 \\ 9 & 0 & 12 & 0 \\ 0 & -15 & 0 & -18 \end{bmatrix} = \begin{bmatrix} 8 & 3 & 4 & 4 \\ 3 & 14 & -14 & 8 \\ 18 & 35 & -16 & 24 \end{bmatrix}$

6. $A + B - 2C = \begin{bmatrix} 16 & 12 & 8 & 20 \\ 24 & 28 & -4 & 16 \\ 36 & 40 & -32 & 12 \end{bmatrix} + \begin{bmatrix} 3 & -2 & 1 & 0 \\ 5 & -1 & 2 & 0 \\ 4 & 3 & -2 & 0 \end{bmatrix} - \begin{bmatrix} 0 & 2 & 0 & 4 \\ 6 & 0 & 8 & 0 \\ 0 & -10 & 0 & -12 \end{bmatrix} = \begin{bmatrix} 19 & 8 & 9 & 16 \\ 23 & 27 & -10 & 16 \\ 40 & 53 & -34 & 24 \end{bmatrix}$

7. $\begin{bmatrix} 3 & 2 \\ 4 & 5 \\ 1 & 0 \end{bmatrix} \times \begin{bmatrix} 1 & 2 \\ 0 & 1 \end{bmatrix} = \begin{bmatrix} 3+0 & 6+2 \\ 4+0 & 8+5 \\ 1+0 & 2+0 \end{bmatrix} = \begin{bmatrix} 3 & 8 \\ 4 & 13 \\ 1 & 2 \end{bmatrix}$

8. $\begin{bmatrix} 1 & -4 \\ 5 & 1 \end{bmatrix} \begin{bmatrix} 3 & 4 & 1 \\ 2 & 5 & 0 \end{bmatrix} =$

$\begin{bmatrix} 3-8 & 4-20 & 1+0 \\ 15+2 & 20+5 & 5+0 \end{bmatrix} = \begin{bmatrix} -5 & -16 & 1 \\ 17 & 25 & 5 \end{bmatrix}$

9. $\begin{bmatrix} 3 \\ 2 \\ -1 \end{bmatrix} [4 \quad 5 \quad 1] = \begin{bmatrix} 12 & 15 & 3 \\ 8 & 10 & 2 \\ -4 & -5 & -1 \end{bmatrix}$

10. $[4 \quad 5 \quad 1] \begin{bmatrix} -2 \\ 7 \\ -3 \end{bmatrix} = [-8 + 35 - 3] = [24]$

11. $\begin{bmatrix} 2 & 3 \\ -4 & -5 \end{bmatrix}^{-1} = \frac{1}{(-10+12)} \begin{bmatrix} -5 & -3 \\ 4 & 2 \end{bmatrix} =$

$\begin{bmatrix} -\frac{5}{2} & -\frac{3}{2} \\ 2 & 1 \end{bmatrix}$

12. $\begin{bmatrix} 2 & -1 \\ 0 & 4 \end{bmatrix}^{-1} = \frac{1}{8} \begin{bmatrix} 4 & 1 \\ 0 & 2 \end{bmatrix} = \begin{bmatrix} \frac{1}{2} & \frac{1}{8} \\ 0 & \frac{1}{4} \end{bmatrix} =$

$\begin{bmatrix} 0.5 & 0.125 \\ 0 & 0.25 \end{bmatrix}$

13. $\begin{bmatrix} -2 & 1 & 0 & \vdots & 1 & 0 & 0 \\ 0 & 4 & 0 & \vdots & 0 & 1 & 0 \\ 1 & 0 & 1 & \vdots & 0 & 0 & 1 \end{bmatrix} \begin{matrix} 2R_3 + R_1 \\ \\ \end{matrix}$

$\begin{bmatrix} 0 & 1 & 2 & \vdots & 1 & 0 & 2 \\ 0 & 4 & 0 & \vdots & 0 & 1 & 0 \\ 1 & 0 & 1 & \vdots & 0 & 0 & 1 \end{bmatrix}$ Swap R_1 and R_3

$\begin{bmatrix} 1 & 0 & 1 & \vdots & 0 & 0 & 1 \\ 0 & 4 & 0 & \vdots & 0 & 1 & 0 \\ 0 & 1 & 2 & \vdots & 1 & 0 & 2 \end{bmatrix} \begin{matrix} \\ \\ -\frac{1}{4}R_2 + R_3 \end{matrix}$

$\begin{bmatrix} 1 & 0 & 1 & \vdots & 0 & 0 & 1 \\ 0 & 4 & 0 & \vdots & 0 & 1 & 0 \\ 0 & 0 & 2 & \vdots & 1 & -\frac{1}{4} & 2 \end{bmatrix} \begin{matrix} -\frac{1}{2}R_3 + R_1 \\ \\ \end{matrix}$

$\begin{bmatrix} 1 & 0 & 0 & \vdots & -\frac{1}{2} & \frac{1}{8} & 0 \\ 0 & 4 & 0 & \vdots & 0 & 1 & 0 \\ 0 & 0 & 2 & \vdots & 1 & -\frac{1}{4} & 2 \end{bmatrix} \begin{matrix} \\ \frac{1}{4}R_2 \\ \frac{1}{2}R_3 \end{matrix}$

$\begin{bmatrix} 1 & 0 & 0 & \vdots & -\frac{1}{2} & \frac{1}{8} & 0 \\ 0 & 1 & 0 & \vdots & 0 & \frac{1}{4} & 0 \\ 0 & 0 & 1 & \vdots & \frac{1}{2} & -\frac{1}{8} & 1 \end{bmatrix}$. The inverse is

$\begin{bmatrix} -\frac{1}{2} & \frac{1}{8} & 0 \\ 0 & \frac{1}{4} & 0 \\ \frac{1}{2} & -\frac{1}{8} & 1 \end{bmatrix}$ or $\begin{bmatrix} -0.5 & 0.125 & 0 \\ 0 & 0.25 & 0 \\ 0.5 & -0.125 & 1 \end{bmatrix}$.

14. $\begin{bmatrix} 2 & 3 & -1 & \vdots & 1 & 0 & 0 \\ 1 & 2 & 1 & \vdots & 0 & 1 & 0 \\ -1 & -1 & 3 & \vdots & 0 & 0 & 1 \end{bmatrix} \begin{matrix} \\ -R_1 + 2R_2 \\ R_2 + R_3 \end{matrix}$

$\begin{bmatrix} 2 & 3 & -1 & \vdots & 1 & 0 & 0 \\ 0 & 1 & 3 & \vdots & -1 & 2 & 0 \\ 0 & 1 & 4 & \vdots & 0 & 1 & 1 \end{bmatrix} \begin{matrix} -3R_2 + R_1 \\ \\ -R_2 + R_3 \end{matrix}$

$\begin{bmatrix} 2 & 0 & -10 & \vdots & 4 & -6 & 0 \\ 0 & 1 & 3 & \vdots & -1 & 2 & 0 \\ 0 & 0 & 1 & \vdots & 0 & -1 & 1 \end{bmatrix} \begin{matrix} 10R_3 + R_1 \\ -3R_3 + R_2 \\ \end{matrix}$

$\begin{bmatrix} 2 & 0 & 0 & \vdots & 14 & -16 & 10 \\ 0 & 1 & 0 & \vdots & -4 & 5 & -3 \\ 0 & 0 & 1 & \vdots & 1 & -1 & 1 \end{bmatrix} \begin{matrix} \frac{1}{2}R_1 \\ \\ \end{matrix}$

$\begin{bmatrix} 1 & 0 & 0 & \vdots & 7 & -8 & 5 \\ 0 & 1 & 0 & \vdots & -4 & 5 & -3 \\ 0 & 0 & 1 & \vdots & 1 & -1 & 1 \end{bmatrix}$. The inverse is

$\begin{bmatrix} 7 & -8 & 5 \\ -4 & 5 & -3 \\ 1 & -1 & 1 \end{bmatrix}$

15. $A = \begin{bmatrix} 12 & 5 \\ 3 & 1 \end{bmatrix}, A^{-1} = -\frac{1}{3} \begin{bmatrix} 1 & -5 \\ -3 & 12 \end{bmatrix}, K = \begin{bmatrix} -2 \\ 1.1 \end{bmatrix}, A^{-1}K = \begin{bmatrix} 2.5 \\ -6.4 \end{bmatrix}, x = 2.5, y = -6.4$

16. $A = \begin{bmatrix} 4 & 1 \\ -3 & 2 \end{bmatrix}, A^{-1} = \frac{1}{11} \begin{bmatrix} 2 & -1 \\ 3 & 4 \end{bmatrix}, K = \begin{bmatrix} -4 \\ 14 \end{bmatrix}, A^{-1}K = \begin{bmatrix} -2 \\ 4 \end{bmatrix}, x = -2, y = 4$

17. $A = \begin{bmatrix} 2 & 3 & 5 \\ -2 & 3 & 5 \\ 5 & -3 & -2 \end{bmatrix}, K = \begin{bmatrix} 20 \\ 12 \\ 9 \end{bmatrix},$

$A^{-1}K = \begin{bmatrix} 2 \\ -3 \\ 5 \end{bmatrix}, x = 2, y = -3, z = 5$

18. $A = \begin{bmatrix} 1 & 1 & 6 \\ -2 & 2 & 4 \\ 3 & 2 & 4 \end{bmatrix}, K = \begin{bmatrix} -3 \\ -2 \\ 14 \end{bmatrix},$

$A^{-1}K = \begin{bmatrix} 3.2 \\ 6.4 \\ -2.1 \end{bmatrix}, x = 3.2, y = 6.4, z = -2.1$

19. $\begin{cases} -9a + T = 75.4 \\ -11a - T = 100.0 \end{cases} A = \begin{bmatrix} -9 & 1 \\ -11 & -1 \end{bmatrix},$

$K = \begin{bmatrix} 75.4 \\ -100.0 \end{bmatrix}, A^{-1}K = \begin{bmatrix} 1.23 \\ 86.47 \end{bmatrix}, a = 1.23,$

$T = 86.47$

20. $\begin{cases} 4P + 8B + 12T = 1872 \\ 2P + 3B + 5T = 771 \\ 7P + 6B + 11T = 1770 \end{cases} A = \begin{bmatrix} 4 & 8 & 12 \\ 2 & 3 & 5 \\ 7 & 6 & 11 \end{bmatrix},$

$K = \begin{bmatrix} 1872 \\ 771 \\ 1770 \end{bmatrix}, A^{-1}K = \begin{bmatrix} 45 \\ 72 \\ 93 \end{bmatrix}$, and so they can

make 45 PCs, 72 BCs and 93 TCs.

21. The given systems can be expressed in matrices as follows:

$$\begin{bmatrix} \frac{1}{2} & \frac{\sqrt{3}}{2} \\ -\frac{\sqrt{3}}{2} & \frac{1}{2} \end{bmatrix} \begin{bmatrix} x \\ y \end{bmatrix} = \begin{bmatrix} x' \\ y' \end{bmatrix} \text{ and}$$

$$\begin{bmatrix} -\frac{1}{2} & \frac{\sqrt{3}}{2} \\ -\frac{\sqrt{3}}{2} & -\frac{1}{2} \end{bmatrix} \begin{bmatrix} x' \\ y' \end{bmatrix} = \begin{bmatrix} x'' \\ y'' \end{bmatrix}. \text{ Hence,}$$

$$\begin{bmatrix} x'' \\ y'' \end{bmatrix} = \begin{bmatrix} -\frac{1}{2} & \frac{\sqrt{3}}{2} \\ -\frac{\sqrt{3}}{2} & -\frac{1}{2} \end{bmatrix} \begin{bmatrix} \frac{1}{2} & \frac{\sqrt{3}}{2} \\ -\frac{\sqrt{3}}{2} & \frac{1}{2} \end{bmatrix} \begin{bmatrix} x \\ y \end{bmatrix} =$$

$$\begin{bmatrix} -1 & 0 \\ 0 & -1 \end{bmatrix} \begin{bmatrix} x \\ y \end{bmatrix} = \begin{bmatrix} -x \\ -y \end{bmatrix}. \text{ Thus, we have shown}$$
that $x'' = -x$ and $y'' = -y$.

22. Since $x'' = -x$ and $y'' = -y$, we have that $x\cos\theta + y\sin\theta = -x$ or $\cos\theta = -1$ and $\sin\theta = 0$. This is a $180°$ rotation.

23. $M = \begin{bmatrix} 1 & -\frac{1}{f_2} \\ 0 & 1 \end{bmatrix} \cdot \begin{bmatrix} 1 & 0 \\ d & 1 \end{bmatrix} \cdot \begin{bmatrix} 1 & -\frac{1}{f_1} \\ 0 & 1 \end{bmatrix} =$

$$\begin{bmatrix} 1 - \frac{d}{f_2} & -\frac{1}{f_2} \\ d & 1 \end{bmatrix} \begin{bmatrix} 1 & -\frac{1}{f_1} \\ 0 & 1 \end{bmatrix} =$$

$$\begin{bmatrix} 1 - \frac{d}{f_2} & -\frac{1}{f_1} + \frac{d}{f_1 f_2} - \frac{1}{f_2} \\ d & -\frac{d}{f_1} + 1 \end{bmatrix} \text{ or}$$

$$M = \begin{bmatrix} 1 - \frac{d}{f_2} & \frac{d}{f_1 f_2} - \frac{1}{f_1} - \frac{1}{f_2} \\ d & 1 - \frac{d}{f_1} \end{bmatrix}.$$

24. $-\frac{1}{f} = \frac{d}{f_1 f_2} - \frac{1}{f_1} - \frac{1}{f_2}$ so $\frac{1}{f} = \frac{1}{f_1} + \frac{1}{f_2} - \frac{d}{f_1 f_2}$

CHAPTER 17 TEST

1. **(a)** $A+B = \begin{bmatrix} 8 & 0 & -4 \\ 16 & -6 & 2 \end{bmatrix} + \begin{bmatrix} -1 & 5 & -3 \\ 3 & 0 & 4 \end{bmatrix} =$

$$\begin{bmatrix} 7 & 5 & -7 \\ 19 & -6 & 6 \end{bmatrix}$$

(b) $3A - 2B = \begin{bmatrix} 24 & 0 & -12 \\ 48 & -18 & 6 \end{bmatrix} -$

$$\begin{bmatrix} -2 & 10 & -6 \\ 6 & 0 & 8 \end{bmatrix} = \begin{bmatrix} 26 & -10 & -6 \\ 42 & -18 & -2 \end{bmatrix}$$

2. $[1 \quad -2 \quad 3] \begin{bmatrix} -4 \\ -6 \\ 8 \end{bmatrix} = [-4 + 12 + 24] = [32]$

3. **(a)** $CD = \begin{bmatrix} 4 & 6 \\ -10 & 4 \end{bmatrix} \cdot \begin{bmatrix} \frac{1}{2} & 0 \\ -\frac{3}{2} & 4 \end{bmatrix} =$

$$\begin{bmatrix} 2 - 9 & 0 + 24 \\ -5 - 6 & 0 + 16 \end{bmatrix} = \begin{bmatrix} -7 & 24 \\ -11 & 16 \end{bmatrix}.$$

(b) $DC = \begin{bmatrix} \frac{1}{2} & 0 \\ -\frac{3}{2} & 4 \end{bmatrix} \begin{bmatrix} 4 & 6 \\ -10 & 4 \end{bmatrix} =$

$$\begin{bmatrix} 2 + 0 & 3 + 0 \\ -6 - 40 & -9 + 16 \end{bmatrix} = \begin{bmatrix} 2 & 3 \\ -46 & 7 \end{bmatrix}.$$

4. **(a)** $E = \begin{bmatrix} 2 & -3 \\ 7 & 9 \end{bmatrix}, E^{-1} = \frac{1}{18+21} \begin{bmatrix} 9 & 3 \\ -7 & 2 \end{bmatrix} \approx$

$$\begin{bmatrix} 0.2308 & 0.0769 \\ -0.1795 & 0.0513 \end{bmatrix}$$

(b) $EX = F$ so $X = E^{-1}F = \begin{bmatrix} -5 \\ 7 \end{bmatrix}.$

5. The coefficient matrix is $A = \begin{bmatrix} 1 & 3 & 1 \\ 2 & 5 & 1 \\ 1 & 2 & 3 \end{bmatrix}$ and the

constant matrix is $K = \begin{bmatrix} -2 \\ -5 \\ 6 \end{bmatrix}$. If $X = \begin{bmatrix} x \\ y \\ z \end{bmatrix}$, we

have the system $AX = K$, so $X = A^{-1}K$. You

can determine that $A^{-1} = \begin{bmatrix} -\frac{13}{3} & \frac{7}{3} & \frac{2}{3} \\ \frac{5}{3} & -\frac{2}{3} & -\frac{1}{3} \\ \frac{1}{3} & -\frac{1}{3} & \frac{1}{3} \end{bmatrix} =$

$\frac{1}{3} \begin{bmatrix} -13 & 7 & 2 \\ 5 & -2 & -1 \\ 1 & -1 & 1 \end{bmatrix}$ and so, $A^{-1}K = \begin{bmatrix} 1 \\ -2 \\ 3 \end{bmatrix}.$

Thus, $x = 1, y = -2, z = 3$.

6. Let the costs be $x, y,$ and z. Then we get the system $\begin{cases} x + y + z = 60 \\ x = y + z \\ y = 2z + 3 \end{cases}$ or $\begin{cases} x + y + z = 60 \\ x - y - z = 0 \\ y - 2z = 3 \end{cases}$. Using matrices we get $\begin{bmatrix} 1 & 1 & 1 \\ 1 & -1 & -1 \\ 0 & 1 & -2 \end{bmatrix}^{-1} \begin{bmatrix} 60 \\ 0 \\ 3 \end{bmatrix} = \begin{bmatrix} 30 \\ 21 \\ 9 \end{bmatrix}$. Thus, we see that $x = 30, y = 21,$ and $z = 9$ and so, the parts cost \$30, \$21, and \$9.

7. By Kirchhoff's laws, we have the following equations

$$I_1 - I_2 - I_3 = 0 \qquad (1)$$
$$6I_1 + 4I_2 = 20 \qquad (2)$$
$$4I_2 - 2I_3 = 10 \qquad (3)$$

This yields the matrix equation

$$\begin{bmatrix} 1 & -1 & -1 \\ 6 & 4 & 0 \\ 0 & 4 & -2 \end{bmatrix}^{-1} \begin{bmatrix} 0 \\ 20 \\ 10 \end{bmatrix} = \begin{bmatrix} \frac{20}{11} \\ \frac{25}{11} \\ -\frac{5}{11} \end{bmatrix}.$$ Thus, we have determined that $I_1 = \frac{20}{11}$ A, $I_2 = \frac{25}{11}$ A, and $I_3 = -\frac{5}{11}$ A.

CHAPTER
18
Higher Degree Equations

≡ 18.1 THE REMAINDER AND FACTOR THEOREM

1. $P(2) = 3 \cdot 2^2 - 2 \cdot 2 + 1 = 12 - 4 + 1 = 9$

2. $P(-1) = 2(-1)^3 - 4(-1) + 5 = -2 + 4 + 5 = 7$

3. $P(-1) = (-1)^4 + (-1)^3 + (-1)^2 - (-1) + 1 = 1 - 1 + 1 + 1 + 1 = 3$

4. $P(-2) = (-2)^4 - 2(-2)^2 + (-2) = 16 - 8 - 2 = 6$

5. $P(3) = 5(3)^3 - 4(3) + 7 = 135 - 12 + 7 = 130$

6. $P(-2) = 7(-2)^4 - 5(-2)^2 + (-2) - 7 = 112 - 20 - 2 - 7 = 83$

7. $P(-1) = (-1)^5 - (-1)^4 + (-1)^3 + (-1)^2 - (-1) + 1 = -1 - 1 - 1 + 1 + 1 + 1 = 0$

8. $P(-3) = 3(-3)^5 + 4(-3)^2 - 3 = -729 + 36 - 3 = -696$

9. $R = P(1) = 1^3 + 2 \cdot 1^2 - 1 - 2 = 1 + 2 - 1 - 2 = 0$

10. $R = P(3) = 3^3 + 2 \cdot 3^2 - 12 \cdot 3 - 9 = 27 + 18 - 36 - 9 = 0$

11. $R = P(3) = 3^3 - 3 \cdot 3^2 + 2 \cdot 3 + 5 = 27 - 27 + 6 + 5 = 11$

12. $R = P(1) = 1^3 - 9 \cdot 1^2 + 23 \cdot 1 - 15 = 1 - 9 + 23 - 15 = 0$

13. $R = P(-2) = 4(-2)^4 + 13(-2)^3 - 13(-2)^2 - 40(-2) + 12 = 64 - 104 - 52 + 80 + 12 = 0$

14. $R = P(7) = 2 \cdot 7^4 - 2 \cdot 7^3 - 6 \cdot 7^2 - 14 \cdot 7 - 7 = 4802 - 686 - 294 - 98 - 7 = 3717$

15. $R = P(5) = 3 \cdot 5^4 - 12 \cdot 5^3 - 60 \cdot 5 + 4 = 1875 - 1500 - 300 + 4 = 79$

16. $R = P\left(\frac{1}{2}\right) = 4 \cdot \left(\frac{1}{2}\right)^3 - 4\left(\frac{1}{2}\right)^2 - 10\left(\frac{1}{2}\right) + 8 = \frac{1}{2} - 1 - 5 + 8 = 2\frac{1}{2}$ or 2.5

17. $P(3) = 3^3 + 2 \cdot 3^2 - 12 \cdot 3 - 9 = 27 + 18 - 36 - 9 = 0$; yes

18. $P(-1) = (-1)^4 - 9(-1)^3 + 18(-1)^2 - 3 = 1 + 9 + 18 - 3 = 25$; no

19. $P(-1) = 2(-1)^5 - 6(-1)^3 + (-1)^2 + 4(-1) - 1 = -2 + 6 + 1 - 4 - 1 = 0$; yes

20. $P(1) = 2(1)^5 - 6(1)^3 + 1^2 + 4 \cdot 1 - 1 = 2 - 6 + 1 + 4 - 1 = 0$; yes

21. $P(-3) = 3(-3)^5 + 3(-3)^4 - 14(-3)^3 + 4(-3)^2 - 24(-3) = -729 + 243 + 378 + 36 + 72 = 0$; yes

22. $P(2) = 3 \cdot 2^5 + 3 \cdot 2^4 - 14 \cdot 2^3 + 4 \cdot 2^2 - 24 \cdot 2 = 96 + 48 - 112 + 16 - 48 = 0$: yes

23. $P\left(\frac{5}{2}\right) = 6 \cdot \left(\frac{5}{2}\right)^4 - 15\left(\frac{5}{2}\right)^3 - 8\left(\frac{5}{2}\right)^2 + 20\left(\frac{5}{2}\right) = 234\frac{3}{8} - 234\frac{3}{8} - 50 + 50 = 0$; yes

24. $P\left(-\frac{3}{5}\right) = 20\left(-\frac{3}{5}\right)^4 + 12\left(-\frac{3}{5}\right)^3 + 10\left(-\frac{3}{5}\right) + 9 = 2.592 - 2.592 - 6 + 9 = 3$; no

25.
```
3 | 1    0   -17    0    75    9
  |      3     9  -24   -72    9
  ----------------------------------
    1    3    -8  -24     3   18
```
So, $Q(x) = x^4 + 3x^3 - 8x^2 - 24x + 3$, $R(x) = 18$

26.
```
-2 | 2    0    0   -1    8    44
   |     -4    8  -16   34   -84
   ---------------------------------
     2   -4    8  -17   42   -40
```
So, $Q(x) = 2x^4 - 4x^3 + 8x^2 - 17x + 42$, $R(x) = -40$

27.
```
-3 | 5    7    0    9
   |     -15   24  -72
   -------------------
     5   -8   24  -63
```
So, $Q(x) = 5x^2 - 8x + 24$; $R(x) = -63$

28.
```
2 | 1    3   -2   -4
  |      2   10   16
  -------------------
    1    5    8   12
```
So, $Q(x) = x^2 + 5x + 8$; $R(x) = 12$

29.

$$\frac{1}{2} \,\big|\, \begin{array}{ccccc} 8 & 0 & -4 & 7 & -2 & 0 \end{array}$$
$$\begin{array}{ccccc} & 4 & 2 & -1 & 3 & \frac{1}{2} \end{array}$$
$$\begin{array}{ccccc} \hline 8 & 4 & -2 & 6 & 1 & \frac{1}{2} \end{array}$$

So, $Q(x) = 8x^4 + 4x^3 - 2x^2 + 6x + 1$ and $R(x) = \frac{1}{2}$

30.

$$-\frac{1}{3} \,\big|\, \begin{array}{cccccc} 9 & 3 & -6 & -2 & 6 & 1 \end{array}$$
$$\begin{array}{cccccc} & -3 & 0 & 2 & 0 & -2 \end{array}$$
$$\begin{array}{cccccc} \hline 9 & 0 & -6 & 0 & 6 & -1 \end{array}$$

So, $Q(x) = 9x^4 - 6x^2 + 6$ and $R(x) = -1$

31.

$$\frac{3}{2} \,\big|\, \begin{array}{ccccc} 4 & -12 & 9 & -8 & 12 \end{array}$$
$$\begin{array}{ccccc} & 6 & -9 & 0 & -12 \end{array}$$
$$\begin{array}{ccccc} \hline 4 & -6 & 0 & -8 & 0 \end{array}$$

Since we used $\frac{2x-3}{2} = x - \frac{3}{2}$, we need to divide the depressed polynomial by 2 also. So, $Q(x) = \frac{4x^3 - 6x^2 - 8}{2} = 2x^3 - 3x^2 - 4$; $R(x) = 0$.

32.

$$\frac{5}{4} \,\big|\, \begin{array}{cccc} 4 & 7 & -3 & -15 \end{array}$$
$$\begin{array}{cccc} & 5 & 15 & 15 \end{array}$$
$$\begin{array}{cccc} \hline 4 & 12 & 12 & 0 \end{array}$$

Since we used $\frac{4x-5}{4} = x = -\frac{5}{4}$, we need to divide the depressed polynomial by 4 also. So, $Q(x) = \frac{4x^2 + 12x + 12}{4} = x^2 + 3x + 3$; $R(x) = 0$

≡ 18.2 ROOTS OF AN EQUATION

1. Using synthetic division, we get

$$1 \,\big|\, \begin{array}{cccc} 5 & 0 & -8 & 3 \end{array}$$
$$\begin{array}{cccc} & 5 & 5 & -3 \end{array}$$
$$\begin{array}{cccc} \hline 5 & 5 & -3 & 0 \end{array}$$

We now use the quadratic equation on the depressed equation $5x^2 + 5x - 3 = 0$, with the results
$$x = \frac{-5 \pm \sqrt{5^2 - 4 \cdot 5 \cdot (-3)}}{2 \cdot 5} = \frac{-5 \pm \sqrt{25 + 60}}{10} = \frac{-5 \pm \sqrt{85}}{10}.$$
The roots are 1, $\frac{-5 + \sqrt{85}}{10}$ and $\frac{-5 - \sqrt{85}}{10}$.

2. Using synthetic division, we get

$$-4 \,\big|\, \begin{array}{cccc} 2 & 5 & -11 & 4 \end{array}$$
$$\begin{array}{cccc} & -8 & 12 & -4 \end{array}$$
$$\begin{array}{cccc} \hline 2 & -3 & 1 & 0 \end{array}$$

The depressed equation $2x^2 - 3x + 1 = 0$ factors into $(2x - 1)(x - 1) = 0$, which has roots $\frac{1}{2}$ and 1. So, the roots of the given equation are -4, 1 and $\frac{1}{2}$.

3. Using synthetic division, we get

$$\frac{1}{3} \,\big|\, \begin{array}{cccc} 9 & -3 & -81 & 27 \end{array}$$
$$\begin{array}{cccc} & 3 & 0 & -27 \end{array}$$
$$\begin{array}{cccc} \hline 9 & 0 & -81 & 0 \end{array}$$

Now, $9x^2 - 81 = 9(x^2 - 9) = 9(x + 3)(x - 3)$ and so the roots are $\frac{1}{3}, 3$, and -3.

4. With synthetic division, we get

$$\frac{2}{5} \,\big|\, \begin{array}{cccc} 10 & -4 & -40 & 16 \end{array}$$
$$\begin{array}{cccc} & 4 & 0 & -16 \end{array}$$
$$\begin{array}{cccc} \hline 10 & 0 & -40 & 0 \end{array}$$

Now, $10x^2 - 40 = 10(x^2 - 4) = 10(x + 2)(x - 2)$. The roots are $\frac{2}{5}$, -2 and 2.

5. Synthetic division and the complex conjugate of r_1, which is $-j$, yields

$$j \,\big|\, \begin{array}{ccccc} 1 & 0 & -3 & 0 & -4 \end{array}$$
$$\begin{array}{ccccc} & j & -1 & -4j & 4 \end{array}$$
$$-j \,\big|\, \begin{array}{ccccc} 1 & j & -4 & -4j & 0 \end{array}$$
$$\begin{array}{ccccc} & -j & 0 & 4j \end{array}$$
$$\begin{array}{ccccc} \hline 1 & 0 & -4 & 0 \end{array}$$

Since $x^2 - 4 = (x + 2)(x - 2)$, we see that the roots are $2, -2, j$, and $-j$.

6. Synthetic division and the complex conjugate of r_1, which is $3j$, yields

$$-3j \,\big|\, \begin{array}{ccccc} 3 & 0 & 6 & 0 & -189 \end{array}$$
$$\begin{array}{ccccc} & -9j & -27 & 63j & 189 \end{array}$$
$$3j \,\big|\, \begin{array}{ccccc} 3 & -9j & -21 & 63j & 0 \end{array}$$
$$\begin{array}{ccccc} & 9j & 0 & -63j \end{array}$$
$$\begin{array}{ccccc} \hline 3 & 0 & -21 & 0 \end{array}$$

Since $3x^2 - 21 = 3(x^2 - 7) = 3(x + \sqrt{7})(x - \sqrt{7})$, the roots are $\sqrt{7}, -\sqrt{7}, 3j$, and $-3j$.

7. Synthetic division and the complex conjugate of r_1, which is $-1 - 2j$, yields

$$-1 + 2j \,\big|\, \begin{array}{ccccc} 1 & 2 & -4 & -18 & -45 \end{array}$$
$$\begin{array}{ccccc} & -1 + 2j & -5 & 9 - 18j & 45 \end{array}$$
$$-1 - 2j \,\big|\, \begin{array}{ccccc} 1 & 1 + 2j & -9 & -9 - 18j & 0 \end{array}$$
$$\begin{array}{ccccc} & -1 - 2j & 0 & 9 + 18j \end{array}$$
$$\begin{array}{ccccc} \hline 1 & 0 & -9 & 0 \end{array}$$

Since $x^2 - 9 = (x + 3)(x - 3)$, the roots are $3, -3, -1 + 2j$, and $-1 - 2j$.

8. Synthetic division and the complex conjugate of r_1, which is $-1 + 2j$, yields

$$
\begin{array}{r|rrrrr}
-1-2j & 2 & 4 & 2 & -16 & -40 \\
 & & -2-4j & -10 & 8+16j & 40 \\
\hline
-1+2j & 2 & 2-4j & -8 & -8+16j & 0 \\
 & & -2+4j & 0 & 8-16j \\
\hline
 & 2 & 0 & -8 & 0
\end{array}
$$

Since $2x^2 - 8 = 2(x^2 - 4) = 2(x + 2)(x - 2)$, the roots are $2, -2, -1 - 2j$, and $-1 + 2j$.

9. Repeated use of synthetic division yields

$$
\begin{array}{r|rrrrr}
2 & 1 & -3 & -3 & 7 & 6 \\
 & & 2 & -2 & -10 & -6 \\
\hline
3 & 1 & -1 & -5 & -3 & 0 \\
 & & 3 & 6 & 3 \\
\hline
 & 1 & 2 & 1 & 0
\end{array}
$$

Since $x^2 + 2x + 1 = (x+1)^2$, the roots are $-1, -1, 2$, and 3.

10. Repeated use of synthetic division yields

$$
\begin{array}{r|rrrrr}
2 & 1 & -3 & -12 & 52 & -48 \\
 & & 2 & -2 & -28 & 48 \\
\hline
-4 & 1 & -1 & -14 & 24 & 0 \\
 & & -4 & 20 & -24 \\
\hline
 & 1 & -5 & 6 & 0
\end{array}
$$

Since $x^2 - 5x + 6 = (x - 2)(x - 3)$, the roots are $2, -4, 2$, and 3.

11. Repeated use of synthetic division yields

$$
\begin{array}{r|rrrrr}
-1 & 3 & 12 & 6 & -12 & -9 \\
 & & -3 & -9 & 3 & 9 \\
\hline
-1 & 3 & 9 & -3 & -9 & 0 \\
 & & -3 & -6 & 9 \\
\hline
 & 3 & 6 & -9 & 0
\end{array}
$$

Since $3x^2 + 6x - 9 = 3(x^2 + 2x - 3) = 3(x + 3)(x - 1)$, the roots are $-1, -1, -3$, and 1.

12. Repeated use of synthetic division yields

$$
\begin{array}{r|rrrrr}
2 & 2 & 6 & -12 & -24 & 16 \\
 & & 4 & 20 & 16 & -16 \\
\hline
-2 & 2 & 10 & 8 & -8 & 0 \\
 & & -4 & -12 & 8 \\
\hline
 & 2 & 6 & -4 & 0
\end{array}
$$

Here $2x^2 + 6x - 4 = 2(x^2 + 3x - 2)$ and using the quadratic formula produces $x = \frac{-3 \pm \sqrt{3^2 - 4 \cdot 1(-2)}}{2} = \frac{-3 \pm \sqrt{9+8}}{2}$. The roots are $2, -2, \frac{-3 + \sqrt{17}}{2}, \frac{-3 - \sqrt{17}}{2}$.

13. Repeated use of synthetic division yields

$$
\begin{array}{r|rrrrr}
-\frac{1}{2} & 6 & 25 & 33 & 1 & -5 \\
 & & -3 & -11 & -11 & 5 \\
\hline
\frac{1}{3} & 6 & 22 & 22 & -10 & 0 \\
 & & 2 & 8 & 10 \\
\hline
 & 6 & 24 & 30 & 0
\end{array}
$$

Here $6x^2 + 24x + 30 = 6(x^2 + 4x + 5)$, so using the quadratic formula, we get $x = \frac{-4 \pm \sqrt{16-20}}{2} = -2 \pm j$. The roots are $-\frac{1}{2}, \frac{1}{3}, -2 + j$, and $-2 - j$.

14. Repeated use of synthetic division yields

$$
\begin{array}{r|rrrrr}
\frac{1}{4} & 12 & -47 & 55 & 9 & -5 \\
 & & 3 & -11 & 11 & 5 \\
\hline
-\frac{1}{3} & 12 & -44 & 44 & 20 & 0 \\
 & & -4 & 16 & -20 \\
\hline
 & 12 & -48 & 60 & 0
\end{array}
$$

Here $12x^2 - 48x + 60 = 12(x^2 - 4x + 5)$, and so $x = \frac{4 \pm \sqrt{16-20}}{2} = 2 \pm j$. The roots are $\frac{1}{4}, -\frac{1}{3}, 2 + j$, and $2 - j$.

15. Repeated use of synthetic division yields

$$
\begin{array}{r|rrrrr}
\frac{2}{3} & 3 & -2 & 0 & -3 & 2 \\
 & & 2 & 0 & 0 & -2 \\
\hline
1 & 3 & 0 & 0 & -3 & 0 \\
 & & 3 & 3 & 3 \\
\hline
 & 3 & 3 & 3 & 0
\end{array}
$$

Here $3x^2 + 3x + 3 = 3(x^2 + x + 1)$, and by using the quadratic formula, we get $x = \frac{-1 \pm \sqrt{1-4}}{2} = -\frac{1}{2} \pm \frac{\sqrt{3}}{2}j$. The roots are $1, \frac{2}{3}, -\frac{1}{2} + \frac{\sqrt{3}}{2}j$, and $-\frac{1}{2} - \frac{\sqrt{3}}{2}j$.

16. Repeated use of synthetic division yields

$$
\begin{array}{r|rrrrr}
2 & 4 & 3 & 0 & -32 & -24 \\
 & & 8 & 22 & 44 & 24 \\
\hline
-\frac{3}{4} & 4 & 11 & 22 & 12 & 0 \\
 & & -3 & -6 & -12 \\
\hline
 & 4 & 8 & 16 & 0
\end{array}
$$

Here $4x^2 + 8x + 16 = 4(x^2 + 2x + 4)$ and use of the quadratic formula yields $x = \frac{-2 \pm \sqrt{4-16}}{2} = -1 \pm j\sqrt{3}$. The roots are $2, -\frac{3}{4}, -1 + j\sqrt{3}$, and $-1 - j\sqrt{3}$.

17. Here we use synthetic division three times, with the following results:

$$
\begin{array}{r|rrrrrr}
1 & 1 & -1 & 1 & -7 & 10 & -4 \\
  &   & 1 & 0 & 1 & -6 & 4 \\
\hline
1 & 1 & 0 & 1 & -6 & 4 & 0 \\
  &   & 1 & 1 & 2 & -4 & \\
\hline
1 & 1 & 1 & 2 & -4 & 0 & \\
  &   & 1 & 2 & 4 & & \\
\hline
  & 1 & 2 & 4 & 0 & &
\end{array}
$$

Here $x^2 + 2x + 4$ has roots $x = \frac{-2 \pm \sqrt{4-16}}{2} = -1 \pm j\sqrt{3}$. The roots are $1, 1, 1, -1 + j\sqrt{3}$, and $-1 - j\sqrt{3}$.

18. Here we use synthetic division three times, with the following results:

$$
\begin{array}{r|rrrrrr}
2 & 1 & -5 & 7 & -2 & 4 & -8 \\
  &   & 2 & -6 & 2 & 0 & 8 \\
\hline
2 & 1 & -3 & 1 & 0 & 4 & 0 \\
  &   & 2 & -2 & -2 & -4 & \\
\hline
2 & 1 & -1 & -1 & -2 & 0 & \\
  &   & 2 & 2 & 2 & & \\
\hline
  & 1 & 1 & 1 & 0 & &
\end{array}
$$

Here $x^2 + x + 1 = \frac{-1 \pm \sqrt{1-4}}{2} = \frac{-1 \pm j\sqrt{3}}{2}$ and the roots are $2, 2, 2, -\frac{1}{2} + \frac{\sqrt{3}}{2}j$, and $-\frac{1}{2} - \frac{\sqrt{3}}{2}j$.

19. Here we use synthetic division three times, with the following results:

$$
\begin{array}{r|rrrrrr}
3 & 3 & -2 & -24 & 1 & 28 & -12 \\
  &   & 9 & 21 & -9 & -24 & 12 \\
\hline
-2 & 3 & 7 & -3 & -8 & 4 & 0 \\
  &   & -6 & -2 & 10 & -4 & \\
\hline
\frac{2}{3} & 3 & 1 & -5 & 2 & 0 & \\
  &   & 2 & 2 & -2 & & \\
\hline
  & 3 & 3 & -3 & 0 & &
\end{array}
$$

Since $3x^2 + 3x - 3 = 3(x^2 + x - 1)$, we find that $x = \frac{-1 \pm \sqrt{1+4}}{2} = \frac{-1 \pm \sqrt{5}}{2}$. The roots are 3, $-2, \frac{2}{3}, \frac{-1+\sqrt{5}}{2}$, and $\frac{-1-\sqrt{5}}{2}$.

20. Repeated use of synthetic division yields

$$
\begin{array}{r|rrrrrr}
-\frac{2}{3} & 3 & -4 & 5 & -18 & -28 & -8 \\
  &   & -2 & 4 & -6 & 16 & 8 \\
\hline
2j & 3 & -6 & 9 & -24 & -12 & 0 \\
  &   & 6j & -12-12j & 24-6j & 12 & \\
\hline
-2j & 3 & -6+6j & -3-12j & -6j & 0 & \\
  &   & -6j & 12j & 6j & & \\
\hline
  & 3 & -6 & -3 & 0 & &
\end{array}
$$

Since $3x^2 - 6x - 3 = 3(x^2 - 2x - 1)$, we see that $x = \frac{2 \pm \sqrt{4+4}}{2} = 1 \pm \sqrt{2}$. The roots are $-\frac{2}{3}, 2j, -2j, 1 + \sqrt{2}$, and $1 - \sqrt{2}$.

21. Use of synthetic division four times yields

$$
\begin{array}{r|rrrrrrr}
j & 1 & -6 & 3 & -60 & -61 & -54 & -63 \\
  &   & j & -1-6j & 6+2j & -2-54j & 54-63j & 63 \\
\hline
-j & 1 & -6+j & 2-6j & -54+2j & -63-54j & -63j & 0 \\
  &   & -j & 6j & -2j & 54j & 63j & \\
\hline
-3j & 1 & -6 & 2 & -54 & -63 & 0 & \\
  &   & -3j & -9+18j & 54+21j & 63 & & \\
\hline
3j & 1 & -6-3j & -7+18j & 21j & 0 & & \\
  &   & 3j & -18j & -21j & & & \\
\hline
  & 1 & -6 & -7 & 0 & & &
\end{array}
$$

Since $x^2 - 6x - 7 = (x + 1)(x - 7)$, the roots are $j, -j, 3j, -3j, -1$, and 7.

22.   Use of synthetic division four times yields

$$
\begin{array}{r|rrrrrrr}
2j & 6 & -19 & 63 & -152 & 216 & -304 & 240 \\
   &   & 12j & -24-38j & 76+78j & -156-152j & 304+120j & -240 \\
\hline
-2j & 6 & -19+12j & 39-38j & -76+78j & 60-152j & 120j & 0 \\
   &   & -12j & 38j & -78j & 152j & 120j & \\
\hline
2j & 6 & -19 & 39 & -76 & 60 & 0 & \\
   &   & 12j & -24-38j & 76+30j & -60 & & \\
\hline
-2j & 6 & -19+12j & 15-38j & 30j & 0 & & \\
   &   & -12j & 38j & -30j & & & \\
\hline
   & 6 & -19 & 15 & 0 & & & \\
\end{array}
$$

Here $6x^2 - 19x + 15 = (3x - 5)(2x - 3)$, and so the roots are $2j, 2j, -2j, -2j, \frac{5}{3}$, and $\frac{3}{2}$.

≡ 18.3 RATIONAL ROOTS

1. $P(x) = x^3 - 3x - 2$ has one sign change, so there is one positive root. $P(-x) = -x^3 + 3x - 2$ has two sign changes so there are either 0 or 2 negative roots. $c = \pm1, \pm2; d = 1$, possible rational roots are ±1 and ±2.

2. $P(x) = x^3 - 2x^2 - x + 2$ has two sign changes so 0 or 2 positive roots. $P(-x) = -x^3 - 2x^2 + x + 2$ has one sign change so one negative root. $c = \pm1, \pm2; d = 1$, so possible rational roots are ±1 and ±2.

3. $P(x) = x^3 - 8x^2 - 17x + 6$ has two sign changes so 0 or 2 positive roots. $P(-x) = -x^3 - 8x^2 + 17x + 6$ has one sign change so 1 negative root. $c = \pm1, \pm2, \pm3, \pm6; d = 1$ so possible rational roots are $\pm1, \pm2, \pm3, \pm6$.

4. $P(x) = x^3 + 2x^2 - x - 2$ has one sign change so 1 positive root. $P(-x) = -x^3 + 2x^2 + x - 2$ has two sign changes so 0 or 2 negative roots. $c = \pm1, \pm2; d = 1$, so possible rational roots are ±1 and ±2.

5. $P(x) = x^3 - 2x^2 - 5x + 6$ has two sign changes so 0 or 2 positive roots. $P(-x) = -x^3 - 2x^2 + 5x + 6$ has one sign change so 1 negative root. $c = \pm1, \pm2, \pm3, \pm6; d = 1$ so possible rational roots are $\pm1, \pm2, \pm3, \pm6$.

6. $P(x) = 3x^3 - x^2 - 2x + 2$ has two sign changes so 0 or 2 positive roots. $P(-x) = 3x^3 - x^2 + 2x + 2$ has one sign change so one negative root. $c = \pm1, \pm2, d = 1, 3$. Possible rational roots are $\pm1, \pm2, \pm\frac{1}{3}, \pm\frac{2}{3}$.

7. $P(x) = 2x^4 - x^3 - 5x^2 + x + 3$ has two sign changes so 0 or 2 positive roots. $P(-x) = 2x^4 + x^3 - 5x - x + 3$ has 2 sign changes so 0 or 2 negative roots.

$c = \pm1, \pm3; d = 1, 2$. Possible rational roots are $\pm1, \pm3, \pm\frac{1}{2}, \pm\frac{3}{2}$.

8. $P(x) = 2x^4 - 7x^3 - 10x^2 + 33x + 18 = 0$ has two sign changes so there are either 0 or 2 positive roots. $P(-x) = 2x^4 + 7x^3 - 10x^2 - 33x + 18$ has two sign changes so there are 0 or 2 negative roots. $c = \pm1, \pm2, \pm3, \pm6, \pm9, \pm18; d = 1, 2$. Possible rational roots are $\pm1, \pm2, \pm3, \pm6, \pm9, \pm18, \pm\frac{1}{2}, \pm\frac{3}{2}, \pm\frac{9}{2}$.

9. $P(x) = 6x^4 - 7x^3 - 13x^2 + 4x + 4$ has two sign changes so 0 or 2 positive roots. $P(-x) = 6x^4 + 7x^3 - 13x^2 - 4x + 4$ has two sign changes so 0 or 2 negative roots. $c = \pm1, \pm2, \pm4; d = 1, 2, 3, 6$. Possible rational roots are $\pm1, \pm2, \pm4, \pm\frac{1}{2}, \pm\frac{1}{3}, \pm23, \pm\frac{4}{3}, \pm\frac{1}{6}$.

10. $P(x) = 5x^6 - x^5 - 5x^4 + 6x^3 - x^2 - 5x + 1 = 0$ has 4 sign changes so 0, 2 or 4 positive roots. $P(-x) = 5x^6 + x^5 - 5x^4 - 6x^3 - x^2 + 5x + 1$ has 2 sign changes so 0 or 2 negative roots. $c = \pm1; d = 1, 5$. Possible rational roots are $\pm1, \pm\frac{1}{5}$.

11.   Possible rational roots are $\pm1, \pm3, \pm5, \pm15$. $P(-x) = -x^3 - 9x^2 - 23x - 15$ so no negative roots

$$
\begin{array}{r|rrrr}
1 & 1 & -9 & 23 & -15 \\
  &   & 1 & -8 & 15 \\
\hline
  & 1 & -8 & 15 & 0 \\
\end{array}
$$

Since $x^2 - 8x + 15 = (x - 3)(x - 5)$, we see that the roots are 1, 3, 5.

12. Possible rational roots are $\pm1, \pm2, \pm3, \pm4, \pm6, \pm12$. It has 0 or 2 positive roots and one negative. Using synthetic division, we see that

$$\begin{array}{r|rrrr} 2 & 1 & -3 & -4 & 12 \\ & & 2 & -2 & -12 \\ \hline & 1 & -1 & -6 & 0 \end{array}$$

Since $x^2 - x - 6 = (x-3)(x+2)$ the roots are 2, 3, and -2.

13. Possible rational roots are ±1 and ±2. Using synthetic division, we obtain

$$\begin{array}{r|rrrr} -1 & 1 & 0 & -3 & -2 \\ & & -1 & 1 & 2 \\ \hline & 1 & -1 & -2 & 0 \end{array}$$

Since $x^2 - x - 2 = (x-2)(x+1)$, the roots are $-1, -1$, and 2.

14. Possible rational roots are $\pm1, \pm2, \pm3, \pm4, \pm6, \pm12$.

$$\begin{array}{r|rrrr} 2 & 1 & -1 & -8 & 12 \\ & & 2 & 2 & -12 \\ \hline & 1 & 1 & -6 & 0 \end{array}$$

Since $x^2 + x - 6 = (x-2)(x+3)$, the roots are 2, 2, and -3.

15. Possible roots are $\pm1, \pm2, \pm3, \pm4, \pm6, \pm12, \pm\frac{1}{2}, \pm\frac{3}{2}, \pm\frac{1}{4}$, and $\pm\frac{3}{4}$. Synthetic division produces

$$\begin{array}{r|rrrr} -\frac{3}{4} & 4 & -5 & 10 & 12 \\ & & -3 & 6 & -12 \\ \hline & 4 & -8 & 16 & 0 \end{array}$$

We see that $4x^2 - 8x + 16 = 4(x^2 - 2x + 4)$ and so $x = \frac{+2 \pm \sqrt{4-16}}{2} = 1 \pm j\sqrt{3}$. Roots are $-\frac{3}{4}, 1 + j\sqrt{3}$, and $1 - j\sqrt{3}$.

16. Possible rational roots are $\pm1, \pm2, \pm3, \pm6, \pm\frac{1}{3}, \pm\frac{2}{3}$. Synthetic division produces

$$\begin{array}{r|rrrr} 3 & 3 & -4 & -17 & 6 \\ & & 9 & 15 & -6 \\ \hline & 3 & 5 & -2 & 0 \end{array}$$

Since $3x^2 + 5x - 2 = (3x-1)(x+2)$, we see that the roots are $3, \frac{1}{3}$, and -2.

17. Possible rational roots are $\pm1, \pm2, \pm3, \pm4, \pm6, \pm12$.

$$\begin{array}{r|rrrrr} 2 & 1 & -4 & 7 & -12 & 12 \\ & & 2 & -4 & 6 & -12 \\ \hline 2 & 1 & -2 & 3 & -6 & 0 \\ & & 2 & 0 & 6 & \\ \hline & 1 & 0 & 3 & 0 & \end{array}$$

Thus, $x^2 + 3 = 0$ and so $x^2 = -3$ and $x = \pm j\sqrt{3}$. The roots are $2, 2, j\sqrt{3}$, and $-j\sqrt{3}$.

18. Possible rational roots are $\pm1, \pm2, \pm\frac{1}{2}, \pm\frac{1}{4}$. There are no positive roots.

$$\begin{array}{r|rrrrr} -\frac{1}{2} & 4 & 4 & 9 & 8 & 2 \\ & & -2 & -1 & -4 & -2 \\ \hline -\frac{1}{2} & 4 & 2 & 8 & 4 & 0 \\ & & -2 & 0 & -4 & \\ \hline & 4 & 0 & 8 & 0 & \end{array}$$

Here $4x^2 + 8 = 4(x^2 + 2)$. Solving $x^2 + 2 = 0$, we get $x^2 = -2$ and so $x = \pm j\sqrt{2}$. The roots are $-\frac{1}{2}, -\frac{1}{2}, j\sqrt{2}$, and $-j\sqrt{2}$.

19. Possible rational roots are $\pm1, \pm2, \pm3, \pm6, \pm\frac{1}{2}, \pm\frac{3}{2}, \pm\frac{1}{4}, \pm\frac{3}{4}$. There are either 0 or 2 positive roots, and 0 or 2 negative roots.

$$\begin{array}{r|rrrrr} -\frac{1}{2} & 4 & -20 & 1 & 18 & 6 \\ & & -2 & 11 & -6 & -6 \\ \hline -\frac{1}{2} & 4 & -22 & 12 & 12 & 0 \\ & & -2 & 12 & -12 & \\ \hline & 4 & -24 & 24 & 0 & \end{array}$$

Since $4x^2 - 24x + 24 = 4(x^2 - 6x + 6)$ we can use the quadratic formula to see that $x = \frac{6 \pm \sqrt{36-24}}{2} = \frac{6 \pm \sqrt{12}}{2} = 3 \pm \sqrt{3}$. Hence, the roots are $-\frac{1}{2}, -\frac{1}{2}, 3 + \sqrt{3}, 3 - \sqrt{3}$.

20. Possible rational roots are $\pm1, \pm3, \pm9, \pm27$. There are 1 or 3 positive roots and 0 or 2 negative roots. Using synthetic division three times, we get

$$\begin{array}{r|rrrrrr} 1 & 1 & 0 & -1 & 27 & 0 & -27 \\ & & 1 & 1 & 0 & 27 & 27 \\ \hline -1 & 1 & 1 & 0 & 27 & 27 & 0 \\ & & -1 & 0 & 0 & -27 & \\ \hline -3 & 1 & 0 & 0 & 27 & 0 & \\ & & -3 & 9 & -27 & & \\ \hline & 1 & -3 & 9 & 0 & & \end{array}$$

$\rightarrow$ no more negative roots

We have $x^2 - 3x + 9 = 0$ and so, $x = \frac{3 \pm \sqrt{9-36}}{2} = \frac{3 \pm 3j\sqrt{3}}{2}$. The roots are $1, -1, -3, \frac{3}{2} + \frac{3j\sqrt{3}}{2}$, and $\frac{3}{2} - \frac{3j\sqrt{3}}{2}$.

21. Possible rational roots are $\pm1, \pm2, \pm5, \pm10$. There are no rational roots.

22. Possible rational roots are $\pm1, \pm3, \pm9, \pm27, \pm\frac{1}{2}$, $\pm\frac{3}{2}, \pm\frac{9}{2}, \pm\frac{27}{2}, \pm\frac{1}{4}, \pm\frac{3}{4}, \pm\frac{9}{4}, \pm\frac{27}{4}, \pm\frac{1}{8}, \pm\frac{3}{8}, \pm\frac{9}{8}, \pm\frac{27}{8}$. $P(-x) = -8x^3 - 36x^2 - 54x - 27$ so there are no negative roots. Using synthetic division we get

$$\frac{3}{2} \,\Big|\; \begin{array}{cccc} 8 & -36 & 54 & -27 \\ & 12 & -36 & 27 \\ \hline 8 & -24 & 18 & 0 \end{array}$$

The depressed equation is $8x^2 - 24x + 18 = 0$. This factors as $2(2x - 3)(2x - 3) = 0$. From the synthetic division and the factoring, we see that the roots are $\frac{3}{2}, \frac{3}{2}$ and $\frac{3}{2}$ again. [Thus, we say that $\frac{3}{2}$ is a triple root.]

23. $P(-x) = -2x^5 - 13x^4 - 26x^3 - 22x^2 - 24x - 9$ so no sign changes so no negative roots. Possible rational roots are $1, 3, 9, \frac{1}{2}, \frac{3}{2}$, and $\frac{9}{2}$. Using synthetic division three times, we get

$$\begin{array}{c|cccccc} 3 & 2 & -13 & 26 & -22 & 24 & -9 \\ & & 6 & -21 & 15 & -21 & 9 \\ \hline 3 & 2 & -7 & 5 & -7 & 3 & 0 \\ & & 6 & -3 & 6 & -3 & \\ \hline \frac{1}{2} & 2 & -1 & 2 & -1 & 0 & \\ & & 1 & 0 & 1 & & \\ \hline & 2 & 0 & 2 & 0 & & \end{array}$$

We have $2x^2 + 2 = 0$ or $2x^2 = -2$ or $x^2 = -1$ and so, $x = \pm j$. The roots are $\frac{1}{2}, 3, 3, j$, and $-j$.

24. Possible rational roots are $\pm1, \pm\frac{1}{2}, \pm\frac{1}{4}, \pm\frac{1}{8}$

$$\frac{1}{2} \,\Big|\; \begin{array}{cccccc} 8 & -4 & 6 & -3 & -2 & 1 \\ & 4 & 0 & 3 & 0 & -1 \\ \hline 8 & 0 & 6 & 0 & -2 & 0 \end{array}$$

Now, $8x^4 + 6x^2 - 2 = (8x^2 - 2)(x^2 + 1)$. We see that $8x^2 - 2 = 0$ or $x^2 = \frac{1}{4}$, which produces $x = \pm\frac{1}{2}$. We also see that $x^2 + 1 = 0$ or $x^2 = -1$, and so $x = \pm j$. The roots are $\frac{1}{2}, \frac{1}{2}, -\frac{1}{2}, j$, and $-j$.

25. Possible rational roots are $\pm1, \pm2, \pm3, \pm5, \pm6, \pm10$, $\pm15, \pm30$. Using synthetic division three times, we get

$$\begin{array}{c|cccccc} 3 & 1 & -5 & 6 & 11 & -43 & 30 \\ & & 3 & -6 & 0 & 33 & -30 \\ \hline 1 & 1 & -2 & 0 & 11 & -10 & 0 \\ & & 1 & -1 & -1 & 10 & \\ \hline -2 & 1 & -1 & -1 & 10 & 0 & \\ & & -2 & 6 & -10 & & \\ \hline & 1 & -3 & 5 & 0 & & \end{array}$$

We have $x^2 - 3x + 5 = 0$ and so, $x = \frac{3 \pm \sqrt{9-20}}{2} = \frac{3 \pm j\sqrt{11}}{2}$. The roots are $1, 3, -2, \frac{3}{2} + \frac{\sqrt{11}}{2}j$, and $\frac{3}{2} - \frac{\sqrt{11}}{2}j$.

26. Possible rational roots are $\pm1, \pm2, \pm4, \pm5, \pm8, \pm20$, $\pm40, \pm\frac{1}{2}, \pm\frac{5}{2}, \pm10$. There are either 0 or 2 positive roots and 1 or 3 negative roots.

$$\frac{5}{2} \,\Big|\; \begin{array}{cccccc} 2 & -1 & -6 & -18 & 4 & 40 \\ & 5 & 10 & 10 & -20 & -40 \\ \hline 2 & 4 & 4 & -8 & -16 & 0 \end{array}$$

Thus, $\frac{5}{2}$ is the only rational root. The other roots are $\pm\sqrt{2}$ and $-1 \pm j\sqrt{3}$ but, at present, we have no way to use algebra to find them.

27. From the given information, we have the equation $(8 - 2x)(10 - 2x)x = 48$ where $0 < x < 4$. Expanding this equation produces $4x^3 - 36x^2 + 80x - 48 = 0$. Dividing both sides of this equation by 4 we get the equation $x^3 - 9x^2 + 20x - 12 = 0$. Possible rational roots are $1, 2$, and 3. (Notice, that because $0 < x < 4$, we can ignore any possible negative roots.) Using synthetic division, we get

$$1 \,\Big|\; \begin{array}{cccc} 1 & -9 & 20 & -12 \\ & 1 & -8 & 12 \\ \hline 1 & -8 & 12 & 0 \end{array}$$

The depressed equation factors: $x^2 - 8x + 12 = (x - 2)(x - 6) = 0$. Solutions to the equation are $1, 2$ and 6. Solutions to the problem are 1 in. and 2 in.

28. If we let x = height, then the width = $x + 2.5$ and length = $x + 6.5$. Multiplying these we see that Volume = $x(x + 2.5)(x + 6.5) = 210$ or $x^3 + 9x^2 + 16.25x - 210 = 0$. Multiplying by 4, we can eliminate the fractional parts, and work with the easier equation $4x^3 + 36x^2 + 65x - 840 = 0$. We know that x must be positive and by trial, we determine that 4 is too large. Since $\frac{7}{2}$ is a possible solution less than 4, we will use synthetic division to test and see if it is a root. Using synthetic division, we get

$$\frac{7}{2} \,\Big|\; \begin{array}{cccc} 4 & 36 & 65 & -840 \\ & 14 & 175 & 840 \\ \hline 4 & 50 & 240 & 0 \end{array}$$

Thus, $\frac{7}{2} = 3.5$ is a solution. Since the original equation has only one sign change, there are no other positive solutions. Height = 3.5 cm, width = 6 cm, length = 10 cm.

29. The volume of a cylinder is $\pi r^2 h$ and the volume of a hemisphere is one-half the volume of a sphere or $\frac{1}{2}\frac{4}{3}\pi r^3 = \frac{2}{3}\pi r^3$. Since the total height of the silo is 34 ft, and the radius or the hemisphere is r ft, the height of the cylinder is $34 - r$ ft. Thus, the total volume of this grain silo is $V = \pi r^2 h + \frac{2}{3}\pi r^3 = \pi r^2(34 - r) + \frac{2}{3}\pi r^3 = 2511\pi$. Expanding, and dividing by π, we obtain $34r^2 - r^3 + \frac{2}{3}r^3 = 2511$. Collecting terms we get $-\frac{1}{3}r^3 + 34r^2 - 2511 = 0$ or $r^3 - 102r^2 + 7533 = 0$. Now, $7533 = 3^5 \cdot 31$ so factors include 1, 3, 9, 27, and 31. Synthetic division shows that 9 is a solution

$$
\begin{array}{r|rrrr}
9 & 1 & -102 & 0 & 7533 \\
  &   & 9 & -837 & -7533 \\
\hline
  & 1 & -93 & -837 & 0
\end{array}
$$

The radius is 9 ft.

30. As in the previous exercise, $V = \pi r^2 h + \frac{4}{3}\pi r^3 = 6\pi r^2 + \frac{4}{3}\pi r^3 = 18\pi$. Dividing this equation by π, we obtain $6r^2 + \frac{4}{3}r^3 + 18$ or $4r^3 + 18r^2 - 54 = 0$. Using synthetic division, we get

$$
\begin{array}{r|rrrr}
\frac{3}{2} & 4 & 18 & 0 & -54 \\
  &   & 6 & 36 & 54 \\
\hline
  & 4 & 24 & 36 & 0
\end{array}
$$

There are no other positive solutions, so the radius is $\frac{3}{2}$ or 1.5 m.

31. Multiplying $\frac{1}{R} + \frac{1}{R+4} + \frac{1}{R+1} = 1$ by the common denominator $R(R + 1)(R + 4)$ yields

$$(R + 1)(R + 4) + R(R+1) + R(R + 4)$$
$$= R(R + 1)(R + 4);$$
$$R^2 + 5R + 4 + R^2 + R + R^2 + 4R$$
$$= R^3 + 5R^2 + 4R;$$
$$3R^2 + 10R + 4 = R^3 + 5R^2 + 4R;$$
$$R^3 + 2R^2 - 6R - 4 = 0$$

Using synthetic division, produces

$$
\begin{array}{r|rrrr}
2 & 1 & 2 & -6 & -4 \\
  &   & 2 & 8 & 4 \\
\hline
  & 1 & 4 & 2 & 0
\end{array}
$$

There are no other positive solutions, and so we have $R = 2, R + 4 = 6$ and $R + 1 = 5$.

32. Multiplying $\frac{1}{R} + \frac{1}{R+4} + \frac{1}{R+9} = \frac{1}{3}$ by the common denominator $3R(R + 4)(R + 9)$ produces

$$3(R + 4)(R + 9) + 3R(R + 9) + 3(R)(R + 4)$$
$$= R(R + 4)(R + 9);$$
$$3(R^2 + 13R + 36) + 3R^2 + 27R + 3R^2 + 12R$$
$$= R(R^2 + 13R + 36);$$
$$9R^2 + 78R + 108 = R^3 + 13R^2 + 36R;$$
$$R^3 + 4R^2 - 42R - 108 = 0$$

Using synthetic division to divide by $x - 9$, we obtain

$$
\begin{array}{r|rrrr}
9 & 1 & 4 & -42 & -108 \\
  &   & 9 & 117 & 675 \\
\hline
  & 1 & 13 & 75 & 576
\end{array}
$$

From this we can see that 9 is not a factor but that it is an upper bound.

$$
\begin{array}{r|rrrr}
6 & 1 & 4 & -42 & -108 \\
  &   & 6 & 60 & 108 \\
\hline
  & 1 & 10 & 18 & 0
\end{array}
$$

This shows that 6 is a root. There are no other postitive solutions, and we get $R = 6, R+4 = 10$, and $R + 9 = 15$.

33. If the side of the square that is cut from each corner has length x, then the volume of the box is given by

$$(1 - 2x)(19 - 2x)x = 210$$
$$(228 - 62x + 4x^2)x = 210$$
$$4x^3 - 62x^2 + 228x - 210 = 0$$
$$2x^3 - 31x^2 + 114x - 105 = 0$$

Now, x must be positive and the realistic domain is $0 < x < 6$. The possible rational solutions include $1, 5, 7, \frac{1}{2}, \frac{5}{2}$. and $\frac{7}{2}$.

$$
\begin{array}{r|rrrr}
\frac{7}{2} & 2 & -31 & 114 & -105 \\
  &   & 7 & -84 & 105 \\
\hline
  & 2 & -24 & 30 & 0
\end{array}
$$

Thus, $\frac{7}{2}$ is a root. The depressed polynomial $2x^2 - 24x + 30$ has roots at $x = \frac{24 \pm \sqrt{(-24)^2 - 4(2)(30)}}{2(2)} = \frac{24 \pm \sqrt{336}}{4} = \frac{24 \pm 4\sqrt{21}}{4} = 6 \pm \sqrt{21}$. There are no other real solutions. Since $6 + \sqrt{21} \approx 10.583$ is not in the realistic domain, it is not a solution. So the only solutions are $x = \frac{5}{2} = 3.5$ cm or $6 - \sqrt{21} \approx 1.417$ cm.

34. $M(d) = 0.1d^4 - 2.2d^3 + 15.2d^2 - 32d$
    $10M(d) = d^4 - 22d^3 + 152d^2 - 320d$
    $\qquad\quad = d(d^3 - 22d^2 + 152d - 320)$
    Thus, 0 is one root. Possible rational roots include 1, 2, 4, 5, 8, 10, 16, ...

$$
\begin{array}{r|rrr}
4 & 1 & -22 & 152 & -320 \\
  &   & 4   & -72 & 320 \\
\hline
8 & 1 & -18 & 80 & 0 \\
  &   & 8   & -80 & \\
\hline
10 & 1 & -10 & 0 & \\
  &   & 10 & & \\
\hline
  & 1 & 0 & &
\end{array}
$$

Solutions are 0, 4, 8 and 10 meters.

 ## 18.4 IRRATIONAL ROOTS

1. $P(x) = x^3 + 5x - 3$, $P(0) = -3$, and $P(1) = 3$ so

a	$P(a)$	b	$P(b)$	$c = a - \dfrac{P(a)}{m}$	$P(c)$
0	−3	1	3	0.5	−0.375
0	−3	0.5	−0.375	0.5714	0.0437
0	−3	0.5714	0.0437	0.5632	−0.0052
0	−3	0.5632	−0.0052	0.5642	0.0006

Thus, $c \approx 0.56$

2. $P(x) = x^3 - 3x + 1$, $P(1) = -1$, and $P(2) = 3$

a	$P(a)$	b	$P(b)$	$c = a - \dfrac{P(a)}{m}$	$P(c)$
1	−1	2	3	1.25	−0.7969
1	−1	1.25	−0.7969	2.2308	5.4

The method fails because 1 is not close enough. We try $P(1.5) = -0.125$, with the following results:

a	$P(a)$	b	$P(b)$	$c = a - \dfrac{P(a)}{m}$	$P(c)$
1.5	−0.125	2	3	1.52	−0.04819
1.5	−0.125	1.52	−0.04819	1.5325	−0.00186
1.5	−0.125	1.5325	0.00186	1.5321	−0.00007

Thus, $c \approx 1.53$.

3. Let $P(x) = x^4 + 2x^3 - 5x^2 + 1$, $P(0) = 1$ and $P(1) = -1$, with the following results:

a	$P(a)$	b	$P(b)$	$c = a - \dfrac{P(a)}{m}$	$P(c)$
0	1	1	-1	0.5	0.0625
0	1	0.5	0.0625	0.5333	-0.379
0	1	0.5333	-0.0379	0.5139	0.0208
0	1	0.5139	0.0208	0.5248	-0.0121
0	1	0.5248	-0.0121	0.5185	0.0068
0	1	0.5185	0.0068	0.5221	-0.0039

Thus, $c \approx 0.52$.

4. Here we let $P(x) = x^4 + 2x^3 - 5x^2 + 1$. We see that $P(1) = -1$ and $P(2) = 13$ and get the following result:

a	$P(a)$	b	$P(b)$	$c = a - \dfrac{P(a)}{m}$	$P(c)$
1	-1	2	13	1.0714	-0.9621
1	-1	1.0714	-0.962	2.883	76.49

The method fails, so we try $a = 1.2$, with the following results:

a	$P(a)$	b	$P(b)$	$c = a - \dfrac{P(a)}{m}$	$P(c)$
1.2	-0.6704	2	13	1.239	-0.5139
1.2	-0.6704	1.239	-0.5139	1.368	0.2662
1.2	-0.6704	1.368	0.2662	1.3203	-0.0740
1.2	-0.6704	1.3203	-0.0740	1.3352	0.02549
1.2	-0.67045	1.3352	0.02549	1.3303	-0.0082
1.2	-0.67045	1.3303	-0.0082	1.3319	0.0027

Thus, $c \approx 1.33$.

5. Let $P(x) = x^3 + x^2 - 7x + 3$, $P(0) = 3$, and $P(1) = -2$. We get

a	$P(a)$	b	$P(b)$	$c = a - \dfrac{P(a)}{m}$	$P(c)$
0	3	1	-2	0.6	-0.624
0	3	0.6	-0.624	0.4967	-0.1076
0	3	0.4967	-0.1076	0.4795	-0.0163
0	3	0.4795	-0.0163	0.4769	-0.0024
0	3	0.4769	-0.0024	0.4765	-0.00036

Thus, $c \approx 0.48$.

6. Here $P(x) = x^3 + x^2 - 7x + 3$, $P(1) = -2$, and $P(2) = 1$. We get

a	$P(a)$	b	$P(b)$	$c = a - \dfrac{P(a)}{m}$	$P(c)$
1	-2	2	1	1.6667	-1.2593
1	-2	1.6667	-1.2593	2.8	13.192

The method fails, so we try $a = 1.8$, with the following results:

a	$P(a)$	b	$P(b)$	$c = a - \dfrac{P(a)}{m}$	$P(c)$
1.8	-0.528	2	1	1.8691	-0.0603
1.8	-0.528	1.8691	-0.0603	1.8780	0.00455
1.8	-0.528	1.8780	0.00455	1.8774	-0.00034

Thus, $c \approx 1.88$.

7. $P(x) = x^4 - x - 2$. Since $P(-1) = 0$, one root is at -1. Using synthetic division we get the depressed equation $D(x) = x^3 - x^2 + x - 2$. Since $D(1) = -1$ and $D(2) = 4$, there is a real root between 1 and 2.

a	$P(a)$	b	$P(b)$	$c = a - \dfrac{P(a)}{m}$	$P(c)$
1	-1	2	4	1.2	-0.512
1.2	-0.512	2	4	1.29	-0.2274
1.29	-0.2274	2	4	1.328	-0.0935
1.328	-0.0935	2	4	1.343	-0.03835
1.343	-0.03835	2	4	1.349	-0.0159
1.349	-0.0159	2	4	1.352	-0.0046

Thus, $c \approx 1.35$ and we have found an approximation of the second real root. The other two roots are complex numbers.

8. $x^5 - 2x^2 + 4 = 0$; Root is -1.10

9. Letting $P(x) = x^4 + x^3 - 2x^2 - 7x - 5$, we see that $P(-1) = 0$, so -1 is a root. Using synthetic division, we obtain

$$
\begin{array}{r|rrrrr}
-1 & 1 & 1 & -2 & -7 & -5 \\
   &   & -1 & 0 & 2 & 5 \\
\hline
   & 1 & 0 & -2 & -5 & 0
\end{array}
$$

This gives the depressed equation $D(x) = x^3 - 2x - 5$. Since $D(2) = -1$ and $D(3) = 16$, we see that there is a root between 2 and 3.

a	$P_1(a)$	b	$P_1(b)$	$c = a - \dfrac{P_1(a)}{m}$	$P_1(c)$
2	-1	3	16	2.0588	-0.39105
2.0588	-0.39105	3	16	2.081	-0.1501
2.081	-0.1501	3	16	2.0895	-0.5622
2.0895	-0.05622	3	16	2.0927	-0.02064

Thus, $c \approx 2.09$ is a second root. The other two roots are complex numbers.

10. Here $P(x) = x^4 - 2x^3 - 3x + 4$ and, since $P(1) = 0$, 1 is a root. Synthetic division gives the depressed equation $D(x) = x^3 - x^2 - x - 4$. Since $D(2) = -2$ and $D(3) = 11$, we know that there is a real root between -2 and 3. Using linear interpolation, we see that this root is approximately 2.24. There are no other real roots.

11. Here $P(x) = 2x^4 + 3x^3 - x^2 - 2x - 2$. $P(1) = 0$ and synthetic division gives the depressed equation $D(x) = 2x^3 + 5x^2 + 4x + 2$. Since $D(-2) = -2$ and $D(-1) = 1$, we see that there is a root between -2 and -1. Linear interpolation shows that this root is approximately -1.66.

12. Here $P(x) = 3x^4 + 3x^3 + x^2 + 4x + 3$. $P(-1) = 0$ and synthetic division gives the depressed equation $D(x) = 3x^3 + x + 3$. Since $D(-1) = -1$ and $D(0) = 3$, we see that there is a root between -1 and 0. Linear interpolation shows that this root is approximately -0.89.

13. Here $P(x) = 2x^5 - 5x^3 + 2x^2 + 4x - 1$. $P(-1) = 0$ and synthetic division gives the depressed equation $D(x) = 2x^4 - 2x^3 - 3x^2 + 5x - 1$. Since $D(-2) =$

25, $D(-1) = -5$, $D(0) = -1$, and $D(1) = 1$, we see that there is a root between -2 and -1 and another root between 0 and 1. Linear interpolation shows that these roots are approximately -1.43 and 0.24.

14. Here $P(x) = x^5 + 3x^4 - 5x^3 - 2x^2 + x + 2$. $P(1) = 0$ and synthetic division gives the depressed equation $D(x) = x^4 + 4x^3 - x^2 - 3x - 2$. Since $D(-5) = 113$, $D(-4) = -6$, $D(1) = -1$, and $D(2) = 36$, we see that there is a root between -5 and -4 and another root between 0 and 1. Linear interpolation shows that these roots are approximately -4.09 and 1.08.

15. If we let $P(x) = 8x^4 + 6x^3 - 15x^2 - 12x - 2$, we can see that there are rational roots at $-\frac{1}{2}$ and $-\frac{1}{4}$. Using synthetic division with these factors, we get the depressed equation $D(x) = 8x^2 - 16 = 8(x^2 - 2)$. This depressed equation has roots at $\pm\sqrt{2}$. Thus, this equation has the four real roots $-\frac{1}{2}, -\frac{1}{4}, -1.41$, and 1.41.

16. If we let $P(x) = 9x^4 + 15x^3 - 20x^2 + 16 = 0$ we can see that there are rational roots at 1 and -2. Using synthetic division, we get the depressed equation $D(x) = 9x^2 + 6x - 8 = (3x+4)(3x-2)$ and so $x = -\frac{4}{3}$ and $x = \frac{2}{3}$. Thus, the roots are $1, -2, -\frac{4}{3}$, and $\frac{2}{3}$.

17. Here we have $P(t) = 150t - 20t + t^3$. We want to know the values of t when $P(t) = 400$. Thus, we want the solution to $P(t) = 150t - 20t^2 + t^3 - 400 = 0$. Since $P(6) = -4$ and $P(7) = 13$, we see that this will be between 6 and 7. Using linear interpolation, we obtain

a	$P(a)$	b	$P(b)$	$c = a - \dfrac{P(a)}{m}$	$P(c)$
6	-4	7	-13	6.23529	0.13759
6	-4	6.23529	0.13759	6.22747	0.002736
6	-4	6.22747	0.002736	6.22731	0.0000546
6	-4	6.22731	0.0000546	6.22731	0.000001

Thus, $c \approx 6.22731$ and so it will take about 6 years 83 days.

18. Let $x =$ thickness of the tank in feet. Then, $h = 12 - 2x$ and $r = 4.5 - x$. Using the formula for the volume of a cylinder, we obtain

$$V = \pi r^2 h = \pi (4.5 - x)^2 (12 - x) = 674$$
$$(4.5 - x)^2 (12 - 2x) = 214.54$$
$$(20.25 - 9x + x^2)(6 - x) = 107.27$$
$$-x^3 + 15x^2 - 74.25x + 121.5 = 107.27$$
$$x^3 - 15x^2 + 74.25x - 14.23 = 0$$

Since $V = -14.23$ when $x = 0$ and $V = 46.02$ when $x = 1$, we have a solution between 0 and 1. Using linear interpolation, we get

a	$V(a)$	b	$V(b)$	$c = a - \dfrac{V(a)}{m}$	$V(c)$
0	-14.23	1	46.02	0.23618	2.4830
0	-14.23	0.23618	2.4830	0.2011	0.10276
0	-14.23	0.2011	0.10276	0.19965	0.00420
0	-14.23	0.19965	0.0042	0.19959	0.00017

Thus, $x \approx 0.1996 \approx 0.20$ ft $= 2.4$ in.

19. **(a)** Since $h = 2r$, we have the following

$$V = \pi (r + 15.5)^2 (2r + 31) = 1\ 000\ 000\pi$$
$$(r^2 + 31r + 240.25)(2r + 31) = 1\ 000\ 000$$
$$2r^3 + 93r^2 + 1\ 441.5r + 7\ 447.75 = 1\ 000\ 000$$
$$2r^3 + 93r^2 + 1\ 441.5r - 992\ 552.25 = 0$$

Using linear interpolation, we get

a	$V(a)$	b	$V(b)$	$c = a - \dfrac{V(a)}{m}$	$V(c)$
63	-32526	64	4919	63.869	-54.17
63	-32526	63.869	-54.177	63.870	0.5983
63	-32526	63.870	0.5983	63.870	-0.0066

Thus, $r \approx 63.870$ cm.

(b) $h = 2r$, $V = \pi(63.87)^2(2 \cdot 63.87) = 521\ 100\pi$ cm$^3 = 1\ 637\ 000$ cm^3.

20. Letting $P(S) = S^3 - 6S^2 - 78S + 108$, we get

a	$P(a)$	b	$P(b)$	$c = a - \dfrac{P(a)}{m}$	$P(c)$
1	25	2	−64	1.2809	0.3472
1	25	1.2809	0.3472	1.2849	−0.00274
1	25	1.2849	− 0.00274	1.2848	0.000022

Thus, $S \approx 1.2848$ psi.

21. R is an even function so all solutions come in pairs. Evaluating R at various points, we see that $R(0.3) = -0.675$; $R(0.5) = 1.536$; $R(0.8) = 0.864$; $R(0.9) = -0.509$; $R(1.4) = -1.278$; $R(1.5) = 0.700$; $R(1.7) = 1.0183$; $R(1.8) = -3.561$; $R(2.2) = -13.699$; and $R(2.3) = 51.874$. By the locator theorem, there are roots between 0.3 and 0.5, between 0.8 and 0.9, between 1.4 and 1.5, between 1.7 and 1.8 as well as between 2.2 and 2.3.

Linear interpolation for the one between 1.7 and 1.8 is as follows

a	$R_1(a)$	b	$R_1(b)$	$c = a - \dfrac{R_1(a)}{m}$	$R_1(c)$
1.7	1.0183	1.8	−3.5614	1.7222	0.3546
1.7	1.0183	1.7222	0.3546	1.7341	−0.0796
1.7	1.0183	1.7341	−0.0796	1.7316	0.01548
1.7	1.0183	1.7316	0.01548	1.7321	−0.0031
1.7	1.0183	1.7321	−0.0031	1.7320	0.00062
1.7	1.0183	1.7320	0.00062	1.7321	−0.0001
1.7	1.0183	1.7321	−0.00012	1.7321	0.000025

The approximate roots are ± 0.3536, ± 0.8654, ± 1.4639, ± 1.7321, and ± 2.2323.

22. See *Computer Programs* in main text.

23. We want to determine the value of t for which $-0.009t^5 + t^3 + 5.5 = 32.6$ or $-0.009t^5 + t^3 - 27.1 = 0$. Using a graphing calculator we estimate one root at 3.095 h or 3 h 5.7 min and the other root is at 10.414 h or about 10 h 24.8 min.

18.5 RATIONAL FUNCTIONS

1. Vertical asymptote: $x = -2$; Horizontal asymptote:
$y = 0$

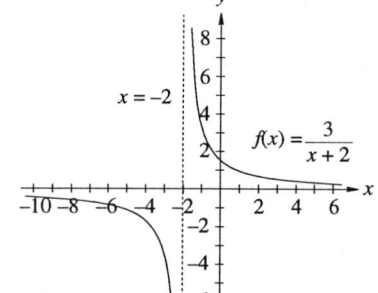

$f(x) = \dfrac{3}{x+2}$

2. Vertical asymptote: $x = 3, x = -3$; Horizontal asymptote: $y = 0$

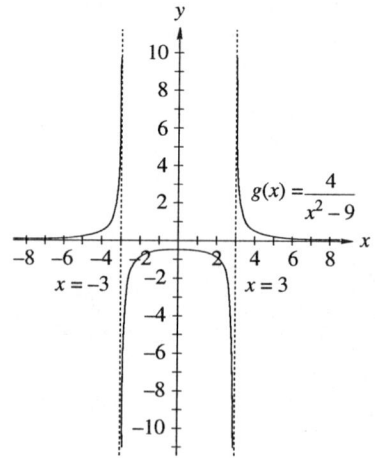

$g(x) = \dfrac{4}{x^2 - 9}$

3. Vertical asymptote: $x = -4$; Horizontal asymptote: $y = 2$

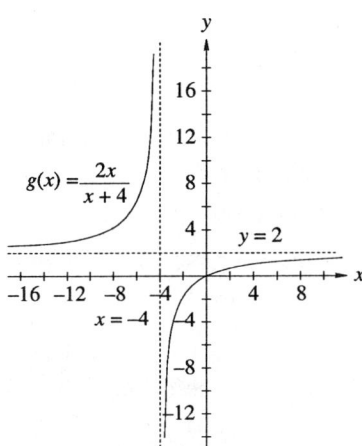

4. $\frac{5x}{x^2+x-6} = \frac{5x}{(x+3)(x-2)}$, Vertical asymptotes: $x = -3$ and $x = 2$; Horizontal asymptote: $y = 0$

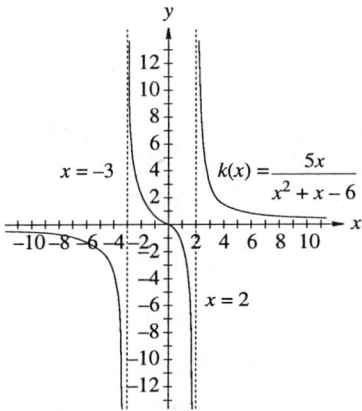

5. $\frac{3x^2}{x^2+6x+8} = \frac{3x^2}{(x+4)(x+2)}$, Vertical asymptotes: $x = -2$ and $x = -4$; Horizontal asymptote: $y = 3$

6. $\frac{x^2-4}{x^2-16} = \frac{(x+2)(x-2)}{(x+4)(x-4)}$; Vertical asymptotes: $x = -4$ and $x = 4$; Horizontal asymptote: $y = 1$

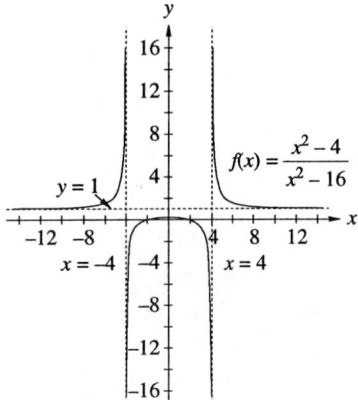

7. $\frac{x(x+1)}{(x+1)(x+2)}$; Vertical asymptote: $x = -2$; Horizontal asymptote: $y = 1$; (Note: hole at $x = -1$)

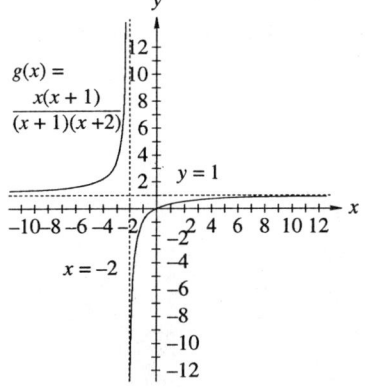

8. $\frac{x^2+1}{2x^2}$; Vertical asymptote: $x = 0$; Horizontal asymptote: $y = \frac{1}{2}$

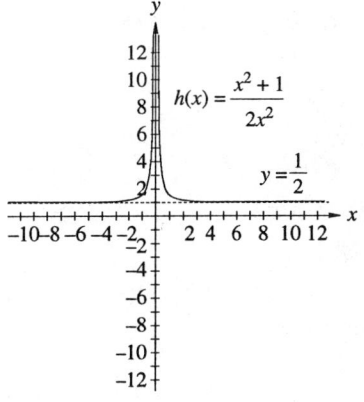

9. $\frac{4x^2-12x-27}{x^2-6x+8} = \frac{(2x-9)(2x+3)}{(x-2)(x-4)}$. Vertical asymptotes: $x = 2$ and $x = 4$; Horizontal asymptote: $y = 4$

10. $\frac{4x^2-9}{2x^2+7x+3} = \frac{(2x+3)(2x-3)}{(2x+1)(x+3)}$. Vertical asymptotes: $x = -\frac{1}{2}$ and $x = -3$; Horizontal asymptote: $y = \frac{4}{2} = 2$

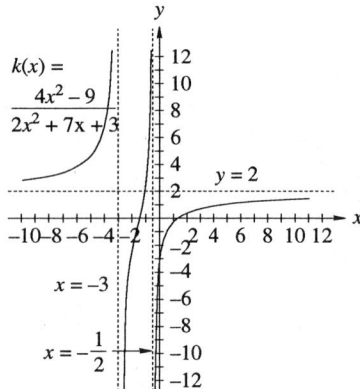

11. $\frac{(x+2)(x-3)}{(x+5)(x+4)} = 0 \Rightarrow (x+2)(x-3) = 0; x = -2$ or $x = 3$

12. $\frac{x^2-2x-1}{x(x+1)} \Rightarrow x^2 - 2x - 1 = 0; \frac{2\pm\sqrt{4+4}}{2} = 1 \pm \sqrt{2}, x = 1 + \sqrt{2}$ or $x = 1 - \sqrt{2}$

13. $\frac{x^2-1}{x^2+2x+1} = \frac{(x+1)(x-1)}{(x+1)^2} = 0; x \neq -1$ so $x = 1$ is the only solution.

14. $\frac{3x-4}{6x^2-5x-4} = 0; \frac{3x-4}{(3x-4)(2x+1)}$; Since x cannot equal $\frac{4}{3}$, there is no solution.

15. The given equation $\frac{4}{x-3} = \frac{2}{x+1}$. Subtracting, we get $\frac{4}{x-3} - \frac{2}{x+1} = 0$. This has a common denominator of $(x-3)(x+1)$. Multiplying the equation by this common denominator produces $\frac{4(x+1)-2(x-3)}{(x-3)(x+1)} = 0$ which simplifies to $\frac{4x+4-2x+6}{(x-3)(x+1)} = 0$ or $\frac{2x+10}{(x-3)(x+1)} = 0$. This has the solution $2x + 10 = 0$ or $x = -5$.

16. Subtracting, we get $\frac{5}{x+7} - \frac{2}{x-8} = 0$ and multiplying this equation by the common denominator $(x + 7)(x - 8)$ produces $\frac{5(x-8)-2(x+7)}{(x+7)(x-8)} = 0$. This can be simplified as $\frac{5x-40-2x-14}{(x+7)(x-8)} = 0$ and it has the solution $3x - 54 = 0$ or $3x = 54$ and so, we see that $x = 18$.

17. Subtracting $\frac{2}{3}$ from both sides of the equation produces $\frac{5}{2x} - \frac{3}{4x} - \frac{2}{3} = 0$. Finding a common denominator, we get $\frac{6\cdot5-3\cdot3-2\cdot4x}{12x} = 0$ or $\frac{30-9-8x}{12x} = 0$. Any solutions will be when $30 - 9 - 8x = 0$ or $21 - 8x = 0$ which means that $x = \frac{21}{8}$.

18. Subtracting 2 produces $\frac{2x-3}{x+5} - 2 = 0$ or $\frac{2x-3}{x+5} - \frac{2(x+5)}{x+5} = 0$ which is equivalent to $\frac{2x-3-2x-10}{x+5} = 0$. This simplifies to $\frac{-13}{x+5}$. Since the numerator can never be 0, there is no solution.

19. Finding that $(x + 2)(x - 2) = x^2 - 4$ is a common denominator for $\frac{3}{x-2} + \frac{5}{x+2} = \frac{20}{x^2-4}$, we get $\frac{3(x+2)}{x^2-4} + \frac{5(x-2)}{x^2-4} - \frac{20}{x^2-4} = 0$ or $\frac{3x+6+5x-10-20}{x^2-4} = 0$. The numerator simplifies to $8x - 24$ and is 0 only when $x = 3$.

20. $\frac{4}{x+3} - \frac{6}{x-3} = \frac{10}{x^2-9}$ is equivalent to $\frac{4(x-3)-6(x+3)-10}{x^2-9} = 0$. The numerator $4x - 12 - 6x - 18 - 10 = 0$, and hence the given equation has a solution, only when $x = -20$.

21. If we let $P(t) = 3t^4 + 2t^3 - 300t - 50$, we see that $P(4.4) = -75.20$ and $P(4.5) = 12.4375$. Using linear interpolation, we get

a	$P(a)$	b	$P(b)$	$c = a - \dfrac{P(a)}{m}$	$P(c)$
4.4	−75.20	4.5	12.4375	4.4858	−.469
4.4	−75.20	4.4858	−.469	4.4863	+0.01787
4.4	−75.20	4.4863	0.01787	4.4863	−0.00068

Thus, $t \approx 4.49$ hours.

22. $C\frac{L}{r^4} = R$ and so $r^4 = \frac{CL}{R}$. Substituting the given values, we see that $r^4 = \frac{1\cdot100}{6.25} = 16$, and thus, $r = \sqrt[4]{16} = 2$ mm.

23. As x gets very large, the fraction $\frac{7{,}250}{x}$ gets close to 0. Hence, the lowest cost is $73 + 0 = 73$ or \$73/set.

24. As t gets very large, $\frac{70t^2}{t^2+15}$ approaches 70. Hence, $W(t) = 65 + 70 = 135$ and the most words per minute that we can expect this student to keyboard is 135 words per miniute.

≡ CHAPTER 18 REVIEW

1. $P(2) = 9\cdot2^2 - 4\cdot2 + 2 = 36 - 8 + 2 = 30$, degree $= 2$

2. $P(3) = 2\cdot3^3 - 4\cdot3^2 + 7 = 54 - 36 + 7 = 25$, degree $= 3$

3. $P(-1) = 4(-1)^3 - 5(-1)^2 + 7(-1) + 3 = -4 - 5 - 7 + 3 = -13$, degree $= 3$

4. $P(-3) = 5(-3)^2 + 6(-3) - 3 = 45 - 18 - 3 = 24$, degree $= 2$

5.
$$\begin{array}{r|rrr} -3 & 4 & -6 & -3 \\ & & -12 & 54 \\ \hline & 4 & -18 & 51 \end{array}$$
So, $Q(x) = 4x - 18; R = 51$

6.
$$\begin{array}{r|rrrr} 1 & 6 & 0 & -2 & 1 \\ & & 6 & 6 & 4 \\ \hline & 6 & 6 & 4 & 5 \end{array}$$
So, $Q(x) = 6x^2 + 6x + 4; R = 5$

7.
$$\begin{array}{r|rrrrr} 1 & 9 & -5 & 2 & -7 & 1 \\ & & 9 & 4 & 6 & -1 \\ \hline & 9 & 4 & 6 & -1 & 0 \end{array}$$
So, $Q(x) = 9x^3 + 4x^2 + 6x - 1; R = 0$

8.
$$\begin{array}{r|rrrr} -2 & 7 & -3 & 0 & -12 \\ & & -14 & 34 & -68 \\ \hline & 7 & -17 & 34 & -80 \end{array}$$
So, $Q(x) = 7x^2 - 17x + 34; R = -80$

9. $x^3 - 3x^2 + 3x - 1 = 0$; possible rational roots are ±1. Using synthetic division, we find
$$\begin{array}{r|rrrr} 1 & 1 & -3 & 3 & -1 \\ & & 1 & -2 & 1 \\ \hline & 1 & -2 & 1 & 0 \end{array}$$
The quotient is $x^2 - 2x + 1 = (x-1)(x-1)$. Thus, the roots are 1, 1, and 1.

10. $x^4 - 5x^3 + 9x^2 - 7x + 2 = 0$; possible roots $\pm1, \pm2$. Using synthetic divison with 1, we find
$$\begin{array}{r|rrrrr} 1 & 1 & -5 & 9 & -7 & 2 \\ & & 1 & -4 & 5 & -2 \\ \hline 1 & 1 & -4 & 5 & -2 & 0 \\ & & 1 & -3 & 2 & \\ \hline & 1 & -3 & 2 & 0 & \end{array}$$
Since $x^2 - 3x + 2 = (x-2)(x-1)$. Thus, the roots are 1, 1, 1, and 2.

11. $x^4 - 6x^2 - 8x - 3 = 0$. Possible roots ±1, and ±3.
$$\begin{array}{r|rrrrr} -1 & 1 & 0 & -6 & -8 & -3 \\ & & -1 & 1 & 5 & 3 \\ \hline -1 & 1 & -1 & -5 & -3 & 0 \\ & & -1 & 2 & 3 & \\ \hline & 1 & -2 & -3 & 0 & \end{array}$$
Since $x^2 - 2x - 3 = (x+1)(x-3)$, the roots are $-1, -1, -1$, and 3.

12. $x^4 + 10x^3 + 25x^2 - 16 = 0$; Possible roots $\pm1, \pm2, \pm4, \pm8, \pm16$
$$\begin{array}{r|rrrrr} -1 & 1 & 10 & 25 & 0 & -16 \\ & & -1 & -9 & -16 & 16 \\ \hline -4 & 1 & 9 & 16 & -16 & 0 \\ & & -4 & -20 & 16 & \\ \hline & 1 & 5 & -4 & 0 & \end{array}$$
Using the quadratic formula on $x^2 + 5x - 4 = 0$, we get $x = \frac{-5\pm\sqrt{25+16}}{2} = \frac{-5\pm\sqrt{41}}{2}$. The roots are $-1, -4, \frac{-5\pm\sqrt{41}}{2}$, and $\frac{-5-\sqrt{41}}{2}$.

13. $x^4 + x^3 - 6x^2 + x + 1 = 0$. Only possible rational roots are ± 1. Neither works

a	$P(a)$	b	$P(b)$	$c = a - \dfrac{P(a)}{m}$	$P(c)$
-3.2	8.4496	-3	-2	-3.038	-0.2580
-3.2	8.4496	-3.038	-0.2580	-3.043	-0.3169
-3.2	8.4496	-3.043	-0.3169	-3.0436	$-.00387$
-3.2	8.4496	-3.0436	-0.00387	-3.0437	$-.00047$
-0.4	-0.3984	-0.3	0.1411	-0.3262	0.01221
-0.4	-0.3984	-0.3262	0.01221	-0.3283	0.00099
-0.4	-0.3984	-0.3283	0.00099	-0.3285	0.00008
-0.4	-3.984	-0.3285	0.00008	-0.3285	0.000006
0.5	0.1875	0.6	-0.2144	0.5467	0.00633
0.5	0.1875	0.5467	0.00633	0.5482	-0.00021
0.5	0.1875	0.5482	-0.00021	0.5482	0.000007
1.8	-0.3104	1.9	1.1311	1.8215	-0.03354
1.8	-0.3104	1.8215	-0.03354	1.8241	0.00121
1.8	-0.3104	1.8241	0.00121	1.8240	-0.00004

The approximate solutions are $-3.044, -0.328, 0.548$, and 1.824.

14. $2x^5 - 3x^4 - 18x^3 + 75x^2 - 104x + 48 = 0$; possible roots include $\pm 1, \pm 2, \pm 3, \pm 4, \pm 6, \pm 8, \pm\frac{1}{2}$, and $\pm\frac{3}{2}$.

$$
\begin{array}{r|rrrrrr}
1 & 2 & -3 & -18 & 75 & -104 & 48 \\
  &   & 2 & -1 & -19 & 56 & -48 \\
\hline
-4 & 2 & -1 & -19 & 56 & -48 & 0 \\
   &   & -8 & 36 & -68 & 48 & \\
\hline
\frac{3}{2} & 2 & -9 & 17 & -12 & 0 & \\
   &   & 3 & -9 & 12 & & \\
\hline
   & 2 & -6 & 8 & 0 & &
\end{array}
$$

Here $2x^2 - 6x + 8 = 2(x^2 - 3x + 4)$. Using the quadratic formula, we get $x = \frac{3\pm\sqrt{9-16}}{2} = \frac{3\pm\sqrt{-7}}{2} = \frac{3}{2}\pm\frac{\sqrt{7}}{2}j$. The roots are $1, -4, \frac{3}{2}, \frac{3}{2}+\frac{\sqrt{7}}{2}j$, and $\frac{3}{2} - \frac{\sqrt{7}}{2}j$.

15. $x^6 + 2x^5 + 3x^4 + 4x^3 + 3x^2 + 2x + 1 = 0$; possible solution is -1.

$$
\begin{array}{r|rrrrrrr}
-1 & 1 & 2 & 3 & 4 & 3 & 2 & 1 \\
   &   & -1 & -1 & -2 & -2 & -1 & -1 \\
\hline
-1 & 1 & 1 & 2 & 2 & 1 & 1 & 0 \\
   &   & -1 & 0 & -2 & 0 & -1 & \\
\hline
   & 1 & 0 & 2 & 0 & 1 & 0 &
\end{array}
$$

Notice that $x^4 + 2x^2 + 1 = (x^2 + 1)^2$ and that $x^2 + 1 = 0 \Rightarrow x^2 = -1$ or $x = \pm j$. This happens twice. The solutions are $-1, -1, j, j, -j$, and $-j$.

16. $x^6 - 2x^5 - 4x^4 + 6x^3 - x^2 + 8x + 4 = 0$. Possible solutions are $\pm 1, \pm 2, \pm 4$

$$
\begin{array}{r|rrrrrrr}
-2 & 1 & -2 & -4 & 6 & -1 & 8 & 4 \\
   &   & -2 & 8 & -8 & 4 & -6 & -4 \\
\hline
2 & 1 & -4 & 4 & -2 & 3 & 2 & 0 \\
  &   & 2 & -4 & 0 & -4 & -2 & \\
\hline
  & 1 & -2 & 0 & -2 & -1 & 0 &
\end{array}
$$

Here $x^4 - 2x^3 - 2x - 1 = (x^2 + 1)(x^2 - 2x - 1)$ and that $x^2 + 1 = 0$ or $x^2 = -1$ and $x = \pm j$. Also, $x^2 - 2x - 1$ and $x = \frac{2\pm\sqrt{4+4}}{2} = 1 \pm \sqrt{2}$. The roots are $-2, 2, 1 + \sqrt{2}, 1 - \sqrt{2}, j, -j$.

17. $\frac{5x}{x^2+3x-4} = \frac{5x}{(x+4)(x-1)}$; Vertical asymptotes: $x = -4$ and $x = 1$; Horizontal asymptote: $y = 0$. A rational equation is 0 only when the numerator is 0, and so $5x = 0$ or $x = 0$, the solution is $x = 0$.

18. $\frac{7x^3}{x^3+1} = \frac{7x^3}{(x+1)(x^2-x+1)}$; Vertical asymptote: $x = -1$; Horizontal asymptote: $y = 7$. A rational equation is 0 only when the numerator is 0, and since $7x^3 = 0 \Rightarrow x = 0$, the solution is $x = 0$.

19. $\frac{4}{x-2} - \frac{5}{x+2} = \frac{4(x+2)-5(x-2)}{(x-2)(x+2)} = \frac{4x+8-5x+10}{(x-2)(x+2)} = \frac{-x+18}{(x-2)(x+2)}$ Vertical asymptote: $x = 2, x = -2$; Horizontal asymptote: $y = 0$. The solution is when $-x + 18 = 0$ or when $x = 18$.

20. $\frac{3}{x-1} + \frac{2}{x+1} - \frac{4}{x^2-1} = \frac{3(x+1)+2(x-1)-4}{(x-1)(x+1)}$
$= \frac{3x+3+2x-2-4}{(x-1)(x+1)} = \frac{5x-3}{(x-1)(x+1)}$. Vertical asymptotes: $x = 1$ and $x = -1$; Horizontal asymptote: $y = 0$. The solution is when $5x - 3 = 0$ or when $x = \frac{3}{5}$.

21. $P(n) = \frac{0.5+0.8(n-1)}{1.5+0.8(n-1)} = \frac{0.8n-0.3}{0.8n+0.7}$. Horizontal asymptote is $\frac{0.8}{0.8} = 1$ or 100%.

22. $N(t) = \frac{5t^2+6t+15}{t^2}$ Horizontal asymptote is $y = \frac{5}{1} = 5$.

CHAPTER 18 TEST

1. $P(x) = 7x^5 - 2x^4 - 5$. Degree is 5

2.
$$P(x) = 4x^3 - 5x^2 + 2x - 8$$
$$P(-2) = 4(-2)^3 - 5(-2)^2 + 2(-2) - 8$$
$$= -32 - 20 - 4 - 8 = -64$$
or

$$\begin{array}{r|rrrr} -2 & 4 & -5 & 2 & -8 \\ & & -8 & 26 & -56 \\ \hline & 4 & -13 & 28 & -64 \end{array}$$

and so, $R = -64$.

3.
$$\begin{array}{r|rrrrr} 6 & 1 & -8 & 3 & 44 & 75 \\ & & 6 & -12 & -54 & -60 \\ \hline & 1 & -2 & -9 & -10 & 15 \end{array}$$

and so, $Q(x) = x^3 - 2x^2 - 9x - 10$ and $R = 15$.

4. $c = \pm 1, \pm 2, d = 1, 3$ and so $\frac{c}{d} = \pm 1, \pm 2, \pm\frac{1}{3}, \pm\frac{2}{3}$.

5.
$$\begin{array}{r|rrrr} 1 & 3 & 5 & -4 & -4 \\ & & 3 & 8 & 4 \\ \hline & 3 & 8 & 4 & 0 \end{array}$$

and $3x^2 + 8x + 4 = (3x + 2)(x + 2)$ and the roots are $1, -2, -\frac{2}{3}$.

6. Possible roots include $\pm 1, \pm 2, \pm 3, \pm 4$, and ± 6.

$$\begin{array}{r|rrrrrr} -1 & 1 & 2 & -4 & -10 & -41 & -72 & -36 \\ & & -1 & -1 & 5 & 5 & 36 & 36 \\ \hline -1 & 1 & 1 & -5 & -5 & -36 & -36 & 0 \\ & & -1 & 0 & 5 & 0 & 36 \\ \hline 3 & 1 & 0 & -5 & 0 & -36 & 0 \\ & & 3 & 9 & 12 & 36 \\ \hline -3 & 1 & 3 & 4 & 12 & 0 \\ & & -3 & 0 & -12 \\ \hline & 1 & 0 & 4 & 0 \end{array}$$

From the quotient we see that

$$x^2 + 4 = 0$$
$$x^2 = -4$$
$$x = \pm 2j$$

and so the roots are $-1, -1, -3, 3, 2j$, and $-2j$.

7. $\frac{3x^2 - 2x - 1}{x^2 + 5x + 6} = \frac{(3x+1)(x-1)}{(x+2)(x+3)}$. Vertical asymptote: $x = -2, x = -3$. Horizontal asymptote: $y = 3$. Solutions: $x = -\frac{1}{3}, x = 1$.

8. The volume of the box is $V = lwh = 175$ or $(12 - 2x)(22 - 2x)x = 175$. Expanding, we get $(4x^2 - 68x + 264)x = 175$ or $4x^3 - 68x^2 + 264x - 175 = 0$. Graphing the left-hand side of this equation, we get roots at about 0.83, 4.50 and 11.66. Since the answer must be less than half of the shortest measurement, this is 6, we obtain 0.8 in. or 4.5 in.

9. We have $t^3 - 22t^2 + 250t = 500$ and so $t^3 - 22t^2 + 250t - 500 = 0$. Graphing, we obtain a root at $t = 2.48$. Thus, after 2.48 yr, the pressure has dropped 500 lb/in.^{2}.

CHAPTER
19

Sequences, Series, and the Binomial Formula

▤ 19.1 SEQUENCES

1. $a_1 = \frac{1}{1+1} = \frac{1}{2}$; $a_2 = \frac{1}{2+1} = \frac{1}{3}$; $a_3 = \frac{1}{3+1} = \frac{1}{4}$; $a_4 = \frac{1}{4+1} = \frac{1}{5}$; $a_5 = \frac{1}{5+1} = \frac{1}{6}$; $a_6 = \frac{1}{6+1} = \frac{1}{7}$

2. $b_1 = \frac{(-1)^1}{1} = -1$; $b_2 = \frac{(-1)^2}{2} = \frac{1}{2}$; $b_3 = \frac{(-1)^3}{3} = -\frac{1}{3}$; $b_4 = \frac{(-1)^4}{4} = \frac{1}{4}$; $b_5 = \frac{(-1)^5}{5} = -\frac{1}{5}$; $b_6 = \frac{(-1)^6}{6} = \frac{1}{6}$

3. $a_1 = (1+1)^2 = 4$; $a_2 = (2+1)^2 = 9$; $a_3 = (3+1)^2 = 16$; $a_4 = (4+1)^2 = 25$; $a_5 = (5+1)^2 = 36$; $a_6 = (6+1)^2 = 49$

4. $a_1 = \frac{1}{1(1+1)} = \frac{1}{2}$; $a_2 = \frac{1}{2(2+1)} = \frac{1}{6}$; $a_3 = \frac{1}{3(3+1)} = \frac{1}{12}$; $a_4 = \frac{1}{4(4+1)} = \frac{1}{20}$; $a_5 = \frac{1}{5(5+1)} = \frac{1}{30}$; $a_6 = \frac{1}{6(6+1)} = \frac{1}{42}$

5. $a_1 = (-1)^1 \cdot 1 = -1$; $a_2 = (-1)^2 \cdot 2 = 2$: $a_3 = (-1)^3 \cdot 3 = -3$; $a_4 = (-1)^4 \cdot 4 = 4$; $a_5 = (-1)^5 \cdot 5 = -5$; $a_6 = (-1)^6 \cdot 6 = 6$

6. $a_1 = \frac{1+(-1)^1}{1+4 \cdot 1} = 0$; $a_2 = \frac{1+(-1)^2}{1+4 \cdot 2} = \frac{2}{9}$; $a_3 = \frac{1+(-1)^3}{1+4 \cdot 3} = 0$; $a_4 = \frac{1+(-1)^4}{1+4 \cdot 4} = \frac{2}{17}$; $a_5 = \frac{1+(-1)^5}{1+4 \cdot 5} = 0$; $a_6 = \frac{1+(-1)^6}{1+4 \cdot 6} = \frac{2}{25}$

7. $a_1 = \frac{2 \cdot 1 - 1}{2 \cdot 1 + 1} = \frac{1}{3}$; $a_2 = \frac{2 \cdot 2 - 1}{2 \cdot 2 + 1} = \frac{3}{5}$; $a_3 = \frac{2 \cdot 3 - 1}{2 \cdot 3 + 1} = \frac{5}{7}$; $a_4 = \frac{2 \cdot 4 - 1}{2 \cdot 4 + 1} = \frac{7}{9}$; $a_5 = \frac{2 \cdot 5 - 1}{2 \cdot 5 + 1} = \frac{9}{11}$; $a_6 = \frac{2 \cdot 6 - 1}{2 \cdot 6 + 1} = \frac{11}{13}$

8. $a_1 = \left(-\frac{1}{2}\right)^1 = -\frac{1}{2}$; $a_2 = \left(-\frac{1}{2}\right)^2 = \frac{1}{4}$; $a_3 = \left(-\frac{1}{2}\right)^3 = -\frac{1}{8}$; $a_4 = \left(-\frac{1}{2}\right)^4 = \frac{1}{16}$; $a_5 = \left(-\frac{1}{2}\right)^5 = -\frac{1}{32}$; $a_6 = \left(-\frac{1}{2}\right)^6 = \frac{1}{64}$

9. $a_1 = \left(\frac{2}{3}\right)^0 = 1$; $a_2 = \left(\frac{2}{3}\right)^1 = \frac{2}{3}$; $a_3 = \left(\frac{2}{3}\right)^2 = \frac{4}{9}$; $a_4 = \left(\frac{2}{3}\right)^3 = \frac{8}{27}$; $a_5 = \left(\frac{2}{3}\right)^4 = \frac{16}{81}$; $a_6 = \left(\frac{2}{3}\right)^5 = \frac{32}{243}$

10. $a_1 = 1 \cos \pi = -1$; $a_2 = 2 \cos 2\pi = 2$; $a_3 = 3 \cos 3\pi = -3$; $a_4 = 4 \cos 4\pi = 4$; $a_5 = 5 \cos 5\pi = -5$; $a_6 = 6 \cos 6\pi = 6$

11. $a_1 = \left(\frac{1-1}{1+1}\right)^2 = 0$; $a_2 = \left(\frac{2-1}{2+1}\right)^2 = \frac{1}{9}$; $a_3 = \left(\frac{3-1}{3+1}\right)^2 = \frac{4}{16} = \frac{1}{4}$; $a_4 = \left(\frac{4-1}{4+1}\right)^2 = \frac{9}{25}$; $a_5 = \left(\frac{5-1}{5+1}\right)^2 = \frac{16}{36} = \frac{4}{9}$; $a_6 = \left(\frac{6-1}{6+1}\right)^2 = \frac{25}{49}$

12. $a_1 = \frac{1^2-1}{1^2+1} = 0$; $a_2 = \frac{2^2-1}{2^2+1} = \frac{3}{5}$; $a_3 = \frac{3^2-1}{3^2+1} = \frac{8}{10} = \frac{4}{5}$; $a_4 = \frac{4^2-1}{4^2+1} = \frac{15}{17}$; $a_5 = \frac{5^2-1}{5^2+1} = \frac{24}{26} = \frac{12}{13}$; $a_6 = \frac{6^2-1}{6^2+1} = \frac{35}{37}$

13. $a_1 = 1$; $a_2 = 2 \cdot 1 = 2$; $a_3 = 3 \cdot 2 = 6$; $a_4 = 4 \cdot 6 = 24$; $a_5 = 5 \cdot 24 = 120$; $a_6 = 6 \cdot 120 = 720$

14. $a_1 = 3$; $a_2 = 3 + 2 = 5$; $a_3 = 5 + 3 = 8$; $a_4 = 8 + 4 = 12$; $a_5 = 12 + 5 = 17$; $a_6 = 17 + 6 = 23$

15. $a_1 = 5$; $a_2 = 5 + 3 = 8$; $a_3 = 8 + 3 = 11$; $a_4 = 11 + 3 = 14$; $a_5 = 14 + 3 = 17$; $a_6 = 17 + 3 = 20$

16. $a_1 = 2$; $a_2 = 2^2 = 4$; $a_3 = 4^3 = 64$; $a_4 = 64^4 = 16,777,216$; $a_5 = 16,777,216^5 = 1.329228 \times 10^{36}$; $a_6 = (1.329228 \times 10^{36})^6 = 5.51565 \times 10^{216}$

17. $a_1 = 2$; $a_2 = 2^1 = 2$; $a_3 = 2^2 = 4$; $a_4 = 4^3 = 64$; $a_5 = 64^4 = 16,777,216$; $a_6 = (16,777,216)^5 = 1.329228 \times 10^{36}$

18. $a_1 = 1$; $a_2 = \frac{-1}{2} \cdot 1 = -\frac{1}{2}$; $a_3 = \frac{-1}{3} \cdot \frac{-1}{2} = \frac{1}{6}$; $a_4 = \frac{-1}{4} \cdot \frac{1}{6} = \frac{-1}{24}$; $a_5 = \frac{-1}{5} \cdot \frac{-1}{24} = \frac{1}{120}$; $a_6 = \frac{-1}{6} \cdot \frac{1}{120} = \frac{-1}{720}$

19. $a_1 = 1; a_2 = 2^1 = 2; a_3 = 3^2 = 9; a_4 = 4^9 = 262{,}144; a_5 = 5^{262{,}144}; a_6 = 6^{(5^{262{,}144}})$

20. $a_1 = \frac{1}{2}; a_2 = \left(\frac{1}{2}\right)^{-2} = 4; a_3 = 4^{-3} = \frac{1}{64}; a_4 = \left(\frac{1}{64}\right)^{-4} = 16{,}777{,}216; a_5 = (16{,}777{,}216)^{-5} = 7.52316385 \times 10^{-37}; a_6 = 5.5256524 \times 10^{216}$

21. $a_1 = 1; a_2 = \frac{1}{2}; a_3 = 1 \cdot \frac{1}{2} = \frac{1}{2}; a_4 = \frac{1}{2} \cdot \frac{1}{2} = \frac{1}{4}; a_5 = \frac{1}{2} \cdot \frac{1}{4} = \frac{1}{8}; a_6 = \frac{1}{4} \cdot \frac{1}{8} = \frac{1}{32}$

22. $a_1 = 1; a_2 = 1; a_3 = 1 + 1 = 2; a_4 = 1 + 2 = 3; a_5 = 2 + 3 = 5; a_6 = 3 + 5 = 8$

23. $a_1 = 0; a_2 = 2; a_3 = 2 - 0 = 2; a_4 = 2 - 2 = 0; a_5 = 0 - 2 = -2; a_6 = -2 - 0 = -2$

24. $a_1 = 1; a_2 = 2; a_3 = 1 \cdot 2 = 2; a_4 = 2 \cdot 2 = 4; a_5 = 2 \cdot 4 = 8; a_6 = 4 \cdot 8 = 32$

25. $a_1 = 25 \text{ ft} - 3 \text{ in.}; a_{25} = 25 \text{ ft} - 3(25) \text{ in.} = 25 \text{ ft} - 75 \text{ in.};$ Since 75 in. $= 6'3'' = 6.25$ ft, $a_{25} = 25 - 6.25 = 18.75$ ft

26. $a_1 = 320{,}000; a_2 = 240{,}000; a_3 = 240{,}000 + 12{,}000; = 252{,}000; a_4 = 252{,}000 + 12{,}000 = 264{,}000; a_n = 240{,}000 + (n - 2) \cdot 12{,}000; n \geq 2; a_{10} = 240{,}000 + (10 - 2)12{,}000 = 240{,}000 + 8 \cdot 12{,}000 = \$336{,}000; a_{20} = 240{,}000 + (20 - 2)12{,}000 = \$456{,}000$

27. $a_1 = 26000, a_2 = 26000 + 1100 = 27{,}100; a_n = 26000 + (n-1) \cdot 1100; a_5 = 26000 + (5-1)1100 = 26000 + 4400 = \$30{,}400; a_{10} = 26000 + 9(1100) = \$35{,}900$

28. $a_1 = 60; a_2 = 60 + 4; a_3 = 60 + 2 \cdot 4; a_{10} = 60 + 9 \cdot 4 = 60 + 36 = 96$ seats

≡ 19.2 ARITHMETIC AND GEOMETRIC SEQUENCES

1. Common difference, so arithmetic: $d = 9 - 1 = 8; a_9 = 1 + (9 - 1)8 = 1 + 64 = 65$

2. Common difference, so arithmetic: $d = -5 - 3 = -8, a_7 = 3 + 6(-8) = 3 - 48 = -45$

3. Common ratio, so geometric: $r = \frac{-4}{8} = -\frac{1}{2}; a_{10} = \left(-\frac{1}{2}\right)^9 \cdot 8 = -\frac{1}{512} \cdot 8 = -\frac{1}{64}$

4. Common ratio, so geometric: $r = \frac{1}{4}; a_8 = \left(\frac{1}{4}\right)^7 \times 4 = \left(\frac{1}{4}\right)^6 = \frac{1}{4096}$

5. Geometric: $r = \frac{4}{-1} = -4; a_6 = (-4)^5 \cdot (-1) = 1024$

6. Arithmetic: $d = 4 - (-1) = 5; a_{12} = -1 + 11 \cdot 5 = 54$

7. Arithmetic: $d = -2 - 3 = -5; a_8 = 3 + 7(-5) = -32$

8. Geometric: $r = \frac{-3}{1} = -3; a_7 = (-3)^6 \cdot 1 = 729$

9. Neither a common difference nor a common ratio so it is neither arithmetic nor geometric

10. Geometric: $r = \sqrt{2}; a_{12} = \left(\sqrt{2}\right)^{11} \cdot 1 = 2^5 \cdot \sqrt{2} = 32\sqrt{2}$

11. Arithmetic: $d = 0.4; a_9 = 0.4 + 8(0.4) = 9(0.4) = 3.6$

12. Neither: it could be $a_n = \sqrt{a_{n-1}}$ so $a_{11} = 4^{(1/2)^{11}} = 2^{(1/2)^{10}} = 2^{1/1024}$

13. Arithmetic: $d = -\frac{1}{6}; a_8 = \frac{2}{3} + 7\left(-\frac{1}{6}\right) = \frac{4}{6} - \frac{7}{6} = -\frac{3}{6} = -\frac{1}{2}$

14. Geometric: $r = \frac{1}{2}; a_8 = \left(\frac{1}{2}\right)^7 \cdot \frac{2}{3} = \frac{1}{64 \cdot 3} = \frac{1}{192}$

15. Geometric: $r = \frac{1}{10}$ or 0.1; $a_6 = (0.1)^5 \cdot 5 = 0.00005$

16. Geometric: $r = -\frac{1}{3}; a_7 = \left(-\frac{1}{3}\right)^6 \cdot 6 = \frac{2}{243}$

17. Arithmetic: $d = 2; a_9 = (1 + \sqrt{3}) + 8(2) = 17 + \sqrt{3}$

18. Geometric: $r = 10; a_{10} = 1.02 \times 10^9$

19. Neither. It looks like an arithmetic sequence with alternating signs, so try $[4.3 - (n-1)1.1](-1)^{n-1}$; $a_6 = 1.2$

20. Geometric: $r = -2; a_9 = (-2)^8 \cdot 3.4 = 870.4$

21. $d = a_6 - a_5 = 24 - 9 = 15; a_5 = a_1 + 4d; 9 = a_1 + 4 \cdot 15; a_1 = 9 - 60 = -51$

22. $a_9 = a_7 + 2d; 15 = 11 + 2d; 2d = 4; d = 2; a_7 = a_1 + 6d; 11 = a_1 + 12; a_1 = -1$

23. $r = \frac{a_5}{a_3} = \frac{3}{9} = \frac{1}{3}; a_4 = \left(\frac{1}{3}\right)^3 \cdot a_1; 9 = \frac{1}{27} \cdot a_1; a_1 = 243$

24. $r = \frac{a_9}{a_3} = \frac{5}{\frac{1}{3}} = 15; a_8 = (15)^7 \cdot a_1; a_1 = \frac{1}{3} \div 15^7; a_1 = 1.9509221 \times 10^{-9}$

25. $81 = 5 + (n-1)4; 76 = 4(n-1); 19 = (n-1); n = 20$

26. $-31 = 20 + (n-1)(-3); -51 = -3(n-1); 17 = n-1; n = 18$

27. $\frac{1}{559,872} = 3\left(\frac{1}{6}\right)^{n-1}; \frac{1}{1,679.616} = \left(\frac{1}{6}\right)^{n-1}; 6^{n-1} = 1,679,619; \ln 6^{n-1} = (n-1)\ln 6 = \ln 1,679,619;$
$n-1 = \frac{\ln 1,679,619}{\ln 6} = 8; n = 9$

28. $\frac{-1}{4,096} = (-4)\left(\frac{1}{2}\right)^{n-1}; \frac{1}{16,384} = \left(\frac{1}{2}\right)^{n-1}; 2^{n-1} = 16,384; (n-1)\ln 2 = \ln 16,384; n-1 = 14; n = 15$

29. $a_1 = -5; a_7 = 4; a_7 = a_1 + (n-1)d; 4 = -5 + 6d; 9 = 6d; d = \frac{3}{2}; a_2 = -5 + \frac{3}{2} = -3\frac{1}{2}$ or -3.5

30. $a_1 = \frac{1}{3}; a_8 = 729; a_8 = a_1 r^{n-1}; 729 = \frac{1}{3}r^7; 2187 = r^7; r = 3; a_3 = \frac{1}{3} \cdot 3^2 = 3$

31. $a_{61} = a_1 + 60d; 8 = \frac{1}{2} + 60d; 7\frac{1}{2} = 60d; d = \frac{7.5}{60} = 0.125 = \frac{1}{8}$ mi

32. $a_6 = a_1\left(\frac{7}{8}\right)^5; a_6 = \left(80 \cdot \frac{7}{8}\right) \cdot \left(\frac{7}{8}\right)^5 = 70 \cdot \left(\frac{7}{8}\right)^5 = 35.9036$ m

33. $a_{10} = a_1 + (10-1)d; a_{10} = 15 + 9(-0.3); a_{10} = 15 - 2.7 = 12.3$ cm$; 0 = 15 + (n-1)(-0.3); -15 = -0.3(n-1); n-1 = 50; n = 51$

34. Geometric sequence; $r = 1 - 0.125 = 0.875; P = 100(1-0.125)^{8.8} = 100(0.875)^{8.8} = 30.88$ kPa

35. **(a)** Geometric sequence: Here $r = (1-0.162) = 0.838$ and $a_1 = 940$, so the concentration is reduced to 100 ppm when $100 = 940(0.838)^{n-1}$. Thus, $a_n = \frac{100}{940} = (0.838)^{n-1}$ and so, $(n-1)\ln(0.838) = \ln\frac{10}{94}$. Dividing, we get $n-1 = \dfrac{\ln\left(\frac{10}{94}\right)}{\ln(0.838)} \approx 12.678$, and hence $n = 13.678$. Thus, the concentration will be reduced to 100 ppm about 13,678 mi from the first monitor or about 14.678 mi from the spill.

(b) Here $a_n = 1.5 = 940(0.838)^{n-1}$ and so $n - 1 = \dfrac{\ln\left(\frac{1.5}{940}\right)}{\ln(0.838)} = 36.44$ which is $n = 37.44$ mi from the monitor so about 38.44 mi from the spill.

36. **(a)** Using $t = 4$ mo, $a_1 = 3.50$, and $a_4 = 18.76$, then $18.76 = 3.50r^3$ and $r^3 = \frac{18.76}{3.50} = 5.36$. Thus, $r = \sqrt[3]{5.36} \approx 1.75$.
(b) December is the 12th month, so we want to determine a_{12}. Using $r = 1.75$, we see that $a_{12} = 3.50 \times 1.75^{12} = 1650$ ppm.

≡ 19.3 SERIES

1. $\sum_{k=1}^{20} 3\left(\frac{1}{2}\right)^k = 3\left(\frac{1}{2}\right)^1 + 3\left(\frac{1}{2}\right)^2 + 3\left(\frac{1}{2}\right)^3 + 3\left(\frac{1}{2}\right)^4 + \cdots = 3\left(\frac{1}{2}\right) + 3\left(\frac{1}{4}\right) + 3\left(\frac{1}{8}\right) + 3\left(\frac{1}{16}\right) \cdots = \frac{3}{2} + \frac{3}{4} + \frac{3}{8} + \frac{3}{16} + \cdots$

2. $\sum_{k=1}^{50} (4 + (n-1)3) = 4 + 7 + 10 + 13 + \cdots$

3. $\sum_{n=0}^{20} (-1)^n \frac{3^n}{n+1} = 1 - \frac{3}{2} + 3 - \frac{27}{4} + \cdots$

4. $\sum_{k=1}^{60} (-1)^k = -1 + 1 - 1 + 1 - \cdots$

5. $\sum_{k=1}^{5} (k-3) = -2 - 1 + 0 + 1 + 2 = 0$

6. $\sum_{k=0}^{4} (2k+1) = 1 + 3 + 5 + 7 + 9 = 25$

7. $\sum_{k=1}^{6} k^2 = 1 + 4 + 9 + 16 + 25 + 36 = 91$

8. $\sum_{k=2}^{8} \frac{k+2}{(k-1)k} = \frac{4}{2} + \frac{5}{6} + \frac{6}{12} + \frac{7}{20} + \frac{3}{30} + \frac{9}{42} + \frac{10}{56} = 4.342837$

9. $\sum_{n=1}^{5} \sum n^3 = 1 + 8 + 27 + 64 + 125 = 225$

10. $\sum_{n=1}^{8} \frac{(-1)^n(n^2+1)}{h} = -\frac{2}{7} + \frac{5}{2} - \frac{10}{3} + \frac{17}{4} - \frac{26}{5} + \frac{37}{6} - \frac{50}{7} + \frac{65}{8} = \frac{2827}{840} \approx 3.36547$

11. $\sum_{n=1}^{8} \frac{2^i}{3i+1} = \frac{8}{10} + \frac{16}{13} + \frac{32}{16} + \frac{64}{19} + \frac{128}{22} + \frac{256}{25} = \frac{1,593,342}{67,925} \approx 23.457$

12. $\sum_{i=0}^{7} \frac{(-1)^{i+1}(i+1)^2}{2i+1} = \frac{-1}{1} + \frac{4}{3} - \frac{9}{5} + \frac{16}{7} - \frac{25}{9} + \frac{36}{11} - \frac{49}{13} + \frac{64}{15} = \frac{244,788}{135,135} \approx 1.811$

13. Aritimetic: $d = 3; a_{10} = 3 + 9 \cdot 3 = 30; S_{10} = \frac{10(3+30)}{2} = 165$

14. Arithmetic; $d = -4; a_{12} = 5 - 11(4) = -39; S_{12} = \frac{12(5-39)}{2} = -204$

15. Geometric: $r = 2; S_8 = 1 \cdot \frac{1-2^8}{1-2} = \frac{-255}{-1} = 255$

16. Geometric: $r = 4; S_9 = \frac{1}{2} \cdot \frac{1-4^9}{1-4} = \frac{1}{2} \cdot \frac{-262143}{-3} = 43690.5$

17. Arithmetic: $d = 4; a_{12} = -6 + 11 \cdot 4 = 38; S_{12} = \frac{12(-6+38)}{2} = 192$

18. Geometric: $r = -\frac{1}{2}; S_{10} = 1 \cdot \frac{1 - \left(\frac{-1}{2}\right)^{10}}{1 + \frac{1}{2}} = \frac{2}{3}\left(1 - \left(\frac{1}{2}\right)^{10}\right) = \frac{341}{512} \approx 0.666016$

19. Arithmetic: $d = 0.25, a_{20} = 0.5 + 19(.25) = 5.25; S_{20} = \frac{20}{2}(0.5 + 5.25) = 57.5$

20. Geometric: $r = 0.4; S_{15} = 0.5\frac{(1-0.4^{15})}{1-0.4} \approx 0.8333324$

21. Geometric: $r = -\frac{1}{3}; S_{16} = \frac{3}{4}\frac{\left(1 - \left(-\frac{1}{3}\right)^{16}\right)}{\left(1 + \frac{1}{3}\right)} = \frac{9}{10}\left(1 - \left(\frac{1}{3}\right)^{16}\right) \approx 0.5625$

22. Geometric: $r = 0.1; S_6 = 0.444444$

23. Arithmetic: $d = 1.2; a_{14} = 0.4 + 13(1.2) = 16; S_{14} = \frac{14}{2}(0.4 + 16) = 114.8$

24. Geometric: $r = \frac{1}{2}; S_{10} = 16 \cdot \frac{1 - \left(\frac{1}{2}\right)^{10}}{1 - \frac{1}{2}} = 32(1 - 0.5^{10}) = 31.96875$

25. Geometric: $r = \sqrt{3}; S_{15} = 2\sqrt{3} \cdot \left(\frac{1-\sqrt{3}^{15}}{1-\sqrt{3}}\right) = 17,920.253$

26. Geometric: $r = 2\sqrt{5}; S_{12} = \sqrt{5} \cdot \left(\frac{1-(2\sqrt{5})^{12}}{1-2\sqrt{5}}\right) = 41,216,228.33$

27. Geometric: $r = 3; S_8 = \left(\frac{1-3^8}{1-3}\right) = 9840$

28. Geometric: $r = -5; S_{10} = 5\left(\frac{1-(-5)^{10}}{1+5}\right) = -8,138,020$

29. First time it hits the ground it has only fallen 80m. After this it rises and falls so $2 \times 80(.8)$. So this is a sequence $80 + 160(.8) + 160(0.8)^2 + 160(0.8)^3 = 160 + 160(0.8) + 160(0.8)^2 + 160(0.8)^3 - 80 = 160\left(\frac{1-0.8^4}{1-0.8}\right) - 80 = 392.32$m when it hits the fourth time. $160\left(\frac{1-0.8^{10}}{1-0.8}\right) - 80 = 634.10$ when it hits the tenth time.

30. Geometric with $a_1 = 50; r = 0.8; S_{10} = 50\left(\frac{1-0.8^{10}}{1-0.8}\right) = 223.156$ cm

31. In a 16 ft span there will be 12 studs 16″ apart. The longest is 8′ and the shortest is 8″ or $\frac{2}{3}$ ft. $S_{12} = \frac{12}{2}\left(8 + \frac{2}{3}\right) = 6\left(8\frac{2}{3}\right) = 52$ ft. This is only half of the roof. The whole roof will be twice that or 104 ft.

32. The rate of 8% compounded quarterly is 2% or 0.02 per quarter, 3 years is 12 quarters, so $A = 2000(1 + 0.02)^{12} = \$2536.48$.

33. The rate of 8% compounded monthly is $\frac{8}{12} = \frac{2}{3}\%$ per month $\approx 0.00\overline{6}$ per month, 3 years have 36 months $A = 2000(1 + 0.00\overline{6})^{36} = \$2540.47$. This is $3.99 more than exercise #32.

34. $A = Pe^{rt} = 2000e^{(0.08 \times 3)} = 2542.50$

35. **(a)** Here the sequence is $\$750,000 + \$320,000 + \$240,000 + \$250,000 + \cdots + \$336,000$. Here $a_1 = 240,000$ is the cost of the 2nd floor, so $a_9 = 336,000$ and $S_9 = \frac{9(240,000+336,000)}{2} = 2,592,000$ is the cost of floors 2–10. So, the cost of a 10-story building is $\$750,000 + \$320,000 + \$2,592,000 = \$3,662,000$.
(b) $\$750,000 + \$320,000 + S_{19} = \$7,682,000$. ($S_{19} = $ cost of floors 2–20 $= \frac{19}{2}(240,000 + 456,000) = 6,612,000$.)

36. This is a geometric series, with $a_1 = 3,000$ and $r = 1 + 5.25\% = 1 + 0.0525 = 1.0525$. The first 3,000 is worth $3,000(1.0525)^{25}$ when re retires. The 3,000 the owner deposits at the beginning of the second year is worth $3,000(1.0525)^{24}$ when the owner retires. This produces the geometric series

$$3,000(1.0525)^{25} + 3,000(1.0525)^{24} + \cdots + 3,000(1.0525)^1$$
$$= 3,000[(1.0525)^{25} + (1.0525)^{24} + \cdots + (1.0525)^1]$$
$$= 3,000[(1.0525)^1 + (1.0525)^2 + \cdots + (1.0525)^{25}]$$

We want the sum of the 25 terms of this sequence, S_{25}. Notice from the last line of the above display, that $a_1 = 3,000(1.0525)^1 = 3,157.50$. Thus,

$$S_{25} = a_1 \frac{1 - r^n}{1 - r}$$
$$= 3,157.50 \left(\frac{1 - (1.0525)^{25}}{1 - (1.0525)} \right)$$
$$= 155,997.90$$

The owner of this automotive repair shop will have \$155,997.90 in the account upon retirement.

≡ 19.4 INFINITE GEOMETRIC SERIES

1. $r = \frac{-1}{2}$ so converges; $S = \dfrac{\frac{1}{2}}{1 - \frac{-1}{2}} = \dfrac{\frac{1}{2}}{\frac{3}{2}} = \frac{1}{3}$

2. $r = \frac{3}{4}$ so converges; $S = \dfrac{1}{1 - \frac{3}{4}} = \dfrac{1}{\frac{1}{4}} = 4$

3. $r = -\frac{2}{3}$ so converges; $S = \dfrac{1}{1 - \frac{-2}{3}} = \dfrac{1}{\frac{5}{3}} = \frac{3}{5}$ or 0.6

4. $r = 1.5$ so diverges

5. $r = \frac{1}{4}$ so converges; $S = \dfrac{3}{1 - \frac{1}{4}} = \dfrac{3}{\frac{3}{4}} = 4$

6. $r = -\frac{1}{4}$ so converges; $S = \dfrac{4}{1 - \frac{-1}{4}} = \dfrac{4}{\frac{5}{4}} = \frac{16}{5}$ or 3.2

7. $r = \frac{1}{10}$ so converges; $S \dfrac{\frac{5}{4}}{1 - \frac{1}{10}} = \dfrac{\frac{5}{4}}{\frac{9}{10}} = \frac{5}{4} \cdot \frac{10}{9} = \frac{50}{36} = \frac{25}{18} = 1.3888\ldots$

8. $r = \frac{1}{4}$ so converges; $S = \dfrac{\frac{3}{4}}{1 - \frac{1}{4}} = \dfrac{\frac{3}{4}}{\frac{3}{4}} = 1$

9. $r = 0.1$ so converges; $S = \frac{0.03}{1 - 0.1} = \frac{0.03}{0.9} = \frac{3}{90} = \frac{1}{30} = 0.03333\ldots$

10. 0.01 so converges; $S = \frac{1.4}{1 - 0.01} = \frac{1.4}{0.99} = \frac{140}{99} = 1.41414\ldots$

11. $r = (-0.8)$ so converges; $S = \frac{1}{1 + 0.8} = \frac{1}{1.8} = \frac{10}{18} = \frac{5}{9} = 0.555\ldots$

12. $r = \frac{3}{4}$ so converges; $= \dfrac{\frac{3}{4}}{1 - \frac{3}{4}} = \dfrac{\frac{3}{4}}{\frac{1}{4}} = 3$

13. $r = \frac{5}{4}$ so diverges

14. $r = -1.1$ so diverges

15. $r = 10$ so diverges

16. $2^{-n} = (2^{-1})^n = \frac{1}{2}^n$; $r = \frac{1}{2}$ so converges $S = \dfrac{\frac{1}{2}}{1 - \frac{1}{2}} = \dfrac{\frac{1}{2}}{\frac{1}{2}} = 1$

17. $r = \frac{2}{3}$ so converges; $S = \dfrac{\frac{2}{3}}{1 - \frac{2}{3}} = \dfrac{\frac{2}{3}}{\frac{1}{3}} = 2$

18. $r = -\frac{2}{3}$ so converges; $S = \dfrac{-4}{1 + \frac{2}{3}} = \dfrac{-4}{\frac{5}{3}} = -\frac{12}{5} = -2.4$

19. $r = \frac{1}{\sqrt{2}}$ so converges; $S = \dfrac{\frac{1}{\sqrt{2}}}{1 - \frac{1}{\sqrt{2}}} = \dfrac{\frac{1}{\sqrt{2}}}{\frac{\sqrt{2}-1}{\sqrt{2}}} = \frac{1}{\sqrt{2}-1} = \sqrt{2} + 1 \approx 2.4142$

20. $r = \frac{\sqrt{3}}{3}$ so converges; $S = \dfrac{\frac{\sqrt{3}}{3}}{1 - \frac{\sqrt{3}}{3}} = \dfrac{\frac{\sqrt{3}}{3}}{3 - \frac{\sqrt{3}}{3}} = $

$\frac{\sqrt{3}}{3 - \sqrt{3}} = \frac{3\sqrt{3}+3}{6} = \frac{\sqrt{3}}{2} + \frac{1}{2} \approx 1.366$

21. $r = \sqrt{5}$ so diverges

22. Think of this as $\displaystyle\sum_{n=1}^{\infty} \left[(\sqrt{7})^{-1}\right]^{n-1} = \sum_{n=1}^{\infty} \left(\frac{1}{\sqrt{7}}\right)^{n-1}$.

Thus, $r = \frac{1}{\sqrt{7}}$ so the series converges; $S = \dfrac{1}{1 - \frac{1}{\sqrt{7}}}$

$= \frac{1}{\frac{\sqrt{7}-1}{\sqrt{7}}} = \frac{\sqrt{7}}{\sqrt{7}-1} \approx 1.6076$

23. $a_1 = 0.4; r = 0.1; S = \frac{0.4}{1-0.1} = \frac{0.4}{0.9} = \frac{4}{9}$

24. $a_1 = 0.7; r = 0.1; S = \frac{0.7}{1-0.1} = \frac{0.7}{0.9} = \frac{7}{9}$

25. $a_1 = 0.57; r = 0.01; S = \frac{0.57}{1-0.01} = \frac{0.57}{0.99} = \frac{57}{99} = \frac{19}{33}$

26. $a_1 = 0.84; r = 0.01; S = \frac{0.84}{1-0.01} = \frac{0.84}{0.99} = \frac{28}{33}$

27. $a_1 = 1.352; r = 0.0001; S = \frac{1.352}{1-0.0001} = \frac{1.352}{0.9999} = $

$\frac{13520}{9999}$

28. $a_1 = 4.123; r = 0.0001; S = \frac{4.123}{1-0.0001} = \frac{4.123}{0.9999} = $

$\frac{41230}{9999}$

29. $6.3021021\ldots = 6.3 + (0.0021 + 0.0000021 \cdots);$

$a_1 = 0.0021; r = 0.001; S = \frac{0.0021}{.999} = \frac{21}{9990} = $

$\frac{7}{3330}; 6.3 + \frac{7}{3330} = \frac{63}{10} + \frac{7}{3330} = \frac{20986}{3330} = \frac{10493}{1665}$

30. $2.1906906\ldots = 2.1 + (0.0906 + 0.0000906 + \ldots);$

$S = \frac{0.0906}{.999} = \frac{906}{9990} = \frac{302}{3330}; 2.1 + \frac{302}{3330} = \frac{21}{10} = $

$\frac{302}{3330} = \frac{6993+302}{3330} = \frac{7295}{3330} = \frac{1459}{666}$

31. After the first bounce it is a geometric series with $a_1 = 8; r = 0.8; S = \frac{8}{1-0.8} = \frac{8}{0.2} = 40$. The total distance is $5 + 40 = 45$ m.

32. Geometric series: $a_1 = 100; r = 0.9; S = \frac{100}{1-0.9} = $ 1000 mm

33. Geometric series with $a_1 = 250; r = \frac{8}{10} = 0.8$ so $S = \frac{250}{1-0.8} = \frac{250}{0.2} = 1250$ rev.

34. Geometric series with $a_1 = 50$ and $r = 0.85$, hence $S = \frac{50}{1-0.85} = \frac{50}{0.15} = 333.33$ cm.

▤ 19.5 THE BINOMIAL THEOREM

1. $(a+1)^4 = 1a^4 + 4a^3 \cdot 1^1 + 6a^2 1^2 + 4a1^3 + 1^4 = a^4 + 4a^3 + 6a^2 + 4a + 1$

2. $(2x+b)^5 = (2x)^5 + 5(2x)^4 b + 10 \cdot (2x)^3 b^2 + 10(2x)^2 b^3 + 5(2x)b^4 + b^5 = 32x^5 + 80x^4 b + 80x^3 b^2 + 40x^2 b^3 + 10xb^4 + b^5$

3. $(3x-1)^4 = (3x)^4 + 4(3x)^3(-1) + 6(3x)^2(-1)^2 + 4(3x)(-1)^3 + (-1)^4 = 81x^4 - 108x^3 + 54x^2 - 12x + 1$

4. $(x-2y)^5 = x^5 + 5x^4(-2y)^1 + 10x^3(-2y)^2 + 10x^2(-2y)^3 + 5x(-2y)^4 + (-2y)^5 = x^5 - 10x^4 y + 40x^3 y^2 - 80x^2 y^3 + 80xy^4 - 32y^5$

5. $\left(\frac{x}{2}+d\right)^6 = \left(\frac{x}{2}\right)^6 + 6\left(\frac{x}{2}\right)^5 d + 15\left(\frac{x}{2}\right)^4 d^2 + 20\left(\frac{x}{2}\right)^3 d^3 + 15\left(\frac{x}{2}\right)^2 d^4 + 6\left(\frac{x}{2}\right)d^5 + d^6 = \frac{x^6}{64} + \frac{3x^5 d}{16} + \frac{15x^4 d^2}{16} + \frac{5x^3 d^3}{2} + \frac{15x^2 d^4}{4} + 3xd^5 + d^6$

6. $\left(xy-\frac{a}{3}\right)^6 = (xy)^6 + 6(xy)^5\left(-\frac{a}{3}\right)^1 + 15(xy)^4\left(-\frac{a}{3}\right)^2 + 20(xy)^3\left(-\frac{a}{3}\right)^3 + 15(xy)^2\left(-\frac{a}{3}\right)^4 + 6(xy)\left(-\frac{a}{3}\right)^5 + $

$\left(-\frac{a}{3}\right)^6 = x^6 y^6 - 2x^5 y^5 a + \frac{5}{3}x^4 y^4 a^2 - \frac{20}{27}x^3 y^3 a^3 + \frac{5}{27}x^2 y^2 a^4 - \frac{2}{81}xya^5 + \frac{1}{729}a^6$

7. $\left(\frac{a}{2}-\frac{4}{b}\right)^5 = \left(\frac{a}{2}\right)^5 + 5\left(\frac{a}{2}\right)^4\left(-\frac{4}{b}\right) + 10\left(\frac{a}{2}\right)^3\left(-\frac{4}{b}\right)^2 + 10\left(\frac{a}{2}\right)^2\left(-\frac{4}{b}\right)^3 + 5\left(\frac{a}{2}\right)\left(-\frac{4}{b}\right)^4 + \left(-\frac{4}{b}\right)^5 = \frac{a^5}{32} - \frac{5a^4}{4b} + 20\frac{a^3}{b^2} - 160\frac{a^2}{b^3} + 640\frac{a}{b^4} - \frac{1024}{b^5}$

8. $\left(\frac{x}{3}+\frac{2}{y}\right)^4 = \left(\frac{x}{3}\right)^4 + 4\left(\frac{x}{3}\right)^3\left(\frac{2}{y}\right) + 6\left(\frac{x}{3}\right)^2\left(\frac{2}{y}\right)^2 + 4\left(\frac{x}{3}\right)\left(\frac{2}{y}\right)^3 + \left(\frac{2}{y}\right)^4 = \frac{x^4}{81} + \frac{8x^3}{27y} + \frac{8x^2}{3y^2} + \frac{32x}{3y^3} + \frac{16}{y^4}$

9. $(a+b)^7 = \binom{7}{0}a^7 + \binom{7}{1}a^6 b + \binom{7}{2}a^5 b^2 + \binom{7}{3}a^4 b^3 + \binom{7}{4}a^3 b^4 + \binom{7}{5}a^2 b^5 + \binom{7}{6}ab^6 + \binom{7}{7}b^7 = a^7 + 7a^6 b + 21a^5 b^2 + 35a^4 b^3 + 35a^3 b^4 + 21a^2 b^5 + 7ab^6 + b^7$

10. $(a+b)^9 = \binom{9}{0}a^9 + \binom{9}{1}a^8 b^1 + \binom{9}{2}a^7 b^2 + \binom{9}{3}a^6 b^3 + \binom{9}{4}a^5 b^4 + \binom{9}{5}a^4 b^5 + \binom{9}{6}a^3 b^6 + \binom{9}{7}a^2 b^7 + \binom{9}{8})ab^8 + b^9 = a^9 + 9a^8 b + 36a^7 b^2 + 84a^6 b^3 + 126a^5 b^4 + 126a^4 b^5 + 84a^3 b^6 + 36a^2 b^7 + 9ab^8 + b^9$

11. $(t - a)^8 = \binom{8}{0}t^8 + \binom{8}{1}t^7(-a)^1 + \binom{8}{2}t^6(-a)^2 +$
$\binom{8}{3}t^5(-a)^3 + \binom{8}{4}t^4(-a)^4 + \binom{8}{5}t^3(-a)^5 +$
$\binom{8}{6}t^2(-a)^6 + \binom{8}{7}t(-a)^7 + \binom{8}{8}(-a)^8 = t^8 - 8t^7a +$
$28t^6a^2 - 56t^5a^3 + 70t^4a^4 - 56t^3a^5 + 28t^2a^6 - 8ta^7 +$
a^8

12. $(z - 2a)^7 = \binom{7}{0}x^7 + \binom{7}{1}x^6(-2a)^1 +$
$\binom{7}{2}x^5(-2a)^2 + \binom{7}{3}x^4(-2a)^3 + \binom{7}{4}x^3(-2a)^4 +$
$\binom{7}{5}x^2(-2a)^5 + \binom{7}{6}x(-2a)^6 + \binom{7}{7}(-2a)^7 = 1x^7 +$
$7x^6(-2a) + 21x^5(4a^2) + 35x^4(-8a^3) +$
$35x^3(-16a^4) + 21x^2(-32a^5) + 7x(64a^6) +$
$1(-128a^7) = x^7 - 14x^6a + 84x^5a^2 - 280x^4a^3 +$
$560x^3a^4 - 672x^2a^5 + 448xa^6 - 128a^7$

13. $(2a - 1)^6 = \binom{6}{0}(2a)^6 + \binom{6}{1}(2a)^5(-1) +$
$\binom{6}{2}(2a)^4(-1)^2 + \binom{6}{3}2a)^3(-1)^3 + \binom{6}{4}(2a)^2(-1)^4 +$
$\binom{6}{5}(2a)(-1)^5 + \binom{6}{6}(-1)^6 = 64a^6 - 192a^5 + 240a^4 -$
$160a^3 + 60a^2 - 12a + 1$

14. $(a^2 - 3)^9 = \binom{9}{0}(a^2)^9 + \binom{9}{1}(a^2)^8(-3) + \binom{9}{2}(a^2)^7(-3)^2$
$+ \binom{9}{3}(a^2)^6(-3)^3 + \binom{9}{4}(a^2)^5(-3)^4 + \binom{9}{5}(a^2)^4(-3)^5 +$
$\binom{9}{6}(a^2)^3(-3)^6 + \binom{9}{7}(a^2)^2(-3)^7 + \binom{9}{8}a^2(-3)^8 +$
$\binom{9}{9}(-3)^9 = a^{18} - 27a^{16} + 324a^{14} - 2268a^{12} +$
$10206a^{10} - 30618a^8 + 61236a^6 - 78732a^4 + 59049a^2$
$- 19683$

15. $\left(x^2y + \frac{a}{2}\right)^7 = \binom{7}{0}\left(x^2y\right)^7 + \binom{7}{1}\left(x^2y\right)^6\left(\frac{a}{2}\right) +$
$\binom{7}{2}\left(x^2y\right)^5\left(\frac{a}{2}\right)^2 + \binom{7}{3}\left(x^2y\right)^4\left(\frac{a}{2}\right)^3\binom{7}{4}\left(x^2y\right)^3\left(\frac{a}{2}\right)^4 +$
$\binom{7}{5}\left(x^2y\right)^2\left(\frac{a}{2}\right)^5 + \binom{7}{6}\left(x^2y\right)\left(\frac{a}{2}\right)^6 + \binom{7}{7}\left(\frac{a}{2}\right)^7 = x^{14}y^7 +$
$\frac{7}{2}x^{12}y^6a + \frac{21}{4}x^{10}y^5a^2 + \frac{35}{8}x^8y^4a^3 + \frac{35}{16}x^6y^3a^4 +$
$\frac{21}{32}x^4y^2a^5 + \frac{7}{64}x^2ya^6 + \frac{a^7}{128}$

16. $\left(\frac{a}{3} - \frac{xy}{2}\right)^6 = \binom{6}{0}\left(\frac{a}{3}\right)^6 + \binom{6}{1}\left(\frac{a}{3}\right)^5\left(-\frac{xy}{2}\right) +$
$\binom{6}{2}\left(\frac{a}{3}\right)^4\left(\frac{-xy}{2}\right)^2 + \binom{6}{3}\left(\frac{a}{3}\right)^3\left(-\frac{xy}{2}\right)^3 + \binom{6}{4}\left(\frac{a}{3}\right)^2 +$
$\binom{6}{5}\left(\frac{a}{3}\right)\left(-\frac{xy}{2}\right)^5 + \binom{6}{6}\left(-\frac{xy}{2}\right)^6 = \frac{a^6}{729} - \frac{a^5xy}{81} +$
$\frac{5a^4x^2y^2}{108} - \frac{5a^3x^3y^3}{54} + \frac{5a^2x^4y^4}{48} - \frac{ax^5y^5}{16} + \frac{x^6y^6}{64}$

17. $(x + y)^{12} = \binom{12}{0}x^{12} + \binom{12}{1}x^{11}y + \binom{12}{2}x^{10}y^2 +$
$\binom{12}{3}x^9y^3 + \cdots = x^{12} + 12x^{11}y + 66x^{10}y^2 + 220x^9y^3$
$+ \cdots$

18. $(x - 5)^{10} = \binom{10}{0}x^{10} + \binom{10}{1}x^9(-5) + \binom{10}{2}x^8(-5)^2 +$
$\binom{10}{3}x^7(-5)^3 + \cdots = x^{10} - 50x^9 + 1125x^8 - 15000x^7$
$+ \cdots$

19. $(1 - a)^{-2} = 1 + (-2)(-a) + \frac{(-2)(-3)}{2!}(-a)^2 +$
$\frac{(-2)(-3)(-4)}{3!}(-a)^3 + \cdots = 1 + 2a + 3a^2 + 4a^3 + \cdots$

20. $(1 + x)^{\frac{1}{2}} = 1 + \frac{1}{2}x + \frac{\frac{1}{2} \cdot \left(-\frac{1}{2}\right)}{2!}x^2 +$
$\frac{\frac{1}{2}\left(-\frac{1}{2}\right)\left(-\frac{3}{2}\right)}{3!}x^3 + \cdots = 1 + \frac{x}{2} - \frac{x^2}{8} + \frac{x^3}{16} + \cdots$

21. $(1 + b)^{-\frac{1}{3}} = 1 + \left(-\frac{1}{3}\right)b + \frac{\left(-\frac{1}{3}\right)\left(-\frac{4}{3}\right)}{2!}b^2 +$
$\frac{\left(-\frac{1}{3}\right)\left(-\frac{4}{3}\right)\left(-\frac{7}{3}\right)}{3!}b^3 + \cdots = 1 - \frac{b}{3} + \frac{2b^2}{9} + \frac{14b^3}{81} + \cdots$

22. $(1 - y)^{-\frac{1}{4}} = 1 + \left(-\frac{1}{4}\right)(-y) + \frac{\left(-\frac{1}{4}\right)\left(-\frac{5}{4}\right)\left(-\frac{9}{4}\right)}{3!}(-y)^3 =$
$1 + \frac{y}{4} + \frac{5y^2}{32} + \frac{15y^3}{128} + \cdots$

23. $(1.1)^4 = (1 + 0.1)^4 = 1 + 4(0.1) + \frac{4 \cdot 3}{2!}(0.1)^2 +$
$\frac{4 \cdot 3 \cdot 2}{3!}(0.1)^3 = 1 + 0.4 + 0.06 + 0.004 = 1.464$

24. $(0.85)^5 = (1 - 0.15)^5 = 1 + 5(-0.15)$
$+ \frac{5 \cdot 4}{2!}(-0.15)^2 + \frac{5 \cdot 4 \cdot 3}{3!}(-0.15)^3 + \frac{5 \cdot 4 \cdot 3 \cdot 2}{4!}(-0.15)^4 =$
$1 - 0.75 + 0.225 - 0.03375 + 0.00253 = 0.444$

25. $\sqrt{1.1} = (1 + 0.1)^{\frac{1}{2}} = 1 + \frac{1}{2}(0.1) + \frac{\frac{1}{2} \cdot \left(-\frac{1}{2}\right)}{2!}(0.1)^2 +$
$\frac{\frac{1}{2}\left(-\frac{1}{2}\right)\left(-\frac{3}{2}\right)}{3!}(0.1)^3 = 1 + 0.05 - 0.00125 + 0.00006 =$
1.049

26. $\sqrt[3]{1.01} = (1 + 0.01)^{\frac{1}{3}} = 1 + \frac{1}{3}(0.01) + \frac{\frac{1}{3}\left(-\frac{2}{3}\right)}{2!}(0.01)^2$
$= 1 + 0.00333 - 0.00001 \approx 1.003$

27. $\sqrt[5]{1.04} = (1 + 0.04)^{\frac{1}{5}} = 1 + \frac{1}{5}(0.04) + \frac{\frac{1}{5}\left(-\frac{4}{5}\right)}{2!}(0.04)^2$
$= 1 + 0.008 - 0.000128 \approx 1.008$

28. $\sqrt[6]{0.98} = (1 - 0.02)^{\frac{1}{6}} = 1 + \frac{1}{6}(-0.02) +$
$\frac{\frac{1}{6}\left(-\frac{5}{6}\right)}{2!}(-0.02)^2 = 1 - 0.0033 - 0.000028 \approx 0.997$

29. $\sqrt[3]{0.95} = (1 - 0.05)^{\frac{1}{3}} = 1 + \frac{1}{3}(-0.05) = 1 - 0.017 \approx 0.983$

30. $\sqrt[4]{0.925} = (1 - 0.075)^{\frac{1}{4}} = 1 + \frac{1}{4}(-0.075) +$
$\frac{\frac{1}{4}\left(-\frac{3}{4}\right)}{2!}(-0.075)^2 = 1 - 0.01875 - 0.000527 \approx$
0.981

31. Counting from 0, the sixth term is $\binom{15}{5}(x^2)^{10}y^5 = 3003x^{20}y^5$

32. Counting from 0, the eighth term is $\binom{20}{7}a^{13}b^7 = 77{,}520a^{13}b^7$

33. Fifth term is $\binom{12}{4}(2x)^8(-y)^4 = 126{,}720x^8y^4$

34. Fourth term is $\binom{10}{3}(3x)^7(-2y)^3 = -2{,}099{,}520x^7y^3$

35. Term involving b^4 is the 5th term: $\binom{14}{4}a^{10}b^4 = 1001a^{10}b^4$

36. Term involving x^5 is the $(15-5+1) = 11$th term: $\binom{15}{10}x^5y^{10} = 3003x^5y^{10}$

37. (a) $\left(1 - \frac{v^2}{c^2}\right)^{-\frac{1}{2}} = 1 + \left(-\frac{1}{2}\right)\left(-\frac{v^2}{c^2}\right) + \frac{\left(-\frac{1}{2}\right)\left(-\frac{3}{2}\right)}{2!}\left(-\frac{v^2}{c^2}\right)^2 = 1 + \frac{v^2}{2c^2} + \frac{3v^4}{8c^4}$

(b) $mc^2 + \frac{mv^2}{2} + \frac{3mv^4}{8c^2}$

38. (a) $\sqrt{1 - \frac{1}{4V^2}} = \left(1 - \frac{1}{4V^2}\right)^{\frac{1}{2}} = 1 - \frac{1}{2}\cdot\frac{1}{4V^2} = 1 - \frac{1}{8V^2}$,

(b) $1 - \frac{v}{c} = 1 - \left(1 - \frac{1}{8V^2}\right) = \frac{1}{8V^2}$,

(c) $V = 100$; $\frac{1}{8V^2} = 1.25 \times 10^{-5}$,

(d) $V = 500$; $\frac{1}{8V^2} = 5 \times 10^{-7}$,

(e) $V = 125{,}000$; $\frac{1}{8V^2} = 8 \times 10^{-12}$,

(f) $V = 2.50 \times 10^5$; $\frac{1}{8V^2} = 2 \times 10^{-12}$

39.
$$(r^2)^{3/2}\left(1 + \frac{\ell^2}{r^2}\right)^{3/2}$$
$$= r^3\left[1 + \frac{3}{2}\cdot\frac{\ell^2}{r^2} + \frac{\frac{3}{2}\cdot\frac{1}{2}}{2!}\left(\frac{\ell^2}{r^2}\right)^2\right]$$
$$= r^3\left[1 + \frac{3\ell^2}{2r^2} + \frac{3\ell^4}{8r^4}\right]$$
$$= r^3 + \frac{3}{2}r\ell^2 + \frac{3\ell^4}{8r}$$

40.
$$\frac{(x+h)^4 - x^4}{h} = \frac{x^4 + 4x^3h + 6x^2h^2 + 4xh^3 + h^4 - x^4}{h}$$
$$= \frac{4x^3h + 6x^2h^2 + 4xh^3 + h^4}{h}$$
$$= 4x^3 + 6x^2h + 4xh^2 + h^3$$

CHAPTER 19 REVIEW

1. $\frac{1}{1+2} = \frac{1}{3}$; $\frac{1}{2+2} = \frac{1}{4}$; $\frac{1}{3+2} = \frac{1}{5}$; $\frac{1}{4+2} = \frac{1}{6}$; $\frac{1}{5+2} = \frac{1}{7}$; $\frac{1}{6+2} = \frac{1}{8}$

2. $\frac{3}{2-1} = 3$; $\frac{3}{4-1} = 1$; $\frac{3}{6-1} = \frac{3}{5}$; $\frac{3}{8-1} = \frac{3}{7}$; $\frac{3}{10-1} = \frac{1}{3}$; $\frac{3}{12-1} = \frac{3}{11}$

3. $\frac{-1}{3}$; $\frac{1}{2\cdot5} = \frac{1}{10}$; $\frac{-1}{3(7)} = \frac{-1}{21}$; $\frac{1}{4(9)} = \frac{1}{36}$; $\frac{-1}{5(11)} = \frac{-1}{55}$; $\frac{1}{6(13)} = \frac{1}{78}$

4. $\frac{1+1}{3-1} = \frac{2}{2} = 1$; $\frac{5}{5} = 1$; $\frac{10}{8} = \frac{5}{4}$; $\frac{17}{11}$; $\frac{26}{14} = \frac{13}{7}$; $\frac{37}{17}$

5. $a_1 = 1$; $a_2 = \frac{1}{2}$; $a_3 = \frac{\frac{1}{2}}{3} = \frac{1}{6}$; $a_4 = \frac{\frac{1}{6}}{4} = \frac{1}{24}$; $a_5 = \frac{\frac{1}{24}}{5} = \frac{1}{120}$; $a_6 = \frac{\frac{1}{120}}{6} = \frac{1}{720}$

6. $a_1 = 5$; $a_2 = 5 - 2 = 3$; $a_3 = 3 - 3 = 0$; $a_4 = 0 - 4 = -4$; $a_5 = -4 - 5 = -9$; $a_6 = -9 - 6 = -15$

7. $a_1 = 1$; $a_2 = 3$; $a_3 = 1 + 3 = 4$; $a_4 = 3 + 4 = 7$; $a_5 = 4 + 7 = 11$; $a_6 = 7 + 11 = 18$

8. $a_1 = 1$; $a_2 = 2$; $a_3 = 1\cdot2 = 2$; $a_4 = 2\cdot2 = 4$; $a_5 = 2\cdot4 = 8$; $a_6 = 4\cdot8 = 32$

9. Arithmetic; $d = -3$, $a_{10} = 10 + 9(-3) = -17$

10. Geometric: $r = \frac{1}{4}$; $a_8 = 8\cdot\left(\frac{1}{4}\right)^7 = \frac{1}{2048}$

11. Arithmetic: $d = 4$; $a_7 = -1 + 6(4) = 23$

12. Neither; $7, 4, 0, -5, -11, -18, -26, -35, -45, = a_9$

13. Geometric: $r = -\frac{1}{6}$; $a_{10} = 3\left(-\frac{1}{6}\right)^9 = \frac{-1}{3,359,232} \approx 2.97687 \times 10^{-7} = 0.0000003$

14. Geometric: $r = 10$; $a_7 = 2.05 \times 10^6 = 2{,}050{,}000$

15. $d = 18 - 12 = 6$; $a_7 = a_1 + 6d$; $12 = a_1 + 6\cdot6$; $a_1 = 12 - 36 = -24$

16. $r = \frac{\frac{9}{7}}{9} = \frac{1}{7}$; $a_8 = a_7\cdot r = \frac{9}{7}\cdot\frac{1}{7} = \frac{9}{49}$

17. (a) $4\cdot\frac{1}{3} + 4\cdot\frac{1}{9} + 4\cdot\frac{1}{27} + 4\cdot\frac{1}{81} = \frac{4}{3} + \frac{4}{9} + \frac{4}{27} + \frac{4}{81}$

(b) $2 + \frac{3}{2} + \frac{4}{3} + \frac{5}{4}$

18. $2 + 3 + 4 + 5 = 14$

19. $-1 + 2 + 5 + 8 + 11 + 14 = 39$

20. $\frac{-1}{3} + \frac{2}{5} - \frac{3}{9} + \frac{4}{17} = \frac{-8}{255} = -0.03137$

21. Arithmetic: $d = 5; a_{10} = 4 + 9 \cdot 5 = 49; S_{10} = \frac{10}{2}(4 + 49) = 5(53) = 265$

22. Arithmetic: $d = -7; a_{12} = 2 + 11(-7) = -75; S_{12} = \frac{12}{2}(2 - 75) = 6(-73) = -438$

23. Geometric: $r = \frac{1}{3}; S_{14} = 1 \cdot \dfrac{1 - \frac{1}{3}^{14}}{1 - \frac{1}{3}} = 1.4999997$

24. Arithmetic: There are two possible ways to look at the sequence. One way is to let $a_1 = \sqrt{5} + 1$ and $d = \sqrt{5} + 1$. Then $a_8 = \sqrt{5} + 1 + (7)(\sqrt{5} + 1) = 8\sqrt{5} + 8$ and $S_8 = \frac{8}{2}(\sqrt{5} + 1 + 8\sqrt{5} + 8) = 4(9\sqrt{5} + 9) = 36\sqrt{5} + 36$. The other way to look at this sequence is to let $a_1 = \sqrt{5}$ and $d = 1 + \sqrt{5}$. Then, $a_8 = \sqrt{5} + 7(1 + \sqrt{5}) = 7 + 8\sqrt{5}$ and $S_8 = \frac{8}{2}(\sqrt{5} + 7 + 8\sqrt{5}) = 4(7 + 9\sqrt{5}) = 28 + 36\sqrt{5}$.

25. Converges: $r = \frac{1}{2}; S = \dfrac{\frac{1}{3}}{1 - \frac{1}{2}} = \frac{2}{3}$

26. Converges: $r = \frac{3}{4}; S = \dfrac{8}{1 - \frac{3}{4}} = 32$

27. Diverges: $r = 10$

28. Converges: $r = -\frac{1}{10}; S = \frac{1.5}{1 + 0.1} = \frac{1.5}{1.1} = \frac{15}{11} \approx 1.3636$

29. Converges: $r = \frac{1}{\sqrt{3}}; S = \dfrac{\frac{1}{\sqrt{3}}}{1 - \frac{1}{\sqrt{3}}} = \dfrac{\frac{1}{\sqrt{3}}}{\frac{\sqrt{3}-1}{\sqrt{3}}} = \frac{1}{\sqrt{3}-1} = \frac{\sqrt{3}+1}{2} \approx 1.3660$

30. Converges: $r = \frac{1}{\sqrt{2}}; a_1 = 1; S = \dfrac{1}{1 - \frac{1}{\sqrt{2}}} = \dfrac{1}{1 - \frac{\sqrt{2}-1}{\sqrt{2}}} = \frac{\sqrt{2}}{\sqrt{2}-1} = \frac{2+\sqrt{2}}{1} \approx 3.4142$

31. $a_1 = 0.185; r = 0.001; S = \frac{0.185}{0.999} = \frac{185}{999} = \frac{5}{27}$

32. Start with $a_1 = 0.01; r = 0.1; S = \frac{0.01}{.9} = \frac{1}{90}; 0.6 + \frac{1}{90} = \frac{6}{10} + \frac{1}{90} = \frac{54}{90} + \frac{1}{90} = \frac{55}{90} = \frac{11}{18}$

33. $(a+2)^5 = a^5 + 5a^4 \cdot 2 + 10a^3 \cdot 2^2 + 10a^2 \cdot 2^3 + 5a^1 \cdot 2^4 + 2^5 = a^5 + 10a^4 + 40a^3 + 80a^2 + 80a + 32$

34. $(3x - y)^6 = (3x)^6 + 6(3x)^5(-y) + 15(3x)^4(-y)^2 + 20(3x)^3(-y)^3 + 15(3x)^2(-y)^4 + 6(3x)(-y)^5 + (-y)^6 = 729x^6 - 1458x^5y + 1215x^4y^2 - 540x^3y^3 + 135x^2y^4 - 18xy^5 + y^6$

35. $\left(\frac{x}{2} - 3y^2\right)^6 = \left(\frac{x}{2}\right)^6 + 6\left(\frac{x}{2}\right)^5(-3y^2) + 15\left(\frac{x}{2}\right)^4(-3y^2)^2 + 20\left(\frac{x}{2}\right)^3(-3y^2)^3 + 15\left(\frac{x}{2}\right)^2(-3y^2)^4 + 6\left(\frac{x}{2}\right)(-3y^2)^5 + (-3y^2)^6 = \frac{x^6}{64} - \frac{9}{16}x^5y^2 + \frac{135}{16}x^4y^4 - \frac{135}{2}x^3y^6 + \frac{1215}{4}x^2y^8 - 729xy^{10} + 729y^{12}$

36. $\left(\frac{2a^3}{5} + \frac{5b}{2}\right)^5 = \left(\frac{2a^3}{5}\right)^5 + 5\left(\frac{2a^3}{5}\right)^4\left(\frac{5b}{2}\right) + 10\left(\frac{2a^3}{5}\right)^3\left(\frac{5b}{2}\right)^2 + 10\left(\frac{2a^3}{5}\right)^2\left(\frac{5b}{2}\right)^3 + 5\left(\frac{2a^3}{5}\right)\left(\frac{5b}{2}\right)^4 + \left(\frac{5b}{2}\right)^5 = \frac{32a^{15}}{3125} + \frac{8a^{12}b}{25} + 4a^9b^2 + 25a^6b^3 + \frac{625a^3b^4}{8} + \frac{3125b^5}{32}$

37. $(2x + y)^{15} = (2x)^{15} + 15(2x)^{14}y + \binom{15}{2}(2x)^{13}y^2 + \binom{15}{3}(2x)^{12}y^3 + \cdots = 32{,}768x^{15} + 245{,}760x^{14}y + 860{,}160x^{13}y^2 + 1{,}863{,}680x^{12}y^3 + \cdots$

38. $(1 + ax^2)^{10} = 1^{10} + 10 \cdot 1^9(ax^2) + \binom{10}{2}1^8(ax^2)^2 + \binom{10}{3}1^7(ax^2)^3 + \cdots = 1 + 10ax^2 + 45a^2x^4 + 120a^3x^6 + \cdots$

39. $(1 - x)^{-5} = 1 - 5(-x) + \frac{(-5)(-6)}{2!}x^2 + \frac{(-5)(-6)(-7)}{3!}(-x)^3 + \cdots = 1 + 5x + 15x^2 + 35x^3 + \cdots$

40. $(1 + b)^{-\frac{1}{4}} = 1 - \frac{1}{4}b + \frac{(-\frac{1}{4})(-\frac{5}{4})}{2!}b^2 + \frac{(-\frac{1}{4})(-\frac{5}{4})(-\frac{9}{4})}{3!}b^3 + \cdots = 1 - \frac{b}{4} + \frac{5b^2}{32} - \frac{15b^3}{128} + \cdots$

41. $\binom{20}{6}x^{14}(2y)^6 = 2{,}480{,}640x^{14}y^6$

42. $\binom{12}{4}a^8(-3b)^4 = 40{,}095a^8b^4$

43. $\sqrt[5]{1.02} = (1 + 0.02)^{\frac{1}{5}} = 1 + \frac{1}{5}(0.02) + \frac{\frac{1}{5}(-\frac{4}{5})}{2!}(0.02)^2 = 1 + 0.004 - 0.000032 \approx 1.004$

44. $\sqrt[7]{0.98} = (1 - 0.02)^{\frac{1}{7}} = 1 - \frac{1}{7}(0.02) \approx 1 - 0.003 = 0.997$

45. $\frac{1}{2}\%$ per month $= 0.005$; 10 years $= 120$ months; $A = 400(1.005)^{120} = \$727.76$

46. Geometric sequence: $a_1 = 20; r = \frac{3}{4}; a_6 = 20 \cdot \left(\frac{3}{4}\right)^5 = 4.75$ cm

47. $S_6 = 20 \cdot \dfrac{1 - \left(\frac{3}{4}\right)^6}{1 - \frac{3}{4}} = 65.76$ cm

48. $S = \dfrac{20}{1 - \frac{3}{4}} = \dfrac{20}{\frac{1}{4}} = 80$ cm

49. First convert 250 ft to inches. Since $250 \times 12 = 3000$, we see that 250 ft $= 3000$ in. $S_n = \frac{n}{2}(a_1 + a_n)$; $a_n = a_1 + (n-1)d$ so $a_n = 10 + (n-1)15 = -5 + 15n$. Substituting, we get $S_n = \frac{n}{2}(10 - 5 + 15n) = \frac{n}{2}(5 + 15n)\frac{5n+15n^2}{2} = 3000$. Hence, $15n^2 + 5n - 6000 = 0$. Dividing by 5, we have $3n^2 + n - 1200 = 0$. The quadratic formula gives answers of -20.167 and 19.834. Since n must be positive, the answer is 19.834 s

50. $a_{10} = 250,000 + 9(40,000) + \$610,000$. $S_{10} = \frac{10}{2}(250,000 + 610,000) = \$4,300,000$

CHAPTER 19 TEST

1. $\frac{5}{-1} = -5; \frac{5}{1} = 5; \frac{5}{3}; \frac{5}{5} = 1; \frac{5}{7}; \frac{5}{9}$

2. $a_1 = -2; a_2 = 3 + 2(-2) = -1; a_3 = 3 + 3(-1) = 0; a_4 = 3 + 4 \cdot 0 = 3; a_5 = 3 + 5 \times 3 = 18; a_6 = 3 + 6 \times 18 = 111$

3. Arithmetic: $d = -2.5; a_{10} = 15 + 9(-2.5) = -7.5$

4. $d = 14 - 10 = 4; a_5 = a_1 + 4d; 10 = a_1 + 4 \cdot 4; a_1 = -6$

5. $\displaystyle\sum_{k=1}^{5}(2k - 1) = 1 + 3 + 5 + 7 + 9 = 25$

6. Converges: $r = -\frac{1}{2}; S = \dfrac{\frac{1}{5}}{1 + \frac{1}{2}} = \dfrac{\frac{1}{5}}{\frac{3}{2}} = \dfrac{2}{15}$

7. $a_1 = 0.435; r = 0.001; S = \dfrac{0.435}{0.999} = \dfrac{435}{999} = \dfrac{145}{333}$

8. $(3x - 2y)^{12} = (3x)^{12} - 12(3x)^{11}(2y) + \binom{12}{2}(3x)^{10}(2y)^2 - \binom{12}{3}(3x)^9(2y)^3 + \cdots = 531,441x^{12} - 4,251,528x^{11}y + 15,588,936x^{10}y^2 - 34,642,080x^9y^3$

9. Arithmetic Series: $a_1 = 2000; d = 1500; a_{12} = 2000 + 11(1500) = 18,500; S_{12} = \frac{12}{2}(2000 + 18,500) = 123,000$ copies

10. Using $a_1 = 1.4$ and $n = 11$, we have that $a_n = a_1 r^{10} = 1.40 \times 1.01^{10} = 1.55$ nA.

CHAPTER

20

Trigonometric Formulas, Identities and Equations

≡ 20.1 BASIC IDENTITIES

1. $\sin^2\theta + \cos^2\theta = 1$; Dividing by $\sin^2\theta$ produces
$\frac{\sin^2\theta}{\sin^2\theta} + \frac{\cos^2\theta}{\sin^2\theta} = \frac{1}{\sin^2\theta}$ or $1 + \cot^2\theta = \csc^2\theta$

2. $\tan x \cot x = 1$

$$\frac{\sin x}{\cos x} \cdot \frac{\cos x}{\sin x} \,\Big|\, 1$$
$$1$$

3. $\sin\theta \sec\theta = \tan\theta$

$$\sin\theta \frac{1}{\cos\theta} \,\Big|\, \tan\theta$$
$$\frac{\sin\theta}{\cos\theta}$$
$$\tan\theta$$

4. $\cos\theta(\tan\theta + \sec\theta) = \sin\theta + 1$

$$\cos\theta\tan\theta + \cos\theta\sec\theta \,\Big|\, \sin\theta + 1$$
$$\cos\theta\frac{\sin\theta}{\cos\theta} + \cos\theta \cdot \frac{1}{\cos\theta}$$
$$\sin\theta + 1$$

5. $\dfrac{\sin\theta}{\cot\theta} = \sec\theta - \cos\theta$

$$\frac{\sin\theta}{\frac{\cos\theta}{\sin\theta}} \,\Big|\, \frac{1}{\cos\theta} - \cos\theta$$
$$\frac{\sin^2\theta}{\cos\theta}$$
$$\frac{1-\cos^2\theta}{\cos\theta}$$
$$\frac{1}{\cos\theta} - \cos\theta$$

6. $\tan x = \dfrac{\sec x}{\csc x}$

$$\tan x \,\Bigg|\, \frac{\frac{1}{\cos x}}{\frac{1}{\sin x}}$$
$$\frac{1}{\cos x} \cdot \frac{\sin x}{1}$$
$$\frac{\sin x}{\cos x}$$
$$\tan x$$

7. $\dfrac{(1 - \sin^2\theta)(1 + \tan^2\theta) = 1}{}$

$$1 + \tan^2\theta - \sin^2\theta - \sin^2\theta\tan^2\theta \,\Big|\, 1$$
$$1 + \frac{\sin^2\theta}{\cos^2\theta} - \sin^2\theta - \sin^2\theta \cdot \frac{\sin^2\theta}{\cos^2\theta}$$
$$\frac{\cos^2\theta}{\cos^2\theta} + \frac{\sin^2\theta}{\cos^2\theta} - \frac{\sin^2\theta\cos^2\theta}{\cos^2\theta} - \frac{\sin^2\theta\sin^2\theta}{\cos^2\theta}$$
$$\frac{\cos^2\theta + \sin^2\theta - \sin^2\theta(\cos^2\theta + \sin^2\theta)}{\cos^2\theta}$$
$$\frac{1-\sin^2\theta}{\cos^2\theta}$$
$$\frac{\cos^2\theta}{\cos^2\theta}$$
$$1$$

8. $\dfrac{\sin A}{\csc A} + \dfrac{\cos A}{\sec A} = 1$

$$\frac{\sin A}{\frac{1}{\sin A}} + \frac{\cos A}{\frac{1}{\cos A}} \,\Bigg|\, 1$$
$$\sin^2 A + \cos^2 A$$
$$1$$

9. $1 - \dfrac{\sin A}{\csc A} = \cos^2 A$

$$1 - \frac{\sin A}{1/\sin A} \,\Big|\, \cos^2 A$$
$$1 - \sin^2 A$$
$$\cos^2 A$$

10. $(1 + \tan\theta)(1 - \tan\theta) = 2 - \sec^2\theta$

$$1 - \tan^2\theta \,\Big|\, 2 - \sec^2\theta$$
$$1 - (\sec^2\theta - 1)$$
$$2 - \sec^2\theta$$

11. $(1 + \cos x)(1 - \cos x) = \sin^2 x$

$$1 - \cos^2 x \,\Big|\, \sin^2 x$$
$$\sin^2 x$$

12. $\dfrac{\sec^4 x - \sec^2 x = \tan^4 x + \tan^2 x}{}$

$$\sec^2 x(\sec^2 x - 1) \,\Big|\, \tan^4 x + \tan^2 x$$
$$\sec^2 x(\tan^2 x)$$
$$(\tan^2 x + 1)(\tan^2 x)$$
$$\tan^4 x + \tan^2 x$$

13. $2\csc\theta = \frac{\sin\theta}{1+\cos\theta} + \frac{1+\cos\theta}{\sin\theta}$

$2\csc\theta$	$\frac{\sin^2\theta}{\sin\theta(1+\cos\theta)} + \frac{(1+\cos\theta)(1+\cos\theta)}{\sin\theta(1+\cos\theta)}$
	$\frac{\sin^2\theta+1+2\cos\theta+\cos^2\theta}{\sin\theta(1+\cos\theta)}$
	$\frac{\sin^2\theta+\cos^2\theta+1+2\cos\theta}{\sin\theta(1+\cos\theta)}$
	$\frac{2+2\cos\theta}{\sin\theta(1+\cos\theta)}$
	$\frac{2(1+\cos\theta)}{\sin\theta(1+\cos\theta)}$
	$\frac{2}{\sin\theta}$
	$2\csc\theta$

14. $\cos x = \sin x\cot x$

$\cos x$	$\sin x\cdot\frac{\cos x}{\sin x}$
	$\cos x$

15. $(\sin\theta+\cos\theta)^2 = 1+2\sin\theta\cos\theta$

$\sin^2\theta+2\sin\theta\cos\theta+\cos^2\theta$	$1+2\sin\theta\cos\theta$
$\sin^2\theta+\cos^2\theta+2\sin\theta\cos\theta$	
$1+2\sin\theta\cos\theta$	

16. $\csc^2 x(1-\cos^2 x) = 1$

$\csc^2 x-\csc^2 x\cos^2 x$	1
$\frac{1}{\sin^2 x}-\frac{\cos^2 x}{\sin^2 x}$	
$\frac{1-\cos^2 x}{\sin^2 x}$	
$\frac{\sin^2 x}{\sin^2 x}$	
1	

17. $\frac{\tan\theta+\cot\theta}{\tan\theta-\cot\theta} = \frac{\tan^2\theta+1}{\tan^2\theta-1}$

$\frac{\frac{\sin\theta}{\cos\theta}+\frac{\cos\theta}{\sin\theta}}{\frac{\sin\theta}{\cos\theta}-\frac{\cos\theta}{\sin\theta}}$	$\frac{\frac{\sin^2\theta}{\cos^2\theta}+\frac{\cos^2\theta}{\cos^2\theta}}{\frac{\sin^2\theta}{\cos^2\theta}-\frac{\cos^2\theta}{\cos^2\theta}}$
$\frac{\frac{\sin^2\theta+\cos^2\theta}{\sin\theta\cos\theta}}{\frac{\sin^2\theta-\cos^2\theta}{\sin\theta\cos\theta}}$	$\frac{\sin^2\theta+\cos^2\theta}{\sin^2\theta-\cos^2\theta}$
$\frac{\sin^2\theta+\cos^2\theta}{\sin^2\theta-\cos^2\theta}$	

18. $\frac{1-\sin x}{\cos x} = \frac{\cos x}{1+\sin x}$

$\frac{1-\sin x}{\cos x}\cdot\frac{1+\sin x}{1+\sin x}$	$\frac{\cos x}{1+\sin x}$
$\frac{1-\sin^2 x}{\cos x(1+\sin x)}$	
$\frac{\cos^2 x}{\cos x(1+\sin x)}$	
$\frac{\cos x}{1+\sin x}$	

19. $\frac{\sec\theta-\csc\theta}{\sec\theta+\csc\theta} = \frac{\tan\theta-1}{\tan\theta+1}$

$\frac{\frac{1}{\cos\theta}-\frac{1}{\sin\theta}}{\frac{1}{\cos\theta}+\frac{1}{\sin\theta}}$	$\frac{\tan\theta-1}{\tan\theta+1}$
$\frac{\left(\frac{1}{\cos\theta}-\frac{1}{\sin\theta}\right)\sin\theta}{\left(\frac{1}{\cos\theta}+\frac{1}{\sin\theta}\right)\sin\theta}$	
$\frac{\frac{\sin\theta}{\cos\theta}-\frac{\sin\theta}{\sin\theta}}{\frac{\sin\theta}{\cos\theta}+\frac{\sin\theta}{\sin\theta}}$	
$\frac{\tan\theta-1}{\tan\theta+1}$	

20. $(\sin A+\cos A)^2 + (\sin A-\cos A)^2 = 2$

$\sin^2 A+2\sin A\cos A+\cos^2 A$ $+\sin^2 A-2\sin A\cos A+\cos^2 A$	2
$\sin^2 A+\cos^2 A+\sin^2 A+\cos^2 A$	
$1+1$	
2	

21. $\tan^2 x\cos^2 x + \cot^2 x\sin^2 x = 1$

$\frac{\sin^2 x}{\cos^2 x}\cdot\frac{\cos^2 x}{1} + \frac{\cos^2 x}{\sin^2 x}\cdot\frac{\sin^2 x}{1}$	1
$\sin^2 x+\cos^2 x$	
1	

22. $\tan\theta + \frac{\cos\theta}{1+\sin\theta} = \sec\theta$

$\frac{\sin\theta}{\cos\theta} + \frac{\cos\theta}{1+\sin\theta}$	$\frac{1}{\cos\theta}$
$\frac{\sin\theta(1+\sin\theta)}{\cos\theta(1+\sin\theta)} + \frac{\cos^2\theta}{\cos\theta(1+\sin\theta)}$	
$\frac{\sin\theta+\sin^2\theta+\cos^2\theta}{\cos\theta(1+\sin\theta)}$	
$\frac{\sin\theta+1}{\cos\theta(1+\sin\theta)}$	
$\frac{1}{\cos\theta}$	

23. $\sec^4 x - \sec^2 x = \tan^2 x\sec^2 x$

$\sec^2 x(\sec^2 x-1)$	$\sec^2 x\tan^2 x$
$\sec^2 x\tan^2 x$	

24. $\cos^2 A - \sin^2 A = 2\cos^2 A - 1$

$\cos^2 A-(1-\cos^2 A)$	$2\cos^2 A-1$
$\cos^2 A-1+\cos^2 A$	
$2\cos^2 A-1$	

25. $\frac{\tan\theta-\sin\theta}{\sin^3\theta} = \frac{\sec\theta}{1+\cos\theta}$

$\frac{\frac{\sin\theta}{\cos\theta}-\sin\theta}{\sin\theta\cdot\sin^2\theta}$	$\frac{\sec\theta}{1+\cos\theta}$
$\frac{\sin\theta\left(\frac{1}{\cos\theta}-1\right)}{\sin\theta(1-\cos^2\theta)}$	
$\frac{\frac{1-\cos\theta}{\cos\theta}}{(1+\cos\theta)(1-\cos\theta)}$	
$\frac{\frac{1}{\cos\theta}}{1+\cos\theta}$	
$\frac{\sec\theta}{1+\cos\theta}$	

26. $\frac{\sin x-\cos x+1}{\sin x+\cos x-1} = \frac{\sin x+1}{\cos x}$

$\frac{\sin x-\cos x+1}{\sin x+\cos x-1}\cdot\frac{\sin x+\cos x+1}{\sin x+\cos x+1}$	$\frac{\sin x+1}{\cos x}$
$\frac{[(\sin x+1)-\cos x][(\sin x+1)+\cos x]}{[(\sin x+\cos x)-1][(\sin x+\cos x)+1]}$	
$\frac{(\sin x+1)^2-\cos^2 x}{(\sin x+\cos x)^2-1}$	
$\frac{\sin^2 x+2\sin x+1-\cos^2 x}{\sin^2 x+2\sin x\cos x+\cos^2 x-1}$	
$\frac{\sin^2 x+2\sin x+1-(1-\sin^2 x)}{2\sin x\cos x}$	
$\frac{2\sin^2 x+2\sin x}{2\sin x\cos x}$	
$\frac{2\sin x(\sin x+1)}{2\sin x\cos x}$	
$\frac{\sin x+1}{\cos x}$	

27.

$$\tan^2\theta\,\csc^2\theta\,\cot^2\theta\,\sin^2\theta = 1$$

$\dfrac{\sin^2\theta}{\cos^2\theta}\cdot\dfrac{1}{\sin^2\theta}\cdot\dfrac{\cos^2\theta}{\sin^2\theta}\cdot\dfrac{\sin^2\theta}{1}$	1
1	

28.

$$\tan x\,\sin x + \cos x = \sec x$$

$\dfrac{\sin x}{\cos x}\cdot\dfrac{\sin x}{1}+\cos x$	$\dfrac{1}{\cos x}$
$\dfrac{\sin^2 x}{\cos x}+\dfrac{\cos^2 x}{\cos x}$	
$\dfrac{\sin^2 x+\cos^2 x}{\cos x}$	
$\dfrac{1}{\cos x}$	

29.

$$\frac{\sec A+\csc A}{\tan A+\cot A} = \sin A + \cos A$$

$\dfrac{\left(\dfrac{1}{\cos A}+\dfrac{1}{\sin A}\right)\sin A\cos A}{\left(\dfrac{\sin A}{\cos A}+\dfrac{\cos A}{\sin A}\right)\sin A\cos A}$	$\sin A + \cos A$
$\dfrac{\sin A+\cos A}{\sin^2 A+\cos^2 A}$	
$\dfrac{\sin A+\cos A}{1}$	
$\sin A + \cos A$	

30.

$$\frac{\sin^3 x+\cos^3 x}{\sin x+\cos x} = 1 - \sin x\cos x$$

$(\sin x + \cos x)\times$	$1-\sin x\cos x$
$\dfrac{(\sin^2 x-\sin x\cos x+\cos^2 x)}{\sin x+\cos x}$	
$\sin^2 x + \cos^2 x - \sin x\cos x$	
$1 - \sin x\cos x$	

31. This identity is true based on the graphs of $y = 2\csc 2x$ and $y = \sec x\csc x$ shown below.

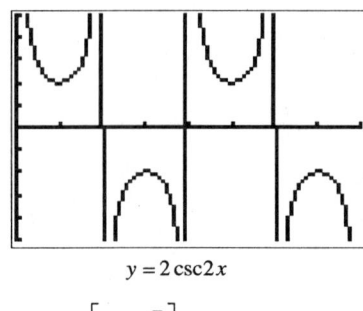

$$y = 2\csc 2x$$

$$\left[0,2\pi,\frac{\pi}{4}\right]\cdot[-5,5,1]$$

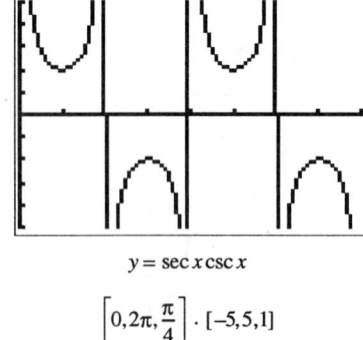

$$y = \sec x\csc x$$

$$\left[0,2\pi,\frac{\pi}{4}\right]\cdot[-5,5,1]$$

32. This identity is true based on the graphs of $y = \cos 2x + 1$ and $y = 2\cos^2 x$ shown below.

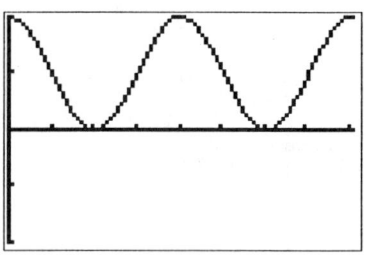

$$y = \cos 2x + 1$$

$$\left[0,2\pi,\frac{\pi}{4}\right]\cdot[-2,2,1]$$

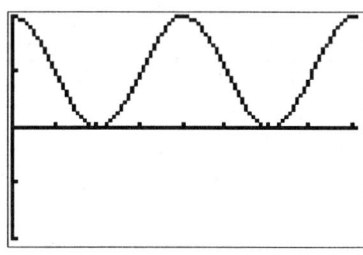

$$y = 2\cos^2 x$$

$$\left[0,2\pi,\frac{\pi}{4}\right]\cdot[-2,2,1]$$

33. This identity is not true based on the graphs of $y = \sin\frac{1}{2}x$ and $y = \frac{1}{2}\sin x$ shown below.

$$y = \sin\tfrac{1}{2}x$$

$$\left[0,2\pi,\frac{\pi}{4}\right]\cdot[-2,2,1]$$

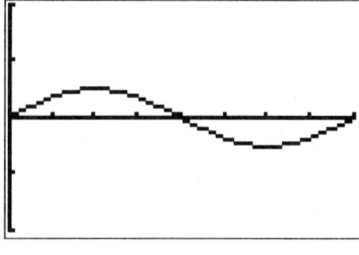

$$y = \tfrac{1}{2}\sin x$$

$$\left[0,2\pi,\frac{\pi}{4}\right]\cdot[-2,2,1]$$

34. This identity is not true based on the graphs of $y = 1 - \cos 2x$ and $y = \sin^2 x$ shown below. However, it appears as if $1 - \cos 2x = 2\sin^2 x$.

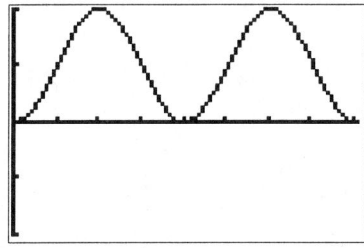

$$y = 1 - \cos 2x$$

$$\left[0, 2\pi, \frac{\pi}{4}\right] \cdot [-2, 2, 1]$$

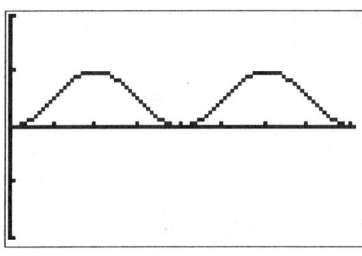

$$y = \sin^2 x$$

$$\left[0, 2\pi, \frac{\pi}{4}\right] \cdot [-2, 2, 1]$$

35. For $2\sin\theta \neq \sin 2\theta$ let $\theta = 90°$. $2\sin 90° \neq \sin(2 \cdot 90°)$, $2 \times 1 \neq \sin 180°$ and $2 \neq 0$.

36. If $A = 60°$, then $\frac{\tan 60°}{2} \neq \tan\left(\frac{60°}{2}\right)$ or $\frac{\sqrt{3}}{2} \neq \tan 30°$ which is equivalent to $\frac{\sqrt{3}}{2} \neq \frac{\sqrt{3}}{3}$

37. If $\theta = \pi$, then we have $\cos \pi^2 \neq (\cos \pi)^2$ or $-0.90269 \neq (-1)^2$, which is equivalent to $-0.90269 \neq 1$.

38. If $x = 60°$ and $y = 30°$, then $\sin(60° - 30°) \neq \sin 60° - \sin 30°$ or $\sin 30° \neq \frac{\sqrt{3}}{2} - \frac{1}{2}$, and so $\frac{1}{2} \neq \frac{\sqrt{3}-1}{2}$.

39. Here we let $x = 120°$. Now, $\sin 120° = \frac{\sqrt{3}}{2}$ and $\frac{\tan 120°}{\sqrt{1+\tan^2 120°}} = \frac{-\sqrt{3}}{\sqrt{1+(-\sqrt{3})^2}}$, which is negative. Hence, we see that $\sin x \neq \frac{\tan x}{\sqrt{1+\tan^2 x}}$ when $x = 120°$.

40. $\frac{(\sin x)(-\sin x) - (\cos x)(\cos x)}{\sin^2 x} = \frac{-\sin^2 x - \cos^2 x}{\sin^2 x} = \frac{-1}{\sin^2 x} = -\csc^2 x$

41. $-2\cot x \csc^2 x = -2\frac{\cos x}{\sin x} \cdot \csc^2 x = -2\cos x \cdot \frac{1}{\sin x} \cdot \csc^2 x = -2\cos x \csc x \csc^2 x = -2\cos x \csc^3 x$

42. $\sin^5 x = \sin^4 x \sin x = (\sin^2 x)^2 \sin x = (1 - \cos^2 x)^2 \sin x = (1 - 2\cos^2 x + \cos^4 x)\sin x$

43. $\frac{(1.2\sin\omega t - 1.6\cos\omega t)^2 + (1.6\sin\omega t + 1.2\cos\omega t)^2}{2L} = \frac{1.2^2\sin^2\omega t - 2(1.2)(1.6)\sin\omega t \cos\omega t + (1.6)^2\cos^2\omega t}{2L} + \frac{(1.6)^2\sin^2\omega t + 2(1.6)(1.2)\sin\omega t \cos\omega t + (1.2)^2\cos^2\omega t}{2L} = \frac{(1.2)^2 + (1.6)^2}{2L} = \frac{1.44 + 2.56}{2L} = \frac{4.00}{2L} = \frac{2.0}{L}$

44. $\frac{2v^2\cos\theta\sin\theta}{g} = \frac{2v^2\sqrt{1-\sin^2\theta}\sin\theta}{g}$, since $\cos\theta \geq 0$.

≣ 20.2 THE SUM AND DIFFERENCE IDENTITIES

1. $\sin 15° = \sin(45° - 30°) = \sin 45° \cos 30° - \cos 45° \sin 30° = \frac{\sqrt{2}}{2} \cdot \frac{\sqrt{3}}{2} - \frac{\sqrt{2}}{2} \cdot \frac{1}{2} = \frac{\sqrt{6}-\sqrt{2}}{4}$

2. $\cos 75° = \cos(45° + 30°) = \cos 45° \cos 30° - \sin 45° \sin 30° = \frac{\sqrt{2}}{2} \cdot \frac{\sqrt{3}}{2} - \frac{\sqrt{2}}{2} \cdot \frac{1}{2} = \frac{\sqrt{6}-\sqrt{2}}{4}$

3. $\sin 120° = \sin(60° + 60°) = \sin 60° \cos 60° + \cos 60° \sin 60° = \frac{\sqrt{3}}{2} \cdot \frac{1}{2} + \frac{1}{2} \cdot \frac{\sqrt{3}}{2} = \frac{\sqrt{3}}{4} + \frac{\sqrt{3}}{4} = \frac{\sqrt{3}}{2}$

4. $\cos(-15°) = \cos(30° - 45°) = \cos 30° \cos 45° + \sin 30° \sin 45° = \frac{\sqrt{3}}{2} \cdot \frac{\sqrt{2}}{2} + \frac{1}{2} \cdot \frac{\sqrt{2}}{2} = \frac{\sqrt{6}+\sqrt{2}}{4}$

5. $\tan 15° = \tan(45° - 30°) = \frac{\tan 45° - \tan 30°}{1 + \tan 45° \tan 30°} = \frac{1 - \frac{\sqrt{3}}{3}}{1 + 1 \cdot \frac{\sqrt{3}}{3}} = \frac{\frac{3-\sqrt{3}}{3}}{\frac{3+\sqrt{3}}{3}} = \frac{3-\sqrt{3}}{3+\sqrt{3}} = \frac{3-\sqrt{3}}{3+\sqrt{3}} \cdot \frac{3-\sqrt{3}}{3-\sqrt{3}} = \frac{9-6\sqrt{3}+3}{9-3} = \frac{12-6\sqrt{3}}{6} = 2 - \sqrt{3}$

6. $\tan 135° = \tan(180° - 45°) = \frac{\tan 180° - \tan 45°}{1 + \tan 180° \tan 45°} = \frac{0-1}{1+0} = -1$

7. $\sin 150° = \sin(180° - 30°) = \sin 180° \cos 30° - \cos 180° \sin 30° = 0 \cdot \frac{\sqrt{3}}{2} - (-1) \cdot \frac{1}{2} = \frac{1}{2}$

8. $\cos 105° = \cos(60° + 45°) = \cos 60° \cos 45° - \sin 60° \sin 45° = \frac{1}{2} \cdot \frac{\sqrt{2}}{2} - \frac{\sqrt{3}}{2} \cdot \frac{\sqrt{2}}{2} = \frac{\sqrt{2}-\sqrt{6}}{4}$

9. $\sin(x+90°) = \sin x \cos 90° + \cos x \sin 90° = \sin x \cdot 0 + \cos x \cdot 1 = \cos x$

10. $\cos(x + \pi) = \cos x \cos \pi - \sin x \sin \pi = \cos x \cdot (-1) - \sin x \cdot 0 = -\cos x$

11. $\cos\left(x + \frac{\pi}{2}\right) = \cos x \cos \frac{\pi}{2} - \sin x \sin \frac{\pi}{2} = \cos x \cdot (0) - \sin x \cdot (1) = -\sin x$

12. $\sin\left(\frac{\pi}{2} - x\right) = \sin \frac{\pi}{2} \cos x - \cos \frac{\pi}{2} \sin x = 1 \cos x - 0 \sin x = \cos x$

13. $\cos(\pi - x) = \cos \pi \cos x + \sin \pi \sin x = -1 \cos x + 0 \sin x = -\cos x$

14. $\tan\left(x - \frac{\pi}{4}\right) = \dfrac{\tan x - \tan \frac{\pi}{4}}{1 + \tan x \tan \frac{\pi}{4}} = \frac{\tan x - 1}{1 + \tan x}$

15. $\sin(180° - x) = \sin 180° \cos x - \cos 180° \sin x = 0 \cos x - (-1) \sin x = \sin x$

16. $\tan(180° + x) = \frac{\tan 180° + \tan x}{1 - \tan 180° \tan x} = \frac{0 + \tan x}{1 - 0 \tan x} = \tan x$

Note: For exercises 17–24, if $\sin \alpha = \frac{3}{4}$, then $\cos \alpha = \sqrt{1 - \frac{9}{16}} = \frac{\sqrt{7}}{4}$ and $\tan \alpha = \frac{3}{\sqrt{7}} = \frac{3\sqrt{7}}{7}$. Similarly, if $\cos \beta = \frac{7}{8}$, then $\sin \beta = \sqrt{1 - \frac{49}{64}} = \frac{\sqrt{15}}{8}$ and $\tan \beta = \frac{\sqrt{15}}{7}$.

17. $\sin(\alpha + \beta) = \sin \alpha \cos \beta + \cos \alpha \sin \beta = \frac{3}{4} \cdot \frac{7}{8} + \frac{\sqrt{7}}{4} \cdot \frac{\sqrt{15}}{8} = \frac{21 + \sqrt{105}}{32} \approx 0.97647$

18. $\cos(\alpha + \beta) = \cos \alpha \cos \beta - \sin \alpha \sin \beta = \frac{\sqrt{7}}{4} \cdot \frac{7}{8} - \frac{3}{4} \cdot \frac{\sqrt{15}}{8} = \frac{7\sqrt{7} - 3\sqrt{15}}{32} \approx 0.21567$

19. $\tan(\alpha + \beta) = \frac{\sin(\alpha + \beta)}{\cos(\alpha + \beta)} = \frac{21 + \sqrt{105}}{7\sqrt{7} - 3\sqrt{15}} = 4.52768$

or $\tan(\alpha + \beta) = \frac{\tan \alpha + \tan \beta}{1 - \tan \alpha - \tan \beta} = \dfrac{\frac{3\sqrt{7}}{7} + \frac{\sqrt{15}}{7}}{1 - \frac{3\sqrt{7}}{7} \cdot \frac{\sqrt{15}}{7}} = \dfrac{\frac{3\sqrt{7}+\sqrt{15}}{7}}{\frac{49 - 3\sqrt{105}}{49}} = \frac{7(3\sqrt{7}+\sqrt{15})}{49 - 3\sqrt{105}} \approx 4.52768$

20. $\sin(\alpha - \beta) = \sin \alpha \cos \beta - \cos \alpha \sin \beta = \frac{3}{4} \cdot \frac{7}{8} - \frac{\sqrt{7}}{4} \cdot \frac{\sqrt{15}}{8} = \frac{21 - \sqrt{105}}{32} \approx 0.33603$

21. $\cos(\alpha - \beta) = \cos \alpha \cos \beta + \sin \alpha \sin \beta = \frac{\sqrt{7}}{4} \cdot \frac{7}{8} + \frac{3}{4} \cdot \frac{\sqrt{15}}{8} = \frac{7\sqrt{7} + 3\sqrt{15}}{32} \approx 0.94185$

22. $\tan(\alpha - \beta) = \frac{\sin(\alpha - \beta)}{\cos(\alpha - \beta)} = \frac{21 - \sqrt{105}}{7\sqrt{7} + 3\sqrt{15}} \approx 0.35678$

or $\tan(\alpha - \beta) = \frac{\tan \alpha - \tan \beta}{1 + \tan \alpha \tan \beta} = \dfrac{\frac{3\sqrt{7}}{7} - \frac{\sqrt{15}}{7}}{1 + \frac{3\sqrt{7}}{7} \cdot \frac{\sqrt{15}}{7}} = \frac{7(3\sqrt{7} - \sqrt{15})}{49 + 3\sqrt{105}} \approx 0.35678$

23. Since both $\sin(\alpha + \beta)$ and $\cos(\alpha + \beta)$ are positive $(\alpha - \beta)$ is in Quadrant I.

24. Since both $\sin(\alpha - \beta)$ and $\cos(\alpha - \beta)$ are positive, $(\alpha - \beta)$ is in Quadrant I.

25. $\sin(\alpha + \beta) = \sin \alpha \cos \beta + \cos \alpha \sin \beta = \frac{3}{4} \cdot \frac{-7}{8} + \frac{-\sqrt{7}}{4} \cdot \frac{-\sqrt{15}}{8} = \frac{-21 + \sqrt{105}}{32} = \frac{\sqrt{105} - 21}{32} \approx -0.33603$

26. $\cos(\alpha + \beta) = \cos \alpha \cos \beta - \sin \alpha \sin \beta = \frac{-\sqrt{7}}{4} \cdot \frac{-7}{8} - \frac{3}{4} \cdot \frac{-\sqrt{15}}{8} = \frac{7\sqrt{7} + 3\sqrt{15}}{32} \approx 0.94185$

27. $\tan(\alpha + \beta) = \frac{\sin(\alpha + \beta)}{\cos(\alpha + \beta)} = \frac{\sqrt{105} - 21}{7\sqrt{7} + 3\sqrt{15}} \approx -0.35678$

or $\tan(\alpha + \beta) = \frac{\tan \alpha + \tan \beta}{1 - \tan \alpha \tan \beta} = \dfrac{\frac{-3\sqrt{7}}{7} + \frac{\sqrt{15}}{7}}{1 - \frac{-3\sqrt{7}}{7} \cdot \frac{\sqrt{15}}{7}} = \frac{7(\sqrt{15} - 3\sqrt{7})}{49 + 3\sqrt{105}} \approx -0.35678$

28. $\sin(\alpha - \beta) = \sin \alpha \cos \beta - \cos \alpha \sin \beta = \frac{3}{4} \cdot \frac{-7}{8} - \frac{-\sqrt{7}}{4} \cdot \frac{-\sqrt{15}}{8} = \frac{-21 - \sqrt{105}}{32} \approx -0.97647$

29. $\sin(\alpha - \beta) = \cos \alpha \cos \beta + \sin \alpha \sin \beta = \frac{-7}{8} \cdot \frac{-\sqrt{7}}{4} + \frac{3}{4} \cdot \frac{-\sqrt{15}}{8} = \frac{7\sqrt{7} - 3\sqrt{15}}{32} \approx 0.21567$

30. $\tan(\alpha - \beta) = \frac{\sin(\alpha - \beta)}{\cos(\alpha - \beta)} = \frac{-(21 + \sqrt{105})}{7\sqrt{7} - 3\sqrt{15}} \approx -4.52768$

or $\tan(\alpha - \beta) = \frac{\tan \alpha - \tan \beta}{1 + \tan \alpha \tan \beta} = \dfrac{\frac{-3\sqrt{7}}{7} - \frac{\sqrt{15}}{7}}{1 + \frac{-3\sqrt{7}}{7} \frac{\sqrt{15}}{7}} = \frac{-7(3\sqrt{7} + \sqrt{15})}{49 - 3\sqrt{105}} \approx -4.52768$

31. Since $\sin(\alpha + \beta)$ is negative and $\cos(\alpha + \beta)$ is positive, $(\alpha + \beta)$ is in Quadrant IV.

32. Since $\sin(\alpha - \beta)$ is negative and $\cos(\alpha - \beta)$ is positive, $(\alpha - \beta)$ is in Quadrant IV.

33. $\sin 47° \cos 13° + \cos 47° \sin 13° = \sin(47° + 13°) = \sin 60° = \frac{\sqrt{3}}{2}$

34. $\sin 47° \sin 13° + \cos 47° \cos 13° = \cos(47° - 13°) = \cos 34°$

35. $\cos 32° \cos 12° - \sin 32° \sin 12° = \cos(32° + 12°) = \cos 44°$

36. $\frac{\tan 40° + \tan 15°}{1 - \tan 40° \tan 15°} = \tan(40° + 15°) = \tan 55°$

37. $\cos(\alpha + \beta)\cos\beta + \sin(\alpha + \beta)\sin\beta = \cos(\alpha + \beta - \beta) = \cos\alpha$

38. $\sin(x-y)\cos y + \cos(x-y)\sin y = \sin(x-y+y) = \sin x$

39. $\cos(x + y)\cos(x - y) - \sin(x + y)\sin(x - y) = \cos(x + y + x - y) = \cos 2x = \cos x \cos x - \sin x \sin x = \cos^2 x - \sin^2 x = \cos 2x$

40. $\sin A \cos(-B) + \cos A \sin(-B) = \sin(A - B)$

41.

$\sin(x + y)\sin(x - y) = \sin^2 x - \sin^2 y$	
$(\sin x \cos y + \cos x \sin y)$ $\cdot (\sin x \cos y - \cos x \sin y)$ $\sin^2 x \cos^2 y - \cos^2 x \sin^2 y$ $\sin^2 x (1 - \sin^2 y)$ $-(1 - \sin^2 x)\sin^2 y$ $\sin^2 x - \sin^2 x \sin^2 y$ $- \sin^2 y + \sin^2 y \sin^2 x$ $\sin^2 x - \sin^2 y$	$\sin^2 x - \sin^2 y$

42. $(\sin A \cos B - \cos A \sin B)^2$

$+ (\cos A \cos B + \sin A \sin B)^2 = 1$	
$[\sin(A - B)]^2 + [\cos(A - B)]^2$ $\sin^2(A - B) + \cos^2(A - B)$ 1	1

43. $\cos\theta = \sin(\theta + 30°) + \cos(\theta + 60°)$

$\cos\theta$	$\sin\theta\cos 30° + \cos\theta\sin 30°$ $+ \cos\theta\cos 60° - \sin\theta\sin 60°$ $\frac{\sqrt{3}}{2}\sin\theta + \frac{1}{2}\cos\theta + \frac{1}{2}\cos\theta - \frac{\sqrt{3}}{2}\sin\theta$ $\cos\theta$

44.

$$\frac{\sin(x+y)}{\cos(x-y)} = \frac{\tan x + \tan y}{1 + \tan x \tan y}$$

$\frac{\sin x \cos y + \cos x \sin y}{\cos x \cos y + \sin x \sin y}$	$\frac{\tan x + \tan y}{1 + \tan x \tan y}$
$\frac{\frac{\sin x \cos y}{\cos x \cos y} + \frac{\cos x \sin y}{\cos x \cos y}}{\frac{\cos x \cos y}{\cos x \cos y} + \frac{\sin x \sin y}{\cos x \cos y}}$	
$\frac{\tan x + \tan y}{1 + \tan x \tan y}$	

45. $\tan x - \tan y = \frac{\sin(x-y)}{\cos x \cos y}$

$\tan x - \tan y$	$\frac{\sin x \cos y - \cos x \sin y}{\cos x \cos y}$
	$\frac{\sin x \cos y}{\cos x \cos y} - \frac{\cos x \sin y}{\cos x \cos y}$
	$\tan x - \tan y$

46.

$$\cos(A + B)\cos(A - B) = 1 - \sin^2 A - \sin^2 B$$

$(\cos A \cos B - \sin A \sin B)$ $\cdot (\cos A \cos B + \sin A \sin B)$ $\cos^2 A \cos^2 B - \sin^2 A \sin^2 B$ $(1 - \sin^2 A)(1 - \sin^2 B)$ $- \sin^2 A \sin^2 B$ $1 - \sin^2 A - \sin^2 B$ $+ \sin^2 A \sin^2 B - \sin^2 A \sin^2 B$ $1 - \sin^2 A - \sin^2 B$	$1 - \sin^2 A - \sin^2 B$

47. Both sides equal 0.8386705679

48. Both sides equal 0.2079116908

49. Both sides equal 14.10141995

50. Both sides equal 0.932039086

51. $\sin(55° + 37°) = 0.99939$; $\sin 55° + \sin 37° = 1.420967$; not equal

52. $\cos(68° - 24°) = 0.7193398$; $\cos 68° - \cos 24° = -0.538939$; not equal

53. $\cos(40° + 35°) = 0.258819$; $\cos 40° + \cos 35° = 1.585196$; not equal

54. $\tan(76° - 37°) = 0.809784$; $\tan 76° - \tan 37° = 3.257227$; not equal

55. $[r_1(\cos\theta_1 + j\sin\theta_1)][r_2(\cos\theta_2 + j\sin\theta_2)] = r_1 r_2[\cos\theta_1\cos\theta_2 + j\cos\theta_2\sin\theta_1 + j\sin\theta_2\cos\theta_1 + j^2\sin\theta_1\sin\theta_2] = r_1 r_2[(\cos\theta_1\cos\theta_2 - \sin\theta_1\sin\theta_2) + j(\sin\theta_1\cos\theta_2 + \cos\theta_1\sin\theta_2)] = r_1 r_2[\cos(\theta_1 + \theta_2) + j\sin(\theta_1 + \theta_2)]$

56. $y_1 + y_2 = A_1\cos(\omega t + \pi) + A_2\cos(\omega t - \pi) = A_1[\cos\omega t\cos\pi - \sin\omega t\sin\pi] + A_2[\cos\omega t\cos\pi + \sin\omega t\sin\pi] = A_1[-\cos\omega t] + A_2[-\cos\omega t] = -[A_1 + A_2]\cos\omega t$

57. $d = \frac{h}{\cos\theta_r} \cdot \sin(\theta_i - \theta_r) = \frac{h}{\cos\theta_r}[\sin\theta_i\cos\theta_r - \cos\theta_i\sin\theta_r] = h\left[\sin\theta_i\cos\theta_r \cdot \frac{1}{\cos\theta_r} - \cos\theta_i\sin\theta_r \cdot \frac{1}{\cos\theta_r}\right] = h[\sin\theta_i - \cos\theta_i\tan\theta_r]$

58. $i(t) = I_p\sin(\theta - \omega t) = 14.8[\sin\theta\cos\omega t - \cos\theta\sin\omega t] = 14.8[\sin 45°\cos\omega t - \cos 45°\sin\omega t] = 14.8\left[\frac{\sqrt{2}}{2}\cos\omega t - \frac{\sqrt{2}}{2}\sin\omega t\right] = 7.4\sqrt{2}[\cos\omega t - \sin\omega t]$

▤ 20.3 THE DOUBLE- AND HALF-ANGLE IDENTITIES

1. $\cos 15° = \cos \frac{30°}{2} = \sqrt{\frac{1+\cos 30°}{2}} = \sqrt{\frac{1+\sqrt{3}/2}{2}} =$
$\sqrt{\frac{2+\sqrt{3}}{4}} \approx 0.96593$

2. $\sin 75° = \sin \frac{150°}{2} = \sqrt{\frac{1-\cos 150°}{2}} = \sqrt{\frac{1+\sqrt{3}/2}{2}} =$
$\sqrt{\frac{2+\sqrt{3}}{4}} \approx 0.96593$

3. $\sin 15° = \sin \frac{30°}{2} = \sqrt{\frac{1-\cos 30°}{2}} = \sqrt{\frac{1-\sqrt{3}/2}{2}} =$
$\sqrt{\frac{2-\sqrt{3}}{4}} \approx 0.25882$

4. $\cos 105° = \cos \frac{210°}{2} = -\sqrt{\frac{1+\cos 210°}{2}} = -\sqrt{\frac{1-\sqrt{3}/2}{2}}$
$= -\sqrt{\frac{2-\sqrt{3}}{4}} \approx -0.25882$

5. $\sin 105° = \sin \frac{210°}{2} = \sqrt{\frac{1-\cos 210°}{2}} = \sqrt{\frac{1+\sqrt{3}/2}{2}} =$
$\sqrt{\frac{2+\sqrt{3}}{4}} \approx 0.96593$

6. Using a double-angle formula and the value of $\cos 105°$ from exercise 4, we obtain $\cos 210° =$
$2(\cos 105°)^2 - 1 = 2\left(\sqrt{\frac{2-\sqrt{3}}{4}}\right)^2 - 1 = \frac{2-\sqrt{3}}{2} -$
$1 = \frac{-\sqrt{3}}{2} \approx -0.86603.$

7. $\cos 7\frac{1}{2}° = \cos \frac{15°}{2} = \sqrt{\frac{1+\cos 15°}{2}} =$
$\sqrt{\frac{1+\sqrt{\frac{2+\sqrt{3}}{4}}}{2}} = \sqrt{\frac{2+\sqrt{2+\sqrt{3}}}{4}} \approx 0.99145$

8. $\tan 15° = \tan \frac{30°}{2} = \sqrt{\frac{1-\cos 30°}{1+\cos 30°}} = \sqrt{\frac{1-\sqrt{3}/2}{1+\sqrt{3}/2}} =$
$\sqrt{\frac{2-\sqrt{3}}{2+\sqrt{3}}} \approx 0.26795$

9. $\tan 22\frac{1}{2}° = \tan \frac{45°}{2} = \sqrt{\frac{1-\cos 45°}{1+\cos 45°}} = \sqrt{\frac{1-\sqrt{2}/2}{1+\sqrt{2}/2}} =$
$\sqrt{\frac{2-\sqrt{2}}{2+\sqrt{2}}} \approx 0.41421$

10. $\cos 67\frac{1}{2}° = \cos \frac{135°}{2} = \sqrt{\frac{1+\cos 135°}{2}} = \sqrt{\frac{1+\frac{-\sqrt{2}}{2}}{2}}$
$= \frac{\sqrt{2-\sqrt{2}}}{2} \approx 0.38268$

11. $\cos 75° = \cos \frac{150°}{2} = \sqrt{\frac{1+\cos 150°}{2}} = \sqrt{\frac{1-\frac{\sqrt{3}}{2}}{2}} =$
$\frac{\sqrt{2-\sqrt{3}}}{2} \approx 0.25882$

12. $\sin 37\frac{1}{2}° = \sin \frac{75°}{2} = \sqrt{\frac{1-\cos 75°}{2}} = \sqrt{\frac{1-\frac{\sqrt{2-\sqrt{3}}}{2}}{2}}$
$= \frac{\sqrt{2-\sqrt{2-\sqrt{3}}}}{2} \approx 0.60876$

13. $\sin 127\frac{1}{2}° = \sin(105° + 22.5°)$. From exercise 5, we have $\sin 105° = \frac{\sqrt{2+\sqrt{3}}}{2}$, and from exercise 4, $\cos 105° = \frac{-\sqrt{2+\sqrt{3}}}{2}$. We determine that
$\sin 22\frac{1}{2}° = \sin \frac{45°}{2} = \sqrt{\frac{1-\frac{\sqrt{2}}{2}}{2}} = \frac{\sqrt{2-\sqrt{2}}}{2}$ and
that $\cos 22\frac{1}{2}° = \cos \frac{45°}{2} = \sqrt{\frac{1+\frac{\sqrt{2}}{2}}{2}} = \frac{\sqrt{2+\sqrt{2}}}{2}$.
Using these values, we find the desired result as follows:
$\sin(105° + 22.5°) = \sin 105° \cos 22.5° +$
$\cos 105° \sin 22.5° = \sin \left(\frac{\sqrt{2+\sqrt{3}}}{2}\right)\left(\frac{\sqrt{2+\sqrt{2}}}{2}\right) +$
$\left(\frac{-\sqrt{2-\sqrt{3}}}{2}\right)\left(\frac{\sqrt{2-\sqrt{2}}}{2}\right)$. This "simplifies" as
$\frac{1}{4}\left[\sqrt{4+2\sqrt{3}+2\sqrt{2}+\sqrt{6}} - \sqrt{4-2\sqrt{2}-2\sqrt{3}+\sqrt{6}}\right] \approx 0.79335$

14. $\tan -15° = \tan \frac{-30°}{2} = \frac{1-\cos(-30°)}{\sin -30°} = \frac{1-\frac{\sqrt{3}}{2}}{-\frac{1}{2}} =$
$\sqrt{3} - 2 \approx -0.26795$

15. $\sin x = \frac{7}{25}; \cos x = -\sqrt{1-\left(\frac{7}{25}\right)^2} = -\frac{24}{25}; \sin 2x =$
$2 \sin x \cos x = 2 \cdot \frac{7}{25} \cdot \frac{-24}{25} = \frac{-336}{625}; \cos 2x = \cos^2 x -$
$\sin^2 x = \left(\frac{24}{25}\right)^2 - \left(\frac{7}{25}\right)^2 = \frac{527}{625}; \tan 2x = \frac{-336}{527};$
$\sin \frac{x}{2} = \sqrt{\frac{1-\cos x}{2}} = \sqrt{\frac{1+\frac{24}{25}}{2}} = \sqrt{\frac{49}{50}} = \frac{7\sqrt{2}}{10};$
$\cos \frac{x}{2} = \sqrt{\frac{1+\cos x}{2}} = \sqrt{\frac{1-\frac{24}{25}}{2}} = \sqrt{\frac{1}{50}} = \frac{\sqrt{2}}{10};$
$\tan \frac{x}{2} = \frac{7\sqrt{2}}{\sqrt{2}} = 7$

16. $\cos x = \frac{8}{17}; \sin x = -\sqrt{1 - \left(\frac{8}{17}\right)^2} = -\frac{15}{17}; \sin 2x =$
$2 \sin x \cos x = 2 \frac{8}{27} \cdot \frac{-15}{17} = \frac{-240}{459}; \cos 2x = \cos^2 x - $
$\sin^2 x = \left(\frac{64}{289}\right) - \left(\frac{225}{289}\right) = \frac{-161}{289}; \tan 2x = \frac{240}{161}; \sin \frac{x}{2} =$
$\sqrt{\frac{1-\cos x}{2}} = \sqrt{\frac{1 - \frac{8}{17}}{2}} = \sqrt{\frac{9}{34}} = \frac{3}{\sqrt{34}}; \cos \frac{x}{2} =$
$-\sqrt{\frac{1+\cos x}{2}} = -\sqrt{\frac{1 + \frac{8}{17}}{2}} = \frac{-5}{\sqrt{34}}; \tan \frac{x}{2} = -\frac{3}{5}$

17. $\sec x = \frac{29}{20}; \cos x = \frac{20}{29}; \sin x = \frac{21}{29}; \sin 2x =$
$2\left(\frac{21}{29}\right)\left(\frac{20}{29}\right) = \frac{840}{841}; \cos 2x = \cos^2 x - \sin^2 x =$
$\frac{400-441}{841} = \frac{-41}{841}; \tan 2x = -\frac{840}{41}; \sin \frac{x}{2} = \sqrt{\frac{1 - \frac{20}{29}}{2}}$
$= \sqrt{\frac{9}{58}} = \frac{3}{\sqrt{58}}; \cos \frac{x}{2} = \sqrt{\frac{1 + \frac{20}{29}}{2}} = \sqrt{\frac{49}{58}} = \frac{7}{\sqrt{58}};$
$\tan \frac{x}{2} = \frac{3}{7}$

18. $\csc x = \frac{-41}{9}; \sin x = -\frac{9}{41}; \cos x = -\frac{40}{41}; \sin 2x =$
$2\left(\frac{-9}{41}\right)\left(\frac{-40}{41}\right) = \frac{720}{1681}; \cos 2x = 1 - 2\sin^2 x = 1 - $
$2\left(\frac{9}{41}\right)^2 = \frac{1519}{1681}; \tan 2x = \frac{720}{1519}; \sin \frac{x}{2} = \sqrt{\frac{1 + \frac{40}{41}}{2}} =$
$\sqrt{\frac{81}{82}} = \frac{9}{\sqrt{82}}; \cos \frac{x}{2} = -\sqrt{\frac{1 - \frac{40}{41}}{2}} = -\sqrt{\frac{1}{82}} =$
$-\frac{1}{\sqrt{82}}; \tan \frac{x}{2} = -9$

19. $\tan x = \frac{35}{12}; \sqrt{u^2 + v^2} = \sqrt{35^2 + 12^2} = 37; \sin x =$
$-\frac{35}{37}; \cos x = -\frac{12}{37}; \sin 2x = 2\left(\frac{-35}{37}\right)\left(-\frac{12}{37}\right) = \frac{840}{1369};$
$\cos 2x = 1 - 2\sin^2 x = 1 - 2\left(\frac{-35}{37}\right)^2 = \frac{-1081}{1369};$
$\tan 2x = -\frac{840}{1081}; \sin \frac{x}{2} = \sqrt{\frac{1 + \frac{12}{37}}{2}} = \sqrt{\frac{49}{74}} =$
$\frac{7}{\sqrt{74}}; \cos \frac{x}{2} = -\sqrt{\frac{1 - \frac{12}{37}}{2}} = -\sqrt{\frac{25}{74}} = \frac{-5}{\sqrt{74}}; \tan \frac{x}{2} =$
$-\frac{7}{5}$

20. $\cot x = \frac{-45}{28}; \sqrt{u^2 + v^2} = 53; \sin x = \frac{28}{53}; \cos x =$
$\frac{-45}{53}; \sin 2x = 2\left(\frac{28}{53}\right)\left(\frac{-45}{53}\right) = \frac{-2520}{2809}; \cos 2x = 2\cos^2$
$x - 1 = \left(\frac{-45}{53}\right)^2 - 1 = \frac{1241}{2809}; \tan 2x = \frac{-2520}{1241}; \sin \frac{x}{2} =$
$\sqrt{\frac{1 + \frac{45}{53}}{2}} = \sqrt{\frac{98}{106}} = \sqrt{\frac{49}{53}} = \frac{7}{\sqrt{53}}; \cos \frac{x}{2} =$
$\sqrt{\frac{1 - \frac{45}{53}}{2}} = \sqrt{\frac{8}{106}} = \sqrt{\frac{4}{53}} = \frac{2}{\sqrt{53}}; \tan \frac{x}{2} = \frac{7}{2}$

21.
$$\cos^2 x = \sin^2 x + \cos 2x$$

$\cos^2 x$	$\sin^2 x + 2\cos^2 x - 1$
	$\sin^2 x + \cos^2 x + \cos^2 x - 1$
	$1 + \cos^2 x - 1$
	$\cos^2 x$

22.
$$\cos^2 4x - \sin^2 4x = \cos 8x$$

$\cos^2 4x - \sin^2 4x$	$\cos(2 \cdot 4x)$
	$\cos^2 4x - \sin^2 4x$

23.
$$\cos 4x = 1 - 8\sin^2 x \cos^2 x$$

$\cos(2 \cdot 2x)$	$1 - 8\sin^2 x \cos^2 x$
$1 - 2\sin^2 2x$	
$1 - 2(2\sin x \cos x)^2$	
$1 - 2(4\sin^2 x \cos^2 x)$	
$1 - 8\sin^2 x \cos^2 x$	

24.
$$\cos 3x = 4\cos^3 x - 3\cos x$$

$\cos(x + 2x)$	$4\cos^3 x - 3\cos x$
$\cos x \cos 2x - \sin x \sin 2x$	
$\cos x(2\cos^2 x - 1)$	
$\quad - \sin x(2\cos x \sin x)$	
$2\cos^3 x - \cos x - 2\sin^2 x \cos x$	
$2\cos^3 x - \cos x - 2(1 - \cos^2 x)\cos x$	
$2\cos^3 x - \cos x - 2\cos x + 2\cos^3 x$	
$4\cos^3 x - 3\cos x$	

25. $\dfrac{1+\tan^2 \alpha}{1-\tan^2 \alpha} = \sec 2\alpha$

$\dfrac{1+\tan^2 \alpha}{1-\tan^2 \alpha}$	$\dfrac{1}{\cos 2\alpha}$
	$\dfrac{1}{\cos^2 \alpha - \sin^2 \alpha}$
	$\dfrac{\cos^2 \alpha + \sin^2 \alpha}{\cos^2 \alpha - \sin^2 \alpha}$
	$\dfrac{\frac{\cos^2 \alpha}{\cos^2 \alpha} + \frac{\sin^2 \alpha}{\cos^2 \alpha}}{\frac{\cos^2 \alpha}{\cos^2 \alpha} - \frac{\sin^2 \alpha}{\cos^2 \alpha}}$
	$\dfrac{1+\tan^2 \alpha}{1-\tan^2 \alpha}$

26. $\sin 2x \cos 2x = \frac{1}{2} \sin 4x$

$\sin 2x \cos 2x$	$\frac{1}{2} \sin(2 \cdot 2x)$
	$\frac{1}{2} \cdot 2 \sin 2x \cos 2x$
	$\sin 2x \cos 2x$

27. $1 - 2\sin^2 3x = \cos 6x$

$1 - 2\sin^2 3x$	$\cos(2 \cdot 3x)$
	$1 - 2\sin^2 3x$

28. $\dfrac{2\tan 3x}{1-\tan^2 3x} = \tan 6x$

$\dfrac{2\tan 3x}{1-\tan^2 3x}$	$\tan(2 \cdot 3x)$
	$\dfrac{2\tan 3x}{1-\tan^2 3x}$

29. $\sin^2 x \cos^2 x = \frac{1}{4} \sin^2 2x$

$\sin^2 x \cos^2 x$	$\frac{1}{4}[2\sin x \cos x]^2$
	$\frac{1}{4}(4\sin^2 x \cos^2 x)$
	$\sin^2 x \cos^2 x$

30. $1 + \tan \beta \tan \frac{\beta}{2} = \dfrac{1}{\cos \beta}$

$1 + \tan \beta \left(\dfrac{\sin \beta}{1+\cos \beta} \right)$	$\dfrac{1}{\cos \beta}$
$1 + \dfrac{\sin \beta}{\cos \beta} \left(\dfrac{\sin \beta}{1+\cos \beta} \right)$	
$1 + \dfrac{\sin^2 \beta}{\cos \beta(1+\cos \beta)}$	
$1 + \dfrac{1-\cos^2 \beta}{\cos \beta(1+\cos \beta)}$	
$1 + \dfrac{(1+\cos \beta)(1-\cos \beta)}{\cos \beta(1+\cos \beta)}$	
$1 + \dfrac{1-\cos \beta}{\cos \beta}$	
$1 + \dfrac{1}{\cos \beta} - \dfrac{\cos \beta}{\cos \beta}$	
$\dfrac{1}{\cos \beta}$	

31.
$$\tan\left(\frac{\alpha+\beta}{2}\right) \cot\left(\frac{\alpha-\beta}{2}\right) = \left(\frac{1-\cos(\alpha+\beta)}{\sin(\alpha+\beta)}\right)\left(\frac{1+\cos(\alpha-\beta)}{\sin(\alpha-\beta)}\right)$$
$$= \left(\frac{1-\cos\alpha\cos\beta+\sin\alpha\sin\beta}{\sin\alpha\cos\beta+\cos\alpha\sin\beta}\right)\left(\frac{1+\cos\alpha\cos\beta+\sin\alpha\sin\beta}{\sin\alpha\cos\beta-\cos\alpha\sin\beta}\right)$$
$$= \frac{1+2\sin\alpha\sin\beta+\sin^2\alpha\sin^2\beta-\cos^2\alpha\cos^2\beta}{\sin^2\alpha\cos^2\beta-\cos^2\alpha\sin^2\beta}.$$

In the denominator, let $\cos^2 \beta = 1 - \sin^2 \beta$ and $\cos^2 \alpha = 1 - \sin^2 \alpha$ to produce

$$\frac{1+2\sin\alpha\sin\beta+\sin^2\alpha\sin^2\beta-(1-\sin^2\alpha)(1-\sin^2\beta)}{\sin^2\alpha(1-\sin^2\beta)-(1-\sin^2\alpha)\sin^2\beta}$$

$$= \frac{1+\alpha\sin\alpha\sin\beta+\sin^2\alpha\sin^2\beta-1+\sin^2\alpha+\sin^2\beta-\sin^2\alpha\sin^2\beta}{\sin^2\alpha-\sin^2\alpha\sin^2\beta-\sin^2\beta+\sin^2\alpha\sin^2\beta}$$

$$= \frac{\sin^2\alpha+2\sin\alpha\sin\beta+\sin^2\beta}{\sin^2\alpha-\sin^2\beta} = \frac{(\sin\alpha+\sin\beta)^2}{\sin^2\alpha-\sin^2\beta}$$

32. $\cot\theta = \frac{1}{2}\left(\cot\frac{\theta}{2} - \tan\frac{\theta}{2}\right)$

$\cot\theta$	$\frac{1}{2}\left(\dfrac{1}{\tan\frac{\theta}{2}} - \tan\frac{\theta}{2}\right)$
	$\frac{1}{2}\left[\dfrac{1+\cos\theta}{\sin\theta} - \dfrac{1-\cos\theta}{\sin\theta}\right]$
	$\frac{1}{2}\left[\dfrac{2\cos\theta}{\sin\theta}\right]$
	$\cot\theta$

33. $\cos 2 \cdot 45° = \cos 90° = 0; 2\cos 45° = 2 \cdot \frac{\sqrt{2}}{2} = \sqrt{2}; 0 \neq \sqrt{2}$

34. $\tan \frac{80°}{2} = \tan 40° \approx 0.8391; \frac{\tan 80°}{2} \approx 2.8356; 0.8391 \neq 2.8356$

35. $\cot 2 \cdot 150° = \cot 300° \approx -0.5774; 2\cot 150° \approx -3.4641; -0.5774 \neq -3.4641$

36. $\sin \frac{210°}{2} = \sin 105° \approx 0.9659; \frac{\sin 210°}{2} = -0.25;$
$0.9659 \neq -0.25$

37. $n = \dfrac{\sin \frac{\alpha+\phi}{2}}{\sin \frac{\alpha}{2}} = \dfrac{\sqrt{\frac{1-\cos(\alpha+\phi)}{2}}}{\sqrt{\frac{1-\cos \alpha}{2}}} = \sqrt{\dfrac{1-\cos(\alpha+\phi)}{1-\cos \alpha}} =$

$\sqrt{\dfrac{1-\cos \alpha \cos \phi + \sin \alpha \sin \phi}{1-\cos \alpha}}$

38. Since $\cot \frac{\alpha}{2} = \dfrac{\cos \frac{\alpha}{2}}{\sin \frac{\alpha}{2}}$, we can write $n = \sqrt{\dfrac{1+\cos \phi}{2}} +$

$\left(\cot \frac{\alpha}{2}\right)\sqrt{\dfrac{1-\cos \phi}{2}} = \sqrt{\dfrac{1+\cos \phi}{2}} + \left(\dfrac{\cos \frac{\alpha}{2}}{\sin \frac{\alpha}{2}}\right)\sqrt{\dfrac{1-\cos \phi}{2}}.$

We know $\sqrt{\dfrac{1+\cos \phi}{2}} = \cos \frac{\phi}{2}$ and $\sqrt{\dfrac{1-\cos \phi}{2}} = \sin \frac{\phi}{2},$

and so we can rewrite the given expression as

$n = \cos \frac{\phi}{2} + \left(\dfrac{\cos \frac{\alpha}{2}}{\sin \frac{\alpha}{2}}\right) \sin \frac{\phi}{2}$. Writing this with

a common denominator of $\sin \frac{\alpha}{2}$, we get

$n = \dfrac{\sin \frac{\alpha}{2} \cos \frac{\phi}{2} + \cos \frac{\alpha}{2} \sin \frac{\phi}{2}}{\sin \frac{\alpha}{2}} = \dfrac{\sin \frac{(\alpha+\phi)}{2}}{\sin \frac{\alpha}{2}}.$

39. $P = V_{max} I_{max} \cos \omega t \sin \omega t = V_{max} I_{max} \frac{1}{2} \cdot 2 \cos \omega t$
$\sin \omega t = \frac{V_{max} I_{max}}{2} \cdot \sin 2\omega t$

40. $A = \sqrt{e^{-2x}(1+\sin 2x)} = \sqrt{e^{-2x}}\sqrt{1+\sin 2x} =$
$e^{-x}\sqrt{1+\sin 2x} = e^{-x}\sqrt{1+2\sin x \cos x} =$
$e^{-x}\sqrt{\sin^2 x + \cos^2 x + 2 \sin x \cos x} =$
$e^{-x}\sqrt{(\sin x + \cos x)^2} = e^{-x}(\sin x + \cos x)$

≡ 20.4 TRIGONOMETRIC EQUATIONS

1. $2\cos \theta = 0; \cos \theta = 0; \theta = 90°, 270°$

2. $2\sin \theta = -1; \sin \theta = -\frac{1}{2}; \theta = 210°, 330°$

3. $\sqrt{3}\tan x = 1; \tan x = \frac{1}{\sqrt{3}}; x = 30°, 210°$

4. $\sqrt{3}\sec x = -2; \sec x = -\frac{2}{\sqrt{3}}$ or $\cos x = -\frac{\sqrt{3}}{2};$
$x = 150°, 210°$

5. $4\sin \theta = -3; \sin \theta = \frac{-3}{4}; \theta \approx 228.59°, 311.41°$

6. $2\cos x = 3; \cos x = \frac{3}{2};$ no solution

7. $4\tan \alpha = 5; \tan \alpha = \frac{5}{4}; \alpha = 51.34°, 231.34°$

8. $3\csc x = 1; \csc x = \frac{1}{3}$ or $\sin x = 3;$ no solution

9. $\cos 2x = -1; 2x = 180°$ or $540°; x = 90°$ or $270°$

10. $\sin 2x = \frac{1}{2}; 2x = 30°, 150°, 390°, 510°; x = 15°,$
$75°, 195°, 255°$

11. $\tan \frac{\theta}{4} = 1; \frac{\theta}{4} = 45°; \theta = 180°$

12. $\cos \frac{\theta}{3} = -1; \frac{\theta}{3} = 180°; \theta = 540°$ is the smallest positive answer. No solutions between $0°$ and $360°$

13. $\sin^2 \alpha = \sin \alpha; \sin^2 \alpha - \sin \alpha = 0; \sin \alpha(\sin \alpha - 1) = 0$, so $\sin \alpha = 0$ or $\sin \alpha = 1; \alpha = 0°, 90°, 180°$

14. $\cos^2 \beta = \frac{1}{2}\cos \beta; \cos^2 \beta - \frac{1}{2}\cos \beta = 0; \cos \beta(\cos \beta - \frac{1}{2}) = 0; \cos \beta = 0$ or $\cos \beta = \frac{1}{2}; \beta = 60°, 90°, 270°, 300°$

15. $\sin x \cos x = 0; \sin x = 0 \cos x = 0; x = 0°, 90°, 180°, 270°$

16. $\frac{\sec \theta}{\csc \theta} = -1; \frac{\sin \theta}{\cos \theta} = -1; \tan \theta = -1; \theta = 135°, 315°$

17. $3\tan^2 x = 1; \tan^2 x = \frac{1}{3}; \tan x = \pm\sqrt{\frac{1}{3}}; x = 30°, 150°, 210°, 330°$

18. $\sec^2 \theta = 2; \sec \theta = \pm\sqrt{2}; \cos \theta = \pm\sqrt{\frac{1}{2}}; \theta = 45°, 135°, 225°, 315°$

19. $4\sin \alpha \cos \alpha = 1; 2(2\sin \alpha \cos \alpha) = 2(\sin 2\alpha) = 1; \sin 2\alpha = \frac{1}{2}; 2\alpha = 30°, 150°, 390°, 510°$ and so, $\alpha = 15°, 75°, 195°, 255°$

20. $\sin^2 \beta = \frac{1}{2}\sin \beta; \sin^2 \beta - \frac{1}{2}\sin \beta = 0; \sin \beta(\sin \beta - \frac{1}{2}) = 0; \sin \beta = 0$ and $\beta = 0°, 180°$ or $\sin \beta = \frac{1}{2}$, and $\beta = 30°, 150°$

21. $\sin \theta - \cos \theta = 0; \sin \theta = \cos \theta; \frac{\sin \theta}{\cos \theta} = 1$ or $\tan \theta = 1$, and so $\theta = 45°, 225°$

22. $\tan \theta = \csc \theta; \frac{\sin \theta}{\cos \theta} = \frac{1}{\sin \theta}; \sin^2 \theta = \cos \theta; 1 - \cos^2 \theta = \cos \theta; \cos^2 \theta + \cos \theta - 1 = 0; \cos \theta = \frac{-1 \pm \sqrt{1+4}}{2} = \frac{1 \pm \sqrt{5}}{2}; \cos \theta \approx 0.618$ or $\cos \theta \approx -1.618; \theta = 51.83°, 308.17°.$ (Note: $\cos \theta = -1.618$ yields no solutions)

23. $\sin 6\theta + \sin 3\theta = 0; 2\sin 3\theta \cos 3\theta + \sin 3\theta = 0;$
$\sin 3\theta(2\cos 3\theta+1) = 0; \sin 3\theta = 0$ or $\cos 3\theta = -\frac{1}{2}$.
If $\sin 3\theta = 0$, then $3\theta = 0°, 180°, 360°, 540°,$
$720°, 900°$; and if $\cos 3\theta = -\frac{1}{2}$, then $3\theta = 120°,$
$240°; 480°; 600°; 840°; 960°$.   Thus, we see that
$\theta = 0°, 60°, 120°, 180°, 240°, 300°, 40°, 80°, 160°,$
$200°, 280°, 320°$

24. $4\tan^2 x = 3\sec^2 x; 4\tan^2 x = 3(\tan^2 x + 1); \tan^2 x$
$= 3; \tan x = \pm\sqrt{3}; x = 60°, 120°, 240°, 300°$

25. $2\cos^2 x - 3\cos 2x = 1; 2\cos^2 x - 3(2\cos^2 x -$
$1) = 1; -4\cos^2 x + 3 = 1; \cos^2 x = \frac{1}{2}; \cos x =$
$\pm\frac{\sqrt{2}}{2}; x = 45°, 135°, 225°, 315°$

26. $\sin^2 4\alpha = \sin 4\alpha + 2; \sin^2 4\alpha - \sin 4\alpha - 2 = 0;$
$(\sin 4\alpha - 2)(\sin 4\alpha + 1) = 0; \sin 4\alpha = 2$ or $\sin 4\alpha =$
$-1; \sin 4\alpha = 2$ yields no solutions. $\sin 4\alpha = -1$
yields $4\alpha = 270°, 630°, 990°, 1350°; \alpha = 67.5°,$
$157.5°, 247.5°, 337.5°$

27. $\sec^2 \theta + \tan \theta = 1; (\tan^2 \theta + 1) + \tan \theta = 1; \tan^2 \theta +$
$\tan \theta = 0; \tan \theta(\tan \theta + 1) = 0; \tan \theta = 0$ or $\tan \theta =$
$-1; \theta = 0°, 180°, 135°, 315°$

28. $\tan 2x + \sec 2x = 1; \frac{\sin 2x}{\cos 2x} + \frac{1}{\cos 2x} = \frac{\cos 2x}{\cos 2x}; \sin 2x +$
$1 = \cos 2x$; squaring both sides yields $\sin^2 2x +$
$2\sin 2x + 1 = \cos^2 2x = 1 - \sin^2 2x; 2\sin^2 2x +$
$2\sin 2x = 0; \sin 2x(\sin 2x + 1) = 0;$ and so $\sin 2x$
$= 0$ or $\sin 2x = -1$. If $\sin 2x = 0$, then $2x =$
$0°, 180°, 360°, 540°$. If $\sin 2x = -1$, then $2x =$
$270°, 630°$ and $x = 0°, 90°, 180°, 270°, 135°, 315°.$
Continue checking for extraneous solutions, the
ones that are solutions are $x = 0°$ and $180°$

29. $\sin 2x = \cos x; 2\sin x \cos x = \cos x; 2\sin x \cos x -$
$\cos x = 0; \cos x(2\sin x - 1) = 0; \cos x = 0$ or
$\sin x = \frac{1}{2}; x = 30°, 90°, 150°, 270°$

30. $\csc^2 \theta - \cot \theta = 1; \cot^2 \theta + 1 - \cot \theta = 1; \cot^2 \theta -$
$\cot \theta = 0; \cot \theta(\cot \theta - 1) = 0$. So $\cot \theta = 0$ or
$\cot \theta = 1$. Thus, $\theta = 45°, 90°, 225°, 270°$

31. $\frac{\sin 35°}{\sin \theta_r} = 1.61; \sin \theta_r = \frac{\sin 35°}{1.61} = 0.356258; \theta_r =$
$20.87°$

32. $\frac{\sin 27°}{\sin \theta_r} = 1.43; \sin \theta_4 = \frac{\sin 27°}{1.43} = 0.317476; \theta_r =$
$18.51°$

33. Since $E = 125\cos(\omega t - \phi)$, we have $60 =$
$125\cos(120\pi t - \frac{\pi}{2})$ so $\cos(120\pi t - \frac{\pi}{2}) = \frac{60}{125}$.
Thus, $120\pi t - \frac{\pi}{2} \approx 1.070\,14$ and so $120\pi t =$
$2.640\,938$. Hence, $t = \frac{2.640\,938}{120\pi} \approx 0.007\,005$ s.

34. Substituting the given values for a, b and A into
$A = a\sin \theta(b + a\cos \theta)$ we get $10 = 3\sin \theta(3.4 +$
$3\cos \theta)$ or $10 = 10.2\sin \theta + 9\sin \theta \cos \theta$. Since
we know that $\sin \theta \cos \theta \approx 0.4675$, we have $10 \approx$
$10.2\sin \theta + 9(0.4675)$ or $10 \approx 10.2\sin \theta + 4.2075$.
Thus, $10.2\sin \theta \approx 10 - 4.2075 = 5.7925$ or $\sin \theta \approx$
$\frac{5.7925}{10.2} \approx 0.5679$ and so $\theta = \sin^{-1} 0.5679 \approx 34.6°$
or $145.4°$. But, $\sin(145.4°)\cos(145.4°) \approx -0.4662$
and so it does not satisfy the stated conditions.
Thus, $\theta \approx 34.6°$.

35. Since $d = \sin \omega t + \frac{1}{2}\sin 2\omega t$ we use a double-angle
formula to get $0 = \sin \omega t + \frac{1}{2}\cdot 2\sin \omega t \cos \omega t$ which
factors as $0 = \sin \omega t(1 + \cos \omega t)$. Thus, $\sin \omega t = 0$
or $\cos \omega t = -1$ and so, $\omega t = 0, \pi$

36. $\frac{a}{b} = \sqrt{\frac{\mu^2 - \sin^2 \theta_i}{\cos^2 \theta_i}}; \frac{14}{10} = \sqrt{\frac{1.333^2 - \sin^2 \theta_i}{\cos^2 \theta_i}}; 1.4^2 =$
$\frac{1.333^2 - \sin^2 \theta_i}{\cos^2 \theta_i}; 1.96\cos^2 \theta_i = 1.333^2 - \sin^2 \theta_i =$
$1.7769 - (1 - \cos^2 \theta_i); 0.96\cos^2 \theta_i = 0.7769;$
$\cos^2 \theta_i = 0.80926; \cos \theta_i = 0.8996; \theta_i = 25.9°$

CHAPTER 20 REVIEW

1.

$\frac{\sin(x+y)}{\cos x \cos y} = \tan x + \tan y$	
$\frac{\sin x \cos y + \cos x \sin y}{\cos x \cos y}$	$\tan x + \tan y$
$\frac{\sin x \cos y}{\cos x \cos y} + \frac{\cos x \sin y}{\cos x \cos y}$	
$\frac{\sin x}{\cos x} + \frac{\sin y}{\cos y}$	
$\tan x + \tan y$	

2.

$(\sin x + \cos x)^2 = 1 + \sin 2x$	
$\sin^2 x + 2\sin x \cos x + \cos^2 x$	$1 + \sin 2x$
$1 + 2\sin x \cos x$	

3.
$$\frac{\sin 3x}{\sin x} - \frac{\cos 3x}{\cos x} = 2$$

$$\frac{\sin(x+2x)}{\sin x} - \frac{\cos(x+2x)}{\cos x} \quad \Big| \quad 2$$

$$\frac{\sin x \cos 2x + \cos x \sin 2x}{\sin x} - \frac{\cos x \cos 2x - \sin x \sin 2x}{\cos x}$$

$$\frac{\sin x(\cos^2 x - \sin^2 x) + \cos x(2\sin x \cos x)}{\sin x}$$

$$- \frac{\cos x(\cos^2 x - \sin^2 x) - \sin x(2\sin x \cos x)}{\cos x}$$

$$\cos^2 x + \sin^2 x + 2\cos^2 x - \cos^2 x + \sin^2 x + 2\sin^2 x$$

$$2\cos^2 x + 2\sin^2 x$$

$$2(\cos^2 x + \sin^2 x)$$

$$2$$

4.
$$\cos(\theta + \phi)\cos(\theta - \phi) = \cos^2\phi - \sin^2\theta$$

$$(\cos\theta\cos\phi - \sin\theta\sin\phi)(\cos\theta\cos\phi + \sin\theta\sin\phi) \quad \Big| \quad \cos^2\theta - \sin^2\theta$$

$$\cos^2\theta\cos^2\phi - \sin^2\theta\sin^2\phi$$

$$(1 - \sin^2\theta)\cos^2\phi - \sin^2\theta(1 - \cos^2\phi)$$

$$\cos^2\phi - \sin^2\theta\cos^2\phi - \sin^2\theta + \sin^2\theta\cos^2\phi$$

$$\cos^2\phi - \sin^2\theta$$

5.
$$\sin(\alpha + \beta)\cos\beta - \cos(\alpha + \beta)\sin\beta = \sin\alpha$$

$$(\sin\alpha\cos\beta + \cos\alpha\sin\beta)\cos\beta$$
$$-(\cos\alpha\cos\beta - \sin\alpha\sin\beta)\sin\beta \quad \Big| \quad \sin\alpha$$
$$\sin\alpha\cos^2\beta + \cos\alpha\sin\beta\cos\beta$$
$$- \cos\alpha\cos\beta\sin\beta + \sin\alpha\sin^2\beta$$
$$\sin\alpha(\cos^2\beta \, \sin^2\beta)$$
$$\sin\alpha$$

6.
$$\tan 2x = \frac{2\cos x}{\csc x - 2\sin x}$$

$$\frac{2\tan x}{1 - \tan^2 x} \quad \Big| \quad \frac{2\cos x}{\csc x - 2\sin x}$$

$$\frac{2\frac{\sin x}{\cos x}}{1 - \frac{\sin^2 x}{\cos^2 x}} \cdot \frac{\frac{\cos^2 x}{\sin x}}{\frac{\cos^2 x}{\sin x}}$$

$$\frac{2\cos x}{\frac{\cos^2 x}{\sin x} - \frac{\sin^2 x}{\sin x}}$$

$$\frac{2\cos x}{\frac{1}{\sin x}(\cos^2 x - \sin^2 x)}$$

$$\frac{2\cos x}{\csc x(1 - 2\sin^2 x)}$$

$$\frac{2\cos x}{\csc x(1 - 2\sin x)}$$

7.
$$\cos^4 x - \sin^4 x = \cos 2x$$

$$(\cos^2 x + \sin^2 x)(\cos^2 x - \sin^2 x) \quad \Big| \quad \cos 2x$$

$$1 \cdot \cos 2x$$

$$\cos 2x$$

8.
$$\frac{\sin 2x - \sin x}{\cos 2x + \cos x} = \tan\frac{x}{2}$$

$$\frac{2\sin x \cos x - \sin x}{2\cos^2 x - 1 + \cos x} \quad \Big| \quad \frac{\sin x}{\cos x + 1}$$

$$\frac{\sin x(2\cos x - 1)}{(2\cos x - 1)(\cos x + 1)}$$

$$\frac{\sin x}{\cos x + 1}$$

9.
$$\sin 3\theta = 2\sin\theta\cos 2\theta + \sin\theta$$

$$\sin(\theta + 2\theta) \quad \Big| \quad 2\sin\theta\cos 2\theta + \sin\theta$$
$$\sin\theta\cos 2\theta + \cos\theta\sin 2\theta$$
$$\sin\theta(1 - 2\sin^2\theta) +$$
$$\cos\theta(2\sin\theta\cos\theta)$$
$$\sin\theta - 2\sin^3\theta + 2\sin\theta\cos^2\theta$$
$$\sin\theta + 2\sin\theta(\cos^2\theta - \sin^2\theta)$$
$$2\sin\theta\cos 2\theta + \sin\theta$$

10.

$$\frac{\cos 3x - \cos 5x}{\sin 3x + \sin 5x} = \tan x$$

$\frac{\cos(4x-x)-\cos(4x+x)}{\sin(4x-x)+\sin(4x+x)}$	$\tan x$
$(\cos 4x \cos x + \sin 4x \sin x) -$	
$\frac{(\cos 4x \cos x - \sin 4x \sin x)}{(\sin 4x \cos x - \cos 4x \sin x)+} \sin 4x \cos x + \cos 4x \sin x$	
$\frac{2 \sin 4x \sin x}{2 \sin 4x \cos x}$	
$\frac{\sin x}{\cos x}$	
$\tan x$	

11. $2 \tan x = -\sqrt{3}$ and so $\tan x = -\frac{\sqrt{3}}{2}$ and $x \approx 139.11°$ or $319.11°$.

12. $3 \sin x = -2; \sin x = \frac{-2}{3}; x = 221.81°, 318.19°$

13. $\cos 2x + \cos x = -1; 2\cos^2 x - 1 + \cos x = -1;$
$2\cos^2 x + \cos x = 0; \cos x(2 \cos x + 1) = 0; \cos x = 0$ or $\cos x = \frac{-1}{2}; x = 90°, 120°, 240°, 270°$

14. $\cos x - \sin 2x - \cos 3x = 0; \cos x - 2 \sin x \cos x -$
$\cos x \cos 2x + \sin x \sin 2x = 0; \cos x - 2 \sin x \cos x$
$-\cos x(1 - 2\sin^2 x) + \sin x(2 \sin x \cos x) = 0; \cos x$
$-2 \sin x \cos x - \cos x + 2 \sin^2 x \cos x + 2 \sin^2 x \cos x$
$= 0; 4 \sin^2 x \cos x - 2 \sin x \cos x = 0; 2 \sin x \cos x$
$(2 \sin x - 1) = 0$. Hence $\sin x = 0$ or $\cos x = 0$ or
$\sin x = \frac{1}{2}$ and so $x = 0°, 30°, 90°, 150°, 180°, 270°$

15. $\sin 4x - 2 \sin 2x = 0; 2 \sin 2x \cos 2x - 2 \sin 2x =$
$0; 2 \sin 2x(\cos 2x - 1) = 0; \sin 2x = 0$ or $\cos 2x =$
$1; 2x = 0°, 180°, 360°, 540°, x = 0°, 90°, 180°,$
$270°$

16. $\sin(30° + x) - \cos(60° + x) = \frac{-\sqrt{3}}{2}; \sin 30° \cos x +$
$\cos 30° \sin x - \cos 60° \cos x + \sin 60° \sin x = \frac{-\sqrt{3}}{2};$
$\frac{1}{2} \cos x + \frac{\sqrt{3}}{2} \sin x - \frac{1}{2} \cos x + \frac{\sqrt{3}}{2} \sin x = \frac{-\sqrt{3}}{2};$
$\sqrt{3} \sin x = \frac{-\sqrt{3}}{2}; \sin x = -\frac{1}{2}; x = 210°, 330°$

17. $2 \sin \theta = \sin 2\theta; 2 \sin \theta = 2 \sin \theta \cos \theta; \sin \theta -$
$\sin \theta \cos \theta = 0; \sin \theta(1 - \cos \theta) = 0$, so $\sin \theta = 0$
or $\cos \theta = 1$ with the result that $\theta = 0; 180°$.

18. $\sin^2 \alpha + 5 \cos^2 \alpha = 3; \sin^2 \alpha + \cos^2 \alpha + 4 \cos^2 \alpha =$
$3; 1 + 4 \cos^2 \alpha = 3; 4 \cos^2 \alpha = 2$, or $\cos^2 \alpha = \frac{1}{2}$.

Thus, we have $\cos \alpha = \pm\frac{\sqrt{2}}{2}$, with the result that
$\alpha = 45°, 135°, 225°, 315°$.

19. $\sin^2 x = 1 + \sin x; \sin^2 x - \sin x = -1 = 0; \sin x$
$= \frac{1\pm\sqrt{1+4}}{2} = \frac{1\pm\sqrt{5}}{2} \approx 1.618$ or -0.618. Since
1.618 is not in the range of the sine function, only
-0.618 yields answers. They are $x = 218.17°$ and
$321.83°$.

20. $\sin \theta - 2 \csc \theta = -1; \sin \theta - \frac{2}{\sin \theta} = -1; \sin^2 \theta - 2 =$
$- \sin \theta; \sin^2 \theta + \sin \theta - 2 = 0; (\sin \theta + 2)(\sin \theta - 1) =$
$0; \sin \theta = -2$ or $\sin \theta = 1$. But, $\sin \theta$ cannot be
-2, so only $\sin \theta = 1$ yields an answer and it is
$\theta = 90°$.

21. $\sin 2x = 2 \sin x \cos x = 2\left(\frac{5}{13}\right)\left(\frac{-12}{13}\right) = \frac{-120}{169}$

22. $\cos \frac{x}{2} = \sqrt{\frac{1+\cos x}{2}} = \sqrt{\frac{1 + \frac{-12}{13}}{2}} = \sqrt{\frac{1}{26}} \approx$
0.196116

23. $\sin \frac{x}{2} = \sqrt{\frac{1-\cos x}{2}} = \sqrt{\frac{1 + \frac{12}{13}}{2}} = \sqrt{\frac{25}{26}} \approx 0.98058$

24. $\tan 2x = \frac{2 \tan x}{1-\tan^2 x} = \frac{2 \cdot \frac{-5}{12}}{1 - \left(\frac{-5}{12}\right)^2} = \frac{\frac{-10}{12}}{\frac{119}{144}} = \frac{-120}{119}$

25. $\sin(x + y) = \sin x \cos y + \cos x \sin y = \frac{-5}{13} \cdot \frac{-15}{17} +$
$\frac{-12}{13} \cdot \frac{8}{17} = \frac{75}{221} - \frac{96}{221} = \frac{-21}{221} \approx -0.09502$

26. $\cos(x - y) = \cos x \cos y + \sin x \sin y = \frac{-12}{13} \cdot \frac{-15}{17} +$
$\frac{-5}{13} \cdot \frac{8}{17} = \frac{180}{221} - \frac{40}{221} = \frac{140}{221} \approx 0.63348$

27. $\cos(x + y) = \frac{-12}{13} \cdot \frac{-15}{17} - \frac{-5}{13} \cdot \frac{8}{17} = \frac{180}{221} + \frac{40}{221} =$
$\frac{220}{221} \approx 0.99548$

28. $\tan(x + y) = \frac{\tan x + \tan y}{1-\tan x \tan y} = \frac{\frac{5}{12} + \frac{-8}{15}}{1 - \frac{5}{12} \cdot \frac{-8}{15}} = \frac{\frac{-21}{180}}{\frac{220}{180}} =$
$\frac{-21}{220} \approx -0.09545$ or $\tan(x + y) = \frac{\sin(x+y)}{\cos(x+y)} =$
$\frac{\frac{-21}{221}}{\frac{220}{221}} = \frac{-21}{220} \approx -0.09545$

29. $\cos(y - x) = \cos y \cos x + \sin y \sin x = \frac{-15}{17} \cdot \frac{-12}{13} +$
$\frac{8}{17} \cdot \frac{-5}{13} = \frac{140}{221} \approx 0.63348$ (Note #26 and #29 are
the same answer.)

30. $\sin(x - y) = \sin x \cos y - \cos x \sin y = \frac{-5}{13} \cdot \frac{-15}{17} -$
$\frac{-12}{13} \cdot \frac{8}{17} = \frac{75}{221} + \frac{96}{221} = \frac{171}{221} \approx 0.77376$

31. We begin with the given equation $a = 5.0(\sin \omega t + \cos 2\omega t)$ and substitute $a = 0$ and a double-angle identity for $\cos \omega t$ with the result that $0 = \sin \omega t + 1 - 2\sin^2 \omega t$ which is equivalent to $2\sin^2 \omega t - \sin \omega t - 1 = 0$. This last equation factors as $(2\sin \omega t + 1)(\sin \omega t - 1) = 0$. Thus, by the zero product rule $\sin \omega t = -\frac{1}{2}$ or $\sin \omega t = 1$, and so $\omega t = 90°, 210°, 330°$.

32. $y = y_1 + y_2 = A_1 \cos \omega_1 t + A_2 \cos \omega_2 t$. If $\alpha = \frac{\omega_1 + \omega_2}{2}$ and $\beta = \frac{\omega_2 + \omega_1}{2}$, then $\alpha - \beta = \omega_1$ and $\alpha + \beta = \omega_2$. Substituting produces $A_1 \cos \omega_1 t = A_1 \cos(\alpha - \beta)t$ and $A_2 \cos \omega_2 t = A_2 \cos(\alpha + \beta)t$, and so $y_1 + y_2 = A_1 \cos(\alpha - \beta)t + A_2 \cos(\alpha + \beta)t$.

33. We want to show that $x = 0.01e^{-6t}(\cos 8t + \sin 8t)$ and $x = \frac{\sqrt{2}}{100}e^{-6t}\cos(8t - \frac{\pi}{4})$ are identically equivalent. Now, using a difference formula, we see that the second equation can be rewritten as $\frac{\sqrt{2}}{100}e^{-6t}\cos(8t - \frac{\pi}{4}) = 0.01e^{-6t}\sqrt{2}[\cos 8t \cos \frac{\pi}{4} + \sin 8t \sin \frac{\pi}{4}] = 0.01e^{-6t}\sqrt{2}[\cos 8t \cdot \frac{1}{\sqrt{2}} + \sin 8t \cdot \frac{1}{\sqrt{2}}] = 0.01e^{-6t}\cos 8t + \sin 8t)$, which is the first equation. Hence the two equations are identical.

CHAPTER 20 TEST

1. Since, $\sin \alpha = \frac{4}{5}$ and α is in Quadrant II, we have $\cos \alpha = -\frac{3}{5}$. Also, since $\cos \beta = -\frac{12}{13}$, and β is in Quadrant III, then $\sin \beta = -\frac{5}{13}$. Thus, $\sin(\alpha + \beta) = \frac{4}{5} \cdot \frac{-12}{13} + \frac{-3}{5} \cdot \frac{-5}{13} = \frac{-48}{65} + \frac{15}{65} = \frac{-33}{65}$.

2. Using the values of $\sin \alpha = \frac{4}{5}, \cos \alpha = -\frac{3}{5}, \cos \beta = -\frac{12}{13}$, and $\sin \beta = -\frac{5}{13}$ from Exercise #1, we see that $\cos(\alpha + \beta) = \frac{-3}{5} \cdot \frac{-12}{13} - \frac{4}{5} \cdot \frac{-5}{13} = \frac{36}{65} + \frac{20}{65} = \frac{56}{65}$.

3. Using the values of $\sin \alpha = \frac{4}{5}, \cos \alpha = -\frac{3}{5}, \cos \beta = -\frac{12}{13}$, and $\sin \beta = -\frac{5}{13}$ from Exercise #1, we see that $\sin(\alpha - \beta) = \frac{4}{5} \cdot \frac{-12}{13} - \frac{-3}{5} \cdot \frac{-5}{13} = \frac{-48}{65} - \frac{15}{65} = \frac{-63}{65}$.

4. Using the values of $\sin \alpha = \frac{4}{5}, \cos \alpha = -\frac{3}{5}, \cos \beta = -\frac{12}{13}$, and $\sin \beta = -\frac{5}{13}$ from Exercise #1, we see that $\cos(\alpha - \beta) = \frac{-3}{5} \cdot \frac{-12}{13} + \frac{4}{5} \cdot \frac{-5}{13} = \frac{36}{65} - \frac{20}{65} = \frac{16}{65}$.

5. Using the values of $\sin \alpha = \frac{4}{5}, \cos \alpha = -\frac{3}{5}$ from Exercise #1, we see that $\sin 2\alpha = 2 \cdot \frac{4}{5} \cdot \frac{-3}{5} = \frac{-24}{25}$.

6. Using the values of $\cos \beta = -\frac{12}{13}$, and $\sin \beta = -\frac{5}{13}$ from Exercise #1, we see that $\cos 2\beta = 2\left(\frac{-12}{13}\right)^2 - 1 = \frac{288}{169} - \frac{169}{169} = \frac{119}{169}$.

7. Using the values of $\sin \alpha = \frac{4}{5}, \cos \alpha = -\frac{3}{5}$ from Exercise #1, we see that $\cos \frac{\alpha}{2} = \sqrt{\frac{1 + \cos \alpha}{2}} = \sqrt{\frac{1 - \frac{3}{5}}{2}} = \sqrt{\frac{8}{10}} = \sqrt{\frac{4}{5}} = \frac{2}{\sqrt{5}}$.

8. Using the values of $\cos \beta = -\frac{12}{13}$, and $\sin \beta = -\frac{5}{13}$ from Exercise #1, we see that $\cos \frac{\beta}{2} = -\sqrt{\frac{1 + \cos \beta}{2}} = -\sqrt{\frac{1 - \frac{12}{13}}{2}} = -\sqrt{\frac{1}{26}} = \frac{-1}{\sqrt{26}}$.

9. Using the double angle formula, $2 \cos x \sin x = \sin 2x$, we see that can be rewritten as $8 \cos 6x \sin 6x = 4 \cdot 2 \sin\left(\frac{12x}{2}\right) \cos\left(\frac{12x}{2}\right) = 4 \sin 12x$.

10. $\tan x = \frac{\sec x}{\csc x}; \frac{\sec x}{\csc x} = \frac{\frac{1}{\cos x}}{\frac{1}{\sin x}} = \frac{\sin x}{\cos x} = \tan x$

11.

$$
\begin{array}{c|c}
\dfrac{\sin x}{1-\cos x} + \dfrac{\sin x}{1+\cos x} & 2\csc x \\[2ex]
\left(\dfrac{\sin x}{1-\cos x}\right)\left(\dfrac{1+\cos x}{1+\cos x}\right) & \\
+\left(\dfrac{\sin x}{1+\cos x}\right)\left(\dfrac{1-\cos x}{1-\cos x}\right) & 2\csc x \\[2ex]
\dfrac{\sin x(1+\cos x)}{1-\cos^2 x} + \dfrac{\sin x(1-\cos x)}{1-\cos^2 x} & \\[2ex]
\dfrac{\sin x(1+\cos x)}{\sin^2 x} + \dfrac{\sin x(1-\cos x)}{\sin^2 x} & \\[2ex]
\dfrac{1+\cos x}{\sin x} + \dfrac{1-\cos x}{\sin x} & \\[2ex]
\dfrac{2}{\sin x} & \\[2ex]
2\csc x &
\end{array}
$$

12. The equation $6\cos^2 x + \cos x = 2$ is equivalent to $6\cos^2 x + \cos x - 2 = 0$. This expression factors as $(3\cos x + 2)(2\cos x - 1) = 0$ and so by the zero-product rule, $3\cos x + 2 = 0$ or $2\cos x - 1 = 0$. If $3\cos x + 2 = 0$, then $\cos x = -\frac{2}{3}$ and $x \approx 131.81°$ or $228.19°$. If $2\cos x - 1 = 0$, then $\cos x = \frac{1}{2}$ and $x = 60°$ or $300°$. In radians the solutions are $\frac{\pi}{3}, \frac{5\pi}{3}, 2.3$, and 3.98.

13. $x = 4r\sin^2\left(\frac{\theta}{2}\right) = 4r\left(\frac{1-\cos\theta}{2}\right) = 2r - 2r\cos\theta = 2r(1 - \cos\theta)$

21

Statistics and Empirical Methods

≡ 21.1 PROBABILITY

1. $\frac{13}{52} = \frac{1}{4}$ **5.** $\frac{12}{52} = \frac{3}{13}$ **9.** $\frac{1}{2} \cdot \frac{26}{51} = \frac{13}{51}$ **13.** $\frac{3}{6} = \frac{1}{2}$

2. $\frac{4}{52} = \frac{1}{13}$ **6.** $\frac{26}{52} = \frac{1}{2}$ **10.** $\frac{3}{13} \cdot \frac{4}{51} = \frac{4}{221}$ **14.** $\frac{3}{6} = \frac{1}{2}$

3. $\frac{4}{52} = \frac{1}{13}$ **7.** $\frac{1}{2} \cdot \frac{1}{2} = \frac{1}{4}$ **11.** $\frac{1}{6}$ **15.** $\frac{1}{6} \cdot \frac{1}{6} = \frac{1}{36}$

4. $\frac{8}{52} = \frac{2}{13}$ **8.** $\frac{1}{2} \cdot \frac{3}{13} = \frac{3}{26}$ **12.** $\frac{2}{6} = \frac{1}{3}$ **16.** $\frac{5}{6}$

17. $\frac{5}{6} \cdot \frac{5}{6} = \frac{25}{36}; P = 1 - \frac{25}{36} = \frac{11}{36}$ **18.** $\frac{1}{6} \cdot \frac{1}{6} \cdot \frac{1}{6} = \frac{1}{216}$

19.

Die 1 \ Die 2	1	2	3	4	5	6
1	(1, 1)	(1, 2)	(1, 3)	(1, 4)	(1, 5)	(1, 6)
2	(2, 1)	(2, 2)	(2, 3)	(2, 4)	(2, 5)	(2, 6)
3	(3, 1)	(3, 2)	(3, 3)	(3, 4)	(3, 5)	(3, 6)
4	(4, 1)	(4, 2)	(4, 3)	(4, 4)	(4, 5)	(4, 6)
5	(5, 1)	(5, 2)	(5, 3)	(5, 4)	(5, 5)	(5, 6)
6	(6, 1)	(6, 2)	(6, 3)	(6, 4)	(6, 6)	(6, 6)

20. $\frac{1}{36}$ **24.** $\frac{2}{36} = \frac{1}{18}$ **28.** $\frac{1}{6} \cdot \frac{1}{6} = \frac{1}{36}$ **31.** $\frac{6}{21} \cdot \frac{6}{21} = \frac{36}{441} = \frac{4}{49}$

21. $\frac{1}{36}$ **25.** $\frac{6}{36} = \frac{1}{6}$ **29.** $\frac{1}{36} \cdot \frac{1}{36} = \frac{1}{1,296}$ **32.** $\frac{3}{21} \cdot \frac{4}{21} = \frac{4}{147}$

22. $\frac{1}{36}$ **26.** $\frac{6+2}{36} = \frac{8}{36} = \frac{2}{9}$ **30.** $\frac{6+5+4+3+2+1}{36} = \frac{21}{36} = \frac{7}{21}$ **33.** $\frac{2+4+6}{21} = \frac{12}{21} = \frac{4}{7}$

23. 0   **27.** $\frac{1+2+3}{36} = \frac{6}{36} = \frac{1}{6}$ **34.** $\frac{6}{21} \cdot \frac{6}{21} = \frac{4}{49}$

35. (a) $.75 \times .75 = 0.5625 = \frac{9}{16}$,

(b) $.25 \times .25 = 0.0625 = \frac{1}{16}$,

(c) $1 - 0.5625 - 0.0625 = 0.375 = \frac{3}{8}$

36. $0.001 \times 0.005 = 0.000005$

37. (a) $0.7 \times 0.8 = 0.56$,
(b) $0.3 \times 0.2 = 0.06$,
(c) $1 - 0.56 - 0.06 = 0.38$

38. $\frac{25}{1000} = 0.025$

39. If one is defective, then the other is not defective. There are two ways this can happen, so the probability is $2(0.025)(1 - 0.025) = 2(0.025)(0.975) = 0.04875$.

40. The probability that it is not defective is $\frac{975}{1000} = 0.975$. The probability that four chosen at random are non-defective is $(0.975)^4 = 0.9037$.

41. The total number of people is 200. Since 20 are blood group B, the probability is $\frac{20}{200} = \frac{1}{10} = 0.10$.

42. $\frac{316}{800} = 0.395$

≡ 21.2 MEASURES OF CENTRAL TENDENCY

1.

Number	2	3	4	6	7
Frequency	2	1	4	2	1

2.

Number	50	52	54	56	58
Frequency	1	2	4	1	2

3.

Number	73	74	77	80	82	83	84	87	89	92	94	96	100
Frequency	1	1	1	1	1	1	2	1	1	1	3	5	1

4.

Number	64	66	76	78	82	85	89	90	93	94	96	98	100
Frequency	3	1	1	2	1	1	2	1	3	1	4	2	1

5.

Interval	1 − 2	3 − 4	5 − 6	7 − 8
Frequency	2	5	2	1

6.

Interval	50 − 52	53 − 55	56 − 58
Frequency	3	4	3

7.

Interval	71 − 75	76 − 80	81 − 85	86 − 90	91 − 95	96 − 100
Frequency	2	2	4	2	4	6

8.

Interval	61 − 65	66 − 70	71 − 75	76 − 80	81 − 85	86 − 90	91 − 95	96 − 100
Frequency	3	1	0	3	2	3	4	7

9.

11.

10.

12.

13.

14.

15.

16.

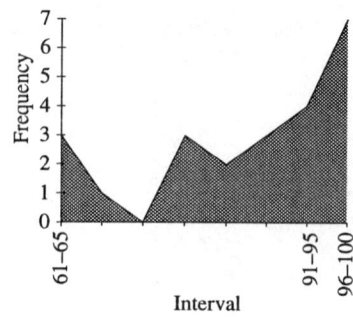

17.   Mean $= \frac{2\times2+3+4\times4+6\times2+7}{10} = \frac{42}{10} = 4.2$; Median $= 4$; Mode $= 4$; $Q_1 = 3$; $Q_2 = 4$; $Q_3 = 6$;

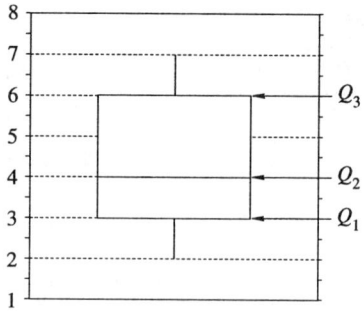

18.   Mean $= \frac{50+52\times2+54\times4+56+58\times2}{10} = \frac{542}{10} = 54.2$; Median $= 54$; Mode $= 54$; $Q_1 = 52$; $Q_2 = 54$; $Q_3 = 56$

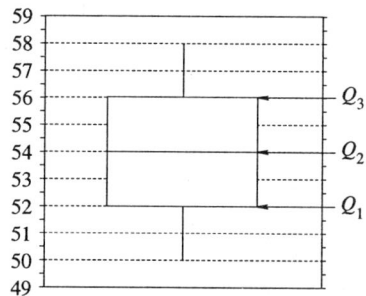

19.   Mean $= \frac{\substack{73+74+77+80+82+83+84\times2+87 \\ +89+92+94\times3+96\times5+100}}{20} = \frac{1767}{20} = 88.35$; Median $= \frac{89+92}{2} = 90.5$; Mode $= 96$; $Q_1 = 82.5$; $Q_2 = 90.5$; $Q_3 = 96$;

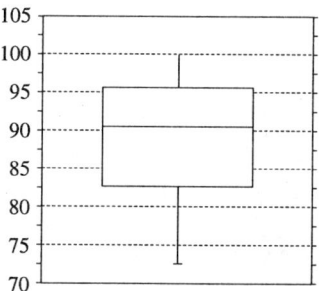

20. Mean $= \dfrac{\substack{64\times3+66+76+78\times2+82+85+89\times2\\+90+93\times3+94+96\times4+98\times2+10}}{23} = 86;$

Median $= 90;$ Mode $= 96; Q_1 = 78; Q_2 = 90;$
$Q_3 = 96$

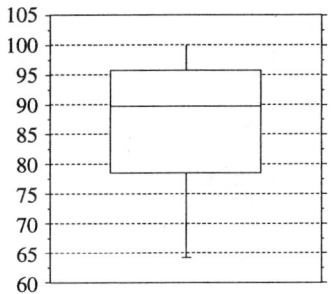

21. See *Computer Programs* in main text.

22.

23.

24.

25.

26.

27.

28. $\overline{x}$ $=$ mean $= \dfrac{\substack{0.01\times1+0.02\times5+0.03\times40+0.04\times50+0.05\times36\\+0.06\times30+0.07\times25+0.08\times10+0.09\times3}}{200};$

$\overline{x} = 0.04865;$ median $= 0.05;$ mode $= 0.04; Q_1 = 0.04; Q_2 = 0.05; Q_3- = 0.06$

29.

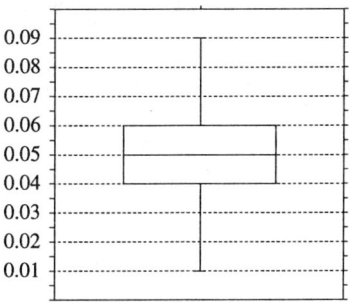

30.

mA	5.24	5.26	5.27	5.28	5.29	5.31	5.34	5.35	5.39	5.42	5.43	5.44	5.45	5.46	5.47
Frequency	1	2	2	1	2	3	4	3	2	2	1	1	1	3	2

31.

32. $\overline{x} = \frac{160.76}{30} = 5.3587$; Median $= 5.345$; mode $= 5.34$

33. $Q_1 = 5.29; Q_2 = 5.345; Q_3 = 5.43$

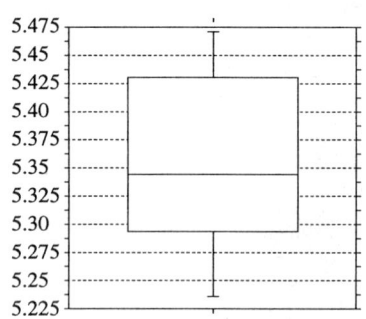

34. **(a)** Sorting the data from low to high, we get

$$52 \quad 54 \quad 55 \quad 57 \quad 59 \quad 59$$
$$62 \quad 63 \quad 64 \quad 65 \quad 67 \quad 72$$

The sum of the data is 729, hence the mean is $\frac{729}{12} = 60.75$. The median is halfway between the 6th and 7th numbers, 59 and 62, so the median is $\frac{59+62}{2} = \frac{121}{2} = 60.5$. The mode is 59. Q_1 is halfway between the 3rd and 4th numbers, 55 and 57, so $Q_1 = \frac{55+57}{2} = \frac{112}{2} = 56$. $Q_2 =$ median $= 60.5$. $Q_3 = \frac{64+65}{2} = \frac{129}{2} = 64.5$.

(b)

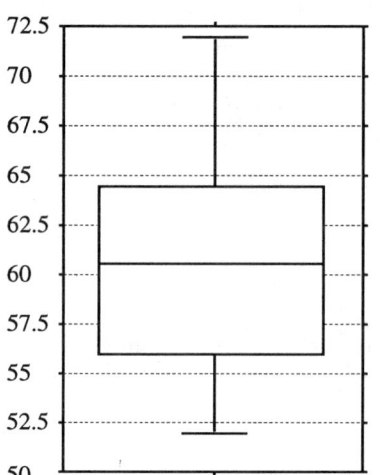

35. **(a)** Mean $= 10.01$, Median $= 10.31$, Mode: None, $Q_1 = 7.44, Q_2 = 10.31, Q_3 = 12.36$

(b)

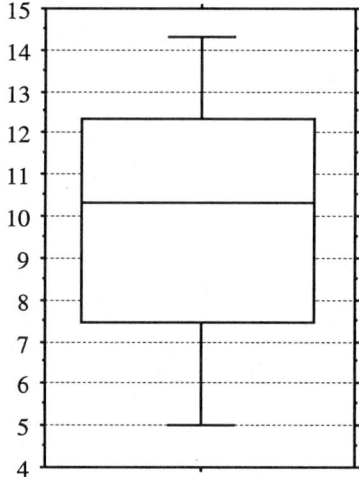

36. Mean $= 659.25$, Median $= 626$, Mode $= 892$, $Q_1 = 528, Q_2 = 626, Q_3 = 800$

37. Mean $= 0.17$, Median $= 0.16$, Mode $= 0.13$, $Q_1 = 0.12, Q_2 = 0.16, Q_3 = 0.22$

38. Mean $= 25{,}905.65$, Median $= 19{,}478$, Mode: None, $Q_1 = 15{,}122, Q_2 = 19{,}478, Q_3 = 38{,}646$

21.3 MEASURES OF DISPERSION

1. $v = 2.84$

3. $v = 67.71$

5. $\sigma_n = 1.69$

7. $\sigma = 8.23$

2. $v = 6.62$

4. $v = 146.63636$

6. $\sigma_n = 2.57$

8. $\sigma = 12.11$

9. The mean is 4.2; $\overline{x} - s = 2.51$; $\overline{x} + s = 5.89$; 5 of the 10 values are within one standard deviation. $\overline{x} - 2s = 0.82$; $\overline{x} + 2s = 7.58$; All 10 values are within two standard deviations.

10. $\overline{x} = 54.2$; $\overline{x} - s = 51.63$; $\overline{x} + s = 56.77$; there are 7 values within one standard deviation $\overline{x} - 2s = 49.06$; $\overline{x} + 2s = 59.34$; there are 10 values within two standard deviations.

11. The mean is 88.35; $\overline{x} - 2s = 80.12$; $\overline{x} + s = 96.58$. There are 15 values within one standard deviation. $\overline{x} - 2s = 71.89$; $\overline{x} + 2s = 104.81$; there are 20 values within two standard deviations.

12. $\overline{x} = 86$; $\overline{x} - s = 73.89$; $\overline{x} + s = 98.11$; there are 18 values within one standard deviation. $\overline{x} - 2s = 61.78$; $\overline{x} + 2s = 110.22$; there are 23 values within two standard deviations.

13. See *Computer Programs* in main text.

14. See *Computer Programs* in main text.

15. (a) The standard deviation is 0.0217.
(b) The mean is 10.0214; $\overline{x} - s = 9.9997$; $\overline{x} + s\,10.0431$; The number of rods within one standard deviation is 390.
(c) $\overline{x} - 2s = 9.978$; $\overline{x} + 2s = 10.0648$; The number of rods within two standard deviations is 480.

16. (a) The standard deviation is 8.89.
(b) The mean is 230.25 so $\overline{x} - s = 221.36$; $\overline{x} + s = 239.14$; There are 72 within one standard deviation.
(c) $\overline{x} - 2s = 212.47$; $\overline{x} + 2s = 248.03$. All 100 values are within two standard deviations.

17. (a) The standard deviation is 0.01636;
(b) $\overline{x} = 0.04865$; $\overline{x} - s = 0.03229$; $\overline{x} + s = 0.06501$; 116 are within one standard deviation.
(c) $\overline{x} - 2s = 0.01593$; $\overline{x} + 2s = 0.08137$; 196 are within two standard deviations.

18. (a) The standard deviation is 0.0740
(b) $\overline{x} = 5.587$ so $\overline{x} - s = 5.2847$; $\overline{x} + s = 5.4340$; there are 17 within one standard deviation.
(c) $\overline{x} - 2s = 5.2107$; $\overline{x} + 2s = 5.5067$; all 30 are within two standard deviations.

19. (a) The mean is 60.75 and the standard deviation is 5.85.

(b) $\overline{x} - s = 60.75 - 5.85 = 54.9$ and $\overline{x} + s = 60.75 + 5.85 = 66.6$. There are 8 data points within one standard deviation of the mean and this is $\frac{8}{12} \approx 66.7\%$.
(c) $\overline{x} - 2s = 60.75 - 2(5.85) = 49.05$ and $\overline{x} + 2s = 60.75 + 2(5.85) = 72.45$. There are 12 data points within two standard deviations of the mean and this is $\frac{12}{12} = 100\%$.

20. (a) The mean is 10.01 and the standard deviation is 2.94.
(b) $\overline{x} - s = 10.01 - 2.94 = 7.07$ and $\overline{x} + s = 10.01 + 2.94 = 12.95$. There are 9 data points within one standard deviation of the mean and this is $\frac{9}{14} \approx 64.3\%$.
(c) $\overline{x} - 2s = 10.01 - 2(2.94) = 4.13$ and $\overline{x} + 2s = 10.01 + 2(2.94) = 15.89$. There are 14 data points within two standard deviations of the mean and this is $\frac{14}{14} = 100\%$.

21. (a) $\overline{x} = 659.25$, $s = 174.67$,
(b) $\overline{x} - s = 659.25 - 174.67 = 484.58$ and $\overline{x} + s = 659.25 + 174.67 = 833.92$. There are 14 data points within one standard deviation of the mean and this is $\frac{14}{24} \approx 58.3\%$,
(c) $\overline{x} - 2s = 659.25 - 2(174.67) = 309.91$ and $\overline{x} + 2s = 659.25 + 2(171.01) = 1008.59$. There are 23 data points within two standard deviations of the mean and this is $\frac{23}{24} \approx 95.8\%$

22. (a) $\overline{x} = 0.169$, $s = 0.053$,
(b) $\overline{x} - s = 0.169 - 0.053 = 0.116$ and $\overline{x} + s = 0.169 + 0.053 = 0.222$. There are 10 data points within one standard deviation of the mean and this is $\frac{10}{15} \approx 66.7\%$,
(c) $\overline{x} - 22 = 0.169 - 2(0.053) = 0.063$ and $\overline{x} + 2s = 0.169 + 2(0.053) = 0.275$. There are 15 data points within two standard deviations of the mean and this is $\frac{15}{15} = 100\%$

23. (a) $\overline{x} = 25,905.65$, $s = 13,280.32$,
(b) $\overline{x} - s = 25,905.65 - 13,280.32 = 12,625.33$ and $\overline{x} + s = 25,905.65 - 13,280.32 = 39,185.97$. There are 23 data points within one standard deviation of the mean and this is $\frac{23}{31} \approx 74.2\%$,
(c) $\overline{x} - 2s = 25,905.65 - 2(13,280.32) = -654.99$ and $\overline{x} + 2s = 25,905.65 + 2(13,280.32) = 52,466.29$. There are 29 data points within two standard deviations of the mean and this is $\frac{29}{31} = 93.5\%$.

24. To find the mean and standard deviations, we use the midpoint of each class. The midpoint of the first class is 709.5.
(a) $\bar{x} = 802.83, s = 49.45$,
(b) $\bar{x} - s = 802.83 - 49.45 = 753.387$ and $\bar{x} + s = 802.83 + 49.45 = 852.28$. There are $5 + 3 + 4 + 7 +$

$5 = 24$ data points within one standard deviation of the mean and this is $\frac{24}{36} \approx 66.7\%$,
(c) $\bar{x} - 2s = 802.83 - 2(49.45) = 703.93$ and $\bar{x} + 2s = 802.83 + 2(49.45) = 901.73$. All 36 data points are within two standard deviations of the mean and this is $\frac{36}{36} = 100\%$.

21.4 FITTING A LINE TO DATA

1. $y = 0.833x - 1.333, r = 0.943$

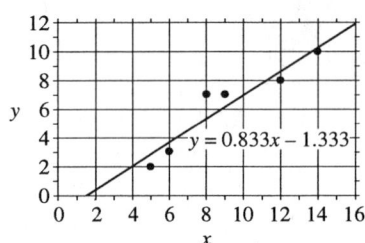

2. $y = 14.55x + 3.85, r = 0.994$

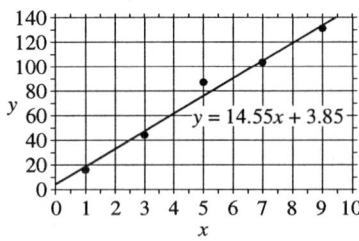

3. $y = 13.55x + 5.1, r = 0.986$

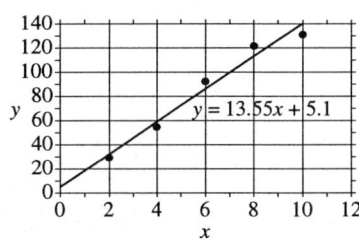

4. $y = 0.175x + 20.974; r = 0.150$

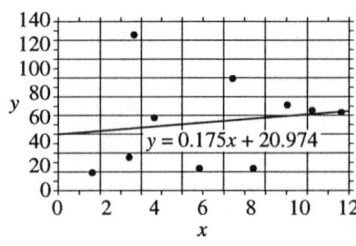

5. $y = 14.955x + 5.379; r = 0.996$

6. $y = 10.33x - 35; r = 0.692$

7. $y = 0.000425x + 0.072; r = 0.987$

8. $y = -0.278x + 33.398; r = -0.784$

9. See *Computer Programs* in main text.

10. $y = -0.00132x + 0.24414, r = -0.49.$ Note that $|r| < 0.75$, so the correlation is poor.

11. $y = -0.26825x + 74.6865, r = -0.70.$ Note that $|r| < 0.75$, so the correlation is poor.

12. $y = -45.4911x + 91,383.3036, r = -0.99.$ Since $|r| > 0.75$, in fact very close to 1, there is a strong correlation.

13. $y = 5.1x - 10,121, r = 0.98.$ Since $|r| > 0.75$, in fact very close to 1, there is a strong correlation.

14. $y = -4.8298x + 48.1106, r = -0.97.$ Since $|r|$ is very close to 1, there is a strong correlation.

≡ 21.5 FITTING NONLINEAR CURVES TO DATA

1. **(a)**

(b) $y = (2.002)(0.988)^x$ Calculator Function

2. **(a)**

(b) $y = 69.541 \times 1.149^x$ Calculator Function

3. **(a)** See the graph below.

(b)

x	5	10	4	10	7	8	8	5	10	5	12	6
y	30	51	26	52	40	43	45	31	52	30	59	36
x^2	25	100	16	100	49	64	64	25	100	25	144	36
x^3	125	1000	64	1000	343	512	512	125	1000	125	1728	216
x^4	625	10000	256	10000	2401	4096	4096	625	10000	625	20736	1296
xy	150	510	104	520	280	344	360	155	520	150	708	216
x^2y	750	5100	416	5200	1960	2752	2880	775	5200	750	8496	1296

$\sum x = 90; \sum x^2 = 748; \sum xy = 4017; \sum y = 495; \sum x^3 = 6750; \sum x^2y = 35,575; \sum x^4 = 64,756.$

Hence:

$$495 = 12a_0 + 90a_1 + 748a_2$$
$$4017 = 90a_0 + 748a_1 + 6750a_2$$
$$35,575 = 748a_0 + 6750a_1 + 64,756a_2$$

Solving yields $a_0 = 4.084; a_1 = 5.848, a_2 = -0.107.$ The equation is $y = 4.084 + 5.848x - 0.107x^2$

(c)

4.   **(a)** See the graph below.

(b)

x	1	2	3	4	5	6	7	8	9	10	$\sum$ 55
y	1.28	1.47	1.00	0.85	0.72	0.63	0.89	0.91	1.05	1.08	9.88
x^2	1	4	9	16	25	36	49	64	81	100	385
x^3	1	8	27	64	125	216	343	512	729	1000	3025
x^4	1	16	81	256	625	1296	2401	4096	6561	10000	25333
xy	1.28	2.94	3.00	3.40	3.60	3.78	6.23	7.28	9.45	10.8	51.76
x^2y	1.28	5.88	9.00	13.6	18	22.68	43.61	58.24	85.05	108	365.34

Hence:

$$9.88 = 10a_0 + 55a_1 + 385a_2$$
$$51.76 = 55a_0 + 385a_1 + 3025a_2$$
$$365.34 = 385a_0 + 3025a_1 + 25333a_2$$

Solving this system of three equations in three variables, yields $a_0 = 1.716, a_1 = -0.309$, and $a_2 = 0.025$. Thus, we have $y = 1.716 - 0.309x + 0.025x^2$.

(c)

5.   **(a)** See the graph below.

(b) To find the least square curve for $y = ae^{-kx}$ we first take natural logarithms giving $\ln y = \ln a - kx$. To use the normal equations of the previous section we need to compute several sums. These are $\sum x = 36; \sum x^2 = 204; \sum \ln y = -4.48183$ and $\sum x \ln y = -25.3578$. Also $n = 8$.

$$\sum \ln y = nb - k \sum x$$
$$\sum x \ln y = b \sum x - k \sum x^2 \qquad \text{or}$$
$$-4.48183 = 8b - 36k$$
$$-25.3578 = 36b - 204k$$

Solving we get $b = -0.004208$ and $k = 0.12356$. Since $b = \ln a; a = e^{-0.004208} = 0.9958$. The final equation is $y = (0.9958)e^{-0.12356x}$. Note: Some calculators have built in functions to compute the least square curve of the form $y = ab^x$. This yields $a = 0.9958$ and $b = 0.883768$. To convert to $y = ae^{-kx}$ find $\ln b = \ln(0.883768) = -0.12356$ or $y = (0.9958)e^{-0.12356x}$.

(c)

6.   **(a)** See the graph below.

(b) To use the least square curve we need the following sums $\sum y = 139; \sum \frac{1}{x^2} = 154.98; \sum \frac{y}{x^2} = 9726$ and $\sum \frac{1}{x^4} = 10{,}820.4$. The normal equations are $139 = 10b + 154.977m; 9726.55 = 154.977b + 10{,}820.4m$. Solving yields $b = -0.039$ and $m = 0.899$. The final equation is $y = \frac{0.899}{x^2} - 0.039$.

(c)

diameter (mm), x

7. **(a)** See the graph below.

(b) As in the previous examples we need sums for the following: $\sum \frac{1}{p}; \sum \frac{1}{p^2}; \sum y;$ and $\sum \frac{y}{p}$. These are 2.0290; 1.5498; 15.2; and 8.1939, respectively. Solving the system $15.2 = 10b + 2.929m$; $8.1939 = 2.9292b + 1.5498m$ yields $b = -0.064$ and $m = 5.41$. The final equation is $y = \frac{5.41}{p} - 0.064$.

(c)

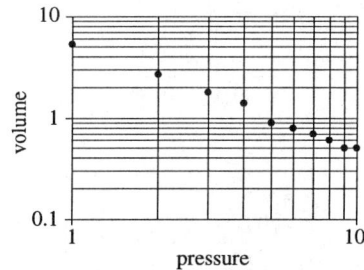

pressure

8. **(c)** $P = 3.46(1.172)^x$;

(d) Shown by dashed curve in Figure.

(a)

(b)

(c) From the graphs in a and b we can see that the best fit will be of the form $y = ab^x$. The equation is $y = 3.46(1.172)^x$.

(d) Shown by dashed curve in Figure 21.0.

CHAPTER 21 REVIEW

1. $\frac{1}{2}$

2. $\frac{2}{54} = \frac{1}{27}$

3. $\frac{10}{16} = \frac{5}{8}$

4. $\frac{48}{52} = \frac{12}{13}$

5. $\frac{2}{6} = \frac{1}{3}$

6. $\frac{1}{36} + \frac{2}{36} + \frac{3}{36} = \frac{6}{36} = \frac{1}{6}$

7. $\frac{1}{6} \times \frac{1}{6} = \frac{1}{36}$

8. $\frac{4}{52} \times \frac{3}{51} = \frac{1}{221}$

9. $\frac{46}{50} = \frac{23}{25} = $ probability that a drive is not defective $\left(\frac{23}{25}\right)^3 = \frac{12167}{15625}$

10. 297.8, 298.9, 299.6, 299.8, 300.1, 300.1, 300.3, 300.4, 300.5, 300.7, 301.2. The median is 300.1

11. $Q_1 = \frac{299.6 + 299.8}{2} = 299.7; Q_2 = 300.1; Q_3 = 300.45$

12.

13. 300.1   **14.** $\bar{x} = 299.945$ **15.** 0.931   **16.** $\frac{8}{11} = 72.7\%$

17. $\frac{10}{11} = 90.9\%$

18. 23, 24, 26, 32, 33, 37, 37, 38, 39, 42, 43, 45, 48, 50, 52;
(a) 37.93,
(b) 38,
(c) $Q_1 = 32.5, Q_2 = 38, Q_3 = 44$;
(d) 8.79

19. (a) See the figure below.
(b) $y = 43.94x + 18.94$,
(c)

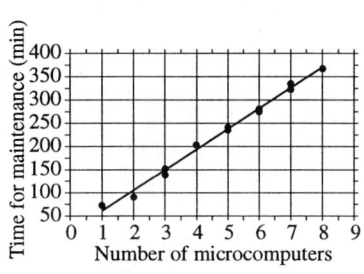

(d) $r = 0.996$

20. (a) See the figure below.
(b) $P = 3983.96V^{-1.405}$ (Note: enter data with V first and P second),

(c)

21. (a) See the figure below.
(b) $n = -0.686p + 83.714, r = -0.9384$
(c) Shown by curve in the figure below.

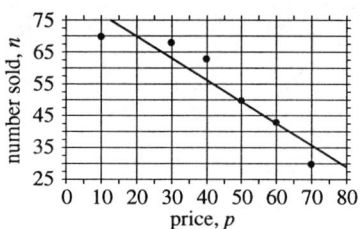

22. To find this equation we need the followings sums $\sum \sqrt{x} = 31.775; \sum x = 110; \sum y = 111.5; \sum \sqrt{x}y = 346.772$. This gives the normal equations $111.5 = 11b + 31.775m, 346.772 = 31.775b + 110m$. Solving this system yields $b = 6.221, m = 1.356$: Thus the equation is $y = 1.356\sqrt{x} + 6.221$.

CHAPTER 21 TEST

1. A deck of 52 cards has 2 red queens, so the probability of drawing a red queen from this deck is $\frac{2}{52} = \frac{1}{26}$.

2. The probability of no defect is $\frac{49}{50}$. To get 5 with no defect, the probability is $\left(\frac{49}{50}\right)^5 \approx 0.904$.

3. 3598.2, 3598.6, 3598.7, 3598.9, 3599.2, 3599.2, 3599.6, 3599.6, 3599.9, 3600.1, 3600.1, 3600.2, 3600.4, 3600.4, 3600.4, 3600.4, 3600.8, 3601.2, 3601.4, 3601.6, median = 3600.1

4. $Q_1 = 3599.2, Q_2 = 3600.1, Q_3 = 3600.4$

5.

6. mode = 3600.4   **7.** mean = 3599.945   **8.** $\sigma = 0.9206, s = 0.9445$

9. $\frac{13}{20} = 65\%$ **10.** $\frac{20}{20} = 100\%$

11. (b) $y = -39.774x + 1063.76$,   (c) See line in graph;   (d) $r = -0.8997$

CHAPTER
22
An Introduction to Calculus

≡ 22.1 THE TANGENT QUESTION

1. $\dfrac{f(3)-f(1)}{3-1} = \dfrac{24-6}{3-1} = \dfrac{18}{2} = 9$

2. $\dfrac{g(7)-g(4)}{7-4} = \dfrac{32-17}{7-4} = \dfrac{15}{3} = 5$

3. $\dfrac{h(-1)-h(-3)}{-1--3} = \dfrac{-6--20}{-1+3} = \dfrac{14}{2} = 7$

4. $\dfrac{j(0)-j(-2)}{0-(-2)} = \dfrac{-5-11}{2} = \dfrac{-16}{2} = -8$

5. $\dfrac{k(-1-k(-2)}{-1--2} = \dfrac{-11--39}{-1+2} = \dfrac{28}{1} = 28$

6. $\dfrac{m(1)-m(0)}{1-0} = \dfrac{-3--1}{1} = \dfrac{-2}{1} = -2$

7. $\dfrac{f(5)-f(2)}{5-2} = \dfrac{1-\frac{5}{2}}{5-2} = \dfrac{\frac{-3}{2}}{3} = -\dfrac{1}{2}$

8. $\dfrac{g(-3)-g(-6)}{-3--6} = \dfrac{-8--2}{-3+6} = \dfrac{-6}{3} = -2$

9. $\dfrac{h(0)-h(1)}{0-1} = \dfrac{0-1}{-1} = 1$

10. $\dfrac{j(b)-j(-2)}{b--2} = \dfrac{b^2+3b+5-3}{b+2} = \dfrac{b^2+3b+2}{b+2} = \dfrac{(b+1)(b+2)}{b+2} = b+1$

11. $\dfrac{k(x_1)-k(1)}{x_1-1} = \dfrac{x_1^2+x_1-7--5}{x_1-1} = \dfrac{x_1^2+x_1-2}{x_1-1} = \dfrac{(x_1+2)(x_1-1)}{x_1-1} = x_1+2$

12. $\dfrac{m(x_1)-m(1)}{x_1-1} = \dfrac{x_1^3+4-5}{x_1-1} = \dfrac{x_1^3-1}{x_1-1} = \dfrac{(x_1-1)(x_1^2+x_1+1)}{x_1-1} = x_1^2+x_1+1$

13. $\dfrac{f(x_1+h)-f(x_1)}{(x_1+h)-x_1} = \dfrac{[(x_1+h)^2+1]-[x_1^2+1]}{h} = \dfrac{x_1^2+2x_1h+h^2+1-x_1^2-1}{h} = \dfrac{2x_1h+h^2}{h} = \dfrac{h(2x_1+h)}{h} = 2x_1+h$

14. $\dfrac{g(x_1+h)-g(x_1)}{x_1+h-x_1} = \dfrac{[(x_1+h)^2+4(x_1+h)-1]-[x_1^2+4x_1-1]}{h} = \dfrac{x_1^2+2x_1h+h^2+4x_1+4h-1-x_1^2-4x_1+1}{h} = \dfrac{2x_1h+h^2+4h}{h} = \dfrac{h(2x_1+4+h)}{h} = 2x_1+4+h$

15. $V(t) = 5t^2 + 4t;$

(a) $\dfrac{V(3)-V(2)}{3-2} = \dfrac{57-28}{1} = 29$ L/min

(b) $\dfrac{V(4)-V(3)}{4-3} = \dfrac{96-57}{1} = 39$ L/min

(c) $\dfrac{V(4)-V(2)}{4-2} = \dfrac{96-28}{2} = \dfrac{68}{2} = 34$ L/min

16. The formula for area is $A = \pi(2t)^2 = 4\pi t^2$, thus
$\overline{A} = \dfrac{A(4)-A(3)}{4-3} = \dfrac{64\pi-36\pi}{1} = 28\pi$ m²/sec

17. **(a)** $\dfrac{f(6)-f(4)}{6-4} = \dfrac{103-43}{2} = \dfrac{60}{2} = 30$

(b) $\dfrac{f(5)-f(4)}{5-4} = \dfrac{70-43}{1} = 27$

(c) $\dfrac{f(4.5)-f(4)}{4.5-4} = \dfrac{55.75-43}{0.5} = \dfrac{12.75}{0.5} = 25.5$

(d) $\dfrac{f(4.25)-f(4)}{4.25-4} = \dfrac{49.1875-43}{0.25} = \dfrac{6.1875}{0.25} = 24.75$

(e) $\dfrac{f(4.1)-f(4)}{4.1-4} = \dfrac{45.43-43}{0.1} = \dfrac{2.43}{0.1} = 24.3$

(f) $\dfrac{f(4.05)-f(4)}{4.05-4} = \dfrac{44.2075-43}{0.05} = \dfrac{1.2075}{0.05} = 24.15$

18. **(a)** $\dfrac{f(2)-f(4)}{2-4} = \dfrac{7-43}{-2} = \dfrac{-36}{-2} = 18$

(b) $\dfrac{f(3)-f(4)}{3-4} = \dfrac{22-43}{-1} = \dfrac{-21}{-1} = 21$

(c) $\dfrac{f(3.5)-f(4)}{3.5-4} = \dfrac{31.75-43}{-0.5} = \dfrac{-11.25}{-0.5} = 22.5$

(d) $\dfrac{f(3.75)-f(4)}{3.75-4} = \dfrac{37.1875-43}{-0.25} = \dfrac{-5.8125}{-0.25} = 23.25$

(e) $\dfrac{f(3.9)-f(4)}{3.9-4} = \dfrac{40.63-43}{-0.1} = \dfrac{-2.37}{-0.1} = 23.7$

(f) $\dfrac{f(3.95)-f(4)}{3.95-4} = \dfrac{41.8075-43}{-0.05} = \dfrac{-1.1925}{-0.05} = 23.85$

19. See *Computer Programs* in main text.

20. **(a)** One possible solution is shown in the figure.

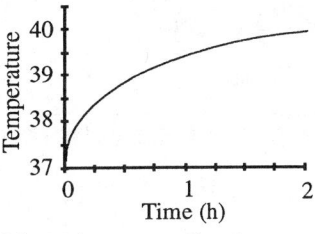

(b) Each tangent line has a positive slope indicating that the temperature is rising. As you move from left to right, the slopes are getting closer to

0, that is, closer to horizontal, showing that the rate of change in the temperature is decreasing.

21. **(a)** Using the formula $d = rt$, the two distances are $d_1 = (20)\frac{15}{60} = 5$ mi and $d_2 = (54)\frac{35}{60} = 31.5$ mi. The total distance is $d = d_1 + d_2 = 5 + 31.5 = 36.5$ mi.

(b) The total time of the trip was 50 min or $\frac{50}{60} = \frac{5}{6}$ h. $v = \frac{d}{t} = \frac{36.5}{5/6} = 43.8$ mph.

22. **(a)** There were 5,100 farms in 1980 and 5,700 farms in 1985. This is an average rate of change of $\frac{5,700-5,100}{5} = \frac{600}{5} = 120$ farms/yr.

(b) There were 2,800 farms in 1990 and 1,300 farms in 1995. This is an average rate of change of $\frac{1,300-2,800}{5} = \frac{-1,500}{5} = -300$ farms/yr. The negative sign indicates that this is a decrease of 300 farms/yr.

(c) Between 1986 and 1988.

(d) In 1986 there were about 5,300 farms and in 1988 there were about 4,100 farms. This is an average change of $\frac{4,100-5,300}{2} = \frac{-1,200}{2} = -600$ farms/yr.

▤ 22.2 THE AREA QUESTION

1. $\Delta x = \frac{b-a}{6} = \frac{7-1}{6} = 1$;

x^*	1.5	2.5	3.5	4.5	5.5	6.5
y^*	6.5	9.5	12.5	15.5	18.5	21.5

$\sum_{i=1}^{6} y_i^* \Delta x = 84$

2. $\Delta x = \frac{1-(-4)}{5} = 1$;

x^*	−3.5	−2.5	−1.5	−0.5	0.5
y^*	21	17	13	9	5

$\sum_{i=1}^{5} y_i^* \Delta x = 65$

3. $\Delta x = \frac{3-0}{6} = \frac{1}{2}$ or 0.5;

x^*	0.25	0.75	1.25	1.75	2.25	2.75
y^*	1.0625	1.5625	2.5625	4.0625	6.0625	8.5625

$\sum_{i=1}^{6} y_i^* \Delta x = 11 \cdot 9375$

4. $\Delta x = \frac{3-1}{4} = \frac{1}{2}$ or 0.5;

x^*	1.25	1.75	2.25	2.75
y^*	2.6875	7.1875	13.1875	20.6875

$\sum_{i=1}^{4} y_i^* \Delta x = 43.75 \times 0.5 = 21.875$

5. $\Delta x = \frac{5-1}{8} = 0.5$;

x^*	1.25	1.75	2.25	2.75	3.25	3.75	4.25	4.75
y^*	5	12.5	22	33.5	47	62.5	80	99.5

$\sum_{i=1}^{8} y_i^* \Delta x = 362(0.5) = 181$

6. $\Delta x = \frac{2--1}{6} = 0.5$;

x^*	−0.75	−0.25	0.25	0.75	1.25	1.75
y^*	1.578125	1.984375	2.015625	2.421875	3.953125	7.359375

$\sum_{i=1}^{6} y_i^* \Delta x = 19.3125(0.5) = 9.65625$

7. $\Delta x = \frac{4--2}{8} = \frac{6}{8} = \frac{3}{4} = 0.75$;

x^*	−1.625	−0.875	−0.125	0.625	1.375	2.125	2.875	3.625
y^*	13.359375	15.234375	15.984375	15.609375	14.109375	11.484375	7.734375	2.859375

$\sum_{i=1}^{8} y_i^* \Delta x = 96.375(0.75) = 72.28125$

8. $\Delta x = \frac{2-0}{10} = \frac{1}{5} = 0.2;$

x^*	0.1	0.3	0.5	0.7	0.9	1.1	1.3	1.5	1.7	1.9
y^*	0.399	1.173	1.875	2.457	2.871	3.069	3.003	2.625	1.887	0.741

$\sum y_i^* \Delta x = 20.1(0.2) = 4.02$

9. $\Delta x = \frac{-1--5}{8} = \frac{1}{2} = 0.5;$

x^*	−4.75	−4.25	−3.75	−3.25	−2.75	−2.25	−1.75	−1.25
y^*	84.1875	69.6875	56.6875	45.1875	35.1875	26.6875	19.6875	14.1875

$\sum y_i^* \Delta x = 351.5(0.5) = 175.75$

10. $\Delta x = \frac{4-2}{8} = \frac{2}{8} = 0.25;$

x^*	2.125	2.375	2.625	2.875	3.125	3.375	3.625	3.875
y^*	17.3909	28.8167	44.4807	65.3206	92.3674	126.7463	169.6760	222.4690

$\sum y_i^* \Delta x = 767.26758(0.25) = 191.8169$ more accurately 191.8168945

11. $\Delta x = 1, \sum y_i^* = 4.05 + 3.85 + 3.9 + 3.75 + 3.65 + 4.0 + 4.2 = 30.85; \sum y_i^* \Delta x = 30.85$

12. $\Delta x = 0.25, \sum y_i^* = 16.4 + 16.605 + 16.575 + 16.4 + 16.335 + 16.27 = 98.585; \sum y_i^* \Delta x = 98.585(0.25) = 24.64625$

13. See *Computer Programs* in main text.

14. For Exercise 9 the answers are:
(a) 175.928889,
(b) 175.972222,
(c) 175.993055,
(d) 175.998264;

and for Exercise 10 you should get:
(a) 192.234094,
(b) 192.335187,
(c) 192.383798,
(d) 192.395945

15. Shading the desired area on a graph of $P'(x)$ indicates that the desired area is a trapezoid with bases of length $P'(0) = 78$ and $P'(40) = 46$ and height 40. The area is $\frac{40}{2}(78 + 40) = 2{,}480$. Thus, the profit for the first 40 coats produced and sold is $2,480.

16. The total length of the interval is 5, so 10 divisions yields $\Delta x = 0.5$. This produces the following table

x^*	0.25	0.75	1.25	1.75	2.25	2.75	3.25	3.75	4.25	4.75
y^*	94.57	276.92	441.13	581.04	695.67	787.45	860.06	917.36	962.73	998.89

The total amount of pollution is $\sum y^* \Delta x = (6{,}615.85)(0.5) = 3{,}307.92$ or 3308 ppm.

17. The total length of the interval is $3 - 1 = 2$, so 8 divisions yields $\Delta x = 0.25$. This produces the following table

x^*	1.125	1.375	1.625	1.875	2.125	2.375	2.625	2.875
y^*	0.475	0.614	0.726	0.791	0.790	0.705	0.517	0.207

The total charge on this capacitor is $\sum y^* \Delta x = (4.825)(0.25) \approx 1.206 \ \Omega$.

18. **(a)** The length of each interval is 10 ft, so $\Delta x = 10$ ft. As in Example 22.9, we approximate the lengths at the midpoints by taking the average of the values at each end of the intervals. The approximate values at the midpoints of these intervals is shown in the following table.

x^*	5	15	25	35	45	55	65	75	85	95
y^*	9	28	40.5	42.5	41	43	45	37	24.5	9.5

Using the values in the above table, we approximate the area as follows:

$$A \approx (y_1^* + y_2^* + y_3^* + y_4^* + y_5^* + y_6^* + y_7^* + y_8^* + y_9^* + y_{10}^*)\Delta x$$
$$= (9 + 28 + 40.5 + 42.5 + 41 + 43 + 45 + 37 + 24.5 + 9.5)(10)$$
$$= (320)(10) = 3200$$

The surface area of the swamp is 3,200 ft^2
(b) The volume of the swamp is obtained by taking the product of its surface area and its average depth. Thus, $S = 3,200$ ft^2 × 6 ft = 19,200 ft^3. Since 27 ft^3 = 1 yd^3, then 19,200 ft^3 = $\frac{19,200}{27} = 711\frac{1}{9} \approx 711.11$ yd^3.

≡ 22.3 LIMITS: AN INTUITIVE APPROACH

1.

x	0.9	0.99	0.999	0.9999	1.0001	1.001	1.01	1.1
$f(x) = 3x$	2.7	2.97	2.997	2.9997	3.0003	3.003	3.03	3.3

$$\lim_{x \to 1} 3x = 3$$

2.

x	2.9	2.99	2.999	2.9999	3.0001	3.001	3.01	3.1
$g(x) = x - 4$	-1.1	-1.01	-1.001	-1.0001	-0.9999	-0.999	-0.99	-0.9

$$\lim_{x \to 3}(x - 4) = -1$$

3.

x	-1.1	-1.01	-1.001	-1.0001	-0.9999	-0.999	-0.99	-0.9
$h(x) = x^2 + 2$	3.21	3.0201	3.002	3.0002	2.9998	2.998	2.9801	2.81

$$\lim_{x \to -1}(x^2 + 2) = 3$$

4.

x	-2.1	-2.01	-2.001	-2.0001	-1.9999	-1.999	-1.99	-1.9
$k(x) = \frac{x^2-4}{x+2}$	-4.1	-4.01	-4.001	-4.0001	-3.9999	-3.999	-3.99	-3.9

$$\lim_{x \to -2} \frac{x^2 - 4}{x + 2} = -4$$

5.

x	-0.1	-0.01	-0.001	-0.0001	0.0001	0.001	0.01	0.1
$f(x) = \frac{\tan x}{x}$	1.0033467	1.0000333	1.0000003	1	1	1.0000003	1.0000333	1.0033467

$$\lim_{x \to 0} \frac{\tan x}{x} = 1$$

6.

x	-0.1	-0.01	-0.001	-0.0001	0.0001	0.001	0.01	0.1
$g(x) = \frac{1-\cos x}{x^2}$	0.49958	0.499996	0.5*	0.5*	0.5*	0.5*	0.499996	0.49958

$$\lim_{x \to 0} \frac{1 - \cos x}{x^2} = 0.5$$

7.

x	0.9	0.99	0.999	0.9999	1.0001	1.001	1.01	1.1
$h(x) = \frac{x}{x-1}$	-9	-99	-999	-9999	10001	1001	101	11

$$\lim_{x \to 1} \frac{x}{x-1} \text{ Does not exist}$$

8.

x	-0.1	-0.01	-0.001	-0.0001	0.0001	0.001	0.01	0.1
$k(x) = \frac{1}{x^2}$	100	10000	10^6	10^8	10^8	10^6	10^4	100

$$\lim_{x \to 0} \frac{1}{x^2} \text{ Does not exist}$$

9. (a) -1, both sides of graph near $x = 0$ converge on -1.
(b) Does not exist, graph goes toward -2 when $x < 2$ and goes toward 1 for $x > 2$;
(c) 4, both sides of graph near $x = 3$ converge on 4.

10. (a) Does not exist, both sides of the graph keep getting larger as x nears -2.
(b) Does not exist, graph goes toward 0 from the left of 0 and keeps getting smaller for $x > 0$;
(c) 2, boths sides of graph near $x = 3$ converge on 2.

11.

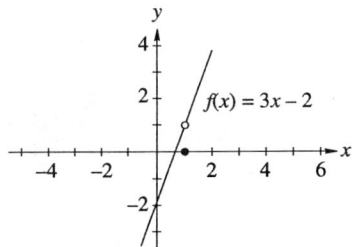

$$\lim_{x \to 1} f(x) = 1$$

12.

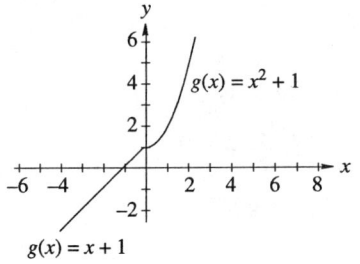

$$\lim_{x \to 0} g(x) = 1$$

13.

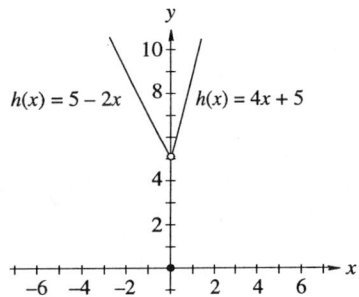

$$\lim_{x \to 0} h(x) = 5$$

14.

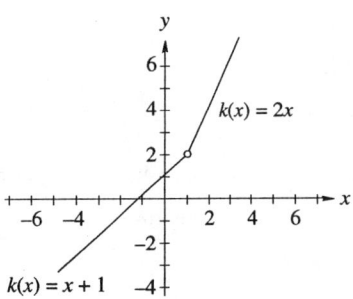

$$\lim_{x \to 1} k(x) = 2$$

15.

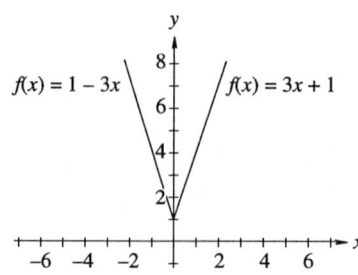

$$\lim_{x \to 0} f(x) = 1$$

16.

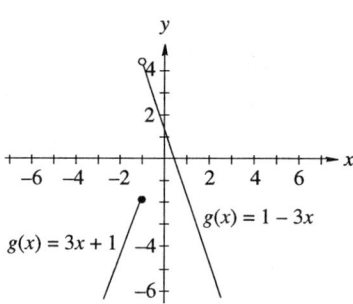

$$\lim_{x \to -1} g(x) \text{ Does not exist}$$

17.

$$\lim_{x \to 0} h(x) = 1$$

18. $\lim_{x \to 0} k(x) = 1$

19. $\lim_{x \to 2}(-15) = -15$

20. $\lim_{x \to 0} x = 0$

21. $\lim_{x \to -4} x = -4$

22. $\lim_{x \to -7} 19 = 19$

23. $\lim_{x \to 8} x = 8$

24. $\lim_{x \to -5}(-9) = -9$

25. 0   **26.** $\frac{5}{3}$ **27.** 4   **28.** 0

≡ 22.4 ONE-SIDED LIMITS

1. (a) $\lim_{x \to 2^-} f(x) = -2$;
    (b) $\lim_{a \to 2^+} f(x) = 1$;
    (c) $\lim_{x \to 3^-} f(x) = 4$;
    (d) $\lim_{x \to 3^+} f(x) = 4$

2. (a) $\lim_{x \to -1^-} g(x) = \infty$, so the limit does not exist;
    (b) $\lim_{x \to -1^+} g(x) = \infty$, so the limit does not exist;
    (c) $\lim_{x \to 0^-} g(x) = 0$;
    (d) $\lim_{x \to 0^+} g(x) = -\infty$, so the limit does not exist;
    (e) $\lim_{x \to \infty} g(x) = 2$;
    (f) $\lim_{x \to -\infty} g(x) = 0$

3. $\lim_{x \to 5^+}(x - 2) = 3$;

x	6	5.5	5.1
$x - 2$	4	3.5	3.1

4. $\lim_{x \to 2^-}(x^2 - 1) = 3$;

x	1	1.5	1.9	1.99
$x^2 - 1$	0	1.25	2.61	2.9601

5. $\lim_{x \to 4^+} \dfrac{x}{x - 4} = \infty$;

x	5	4.5	4.1	4.01
$\frac{x}{x-4}$	5	9	41	401

6. $\lim_{x \to 3^-} \dfrac{2x}{x - 3} = -\infty$;

x	2	2.5	2.9	2.99
$\frac{2x}{x-3}$	-4	-10	-58	-598

7. $\displaystyle\lim_{x\to-2^+}\frac{x^2-4}{x+2}=-4;$

x	-1	-1.5	-1.9	-1.99
$\frac{x^2-4}{x+2}$	-3	-3.5	-3.9	-3.99

8. $\displaystyle\lim_{x\to-5^-}\frac{x^2+25}{x+5}=-\infty;$

x	-6	-5.5	-5.1
$\frac{x^2+25}{x+5}$	-61	-110.5	-510.1

9. $\displaystyle\lim_{x\to1^-}\frac{|x-1|}{x-1}=-1;$

x	0	0.5	0.9
$\frac{|x-1|}{x-1}$	-1	-1	-1

10. $\displaystyle\lim_{x\to1^+}\frac{|x-1|}{x-1}=1$

x	1.1	1.5	2
$\frac{|x-1|}{x-1}$	1	1	1

11. $\displaystyle\lim_{x\to2^-}\frac{|x-2|}{x-2}=-1;$

x	1	1.5	1.9
$\frac{|x-2|}{x-2}$	-1	-1	-1

12. $\displaystyle\lim_{x\to2^+}\frac{|x-2|}{x-2}=1;$

x	3	2.5	2.1
$\frac{|x-2|}{x-2}$	1	1	1

19. 2

20. 2.5

21. 0

22. 0

13. $\displaystyle\lim_{x\to2^+}\sqrt{x-2}=0;$

x	3	2.5	2.1	2.01
$\sqrt{x-2}$	1	0.707	0.316	0.1

14. $\displaystyle\lim_{x\to-4^+}\sqrt{x+4}=0;$

x	-3	-3.5	-3.9	-3.99
$\sqrt{x+4}$	1	0.707	0.316	0.1

15. $\displaystyle\lim_{x\to\infty}\frac{9}{3x}=0;$

x	10	100	1000
$\frac{9}{3x}$	0.3	0.03	0.003

16. $\displaystyle\lim_{x\to-\infty}\frac{-8}{4x}=0;$

x	-10	-100	-1000
$\frac{-8}{4x}$	0.2	0.02	0.002

17. $\displaystyle\lim_{x\to-\infty}\frac{1}{2x+1}=0;$

x	-10	-100	-1000
$\frac{1}{2x+1}$	-0.0526	-0.00503	-0.0005

18. $\displaystyle\lim_{x\to\infty}\frac{-5}{9x-4}=0;$

x	0	10	100
$\frac{-5}{9x-4}$	$\frac{5}{4}$	-0.0581	-0.00558

23. $-\infty$

24. 0

25. (a) $C(0.5)=\frac{0.25(0.5)}{(0.5)^2+5}=0.02381$

(b) $C(2)=\frac{0.25(2)}{(2)^2+5}=0.0556$

(c) $\displaystyle\lim_{t\to\infty}C(t)=\lim_{t\to\infty}\frac{0.25t}{t^2+5}=\frac{\frac{0.25t}{t^2}}{\frac{t^2}{t^2}+\frac{5}{t^2}}=$

$\displaystyle\lim_{t\to\infty}\frac{\frac{0.25}{t}}{1+\frac{5}{t^2}}=\lim_{t\to\infty}\frac{0}{1+0}=0$

26. $\displaystyle\lim_{t\to\infty}\left(65+\frac{70t^2}{t^2+15}\right)=65+\lim_{t\to\infty}\left(\frac{70t^2}{t^2+15}\right)=$

$65+\displaystyle\lim_{t\to\infty}\left(\frac{\frac{70t^2}{t^2}}{\frac{t^2}{t^2}+\frac{15}{t^2}}\right)=65+\lim_{t\to\infty}\left(\frac{70}{1+\frac{15}{t^2}}\right)=$

$65+\left(\dfrac{70}{1+0}\right)=65+70=135.$ The most words per minute that we can expect this student to keyboard is 135 words per minute.

27. (a) $\displaystyle\lim_{t\to3^-}C(t)=4.75$ since $t<3$
(b) $\displaystyle\lim_{x\to3^+}C(t)=0.65(3)+2.80=4.75$ since $t>3$.

(c) $\lim\limits_{x\to3} C(t) = 4.75$ since $\lim\limits_{t\to3^-} C(t) = \lim\limits_{x\to3^+} C(t) = 4.75.$

28. (a) $\lim\limits_{x\to12,500^-} T(x) = 0.04(12,500) = \$500$

(b) $\lim\limits_{x\to12,500^+} T(x) = 0.03(12,500) + 125 = \$500$

(c) $\lim\limits_{x\to12,500} T(x) = \$500$ since $\lim\limits_{x\to12,500^-} T(x) = \lim\limits_{x\to12,500^+} T(x) = \$500.$

(d) $\lim\limits_{x\to35,000^-} T(x) = 0.03(35,000) + 125 = \$1,175$

(e) $\lim\limits_{x\to35,000^+} T(x) = 0.02(35,000) + 465 = \$1,165$

(f) $\lim\limits_{x\to35,000} T(x)$ does not exist since $\lim\limits_{x\to35,000^-} T(x) \neq \lim\limits_{x\to35,000^+} T(x).$

≡ 22.5 ALGEBRAIC TECHNIQUES FOR FINDING LIMITS

1. $\lim\limits_{x\to5} 8 = 8$

2. $\lim\limits_{x\to-4} x = -4$

3. $\lim\limits_{x\to8}(x-3) = 8-3 = 5$

4. $\lim\limits_{x\to-7}(x+4) = -7+4 = -3$

5. $\lim\limits_{x\to-6}(x+7) = -6+7 = 1$

6. $\lim\limits_{x\to6}(9-x) = 9-6 = 3$

7. $\lim\limits_{x\to2} 3x = 3\cdot2 = 6$

8. $\lim\limits_{t\to-3} 5t = 5(-3) = -15$

9. $\lim\limits_{x\to6}\dfrac{2}{3}x + 5 = \dfrac{2}{3}\cdot6 + 5 = 4 + 5 = 9$

10. $\lim\limits_{x\to4}(\dfrac{3}{2}x - 2) = \dfrac{3}{2}\cdot4 - 2 = 6 - 2 = 4$

11. $\lim\limits_{p\to3}\dfrac{p^2+6}{p} = \dfrac{3^2+6}{3} = \dfrac{15}{3} = 5$

12. $\lim\limits_{x\to-2}\dfrac{x^2-6}{x} = \dfrac{(-2)^2-6}{-2} = \dfrac{-2}{-2} = 1$

13. $\lim\limits_{s\to5}\dfrac{3s^2+5}{2s} = \dfrac{3\cdot5^2+5}{2\cdot5} = \dfrac{80}{10} = 8$

14. $\lim\limits_{x\to6}\dfrac{2x^2-3}{3-x} = \dfrac{2\cdot6^2-3}{3-6} = \dfrac{69}{-3} = -23$

15. $\lim\limits_{x\to2}(x+1)^3 = (2+1)^3 = 3^3 = 27$

16. $\lim\limits_{y\to5}(y-1)^3 = (5-1)^3 = 4^3 = 64$

17. $\lim\limits_{x\to-1}(2x^2-1)^2 = (2(-1)^2-1)^2 = 1^2 = 1$

18. $\lim\limits_{t\to4}(3t^2+2)^2 = (3\cdot4^2+2)^2 = 50^2 = 2500$

19. $\lim\limits_{x\to3}\sqrt{x+3} = \sqrt{3+3} = \sqrt{6}$

20. $\lim\limits_{x\to-6}\sqrt{6-7x} = \sqrt{6-7(-6)} = \sqrt{6+42} = \sqrt{48} = 4\sqrt{3}$

21. $\lim\limits_{x\to2}\dfrac{x^2-2x}{x-2} = \lim\limits_{x\to2}\dfrac{x(x-2)}{x-2} = \lim\limits_{x\to2}x = 2$

22. $\lim\limits_{w\to3}\dfrac{w^2-1}{w+1} = \dfrac{9-1}{3+1} = \dfrac{8}{4} = 2$

23. $\lim\limits_{x\to-1}\dfrac{x^2-1}{x+1} = \lim\limits_{x\to-1}\dfrac{(x+1)(x-1)}{x+1} = \lim\limits_{x\to-1}(x-1) = -2$

24. $\lim\limits_{y\to3}\dfrac{y^2-7y+12}{y-3} = \lim\limits_{y\to3}\dfrac{(y-3)(y-4)}{y-3} = \lim\limits_{y\to3}(y-4) = -1$

25. $\lim\limits_{x\to-4}\dfrac{x^2+2x-8}{x^2+5x+4} = \lim\limits_{x\to-4}\dfrac{(x+4)(x-2)}{(x+4)(x+1)} = \lim\limits_{x\to-4}\dfrac{x-2}{x+1} = \dfrac{-6}{-3} = 2$

26. $\lim\limits_{x\to\frac{1}{2}}\dfrac{2x^2+5x-3}{4x^2-2x} = \lim\limits_{x\to\frac{1}{2}}\dfrac{(x+3)(2x-1)}{2x(2x-1)} = \lim\limits_{x\to\frac{1}{2}}\dfrac{x+3}{2x} = 3\dfrac{1}{2} \text{ or } \dfrac{7}{2}$

27. $\lim\limits_{t\to-7}\dfrac{2t^2+15t+7}{t^2+5t-14} = \lim\limits_{t\to-7}\dfrac{(2t+1)(t+7)}{(t-2)(t+7)} = \lim\limits_{t\to-7}\dfrac{2t+1}{t-2} = \dfrac{-14+1}{-7-2} = \dfrac{-13}{-9} = \dfrac{13}{9}$

28. $\lim\limits_{x\to0}\dfrac{(x+3)^2-9}{x} = \lim\limits_{x\to0}\dfrac{x^2+6x+9-9}{x} = \lim\limits_{x\to0}(x+6) = 6$

29. $\lim\limits_{h\to0}\dfrac{(4+h)^2-4^2}{h} = \lim\limits_{h\to0}\dfrac{16+8h+h^2-16}{h} = \lim\limits_{h\to0}(8+h) = 8$

30. $\lim\limits_{x\to c}\dfrac{x^3-c^3}{x-c} = \lim\limits_{x\to c}\dfrac{(x-c)(x^2+xc+c^2)}{x-c} = \lim\limits_{x\to c}(x^2+xc+c^2) = c^2+c^2+c^2 = 3c^2$

31. $\lim\limits_{x\to\infty} \dfrac{2x^2+4}{5x^2+3x+2} =$

$\lim\limits_{x\to\infty} \dfrac{2+\frac{4}{x^2}}{5+\frac{3}{x}+\frac{2}{x^2}} x^2 = \dfrac{2}{5}$

32. $\lim\limits_{x\to\infty} \dfrac{6x^3+9x+1}{2x^3+4x+8} = \lim\limits_{x\to\infty} \dfrac{6+\frac{9}{x^2}+\frac{1}{x^3}}{2+\frac{4}{x^2}+\frac{8}{x^3}} =$

$\dfrac{6}{2} = 3$

33. $\lim\limits_{x\to\infty} \dfrac{7x^4+5x^2}{x^5+2} = \lim\limits_{x\to\infty} \dfrac{\frac{7}{x}+\frac{5}{x^3}}{1+\frac{2}{x^5}} = \dfrac{0}{1} = 0$

34. $\lim\limits_{x\to\infty} \dfrac{4x^5+3x}{2x^4+1} = \lim\limits_{x\to\infty} \dfrac{4x+\frac{3}{x^3}}{2+\frac{1}{x^4}} = \lim\limits_{x\to\infty} 2x = \infty$

or does not exist.

35. **(a)** This is true for any two functions f and g where $\lim\limits_{x\to1} f(x) = 0$ and $\lim\limits_{x\to1} g(x) = \infty$. For example, $f(x) = x-1$ and $g(x) = \frac{1}{(x-1)^2}$.

(b) One such example is $f(x) = (x-1)^3$ and $g(x) = \frac{1}{(x-1)^2}$.

(c) One such example is $f(x) = 5(x-1)^2$ and $g(x) = \frac{1}{(x-1)^2}$.

36. One such example is $f(x) = \frac{1}{x-2}$ and $g(x) = -\frac{1}{x-2}$. Here $f(x)+g(x) = \frac{1}{x-2} - \frac{1}{x-2} = 0$. Thus, $\lim\limits_{x\to2}[f(x)+g(x)] = 0$.

▤ 22.6 CONTINUITY

1. $f(x) = \begin{cases} 3x+1 & \text{if } x \le 4 \\ x^2-3 & \text{if } x > 4 \end{cases}$ $c=4=$

Continuous

2. Not continuous, $\lim\limits_{x\to1^-} g(x) = 1$; $\lim_{x\to1^-} g(x) = 2$

3. Not continuous, $\lim\limits_{x\to1} h(x) = 4 \ne 3 = h(1)$

4. Continuous

5. Not continuous, $k(2)$ is not defined

6. Not continuous, $\lim\limits_{x\to2} f(x) = 9 \ne f(2) = 0$

7. Not continuous, $\lim\limits_{x\to-2^-} g(x) = 4$ and $\lim\limits_{x\to-2^+} g(x) = 10$. Since $\lim\limits_{x\to-2^-} g(x) \ne \lim\limits_{x\to-2^+} g(x)$, the limit does not exist.

8. Not continuous, $\lim\limits_{x\to-2^-} h(x) = 4 \ne -4 = \lim_{x\to-2^+} h(x)$

9. Not continuous, $\lim\limits_{x\to2} j(x) = 4 \ne 5 = j(2)$

10. Not continuous, $k(1)$ is not defined

11. None

12. 5

13. -1

14. None

15. $-3, 2$

16. $-2, -3$

17. None

18. 0

19. 2

20. 0, 5

21. $f(x) = \frac{x}{x+3}; (-\infty, -3), (-3, \infty)$

22. $g(x) = \frac{x}{x^2-1}; (-\infty, -1), (-1, 1), (1, \infty)$

23. $h(x) = \frac{x+2}{x^2-4}; (-\infty, -2), (-2, 2), (2, \infty)$

24. $j(x) = \frac{x^2+3x-4}{x+4}; (-\infty, -4), (-4, \infty)$

25. $k(x) = \sqrt{x+3}; [-3, \infty)$

26. $f(x) = \sqrt{5-x}; (-\infty, 5]$

27. $g(x) = \sqrt{x^2-9}; (-\infty, -3), [3, \infty)$

28. $h(x) = \sqrt{x^2+4x+4}; (-\infty, \infty)$

29. $j(x) = \frac{1}{x}; (-\infty, 0), (0, \infty)$

30. $k(x) = \frac{1}{\sqrt{x}}; (0, \infty)$

31. **(a)** Yes, because when $r = R$, $\frac{GMr}{R^3} = \frac{GMr}{r^3} = \frac{GM}{r^2}$ and the two formulas connect.

(b)

32. Yes, since when $t = 3, 0.85(3) + 5.95 = 2.55 + 5.95 = 8.50$ and the two formulas connect.

33. **(a)** $\lim\limits_{x \to 23,900^-} T(x) = 0.15(23,900) = \$3,585$

　　(b) $\lim\limits_{x \to 23,900^+} T(x) = 0.28(23,900) - 3,107 = \$3,585$

(c) $\lim\limits_{x \to 23,900} T(x) = \$3,585$ since $\lim\limits_{x \to 23,900^-} T(x) = \lim\limits_{x \to 23,900^+} T(x)$

(d) Yes

(e) $\lim\limits_{x \to 61,650^-} T(x) = 0.28(61,650) - 3,107 = \$14,155$

(f) $\lim\limits_{x \to 61,650^+} T(x) = 0.33(61,650) - 6,189.50 = \$14,155$

(g) $\lim\limits_{x \to 61,650} T(x) = 14,155$ because $\lim\limits_{x \to 61,650^-} T(x) = \lim\limits_{x \to 61,650^+} T(x)$

(h) Yes

34. **(a)** $\lim\limits_{x \to 24^-} C(x) = 0.47(24) + 2.95 = \$14.23$

(b) $\lim\limits_{x \to 24^+} C(x) = 0.89(24) - 7.13 = \$14.23$

(c) $\lim\limits_{x \to 24} C(x) = \$14.23$ since $\lim\limits_{x \to 24^-} C(x) = \lim_{x \to 24^+} C(x)$

(d) Yes

CHAPTER 22 REVIEW

1. $\overline{m} = \dfrac{f(2) - f(0)}{2 - 0} = \dfrac{-3 - -7}{2} = \dfrac{4}{2} = 2$

2. $\overline{m} = \dfrac{g(-1) - g(-3)}{-1 - -3} = \dfrac{1 - -7}{-1 + 3} = \dfrac{8}{2} = 4$

3. $\overline{m} = \dfrac{h(-2) - h(-3)}{-2 - -3} = \dfrac{9 - 19}{-2 + 3} = \dfrac{-10}{1} = -10$

4. $\overline{m} = \dfrac{j\left(\frac{1}{2}\right) - j(0)}{\frac{1}{2} - 0} = \dfrac{\frac{8}{9/2} - 2}{\frac{1}{2}} = \dfrac{\frac{16}{9} - 2}{\frac{1}{2}} = \dfrac{\frac{-2}{9}}{\frac{1}{2}} = \dfrac{-4}{9}$

5. $\overline{m} = \dfrac{k(b) - k(-1)}{b - -1} = \dfrac{b^2 + 2b - (-1)}{b + 1} = \dfrac{b^2 + 2b + 1}{b + 1} = b + 1$

6. $\overline{m} = \dfrac{m(x_1) - m(6)}{x_1 - 6} = \dfrac{x_1^2 - 5x_1 - 6}{x_1 - 6} = \dfrac{(x_1 - 6)(x_1 + 1)}{x_1 - 6} = x_1 + 1$

7. $\Delta x = \dfrac{-2 + 5}{6} = \dfrac{3}{6} = 0.5;$

x^*	-4.75	-4.25	-3.75	-3.25	-2.75	-2.25
y^*	12	10	8	6	4	2

$\sum y_i^* \Delta x = 42(0.5) = 21$

8. $\Delta x = \dfrac{4 - 0}{8} = 0.5;$

x^*	0.25	0.75	1.25	1.75	2.25	2.75	3.25	3.75
y^*	5.125	6.125	8.125	11.125	15.125	20.125	26.125	33.125

$\sum y_i^* \Delta x = 125(0.5) = 62.5$

9. $\Delta x = \dfrac{4 - 2}{8} = \dfrac{2}{8} = 0.25;$

x^*	2.125	2.375	2.625	2.875	3.125	3.375	3.625	3.875
y^*	16.796875	20.671875	24.921875	29.546875	34.546875	39.921875	45.671875	51.796875

$\sum y_i^* \Delta x = 263.875(0.25) = 65.96875$

10. $\Delta x = \frac{1-(-1)}{8} = \frac{2}{8} = 0.25;$

x^*	-0.875	-0.625	-0.375	-0.125	0.125	0.375	0.625	0.875
y^*	3.46875	4.21875	4.71875	4.96875	4.96875	4.71875	4.21875	3.46875

$\sum y_i^* \Delta x = 34.75(0.25) = 8.6875$

11. $\lim\limits_{x \to 1}(3x^2 + 2x - 5) = 3 + 2 - 5 = 0$

12. $\lim\limits_{x \to -2} \dfrac{3x^2 - 12}{x + 2} = \lim\limits_{x \to -2} = \dfrac{3(x + 2)(x - 2)}{x + 2} = $
$\lim\limits_{x \to -2} 3(x - 2) = 3(-2 - 2) = -12$

13. $\lim\limits_{x \to 0} \dfrac{3x^2 - 12}{x - 2} = \dfrac{-12}{-2} = 6$

14. $\lim\limits_{x \to -3} \dfrac{x^2 - 9}{3x + 9} = \lim\limits_{x \to -3} \dfrac{(x + 3)(x - 3)}{3(x + 3)} = $
$\lim\limits_{x \to -3} \dfrac{x - 3}{3} = \dfrac{-6}{3} = -2$

15. $\lim\limits_{x \to \infty} \dfrac{6x^2 + 3}{2x^2} = \lim\limits_{x \to \infty} \dfrac{6 + \frac{3}{x^2}}{2} = \dfrac{6}{2} = 3$

16. $\lim\limits_{x \to \infty} \dfrac{9x^3 + 2x - 1}{3x^2 + x + 1} = \lim\limits_{x \to \infty} \dfrac{9x + \frac{2}{x} - \frac{1}{x^2}}{3 + \frac{1}{x} + \frac{1}{x^2}} = $
$\lim\limits_{x \to \infty} \dfrac{9}{3}x = \infty$ or does not exist

17. $\lim\limits_{x \to 5^+} \sqrt{x - 5} = \sqrt{5 - 5} = \sqrt{0} = 0$

18. $\lim\limits_{x \to 4^-} \dfrac{x + 4}{x^2 - 16} = \lim\limits_{x \to 4^-} \dfrac{1}{x - 4} = -\infty$ or does not exist

19. $\lim\limits_{x \to \infty} \dfrac{x + 4}{2x^2 + 1} = \lim\limits_{x \to \infty} \dfrac{\frac{1}{x} + \frac{4}{x^2}}{2 + \frac{1}{x^2}} = \dfrac{0}{2} = 0$

20. Since both $\lim\limits_{x \to 1^-} f(x) = 0$ and $\lim\limits_{x \to 1^+} f(x) = 0$, we have $\lim\limits_{x \to 1} f(x) = 0$.

21. At $x = 5$ **22.** At $x = 0$ **23.** At $x = -3$ **24.** At $x = -1$ and 3

25. x cannot equal -7 so $(-\infty, -7), (-7, \infty)$

26. x cannot equal -3 or -2 so $(-\infty, -3), (-3, -2), (-2, \infty)$

27. $x^2 - 25$ must be greater than 0 so $x^2 > 25$; or $x < -5$ or $x > 5; (-\infty, -5), (5, \infty)$

28. $3 + x > 0$ and $3 - x \geq 0$ so $x > -3$ and $x \leq 3$; $(-3, 3]$

CHAPTER 22 TEST

1. $\overline{m} = \dfrac{f(1) - f(0)}{1 - 0} = \dfrac{-1 - (-3)}{1 - 0} = 2$

2. Here $\Delta x = \frac{5-1}{4} = 1$, so the area is $\sum\limits_{i=1}^{4} g(x_i^*)\Delta x = $
$[g(1.5) + g(2.5) + g(3.5) + g(4.5)]\Delta x = (3.75 + 13.75 + 29.75 + 51.75)(1) = 99.$

3. $\lim\limits_{x \to 2}(5x^2 - 7) = \left(\lim\limits_{x \to 2} 5\right)\left(\lim\limits_{x \to 2} x^2\right) - \lim\limits_{x \to 2} 7 = 5(2^2) - 7 = 5 \cdot 4 - 7 = 20 - 7 = 13$

4. $\lim\limits_{x \to 2}\left(\dfrac{x^2 - 4}{3x - 6}\right) = \lim\limits_{x \to 2}\left(\dfrac{(x + 2)(x - 2)}{3(x - 2)}\right) = $
$\lim\limits_{x \to 2}\left(\dfrac{x + 2}{3}\right) = \dfrac{4}{3}$

5. $\lim\limits_{x \to 1^+} \begin{cases} x^3 - 1 & \text{if } x < 1 \\ 4 & \text{if } x = 1 = \lim\limits_{x \to 1^+}(1 + x) = 1 + 1 = 2 \\ 1 + x & \text{if } x > 1 \end{cases}$

6. $\lim\limits_{x \to \infty} \dfrac{5x^3 - 4x + 1}{7x^3 - 2} = \lim\limits_{x \to \infty} \dfrac{\frac{5x^3 - 4x + 1}{x^3}}{\frac{7x^3 - 2}{x^3}} = $
$\lim\limits_{x \to \infty} \dfrac{\frac{5x^3}{x^3} - \frac{4x}{x^3} + \frac{1}{x^3}}{\frac{7x^3}{x^3} - \frac{2}{x^3}} = \lim\limits_{x \to \infty} \dfrac{5 - \frac{4}{x^2} + \frac{1}{x^3}}{7 - \frac{2}{x^3}} = $
$\dfrac{5 - 0 + 0}{7 - 0} = \dfrac{5}{7}$

7. $\lim\limits_{x \to 2^-} h(x) = \lim\limits_{x \to 2^-}(x + 7) = 9; \lim\limits_{x \to 2^+} h(x) = \lim\limits_{x \to 2^+} \frac{1}{x^2 - 9} = -\frac{1}{5}$. Since $\lim\limits_{x \to 2^-} h(x) \neq \lim\limits_{x \to 2^+} h(x)$, $x = 2$ is a point of discontinuity. Also, $\frac{1}{x^2 - 9} = \frac{1}{(x + 3)(x - 3)}$, so h is not defined for $x = 3$, and 3 is a point of discontinuity.

8. The only possible trouble points are where the denominator is 0. The denominator of $j(x) = \frac{x^2-9}{x^2-x-12} = \frac{x^2-9}{(x+3)(x-4)}$ is 0 when $x^2 - x - 12 = (x+3)(x-4) = 0$ or when $x = -3$ or $x = 4$. So, these are the only two points of discontinuity and j is continuous over the intervals $(-\infty, -3), (-3, 4), (4, \infty)$.

CHAPTER
23
The Derivative

☰ 23.1 THE TANGENT QUESTION AND THE DERIVATIVE

1. Step 1: $f(x+h) = 3(x+h) + 2 = 3x + 3h + 2$.
Step 2: $f(x+h) - f(x) = (3+3h+2) - (3x+2) = 3h$.
Step 3: $\frac{3h}{h} = 3$.
Step 4: $\lim_{h\to 0} 3 = 3 : f'(x) = 3$

2. Step 1: $g(x+h) = 5(x+h) - 7 = 5x + 5h - 7$.
Step 2: $g(x+h) - g(x) = (5x+5h-7) - (5x-7) = 5h$.
Step 3: $\frac{5h}{h} = 5$.
Step 4: $\lim_{h\to 0} 5 = 5 : g'(x) = 5$

3. Step 1: $f(x+h) = 4 - 3(x+h) = 4 - 3x - 3h$.
Step 2: $f(x+h) - f(x) = (4-3x-3h) - (4-3x) = -3h$.
Step 3: $\frac{-3h}{h} = -3$.
Step 4: $\lim_{h\to 0} -3 = -3 : f'(x) = -3$

4. Step 1: $g(t+h) = 6 - 4(t+h) = 6 - 4t - 4h$.
Step 2: $g(t+h) - g(t) = (6-4t-4h) - (6-4t) = -4h$.
Step 3: $\frac{-4h}{h} = -4$.
Step 4: $\lim_{h\to 0} -4 = -4 : g'(t) = -4$

5. Step 1: $3(x+h)^2 = 3(x^2 + 2xh + h^2) = 3x^2 + 6xh + 3h^2$.
Step 2: $(3x^2 + 6xh + 3h^2) - 3x^2 = 6xh + 3h^2$.
Step 3: $\frac{6xh+3h^2}{h} = 6x + 3h$.
Step 4: $\lim_{h\to 0}(6x + 3h) = 6x : y' = 6x$

6. Step 1: $-5(t+h)^2 = -5(t^2 + 2th + h^2) = -5t^2 - 10th - 5h^2$.
Step 2: $(-5t^2 - 10th - 5h^2) - (-5t^2) = -10th - 5h^2$.
Step 3: $\frac{-10th-5h^2}{h} = -10t - 5h$
Step 4: $\lim_{h\to 0}(-10t - 5h) = -10t : y' = -10t$

7. Step 1: $2(x+h)^2 + 5 = 2(x^2 + 2xh + h^2) + 5 = 2x^2 + 4xh + 2h^2 + 5$.

Step 2: $(2x^2 + 4xh + 2h^2 + 5) - (2x^2 + 5) = 4xh + 2h^2$.
Step 3: $\frac{4xh+2h^2}{h} = 4x + 2h$.
Step 4: $\lim_{h\to 0}(4x + 2h) = 4x : y' = 4x$

8. Step 1: $j(x+h) = 3(x+h)^2 - (x+h) + 2 = 3(x^2 + 2xh + h^2) - (x+h) + 2 = 3x^2 + 6xh + 3h^2 - x - h + 2$.
Step 2: $(3x^2 + 6xh + 3h^2 - x - h + 2) - (3x^2 - x + 2) = 6xh + 3h^2 - h$.
Step 3: $\frac{6xh+3h^2-h}{h} = 6x - 1 + 3h$.
Step 4: $\lim_{h\to 0}(6x - 1 + 3h) = 6x - 1 : j' = 6x - 1$

9. Step 1: $k(x+h) = 3 - 4(x+h)^2 = 3 - 4(x^2 + 2xh + h^2) = 3 - 4x^2 - 8xh - 4h^2$.
Step 2: $(3 - 4x^2 - 8xh - 4h^2) - (3 - 4x^2) = -8xh - 4h^2$
Step 3: $\frac{-8xh-4h^2}{h} = -8x - 4h$
Step 4: $\lim_{h\to 0}(-8x - 4h) = -8x : k'(x) = -8x$

10. Step 1: $g(x+h) = 2(x+h) - 7(x+h)^2 = 2x + 2h - 7x^2 - 14xh - 7h^2$.
Step 2: $(2x+2h-7x^2-14xh-7h^2) - (2x-7x^2) = 2h - 14xh - 7h^2$.
Step 3: $\frac{2h-14xh-7h^2}{h} = 2 - 14x - 7h$.
Step 4: $\lim_{h\to 0}(2 - 14x - 7h) = 2 - 14x : g'(x) = 2 - 14x$

11. Step 1: $-4(x+h)^3 = -4(x^3 + 3x^2h + 3xh^2 + h^3) = -4x^3 - 12x^2h - 12xh^2 - 4h^3$.
Step 2: $(-4x^3 - 12x^2h - 12xh^2 - 4h^3) - (-4x^3) = -12x^2h - 12xh^2 - 4h^3$.
Step 3: $\frac{12x^2h-12xh^2-4h^3}{h} = -12x^2 - 12xh - 4h^2$.
Step 4: $\lim_{h\to 0} -12x^2 - 12xh - 4h^2 = -12x^2 : y' = -12x^2$

12. Step 1: $(x+h)^2 - 2(x+h)^3 = x^2 + 2xh + h^2 - 2x^3 - 6x^2h - 6xh^2 - 2h^3$.
Step 2: $(x^2 + 2xh + h^2 - 2x^3 - 6x^2h - 6xh^2 - 2h^3) - (x^2 - 2x^3) = 2xh + h^2 - 6x^2h - 6xh^2 - 2h^3$.

Step 3: $\frac{2xh+h^2-6x^2h-6xh^2-2h^3}{h} = 2x + h - 6x^2 - 6xh - 2h^2$.

Step 4: $\lim_{h\to 0}(2x + h - 6x^2 - 6xh - 2h^2) = 2x - 6x^2 : y' = 2x - 6x^2$

13. Step 1: $(t + h) - 3(t + h)^3 = t + h - 3t^3 - 9t^2h - 9th^2 - 3h^3$.

Step 2: $(t+h-3t^3-9t^2h-9th^2-3h^3)-(t-3t^3) = h - 9t^2h - 9th^2 - 3h^3$.

Step 3: $\frac{h-9t^2h-9th^2-3h^3}{h} = 1 - 9t^2 - 9th - 3h^2$.

Step 4: $\lim_{h\to 0}(1 - 9t^2 - 9th - 3h^2) = 1 - 9t^2 : y' = 1 - 9t^2$.

14. Step 1: $f(x + h) = \frac{2(x+h)+1}{3} = \frac{2x+2h+1}{3}$.

Step 2: $\frac{2x+2h+1}{3} - \frac{2x+1}{3} = \frac{(2x+2h+1)-(2x+1)}{3} = \frac{2h}{3}$.

Step 3: $\frac{\frac{2h}{3}}{h} = \frac{2}{3}$.

Step 4: $\lim_{h\to 0}\frac{2}{3} = \frac{2}{3} : f'(x) = \frac{2}{3}$

15. Step 1: $f(x + h) = \frac{1}{x+h+1}$.

Step 2: $\frac{1}{x+h+1} - \frac{1}{x+1} = \frac{(x+1)-(x+h+1)}{(x+1)(x+h+1)} = \frac{-h}{(x+1)(x+h+1)}$.

Step 3: $\frac{\frac{-h}{(x+1)(x+h+1)}}{h} = \frac{-1}{(x+1)(x+h+1)}$.

Step 4: $\lim_{h\to 0}\frac{-1}{(x+1)(x+h+1)} = \frac{-1}{(x+1)^2} : f'(x) = \frac{-1}{(x+1)^2}$

16. Step 1: $g(x + h) = \frac{5}{2(x+h)-3} = \frac{5}{2x+2h-3}$.

Step 2: $\frac{5}{2x+2h-3} - \frac{5}{2x-3} = \frac{5(2x-3)-5(2x+2h-3)}{(2x+2h-3)(2x-3)} = \frac{10x-15-10x-10h+15}{(2x+2h-3)(2x-3)} = \frac{-10h}{(2x+2h-3)(2x-3)}$.

Step 3: $\frac{\frac{-10h}{(2x+2h-3)(2x-3)}}{h} = \frac{-10}{(2x+2h-3)(2x-3)}$.

Step 4: $\lim_{h\to 0}\frac{-10}{(2x+2h-3)(2x-3)} = \frac{-10}{(2x-3)^2} : g'(x) = \frac{-10}{(2x-3)^2}$

17. Step 1: $j(x + h) = \frac{4}{1-(x+h)^2}$.

Step 2: $\frac{4}{1-(x+h)^2} - \frac{4}{1-x^2} = \frac{4(1-x^2)-4[1-(x+h)^2]}{[1-(x+h)^2](1-x^2)} = \frac{4-4x^2-4+4x^2+8xh+4h^2}{[1-(x+h)^2](1-x^2)} = \frac{8xh+4h^2}{[1-(x+h)^2](1-x^2)}$.

Step 3: $\frac{\frac{8xh+4h^2}{[1-(x+h)^2](1-x^2)}}{h} = \frac{8x+4h}{[1-(x+h^2](1-x^2)}$.

Step 4: $\lim_{h\to 0}\frac{8x+4h}{[1-(x+h)^2](1-x^2)} = \frac{8x+4\cdot 0}{[1-(x+0)^2](1-x^2)} = \frac{8x}{(1-x^2)^2} : j'(x) = \frac{8x}{(1-x^2)^2}$

18. Step 1: $f'(x + h) = \frac{9}{(x+h)^2+2(x+h)}$.

Step 2: $\frac{9}{(x+h)^2+2(x+h)} - \frac{9}{x^2+2x} = \frac{9(x^2+2x)-9[(x+h)^2+2(x+h)]}{[(x+h)^2+2(x+h)](x^2+2x)} = \frac{9x^2+18x-9x^2-18xh-9h^2-18x-18h}{[(x+h)^2+2(x+h)](x^2+2x)} = \frac{-18xh-9h^2-18h}{[(x+h)^2+2(x+h)](x^2+2x)}$.

Step 3: $\frac{18x-9h-18}{[(x+h)^2+2(x+h)](x^2+2x)}$.

Step 4: $\lim_{h\to 0}\frac{-18x-9h-18}{[(x+h)^2+2(x+h)](x^2+2x)} = \frac{-18x-18}{[x^2+2x]^2} = \frac{-18(x+1)}{[x^2+2x]^2}$ or $\frac{-9(2x+2)}{[x^2+2x]^2} : f'(x) = \frac{-9(2x+2)}{(x^2+2x)^2}$

19. Step 1: $16(t + h)^2 - 6(t + h) + 3 = 16t^2 + 32th + 16h^2 - 6t - 6h + 3$.

Step 2: $(16t^2+32th+16h^2-6t-6h+3)-(16t^2-6t + 3) = 32th + 16h^2 - 6h$

Step 3: $\frac{32th+16h^2-6h}{h} = 32t + 16h - 6$

Step 4: $\lim_{h\to 0}(32t+16h-6) = 32t-6 : s' = 32t-6$

20. Step 1: $(t + h)^3 - 4(t + h)^2 + 5(t + h) - 2 = t^3 + 3t^2h + 3th^2 + h^3 - 4(t^2 + 2th + h^2) + 5(t + h) - 2 = t^3 + 3t^2h + 3th^2 + h^3 - 4t^2 - 8th - 4h^2 + 5t + 5h - 2$

Step 2: $(t^3 + 3t^2h + 3th^2 + h^3 - 4t^2 - 8th - 4h^2 + 5t + 5h - 2) - (t^3 - 4t^2 + 5t - 2) = 3t^2h + 3th^2 + h^3 - 8th - 4h^2 + 5h$

Step 3: $\frac{3t^2h+3th^2+h^3-8th-4h^2+5h}{h} = 3t^2 + 3th + h^2 - 8t - 4h + 5$

Step 4: $\lim_{h\to 0}(3t^2 + 3th + h^2 - 8t - 4h + 5) = 3t^2 - 8t + 5 : q' = 3t^2 - 8t + 5$

21. Step 1: $\sqrt{(x + h)^2 + 4}$

Step 2: $\sqrt{(x + h)^2 + 4} - \sqrt{x^2 + 4}$.

Step 3: $\frac{\sqrt{(x+h)^2+4}-\sqrt{x^2+4}}{h}$.

$\frac{\sqrt{(x+h)^2+4}+\sqrt{x^2+4}}{\sqrt{(x+h)^2+4}+\sqrt{x^2+4}} = \frac{((x+h)^2+4)-(x^2+4)}{h[\sqrt{(x+h)^2+4}+\sqrt{x^2+4}]} = \frac{(x^2+2xh+h^2+4)-(x^2+4)}{h[\sqrt{(x+h)^2+4}+\sqrt{x^2+4}]} = \frac{2xh+h^2}{h[\sqrt{(x+h)^2+4}+\sqrt{x^2+4}]} = \frac{2x+h}{\sqrt{(x+h)^2+4}+\sqrt{x^2+4}}$

Step 4: $\lim_{h\to 0}\frac{2x+h}{h[\sqrt{(x+h)^2+4}+\sqrt{x^2+4}]} = \frac{2x}{2\sqrt{x^2+4}} = \frac{x}{\sqrt{x^2+4}} : y' = \frac{x}{\sqrt{x^2+4}}$

22. Step 1: $\sqrt{5 - (x + h)} = \sqrt{5 - x - h}$

Step 2: $\sqrt{5 - x - h} - \sqrt{5 - x}$

Step 3: $\frac{\sqrt{5-x-h}-\sqrt{5-x}}{h} \cdot \frac{\sqrt{5-x-h}+\sqrt{5-x}}{\sqrt{5-x-h}+\sqrt{5-x}} = \frac{(5-x-h)-(5-x)}{h[\sqrt{5-x-h}+\sqrt{5-x}]} = \frac{-h}{h[\sqrt{5-x-h}+\sqrt{5-x}]} = \frac{-1}{\sqrt{5-x-h}+\sqrt{5-x}}$

Step 4: $\lim\limits_{h \to 0} \frac{-1}{\sqrt{5-x-h}+\sqrt{5-x}} = \frac{-1}{2\sqrt{5-x}} : y' \frac{-1}{2\sqrt{5-x}}$

23.   Step 1: $\sqrt{2(x+h)-(x+h)^2}$

Step 2: $\sqrt{2(x+h)-(x+h)^2} - \sqrt{2x-x^2}$

Step 3: $\dfrac{\sqrt{2(x+h)-(x+h)^2}-\sqrt{2x-x^2}}{h} \cdot$

$\dfrac{\sqrt{2(x+h)-(x+h)^2}+\sqrt{2x-x^2}}{\sqrt{2(x+h)-(x+h)^2}+\sqrt{2x-x^2}} =$

$\dfrac{[2(x+h)-(x+h)^2]-[2x-x^2]}{h[\sqrt{2(x+h)-(x+h)^2}+\sqrt{2x-x^2}]} =$

$\dfrac{2h-2xh-h^2}{h[\sqrt{2(x+h)-(x+h)^2}+\sqrt{2x-x^2}]} =$

$\dfrac{2-2x-h}{\sqrt{2(x+h)-(x+h)^2}+\sqrt{2x-x^2}}$

Step 4: $\lim\limits_{h \to 0} \dfrac{2-2x-h}{\sqrt{2(x+h)-(x+h)^2}+\sqrt{2x-x^2}}$

$= \dfrac{2-2x}{2\sqrt{2x-x^2}} = \dfrac{1-x}{\sqrt{2x-x^2}}$

$y' = \dfrac{1-x}{\sqrt{2x-x^2}}$

24.   Step 1: $\sqrt{5(x+h)-4(x+h)^3}$.

Step 2: $\sqrt{5x+5h-4x^3-12x^2h-12xh^2-4h^3}$
$- \sqrt{5x-4x^3} = \sqrt{E_1}-\sqrt{E_2}$

Step 3: $\dfrac{\sqrt{E_1}-\sqrt{E_2}}{h} \cdot \dfrac{\sqrt{E_1}+\sqrt{E_2}}{\sqrt{E_1}+\sqrt{E_2}} = \dfrac{E_1-E_2}{h[\sqrt{E_1}+\sqrt{E_2}]} =$

$\dfrac{5x+5h-4x^3-12x^2h-12xh^2-4h^3-5x+4x^3}{h[\sqrt{E_1}+\sqrt{E_2}]}$

$= \dfrac{5h-12x^2h-12xh^2-4h^3}{h[\sqrt{E_1}+\sqrt{E_2}]} = \dfrac{5-12x^2-12xh^2-4h^2}{[\sqrt{E_1}+\sqrt{E_2}]}$

Step 4: $\lim\limits_{h \to 0} \dfrac{5-12x^2-12xh-4h^2}{\sqrt{5(x+h)-4(x+h)^3}+\sqrt{5x-4x^3}} =$

$\dfrac{5-12x^2}{2\sqrt{5x-4x^3}} : y' = \dfrac{5-12x^2}{2\sqrt{5x-4x^3}}$

25. $f(x) = 3x+7; f'(x) = 3; f'(2) = 3; f'(9) = 3$

26. $g(x) = x^2-2; g'(x) = 2x; g'(2) = 4; g'(9) = 18$

27. $s(t) = 16t^2+2t; s'(t) = 32t+2; s'(3) = 32 \cdot 3+2 = 98; s'(2) = 32 \cdot 2 + 2 = 66; s'(0) = 32 \cdot 0+2 = 2$

28. $s(t) = 2t^2-1; s'(t) = 4t; s'(1) = 4; s'(5) = 20$

29. $q(t) = t^3-4t^2; q'(t) = 3t^2-8t; q'(0) = 0; q'(3) = 3 \cdot 3^2-8 \cdot 3 = 27-24 = 3.$

30. $j(x) = \frac{1}{\sqrt{x}}$; Since this function is unlike any in exercises 1–24, its derivative will be evaluated using the four-step process.

Step 1: $\frac{1}{\sqrt{x+h}}$

Step 2: $\frac{1}{\sqrt{x+h}} - \frac{1}{\sqrt{x}} = \frac{\sqrt{x}-\sqrt{x+h}}{\sqrt{x+h}\sqrt{x}}$

Step 3: $\dfrac{\sqrt{x}-\sqrt{x+h}}{h(\sqrt{x+h}\sqrt{x})} \cdot \dfrac{\sqrt{x}+\sqrt{x+h}}{\sqrt{x}+\sqrt{x+h}} \quad \dfrac{x-(x+h)}{h(\sqrt{x+h}\sqrt{x})(\sqrt{x}+\sqrt{x+h})}$

$= \dfrac{-1}{h(\sqrt{x+h}\sqrt{x})(\sqrt{x}+\sqrt{x+h}]}$

Step 4: $\lim\limits_{h \to 0} \dfrac{-1}{\sqrt{x+h}\sqrt{x}(\sqrt{x}+\sqrt{x+h})} = \dfrac{-1}{\sqrt{x}\sqrt{x}(\sqrt{x}+\sqrt{x})}$

$= \dfrac{-1}{2\sqrt{x^3}}$

$j'(x) = \dfrac{-1}{2\sqrt{x^3}}$

$j'(4) = \dfrac{-1}{2\sqrt{4^3}} = \dfrac{-1}{2 \cdot 2^3} = -\dfrac{1}{16}$

$j'(9) = \dfrac{-1}{2\sqrt{9^3}} = \dfrac{-1}{2 \cdot 3^3} = \dfrac{-1}{2 \cdot 27} = -\dfrac{1}{54}$

31. $k(x) = \frac{1}{x^2-4}; k'(x) = \frac{-2x}{(x^2-4)^2}; k'(5) = \frac{-10}{21^2} = \frac{-10}{441}; k'(-5) = \frac{-2(-5)}{((-5)^2-4)^2} = \frac{10}{(25-4)^2} = \frac{10}{21^2} = \frac{10}{441}.$

32. $f(x) = \frac{4}{x^2+1}; f'(x) = \frac{-8x}{(x^2+1)^2}; f'(0) = 0; f'(3) = \frac{-8 \cdot 3}{(3^2+1)^2} = \frac{-24}{10^2} = \frac{-24}{100} = \frac{-6}{25}; f'(-5) = \frac{-8(-5)}{(-5^2+1)^2} = \frac{40}{26^2} = \frac{40}{676} = \frac{10}{169}$

33. $j'(t) = \sqrt{5t-t^2}; j'(t) = \frac{5-2t}{2\sqrt{5t-t^2}}; j'(1) = \frac{5-2}{2\sqrt{5-1}} = \frac{3}{2\sqrt{4}} = \frac{3}{4}; j'(4) = \frac{5-8}{2\sqrt{20-16}} = \frac{-3}{2\sqrt{4}} = \frac{-3}{4} \text{ or } -\frac{3}{4}.$

34. $g(x) = \frac{1}{\sqrt{x+4}}; g'(x) = \frac{-1}{2\sqrt{x+4}^3}; g'(0) = \frac{-1}{2\sqrt{4}^3} = \frac{-1}{16}; g'(5) = \frac{-1}{2\sqrt{5+4}^3} = \frac{-1}{2\sqrt{9}^3} = \frac{-1}{2 \cdot 27} = \frac{-1}{54}; g'(-2) = \frac{-1}{2\sqrt{-2+4}^2} = \frac{-1}{2\sqrt{x^3}} = \frac{-1}{4\sqrt{2}}$

35. $f(x) = \dfrac{5}{\sqrt{25-x^2}}$

$f'(x) = \dfrac{-5(-2x)}{2\sqrt{25-x^2}^3} = \dfrac{5x}{\sqrt{25-x^2}^3}$

$f(0) = 0; f'(3) = \dfrac{15}{\sqrt{25-9}^3} = \dfrac{15}{\sqrt{16}^3}$

$= \dfrac{15}{4^3} = \dfrac{15}{64};$

$f'(-4) = \dfrac{-20}{\sqrt{25-16}^3} = \dfrac{-20}{3^3} = \dfrac{-20}{27}$

or $-\dfrac{20}{27}$

The four-step process for finding $f'(x)$ is as follows:

Step 1: $f(x+h) = \dfrac{5}{\sqrt{25-(x+h)^2}}$

Step 2: $\dfrac{5}{\sqrt{25-(x+h)^2}} - \dfrac{5}{\sqrt{25-x^2}} =$

$\dfrac{5\sqrt{25-x^2}-5\sqrt{25-(x+h)^2}}{\sqrt{25-(x+h)^2} \cdot \sqrt{25-x^2}} \times$

$\dfrac{5\sqrt{25-x^2}+5\sqrt{25-(x+h)^2}}{5\sqrt{25-x^2}+5\sqrt{25-(x+h)^2}} =$

$\dfrac{5(25-x^2)-(25-x^2-2xh-h^2)}{\sqrt{E_1}\sqrt{E_2}(\sqrt{E_2}+\sqrt{E_1})} =$

$\dfrac{5(2xh+h^2)}{\sqrt{E_2}\sqrt{E_2}(\sqrt{E_2}+\sqrt{E_1})}$

Step 3: $\dfrac{10x+5h}{\sqrt{E_1}\sqrt{E_2}(\sqrt{E_2}+\sqrt{E_1})}$

Step 4: $\lim\limits_{h \to 0} \dfrac{10x+5h}{\sqrt{25-(x+h)^2}\sqrt{25-x^2} \times (\sqrt{25-(x+h)^2}+\sqrt{25-x^2})}$

$= \dfrac{10x}{2\sqrt{25-x^2}^3} = \dfrac{5x}{\sqrt{25-x^2}^3}$

36. First find $h'(x)$ using the four step process.

Step 1: $h(x+c) = \sqrt{(x+c)^2 + 4(x+c) + 9} = \sqrt{x^2 + 2xc + c^2 + 4x + 4c + 9}$

Step 2: $h(x+c) - h(x) =$
$\sqrt{x^2 + 2xc + c^2 + 4x + 4c + 9} - \sqrt{x^2 + 4x + 9}$

Step 3: $\frac{h(x+c) - h(x)}{c} =$

$\frac{\sqrt{x^2+2xc+c^2+4x+4c+9} - \sqrt{x^2+4x+9}}{c} =$

$\frac{\sqrt{x^2+2xc+c^2+4x+4c+9} - \sqrt{x^2+4x+9}}{c} \cdot$

$\frac{\sqrt{x^2+2xc+c^2+4x+4c+9} + \sqrt{x^2+4x+9}}{\sqrt{x^2+2xc+c^2+4x+4c+9} + \sqrt{x^2+4x+9}} =$

$\frac{(x^2+2xc+c^2+4x+4c+9) - (x^2+4x+9)}{c[\sqrt{x^2+2xc+c^2+4x+4c+9} + \sqrt{x^2+4x+9}]} =$

$\frac{2xc+c^2+4c}{c[\sqrt{x^2+2xc+c^2+4x+4c+9} + \sqrt{x^2+4x+9}]} =$

$\frac{2x+c+4}{\sqrt{x^2+2xc+c^2+4x+4c+9} + \sqrt{x^2+4x+9}}$

Step 4: $\lim\limits_{c \to 0} \frac{2x+c+4}{\sqrt{x^2+2xc+c^2+4x+4c+9} + \sqrt{x^2+4x+9}} = \frac{2x+4}{2\sqrt{x^2+4x+9}}$

Now, we will evaluate $h'(x)$ at $x = -4, x = 0$, and $x = 4$.

$h'(-4) = \dfrac{2(-4)+4}{2\sqrt{(-4)^2+4(-4)+9}}$

$= \dfrac{-8+4}{2\sqrt{16-16+9}} = \dfrac{-4}{2\sqrt{9}} = \dfrac{-2}{3} = -\dfrac{2}{3}$

$h'(0) = \dfrac{2(0)+4}{2\sqrt{(0)^2+4(0)+9}}$

$= \dfrac{0+4}{2\sqrt{0+0+9}} = \dfrac{4}{2\sqrt{9}} = \dfrac{2}{3}$

$h'(4) = \dfrac{2(4)+4}{2\sqrt{(4)^2+4(4)+9}}$

$= \dfrac{8+4}{2\sqrt{16+16+9}} = \dfrac{12}{2\sqrt{41}} = \dfrac{6}{\sqrt{41}}$

$\equiv$ 23.2 DERIVATIVES OF POLYNOMIALS

1. $f(x) = 19; f'(x) = 0$

2. $g(x) = -4; g'(x) = 0$

3. $h(x) = 7x - 5; h'(x) = 7$

4. $j(x) = -3x + 7; j'(x) = -3$

5. $k(x) = 9x^2; k'(x) = 9(2x) = 18x$

6. $m(x) = 15x^3; m'(x) = 15(3x^2) = 45x^2$

7. $f(x) = \frac{1}{3}x^{15}; f'(x) = \frac{1}{3}(15x^{14}) = 5x^{14}$

8. $g(x) = \frac{2}{5}x^{10}; g'(x) = \frac{2}{5}(10x^9) = 4x^9$

9. $h(x) = 5x^{-2}; h'(x) = 5(-2x^{-3}) = -10x^{-3}$

10. $j(x) = -3x^{-4}; j'(x) = -3(-4x^{-5}) = 12x^{-5}$

11. $k(x) = -\frac{2}{3}x^{3/2}; k'(x) = -\frac{2}{3}(\frac{3}{2}x^{1/2}) = -x^{1/2}$

12. $j(x) = \frac{4\sqrt{x}}{3} = \frac{4}{3}x^{1/2}; j'(x) = \frac{4}{3} \cdot \frac{1}{2}x^{-1/2} = \frac{2}{3}x^{-1/2}$ or $\frac{2}{3\sqrt{x}}$

13. $h(x) = 4\sqrt{3x} = 4\sqrt{3}x^{1/2}; h'(x) = 4\sqrt{3} \cdot \frac{1}{2}x^{-1/2} = 2\sqrt{3}x^{-1/2}$ or $\frac{2\sqrt{3}}{\sqrt{x}}$

14. $m(x) = \frac{3}{4}x^{-5/3}; m'(x) = \frac{3}{4}(-\frac{5}{3}x^{-8/3}) = -\frac{5}{4}x^{-8/3}$

15. $g(x) = \frac{3}{\sqrt[3]{x}} = 3x^{-1/3}; g(x) = 3(-\frac{1}{3}x^{-4/3}) = -x^{-4/3}$ or $\frac{-1}{\sqrt[3]{x^4}} = \frac{-1}{x\sqrt[3]{x}}$

16. $f(x) = 4x^3 - 5x^2 + 2; f'(x) = 12x^2 - 10x$

17. $g(x) = 9x^2 + 3x - 4; g'(x) = 18x + 3$

18. $n(x) = 7x^7 - 5x^5; n'(x) = 49x^6 - 25x^4$

19. $h(x) = \frac{1}{3}x^3 + \frac{1}{2}x^2 - 5x + \frac{2}{3}x^{-2} + 5; h'(x) = x^2 + x - 5 - \frac{4}{3}x^{-3}$

20. $j(x) = \frac{3}{4}x^4 - \frac{2}{3}x^3 + \frac{5}{4}x^2 - \frac{1}{2}x + 2x^{-3} + x^{-4} + 7; j'(x) = 3x^3 - 2x^2 + \frac{5}{2}x - \frac{1}{2} - 6x^{-4} - 4x^{-5}$

21. $f(x) = \frac{4}{5}x^5 - \frac{3}{2}x^3 - 7x^0 + \frac{1}{2}x^{-2} - \frac{2}{3}x^{-3}; f'(x) = 4x^4 - \frac{9}{2}x^2 - x^{-3} + 2x^{-4}$

22. $k(x) = \sqrt{2}x^5 - \sqrt{3}x^3 + 3x^3 + \pi; k'(x) = 5\sqrt{2}x^4 - 3\sqrt{3}x^2 + 3$

23. $l(x) = 3\sqrt{3}x^4 + 3\sqrt{5}x + \sqrt{4x^4} = 3\sqrt{3}x^4 + 3\sqrt{5}x^{1/2} + 2x^2; l'(x) = 12\sqrt{3}x^3 + \frac{3\sqrt{5}}{2}x^{-1/2} + 4x = 12\sqrt{3}x^3 + \frac{3\sqrt{5}}{2\sqrt{x}} + 4x$

24. $m(x) = \sqrt{7}x^6 - 2x^3 + 5\sqrt{3}x^2 - \sqrt{7}; m'(x) = 6\sqrt{7}x^5 - 6x^2 + 10\sqrt{3}x$

25. $s(t) = 16t^2 - 32t + 5; s'(t) = 32t - 32$

26. $q(t) = 4.3t^3 - 2.7t + 3.0; q'(t) = 12.9t^2 - 2.7$

27. $\alpha(t) = 30 - 4.0t^2 + 2t^{1/2}; \alpha'(t) = -8.0t + t^{-1/2}$

28. $s(t) = \frac{5}{3}t^4 + 7\sqrt{2}t^3 - 8t^{2/3} + 7t^{-1/2}$; $s'(t) = \frac{20}{3}t^3 + 21\sqrt{2}t^2 - \frac{16}{3}t^{-1/3} - \frac{7}{2}t^{-3/2}$

29. First, rewrite $q(t) = \sqrt{6t} - 4t\sqrt{3t} + \sqrt[3]{t^2}$ as $q(t) = \sqrt{6}t^{1/2} - 4\sqrt{3}t^{3/2} + t^{2/3}$. Then, differentiating produces $q'(t) = \frac{\sqrt{6}}{2}t^{-1/2} - 6\sqrt{3}t^{1/2} + \frac{2}{3}t^{-1/3}$.

30. First, rewrite $f(x) = 3x\sqrt{x}$ as $f(x) = 3x^{3/2}$. Then, differentiating produces $f'(x) = \frac{9}{2}x^{1/2} = \frac{9}{2}\sqrt{x}$.

31. Rewrite $g(x) = 5x^2\sqrt[3]{x}$ as $g(x) = 5x^{2\frac{1}{3}}$ or $g(x) = 5x^{7/3}$. Then, differentiating gives $g'(x) = 5 \cdot \frac{7}{3}x^{4/3} = \frac{35}{3}x^{4/3}$ or $\frac{35}{3}x\sqrt[3]{x}$.

32. Rewrite $j(x) = \frac{3x^3}{\sqrt[3]{x}}$ as $j(x) = 3x^{3-\frac{1}{3}} = 3x^{8/3}$. Then, differentiating gives $j'(x) = 8x^{5/3}$.

33. Here, we rewrite $h(x) = \frac{4}{x^2} = 4x^{-2}$. Then, differentiation produces $h'(x) = -8x^{-3} = \frac{-8}{x^3}$

34. Rewrite $j'(x) = 7x^2 + 2x\sqrt{x} + \frac{5}{x^3}$ as $j(x) = 7x^2 + 2x^{3/2} + 5x^{-3}$. Then, differentiating gives $j'(x) = 14x + 3x^{1/2} - 15x^{-4} = 14x + 3\sqrt{x} - \frac{15}{x^4}$.

35. Rewrite $K(x) = \frac{4\sqrt[3]{x}}{\sqrt{x}}$ as $K(x) = 4x^{1/3-1/2} = 4x^{-1/6}$. Then, differentiating produces $K'(x) = -\frac{2}{3}x^{-7/6}$

36. Here $M(x) = \frac{3\sqrt{x^3}}{\sqrt{3x}} = \frac{3}{\sqrt{3}}x^{3/2-1/2} = \sqrt{3}x^{2/2} = \sqrt{3}x$, so $M'(x) = \sqrt{3}$.

37.
$$f(x) = \frac{4\sqrt{3x} + 3x^2 - 7\sqrt{x^3} + 2x^{-1}}{\sqrt{x}}$$
$$= \frac{4\sqrt{3x}}{\sqrt{x}} + \frac{3x^2}{\sqrt{x}} - \frac{7\sqrt{x^3}}{\sqrt{x}} + \frac{2x^{-1}}{\sqrt{x}}$$
$$= 4\sqrt{3} + 3x^{3/2} - 7x + 2x^{-3/2}$$
$$f'(x) = \frac{9}{2}x^{1/2} - 7 - 3x^{-5/2}$$

38.
$$F(x) = f(x) - g(x)$$
$$F'(x) = \lim_{h \to 0} \frac{F(x+h) - F(x)}{h}$$
$$= \lim_{h \to 0} \frac{[f(x+h) - g(x+h)] - [f(x) - g(x)]}{h}$$

$$= \lim_{h \to 0} \frac{[f(x+h) - f(x)] - [g(x+h) - g(x)]}{h}$$
$$= \lim_{h \to 0} \frac{f(x+h) - f(x)}{h}$$
$$- \lim_{h \to 0} \frac{g(x+h) - g(x)}{h}$$
$$= f'(x) - g'(x).$$
Hence: $F'(x) = f'(x) - g'(x)$.

39. Here we have $f(x) = x^3 + 3x^2$, and so $f'(x) = 3x^2 + 6x$. Now, $f'(x) = 0$ means that $3x^2 + 6x = 0$ or $3x(x+2) = 0$. Thus, by the zero product principle, $x = 0$ or $x = -2$.

40. Here we have $\theta(t) = 3t^3 - t$, and so $\theta'(t) = 9t^2 - 1$. Since $\theta'(t) = 0$ is the same as $9t^2 - 1 = 0$, then $9t^2 = 1$, or $t^2 = \frac{1}{9}$, and so $t = \pm\frac{1}{3}$. Thus, $t = -\frac{1}{3}$ or $t = \frac{1}{3}$.

41. We are given $g(x) = 2x^3 + x^2 - 4x$ and determine that $g'(x) = 6x^2 + 2x - 4$. Now $g'(x) = 0$ is the same as $6x^2 + 2x - 4 = 0$, or $2(3x^2 + x - 2) = 0$. This factors as $2(3x - 2)(x + 1) = 0$ which, by the zero product principle, has solutions when $x = \frac{2}{3}$ or $x = -1$.

42. Since $\omega(t) = 4.0t^2 - 2.0t^{1/2}$, then $\omega'(t) = 8.0t - t^{-1/2}$. Now, $\omega'(t) = 0$ means that $8.0t - t^{-1/2} = 0$, or $8t = \frac{1}{\sqrt{t}}$, or $t^{3/2} = \frac{1}{8}$ and so $t = \left(\frac{1}{8}\right)^{\frac{2}{3}} = (1/2)^2 = \frac{1}{4}$.

43. Since $j(t) = 4t^3 + \frac{1}{t}$, then $j'(t) = 12t^2 - t^{-2} = 12t^2 - \frac{1}{t^2}$. Now, $j'(t) = 0$, means that $12t^2 - \frac{1}{t^2} = 0$ or $12t^2 = \frac{1}{t^2}$ and so $12t^4 = 1$, or $t^4 = \frac{1}{12}$. Hence, $t = \sqrt[4]{\frac{1}{12}} \approx 0.537285$.

44. Rewriting h as $h(x) = 4 + x^{-1} - 8x^{-3}$ and differentiating, we get $h'(x) = -x^{-2} + 24x^{-4} = -\frac{1}{x^2} + \frac{8}{x^4}$. Setting this equal to 0 produces $-\frac{1}{x^2} + \frac{8}{x^4} = 0$ or $x^2 - 24 = 0$. Solving this equation, we get $x^2 = 24$ and so, $x = \pm\sqrt{24} = \pm 2\sqrt{6}$.

45. (a) $P(6) = 146{,}000 + 64(6) - 0.04(6)^2 = 146{,}382$
(b) Since the rate of change at time t is given by $P'(t) = 64 - 0.08t$, then the rate of change in the city's population six months from now is $P'(6) = 64 - 0.08(6) = 63.52$. This represents a 63.52 increase in the population per month.

46. (a) $R(30) = 120(30) - (30)^2 = 2{,}700$. The sale of 30 telephones will produce revenue of $2,700.

(b) The rate of change in the revenue is given by $R'(c) = 120 - 2c$. Thus, the rate of change in the revenue when $c = 30$ is $R'(30) = 120 - 2(30) = 60$. This represents a \$60 increase in revenue per sale.

47. **(a)** The distance the ball has traveled when $t = 4$ s is given by $s(4) = 15(4) + 7(4)^2 = 172$. The ball has traveled 172 ft when $t = 4$ s.
(b) The velocity (rate of change) of the ball is given by $s'(t) = 15 + 14t$. Thus, $s'(4) = 15 + 14(4) = 71$ and the velocity of the ball when $t = 4$ s is 71 ft/s.

48. **(a)** The number of gallons of oil left in the tank 20 min after the drain is opened is $V(20) = 75,000 - 3,000(20) + 30(20)^2 = 27,000$ gal.
(b) The average rate that the oil drains during the first 20 min is $\frac{V(20)-V(0)}{20} = \frac{27,000-75,000}{20} = -2,4000$. The negative sign indicates that the amount of oil is decreasing at an average rate of 2,400 gal/min.
(c) When $t = 20$ the oil is draining at $V'(20) = -3,000 + 30(20) = -1,800$ gal/min.

49. **(a)** The city's population six years from now will be $P(6) = 125,000 + 1,500(6) - 36\sqrt{6} \approx 133,912$ people.
(b) We first determine $P'(t)$ by rewriting $P(t) = 125,000 + 1,500t - 36\sqrt{t}$ as $P(t) = 125,000 + 1,500t - 36t^{1/2}$ and then differentiating. Thus, we get $P'(t) = 1,500 - 36\frac{1}{2}t^{-1/2} = 1,500 - \frac{18}{\sqrt{t}}$. Six years from now the rate at which the city's population is growing will be $P'(6) = 1,500 - \frac{18}{\sqrt{6}} \approx 1,493$ people per year.
(c) The city's population ten years from now will be $P(10) = 125,000 + 1,500(10) - 36\sqrt{10} \approx 139,886$ people.
(d) The rate at which the city's population is growing ten years from now is $P'(10) = 1,500 - \frac{18}{\sqrt{10}} \approx 1,494$ people per year.

50. $C'(n) = 25 - 0.8n$, where $0 \le n \le 60$.
(a) $C'(10) = 25 - 0.8(10) = \$17/\text{unit}$
(b) $C'(30) = 25 - 0.8(30) = \$1/\text{unit}$
(c) $C'(50) = 25 - 0.8(50) = \$-15/\text{unit}$

51. Since $P'(i) = 2 \cdot 25i = 50i$, then $P'(2.4) = 50(2.4) = 120$ W/A.

52. Since $S'(T) = -0.00017T$, we have $S'(135) = -0.00017(135) = -0.02295$ lb/°F.

53. **(a)** $P(n) = R(n) - C(n) = (78n - 0.025n^2) - (9,800 + 22.5n) = 55.5n - 0.025n^2 - 9,800$
(b) $P'(n) = 55.5 - 0.05n$
(c) $P'(150) = 55.5 - 0.05(150) = 55.5 - 2.5 = \$53/\text{telephone}$

54. $V'(t) = 2 \cdot 0.04t = 0.08t$ in V/s

55. **(a)** When the rocket is first launched $t = 0$, so the rocket's altitude at launch is $h(0) = -0^4 + 12(0)^3 - 16(0)^2 + 605 = 605$ ft above sea level.
(b) The rocket will fall into the ocean when $h(t) = 0$. That is, when $-t^4 + 12t^3 - 16t^2 + 605 = 0$. Using a graphing utility, we obtain $t = 11$ s.
(c) The rocket's altitude at $t = 2$ s is $h(2) = -(2)^4 + 12(2)^3 - 16(2)^2 + 605 + 621$ ft. The rocket's velocity at any time t is given by $h'(t)$. Differentiating h, we see that $v(t) = h'(t) = -4t^3 + 36t^2 - 32t$. Thus, the rocket's velocity at $t = 2$ s is $v(2) = -4(2)^3 + 36(2)^2 - 32(2) = 48$ ft/s.
(d) The rocket's altitude at $t = 4$ s is $h(4) = -(4)^4 + 12(4)^3 - 16(4)^2 + 605 = 861$ ft. The rocket's velocity at $t = 4$ s is $v(4) = -4(4)^3 + 36(4)^2 - 32(4) = 192$ ft/s.
(e) The rocket's altitude at $t = 6$ s is $h(6) = -(6)^4 + 12(6)^3 - 16(6)^2 + 605 = 1,325$ ft. The rocket's velocity at $t = 6$ s is $v(6) = -4(6)^3 + 36(6)^2 - 32(6) = 240$ ft/s.
(f) The rocket's altitude at $t = 10$ s is $v(10) = -(10)^4 + 12(10)^3 - 16(10)^2 + 605 = 1,005$ ft. The rocket's velocity at $t = 10$ s is $v(10) = -4(10)^3 + 36(10)^2 - 32(10) = -720$ ft/s.
(g) The rocket's velocity when it falls into the ocean, at $t = 11$ s is $v(11) = -4(11)^3 + 36(11)^2 - 32(11) = -1,320$ ft/s.

56. **(a)** $v(t) = s'(t) = 84.5 - \frac{1}{2}t^2$ in m/min.
(b) The engines were put in reverse at $t = 0$, so $s'(0) = 84.5 - \frac{1}{2}(0)^2 = 84.5$ m/min. Converting this to km/h, we have 84.5 m/min $= \frac{84.5m}{1 \text{ min}} \cdot \frac{1km}{1000m} \cdot \frac{60 \text{ min}}{1 \text{ hr}} = 5.07$ km/h.
(c) $v(t) = 0$ when $84.5 - \frac{1}{2}t^2 = 0$ or when $t^2 = 169$. Taking the square root, we get $t = 13$. Thus, it will take 13 min for the supertanker to stop.
(d) $s(13) = 84.5(13) - \frac{1}{6}(13)^3 = 732.3$ m.

≡ 23.3 DERIVATIVES OF PRODUCT AND QUOTIENTS

1. $f'(x) = (3x+1)(2) + (3)(2x-7) = 6x + 2 + 6x - 21 = 12x - 19$

2. $g'(x) = (6x-2)(-4) + (6)(5-4x) = -24x + 8 + 30 - 24x = -48x + 38$

3. $h'(x) = (2x^2 + x - 1)(3) + (4x+1)(3x-5) = 6x^2 + 3x - 3 + 12x^2 - 17x - 5 = 18x^2 - 14x - 8$

4. $k'(x) = (4x^3-1)(2x-7) + (12x^2)(x^2-7x) = 8x^4 - 28x^3 - 2x + 7 + 12x^4 - 84x^3 = 20x^4 - 112x^3 - 2x + 7$

5. $j'(x) = (4-3x)(12x+6) + (-3)(6x^2 + 6x - 4)$
$= 48x + 24 - 36x^2 - 18x - 18x^2 - 18x + 12$
$= -54x^2 + 12x + 36$

6. $s'(t) = (6-4t)(6t-7) + (-4)(3t^2 - 7t)$
$= 36t - 42 - 24t^2 + 28t - 12t^2 + 28t$
$= -36t^2 + 92t - 42$

7. $q(t) = 4(t^2 - 3t)(t^2 - 3t)$
$q'(t) = 4[(t^2 - 3t)(2t-3) + (2t-3)(t^2 - 3t)]$
$= 4[2t^3 - 9t^2 + 9t + 2t^3 - 9t^2 + 9t]$
$= 16t^3 - 72t^2 + 72t$
(Note: we learn a better way to do this problem in the next unit).

8. $r'(p) = (4p - 3p^2)(2) + (4-6p)(2p-4)$
$= 8p - 6p^2 + 8p - 16 - 12p^2 + 24p$
$= -18p^2 + 40p - 16$

9. $f'(w) = (3w^3 - 4w^2 + 2w - 5)(2w + w^{-2})$
$\quad + (9w^2 - 8w + 2)(w^2 - w^{-1})$
$= 6w^4 - 8w^3 + 4w^2 - 10w + 3w - 4$
$\quad + 2w^{-1} - 5w^{-2} + 9w^4 - 8w^3$
$\quad + 2w^2 - 9w + 8 - 2w^{-1}$
$= 15w^4 - 16w^3 + 6w^2 - 16w + 4 - 5w^{-2}$

10. $g'(r) = (2r^4 - 4r^2 + 2)(2r+2)$
$\quad + (8r^3 - 8r + 2)(r^2 + 2r - 1)$
$= 4r^5 + 4r^4 - 8r^3 - 8r^2 + 4r^2$
$\quad + 4r + 8r^5 + 16r^4 - 8r^3$
$\quad - 8r^3 - 16r^2 + 8r + 2r^2 + 4r - 2$
$= 12r^5 + 20r^4 - 24r^3 - 18r^2 + 16r - 2$

11. $f'(x) = \frac{4(2x+3) - (4x-1)(2)}{(2x+3)^2} = \frac{8x+12-8x+2}{(2x+3)^2} = \frac{14}{(2x+3)^2}$

12. $h'(y) = (y^3 - 1)[(y^2 - 1) \cdot 1 + (2y)(y-1)]$
$\quad + 3y^2[(y^2 - 1)(y-1)]$
$= (y^3 - 1)[y^2 - 1 + 2y^2 - 2y]$
$\quad + 3y^2[y^3 - y^2 - y + 1]$

$= (y^3 - 1)(3y^2 - 2y - 1) + 3y^2(y^3 - y^2 - y + 1)$
$= (3y^5 - 2y^4 - y^3 - 3y^2 + 2y + 1 + 3y^5 - 3y^4 - 3y^3 + 3y^2$
$= 6y^5 - 5y^4 - 4y^3 + 2y + 1$

13. $g'(x) = \frac{(4x-1)(18x) - (9x^2 + 2)(4)}{(4x-1)^2}$
$= \frac{72x^2 - 18x - 36x^2 - 8}{(4x-1)^2}$
$= \frac{36x^2 - 18x - 8}{(4x-1)^2}$

14. $h'(x) = 8x - \frac{x^2 \cdot 0 - 4 \cdot 2x}{x^4} = 8x + \frac{8}{x^3}$

15. $j'(s) = 12s - \frac{-12s}{36s^4} = 12s + \frac{1}{3s^3}$

16. $K'(s) = \frac{(5s - 2s^2)(-4) - (3 - 4s)(5 - 4s)}{(5s - 2s^2)^2}$
$= \frac{-20s + 8s^2 - 15 + 32s - 16s^2}{(5s - 2s^2)^2}$
$= \frac{-8s^2 + 12s - 15}{(5s - 2s^2)^2}$

17. $f'(t) = 12t^2 - \frac{(t-2)2 - 2t}{(t-2)^2} = 12t^2 + \frac{4}{(t-2)^2}$

18. $j'(x) = \frac{(1-2x)2 - (1+2x)(-2)}{(1-2x)^2}$
$= \frac{2 - 4x + 2 + 4x}{(1-2x)^2}$
$= \frac{4}{(1-2x)^2}$

19. $H'(x) = \frac{(x-1)(3x^2) - (x^3 - 1)}{(x-1)^2}$
$= \frac{3x^3 - 3x^2 - x^3 + 1}{(x-1)^2}$
$= \frac{2x^3 - 3x^2 + 1}{(x-1)^2}$
$= \frac{(2x+1)(x-1)(x-1)}{(x-1)^2}$
$= 2x + 1$
(Note: $H(x) = \frac{x^3 - 1}{x-1} = \frac{(x-1)(x^2 + x + 1)}{x-1} = x^2 + x + 1$
and so, $H'(x) = 2x + 1$.)

20. $k'(x) = \frac{3x(4x^3) - (x^4 + 4)3}{9x^2} = \frac{12x^4 - 3x^4 - 12}{9x^2}$
$= \frac{9x^4 - 12}{9x^2} = \frac{3x^4 - 4}{3x^2} = x^2 - \frac{4}{3x^2}$

21. $f'(\phi) = \dfrac{(3\phi^2 - 1)(2\phi) - (\phi^2)(6\phi)}{(3\phi^2 - 1)^2}$

$= \dfrac{6\phi^3 - 2\phi - 6\phi^3}{(3\phi^2 - 1)^2}$

$= \dfrac{-2\phi}{(3\phi^2 - 1)^2}$

22. $L'(t) = \dfrac{(t^2 - 4t - 4)(2t + 3) - (t^2 + 3t + 2)(2t - 4)}{(t^2 - 4t - 4)^2}$

$= \dfrac{2t^3 - 5t^2 - 20t - 12 - 2t^3 - 2t^2 + 8t + 8}{(t^2 - 4t - 4)^2}$

$= \dfrac{-7t - 12t - 4}{(t^2 - 4t - 4)^2} = \dfrac{-(7t^2 + 12t + 4)}{(t^2 - 4t - 4)^2}$

23. $h'(x) = \dfrac{(x^3 - 3x - 1)(9x^2 - 1) - (3x^3 - x + 1)(3x^2 - 3)}{(x^3 - 3x - 1)^2}$

$= \dfrac{(9x^5 - 28x^3 - 9x^2 + 3x + 1) - (9x^5 - 12x^3 + 3x^2 + 3x - 3)}{(x^3 - 3x - 1)^2}$

$= \dfrac{-16x^3 - 12x^2 + 4}{(x^3 - 3x - 1)^2}$

24. $m'(x) = \dfrac{(x^2 - 5x + 6)(2x + 5) - (x^2 + 5x + 6)(2x - 5)}{(x^2 - 5x + 6)^2}$

$= \dfrac{(2x^3 - 5x^2 - 13x + 30) - (2x^3 + 5x^2 - 13x - 30)}{(x^2 - 5x + 6)^2}$

$= \dfrac{-10x^2 + 60}{(x^2 - 5x + 6)^2}$

25. $f'(t) = \dfrac{t^{1/3}(6t - 1) - (3t^2 - t - 1)(\frac{1}{3}t^{-2/3})}{t^{2/3}}$

$\cdot \dfrac{t^{2/3}}{t^{2/3}} = \dfrac{t(6t - 1) - (3t^2 - t - 1)\frac{1}{3}}{t^{4/3}}$

$= \dfrac{6t^2 - t - t^2 + \frac{1}{3}t + \frac{1}{3}}{t^{4/3}}$

$= \dfrac{5t^2 - \frac{2}{3}t + \frac{1}{3}}{t^{4/3}} = \dfrac{15t^2 - 2t + 1}{3t^{4/3}}$

26. $F'(x) = \dfrac{\substack{(4x^{1/2} - 2)(3x^{1/2} - 2x^{-1/2}) \\ -(2x^{3/2} - 4x^{1/2})(2x^{-1/2})}}{(4x^{1/2} - 2)^2}$

$= \dfrac{12x - 6x^{1/2} - 8 + 4x^{-1/2} - 4x + 8}{(4x^{1/2} - 2)^2}$

$= \dfrac{8x - 6x^{1/2} + 4x^{-1/2}}{(4x^{1/2} - 2)^2}$

27. $g'(w) = 24w + \dfrac{(w + 1) - (w - 1)}{(w + 1)^2} = 24w + \dfrac{2}{(w + 1)^2}$

28. $j'(t) = \dfrac{(2t + 1)(2t) - (t^2 + 1)2}{(2t + 1)^2}$

$+ \dfrac{(2t + 1) - (t - 1)2}{(2t + 1)^2}$

$= \dfrac{4t^2 + 2t - 2t^2 - 2 + 2t + 1 - 2t + 2}{(2t + 1)^2}$

$= \dfrac{2t^2 + 2t + 1}{(2t + 1)^2}$

29. $h'(x) = \dfrac{(x^2 + 1)0 - 1(2x)}{(x^2 + 1)^2} = \dfrac{-2x}{(x^2 + 1)^2}$

30. $k'(x) = \dfrac{-5(3x^2)}{(x^3 - 1)^2} = \dfrac{-15x^2}{(x^3 - 1)^2}$

31. $H'(x) = \dfrac{\substack{(2 - x^2)(3x - x^3)(-3x^2) - \\ (4 - x^3)[(-2x)(3x - x^3) + (2 - x^2)(3 - 3x^2)]}}{[(2 - x^2)(3x - x^3)]^2}$

$= \dfrac{\substack{(6x - 5x^3 + x^5)(-3x^2) \\ -(4 - x^3)[-6x^2 + 2x^4 + 6 - 9x^2 + 3x^4]}}{[(2 - x^2)(3x - x^3)]^2}$

$= \dfrac{-18x^3 + 15x^5 - 3x^7 - (4 - x^3)[5x^4 - 15x^2 + 6]}{(2 - x^2)(3x - x^3)^2}$

$= \dfrac{-18x^3 + 15x^5 - 3x^7 - 20x^4 + 60x^2 - 24 + 5x^7 - 15x^5 + 6x^3}{(2 - x^2)^2(3x - x^3)^2}$

$= \dfrac{2x^7 - 20x^4 - 12x^3 + 60x^2 - 24}{(2 - x^2)^2(3x - x^3)^2}$

32. $L'(x) = \dfrac{\substack{(7x - 1)[(2x + 1)(6x - 4) + (2)(3x^2 - 4x)] \\ -(2x + 1)(3x^2 - 4x)\cdot 7}}{(7x - 1)^2}$

$= \dfrac{\substack{(7x - 1)[12x^2 - 2x - 4 + 6x^2 - 8x] \\ -[6x^3 - 5x^2 - 4x]\cdot 7}}{(7x - 1)^2}$

$= \dfrac{(7x - 1)[18x^2 - 10x - 4] - [6x^3 - 5x^2 - 4x]\cdot 7}{(7x - 1)^2}$

$= \dfrac{126x^3 - 88x^2 - 18x + 4 - 42x^3 + 35x^2 + 28x}{(7x - 1)^2}$

$= \dfrac{84x^3 - 53x^2 + 10x + 4}{(7x - 1)^2}$

33. $n(s) = \dfrac{2s^3}{s^3 - s^2 - s + 1}$

$n'(s) = \dfrac{(s^3 - s^2 - s + 1)(6s^2) - 2s^3(3s^2 - 2s - 1)}{(s^3 - s^2 - s + 1)^2}$

$= \dfrac{6s^5 - 6s^4 - 6s^3 + 6s^2 - 6s^5 + 4s^4 + 2s^3}{(s^3 - s^2 - s + 1)^2}$

$= \dfrac{-2s^4 - 4s^3 + 6s^2}{(s^2 - 1)^2(s - 1)^2}$

34. $m(z) = \dfrac{z^3 - 5z^2 + 7z - 35}{z^2 + 3z};$

$m'(z) = \dfrac{\substack{(z^2 + 3z)(3z^2 - 10z + 7) \\ -(z^3 - 5z^2 + 7z - 35)(2z + 3)}}{(z^2 + 3z)^2}$

$= \dfrac{3z^4 - z^3 - 23z^2 + 21z - 2z^4 + 7z^3 + z^2 + 49z + 105}{(z^2 + 3z)^2}$

$= \dfrac{z^4 + 6z^3 - 22z^2 + 70z + 105}{(z^2 + 3z)^2}$

35. $y = \dfrac{(2x - 3)(x^2 - 4x + 1)}{3x^3 + 1}$

$= \dfrac{2x^3 - 11x^2 + 14x - 3}{3x^3 + 1}$

$$y' = \frac{(3x^3+1)(6x^2-22x+14)}{-(2x^3-11x^2+14x-3)(9x^2)}$$
$$= \frac{\begin{matrix}18x^5-66x^4+42x^3+6x^2- \\ 22x+14-18x^5+99x^4-126x^3+27x^2\end{matrix}}{(3x^3+1)^2}$$
$$= \frac{33x^4-84x^3+33x^2-22x+14}{(3x^3+1)^2}$$

36. $y = \dfrac{5x}{5-x}(25-x^2) = \dfrac{5x(5+x)(5-x)}{5-x}$
$= 25x + 5x^2$
$y' = 25 + 10x$

37. $y = \dfrac{(t^2-1)(t-1)}{(2t+1)^2} = \dfrac{t^3-t^2-t+1}{4t^2+4t+1}$
$$y' = \frac{\begin{matrix}(4t^2+4t+1)(3t^2-2t-1)\\-(t^3-t^2-t+1)(8t+4)\end{matrix}}{(4t^2+4t+1)^2} = \frac{\begin{matrix}12t^4+4t^3-9t^2-6t-1\\-8t^4+4t^3+12t^2-4t-4\end{matrix}}{(4t^2+4t+1)^2}$$
$$= \frac{4t^4+8t^3+3t^2-10t-5}{(2t+1)^4}$$
$$= \frac{(2t+1)(2t^3+3t^2-5)}{(2t+1)^4} = \frac{2t^3+3t^2-5}{(2t+1)^3}$$

38. $y = \left(\dfrac{1-x}{x}\right)(1-x^2)$
$$= \frac{1-x^2-x+x^3}{x} = \frac{x^3-x^2-x+1}{x}$$
$$= x^2 - x - 1 + \frac{1}{x}$$
$$y' = 2x - 1 + \frac{-1}{x^2} = 2x - 1 - \frac{1}{x^2}$$

39. $y' = (3x^2+2x-1)(3x^2-1)+(6x+2)(x^3-x+1)$. At $x=1, y' = (3+2-1)(3-1)+(6+2)(1-1+1) = 4\cdot 2 + 8\cdot 1 = 8+8 = 16$. The slope at (1,4) is 16.

40. $y' = \dfrac{(x^2+1)(3x^2)-x^3(2x)}{(x^2+1)^2}$. At $x=-1, y' = \dfrac{(1+1)(3)-(-1)(-2)}{(1+1)^2} = \dfrac{6-2}{4} = \dfrac{4}{4} = 1$. The slope at $\left(-1,-\frac{1}{2}\right)$ is 1.

41. $y' = \dfrac{-8}{(x+1)^2}$; at $x=3, y' = \dfrac{-8}{4^2} = -\dfrac{1}{2}$. Slope of the tangent is $-\frac{1}{2}$. Equation of the tangent is $y-2 = -\frac{1}{2}(x-3)$ or $2y+x = 7$. The slope of the normal is 2; the equation of the normal is $y-2 = 2(x-3)$ or $2x-4$.

42. $y' = \dfrac{5x(4)-(4x+1)\cdot 5}{25x^2}$; at $x=1, y' = \dfrac{20-25}{25} = -\dfrac{1}{5}$. The equation of the tangent line is $y-1 = -\frac{1}{5}(x-1)$ or $5y+x = 6$. The equation of the normal line is $y-1 = 5(x-1)$ or $y-5x = -4$.

43. $y' = \dfrac{(2x+1)5-(5x+4)2}{(2x+1)^2}$; at $x=1, y' = \dfrac{15-18}{9} = -\dfrac{1}{3}$. The equation of the tangent is $y-3 = -\frac{1}{3}(x-1)$

or $3y + x = 10$. The equation of the normal is $y - 3 = 3(x-1)$ or $y - 3x = 0$.

44. $y' = (x^2-3x-6)(2)+(2x-3)(2x-6)$; at $x=-1$, $y' = (1+3-6)(2)+(-5)(-8) = -4+40 = 36$. The equation of the tangent line is $y-16 = 36(x+1)$ or $y - 36x = 52$. The equation of the normal line is $y-16 = -\frac{1}{36}(x+1)$ or $36y + x = 575$.

45. $y = \dfrac{9x+2}{x^3+4x^2}$. $y' = \dfrac{(x^3+4x^2)9-(9x+2)(3x^2+8x)}{(x^3+4x^2)^2}$ at $x=-2, y' = \dfrac{(-8+16)9-(-16)(12-16)}{(-8+16)^2} = \dfrac{72-64}{64} = \dfrac{8}{64} = \dfrac{1}{8}$. Tangent: $y+2 = \frac{1}{8}(x+2)$ or $8y-x = -14$. Normal: $y+2 = -8(x+2)$ or $y+8x = -18$.

46. $y' = (x^3+5x^2-1)(2x-2)+(3x^2+10x)(x^2-2x)$ and at $x=1$, we see that $y' = (1+5-1)(2-2)+(3+10)(1-2) = -13$. The equation of the line with slope $m=-13$ and through $(1,-5)$ is $y-(-5) = -13(x-1)$ or $y+5 = -13(x-1)$ or $y = -13x+8$. The slope of the normal line is $-\frac{1}{-13} = \frac{1}{13}$ and its equation is $y-(-5) = \frac{1}{13}(x-1)$ or $y+5 = \frac{1}{13}(x-1)$ or $x - 13y = 66$.

47. First, find $P'(t) = (0.8t-6)(0.5)+(0.8)(0.5t+9) = (0.4t-3)+(0.4t+7.2) = 0.8t+4.2$. Then, $P'(5) = 0.8(5)+4.2 = 8.2$. Thus, in 5 years the population will be growing at a rate of 8.2 thousand people per year or 8,200 people per year.

48. $P'(n) = (4-0.2n)(2.6)+(-0.2n)(2.6n+7) = (10.4 = 0.52n)+(-0.52n-1.4) = 9-1.04n$. Thus, when $n=7$, the marginal profit is $P'(7) = 9-1.04(7) = 9-1.04(7) = 1.72$ hundred dollars/item = \$172/item.

49. $V' = \dfrac{(R+60)110-110R(1)}{(R+60)^2} = \dfrac{110R+6600-110R}{(R+60)^2} = \dfrac{6600}{(R+60)^2}$

50. First, rewrite $Z = \dfrac{(1+8R)^2+8R}{8(1+8R)} = \dfrac{(1+8R)^2}{8(1+8R)} + \dfrac{8R}{8(1+8R)} = \frac{1}{8}(1+8R) + \dfrac{R}{1+8R}$. Then $Z' = \frac{1}{8}(8) + \dfrac{(1+8R)(1)-R(8)}{(1+8R)^2} = 1 + \dfrac{1+8R-8R}{(1+8R)^2} = 1 + \dfrac{1}{(1+8R)^2} = \dfrac{(1+8R)}{(1+8R)^2} + \dfrac{1}{(1+8R)^2} = \dfrac{(1+8R)^2+1}{(1+8R)^2} = \dfrac{1+16R+64R^2+1}{(1+8R)^2} = \dfrac{2+16R+64R^2}{(1+8R)^2}$.

51. **(a)** $N'(t) = \dfrac{(3\sqrt{t}+2)(2t-2)-(t^2-2t+1)(\frac{3}{2}t^{-1/2})}{(3\sqrt{t}+2)^2}$. Multiply this result by $\dfrac{2\sqrt{t}}{2\sqrt{t}}$ to get

$$N'(t) = \frac{2\sqrt{t}(3\sqrt{t}+2)(2t-2)-2\sqrt{t}(t^2-2t+1)(\frac{3}{2}t^{-1/2})}{2\sqrt{t}(3\sqrt{t}+2)^2}$$
$$= \frac{(6t+4\sqrt{t})(2t-2)-3(t^2-2t+1)}{2(3\sqrt{t}+2)^2\sqrt{t}}$$

$$= \frac{(12t^2+8t\sqrt{t}-12t-8\sqrt{t})-3t^2+6t-3}{2(3\sqrt{t}+2)^2\sqrt{t}}$$

$$= \frac{9t^2+8t^{3/2}-6t-8\sqrt{t}-3}{2(3\sqrt{t}+2)^2\sqrt{t}}$$

(b) $N'(5) = \frac{9(5)^2+8(5)^{3/2}-6(5)-8\sqrt{5}-3}{2(3\sqrt{5}+2)^2\sqrt{5}} \approx 0.777$ thousand bacteria/hr $= 777$ bacteria/hr.

52. $\frac{ds}{ds'} = \frac{(s'-f)f-s'f(1)}{(s'-f)^2} = \frac{s'f-f^2-s'f}{(s'-f)^2} = \frac{-f^2}{(s'-f)^2}$

☰ 23.4 DERIVATIVES OF COMPOSITE FUNCTIONS

1. $f'(x) = 4(3x-6)^3(3) = 12(3x-6)^3$

2. $g'(x) = 3(7-2x)^2(-2) = -6(7-2x)^2$

3. $h'(x) = -5(5x-7)^{-6}(5) = -25(5x-7)^{-6}$

4. $k'(x) = -3(9x+5)^{-4}(9) = -27(9x+5)^{-4}$

5. $f'(x) = 4(x^2+3x)^3(2x+3) = 4(2x+3)(x^2+3x)^3$

6. $h'(x) = \frac{1}{2}(4x^2+7)^{-1/2}(8x) = 4x(4x^2+7)^{-1/2}$ or
$$\frac{4x}{\sqrt{4x^2+7}}$$

7. $H(x) = (4x^2+7)^{-1/2}$, so $H'(x) = -\frac{1}{2}(4x^2+7)^{-3/2}(8x) = -4x(4x^2+7)^{-3/2}$ or $\frac{-4x}{\sqrt{4x^2+7}^3}$.

8. $g'(x) = 6(x^7-9)^5(7x^6) = 42x^6(x^7-9)^5$

9. $s'(t) = 3(t^4-t^3+2)^2(4t^3-3t^2)$

10. $g'(t) = 10(3t^5+2t^3-t)^9(15t^4+6t^2-1)$

11. $f'(u) = 3(u^3+2u^{-4})^2(3u^2-8u^{-5})$

12. $f'(u) = 2(4u^2-3u^{-1})(8u+3u^{-2})$

13. $g'(x) = 3[(x-2)(3x^2-x)]^2$
$$\times [(x-2)(6x-1)+3x^2-x]$$
$$= 3[(x-2)(3x^2-x)]^2$$
$$\times [6x^2-13x+2+3x^2-x]$$
$$= 3[(x-2)(3x^2-x)]^2[9x^2-14x+2]$$

14. $j(t) = 4t^3(t^2-4)^{1/2}$
$$j'(t) = 4t^3\left[\frac{1}{2}(t^2-4)^{-1/2}(2t)\right]$$
$$+ (t^2-4)^{1/2}(12t^2)$$
$$= \frac{4t^4}{\sqrt{t^2-4}} + 12t^2\sqrt{t^2-4}$$

15. $h'(v) = (v^2+1)^2[3(2v-5)^2(2)]$
$$+ (2v-5)^3[2(v^2+1)2v]$$
$$= 6(v^2+1)^2(2v-5)^2$$
$$+ 4v(v^2+1)(2v-5)^3$$

16. $g'(x) = 3[(x^2-4x)(2x^3-7)]^2[(x^2-4x)(6x^2)+(2x-4)(2x^3-7)] = 3[(x^2-4x)(2x^3-7)]^2[6x^4-$

$24x^3+4x^4-8x^3-14x+28] = 3[(x^2-4x)(2x^3-7)]^2[10x^4-32x^3-14x+28]$

17. $h'(x) = 4[(x^3-6x^2)(5x-6x^2+x^3)]^3[x^3-6x^2)(5-12x+3x^2)+(3x^2-12x)(5x-6x^2+x^3)] = 4[(x^3-6x^2)(5x-6x^2+x^3)]^3[3x^5-30x^4+77x^3-30x^2+3x^5-30x^4+87x^3-60x^2] = 4[(x^3-6x^2)(5x-6x^2+x^3)]^3(6x^5-60x^4+164x^3-90x^2)$

18. $j'(w) = (w^3+2w)^4(-2)(3w-5)^{-3}(3) + 4(w^3+2w)^3(3w^2+2)\times(3w-5)^{-2} = -6(w^3+2w)^4(3w-5)^{-3} + 4(w^3+2w)^3(3w^2+2)(3w-5)^{-2}$

19. $k'(x) = (3x^2+2)^3 \cdot \frac{1}{2}(x^3-7x)^{-1/2}(3x^2-7)$
$$+ 3(3x^2+2)^2(6x)(x^3-7x)^{1/2}$$
$$= \frac{(3x^2+2)^3(3x^2-7)}{2\sqrt{x^3-7x}}$$
$$+ 18x(3x^2+2)^2\sqrt{x^3-7x}$$
$$= \frac{(3x^2+2)^3(3x^2-7)}{2\sqrt{x^3-7x}}$$
$$+ \frac{36x(3x^2+2)^2(x^3-7x)}{2\cdot\sqrt{x^3-7x}}$$
$$= \frac{(3x^2+2)^2[(3x^2+2)(3x^2-7)+36x(x^3-7x)]}{2\sqrt{x^3-7x}}$$
$$= \frac{(3x^2+2)^2[9x^4-15x^2-14+36x^4-252x^2]}{2\sqrt{x^3-7x}}$$
$$= \frac{(3x^2+2)^2[45x^4-267x^2-14]}{2\sqrt{x^3-7x}}$$

20. $h'(x) = 4\left(\frac{3x+4}{2x^2-1}\right)^3$
$$\cdot \frac{(2x^2-1)(3)-(3x+4)(4x)}{(2x^2-1)^2}$$
$$= \frac{4(3x+4)^3[6x^2-3-12x^2-16x]}{(2x^2-1)^5}$$
$$= \frac{4(3x+4)^3(-6x^2-16x-3)}{(2x^2-1)^5}$$
$$= \frac{-4(3x+4)^3(6x^2+16x+3)}{(2x^2-1)^5}$$

21. $g'(v) = \dfrac{(v^3-9)^2 \cdot 3(v^2-4v)^2(2v-4)-(v^2-4v)^3 2(v^3-9)(3v^2)}{(v^3-9)^4}$

$= \dfrac{3(v^3-9)(v^2-4v)^2(2v-4)-6v^2(v^2-4v)^3}{(v^3-9)^4}$

$= \dfrac{(v^2-4v)^2[3(v^3-9)(2v-4)-6v^2(v^2-4v)]}{(v^3-9)^3}$

$= \dfrac{(v^2-4v)^2[6v^4-12v^3-54v+108-6v^4+24v^3]}{(v^3-9)^3}$

$= \dfrac{(v^2-4v)^2[12v^3-54v+108]}{(v^3-9)^3}$

22. $f'(u) = 5\left(\dfrac{4u^2+5}{6u^3-3u}\right)^4$

$\times \left(\dfrac{(6u^3-3u)(8u)-(4u^2+5)(18u^2-3)}{(6u^3-3u)^2}\right)$

$= \dfrac{5(4u^2+5)^4[48u^4-24u^2-72u^4-78u^2+15]}{(6u^3-3u)^6}$

$= \dfrac{5(4u^2+5)^4(-24u^4-102u^2+15)}{(6u^3-3u)^6}$

$= \dfrac{-5(4u^2+5)^4(24u^4+102u^2-15)}{(6u^3-3u)^6}$

23. $f'(u) = 6u^5; \ u' = 5x^4$

$y' = 6u^5 \cdot 5x^4 = 6(x^5+4)^5 \cdot 5x^4$

$= 30x^4(x^5+4)^5$

24. $f'(u) = 3u^2; \ u' = (6x^2-5)$

$y' = 3u^2(6x^2-5) = 3(2x^3-5x)^2(6x^2-5)$

25. $f'(u) = \dfrac{1}{2}u^{-1/2}; \ u' = 8x$

$y' = \dfrac{1}{2}u^{-1/2} \cdot 8x = 4x(4x^2-5)^{-1/2}$

26. $f'(u) = 20u^4; \ u' = \dfrac{1}{2}(3x^2+5x)^{-1/2}(6x+5)$

$y' = 20u^4\left(\dfrac{1}{2}(3x^2+5x)^{-1/2}(6x+5)\right)$

$= 10\left(\sqrt{3x^2+5x}\right)^4(3x^2+5x)^{-1/2}(6x+5)$

$= 10(3x^2+5x)^{3/2}(6x+5)$

27. $f'(u) = \dfrac{2}{3}u^{-1/3}; \ u' = 21x^2-9$

$y' = \dfrac{2}{3}(7x^3-9x)^{-1/3}(21x^2-9)$

$= 2(7x^2-3)(7x^3-9x)^{-1/3}$

28. $f'(u) = 2(u+1); \ u' = -2x^{-2}$

$y' = 2(u+1)(-2x^{-2}) = 2\left(\dfrac{2}{x}+1\right)(-2x^{-2})$

$= \dfrac{-4}{x^2}\left(\dfrac{2}{x}+1\right)$

29. $f'(u) = 3(u^2+1)^2(2u); \ u' = -(x+3)^{-2}$

$y' = 3(u^2+1)^2(2u)(-(x+3)^{-2})$

$= 3\left[\dfrac{1}{(x+3)^2}+1\right]^2\left[\dfrac{2}{x+3}\right]\left[\dfrac{-1}{(x+3)^2}\right]$

$= \dfrac{-6}{(x+3)^3}\left[\dfrac{1}{(x+3)^2}+1\right]^2$

30. $f'(u) = 3u^2-4; \ u' = 4x^3$

$y' = (3u^2-4)(4x^3) = [3(x^4+5)^2-4](4x^3)$

31. $f'(u) = \dfrac{1}{3}(u^2-2u)^{-2/3}(2u-2); \ u' = 3x^2$

$y' = \dfrac{1}{3}\left[(x^3+4)^2-2(x^3+4)\right]^{-2/3}$

$\times [2(x^3+4)-2] \cdot 3x^2$

$= 2x^2[(x^3+4)^2-2(x^3+4)]^{-2/3}(x^3+3)$

32. $f'(u) = 6u^2-8; \ u' = 12x-5$

$y' = [6(6x^2-5x-1)^2-8](12x-5)$

33. $g'(x) = 16x^3-3; \ x' = 6t-4$

$y'(t) = [16x^3-3](6t-4)$

$= [16(3t^2-4t)^3-3](6t-4)$

34. $f'(u) = 15u^2-14u-5; \ u' = 12t-8$

$y' = [15(6t^2-8t)^2-14(6t^2-8t)-5]$

$\times (12t-8)$

35. $y' = 6(9x^2+4x)^5(18x+4)$

36. $y' = 5((2x^2-5x+1)^{2/3}(4x-5)$

37. $y' = 10(11x^5-2x+1)^9(55x^4-2)$

38. $y' = -32(x^3-5x+2)^{-5}(3x^2-5)$

$= \dfrac{-32(3x^2-5)}{(x^3-5x+2)^5}$

39. $y = 7(9x^2-4)^{-8}$

$y' = -56(9x^2-4)^{-9}(18x)$

$= \dfrac{-56(18x)}{(9x^2-4)^9}$

40. $y = (2x^5-x^3)^{4/3}$

$y' = \dfrac{4}{3}(2x^5-x^3)^{1/3}(10x^4-3x^2)$

$= \dfrac{4}{3} \cdot \sqrt[3]{2x^5-x^3}(10x^4-3x^2)$

41. $y = (2x^2-5)^{3/4}$

$y' = \dfrac{3}{4}(2x^2-5)^{1/4}(4x)$

$= \dfrac{3x}{\sqrt[4]{2x^2-5}}$

42. $y = (7x - 4x^3)^{-2/5}$

$$y' = -\frac{2}{5}(7x - 4x^3)^{-7/5}(7 - 12x^2)$$

$$= \frac{-2(7 - 12x^2)}{5\sqrt[5]{(7x - 4x^3)^7}}$$

43. $\dfrac{dy}{dx} = \dfrac{dy}{du} \cdot \dfrac{du}{dv} \cdot \dfrac{dv}{dx}; \dfrac{dy}{du} = 4u^3; \dfrac{du}{dv} = 6v^2;$

$$\frac{dv}{dx} = -8x^{-3} = \frac{-8}{x^3}$$

$$\frac{dy}{dx} = (4u^3)(6v^2)\left(\frac{-8}{x^3}\right)$$

$$= (4(2v^3 - 1)^3)(6v^2)\left(\frac{-8}{x^3}\right)$$

$$= 4\left(2 \cdot \left(\frac{4}{x^2}\right)^3 - 1\right)^3\left(6\left(\frac{4}{x^2}\right)^2\right)\left(\frac{-8}{x^3}\right)$$

$$= 4\left(\frac{128}{x^6} - 1\right)^3\left(\frac{96}{x^4}\right)\left(\frac{-8}{x^3}\right)$$

$$= \frac{-3072}{x^7}\left(\frac{128}{x^6} - 1\right)^3$$

44. $\dfrac{dy}{du} = 8u - 1; \dfrac{du}{dx} = 3x^2; \dfrac{dx}{dt} = 6$

$$\frac{dy}{dt} = (8u - 1)(3x^2)(6) = [(8)(x^3 - 8) - 1](3x^2)(6)$$

$$= 18x^2(8x^3 - 65)$$

$$= 18(6t + 4)^2[8(6t + 4)^3 - 65]$$

45. $\dfrac{dy}{du} = 6u; \dfrac{du}{dv} = -4v^{-2}; \dfrac{dv}{dx} = 5x^4$

$$\frac{dy}{dx} = (6u)\left(\frac{-4}{v^2}\right)(5x^4) = 6\left(\frac{4}{v}\right)\left(\frac{-4}{v^2}\right)(5x^4)$$

$$= \frac{-96}{v^3}(5x^4) = \frac{-96}{(x^5)^3} \cdot 5x^4$$

$$= \frac{-480x^4}{x^{15}} = \frac{-480}{x^{11}}$$

46. $y = (x - 5)^{2/3}; \; y' = \frac{2}{3}(x - 5)^{-1/3};$ at $x = 4,$

$y' = \frac{2}{3}(4 - 5)^{-1/3} = -\frac{2}{3}.$ Tangent equation is

$y - 1 = -\frac{2}{3}(x - 4); 3y - 3 = -2x + 8; 2x + 3y = 11$

47. $y' = 3(4x^2 - 18)^2(8x);$ at $x = 2, y' = 3(4 \cdot 2^2 - 18)^2(8 \cdot 2) = 3(-2)^2(16) = 192.$ Tangent equation is $y + 8 = 192(x - 2)$ or $y = 192x - 392$ or $192x - y = 392.$

48. Here $y = 4(25 - x^2)^{-1/2},$ so $y' = -2(25 - x^2)^{-3/2} \times (-2x).$ Thus, at $x = -3, y' = -2(25 - 9)^{-3/2} \times (-2(-3)) = -12(16)^{-3/2} = -\frac{12}{64} = -\frac{3}{16}.$ The slope of the normal is $\frac{16}{3}.$ The equation of the normal is $y - 1 = \frac{16}{3}(x + 3); 3y - 3 = 16x + 48$ or $16x - 3y = -51.$

49. **(a)** $P = E' = \frac{dE}{dt} = 3 \cdot 6(1 + 4t^2)^2(8t) = 144(1 + 4t^2)^2 t$ W

(b) $144(1 + 4t^2)^2 t$ W$= 144(1 + 4t^2)^2 t$ W$\cdot \frac{1 \text{ kW}}{1000 \text{ W}} = \frac{144(1 + 4t^2)^2 t}{1000} = 0.144(1 + 4t^2)^2 t$ kW

(c) $P(2.5) = 0.144(1 + 4(2.5)^2)^2(2.5) = 243.36$ kW

50. **(a)** $P'(t) = 2 \cdot 10(45 + 3.5t)(3.5) - 1750 = 3150 + 245t - 1750 = 1,400 + 245t$ people/year

(b) $P'(5) = 1,400 + 245(5) = 2,625$ people/year

51. **(a)** Rewrite $P(t) = \frac{250t^2}{\sqrt{t^2 + 15}} = \frac{250t^2}{(t^2 + 15)^{1/2}}.$ Then differentiating, we get

$$P'(t) = \frac{(t^2 + 15)^{1/2}(500t) - (250t^2)\frac{1}{2}(t^2 + 15)^{-1/2}(2t)}{\left((t^2 + 15)^{1/2}\right)^2}$$

$$= \frac{(t^2 + 15)^{1/2}(500t) - (250t^2)\frac{1}{2}(t^2 + 15)^{-1/2}(2t)}{\left((t^2 + 15)^{1/2}\right)^2}$$

$$\cdot \frac{(t^2 + 15)^{1/2}}{(t^2 + 15)^{1/2}}$$

$$= \frac{(t^2 + 15)(500t) - (250t^2)\frac{1}{2}(2t)}{(t^2 + 15)^{3/2}}$$

$$= \frac{500t^3 + 7500t - 250t^3}{(t^2 + 15)^{3/2}} = \frac{250t^3 + 7500t}{(t^2 + 15)^{3/2}}$$

$$= \frac{250t(t^2 + 30)}{(t^2 + 15)^{3/2}}$$

(b) When $t = 5.0,$ we have $P'(5.0) = \frac{250(5.0)((5.0)^2 + 30)}{((5.0)^2 + 15)^{3/2}}$

$= \frac{1250(55)}{(40)^{3/2}} \approx 271.8$ ppm/day.

52. **(a)** To find the rate of change in the level of pollution we need to differentiate $P(t) = \frac{0.6\sqrt{8t^2 + 11t + 60}}{(t + 1)^2}.$ Using the quotient rule, we obtain $P'(t) =$

$$\frac{(t + 1)^2\left[0.6 \cdot \frac{1}{2}(8t^2 + 11t + 60)^{-1/2}\right](16t + 11) - 0.6(8t^2 + 11t + 60)2(t + 1)}{(t + 1)^4}$$

Multiplying by $\frac{(8t^2 + 11t + 60)^{1/2}}{(8t^2 + 11t + 60)^{1/2}}$ and simplifying, produces

$$P'(t) = \frac{(t + 1)^2(0.3)(16t + 11) - (0.6(8t^2 + 11t + 60)2(t + 1)}{(8t^2 + 11t + 60)^{1/2}(t + 1)^4}$$

$$= \frac{0.3(t + 1)(16t + 11) - 1.2(8t^2 + 11t + 60)}{(8t^2 + 11t + 60)^{1/2}(t + 1)^3}$$

$$= \frac{0.3(16t^2 + 27t + 11) - 1.2(8t^2 + 11t + 60)}{\sqrt{8t^2 + 11t + 60}(t + 1)^3}$$

$$= \frac{4.8t^2 + 8.1t + 3.3 - 9.6t^2 - 13.2t - 72}{\sqrt{8t^2 + 11t + 60}(t + 1)^3}$$

$$= \frac{-4.8t^2 - 5.1t - 68.7}{\sqrt{8t^2 + 11t + 60}(t + 1)^3}$$

(b) $P'(10.0) = \frac{-4.8(10)^2 - 5.1(10) - 68.7}{\sqrt{8(10)^2 + 11(10) + 60}[(10) + 1]^3} = -0.0145$ ppm/year

53. $R' = 2\left(\frac{n-1}{n+1}\right)^1 \left[\frac{(n+1)-(n-1)}{(n+1)^2}\right] = 2\frac{(n-1)(2)}{(n+1)^3} = \frac{4n-4}{(n+1)^3}$

54. **(a)** First, rewrite $q = 5t^2 + \sqrt{2-5t}$ as $q = 5t^2 + (2-5t)^{1/2}$ and then differentiate. $I = q' = 10t + \frac{1}{2}(2-5t)^{-1/2}(-5) = 10t + \frac{-5}{2}(2-5t)^{-1/2} = 10t - \frac{5}{2\sqrt{2-5t}} = \frac{10t(2\sqrt{2-5t})-5}{2\sqrt{2-5t}} = \frac{20t\sqrt{2-5t}-5}{2\sqrt{2-5t}}$

(b) $I(0.25) = \frac{20(0.25)\sqrt{2-5(0.25)}-5}{2\sqrt{2-5(0.25)}} = -0.387$ A

55. **(a)** Substituting the given information into the formula for the current produces $I = \frac{V}{\sqrt{R^2+X_C^2}} = \frac{120}{\sqrt{30^2+X_C^2}} = \frac{120}{\sqrt{900+X_C^2}} = 120(900 + X_C^2)^{-1/2}$. Differentiating, we get $I' = -\frac{1}{2}(120)(900+X_C^2)^{-3/2} \times (2X_C) = -\frac{120X_C}{(900+X_C^2)^{3/2}}$.

(b) $I'(25) = -\frac{120(25)}{(900+(25)^2)^{3/2}} = -0.0504$ A/Ω

56. **(a)** We are given $L = 1 + 0.25x + 0.001x^2$ and $x = \frac{200}{\sqrt{9-0.4t}} = 200(9-0.4t)^{-1/2}$. We want to find $\frac{dL}{dt}$. Using the chain rule, we have

$$\frac{dL}{dt} = \frac{dL}{dx}\frac{dx}{dt}$$
$$= (0.25 + 0.002x)\left(-\frac{1}{2}\right)(200)$$
$$\times (9 - 0.4t)^{-3/2}(-0.4)$$
$$= \left[0.25 + 0.002\left(\frac{200}{\sqrt{9-0.4t}}\right)\right](40)$$
$$\times (9 - 0.4t)^{-3/2}$$
$$= \left[\frac{0.25\sqrt{9-0.4t}+0.4}{\sqrt{9-0.4t}}\right]\frac{40}{(9-0.4t)^{3/2}}$$
$$= \frac{10\sqrt{9-0.4t}+16}{(9-0.4t)^2}$$

(b) $L'(5) = \frac{10\sqrt{9-0.4(5)}+16}{(9-0.4(5))^2} \approx 0.866$ ppm/year

(c) $L'(10) = \frac{10\sqrt{9-0.4(10)}+16}{(9-0.4(10))^2} \approx 1.534$ ppm/year

23.5 IMPLICIT DIFFERENTIATION

1. $\frac{d}{dx}(4x + 5y) = \frac{d}{dx}(0)$ gives $4 + 5\frac{dy}{dx} = 0$, and so $\frac{dy}{dx} = -\frac{4}{5}$.

2. $\frac{d}{dx}(6x - 7y) = \frac{d}{dx}(3)$ gives $6 - 7\frac{dy}{dx} = 0$, and so $\frac{dy}{dx} = \frac{6}{7}$.

3. $\frac{d}{dx}(x - y^2) = \frac{d}{dx}(4)$ yields $1 - 2y\frac{dy}{dx} = 0$. Thus, $\frac{dy}{dx} = \frac{1}{2y}$.

4. $\frac{d}{dx}(x^2 + y^2) = \frac{d}{dx}(9)$ produces $2x + 2y\frac{dy}{dx} = 0$. Thus, $\frac{dy}{dx} = -\frac{x}{y}$.

5. Here we see that $\frac{d}{dx}(x^2 - y^2) = \frac{d}{dx}(16)$ produces $2x - 2yy' = 0$. Solving for y', we get $2yy' = 2x$ or $y' = \frac{2x}{2y} = \frac{x}{y}$.

6. Here $\frac{d}{dx}(x^2 - 2y^2) = \frac{d}{dx}(2x)$ gives $2x - 4yy' = 2$. Solving for y', we get $y' = \frac{2-2x}{-4y}$ or $y' = \frac{x-1}{2y}$.

7. $\frac{d}{dx}(9x^2 + 16y^2) = \frac{d}{dx}(144)$ means that $18x + 32yy' = 0$. Solving for y', we get $y' = \frac{-18x}{32y} = \frac{-9x}{16y}$.

8. Here $\frac{d}{dx}(16x^2 - 9y^2) = \frac{d}{dx}(144)$ and so $32x - 18yy' = 0$. Solving for y', we get $y' = \frac{-32x}{-18y} = \frac{16x}{9y}$.

9. $\frac{d}{dx}(4x^2 - y^2) = \frac{d}{dx}4y$ produces $8x - 2yy' = 4y'$ or $8x = 4y' + 2yy' = y'(4 + 2y)$. Solving for y', we get $y' = \frac{8x}{4+2y} = \frac{4x}{2+y}$.

10. Here, we see that $\frac{d}{dx}(x^2y + xy^2) = \frac{d}{dx}x$ gives $2xy + x^2y' + y^2 + 2xyy' = 1$ or $x^2y' + 2xyy' = 1-2xy-y^2$. Solving for y', we get $y' = \frac{1-2xy-y^2}{x^2+2xy}$.

11. Here we have $\frac{d}{dx}(xy + 5xy^2) = \frac{d}{dx}x^2$ or $xy' + y + 5y^2 + 10xyy' = 2x$ which is the same as $xy' + 10xyy' = 2x - y - 5y^2$. Solving for y', we get $y' = \frac{2x-y-5y^2}{x+10xy}$.

12. $\frac{d}{dx}(x^2y - 3xy^2) = \frac{d}{dx}y^3$ produces $2xy + x^2y' - 3y^2 - 6xyy' = 3y^2y'$ or $2xy - 3y^2 = 3y^2y' - x^2y' + 6xyy'$. Solving for y', we get $y' = \frac{2xy-3y^2}{3y^2-x^2+6xy}$ or $\frac{3y^2-2xy}{x^2-6xy-3y^2}$.

13. Here, $\frac{d}{dx}(x^2 + 4xy + y^2) = \frac{d}{dx}y$ gives $2x + 4y + 4xy' + 2yy' = y'$ or $4xy' + 2yy' - y' = -2x - 4y$, and so we get $y' = \frac{-2x-4y}{4x+2y-1}$ or $\frac{2x+4y}{1-2y-4x}$.

14. $\frac{d}{dx}(x^3 y - 3xy^2) = \frac{d}{dx}2x$ produces $3x^2 y + x^3 y' - 3y^2 - 6xyy' = 2$ or $x^3 y' - 6xyy' = 2 + 3y^2 - 3x^2 y$. Thus, $y' = \frac{2+3y^2-3x^2y}{x^3-6xy}$

15. Here, $\frac{d}{dx}(x + 5x^2 - 10y^2) = \frac{d}{dx}3y$ gives $1 + 10x - 20yy' = 3y'$ or $1 + 10x = 3y' + 20yy'$. Solving for y', we get $y' = \frac{1+10x}{2+20y}$.

16. Here $\frac{d}{dx}(x^2 - 2xy + y^2) = \frac{d}{dx}x$ yields $2x - 2y - 2xy' + 2yy' = 1$ or $2yy' - 2xy' = 1 + 2y - 2x$. Thus, $y' = \frac{1+2y-2x}{2y-2x}$.

17. $\frac{d}{dx}(4x^2 + y^3) = \frac{d}{dx}9$ yields $8x + 3y^2 y' = 0$, and so $y' = \frac{-8x}{3y^2}$.

18. $\frac{d}{dx}(y^4 - 9x^2) = \frac{d}{dx}9$ gives $4y^3 y' - 18x = 0$, and so $y' = \frac{18x}{4y^3} = \frac{9x}{2y^3}$.

19. Here $\frac{d}{dx}(6y^3 + x^3) = \frac{d}{dx}(xy)$ produces $18y^2 y' + 3x^2 = xy' + y$ or $18y^2 y' - xy' = y - 3x^2$. Solving for y', we get $y' = \frac{y-3x^2}{18y^2-x}$.

20. $\frac{d}{dx}(x^2 y) = \frac{d}{dx}(7y)$ produces $2xy + x^2 y' = 7y'$ or $x^2 y' - 7y' = -2xy$. Solving for y', we get $y' = \frac{2xy}{7-x^2}$.

21. $\frac{d}{dx}(x^2 y) = \frac{d}{dx}(x + 1)$ yields $2xy + x^2 y' = 1$ or $x^2 y' = 1 - 2xy$. Solving for y', we get $y' = \frac{1-2xy}{x^2}$.

22. $\frac{d}{dx}(x^2 + y^{-1}) = \frac{d}{dx}(2x)$ yields $2x - y^{-2} y' = 2$. This simplifies to $\frac{y'}{y^2} = 2x - 2$ and so $y' = y^2(2x - 2)$.

23. $\frac{d}{dx}(x^{-1} + y^{-1}) = \frac{d}{dx}16$ produces $-x^{-2} - y^{-2} y' = -0$. Solving for y', we get $y' = \frac{x^{-2}}{-y^{-2}} = -\frac{y^2}{x^2}$.

24. $\frac{d}{dx}(x^{-2} - y^{-2}) = \frac{d}{dx}(xy)$, and so $-2x^{-3} + 2y^{-3} y' = y + xy'$ which can be rewritten as $2y^{-3} y' - xy' = y + 2x^{-3}$. Solving for y', we get $y' = \frac{y+2x^{-3}}{2y^{-3}-x} = \frac{y+\frac{2}{x^3}}{\frac{2}{y^3}-x} = \frac{x^3 y^4 + 2y^3}{2x^3 - x^4 y^3}$.

25. $\frac{d}{dx}\left(\frac{3}{x+1} + x^2 y\right) = \frac{d}{dx}5$ produces $-3(x + 1)^{-2} + 2xy + x^2 y' = 0$. Solving for y', we get $y' = \frac{3(x+1)^{-2}-2xy}{x^2} = \frac{\frac{3}{(x+1)^2}-2xy}{x^2} = \frac{3 - 2xy(x + 1)^2}{x^2(x + 1)^2}$.

26. Here $\frac{d}{dx}(y^2) = \frac{d}{dx}\left(\frac{x}{y+1}\right)$ produces $2yy' = \frac{(y+1)-xy'}{(y+1)^2}$, or $y' \cdot 2y(y + 1)^2 = y + 1 - xy'$, or $y'[2y(y + 1)^2 + x] = y + 1$. Solving for y', we get $y' = \frac{y+1}{2y(y+1)^2+x}$.

27. $\frac{d}{dx}(xy + \frac{y}{x}) = \frac{d}{dx}x$ gives $y + xy' + \frac{xy'-y}{x^2} = 1$, or $yx^2 + x^3 y' + xy' - y = x^2$ which can be written as $x^3 y' + xy' = x^2 + y - yx^2$. Solving for y', we get $y' = \frac{x^2+y-yx^2}{x^3+x}$.

28. Here $\frac{d}{dx}(x^3 - 8x^2 y^2 + y) = \frac{d}{dx}9x$ produces $3x^2 - 8x^2(2yy') - 16xy^2 + y' = 9$, or $y' - 16x^2 yy' = 9 + 16xy^2 - 3x^2$. Thus, $y' = \frac{9+16xy^2-3x^2}{1-16x^2y}$.

29. Here, $\frac{d}{dx}(x^4 - 6x^2 y^2 + y^2) = \frac{d}{dx}(10x)$ yields $4x^3 - 6x^2(2yy') - 12xy^2 + 2yy' = 10$, or $2yy' - 12x^2 yy' = 10 + 12xy^2 - 4x^3$, and so $y' = \frac{10+12xy^2-4x^3}{2y-12x^2y} = \frac{5+6xy^2-2x^3}{y-6x^2y}$.

30. Here, $\frac{d}{dx}(x + 2y)^2 = \frac{d}{dx}4x$ produces $2(x + 2y)(1 + 2y') = 4$, or $(x + 2y)(1 + 2y') = 2$, or $1 + 2y' = \frac{2}{x+2y}$. Thus, $2y' = \frac{2}{x+2y} - 1$, and so $y' = \frac{1}{x+2y} - \frac{1}{2} = \frac{2-x-2y}{(x+2y)2}$.

31. $\frac{d}{dx}(x^2 + y^2)^{1/2} = \frac{d}{dx}\left(\frac{2y}{x}\right)$ gives $\frac{1}{2}(x^2 + y^2)^{-1/2}(2x + 2yy') = \frac{x\cdot 2y'-2y}{x^2}$, or $x^2(2x + 2yy') = 2\sqrt{x^2 + y^2}(2xy' - 2y)$, or $2x^3 + 2x^2 yy' = 4x\sqrt{x^2 + y^2}y' - 4y\sqrt{x^2 + y^2}$, and so $2x^2 yy' - 4x\sqrt{x^2 + y^2}y' = -4y\sqrt{x^2 + y^2} - 2x^3$. Solving for y', we get $y' = \frac{4y\sqrt{x^2+y^2}+2x^3}{4x\sqrt{x^2+y^2}-2x^2y} = \frac{2y\sqrt{x^2+y^2}+x^3}{2x\sqrt{x^2+y^2}-x^2y}$.

32. Here, $\frac{d}{dx}(x + xy)^{1/2} = \frac{d}{dx}(x^2 y^2)$ gives $\frac{1}{2}(x + xy)^{-1/2}(1 + xy' + y) = 2xy^2 + 2x^2 yy'$, or $1 + xy' + y = 2\sqrt{x + xy}(2xy^2 + 2x^2 yy')$ which is the same as $1 + xy' + y = 4xy^2\sqrt{x + xy} + 4x^2 yy'\sqrt{x + xy}$ or $xy' - 4x^2 yy'\sqrt{x + xy} = 4xy^2\sqrt{x + xy} - y - 1$. Solving for y', we get $y' = \frac{4xy^2\sqrt{x+xy}-y-1}{x-4x^2y\sqrt{x+xy}}$.

33. $\frac{d}{dx}(x^2 + y^2)^3 = \frac{d}{dx}y$ produces $3(x^2 + y^2)^2(2x + 2yy') = y'$, or $6x(x^2 + y^2)^2 + 6yy'(x^2 + y^2)^2 = y'$, or $6x(x^2 + y^2)^2 = y' - 6yy'(x^2 + y^2)^2$. Solving for y', we get $y' = \frac{6x(x^2+y^2)^2}{1-6y(x^2+y^2)^2}$.

34. Here, $\frac{d}{dx}(x^2y+4)^2 = \frac{d}{dx}(3x)$ gives $2(x^2y+4)(2xy + x^2y') = 3$, or $(2x^2y + 8)(2xy + x^2y') = 3$, and so $4x^3y^2 + 2x^4yy' + 16xy + 8x^2y' = 3$, or $2x^4yy' + 8x^2y' = 3 - 4x^3y^2 - 16xy$. Solving for y', we get $y' = \frac{3-4x^3y^2-16xy}{2x^4y+8x^2} = \frac{3-4xy(x^2y+4)}{2x^2(x^2y+4)}$.

35. $\frac{d}{dx}(x^2 + y^2) = \frac{d}{dx}25$ give $2x + 2yy' = 0$ and so $y' = \frac{-2x}{2y} = \frac{-x}{y}$. Thus, the slope of the tangent to the circle at the point $(3, -4)$ is $m = \frac{-3}{-4} = \frac{3}{4}$. Using the point-slope form for the equation of a line, we see that the tangent line is $y+4 = \frac{3}{4}(x-3)$ or $4y + 16 = 3x - 9$ or $4y - 3x = -25$.

36. $\frac{d}{dx}(x^2 + y^2 - 2x - 4y - 20) = \frac{d}{dx}0$ produces $2x + 2yy' - 2 - 4y' = 0$, or $2yy' - 4y' = 2 - 2x$, and so $y' = \frac{2-2x}{2y-4} = \frac{1-x}{y-2}$. Thus, the slope of the tangent to the circle at the point $(-2, 6)$ is $m = \frac{1+2}{6-2} = \frac{3}{4}$ and the slope of the normal is $-\frac{4}{3}$. Hence, the equation of the normal is $y - 6 = -\frac{4}{3}(x + 2)$ or $3y - 18 = -4x - 8$ or $4x + 3y = 10$.

37. Here, $\frac{d}{dx}(9x^2 + 16y^2 - 100) = \frac{d}{dx}0$ gives $18x + 32yy' = \frac{d}{dx}0$, and so $y' = \frac{-18x}{32y} = \frac{-9x}{16y}$. Thus, the slope of the tangent to the ellipse at the point $(-2, 2)$ is $m = \frac{-9(-2)}{16(2)} = \frac{+18}{32} = \frac{9}{16}$. Hence, the tangent line is $y - 2 = \frac{9}{16}(x + 2)$ or $16y - 32 = 9x + 18$ or $16y - 9x = 50$ or $9x - 16y + 50 = 0$.

38. $\frac{d}{dx}y^2 = \frac{d}{dx}(-16x)$, or $2yy' = -16$, and so $y' = \frac{-16}{2y} = \frac{-8}{y}$. Thus, the slope of the tangent to the parabola at the point $(-1, 4)$ is $m = \frac{-8}{4} = -2$ and the slope of the normal is $\frac{1}{2}$.

39. $\frac{d}{dx}(12x^2 - 16y^2 - 192) = \frac{d}{dx}0$ gives $24x - 32yy' = 0$ and so $y' = \frac{24x}{32y} = \frac{3x}{4y}$. Thus, the slope of the tangent to the hyperbola at the point $(-8, -6)$ is $m = \frac{3(-8)}{4(-6)} = \frac{-24}{-24} = 1$ and the slope of the normal is -1.

40. Differentiating $s^3 - 4st + 2t^3 - 5t = 0$ with respect to t produces $3s^2\frac{ds}{st} - 4\left(\frac{ds}{st}t + s\right) + 6t^2 - 5 = 0$ or $3s^2\frac{ds}{st} - 4t\frac{ds}{st} - 4s + 6t^2 - 5 = 0$. Collecting terms

results in $\left(3s^2 - 4t\right)\frac{ds}{st} = 4s - 6t^2 + 5$. Dividing, we get the desired result $\frac{ds}{dt} = \frac{4s-6t^2+5}{3s^2-4t}$.

41. **(a)** Rewrite the given equation $2s^2 + \sqrt{st} - 3t = 0$ as $2s^2 + (st)^{1/2} - 3t = 0$. Differentiating produces $4s\frac{ds}{dt} + \frac{1}{2}(st)^{-1/2}\left(t\frac{ds}{dt} + s\right) - 3 = 0$ or $4s\frac{ds}{dt} + \frac{t}{2(st)^{1/2}}\frac{ds}{dt} + \frac{s}{2(st)^{1/2}} - 3 = 0$. Collecting terms yields $\left(4s + \frac{t}{2(st)^{1/2}}\right)\frac{ds}{dt} = 3 - \frac{s}{2(st)^{1/2}}$ or $\left(\frac{8s(st)^{1/2}+t}{2(st)^{1/2}}\right)\frac{ds}{dt} = \frac{6(st)^{1/2}-s}{2(st)^{1/2}}$. Dividing produces the answer $\frac{ds}{dt} = \frac{6(st)^{1/2}-s}{8s(st)^{1/2}+t} = \frac{6\sqrt{st}-s}{8s\sqrt{st}+t}$.

(b) When $t = 2$ the automobile's position can be determined by solving $2s^2 + \sqrt{2(2)} - 3(2) = 0$ or $2s^2 + \sqrt{2}s - 6 = 0$ for s. A graph yields $s \approx 1.464$ mi.

(c) Substituting $t = 2$ and $s = 1.464$ into $\frac{ds}{dt} = \frac{6\sqrt{st}-s}{8s\sqrt{st}+t}$ produces $\frac{ds}{dt} = \frac{6\sqrt{1.464(2)}-1.464}{11.712\sqrt{2.928}+2} = \frac{6\sqrt{2.928}-1.464}{11.712\sqrt{2.928}+2} \approx 0.399$ mi/min $= 23.94$ mi/hr.

42. If the same level of production is maintained, then $P' = 0$. Differentiating yields $P' = 20\left[\frac{1}{5}x^{-4/5}\frac{dx}{dt}y^{4/5} + x^{1/5}\frac{dy}{dt}\right] = 0$. Dividing by 20, produces $\frac{1}{5}x^{-4/5}\frac{dx}{dt}y^{4/5} + x^{1/5}\frac{4}{5}y^{-1/5}\frac{dy}{dt} = 0$ and multiplying by $y^{1/5}x^{4/5}$, the equation simplifies to $\frac{1}{5}y\frac{dx}{dt} + \frac{4}{5}x\frac{dy}{dt} = 0$ and multiplying by 5, the equation becomes $y\frac{dx}{dt} + 4x\frac{dy}{dt} = 0$. When $x = 32$, an increase of 25% means $\frac{dx}{dt} = 8$. Substituting these values into the equation it becomes $8y + 4x\frac{dy}{dt} = 0$ and solving for $\frac{dy}{dt} = \frac{-8y}{4x} = \frac{-2y}{x}$. Thus, when $y = 243$ and $x = 32$, the change in units of capital needed to maintain the same level of production is $\frac{dy}{dt} = \frac{-2(243)}{32} = -15.1875$ units per month.

43. **(a)** Differentiating produces $1.4y\frac{dy}{dx} = 12+y+x\frac{dy}{dx}$ and so $(1.4y-x)\frac{dy}{dx} = 12+y$. Dividing by $1.4y-x$ gives the result $\frac{dy}{dx} = \frac{12+y}{1.4y-x}$. When $x = 10$ and $y = 25$, we have $\frac{dy}{dx} = \frac{12+25}{1.4(25)-10} = \frac{37}{25} = 1.48$.

(b) The store should sell 1.48 television sets for each digital satellite that is sold.

44. **(a)** Differentiating produces $48 = 12\frac{dy}{dx} + 0.02xy + 0.01x^2\frac{dy}{dx}$. Collecting terms yields $(12+0.01x^2)\frac{dy}{dx} = 48 - 0.02xy$. Dividing, gives $\frac{dy}{dx} = \frac{48-0.02xy}{12+0.01x^2}$.

When $x = 10$ and $y = 40$, the derivative becomes

$\frac{dy}{dx} = \frac{48 - 0.02(10)(40)}{12 + 0.01(10)^2} = \frac{40}{13} \approx 3.08$.

(b) The store should sell about 3.08 shirts for each pair of trousers that is sold.

≡ 23.6 HIGHER ORDER DERIVATIVES

1. $y = 4x^3 - 6x^2 + 3x - 10$
$y' = 12x^2 - 12x + 3$
$y'' = 24x - 12$

2. $y = 9x^4 + x^3 - 10;$
$y' = 36x^3 + 3x^2;$
$y'' = 108x^2 + 6x;$
$y''' = 216x + 6$

3. $y = 7x^5 - 3x^3 + x;$
$y' = 35x^4 - 9x^2 + 1;$
$y'' = 140x^3 - 18x;$
$y''' = 420x^2 - 18;$
$y^{(4)} = 840x$

4. $f(x) = 10x - 1;$
$f'(x) = 10;$
$f''(x) = 0$

5. $f(x) = x + x^{-1};$
$f'(x) = 1 - x^{-2};$
$f''(x) = 2x^{-3} = \frac{2}{x^3}$

6. $f(t) = t^3 - t^{-2};$
$f'(t) = 3t^2 + 2t^{-3};$
$f''(t) = 6t - 6t^{-4};$
$f'''(t) = 6 + 24t^{-5}$

7. $f(t) = t^2 + \frac{1}{t+1};$
$f'(t) = 2t - \frac{1}{(t+1)^2};$
$f''(t) = 2 + \frac{2}{(t+1)^3}$

8. $h(x) = (x^2 + 2)(x - 1) = x^3 - x^2 + 2x - 2$
$h'(x) = 3x^2 - 2x + 2;$
$h''(x) = 6x - 2$

9. $g(u) = \frac{u - 1}{u^2};$
$g'(u) = \frac{u^2 - (u-1)2u}{u^4} = \frac{u^2 - 2u^2 + 2u}{u^4}$
$= \frac{2u - u^2}{u^4} = \frac{2 - u}{u^3}$

$g''(u) = \frac{u^3(-1) - (2-u)3u^2}{u^6}$
$= \frac{-u^3 - 6u^2 + 3u^3}{u^6}$
$= \frac{2u^3 - 6u^2}{u^6} = \frac{2u - 6}{u^4}$
$g'''(u) = \frac{u^4(2) - (2u - 6)4u^3}{u^8} = \frac{2u - 8u + 24}{u^5}$
$= \frac{-6u + 24}{u^5}$

10. $y = 3(x^2 + 1)^{1/3}; \frac{dy}{dx} = (x^2 + 1)^{-2/3}(2x)$

$\frac{d^2y}{dx^2} = 2(x^2 + 1)^{-2/3} - \frac{2}{3}(2x)(x^2 + 1)^{-5/3}(2x)$

$= 2(x^2 + 1)^{-2/3} - \frac{8x^2}{3}(x^2 + 1)^{-5/3}$

$= \frac{2}{\sqrt[3]{x^2 + 1}} - \frac{8x^2}{3(\sqrt[3]{x^2 + 1})^5}$

$= \frac{6(x^2 + 1) - 8x^2}{3(\sqrt[3]{x^2 + 1})^5} = \frac{-2x^2 + 6}{3(\sqrt[3]{x^2 + 1})^5}$

11. $y = (x + 1)^{1/2}$
$\frac{dy}{dx} = \frac{1}{2}(x + 1)^{-1/2}$
$\frac{d^2y}{dx^2} = -\frac{1}{4}(x + 1)^{-3/2} = \frac{-1}{4\sqrt{x+1}^3}$

12. $y = (4x - x^2)^2;$
$\frac{dy}{dx} = 2(4x - x^2)(4 - 2x)$
$= 4(2 - x)(4x - x^2);$
$\frac{d^2y}{dx^2} = 4(-1)(4x - x^2) + 4(2 - x)(4 - 2x)$
$= -4(4x - x^2) + 4(2 - x)(4 - 2x)$
$= -16x + 4x^2 + 8x^2 - 32x + 32$
$= 12x^2 - 48x + 32$

Note: An easier method would be to first expand $y = (4x - x^2)^2$, to get $y = 16x^2 - 8x^3 + x^4$; $y' = 32x - 24x^2 + 4x^3$; $y'' = 12x^2 - 48x + 32$.

13. $y = (2x + 1)^3 = 8x^3 + 12x^2 + 6x + 1$
$\frac{dy}{dx} = 24x^2 + 24x + 6; \frac{d^2y}{dx^2} = 48x + 24$
$\frac{d^3y}{dx^3} = 48$

14. $y = x^4 - 3x^2 + 2$

$\dfrac{dy}{dx} = 4x^3 - 6x$;

$\dfrac{d^2y}{dx^2} = 12x^2 - 6$

$\dfrac{d^3y}{dx^3} = 24x$

15. $y = x^{-2/3}; \dfrac{dy}{dx} = -\dfrac{2}{3}x^{-5/3}; \dfrac{d^2y}{dx^2} = \dfrac{10}{9}x^{-8/3}$

16.   Since $y = \dfrac{x^3}{\sqrt[3]{x}} = x^{8/3}$, we have $\dfrac{dy}{dx} = \dfrac{8}{3}x^{5/3}$. The

second derivative is $\dfrac{d^2y}{dx^2} = \dfrac{40}{9}x^{2/3}$; and the third

derivative is $\dfrac{d^3y}{dx^3} = \dfrac{80}{27}x^{-1/3} = \dfrac{80}{27\sqrt[3]{x}}$.

17. $f(x) = x(x+1)^{-1}$;

$D_x f(x) = 1(x+1)^{-1} - x(x+1)^{-2}$

$D_x^2 f(x) = -(x+1)^{-2} - (x+1)^{-2}$
$\qquad\qquad + 2x(x+1)^{-3}$
$\qquad\quad = -2(x+1)^{-2} + 2x(x+1)^{-3}$

$D_x^3 f(x) = 4(x+1)^{-3} + 2(x+1)^{-3}$
$\qquad\qquad - 6x(x+1)^{-4}$
$\qquad\quad = 6(x+1)^{-3} - 6x(x+1)^{-4}$

$D_x^4 f(x) = -18(x+1)^{-4} - 6(x+1)^{-4}$
$\qquad\qquad + 24x(x+1)^{-5}$
$\qquad\quad = -24(x+1)^{-4} + 24x(x+1)^{-5}$
$\qquad\quad = -24(x+1)(x+1)^{-5} + 24x(x+1)^{-5}$
$\qquad\quad = -24x(x+1)^{-5} - 24(x+1)^{-5}$
$\qquad\qquad + 24x(x+1)^{-5}$
$\qquad\quad = -24(x+1)^{-5}$

18. $f(t) = (t+1)(t+2)^{-1}$;

$D_t f(t) = (t+1)(-1)(t+2)^{-2} + (t+2)^{-1}$
$\qquad\quad = -(t+1)(t+2)^{-2} + (t+2)^{-1}$;

$D_t^2 f(t) = -(t+2)^{-2} + 2(t+1)(t+2)^{-3}$
$\qquad\qquad - (t+2)^{-2}$
$\qquad\quad = -2(t+2)^{-2} + 2(t+1)(t+2)^{-3}$
$\qquad\quad = -2(t+2)(t+2)^{-3} + 2(t+1)(t+2)^{-3}$
$\qquad\quad = (-2t-4)(t+2)^{-3} + (2t+2)(t+2)^{-3}$
$\qquad\quad = -2(t+2)^{-3}$

$\qquad\quad = D_t^3 f(t) = 6(t+2)^{-4} = \dfrac{6}{(t+2)^4}$

19. $f(w) = (3w^2+1)^{-2}$;

$D_w f(w) = -2(3w^2+1)^{-3}(6w)$
$\qquad\qquad = -12w(3w^2+1)^{-3}$

$D_w^2 f(w) = -12(3w^2+1)^{-3}$
$\qquad\qquad + 36w(3w^2+1)^{-4}(6w)$
$\qquad\qquad = 216w^2(3w^2+1)^{-4} - 12(3w^2+1)^{-3}$

$\qquad\qquad = 216w^2(3w^2+1)^{-4}$
$\qquad\qquad\quad - 12(3w^2+1)(3w^2+1)^{-4}$
$\qquad\qquad = (216w^2 - 36w^2 - 12)(3w^2+1)^{-4}$
$\qquad\qquad = (180w^2 - 12)(3w^2+1)^{-4}$

20. $\dfrac{d}{dx}(x^4 - 3x^2 + 1) = 4x^3 - 6x$;

$\dfrac{d^2}{dx^2}(x^4 - 3x^2 + 1) = 12x^2 - 6$;

$\dfrac{d^3}{dx^3}(x^4 - 3x^2 + 1) = 24x$;

$\dfrac{d^4}{dx^4}(x^4 - 3x^2 + 1) = 24$;

$\dfrac{d^5}{dx^5}(x^4 - 3x^2 + 1) = 0$

21. $\dfrac{d}{dx}(x^7 - 4x^5 - x^3) = 7x^6 - 20x^4 - 3x^2$;

$\dfrac{d^2}{dx^2}(x^7 - 4x^5 - x^3) = 42x^5 - 80x^3 - 6x$;

$\dfrac{d^3}{dx^3}(x^7 - 4x^5 - x^3) = 210x^4 - 240x^2 - 6$;

$\dfrac{d^4}{dx^4}(x^7 - 4x^5 - x^3) = 840x^3 - 480x$

22. $\dfrac{d}{dx} = 8x^7 - 6x$;

$\dfrac{d^2}{dx^2} = 8 \cdot 7 \cdot x^6 - 6$;

$\dfrac{d^3}{dx^3} = 8 \cdot 7 \cdot 6x^5; \dfrac{d^4}{dx^4} = 8 \cdot 7 \cdot 6 \cdot 5x^4; \ldots$;

$\dfrac{d^8}{dx^8} = 8! = 40{,}320$

23. $\dfrac{d}{dx}(x - x^{-2}) = 1 + 2x^{-3}$;

$\dfrac{d^2}{dx^2}(x - x^{-2}) = -6x^{-4}$;

$\dfrac{d^3}{dx^3}(x - x^{-2}) = 24x^{-5}$

24. $\dfrac{d}{dx}\left(\dfrac{x^2}{2x+3}\right)$

$\qquad = \dfrac{(2x+3)2x - x^2 \cdot 2}{(2x+3)^2} = \dfrac{4x^2 + 6x - 2x^2}{(2x+3)^2}$

$\qquad = \dfrac{2x^2 + 6x}{(2x+3)^2}$;

$$\frac{d^2}{dx^2}\left(\frac{x^2}{2x+3}\right)$$

$$= \frac{d}{dx}\left(\frac{2x^2+6x}{(2x+3)^2}\right)$$

$$= \frac{(2x+3)^2(4x+6)-(2x^2+6x)(2)(2x+3)2}{(2x+3)^4}$$

$$= \frac{(2x+3)(4x+6)-4(2x^2+6x)}{(2x+3)^3}$$

$$= \frac{8x^2+24x+18-8x^2-24x}{(2x+3)^3}$$

$$= \frac{18}{(2x+3)^3}$$

25. $\dfrac{d}{dx}\left(\dfrac{x^3}{3x-1}\right)$

$$= \frac{(3x-1)(3x^2)-x^3(3)}{(3x-1)^2} = \frac{9x^3-3x^2-3x^3}{(3x-1)^2}$$

$$= \frac{6x^3-3x^2}{(3x-1)^2};$$

$$\frac{d}{dx}\left(\frac{6x^3-3x^2}{(3x-1)^2}\right)$$

$$= \frac{(3x-1)^2(18x^2-6x)-(6x^3-3x^2)(2)(3x-1)(3)}{(3x-1)^4}$$

$$= \frac{(3x-1)(18x^2-6x)-6(6x^3-3x^2)}{(3x-1)^3}$$

$$= \frac{54x^3-36x^2+6x-36x^3+18x^2}{(3x-1)^3}$$

$$= \frac{18x^3-18x^2+6x}{(3x-1)^3} = \frac{6x(3x^2-3x+1)}{(3x-1)^3}$$

26. $\dfrac{d}{dx}[x^3+x^2-2x] = 3x^2+2x-2$

$$\frac{d}{dx}(3x^2+2x-2) = 6x+2$$

27. $\dfrac{d}{dx}(x(x+1)^{-1/2})$

$$= x\left(-\frac{1}{2}\right)(x+1)^{-3/2}+(x+1)^{-1/2}$$

$$= -\frac{x}{2}(x+1)^{-3/2}+(x+1)^{-1/2}$$

$$\frac{d^2}{dx^2}[x(x+1)^{-1/2}]$$

$$= -\frac{x}{2}(-3/2)(x+1)^{-5/2}-\frac{1}{2}(x+1)^{-3/2}$$

$$\quad -\frac{1}{2}(x+1)^{-3/2}$$

$$= \frac{3}{4}x(x+1)^{-5/2}-(x+1)^{-3/2}$$

$$= \frac{3}{4}x(x+1)^{-5/2}-(x+1)(x+1)^{-5/2}$$

$$= \frac{-x-4}{4(x+1)^{5/2}}$$

28. $2x+2yy'=0; y'=-\dfrac{x}{y};$

$$y'' = \frac{-y+xy'}{y^2} = \frac{-y+x\left(-\frac{x}{y}\right)}{y^2}$$

$$= \frac{\frac{-y^2}{y}-\frac{x^2}{y}}{y^2} = -\frac{x^2+y^2}{y^3} = \frac{-9}{y^3}$$

(Note: $x^2+y^2=9$)

29. $2x-8yy'=1;$

$$y' = \frac{1-2x}{-8y} = \frac{2x-1}{8y}$$

$$y'' = \frac{8y(2)-(2x-1)8y'}{64y^2}$$

$$= \frac{2y-(2x-1)\left(\frac{2x-1}{8y}\right)}{8y^2}$$

$$= \frac{\frac{16y^2}{8y}-\frac{(2x-1)^2}{8y}}{8y^2} = \frac{16y^2-(2x-1)^2}{64y^3}$$

$$= \frac{16y^2-4x^2+4x-1}{64y^3}$$

$$= \frac{4(4y^2-x^2)+4x-1}{64y^3}$$

$$= \frac{4(-x)+4x-1}{64y^3} = \frac{-1}{64y^3}$$

(Note: In the second-to-last step, we used the fact that since $x^2-4y^2=x$, then $4y^2-x^2=-x$.)

30. $2x-2yy'=0;$

$$y' = \frac{x}{y}$$

$$y'' = \frac{y-xy'}{y^2} = \frac{y-x\cdot\frac{x}{y}}{y^2} = \frac{\frac{y^2}{y}-\frac{x^2}{y}}{y^2}$$

$$= \frac{y^2-x^2}{y^3} = -\frac{(x^2-y^2)}{y^3} = \frac{-16}{y^3}$$

(Note: In the last step, remember that $x^2-y^2=16$.)

31. Differentiating, we get $x(2yy')+y^2+4=y'$. Then, solving for y' produces
$$2xyy'-y'=-y^2-4;$$

$$y' = \frac{y^2+4}{1-2xy}$$

$$y'' = \frac{(1-2xy)(2yy')-(y^2+4)[-2xy'-2y]}{(1-2xy)^2}$$

$$= \frac{(1-2xy)\left(2y\cdot\frac{y^2+4}{1-2xy}\right)+(y^2+4)\left(2x\cdot\frac{y^2+4}{1-2xy}+2y\right)}{(1-2xy)^2}$$

$$= \frac{(2y^3+8y)+(y^2+4)\left[\frac{2xy^2+8x}{1-2xy}+\frac{2y(1-2xy)}{1-2xy}\right]}{(1-2xy)^2}$$

$$= \frac{(1-2xy)2y(y^2+4)+(y^2+4)(8x+2y-2xy^2)}{(1-2xy)^3}$$

$$= \frac{(y^2+4)[2y-4xy^2+8x+2y-2xy^2]}{(1-2xy)^3}$$
$$= \frac{(y^2+4)(4y-6xy^2+8x)}{(1-2xy)^3}$$

32. $x(2yy')+y^2-5=4y'$;
$2xyy'-4y'=5-y^2$
$$y'=\frac{5-y^2}{2xy-4} \text{ or } \frac{y^2-5}{4-2xy}$$
$$y''=\frac{(4-2xy)(2yy')-(y^2-5)(-2xy'-2y)}{(4-2xy)^2}$$
$$=\frac{(4-2xy)\left(2y\cdot\frac{y^2-5}{4-2xy}\right)+(y^2-5)(2x\cdot\frac{y^2-5}{4-2xy}+2y)}{(4-2xy)^2}$$
$$=\frac{(y^2-5)(8y-4xy^2)+(y^2-5)[2xy^2-10x+8y-4xy^2]}{(4-2xy)^3}$$
$$=\frac{(y^2-5)(16y-6xy^2-10x)}{(4-2xy)^3}$$

34. $xy'+y+2yy'=4x$;
$xy'+2yy'=4x-y$
$$y'=\frac{4x-y}{x+2y}$$
$$y''=\frac{(x+2y)(4-y')-(4x-y)(1+2y')}{(x+2y)^2}$$
$$=\frac{f(x+2y)\left(4-\frac{4x-y}{x+2y}\right)-(4x-y)\left(1+2\cdot\frac{4x-y}{x+y}\right)(x+2y)^2}$$
$$=\frac{(x+2y)\left[\frac{4x+8y-4x+y}{x+2y}\right]-(4x-y)\left[\frac{x+2y+8x-2y}{x+2y}\right]}{(x+2y)^2}$$
$$=\frac{(x+2y)9y-(4x-y)9x}{(x+2y)^3}$$
$$=\frac{9xy+18y^2-36x^2+9xy}{(x+2y)^3}$$
$$=\frac{18y^2+18xy-36x^2}{(x+2y)^3}$$
$$=\frac{18(y^2+xy-2x^2)}{(x+2y)^3}$$

33. $x^2y'+2xy=0$;
$$y'=\frac{-2xy}{x^2}=-\frac{2y}{x}$$
$$y''=\frac{x(-2y')+(2y\cdot1)}{x^2}$$
$$=\frac{-2x\cdot\left(\frac{-2y}{x}\right)+2y}{x^2}$$
$$=\frac{4y+2y}{x^2}=\frac{6y}{x^2}$$

35. (a) $N'(t)=10t-0.08\left(\frac{5}{2}\right)t^{3/2}=10t-0.2t^{3/2}$,
(b) $N''(t)=10-0.2\left(\frac{3}{2}\right)t^{1/2}=10-0.3t^{1/2}$

36. (a) $T'(t)=-0.04(3)t^2+1.16(2)t-9.3=-0.12t^2+2.32t-9.3$,
(b) $T''(t)=-0.12(2)t+2.32=-0.24t+2.32$

37. (a) $h'(t)=120-24t$,
(b) $h''(t)=-24$

38. (a) $s'(t)=3(2)t-\frac{1}{60}(3t^2)=6t-\frac{1}{20}t^2$ or $6t-0.05t^2$,
(b) $s''(t)=6-\frac{1}{10}t=6-0.1t$

CHAPTER 23 REVIEW

1. Step 1: $2(x+h)^2-(x+h)=2x^2+4xh+2h^2-x-h$
Step 2: $[2x^2+4xh+2h^2-x-h]-[2x^2-x]=4xh+2h^2-h$
Step 3: $\frac{4xh+2h^2-h}{h}=4x-1+2h$.
Step 4: $\lim_{h\to0}(4x-1+2h)=4x-1=f'(x)$

2. Step 1: $(x+h)^3-1=x^3+3x^2h+3xh^2+h^3-1$
Step 2: $[x^3+3x^2h+3xh^2+h^3-1]-[x^3-1]=3x^2h+3xh^2+h^3$
Step 3: $\frac{3x^2h+3xh^2+h^3}{h}=3x^2+3xh+h^2$.
Step 4: $\lim_{h\to0}(3x^2+3xh+h^2)=3x^2=g'(x)$

3. Step 1: $f(x+h) = (x+h) - 2(x+h)^3 =$
$x + h - 2x^3 - 6x^2h - 6xh^2 - 2h^3$
Step 2: $[x + h - 2x^3 - 6x^2h - 6xh^2 - 2h^3] -$
$(x - 2x^3) = h - 6x^2h - 6xh^2 - 2h^3$
Step 3: $\frac{h - 6x^2h - 6xh^2 - 2h^3}{h} = 1 - 6x^2 - 6xh - 2h^2.$
Step 4: $\lim_{h \to 0}(1 - 6x^2 - 6xh - 2h^2) = 1 - 6x^2$
$f'(x) = 1 - 6x^2$

4. Step 1: $k(x+h) = \frac{4}{(x+h)^2}$
Step 2: $\frac{4}{(x+h)^2} - \frac{4}{x^2} = \frac{4x^2 - 4(x+h)^2}{x^2(x+h)^2}$
$= \frac{4x^2 - 4x^2 - 8xh - 4h^2}{x^2(x+h)^2} = \frac{-8xh - 4h^2}{x^2(x+h)^2}$
Step 3: $\frac{-8xh - 4h^2}{x^2(x+h)^2} \cdot \frac{1}{h} = \frac{-8x - 4h}{x^2(x+h)^2}$
Step 4: $\lim_{h \to 0} \frac{-8x - 4h}{x^2(x+h)^2} = \frac{-8x - 0}{x^2(x+0)^2} = \frac{-8x}{x^4} = \frac{-8}{x^3}$
$k'(x) = \frac{-8}{x^3}$

5. Step 1: $m(x+h) = \sqrt{x+h}$
Step 2: $m(x+h) - m(x) = \sqrt{x+h} - \sqrt{x}.$
Step 3: $\frac{\sqrt{x+h} - \sqrt{x}}{h} \cdot \frac{\sqrt{x+h} + \sqrt{x}}{\sqrt{x+h} + \sqrt{x}} = \frac{x+h-x}{h(\sqrt{x+h} + \sqrt{x})} =$
$\frac{h}{h(\sqrt{x+h} + \sqrt{x})} = \frac{1}{\sqrt{x+h} + \sqrt{x}}$
Step 4: $\lim_{h \to 0} \frac{1}{\sqrt{x+h} + \sqrt{x}} = \frac{1}{\sqrt{x} + \sqrt{x}} = \frac{1}{2\sqrt{x}}$
$m'(x) = \frac{1}{2\sqrt{x}}$

6. Step 1: $j(t+h) = \sqrt{t+h+4}$
Step 2: $\sqrt{t+h+4} - \sqrt{t+4}$
Step 3: $\frac{\sqrt{t+h+4} - \sqrt{t+4}}{h} \cdot \frac{\sqrt{t+h+4} + \sqrt{t+4}}{\sqrt{t+h+4} + \sqrt{t+4}}$
$= \frac{(t+h+4) - (t+4)}{h(\sqrt{t+h+4} + \sqrt{t+4})} = \frac{1}{\sqrt{t+h+4} + \sqrt{t+4}}$
Step 4: $\lim_{h \to 0} \frac{1}{\sqrt{t+h+4} + \sqrt{t+4}} = \frac{1}{\sqrt{t+4} + \sqrt{t+4}} = \frac{1}{2\sqrt{t+4}}$
$j'(t) = \frac{1}{2\sqrt{t+4}}$

7. $f'(x) = 6x + 2$

8. $g'(x) = -20x^{-6} + 2$

9. $H'(x) = 15x^4 - 12x^{-4} + \frac{3}{2}x^{-1/2} = 15x^4 - \frac{12}{x^4} + \frac{3}{2\sqrt{x}}$

10. $y' = \frac{(x^2+1)(0) - 1(2x)}{(x^2+1)^2} = \frac{-2x}{(x^2+1)^2}$

11. $y' = \frac{(x+3)3 - 3x \cdot 1}{(x+3)^2} = \frac{3x + 9 - 3x}{(x+3)^2} = \frac{9}{(x+3)^2}$

12. Method 1: Expand the given function as $y = (2x+1)(x^2+3) = 2x^3 + x^2 + 6x + 3$; and the differentiating produces $y' = 6x^2 + 2x + 6$. Method 2: Use the Product rule with the result: $y' = (2x+1)(2x) + 2(x^2+3) = 4x^2 + 2x + 2x^2 + 6 = 6x^2 + 2x + 6$.

13. $y' = (x-4)(3x^2 - 3) + (1)(x^2 - 3x)$
$= 3x^3 - 12x^2 - 3x + 12 + x^3 - 3x$
$= 4x^3 - 12x^2 - 6x + 12$

14. $F'(x) = \dfrac{x \cdot \frac{1}{2}(x+1)^{-1/2} - \sqrt{x+1}}{x^2}$
$= \dfrac{\frac{x}{2\sqrt{x+1}} - \sqrt{x+1}}{x^2} = \dfrac{\frac{x}{2\sqrt{x+1}} - \frac{2(x+1)}{2\sqrt{x+1}}}{x^2}$
$= \dfrac{\frac{-x-2}{2\sqrt{x+1}}}{x^2} = \dfrac{-x-2}{2x^2\sqrt{x+1}}$

15. $g'(x) = \dfrac{\sqrt{x+1}(1) - (x+1) \cdot \frac{1}{2}(x+1)^{-1/2}}{\sqrt{x+1}^2}$
$= \dfrac{\sqrt{x+1} - \frac{(x+1)}{2\sqrt{x+1}}}{x+1} = \dfrac{\frac{2(x+1)}{2\sqrt{x+1}} - \frac{(x+1)}{2\sqrt{x+1}}}{x+1}$
$= \dfrac{\frac{x+1}{2\sqrt{x+1}}}{x+1} = \dfrac{1}{2\sqrt{x+1}}$

16. $G'(x) = \dfrac{\sqrt{x+1}(2x) - x^2 \cdot \frac{1}{2}(x+1)^{-1/2}}{x+1}$
$= \dfrac{\frac{2(x+1)2x}{2\sqrt{x+1}} - \frac{x^2}{2\sqrt{x+1}}}{x+1} = \dfrac{4x^2 + 4x - x^2}{2\sqrt{x+1}(x+1)}$
$= \dfrac{3x^2 + 4x}{2(x+1)^{3/2}}$

17. $h'(x) = (x^2 + 4)[(x-2)(3x^2) + (x^3 - 2)]$
$+ (2x)(x-2)(x^3 - 2)$
$= (x^2 + 4)[3x^3 - 6x^2 + x^3 - 2]$
$+ (2x^2 - 4x)(x^3 - 2)$
$= (x^2 + 4)(4x^3 - 6x^2 - 2)$
$+ (2x^2 - 4x)(x^3 - 2)$
$= 4x^5 - 6x^4 + 16x^3 - 26x^2 - 8$
$+ 2x^5 - 4x^4 - 4x^2 + 8x$
$= 6x^5 - 10x^4 + 16x^3 - 30x^2 + 8x - 8$

18. $f'(x) = (3x^2 - 4x)(15x^2 + 2)$
$+ (6x - 4)(5x^3 + 2x - 1)$
$= 45x^4 - 60x^3 + 6x^2 - 8x$
$+ 30x^4 - 20x^3 + 12x^2 - 14x + 4$
$= 75x^4 - 80x^3 + 18x^2 - 22x + 4$

19. $F'(x) = \frac{1}{2}[(x^3 + 7)(x + 5)]^{-1/2}$
$\cdot [(x^3 + 7) + 3x^2(x + 5)]$
$= \dfrac{4x^3 + 15x^2 + 7}{2\sqrt{(x^3 + 7)(x + 5)}}$

20. $g'(x) = 3(x + 5)^2$

21. $y' = (x^2 + 5x)^2(-2)(x^3 + 1)^{-3}(3x^2)$
$\qquad + 2(x^2 + 5x)(2x + 5)(x^3 + 1)^{-2}$

$\qquad = \dfrac{-6x^2(x^2 + 5x)^2}{(x^3 + 1)^3} + \dfrac{2(x^2 + 5x)(2x + 5)}{(x^3 + 1)^2}$

$\qquad = \dfrac{-6x^2(x^2+5x)^2 + 2(x^2+5x)(2x+5)(x^3+1)^3}{(x^3 + 1)^3}$

$\qquad = \dfrac{(x^2+5x)[(-6x^4-30x^3) + 2(2x^4+5x^3+2x+5)]}{(x^3 + 1)^3}$

$\qquad = \dfrac{(x^2+5x)(-6x^4-30x^3+4x^4+10x^3+4x+10)}{(x^3 + 1)^3}$

$\qquad = \dfrac{(x^2 + 5x)(-2x^4 - 20x^3 + 4x + 10)}{(x^3 + 1)^3}$

22. $h'(x) = 4(x^2 + 7x - 1)^3(2x + 7)$

23. $y' = (x + 1)^2 3(x - 1)^2 + 2(x + 1)(x - 1)^3$
$\qquad = (x + 1)(x - 1)^2[3(x + 1) + 2(x - 1)]$
$\qquad = (x + 1)(x - 1)^2[5x + 1]$

24. $y' = (x + 4)^3 \frac{1}{2}(x - 1)^{-1/2} + 3(x + 4)^2\sqrt{x - 1}$

25. Here $y = 5(x^2 + 1)^{-3}$, and so $y' = -15(x^2 + 1)^{-4}(2x) = \frac{-30x}{(x^2+1)^4}$.

26. $y' = \dfrac{(3x + 4)^2 \cdot 4 - 4x \cdot 2(3x + 4) \cdot 3}{(3x + 4)^4}$

$\qquad = \dfrac{4(3x + 4) - 24x}{(3x + 4)^3} = \dfrac{-12x + 16}{(3x + 4)^3}$

27. Method 1: Using the Quotient rule produces $y' =$

$\qquad \dfrac{(x^2 - 2)^{3/2}(2x) - x^2 \cdot \frac{3}{2}(x^2 - 2)^{1/2} \cdot 2x}{\left(\sqrt{(x^2 - 2)^3}\right)^2}$

$\qquad = \dfrac{(x^2 - 2)^{3/2}(2x) - x^2 \cdot \frac{3}{2}(x^2 - 2)^{1/2} \cdot 2x}{(x^2 - 2)^3}$

$\qquad = \frac{(x^2-2)2x-3x^3}{(x^2-2)^{5/2}} = \frac{2x^3-4x-3x^3}{(x^2-2)^{5/2}} = \frac{-x^3-4x}{(x^2-2)^{5/2}}$
Method 2: Rewriting as a product (using negative exponents), we obtain $y = x^2(x^2 - 2)^{-3/2}$, and then, using the Product rule produces $y' = x^2(-\frac{3}{2})(x^2 - 2)^{-5/2}(2x) + 2x(x^2 - 2)^{-3/2} = (x^2 - 2)^{-5/2}[-3x^3 + 2x(x^2 - 2)] = (x^2 - 2)^{-5/2}[-x^3 - 4x]$.

28. $f'(t) = \dfrac{(t^2 - 2)(12t^2) - (4t^3 - 3)(2t)}{(t^2 - 2)^2}$

$\qquad = \dfrac{12t^4 - 24t^2 - 8t^4 + 6t}{(t^2 - 2)^2} = \dfrac{4t^4 - 24t^2 + 6t}{(t^2 - 2)^2}$

29. $g'(u) = -4(u^2 + \frac{1}{u} - \frac{4}{u^2})^{-5}(2u - \frac{1}{u^2} + \frac{8}{u^3})$

30. $R(t) = (t^2 - 2t + 1)^{1/2}(t^2 + 1)^{1/3}$

$\qquad R'(t) = \dfrac{1}{3}(t^2 + 1)^{-2/3} \cdot (2t)(t^2 - 2t + 1)^{1/2}$

$\qquad + \dfrac{1}{2}(t^2 - 2t + 1)^{-1/2}(2t - 2)(t^2 + 1)^{1/3}$

$\qquad = \dfrac{2t}{3}(t^2 - 2t + 1)^{1/2}(t^2 + 1)^{-2/3}$

$\qquad + (t - 1)(t^2 + 1)^{1/3}(t^2 - 2t + 1)^{-1/2}$

31. Here we have $\frac{dy}{du} = 3u^2 - 1$ and $\frac{du}{dx} = 2x$. Applying the Chain rule, we get

$\qquad \dfrac{dy}{dx} = (3u^2 - 1)(2x) = [3(x^2 + 1)^2 - 1](2x)$

$\qquad = (3x^4 + 6x^2 + 3 - 1)(2x)$

$\qquad = (3x^4 + 6x^2 + 2)(2x)$

$\qquad = 6x^5 + 12x^3 + 4x$

32. Here $\frac{dy}{du} = 3(u + 4)^2$ and $\frac{du}{dt} = 2$, so if we apply the Chain rule, we obtain

$\qquad \dfrac{dy}{dt} = 3(u + 4)^2 \cdot 2 = 6(2t - 5 + 4)^2$

$\qquad = 6(2t - 1)^2 = 6(4t^2 - 4t + 1)$

$\qquad = 24t^2 - 24t + 6$

33. $\dfrac{d}{dx}(x^2 + 4xy + 4y^2) = \dfrac{d}{dx}16;$

$\qquad 2x + 4xy' + 4y + 8yy' = 0$

$\qquad\qquad (4x + 8y)y' = -2x - 4y$

$\qquad\qquad y' = -\dfrac{2x + 4y}{4x + 8y} = -\dfrac{1}{2}$

34. $3x^2 - 2xy' - 2y = 2yy';$
$\qquad (-2x - 2y)y' = 2y - 3x^2;$
$\qquad\qquad y' = \dfrac{3x^2 - 2y}{2x + 2y}$

35. $2xy' + 2y - 2x = 2yy'x + y^2;$
$\qquad (2x - 2xy)y' = y^2 + 2x - 2y;$
$\qquad\qquad y' = \dfrac{y^2 + 2x - 2y}{2x - 2xy}$ or $\dfrac{2y - 2x - y^2}{2xy - 2x}$

36. $x^2y' + 2xy + y3x^2 + y'x^3 = 0;$
$\qquad (x^2 + x^3)y' = -2xy - 3x^2y;$
$\qquad\qquad y' = -\dfrac{2xy + 3x^2y}{x^2 + x^3}$
$\qquad\qquad = -\dfrac{2y + 3xy}{x + x^2}$

37. $x^2 3y^2 y' + 2xy^3 - \dfrac{y'}{y^2} = 1;$

$$\left(3x^2 y^2 - \dfrac{1}{y^2}\right) y' = 1 - 2xy^3$$

$$(3x^2 y^4 - 1)y' = y^2 - 2xy^5$$

$$y' = \dfrac{y^2 - 2xy^5}{3x^2 y^4 - 1}$$

38. $8x + x2yy' + y^2 - y'y^{-2} = 2$

$$8xy^2 + 2xy^3 y' + y^4 - y' = 2y^2$$

$$(2xy^3 - 1)y' = 2y^2 - 8xy^2 - y^4$$

$$y' = \dfrac{2y^2 - 8xy^2 - y^4}{2xy^3 - 1}$$

39. $f(x) = x^7 + 2x^4 + 3x;$

$f'(x) = 7x^6 + 8x^3 + 3$

$f''(x) = 42x^5 + 24x^2$

40. $h(x) = x^{1/2} - x^{-1/2};$

$h'(x) = \dfrac{1}{2}x^{-1/2} + \dfrac{1}{2}x^{-3/2}$

$h''(x) = -\dfrac{1}{4}x^{-3/2} - \dfrac{3}{4}x^{-5/2};$

$h'''(x) = \dfrac{3}{8}x^{-5/2} + \dfrac{15}{8}x^{-7/2}$

41. $g(x) = 4x^3 - 5x + 2x^{-1};$

$g'(x) = 12x^2 - 5 - 2x^{-2}$

$g''(x) = 24x + 4x^{-3}$

$g'''(x) = 24 - 12x^{-4}$ or $24 - \dfrac{12}{x^4}$

42. $\dfrac{dy}{dx} = 20x^4 + 6x^{-4}; \dfrac{d^2 y}{dx^2} = 80x^3 - 24x^{-5}$

43. $\dfrac{dy}{dx} = 21x^2 - 21x^{-4};$

$\dfrac{d^2 y}{dx^2} = 42x + 84x^{-5}$

$\dfrac{d^3 y}{dx^3} = 42 - 420x^{-6}$

44. $\dfrac{dy}{dx} = -3x^{-4} - 4x^{-2};$

$\dfrac{d^2 y}{dx^2} = 12x^{-5} + 8x^{-3}$

$\dfrac{d^3 y}{dx^3} = -60x^{-6} - 24x^{-4}$

45. $D_x f(x) = 15x^2 + \dfrac{1}{2}x^{-1/2};$

$D_x^2 f(x) = 30x - \dfrac{1}{4}x^{-3/2}$

$D_x^3 f(x) = 30 + \dfrac{3}{8}x^{-5/2}$

46. $\dfrac{d}{dx}g(x) = 3(x^2 + 1)^2(2x) = 6x(x^2 + 1)^2$

$\dfrac{d^2}{dx^2}g(x) = 12x(x^2 + 1)(2x) + 6(x^2 + 1)^2$

$$= 24x^4 + 24x^2 + 6x^4 + 12x^2 + 6$$

$$= 30x^4 + 36x^2 + 6$$

$\dfrac{d^3}{dx^3}g(x) = 120x^3 + 72x$

47. Using implicit differentiation, we get $x^2 y' + 2xy = y' + 2$ or $(x^2 - 1)y' = 2 - 2xy$. Solving for y', we obtain $y' = \dfrac{2-2xy}{x^2-1}$. Differentiating again yields the following:

$$y'' = \dfrac{(x^2 - 1)(-2xy' - 2y) - (2 - 2xy)2x}{(x^2 - 1)^2} =$$

$$\dfrac{(x^2 - 1)\left(-2x \cdot \frac{2-2xy}{x^2-1} - 2y\right) - (2 - 2xy)2x}{(x^2 - 1)^2} =$$

$$\dfrac{-2x(2-2xy)-(x^2-1)2y-2x(2-2xy)}{(x^2 - 1)^2} =$$

$$\dfrac{-4x(2 - 2xy) - (x^2 - 1)2y}{(x^2 - 1)^2} =$$

$$\dfrac{-8x + 8x^2 y - 2x^2 y + 2y}{(x^2 - 1)^2} = \dfrac{6x^2 y - 8x + 2y}{(x^2 - 1)^2}$$

48.
$$2yy' = 2xyy' + y^2 + 3x^2$$

$$(2y - 2xy)y' = y^2 + 3x^2$$

$$y' = \dfrac{y^2 + 3x^2}{2y - 2xy}$$

49. Using implicit differentiation, we get $xy' + y = 2x + 2yy'$ or $(x - 2y)y' = 2x - y$. Solving for y', we obtain $y' = \dfrac{2x-y}{x-2y}$. Differentiating again yields the following:

$$y'' = \dfrac{(x - 2y)(2 - y') - (2x - y)(1 - 2y')}{(x - 2y)^2} =$$

$$\dfrac{2x-xy'-4y+2yy'-2x+4xy'+y-2yy'}{(x - 2y)^2} =$$

$$\dfrac{3xy' - 3y}{(x - 2y)^2} = \dfrac{3x\left(\frac{2x-y}{x-2y}\right) - 3y}{(x - 2y)^2} =$$

$$\dfrac{3x(2x - y) - 3y(x - 2y)}{(x - 2y)^3} =$$

$$\dfrac{6x^2 - 3xy - 3xy + 6y^2}{(x - 2y)^3} = \dfrac{6x^2 - 6xy + 6y^2}{(x - 2y)^3}$$

50. $y' = 8x - 9;$ at $x = 3, y' = 8 \cdot 3 - 9 = 24 - 9 = 15$. The slope of the tangent is 15 and the slope of the normal is $-\dfrac{1}{15}$.

51. $f'(x) = 21x^2 - 3;$ at $x = -1, f'(-1) = 21 - 3 = 18$. The slope of the tangent is 18.

52. $8x + 18yy' = 0; y' = \frac{-8x}{18y} = \frac{-4x}{9y}$; at $(-1, 2)$,

$y' = \frac{-4(-1)}{9 \cdot 2} = \frac{4}{18} = \frac{2}{9}$. The slope of the tangent

is $\frac{2}{9}$.

53. Finding y' by implicit differentiation, we have
$$8x + 6yy' - 5y - 5xy' - 4 + 10y' = 0$$
$$(6y - 5x + 10)y' = 5y + 4 - 8x$$
$$y' = \frac{5y + 4 - 8x}{6y - 5x + 10}$$

At $(2, -1)$, we see that
$$y' = \frac{5(-1) + 4 - 8(2)}{6(-1) - 5 \cdot 2 + 10}$$
$$= \frac{-5 + 4 - 16}{-6 - 10 + 10} = \frac{-17}{-6} = \frac{17}{6}$$

The slope of the tangent is $\frac{17}{6}$ and the slope of the

normal is $-\frac{6}{17}$.

54. $f'(x) = 3x^2 - 3; 3x^2 - 3 = 0 \Rightarrow x^2 = 1 \Rightarrow x = \pm 1$

55. $g'(x) = \frac{(x+1)2x - x^2 \cdot 1}{(x+1)^2} = \frac{2x^2 + 2x - x^2}{(x+1)^2}$

$= \frac{x^2 + 2x}{(x+1)^2}$

$g'(x) = 0$ when $x^2 + 2x = 0$ so
$x(x + 2) = 0 \Rightarrow x = 0$ or $x = -2$

56. $y' = 2x$, at $(-1, 5), y' = -2$. Equation of tangent is $y - 5 = -2(x+1)$ or $y = -2x + 3$ or $2x + y - 3 = 0$

57. Finding y' by implicit differentiation, we have
$8x + 10yy' = 0$;
$$y' = \frac{-4x}{5y}$$

At $(-2, 2)$, we see that $y' = \frac{-4(-2)}{5(2)} = \frac{4}{5}$.

Slope of the normal is $-\frac{5}{4}$. Equation of the normal

is $y - 2 = -\frac{5}{4}(x + 2)$ or $4y - 8 = -5x - 10$ or
$5x + 4y + 2 = 0$

58. $y' = \frac{1}{2}(x + 4)^{-1/2} = \frac{1}{2\sqrt{x+4}}$. This is never equal to zero so the answer is none.

59. $f(x) = 3x^4 - 6x^2 + 2; f'(x) = 12x^3 - 12x$
$f''(x) = 36x^2 - 12$

Now, $36x^2 - 12 = 0 \Rightarrow x^2 = \frac{1}{3}$ and so, $x = \pm\sqrt{\frac{1}{3}} = \frac{\pm\sqrt{3}}{3}$

60. $y' = 3x^2 - 18x; y'' = 6x - 18$. Now, $y'' = 0$ means that $6x - 18 = 0 \Rightarrow x - 3 = 0$ or $x = 3$.

CHAPTER 23 TEST

1. $f'(x) = 15x^2$

2. $g'(x) = -12x^{-5} + 10x$

3. Rewrite as $h(x) = (x^6 - 2)^{1/3}$ and, using the Chain Rule, we obtain $h'(x) = \frac{1}{3}(x^6 - 2)^{-2/3}(6x^5) = 2x^5(x^6 - 2)^{-2/3}$

4. Using the Product rule, we get $j'(x) = (2x + 1)(6x - 1) + (3x^2 - x)(2) = 12x^2 + 4x - 1 + 6x^2 - 2x = 18x^2 + 2x - 1$

5. Using the Quotient rule produces
$$k'(x) = \frac{(4x^2+1)^{1/2}(2) - (2x+1)\frac{1}{2}(4x^2+1)^{-1/2}(8x)}{[(4x^2+1)^{1/2}]^2}$$
$$= \frac{(4x^2+1)^{-1/2}[(4x^2+1)(2) - (2x+1)4x]}{4x^2+1}$$
$$= \frac{8x^2 + 2 - 4x(2x+1)}{(4x^2+1)^{3/2}} = \frac{2 - 4x}{(4x^2+1)^{3/2}}$$

6. $f'(x) = 20x^3 + 12x^{-4}; f''(x) = 60x^2 - 48x^{-5};$
$f'''(x) = 120x + 240x^{-6}$

7. Using implicit differentiation, we obtain $2y + 2xy' + 3y^2xy' + y^3 = 3x^2$. Rewriting, we get $(2x + 3y^2x)y' = 3x^2 - 2y - y^3$ and so $y' = \frac{3x^2 - 2y - y^3}{2x + 3y^2x}$.

8. Taking the derivative of g we obtain $g'(x) = 12x^2 - 4x$. When $x = 1, g'(1) = 8$, and so the slope of the tangent to the curve at $(1, 6)$ is 8. Using the point-slope form for the equation of a line, we see the equation of the tangent line is $y - 6 = 8(x - 1)$ or $y = 8x - 2$.

9. $v(t) = s'(t) = 4.5(2t) + 27 = 9t + 27$

10. (a) Using the quotient rule produces
$P'(t) = \frac{(2\sqrt{t}+7)(2t-5) - (t^2-5t)(t^{-1/2})}{(2\sqrt{t}+7)^2}$

Multiplying by $\frac{\sqrt{t}}{\sqrt{t}}$, we obtain
$$P'(t) = \frac{(2t + 7\sqrt{t})(2t - 5) - (t^2 - 5t)}{(2\sqrt{t} + 7)^2\sqrt{t}}$$
$$= \frac{4t^2 + 14t^{3/2} - 10t - 35\sqrt{t} - t^2 + 5t}{(2\sqrt{t} + 7)^2\sqrt{t}}$$
$$= \frac{3t^2 + 14t^{3/2} - 5t - 35\sqrt{t}}{(2\sqrt{t} + 7)^2\sqrt{t}}$$

(b) $P'(5) = \frac{3(5)^2 + 14(5)^{3/2} - 5(5) - 35\sqrt{(5)}}{(2\sqrt{(5)} + 7)^2\sqrt{(5)}} \approx 0.4358$ thousand bacteria/ hour = 435.8 bacteria/hour.

CHAPTER

24

Applications of Derivatives

≡ **24.1** RATES OF CHANGE

1. $s(t) = 3t^2 - 12t + 5; v(t) = s'(t) = 6t - 12; a(t) = s''(t) = 6$

t	0	1	2	3	4	5
$s(t)$	5	−4	−7	−4	5	20
$v(t)$	−12	−6	0	6	12	18
$a(t)$	6	6	6	6	6	6

2. $s(t) = 3t^2 - 18t + 10; v(t) = s'(t) = 6t - 18; a(t) = s''(t) = 6$

t	0	1	2	3	4
$s(t)$	10	−5	−14	−17	−14
$v(t)$	−18	−12	−6	0	6
$a(t)$	6	6	6	6	6

3. $s(t) = t^3 - 12t + 2; v(t) = s'(t) = 3t^2 - 12; a(t) = s''(t) = 6t$

t	−4	−3	−2	−1	0	1	2	3	4
$s(t)$	−14	11	18	13	2	−9	−14	−7	18
$v(t)$	36	15	0	−9	−12	−9	0	15	36
$a(t)$	−24	−18	−12	−6	0	6	12	18	24

4. $s(t) = t^3 - 27t + 20; v(t) = 3t^2 - 27; a(t) = s''(t) = 6t$

t	−4	−3	−2	−1	0	1	2	3	4
$s(t)$	64	74	66	46	20	−6	−26	−34	−24
$v(t)$	21	0	−15	−24	−27	−24	−15	0	21
$a(t)$	−24	−18	−12	−6	0	6	12	18	24

5. $s(t) = t + 4/t; v(t) = 1 - 4/t^2; a(t) = 8/t^3$

t	1	2	3	4
$s(t)$	5	4	4.3333	5
$v(t)$	-3	0	0.5555	0.75
$a(t)$	8	1	0.2963	0.125

6. $s(t) = t + 8/t; v(t) = 1 - 8/t^2; a(t) = 16/t^3$

t	1	2	3	4
$s(t)$	9	6	5.6667	6
$v(t)$	-7	-1	0.1111	0.5
$a(t)$	16	2	0.5926	0.25

7. $s(t) = 2\sqrt{t} + \frac{1}{\sqrt{t}} = 2t^{1/2} + t^{-1/2}; v(t) = t^{-1/2} - \frac{1}{2}t^{-3/2}; a(t) = -\frac{1}{2}t^{-3/2} + \frac{3}{4}t^{-5/2}$

t	1	2	3	4
$s(t)$	3	3.5355	4.0415	4.5
$v(t)$	0.5	0.5303	0.4811	0.4375
$a(t)$	0.25	-0.0442	-0.0481	-0.0391

8. $s(t) = t^3 - 8; v(t) = 3t^2; a(t) = 6t$

t	0	1	2	3	4
$s(t)$	-8	-7	0	19	56
$v(t)$	0	3	12	27	48
$a(t)$	0	6	12	18	24

9. **(a)** The object hits the ground when $s(t) = 0$. Solving $144t - 16t^2 = 0$ we get $t(144 - 16t) = 0$ so $t = 0$ or $t = 9$. The object is fired at $t = 0$ and hits the ground at $t = 9$ s.
(b) $v(t) = 144 - 32t; v(9) = 144 - 32 \cdot 9 = -144$ ft/s; $a(t) = -32; a(9) = -32$ ft/s^2.
(c) $s(0) = 0$ ft = height when fired.

10. **(a)** Solving for t when $250 + 256t - 16t^2 = 0$. The quadratic formula produces $t = -0.9233$ or $t = 16.9233$. Since t cannot be negative it hits the ground at 16.9233 s.
(b) $v(t) = 256 - 32t; v(16.9233) = -285.55$ ft/s; $a(t) = -32; a(16.9233) = -32$ ft/s^2.
(c) $s(0) = 250$ ft = height when fired.

11. **(a)** $s(t) = 29.4t - 4.9t^2 = 0 \Rightarrow t(29.4 - 4.9t) = 0$. By the zero product property, $t = 0$ or $t = \frac{29.4}{4.9} = 6$. Since t must be greater than 0, it hits the ground at 6 s.
(b) $v(t) = 29.4 - 9.8t; v(6) = -29.4$ m/s; $a(t) = -9.8; a(6) = -9.8$ m/s^2.
(c) $s(0) = 0$ m is the initial height.

12. **(a)** $s(t) = 180 + 98t - 4.9t^2 = 0$. Solving for t using the quadratic formula yields $t = -1.6934$ and $t = 21.6934$. Since $t > 0$ we get 21.6934 s.
(b) $v(t) = 98 - 9.8t$ means that $v(21.6934) = -114.595$ m/s, and $a(t) = -9.8$ gives $a(21.6934) = -9.8$ m/s^2.
(c) $s(0) = 180$ m is the initial height.

13. **(a)** The rate of change of $F(s)$ is $\frac{dF}{ds} = F'(s) = -100/s^3$ dynes/cm,
(b) $F'(3) = -\frac{100}{27}$ dynes/cm

14. **(a)** $i = \frac{dq}{dt} = 10 - 4.0t$,
(b) $q(0) = 25$ C,
(c) $i(0) = 10$ A; $i(4) = 10 - 16 - 6$ A

15. **(a)** $\frac{dF}{dC} = \frac{9}{5}$,
(b) $\frac{dC}{dF} = \frac{d}{dF}\left(\frac{5}{9}(F - 32)\right) = \frac{5}{9}$,
(c) $\frac{dC}{dF}$ at $C = 20°$ is $\frac{5}{9}$

16. **(a)** Since the area of a circle is related to (is a function of) its radius, we can think of $A = \pi r^2$ as $A(r) = \pi r^2$ and thus, the rate of change of the area with respect to the radius is $\frac{dA}{dr} = A'(r) = 2\pi r$ cm^2/cm;
(b) at $r = 1.5$, $A'(1.5) = 2\pi(1.5) = 3\pi \approx 9.4248$ cm^2/cm;
(c) at $r = 3$, we have $A'(3) = 2\pi \cdot 3 = 6\pi \approx 18.8496$ cm^2/cm.

17. (a) The rate of change in the quantity of water is $\frac{dQ}{dt} = Q'(t) = -50 + t$ gal/min,

(b) $Q'(4) = -50 + 4 = -46$ gal/min,

(c) $Q'(8) = -50 + 8 = -42$ gal/min.

18. (a) The rate of change of the distance with respect to time is called the velocity. Thus, $v(t) = s'(t) = 10t + 2$ mm/s;

(b) The acceleration is the rate of change of the velocity, and so $a(t) = v'(t) = s''(t) = 10$ mm/s^2;

(c) $s(3) = 5 \cdot 3^2 + 3 \cdot 2 = 45 + 6 = 51$ mm; $v(3) = 10 \cdot 3 + 2 = 32$ mm/s; $a(3) = 10$ mm/s^2;

(d) $10t + 2 = 46$ or $10t = 44$, and so $t = 4.4$ s.

19. (a) The velocity, v, at any time t is given by $v(t) = s'(t) = 8.0 - 4.0t$. The initial velocity is when $t = 0$, and so the initial velocity is $v(0) = 8.0$ m/s.

(b) When the ball stops rolling its velocity is 0. Thus, we want the solution to $v(t) = 0$ or $8.0 - 4.0t = 0$ or $t = 2$ s. When $t = 2$, the ball's position is $s(2) = 8.0 \cdot 2 - 2.0 \cdot 2^2 = 16 - 8 = 8$ m;

(c) $s(t) = 0$ means that $8.0t - 2.0t^2 = 0$ and so $t(8 - 2t) = 0$. By the zero product property, this means that $t = 0$ or $t = 4$. It will take 4.0 s to return to its initial position.

20. The area of a circle is $A(r) = \pi r^2$ and its circumference $C = 2\pi r$. So, $r = \frac{C}{2\pi}$ and substitution produces the area of a circle as a function of its circumference $A(C) = \frac{C^2}{4\pi}$. The rate of change of the area with respect to its circumference is $A'(C)$ or $\frac{dA}{dC} = \frac{C}{2\pi}$.

21. (a) $v = -N\frac{d\phi}{dt} = -50\frac{d(2.4t^{1/2} - 0.4t^2)}{dt} = -50$ $(1.2t^{-1/2} - 0.8t); v(t) = -60t^{-1/2} + 40t$;

(b) $v(1.0) = \frac{-60}{\sqrt{1}} + 40 \cdot 1 = -60 + 40 = -20$ V and $v(9.0) = \frac{-60}{\sqrt{9}} + 40 \cdot 9 = \frac{-60}{3} + 360 = 340$ V.

22. (a) The formula for the volume of a cone is $V = \frac{1}{3}\pi r^2 h$. We are given $h = 2r$, so by substituting, we can write the volume as a function of the radius $V(r) = \frac{1}{3}\pi r^2 \cdot 2r = \frac{2}{3}\pi r^3$. Thus, $\frac{dV}{dr} = 2\pi r^2$;

(b) $\frac{dV}{dr}(24.3) = 2\pi(24.3)^2 \approx 3710$ mm^2.

23. $A = \pi r^2$ and $r = 1.7t$ m so $A = 1.7^2\pi t^2 = 2.89\pi t^2$ m^2. Thus, $\frac{dA}{dt} = 5.78\pi t$ and when $t = 8$, the area is increasing at the rate of 46.24π m^2/s.

24. (a) $i = \frac{V}{r} = \frac{t^2 - 1.2}{3.0t^{1/2}}$;

(b) In (a), we expressed i as a function of t, thus the derivative is

$$\frac{di}{dt} = \frac{3.0t^{1/2}(2t) - (t^2 - 1.2)\frac{3}{2}t^{-1/2}}{9t}$$

$$= \frac{6t^{3/2} - \frac{3}{2}t^{3/2} + 1.8t^{-1/2}}{9t}$$

$$= \frac{\frac{9}{2}t^{3/2}}{9t} + \frac{1.8t^{-1/2}}{9t}$$

$$= \frac{1}{2}t^{1/2} + 0.2t^{-3/2}$$

$$= \frac{t^{1/2}}{2.0} + \frac{1}{5.0t^{3/2}}$$

(c) At $t = 2.4$ s this is $\frac{\sqrt{2.4}}{2.0} + \frac{1}{5.0\sqrt{2.4^3}} \approx 0.83$ A/s.

25. (a) $P = \frac{k}{V}$ or $5.21 = \frac{k}{1.23}$ so $k = 6.4083 \approx 6.41$. Hence, $P = \frac{6.41}{V}$;

(b) $\frac{dP}{dV} = -6.41V^{-2}$ or $\frac{-6.41}{V^2}$.

26. (a) $\omega(t) = \theta'(t) = 3.4 - 2.4t$ rad/s;

(b) $\alpha(t) = \theta''(t) = -2.4$ rad/s^2;

(c) $\omega(0) = 3.4$ rad/s; $\alpha(0) = -2.4$ rad/s^2

27. (a) $v(t) = s'(t) = 6t - 0.05t^2$,

(b) $a(t) = v'(t) = s''(t) = 6 - 0.1t$,

(c) $|v(t)| = |6t - 0.05t^2|$

28. (a) $v(t) = h'(t) = 120 - 24t$,

(b) $a(t) = v'(t) = h''(t) = -48$

≡ 24.2 EXTREMA AND THE FIRST DERIVATIVE TEST

1. $f'(x) = 2x - 8$, so we have $2x - 8 = 0$, and $x = 4$ is the critical value. On the interval $(-\infty, 4)$, $f'(0) = -8$ so decreasing. On the interval $(4, \infty)$, $f'(5) = 10 - 8 = 2$, so f is increasing. No maximums; minimum at $x = 4$, $f(4) = 16 - 32 = -16$.

2. $g'(x) = 10 - 2x$, so we have $10 - 2x = 0$, and $x = 5$ is the critical value. On the interval $(-\infty, 5)$, $g'(0)$ is positive, so g is increasing. On the interval $(5, \infty)$, $g'(6)$ is negative, and so g is decreasing. Maximum at $x = 5$; $g(5) = 50 - 25 = 25$. No minimum.

3. $m'(x) = -2x$; $-2x = 0 \Rightarrow 0$ is the critical value. On the interval $(-\infty, 0)$, $m'(-1)$ is positive so increasing. On the interval $(0, \infty)$, $m'(1)$ is negative so decreasing. Maximum at $x = 0$; $m(0) = 16$. No minimum.

4. $h'(x) = 6x - 12; 6x - 12 = 0 \Rightarrow x = 2$ is the critical value. On $(-\infty, 2), h'(0)$ is negative so decreasing. On the interval $(2, \infty), h'(3)$ is positive so increasing. No maximum. Minimum at $x = 2; h(2) = -2$.

5. $j'(x) = 16 - 8x; 16 - 8x = 0 \Rightarrow x = 2$ is the critical value. On $(-\infty, 2), j'(0)$ is positive so increasing. On the interval $(2, \infty), j'(3)$ is negative so decreasing. Maximum at $x = 2; j(2) = 31$. Minimum: none.

6. $k'(x) = 2x - 4; 2x - 4 = 0 \Rightarrow x = 2$ is the critical value. On the interval $(-\infty, 2), k'(0)$ is negative so decreasing. On the interval $(2, \infty), k'(3)$ is positive so increasing. Maximum: none. Minimum: $k(2) = 7$.

7. $f'(x) = 8x - 16; x = 2$ is the critical value. On the interval $(-\infty, 2), f'(0)$ is negative so decreasing. On the interval $(2, \infty), f'(3)$ is positive so increasing. Maximum: none; Minimum: $f(2) = -15$.

8. $s'(t) = -2t + 7$ and $-2t + 7 = 0 \Rightarrow t = \frac{7}{2}$ is the critical value. On the interval $(-\infty, \frac{7}{2}), s'(0)$ is positive so increasing. On the interval $(\frac{7}{2}, \infty), s'(4)$ is negative so decreasing. Maximum: $s(\frac{7}{2}) = \frac{65}{4} = 16.25$. Minimum: none.

9. Here $q'(t) = 3t^2 - 3$ and so, $3t^2 - 3 = 3(t + 1)(t - 1) = 0$. Thus, by the zero product property, -1 and 1 are the critical values. On $(-\infty, -1), q'(-2)$ is positive so increasing. On the interval $(-1, 1)$, $q'(0)$ is negative so decreasing. On the interval $(1, \infty), q'(2)$ is positive so increasing. Maximum: $q(-1) = 2$. Minimum: $q(1) = -2$.

10. $v'(t) = 21 + 18t - 3t^2; -3t^2 + 18t + 21 = -3(t^2 - 6t - 7) = -3(t - 7)(t + 1) \Rightarrow -1, 7$ are critical values. On the interval $(-\infty, -1), v'(-2)$ is negative so decreasing. On the interval $(-1, 7), v'(0)$ is positive so increasing. On the interval $(7, \infty), v'(8)$ is negative so decreasing. Maximum: $v(7) = 245$; Minimum: $v(-1) = -11$.

11. The derivative is $y' = x^3 - 36x$. By factoring we can determine when the derivative is 0, thus $x^3 - 36x = x(x + 6)(x - 6) = 0$ and so, the critical values are at $-6, 0,$ and 6. On the interval $(-\infty, -6), y'$ is negative so decreasing. On the interval $(-6, 0), y'$ is positive so increasing. On the interval $(0, 6), y'$ is negative so decreasing. On the interval $(6, \infty), y'$ is positive so increasing. Maximum: $y(0) = 16$. Minimum: $y(-6) = y(6) = -308$.

12. $t' = -25w + w^3; w^3 - 25w = w(w + 5)(w - 5) = 0$, and so the critical values are at $-5, 0$ and 5. On the interval $(-\infty, -5), t'$ is negative so decreasing. On $(-5, 0), t'$ is positive so increasing. On the interval $(0, 5), t'$ is negative so decreasing. On the interval $(5, \infty), t'$ is positive so increasing. Maximum at 0 of 3.2; Minimum at both ± 5 of -153.05.

13. $f'(z) = 3(z + 4)^2$, so $3(z + 4)^2 = 0$ when $x = -4$. Thus, -4 is only critical value. On the interval $(-\infty, -4), f'(z)$ is positive so increasing. On $(-4, \infty), f'(z)$ is positive so increasing. Hence $f(z)$ is always increasing and there are no extrema.

14. $g'(x) = 2(2x^2 - 8)(4x) = 8x(2x^2 - 8) = 0$, and so the critical values are at $x = -2, 0, 2$. On the interval $(-\infty, -2), g'(x)$ is negative so g is decreasing. On the interval $(-2, 0), g'(x)$ is positive so g is increasing. On the interval $(0, 2), g'(x)$ is negative so g decreasing. On the interval $(2, \infty), g'(x)$ is positive, so g is increasing. Maximum: $g(0) = 64$; Minimum: $g(-2) = g(2) = 0$.

15. $g'(v) = 12v^3 - 12v^2 - 24v = 12v(v^2 - v - 2) = 12v(v - 2)(v + 1)$, so the critical values are at -1, 0, and 2. On the interval $(-\infty, -1), g'(v)$ is negative so decreasing. On the interval $(-1, 0), g'(v)$ is positive so increasing. On $(0, 2), g'(v)$ is negative so decreasing. On the interval $(2, \infty), g'(v)$ is positive so increasing. Maximum: $g(0) = 12$. Minima: $g(-1) = 7, g(2) = -20$.

16. Rewriting the given function as $y = 5x^{-1}$, we find that its derivative is $y' = -5x^{-2} = \frac{-5}{x^2}$. Since y is undefined at $x = 0$, 0 is *not* a critical value. On the interval $(-\infty, 0), y'$ is negative so y is decreasing. On the interval $(0, \infty), y'$ is negative so y is decreasing. There are no extrema.

17. $s'(t) = \frac{-6}{(t+1)^2}$. Since s is undefined at $t = -1, -1$ is *not* a critical value. On the interval $(-\infty, -1)$, $s'(t)$ is negative, so s decreasing over this interval. On the interval $(-1, \infty), s'(t)$ is negative so s is decreasing over this interval. $s(t)$ is undefined at -1. There are no extrema.

18. We see that $x = 0$ is not in the domain of h (or h'). $h'(x) = 1 - \frac{1}{x^2} = \frac{x^2 - 1}{x^2}$. Hence, $x^2 - 1 = 0 \Rightarrow x = \pm 1$ are critical values. On the interval $(-\infty, -1), h'(x)$ is positive so h increasing over this interval. On the interval $(-1, 0), h'(x)$ is negative so h is decreasing over this interval. On the interval $(0, 1), h'(x)$ is negative so h decreasing. On the interval $(1, \infty), h'(x)$ is positive so h

is increasing on this interval. $h(0)$ is undefined. Maximum: $h(-1) = -2$; Minimum: $h(1) = 2$.

19. $g(x)$ is not defined at $x = 0$. Thus, $g'(x) = 2x + \frac{2}{x^2} = \frac{2x^3+2}{x^2}$ which is 0 when $2x^3 + 2 = 0$ or $x = -1$. The critical value is -1. On the interval $(-\infty, -1)$, $g'(x)$ is negative so g is decreasing. On the interval $(-1, 0)$, $g'(x)$ is positive so g is increasing. On the interval $(0, \infty)$, $g'(x)$ is positive so g is increasing. Maximum: none; Minimum: $g(-1) = 3$.

20. $j(x)$ is not defined at -3. $j'(x) = 1 - \frac{1}{(x+3)^2} = \frac{(x+3)^2-1}{(x+3)^2} = \frac{x^2+6x+8}{(x+3)^2}$. Thus, $j'(x) = 0$, when $x^2 + 6x + 8 = (x+4)(x+2) = 0$, and so critical values are $x = -4$ and $x = -2$. On the interval $(-\infty, -4)$, $j'(x)$ is positive so $j(x)$ is increasing. On the interval $(-4, -3)$, $j'(x)$ is negative so $j(x)$ is decreasing. On $(-3, -2)$, $j'(x)$ is negative, so $j(x)$ decreasing. On the interval $(-2, \infty)$, $j'(x)$ is positive so $j(x)$ increasing. Maximum: $j(-4) = -5$; Minimum: $j(-2) = -1$.

21. $k(x)$ is undefined at $x = 1$. $k'(x) = \frac{(1-x)2x+(x^2)}{(1-x)^2} = \frac{2x-x^2}{(1-x)^2}$. Critical values are at 0 and 2. On the interval $(-\infty, 0)$, $k'(x)$ is negative so $k(x)$ decreasing. On the interval $(0, 1)$, $k'(x)$ is positive, so $k(x)$ increasing. On the interval $(1, 2)$, $k'(x)$ is positive, so $k(x)$ increasing. On the interval $(2, \infty)$, $k'(x)$ is negative, so $k(x)$ is decreasing. Maximim: $k(2) = -4$; Minimum: $k(0) = 0$.

22. Here $m'(b) = \frac{2}{\sqrt{b}}$ an a critical number is $b = 0$. Domain of m is $[0, \infty)$. On the interval $(0, \infty)$, $m'(b)$ is positive so $m(b)$ is increasing. Maximum: none; Minimum, $m(0) = 0$.

23. The domain of f is all t when $1 - t^2 \geq 0$ or $[-1, 1]$. The derivative is

$$f'(t) = \sqrt{1 - t^2} - t \cdot \frac{1}{2}(1 - t^2)^{-1/2}(2t)$$

$$= \sqrt{1 - t^2} - \frac{t^2}{\sqrt{1 - t^2}}$$

$$= \frac{1 - t^2 - t^2}{1\sqrt{1 - t^2}}$$

$$= \frac{1 - 2t^2}{\sqrt{1 - t^2}}$$

Critical values are when $1 - 2t^2 = 0$, or $t = \pm\frac{\sqrt{2}}{2}$ and when $1 - t^2 = 0$ or $t = \pm 1$. The domain of $f(t)$ is $[-1, 1]$. On the interval $(-1, -\frac{\sqrt{2}}{2})$, $f'(t)$ is negative, so $f(t)$ decreasing. On the interval $(-\frac{\sqrt{2}}{2}, \frac{\sqrt{2}}{2})$, we see that $f'(t)$ is positive, so $f(t)$

increasing. On the interval $(\frac{\sqrt{2}}{2}, 1)$, $f'(t)$ is negative, so $f(x)$ decreasing. Maximum: $f(-1) = 0$ and $f(\frac{\sqrt{2}}{2}) = 0.5$; Minimum: $f(-\frac{\sqrt{2}}{2}) = -0.5$ and $f(1) = 0$.

24. $g'(s) = 2(s+2)(s-1)^{2/3} + \frac{2}{3}(s-1)^{-1/3}(s+2)^2$

$$= \frac{2(s+2)(s-1)^{2/3}}{1} + \frac{2(s+2)^2}{3(s-1)^{1/3}}$$

$$= \frac{3 \cdot 2(s+2)(s-1) + 2(s+2)^2}{3(s-1)^{1/3}}$$

$$= \frac{6(s^2+s-2) + 2(s^2+4s+4)}{3\sqrt[3]{s-1}}$$

$$= \frac{8s^2 + 14s - 4}{3\sqrt[3]{s-1}}$$

$$= \frac{2(4s^2+7s-2)}{3\sqrt[3]{s-1}} = \frac{2(4s-1)(s+2)}{3\sqrt[3]{s-1}}.$$

The critical values are $\frac{1}{4}$, -2, and 1. On the interval $(\infty, -2)$, we see that $g'(s)$ is negative so g is decreasing. On $(-2, \frac{1}{4})$, $g'(s)$ is positive so g is increasing. On the interval $(\frac{1}{4}, 1)$, $g'(s)$ is negative, so g is decreasing. On $(1, \infty)$, $g'(s)$ is positive so g is increasing. Maximum: $g(\frac{1}{4}) = 4.179$; Minimum: $g(-2) = 0; g(1) = 0$.

25. $y' = (x-1)^{2/3} + \frac{2}{3}(x-1)^{-1/3} \cdot x$

$$= (x-1)^{2/3} + \frac{2x}{3(x-1)^{1/3}}$$

$$= \frac{3(x-1) + 2x}{3(x-1)^{1/3}}$$

$$= \frac{5x-3}{3(x-1)^{2/3}}$$

The critical values are 0.6 and 1. On the interval $(-\infty, 0.6)$, y' is positive so y is increasing. On the interval $(0.6, 1)$, y' is negative so y is decreasing. On the interval $(1, \infty)$, y' is positive so y is increasing. Maximum: $(0.6, 0.326)$; Minimum: $(1, 0)$.

26. The derivative is $f'(x) = 3x^{-2/3} - 4 = \frac{3}{x^{2/3}} - 4 = \frac{3-4x^{2/3}}{x^{2/3}}$. The derivative is 0 when the numerator is 0, or when $3 - 4x^{2/3} = 0$, and so $4x^{2/3} = 3$, or $2x^{1/3} = \pm\sqrt{3}$, or $x^{1/3} = \pm\frac{\sqrt{3}}{2}$, and so $x = \pm(\frac{\sqrt{3}}{2})^3 = \frac{\pm3\sqrt{3}}{8}$. The derivative is undefined when the denominator is 0, or when $x^{2/3} = 0$, or $x = 0$. Thus, the critical values are 0 and $\pm\frac{3\sqrt{3}}{8}$. On the interval $(-\infty, \frac{-3\sqrt{3}}{8})$, f' is negative so f is decreasing. On the interval $(\frac{-3\sqrt{3}}{8}, 0)$, f' is positive, so f increasing. On the interval $\left(0, \frac{3\sqrt{3}}{8},\right)$, f' is positive, so f increasing. On the interval $\left(\frac{3\sqrt{3}}{8}, \infty\right)$, f' is negative, so f decreasing. Maximum: $f\left(\frac{3\sqrt{3}}{8}\right) = 5.19615$; Minimum: $f\left(\frac{-3\sqrt{3}}{9}\right) = -5.19615$.

27. (a) $s(t) = 250 + 256t - 16t^2$; $s(t) = 256 - 32t$; Critical value when $256 - 32t = 0$ or $t = 8$. On the interval $[0, 8)$, $s'(t)$ is positive so increasing. On the interval $(8, \infty)$, $s'(t)$ is negative, so s is decreasing. Maximum at 8 s
(b) $s(8) = 1274$ ft.

28. $s(t) = 180 + 98t - 4.9t^2$; $s'(t) = 98 - 9.8t$; The critical value is 10 and this gives a maximum
(a) 10 s
(b) $s(10) = 670$ m.

29. $P(t) = 4.7(t - 2.0)^2$; $P'(t) = 9.4(t - 2.0)$
(a) The critical value is 2.0 s and this will give the lowest power $P(2) = 0.0$ W
(b) This occurs at 2.0 s.

30. $i(t) = 4.8t - 1.2t^2$; $i'(t) = 4.8 - 2.4t$; the critical value is $t = 2$.
(a) The current is maximum at $t = 2.0$ s.
(b) $i(2) = 4.8$ A.

31. $R = \dfrac{9R_2}{9 + R_2}$;
$R' = \dfrac{(9 + R_2)9 - 9R_2(1)}{(9 + R_2)^2} = \dfrac{81 + 9R_2 - 9R_2}{(9 + R_2)^2}$
$= \dfrac{81}{(9 + R_2)^2}$.
(a) the only critical value is -9 but R_2 must be greater than or equal to 0. R' is always positive hence R is always increasing. There is no maximum and the minimum value ocurs at the minimum value for R_2 which is 0.
(b) The mimimum value for R is 0.

32. $x(t) = 8t^3 - t^4$; $x'(t) = 24t^2 - 4t^3$; $a(t) = x''(t) = 48t - 12t^2$; $a'(t) = 48 - 24t$. The critical value for

$a(t)$ is 2 which yields a maximum $v(2) = x'(2) = 24 \cdot 2^2 - 4 \cdot 2^3 = 64$

33. (a) Using the product rule, we get $P'(n) = (4 - 0.2n)(2.5) + (-0.2)(2.5n + 7) = 10 - 0.5n - 0.5n - 1.4 = -n + 8.6$
(b) The first derivative is 0 when $n = 8.6$. Since $P'(n) > 0$ for $n < 8.6$ and $P'(n) < 0$ for $n > 8.6$, P has a maximum at $n = 8.6$. Thus, 8.6 items must be sold to make the most profit.
(c) The maximum profit is $P(8.6) = [4 - 0.2(8.6)][2.5(8.6) + 7] = 64.98$ hundreds of dollars or $64.98 \times 100 = \$6,498$.

34. (a) $E' = \frac{dE}{dt} = 0.03(6t - t^2)^2(6 - 2t)$ W. The critical values are at $t = 0$, 3, and 6. The first derivative changes sign from positive to negative only at $t = 3$. Hence, $t = 3$ yields a maximum value of E.
(b) When $t = 3$, we see that $E(3) = 0.01[6(3) - (3)^2]^3 = 7.29$. The maximum energy output is 7.29 J.
(c) The first derivative changes sign from negative to positive at $t = 0$ and $t = 6$. Hence, both $t = 0$ and $t = 6$ yield minimum energy outputs (of 0 J) for E.

35. (a) $N'(t) = 2t^{-1/2} + 3t^{1/2} - 2t^{3/2}$
(b) We want to solve $2t^{-1/2} + 3t^{1/2} - 2t^{3/2} = 0$. Multiplying both sides by $\sqrt{t} = t^{1/2}$ we obtain $2 + 3t - 2t^2 = 0$. This yields critical values of $t = -0.5$ and $t = 2$. Since, $0 \le t \le 6$, the only critical value to consider is when $t = 2$. The first derivative test gives us as a maximum $t = 2$ mo.

36. (a) $T'(t) = -0.12t^2 + 2.52t - 9.3$
(b) Setting $T'(t) = 0$, the quadratic formula yields critical values at $t \approx 4.777$ and $t \approx 16.223$. The first derivative test yields a maximum at $t \approx 16.223$ or about 4:13.4 P.M.
(c) As in (b), the minimum is at $t \approx 4.777$ or about 4:46.6 A.M.

$\equiv$ 24.3 CONCAVITY AND THE SECOND DERIVATIVE TEST

1. Since $f(x) = -3x^2 + 12x$, $f'(x) = -6 + 12 = 0$ when $x = 2$. $f''(x) = -6$ is always negative, so f is concave down over the interval $(-\infty, \infty)$. Maximum at (2, 12).

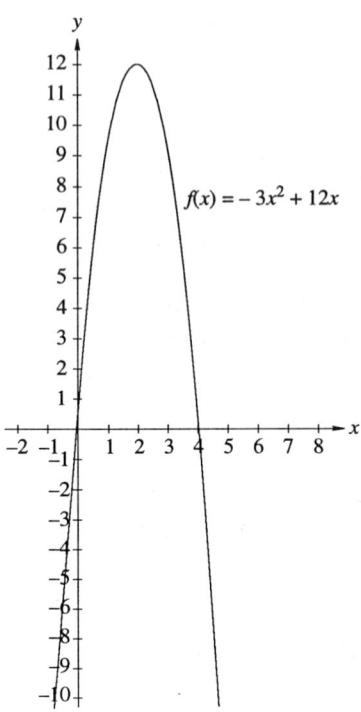

2. $g(x) = 3x^2 - 6x + 5; g'(x) = 6x - 6 = 0$ when $x = 1$. $g''(x) = 6$ always positive so concave up $(-\infty, \infty)$. Minimum at $(1, 2)$.

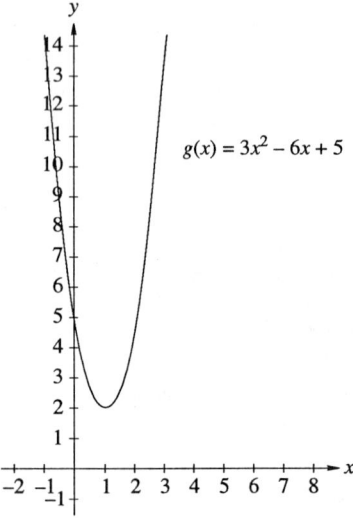

3. $g(x) = x^3 - 12x + 12; g'(x) = 3x^2 - 12 = 0$ when $x = \pm 2$; and $g''(x) = 6x = 0$ when $x = 0$. Concave up: $(0, \infty)$; Concave down: $(-\infty, 0)$. Maximum: $(-2, 28)$; Minimum: $(2, -4)$; Inflection Point: $(0, 12)$.

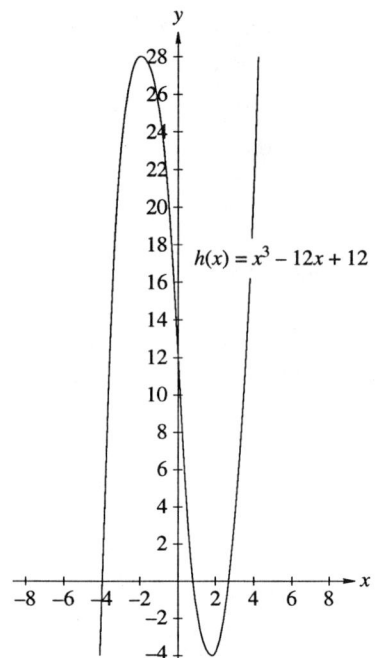

4. Since $f(x) = x^3 + 3x^2 - 5$, then $f'(x) = 3x^2 + 6x$ and $f''(x) = 6x + 6$. We have $f'(0) = 0$, $f'(-2) = 0$, and $f''(-1) = 0$. Hence, f is concave up on the interval $(-1, \infty)$; Concave down over $(-\infty, -1)$. Maximum: $(-2, -1)$; Minimum: $(0, -5)$; Inflection point: $(-1, -3)$.

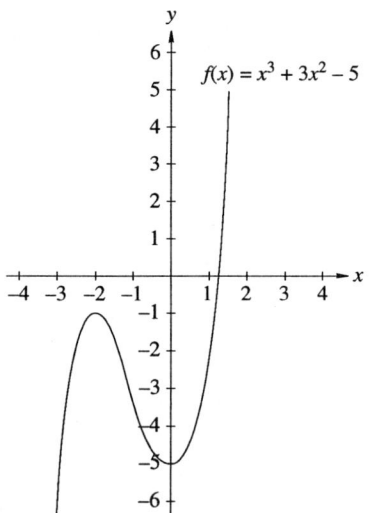

5. $j(x) = x^3 - 6x^2 + 12x - 4; j'(x) = 3x^2 - 12x + 12 = 3(x-2)^2 = 0$ when $x = 2$; $j''(x) = 6x - 12 = 0$ when $x = 2$. Inflection point at $(2, 4)$. No extrema. Concave up: $(2, \infty)$; Concave down: $(-\infty, 2)$.

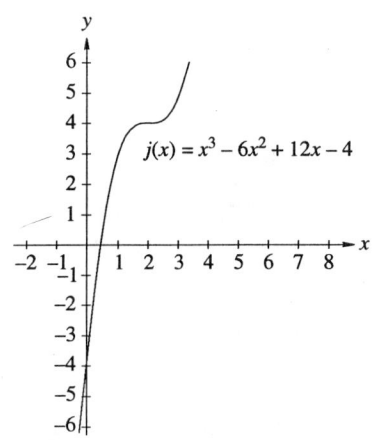

6. $k(x) = x^3 - 6x^2 - 20; k'(x) = 3x^2 - 12x = 3x(x-4) = 0$ when $x = 0$ or 4; $k''(x) = 6x - 12 = 0$ when $x = 2$. Concave up: $(2, \infty)$; Concave down: $(-\infty, 2)$ Inflection point: $(2, -36)$; Maximum: $(0, -20)$; Minimum: $(4, -52)$.

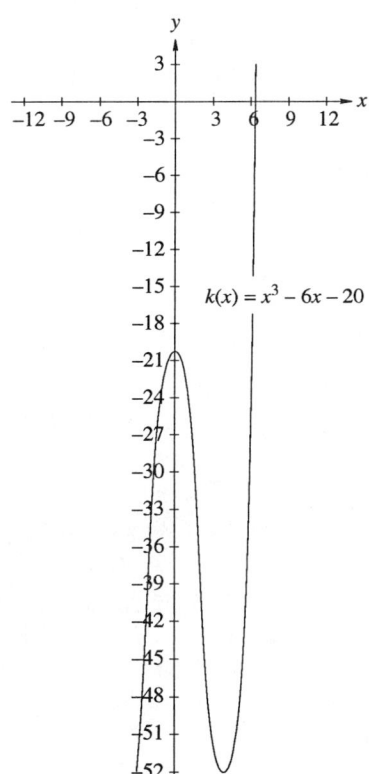

7. $h(x) = -2x^3 + 3x^2 - 12x + 5; h'(x) = -6x^2 + 6x - 12 = -6(x^2 - x + 2)$ never equals 0 since the discriminant $= 1 - 8 = -7$ is negative; hence no extrema. $h''(x) = -12x + 6 = 0$ when $x = \frac{1}{2}$.

Concave up: $(-\infty, \frac{1}{2})$; Concave down: $(\frac{1}{2}, \infty)$;

Inflection point: $(\frac{1}{2}, -\frac{1}{2})$.

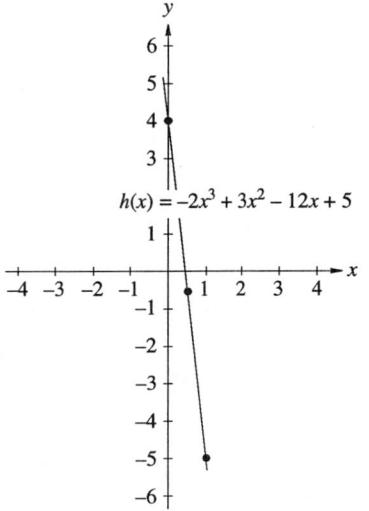

8. $m(x) = x^4 - 4x^3 + 15; m'(x) = 4x^3 - 12x^2 = 4x^2(x - 3) = 0$ when $x = 0$ or $x = 3$; $m''(x) = 12x^2 - 24x = 12x(x - 2) = 0$ when $x = 0$ or $x = 2$. Concave up: $(-\infty, 0)$ and $(2, \infty)$: Concave down: $(0, 2)$; Inflection points: $(0, 15)$ and $(2, -1)$; Minimum: $(3, -12)$, Maximum: none.

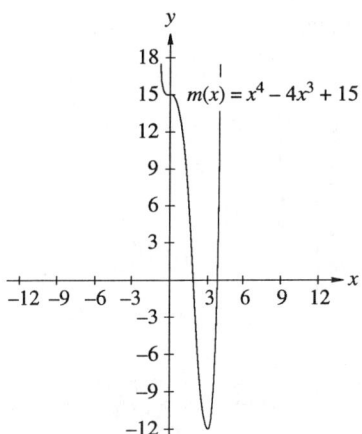

9. $j(x) = 4x^2 - \frac{4}{3}x^3 - x^4; j'(x) = 8x - 4x^2 - 4x^3 = -4x(x^2 + x - 2) = -4x(x + 2)(x - 1) = 0$ when $x = 0, -2$ and 1; $j''(x) = 8 - 8x - 12x^2 = -4(3x^2 + 2x - 2) = 0$ when $x = -1.22$ and 0.55. Concave up: $(-1.22, 0.55)$; Concave down $(-\infty, -1.22)$ and $(0.55, \infty)$; Inflection points: $(-1.22, 6.16)$ and $(0.55, 0.9)$; Maximum: $(-2, 10.67)$ and $(1, 1.67)$; Minimum: $(0,0)$

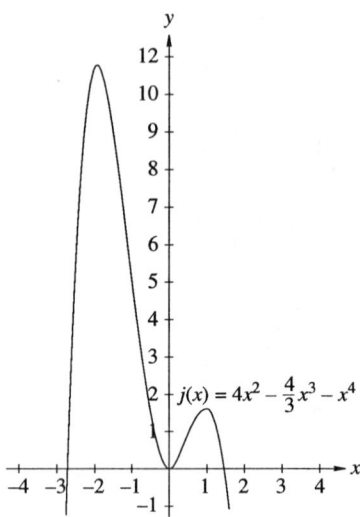

10. $f(x) = 3x^4 - 4x^3 + 8$; $f'(x) = 12x^3 - 12x^2 = 12x^2(x-1) = 0$ when $x = 0$ or $x = 1$; $f''(x) = 36x^2 - 24x = 12x(3x-2) = 0$ when $x = 0$ or $x = \frac{2}{3}$. Thus, f is concave up on the interval $(-\infty, 0)$ and $(\frac{2}{3}, \infty)$ and concave down over $(0, \frac{2}{3})$. Inflection points are $(0, 8)$ and $(\frac{2}{3}, 7\frac{11}{27})$; Maximum: none; Minimum: $(1, 7)$.

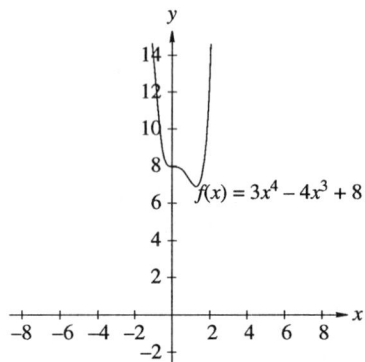

11. $h(x) = (1-x)^3$; $h'(x) = -3(1-x)^2 = 0$ when $x = 1$; $h''(x) = 6(1-x) = 0$ when $x = 1$; Concave down: $(1, \infty)$; Concave up: $(-\infty, 1)$; No extrema. Inflection point: $(1,0)$.

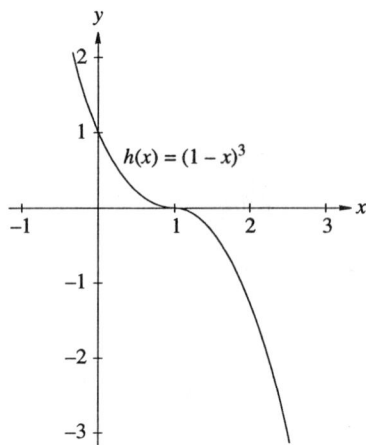

12. $g(x) = 3x^5 - 5x^3$; $g'(x) = 15x^4 - 15x^2 = 15x^2(x^2 - 1) = 0$ when $x = 0$ and $x = \pm 1$; $g''(x) = 60x^3 - 30x = 30x(2x^2 - 1) = 0$ when $x = 0$ and $x = \pm\frac{\sqrt{2}}{2}$; Concave up: $\left(-\frac{\sqrt{2}}{2}, 0\right)$ and $\left(\frac{\sqrt{2}}{2}, \infty\right)$; Concave down: $\left(-\infty, -\frac{\sqrt{2}}{2}\right)$ and $\left(0, \frac{\sqrt{2}}{2}\right)$; Inflection points: $\left(\frac{\sqrt{2}}{2}, 1.2374\right)$, $(0,0)$ and $\left(\frac{\sqrt{2}}{2}, -1.2374\right)$; Maximum: $(-1, 2)$; Minimum: $(1, -2)$;

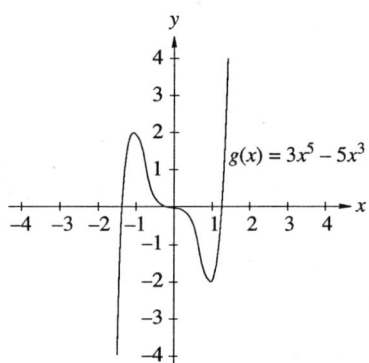

13. Since $h(x) = \sqrt[3]{x} + 2 = x^{1/3} + 2$, then $h'(x) = \frac{1}{3}x^{-2/3}$ and $h''(x) = -\frac{2}{9}x^{-5/3}$. We see that h is defined at $x = 0$, but h' is undefined at $x = 0$, so there is a critical value at $x = 0$; also $h''(x)$ is undefined at $x = 0$. Concave up: $(-\infty, 0)$; Concave down: $(0, \infty)$; Inflection point: $(0, 2)$; No extrema.

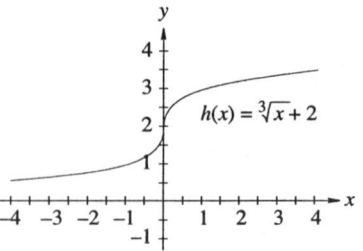

14. Here $j(x) = x - x^{2/3}$, so $j'(x) = 1 - = \frac{2}{3}x^{-1/3} = 1 - \frac{2}{3\sqrt[3]{x}} = \frac{3\sqrt[3]{x}-2}{3\sqrt[3]{x}}$. Critical values at $x = 0$ and $x = \frac{8}{27}$. $j''(x) = \frac{2}{9}x^{-4/3}$ which is undefined at $x = 0$. Concave up: $(-\infty, 0)$ and $(0, \infty)$; Increasing from $(-\infty, 0)$ and from $\left(\frac{8}{27}, \infty\right)$; decreasing from $\left(0, \frac{8}{27}\right)$. Maximum: $(0,0)$, Minimum: $\left(\frac{8}{27}, \frac{-4}{27}\right)$

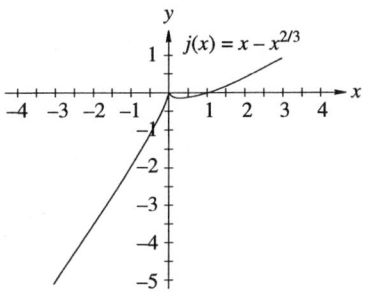

15. $k(x) = x + \frac{4}{x}$ has the derivative $k'(x) = 1 - \frac{4}{x^2} = \frac{x^2-4}{x^2}$. Now, $k'(x) = 0$ when $x = \pm 2$ and k' undefined at $x = 0$, so critical values are at $x = -2$ and $x = 2$. $k''(x) = \frac{8}{x^3}$ which is undefined at $x = 0$; Concave up: $(0, \infty)$; Concave down: $(-\infty, 0)$; Maximum: $(-2, -4)$; Minimum: $(2, 4)$; Vertical Asymptote at $x = 0$; No inflection point

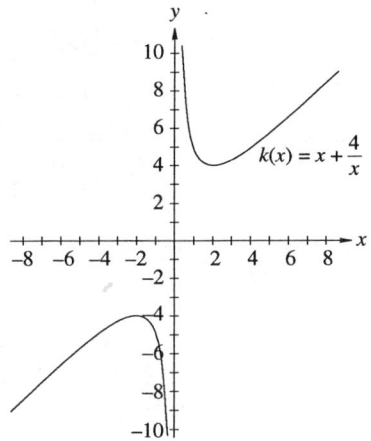

16. $m(x) = x - \frac{9}{x}$ has the derivative $m'(x) = 1 - \frac{9}{x^2}$ which is undefined at $x = 0$. But $x = 0$ is *not* a critical value, since 0 is not in the domain of m. The second derivative is $m''(x) = -\frac{18}{x^3}$, which is also undefined at $x = 0$; Concave up: $(-\infty, 0)$; Concave down: $(0, \infty)$. No extrema; Vertical Asymptote at $x = 0$; No inflection point

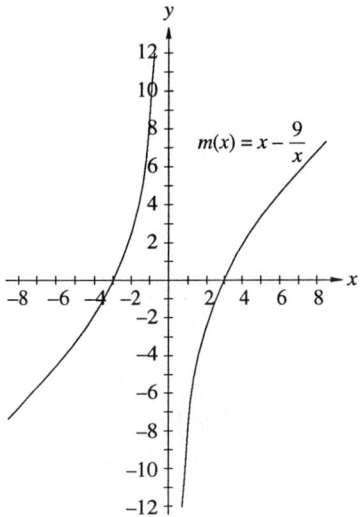

17. $n(x) = x^2 - \frac{2}{x}$ has the derivative $n'(x) = 2x + \frac{2}{x^2} = \frac{2x^3+2}{x^2}$. Critical value at -1. The second derivative is $n''(x) = 2 - \frac{4}{x^3} = \frac{2x^3-4}{x^3}$; this is undefined at 0 and is zero at $x = \sqrt[3]{2}$. Concave up $(-\infty, 0)$ and $\left(\sqrt[3]{2}, \infty\right)$, Concave down $\left(0, \sqrt[3]{2}\right)$. Maximum: none; Minimum: $(-1, 3)$; Vertical Asymptote: $x = 0$; Inflection point: $\left(\sqrt[3]{2}, 0\right)$.

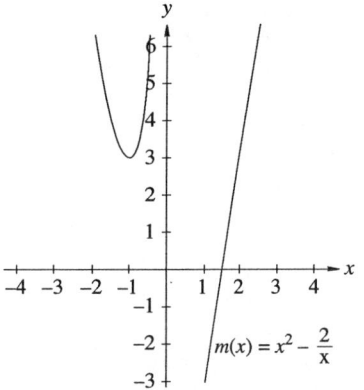

18. $f(x) = \frac{1}{x-2} = (x-2)^{-1}$; $f'(x) = -(x-2)^{-2}$ which is undefined at $x = 2$. $f''(x) = 2(x-2)^{-3}$ which is also undefined at $x = 2$. Concave up: $(2, \infty)$; Concave down: $(-\infty, 2)$ No extrema; Vertical Asymptote: $x = 2$; No inflection point.

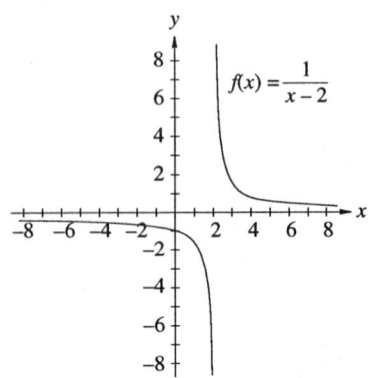

19. $g(x) = \frac{x^3}{3} - \frac{x^2}{2} - 6x$ has the derivative $g'(x) = x^2 - x - 6 = (x-3)(x+2)$, and so $g'(x) = 0$ when $x = -2$ or $x = 3$. The second derivative $g''(x) = 2x - 1 = 0$ when $x = \frac{1}{2}$. Concave up: $(\frac{1}{2}, \infty)$; Concave down: $(-\infty, \frac{1}{2})$. Maximum: $(-2, 7\frac{1}{3})$; Minimum: $(3, -13\frac{1}{2})$; Inflection point: $(\frac{1}{2}, -3.0833)$.

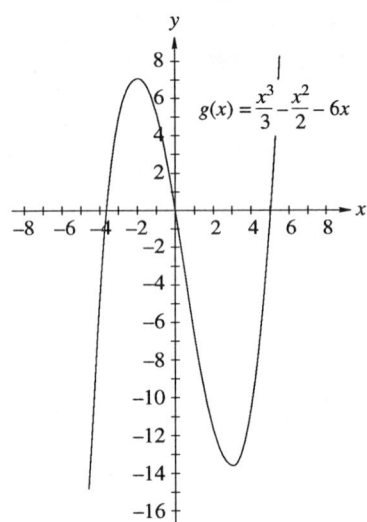

20. $h(x) = x^4 - 32x + 48$; $h'(x) = 4x^3 - 32 = 4(x^3 - 8) = 0$ when $x = 2$; $h''(x) = 12x^2 = 0$ when $x = 0$. Concave up $(-\infty, \infty)$. Minimum: $(2, 0)$; no inflection points

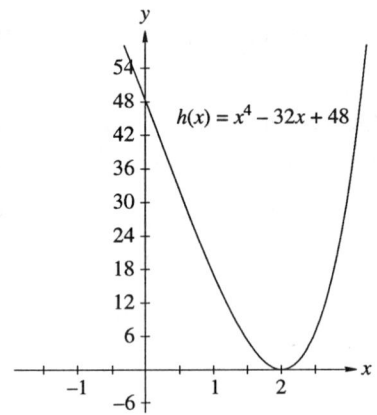

21. $j(x) = 2x^{1/2} - x$ has the derivative $j'(x) = x^{-1/2} - 1$, and $j'(x) = 0$ when $x = 1$ and j' is undefined at $x = 0$. The second derivative is $j''(x) = -\frac{1}{2}x^{-3/2}$ which is undefined at $x = 0$; Concave up: nowhere; Concave down: $(0, \infty)$. Maximum: $(1, 1)$; Minimum: $(0, 0)$; Inflection points: none.

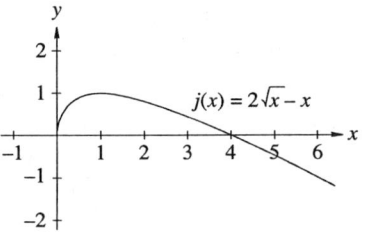

22. $k(x) = -(x^2 - 4)^{1/2}$, so $k'(x) = -\frac{1}{2}(x^2 - 4)^{-1/2}(2x) = \frac{-x}{\sqrt{x^2-4}}$. Since the domain of $k(x)$ is $(-\infty, 2)$ and $(2, \infty)$ the only critical values are -2 and 2. $k''(x) = -(x^2 - 4)^{-1/2} + (x^2 - 4)^{-3/2}x^2 = \frac{-(x^2-4)+x^2}{(x^2-4)^3} = \frac{4}{(x^2-4)^3}$. On the domain of $k(x)$, we see that k'' is positive, so Concave up on the intervals $(-\infty, -2)$ and $(2, \infty)$. We also see that $k'(x)$ is positive on $(-\infty, -2)$ so is increasing and yields a maximum at $(-2, 0)$. $k'(x)$ is negative on $(2, \infty)$ so decreasing and yields a maximum at $(2, 0)$ also. No inflection points.

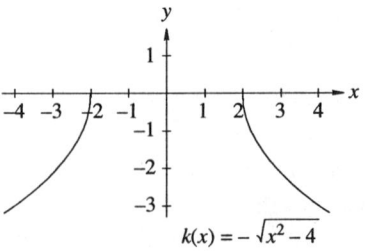

23. $g(x) = 2 + x^{2/3}$, so $g'(x) = \frac{2}{3}x^{-1/3}$. Since g is defined at $x = 0$ and g' is undefined at $x = 0$, 0 is a critical value. $g''(x) = \frac{-2}{9}x^{-4/3}$ is undefined at 0 but negative everywhere else, so concave down over $(-\infty, 0)$ and $(0, \infty)$ and $g'(x) > 0$ on $(0, \infty)$ and $g'(x) < 0$ on $(-\infty, 0)$, so Minimum value: $(0, 2)$. No inflection points.

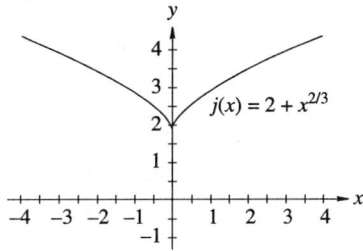

24. $h(x) = 2x - 3x^{2/3}; h'(x) = 2 - 2x^{-1/3} = \frac{2x^{-1/3}-2}{x^{1/3}}$.

Critical values at 1 and 0. $h''(x) = \frac{2}{3}x^{-4/3}$, which is undefined at $x = 0$ and positive everywhere else. Hence, concave up on the intervals $(-\infty, 0)$ and $(0, \infty)$. Maximum: $(0, 0)$; Minimum: $(1, -1)$. No inflection points.

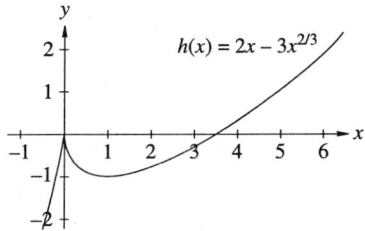

25. $f(x) = x^{2/3}(x - 10) = x^{5/3} - 10x^{2/3}$

$$f'(x) = \frac{5}{3}x^{2/3} - \frac{20}{3}x^{-1/3}$$
$$= \frac{5x^{2/3}}{3} - \frac{20}{3x^{1/3}}$$
$$= \frac{5x - 20}{3x^{1/3}}$$

Critical values are at 0 and 4. $f''(x) = \frac{10}{9}x^{-1/3} + \frac{20}{9}x^{-4/3} = \frac{10x+20}{9x^{4/3}} = 0$ when $x = -2$, undefined at $x = 0$. Concave down $(-\infty, -2)$. Concave up $(-2, 0)$ and $(0, \infty)$. Maximum: $(0,0)$; Minimum: $(4, -15.119)$. Inflection point: $(-2, -19.049)$.

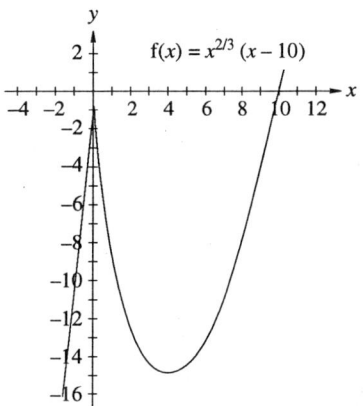

26. $g(x) = \frac{2}{x^2-4}$; Domain excludes $x = \pm 2$. Vertical Asymptotes at $x = 2$, $x = -2$. Now, $g'(x) = -2(x^2 - 4)^{-2}(2x) = \frac{-4x}{(x^2-4)^2}$. Critical value at 0. $g''(x) = 8x(x^2-4)^{-3}(2x) - 4(x^2-4)^{-2} = \frac{16x^2}{(x^2-4)^3} - \frac{4}{(x^2-4)^2} = \frac{16x^2-4x^2+16}{(x^2-4)^3} = \frac{12x^2+16}{(x^2-4)^3}$. We can see that g'' is undefined at ± 2. Concave up: $(-\infty, -2)$ and $(2, \infty)$; Concave down: $(-2, 2)$; Maximum: $(0, -\frac{1}{2})$. No inflection points.

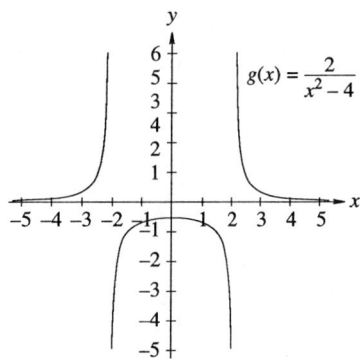

27. $h(x) = x(x^2 + 1)^{-1}$
$$h'(x) = (x^2 + 1)^{-1} - 2x^2(x^2 + 1)^{-2}$$
$$= \frac{1}{x^2 + 1} - \frac{2x^2}{(x^2 + 1)^2} = \frac{x^2 + 1 - 2x^2}{(x^2 + 1)^2}$$
$$= \frac{1 - x^2}{(x^2 + 1)^2}$$

Critical values at $x = \pm 1$.
$$h''(x) = \frac{(x^2 + 1)^2(-2x) - (1 - x^2)(x^2 + 1)(2x)2}{(x^2 + 1)^4}$$
$$= \frac{-2x^3 - 2x - 4x + 4x^3}{(x^2 + 1)^3}$$
$$= \frac{2x^3 - 6x}{(x^2 + 1)^3}$$

You can determine that $h''(x) = 0$ when $x = 0$ and $x = \pm\sqrt{3}$. Thus h is concave up over the intervals $(-\sqrt{3}, 0)$ and $(\sqrt{3}, \infty)$ and h is concave down over the intervals $(-\infty, -\sqrt{3})$ and $(0, \sqrt{3})$.

Maximum: $(1, \frac{1}{2})$; Minimum: $(-1, -\frac{1}{2})$; Inflection point: $\left(-\sqrt{3}, \frac{-\sqrt{3}}{4}\right), (0,0)$, and $\left(\sqrt{3}, \frac{\sqrt{3}}{4}\right)$.

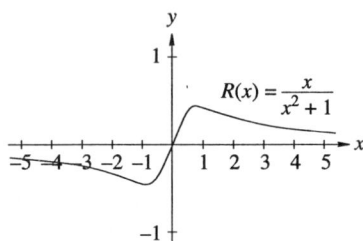

28. $j(x) = x^{2/3}(x-4)^{-1}$

$j'(x) = \frac{2}{3}x^{-1/3}(x-4)^{-1} - (x-4)^{-2}x^{2/3}$

$= \frac{2}{3x^{1/3}(x-4)} - \frac{x^{2/3}}{(x-4)^2} = \frac{2(x-4) - 3x}{3x^{1/3}(x-4)^2}$

$= \frac{-x - 8}{3x^{1/3}(x-4)^2}$.

Critical values at -8 and 0. j is undefined at $x = 4$, so there is a vertical asymptote at $x = 4$.

$j''(x) = \frac{3x^{1/3}(x-4)^2(-1) + (x+8)\left[x^{-2/3}(x-4)^2 + 6x^{1/3}(x-4)\right]}{9x^{2/3}(x-4)^4}$

$= \frac{-3x^{1/3}(x-4)^2 + (x+8)(x-4)\left[x^{1/3} - 4x^{-2/3} + 6x^{1/3}\right]}{9x^{2/3}(x-4)^4}$

$= \frac{-3x(x-4) + (x+8)(7x-4)}{9x^{4/3}(x-4)^3}$

$= \frac{-3x^2 + 12x + 7x^2 + 52x - 32}{9x^{4/3}(x-4)^3}$

$= \frac{4x^2 + 64x - 32}{9x^{4/3}(x-4)^3} = \frac{4(x^2 + 16x - 8)}{9x^{4/3}(x-4)^3}$

j'' is undefined at 0 and 4 and zero at $x \approx -16.485$ and $x \approx 0.485$. Thus, j is concave up over the intervals $(-16.485, 0), (0, 0.4585)$, and $(4, \infty)$ and concave down over $(-\infty, -16.485)$ and $(0.485, 4)$.

Maximum: $(0,0)$; Minimum: $\left(-8, -\frac{1}{3}\right)$; Inflection points: $(-16.485, -0.316)$ and $(0.485, -0.176)$.

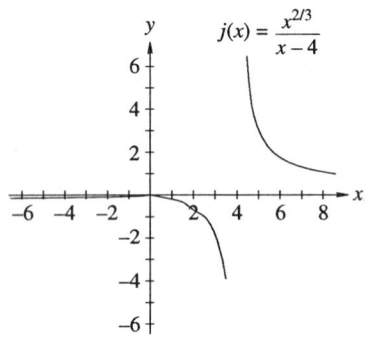

29. To find the potential inflection points, we find $F''(t)$ and then solve $F''(t) = 0$. Differentiating produces $F'(t) = 1.242t - 0.027t^2$ and $F''(t) = 1.242 - 0.054t$. Solving $1.242 - 0.054t = 0$ yields $t = 23$. Since $F''(t)$ changes from concave upward to concave downward at $t = 23$, this is an inflection point. Hence, the best time to harvest is during the 23rd day.

30. First, we find $S''(x)$ and then solve $S''(x) = 0$. Differentiating produces $S'(x) = 4.2x - 0.3x^2$ and $S''(x) = 4.2 - 0.6x$. Solving $4.2 - 0.6x = 0$ yields $x = 7$ as the only possible inflection point. Since S'' changes from positive to negative at $x = 7$ we see that it is an inflection point. Thus, the point of diminishing returns for the sales function $S(x) = 97 + 2.1x^2 - 0.1x^3$ is when $x = 7$, that is, when \$7,000 is spent on advertising.

31. (a) To find the maximum of $N(t)$, we set $N'(t) = 0$ to determine the critical values of N. Differentiating produces $N'(t) = 6t - 2t^{3/2}$ and $6t - 2t^{3/2} = 2t(3 - t^{1/2}) = 0$ yields critical values at $t = 0$ and $t = 9$. Since $N''(t) = 6 - 3t^{1/2}$ is positive at $t = 0$, N has a relative minimum at $t = 0$. Since $N''(9) < 0$, we see that N has a relative maximum at $t = 9$. Evaluating $N(9)$ we see that $N(9) = 116$. The maximum is 116 crimes in the ninth month.
(b) Solving $N''(t) = 0$ we get $3\sqrt{t} = 6$ or $\sqrt{t} = 2$ and so, $t = 4$. The maximum rate of increase is $N'(4)6(4) - 2(4^{3/2}) = 8$ crimes/mo.

32. (a) First, find $T'(t)$ and $T''(t)$. Differentiating produces $T'(t) = -0.12t^2 + 2.52t - 9.3$ and $T''(t) = -0.24t + 2.52$. Solving $T'(t) = 0$ we find the critical values $t \approx 4.777$ and $t \approx 16.223$. Since $T''(4.777) > 0$, we know that $t \approx 4.777$ is a relative minimum and since $T''(16.223) < 0$, we know that $t \approx 16.223$ is a relative maximum. Evaluating T at each of these values for t, we see that the maximum temperature is $T(16.223) = 65.95°F$ and the minimum is $T(4.777) = 35.97°F$.
(b) Setting $T''(t) = 0$, or $-0.24t + 2.52 = 0$, yields $t = 10.5$. At 10.5 hours, the rate of increase is at a maximum: $T'(10.5) = 3.93°F/hour$.

≡ 24.4 APPLIED EXTREMA PROBLEMS

1. $A = \ell \cdot w; \ell + 2w = 10{,}000 \Rightarrow \ell = 10{,}000 - 2w$. Substituting into the area formula yields $A = (10{,}000-2w)w = 10{,}000w-2w^2$, so $A' = 10{,}000 - 4w = 0 \Rightarrow w = 2{,}500$. $A'' = -4 < 0$ so $w = 2{,}500$ yields a maximum area, and $\ell = 10{,}000 - 5000 = 5000; A = 5{,}000 \times 2{,}500 = 12{,}500{,}000$ ft^2.

2. This time $2\ell + 2w = 10{,}000 \Rightarrow \ell + w = 5{,}000 \Rightarrow \ell = 5{,}000 - w$. Substituting into the area formula, we have $A = (5{,}000 - w) = 5000w - w^2; A' = 5000 - 2w = 0 \Rightarrow w = 2{,}500$. $\ell = 5000 - 2500 = 2{,}500$. $A = 2{,}500 \times 2{,}500 = 6{,}250{,}000$ ft^2.

3. **(a)** Let x be the length of the side along the river, as well as the side opposite. Then the length of the other two sides is $6{,}000{,}000/x$. The cost $C(x) = 2x + x + 2(6{,}000{,}000/x)$, so $C(x) = 3x + \frac{12{,}000{,}000}{x}$.

$C'(x) = 3 - \frac{12{,}000{,}000}{x^2} = 0 \Rightarrow 3x^2 = 12{,}000{,}000 \Rightarrow x^2 = 4{,}000{,}000 \Rightarrow x = 2{,}000$. The other sides are 3,000 ft. The dimensions are 2,000ft. $\times$ 3,000 ft.
(b) The cost is $3(2000) + 2(3{,}000) = \$12{,}000$.

4. Let x be length of a side of square and $200 - 4x$ be the circumference of the circle. The radius of the circle is $\frac{100-2x}{\pi}$. The total area for the square and circle is $A = x^2 + \pi\left(\frac{100-2x}{\pi}\right)^2 = x^2 + \frac{1}{\pi}(100 - 2x)^2$ and $A' = 2x + \frac{2}{\pi}(100 - 2x)(-2) = 2x - \frac{4}{\pi}(100 - 2x) = 2x - \frac{400}{\pi} + \frac{8}{\pi}x = \left(2 + \frac{8}{\pi}\right)x - \frac{400}{\pi}$. We see that $A'(x) = 0$ when $x = \frac{200}{\pi+4} \approx 28.005$. When $x < 28.005$, $A'(x) < 0$ and when $x > 28.005$, $A'(x) > 0$ and so, when $x \approx 28.005$, the area is a minimum. Thus, the fencing should not be cut and it all should be used to form a circular fence.

5. After cutting out the squares, the resulting length and width of the box will both be $(24 - 2x)$. The height will be $x; V = \ell \cdot w \cdot h = (24 - 2x)^2 x = 4x^3 - 96x^2 + 576x; V' = 12x^2 - 192x + 576 = 12(x^2 - 16x + 48) = 12(x-4)(x-12)$. Now, $V' = 0$ when $x = 4$ or 12. $V'' = 24x - 192; V''(4) < 0$ so $V(4)$ is maximum. $V''(12) > 0$ so $V(12)$ is minimum. $V(4) = 4 \cdot 4^3 - 96 \cdot 4^2 + 576 \cdot 4 = 1024$ cm^3. The length of the side of a removed square should be 4 cm.

6. The bottom center of the building can be placed on the origin of a coordinate system. Then the formula for the edge of the building is $y = \sqrt{100 - x^2}$. If $2x$ is the length of the sign, then $\sqrt{100 - x^2}$ will be the height. Its area is given by the formula $A = 2x\sqrt{100 - x^2}$.

$A' = 2(100 - x^2)^{1/2} - (100 - x^2)^{-1/2} \cdot 2x^2$

$= 2(100 - x^2)^{1/2} - \frac{2x^2}{(100 - x^2)^{1/2}}$

$= \frac{2(100 - x^2)}{(100 - x^2)^{1/2}} - \frac{2x^2}{(100 - x^2)^{1/2}}$

$= \frac{200 - 4x^2}{(100 - x^2)^{1/2}}$

$200 - 4x^2 = 0$, or $4x^2 = 200$, or $x^2 = 50$, and so $x = \pm\sqrt{50} \approx \pm 7.071$. Since x cannot be negative, 7.071 is the answer we seek. Length of the sign is $2x = 14.14$ m. The height is $\sqrt{100 - x^2} = \sqrt{50} = 7.07$ m.

7. First solve $9x^2 + 16y^2 = 144$ for y. This yields $y = \sqrt{9 - \frac{9}{16}x^2}$. The area of the rectangle is $A = xy$, or after substituting for y, can be written as $A = x\sqrt{9 - \frac{9}{16}x^2}$. $A' = \left(9 - \frac{9}{16}x^2\right)^{1/2} + x \cdot \frac{1}{2}\left(9 - \frac{9}{16}x^2\right)^{-1/2}\left(\frac{-18}{16}x\right) = \frac{9 - \frac{9}{16}x^2 - \frac{9}{16}x^2}{\left(9 - \frac{9}{16}x^2\right)^{1/2}} = \frac{9 - \frac{9}{8}x^2}{\left(9 - \frac{9}{16}x^2\right)^{1/2}}$. Critical values are when $9 - \frac{9}{8}x^2 = 0$, or $x^2 = 8$, or $x = 2\sqrt{2}$, and when $9 - \frac{9}{16}x^2 = 0$ or $x = 4$. When $x = 4$, $y = 0$ so $A = 0$ which is the minimum. When $x = 2\sqrt{2}$, we see that $y = \sqrt{9 - \frac{9}{2}} = \sqrt{\frac{9}{2}} = \frac{3}{\sqrt{2}} = \frac{3\sqrt{2}}{2}$ and the area is $A = 2\sqrt{2} \cdot \frac{3\sqrt{2}}{2} = 6$, which is a maximum. The dimensions are $2\sqrt{2} \approx 2.83$ units horizontally and $\frac{3}{2}\sqrt{2} \approx 2.12$ units vertically.

8. Similar to exercise #5, the length will be $15 - 2x$, the width $8 - 2x$ and the height x. $V = \ell \cdot w \cdot h = (15 - 2x)(8 - 2x)x = 120x - 46x^2 + 4x^3$. $V' = 12x^2 - 92x + 120$ and solving yields $x = \frac{5}{3}$ or $x = 6$. Since $0 \le x \le 4$, the answer is $\frac{5}{3}$ in (Note: $V'' = 24x - 92 < 0$ when $x = \frac{5}{3} \Rightarrow$ Concave down $\Rightarrow$ is maximum.)

9. Let x be the length and width of the box and $y = \frac{32}{x^2}$ be the height. The material used is proportional to the area of the 4 sides plus the base. $A = x^2 + 4 \cdot x \cdot \frac{32}{x^2} = x^2 + \frac{128}{x}$, and so $A' = 2x - \frac{128}{x^2} = \frac{2x^3 - 128}{x^2}$. Solving $2x^3 - 128 = 0$ yields $x^3 = 64$ or $x = 4$. $A'' = 2 + \frac{256}{x^3}$ is positive at $x = 4$, so this yields a minimum. The height is $\frac{32}{16} = 2$. The dimensions are $4'' \times 4'' \times 2''$.

10. This time the area is $A = 2x^2 + \frac{128}{x}$, so $A' = 4x - \frac{128}{x^2} = \frac{4x^3-128}{x^2}$. Solving $4x^3 - 128 = 0 \Rightarrow x^3 = 32 \Rightarrow x = \sqrt[3]{32} \approx 3.175$. Height is $\frac{32}{\sqrt[3]{32^2}} = \sqrt[3]{32} \approx 3.175$. The box is a cube $\sqrt[3]{32} \approx 3.175$ in. on each side.

11. $A = 2\pi r^2 + 2\pi rh = 480 \Rightarrow h = \frac{480-2\pi r^2}{2\pi r} = \frac{240-\pi r^2}{\pi r}$, so the volume is $V = \pi r^2 h = \pi r^2 \left(\frac{240-\pi r^2}{\pi r}\right) = 240r - \pi r^3$, which has the derivative $V' = 240 - 3\pi r^2$. Setting $240 - 3\pi r^2 = 0$ and solving for r yields $\pi r^2 = 80$ or $r = \sqrt{\frac{80}{\pi}} \approx 5.05$. $h = 10.09$. Radius $= 5.05$ cm; height $= 10.09$ cm.

12. Here $s(t) = \sqrt{t - 9t^2}$ has the derivative $s'(t) = \frac{1}{2}(t - 9t^2)^{-1/2}(1 - 18t) = \frac{1-18t}{2\sqrt{t-9t^2}}$. We see that $s'(t) = 0$, when $1 - 18t = 0 \Rightarrow t = \frac{1}{18}$. The other critical values are 0 and $\frac{1}{9}$ which yield a speed of 0. $s(\frac{1}{18}) = 0.16667$ m/min.

13. $A = \ell \cdot w = 580$, and solving for ℓ, we get $\ell = \frac{580}{w}$. The printing area of the page is $P = (\ell - 5)(w - 6) = \left(\frac{580}{w} - 5\right)(w - 6) = 580 - \frac{3480}{w} - 5w + 30 = 610 - 5w - \frac{3480}{w}$. In order to determine when the printed area of the page is the largest, we take the derivative, with the result $P' = -5 + \frac{3480}{w^2}$. Setting the derivative equal to 0, we have $-5 + \frac{3480}{w^2} = 0$ or $w^2 = \frac{3480}{5} = 696$ and so $w = \sqrt{696} = 26.38$ and $\ell = \frac{580}{w} = 21.98$. The largest printed area will have a length of about 21.98 cm and a width of about 26.38 cm.

14. $d = \sqrt{(x - 5)^2 + (y - 1)^2}$, so $d^2 = (x-5)^2 + (y-1)^2$. The distance is minimum when d^2 is minimum. $d^{2'} = 2(x - 5) + 2(y - 1)y'; y = 2x^2; y' = 4x; d^{2'} = 2(x - 5) + 2(2x^2 - 1)(4x) = 2x - 10 + 16x^3 - 8x = 16x^3 - 6x - 10 = 2(8x^3 - 3x - 5) = 2(x-1)(8x^2+8x+5)$. The only critical number is 1. $d^{2''} = 48x^2 - 6 > 0$ when $x = 1$ so this yields a minimum $y = 2 \cdot 1^2 = 2$. The point is $(1, 2)$.

15. $r = 15$. Placing the center of the log at the origin yields the equation $x^2 + y^2 = 225$ and so, $y = \sqrt{225 - x^2}, d = 2y$, and $w = 2x$. Thus, the strength is $S = wd^2 = 2x\left(2\sqrt{225 - x^2}\right)^2 =$

$8x(225 - x^2) = 1800x - 8x^3$. To find when the strength is the most, we take the derivative, getting $S' = 1800 - 24x^2$, then determine the critical values by finding when the derivative is 0. Thus, $1800 - 24x^2 = 0 \Rightarrow 24x^2 = 1800 \Rightarrow x^2 = 75$, and so $x = 5\sqrt{3}$. As a result, the width is $w = 2x = 10\sqrt{3} \approx 17.32$ in. and the depth is $d = 2y = 2\sqrt{225 - 75} = 2\sqrt{150} = 24.49$ in.

(b) This time our equation is $x^2 + y^2 = r^2$ so $y = \sqrt{r^2 - x^2}$ and so the strength is $S = 2x (2\sqrt{r^2 - x^2})^2 = 8x(r^2-x^2) = 8xr^2 - 8x^3$. Taking the derivative, we obtain $S' = 8r^2 - 24x^2$, and to find when the derivative is 0, we get $8r^2 - 24x^2 = 0 \Rightarrow 24x^2 = 8r^2 \Rightarrow x^2 = \frac{1}{3}r^2$.

Thus, $x = \sqrt{\frac{r^2}{3}} = \frac{r}{\sqrt{3}} = \frac{r\sqrt{3}}{3}$. For the width we have $w = 2x = 2r\sqrt{3}/3$ and the depth is $d = 2y = 2\sqrt{r^2 - \frac{r^2}{3}} = 2\sqrt{\frac{2r^2}{3}} = \frac{2r\sqrt{6}}{3}$.

16. $A = (s(s - a)(s - b)(s - c))^{1/2}$

$= \left[\frac{1}{2}(300 + c)\left(\frac{300 + c}{x} - 150\right) \times \left(\frac{300 + c}{2} - 150\right)\left(\frac{300 + c}{2}\right)\right]^{1/2}$

$= \left[\left(150 + \frac{c}{2}\right)\left(\frac{c}{2}\right)\left(\frac{c}{2}\right)\left(150 - \frac{c}{2}\right)\right]^{1/2}$

$= \left[\left(22500 - \frac{c^2}{4}\right)\left(\frac{c^2}{4}\right)\right]^{1/2}$

$= \left[5625c^2 - \frac{c^4}{16}\right]^{1/2}$

$A' = \frac{1}{2}\frac{11250c - \frac{c^3}{4}}{\sqrt{5625c^2 - \frac{c^4}{16}}}$

Critical values are when the numerator is 0 or the denominator is 0. Now, if $11250c - \frac{c^3}{4} = 0$, then $\frac{c}{4}(45000 - c^2) = 0$, and we have $c = 0$, $c = \pm\sqrt{45000} = \pm 212.13$. The denominator yields $5625 - \frac{c^2}{16} = 0$, or $c = \sqrt{90000}$ and $c = 300$. Since c cannot be negative, 0 and 300 yield an area of 0 which is a minimum, so 212.13 gives maximum. Use the Law of Cosines to determine θ. Then, $\cos\theta = \frac{150^2 + 150^2 - 212.13^2}{2 \cdot 150 \cdot 150} = 0$ and $\theta = \cos^{-1} 0 = 90°$. The angle that gives the maximum capacity is 90°.

17. $A = \ell \cdot w; \ell = 320 - 2w$
$A = (320 - 2w)w = 320w - 2w^2$
$A' = 320 - 4w = 0 \Rightarrow w = 80\,\text{mm}$
$\ell = 320 - 2 \cdot 80 = 320 - 160 = 160\,\text{mm}$

The gutter is 80 mm deep and 160 mm wide.

18. **(a)** $I = \dfrac{E}{\sqrt{R^2 + X^2}} = E(R^2 + X^2)^{-1/2}$

$I' = \dfrac{-1}{2}E(R^2 + X^2)^{-3/2}(2X)$

$\quad = \dfrac{-XE}{(R^2 + X^2)^{3/2}}$

The only critical value is 0. Since $I' < 0, I$ is decreasing and this yields a maximum.

(b) Maximum $= \dfrac{E}{\sqrt{R^2}} = \dfrac{E}{R}$.

19. $P = VI - RI^2 = 10I - 4I^2$ and so $P' = 10 - 8I$. Critical value is $\frac{5}{4} = 1.25$. $P'' = -8$ so critical value yields a maximum. Maximum power given by current of 1.25 A.

20. $v(x) = 5x(a - x) = 5ax - 5x^2$

$v' = 5a - 10x = 0 \Rightarrow x = \dfrac{a}{2}$

$v'' = -10$

so the reaction is a maximum at $x = \frac{a}{2}$.

21. $C(A) = bA + \frac{c}{A}; C' = b - \frac{c}{A^2} = 0$ when $A = \sqrt{c/b}; C'' = 2c/A^3 > 0$, so minimum cost at $A = \sqrt{c/b}$.

22. Let $\ell =$ height of rectangular part of window; $w = d =$ width. Then,

$A = \ell \cdot w = 2$ so $w = \dfrac{2}{\ell} = d$

$r = \dfrac{1}{2}w = \dfrac{1}{\ell}$

Perimeter of window $= 2\ell + w + \pi r = 2\ell + w + \pi r = 2\ell + \frac{2}{\ell} + \frac{\pi}{\ell} = 2\ell + \frac{2+\pi}{\ell}$. $P' = 2 - (2 + \pi)/\ell^2 = 0$ and $\frac{2+\pi}{\ell^2} = 2; 2\ell^2 = 2 + \pi; \ell^2 = \frac{2+\pi}{2}; \ell = \sqrt{\frac{2+\pi}{2}} \approx 1.6; d = \frac{2}{1.6} = 1.247 \approx 1.25$ m.

23. $\ell =$ height; $w = d =$ width $= 2r$. Thus,

$P = 8 = 2\ell + w + \dfrac{\pi w}{2} = 2\ell + \dfrac{2 + \pi}{2}w$

$2\ell = 8 - \dfrac{2 + \pi}{2}w; \ell = 4 - \dfrac{2 + \pi}{4}w$

$A = \ell \cdot w + \dfrac{1}{2}\pi\left(\dfrac{w}{2}\right)^2$

$\quad = \left(4 - \dfrac{2 + \pi}{4} \cdot w\right)w + \dfrac{\pi}{8}w^2$

$\quad = 4w - \dfrac{2 + \pi}{4}w^2 + \dfrac{\pi}{8}w^2 = 4w - \dfrac{4 + \pi}{8}w^2$

$A' = 4 - \dfrac{4 + \pi}{4}w = 0; (4 + \pi)w = 16;$

$w = \dfrac{16}{4 + \pi} = 2.24$ m.

The diameter is 2.24 m.

24. **(a)** Time $= \frac{800}{s}$. Total cost is

$C = \left(0.35 + \dfrac{s}{300}\right)800 + 12\left(\dfrac{800}{s}\right).$

$C = 800\left(0.35 + \dfrac{s}{300} + \dfrac{12}{s}\right).$

$C' = 800\left(\dfrac{1}{300} - \dfrac{12}{s^2}\right) = 0.$

This is 0 when $\frac{s^2}{12} = 300$, or $s^2 = 300 \times 12 = 3600$, and so $s = \sqrt{3600} = 60$ mph yields the least cost.

(b) $C = \left(0.35 + \frac{60}{300} + \frac{12}{60}\right)800 = (0.35 + 0.2 + 0.2)800 = (0.75)800 = \$600$ is the least cost.

25. **(a)** $P = I^2R = (4.5 - t)^2(25.4)$

$P' = 2(4.5 - t)(25.4)(-1) = 0$

when $t = 4.5s$. P'' is always positive so this is a minimum.

(b) $P = (4.5 - 4.5)^2(25.4) = 0$ W.

26. If $i(t)$ is the current, then

$q = -4.36t^2 + 2.14t^3$

$i(t) = \dfrac{dq}{dt} = -8.72t + 6.42t^2$

$i'(t) = -8.72 + 12.84t$

$i'(t) = 0$ when $t = 0.679$ s. Thus, $i(0.679) = -2.96$ A.

27. Let $x =$ distance from less intense light and $8 - x =$ distance from more intense light. Light intensity $= \frac{1}{x^2} + \frac{4}{(8-x)^2}$. Taking the derivative, we get $L' = \frac{-2}{x^3} + \frac{8}{(8-x)^3}$. Critical values will be when $\frac{(8-x)^3}{8} = \frac{x^3}{2}$, or $(8 - x)^3 = 4x^3$, or $512 - 192x + 24x^2 - x^3 = 4x^3$, or when $512 - 192x + 24x^2 - 5x^3 = 0$. This has one solution between 0 and 8. It is about 3.1 m. Thus, the desired answer is that the light is the least about 3.1 m from the less intense light.

28. $D = av^2 + \dfrac{b}{v^2}$

$D' = 2av - 2\dfrac{b}{v^3} = 0$

$2av = \dfrac{2b}{v^3}$

$v^4 = \dfrac{b}{a}$

$v = \sqrt[4]{b/a}$

(Note: $D'' = 2a + 6\frac{b}{v^4} > 0$, so we have a minimum.)

29. Here, we have $Y = k(2x^4 - 5Lx^3 + 3L^2x^2)$, and so $Y' = k(8x^3 - 15Lx^2 + 6L^2x) = 0$ or

$x(8x^2 - 15Lx + 6L^2) = 0$. This yields critical values when $x = 0$, $0.5785L$, and $1.2965L$. Now, $Y'' = k(24x^2 - 30Lx + 6L^2)$ and, since $L > 0$, we find that $Y''(0) > 0, Y''(0.5785L) < 0; Y''(1.2965L) > 0$. Thus, the maximum deflection occurs at $x = 0.5785L$.

30. $E = \dfrac{T(1 - T\mu)}{T + \mu} = \dfrac{T - T^2\mu}{T + \mu}$;

$E' = \dfrac{(T + \mu)(1 - 2T\mu) - (T - T^2\mu)1}{(T + \mu)^2}$

$= \dfrac{T + \mu - 2T^2\mu - 2T\mu^2 - T + T^2\mu}{(T + \mu)^2}$

$= \dfrac{\mu - T^2\mu - 2T\mu^2}{(T + \mu)^2} = \dfrac{\mu(-T^2 - 2T\mu + 1)}{(T + \mu)^2}$

Setting the numerator equal to 0, we have $-T^2 - 2T\mu + 1 = 0; T^2 + 2\mu T - 1 = 0$ and so $T = \dfrac{-2\mu + \sqrt{4\mu^2 + 4}}{2} = -\mu + \sqrt{\mu^2 + 1}$ is the critical value that yields the angle of greatest efficiency.

31. Let d represent the diameter (or width) and h the height of the vertical walls of the tunnel. Then the area of the opening is $A = hd + \frac{1}{2}\pi\left(\frac{d}{2}\right)^2 = hd + \frac{\pi d^2}{8}$, or $hd = A - \frac{\pi d^2}{8}$ and so $h = \frac{A}{d} - \frac{\pi d}{8}$ where A is the constant area. Now, the cost, C, is given by $C = 2h + d + \frac{1}{2}\pi d \cdot 4$. We begin by substituting the above value of h to get the cost as a function of just one variable, d. Thus, we have

$C = 2h + d + 2\pi d = 2\left(\dfrac{A}{d} - \dfrac{\pi d}{8}\right) + (2\pi + 1)d$

$= \dfrac{2A}{d} - \dfrac{\pi d}{4} + (2\pi + 1)d$

To determine when the cost is a minimum, we take the derivative, getting

$C' = \dfrac{-2A}{d^2} - \dfrac{\pi}{4} + 2\pi + 1$

$= \dfrac{-2A}{d^2} + \dfrac{7\pi}{4} + 1$

Setting this equal to 0, we get $\frac{2A}{d^2} = \frac{7\pi + 4}{4}$. Substituting for A, we have

$\dfrac{2\left(hd + \frac{\pi d^2}{8}\right)}{d^2} = \dfrac{7\pi + 4}{4}$

or $\frac{hd}{d^2} + \frac{\pi d^2}{8d^2} = \frac{7\pi + 4}{8}$ which produces $\frac{h}{d} = \frac{7\pi + 4}{8} - \frac{\pi}{8} = \frac{6\pi + 4}{8} = \frac{3\pi + 2}{4}$.

32. Let x be the distance down the shore from P. Then, by the Pythagorean theorem, the distance that must be rowed is $\sqrt{x^2 + 4^2}$. The distance that will be walked is $20 - x$. The total time is then given by the equation $T = \frac{\sqrt{x^2 + 4^2}}{2.5} + \frac{20 - x}{4.5}$.

To find the shortest time, we find T' and set it equal to 0. Thus,

$T' = \dfrac{1}{2.5} \cdot \dfrac{1}{2} \dfrac{2x}{\sqrt{x^2 + 4^2}} - \dfrac{1}{4.5}$

$= \dfrac{x}{2.5\sqrt{x^2 + 4^2}} - \dfrac{1}{4.5}$

Setting this equal to 0, we have $\frac{x}{2.5\sqrt{x^2 + 4^2}} = \frac{1}{4.5}$ or $4.5x = 2.5\sqrt{x^2 + 4^2}$ or $1.8x = \sqrt{x^2 + 4^2}$. Squaring both sides, we get $3.24x^2 = x^2 + 16$, or $2.24x^2 = 16$, or $x^2 \approx 7.14286$, and so $x \approx \pm 2.6726$. Since the distance must be positive, the answer is 2.673 km from P. Checking the critical value of $x = 2.673$ with the second derivative, we find T'' as

$T'' = \dfrac{1}{2.5} \dfrac{\sqrt{x^2 + 4^2} - x\frac{1}{2}\frac{2x}{\sqrt{x^2 + 4^2}}}{x^2 + 4^2}$

$= \dfrac{1}{2.5} \dfrac{x^2 + 16 - x^2}{(x^2 + 16)^{3/2}}$

$= \dfrac{16}{2.5}(x^2 + 16)^{-3/2}$

This is always positive, so the critical value is a minimum.

33. Draw a line from the oil rig perpendicular to the shore at point P. This line represents the 5 mi from the rig to the shore. Let x be the distance from the connection to P. Then the pipe must go a distance of $\sqrt{5^2 + x^2}$ mi underwater and $12 - x$ mi along the shore. The total cost is $C(x) = 1200\sqrt{25 + x^2} + 600(12 - x) = 600\left(2\sqrt{25 + x^2} + 12 - x\right)$.

Finding the derivative, we get $C'(x) = 600\left[\dfrac{2x}{\sqrt{25 + x^2}} - 1\right]$. Setting $C'(x) = 0$, we obtain $\frac{2x}{\sqrt{25 + x^2}} - 1 = 0$ or $\frac{2x}{\sqrt{25 + x^2}} = 1$ or $2x = \sqrt{25 + x^2}$. Squaring both sides, we get $4x^2 = 25 + x^2$ or $3x^2 = 25$. Hence $x = +\frac{5}{\sqrt{3}}$. From the given information, we know $x \geq 0$, and so, $x = \pm\frac{5}{\sqrt{3}} = \frac{5}{3}\sqrt{3}$. The connection should be $\frac{5}{3}\sqrt{3}$ mi from P or $12 - \frac{5}{3}\sqrt{3} \approx 9.11$ mi from the storage tank.

34. (a) Let the bottom dimensions be x by x and so, the area of the bottom is x^2 and the height of the crate is $36/x^2$. The cost is $C(x) = 1.50x^2 + 1.00x^2 + 4(0.75)x(36/x^2)$. This simplifies to $C(x) = 2.5x^2 + \frac{108}{x}$. The derivative is $C'(x) = 5x - \frac{108}{x^2}$. Setting this equal to 0, we obtain $5x - \frac{108}{x^2} = 0$ or $5x = \frac{108}{x^2}$ or $5x^3 = 108$ and so, $x^3 = \frac{108}{5}$ and

$x = \sqrt[3]{\frac{108}{5}} \approx 2.785$. The height is $36/x^2 \approx 4.641$. The dimensions are $2.785 \times 2.785 \times 4.641$ in ft.

(b) The minimum cost is $C(2.785) = 2.5(2.785)^2 + \frac{108}{2.785} \approx \$58.17$.

35. **(a)** Let x represent the length of the interior partition wall. Then, there are two exterior walls of length x and two of length $4200/x$. The total cost is $C(x) = 2(177)x + 2(175)(4200/x) + 115x$. This simplifies to $C(x) = 465x + \frac{1,470,000}{x}$. The derivative is $C'(x) = 465 - \frac{1,470,000}{x^2}$. Setting $C'(x) = 0$ we obtain $465 - \frac{1,470,000}{x^2} = 0$ or $465 = \frac{1,470,000}{x^2}$. Hence $x^2 = \frac{1,470,000}{465}$ and $x \approx 56.225$ ft. The other length is $4200/x \approx 74.700$. The dimensions that will minimize costs are about 56.2×74.7 ft.

(b) The minimum cost is $C(56.225) = 465(56.225) + \frac{1,470,000}{56.225} \approx \$52,290$.

36. Let x represent the length of one side of the square and h the other dimension. If we assume the girth to be about a square, then $4x + h = 108$ and $h = 108 - 4x$. The volume is $V = x^2 h = x^2(108 - 4x) = 108x^2 - 4x^3$ and $V' = 216x - 12x^2$. Setting $V' = 0$ gives $x = 18$. Thus, one such box measures 18 in. $\times$ 18 in. $\times$ 36 in.

If we assume the girth to be about a non-square side of the box, then $2x + 2h + x = 108$ or $3x + 2h = 108$. Solving for h, we obtain $h = 54 - 1.5x$. The volume is $V = x^2 h = x^2(54 - 1.5x) = 54x^2 - 1.5x^3$ and $V' = 108x - 4.5x^2$. Solving $V' = 0$ gives $x = 24$. Hence, the dimensions of the other box is 24 in. $\times$ 24 in. $\times$ 18 in.

≡ 24.5 RELATED RATES

1. As in example 24.19 we have variables $t, r, S,$ and V with formulas $S = 4\pi r^2$ and $V = \frac{4}{3}\pi r^3$. $\frac{dS}{dt} = 8\pi r \frac{dr}{dt}$; $\frac{dV}{dt} = 4\pi r^2 \frac{dr}{dt}$; and $\frac{dV}{dt} = 20$ L/min $= 20,000$ cm³/min; $20,000 = 4\pi r^2 \frac{dr}{dt} \Rightarrow \frac{dr}{dt} = \frac{5,000}{\pi r^2}$. This yields $\frac{dS}{dt} = 8\pi r \cdot \frac{5,000}{\pi r^2} = \frac{40,000}{r}$. When $r = 300$ cm, then $\frac{dS}{dt} = \frac{40,000}{300} = 133\frac{1}{3}$ cm²/min.

2. The variables are $t, r, h,$ and V with $h = 2r$, $V = \frac{1}{3}\pi r^2 h = \frac{1}{3}\pi r^2 (2r) = \frac{2}{3}\pi r^3$. $\frac{dV}{dt} = 2\pi r^2 \frac{dr}{dt}$; $\frac{dV}{dt}$ is given as 10 ft³/s and h is 4 ft. When $h = 4$; $r = 2$ ft. Substituting we get $10 = 2\pi(2)^2 \frac{dr}{dt}$ or $\frac{dr}{dt} = \frac{10}{8\pi} = \frac{5}{4\pi}$ ft/s ≈ 0.40 ft/s.

3. In this problem we are not concerned about the volume, only the radius and height. These are related by the equation $(24 - h)^2 + r^2 = 24^2$ or $(24^2 - 48h + h^2) + r^2 = 24^2$ or $-48h + h^2 = -r^2$. Taking the derivatives with respect to time we get

$$-48\frac{dh}{dt} + 2h\frac{dh}{dt} = -2r\frac{dr}{dt}$$

and thus

$$\frac{dr}{dt} = \frac{24 - h}{r}\frac{dh}{dt}$$

When $h = 8$, then $r^2 = 24^2 - (24 - 8)^2 = 24^2 - 16^2 = 576 - 256$, so $r^2 = 320$ and $r =$

$\sqrt{320} = 8\sqrt{5}$ ft. We were given $\frac{dh}{dt} = 2$ ft/min . Substituting we get

$$\frac{dr}{dt} = \frac{24 - 8}{8\sqrt{5}}(2)$$
$$= \frac{32}{8\sqrt{5}} = \frac{4}{\sqrt{5}} = \frac{4\sqrt{5}}{5} \approx 1.789 \text{ ft/min}$$

4. $V = \pi r^2 h = 10$ m³ $= 10$. Taking derivatives with respect to time yields $2\pi r h \frac{dr}{dt} + \pi r^2 \frac{dh}{dt} = 0$ or $2h\frac{dr}{dt} + r\frac{dh}{dt} = 0$. Solving for $\frac{dr}{dt}$, we obtain $2h\frac{dr}{dt} = -r\frac{dh}{dt}$, or $\frac{dr}{dt} = \frac{-r}{2h}\frac{dh}{dt}$. When $r = 12$ m, then $h = \frac{10}{\pi 12^2} \approx 0.02210485$, so we get $\frac{dr}{dt} = \frac{-12}{2(0.0221)} \cdot (-1)$ mm/h $= 271.4$ mm/h $= 0.2714$ m/h.

5. Let x be the distance the helicopter flies after passing over you and y the distance from you to the helicopter. x and y are related by the equation $x^2 + (0.75)^2 = y^2$. Taking derivatives with respect to time yields $2x\frac{dx}{dt} + 0 = 2y\frac{dy}{dt}$ or $\frac{dy}{dt} = \frac{x}{y}\frac{dx}{dt}$. After 1 minute, $x = 3$ km and $y = 3.0923$ km. Substituting yields $\frac{dy}{dt} = \frac{3}{3.0923} \cdot 3 = 2.9104$ km/min $= 174.6$ km/h.

6. In this problem the width of the water is a constant of 10 m but the length and depth vary. They are related by the equation $\ell = 5h$ when $1 \leq h \leq 4$. The volume of the pool is $V = \frac{1}{2}\ell w h$, and so $V = \frac{1}{2}(5h) \cdot 10 \cdot h$ or $V = 25h^2$. Taking the derivative, we have $\frac{dV}{dt} = 50 \cdot h \cdot \frac{dh}{dt}$. Since we are given $\frac{dV}{dt} = 50$ L/min $= 50$ L/min $\cdot \frac{1\,m^3}{1000\,L} = 0.05$ m^3/min, substituting, we get $0.05 = 50 \cdot 2\frac{dh}{dt} \Rightarrow$ $\frac{dh}{dt} = \frac{0.05}{100} = 0.0005$ m/min.

7. Using the formula $V = \frac{1}{3}\pi h^2(3R - h)$ with $R = 5$ we get

$$V = \frac{1}{3}\pi h^2(15 - h) = 5\pi h^2 - \frac{1}{3}\pi h^3$$

$$\frac{dV}{dt} = 10\pi h \frac{dh}{dt} - \pi h^2 \frac{dh}{dt} = (10\pi h - \pi h^2)\frac{dh}{dt}$$

$$\frac{dh}{dt} = \frac{dV/dt}{10\pi h - \pi h^2}$$

We are given $\frac{dV}{dt} = 10$ ft^3/min, so $\frac{dh}{dt} = \frac{10}{10\pi h - \pi h^2}$ $= \frac{10}{(10h - h^2)\pi}$.

(a) When $h = 2$, we get $\frac{dh}{dt} = \frac{10}{(10 \cdot 2 - 2^2)\pi} = \frac{10}{(20-4)\pi} = \frac{10}{16\pi} = \frac{5}{8\pi} \approx 0.1989$ ft/min

(b) When $h = 3$, then $\frac{dh}{dt} = \frac{10}{(30-9)\pi} = \frac{10}{21\pi} \approx$ 0.1516 ft/min

(c) When $h = 4$, we get $\frac{dh}{dt} = \frac{10}{(40-16)\pi} = \frac{10}{24\pi} = \frac{5}{12\pi} \approx 0.1326$ ft/min

8. As in Exercise #7, $\frac{dh}{dt} = \frac{dV/dt}{(10h - h^2)\pi}$, and since $\frac{dV}{dt} = -45$ ft^3/min, we have $\frac{dh}{dt} = \frac{-45}{(10h - h^2)\pi}$.

(a) When $h = 4$, we have $\frac{dh}{dt} = \frac{-45}{(40-16)\pi} = \frac{-45}{24\pi} \approx$ -0.5968 ft/min

(b) When $h = 2$, then $\frac{dh}{dt} = \frac{-45}{(20-4)\pi} = \frac{-45}{16\pi} \approx$ -0.8952 ft/min.

9. Let h be the height of the surface of the water. Since the trough is an equilateral triangle we know that the width of the surface and the height are related by the equation $w = \frac{2\sqrt{3}h}{3}$. The volume of the trough is $V = \frac{1}{2}wh \cdot 10 = \frac{10\sqrt{3}h^2}{3}$ and $\frac{dV}{dt} = 25.0$ L/min $= 25$ L/min $\cdot \frac{1\,m^3}{1000\,L} = 0.025$ m^3/min; $\frac{dV}{dt} = \frac{20\sqrt{3}}{3}h\frac{dh}{dt} \Rightarrow \frac{dh}{dt} = \frac{3dV/dt}{20\sqrt{3}h}$. When $h = 0.175$, we have $\frac{dh}{dt} = \dfrac{0.025}{\frac{20\sqrt{3}}{3}(0.175)} \approx 0.0124$ m/min $= 12.4$ mm/min.

10. $A = \frac{1}{2}bh$. Differentiating, with respect to t, we have $\frac{dA}{dt} = \frac{1}{2}b\frac{dh}{dt} + \frac{1}{2}h\frac{db}{dt}$. Now, substituting the given information yields $\frac{dA}{dt} = \frac{1}{2}(262.5 \cdot (-4.00) +$ $(137.5) \cdot (5.00)) = -181.25$ cm^2/s.

11. We start with the formulas $V = \frac{4}{3}\pi r^3$ and $S = 4\pi r^2$. Differentiating V and setting it equal to 60, we get $\frac{dV}{dt} = 4\pi r^2 \frac{dr}{dt} = 60 \Rightarrow \frac{dr}{dt} = \frac{60}{4\pi r^2} = \frac{15}{\pi r^2}$. Now, differentiating S, we find $\frac{dS}{dt} = 8\pi r \frac{dr}{dt} = 8\pi r \cdot \frac{15}{\pi r^2} = \frac{120}{r}$. When $r = 8$ cm, $\frac{dS}{dt} = \frac{120}{8} = 15$ cm^2/s.

12. $LA = \frac{1}{2} \cdot 2\pi r \cdot \ell$ where ℓ is the slant height. Thus, $\ell = \sqrt{r^2 + h^2}$ so the lateral area is $LA = \pi r\sqrt{r^2 + h^2}$. From the given information, we see that $\frac{r}{h} = \frac{2.43}{4.27} \Rightarrow r = 0.57h$. Substituting, we get $LA = \pi(0.57h)\sqrt{(0.57h)^2 + h^2} = \pi(0.57h)$ $(1.15h); LA = 0.656\pi h^2$. Differentiating and substituting the given values produces $\frac{dLA}{dt} = 1.312$ $\pi h\frac{dh}{dt} = 1.312(\pi)(4.27)(0.425) = 7.48$ m^2/h.

13. We have $V = \frac{4}{3}\pi r^3 + \pi r^2 h$ and $h = 4r$, so $V = \frac{4}{3}\pi r^3 + 4\pi r^3 = \frac{16}{3}\pi r^3$. Differentiating, we obtain $\frac{dV}{dt} = 16\pi r^2 \frac{dr}{dt} = 60 \Rightarrow \frac{dr}{dt} = \frac{60}{16\pi r^2} = \frac{15}{4\pi r^2}$. The formula for the surface area is $S = 4\pi r^2 + 2\pi rh = 4\pi r^2 + 2\pi r(4r) = 4\pi r^2 + 8\pi r^2 = 12\pi r^2$. Differentiating this, we get $\frac{dS}{dt} = 24\pi r\frac{dr}{dt} = \frac{24\pi r \cdot 15}{4\pi r^2} = \frac{90}{r} = \frac{90}{8} = 11.25$ cm^2/s.

14. $A = \pi r^2; \frac{dA}{dt} = 2\pi r\frac{dr}{dt} = 2\pi(20)(0.2) = 8\pi \approx 25.13$ mm^2/s.

15. $A = s^2; \frac{dA}{dt} = 2s\frac{ds}{dt} = 2(40) \cdot (0.3) = 24$ mm^2/s.

16. As in example 24.21 $PV^{1.4} = k$, so $k = 60 \times 100^{1.4} = 37857. V^{1.4} = kP^{-1}; 1.4V^{0.4}\frac{dV}{dt} = -kP^{-2}\frac{dP}{dt}$. Solving for dV/dt, we obtain $\frac{dV}{dt} = \frac{-kP^{-2}}{1.4V^{0.4}}\frac{dP}{dt} = -\frac{37857}{1.4}(100)^{-0.4}(4)(60)^{-2} = -4.76$ cm^2/s.

17. Here $PV^{1.2} = 400$, so $P = 400V^{-1.2}. \frac{dP}{dt} = -1.2(400)V^{-2.2}\frac{dV}{dt} = \frac{-480}{V^{2.2}}\frac{dV}{dt}$. We are given that $\frac{dP}{dt} = 0.1P = 0.1(400V^{-1.2}) = 40V^{-1.2}$. Thus, $40V^{-1.2} = \frac{-480}{V^{2.2}}\frac{dV}{dt}$. Solving for dV/dt, we get $\frac{dV}{dt} = \frac{40V^{-1.2} \times V^{2.2}}{-480} = -0.0833V = -8.3\%$ of the volume per hour.

18. The height of the ballon, h, and the distance from the ballon to the observer, s, are related by the following equation $s^2 = h^2 + 500^2$. Taking the derivative with respect to t, we get $2s\frac{ds}{dt} = 2h\frac{dh}{dt}$

so $\frac{ds}{dt} = \frac{h}{s}\frac{dh}{dt}$. When $h = 2000$, $s = \sqrt{2000^2 + 500^2} = 2061.6$. Hence, $\frac{ds}{dt} = \frac{2000}{2061} \cdot 5 = 4.85$ ft/s.

19. Let a = altitude of the rocket and s be the distance from the radar station to the rocket. Then $a^2 + 2.430^2 = s^2$. Taking the derivatives, we obtain

$$2a\frac{da}{dt} = 2s\frac{ds}{dt} \Rightarrow \frac{da}{dt} = \frac{s}{a}\frac{ds}{dt}$$

When $a = 3.750, s = \sqrt{3.750^2 + 2.430^2} \approx 4.468$, and so $\frac{da}{dt} = \frac{4.468}{3.750}(325.0 \text{ m/s}) = 387.2$ m/s or 1394 km/hr.

20. $PV = k$ so $k = 20 \cdot 50 = 1000$ in./lb; $V = kP^{-1}$; $\frac{dV}{dt} = -kP^{-2}\frac{dP}{dt} = \frac{-k}{P^2}\frac{dP}{dt} = \frac{-1000}{20^2} \cdot (1.5) = -3.75$ ft³/min.

21. **(a)** The derivative of $v^2 = 1200 - 36.0s$ is $2v\frac{dv}{dt} = -36.0\frac{ds}{dt}$. Now, $\frac{dv}{dt} = a$ and $\frac{ds}{dt} = v$, so we get $2va = -36.0v$ or $a = -18.0$ m/s².
(b) The acceleration is constant so when $s = 2.45$ m, then a is -18.0 m/s².

22. Differentiating $h^2 + x^2 = 10^2$ we obtain $2h\frac{dh}{dt} + 2x\frac{dx}{dt} = 0$ and so $\frac{dh}{dt} = \frac{-x}{h}\frac{dx}{dt} = \frac{-x}{h}(0.5 \text{ m/s})$.
(a) When $x = 2, h = \sqrt{100 - 2^2} = \sqrt{96}$ and $\frac{dh}{dt} = \frac{-2}{\sqrt{96}}(0.5) = -0.1021$ m/s.
(b) When $x = 3, h = \sqrt{100 - 9} = \sqrt{91}$, and $\frac{dh}{dt} = \frac{-3}{\sqrt{91}}(0.5) = -0.1572$ m/s.
(c) When $x = 4, h = \sqrt{100 - 16} = \sqrt{84}$, and $\frac{dh}{dt} = \frac{-4}{\sqrt{84}}(0.5) = -0.2182$ m/s.
(d) When $x = 6, h = \sqrt{100 - 36} = \sqrt{64} = 8$, and $\frac{dh}{dt}\frac{-6}{8}(0.5) = -0.375$ m/s.

23. **(a)** We let A be at the origin and B at $(8, 0)$. The location of the ship at 8 mi from A means that the ship is on the circle $x^2 + y^2 = 64$; and the fact that the ship is 6 mi from B means that it is on the circle $(x - 8)^2 + y^2 = 36$. Solving the first equation for $y^2 = 64 - x^2$ and substituting into the second gives $(x - 8)^2 + (64 - x^2) = 36$ or $x^2 - 16x + 64 + 64 - x^2 = 36$, or $-16x = -92$, or $x = 5.75$ and so, $y = \pm\sqrt{64 - 5.75^2} = \pm 5.56$. The ship is 5.75 mi east of A and 5.56 mi north or south of A.
(b) Using the same setup as (a), the velocity vector from A has magnitude 14 and direction $\tan^{-1}\left(\frac{5.56}{5.75}\right)$

$= 44.0°$. The velocity vector from B has magnitude 2 and direction $\tan^{-1}\left(\frac{5.56}{2.25}\right) = 112.0°$. Using component vectors to add gives

$A_x =$	$14\cos 44° =$	10.07	$A_y =$	$14\sin 44° =$	9.73
$B_x =$	$2\cos 112° =$	-0.75	$B_y =$	$2\sin 112° =$	1.85
$R_x =$		9.32	$R_y =$		11.58

Thus, $|R| = \sqrt{9.32^2 + 11.58^2} = 14.86$ mph.
(c) $\theta = 51.2°$ north of east or 38.8° east of north.

24. **(a)** The first car's position t hours after the second car leaves is $y = 40 + 80t$. The second car's position is $x = -100t$. The distance between them is $s^2 = x^2 + y^2$.

$$2s\frac{ds}{dt} = 2x\frac{dx}{dt} + 2y\frac{dy}{dt} \quad \text{or} \quad \frac{ds}{dt} = \frac{xx' + yy'}{s}$$

When $t = 1$, then $x = -100$ and $y = 120$. We are given $x' = -100$ and $y' = 80$. Thus, $s = \sqrt{100^2 + 120^2} = 156.2$; $\frac{ds}{dt} = \frac{(-100)(-100)+120\cdot80}{156.2} = \frac{19600}{156.2} = 125.5$ km/h.
(b) When $t = 2$; $x = -200$; $x' = -100$; $y = 200, y' = 80$; $s = \sqrt{200^2 + 200^2} = 200\sqrt{2} = 282.8$, and so, $\frac{ds}{dt} = \frac{(-200)(-100) + (200)(80)}{282.8} = 127.3$ km/h.

25. **(a)** Let x be the distance from the person to the base of the light and y be the distance from the light to the top of the shadow. Then we have similar triangles so $\frac{y}{20} = \frac{y-x}{6}$. Simplifying: $6y = 20y - 20x$ or $-14y = -20x$ or $y = \frac{10}{7}x$, we get $\frac{dy}{dt} = \frac{10}{7} dx/dt = \frac{10}{7}(5) = \frac{50}{7} \approx 7.143$ ft/s.
(b) Same as (a) $\frac{50}{7} = 7.143$ ft/s.
(c) Same as (a) 7.143 ft/s.

26. Let s be the distance from the truck to the pulley, x the horizontal distance, and y the height of the weight above the ground. Then, $s^2 = x^2 + 18^2$ and $(20 - y) + s = 50$, so $s = 50 - 20 + y = 30 + y$ and $s' = y'$, which means that $2ss' = 2xx'$ and so, $s' = \frac{x}{s}x'$. When the weight is 10 ft above the ground, then $s = 30 + 10 = 40$ and $x = \sqrt{40^2 - 18^2} = 35.721$, so $s' = \frac{35.721}{40}(8) = 7.144$ ft/s. $y' = s' = 7.144$ ft/s.

27. Here $R = 6 + 0.008T^2$, so $\frac{dR}{dt} = 0.016T\frac{dt}{dt} = 0.016(40)(0.01) = 0.0064$ Ω/s.

28. The first car's position is $x = 80t$ and the second is $y = 40t$. The distance between the two cars is $d^2 = x^2 + y^2$ so $2dd' = 2xx' + 2yy'$ or $d' = \frac{xx'+yy'}{d}$

(a) After 3 min or 0.05 h, $x = 30(0.05) = 1.5$ mi; $x' = 30; y = 40(0.05) = 2$ mi and $d = \sqrt{1.5^2 + 2^2} = 2.5; d' = \frac{1.5(30)+2(40)}{2.5} = 50$ mph.

(b) After 6 min or 0.10 h, $x = 3$ mi; $x' = 30; y = 4$ mi; $y' = 40; d = \sqrt{3^2 + 4^2} = 5; d' = \frac{3(30)+4(40)}{5} = 50$ mph.

29. $Z = \frac{RX}{R+X}$. If R is constant 3 then $Z = \frac{3X}{3+X}$ and $Z' = \frac{(3+X)3X'-3X(X')}{(3+X)^2} = \frac{9X'}{(3+X)^2}$, or $Z' = \frac{9(1.45)}{(3+1.05)^2} = 0.80\,\Omega/\text{min}$.

30. $Z^2 = R^2 + X^2 = R^2 + 12^2$, and differentiating gives $2ZZ' = 2RR'$ or $Z' = \frac{R}{Z}R'$. When $R = 6$ and $X = 12, Z = \sqrt{6^2 + 12^2} = 6\sqrt{5}$ so $Z' = \frac{6}{6\sqrt{5}}(3) = \frac{3\sqrt{5}}{5} \approx 1.3416\,\Omega/\text{s}$.

31. $R = \frac{1}{R_1} + \frac{1}{R_2} = R_1^{-1} + R_2^{-1}; R' = \frac{-R_1'}{R_1^2} + \frac{-R_2'}{R_2^2} = \frac{-0.5}{4^2} + \frac{-0.4}{5^2} = -0.04725\,\Omega/\text{s}$. Thus, R is decreasing at the rate of $0.04725\,\Omega/\text{s}$.

32. $P = RI^2 = 80.00I^2$, so $P' = 160.00II' = 160.00(2.5)(0.24) = 96.00$ W/s.

33. $R = 35.0 + 0.0174T^2$, so $R' = 0.0348TT'$, and after substituting the given values, we have $R' = 0.0348(47.0)(-1.25) = -2.04\,\Omega/\text{min}$.

34. $v = 331\left(\frac{T}{273}\right)^{1/2}$ and $C + 273 = K$ so $C' = K'$. When $C = -30°$, we have $K = 243°. C' = -5.8 = K' = T'; v' = \frac{1}{2}\left(\frac{331}{273}\right)\left(\frac{T}{273}\right)^{-1/2}T' = \frac{1}{2}\left(\frac{331}{273}\right)\left(\frac{273}{243}\right)^{1/2}(-5.8) = -3.727$ m/s.

35. **(a)** $f = f_s\frac{v_L}{v_L-v_s}; f_s = 200; v_L = 343, v_s = 29; f = 200\left(\frac{343}{343-29}\right) = 218.47$ Hz.

(b) Here f_s and v_L are constants. We want to find f' and, in this case, we will write f as $f = f_s v_L(v_L - v_s)^{-1}$. Here v_s is the only variable, so $f' = f_s v_L(v_L - v_s)^{-2} = \frac{f_s v_L}{(v_L-v_s)^2}$. Substituting the given values, we obtain $f' = \frac{200(343)}{(343-29)^2} \approx 20.17$ Hz/s.

36. **(a)** Substituting $k = 475$ in $V = k\left(R^2 - r^2\right)$ produces $V = 475\left(R^2 - r^2\right) = 475R^2 - 475r^2$. Then, $\frac{dV}{dt} = 2 \cdot 475R\frac{dR}{dt} = 950R\frac{dR}{dt}$.

(b) $\frac{dV}{dt} = 950(0.015)(-0.001)$

$= -0.01425$mm/min

37. Substituting $k = 637.5$ in $V = k\left(R^2 - r^2\right)$ produces $V = 637.5\left(R^2 - r^2\right) = 637.5R^2 - 637.5r^2$. Then, $\frac{dV}{dt} = 2 \cdot 637.5R\frac{dR}{dt} = 1275R\frac{dR}{dt}$. Evaluating this derivative using the given data we have $\frac{dV}{dt} = 1275(0.0250)(0.002) = 0.06375$ mm/min.

38. Differentiating $C = 0.1(T - 47)^2 + 95$, produces $\frac{dC}{dt} = 0.2(T - 47)\frac{dT}{dt}$. Evaluating this using the given data we obtain $\frac{dC}{dt} = 0.2(T - 47)\frac{dT}{dt} = 0.2(51 - 47)(7) = 5.6$ crimes/mo.

39. Substituting the given values of $R = 7.5\,\Omega$ into $Z = \frac{RX}{R+X}$ we obtain $Z = \frac{7.5X}{7.5+X}$. Differentiating produces $\frac{dZ}{dt} = \frac{(7.5+X)7.5-7.5X}{(7.5+X)^2}\frac{dX}{dt} = \frac{7.5^2}{(7.5+X)^2}\frac{dX}{dt}$. Evaluating this when $X = 4.0\,\Omega$ and $\frac{dX}{dt} = -2.5$ Ω/s, we obtain $\frac{dZ}{dt} = \frac{7.5^2}{(7.5+4.0)^2}(-2.5) = -1.06\,\Omega/\text{s}$.

40. Differentiating the formula for the area of a circle $A = \pi r^2$ with respect to t, we obtain $\frac{dA}{dt} = 2\pi r\frac{dr}{dt}$. Evaluating this derivative at the given data produces $\frac{dA}{dt} = 2\pi(225)(0.9) = 405\pi$ m²/s $\approx 1\,272.3$ m²/s.

41. As in Exercise 39, $\frac{dZ}{dt} = \frac{7.5^2}{(7.5+X)^2}\frac{dX}{dt}$, so $\frac{dX}{dt} = \frac{dZ}{dt}\Big/\frac{7.5^2}{(7.5+X)^2}$. Substituting, we get $\frac{dX}{dt} = 2.0\Big/\frac{7.5^2}{(7.5+3.5)^2} = 4.30\,\Omega/\text{s}$.

42. The oil slick is in the shape of a right circular cylinder on top of a circular cone. The height of the cylinder is 0.5 cm and its radius is r. The cone has radius r and height 4.5 cm. The volume of the oil slick at time t, in min, is $V = \pi r^2 h + \frac{1}{3}\pi r^2 h = \pi r^2(0.5) + \frac{1}{3}\pi r^2(4.5) = 2\pi r^2$. Thus, $\frac{dV}{dt} = 4\pi r\frac{dr}{dt}$ and so, $\frac{dr}{dt} = \frac{dV}{dt}\frac{1}{4\pi r}$. We are given $\frac{dV}{dt} = 5\,000$ L/min $= 5\,000\,000$ cm³/min. After 4 h, the volume of the spill is $V = 5\,000$ L/min $\times 60$ min/h $\times 4$ h $= 1\,200\,000$ L. At this time $r = \sqrt{\frac{1\,200\,000}{2\pi}} = \sqrt{\frac{600\,000}{\pi}}$ cm. Substituting, we obtain $\frac{dr}{dt} = 5\,000\,000\left(\frac{1}{4\pi\sqrt{\frac{600\,000}{\pi}}}\right) = \frac{1250\sqrt{15}}{3\sqrt{\pi}} \approx 910.5$ cm/min.

43. Since the cross-section of the pipe is circular its area is given by $A = \pi r^2$ and so, $\frac{dA}{dt} = 2\pi r \frac{dr}{dt}$. Evaluating this derivative at the given data produces $\frac{dA}{dt} = 2\pi(0.5)(-0.1) = -0.1\pi$ in.^{2}/yr. Thus, the cross-sectional area of the pipe's opening is decreasing at the rate of 0.1π in.^{2}/yr ≈ 0.314 in.^{2}/yr.

44. The formula for the volume of the frustum of a square pyramid is $V = \frac{1}{3}h\left(B_1 + B_2 + \sqrt{B_1 B_2}\right)$. In this problem, $B_1 = 3^2 = 9$, h is changing and $B_2 = \left(3 - \frac{1}{2}h\right)^2$. Hence, we obtain $V =$

$$\frac{1}{3}h\left(9 + \left(3 - \frac{1}{2}h\right)^2 + \sqrt{9\left(3 - \frac{1}{2}h\right)^2}\right) =$$

$$\frac{1}{3}h\left(9 + 9 - 3h + \frac{1}{4}h^2 + 9 - \frac{3}{2}h\right) =$$

$$\frac{1}{3}h\left(27 - \frac{9}{2}h + \frac{1}{4}h^2\right) = 9h - \frac{3}{2}h^2 + \frac{1}{12}h^3.$$ Differentiating, we obtain $\frac{dV}{dt} = \left(9 - 3h + \frac{h^2}{4}\right)\frac{dh}{dt}$. Substituting the given information produces $0.01 = \left(9 - 3 + \frac{1}{4}\right)\frac{dh}{dt} = \frac{25}{4}\frac{dh}{dt}$. Hence, $\frac{dh}{dt} = \frac{0.01}{25/4} = \frac{0.01}{6.25} = 0.0016$ m/s.

≡ 24.6 NEWTON'S METHOD

1. See *Computer Programs* in main text.

2. $x^3 + 5x - 8 = 0$; $P(x) = x^3 + 5x - 8$; $P'(x) = 3x^2 + 5$. There is a solution near 1. Using Newton's method, we obtain the following results:

i	x_i	$P(x_i)$	$P'(x_i)$	$x_{i+1} = x_i - \dfrac{P(x_i)}{P'(x_i)}$	$P(x_{i+1})$
1	1	-2	8	1.25	0.2031
2	1.25	0.2031	9.6875	1.2290	0.0013
3	1.229	0.0013	9.5313	1.2289	0.0004
4	1.2289	0.0004	9.5306	1.22885	-0.000002

Hence, $x \approx 1.2289$

3. $x^3 - 4x + 2 = 0$; $P(x) = x^3 - 4x + 2$; $P'(x) = 3x^2 - 4$. There are three solutions; one each near -2; 0.5, and 1.5. Newton's method produces the following results for each solution.

i	x_i	$P(x_i)$	$P'(x_i)$	$x_{i+1} = x_i - \dfrac{P(x_i)}{P'(x_i)}$	$P(x_{i+1})$
1	-2	2	8	-2.25	-0.3906
2	-2.25	-0.3906	11.1875	-2.2151	-0.0084
3	-2.2151	-0.0084	10.7200	-2.2143	2.11×10^{-4}

Hence, one solution is $x \approx -2.2143$.

i	x_i	$P(x_i)$	$P'(x_i)$	$x_{i+1} = x_i - \dfrac{P(x_i)}{P'(x_i)}$	$P(x_{i+1})$
1	0.5	0.125	-3.25	0.5385	0.0022
2	0.5385	0.0022	-3.1301	0.5392	-3.48×10^{-5}

Another solution is $x \approx 0.5392$.

i	x_i	$P(x_i)$	$P'(x_i)$	$x_{i+1} = x_i - \dfrac{P(x_i)}{P'(x_i)}$	$P(x_{i+1})$
1	1.5	-0.625	2.75	1.7273	0.2443
2	1.7273	.2443	4.9507	1.6780	0.0127
3	1.6780	0.127	4.4471	1.6751	-1.36×10^{-4}

And the final solution is $x \approx 1.6751$.

4. Let $x^4 - x - 4 = P(x)$, so $P'(x) = 4x^3 - 1$. Solutions are near -1.2 and 1.4. Newton's method produces the following results for each solution.

i	x_i	$P(x_i)$	$P'(x_i)$	$x_{i+1} = x_i - \dfrac{P(x_i)}{P'(x_i)}$	$P(x_{i+1})$
1	-1.2	-0.7264	-7.912	-1.2918	0.0766
2	-1.2918	0.0766	-9.6228	-1.2838	1.74×10^{-4}

Hence, one solution is $x \approx -1.2838$.

i	x_i	$P(x_i)$	$P'(x_i)$	$x_{i+1} = x_i - \dfrac{P(x_i)}{P'(x_i)}$	$P(x_{i+1})$
1	1.4	-1.5584	9.9760	1.5562	0.3088
2	1.5562	0.3088	14.0750	1.5343	0.0074
3	1.5343	0.0074	13.4474	1.5337	-6.9×10^{-4}

And the other solution is $x \approx 1.5337$.

5. $P(x) = x^4 - 2x^3 - 3x + 2$ and $P'(x) = 4x^3 - 6x^2 - 3$. Solutions are near 0.6 and 2.3. Newton's method produces the following results for each solution.

i	x_i	$P(x_i)$	$P'(x_i)$	$x_{i+1} = x_i - \dfrac{P(x_i)}{P'(x_i)}$	$P(x_{i+1})$
1	0.6	-0.1024	-4.2960	0.5762	-9.76×10^{-4}
2	0.5762	-9.76×10^{-4}	-4.2268	0.5760	-1.3×10^{-4}

Hence, one solution is $x \approx 0.5760$.

i	x_i	$P(x_i)$	$P'(x_i)$	$x_{i+1} = x_i - \dfrac{P(x_i)}{P'(x_i)}$	$P(x_{i+1})$
1	2.3	-1.2499	13.9280	2.3897	0.1490
2	2.3897	0.1490	17.3231	2.3811	0.0015
3	2.3811	0.0015	16.9821	2.3810	-1.6×10^{-4}

And the other solution is $x \approx 2.3810$.

6. $x^4 = 125 \Rightarrow x^4 - 125 = 0$ so let $P(x) = x^4 - 125$, and then $P'(x) = 4x^3$. Since $P(x)$ is an even function, solutions will be in positive and negative pairs. One solution is near 3.3. Newton's method will show that $x \approx -3.3437$ and 3.3437.

7. $x^3 = 91 \Rightarrow x^3 - 91 = 0$; $P(x) = x^3 - 91$; $P'(x) = 3x^2$. A solution is near 4.5. Using Newton's method, you can show that $x \approx 4.4979$.

8. $x^5 = 111 \Rightarrow x^5 - 111 = 0$ so $P(x) = x^5 - 111$ and $P'(x) = 5x^4$. A solution is near 2.5. Using Newton's method, you can show that $x \approx 2.5649$.

9. (a) The volume is given by $V = \pi r^2 h$ and the total surface area is $A = 2\pi rh + \pi r^2$. Then $2V = 2\pi r^2 h$ and so,

$$2V - 2\pi r^2 h = 0.$$

From the area formula, we see that $A - \pi r^2 = 2\pi rh$ and so, $Ar - \pi r^3 = 2\pi r^2 h$. Substituting this for $2\pi r^2 h$ in the displayed equation produces the desired result

$$2V - (Ar - \pi r^3) = 2V - Ar + \pi r^3 = 0$$
$$2V + \pi r^3 - Ar = 0.$$

(b) Substituting 10 000 for V and 2 400 for A, we get

$$2(10000) + \pi r^3 - (2400)r = 0$$
$$\pi r^3 - 2400r + 20000 = 0.$$

Using Newton's method we obtain solution at $r_1 = 9.4315$ cm, $r_2 = 21.6893$ cm

10. In this problem, we need to determine a formula for the cross-sectional area of the trough. The area of the sector is $A = \frac{1}{2}r^2\theta$. Since the arc length is

a constant 0.7 m and $S = r\theta$, we get $r = 0.7/\theta$. Hence, the area of the sector is $\frac{1}{2}\left(\frac{0.7}{\theta^2}\right)\theta = \frac{0.245}{\theta}$.

The area of the trough is the sector less the triangle. The height of the triangle is $r\cos\frac{\theta}{2}$ and the base is $2r\sin\frac{\theta}{2}$. Hence, its area is $\frac{1}{2}r\cos\frac{\theta}{2}2r\sin\frac{\theta}{2}$ $= \frac{1}{2}r^2\left(2\sin\frac{\theta}{2}\cos\frac{\theta}{2}\right) = \frac{1}{2}r^2\sin\theta$.

Substituting again for r, the area of the triangle is $\frac{1}{2}\left(\frac{0.7}{\theta}\right)^2\sin\theta = \frac{0.245}{\theta^2}\sin\theta$. The area of the trough is $\frac{0.245}{\theta} - \frac{0.245}{\theta^2}\sin\theta$. The volume is $20\left(\frac{0.245}{\theta} - \frac{0.245}{\theta^2}\sin\theta\right) = 4.9\left(\frac{1}{\theta} - \frac{\sin\theta}{\theta^2}\right) = 0.85$.

Thus, $\frac{1}{\theta} - \frac{\sin\theta}{\theta^2} \approx 0.173469$. Multiplying by θ^2 we get $\theta - \sin\theta - \sin\theta - 0.173469\theta^2 = 0$. By Newton's method, $\theta = 1.1066$ rad $\approx 63.40°$.

24.7 DIFFERENTIALS

1. $dy = f'(x)dx = (4x^3 - 2x)dx$

2. $dy = g'(x)dx = (6x + 2)dx$

3. $dy = (15x^2 - 2x + 1)dx$

4. $dy = 3(x^2 - 4)^2(2x)dx = 6x(x^2 - 4)^2 dx$

5. $dy = \frac{1}{3}(4 - 2x)^{-2/3}(-2)dx = \frac{-2}{3}(4 - 2x)^{-2/3}dx$

6. $dy = \frac{(x-4)-x}{(x-4)^2}dx = \frac{-4}{(x-4)^2}dx$ or $-4(x - 4)^{-2}dx$

7. $\Delta y = ((3.2)^2 - 3.2) - (3^2 - 3) = 7.04 - 6 = 1.04$, $dy = (2x - 1)dx = 5(0.2) = 1.0$

8. $\Delta y = f(2.1) - f(2) = 11.361 - 10 = 1.361$, $dy = f'(2)dx = 13(0.1) = 1.3$

9. $\Delta y = f(5.15) - f(5) = 59.9675 - 56 = 3.9675$, $dy = f'(5)(0.15) = 26(0.15) = 3.9$

10. $\Delta y = g(4.2) - g(4.1) = 311.1696 - 282.5761 = 28.5935$, $dy = g'(4.1)0.1 = 27.5684$

11. $\Delta y = h(3.2) - h(3) = 0.0305175781 - 0.037037037 = -0.00651946$, $dy = h'(3)(0.2) = -0.037037037(0.2) = -0.00740741$

12. $\Delta y = j(0.6) - j(0.5) = 0.375 - 0.333333 = 0.0416667$, $dy = j'(0.5)(0.1) = 0.444444(0.1) = 0.0444444$

13. **(a)** $V = \frac{1}{3}\pi r^2 h = \frac{1}{3}\pi(2)^2 \cdot 4 = \frac{16\pi}{3} = 16.755$ m^3

(b) $dV = \frac{2}{3}\pi rh\,dr = \frac{2}{3}(\pi \cdot 2 \cdot 4)(0.01) = \frac{0.16\pi}{3} = \frac{16}{300}\pi \approx 0.16755$ m^3

(c) Relative error $= \frac{dV}{V} = \frac{\frac{16\pi}{300}}{16\frac{\pi}{3}} = \frac{1}{100}$; $\frac{1}{100} = 0.01 = 1\%$.

14. **(a)** $A = \pi r^2$; $dA = 2\pi r\,dr = 2\pi(200)(1.5) = 600\pi$ mm^2

(b) $\frac{dr}{r} = \frac{1.5}{200} = 0.0075 = 0.75\%$

(c) $\frac{dA}{A} = \frac{600\pi}{\pi(200)^2} = \frac{6}{400} = 0.015 = 1.5\%$

15. **(a)** $A = s^2$; $dA = 2s\,ds = 2(24)(0.02) = 0.96$ in^2

(b) $\frac{dA}{A} = \frac{0.96}{24^2} = 0.00167 = 0.167\%$

16. $V = s^3$; $dV = 3s^2 ds = 3(1.452)^2(0.0005) = 0.00316$ m^3 = 3162 cm^3 (Note: 0.5 mm = 0.0005 m and 1 m^3 = $(100\text{ cm})^3$).

17. $S = 2\pi r^2$; $dS = 4\pi r\,dr = 4\pi(100)(0.01) = 4\pi$ m^2. (Note: 10 mm = 0.01 m.)

18. $\frac{dS}{S} = \frac{4\pi r\,dr}{2\pi r^2} = \frac{2dr}{r} = 0.01\% = 0.0001$, $dr = \frac{0.0001r}{2} = \frac{0.01}{2} = 0.005$ m = 5 mm.

19. **(a)** The volume of a sphere is $V = \frac{4}{3}\pi r^3$. Since we are given $d = 12.4$ m, we see that $r = 6.2$ m, and $V = \frac{4}{3}\pi(6.2)^3 \approx 317.8\pi$. Taking the derivative of the formula for the volume, we obtain $dV = 4\pi r^2 dr = 4\pi(6.2)^2(\pm 0.05) = \pm 7.69\pi$ m$^3 \approx \pm 24.2$ m^3

(b) $\frac{dV}{V} = \frac{7.69\pi}{317.8\pi} = 0.024 = \pm 2.4\%$.

20. **(a)** $dV = 4\pi r^2 dr = 4\pi(6.20)^2(0.005) = \pm 0.7688\pi \approx 2.42$ m^3

(b) $\frac{dV}{V} = \frac{0.7688\pi}{317.8\pi} = \pm 0.00242 = \pm 0.242\%$.

21. $R = \sigma T^4$, so $dR = 4\sigma T^3 dT$ and $\frac{dT}{T} = 0.02$; $\frac{dR}{R} = \frac{4\sigma T^3 dT}{\sigma T^4} = 4\frac{dT}{T} = 4(0.02) = 0.08$.

22. Here $R = 250r^{-2}$ and $\frac{dr}{r} = \pm 0.4\%$.

(a) $dR = -500r^{-3}dr = \frac{-500}{r^2} \cdot \frac{dr}{r} = \frac{-500}{r^2} \cdot (\pm 0.4\%) = \frac{\pm 200\%}{r^2} = \pm \frac{2}{r^2}\,\Omega$.

(b) $\frac{dR}{R} = \frac{\pm \frac{2}{r^2}}{\frac{250}{r^2}} = \frac{\pm 2}{250} = \pm 0.008 = \pm 0.8\%$.

23. $V = IR; 30 = IR; I = 30R^{-1}; dI = -30R^{-2}dR = -30(10)^{-2}(0.1) = -0.03$ A

24. $R = \frac{20R_v}{20+R_v}; dR = \frac{(20+R_v)20(dR_v)-20R_v(dR_v)}{(20+R_v)^2} = \frac{400 dR_v}{(20+R_v)^2} = \frac{400(0.1)}{(20+10)^2} = \frac{40}{900} = 0.0444\,\Omega$.

25. $R = 35.0 + 0.0174T^2; dR = 0.0348T \cdot dT = 0.0348(125)(\pm 0.5) = \pm 2.175\,\Omega$.

26. Since $1\text{ dm}^3 = 1$ L, we will convert all units to decimeters. Here $V = \frac{4}{3}\pi r^3$, and so, $dV = 4\pi r^2 dr$. Since 20 m $= 200$ dm and 0.5 mm $= 0.005$ dm, we have $dV = 4\pi(200)^2(0.005) \approx 2\,531\text{ dm}^3 = 2513$ L.

27. $S = 4\pi r^2; \frac{dS}{S} = \frac{8\pi r dr}{4\pi r^2} = 2\frac{dr}{r} = 2(0.5\%) = 1.0\%$

28. **(a)** $V = \frac{4}{3}\pi r^3 = \frac{4}{3}\pi(8)^3 = 2144.7\text{ mm}^3$, mass $= (2144.7)(0.00794)$ g $= 17.03$ g,
(b) $dV = 4\pi r^2 dr = 4\pi \cdot 8^2 \cdot (.005 \times 8) = 32.17$, so the mass is $32.17(0.00794 = 0.255$ g. (Note: $\frac{dr}{r} = 0.5\% = 0.005$ so $dr = 0.005r = 0.005 \times 8$.)

29. A ball bearing is spherical and its volume is given by $V = \frac{4}{3}\pi r^3$. Differentiating produces $\frac{dV}{dr} = 4\pi r^2$ or $dV = 4\pi r^2 dr$. Substituting the given values yields $dV = 4\pi(0.6)^2(\pm 0.015) = \pm 0.0678$ or $\pm 0.07\text{ mm}^3$.

30. Differentiating $C(x) = \frac{5x}{9+x^2}$ we get $C'(x) = \frac{dC}{dx} = \frac{(9+x^2)5-5x(2x)}{(9+x^2)^2} = \frac{45+5x^2-10x^2}{(9+x^2)^2}$. Thus, $dC = \frac{45-5x^2}{(9+x^2)^2}dx$.
(a) Evaluating dC when $x = 1$ and $dx = 1.5 - 1.0 = 0.5$ produces

$$dC = \left(\frac{45-5(1)^2}{(9+(1)^2)^2}\right)(0.5) = \left(\frac{40}{100}\right)(0.5) = 0.2$$

Thus, the concentration of this drug in the bloodstream increases 0.2 during the period from 1 h to 1.5 h after being administered.
(b) Evaluating dC when $x = 2$ and $dx = 2.25 - 2.0 = 0.25$ produces

$$dC = \left(\frac{45-5(2)^2}{(9+(2)^2)^2}\right)(0.25) = \left(\frac{25}{169}\right)(0.25) \approx 0.037$$

Thus, during the 15 min period from 2 h to 2.25 h after being administered, the concentration of this drug in the bloodstream increases 0.04.
(c) Evaluating dC when $x = 3$ and $dx = 3.25 - 3.0 = 0.25$ produces

$$dC = \left(\frac{45-5(3)^2}{(9+(3)^2)^2}\right)(0.25) = \left(\frac{0}{324}\right)(0.25) = 0$$

Thus, during the 15 min period from 3 h to 3.25 h after being administered, the concentration of this drug in the bloodstream does not increase.

31. The volume of a sphere is given by $V = \frac{4}{3}\pi r^3$. Differentiating produces $\frac{dV}{dr} = 4\pi r^2$ or $dV = 4\pi r^2 dr$. Substituting the given values yields $dV = 4\pi(15)^2(0.5) = 450\pi \approx 1414\text{ mm}^3$.

32. The area of a circle is $A = \pi r^2$ and so $dA = 2\pi r\, dr$. Here $dr = 2.3 - 2.1 = 0.2$ km. Using $r = 2.1$ km produces $dA = 2\pi(2.1)(0.2) = 0.84\pi \approx 2.639\text{ km}^2$.

33. Here $dR = (0.04n + 0.00045n^2)dn$. Using $n = 500$ and $dn = 1$ produces $dR = (0.04(500) + 0.00045(500)^2)(1) = 132.5$. Revenue will increase by \$132.50 from the sale of one more computer when 500 have been sold.

34. The volume of a sphere is given by $V = \frac{4}{3}\pi r^3$. Differentiating produces $\frac{dV}{dr} = 4\pi r^2$ or $dV = 4\pi r^2 dr$. Substituting $r = 20.0$ and $dr = 21.5 - 20.0 = 1.5$ yields $dV = 4\pi(20.0)^2(1.5) = 2400\pi \approx 7539.8$. The volume of the weather balloon increases by $2,400\pi \approx 7540\text{ in.}^3$ as the radius changes from 20.0 in. to 21.5 in.

35. The volume of a right circular cylinder is given by $V = \pi r^2 h$. Differentiating with respect to r produces $\frac{dV}{dr} = 2\pi rh$ or $dV = 2\pi rh\, dr$. We are given $h = 3.2$, $r = \frac{1}{2}(30.0) = 15.0$, and $dr = 0.2$. Substituting these values produces $dV = 2\pi(15.0)(3.2)(0.2) = 19.2\pi \approx 60.3\text{ cm}^3$.

36. Since we are differentiating with respect to T, we can treat L and μ as constants. Thus, we can rewrite the formula for f as $f = \frac{1}{2L}\sqrt{\frac{T}{\mu}} = \frac{1}{2L}\left(\frac{T}{\mu}\right)^{1/2} = \frac{1}{2L\sqrt{\mu}}T^{1/2}$. Differentiating f with respect to T produces $df = \left(\frac{1}{2}\right)\frac{1}{2L\sqrt{\mu}}T^{-1/2}dT = \frac{1}{4L\sqrt{\mu}}T^{-1/2}dT$. The change in the tension is $dT =$

$1\% = 0.01 \times 620 = 6.2$. Substituting this value for dT along with $L = 1.15$ m, $\mu = 0.0036$ kg/m, and $T = 620$ N, we get $df = \frac{1}{4(1.15)\sqrt{0.0036}}(620)^{-1/2}6.2 \approx 0.902$. The change in frequency is about 0.902 Hz.

24.8 ANTIDERIVATIVES

1. $f(x) = 7; F(x) = 7x + C$

2. $g(x) = 4x; G(x) = 4 \cdot \frac{1}{2}x^2 + C = 2x^2 + C$

3. $h(x) = 2 - 3x^2; H(x) = 2x - 3\left(\frac{1}{3}\right)x^3 + C = 2x - x^3 + C$

4. $j(x) = x^2 - 3x + 5; J(x) = \frac{1}{3}x^3 - 3 \cdot \frac{1}{2}x^2 + 5x + C = \frac{1}{3}x^3 - \frac{3}{2}x^2 + 5x + C$

5. $k(x) = 4x^3 - 3x^2 + 2x + 9; K(x) = 4 \cdot \frac{1}{4}x^4 - 3 \cdot \frac{1}{3}x^3 + 2 \cdot \frac{1}{2}x^2 + 9x + C = x^4 - x^3 + x^2 + 9x + C$

6. $m(x) = \frac{4}{5}x^3 - 6x^2; M(x) = \frac{4}{5} \cdot \frac{1}{4}x^4 - 6 \cdot \frac{1}{3}x^3 + C = \frac{1}{5}x^4 - 2x^3 + C$

7. $f(x) = x^{-4} + 2x^{-3} - x^{-2} + 5; F(x) = \frac{1}{-3}x^{-3} + 2 \cdot \frac{1}{-2}x^{-2} - \frac{1}{-1}x^{-1} + 5x + C = \frac{-1}{3}x^{-3} - x^{-2} + x^{-1} + 5x + C$

8. $g(x) = \frac{1}{x^3} + 2 = x^{-3} + 2; G(x) = \frac{1}{-2}x^{-2} + 2x + C = -\frac{1}{2}x^{-2} + 2x + C$

9. $h(x) = \sqrt{x} + x - \frac{1}{x^2} = x^{1/2} + x - x^{-2}; H(x) = \frac{1}{3/2}x^{3/2} + \frac{1}{2}x^2 - \frac{1}{-1}x^{-1} + C = \frac{2}{3}x^{3/2} + \frac{1}{2}x^2 + x^{-1} + C$

10. $j(x) = 3x^2 - x^{-1/3} - 4x^{-2}; J(x) = 3 \cdot \frac{1}{3}x^3 - \frac{1}{2/3}x^{2/3} - 4 \cdot \frac{1}{-1}x^{-1} + C = x^3 - \frac{3}{2}x^{2/3} + 4x^{-1} + C$

11. $k(x) = 2x^{-1.4} + 3.5x^{-6} + 1.2x^{-1.3}; K(x) = 2 \cdot \frac{1}{-.4}x^{-.4} + 3.5 \cdot \frac{1}{-5}x^{-5} + 1.2 \cdot \frac{1}{-0.3}x^{-.3} + C = -5x^{-0.4} - 0.7x^{-5} - 4x^{-0.3} + C$

12. $m(x) = 2x^{-1/2} + 3x^{5/2}; M(x) = 2 \cdot \frac{1}{1/2}x^{1/2} + 3 \cdot \frac{1}{7/2}x^{7/2} + C = 4x^{1/2} + \frac{6}{7}x^{7/2} + C$

13. $s(t) = t^2 + 2t; S(t) = \frac{1}{3}t^3 + t^2 + C$

14. $a(t) = \frac{2}{3}t - \frac{3}{4}t^2 + t^{3/2}; A(t) = v(t) = \frac{2}{3} \cdot \frac{1}{2}t^2 - \frac{3}{4} \cdot \frac{1}{3}t^3 + \frac{1}{5/2}t^{5/2} = \frac{1}{3}t^2 - \frac{1}{4}t^3 + \frac{2}{5}t^{5/2} + C$

15. $v(t) = 42t - 5; V(t) = \frac{42}{2}t^2 - 5t + C = 21t^2 - 5t + C$

16. $p(v) = \frac{1}{3}v^3 - 3v^{-4}; P(v) = \frac{1}{3} \cdot \frac{1}{4}v^4 - 3 \cdot \frac{1}{-3}v^{-3} + C = \frac{1}{12}v^4 + v^{-3} + C$

17. $a(t) = -32$ ft/s^2; $v(t) = A(t) = -32 \cdot t + C_1$. When the ball is dropped at $t = 0$, we have $v(0) = 0$, so we can see that $C_1 = 0$. Hence, $v(t) = -32t$. Now, $s(t) = V(t) = -16t^2 + C_2$. When $t = 0$, we are given that $s(0) = 600$, so $C_2 = 600$ and we have $s(t) = -16t^2 + 600$.
(a) When the ball hits the ground, $s(t) = 0$. Solving $s(t) = 0$ for t, we have $16t^2 = 600$ or $t^2 = \frac{600}{16}$ and so the ball hits the ground after $t = \sqrt{\frac{600}{16}} \approx 6.12$ s
(b) $v(6.12) = -32(6.12) = -195.96$ ft/s

18. $a(t) = -9.8$ m/s^2; $v(t) = -9.8t + C$. Since at the time the ball is dropped ($t = 0$), the velocity is 0, we have $C = 0$, and so $v(t) = -9.8t$. Thus, we find that $s(t) = -4.9t^2 + C_2$. When $t = 0, s = 140$ m, and so $s(t) = -4.9t^2 + 140$.
(a) $s(t) = 0$ when $4.9t^2 = 140$ or $t \approx 5.345$ s.
(b) $v(5.345) = -52.38$ m/s

19. $a(t) = -32; v(t) = -32t + C$. Since the ball was thrown down at $t = 0$ at 25 ft/s, we have $C = -25$ and thus, $v(t) = -32t - 25$.
(a) $v(t) = -185 = -32t - 25 \Rightarrow 32t = 160 \Rightarrow t = 5$ s.
(b) $s(t) = -16t^2 - 25t + C, s(5) = 0 = -16 \cdot 5^2 - 25 \cdot 5 + C, C = 16 \cdot 5^2 + 25 \cdot 5 = 525$. The building is 525 feet high.

20. $a(t) = -9.8$ m/s^2; $v(t) = -9.8t + C$; when $t = 0, v = -12.1$ so $C = -12.1$. This gives $v(t) = -9.8t - 12.1$.
(a) $v(t) = -9.8t - 12.1 = -66 \Rightarrow 9.8t = 53.9; t = 5.5$ s

(b) $s(t) = -4.9t^2 - 12.1t + C; s(5.5) = -4.9(5.5)^2$ $-12.1(5.5) + C = 0$, so $C = 4.9(5.5)^2 + 12.1(5.5) =$ 214.775 m

21. $a(t) = -32$ and $v(t) = -32t + 160$.
(a) Since $v(t) = -32t + 160 = -384$, then $-32t = -544$ and $t = 17$ s.
(b) $s(t) = -16t^2 + 160t + C$, so $s(17) = -16 \cdot 17^2 + 160 \cdot 17 + C = 0$, thus $C = 4.9(16.837)^2 - 45(16.837) = 631.41$ m $= 1904$ ft.

22. $a(t) = -9.8; v(t) = -9.8t + 45$.
(a) $-9.8t + 45 = -120$ or $-9.8t = -165$ and so $t = 16.837$ s.
(b) $s(t) = -4.9t^2 + 45t + C$, and $s(16.837) = -4.9(16.837)^2 + 45(16.837) + C = 0, C = 4.9 \times (16.837)^2 - 45(16.837) = 631.41$ m

23. $a(t) = 3.2; v(t) = 3.2t + 0; 3.2t = 32 \Rightarrow t = 10$ s, $s(t) = 1.6t^2, s(10) = 1.6(10)^2 = 160$ m

24. 45 mph $= 45\frac{\text{mi}}{\text{hr}} \times 5280$ ft/mi $\times \frac{1\text{hr}}{3600\text{s}} = 66$ ft/s
(a) $a(t) = -20$ ft/s^2; $v(t) = -20t + 66 = 0 \Rightarrow t = \frac{66}{20} = 3.3$ s
(b) $s(t) = -10t^2 + 66t; s(3.3) = -10(3.3)^2 + 66 \cdot 3.3 = 108.9$ ft

25. $v(t) = a \cdot t + 50 \Rightarrow a = \frac{-50}{t}, s(t) = \frac{a}{2}t^2 + 50t + 500 = 0, \frac{-50}{2t}t^2 + 50t = -500$. This simplifies to $-25t + 50t = 500$, or $25t = 500$, or $t = 20$ s. Finally, $a = \frac{-50}{20} = -2.5$ m/s^2.

26. **(a)** $\alpha(t) = 4 - t, \omega(t) = 4t - \frac{1}{2}t^2 + C, \omega(0) = 10 = 4t - \frac{1}{2}t^2 + C$ so $C = 10, \omega(t) = 10 + 4t - \frac{1}{2}t^2, \theta(t) = 10t + 2t^2 - \frac{1}{6}t^3 + C, \theta(0) = 0$ so $C = 0$, $\theta(t) = 10t + 2t^2 - \frac{1}{6}t^3$
(b) The wheel begins to turn in the opposite direction when $\omega(t) = 0$. This is at -2 and 10 seconds. Since $t \geq 0, t = 10$ and $\theta(10) = 133.3$ rad.
(c) $\omega(t) = -14 = 10 + 4t - \frac{1}{2}t^2, 0 = 24 + 4t - \frac{1}{2}t^2 \Rightarrow t^2 - 8t - 48 = 0, (t - 12)(t + 4) = 0; t = 12$ or $t = -4$. In this problem time cannot be negative, so $t = 12$ s.

27. **(a)** $\theta(t) = 12.4t - \frac{3.4}{2}t^2 + \frac{0.30}{3}t^3 =$ antiderivative of $\omega(t) = 12.4t - 1.7t^2 + 0.10t^3; \theta(3) = 24.6$ rad $= \frac{24.6}{2\pi} = 3.9$ rev.
(b) $\alpha(t) = \omega'(t) = -3.4 + 0.6t = -1, 0.6t = 2.4, t = 4$ s, $\theta(4) = 28.8$ rad $= \frac{28.8}{2\pi} = 4.6$ rev.

28. $a(t) = kt; v(t) = \frac{k}{2}t^2 + C$. Since the ball starts

from rest, let $C = 0$. $s(t) = \frac{k}{6}t^3; \frac{k}{6}4^3 = 24; k = \frac{24 \cdot 6}{4^3} = 2.25, s(t) = \frac{2.25}{6}t^3 = 0.375t^3; s(t) = 0.375t^3$

29. **(a)** $i = 4.4t - 2.1t^2; q = 2.2t^2 - 0.7t^3 + C = 2.2t^2 - 0.7t^3 + 5$
(b) $q(3.2) \approx 4.59 \approx 4.6$ C

30. $i(t) = 6\sqrt{t} = 6t^{1/2}, q(t) = 6 \cdot \frac{1}{3/2}t^{3/2} + C = 4t^{3/2} + C$. Now, finding C, we have $q(0.16) = 4(0.16)^{3/2} + C = 0.347, C = 0.347 - 4(0.16)^{3/2} = 0.091$. Hence, we get $q(t) = 4t^{3/2} + 0.091, q(0.25) = 4(0.25)^{3/2} + 0.091 = 0.591$ C

31. $v(t) = -N\phi'(t)$
(a) $-200\phi'(t) = 2t - 4t^{1/3}, \phi'(t) = -\frac{t}{100} + \frac{1}{50}t^{1/3}$, $\phi(t) = \frac{-t^2}{200} + \frac{3}{200}t^{4/3} + C, C = 0.02; \phi(t) = -0.005t^2 + 0.015t^{4/3} + 0.02$, or $-\frac{1}{200}(t^2 - 3t^{4/3} - 4)$
(b) $\phi(0.729) = -\frac{1}{200}(0.729^2 - 3(0.729)^{4/3} - 4) = 0.027$ Wb

32. $m = 4x + 2 = y'; y =$ antiderivative of $4x + 2$, so $y = 2x^2 + 2x + C; 5 = 2(-1)^2 + 2(-1) + C$, so $C = 5$ and we have $y = 2x^2 + 2x + 5$.

33. $m = \sqrt{x}$ so $y' = \frac{-1}{\sqrt{x}} = -x^{-1/2}, y = -\frac{1}{1/2}x^{1/2} = -2x^{1/2} + C$, and $-3 = -2(4)^{1/2} + C$ gives us that $C = -3 + 4 = 1$. Hence, $f(x) = -2x^{1/2} + 1 = -2\sqrt{x} + 1$.

34. **(a)** The antiderivative is $N(t) = 175t + t^2 - \frac{0.2}{3}t^3 + C$.
(b) Since $N(0) = 175t + t^2 - \frac{0.2}{3}t^3 + C = 0$, we see that $C = 0$ and so, $N(t) = 175t + t^2 - \frac{0.2}{3}t^3$. Thus, $N(8) = 175(8) + 8^2 - \frac{0.2}{3}(8)^3 \approx 1429.86$. Thus, for an 8-hour day, the machine produces 1,429 computer cases (it does not finish producing the last case).

35. Rewriting $\frac{dT}{dx} = -\frac{5750}{(x+1)^3}$ as $\frac{dT}{dx} = -5750(x + 1)^{-3}$. The antiderivative is $T(x) = -\frac{5750}{-2}(x + 1)^{-2} + C = 2875(x + 1)^{-2} + C$. Since $T(0) = 2875(0 + 1)^{-2} + C = 2875 + C = 2900$, we have $C = 25$. Thus, $T = 2875(x + 1)^{-2} + 25$.

36. Rewriting $\frac{dN}{dt} = -\frac{870}{t^2} - 210$ as $\frac{dN}{dt} = -870t^{-2} - 210$, we see that the antiderivative is $N(t) = \frac{-870}{-1} \times t^{-1} - 210t + C = 870t^{-1} - 210t + C$. Since $N(1) = 870(1)^{-1} - 210(1) + C = 660 + C = 7320$, we find that $C = 6660$. Hence, $N(t) = 870t^{-1} - 210t + 6660$.

CHAPTER 24 REVIEW

1. $f(x) = x^4 - x^2$; $f'(x) = 4x^3 - 2x = 2x(2x^2 - 1) = 0$ when $x = 0$ and $\pm\frac{\sqrt{2}}{2}$. These are the critical values. $f''(x) = 12x^2 - 2$; $f''(0) < 0$, so there is a maximum at $(0, 0)$. $f''\left(\frac{\sqrt{2}}{2}\right) = f''\left(-\frac{\sqrt{2}}{2}\right) > 0$, so there are minimums at $\left(\pm\frac{\sqrt{2}}{2}, -\frac{1}{4}\right)$. $f''(x) = 12x^2 - 2 = 0$ when $x = \pm\frac{1}{\sqrt{6}} = \frac{\pm\sqrt{6}}{6}$. Inflection points are at $\left(\frac{\pm\sqrt{6}}{6}, -0.1389\right)$. Concave up over the intervals $\left(-\infty, \frac{-\sqrt{6}}{6}\right)$ and $\left(\frac{\sqrt{6}}{6}, \infty\right)$. Concave down over $\left(\frac{-\sqrt{6}}{6}, \frac{\sqrt{6}}{6}\right)$.

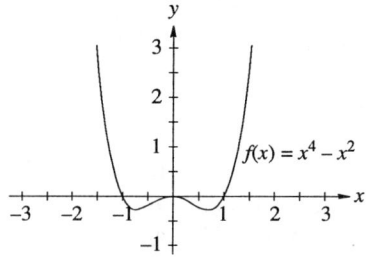

2. $g(x) = x^4 - 32x$; $g'(x) = 4x^3 - 32 = 4(x^3 - 8)$. Critical value is 2. $g''(x) = 12x^2$ is always ≥ 0 so there is a minimum at $(2, -48)$. There are no maxima. $g''(x) = 0$ when $x = 0$. Since $g''(x)$ is never negative there is no inflection point and $g(x)$ is concave up over the interval $(-\infty, \infty)$.

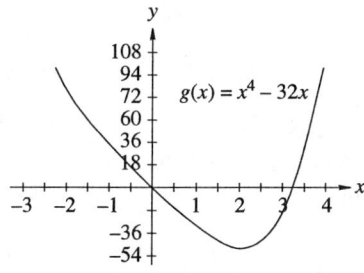

3. $h(x) = 18x^2 - x^4$, so $h'(x) = 36x - 4x^3 = 4x(9 - x^2)$. Critical values are at 0 and ± 3. $h''(x) = 36 - 12x^2 = 12(3 - x^2)$. Now, $h''(x) > 0$, so there is a minimum at $(0, 0)$; and $h''(\pm 3) < 0$, so there is a maximum at $(\pm 3, 81)$. Inflection points at $\left(\pm\frac{\sqrt{3}}{3}, 5.889\right)$. Concave up $\left(\frac{-\sqrt{3}}{3}, \frac{\sqrt{3}}{3}\right)$. Concave down $\left(-\infty, \frac{-\sqrt{3}}{3}\right)$, and $\left(\frac{\sqrt{3}}{3}, \infty\right)$.

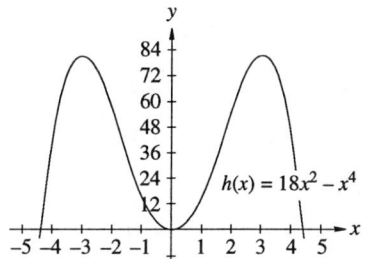

4. $j(x) = 3x^4 - 4x^3 + 1$; $j'(x) = 12x^3 - 12x^2 = 12x^2(x - 1)$. Critical values at $0, 1$. $j''(x) = 36x^2 - 24x$; $j''(0) = 0$ so the Second Derivative Test fails. $j''(1) > 0$ so minimum at $(1, 0)$. Use the First Derivative Test to determine if the critical value 0 yields a maximum or minimum. On the interval $(-\infty, 0)$; $j' < 0$. Likewise, on $(0, 1)$, $j' < 0$ so also decreasing. No maxima. $j''(x)36x^2 - 24x = 12x(3x - 2)$. Inflection points at $(0, 1)$ and $\left(\frac{2}{3}, 0.4074\right)$. Concave up $(-\infty, 0)$ and $\left(\frac{2}{3}, \infty\right)$. Concave down $\left(0, \frac{2}{3}\right)$.

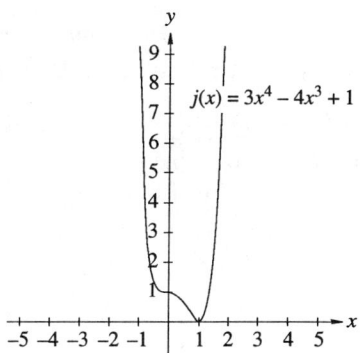

5. $k(x) = \sqrt[5]{x} = x^{1/5}$; $k'(x) = \frac{1}{5}x^{-4/5} = \frac{1}{5} \cdot \frac{1}{\sqrt[5]{x^4}}$. Critical value at 0, but since $k'(x)$ is always positive or undefined there are no extrema. $k''(x) = -\frac{4}{25}x^{-8/5}$. This is undefined at 0. On the interval $(-\infty, 0)$, $k'' > 0$ so Concave up. On the interval $(0, \infty)$, $k'' < 0$ so concave down. Inflection point: $(0, 0)$.

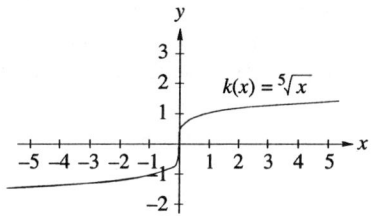

6. Here $f(x) = x\sqrt[3]{4-x}$, so $f'(x) = \sqrt[3]{4-x} + x \cdot \frac{1}{3}(4-x)^{-2/3}(-1) = \sqrt[3]{4-x} - \frac{x}{3\left(\sqrt[3]{4-x}\right)^2} = $

$\frac{3(4-x)-x}{3\left(\sqrt[3]{4-x}\right)^2} = \frac{12-4x}{3\left(\sqrt[3]{4-x}\right)^2}$. Critical values are at 3 and 4. $f''(x) = \frac{3\sqrt[3]{4-x}^2(-4)-(12-4x)2(4-x)^{-1/3}(-1)}{9\left(\sqrt[3]{4-x}\right)^4} = $

$\frac{3(4-x)(-4)+(12-4x)2}{9\left(\sqrt[3]{4-x}\right)^5} = \frac{-48+12x+24-8x}{9\left(\sqrt[3]{4-x}\right)^5} = \frac{4x-24}{9\left(\sqrt[3]{4-x}\right)^5}$.

The second derivative is undefined at 4 and zero at 6. Concave down over $(-\infty, 4)$ and $(6, \infty)$; concave up over $(4, 6)$. Inflection points at $(4, 0)$ and $(6, -7.560)$. Maximum at $(3, 3)$, no minimum.

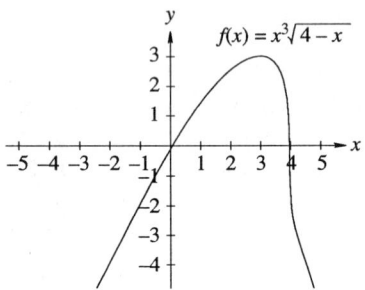

7. $g(x) = \frac{x^2+1}{x^2-4}; g'(x) = \frac{(x^2-4)2x-(x^2+1)(2x)}{(x^2-4)^2} = \frac{-10x}{(x^2-4)^2}$. Critical values are at 0 and ± 2. Vertical asymptotes at $x = 2$ and $x = -2$. Horizontal asymptote at $y = 1$. $g''(x) = \frac{(x^2-4)^2(-10)+10x(2)(x^2-4)(2x)}{(x^2-4)^4} = \frac{(x^2-4)(-10)+40x^2}{(x^2-4)^3} = \frac{30x^2+40}{(x^2-4)^3}$. $g''(x)$ is undefined at $x = \pm 2$. Concave up over the intervals $(-\infty, -2)$ and $(2, \infty)$. Concave down over $(-2, 2)$. Maximum at the point $(0, -\frac{1}{4})$, no minimum. No inflection points.

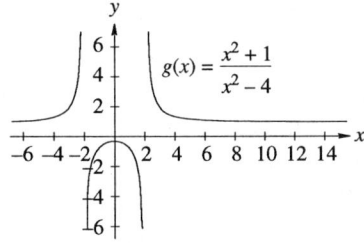

8. $h(x) = \frac{x-1}{x+2}; h'(x) = \frac{(x+2)-(x-1)}{(x+2)^2} = \frac{3}{(x+2)^2}$. Critical value $x = -2$; Vertical asymptote: $x = -2$, horizontal asymptote : $y = 1$. No extrema. $h''(x) = -6(x+2)^{-3}$ undefined at $x = -2$. Concave up: $(-\infty, -2)$; Concave down $(-2, \infty)$; No inflection points.

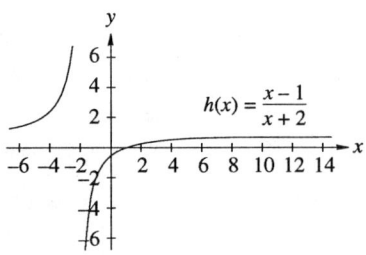

9. $j(x) = \frac{x}{x^2+x-2}; j'(x) = \frac{(x^2+x-2)-x(2x+1)}{(x^2+x-2)^2} = \frac{x^2+x-2-2x^2-x}{(x^2+x-2)^2} = \frac{-x^2-2}{(x^2+x-2)^2}$. Since $j(x)$ is undefined at $x = -2$ and 1, there are no critical values. No extrema. Vertical asymptotes: $x = -2$ and $x = 1$; Horizontal asymptote: $y = 0$.

$j''(x) = \frac{(x^2+x-2)^2(-2x)+(x^2+2)(2)(x^2+x-2)(2x+1)}{(x^2+x-2)^4}$

$= \frac{(x^2+x-2)(-2x)+2(x^2+2)(2x+1)}{(x^2+x-2)^3}$

$= \frac{(-2x^3-2x^2+4x)+(4x^3+2x^2+8x+4)}{(x^2+x-2)^3}$

$= \frac{2x^3+12x+4}{(x^2+x-2)^3}$

Using Newton's method to solve $2x^3+12x+4 = 0$ we get an inflection point when $x = -0.3275$. Concave up: $(-2, -0.3275)$ and $(1, \infty)$, Concave down: $(-\infty, -2)$ and $(-0.3275, 1)$.

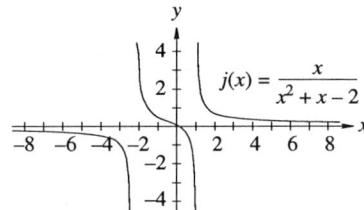

10. $k(x) = \frac{x^3}{x^2-9}; k'(x) = \frac{(x^2-9)3x^2-x^3(2x)}{(x^2-9)^2} = \frac{3x^4-27x^2-2x^4}{(x^2-9)^2} = \frac{x^4-27x^2}{(x^2-9)^2}$. Critical values at 0 and $\pm 3\sqrt{3}$. Vertical asymptotes at $x = \pm 3$. No horizontal asymptote but there is a slant asymptote $y = x$.

$k''(x) = \frac{(x^2-9)(4x^3-54x)-(x^4-27x^2)(2)(x^2-9)(2x)}{(x^2-9)^4}$

$= \frac{(x^2-9)(4x^3-54x)-4x(x^4-27x^2)}{(x^2-9)^3}$

$= \frac{4x^5-90x^3+486x-4x^5+108x^3}{(x^2-9)^3}$

$= \frac{18x^3+486x}{(x^2-9)^3}$

This is undefined at $x = \pm 3$ and zero at $x = 0$. Inflection point: $(0, 0)$; Maximum: $(-3\sqrt{3}, -7.794)$; Minimum: $(3\sqrt{3}, 7.794)$; Concave up: $(-3, 0)$; $(3, \infty)$; Concave down: $(-\infty, -3)$; $(0, 3)$

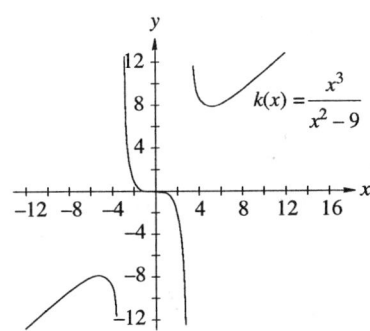

$$k(x) = \frac{x^3}{x^2 - 9}$$

11. $s(t) = t^3 - 9t^2 + t, v(t) = s'(t) = 3t^2 - 18t + 1; a(t) = v'(t) = 6t - 18, a'(t) = 6; v(t) = 0 = 3t^2 - 18t + 1 = 0$ when $t = 0.056$ and 5.944. Distance: maximum at $(0.056, 0.028)$; minimum $(-4, -212)$ and $(4, -76); a(t) = 0$ at $t = 3$. Velocity: minimum $(3, -26)$; maxima at $(-4, 121)$ and $(4, -23)$. Acceleration is always increasing. Minimum: $(-4, -42)$, Maximum: $(4, 6)$.

12. $s(t) = t + 4t^{-1}; v(t) = 1 - 4t^{-2}; a(t) = 8t^{-3}; v(t) = 0 = 1 - \frac{4}{t^2} \Rightarrow \frac{4}{t^2} = 1; t^2 = 4$ or $t = 2$; distance: Maximum: $(4, 5)$; Minimum: $(1, 5), (2, 4); a(t) = 0 = 8t^{-3}$ is never 0. $v(1) = -3; v(4) = 0.75$. Velocity: Maximum: $(4, 0.75)$; Minimum: $(1, -3)$. Acceleration is always decreasing on $[1, 4]$, so maximum $(1, 8)$; minimum $(4, 0.125)$.

13. $f(x) = 3x^2 - 4x; F(x) = 3 \cdot \frac{1}{3}x^3 - 4 \cdot \frac{1}{2}x^2 + C; F(x) = x^3 - 2x^2 + C$

14. $g(x) = 2x^{-3} + 5x^6; G(x) = 2 \cdot \frac{1}{-2}x^{-2} + 5 \cdot \frac{1}{7}x^7 + C; G(x) = -x^{-2} + \frac{5}{7}x^7 + C$

15. $h(x) = x^{1/2} + x^2 - 3x^{-2}; H(x) = \frac{1}{3/2}x^{3/2} + \frac{1}{3}x^3 - 3 \cdot \frac{1}{-1}x^{-1} + C; H(x) = \frac{2}{3}x^{3/2} + \frac{1}{3}x^3 + 3x^{-1} + C$

16. $s(t) = 4.9t^2 - 3.6t + 14; S(t) = 4.9 \cdot \frac{1}{3}t^3 - 3.6 \cdot \frac{1}{2}t^2 + 14t + C; S(t) = 1.63t^3 - 1.8t^2 + 14t + C$

17. (a) $s(t) = 288t - 16t^2; v(t) = s'(t) = 288 - 32t$ ft/s; $a(t) = v'(t) = -32$ ft/s^2
(b) $s(4) = 896$ ft; $v(4) = 160$ ft/s; $a(4) = -32$ft/s^2
(c) Maximum height is when $v(t) = 0; 288 - 32t = 0 \Rightarrow 32t = 288 \Rightarrow t = 9; s(9) = 1296$ ft.
(d) at $t = 9$ s.
(e) $s(t) = 0 = 288t - 16t^2 = t(288 - 16t); t = 288/16 = 18$ s.
(f) $v(18) = -288$ ft/s

18. (a) $s(t) = 15t^2 + 5; v(t) = s'(t) = 30t$ cm/s
(b) $v(2) = 60$ cm/s
(c) $v(t) = 30t = 105 \Rightarrow t = 3.5$ s.

19. The total cost is $C = (1 + 0.00058v^{3/2}) \times 1000 + 25\left(\frac{1000}{v}\right) = (1000 + 0.58v^{3/2}) + \left(\frac{25,000}{v}\right)$. To find the minimum cost, we find the derivative and set it equal to 0. Thus, $C' = \frac{3}{2}(0.58)v^{1/2} - \frac{25,000}{v^2} = 0$. Hence, $0.87v^{5/2} - 25,000 = 0$ or $v^{5/2} = \frac{25,000}{0.87} \approx 28,735$ and $v = (28,735)^{2/5} \approx 60.7$. Checking the second derivative, we have $C'' = 0.435v^{-1/2} + 50,000v^{-3}$, and we see that $C''(60.7) > 0$, so, by the second derivative test, $v \approx 60.7$ is a minimum. Thus, the trucker should drive at 60.7 mph in order to minimize the cost of the trip.

20. $V = \pi r^2 h = 246\pi \Rightarrow h = \frac{246}{r^2}$. Area of top and bottom $= 2\pi r^2$. Area of side $= 2\pi rh = 2\pi r \cdot \frac{246}{r^2} = 492\frac{\pi}{r}$. Cost is $C(r) = 2\pi r^2 + 3 \cdot 492\frac{\pi}{r} = 2\pi r^2 + 1476\pi r^{-1}$ and has the derivative $C'(r) = 4\pi r - 1476\pi r^{-2} = 0 \Rightarrow r - \frac{369}{r^2} = 0$, or $r = \frac{369}{r^2}$, or $r^3 = 369$. Thus, $r = \sqrt[3]{369} \approx 7.17258$ ft; $h = \frac{246}{7.17258^2} = 4.78$ ft.

21. $A = \ell \cdot w; 2\ell + 2w = 1200$ m; $\ell + w = 600; w = 600 - \ell; A = \ell(600 - \ell) = 600\ell - \ell^2; A' = 600 - 2\ell \Rightarrow \ell = 300$ m; $w = 600 - 300 = 300$ m; $A = (300 \times 300) = 90000$ m^2

22. Since the radius of the tank is 8 ft and the height is 11 ft, the tank is more than half full. The height left is $16 - 11 = 5$ ft and the volume remaining is given by the formula $V = \frac{1}{3}\pi h^2(3R - h)$ where h is the height left and $R = 8$ ft. $V = \frac{1}{3}\pi h^2(24) - \frac{1}{3}\pi h^3; \frac{dV}{dt} = (16\pi h - \pi h^2)dh/dt$ and $\frac{dV}{dt} = 20$ gal/min $= 20(0.13358)$ ft^3/min; $\frac{dh}{dt} = \frac{20(0.13358)}{[16\pi \cdot 5 - \pi \cdot 5^2]} = \frac{2.6716}{55\pi} = 0.01546$ ft/min.

23. $V = \frac{4}{3}\pi r^3; dV = 4\pi r^2 dr; dV = 4\pi(1.5)^2(0.01) = 0.2827$ m^3; $A = \pi r^2; dA = 8\pi r dr = 8\pi(1.5)(0.01) \approx 0.3770$ m^2

24. The largest capacity is when the cross-sectional area is the largest. $A = \ell \cdot w; \ell + 2w = 380$ mm; $\ell = 380 - 2w; A = (380 - 2w)w = 380w - 2w^2; A' = 380 - 4w$; maximum when $A' = 0; 4w = 380; w = 95; \ell = 380 - 2 \cdot 95 = 190$. The gutter should measure 95 mm $\times$ 190 mm.

25. $V = 4.5T + 0.0003T^3; \Delta v \approx dv = (4.5 + 0.0009T^2)dT = (4.5 + 0.0009 \cdot 100^2)(1) = 13.5$ V

26. The distance between the ships is related by the equation $d^2 = x^2 + y^2$ so $2dd' = 2xx' + 2yy'$

or $d' = \frac{xx'+yy'}{d}$. After 3 hours, we have $x = 12 \cdot 3 = 36, y = 5 \cdot 3 = 15, d = \sqrt{36^2 + 15^2} = 39$, and $d' = \frac{36 \cdot 12 + 15 \cdot 5}{39} = 13$ km/h.

27. $V = \pi r^2 h = 20$ kL $= 20\,000$ L $= 20\,000$ dm$^3 = 20$ m^3; $h = \frac{20}{\pi r^2}$; $A = 2\pi r^2 + 2\pi rh = 2\pi r^2 + 2\pi r\left(\frac{20}{\pi r^2}\right)$. $\frac{dA}{dr} = 4\pi r - 40r^{-2}$; minimized when

$\frac{dA}{dr} = 0$ or $40r^{-2} = 4\pi r$; $\frac{10}{\pi} = r^3$; $r = \sqrt[3]{\frac{10}{\pi}} \approx$

1.4710 m; $h = \frac{20}{\pi r^2} = 2.942$ m.

28. Distance in the desert is $\sqrt{8^2 + x^2} = \sqrt{64 + x^2}$. Distance on the road is $10 - x$. Total time $=$

$\frac{\sqrt{64+x^2}}{15} + \frac{10-x}{55}$; $\frac{dt}{dx} = \frac{\frac{1}{2}(64 + x^2)^{-1/2}(2x)}{15} - \frac{1}{55}$.

Set $\frac{dt}{dx} = 0$ and solve $\frac{x}{15\sqrt{64+x^2}} = \frac{1}{55}$, or $55x = 15\sqrt{64 + x^2}$, and so $\frac{11}{3}x = \sqrt{64 + x^2}$, or $\frac{121}{9}x^2 = 64 + x^2$; $\frac{112}{9}x^2 = 64$; $112x^2 = 576$; $x^2 = 5.14287$ and we get $x = 2.2678$; 2.2678 mi from A.

29. **(a)** $a = -32$ ft/s^2; $v = -32t$; $s(t) = -16t^2 + C$; $s(7) = 0 = -16 \cdot 7^2 + C$; $C = 16 \cdot 7^2 = 16 \cdot 49 = 784$ ft. The building is 784 ft tall.
(b) $v(7) = -32 \cdot 7 = -224$ ft/s; Speed $= |v| = |-224| = 224$ ft/s.

CHAPTER 24 TEST

1. $f(x) = x^4 - 8x^2$; $f'(x) = 4x^3 - 16x = 4x(x^2 - 4)$
(a) Critical values are $0, \pm2$.

(b) $f''(x) = 12x^2 - 16$; $x = \pm\sqrt{\frac{16}{12}} = \pm\sqrt{4/3} = \frac{\pm2\sqrt{3}}{3}$ are inflection points. Maximum: $(0, 0)$; Minima: $(-2, -16), (2, 16)$

(c) Concave up: $\left(-\infty, \frac{-2\sqrt{3}}{3}\right), \left(\frac{2\sqrt{3}}{3}, \infty\right)$

(d) Concave down $\left(\frac{-2\sqrt{3}}{3}, \frac{2\sqrt{3}}{3}\right)$

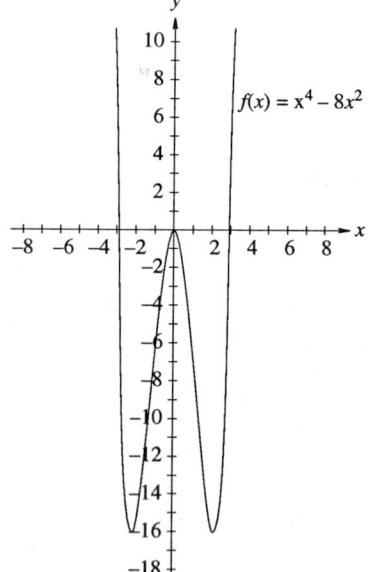

2. $g(x) = \frac{x^2}{9-x^2}$; $g'(x) = \frac{(9-x^2)(2x)-x^2(-2x)}{(9-x^2)^2} = \frac{18x-2x^3+2x^3}{(9-x^2)^2} = \frac{18x}{(9-x^2)^2}$.
(a) Critical value: $x = 0$. Vertical asymptote at $x = 3, x = -3$
(b) $g''(x) = \frac{(9-x^2)^2(18)-18x(2)(9-x^2)(-2x)}{(9-x^2)^4} = \frac{18(9-x^2)+18\cdot4x^2}{(9-x^2)^3} = \frac{54x^2+162}{(9-x^2)^3}$. The second derivative is undefined at ±3. Maximum: none; Minimum: $(0, 0)$,
(c) Concave up: $(-3, 3)$,
(d) Concave down: $(-\infty, -3), (3, \infty)$,
(e) Vertical asymptotes: $x = \pm3$. Horizontal asymptote: $y = -1$

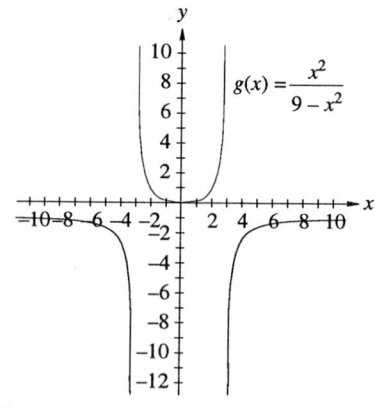

3. $x^2 - 7 = 0; P(x) = x^2 - 7; P'(x) = 2x$

i	x_i	$P(x_i)$	$P'(x_i)$	$x_{i+1} = x_i - \dfrac{P(x_i)}{P'(x_i)}$	$P(x_{i+1})$
1	3	2	6	2.6667	0.1113
2	2.6667	0.1113	5.3333	2.6458	2.58×10^{-4}
3	2.6458	2.58×10^{-4}	5.2916	2.64575	-6.9×10^{-6}

Hence, $x \approx 2.646$

4. $f(x) = 5x^4 - 7x; F(x) = 5 \cdot \frac{1}{5}x^5 - 7 \cdot \frac{1}{2}x^2 + C; F(x) = x^5 - \frac{7}{2}x^2 + C$

5. $s(t) = t^3 - 12t^2 + 5$
(a) $v(t) = s'(t) = 3t^2 - 24t; a(t) = v'(t) = 6t - 24$
(b) Extrema for $s(t)$ are when $v(t) = 0$ that is at 0 and 8; $s(-2) = -51; s(0) = 5; s(4) = -123$. Position: Maximum at $(0, 5)$; Minima at $(-2, -51)$ and $(4, -123)$; Extrema for velocity when $a(t) = 0$ or $t = 4; v(-2) = 60; v(4) = -48$; Velocity: Maximum at $(-2, 60)$, Minimum at $(4, -48)$. Acceleration is always increasing on the interval $[-2, 4]$, so Maximum: $(4, 0)$, Minimum: $(-2, -36)$

6. Let ℓ and w be the dimensions of the printed part. Then $\ell \cdot w = 24$ so $\ell = \frac{24}{w}$. The dimensions of the page are $\ell + 3$ and $w + 2$. Area of the page is $(\ell+3)(w+2) = (\frac{24}{w}+3)(w+2) = 24+\frac{48}{w}+3w+6 = 3w + \frac{48}{w} + 30; A' = 3 - 48/w^2 = 0; 48/w^2 = 3; 16 = w^2; w = 4$. The width is 4 in. and the length is 6 in. for the printed portion. The page is 6 in. wide by 9 in. long.

7. Making the necessary substitutions, we get $800 = \pi r^2(20) + \frac{2}{3}\pi r^3$ or $\frac{2}{3}\pi r^3 + \pi r^2(20) - 800 = 0$. Using Newton's method produces $r \approx 3.38$ m.

CHAPTER
25
Integration

≡ 25.1 THE AREA QUESTION AND THE INTEGRAL

1. $\int_1^3 x\,dx$. The interval $[1,3]$ has length 2 so each subinterval is length $\frac{2}{n}$. Using the right end point of each subinterval, $x_i = 1 + \frac{2i}{n} = \frac{n+2i}{h}$. Since $f(x) = x; f(x_i) = x_i = \frac{n+2i}{n}$

$$\sum_{i=1}^{n} f(x_i)nx = \sum_{i=1}^{n} \left(\frac{n+2i}{n}\right) \cdot \frac{2}{n}$$
$$= \sum_{i=1}^{n} \left(\frac{2}{n} + \frac{4i}{n^2}\right)$$
$$= \frac{2}{n}\sum_{i=1}^{n} 1 + \frac{4}{n^2}\sum_{i=1}^{n} i$$
$$= \frac{2}{n}\cdot n + \frac{4}{n^2}\left(\frac{n^2+n}{2}\right)$$
$$= 2 + 2 + \frac{2}{n} = 4 + \frac{2}{n}$$
$$\lim_{n\to\infty}\left(4 + \frac{2}{n}\right) = 4$$

2. $\int_0^2 2x\,dx; \Delta x = \frac{2}{n}; x_i = \frac{2i}{n}; f(x_i) = 2x_i = \frac{4i}{n};$

$$\sum_{i=1}^{n} f(x_i)\Delta x = \sum_{i=1}^{n} \frac{4i}{n}\cdot\frac{2}{n} = \frac{8}{n^2}\sum_{i1}^{n} i$$
$$= \frac{8}{n^2}\left(\frac{n^2+n}{2}\right) = 4 + \frac{4}{n};$$
$$\lim_{n\to\infty}\left(4 + \frac{4}{n}\right) = 4$$

3. $\int_0^4 (3x+1)dx; \Delta x = \frac{4}{n}; x_i = \frac{4i}{n}; f(x_i) = 3\left(\frac{4i}{n}\right) + 1 = \frac{12i}{n} + 1;$

$$\sum_{i=1}^{n}\left(\frac{12i}{n}+1\right)\cdot\frac{4}{n} = \frac{48}{n^2}\sum_{i=1}^{n} i + \frac{4}{n}\sum_{i=1}^{n} 1$$
$$= \frac{48}{n^2}\left(\frac{n^2+n}{2}\right) + \frac{4}{n}\cdot n$$
$$= 24 + \frac{18}{n} + 4;$$
$$\lim_{n\to\infty}\left(28 + \frac{18}{n}\right) = 28$$

4. $\int_1^3 (2x+2)\,dx; \Delta x = \frac{2}{n}; x_i = 1 + \frac{2i}{n}; f(x_i) = 2(1+\frac{2i}{n}) + 2 = 4 + \frac{4i}{n}$

$$\sum_{i=1}^{n}\left(4+\frac{4i}{n}\right)\frac{2}{n} = \sum_{i=1}^{n}\left(\frac{8}{n}+\frac{8i}{n^2}\right)$$
$$= \frac{8}{n}\sum_{i=1}^{n} 1 + \frac{8}{n}\sum_{i=1}^{n} i$$
$$= \frac{8}{n}n + \frac{8}{n^2}\left(\frac{n^2+n}{2}\right)$$
$$= 8 + 4 + \frac{4}{n};$$
$$\lim_{n\to\infty}\left(12 + \frac{4}{n}\right) = 12$$

5. $\int_1^3 (2x-2)\,dx; \Delta x = \frac{2}{n}; x_i = 1 + \frac{2i}{n}; f(x_i) = 2\left(1+\frac{2i}{n}\right) - 2 = \frac{4i}{n};$

$$\sum_{i=1}^{n}\frac{4i}{n}\cdot\frac{2}{n} = \frac{8}{n^2}\sum_{i=1}^{n} = \frac{8}{n^2}\left(\frac{n^2+n}{2}\right) = 4 + \frac{4}{n};$$
$$\lim_{n\to\infty}\left(4 + \frac{4}{n}\right) = 4$$

6. $\int_0^4 (1-3x)\,dx; \Delta x = \frac{4}{n}; x_n = \frac{4i}{n}; f(x_i) = 1 - 3\cdot\frac{4i}{n} = 1 - \frac{12i}{n};$

$$\sum_{i=1}^{n}\left(1-\frac{12i}{n}\right)\frac{4}{n} = \sum_{i=1}^{n}\left(\frac{4}{n}-\frac{48i}{n^2}\right)$$
$$= \frac{4}{n}\sum_{i=1}^{n} 1 - \frac{48}{n^2}\sum_{i=1}^{n} i$$
$$= \frac{4}{n}\cdot n - \frac{48}{n^2}\cdot\left(\frac{n^2+n}{2}\right)$$
$$= 4 - 24 - \frac{2}{n} = -20 - \frac{24}{n};$$
$$\lim_{n\to\infty}\left(-20 - \frac{24}{n}\right) = -20$$

7. $\int_1^4 (3-2x)\,dx$; $\Delta x = \frac{3}{n}$; $x_i = 1 + \frac{3i}{n}$; $f(x_i) = 3 - 2\left(1 + \frac{3i}{n}\right) = 3 - 2 - \frac{6i}{n} = 1 - \frac{6i}{n}$;

$$\sum_{i=1}^n \left(1 - \frac{6i}{n}\right)\frac{3}{n} = \sum_{i=1}^n \left(\frac{3}{n} - \frac{18i}{n^2}\right)$$
$$= \frac{3}{n}\sum_{i=1}^n 1 - \frac{18}{n^2}\sum_{i=1}^n i$$
$$= \frac{3}{n}\cdot n - \frac{18}{n^2}\left(\frac{n^2+n}{2}\right)$$
$$= 3 - 9 - \frac{9}{n} = -6 - \frac{9}{n};$$

$$\lim_{n\to\infty}\left(-6 - \frac{9}{n}\right) = -6$$

8. $\int_0^4 (x^2)\,dx$; $\Delta x = \frac{4}{n}$; $x_i = \frac{4i}{n}$; $f(x_i) = \left(\frac{4i}{n}\right)^2 = \frac{16i^2}{n^2}$;

$$\sum_{i=1}^n \frac{16i^2}{n^2}\cdot\frac{4}{n} = \frac{64}{n^3}\sum_{i=1}^n i^2$$
$$= \frac{64}{n^3}\left(\frac{n(n+1)(2n+1)}{6}\right)$$
$$= \frac{64}{n^3}\left(\frac{2n^3+3n^2+n}{6}\right)$$
$$= \frac{64}{3} + \frac{32}{n} + \frac{32}{3n^2}$$

$$\lim_{n\to\infty}\left(\frac{64}{3} + \frac{32}{n} + \frac{32}{3n^2}\right) = \frac{64}{3}$$

9. $\int_0^2 (x^2-1)\,dx$; $\Delta x = \frac{2}{n}$; $x_i = \frac{2i}{n}$; $f(x_i) = \left(\frac{2i}{n}\right)^2 - 1 = \frac{4i^2}{n^2} - 1$;

$$\sum_{i=1}^n \left(\frac{4i^2}{n^2} - 1\right)\cdot\frac{2}{n} = \sum_{i=1}^n \frac{8i^2}{n^3} - \sum_{i=1}^n \frac{2}{n}$$
$$= \frac{8}{n^3}\sum_{i=1}^n i^2 - \frac{2}{n}\sum_{i=1}^n 1$$
$$= \frac{8}{n^3}\left(\frac{n^3}{3} + \frac{n^2}{2} + \frac{n}{6}\right) - \frac{2}{n}\cdot n$$
$$= \frac{8}{3} + \frac{4}{n} + \frac{4}{3n^2} - 2$$
$$= \frac{2}{3} + \frac{4}{n} + \frac{4}{3n^2};$$

$$\lim_{n\to\infty}\left(\frac{2}{3} + \frac{4}{n} = \frac{4}{3n^2}\right) = \frac{2}{3}$$

10. $\int_0^4 (x^2+2)\,dx$; $\Delta x = \frac{4}{n}$; $x_i = \frac{4i}{n}$; $f(x_i) = \left(\frac{4i}{n}\right)^2 + 2 = \frac{16i^2}{n^2} + 2$;

$$\sum_{i=1}^n \left(\frac{16i^2}{n^2} + 2\right)\frac{4}{n} = \frac{64}{n^3}\sum_{i=1}^n i^2 + \frac{8}{n}\sum_{i=1}^n 1$$
$$= \frac{64}{n^3}\left(\frac{n^3}{3} + \frac{n^2}{2} + \frac{n}{6}\right)$$
$$+ \frac{8}{n}\cdot n$$
$$= \frac{64}{3} + \frac{32}{n} + \frac{32}{3n^2} + 8$$
$$= \frac{88}{3} + \frac{32}{n} + \frac{32}{3n^2};$$

$$\lim_{n\to\infty}\left(\frac{88}{3} + \frac{32}{n} + \frac{32}{3n^2}\right) = \frac{88}{3}$$

11. $\int_0^2 (2x^2+1)\,dx$; $\Delta x = \frac{2}{n}$; $x_i = \frac{2i}{n}$; $f(x_i) = 2\cdot\left(\frac{2i}{n}\right)^2 + 1 = \frac{8i^2}{n} + 1$;

$$\sum_{i=1}^n \left(\frac{8i^2}{n^2} + 1\right)\cdot\frac{2}{n} = \sum_{i=1}^n \left(\frac{16i^2}{n^3} + \frac{2}{n}\right)$$
$$= \frac{16i}{n^3}\sum_{i=1}^n i^2 + \frac{2}{n}\sum_{i=1}^n 1$$
$$= \frac{16}{n^3}\left(\frac{n^3}{3} + \frac{n^3}{2} + \frac{n}{6}\right)$$
$$+ \frac{2}{n}(n)$$
$$= \frac{16}{3} + \frac{8}{n} + \frac{8}{3n^2} + 2$$
$$= \frac{22}{3} + \frac{8}{n} + \frac{8}{3n^2};$$

$$\lim_{n\to\infty}\left(\frac{22}{3} + \frac{8}{n} + \frac{8}{3n^2}\right) = \frac{22}{3}.$$

12. $\int_0^4 (2-x^2)\,dx$; $\Delta x = \frac{4}{n}$; $x_i = \frac{4i}{n}$; $f(x_i) = 2 - \left(\frac{4i}{n}\right)^2 = 2 - \frac{16i^2}{n^2}$;

$$\sum_{i=1}^n \left(2 - \frac{16i^2}{n^2}\right)\frac{4}{n} = \sum_{i=1}^n \left(\frac{8}{n} - \frac{64i^2}{n^3}\right)$$
$$= \frac{8}{n}\cdot n - \frac{64}{n^3}$$
$$\times \left(\frac{n^3}{3} + \frac{n^2}{2} + \frac{n}{b}\right)$$
$$= 8 - \frac{64}{3} - \frac{32}{n} - \frac{32}{3n^2};$$

$$\lim_{n\to\infty}\left(8 - \frac{64}{3} - \frac{32}{n} - \frac{32}{3n^2}\right) = 8 - \frac{64}{3} = -\frac{40}{3}$$

13. $\int_1^3 x^3\,dx$; $\Delta x = \frac{2}{n}$; $x_i = 1 + \frac{2i}{n}$; $f(x_i) = \left(1 + \frac{2i}{n}\right)^3 = 1 + \frac{6i}{n} + \frac{12i^2}{n^2} + \frac{8i^3}{n^3}$;

$$\sum_{i=1}^{n}\left(1+\frac{6i}{n}+\frac{12i^2}{n^2}+\frac{8i^3}{n^3}\right)\frac{2}{n}$$

$$=\sum_{i=1}^{n}\left(\frac{2}{n}+\frac{12i^2}{n^2}+\frac{24i^2}{n^3}+\frac{16i^3}{n^4}\right)$$

$$=\frac{2}{n}\cdot n+\frac{12}{n^2}\left(\frac{n^2+n}{2}\right)+\frac{24}{n^3}$$

$$\times\left(\frac{n^3}{3}+\frac{n^2}{2}+\frac{n}{6}\right)$$

$$+\frac{16}{n^4}\left(\frac{n^4}{4}+\frac{n^3}{2}+\frac{n^2}{4}\right)$$

$$=2+6+8+4+\frac{6}{n}+\frac{12}{n}+\frac{4}{n^2}+\frac{8}{n}+\frac{4}{n^2};$$

$$\lim_{n\to\infty}\left(20+\frac{26}{n}+\frac{8}{n^2}\right)=20$$

14. $\int_{1}^{3}(1-x^3)dx;\ \Delta x=\frac{2}{n};\ x_i=1+\frac{2i}{n};\ f(x_i)=1-$
$\left[1+\frac{2i}{n}\right]^3=1-\left[1+\frac{6i}{n}+\frac{12i^2}{n^2}+\frac{8i^3}{n^3}\right];$

$$\sum_{i=1}^{n}-\left[\frac{6i}{n}+\frac{12i^2}{n^2}+\frac{8i^3}{n^3}\right]\frac{2}{n}$$

$$=-\left[\frac{12}{n^2}\cdot\left(\frac{n^2+n}{2}\right)+\frac{24}{n^3}\left(\frac{n^3}{3}+\frac{n^2}{2}+\frac{n}{6}\right)\right.$$

$$\left.+\frac{16}{n^4}\left(\frac{n^4}{4}+\frac{n^3}{2}+\frac{n^2}{4}\right)\right]$$

$$\lim_{n\to\infty}-\left[\frac{12}{n^2}\cdot\left(\frac{n^2+n}{2}\right)+\frac{24}{n^3}\left(\frac{n^3}{3}+\frac{n^2}{2}+\frac{n}{6}\right)+\right.$$
$$\left.\frac{16}{n^4}\left(\frac{n^4}{4}+\frac{n^3}{2}+\frac{n^2}{4}\right)\right]=-[6+8+4]=-18$$

15. $\int_{0}^{4}(x^3+2)dx;\ \Delta x=\frac{4}{n};\ x_i=\frac{4i}{n};\ f(x_i)=\left(\frac{4i}{n}\right)^3+$
$2=\frac{64i^3}{n^3}+2;$

$$\sum_{i=1}^{n}\left(\frac{64i^3}{n^3}+2\right)\frac{4}{n}=\sum_{i=1}^{n}\left(\frac{256i^3}{n^4}+\frac{8}{n}\right)$$

$$=\frac{256}{n^4}\left[\frac{n^4}{4}+\frac{n^3}{2}+\frac{n^2}{2}\right]$$

$$+\frac{8}{n}\cdot n$$

$$\lim_{n\to\infty}\left[\frac{256}{n^4}\left[\frac{n^4}{4}+\frac{n^3}{2}+\frac{n^2}{2}\right]+\frac{8}{n}\cdot n\right]=64+8=72$$

16. $\int_{0}^{4}(2x^3-1)dx;\ \Delta x=\frac{3}{n};\ x_i=\frac{3i}{n};\ f(x_i)=2\cdot$
$\left(\frac{3i}{b}\right)^3-1=\frac{54i^3}{n^3}-1;$

$$\sum_{i=1}^{n}\left(\frac{54i^3}{n}-1\right)\frac{3}{n}=\sum_{i=1}^{n}\left(\frac{162i^3}{n}-\frac{3}{n}\right)$$

$$=\frac{162}{n^4}\left(\frac{n^4}{4}+\frac{n^3}{2}+\frac{n^2}{2}\right)$$

$$-\frac{3}{n}\cdot n$$

$$\lim_{n\to\infty}\left[\frac{162}{n^4}\left[\frac{n^4}{4}+\frac{n^3}{2}+\frac{n^2}{2}\right]+\frac{3}{n}\cdot n\right]=40.5-3=$$
37.5 or $\frac{75}{2}$.

17. $\int_{0}^{2}(x^2+x)dx;\ \Delta x=\frac{2}{n};\ x_i=\frac{2i}{n};\ f(x_i)=\left(\frac{2i}{n}\right)^2+$
$\frac{2i}{n}=\frac{4i^2}{n^2}+\frac{2i}{n};$

$$\sum_{i=1}^{n}\left(\frac{4i^2}{n^2}+\frac{2i}{n}\right)\frac{2}{n}=\sum_{i=1}^{n}\left(\frac{8i^2}{n^3}+\frac{4i}{n^2}\right)$$

$$=\frac{8}{n^3}\left(\frac{n^3}{2}+\frac{n^2}{2}+\frac{n}{6}\right)$$

$$+\frac{4}{n^2}\left(\frac{n^2}{2}+\frac{n}{2}\right)$$

$$\lim_{n\to\infty}\left[\frac{8}{n^3}\left(\frac{n^3}{3}+\frac{n^2}{2}+\frac{n}{6}\right)+\frac{4}{n^2}\left(\frac{n^2}{2}+\frac{n}{2}\right)\right]=\frac{8}{3}+2=\frac{14}{3}$$

18. $\int_{0}^{1}x^4\,dx;\ \Delta x=\frac{1}{n};\ x_i=\frac{i}{n};\ f(x_i)=\frac{i^4}{n^4};$

$$\sum_{i=1}^{n}\frac{i^4}{n^4}\cdot\frac{1}{n}$$

$$=\frac{1}{n^5}\sum_{i=1}^{n}i^4$$

$$=\frac{1}{n^5}\left[\frac{n(n+1)(2n+1)(3n^2+3n-1)}{30}\right]$$

$$=\frac{1}{n^5}\left[\left(\frac{6n^5+15n^4+10n^3-n}{30}\right)\right]$$

$$=\frac{1}{n^5}\left[\left(\frac{n^5}{5}+\frac{15n^4}{2}+\frac{10n^3}{3}-\frac{n}{30}\right)\right]$$

$$\lim_{n\to\infty}\frac{1}{n^5}\left[\left(\frac{n^5}{5}+\frac{15n^4}{2}+\frac{10n^3}{3}-\frac{n}{30}\right)\right]=\frac{1}{5}$$

☰ 25.2 THE FUNDAMENTAL THEOREM OF CALCULUS

1. $\int_{0}^{5}6\,dx=6x\big|_{0}^{5}=6\cdot5-6\cdot0=30$

2. $\int_{1}^{2}(x-6)dx=\frac{1}{2}x^2-6x\big|_{1}^{2}=\left(\frac{1}{2}\cdot2^2-6\cdot2\right)-\left(\frac{1}{2}\cdot\right.$
$\left.1^2-6\cdot1\right)=2-12-\frac{1}{2}+6=-\frac{9}{2}$

3. $\int_{1}^{5}2x\,dx=x^2\big|_{1}^{5}=5^2-1^2=25-1=24$

4. $\int_{-1}^{2}(6-2x)dx=6x-x^2\big|_{-1}^{2}=(6\cdot2-2^2)-$
$(6(-1)-(-1)^2)=12-4+6+1=15$

5. $\int_{-1}^{3}\frac{5}{3}x^3\,dx=\frac{5}{12}x^4\big|_{-1}^{3}=\frac{5}{12}\cdot3^4-\frac{51}{2}\cdot(-1)^4=$
$\frac{5\cdot81}{12}-\frac{5}{12}=\frac{400}{12}=\frac{100}{3}$

6. $\int_{0}^{4}4x^2\,dx=\frac{4}{3}x^3\big|_{0}^{5}=\frac{4}{3}\cdot5^3-\frac{4}{3}\cdot0^3=\frac{4}{3}\cdot125=\frac{500}{3}$

7. $\int_{1}^{2}\frac{8}{5}x^3\,dx=\frac{2}{5}x^4\big|_{1}^{2}=\frac{2}{5}\cdot2^4-\frac{2}{5}\cdot1^4=\frac{32}{5}-\frac{2}{5}=$
$\frac{30}{5}=6$

8. $\int_{-2}^{2} 6x\, dx = 3x^2 \big|_{-2}^{2} = 3 \cdot 2^2 - 3 \cdot (-2)^2 = 12 - 12 = 0$

9. $\int_{0}^{2}(4x+5)dx = 2x^2 + 5x \big|_{0}^{2} = (2 \cdot 2^2 + 5 \cdot 2) - (2 \cdot 0^2 + 5 \cdot 0) = 8 + 10 = 18$

10. $\int_{1}^{3}(-4)dx = (-4x)\big|_{1}^{3} = -4 \cdot 3 - (-4 \cdot 1) = -12 + 4 = -8$

11. $\int_{2}^{0}(4x+5)dx = 2x^2 + 5x \big|_{2}^{0} = (2 \cdot 0^2 + 5 \cdot 0) - (2 \cdot 2^2 + 5 \cdot 2) = 0 - 18 = -18$

12. $\int_{-1}^{2}(2t - t^2)dx = t^2 - \frac{1}{3}t^3 \big|_{-1}^{2} = \left(2^2 - \frac{1}{3} \cdot 2^3\right) - \left((-1)^2 - \frac{1}{3}(-1)^3\right) = \left(4 - \frac{8}{3}\right) - \left(1 + \frac{1}{3}\right) = 0$

13. $\int_{1}^{3}(-4x)dx = -2x^2 \big|_{1}^{3} = -2 \cdot 3^2 - (-2 \cdot 1^2) = -2 \cdot 9 - (-2 \cdot 1) = -18 + 2 = -16$

14. $\int_{2}^{5}(5w^2 + 2w)dw = \frac{5}{3}w^3 + w^2 \big|_{2}^{5} = \left(\frac{5}{3} \cdot 5^3 + 5^2\right) - \left(\frac{5}{3} \cdot 2^3 + 2^2\right) = \left(\frac{625}{3} + 25\right) - \left(\frac{40}{3} + 4\right) = \frac{585}{3} + 21 = 195 + 21 = 216$

15. $\int_{0}^{3}(t^2 + 2t)dt = \frac{1}{3}t^3 + t^2 \big|_{0}^{3} = \frac{1}{3} \cdot 3^3 + 3^2 - 0 = 18$

16. $\int_{-2}^{0}(5 - t^3)dt = 5t - \frac{t^4}{4} \big|_{-2}^{0} = 0 - \left(5(-2) - \frac{(-2)^4}{4}\right) = -(-10 - 4) = 14$

17. $\int_{-1}^{3}(3y^2 - 2y)dy = y^3 - y^2 \big|_{-1}^{3} = (3^3 - 3^2) - ((-1)^3 - (-1)^2) = 18 + 2 = 20$

18. $\int_{0}^{4}(s^2 - 4s + 16)ds = \frac{s^3}{3} - 2s^2 + 16s \big|_{0}^{4} = \frac{4^3}{3} - 2 \cdot 4^2 + 16 \cdot 4 - 0 = \frac{64}{3} - 32 + 64 = \frac{64}{3} + 32 = \frac{160}{3}$

19. $\int_{1}^{4}(4 - 2w^2)dw = 4w - \frac{2}{3}w^3 \big|_{1}^{4} = \left(4 \cdot 4 - \frac{2}{3} \cdot 4^3\right) - \left(4 \cdot 1 - \frac{2}{3} \cdot 1^3\right) = \left(16 - \frac{128}{3}\right) - \left(4 - \frac{2}{3}\right) = 12 - \frac{126}{3} = 12 - 42 = -30$

20. $\int_{-2}^{0} -dt = -t \big|_{-2}^{0} = 0 - (-(-2)) = -2$

21. $\int_{2}^{3}(x^2 + 2x + 1)dx = \frac{x^3}{3} + x^2 + x \big|_{2}^{3} = \left(\frac{3^3}{3} + 3^2 + 3\right) - \left(\frac{2^3}{3} + 2^2 + 2\right) = 21 - \frac{26}{3} = \frac{37}{3}$

22. $\int_{-3}^{0}(3y - y^3)dy = \frac{3}{2}y^2 - \frac{y^4}{4} \big|_{-3}^{0} = 0 - \left(\frac{3}{2}(-3)^2 - \frac{(-3)^4}{4}\right) = -\left(\frac{27}{2} - \frac{81}{4}\right) = \frac{81}{4} - \frac{54}{4} = \frac{27}{4}$

23. $\int_{3}^{2}(x^2 + 2x + 1)dx = \frac{x^3}{3} + x^2 + x \big|_{3}^{2} = \left(\frac{2^3}{3} + 2^2 + 2\right) - \left(\frac{3^3}{3} + 3^2 + 3\right) = \frac{26}{3} - 21 = -\frac{37}{3}$

24. $\int_{1}^{4}\left(\sqrt{x} - \frac{1}{\sqrt{x}}\right)dx = \int_{1}^{4}(x^{1/2} - x^{-1/2})dx = \frac{2}{3}x^{3/2} - 2x^{1/2} \big|_{1}^{4} = \left(\frac{2}{3} \cdot 4^{3/2} - 2 \cdot 4^{1/2}\right) - \left(\frac{2}{3} \cdot 1^{3/2} - 2 \cdot 1^{1/2}\right) = \left(\frac{16}{3} - 4\right) - \left(\frac{2}{3} - 2\right) = \frac{14}{3} - 2 = \frac{8}{3}$

25. $\int_{6}^{8} dx = x \big|_{6}^{8} = 8 - 6 = 2$

26. $\int_{1/2}^{1} \frac{4}{x^2}dx = 4 \int_{1/2}^{1} x^{-2}dx = 4 \cdot -x^{-1} \big|_{1/2}^{1} = -4 \cdot 1 + 4 \cdot \frac{2}{1} = -4 + 8 = 4$

27. $\int_{-4}^{-2}(3y^2 - y - 1)dy = y^3 - \frac{1}{2}y^2 - y \big|_{-4}^{-2} = \left((-2)^3 - \frac{1}{2}(-2)^2 - (-2)\right) - \left((-4)^3 - \frac{1}{2}(-4)^2 - (-4)\right) = (-8 - 2 + 2) - (-64 - 8 + 4) = -8 - (-68) = 60$

28. $\int_{1}^{4}\left(\frac{3}{x^3} - \frac{2}{x^2}\right)dx = \int_{1}^{4}(3x^{-3} - 2x^{-2})dx = -\frac{3}{2}x^{-2} + 2x^{-1} \big|_{1}^{4} = \left(\left(-\frac{3}{2}\right)4^{-2} + 2 \cdot 4^{-1}\right) - \left(-\frac{3}{2}1^{-2} + 2 \cdot 1^{-1}\right) = \left(\frac{-3}{32} + \frac{1}{2}\right) - \left(-\frac{3}{2} + 2\right) = -\frac{3}{32} + \frac{1}{2} + \frac{3}{2} - 2 = -\frac{3}{32}$

29. $\int_{2}^{4}(x^2 - 2x + 1)dx = \frac{x^3}{3} - x^2 + x \big|_{2}^{4} = \left(\frac{4^3}{3} - 4^2 + 4\right) - \left(\frac{2^3}{3} - 2^2 + 2\right) = \left(\frac{64}{3} - 16 + 4\right) - \left(\frac{8}{3} - 4 + 2\right) = \frac{64}{3} - 16 + 4 - \frac{8}{3} + 4 - 2 = \frac{56}{3} - 10 = \frac{26}{3}$

30. $\int_{0}^{4} \sqrt{x^3}\, dx = \int_{0}^{4} x^{3/2}dx = \frac{2}{5}x^{5/2} \big|_{0}^{4} = \frac{2}{5}4^{5/2} - 0 = \frac{2}{5} \cdot 32 = \frac{64}{5}$

31. $\int_{1}^{2} -3x^{-4}dx = x^{-3} \big|_{1}^{2} = \frac{1}{8} - 1 = -\frac{7}{8}$

32. $\int_{0}^{1} 3x^{5/6}dx = 3 \cdot \frac{6}{11}x^{11/6} \big|_{0}^{1} = \frac{18}{11}$

33. $\int_{-2}^{-1} \frac{t^{-2}}{3}dx = -\frac{1}{3}t^{-1} \big|_{-2}^{-1} = \frac{1}{3} - \frac{1}{6} = \frac{1}{6}$

34. $\int_{0}^{2}(x^4 - x^3)dx = \frac{1}{5}x^5 - \frac{1}{4}x^4 \big|_{0}^{2} = \frac{1}{5} \cdot 2^5 - \frac{1}{4} \cdot 2^4 - 0 = \frac{32}{5} - \frac{16}{4} = \frac{32}{5} - 4 = \frac{12}{5}$

35. $\int_{0}^{2} \sqrt{x}\, dx = \int_{0}^{2} x^{1/2}dx = \frac{2}{3}x^{3/2} \big|_{0}^{2} = \frac{2}{3} \cdot 2^{3/2} - 0 = \frac{4}{3}\sqrt{2} \approx 1.8856$

36. $\int_{1}^{3} x\sqrt[3]{x}\, dx = \int_{1}^{3} x^{4/3}dx = \frac{3}{7}x^{7/3} \big|_{1}^{3} = \frac{3}{7} \cdot 3^{7/3} - \frac{3}{7} \approx 5.1344$

37. $\int_{1/2}^{3}\left(\frac{1}{\sqrt{x}} - 2\right)dx = \int_{1/2}^{3}(x^{-1/2} - 2)dx = 2x^{1/2} - 2x \big|_{1/2}^{3} = \left(2\sqrt{3} - 2 \cdot 3\right) - \left(2 \cdot \sqrt{1/2} - 1\right) = 2\sqrt{3} - 6 - \sqrt{2} + 1 = 2\sqrt{3} - \sqrt{2} - 5 \approx -2.9501$

38. To use the Fundamental Theorem of Calculus, the function must be continuous over a closed interval. This function, $f(x) = x^{-2}$ is not continuous at $x = 0$.

39. Since velocity is the derivative of displacement, the displacement is the integral of the velocity.

$\int_0^4 (4.00t - 1.00t^2)dt = 2.00t^2 - \frac{1}{3}t^3\big|_0^4 = 2 \cdot 4^2 - \frac{1}{3} \cdot 4^3 - 0 = 32 - \frac{64}{3} = 10.7$ m

40. Angular displacement is the integral of angular velocity. Hence, we have $\int_1^3 (1.00t^3 - 27.0t)dt = \frac{1}{4}t^4 - \frac{27}{2}t^2\big|_1^3 = \left(\frac{3^4}{4} - \frac{27 \cdot 3^2}{2}\right) - \left(\frac{1}{4} - \frac{27}{2}\right) = -88.0$ rad

41. **(a)** $a(t) = -9.8$ m/s^2; $v(t) =$ antiderivative of $a(t)$; $v(t) = -9.8t + C_1$; at $t = 0; v(t) = -10$ so $v(t) = -9.8t - 10; v(3) = -9.8 \cdot 3 - 10 = -39.4$ m/s.
The negative means its velocity is downward. Its speed is $|-39.4| = 39.4$ m/s
(b) Displacement is the integeral of the velocity. Hence $\int_0^3 (-9.8t - 10)dt = -4.9t^2 - 10t\big|_0^3 = -4.9(3^2) - 10.3 - 0 = -74.1$. The negative indicates a downward movement. Distance is $|-74.1| = 74.1$ m.
(c) $4.9t^2 - 10t = 180; 4.9t^2 + 10t - 180 = 0$. Using the quadratic formula, $t = -7.167$ or $t = 5.1258$. Since t must be positive, $t = 5.1258$ s.

42. $v(t) = \int a(t)dt = \int -32\, dt = -32t + C_1$. $s(t) = \int v(t)dt = -16t^2 + C_1t + C_2$. Since the sandbag is dropped from an elevation of 840 ft, $s(0) = 840$ and so, $C_2 = 840$, which means that $s(t) = -16t^2 + C_1t + 840$. We also have $s(8) = 0 = -16 \cdot 8^2 + C_1 \cdot 8 + 840 = -1024 + C_18 + 840$ and so, $8C_1 = 184$, or $C_1 = \frac{184}{8} = 23$ ft/s.

43. **(a)** Acceleration is change in velocity divided by time so $a = (10,500 - 500)/0.01 = 10,000/0.01 = 1\,000\,000$ m/s^2.
(b) $v = \int a(t)dt = 1\,000\,000t + 500$; $s = \int_0^{0.01} v(t) = 500\,000t^2 + 500t\big|_0^{0.01} = 50 + 5 = 55$ m.

44. We have $i = 1.00t + 1.00\sqrt{t}$. Since charge is the integral of current, we have $q = \int_1^9 (t + t^{1/2})dt = \frac{t^2}{2} + \frac{2}{3}t^{3/2}\big|_1^9 = \left(\frac{9^2}{2} + \frac{2}{3} \cdot 9^{3/2}\right) - \left(\frac{1}{2} + \frac{2}{3}\right) = \frac{81}{2} + \frac{54}{3} - \frac{1}{2} - \frac{2}{3} = \frac{80}{2} + \frac{52}{3} \approx 57.3$ C

45. $\int_0^9 N'(t)\,dt = \int_0^9 25 - 6t^{1/2}\,dt = 25t - 4t^{3/2}\big|_0^9 = 25(9) - 4(9)^{3/2} = 225 - 108 = 117$. There will be an increase of 117 moose.

46. **(a)** $N(t) = \int (175 + 2t - 0.2t^2)\,dt = 175t + t^2 - \frac{0.2}{3}t^3 + C$. Since $N(0) = 175(0) + (0)^2 - \frac{0.2}{3}(0)^3 + C = C = 0$, we have $N(t) = 175t + t^2 - \frac{0.2}{3}t^3$.
(b) The number of computer cases the machine produces in the first 4 hours of the day is given by $\int_0^4 N(t)\,dt = \int_0^4 (175 + 2t - 0.2t^2)\,dt = 175t + t^2 - \frac{0.2}{3}t^3\big|_0^4 = 711.7$ or about 712 cases.
(c) $\int_4^8 N(t)\,dt = \int_4^8 (175 + 2t - 0.2t^2)\,dt = 175t + t^2 - \frac{0.2}{3}t^3\big|_4^8 \approx 1429.9 - 711.7 = 718.2$ or about 718 cases.

47. **(a)** $\int_1^2 (0.5 + 4t^{-3})\,dt = 0.5t - 2t^{-2}\big|_1^2 = \left[0.5(2) - 2(2)^{-2}\right] - \left[0.5(1) - 2(1)^{-2}\right] = [1.0 - 0.5] - [0.5 - 2] = 2$. The tree will grow 2 ft in the second year after it is 20 years old.
(b) $\int_4^5 (0.5 + 4t^{-3})\,dt = 0.5t - 2t^{-2}\big|_4^5 = \left[0.5(5) - 2(5)^{-2}\right] - \left[0.5(4) - 2(4)^{-2}\right] = [2.5 - 0.8] - [2.0 - 0.125] = 0.545$. The tree will grow 0.545 ft in the fifth year after it is 20 years old.
(c) $\int_1^5 (0.5 + 4t^{-3})\,dt = 0.5t - 2t^{-2}\big|_1^5 = \left[0.5(5) - 2(5)^{-2}\right] - \left[0.5(1) - 2(1)^{-2}\right] = [2.5 - 0.8] - [0.5 - 2] = 3.92$. The tree will grow 3.92 ft in years 1 through 5 after it is 20 years old.

48. The total concentration of the pollutant from $t = 0$ to $t = x$ is $\int_0^x P'(t)\,dt = \int_0^x (91t^{5/2})\,dt = 91\left(\frac{2}{7}\right)t^{7/2}\big|_0^x = 26x^{7/2}$. We set this equal to 5720 and solve for x.
$26x^{7/2} = 5720$
$x^{7/2} = 220$
$x = 220^{2/7} = 4.67$
It will take about 4.67 yr or 4 yr 8 mo.

≡ 25.3 THE INDEFINITE INTEGRAL

1. $\int 9\,dx + 9x + C$

2. $\int 3x\,dx = \frac{3}{2}x^2 + C$

3. $\int 6x^2 dx = 6 \cdot \frac{1}{3}x^3 + C = 2x^3 + C$

4. $\int \sqrt{x}\,dx = \int x^{1/2}dx = \frac{2}{3}x^{3/2} + C$

5. $\int x^2\sqrt{x}\,dx = \int x^{5/2}dx = \frac{2}{7}x^{7/2} + C$

6. $\int x^{-2/3}dx = 3x^{1/3} + C$

7. $\int (t^3 + 1)dt = \frac{1}{4}t^4 + t + C$

8. $\int (2 - 4s^3)ds = 2s - s^4 + C$.

9. $\int \left(y^2 + 4y - 3\right)dy = \frac{1}{3}y^3 + 2y^2 - 3y + C.$

10. $\int \left(4\sqrt{y} + \frac{1}{4\sqrt{y}}\right) = 4\int y^{1/2} + \frac{1}{4}\int y^{-1/2}dy = 4 \cdot \frac{2}{3}y^{3/2} + \frac{1}{4} \cdot 2y^{1/2} + C = \frac{8}{3}y^{3/2} + \frac{1}{2}y^{1/2} + C.$

11. $\int \left(\sqrt[3]{x} - \frac{3}{\sqrt[3]{x^2}}\right)dx = \int \left(x^{1/3} - 3x^{-2/3}\right)dx = \frac{3}{4}x^{4/3} - 9x^{1/3} + C = \frac{3x\sqrt[3]{x}}{4} - 9\sqrt[3]{x} + C.$

12. $\int \left(x^{-3} - 3x^{-1/2} + \frac{1}{3}x^{1/2}\right)dx = -\frac{1}{2}x^{-2} - 3 \cdot 2x^{1/2} + \frac{1}{3} + \frac{2}{3}x^{3/2} + C = \frac{-1}{2x^2} - 6\sqrt{x} + \frac{2}{9}x\sqrt{x} + C.$

13. $\int \left(x^2 + 3\right)2x\,dx.$ Let $u = x^2 + 3$, then $du = 2x\,dx.$ Then, substituting we get $\int u\,du = \frac{1}{2}u^2 + C = \frac{1}{2}\left(x^2 + 3\right)^2 + C.$

14. $\int \left(3x^2 - 5\right)6x\,dx.$ Let $u = 3x^2 - 5$, then $du = 6x\,dx.$ Substituting, we see that $\int u\,du = \frac{1}{2}u^2 + C = \frac{1}{2}\left(3x^2 - 5\right)^2 + C.$

15. $\int \left(4 - 2x^2\right)4x\,dx.$ Let $u = 4 - 2x^2$, then $du = -4x\,dx.$ Substituting, we can rewrite the given integral as $\int \left(4 - 2x^2\right)4x\,dx = -\int u\,du = -\frac{1}{2}u^2 + C = -\frac{1}{2}\left(4 - 2x^2\right)^2 + C.$

16. $\int \left(2x^3 + 1\right)^4 6x^2dx.$ Let $u = 3x^2 + 1$, then $du = 6x^2dx.$ Substituting, we have $\int u^4du = \frac{1}{5}u^5 + C = \frac{1}{5}\left(2x^3 + 1\right)^5 + C.$

17. $\int \left(3 - x^2\right)^3 2x\,dx.$ Let $u = 3 - x^2$, then $du = -2x\,dx$ or $-du = 2x\,dx.$ Substituting, we have $-\int u^3du = -\frac{1}{4}u^4 + C = -\frac{1}{4}\left(3 - x^2\right)^4 + C.$

18. $\int \left(x^2 - 3\right)^4 x\,dx.$ Let $u = x^2 - 3$, then $du = 2x\,dx$ or $x\,dx = \frac{1}{2}du.$ Substituting, produces $\frac{1}{2}\int u^4du = \frac{1}{2} \cdot \frac{1}{5}u^5 + C = \frac{1}{10}\left(x^2 - 3\right)^5 + C.$

19. $\int \left(x^2 + 4\right)^{1/2}x\,dx.$ Let $u = x^2 + 4$, then $du = 2x\,dx$ or $x\,dx = \frac{1}{2}du.$ Substituting, we obtain $\frac{1}{2}\int u^{1/2}du = \frac{1}{2} \cdot \frac{2}{3}u^{3/2} + C = \frac{1}{3}\left(x^2 + 4\right)^{3/2} + C.$

20. $\int \left(x^3 + 1\right)^5 x^2\,dx.$ Let $u = x^3 + 1$, then $du = 3x^2dx$ or $x^2dx = \frac{1}{3}du.$ Substituting, we see that $\frac{1}{3}\int u^5du = \frac{1}{3} \cdot \frac{1}{6}u^6 + C = \frac{1}{18}\left(x^3 + 1\right)^6 + C.$

21. $\int \frac{(\sqrt{x}-1)^3}{\sqrt{x}}dx = \int (\sqrt{x} - 1)^3 \cdot x^{-1/2}dx.$ Let $u = \sqrt{x} - 1$, then $du = \frac{1}{2}x^{-1/2}$ or $2\,du = x^{-1/2}dx.$ Substituting, we have $2\int u^3du = 2 \cdot \frac{1}{4}u^4 + C = \frac{1}{2}(\sqrt{x} - 1)^4 + C.$

22. $\int \left(x^4 - 5\right)^7 x^3dx.$ Let $u = x^4 - 5$, then $du = 4x^3dx$ or $x^3dx = \frac{1}{4}du.$ Substituting, we have $\frac{1}{4}\int u^7du = \frac{1}{4} \cdot \frac{1}{8}u^8 + C = \frac{1}{32}\left(x^4 - 5\right)^8 + C.$

23. $\int \left(3x^3 + 1\right)^4 x^2dx.$ Let $u = 3x^3 + 1$, then $du = 9x^2dx$ or $x^2dx = \frac{1}{9}du.$ Substituting, we obtain $\frac{1}{9}\int u^4du = \frac{1}{9} \cdot \frac{1}{5}u^5 + C = \frac{1}{45}\left(3x^3 + 1\right)^5 + C.$

24. $\int \left(3 - 6x^2\right)^{3/2}2x\,dx.$ Let $u = 3 - 6x^2$, then $du = -12x\,dx$, or $2x\,dx = -\frac{1}{6}du.$ Substituting, we see that $-\frac{1}{6}\int u^{3/2}du = -\frac{1}{6} \cdot \frac{2}{5}u^{5/2} + C = -\frac{1}{15}\left(3 - 6x^2\right)^{5/2} + C.$

25. $\int \frac{x\,dx}{\sqrt{x^2+3}} = \int \left(x^2 + 3\right)^{-1/2}x\,dx.$ Let $u = x^2 + 3$, then $du = 2x\,dx$ or $x\,dx = \frac{1}{2}du.$ Substituting produces $\frac{1}{2}\int u^{-1/2}du = \frac{1}{2} \cdot 2u^{1/2} + C = \left(x^2 + 3\right)^{1/2} + C = \sqrt{x^2 + 3} + C.$

26. $\int \frac{x\,dx}{\sqrt{4x^2-1}} = \int \left(4x^2 - 1\right)^{-3/2}x\,dx.$ Let $u = 4x^2 - 1$, then $du = 8x\,dx$ and so $\frac{1}{8}du = x\,dx.$ Substituting, we have $\frac{1}{8}\int u^{-3/2}du = \frac{1}{8}(-2)u^{-1/2} + C = -\frac{1}{4}\left(4x^2 - 1\right)^{-1/2} + C = \frac{-1}{4\sqrt{4x^2-1}} + C = -\frac{1}{4}\left(4x^2 - 1\right)^{-1/2} + C.$

27. $\int \frac{x\,dx}{\left(x^2+3\right)^3} = \int \left(x^2 + 3\right)^{-3}x\,dx.$ Let $u = x^2 + 3$, then $du = 2x\,dx$, or $x\,du = \frac{1}{2}du.$ Substitution yields $\frac{1}{2}\int u^{-3}du = \frac{1}{2} \cdot \frac{-1}{2}u^{-2} + C = \frac{-1}{4}\left(x^2 + 3\right)^{-2} + C = \frac{-1}{4\left(x^2+3\right)^2} + C = -\frac{1}{4}\left(x^2 + 3\right)^{-2} + C.$

28. $\int \left(x^2 + 3\right)^2 dx.$ We cannot use substitution, so we expand $\left(x^2 + 3\right)^2$ to get $\int \left(x^4 + 6x^2 + 9\right)dx = \frac{1}{5}x^5 + 2x^3 + 9x + C.$

29. $\int \left(3x^2 - 1\right)^2 dx.$ We cannot use substitution so we expand to get $\int \left(9x^4 - 6x^2 + 1\right)dx = \frac{9}{5}x^5 - 2x^3 + x + C.$

30. $\int \frac{x^2-1}{x+1}dx = \int \frac{(x+1)(x-1)}{x+1}dx = \int (x-1)dx = \frac{1}{2}x^2 - x + C, \ x \neq -1.$

31. $\int \frac{x^2-x-6}{x+1}dx = \int \frac{(x-3)(x+2)}{x+2}dx = \int (x-3)dx = \frac{1}{2}x^2 - 3x + C, \ x \neq -2.$

32. $\int \frac{3x\,dx}{(x^2+4)^5}$. Let $u = x^2+4$, and then $du = 2x\,dx$, or $3x\,dx = \frac{3}{2}du$. Substitution yields $\frac{3}{2}\int u^{-5}du = \frac{3}{2}\cdot\frac{-1}{4}u^{-4} + C = -\frac{3}{8}(x^2+4)^{-4} + C.$

33. $\int \frac{5x\,dx}{(x^2-1)^{1/3}}$. Let $u = x^2-1$, then $du = 2x\,dx$ or $5x\,dx = \frac{5}{2}du$. Substitution produces $\frac{5}{2}\int u^{-1/3}du = \frac{5}{2}\cdot\frac{3}{2}u^{2/3} + C = \frac{15}{4}(x^2-1)^{2/3} + C.$

34. $\int (4-x)^{-1/2}dx$. Let $u = 4-x$, and then $du = -dx$. Substituting we see that $-\int (u)^{-1/2}du = -\frac{2}{1}u^{1/2} + C = -2(4-x)^{1/2} + C.$

35. $\int (1+3x^2)^2 x\,dx$. Let $u = 1+3x^2$, and then $du = 6x\,dx$, or $x\,dx = \frac{1}{6}du$. Substitution produces $\frac{1}{6}\int u^2 du = \frac{1}{6}\cdot\frac{1}{3}u^3 + C = \frac{1}{18}(1+3x^2)^3 + C.$

36. $\int 2x\sqrt{5x^2+3}\,dx$. Let $u = 5x^2+3$, then $du = 10x\,dx$, or $\frac{1}{5}du = 2x\,dx$. Substitution produces $\frac{1}{5}\int u^{1/2}du = \frac{1}{5}\cdot\frac{2}{3}u^{3/2} + C = \frac{2}{15}(5x^2+3)^{3/2} + C.$

37. $\int 3x^2(4x^3-5)^{2/3}dx$. Let $u = 4x^3-5$, then $du = 12x^2dx$, and we have $\frac{1}{4}du = 3x^2dx$. Thus, $\frac{1}{4}\int u^{2/3}du = \frac{1}{4}\cdot\frac{3}{5}u^{5/3} + C = \frac{3}{20}(4x^3-5)^{5/3} + C.$

38. $\int \frac{2x^2dx}{(5-3x^3)^{2/3}}$. Let $u = 5-3x^3$, and $du = -9x^2dx$; $2x^2dx = -\frac{2}{9}du$. Thus, $-\frac{2}{9}\int u^{-2/3}du = -\frac{2}{9}\cdot\frac{3}{1}u^{1/3} + C = -\frac{2}{3}(5-3x^3)^{1/3} + C.$

39. $\int (1+3x^2)^2 dx$. We cannot use substitution so we expand to get $\int (1+6x^2+9x^4)dx = x + 2x^3 + \frac{9}{5}x^5 + C.$

40. $\int (2x^2+4x)^3(4x+4)dx$. Let $u = 2x^2+4x$, and then $du = (4x+4)dx$. Hence, $\int u^3 du = \frac{1}{4}u^4 + C = \frac{1}{4}(2x^2+4x)^4 + C.$

41. $\int \sqrt{4x^2+2x}(4x+1)dx$. Let $du = 4x^2+2x$, and then $du = (8x+2)dx$, or $\frac{1}{2}du = (4x+1)dx$. Thus,

substituting results in $\frac{1}{2}\int u^{1/2}du = \frac{1}{2}\cdot\frac{2}{3}u^{3/2} + C = \frac{1}{3}(4x^2+2x)^{3/2} + C.$

42. $\int (x^2+x-5)^4(2x+1)dx$. Let $u = x^2+x-5$, then $du = (2x+1)dx$. Substituting, we obtain $\int u^4 du = \frac{1}{5}u^5 + C = \frac{1}{5}(x^2+x-5)^5 + C.$

43. $\int \frac{x-1}{(x-1)^3}dx = \int (x-1)^{-2}dx = -1(x-1)^{-1} + C = \frac{-1}{x-1} + C$

44. $\int \frac{2x+1}{(x^2+x-5)^3}dx$. Letting $u = x^2+x-5$, we obtain $du = (2x+1)dx$. Substitution yields $\int u^{-3}du = -\frac{1}{2}u^{-2} + C = -\frac{1}{2}(x^2+x-5)^{-2} + C = \frac{-1}{2(x^2+x-5)^2} + C.$

45. $\int (2x^4-3)^2 8x\,dx$. Let $u = 2x^4-3$ and $du = 8x^3$. We do not have x^3 in the original integrand, so substitution will not work. Expand the indicated binomial to get $\int (4x^8-12x^4+9)8x\,dx = \int (32x^9-96x^5+72x)dx = \frac{32}{10}x^{10} - \frac{96}{6}x^6 + \frac{72}{2}x^2 + C = \frac{16}{5}x^{10} - 16x^6 + 36x^2 + C.$

46. $\int \frac{x+3}{(x^2+6x)^{1/3}}dx$. Let $u = x^2+6x$, then $du = (2x+6)dx$, and we have $\frac{1}{2}du = (x+3)dx$. Thus, $\frac{1}{2}\int u^{-1/3}du = \frac{1}{2}\cdot\frac{3}{2}u^{2/3} + = \frac{3}{4}(x^2+6x)^{2/3} + C$

47. $\int (x^3-3x)^{2/3}(x^2-1)dx$. Let $u = x^3-3x$, then $du = (3x^2-3)dx$, or $\frac{1}{3}du = (x^2-1)dx$; $\frac{1}{3}\int u^{2/3}du = \frac{1}{3}\cdot\frac{3}{5}u^{5/3} + C = \frac{1}{5}(x^3-3x)^{5/3} + C$

48. $\int (x^3-2)(x^4-8x)^{1/2}dx$. Let $u = x^4-8x$, then $du = (4x^3-8)dx$, or $\frac{1}{4}du = (x^3-2)dx$. Thus, $\int (x^3-2)(x^4-8x)^{1/2}dx = \frac{1}{4}\int u^{1/2}du = \frac{1}{4}\cdot\frac{2}{3}u^{3/2} + C = \frac{1}{6}(x^4-8x)^{3/2} + C.$

49. $\int (3x+1)^3 dx$. Let $u = 3x+1$, then $du = 3\,dx$, or $\frac{1}{3}du = dx$. Hence, we obtain $\frac{1}{3}\int u^3 du = \frac{1}{3}\cdot\frac{1}{4}u^4 + C = \frac{1}{12}(3x+1)^4 + C.$

50. $\int 7x(x^3-2)^2 dx$. For substitution to work, we must have an x^2 term. We do not, so expand $\int 7x(x^6-4x^3+4)dx = \int (7x^7-28x^4+28x)dx = \frac{7}{8}x^8 - \frac{28}{5}x^5 + 14x^2 + C.$

51. $\int \left(2x^3 - 1\right)^4 \sqrt{x^4 - 2x}\, dx$. Let $u = x^4 - 2x$, then $du = \left(4x^3 - 2\right)dx$, or $\frac{1}{3}du = \left(2x^3 - 1\right)dx$. Substitution results in $\frac{1}{2}\int u^{1/4}du = \frac{1}{2} \cdot \frac{1}{5}u^{5/4} + C = \frac{2}{5}\left(x^4 - 2x\right)^{5/4} + C$.

52. $\int \frac{x^2}{\left(4-x^3\right)^{3/2}}dx$. Setting $u = 4 - x^3$, we have $du = -3x^2 dx$, or $x^2 dx = -\frac{1}{3}du$. Thus, $-\frac{1}{3}\int u^{-3/2}du = -\frac{1}{3} \cdot \frac{-2}{1}u^{-1/2} + C = \frac{2}{3}\left(4 - x^3\right)^{-1/2} + C$ or $\frac{2}{3\sqrt{4-x^3}} + C$.

53. $\int (t+7)^{1/2}dt = \frac{2}{3}(t+7)^{3/2} + C$

54. $\int \frac{x^4-1}{x-1}dx = \int \frac{(x-1)(x+1)\left(x^2+1\right)}{x-1}dx = \int \left(x^3 + x^2 + x + 1\right)dx = \frac{x^4}{4} + \frac{x^3}{3} + \frac{x^2}{2} + x + C$, $x \neq 1$.

55. $\int \frac{y^3-8}{y-2}dy = \int \frac{(y-2)\left(y^2+2y+4\right)}{y-2}dy = \int \left(y^2 + 2y + 4\right)dy = \frac{1}{3}y^3 + y^2 + 4y + C$, $y \neq 2$.

56. $\int (6x - 7)^8 dx$. Let $u = 6x - 7$, then $du = 6\, dx$; $\frac{1}{6}du = dx$. Hence, we obtain $\frac{1}{6}\int u^8 du = \frac{1}{6} \cdot \frac{u^9}{9} + C = \frac{1}{54}(6x - 7)^9 + C$.

57. $\int_0^2 \left(x^2 - 4\right)2x\, dx$. Let $u = x^2 - 4$, then $du = 2x\, dx$. For the limits of integration, we have $x = 0 \Rightarrow u = -4$ and $x = 2 \Rightarrow u = 0$. Thus, $\int_{-4}^0 u\, du = \frac{u^2}{2}\Big|_{-4}^0 = 0 - \frac{16}{2} = -8$.

58. $\int_0^4 x\left(4x^2 + 2\right)^3 dx$. Let $u = 4x^2 + 2$, then $du = 8x\, dx$, or $\frac{1}{8}du = x\, dx$. For the limits of integration, we have $x = 0 \Rightarrow u = 2$ and $x = 4 \Rightarrow u = 66$. Thus, by substitution, we have $\frac{1}{8}\int_2^{66} u^3 du = \frac{1}{8} \cdot \frac{1}{4}u^4\Big|_2^{66} = \frac{1}{32}\left(66^4 - 2^4\right) = \frac{18974720}{32} = 592960$.

59. $\int_1^2 6x\left(3x^2 - 7\right)dx$. Letting $u = 3x^2 - 7$, we obtain $du = 6x\, dx$ and $x = 1 \Rightarrow u = -4$, and $x = 2 \Rightarrow u = 5$. Substitution produces $\int_{-4}^5 u\, du = \frac{1}{2}u^2\Big|_{-4}^5 = \frac{25}{2} - \frac{16}{2} = \frac{9}{2}$.

60. $\int_1^3 2x^2\left(3x^3 - 1\right)^2 dx$. Letting $u = 3x^3 - 1$, we see that $du = 9x^2 dx$, or $\frac{2}{9}du = 2x^2 dx$. For the limits of integration, we have $x = 1 \Rightarrow u = 2$ and $x = 3 \Rightarrow u = 80$. Thus, the integral becomes $\frac{2}{9}\int_2^{80} u^2 du = \frac{2}{9} \cdot \frac{u^3}{3}\Big|_2^{80} = \frac{2}{27}\left(80^3 - 2^3\right) = 37{,}925\frac{1}{3}$.

61. $\int_0^2 2x\left(20 - 3x^2\right)dx$. Let $u = 20 - 3x^2$, then $du = -6x\, dx$, and so $-\frac{1}{3}du = 2x\, dx$. Here, we see that the limits of integration can be changed as $x = 0 \Rightarrow u = 20$ and $x = 2 \Rightarrow u = 8$. Substitution produces $-\frac{1}{3}\int_{20}^8 u\, du = -\frac{1}{3} \cdot \frac{1}{2}u^2\Big|_{20}^8 = -\frac{1}{6}\left(8^2 - 20^2\right) = 56$.

62. $\int_0^5 x\left(x^2 + 144\right)^{1/2}dx$. Let $u = x^2 + 144$, then $du = 2x\, dx$, and so $\frac{1}{2}du = x\, dx$. Here we see that $x = 0 \Rightarrow u = 144$ and $x = 5 \Rightarrow u = 169$. Using all these substitutions, you get $\frac{1}{2}\int_{144}^{169} u^{1/2}du = \frac{1}{2} \cdot \frac{2}{3}u^{3/2}\Big|_{144}^{169} = \frac{1}{3}\left(169^{3/2} - 144^{3/2}\right) = \frac{1}{3}\left(13^3 - 12^3\right) = \frac{1}{3}(2197 - 1728) = \frac{469}{3} = 156\frac{1}{3}$.

63. $\int_0^3 \left(x^3 + 2\right)x^2 dx$. Let $u = x^3 + 2$, then $du = 3x^2 dx$, or $\frac{1}{3}du = x^2 dx$. Here, we see that $x = 0 \Rightarrow u = 2$ and $x = 3 \Rightarrow u = 29$. Substitution yields $\frac{1}{3}\int_2^{29} u\, du = \frac{1}{3} \cdot \frac{1}{2}u^2\Big|_2^{29} = \frac{1}{6}\left(29^2 - 2^2\right) = 139.5$.

64. $\int_1^2 \frac{dx}{(x+4)^2}$. Let $u = x+4$, then $du = dx$. Here $x = 1 \Rightarrow u = 5$; and $x = 2 \Rightarrow u = 6$. Substitution produces $\int_5^6 u^{-2}du = -u^{-1}\Big|_5^6 = -\frac{1}{6} + \frac{1}{5} = \frac{1}{30}$

65. $\int_1^4 \left(3x^2 - 2\right)^4 x\, dx$. Let $u = 3x^2 - 2$, then $du = 6x\, dx$, or $\frac{1}{6}du = x\, dx$ and we see that $x = 1 \Rightarrow u = 1$ and $x = 4 \Rightarrow 4 = 46$. Substitution produces $\frac{1}{6}\int_1^{46} u^4 du = \frac{1}{6} \cdot \frac{1}{5}u^5\Big|_1^{46} = \frac{1}{30}\left(46^5 - 1^5\right) = 6{,}865{,}432.5$

66. $\int_{-1}^0 \left(6x^2 - 1\right)^3 x\, dx$. Let $u = 6x^2 - 1$, then $du = 12x\, dx$, or $\frac{1}{12}du = x\, dx$ and we see that $x = -1 \Rightarrow u = 5$ and $x = 0 \Rightarrow u = -1$. Substitution produces $\frac{1}{12}\int_5^{-1} u^3 du = \frac{1}{12} \cdot \frac{1}{4}u^4\Big|_5^{-1} = \frac{1}{48}\left((-1)^4 - 5^4\right) = \frac{1}{48}(-624) = -13$.

67. $L(x) = \int L'(x)dx = \int (0.3 + 0.002x)dx = 0.3x + 0.001x^2 + C$. When the population is 20,000 people, $x = 20$ and we are given $L(20) = 7.2$ or $0.3(20) + 0.001(20)^2 + C = 6.4 + C = 7.2$. Thus, $C = 7.2 - 6.4 = 0.8$ and $L(x) = 0.3x + 0.001x^2 + 0.8$

68. $N(t) = \int N'(t)dt = \int \left(-\frac{750}{t^2} - 120\right)dt = 750t^{-1} - 120t + C$. We are given $N(1) = 7320 = 750(1)^{-1} - 120(1) + C = 630 + C$. Thus, $C = 7320 - 630 = 6690$ and we conclude that $N(t) = 750t^{-1} - 120t + 6690$.

69. $P(x) = \int P'(x)dx = \int (4x+35)dx = 2x^2 + 35x + C$. We are given $P(0) = -150$ and so, $C = -150$. Thus, $P(x) = 2x^2 + 35x - 150$

70. $P(t) = \int_0^8 P'(t)dt$

$= \int_0^8 \left(\frac{1800}{(18-0.5t)^{1/2}} \right) dt$

$= \int_0^8 \left[1800(18-0.5t)^{-1/2} \right] dt$

$= 1800 \int_0^8 \left[(18-0.5t)^{-1/2} \right] dt$

Let $u = 18 - 0.5t$, and then $du = -0.5\,dt$ and $dt = -2\,du$. When $t = 0$, $u = 18$ and when $t = 8$, $u = 14$. With these substitutions, the integral becomes

$P(t) = 1800 \int_0^8 \left[(18-0.5t)^{-1/2} \right] dt$

$= 1800 \int_{18}^{14} u^{-1/2}(-2\,du)$

$= -3600 \int_{18}^{14} u^{-1/2}du$

$= -7200 u^{1/2} \Big|_{18}^{14}$

$= -7200 \left[(14)^{1/2} - (18)^{1/2} \right] \approx 3607$

During the first 8 hr, there will be an increase of about 3607 bacteria.

71. $T(x) = \int \left(-\frac{5750}{(x+1)^3} \right) dx$

$= \int -5750(x+1)^{-3}dx$

$= \frac{-5750}{-2}(x+1)^{-2} + K$

$= 2875(x+1)^{-2} + K$

Since $T(0) = 2900°$C, then $2875 + K = 2900$ and $K = 25$. Hence, $T(x) = 2875(x+1)^{-2} + 25$.

72. $q(t) = \int I\,dt = \int 0.005\,dt = 0.005t + C$. We are given that the capacitor is discharged at $t = 0$, and so $t(0) = C = 0$. Thus, $q(t) = 0.005t$.

≡ 25.4 THE AREA BETWEEN TWO CURVES

1.

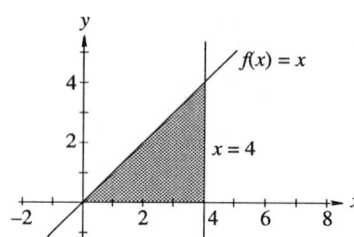

From the graph we can determine that all the area is above the x-axis so we only need to integrate from 0 to 4. $\int_0^4 x\,dx = \frac{x^2}{2}\Big|_0^4 = \frac{4^2}{2} - \frac{0^2}{2} = 8$

2.

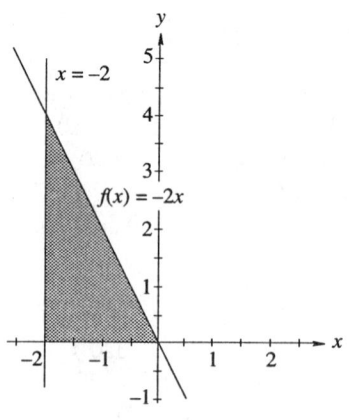

From the graph we can determine that all the area is above the x-axis so we need only integrate from -2 to 0. $\int_{-2}^0 -2x\,dx = -x^2\Big|_{-2}^0 = -0^2 + (-2)^2 = 4$

3.

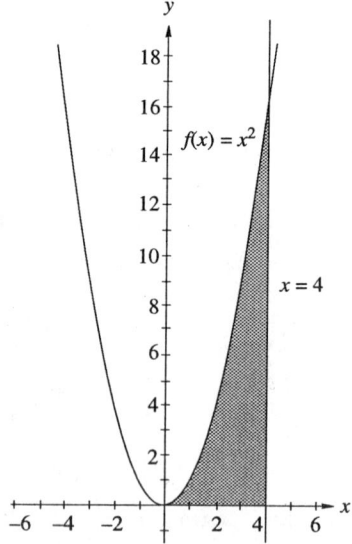

All the area is above the x-axis $\int_0^4 x^2 = \frac{x^3}{3}\Big|_0^4 = \frac{4^3}{3} - \frac{0^3}{3} = \frac{64}{3}$

4.

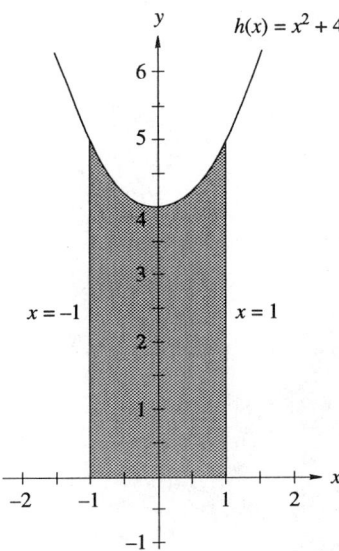

All the area above the x-axis $\int_{-1}^{1}\left(x^2+4\right)dx =$

$\frac{x^3}{3}+4x\Big|_{-1}^{1} = \left(\frac{1}{3}+4\right)-\left(-\frac{1}{3}-4\right) = 8\frac{2}{3}$.

5.

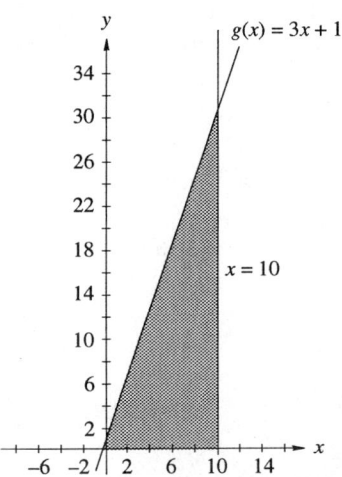

All the area is above the x-axis. We can determine that $3x+1$ intersects the x-axis at $-\frac{1}{3}$. Hence, the desired area is

$$\int_{-1/3}^{10}(3x+1)dx = \frac{3}{2}x^2+x\Big|_{-1/3}^{10}$$

$$= \left(\frac{300}{2}+10\right)-\left(\frac{1}{6}-\frac{1}{3}\right)$$

$$= 160\frac{1}{6}$$

6.

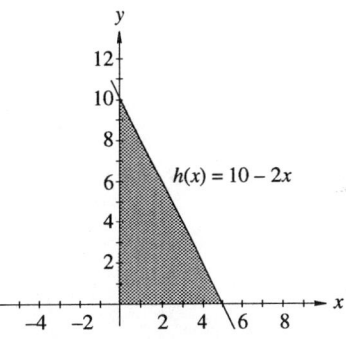

All the area is above the x-axis. We can see that $h(x)$ intersects the x-axis at 5. $\int_{0}^{5}(10-2x)dx = 10x-x^2\Big|_0^5 = 50-25 = 25$

7.

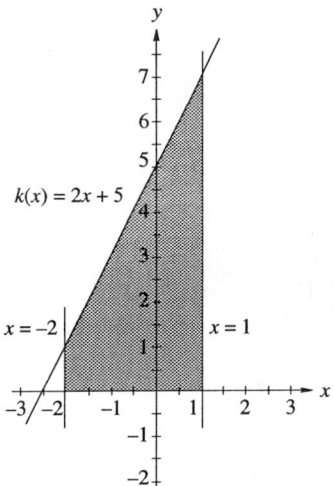

Between -2 and 1, $k(x)$ is above the x-axis. Thus, the desired area is $\int_{-2}^{1}(2x+5)dx = x^2+5x\Big|_{-2}^{1} = (1+5)-(4-10) = 6+6 = 12$.

8.

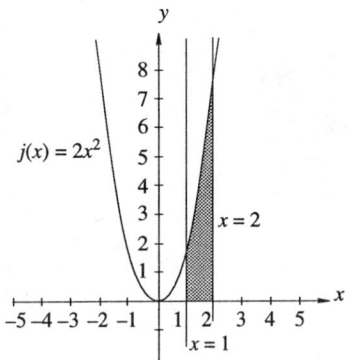

Between 1 and 2, j is above the x-axis. Hence the desired area is $\int_{1}^{2}2x^2dx = \frac{2x^3}{3}\Big|_1^2 = \frac{16}{3}-\frac{2}{3} = \frac{14}{3}$.

9.

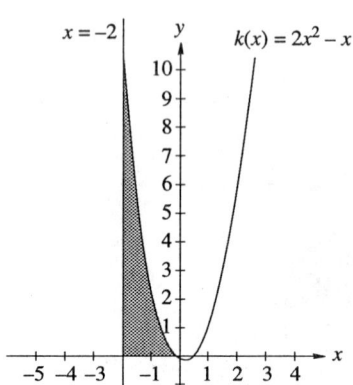

Between -2 and 0, we can see that k is above the x-axis. Thus, the desired area is $\int_{-2}^{0} (2x^2 - x)\,dx = \frac{2x^3}{3} - \frac{x^2}{2}\Big|_{-2}^{0} = 0 - \left(\frac{-16}{3} - \frac{4}{2}\right) = 7\frac{1}{3} = \frac{22}{3}$.

10.

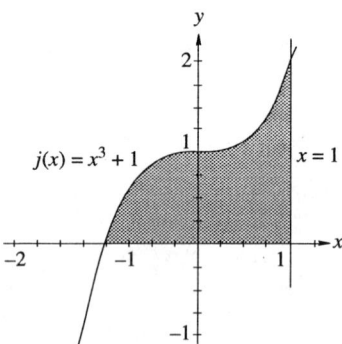

Between -1 and 1, we see that j is above the x-axis. As a result, the desired area is $\int_{-1}^{1} (x^3 + 1)\,dx = \frac{x^4}{4} + x\Big|_{-1}^{1} = \left(\frac{1}{4} + 1\right) - \left(\frac{1}{4} - 1\right) = 2$.

11.

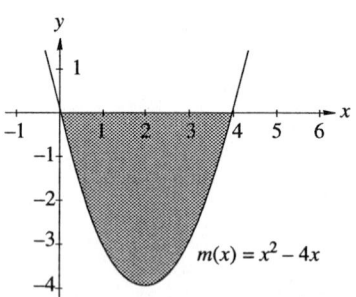

From the graph we can determine that m intersects the x-axis at 0 and 4 and is below the axis.

Hence we take the absolute value of the integral.

$$\left| \int_{0}^{4} (x^2 - 4x)\,dx \right| = \left| \left[\frac{x^3}{3} - 2x^2 \right]_{0}^{4} \right| = \left| \frac{64}{3} - 32 \right|$$

$$= \left| -10\frac{2}{3} \right| = 10\frac{2}{3} \text{ or } \frac{32}{3}$$

12.

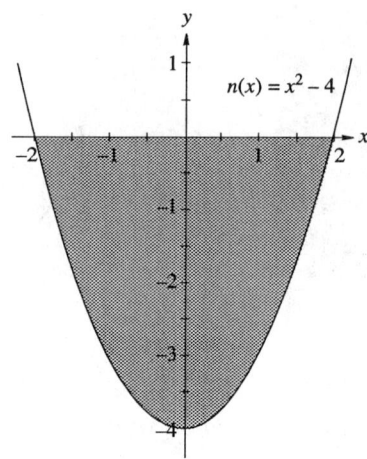

From the graph we determine that n intersects the x-axis at -2 and 2 and the area is below the axis between these points. We need the absolute value of the integral or its negative.

$$\left| \int_{-2}^{2} (x^2 - 4)\,dx \right|$$

$$= -\int_{-2}^{2} (x^2 - 4)\,dx = -\left(\frac{x^3}{3} - 4x \right)\Big|_{-2}^{2}$$

$$= -\left[\left(\frac{8}{3} - 8 \right) - \left(\frac{-8}{3} + 8 \right) \right]$$

$$= -\left[\frac{16}{3} - 16 \right] = -\left[\frac{-32}{3} \right] = \frac{32}{3} = 10\frac{2}{3}$$

13.

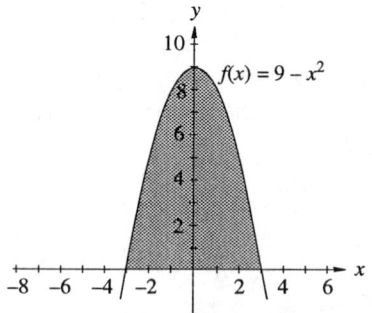

The area is above the x-axis and f intersects the x-axis at -3 and 3. Thus, we have

$$\int_{-3}^{3} (9 - x^2)\,dx = 9x - \frac{x^3}{3}\Big|_{-3}^{3}$$

$$= (27 - 9) - (-27 + 9) = 36$$

14.

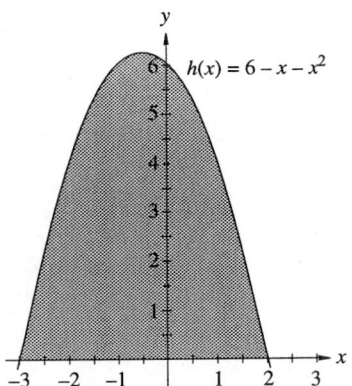

The area is above the x-axis and h intersects the axis at -3 and 2.

$$\int_{-3}^{2} (6 - x - x^2)\,dx = 6x - \frac{x^2}{2} - \frac{x^3}{3}\Big|_{-3}^{2}$$

$$= \left(12 - 2 - \frac{8}{3}\right)$$

$$- \left(-18 - \frac{9}{2} + 9\right)$$

$$= 20\frac{5}{6}$$

15.

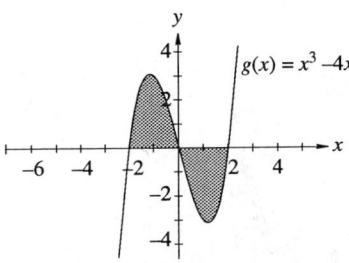

The graph is above the x-axis from -2 to 0 and below the x-axis from 0 to 2.

$$\int_{-2}^{0} (x^3 - 4x)\,dx - \int_{0}^{2} (x^3 - 4x)\,dx$$

$$= \left(\frac{x^4}{4} - 2x^2\right)\Big|_{-2}^{0} - \left(\frac{x^4}{4} - 2x^2\right)\Big|_{0}^{2}$$

$$= 0 - \left(\frac{16}{4} - 8\right) - \left(\frac{16}{4} - 8\right) + 0 = 8$$

16.

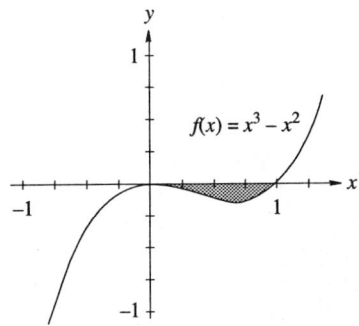

The graph is below the axis from 0 to 1, the two places it intersects the x-axis.

$$-\int_{0}^{1} (x^3 - x^2)\,dx = -\left(\frac{x^4}{4} - \frac{x^3}{3}\right)\Big|_{0}^{1}$$

$$= -\left(\frac{1}{4} - \frac{1}{3}\right) = \frac{1}{12}.$$

17.

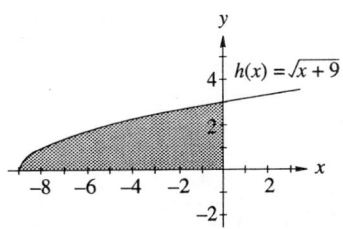

The graph is above the x-axis and intersects at -9. Thus, the desired area is

$$\int_{-9}^{0} \sqrt{x + 9}\,dx = \frac{2}{3}(x + 9)^{3/2}\Big|_{-9}^{0}$$

$$= \frac{2}{3}\left[(0 + 9)^{3/2} - (-9 + 9)^{3/2}\right]$$

$$= \frac{2}{3} \cdot 27 = 18.$$

18.

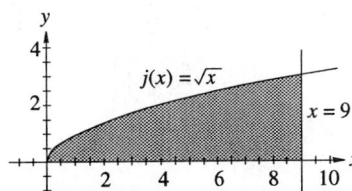

The graph is above the x-axis and intersects the axis at 0. Thus, the desired area is

$$\int_{0}^{9} \sqrt{x}\,dx = \frac{2}{3}x^{3/2}\Big|_{0}^{9} = \frac{2}{3} \cdot 9^{3/2} = 18.$$

19.

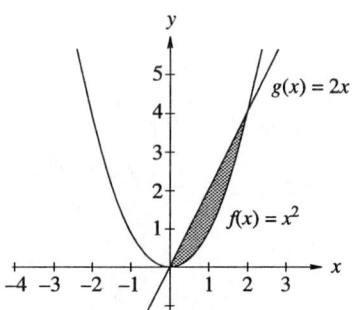

From the graph we determine that $g(x) \geq f(x)$ on $[0, 2]$. Hence, the desired area is

$$\int_0^2 \left(2x - x^2\right) dx = x^2 - \frac{x^3}{3}\bigg|_0^2 = 4 - \frac{8}{3} = \frac{4}{3}.$$

20.

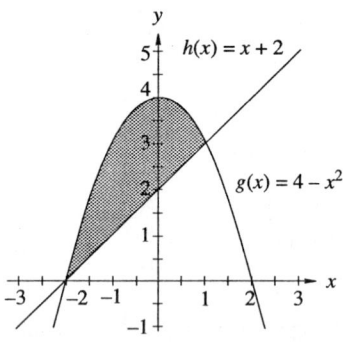

The graphs intersect at $x = -2$ and $x = 1$. Since $g(x) \geq h(x)$, the desired area is

$$\int_{-2}^1 (g(x) - h(x))dx$$

$$= \int_{-2}^1 \left((4 - x^2) - (x + 2)\right) dx$$

$$= \int_{-2}^1 \left(2 - x - x^2\right) dx = 2x - \frac{x^2}{2} - \frac{x^3}{3}\bigg|_{-2}^1$$

$$= \left(2 - \frac{1}{2} - \frac{1}{3}\right) - \left(-4 - 2 + \frac{8}{3}\right)$$

$$= 2 - \frac{1}{2} - \frac{1}{3} + 4 + 2 - \frac{8}{4} = 4\frac{1}{2}$$

21.

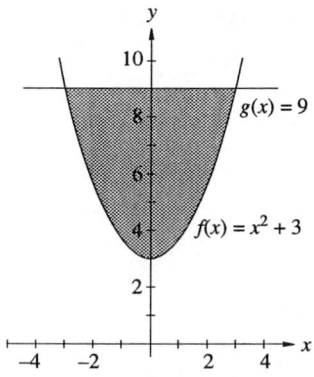

f and g intersect at $(\pm\sqrt{6}, 9)$. On the interval $[-\sqrt{6}, \sqrt{6}]$, we see that $g(x) \geq f(x)$. Thus, the desired area is

$$\int_{-\sqrt{6}}^{\sqrt{6}} \left(9 - \left(x^2 + 3\right)\right) dx$$

$$= \int_{-\sqrt{6}}^{\sqrt{6}} \left(6 - x^2\right) dx = 6x - \frac{x^3}{3}\bigg|_{-\sqrt{6}}^{\sqrt{6}}$$

$$= \left(6\sqrt{6} - \frac{\sqrt{6}^3}{3}\right) - \left(-6\sqrt{6} + \frac{\sqrt{6}^3}{3}\right)$$

$$= 12\sqrt{6} - 4\sqrt{6} = 8\sqrt{6} \approx 19.596$$

22.

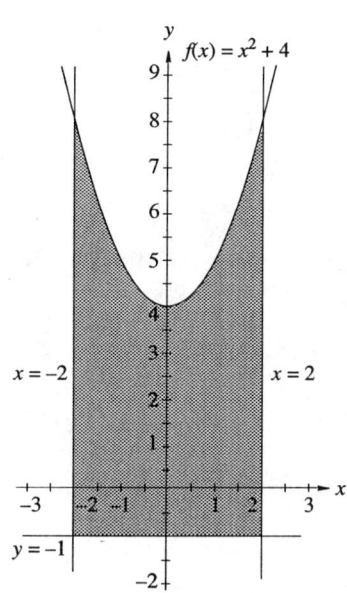

From the figure we determine that we need to integrate from -2 to 2 and that the needed integrand is $f(x) - (-1)$.

$$\int_{-2}^{2} \left(x^2 + 4 - (-1)\right) dx$$

$$= \int_{-2}^{2} \left(x^2 + 5\right) dx$$

$$= \frac{x^3}{3} + 5x \bigg|_{-2}^{2} = \left(\frac{8}{3} + 10\right) - \left(\frac{-8}{3} - 10\right)$$

$$= 25\frac{1}{3}$$

23.

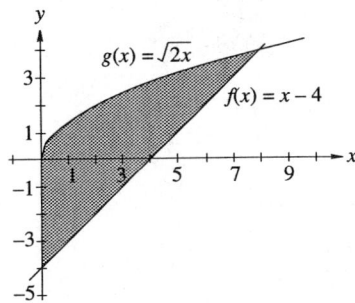

f and g intersect at $(8,4)$. g is defined on $[0, \infty)$. On $[0, 8]$, we see that $g(x) \geq f(x)$.

$$\int_{0}^{8} (g(x) - f(x)) dx$$

$$= \int_{0}^{8} (\sqrt{2x} - x + 4) dx$$

$$= \int_{0}^{8} \sqrt{2x}\, dx + \int_{0}^{8} (4 - x) dx$$

For the first integral let $u = 2x$, then $du = 2\,dx$, so $dx = \frac{1}{2} du$ and the limits of integration become $x = 0 \Rightarrow u = 0$ and $x = 8 \Rightarrow u = 16$. Thus, by substitution into the first integral, we get

$$\frac{1}{2} \int_{0}^{16} \sqrt{u}\, du + \int_{0}^{8} (4 - x) dx$$

$$= \frac{1}{2} \frac{2}{3} u^{3/2} \bigg|_{0}^{16} + \left(4x - \frac{x^2}{2}\right) \bigg|_{0}^{8}$$

$$= \left(\frac{64}{3} - 0\right) + \left(32 - \frac{64}{2}\right) = 21\frac{1}{3} = 21.333$$

24.

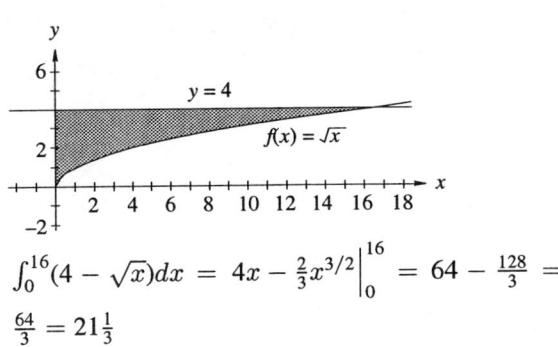

$\int_{0}^{16} (4 - \sqrt{x}) dx = 4x - \frac{2}{3} x^{3/2} \bigg|_{0}^{16} = 64 - \frac{128}{3} = \frac{64}{3} = 21\frac{1}{3}$

25.

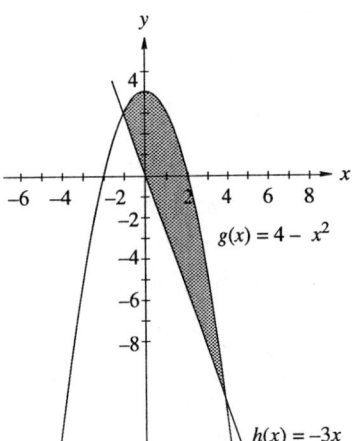

f and g intersect at $(-1, \ 3)$ and $(4, \ -12)$

$$\int_{-1}^{4} \left((4 - x^2) - (-3x)\right) dx$$

$$= \int_{-1}^{4} (4 + 3x - x^2) dx$$

$$= 4x + \frac{3x^2}{2} - \frac{x^3}{3} \bigg|_{-1}^{4}$$

$$= \left(16 + 24 - \frac{64}{3}\right) - \left(-4 + \frac{3}{2} + \frac{1}{3}\right)$$

$$= 44 - \frac{3}{2} - \frac{65}{3} = 20.8333$$

26.

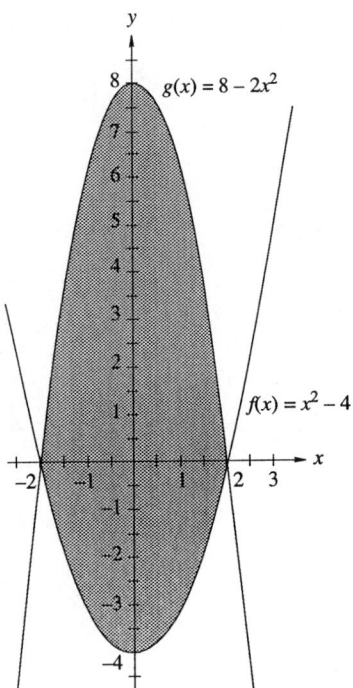

f and g intersect at $(\pm 2, \ 0)$. On $[-2, \ 2]$, $g(x) \geq f(x)$.

$$\int_{-2}^{2} \left[(9 - 2x^2) - (x^2 - 4) \right] dx$$

$$= \int_{-2}^{2} (12 - 3x^2) \, dx$$

$$= 12x - x^3 \Big|_{-2}^{2} = (24 - 8) - (-24 + 8) = 32$$

27.

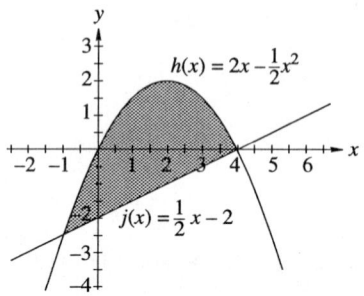

h and j intersect at $\left(-1, \ -2\frac{1}{3}\right)$ and $(4, \ 0)$

$$\int_{-1}^{4} (h(x) - j(x) \, dx$$

$$= \int_{-1}^{4} \left(\frac{3}{2}x - \frac{1}{2}x^2 + 2 \right) dx$$

$$= \frac{3}{4}x^2 - \frac{1}{6}x^3 + 2x \Big|_{-1}^{4}$$

$$= \left(12 - \frac{64}{6} + 8 \right) - \left(\frac{3}{4} + \frac{1}{6} - 2 \right)$$

$$= 10.417$$

28.

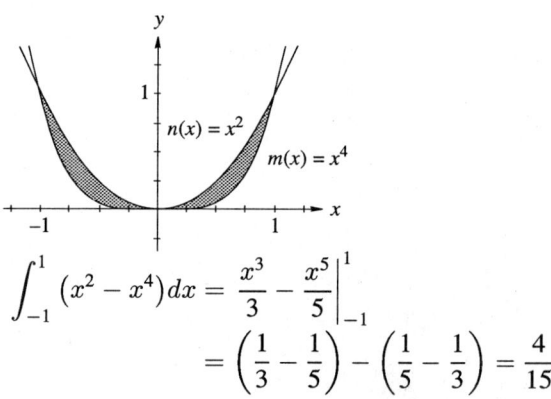

$$\int_{-1}^{1} (x^2 - x^4) \, dx = \frac{x^3}{3} - \frac{x^5}{5} \Big|_{-1}^{1}$$

$$= \left(\frac{1}{3} - \frac{1}{5} \right) - \left(\frac{1}{5} - \frac{1}{3} \right) = \frac{4}{15}$$

29.

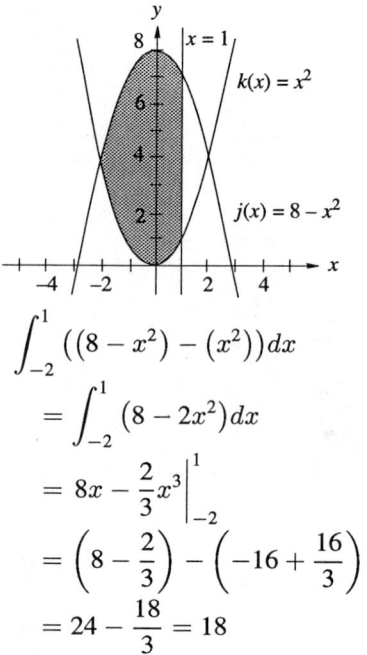

$$\int_{-2}^{1} \left((8 - x^2) - (x^2) \right) dx$$

$$= \int_{-2}^{1} (8 - 2x^2) \, dx$$

$$= 8x - \frac{2}{3}x^3 \Big|_{-2}^{1}$$

$$= \left(8 - \frac{2}{3} \right) - \left(-16 + \frac{16}{3} \right)$$

$$= 24 - \frac{18}{3} = 18$$

30.

$$\int_{-2}^{0} 4 \, dx + \int_{0}^{4} (4 - 2\sqrt{x}) \, dx$$

$$= 4x \Big|_{-2}^{0} + \left(4x - \frac{4}{3}x^{3/2} \right) \Big|_{0}^{4}$$

$$= 8 + \left(16 + \frac{32}{3} \right) = 24 - \frac{32}{3} = \frac{40}{3} = 13\frac{1}{3}$$

31.

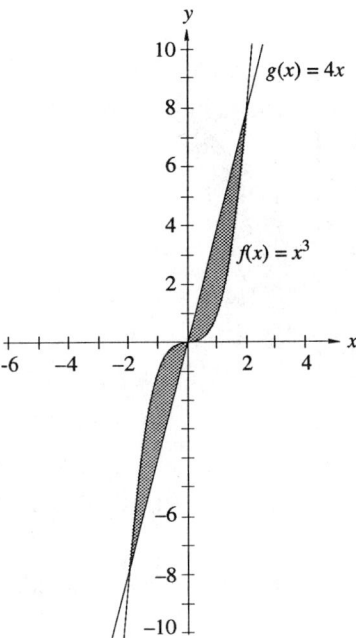

$$\int_{-2}^{0} \left(x^3 - 4x\right)dx + \int_{0}^{2} \left(4x - x^3\right)dx$$

$$= \left(\frac{x^4}{4} - 2x^2\right)\Big|_{-2}^{0} + \left(2x^2 - \frac{x^4}{4}\right)\Big|_{0}^{2}$$

$$= 0 - \left(\frac{16}{4} - 18\right) + \left(8 - \frac{16}{4}\right) - 0 = 8$$

32.

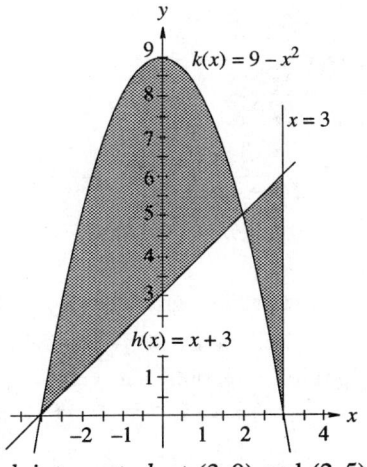

h intersects k at $(3, 0)$ and $(2, 5)$

$$\int_{-3}^{2} \left[(9 - x^2) - (x + 3)\right]dx$$

$$\quad + \int_{2}^{3} \left[(x + 3) - (9 - x^2)\right]dx$$

$$= \int_{-3}^{2} \left(6 - x - x^2\right)dx + \int_{2}^{3} \left(x^2 + x - 6\right)dx$$

$$= \left(6x - \frac{x^2}{2} - \frac{x^3}{3}\right)\Big|_{-3}^{2} + \left(\frac{x^3}{3} + \frac{x^2}{2} - 6x\right)\Big|_{2}^{3}$$

$$= \left(12 - 2 - \frac{8}{3}\right) - \left(-18 - \frac{9}{2} +\right)$$

$$\quad + \left(9 + \frac{9}{2} - 18\right) - \left(\frac{8}{3} + 2 - 12\right)$$

$$= 12 - 2 - \frac{8}{3} + 18 + \frac{9}{2} - 9 + 9$$

$$\quad + \frac{9}{12} - 18 - \frac{8}{3} - 2 + 12$$

$$= 24 - 4 + 9 - \frac{16}{3} = 29 - \frac{16}{3} = 23\frac{2}{3}$$

33.

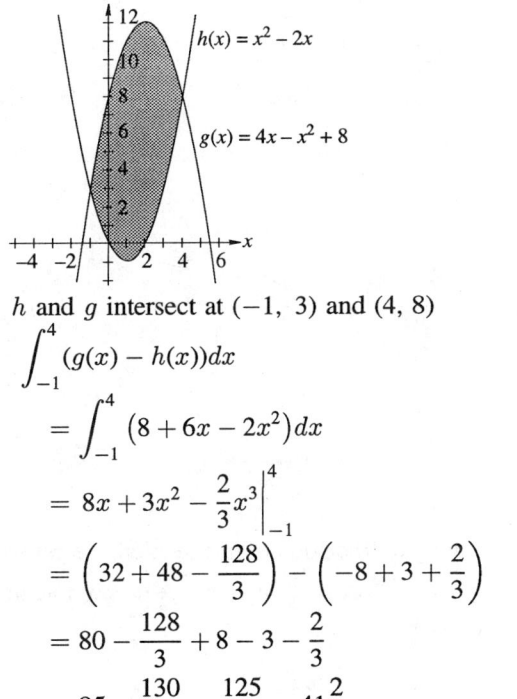

h and g intersect at $(-1, 3)$ and $(4, 8)$

$$\int_{-1}^{4} (g(x) - h(x))dx$$

$$= \int_{-1}^{4} \left(8 + 6x - 2x^2\right)dx$$

$$= 8x + 3x^2 - \frac{2}{3}x^3\Big|_{-1}^{4}$$

$$= \left(32 + 48 - \frac{128}{3}\right) - \left(-8 + 3 + \frac{2}{3}\right)$$

$$= 80 - \frac{128}{3} + 8 - 3 - \frac{2}{3}$$

$$= 85 - \frac{130}{3} = \frac{125}{3} = 41\frac{2}{3}$$

34.

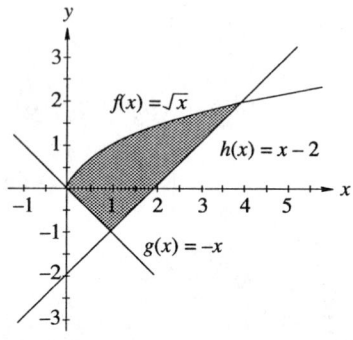

f is always the top graph. From 0 to 1, g is the bottom; from 1 to 4, h is the bottom.

$$\int_0^1 (\sqrt{x} + x)dx + \int_1^4 (\sqrt{x} - x + 2)dx$$

$$= \left(\frac{2}{3}x^{3/2} + \frac{x^2}{2}\right)\Big|_0^1 + \left(\frac{2}{3}x^{3/2} - \frac{x^2}{2} + 2x\right)\Big|_1^4$$

$$= \left(\frac{2}{3} + \frac{1}{2}\right) - 0 + \left(\frac{16}{3} - \frac{16}{2} + 8\right)$$
$$\quad - \left(\frac{2}{3} - \frac{1}{2} + 2\right)$$

$$= \frac{2}{3} + \frac{1}{2} + \frac{16}{3} - 8 + 8 - \frac{2}{3} + \frac{1}{2} - 2$$

$$= \frac{16}{3} - 1 = 4\frac{1}{3}$$

35.

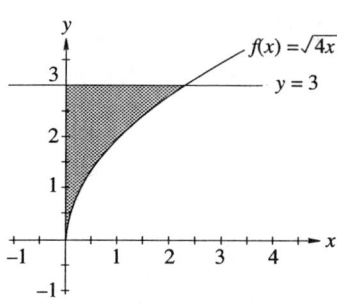

Solving for the point of intersection, we get $\sqrt{4x} = 2\sqrt{x} = 3$; $\sqrt{x} = \frac{3}{2}$, and $x = \frac{9}{4}$. Now integrating, we have

$$\int_0^{9/4} (3 - 2\sqrt{x})dx$$

$$= 3x - \frac{4}{3}x^{3/2}\Big|_0^{9/4}$$

$$= \frac{27}{4} - \frac{4}{3} \cdot \frac{27}{8} = \frac{27}{4} - \frac{27}{6} = 2.25$$

36.

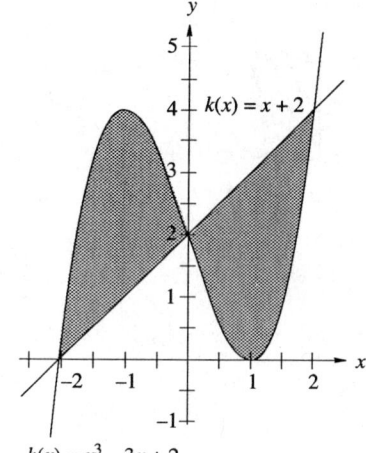

$h(x) = x^3 - 3x + 2$

h and k intersect at $(-2, 0)$, $(0, 2)$ and $(2, 4)$. Thus, the desired area is

$$\int_{-2}^0 (h(x) - k(x))dx + \int_0^2 (k(x) - h(x))dx$$

$$= \int_{-1}^0 (x^3 - 4x)dx + \int_0^2 (4x - x^3)dx$$

$$= \left(\frac{x^4}{4} - 2x^2\right)\Big|_{-2}^0 + \left(2x^2 - \frac{x^4}{4}\right)\Big|_0^2$$

$$= 0 - (-4 + 8) + (8 - 4) - 0 = 8$$

37.

$\int_0^4 (\sqrt{16 - 2x} - \sqrt{16 - 4x}\, dx + \int_4^8 \sqrt{16 - 2x}\, dx = \int_0^4 \sqrt{16 - 2x}\, dx - \int_0^4 \sqrt{16 - 4x}\, dx + \int_4^8 \sqrt{16 - 2x}\, dx.$

For the first and third integrals let $u = 16 - 2x$, then $du = -2\,dx$ or $dx = -\frac{1}{2}du$. For the limits of integration, we obtain $x = 0 \Rightarrow u = 16$; $x = 4 \Rightarrow u = 8$ and $x = 8 \Rightarrow u = 0$.

$$-\frac{1}{2}\int_{16}^8 u^{1/2}du = -\frac{1}{2} \cdot \frac{2}{3}u^{3/2}\Big|_{16}^8$$

$$= -\frac{1}{3}\left(8^{3/2} - 64\right)$$

$$-\frac{1}{2}\int_8^0 u^{1/2}dx = \frac{1}{2} \cdot \frac{2}{3}u^{3/2}\Big|_8^0$$

$$= -\frac{1}{3}\left(0 - 8^{3/2}\right)$$

The sum of these two integrals is $\frac{64}{3}$. For the second integral let $u = 16 - 4x$, and then $du = -4\,dx$ or $dx = -\frac{1}{4}du$. Here the limits of integration are $x = 0 \Rightarrow u = 16$ and $x = 4 \Rightarrow u = 0$.

$$-\frac{1}{4}\int_{16}^{0} u^{1/2}\,du = -\frac{1}{4}\cdot\frac{2}{3}u^{3/2}\Big|_{16}^{0}$$
$$= 0 + \frac{1}{6}\cdot 64 = \frac{64}{6}$$

For the final answer, we subtract this last answer from the earlier result, and obtain $\frac{64}{3} - \frac{64}{6} = \frac{64}{6} = \frac{32}{3} = 10\frac{2}{3} = 10.667$.

38.

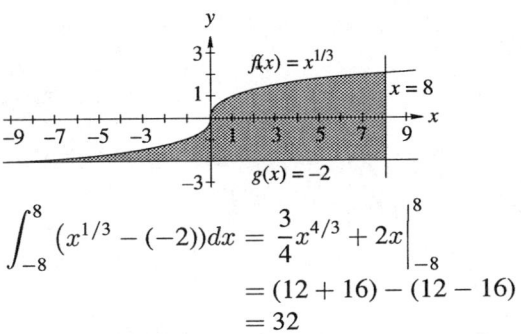

$$\int_{-8}^{8}\left(x^{1/3} - (-2)\right)dx = \frac{3}{4}x^{4/3} + 2x\Big|_{-8}^{8}$$
$$= (12 + 16) - (12 - 16)$$
$$= 32$$

39.

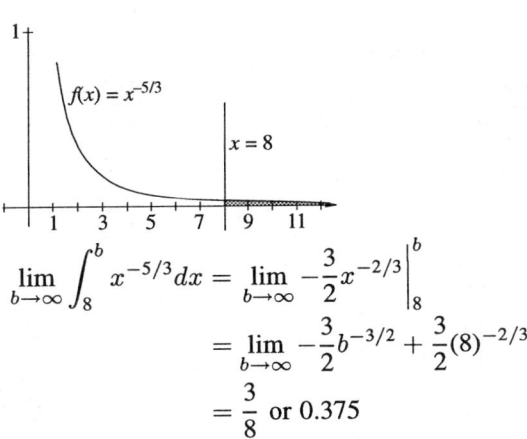

$$\lim_{b\to\infty}\int_{8}^{b} x^{-5/3}\,dx = \lim_{b\to\infty} -\frac{3}{2}x^{-2/3}\Big|_{8}^{b}$$
$$= \lim_{b\to\infty} -\frac{3}{2}b^{-3/2} + \frac{3}{2}(8)^{-2/3}$$
$$= \frac{3}{8} \text{ or } 0.375$$

40.

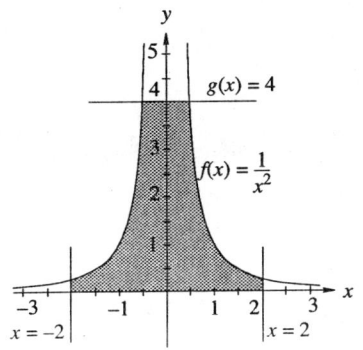

The graphs of f and g intersect when $\frac{1}{x^2} = 4$, or when $x^2 = \frac{1}{4} \Rightarrow x = \pm\frac{1}{2}$. Thus, the desired area is

$$\int_{-2}^{-1/2}\frac{1}{x^2}\,dx + \int_{-1/2}^{1/2} 4\,dx + \int_{1/2}^{2}\frac{1}{x^2}\,dx$$
$$= -x^{-1}\Big|_{-2}^{-1/2} + 4x\Big|_{-1/2}^{1/2} + -x^{-1}\Big|_{1/2}^{2}$$
$$= 2 - \frac{1}{2} + 2 - (-2) - \frac{1}{2} + 2 = 7$$

41.

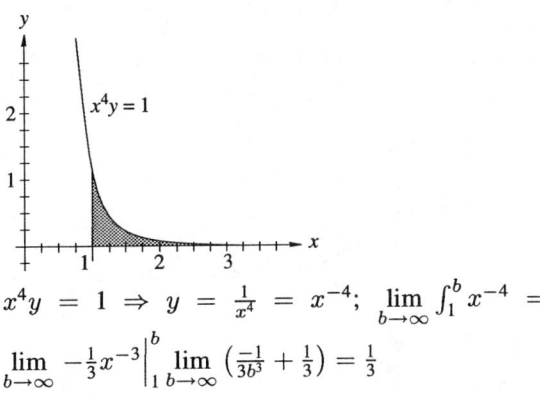

$x^4 y = 1 \Rightarrow y = \frac{1}{x^4} = x^{-4}$; $\displaystyle\lim_{b\to\infty}\int_{1}^{b} x^{-4} = \lim_{b\to\infty} -\frac{1}{3}x^{-3}\Big|_{1}^{b} \lim_{b\to\infty}\left(\frac{-1}{3b^3} + \frac{1}{3}\right) = \frac{1}{3}$

42.

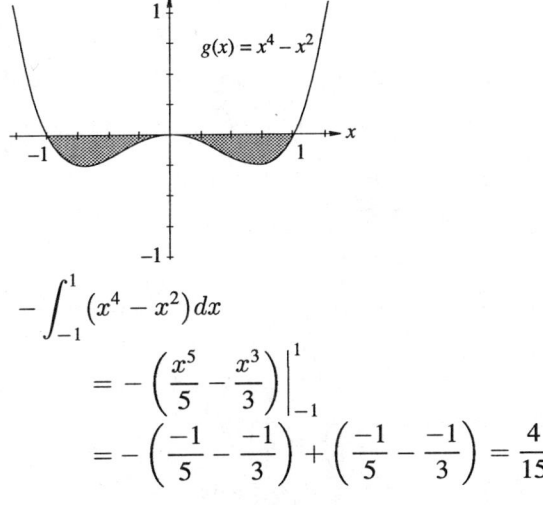

$$-\int_{-1}^{1}\left(x^4 - x^2\right)dx$$
$$= -\left(\frac{x^5}{5} - \frac{x^3}{3}\right)\Big|_{-1}^{1}$$
$$= -\left(\frac{-1}{5} - \frac{-1}{3}\right) + \left(\frac{-1}{5} - \frac{-1}{3}\right) = \frac{4}{15}$$

43. The area of the blue is $\int_{0}^{1}\left(\sqrt{x} - x^3\right)dx = \frac{2}{3}x^{3/2} - \frac{x^4}{4}\Big|_{0}^{1} = \frac{2}{3} - \frac{1}{4} = \frac{5}{12}$. The total area is 1 so the area of the white is $1 - \frac{5}{12} = \frac{7}{12}$. More white enamel is needed.

44. (a) $S(x) - C(x) = \left(-x^2 + 11x + 15\right) - \left(\frac{5}{4}x^2\right) = -\frac{9}{4}x^2 + 11x + 15$. Using the quadratic formula,

$x = 6$ or $x = -\frac{10}{9}$. Since $x > 0$, the answer is 6 yr.

(b) $\int_0^6 \left(-\frac{9}{4}x^2 + 11x + 15\right)dx = -\frac{3}{4}x^3 + \frac{11}{2}x^2 + 15x\big|_0^6 = 126$ or \$126,000,000.

45. The graphs intersect at $(-2, 9)$ and $(2, 9)$. For the cross-sectional area of the trough, we use $\int_{-2}^2 (9 + 2x^2 - x^4)dx = 2\int_0^2 (9 + 2x^2 - x^4)dx = 2 \cdot 9x + \frac{2}{3}x^3 - \frac{1}{5}x^5\big|_0^2 = \frac{508}{15} \approx 33.87$ cm^2. The volume is $\frac{508}{15} \times 1\,000 = 33\,866\frac{2}{3} \approx 33\,866.67$ cm^3.

46. The graphs intersect at $(0, 0)$ and $(8, 16)$. $A = \int_0^8 \left(10x - x^2 - \frac{1}{4}x^2\right)dx = \int_0^8 \left(10x - \frac{5}{4}x^2\right)dx = 5x^2 - \frac{5}{12}x^3\big|_0^8 = 106\frac{2}{3} \approx 106.67$ cm^2.

47. $\int_{0.6}^{2.4} \left(\frac{3000}{V^{1.33}} - \frac{1000}{V^{1.33}}\right)dV = \int_{0.6}^{2.4} 2000V^{-1.33}dV = \frac{2000}{-0.33}V^{-0.33}\big|_{0.6}^{2.4} = 2633.5$ or about 2630 Btu.

48. $W = \int_{0.6}^{1.0} \left(2044.5 - \frac{1000}{V^{1.4}}\right)dV$
$+ \int_{1.0}^{2.6} \left(\frac{2000}{V^{1.4}} - \frac{1000}{V^{1.4}}\right)$
$= [2044.5V + 2500V^{-0.4}]_{0.6}^{1.0}$
$+ [-5000V^{-0.4} + 2500V^{-0.4}]_{1.0}^{2.6}$
$\approx [(2044.5 + 2500) - (1226.7 + 3066.76)]$
$+ [(-3411.78 + 1705.89)$
$- (-5000 + 2500)]$
$= 1045.15$
The work done is about 1045 Btu.

49. $\omega = \int_0^{10} \alpha\, dt = \int_0^{10} \frac{100}{\sqrt{(t+1)^3}}dt =$
$\frac{-200}{\sqrt{t+1}}\big|_0^{10} = -200\left(\frac{1}{\sqrt{11}} - 1\right) \approx 139.7$ rad/s.

50. $V = l\int_a^\infty \frac{x}{(R^2+x^2)^{3/2}}dx = \frac{1}{2}l\int_a^\infty \frac{2x}{(R^2+x^2)^{3/2}}dx =$
$\frac{1}{2}l\int_a^\infty (2x)(R^2 + x^2)^{-3/2}dx =$
$\frac{1}{2}l\left(\frac{-2}{1}\right)(R^2 + x^2)^{-1/2}\big|_a^\infty = \frac{-l}{\sqrt{R^2+a^2}}$.

≡ 25.5 NUMERICAL INTEGRATION

1. $\int_1^2 x^4 dx$; $n = 6$; $\Delta x = \frac{1}{6}$

(a) $\frac{1}{2}\left[1^4 + 2\cdot\left(\frac{7}{6}\right)^4 + 2\left(\frac{8}{6}\right)^4 + 2\left(\frac{9}{6}\right)^4 + 2\left(\frac{10}{6}\right)^4 + 2\left(\frac{11}{6}\right)^4 + 2^4\right] \cdot \frac{1}{6} = \frac{1}{12}(75.17746914) = 6.2647891$

(b) $\frac{1}{3}\left[1^4 + 4\cdot\left(\frac{7}{6}\right)^4 + 2\left(\frac{8}{6}\right)^4 + 4\left(\frac{9}{6}\right)^4 + 2\left(\frac{10}{6}\right)^4 + 4\left(\frac{11}{6}\right)^4 + 2^4\right] \cdot \frac{1}{6} = \frac{1}{18}(111.6018519) = 6.20010286$

(c) $\int_1^2 x^4 dx = \frac{x^4}{5}\big|_1^2 = \frac{32}{5} - \frac{1}{5} = \frac{31}{5} = 6.2$

2. $\int_1^2 2x^5 dx$; $n = 8$; $\Delta x = \frac{1}{4}$

(a) $\frac{1}{2}\left[0 + 2\cdot2\left(\frac{1}{4}\right)^5 + 2\cdot2\left(\frac{1}{2}\right)^5 + 4\left(\frac{3}{4}\right)^5 + 4(1)^5 + 4\left(\frac{5}{4}\right)^5 + 4\left(\frac{3}{2}\right)^5 + 4\left(\frac{7}{4}\right)^5 + 2(2)^5\right] \cdot \frac{1}{4} = \frac{1}{8}(177.3125) = 22.1640625$

(b) $\frac{1}{3}\left[0 + 4\cdot2\left(\frac{1}{4}\right)^5 + 2\cdot2\left(\frac{1}{2}\right)^5 + 4\cdot2\cdot\left(\frac{3}{4}\right)^5 + 2\cdot2\cdot1^5 + 4\cdot2\left(\frac{5}{4}\right)^5 + 2\cdot2\left(\frac{3}{2}\right)^5 + 4\cdot2\cdot\left(\frac{7}{4}\right)^5 + 2(2)^5\right] \cdot \frac{1}{4} = \frac{1}{12}(256.125) = 21.34375$

(c) $\int_1^2 2x^5 dx = \frac{1}{3}x^6\big|_0^2 = \frac{2^6}{3} = \frac{64}{3} = 21\frac{1}{3}$.

3. $\int_0^1 \sqrt{x}\, dx$; $n = 4$; $\Delta x = \frac{1}{4}$

(a) $\frac{1}{2}\left(0 + 2\sqrt{\frac{1}{4}} + 2\sqrt{1/2} + 2\sqrt{\frac{3}{4}} + 1\right) \cdot \frac{1}{4} = \frac{1}{8}(5.14626437) = 0.643283$

(b) $\frac{1}{3}\left(0 + 4\sqrt{\frac{1}{4}} + 2\sqrt{1/2}\sqrt{\frac{3}{4}} + 1\right) \cdot \frac{1}{4} = \frac{1}{12}(7.878315178) = 0.6565263$

(c) $\int_0^1 x^{1/2}dx = \frac{2}{3}x^{3/2}\big|_0^1 = \frac{2}{3} = 0.6667$

4. $\int_1^2 x^{1/3}dx$
(a) 1.139241,
(b) 1.139875,
(c) $\int_1^2 x^{1/3}dx = \frac{3}{4}x^{\frac{4}{3}}\big|_1^2 = \frac{3}{4} \cdot 2^{4/3} - \frac{3}{4} = 1.139882$

5. $\int_1^5 \frac{dx}{x}$, $n = 8$
(a) 1.628968,
(b) 1.610847

6. $\int_1^4 \frac{dx}{x^2}$; $n = 6$
(a) 0.789219,
(b) 0.754838, actual value $\int_1^4 x^{-2}dx = -x^{-1}\big|_1^4 = -\frac{1}{4} + 1 = 0.75$

7. $\int_1^5 \frac{dx}{1+x}$, $n = 8$
(a) 1.1032107,
(b) 1.098726

8. $\int_0^2 x^\pi dx$; $n = 4$
(a) 4.550118,
(b) 4.262661 actual $\frac{1}{\pi+1} x^{\pi+1}\big|_0^2 = \frac{2^{\pi+1}}{\pi+1} =$
4.261635

9. $\int_0^1 \frac{4}{1+x^2} dx$, $n = 10$
(a) 3.139926;
(b) 3.1415926 (Note: The actual value is π.)

10. $\int_0^2 \sqrt{1+x^3} dx$, $n = 8$
(a) 6.521610,
(b) 6.391210

11. $\int_0^2 \sqrt{1+x^3} dx$; $n = 8$
(a) 3.251744,
(b) 3.241238

12. $\int_0^{0.2} \sin x \, dx$; $n = 4$
(a) 0.019929,
(b) 0.019933

13. $\int_{-2}^2 \sqrt{x+4} \, dx$, $n = 8$
(a) 7.909233,
(b) 7.912321

14. $\int_1^3 \ln x \, dx$, $n = 8$
(a) 1.292375,
(b) 1.295798

15. $\int_0^4 \frac{\sqrt{x}}{1+\sqrt{x}} dx$; $n = 8$
(a) 2.141030,
(b) 2.170342

16. $\int_0^1 \frac{e^x + e^{-x}}{2} dx$; $n = 4$
(a) 1.181316,
(b) 1.175227

17. (a) $\Delta x = 1$; $\frac{1}{2}(4.2 + 2(3.9) + 2(3.8) + 2(4.0) +$
$2(3.5) + 2(3.4) + 2(3.9) + 2(4.1) + 4.3) = \frac{1}{2}(61.7) =$
30.85
(b) $\frac{1}{3}(4.2 + 4(3.9) + 2(3.8) + 4(4.0) + 2(3.5) + 4(3.4) +$
$2(3.9) + 4(4.1) + 4.3) = \frac{1}{3}(92.5) = 30.83333$

18. $\Delta x = 0.1$
(a) $\frac{1}{2}(4.32 + 2(4.57) + 2(5.14) + 2(5.78) + 2(6.84) +$
$2(6.62) + 6.51)(0.1) = \frac{1}{2}(68.73)(0.1) = 3.4365$
(b) $\frac{1}{3}(4.32 + 4(4.57) + 2(5.14) + 4(5.78) + 2(6.84) +$
$4(6.62) + 6.51)(0.1) = \frac{1}{3}(102.67)(0.1) = 3.4223$

19. (a) $\Delta x = 0.25$; $\frac{1}{2}(16.32 + 2(16.48) + 2(16.73) +$
$2(16.42) + 2(16.38) + 2(16.29) + 16.25)(0.25) =$
$\frac{1}{2}(197.17)(0.25) = 24.64625$

(b) $\frac{1}{3}(16.32 + 4(16.48) + 2(16.73) + 4(16.42) +$
$2(16.38) + 4(16.29) + (16.25)(0.25)$
$= \frac{1}{3}(295.55)(0.25) = 24.629167$

20. $\Delta x = 5$
(a) $\frac{1}{2}(0 + 2 \cdot 21 + 2 \cdot 18 + 2 \cdot 17 + 2 \cdot 18 + 2 \cdot 14 + 0) \cdot 5 =$
$\frac{5}{2}(142) = 355 \text{ m}^2$

(b) $\frac{1}{3}(0 + 4 \cdot 21 + 2 \cdot 18 + 4 \cdot 17 + 2 \cdot 18 + 4 \cdot 14 + 0) \cdot 5 =$
$\frac{5}{3}(280) = 466.7 \text{ m}^2$

21. $\int_1^2 4.0e^{-10t} - 4.0e^{-20t} dt$; $\Delta t = \frac{1}{4}$
(a) $2.676 \times 10^{-5} C = 26.76 \ \mu C$
(b) 20.34 μC

22. $\Delta x = 5$ milliseconds moving from 10 to 60.
(a) $\frac{1}{2}[0.3 + 2(0.7) + 2(0.8) + 2(0.7) + 2(0.8) +$
$2(0.7) + 2(0.5) + 2(0.4) + 2(0.2) + 2(0.2) + 0.1](5) =$
$\frac{5}{2}(10.8) = 27$

(b) $\frac{5}{3}[0.3 + 4(0.7) + 2(0.8) + 4(0.7) + 2(0.8) +$
$4(0.7) + 2(0.5) + 4(0.4) + 2(0.2) + 4(0.2) + 0.1] =$
$\frac{5}{3}(16.2) = 27$

23. See *Computer Programs* in main text.

24. With $n = 10$m the interval and points are at 0, 60,
120, . . . ,600.
$$\int_0^{600} R'(x) dx$$
$$= \frac{60}{2}[R'(0) + 2R'(60) + 2R'(120) + \cdots$$
$$+ 2R'(540) + R'(600)]$$
$$\approx 30[0 + 2(152.33) + 2(169.46) + 2(163.26)$$
$$+ 2(148.29) + 2(130.42) + 2(112.38)$$
$$+ 2(95.49) + 2(80.30) + 2(67.00) + 55.55]$$
$$= 68,802.3$$
The revenue from the sale of the first 600 computers is about \$68,802.30.

25. Trapezoidal:
$$A = \frac{100}{2}[300 + 2(300) + 2(350) + 2(400)$$
$$+ 2(500) + 2(600) + 2(500) + 2(600)$$
$$+ 2(550) + 2(650) + 600]$$
$$= 490,000 \text{ ft}^2 \approx 11.264 \text{ acres}$$

Simpson:

$$A = \frac{100}{3}[300 + 4(300) + 2(350) + 4(400)$$
$$+ 2(500) + 4(600) + 2(500)$$
$$+ 4(600) + 2(550) + 4(650) + 600]$$
$$= 496{,}667 \text{ ft}^2 \approx 11.417 \text{ acres}$$

26. Here, $\Delta x = \frac{9-1}{8} = 1$.

Trapezoidal:

$$\text{Reaction} = \frac{1}{2}[r(1) + 2r(2) + 2r(3) + 2r(4)$$
$$+ 2r(5) + 2r(6) + 2r(7) + 2r(8) + 2r(9)]$$
$$= \frac{1}{2}[1.3679 + 2(0.5183) + 2(0.3335)$$
$$+ 2(0.2500) + 2(0.2000) + 2(0.1667)$$

$$+ 2(0.1429) + 2(0.1250) + 1111]$$
$$= 2.47585$$

Simpson:

$$\text{Reaction} = \frac{1}{3}[r(1) + 4r(2) + 2r(3) + 4r(4)$$
$$+ 2r(5) + 4r(6) + 2r(7) + 4r(8)$$
$$+ r(9)]$$
$$= \frac{1}{3}[1.3679 + 4(0.5183) + 2(0.3335)$$
$$+ 4(0.2500) + 2(0.2000) + 4(0.1667)$$
$$+ 2(0.1429) + 4(0.1250) + 0.1111]$$
$$= 2.35723$$

CHAPTER 25 REVIEW

1. $\int x^5 dx = \frac{1}{6}x^6 + C$

2. $\int(\sqrt[3]{x}+4)dx = \int\left(x^{1/3}+4\right)dx = \frac{3}{4}x^{4/3}+4x+C$

3. $\int_0^3 \left(x^2 + 4x - 6\right)dx = \frac{1}{3}x^3 + 2x^2 - 6x\Big|_0^3 = 9 + 18 - 18 = 9$

4. $\int_1^4 (x+4)^2 dx = \int_1^4 \left(x^2 + 8x + 16\right)dx = \frac{1}{3}x^3 + 4x^2 + 16x\Big|_1^4 = \left(\frac{64}{3}+64+64\right) - \left(\frac{1}{3}+4+16\right) = 129$

5. $\int_1^8 \frac{dt}{\sqrt[3]{t}} = \int_1^8 t^{-1/3}dt = \frac{3}{2}t^{2/3}\Big|_1^8 = 6 - \frac{3}{2} = \frac{9}{2}$

6. $\int_1^4 (\sqrt{x}+4)dx = \frac{2}{3}x^{3/2} + 4x\Big|_1^4 = \left(\frac{16}{3}+16\right) - \left(\frac{2}{3}+4\right) = \frac{14}{3} + 12 = 16\frac{2}{3} = \frac{50}{3}$

7. $\int \frac{dt}{(t+5)^4} = \int(t + 5)^{-4}dt = -\frac{1}{3}(t + 5)^{-3} + C$

8. $\int \frac{x\,dx}{\sqrt[3]{x^2+4}} = \int \left(x^2 + 4\right)^{-1/3}x\,dx$. Let $u = x^2 + 4$, and then $du = 2x\,dx$, or $\frac{1}{2}du = x\,dx$. Substitution produces $\frac{1}{2}\int u^{-1/3}du = \frac{1}{2}\cdot\frac{3}{2}u^{2/3} + C = \frac{3}{4}\left(x^2 + 4\right)^{2/3} + C$.

9. $\int_1^8 x\sqrt[3]{x}\,dx = \int_1^8 x^{4/3}dx = \frac{3}{7}x^{7/3}\Big|_1^8 = \frac{3}{7}\cdot 2^7 - \frac{3}{7} = \frac{381}{7} \approx 54.429$.

10. $\int_1^4 \frac{4\,dx}{3\sqrt{x^3}} = \frac{4}{3}\int_1^4 x^{-3/2}dx = \frac{4}{3}\cdot(-2)x^{-1/2}\Big|_1^4 = -\frac{8}{3}\left(\frac{1}{2} - 1\right) = -\frac{8}{3}\cdot\frac{-1}{2} = \frac{4}{3}$

11. $\int \left(u^4 - u^{-4}\right)du = \frac{1}{5}u^5 + \frac{1}{3}u^{-3} + C$

12. $\int \frac{x^4-1}{x^3}dx = \int \left(x - x^{-3}\right)dx = \frac{x^2}{2} + \frac{1}{2}x^{-2} + C = \frac{x^2}{2} + \frac{1}{2x^2} + C$

13. $\int 2x\sqrt{x^2 - 5}\,dx$. Let $u = x^2 - 5$, then $du = 2x\,dx$. Thus, $\int u^{1/2}du = \frac{2}{3}u^{3/2} + C = \frac{2}{3}\left(x^2 - 5\right)^{3/2} + C$

14. $\int_0^4 x\sqrt{x^2 + 9}\,dx$. Let $u = x^2 + 9$, then $du = 2x\,dx$, or $\frac{1}{2}du = x\,dx$. For the limits of integration, we have $x = 0 \Rightarrow u = 9$ and $x = 4 \Rightarrow u = 25$. Thus $\frac{1}{2}\int_9^{25} u^{1/2}du = \frac{1}{2}\cdot\frac{2}{3}u^{3/2}\Big|_9^{25} = \frac{1}{3}(125 - 27) = \frac{1}{3}(98) = \frac{98}{3} = 32\frac{2}{3}$.

15. $\int x\sqrt[3]{6 - x^2}dx$. Let $u = 6-x^2$, then $du = -2x\,dx$, or $-\frac{1}{2}du = x\,dx$. Substitution yields $-\frac{1}{2}\int u^{1/3}du = -\frac{1}{2}\cdot\frac{3}{4}u^{4/3} + C = \frac{-3}{8}\left(6 - x^2\right)^{4/3} + C$.

16. $\int 3x\left(4 - x^2\right)^3 dx$. Let $u = 4 - x^2$, then $du = -2x\,dx$, or $3x\,dx = -\frac{3}{2}du$. Substituting, we have $-\frac{3}{2}\int u^3 du = -\frac{3}{2}\cdot\frac{1}{4}u^4 + C = \frac{-3}{8}\left(4 - x^2\right)^4 + C$.

17. $\int_{-1}^2 \left(x^2 + 4\right)\left(x^3 - 3\right)dx = \int_{-1}^2 \left(x^5 + 4x^3 - 3x^2 - 12\right)dx = \frac{x^6}{6} + x^4 - x^3 - 12x\Big|_{-1}^2 = \left(\frac{64}{6} + 16 - 8 - 24\right) - \left(\frac{1}{6} + 1 + 1 + 12\right) = -\frac{39}{2}$

18. $\int_1^2 x(x+2)^2\,dx = \int_1^2 \left(x^3+4x^2+4x\right)dx = \frac{x^4}{4}+\frac{4}{3}x^3$

$+ 2x^2\Big|_1^2 = \left(4+\frac{32}{3}+8\right) - \left(\frac{1}{4}+\frac{4}{3}+2\right) = 19\frac{1}{12}$

19. $\int \frac{y^2 dy}{\left(y^3-5\right)^{2/3}}$; Let $u = y^3 - 5$, then $du = 3y^2 dy$, or

$\frac{1}{3}du = y^3 dy$. Substitution produces $\frac{1}{3}\int u^{-2/3}du = $

$\frac{1}{3}\cdot\frac{3}{1}u^{1/3} + C = \left(y^3 - 5\right)^{1/3} + C.$

20. $\int \frac{5-6x+x^2}{1-x}dx = \int \frac{(5-x)(1-x)}{1-x}dx = \int(5-x)dx = $

$5x - \frac{x^2}{2} + C,\ x \neq 1$

21. $\int_0^1 \left(1-x^2\right)^3 dx = \int_0^1 \left(1-3x^2+3x^4-x^6\right)dx = $

$\left(x - x^3 + \frac{3}{5}x^5 - \frac{1}{7}x^7\right)\Big|_0^1 = 1-1+\frac{3}{5}-\frac{1}{7} = \frac{21}{35} - $

$\frac{5}{35} = \frac{16}{35}$

22. $\int \frac{x+2}{\left(x^2+4x\right)}dx$. Let $u = x^2 + 4x$, then $du = (2x + $

$4)dx$, or $\frac{1}{2}du = (x+2)dx$. Thus, $\frac{1}{2}\int u^{-2}du = \frac{1}{2}\cdot$

$(-1)u^{-1}+C = -\frac{1}{2}\left(x^2+4x\right)^{-1}+C$ or $\frac{-1}{2\left(x^2+4x\right)} + $

$C.$

23. $\int \left(\sqrt{2x}-\frac{1}{\sqrt{2x}}\right)dx = \sqrt{2}\cdot\frac{2}{3}x^{3/2} - \frac{1}{\sqrt{2}}\cdot\frac{2}{1}x^{1/2}+C = $

$\frac{2\sqrt{2}}{3}x^{3/2} - \sqrt{2}x^{1/2} + C = \frac{1}{3}(2x)^{3/2} - (2x)^{1/2} + C$

24. $\int \left(x^3 - 4x\right)\left(3x^2 - 4\right)dx$. Let $u = x^3 - 4x$, then

$du = \left(3x^2 - 4\right)dx$. Thus, $\int u\,du = \frac{u^2}{2} + C = $

$\frac{1}{2}\left(x^3 - 4x\right)^2 + C.$

25. $\int_1^4 \frac{9-t^3}{3t^2}dt = \int_1^4 \left(3t^{-2} - \frac{1}{3}t\right)dt = \left(-3t^{-1} - \frac{1}{3}\cdot\right.$

$\left.\frac{1}{2}t^2\right)\Big|_1^4 = \left(\frac{-3}{4} - \frac{8}{3}\right) - \left(-3 - \frac{1}{2}\right) = -\frac{1}{4}$

26. $\int_0^1 \sqrt{x}(3\sqrt{x} - 5)dx = \int_0^1 (3x - 5\sqrt{x})dx = \frac{3}{2}x^2 - $

$5\cdot\frac{2}{3}x^{3/2}\Big|_0^1 = \frac{3}{2} - \frac{10}{3} = -\frac{11}{6}$

27. $\int_{-1}^8 \frac{4\sqrt[3]{u}\,du}{\left(1+u^{4/3}\right)^3}$. Let $v = 1 + u^{4/3}$, then $dv = $

$\frac{4}{3}u^{1/3}du$, or $3dv = 4u^{1/3}du$. The limits of integra-

tion become $u = -1 \Rightarrow v = 2$ and $u = 8 \Rightarrow v = $

17. Thus, we have $3\int_2^{17}v^{-3}dv = 3\cdot\frac{-1}{2}v^{-2}\Big|_2^{17} = $

$\frac{-3}{2\cdot17^2} + \frac{3}{2\cdot2^2} \approx -0.00519 + 0.375 = 0.36981.$

28. $\int_1^2 \frac{3x^3-2x^{-3}}{x^2}dx = \int_1^2 \left(3x - 2x^{-5}\right)dx = \left(\frac{3}{2}x^2 + \right.$

$\left.\frac{1}{2}x^{-4}\right)\Big|_1^2 = \left(6 + \frac{1}{32}\right) - \left(\frac{3}{2} + \frac{1}{2}\right) = 6\frac{1}{32} - 2 = 4\frac{1}{32}$

29. $\int_1^\infty \frac{dx}{x^4} = \lim_{b\to\infty}\int_1^b x^{-4}dx = \lim_{b\to\infty}\left(\frac{-1}{3}x^{-3}\right)\Big|_1^b = $

$\lim_{b\to\infty}\left(\frac{1}{3b^3} + \frac{1}{3}\right) = \frac{1}{3}$

30. $\int_0^\infty \frac{4x\,dx}{\left(3x^2+1\right)^5}$; Let $u = 3x^2 + 1$, then $du = 6x\,dx$,

or $4x\,dx = \frac{2}{3}du$. For the limits of integration, we

obtain $x = 0 \Rightarrow u = 1$ and $x \to \infty \Rightarrow$

$u \to \infty$. Thus, $\frac{2}{3}\int_1^\infty u^{-5}du = \frac{2}{3}\left(-\frac{1}{4}\right)u^{-4}\Big|_1^\infty = $

$\lim_{b\to\infty}\left(\frac{-1}{6b^4} + \frac{1}{6}\right) = \frac{1}{6}.$

31. $f(x) = 4x - 6 = 0;\ 4x = 6;\ x = \frac{3}{2}$. Thus,

$-\int_{-3}^{3/2}(4x-6)dx + \int_{3/2}^4(4x-6)dx = -\left(2x^2 - \right.$

$\left.6x\right)\Big|_{-3}^{3/2} + \left(2x^2 - 6x\right)\Big|_{3/2}^4 = -\left(\frac{9}{2} - 9\right) + (18 + $

$18) + (32 - 24) - \left(\frac{9}{2} - 9\right) = \frac{9}{2} + 36 + 8 + \frac{9}{2} = 53.$

32. $g(x) = 2 - x - x^2 = (2+x)(1-x)$. This crosses

the x-axis at -2 and 1. $\int_{-2}^1 \left(2 - x - x^2\right)dx = $

$2x - \frac{x^2}{2} - \frac{x^3}{3}\Big|_{-2}^1 = \left(2 - \frac{1}{2} - \frac{1}{3}\right) - \left(-4 - 2 + \frac{8}{3}\right) = $

$2 - \frac{5}{6} + 6 - \frac{8}{3} = 8 - \frac{21}{6} = 8 - \frac{7}{2} = \frac{9}{2} = 4\frac{1}{2}$ or $4.5.$

33.

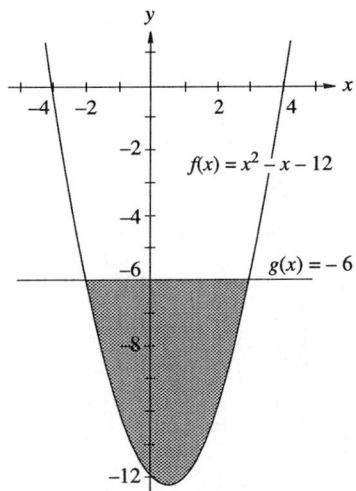

$f(x) = x^2 - x - 12$

$g(x) = -6$

The graphs will intersect when $f(x) = g(x)$ or

when $x^2 - x - 12 = -6$. This can be rewritten as

$x^2 - x - 6 = 0$ which factors as $(x-3)(x+2) = $

0. The graphs intersect at $x = -2$ and $x = 3$.

Thus, the desired area is $\left|\int_{-2}^3 \left(x^3 - x - 6\right)dx\right| = $

$\left|\frac{x^3}{3} - \frac{x^2}{2} - 6x\Big|_{-2}^3\right| = \left|\left(9 - \frac{9}{2} - 18\right) - \left(\frac{-8}{3} - 2 + \right.\right.$

$\left.\left.12\right)\right| = \left|-\frac{125}{6}\right| = \frac{125}{6}.$

34. $h(x) = \sqrt{x+2} = 0$ at $x = -2$. Thus,

$\int_{-2}^7 \sqrt{x+2}\,dx = \frac{2}{3}(x+2)^{3/2}\Big|_{-2}^7 = \frac{2\cdot27}{3} - 0 = 18.$

35. $4x^3 = x \Rightarrow 4x^3 - x = 0 \Rightarrow x(4x^2 - 1) = 0 \Rightarrow$
$x = 0, -\frac{1}{2}, \frac{1}{2}.$

$$\int_{-1/2}^{0} (4x^2 - x)\,dx + \int_{0}^{1/2} (x - 4x^3)\,dx$$

$$= x^4 - \frac{x^2}{2}\Big|_{-1/2}^{0} + \frac{x^2}{2} - x^4\Big|_{0}^{1/2}$$

$$= -\left(\frac{1}{16} - \frac{1}{8}\right) + \left(\frac{1}{8} - \frac{1}{16}\right) = \frac{1}{8}$$

36.

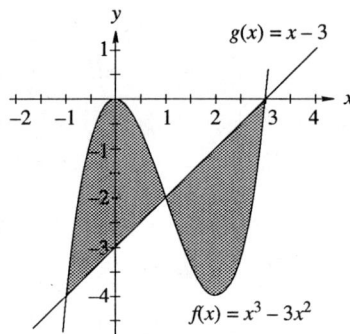

$g(x) = x - 3$

$f(x) = x^3 - 3x^2$

The graphs intersect when $f(x) = g(x)$ or $x^3 - 3x^2 = x - 3$. This can be rewritten as $x^2(x - 3) = (x - 3)$, or $x^2(x - 3) - (x - 3) = 0$, or $(x^2 - 1)(x - 3) = 0$. Thus, we see that the graphs intersect at -1, 1 and 3. The desired area is

$$\int_{-1}^{1} (x^3 - 3x^2 - x + 3)\,dx$$

$$+ \int_{1}^{3} (x - 3 - x^3 + 3x^2)\,dx$$

$$= \left(\frac{x^4}{4} - x^3 - \frac{x^2}{2} + 3x\right)\Big|_{-1}^{1}$$

$$+ \left(\frac{x^2}{2} - 3x - \frac{x^4}{4} + x^3\right)\Big|_{1}^{3}$$

$$= \left(\frac{1}{4} - 1 - \frac{1}{2} + 3\right) - \left(\frac{1}{4} + 1 - \frac{1}{2} - 3\right)$$

$$+ \left(\frac{9}{2} - 9 - \frac{81}{4} + 27\right) - \left(\frac{1}{2} - 3 - \frac{1}{4} + 1\right)$$

$$= \left(1\frac{3}{4}\right) - \left(-2\frac{1}{4}\right) + \left(2 + \frac{1}{4}\right) - \left(-1\frac{3}{4}\right)$$

$$= 8$$

37. $y^2 + x = 0$; $y = 2x + 1$; $(2x+1)^2 + x = 0$; $4x^2 + 4x + 1 + x = 0$; $4x^2 + 5x + 1 = 0$; $(4x+1)(x+1)$. The graphs intersect at $(-1, -2)$ and $\left(-\frac{1}{4}, \frac{1}{2}\right)$

$$\int_{-1}^{-1/4} (2x + 1 + \sqrt{-x})\,dx$$

$$+ \int_{-1/4}^{0} 2\sqrt{-x}\,dx$$

$$= \left(x^2 + x - \frac{2}{3}(-x)^{3/2}\right)\Big|_{-1}^{-1/4}$$

$$+ \left(-2 \cdot \frac{2}{3}(-x)^{3/2}\right)\Big|_{-1/4}^{0}$$

$$= \left(\frac{1}{16} - \frac{1}{4} - \frac{1}{12}\right) - \left(1 - 1 - \frac{2}{4}\right) + 0 + \frac{1}{6}$$

$$= \frac{9}{16} = 0.5625$$

38. $2x^3 = x^4$; $x^4 - 2x^3 = 0$; $x^3(x - 2) = 0$ they intersect at $(0, 0)$ and $(2, 16)$ $\int_{0}^{2} (2x^3 - x^4)\,dx =$
$\frac{1}{2}x^4 - \frac{x^5}{5}\Big|_{0}^{2} = 8 - \frac{32}{5} = \frac{8}{5} = 1\frac{3}{5} = 1.6$

39. $\int_{0}^{1} \sqrt{1 - x^2}\,dx$; $n = 10$
(a) 0.776130,
(b) 0.781752

40. $\int_{1}^{4} \frac{1}{3x}\,dx$; $n = 6$
(a) 0.468452,
(b) 0.462566

41. $\int_{2}^{2.5} \sqrt{x^3 - 1}\,dx$; $n = 10$
(a) 1.613695,
(b) 1.613656

42. $\int_{-1}^{4} \sqrt{1 + x^5}\,dx$; $n = 10$
(a) 38.795198,
(b) 38.464568

43. $\int_{0}^{1} (e^x - e^{-x})\,dx$; $n = 4$
(a) 1.091812,
(b) 1.086185

44. (a) $\frac{1}{2}(4.25 + 2(5.72) + 2(5.13) + 2(3.19) + 2(2.10) + 2(0.15) + 2(1.65) + 2(3.10) + 3.70)(0.25) = \frac{1}{8}(50.03)$
$= 6.25375$
(b) $\frac{1}{3}(4.25 + 4(5.72) + 2(5.13) + 4(3.19) + 2(2.10) + 4(0.15) + 2(1.65) + 4(3.10) + 3.70)(0.25) = \frac{1}{12}(74.35)$
$= 6.195833$

45. (a) $\int_{0}^{15} \left(t + \frac{1}{\sqrt{1+t}}\right)dt = \frac{t^2}{2} + 2\sqrt{1 + t}\Big|_{0}^{15} = \frac{225}{2} + 2\sqrt{16} - 2 = 112.5 + 8 - 2 = 118.5$ m
(b) The acceleration is given by $a(t) = v'(t) = 1 - \frac{1}{2}(1 + t)^{-3/2}$. Thus, the acceleration at $t = 15$ is $a(15) = 1 - \frac{1}{2}(16)^{-3/2} = 1 - \frac{1}{128} = \frac{127}{128} \approx 0.992$ m/s^2.

46. $\omega = \int_3^5 10t^3 dt = \frac{10}{4}t^4 \Big|_0^5 = \frac{5}{2}(625-81) = \frac{5}{2}(544) = 1360$ J

47. $2.0 - 1.0t^2 = -2\frac{di}{dt} \Rightarrow \frac{di}{dt} = \frac{1}{2}t^2 - 1$. Hence, $i = \int_1^6 \left(\frac{1}{2}t^2 - 1\right)dt = \left(\frac{1}{6}t^3 - t\right)\Big|_1^6 = (36 - 6) - \left(\frac{1}{6} - 1\right) = 30 + \frac{5}{6} = 30\frac{5}{6} \approx 30.8$ A.

48. **(a)** When the car stops its velocity is 0. Hence, we want to know when $v(t) = 88 - 4t = 0$, which is the same as $4t = 88$, and so $t = 22$ s

(b) The distance the car travels is given by $s(t) = \int v(t) = 88t - 2t^2$, and so, the distance it travels in those 22 s is $s(22) = 88 \cdot 22 - 2 \cdot 22^2 = 968$ ft.

CHAPTER 25 TEST

1. $\int_0^1 2x \, dx = x^2 \Big|_0^1 = 1 - 0 = 1.$

2. $\int_1^4 (x+1)(x^2+1)dx = \int_1^4 (x^3 + x^2 + x + 1)dx = \left(\frac{x^4}{4} + \frac{x^3}{3} + \frac{x^2}{2} + x\right)\Big|_1^4 = (64 + \frac{64}{3} + 8 + 4) - (\frac{1}{4} + \frac{1}{3} + \frac{1}{2} + 1) = 95.25.$

3. $\int (x^2 + 3)^2 dx = \int (x^4 + 6x^2 + 9)dx = \frac{1}{5}x^5 + 2x^3 + 9x + C.$

4. $\int 4(x^4 + 7)^5 x^3 dx$; Let $u = x^4 + 7$, then $du = 4x^3 dx$. Thus, $\int u^5 du = \frac{1}{6}u^6 + C = \frac{1}{6}(x^4 + 7)^6 + C.$

5. $\int \frac{x^2+2}{x^2} dx = \int (1 + 2x^{-2})dx = x - 2x^{-1} + C.$

6. $\int (x^2 + 8x)^3 (x+4)dx$. Let $u = x^2 + 8x$, then $du = (2x+8)dx$, or $\frac{1}{2}du = (x+4)dx$. Substitution yields $\frac{1}{2}\int u^3 du = \frac{1}{2} \cdot \frac{1}{4}u^4 + C = \frac{1}{8}(x^2 + 8x)^4 + C.$

7. $\int_1^2 \left(3x^2\sqrt{x^3 + x} + \sqrt{x^3 + x}\right)dx = \int_1^2 \sqrt{x^3 + x} \times (3x^2 + 1)dx$. Let $u = x^3 + x$, then $du = (3x^2 + 1)dx$. When $x = 1 \Rightarrow u = 2$ and when $x = 2 \Rightarrow u = 10$. Thus, with these substitutions we

have $\int_2^{10} \sqrt{u} \, du = \frac{2}{3}u^{3/2}\Big|_1^{10} = \frac{2}{3}\left(10^{3/2} - 2^{3/2}\right) \approx 19.196.$

8. $4x^3 - 4 = 4(x^3 - 1) = 4(x-1)(x^2 + x + 1)$. This crosses the x-axis at $x = 1$. $\int_{-1}^1 (4 - 4x^3)dx + \int_1^2 (4x^3 - 4)dx = (4x - x^4)\Big|_{-1}^1 + (x^4 - 4x)\Big|_1^2 = (4-1)-(-4-1)+(16-8)-(1-4) = 3+5+8+3 = 19.$

9. The graphs intersect at $(-3, 7)$ and $(2, 2)$. Thus, the desired area is $\int_{-3}^2 \left[(4 - x) - (x^2 - 2)\right]dx = \int_{-3}^2 (6 - x - x^2)dx = \left(6x - \frac{x^2}{2} - \frac{x^3}{3}\right)\Big|_{-3}^2 = (12 - 2 - \frac{8}{3}) - (-18 - \frac{9}{2} + 9) = \frac{125}{6} \approx 20.833.$

10. $\int_2^4 \frac{1}{x}dx$; $n = 8$; $\Delta x = \frac{2}{8} = \frac{1}{4}$. The trapezoidal rule yields $\frac{1}{2}\left(\frac{1}{2} + \frac{2}{2.25} + \frac{2}{2.5} + \frac{2}{2.75} + \frac{2}{3} + \frac{2}{3.25} + \frac{2}{3.5} + \frac{2}{3.75} + \frac{1}{4}\right) \cdot \frac{1}{4} = \frac{1}{8}(5.55297) = 0.694122$. Simpson's rule produces $\frac{1}{3}\left(\frac{1}{2} + \frac{4}{2.25} + \frac{2}{2.5} + \frac{4}{2.75} + \frac{2}{3} + \frac{4}{3.25} + \frac{2}{3.5} + \frac{4}{3.75} + \frac{1}{4}\right) \cdot \frac{1}{4} = \frac{1}{12}(8.31785) = 0.693154$

11. $\int_0^1 0.0004\left(t^{-1/2} + t^{1/2}\right)dt = 0.0004\left[2t^{1/2} + \frac{2}{3}t^{3/2}\right]_0^1 = 0.0004\left[\left(2 + \frac{2}{3}\right) - (0 + 0)\right] \approx 0.00107$ C

CHAPTER
26
Applications of Integration

≡ 26.1 AVERAGE VALUES AND OTHER ANTIDERIVATIVE APPLICATIONS

1. $\bar{y} = \frac{1}{1} \int_0^1 x^2 dx = \frac{1}{3} x^3 \big|_0^1 = \frac{1}{3}$

$f_{rms} = \sqrt{\frac{1}{1} \int_0^1 x^4 dx} = \sqrt{\frac{1}{5} x^5 \big|_0^1} = \sqrt{\frac{1}{5}} = \frac{\sqrt{5}}{5} \approx$ 0.4472

2. $\bar{y} = \frac{1}{4-2} \int_2^4 3x^2 dx = \frac{1}{2} [x^3]_2^4 = \frac{1}{2}(64 - 8) = 28$

$g_{rms} = \sqrt{\frac{1}{2} \int_2^4 9x^4 dx} = \sqrt{\frac{9}{2} \left(\frac{x^5}{5}\right) \big|_2^4} =$

$\sqrt{\frac{9}{2}\left(\frac{1024-32}{5}\right)} = \sqrt{892.8} = 29.8798$

3. $\bar{y} = \frac{1}{3-1} \int_1^3 (x^2 - 1) dx = \frac{1}{2} \left[\frac{x^3}{3} - x\right]_1^3 = \frac{1}{2}[(9 - 3) - (\frac{1}{3} - 1)] = \frac{1}{2}[\frac{20}{3}] = \frac{10}{3}$

$h_{rms} = \sqrt{\frac{1}{2} \int_1^3 (x^2 - 1)^2 dx} =$

$\sqrt{\frac{1}{2} \int_1^3 (x^4 - 2x^2 + 1) dx} = \sqrt{\frac{1}{2} \left[\frac{x^5}{5} - \frac{2x^3}{3} + x\right]_1^3}$

$= \sqrt{\frac{1}{2}\left[\left(\frac{243}{5} - \frac{54}{3} + 3\right) - \left(\frac{1}{5} - \frac{2}{3} + 1\right)\right]} = \sqrt{\frac{248}{15}}$

$= \sqrt{16.53} \approx 4.0661$

4. $\bar{y} = \frac{1}{3+3} \int_{-3}^3 (9 - x^2) dx = \frac{1}{6} \left[9x - \frac{x^3}{3}\right]_{-3}^3 = \frac{1}{6}[(27 - 9) - (-27 + 9)] = \frac{1}{6}(36) = 6.$

$j_{rms} = \sqrt{\frac{1}{6} \int_{-3}^3 (9 - x^2)^2 dx} =$

$\sqrt{\frac{1}{6} \int_{-3}^3 (81 - 18x^2 + x^4) dx} =$

$\sqrt{\frac{1}{6}\left[81x - 6x^3 - \frac{x^5}{5}\right]_1^3} = \sqrt{43.2} \approx 6.5727$

5. $\bar{y} = \int_0^5 \frac{1}{5}\sqrt{x+4}\, dx = \frac{1}{5}\left[\frac{2}{3}(x+4)^{3/2}\right]_0^5$

$= \frac{2}{15}[(27 - 8)] = \frac{38}{15}.$

$f_{rms} = \sqrt{\frac{1}{5} \int_0^5 (x + 4)\, dx} =$

$\sqrt{\frac{1}{5}\left[\frac{x^2}{2} + 4x\right]_0^5} = \sqrt{\frac{1}{5}\left[\frac{25}{2} + 20\right]} = \sqrt{\frac{5}{2} + 4} =$

$\sqrt{\frac{13}{2}} = 2.5495$

6. $\bar{y} - \frac{1}{3-1} \int_1^3 (x^2 - 1) dx = \frac{1}{2}\left[\frac{x^4}{4} - x\right]_1^3 =$

$\frac{1}{2}\left[\frac{81}{4} - 3 - \frac{1}{4} + 1\right] = \frac{1}{2}[18] = 9.$

$k_{rms} = \sqrt{\frac{1}{2} \int_1^3 (x^6 - 2x^3 + 1)\, dx} =$

$\sqrt{\frac{1}{2}\left[\frac{x^7}{7} - \frac{x^4}{2} + x\right]_1^3} =$

$\sqrt{\frac{1}{2}\left[\frac{2187}{7} - \frac{81}{2} + 3 - \frac{1}{7} + \frac{1}{2} - 1\right]} = \sqrt{\frac{960}{7}} =$

$\sqrt{137.1428} \approx 11.7108$

7. $\bar{y} = \frac{1}{4} \int_0^4 (t^2 + 4) dt = \frac{1}{4}\left[\frac{t^3}{3} + 4t\right]_0^4 = \frac{1}{4}\left[\frac{64}{3} + 16\right] = \frac{16}{3} + 4 = 9\frac{1}{3} = \frac{28}{3}$

$f_{rms} = \sqrt{\frac{1}{4} \int_0^4 (t^4 + 8t^2 + 16) dt} =$

$\sqrt{\frac{1}{4}\left[\frac{t^5}{5} + \frac{8t^3}{3} + 16t\right]_0^4} =$

$\sqrt{\frac{1}{4}\left[\frac{1024}{5} + \frac{512}{3} + 64\right]} = \sqrt{\frac{1648}{15}} = \sqrt{109.8667} \approx$ 10.4817

8. $\bar{y} = \frac{1}{4-2} \int_2^4 (16t^2 - 25) dt = \frac{1}{2}\left[\frac{16t^3}{3} - 25t\right]_2^4 =$

$\frac{1}{2}\left(\frac{1024}{3} - 100\right) - \left(\frac{128}{3} - 50\right) = \frac{1}{2}\left[248\frac{2}{3}\right] = 124.333$

$g_{rms} = \sqrt{\frac{1}{2} \int_2^4 (256t^4 - 800t^2 + 625) dt} =$

$\sqrt{\frac{1}{2}\left[\frac{256t^5}{5} + \frac{800t^3}{3} + 625t\right]_0^4} = \sqrt{18553.53} =$ 136.2114

9. $\bar{y} = \frac{1}{1} \int_1^2 (6x - x^2) dx = \left[3x^2 - \frac{x^3}{3}\right]_1^2$

$= \left(12 - \frac{8}{3}\right) - \left(3 - \frac{1}{3}\right) = \frac{20}{3}$

$f_{rms} = \sqrt{\int_1^2 (36x^2 - 12x^3 + x^4) dx} =$

$\sqrt{\left[12x^3 - 3x^4 + \frac{x^5}{5}\right]_1^2} = \sqrt{45.2} \approx 6.7231$

10. $\bar{y} = \frac{1}{3}\int_0^3 (16t^2 - 8t + 1)dt = \frac{1}{3}\left[\frac{16t^2}{3} - 4t^2 + t\right]_0^3 = \frac{1}{3}[144 - 36 + 3] = 37$

$h_{\text{rms}} =$

$\sqrt{\frac{1}{3}\int_0^3 (256t^4 - 256t^3 + 96t^2 - 16t + 1)dt} = $

$\sqrt{\frac{1}{3}\left[\frac{256t^5}{5} - 64t^4 + 32t^3 - 8t^2 + t\right]_0^3} = \sqrt{2684.2} \approx 51.8093$

11. $\bar{y} = \frac{1}{2.5}\int_0^{2.5} (4.9t^2 - 2.8t + 4)dt = \frac{1}{2.5}\left[\frac{4.9t^3}{3} - 1.4t^2 + 4t\right]_0^{2.5} = \frac{1}{2.5}(26.7708) = 10.7083$

$j_{\text{rms}} = \sqrt{\frac{1}{2.5}\int_0^{2.5} (4.9t^2 - 2.8t + 4)^2 dt}$

$= \sqrt{\frac{1}{2.5}(415.9765)} = 12.8992$

12. $\bar{y} = \frac{1}{1.6}\int_0^{1.6} (4.9t^2 - 4.2t + 9)\,dt = \frac{1}{1.6}(15.71413) = 9.8213$

$k_{\text{rms}} = \sqrt{\frac{1}{1.6}\int_0^{1.6} (4.9t^2 - 4.2t^3 + 9)^2 dt} = \sqrt{\frac{1}{1.6}(160.25496)} = 10.0080$

13. $\frac{1}{90}\int_0^{90} 0.0002(4991 + 366x - x^2)dx = \frac{0.0002}{90}\left[4991x + 183x^2 - \frac{x^3}{3}\right]_0^{90} = \frac{0.0002}{90}[1688490] = 3.7522$ cm

for 90 days. $\frac{0.0002}{365}\int_0^{365} (4991 + 366x - x^2)dx = \frac{0.0002}{365}\left[4991x + 183x^2 - \frac{x^3}{3}\right]_0^{365} =$

$\frac{0.0002}{365}(9992848) = 5.4755$ cm for the year.

14. $\frac{1}{4}\int_0^4 (1 + x - \sqrt{x})dx = \frac{1}{4}\left[x + \frac{x^2}{2} - \frac{2}{3}x^{3/2}\right]\Big|_0^4 = \frac{1}{4}\left[4 + 8 - \frac{16}{3}\right] = 1\frac{2}{3} = 1.6667$ kg

15. $a(t) = -32$; $v(t) = -32t$; $s(t) = -16t^2 + 360$. Setting $s(t) = 0$, and solving for t, we obtain $-16t^2 + 360 = 0$, or $16t^2 = 360$, and so $t = 4.743416$. Thus, $\bar{s} = \frac{1}{4.74}\int_0^{4.74} (-16t^2 + 360)dt = \frac{1}{4.74}\left[-\frac{16}{3}t^3 + 360t\right]_0^{4.74} = 240$ ft. The average velocity is $\bar{v} = \frac{1}{4.74}\int_0^{4.74}(-32t)dt = \left[\frac{-16t^2}{4.74}\right]_0^{4.74} = -75.8947$ ft/s.

16. $V(t) = 200(100 - t^2)$. $\bar{V} = \frac{200}{10}\int_0^{10} (100 - t^2)dt = 20\left[100t - \frac{t^3}{3}\right]_0^{10} = 20\left(1000 - \frac{1000}{3}\right) = 13333.33$ L

17. (a) $\frac{1}{6}\int_0^6 (60 + 4t - t^{2/3})dt = \frac{1}{6}\left[60t + 2t^2 - \frac{3}{5}t^{5/3}\right]_0^6 = 60 + 2\cdot 6 - \frac{3}{5}6^{2/3} = 70.018$ or $70°$F

(b) $\frac{1}{6}\int_3^9 (60 + 4t - t^{2/3})dt = \frac{1}{6}\left[60t + 2t^2 - \frac{3}{5}t^{5/3}\right]_3^9 = \frac{1}{6}[484.3797] = 80.7°$F

18. $\frac{1}{30}\int_0^{30} (900 - 15\sqrt{120t})dt = \frac{1}{30}\left[900t - 10\sqrt{120}t^{3/2}\right]_0^{30} = \frac{1}{30}[27000 - 18000] = \frac{1}{30}[9{,}000]$ $= 300$ cases

19. $\frac{1}{3}\int_0^3 (1.0 + \sqrt{t})dt = \frac{1}{3}\left[\frac{t^2}{2} + \frac{2}{3}t^{3/2}\right]_0^3 = \frac{3}{2} + \frac{2}{3}\cdot\sqrt{3} \approx 2.6547$ A.

20. $i_{\text{eff}} = \sqrt{\frac{1}{3}\int_0^3 (t + \sqrt{t})^2 dt}$

$= \sqrt{\frac{1}{3}\int_2^3 (t^2 + 2t^{3/2} + t)dt}$

$= \sqrt{\frac{1}{3}\left[\frac{t^3}{3} + \frac{4}{5}t^{5/2} + \frac{t^2}{2}\right]_0^3}$

$= \sqrt{3 + \cdot\frac{4}{5}\cdot 3^{3/2} + \frac{3}{2}} = \sqrt{8.6569} = 2.9423$ A.

21. $\frac{1}{4}\int_0^4 4t\sqrt{t^2 + 1} = \int_0^4 t\sqrt{t^2 + 1}$. Let $u = t^2 + 1$ and then, $du = 2t\,dt$. When $t = 0 \Rightarrow u = 1$, and when $t = 4 \Rightarrow u = 17$. Making these substitutions, we have $\frac{1}{2}\int_1^{17} \sqrt{u}\,du = \frac{1}{2}\cdot\frac{2}{3}u^{3/2}\big|_1^{17} = \frac{1}{3}(69.092796) = 23.0309$ A.

22. $i_{\text{eff}} = \sqrt{\frac{1}{4}\int_0^4 \left[(4t)^2(t^2 + 1)\right]dt} =$

$\sqrt{4\int_0^4 (t^4 + t^2)dt} = 2\cdot\sqrt{\frac{t^5}{5} + \frac{t^3}{3}\big|_0^4} = 2\cdot$

$\sqrt{\frac{1024}{5} + \frac{64}{3}} = 30.0755$ A

23. $P = (i_{\text{eff}})^2 \cdot R = (30.0755)^2 \cdot 30 = 27136$ W

24. $i_{\text{eff}} = \sqrt{\frac{1}{0.25}\int_0^{0.25} (t^3 - 2t^2)^2 dt} =$

$\sqrt{4\int_0^{0.25} (t^6 - 4t^5 + 4t^4)dt} =$

$2\sqrt{\frac{1}{7}t^7 - \frac{2}{3}t^6 + \frac{4}{5}t^5\big|_0^{0.25}} = 2\sqrt{0.000\,627\,2}$. Thus, $(i_{\text{eff}})^2 = 4(0.000\,627\,2) \approx 0.002\,509$ and so, $P = i_{\text{eff}}^2 \cdot R = 0.002\,509 \cdot 5 = 0.012\,545 \approx 12.5$ mW.

25. We have $v = 4\sqrt{t} - 2t = 2(2\sqrt{t} - t)$.

(a) So, $v_{\text{eff}} = \sqrt{\frac{1}{6}\int_0^6 \left[2(2\sqrt{t} - t)\right]^2 dt} =$

$\sqrt{\frac{2}{3}\int_0^6 (4t - 4t^{3/2} + t^2)dt} =$

$\sqrt{\frac{1}{3}\left[2t^2 - \frac{8}{5}t^{5/2} + \frac{1}{3}t^3\right]_0^6} =$

$\sqrt{\frac{2}{3}\left(72 - \frac{288}{5}\sqrt{6} + 72\right)} = \sqrt{96 - \frac{192}{5}\sqrt{6}}$

$\approx \sqrt{1.939593877} \approx 1.3927$. So, the effective voltage is about 1.3927 V.

(b) The average power is $P = i_{eff}^2 \cdot R = \left(\sqrt{1.939593877}\right)^2 (10) \approx 19.40$. Hence, the average power is about 19.40 W.

26. $V = 8.0\sqrt{t} - 16\sqrt{t^3} = 8\left(\sqrt{t} - 2\sqrt{t^3}\right) = 8\sqrt{t}(1 - 2t)$ and is greater than 0 from $t = 0$ to $t = \frac{1}{2}$ s. $\left(V_{rms}\right)^2 = 2\int_0^{1/2} \left[8\left(\sqrt{t} - 2\sqrt{t^3}\right)\right]^2 dt = 2 \cdot 64\int_0^{1/2}\left(t - 4t^2 + 4t^3\right)dt = 128\left[\frac{t^2}{2} - \frac{4}{3}t^3 + t^4\right]_0^{1/2} = 128[0.02083] = 2.6667$. Hence, $P = \left(V_{rms}\right)^2 R = 2.6667 \times 9.75 = 26.0$ W.

27. $q = \int_0^{0.1} 2t\, dt = t^2 \big|_0^{0.1} = 0.01$ C is the increase. The new charge is $0.01 + 0.01 = 0.02$ C.

28. $V_c = \frac{1}{C}\int i\, dt = \frac{1}{80\times 10^{-6}}\int 0.04t^3 dt = \frac{1}{80\times 10^{-6}}\left(0.01t^4\right) + k$; $k = 100$, so $V_c = 125t^4 +$

29. $100\ 225 = 125t^5 + 100\ 125 = 125t$, $t^4 = 1$, so $t = 1$ s.

29. $V = 2t + 1$, $i = 0.03t$ W$= \int P\, dt = \int Vi\, dt = \int(2t + 1)(0.03t)dt = \int_0^{50}\left(0.06t^2 + 0.03t\right)dt = \left[0.02t^2 + \frac{0.03}{2}t^2\right]_0^{50} = 2537.5$ J

30. $V = \frac{1}{C}\int_0^{0.001} 0.2\, dt = \frac{1}{90\times 10^{-6}} \cdot 0.0002 = 2.222$ V

31. $V = \frac{1}{C}\int 0.04t\, dt = \frac{1}{C}0.2t^2 + k$, so $V_c = \frac{0.2t^2}{7.5\times 10^{-6}}$. At $t = 0.005$, $V = \frac{0.2(0.005)^2}{7.5\times 10^{-6}} = 0.6667$ V.

32. $q = \int i\, dt$; $q = \int_0^3 \sqrt[3]{1 + 5t}\, dt$. Let $u = 1 + 5t$, then $du = 5\, dt$. When $t = 0$, then $u = 1$ and when $t = 3$, then $u = 16$. Substituting, we get $\frac{1}{5}\int_1^{16} u^{1/3} du = \frac{1}{5}\cdot\frac{3}{4}u^{4/3}\big|_1^{16} = \frac{3}{20}\left[16^{4/3} - 1^{4/3}\right] = 5.8976$ C.

≣ ## 26.2 VOLUMES OF REVOLUTION: DISK AND WASHER METHODS

1. $\pi\int_1^5 (4x)^2 dx = \frac{16\pi}{3}r^3 \big|_1^5 = \frac{16\pi}{3}(124) = 661.33\pi \approx 2077.64$.

2. $\pi\int_0^2 x^4 dx = \frac{\pi}{5}x^5 \big|_0^2 = \frac{32\pi}{5} \approx 20.11$

3. $\pi\int_0^4 x^2 dx = \frac{\pi}{3}x^3 \big|_0^4 = \frac{64\pi}{3} \approx 67.02$

4. $\pi\int_0^6 (6 - x)^2 dx = \pi\int_0^6 \left(36 - 12x + x^2\right)dx = \pi\left[36x - 6x^2 + \frac{x^3}{3}\right]_0^6 = 72\pi \approx 226.19$

5. $\pi\int_1^7 (x+1)dx = \pi\left[\frac{x^2}{2} + x\right]_1^7 = \pi\left[\frac{49}{2} + 7 - \frac{1}{2} - 1\right] = 30\pi \approx 94.25$

6. $\pi\int_1^2 \left(x^6 - 1\right)dx = \pi\left[\frac{x^7}{7} - x\right]_1^2 = \pi\left[\frac{128}{7} - 2 - \frac{1}{7} + 1\right] = \frac{120\pi}{7}$

7. $y = x^2 \Rightarrow x = \sqrt{y}$; $\pi\int_0^4 \sqrt{y}^2 dy = \pi\frac{y^2}{2}\big|_0^4 = 8\pi$

8. $x + y = 5 \Rightarrow x = 5 - y$; $\pi\int_0^5(5 - y)^2 dy = \pi\int_0^5\left(25 - 10y + y^2\right)dy = \pi\left[25y - 5y^2 + \frac{y^3}{3}\right]_0^5 = \frac{125\pi}{3}$

9. The two curves intersect at $(2, 2)$. Thus, $V = \pi\int_0^2\left((4 - x)^2 - x^2\right)dx = \pi\int_0^2\left(16 - 8x + x^2 - x^2\right)dx = \pi\left[16x - 4x^2\right]_0^2 = 16\pi$

10. $y = \sqrt[3]{x} \Rightarrow x = y^3$ and so $V = \pi\int_0^8 y^6 dy = \frac{y^7}{7}\big|_0^8 = \frac{8^7\pi}{7} = \frac{2\ 097\ 152\pi}{7}$.

11. $y = \sqrt[3]{x} \Rightarrow x = y^3$ and so $V\pi\int_0^2\left(8^2 - y^{3\cdot 2}\right)dy = \pi\left[64y - \frac{y^7}{7}\right]_0^2 = \pi\left[128 - \frac{128}{7}\right] = \frac{768\pi}{7}$.

12. $y = \sqrt{x} \Rightarrow x = y^2$; $y = x^2 \Rightarrow x = \sqrt{y}$. The curves intersect at $(1, 1)$. Thus, $\pi\int_0^1\left(y - y^4\right)dy = \pi\left[\frac{y^2}{2} - \frac{y^5}{5}\right]_0^1 = \left[\frac{1}{2} - \frac{1}{5}\right] = \frac{3\pi}{10}$

13. They intersect at $(\pm 2,\ 0)$ with $8 - 2x^2$ on the outside. Thus, we have $\pi\int_{-2}^2\left(\left(8 - 2x^2\right)^2 - \left(4 - x^2\right)^2\right)dx = \pi\int_{-2}^2\left[\left(64 - 32x^2 + 4x^4\right) - \left(16 - 8x^2 + x^4\right)\right]dx = \pi\int_{-2}^2\left(48 - 24x^2 + 3x^4\right)dx = \pi\left[48x - 8x^3 + \frac{3}{5}x^5\right]_{-2}^2 = 102.4\pi$ or $\frac{512\pi}{5}$

14. $y = x^2 \Rightarrow x = \sqrt[2]{y}$; $y = x^3 \Rightarrow x = \sqrt[3]{y}$. Thus, we have $\pi\int_0^1\left(\sqrt[3]{y}^2 - \sqrt[2]{y}^2\right)dy = \pi\left(\frac{3}{5}y^{5/3} - \frac{y^2}{2}\right)\big|_0^1 = \pi\left(\frac{3}{5} - \frac{1}{2}\right) = \frac{\pi}{10}$

15. $\pi\int_0^1\left(x^4 - x^6\right)dx = \pi\left[\frac{x^5}{5} - \frac{x^7}{7}\right]_0^1 = \frac{2\pi}{35}$

16. They intersect at $(0, 0)$ and $(4, 16)$. Thus, we have $\pi\int_0^4\left((4x)^2 - x^4\right)dx = \pi\left[\frac{16x^3}{3} - \frac{x^5}{5}\right]_0^4 = \frac{2048\pi}{15} = 428.93$

17. $x = \sqrt{y}$; $x = \frac{y}{4}$. Hence, we obtain $\pi\int_0^{16}\left(\sqrt{y}^2 - \left(\frac{y}{4}\right)^2\right)dy = \pi\left[\frac{y^2}{2} - \frac{y^3}{48}\right]_0^{16} = \pi\left(\frac{256}{2} - \frac{256}{3}\right) = \pi\left(\frac{256}{6}\right) = \frac{128\pi}{3}$

18. $\pi \int_0^2 \sqrt{8 - x^3}^2 \, dx = \pi \left[8x - \frac{x^4}{4}\right]_0^2 = \pi\left(16 - \frac{16}{4}\right) = 12\pi$

19. $y = \sqrt{8 - x^2}$; $y^2 = 8 - x^2$; $x^2 = 8 - y^2$. So, $x = \sqrt{8 - y^2}$. Now, we obtain $\pi \int_0^{\sqrt{8}} \left(8 - y^2\right) dt = \pi\left[8y - \frac{y^3}{3}\right]_0^{\sqrt{8}} = \pi\left[8\sqrt{8} - \frac{8\sqrt{8}}{3}\right] = \pi\left[16\sqrt{2} - \frac{16\sqrt{2}}{3}\right] = \frac{32\sqrt{2}\pi}{3}$

20. Take the line $y = \frac{r}{h}x$. This line rotated about the x-axis forms a cone of height h and radius r. $V = \pi \int_0^h \frac{r^2}{h^2} x^2 dx = \pi\left[\frac{r^2}{h^2}\frac{x^3}{3}\right]_0^h = \pi \cdot \frac{r^2}{h^2} \cdot \frac{h^3}{3} = \frac{1}{3}\pi r^2 h$

21. Use the equation for the top of a semicircle $y = \sqrt{r^2 - x^2}$ rotated about the x-axis.

$$V = \pi \int_{-r}^r \sqrt{r^2 - x^2}\,dx = \pi \int_{-r}^r \left(r^2 - x^2\right) dx$$

$$= \pi\left[r^2 x - \frac{x^3}{3}\right]_{-r}^r = \pi\left[r^3 - \frac{r^3}{3} + r^3 - \frac{r^3}{3}\right]$$

$$= \frac{4}{3}\pi r^3$$

22. $\frac{x^2}{a^2} + \frac{y^2}{b^2} = 1$; $\frac{y^2}{b^2} = 1 - \frac{x^2}{a^2} = \frac{a^2 - x^2}{a^2}$, so $y^2 = \frac{a^2 b^2 - b^2 x^2}{a^2}$, and we have $y = \sqrt{\frac{a^2 b^2 - b^2 x^2}{a^2}}$. The volume is $\pi \int_{-a}^a \left(\frac{a^2 b^2}{a^2} - \frac{b^2}{a^2} x^2\right) dx = \pi\left[b^2 x - \frac{b^2 x^3}{3a^2}\right]_{-a}^a = \pi\left[b^2 a - \frac{b^2 a}{3} + b^2 a - \frac{b^2 a}{3}\right] = \frac{4}{3}ab^2\pi$ or $\frac{4b^2 a\pi}{3}$. (Note: if $a = b = r$ we have the formula for the volume of a circle.)

23. (a) The equation must be of the form $y = mx^2$ assuming it passes through the origin. It must also contain the point $(1.5, 1)$ so $1 = m(1.5)^2$, $1 = m(2.25)$ or $m = \frac{1}{2.25} = \frac{4}{9}$. The equation is $y = \frac{4}{9}x^2$ of $x = \sqrt{\frac{9}{4}y} = \frac{3}{2}\sqrt{y}$.

(b) To find its volume rotate about the y-axis we evaluate $\pi \int_0^1 \sqrt{\frac{9}{4}y}^2 \, dy = \pi \int_0^1 \frac{9}{4}y\,dy = \pi\left[\frac{9}{4} \cdot \frac{y^2}{2}\right]_0^1 = \frac{9\pi}{8}$ m$^2 = 1.125\pi$ m^2.

24. $\frac{dV}{dy} = V' = \frac{9\pi}{4}y$; $dV = \frac{9\pi}{4}y\,dy = \frac{9\pi}{4} \cdot 1 \cdot 1(0.004) = 0.009\pi$, and $\Delta v = v'\,dy = \frac{9\pi}{4}y\,dy = \frac{9\pi}{4} \cdot 1(0.004) = 0.009\pi \approx 0.028\,27$ m^3.

25. $\pi \int_{-442}^{123} \sqrt{(147^2 + 0.16y^2)^2}\,dy = \pi \int_{-422}^{123} \left(147^2 + 0.16y^2\right) dy = \pi\left[21609y + \frac{0.16y^3}{3}\right]_{-442}^{123} = 16913711\pi$ ft$^2 \approx 53135993$ ft^3.

26. $\pi \int_{-442}^{123} \left(146.5^2 + 0.16y^2\right) dy = \pi\left[(146.5)^5 \cdot y + \frac{10.16y^3}{3}\right]_{-442}^{123} = 16{,}830{,}798\pi$ ft^3 or $52{,}875{,}552$ ft^3. The volume is $(16{,}913{,}711 - 16{,}830{,}798)\pi$ ft$^3 = 82{,}913\pi$ or $260{,}479$ ft^3

27. Use the equation for a circle $x^2 + y^2 = 18^2$ or $x = \sqrt{18^2 - y^2}$

(a) $\pi \int_{-18}^{-14} \left(18^2 - y^2\right) dy = \left[18^2 y - \frac{y^3}{3}\right]_{-18}^{-14} = 226\frac{2}{3}\pi$ or 837.758 m^3

(b) $\pi \int_{-18}^6 \left(18^2 - y^2\right) dy = \pi\left[18^2 y - \frac{y^3}{3}\right]_{-18}^6 = 5760\pi \approx 18\,095.574$ m^3

28. Using the washer method, we got $m = \rho\pi \int_0^4 \left[(x^2 + 5)^2 - 2^2\right] dx = \rho\pi \int_0^4 \left(x^4 + 10x^2 + 21\right) dx = \rho\pi\left[\frac{1}{5}x^5 + \frac{10}{3}x^3 + 21x\right]_0^4 = 502.13\rho\pi$. Since $\rho = 0.016$ kg/cm^3, we have $m = (502.13$ cm$^3) \times (0.016$ kg/cm$^3)\pi \approx 25.24$ kg.

≡ 26.3 VOLUMES OF REVOLUTION: SHELL METHOD

1. $2\pi \int_0^2 x\left(x^2 - 0\right) dx = 2\pi \int_0^2 x^3 dx = 2\pi\left[\frac{x^4}{4}\right]_0^2 = 8\pi$

2. $x = y^3 \Rightarrow y = \sqrt[3]{x} = x^{1/3}$. Hence we have $= 2\pi \int_0^8 x \cdot x^{1/3} dx = 2\pi \int_0^8 x^{4/3} dx = 2\pi\left[\frac{3}{7}x^{7/3}\right]_0^8 = \frac{768\pi}{7}$

3. When $y = 8$ we see that $x = \pm 3$, we will rotate the portion in the first quadrant. Thus, $2\pi \int_1^3 x(8 - x^2 + 1) dx = 2\pi \int_1^3 \left(9x - x^3\right) dx = 2\pi\left[\frac{9x^2}{2} - \frac{x^4}{4}\right]_1^3 = 2\pi\left[\frac{81}{4} - \frac{17}{4}\right] = 32\pi$

4. $2\pi \int_0^2 x\left(x^3 + x\right) dx = 2\pi \int_0^2 \left(x^4 + x^2\right) dx = 2\pi\left[\frac{x^5}{5} + \frac{x^3}{3}\right]_0^2 = 2\pi\left[\frac{32}{5} + \frac{8}{3}\right] = \frac{272\pi}{15}$

5. $2\pi \int_0^5 x\left(25 - x^2\right) dx = 2\pi \int_0^5 \left(25x - x^3\right) dx = 2\pi\left[\frac{25x^2}{2} - \frac{x^4}{4}\right]_0^5 = \frac{625\pi}{2}$

6. $x^2 = 8 - x^2$, or $2x^2 = 8$, or $x^2 = 4$, so $x = \pm 2$. Now we can solve this using $2\pi \int_0^2 x\left(8 - x^2 - x^2\right) dx = 2\pi \int_0^2 \left(8x - 2x^3\right) dx = 2\pi\left[4x^2 - \frac{x^4}{2}\right]_0^2 = 16\pi$

7. $y = x^3 \Rightarrow x = \sqrt[3]{y} = y^{1/3}$. Hence, we have
$2\pi \int_0^8 y(y^{1/3} - 0)\,dy = 2\pi \int_0^8 y^{4/3}\,dy = 2\pi \left[\frac{3}{7}y^{7/3}\right]_0^8$
$= \frac{768\pi}{7}$

8. $y = x^4 \Rightarrow x = y^{\frac{1}{4}}$; $y = x^3 \Rightarrow x = y^{1/3}$.
They intersect at $(0, 0)$ and $(1, 1)$, so we have
$2\pi \int_0^1 y(y^{1/4} - y^{1/3})\,dy = 2\pi \int_0^1 (y^{5/4} - y^{4/3})\,dy =$
$2\pi \left[\frac{4}{9}y^{9/4} - \frac{3}{7}y^{7/3}\right] = 2\pi\left(\frac{1}{63}\right) = \frac{2\pi}{63}$

9. $y = \sqrt[3]{x} \Rightarrow x = y^3$; $2\pi \int_0^2 y(y^3)\,dy = 2\pi \int_0^2 y^4\,dy$
$= 2\pi \left[\frac{y^5}{5}\right]^2 = \frac{64\pi}{5}$

10. $x = y^{2/3}$; $2\pi \int_0^8 y \cdot y^{2/3}\,dy = 2\pi \left[\frac{3}{8}y^{8/3}\right]_0^8 = 192\pi$

11. $2\pi \int_0^1 x(x^{1/2} - x^{3/2})\,dx = 2\pi \int_0^1 (x^{3/2} - x^{5/2})\,dx =$
$2\pi \left[\frac{2}{5}x^{5/2} - \frac{2}{7}x^{7/2}\right]_0^1 = \frac{8\pi}{35}$

12. $x + 2y = 3 \Rightarrow x = 3 - 2y$; $x = y$; $y = 0$.
These graphs intersect at $(1, 1)$. $2\pi \int_0^1 y[(3 - 2y) -$
$y]\,dy = 2\pi \int_0^1 (3y - 3y^2)\,dy = 2\pi\left[\frac{3}{2}y^2 - y^3\right]_0^1 =$
$2\pi(0.5) = \pi$.

13. $y = 2x^3 \Rightarrow x = \sqrt[3]{\frac{y}{2}}$; $y^2 = 4x \Rightarrow x = \frac{1}{4}y^2$.
These graphs intersect at $(0, 0)$ and $(1, 2)$.
$2\pi \int_0^2 y\left[\left(\frac{y}{2}\right)^{1/3} - \frac{y^2}{4}\right]dy$
$= 2\pi \int_0^2 \left[\frac{1}{\sqrt[3]{2}}y^{4/3} - \frac{y^3}{4}\right]dy$
$= 2\pi \left[\frac{1}{\sqrt[3]{2}}\frac{3}{7}y^{7/3} - \frac{y^4}{16}\right]_0^2$
$= 2\pi \left[\frac{12}{7} - 1\right] = \frac{10\pi}{7}$

14. They intersect at $(0, 0)$ and $(0, 2)$
$2\pi \int_0^2 y(4y - y^3)\,dy = 2\pi \int_0^2 (4y^2 - y^4)\,dy$
$= 2\pi \left[\frac{4y^3}{3} - \frac{y^5}{5}\right]_0^2$
$= \left(\frac{64}{3} - \frac{64}{5}\right)\pi = \frac{128\pi}{15}$

15. They intersect at $(0, 0)$ and $(2, 4)$. Rotating about the line $y = 6$ means we must integrate with respect to dy. $y = x^2 \Rightarrow x = \sqrt{y}$ and $y = 2x \Rightarrow$
$x = \frac{1}{2}y$
$2\pi \int_0^4 (6 - y)\left(y^{1/2} - \frac{y}{2}\right)dy$
$= 2\pi \int_0^4 \left(6y^{1/2} - 3y - y^{3/2} + \frac{y^2}{2}\right)dy$

$= 2\pi \left[4y^{3/2} - \frac{3y^2}{2} - \frac{2}{5}y^{5/2} + \frac{y^3}{6}\right]_0^4$
$= 2\pi \left[32 - 24 - \frac{64}{5} + \frac{32}{3}\right]$
$= \frac{176}{15}\pi \approx 11.7333\pi$

16. $y = 2 - x$ and $y = x^2$ intersect at $(-2, 4)$ and $(1, 1)$. This region is bounded on the left by $y = 2 - x$ or $x = 2 - y$ and on the right by $y = x^2$ or $x = -\sqrt{y}$.
$2\pi \int_1^4 y[(2 - y) - (-\sqrt{y})]\,dy$
$= 2\pi \int_1^4 (2y - y^2 + y\sqrt{y})\,dy$
$= 2\pi \left[y^2 - \frac{y^3}{3} + \frac{2}{5}y^{5/2}\right]_1^4$
$= 2\pi \left[16 - \frac{64}{3} + \frac{64}{5} - 1 + \frac{1}{3} - \frac{2}{5}\right]$
$= \frac{64\pi}{5}$

17. $2\pi \int_{-2}^1 (1 - x)(2 - x - x^2)\,dx$
$= 2\pi \int_{-2}^1 (2 - 3x + x^3)\,dx$
$= 2\pi \left[2x - \frac{3}{2}x^2 + \frac{x^4}{4}\right]_{-2}^1$
$= 2\pi \left[2 - \frac{3}{2} + \frac{1}{4} + 4 + 6 - 4\right]$
$= 2\pi \left(6\frac{3}{4}\right) = 2\pi \left(\frac{27}{4}\right) = \frac{27\pi}{2}$

18. $2\pi \int_{-2}^2 (4 - x)(4 - x^2)\,dx$
$= 2\pi \int_{-2}^2 (16 - 4x - 4x^2 + x^3)\,dx$
$= 2\pi \left[16x - 2x^2 - \frac{4}{3}x^3 + \frac{x^4}{4}\right]_{-2}^2$
$= 2\pi \left[\left(32 - 8 - \frac{32}{3} + 4\right)\right.$
$\left. - \left(-32 - 8 + \frac{32}{3} + 4\right)\right]$
$= 2\pi \left[32 - 8 - \frac{32}{3} + 4 + 32\right.$
$\left. + 8 - \frac{32}{3} - 4\right]$
$= 2\pi \left[\frac{138}{3}\right] = \frac{256\pi}{3}$

19. Start with the equation $x = \pm\sqrt{25 = y^2}$ rotated about the x axis. We will take the shells starting with $y = 2$ to $y = 5$. The height of a shell is $2\sqrt{25 - y^2}$. $2\pi \int_2^5 y(2\sqrt{25 - y^2})\,dy$; Let $u = 25 - y^2$; $du = -2y\,dy$; $y = 2 \Rightarrow u = 21$; $y = 5 \Rightarrow u = 0$. Substituting, we have

$$2\pi \int_{21}^0 -u^{1/2}\,du = 2\pi\left[-\frac{2}{3}u^{3/2}\right]_{21}^0$$

$$= 2\pi\left(\frac{2}{3}\cdot 21^{3/2}\right) = 28\pi\sqrt{21}$$

$$= 403.1$$

By the washer method:

$$\pi \int_{-\sqrt{21}}^{\sqrt{21}} \left((25 - y^2) - 4\right)dy$$

$$= \pi \int_{-\sqrt{21}}^{\sqrt{21}} (21 - y^2)\,dy$$

$$= \pi\left[(21y - \frac{y^3}{3}\right]_{-\sqrt{21}}^{\sqrt{21}}$$

$$= \pi(21\sqrt{21} - 7\sqrt{21}) \times 2 = 28\pi\sqrt{21}$$

20. (a) First we obtain the value of y when $x = 7$, using $y = -\sqrt{100 - x^2}$; $= \sqrt{10^2 - 7^2} = 100 - 49 = \sqrt{51}$. Now, integrating, we have

$$2\pi \int_0^{\sqrt{51}} x\left(\sqrt{100 - x^2} - 7\right)dx$$

$$= 2\pi \int_0^{\sqrt{51}} \left(x\sqrt{100 - x^2} - 7x\right)dx$$

$$= 2\pi \int_0^{\sqrt{51}} x\sqrt{100 - x^2}$$

$$- 2\pi \int_0^{\sqrt{51}} 7x\,dx$$

For the first integral, let $u = 100 - x^2$; $du = -2x\,dx$ so $-\frac{1}{2}du = x\,dx$. When $x = 0$, then $u = 100$ and when $x = \sqrt{51}$, then $u = 49.1$. Substituting, we obtain

$$2\pi \cdot \frac{-1}{2}\int_{100}^{49} u^{1/2}\,du = -\pi \cdot \frac{2}{3}u^{\frac{3}{2}}\Big|_{100}^{49}$$

$$= -\frac{2\pi}{3}[7^3 - 10^3] = 483\pi$$

The second integral is equal to $2\pi\left[\frac{7}{2}x^2\right]_0^{\sqrt{51}} = 7 \cdot 51\pi = 357\pi$. Substituting these two answers, we have $438\pi - 357\pi = 81\pi$ m^3

(b) $81\pi \cdot 1000 = 254\ 469$ kg

21. The graph intersects the x-axis at 2 and 4.5.

$$V = 2\pi \int_2^{4.5} x\left[-(x - 3.25)^4 + \left(\frac{5}{4}\right)^4\right]dx$$

$$= 2\pi \int_2^{4.5} \left(-x^5 + 13x^4 - 63.375x^3\right.$$

$$+ 137.3125x^2 - 109.125x\Big)dx$$

$$= 2\pi\left[-\frac{1}{6}x^6 + 2.6x^5 - 15.84375x^4\right.$$

$$\left. + \frac{2197}{48}x^3 - \frac{873}{16}x^2\right]_2^{4.5}$$

$$\approx 2\pi\left[-17.18 - (-33.05)\right]$$

$$\approx 31.74\pi \approx 99.7$$

The volume of the pan is 99.7 in.^{3}.

22. This is a constant thickness of 0.1 cm. Hence, $V = 2\pi \int_0^2 0.1y\,dy = 2\pi 0.05y^2\big|_0^2 = 2\pi(0.2) = 0.4\pi \approx 1.2566$ cm^3

26.4 ARC LENGTH AND SURFACE AREA

1. $y = \frac{1}{3}(x^2 + 2)^{3/2}$; $y' = \frac{1}{2}(x^2 + 2)^{1/2} \cdot 2x = x(x^2 + 2)^{1/2}$

$$\int_0^3 \sqrt{1 + \left[x(x^2 + 2)^{1/2}\right]}dx$$

$$= \int_0^3 \sqrt{1 + x^2(x^2 + 2)}dx$$

$$= \int_0^3 \sqrt{1 + x^4 + 2x^2}\,dx$$

$$= \int_0^3 \sqrt{x^4 + 2x^2 + 1}\,dx$$

$$= \int_0^3 (x^2 + 1) = \frac{x^3}{3} + x\big|_0^3 = 9 + 3 = 12$$

2. $y = x^{3/2}$; $y' = \frac{3}{2}x^{1/2}$

$$\int_0^8 \sqrt{1 + \left(\frac{3}{2}x^{1/2}\right)} = \int_0^8 \sqrt{1 + \frac{9x}{4}}dx$$

$$= \frac{4}{9}\int_0^8 \frac{9}{4}\sqrt{1 + \frac{9}{4}x}/,dx$$

$$= \frac{4}{9}\cdot\frac{2}{3}\left(1 + \frac{9}{4}x\right)^{3/2}\Big|_0^8$$

$$= \frac{8}{27}\left[19^{3/2} + 1\right] = 24.8353$$

3. $9x^2 = 4y^3$, or $x^2 = \frac{4}{9}y^3$, or $x = \frac{2}{3}y^{3/2}$, and so $x' = y^{1/2}$

$$\int_0^3 \sqrt{1+\left(y^{1/2}\right)^2}\,dy = \int_0^3 \sqrt{1+y}\,dy$$
$$= \frac{2}{3}(1+y)^{3/2}\,\Big|_0^3$$
$$= \frac{2}{3}\left[4^{3/2}-1^{3/2}\right]$$
$$= \frac{2}{3}[8-1] = \frac{14}{3}$$

4. $y = \frac{x^4}{4} + \frac{1}{8}x^{-2}$ and $y' = x^3 - \frac{1}{4}x^{-3}$

$$\int_1^2 \sqrt{1+\left(x^3 - \frac{1}{4x^3}\right)^2}\,dx$$
$$= \int_1^2 \sqrt{1+x^6 - \frac{1}{2} + \frac{1}{16x^{65}}}\,dx$$
$$= \int_0^2 \sqrt{x^6 + \frac{1}{2} + \frac{1}{16x^6}} = \int_1^2 \sqrt{\left(x^3 + \frac{1}{4x^3}\right)}$$
$$= \int_1^2 \left(x^3 + \frac{1}{4}x^{-3}\right)dx = \left(\frac{x^4}{4} - \frac{1}{8}x^{-2}\right)\Big|_1^2$$
$$= 4 - \frac{1}{32} - \frac{1}{4} + \frac{1}{8} = \frac{123}{32} = 3.84375$$

5. $y = \frac{x^3}{6} + \frac{1}{2x}$, so $y' = \frac{x^2}{2} - \frac{1}{2x^2}$

$$\int_1^3 \sqrt{1+\left[\frac{x^2}{2} - \frac{1}{2x^2}\right]^2}\,dx$$
$$= \int_1^3 \sqrt{1+\left(\frac{x^4}{4} - \frac{1}{2} + \frac{1}{4x^4}\right)}\,dx$$
$$= \int_1^3 \sqrt{\frac{x^4}{4} + \frac{1}{2} + \frac{1}{4x^4}}\,dx$$
$$= \int_1^3 \left(\frac{x^2}{2} + \frac{1}{2x^2}\right)dx$$
$$= \frac{x^3}{6} - \frac{1}{2x}\Big|_1^3 = \frac{9}{2} - \frac{1}{6} - \frac{1}{6} + \frac{1}{2}$$
$$= 5 - \frac{1}{3} = 4\frac{2}{3} = \frac{14}{3}$$

6. $x = \frac{y^3}{3} + \frac{1}{4y}$, so $x' = y^2 - \frac{1}{4y^2}$

$$\int_1^3 \sqrt{1+\left(y^2 - \frac{1}{4y^2}\right)^{2}}\,dy$$
$$= \int_1^3 \sqrt{1+y^4 - \frac{1}{2} + \frac{1}{16y^4}}\,dy$$
$$= \int_1^3 \sqrt{\left(y^4 + \frac{1}{2} + \frac{1}{16y^4}\right)}\,dy$$
$$= \int_1^3 \sqrt{\left(y^2 + \frac{1}{4y^2}\right)}\,dy$$

$$= \int_1^3 \left(y^2 + \frac{1}{4}y^{-2}\right)dy = \frac{y^3}{3} - \frac{y^{-1}}{4}\Big|_1^3$$
$$= \left(9 - \frac{1}{12} - \frac{1}{3} + \frac{1}{4}\right)$$
$$= \frac{108-1-4+3}{12} = \frac{106}{12} = \frac{53}{6}$$

7. $2\pi \int_0^2 4x\sqrt{1+4^2}\,dx = 8\pi\sqrt{17}\int_0^2 x\,dx = 8\pi\sqrt{17}\left[\frac{x^2}{2}\right]_0^2 = 16\pi\sqrt{17}$

8. $2\pi \int_0^8 \frac{12}{5}x\sqrt{1+\left(\frac{12}{5}\right)^2}\,dx = \pi 2 \cdot \frac{12}{5} \cdot \frac{13}{5} \int_0^8 x\,dx = \frac{312\pi}{25}\frac{x^2}{2}\Big|_0^8 = \frac{9984\pi}{25} = 399.36\pi$

9. $y = x$; $y' = 1$; $2\pi \int_0^2 x\sqrt{1+1^2}\,dx = 2\sqrt{2}\pi\frac{x^2}{2}\Big|_0^2$
$$= 4\pi\sqrt{2}$$

10. $y^2 = 4x$, so $y = 2x^{1/2}$; $y' = x^{-1/2} = \frac{1}{\sqrt{x}}$; $\left(\frac{1}{\sqrt{x}}\right)^2 = \frac{1}{x}$

$$2\pi \int_0^8 2x^{1/2}\sqrt{1+\frac{1}{x}}\,dx = 4\pi \int_0^8 \sqrt{x}\sqrt{\frac{x+1}{x}}\,dx$$
$$= 4\pi \int_0^8 \sqrt{x+1}\,dx$$
$$= \frac{2}{3}4\pi(x+1)^{3/2}\Big|_0^8$$
$$= \frac{8\pi}{3}[27-1] = \frac{208\pi}{3}$$

11. $x = 4y \Rightarrow y = \frac{x}{4}$, $y = 1 \Rightarrow x = 4$, and $y = 3 \Rightarrow x = 12$. We have $y' = \frac{1}{4}$, and so, the surface area is

$$2\pi \int_4^{12} \frac{x}{4}\sqrt{1+\frac{1}{16}}\,dx = \frac{\pi\sqrt{17}}{8} \int_4^{12} x\,dx$$
$$= \frac{\pi\sqrt{17}}{8}\left[\frac{x^2}{2}\right]_4^{12}$$
$$= \frac{\pi\sqrt{17}}{8}[72-8]$$
$$= 8\pi\sqrt{17} \approx 103.62$$

12. Since $x = \sqrt{y} = y^{1/2}$, then $x' = \frac{1}{2}y^{-1/2} = \frac{1}{2\sqrt{y}}$ and $x'^2 = \frac{1}{4y}$. Thus, the surface area is

$$2\pi \int_0^6 \sqrt{y}\sqrt{1+\frac{1}{4y}}\,dy$$
$$= 2\pi \int_0^6 \sqrt{y}\sqrt{\frac{4y+1}{4y}}\,dy$$

$$= 2\pi \int_0^6 \frac{\sqrt{4y+1}}{2}\, dy = \pi \int_0^6 \sqrt{4y+1}\, dy$$

$$= \pi \cdot \frac{1}{4} \cdot \frac{2}{3}(4y+1)^{3/2}\,\big|_0^6$$

$$= \frac{\pi}{6}\left(25^{3/2} - 1^{3/2}\right) = \frac{62\pi}{3}$$

13. Since $y = \frac{2}{75}x^{3/2}$, then $y' = \frac{1}{25}x^{1/2}$. Place the origin at the point where the cable meets the ground, then the length of the cable is given by $L = \int_0^{225} \sqrt{1 + \left(\frac{1}{25}x^{1/2}\right)^2}\, dy = \int_0^{225} \sqrt{1 + \left(\frac{1}{25}\right)^2 x}\, dy$
$= \int_0^{225} \left(1 + \frac{1}{625}x\right)^{1/2} dy = 625 \cdot \frac{2}{3}\left(1 + \frac{1}{625}x\right)^{3/2}\big|_0^{225}$

26.5 CENTROIDS

1. $\overline{x} = \frac{4 \cdot 5 + 6(-3)}{4+6} = \frac{20-18}{10} = \frac{1}{5}$; $\overline{y} = \frac{4 \cdot 2 + 6 \cdot 7}{4+6} = \frac{8+42}{10} = 5$. The centroid is at $(\overline{x},\ \overline{y}) = (0.2,\ 5)$.

2. $\overline{x} = \frac{3(-1)+4(-2)+13(1)}{3+4+13} = \frac{-3-8+13}{20} = \frac{1}{10} = 0.1$; $\overline{y} = \frac{3(4)+4(-5)+13(2)}{3+4+13} = \frac{12-20+26}{20} = \frac{18}{20} = 0.9$. The centroid is at $(\overline{x},\ \overline{y}) = (0.1,\ 0.9)$.

3. $\overline{x} = \frac{2 \cdot 1 + 3(-1)+5(6)}{2+3+5} = \frac{2-3+30}{10} = 2.9$; $\overline{y} = \frac{2 \cdot 5 + 3(4)+5(-4)}{2+3+5} = \frac{2}{10} = 0.2$. The centroid is at $(2.9,\ 0.2)$.

4. $\overline{x} = \frac{1 \cdot 1 + 2 \cdot 2 + 3(-3)+4(-4)}{1+2+3+4} = \frac{1+4-9-16}{10} = \frac{-20}{10} = -2$; $\overline{y} = \frac{1 \cdot 1 + 2(-2)+3 \cdot 3 + 4(-4)}{10} = \frac{1-4+9-16}{10} = \frac{-10}{10} = -1$. The centroid is at $(-2,\ -1)$.

5. The left rectangle has center at $\left(-\frac{1}{2},\ 2\right)$ and area 4. The right rectangle has center at $(2,\ 1)$ and area 8. $\overline{x} = \dfrac{4\left(-\frac{1}{2}\right)+8(2)}{4+8} = \frac{-2+16}{12} = \frac{14}{12} = \frac{7}{6}$; $\overline{y} = \frac{4 \cdot 2 + 8 \cdot 1}{12} = \frac{8+8}{12} = \frac{16}{12} = \frac{4}{3}$. The centroid is at $\left(\frac{7}{6},\ \frac{4}{3}\right)$.

6. The upper rectangle has center at $(3,\ 1)$ and area 12. The lower rectangle has center at $\left(\frac{1}{2},\ -\frac{1}{2}\right)$ and area 7. $\overline{x} = \frac{12(3)+7(1/2)}{12+7} = \dfrac{36+3\frac{1}{2}}{19} = \frac{79}{38}$; $\overline{y} = \dfrac{12 - \frac{7}{2}}{19} = \frac{17}{38}$. The centroid is at $\left(\frac{79}{38},\ \frac{17}{38}\right)$.

7. The upper left rectangle has center $(-3.5,\ 1)$ and area 6. Lower rectangle has center $(-1,\ -1)$ and area 16. The upper right rectangle has center

$\approx 660.841\ 2147 - 416.666\ 6667 = 244.174\ 548$ or approximately 244.17 m.

14. Here $y = (4-x)^3$, so $y' = -3(4-x)^2$. Thus, we find that the surface area is $S = 2\pi \int_2^4 (4-x)^3 \sqrt{1 + \left[-3(4-x)^2\right]^2}\, dx = 2\pi \int_2^4 (4-x)^3 \sqrt{1 + 9(4-x)^4}\, dx$. Let $u = (4-x)^4$, and then $du = -4(4-x)^3 dx$. When $x = 4$ we see that $u = 0$, and when $x = 2$, we find that $u = 16$. Substituting these values in S, we obtain $S = -\frac{\pi}{2}\int_{16}^0 \sqrt{1+9u}\, du = -\frac{\pi}{2}\int_{16}^0 (1+9u)^{1/2}\, du = -\frac{\pi}{2} \cdot \frac{2}{3} \cdot \frac{1}{9}(1+9y)^{3/2}\big|_{16}^0 = -\frac{\pi}{27}(1-576) = \frac{575}{27}\pi \approx 66.904$ in.^{2}.

$(2.5,\ 1.5)$ and area 15. $\overline{x} = \frac{6(-3.5)+16(-1)+15(2.5)}{6+16+15} = \frac{0.5}{37} = \frac{1}{74}$; $\overline{y} = \frac{6-16+15 \cdot (1.5)}{37} = \frac{12.5}{37} = \frac{25}{74}$. The centroid is at $\left(\frac{1}{74},\ \frac{25}{74}\right)$.

8. Divide the region into several rectangles giving areas and centers similar to the following: 4, $(-2, 3.5)$; 10, $(-3.0.5)$; 16, $(0,0)$; 2, $(3, -1.5)$; 8, $(4, 1)$; and 2, $(5.5, 3)$.

$$\overline{x} = \frac{4(-2)+10(-3)+16(0)+2(3)+8(4)+2(5.5)}{4+10+16+2+8+2} = \frac{11}{42}$$

$$\overline{y} = \frac{4(3.5)+10(0.5)+16(0)+2(-1.5)+8(1)+2(3)}{42} = \frac{30}{42} = \frac{5}{7}$$

The centroid is at $\left(\frac{11}{42},\ \frac{5}{7}\right)$.

9. $y = 2x + 3[0,\ 3]$

$$M_y = \int_0^3 x(2x+3)\, dx = \int_0^3 \left(2x^2 + 3x\right) dx$$

$$= \frac{2x^3}{3} + \frac{3x^2}{2}\,\bigg|_0^3 = 18 + \frac{27}{2} = \frac{63}{2}$$

$$M_x = \frac{1}{2}\int_0^3 (2x+3)^2\, dx$$

$$= \frac{1}{2}\int_0^3 \left(4x^2 + 12x + 9\right) dx$$

$$= \frac{1}{2}\left[\frac{4x^3}{3} + 6x^2 + 9x\right]_0^3 = \frac{1}{2}[36 + 54 + 27]$$

$$= \frac{117}{2}$$

$$m = \int_0^3 (2x+3)\, dx = x^2 + 3x\,\big|_0^3 = 9 + 9 = 18$$

$$\overline{x} = \frac{63}{36} = \frac{7}{4} = 1.75$$

$$\overline{y} = \frac{117}{36} = \frac{13}{4} = 3.25$$

The centroid is $(1.75, 3.25)$.

10. $y = x^3 [0, \ 2]$

$$M_y = \int_0^2 x \cdot x^3 dx = \frac{x^5}{5}\bigg|_0^2 = \frac{32}{5}$$

$$M_x = \frac{1}{2}\int_0^2 x^6 dx = \frac{1}{2} \cdot \frac{x^7}{7}\bigg|_0^2 = \frac{64}{7}$$

$$m = \int_0^2 x^3 dx = \frac{x^4}{4}\bigg|_0^2 = 4$$

$$\bar{x} = \frac{32}{20} = \frac{8}{5}$$

$$\bar{y} = \frac{64}{7.4} = \frac{16}{7}$$

The centroid is $\left(\frac{8}{5}, \ \frac{16}{7}\right)$.

11. $y = x^{1/3} [0, \ 8]$

$$M_y = \int_0^8 x \cdot x^{2/3} dx = \int_0^8 x^{4/3} dx$$

$$= \frac{3}{7}x^{7/3}\,|_0^8 = \frac{384}{7}$$

$$M_x = \frac{1}{2}\int_0^8 x^{2/3} dx = \frac{1}{2} \cdot \frac{3}{5}x^{5/3}\,|_0^8 = \frac{48}{5}$$

$$m = \int_0^8 x^{1/3} dx = \frac{3}{4} \cdot x^{4/3}\,|_0^8 = 12$$

$$\bar{x} = \frac{384}{7 \cdot 12} = \frac{32}{7}$$

$$\bar{y} = \frac{48}{5 \cdot 12} = \frac{4}{5}.$$

The centroid is $\left(\frac{32}{7}, \ \frac{4}{5}\right)$.

12. $y = x^4 [-1, \ 2]$

$$M_y = \int_{-1}^2 x \cdot x^4 dx = \int_{-1}^2 x^5 dx$$

$$= \frac{x^6}{6}\bigg|_{-1}^2 = \frac{69 - 1}{6} = \frac{63}{6} = \frac{21}{2}$$

$$M_x = \frac{1}{2}\int_{-1}^2 x^8 dx = \frac{1}{2} \cdot \frac{x^9}{9}\bigg|_{-1}^2$$

$$= \frac{513}{18} = \frac{171}{6} = \frac{57}{2}$$

$$m = \int_{-1}^2 x^4 dx = \frac{x^5}{5}\bigg|_{-1}^2 = \frac{32 + 1}{5} = \frac{33}{5};$$

$$\bar{x} = \frac{21}{2} \cdot \frac{5}{33} = \frac{35}{22}$$

$$\bar{y} = \frac{57}{2} \cdot \frac{5}{33} = \frac{285}{66} = \frac{95}{22}.$$

The centroid is $\left(\frac{35}{22}, \ \frac{95}{22}\right)$.

13. $y = \sqrt{x + 4}$ on $[0, 5]$; $M_y = \int_0^5 x\sqrt{x + 4}\, dx$; $u = x + 4$; $du = dx$; $x = u - 4\, x = 0 \Rightarrow u = 4$ and $x = 5 \Rightarrow u = 9$. Substituting, we obtain

$$\int_4^9 (u - 4)\sqrt{u}\, du$$

$$= \int_4^9 \left(u^{3/2} - 4u^{1/2}\right) du$$

$$= \left[\frac{2}{5}u^{5/2} - 4 \cdot \frac{2}{3}u^{3/2}\right]_4^9$$

$$= \frac{2 \cdot 243}{5} - 72 - \frac{64}{5} + \frac{64}{3}$$

$$= \frac{1458 - 1080 - 192 + 320}{15} = \frac{506}{15}$$

$$M_x = \frac{1}{2}\int_0^5 (x + 4) dx = \frac{1}{2}\left[\frac{x^2}{2} + 4x\right]_0^5$$

$$= \frac{1}{2}\left[\frac{25}{2} + 20\right] = \frac{65}{4}$$

$$m = \int_0^5 \sqrt{x + 4}\, dx = \frac{2}{3}(x + 4)^{3/2}\,|_0^5$$

$$= \frac{2}{3}[27 - 8] = \frac{38}{3}$$

$$\bar{x} = \frac{506}{15} \cdot \frac{3}{38} = \frac{253}{95}$$

$$\bar{y} = \frac{65}{4} \cdot \frac{3}{38} = \frac{195}{152}.$$

The centroid is $\left(\frac{253}{95}, \ \frac{195}{152}\right)$.

14. $y = x^2 + 16; \ [0, \ 4]$

$$M_y = \int_0^4 x\left(x^2 + 16\right) dx = \left[\frac{x^4}{4} + 8x^2\right]_0^4$$

$$= 64 + 128 = 192$$

$$M_x = \frac{1}{2}\int_0^4 \left(x^2 + 16\right)^2 dx$$

$$= \frac{1}{2}\int_0^4 \left(x^4 + 32x^2 + 256\right) dx$$

$$= \frac{1}{2}\left[\frac{x^5}{5} + \frac{32x^3}{3} + 256x\right]_0^4$$

$$= \frac{1}{2}\left[\frac{1024}{5} + \frac{2048}{3} + 1024\right]$$

$$= \frac{512}{5} + \frac{1024}{3} + 512$$

$$= \frac{1536 + 5120 + 7680}{15} = \frac{14336}{15}$$

$$m = \int_0^4 \left(x^2 + 16\right) dx = \left[\frac{x^3}{3} + 16x\right]_0^4$$

$$= \frac{64}{3} + 64 = \frac{256}{3}$$

$$\bar{x} = 192 \cdot \frac{3}{256} = \frac{9}{4} = 2.25$$

$$\bar{y} = \frac{14336}{15} \cdot \frac{3}{256} = \frac{56}{5} = 11.2$$

The centroid is $(2.25, 11.2)$.

15. $M_y = \int_0^2 x(4x - x^2)\,dx = \int_0^2 (4x^2 - x^3)\,dx$

$= \left[\dfrac{4x^3}{3} - \dfrac{x^4}{4}\right]_0^2 = \dfrac{32}{3} - 4 = \dfrac{20}{3}$

$M_x = \dfrac{1}{2}\int_0^2 \left[(4x)^2 - (x^2)^2\right]\,dx$

$= \dfrac{1}{2}\int_0^2 (16x^2 - x^4)\,dx$

$= \dfrac{1}{2}\left[\dfrac{16x^3}{3} - \dfrac{x^5}{5}\right]_0^2 = \dfrac{1}{2}\left[\dfrac{128}{3} - \dfrac{32}{5}\right]$

$= \dfrac{64}{3} - \dfrac{16}{5} = \dfrac{272}{15}$

$m = \int_0^2 (4x - x^2)\,dx = \left[2x^2 - \dfrac{x^3}{3}\right]_0^2$

$= 8 - \dfrac{8}{3} = \dfrac{16}{3}$

$\bar{x} = \dfrac{20}{3}\cdot\dfrac{3}{16} = \dfrac{5}{4}$

$\bar{y} = \dfrac{272}{15}\cdot\dfrac{3}{16} = \dfrac{17}{5}.$

The centroid is $\left(\dfrac{5}{4},\ \dfrac{17}{5}\right)$.

16. $M_y = \int_0^1 x(2x - x^3)\,dx = \int_0^1 (2x^2 - x^4)\,dx$

$= \left[\dfrac{2x^3}{3} - \dfrac{x^5}{5}\right]_0^1 = \dfrac{2}{3} - \dfrac{1}{5} = \dfrac{7}{15}$

$M_x = \dfrac{1}{2}\int_0^1 \left[(2x)^2 - (x^3)^2\right]\,dx$

$= \dfrac{1}{2}\int_0^1 (4x^2 - x^6)\,dx$

$= \left[\dfrac{2x^5}{5} - \dfrac{x^7}{14}\right]_0^1 = \dfrac{2}{5} - \dfrac{1}{14} = \dfrac{23}{70}$

$m = \int_0^1 (2x - x^3)\,dx = \left[x^2 - \dfrac{x^4}{4}\right]_0^1$

$= 1 - \dfrac{1}{4} = \dfrac{3}{4}$

$\bar{x} = \dfrac{7}{15}\cdot\dfrac{4}{3} = \dfrac{28}{45}$

$\bar{y} = \dfrac{23}{70}\cdot\dfrac{4}{3} = \dfrac{92}{210} = \dfrac{46}{105}.$

The centroid is $\left(\dfrac{28}{45},\ \dfrac{46}{105}\right)$.

17. The graphs intersect at $(-2, 4)$ and $(1, 1)$

$M_y = \int_{-2}^1 x(2 - x - x^2)\,dx$

$= \int_{-2}^1 (2x - x^2 - x^3)\,dx$

$= \left[x^2 - \dfrac{x^3}{3} - \dfrac{x^4}{4}\right]_{-2}^1$

$= 1 - \dfrac{1}{3} - \dfrac{1}{4} - 4 - \dfrac{8}{4} + 4 = -\dfrac{9}{4}$

$M_x = \dfrac{1}{2}\int_{-2}^1 \left((2 - x)^2 - (x^2)^2\right)\,dx$

$= \dfrac{1}{2}\int_{-2}^1 (4 - 4x + x^2 - x^4)\,dx$

$= \dfrac{1}{2}\left[4x - 2x^2 + \dfrac{4^3}{3} - \dfrac{x^5}{5}\right]_{-2}^1$

$= \dfrac{1}{2}\left[4 - 2 + \dfrac{1}{3} - \dfrac{1}{5} + 8 + 8 + \dfrac{8}{4} - \dfrac{32}{5}\right]$

$= \dfrac{1}{2}\left[18 + 3 - \dfrac{33}{5}\right] = \dfrac{36}{5}$

$m = \int_{-2}^1 (2 - x - x^2)\,dx = \left[2x - \dfrac{x^2}{2} - \dfrac{x^3}{3}\right]_{-2}^1$

$= 2 - \dfrac{1}{2} - \dfrac{1}{3} + 4 + 2 - \dfrac{8}{3} = 8 - \dfrac{1}{2} - 3$

$= 4\dfrac{1}{2} = \dfrac{9}{2}$

$\bar{x} = -\dfrac{9}{4}\cdot\dfrac{2}{9} = -\dfrac{1}{2}$

$\bar{y} = \dfrac{36}{5}\cdot\dfrac{2}{9} = \dfrac{8}{5}.$

The centroid is $\left(-\dfrac{1}{2},\ \dfrac{8}{5}\right)$.

18. The graphs intersect at $(0, 0)$ and $(1, 1)$

$M_y = \int_0^1 x(x - x^{3/2})\,dx = \int_0^1 (x^2 - x^{5/2})\,dx$

$= \left[\dfrac{x^3}{3} - \dfrac{2}{7}x^{7/2}\right]_0^1 = \dfrac{1}{3} - \dfrac{2}{7} = \dfrac{1}{21}$

$M_x = \dfrac{1}{2}\int_0^1 (x^2 - x^3)\,dx = \dfrac{1}{2}\left[\dfrac{x^3}{3} - \dfrac{x^4}{4}\right]_0^1$

$= \dfrac{1}{2}\left(\dfrac{1}{3} - \dfrac{1}{4}\right) = \dfrac{1}{24}$

$m = \int_0^1 (x - x^{3/2})\,dx = \left[\dfrac{x^2}{2} - \dfrac{2}{5}x^{5/2}\right]_0^1$

$= \dfrac{1}{2} - \dfrac{2}{5} = \dfrac{1}{10}$

$\bar{x} = \dfrac{1}{21}\cdot\dfrac{10}{1} = \dfrac{10}{21}$

$\bar{y} = \dfrac{1}{24}\cdot\dfrac{10}{1} = \dfrac{10}{24} = \dfrac{5}{24}.$

The centroid is $\left(\dfrac{10}{21},\ \dfrac{5}{12}\right)$

19. The graphs intersect at $(\pm 3,\ 9)$

$$M_y = \int_{-3}^{3} x\left(18 - x^2 - x^2\right)dx$$

$$= \int_{-3}^{3} \left(18x - 2x^3\right)dx = 0$$

$$M_x = \frac{1}{2}\int_{-3}^{3} \left(\left(18 - x^2\right)^2 - \left(x^2\right)^2\right)dx$$

$$= \frac{1}{2}\int_{-3}^{3} \left(324 - 36x^2 + x^4 - x^4\right)dx$$

$$= \frac{1}{2}\left[324x - 12x^3\right]_{-3}^{3}$$

$$= \frac{1}{2}[972 - 324 + 972 - 324] = 648$$

$$m = \int_{-3}^{3} \left(18 - x^2 - x^2\right)dx = \left[18x - \frac{2x^3}{3}\right]_{-3}^{3}$$

$$= (54 - 18 + 54 - 18) = 72$$

$$\bar{x} = 0$$

$$\bar{y} = \frac{648}{72} = 9$$

The centroid is $(0,\ 9)$.

20. The intersect at $(0,\ 0)$ and $(1,\ 1)$

$$M_y = \int_{0}^{1} x(\sqrt{x} - x)dx = \int_{0}^{1} \left(x^{3/2} - x^2\right)dx$$

$$= \left[\frac{2}{5}x^{5/2} - \frac{x^3}{3}\right]_{0}^{1}$$

$$= \frac{2}{5} - \frac{1}{3} = \frac{1}{15}$$

$$M_x = \frac{1}{2}\int_{0}^{1} \left(x - x^2\right)dx = \frac{1}{2}\left[\frac{x^2}{2} - \frac{x^3}{3}\right]_{0}^{1} = \frac{1}{12}$$

$$m = \int_{0}^{1} (\sqrt{x} - x)dx = \left[\frac{2}{3}x^{3/2} - \frac{x^2}{2}\right]_{0}^{1}$$

$$= \frac{2}{3} - \frac{1}{2} = \frac{1}{6}$$

$$\bar{x} = \frac{1}{15} \cdot \frac{6}{1} = \frac{6}{15} = \frac{2}{5}$$

$$\bar{y} = \frac{1}{12} \cdot \frac{6}{1} = \frac{1}{2}.$$

The centroid is $\left(\frac{2}{5},\ \frac{1}{2}\right)$.

21. They intersect at $(-4,\ -4)$ and $(3,\ 3)$

$$M_y = \int_{-4}^{3} x\left(12 - x^2 - x\right)dx$$

$$= \int_{-4}^{3} \left(12x - x^3 - x^2\right)dx$$

$$= \left[6x^2 - \frac{x^4}{4} - \frac{x^3}{3}\right]_{-4}^{3}$$

$$= 54 - \frac{81}{4} - 9 - 96 + 64 - \frac{64}{3}$$

$$= 13 - \frac{81}{4} - \frac{64}{3} = -\frac{343}{12}$$

$$M_x = \frac{1}{2}\int_{-4}^{3} \left(12 - x^2\right)^2 - (x)^2 dx$$

$$= \frac{1}{2}\int_{-4}^{3} \left(144 - 24x^2 + x^4 - x^3\right)dx$$

$$\times \frac{1}{2}\left[144x - \frac{25x^3}{3} + \frac{x^5}{5}\right]_{-4}^{3}$$

$$= \frac{1}{2}\left[432 - 225 + \frac{243}{5} + 576\right.$$

$$\left. - \frac{1600}{3} + \frac{1024}{5}\right] = \frac{3773}{15}$$

$$m = \int_{-4}^{3} \left(12 - x^2 - x\right)dx$$

$$= \left[12x - \frac{x^3}{3} - \frac{x^3}{2}\right]_{-4}^{3}$$

$$= 36 - 9 - \frac{9}{2} + 48 - \frac{64}{3} + 8 = 57\frac{1}{6} = \frac{343}{6}$$

$$\bar{x} = -\frac{343}{12} \times \frac{6}{343} = -\frac{1}{2} = -0.5$$

$$\bar{y} = \frac{3773}{15} \cdot \frac{6}{343} = 4.4$$

The centroid is $(-0.5,\ 4.4)$.

22. $y = x^2$ and $y = x^3$ intersect at $(0,\ 0)$ and $(1,\ 1)$.

$$M_y = \int_{0}^{1} x\left(x^2 - x^3\right)dx = \left[\frac{x^4}{4} - \frac{x^5}{5}\right]_{0}^{1} = \frac{1}{4} - \frac{1}{5}$$

$$= \frac{1}{20}$$

$$M_x = \frac{1}{2}\int_{0}^{1} \left(x^4 - x^7\right)dx = \frac{1}{2}\left[\frac{x^5}{5} - \frac{x^7}{7}\right]_{0}^{1}$$

$$= \frac{1}{2}\left(\frac{1}{5} - \frac{1}{7}\right) = \frac{2}{23} \cdot \frac{1}{2} = \frac{1}{35}$$

$$m = \int_{0}^{1} \left(x^2 - x^3\right)dx = \left[\frac{x^3}{3} - \frac{x^4}{4}\right]_{0}^{1} = \frac{1}{3} - \frac{1}{4}$$

$$= \frac{1}{12}$$

$$\bar{x} = \frac{1}{20} \cdot \frac{12}{1} = \frac{3}{5}$$

$$\bar{y} = \frac{1}{35} \cdot \frac{12}{1} = \frac{12}{35}$$

The centroid is $\left(\frac{3}{5},\ \frac{12}{35}\right)$.

23. $y = x^3$;

$$M_y = \pi\int_{0}^{2} x\left(x^2\right)dx = \pi\left.\frac{x^3}{8}\right|_{0}^{2} = 32\pi$$

$$m = \pi\int_{0}^{2} \left(x^3\right)^2 = \pi \cdot \left.\frac{x^7}{7}\right|_{0}^{2} = \frac{128\pi}{7}.$$

Thus, we find that $\bar{x} = \dfrac{32\pi}{\frac{128\pi}{7}} = 32\pi\left(\frac{7}{128\pi}\right) = \frac{7}{4}$

and $\bar{y} = 0$ since it's rotated about the x-axis. The centroid is $\left(\frac{7}{4},\ 0\right)$.

24. $y = x^3 \Rightarrow x = y^{1/3}$

$$M_x = \pi \int_0^8 y\left(y^{2/3}\right) dy = \pi \frac{3}{8} y^{8/3} \Big|_0^8 = 96\pi$$

$$m = \pi \int_0^8 y^{2/3} dy = \pi \cdot \frac{3}{5} y^{5/3} \Big|_0^8 = \frac{96\pi}{5}$$

Thus, we see that $\overline{x} = 0$ since it is rotated about the y-axis, and $\overline{y} = \dfrac{96\pi}{96\frac{\pi}{5}} = 96\pi\left(\frac{5}{96\pi}\right) = 5$. The centroid is (0, 5).

25. $y = x^4 \Rightarrow x = y^{1/4}$

$$M_x = \pi \int_0^1 y \cdot \left(y^{1/4}\right)^2 dy = \pi \frac{2}{5} \cdot y^{5/2} \Big|_0^1 = \frac{2\pi}{5}$$

$$m = \pi \int_0^1 \left(y^{1/4}\right)^2 dy = \pi \cdot \frac{2}{3} y^{3/2} \Big|_0^1 = \frac{2\pi}{3}$$

$$\overline{x} = 0$$

$$\overline{y} = \frac{\frac{2\pi}{5}}{\frac{2\pi}{3}} = \frac{2\pi}{5} \cdot \frac{3}{2\pi} = \frac{3}{5}$$

The centroid is $\left(0, \frac{3}{5}\right)$.

26. $y = (x-1)^2 \Rightarrow x - 1 = \pm y^{1/2} \Rightarrow x = 1 \pm \sqrt{y}$

$$M_x = \pi \int_0^1 y\left[(1+\sqrt{y})^2 - (1-\sqrt{y})^2\right] dy$$

$$= \pi \int_0^1 y[1 + 2\sqrt{y} + y - 1 + 2\sqrt{y} - y] dy$$

$$= \pi \int_0^1 4y^{3/2} dy$$

$$= \pi \left[4 \cdot \frac{2}{5} y^{5/2}\right]_0^1 = \frac{8\pi}{5}$$

$$m = \pi \int_0^1 4y^{1/2} dy = \pi 4 \cdot \frac{2}{3} y^{3/2} \Big|_0^1 = \frac{8\pi}{3}$$

$$\overline{x} = 0$$

$$\overline{y} = \frac{\frac{8\pi}{5}}{\frac{8\pi}{3}} = \frac{8\pi}{5} \cdot \frac{3}{8\pi} = \frac{3}{5}.$$

The centroid is $\left(0, \frac{3}{5}\right)$.

27. $x + y = 6 \Rightarrow y = 6 - x;\ x = 3$

$$M_y = \pi \int_0^3 x\left[(6-x)^2\right] dx$$

$$= \pi \int_0^3 \left(36x - 12x^2 + x^3\right) dx$$

$$= \pi \left[18x^2 - 4x^3 + \frac{x^4}{4}\right]_0^3 = 74.25\pi = \frac{297}{4}\pi$$

$$m = \pi \int_0^3 (6-x)^2 dx = \pi \int_0^3 \left(36 - 12x + x^2\right) dx$$

$$= \pi \left(36x - 6x^2 + \frac{x^3}{3}\right) \Big|_0^3 = 63\pi$$

$$\overline{x} = \frac{297}{4} \cdot \frac{1}{63} = \frac{33}{28}$$

$$\overline{y} = 0$$

The centroid is $\left(\frac{33}{28}, 0\right)$.

28. $y = 8 - x$

$$M_y = \pi \int_4^8 x(8-x)^2 dx$$

$$= \pi \int_4^8 \left(64x - 16x^2 + x^3\right) dx$$

$$= \pi \left(32x^2 - \frac{16}{3}x^3 + \frac{1}{4}x^4\right) \Big|_0^4 = 106\frac{2}{3}$$

$$m = \pi \int_4^8 (8-x)^2 dx = -\frac{1}{3}(8-x)^3 \Big|_4^8 = 21\frac{1}{3}$$

$$\overline{x} = 5$$

$$\overline{y} = 0$$

The centroid is (5, 0).

29. $M_x = \pi \int_0^4 y \cdot 4^2 dy + \pi \int_4^8 y(8-y)^2 dy$

$$= \pi \left[128 + 106\frac{2}{3}\right] = 234\frac{2}{3}\pi$$

$$m = \pi \left[\int_0^4 16\, dy + \int_4^8 (8-y)^2 dy\right]$$

$$= \pi \left[64 + 21\frac{1}{3}\right] = 85\frac{1}{3}\pi$$

$$\overline{x} = 0$$

$$\overline{y} = \frac{234\frac{2}{3}}{85\frac{1}{3}} = 2.75$$

The centroid is (0, 2.75).

30. $y = \sqrt{16 - x^2}$

$$M_y = \pi \int_0^4 x\sqrt{16-x^2}^2 dx = \pi \int_0^4 16x - x^2 dx$$

$$= \pi \left[8x^2 - \frac{x^4}{4}\right]_0^4 = \pi(128 - 64) = 64\pi$$

$$m = \pi \int_0^4 \left(16 - x^2\right) dx = \pi \left[16x - \frac{x^3}{3}\right]_0^4$$

$$= \frac{128\pi}{3}$$

$$\overline{x} = \frac{64\pi}{\frac{128\pi}{3}} = \frac{3}{2} \text{ or } 1.5$$

$$\overline{y} = 0$$

The centroid is (1.5, 0).

31. $x^2 - y^2 = 1;\ -y^2 = 1 - x^2;\ y^2 = x^2 - 1;\ y = \sqrt{x^2 - 1}$

$$M_y = \pi \int_1^3 x(x^2 - 1)\,dx = \pi \left[\frac{x^4}{4} - \frac{x^2}{2} \right]_1^3$$

$$= \pi \left[\frac{81}{4} - \frac{9}{2} - \frac{1}{4} + \frac{1}{2} \right] = \pi[20 - 4] = 16\pi$$

$$m = \pi \int_1^3 (x^2 - 1)\,dx = \pi \left[\frac{x^3}{3} - x \right]_1^3$$

$$= \pi \left(9 - 3 - \frac{1}{3} + 1 \right) = \pi 6 \frac{2}{3} \text{ or } \frac{20}{3}\pi$$

$$\overline{x} = \frac{16}{\frac{20}{3}} = \frac{4 \cdot 3}{5} = \frac{12}{5}.$$

The centroid is $\left(\frac{12}{5}, 0 \right)$ or (2.4, 0).

32. $x^2 - y^2 = 1 \Rightarrow x = \sqrt{y^2 + 1}$ and $x = 3 \Rightarrow y = \sqrt{8}$

$$M_x = \pi \int_0^{\sqrt{8}} y(3^2 - y^2 - 1)\,dy$$

$$= \pi \int_0^{\sqrt{8}} y(8 - y^2)\,dy$$

$$= \pi \left[4y^2 - \frac{y^4}{4} \right]_0^{\sqrt{8}} = \pi[32 - 16] = 16\pi$$

$$m = \pi \int_0^{\sqrt{8}} (9 - y^2 - 1)\,dy = \pi \left[8y - \frac{y^3}{3} \right]_0^{\sqrt{8}}$$

$$= \left[8\sqrt{8} - \frac{8\sqrt{8}}{3} \right] = \frac{\pi 16\sqrt{8}}{3} = \frac{32\pi\sqrt{2}}{3}$$

$$\overline{x} = 0$$

$$\overline{y} = \frac{16\pi}{\frac{32\pi\sqrt{2}}{3}} = \frac{3\sqrt{2}}{4} = 1.0607$$

The centroid is (0, 1.0607).

33. $y = x^2$; $y = \sqrt{x}$

$$M_y = \pi \int_0^1 x[x - x^4]\,dx = \pi \left[\frac{x^3}{3} - \frac{x^6}{6} \right]_0^1 = \frac{\pi}{6}$$

$$m = \pi \int_0^1 x - x^4\,dx = \pi \left[\frac{x^2}{2} - \frac{x^5}{5} \right] = \frac{3\pi}{10}$$

$$\overline{x} = \frac{\frac{\pi}{6}}{\frac{3\pi}{10}} = \frac{10}{18} = \frac{5}{9}.$$

The centroid is $\left(\frac{5}{9}, 0 \right)$.

34. These two curves intersect at (0, 0) and (1, 1).
$y = x^4 \Rightarrow x = y^{1/4}$; $y = x^2 \Rightarrow x = y^{1/2}$.

$$M_x = \pi \int_0^1 y(y^{1/2} - y)\,dy = \pi \left[\frac{2}{5}y^{5/2} - \frac{y^3}{3} \right]_0^1$$

$$= \pi \left[\frac{2}{5} - \frac{1}{3} \right] = \frac{\pi}{15}$$

$$m = \pi \int_0^1 (y^{1/2} - y)\,dy = \pi \left[\frac{2}{3}y^{3/2} - \frac{y^2}{2} \right]_0^1$$

$$= \pi \left[\frac{2}{3} - \frac{1}{2} \right] = \frac{\pi}{6}$$

$$\overline{y} = \frac{\frac{1}{15}}{\frac{1}{6}} = \frac{6}{15} = \frac{2}{5}$$

The centroid is $\left(0, \frac{2}{5} \right)$.

35.   (a) Make two rectangles by cutting off the top piece. Then the bottom has center at $\left(50, \frac{1}{2} \right)$ and area 100 cm². The left piece has center at $\left(\frac{1}{2}, \frac{101}{2} \right)$ and area 99 cm².

$$\overline{x} = \frac{100 \times 50 + 99 \cdot \frac{1}{2}}{100 + 99} = \frac{5049.5}{199} \approx 25.374$$

$$\overline{y} = 100 \times \frac{1}{2} + 101 \left(\frac{99}{2} \right) = \frac{5049.5}{199} \approx 25.374$$

The centroid of the frame is (25.374, 25.374).

(b) The bottom area is $h \cdot w$ with center $\left(\frac{h}{2}, \frac{w}{2} \right)$ and the left area is $(h - w)$, center $\left(\frac{w}{2}, \frac{h+w}{2} \right)$

$$\overline{x} = \frac{hw\left(\frac{h}{2} \right) + w(h - w) \cdot \frac{w}{2}}{hw + (h - w)w} = \frac{\frac{h^2w}{2} + \frac{hw^2}{2} - \frac{w^3}{2}}{2wh - w^2}$$

$$= \frac{h^2 + hw - w^2}{2(2h - w)}.$$

$$\overline{y} = \frac{\left(\frac{hw \cdot w}{2} \right) + (h - w)w\left(\frac{h+w}{2} \right)}{w(2h - w)}$$

$$= \frac{hw + h^2 - w^2}{2(2h - w)}$$

The centroid is $\left(\frac{h^2 + hw - w^2}{2(2h - w)}, \frac{h^2 + hw - w^2}{2(2h - w)} \right)$.

36.   (a) Use the center of the bottom of the cone as the origin and the y-axis as the axis of the cone. Then the cone is formed by rotating the line $x = 7 - \frac{4.5}{28}y$ or $x = 7 - \frac{45}{280}y = 7 - \frac{9}{56}y$.

$$M_x = \pi \int_0^{28} y \left(7 - \frac{9}{56}y \right)^2 dy$$

$$= \pi \int y \left(49 - \frac{9}{4}y + \frac{81}{3136}y^2 \right)$$

$$= \pi \int_0^{28} \left(49y - \frac{9}{4}y^2 + \frac{81}{3136}y^3 \right) dy$$

$$= \pi \left[\frac{49y^2}{2} - \frac{3y^3}{4} + \frac{81y^4}{12544} \right]_0^{28} = 6713\pi$$

$$m = \pi \int_0^{28} \left(7 - \frac{9}{56}y \right)^2 dy$$

$$= \pi \int_0^{28} \left(49 - \frac{9}{4}y + \frac{81}{3136}y^2 \right) dy$$

$$= \pi \left[49y - \frac{9y^2}{8} + \frac{27y^3}{3136} \right]_0^{28} = 679\pi$$

$\overline{y} = \frac{6713}{679} = 9.8866$ inches from the top of the base. The centroid is (0, 9.8866).

(b) The bottom has volume of $18^2 = 324$ in.^{2} and centroid $\frac{1}{2}$ in. below the top of the base.

$$\bar{y} = \frac{9.8866 \times 679\pi - \frac{1}{2}(324)}{679\pi + 324}$$

$$= \frac{21089.52 - 162}{2457.14} = \frac{20{,}927.52}{2457.14} \approx 8.517$$

The centroid is 8.517 in. above the top of the base.

37. **(a)** $m = \int_0^{40}(5x+1)dx = \left[\frac{5x^2}{2} + x\right]_0^{40} = 4040$ g

(b) $M_y = \int_0^{40} x(5x+1)dx = \left[\frac{5x^3}{3} + \frac{x^3}{2}\right]_0^{40} = 107{,}467$; $\bar{x} = \frac{107{,}467}{4040} = 26.6$ cm from the less dense end.

38. **(a)** $m = \int_0^8(x+5)dx = \left[\frac{x^2}{2} + 5x\right]_0^8 = 32+40 = 72$ kg

(b) $M_y = \int_0^8 x(x+5)dx = \left[\frac{x^3}{3} + \frac{5x^2}{2}\right]_0^8 = 330.67$; $\bar{x} = \frac{330.67}{72} = 4.59$ m from the one end.

26.6 MOMENTS OF INERTIA

1. $I_y = (3 \cdot 4^2) + 5(-3)^2 = 48 + 45 = 93$
$I_x = 3(0)^2 + 5(0)^2 = 0$
$r_y = \sqrt{\frac{93}{3}} = 3.41$; $r_x = 0$; $r_0 = \sqrt{r_x^2 + r_y^2} = 3.4095$

2. $I_y = 6 \cdot 0^2 + 3 \cdot 0^2 + 1 \cdot 0^2 = 0$
$I_x = 6 \cdot 2^2 + 3 \cdot (-5)^2 + 1 \cdot 4^2 = 115$
$r_y = 0$; $r_x = \sqrt{\frac{115}{10}} = 3.39$; $r_0\sqrt{r_x^2 + r_y^2} = 3.3912$

3. $I_y = 4 \cdot 2^2 + 3 \cdot 1^2 = 19$; $r_y = \sqrt{\frac{19}{7}} = 1.6475$
$I_x = 4 \cdot 1^2 + 3 \cdot 4^2 = 52$; $r_x = \sqrt{\frac{52}{7}} = 2.7255$; $r_0 = \sqrt{\frac{19+52}{7}} = 3.1847$

4. $I_y = 3 \cdot 2^2 + 4 \cdot 2^2 = 3 \cdot 4^2 = 76$
$I_x = 3 \cdot 3^2 + 4 \cdot 4^2 + 3 \cdot 5^2 = 166$
$r_0 = \sqrt{\frac{76 + 166}{10}} = 4.9193$

5. $I_y = \rho\int_0^{\sqrt{2}} x^2 \cdot (2 - x^2)\, dx = \rho\left[\frac{2x^3}{3} - \frac{x^5}{5}\right]_0^{\sqrt{2}}$
$= \rho\left[\frac{4\sqrt{2}}{3} - \frac{4\sqrt{2}}{5}\right] = \rho \cdot \frac{8\sqrt{2}}{15} \approx 0.7542\rho$
$m = \rho\int_0^{\sqrt{2}}(2-x^2)dx = \rho\left[2x - \frac{x^3}{3}\right]_0^{\sqrt{2}}$
$= \rho\left(2\sqrt{2} - \frac{2\sqrt{2}}{3}\right)$
$= \rho \cdot \frac{4\sqrt{2}}{3}$; $r_y = \sqrt{\frac{\frac{8}{15}}{4/3}} = \sqrt{\frac{2}{5}} = 0.6325$

6. $I_x = \rho\int_0^2 y^2(\sqrt{y})dy = \rho\left[\frac{2}{7}y^{7/2}\right]_0^3 = \frac{16\sqrt{2}}{7}\rho$
$= 3.2325\rho$
$m = \frac{4\sqrt{2}}{3}\rho$
$r_x = \sqrt{\frac{\frac{16\sqrt{2}}{7}}{\frac{4\sqrt{2}}{3}}} = \sqrt{\frac{12}{7}} = 1.3093$

7. $I_x = \rho\int_0^5 y^2 \cdot 3\, dy = \rho y^3\,|_0^5 = 125\rho$
$m = \rho\int_0^5 3\, dx = 15$
$r_x = \sqrt{\frac{125}{15}} = \sqrt{\frac{25}{3}} \approx 2.8868$

8. $I_y = \rho\int_0^3 x^2 \cdot 5\, dx = \rho\frac{5x^3}{3}\Big|_0^3 = 45\rho$
$m = 15$
$r_y = \sqrt{\frac{45}{15}} = \sqrt{3} \approx 1.7321$

9. $I_x = \rho\int_0^1 y^2\left(y^{1/3} - y^{1/2}\right)dy$
$= \rho\left[\frac{3}{10}y^{10/3} - \frac{2}{7}y^{7/2}\right]_0^1$
$= \frac{\rho}{70} \approx 0.014286\rho$
$m = \rho\int_0^1(x^2 - x^3)dx = \rho\left[\frac{x^3}{3} - \frac{x^4}{4}\right]_0^1 = \frac{1}{12}\rho$
$r_x = \sqrt{\frac{\frac{1}{70}}{\frac{1}{12}}} = \sqrt{\frac{12}{70}} \approx 0.4140$

10. $I_y = \rho \displaystyle\int_0^1 x^2(x^2 - x^3)dx = \rho\left[\dfrac{x^5}{5} - \dfrac{x^6}{6}\right] = \dfrac{\rho}{30}$

$\approx 0.0333\rho$

$m = \dfrac{\rho}{12};\ r_y = \sqrt{\dfrac{\frac{1}{30}}{\frac{1}{12}}} = \sqrt{\dfrac{12}{30}} = \sqrt{\dfrac{2}{5}} \approx 0.6325$

11. $I_y = 5\displaystyle\int_0^4 x^2(4x - x^2)dx = 5\left[x^4 - \dfrac{x^5}{5}\right]_0^4$

$= 5\left[256 - \dfrac{1024}{5}\right] = 256\ \text{g}\cdot\text{cm}^2$

$m = 5\displaystyle\int_0^4 (4x - x^2)dx = 5\left[2x^2 - \dfrac{x^3}{3}\right]_0^4$

$= 5\left[32 - \dfrac{64}{3}\right]$

$= 53.333;\ r_y = \sqrt{\dfrac{256}{53.333}} \approx 2.1909\ \text{cm}$

12. $I_x = 4\displaystyle\int_0^4 y^2(y^{3/2})dy = 4\int_0^4 y^{7/2}dy$

$= 4\left[\dfrac{2}{9}y^{9/2}\right]_0^4 = \dfrac{4096}{9} = 455.11\ \text{g}\cdot\text{cm}^2$

$m = 4\displaystyle\int_0^8 x^{2/3}dx = 4\left[\dfrac{3}{5}x^{5/3}\right]_0^8 = \dfrac{32\cdot3\cdot4}{5}$

$= 51.2\ \text{g}$

$r_x = \sqrt{\dfrac{455.1111}{51.2}} \approx 2.9814\ \text{cm}$

13. $I_y = 8\displaystyle\int_1^2 x^2\left(x^2 - \dfrac{1}{x^2}\right)dx = 8\left[\dfrac{x^5}{5} - x\right]_1^2$

$= 8\left[\dfrac{32}{5} - 2 - \dfrac{1}{5} + 1\right] = 8\left[\dfrac{26}{5}\right] = 41.6\ \text{g}\cdot\text{cm}^2$

$m = 8\displaystyle\int_1^2 \left(x^2 - \dfrac{1}{x^2}\right)dx = 8\left[\dfrac{x^3}{3} + \dfrac{1}{x}\right]_1^2$

$= 8\left[\dfrac{8}{3} + \dfrac{1}{2} - \dfrac{1}{3} - 1\right] = 14\dfrac{2}{3}\ \text{g}$

$r_y = \sqrt{\dfrac{41.6}{14.6667}} \approx 1.6842\ \text{cm}$

14. $I_x = 8\displaystyle\int_0^1 y^2(2 - y^{-1/2})dy$

$\quad + 8\displaystyle\int_1^4 y^2(2 - y^{1/2})dy$

$= 8\left[\dfrac{2y^3}{3} - \dfrac{2}{5}y^{5/2}\right]_0^1 + 8\left[\dfrac{2y^3}{3} - \dfrac{2}{7}y^{7/2}\right]_1^4$

$= 8\left[\dfrac{2}{3} - \dfrac{2}{5}\right] + 8\left[\dfrac{128}{3} - \dfrac{256}{7} - \dfrac{2}{3} + \dfrac{2}{7}\right]$

$= 8\left[\dfrac{128}{3} - \dfrac{2}{5} - \dfrac{254}{7}\right] = 47.8467\ \text{g}\cdot\text{cm}^2$

$r_x = \sqrt{\dfrac{47.8476}{14.6666}} = 1.8062\ \text{cm}$

15. $I_y = 2\pi\rho\displaystyle\int_0^{\sqrt{2}} x^3[2 - x^2]dx$

$= 2\pi\rho\left[\dfrac{2x^4}{4} - \dfrac{x^6}{6}\right]_0^{\sqrt{2}}$

$= 2\pi\rho\left[2 - \dfrac{8}{6}\right] = \dfrac{4\pi\rho}{3}$

$m = 2\pi\rho\displaystyle\int_0^{\sqrt{2}} x(2 - x^2)dx = 2\pi\rho\left[x^2 - \dfrac{x^4}{4}\right]_0^{\sqrt{2}}$

$= 2\pi\rho$

$r_y = \sqrt{\dfrac{4/3}{2}} = \sqrt{2/3} \approx 0.8165$

16. $I_x = 2\pi\rho\displaystyle\int_0^2 y^3\left[y^{1/2}\right]dy = 2\pi\rho\left[\dfrac{2}{9}y^{9/2}\right]_0^2$

$= 2\pi\rho\cdot\dfrac{2}{9}\cdot16\sqrt{2} = \dfrac{64\pi\rho\sqrt{2}}{9} \approx 31.5938\rho$

$m = 2\pi\rho\displaystyle\int_0^2 y^{3/2}dy = 2\pi\rho\left[\dfrac{2}{5}y^{5/2}\right]_0^2 = \dfrac{16\pi\sqrt{2}\rho}{5}$

$r_x = \sqrt{\dfrac{\frac{64\sqrt{2}}{2}}{\frac{16\sqrt{2}}{5}}} = \sqrt{\dfrac{20}{9}} = \dfrac{2}{3}\sqrt{5} \approx 1.4907$

17. $I_y = 2\pi\rho\displaystyle\int_0^4 x^3(4x - x^2)dx = 2\pi\rho\left[\dfrac{4x^5}{5} - \dfrac{x^6}{6}\right]_0^4$

$= 2\pi\rho\left(\dfrac{4096}{30}\right) \approx 857.86\rho\ \text{or}\ 273.07\pi\rho$

$m = 2\pi\rho\displaystyle\int_0^4 x(4x - x^2)dx = 2\pi\rho\left[\dfrac{4x^3}{3} - \dfrac{x^4}{4}\right]_0^4$

$= \dfrac{128\pi\rho}{3}$

$\approx 134.04\rho\ \text{or}\ 42.667\pi\rho$

$r_y = \sqrt{\dfrac{273.07}{42.667}} = 2.5298$

18. $I_x = 2\pi\rho\displaystyle\int_0^{16} y^3\left(y^{1/2} - \dfrac{1}{4}y\right)dy$

$= 2\pi\rho\displaystyle\int_0^{16} \left(y^{7/2} - \dfrac{1}{4}y^4\right)dy$

$= 2\pi\rho\left[\dfrac{2}{9}y^{9/2} - \dfrac{4y^5}{5}\right]_0^{16} \approx 11{,}650.84\pi\rho$

$m = 2\pi\rho\displaystyle\int_0^{16} y\left(y^{1/2} - \dfrac{1}{4}y\right)dy$

$= 2\pi\rho\displaystyle\int_0^{16} \left(y^{3/2} - \dfrac{1}{4}y^2\right)dy$

$$= 2\pi\rho\left[\frac{2}{5}y^{5/2} - \frac{1}{12}y^3\right]_0^{16} \approx 136.53\pi\rho$$

$$r_x \approx \sqrt{\frac{11{,}650.84}{136.53}} = 9.24$$

19. $I_x = 2\pi\rho\displaystyle\int_0^6 y^3 \cdot 4\,dy = 2\pi\rho y^4\,\big|_0^6 = 2592\pi\rho$

$m = 2\pi\rho\displaystyle\int_0^6 y \cdot 4\,dx = 2\pi\rho\left[2y^2\right]_0^6 = 144\pi\rho$

$r_x = \sqrt{\dfrac{2592}{144}} \approx 4.2426$

20. $I_y = 2\pi\rho\displaystyle\int_0^4 x^3 \cdot 6\,dx = 2\pi\rho\left[\frac{6x^4}{4}\right]_0^4 = 768\pi\rho$

$m = 2\pi\rho\displaystyle\int_0^4 x \cdot 6\,dx = 2\pi\rho\left[3x^2\right]_0^4 = 96\pi\rho$

$r_y = \sqrt{\dfrac{768}{96}} = \sqrt{8} \approx 2.8284$

21. $I_y = 2\pi \cdot 3\displaystyle\int_1^4 x^3\big(4x - x^2(4 - x)\big)\,dx$

$= 6\pi\displaystyle\int_1^4 x^3\big(5x - x^2 - 4\big)\,dx$

$= 6\pi\displaystyle\int_1^4 \big(5x^4 - x^5 - 4x^3\big)\,dx$

$= 6\pi\left[x^5 - \dfrac{x^6}{6} - x^4\right]_1^4$

$= 6\pi(85.5) = 513\pi\ \text{g}\cdot\text{cm}$

$m = 6\pi\displaystyle\int_1^4 x\big(5x - x^2 - 4\big)\,dx$

$= 6\pi\displaystyle\int_1^4 \big(5x^2 - x^3 - 4x\big)\,dx$

$= 6\pi\left[\dfrac{5x^3}{3} - \dfrac{x^3}{4} - 2x^2\right]_1^4 = 67.5\pi\ \text{g}$

$r_y = \sqrt{\dfrac{513}{67.5}} = 2.7568\ \text{cm}$

22. $y = \sqrt{4 - x} \Rightarrow y^2 = 4 - x \Rightarrow x = 4 - y^2$

$x + 2y = 4 \Rightarrow x = 4 - 2y$

$I_x = 2\pi \cdot 5\displaystyle\int_0^2 y^3\big(4 - y^2 - (4 - 2y)\big)\,dy$

$= 10\pi\displaystyle\int_0^2 y^3\big(2y - y^2\big)\,dy$

$= 10\pi\displaystyle\int_0^2 \big(2y^4 - y^5\big)\,dy$

$= 10\pi\left[\dfrac{2y^5}{5} - \dfrac{y^6}{6}\right]_0^2 = 21.333\pi\ \text{g}\cdot\text{cm}^2$

$m = 10\pi\displaystyle\int_0^2 y\big(2y - y^2\big)$

$= 10\pi\displaystyle\int_0^2 \big(2y^2 - y^3\big)\,dy$

$= 10\pi\left[\dfrac{2y^3}{3} - \dfrac{y^4}{4}\right]_0^2 = 13.333\ \text{g}$

$r_x = \sqrt{\dfrac{21.333}{13.333}} = \sqrt{1.6} = 1.2649\ \text{cm}$

23. $I_y = 2\pi \cdot 5\displaystyle\int_1^2 x^3\big(x^3 - x^{-1}\big)\,dx$

$= 10\pi\displaystyle\int_1^2 \big(x^6 - x^2\big)\,dx = 10\pi\left[\dfrac{x^7}{7} - \dfrac{x^3}{3}\right]_1^2$

$= 158.0952\pi\ \text{g}\cdot\text{cm}^2$

$m = 10\pi\displaystyle\int_1^2 x\big(x^3 - x^{-1}\big)\,dx$

$= 10\pi\displaystyle\int_1^2 \big(x^4 - 1\big)\,dx$

$= 10\pi\left[\dfrac{x^5}{5} - x\right]_1^2 = 52\pi\ \text{g}$

$r_y = \sqrt{\dfrac{158.0952}{52}} = 1.7436\ \text{cm}$

24. $I_x = 10\pi\displaystyle\int_{1/2}^1 y^3\big(2 - y^{-1}\big)\,dy$

$\qquad + 10\pi\displaystyle\int_1^8 y^3\big(2 - y^{1/3}\big)\,dy$

$= 10\pi\displaystyle\int_{1/2}^1 \big(2y^3 - y^2\big)\,dy$

$\qquad + 10\pi\displaystyle\int_1^8 \big(2y^3 - y^{10/3}\big)\,dy$

$= 10\pi\left[\dfrac{y^4}{2} - \dfrac{y^3}{3}\right]_{1/2}^1 + 10\pi\left[\dfrac{y^4}{2} - \dfrac{3}{13}y^{\frac{33}{3}}\right]_1^8$

$= 1574.4631\pi\ \text{g}\cdot\text{cm}^2$

$m = 10\pi\displaystyle\int_{1/2}^1 y\big(2 - y^{-1}\big)\,dy$

$\qquad + 10\pi\displaystyle\int_1^8 y\big(2 - y^{1/3}\big)\,dy$

$= 10\pi\left[y^2 - y\right]_{1/2}^1 + 10\pi\left[y^2 - \dfrac{3}{7}y^{7/3}\right]_1^8$

$= 88.2143\pi\ \text{g}$

$r_x = \sqrt{\dfrac{1574.4631}{88.2143}} = 4.2247\ \text{cm}$

26.7 WORK AND FLUID PRESSURE

1. $F = kx$; $3 = k(1/3)^1$ yields $k = 9$. Hence, $F = 9x$; 10 in. $= \frac{5}{6}$ ft. Work $= \int_0^{5/6} 9x \, dx = \frac{9}{2}x^2\big|_0^{5/6} = \frac{9}{2} \cdot \frac{25}{36} = \frac{25}{8} = 3.125$ ft·lb

2. $\int_1^2 9x \, dx = \frac{9}{2}x^2\big|_1^2 = \frac{9}{2}(4-1) = \frac{27}{2} = 13.5$ ft·lb.

3. 20 cm $= 0.2$ m; $6 = k(0.2)$; so $k = 30$ and we get $F = 30x$. Thus, work $= \int_0^1 30x = \frac{30x^2}{2}\big|_0^1 = 15$ J

4. $\int_{0.1}^{0.2} 30x = 15x^2\big|_{0.1}^{0.2} = 15(0.04 - 0.01) = 0.45$ J

5. $\int_0^{0.1} kx \, dx = \frac{k}{2}x^2\big|_0^{0.1} = \frac{k(0.01)}{2} = 0.20$. Hence, $k = \frac{0.4}{0.01} = 40$; $F = kx = 40 \cdot (0.1) = 4$N

6. $F = kx$; $5 = k\left(\frac{1}{6}\right)$, so $k = 30$. Work $= \int_{1/12}^{4/12} 30x \, dx = 15x^2\big|_{1/12}^{1/3} = 15\left(\frac{1}{9} - \frac{1}{144}\right) = 15\left(\frac{15}{144}\right) = \frac{225}{144}$ ft·lb or 1.5625 ft·lb

7. $\int_0^b 9x \, dx = \frac{9}{2}x^2\big|_0^b = \frac{9}{2}b^2$; $\frac{9}{2}b^2 = 105$; $b^2 = 23.333$; $b = 4.8305$ ft

8. The weight of a section is $2\,dy$; distance moved is $50-y$. Work $= \int_0^{20} 2(50-y)dy = 2\big[50y - \frac{y^2}{2}\big]_0^{20} = 2[1000 - 200] = 2(800) = 1600$ ft·lb.

9. $\int_0^{30} 2.5(30-y)dy = 2.5\big[30y - \frac{y^2}{2}\big]_0^{30} = 2.5(450) = 1125$ J

10. The mass $= 4.7$ kg/m; weight $= 4.7 \times 9.8 = 46.06$ N. A mass of 87 kg has a weight of $87 \times 9.8 = 852.6$ N. The work in lifting the chain is $\int_0^{10} 46.06(15-y)dy = 46.06\big[15 - \frac{y^2}{2}\big]_0^{10} = 4606$ J. The work in lifting the load is $852.6 \times 10 = 8526$ J. The total work is $4606 + 8526 = 13,132$ J or about 13 kJ.

11. For the first 12 feet; each section will have a weight of $0.75\,dy$. $\int_0^{12} 0.75(12-y)dy = 0.75\big(12y - \frac{y^2}{2}\big)_0^{12} = \frac{3}{4}(72) = 54$ ft·lb. The total weight of the chain is $12 \cdot \frac{3}{4} = 9$ lb. The work in lifting the last 8 feet is 72 ft·lb. The total work is $54 + 72 = 126$ ft·lb.

12. For the first 5 meters; each section has a weight of $0.4 \times 0.8\Delta y = 3.92\Delta y$ N. The work in raising one end 5 meters is $\int_0^5 3.92(5-y)dy = 3.92\big(5y - \frac{y^2}{2}\big)\big|_0^5 = 3.92 \times 12.5 = 49$ N·m $= 49$ J. The total weight of the chain is $5(3.92) = 19.6$ N. The work

in lifting the chain the last 5 meters is $19.6 \times 5 = 98$ J. The total work done is $49 + 98 = 147$ J.

13. When the tank is x meters above the ground it has taken $2x$ minutes to get there and has used $20 \cdot 2x = 40x$ L. Hence the volume of the tank is $(1000 - 40x)$ L. Water has a density of 1 kg/L so its mass is $(1000 - 40x)$kg. The total mass is $(1047 - 40x)$ kg so its weight is $9.8(1047 - 40x)$. $W = \int_0^{20} 9.8(1047 - 40x)dx = 9.8\big[1047x - 20x^2\big]_0^{20} = 9.8(12940) = 126,812$ J or 127 kJ.

14. 1 cm $= 0.01$ m; 3 mm $= 0.003$ m
$$W = \int_{0.003}^{0.01} 8.988 \times 10^9 \cdot 5 \times 10^{-7}$$
$$\times (-2) \times 10^{-7} \cdot \frac{1}{r^2}dr$$
$$= -8.988 \times 10^{-4}\left[\frac{-1}{r}\right]_{0.003}^{0.01}$$
$$= -8.988 \times 10^{-4}(233.33)$$
$$= -2.0972 \times 10^{-1} = -0.20972 \text{ J}$$

15. $$W = \int_{0.05}^{10 \times 10^{-9}} 8.988 \times 4 \times 10^{-19} \times 5$$
$$\times 10^{-8}\frac{1}{r^2}dr$$
$$= 1.7976 \times 10^{-16}\left[\frac{-1}{r}\right]_{0.05}^{10^{-8}}$$
$$= 1.7976 \times 10^{-16} \times 10^8$$
$$= 1.7976 \times 10^{-8} \text{ J} = 179.76 \ \mu\text{J}$$

16. $W = \int_{5.3 \times 10^{-7}}^{10^{-3}} 8.988 \times 1.6 \times 10^{-18} \times (-1.6) \times 10^{-19} \cdot \frac{1}{r^2}dr = 2.3009 \times 10^{-27}\big[\frac{-1}{r}\big]_{5.3 \times 10^{-7}}^{10^{-3}} = -2.3009 \times 10^{-27}(1.8858 \times 10^6) = -4.3391 \times 10^{-21}$ J

17. $W = \int F(r)dr = \int_{0.01}^1 \frac{k}{r^2} = -k\big[\frac{1}{r}\big]_{0.01}^1 = -k[1 - 100] = 99k$ J.

18. $W = \int_1^{20} \frac{k}{r^2}dr = \frac{-k}{r}\big|_1^{20} = -k\big[\frac{1}{20} - 1\big] = \frac{19k}{20}$ ft·lb.

19. $W = \int_{4000}^{5000} \frac{k}{r^2} = -k\big[\frac{1}{r}\big]_{4000}^{5000} = k\big[5 \times 10^{-5}\big]$. Since $k = 16 \times 10^9$, we get $W = 16 \times 10^9 \cdot 5 \times 10^{-5} = 80,000$ mi·lb $= 4.224 \times 10^9$ ft·lb.

20. $\int_0^8 (54.8)(16\pi)x \, dx = 876.8\pi\frac{x^2}{2}\big|_0^8 = 28057.6\pi \approx 88145.6$ ft·lb.

21. $\int_0^8 880(4\pi)x \, dx = 1.264\pi \times 10^5$ kg·m. Multiplying by the gravitational constant to get joules, we have $1.264\pi \times 10^5 \times 9.8 = 1.23872\pi \times 10^6$ J $= 3.89155 \times 10^6$ J

22. $\int_0^2 70 \cdot 9.8 \cdot 10^3 \cdot x \, dx = 6.86 \times 10^5 \frac{x^2}{2} \big|_0^2 = 1.372 \times 10^6 \text{J} = 1.372$ MJ. (Note: 1 m^3 of water has a mass of 10^3 kg and a weight of 9.8×10^3 N.)

23. $\int_0^{1.5} 6.86 \times 10^5 (2-x) dx = 6.86 \times 10^5 \left[2x - \frac{x^2}{2}\right]_0^{1.5} = 6.86 \times 10^5 \times 1875 = 1.28625$ MJ $= 1{,}286{,}250$ J

24. $\int_0^{10} \rho(9\pi)(x) dx + 9\pi\rho 10 \int_0^5 1 dx = 9\pi\rho\frac{x^2}{2}\big|_0^{10} + 450\pi\rho = 450\pi\rho + 450\pi\rho = 900\pi\rho = 900 \cdot 54.8\pi = 49{,}320\pi = 1.549 \times 10^5$ ft · lb.

25. Position the tank so that the bottom is at (0, 0). Then the radius will be $\frac{2}{3}y$ where y is the height. The volume will be $\pi\left(\frac{2}{3}y\right)^2 dy$; $\rho = 847$ kg/m^3 = 8300.6 N/m^3. $W = 8300.6\pi \int_0^{3.6} \left(\frac{2}{3}y\right)^2 (18.6 - y) dy = 8300.6\pi \cdot \left[\frac{4}{9}(18.6)\frac{y^3}{3} - \frac{4}{9} \cdot \frac{y^4}{4}\right]_0^{3.6} = 8300.6\pi[109.9] = 912\,236\pi$ J $= 2\,866\,000$ J $= 2.866$ MJ

26. $r = \sqrt{16 - y^2} = r^2 = (16 - y^2)$; distance raised is $(-y) W = \int_{-4}^0 (1000 \cdot 9.8)\pi(16 - y^2)(-y) dy = 9800\pi\left[-8y^2 + \frac{y^4}{4}\right]_{-4}^0 = 9800\pi[64] = 627200\pi$ J.

27. $W = 9800\pi \int_{-2}^0 (16 - y^2)(-y) dy = 9800\pi\left[-8y^2 + \frac{y^4}{4}\right]_{-2}^0 = 9800\pi[32 - 4] = 9800\pi[28] = 274400\pi$ J

28. Using the same set-up as Example 26.40 we get $9800\pi \int_1^4 (16 - x^2)(14 - x) dx$ where x is the distance from the top of the tank. This is equal to $9800\pi \int_1^4 (224 - 16x - 14x^2 + x^3) dx = 9800\pi\left[224x - 8x^2 - \frac{14x^3}{3} + \frac{x^4}{4}\right]_1^4 = 9800\pi(321.75) = 3\,153\,150\pi$ J.

29. (a) $P = \rho gh = 1000 \cdot 9.8 \cdot 2 = 19{,}600$ N/m^2
(b) $F = PA = 19{,}600 \times 10 \times 7 = 1372000$ N

30. (a) $P = \rho gh = 54.8 \cdot 10 = 548$ lb/ft^2;
(b) $F = PA = 548 \cdot 9\pi = 4932\pi = 15494$ lb.

31. The area of the short side is 7 m $\times$ 2m. So, we get $= 9800 \int_0^2 h(y) \cdot L(y) dy - 9800 \int_0^2 (2 - y) \cdot 7 dy = 9800\left[14y - \frac{7y^2}{2}\right]_0^0 = 9800 \cdot 14 = 137200$ N.

32. Long side $9800 \int_0^2 (2-y) \cdot 10 = 9800\left[20y - 5y^2\right]_0^2 = 9800 \cdot 20 = 196\,000$ N

33. $9800 \int_0^2 (2 - y)\left(3 - \frac{3}{2}y\right) dy = 9800 \int_0^2 \left(6 - 6y + \frac{3}{2}y^2\right) dy = 9800\left[6y - 3y^2 + \frac{y^3}{2}\right]_0^2 = 9800[4] = 39\,200$ N

34. $\rho g \int_0^1 h(40 dh + \rho g \int_0^2 (1 + h)(40 - 20h) dh$
$= 9800 \int_0^1 40h + 9800 \int_0^2 \left(40 + 20h - 20h^2\right) dh$
$= 9800\left[20h^2\right]_0^1 + 9800\left[40h + 10h^2 - \frac{20h^3}{3}\right]_0^2$
$= 9800[20] + 9800[66.67] = 9800[86.67]$
$= 849300$ N $= 849$ kN

35. $r = 1.6$ m; $\rho = 680$ kg/m^3; $\rho g = 680 \times 9.8 = 6664$ N/m^3. Thus, we have $\rho g \int_{-1.6}^0 (1 \cdot 6 + y)(2\sqrt{2.56 - y^2}) dy = \rho g \cdot 3.2 \int_{-1.6}^0 \sqrt{2.56 - y^2} - \rho g \int_{-1.6}^0 -2y\sqrt{2.56 - y^2} dy$. As in example 26.43. The left integral will be $\frac{1}{4}$ the area of a circle so $\frac{1}{4} \cdot \pi(1.6)^2 = 2.0106$. The right integral is $\frac{2}{3}(2.56 - y^2)^{3/2} \big|_{-1.6}^0 = 4.096 \cdot \frac{2}{3} = 2.7307$; $\rho g[3.2 \times 2.0106 - 2.7307] = \rho g[3.7032] = 6664(3.7032) = 24\,578$ N.

36. $x^2 + y^2 = 16$. $2\rho \int_{-4}^0 (4 + y)\sqrt{16 - y^2} dy = 2\rho \int_{-4}^0 \sqrt{16 - y^2} - 2\rho \int_{-4}^0 -y\sqrt{16 - y^2} dy = 8\rho \int_{-4}^0 \sqrt{16 - y^2} dy - \rho \int_{-4}^0 -2y\sqrt{16y^2} d6$. The left integral is $\frac{1}{4}$ the area of a circle of radius 4 so 4π. The right integral is $\frac{2}{3}(16 - y^2)^{3/2} \big|_{-4}^0 = \frac{128}{3}$. Putting this together $8\rho \cdot 4\pi - \rho \cdot \frac{128}{3} = 42.5(57.86) = 2{,}459.2$ lb.

CHAPTER 26 REVIEW

1. $\bar{y} = \frac{1}{5} \int_1^6 4x \, dx = \frac{1}{5} 2x^2 \big|_1^6 = \frac{1}{5}(72 - 2)$
$= \frac{20}{5} = 14$
$f_{\text{rms}} = \sqrt{\frac{1}{5} \int_1^6 16x \, dx} = \sqrt{\frac{1}{5}\left[\frac{16x^3}{3}\right]_1^6}$
$= \sqrt{229.\overline{3}} = 15.1438$

2. $\bar{y} = \frac{1}{2} \int_0^2 x^3 \, dx = \frac{1}{2}\frac{x^4}{4}\big|_0^2 = \frac{16}{8} = 2$
$y_{\text{rms}} = \sqrt{\frac{1}{2} \int_0^2 x^6 \, dx} = \sqrt{\frac{1}{2}\frac{x^7}{7}}\big|_0^7$
$= \sqrt{9.142857} = 3.0237$

3.
$$\bar{y} = \frac{1}{2}\int_2^4 (x^2 - 4)\,dx = \frac{1}{2}\left[\frac{x^3}{3} - 4x\right]_2^4$$
$$= \frac{1}{2} \cdot \frac{32}{3} = \frac{16}{3}$$

$$h_{\text{rms}} = \sqrt{\frac{1}{2}\int_2^4 (x^4 - 8x^2 + 16)\,dx}$$
$$= \sqrt{\frac{1}{2}\left[\frac{x^5}{5} - \frac{8x^3}{3} + 16x\right]_2^4}$$
$$= \sqrt{40.5\overline{3}} = 6.3666$$

4.
$$\bar{y} = \frac{1}{2.5}\int_0^{2.5} (4.9t^2 - 2.8t - 4)\,dt = \frac{6.7708\overline{3}}{2.5}$$
$$= 2.7083$$

$$k_{\text{rms}} = \sqrt{\frac{1}{2.5}\int_0^{2.5} (4.9t^2 - 2.8t - 4)^2\,dt}$$
$$= \sqrt{\frac{147.6432}{2.5}}$$
$$= \sqrt{59.0573} = 7.6849$$

5.
$$V = \pi \int_1^5 (6x)^2\,dx = 36\pi \cdot \frac{x^3}{3}\bigg|_1^5$$
$$= 12\pi[125 - 1] = 1488\pi$$

$$M_y = \pi \int_1^5 x(6x)^2 = 36\pi \int_1^5 x^3\,dx = 36\pi\left[\frac{x^4}{4}\right]_1^5$$
$$= 9\pi[625 - 1] = 5616\pi$$

$$\bar{x} = \frac{5676}{1488} = 3.7742$$
The centroid is (3.7742, 0).

6. $y = -x \Rightarrow x = -y$. Hence,
$$V = \pi \int_1^4 (-y)^2\,dy = \pi\frac{y^3}{3}\bigg|_1^4 = \frac{\pi}{3}(64 - 1) = 2\pi$$
$$M_x = \pi \int_1^4 \pi y^3\,dy = \pi\frac{y^4}{4}\bigg|_1^4 = \pi\left(64 - \frac{1}{4}\right)$$
$$= 63.75\pi$$
$$\bar{y} = \frac{63.75}{21} = 3.0357$$
The centroid is (0, 3.0357).

7.
$$V = \pi \int_1^2 (x^4)^2\,dx = \pi\frac{x^9}{9}\bigg|_1^2 = \frac{\pi}{9}(512 - 1)$$
$$= \frac{511\pi}{9}$$

$$M_y = \pi \int_1^2 x^9\,dx = \pi\frac{x^{10}}{10}\bigg|_1^2 = \frac{\pi}{10}(1024 - 1)$$
$$= \frac{1023\pi}{10}$$

$$\bar{x} = \frac{1023}{10} \times \frac{9}{511} = 1.8018$$
The centroid is (1.8018, 0).

8.
$$V = \pi \int_0^1 (x^{2/3})^2\,dx = \pi\frac{3}{7}x^{7/3}\big|_0^1 = \frac{3\pi}{7}$$
$$M_y = \pi \int_0^1 x(x^{4/3})\,dx = \pi\frac{3}{10}x^{10/3} = \frac{3\pi}{10}$$
$$\bar{x} = \frac{3}{10} \cdot \frac{7}{3} = \frac{7}{10} \text{ or } 0.7$$
The centroid is (0.7, 0).

9.
$$V = \pi \int_0^1 (y^{2/3} - y^6)\,dy = \pi\left[\frac{3}{5}y^{5/3} - \frac{y^7}{7}\right]_0^1$$
$$= \frac{16\pi}{35} \approx 0.4571\pi$$

$$M_x = \pi \int_0^1 (y^{5/3} - y^7)\,dy = \pi\left[\frac{3}{8}y^{8/3} - \frac{y^8}{3}\right]_0^1$$
$$= \frac{\pi}{4}$$

$$\bar{y} = \frac{1}{4} \cdot \frac{35}{16} = \frac{35}{64} \approx 0.5469$$
The centroid is (0, 0.5469).

10.
$$V = \pi \int_0^1 (x^{2/3} - x^6)\,dx = \frac{16\pi}{35}$$

$$M_y = \pi \int_0^1 x(x^{2/3} - x^6) = \frac{\pi}{4}$$

$$\bar{x} = \frac{1}{4} \cdot \frac{35}{16} = \frac{35}{64} \approx 0.5469$$
The centroid is (0.5469, 0).

11. Here $y = x^2 \Rightarrow x = y^{1/2}$, $y = 9 - x^2 \Rightarrow x = \sqrt{9 - y}$, and $x^2 = 9 - x^2 \Rightarrow 2x^2 = 9$. Hence $x = \pm\frac{3}{\sqrt{2}}$ and $y = \frac{9}{2} = 4.5$.

$$V = \pi \int_0^{4.5} y\,dy + \pi \int_{4.5}^9 (9 - y)\,dy$$
$$= \pi\frac{y^2}{2}\big|_0^{4.5} + \pi\left[9y - \frac{y^2}{2}\right]_{4.5}^9 = 20.25\pi$$
$$M_x = \pi \int_0^{4.5} y^2\,dy + \pi \int_{4.5}^9 (9y - y^2)\,dy$$
$$= \pi\left(\left[\frac{y^3}{3}\right]_0^{4.5} + \left[\frac{9y^2}{2} - \frac{y^3}{3}\right]_{4.5}^9\right) = 91.125\pi$$

$$\bar{y} = \frac{91.125}{20.25} = 4.5$$
The centroid is (0, 4.5). (Note: this is the answer you would expect due to the symmetry of the curves.)

12. Curves intersect at (0, 0) and (2, 4). Here $x = y^{1/3}$ and $x = \frac{y}{4}$.

$$V = \pi \int_0^4 \left(5 - \frac{y}{4}\right)^2 - \left(5 - y^{1/3}\right)^2 dy$$

$$= \pi \int_0^4 \left[\left(25 - \frac{5}{2}y + \frac{y^2}{16}\right)\right.$$

$$\left. - \left(25 - 10y^{1/3} + y^{2/3}\right)\right] dy$$

$$= \pi \int_0^4 \left(-\frac{5}{2}y + \frac{y^2}{16} + 10y^{1/3} - y^{2/3}\right) dy$$

$$= \pi \left[\frac{-5y^2}{4} + \frac{y^3}{48} + \frac{15}{2}y^{4/3} - \frac{3}{5}y^{5/3}\right]_0^4$$

$$= \pi[22.9077]$$

$$M_x = \pi \int_0^4 \left(-\frac{5}{2}y^2 + \frac{y^3}{16} + 10y^{4/3} - y^{5/3}\right) dy$$

$$= \pi \left[-\frac{5y^3}{6} + \frac{y^3}{64} + \frac{30}{7}y^{7/3} - \frac{3}{8}y^{8/3}\right]_0^4$$

$$= 44.3980\pi$$

$$\overline{y} = \frac{44.3980}{22.9077} = 1.93812$$

Since it's rotated about $x = 5$, then $\overline{x} = 5$. The centroid is $(5, 1.0169)$.

13. $x = y^{3/2}$; $x' = \frac{3}{2}y^{1/2}$; $x'^2 = \frac{9}{4}y$

$$L = \int_0^4 \sqrt{1 + x'^2}\, dy = \int_0^4 \sqrt{1 + \frac{9}{4}y}\, dy$$

$$= \frac{4}{9} \int_0^4 \frac{9}{4} \sqrt{1 + \frac{9}{4}y}\, dy$$

$$= \frac{4}{9} \cdot \frac{2}{3} \left(1 + \frac{9}{4}y\right)^{3/2} \Big|_0^4 = \frac{8}{27}\left(10^{3/2} - 1\right)$$

$$= 9.0734$$

14. $y = \left(8x^3\right)^{1/2}$;

$$y' = \frac{1}{2}\left(8x^3\right)^{-1/2} \cdot 24x^2 = \frac{24x^{1/2}}{4\sqrt{2}} \cdot 24x^2$$

$$= \frac{6x^{3/2}}{\sqrt{2}} = 3\sqrt{2}x^{3/2}$$

$$L = \int_0^2 \sqrt{1 + \left(3\sqrt{2}x^{1/2}\right)^2}\, dx$$

$$= \int_0^2 \sqrt{1 + 18x}\, dx$$

$$= \frac{1}{18} \cdot \frac{2}{3}(1 + 18x)^{3/2} \Big|_0^2 = 8.2986$$

15. (a) Here $y = \frac{x^4}{4} + \frac{1}{8x^2}$ and $y' = x^3 - \frac{1}{4x^3}$.

$$L = \int_1^3 \sqrt{1 + \left(x^3 - \frac{1}{4x^3}\right)^2}\, dx$$

$$= \int_1^3 \sqrt{1 + \left(x^6 - \frac{1}{2} + \frac{1}{16x^6}\right)}\, dx$$

$$= \int_1^3 \sqrt{x^6 + \frac{1}{2} + \frac{1}{16x^6}}\, dx$$

$$= \int_1^3 \left(x^3 + \frac{1}{4x^3}\right) dx$$

$$= x^4 - \frac{1}{4} \cdot \frac{1}{2}x^{-2} \Big|_1^3 = \frac{81}{4}$$

$$- \frac{1}{72} - \frac{1}{4} + \frac{1}{8} = 20.111$$

(b) $S = 2\pi \int_1^3 \left(\frac{x^4}{4} + \frac{1}{8x^2}\right)$

$$\times \sqrt{1 + \left(x^6 - \frac{1}{2} + \frac{1}{16x^2}\right)}\, dx$$

$$= 2\pi \int_1^3 \left(\frac{x^4}{4} + \frac{1}{8x^3}\right)\left(x^3 + \frac{1}{4x^3}\right) dx$$

$$= 2\pi \int_1^3 \left(\frac{x^7}{4} + \frac{x}{6} + \frac{1}{8} + \frac{1}{32x^6}\right) dx$$

$$= 2\pi \left[\frac{x^8}{32} + \frac{x^2}{32} + \frac{x}{8} - \frac{1}{32}\frac{1}{5}x^{-5}\right]_1^3$$

$$= 2\pi[205.5062] = 411.0124\pi$$

16. (a) $x = \frac{1}{6}y^6 + \frac{1}{16}y^{-4}$; $x' = y^5 - \frac{1}{4}y^{-5}$; $x'^2 =$
$$\left(y^{10} - \frac{1}{2} + \frac{1}{16}y^{-10}\right)$$

$$L = \int_1^2 \sqrt{1 + \left(y^{10} - \frac{1}{2} + \frac{1}{16}y^{-10}\right)}\, dy$$

$$= \int_1^2 \left(y^5 + \frac{1}{4}y^{-5}\right) dy$$

$$= \left(\frac{y^6}{6} - \frac{1}{16}y^{-4}\right)\Big|_1^2 = 10.5586$$

$$S = 2\pi \int_1^2 \left(\frac{y^6}{6} + \frac{1}{16y^4}\right)\left(y^5 + \frac{1}{4y^5}\right) dy$$

$$= 2\pi \int_1^2 \left(\frac{y^{11}}{6} + \frac{y}{24} + \frac{y}{16} + \frac{1}{64y^9}\right) dy$$

$$= 2\pi \left[\frac{y^{12}}{72} + \frac{y^2}{48} + \frac{y^2}{32} - \frac{1}{512y^8}\right]_1^2$$

$$= 2\pi[57.08332] = 114.0664\pi$$

17. $F = kx$; $5 = k(0.2)$; $k = 25$; $W = \int_0^{0.6} 25x\, dx =$
$\frac{25x^2}{2} \Big|_0^{0.6} = 4.5 \text{ N} \cdot \text{m} = 4.5 \text{ J}$

18. $V = \frac{1}{60 \times 10^{-6}} \int_0^{0.001} 0.40\, dt = \frac{1}{60 \times 10^{-6}}0.40t \Big|_0^{0.001} =$
$\frac{1}{60 \times 10^{-6}}(0.004) = 6.6667 \text{ V}$

19. $I_{avg} = \frac{1}{2} \int_0^2 \left(4t - t^3\right) dt = \frac{1}{2}\left[2t^2 - \frac{t^4}{4}\right]_0^2 = \frac{1}{2}[8 - 4] = 2 \text{ A}$

20. $a(t) = -32$; $v(t) = -32t$; $s(t) = -16t^2 + 555$;
$s(t) = 0 \Rightarrow 16t^2 = 555 \Rightarrow t = 5.8896$.

$$\bar{s} = \frac{1}{5.8896} \int_0^{5.8896} (-16t^2 + 555)dt$$

$$= \frac{1}{5.8896} \left[\frac{-16t^3}{3} + 555t \right]_0^{5.8896}$$

$$\bar{s} = 370 \text{ ft}$$

$$\bar{v} = \frac{1}{5.8896} \int_0^{5.8896} (-32t)dt$$

$$= \frac{1}{5.8896} \left[-16t^2 \right]_0^{5.8896} = -94.2338 \text{ft/s}$$

21. $W = \int_0^{30} [1000 + 5(50 - x)]dx$

$$= \int_0^{30} (1250 - 5x)dx$$

$$= \left[1250x - \frac{5x^2}{2} \right]_0^{30} = 35250 \text{ ft} \cdot \text{lb}$$

22. $h(y) = (100 - y);\ L(y) = 200 + 2y$

$P = 62.4 \int_0^{100} (100 - y)(200 + 2y)dy$

$$= 62.4 \int_0^{100} (20000 - 2y^2)dy$$

$$= 64.2 \left[20000y - \frac{2}{3}y^3 \right]_0^{100} = 62.4(1.333 \times 10^6)$$

$$= 8.32 \times 10^7 \text{ lb} = 4.16 \times 10^4 \text{ tons}$$

23. The top of the tank is at $x = 0$ and the bottom at
$x = 16$ $W = 880 \int_0^{16} 36\pi \cdot x\, dx = 31680\pi \frac{x^2}{2} \big|_0^{16} =$
$4\ 055\ 040\pi$ kg $\cdot$ m $= 39\ 739\ 392\pi$ J

24. $62.4\pi \int_0^{16} \left(6 - \frac{3x}{8}\right)^2 x\, dx$

$$= 62.4\pi \int_0^{10} \left(36 - \frac{9}{2}x + \frac{9}{64}x^2\right) x\, dx$$

$$= 62.4\pi \int_0^{10} \left(36x - \frac{9}{2}x^2 + \frac{9}{64}x^3\right) dx$$

$$= 62.4\pi \left[18x^2 - \frac{3}{2}x^3 + \frac{9}{256}x^4 \right]_0^{16}$$

$$= 62.4\pi(768) = 4.79232 \times 10^4 \pi \text{ ft} \cdot \text{lb}$$

CHAPTER 26 TEST

1. $\bar{y} = \frac{1}{5-1} \int_1^5 (3x^2 + 1)dx = \frac{1}{4}(x^3 + x)\big|_1^5 = \frac{1}{4}(130 - 2) = 32$

2. $g_{rms} = \sqrt{\frac{1}{3-1} \int_1^3 (x^3 - 1)^2 dx} =$

$\sqrt{\frac{1}{2} \int_1^3 (x^6 - 2x^3 + 1)dx} = \sqrt{\frac{1}{2} \left[\frac{1}{7}x^7 - \frac{1}{2}x^4 + x\right]_1^3}$
$\approx \sqrt{137.1429} \approx 11.71$

3. $V = \pi \int_0^7 [f(x)]^2 dx = \pi \int_0^7 \sqrt{9x+1}^2 dx =$

$\pi \int_0^7 (9x + 1)dx = \pi \left[\frac{9}{2}x^2 + x\right]_0^7 = 227.5\pi$

4. We begin by determining the moment: $M_y = \pi \int_0^4 \rho x (x^2 + 1)^2 dx = \rho\pi \int_0^4 (x^5 + 2x^3 + x)dx = \rho\pi \left[\frac{1}{6}x^6 + \frac{1}{2}x^4 + \frac{1}{2}x^2\right]_0^4 = \frac{2456}{3}\rho\pi$. Now we find the mass, $m = \rho V$, from $V == \pi \int_0^4 (x^2 + 1)^2 dx = \pi \int_0^4 (x^4 + 2x^2 + 1)dx = \pi \left[\frac{1}{5}x^5 + \frac{2}{3}x^3 + x\right]_0^4 = \frac{7544}{30}\pi$. so, $m = \frac{7544}{30}\rho\pi$ and $\bar{x} = \frac{M_y}{m} = \left(\frac{2456}{3}\rho\pi\right)/\left(\frac{7544}{30}\rho\pi\right) = \frac{24560}{7544} \approx 3.26$. Thus, the centroid is at $(3.26, 0)$

5. Since $y' = \frac{3}{2}x^{1/2}$, the arc length

$L = \int_1^6 \sqrt{1 + \left(\frac{3}{2}x^{1/2}\right)^2}\, dx = \int_1^6 \sqrt{1 + \frac{9}{4}x}\, dx = \frac{1}{2} \int_1^6 \sqrt{4 + 9x}\, dx$. Let $u = 4 + 9x$ and then $du = 9dx$, and if $x = 1$, $u = 13$ and if $x = 6$, $u = 58$, so $L = \frac{1}{18} \int_{13}^{58} u^{1/2} du = \frac{1}{27} u^{3/2}\big]_{13}^{58} \approx 14.6238$.

6. Here $y' = \frac{1}{2}x^2$ and so, the surface area $S =$

$\int_1^3 2\pi\left(\frac{1}{6}x^3\right)\sqrt{1 + \left(\frac{1}{2}x^2\right)^2}\, dx =$

$\frac{\pi}{3} \int_1^3 x^3 \sqrt{1 + \frac{1}{4}x^4}\, dx = \frac{4\pi}{9}\left(1 + \frac{1}{4}x^4\right)^{3/2}\big|_1^3 \approx 42.9156\pi \approx 134.82$

7. $W = \int_0^{10} 25\pi\rho(y + 4)dy = 25\pi\rho \int_0^{10} (y + 4)dy = 25\pi\rho\left[\frac{1}{2}y^2 + 4y\right]_0^{10} = 25\pi\rho(50 + 40) = 2250\pi\rho = 2250\ 000\pi$ J

8. $F = \rho \int_5^8 y2(8 - y)dy = \rho \int_5^8 (16y - 2y^2)dy = \rho\left[8y^2 - \frac{2}{3}y^3\right]_5^8 \approx 54\rho$. Since the density of water is $\rho = 62.5$ lb, we have a force of $F = 54(62.5) = 3375$.

27

Derivatives of Transcendental Functions

≡ 27.1 DERIVATIVES OF THE SINE AND COSINE FUNCTIONS

1. $y = \sin 3x;\ y' = 3\cos 3x$

2. $y = \cos 4x;\ y' = -4\sin 4x$

3. $y = 3\cos 2x;\ y' = 3(-\sin 2x)\cdot 2 = -6\sin 2x$

4. $y = 5\sin 6x;\ y' = 5(\cos 6x)\cdot 6 = 30\cos 6x$

5. $y = \sin\left(x^2 + 1\right);\ y' = 2x\cos\left(x^2 + 1\right)$

6. $y = \cos\left(x^3 - 5\right);\ y' = -\sin\left(x^3 - 5\right)\cdot 3x^2 = -3x^2\sin\left(x^3 - 5\right)$

7. $y = 4\sin^2 3x = 4(\sin 3x)^2;\ y' = 4\cdot 2(\sin 3x)\times (\cos 3x)(3) = 24\sin 3x\cos 3x$

8. $y = 5\cos^2 2x = 5(\cos 2x)^3;\ y' = 5\cdot 3(\cos 2x)^2\times(-\sin 2x)(2) = -30\cos^2 2x\sin 2x$

9. $y = \cos\left(3x^2 - 2\right);\ y' = -6x\sin\left(3x^2 - 2\right)$

10. $y = \sin\left(2x^3 + 1\right);\ y' = 6x^2\cos\left(2x^3 + 1\right)$

11. $y = \sin\sqrt{x} = \sin x^{1/2};\ y' = \cos x^{1/2}\cdot\frac{1}{2}x^{-1/2} = \frac{\cos\sqrt{x}}{2\sqrt{x}}$

12. $y = \cos x^{3/2};\ y' = -\sin x^{3/2}\cdot\frac{3}{2}x^{1/2} = -\frac{3}{2}x\sin x^{3/2}$

13. $y = \cos\sqrt{2x^3 - 4};\ y' = -\sin\sqrt{2x^3 - 4}\times\left(\frac{1}{2}\left(2x^3 - 4\right)^{-1/2}\left(6x^2\right)\right) = \frac{-3x^2\sin\sqrt{2x^3 - 4}}{\sqrt{2x^3 - 4}}$

14. $y = \sin^2\sqrt{x + 1};\ y' = 2\sin^2\cos\sqrt{x + 1}\cdot\frac{1}{2}(x + 1)^{-1/2} = \frac{\sin\sqrt{x+1}\cos\sqrt{x+1}}{\sqrt{x+1}}$

15. $y = x^2 + \sin^2 x;\ y' = 2x + 2\sin x\cos x$

16. $y = \cos x + \sin x;\ y' = -\sin x + \cos x = \cos x - \sin x$

17. $y = \sin x\cos x;\ y' = \sin x(-\sin x) + \cos x\cos x = \cos^2 x - \sin^2 x = \cos 2x$

18. $y = (\sin x - \cos x)^2;\ y' = 2(\sin x - \cos x)(\cos x + \sin x) = 2\left(\sin^2 x - \cos^2 x\right) = -2\cos 2x$

19. $y = \dfrac{2\cos x}{\sin 2x};$

$y' = \dfrac{\sin 2x(-2\sin x) - 2\cos x\cos 2x\cdot 2}{\sin^2 2x}$

$= \dfrac{-2\sin x\sin 2x - 4\cos x\cos 2x}{\sin^2 2x}$

$= \dfrac{-2\sin x(2\sin x\cos x) - 4\cos x\left(2\cos^2 x - 1\right)}{(2\sin x\cos x)^2}$

$= \dfrac{-4\cos x\left(\sin^2 x + 2\cos^2 x - 1\right)}{4\sin^2 x\cos^2 x}$

$= \dfrac{-4\cos x\left(\sin^2 x + \cos^2 x + \cos^2 x - 1\right)}{4\sin^2 x\cos^2 x}$

$= \dfrac{-4\cos x\left(\cos^2 x\right)}{4\sin^2 x\cos^2 x} = \dfrac{-\cos x}{\sin^2 x}$

20. $y = \frac{\sin 2x}{2x};\ y' = \frac{2x\cos 2x\cdot 2 - \sin 2x\cdot 2}{4x^2} = \frac{4x\cos 2x - 2\sin 2x}{4x^2}$

$= \frac{2x\cos 2x - \sin 2x}{2x^2}$

21. $y = x^2\sin x;\ y' = 2x\sin x + x^2\cos x$

22. $y = x^3\cos x;\ y' = 3x^2\cos x - x^3\sin x$

23. $y = \sqrt{x}\sin x;\ y' = \frac{1}{2}x^{-1/2}\sin x + \sqrt{x}\cos x$

24. $y = \frac{\sin^2 x}{x};\ y' = \frac{3x\sin^2 x\cos x - \sin^3 x}{x^2} = \frac{\sin^2 x(3x\cos x - \sin x)}{x^2}$

25. $y = (\sin 2x)(\cos 3x);\ y' = \sin 2x(-3\sin 3x) + 2\cos 2x\cos 3x = 2\cos 2x\cos 3x - 3\sin 2x\sin 3x$

26. $y = x^2\cos\left(3x^2 - 1\right);\ y' = 2x\cos\left(3x^2 - 1\right) + x^2\left(-\sin\left(3x^2 - 1\right)\right)(6x) = 2x\cos\left(3x^2 - 1\right) - 6x^3\sin\left(3x^2 - 1\right)$

27. $y = \sin^3\left(x^4\right);\ y' = 3\sin^2\left(x^4\right)\cos\left(x^4\right)\cdot\left(4x^3\right) = 12x^3\sin^2\left(x^4\right)\cos\left(x^4\right)$

28. $y = \sqrt{\cos\sqrt{x}} = (\cos\sqrt{x})^{\frac{1}{2}};\ y' = \frac{1}{2}(\cos\sqrt{x})^{-1/2}\times(-\sin\sqrt{x})\left(\frac{1}{2}x^{-1/2}\right) = \frac{-\sin\sqrt{x}}{4\sqrt{x}\sqrt{\cos\sqrt{x}}}$

29. $y = \sin^2 x + \cos^2 x = 1$; $y' = 0$

30. $y = \sin^2 x - 2\cos^2 x$; $y' = 2\sin x \cos x + 4\cos x \sin x = 6\sin x \cos x = 3\sin 2x$

31. $f(x) = \sin x$; $f'(x) = \cos x$; $f''(x) = -\sin x$

32. $g(x) = \cos x$; $g'(x) = -\sin x$; $g''(x) = -\cos x$

33. $y = \sin x$; $y' = \cos x$; $y'' = -\sin x$; $y''' = -\cos x$

34. $y = \cos x$; $y' = -\sin x$; $y'' = -\cos x$; $y''' = \sin x$

35. $y = \sin x$; $y' = \cos x$; $y'' = -\sin x$; and $y''' = -\cos x$; $y^{(4)} = \sin x$, so $\frac{d^4}{dx^4}(\sin x) = \sin x$

36. $y = \cos x$; $y' = -\sin x$; $y'' = -\cos x$; $y''' = \sin x$, and $y^{(4)} = \cos x$, so $\frac{d^4}{dx^4}(\cos x) = \cos x$.

37. **(a)** When $x = \frac{\pi}{4}$, then $f(x) = \sqrt{2}\cos\left(\frac{\pi}{4}\right) = \sqrt{2}\left(\frac{\sqrt{2}}{2}\right) = 1$. The line is tangent at the point $\left(\frac{\pi}{4}, 1\right)$. The slope of the tangent line at this point is $f'\left(\frac{\pi}{4}\right)$. Since $f'(x) = -\sqrt{2}\sin x$, the slope of the tangent line is $-\sqrt{2}\sin\left(\frac{\pi}{4}\right) = -1$. Using the slope-intercept form for the equation of a line, the equation of the tangent line is
$$y - y_0 = m(x - x_0)$$
$$y - 1 = -1\left(x - \frac{\pi}{4}\right)$$
$$= -x + \frac{\pi}{4}$$
$$y = -x + \frac{\pi}{4}$$
(b)

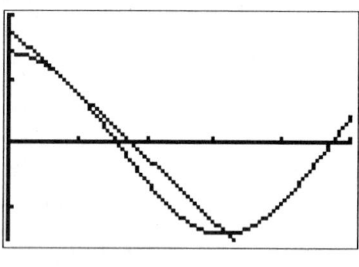

[0, 5, 1] × [−1.5, 2, 1]

38. **(a)** When $x = \frac{2\pi}{3}$, then $g(x) = 2\cos^2\left(\frac{2\pi}{3}\right) = 2\left[\cos\left(\frac{2\pi}{3}\right)\right]^2 = 2\left(-\frac{1}{2}\right)^2 = \frac{1}{2}$. The point of tangency is $\left(\frac{2\pi}{3}, \frac{1}{2}\right)$. The slope of the tangent line at this point is $g'\left(\frac{2\pi}{3}\right)$. Since $g'(x) = -4\cos x \sin x$, then $g'\left(\frac{2\pi}{3}\right) = -4\left(-\frac{1}{2}\right)\left(\frac{\sqrt{3}}{2}\right) = \sqrt{3}$. Using the

slope-intercept form for the equation of a line, the equation of the tangent line is
$$y - y_0 = m(x - x_0)$$
$$y - \frac{1}{2} = \sqrt{3}\left(x - \frac{2\pi}{3}\right)$$
$$y = \sqrt{3}x + \frac{1}{2} - \frac{2\pi}{\sqrt{3}}$$
(b)

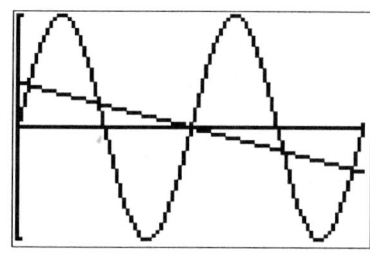

$\left[0, \pi, \frac{\pi}{6}\right] \cdot [-1, 2.2, 1]$

39. **(a)** The slope of the tangent line to h at this point is $h'\left(\frac{\pi}{2}\right)$. Since $h'(x) = 4\cos(4x)$, we see that $h'\left(\frac{\pi}{2}\right) = 4\cos(2\pi) = 4$. Thus, the slope of the normal line to h at this point is $-\frac{1}{4}$ and its equation is $y - y_0 = m(x - x_0)$
$$y - 0 = -\frac{1}{4}\left(x - \frac{\pi}{2}\right)$$
$$y = -\frac{1}{4}x + \frac{\pi}{8}$$
(b)

$\left[0, \pi, \frac{\pi}{4}\right] \cdot [-1, 1, 1]$

40. **(a)** Since $j'(x) = \sin x(-2\sin 2x) + \cos x \cos 2x$, we see that $j'\left(\frac{\pi}{4}\right) = -\frac{\sqrt{2}}{2} \cdot 2 + \frac{\sqrt{2}}{2} \cdot 0 = -\sqrt{2}$. Thus, the slope of the normal line to j at this point is $\frac{1}{\sqrt{2}}$ and its equation is $y = \frac{1}{\sqrt{2}}\left(x - \frac{\pi}{4}\right)$ or $y = -\frac{1}{\sqrt{2}}x + \frac{\pi}{4\sqrt{2}}$.

(b)

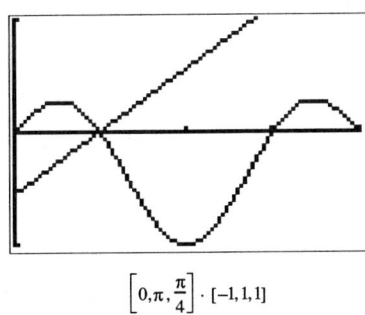

$$\left[0, \pi, \frac{\pi}{4}\right] \cdot [-1, 1, 1]$$

41. As in Example 27.7, $V' = 45{,}240 \cos\left[377(1.571)\right]$
$= -3470.332$ or about -3470 V.

42. The velocity function is $v = s' = (-4.8)75 \sin 75t$. When $t = 0.01$ s, $v(0.01) = (-4.8)75 \sin\left[75(0.01)\right]$ $= (-4.8)75 \sin(0.75) = -245.4$ cm/s. The acceleration function is $a = s''(-4.8)75^2 \cos 75t$. When $t = 0.01$ s, $a(0.01) = (4.8)75^2 \cos\left[75(0.01)\right]$ $= (-4.8)75^2 \cos(0.75) = -19\,755.6 \text{ cm}^2/\text{s}$

43. The current to the capacitor is given by $I = q'(t) = -0.25 \sin(t - 1.45)$. When $t = 7.6$ s, the current is $I = q'(7.6) = -0.25 \sin(7.6 - 1.45) = -0.25 \sin(6.15) = 0.033$ A.

44. The rate at which l changes with respect to θ is given by $l' = 62 \sin 2\theta$. When $\theta = \frac{\pi}{3}$, $l' = 62 \sin\left[2\left(\frac{\pi}{3}\right)\right] = 62 \sin\left(\frac{2\pi}{3}\right) = 53.7$ ft/rad.

45. $I'(\theta) = -2M \cos(\theta) \sin(\theta)$; $I'\left(\frac{\pi}{3}\right) = -\frac{\sqrt{3}}{2} M \approx -0.866M$ cd/rad

46. Notice here that, I_{L_r} is being differentiated with respect to θ_i, and so both μ_r and I_{L_i} are treated as constants. Thus, $\frac{dI_{L_r}}{d\theta_i} = I_{L_r}'(\theta_i) = I_{L_i}(1 - \mu_r) \sin\theta_i$. When $\mu_r = 0.6$ and $I_{L_i} = 100$ cd, we have $\frac{dI_{L_r}}{d\theta_i} = I_{L_r}'(\theta_i) = 100(1 - 0.6) \sin\theta_i = 40 \sin\theta_i$. The rate at which I_{L_r} changes with respect to θ_i when $\theta_i = \frac{\pi}{6}$ is given by $I_{L_r}'\left(\frac{\pi}{6}\right) = 40 \sin\left(\frac{\pi}{6}\right) = 20$ cd/rad.

47. To find the maximum and minimum, we first find the critical values. $I'(t) = 2 \cos t - \sin t = 0$ or $2 \cos t = \sin t$ and so, $\frac{\sin t}{\cos t} = \tan t = 2$. This occurs when $t \approx 1.107$ and $t \approx 4.249$. The maximum is at $I(1.107) = 2 \sin(1.107) + \cos(1.107) \approx 2.24$ A. The minimum is $I(4.249) = 2 \sin(4.249) + \cos(4.249) \approx -2.24$ A.

48. $B' = \dfrac{dB}{d\theta}$

$$= \frac{\left[1 + \cos^2(2\theta)\right](2\cos\theta) - 2\sin\theta\left[2\cos(2\theta)\right]\left[-2\sin(2\theta)\right]}{\left[1 + \cos^2(2\theta)\right]^2}$$

$$= \frac{\left[1 + \cos^2(2\theta)\right](2\cos\theta) + 8\sin\theta \sin(2\theta) \cos(2\theta)}{\left[1 + \cos^2(2\theta)\right]^2}$$

When $\theta = \frac{\pi}{6}$, we obtain $B' \approx 2.494$ T/rad.

27.2 DERIVATIVES OF THE OTHER TRIGONOMETRIC FUNCTIONS

1. $y = \tan^2 \sqrt{x}$; $y' = 2 \tan\sqrt{x} \cdot \sec^2\sqrt{x} \cdot \frac{1}{2}x^{-1/2} = \dfrac{\tan\sqrt{x} \sec^2\sqrt{x}}{\sqrt{x}}$

2. $y = \tan 4x$; $y' = 4 \sec^2 4x$

3. $y = \sec 5x$; $y' = 5 \sec 5x \tan 5x$

4. $y = \cot(1 + 2x)$; $y' = -2 \csc^2(1 + 2x)$

5. $y = \csc(2x - 1)$; $y' = -2 \csc(2x - 1) \cot(2x - 1)$

6. $y = \tan\frac{x}{2}$; $y' = \frac{1}{2} \sec^2 \frac{x}{2}$

7. $y = \sin x \tan x$; $y' = \sin x \sec^2 x + \tan x \cos x = \sin x \sec^2 x + \sin x = \sin x\left(\sec^2 x + 1\right)$

8. $y = \sin x \cot x$; $y' = -\sin x \csc^2 x + \cot x \cos x = \dfrac{-1}{\sin x} + \dfrac{\cos^2 x}{\sin x} = \dfrac{\cos^2 x - 1}{\sin x} = \dfrac{-\sin^2 x}{\sin x} = -\sin x$

9. $y = \tan^3 x$; $y' = 3 \tan^2 x \cdot \sec^2 x$

10. $y = \cot^3 4x$; $y' = 3 \cot^2 4x \cdot \left(-\csc^2 4x\right) \cdot 4x = -12x \cot^2 4x \csc^2 4x$

11. $y = \sec^4\left(x^2\right)$; $y' = 4 \sec^3\left(x^2\right)\left(\sec\left(x^2\right)\tan\left(x^2\right)\right)(2x) = 8x \sec^4\left(x^2\right)\tan\left(x^2\right)$

12. $y = \csc^3 \sqrt{x^3} = \left[\csc\left(x^{3/2}\right)\right]^3$; $y' = 3\left[\csc^2\left(x^{3/2}\right)\right]\left[-\csc\left(x^{3/2}\right)\cot\left(x^{3/2}\right)\right]\left(\frac{3}{2}x^{1/2}\right) = -\frac{9}{2}x^{1/2}\csc^3\left(x^{3/2}\right)\cot\left(x^{3/2}\right)$

13. $y = \tan x \cot x$; $y' = \tan x\left(-\csc^2 x\right) + \cot x \times \left(\sec^2 x\right) = \sec^2 x \cot x - \csc^2 x \tan x$

14. $y = \sec x \csc x$; $y' = \sec x \tan x \csc x - \sec x \csc x \cot x = \sec x \cdot \dfrac{\sin x}{\cos x} \cdot \dfrac{1}{\sin x} - \dfrac{1}{\cos x} \cdot \dfrac{\cos x}{\sin} \csc x = \sec^2 x - \csc^2 x$

15. $y = \sin^2 x \cot x$; $y' = 2\sin x \cos x \cot x + \sin^2 x \times$
$\left(-\csc^2 x\right) = 2\sin x \cos x \cdot \frac{\cos x}{\sin x} - \sin^2 x \cdot \frac{1}{\sin^2 x} =$
$2\cos^2 x - 1 = \cos 2x$

16. $y = \sin 5x^2$; $y' = \cos 5x^2(10x) = 10x \cos 5x^2$

17. $y = \tan \frac{1}{x}$; $y' = \sec^2 \frac{1}{x} \cdot \left(-x^{-2}\right) = -\frac{1}{x^2}\sec^2 \frac{1}{x}$

18. $y = \cot\sqrt{x^2 + 1}$;
$y' = -\csc^2 \sqrt{x^2 + 1}\left(\frac{1}{2}(x^2 + 1)^{-1/2}\right) 2x$
$= \frac{-x}{\sqrt{x^2 + 1}}\csc\sqrt{x^2 + 1}$

19. $y = \sqrt{1 + \tan x^2} = \left(1 + \tan x^2\right)^{1/2}$; $y' = \frac{1}{2}\left(1 + \tan x^2\right)^{-1/2}\left(\sec^2 x^2\right)(2x) = \frac{x \sec^2 x^2}{\sqrt{1 + \tan x^2}}$

20. $y = \sqrt{\tan x + \cot x} = (\tan x + \cot x)^{1/2}$; $y' = \frac{1}{2}(\tan x + \cot x)^{-1/2}\left(\sec^2 x - \csc^2 x\right) = \frac{\sec^2 - \csc^2 x}{2\sqrt{\tan x + \cot x}}$

21. $y = \frac{\tan x}{1 + \sec x}$;
$y' = \frac{(1 + \sec x)\sec^2 x - \tan x(\sec x \tan x)}{(1 + \sec x)^2}$
$= \frac{\sec^2 x + \sec^3 x - \sec x \tan^2 x}{(1 + \sec x)^2}$
$= \frac{\sec x\left[\sec x + \sec^2 x - \tan^2 x\right]}{(1 + \sec x)^2}$
$= \frac{\sec x[\sec x + 1]}{(1 + \sec x)^2} = \frac{\sec x}{1 + \sec x}$

22. $y(\csc x + \cot x)^3$; $y' =$
$3(\csc x + \cot x)^2(-\csc x \cot x - \csc^2 x) =$
$-3\csc x(\csc x + \cot x)^3$

23. Here $y = \frac{\cot x}{1 - \csc x}$, so $y' =$
$\frac{(1 - \csc x)\left(-\csc^2 x\right) - \cot x(\csc x \cot x)}{(1 - \csc x)^2} =$
$\frac{-\csc^2 x + \csc^3 x - \csc x \cot^2 x}{(1 - \csc x)^2} = \frac{\csc x\left(\csc^2 x - \cot^2 x - \csc x\right)}{(1 - \csc x)^2} =$
$\frac{\csc x(1 - \csc x)}{(1 - \csc x)^2} = \frac{\csc x}{1 - \csc x}$

24. Here $y = \sqrt{x} + \tan\sqrt{x}$, so $y' = \frac{1}{2\sqrt{x}} + \frac{\sec^2 \sqrt{x}}{2\sqrt{x}} = \frac{1}{2\sqrt{x}}\left(1 + \sec^2 \sqrt{x}\right)$

25. $y = (\csc x + 2\tan x)^3$; $y' = 3(\csc x + 2\tan x)^2 \times$
$\left(-\csc x \cot x + 2\sec^2 x\right) = 3(\csc x + 2\tan x)^2 \times$
$\left(2\sec^2 x - \csc x \cot x\right)$

26. $y = \left(1 + \cot^2 x\right)^{1/2}$; $y' = \frac{1}{2}\left(1 + \cot^2 x\right)^{-1/2} \times$
$(2\cot x)(-\csc^2 x) = \frac{-\cot x \csc^2 x}{\sqrt{1 + \cot^2 x}} = \frac{-\cot x \csc^2 x}{\csc x} =$

$-\cot x \csc x$ or $y = \sqrt{1 + \cot^2 x} = \csc x$; $y' = -\csc x \cot x$

27. $y = (\tan 2x)^{3/5}$; $y' = \frac{3}{5}(\tan 2x)^{-2/5}\left(\sec^2 2x\right)(2) = \frac{6}{5}\sec^2 2x(\tan 2x)^{-2/5}$

28. $y = x \csc x$; $y' = x(-\csc x \cot x) + \csc x = \csc x(1 - x \cot x)$

29. Here $y = \frac{\sec x}{1 + \tan x}$, and so $y' =$
$\frac{(1 + \tan x)(\sec x \tan x) - \sec^3 x}{(1 + \tan x)^2} = \frac{\sec x\left(\tan x + \tan^2 x - \sec^2 x\right)}{(1 + \tan x)^2} =$
$\frac{\sec x(\tan x - 1)}{(1 + \tan x)^2}$

30. $y = \frac{\cos x}{1 + \sec^2 x}$;
$y' = \frac{\left(1 + \sec^2 x\right)(-\sin x) - \cos x\left(2\sec^2 x \tan x\right)}{\left(1 + \sec^2 x\right)^2}$
$= \frac{-\sin x - \sin x \sec^2 x - 2\sec^2 x \sin x}{\left(1 + \sec^2\right)^2}$
$= \frac{-\sin x\left(1 + 3\sec^2 x\right)}{\left(1 + \sec^2 x\right)^2}$

31. $\frac{d}{dx}\cot u = \frac{d}{dx}\left(\frac{\cos u}{\sin u}\right) =$
$\frac{\sin u \frac{d}{dx}\cos u - \cos u^2 \sin u}{\sin^2 u} =$
$\frac{\sin u(-\sin u)\frac{du}{dx} - \cos u \cos u \frac{du}{dx}}{\sin^2 u} =$
$\frac{\left(-\sin^2 u - \cos^2 u\right)\frac{du}{dx}}{\sin^2 u} = -\csc^2 u \frac{du}{dx}$

32. $\frac{d}{dx}\csc u = \frac{d}{du}\frac{1}{\sin x} = \frac{d}{dx}(\sin u)^{-1} = -1(\sin u)^{-2}\frac{d}{dx}$
$\sin u = -1(\sin u)^{-2}\cos u \frac{du}{dx} = -\frac{1}{\sin u} \cdot \frac{\cos u}{\sin u}\frac{du}{dx} =$
$-\csc u \cot u \frac{du}{dx}$

33. $y = \tan 2x$; $y' = 2\sec^2 2x = 2(\sec 2x)^2$; $y'' = 4(\sec 2x)(\sec 2x \tan 2x)2 = 8\sec^2 2x \tan 2x$

34. $y = \sec 3x$; $y' = 3\sec 3x \tan 3x$; $y'' = 3\sec 3x \times \left(\sec^2 3x \cdot 3\right) + 3\tan 3x(\sec 3x \tan 3x \cdot 3) = 9\sec 3x\left(\sec^2 3x + \tan^2 3x\right)$

35. $y = x\tan x$; $y' = x\left(\sec^2 x\right) + \tan x$; $y'' = x(2\sec x \cdot \sec x \tan x) + \sec^2 x + \sec^2 x = 2x\sec^2 x \tan x + 2\sec^2 x = 2\sec^2 x(x\tan x + 1)$

36. Here $y = \frac{\cos x}{x}$, and so $y' = \frac{x(-\sin x) - \cos x \cdot 1}{x^2} = \frac{-x\sin x - \cos x}{x^2}$; $y'' =$

$$\frac{x^2(-x\cos x-\sin x+\sin x)+2x(x\sin x+\cos)}{x^4}=$$

$$\frac{-x^2\cos x+2x\sin x+2\cos x}{x^3}\text{ or }\frac{2\cos x+2x\sin x-x^2\cos x}{x^3}$$

37. Since $y=x\sin y$, then $y'=x(\cos y)y'+\sin y$, or $y'-x(\cos y)y'=\sin y$, or $y'(1-x\cos y)=\sin y$ and so $y'=\frac{\sin y}{1-x\cos y}$.

38. Since $\sin xy+xy=0$, then $\cos xy(xy'+y)+(xy'+y)=0$, or $(xy'+y)[\cos xy+1]=0\Rightarrow xy'+y=0$, and so $y'=\frac{-y}{x}$.

39. Here $x+y=\sin(x+y)$, and differentiating, we obtain $1+y'=\cos(x+y)[1+y']=\cos(x+y)+y'\cos(x+y)$, or $y'-y'\cos(x+y)=\cos(x+y)-1$, and so $y'=\frac{\cos(x+y)-1}{1-\cos(x+y)}$ or $\frac{1-\cos(x+y)}{\cos(x+y)-1}$.

40. Differentiating $y^2=\tan 4xy$ we get $2yy'=\sec^2 4xy(4xy'+4y)$, or $2yy'=4xy'\sec^2 4xy+4y\sec^2 4xy$, or $y'(2y-4x\sec^2 4xy)=4y\sec^2 4xy$, and so $y'=\frac{2y\sec^2 4xy}{y-2x\sec^2 4xy}$.

41. $n'=-\csc\theta\cot\theta=-\csc\left(\frac{\pi}{8}\right)\cot\left(\frac{\pi}{8}\right)=-6.31$

42. The relationship between θ and x is given by $\cot\theta=\frac{x}{h}$. Taking the derivative of both sides with respect to t, we get $-\csc^2\theta\frac{d\theta}{dt}=\frac{1}{h}\frac{dx}{dt}$ or $\frac{d\theta}{dt}=-\frac{\sin^2\theta}{h}\frac{dx}{dt}$. Now, 360 mph = 528 ft/s, and so $dx/dt=-528$

ft/s (negative because the airplane is approaching the tower). Putting this all together, produces $\frac{d\theta}{dt}=-\frac{\sin^2 37°}{6250\,\text{ft}}(-528\,\text{ft/s})=0.0306$ rad/s.

43. Differentiating $I=16\cot^2(2\theta)$, produces
$$I'=2\left[16\cot(2\theta)\right]\left[-\csc^2(2\theta)\right](2)$$
$$=-64\cot(2\theta)\csc^2(2\theta)$$
When $\theta=\frac{7\pi}{16}$, then $I'\left(\frac{7\pi}{16}\right)=$
$$-64\cot\left[2\left(\frac{7\pi}{16}\right)\right]\csc^2\left[2\left(\frac{7\pi}{16}\right)\right]=$$
$$-64\cot\left(\frac{7\pi}{8}\right)\csc^2\left(\frac{7\pi}{8}\right)\approx 1055.05\text{ mV/rad.}$$

44. The velocity is found by differentiating s, as
$$v=s'=\frac{ds}{dt}=16\tan t^2+8t^2\left(\sec^2 t^2\right)(2t)$$
$$=16t\tan t^2+16t^3\sec^2 t^2$$
When $t=0.8$ s, the velocity is $v(0.8)=16(0.8)\tan(0.8)^2+16(0.8)^3\sec^2(0.8)^2=22.26$ m/s.

Differentiating the velocity, the acceleration is
$$a=v'=\frac{dv}{dt}=16\tan t^2+16t\left(\sec^2 t^2\right)(2t)$$
$$+48t^2\sec^2 t^2$$
$$+\left(2\sec t^2\right)\left(\sec t^2\right)\left(\tan t^2\right)(2t)\left(16t^3\right)$$
$$=16\tan t^2+32t^2\sec^2 t^2+48t^2\sec^2 t^2$$
$$+64t^4\sec^2 t^2\tan t^2$$
$$=16\tan t^2+80t^2\sec^2 t^2+64t^4\sec^2 t^2\tan t^2$$
When $t=0.8$ s, the acceleration is $a(0.8)=16\tan(0.8)^2+80(0.8)^2\sec^2(0.8)^2+64(0.8)^4\sec^2(0.8)^2\tan(0.8)^2=121.83$ m/s^2.

≡ 27.3 DERIVATIVES OF INVERSE TRIGONOMETRIC FUNCTIONS

1. $y=\sin^{-1}2x$, so $y'=\frac{2}{\sqrt{1-4x^2}}$.

2. $y=\cos^{-1}4x$, so $y'=\frac{-4}{\sqrt{1-16x^2}}$.

3. $y=\tan^{-1}\frac{x}{2}$, so $y'=\frac{\frac{1}{2}}{1+\frac{x^2}{4}}=\frac{2}{4+x^2}$.

4. $y=\sin^{-1}\left(\frac{x}{3}\right)$, so $\frac{y'-\frac{1}{3}}{\sqrt{1-\frac{x^3}{9}}}=\frac{\frac{1}{3}}{\frac{\sqrt{9-x^2}}{3}}=\frac{1}{\sqrt{9-x^2}}$

5. $y=\cos^{-1}\left(1-x^2\right)$ and $y'=\frac{2x}{\sqrt{1-\left(1-x^2\right)^2}}=$

$\frac{2x}{\sqrt{2x^2-x^4}}$

6. $y=\sin^{-1}\sqrt{x}$, and $y'=\frac{\frac{1}{2\sqrt{x}}}{\sqrt{1-\sqrt{x}^2}}=\frac{1}{2\sqrt{x}\sqrt{1-x}}=$

$\frac{1}{2\sqrt{x-x^2}}$

7. $y=\sin^{-1}(1-2x)$, so $y'=\frac{-2}{\sqrt{1-(1-2x)^2}}=\frac{-2}{\sqrt{4x-4x^2}}$

$=\frac{-1}{\sqrt{x-x^2}}$

8. $y=\tan^{-1}(4x-1)$; $y'=\frac{4}{1+(4x-1)^2}=\frac{4}{16x^2-8x+2}=$

$\frac{2}{8x^2-4x+1}$ or $\frac{2}{1-4x+8x^2}$

9. $y=\cos^{-1}\left(x^3-x\right)$; $y'=\frac{1-3x^2}{\sqrt{1-\left(x^3-x\right)^2}}=$

$\frac{1-3x^2}{\sqrt{1-x^2+6x^4-x^6}}$

10. $y = \sin^{-1}(x-3)^2$; $y' = \frac{2(x-3)}{\sqrt{1-(x-3)^4}}$

11. $y = \sec^{-1}(4x+2)$; $y' = \frac{4}{(4x+2)\sqrt{(4x+2)^2-1}} = \frac{2}{(2x+1)\sqrt{(4x+2)^2-1}}$

12. $y = \csc^{-1}(3x+1)$; $y' = \frac{-3}{(3x+1)\sqrt{(3x+1)^2-1}}$

13. $y = x\sin^{-1} x$; $y' = x \cdot \frac{1}{\sqrt{1-x^2}} + \sin^{-1} x$

14. $y = \sin^{-1}\sqrt{x+1}$; $y' = \frac{\frac{1}{2\sqrt{x+1}}}{\sqrt{1-\left(\sqrt{x+1}\right)^2}} = \frac{1}{2\sqrt{x+1}\sqrt{-x}}$

15. $y = x^2\cos^{-1} x$; $y' = 2x\cos^{-1} x - \frac{x^2}{\sqrt{1-x^2}}$

16. $y = \cos^{-1}\sqrt{1-x^2}$; $y' = \frac{-\frac{1}{2}(1-x^2)^{-1/2}(-2x)}{\sqrt{1-(1-x^2)}} = \frac{x(1-x^2)^{-1/2}}{\sqrt{x^2}} = \frac{1}{\sqrt{1-x^2}}$

17. Since $y = \tan^{-1}\left(\frac{2x-1}{2x}\right)$, then $y' = \frac{1}{1+\left(\frac{2x-1}{2x}\right)^2} \times$
$\left(\frac{2x\cdot 2-(2x-1)2}{4x^2}\right) = \frac{1}{\frac{4x^2+4x^2-4x+1}{4x^2}}\left(\frac{4x-4x+2}{4x^2}\right) =$
$\frac{2}{8x^2-4x+1}$

18. $y = \csc^{-1}\sqrt{x}$; $y' = \frac{-1}{\sqrt{x}\sqrt{x-1}} \cdot \frac{1}{2\sqrt{x}} = \frac{-1}{2x\sqrt{x-1}}$

19. $y = x\tan^{-1}(x+1)$; $y' = \tan^{-1}(x+1) + \frac{x}{1+(x+1)^2}$

20. $y = (1+\tan^{-1} x)^2$; $y' = 2(1+\tan^{-1} x)\left(\frac{1}{1+x^2}\right) = \frac{2(1+\tan^{-1} x)}{1+x^2}$

21. Here we are given $y = \sqrt{\sin^{-1}(1-x^2)}$, and so
$y' = \frac{1}{2}\left(\sin^{-1}(1-x^2)\right)^{-1/2}\left(\frac{-2x}{\sqrt{1-(1-x^2)^2}}\right) =$
$\frac{-x}{\sqrt{\sin^{-1}(1-x^2)}\sqrt{1-(1-x^2)^2}} = \frac{-x}{\sqrt{\sin^{-1}(1-x^2)}\sqrt{2x^2-x^4}}$
$= \frac{-1}{\sqrt{\sin^{-1}(1-x^2)}\sqrt{2-x^2}}.$

22. $y = (1-x^2)^{1/2}\sin^{-1} x$; $y' = \frac{1}{2}(1-x^2)^{-1/2} \times$
$(-2x)\sin^{-1} x + (1-x^2)^{1/2} \cdot \frac{1}{\sqrt{1-x^2}} = \frac{-\sin^{-1} x}{\sqrt{1-x^2}} + 1$
or $1 - \frac{\sin^{-1} x}{\sqrt{1-x^2}}$

23. $y = x\cot^{-1}(1+x^2)$; $y' = \cot^{-1}(1+x^2) +$
$x\frac{-1}{1+(1+x^2)^2} \cdot 2x = \cot^{-1}(1+x^2) - \frac{2x^2}{2+2x^2+x^4}$

24. Here $y = \sin^{-1}(fx - 1x + 1)$, and so $y' =$
$\frac{1}{\sqrt{1-\frac{(x-1)^2}{(x+1)^2}}} \cdot \frac{(x+1)-(x-1)}{(x+1)^2} = \frac{1}{\frac{\sqrt{(x+1)^2-(x-1)^2}}{x+1}} \cdot$
$\frac{2}{(x+1)^2} = \frac{2}{(x+1)\sqrt{4x}} = \frac{1}{(x+1)\sqrt{x}}$

25. Since $y = \frac{\sin^{-1} x}{x}$, then $y' = \frac{x\left(\frac{1}{\sqrt{1-x^2}}\right) - \sin^{-1} x}{x^2}$
$= \frac{1}{x\sqrt{1-x^2}} - \frac{\sin^{-1} x}{x^2} = \frac{1}{x}\left(\frac{1}{\sqrt{1-x^2}} - \frac{\sin^{-1} x}{x}\right).$

26. $y = \frac{\arccos x^2}{x}$; $y' = \frac{x \cdot \frac{-1}{\sqrt{1-x^4}} \cdot 2x - \arccos x^2}{x^2} =$
$\frac{-2}{\sqrt{1-x^4}} - \frac{\arccos x^2}{x^2}$

27. $y = (\arcsin x)^{1/3}$; $y' = \frac{1}{3}(\arcsin x)^{-2/3}\left(\frac{1}{\sqrt{1-x^2}}\right)$
$= \frac{1}{3\sqrt{1-x^2}(\arcsin x)^{2/3}}$

28. $y = \left(\arctan x^2\right)^{1/3}$; $y' = \frac{1}{3}\left(\arctan x^2\right)^{-2/3} \times$
$\left(\frac{1}{1+x^4}\right)2x = \frac{2x}{3(1+x^4)\left(\arctan x^2\right)^{2/3}}$

29. Let $y = \cos^{-1} u$ or $u = \cos y$. Hence $\frac{du}{dx} = \frac{d}{dx}\cos y = -\sin y\frac{dy}{dx}$ so $\frac{dy}{dx} = \frac{-1}{\sin y}\frac{du}{dx}$. Since $\sin^2 y + \cos^2 y = 1$ we have $\sin y = \sqrt{1-\cos^2 y} = \sqrt{1-u^2}$, and so $\frac{dy}{dx} = \frac{d}{dx}\cos^{-1} u = \frac{-1}{\sqrt{1-u^2}}\frac{du}{dx}$.

30. Let $y = \cot^{-1} u$ and then, $u = \cot y$. Thus, we have $\frac{du}{dx} = -\csc^2 y\frac{dy}{dx} \Rightarrow \frac{dy}{dx} = \frac{-1}{\csc^2 y}\frac{du}{dx}$. By the Pythagorean identities we have $\csc^2 = 1+\cot^2 y = 1 + u^2$, and so $\frac{dy}{dx} = \frac{-1}{1+u^2}\frac{du}{dx}$ or $\frac{d}{dx}\cot^{-1} u = \frac{-1}{1+u^2}\frac{du}{dx}$.

31. In this problem, let P be a point on the track directly in front of the camera and x the distance the car has traveled past P, and θ the angle from P to the camera to x. Then $\theta = \tan^{-1}\left(\frac{x}{42}\right)$ and $\frac{d\theta}{dt} = \frac{1}{1+\left(\frac{x}{42}\right)^2}\frac{1}{42}\frac{dx}{dt} = \frac{42}{42^2+x^2}\frac{dx}{dt}$. When $\theta = 15°$, then $\theta = \tan^{-1}\left(\frac{x}{42}\right)$ or $\tan\theta = \frac{x}{42}$ and $x = 42\tan 15° \approx 11.25387$. Substituting this value of x and the given information that

$\frac{dx}{dt} = 320$ ft/s in the formula for $\frac{d\theta}{dt}$ produces $\frac{d\theta}{dt} = \frac{42}{42^2+(11.25387)^2}(320) \approx 7.1087$ rad/s.

32. We are given $h = 7920$ ft $= 1.5$ mi, $x = 6$ mi, and $dx/dt = -270.0$ mi/s (negative because it is approaching tower). Since $\tan\theta = \frac{h}{x} = \frac{1.5}{x}$, we have

$\theta = \arctan\left(\frac{1.5}{x}\right)$. Thus, $\frac{d\theta}{dt} = \frac{d}{dt}\arctan\left(\frac{1.5}{x}\right) = \frac{1}{1+\frac{2.25}{x^2}}\left(-\frac{1.5}{x^2}\right)\frac{dx}{dt} = -\frac{1.5}{x^2+2.25}\frac{dx}{dt}$. Substituting the given values for x and θ produces $\frac{d\theta}{dt} = -\frac{1.5}{6^2+2.25} \times (-270) \approx 10.59$ rad/h ≈ 0.1765 rad/min ≈ 0.0029 rad/s $\approx 0.1685°$/s.

27.4 APPLICATIONS

1. If $y = \sin x$, then $y' = \cos x$. Since $\cos x > 0$ on $\left(-\frac{\pi}{2}, \frac{\pi}{2}\right)$, $\sin x$ is increasing on $\left[-\frac{\pi}{2}, \frac{\pi}{2}\right]$.

2. If $y = \tan x$, $y' = \sec^2 x$. Since $\sec^2 x \geq 0$ everywhere it is defined, $\tan x$ is increasing for all values where it is defined.

3. If $y = \arccos x$, then $y' = \frac{-1}{\sqrt{1-x^2}}$ when $|x| < 1$. Hence $y' < 0$ which means that the $\arccos x$ is decreasing for $|x| \leq 1$.

4. If $y = \arctan x$, then $y' = \frac{1}{1+x^2}$ which is always positive. As a result, we see that $\arctan x$ is always increasing.

5. $f(x) = 2\sin x + \cos 2x$; $f'(x) = 2\cos x - 2\sin 2x$. The critical values occur when $f'(x) = 0$. Hence, we have $2\cos x - 2\sin 2x = 0$; $2\cos x - 4\sin x\cos x = 0$. Factoring, we obtain $2\cos x(1 - 2\sin x) = 0$. If $\cos x = 0 \Rightarrow x = \frac{\pi}{2}, \frac{3\pi}{2}$ and if $1 - 2\sin x = 0 \Rightarrow \sin x = \frac{1}{2} \Rightarrow x = \frac{\pi}{6}, \frac{5\pi}{6}$. Differentiating again, we get $f''(x) = -2\sin x - 4\cos 2x = 0$, or $-2(\sin x + 2(1 - 2\sin^2 x)) = 0$, or $-2(\sin x + 2 - 4\sin^2 x) = 0$, and so $2(4\sin^2 x - \sin x - 2) = 0$. The quadratic formula yields $\sin x = -0.59307$ or $\sin x = 0.84307$. Hence, $x = 3.7765$, $x = 5.6483$, $x = 1.0030$, or $x = 2.1386$. Maxima are at $\left(\frac{\pi}{6}, 1.5\right)$, $\left(\frac{5\pi}{6}, 1.5\right)$, minima are at $\left(\frac{\pi}{2}, 1\right)$ and $\left(\frac{3\pi}{2}, -3\right)$. Inflection points are at $(1.0030, 1.2646)$, $(2.1386, 1.2646)$, $(3.7765, -0.8896)$ and $(5.6483, -0.8896)$.

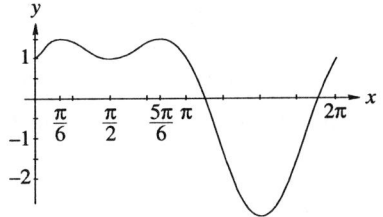

6. $g(x) = 2\cos x + \sin 2x$. To find the critical values, we take the derivative and set it equal to 0. $g'(x) = -2\sin x + 2\cos 2x = 0 \Rightarrow -2\sin x + 2(1 - 2\sin^2 x) = 0$ or $-4\sin^2 x - 2\sin x + 2 = 0$. Factoring, we have $(-4\sin x + 2)(\sin x + 1) = 0$. This gives $\sin x = -1$ or $\sin x = \frac{1}{2}$. Hence $x = \frac{3\pi}{2}$ of $x = \frac{\pi}{6}$ or $\frac{5\pi}{6}$. Taking the second derivative and setting it equal to 0 produces $g''(x) = -2\cos x - 4\sin 2x = -2\cos x - 4(2\sin x\cos x) = -2\cos x(1 + 4\sin x) = 0$, or $\cos x = 0 \Rightarrow x = \frac{\pi}{2}$ and so, $x = \frac{3\pi}{2}$ or $\sin x = -\frac{1}{4}$, and so, $x = 3.3943$ or $x = 6.0305$. As a result, the maximum is at $\left(\frac{\pi}{6}, 2.5981\right)$, the minimum at $\left(\frac{5\pi}{6}, -2.5981\right)$, and the inflection points are at $\left(\frac{\pi}{2}, 0\right)$, $(3.3943, -1.4523)$, $\left(\frac{3\pi}{2}, 0\right)$, and $(6.0305, 1.4524)$.

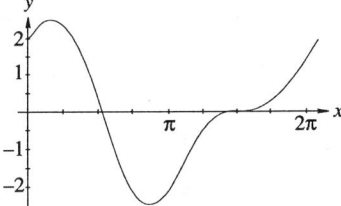

7. $h(x) = \cos x - \sin x$. To find the critical values, we take the derivative and set it equal to 0. Thus, $h'(x) = -\sin x - \cos x = 0$ and so, $-\sin x = \cos x$, or $\tan x = -1$, which means that $x = \frac{3\pi}{4}$ or $\frac{7\pi}{4}$. The second derivative is $h''(x) = -\cos x + \sin x$ and setting it equal to 0 produces $\sin x = \cos x$ or $\tan x = 1$. Thus, $x = \frac{\pi}{4}$ or $x = \frac{5\pi}{4}$. As a result, the maxima are $\left(\frac{7\pi}{4}, \sqrt{3}\right)$; the minimum is $\left(\frac{3\pi}{4}, -\sqrt{3}\right)$, and the inflection points are $\left(\frac{\pi}{4}, 0\right)$ and $\left(\frac{5\pi}{4}, 0\right)$.

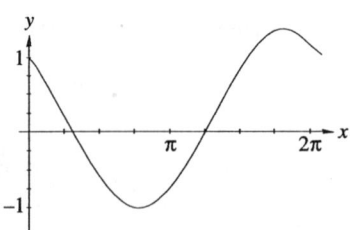

8. $j(x) = \cos^2 x - \sin x$. Taking the derivative and setting it equal to 0, we have $j'(x) = -2 \cos x \sin x - \cos x = -\cos x(2 \sin x + 1) = 0$. This yields $\cos x = 0$ which produces the critical values $x = \frac{\pi}{2}$ and $\frac{3\pi}{2}$ or $\sin x - \frac{1}{2}$ gives the critical values $x = \frac{7\pi}{6}$ and $\frac{11\pi}{6}$. For inflection points take the second derivative $j''(x) = -\cos x(2 \cos x) + \sin(2 \sin x + 1) = -2 \cos^2 x + 2 \sin^2 x + \sin x = -2(1 - 2 \sin^2 x) + 2 \sin^2 x + \sin x = -2 + 6 \sin^2 x + \sin x = 6 \sin^2 x + \sin x - 2 = (3 \sin x + 2)(2 \sin x - 1)$. Setting the second derivative equal to 0 and solving, we obtain $\sin x = \frac{1}{2} \Rightarrow x = \frac{\pi}{6}$ or $\frac{5\pi}{6}$ or $\sin x = -\frac{2}{3} \Rightarrow x = 3.8713$ or $x = 5.5535$. From the above information, we obtain maxima at $\left(\frac{7\pi}{6}, 1.25\right)$ and $\left(\frac{11\pi}{6}, 1.25\right)$; minima at $\left(\frac{\pi}{2}, -1\right)$ and $\left(\frac{3\pi}{2}, 1\right)$; and inflection points at $\left(\frac{\pi}{6}, 0\right)$, $\left(\frac{5\pi}{6}, 0\right)$, (3.8713, 1.2222), and (5.5535, 1.2222).

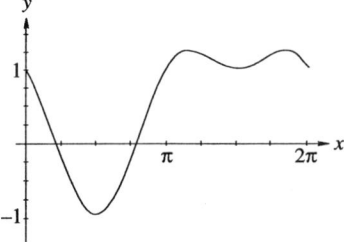

9. $k(x) = \cos^2 x + \sin x$. To find the critical values, we take the derivative and set it equal to 0. $k'(x) = -2 \cos x \sin x + \cos x = \cos x(-2 \sin x + 1) = 0$. This yields $\cos = 0$ which means that there are critical values at $x = \frac{\pi}{2}$ and $x = \frac{3\pi}{2}$. We also get $\sin x = \frac{1}{2}$, and so there are additional critical values at $x = \frac{\pi}{6}$ and $x = \frac{5\pi}{6}$. For inflection points, we take the second derivatives and set them equal to 0: $k''(x) = \cos x(-2 \cos x) - \sin x(-2 \sin x + 1) = -2 \cos^2 x + 2 \sin^2 x - \sin x = -2(1 - 2 \sin^2 x) + 2 \sin^2 x - \sin x = 6 \sin^2 - \sin x - 2 = (3 \sin x - 2)(2 \sin x + 1) = 0$. Hence, $\sin x = \frac{2}{3} \Rightarrow x = 0.7297$ or $x = 2.4119$, or $\sin x = -\frac{1}{2} \Rightarrow x = \frac{7\pi}{6}$

or $\frac{11\pi}{6}$. Thus, we see that this function has the following maxima: $\left(\frac{\pi}{6}, 1.25\right)$ and $\left(\frac{5\pi}{6}, 1.25\right)$; minima: $\left(\frac{\pi}{2}, 1\right)$ and $\left(\frac{3\pi}{2}, -1\right)$; and inflection points: (0.7297, 1.2222), (2.4119, 1.2222), $\left(\frac{7\pi}{6}, 0\right)$ and $\left(\frac{11\pi}{6}, 0\right)$.

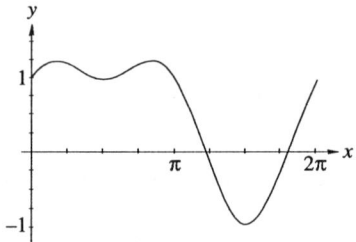

10. $m(x) = \sec x - \tan x$. For critical values $m'(x) = \sec x \tan x - \sec^2 x = \sec x(\tan x - \sec x) = 0$. Hence, either $\sec x = 0$ which yields no solutions or $\tan x = \sec x$, or $\frac{\sin x}{\cos x} = \frac{1}{\cos x} \Rightarrow \sin x = 1$ or $x = \frac{\pi}{2}$. This is not a solution since both sec and tan are undefined at $\frac{\pi}{2}$. Hence, there are not critical values and m has no maxima or minima. To determine inflection points we take the second derivative and set it equal to 0. $m''(x) = \sec x \tan^2 x + \sec^3 x - 2 \sec^2 x \tan x = \sec x(\tan^2 x + \sec^2 x - 2 \sec \tan x) = \sec x(\tan x - \sec x)^2 = 0$. Once again undefined at $\frac{\pi}{2}$ and $\frac{3\pi}{2}$. There is an apparent inflection point at $\left(\frac{\pi}{2}, 0\right)$, but this is actually an undefined point and hence a hole. Thus, there are no inflection points. Concave up $\left(\frac{-\pi}{2}, \frac{\pi}{2}\right)$ Concave down $\left(\frac{-\pi}{2}, \frac{3\pi}{2}\right)$.

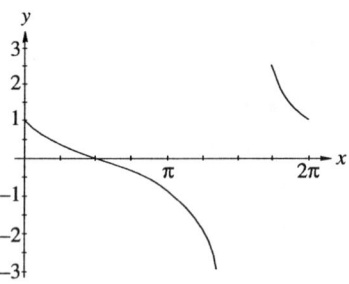

11. $y = \sin x$, at $x = \frac{\pi}{6}$, we get $y = \sin \frac{\pi}{6} = 0.5$. Thus, we want to find the line tangent at this point $\left(\frac{\pi}{6}, 0.5\right)$. To find the slope of the tangent line we take the derivative $y' = \cos x$ and evaluate it when $x = \frac{\pi}{6}$ to get $\cos \frac{\pi}{6} = \frac{\sqrt{3}}{2}$. Using the point-slope form for the equation of the line, we obtain the desired equation: $y - 0.5 = \frac{\sqrt{3}}{2}\left(x - \frac{\pi}{6}\right)$.

12. $y = \sin x$, at $x = \frac{\pi}{6}$, we get $y = \sin \frac{\pi}{6} = 0.5$. Thus, we want to find the line normal at the point $\left(\frac{\pi}{6}, 0.5\right)$. As in Exercise #11, the slope of the tangent line when $x = \frac{\pi}{6}$ is $\frac{\sqrt{3}}{2}$, and so the slope of the normal line is $-\frac{2}{\sqrt{3}}$. Using the point-slope form for the equation of the line, we obtain the desired equation: $y - 0.5 = \frac{-2}{\sqrt{3}}\left(x - \frac{\pi}{6}\right)$ or $y - 0.5 = \frac{-2\sqrt{3}}{3}\left(x - \frac{\pi}{6}\right)$.

13. $y = \tan^3 x$; at $x = \frac{\pi}{4}$, we get $y = 1$, so point is $\left(\frac{\pi}{4}, 1\right)$. To find the slope, we take the derivative $y' = 3\tan x\left(\sec^2 x\right)$ and at $\frac{\pi}{4}$, we get $y' = 3 \cdot 1^2(\sqrt{2})^2 = 6$. Hence, the equation of the tangent line is $y - 1 = 6\left(x - \frac{\pi}{4}\right)$.

14. As in Exercise #13, the point is $\left(\frac{\pi}{4}, 1\right)$, but the slope of the normal line is $-\frac{1}{6}$. Thus, the equation of the normal line is $y - 1 = -\frac{1}{6}\left(x - \frac{\pi}{4}\right)$.

15. The horizontal distance from the light to the plane is 8000 ft. The velocity is $\frac{dx}{dt} = 400$ mi/h $= 400 \cdot \frac{\text{mile}}{\text{hr}} \cdot \frac{5280 \text{ ft}}{1 \text{ mi}} \cdot \frac{1 \text{ hr}}{3600 \text{ s}} = 586.7$ ft/s; $\tan\theta = \frac{x}{6000}$. Taking derivatives we have $\sec^2\theta\frac{d\theta}{dt} = \frac{1}{6000}\frac{dx}{dt}$. Also, $\tan\theta = \frac{8000}{6000} = \frac{4}{3}$, and so $\theta = 0.9273$ rad. Hence, $\frac{d\theta}{dt} = \frac{dx/dt}{6000\sec^2\theta} = \frac{(dx/dt)\cos^2\theta}{6000} = \frac{(586.7)\cos^2(0.9273)}{6000} = 0.0352$ rad/s.

16. 300 m $= 0.3$ km; 400 m $= 0.4$ km. The ship is $\sqrt{400^2 - 300^2} = 264.6$ m $= 0.2646$ km from its closest point. $\tan\theta = \frac{x}{0.3}$. Taking derivatives, we have $\sec^2\theta\frac{d\theta}{dt} = \frac{1}{0.3}dx/dt$. We know $\cos\theta = \frac{3}{4}$, and so $\sec\theta = \frac{4}{3}$. Substituting, we obtain $\frac{d\theta}{dt} = \frac{10}{0.3\left(\frac{4}{3}\right)^2} = 18.75$ rad/hr $= 0.0052$ rad/s. (Note: $\frac{dx}{dt} = $ velocity $= 10$ km/h).

17. $\frac{d\theta}{dt} = 1$ rpm $= 2\pi$ rad/min $= 0.10472$ rad/s. $\tan\theta = \frac{x}{500}$. Taking derivatives we have $\sec^2\theta\frac{d\theta}{dt} = \frac{1}{500}\frac{dx}{dt}$ or $\frac{dx}{dt} = \sec^2\theta \cdot \frac{d\theta}{dt} \cdot 500 = \left(\sec^2\frac{\pi}{4}\right)(0.10472)(500) = 104.72$ m/s.

18. $P_{\text{app}} = P\sec\theta$; $P'_{\text{app}} = P\sec\theta\tan\theta\theta' = 15\sec\frac{\pi}{3}\tan\frac{\pi}{3}(0.6) = 31.1769$ W/min

19. $\frac{d\theta}{dt} - 30$ rpm $= 3.1416$ rad/s. We know that $\tan\theta = \frac{x}{8}$, and taking derivatives of both sides we have $\sec^2\theta\frac{d\theta}{dt} = \frac{1}{8}\frac{dx}{dt}$. Hence, $\frac{dx}{dt} = \left(8\sec^2 40°\right) \times (3.1416) = 42.8284$ mi/s.

20. **(a)** $x(t) = 2 + 0.75\cos 2\pi t$; $x(0) = 2.75$ m; $x\left(\frac{1}{2}\right) = 1.245$ m; $x(1) = 2.75$ m
(b) $v(t) = x'(t) = -1.5\pi\sin 2\pi t$
(c) $v(0) = 0$ m/s; $v\left(\frac{1}{2}\right) = 0$ m/s; $v(1) = 0$ m/s
(d) $a(t) = v'(t) = -3.0\pi^2\cos 2\pi t$
(e) Velocity is a maximum when $a(t) = 0$. Hence $-3.0\pi^2\cos 2\pi t = 0$; or $\cos 2\pi t = 0 \Rightarrow 2\pi t = \frac{\pi}{2} + k\pi \Rightarrow t = \frac{1}{4} + \frac{k}{2}$: $k = 0, 1, 2\ldots$; Using the second derivative test, we have $a'(t) = 6.0\pi^3\sin 2\pi t$. At $= \frac{1}{4} + \frac{k}{2}$, with k is even, we see that $a'(t)$ is positive, so $v(t)$ is minimum. When k is odd, $a'(t)$ is negative so $v(t)$ is maximum. Answer: $t = \frac{1}{4} + \frac{k}{2}$; $k = 1, 3, 5\ldots$ or $t = \frac{4n+3}{4}$, $n = 0, 1, 2, \ldots$.

21. **(a)** $\overline{Q} = 1.33(2)^{5/2}\tan\frac{\theta}{2} = 7.523\tan\frac{\theta}{2}$; $\overline{Q}' = \frac{1}{2}(7.5236)\sec^2\frac{\theta}{2}$
(b) $Q'(30) = 3.7618\sec^2 15° = 3.7518(1.0353)^2 = 4.0319$ m^3/s
(c) $\overline{Q}(40) = 3.7168\sec^2 20° = 4.2602$
(d) $\overline{Q}' = k\sec^2\frac{\theta}{2}$; critical values at $\frac{\theta}{2} = 90°$ or $\theta = 180°$. $\overline{Q}$ is increasing so maximum at $\theta = 180°$.

22. $\cos\theta = \frac{12}{s} = 12s^{-1}$. Taking derivatives we have $-\sin\theta\frac{d\theta}{dt} = -12s^{-2}\frac{ds}{dt}$. When $s = 20$, $\cos\theta = \frac{12}{20} \Rightarrow \theta = 0.9273$ rad. $\frac{d\theta}{dt} = \frac{12 \times (0.4)}{\sin(0.9273)20^2} = 0.015$ rad/min smaller.

23. $I = \sin^2 3t$; $I' = 6\sin 3t\cos 3t$; $I'(1.5) = 6(\sin 4.5)(\cos 4.5) = 1.2364$ A/s

24. Using $\tan(\theta + \phi) = \frac{6.5}{x}$, with $\tan(\phi) = \frac{2.5}{x}$, then $\tan\theta = \tan\left[(\theta + \phi) - \phi\right] = \frac{\tan(\theta+\phi) - \tan\phi}{1 + \tan(\theta+\phi)\tan\phi} = \frac{\frac{6.5}{x} - \frac{2.5}{x}}{1 + \frac{6.5}{x} \cdot \frac{2.5}{x}} = \frac{\frac{9}{x}}{1 + \frac{16.25}{x^2}} = \frac{9x}{x^2 + 16.25}$. Hence $\theta = \arctan\frac{9x}{x^2 + 162.5}$. Taking derivatives we have $\frac{d\theta}{dx} =$

$$\frac{1}{1+\left(\frac{9x}{x^2+16.25}\right)^2}\cdot\frac{d}{dx}\left(\frac{9x}{x^2+16.25}\right)=$$

$$\frac{(x^2+16.25)^2}{x^4+32.5x^2+18x^2+264.0625}\cdot\frac{(x^2+16.25)9-9x(2x)}{(x^2+16.25)^2}=$$

$$\frac{146.25-9x^2}{x^4+50.5x^2+264.0625}.$$ Now, $\frac{d\theta}{dx}=0$ when $9x^2=$ 146.25 or $x=4.0311$. By the first derivative test this yields a maximum. Answer: 4.0311 ft.

25. $x(t)=\frac{1}{4}\cos 2t$; $v(t)=-\frac{1}{2}\sin 2t$; $a(t)=v(t)=$ $-\cos 2t$; $a(t)=0$ when $2t=\frac{\pi}{2}+2k\pi$ or $\frac{3\pi}{2}+2k\pi$. By the first derivative test the maximums occur at $2t+\frac{3\pi}{2}+2k\pi$ or $t=\frac{3\pi}{4}+k\pi$ or $t=\frac{3\pi+4k\pi}{4}$, $k=$ 0, 1, 2, 3,.... Thus, $v\left(\frac{3\pi}{4}\right)=-\frac{1}{2}\sin\left(\frac{3\pi}{2}\right)=$ $-\frac{1}{2}(-1)=\frac{1}{2}$ m/s = 0.5 m/s.

26. $A=\frac{1}{2}(b_1+b_2)h$; $b_1=10$; $b_2=10+20\sin\theta$; $h=$ $10\cos\theta\left(0\le\theta\le\frac{\pi}{2}\right)$. Hence, we have $A=$ $\frac{1}{2}(10+10+20\sin\theta)(10\cos\theta)=5\cos\theta\times$ $(20+20\sin\theta)=100\cos\theta(1+\sin\theta)$. Taking the derivative, we get $A'=100\cos\theta(+\cos\theta)-$ $100\sin\theta(1+\sin\theta)=100\left(\cos^2\theta-\sin\theta-\sin^2\theta\right)$ and setting $A'=0$ we obtain $\cos^2\theta-\sin\theta-$ $\sin^2\theta=0$; or $1-\sin^2\theta-\sin\theta-\sin^2\theta=0$.

Hence, $1-\sin\theta-2\sin^2\theta=0$ and factoring, we have $(1-2\sin\theta)(1+\sin\theta)=0$. Thus, $\sin\pi=$ $\frac{1}{2}\Rightarrow\theta=\frac{\pi}{6}$; or $\sin\theta=-1\Rightarrow\theta=\pi$ which is out of range. Hence the only critical value is $\frac{\pi}{6}$. By the first derivative test, this is a maximum.

Answer: $\theta=\frac{\pi}{6}$ of 30°.

27. $f(x)=\cos x+x-2$; $f'(x)=-\sin x+1$. Use first $x=3$.

n	x_n	$f(x_n)$	x_{n+1}	$f(x_{n+1})$
1	3	0.0100	2.98835	6.716×10^{-5}
2	2.98835	6.716×10^{-5}	2.9883	3.105×10^{-9}

28. $f(x)=\sin x-x^2$; $f'(x)=\cos x-2x$. $x=0$ is one answer. Using Newton's method to find the other, $x=0.8767$ more exactly 0.8767262154.

29. $g(x)=\sin x+\frac{x^2}{4}-2x$; $g'(x)=\cos x+\frac{3}{4}x^2-$ 2; $x=0$ is one answer. By Newton's method $x=$ 2.71995214. Since $g(x)$ is odd, $x=-2.71995214$ is also an answer.

30. $h(x)=\cos x+x^2-3$; $h'(x)=-\sin x+2x$. Since $h(x)$ is even, roots come in pairs, so $x=$ ±1.795116642.

▤ 27.5 DERIVATIVES OF LOGARITHMIC FUNCTIONS

1. Since $y=\log 5x$, its derivative is $y'=\frac{1}{5x}\cdot\log e\cdot$ $5=\frac{1}{x}\cdot\frac{1}{\ln 10}$.

2. Here $y=\log_5 x^3$, and its derivative is $y'=\frac{3x^2}{x^3}\cdot$ $\log_5 e=\frac{3}{x}\cdot\frac{1}{\ln 5}$.

3. $y=\ln\left(x^2+4x\right)$; $y'=\frac{2x+4}{x^2+4x}$

4. $y=\ln\sqrt{x^3+4x}$; $y'=\frac{1}{\sqrt{x^2+4x}}\cdot\frac{1}{2}(x^2+4x)^{-1/2}(2x$ $+4)=\frac{x+2}{x^2+4x}$

5. $y=\sqrt{\ln\left(x^2+4x\right)}$; $y'=\frac{1}{2}\left(\ln\left(x^2+4x\right)\right)^{-1/2}\cdot$ $\frac{2x+4}{x^2+4x}=\frac{x+2}{\sqrt{\ln\left(x^2+4x\right)\cdot\left(x^2+4x\right)}}$

6. $y=\sqrt{\ln x^2+4x}$; $y'=\frac{1}{2}\left(\ln x^2+4x\right)^{-1/2}\cdot\left[\frac{2x}{x^2}+$ $4\right]=\left(\frac{1}{x}+2\right)\left(\ln x^2+4x\right)^{-1/2}$

7. Here $y=\ln\frac{1}{x}$ and its derivative is $y'=\frac{1}{1/x}\cdot$ $\left(-1x^{-2}\right)=-\frac{1}{x}$. By an alternate method: $y=$ $\ln x^{-1}=\ln x$, so $y'=-1\frac{1}{x}=\frac{-1}{x}$.

8. $y=\ln\cos x$; $y'=\frac{1}{\cos x}(-\sin x)=-\tan x$

9. $y=\ln\tan x$; $y'=\frac{1}{\tan x}\left(\sec^2 x\right)=\sec x\csc x$

10. $y=\log_4\sqrt{x}$; $y'=\frac{1}{\sqrt{x}}\cdot\frac{1}{\ln 4}\cdot x^{-1/2}=\frac{1}{2x\ln 4}$

11. $y=\frac{\ln x}{x}$; $y'=\frac{x\cdot\frac{1}{x}-\ln x}{x^2}=\frac{1-\ln x}{x^2}$

12. $y=\ln\left(\frac{x}{1+x}\right)$; $y'=\frac{1+x}{x}\cdot\left(\frac{1+x-x}{(1+x)^2}\right)=\frac{1}{x(1+x)}$

13. $y=\left[\ln\left(\frac{1+x}{1-x}\right)\right]^{-1/2}$; $y'=\frac{1}{2}\left[\ln\left(\frac{1+x}{1-x}\right)\right]^{-1/2}\cdot\left(\frac{1-x}{1+x}\right)$ $\cdot\left(\frac{(1-x)+(1+x)}{(1-x)^2}\right)=\frac{1}{2}\left[\ln\left(\frac{1+x}{1-x}\right)\right]^{-1/2}\left(\frac{1-x}{1+x}\right)\left(\frac{2}{(1-x)^2}\right)$ $=\frac{1}{(1+x)(1-x)}\cdot\left[\ln\left(\frac{1+x}{1-x}\right)\right]^{-1/2}=$ $\frac{1}{1-x^2}\left[\ln\left(\frac{1+x}{1-x}\right)\right]^{-1/2}$

14. $y = \frac{1}{2}\ln\left(\frac{1+x^2}{1-x^2}\right)$; $y' = \frac{1}{2}\left(\frac{1-x^2}{1+x^2}\right) \cdot$

$\left(\frac{\left(1-x^2\right)2x+2x\left(1+x^2\right)}{\left(1-x^2\right)^2}\right) = \frac{1}{2}\left(\frac{1}{1+x^2}\right) \times \left(\frac{4x}{1-x^2}\right) =$

$\frac{2x}{1-x^4}$

15. $y = \ln\frac{4x^3}{\sqrt{x^2+4}}$; $y' = \frac{\sqrt{x^2+4}}{4x^3} \cdot$

$\left(\dfrac{\sqrt{x^2+4}\cdot 12x^2 - 4x^3 \cdot \frac{1}{2}\left(x^2+4\right)^{-1/2} \cdot 2x}{x^2+4}\right) =$

$\frac{\sqrt{x^2+4}}{4x^3} \cdot \frac{\left(x^2+4\right)\cdot 12x^2 - 4x^4}{\left(x^2+4\right)^{3/2}} = \frac{1}{4x^3}\cdot \frac{4x^2\left[\left(x^2+4\right)\cdot 3 - x^2\right]}{x^2+4} =$

$\frac{2x^2+12}{x\left(x^2+4\right)}$

16. $y = \frac{\ln x}{x^3}$; $y' = \frac{x^3 \cdot \frac{1}{x} - \ln x\left(3x^2\right)}{x^6} = \frac{x^2 - 3x^2\ln x}{x^6} =$

$\frac{1-3\ln x}{x^4}$

17. $y = (\ln x)^2$; $y' = 2\ln x \cdot \frac{1}{x} = \frac{2\ln x}{x}$

18. $y = \sin(\ln x)$; $y' = \cos(\ln x) \cdot \frac{1}{x} = \frac{\cos(\ln x)}{x}$ or $\frac{1}{x}\cos(\ln x)$

19. $y = \ln(\ln x)$; $y' = \frac{1}{\ln x} \cdot \frac{1}{x} = \frac{1}{x\ln x}$

20. $y = \frac{x}{\ln x}$; $y' = \frac{\ln x - x \cdot \frac{1}{x}}{(\ln x)^2} = \frac{\ln x - 1}{(\ln x)^2}$

21. Since $y = \ln x$, then $y' = \frac{1}{x} = x^{-1}$, and $y'' = -1x^{-2} = \frac{-1}{x^2}$.

22. Here $y = x\ln x$, and $y' = 1 + \ln x$, so $y'' = \frac{1}{x}$.

23. $y = \frac{1}{x}\ln x$; $y' = \frac{1}{x} \cdot \frac{1}{x} + \ln x \cdot \left(-1x^{-2}\right) =$

$\frac{1}{x^2} - \frac{\ln x}{x^2} = \frac{1-\ln x}{x^2}$; $y'' = \dfrac{x^2\left(-\frac{1}{x}\right) - (1-\ln x)2x}{x^4}$

$= \frac{-x - 2x(1-\ln x)}{x^4} = \frac{-1-2+2\ln x}{x^3} = \frac{2\ln x - 3}{x^3} = \frac{\ln x^2 - 3}{x^3}$

24. $y = \sqrt{\ln x} = (\ln x)^{1/2}$; $y' = \frac{1}{2}(\ln x)^{-1/2} \cdot \frac{1}{x} =$

$\frac{1}{2x(\ln x)^{1/2}}$; $y'' = -1\left(2x(\ln x)^{1/2}\right)^{-2}\left(2(\ln x)^{1/2} + \right.$

$\left. 2x \cdot \frac{1}{x} \cdot \frac{1}{2}(\ln x)^{-1/2}\right)\frac{-1}{4x^2\ln x} \cdot \frac{2(\ln x)^{1/2} + (\ln x)^{-1/2}}{1} =$

$\frac{-1}{4x^2(\ln x)^{3/2}} \cdot \frac{2\ln +1}{1} = -\frac{2\ln x + 1}{4x^2(\ln x)^{3/2}}$

25. $y = \ln\sqrt{\frac{\sin^2 x}{x^3}} = \frac{1}{2}\ln\frac{\sin^2 x}{x^3}$; $y' = \frac{1}{2} \cdot \frac{x^3}{\sin^2 x} \cdot$

$\frac{2x^3\sin x\cos x - 3x^2\sin^2 x}{x^6} = \frac{2x\cos x - 3\sin x}{2\sin x \cdot x} = \cot x -$

$\frac{3}{2x}$; $y'' = -\csc^2 x - \frac{3}{2}\cdot\frac{(-1)}{x^2} = \frac{3}{2x^2} - \csc^2 x$

26. $y = \ln\sqrt[3]{\frac{\sin^3 x}{x}} = \frac{1}{3}\ln\left(\frac{\sin^3 x}{x}\right)$; $y' = \frac{1}{3} \cdot \frac{x}{\sin^3 x} \cdot$

$\frac{x\cdot 3\sin^2 x\cdot\cos x - \sin^3 x}{x^2} = \frac{1}{3} \cdot \frac{3x\cos x - \sin x}{x\sin x} = \cot x -$

$\frac{1}{3x}$; $y'' - \csc^2 x + \frac{1}{3x^2} = \frac{1}{3x^2} - \csc^2 x$

27. $y = \sqrt{\ln x^2} = \left(\ln x^2\right)^{1/2}$; $y' = \frac{1}{2}\left(\ln x^2\right)^{-1/2} \cdot \frac{1}{x^2} \cdot$

$2x = \frac{1}{x\left(\ln x^2\right)^{1/2}}$; $y'' = \frac{-1}{x^2\ln x^2}\left(x \cdot \frac{1}{2}\left(\ln x^2\right)^{-1/2} \cdot\right.$

$\left.\frac{2}{x} + \left(\ln x^2\right)^{1/2}\right) = \frac{-1}{x^2\ln x^2}\left(\frac{1}{\left(\ln x^2\right)^{1/2}} + \left(\ln x^2\right)^{1/2}\right)$

$= -\frac{1+\ln x^2}{x^2\left(\ln x^2\right)^{3/2}}$

28. Here $y = \ln(x + \cos x)$, and $y' = \frac{1}{x+\cos x} \cdot$

$|1 - \sin x| = \frac{1-\sin x}{x+\cos x}$, which means that $y'' =$

$\frac{(x+\cos x)(-\cos x)-(1-\sin x)(1-\sin x)}{(x+\cos x)^2} =$

$\frac{-x\cos x-\cos^2 x-1+2\sin x-\sin^2 x}{(x+\cos x)^2} = \frac{2\sin x-x\cos x-2}{(x+\cos x)^2}$.

29. Since $y = x^2\ln(\sin 2x)$, then $y' = 2x\ln(\sin 2x) +$

$\frac{2x^2}{\sin 2x}\cos 2x = 2x\ln(\sin 2x) + 2x^2\cot 2x$, and so

$y'' = 2\ln(\sin 2x) + 2x\cdot\frac{1}{\sin 2x}\cdot\cos 2x\cdot 2 - 2x^2\csc^2 2x\cdot$

$2 + 4x\cot 2x = 2\ln(\sin 2x) + 4x\cot 2x - 4x^2\csc^2 2x$

$+4x\cot 2x = 2\ln(\sin 2x) + 8x\cot 2x - 4x^2\csc^2 2x$.

30. $y = (\cos x)(\ln x)$; $y' = \cos x \cdot \frac{1}{x} + \ln x(-\sin x) =$

$\frac{1}{x}\cos x - (\sin x)(\ln x)$; $y'' = \frac{-1}{x}\sin x - \frac{1}{x^2}\cos x -$

$\sin x \cdot \frac{1}{x} - (\cos x)(\ln x) = -\frac{2}{x}\sin x - \frac{1}{x^2}\cos x -$

$(\cos x)(\ln x) = -\frac{2x\sin x+\cos x+x^2(\cos x)(\ln x)}{x^2}$

31. (a) $N'(t) = -5\ln(0.045t) - \frac{0.225t}{0.045t} - 0.5 =$
$-5\ln(0.045t) - 5.5$. Solving $-5\ln(0.045t) - 5.5 =$
0, or $\ln(0.045t) = -1.1$, we have $0.045t = e^{-1.1}$
and so, $t = \frac{1}{0.045}e^{-1.1} \approx 7.397$ h. The maximum
number of medflies will be $N(7.397) \approx 108.96 \approx$
109.
(b) Graphing $N(t)$, we see that it is 0 at about 31.67
h, so the medflies will be eradicated within the 36
h period.

32. (a) $\frac{ds}{dt} = -2kt\ln t - kt^2\left(\frac{1}{t}\right) = -kt(2\ln t + 1)$
(b) Finding the critical value, we get $t = 0$ or
$2\ln t + 1 = 0$. Hence $\ln t = -\frac{1}{2}$ and $t = e^{-1/2}$.
$s(0)$ is undefined and $s\left(e^{-1/2}\right) =$
$-k\left(e^{-1/2}\right)^2\ln\left(e^{-1/2}\right) = -\frac{k}{e}\left(-\frac{1}{2}\right) = \frac{k}{2e}$.

33. (a) In order to find the critical values, we determine when $P'(t) = -0.008\ln(0.025t) - 0.008 = 0$.

The maximum value is when $\ln(0.025t) + 1 = 0$ or when $t = 40e^{-1} \approx 14.72$ h

(b) $P(14.72) = 0.014 - 0.008(14.72)\ln \times$
$[0.025(14.72)] = 0.014 - 0.11776\ln(0.368) \approx$
0.132 ppm.

34. Rewriting $C = \frac{kL}{\ln(r_b/r_a)} = kL\left[\ln\left(\frac{r_b}{r_a}\right)\right]^{-1} =$
$kL\left[\ln\{(r_b)(r_a)^{-1}\}\right]^{-1}$. Treating k, L, and r_b as constants, and differentiating, we obtain

$$\frac{dC}{dr_a} = -kL\left[\ln\{(r_b)(r_a)^{-1}\}\right]^{-2}\frac{1}{(r_b)(r_a)^{-1}}(-1)$$

$$\times \left[(r_b)(r_a)^{-2}\right]$$
$$= kL\left[\ln\left(\frac{r_b}{r_a}\right)\right]^{-2}\frac{r_a}{r_b}\left[\frac{r_b}{(r_a)^2}\right]$$
$$= \frac{kL}{\left[\ln\left(\frac{r_b}{r_a}\right)\right]^2(r_a)}$$

Substituting, $r_b = 1.5$ cm, $r_a = 1.0$ cm, $L = 6.0$ m $= 600$ cm, and $k = 55.6$ pF/m $= 0.556$ pF/cm, produces $\frac{dC}{dr_a} \approx 2029.2$ pF/cm.

≡ 27.6 DERIVATIVES OF EXPONENTIAL FUNCTIONS

1. $y = 4^x$; $y' = 4^x \cdot 1 \cdot \ln 4 = 4^x \ln 4$

2. $y = e^{x^2}$; $y' = e^{x^2} \cdot 2x = 2xe^{x^2}$

3. $y = 5^{\sqrt{x}}$; $y' = 5^{\sqrt{x}} \cdot \frac{1}{2\sqrt{x}} \cdot \ln 5 = \frac{5^{\sqrt{x}}}{2\sqrt{x}}\ln 5$

4. $y = 6^{x^2+x}$; $y' = 6^{x^2+x}(2x+1)\ln 6 = (2x+1)6^{x^2+x} \cdot \ln 6$

5. $y = 2^{\sin x}$; $y' = 2^{\sin x} \cdot \cos x \cdot \ln 2 = \cos x \cdot 2^{\sin x} \cdot \ln 2$

6. $y = e^{5x+3}$; $y' = e^{5x+3} = 5e^{5x+3}$

7. $y = e^{x^2+x}$; $y' = (2x+1)e^{x^2+x}$

8. $y = e^{2x}$; $y' = 2e^{2x}$

9. $y = 4^{x^4}$; $y' = 4^{x^4} \cdot 4x^3 \cdot \ln 4 = 4x^3 4^{x^4}\ln 4$

10. $y = x + e^x$; $y' = 1 + e^x$

11. $y = \frac{e^x}{x^2}$; $y' = \frac{x^2 e^x - e^x(2x)}{x^4} = \frac{e^x(x-2)}{x^3}$

12. $y = x^3 e^x$; $y' = 3x^2 e^x + x^3 e^x = x^2 e^x(3+x)$

13. $y = \frac{1+e^x}{x^2}$; $y' = \frac{x^2(e^x) - (1+e^x)2x}{x^4} = \frac{xe^x - 2 - 2e^x}{x^3} = \frac{e^x(x-2)-2}{x^3}$

14. $y = \frac{1+e^x}{e^x}$; $y' = \frac{e^x(e^x) - (1+e^x)e^x}{(e^x)^2} = \frac{e^{2x}-e^x-e^{2x}}{e^{2x}} = \frac{-e^x}{e^{2x}} = \frac{-1}{e^x} = -e^{-x}$

15. $y = e^{\tan x}$; $y' = e^{\tan x} \cdot \sec^2 x = \sec^2 x \cdot e^{\tan x}$

16. $y = e^{\cos 3x}$; $y' = e^{\cos 3x}(-\sin 3x)(3) = -3\sin 3xe^{\cos 3x}$

17. $y = \sin e^{x^2}$; $y' = \cos e^{x^2} \cdot e^{x^2} \cdot 2x = 2xee^{x^2}\cos e^{x^2}$

18. $y = \cos 2^{x^x}$; $y' = -\sin 2^{x^x} \cdot 2^{x^x} \cdot 2x \cdot \ln 2 = -2x \cdot 2^{x^x} \cdot \sin 2^{x^x}\ln 2$

19. $y = 3^x(x^3-1)$; $y' = 3^x(3x^2) + (x^3-1)3^x \cdot \ln 3 = 3x^2 3^x + (x^3-1)3^x\ln 3$

20. $y = e^{x^2-\ln 2}$; $y' = e^{x^2-\ln x}\left(2x - \frac{1}{x}\right) = \left(2x - \frac{1}{x}\right)e^{x^2-\ln x}$

21. $y = x^{\cos x}$. Using logarithmic differentiation, we obtain $\ln y = \ln x^{\cos x}$ or $\ln y = \cos x \ln x$, or $\frac{1}{y}y' = \cos x \cdot \frac{1}{x} + (-\sin x)\ln x$, and so $y' = y\left[\frac{1}{x}\cos x - \sin x \cdot \ln x\right]$, or $y' = x^{\cos x} \cdot \left[\frac{\cos x}{x} - (\sin x)(\ln x)\right]$.

22. $y = x^{x^3}$; Use logarithmic differentiation. $\ln y = x^3\ln x$; $\frac{1}{y}y' = x^3 \cdot \frac{1}{x} + 3x^2\ln x$; $y' = y[x^2 + 3x^2\ln x]$; $y = x^{x^3}[x^2 + 3x^x\ln x] = x^{x^3} \cdot x^2[1 + 3\ln x] = x^{x^3+2}[1 + 3\ln x]$

23. $y = (\sin x)^x$; Use logarithmic differentiation. Then, $\ln y = x\ln(\sin x)$, or $\frac{1}{y}y' = x \cdot \frac{1}{\sin x} \cdot \cos x + \ln(\sin x)$, and so $y' = y[x\cot x + \ln(\sin x)]$ or $y' = (\sin x)^x[x\cot x + \ln(\sin x)]$.

24. $y = e^{\arctan x}$; $y' = e^{\arctan x} \cdot \frac{1}{1+x^2} = \frac{e^{\arctan x}}{1+x^2}$

25. $y = \ln\sin e^{3x}$; $y' = \frac{1}{\sin e^{3x}}\cos e^{3x} \cdot e^{3x} \cdot 3 = \frac{3e^{3x}\cos e^{3x}}{\sin e^{3x}} = 3e^{3x}\cot e^{3x}$

26. $y = e^{\sin x}\ln\sqrt{x}$; Use the product rule with the chain rule. $y' = e^{\sin x} \cdot \frac{d}{dx}\ln\sqrt{x} + \ln\sqrt{x}\frac{d}{dx}e^{\sin x} = e^{\sin x} \cdot \frac{1}{\sqrt{x}} \cdot \frac{1}{2\sqrt{x}} + \ln\sqrt{x} \cdot e^{\sin x} \cdot \cos x = e^{\sin x}\left[\frac{1}{2x} + \cos x\ln\sqrt{x}\right]$

27. $y = e^y + y + x$; this is equivalent to $0 = e^y + x$. By implicit differentiation we obtain $0 = e^y y' + 1$ or $e^y y' = -1$ so $y' = \frac{-1}{e^y}$ or $y' = -e^{-y}$

28. $ye^x - xe^y = 1$; $(ye^x + y'e^x) - (e^y + xe^y y') = 0$; $y'e^x - xe^y y' = e^y - ye^x$; $y' = \frac{e^y - ye^x}{e^x - xe^y}$

29. $\frac{d}{dx}(\cosh x) = \frac{d}{dx}\left(\frac{e^x+e^{-x}}{2}\right) = \frac{1}{2}\left(e^x + e^{-x}(-1)\right) =$
$\frac{e^x-e^{-x}}{2} = \sinh x$

30. $\cosh^2 x - \sinh^2 x = \left(\frac{e^x+e^{-x}}{2}\right)^2 - \left(\frac{e^x-e^{-x}}{2}\right)^2 =$
$\frac{e^{2x}+2+e^{-2x}}{4} - \frac{e^{2x}-2+e^{-2x}}{4} = \frac{2}{4} + \frac{2}{4} = 1$

31. $\cosh x + \sin hx = \left(\frac{e^x+e^{-x}}{2}\right) + \left(\frac{e^x-e^{-x}}{2}\right) = \frac{2e^x}{2} =$
e^x

32. $\sinh 2x = \frac{e^{2x}-e^{-2x}}{2} = \frac{\left(e^x+e^{-x}\right)\left(e^x-e^x\right)}{2} = 2 \cdot$
$\left(\frac{e^x-e^{-x}}{2}\right)\left(\frac{e^x+e^{-x}}{2}\right) = 2\sinh x \cosh x$

33. $y = e^{\sinh 3x};\ y' = e^{\sinh 3x} \cdot \cosh 3x \cdot 3 =$
$3(\cosh 3x)e^{\sinh 3x}$

34. $f(x) = \sinh(2x + 3);\ f'(x) = \cosh(2x + 3) \cdot 2 =$
$2\cosh(2 + 3)$

35. $g(x) = \tanh\sqrt{x^3 + 4};\ g'(x) = \text{sech}^2\sqrt{x^3 + 4} \cdot$
$\frac{1}{2}\left(x^3 + 4\right)^{-1/2} \cdot 3x^2 = \frac{3x^2}{2\sqrt{x^3+4}} \cdot \text{sech}^2\sqrt{x^3 + 4}$

36. $h(x) = \sinh\left[\ln\left(x^2 + 3x\right)\right];\ h'(x) = \cos h\left[\ln\left(x^2 + \right.\right.$
$\left.\left. 3x\right)\right] \cdot \frac{1}{x^2+3x} \cdot (2x + 3) = \frac{(2x+3)\cosh\ln\left(x^2+3x\right)}{x^2+3x}$

37. $j(x) = \cosh^3\left(x^4 + \sin x\right);\ j'(x) = 3\cosh^2\left(x^4 + \right.$
$\left.\sin x\right) \cdot \sinh\left(x^4 + \sin x\right)\left(4x^3 + \cos x\right) = 3\left(4x^3 + \right.$
$\left.\cos x\right)\cosh^2\left(x^4 + \sin x\right)\sinh\left(x^4 + \sin x\right)$

38. $k(x) = \tanh^2 x + \text{sech}^2 x$; by one of the hyperbolic identities, $k(x) = 1$ so $k'(x) = 0$

39. Here $f(x) = \sinh^{-1} 7x$, so $f'(x) = \frac{1}{\sqrt{(7x)^2+1}} \cdot 7 =$
$\frac{7}{\sqrt{49x^2+1}}.$

40. Since $g(x) = \cosh^{-1}\sqrt{x}$, we have $g'(x) = \frac{1}{\sqrt{x^2-1}} \cdot$
$\frac{1}{2\sqrt{x}} = \frac{1}{2\sqrt{x-1}\sqrt{x}}.$

41. $j(x) = x\sinh^{-1}\frac{1}{x}$. Using the product rule with the chain rule we obtain $j'(x) = \sinh^{-1}\frac{1}{x} + x \cdot$
$\frac{1}{\sqrt{\left(\frac{1}{x}\right)^2 + 1}} \cdot \frac{-1}{x^2} = \sinh^{-1}\frac{1}{x} - \frac{1}{x\sqrt{\frac{1}{x^2} + 1}} =$

$\sinh^{-1}\frac{1}{x} - \frac{|x|}{x\sqrt{\frac{1+x^2}{x^2}}} = \sinh^{-1}\frac{1}{x} - \frac{|x|}{x\sqrt{1+x^2}}.$ Recall that $\sqrt{x^2} = |x|$.

42. $k(x) = \tanh^{-1}(\sin x);\ k'(x) = \frac{1}{1-\sin^2 x} \cdot \cos x =$
$\frac{\cos x}{\cos^2 x} = \frac{1}{\cos x} = \sec x$

43. The rate at which N is growing is $N'(t) = 615\left(-e^{0.02t}\right)(-0.02) = 12.3e^{-0.02t}$. After 7 days, N is growing at the rate of $N'(7) = 12.3e^{-0.14} = 10.69$ people/day.

44. (a) $N(0) = \frac{10,000}{1+99} = 100$ bacteria

(b) $N'(t) = \frac{-10,000(99)(-0.14)e^{-0.14t}}{\left(1+99e^{-0.14t}\right)^2} = \frac{138,600e^{-0.14t}}{\left(1+99e^{-0.14t}\right)^2}$
bacteria/hour

(c) $N'(t) \approx 27.35 \approx 27$ bacteria/hour

(d) $N''(t) = 138,600 \times$
$\left(\frac{\left(1+99e^{-0.14t}\right)^2(-0.14)e^{-0.14t}-e^{-0.14t} \times}{\left(1+99e^{-0.14t}\right)^4}\right) \times$
$\frac{2\left(1+99e^{-0.14t}\right)(99)(-0.14)e^{-0.1}}{}$

$= 138,600\left(\frac{-0.14e^{-0.14t}\left(1-99e^{-0.14t}\right)}{\left(1+99e^{-0.14t}\right)^3}\right).$ Setting the numerator equal to 0, and simplifying, we get $e^{0.14t} = 99$ or $0.14t = \ln 99$ and so $N''(t) = 0$ when $t = \frac{\ln 99}{0.14} \approx 32.8$ h.

(e) $N'(32.8) \approx 350$ bacteria/hour

45. (a) Using the product rule, we obtain $I = q'(t) = -5e^{-t}\cos 2.5t - 12.5e^{-t}\sin 2.5t$,

(b) $q'(0.45) \approx -8.57$ A

46. (a) $\frac{dP}{dh} = 14.7(-0.0003655)e^{-0.0003655h} = -0.00537285e^{-0.0003655h}$ in lb/in.^{2}/ft

(b) $P'(35,000) = -1.4945 \times 10^{-8}$ lb/in.^{2}/ft

47. (a) $N'(t) = 10^6(0.1)\text{sech}^2(0.1t) = 10,000\,\text{sech}^2(0.1t)$,

(b) $N'(6.93) \approx 6401$ bacteria/h,

(c) $N(6.93) = 699,906 \approx 7 \times 10^5$ bacteria

48. $\frac{dC}{dx} = Aq\cosh(qx) + Bq\sinh(qx)$ J/cm^2

≣ 27.7 APPLICATIONS

1. $y = x \ln x$. The domain is $(0, \infty)$. $y' = x \cdot \frac{1}{x} + 1 \ln x = 1 + \ln x$. Setting $y' = 0$ we obtain the critical values $1 + \ln x = 0$; $\ln x = -1$, $x = e^{-1} = 0.3679$; $y'' = \frac{1}{x}$; since $x > 0$, y'' is always positive so y is always concave up. This also means that the critical value yields a minimum. Summary: maxima: none, minima: $(e^{-1}, -e^{-1})$ or $(0.3679 - 0.3679)$; inflection points: none.

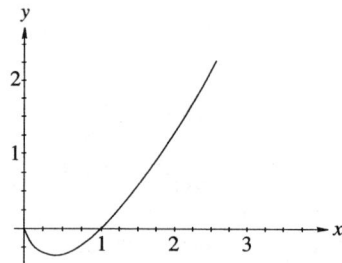

2. $y = \frac{\ln x}{x}$. The domain is $(0, \infty)$. We determine that $y' = \frac{1 - \ln x}{x^2}$; and the denominative cannot be zero since the domain is $(0, \infty)$. The critical values occur when $1 - \ln x = 0$ or $x = e$. Now, $y'' = \frac{x^2 \left(-\frac{1}{2}\right) - (1 - \ln x) 2x}{x^4} = \frac{-1 - 2 + 2 \ln x}{x^3} = \frac{2 \ln x - 3}{x^3}$.

Inflection point when $2 \ln x - 3 = 0$ or $x = e^{3/2}$. At $x = e$, we see that $y'' < 0$, so the critical value yields a maximum. Summary: Maxima: $\left(e, \frac{1}{e}\right)$, minima: none; inflection point: $\left(e^{3/2}, \frac{3}{2e^{3/2}}\right)$.

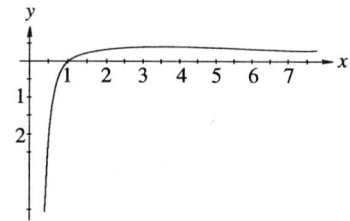

3. $y = xe^x$; $y' = xe^x + e^x = e^x(x + 1)$. The critical values occur when $y' = 0$. Since e^x cannot be 0, the only one is when $x + 1 = 0$ or $x = -1$. $y'' = e^x(1) + (x + 1)e^x = 2e^x + xe^x = e^x(2 + x)$. Inflection point is when $x = -2$; $y''(-1) > 0$ so at $x = -1$ there is a minimum. Summary: Maxima; none; minimum: $\left(-1, -e^{-1}\right)$; inflection point: $\left(-2, -2e^{-2}\right)$

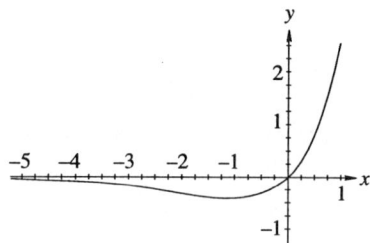

4. $y = x^2 e^x$; $y' = x^2 e^x + 2xe^x = e^x(x^2 + 2x)$. Critical values occur when $x^2 + 2x = $ or at $x = 0$ and $x = -2$. $y'' = e^x(2x + 2) + (x^2 + 2x)e^2 = e^x[x^2 + 4x + 2]$ Inflection points may occur when $x^2 + 4x + 2 = 0$. The quadratic formula yields $x = -2 + \sqrt{2}$. When $x = -2$; $y'' < 0$ so y is maximum at $x = -2$. When $x = 0$; $y'' > 0$ so y is minimum at $x = 0$. Summary: Maxima: $\left(-2, 4e^{-2}\right)$; minima: $(0, 0)$; Inflection points: $(-2 - \sqrt{2}, 0.3835370)$ and $(-2 + \sqrt{2}, 0.1910182)$.

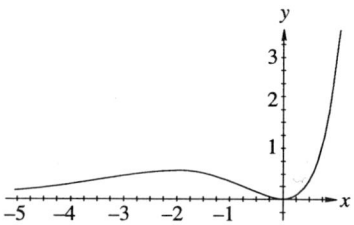

5. Here $y = \ln \frac{1}{x^2 + 1}$, and so $y' = \frac{x^2 + 1}{1} \cdot \frac{-2x}{\left(x^2 + 1\right)^2} = \frac{-2x}{x^2 + 1}$. Since the denominator cannot be 0, the only critical value is when $x = 0$. $y'' = \frac{\left(x^2 + 1\right)(-2) + 2x(2x)}{\left(x^2 + 1\right)^2} = \frac{-2x^2 - 2 + 4x^2}{\left(x^2 + 1\right)^2} = \frac{2x^2 - 2}{\left(x^2 + 1\right)^2}$. Inflection points when $x^2 - 1 = 0$ or at $x = \pm 1$. At $x = 0$; $y'' < 0$ so y is a maximum. Summary: maxima; $(0, 0)$; minima: none; inflection points: $\left(-1, \ln \frac{1}{2}\right)$; $\left(1, \ln \frac{1}{2}\right)$

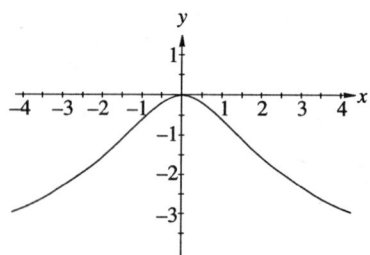

6. $y = \frac{4\ln^2 x}{x}$; the domain is $(0, \infty)$. $y' =$

$\frac{x \cdot 8\ln x \cdot \frac{1}{x} - 4\ln^2 x}{x^2} = \frac{4\ln x(2 - \ln x)}{x^2}$. Critical values occur when $\ln x = 0$ or $x = 1$ and when $2 - \ln x = 0$ or $x = e^2$. $y'' =$

$\frac{x^2\left[\frac{8}{x} - \frac{8\ln x}{x}\right] - 2x \cdot 4\ln x(2 - \ln x)}{x^4} =$

$\frac{8 - 8\ln x - 8\ln x(2 - \ln x)}{x^3} = \frac{8 - 24\ln x + 8\ln^2 x}{x^3} = \frac{8}{x^3}\left(1 - 3\ln x + \ln^2 x\right)$. The inflections occur when $\ln^2 x - 3\ln x + 1 = 0$ or, by the quadratic formula, when $\ln x = 0.381966$ or $\ln x = 2.618034$. This yields $x = 1.46516$ or $x = 13.70875$. At $x = 1$; $y'' > 0$ so y is a minimum. At $x = e^2$, we see that $y'' < 0$, so y is a maximum. Summary: Maxima: $\left(e^2, \ 2.16536\right)$; minima $(1, 0)$; Inflection points: $(1.46516, 0.39831)$ and $(13.70875, 1.99992)$.

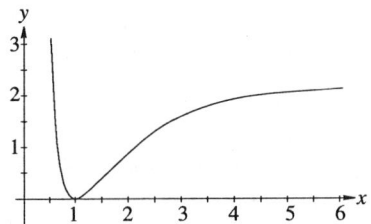

7. $y = \sinh x$; $y' = \cosh x = \frac{e^x + e^{-x}}{2}$, which is never 0. $y'' = \sinh x = \frac{e^x - e^{-x}}{2}$, which is 0 when $e^x = e^{-x}$ or when $x = 0$. Inflection point at $x = 0$. Summary: Maxima: none, minima: none, inflection point $(0, 0)$

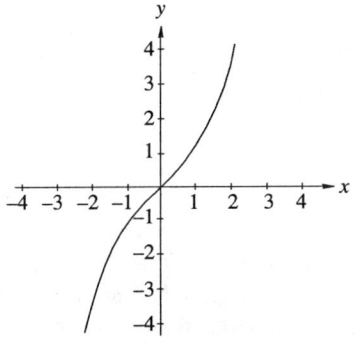

8. Since $y = \cosh x$, its derivative is $y' = \sinh x$; which is 0 when $x = 0$. The second derivative is $y'' = \cosh x$, which is never 0. At $x = 0$, $y'' > 0$, so minimum. Summary: maxima: none; minima: $(0, 1)$; inflection points: none.

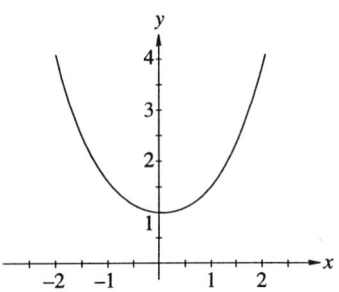

9. $y = e^{-x}\cos x$; $-2\pi \le x \le 2\pi$; $y' = -e^{-x}\sin x - \cos x e^{-x} = -e^{-x}(\sin x + \cos x)$. Critical values occur when $\sin x + \cos x = 0$ or $\sin x = -\cos x$ or $\tan x = -1$. This is at $-\frac{5\pi}{4}$, $-\frac{\pi}{4}$, $\frac{3\pi}{4}$, and $\frac{7\pi}{4}$. $y'' = -e^{-x}(\cos x - \sin x) + e^x(\sin x + \cos x) = e^{-x}(-\cos x + \sin x + \sin x + \cos x) = 2e^{-x}\sin x$. Inflection points occur when $\sin x = 0$ or $x = -2\pi, \ -\pi, \ 0, \ \pi, \ 2\pi$. At $x = -\frac{5\pi}{4}$, $y'' > 0$ so y is minima. At $x = -\frac{\pi}{4}$, $y'' < 0$ so y is maxima. At $x = \frac{3\pi}{4}$, $y'' > 0$ so y is minima. At $x = \frac{7\pi}{4}$, $y'' < 0$ so y is maxima. Summary: maxima $\left(\frac{-\pi}{4}, \ 1.5509\right)$, $\left(\frac{7\pi}{4}, \ 0.002896\right)$, minima at $\left(\frac{-5\pi}{4}, \ -35.8885\right)$ and $\left(\frac{3\pi}{4}, \ 0.06702\right)$; Inflection points at $(-2\pi, 535.49)$, $(-\pi, -23.1407)$, $(0, 1)$, $(\pi, -0.0432)$ and $(2\pi, 0.001867)$.

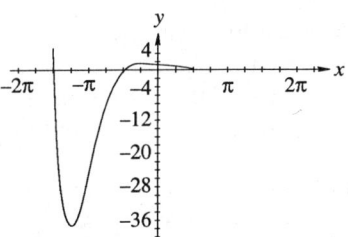

10. $y = \frac{\sin x}{\ln(x+2)}$; $0 \le x \le 4\pi$; $y' = \frac{\ln(x+2)\cos x - \frac{\sin x}{x+2}}{\ln^2(x+2)}$. This is 0 when $\ln(x+2)\cos x - \frac{\sin x}{x+2} = 0$. We can solve this using Newton's method. $f(x) = y' = \ln(x+2)\cos x - \frac{\sin x}{x+2}$; $f'(x) = y'' = -\ln(x+2)\sin x + \frac{1}{x+2}\cos x + \frac{(x+2)\cos x + \sin x}{(x+2)^2}$; $x = 1.3256$, 4.6329, 7.8094 or 10.9655. Using a computer algebraic system, we can determine the inflection points occur at 2.1469, 5.1439, 8.2376, and 11.3605. Summary: maxima: $(1.3256, 0.8073)$, $(7.8094, 0.4375)$; minima: $(4.6329, 0.5269)$, $(10.9655, -0.3901)$ inflection points: $(2.1469, 0.5896)$, $(5.1439, -0.4620)$, $(8.2376, 0.3987)$, $(11.3605, -0.3604)$

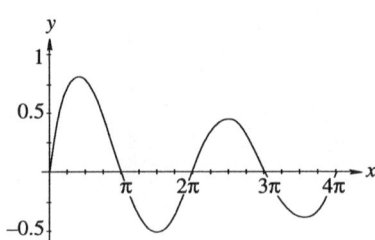

11. $y = x^3 \ln x$; $y' = 3x^2 \ln x + x^2$; at $x = 1$, $y' = 3 \cdot 1^2 \cdot \ln 1 + 1^2 = 1^2 = 1$ so the slope is 1. $y - 0 = 1(x - 1)$ or $y = x - 1$

12. If the slope in Exercise #1 is 1, then the normal is -1 so we get $y = -1(x - 1)$ or $y = -x + 1$.

13. $y = x^2 e^x$; $y' = 2xe^x + x^2 e^x$; at $x = 1$; $y' = 2e + e = 3e$. The equation of the line is $y - e = 3e(x - 1)$

14. The normal has slope $-\frac{1}{3e}$. Its equation is $y - e = -\frac{1}{3e}(x - 1)$

15. (a) $s(t) = e^{-3t}$; $v(t) = s'(t) = -3e^{-3t}$,
(b) $v'(t) = 9e^{-3t}$ which is never undefined or 0 so velocity is never maximum.

16 $s(t) = t^2 + 4\ln(t + 1)$
(a) $v(t) = s'(t) = 2t + \frac{4}{t+1}$
(b) $a(t) = v'(t) = 2 - \frac{4}{(t+1)^2} = \frac{2(t+1)^2 - 4}{(t+1)^2}$
(c) $a(t) = 0$ when $2(t + 1)^2 - 4 = 0$ or $t + 1 = \pm\sqrt{2}$ or $t = -1 \pm \sqrt{2}$; $a(t) < 0$ when $-1 - \sqrt{2} < t < -1 + \sqrt{2}$. This means the v will be a maximum at $= -1 - \sqrt{2}$ and minimum at $t = -1 + \sqrt{2}$
(d) This minimum is $v(-1+\sqrt{2}) = 2(-1+\sqrt{2}) + \frac{4}{\sqrt{2}} = -2 + 2\sqrt{2} + 2\sqrt{2} = 4\sqrt{2} - 2 \approx 3.6569$ units/s

17. $s(t) = \sin e^t$;
(a) $v(t) = s'(t) = \cos e^t \cdot e^t$; $v(t) = e^t \cos e^t$; $a(t) = v'(t) = e^t \cos e^t - \sin e^t (e^{2t})$ or $a(t) = e^t[\cos e^t - e^t \sin e^t]$
(b) $v(t) = 0$ when $\cos e^t = 0$ or $e^t = \frac{\pi}{2}$; this gives $t = \ln \frac{\pi}{2} \approx 0.4516$ s; $a(\ln \frac{\pi}{2}) = \frac{\pi}{2}[0 - $

$\frac{\pi}{2} \sin \frac{\pi}{2}] = -\left(\frac{\pi}{2}\right)^2$; $-\left(\frac{\pi}{2}\right)^2 \approx -2.4674$ units/ss.

18. $x = 5000e^{0.5t}$; the growth rate is the derivative of x or $x' = 2500e^{0.5t}$ bacteria/h.

19. $\rho(x) = xe^{-x^{2/3}}$; this will have a maximum when $\rho'(x) = 0$. $\rho'(x) = x \cdot e_x^{2/3} \cdot \left(-\frac{2}{3}x^{-1/3}\right) + e^{-x^{2/3}} = e^{-x^{2/3}}\left(1 - \frac{2}{3}x^{2/3}\right) = 0$ when $\frac{2}{3}x^{\frac{2}{3}} = 1$ or $x^{2/3} = \frac{3}{2}$ or $x = \left(\frac{3}{2}\right)^{3/2} \approx 1.8371$ m. We can use the first derivative test to verify that this is a maximum.

20. $P(x) = 10^4 e^{-0.0012x}$. The rate of change is $P'(x)$; $P'(x) = -0.0012 \times 10^4 e^{-0.0012x} = -12e^{-0.0012x}$ kg/m^2

21. $i = 1 - e^{-t^2/2L}$; $i = \frac{t}{L}e^{-t^2/2L}$; The critical value is when $t = 0$. When $t < 0$, $i' < 0$ and when $t > 0$, $i' > 0$ so this yields a maximum.

22. $Q(t) = e^{-6t}(4\cos 8t + 3\sin 8t) - 0.4\cos 10t$.
(a) $Q'(t) = e^{-6t}(-4\sin 8t \cdot 87 + 3\cos 85 \cdot 8) - 6e^{-6t}(4\cos 8t + 3\sin 8t) + 4\sin 10t = e^{-6t} \times (-50\sin 8t) + 4\sin 10t = 4\sin 10t - 50e^{-6t}\sin 8t$
(b) $Q'\left(\frac{\pi}{16}\right) = -11.69768$

23. $P(t) = 100e^{-0.015t}$; $P'(t) = -1.5e^{-0.015t}$; $P'(50) = -1.5e^{-0.015\cdot50} = -0.7085$ W/day

24. $T = 10 + 15e^{-0.875t}$; $T' = -0.875 \cdot 15 \cdot e^{-0.875t} = -13.125e^{-0.875t}$; $T'(30) = -13.125e^{-0.875\cdot30} = -5.2224 \times 15^{-15}$°C/min

25. $f(x) = x - \ln x - 2$; $f'(x) = 1 - \frac{1}{x}$. The roots are 0.1586; 3.1462.

26. $g(x) = \ln x - \sin x$; $g'(x) = \frac{1}{x} - \cos x$. The root is 2.2191.

27. $h(x) = e^x \cdot \cos x$; $h'(x) = e^x \cos x - e^x \sin x = e^x(\cos x - \sin x)$. Newton's method yields 1.570796327 and several other answers. Reexamining $h(x)$ we can see that e^x is never 0 and $\cos x = 0$ when $x = \frac{\pi}{2} + k\pi$ where k is an integer.

28. $j(x) = e^x + 2x - 2$; $j'(x) = e^x + 2$. The root is 0.3149.

≡ CHAPTER 27 REVIEW

1. $y = \sin 2x + \cos 3x$; $y' = \cos 2x \cdot 2 - \sin 3x \cdot 3 = 2\cos 2x - 3\sin 3x$

2. $y = \tan 3x^2$; $y' = \sec^2 3x^2 \cdot 6x = 6x\sec^2 3x^2$

3. $y = \tan^2 3x$; $y' = 2\tan 3x \cdot \sec^2 3x \cdot 3 = 6\tan 3x \sec^2 3x$

4. $y = \sqrt{\sin 2x} = (\sin 2x)^{1/2}$; $y' = \frac{1}{2}(\sin 2x)^{-1/2} \cdot (\cos 2x) \cdot = \frac{\cos 2x}{\sqrt{\sin 2x}} = \frac{\cos 2x\sqrt{\sin 2x}}{\sin 2x} = \cot 2x\sqrt{\sin 2x}$

5. $y = x^2 \sin x$. Using the product rule we obtain $y' = x^2 \cos x + 2x\sin x = x(x\cos x + 2\sin x)$

6. $y = \frac{\cos 3x}{3x}$. Using the quotient rule we obtain $y' = \frac{3x(-\sin 3x)3 - \cos 3x \cdot 3}{9x^2} = \frac{-3x\sin x - \cos 3x}{3x^2}$

7. $y = \sin^{-1}(3x - 2)$; $y' = \frac{1}{\sqrt{1-(3x-2)^2}} \cdot 3 = \frac{3}{\sqrt{12x - 9x^2 - 3}}$

8. $y = \cos^{-1} x^2$; $y' = \frac{-1}{\sqrt{1-x^4}} \cdot 2x = \frac{-2x}{\sqrt{1-x^4}}$

9. $y = \arctan 3x^2$; $y' = \frac{1}{1+9x^4} \cdot 6x = \frac{6x}{1+9x^4}$

10. $y = \arctan\left(\frac{1+x}{1-x}\right)$; $y' = \frac{1}{1 + \left(\frac{1+x}{1-x}\right)^2} \cdot \frac{(1-x)1 + (1+x)}{(1-x)^2}$
$= \frac{2}{(1-x)^2 + (1+x)^2} = \frac{1}{1+x^2}$

11. Here $y = \arcsin\sqrt{1-x^2}$, and so $y' = \frac{1}{\sqrt{1-(1-x)^2}} \cdot$
$\frac{1}{2}\left(1 - x^2\right)^{-1/2} \cdot (-2x) = \frac{-x}{\sqrt{x^2}\sqrt{1-x^2}} = \frac{-x}{|x|\sqrt{1-x^2}}$.
If $x > 0$ then $y' = \frac{-1}{\sqrt{1-x^2}}$, $x < 0$, $y' = \frac{1}{\sqrt{1-x^2}}$

12. Here $y = \arccos\sqrt{-3+4x-x^2}$, and so $y' = \frac{-1}{\sqrt{1-(-3+4x-x^2)}} \cdot \frac{1}{2}\left(-3+4x-x^2\right)^{-1/2}(4-2x) = \frac{-1}{\sqrt{x^2-4x+4}} \cdot \frac{2-x}{\sqrt{-3+4x-x^2}} = \frac{1}{|x-2|} \cdot \frac{2-x}{\sqrt{-3+4x-x^2}}$. If $x - 2 > 0$ then $y' = \frac{1}{-3+4x-x^2}$.

13. $y = \log_4\left(3x^2 - 5\right)$; $y' = \frac{1}{3x^2-5}(6x) \cdot \frac{1}{\ln 4} = \frac{6x}{3x^2-5} \cdot \frac{1}{\ln 4}$

14. $y = \ln(x+4)^3 = 3\ln(x+4)$; $y' = \frac{3}{x+4}$

15. $y = \ln^2\left(x^3 + 4\right)$; $y' = \frac{2\ln\left(x^3+4\right)}{x^3+4} \cdot 3x^2 = \frac{6x^2\ln\left(x^3+4\right)}{x^3+4}$

16. $y = \ln\cos x$; $y' = \frac{1}{\cos x} \cdot (-\sin x) = -\tan x$

17. $y = \ln\frac{x^3}{(4x-3)^2}$; $y' = \frac{(4x-3)^2}{x^3} \cdot \frac{(4x-3)^2 \cdot 3x^2 - x^3 \cdot 2 \cdot (4x-3) \cdot 4}{(4x-3)^4}$
$= \frac{(4x-3)^2}{x^3} \cdot \frac{3x^2(4x-3)^2 - 8x^3(4x-3)}{(4x-3)^4} = \frac{3(4x-3) - 8x}{x(4x-3)} =$
$\frac{4x-9}{x(4x-3)}$

18. $y = \ln\arctan x$; $y' = \frac{1}{\arctan x} \cdot \frac{1}{1+x^2} = \frac{1}{\left(1+x^2\right)\arctan x}$

19. $y = e^{4x^2}$; $y' = 8xe^{4x^2}$

20. $y = e^{\ln x^2} = x^2$; $y' = 2x$

21. $y = e^{\sin x^2}$; $y' = 2x\cos x^2 e^{\sin x^2}$

22. $y = exe^{x^2} + \sin x$; $y' = xe^{x^2} \cdot 2x + e^{x^2} + \cos x = e^{x^2}\left(2x^2 + 1\right) + \cos x$

23. $y = e^{-x}\sin 5x$; $y' = e^{-x}\cos 5x \cdot 5 + (-1)e^{-x}\sin 5x = e^{-x}(5\cos 5x - \sin 5x)$

24. $y = e^{-x}\ln x$; $y' = e^{-x} \cdot \frac{1}{x} + e^{-x}(-1)\ln x = e^{-x}\left(\frac{1}{x} - \ln x\right)$

25. $y = x^2 e^x$; $y' = 2xe^x + x^2 e^2 = e^x\left(x^2 + 2x\right)$. The critical values occur when $x^2 + 2x = 0$ or $x = 0$ and $x = -2$. For the inflection points, we find $y'' = 2e^x + 2xe^x + x^2 e^x + 2xe^x = e^x\left(x^2 + 4x + 2\right)$. Inflection points occur when $x^2 + 4x + 2 = 0$. The quadratic formula yields $x = -2 \pm \sqrt{2}$. When $x = -2$, then $y'' < 0$, and so y is a maximum. When $x = 0$, then $y'' > 0$, and so y is a minimum. Summary: Maxima: $(-2, 0.5413)$; minima: $(0, 0)$; inflection points: $(-2-\sqrt{2}, 0.3835)$ and $(-2+\sqrt{2}, 0.1910)$.

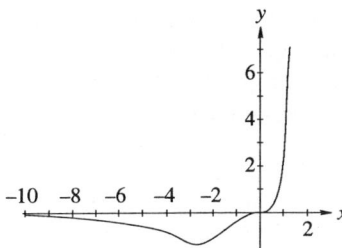

26. $y = 4e^{-9x^2}$. Finding y', we have $y' = (-18x) \times 4e^{-9x^2} = -72xe^{-9x^2}$. The only critical value is 0. Next, we find $y'' = -72e^{-9x^2} + 72x(18x)e^{-9x^2} = -72e^{-9x^2} + 1296x^2 e^{-9x^2}$. Setting this equal to 0 we find $1296x^2 - 72 = 0$, or $18x^2 = 1$, or $x = \pm\frac{1}{\sqrt{18}} = \frac{\pm\sqrt{2}}{6}$. Summary: maxima: $(0, 4)$; minima: none; inflection points: $\left(\pm\frac{\sqrt{2}}{6}, 2.4261\right)$

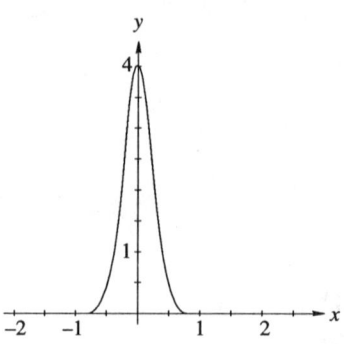

27. $y = e^{-x/2} \sin 2x$ on $[0, \ 2\pi]$; $y' = -\frac{1}{2}e^{-x/2}\sin 2x$ $+2\cos 2x e^{-x/2} = e^{-x/2}\left(2\cos 2x - \frac{1}{2}\sin 2x\right)$. Since $e^{-\frac{1}{2}x}$ never equals 0, the critical values occur when $2\cos 2x - \frac{1}{2}\sin 2x = 0$. Solving we get $2\cos 2x = \frac{1}{2}\sin 2x$ or $\tan 2x = 4$; so $2x = 1.325817, \ 4.45741,$ 7.6090, and 10.75060, and as a result $x = -.6629,$ 2.2337, 2.38045, 5.3753. Taking the second derivative, we obtain $y'' = -\frac{1}{2}e^{-x/2}\left(2\cos 2x - \frac{1}{2}\sin 2x\right) + e^{-x/2}(-4\sin 2x - \cos 2x) = e^{-x/2} \times$ $\left(-\cos 2x + \frac{1}{4}\sin 2x - 4\sin 2x - \cos 2x\right) = e^{-x/2} \times$ $\left(-2\cos 2x - \frac{15}{4}\sin 2x\right)$. Solving $2\cos 2x + \frac{15}{4}\sin 2x = 0$, we obtain $\frac{15}{4}\sin 2x = -2\cos 2x$, or $\frac{\sin 2x}{\cos 2x} = \frac{-2}{\frac{15}{4}}$, or $\tan 2x = -\frac{8}{14}$ which means that $2x = $ $2.6516, 5.7932,\ 8.9348,$ and 12.0764, and as a result, that $x = 1.3258, \ 2.8966, 4.4675, \ 6.0382.$ Summary: maxima: $(0.6629, 0.6964)$, $(3.8045,$ $0.1448)$; minima: $(2.2337, \ -0.3175)$, $(5.3753,$ $-0.0660)$; inflection points: $(1.3258, 0.2425)$, $(2.8966, \ -0.1106)$, $(4.4674, 0.0504)$, and $(6.0382, \ -0.0230)$

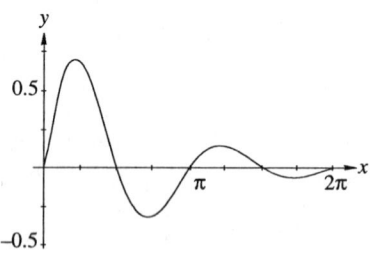

28. $y = e\sin 3x - e^{-1}\cos x$. Noting the e and e^{-1} are constants, we find $y' = e\cos x + e^{-1}\sin x$. Setting y' equal to 0, we have $e\cos x + e^{-1}\sin x = 0$, or $e^{-1}\sin x = -e\cos x$, or $\tan x = -e^2$. Hence, $x = 1.7053$ or 4.8469 are the critical values. Now, $y'' = -e\sin x + e^{-1}\cos x$. Setting this equal to

0 we have $\tan x = e^{-2}$ or $x = 0.1345$ or 3.2761. Thus, we have a maxima at $(1.7053, 2.7431)$, minima at $(4.8469, \ -2.7431)$, and inflection points at $(1.4363, 0)$ and $(3.2761, 0)$.

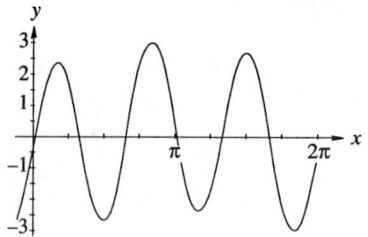

29. With $y = x^3 \ln x$, we get $y' = 3x^2 \ln x + x^2 = x^2(3\ln + 1)$. Solving $3\ln x + 1 = 0$, we obtain $\ln x = -\frac{1}{3}$ or $e^{-1/3} \approx 0.7165$. For the second derivative, we have $y'' = 6x\ln x + 3x + 2x = 6x\ln x + 5x = x(6\ln x + 5)$. Solving $6\ln x + 5 = 0$, we find $\ln x = -\frac{5}{6}$ or $x = e^{-\frac{5}{6}} \approx 0.4346$. At $x = -0.7165$, y'' is positive so y has a minimum at this point. Summary: Maxima: none; minima; $(0.7165, \ -0.1226)$; inflection point $(0.4346,$ $0.0684)$

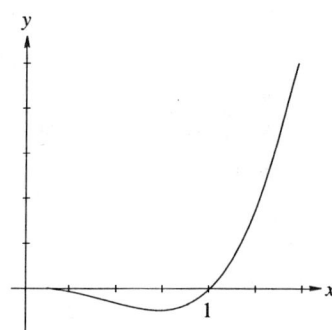

30. $y = 2\arccos\frac{x}{4}$; $y' = \frac{1}{2}\cdot\dfrac{-1}{\sqrt{1 - \frac{x^2}{16}}} = \frac{1}{2}\cdot\dfrac{-1}{\sqrt{\frac{16-x^2}{16}}} = \dfrac{-2}{\sqrt{16-x^2}}$. This is undefined at -4 and 4 which are the ends of the domain. $y(-4) = 2\pi$, $y(4) = 0$, $y' = -2(16-x)^{-1/2}$; $y'' = \left(16-x^2\right)^{-3/2}(-2x)$. This is 0 when $x = 0$ Summary: Maxima: $(-4, \ 2\pi)$; minima: $(4, 0)$; inflection point $(0, \ \pi)$

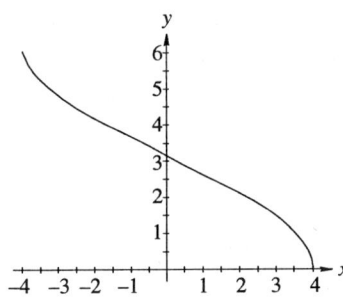

31. $y = \arctan x$; $y' = \frac{1}{1+x^2}$; $y'(1) = \frac{1}{1+1} = \frac{1}{2}$

(a) The slope of the tangent is $\frac{1}{2}$ and its equation

is $y = \frac{\pi}{4} = \frac{1}{2}(x-1)$,

(b) The slope of the normal is -2 and its equation

is $y = \frac{\pi}{4} = -2(x-1)$

32. $y = \sin^3 x$. The first derivative is $y' = 3\sin^2 x \cos x$ and so $y'\left(\frac{\pi}{4}\right) = 3 \cdot \left(\frac{\sqrt{2}}{2}\right)^2 \cdot \frac{\sqrt{2}}{2} = \frac{3\sqrt{2}}{4}$.

(a) The slope of the tangent line is $\frac{3\sqrt{2}}{4}$ and its

equation is $y = \frac{\sqrt{2}}{4} = \frac{3\sqrt{2}}{4}\left(x - \frac{\pi}{4}\right)$.

(b) The slope of the normal is $-\frac{4}{3\sqrt{2}} = \frac{-2\sqrt{2}}{3}$ and

its equation is $y - \frac{\sqrt{2}}{4} = -\frac{2\sqrt{2}}{3}\left(x - \frac{\pi}{4}\right)$.

33. $y = \ln 3x$ has derivative $y' = \frac{1}{x}$ and, when $x = 1$, the derivative is $y'(1) = 1$.

(a) The slope of the tangent at the point $(1, 0)$ is 1 and its equation is $y = x - 1$.

(b) The slope of the normal line at the point $(1, 0)$ is -1 and its equation is $y = -x + 1$ or $y = 1 - x$.

34. $y = \frac{d^x}{x}$ has the derivative $y' = \frac{xe^x - e^x}{x^2}$ and, when $x = 1$, the derivative is $y'(1) = \frac{1e - e}{1^2} = 0$.

(a) The slope of the tangent at $(1, e)$ is 0 and its equation is $y = e$.

(b) The slope of the normal at $(1, e)$ is undefined, which means that the normal line is a vertical line. Its equation is $x = 1$.

35. $y = \ln(\sin x)$ has the derivative $y' = \frac{1}{\sin x} \cdot \cos x = \tan x$ and, when $x = \frac{\pi}{4}$, the derivative is $y'\left(\frac{\pi}{4}\right) = \tan\left(\frac{\pi}{4}\right) = 1$.

(a) The slope of the tangent is 1 and its equation is $y + \ln\sqrt{2} = x - \frac{\pi}{4}$.

(b) The slope of the normal is -1 and its equation is $y + \ln\sqrt{2} = \frac{\pi}{4} - x$.

36. $y = \sin(\ln x)$ has the derivative $y' = \cos(\ln x) \cdot \frac{1}{x}$ and, when $x = 1$, the derivative is $y'(1) = \cos(\ln 1) \cdot \frac{1}{1} = \cos 0 = 1$.

(a) The slope of the tangent is 1 and its equation is $y = x - 1$.

(b) The slope of the normal is -1 and its equation is $y = 1 - x$.

37. $T = 75e^{-0.05109t} + 29$

(a) $T(10) = 75e^{-0.5109} + 29 = 64.9967°C$,

(b) $T' = (-0.05109)75e^{-0.05109t}$ and $T'(10) = (-0.05109)75e^{-0.5109} = -2.2989°C$

38. $s(t) = 10{,}000 - \frac{400}{3}\ln(\cosh 1.6t)$

(a) $v(t) = s'(t) = \sqrt{\frac{400}{3}} \cdot \frac{1}{\cosh 1.6t} \cdot \sinh 1.6t \cdot 1.6 = \frac{-640}{3}\tanh 1.6t$

(b) $v(60) = \frac{-640}{3}\tan 1.6(60) = -213.33$ ft/s

39. Using the same set-up as Example 27.20 we have $\tan(\theta + \phi) = \frac{48}{x}$ and $\tan\phi = \frac{8}{x}$. Writing $\tan\theta = \tan\left[(\theta + \phi) - \phi\right]$, we obtain $\tan\theta = \tan\left[(\theta + \phi) - \phi\right] = \frac{\tan(\theta+\phi) - \tan\phi}{1 + \tan(\theta+\phi)\tan\phi} = \frac{\frac{48}{x} - \frac{8}{x}}{1 + \frac{48}{x} \cdot \frac{8}{x}} = \frac{\frac{40}{x}}{1 + \frac{384}{x^2}} = \frac{40x}{x^2 + 384}$. Hence, $\theta = \arctan\left(\frac{40x}{x^2+384}\right)$ and $\theta' = \frac{1}{1 + \left(\frac{40x}{x^2+384}\right)^2} \cdot \frac{\left(x^2+384\right)40 - 40x(2x)}{\left(x^2+384\right)^2} = \frac{\left(x^2+384\right)^2}{\left(x^2+384\right)^2 + (40x)^2} \cdot \frac{40x^2 + 15360 - 80x^2}{\left(x^2+384\right)^2} = \frac{15360 - 40x^2}{\left(x^2+384\right)^2 + (40x)^2}$. The derivative is 0 when $15360 - 40x^2 = 0$ or $40x^2 = 15360$, or $x^2 = 384$, which is $x = 19.5959$. By the first derivative test this is a maximum. Answer 19.5959 ft.

40. $v(t) = 2$ m/s. Letting y represent the height of the balloon above the ground, we have $\tan\theta = \frac{y}{100}$. Taking the derivative of $\tan\theta = \frac{y}{100}$, we obtain $\sec^2\theta\frac{d\theta}{dt} = \frac{1}{100}\frac{dy}{dt}$ and so, $\frac{d\theta}{dt} = \frac{1}{dt} \cdot \frac{dy}{dt} \cdot \frac{1}{\sec^2\theta} = \frac{\cos^2\theta}{100}\frac{dy}{dt}$. Since the balloon is rising at the rate of 2 m/s, we have $\frac{dy}{dt} = 2$ m/s. Thus, when the balloon is 30 m above the ground, $y = 30$, and we see that we obtain $\cos\theta = \frac{100}{\sqrt{100^2 + 30^2}} = 0.957826$ and the rate of increase of the angle of inclination of the observer's line of sight when the balloon is 30 m high is $\frac{d\theta}{dt} = \frac{(0.957826)^2 \cdot 2}{100} = 0.0183$ rad/s.

CHAPTER 27 TEST

1. $f'(x) = \cos 7x \frac{d}{dx}(7x) = 7\cos 7x$

2. $g'(x) = \sec^2\left(3x^2 + 2x\right)\left[\frac{d}{dx}\left(3x^2 + 2x\right)\right] = \left[\sec^2\left(3x^2 + 2x\right)\right](6x + 2) = (6x + 2)\sec^2\left(3x^2 + 2x\right)$

3. $h'(x) = e^{2x}\frac{d}{dx}(2x) = 2e^{2x}$

4. $j'(x) = \frac{1}{5x^2}\frac{d}{dx}\left(5x^2\right) = \frac{1}{5x^2}(10x) = \frac{10x}{5x^2} = \frac{2}{x}$

5. $k'(x) = \dfrac{1}{\sqrt{1 - \left(e^{x^2}\right)^2}}\frac{d}{dx}\left(e^{x^2}\right) = \dfrac{1}{\sqrt{1 - e^{2x^2}}}\left(e^{x^2}\right)\frac{d}{dx}\times \left(x^2\right) = \dfrac{2xe^{x^2}}{\sqrt{1 - e^{2x^2}}}$

6. Using the quotient rule: $f'(x) = \dfrac{\left(1 + e^{2x}\right)4\cos 4x - \sin 4x\left(2e^{2x}\right)}{\left(1 + e^{2x}\right)^2}$

7. Using the product rule: $g'(x) = \left(7x^2 + 3x\right)\frac{d}{dx}\times \left(\tan^2 5x\right)\frac{d}{dx}\left(7x^2 + 3x\right)$. Then, by using the chain rule we obtain $g'(x) = \left(7x^2 + 3x\right)(2\tan 5x)\times \left(\sec^2 5x\right)5 + \left(\tan^2 5x\right)(14x + 3) = (\tan 5x)\left[10\left(7x^2 + 3x\right)\left(\sec^2 5x\right) + (14x + 3)(\tan 5x)\right]$.

8. Using the product rule: $h'(x) = e^{\sin x}\frac{d}{dx}(\ln \sqrt{x}) + (\ln \sqrt{x})\frac{d}{dx}\left(e^{\sin x}\right)$. We simplify this by using the chain rule: $h'(x) = e^{\sin x}\frac{1}{\sqrt{x}}\times \frac{d}{dx}\sqrt{x} + (\ln \sqrt{x})\times e^{\sin x}\frac{d}{dx}(\sin x) = e^{\sin x}\frac{1}{\sqrt{x}}\frac{1}{2\sqrt{x}} + (\ln \sqrt{x})e^{\sin x}(\cos x) = e^{\sin x}\left(\frac{1}{2x} + \cos x \ln \sqrt{x}\right)$.

9. The current, i, at a particular time is found by taking the derivative of the charge. So, $i(t) = q'(t) = -3e^{-t}\cos 2.5t - 7.5e^{-t}\sin 2.5t$. At $t = 0.65$ sec, $i(0.65) \approx -3.82$ A.

10. The velocity at time t, $v(t)$, is the derivative of the position function. So, $v(t) = x'(t) = 1.6 \times (\tan 5.0t)^{-1/2}\left(5\sec^2 5.0t\right) = 8\frac{\sec^2 5.0t}{\sqrt{\tan 5.0t}}$. When $t = 1.5$, we see that $v(1.5) = 8\frac{\sec^2 7.5}{\sqrt{\tan 7.5}} \approx 40.4743$. The particle is at 40.4743 m.

CHAPTER
28
Techniques of Integration

☰ 28.1 THE GENERAL POWER FORMULA

1. $\int \sin^3 x \cos x \, dx$. Let $u = \sin x$ and $du = \cos x \, dx$, then $\int u^3 du = \frac{u^4}{4} + C = \frac{1}{4}\sin^4 x + C$.

2. $\int \cos^4 x \sin x \, dx$. If you let $u = \cos x$ and $du = -\sin x \, dx$, then you get $-\int u^4 du = -\frac{1}{5}u^5 + C = -\frac{1}{5}\cos^5 x + C$.

3. $\int \sqrt{\sin^3 x} \cos x \, dx$. Let $u = \sin x$ and $du = \cos x \, dx$, so we have $\int u^{3/2} du = \frac{2}{5}u^{5/2} + C = \frac{2}{5}\sin^{5/2} x + C$.

4. $\int \sin 2x \cos 2x \, dx$. Let $u = \sin 2x$ and $du = 2\cos 2x \, dx$, so we have $\cos 2x \, dx = \frac{1}{2}du$. Substituting, we have $\frac{1}{2}\int u \, du = \frac{1}{4}u^2 + C = \frac{1}{4}\sin^2 2x + C$.

5. $\int x \sec^2 x^2 dx$. If you let $u = x^2$ and $du = 2x \, dx$, or $\frac{1}{2}du = x \, dx$, then you get $\frac{1}{2}\int \sec^2 u \, du = \frac{1}{2}\tan u + C = \frac{1}{2}\tan x^2 + C$.

6. $\int \cos^3 x \sin x \, dx$. If you let $u = \cos x$ and $du = -\sin x \, dx$, then you get $-\int u^3 du = -\frac{1}{4}u^4 + C = -\frac{1}{4}\cos^4 x + C$.

7. $\int \sec^2 x \tan x \, dx$. If you let $u = \tan x$ and $du = \sec^2 x \, dx$, then you get $\int u \, du = \frac{1}{2}u^2 + C = \frac{1}{2}\tan^2 x + C$ or rewrite the original integral as $\int \sec^2 x \tan x \, dx = \int \sec x(\sec x \tan x \, dx)$ and let $u = \sec x$ and $du = \sec x \tan x \, dx$. Thus, $\int u \, du = \frac{1}{2}u^2 + C = \frac{1}{2}\sec^2 x + C$.

8. $\int \frac{e^{2x}}{(1+e^{2x})^{1/2}}dx$. Here $u = 1 + e^{2x}$ and $du = 2e^{2x}dx$, so we have $\frac{1}{2}du = e^{2x}dx$. Thus, we rewrite the given integral as $\frac{1}{2}\int(u)^{1/2}du = \frac{1}{2} \cdot 2u^{1/2} + C = u^{1/2} + C = (1+e^{2x})^{1/2} + C$.

9. $\int \frac{\arccos x}{\sqrt{1-x^2}}dx$. Let $u = \arccos x$ and then $du = \frac{-1}{\sqrt{1-x^2}}dx$, with the result $-\int u \, du = \frac{-1}{2}u^2 + C = \frac{-1}{2}(\arccos x)^2 + C$.

10. $\int \frac{\arctan x}{1+x^2}dx$. Let $u = \arctan x$ and then $du = \frac{1}{1+x^2}dx$, with the result $\int u \, du = \frac{1}{2}u^2 + C = \frac{1}{2}(\arctan x)^2 + C$.

11. $\int \frac{1}{x}\sqrt{\frac{\operatorname{arccsc} x}{x^2-1}}dx$. Let $u = \operatorname{arccsc} x$ and $du = \frac{-1}{x\sqrt{x^2-1}}dx$, with the result $-\int u^{1/2}du = -\frac{2}{3}u^{3/2} + C = -\frac{2}{3}(\operatorname{arccsc} x)^{3/2} + C$.

12. $\int \frac{\arctan 6x}{1+36x^2}dx$. Let $u = \arctan 6x$ and $du = 6 \cdot \frac{1}{1+36x^2}dx$, so $\frac{1}{6}du = \frac{1}{1+36x^2}dx$. Substitution yields $\frac{1}{6}\int u \, du = \frac{1}{6} \cdot \frac{1}{2}u^2 + C = \frac{1}{12}(\arctan 6x)^2 + C$.

13. $\int \frac{(\arcsin 2x)^3}{\sqrt{1-4x^2}}dx$. Let $u = \arcsin 2x$ and $du = 2 \cdot \frac{1}{\sqrt{1-4x^2}}dx$, so $\frac{1}{2}du = \frac{dx}{\sqrt{1-4x^2}}$. Substitution yields $\frac{1}{2}\int u^3 du = \frac{1}{2} \cdot \frac{1}{4}u^4 + C = \frac{1}{8}(\arcsin 2x)^4 + C$.

14. $\int \frac{(\ln x)^3}{x}dx$. Let $u = \ln x$ and $du = \frac{dx}{x}$, and then $\int u^3 du = \frac{1}{4}u^4 + C = \frac{1}{4}(\ln x)^4 + C$.

15. $\int \frac{\left[\ln(x+4)\right]^4}{x+4}dx$. Let $u = \ln(x+4)$ and $du = \frac{dx}{x+4}$, then $\int u^4 du = \frac{1}{5}u^5 + C = \frac{1}{5}\left[\ln(x+4)\right]^5 + C$.

16. $\int \frac{[5+2\ln x]^3}{x}dx$. If you let $u = 5 + 2\ln x$ and $du =$ $\frac{2dx}{x}$, then $\frac{1}{2}du = \frac{dx}{x}$. Thus, we have $\frac{1}{2}\int u^3 du = \frac{1}{2}\cdot\frac{1}{4}u^4 + C = \frac{1}{8}[5+2\ln x]^4 + C$.

17. $\int \frac{e^x dx}{(e^x+4)^2}$. Let $u = e^x + 4$, and $du = e^x\ dx$, then you have $\int u^{-2}du = -1u^{-1}+C = -(e^x+4)^{-1}+C$.

18. $\int \left(e^x + e^{-x}\right)^{1/3}\left(e^x - e^{-x}\right)dx$. Let $u = e^x + e^{-x}$ and $du = \left(e^x - e^{-x}\right)dx$. Then we have $\int u^{1/3}du = \frac{3}{4}u^{4/3} + C = \frac{3}{4}\left(e^x + e^{-x}\right)^{4/3} + C$.

19. $\int \left(e^{2x} - 1\right)^3 e^{2x}dx$. Let $u = e^{2x} - 1$ and $du = 2e^{2x}dx$ so $\frac{1}{2}du = e^{2x}dx$. Then, $\frac{1}{2}\int u^3 du = \frac{1}{2}\cdot\frac{1}{4}u^4 + C = \frac{1}{8}\left(e^{2x} - 1\right)^4 + C$.

20. $\int \left(3e^{3x} + 1\right)^{2/3}e^{3x}dx$. Let $u = 3e^{3x} + 1$ and $du = 9e^{3x}dx$ so $\frac{1}{9}du = e^{3x}dx$. Then we have $\frac{1}{9}\int u^{2/3}du = \frac{1}{9}\cdot\frac{3}{5}u^{5/3}+C = \frac{1}{15}\left(3e^{3x}+1\right)^{5/3}+C$.

21. $\int_0^{\pi/4}\sin^3 x\cos x\ dx$. If you let $u = \sin x$ and $du = \cos x\ dx$, then you have $\int u^3 du = \frac{1}{4}u^4$. Substituting for u produces $\frac{1}{4}\sin^4 x\Big|_0^{\pi/4} = \frac{1}{4}\Big[\sin^4\frac{\pi}{4} - \sin^4 0\Big] = \frac{1}{4}\Big[\left(\frac{\sqrt{2}}{2}\right)^4 = 0^4\Big] = \frac{1}{4}\cdot\frac{1}{4} = \frac{1}{16}$.

22. $\int_{\pi/4}^{\pi/3}\tan^3 x\sec^2 x\ dx$. Let $u = \tan x$ and $du = \sec^2 x\ dx$, which leads to $\int u^3 du = \frac{1}{4}u^4$. Substituting for u produces $\frac{1}{4}\tan^4 x\Big|_{\pi/4}^{\pi/3} = \frac{1}{4}\Big[\tan^4\frac{\pi}{3} - \tan^4\frac{\pi}{4}\Big] = \frac{1}{4}[\sqrt{3}^4 - 1^4] = \frac{1}{4}[9 - 1] = 2$.

23. $\int_1^2 \frac{[\ln(x^2+1)]^3}{x^2+1}dx$. Let $u = \ln\left(x^2 + 1\right)$ and $du = \frac{2x}{x^2+1}dx$ and so $\frac{1}{2}du = \frac{x}{x^2+1}dx$. Thus, we get $\frac{1}{2}\int u^3 du = \frac{1}{8}u^4$ or $\frac{1}{8}\ln^4\left(x^2 + 1\right)\Big|_1^2 = \frac{1}{8}\left[\ln^4 5 - \ln^4 2\right] \approx 0.8098$.

24. $\int_1^{\sqrt{3}}\frac{(\arctan x)^3}{1+x^2}dx$. Let $u = \arctan x$ and $du = \frac{dx}{1+x^2}$. Then, we get $\int u^3 du = \frac{1}{4}u^4$ or $\frac{1}{4}(\arctan x)^4\Big|_1^{\sqrt{3}} = \frac{1}{4}\Big[\left(\frac{\pi}{3}\right)^4 - \left(\frac{\pi}{4}\right)^4\Big] \approx 0.2055$.

25. $\int_1^2 \frac{e^{2x}dx}{(e^{2x}-1)^2}$. Let $u = e^{2x} - 1$ and $du = 2e^{2x}dx$ so $\frac{1}{2}du = e^{2x}dx$. Then, $\frac{1}{2}\int u^{-2}du = -\frac{1}{2}u^{-1}$ or $-\frac{1}{2}\left(e^{2x} - 1\right)^{-1} = \Big|_1^2 = -\frac{1}{2}\left(\left(e^4 - 1\right)^{-1} - \left(e^2 - 1\right)^{-1}\right) \approx 0.0689$.

26. $\int_0^2 8\left(e^{2x} - 1\right)^3 e^{2x}dx$. Let $u = e^{2x} - 1$ and $du = 2e^{2x}dx$, so $4du = 8e^{2x}dx$. This leads to $4\int u^3 du = 4\cdot\frac{1}{4}u^4$. Substituting for u produces $\int_0^2 8\left(e^{2x} - 1\right)^3 e^{2x}dx = \left(e^{2x}-1\right)^4\Big|_0^2 = \Big[\left(e^4 - 1\right) - \left(e^0 - 1\right)^4\Big] = \left(e^4 - 1\right)^4 = 8.25276\times 10^6$.

27. $\int_0^\pi \sin^2 x\cos x\ dx$. Let $u = \sin x$ and $du = \cos x\times dx$, and so $\int u^2 du = \frac{1}{3}u^3$. The curve is above the x axis from 0 to $\frac{\pi}{2}$ and below from $\frac{\pi}{2}$ to π. Thus, the area is $\frac{1}{3}\sin^3 x\Big|_0^{\pi/2} - \frac{1}{3}\sin^3 x\Big|_{\pi/2}^\pi = \frac{1}{3}\Big(\sin^3\frac{\pi}{2} - \sin^3 0 - \sin^3\pi + \sin^3\frac{\pi}{2}\Big) = \frac{1}{3}(1 - 0 - 0 + 1) = \frac{2}{3}$.

28. $\int \left(e^{3x} + 1\right)e^{3x}dx$. Let $u = e^{3x} + 1$ and $du = 3e^{3x}dx$. Then, $\frac{1}{3}\int u\ du = \frac{1}{6}u^2$. Substituting for u gives $\frac{1}{6}\left(e^{3x} + 1\right)^2\Big|_0^1 = \frac{1}{6}\Big[\left(e^3 + 1\right)^2 - 2^2\Big] = \frac{1}{6}[444.599 - 4] \approx 73.43331$.

29. $\int_{0.7}^3 e^{-t}\left(e^{-t} - 2\right)^2 dt$. Let $u = e^{-t} - 2$ and $du = -e^{-t}dt$, and so $-\int u^2 du = -\frac{1}{3}u^3$. Substituting for u we obtain $\int_{0.7}^3 e^{-t}\left(e^{-t}-2\right)^2 dt = -\frac{1}{3}\Big[\left(e^{-t}-2\right)^3\Big]_{0.7}^3 = -\frac{1}{3}\Big[\left(e^{-3} - 2\right)^3 - \left(e^{-0.7} - 2\right)^3\Big] \approx 1.3397$ C.

30. $\int_{0.25}^{2.50} e^{-t}\left(e^{-t} - 5\right)^2 dt$. Substituting $u = e^{-t} - 5$ and $du = -e^{-t}dt$, produces $-\int u^2 du = -\frac{1}{3}u^3$, which leads to $\int_{0.25}^{2.50} e^{-t}\left(e^{-t} - 5\right)^2 dt = -\frac{1}{3}\Big[\left(e^{-t} - 5\right)^3\Big]_{0.25}^{2.50} = -\frac{1}{3}\Big[\left(e^{-2.50} - 5\right)^3 - \left(e^{-0.25} - 5\right)^3\Big] \approx 14.5762$ C.

31. **(a)** $N(t) = \int -\frac{1020}{t^2} - 340 dt = \frac{1020}{t} - 340t + C$. Since $N(1) = 1020 - 340 + C = 9750$ and so, $C = 9070$. Thus, $N(t) = \frac{1020}{t} - 340t + 9070$.

(b) $N(10) = \frac{1020}{10} - 340(10) + 9070 = 102 - 3400 + 9070 = 5772$.

32. **(a)** $M(t) = \int \sqrt{3t^2 + 18t}(6t + 18)dt = \int \left(3t^2 + 18t\right)^{1/2}(6t + 18)dt = \frac{2}{3}\left(3t^2 + 18t\right)^{3/2} + C$. Since $M(3) = \frac{2}{3}\left[3(3)^2 + 18(3)\right]^{3/2} + C = 486 + C = 945$, then $C = 459$ and $M(t) = \frac{2}{3}\left(3t^2 + 18t\right)^{3/2} + 459$.

(b) $M(t) = 7200$ means that

$$\frac{2}{3}(3t^2 + 18t)^{3/2} + 459 = 7200$$

$$\frac{2}{3}(3t^2 + 18t)^{3/2} = 6741$$

$$(3t^2 + 18t)^{3/2} = 10{,}111.5$$
$$3t^2 + 18t = 10{,}111.5^{2/3} \approx 467.6$$
$$3t^2 + 18t - 467.6 = 0$$

By the quadratic formula, $t \approx -15.84$ or 9.84. Since $t \geq 0$, the machine should be replaced in about 9.84 yr or 9 yr 10 mo.

28.2 BASIC LOGARITHMIC AND EXPONENTIAL INTEGRALS

1. $\int \frac{dx}{4x+1}$. Let $u = 4x + 1$ and $du = 4\,dx$, so $\frac{1}{4}du = dx$. Then, $\frac{1}{4}\int \frac{du}{u} = \frac{1}{4}|u| + C = \frac{1}{4}\ln|4x+1| + C$.

2. $\int \frac{dx}{9x-5}$. Let $u = 9x - 5$ and $du = 9\,dx$ and $\frac{1}{9}du = dx$. Then, $\frac{1}{9}\int \frac{du}{u} = \frac{1}{9}\ln|u| + C = \frac{1}{9}\ln|9x-5| + C$.

3. $\int e^{-3x}dx$. Let $u = -3x$ and $du = -3\,dx$, or $-\frac{1}{3}du = dx$, which leads to $-\frac{1}{3}\int e^u du = -\frac{1}{3}e^u + C = -\frac{1}{3}e^{-3x} + C$.

4. $\int e^{6x}dx$. Here let $u = 6x$ and $du = 6\,dx$ or $\frac{1}{6}du = dx$. This leads to $\frac{1}{6}\int e^u\,du = \frac{1}{6}e^u + C = \frac{1}{6}e^{6x} + C$.

5. $\int \frac{\sin x}{\cos x}dx$. Let $u = \cos x$ and $du = -\sin x\,dx$, then you have $-\int \frac{du}{u} = -\ln|u| + C = -\ln|\cos x| + C$.

6. $\int \frac{10x+3}{5x^2+3x-7}dx$. Here $u = 5x^2 + 3x - 7$ and $du = (10x+3)dx$, and so $\int \frac{du}{u} = \ln|u| + C = \ln|5x^2 + 3x - 7| + C$.

7. $\int \frac{2\sec^2 x}{\tan x}$. If you let $u = \tan x$ and $du = \sec^2 x\,dx$, then $2\,du = 2\sec^2 x\,dx$. Substituting we get $2\int \frac{du}{u} = 2\ln|u| + C = 2\ln|\tan x| + C$.

8. $\int \frac{2e^x}{e^x+4}dx = 2\int \frac{e^x}{e^x+4}dx$. If you let $u = e^x + 4$, then $du = e^x\,dx$. Substituting we get $2\int \frac{du}{u} = 2\ln|u| + C = 2\ln|e^x + 4| + C$. Since $e^x > 0$ the absolute value signs are not needed. The final answer is $2\ln(e^x + 4) + C$.

9. $\int 4^x dx = \frac{4^x}{\ln 4} + C$.

10. $\int (e^x - e^{-x})dx = \int e^x dx + \int -e^{-x}dx = e^x + e^{-x} + C$. (Note: for the second integral let $u = x$ then $du = -dx$).

11. $\int (e^x + e^{-x})dx = \int e^x dx - \int -e^{-x}dx = e^x - e^{-x} + C$. (Note: for the second integral let $u = -x$, then $du = -dx$)

12. $\int \sec^2 x e^{\tan x}dx$. Here you should let $u = \tan x$, and $du = \sec^2 x\,dx$. Then, you have $\int e^u du = e^u + C = e^{\tan x} + C$.

13. $\int \frac{e^{2x}}{1+e^{2x}}dx$. Let $u = 1 + e^{2x}$, so $du = e^{2x}2\,dx$ or $\frac{1}{2}du = e^{2x}dx$. Substituting we get $\frac{1}{2}\int \frac{du}{u} = \frac{1}{2}\ln|u| + C = \frac{1}{2}\ln(1+e^{2x}) + C = \ln\sqrt{1+e^{2x}} + C$.

14. $\int \frac{x}{1+x^2}dx$. Here $u = 1 + x^2$ and $du = 2x\,dx$, so $\frac{1}{2}du = x\,dx$. Substituting we get $\frac{1}{2}\int \frac{du}{u} = \frac{1}{2}\ln|u| + C = \frac{1}{2}\ln(1+x^2) + C$ or $\ln\sqrt{1+x^2} + C$.

15. $\int \frac{\ln(1/x)}{x}dx$. Let $u = \ln\frac{1}{x} = \ln(x)^{-1} = -\ln x$, and then $du = -\frac{1}{x}dx$. Substituting we get $-\int u\,du = -\frac{1}{2}u^2 + C = -\frac{1}{2}\left[\ln\left(\frac{1}{x}\right)\right]^2 + C$.

16. $\int \frac{e^{\sqrt{x}}}{8\sqrt{x}}dx$. Let $u = \sqrt{x} = x^{1/2}$; $du = \frac{1}{2}x^{-1/2}dx = \frac{1}{2\sqrt{x}}dx$ so $\frac{1}{4}du = \frac{1}{8\sqrt{x}}dx$. Substituting we get $\frac{1}{4}\int e^u du = \frac{1}{4}e^u + C = \frac{1}{4}e^{\sqrt{x}} + C$.

17. $\int \frac{e^{\sqrt[3]{x}}}{x^{2/3}}dx$. Let $u = \sqrt[3]{x} = x^{1/3}$; $du = \frac{1}{3}x^{-2/3}dx$ so $3\,du = \frac{1}{x^{2/3}}dx$. Substituting we get $3\int e^u du = 3e^u + C = 3e^{\sqrt[3]{x}} + C$.

18. $\int \frac{(\ln x)^{3/4}}{x}dx$. Let $u = \ln x$ and $du = \frac{1}{x}dx$. Substituting we get $\int u^{3/4}du = \frac{4}{7}u^{7/4} + C = \frac{4}{7}(\ln x)^{7/4} + C$.

19. $\int_0^4 \frac{x}{x^2+1}dx$. Let $u = x^2 + 1$ and $du = 2x\,dx$ or $\frac{1}{2}du = x\,dx$. Substituting we get $\frac{1}{2}\int \frac{du}{u} = \frac{1}{2}\ln|u| = \frac{1}{2}\ln(x^2 + 1)$. This leads to $\frac{1}{2}\ln(x^2 + 1)\big|_0^4 = \frac{1}{2}\big[\ln(17) - \ln(1)\big] = \frac{1}{2}\ln 17 \approx 1.4166$.

20. $\int_0^1 \frac{dx}{e^{2x}} = \int_0^1 e^{-2x}dx$. Let $u = -2x$ and $du = -2\,dx$ or $-\frac{1}{2}du = dx$. Substituting we get $-\frac{1}{2}\int e^u du = -\frac{1}{2}e^u = -\frac{1}{2}e^{-2x}$. Evaluating from 0 to 1 we have $-\frac{1}{2}e^{-2x}\big|_0^1 = -\frac{1}{2}\left[e^{-2} - e^0\right] = -\frac{1}{2}\left[e^{-2} - 1\right] = \frac{1}{2}\left[1 - e^{-2}\right] \approx 0.4323$.

21. $\int_1^3 \frac{1}{\ln e^x}dx$. Recall that $\ln e^x = x$ so $\int_1^3 \frac{1}{x}dx = \ln|x|\,\big|_1^3 = \ln 3 - \ln 1 = \ln 3 \approx 1.0986$.

22. $\int_0^3 \frac{x^2}{x^3+1}dx$. Let $u = x^3 + 1$ and $du = 3x^2 dx$ or $\frac{1}{3}du = x^2 dx$. Thus, we have $\frac{1}{3}\int \frac{1}{u}du = \frac{1}{3}\ln|u|$. Evaluating produces $\frac{1}{3}\ln\left(x^3 + 1\right)\big|_0^3 = \frac{1}{3}(\ln 28 - \ln 1) = \frac{1}{3}\ln 28 \approx 1.1107$.

23. $\int_0^1 x^3 e^{x^4} dx$. Let $u = x^4$ and $du = 4x^3 dx$ or $\frac{1}{4}du = x^3 dx$. Substituting produces $\frac{1}{4}\int e^u du = \frac{1}{4}e^u = \frac{1}{4}e^{x^4}$ which leads to $\frac{1}{4}e^{x^4}\big|_0^1 = \frac{1}{4}\left(e^1 - e^0\right) = \frac{1}{4}(e - 1) \approx 0.4296$.

24. $\int_{\pi/4}^{\pi/3} \frac{dx}{(x^2+1)\arctan x}$. Let $u = \arctan x$, then $du = \frac{1}{x^2+1}dx$. Substituting we get $\int \frac{1}{u}du = \ln|u| = \ln|\arctan x|\,\big|_{\pi/4}^{\pi/3} = \ln \arctan \frac{\pi}{3} - \ln \arctan \frac{\pi}{4} = 0.19417$.

25. $\int_2^4 \left(\frac{1}{5}\right)^{x/2} dx$. Let $u = \frac{x}{2}$ and $du = \frac{1}{2}dx$. Substituting, we get $2\int \left(\frac{1}{5}\right)^u du = 2 \cdot \frac{\left(\frac{1}{5}\right)^u}{\ln \frac{1}{5}} = \frac{2\left(\frac{1}{5}\right)^{x/2}}{\ln \frac{1}{5}}$. Evaluating, we have $\frac{2\left(\frac{1}{5}\right)^{x/2}}{\ln \frac{1}{5}}\bigg|_2^4 = \frac{2}{\ln \frac{1}{5}}\left[\left(\frac{1}{5}\right)^2 - \frac{1}{5}\right] = 0.1998$.

26. $\int_{\ln 2}^{\ln 5} \frac{e^x}{e^x+1}dx$. Let $u = e^x + 1$ and $du = e^x dx$. Hence, we have $\int \frac{1}{u}du = \ln|u| = \ln\left(e^x + 1\right)$. Evaluating, we get $\ln\left(e^x + 1\right)\big|_{\ln 2}^{\ln 5} = \ln\left(e^{\ln 5} + 1\right) - \ln\left(e^{\ln 2} + 1\right) = \ln(5+1) - \ln(2+1) = \ln 6 - \ln 3 = \ln \frac{6}{3} = \ln 2 \approx 0.6931$.

27. Since $e^{x+1} > x^2$ on $[0, 1]$, the area we want is $\int_0^1 \left(e^{x+1} - x^2\right)dx = \int_0^1 e^{x+1} - \int_0^1 x^2 dx = e^{x+1}\big|_0^1 - \frac{x^3}{3}\big|_0^1 = e^2 - e^1 - \frac{1^3}{3} = e^2 - e - \frac{1}{3} \approx 4.3374$.

28. This area is $\int_0^4 e^{-x}dx = -e^{-x}\big|_0^4 = -e^{-4} + e^0 = 1 - e^{-4} = 0.9817$.

29. **(a)** $s(t) = \int \left(394e^{-0.025t} - 384\right)dt = -15760e^{-0.025t} - 384t + C$. At $t = 0$, the object is on the ground, so $s(0) = 0$ and we see that, $C = 15,760$. Thus, the position function is $s(t) = -15,760e^{-0.025t} - 384t + 15,760$;
(b) The object reaches its maximum height when $v(t) = 0$ or when $e^{-0.025t} = \frac{384}{394}$. Taking the natural logarithm of both sides we get $-0.025t = \ln\left(\frac{384}{394}\right) = \ln 384 - \ln 394$. So, $t = \frac{\ln 384 - \ln 394}{-0.025} \approx 1.028$;
(c) $s(1.028) = 5.1196$.

30. The average value is $\frac{1}{2}\int_0^2 e^{(-2/3)t}dt$. Let $u = -\frac{2}{3}t$, then $du = -\frac{2}{3}dt$ or $-\frac{3}{2}du = dt$. Then, $\frac{1}{2} \cdot \frac{-3}{2}\int e^u du = \frac{-3}{4}e^{(-2/3)t}$. Hence, $-\frac{3}{4}e^{-(2/3)t}\big|_0^2 = -\frac{3}{4}\left[e^{-4/3} - e^0\right] = \frac{3}{4}\left[1 - e^{-4/3}\right] = 0.5523$.

31. Using the disc method this volume is $\pi \int_0^2 \left(e^{-x}\right)^2 dx = \pi \int_0^2 e^{-2x}dx$. Let $u = -2x$, then $du = -2\,dx$ or $-\frac{1}{2}du = dx$. Thus, the integral becomes $\pi \cdot -\frac{1}{2}\int e^u du = \frac{-\pi}{2}e^u = -\frac{\pi}{2}e^{-2x}$. Evaluating, we get $\frac{-\pi}{2}e^{-2x}\big|_0^2 = \frac{-\pi}{2}\left(e^{-4} - e^0\right) = \frac{\pi}{2}\left(1 - e^{-4}\right) \approx 1.5420$.

32. **(a)** $v = 200e^{-50t}$. The average voltage is $\frac{1}{0.02} \times \int_0^{0.02} 200e^{-50t}dt = \frac{200}{0.02} \cdot \left(\frac{-1}{50}\right)e^{-50t}\big|_0^{0.02} = -200e^{-50t}\big|_0^{0.02} = -200\left(e^{-1} - e^0\right) = 200\left(1 - e^{-1}\right) = 126.4241$ V
(b) The effective voltage from $t = 0$ to $t = 0.02$ s is

$$v_{\text{eff}} = \sqrt{\frac{1}{0.02}\int_0^{0.02} \left(200e^{-50t}\right)^2 dt}$$

$$= \sqrt{2,000,000 \int_0^{0.02} e^{-100t}dt}$$

$$= \sqrt{(2,000,000)\left(\frac{-1}{100}\right)e^{-100t}\big|_0^{0.02}}$$

$$= \sqrt{(2,000,000)\left(\frac{-1}{100}\right)\left(e^{-2t} - e^0\right)}$$

$$= \sqrt{(20,000)\left(1 - e^{-2}\right)}$$

$$\approx \sqrt{17,293.29434} \approx 131.5040$$

The effective voltage is about 131.5040 V.

33. Since $\frac{dV}{ds} = -0.45e^{-0.15s}$, we see that $V = \int -0.45e^{-0.15s}ds = \frac{-0.45}{-0.15}e^{-0.15s} + C = 3e^{-0.15s} + C$. Hence, $V(0) = 3 = 3e^0 + C = 3 + C$ so $C = 0$ and $V(s) = 3e^{-0.15s}$. Also, $3e^{-0.15s} = 1.5$

yields $e^{-0.15s} = \frac{1}{2}$. Taking the natural logarithm of both sides we get $-0.15s = \ln\frac{1}{2}$ or $s = \ln(0.5) \div (-0.15) = 4.6210$ mi.

34. Since $P = \frac{k}{V}$, we can solve for $k = PV = 500 \cdot 0.200 = 100$. Thus, the work done by the gas is given by $W = \int_{V_0}^{V_1} \frac{k}{V} dV = \int_{0.200}^{0.800} \frac{100}{V} dV = 100 \ln|V| \big|_{0.2}^{0.8} = 100(\ln 0.8 - \ln 0.2) \approx 138.6294\,\text{N} \cdot \text{m} = 138.6$ J.

35. First we find $k = PV = 20 \times 0.4 = 8$. Then, the work done by the gas is $W = \int_{0.400}^{1.500} \frac{8}{V} dN = 8(\ln 1.5 - \ln 0.4) \approx 10.574\,\text{ft} \cdot \text{lb}$.

36. $T(x) = \int \frac{4}{2x+9} dx = 2\int \frac{2}{2x+9} dx = 2\ln(2x+9) + C$. Since $T(1) = 2\ln(11) + C = 2.4$, then $C = 2.4 - 2\ln 11 \approx -2.4$. Thus, $T(x) = 2\ln(2x+9) - 2.4$.

37. $F(x) = \int_0^x 12te^{-3t^2} dt = -2\int_0^x (-6t)e^{-3t^2} dt = -2e^{-3t^2}\big|_0^x = -2e^{-3x^2} - \left(-2e^{-3\cdot 0^2}\right) = -2e^{-3x^2} + 2 = 2 - 2e^{-3x^2}$.

38. $C(t) = \int C'(t) dt = \int 2^{-t} dt = -\frac{1}{\ln 2} 2^{-t} + K$. Since $C(0) = \frac{-1}{\ln 2} + K = 1$, we see that $K = 1 + \frac{1}{\ln 2}$. Hence, $C(t) = 1 + \frac{1 - 2^{-t}}{\ln 2}\mu\text{g}$.

28.3 BASIC TRIGONOMETRIC AND HYPERBOLIC INTEGRALS

1. $\int \sin\frac{x}{2} dx$. Let $u = \frac{x}{2}$ and $du = \frac{1}{2} dx$ gives $2\int \sin u\, du = -2\cos u + C = -2\cos\frac{x}{2} + C$.

2. $\int \cos 3x\, dx$. Let $u = 3x$, then $du = 3\, dx$ and $\frac{1}{3} du = dx$. This gives $\frac{1}{3}\int \cos u\, du = \frac{1}{3}\sin u + C = \frac{1}{3}\sin 3x + C$.

3. $\int \tan 5x\, dx$. Let $u = 5x$, then $du = 5\, dx$ and $\frac{1}{5} du = dx$. This gives $\frac{1}{5}\int \tan u\, du = -\frac{1}{5}\ln|\cos u| + C = -\frac{1}{5}\ln|\cos 5x| + C$ or $\frac{1}{5}\ln|\sec 5x| + C$.

4. $\int x\csc x^2 dx$. Let $u = x^2$ then $du = 2x\, dx$ and $\frac{1}{2} du = x\, dx$. This gives $\frac{1}{2}\int \csc u\, du = \frac{1}{2}\ln|\csc u - \cot u| + C = \frac{1}{2}\ln|\csc x^2 - \cot x^2| + C$.

5. $\int \frac{1}{\cos x} dx = \int \sec x\, dx = \ln|\sec x + \tan x| + C$

6. $\int \frac{1}{\sin x} dx = \int \csc x\, dx = \ln|\csc x - \cot x| + C$

7. $\int \frac{1}{\cos^2 x} dx = \int \sec^2 x\, dx = \tan x + C$

8. $\int \frac{\csc x}{\sin x} dx = \int \csc^2 x\, dx = -\cot x + C$

9. $\int \frac{\tan\sqrt{x}}{\sqrt{x}} dx$. Let $u = \sqrt{x}$ then $du = \frac{1}{2\sqrt{x}} dx$ and $2du = \frac{1}{\sqrt{x}} dx$. Substituting produces $2\int \tan u\, du = -2\ln|\cos u| + C = -2\ln|\cos\sqrt{x}| + C$ or $2\ln|\sec\sqrt{x}| + C$.

10. $\int \tan^2 6x\, dx = \int \left(\sec^2 6x - 1\right) dx = \frac{1}{6}\tan 6x - x + C$.

11. $\int \sin(4x - 1) dx = -\frac{1}{4}\cos(4x - 1) + C$

12. $\int \frac{\sin x + \cos x}{\cos x} dx = \int(\tan x + 1) dx = \ln|\sec x| + x + C$ or $x - \ln|\cos x| + C$.

13. $\int \frac{\sin x}{\cos^2 x} dx$. Let $u = \cos x$ and $du = -\sin x\, dx$. Substituting, we obtain $-\int u^{-2} du = u^{-1} + C = (\cos x)^{-1} + C = \sec x + C$.

14. $\int \frac{1 + \sin x}{\cos x} dx = \int \left(\frac{1}{\cos x} + \tan x\right) dx = \int(\sec x + \tan x) dx = \ln|\sec x + \tan x| + \ln|\sec x| + C$.

15. $\int \frac{\sec 3x\tan 3x}{5 + 2\sec 3x} dx$. Let $u = 5 + 2\sec 3x$, and then $du = 6\sec 3x\tan 3x\, dx$ or $\frac{1}{6} du = \sec 3x\tan 3x\, dx$. This yields $\frac{1}{6}\int \frac{1}{u} du = \frac{1}{6}\ln|u| + C = \frac{1}{6}\ln|5 + 2\sec 3x| + C$.

16. $\int \cos^2 x\sin x\, dx$. Let $u = \cos x$ and $du = -\sin x\, dx$. Substituting yields $-\int u^2 du = -\frac{1}{3}u^3 + C = -\frac{1}{3}\cos^3 x + C$.

17. $\int \sec^4 3x\tan 3x\, dx = \int \sec^3 3x\sec 3x\tan 3x\, dx$. If you let $u = \sec 3x$, then $du = 3\sec 3x\tan 3x\, dx$. This leads to $\frac{1}{3}\int u^3 du = \frac{1}{12}u^4 + C = \frac{1}{12}\sec^4 3x + C$.

18. $\int \tan 3x\sec^2 3x\, dx = \int \sec 3x\tan 3x\sec 3x\, dx$. If you let $u = \sec 3x$, then $du = 3\sec 3x\tan 3x\, dx$. Hence, $\frac{1}{3}\int u\, du = \frac{1}{6}u^2 + C = \frac{1}{6}\sec^2 3x + C$. An alternate method is to let $u = \tan 3x$ and $du = 3\sec^2 3x\, dx$, and so $\frac{1}{6}\int u\, du = \frac{1}{6}u^2 + C = \frac{1}{6}\tan^2 3x + C$.

19. $\int \frac{3\,dx}{\tan 3x} = \int 3\cot(3x)dx = \ln|\sin 3x| + C$

20. $\int \frac{5\,dx}{\sin^2 5x} = \int 5\csc^2 5x\ dx = -\cot 5x + C$

21. $\int (\sec x + 2)^2 dx = \int \left(\sec^2 x + 4\sec x + 4\right)dx = \tan x + 4\ln|\sec x + \tan x| + 4x + C$

22. $\int (\tan x + 3)^2 dx = \int \left(\tan^2 x + 6\tan x + 9\right)dx = \int \left(\sec^2 x - 1 + 6\tan x + 9\right)dx = \tan x - x + 6\ln|\sec x| + 9x + C = \tan x + 6\ln|\sec x| + 8x + C$

23. $\int_0^{\pi/2} (\sec 0.5x + 5)^2 dx = \int_0^{\pi/2}$
$\times \left(\sec^2 0.5x + 10\sec 0.5x + 25\right)dx$
$= \left[2\tan 0.5x + 20\ln|\sec 0.5x + \tan 0.5x| + 25x\right]_0^{\pi/2}$
$= \left[2\tan\left(\frac{\pi}{4}\right) + 20\ln\left|\sec\left(\frac{\pi}{4}\right) + \tan\left(\frac{\pi}{4}\right)\right| + 25\left(\frac{\pi}{2}\right)\right]$
$- \left[2\tan 0 + 20\ln|\sec 0 + \tan 0| + 25(0)\right]$
$= \left[2(1) + 20\ln\left|\sqrt{2} + 1\right| + \frac{25\pi}{2}\right]$
$- \left[2(0) + 20\ln|1 + 0| + 25(0)\right]$
$= 2 + 20\ln\left|\sqrt{2} + 1\right| + \frac{25\pi}{2} \approx 58.897$

24. $\int_{\pi/12}^{\pi/6} (\cot 3x - 4)^2 dx$
$= \int_{\pi/12}^{\pi/6} \left(\cot^2 3x - 8\cot 3x + 16\right) dx$
$= \left[-\frac{1}{3}\cot 3x - \frac{8}{3}\ln|\sin 3x| + 15x\right]_{\pi/12}^{\pi/6}$
$= \left[-\frac{1}{3}(0) - \frac{8}{3}\ln|1| + 15\left(\frac{\pi}{6}\right)\right]$
$- \left[-\frac{1}{3}1 - \frac{8}{3}\ln\left|\frac{\sqrt{2}}{2}\right| + 15\left(\frac{\pi}{12}\right)\right]$
$= 15\left(\frac{\pi}{12}\right) + \frac{1}{3} + \frac{8}{3}\ln\left|\frac{\sqrt{2}}{2}\right| \approx 6.1459$

25. $\int \frac{1+\cos 4x}{\sin^2 4x} dx = \int \left(\frac{1}{\sin^2 4x} + \frac{\cos 4x}{\sin^2 4x}\right)dx = \int \csc^2 4x\times$
$dx + \int \frac{\cos 4x}{\sin^2 4x} dx$. Now, the first integral is $\int \csc^2 4x$
$= \frac{-1}{4}\cot^2 4x + C$. For the second integral, let
$u = \sin 4x$ then $du = 4\cos 4x\ dx$ or $\frac{1}{4}du = \cos 4x\ dx$. Substitution yields $\frac{1}{4}\int u^{-2}du = \frac{1}{4}\cdot$
$-1u^{-1} + C = -\frac{1}{4}(\sin 4x)^{-1} + C = -\frac{1}{4}\csc 4x + C$. Putting these together we get $-\frac{1}{4}\cot^2 4x -$

$\frac{1}{4}\csc 4x + C = -\frac{1}{4}\left[\cot^2 4x + \csc 4x\right] + C$ or
$-\frac{1}{4}\left(\frac{\cos 4x + 1}{\sin 4x}\right) + C.$

26. $\int x\csc^2 x^2 dx$. Let $u = x^2$ and $du = 2x\ dx$ or $\frac{1}{2}du = x\ dx$. Then, you get $\frac{1}{2}\int \csc^2 u\ du = \frac{-1}{2}\cot u + C = -\frac{1}{2}\cot x^2 + C.$

27. $\int \tan\frac{x}{4}dx = 4\ln\left|\sec\frac{x}{4}\right| + C$

28. $\int x\sec^2\left(x^2+1\right)\tan\left(x^2+1\right)dx$. Let $u = \sec\left(x^2+1\right)$ then $du = 2x\sec\left(x^2+1\right)\tan\left(x^2+1\right)dx$. Then, you get $\frac{1}{2}\int u\ du = \frac{1}{4}u^2 + C = \frac{1}{4}\sec^2\left(x^2+1\right) + C$ or let $u = \tan\left(x^2 + 1\right)$ which yields the answer $\frac{1}{4}\tan^2\left(x^2 + 1\right) + C.$

29. Using trig identities, we get $\int_{\pi/4}^{\pi/2} \frac{1+\cot^2 x}{\csc^2 x}dx = \int_{\pi/4}^{\pi/2} \left(\frac{1}{\csc^2 x}+\cos^2 x\right)dx = \int_{\pi/4}^{\pi/2} \left(\sin^2 x+\cos^2 x\right)dx = \int_{\pi/4}^{\pi/2} dx = x\big|_{\pi/4}^{\pi/2} = \frac{\pi}{2} - \frac{\pi}{4} = \frac{\pi}{4}.$

30. $\int_0^{\pi/2} \frac{\cos x}{1+\sin x}dx$. Let $u = 1 + \sin x$ and $du = \cos x\ dx$. This leads to $\int \frac{1}{u}du = \ln|u| = \ln|1 + \sin x|\big|_0^{\pi/2} = \ln|2| - \ln|1| = \ln 2 \approx 0.6931.$

31. $\int_{\pi/4}^{\pi/2} \frac{\csc\sqrt{x}\cot\sqrt{x}}{\sqrt{x}}dx$. Let $u = \sqrt{x}$ and $du = \frac{1}{2\sqrt{x}}$. Then, $2\int \csc u\cot u\ du = -2\csc u$. Evaluating, we obtain $-2\csc\sqrt{x}\big|_{\pi/4}^{\pi/2} = 0.4765.$

32. $\int_0^{\pi/8} \sec^5 2x\tan 2x\ dx = \int_0^{\pi/8} \sec^4 2x\sec 2x\tan 2x\ dx$. Let $u = \sec 2x$ and $du = 2\sec 2x\tan 2x$. Dividing by 2 and substituting gives $\frac{1}{2}\int u^4 du = \frac{1}{10}u^5$ or $\frac{1}{10}\sec^5 2x\big|_0^{\pi/8} = \frac{1}{10}\left(\sec^5\frac{\pi}{4} - \sec^5 0\right) = \frac{1}{10}\left(\sqrt{2}^5 - 1^5\right) = \frac{1}{10}(4\sqrt{2} - 1) \approx 0.4657.$

33. Using a Pythagorean trigonometric substitution gives the result $\int_{\pi/6}^{\pi/2} \frac{\cos^2 x}{\sin x}dx = \int_{\pi/6}^{\pi/2} \frac{1-\sin^2 x}{\sin x}dx = \int_{\pi/6}^{\pi/2} \times \left(\frac{1}{\sin x} - \sin x\right)dx = \int_{\pi/6}^{\pi/2}(\csc x - \sin x)dx = \left[\ln|\csc x - \cot x| + \cos x\right]_{\pi/6}^{\pi/2} = \ln|1 - 0| + 0 - \ln|2 - \sqrt{3}| - \frac{\sqrt{3}}{2} = 0.4509.$

34. $\int_0^{5\pi/12} \frac{\sec^2 x}{2\tan x+4}dx$; Let $u = 2\tan x + 4$, and then $du = 2\sec^2 x\ dx$ or $\frac{1}{2}du = \sec^2 x\ dx$. Substituting yields $\frac{1}{2}\int \frac{1}{u}du = \frac{1}{2}\ln|u| = \frac{1}{2}\ln|2\tan x + 4|\big|_0^{5\pi/12} = \frac{1}{2}\ln|2\tan\frac{5\pi}{12} + 4| - \frac{1}{2}\ln|4| = 0.5265.$

35. $\int x \sinh x^2 \, dx$. Let $u = x^2$ and $du = 2x \, dx$ or $\frac{1}{2} du = x \, dx$. Substituting we get $\frac{1}{2} \int \sinh u \, du = \frac{1}{2} \cosh u + C = \frac{1}{2} \cosh x^2 + C$

36. $\int \frac{\sinh x}{1 + \cosh x} dx$. Let $u = 1 + \cosh x$ and then $du = \sinh x \, dx$. This leads to $\int \frac{1}{u} du = \ln|u| + C = \ln(1 + \cosh x) + C$. Absolute value signs are not needed as $1 + \cosh x$ is never negative.

37. $\int 3x \cosh x^2 \sqrt{\sinh x^2} dx$. Let $u = \sinh x^2$, then $du = 2x \cosh x^2 dx$ or $\frac{3}{2} du = 3x \cosh x^2 dx$. Substituting we get $\frac{3}{2} \int \sqrt{u} \, du = \frac{3}{2} \cdot \frac{2}{3} u^{3/2} + C = \left(\sinh x^2 \right)^{3/2} + C$

38. $\int \operatorname{sech}^2 5x \, dx = \frac{1}{5} \tanh 5x + C$

39. $\int \sinh^3 x \cdot \cosh^2 x \, dx$. Recall that $\sinh^3 x = \sinh x \sinh^2 x = \sinh x \left(\cosh^2 x - 1 \right)$, and then we have $\int \sinh x \left(\cosh^2 x - 1 \right) \cosh^2 x \, dx = \int \cosh^4 x \sinh x \, dx - \int \cosh^2 x \sinh x \, dx = \frac{1}{5} \cosh^5 x - \frac{1}{3} \cosh^3 x + C$.

40. $\int \tanh 3x \operatorname{sech} 3x \, dx = -\frac{1}{3} \operatorname{sech} 3x + C$

41. We begin by rewriting $\cot u$ as $\frac{\cos u}{\sin u}$. Here, if we let $v = \sin u$, then $dv = \cos u \, du$, and the integral is $\int \cot u \, du = \int \frac{dv}{v} = \ln|v| + C$. Back substitution gives the desired result: $\int \cot u \, du = \ln|\sin u| + C$.

42. We will follow a procedure similar to the one we followed prior to Example 28.18. Begin by multiplying the integrand by $\frac{\csc u - \cot u}{\csc u - \cot u}$. This produces $\int \csc u \, du = \int \csc u \left(\frac{\csc u - \cot u}{\csc u - \cot u} \right) du = \int \frac{\csc u (\csc u - \cot u)}{\csc u - \cot u} du = \int \frac{\csc^2 u - \csc u \cot u}{\csc u - \cot u} du$. Let $v = \csc u - \cot u$. Then, $dv = -\csc u \cot u + \csc^2 u) du$ and we can write the original integral as $\int \csc u \, du = \int \frac{dv}{v} = \ln|v| + C$. Back substitution produces the result we wanted to show, namely that $\int \csc u \, du = \ln|\csc u - \cot u| + C$.

43. For $0 \le x \le \frac{\pi}{2}$, we see that $\sin 2x \ge 0$, so the area is the integral $\int_0^{\pi/2} \sin 2x \, dx = \frac{-1}{2} \cos 2x \big|_0^{\pi/2} = -\frac{1}{2}(\cos \pi - \cos 0) = \frac{-1}{2}(-1 - 1) = 1$.

44. For $0 \le x \le \frac{\pi}{4}$, we see that $\sec x \ge x$, so the area is $\int_0^{\pi/4}(\sec x - x) dx = \left(\ln|\sec x + \tan x| - \right.$

$\left. \frac{x^2}{2} \right) \big|_0^{\pi/4} = \ln(\sqrt{2} + 1) - \frac{(\pi/4)^2}{2} - \ln(1 + 0) - 0 \approx 0.5729$.

45. The average value is $\frac{1}{\pi/4} \int_0^{\pi/4} \tan x \, dx = \frac{4}{\pi} \ln|\sec x| \big|_0^{\pi/4} = \frac{4}{\pi}(\ln\sqrt{2} - \ln 1) = \frac{4 \ln\sqrt{2}}{\pi} \approx 0.4413$.

46. Since $v(t) = 5 \sin 2t$, then $s(t) = \int v(t) dt = \int 5 \sin 2t \, dt = \frac{-5}{2} \cos 2t + C$. Since $s(0) = 4$, we can solve for C. $C = 4 + \frac{5}{2} \cos 0 = 4 + \frac{5}{2} = \frac{13}{2}$. Thus, we see that $s(t) = -\frac{5}{2} \cos 2t + \frac{13}{2}$.

47. $E = \frac{1}{1-0} \int_0^1 5 \sin 4t \, dt = -\frac{5}{4} \cos 4t \big|_0^1 = -\frac{5}{4}(-0.6536 - 1) \approx 2.067$ V.

48. The time to complete one cycle is $\frac{2\pi}{377}$ sec, so the average rate at which heat is produced is $\overline{P} = \frac{1}{\frac{2\pi}{377} - 0} \int_0^{3\pi/277} 40(6.5 \sin 377t)^2 dt = \frac{377}{2\pi} \times \int_0^{2\pi/377} 1690 \sin^2 377t \, dt$. Using the half-angle identity, $\sin \frac{\alpha}{2} = \pm\sqrt{\frac{1 - \cos \alpha}{2}}$, we see that $\sin^2 377t = \frac{1 - \cos 745t}{2}$. Thus, we get $\frac{377}{2\pi} \int_0^{2\pi/377} 845(1 - \cos 754t) dt = \frac{(377)(845)}{2\pi} \int_0^{2\pi/377}(1 - \cos 754t) dt = \frac{(377)(845)}{2\pi}(t - \sin 754t) \big|_0^{2\pi/377} = \frac{(377)(845)}{2\pi} \times \left(\frac{2\pi}{377} - 0 \right) = 845$ W.

49. Here $y = 12 \cosh \frac{x}{12}$ and so $y' = \sinh \frac{x}{12}$. The length of the cable is the arc length $= \int_{-18}^{18} \sqrt{1 + \sinh^2 \frac{x}{12}} \, dx = \int_{-18}^{18} \cosh \frac{x}{12} dx = 12 \sinh \frac{x}{12} \big|_{-18}^{18} = 12 \sinh \frac{3}{2} - 12 \sinh \frac{-3}{2} = 51.1027$ m.

50. The center of the towers is 0 so the two towers are at -60 and 60. $y = 60 \cosh \frac{x}{60}$; $y' = \sinh \frac{x}{60}$. The arc length is $\int_{-60}^{60} \sqrt{1 + \sinh^2 \frac{x}{60}} dx = \int_{-60}^{60} \cosh \frac{x}{60} dx = 60 \sinh \frac{x}{60} \big|_{-60}^{60} = 60 \sinh 1 - 60 \sinh(-1) = 141.02$ ft.

51. The center of the towers is at 0 and the towers are at -150 and 150. We are given $y = 80 \cosh \frac{x}{80}$ and differentiating, we obtain $y' = \sinh \frac{x}{80}$. The arc length is $\int_{-150}^{150} \sqrt{1 + \sinh^2 \frac{x}{80}} dx = \int_{-150}^{150} \cosh \frac{x}{80} dx = 80 \left[\sinh \frac{x}{80} \right]_{-150}^{150} = 509.397$ ft.

52. **(a)** The maximum will occur at the center or when $x = 0$. $(-\cosh x)' = -\sinh x = 0$ when $x = 0$. Hence, $y(0) = -127.7\cosh 0 + 757.7 = 630$ ft.
(b) $315 - (-315) = 630$ ft.
(c) Area $= \int_{-315}^{315} \left[-127.7\cosh\left(\frac{x}{127.7}\right) + 757.7\right]dx$
$= -(127.7)^2\sinh\left(\frac{x}{127.7}\right) + 757.7x\big|_{-315}^{315} =$
$286{,}574$ ft^2.

(d) To find the arc length we first find y' as $y' = -\sinh\left(\frac{x}{127.7}\right)$. Substituting this in the arc length formula produces $\int_{-315}^{315}\sqrt{1 + \left[-\sinh\left(\frac{x}{127.7}\right)\right]^2}\,dx$
$= \int_{-315}^{315}\cosh\left(\frac{x}{127.7}\right)dx = 127.7\sinh\left(\frac{x}{127.7}\right)\big|_{-315}^{315} =$
1493.94 ft.

≡ 28.4 MORE TRIGONOMETRIC INTEGRALS

1. $\int \sin^2 3x \cos 3x\ dx$. Legt $u = \sin 3x$ and $du = 3\cos 3x\ dx$. Then, $\frac{1}{3}\int u^2 du = \frac{1}{9}u^3 + C = \frac{1}{9}\sin^3 3x + C$.

2. $\int \sin 2x \cos^2 2x\ dx$. Let $u = \cos 2x$ and $du = -2\sin 2x\ dx$. Then, $-\frac{1}{2}\int u^2 du = -\frac{1}{6}u^3 + C = -\frac{1}{6}\cos^3 2x + C$.

3. $\int \cos x \sin^3 x\ dx = \frac{1}{4}\sin^4 x + C$.

4. $\int \sin^3 4x \cos^2 4x\ dx = \int \sin 4x(1 - \cos^2 4x)\cos^2 4x\ dx$. Let $u = \cos 4x$ and $du = -4\sin 4x\ dx$. Then,

$-\frac{1}{4}\int(1-u^2)u^2 du = -\frac{1}{4}\int u^2 - u^4 du = -\frac{1}{4}\left(\frac{1}{3}u^3 - \frac{1}{5}u^5\right) + C = \frac{1}{20}\cos^5 4x - \frac{1}{12}\cos^3 4x + C$

5. $\int \cos 5x \sin^4 5x\ dx = \frac{1}{25}\sin^5 5x + C$

6. $\int \cos^2 y \sin^5 y\ dy = \int \cos^2 y(1 - \cos^2 y)^2\sin y\ dy$. Let $u = \cos y$ and $du = -\sin y\ dy$, and we get
$-\int u^2(1 - u^2)^2 du = -\int(u^2 - 2u^4 + u^6)du = -\frac{1}{3}u^3 + \frac{2}{5}u^5 - \frac{1}{7}u^7 + C = -\frac{1}{3}\cos^3 y + \frac{2}{5}\cos^5 y - \frac{1}{7}\cos^7 y + C$.

7.
$$\int \sin^2 x\cos^4 x\ dx = \int\left(\frac{1 - \cos 2x}{2}\right)\left(\frac{1 + \cos 2x}{2}\right)^2 dx = \frac{1}{8}\int(1 - \cos 2x)(1 + 2\cos 2x + \cos^2 2x)dx$$
$$= \frac{1}{8}\int\left(1 + \cos 2x - \cos^2 2x - \cos^3 2x\right)dx$$
$$= \frac{1}{8}\int\left(1 + \cos 2x - \cos^3 2x\right) - \frac{1}{8}\int\cos^2 2x\ dx$$
$$= \frac{1}{8}x - \frac{1}{16}\sin 2x - \frac{1}{8}\int(1 - \sin^2 2x)\cos 2x\ dx - \frac{1}{8}\int\frac{1 + \cos 4x}{2}dx$$
$$= \frac{1}{8}x - \frac{1}{16}\sin 2x + \frac{1}{16}\sin 2x + \frac{1}{48}\sin^3 2x - \frac{1}{16}x + \frac{1}{16}\frac{1}{4}\sin 4x + C$$
$$= \frac{1}{16}x + \frac{1}{48}\sin^3 2x + \frac{1}{64}\sin 4x + C$$

8.
$$\int \sin^8 x\ dx = \int(\sin^2 x)^4 dx = \int\left(\frac{1 - \cos 2x}{2}\right)^4 dx = \frac{1}{16}\int(1 - \cos 2x)^4 dx$$
$$= \frac{1}{16}\int\left(1 - 4\cos 2x + 6\cos^2 2x - 4\cos^3 2x + \cos^4 2x\right)dx$$
$$= \frac{1}{16}\int 1 - 4\cos 2x + 3(1 + \cos 4x) - 4(1 - \sin^2 2x)\cos 2x + \left(\frac{1 + \cos 4x}{2}\right)^2 dx$$
$$= \frac{1}{16}\int\left(1 - 4\cos 2x + 3 + 3\cos 4x - 4\cos 2x + 4\cos 2x\sin^2 2x\right.$$
$$\left. + \frac{1}{4} + \frac{1}{2}\cos 4x + \frac{1}{4}\cos^2 4x\right)dx$$
$$= \frac{1}{16}\int\frac{17}{4} - 8\cos 2x + \frac{7}{2}\cos 4x + 4\cos 2x\sin^2 2x + \frac{1}{4}\left(\frac{1 + \cos 8x}{2}\right)dx$$

$$= \frac{1}{16} \int \left(\frac{35}{8} - 8\cos 2x + \frac{7}{2}\cos 4x + 4\cos 2x \sin^2 2x + \frac{1}{8}\cos 8x \right) dx$$

$$= \frac{1}{16} \left(\frac{35}{8}x - 4\sin 2x + \frac{7}{8}\sin 4x + \frac{2}{3}\sin^3 2x + \frac{1}{64}\sin 8x \right) + C$$

9. $\displaystyle\int \cos^6 3x\, dx = \int \left(\frac{1 + \cos 6x}{2} \right)^3 dx = \frac{1}{8} \int \left(1 + 3\cos 6x + 3\cos^2 6x + \cos^3 6x \right) dx$

$$= \frac{1}{8} \int \left[1 + 3\cos 6x + \frac{3}{2}(1 + \cos 12x) + \left(1 - \sin^2 6x \right)\cos 6x \right] dx$$

$$= \frac{1}{8} \int \left[1 + 3\cos 6x + \frac{3}{2} + \frac{3}{2}\cos 12x + \cos 6x - \sin^2 6x\cos 6x \right] dx$$

$$= \frac{1}{8} \int \left[\frac{5}{2} + 4\cos 6x + \frac{3}{2}\cos 12x - \sin^2 6x\cos 6x \right] dx$$

$$= \frac{1}{8} \left[\frac{5x}{2} + \frac{2}{3}\sin 6x + \frac{1}{8}\sin 12x - \frac{1}{18}\sin^3 6x \right] + C$$

$$= \frac{5x}{16} + \frac{1}{12}\sin 6x + \frac{1}{64}\sin 12x - \frac{1}{144}\sin^3 6x + C$$

10. $\int \frac{\cos x}{\sin^3 x}\, dx$. Let $u = \sin x$ and $du = \cos x\, dx$, then $\int \frac{du}{u^3} = -\frac{1}{2}u^{-2} + C = \frac{-1}{2\sin^2 x} + C.$

11. $\displaystyle\int \sin^2 2\theta \cos^4 2\theta\, d\theta = \int \left(\frac{1 - \cos 4\theta}{2} \right)\left(\frac{1 + \cos 2\theta}{2} \right)^2 d\theta = \frac{1}{8} \int \left(1 + \cos 4\theta - \cos^2 4\theta - \cos^3 4\theta \right) d\theta$

$$= \frac{1}{8} \int \left[1 + \cos 4\theta - \frac{1}{2}(1 + \cos 8\theta) - \left(1 - \sin^2 4\theta \right)\cos 4\theta \right] d\theta$$

$$= \frac{1}{8} \int \left[1 + \cos 4\theta - \frac{1}{2} - \frac{1}{2}\cos 8\theta - \cos 4\theta + \sin^2 4\theta \cos 4\theta \right] d\theta$$

$$= \frac{1}{8} \int \left[\frac{1}{2} - \frac{1}{2}\cos 8\theta + \sin^2 4\theta \cos 4\theta \right] d\theta = \frac{1}{8} \left[\frac{\theta}{2} - \frac{1}{16}\sin 8\theta + \frac{1}{12}\sin^3 4\theta \right] + C$$

$$= \frac{\theta}{16} - \frac{1}{128}\sin 8\theta + \frac{1}{96}\sin^3 4\theta + C$$

12. $\displaystyle\int \sin^4 3\theta \cos^2 3\theta\, d\theta = \int \left(\frac{1 - \cos 6\theta}{2} \right)^2\left(\frac{1 + \cos 6\theta}{2} \right) d\theta = \frac{1}{8} \int \left(1 - 2\cos 6\theta + \cos^2 6\theta \right)(1 + \cos 6\theta) d\theta$

$$= \frac{1}{8} \int \left(1 - \cos 6\theta - \cos^2 6\theta + \cos^3 6\theta \right) d\theta$$

$$= \frac{1}{8} \int \left[1 - \cos 6\theta - \frac{1}{2}(1 + \cos 12\theta) + \left(1 - \sin^2 6\theta \right)\cos 6\theta \right] d\theta$$

$$= \frac{1}{8} \int \left[\frac{1}{2} - \cos 6\theta - \frac{1}{2}\cos 12\theta + \cos 6\theta - \sin^2 6\theta \cos 6\theta \right] d\theta$$

$$= \frac{1}{8} \int \left[1/2 - \frac{1}{2}\cos 12\theta - \sin^2 6\theta \cos 6\theta \right] d\theta = \frac{1}{8} \left[\frac{\theta}{2} - \frac{1}{24}\sin 12\theta - \frac{1}{18}\sin^3 6\theta \right] + C$$

$$= \frac{\theta}{16} - \frac{1}{192}\sin 12\theta - \frac{1}{144}\sin^3 6\theta + C$$

13. $\int \sec^2 x \tan^2 x\, dx$. Let $u = \tan x$ and then $du = \sec^2 x\, dx$. Hence, we get $\int u^2 du = \frac{1}{3}u^3 + C = \frac{1}{3}\tan^3 x + C.$

14. $\int \sec^4 y \tan^3 y\, dy = \int \sec^2 y \left(\sec^2 y \right)\tan^3 y\, dy = \int \left[\sec^2 y \left(1 + \tan^2 y \right)\tan^3 y \right] dy = \int \sec^2 y \left[\tan^3 y + \tan^5 y \right] dy$. Let $u = \tan y$ and then $du = \sec^2 y\, dy$,

and you have $\int \left(u^3 + u^5 \right)du = \frac{1}{4}u^4 + \frac{1}{6}u^6 + C = \frac{1}{4}\tan^4 y + \frac{1}{6}\tan^6 y + C.$

15. First factor $\int \csc^4 x \cot x\, dx = \int \csc^3 x \times (\csc x \cot x\, dx)$. Then let $u = \csc x$ and $du = -\csc x \cot x\, dx$. These produce $-\int u^3 du = -\frac{1}{4}u^4 + C = -\frac{1}{4}\csc^4 x + C_1$ or $\int \csc^4 x \cot x\, dx = \int \csc^2 x \left(1 + \cot^2 x \right)\cot x\, dx = \int \left(\cot^3 x + \cot x \right) \times$

$\csc^2 x \, dx$. Let $u = \cot x$ and $du = -\csc^2 x \, dx$, and then you have $-\int (u^3 + u)\,du = -\frac{1}{4}u^4 - \frac{1}{2}u^2 + C = -\frac{1}{4}\cot^4 x - \frac{1}{2}\cot^2 x + C_2$ where $C_2 = C_1 - 0.25$.

16. $\int \frac{\sin^2 \theta}{\cos^4 \theta}d\theta = \int \sec^2 \theta \tan^2 \theta \, d\theta$. Let $u = \tan\theta$ and $du = \sec^2\theta \, d\theta$, then $\int u^2 du = \frac{1}{3}u^3 + C = \frac{1}{3}\tan^3\theta + C$.

17. We first factor the integrand and use a Pythagorean identity: $\int \sin^{1/2} 3\theta \cos^3 3\theta d\theta = \int \sin^{1/2} 3\theta (1 - \sin^2 3\theta) \cos 3\theta \, d\theta$. Then multiply and integrate, with the result $\int (\sin^{1/2} 3\theta - \sin^{5/2} 3\theta) \cos 3\theta d\theta = \frac{1}{3} \cdot \frac{2}{3}\sin^{3/2} 3\theta - \frac{1}{3} \cdot \frac{2}{7}\sin^{7/2} 3\theta + C = \frac{2}{9}\sin^{3/2} 3\theta - \frac{2}{21}\sin^{7/2} 3\theta + C$.

18. The given integral $\int \csc^6 2x \cot^2 2x \, dx$ can be written as $\int \csc^4 2x \cot^2 2x \csc^2 2x \, dx$ or, using a Pythagorean identity, as $\int (1 + \cot^2 2x)^2 \cot^2 2x \csc^2 2x \, dx$. This expands to $\int (1 + 2\cot^2 2x + \cot^4 2x) \times \cot^2 2x \csc^2 2x \, dx$ or $\int (\cot^2 2x + 2\cot^4 2x + \cot^6 2x) \csc^2 2x \, dx$. Let $u = \cot 2x$ and $du = -2\csc^2 2x \, dx$ and we rewrite the integral as $-\frac{1}{2} \int (u^2 + 2u^4 + u^6)\,du = -(\frac{1}{6}u^3 + \frac{1}{5}u^5 + \frac{1}{14}u^7) + C = -(\frac{1}{6}\cot^3 2x + \frac{1}{5}\cot^5 2x + \frac{1}{14}\cot^7 2x) + C$.

19. $\int \csc x \cot^3 x \, dx = \int (\csc^2 x - 1) \csc x \cot x \, dx = -\frac{1}{3}\csc^3 x + \csc x + C$.

20. $\int \sec t \tan^5 t \, dt = \int \sec t \tan t (\tan^4 t)\,dt = \int \sec t \tan t (\tan^2 t)^2 dt = \int \sec t \tan t (\sec^2 - 1t)^2 dt = \int \sec t \tan t (\sec^4 + 2\sec^2 t + 1)\,dt = \frac{1}{5}\sec^5 t - \frac{2}{3}\sec^3 t + \sec t + C$.

21. $\int \tan^3 x \sec^2 x \, dx = \frac{1}{4}\tan^4 x + C$.

22. $\int \cot 2x \csc^4 2x \, dx$. Let $u = \csc 2x$ and then $du = -2\csc 2x \cot 2x \, dx$. Thus, we then integrate $-\frac{1}{2} \int u^3 du = -\frac{1}{8}u^4 + C = -\frac{1}{8}\csc^4 2x + C_1$ or $\int \cot 2x \csc^4 2x = \int \cot 2x (\cot^2 2x + 1) \csc^2 2x \, dx = \int (\cot^2 2x + \cot 2x) \csc^2 2x \, dx = -\frac{1}{8}\cot^4 2x - \frac{1}{4}\cot^2 2x + C_2$ where $C_2 = C_1 - \frac{1}{8}$.

23. $\int \tan^6 x \sec^2 x \, dx = \frac{1}{7}\tan^7 x + C$

24. $\int \tan^5 x \, dx = \int (\sec^2 x - 1)^2 \tan x \, dx = \int (\sec^4 x - 2\sec^2 x + 1) \tan x \, dx = \int \sec^4 x \tan x \, dx + 2\int \sec^2 x \tan x \, dx + \int \tan x \, dx = \frac{1}{4}\sec^4 x - \sec^2 x +$

$\ln |\sec x| + C$. An alternate method is $\int \tan^5 x \, dx = \int \tan^3 x (\sec^2 x - 1)\,dx$ which can be written as $\int \tan^3 x \sec^2 x - \int \tan x (\sec^2 - 1)\,dx = \frac{1}{4}\tan^4 x - \frac{1}{2}\tan^2 x + \ln |\sec x| + C$.

25. $\int \cot^6 x \, dx = \int \cot^4 x (\csc^2 x - 1)\,dx = \int \cot^4 x \csc^2 x \, dx - \int \cot^2 x (\csc^2 x - 1)\,dx = \int \cot^4 x \csc^2 x \, dx - \int \cot^2 x \csc^2 x \, dx + \int \cot^2 x \, dx = -\frac{1}{5}\cot^5 x + \frac{1}{3}\cot^3 x + \int (\csc^2 x - 1)\,dx = -\frac{1}{5}\cot^5 x + \frac{1}{3}\cot^3 x - \cot x - x + C$. Recall: $\cot^2 x = \csc^2 x - 1$.

26. $\int \sec^6 2\theta d\theta = \int \sec^2 2\theta (\tan^2 2\theta + 1)^2 d\theta = \int \sec^2 2\theta (\tan^4 2\theta + 2\tan^2 2\theta + 1)\,d\theta = \frac{1}{10}\tan^5 2\theta + \frac{1}{3}\tan^3 2\theta + \frac{1}{2}\tan 2\theta + C$

27. $\int_0^{\pi/4} \tan^2 x \, dx = \int_0^{\pi/4} (\sec^2 x - 1)\,dx = \tan x - x|_0^{\pi/4} = 1 - \frac{\pi}{4} - 0 = 1 - \frac{\pi}{4}$

28. $\tan^3 2t$ is not defined at $\frac{\pi}{4}$ and so, $\tan^3 2t$ is not continuous on the interval $[\pi/6, \pi/3]$. Hence, this integral does not exist.

29. $\int_0^{\pi/2} \sin^4 x \, dx = \int_0^{\pi/2} (\frac{1 - \cos 2x}{2})^2 dx$ by a half-angle identity. This expands as $\frac{1}{4} \int_0^{\pi/2} (1 - 2\cos 2x + \cos^2 2x)\,dx$ which becomes, by another half-angle identity, $\frac{1}{4} \int_0^{\pi/2} (1 - 2\cos 2x + \frac{1 + \cos 4x}{2})\,dx$ or $\frac{1}{4} \int_0^{\pi/2} (\frac{3}{2} - 2\cos 2x + \frac{\cos 4x}{2})\,dx = \frac{1}{4}[\frac{3x}{2} - \sin 2x + \frac{1}{8}\sin 4x]_0^{\pi/2} = \frac{1}{8}[\frac{3\pi}{4} - \sin \pi + \frac{1}{8}\sin 2\pi] = \frac{1}{4}[\frac{3\pi}{4} - 0 + 0] = \frac{3\pi}{16}$.

30. $\int_0^{\pi/6} \sec^3 2x \tan 2x \, dx = \frac{1}{2} \cdot \frac{1}{3}\sec^3 2x|_0^{\pi/6} = \frac{1}{6}(\sec^3 \frac{\pi}{3} - \sec^3 0) = \frac{1}{6}(8 - 1) = \frac{7}{6}$

31. Using the disc method, you get $\pi \int_0^{\pi/3} \cos^2 x \, dx = \pi \int_0^{\pi/3} \frac{1 + \cos 2x}{2}\,dx = \frac{\pi}{2}[x + \frac{1}{2}\sin 2x]_0^{\pi/3} = \frac{\pi}{2}[\frac{\pi}{3} + \frac{\sqrt{3}}{4}] \approx 2.3251$.

32. $V = 80\sin 120\pi t$. One period is $\frac{2\pi}{120\pi} = \frac{1}{60}$. Thus, the root mean square of the voltage for one period is $\sqrt{\frac{1}{1/60} \int_0^{1/60} (80\sin 120\pi t)^2 dt} = 80\sqrt{60 \int_0^{1/60} \sin^2 120\pi t \, dt} = 80\sqrt{60 \int_0^{1/60} \frac{1 - \cos 240\pi t}{2}dt} =$

$$80\sqrt{60\left[\tfrac{1}{2}t - \tfrac{\sin 240\pi t}{480\pi}\right]_0^{1/60}} =$$

$$80\sqrt{60\left(\tfrac{1}{120}\right)} = 80\sqrt{1/2} = 80\tfrac{\sqrt{2}}{2} = 40\sqrt{2} \approx$$
56.5685 V.

33. **(a)** $I = \dfrac{mr^2}{4}\displaystyle\int_0^{2\pi} \sin^2\theta\; d\theta$

$$= \dfrac{mr^2}{4}\cdot\dfrac{1}{2}\int_0^{2\pi}(1-\cos 2\theta)d\theta$$

$$= \dfrac{mr^2}{8}\left[\theta - \dfrac{1}{2}\sin 2\theta\right]_0^{2\pi}$$

$$= \dfrac{mr^2}{8}\left[(2\pi - 0) - (0-0)\right] = \dfrac{mr^2\pi}{4}$$

(b) Substituting $r = 0.1$ m and $m = 12.4$ kg into
$I = \frac{mr^2\pi}{4}$ produces $I = \frac{(12.4)(0.1)^2\pi}{4} \approx 0.0974\,\text{kg}\cdot\text{m}^2$.

34. $R = \displaystyle\int \cos^3\theta\; d\theta = \int(1-\sin^2\theta)\cos\theta\; d\theta =$

$$\int(1-u^2)du = u - \dfrac{1}{3}u^3 + C = \sin\theta - \dfrac{1}{3}\sin^3\theta + C$$

35. $i_{\text{eff}} = \sqrt{\dfrac{1}{0.5-0}\displaystyle\int_0^{0.5}(2\sin t\sqrt{\cos t})^2 dt}$

$$= \sqrt{8\int_0^{0.5}\left(\sin^2 t\cos t\right)dt}$$

$$= \sqrt{\dfrac{8}{3}\sin^3 t\Big|_0^{0.5}}$$

$$= \sqrt{\dfrac{8}{3}\sin^3 0.5} \approx 0.5421$$

The effective current from $t = 0$ to $t = 0.5$ s is about 0.5421 A.

36. The rms of the current is the same as the effective current.

$$i_{\text{rms}} = \sqrt{\dfrac{1}{\pi-0}\int_0^\pi (4\sin t\cos t)^2 dt} = \sqrt{\dfrac{16}{\pi}\int_0^\pi\left(\sin^2 t\cos^2 t\right)dt} = \sqrt{\dfrac{16}{\pi}\int_0^\pi\left[\sin^2 t\left(1-\sin^2 t\right)\right]dt}$$

$$= \sqrt{\dfrac{16}{\pi}\int_0^\pi\left(\sin^2 t - \sin^4 t\right)dt} = \sqrt{\dfrac{16}{\pi}\int_0^\pi\left[\left(\dfrac{1-\cos 2t}{2}\right) - \left(\dfrac{1-\cos 2t}{2}\right)^2\right]dt}$$

$$= \sqrt{\dfrac{16}{\pi}\int_0^\pi\left[\left(\dfrac{1}{2} - \dfrac{\cos 2t}{2}\right) - \left(\dfrac{1-2\cos 2t + \cos^2 2t}{4}\right)\right]dt}$$

$$= \sqrt{\dfrac{16}{\pi}\int_0^\pi\left[\left(\dfrac{1}{2} - \dfrac{\cos 2t}{2}\right) - \left(\dfrac{1}{4} - \dfrac{\cos 2t}{2} + \dfrac{\cos^2 2t}{4}\right)\right]dt}$$

$$= \sqrt{\dfrac{16}{\pi}\int_0^\pi\left[\left(\dfrac{1}{4} - \dfrac{\cos^2 2t}{4}\right)\right]dt} = \sqrt{\dfrac{16}{\pi}\int_0^\pi\left[\left(\dfrac{1}{4} - \dfrac{1-\cos 4t}{8}\right)\right]dt} = \sqrt{\dfrac{16}{\pi}\int_0^\pi\left[\left(\dfrac{1}{8} + \dfrac{\cos 4t}{8}\right)\right]dt}$$

$$= \sqrt{\dfrac{2}{\pi}\int_0^\pi\left[(1+\cos 4t)\right]dt} = \sqrt{\dfrac{2}{\pi}\left[t + \dfrac{1}{4}\sin 4t\right]_0^\pi} = \sqrt{\dfrac{2}{\pi}(\pi)} = \sqrt{2}$$

The rms of the current from $t = 0$ to $t = \pi$ s is $\sqrt{2}$ A.

≡ 28.5 INTEGRALS RELATED TO INVERSE TRIGONOMETRIC AND INVERSE HYPERBOLIC FUNCTIONS

1. $\displaystyle\int \dfrac{dx}{\sqrt{4-x^2}} = \arcsin\dfrac{x}{2} + C$

2. $\displaystyle\int \dfrac{dx}{\sqrt{1-9x^2}}$. Let $u = 3x$ and $du = 3\,dx$ so $\frac{1}{3}du = dx$. Substituting, we have $\frac{1}{3}\int\dfrac{du}{\sqrt{1-u^2}} = \frac{1}{3}\arcsin u + C = \frac{1}{3}\arcsin 3x + C$.

3. $\displaystyle\int \dfrac{dx}{\sqrt{4-9x^2}}$. Let $u = 3x$ and $du = 3\,dx$ so $\frac{1}{3}du = dx$. Substituting, we have $\frac{1}{3}\int\dfrac{du}{\sqrt{4-u^2}} = \frac{1}{3}\arcsin\frac{u}{2} + C = \frac{1}{3}\arcsin\frac{3x}{2} + C$.

4. $\displaystyle\int \dfrac{8\,dx}{1+16x^2}$. Let $u = 4x$ and then $du = 4\,dx$ so $2\,du = 8\,dx$. Thus, $2\int\dfrac{du}{1+u^2} = 2\cdot\arctan u + C = 2\arctan 4x + C$.

5. $\displaystyle\int \dfrac{-x\,dx}{\sqrt{9-x^2}}$. Let $u = 9 - x^2$ and $du = -2x\,dx$ so $\frac{1}{2}du = -x\,dx$. Then, $\frac{1}{2}\int u^{-1/2}du = \frac{1}{2}\cdot\frac{2}{1}u^{1/2} + C = \sqrt{9-x^2} + C$.

6. $\displaystyle\int \dfrac{dx}{x\sqrt{x^2-16}} = \frac{1}{4}\arcsec\frac{x}{4} + C$.

7. $\int \frac{dx}{3x\sqrt{9x^2-4}}$. Let $u = 3x$, then $du = 3\,dx$, so $\frac{1}{3}du = dx$. Thus, we can rewrite the integral as $\frac{1}{3}\int \frac{du}{u\sqrt{u^2-4}} = \frac{1}{3}\cdot\frac{1}{2}\operatorname{arcsec}\frac{u}{2}+C = \frac{1}{6}\operatorname{arcsec}\frac{3x}{2}+C$.

8. $\int \frac{x\,dx}{\sqrt{1-25x^2}}$. Here $u = 1-25x^2$ and $du = -50x\,dx$ and the integral becomes $-\frac{1}{50}\int u^{-1/2}du = -\frac{1}{50}\cdot\frac{2}{1}u^{1/2}+C = -\frac{1}{25}\sqrt{1-25x^2}+C$.

9. $\int \frac{dx}{1+(3-x)^2}$. Here $u = 3-x$ and $du = -dx$ and so we get $-\int \frac{du}{1+u^2} = -\arctan u + C = -\arctan(3-x)+C$.

10. $\int \frac{dx}{(2x+1)\sqrt{(2x+1)^2-4}}$. Let $u = 2x+1$ and $du = 2\,dx$, then $\frac{1}{2}du = dx$. $\frac{1}{2}\int \frac{du}{u\sqrt{u^2-4}} = \frac{1}{2}\cdot\frac{1}{2}\operatorname{arcsec}\frac{u}{2} + C = \frac{1}{4}\operatorname{arcsec}\frac{2x+1}{2}+C$

11. $\int \frac{4x-6}{4x^2+25}dx = \int \frac{4x}{4x^2+25}dx - \int \frac{6}{4x^2+25}dx$. In the first integral let $u = 4x^2+25$; $du = 8x\,dx$ and we get by substitution $2\int \frac{du}{u} = 2\ln|u| = 2\ln(4x^2+25)$. The second integral is $-6\int \frac{dx}{4x^2+25} = -6\cdot\frac{1}{10}\arctan\frac{2x}{5}$. Putting these together we have $2\ln\times(4x^2+25) - \frac{3}{5}\arctan\frac{2x}{5}+C$

12. $\int \frac{dx}{(x+3)\sqrt{(x+3)^2-1}} = \operatorname{arcsec}(x+3)+C$.

13. Completing the square, we obtain $\int \frac{dx}{x^2+6x+10} = \int \frac{dx}{(x+3)^2+1} = \arctan(x+3)+C$.

14. We begin by completing the square on the expression under the radical. $\int \frac{dx}{\sqrt{-4x^2+4x+15}} + \int \frac{dx}{\sqrt{-(4x^2-4x+1)+15+1}} = \int \frac{dx}{\sqrt{16-(2x-1)^2}}$. Now let $u = 2x-1$ and $du = 2\,dx$ and you get $\frac{1}{2}\int \frac{du}{\sqrt{16-u^2}} = \frac{1}{2}\cdot\arcsin\frac{u}{2}+C = \frac{1}{2}\arcsin\frac{2x-1}{2}+C$.

15. $\int \frac{x\,dx}{1+x^4}$; Let $u = x^2$ and then $du = 2x\,dx$. Substituting, we obtain $\frac{1}{2}\int \frac{du}{1+u^2} = \frac{1}{2}\cdot\arctan u + C = \frac{1}{2}\arctan x^2 + C$.

16. $\int \frac{\sec^2 x\,dx}{4+\tan^2 x}$. Let $u = \tan x$ and $du = \sec^2 x\,dx$. Substituting, produces $\int \frac{du}{4+u^2} = \frac{1}{2}\arctan\frac{u}{2} + C = \frac{1}{2}\arctan\left(\frac{\tan x}{2}\right) + C$.

17. $\int \frac{\sin x\,dx}{\sqrt{1-\cos^2 x}}$. Let $u = \cos x$ and $du = -\sin x\,dx$. Substituting produces $-\int \frac{du}{\sqrt{1-u^2}} = -\arcsin u + C = -\arcsin(\cos x) + C = -\left(\frac{\pi}{2}-x\right) + C = x - \frac{\pi}{2} + C$.

18. $\int \frac{x-5}{\sqrt{x^2-10x+16}}dx$. Let $u = x^2 - 10x + 16$; $du = (2x-10)dx$ or $\frac{1}{2}du = (x-5)dx$. Hence, we have $\frac{1}{2}\int \frac{du}{\sqrt{u}} = \frac{1}{2}\cdot\frac{2}{1}u^{1/2}+C = \sqrt{x^2-10x+16}+C$.

19. $\int_0^{\pi/4} \frac{\cos x\,dx}{1+\sin^2 x} = \arctan(\sin x)\Big|_0^{\pi/4} = \arctan\left(\sin\frac{\pi}{4}\right) - \arctan(\sin 0) = \arctan\frac{\sqrt{2}}{2} - \arctan 0 = 0.6155 - 0 = 0.6155$.

20. Completing the square, we have $\int_3^4 \frac{dx}{\sqrt{-x^2+8x-15}} = \int_3^4 \frac{dx}{\sqrt{1-(x^2-8x-16)}} = \int_3^4 \frac{dx}{\sqrt{1-(x-4)^2}} = \arcsin(x-4)\Big|_3^4 = \arcsin 0 - \arcsin(-1) = 0 - \frac{-\pi}{2} = \frac{\pi}{2}$.

21. $\int \frac{dx}{\sqrt{x^2-25}} = \cosh^{-1}\frac{x}{5} + C$.

22. $\int \frac{dx}{\sqrt{(x-3)^2+16}} = \sinh^{-1}\left(\frac{x-3}{4}\right) + C$.

23. $\int \frac{dx}{\sqrt{25+9x^2}}$. Let $u = 3x$ and $du = 3\,dx$ and so $\frac{1}{3}du = dx$. Substituting produces $\int \frac{du}{\sqrt{25+u^2}} = \frac{1}{3}\sinh^{-1}\frac{u}{5} + C = \frac{1}{3}\sinh^{-1}\frac{3x}{5} + C$.

24. $\int \frac{e^{2x}}{25-e^{4x}}dx$. Let $u = e^{2x}$ and then $du = 2e^{2x}dx$. Substituting, we obtain $\frac{1}{2}\int \frac{du}{25-u^2} = \frac{1}{2}\cdot\frac{1}{5}\tanh^{-1}\frac{u}{5} + C = \frac{1}{10}\tanh^{-1}\frac{e^{2x}}{5} + C$.

25. $\int_0^{0.5} \frac{dx}{\sqrt{1-x^2}} = \arcsin\Big|_0^{0.5} = \arcsin 0.5 - \arcsin 0 = \frac{\pi}{6} - 0 = \frac{\pi}{6} \approx 0.5236$

26. Let $u = 1+x^2$ and $du = 2x\,dx$. Then, $\int_0^{\sqrt{3}} \frac{x\,dx}{\sqrt{1+x^2}} = \frac{1}{2}\int_0^{\sqrt{3}} \frac{2x\,dx}{\sqrt{1+x^2}} = \frac{1}{2}\int_1^4 \frac{du}{\sqrt{u}} = \frac{1}{2}\int_1^4 u^{-1/2}du = u^{1/2}\Big|_1^4 = 2 - 1 = 1$

27. $\int_1^5 \frac{dx}{x^2-4x+13} = \int_1^5 \frac{dx}{(x^2-4x+4)+9} = \int_1^5 \frac{dx}{3^2+(x-2)^2} = \frac{1}{3}\arctan\frac{(x-2)}{3}\Big|_1^5 = \frac{1}{3}\left[\arctan 1 - \arctan\left(\frac{-1}{3}\right)\right] = \frac{1}{3}\left(\frac{\pi}{4} + 0.32175\right) \approx 0.36905$

28. $\int_{-5}^{-1} \frac{dx}{\sqrt{-x^2-7x-6}} = \int_{-5}^{-1} \frac{dx}{\sqrt{\frac{25}{4} - \left(x^2 + 7x + \frac{49}{4}\right)}} =$

$\int_{-5}^{-1} \frac{dx}{\sqrt{\left(\frac{5}{2}\right)^2 - \left(x + \frac{7}{2}\right)^2}} = \arcsin\left(\frac{x + \frac{7}{2}}{5/2}\right)\Big|_{-5}^{-1} =$

$\arcsin\left(\frac{2x+7}{5}\right)\Big|_{-5}^{-1} = \arcsin 1 - \arcsin\left(\frac{-3}{5}\right) \approx$
2.2143

29. Let $u = 4\tan x$ and $du = 4\sec^2 x\, dx$. Then,
$\int_0^{\pi/6} \frac{\sec^2 x\, dx}{1+16\tan^2 x} = \frac{1}{4}\int_0^{\pi/6} \frac{4\sec^2 x\, dx}{1+(4\tan x)^2} = \frac{1}{4}\int_0^{4\sqrt{3}/3} \frac{du}{1+u^2}$
$= \frac{1}{4}\arctan u\Big|_0^{4\sqrt{3}/3} = \frac{1}{4}\left[\arctan\left(\frac{4\sqrt{3}}{3}\right) - \arctan 0\right] =$
$\frac{1}{4}\arctan\left(\frac{4\sqrt{3}}{3}\right) \approx 0.2905$

30. Let $u = x^2$ and $du = 2x\, dx$ $\int_0^1 \frac{x\, dx}{\sqrt{1+x^4}} =$
$\frac{1}{2}\int_0^1 \frac{2x\, dx}{\sqrt{1+x^4}} = \frac{1}{2}\int_0^1 \frac{1}{\sqrt{1+u^2}}du = \frac{1}{2}\sinh^{-1} u\Big|_0^1 =$
$\frac{1}{2}\left(\sinh^{-1} 1 - \sinh^{-1} 0\right) \approx 0.44069$

31. $\int_0^1 \frac{1}{1+x^2}dx = \arctan x\Big|_0^1 = \arctan 1 - \arctan 0 =$
$\frac{\pi}{4} - 0 = \frac{\pi}{4}$.

32. $\int_{-1/4}^{1/4} \frac{1}{\sqrt{1-4x^2}}dx = \frac{1}{2}\arcsin 2x\Big|_{-1/4}^{1/4} = \frac{1}{2}\left[\arcsin - \right.$
$\left.\arcsin\left(-\frac{1}{2}\right)\right] = \frac{1}{2}\left[\frac{\pi}{6} - \frac{-\pi}{6}\right] = \frac{\pi}{6}$.

33. These two graphs intersect at $(0, 1)$ and, from 0 to 1, e^x is always larger. Using the washer method
we get $\pi \int_0^1 \left[(e^x)^2 - \left(\frac{1}{\sqrt{x^2 - 1}}\right)^2\right] dx$
$= \pi \int_0^1 \left(e^{2x} - \frac{1}{x^2 - 1}\right) dx$
$= \pi \int_0^1 e^{2x} dx - \pi \int_0^1 \frac{1}{x^2 + 1}dx$
$= \frac{\pi}{2}e^{2x}\Big|_0^1 - \pi \arctan x\Big|_0^1$
$= \frac{\pi}{2}\left(e^2 - 1\right) - \pi(\arctan 1 - \arctan 0)$
$= \frac{\pi}{2}\left(e^2 - 1\right) - \pi\left(\frac{\pi}{4} - 0\right)$
$= \frac{\pi}{2}\left(e^2 - 1 - \frac{\pi}{2}\right) \approx 7.5685$

34. $M_y = \int_1^2 x\left(\frac{1}{1+x^4}\right)dx = \int_1^2 \frac{x\, dx}{1+x^4}$. Let $u = x^2$
and $du = 2x\, dx$ or $\frac{1}{2}du = x\, dx$. Substituting
we get $\frac{1}{2}\int \frac{du}{1+u^2} = \frac{1}{2}\cdot \tan^{-1} u = \frac{1}{2}\tan^{-1} x^2\Big|_1^2 =$

$\frac{1}{2}\left[\tan^{-1} 4 - \tan^{-1} 1\right] = \frac{1}{2}[1.3258 - 0.7854] =$
0.2702

35.
$$y = \sqrt{4 - x^2}$$
$$y' = \frac{1}{2}(4 - x)^{-1/2} \cdot (-2x)$$
$$= -\frac{x}{\sqrt{4 - x^2}}$$
$$y'^2 = \frac{x^2}{4 - x^2}$$
$$\int_{-1}^1 \sqrt{1 + (y')^2} = \int_{-1}^1 \sqrt{1 + \frac{x^2}{4 - x^2}}dx$$
$$= \int_{-1}^1 \sqrt{\frac{4 - x^2 + x^2}{4 - x^2}}dx$$
$$= \int_{-1}^1 \sqrt{\frac{4}{4 - x^2}}dx$$
$$= \int_{-1}^1 \frac{2\, dx}{\sqrt{4 - x^2}}$$
$$= 2\arcsin\frac{x}{2}\Big|_{-1}^1$$
$$= 2\left[\arcsin\frac{1}{2} - \arcsin\frac{-1}{2}\right]$$
$$= 2\left[\frac{\pi}{6} - \frac{-\pi}{6}\right] = \frac{2\pi}{3}$$

36. $m = \int_0^1 \frac{4}{\sqrt{4 - x^2}}$
$= 4\sin^{-1}\frac{x}{2}\Big|_0^1$
$= 4\left(\frac{\pi}{6} - 0\right) = \frac{2\pi}{3} \approx 2.0944$
$M_x = \frac{1}{2}\int_0^1 \left(\frac{4}{\sqrt{4 - x^2}}\right)^2 dx$
$= 8\int_0^1 \frac{1}{4 - x^2}dx$
$= 8 \cdot \frac{1}{4}\ln\left|\frac{x + 2}{x - 2}\right|\Big|_0^1$
$= 2[\ln 3 - \ln -1] = 2\ln 3 \approx 2.1972$
$M_y = \int_0^1 x\left(\frac{4}{\sqrt{4 - x^2}}\right)dx$.

Let $u = 4 - x^2$; $du = -2x\, dx$ so $-2\, du = 4x\, dx$.
When $x - 0$, then $u = 4 - 0^2 = 4$, and when $x = 1$,
then $u = 4 - 1^2 = 3$. Substituting we get
$-2\int_0^1 \frac{du}{\sqrt{u}} = -2\int_4^3 u^{-1/2}du = -2 \cdot 2u^{1/2}\Big|_4^3$
$= -4\left(\sqrt{3} - \sqrt{4}\right) = 1.0718$
$\bar{x} = \frac{M_y}{m} = \frac{1.0718}{2.0944} = 0.51174$
$\bar{y} = \frac{M_x}{m} = \frac{2.1972}{2.0944} = 1.0491$.

37.

$$f'(x) = -\frac{1}{x\sqrt{1-x^2}} + \frac{x}{\sqrt{1-x^2}}$$

$$f(x) = \int f'(x)dx = \int \frac{-1}{x\sqrt{1-x^2}}dx$$

$$+ \int \frac{x}{\sqrt{1-x^2}}dx$$

$$= -\operatorname{sech}^{-1} x + \int \frac{x\,dx}{\sqrt{1-x^2}}.$$

For the second integral let $u = 1 - x^2$, then $du = -2x\,dx$ or $-\frac{1}{2}du = x\,dx$. These substitutions make $\int \frac{x\,dx}{\sqrt{1-x^2}} = -\frac{1}{2}\int \frac{du}{\sqrt{u}} = -\frac{1}{2}\frac{2}{1}u^{1/2} = -\sqrt{u} =$

$-\sqrt{1-x^2}$. Putting these together we get $f(x) = -\operatorname{sech}^{-1} x - \sqrt{1-x^2} + C$. Since $f(1) = 0$ we have $C = \operatorname{sech}^{-1} 1 + \sqrt{1-1^2} = \operatorname{sech}^{-1} 1$. Thus, the desired equation is $y = -\operatorname{sech}^{-1} x - \sqrt{1-x^2} + \operatorname{sech}^{-1} 1$.

38. $s(t) = \int v(t)dt = \int 4t - \frac{12}{4-t^2}dt = 2t^2 - \frac{12}{2}\tanh^{-1}\left(\frac{t}{2}\right) + C$. Since $s(0) = 0$, we have $C = 0$, so $s(t) = 2t^2 - 6\tanh^{-1}\left(\frac{1}{2}t\right)$.

28.6 TRIGONOMETRIC SUBSTITUTION

1. $\int \frac{x}{\sqrt{9-x^2}}dx$. This can be done with trigonometric substitutions but it is much easier to do the following. Let $u = 9 - x^2$; $du = -2x\,dx$ so $x\,dx = -\frac{1}{2}du$. Substituting yields $-\frac{1}{2}\int \frac{du}{\sqrt{u}} = -\frac{1}{2}\frac{2}{1}u^{1/2} + C = -\sqrt{9-x^2} + C$.

2. $\int \frac{dx}{\sqrt{9-x^2}} = \arcsin\frac{x}{3} + C.$

3. $\int \frac{x^2}{\sqrt{9-x^2}}dx$. Let $x = 3\sin\theta$ and $dx = 3\cos\theta\,d\theta$.

Substituting produces

$$\int \frac{(3\sin\theta)^2 3\cos\theta\,d\theta}{\sqrt{9 - 9\sin^2\theta}} = \int 9\sin^2\theta\,d\theta$$

$$= 9\int \frac{1 - \cos 2\theta}{2}$$

$$= \frac{9}{2}\int 1\,d\theta - \frac{9}{2}\int \cos 2\theta\,d\theta$$

$$= \frac{9}{2}\theta - \frac{9}{2}\cdot\frac{1}{2}\sin 2\theta + C.$$

Since $x = 3\sin\theta$, we have $\sin\theta = \frac{x}{3}$ and $\theta = \sin^{-1}\frac{x}{3}$. Hence $\sin 2\theta = 2\sin\theta\cos\theta$, which yields $\sin 2\theta = 2\cdot\frac{x}{3}\cdot\frac{\sqrt{9-x^2}}{3} = \frac{2}{9}x\sqrt{9-x^2}$. Thus, the final answer is $\frac{9}{2}\arcsin\frac{x}{3} - \frac{1}{2}x\sqrt{9-x^2} + C$.

4. $\int \left(9-x^2\right)^{3/2}dx$. Let $x = 3\sin\theta$ and $dx = 3\cos\theta\,d\theta$. This produces

$$\int \left(9 - 9\sin^2\theta\right)^{3/2} 3\cos\theta\,d\theta)$$

$$= \int \left(9\cos^2\theta\right)^{3/2} 3\cos\theta\,d\theta$$

$$= 81\int \cos^4\theta\,d\theta = 81\int \left(\frac{1+\cos 2\theta}{2}\right)^2 d\theta$$

$$= \frac{81}{4}\int \left(1 + 2\cos 2\theta + \cos^2 2\theta\right)d\theta$$

$$= \frac{81}{4}\int \left[1 + 2\cos 2\theta + \frac{1}{2}(1+\cos 4\theta)\right]d\theta$$

$$= \frac{81}{4}\int \left[\frac{3}{2} + 2\cos 2\theta + \frac{1}{2}\cos 4\theta\right]d\theta$$

$$= \frac{81}{4}\left[\frac{3}{2}\theta + \sin 2\theta + \frac{1}{8}\sin 4\theta\right] + C$$

Since $\theta = \sin^{-1}\frac{x}{3}$, then $\sin\theta = \frac{x}{3}$ and $\cos\theta = \frac{\sqrt{9-x^2}}{3}$, and we have $\sin 2\theta = 2\sin\theta\cos\theta = 2\cdot\frac{x}{3}\cdot\frac{\sqrt{9-x^2}}{3} = \frac{2x\sqrt{9-x^2}}{9}$ $\sin 4\theta = 2\sin 2\theta\cos 2\theta = 2\sin\theta\cos\theta\left(1 - 2\sin^2\theta\right)$

$$= 2\cdot\frac{x}{3}\cdot\frac{\sqrt{9-x^2}}{3}\left[1 - 2\left(\frac{x}{3}\right)^2\right]$$

$$= \frac{2x\sqrt{9-x^2}}{9}\left(1 - \frac{2x^2}{9}\right).$$

Back substitution yields

$$\frac{81}{4}\left[\frac{3}{2}\sin^{-1}\frac{x}{3} + \frac{2x\sqrt{9-x^2}}{9}\right.$$

$$\left. + \frac{1}{8}\left(\frac{2x\sqrt{9-x^2}}{9}\right)\left(1 - \frac{2x^2}{9}\right)\right]$$

$$= \frac{243}{8}\sin^{-1}\frac{x}{3} + \left(\frac{81}{4}\right)\frac{2x\sqrt{9-x^2}}{9}$$

$$\times \left(\frac{9}{8} - \frac{2x^2}{72}\right)$$

$$= \frac{243}{8}\sin^{-1}\frac{x}{3} + \left(\frac{9}{2}\right)x\sqrt{9-x^2}\left(\frac{9}{8} - \frac{2x^2}{72}\right)$$

$$= \frac{243}{8}\sin^{-1}\frac{x}{3} + x\sqrt{9-x^2}\left(\frac{81}{16} - \frac{x^2}{8}\right)$$

5. $\int x^3\sqrt{9-x^2}\,dx$. Again, this can be done using trig substitutions but the following method is easier. Let $u = 9-x^2$ and $du = -2x\,dx$, and so $x^2 = 9 - u$ and $-\frac{1}{2}du = x\,dx$. Thus, $\int x^3\sqrt{9-x^2} = \int x^2\sqrt{9-x^2}x\,dx$. Now substitute

$$\int(9-u)\sqrt{u}\left(-\frac{1}{2}du\right)$$

$$= -\frac{1}{2}\int\left(9\sqrt{u} - u^{3/2}\right)du$$

$$= -\frac{1}{2}\cdot 9 \cdot \frac{2}{3}u^{3/2} + \frac{1}{2}\frac{2}{5}u^{5/2} + C$$

$$= -3\left(9-x^2\right)^{3/2} + \frac{1}{5}\left(9-x^2\right)^{5/2} + C$$

$$= -3\left(9-x^2\right)^{3/2} + \frac{1}{5}\left(9-x^2\right)\left(9-x^2\right)^{3/2} + C$$

$$= -\frac{15}{5}\left(9-x^2\right)^{3/2}$$

$$\quad + \frac{1}{5}\left(9-x^2\right)\left(9-x^2\right)^{3/2} + C$$

$$= -\frac{1}{5}\left(9-x^2\right)^{3/2}\left(6+x^2\right) + C.$$

6. $\int \frac{x}{\sqrt{x^2-9}}\,dx$: Let $u = x^2 - 9$; $du = 2x\,dx$ or $x\,dx = \frac{1}{2}du$. Substituting yields $\frac{1}{2}\int\frac{du}{\sqrt{u}} = \frac{1}{2}\cdot\frac{2}{1}u^{1/2} + C = \sqrt{x^2-9} + C.$

7. $\int \frac{dx}{\sqrt{x^2-9}}$. Let $x = 3\sec\theta$ and then $dx = 3\sec\theta\times\tan\theta\,d\theta$. Substitution into the given integral yields $\int\frac{3\sec\theta\tan\theta}{\sqrt{9\sec^2\theta-9}}d\theta = \int\frac{3\sec\theta\tan\theta\,d\theta}{3\tan\theta} = \int\sec\theta\,d\theta = \ln|\sec\theta + \tan\theta| + C$. Since $\sec\theta = \frac{x}{3}$ and $\tan\theta = \frac{\sqrt{x^2-9}}{3}$, the answer is $\ln\left|\frac{x}{3} + \frac{\sqrt{x^2-9}}{3}\right| + C = \ln\frac{1}{3}\left|x + \sqrt{x^2-9}\right| + C = \ln\frac{1}{3} + \ln\left|x + \sqrt{x^2-9}\right| + C$ or $\ln\left|x + \sqrt{x^2-9}\right| + k$ where $k = C + \ln\frac{1}{3} = C - \ln 3$.

8. $\int \frac{x^3}{\sqrt{x^2-9}}\,dx$. Once again we use a substitution but do not need trigonometric functions. Let $u = x^2 - 9$; $du = 2x\,dx$ or $\frac{1}{2}du = x\,dx$ also $u = x^2 - 9$ yields $x^2 = u + 9$

$$\int\frac{x^3}{\sqrt{x^2-9}}dx = \int\frac{x^2x\,dx}{\sqrt{x^2-9}} = \frac{1}{2}\int\frac{(u+9)du}{\sqrt{u}}$$

$$= \frac{1}{2}\int\left(\sqrt{u} + \frac{9}{\sqrt{u}}\right)du$$

$$= \frac{1}{2}\cdot\frac{2}{3}u^{3/2} + \frac{1}{2}\cdot 9\cdot\frac{2}{1}u^{1/2} + C$$

$$= \frac{1}{3}\left(x^2-9\right)^{3/2}$$

$$\quad + 9\left(x^2 - +9\right)^{1/2} + C$$

$$= \frac{1}{2}\left(x^2-9\right)\sqrt{x^2-9}$$

$$\quad + 9\sqrt{x^2-9} + C$$

$$= \left(\frac{x^2}{3} + 6\right)\sqrt{x^2-9} + C$$

9. $\int\left(x^2-9\right)^{1/2}dx = \int\sqrt{x^2-9}\,dx$. Let $x = 3\sec\theta$, then $dx = 3\sec\theta\tan\theta\,d\theta$. Substituting yields $\int\sqrt{9\sec^2\theta-9}\cdot 3\sec\theta\tan\theta\,d\theta = \int 9\tan^2\theta\sec\times\theta\,d\theta = 9\int\left(\sec^2\theta-1\right)\sec\theta\,d\theta = 9\int\left(\sec^3\theta - \sec\theta\right)d\theta = 9\int\sec^2\theta\,d\theta - 9\int\sec\theta\,d\theta$. The first integral is example 20.50 and the second is a formula. They yield $9\left[\frac{1}{2}\sec\theta\tan\theta + \frac{1}{2}\ln|\sec\theta + \tan\theta| - \ln|\sec\theta + \tan\theta|\right] = \frac{9}{2}\sec\theta\tan\theta - \frac{9}{2}\ln|\sec\theta + \tan\theta|$. Since $\sec\theta = \frac{x}{3}$ and $\tan\theta = \frac{\sqrt{x^2-9}}{3}$, we get $\int\left(x^2-9\right)^{1/2}dx = \frac{9}{2}\cdot\frac{x}{3}\cdot\frac{\sqrt{x^2-9}}{3} - \frac{9}{2}\ln\left|\frac{x}{3} + \frac{\sqrt{x^2-9}}{3}\right| + K = \frac{1}{2}\left[x\sqrt{x^2-9} - 9\ln\left|x + \sqrt{x^2-9}\right|\right] + C$ where $C = K - \frac{9}{2}\ln 3$.

10. $\int x^3\sqrt{x^2-9}\,dx$; Let $u = x^2-9$; $du = 2x\,dx$ so $x\,dx = \frac{1}{2}du$ and $x^2 = u + 9$. Substituting, we get $\frac{1}{2}\int(u+9)\sqrt{u}\,du = \frac{1}{2}\int\left(u^{3/2} + 9u^{1/2}\right)du = \frac{1}{2}\frac{2}{5}u^{5/2} + \frac{9}{2}\frac{2}{3}u^{3/2} + C = \frac{1}{5}\left(x^2-9\right)^{5/2} + 3\left(x^2-9\right)^{3/2} = \left(\frac{6+x^2}{5}\right)\left(x^2-9\right)^{3/2} + C.$

11. $\int\frac{x}{\sqrt{x^2+9}}\,dx$. Letting $u = x^2 + 9$ and $du = 2x\,dx$, produces $\frac{1}{2}\int\frac{du}{\sqrt{u}} = \frac{1}{2}\cdot\frac{2}{1}u^{1/2} + C = \sqrt{x^2+9} + C.$

12. $\int\frac{dx}{\sqrt{x^2+9}}$. Let $x = 3\tan\theta$ and $dx = 3\sec^2\theta\,d\theta$. This gives $\int\frac{3\sec^2\theta\,d\theta}{\sqrt{9\tan^2\theta+9}} = \int\frac{3\sec^2\theta\,d\theta}{3\sec\theta} = \int\sec\theta\,d\theta = \ln|\sec\theta + \tan\theta| + C = \ln\left|\frac{\sqrt{x^2+9}}{3} + \frac{x}{3}\right| + C = \ln\left|\sqrt{x^2+9} + x\right| + K$ where $K = C - \ln 3$.

13. $\int\frac{dx}{x^2+9} = \frac{1}{3}\arctan\frac{x}{3} + C.$

14. $\int\frac{x^2}{x^2+9}\,dx$. Let $x = 3\tan\theta$ and $d\theta = 3\sec^2\theta\,d\theta$. Substituting this into the given integral produces $\int\frac{9\tan^2\theta}{9\tan^2\theta+9}\cdot 3\sec^2\theta\,d\theta = \int\frac{9\tan^2\theta}{9\sec^2\theta}3\sec^2\theta\,d\theta = \int 3\tan^2\theta\,d\theta = 3\int\left(\sec^2\theta - 1\right)d\theta = 3\int\sec^2\theta - 3\int d\theta = 3\tan\theta - 3\theta + C = 3\frac{x}{3} - 3\tan^{-1}\frac{x}{3} + C = x - 3\tan^{-1}\frac{x}{3} + C.$

15. $\int x^3\sqrt{9+x^2}\,dx$. Let $u = 9 + x^2$ and $du = 2x\,dx$ or $\frac{1}{2}du = x\,dx$. Thus, $x^2 = u - 9$. Substituting these in the given integral yields $\frac{1}{2}\int(u - 9)\sqrt{u}\,du = \frac{1}{2}\int\left(u^{3/2} - 9u^{1/2}\right)du = \frac{1}{2}\cdot\frac{2}{5}u^{5/2} - \frac{9}{2}\cdot\frac{2}{3}u^{3/2} + C = \frac{1}{5}u^{5/2} - 3u^{3/2} + C = \frac{1}{5}uu^{3/2} - 3u^{3/2} = \frac{1}{5}(9 + x^2)(9 + x^2)^{3/2} - 3(9 + x^2)^{3/2} = \left(\frac{x^2-6}{5}\right)(9 + x^2)^{3/2}$.

16. $\int\frac{\sqrt{1-x^2}}{x^2}\,dx$. Let $x = \sin\theta$ and $dx = \cos\theta\,d\theta$. Substitution yields $\int\frac{\sqrt{1-\sin^2\theta}}{\sin^2\theta}\cdot\cos\theta\,d\theta = \int\frac{\cos^2\theta}{\sin^2\theta}\,d\theta = \int\cot^2\theta\,d\theta = \int\left(\csc^2\theta - 1\right)d\theta = -\cot\theta - \theta + C;\ \theta = \sin^{-1}x;\ \cot\theta = \frac{\sqrt{1-x^2}}{x}$. The final answer is $-\frac{\sqrt{1-x^2}}{x} - \sin^{-1}x + C$.

17. $\int\frac{\sqrt{4-3x^2}}{x^4}\,dx$. Let $\sqrt{3}x = 2\sin\theta$, so $x = \frac{2}{\sqrt{3}}\sin\theta$ and $dx = \frac{2}{\sqrt{3}}\cos\theta\,d\theta$. Substitution produces

$$\int\frac{\sqrt{4 - 4\sin^2\theta}}{\left(\frac{2}{\sqrt{3}}\sin\theta\right)^4}\frac{2}{\sqrt{3}}\cos\theta\,d\theta$$

$$= \int\frac{2\cos\theta\frac{2}{\sqrt{3}}\cos\theta\,d\theta}{\frac{16}{9}\sin^4\theta}$$

$$= \frac{4}{\sqrt{3}}\cdot\frac{9}{16}\int\frac{\cos^2\theta}{\sin^4\theta}\,d\theta$$

$$= \frac{3\sqrt{3}}{4}\int\csc^2\theta\cot^2\theta\,d\theta$$

$$= \frac{-3\sqrt{3}}{4}\cdot\frac{1}{3}\cot^3\theta + C$$

$$= \frac{-\sqrt{3}}{4}\left(\frac{\sqrt{4-3x^2}}{\sqrt{3}x}\right)^3 + C$$

$$= -\frac{1}{12}\frac{(\sqrt{4-3x^2})^3}{x^3} + C$$

$$= -\frac{1}{12}\frac{(4-3x^2)^{3/2}}{x^3} + C$$

18. $\int\frac{\sqrt{4x^2-9}}{x^3}\,dx$. Let $2x = 3\sec\theta$, then $x = \frac{3}{2}\sec\theta$ and $dx = \frac{3}{2}\sec\theta\tan\theta\,d\theta$. Substituting we get

$$\int\frac{\sqrt{9\sec^2\theta - 9}}{\frac{27}{8}\sec^2\theta}\frac{3}{2}\sec\theta\tan\theta\,d\theta$$

$$= \int\frac{8\cdot3\tan\theta}{27\sec^2\theta}\frac{3}{2}\sec\theta\tan\theta\,d\theta$$

$$= \int\frac{8\tan^2\theta\,d\theta}{6\sec^2\theta} = \frac{4}{3}\int\frac{\sec^2\theta - 1}{\sec^2\theta}\,d\theta$$

$$= \frac{4}{3}\int\left(1 - \frac{1}{\sec^2\theta}\right)d\theta = \frac{4}{3}\int\left(1 - \cos^2\theta\right)d\theta$$

$$= \frac{4}{3}\int\sin^2\theta\,d\theta = \frac{4}{3}\int\frac{1 - \cos2\theta}{2}\,d\theta$$

$$= \frac{4}{3}\cdot\frac{1}{2}\theta - \frac{4}{3}\cdot\frac{1}{4}\sin2\theta + C$$

$$= \frac{2}{3}\theta - \frac{\sin2\theta}{3} + C$$

$$= \frac{2}{3}\sec^{-1}\frac{2x}{3} - \frac{2\sin\theta\cos\theta}{3} + C$$

$$= \frac{2}{3}\sec^{-1}\frac{2x}{3} - \frac{2}{3}\frac{\sqrt{4x^2 - 9}}{2x}\cdot\frac{3}{2x} + C$$

$$= \frac{2}{3}\sec^{-1}\frac{2x}{3} - \frac{\sqrt{4x^2 - 9}}{2x^2} + C$$

19. $\int\frac{x^3}{\sqrt{9x^2+4}}\,dx$; $u = 3x$; $3x = 2\tan\theta$ and so $x = \frac{2}{3}\tan\theta$; and $dx = \frac{2}{3}\sec^2\theta\,d\theta$. Substituting these values in the given integral produces

$$\int\frac{\left(\frac{2}{3}\tan\theta\right)^3}{\sqrt{(2\tan\theta)^2 + 4}}\frac{2}{3}\sec^2\theta\,d\theta$$

$$= \int\frac{\frac{16}{81}\tan^3\theta\sec^2\theta}{2\sec\theta}\,d\theta$$

$$= \frac{8}{81}\int\tan^3\theta\sec\theta\,d\theta$$

$$= \frac{8}{81}\int\tan\theta\left(\sec^2\theta - 1\right)\sec\theta\,d\theta$$

$$= \frac{8}{81}\int\left(\tan\sec^3\theta - \tan\theta\sec\theta\right)d\theta$$

$$= \frac{8}{81}\left(\frac{1}{3}\sec^3\theta - \sec\theta\right) + C$$

$$= \frac{8}{81}\left[\frac{1}{3}\left(\frac{\sqrt{9x^2+4}}{2}\right)^3 - \left(\frac{\sqrt{9x^2+4}}{2}\right)\right] + C$$

$$= \frac{1}{243}\sqrt{(9x^2 + 4)^3} - \frac{4}{81}\sqrt{9x^2 + 4} + C$$

20. $\int\frac{dx}{(3x^2+6)^{3/2}} = \frac{1}{3^{3/2}}\int\frac{dx}{\sqrt{x^2+2}^3}$. Let $x = \sqrt{2}\tan\theta$ and $dx = \sqrt{2}\sec^2\theta$. Substituting we get $\frac{1}{3\sqrt{3}}\cdot\int\frac{\sqrt{2}\sec^2\theta}{\sqrt{(2\tan^2\theta+2)}^3}\,d\theta = \frac{1}{3\sqrt{3}}\int\frac{\sqrt{2}\sec^2\theta}{(\sqrt{2}\sec\theta)^3}\,d\theta = \frac{1}{3\sqrt{3}}\cdot\int\frac{1}{2\sec\theta}\,d\theta = \frac{1}{6\sqrt{3}}\int\cos\theta\,d\theta = \frac{1}{6\sqrt{3}}\sin\theta + C = \frac{1}{6\sqrt{3}}\cdot\frac{x}{\sqrt{x^2+2}} + C = \frac{x\sqrt{3}}{6\sqrt{3x^2+6}} + C$.

21. $\int\frac{\sqrt{x^2+1}}{x^2}\,dx$. Let $x = \tan\theta$ and $dx = \sec^2\theta$. Then $\int\frac{\sqrt{\tan^2\theta + 1}}{\tan^2\theta}\sec^2\theta\,d\theta = \int\frac{\sec\theta}{\tan^2\theta}\sec^2\theta\,d\theta$

$$= \int \frac{\sec^3 \theta}{\tan^2 \theta} d\theta = \int \frac{\left(1 + \tan^2 \theta\right) \sec \theta}{\tan^2 \theta} d\theta$$

$$= \int \frac{\sec \theta}{\tan^2 \theta} + \int \sec \theta \, d\theta$$

$$= \int \frac{\frac{1}{\cos \theta}}{\frac{\sin^2 \theta}{\cos^2 \theta}} d\theta + \int \sec \theta \, d\theta$$

$$= \int \frac{\cos \theta}{\sin^2 \theta} d\theta + \int \sec \theta \, d\theta$$

$$= -(\sin \theta)^{-1} + \ln |\sec \theta + \tan \theta| + C$$

$$= -\frac{\sqrt{x^2 + 1}}{x} + \ln \left(\sqrt{x^2 + 1} + x\right) + C$$

22. $\int \frac{dx}{\sqrt{4-(x-1)^2}} dx = \arcsin\left(\frac{x-1}{2}\right) + C$

23. $\int \sqrt{4 - (x - 1)^2} dx$. Here we have $a = 2$ and $u = x - 1$. Let $x - 1 = 2 \sin \theta$, then $x = 2 \sin \theta + 1$ and $dx = 2 \cos \theta \, d\theta$. Substitution yields $\int \sqrt{4 - 4 \sin^2 \theta}(2 \cos \theta) \, d\theta = \int (2 \cos \theta)(2 \cos \theta) \times$

$d\theta = 4 \int \cos^2 \theta \, d\theta = 4 \int \frac{1 + \cos 2\theta}{2} d\theta = 2 \int (1 + \cos 2\theta) \, d\theta + 2\theta + \sin 2\theta + C$. Thus $\theta = \sin^{-1} \frac{x-1}{2}$ and using the identity $\sin 2\theta = 2 \sin \theta \cos \theta = 2 \cdot \frac{x-1}{2} \cdot \frac{\sqrt{4-(x-1)^2}}{2}$ so the answer is $2 \sin^{-1} \left(\frac{x-1}{2}\right) + \frac{(x-1)\sqrt{4-(x-1)^2}}{2} + C$

24. $\int \frac{dx}{\sqrt{4-(x-1)^2}}$. Let $x - 1 = 2 \tan \theta$, then $dx = 2 \sec^2 \theta \, d\theta$. Substituting produces

$$\int \frac{2 \sec^2 \theta \, d\theta}{\sqrt{4 + 4 \tan^2 \theta}} = \int \frac{2 \sec^2 \theta \, d\theta}{2 \sec \theta} = \int \sec \theta \, d\theta$$

$$= \ln \left| \sec \theta + \tan \theta \right| + C$$

$$= \left| \frac{\sqrt{4 + (x - 1)^2}}{2} \right.$$

$$\left. + \frac{x - 1}{2} \right| + C$$

25. $\int \frac{dx}{\sqrt{(x-1)^2-4}} dx$. Let $x - 1 = 2 \sec \theta$ and $dx = 2 \sec \theta \tan \theta \, d\theta$. Then, you get

$$\int \frac{\sec \theta \tan \theta \, d\theta}{\sqrt{(2 \sec \theta)^2 - 4}} = \int \frac{2 \sec \theta \tan \theta \, d\theta}{2 \tan \theta}$$

$$= \int \sec \theta \, d\theta$$

$$= \ln | \sec \theta + \tan \theta | + C$$

$$= \ln \left| \frac{x - 1}{2} \right.$$

$$\left. + \frac{\sqrt{(x - 1)^2 - 4}}{2} \right| + C$$

$$= \ln \frac{1}{2} \left| x - 1 \right.$$

$$\left. + \sqrt{(x - 1)^2 - 4} \right| + C$$

$$= \ln \left| x - 1 \right.$$

$$\left. + \sqrt{(x - 1)^2 - 4} \right| + K$$

where $K = C - \ln 2$.

26. Completing the square gives $x^2 - 2x + 5 = x^2 - 2x + 1 + 4 = (x - 1)^2 + 4$. Thus, the given integral can be written as $\int \sqrt{x^2 - 2x + 5} dx = \int \sqrt{(x - 1)^2 + 4} dx$. If we let $x - 1 = 2 \tan \theta$ then $dx = 2 \sec^2 \theta \, d\theta$. Substitution yields

$\int \sqrt{(2 \tan \theta)^2 + 4}(2 \sec^2 \theta) \, d\theta = \int (2 \sec \theta) \times (2 \sec^2 \theta) \, d\theta = 4 \int \sec^3 \theta \, d\theta = 4 [\frac{1}{2} \sec \theta \tan \theta + \frac{1}{2} \ln | \sec \theta + \tan \theta |] + C = 2 \cdot \frac{\sqrt{x^2 - 2x + 5}}{2} \cdot \frac{x-1}{2} + 2 \ln \left| \frac{\sqrt{x^2 - 2x + 5}}{2} + \frac{x-1}{2} \right| + C = \frac{1}{2}\sqrt{x^2 - 2x + 5}(x - 1) + 2 \ln \left| \sqrt{x^2 - 2x + 5} + x - 1 \right| + K$ where $K = C - \ln 2$.

27. Completing the square produces $x^2 - 2x - 3 = x^2 - 2x + 1 - 4 = (x - 1)^2 - 4$. Thus $\int \sqrt{x^2 - 2x - 3} \, dx = \int \sqrt{(x - 1)^2 - 4} \, dx$. Let $x - 1 = 2 \sec \theta$, then $dx = 2 \sec \theta \tan \theta \, d\theta$. Substitution gives

$$\int \sqrt{(2 \sec \theta)^2 - 4}(2 \sec \theta \tan \theta) \, d\theta$$

$$= \int 2 \tan \theta \cdot 2 \sec \theta \tan \theta \, d\theta$$

$$= 4 \int \tan^2 \theta \sec \theta \, d\theta$$

$$= 4 \int \left(\sec^2 \theta - 1\right) \sec \theta \, d\theta$$

$$= 4 \int \sec^3 \theta \, d\theta - 4 \int \sec \theta \, d\theta$$

$$= 4 \cdot \left[\frac{1}{2} \sec \theta \tan \theta + \frac{1}{2} \ln | \sec \theta + \tan | \right.$$

$$\left. - 4 \ln | \sec \theta + \tan \theta| \right] + C$$

$$= 4 \left[\frac{1}{2} \sec \theta \tan \theta \right.$$

$$\left. - \frac{1}{2} \ln | \sec \theta + \tan \theta| \right] + C$$

$$= 2 \cdot \frac{x - 1}{2} \cdot \frac{\sqrt{x^2 - 2x - 3}}{2}$$

$$- 2 \ln \left| \frac{x - 1}{2} + \frac{\sqrt{x^2 - 2x - 3}}{2} \right| + C$$

$$= \frac{1}{2}(x - 1)\sqrt{x^2 - 2x - 3}$$

$$- 2 \ln \left| x - 1 + \sqrt{x^2 - 2x - 3} \right| + K$$

where $K = C + \ln 2$.

28. $\int \dfrac{dx}{\left(x^2-2x+10\right)^{3/2}} = \int \dfrac{dx}{\left[\left(x^2-2x+1\right)+9\right]^{3/2}} =$

$\int \dfrac{dx}{\left[(x-1)^2+9\right]^{3/2}}.$ Here $u = x-1$ and $a = 3$.

Let $x-1 = 3\tan\theta$ and $dx = 3\sec^2\theta\, d\theta$. Then,

$\int \dfrac{3\sec^2\theta\, d\theta}{\left[(3\tan\theta)^2+9\right]^{3/2}} = \int \dfrac{3\sec^2\theta\, d\theta}{27\sec^3\theta} = \int \dfrac{d\theta}{9\sec\theta} = \frac{1}{9}\cdot$

$\int \cos\theta = \frac{1}{9}\sin\theta + C = \frac{1}{9}\left(\dfrac{x-1}{\sqrt{x^2-2x+10}}\right) + C.$

29. $\int \dfrac{dx}{(x-3)\sqrt{x^2-6x+25}} = \int \dfrac{dx}{(x-3)\sqrt{(x-3)^2+16}}.$ Here $u = $
$x-3$ and $a = 4$. Let $x-3 = 4\tan\theta$, so $dx = $
$4\sec^2\theta\, d\theta.$ Then, $\int \dfrac{4\sec^2\theta\, d\theta}{4\tan\theta\sqrt{(4\tan\theta)^2+16}} = $

$\int \dfrac{4\sec^2\theta\, d\theta}{4\tan\theta 4\sec\theta} = \frac{1}{4}\int \dfrac{\sec\theta}{\tan\theta}\, d\theta = \frac{1}{4}\int \dfrac{1}{\sin\theta} = \frac{1}{4}\cdot$

$\int \csc\theta\, d\theta = \frac{1}{4}\ln|\csc\theta - \cot\theta| + C = $

$\frac{1}{4}\ln\left|\dfrac{\sqrt{x^2-6x+25}}{x-3} - \dfrac{4}{x-3}\right| + C = $

$\frac{1}{4}\ln\left|\dfrac{\sqrt{x^2-6x+25}-4}{x-3}\right| + C.$

30. $\int \dfrac{x-3}{\sqrt{x^2-6x+25}}dx.$ Let $u = x^2 - 6x + 25$, and then

$du = (2x-6)dx$ or $\frac{1}{2}du = (x-3)dx.$ Substi-

tution yields $\frac{1}{2}\int \dfrac{du}{\sqrt{u}} = \frac{1}{2}\cdot\frac{2}{1}u^{1/2} = u^{1/2} + C = $

$\sqrt{x^2-6x+25} + C.$

31. Using the trigonometric substitution $x = 3\sin\theta$,
produces $\sqrt{9-x^2} = 3\cos\theta$ and $dx = 3\cos\theta\, d\theta.$
Substituting these values, we obtain

$\displaystyle\int_1^3 \dfrac{\sqrt{9-x^2}}{x^2}dx = \int_{x=1}^{x=3} \dfrac{3\cos\theta}{(3\sin\theta)^2}3\cos\theta\, d\theta$

$= \displaystyle\int_{x=1}^{x=3} \cot^2\theta\, d\theta$

$= \displaystyle\int_{x=1}^{x=3} \left(\csc^2\theta - 1\right) d\theta$

$= \left[-\cot\theta - \theta\right]_{x=1}^{x=3}$

$= \left[-\dfrac{\sqrt{9-x^2}}{x} - \arcsin\left(\dfrac{x}{3}\right)\right]_1^3$

$= \left(-0 - \dfrac{\pi}{2}\right)$

$\quad - (-2.82843 - 0.33984)$

≈ 1.59747

32. Let $x = \sqrt{8}\sin\theta$, so $\theta = \arcsin\left(\frac{x}{\sqrt{8}}\right)$, $dx = $
$\sqrt{8}\cos\theta\, d\theta$ and $\sqrt{8-x^2} = \sqrt{8}\cos\theta.$ Substituting in the given integral produces

$\displaystyle\int_1^2 \dfrac{dx}{\left(8-x^2\right)^{3/2}} = \int_{x=1}^{x=2} \dfrac{\sqrt{8}\cos\theta\, d\theta}{(\sqrt{8}\cos\theta)^3}$

$= \displaystyle\int_{x=1}^{x=2} \dfrac{d\theta}{(\sqrt{8}\cos\theta)^2}$

$= \dfrac{1}{8}\displaystyle\int_{x=1}^{x=2} \sec^2\theta\, d\theta$

$= \dfrac{1}{8}\tan\theta\,\Big|_{x=1}^{x=2}$

$= \dfrac{1}{8}\dfrac{x}{\sqrt{8-x^2}}\bigg|_1^2$

$= \dfrac{1}{8}\left[\dfrac{2}{2} - \dfrac{1}{\sqrt{7}}\right] = \dfrac{1}{8} - \dfrac{1}{8\sqrt{7}}$

≈ 0.0778

33. Let $x = 2\sec\theta$, $\theta = \sec^{1-}\left(\frac{\pi}{2}\right)$, and $dx = $
$2\sec\theta\tan\theta\, d\theta$ so that $\sqrt{x^2-4} = 2\tan\theta.$ Then

$\displaystyle\int_4^8 \dfrac{dx}{x\sqrt{x^2-4}} = \int_{x=4}^{x=8} \dfrac{2\sec\theta\tan\theta\, d\theta}{2\sec\theta 2\tan\theta}$

$= \dfrac{1}{2}\displaystyle\int_{x=4}^{x=8} d\theta = \dfrac{1}{2}\theta\,\bigg|_{x=4}^{x=8}$

$= \dfrac{1}{2}\sec^{-1}\left(\dfrac{x}{2}\right)\bigg|_4^8$

$= \dfrac{1}{2}\left(\sec^{-1}4 - \sec^{-1}2\right)$

≈ 0.1355

34. Let $x = 2\tan\theta$, $dx = 2\sec^2\theta$, $\theta = \tan^{-1}\left(\frac{\pi}{2}\right)$,
and then $\sqrt{x^2+4} = 2\sec\theta.$ Substituting pro-
duces

$\displaystyle\int_1^4 \dfrac{dx}{x^4\sqrt{x^2+4}} = \int_{x=1}^{x=4} \dfrac{2\sec^2\theta}{(2\tan\theta)^4 2\sec\theta}d\theta$

$= \dfrac{1}{16}\displaystyle\int_{x=1}^{x=4} \dfrac{\sec\theta}{\tan^4\theta}d\theta$

$= \dfrac{1}{16}\displaystyle\int_{x=1}^{x=4} \dfrac{\cos^3\theta}{\sin^4\theta}d\theta$

$= \dfrac{1}{16}\displaystyle\int_{x=1}^{x=4} \cot^3\theta\csc\theta\, d\theta$

$= \dfrac{1}{16}\displaystyle\int_{x=1}^{x=4} \cot^2\theta(\cot\theta\csc\theta)\, d\theta$

$= \dfrac{1}{16}\displaystyle\int_{x=1}^{x=4} \left(\csc^2\theta - 1\right)$
$\quad \times (\cot\theta\csc\theta)\, d\theta$

$= \dfrac{1}{16}\left[-\dfrac{1}{3}\csc^3\theta + \csc\theta\right]_{x=1}^{x=4}$

$= \left[\dfrac{\csc\theta}{16} - \dfrac{\csc^3\theta}{48}\right]_{x=1}^{x=4}$

$= \left[\dfrac{\sqrt{x^2+4}}{16x} - \dfrac{\left(x^2+4\right)^{3/2}}{48x^3}\right]_1^4$

$$= \left[\frac{\sqrt{20}}{64} - \frac{(20)^{3/2}}{3072}\right]$$

$$- \left[\frac{\sqrt{5}}{16} - \frac{(5)^{3/2}}{48}\right]$$

$$= \frac{2\sqrt{5}}{64} - \frac{40\sqrt{5}}{3072} - \frac{\sqrt{5}}{16} + \frac{5\sqrt{5}}{48}$$

$$= \frac{23\sqrt{5}}{384} \approx 0.1339$$

35. Let $x = 6\sin\theta$, $dx = 6\cos\theta\, d\theta$, $\sqrt{36 - x^2} = 6\cos\theta$, and $\theta = \arcsin\left(\frac{\pi}{6}\right)$. Thus, when $x = 0$, $\theta = 0$ and when $x = 6$, $\theta = \frac{\pi}{2}$. Then

$$\int_0^6 x^2\sqrt{36 - x^2}\,dx = \int_0^{\pi/2} (6\sin\theta)^2$$
$$\times x(6\cos\theta)(6\cos\theta)\,d\theta$$

$$= 1296\int_0^{\pi/2} \sin^2\theta\cos^2\theta\,d\theta$$

$$= 324\int_0^{\pi/2} \sin^2 2\theta\, d\theta$$

$$= 162\int_0^{\pi/2} (1 - \cos 2\theta)\,d\theta$$

$$= 162\left[\theta - \frac{1}{2}\sin 2\theta\right]_0^{\pi/2}$$

$$= 81\pi \approx 254.4690$$

36. Let $x = 3\sin\theta$, $dx = 3\cos\theta$, and $\theta = \arcsin\left(\frac{\pi}{3}\right)$. Then, when $x = 0$, $\theta = 0$ and when $x = 1.5\sqrt{3}$, $\theta = \frac{\pi}{3}$. Substituting produces

$$\int_0^{1.5\sqrt{3}} \frac{x^2\,dx}{\sqrt{9 - x^2}} = \int_0^{\pi/3} \frac{(3\sin\theta)^2}{3\cos\theta} 3\cos\theta\, d\theta$$

$$= 9\int_0^{\pi/3} \sin^2\theta\, d\theta$$

$$= \frac{9}{2}\int_0^{\pi/3} (1 - \cos 2\theta)\,d\theta$$

$$= \frac{9}{2}\left[\theta - \frac{1}{2}\sin 2\theta\right]_0^{\pi/3}$$

$$= \frac{9}{2}\left[\frac{\pi}{3} - \frac{1}{2}\frac{\sqrt{3}}{2}\right]$$

$$= \frac{3\pi}{2} - \frac{9\sqrt{3}}{8} \approx 2.76383$$

37. The area under $\frac{x^3}{\sqrt{16 - x^2}}$ from 0 to 3 is $\int_0^3 \frac{x^2\,dx}{\sqrt{16 - x^2}}$. Let $x = 4\sin\theta$, then $dx = 4\cos\theta\, d\theta$. Substituting gives

$$\int \frac{(4\sin\theta)^3(4\cos\theta)\,d\theta}{\sqrt{16 - (4\sin\theta)^2}}$$

$$= \int \frac{64\sin^3\theta\, 4\cos\theta}{4\cos\theta}$$

$$= 64\int \sin^3\theta\, d\theta = 64\int (1 - \cos^2\theta)\sin\theta\, d\theta$$

$$= 64\left[-\cos\theta + \frac{1}{3}\cos^3\theta\right]$$

$$= 64\left[\frac{1}{3}\left(\frac{\sqrt{16 - x^2}}{4}\right)^3 - \frac{\sqrt{16 - x^2}}{4}\right]_0^3$$

$$= \frac{1}{3}\sqrt{7}^3 - 16\sqrt{7} - \frac{1}{3}\cdot 4^3 + 16\cdot 4$$

$$= \frac{1}{3}(7\sqrt{7} - 48\sqrt{7} - 64 + 192)$$

$$= \frac{1}{3}(128 - 41\sqrt{7}) \approx 6.5080654$$

38. $y = 10x - x^2$ is above the x-axis between 0 and 10 and $y' = 10 - 2x$. The arc length is $\int_0^{10}\sqrt{1 + (10 - 2x)^2}\,dx = \int_0^{10}\sqrt{1 + (2x - 10)^2}\,dx$. Let $u = 2x - 10 = \tan\theta$; $x - 5 = \frac{1}{2}\tan\theta$ so $dx = \frac{1}{2}\sec^2\theta\, d\theta$. Thus, $\int \sqrt{1 + \tan^2\theta}\cdot\frac{1}{2}\sec^2\theta\, d\theta = \int\sec\theta\cdot\frac{1}{2}\sec^2\theta\, d\theta = \frac{1}{2}\int\sec^3\theta\, d\theta$. As in Example 28.50, this is $\frac{1}{2}\left[\frac{1}{2}\sec\theta\tan\theta + \frac{1}{2}\ln|\sec\theta + \tan\theta| = \frac{1}{4}\left[\sqrt{1 + (2x - 10)^2}\cdot(2x - 10) + \ln|\sqrt{1 + (2x - 10)^2} + 2x - 10]_0^{10} = \frac{1}{4}[10\sqrt{101} + 10\sqrt{101} + \ln(\sqrt{101} + 10) - \ln(\sqrt{101} - 10) = 5\sqrt{101} + \frac{1}{4}\ln\left(\frac{\sqrt{101} + 10}{\sqrt{101} - 10}\right) \approx 51.7485$.

39. The area described is $\int_0^4 \sqrt{9 + x^2}\,dx$. Let $x = 3\tan\theta$ and $dx = 3\sec^2\theta\, d\theta$. Then, $\int\sqrt{9 + 9\tan^2\theta}\times(3\sec^2\theta)\,d\theta = \int(3\sec\theta)(3\sec^2\theta)\,d\theta = 9\int\sec^3\theta\, d\theta = 9\left[\frac{1}{2}\sec\theta\tan\theta + \frac{1}{2}\ln|\sec\theta + \tan\theta|\right.$

$$= \frac{9}{2}\left[\frac{\sqrt{9 + x^2}}{3}\cdot\frac{x}{3}\right] + \ln\left[\frac{\sqrt{9 + x^2}}{3} + \frac{x}{3}\right]\Big|_0^4 \text{ or, we get,}$$

$\frac{9}{2}\left[\frac{5}{3}\cdot\frac{4}{3} + \ln|\frac{5}{3} + \frac{4}{3}| - \frac{3}{3}\cdot\frac{0}{3} - \ln|\frac{3}{3}|\right] = \frac{9}{2}\left[\frac{20}{9} + \ln|3| - 0 - 0\right] = 5 + \frac{9}{2}\ln 3$ or $5 + \ln 3^{9/2} \approx 9.9438$.

40. Since $\rho(x) = \sqrt{108 - 3x^2}$, then the mass is $m = \int_0^6\sqrt{108 - 3x^2}\,dx = \sqrt{3}\int_0^6\sqrt{36 - x^2}\,dx$. Let $x = 6\sin\theta$ and $dx = 6\cos\theta\, d\theta$. Substituting these into the original integral gives $\sqrt{3}\int\sqrt{36 - 36\sin^2\theta}\cdot 6\cos\theta\, d\theta = \sqrt{3}\int(6\cos\theta)(6\cos\theta)\,d\theta = 36\sqrt{3}\times\int\cos^2\theta\, d\theta = 36\sqrt{3}\int\frac{1 + \cos 2\theta}{2}\,d\theta = 18\sqrt{3}[\theta + \frac{1}{2}\sin 2\theta] = 18\sqrt{3}[\theta + \sin\theta\cos\theta] = 18\sqrt{3}\left[\sin^{-1}\frac{x}{6} + \frac{x}{6}\cdot\frac{\sqrt{36 - x^2}}{6}\right]_0^6 = 18\sqrt{3}[\frac{\pi}{2} + 0 - 0 - 0] = 9\sqrt{3}\pi \approx 48.9726$.

41. To find the desired area we first solve $\frac{x^2}{25} + \frac{y^2}{4} = 1$

for y : $\frac{y^2}{4} = 1 - \frac{x^2}{25}$ so $y^2 = 4 - \frac{4x^2}{25}$ or $y^2 = \frac{100 - 4x^2}{25}$

and we get $y = \sqrt{\frac{100 - 4x^2}{25}} = \frac{2}{5}\sqrt{25 - x^2}$. We also

need $y' = \frac{2}{5}(25 - x^2)^{-1/2} \cdot \frac{1}{2} \cdot 2x = \frac{2}{5}x(25 - $

$x^2)^{-1/2} = \frac{2x}{5\sqrt{25-x^2}}$. Thus, $y'^2 = \frac{4x^2}{25(25-x^2)}$ and

so

$$S = \int_{-5}^{5} 2\pi y \sqrt{1 + y'^2}\, dx$$

$$= \int_{-5}^{5} 2\pi \left(\frac{2}{5}\sqrt{25 - x^2}\right)\sqrt{1 + \frac{4x^2}{25(25 - x^2)}}\, dx$$

$$= \int_{-5}^{5} 2\pi \cdot \frac{2}{5}\sqrt{25 - x^2}\sqrt{\frac{625 - 21x^2}{25(25 - x^2)}}\, dx$$

$$= \int_{-5}^{5} 2\pi \cdot \frac{2}{5}\sqrt{25 - x^2} \cdot \frac{1}{5}\frac{\sqrt{625 - 21x^2}}{\sqrt{25 - x^2}}\, dx$$

$$= \frac{4\pi}{25}\int_{-5}^{5} \sqrt{625 - 21x^2}\, dx.$$

If you let $\sqrt{21}x = 25\sin\theta$, then $dx = \frac{25}{\sqrt{21}}\cos\theta\, d\theta$.
Substitution gives

$$\frac{4\pi}{25}\int \sqrt{625 - 625\sin^2\theta} \cdot \frac{25}{\sqrt{21}}\cos\theta\, d\theta$$

$$= \frac{4\pi}{25}\int 25\cos\theta \cdot \frac{25}{\sqrt{21}}\cos\theta\, d\theta$$

$$= \frac{100\pi}{25}\int \cos^2\theta\, d\theta = \frac{100\pi}{\sqrt{21}}\int \frac{1 + \cos 2\theta}{2}\, d\theta$$

$$= \frac{50\pi}{\sqrt{21}}\left[\theta + \frac{1}{2}\sin 2\theta\right] = \frac{50\pi}{\sqrt{21}}[\theta + \sin\theta\cos\theta]$$

$$= \frac{50\pi}{\sqrt{21}}\left[\sin^{-1}\frac{\sqrt{21}x}{25} + \frac{\sqrt{21}x}{25}\right.$$

$$\left.\times \frac{\sqrt{625 - 21x^2}}{25}\right]_{-5}^{5}$$

$$= \frac{50\pi}{\sqrt{21}}\left[\sin^{-1}\frac{\sqrt{21}}{5}\right.$$

$$\left.+ \frac{\sqrt{21}\cdot 5\sqrt{625 - 21\cdot 25}}{625}\right] (2)$$

$$= \frac{100\pi}{\sqrt{21}}\left[\sin^{-1}\frac{\sqrt{21}}{5} + \frac{\sqrt{21}\cdot 50}{625}\right]$$

$$= \frac{100\pi \cdot 50}{625} + \frac{100\pi}{\sqrt{21}}\sin^{-1}\frac{\sqrt{21}}{5}$$

$$= 8\pi + \frac{100\pi}{\sqrt{21}}\sin^{-1}\frac{\sqrt{21}}{5} \approx 104.6073.$$

42. Charge is the integral of the current. As a result,
we have $\int_0^1 t\sqrt{t + 2}\, dt$. Let $u = t + 2$ with $du =$
dt and $t = u - 2$. Substituting we get $\int (u -$
$2)\sqrt{u} = \int u^{3/2} - 2u^{1/2}\, du = \frac{2}{5}u^{5/2} - \frac{4}{3}u^{3/2} =$

$\frac{2}{5}(t + 2)^{5/2} - \frac{4}{3}(t + 2)^{3/2}\big|_0^1 = \frac{2}{5}(3)^{5/2} - \frac{4}{3}(3)^{3/2} -$
$\frac{2}{5}(2)^{5/2} + \frac{4}{3}(2)^{3/2} \approx 0.8157.$

43. Let $u = t = \tan\theta$ and $du = dt = \sec^2\theta\, d\theta$. Then

$$\int \frac{\sqrt{t^2 + 1}}{9t^2}\, dt = \int \frac{\sqrt{\tan^2\theta + 1}}{9\tan^2\theta}\sec^2\theta\, d\theta$$

$$= \frac{1}{9}\int \frac{\sec^3\theta}{\tan^2\theta}\, d\theta = \frac{1}{9}\int \csc^2\theta\sec\theta\, d\theta$$

$$= \frac{1}{9}\int (1 + \cot^2\theta)\sec\theta\, d\theta$$

$$= \frac{1}{9}\int \left(\sec\theta + \frac{\cos\theta}{\sin^2\theta}\right)d\theta$$

$$= \frac{1}{9}\left[\ln|\sec\theta + \tan\theta| - (\sin\theta)^{-1}\right]$$

$$= \frac{1}{9}\left[\ln\left(\sqrt{1 + t^2} + t\right) - \frac{\sqrt{1 + t^2}}{t}\right]$$

So from $t = 0.5$ to $t = 1$, we have $i = \frac{1}{9} \times$

$$\left[\ln\left(\sqrt{1 + t^2} + t\right) - \frac{\sqrt{1+t^2}}{t}\right]_{0.5}^{1} \approx 0.1829 \text{ A}$$

44. Since the tank is a cylinder of height 10 ft, the
volume will be $V = 10B$ where B is the area of
the cross-section of oil in the tank. For a depth of
2.5 ft, we integrate with respect to t from -3 to
-0.5. $B = 2\int_{-3}^{-0.5}\sqrt{x}\, dy = 2\int_{-3}^{-0.5}2\sqrt{9 - y^2}\, dy$
$= 4\int_{-3}^{-0.5}\sqrt{9 - y^2}\, dy$. Let $y = 3\sin\theta$ and $dy = $
$3\cos\theta$. Then the integral becomes

$$B = 4\int_{-3}^{-0.5}\sqrt{9 - y^2}\, dy$$

$$= 4\int_{y=-3}^{y=-0.5}(3\cos\theta)(3\cos\theta)\, d\theta$$

$$= 36\int_{y=-3}^{y=-0.5}\cos^2\theta\, d\theta$$

$$= 18\int_{y=-3}^{y=-0.5}(1 + \cos 2\theta)\, d\theta$$

$$= 18\left[\theta + \frac{1}{2}\sin 2\theta\right]_{y=-3}^{y=-0.5}$$

$$= 18[\theta + \sin\theta\cos\theta]_{y=-3}^{y=-0.5}$$

Substituting $\sin\theta = \frac{y}{3}, \theta = \arcsin\frac{y}{3}$, and $\cos\theta = $
$\frac{\sqrt{9 - y^2}}{3}$ produces

$$B = 18\arcsin\frac{y}{3} + 18\left(\frac{y}{3}\right)\left(\frac{\sqrt{9 - y^2}}{3}\right)\Big|_{-3}^{-0.5}$$

$$= 18\arcsin\frac{y}{3} + 2y\sqrt{9 - y^2}\Big|_{-3}^{-0.5}$$

$$= \left(18\arcsin\frac{-0.5}{3} + 2(-0.5)\sqrt{9 - (-0.5)^2}\right)$$

$$- \left(18\arcsin\frac{-3}{3} + 2(-3)\sqrt{9 - (-3)^2}\right)$$

$\approx (-3.01407 - 2.95804) - (-9\pi + 0)$

$= 9\pi - 5.97211 \approx 22.30223$

Thus, the volume of oil in the tank is $10B = 223.022$ ft^3.

28.7 INTEGRATION BY PARTS

1. $\int x \ln x \, dx$. Completing the table, we get

$u = \ln x$	$v = \frac{1}{2}x^2$
$du = \frac{1}{x}dx$	$dv = x\,dx$

Thus, using the table, we obtain $\int x \ln x \, dx = \frac{1}{2}x^2 \ln x - \int \frac{1}{2}x^2 \cdot \frac{1}{x}dx = \frac{1}{2}x^2 \ln x - \frac{1}{2}\int x \, dx = \frac{1}{2}x^2 \ln x - \frac{1}{4}x^2 + C$ or $\frac{1}{2}x^2(\ln x - \frac{1}{2}) + C$.

2. $\int \tan^{-1} x \, dx$

$u = \tan^{-1} x$	$v = x$
$du = \frac{1}{1+x^2}dx$	$dv = dx$

Thus, $\int \tan^{-1} x \, dx = x \tan^{-1} x - \int \frac{x\,dx}{1+x^2} = x \tan^{-1} x - \frac{1}{2}\ln(1 + x^2) + C$.

3. $\int x \sin 2x \, dx$

$u = x$	$v = -\frac{1}{2}\cos 2x$
$du = dx$	$dv = \sin 2x \, dx$

Using the table, we see that $\int x \sin 2x \, dx = -\frac{1}{2}x \cos 2x - \int -\frac{1}{2}\cos 2x \, dx = -\frac{1}{2}x \cos 2x + \frac{1}{4}\sin 2x + C$.

4. $\int x^5 \ln x \, dx$

$u = \ln x$	$v = \frac{1}{6}x^6$
$du = \frac{1}{x}dx$	$dv = x^5 \, dx$

Thus, we have $\int x^5 \ln x \, dx = \frac{1}{6}x^6 \ln x - \int \frac{1}{6}x^5 \, dx = \frac{1}{6}x^6 \ln x - \frac{1}{36}x^6 + C = \frac{1}{6}x^6 \left[\ln x - \frac{1}{6}\right] + C$.

5. $\int x^2 \ln x \, dx$

$u = \ln x$	$v = \frac{1}{3}x^3$
$du = \frac{1}{x}dx$	$du = x^2 \, dx$

Thus, from the table, we see that $\int x^2 \ln x \, dx = \frac{1}{3}x^3 \ln x - \int \frac{1}{3}x^3 \cdot \frac{1}{x}dx = \frac{1}{3}x^3 \ln x - \frac{1}{9}x^3 + C$.

6. $\int x\sqrt{1+x} \, dx$. Let $u = 1 + x$, then $du = dx$ and $x = u - 1$. We can rewrite the given integral as $\int (u - 1)\sqrt{u}\,du = \int \left(u^{3/2} - u^{1/2}\right)du = \frac{2}{5}u^{5/2} - \frac{2}{3}u^{3/2} + C = \frac{2}{5}(1 + x)^{5/2} - \frac{2}{3}(1 + x)^{3/2} + C =$

$\frac{2}{5}(1 + x)(1 + x)^{3/2} - \frac{2}{3}(1 + x)^{3/2} + C = \frac{2}{15}(3x - 2)(1 + x)^{3/2} + C$.

7. $\int x \sin^{-1} x^2 \, dx$.

$u = \sin^{-1} x^2$	$v = \frac{1}{2}x^2$
$du = \frac{2x}{\sqrt{1-x^2}}dx$	$dv = x \, dx$

Hence, $\int x \sin^{-1} x^2 dx = \frac{1}{2}x^2 \sin^{-1} x^2 - \int \frac{x^3}{\sqrt{1-x^2}}dx$. For the second integral let $w = 1 - x^4, dw = -4x^3 dx$ or $\frac{1}{4}dw = -x^3 dx$. Substituting, we get $\frac{1}{2}x^2 \sin^{-1} x^2 - \int \frac{x^3}{1-x^2} \, dx = \frac{1}{2}x^2 \sin^{-1} x^2 - \frac{1}{4}\int \frac{dw}{\sqrt{w}} = \frac{1}{2}x^2 \sin^{-1} x^2 - \frac{1}{4}\int w^{-1/2}dw = \frac{1}{2}x^2 \sin^{-1} x^2 + \frac{1}{2}w^{1/2} + C = \frac{1}{2}x^2 \sin^{-1} x^2 + \frac{1}{2}(1 - x^4)^{1/2} + C$. The final answer is $\frac{1}{2}x^2 \sin^{-1} x^2 + \frac{1}{2}(1 - x^4)^{1/2} + C = \frac{1}{2}\left[x^2 \sin^{-1} x^2 + \sqrt{1 - x^4}\right] + C$.

8. $\int x^2 \sin x \, dx$

$u = x^2$	$v = -\cos x$
$du = 2x \, dx$	$dv = \sin x \, dx$

From the table, we see that $\int x^2 \sin x \, dx = -x^2 \cos x + \int 2x \cos x \, dx$. We now use a second table.

$u = 2x$	$v = \sin x$
$du = 2\,dx$	$dv = \cos x \, dx$

Hence, $-x^2 \cos x + \int 2x \cos x \, dx = -x^2 \cos x + 2x \sin x - \int 2 \sin x \, dx = -x^2 \cos x + 2x \sin x - 2\cos x + C = 2x \sin x - (x^2 + 2)\cos x + C$.

9. $\int x^3 e^{2x} dx$

$u = x^3$	$v = \frac{1}{2}e^{2x}$
$du = 3x^2 dx$	$dv = e^{2x} dx$

So, $\int x^3 e^{2x} \, dx = \frac{1}{2}x^3 e^{2x} - \frac{3}{2}\int x^2 e^{2x} dx$. Using a second table produces

$u = x^2$	$v = \frac{1}{2}e^{2x}$
$du = 2x \, dx$	$dv = e^{2x} dx$

Combined with the result from the first table, this produces $\frac{1}{2}x^3 e^{2x} - \frac{3}{2}\int x^2 e^{2x}\,dx = \frac{1}{2}x^3 e^{2x} - \frac{3}{2}\left[\frac{1}{2}x^2 e^{2x} - \int xe^{2x}\,dx\right]$. A third table produces

$u = x$	$v = \frac{1}{2}e^{2x}$
$du = dx$	$dv = e^{2x}\,dx$

When combined with the previous result, this yields
$\frac{1}{2}x^3 e^{2x} - \frac{3}{2}\left[\frac{1}{2}x^2 e^{2x} - \int xe^{2x}\,dx\right] = \frac{1}{2}x^3 e^{2x} - \frac{3}{4}x^2 e^{2x}$
$+ \frac{3}{4}xe^{2x} - \int \frac{1}{2}e^{2x}\,dx = \frac{1}{2}x^3 e^{2x} - \frac{3}{4}x^2 e^{2x} + \frac{3}{4}xe^{2x} -$
$\frac{1}{4}e^{2x} + C = e^{2x}\left(\frac{1}{2}x^3 - \frac{3}{4}x^2 + \frac{3}{4}x - \frac{1}{4}\right) + C$

10. $\int x \tan^{-1} x\,dx$

$u = \tan^{-1} x$	$v = \frac{1}{2}x^2$
$du = \frac{1}{1+x^2}\,dx$	$dv = x\,dx$

$\int x \tan^{-1} x\,dx = \frac{1}{2}x^2 \tan^{-1} x - \int \frac{1}{2}x^2 \frac{1}{1+x^2}\,dx$. Now in the second integral let $x = \tan\theta$ and $dx = \sec\theta\,d\theta$. Then, we have $-\int \frac{1}{2}x^2 \frac{1}{1+x^2}\,dx = -\frac{1}{2} \times$
$\int \frac{\tan^2\theta}{1+\tan^2\theta} \cdot \sec^2\theta\,d\theta = -\frac{1}{2}\int \frac{\tan^2\theta}{\sec^2\theta} \cdot \sec^2\theta\,d\theta =$
$-\frac{1}{2}\int \tan^2\theta\,d\theta = -\frac{1}{2}\int(\sec^2\theta - 1)d\theta = \frac{1}{2}\int(1 - \sec^2\theta)d\theta = \frac{1}{2}\theta - \frac{1}{2}\tan\theta + C = \frac{1}{2}\tan^{-1}x - \frac{1}{2}x + C$.
Thus, the final answer is $\frac{1}{2}x^2 \tan^{-1} x + \frac{1}{2}\tan^{-1} x - \frac{1}{2}x + C = \frac{1}{2}\tan^{-1}x(x^2 + 1) - \frac{1}{2}x + C$.

11. $\int x\sqrt{4x+1}\,dx$. Let $u = 4x + 1$ and $du = 4\,dx$. So, $\frac{u-1}{4} = x$ and $\frac{1}{4}du = dx$. Thus, the given integral is $\int \frac{u-1}{4} \cdot \sqrt{u} \cdot \frac{1}{4}dx = \frac{1}{16}\int u^{3/2} - u^{1/2}du =$
$\frac{1}{16}\left[\frac{2}{5}u^{5/2} - \frac{2}{3}u^{3/2}\right] + C = \frac{1}{16}\left[\frac{2}{5}(4x+1)^{5/2} - \frac{2}{3}(4x+1)^{3/2}\right] + C = \frac{1}{16}\left[\frac{2}{5}(4x+1)(4x+1)^{3/2} - \frac{2}{3}(4x+1)^{3/2}\right] + C = \frac{1}{16}\left[\left(\frac{8x}{5} - \frac{4}{15}\right)(4x+1)^{3/2}\right] + C =$
$\frac{1}{16}\left[\frac{24x-4}{15}(4x+1)^{3/2}\right] + C = \frac{1}{60}(6x-1)(4x+1)^{3/2} + C$.

12. $\int x^2 \cos 3x\,dx$

$u = x^2$	$v = \frac{1}{3}\sin 3x$
$du = 2x\,dx$	$dv = \cos 3x\,dx$

Using the table, we see that $\int x^2 \cos 3x\,dx = \frac{1}{3}x^2 \sin 3x - \int \frac{2}{3}x \sin 3x$. We again use integration by parts to determine the second integral.

$u = \frac{2}{3}x$	$v = -\frac{1}{3}\cos 3x$
$du = \frac{2}{3}dx$	$dv = \sin 3x\,dx$

Thus, $\frac{1}{3}x^2 \sin 3x - \int \frac{2}{3}x \sin 3x = \frac{1}{3}x^2 \sin 3x + \frac{2}{9}x \cos 3x + \int -\frac{2}{9}\cos 3x\,dx = \frac{1}{3}x^2 \sin 3x + \frac{2}{9}x \cos 3x - \frac{2}{27}\sin 3x + C$ or $\frac{1}{9}\left[2x\cos 3x + (3x^2 - \frac{2}{3})\sin 3x\right] + C$.

13. $\int x^2 e^{x/4}\,dx$

$u = x^2$	$v = 4e^{x/4}$
$du = 2x\,dx$	$dv = e^{x/4}\,dx$

Thus, $\int x^2 e^{x/4}\,dx = 4x^2 e^{x/4} - \int 8xe^{x/4}$. Using integration by parts a second time, we get the following table:

$u = 8x$	$v = 4e^{x/4}$
$du = 8\,dx$	$dv = e^{x/4}\,dx$

Hence, we get $4x^2 e^{x/4} - \int 8xe^{x/4} = 4x^2 e^{x/4} - 32xe^{x/4} + \int 32e^{x/4}\,dx = 4x^2 e^{x/4} - 32xe^{x/4} + 128e^{x/4} + C = e^{x/4}(4x^2 - 32x + 128) + C$.

14. $\int e^{4x} \sin 2x\,dx$

$u = e^{4x}$	$v = -\frac{1}{2}\cos 2x$
$du = 4e^{4x}$	$dv = \sin 2x\,dx$

This produces $\int e^{4x} \sin 2x\,dx = -\frac{1}{2}e^{4x}\cos 2x + 2\int e^{4x}\cos 2x$. But, we need to use integration by parts a second time. Here we get the table

$u = 2e^{4x}$	$v = \frac{1}{2}\sin 2x$
$du = 8e^{4x}$	$dv = \cos 2x\,dx$

Thus, we have $-\frac{1}{2}e^{4x}\cos 2x + 2\int e^{4x}\cos 2x = -\frac{1}{2}e^{4x}\cos 2x + e^{4x}\sin 2x - 4\int e^{4x}\sin 2x$. So, $5\int e^{4x}\sin 2x\,dx = e^{4x}\sin 2x - \frac{1}{2}e^{4x}\cos 2x + C$ or $\int e^{4x}\sin 2x\,dx = \frac{1}{10}e^{4x}[2\sin 2x - \cos 2x] + C$.

15. $\int e^x \cos x\,dx$

$u = e^x$	$v = \sin x$
$du = e^x\,dx$	$dv = \cos x\,dx$

This results in $\int e^x \cos x\,dx = e^x \sin x - \int e^x \sin x\,dx$. Using integration by parts again produces

$u = e^x$	$v = -\cos x$
$du = -e^x\,dx$	$dv = \sin x\,dx$

Combining this with the previous result yields $e^x \sin x - \int e^x \sin x\,dx = e^x \sin x + e^x \cos x - \int e^x \cos x\,dx$. Hence $2\int e^x \cos x\,dx = e^x \sin x + e^x \cos x$ or $\int e^x \cos x\,dx = \frac{1}{2}e^x(\sin x + \cos x) + C$.

16. $\int x^4 e^{2x} dx$

$u = x^4$	$v = \frac{1}{2}e^{2x}$
$du = 4x^3 dx$	$dv = e^{2x} dx$

Thus, $\int x^4 e^{2x} dx = x^4 \cdot \frac{1}{2}e^{2x} - \int 2x^3 e^{2x}$. From Exercise #9, we know that

$$\int x^3 e^{2x} dx = e^{2x}\left(\frac{1}{2}x^3 - \frac{3}{4}x^2 + \frac{3}{4}x - \frac{3}{8}\right) + C$$

Using this result, we have

$$\int x^4 e^{2x} dx$$

$$= x^4 \cdot \frac{1}{2}e^{2x} - 2\int x^3 e^{2x}$$

$$= x^4 \cdot \frac{1}{2}e^{2x}$$

$$\quad - 2\left[e^{2x}\left(\frac{1}{2}x^3 - \frac{3}{4}x^2 + \frac{3}{4}x - \frac{3}{8}\right)\right] + C$$

$$= e^{2x}\left(\frac{1}{2}x^4 - x^3 + \frac{3}{2}x^2 - \frac{3}{2}x + \frac{3}{4}\right) + C$$

17. Rewrite the given integral as $\int x^3 \cos x^2 dx = \int \frac{x^2}{2} \cdot 2x \cos x^2 dx$, and then use integration by parts.

$u = \frac{x^2}{2}$	$v = \sin x^2$
$du = x\, dx$	$dv = 2x \cos x^2 dx$

So the integral equals $\frac{x^2}{2}\sin x^2 - \int x \sin x^2 = \frac{x^2}{2}\sin x^2 + \frac{1}{2}\cos x^2 + C$

18. $\int x^2 e^{-x} dx$

$u = x^2$	$v = -e^{-x}$
$du = 2x\, dx$	$dv = e^{-x} dx$

Hence $\int x^2 e^{-x} dx = -x^2 e^{-x} + \int 2xe^{-x} dx$. Integrating by parts again,

$u = 2x$	$v = -e^{-x}$
$du = 2\, dx$	$dv = e^{-x} dx$

Thus, $-x^2 e^{-x} - 2xe^{-x} + \int 2e^{-x} dx = -x^2 e^{-x} - 2xe^{-x} - 2e^{-x} + C = -e^{-x}(x^2 + 2x + 2) + C$.

19. $\int x^3 e^{-x} dx$

$u = x^3$	$v = -e^{-x}$
$du = 3x^2 dx$	$dv = e^{-x} dx$

Thus, we obtain $\int x^3 e^{-x} dx = -x^3 e^{-x} + \int 3x^2 e^{-x}$. Using integration by parts a second time, we have the following table.

$u = 3x^2$	$v = -e^{-x}$
$du = 6x\, dx$	$dv = e^{-x} dx$

Combined with the earlier result, this produces $-x^3 e^{-x} + \int 3x^2 e^{-x} = -x^3 e^{-x} - 3x^2 e^{-x} + \int 6xe^{-x}$. We now use integration by parts a third time.

$u = 6x$	$v = -e^{-x}$
$du = 6\, dx$	$dv = e^{-x} dx$

Hence, $-x^3 e^{-x} - 3x^2 e^{-x} + \int 6xe^{-x} = -x^3 e^{-x} - 3x^2 e^{-x} - 6xe^{-x} + \int 6e^{-x} = -x^3 e^{-x} - 3x^2 e^{-x} - 6xe^{-x} - 6e^{-x} + C = -e^{-x}(x^3 + 3x^2 + 6x + 6) + C$

20. $\int \sin^2 x\, dx = \int \frac{1 - \cos 2x}{2} dx = \frac{1}{2}x - \frac{1}{4}\sin 2x + C$ or $= \frac{1}{2}x - \frac{1}{2}\sin x \cos x + C$.

21. $\int e^{-x} \cos x\, dx$

$u = e^{-x}$	$v = \sin x$
$du = -e^{-x} dx$	$dv = \cos x\, dx$

Thus, $\int e^{-x} \cos x\, dx = e^{-x} \sin x + \int \sin x e^{-x} dx$. Applying integration by parts again, we have the following table.

$u = e^{-x}$	$v = -\cos x$
$du = -e^{-x} dx$	$dv = \sin x\, dx$

Hence, $e^{-x} \sin x + \int \sin x e^{-x} dx = e^{-x} \sin x - e^{-x} \cos x - \int e^{-x} \cos x\, dx$ which produces $2\int e^{-x} \cos x\, dx = e^{-x} \sin x - e^{-x} \cos x$ and means that $\int e^{-x} \cos x = \frac{1}{2}e^{-x}(\sin x - \cos x) + C$.

22. $\int x4^x dx$

$u = x$	$v = \frac{4^x}{\ln 4}$
$du = dx$	$dv = 4^x dx$

Thus, $\int x4^x dx = x \cdot \frac{1}{\ln 4}4^x - \int \frac{4^x}{\ln 4} dx = \frac{x4^x}{\ln 4} - \frac{1}{\ln 4} \cdot \frac{1}{\ln 4}4^x + C = \frac{4^x}{\ln 4}\left(x - \frac{1}{\ln 4}\right) + C$.

23. $\int x \cos 4x\, dx$

$u = x$	$v = \frac{1}{4}\sin 4x$
$du = dx$	$dv = \cos 4x\, dx$

Hence, $\int x \cos 4x\, dx = \frac{x}{4}\sin 4x - \frac{1}{4}\int \sin 4x\, dx = \frac{x}{4}\sin 4x + \frac{1}{16}\cos 4x + C$.

24. $\int e^{4x} \cos 2x\, dx$

$u = e^{4x}$	$v = \frac{1}{2}\sin 2x$
$du = 4e^{4x}$	$dv = \cos 2x\, dx$

Which means $\int e^{4x} \cos 2x\, dx = \frac{1}{2}e^{4x} \sin 2x - \int 2e^{4x} \sin 2x\, dx$. Using integration by parts a second time, we get

$u = 2e^{4x}$ | $v = -\frac{1}{2}\cos 2x$

$du = 8e^{4x}$ | $dv = \sin 2x\ dx$

Combined with the earlier results, we get $\frac{1}{2}e^{4x}\sin 2x$

$-\int 2e^{4x}\sin 2x\ dx = \frac{1}{2}e^{4x}\sin 2x + e^{4x}\cos 2x -$

$4\int e^{4x}\cos 2x\ dx$. This leads to $5\int e^{4x}\cos 2x\ dx =$

$\frac{1}{2}e^{4x}\sin 2x + e^{4x}\cos 2x$ and so, the desired result

is $\int e^{4x}\cos 2x\ dx = \frac{1}{10}e^{4x}(\sin 2x + 2\cos 2x) + C$.

25. Using the disc method, we see that $V = \pi\int_0^{\pi/2}\times$
$\cos^2 x\ dx = \pi\int_0^{\pi/2}\frac{1+\cos 2x}{2} = \pi\left[\frac{1}{2}x + \frac{1}{4}\sin 2x\right]_0^{\pi/2}$
$= \pi\left(\frac{\pi}{4} + 0 - 0 - 0\right) = \frac{\pi^2}{4}$.

26. To find this volume, the disc method will lead to an integral we haven't worked yet, hence we use the shell method, with $V = 2\pi\int_0^{\pi/2}x\cos x\ dx$. This requires integration by parts, with the following table:

$u = x$ | $v = \sin x$

$du = dx$ | $dv = \cos x\ dx$

Hence, $V = 2\pi\int_0^{\pi/2}x\cos x\ dx = 2\pi[x\sin x -$
$\int \sin x\ dx] = 2\pi[x\sin x + \cos x]_0^{\pi/2} = 2\pi\left[\frac{\pi}{2} + 0 - 0 - 1\right] = \pi^2 - 2\pi = 3.5864$.

27. Charge is the integral of current. Hence we must find $\int 4t\sin 2t\ dt$. Use the following integration by parts table.

$u = 4t$ | $v = -\frac{1}{2}\cos 2t$

$du = 4\ dt$ | $dv = \sin 2t\ dt$

Thus, $\int 4t\sin 2t\ dt = -2t\cos 2t + \int 2\cos 2t\ dt =$
$-2t\cos 2t + \sin 2t$. Evaluated from 0 to 4 we have
$[-8\cos 8 + \sin 8] = 2.1534$ C.

28. The total mass is given by $m = \int_0^6 xe^{-x}dx$. Use the following integration by parts table.

$u = x$ | $v = -e^{-x}$

$du = dx$ | $dv = e^{-x}dx$

Hence $m = \int_0^6 xe^{-x}dx = -xe^{-x} + \int e^{-x}dx =$
$\left[-xe^{-x} - e^{-x}\right]_0^6 = -6x^{-6} - e^{-6} + 0 + e^0 = 0.9826$ kg.

29. $M_y = \int_{-1}^2 x\cdot x^2 e^x dx = \int_{-1}^2 x^3 e^x dx$. From Exercise #9, we have $\int x^3 e^{2x}dx = e^{2x}\left(\frac{1}{2}x^3 - \frac{3}{4}x^2 + \frac{3}{4}x - \frac{3}{8}\right) + C$. Hence, we see that

$M_y = \int_{-1}^2 x\cdot x^2 e^x dx = e^x\left[x^3 - 3x^2 + 6x - 6\right]_{-1}^2$
$= 14.7781 + 5.8861 = 20.6642$

$M_x = \frac{1}{2}\int_{-1}^2 \left(x^2 e^x\right)^2 dx = \frac{1}{2}\int_{-1}^2 x^4 e^{2x}dx$. From Exercise #16, we know

$\int x^4 e^{2x}dx$

$= e^{2x}\left(\frac{1}{2}x^4 - x^3 + \frac{3}{2}x^2 - \frac{3}{2}x + \frac{3}{4}\right) + C$

Thus,

$M_x = \frac{1}{2}\int_{-1}^2 \left(x^2 e^x\right)^2 dx = \frac{1}{2}\int_{-1}^2 x^4 e^{2x}dx$

$= \frac{1}{8}e^{2x}\left[2x^4 - 4x^2 + 6x^2 - 6x + 3\right]_{-1}^2$

$= \frac{15}{8}e^4 - \frac{21}{8}e^{-2} = 102.0163$

The mass is $m = \int_{-1}^2 x^2 e^x dx$. Once again, we need to use integration by parts.

$u = x^2$ | $v = e^x$

$du = 2x\ dx$ | $dv = e^x dx$

So, $\int x^2 e^x dx = x^2 e^x - \int 2xe^x dx$. A second application of integration by parts produces

$u = 2x$ | $v = e^x$

$du = 2\ dx$ | $dv = e^x dx$

With the result that $\int x^2 e^x dx = x^2 e^x - 2xe^x + \int 2e^x dx = x^2 e^x - 2xe^x + 2e^x$. Thus, $m = \int_{-1}^2 x^2 e^x dx = \left[x^2 e^x - 2xe^x + 2e^x\right]_{-1}^2 = \left[e^x\left(x^2 - 2x + 2\right)\right]_{-1}^2 = 12.9378$. Using the above result, we see that $\overline{x} = \frac{M_y}{m} = \frac{20.6642}{12.9387} = 1.5971$ and $\overline{y} = \frac{M_x}{m} = \frac{102.0163}{12.9387} = 7.8846$.

30. $\sqrt{\frac{1}{1}\int_0^1 \left[\sqrt{\cos^{-1}x}\right]^2 dx}$

$= \sqrt{\int_0^1 \cos^{-1}x\ dx}$

$= \sqrt{\left[x\cos^{-1}x - \sqrt{1-x^2}\right]_0^1}$

$= \sqrt{\left[1\cdot 0 - \sqrt{0} - 0\cdot 1 + \sqrt{1}\right]}$

$= \sqrt{1} = 1$

31. Work is the integral of force so we get $W = \int x^3\cos x\ dx$. Using integration by parts, we have

$u = x^3$ | $v = \sin x$

$du = 3x^2 dx$ | $dv = \cos x\ dx$

Thus, $\int x^3\cos x\ dx = x^3\sin x - \int 3x^2\sin x\ dx$. Using integration by parts again yields

$u = -3x^2 \quad | \quad v = -\cos x$

$du = -6x\,dx \quad | \quad dv = \sin x\,dx$

Hence, $\int x^3 \cos x\,dx = x^3 \sin x + 3x^2 \cos x - \int 6x \cos x$. We need integration by parts one more time.

$u = -6x \quad | \quad v = \sin x$

$du = -6\,dx \quad | \quad dv = \cos x\,dx$

$\int x^3 \cos x\,dx$

$= x^3 \sin x + 3x^2 \cos x - 6x \sin x$

$\quad + \int 6 \sin x\,dx$

$= x^3 \sin x + 3x^2 \cos x - 6x \sin x + 6\cos x\Big|_0^{\pi/2}$

$= \left(\dfrac{\pi}{2}\right)^3 + 0 - 6\dfrac{\pi}{2} - 0 - 0 + 0 - 0 - 6$

$= \left(\dfrac{\pi}{2}\right)^3 - 3\pi + 6 \approx 0.4510$

32. Since v is always positive we need only integrate $v(t)$ from 0 to 1. Expanding, we obtain $\int t^5\big(1 - t^2\big)^{1/2}dt = \int t^3 \cdot t^2\big(1 - t^3\big)^{1/2}dt$. Let $u = 1 - t^3$, then $t^3 = 1 - u$ and $du = -3t^2 dt$ or $-\frac{1}{3}du = t^2 dt$. Substituting we have $\int (1-u)u^{\frac{1}{2}} \cdot (-\frac{1}{3})du = -\frac{1}{3}\int \big(u^{1/2} - u^{3/2}\big)du = -\frac{1}{3}\big[\frac{2}{3}u^{3/2} - \frac{2}{5}u^{5/2}\big] = -\frac{2}{9}u^{3/2} + \frac{2}{15}u^{5/2} = \big[-\frac{2}{9}\big(1-t^3\big) + \frac{2}{15}\big(1-t^3\big)\big]_0^1 = \frac{2}{9} - \frac{2}{15} = \frac{10-6}{45} = \frac{4}{45}.$

33. As in example 28.54 set up the coordinate axis so that $(0, 0)$ is at the center of the road between the towers. Since it is a parabola the equation of the main cable fits the form $y = 4px^2$ and contains the point $(500, 200)$. Thus, $200 = 4p \cdot (500)$ or $p = \frac{2}{10000}$. The equation of the cable is thus $y = \frac{1}{1250}x^2$. Differentiating we get $y' = \frac{1}{625}x$. The length of the cable is $\int_{-500}^{500}\sqrt{1 + \left(\frac{x}{625}\right)^2}dx$. Evaluate this integral as follows: $\int \sqrt{1 + \left(\frac{x}{625}\right)^2} = \frac{1}{625}\int\sqrt{625^2 + x^2}$. Let $x = 625\tan\theta$ and $dx = 625\sec^2\theta\,d\theta$. This leads to $\frac{1}{625}\int\sqrt{625^2 + 25^2\tan^2\theta} \cdot 625\sec\theta\,d\theta = \frac{1}{625}\int 625\sec\theta \cdot 625\sec^2\theta\,d\theta = 625\int \sec^3\theta\,d\theta$. Using Example 28.50, we get $\frac{625}{2}\big[\sec\theta\tan\theta + \ln|\sec\theta + \tan\theta|\big] = \frac{625}{2}\left[\frac{\sqrt{625+x^2}}{625} + \frac{x}{625} + \ln\left|\frac{\sqrt{625+x^2}}{625} + \frac{x}{625}\right|\right]_{-500}^{500} =$

$625(1.02445 + \ln(2.08062) = 625(1.02445 + 0.73267)$. Thus, the length of the main cable is $625(1.75712) = 1098$ ft.

34. **(a)** $r(t) = \int r'(t)dt = \int 3t^2 e^{-t}dt$. Integrating by parts, we use the following

$u = 3t^2 \quad | \quad v = -e^{-t}$

$du = 6t\,dt \quad | \quad dv = e^t dt$

and so, $r(t) \int r'(t)dt = -3t^2 e^{-t} + \int 6te^{-t}dt$. Integrating by parts a second time, we use the following table

$u = 6t \quad | \quad v = -e^{-t}$

$du = 6\,dt \quad | \quad dv = e^{-t}dt$

and we get $r(t) \int r'(t)dt = -3t^2 e^{-t} + \int 6te^{-t}dt = -3t^2 e^{-t} - 6te^{-t} + \int 6\,dt = -3t^2 e^{-t} - 6te^{-t} - 6e^{-t} + C$. Since $r(0) = 0$, we have $r(0) = -3(0)^2 e^{-0} - 6(0)e^{-0} - 6e^{-0} + C = -6 + C = 0$ and so, $C = 6$. Hence, $r(t) = -3t^2 e^{-t} - 6te^{-t} - 6e^{-t} + 6$.

(b) The total reaction to the drug for the first 8 hours is $r(8) = -3(8)^2 e^{-8} - 6(8)e^{-8} - 6e^{-8} + 6 = 11.9$.

35. **(a)** $F = 2\pi\int_0^R r \cdot P(r)dr = 2\pi\int_0^R r \cdot P_0 e^{-kr}dr$. Integrating by parts, we have

$u = P_o r \quad | \quad v = -\frac{1}{k}e^{-kr}$

$du = P_0 dr \quad | \quad dv = e^{-kr}dr$

and so,

$F = 2\pi\int_0^R r \cdot P_0 e^{-kr}dr$

$= -2\pi\frac{1}{k}P_0 re^{-kr}\Big|_0^R + 2\pi\int_0^R \frac{P_0}{k}e^{-kr}dr$

$= \left[2\pi\frac{1}{k}P_0 re^{-kr} - 2\pi\frac{P_0}{k^2}d^{-kr}\right]_0^R$

$= \frac{-2\pi P_0 Rke^{-kR} - 2\pi P_0 e^{-kR}}{k^2} + \frac{2\pi P_0}{k^2}$

$= -\frac{e^{-kR}2\pi(kR+1)P_0}{k^2} + \frac{2\pi P_0}{k^2}$

(b) $T = 2\pi\mu\int_0^R r^2 \cdot P(r)dr = 2\pi\mu\int_0^R r^2 \times P_0 e^{-kr}dr = 2\pi\mu P_0\int_0^R r^2 \cdot e^{-kr}dr$. Integrating $\int_0^R r^2 \cdot e^{-kr}dr$ we will use integration by parts twice. The first time produces

$u = r^2 \quad | \quad v = -\frac{1}{k}e^{-kr}$

$du = 2r\,dr \quad | \quad dv = e^{-kr}dr$

and we get $\int r^2 \cdot e^{-kr}dr = -\frac{r^2}{k}e^{-kr} + \int \frac{2}{k}re^{-kr}dr$. The second time we use integration by parts, we have

$u = \frac{2}{k}r \quad \bigg| \quad v = -\frac{1}{k}e^{-kr}$

$du = \frac{2}{k}dr \quad \bigg| \quad dv = e^{-kr}dr$

and we get

$$\int r^2 \cdot e^{-kr}dr = -\frac{r^2}{k}e^{-kr} - \frac{2r}{k^2} + \int \frac{2}{k^2}e^{-kr}dr$$

$$= -\frac{r^2}{k}e^{-kr} - \frac{2r}{k^2} - \frac{2}{k^3}e^{-kr} + C$$

Putting the constant multiples and the limits of integration back in, the result is

$$T = 2\pi\mu \int_0^R r^2 \cdot P(r)dr$$

$$= 2\pi\mu \left[-\frac{r^2}{k}e^{-kr} - \frac{2r}{k^2} - \frac{2}{k^3}e^{-kr} \right]_0^R$$

$$= -2\pi\mu \left[\frac{1}{k^3}e^{-kr}\left(k^2r^2 + 2kr + 2\right) \right]_0^R$$

$$= -2\pi\mu \left[\frac{1}{k^3}e^{-kR}\left(k^2R^2 + 2kR + 2\right) \right]$$

$$+ 2\pi\mu \left[\frac{1}{k^3}e^{-0}\left(k^2 \cdot 0^2 + 2k \cdot 0 + 2\right) \right]$$

$$= -2\pi\mu \left[\frac{1}{k^3}e^{-kR}\left(k^2R^2 + 2kR + 2\right) \right]$$

$$+ 4\pi\mu \left(\frac{1}{k^3} \right)$$

36.   (a) Integrating by parts, we obtain

$u = 10t \quad \bigg| \quad v = -10e^{-0.1t}$

$du = 10\,dt \quad \bigg| \quad dv = e^{-0.1t}dt$

and so,

$$\int 10te^{-0.1t}dt = -100e^{-0.1t} + 100\int e^{-0.1t}dt$$

$$= -100te^{-0.1t} - 1000e^{-0.1t} + C$$

$$P(t) = -100te^{-0.1t}$$

$$- 1000e^{-0.1t} + 1000$$

Notice the $P(0) = 0$ implies that $C = 1000$.

(b) $P(12) = -100(12)e^{-0.1(12)} - 1000e^{-0.1(12)} + 1000 = -120e^{-1.2} - 1000e^{-1.2} + 1000 \approx 337.373$ thousands of barrels

(c) The total production in the second year of operation is $P(24) - P(12) = -100(24)e^{-0.1(24)} - 1000e^{-0.1(24)} + 1000 - 337.373 \approx 691.559 - 337.373 = 354.186$ thousands of barrels.

37. $\overline{C} = \int_0^6 \frac{40\ln(t+2)}{(t+2)^2}dt = \frac{20}{3}\int_0^6 \frac{\ln(t+2)}{(t+2)^2}dt$.   To determine the integral, we use integration by parts

$u = \ln(t+2) \quad \bigg| \quad v = -(t+2)^{-1}$

$du = \frac{1}{t+2}dt \quad \bigg| \quad dv = (t+2)^{-2}dt$

and so, $\int \frac{\ln(t+2)}{(t+2)^2}dt = \frac{-\ln(t+2)}{t+2}\int \frac{1}{(t+2)^2}dt = \frac{-\ln(t+2)}{t+2}$

$-\frac{1}{t+2}$. Combining the pieces, we get $\overline{C} = -\frac{20}{3} \times$

$\left[\frac{\ln(t+2)+1}{t+2} \right]_0^6 = -\frac{20}{3}\left[\frac{\ln 8 + 1}{8} - \frac{\ln 2 + 1}{2} \right] \approx 3.0776$ ppm

38.   As in Example 28.49, for $\int e^{kt}\sin\omega t\,dt$, we use integration by parts

$u = e^{kt} \quad \bigg| \quad v = -\frac{1}{\omega}\cos\omega t$

$du = ke^{kt}dt \quad \bigg| \quad dv = \sin\omega t\,dt$

and so, $\int e^{kt}\sin\omega t\,dt = -\frac{1}{\omega}e^{kt}\cos\omega t + \frac{1}{\omega}\int ke^{kt}\cos\omega t\,dt$. Using integration by parts a second time produces

$u = ke^{kt} \quad \bigg| \quad v = -\frac{1}{\omega^2}\sin\omega t$

$du = k^2e^{kt}dt \quad \bigg| \quad dv = \frac{1}{\omega}\cos\omega t\,dt$

and so,

$$\int e^{kt}\sin\omega t\,dt = -\frac{1}{\omega}e^{kt}\cos\omega t + \frac{1}{\omega}\int ke^{kt}\cos\omega t\,dt$$

$$= -\frac{1}{\omega}e^{kt}\cos\omega t + k\frac{1}{\omega^2}e^{kt}\sin\omega - \frac{k^2}{\omega^2}\int e^{kt}\sin\omega t$$

$$\int e^{kt}\sin\omega t\,dt + \frac{k^2}{\omega^2}\int e^{kt}\sin\omega t = -\frac{1}{\omega}e^{kt}\cos\omega t + k\frac{1}{\omega^2}e^{kt}\sin\omega$$

$$\left(1 + \frac{k^2}{\omega^2} \right)\int e^{kt}\sin\omega t\,dt = -\frac{1}{\omega}e^{kt}\cos\omega t + k\frac{1}{\omega^2}e^{kt}\sin\omega$$

$$\left(\frac{\omega^2 + k^2}{\omega^2} \right)\int e^{kt}\sin\omega t\,dt = -\frac{1}{\omega}e^{kt}\cos\omega t + k\frac{1}{\omega^2}e^{kt}\sin\omega$$

and so,

$$\int e^{kt}\sin\omega t\ dt = \left(\frac{\omega^2}{\omega^2+k^2}\right)\left[-\frac{1}{\omega}e^{kt}\cos\omega t + k\frac{1}{\omega^2}e^{kt}\sin\omega\right]$$

$$= \frac{-\omega e^{kt}\cos\omega t + k e^{kt}\sin\omega}{\omega^2+k^2}$$

As a result, we see that $i(t) = \frac{V}{L}\int e^{kt}\sin\omega t\ dt = \frac{V}{L}\left(-\frac{\omega e^{kt}\cos(\omega t)}{k^2+\omega^2}+\frac{k e^{kt}\sin(\omega t)}{k^2+\omega^2}\right)+C.$

≡ 28.8 USING INTEGRATION TABLES

1. $\int(1+\tan 3x)^2 dx = \int\left(1+2\tan 3x+\tan^2 3x\right)dx$ by formulas #12 and #58 we get $x+2\cdot\frac{1}{3}\ln|\sec 3x|+\frac{1}{3}(\tan 3x-3x)+C = x+\frac{2}{3}\ln|\sec 3x|+\frac{1}{3}\tan 3x-x+C = \frac{1}{3}(2\ln|\sec 3x|+\tan 3x)+C$

2. $\int\frac{\sqrt{16+25x^2}}{x}dx.$ Let $u=5x$, $du=5\ dx$, so $x=\frac{u}{5}$ and $dx=\frac{du}{5}$. Let $a=4$ and use formula #32. $\int\frac{\sqrt{4^2+u^2}}{\frac{4}{5}}\cdot\frac{1}{5}du = \int\frac{\sqrt{4^2+u^2}}{u}du = \sqrt{u^2+4^2}-4\ln\left|\frac{4+\sqrt{u^2+4^2}}{u}\right|+C = \sqrt{26x^2+16}-4\ln\left|\frac{4+\sqrt{25x^2+16}}{5x}\right|+C.$

3. $\int\frac{dx}{x(4x-3)} = \int\frac{dx}{x(-3+4x)}$ Let $u=x$, $a=-3$, and $du=dx$. By formula #48 this is $\frac{1}{a}\ln\left|\frac{u}{a+bu}\right|+C = \frac{-1}{3}\ln\left|\frac{x}{4x-3}\right|+C.$

4. $\int\frac{5x}{3+7x}dx = 5\int\frac{x}{3+7x}dx.$ Using formula #46 we obtain $5\left(\frac{1}{7^2}(3+7x-3\ln|3+7x|)\right)+C = \frac{5}{49}(3+7x-3\ln|3+7x|)+C.$

5. $\int\frac{x^2 dx}{x^6\sqrt{16+x^6}}.$ Let $u=x^3$ and $du=3x^2 dx$ or $\frac{1}{3}du=3x^2 dx.$ Substituting yields $\frac{1}{3}\int\frac{du}{u^2\sqrt{4^2+u^2}}$ By formula #39 this is $\frac{1}{3}\left(\frac{-\sqrt{16+u^2}}{16u}\right)+C = \frac{1}{3}\left(\frac{-\sqrt{16+x^6}}{16x^3}\right)+C = -\frac{\sqrt{16+x^6}}{48x^3}+C.$

6. $\int(25-4x^2)^{3/2}dx.$ Let $u=2x$ and then $du=2x$, or $\frac{1}{2}du=dx.$ Substitution yields $\frac{1}{2}\int(25-u^2)^{3/2}.$ By formula #28 we get $\frac{1}{2}\left(\frac{u}{8}(2u^2-5\cdot5^2)\sqrt{25-u^2}+\frac{3\cdot5^4}{8}\sin^{-1}\frac{u}{5}\right)+C = \frac{1}{2}\left(\frac{2x}{8}(8x^2-125)\sqrt{25-4x^2}+\frac{1875}{8}\sin^{-1}\frac{2x}{5}\right)+C = \frac{x}{8}(8x^2-125)\sqrt{25-4x^2}+\frac{1875}{16}\sin^{-1}\frac{2x}{5}+C.$

7. $\int\frac{x^2}{\sqrt{9-x^2}}dx.$ By formula #25 this is $-\frac{x}{2}\sqrt{9-x^2}+\frac{9}{2}\sin^{-1}\frac{x}{3}+C.$

8. $\int x^2\sqrt{9x^2-49}\ dx$ Let $u=3x$ and $du=3\ dx$ or $x=\frac{1}{3}u$ and $x^2=\frac{1}{9}u^2$ and $\frac{1}{3}du=dx.$ Substitution yields $\int\frac{1}{9}u^2\sqrt{u^2-7^2}\cdot\frac{1}{3}du = \frac{1}{27}\int u^2\sqrt{u^2-7^2}du.$ By formula #31, we have $\frac{1}{27}\left(\frac{u}{8}(2u^2-7^2)\times\sqrt{u^2-7^2}-\frac{7^4}{8}\ln|u+\sqrt{u^2-7^2}|\right)+C = \frac{1}{27}\times\left(\frac{3x}{8}(18^2-49)\sqrt{9x^2-49}-\frac{2401}{8}\ln|3x+\sqrt{9x^2-49}|\right)+C.$

9. $\int\cos^3 x\ dx.$ By formula #65 this is $\frac{1}{3}\cos^2 x\sin x+\frac{2}{3}\int\cos x\ dx = \frac{1}{3}\cos^2 x\sin x+\frac{2}{3}\sin x+C = \frac{1}{3}(1-\sin^2 x)\sin x+\frac{2}{3}\sin x+C = \sin x-\frac{1}{3}\sin^3 x+C.$ Alternate solution: $\int\cos^3 x\ dx = \int(1-\sin^2 x)\cos x\ dx = \int\cos x\ dx-\int\sin^2 x\cos x\ dx = \sin x-\frac{1}{3}\sin^3 x+C.$

10. $\int\sin 5x\sin 2x\ dx.$ By formula #70 this is $\frac{\sin 3x}{6}-\frac{\sin 7x}{14}+C$

11. $\int\sin^6 3x\ dx.$ Let $u=3x$, then $du=3\ dx$ or $\frac{1}{3}du=dx.$ Substituting we get $\frac{1}{3}\int\sin^6 u\ du.$ By formula #64 this is $\frac{1}{3}\left[-\frac{1}{6}\sin^5 u\cos u+\frac{5}{6}\int\sin^4 u\ du\right].$ Using formula #64 once more, you obtain $\frac{1}{3}\left[-\frac{1}{6}\sin^5 u\cos u+\frac{5}{6}\left(-\frac{1}{4}\sin^3 u\cos u+\frac{3}{4}\int\sin^2\times u\ du\right)\right].$ By formula #56 we get $\frac{1}{3}\left[-\frac{1}{6}\sin^5 u\cos u+\frac{5}{6}\left(-\frac{1}{4}\sin^3 u\cos u+\frac{3}{4}\{\frac{1}{2}u-\frac{1}{2}\sin u\cos u\}\right)\right]+C = -\frac{1}{18}\sin^5 u\cos u-\frac{5}{72}\sin^3 u\cos u-\frac{5}{48}\sin u\cos u+\frac{5}{48}u+C = -\frac{1}{18}\sin^5 3x\cos 3x-\frac{5}{72}\sin^3 3x\cos 3x-\frac{5}{48}\sin 3x\cos 3x+\frac{5}{48}x+C.$

12. $\int e^{-6x} \sin 10x \; dx$: By formula #87 we have
$\frac{e^{-6x}}{36+100}(-6\sin 10x - 10\cos 10x) + C =$
$\frac{1}{136}e^{-6x}(-6\sin 10x - 10\cos 10x) + C.$

13. $\int e^{10x} \cos 6x \; dx$: By formula #88 we have this is
$\frac{e^{10x}}{10^2+6^2}(10\cos 6x+6\sin 6x)+C = \frac{1}{136}e^{10x}(10\cos 6x$
$+ 6\sin 6x) + C.$

14. $\int x^2 \tan^{-1} x \; dx$. By formula #83 we get $\frac{1}{2+1} \times$
$[x^{2+1}\tan^{-1}x - \int \frac{x^2+1}{1+x^2}] = \frac{1}{3}[x^3\tan^{-1}x - \int \frac{x^3}{1+x^2}].$
Now for the second integral let $u = 1 + x$ with
$du = 2x \; dx$ and $x^2 = u - 1$. Substituting we
have $\int \frac{x^3}{1+x^2}dx = \int \frac{u-1}{u}du = \int 1du - \int u^{-1}du =$
$u - \ln u + C = 1 + x^2 - \ln(1 + x^2) + C.$ Putting
it all together we get $\frac{1}{3}(x^3 \tan^{-1} x - x^2 + \ln(1 +$
$x^2)) + C.$

15. $\int x^7 \ln x \; dx$. By formula #90 we obtain $x^8(\frac{1}{8}\ln x -$
$\frac{1}{64}) + C$

16. $\int \frac{\sqrt{7-9x^2}}{x}dx$. Let $u = 3x$ and $du = 3 \; dx$, then
$x = \frac{1}{3}du$ and we have $\int \frac{\sqrt{7-u^2}}{u}du$. Using formula
#23 and replacing u with $3x$ we get $\sqrt{7 - 9x^2} -$
$\sqrt{7}\ln\left|\frac{\sqrt{7}+\sqrt{7-9x^2}}{3x}\right| + C.$

17. $\int \arcsin 4x \; dx$. Letting $u = 4x$ and $du = 4 \; dx$
and then substituting we get $\frac{1}{4}\int \arcsin u \; du$. By
formula #78, $\int \arcsin 4x \; dx = \frac{1}{4}[(4x\sin^{-1}4x +$
$\sqrt{1 - (4x)^2}] + C = \frac{1}{4}[4x\sin^{-1}4x + \sqrt{1 - 16x^2}]$
$+ C.$

18. $\int \frac{dx}{(4-3x^2)^{3/2}}$. Let $u = \sqrt{3}x$, then $du\sqrt{3} \; dx$ and
substituting produces $\frac{1}{\sqrt{3}}\int \frac{du}{(2^2-u^2)^{3/2}}$. By formula
#29 this is $\frac{1}{\sqrt{3}}\left[\frac{u}{4\sqrt{4-u^2}}\right] + C = \frac{1}{\sqrt{3}} \cdot \left[\frac{\sqrt{3}x}{4\sqrt{4-3x^2}}\right] +$
$C = \frac{x}{4\sqrt{4-3x^2}} + C.$

19. $\int \frac{\sqrt{9+x^2}}{x}dx$. By formula #32 this is $\sqrt{9 + x^2} -$
$3\ln\left|\frac{3+\sqrt{9+x^2}}{x}\right| + C.$

20. $\int e^{\sin x} \sin x \cos x \; dx$. Let $u = \sin x$, then $du = $
$\cos x \; dx$. Substituting we get $\int ue^u du$. By formula #84 this is $\frac{1}{1}(1u - 1)e^u + C = (\sin x -$
$1)e^{\sin x} + C.$

21. $\int x^3 e^{2x}dx$. By formula #86 we get $\frac{1}{2}x^3e^{2x} -$
$\frac{3}{2}\int x^2 e^{2x}dx$. Repeating this process, we get $\frac{1}{2}x^3e^{2x}$
$- \frac{3}{2} \cdot \frac{1}{2}x^2e^{2x} + \frac{3}{2} \cdot \frac{2}{2}\int xe^{2x}dx$ or $\frac{1}{2}x^3e^{2x} - \frac{3}{4}x^2e^{2x} +$
$\frac{3}{2}\int xe^{2x}dx$. Repeating again produces $\frac{1}{2}x^3e^{2x} -$
$\frac{3}{4}x^2e^{2x} + \frac{3}{4}xe^{2x} - \frac{3}{2}\frac{1}{2}\int e^{2x}dx = \frac{1}{2}x^3e^{2x} - \frac{3}{4}x^2e^{2x} +$
$\frac{3}{4}xe^{2x} - \frac{3}{8}e^{2x} + C = \frac{1}{8}e^{2x}(4x^3 - 6x^2 + 6x - 3) + C.$

22. $\int x^4 \ln 2x \; dx$. Let $u = 2x$ then $u^4 = 16x^4$ or $x^4 =$
$\frac{1}{16}u^4$ and $du = 2 \; dx$ or $dx = \frac{1}{2}du$. Substituting
we get $\frac{1}{32}\int u^4 \ln u \; du$. By formula #90 this is
$\frac{1}{32}[u^5(\frac{\ln u}{5} - \frac{1}{5^2}) + C = \frac{1}{32}((2x)^5 \cdot (\frac{\ln 2x}{5} - \frac{1}{25})) +$
$C = \frac{1}{5}x^5(\ln 2x - \frac{1}{5}) + C.$

23. $\int x^3 \sin 2x \; dx$. Let $u = 2x$, then $u^3 = 8x^3$ or
$x^3 = \frac{1}{8}u^3$, and also, $du = 2dx$ or $dx = \frac{1}{2}du$. Substitution produces $\frac{1}{16}\int u^3 \sin u \; du$. By formula
#75 this is $\frac{1}{16}[-u^3\cos u + 3\int u^2 \cos u \; du]$. Now
using formula #76, on the integral in this result, we
obtain $\frac{1}{16}[-u^3\cos u + 3(u^2\sin u - 2\int u\sin u \; du)]$.
Finally, using formula #75 one more time produces
the result $\frac{1}{16}[-u^3\cos u + 3u^2\sin u + 6u\cos u -$
$6\int \cos u]$ or $\frac{1}{16}[-u^3\cos u + 3u^2\sin u + 6u\cos u -$
$6\sin u] + C.$ Back substituting for u produces
$\int x^3 \sin 2x \; dx = \frac{1}{16}[-8x^3\cos 2x + 12x^2\sin 2x +$
$12x\cos 2x - 6\sin 2x] + C = -\frac{1}{2}x^3\cos 2x + \frac{3}{4}x^2 \times$
$\sin 2x + \frac{3}{4}x\cos 2x - \frac{3}{8}\sin 2x + C.$

24. $\int \frac{\sqrt{\tan^2 2x - 9}}{\cos^2 2x}dx$. Let $u = \tan 2x$, then $du =$
$2\sec^2 2x \; dx$ or $\frac{1}{2}du = \sec^2 2x \; dx = \frac{1}{\cos^2 2x}dx$.
Substituting we get $\frac{1}{2}\int \sqrt{u^2 - 3^2}du$. Using for-
mula #30, we obtain $\frac{1}{2}[\frac{u}{2}\sqrt{u^2 - 3^2} - \frac{3^2}{2}\ln|u +$
$\sqrt{u^2 - 3^2}|] + C.$ Back substituting for u, we get
$\frac{1}{2}[\frac{\tan 2x}{2}\sqrt{\tan^2 2x - 9} - \frac{9}{2}\ln|\tan 2x + \sqrt{\tan^2 2x - 9}|]$
$+ C.$

25. $V = \frac{1}{2.5}\int_0^{2.5} t^2 e^{5t}dt$. Using integration form #85
produces
$V = \frac{1}{2.5}[\frac{e^{5t}}{125}(25t^2 - 10t + 2)]_0^{2.5}$
$= \frac{1}{2.5}(286047.5315) \approx 114419.0126$

26. **(a)** $R = \frac{1}{2\pi}\int_0^n te^{0.25t}dt$. By integration form #84,

$$R(n) = \frac{1}{2\pi}\int_0^n te^{0.25t}dt$$

$$= \frac{1}{2\pi}\left[4te^{0.25t} - 16e^{0.25t}\right]_0^n$$

$$= \frac{1}{2\pi}\left[(n-4)4e^{0.25t} + 16\right]$$

$$= \frac{2}{\pi}\left[(n-4)e^{0.25t} + 4\right] \text{ revolutions}$$

(b) $R(10) = \frac{2}{\pi}\left[(10-4)e^{0.25t} + 4\right] = \frac{2}{\pi}\left[6e^{0.25t} + 4\right] \approx 49.08$ revolutions

27. $W = \int F\ dx = \int_0^2 \frac{1}{49-9x^2}dx = \frac{1}{2\cdot7}\ln\left|\frac{3x+7}{3x-7}\right|_0^2$ by integration form #19. This evaluates as $\frac{1}{14}(\ln 13 - \ln 1) = \frac{1}{14}\ln 13 \approx 0.1832$ N · m.

28. $V = \frac{1}{C}\int I\ dt = \frac{1}{0.02}\int \frac{dt}{t^2+t} = 50\left[\ln\left|\frac{t}{1+t}\right|\right] + C$ by integration form #48. When $t = 1$, $V = 50\ln(\frac{1}{2}) + C = 0$, and so $C = 0 - 50\ln(\frac{1}{2}) = 50\ln 2$. Hence, $V = 50\left[\ln\left|\frac{t}{1+t}\right|\right] + 50\ln 2 = 50.0\ln(t) - 50.0\ln(t + 1) + 50\ln 2$.

CHAPTER 28 REVIEW

1. $\int xe^{3x}dx$. Using integration by parts with the table

$u = x$	$v = \frac{1}{3}e^{3x}$
$du = dx$	$dv = e^{3x}dx$

We have $\int xe^{3x}dx = \frac{1}{3}xe^{3x} - \frac{1}{3}\int e^{3x}dx = \frac{1}{3}xe^{3x} - \frac{1}{9}e^{3x} + C = \frac{1}{9}e^{3x}(3x - 1) + C$.

2. $\int \sin^3 x\ \cos^2 x\ dx = \int \sin x(1-\cos^2 x)\cos^2 x\ dx = \int \sin x(\cos^2 x - \cos^4)dx = -\frac{1}{3}\cos^3 x + \frac{1}{5}\cos^5 x + C$.

3. $\int \frac{x}{\sqrt{25-x^2}}dx$. Let $x = 5\sin\theta$, then $dx = 5\cos\theta\ d\theta$. Thus, substitution produces $\int \frac{5\sin\theta}{\sqrt{25-5\sin^2\theta}}\cdot5\cos\theta\ d\theta = \int \frac{5\sin\theta\cdot5\cos\theta\ d\theta}{5\cos\theta} = 5\int \sin\theta\ d\theta = -5\cos\theta + C = 5\cdot\frac{\sqrt{25-x^2}}{5} + C = -\sqrt{25-x^2} + C$. Alternate solution: Let $u = 25 - x^2$, then $du = -2x\ dx$ or $-\frac{1}{2}du = x\ dx$. Substituting, we obtain

$$-\frac{1}{2}\int \frac{du}{\sqrt{u}} = -\frac{1}{2}\int u^{-1/2}du = -\frac{1}{2}\cdot\frac{2}{1}u^{1/2} + C$$

$$= -\sqrt{25 - x^2} + C$$

4. $\int x^3\sqrt{25-x^2}dx$ Let $u = 25 - x^2$ or $x^2 = 25 - u$, and $du = -2x\ dx$ or $-\frac{1}{2}du = x\ dx$. Substituting we get $-\frac{1}{2}\int(25 - u)\sqrt{u}\cdot du = -\frac{1}{2}\int 25\sqrt{u}\ du + \frac{1}{2}\int u^{3/2}du = -\frac{25}{2}\cdot\frac{2}{3}u^{3/2} + \frac{1}{2}\frac{2}{5}u^{5/2} + C = -\frac{25}{3}(25 - x^2)^{3/2} + \frac{1}{5}(25 - x^2)^{5/2} + C = -\frac{25}{3}(25 - x^2)^{3/2} + \frac{1}{5}(25 - x^2)(25 - x^2)^{3/2} + C = (5 - \frac{25}{3} - \frac{x^2}{5})(25 - x^2)^{3/2} + C = (-\frac{1}{5}x^2 - \frac{10}{3})(25 - x^2) + C$.

5. $\int \sin^4 2x\cos 2x\ dx$. Let $u = \sin 2x$ then $du = 2\cos 2x\ dx$ or $\frac{1}{2}du = \cos 2x\ dx$. Substitution produces $\frac{1}{2}\int u^4 du = \frac{1}{2}\cdot\frac{1}{5}u^5 + C = \frac{1}{10}\sin^5 2x + C$.

6. $\int 2(e^x - e^{-x})dx = 2\int e^x dx - 2\int e^{-x}dx = 2e^x + 2e^{-x} + C$.

7. $\int \frac{dx}{9x+5}$. Let $u = 9x + 5$ and $du = 9\ dx$ or $\frac{1}{9}du = dx$. Then $\frac{1}{9}\int \frac{du}{u} = \frac{1}{9}\ln|u| = \frac{1}{9}\ln|9x + 5| + C$.

8. $\int \tan^2 8x\ dx = \int \frac{\sin^2 8x}{\cos^2 8x}dx = \int \frac{1-\cos^2 8x}{\cos^2 8x}dx = \int \frac{1}{\cos^2 8x}dx - \int 1\ dx = \frac{1}{8}\tan 8x - x + C$.

9. $\int \sin(7x + 2)dx$. Let $u = 7x + 2$ and $du = 7\ dx$ or $\frac{1}{7}du = dx$. Then, substitution produces $\frac{1}{7}\int \sin u\ du = \frac{1}{7}(-\cos u) + C = -\frac{1}{7}\cos(7x+2) + C$.

10. $\int \sin^3 2x\cos 2x\ dx$. Let $u = \sin 2x$ and then $du = 2\cos 2x\ dx$. Substitution produces $\frac{1}{2}\int u^3 du = \frac{1}{2}\cdot\frac{1}{4}u^4 + C = \frac{1}{8}\sin^4 2x + C$.

11. $\int \sin^5 3x\cos^2 3x\ dx = \int \sin 3x(\sin^4 3x)\cos^2 3x\ dx = \int \sin 3x(1-\cos^2 3x)^2\cos^2 3x\ dx$. This expands as $\int (\cos^2 3x - 2\cos^4 3x + \cos^6 3x)\sin 3x\ dx$. Let $u = \cos 3x$ and $du = -3\sin 3x\ dx$ or $-\frac{1}{3}du = \sin 3x\ dx$. Substituting we get $-\frac{1}{3}\int (u^2 - 2u^4 + u^6)du = -\frac{1}{3}\left[\frac{1}{3}u^3 - \frac{2}{5}u^5 + \frac{1}{7}u^7\right] + C = -\frac{1}{9}\cos^3 3x + \frac{2}{15}\cos^5 3x - \frac{1}{21}\cos^7 3x + C$.

12. $\int \frac{dx}{x^2+4x+20} = \int \frac{dx}{(x+2)^2+16}$. Let $u = x+2$, $a = 4$, and $du = dx$. Substituting, we get $\int \frac{du}{u^2+4^2} = \frac{1}{4}\tan^{-1}\left(\frac{u}{4}\right) + C = \frac{1}{4}\tan^{-1}\left(\frac{x+2}{4}\right) + C$.

13. $\int \frac{dx}{(4x^2+49)^{3/2}}$. Let $2x = 7\tan\theta$, or $x = \frac{7}{2}\tan\theta$ and then $dx = \frac{7}{2}\sec^2\theta\, d\theta$. Substituting yields $\frac{7}{2}\int \frac{\sec^2\theta\, d\theta}{((7\tan\theta)^2+49)^{3/2}} = \frac{7}{2}\cdot\int \frac{\sec^2\theta\, d\theta}{7^3\sec^3\theta} = \frac{1}{98}\int \frac{1}{\sec\theta}\, d\theta = \frac{1}{98}\int \cos\theta\, d\theta = \frac{1}{98}\cdot\sin\theta + C = \frac{1}{98}\cdot\frac{2x}{\sqrt{4x^2+49}} + C = \frac{x}{49\sqrt{4x^2+49}} + C$.

14. $\int x^3 \ln x\, dx$. Using integration by parts with the table

$u = \ln x$	$v = \frac{1}{4}x^4$
$du = \frac{1}{x}dx$	$dv = x^3 dx$

We obtain $\int x^3 \ln x\, dx = \frac{1}{4}x^4 \ln x - \int \frac{1}{4}x^3 dx = \frac{1}{4}x^4 \ln x - \frac{1}{16}x^4 + C = x^4\left(\frac{1}{4}\ln x - \frac{1}{16}\right) + C$.

15. $\int x^2 e^{x^3} dx$. Let $u = x^3$ and $du = 3x^2 dx$ or $\frac{1}{3}du = x^2 dx$. Substituting, we get $\frac{1}{3}\int e^u du = \frac{1}{3}e^u + C = \frac{1}{3}e^{x^3} + C$.

16. $\int \frac{dx}{\sqrt{4x^2+49}}$. Let $2x = 7\tan\theta$ or $x = \frac{7}{2}\tan\theta$ and then $dx = \frac{7}{2}\sec^2\theta\, d\theta$. Substituting, we get $\frac{7}{2}\int \frac{\sec^2\theta\, d\theta}{\sqrt{40\tan^2\theta+49}} = \frac{1}{2}\int \frac{\sec^2\theta\, d\theta}{\sec\theta} = \frac{1}{2}\int \sec\theta\, d\theta = \frac{1}{2}\ln|\sec\theta + \tan\theta| + K = \frac{1}{2}\ln\left|\frac{\sqrt{4x^2+49}}{7} + \frac{2x}{7}\right| + K = \frac{1}{2}\ln|\sqrt{4x^2+49} + 2x| - \frac{1}{2}\ln 7 + K = \frac{1}{2}\ln|\sqrt{4x^2+49} + 2x| + C$ where $C = K - \frac{1}{2}\ln 7$.

17. $\int \frac{x\, dx}{\sqrt{4x^2+49}}$. Let $u = 4x^2+49$ and $du = 8x\, dx$ or $\frac{1}{8}du = x\, dx$. Substituting we get $\frac{1}{8}\int \frac{du}{\sqrt{u}} = \frac{1}{8}\int u^{-\frac{1}{2}}du = \frac{1}{8}\cdot\frac{2}{1}u^{1/2} + C = \frac{1}{4}\sqrt{4x^2+49} + C$.

18. $\int \cot 5x\, \csc^4 5x\, dx = \int \cot 5x\, \csc 5x\, \csc^3 5x\, dx$. Let $u = \csc 5x$ and $du = -5\cot 5x\, \csc 5x\, dx$. Substituting we get $-\frac{1}{5}\int u^3 du = -\frac{1}{5}\cdot\frac{1}{4}u^4 + C = -\frac{1}{20}\csc^4 5x + C$ or $-\frac{1}{20}\cdot\sin^{-4}5x + C$.

19. $\int \frac{\sec 4x\, \tan 4x}{9+2\sec 4x}dx$. Let $u = 9+2\sec 4x$ and $du = 8\sec 4x\, \tan 4x\, dx$. Substituting we get $\frac{1}{8}\int \frac{du}{u} = \frac{1}{8}\ln|u| + C = \frac{1}{8}\ln|9+2\sec 4x| + C$.

20. $\int \sin^6 \frac{3x}{2}dx$. We begin by rewriting this as a cubic expression, then using a half-angle trig identity, and finally expanding the rewritten expression.

$$\int \sin^6 \frac{3x}{2}dx$$
$$= \int \left(\sin^2 \frac{3x}{2}\right)^3 dx$$
$$= \int \left(\frac{1-\cos 3x}{2}\right)^3 dx$$
$$= \frac{1}{8}\int \left(1 - 3\cos 3x + 3\cos^2 3x - \cos^3 3x\right)dx$$
$$= \frac{1}{8}x - \frac{1}{8}\sin 3x$$
$$+ \frac{3}{8}\int \cos^2 3x - \frac{1}{8}\int \cos^3 3x\, dx$$

Now $= \frac{3}{8}\int \cos^2 3x\, dx = \frac{3}{8}\int \left(\frac{1+\cos 6x}{2}\right)dx = \frac{3}{16}\int(1 + \cos 6x)dx = \frac{3}{16}x + \frac{3}{96}\sin 6x = \frac{3}{16}x + \frac{1}{32}\sin 6x$. And also, $-\frac{1}{8}\int \cos^3 3x\, dx = -\frac{1}{8}\int \left(1 - \sin^2 3x\right)\cos 3x\, dx = -\frac{1}{8}\cdot\frac{1}{3}\sin 3x + \frac{1}{8}\cdot\frac{1}{9}\sin^3 3x + C$. Putting this all together we get $\frac{1}{8}x - \frac{1}{8}\sin 3x + \frac{3}{16}x + \frac{1}{32}\sin 6x - \frac{1}{24}\sin 3x + \frac{1}{72}\sin^3 3x + C = \frac{5}{16}x - \frac{5}{24}\sin 3x + \frac{1}{32}\sin 6x + \frac{1}{72}\sin^3 3x + C$.

21. $\int \frac{[\ln(2x+1)]^5}{2x+1}dx$. Let $u = \ln(2x+1)$ and then $du = \frac{2dx}{2x+1}$. Substituting yields $\frac{1}{2}\int u^5 du = \frac{1}{12}u^6 + C = \frac{1}{12}[\ln(2x+1)]^6 + C$.

22. $\int \frac{\arctan 7x}{1+49x^2}dx$. Let $u = \arctan 7x$ and $du = \frac{7dx}{1+49x^2}$. Substitution produces $\frac{1}{7}\int u\, du = \frac{1}{7}\cdot\frac{1}{2}u^2 + C = \frac{1}{14}(\arctan 7x)^2 + C$.

23. $\int \frac{x^2}{x^3+4}dx$. Let $u = x^3 + 4$ and $du = 3x^2 dx$ or $\frac{1}{3}du = x^2 dx$. Now substituting we get $\frac{1}{3}\int \frac{du}{u} = \frac{1}{3}\ln|u| + C = \frac{1}{3}\ln|x^3 + 4| + C$.

24. $\int 4x^3 e^{x^4} dx$. Let $u = x^4$ and $du = 4x^3 dx$. Substituting yields $\int e^u du = e^u + C = e^{x^4} + C$.

25. $\int \tan \frac{x}{5}dx = 5\ln|\sec \frac{x}{5}| + C$ or $-5\ln|\cos \frac{x}{5}| + C$.

26. $\int x \sin^3 x\, dx = \int x(1 - \cos^2 x)\sin x\, dx = \int x \sin x\, dx - \int x \cos^2 x \sin x\, dx$. For the first integral we use integration by parts with

$u = x$	$v = -\cos x$
$du = dx$	$dv = \sin x\, dx$

Then, $\int x \sin x \; dx = -x \cos x + \int \cos x \; dx = -x \cos x + \sin x + C$. For the second integral by parts

$u = x$	$v = \frac{1}{3} \cos^3 x$
$du = dx$	$dv = -\cos^2 x \sin x \; dx$

Thus, $\int x (-\cos^2 x \sin x) dx = \frac{x}{3} \cos^3 x - \frac{1}{3} \int \times \cos^3 x \; dx = \frac{x}{3} \cos^3 x - \frac{1}{3} \int (1 - \sin^2 x) \cos x \; dx = \frac{x}{3} \cos^3 x - \frac{1}{3} \sin x + \frac{1}{9} \sin^3 x + C$. Putting these together we have $\int x \sin^3 x \; dx = -x \cos x + \sin x - \left(\frac{x}{3} \cos^3 x - \frac{1}{3} \sin x + \frac{1}{9} \sin^3 x \right) + C = -x \cos x + \sin x - \frac{x}{3} \cos^3 x + \frac{1}{3} \sin x - \frac{1}{9} \sin^3 x + C = -x \cos x + \frac{x}{3} \cos^3 x + \frac{4}{3} \sin x - \frac{1}{9} \sin^3 x + C$.

27. $\int e^{\cos x} \sin x \; dx = -e^{\cos x} + C$

28. $\int x^4 e^{-x} dx$. We use integration by parts with

$u = x^4$	$v = -e^{-x}$
$du = 4x^3 dx$	$dv = e^{-x} dx$

Hence, $\int x^4 e^{-x} dx = x^4 (-e^{-x}) + \int 4x^3 e^{-x} dx$. Using integration by parts a second time produces

$u = 4x^3$	$v = -e^{-x}$
$du = 12x^2 dx$	$dv = e^{-x} dx$

Thus, $x^4 (-e^{-x}) + \int 4x^3 e^{-x} dx = -x^4 e^{-x} - 4x^3 e^{-x} + \int 12x^2 e^{-x}$. Using integration by parts a third time, we have

$u = 12x^2$	$v = -e^{-x}$
$du = 24x \; dx$	$dv = e^{-x} dx$

This produces $-x^4 e^{-x} - 4x^3 e^{-x} - 12x^2 e^{-x} - 24x e^{-x} + \int 24x e^{-x} dx$. Finally, using integration by parts a fourth time, we obtain

$u = 24x$	$v = -e^{-x}$
$du = 24 \; dx$	$dv = e^{-x} dx$

$-x^4 - e^{-x} - 4x^3 e^{-x} - 12x^2 e^{-x} - 24x e^{-x} + \int 24 e^{-x} dx = -x^4 e^{-x} - 4x^3 e^{-x} - 12x^2 e^{-x} - 24x e^{-x} - 24 e^{-x} + C = e^{-x} (-x^4 - 4x^3 - 12x^2 - 24x - 24) + C$.

29. $\int \frac{\cos x}{\sin^2 x + 9} dx$. Let $u = \sin x$ and $du = \cos x \; dx$. Substituting we get $\int \frac{du}{u^2 + 9} = \frac{1}{3} \tan^{-1} \frac{u}{3} + C = \frac{1}{3} \tan^{-1} \left(\frac{\sin x}{3} \right) + C$.

30. $\int \frac{e^x}{e^x + 16} dx$. Let $u = e^x + 16$ and $du = e^x dx$. Substituting yields $\int \frac{du}{u} = \ln |u| + C = \ln |e^x + 16| + C = \ln (e^x + 16) + C$.

31. Let $u = x^2$ and $du = 2x \; dx$. Then $\int_1^e x^3 \ln x^2 dx = \frac{1}{2} \int_1^e x^2 \ln x^2 2x \; dx$ and by integration formula #90,

$$\frac{1}{2} \int_1^e x^2 \ln x^2 2x \; dx = \frac{1}{2} \left(x^2 \right)^2 \left(\frac{\ln x^2}{2} - \frac{1}{4} \right) \Big|_1^e$$
$$= \frac{e^4}{2} \left(\frac{\ln e^2}{2} - \frac{1}{4} \right)$$
$$- \frac{1}{2} \left(0 - \frac{1}{4} \right)$$
$$= \frac{e^4}{2} \left(1 - \frac{1}{4} \right) + \frac{1}{8}$$
$$= \frac{3e^4}{8} + \frac{1}{8} \approx 20.5993$$

32. $\int e^{8x} \cos 2x \; dx$. Use integration by parts with $u = e^{8x}$ and $dv = \cos 2x \; dx$. Then $v = \frac{1}{2} \sin 2x$ and $du = 8e^{8x} dx$. This produces $\int e^{8x} \cos 2x \; dx = \frac{1}{2} e^{8x} \sin 2x - \int 4 e^{8x} \sin 2x \; dx$. Use integration by parts again, with

$u = 2e^{8x}$	$v = -\cos 2x$
$du = 16 e^{8x} dx$	$dv = 2 \sin 2x \; dx$

Thus we have $\int e^{8x} \cos 2x \; dx = \frac{1}{2} e^{8x} \sin 2x - \int 4 e^{8x} \sin 2x = \frac{1}{2} e^{8x} \sin 2x + 2e^{8x} \cos 2x - 16 \times \int e^{8x} \cos 2x$. Adding $16 \int e^{8x} \cos 2x$ to both sides of this last equation, we obtain $17 \int e^{8x} \cos 2x \; dx = \frac{1}{2} e^{8x} \sin 2x + 2e^{8x} \cos 2x$, and, dividing by 17 produces the desired integral $\int e^{8x} \cos 2x \; dx = \frac{1}{34} e^{8x} \sin 2x + \frac{2}{17} e^{8x} \cos 2x + C = \frac{1}{34} e^{8x} (\sin 2x + 4 \cos 2x) + C$.

33. $\int \sin^{1/3} 4x \cos^5 4x \; dx = \int \sin^{1/3} 4x \left(\cos^2 4x \right)^2 \times \cos 4x \; dx = \int \sin^{1/3} 4x \left(1 - \sin^2 4x \right)^2 \cos 4x \; dx = \int \left(\sin^{1/3} 4x - 2 \sin^{7/3} 4x + \sin^{13/3} 4x \right) \cos 4x \; dx = \frac{1}{4} \left(\frac{3}{4} \sin^{4/3} 4x - 2 \cdot \frac{1}{10} \sin^{10/3} 4x + \frac{3}{16} \sin^{16/3} 4x \right) + C = \frac{3}{320} \sin^{4/3} 4x (20 - 16 \sin^2 4x + 5 \sin^4 4x) + C$.

34. $\int \frac{\sin^3 x \; dx}{\cos^4 x} = \int \frac{\sin^3 x}{\cos^3 x} \cdot \frac{dx}{\cos x} = \int \tan^3 x \sec x \; dx = \int \tan x (\sec^2 x - 1) \sec x \; dx = \int \sec^2 x \times (\tan x \sec x \; dx) - \int \tan x \sec x \; dx = \frac{1}{3} \sec^3 x - \sec x + C$.

35. $\int \frac{\cos x \, dx}{\sqrt{16-4\sin^2 x}}$. Let $u = 2\sin x$ and $du = 2\cos x \, dx$ or $\frac{1}{2}du = \cos x \, dx$. Substituting we get $\frac{1}{2}\int \frac{du}{\sqrt{4^2-u^2}}$ $= \frac{1}{2}\sin \frac{u}{4} + C = \frac{1}{2}\sin^{-1}\left(\frac{2\sin x}{4}\right) + C = \frac{1}{2}\sin^{-1}\left(\frac{\sin x}{2}\right) + C.$

36. $\int \frac{e^{5x}}{4-e^{5x}}dx$. Let $u = 4 - e^{5x}$ and $du = -5e^{5x}dx$ or $-\frac{1}{5}du = e^{5x}dx$. Substituting we get $-\frac{1}{5}\int \frac{du}{u} = -\ln|u| + C = -\frac{1}{5}\ln|4 - e^{5x}| + C.$

37. $\int x^5 e^{x^2} dx$. Using integration by parts, we have the following table.

$u = x^4$	$v = \frac{1}{2}e^{x^2}$
$du = 4x^3 dx$	$dv = xe^{x^2}dx$

This produces $\int x^5 e^{x^2}dx = \frac{1}{2}e^{x^2}x^4 - 2\int x^3 e^{x^2}dx$. Using integration by parts again with $u = -2x^2$ and $dv = xe^{x^2}dx$, we get $v = \frac{1}{2}e^{x^2}$ and $du = -4x\,dx$. Thus, $\frac{1}{2}e^{x^2}x^4 - 2\int x^3 e^{x^2}dx = \frac{1}{2}e^{x^2}x^4 - e^{x^2}\cdot x^2 + 2\int xe^{x^2} = \frac{1}{2}e^{x^2}x^4 - e^{x^2}x^2 + e^{x^2} + C = e^{x^2}\left(\frac{x^4}{2} - x^2 - 1\right) + C = \frac{e^{x^2}}{2}\left(x^4 - 2x^2 + 2\right) + C.$

38. $\int \frac{e^{3x}}{\left(e^{3x}-1\right)^2}dx$. Let $u = e^{3x} - 1$ and $du = 3e^{3x}dx$ so $\frac{1}{3}du = e^{3x}dx$. Substituting we get $\frac{1}{3}\int \frac{du}{u^2} = -\frac{1}{3}u^{-1} + C = -\frac{1}{3}(e^{3x} - 1)^{-1} + C$

39. $\int \frac{(\arctan 2x)^4}{1+4x^2}dx$. Let $u = \arctan 2x$, then $du = \frac{2}{1+4x^2}dx$ and substituting gives $\frac{1}{2}\int u^4 du = \frac{1}{10}u^5 + C = \frac{1}{10}(\arctan 2x)^5 + C.$

40. $\int \frac{\sec^2 5x \, dx}{2\tan 5x + 9}$. Let $u = 2\tan 5x + 9$, and then $du = 10\sec^2 5x \, dx$ or $\frac{1}{10}du = \sec^2 5x \, dx$. Substituting we get $\frac{1}{10}\int \frac{du}{u} = \frac{1}{10}\ln|u| + C = \frac{1}{10}\ln|2\tan 5x + 9| + C$

41. Rewrite $\int_0^{\ln 49} \sqrt{9 + e^x}dx = 2\int_0^{\ln 49} \frac{\sqrt{9+e^x}}{e^{x/2}}\frac{e^{x/2}}{2}dx.$

Let $u = e^{x/2}, du = \frac{1}{2}e^{x/2}dx$, and $a = 3$ and this fits integration formula #32. Remember that $\ln 49 = \ln 7^2 = 2\ln 7$, and so, evaluating $e^{x/2}$ when $x = \ln 49$ produces $e^{(\ln 49)/2} = e^{(2\ln 7)/2} = e^{\ln 7} = 7.$

$\int_0^{\ln 49} \sqrt{9 + e^x}dx$
$= 2\left[\sqrt{9 + e^{x/2}} - 3\ln\left|\frac{3 + \sqrt{9 + e^{x/2}}}{e^{x/2}}\right|\right]\Big|_0^{\ln 49}$
$= 2\left[\sqrt{9+7} - 3\ln\left|\frac{3 + \sqrt{9+7}}{7}\right|\right] - \left[\sqrt{9+1} - 3\ln\left|\frac{3 + \sqrt{9+1}}{1}\right|\right]$
$= 2\left[4 - 3\ln|1|\right] - 2\left[\sqrt{10} - 3\ln\left|3 + \sqrt{10}\right|\right]$
$= 8 - 2\sqrt{10} + 6\ln\left|3 + \sqrt{10}\right| \approx 12.5861$

42. Using integration by parts, with

$u = \arcsin\left(\frac{x}{2}\right)$	$v = x$
$du = \frac{1}{\sqrt{4-x^2}}dx$	$dv = dx$

Then, we obtain
$\int_0^1 \arcsin\left(\frac{x}{2}\right)dx$
$= x\arcsin\left(\frac{x}{2}\right) - \int \frac{x}{\sqrt{4-x^2}}dx$
$= x\arcsin\left(\frac{x}{2}\right) + \frac{1}{2}\int \frac{2x}{\sqrt{4-x^2}}dx$
$= \left[x\arcsin\left(\frac{x}{2}\right) + \sqrt{4-x^2}\right]_0^1$
$= \arcsin\left(\frac{1}{2}\right) + \sqrt{3} - \arcsin 0 - \sqrt{4}$
$= \frac{\pi}{6} + \sqrt{3} - 2 \approx 0.2556$

43. Using integration by parts, with

$u = x^2$	$v = \sqrt{9+x^2}$
$du = 2x\,dx$	$dv = \frac{x}{\sqrt{9+x^2}}dx$

Then, we obtain
$\int_0^3 \frac{x^3 dx}{\sqrt{9+x^2}}$
$= x^2\sqrt{9+x^2}\Big]_0^3 - \int_0^3 2x\sqrt{9+x^2}dx$
$= \left[x^2\sqrt{9+x^2} - \frac{2}{3}(9+x^2)^{3/2}\right]_0^3$
$= 9\sqrt{18} - \frac{2}{3}(18)^{3/2} - 0 + \frac{2}{3}(9)^{3/2}$
$= 27\sqrt{2} - \frac{2}{3}(3\sqrt{2})^3 + \frac{2}{3}(27)$
$= 27\sqrt{2} - 36\sqrt{2} + 18$
$= 18 - 9\sqrt{2} \approx 5.2721$

44. $\displaystyle\int_0^{\pi/2} \sin^3 x \cos^3 x \ dx$

$$= \int_0^{\pi/2} \sin^3 x(1 - \sin^2 x)\cos x \ dx$$

$$= \int_0^{\pi/2} (\sin^3 x - \sin^5 x)\cos x \ dx$$

$$= \left[\frac{1}{4}\sin^4 x - \frac{1}{6}\sin^6 x\right]_0^{\pi/2}$$

$$= \frac{1}{4} - \frac{1}{6} = \frac{1}{12} \approx 0.0833$$

≡ CHAPTER 28 TEST

1. Let $u = 9 - e^x$, then $du = -e^x\ dx$ and $\int \frac{e^x\ dx}{\sqrt{9-e^x}} = -\int u^{-1/2}du = -2u^{1/2} + C = -2\sqrt{9 - e^x} + C$

2. $\int \tan^2 4x \cos^2 4x \ dx = \int \frac{\sin^2 4x}{\cos^2 4x} \cdot \cos^4 4x \ dx = \int \sin^2 4x \cos^2 4x \ dx = \int \sin^2 4x(1 - \sin^2 4x)dx = \int \sin^2 4x \ dx - \int \sin^4 4x \ dx = \int \frac{1-\cos 8x}{2}dx - \int\left(\frac{1-\cos 8x}{2}\right)^2 dx = \frac{1}{2}x - \frac{1}{16}\sin 8x - \frac{1}{4}\int(1 - 2\cos 8x + \cos^2 8x)dx = \frac{1}{2}x - \frac{1}{10}\sin 8x - \frac{1}{4}x + \frac{1}{16}\sin 8x - \frac{1}{4}\int\left(\frac{1+\cos 16x}{2}\right)dx = \frac{1}{4}x - \frac{1}{8}x - \frac{1}{128}\sin 16x + C = \frac{1}{8}x - \frac{1}{128}\sin 16x + C.$

3. $\int \frac{5\ dx}{x^2+1} = 5\tan^{-1} x + C$

4. Let $u = x^2 + 1$, then $du = 2x\ dx$ and so $\int \frac{4x\ dx}{(x^2+1)^3} = 2\int u^{-3}du = -u^{-2} + C = \frac{-1}{(x^2+1)^2} + C.$

5. Use integration by parts with $u = x$ and $dv = e^{4x}dx$. Then $du = dx$ and $v = \frac{1}{4}e^{4x}$ and so $\int xe^{4x}dx = uv - \int v\ du = \frac{1}{4}xe^{4x} - \int \frac{1}{4}e^{4x}dx = \frac{1}{4}xe^{4x} - \frac{1}{16}e^{4x} + C = \frac{1}{16}e^{4x}(4x - 1) + C.$

6. Let $u = \tan x$ and then $du = \sec^2 x\ dx = \frac{1}{\cos^2 x}dx$ and $\int \frac{e^{\tan x}}{\cos^2 x}dx = \int e^u du = e^u + C = e^{\tan x} + C.$

7. Using the disk method, we have $V = \pi \times \int_0^1 \left(\sqrt{x}e^x\right)^2 dx = \pi \int_0^1 xe^{2x}dx = \frac{\pi}{4}e^{2x}(2x-1)\big]_0^1 = \frac{\pi}{4}(e^2 + 1).$

8. The arc length L uses $(y')^2 = (2x)^2$ and so, $L = \int_0^2 \sqrt{1 + (2x)^2}dx = \int_0^2 \sqrt{1 + 4x^2}dx$. From Formula 30 in Appendix C, we see that $L \times \int_0^2 \sqrt{1 + 4x^2}dx = \frac{2x}{2}\sqrt{1 + 4x^2} + \frac{1}{2}\ln\left|2x + \sqrt{1 + 4x^2}\right|\big]_0^2 = 2\sqrt{17} + \frac{1}{2}\ln\left|4 + \sqrt{17}\right| \approx 9.2936.$

CHAPTER
29

Parametric Equations and Polar Coordinates

≡ 29.1 DERIVATIVES OF PARAMETRIC EQUATIONS

1. $x = t^2 + t$, $y = t + 1$. Thus, $\frac{dx}{dt} = 2t + 1$ and $\frac{dy}{dt} = 1$, which means that $\frac{dy}{dx} = \frac{1}{2t+1}$. Next, $\frac{d}{dt}\left(\frac{dy}{dx}\right) = \frac{d}{dt}(2t+1)^{-1} = \frac{-2}{(2t+1)^2}$ and we get $\frac{d^2y}{dx^2} = \frac{-2}{(2t+1)^3}$. Critical values occur when $\frac{dx}{dy} = 0$ or is undefined. The derivative $\frac{dy}{dx}$ is undefined at $t = -\frac{1}{2}$. The second derivative is also undefined. There are no extrema. Inflection point at $t = -\frac{1}{2}$: $\left(-\frac{1}{4}, \frac{1}{2}\right)$.

2. Here $x = t + 3$ and $y = t^2 + t$ and so the first derivative is $\frac{dy}{dx} = \frac{dy/dt}{dx/dt} = \frac{2t+1}{1} = 2t + 1$ and the second derivative is $\frac{d^2y}{dx^2} = \frac{\frac{d}{dt}\left(\frac{dy}{dx}\right)}{\frac{dx}{dt}} = \frac{2}{1} = 2$.

   Critical value at $t = -\frac{1}{2}$. Since the second derivative is positive, this is a minimum. There are no inflection points. Minimum at $t = -\frac{1}{2}$: $\left(\frac{5}{2}, -\frac{1}{4}\right)$

3. Here $x = t^2 - 6t + 12$ and $y = t + 4$, and so, $\frac{dy}{dx} = \frac{dy/dt}{dx/dt} = \frac{1}{2t-6}$. Critical value at $t = 3$. The second derivative is $\frac{d^2y}{dx^2} = \frac{\frac{d}{dt}\left(\frac{dy}{dx}\right)}{\frac{dx}{dt}} = \frac{\frac{-2}{(2t-6)^2}}{2t-6} = \frac{-2}{(2t-6)^3}$.
   Inflection point at $t = 3$: $(3, 7)$. No extrema.

4. Since $x = t^2 + 6t + 12$ and $y = t + 4$, the first derivative is $\frac{dy}{dx} = \frac{dy/dt}{dx/dt} = \frac{1}{2t+6}$. Critical value at $t = -3$. The second derivative is $\frac{d^2y}{dx^2} = \frac{\frac{d}{dt}\left(\frac{dy}{dx}\right)}{\frac{dx}{dt}} = \frac{\frac{-2}{(2t+6)^2}}{2t+6} = \frac{-2}{(2t+6)^3}$. Inflection point at $t = -3$: $(3, 1)$

5. Here $x = t^2 + t$ and $y = t^2 - t$ which means that the first derivative is $\frac{dy}{dx} = \frac{dy/dt}{dx/dt} = \frac{2t-1}{2t+1}$.

Critical values are at $t = \frac{1}{2}, -\frac{1}{2}$. The second derivative is $\frac{d^2y}{dx^2} = \frac{\frac{d}{dt}\left(\frac{dy}{dx}\right)}{\frac{dx}{dt}} = \frac{\frac{(2t+1)2-(2t-1)2}{(2t+1)^2}}{2t+1} = \frac{4}{(2t+1)^3}$. Inflection point at $t = -\frac{1}{2}$: $\left(-\frac{1}{4}, \frac{3}{4}\right)$. Minimum at $t = \frac{1}{2}$: $\left(\frac{3}{4}, -\frac{1}{4}\right)$. It is a minimum since the second derivative is positive.

6. Since $x = 4t + 1$ and $y = 9t^2$, the first derivative is $\frac{dy}{dx} = \frac{dy/dt}{dx/dt} = \frac{18t}{4} = \frac{9}{2}t$. Critical value at $t = 0$. The second derivative is $\frac{d^2y}{dx^2} = \frac{\frac{d}{dt}\left(\frac{dy}{dx}\right)}{\frac{dx}{dt}} = \frac{\frac{18}{4}}{4} + \frac{18}{16} = \frac{9}{8}$. There are no inflection points. Since the second derivative is positive the critical value yields a minimum. Minimum at $t = 0$: $(1, 0)$.

7. Here $x = 3 + 4\cos t$ and $y = -1 + \cos t$ which means that the first derivative is $\frac{dy}{dx} = \frac{dy/dt}{dx/dt} = \frac{-\sin t}{-4\sin t} = \frac{1}{4}$. The second derivative is $\frac{d^2y}{dx^2} = 0$ since $\frac{dy}{dx}$ is a constant. This is segment where the point (x, y) moves between a minimum when t is an odd multiple of π giving $(-1, -2)$, and a maximum when t is an even multiple of π giving $(7, 0)$.

8. Here $x = 3 + 4\cos t$ and $y = 1 - \sin t$, which means that the first derivative is $\frac{dy}{dx} = \frac{dy/dt}{dx/dt} = \frac{-\cos t}{-4\sin t} = \frac{1}{4}\cot t$. The second derivative is $\frac{d^2y}{dx^2} = \frac{\frac{d}{dt}\left(\frac{dy}{dx}\right)}{\frac{dt}{dt}} = \frac{-\frac{1}{4}\csc^2 t}{-4\sin t} = \frac{1}{16}\csc^3 t$. The critical values occur

when $\cos t = 0$, or when $t = \frac{\pi}{2} + 2k\pi$ yields a minimum $(3, 0)$, as $\frac{d^2y}{dx^2} > 0$. The values of $t = \frac{-\pi}{2} + 2k\pi$ yield a maximum at $(3, 2)$ since $\frac{d^2y}{dx^2} < 0$. Inflection points occur when $\sin t = 0$ or at multiples of π: $t = 0 + 2k\pi$ yields $(7, 1)$; $t = \pi + 2k\pi$ yields $(-1, 1)$. (Note: This curve is an ellipse with major axis between $(-1, 1)$ and $(7, 1)$ and minor axis between $(3, 0)$, and $(3, 2)$.)

9. Since $x = 2 + \sin t$ and $y = -1 + \cos t$, the first derivative is $\frac{dy}{dx} = \frac{dy/dt}{dx/dt} = \frac{-\sin t}{\cos t} = -\tan t$. The second derivative is $\frac{d^2y}{dx^2} = \frac{\frac{d}{dt}\left(\frac{dy}{dx}\right)}{\frac{dx}{dt}} = \frac{-\sec^2 t}{\cos t} = -\sec^3 t$. Critical values occur when $\sin t = 0$. At $t = 2k\pi$, the second derivative is negative so maximum $(2, 0)$. At $t = \pi + 2k\pi$, y'' is positive so we have a minimum $(2, -2)$. Inflection points are at $t = \frac{\pi}{2} + 2k\pi$ and $-\frac{\pi}{2} + 2k\pi$. These are $(3, -1)$ and $(1, -1)$. (Note: this is a circle with vertical diameter between $(2, 0)$ and $(2, -2)$ and horizontal diameter between $(3, -1)$ and $(1, -1)$).

10. Since $x = -2 + 4e^t$ and $y = 3 + 2e^{-t}$, the first derivative is $dy/dx = \frac{dy/dt}{dx/dt} = \frac{-2e^{-t}}{4e^t} = -\frac{1}{2}e^{-2t}$ or $\frac{-1}{2e^{2t}}$. The second derivative is $\frac{d^2y}{dx^2} = \frac{\frac{d}{dt}\left(\frac{dy}{dx}\right)}{\frac{dx}{dt}} = \frac{e^{-2t}}{4e^t} = \frac{1}{4e^{3t}}$. Since $e^t \neq 0$ there are no extrema or inflection points.

11. Here $x = 2t - 1$ and $y = 4t^2 - 2t$ and the first derivative is $\frac{dy}{dx} = \frac{dy/dt}{dx/dt} = \frac{8t-2}{2}$. At $t = 1$, the slope is $m = \frac{dy}{dx} = \frac{8-2}{2} = 3$ and the desired point is $(x, y) = (1, 2)$. Thus, the equation of the tangent is $y - 2 = 3(x - 1) \Rightarrow y = 3x - 1$ and the equation of the normal line is $(y - 2) = -\frac{1}{3}(x - 1) \Rightarrow y = -\frac{1}{3}x + 2\frac{1}{3}$ or $3y + x = 7$.

12. $x = t - 4$, $y = t^3 + 2t^2 - 5t - 2$; $t = 1$. The desired point is $(x, y) = (-3, -4)$. The derivative is $-\frac{dy}{dx} = \frac{3t^2+4t-5}{1}$ and when $t = 1$, the slope is $m = 2$. The requested equations of the lines are: Tangent: $y + 4 = 2(x + 3)$ or $y = 2x + 2$; Normal: $y + 4 = -\frac{1}{2}(x + 3)$ or $2y + x = -11$.

13. $x = t^3$, $y = t^2$, $t = -3$. Thus, $(x, y) = (-27, 9)$. The derivative is $\frac{dy}{dx} = \frac{2t}{3t^2} = \frac{2}{3t}$; at $t = -3$, $m =$

$-\frac{2}{9}$. Tangent: $y - 9 = -\frac{2}{9}(x + 27)$ or $9y + 2x = 27$; Normal: $y - 9 = \frac{9}{2}(x + 27)$ or $2y - 9x = 261$.

14. $x = 2\cos t$, $y = 3\sin t$, $t = \frac{\pi}{4}$, and so $(x, y) = \left(\sqrt{2}, \frac{3\sqrt{2}}{2}\right)$. The derivative is $\frac{dy}{dx} = \frac{3\cos t}{-2\sin t} = -\frac{3}{2}\tan t$ and so, at $\frac{\pi}{4}$, the slope is $m = -\frac{3}{2}$. Tangent: $y - \frac{3\sqrt{2}}{2} = -\frac{3}{2}(x - \sqrt{2})$ or $2y + 3x = 6\sqrt{2}$; Normal: $y - \frac{3\sqrt{2}}{2} = \frac{2}{3}(x - \sqrt{2})$ or $3y - 2x = \frac{5\sqrt{2}}{2}$.

15. $x = 2 + \cos t$, $y = 2\sin t$, $t = \frac{\pi}{2}$, $(x, y) = (2, 2)$. Thus, $\frac{dy}{dx} = \frac{2\cos t}{-\sin t} = -2\cot t$. At $t = \frac{\pi}{2}$, $m = 0$ which means that the desired lines are Tangent: $y = 2$ and Normal: $x = 2$.

16. $x = e^t + 1$, $y = e^t + e^{-t}$, $t = 1$, so $(x, y) = \left(e + 1, e + \frac{1}{e}\right)$ and the derivative is $\frac{dy}{dx} = \frac{e^t - e^{-t}}{e^t}$. At $t = 1$, we determine that $m = \frac{e - e^{-1}}{e} = \frac{e^2-1}{e^2}$. Tangent: $y = \frac{e^2+1}{e} = \frac{e^2-1}{e^2}(x - (e+1))$. Multiplying by e^2, we obtain $e^2 y - e^3 - e = (e^2 - 1)x - e^3 - e^2 + e + 1$ or $e^2 y - (e^2 - 1)x = -e^2 + 2e + 1$ or about $7.389y - 6.389x = -0.9525$. Normal: $y - \frac{e^2+1}{e} = \frac{e^2}{1-e^2}[x - (e+1)]$ or $(1 - e^2)y - \left(\frac{1-e^4}{e}\right) = e^2 x - e^3 - e^2$, or $(1 - e^2)y - e^2 x = -e^2 + e^{-1} - 2e^3$, or $(e^2 - 1)y + e^2 x = e^2 - e^{-1} + 2e^3$. This is about $6.389y + 7.389x = 47.192$.

17. $x = t + 3$, $y = t^2 - 4t$. The derivative is $\frac{dy}{dx} = \frac{dy/dt}{dx/dt} = \frac{2t-4}{2}$. Horizontal tangent when $2t - 4 = 0$ or $t = 2$: $(5, -4)$. Vertical tangents: none.

18. $x = t - 4$, $y = (t^2 + t)^2$. The derivative is $\frac{dy}{dt} = 2(t^2 + t)(2t + 1) = 2t(t + 1)(2t + 1)$. Horizontal tangents are at $t = 0$: $(-4, 0)$, $t = -1$: $(-5, 0)$, and $t = -\frac{1}{2}$: $\left(-4\frac{1}{2}, \frac{1}{16}\right)$. Since $\frac{dx}{dt} = 1$ is never 0 there are no vertical tangents.

19. $x = 3\cos t$, $y = 5\sin t$. To find any horizontal tangents, we have $\frac{dy}{dt} = 5\cos t$ and $5\cos t = 0$, when $t = \frac{\pi}{2} + 2k\pi$: $(0, 5)$ and $t = -\frac{\pi}{2} + 2k\pi$: $(0, -5)$. To find vertical tangents, we take $\frac{dx}{dt} = -3\sin t$; $-3\sin t = 0$ when $t = 2k\pi$: $(3, 0)$ and $t = \pi + 2k\pi$: $(-3, 0)$.

20. $x = t^2 + 1$, $y = \cos t$. To find horizontal tangents, we take $\frac{dy}{dt}$ and set it equal to 0, with the result that $\frac{dy}{dt} = -\sin t = 0$ when $t = 2k\pi$, $k \neq$

0; $(4\pi^2 + 1, \ 1)$, $(16\pi^2 + 1, \ 1)$, etc. or $t = \pi + 2k\pi$: $(\pi^2 + 1, \ -1)$, $(9\pi^2, \ -1)$, etc. To find the vertical tangents, we use $\frac{dx}{dt} = 2t = 0$ when $t = 0$: $\frac{dy/dt}{dx/dt} = \frac{-\sin t}{2t}$. Since $\lim_{t \to 0} \frac{-\sin t}{2t} = -\frac{1}{2}$ there are no vertical tangents.

21. $s_x(t) = 3780t \cos\frac{4\pi}{15}$; $s_y(t) = 3780t \sin\frac{4\pi}{15} - 16t^2$.

(a) $s_x(5) = 3780 \cdot 5 \cdot \cos\frac{4\pi}{15} \approx 12{,}646.57$ ft;

$\quad s_y(5) = 3780 \cdot 5 \cdot \sin\frac{4\pi}{15} - 16 \cdot 5^2$

$\quad\quad \approx 13{,}645.44$ ft.

(b) $v_x(5) = 3780 \cos\frac{4\pi}{15} = 2529.31$ ft/sec, and

$\quad v_y(5) = 3780 \sin\frac{4\pi}{15} - 32 \cdot 5 = 2649.09$ ft/sec

(c) $a_x(t) = 0$;

$\quad a_y(t) = -32$ ft/sec^2

22. $s_x(t) = 1250t \cos\frac{3\pi}{11}$; $s_y(t) = 1250t \sin\frac{3\pi}{11} - 4.9t^2$

(a) $s_x(8) = 1250 \cdot 8 \cdot \cos\frac{3\pi}{11} = 6548.61$ m and

$\quad s_y(8) = 1250 \cdot 8 \sin\frac{3\pi}{11} - 4.9 \cdot 8^2 = 7243.90$ m

(b) $v_x(t) = \frac{ds_x(t)}{dt} = 1250 \cos\frac{3\pi}{11}$, so $v_x(8) =$

$1250 \cos\frac{3\pi}{11} \approx 818.58$ m/s, and $v_y(t) = \frac{ds_y(t)}{dt} =$

$1250 \sin\frac{3\pi}{11} - 9.8t$, so $v_y(8) = 1250 \sin\frac{3\pi}{11} - 9.8 \cdot 8 \approx 866.29$ m/s

(c) $a_x(t) = 0$, so $a_x(8) = 0$, and $a_y(t) = -9.8$, so $a_y(8) = -9.8$ m/s^2

23. (a) When $t = 3.0$, then $x = \frac{100}{\sqrt{3^2+1}} = \frac{100}{\sqrt{10}} =$

$10\sqrt{10}$ and $y = \frac{100(3)}{\sqrt{3^2+1}} = \frac{300}{\sqrt{10}} = 30\sqrt{10}$. Using these two values of x and y, produces magnitude $= \sqrt{x^2 + y^2} = \sqrt{(10\sqrt{10})^2 + (30\sqrt{10})^2} = \sqrt{1000 + 9000} = \sqrt{10{,}000} = 100$ km

$\quad$ The direction is $\theta = \tan^{-1}\left(\frac{y}{x}\right) = \tan^{-1}\frac{30\sqrt{10}}{10\sqrt{10}}$

$= \tan^{-1} 3 = 71.6°$

(b) The horizontal component of this electron's velocity is $v_x = \frac{dx}{dt} = 100(t^2 + 1)^{-3/2}\left(-\frac{1}{2}\right)(2t) = \frac{-100t}{(t^2+1)^{3/2}}$. The vertical component of this electron's velocity is

$v_y = \frac{dy}{dt}$

$= \dfrac{100(t^2 + 1)^{1/2} - \frac{1}{2}(t^2 + 1)^{-1/2}(2t)(100t)}{t^2 + 1}$

$= \dfrac{100(t^2 + 1) - 100t^2}{(t^2 + 1)^{3/2}} = \dfrac{100}{(t^2 + 1)^{3/2}}$

(c) When $t = 3.0$, then $v_x = \frac{-100(3)}{(3^2+1)^{3/2}} = \frac{-300}{(10)^{3/2}} \approx$

-9.487 and $v_y = \frac{100}{(3^2+1)^{3/2}} = \frac{100}{(10)^{3/2}} \approx 3.162$. Using these two values of v_x and v_y, produces a magnitude of

$$\sqrt{(-9.487)^2 + (3.162)^2} = 10.0 \text{ km/s}$$

and the direction is $\theta = \tan^{-1}\left(\frac{3.162}{-9.487}\right) = -18.4°$.

(d) The horizontal component of this electron's acceleration is

$a_x = \dfrac{-100(t^2 + 1)^{3/2} + 100t\left(\frac{3}{2}\right)(t^2 + 1)^{1/2}(2t)}{(t^2 + 1)^3}$

$\quad = \dfrac{-100(t^2 + 1) + 300t^2}{(t^2 + 1)^{5/2}} = \dfrac{200t^2 - 100}{(t^2 + 1)^{5/2}}$

The vertical component of this electron's acceleration is $a_y = 100\left(-\frac{3}{2}\right)(t^2+1)^{-5/2}(2t) = \frac{-300t}{(t^2+1)^{5/2}}$.

(e) When $t = 3.0$, then $a_x = \frac{200(3^2) - 100}{(3^2+1)^{5/2}} \approx 5.376$

and $a_y = \frac{-300(3)}{(3^2+1)^{5/2}} \approx -2.846$. Using these two values of a_x and a_y, produces a magnitude of $\sqrt{5.376^2 + (-2.846)^2} \approx 6.083$ km/s^2 and a direction of $\theta = \tan^{-1}\left(\frac{-2.846}{5.376}\right) \approx -27.9°$.

24. (a) $v_x = dx/dt = V\cos(\alpha) = 44\cos(\alpha)$ ft/s and $v_y = dy/dt = V\sin(\alpha) - gt = 44\sin(\alpha) - 32.0t$ ft/s

(b) When $\alpha = 30°$, $v_x = 44\cos 30° = 44\left(\frac{\sqrt{3}}{2}\right) = 22\sqrt{3} \approx 38.1$ ft/s and $v_y = 44\sin 30° - 32(1) = 22 - 32 = -10$ ft/s. As a result, the magnitude is $\sqrt{(22\sqrt{3})^2 + (-10)^2} = \sqrt{1552} \approx 39.4$ ft/s and the direction is $\theta = \tan^{-1}\left(\frac{-10}{22\sqrt{3}}\right) \approx -14.7°$.

$\quad$ When $\alpha = 45°$, $v_x = 44\cos 45° = 22\sqrt{2} \approx 31.11$ ft/s and $v_y = 44\sin 45° - 32 = 22\sqrt{2} - 32 \approx -0.89$ ft/s. As a result, the magnitude is $\sqrt{v_x^2 + v_y^2} \approx 31.1$ ft/s and the direction is $\theta = \tan^{-1}\left(\frac{-0.89}{31.11}\right) \approx -1.6°$.

$\quad$ When $\alpha = 50°$, $v_x = 44\cos 50° \approx 28.3$ ft/s and $v_y = 44\sin 50° - 32 \approx 1.71$ ft/s. As a result, the magnitude is $\sqrt{v_x^2 + v_y^2} \approx 28.28$ ft/s and the direction is $\theta = \tan^{-1}\left(\frac{1.71}{28.28}\right) = 3.46° \approx 3.5°$.

25. (a) The horizontal component of this spacecraft's velocity is $v_x = \frac{dx}{dt} = 10\left(\frac{1}{2}\right)(t^4 + 1)^{-1/2} \cdot 4t^3 = \frac{20t^3}{\sqrt{t^4+1}}$. The vertical component of this spacecraft's velocity is $v_y = \frac{dy}{dt} = \left(\frac{3}{2}\right)40t^{1/2} = 60\sqrt{t}$.

(b) The magnitude of this spacecraft's velocity at

time t is $\sqrt{v_x^2 + v_y^2} = \sqrt{\left(\frac{20t^3}{\sqrt{t^4+1}}\right)^2 + \left(60\sqrt{t}\right)^2} =$

$20\sqrt{\frac{t^6}{t^4+1} + 9t}$ m/s and the direction is

$$\theta = \tan^{-1}\left(\frac{v_y}{v_x}\right) = \tan^{-1}\left(\frac{60\sqrt{t}}{\frac{20t^3}{\sqrt{t^4+1}}}\right)$$

$= \tan^{-1}\left(\frac{60\sqrt{t}\sqrt{t^4+1}}{20t^3}\right) = \tan^{-1}\left(\frac{3\sqrt{t^4+1}}{t^{5/2}}\right).$

(c) When $t = 5.0$ s, the magnitude is $20 \times$

$\sqrt{\frac{5^6}{5^4+1} + 9(5)} \approx 167.28$ m/s and direction is $\theta =$

$\tan^{-1}\left(\frac{3\sqrt{5^4+1}}{5^{5/2}}\right) \approx 53.3°.$

26. **(a)** The horizontal component of this spacecraft's acceleration is

$$a_x = \frac{60t^2\sqrt{t^4+1} - \frac{1}{2}\left(t^4+1\right)^{-1/2}\left(4t^3\right)\left(20t^3\right)}{t^4+1}$$

$$= \frac{60t^2\left(t^4+1\right) - 40t^6}{\left(t^4+1\right)^{3/2}} = \frac{20t^6 + 60t^2}{\left(t^4+1\right)^{3/2}}$$

$$= \frac{20t^2\left(t^4+3\right)}{\left(t^4+1\right)^{3/2}} \text{ ft/s}^2$$

The vertical component of this spacecraft's acceleration is $a_y = 60\left(\frac{1}{2}\right)t^{-1/2} = \frac{30}{\sqrt{t}}$.

(b) magnitude: $\sqrt{\left(\frac{20t^2\left(t^4+3\right)}{\left(t^4+1\right)^{3/2}}\right)^2 + \frac{900}{t}} = \sqrt{\frac{400t^4\left(t^4+3\right)^2}{\left(t^4+1\right)^3} + \frac{900}{t}}$ m/s^2, direction: $\theta = \tan^{-1}\left(\frac{3\left(t^4+1\right)^{3/2}}{2t^{5/2}\left(t^4+3\right)}\right)$,

(c) When $t = 5.0$ s, the magnitude is $\sqrt{\frac{(400)5^4\left(5^4+3\right)^2}{\left(5^4+1\right)^3} + \frac{900}{5}} \approx 24.1$ m/s^2, direction: $\theta = \tan^{-1}\left(\frac{939\sqrt{626}\sqrt{5}}{78500}\right) \approx$

$\tan^{-1}(0.669219) \approx 33.8°$

≡ 29.2 DIFFERENTIATION IN POLAR COORDINATES

1. Since $r = 3\sin\theta$, then $r' = 3\cos\theta$ which leads

to $\frac{dy}{dx} = \frac{r'\sin\theta + r\cos\theta}{r'\cos\theta - r\sin\theta} = \frac{3\cos\theta\sin\theta + 3\sin\theta\cos\theta}{3\cos\theta\cos\theta - 3\sin\theta\sin\theta} =$

$\frac{2\cos\theta\sin\theta}{\cos^2\theta - \sin^2\theta} = \frac{\sin 2\theta}{\cos 2\theta} = \tan 2\theta$

2. Since $r = -2\cos\theta$, then $r' = 2\sin\theta$ and so $\frac{dy}{dx} =$

$\frac{r'\sin\theta + r\cos\theta}{r'\cos\theta - r\sin\theta} = \frac{2\sin\theta\sin\theta - 2\cos\theta\cos\theta}{2\sin\theta\cos\theta + 2\cos\theta\sin\theta}$

$= \frac{\sin^2\theta - \cos^2\theta}{2\sin\theta\cos\theta} = \frac{-\cos 2\theta}{\sin 2\theta} = -\cot 2\theta$

3. Since $r = 1+\cos\theta$, then $r' = -\sin\theta$ and we have

$\frac{dy}{dx} = \frac{r'\sin\theta + r\cos\theta}{r'\cos\theta - r\sin\theta} = \frac{-\sin\theta\sin\theta + (1+\cos\theta)\cos\theta}{-\sin\theta\cos\theta - (1+\cos\theta)\sin\theta} =$

$\frac{-\sin^2\theta + \cos\theta + \cos^2\theta}{-2\sin\theta\cos\theta - \sin\theta} = \frac{\cos\theta + \cos 2\theta}{-\sin 2\theta - \sin\theta} = -\frac{\cos 2\theta + \cos\theta}{\sin 2\theta + \sin\theta}.$

4. Since $r = \cos 3\theta$, then $r' = -3\sin 3\theta$ and we have

$\frac{dy}{dx} = \frac{r'\sin\theta + r\cos\theta}{r'\cos\theta - r\sin\theta} = \frac{-3\sin 3\theta\sin\theta + \cos 3\theta\cos\theta}{-3\sin 3\theta\cos\theta - \cos 3\theta\sin\theta} =$

$\frac{3\sin 3\theta\sin\theta - \cos 3\theta\cos\theta}{3\sin 3\theta\cos\theta + \cos 3\theta\sin\theta}.$

5. We are given $r = 1+\cos 3\theta$, and so $r' = -3\sin 3\theta$,

which leads to $\frac{dy}{dx} = \frac{-3\sin 3\theta\sin\theta + (1+\cos 3\theta)\cos\theta}{-3\sin 3\theta\cos\theta - (1+\cos 3\theta)\sin\theta} =$

$\frac{3\sin 3\theta\sin\theta - \cos 3\theta\cos\theta - \cos\theta}{3\sin 3\theta\cos\theta + \cos 3\theta\sin\theta + \sin\theta}.$

6. Since $r = \csc\theta$, we have $r' = -\csc\theta\cot\theta$, and

so the desired derivative is $\frac{dy}{dx} =$

$\frac{-\csc\theta\cot\theta\sin\theta + \csc\theta\cos\theta}{-\csc\theta\cot\theta\cos\theta - \csc\theta\sin\theta} = \frac{-\cot\theta + \cot\theta}{-\cot^2\theta - 1} = 0$

7. Here $r = \tan\theta$, which means that $r' = \sec^2\theta$, and

so the desired derivative is $\frac{dy}{dx} = \frac{\sec^2\theta\sin\theta + \tan\theta\cos\theta}{\sec^2\theta\cos\theta - \tan\theta\sin\theta}$

$= \frac{\tan\theta\sec\theta + \tan\theta\cos\theta}{\sec\theta - \tan\theta\sin\theta} \cdot \frac{\cos\theta}{\cos\theta} = \frac{\tan\theta + \tan\theta\cos^2\theta}{1 - \sin^2\theta} =$

$\tan\left(\frac{1+\cos^2\theta}{\cos^2\theta}\right) = \tan\theta\left(\sec^2\theta + 1\right).$

8. $r = e^\theta$, $r' = e^\theta$; $\frac{dy}{dx} = \frac{e^\theta\sin\theta + e^\theta\cos\theta}{e^\theta\cos\theta - e^\theta\sin\theta} = \frac{\sin\theta + \cos\theta}{\cos\theta - \sin\theta}$

9. $r = \sin\theta$, $r' = \cos\theta$; $\frac{dy}{dx} = \frac{\cos\theta\sin\theta + \sin\theta\cos\theta}{\cos\theta\cos\theta - \sin\theta\sin\theta} =$

$\frac{2\sin\theta\cos\theta}{\cos^2\theta - \sin^2\theta} = \frac{\sin 2\theta}{\cos 2\theta} = \tan 2\theta.$ At the point $\left(1, \frac{\pi}{2}\right)$,

$m = \tan 2 \cdot \frac{\pi}{2} = \tan\pi = 0.$ Thus, $x = r\cos\theta =$

$1\cos\frac{\pi}{2} = 0$ and $y = r\sin\theta = 1\sin\frac{\pi}{2} = 1.$ Tangent: $y = 1$; Normal: $x = 0$

10. $r = 2\cos\theta$, $\left(\sqrt{2},\ \frac{\pi}{4}\right)$, $r' = -2\sin\theta$; $\frac{dy}{dx} =$

$\frac{-2\sin\theta\sin\theta + 2\cos\theta\cos\theta}{-2\sin\theta\cos\theta - 2\cos\theta\sin\theta} = \frac{\sin^2\theta - \cos^2\theta}{2\sin\theta\cos\theta} = -\cot 2\theta$. At

$\left(\sqrt{2},\ \frac{\pi}{4}\right)$, we find $m = -\cot\left(2\cdot\frac{\pi}{4}\right) = -\cot\frac{\pi}{2} =$

0. Thus, $x = r\cos\theta = \sqrt{2}\cdot\cos\frac{\pi}{4} = \sqrt{2}\cdot\frac{\sqrt{2}}{2} = 1$,

and $y = r\sin\theta = \sqrt{2}\cdot\sin\frac{\pi}{4} = 1$. Tangent: $y = 1$,

Normal: $x = 1$

11. $r = 5\sin 3\theta$, $\left(\frac{5}{\sqrt{2}},\ \frac{\pi}{12}\right)$, $r' = 15\cos 3\theta$; $\frac{dy}{dx} =$

$\frac{15\cos 3\theta\sin\theta + 5\sin 3\theta\cos\theta}{15\cos 3\theta\cos\theta - 5\sin 3\theta\sin\theta}$. When $\theta = \frac{\pi}{12}$, then $m =$

$\frac{dy}{dx} \approx 0.660254 \approx 0.6603$. Thus, $x = \frac{5}{\sqrt{2}}\cos\frac{\pi}{12} =$

3.4151 and $y = \frac{5}{\sqrt{2}}\sin\frac{\pi}{12} = 0.9151$. Tangent:

$y = -0.9151 = 0.6603(x - 3.4151)$. Normal:

$y - 0.9151 = -1.5146(x - 3.4151)$

12. $r = 2 - 3\sin\theta$, $\left(\frac{1}{2},\ \frac{5\pi}{6}\right)$; $r' = -3\cos\theta$; $\frac{dy}{dx} =$

$\frac{-3\cos\theta\sin\theta + (2 - 3\sin\theta)\cos\theta}{-3\cos\theta\cos\theta - (2 - 3\sin\theta)\sin\theta}$. At $\theta = \frac{5\pi}{6}$, we find that

$m = \frac{dy}{dx} = \frac{-3\left(-\frac{\sqrt{3}}{2}\right)\frac{1}{2} + \left(2 - 3\cdot\frac{1}{2}\right)\cdot\frac{-\sqrt{3}}{2}}{-3\cdot\frac{-\sqrt{3}}{2}\cdot\frac{-\sqrt{3}}{2} - \left(2 - 3\cdot\frac{1}{2}\right)\frac{1}{2}} =$

$\frac{\frac{3\sqrt{2}}{4} + \frac{-\sqrt{3}}{4}}{\frac{-9}{4} - \frac{1}{4}} = \frac{2\sqrt{3}}{-10} = \frac{-\sqrt{3}}{5} \approx -0.3464$. Thus,

we get $x = r\cos\theta = \frac{1}{2}\cdot\frac{-\sqrt{3}}{2} = \frac{-\sqrt{3}}{4}$ and $y =$

$r\sin\theta = \frac{1}{2}\cdot\frac{1}{2} = \frac{1}{4}$. As a result, Tangent: $y - \frac{1}{4} =$

$\frac{-\sqrt{3}}{5}\left(x + \frac{\sqrt{3}}{4}\right) = 4y + \frac{\sqrt{3}}{5}x = \frac{2}{5}$, or $20y + \sqrt{3}x = 2$

or about $20y + 1.732x = 2$, or $y = -0.25 =$

$-0.3464(x + 1.4330)$. Normal: $y - \frac{1}{4} = \frac{5}{\sqrt{3}}\left(x + \right.$

$\left.\frac{\sqrt{3}}{5}\right)$ or $4y - 1 = \frac{20}{\sqrt{3}}x + 5$ or $4y - \frac{20}{\sqrt{3}}x = 6$, or

$y - 0.24 = 2.8858(x + 0.433)$, or $4y - 11.547x = 6$.

13. We are given $r = 6\sin^2\theta$ and $\left(4.5,\ \frac{2\pi}{3}\right)$, and so $r' = 12\sin\theta\cos\theta$ which leads to the derivative $\frac{dy}{dx} = \frac{12\sin\theta\cos\theta\sin\theta + 6\sin^2\theta\cos\theta}{12\sin\theta\cos\theta\cos\theta - 6\sin^2\theta\sin\theta} = \frac{3\sin^2\theta\cos\theta}{\sin\theta\left(2\cos^2\theta - \sin^2\theta\right)}$

$= \frac{3\sin\theta\cos\theta}{2\cos^2\theta - \sin^2\theta}$. At $\theta = \frac{2\pi}{3}$ we get $m = \frac{3\cdot\frac{\sqrt{3}}{2}\cdot\frac{-1}{2}}{2\cdot\frac{1}{4} - \frac{3}{4}}$

$= \frac{\frac{-3\sqrt{3}}{4}}{-\frac{1}{4}} = 3\sqrt{3} \approx 5.1962$. At $\left(4.5,\ \frac{2\pi}{3}\right)$, we find

that $x = 4.5\cos\frac{2\pi}{3} = \frac{-9}{4}$ and $y = 4.5\sin\frac{2\pi}{3} =$

$\frac{9\sqrt{3}}{4}$. Tangent: $y - \frac{9\sqrt{3}}{4} = 3\sqrt{3}\left(x + \frac{9}{4}\right)$ or $4y -$

$9\sqrt{3} = 12\sqrt{3}x + 27\sqrt{3}$ or $4y - 12\sqrt{3}x = 36\sqrt{3}$ or

$y - 3\sqrt{3}x = 9\sqrt{3}$. Normal: $y - \frac{9\sqrt{3}}{4} = \frac{-1}{3\sqrt{3}}\left(x + \frac{9}{4}\right)$,

or $4y - 9\sqrt{3} = \frac{-4}{3\sqrt{3}}x - \frac{9}{3\sqrt{3}}$, or $4y + \frac{4}{3\sqrt{3}}x = 8\sqrt{3}$,

or $y + \frac{1}{3\sqrt{3}}x = 2\sqrt{3}$, or $9y + \sqrt{3}x = 18\sqrt{3}$.

14. $r = 2 + 3\sec\theta$; $\left(-4,\ \frac{2\pi}{3}\right)$, and so $r' = 3\sec\theta\tan\theta$, which leads to

$\frac{dy}{dx} = \frac{3\sec\theta\tan\theta\sin\theta + (2 + 3\sec\theta)\cos\theta}{3\sec\theta\tan\theta\cos\theta - (2 + 3\sec\theta)\sin\theta} =$

$\frac{3\tan^2\theta + 2\cos\theta + 3}{3\tan\theta - 2\sin\theta - 3\tan\theta} = \frac{3\tan^2\theta + 2\cos\theta + 3}{-2\sin\theta}$. So, $\theta = \frac{2\pi}{3}$,

yields $m = \frac{3(-\sqrt{3})^2 + 2\left(\frac{-1}{2}\right) + 3}{-2\left(\frac{\sqrt{3}}{2}\right)} = \frac{9 - 1 + 3}{-\sqrt{3}} =$

$\frac{11}{-\sqrt{3}} = \frac{-11}{3}\sqrt{3}$. As a result, we find that $x =$

$-4\cos\frac{2\pi}{3} = -4\cdot\frac{-1}{2} = 2$ and $y = -4\sin\frac{2\pi}{3} =$

$-4\frac{\sqrt{3}}{2} = -2\sqrt{3}$. Tangent: $y + 2\sqrt{3} = \frac{-11\sqrt{3}}{3}(x -$

2), or $3y + 11\sqrt{3}x = 16\sqrt{3}$; $y + 6.3509x = 9.2376$.

Normal: $y + 2\sqrt{3} = \frac{\sqrt{3}}{11}(x - 2)$, or $11y + 22\sqrt{3} =$

$\sqrt{3}x - 2\sqrt{3}$, or $11y - \sqrt{3}x = -24\sqrt{3}$, or $y -$

$0.1574x = -3.7790$.

15. $r = e^\theta$; $\left(2.8497,\ \frac{\pi}{3}\right)$; By exercise #8, we see

that $\frac{dy}{dx} = \frac{\sin\theta + \cos\theta}{\cos\theta - \sin\theta}$, and at $\frac{\pi}{3}$, we obtain $m =$

$\frac{\frac{\sqrt{3}}{2} + \frac{1}{2}}{\frac{1}{2} - \frac{\sqrt{3}}{2}} = \frac{\sqrt{3} + 1}{1 - \sqrt{3}} = \frac{(\sqrt{3} + 1)(\sqrt{3} + 1)}{(1 - \sqrt{3})(1 + \sqrt{3})} = \frac{4 + 2\sqrt{3}}{-2} =$

$-2 - \sqrt{3}$. Thus, $x = 2.8497\cdot\cos\frac{\pi}{3} = 1.4249$

and $y = 2.8497\cdot\sin\frac{\pi}{3} = 2.4679$. Tangent: $y -$

$2.4679 = -(2 + \sqrt{3})(x - 1.4249)$, or $y - 2.4697 =$

$-3.7321x + 5.3178$, or $y + 3.7321x = 7.7875$.

Normal: $y - 2.4679 = \frac{1}{2 + \sqrt{3}}(x - 1.4249)$, or $y -$

$2.4679 = 0.2679x - 0.3818$, or $y - 0.2679x =$

2.0861.

16. $r = \tan\theta$, $\left(-1,\ \frac{3\pi}{4}\right)$. By exercise #7, $\frac{dy}{dx} =$

$\left(\sec^2\theta + 1\right)\tan\theta$. At $\theta = \frac{3\pi}{4}$, we obtain $m =$

$(2 + 1)\cdot(-1) = -3$. Thus, $x = -1\cos\frac{3\pi}{4} = \frac{\sqrt{2}}{2}$

and $y = -1\sin\frac{3\pi}{4} = -\frac{\sqrt{2}}{2}$. Tangent: $y + \frac{\sqrt{2}}{2} =$

$-3\left(x - \frac{\sqrt{2}}{2}\right)$, or $2y + \sqrt{2} = -6x + 3\sqrt{2}$, or

$2y + 6x = 2\sqrt{2}$, or $y + 3x = \sqrt{2}$. Normal:

$y + \frac{\sqrt{2}}{2} = \frac{1}{3}\left(x - \frac{\sqrt{2}}{2}\right)$, or $6y + 3\sqrt{2} = 2x - \sqrt{2}$,

or $6y - 2x = -4\sqrt{2}$, or $3y - x = -2\sqrt{2}$.

17. Since $r = 3\cos 2\theta$ we have $r' = -6\sin 2\theta$, which gives the derivative $\frac{dy}{dx} = \frac{-6\sin 2\theta\sin\theta + 3\cos 2\theta\cos\theta}{-6\sin 2\theta\cos\theta - 3\cos 2\theta\sin\theta} =$

$\frac{2\sin 2\theta\sin\theta - \cos 2\theta\cos\theta}{2\sin 2\theta\cos\theta + \cos 2\theta\cos\theta}$. If we set the numerator equal to 0 we get $2\sin 2\theta\sin\theta - \cos 2\theta\cos\theta = 0$ which expands to $2(2\sin\theta\cos\theta)\sin\theta - \left(1 - 2\sin^2\theta\right)\cos\theta$

$= 0$; $\cos\theta\left(4\sin^2\theta - 1 + 2\sin^2\theta\right) = 0$. Which simplifies to $\cos\theta\left(6\sin^2\theta - 1\right) = 0$. Hence either $\cos\theta = 0$ or $\sin^2\theta = \frac{1}{6}$ or $\sin\theta = \pm\frac{\sqrt{6}}{6}$. This gives critical values of $\theta = \frac{\pi}{2}$, $\frac{3\pi}{2}$ and $\theta = 0.4205$, 2.7211, 3.5621, and 5.8627.

Set the denominator to 0 and solve $2\sin 2\theta \times \cos\theta + \cos 2\theta\sin\theta = 0$. Using an expansion identity for $2\sin\alpha\cos\alpha$, we get $4\sin\theta\cos\theta\cos\theta + \left(2\cos^2\theta - 1\right)\sin\theta = 0$. Factoring, we get $\sin\theta \times \left[4\cos^2\theta + 2\cos^2\theta - 1\right] = 0$, so $\sin\theta = 0$ or $\cos^2\theta = \frac{1}{6}$ or $\cos\theta = \pm\frac{\sqrt{6}}{6}$. This gives critical values of 0, π, 1.1503, 1.9913, 4.2918, and 5.1329.

The graph of $r = 3\cos 2\theta$ is a four leaf rose. The first set of critical values gives the locations of horizontal tangents and the second set gives the locations of vertical tangents. Horizontal tangents at $\left(-3, \frac{\pi}{2}\right)$, $\left(-3, \frac{3\pi}{2}\right)$, $(2, 0.4205)$, $(2, 2.7211)$, $(2, 3.5621)$, and $(2, 5.8627)$. Vertical tangents at $(3, 0)$, $(3, \pi)$, $(-2, 1.1503)$, $(-2, 1.9913)$, $(-2, 4.2918)$, and $(-2, 5.1329)$.

18. $r = 1 - \cos\theta$; $r' = \sin\theta$, and so
$$\frac{dy}{dx} = \frac{\sin\theta\sin\theta + (1 - \cos\theta)\cos\theta}{\sin\theta\cos\theta - (1 - \cos)\sin\theta}.$$
The numerator is 0 when $\sin^2\theta + \cos\theta - \cos^2\theta = 0$ or $\left(1 - \cos^2\theta\right) + \cos\theta - \cos^2\theta = 0$ or $1 + \cos\theta - 2\cos^2\theta = 0$, which factors as $(1 + 2\cos\theta)(1 - \cos\theta) = 0$. So $\cos\theta = 1$ or $\cos\theta = -\frac{1}{2}$. This gives critical values of $\theta = 0$, $\frac{2\pi}{3}$, and $\frac{4\pi}{3}$. The denominator is 0 when $\sin\theta\cos\theta - (1 - \cos\theta)\sin\theta = 0$; $\sin\theta(\cos\theta - 1 + \cos\theta) = 0$; $\sin\theta = 0$ or $\cos\theta = \frac{1}{2}$. This gives critical values of 0, π, $\frac{\pi}{3}$ and $\frac{5\pi}{3}$. This shape is a cardioid with horizontal tangents at $\theta = \frac{2\pi}{3}$ and $\frac{4\pi}{3}$ and vertical tangents at $\frac{\pi}{3}$, π, and $\frac{5\pi}{3}$. At $\theta = 0$, there is no tangent. Horizontal tangents at $\left(1.5, \frac{3\pi}{3}\right)$, $\left(1.5, \frac{4\pi}{3}\right)$. Vertical tangents at $\left(0.5, \frac{\pi}{3}\right)$, $(2, \pi)$, and $\left(0.5, \frac{5\pi}{3}\right)$.

19. $r = 1 + 2\cos\theta$, $r' = -2\sin\theta$. Hence, we obtain
$$\frac{dy}{dx} = \frac{-2\sin\theta\sin\theta + (1 + 2\cos\theta)\cos\theta}{-2\sin\theta\cos\theta - (1 + 2\cos\theta)\sin\theta}$$
$$= \frac{-2\sin^2\theta + \cos^2\theta + 2\cos\theta}{-2\sin\theta\cos\theta - \sin\theta + 2\cos\theta\sin\theta}$$
$$= \frac{-2\left(1 - \cos^2\theta\right) + \cos\theta + 2\cos^2\theta}{-4\sin\theta\cos\theta - \sin\theta}$$
$$= \frac{4\cos^2\theta + \cos\theta - 2}{-\sin\theta(4\cos\theta + 1)}$$

The numerator equals 0 when $4\cos^2\theta + \cos\theta - 2 = 0$ or $\cos\theta = (-1 \pm \sqrt{33})8$ or when $\theta \approx 0.9359$, 5.3476, 2.5738, or 3.7094. The denominator equals 0 when $\sin\theta = 0$ or $\cos\theta = -\frac{1}{4}$ which yields $\theta = 0$, π, 1.8234, or 4.4597. This is a limaçon. Horizontal tangents are at $(2.1861, 0.9359)$, $(-0.6861, 2.5738)$, $(-0.6861, 3.7094)$, and $(2.1861, 5.3476)$. Vertical tangents are at $(3, 0)$, $(0.5, 1.8234)$, $(-1, \pi)$, and $(0.5, 4.4597)$.

20. $r = \sin^2\theta$, $r' = 2\sin\theta\cos\theta$. Hence, we obtain
$$\frac{dy}{dx} = \frac{2\sin\theta\cos\theta\sin\theta + \sin^2\theta\cos\theta}{2\sin\theta\cos\theta\cos\theta - \sin^2\theta\sin\theta}$$
$$= \frac{2\sin^2\theta\cos\theta + \sin^2\theta\cos\theta}{2\sin\theta\cos^2\theta - \sin^3\theta}$$
$$= \frac{3\sin^2\theta\cos\theta}{2\cos^2\theta - \sin^2\theta}$$
The numerator equals 0 when $\sin\theta = 0$ or $\cos\theta = 0$. These will be when $\theta = 0$, $\frac{\pi}{2}$, π, and $\frac{3\pi}{2}$. The denominator equals 0 when $2\cos^2\theta - \sin^2\theta = 0$ or $2\cos^2\theta = \sin^2\theta$, or $2 = \tan^2\theta$. This will be when $\theta = 0.9553$, 2.1863, 4.0969, or 5.3279. Horizontal tangents occur at $(0, 0)$, $\left(1, \frac{\pi}{2}\right)$, and $\left(1, \frac{3\pi}{2}\right)$. Vertical tangents are at $\left(\frac{2}{3}, 0.9553\right)$, $\left(\frac{2}{3}, 2.1863\right)$, $\left(\frac{2}{3}, 4.0969\right)$, and $\left(\frac{2}{3}, 5.3279\right)$. (Note: $(0, \pi)$ is the same point as $(0, 0)$.)

21. (a) $\frac{dr}{d\theta} = \frac{10(3\sin\theta)}{(5 + 3\cos\theta)^2} = \frac{30\sin\theta}{(5 + 3\cos\theta)^2}$
(b) Critical values are 0, π, and 2π. Since r is increasing on the interval $(0, \pi)$ and decreasing on $(\pi, 2\pi)$, r has a maximum at $\theta = \pi$.

22. (a) $x = \frac{10\cos\theta}{5 + 3\cos\theta}$, $y = \frac{10\sin\theta}{5 + 3\cos\theta}$
(b) $\frac{dy}{dx} = \frac{r'\sin\theta + r\cos\theta}{r'\cos\theta + r\sin\theta}$
$$= \frac{\frac{30\sin^2\theta}{(5 + 3\cos\theta)^2} + \frac{10\cos\theta}{(5 + 3\cos\theta)}}{\frac{30\sin\theta\cos\theta}{(5 + 3\cos\theta)^2} - \frac{10\sin\theta}{(5 + 3\cos\theta)}}$$
$$= \frac{30\sin^2\theta + 10\cos\theta(5 + 3\cos\theta)}{30\sin\theta\cos\theta - 10\sin\theta(5 + 3\cos\theta)}$$
$$= \frac{30\sin^2\theta + 50\cos\theta + 30\cos^2\theta}{30\sin\theta\cos\theta - 50\sin\theta - 30\sin\theta\cos\theta}$$
$$= \frac{30 + 50\cos\theta}{-50\sin\theta} = -\frac{5\cos\theta + 3}{5\sin\theta}$$
(c) Extrema are (a) when $3 + 5\cos\theta = 0$ or $\theta = \cos\left(-\frac{3}{5}\right) \approx 2.21$ or 4.07 radians, or (b) when $\sin\theta = 0$ which is at $\theta = 0$, π. The slope is undefined at $\theta = 0$ and $\theta = \pi$ so there are vertical tangent lines at $(1.25, 0)$ and $(5, \pi)$. The slope is 0 at $\theta = 2.21$ and $\theta = 4.07$ and there are horizontal tangent lines at $(3.125, 2.21)$ and $(3.125, 4.07)$.

29.3 ARC LENGTH AND SURFACE AREA REVISITED

1. $x = 4t^3$; $y = 3t^2$ so $\frac{dx}{dt} = 12t^2$ and $\frac{dy}{dt} = 6t$.

The arc length is $L = \int_0^1 \sqrt{(12t^2)^2 + (6t)^2}\,dt = \int_0^1 \sqrt{144t^4 + 36t^2}\,dt = \int_0^1 6t\sqrt{4t^2 + 1}\,dt$. Let $u = 4t^2 + 1$ and $du = 8t\,dt$. Substituting, we get $\frac{6}{8} \int \sqrt{u}\,du = \frac{3}{4}\frac{2}{3}u^{3/2} = \frac{1}{2}\left(4t^2 + 1\right)^{3/2}\Big|_0^1 = \frac{1}{2}\left[5^{3/2} - 1^{3/2}\right] \approx \frac{1}{2}(10.1803) = 5.0902$.

2. $x = \sin t$; $\frac{dx}{dt} = \cos t$; $y = \cos t$; $\frac{dy}{dt} = -\sin t$
The arc length is $L = \int_0^\pi \sqrt{(\cos t)^2 + (-\sin t)^2}\,dt = \int_0^\pi \sqrt{\cos^2 t + \sin^2 t}\,dt = \int_0^\pi 1\,dt = t\big|_0^\pi = \pi$.

3. $x = 3\cos t$; $\frac{dx}{dt} = -3\sin t$; $y = 3\sin t$; $\frac{dy}{dt} = 3\cos t$. The arc length is
$L = \int_0^{2\pi} \sqrt{(-3\sin t)^2 + (3\cos t)^2}\,dt = \int_0^{2\pi} \sqrt{9\sin^2 t + 9\cos^2 t}\,dt = \int_0^{2\pi} 3\,dt = 3t\big|_0^{2\pi} = 6\pi$. (Note: this curve is a circle of radius 3.)

4. $x = \cos^3 t$; $\frac{dx}{dt} = -3\cos^2 t\sin t$; $y = \sin^3 t$; $\frac{dy}{dt} = 3\sin^2 t\cos t$. The arc length is
$L = \int_0^{\pi/2} \sqrt{\left(-3\cos^2 t\sin t\right)^2 + \left(3\sin^2 t\cos t\right)^2}\,dt$, or $L = \int_0^{\pi/2} \sqrt{9\cos^4 t\sin^2 t + 9\sin^4 t\cos^2 t}\,dt$, or $L = \int_0^{\pi/2} 3\sqrt{\cos^2 t\sin^2 t\left(\cos^2 t + \sin^2 t\right)}\,dt$, or $L = \int_0^{\pi/2} 3\cos t\sin t\,dt$. Let $u = \sin t$; $du = \cos t\,dt$ so by substituting we get $3\int u\,du = 3 \cdot \frac{1}{2}u^2 = \left[\frac{3}{2}\sin^2 t\right]_0^{\pi/2} = \frac{3}{2}$.

5. $x = \cos t + t\sin t$; $\frac{dx}{dt} = -\sin t + t\cos t + \sin t = t\cos t$. $y = \sin t - t\cos t$; $\frac{dy}{dt} = \cos t + t\sin t - \cos t = t\sin t$. The arc length is
$L = \int_0^\pi \sqrt{(t\cos t)^2 + (t\sin t)^2}\,dt = \int_0^\pi \sqrt{t^2\cos^2 t + t^2\sin^2 t}\,dt = \int_0^\pi t\,dt = \frac{t^2}{2}\Big|_0^\pi = \frac{1}{2}\pi^2$.

6. $r = \theta$; $r' = 1$; The arc length is
$L = \int_0^{\pi/2} \sqrt{r^2 + r'^2}\,d\theta = \int_0^{\pi/2} \sqrt{\theta^2 + 1}\,d\theta$. By formula #30 in Appendix C, this is
$\left[\frac{\theta}{2}\sqrt{\theta^2 + 1} + \frac{1}{2}\ln\left|\theta + \sqrt{\theta^2 + 1}\right|\right]_0^{\pi/2} = \frac{\pi}{4}\sqrt{\frac{\pi^2}{4} + 1}$
$+ \frac{1}{2}\ln\left|\frac{\pi}{2} + \sqrt{\frac{\pi^2}{4} + 1}\right| \approx 2.0792$.

7. Since $r = 1 + \cos\theta$ we have $r' = -\sin\theta$. Thus, the arc length is
$L = \int_0^\pi \sqrt{(1 + \cos\theta)^2 + (-\sin\theta)^2}\,d\theta =$

$\int_0^\pi \sqrt{1 + 2\cos\theta + \cos^2\theta + \sin^2\theta}\,d\theta = \int_0^\pi \sqrt{2 + 2\cos\theta}\,d\theta = \sqrt{2}\int_0^\pi \sqrt{1 + \cos\theta}\,d\theta$. Since $\cos^2\frac{\theta}{2} = \frac{1 + \cos\theta}{2}$ we get $\sqrt{1 + \cos\theta} = \sqrt{2}\cos\frac{\theta}{2}$. Now our integral is $\sqrt{2}\cos\int_2^\pi \sqrt{2}\cos\frac{\theta}{2}\,d\theta = 2\int_0^\pi \cos\frac{\theta}{2}\,d\theta$. Letting $u = \frac{\theta}{2}$ we have $du = \frac{1}{2}d\theta$ or $2\,du = d\theta$, and so, we get by substitution, $4\int \cos u\,du = 4\sin u = 4\sin\frac{\theta}{2}\Big|_0^\pi = 4\left(\sin\frac{\pi}{2} - \sin 0\right) = 4$.

8. Here $r = \cos^2\theta$ and so $r' = -2\cos\theta\sin\theta$. Thus, the arc length is given by
$L = \int_0^{\pi/2} \sqrt{\cos^4\theta + (-2\cos\theta\sin\theta)^2}\,d\theta$
$= \int_0^{\pi/2} \sqrt{\cos^4\theta + 4\cos^2\theta\sin^2\theta}\,d\theta$
$= \int_0^{\pi/2} \cos\theta\sqrt{\cos^4\theta + 4\sin^2\theta}\,d\theta$
$= \int_0^{\pi/2} \cos\theta\sqrt{1 + 3\sin^2\theta}\,d\theta$.
Letting $u = \sqrt{3}\sin\theta$, we get $du = \sqrt{3}\cos\theta\,d\theta$. Substituting produces $L = \frac{1}{\sqrt{3}}\int \sqrt{1 + u^2}\,du$. By formula #30 in Appendix C this is
$\frac{1}{\sqrt{3}}\left[\frac{u}{2}\sqrt{1 + u^2} + \frac{1}{2}\ln\left|u + \sqrt{1 + u^2}\right|\right]$
$= \frac{1}{\sqrt{3}}\left[\frac{3\sin\theta}{2}\sqrt{1 + 3\sin^2\theta}\right.$
$\left.+ \frac{1}{2}\ln\left|3\sin\theta + \sqrt{1 + 3\sin^2\theta}\right|\right]_0^{\pi/2}$
$= \frac{1}{\sqrt{3}}\left[\frac{\sqrt{3}}{2} \cdot \sqrt{4} + \frac{1}{2}\ln(\sqrt{3} + \sqrt{4})\right]$
$= \frac{1}{\sqrt{3}}\left[\sqrt{3} + \frac{1}{2}\ln(2 + \sqrt{3})\right]$
$= 1 + \frac{\sqrt{3}}{6}\ln(2 + \sqrt{3}) \approx 1.3802$.

9. Here $r = \sin^2\theta$, and so $r' = 2\sin\theta\cos\theta = \sin 2\theta$. The arc length is
$L = \int_0^{\pi/2} \sqrt{\sin^4\theta + (2\sin\theta\cos\theta)^2}\,d\theta$
$= \int_0^{\pi/2} \sqrt{\sin^4\theta + 4\sin^2\theta\cos^2\theta}\,d\theta$
$= \int_0^{\pi/2} \sin\theta\sqrt{\sin^2\theta + 4\cos^2\theta}\,d\theta$
$= \int_0^{\pi/2} \sin\theta\sqrt{1 + 3\cos^2\theta}\,d\theta$.

Letting $u = \sqrt{3}\cos\theta$, then $du = -\sqrt{3}\sin\theta$ and substituting we get $-\frac{1}{\sqrt{3}}\int\sqrt{1+u^2}\,du$. By formula #30 in Appendix C we get

$-\frac{1}{\sqrt{3}}\left[\frac{u}{2}\sqrt{1+u^2}+\frac{1}{2}\ln\left|u+\sqrt{1+u^2}\right|\right] =$

$-\frac{1}{\sqrt{3}}\left[\frac{\sqrt{3}\cos\theta}{2}\sqrt{1+3\cos^2\theta}+\frac{1}{2}\ln\left|\sqrt{3}\cos\theta+\right.\right.$

$\left.\left.\sqrt{1+3\cos^2\theta}\right|\right]\Big|_0^{\pi/2} = -\frac{1}{\sqrt{3}}\left(-\sqrt{3}-\frac{1}{2}\ln(\sqrt{3}+\right.$

$\left.2)\right) = 1+\frac{\sqrt{3}}{6}\ln(2+\sqrt{3}) \approx 1.3802.$

10. Since $r = e^{\theta/2}$, then $r' = \frac{1}{2}e^{\theta/2}$. The arc length

is $L = \int_0^4 \sqrt{\left(e^{\theta/2}\right)^2+\left(\frac{1}{2}e^{\theta/2}\right)^2}\,d\theta =$

$\int_0^4\sqrt{e^\theta+\frac{1}{4}e^\theta}\,d\theta$. Thus, we have $L = \int_0^4\sqrt{\frac{5e^\theta}{4}}\,d\theta$

$= \frac{\sqrt{5}}{2}\int_0^4\sqrt{e^\theta}\,d\theta = \frac{\sqrt{5}}{2}\int_0^4 e^{\theta/2}\,d\theta$. Letting $u =$

$\frac{\theta}{2}$, then $du = \frac{1}{2}d\theta$, so we get $\frac{\sqrt{5}}{2}\cdot 2\int e^u\,du =$

$\sqrt{5}e^u = \sqrt{5}e^{\theta/2}\Big|_0^4 = \sqrt{5}\left(e^2-e^0\right) = \sqrt{5}\left(e^2-1\right) =$

$14.2864.$

11. $x = t+4$, so $\frac{dx}{dt} = 1$ and $y = t^3$, so $\frac{dy}{dt} = 3t^2$. Thus, the area of the surface of revolution is $S =$

$2\pi\int_0^2 t^3\sqrt{1+\left(3t^2\right)^2}\,dt = 2\pi\int_0^2 t^3\sqrt{1+9t^4}\,dt$. Let $u = 1+9t^4$, and then $du = 36t^3\,dt$. Substituting

we get $\frac{2\pi}{36}\int\sqrt{u}\,du = \frac{\pi}{18}\cdot\frac{2}{3}u^{3/2} = \frac{\pi}{27}\left(1+9t^4\right)^{3/2}\Big|_0^2$

$= \frac{\pi}{27}\left(145^{3/2}-1\right) \approx 203.0436.$

12. Here $x = t$, so $\frac{dx}{dt} = 1$ and $y = 4-t^2$, so $\frac{dy}{dt} = -2t$. Thus, the area of the surface of revolution is $S = 2\pi\int_0^2 t\sqrt{1+\left(4t^2\right)}\,dt$. Let $u = 1+4t^2$, and then $du = 8t\,dt$. Substituting we get $2\pi\cdot\frac{1}{8}\int\sqrt{u}\,du = \frac{\pi}{4}\frac{2}{3}u^{3/2} = \frac{\pi}{6}\left(1+4t^2\right)^{3/2}\Big|_0^2 = \frac{\pi}{6}\left(17^{3/2}-1\right) \approx 36.1769.$

13. Here $x = \cos t$, so $\frac{dx}{dt} = -\sin t$ and $y = \sin t$, so $\frac{dy}{dt} = \cos t$. Thus,

$S = 2\pi\int_0^{\pi/2}\sin t\sqrt{(-\sin t)^2+(\cos t)^2}\,dt$

$= 2\pi\int_0^{\pi/2}\sin t\sqrt{\sin^2 t+\cos^2 t}\,dt$

$= 2\pi\int_0^{\pi/2}\sin t\,dt$

$= 2\pi(-\cos t)\Big|_0^{\pi/2} = 2\pi(0+1) = 2\pi.$

14. $x = 1+\sin t$; $\frac{dx}{dt} = \cos t$; $y = \cos t$; $\frac{dy}{dt} = -\sin t$. Thus,

$S = 2\pi\int_0^{\pi/2}(1+\sin t)\sqrt{\cos^2 t+\sin^2 t}\,dt$

$= 2\pi\int_0^{\pi/2}(1+\sin t)dt = 2\pi[t-\cos t]\Big|_0^{\pi/2}$

$= 2\pi\left[\frac{\pi}{2}-0-0+1\right] = 2\pi\left[\frac{\pi}{2}+1\right]$

$= \pi^2+2\pi \approx 16.1528$

15. $x = 1+\sin t$, $\frac{dy}{dt} = \cos t$; $y = \cos t$; $\frac{dy}{dt} = -\sin t$. Thus, the area is

$S = 2\pi\int_0^{\pi/2}\cos t\sqrt{\cos^2 t+\sin^2 t}\,dt =$

$2\pi\int_0^{\pi/2}\cos t\,dt = 2\pi[\sin t]_0^{\pi/2} = 2\pi.$

16. $r = \sin\theta$ and so $\frac{dr}{d\theta} = \cos\theta$. Thus,

$S = 2\pi\int_0^{\pi/2} r\sin\theta\sqrt{\sin^2\theta+\cos^2\theta}\,d\theta$

$= 2\pi\int_0^{\pi/2}\sin\theta\sin\theta\cdot 1\,d\theta$

$= 2\pi\int_0^{\pi/2}\sin^2\theta\,d\theta = 2\pi\int_0^{\pi/2}\frac{1-\cos 2\theta}{2}\,d\theta$

$= \pi\int_0^{\pi/2}(1-\cos 2\theta)d\theta$

$= \pi\left(\theta-\frac{1}{2}\sin 2\theta\right)\Big|_0^{\pi/2} \pi\left(\frac{\pi}{2}-0-0-0\right)$

$= \frac{\pi^2}{2}.$

17. $r = 1+\cos\theta$; $\frac{dr}{d\theta} = -\sin\theta$. Thus, the area of the surface of revolution is

$S = 2\pi\int_0^\pi r\sin\theta\sqrt{(1+\cos t)^2+(-\sin\theta)^2}\,d\theta$

$= 2\pi\int_0^\pi(1+\cos\theta)\sin\theta$

$\times\sqrt{1+2\cos\theta+\cos^2\theta+\sin^2\theta}\,d\theta$

$= 2\pi\int_0^\pi(1+\cos\theta)\sin\theta\sqrt{2+2\cos\theta}\,d\theta$

$= 2\sqrt{2}\pi\int_0^\pi(1+\cos\theta)^{3/2}\sin\theta\,d\theta$

Let $u = 1+\cos\theta$, then $du = -\sin\theta\,d\theta$, and we get $-2\sqrt{2}\pi\int u^{3/2}du = -2\sqrt{2}\pi\cdot\frac{2}{5}u^{5/2} =$

$\frac{-4\sqrt{2}\pi}{5}(1+\cos\theta)^{5/2}\Big|_0^\pi = \frac{-4\sqrt{2}\pi}{5}\left((1-1)^{5/2}-(1-\right.$

$\left.1)^{5/2}\right) = \frac{-4\sqrt{2}\pi}{5}(2)^{5/2} = \frac{2^5}{5}\pi = \frac{32}{5}\pi \approx 20.1062.$

18. $r = e^{\theta/2}$; $\frac{dr}{d\theta} = \frac{1}{2}e^{\theta/2}$

$$S = 2\pi \int_0^\pi e^{\theta/2} \sin\theta \sqrt{\left(e^{\theta/2}\right)^2 + \left(\frac{1}{2}e^{\theta/2}\right)^2}\, d\theta$$

$$= 2\pi \int_0^\pi e^{\theta/2} \sin\theta \sqrt{e^\theta + \frac{1}{4}e^\theta}\, d\theta$$

$$= 2\pi \int_0^\pi e^{\theta/2} \sin\theta \left(\frac{\sqrt{5}}{2}e^{\theta/2}\right) d\theta$$

$$= \sqrt{5}\pi \int_0^\pi e^\theta \sin\theta\, d\theta.$$

By formula #87 of Appendix C, this is
$\sqrt{5}\pi\left[\frac{e^\theta}{2}(\sin\theta - \cos\theta)\right]_0^\pi = \frac{1}{2}\sqrt{5}\pi\left[e^\pi(1) - e^0(-1)\right] = \frac{1}{2}\sqrt{5}\pi\left(e^\pi + 1\right) \approx 84.7919$.

19. **(a)** $L = \int_0^{8\pi} \sqrt{r^2 + \left(\frac{dr}{d\theta}\right)^2}\, d\theta = \int_0^{8\pi} \sqrt{(b\theta)^2 + b^2}\, d\theta$
$= b \int_0^{8\pi} \sqrt{\theta^2 + 1}\, d\theta$. Integrating by Formula 30,

Appendix C, we obtain $L = b\left(\frac{1}{2}\right)\left[\theta\sqrt{\theta^2 + 1} + \ln(\theta + \sqrt{\theta^2 + 1})\right]_0^{8\pi}$ Here $d = 1$ cm, and so, $b = \frac{1}{2\pi}$. Substituting this value for b, we get $L \approx 50.56$ cm, **(b)** $50.56/5 = 10.11 = 10^+$ h

20. **(a)** $L = \int_{2\pi}^{8\pi} \sqrt{r^2 + \left(\frac{dr}{d\theta}\right)^2}\, d\theta$

$$= \int_{2\pi}^{8\pi} \sqrt{\left(\frac{\theta}{2\pi}\right)^2 + \left(\frac{1}{2\pi}\right)^2}\, d\theta$$

$$= \frac{1}{2\pi}\int_{2\pi}^{8\pi} \sqrt{\theta^2 + 1}\, d\theta$$

$$= \frac{1}{4\pi}\left[\theta\sqrt{\theta^2 + 1} + \ln\left|\theta + \sqrt{\theta^2 + 1}\right|\right]_{2\pi}^{8\pi}$$

$$\approx 47.233973$$

This coil will be about 47.2 cm $= 0.472$ m long.
(b) $P = \frac{V^2}{R} = \frac{110^2}{4(0.472)} \approx 6\ 408.9$ W

29.4 INTERSECTION OF GRAPHS OF POLAR COORDINATES

1. $r = 3\theta$, $r = \frac{\pi}{2}$: By direct substitution we get $\frac{\pi}{2} = 3\theta$ or $\theta = \frac{\pi}{6}$. This gives the one point of intersection $\left(\frac{\pi}{2}, \frac{\pi}{6}\right)$

2. $r = \frac{\theta}{4}$; $r = \frac{\pi}{6}$: By direct substitution we obtain the equation $\frac{\pi}{6} = \frac{\theta}{4}$ or $\theta = \frac{2\pi}{3}$. This gives the one point of intersection $\left(\frac{\pi}{6}, \frac{2\pi}{3}\right)$.

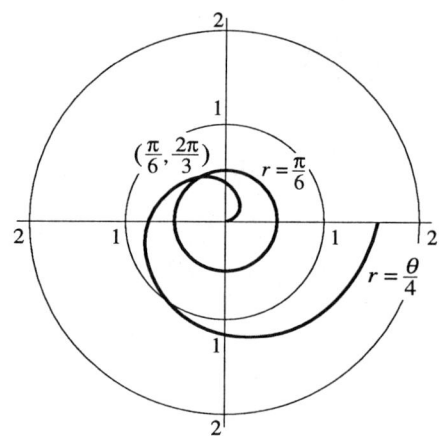

3. $r = \frac{1}{2}$; $r = \cos\theta$: By direct substitution we get $\cos\theta = \frac{1}{2}$ or $\theta = \frac{\pi}{3}$ and $\frac{5\pi}{3}$. This gives the two points of intersection $\left(\frac{1}{2}, \frac{\pi}{3}\right)$ and $\left(\frac{1}{2}, \frac{5\pi}{3}\right)$

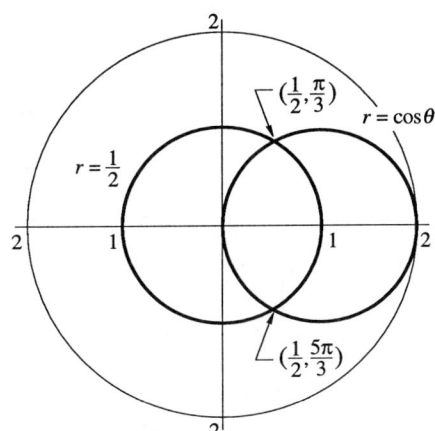

4. $r = 2 - 2\sin\theta$, $r = 2 - 2\cos\theta$: Direct substitution yields $2 - 2\sin\theta = 2 - 2\cos\theta$ or $\sin\theta = \cos\theta$. This is equivalent to $\tan\theta = 1$ or $\theta = \frac{\pi}{4}$ and $\frac{5\pi}{4}$. Hence the two points $\left(2 - \sqrt{2}, \frac{\pi}{4}\right)$ and $\left(2 + \sqrt{2}, \frac{5\pi}{4}\right)$. Looking at the graph we can see a third point, the pole. $(0, 0)$ also checks.

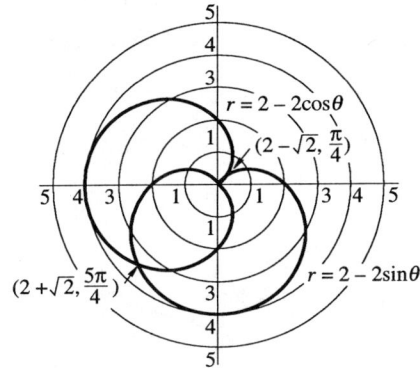

5. $r = 1 - \sin\theta$; $r = 1 + \cos\theta$: Direct substitution yields $1 - \sin\theta = 1 + \cos\theta$ or $-\sin\theta = \cos\theta$. This is equivalent to $\tan\theta = -1$. Hence we get $\theta = \frac{3\pi}{4}$ and $\frac{7\pi}{4}$ and the points $\left(1 - \frac{\sqrt{2}}{2}, \frac{3\pi}{4}\right)$ and $\left(1 + \frac{\sqrt{2}}{2}, \frac{7\pi}{4}\right)$. Looking at the graph we also see that the pole $(0, 0)$ is a common point.

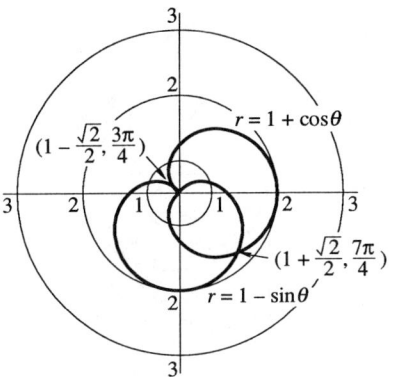

6. $r = -4 + 4\cos\theta$; $r = -4 + 4\sin\theta$. By direct substitution we get $-4 + 4\cos\theta = -4 + 4\sin\theta$ or $\cos\theta = \sin\theta$. This is true at $\theta = \frac{\pi}{4}$ and $\frac{5\pi}{4}$ giving the point $\left(-4 + 2\sqrt{2}, \frac{\pi}{4}\right)$ and $\left(-4 - 2\sqrt{2}, \frac{5\pi}{4}\right)$. Inspection of the graph and checking also gives the pole $(0, 0)$.

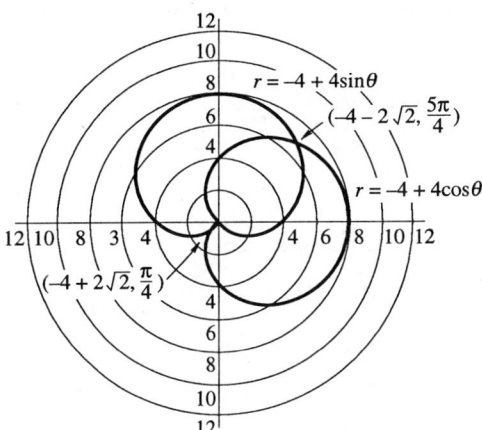

7. $r = \sqrt{3}$; $r = 2\cos\theta$. Direct substitution yields $2\cos\theta = \sqrt{3}$ or $\cos\theta = \frac{\sqrt{3}}{2}$. Hence we get $\theta = \frac{\pi}{6}, \frac{11\pi}{6}$ and the two points $\left(\sqrt{3}, \frac{\pi}{6}\right)$ and $\left(\sqrt{3}, \frac{11\pi}{6}\right)$.

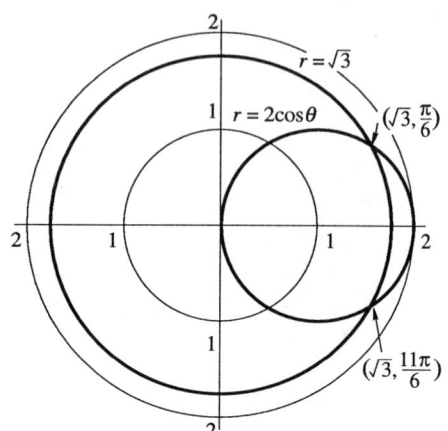

8. $r = \sin 2\theta$; $r = \sin\theta$. Direct substitution gives $\sin 2\theta = \sin\theta$. Using identities we get $2\sin\theta\cos\theta = \sin\theta$ or $2\sin\theta\cos\theta - \sin\theta = 0$. Factoring we get $\sin\theta(2\cos\theta - 1) = 0$. Hence $\sin\theta = 0$ and $\cos\theta = \frac{1}{2}$ or $\theta = 0, \pi$, or $\frac{5\pi}{3}$. This gives four points $(0, 0)$, $\left(\frac{\sqrt{3}}{2}, \frac{\pi}{3}\right)$, $(0, \pi)$, and $\left(\frac{-\sqrt{3}}{2}, \frac{5\pi}{3}\right)$. Since $(0, 0)$ and $(0, \pi)$ are the same we have only three points of intersection.

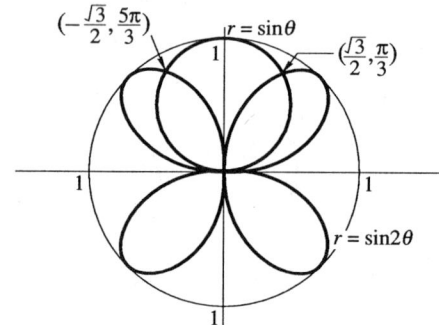

9. $r = \sin 2\theta$; $r = \sqrt{2}\sin\theta$: Direct substitution yields $\sin 2\theta = \sqrt{2}\sin\theta$ or using a trig identity $2\sin\theta\cos\theta = \sqrt{2}\sin\theta$. Hence $2\sin\theta\cos\theta - \sqrt{2}\sin\theta = 0$ or $\sin\theta(\cos\theta - \sqrt{2}) = 0$. Hence, $\sin\theta = 0 \Rightarrow \theta = 0$ or π and $\cos\theta = \frac{\sqrt{2}}{2} \Rightarrow \frac{\pi}{4}, \frac{7\pi}{4}$. Now, $\theta = 0$ and $\theta = \pi$ yield the same point $(0, 0)$, and so we have three solutions $(0, 0)$, $\left(1, \frac{\pi}{4}\right)$, and $\left(-1, \frac{7\pi}{4}\right)$.

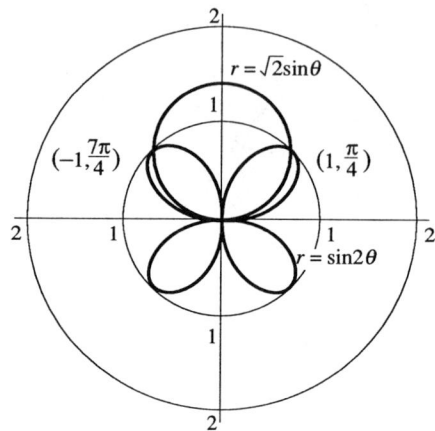

10. $r = 2 + 2\cos\theta$; $r = \frac{1}{1-\cos\theta}$. Direct substitution yields $2 + 2\cos\theta = \frac{1}{1-\cos\theta}$. Multiplying both sides by $1 - \cos\theta$ we get $2 - 2\cos^2\theta = 1$. Hence, $-2\cos^2\theta = -1$ or $\cos^2\theta = \frac{1}{2}$ and $\cos\theta = \pm\frac{\sqrt{2}}{2}$. We now have $\theta = \frac{\pi}{4}, \frac{3\pi}{4}, \frac{5\pi}{4}$, and $\frac{7\pi}{4}$ and the points $\left(2+\sqrt{2}, \frac{\pi}{4}\right)$, $\left(2-\sqrt{2}, \frac{3\pi}{4}\right)$, $\left(2-\sqrt{2}, \frac{5\pi}{4}\right)$, and $\left(2 + \sqrt{2}, \frac{7\pi}{4}\right)$.

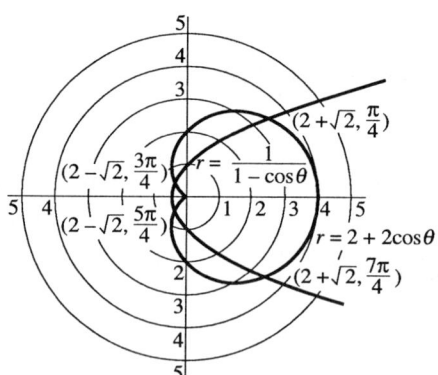

11. $r = 1 - \sin\theta$, $r = \frac{1}{1-\sin\theta}$. Direct substitution gives $1 - \sin\theta = \frac{1}{1-\sin\theta}$. Multiplying both sides by $1 - \sin\theta$ we get $(1-\sin\theta)^2 = 1$ or $1 - \sin\theta = \pm 1$; $-\sin\theta = -1 \pm 1$ or $\sin\theta = 1 \pm 1$; $\sin\theta = 0$ or 2. Since 2 is not in the range of the sine function, $\theta = 0$ or π. This gives the two points $(1, 0)$ and $(1, \pi)$.

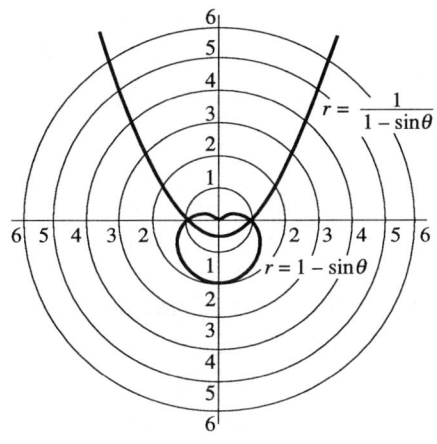

12. $r = \sin 2\theta$; $r = \cos 2\theta$: Direct substitution yields $\sin 2\theta = \cos 2\theta$ or $\frac{\sin 2\theta}{\cos 2\theta} = \tan 2\theta = 1$. From 0 to 4π, $2\theta = \frac{\pi}{4}$, $\frac{5\pi}{4}$, $\frac{9\pi}{4}$, and $\frac{13\pi}{4}$ so θ is $\frac{\pi}{8}$, $\frac{5\pi}{8}$, $\frac{9\pi}{8}$, or $\frac{13\pi}{8}$. Hence the points $\left(\frac{\sqrt{2}}{2}, \frac{\pi}{8}\right)$, $\left(-\frac{\sqrt{2}}{2}, \frac{5\pi}{8}\right)$, $\left(\frac{\sqrt{2}}{2}, \frac{9\pi}{8}\right)$, and $\left(-\frac{\sqrt{2}}{2}, \frac{13\pi}{8}\right)$. Inspecting the graph we see four more points of intersection. By symmetry these are $\left(\frac{\sqrt{2}}{2}, \frac{3\pi}{8}\right)$, $\left(-\frac{\sqrt{2}}{2}, \frac{5\pi}{8}\right)$, $\left(\frac{\sqrt{2}}{2}, \frac{11\pi}{8}\right)$, and $\left(-\frac{\sqrt{2}}{2}, \frac{15\pi}{8}\right)$.

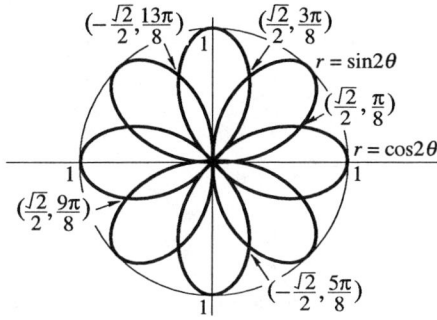

≡ 29.5 AREA IN POLAR COORDINATES

1. $r = 4\sin\theta$. This curve is complete when θ goes from 0 to π so the area is $\frac{1}{2}\int_0^\pi (4\sin\theta)d\theta = \frac{1}{2}\int_0^\pi 16\sin^2\theta\, d\theta = \frac{1}{2}\int_0^\pi 16\frac{1-\cos 2\theta}{2}d\theta = 4\int_0^\pi (1-\cos 2\theta) \times d\theta = 4\left[\theta - \frac{1}{2}\sin\theta\right]_0^\pi = 4[\pi - 0 - 0 + 0] = 4\pi$

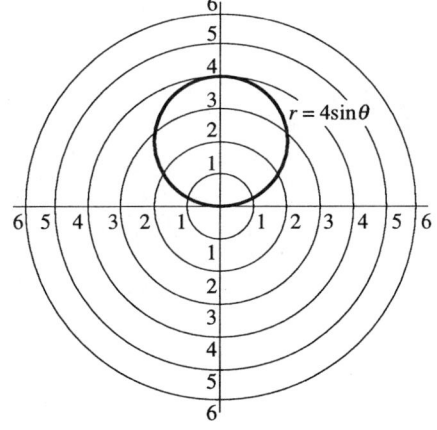

2. $r = 5\cos\theta$. This curve is completed from $\theta = 0$ to $\theta = \pi$, so the area is $\frac{1}{2}\int_0^\pi (5\cos\theta)^2 d\theta = \frac{1}{2}\int_0^\pi 25\cos^2\theta\, d\theta = \frac{25}{2}\int_0^\pi \frac{1+\cos 2\theta}{2}d\theta = \frac{25}{2}\left[\frac{1}{2}\theta + \frac{1}{4}\sin 2\theta\right]_0^\pi = \frac{25}{2}\left[\frac{1}{2}\pi + 0 - 0 - 0\right] = \frac{25}{4}\pi$

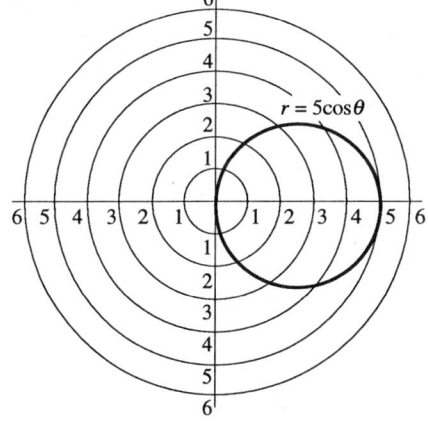

3. $r = 1 - \cos\theta$. For this curve we need to integrate from 0 to 2π with the result $\frac{1}{2}\int_0^{2\pi}(1-\cos\theta)^2\,d\theta =$ $\frac{1}{2}\int_0^{2\pi}(1-2\cos\theta+\cos^2\theta)d\theta = \frac{1}{2}\int_0^{2\pi}\left(1-2\cos\theta+\frac{1+\cos 2\theta}{2}\right)d\theta = \frac{1}{2}\left[\frac{3}{2}\theta-2\sin\theta+\frac{1}{4}\sin 2\theta\right]_0^{2\pi} = \frac{1}{2}[3\pi-0+0-0+0-0] = \frac{3}{2}\pi$

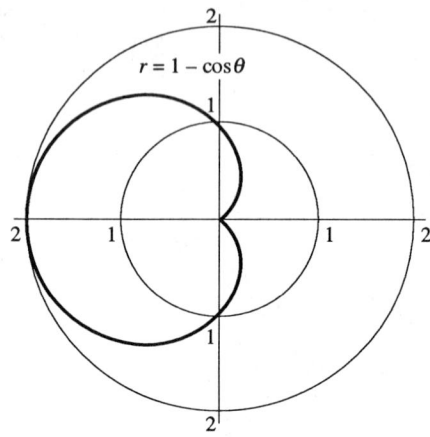

4. $r = 4 + 4\sin\theta$. Integrating from 0 to 2π getting

$\dfrac{1}{2}\displaystyle\int_0^{4\pi}(4+4\sin\theta)^2\,d\theta$

$= \dfrac{1}{2}\displaystyle\int_0^{2\pi}16(1+2\sin\theta+\sin^2\theta)\,d\theta$

$= \dfrac{1}{2}\displaystyle\int_0^{2\pi}16\left(1+2\sin\theta+\frac{1-\cos 2\theta}{2}\right)d\theta$

$= \dfrac{1}{2}\displaystyle\int_0^{2\pi}(24+32\sin\theta-8\cos 2\theta)d\theta$

$= \dfrac{1}{2}[24\theta-32\cos\theta-4\sin 2\theta]_0^{2\pi}$

$= \dfrac{1}{2}[48\pi-32-0-0-(-32)+0] = 24\pi$

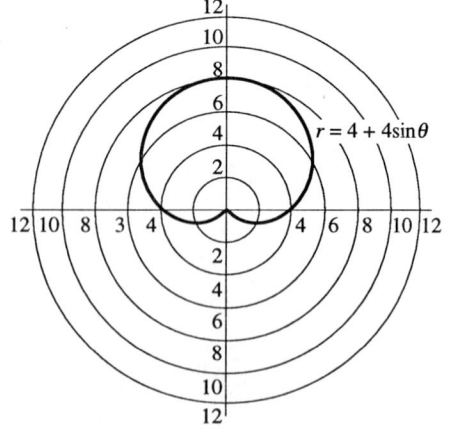

5. $r = \sin 2\theta$. This curve is a four leaf rose. By symmetry we can get the area of one leaf and multiply by 4. The result is $4\cdot\frac{1}{2}\int_0^{\pi/2}(\sin 2\theta)^2\,d\theta = 2\int_0^{\pi/2}\sin^2 2\theta\,d\theta = 2\int_0^{\pi/2}\frac{1-\cos 4\theta}{2}\,d\theta = \theta - \frac{1}{4}\sin 4\theta\Big|_0^{\pi/2} = \frac{\pi}{2}-0 = \frac{\pi}{2}$

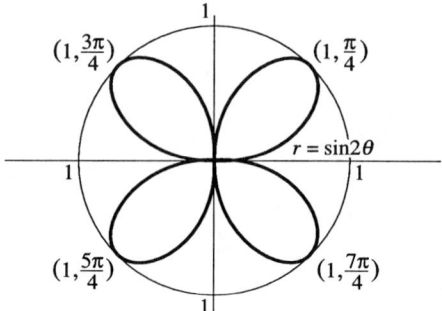

6. $r = \cos 3\theta$. This is a three leaf rose and we can get the area of the right leaf by integrating from $-\frac{\pi}{6}$ to $\frac{\pi}{6}$ and the whole area by multiplying this answer by 3. The result is $3\cdot\frac{1}{2}\int_{-\pi/6}^{\pi/6}(\cos 3\theta)^2\,d\theta = \frac{3}{2}\int_{-\pi/6}^{\pi/6}\cos^2 3\theta\cos\theta\,d\theta = \frac{3}{2}\int_{-\pi/6}^{\pi/6}\frac{1+\cos 6\theta}{2}\,d\theta = \frac{3}{2}\left[\frac{1}{2}\theta+\frac{1}{12}\sin 6\theta\right]_{-\pi/6}^{\pi/6} = \frac{3}{2}\cdot\left[\frac{\pi}{12}+0-\left(-\frac{\pi}{12}\right)-0\right] = \frac{3}{2}\cdot\frac{\pi}{6} = \frac{\pi}{4}.$

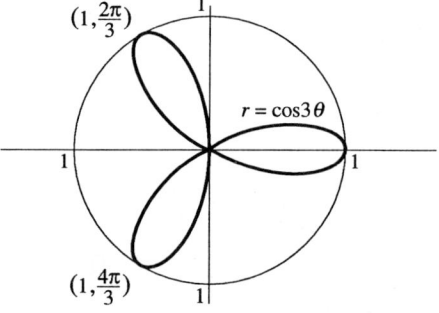

7. $r = 4 + \cos\theta$. Integrating from 0 to 2π, the area is

$$\frac{1}{2}\int_0^{2\pi}(4+\cos\theta)d\theta$$

$$=\frac{1}{2}\int_0^{2\pi}(16+8\cos\theta+\cos^2\theta)d\theta$$

$$=\frac{1}{2}\int_0^{2\pi}\left(16+8\cos\theta+\frac{1+\cos2\theta}{2}\right)d\theta$$

$$=\frac{1}{2}\left[16\theta+8\sin\theta+\frac{1}{2}\theta+\frac{1}{2}\sin2\theta\right]_0^{2\pi}$$

$$=\frac{1}{2}[32\pi+0+\pi+0]=\frac{33}{2}\pi$$

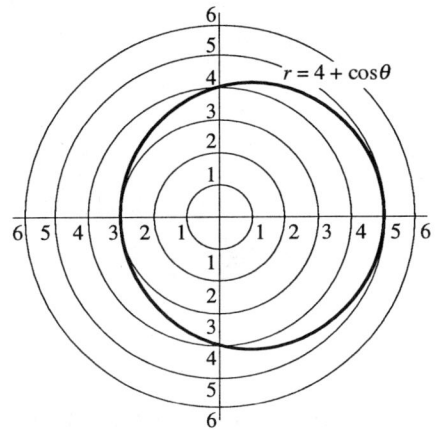

$r = 4 + \cos\theta$

8. $r = 4+3\cos\theta$. By symmetry we can get the total area by doubling the area from $\theta = 0$ to $\theta = \pi$.

$2\cdot\frac{1}{2}\int_0^{\pi}(4+3\cos\theta)^2d\theta = \int_0^{\pi}\left(16+24\cos\theta+9\cos^2\theta\right)d\theta = \int_0^{\pi}\left(16+24\cos\theta+9\left(\frac{1+\cos2\theta}{2}\right)\right)d\theta$

$=\left[16\theta+24\sin\theta+\frac{9}{2}\theta+\frac{9}{4}\sin2\theta\right]_0^{\pi}=16\pi+\frac{9}{2}\pi=\frac{41}{2}\pi.$

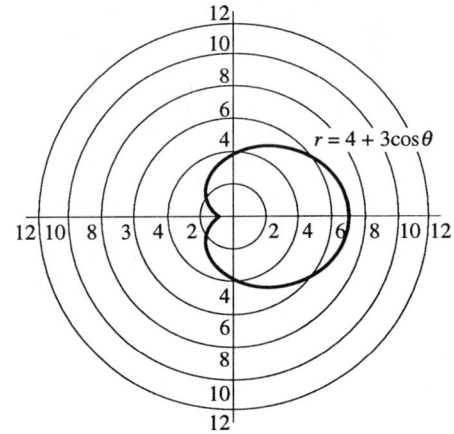

$r = 4 + 3\cos\theta$

9. $r^2 = \sin\theta \Rightarrow r = \pm\sqrt{\sin\theta}$; $r = \sqrt{\sin\theta}$ gives the top half of the curve with $0 \le \theta \le \pi$. The bottom half is $r = -\sqrt{\sin\theta}$. By symmetry we can integrate $\sin\theta$ from 0 to π and double the answer. $2\cdot\frac{1}{2}\int_0^{\pi}(\sqrt{\sin\theta})^2d\theta = \int_0^{\pi}\sin\theta\,d\theta = -\cos\theta|_0^{\pi} = -[(-1)-1] = 2.$

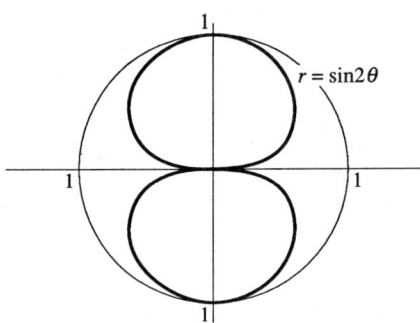

$r = \sin2\theta$

10. $r^2 = 4\cos2\theta$ or $r = \pm\sqrt{4\cos2\theta}$. This curve is undefined when $\cos2\theta < 0$ or $\frac{\pi}{4} < \frac{3\pi}{4}$ and $\frac{5\pi}{4} < \theta < \frac{7\pi}{4}$. By symmetry we can integrate from 0 to $\frac{\pi}{4}$ and multiply the answer by 4. This produces $4\cdot\frac{1}{2}\int_0^{\pi/4}\sqrt{4\cos2\theta}^2d\theta = 2\int_0^{\pi/4}4\cos2\theta\,d\theta = 2\cdot2\sin2\theta|_0^{\pi/4} = 4\cdot1 = 4.$

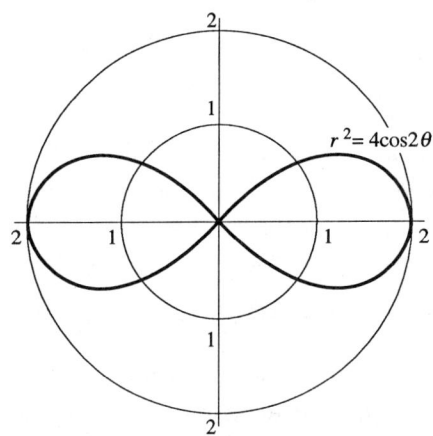

$r^2 = 4\cos2\theta$

11. $\frac{1}{2}\int_0^{\pi/4}(3\cos\theta)^2d\theta = \frac{1}{2}\int_0^{\pi/4}9\cos^2\theta\,d\theta = \frac{9}{2}\int_0^{\pi/4}\frac{1+\cos2\theta}{2}d\theta = \frac{9}{2}\left[\frac{\theta}{2}+\frac{\sin2\theta}{4}\right]_0^{\pi/4} = \frac{9}{3}\left[\frac{\pi}{8}+\frac{1}{4}\right] = \frac{9}{16}[\pi+2] \approx 2.8921.$

12. $\frac{1}{2}\int_0^{\pi/3}(4\cos\theta)^2d\theta = \frac{1}{2}\int_0^{\pi/3}16\cos^2\theta\,d\theta = \frac{1}{2}\int_0^{\pi/3}8(1-\cos2\theta)d\theta = \frac{1}{2}[8\theta-4\sin2\theta]_0^{\pi/3} = \frac{1}{2}\left[\frac{8\pi}{3}-4\cdot\frac{\sqrt{3}}{2}\right] = \frac{4\pi}{3}-\sqrt{3} \approx 2.4567.$

13. $\frac{1}{2}\int_0^{\pi/4}(\cos 2\theta)^2 d\theta = \frac{1}{2}\int_0^{\pi/4}\cos^2\theta\, d\theta =$
$\frac{1}{2}\int_0^{\pi/4}\frac{1+\cos 4\theta}{2}d\theta = \frac{1}{2}\left[\frac{1}{2}\theta + \frac{1}{4}\sin 4\theta\right]_0^{\pi/4} = \frac{1}{2}\left[\frac{\pi}{8}\right] =$
$\frac{\pi}{16} \approx 0.1963.$

14. $\frac{1}{2}\int_0^{\pi/8}(\sin 4\theta)^2 d\theta = \frac{1}{2}\int_0^{\pi/8}\frac{1-\cos 8\theta}{2}\theta\, d\theta = \frac{1}{2}\left[\frac{1}{2}\theta - \frac{\sin 8\theta}{4}\right]_0^{\pi/8} = \frac{\pi}{32} \approx 0.0982.$

15. $\frac{1}{2}\int_0^{\pi/2}\left(e^{2\theta}\right)^2 d\theta = \frac{1}{2}\int_0^{\pi/2}e^{4\theta}d\theta = \frac{1}{2}\left[\frac{1}{4}e^{4\theta}\right]_0^{\pi/2} =$
$\frac{1}{8}\left[e^{2\pi}-1\right] \approx 66.8115$

16. $\frac{1}{2}\int_0^{\pi/6}(5\theta)^2 d\theta = \frac{1}{2}\int_0^{\pi/6}25\theta^2 d\theta = \frac{25}{2}\cdot\frac{1}{3}\theta^3\Big|_0^{\pi/6} =$
$\frac{25}{6}\left(\frac{\pi}{6}\right)^3 = \frac{25\pi^3}{1296} \approx 0.5981$

17. To get the area A of the region enclosed by one loop of $r = 4\cos 2\theta$ we can integrate from $-\frac{\pi}{4}$ to $\frac{\pi}{4}$ with the result $A = \frac{1}{2}\int_{-\pi/4}^{\pi/4}(4\cos 2\theta)^2 d\theta =$
$8\int_{-\pi/4}^{\pi/4}\frac{1+\cos 4\theta}{2}d\theta = 8\left[\frac{1}{2}\theta + \frac{\sin 4\theta}{8}\right]_{-\pi/4}^{\pi/4} = 8\left[\frac{\pi}{8} + \frac{\pi}{8}\right] = 2\pi.$

18. $r^2 = 4\cos 2\theta;\ r = \sqrt{4\cos 2\theta}$. The area of the region enclosed by one loop of this curve can be obtained by integrating from $-\frac{\pi}{4}$ to $\frac{\pi}{4}$. If A is the area, we obtain $\frac{1}{2}\int_{-\pi/4}^{\pi/4}\sqrt{4\cos 2\theta}^2 d\theta =$
$2\int_{-\pi/4}^{\pi/4}\cos 2\theta\, d\theta = \sin 2\theta\big|_{-\pi/4}^{\pi/4} = 1+1 = 2.$

19. $r = 2\sin 3\theta$, and to get the area of the region enclosed by one loop if we integrate from 0 to $\frac{\pi}{3}$. Thus, if A is the area, we have $A =$
$\frac{1}{2}\int_0^{\pi/3}(2\sin 3\theta)^2 d\theta = \int_0^{\pi/3}2\sin 3\theta\, d\theta = \int_0^{\pi/3}(1-\cos 6\theta)d\theta = \left[\theta - \frac{\sin 6\theta}{6}\right]_0^{\pi/3} = \frac{\pi}{3}.$

20. $r = \sin 6\theta$. To get the area of the region enclosed by one loop we integrate from 0 to $\frac{\pi}{6}$. The result is that the area A is $A = \frac{1}{2}\int_0^{\pi/6}(\sin 6\theta)^2 d\theta =$
$\frac{1}{2}\int_0^{\pi/6}\frac{1-\cos 6\theta}{2}d\theta = \left[\frac{1}{4}\theta - \frac{\sin 6\theta}{24}\right]_0^{\pi/6} = \frac{\pi}{24}.$

21. Looking at the figure 29.23S we can determine that the graphs intersect at $\left(1, -\frac{\pi}{2}\right)$ and $\left(1, \frac{\pi}{2}\right)$ so we need to integrate as follows: $\frac{1}{2}\int_{-\pi/2}^{\pi/2}\big((1+\cos\theta)^2 - 1^2\big)d\theta = \frac{1}{2}\int_{-\pi/2}^{\pi/2}\left(1 + 2\cos\theta + \cos^2\theta - 1\right)d\theta = \frac{1}{2}\int_{-\pi/2}^{\pi/2}\left(2\cos\theta + \frac{1+\cos 2\theta}{2}\right)d\theta = \left[\sin\theta + \frac{1}{4}\theta + \frac{1}{8}\sin 2\theta\right]_{-\pi/2}^{\pi/2} = \left(1+\frac{\pi}{8}+0\right) - \left(-1-\frac{\pi}{8}+0\right) = 2 + \frac{\pi}{4} \approx 2.7854.$

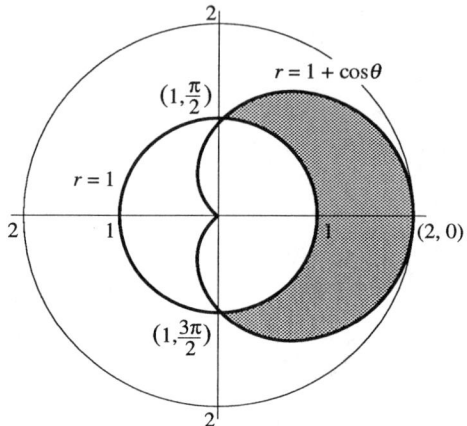

22. For this area we integrate from $\frac{\pi}{6}$ to $\frac{5\pi}{6}$. Thus,
$A = \frac{1}{2}\int_{\pi/6}^{5\pi/6}\big((2\sin\theta)^2 - 1^2\big)d\theta = \frac{1}{2}\int_{\pi/6}^{5\pi/6}\left(4\sin^2\theta - 1\right)d\theta = \frac{1}{2}\int_{\pi/6}^{5\pi/6}\left(2(1-\cos 2\theta) - 1\right)d\theta = \frac{1}{2}\int_{\pi/6}^{5\pi/6}(1-\cos 2\theta)d\theta = \left[\frac{1}{2}\theta - \frac{\sin 2\theta}{4}\right]_{\pi/6}^{5\pi/6} = \frac{\pi}{12} + \frac{\sqrt{3}}{8} - \frac{\pi}{12} + \frac{\sqrt{3}}{8} = \frac{\pi}{3} + \frac{\sqrt{3}}{4} \approx 1.4802.$

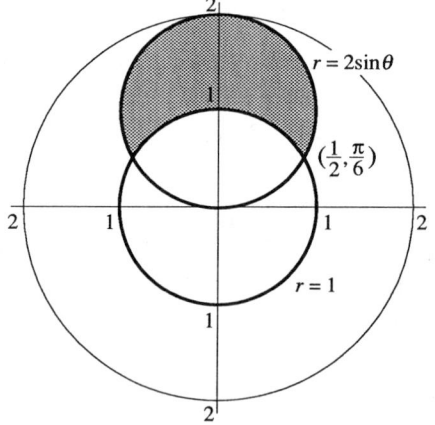

23. First we need to solve for θ. Using direct substitution we get $1 - \sin\theta = 2\cos\theta$. Squaring both sides we have $1 - 2\sin\theta + \sin^2\theta = 4\cos^2\theta$. Since $\cos^2\theta = 1 - \sin^2\theta$ we have $1 - 2\sin\theta + \sin^2\theta = 4 - 4\sin^2\theta$ or $0 = 3 + 2\sin\theta - 5\sin^2\theta$. This factors into $(3 + 5\sin\theta)(1 - \sin\theta) = 0$. Hence $\sin\theta = 1$

or $\sin\theta = -\frac{3}{5}$, and so $\theta = \frac{\pi}{2}$, $\frac{3\pi}{2}$, 3.78509, or 5.63968. Checking we see that $\frac{\pi}{2}$ and 5.63968 are the answers we need. Let $5.63968 = T$. Then,

$$\frac{1}{2}\int_{\pi/6}^{T}\left((1-\sin\theta)^2 - (2\cos\theta)^2\right)d\theta$$

$$= \frac{1}{2}\int_{\pi/2}^{T}\left(1 - 2\sin\theta + \sin^2\theta - 4\cos^2\theta\right)d\theta$$

$$= \frac{1}{2}\int_{\pi/2}^{T}\left(1 - 2\sin\theta + \sin^2\theta - 4 + 4\sin^2\theta\right)d\theta$$

$$= \frac{1}{2}\int_{\pi/2}^{T}\left[-3 - 2\sin\theta + 5\left(\frac{1-\cos 2\theta}{2}\right)\right]d\theta$$

$$= \frac{1}{2}\left[-3\theta + 2\cos\theta + \frac{5}{2}\theta - \frac{5\sin 2\theta}{4}\right]_{\pi/2}^{T}$$

$$= \left[-\frac{1}{4}\theta + \cos\theta - \frac{5\sin 2\theta}{8}\right]_{\pi/2}^{T}$$

$$\approx -0.009921 + 0.392699 = 0.382778$$

So, the desired area is about 0.3828.

24. These two curves intersect at $\theta = \frac{\pi}{6}$, $\frac{5\pi}{6}$, $\frac{7\pi}{6}$, and $\frac{11\pi}{6}$. By symmetry we can integrate from $\frac{\pi}{6}$ to $\frac{5\pi}{6}$ and double the answer. This gives $2 \cdot \frac{1}{2}\int_{\pi/6}^{5\pi/6}\left[\left(\sqrt{8\sin\theta}\right)^2 - 2^2\right]d\theta = \int_{\pi/6}^{5\pi/6}(8\sin\theta - 4)d\theta$
$= [-8\cos\theta - 4\theta]_{\pi/6}^{5\pi/6} = -8\left(-\frac{\sqrt{3}}{2}\right) - \frac{20\pi}{6} + 8\frac{\sqrt{3}}{2} + \frac{4\pi}{6} = 8\sqrt{3} - \frac{8\pi}{3} \approx 5.4788$.

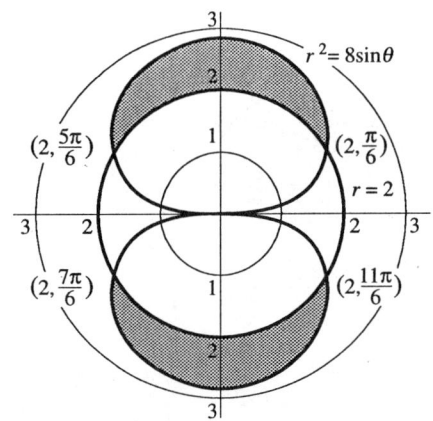

25. $\frac{1}{2}\int_{0}^{\pi/2}\left(\cos^2\theta - (1-\sin\theta)^2\right)d\theta = \frac{1}{2}\int_{0}^{\pi/2}\left(\cos^2\theta - 1 + 2\sin\theta - \sin^2\theta\right)d\theta = \int_{0}^{\pi/2}(\cos 2\theta - 1 + 2\sin\theta)d\theta$
$= \frac{1}{2}\left[-\theta + \frac{\sin 2\theta}{2} - 2\cos\theta\right]_{0}^{\pi/2} = \frac{1}{2}\left[-\frac{\pi}{2} + 0 - 0 - 0 - 0 + 2\right] = -\frac{\pi}{4} + 1 \approx 0.2146$

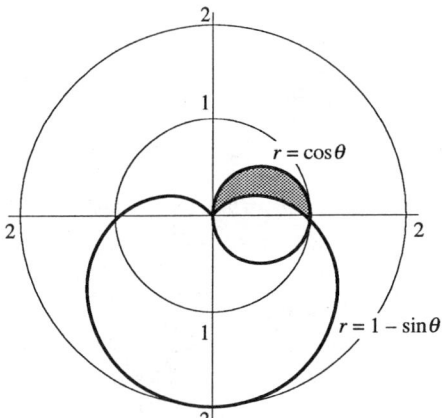

26. The area is $\frac{1}{2}\int_{0}^{\pi/3}\sin^2\theta\,d\theta + \frac{1}{2}\int_{\pi/3}^{\pi/2}3\cos^2\theta\,d\theta = \frac{1}{2}\int_{0}^{\pi/3}\left(\frac{1-\cos 2\theta}{2}\right)d\theta + \frac{1}{2}\int_{\pi/3}^{\pi/2}3\left(\frac{1+\cos 2\theta}{2}\right)d\theta = \left[\frac{1}{4}\theta - \frac{1}{8}\sin 2\theta\right]_{0}^{\pi/3} + \left[\frac{3}{4}\theta + \frac{3}{8}\sin 2\theta\right]_{\pi/3}^{\pi/2} = \left(\frac{\pi}{12} - \frac{\sqrt{3}}{16}\right) + \left(\frac{3\pi}{8} - \frac{\pi}{4} - \frac{3\sqrt{3}}{16}\right) = \left(\frac{5\pi}{24} - \frac{\sqrt{3}}{4}\right) \approx 0.2215$.

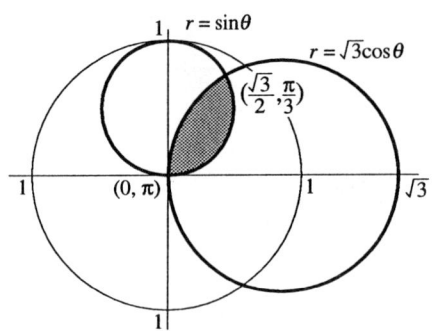

CHAPTER 29 REVIEW

1. $x = t^2$, $y = t^3 + t$.

(a) $\frac{dy}{dx} = \frac{dy/dt}{dx/dt} = \frac{3t^2+1}{2t}$; $\frac{d^2y}{dx^2} = \frac{2t(6t)-\left(3gt^2+1\right)2}{(2t)^3} =$

$\frac{12t^2-6t^2-2}{(2t)^3} = \frac{6t^2-2}{2\left(4t^3\right)} = \frac{3t^2-1}{4t^3}$

(b) At $t = 1$, $\frac{dy}{dx} = \frac{3+1}{2} = 2$.

2. **(a)** $\frac{dy}{dx} = \frac{dy/dt}{dx/dt} = \frac{-2t^{-2}}{-1t^{-2}} = 2$. Since $\frac{dy}{dx}$ is a

constant, $\frac{d^2y}{dx^2} = 0$,

(b) At $t = 1$, $\frac{dy}{dx} = 2$.

3. **(a)** $\frac{dy}{dx} = \frac{\cos t}{2\cos t(-\sin t)} = -\frac{1}{2}\csc t$, and the second

derivative is $\frac{d^2y}{dx^2} = \frac{\frac{1}{2}\csc t\cot t}{-2\cos t\sin t} = -\frac{1}{4}\csc^3 t$

(b) At $t = 1$, $\frac{dy}{dx} = -\frac{1}{2}\csc 1 \approx -0.5942$

4. **(a)** $\frac{dy}{dx} = \frac{-e^{-t}}{\frac{1}{2}t^{-1/2}} = -2\sqrt{t}e^{-t}$; $\frac{d^2y}{dx^2} =$

$\frac{2\sqrt{t}e^{-t} - t^{-1/2}e^{-t}}{\frac{1}{2}t^{-1/2}} = 4te^{-t} - 2e^{-t} = (4t-2)e^{-t}$

(b) At $t = 1$, $\frac{dy}{dx} = -2e^{-1} \approx -0.7358$

5. **(a)** $\frac{dy}{dx} = \frac{-2\sin\theta\sin\theta + (3+2\cos\theta)\cos\theta}{-2\sin\theta\cos\theta - (3+2\cos\theta)\sin\theta}$

$= \frac{-2\sin^2\theta + 3\cos\theta + 2\cos^2\theta}{-2\sin\theta\cos\theta - 3\sin\theta - 2\cos\theta\sin\theta}$

$= \frac{2\left(\cos^2\theta - \sin^2\theta\right) + 3\cos\theta}{-4\sin\theta\cos\theta - 3\sin\theta}$

$= \frac{2\cos 2\theta + 3\cos\theta}{-2\sin 2\theta - 3\sin\theta}$

$= -\frac{2\cos 2\theta + 3\cos\theta}{2\sin 2\theta + 3\sin\theta}$

(b) If $\theta = \frac{\pi}{4}$, then

$\frac{dy}{dx} = -\frac{2\cos\frac{\pi}{2} + 3\cos\frac{\pi}{4}}{2\sin\frac{\pi}{2} + 3\sin\frac{\pi}{4}}$

$= -\frac{2\cdot 0 + \frac{3\sqrt{2}}{2}}{2\cdot 1 + \frac{3\sqrt{2}}{2}} = -\frac{3\sqrt{2}}{4 + 3\sqrt{2}} \approx -0.5147$.

6. **(a)**

$\frac{dy}{dx} = \frac{-12\sin 2\theta\sin\theta + 6\cos 2\theta\cos\theta}{-12\sin 2\theta\cos\theta - 6\cos 2\theta\sin\theta}$

$= \frac{2\sin 2\theta\sin\theta - \cos 2\theta\cos\theta}{2\sin 2\theta\cos\theta + \cos 2\theta\sin\theta}$

$= \frac{2(2\sin\theta\cos\theta)\sin\theta - \left(\cos^2\theta - \sin^2\theta\right)\cos\theta}{2(\sin\theta\cos\theta)\cos\theta + \left(\cos^2\theta - \sin^2\theta\right)\sin\theta}$

$= \frac{\cos\theta\left(4\sin^2\theta - \cos^2\theta + \sin^2\theta\right)}{\sin\theta\left(4\cos^2\theta + \cos^2\theta - \sin^2\theta\right)}$

$= \frac{\cos\theta\left(5\sin^2\theta - \cos^2\theta\right)}{\sin\theta\left(5\cos^2\theta - \sin^2\theta\right)}$

(b) At $\theta = \frac{\pi}{4}$, $\frac{dy}{dx}$ is $\frac{\cos\frac{\pi}{4}\left(5\sin^2\frac{\pi}{4} - \cos^2\frac{\pi}{4}\right)}{\sin\frac{\pi}{4}\left(5\cos^2\frac{\pi}{4} - \sin^2\frac{\pi}{4}\right)} = 1$

7. $r^2 = \sin\theta$; $r = (\sin\theta)^{1/2}$; $r' = \frac{1}{2}(\sin\theta)^{-1/2}\cos\theta$

(a) $\frac{dy}{dx} = \frac{\frac{1}{2}\sin^{-1/2}\theta\cos\theta\sin\theta + \sin^{1/2}\theta\cos\theta}{\frac{1}{2}\sin^{-1/2}\theta\cos\theta\cos\theta - \sin^{1/2}\theta\sin\theta}$

$= \frac{\cos\theta\sin\theta + 2\sin\theta\cos\theta}{\cos^2\theta - 2\sin^2\theta}$

$= \frac{3\cos\theta\sin\theta}{\cos^2\theta - 2\sin^2\theta}$

(b) At $\theta = \frac{\pi}{4}$ we have $\frac{dy}{dx} = \frac{3\cdot\frac{\sqrt{2}}{2}\frac{\sqrt{2}}{2}}{\left(\frac{\sqrt{2}}{2}\right)^2 - 2\left(\frac{\sqrt{2}}{2}\right)^2} =$

$\frac{\frac{3}{2}}{-\frac{1}{2}} = -3$

8. **(a)** If $r = \frac{8}{3+\cos\theta} = 8(3+\cos\theta)^{-1}$, then $r' =$

$-8(3+\cos\theta)^{-2}(-\sin\theta) = \frac{8\sin\theta}{(3+\cos\theta)^2}$. So,

$\frac{dy}{dx} = \frac{\frac{8\sin\theta}{(3+\cos\theta)^2}\sin\theta + \frac{8}{3+\cos\theta}\cos\theta}{\frac{8\sin\theta}{(3+\cos\theta)^2}\cos\theta - \frac{8}{3+\cos\theta}\sin\theta}$

$= \frac{8\sin^2\theta + 8\cos\theta(3+\cos\theta)}{8\sin\theta\cos\theta - 8\sin\theta(3+\cos\theta)}$

$= \frac{\sin^2\theta + 3\cos\theta + \cos^2\theta}{\sin\theta\cos\theta - 3\sin\theta - \sin\theta\cos\theta}$

$= \frac{1 + 3\cos\theta}{-3\sin\theta}$.

(b) At $\theta = \frac{\pi}{4}$, then $\frac{dy}{dx} = \frac{1 + 3\cdot\frac{\sqrt{2}}{2}}{-3\cdot\frac{\sqrt{2}}{2}} \approx -1.4714$.

9. $\frac{dx}{dt} = 2t$; $\frac{dy}{dt} = 3t^2$ and $L = \int_0^2 \sqrt{(2t)^2 + \left(3t^2\right)^2}dt$

$= \int_0^2 \sqrt{4t^2 + 9t^4}dt = \int_0^2 t\sqrt{4 + 9t^2}dt$. Let $u = 4+9t^2$, and then $du = 18t\,dt$, so we get $\frac{1}{18}\int u^{1/2}du$

$= \frac{1}{16}\cdot\frac{2}{5}u^{3/2} = \frac{1}{27}\left(4+9t^2\right)^{3/2}\Big|_0^2 = \frac{1}{27}\left(40^{3/2} - 4^{3/2}\right) \approx 9.0734$.

10. $\frac{dx}{dt} = 2t$; $\frac{dy}{dt} = 2$, Thus, $L = \int_0^3 \sqrt{(2t)^2 + 2^2}dt = \int_0^3 \sqrt{4t^2 + 4}\,dt = 2\int_0^3 \sqrt{t^2 + 1}\,dt =$

$2\left[\frac{t}{2}\sqrt{t^2+1}+\frac{1}{2}\ln\left|t+\sqrt{t^2+1}\right|\right]\Big|_0^3 =$

$\left(3\sqrt{10}+\ln\left|3+\sqrt{10}\right|\right)\approx 11.3053.$

11. $\frac{dx}{dt}=4\cos t;\ \frac{dy}{dt}=(-4\sin t)$, and as a result, $L=\int_0^{\pi/2}\sqrt{(4\cos t)^2+(-4\sin t)^2}dt=\int_0^{\pi/2}4\,dt=4t\big|_0^{\pi/2}=2\pi.$

12. $r=e^{2\theta};\ r'=2e^{2\theta}$, then $L=$
$\int_0^\pi\sqrt{\left(e^{2\theta}\right)^2+\left(2e^{2\theta}\right)}d\theta=\int_0^\pi\sqrt{e^{4\theta}+4e^{4\theta}}d\theta=$
$\sqrt{5}\int_0^\pi e^{2\theta}=\frac{\sqrt{5}}{2}e^{2\theta}\Big|_0^\pi=\frac{\sqrt{5}}{2}(e^{2\pi}-1)\approx 597.5798.$

13. Since $r=\cos^2\left(\frac{\theta}{2}\right)$, we have $r'=$
$2\cos\frac{\theta}{2}\left(-\frac{1}{2}\sin\frac{\theta}{2}\right)=-\cos\frac{\theta}{2}\sin\frac{\theta}{2}$. Thus,
$L=\int_0^\pi\sqrt{\cos^2\frac{\theta}{2}+\cos^2\frac{\theta}{2}\sin^2\frac{\theta}{2}}d\theta=\int_0^\pi\cos\frac{\theta}{2}d\theta$
$=2\sin\frac{\theta}{2}\Big|_0^\pi=2\cdot\sin\frac{\pi}{2}=2.$

14. $r=1-\cos\theta$ so $r'=\sin\theta$
$L=\int_{-\pi}^0\sqrt{(1-\cos\theta)^2+\sin^2\theta}\,d\theta$
$=\int_{-\pi}^0\sqrt{1-2\cos\theta+\cos^2\theta+\sin^2\theta}\,d\theta$
$=\int_{-\pi}^0\sqrt{2-2\cos\theta}=\int_{-\pi}^0 2\sqrt{\frac{1-\cos\theta}{2}}d\theta$
$=2\int_{-\pi}^0\sin\frac{\theta}{2}d\theta=-4\cos\frac{\theta}{2}\Big|_{-\pi}^0=4$

15. $\frac{dx}{dt}=2t;\ \frac{dy}{dt}=1$
$S=2\pi\int_0^2 t\sqrt{(2t)^2+1^2}dt$
$=2\pi\int_0^2 t\sqrt{4t^2+1}\,dt$

Let $u=4t^2+1$, and then $du=8t\,dt$ or $\frac{1}{8}du=t\,dt$. Substituting we get $\frac{1}{4}\pi\int\sqrt{u}\,du=\frac{\pi}{4}\cdot\frac{2}{3}u^{3/2}=$
$\frac{\pi}{6}\left(4t^2+1\right)^{3/2}\Big|_0^2=\frac{\pi}{6}(17^{3/2}-1)\approx 36.1769.$

16. $\frac{dx}{dt}=2t$ and $\frac{dy}{dt}=(1-t^2)$.
$S=2\pi\int_0^1\left(t-\frac{t^3}{3}\right)\sqrt{(2t)^2+\left(1-t^2\right)^2}dt$
$=2\pi\int_0^1\left(t-\frac{t^3}{3}\right)\sqrt{4t^2+1-2t^2+t^4}dt$
$=2\pi\int_0^1\left(t-\frac{t^3}{3}\right)\sqrt{t^4+2t^2+1}dt$
$=2\pi\int_0^1\left(t-\frac{t^3}{3}\right)(t^2+1)dt$
$=2\pi\int_0^1\left(t^3-\frac{t^3}{3}-\frac{t^5}{3}+t\right)dt$
$=2\pi\left[\frac{2}{3}\cdot\frac{1}{4}t^4-\frac{1}{3}\frac{1}{6}t^6+\frac{t^2}{2}\right]_0^1$
$=2\pi\left(\frac{1}{6}-\frac{1}{18}+\frac{1}{2}\right)$
$=2\pi\left(\frac{3-1+9}{18}\right)=\frac{11\pi}{9}\approx 3.8397$

17. Here $\frac{dx}{dt}=2t$ and $\frac{dy}{dt}=3t^2$, so the surface is
$S=2\pi\int_0^2 t^2\sqrt{(2t)^2+(3+2)^2}dt=2\pi\int_0^2 t^2\times$
$\sqrt{4t^2+9t^4}dt=2\pi\int_0^2 t^3\sqrt{4+9t^2}dt$. Let $u=4+9t^2$, then $t^2=\frac{u-4}{9}$ and $du=18t\,dt$ or $\frac{1}{18}du=t\,dt$. Substituting, we get $2\pi\cdot\frac{1}{18}\int\left(\frac{u-4}{9}\right)u^{1/2}du=$
$\frac{\pi}{81}\int(u^{3/2}-4u^{1/2})du=\frac{\pi}{81}\left(\frac{2}{5}u^{5/2}-4\frac{2}{3}u^{3/2}\right)=$
$\frac{\pi}{81}\left[\frac{2}{5}(4+9t^2)^{5/2}-\frac{8}{3}(u+9t^2)^{3/2}\right]_0^2=\frac{\pi}{81}\left[\frac{2}{5}40^{5/2}-\frac{8}{3}40^{3/2}-\frac{64}{5}+\frac{64}{3}\right]\approx 41.7485\pi\approx 131.1568.$

18. Since $x=e^t\sin t$, then $dx=\left(e^t+\cos t+e^t\sin t\right)dt$ and since $y=e^t\cos$, then $dy=\left(e^t\cos t-e^t\sin t\right)dt$. Thus, the desired surface area is given by
$S=2\pi\int_0^{\pi/2}e^t\sin t\sqrt{\left(e^t\cos t+e^t\sin t\right)^2+\left(e^t\cos t-e^t\sin t\right)^2}dt$
$=2\pi\int_0^{\pi/2}e^t\sin t\sqrt{2e^{2t}\cos^2 t+2e^{2t}\sin^2 t}\,dt$
$=2\pi\int_0^{\pi/2}e^t\sin t\sqrt{2}e^t dt=2\sqrt{2}\pi\int_0^{\pi/2}e^{2t}\sin t$

By formula #87, Appendix C, this integral is $2\sqrt{2}\pi\left(\frac{e^{2t}}{2^2+1^2}(2\sin t-\cos t)\right)\Big|_0^{\pi/2}=2\sqrt{2}\pi\left(\frac{e^\pi}{5}\cdot 2+\frac{1}{5}\right)=\frac{2\sqrt{2}\pi}{5}\left(2e^\pi+1\right)\approx 84.0263.$

19. $r=6\sin\theta$ and $r'=6\cos\theta;\ y=6\sin^2\theta$. Hence, we find that the desired surface area is
$S=2\pi\int_0^\pi\sqrt{36\sin^2\theta+36\cos^2\theta}\left(6\sin^2 6\right)dt=$

$2\pi\int_0^\pi 36\sin^2\theta\,d\theta=36\pi\int_0^\pi(1-\cos 2\theta)d\theta=$
$36\pi\left[\theta-\frac{\sin 2\theta}{2}\right]_0^\pi=36\pi^2\approx 355.3058.$

20. Here $r = 4 + 4\cos\theta$, $r' = -4\sin\theta$, and $y = 4\sin\theta + 4\sin\theta\cos\theta$. Hence, the desired surface area is

$$S = 2\pi \int_0^\pi (4\sin\theta + 4\sin\theta\cos\theta)$$
$$\times \sqrt{(4+4\cos\theta)^2 + (4\sin\theta)^2}\,d\theta$$
$$= 2\pi \int_0^\pi (4\sin\theta + 4\sin\theta\cos\theta)$$
$$\times 4\sqrt{1 + 2\cos\theta + \cos^2\theta + \sin^2\theta}\,d\theta$$
$$= 32\pi \int_0^\pi (\sin\theta + \sin\theta\cos\theta)\sqrt{2 + 2\cos\theta}\,d\theta$$
$$= 32\sqrt{2}\pi \int_0^\pi (1+\cos\theta)\sqrt{1+\cos\theta}\sin\theta\,d\theta$$
$$= 32\sqrt{2}\pi \int_0^\pi (1+\cos\theta)^{3/2}\sin\theta\,d\theta$$

Let $u = 1 + \cos\theta$ and $du = -\sin\theta$. Substituting, we get $-32\sqrt{2}\pi\left[\int u^{3/2}du\right] = -32\sqrt{2}\pi\left[\frac{2}{5}u^{5/2}\right] = -32\sqrt{2}\pi\left[\frac{2}{5}(1+\cos\theta)^{5/2}\right]_0^\pi = \frac{-64\sqrt{2}}{5}(0 - 2^{5/2}) = \frac{64\sqrt{2}\pi}{5}(4\sqrt{2}) = \frac{2^9\pi}{5} \approx 102.4\pi \approx 321.6991$.

21. By direct substitution, $4 = 4 + 4\sin\theta$; $0 = \sin\theta$ so $\theta = 0$ or π; $(4, 0)$, $(4, \pi)$.

22. By direct substitution, $3 = 6\sin\theta$; $\sin\theta = \frac{1}{2}$; $\theta = \frac{\pi}{6}$, $\frac{5\pi}{6}$; $\left(3, \frac{\pi}{6}\right)$, $\left(3, \frac{5\pi}{6}\right)$.

23. By direct substitution, $4 + 4\cos\theta = 2$; $4\cos\theta = -2$; $\cos\theta = -\frac{1}{2}$; $\theta = \frac{2\pi}{3}$, $\frac{4\pi}{3}$. $\left(2, \frac{2\pi}{3}\right)$, $\left(2, \frac{4\pi}{3}\right)$.

24. By direct substitution, $2\sin 2\theta = 1$; $\sin 2\theta = \frac{1}{2}$, and so $2\theta = \frac{\pi}{6}$, $\frac{5\pi}{6}$, $\frac{13\pi}{6}$, and $\frac{17\pi}{6}$. Thus, $\theta = \frac{\pi}{12}$, $\frac{5\pi}{12}$, $\frac{13\pi}{12}$, and $\frac{17\pi}{12}$. Graphing yields symmetric points at $\frac{-\pi}{12}$, $\frac{-5\pi}{12}$, $\frac{-13\pi}{12}$, and $\frac{-17\pi}{12}$. The resulting points are $\left(1, \frac{\pi}{12}\right)$, $\left(1, \frac{5\pi}{12}\right)$, etc.

25. The area is $A = \frac{1}{2}\int_0^{2\pi}(1+\cos\theta)^2 d\theta = \frac{1}{2}\int_0^{2\pi}\left(1 + 2\cos\theta + \cos^2\theta\right)d\theta = \frac{1}{2}\int_0^{2\pi}\left(1 + 2\cos\theta + \frac{1+\cos\theta}{2}\right)d\theta = \frac{1}{2}\left[\frac{3}{2}\theta + 2\sin\theta + \frac{\sin 2\theta}{4}\right]_0^{3\pi} = \frac{3\pi}{2}$.

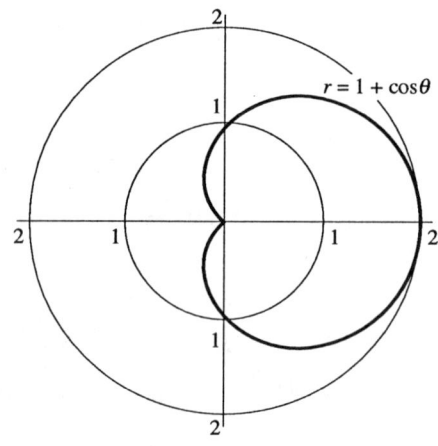

$r = 1 + \cos\theta$

26. The area is $A = \frac{1}{2}\int_0^{2\pi}(4+4\sin\theta)^2 d\theta = 8\int_0^{2\pi}(1 + 2\sin\theta+\sin^2\theta)d\theta = 8\int_0^{2\pi}\left(1+2\sin\theta+\frac{1-\cos\theta}{2}\right)d\theta = 8\left[\frac{3}{2}\theta - 2\cos\theta - \frac{\sin 2\theta}{4}\right]_0^{2\pi} = 24\pi - 16 + 16 = 24\pi$.

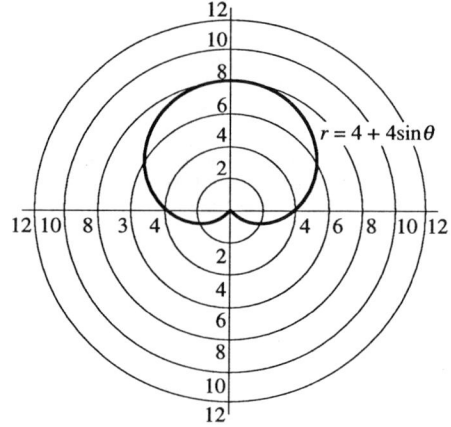

$r = 4 + 4\sin\theta$

27. $A = \frac{1}{2}\int_0^\pi \left((4+4\sin\theta)^2 - 4^2\right)d\theta = \frac{1}{2}\int_0^\pi \left(16 + 32\sin\theta+16\sin^2\theta - 16\right)d\theta = \frac{1}{2}\int_0^\pi \left[32\sin\theta + 8(1 - \cos 2\theta)\right]d\theta = \frac{1}{2}[8\theta - 32\cos\theta - 4\sin 2\theta]_0^\pi = 4\pi + 32 \approx 44.5664$.

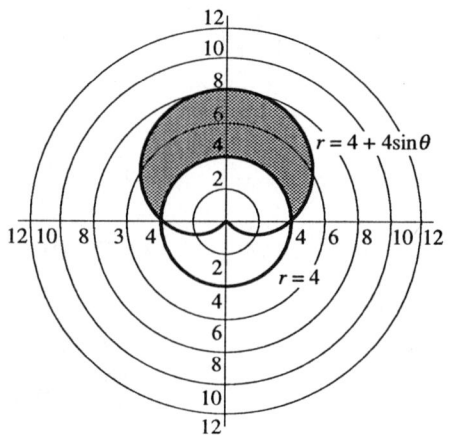

$r = 4 + 4\sin\theta$

$r = 4$

28. The area is $A = \frac{1}{2}\int_{\pi/6}^{5\pi/6}\left[(6\sin\theta)^2 - 3^2\right]d\theta = $
$\frac{1}{2}\int_{\pi/6}^{5\pi/6}\left(36\sin^2\theta - 9\right)d\theta = \frac{1}{2}\int_{\pi/6}^{5\pi/6}\left[18(1-\cos 2\theta)\right.$
$\left. - 9\right]d\theta = \frac{1}{2}\left[9\theta - 9\sin 2\theta\right]_{\pi/6}^{5\pi/6} = \frac{9}{2}\left[\frac{5\pi}{6} + \frac{\sqrt{3}}{2} - \frac{\pi}{6} + \right.$
$\left.\frac{\sqrt{3}}{2}\right] = \frac{9}{2}\left[\frac{2\pi}{3} + \sqrt{3}\right] = 3\pi + \frac{9\sqrt{3}}{2} \approx 17.2190.$

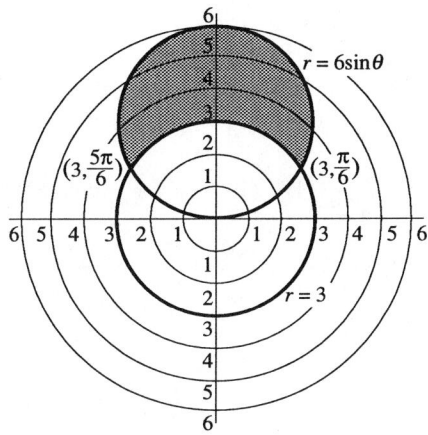

29. $\frac{1}{2}\int_{\pi/2}^{\pi}\left[(\sin\theta)^2 - (1+\cos\theta)^2\right]d\theta = \frac{1}{2}\int_{\pi/2}^{\pi}\left[\sin^2\theta - \right.$
$\left.(1 + 2\cos\theta + \cos^2\theta)\right]d\theta = \frac{1}{2}\int_{\pi/2}^{\pi}\left[\sin^2\theta - \cos^2\theta - \right.$
$\left. 1 - 2\cos\theta\right]d\theta = \frac{1}{2}\int_{\pi/2}^{\pi}[-\cos 2\theta - 1 - 2\cos\theta]d\theta.$
Evaluating this integral, we obtain $\frac{1}{2}\left[-\frac{\sin 2\theta}{2} - 0 - \right.$
$\left. 2\sin\theta\right]_{\pi/2}^{\pi} = \frac{1}{2}\left[-\pi + \frac{\pi}{2} + 2\right] = 1 - \frac{\pi}{4} \approx 0.2146.$

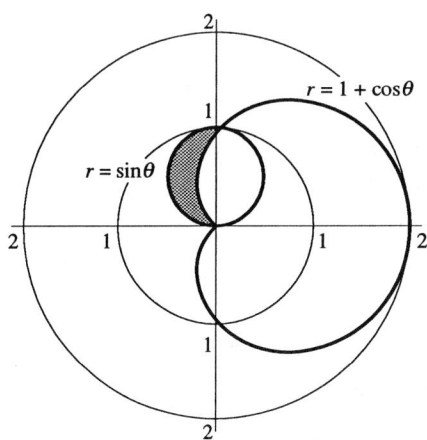

30. In problem #24 we solved for θ. We will use the first pair and multiply the answer by 4. $A = 4 \cdot \frac{1}{2}\int_{\pi/12}^{5\pi/12}\left[(2\sin 2\theta)^2 - 1^2\right]d\theta = 2\int_{\pi/12}^{5\pi/12}\left[4\sin^2 2\theta - \right.$
$\left. 1\right]d\theta = 2\int_{\pi/12}^{5\pi/12}[2(1 - \cos 4\theta) - 1]d\theta = 2\left[\theta - \right.$
$\left.\frac{\sin 4\theta}{2}\right]_{\pi/12}^{5\pi/12} = 2\left[\frac{5\pi}{12} + \frac{\sqrt{3}}{4} - \frac{\pi}{4} + \frac{\sqrt{3}}{4}\right] = \frac{2\pi}{3} + \sqrt{3} \approx $
$3.8264.$

CHAPTER 29 TEST

1. $\frac{dy}{dx} = \frac{dy/dt}{dx/dt} = \frac{2t+5}{3t^2}$; $\frac{d^2y}{dx^2} = \frac{\frac{d}{dt}\left(\frac{dy}{dx}\right)}{dx/dt} = $
$\frac{\left[(3t^2)(2) - (2t+5)(6t)\right]/(3t^2)^2}{3t^2} = \frac{-6t^2 - 30t}{(3t^2)^3} = \frac{-2t-10}{9t^5}.$

2. The first derivative is $\frac{dy}{dx} = \frac{dy/dt}{dx/dt} = \frac{-2\cos t\sin t}{\cos t} = $
$-2\sin t$. When $t = \frac{\pi}{4}$, then $\frac{dy}{dx} = -\sqrt{2}$, so the slope is $-\sqrt{2}$. The tangent line is $y - \frac{1}{2} = \sqrt{2}\left(x - \frac{\sqrt{2}}{2}\right)$ or $y - \sqrt{2}x + \frac{3}{2}$.

3. We will use $\frac{dy}{dx} = \frac{r'\sin\theta + r\cos\theta}{r'\cos\theta - r\sin\theta}$. Differentiating $r = 4 - 2\sin\theta$ we obtain $r' = \frac{dr}{d\theta} = -2\cos\theta$ and so
$\frac{dy}{dx} = \frac{-2\cos\theta\sin\theta + (4 - 2\sin\theta)\cos\theta}{-2\cos\theta\cos\theta - (4 - 2\sin\theta)\sin\theta}$
$= \frac{-4\sin\theta\cos\theta + 4\cos\theta}{-2\cos^2\theta - 4\sin\theta + 2\sin^2\theta}$
$= \frac{4\cos\theta - 4\sin\theta\cos\theta}{-2(1 - \sin^2\theta) - 4\sin\theta + 2\sin^2\theta}$

$= \frac{4\cos\theta(1 - \sin\theta)}{-2 + 2\sin^2\theta - 4\sin\theta + 2\sin^2\theta}$
$= \frac{4\cos\theta(1 - \sin\theta)}{4\sin^2\theta - 4\sin\theta - 2}$
$= \frac{2\cos\theta(1 - \sin\theta)}{2\sin^2\theta - 2\sin\theta - 1}$

4. We find that $r' = 24\cos 3\theta$, so $\frac{dy}{dx} = \frac{r'\sin\theta + r\cos\theta}{r'\cos\theta - r\sin\theta}$
$= \frac{24\cos 3\theta\sin\theta + 8\sin 3\theta\cos\theta}{24\cos 3\theta\cos\theta - 8\sin 3\theta\sin\theta}$. When $\theta = \frac{\pi}{4}$, we see that $\frac{dy}{dx} = \frac{1}{2}$, so the slope of the normal line at this point is -2. Using the relationships that $x = r\cos\theta$ and $y = r\sin\theta$, we see that the polar coordinate $\left(4\sqrt{2}, \frac{\pi}{4}\right)$ has the rectangular coordinate $(4, 4)$ and so the equation of the normal line is $y - 4 = -2(x - 4)$ or $y = -2x + 12$.

5. $L = \int_0^2\sqrt{\left(e^t\right)^2 + \left(2e^t\right)^2}\,dt = \int_0^2\sqrt{e^{2t} + 4e^{2t}}\,dt = \int_0^2\sqrt{5}e^t\,dt = \sqrt{5}e^t\big|_0^2 = \sqrt{5}\left(e^2 - 1\right)$

6.   Here $f(\theta) = 4\sin\theta$, $f'(\theta) = 4\cos\theta$, and $y = 4\sin^2\theta$. So the surface area is

$$S = 2\pi \int_0^{\pi/2} \left(4\sin^2\theta\right)\sqrt{(4\sin\theta)^2 + (4\cos\theta)^2}\,d\theta$$

$$= 8\pi \int_0^{\pi/2} \left(\sin^2\theta\right)\sqrt{16\sin^\theta + 16\cos^2\theta}\,d\theta$$

$$= 32\pi \int_0^{\pi/2} \left(\sin^2\theta\right)d\theta$$

$$= 32\pi \left(\frac{1}{2} - \frac{1}{4}\sin 2\theta\right)\Bigg|_0^{\pi/2}$$

(by Formula 56 in Appendix C). Evaluating this we obtain $32\pi\left(\frac{\pi}{4}\right) = 8\pi^2 \approx 78.9568$

7.   Setting these equations equal to each other, we obtain $4\cos\theta = 1 - \cos\theta$ or $5\cos\theta = 1$ and

so the points of intersection are when $\cos\theta = \frac{1}{5} = 0.2$ or when $\theta \approx 1.3694$ and -1.3694 and the points of intersection are $(0.2,\ 1.3694)$ and $(0.2,\ -1.3694) = (0.2,\ 4.9137)$

8.   The points of intersection are $-\frac{\pi}{2}$ and $\frac{\pi}{2}$ so

$$A = \int_{-\pi/2}^{\pi/2} \frac{1}{2}\left(r_2^2 - r_1^2\right)d\theta$$

$$= \int_{-\pi/2}^{\pi/2} \frac{1}{2}\left[\left(1^2 - (1 - \cos\theta)^2\right)\right]d\theta$$

$$= \int_{-\pi/2}^{\pi/2} \frac{1}{2}\left[\left(1 - (1 - 2\cos\theta + \cos^2\theta)\right)\right]d\theta$$

$$= \frac{1}{2}\left[2\sin\theta - \frac{\sin 2\theta}{4} - \frac{\theta}{2}\right]_{-\pi/2}^{\pi/2} = 2 - \frac{\pi}{4}$$

CHAPTER

30

Partial Derivatives and Multiple Integrals

≡ 30.1 FUNCTIONS IN TWO VARIABLES

1. $f(x, y) = 3x + 4y - xy$:
  (a) $f(1, 0) = 3 \cdot 1 = 3$;
  (b) $f(0, 1) = 4 \cdot 1 = 4$;
  (c) $f(2, 1) = 3 \cdot 2 + 4 \cdot 1 - 2 \cdot 1 = 6 + 4 - 2 = 8$;
  (d) $f(x + h, y) = 3(x + h) + 4y - (x + h)y = 3x + 3h + 4y - xy - hy$;
  (e) $f(x, y + h) = 3x + 4(y + h) - x(y + h) = 3x + 4y + 4h - xy - xh$

2. $g(x, y) = x^2 y + \sin x - \cos y$:
  (a) $g\left(\frac{\pi}{2}, \pi\right) = \left(\frac{\pi}{2}\right)^2 \pi + \sin \frac{\pi}{2} - \cos \pi = \frac{\pi^3}{4} + 2 \approx 9.7516$;
  (b) $g\left(\pi, \frac{\pi}{2}\right) = \pi^2 \cdot \frac{\pi}{2} + \sin \pi - \cos \frac{\pi}{2} = \frac{\pi^3}{2} \approx 15.5031$;
  (c) $g\left(\frac{\pi}{2}, \frac{\pi}{4}\right) = \left(\frac{\pi}{2}\right)^2 + \frac{\pi}{4} + \sin \frac{\pi}{2} - \cos \frac{\pi}{4} = \frac{\pi^3}{16} + 1 - \frac{\sqrt{2}}{2} \approx 2.2308$;
  (d) $g(x + h, y) = (x + h)^2 y + \sin(x + h) - \cos y = x^2 y + 2xhy + h^2 y + \sin(x + h) - \cos y$;
  (e) $g(x, y + h) = x^2(y + h) + \sin x - \cos(y + h) = x^2 y + x^2 h + \sin x - \cos(y + h)$

3. $j(x, y) = \sqrt{xy} - x + \frac{4}{y}$:
  (a) $j(-1, -2) = \sqrt{2} + 1 + \frac{4}{-2} = \sqrt{2} - 1 \approx 0.4142$;
  (b) $j(-1, -4) = \sqrt{4} + 1 - 1 = 2$;
  (c) $j(4, 1) = \sqrt{4} - 4 + 4 = 2$;
  (d) $j(x + h, y) = \sqrt{(x + h)y} - (x + h) + \frac{4}{y}$;
  (e) $j(x, y + h) = \sqrt{x(y + h)} - x + \frac{4}{y + h}$

4. $k(x, y) = e^x + e^{xy} - y^2$:
  (a) $k(1, 2) = e^1 + e^2 - 4$;
  (b) $k(2, 1) = e^2 + e^2 - 1 = 2e^2 - 1$;
  (c) $k(2, 3) = e^2 + e^6 - 9$;
  (d) $k(x + h, y) = e^{x+h} + e^{(x+h)y} - y^2$;
  (e) $k(x, y + h) = e^x + e^{x(y+h)} - (y + h)^2$

5. $f(x, y) = \frac{2xy - x^2}{y - x}$:
  (a) $f(1, 0) = \frac{0 - 1}{-1} = 1$;
  (b) $f(0, 1) = \frac{0}{1} = 0$;
  (c) $f(2, 1) = \frac{2 \cdot 2 - 2^2}{1 - 2} = 0$;
  (d) $f(1, 2) = \frac{4 - 1}{1} = 3$;
  (e) the domain is all real ordered pairs (x, y) such that $x \neq y$.

6. $g(x, y) = \frac{\ln(x+y)}{y}$:
  (a) $g(1, 1) = \frac{\ln(2)}{1} = \ln 2$;
  (b) $g(2, 1) = \frac{\ln(2+1)}{1} = \ln 3$
  (c) $g(3, 6) = \frac{\ln(3+6)}{6} = \frac{1}{6} \ln 9$;
  (d) $g(4, -3) = \frac{\ln(4-3)}{-3} = \frac{\ln 1}{-3} = 0$;
  (e) the domain is the set of all real ordered pairs (x, y) such that $x + y > 0$.

7. $V = \frac{1}{3} \pi r^2 h$

8. $V = Bh$, where B is the area of the base.

9. $A = 2\pi rh + 2\pi r^2$

10. $A = \pi rs + \pi r^2 = \pi r(s + r)$

11. The area of the storage tank is $A = 2\pi r^2 + 2\pi rh$. For the cost we have $C_{\text{bottom}} + C_{\text{top}} = 200\left(\pi r^2\right) + 200\left(\pi r^2\right) = 400\pi r^2$ and $C_{\text{side}} = 1000(2\pi rh) = 2000\pi rh$. Thus, the total cost is $C_{\text{bottom}} + C_{\text{top}} + C_{\text{side}} = 400\pi r^2 + 2000\pi rh = 400\pi r(r + 5h)$.

12. The area of the sewage tank is $A = \pi rs + \pi r^2$. For the cost of the base we have $C_{\text{base}} = 1500\pi r^2$ and for the side the cost is $C_{\text{side}} = 200\pi rs$. Thus, the total cost is $C_{\text{base}} + C_{\text{side}} = 1500\pi r^2 + 200\pi rs = 100\pi r(15r + 2s)$.

13. $v(R, r) = c\left(R^2 - r^2\right)$; $v(0.0075, 0.0045) = \left(0.0075^2 - 0.0045^2\right) = 3.6 \times 10^{-5} = 0.000\,036$ cm/min

14. $S(a, V, d) = \frac{aV}{0.51d^2}$

(a) $S\left(0.62, 2.4 \times 10^6, 100\right) = \frac{0.62 \times 2.4 \times 10^6}{0.51 \times 100^2} = 291.8$ mph,

(b) $S\left(0.62, 2.4 \times 10^6, 200\right) = \frac{0.62 \times 2.4 \times 10^6}{0.51 \times 200^2} = 72.9$ mph.

15. **(a)** $T = 20°F$ and $v = 200$ mph so we use the following part of the formula: $WCI = 91.4 - \frac{(10.45 + 6.69\sqrt{v} - 0.447v)(91.4 - F)}{22} = -10.6$ or $-11°F$,

(b) $T = 10°$, $v = 20$ mph, using the same formula as in part (a) we get the $WCI = -25°F$,

(c) Since $v = 4$ mph, $WCI = F = 10°F$,

(d) As in parts (a) and (b) $WCI = -15°F$

16. $ATI(F, RH) = 0.885F - 78.7RH + 1.20F \cdot RH + 2.70$:

(a) $ATI(95, 0.30) = 97.365 \approx 97°F$

(b) $ATI(95, 0.6) = 107.955 \approx 108°F$

(c) $ATI(95, 9.90) = 118.545 \approx 119°F$

(d) $ATI(25, 0.75) = -11.7 \approx -12°$

17. **(a)** The volume is the product of the box's length, width, and height, or $V(x, y, z) = xyz$.

(b) The amount of the material needed to construct the box is the same as the surface area of the box including the partition, $S(x, y, z) = xy + 2xz + 3yz$.

18. $C(x, y) = (2x)(24) + (2y)(24) + 18.50x + 18.50y = 66.50x + 66.50y$

19. $S(2400, \ 65) = (0.000\ 02)(2,400)(65)^2 = 202.8$ ft

20. **(a)** $t(75, \ 67) = \frac{33(75)}{67 + 33} = \frac{2475}{100} = 24.75$ min $= 24$ min 45 s

(b) $t(60, \ 32) = \frac{33(60)}{32 + 33} = \frac{1980}{65} = \frac{396}{13} \approx 30.4$ min (Note: the fraction has been rounded down for safety reasons.)

21. **(a)** $A(84, 1.80) = 0.18215\left(84^{0.425}\right)\left(1.80^{0.725}\right) \approx 1.834$ m^2

(b) For this, we need to compute a new constant. Since 1 kg $\approx$ 2.2 lb, then $10^{0.425} \approx 2.2^{0.425} \approx 1.398$ and since 1 m $= 39.37$ in., we have $10^{0.725} = 39.37^{0.725} \approx 14.338$. We also need to convert m^2 to in.^{2}, and 1 m$^2 = 39.37^2$ in.$^2 \approx 1550$ in.^{2}. The new constant is $\frac{0.18215 \times 1550}{1.398 \times 14.338} \approx 14.085$. Hence, the formula $A(w, h) = 14.085w^{0.425}h^{0.725}$ can determine the surface area in in.^{2} of a person based on the person's weight, w, in lb and height, h, in in.

(c) $A(185, 71) = 14.085\left(185^{0.425}\right)\left(71^{0.725}\right) \approx 2,847.5$ in.^{2}. (Note that this answer is almost the same as part (a); that is, 1.834 m$^2 \approx 2,847.5$ in.^{2}.)

22. **(a)** $S(4, 3) = 5\left(4^2\right)\left(3^3\right) = 5(16)(27) = 2160$ or sales of \$2,160,000.

(b) $S(3, 4) = 5\left(3^2\right)\left(4^3\right) = 5(9)(64) = 2880$ or sales of \$2,880,000.

23. **(a)** Here $x = 300$ units of labor and $y = 50$ units of capital, we have $P(300, 50) = 400\left(300^{0.35}\right) \times \left(50^{0.65}\right) \approx 37,444$.

(b) Here $x = 600$ and $y = 100$, and we have $P(600, 100) = 400\left(600^{0.35}\right)\left(100^{0.65}\right) \approx 74,888$.

24. **(a)** Here $x = 80$ acres of grass and $y = 20$ acres of alfalfa, we have $C(80, 20) = 9(80) + 6(20) - 7 = 833$.

(b) Here $x = 60$ and $y = 40$, and we have $C(60, 40) = 9(60) + 6(40) - 7 = 773$.

≡ 30.2 SURFACES IN THREE DIMENSIONS

1. $x + 2y + 3z - 6 = 0$: A plane whose intercepts are (6, 0, 0), (0, 3, 0) and (0, 0, 2).

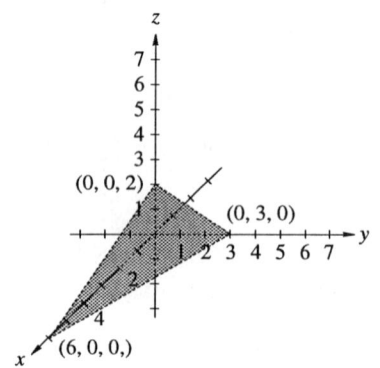

2. $3x + 2y - z - 6 = 0$: A plane whose intercepts are (2, 0, 0), (0, 3, 0) and (0, 0, -6).

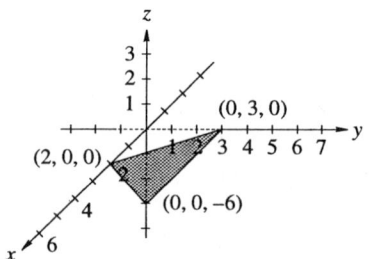

3. $x = 8y^2$: A parabolic cylinder perpendicular to the xy-plane.

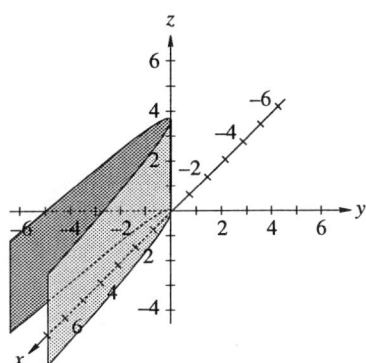

4. $x^2 + y^2 + z^2 = 9$: A sphere of radius 3.

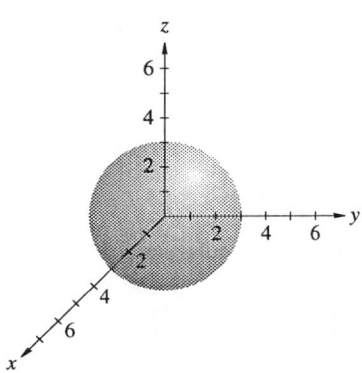

5. $x^2 + 2y^2 + z^2 = 4 \Rightarrow \frac{x^2}{4} + \frac{z^2}{4} = 1$: Ellipsoid with intercepts $(\pm2,\ 0,\ 0)$, $(0,\ \pm\sqrt{2},\ 0)$ and $(0,\ 0,\ \pm2)$.

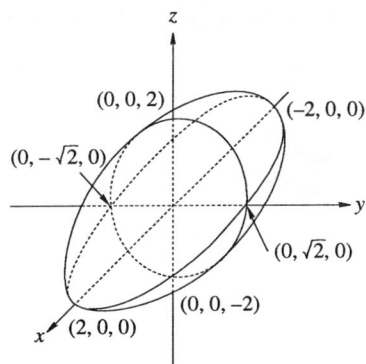

6. $4x^2 + 9y^2 = 36 \Rightarrow \frac{x^2}{9} + \frac{y^2}{4} = 1$: Elliptic cylinder perpendicular to the xy-plane with intercepts $(\pm3,\ 0,\ 0)$ and $(0,\ \pm2,\ 0)$.

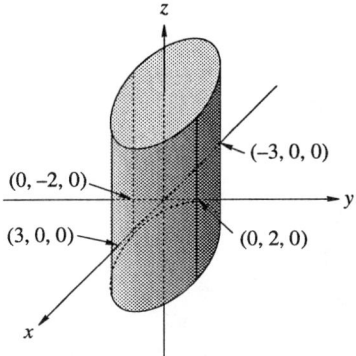

7. $2x^2 + y^2 + 2z^2 = 8 \Rightarrow \frac{x^2}{4} + \frac{y^2}{8} + \frac{z^2}{4} = 1$: Ellipsoid with intercepts $(\pm2,\ 0,\ 0)$, $(0,\ \pm2\sqrt{2},\ 0)$, and $(0,\ 0,\ \pm2)$.

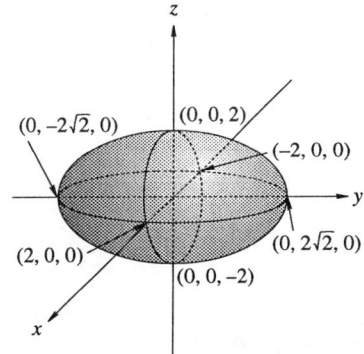

8. $4x^2 + y^2 - z^2 = 1$: Hyperboloid of one sheet with intercepts $(\pm\frac{1}{2},\ 0,\ 0)$, $(0, \pm1,\ 0)$.

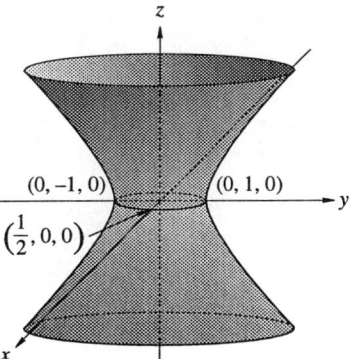

9. $4x^2 + y^2 - z^2 = -1$: Hyperboloid of two sheets with intercepts $(0, 0, \pm 1)$.

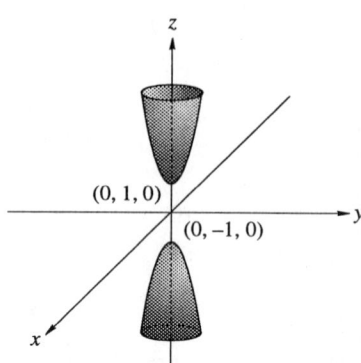

10. $4x^2 - 9y^2 = 1$: Hyperbolic cylinder with intercepts $\left(\pm\frac{1}{2}, 0, 0\right)$.

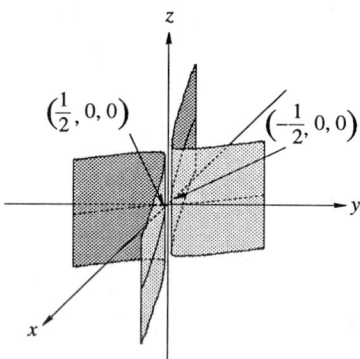

11. $3x - 4y - 6z = 12$: Plane with intercepts $(4, 0, 0)$, $(0, -3, 0)$, and $(0, 0, -2)$.

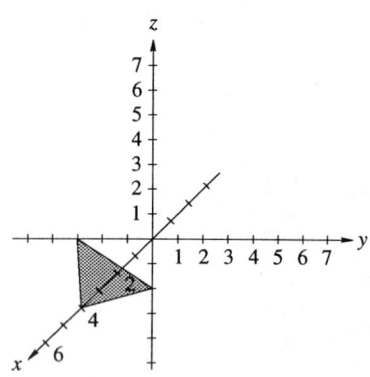

12. $xy = 4$: Hyperbolic cylinder whose asymptote planes are the xz-plane with the yz-plane.

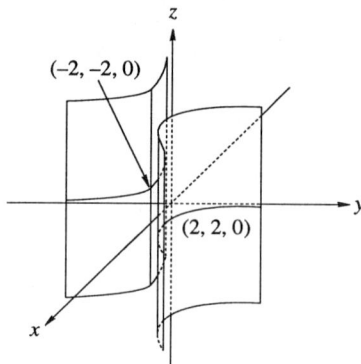

13. $z = x^2 + 4y^2 - 16$: This is an elliptic paraboloid with a z-translation of -16. The intercepts are $(\pm 4, 0, 0)$, $(0, \pm 2, 0)$, and $(0, 0, -16)$.

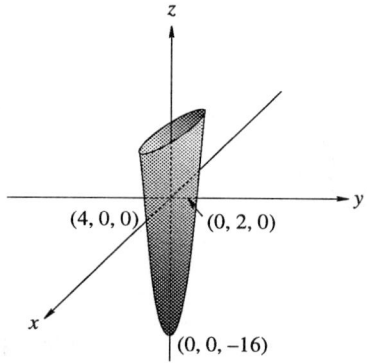

14. $2x^2 + y^2 - 2z^2 = -8 \Rightarrow \frac{x^2}{4} + \frac{y^2}{8} - \frac{z^2}{4} = -1$: Hyperboloid of two sheets with intercepts at $(0, 0, \pm 2)$.

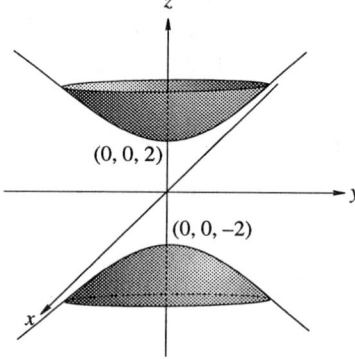

15. $36z = 4x^2 + 9y^2 \Rightarrow z = \frac{x^2}{9} + \frac{y^2}{4}$: Elliptic paraboloid with vertex $(0, 0, 0)$. Four other points are $(\pm3, 0, 1)$ and $(0, \pm2, 1)$.

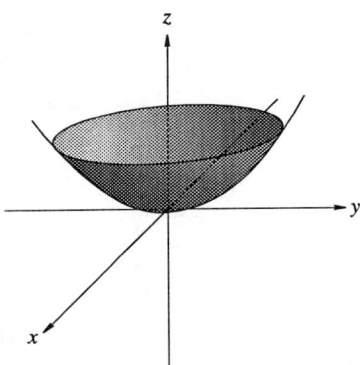

16. $36z = 4x^2 + 9y^2 \Rightarrow z = \frac{x^2}{9} + \frac{y^2}{4}$: Elliptic Cone with vertex at the origin. Eight other points are $(\pm3, 0, \pm1)$ and $(0, \pm2, \pm1)$.

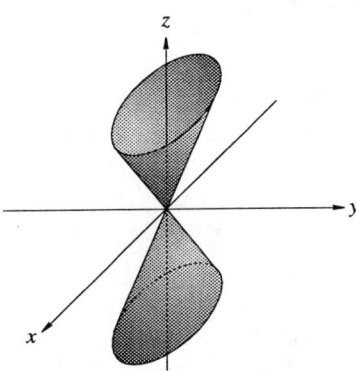

17. $36z = 4x^2 - 9y^2 \Rightarrow z = \frac{x^2}{9} - \frac{y^2}{4}$: Hyperbolic paraboloid with saddle point $(0, 0, 0)$. Four other points are $(\pm3, 0, 1)$ and $(0, \pm2, -1)$.

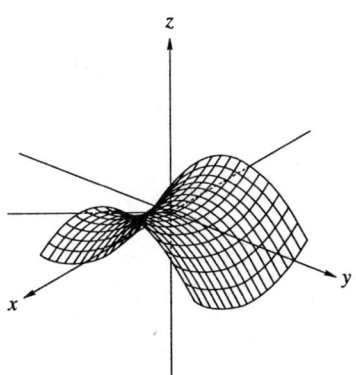

18. $36z^2 = 1 - 4x^2 - 9y^2 \Rightarrow 4x^2 + 9y^2 + 36z^2 = 1$: Ellipsoid with intercepts $(\pm\frac{1}{2}, 0, 0)$, $(0, \pm\frac{1}{3}, 0)$, and $(0, 0, \pm\frac{1}{6})$.

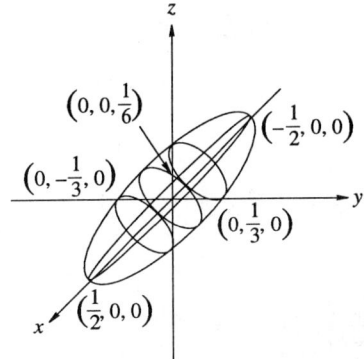

19. $2x = y^2 \Rightarrow x = \frac{1}{2}y^2$: Parabolic cylinder perpendicular to the xy-plane.

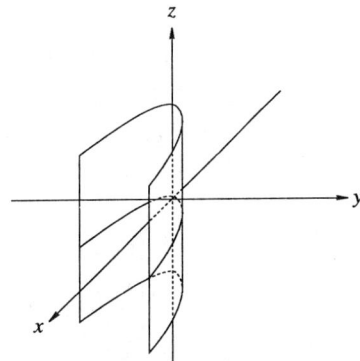

20. $3x + 2y - 3z = 6$: Plane with intercepts $(2, 0, 0)$, $(0, 3, 0)$ and $(0, 0, -2)$.

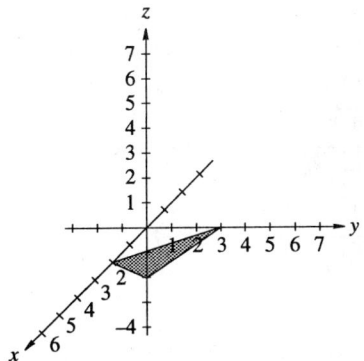

30.3 PARTIAL DERIVATIVES

1. $f(x, y) = x^2y + y^2x$; $f_x = 2xy + y^2$; $f_y = x^2 + 2xy$.

2. $f(x, y) = x^2y + 7y^2$; $f_x = 2xy$; $f_y = x^2 + 14y$

3. $f(x, y) = 3x^2 + 6xy^3$; $f_x = 6x + 6y^3$; $f_y = 18xy^2$

4. $f(x, y) = \frac{x}{y^2} = xy^{-2}$; $f_x = y^{-2} = \frac{1}{y^2}$; $f_y = -2xy^{-3} = \frac{-2x}{y^3}$

5. $f(x, y) = \frac{x^2+y^2}{y} = \frac{x^2}{y} + y$; $f_x = \frac{2x}{y}$; $f_y = \frac{-x^2}{y^2} + 1 = \frac{y^2-x^2}{y^2}$

6. $f(x, y) = \frac{x+y}{\sqrt{xy}}$. So, $f_x = \frac{\sqrt{xy}\cdot 1 - (x+y)(xy)^{-1/2}y}{xy} = \frac{xy - (x+y)y}{(xy)^{3/2}} = \frac{-y^2}{(xy)^{3/2}} = \frac{-y}{x\sqrt{xy}}$, and $f_y = \frac{\sqrt{xy}\cdot 1(x+y)(xy)^{-1/2}\cdot x}{(xy)} = \frac{xy - (x+y)x}{(xy)^{3/2}} = -\frac{x^2}{(xy)^{3/2}} = \frac{-x}{y\sqrt{xy}}$.

7. $f(x, y) = e^y \cos x + e^x \sin y$; $f_x = -e^y \sin x + e^x \sin y$; $f_y = e^y \cos x + e^x \cos y$

8. $f(x, y) = x^2 \cos y + y^2 \cos x$; $f_x = 2x \cos y - y^2 \sin x$; $f_y = -x^2 \sin y + 2y \cos x = 2y \cos x - x^2 \sin y$

9. $f(x, y) = e^{2x+3y}$; $f_x = 2e^{2x+3y}$; $f_y = 3e^{2x+3y}$

10. $f(x, y) = e^{x^2+y^2}$; $f_x = 2xe^{x^2+y^2}$; $f_y = 2ye^{x^2+y^2}$

11. $f(x, y) = \sin\left(x^2y^3\right)$; $f_x = 2xy^3 \cos\left(x^2y^3\right)$; $f_y = 3x^2y^2 \cos\left(x^2y^3\right)$

12. $f(x, y) = \ln\left(x^2y^3\right)$; $f_x = \frac{2xy^3}{x^2y^3} = \frac{2}{x}$; $f_y = \frac{3y^2x^2}{x^2y^3} = \frac{3}{y}$

13. $f(x, y) = \ln\sqrt{x^2 + y^2}$. So, $f_x = \frac{\frac{1}{2}\left(x^2 + y^3\right)^{-1/2} \cdot 2x}{\sqrt{x^2 + y^2}} = \frac{x}{x^2+y^2}$ and $f_y = \frac{\frac{1}{2}\left(x^2 + y^2\right)^{-1/2}(2y)}{\sqrt{x^2 + y^2}} = \frac{y}{x^2+y^2}$.

14. $f(x, y) = \tan{-1}\frac{y}{x}$; $f_x = \frac{-yx^{-2}}{1 + \left(\frac{y}{x}\right)^2} = \frac{\frac{-y}{x^2}}{1 + \frac{y^2}{x^2}} = \frac{-y}{x^2+y^2}$; $f_y = \frac{\frac{1}{x}}{1 + \frac{y^2}{x^2}} = \frac{x}{x^2+y^2}$

15. $f(x, y) = \sin^2(3xy)$; $f_x = 2\sin(3xy) \cdot \cos(3xy) \cdot 3y = 6y\sin(3xy)\cos(3xy)$; $f_y = 2\sin 3xy \cos 3xy \cdot 3x = 6x\sin(3xy)\cos(3xy)$

16. $f(x, y) = e^{\sqrt{x^2+y^2}}$; $f_x = \frac{1}{2}\cdot\left(x^2 + y^2\right)^{-1/2}(2x) \times e^{\sqrt{x^2+y^2}} = \frac{x}{\sqrt{x^2+y^2}}e^{\sqrt{x^2+y^2}}$; $f_y = \frac{y}{\sqrt{x^2+y^2}}e^{\sqrt{x^2+y^2}}$

17. $f(x, y) = x^3y^2 - xy^5$; $f_x = 3x^2y^2 - y^5$; $f_y = 2x^3y - 5xy^4$; $f_{xx} = 6xy^2$; $f_{yx} = 6x^2y - 5y^4$; $f_{yy} = 2x^3 - 20xy^3$; $f_{xy} = 6x^2y - 5y^4$

18. $f(x, y) = \sin xy^3$; $f_x = y^3 \cos xy^3$; $f_y = 3y^2x \times \cos xy^3$; $f_{xx} = -y^6 \sin xy^3$; $f_{yx} = 3y^2 \cos xy^3 - 3xy^5 \sin xy^3$; $f_{yy} = 6yx \cos xy^3 - 9x^2y^3 \sin xy^3$; $f_{xy} = 3y^2 \cos xy^3 - 3xy^5 \sin xy^3$

19. $f(x, y) = 3e^{xy^3}$; $f_x = 3y^3e^{xy^3}$; $f_y = 9xy^2e^{xy^3}$; $f_{xx} = 3y^6e^{xy^3}$; $f_{yx} = f_{xy} = 9e^{xy^3}y^2\left(1 + xy^3\right)$; $f_{yy} = 27x^2y^4e^{xy^3} + 18xye^{ey^3} = 9xye^{xy^3}\left(3xy^3 + 2\right)$

20. $f(x, y) = \sin\left(x + y^2\right)$; $f_x = \cos\left(x + y^2\right)$; $f_y = 2y\cos\left(x + y^2\right)$; $f_{xx} = -\sin(x+y^2)$; $f_{yx} = f_{xy} = -2y\sin(x + y^2)$; $f_{yy} = 2\cos\left(x + y^2\right) - 4y^2 \sin\left(x + y^2\right)$

21. $f(x, y) = \frac{2x}{y^5}$; $f_x = \frac{2}{y^5}$; $f_y = \frac{-10x}{y^6}$; $f_{xx} = 0$; $f_{yx} = f_{xy} = \frac{-10}{y^6}$; $f_{yy} = \frac{60x}{y^7}$

22. $f(x, y) = e^x \tan y$; $f_x = e^x \tan y$; $f_y = e^x \sec^2 y$; $f_{xx} = e^x \tan y$; $f_{yx} = e^x \sec^2 x = f_{xy}$; $f_{yy} = 2e^x \sec^2 y \tan y$

23. $f(x, y) = \ln\left(x^3y^5\right)$; $f_x = \frac{3}{x}$; $f_y = \frac{5}{y}$; $f_{xx} = \frac{-3}{x^2}$; $f_{yx} = f_{xy} = 0$; $f_{yy} = \frac{-5}{y^2}$

24. $f(x, y) = \sqrt{x^2y + 3y^2}$; $f_x = xy\left(x^2y + 3y^2\right)^{-1/2} = \frac{xy}{\sqrt{x^2y+3y^2}}$; $f_y = \frac{1}{2}\left(x^2 + 6y\right)\left(x^2y + 3y^2\right)^{-1/2}$;

$f_{xx} = \frac{\sqrt{x^2y + 3y^2}\cdot y - xy\cdot xy\left(x^2y + 3y^2\right)^{-1/2}}{x^2y + 3y^2}$

$= \frac{\left(x^2y + 3y^2\right)y - x^2y^2}{\left(x^3y + 3y^2\right)^{3/2}}$

$= \frac{x^2y^2 + 3y^3 - x^2y^2}{\left(x^2 + 3y^2\right)^{3/2}}$

$= \frac{3y^2}{\left(x^2 + 3y^2\right)^{3/2}}$

$$f_{yx} = f_{xy}$$

$$= \frac{\sqrt{x^2y+3y^2}\,x - \frac{1}{2}xy(x^2+6y)(x^2+3y^2)^{-1/2}}{x^2y+3y^2}$$

$$= \frac{(x^2y+3y^2)\,x - \frac{1}{2}x^3y - 3xy^2}{(x^2y+3y^2)^{3/2}}$$

$$= \frac{x^3y}{2(x^2y+3y^2)^{3/2}}$$

$$f_{yy} = \frac{2\sqrt{x^2y+3y^2}\cdot 6 - (x^2+6y)(x^2+3y^2)^{-1/2}}{4(x^2y+3y^2)}$$

$$= \frac{12(x^2y+3y^2) - (x^4+12x^2y+36y^2)}{4(x^2y+3y^2)^{3/2}}$$

$$= \frac{12x^2y+36y^2 - x^4 - 12x^2y - 36y^2}{4(x^2+3y^2)^{3/2}}$$

$$= \frac{-x^4}{4(x^2+3y^2)^{3/2}}$$

25. $P = \frac{20 \cdot R \cdot 295}{V}$; $P_V = \frac{-5900R}{V^2}$ at $V = 3$; $P_V = \frac{-5900R}{9}$ kg/m^2

26. $T = \frac{100}{\ln 2}\ln(x^2+y^2)$; $T_x = \frac{100 \cdot 2x}{(x^2+y^2)\ln 2} = \frac{200x}{(x^2+y^2)\ln 2}$; T_x at $(1, 0)$ is $\frac{200}{\ln 2} \approx 288.539$, and at $(0, 1)$, $T_x = 0$.

27. $V = \frac{82.06T}{P}$; $V_P = \frac{-82.06T}{P^2}$ at $T = 300°$, $P = 5$ atm; $V = \frac{-82.06 \times 300}{5^2} = -984.72$ cm^3/atm

28. $V_T = \frac{82.06}{P}$. At $P = 5$, then $V_T = \frac{82.06}{5} = 16.412$ cm^3/°K.

29. $z = 16 - x^2 - y^2$; $z_y = -2y$. At $(1, 3, 6)$, $z_y = -2 \cdot 3 = -6$.

30. $PV = nRT$; $V = \frac{nRT}{P}$; $\frac{\partial V}{\partial T} = \frac{nR}{P}$; $T = \frac{PV}{nR}$; $\frac{\partial T}{\partial P}\cdot\frac{V}{nR}$; $P = \frac{nRT}{V}$; $\frac{\partial P}{\partial V} = \frac{-nRT}{V^2}$; $\left(\frac{\partial V}{\partial T}\right)\left(\frac{\partial T}{\partial P}\right) \times \left(\frac{\partial P}{\partial V}\right) = \frac{nR}{P}\cdot\frac{PV}{nR}\cdot\frac{-nRT}{V^2} = \frac{-nRT}{V} = \frac{-nRT}{\frac{nRT}{P}} = -P$

31. **(a)** $\frac{dR}{dL} = \frac{k}{r^4}$,

(b) Rewriting the formula as $R = kLr^{-4}$, we have $\frac{dR}{dr} = -4kLr^{-5} = -\frac{4kL}{r^5}$.

32. **(a)** $P_x(x, y) = (0.3)(65)x^{-0.7}y^{0.7} = 19.5x^{-0.7}y^{0.7}$

(b) $P_x(1500, 5400) = 19.5\left(1500^{-0.7}\right)\left(5400^{0.7}\right) = 47.80$

(c) $P_y(x, y) = (0.7)(65)x^{0.3}y^{-0.3} = 45.5x^{0.3}y^{-0.3}$

(d) $P_y(1500, 5400) = 45.5\left(1500^{0.3}\right)\left(5400^{-0.3}\right) = 30.98$

(e) Since $P_x(1500, 5400) = 47.80 > 30.98 = P_y(1500, 5400)$, the management should encourage the increase in labor.

33. **(a)** $A_w(w, h) = 5.9861w^{-0.575}h^{0.725}$ in.^{2}/lb; $A_h(w, h) = 10.2116w^{0.425}h^{-0.275}$ in.^{2}/in.

(b) Substituting in the formulas from (a) produces $A_w(185, 71) = 5.9861\left(185^{-0.575}\right)\left(71^{0.725}\right) = 6.54$ in.^{2}/lb and $A_h(185, 71) = 10.2116\left(185^{0.425}\right) \times \left(71^{-0.275}\right) = 29.08$ in.^{2}/in.

(c) $A_w(185, 71) = 6.54$ in.^{2}/lb means that for a 185 lb person 5'11'' tall, the rate of change in surface area is 6.54 in.^{2} for each lb gained in weight if the height is fixed; $A_h(185, 71) = 29.07$ in.^{2}/in. means that for a 185 lb person 5'11'' tall, the rate of change in surface area is 29.07 in.^{2} for each in. gained in height if the weight remains the same.

34. **(a)** $A_w(w, h) = (0.18215)(0.425)w^{-0.575}h^{0.725} \approx 0.077414w^{-0.575}h^{0.725}$ and $A_h(w, h) = (0.18215) \times (0.725)w^{0.425}h^{-0.275} \approx 0.13206w^{0.425}h^{-0.275}$

(b) $A_w(84, 1.80) = 0.077414\left(84^{-0.575}\right)\left(1.80^{0.725}\right) \approx 0.00928$ m^2/kg and $A_h(84, 1.80) = 0.13206 \times \left(84^{0.425}\right)\left(1.80^{-0.275}\right) \approx 0.7386$ m^2/m

35. **(a)** $S_w(w, r) = 0.000\,02r^2$ ft/lb,
(b) $S_w(2400, 65) = 0.00002\left(65^2\right) = 0.0845$ ft/lb,
(c) $S_r(w, r) = 0.000\,04wr$ ft/mph,
(d) $S_r(2400, 65) = 0.00004(2400)(65) = 6.24$ ft/mph,
(e) about 6.24 ft

36. **(a)** $WCI(30, -4F) = 91.4 - \frac{\left[10.45+6.69\sqrt{30}-0.447(30)\right][91.4-(-4)]}{22} \approx -54.7°$F

(b) $WCI_v(v, F)$
$$= -\frac{\left[\left(\frac{1}{2}\right)6.69v^{-1/2} - 0.447\right](91.4 - F)}{22}$$
$$= -\left(\frac{3.345}{\sqrt{v}} - 0.447\right)\left(\frac{91.4 - F}{22}\right)$$
$$= \left(\frac{3.345}{\sqrt{v}} - 0.447\right)\left(\frac{F - 91.4}{22}\right)$$

(c) $WCI_v(30, -4) = \left(\frac{3.345}{\sqrt{30}} - 0.447\right)\left(\frac{-4-91.4}{22}\right) \approx -0.71$

(d) $WCI_F(v, F) = \frac{10.45+6.69\sqrt{v}-0.447v}{22} \approx 0.475 + 0.304\sqrt{v} - 0.0203v$

(e) $WCI_F(30, -4) \approx 0.475 + 0.304\sqrt{30} - 0.0203(30) \approx 1.53$

(f) If you assume that feeling warmer means that you are more comfortable, then an increase of 1 mph in wind velocity would have less effect than an increase of 1°F in temperature. Hence, you would feel more comfortable if the wind velocity increased 1 mph.

30.4 SOME APPLICATIONS OF PARTIAL DERIVATIVES

1. $z = x^2 + y^2$; $z_x = 2x$; $z_y = 2y$; $dz = 2x\,dx + 2y\,dy$

2. $z = 2x^2 + xy - y^2$; $z_x = 4x + y$; $z_y = x - 2y$; $dz = (4x + y)dx + (x - 2y)dy$

3. $z = 3x^2 + 4xy - 2y^3$; $z_x = 6x + 4y$; $z_y = 4x - 6y^2$; $dz = (6x + 4y)dx + (4x - 6y^2)\,dy$

4. $z = xye^{x+y}$; $z_x = (xy + y)e^{x+y}$; $z_y = (xy + x)e^{x+y}$; $dz = (1 + x)ye^{x+y}dx + (1 + y)xe^{x+y}dy$

5. $z = \arctan\left(\frac{y}{x}\right)$; $z_x = \dfrac{-\frac{y}{x^2}}{1 + \frac{y^2}{x^2}} = \dfrac{-y}{x^2+y^2}$; $z_y = \dfrac{\frac{1}{x}}{1 + \frac{y^2}{x^2}} = \dfrac{x}{x^2+y^2}$; $dz = \dfrac{-y}{x^2+y^2}\,dx + \dfrac{x}{x^2+y^2}\,dy$

6. $z = \ln\left(x^2 + y^2\right)$; $z_x = \frac{2x}{x^2+y^2}$; $z_y = \frac{2y}{x^2+y^2}$; $dz = \frac{2x}{x^2+y^2}\,dx + \frac{2y}{x^2+y^2}\,dy$

7. $z = x\tan yx$; $z_x = \tan yx + xy\sec^2 yx$; $z_y = x^2\sec^2 yx$; $dz = \left(\tan yx + xy\sec^2 yx\right)dx + x^2\sec^2 yx\,dy$

8. $z = \ln\left(\frac{x}{y}\right)$; $z_x = \frac{1/y}{x/y} = \frac{1}{x}$; $z_y = \frac{-x/y^2}{x/y} = -\frac{1}{y}$; $dz = \frac{1}{x}dx - \frac{1}{y}dy$

9. $z = x^2 + y^2$; $x = te^t$; $y = t^2e^t$; $z_x = 2x$; $z_y = 2y$; $\frac{dx}{dt} = (t + 1)e^t$; $\frac{dy}{dt} = \left(t^2 + 2t\right)e^t$; $\frac{dz}{dt} = 2x(t + 1)e^t + 2y\left(t^2 + 2t\right)e^t = 2\left(te^t\right)(t + 1)e^t + 2\left(t^2e^t\right)\left(t^2 + 2t\right)e^t = (t + 1)2te^{2t} + \left(t^2 + 2t\right)2t^2e^{2t} = 2te^{2t}\left[1 + t + 2t^2 = t^3\right]$

10. $z = \frac{1}{x^2+y^2}$; $x = \cos 2t$; $y = \sin 2t$; $z_x = -1 \times \left(x^2 + y^2\right)^{-2} \cdot 2x = \frac{-2x}{(x^2+y^2)^2}$; $z_y = \frac{-2y}{(x^2+y^2)^2}$; $\frac{dx}{dt} = -2\sin 2t$; $\frac{dy}{dt} = 2\cos 2t$; $\frac{dz}{dt} = \frac{-2x(-2\sin 2t)}{(x^2+y^2)^2} + \frac{-2y(2\cos 2t)}{(x^2+y^2)^2} = \frac{4x\sin 2t - 4y\cos 2t}{(x^2+y^2)^2} = \frac{4(\cos 2t)\sin 2t - 4(\sin 2t)\cos 2t}{(x^2+y^2)^2} = 0$

11. $z = e^u\sin v$; $u = \sqrt{t}$; $v = \pi t$; $z_u = e^u\sin v$; $z_v = e^u\cos v$; $\frac{du}{dt} = \frac{1}{2\sqrt{t}}$; $\frac{dv}{dt} = \pi$; $\frac{dz}{dt} = e^u\sin v \cdot \frac{1}{2\sqrt{t}} + e^u\cos v \cdot \pi = \frac{e^{\sqrt{t}}\sin \pi t}{2\sqrt{t}} + \pi e^{\sqrt{t}}\cos \pi t$

12. $z = x^3 - y$; $x = te^{-t}$; $y = \sin t$; $z_x = 3x^2$; $z_y = -1$; $\frac{dx}{dt} = (1 - t)e^{-t}$; $\frac{dy}{dt} = \cos t$; $\frac{dz}{dt} = 3x^2(1 - t)e^{-t} + (-1)\cos t = 3t^2e^{-2t}(1 - t)e^{-t} - \cos t = 3t^2e^{-3t}(1 - t) - \cos t$

13. $z = x^2 + 4y^2 + x + 8y - 1$; $z_x = 2x + 1$; $z_y = 8y + 8$; $z_x = 0 \Rightarrow x = -\frac{1}{2}$; $z_y = 0 \Rightarrow y = -1$; $z_{xx} = 2$; $z_{yx} = 0$; $z_{xy} = 0$; $z_{yy} = 8$; $\Delta = \begin{vmatrix} z_{xx} & z_{xy} \\ z_{yx} & z_{yy} \end{vmatrix} = \begin{vmatrix} 2 & 0 \\ 0 & 8 \end{vmatrix} = 16 > 0$. Since z_{xx} is $2 > 0$ we have a local minimum at $\left(-\frac{1}{2}, -1\right)$.

14. $z = x^2 + y^2 - 2x + 4y + 2$; $z_x = 2x - 2 = 0 \Rightarrow x = 1$; $z_y = 2y + 4 = 0 \Rightarrow y = -2$; $z_{xx} = 2$; $z_{xy} = z_{yx} = 0$; $z_{yy} = 2$; $\Delta = \begin{vmatrix} 2 & 0 \\ 0 & 2 \end{vmatrix} = 4 > 0$. Since $z_{yy} = 2$ we have a local minimum at $(1, -2)$.

15. $z = 20 + 12x - 12y - 3x^2 - 2y^2$; $z_x = 12 - 6x = 0 \Rightarrow x = 2$; $z_y = -12 - 4y = 0 \Rightarrow y = -3$; $z_{xx} = -6$; $z_{xy} = z_{yx} = 0$; $z_{yy} = -4$; $\Delta = \begin{vmatrix} -6 & 0 \\ 0 & -4 \end{vmatrix} = 24 > 0$. Since $z_{xx} = 2 > 0$ we have a local minimum at $(2, -3)$.

16. $z = xy + 3x - 2y + 4$; $z_x = y + 3 = 0 \Rightarrow y = -3$; $z_y = x - 2 = 0 \Rightarrow x = 2$; $z_{xx} = 0$; $z_{xy} = z_{yx} = 1$; $z_{yy} = 0$; $\Delta = \begin{vmatrix} 0 & 1 \\ 1 & 0 \end{vmatrix} = -1 < 0$, so there is a saddle point at $(2, -3)$.

17. $z = x^2 + 2xy - y^2$; $z_x = 2x + 2y$; $z_y = 2x - 2y$; $2x + 2y = 0$ and $2x - 2y = 0 \Rightarrow x = 0$ and $y = 0$; $z_{xx} = 2$; $z_{xy} = z_{yx} = 2$; $z_{yy} = -2$. $\Delta \begin{vmatrix} 2 & 2 \\ 2 & -2 \end{vmatrix} = -8 < 0$, so saddle point at $(0, 0)$.

18. $z = x^2 - 3xy - y^2$; $z_x = 2x - 3y$; $z_y = -3x - 2y$; $2x - 3y = 0$ and $-3xy - 2y = 0 \Rightarrow x$ and y both 0. $z_{xx} = 2$; $z_{xy} = z_{yx} = -3$; $z_{yy} = -2$. $\Delta = \begin{vmatrix} 2 & -3 \\ -3 & -2 \end{vmatrix} = -4 - 9 = -13 < 0$, so we have a saddle point at $(0, 0)$.

19. $z = x^3 + x^2y + y^2$; $z_x = 3x^2 + 2xy$; $z_y = x^2 + 2y$; $3x^2 + 2xy = 0$, and $x^2 + 2y = 0$. Solving for $2y$ in the second equation we get $2y = -x^2$. Substituting into the first equation we have $3x^2 - x^3 = 0$ or $x^2(3 - x) = 0$ so $x = 0$ or $x = 3$. Back substituting we get the ordered pairs $(0, 0)$ and $\left(3, -\frac{9}{2}\right)$. $z_{xx} = 6x + 2y$; $z_{xy} = z_{yx} = 2x$; $z_{yy} = 2$. $\Delta = \begin{vmatrix} 6x + 2y & 2x \\ 2x & 2 \end{vmatrix} = 12x + 4y - 4x^2$. At $(0, 0)$ the test fails. At $\left(3, -\frac{9}{2}\right)$, $\Delta = 36 - 18 - 81 < 0$, so there is a saddle point at $\left(3, -\frac{9}{2}\right)$.

20. $z = x^3 + y^3 + 3xy$; $z_x = 3x^2 + 3y$; $z_y = 3y^2 + 3x$. Setting $3x^2 + 3y = 0$ and $3y^2 + 3x = 0$ we solve for y in the first equation getting $y = -x^2$. Substituting this into the second equation yields $3(-x^2)^2 + 3x = 0$ or $x^4 + x = 0$. Hence $x = 0$ or -1. Back substituting we have $y = 0$ or $y = -1$ or the ordered pairs (0, 0) and $(-1, -1)$. $z_{xx} = 6x$; $z_{xy} = z_{yx} = 3$; $z_{yy} = 6y$; $\Delta = \begin{vmatrix} 6x & 3 \\ 3 & 6y \end{vmatrix} = 36xy - 9$; at (0, 0) $\Delta = -9 < 0$ so we have a saddle point at (0, 0). At $(-1, -1)$, $\Delta = 36 - 9 > 0$; $z_{xx} = 6(-1) = -6 < 0$ and we have a local maximum at $(-1, -1)$.

21. $A = bh$; $A_b = h$; $A_h = b$; $dA = hdb + bdh = 25(0.1) + 10(0.1) = 35$ cm^2

22. **(a)** The volume of a right circular cone is $V = \frac{1}{3}\pi r^2 h$. Differentiating with respect to r, we get $V_r = \frac{2}{3}\pi rh$, and with respect to h, produces $V_h = \frac{1}{3}\pi r^2$. Thus, $dV = V_r dr + V_h dh = \frac{2}{3}\pi rh dr + \frac{1}{3}\pi r^2 dh = \frac{2}{3}\pi \cdot 4 \cdot 12 \cdot \frac{1}{16} + \frac{1}{3}\pi(4)^2 \cdot \frac{1}{16} = 2\pi + \frac{\pi}{3} = \frac{7\pi}{3}$ in.^{2}
(b) The total surface area is $A = \pi r^2 + \pi r\ell$, where $\ell = \sqrt{h^2 + r^2} = \sqrt{12^2 + 4^2} = 4\sqrt{10}$. Differentiating with respect to r, yields $A_r = 2\pi r + \pi\sqrt{h^2 + r^2} + \frac{1}{2} \cdot \pi r(2r)\left(h^2 + r^2\right)^{-1/2} = 2\pi r + \pi\sqrt{h^2 + r^2} + \frac{\pi r^2}{\sqrt{h^2 + r^2}}$. Differentiating with respect to h, gives $A_h = \frac{1}{2}\pi r(2h)\left(h^2 + r^2\right)^{-1/2} = \frac{\pi rh}{\sqrt{h^2 + r^2}}$.
Thus we have $dA = A_r dr + A_h dh = \frac{1}{16}\left(8\pi + 4 \times \sqrt{10}\pi + \frac{16\pi}{8\sqrt{10}} + \frac{48\pi}{4\sqrt{10}}\right) = \frac{\pi}{16}\left(8 + 4\sqrt{10} + \frac{2\sqrt{10}}{5} + \frac{6\sqrt{10}}{5}\right) = \frac{\pi}{16}\left(8 + \frac{28\sqrt{10}}{5}\right) = \frac{\pi}{2} + \frac{7\sqrt{10}\pi}{20}$ in.^{2}

23. $R_{R_1} = \frac{(R_1 + R_2)R_2 + R_1 R_2}{(R_1 + R_2)^2}$; $R_{R_2} = \frac{(R_1 + R_2)R_1 + R_1 R_2}{(R_1 + R_2)^2} dR = R_{R_1} dR_1 + R_{R_2} dR_2 = \frac{(500)400 + 40000}{(500)^2} 1 + \frac{500 \cdot 100 + 40000}{500^2} 4 = 0.96 + 1.44 = 2.4\Omega$. (Note $dR_1 = 0.01 \times 100 = 1\Omega$; $dR_2 = 0.01 \times 400 = 4\Omega$.)

24. $PV = kT$; $P = \frac{kT}{V} dP = P_V dV + P_T dT = \frac{-kT}{V^2}$. $(dV) + \frac{k}{V}(dT)$. $T = 100°F = (100 - 32)\frac{5}{9} + 273°K = 316.3°K$. $T = 112°F = (112 - 32)\frac{5}{9} + 273°K = 317.4°K$. $dT = 1.1°K$; $dV = 70 - 66 = 4$ in.^{3}; $k = \frac{PV}{T} = \frac{0.5 \times 66}{316.3} = 0.0104$; $dP = \frac{-0.0104 \cdot 316.3}{66^2} \times 4 + \frac{0.0104}{66} \times 1.1 = -0.00285$ psi.

25. $PV = kT$ or $T = \frac{PV}{k}$. $\frac{dT}{dt} = \frac{\partial T}{\partial P}\frac{dP}{dt} + \frac{\partial T}{\partial V}\frac{dV}{dt} = \frac{V}{k} \cdot \frac{dP}{dt} + \frac{P}{k}\frac{dV}{dt} = \frac{1}{k}\left(V\frac{dP}{dt} + P \cdot \frac{dV}{dt}\right)$.

26. $\frac{dT}{dt} = \frac{1}{0.4}(66 \cdot 0.05 - 0.5 \times 1) = 7$ in · lb/min.

27. $W = \frac{Li^2}{2}$; $dw = \frac{\partial W}{\partial L} dL + \frac{\partial w}{\partial i} di = \frac{i^2}{2}(0.02) + Li(-0.1) = \frac{1^2}{2}(0.02) + 30(1)(-0.1) = 0.01 - 0.3 = -0.29$.

28. $A = xy$; $\frac{dA}{dt} = \frac{\partial A}{\partial x}\frac{dx}{dt} + \frac{\partial A}{\partial y}\frac{dy}{dt} = y\frac{dx}{dt} + x\frac{dy}{dt} = 7 \cdot 2 + 6 \cdot 3 = 14 + 18 = 32$ cm^2/min.

29. $V = \ell \cdot w \cdot h$; $h = \frac{500}{\ell w}$. $A = \ell w + 2\ell \cdot h + 2\ell h = \ell w + 2 \cdot \frac{500}{w} + 2 \cdot \frac{500}{\ell}$. $A_\ell - w - \frac{1000}{w^2}$; $A_w = \ell - \frac{1000}{\ell^2}$; $A_\ell = A_w = 0$ so $w - \frac{1000}{\ell^2} = 0$ and $\ell - \frac{1000}{w^2} = 0$. Hence $\ell = \frac{1000}{w^2}$ and substituting we get $w - \frac{1000}{\left(\frac{1000}{w^2}\right)^2} = 0$ or $w - \frac{w^4}{1000} = 0$. Now $w\left(1 - \frac{w}{1000}\right) = 0$ so $w = 0$ or $w^3 = 1000 \Rightarrow w = 10$. But, w cannot be 0, so $w = 10$ and $\ell = \frac{1000}{10^2} = 10$; $h = \frac{500}{10 \times 10} = 5$. To conclude that this is a minimum we compute $A_{\ell\ell} = \frac{2000}{\ell^3}$; $A_{ww} = \frac{2000}{w^3}$. $A_{w\ell} = A_{\ell w} = 1$; $\Delta \begin{vmatrix} \frac{2000}{10^3} & 1 \\ 1 & \frac{2000}{10^3} \end{vmatrix} = 4 - 1 = 3 > 0$. $A_{\ell\ell} = 2 > 0$ so these dimensions yield a minimum; $h = 5$ cm, $\ell = 10$ cm, $w = 10$ cm.

30. In this case $A = 2(\ell w + \ell h + hw) = 2\left(\ell w + \frac{500}{w} + \frac{500}{\ell}\right)$. $A_\ell = 2\left(w - \frac{500}{\ell^2}\right)$, $A_w = 2\left(\ell - \frac{500}{w^2}\right)$. Setting these both equal to zero and solving we have $w = \frac{500}{\ell^2}$ and substituting $\ell - \frac{500}{\left(\frac{500}{\ell^2}\right)^2} = 0$ or $\ell - \frac{\ell^4}{500} = 0$. Hence, $\ell = 0$ or $\ell = \sqrt[3]{500} = 5\sqrt[3]{4}$. Back substituting we get $h = w = \ell = 5\sqrt[3]{4} \approx 7.937$ cm. As in exercise #29, this yields a minimum.

31. **(a)** $V = \ell wh = 4$ or $h = \frac{4}{\ell w}$. Cost $= 1.25\ell w + 1.50(2\ell h + 2\ell w) = 1.25\ell w + 3\left(\frac{4}{w} + \frac{4}{\ell}\right) = 1.25\ell w + 12\left(\frac{1}{w} + \frac{1}{\ell}\right)$. $C_\ell = 1.25w - \frac{12}{\ell^2}$, $C_w = 1.25\ell - \frac{12}{w^2}$. Setting these both equal to zero we get $w = \frac{9.6}{\ell^2}$. Substituting, we have $1.25\ell - \frac{12}{\left(\frac{9.6}{\ell^2}\right)^2} = 0$ or $1.25\ell - 0.1302\ell^4 = 0$, or $\ell\left(1.25 - 0.1302\ell^3\right) = 0$, so $\ell = 0$ or $\ell = \sqrt[3]{\frac{1.25}{0.1302}} = 2.125$. Since ℓ cannot be 0, we have $\ell = 2.125$ ft, $w = 2.125$ ft and $h = \frac{4}{(2.125)^2} = 0.886$ ft. This will yield a minimum cost.

(b) $C = 1.25(2.125)^2 + 6(2.125 \times 0.886) = \$16.94$

32. $T = x^2 + 2y^2 - x$; $T_x = 2x - 1$; $T_y = 4y$; $T_y = 0$; and $T_x = 0$ yields the point $\left(\frac{1}{2}, 0\right)$. Thus, $T_{xx} = 2$, $T_{xy} = T_{yx} = 0$, $T_{yy} = 4$; $\Delta = \begin{vmatrix} 2 & 0 \\ 0 & 4 \end{vmatrix} = 8 > 0$; $T_{xx} = 2$ so minimum. $T\left(\frac{1}{2}, 0\right) = \frac{1}{2} + 2 \cdot 0 - \frac{1}{2} = -\frac{1}{4} = -0.25$.

33. We begin by differentiating $R(x, y)$ with respect to x and with respect to y, with the following results.
$$R_x(x, y) = 15y(2 - x - 4y) + 15xy(-1)$$
$$= 30y - 15xy - 60y^2 - 15xy$$
$$= 30y - 30xy - 60y^2$$
$$= 30\left(y - xy - 2y^2\right)$$
$$R_y(x, y) = 15x(2 - x - 4y) + 15xy(-4)$$
$$= 30x - 15x^2 - 60xy - 60xy$$
$$= 30x - 15x^2 - 120xy$$
$$= 15\left(2x - x^2 - 8xy\right)$$
Setting $R_x = 0$, we obtain $y - xy - 2y^2 = 0$ or $xy = y - 2y^2$ or $x = 1 - 2y$, if $y \neq 0$. Substituting this value into R_y and setting it equal to 0, we obtain $x = \frac{1}{2}(2y + 1) = \frac{2y+1}{2}$. Substituting this value of x into R_y and setting it equal to 0, we obtain
$$2(1 - 2y) - (1 - 2y)^2 - 8(1 - 2y)y = 0$$
$$2 - 4y - 1 + 4y - 4y^2 - 8y + 16y^2 = 0$$
$$12y^2 - 8y + 1 = 0$$
$$(6y + 1)(2y - 1) = 0$$
Hence, $y = \frac{1}{6}$ or $y = \frac{1}{2}$. Using back substitution produces $x = 1 - 2\left(\frac{1}{6}\right) = \frac{2}{3}$ or $x = 1 - 2\left(\frac{1}{2}\right) = 0$. Since $x > 0$, the maximum reaction occurs when $x = \frac{2}{3}$ and $y = \frac{1}{6}$. This produces a reaction of $15\left(\frac{2}{3} \cdot \frac{1}{6}\right)\left(2 - \frac{2}{3} - 4 \cdot \frac{1}{6}\right) = \frac{10}{9}$.

34. **(a)** Differentiating produces $C_x(x, y) = 4x - 2y$ and $C_y(x, y) = 6y - 2x - 125$. Setting $C_x(x, y) = 0$, we determine that $x = \frac{1}{2}y$. Substituting this value of x into $C_y(x, y)$, we obtain $6y - y - 125$ or $5y - 125$. Setting this equal to 0, we get $y = 25$ or 25 units of packing tape.

Back substitution gives $x = \frac{25}{2}$ or $12\frac{1}{2}$ units of electrical tape. Thus, 12.5 units of electrical

tape and 25 units of packing tape minimize the cost.

(b) The minimum cost would be determined by substituting $x = 12.5$ and $y = 25$ in the given formula
$$C(12.5, \ 25) = 2(12.5)^2 + 3(25)^2 - 2(12.5)(25)$$
$$+ 2(12.5) - 125(25) + 4500$$
$$= 2962.5$$
Thus, the minimum total cost would be \$2962.50.

35. Rewriting $V(x)$ produces
$$V(x) = \frac{Q}{4\pi\epsilon_0}\left(x^2 + a^2\right)^{-1/2}.$$ Differentiating this version, we get
$$E_x = -\frac{\partial V}{\partial x} = -\frac{Q}{4\pi\epsilon_0}\left(-\frac{1}{2}\right)\left(x^2 + a^2\right)^{-3/2}(2x)$$
$$= \frac{Qx}{4\pi\epsilon_0\left(x^2 + a^2\right)^{3/2}}$$

36. Rewriting $V(x)$, produces
$$V(x) = \frac{1}{4\pi\epsilon_0}\frac{Q}{2a}\left[\ln\left(\sqrt{a^2 + x^2} + a\right)\right.$$
$$\left. - \ln\left(\sqrt{a^2 + x^2} - a\right)\right]$$
$$= \frac{Q}{8a\pi\epsilon_0}\left[\ln\left(\left(a^2 + x^2\right)^{1/2} + a\right)\right.$$
$$\left. - \ln\left(\left(a^2 + x^2\right)^{1/2} - a\right)\right]$$
Differentiating this version, we get
$$E_x = -\frac{\partial V}{\partial x} = -\frac{Q}{8a\pi\epsilon_0}\left[\frac{\frac{1}{2}\left(a^2 + x^2\right)^{-1/2}(2x)}{\left(a^2 + x^2\right)^{1/2} + a}\right.$$
$$\left. - \frac{\frac{1}{2}\left(a^2 + x^2\right)^{-1/2}(2x)}{\left(a^2 + x^2\right)^{1/2} - a}\right]$$
$$= -\frac{Q}{8a\pi\epsilon_0}\left[\frac{x\left(a^2 + x^2\right)^{-1/2}}{\left(a^2 + x^2\right)^{1/2} + a}\right.$$
$$\left. - \frac{x\left(a^2 + x^2\right)^{-1/2}}{\left(a^2 + x^2\right)^{1/2} - a}\right]$$
$$= -\frac{Q}{8a\pi\epsilon_0}x\left(a^2 + x^2\right)^{-1/2} \times$$
$$\left[\frac{\left(a^2 + x^2\right)^{1/2} - a}{\left[\left(a^2 + x^2\right)^{1/2} + a\right]\left[\left(a^2 + x^2\right)^{1/2} - a\right]}\right.$$
$$\left. - \frac{\left(a^2 + x^2\right)^{1/2} + a}{\left[\left(a^2 + x^2\right)^{1/2} - a\right]\left[\left(a^2 + x^2\right)^{1/2} + a\right]}\right]$$
$$= -\frac{Q}{8a\pi\epsilon_0}x\left(a^2 + x^2\right)^{-1/2}$$

$$\times \left[\frac{-2a}{a^2 + x^2 - a^2}\right]$$

$$= -\frac{Q}{8a\pi\epsilon_0}x\left(a^2 + x^2\right)^{-1/2}\left[\frac{-2a}{x^2}\right]$$

$$= \frac{Q}{4\pi\epsilon_0}\left[\frac{1}{x\left(a^2 + x^2\right)^{1/2}}\right]$$

30.5 MULTIPLE INTEGRALS

1. $\int_0^1 \int_0^{x^2} xy\, dy\, dx = \int_0^1 \left[\frac{1}{2}xy^2\right]_0^{x^2} dx = \int_0^1 \frac{1}{2}x^5 dx = \left[\frac{1}{12}x^6\right]_0^1 = \frac{1}{12}$

2. $\int_{-1}^1 \int_{-2}^2 (2xy - 3y^2)\, dy\, dx = \int_{-1}^1 \left[xy^2 - y^3\right]_{-2}^2 \times dx = \int_{-2}^1 [(4x - 8) - (4x + 8)]\, dx = \int_{-1}^1 (-16)dx = -16x|_{-1}^1 = (-16) - (+16) = -32$

3. $\int_0^\pi \int_{-\pi/2}^{\pi/2} \sin x \cos y\, dy\, dx = \int_0^\pi [-\sin x \sin y]_{-\pi/2}^{\pi/2} \times dx = \int_0^\pi [-\sin x(1 + 1)]\, dx = -2\cos x|_0^\pi = 2 + 2 = 4$

4. $\int_0^1 \int_0^2 \sqrt{x + y}\, dy\, dx = \int_0^1 \left[\frac{2}{3}(x + y)^{3/2}\right]_0^2 dx = \frac{2}{3}\int_0^1 \left[(x + 2)^{3/2} - x^{3/2}\right] dx = \frac{4}{15}\left[(x + 2)^{5/2} - x^{5/2}\right]_0^1 = \frac{4}{15}\left[(3)^{5/2} - 1 - 2^{5/2}\right] \approx 2.3818$

5. $\int_0^{\ln 2} \int_0^{\ln 5} e^{x+y} dy\, dx = \int_0^{\ln 2} \left[e^{x+y}\right]_0^{\ln 5} dx = \int_0^{\ln 2} e^x(5 - 1) = 4e^x|_0^{\ln 2} = 4(2 - 1) = 4$

6. $\int_0^1 \int_y^{y^2} (x + y)dx\, dy = \int_0^1 \left[\frac{x^2}{2} + xy\right]_y^{y^2} dy = \int_0^1 \left(\frac{y^4}{2} + y^3 - \frac{y^2}{2} - y^2\right) dy = \left[\frac{y^5}{10} + \frac{y^4}{4} - \frac{y^3}{2}\right]_0^1 = \frac{1}{10} + \frac{1}{4} - \frac{1}{2} = \frac{2+5-10}{20} = \frac{-3}{20}$

7. $\int_0^1 \int_x^{\sqrt{x}} (x+y)dy\, dx = \int_0^1 \left[xy + \frac{y^2}{2}\right]_x^{\sqrt{x}} dx = \int_0^1 \times \left(x^{3/2} - \frac{x}{2} - x^2 - \frac{x^2}{2}\right) dx = \left[\frac{2}{5}x^{5/2} + \frac{x^2}{4} - \frac{x^3}{2}\right]_0^1 = \frac{2}{5} + \frac{1}{4} - \frac{1}{2} = \frac{8+5-10}{20} = \frac{3}{20}$

8. $\int_0^4 \int_{-\sqrt{2x}}^{\sqrt{2x}} (2x+y)dy\, dx = \int_0^4 \left(2xy + \frac{y^2}{2}\right)_{-\sqrt{2x}}^{\sqrt{2x}} dx = \int_0^4 \left(2\sqrt{2x}^{3/2} + x + 2\sqrt{2x}^{3/2} - x\right) dx = \int_0^4 4\sqrt{2x}^{3/2} dx = \frac{8\sqrt{2}}{5}x^{5/2}|_0^4 = \frac{8\sqrt{2}\cdot 32}{5} \approx 72.4077$

9. $\int_0^1 \int_{x^3}^x (y - x)dy\, dx = \int_0^1 \left[\frac{y^2}{2} - xy\right]_{x^3}^x dx = \int_0^1 \times \left(\frac{x^2}{2} - x^2 - \frac{x^6}{2} + x^4\right) dx = \left[-\frac{x^3}{6} - \frac{x^7}{14} + \frac{x^5}{5}\right]_0^1 = -\frac{1}{6} - \frac{1}{14} + \frac{1}{5} = \frac{-35-15+42}{210} = \frac{-8}{210} = \frac{-4}{105}$

10. $\int_0^4 \int_y^{\sqrt{y}} (3x + 2y)dx\, dy = \int_0^4 \left[\frac{3x^2}{2} + 2xy\right]_y^{\sqrt{y}} dy = \int_0^4 \left(\frac{3y}{2} + 2y^{3/2} - \frac{3y^2}{2} - 2y^2\right) dy = \int_0^4 \left(\frac{3y}{2} + 2y^{3/2} - \frac{7y^2}{2}\right) dy = \left[\frac{3y^2}{4} + \frac{4}{5}y^{5/2} - \frac{7}{6}y^3\right]_0^4 = 3\cdot 4 + \frac{4}{5}\cdot 4^{5/2} - \frac{7\cdot 64}{6} = 12 + \frac{128}{5} - \frac{7\cdot 64}{6} = \frac{-1112}{30} = \frac{-556}{15} \approx -37.0667$

11. $\int_0^2 \int_0^3 (xy+x-y)dy\, dx = \int_0^2 \left(\frac{xy^2}{2} + xy - \frac{y^2}{2}\right)_0^3 dx = \int_0^2 \left(\frac{9}{2}x + 3x - \frac{9}{2}\right) dx = \int_0^2 \left(\frac{15}{2}x - \frac{9}{x}\right) dx = \left[\frac{15x^2}{4} - \frac{9x}{2}\right]_0^2 = 15 - 9 = 6$

12. $\int_0^1 \int_0^x xy\, dy\, dx = \int_0^1 \frac{xy^2}{2}|_0^x dx = \int_0^1 \frac{x^3}{2} dx = \frac{x^4}{8}|_0^1 = \frac{1}{8}$

13.
$$\int_0^1 \int_0^{1-x} (x^2y + xy^2)\, dy\, dx$$
$$= \int_0^1 \left[\frac{x^2y^2}{2} + \frac{xy^3}{3}\right]_0^{1-x} dy$$
$$= \int_0^1 \left(\frac{x^2(1 - x)^2}{2} + \frac{x(1 - x)^3}{3}\right) dx$$
$$= \int_0^1 \left(\frac{x^2 - 2x^3 + x^4}{2}\right.$$
$$\left. + \frac{x - 3x^2 + 3x^2 - x^4}{3}\right) dx$$
$$= \int_0^1 \left(\frac{x}{3} - \frac{x^2}{2} + \frac{x^4}{6}\right) dx$$
$$= \left[\frac{x^2}{6} - \frac{x^3}{6} + \frac{x^5}{30}\right]_0^1 = \frac{1}{30}$$

14. $\int_0^2 \int_0^{x^2} \left(x^2 - y^2\right) dy \, dx = \int_0^2 \left(x^2 y - \frac{y^3}{3}\right)\Big|_0^{x^2} =$
$\int_0^2 \left[x^4 - \frac{x^6}{3}\right] dx = \left[\frac{x^5}{5} - \frac{x^7}{21}\right]_0^2 = \frac{32}{5} - \frac{128}{21} = \frac{32}{105} \approx$
$0.3048.$

15. $\int_0^1 \int_{y^2}^{y}(xy + 1)dx \, dy = \int_0^1 \left[\frac{x^2 y}{2} + x\right]_{y^2}^{y} dy = \int_0^1 \times$
$\left(\frac{y^3}{2} + y - \frac{y^5}{2} - y^2\right) dy = \left[\frac{y^4}{8} + \frac{y^2}{2} - \frac{y^6}{12} - \frac{y^3}{3}\right]_0^1 =$
$\frac{1}{8} + \frac{1}{2} - \frac{1}{12} - \frac{1}{3} =$
$\frac{3 + 12 - 2 - 8}{24} = \frac{5}{24}.$

16. $\int_1^2 \int_y^{y^2} \frac{x}{y} dx \, dy = \int_1^2 \frac{x^2}{2y}\Big|_y^{y^2} dy = \int_1^2 \left(\frac{y^3}{2} - \frac{y}{2}\right) dy =$
$\left[\frac{y^4}{8} - \frac{y^2}{4}\right]_1^2 = \frac{16}{8} - 1 - \frac{1}{8} + \frac{1}{4} = \frac{16 - 8 - 1 + 2}{8} = \frac{9}{8}.$

17. For this region we can integrate with respect to y
first from x to 1 and then with respect to x from 0
to 1. $\int_0^1 \int_x^1 \left(x^2 - y^2\right) dy \, dx = \int_0^1 \left[x^2 y - \frac{y^3}{3}\right]_x^1 dx$
$= \int_0^1 \left(x^2 - \frac{1}{3} - x^3 + \frac{x^3}{3}\right) dx = \int_0^1 \left(x^2 + \frac{2x^3}{3} - \frac{1}{3}\right)$
$\times dx = \left[\frac{x^3}{3} + \frac{x^4}{6} - \frac{x}{3}\right]_0^1 = \frac{1}{6}.$

18. $\int_0^2 \int_0^{x^2} x^2 y \, dy \, dx = \int_0^2 \frac{x^2 y^2}{2}\Big|_0^{x^2} dx = \int_0^2 \frac{x^6}{2} dx =$
$\frac{x^7}{14}\Big|_0^2 = \frac{128}{14} = \frac{64}{7}.$

19. $\int_0^8 \int_0^{\sqrt[3]{y}} \left(x^3 - y^3\right) dx \, dy = \int_0^8 \left(\frac{x^4}{4} - y^3 x\right)\Big|_0^{\sqrt[3]{y}} dy$
$= \int_0^8 \left(\frac{y^{4/3}}{4} - y^{10/3}\right) dy = \left[\frac{3 y^{7/3}}{28} - \frac{3 y^{13/3}}{13}\right]_0^8 = \frac{3 \cdot 2^7}{28} -$
$\frac{3 \cdot 2^{13}}{13} = -1876.7473.$

20. $\int_0^1 \int_{x^4}^{x} xy \, dy \, dx = \int_0^1 \frac{xy^2}{2}\Big|_{x^4}^{x} dx = \int_0^1 \left(\frac{x^3}{2} - \frac{x^9}{2}\right) \times$
$dx = \frac{x^4}{8} - \frac{x^{10}}{20}\Big|_0^1 = \frac{1}{8} - \frac{1}{20} = \frac{5 - 2}{40} = \frac{3}{40}$

21. $\int_1^8 \int_{8/x}^{9-x} (x + y)dy \, dx$
$= \int_1^8 \left[xy + \frac{y^2}{2}\right]_{8/x}^{9-x} dx$
$= \int_1^8 \left(x(9 - x) + \frac{(9-x)^2}{2} - 8 - \frac{32}{x^2}\right) dx$
$= \int_1^8 \left(9x - x^2 + \frac{81}{2} - 9x\right.$
$\left. + \frac{x^2}{2} - 8 - \frac{32}{x^2}\right) dx$
$= \int_1^8 \left(\frac{-x^2}{2} - \frac{32}{x^2} + \frac{65}{2}\right) dx$
$= \frac{-x^3}{6} + \frac{32}{x} + \frac{65x}{2}\Big|_1^8$

$= \frac{-8^3}{6} + \frac{32}{8} + \frac{65 \cdot 8}{2} + \frac{1}{6} - 32 - \frac{65}{2}$
$= \frac{343}{3} = 114\frac{1}{3}$

22. $\int_1^4 \int_0^5 \left(x^2 + y^2\right) dy \, dx = \int_1^4 \left[x^2 + \frac{y^3}{3}\right]_0^5 dx =$
$\int_1^4 \left(5x^2 + \frac{125}{3}\right) dx = \left[\frac{5x^3}{3} + \frac{125x}{3}\right]_1^4 =$
$\left(\frac{5 \cdot 64}{3} + \frac{126 \cdot 4}{3}\right) - \left(\frac{5}{3} + \frac{125}{3}\right) = \frac{1279}{3} = 230$

23. $\int_0^1 \int_{x^3}^{x} (x + y)^2 dy \, dx$
$= \int_0^1 \int_{x^3}^{x} (x^2 + 2xy + y^2) \, dy \, dx$
$= \int_0^1 \left[x^2 y + xy^2 + \frac{y^3}{3}\right]_{x^3}^{x} dx$
$= \int_0^1 \left[\left(x^3 + x^3 + \frac{x^3}{3}\right)\right.$
$\left. - \left(x^5 + x^7 + \frac{x^9}{3}\right)\right]_{x^3}^{x} dx$
$= \int_0^1 \left[\frac{7}{3}x^3 - x^5 - x^7 - \frac{1}{3}x^9\right] dx$
$= \left[\frac{7}{12}x^4 - \frac{1}{6}x^6 - \frac{1}{8}x^8 - \frac{1}{30}x^{10}\right]_0^1$
$= \frac{7}{12} - \frac{1}{6} - \frac{1}{8} - \frac{1}{30}$
$= \frac{60 + 10 - 20 - 15 - 4}{120} = \frac{31}{120} \approx 0.2583$

24. $x = y^2$ and $x = y + 4$ intersect when $y + 4 = y^2$
or $y^2 - y - 4 = 0$. By the quadratic formula
$y = \frac{1 \pm \sqrt{17}}{2} \approx -1.5616$ and $= 2.5616$. Thus,
the desired integral is $\int_{-1.5616}^{2.5616} \int_{y^2}^{y+4} (xy)dx \, dy =$
$\int_{-1.5616}^{2.5616} \frac{x^2 y}{2}\Big|_{y^2}^{y+4} dy = \frac{1}{2} \int_{-1.5616}^{2.5616} \left[(y + 4)^2 y - y^5\right] \times$
$dy = \frac{1}{2} \int_{-1.5616}^{2.5616} \left(y^3 + 8y^2 + 16y - y^5\right) dy = \frac{1}{2} \times$
$\left[\frac{1}{4}y^4 + \frac{8}{3}y^3 + 8y^2 - \frac{1}{6}y^6\right]_{-1.5616}^{2.5616} \approx \frac{1}{2}(60.9931 -$
$8.4235) = 26.2848.$

25. The desired area is symmetric about the x-axis,
so we need only to integrate the y-values from
0 to 3, and double that answer. Thus, we have
$\int_{-3}^3 \int_0^{y^2 - 9} xy \, dx \, dy = 2 \int_0^3 \int_0^{y^2 - 9} xy \, dx \, dy =$
$2 \int_0^3 \frac{x^2 y}{2}\Big|_0^{y^2 - 9} dy = \int_0^3 \left(y^2 - 9\right)^2 y \, dy = \int_0^3 (y^5 -$
$18y^3 + 81y) \, dy = \left[\frac{1}{6}y^6 - \frac{9}{2}y^4 + \frac{81}{2}y^2\right]_0^3 = 121.5.$

26. The area is $\int_0^1 \int_{x^2}^1 \sqrt{xy}\, dy\, dx = \int_0^1 \frac{2}{3}\sqrt{x}y^{3/2}\big|_{x^2}^1 dx =$

$\frac{2}{3}\int_0^1 (x^{1/2} - x^{7/2})dx = \frac{2}{3}\left[\frac{2}{3}x^{3/2} - \frac{2}{9}x^{9/2}\right]_0^1$

$= \frac{2}{3}\left[\frac{2}{3} - \frac{2}{9}\right] = \frac{2}{3}\left[\frac{6-2}{9}\right] = \frac{2}{3}\cdot\frac{4}{9} = \frac{2}{27}$. (Note: $\sqrt{xy}$ is undefined if $xy < 0$ so the region can only be in the first or third quadrants.) The region described is in the second and first. We can only use the first. Hence $\le x \le 1$ and $x^2 \le y \le 1$).

27. $\overline{f} = \frac{1}{6}\int_0^2 \int_0^3 (x^2 + y^2)\, dy\, dx$

$= \frac{1}{6}\int_0^2 \left[\left(x^2 y + \frac{1}{3}y^3\right)\right]_0^3 dx$

$= \frac{1}{6}\int_0^2 \left[3x^2 + 9\right] dx$

$= \frac{1}{6}\left[x^3 + 9x\right]_0^2$

$= \frac{1}{6}[8 + 18] = \frac{26}{6} = \frac{13}{3}$

28. $\overline{f} = \frac{1}{4\cdot 2}\int_1^5 \int_{-1}^1 (x + y)^2 dy\, dx$

$= \frac{1}{8}\int_1^5 \int_{-1}^1 (x^2 + 2xy + y^2)\, dy\, dx$

$= \frac{1}{8}\int_1^5 \left[x^2 y + xy^2 + \frac{1}{3}y^3\right]_{-1}^1 dx$

$= \frac{1}{8}\int_1^5 \left(2x^2 + \frac{2}{3}\right) dx$

$= \frac{1}{8}\left[\frac{2}{3}x^3 + \frac{2}{3}x\right]_1^5$

$= \frac{1}{8}\left[\frac{250 + 10 - 2 - 2}{3}\right] = \frac{32}{3}$

29. $\overline{f} = \frac{1}{3\cdot 5}\int_1^4 \int_2^7 \frac{x}{y}\, dy\, dx$

$= \frac{1}{15}\int_1^4 [x\ln y]_2^7\, dx$

$= \frac{1}{15}[\ln 7 - \ln 2]\int_1^4 x\, dx$

$= \frac{1}{15}\left(\ln\frac{7}{2}\right)\left[\frac{1}{2}x^2\right]_1^4$

$= \frac{1}{30}\left(\ln\frac{7}{2}\right)[16 - 1] = \frac{1}{2}\ln\frac{7}{2} \approx 0.6264$

30. $\overline{f} = \frac{1}{1\cdot 1}\int_1^2 \int_2^3 e^{2x+y} dy\, dx$

$= \int_1^2 \int_2^3 e^{2x}e^y dy\, dx$

$= \int_1^2 [e^{2x}e^y]_2^3\, dx$

$= (e^3 - e^2)\int_1^2 e^{2x}dx$

$= (e^3 - e^2)\left[\frac{1}{2}e^{2x}\right]_1^2$

$= \frac{1}{2}(e^3 - e^2)(e^4 - e^2)$

$= \frac{1}{2}(e^7 - e^6 - e^5 + e^4) \approx 299.6947$

31. Solving $x + 2y + 2 = 4$ for z we get $z = 4 - x - 2y$. The x-intercept is 4 and the trace in the xy plane is $y = 2 - \frac{1}{2}x$. Thus the volume is

$V = \int_0^4 \int_0^{2-\frac{1}{2}x}(4 - x - 2y)\, dy\, dx$

$= \int_0^4 [4y - xy - y^2]_0^{2-\frac{1}{2}x} dx$

$= \int_0^4 \left[4\left(2 - \frac{1}{2}x\right) - x\left(2 - \frac{1}{2}x\right)\right.$
$\left. - \left(2 - \frac{1}{2}x\right)^2\right] dx$

$= \int_0^4 \left(8 - 2x - 2x + \frac{1}{2}x^2 - 4 + 2x - \frac{1}{4}x^2\right) dx$

$= \int_0^4 \left(4 - 2x + \frac{1}{4}x^2\right) dx$

$= \left[4x - x^2 + \frac{1}{12}x^3\right]_0^4$

$= 16 - 16 + \frac{1}{12}64 = \frac{16}{3} = 5\frac{1}{3}$

32. Solving $x^2 + z^2 = 4$ for z we have $z = \pm\sqrt{4 - x^2}$. To get the volume we integrate from 0 to 2 and double the answer. Thus, $V = 2\int_0^2 \int_0^2 \sqrt{4 - x^2} \times dx\, dy = 2\int_0^2 \left(\frac{x}{4}\sqrt{4 - x^2} + 2\sin^{-1}\frac{x}{2}\right)_0^2 = 2\int_0^2 \times (2\sin^{-1} 1 - 2\sin^{-1} 0)\, dy = 2\int_0^2 2\pi = 4\pi y|_0^2 = 8\pi$.

33. As in problem #28, $z = \pm\sqrt{2 - x^2}$. We can integrate from 0 to $\sqrt{2}$ and double the answer. Thus, $V = 2\int_0^{\sqrt{2}} \int_0^x \sqrt{2 - x^2}dy\, dx = 2\int_0^{\sqrt{2}}\sqrt{2 - x^2} \cdot x\, dx$. Let $u = 2 - x^2$, and $du = -2x\, dx$. Substituting, we get $V = -\int_0^{\sqrt{2}} u^{1/2}du = -\frac{2}{3}u^{3/2} = -\frac{2}{3}\sqrt{2 - x^2}^3|_0^{\sqrt{2}} = \frac{2}{3}\sqrt{2}^3 = \frac{4\sqrt{2}}{3} \approx 1.8856$.

34. $\int_0^\infty \int_0^\infty e^{-(x+y)}dy\, dx = \int_0^\infty -e^{-(x+y)}|_0^\infty dx = \int_0^\infty e^{-x}dx = -e^{-x}|_0^\infty = e^0 = 1$.

35. C_1 and C_2 are independent operations. Therefore, we will figure the averages separately and then add the results.

$$\overline{C_1} = \frac{1}{100} \int_{300}^{400} \left(0.2x^2 + 40x + 3600\right) dx$$

$$= \frac{1}{100} \left[\frac{0.2}{3}x^3 + \frac{40}{2}x^2 + 3600x\right]_{300}^{400}$$

$$= 42{,}266.7$$

$$\overline{C_2} = \frac{1}{200} \int_{350}^{550} \left(0.4y^2 + 24y + 6{,}000\right) dy$$

$$= \frac{1}{200} \left[\frac{0.4}{3}y^3 + 12y^2 + 6{,}000y\right]_{350}^{550}$$

$$= 99{,}133.3$$

Adding produces $C_1 + C_2 = 42{,}266.7 + 99{,}133.3 = 141{,}400$. The average cost of $C_1(x) + C_2(y)$ is \$141,400.

36.
$$\overline{N} = \frac{1}{10} \int_{10}^{20} \int_{10}^{20} x^{0.75} y^{0.25} dy\ dx$$

$$= \frac{1}{10} \frac{1}{1.25} \int_{10}^{20} \left[x^{0.75} y^{1.25}\right]_1^2 dx$$

$$= \frac{1}{10} \frac{2^{1.25} - 1}{1.25} \int_{10}^{20} x^{0.75} dx$$

$$= \frac{1}{10} \frac{2^{1.25} - 1}{1.25} \frac{1}{1.75} \left[x^{1.75}\right]_{10}^{20}$$

$$= \frac{1}{10} \frac{2^{1.25} - 1}{1.25} \frac{1}{1.75} \left[20^{1.75} - 10^{1.75}\right] \approx 8.375$$

Thus, the average number of units produced for these ranges of labor-hour and capital expenditures is 8.375 thousand or 8,375 units.

37. **(a)** By the Pythagorean theorem, $d^2 = x^2 + y^2$, so $C(x, y) = 100 - 15\left(x^2 + y^2\right) = 100 - 15x^2 - 15y^2$.

(b) Since the plant is in the center of town, we integrate for $-1 \le x \le 1$ and $-2 \le y \le 2$.

$$\overline{C} = \frac{1}{4 \cdot 2} \int_{-2}^{2} \int_{-1}^{1} \left(100 - 15x^2 - 15y^2\right) dx\ dy$$

$$= \frac{1}{8} \int_{-2}^{2} \left[100x - 5x^3 - 15xy^2\right]_{-1}^{1} dy$$

$$= \frac{1}{8} \int_{-2}^{2} \left[200 - 10 - 30y^2\right] dy$$

$$= \frac{1}{8} \int_{-2}^{2} \left[190 - 30y^2\right] dy$$

$$= \frac{1}{8} \left[190y - 10y^3\right]_{-2}^{2}$$

$$= \frac{1}{8}[760 - 160] = \frac{1}{8}[600] = 75$$

Thus, the average concentration of particulate matter throughout the town is 75 ppm.

38.
$$\overline{L} = \frac{1}{1000 \cdot 10} \int_{2000}^{3000} \int_{50}^{60} 0.000\ 02xy^2 dy\ dx$$

$$= 2 \times 10^{-9} \int_{2000}^{3000} \int_{50}^{60} xy^2 dy\ dx$$

$$= 2 \times 10^{-9} \int_{2000}^{3000} \left[\frac{1}{3}xy^3\right]_{50}^{60} dx$$

$$= \frac{182{,}000}{3 \times 10^9} \int_{2000}^{3000} x\ dx$$

$$= \frac{182{,}000}{3 \times 10^9} \left[\frac{1}{2}x^2\right]_{2000}^{3000}$$

$$= \frac{91{,}000}{3 \times 10^9}(5{,}000{,}000) \approx 151.67$$

The car will travel about 151.67 ft.

≡ 30.6 CYLINDRICAL AND SPHERICAL COORDINATES

1. Here $x = 2\cos\frac{\pi}{4} = \sqrt{2}$ and $y = 2\sin\frac{\pi}{4} = \sqrt{2}$. Thus, the cylindrical coordinate $(2,\ \pi/4,\ 2)$ has the equivalent rectangular coordinate $(\sqrt{2},\ \sqrt{2},\ 2)$.

2. Here $x = 3\cos\frac{2\pi}{3} = -\frac{3}{2}$ and $y = 3\sin\frac{2\pi}{3} = \frac{3\sqrt{3}}{2}$. Thus, the cylindrical coordinate $(3,\ 2\pi/3,\ -2)$ has the equivalent rectangular coordinate $\left(-\frac{3}{2},\ \frac{3\sqrt{3}}{2},\ -2\right)$ or $(-1.5,\ 2.598,\ -2)$.

3. $x = 2\cos 0 = 2$, $y = 2\sin 0 = 0$. Thus, the cylindrical coordinate $(2,\ 0,\ 4)$ has the equivalent rectangular coordinate $(2,\ 0,\ 4)$.

4. $x = y = 0$. Thus, the cylindrical coordinate $(0,\ 5\pi/4,\ -5)$ has the equivalent rectangular coordinate $(0,\ 0,\ -5)$.

5. Here $x = 5\cos\frac{5\pi}{4} = \frac{-5\sqrt{2}}{2}$ and $y = 5\sin\frac{5\pi}{4} = -\frac{5\sqrt{2}}{2}$, so the cylindrical coordinate $(5,\ 5\pi/4,\ 0)$ has the equivalent rectangular coordinate $\left(\frac{-5\sqrt{2}}{2},\ \frac{-5\sqrt{2}}{2},\ 0\right)$ or $(-3.5355,\ -3.5355,\ 0)$.

6. Here $x = 4\cos\frac{5\pi}{3} = 2$ and $y = 4\sin\frac{5\pi}{3} = -2\sqrt{3}$, which means that the cylindrical coordinate $(4,\ 5\pi/3,\ 7)$ has the equivalent rectangular coordinate $(2,\ -2\sqrt{3},\ 7)$ or $(2,\ -3.4641,\ 7)$.

7. Here $r = \sqrt{x^2 + y^2} = \sqrt{2^2 + 2^2} = \sqrt{8} = 2\sqrt{2}$ and $\theta = \tan^{-1}\frac{y}{x} = \tan^{-1}\frac{2}{2} = \tan^{-1} 1 = \frac{\pi}{4}$. This means that the rectangular coordinates (2, 2, 5) have cylindrical coordinates $\left(2\sqrt{2}, \frac{\pi}{4}, 5\right) = $ (2.8284, 0.7854, 5).

8. Here $r = \sqrt{3^2 + (-4)^2} = 5$ and $\theta = \tan^{-1}\frac{-4}{3} = 5.3559$. This means that the rectangular coordinates (3, −4, −5) have cylindrical coordinates (5, 5.3559, −5).

9. Here $r = \sqrt{(-4)^2 + 3^2} = 5$ and $\theta = \tan^{-1}\frac{-3}{4} = 2.4981$. This means that the rectangular coordinates (−4, 3, 2) have cylindrical coordinates (5, 2.4981, 2).

10. Here $r = \sqrt{(-1)^2 + (-\sqrt{3})^2} = \sqrt{1 + 3} = 2$ and $\theta = \tan^{-1}\frac{-\sqrt{3}}{-1} = 4.1888$. Thus, the rectangular coordinates $(-1, -\sqrt{3}, 4)$ have cylindrical coordinates (2, 4.1888, 4) or $\left(2, \frac{4\pi}{3}, 4\right)$.

11. Here $r = \sqrt{12^2 + (-5)^2} = 13$ and $\theta = \tan^{-1}\frac{-5}{12} = 5.8884$. Thus, the rectangular coordinates (12, −5, −3) have cylindrical coordinates (13, 5.8884, −3)

12. Here $r = \sqrt{6^2 + 8^2} = 10$ and $\theta = \tan^{-1}\frac{8}{6} = 0.9273$. Thus, the rectangular coordinates (6, 8, −4) have cylindrical coordinates (10, 0.9273, −4)

13. Here $x = 4\sin\frac{\pi}{6}\cos\frac{\pi}{4} = 4 \cdot \frac{1}{2} \cdot \frac{\sqrt{2}}{2} = \sqrt{2}$ and $y = 4\sin\frac{\pi}{6}\frac{\pi}{4} = 4 \cdot \frac{1}{2} \cdot \frac{\sqrt{2}}{2} = \sqrt{2}$ and $z = 4\cos\frac{\pi}{6} = 4 \cdot \frac{3}{2} = 2\sqrt{3}$. Thus, the spherical coordinate $\left(4, \frac{\pi}{4}, \frac{\pi}{6}\right)$ has the equivalent rectangular coordinate $(\sqrt{2}, \sqrt{2}, 2\sqrt{3}) \approx (1.4142, 1.4142, 3.4641)$

14. Here $x = \sin\frac{\pi}{2}\cos\frac{\pi}{6} = 1 \cdot \frac{1}{2} \cdot \frac{\sqrt{3}}{2} = \frac{\sqrt{3}}{2}$ and $y = \sin\frac{\pi}{2}\sin\frac{\pi}{6} = \frac{1}{2}$ and $z = \cos\frac{\pi}{2} = 0$. Thus, the spherical coordinate $\left(1, \frac{\pi}{6}, \frac{\pi}{2}\right)$ has the equivalent rectangular coordinate $\left(\frac{\sqrt{3}}{3}, \frac{1}{2}, 0\right) = $ (0.8660, 0.5, 0).

15. Here $x = 3\sin\frac{5\pi}{3}\cos\frac{\pi}{2} = 3 \cdot \frac{-\sqrt{3}}{2} \cdot 0 = 0$, $y = \sin\frac{5\pi}{3}\sin\frac{\pi}{2} = \frac{-3\sqrt{3}}{2}$, and $z = 3\cos\frac{5\pi}{3} = \frac{3}{2}$. Thus, the spherical coordinate $\left(3, \frac{\pi}{2}, \frac{5\pi}{3}\right)$ has the equivalent rectangular coordinate $\left(0, \frac{-3\sqrt{3}}{2}, \frac{3}{2}\right) = $ (0, −2.5981, 1.5).

16. Here $x = 2\sin\frac{3\pi}{4}\cos\frac{5\pi}{6} = 2 \cdot \frac{-\sqrt{2}}{2} \cdot \frac{-\sqrt{3}}{2} = \frac{-\sqrt{6}}{2}$, $y = 2\sin\frac{3\pi}{4}\sin\frac{5\pi}{6} = 2 \cdot \frac{\sqrt{2}}{2} \cdot \frac{1}{2} = \frac{\sqrt{2}}{2}$, and $z = 2\cos\frac{3\pi}{4} = -\sqrt{2}$. Thus, the spherical coordinate $\left(2, \frac{5\pi}{6}, \frac{3\pi}{4}\right)$ has the equivalent rectangular coordinate $\left(\frac{-\sqrt{6}}{2}, \frac{\sqrt{2}}{2}, -\sqrt{2}\right) = $ (−1.2247, 0.7071, −1.4142).

17. Here $x = 5\sin\frac{2\pi}{3}\cos\frac{7\pi}{6} = 5 \cdot \frac{-\sqrt{3}}{2} \cdot \frac{-\sqrt{3}}{2} = \frac{-15}{4} = -3.75$, $y = 5\sin\frac{2\pi}{3}\sin\frac{7\pi}{6} = 5 \cdot \frac{\sqrt{3}}{2} \cdot \frac{-1}{2} = \frac{-5\sqrt{3}}{4} \approx -2.1651$, and $z = 5\cos\frac{2\pi}{3} = 5 \cdot -\frac{1}{2} = -2.5$. Thus, the spherical coordinate $\left(5, \frac{7\pi}{6}, \frac{2\pi}{3}\right)$ has the equivalent rectangular coordinate (−3.75, −2.1651, −2.5).

18. Here $x = 5\sin\frac{\pi}{4}\cos\frac{5\pi}{3} = 5 \cdot \frac{\sqrt{2}}{2} \cdot \frac{1}{2} = \frac{5\sqrt{2}}{4} = 1.7678$, $y = 5\sin\frac{\pi}{4}\sin\frac{5\pi}{3} = 5 \cdot \frac{\sqrt{2}}{2} \cdot \frac{-\sqrt{3}}{2} = \frac{-5\sqrt{6}}{4} \approx -3.0619$, and $z = 5\cos\frac{\pi}{4} = \frac{5\sqrt{2}}{2} = 3.5355$. Thus, the spherical coordinate $\left(5, \frac{5\pi}{3}, \frac{\pi}{4}\right)$ has the equivalent rectangular coordinate (1.7678, −3.0619, 3.5355).

19. Here $\rho = \sqrt{4^2 + 3^2} = 5$, $\theta = \tan^{-1}\frac{3}{4} = 0.6435$, and $\phi = \cos^{-1}\frac{0}{5} = \frac{\pi}{2} \approx 1.5708$. Thus, the rectangular coordinate (4, 3, 0) has the spherical coordinate $\left(5, 0.6435, \frac{\pi}{2}\right)$.

20. Here $\rho = \sqrt{4^2 + \sqrt{5}^2 + (-2)^3} = \sqrt{16 + 5 + 4} = 5$, $\theta = \tan^{-1}\frac{\sqrt{5}}{4} = 0.5097$, and $\phi = \cos^{-1}\frac{-2}{5} \approx 1.9823$. Thus, the rectangular coordinate $(4, \sqrt{5}, -2)$ has the spherical coordinate (5, 0.5097, 1.9823).

21. Here $\rho = \sqrt{2^2 + 1^2 + 2^2} = 3$, $\theta = \tan^{-1}\frac{1}{2} = 0.4636$, and $\phi = \cos^{-1}\frac{-2}{3} = 2.3005$. Thus, the rectangular coordinate (2, 1, −2) has the spherical coordinate (3, 0.4636, 2.3005).

22. Here $\rho = \sqrt{4^2 + (-\sqrt{3})^2 + 2^2} = \sqrt{16 + 3 + 4} = \sqrt{23} = 4.7958$, $\theta = \tan^{-1}\frac{-\sqrt{3}}{4} = 2\pi = -0.408637 = 5.874547$, and $\phi = \cos^{-1}\frac{2}{\sqrt{23}} = 1.1406$. Thus, the rectangular coordinate $(4, -\sqrt{3}, 2)$ has the spherical coordinate (4.7958, 5.8745, 1.1406).

23. Here $\rho = \sqrt{1^2 + 1^2 + \sqrt{2}^2} = 2$, $\theta = \tan^{-1}\frac{1}{1} = \frac{\pi}{4} = 0.7854$, and $\phi = \cos^{-1}\frac{\sqrt{2}}{2} = \frac{\pi}{4}$. Thus, the rectangular coordinate $(1, 1, \sqrt{2})$ has the spherical coordinate $\left(2, \frac{\pi}{4}, \frac{\pi}{4}\right) = (2, \ 0.7854, \ 0.7854)$.

24. Here $\rho = \sqrt{3+3+3} = 3$, $\theta = \tan^{-1}\frac{\sqrt{3}}{-\sqrt{3}} = \frac{3\pi}{4} = 2.3562$, and $\phi = \cos^{-1}\frac{-\sqrt{3}}{3} = 2.1863$. Thus, the rectangular coordinate $(-\sqrt{3}, \sqrt{3}, -\sqrt{3})$ has that spherical coordinate $\left(3, \frac{3\pi}{4}, 2.1863\right)$.

25. The graph is a line that makes an angle of $\frac{\pi}{4}$ with the z-axis and whose image on the xy-plane makes an angle of $\frac{\pi}{4}$ with the x-axis.

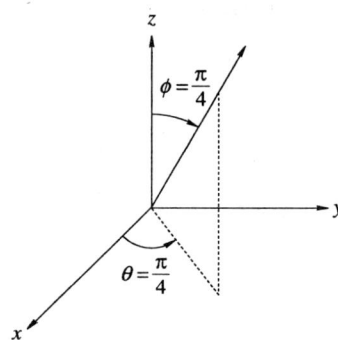

26. A plane through the z-axis and that makes an angle of $\frac{3\pi}{4}$ with the x-axis.

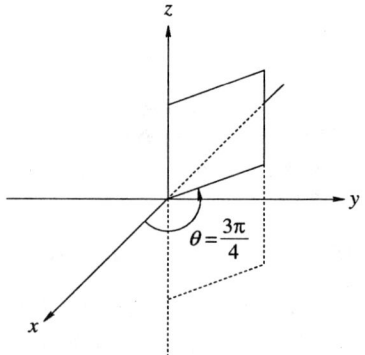

27. The region between the two spheres $\rho = 3$ and $\rho = 5$.

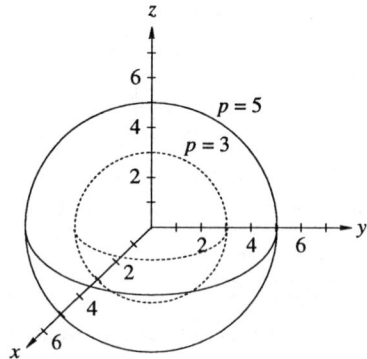

28. A circle of radius 5 in the plane formed by $\theta = \pi/3$.

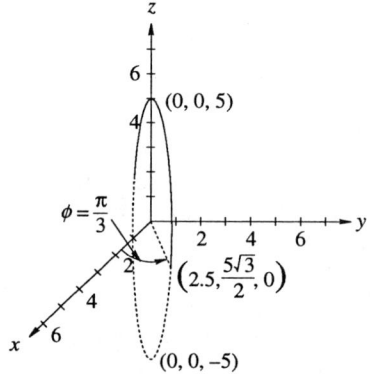

29. A cone shaped figure with spherical base.

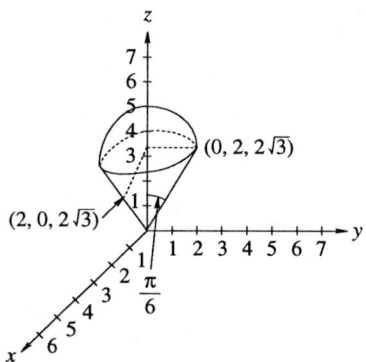

30. A wedge that is a section of a sphere of radius 4 on top of a section of a cone

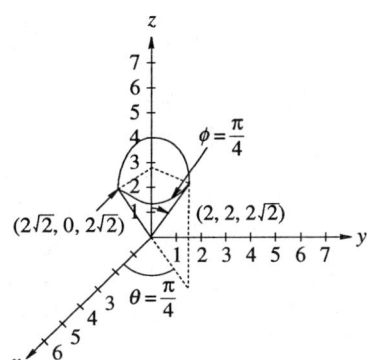

31. See *Computer Programs* in the main text.

30.7 MOMENTS AND CENTROIDS

1. $A = \int_0^3 \int_0^{2x+3} dy\, dx = \int_0^3 y|_0^{2x+3}dx = \int_0^3 (2x + 3)dx = x^2 + 3x|_0^3 = 18.$ $m = \rho A = 18\rho.$
$M_x = \int_0^3 \int_0^{2x+3} y\rho\, dy\, dx = \rho \int_0^3 \frac{y^2}{2}|_0^{2x+3}dx = \frac{1}{2}\rho \int_0^3 (2x+3)^2 dx = \frac{1}{2}\rho \int_0^3 \left(4x^2 + 12x + 9\right) dx = \frac{1}{2}\rho \left[\frac{4}{3}x^3 + 6x^2 + 9x\right]_0^3 = 58.5\rho.$ $M_y = \int_0^3 \int_0^{2x+3} \times x\rho\, dy\, dx = \rho \int_0^3 x[y]_0^{2x+3}dx = \rho \int_0^3 \left(2x^2 + 3x\right) dx = \rho \left[\frac{2}{3}x^3 + \frac{3}{2}x^2\right]_0^3 = 31.5\rho.$ $\overline{x} = \frac{M_y}{m} = \frac{31.5\rho}{18\rho} = 1.75; \overline{y} = \frac{M_y}{m} = \frac{58.5\rho}{18\rho} = 3.25$

2. $m = \rho \int_0^2 \int_0^{x^3} dy\, dx = \rho \int_0^2 x^3 dx = \rho\frac{x^4}{4}|_0^2 = 4\rho.$ $M_x = \rho \int_0^2 \int_0^{x^3} y\, dy\, dx = \rho \int_0^2 \frac{y^2}{2}|_0^{x^3} dx = \frac{\rho}{2} \int_0^2 x^6 dx = \frac{\rho}{2} \cdot \frac{x^7}{7}|_0^2 = \frac{64}{7}\rho \approx 9.142857\rho.$ $M_y = \rho \int_0^2 \int_0^{x^3} x\, dy\, dx = \rho \int_0^2 xy|_0^{x^3} dx = \int_0^2 x^4 dx = \frac{\rho}{2} \cdot \frac{x^5}{5}|_0^2 = \frac{32}{5}\rho = 6.4\rho;$ $\overline{x} = \frac{M_y}{m} = \frac{6.4}{4} = 1.6;$

$\overline{y} = \frac{M_x}{m} = \frac{\frac{64}{7}}{4} = \frac{16}{7} \approx 2.2857$

3. $m = \rho \int_0^8 \int_0^{x^{1/3}} dy\, dx = \rho \int_0^8 x^{1/3}dx = \rho\frac{3}{4}x^{4/3}|_0^8 = 12\rho.$ $M_x = \rho \int_0^8 \int_0^{x^{1/3}} y\, dy\, dx = \rho \int_0^8 \frac{y^2}{2}|_0^{x^{1/3}} dx = \rho \int_0^8 \frac{1}{2}x^{2/3}dx = \rho \left[\frac{1}{2} \cdot \frac{3}{5}x^{5/3}\right]_0^8 = 9.6\rho.$ $M_y = \rho \int_0^8 \int_0^{x^{1/3}} x\, dy\, dx = \rho \int_0^8 x^{4/3}dx = \rho\frac{3}{7}x^{7/3}|_0^8 = 54.8571.$ $\overline{x} = \frac{M_y}{m} = \frac{32}{7} = 4.5714; \overline{y} = \frac{M_x}{m} = 0.8$

4. $m = \rho \int_{-1}^2 \int_0^{x^4} dy\, dx = \rho \int_{-1}^2 x^4 dx = \rho\frac{x^5}{5}|_{-1}^2 = 6.6\rho.$ $M_x = \rho \int_{-1}^2 \int_0^{x^4} y\, dy\, dx = \rho \int_{-1}^2 \frac{y^2}{2}|_0^{x^4} dx = \rho \int_{-1}^2 \frac{x^8}{2}dx = \rho\frac{x^9}{18}|_{-1}^2 = 28.5\rho.$ $M_y = \rho \int_{-1}^2 \int_0^{x^4} \times x\, dy\, dx = \rho \int_{-1}^2 x^5 dx = \rho\frac{x^6}{6}|_{-1}^2 = 10.5\rho.$ $\overline{x} = \frac{M_y}{m} = \frac{35}{22} = 1.5909; \overline{y} = \frac{M_x}{m} = \frac{95}{22} = 4.3182.$

5. $m = \rho \int_0^5 \int_0^{\sqrt{x+4}} dy\, dx = \rho \int_0^5 \sqrt{x + 4}\, dx = \rho \cdot \frac{2}{3}(x + 4)^{3/2}|_0^5 = \left(18 - \frac{16}{3}\right)\rho = \frac{38}{3}\rho = 12.6667\rho.$ $M_x = \rho \int_0^5 \int_0^{\sqrt{x+4}} y\, dy\, dx = \rho \int_0^5 \frac{y^2}{2}|_0^{\sqrt{x+4}}dx = \rho \int_0^5 \frac{x+4}{2}dx = \rho \left[\frac{x^2}{4} + 2x\right]_0^5 = \frac{65}{4}\rho = 16.25\rho.$ $M_y = \rho \int_0^5 \int_0^{\sqrt{x+4}} x\, dy\, dx = \rho \int_0^5 xy|_0^{\sqrt{x+4}}dx = \rho \int_0^5 x\sqrt{x + 4}\, dx.$ Let $u = x + 4$ and $du = dx.$ Then $x = u - 4,$ $x = 0 \Rightarrow u = 4,$ and $x = 5 \Rightarrow u = 9.$ Substituting, we obtain $\rho \int_4^9 (u - u)\sqrt{4}\, du = \rho \int_4^9 \left(u^{3/2} - 4u^{1/2}\right) = \rho \left[\frac{2}{5}u^{5/2} - \frac{8}{3}u^{3/2}\right]_4^9 = 33.7333\rho.$ $\overline{x} = \frac{M_y}{m} = 2.6632, \overline{y} = \frac{M_x}{m} = 1.2829.$

6. $m = \rho \int_0^4 \int_0^{\sqrt{x^2+16}} dy\, dx = \rho \int_0^4 \sqrt{x^2 + 16}\, dx =.$ By Formula #30, in Appendix C, we have $m = \rho \left[\frac{x}{2}\sqrt{x^2 + 16} + 8\ln|x + \sqrt{x^2 + 16}|\right]_0^4 = (29.4551 - 11.0904)\rho = 18.3647\rho.$ $M_x = \rho \int_0^4 \times \int_0^{\sqrt{x^2+16}} y\, dy\, dx = \rho \int_0^4 \frac{y^2}{2}|_0^{\sqrt{x^2+16}}dx = \rho \int_0^4 \times \frac{x^2+16}{2}dx = \rho \left[\frac{x^3}{6} + 8x\right]_0^4 = 42.6667\rho.$ $M_y =$

$\rho \int_0^4 \int_0^{\sqrt{x^2+16}} x \, dy \, dx = \rho \int_0^4 \left(x\sqrt{x^2+16} \right) dx.$
Let $u = x^2 + 16$, and then $du = 2x \, dx$. When $x = 0 \Rightarrow u = 16$, and when $x = 4 \Rightarrow u = 32$. Substitution yields $\frac{1}{2}\rho \int_{16}^{32} \sqrt{u} \, du = \frac{1}{2}\rho \frac{2}{3} u^{3/2}\big|_{16}^{32} = \frac{1}{3}\rho \left[32^{3/2} - 16^{3/2} \right] = 39.0064\rho.$ Thus, $\bar{x} = \frac{M_y}{m} = 2.1240$, $\bar{y} = \frac{M_x}{m} = 2.3233.$

7. $m = \rho \int_0^2 \int_{x^2}^{4x} dy \, dx = \rho \int_0^2 \left(4x - x^2 \right) dx = \left(2x^2 - \frac{x^3}{3} \right)\big|_0^2 \rho = \left(8 - \frac{8}{3} \right)\rho = \frac{16}{3}\rho = 5.3333\rho.$
$M_x = \rho \int_0^2 \int_{x^2}^{4x} y \, dy \, dx = \rho \int_0^2 \frac{y^2}{2}\big|_{x^2}^{4x} dx = \rho \int_0^2 \times \left(8x^2 - \frac{x^4}{2} \right) dx = \rho \left[\frac{8}{3}x^3 - \frac{x^5}{10} \right]_0^2 = 18.1333\rho.$
$M_y = \rho \int_0^2 \int_{x^2}^{4x} x \, dy \, dx = \rho \int_0^2 x \left(4x - x^2 \right) dx = \rho \int_0^2 \left(4x^2 - x^3 \right) dx = \rho \left[\frac{4}{3}x^3 - \frac{x^4}{4} \right]_0^2 = 6.6667 = 6\frac{2}{3}\rho.$ $\bar{x} = \frac{M_y}{m} = 1.25.$ $\bar{y} = \frac{M_x}{m} = 3.4$

8. $m = \rho \int_0^2 \int_{x^3}^{8x} dy \, dx = \rho \int_0^2 \left(8x - x^3 \right) dx = \rho \left[4x^2 - \frac{x^4}{4} \right]_0^2 = 12\rho.$ $M_x = \rho \int_0^2 \int_{x^3}^{8x} y \, dy \, dx = \rho \int_0^2 \frac{y^2}{2}\big|_{x^3}^{8x} dx = \rho \int_0^2 \left(32x^2 - \frac{x^6}{2} \right) dx = \rho \left[\frac{32}{3}x^3 - \frac{x^7}{14} \right]_0^2 = 76.1905\rho.$ $M_y = \rho \int_0^2 \int_{x^3}^{8x} x \times dy \, dx = \rho \int_0^2 \left(8x^2 - x^4 \right) dx = \rho \left[\frac{8}{3}x^3 - \frac{x^5}{5} \right]_0^2 = 14.9333\rho.$ Thus, $\bar{x} = \frac{M_y}{m} = 1.2444$, $\bar{y} = \frac{M_x}{m} = 6.3492.$

9. $m = \rho = \int_0^1 \int_{x^{3/2}}^{x} dy \, dx = \rho \int_0^1 \left(x - x^{3/2} \right) dx = \rho \left[\frac{x^2}{2} - \frac{2}{5}x^{5/2} \right]_0^1 = \rho \left(\frac{1}{2} - \frac{2}{5} \right) = \frac{1}{10}\rho.$ Thus, we have $M_x = \rho \int_0^1 \int_{x^{3/2}}^{x} y \, dy \, dx = \rho \int_0^1 \frac{y^2}{2}\big|_{x^{3/2}}^{x} dx = \rho \int_0^1 \left(\frac{x^2}{2} - \frac{x^3}{2} \right) dx = \rho \left[\frac{x^3}{6} - \frac{x^4}{8} \right]_0^1 = \frac{1}{24}.$ We also have $M_y = \rho = \int_0^1 \int_{x^{3/2}}^{x} x \, dy \, dx = \rho \int_0^1 \left(x^2 - x^{5/2} \right) dx = \rho \left[\frac{x^3}{3} - \frac{2}{7}x^{7/2} \right]_0^1 = \left(\frac{1}{3} - \frac{2}{7} \right)\rho = \frac{1}{21}\rho.$ Hence, $\bar{x} = \frac{M_y}{m} = \frac{1/21}{1/10} = \frac{10}{21} \approx 0.4762$, and $\bar{y} = \frac{M_x}{m} = \frac{1/24}{1/10} = \frac{5}{12} \approx 0.4167.$

10. $m = \int_{-3}^3 \int_{x^2}^{18-x^2} \rho \, dy \, dx = \rho \int_{-3}^3 \left(18 - 2x^2 \right) dx = \rho \left(18x - \frac{2}{3}x^3 \right)\big|_{-3}^3 = 72\rho.$ $M_x = \rho \int_{-3}^3 \int_{x^2}^{18-x^2} \times y \, dy \, dx = \rho \int_{-3}^3 \frac{y^2}{2}\big|_{x^2}^{18-x^2} dx = \frac{\rho}{2} \left(18 - x^2 \right)^2 - \left(x^2 \right)^2 dx = \frac{\rho}{2} \int_{-3}^3 324 - 36x^2 \, dx = \frac{\rho}{2} (324x - 12x^3)\big|_{-3}^3 = 648\rho.$ $M_y = \rho \int_{-3}^3 \int_{x^2}^{18-x^2} x \, dy \, dx = \rho \int_{-3}^3 \left(18x - 2x^3 \right) dx = 0.$ $\bar{x} = 0$, $\bar{y} = 9$

11. $y = x$ and $y = 12 - x^2$ intersect at $x = -4$ and $x = 3$. First, we determine that $m = \rho \int_{-4}^3 \int_x^{12-x^2} \times dy \, dx = \rho \int_{-4}^3 \left(12 - x - x^2 \right) dx = \rho \left(12x - \frac{x^2}{2} - \frac{x^3}{3} \right)_{-4}^3 = 57\frac{1}{6}\rho.$ Then, we have $M_x = \rho \int_{-4}^3 \int_x^{12-x^2} \times y \, dy \, dx = \rho \int_{-4}^3 \frac{y^2}{2}\big|_x^{12-x^2} dx = \frac{\rho}{2} \int_{-4}^3 (144 - 24x^2 + x^4 - x^2) dx = \frac{\rho}{2} \left[144x - \frac{25x^3}{3} + \frac{x^5}{5} \right]_{-4}^3 = \frac{(260.1+276.8)}{2}\rho = 251.5333\rho.$ We also have $M_y = \rho \int_{-4}^3 \int_x^{12-x^2} x \, dy \, dx = \rho \int_{-4}^3 \left[12x - x^3 - x^2 \right] dx = \rho \left[6x^2 - \frac{x^3}{3} - \frac{x^4}{4} \right]_{-4}^3 = 28.5833\rho.$ Hence, $\bar{x} = \frac{-28.5833}{57.1667} = -0.5$ and $\bar{y} = \frac{251.5333}{57.1667} = 4.4.$

12. $m = \rho \int_0^1 \int_{x^3}^{x^2} dy \, dx = \rho \int_0^1 \left(x^2 - x^3 \right) dx = \rho \times \left[\frac{x^3}{3} - \frac{x^4}{4} \right]_0^1 = \frac{\rho}{12}.$ $M_x = \rho \int_0^1 \int_{x^3}^{x^2} y \, dy \, dx = \rho \int_0^1 \frac{y^2}{2}\big|_{x^3}^{x^2} dx = \frac{\rho}{2} \int_0^1 \left(x^4 - x^6 \right) dx = \frac{\rho}{2} \left[\frac{x^5}{5} - \frac{x^9}{7} \right]_0^1 = \frac{\rho}{35}.$ $M_y = \rho \int_0^1 \int_{x^3}^{x^2} x \, dy \, dx = \rho \int_0^1 \left(x^3 - x^4 \right) dx = \rho \left[\frac{x^4}{4} - \frac{x^5}{5} \right]_0^1 = \frac{\rho}{20}.$ $\bar{x} = \frac{12}{20} = \frac{3}{5} = 0.6.$ $\bar{y} = \frac{12}{35} \approx 0.3429$

13. Here $I_y = \rho \int_0^{\sqrt{2}} \int_{x^2}^2 x^2 \, dy \, dx = \rho \int_0^{\sqrt{2}} x^2 y\big|_{x^2}^2 dx = \rho \int_0^{\sqrt{2}} \left(2x^2 - x^4 \right) dx = \rho \left[\frac{2x^3}{3} - \frac{x^5}{5} \right]_0^{\sqrt{2}} = \left(\frac{4}{3}\sqrt{2} - \frac{4}{5}\sqrt{2} \right)\rho = \frac{8\sqrt{2}}{15}\rho = 0.7542\rho.$ We also find that $m = \rho \int_0^{\sqrt{2}} \int_{x^2}^2 dy \, dx = \rho \int_0^{\sqrt{2}} \left(2 - x^2 \right) dx = \rho \left[2x - \frac{x^3}{3} \right]_0^{\sqrt{2}} = \rho \frac{4\sqrt{2}}{3}.$ Hence, $r_y = \sqrt{\frac{\frac{8}{15}}{\frac{4}{3}}} = \sqrt{\frac{2}{5}} \approx 0.6325.$

14. $I_x = \rho \int_0^{\sqrt{2}} \int_{x^2}^2 y^2 \, dy \, dx = \rho \int_0^{\sqrt{2}} \frac{y^3}{3}\big|_{x^2}^2 dx = \rho \times \int_0^{\sqrt{2}} \left[\frac{8}{3} - \frac{x^6}{3} \right] dx = \rho \left[\frac{8x}{3} - \frac{x^7}{21} \right]_0^{\sqrt{2}} = \left(\frac{8\sqrt{2}}{3} - \frac{8\sqrt{2}}{21} \right)\rho = \frac{42\sqrt{2}}{21}\rho = \frac{16\sqrt{2}}{7}\rho.$ $r_x = \sqrt{\frac{\frac{16}{7}}{\frac{4}{3}}} = \sqrt{\frac{12}{7}} \approx 1.3093.$

15. $m = \rho \int_0^3 \int_0^5 dy \, dx = \rho \int_0^3 5 \, dx = 15;$ $I_x = \rho \int_0^3 \int_0^5 y^2 dy \, dx = \rho \int_0^3 \frac{y^3}{3}\big|_0^5 dx = \rho \int_0^3 \frac{125}{3} \, dx = 125\rho;$ $r_x = \sqrt{\frac{125}{15}} = \sqrt{\frac{25}{3}} \approx 2.8868.$

16. $I_y = \rho \int_0^3 \int_0^5 x^2 dy\ dx = \rho \int_0^3 5x^2 dx = \rho \left[\frac{5x^3}{3} \right]_0^3 =$

$45\rho;\ r_y = \sqrt{\frac{45}{15}} = \sqrt{3} \approx 1.7321.$

17. $m = \rho \int_0^4 \int_0^{4x-x^2} dy\ dx = \rho \int_0^4 4x - x^2\ dx =$

$\rho \left[2x^2 - \frac{x^3}{3} \right]_0^4 = \left(32 - \frac{64}{3} \right) \rho = \frac{32}{3}\rho;\ I_x = \rho \int_0^4 \times$

$\int_0^{4x-x^2} y^2 dy\ dx = \rho \int_0^4 \frac{y^3}{3} \Big|_0^{4x-x^2} dx = \frac{\rho}{3} \int_0^4 (64x -$

$48x^4 + 12x^5 - x^6)\ dx = \frac{\rho}{3} \Big[16x^4 - \frac{48}{5}x^5 - 2x^6 -$

$\frac{x^6}{5} \Big]_0^4 \approx 39.010$ or $\frac{4096}{105};\ r_x = \sqrt{\frac{39.010}{10.667}} = 1.9124.$

18. $I_y = \rho \int_0^4 \int_0^{4x-x^2} x^2 dy\ dx = \rho \int_0^4 (4x^3 - x^4)\ dx =$

$\rho \left[x^4 - \frac{x^5}{5} \right]_0^4 = 51.2\rho;\ r_y = \sqrt{\frac{51.2}{10.667}} = 2.1909$

19. $m = \rho \int_1^2 \int_{1/x}^{x^2} dy\ dx = \rho \int_1^2 [x^2 - x^1]\ dx = \rho \times$

$\left[\frac{x^3}{3} - \ln x \right]_1^2 = 1.6402\rho;\ I_y = \rho \int_1^2 \int_{1/x}^{x^2} x^2 dy\ dx =$

$\rho \int_1^2 (x^4 - x)\ dx = \left[\frac{x^5}{5} - \frac{x^2}{2} \right]_1^2 \rho \left(\frac{3^2}{5} - 2 - \frac{1}{5} + \frac{1}{2} \right)$

$\times \rho = 4.7\rho;\ r_y = \sqrt{\frac{4.7}{1.6402}} \approx 1.6928$

20. $I_x = \rho \int_1^2 \int_{1/x}^{x^2} y^2 dy\ dx = \rho \int_1^2 \frac{y^3}{3} \Big|_{1/x}^{x^2} dx = \rho \int_1^2 \times$

$\left(\frac{x^6}{3} - \frac{1}{3x^3} \right) dx = \rho \left[\frac{x^7}{21} = \frac{1}{6x^2} \right]_1^2 = 5.9226\rho;\ r_x =$

$\sqrt{\frac{5.9226}{1.6402}} \approx 1.90024.$

21. **(a)** $I_0 = \rho \int_0^{\sqrt{2}} \int_{x^2}^2 (x^2 + y^2)\ dy\ dx$

$= \rho \int_0^{\sqrt{2}} \left[x^2 y + \frac{1}{3} y^3 \right]_{x^2}^2 dx$

$= \rho \int_0^{\sqrt{2}} \left[2x^2 + \frac{8}{3} - x^4 - \frac{1}{3} x^6 \right] dx$

$= \rho \left[\frac{2}{3} x^3 + \frac{8}{3} x - \frac{1}{5} x^5 - \frac{1}{21} x^7 \right]_0^{\sqrt{2}}$

$= \rho \left[\frac{4\sqrt{2}}{3} + \frac{8\sqrt{2}}{3} - \frac{4\sqrt{2}}{5} - \frac{8\sqrt{2}}{21} \right]$

$= \frac{296\sqrt{2}}{105} \rho$

(b) As in Exercise 13, $m = \frac{4\sqrt{2}}{3}\rho$, so $r_0 \sqrt{\frac{I_0}{m}} =$

$\sqrt{\frac{296\sqrt{2}}{105} \cdot \frac{3}{4\sqrt{2}}} = \sqrt{\frac{74}{35}} \approx 1.45.$

22. **(a)** $I_0 = \rho \int_0^3 \int_0^5 (x^2 + y^2)\ dy\ dx$

$= \rho \int_0^3 \left[x^2 y + \frac{1}{3} y^3 \right]_0^5 dx$

$= \rho \int_0^3 \left[5x^2 + \frac{125}{3} x \right] dx$

$= \rho \int_0^3 \left[\frac{5}{3} x^2 + \frac{125}{3} x \right]_0^3$

$= \rho[45 + 125] = 170\rho$

(b) As in Exercise 15, $m = 15\rho$, so $r_0 = \sqrt{\frac{I_0}{m}} =$

$\sqrt{\frac{170}{15}} = \sqrt{\frac{34}{3}} \approx 3.37$

23. **(a)** $I_0 = \rho \int_0^4 \int_0^{4x-x^2} (x^2 + y^2)\ dy\ dx$

$= \rho \int_0^4 \left[x^2 y + \frac{1}{3} y^3 \right]_0^{4x-x^2} dx$

$= \rho \int_0^4 \left[x^2 (4x - x^2) + \frac{1}{3} (4x - x^2)^3 \right] dx$

$= \rho \int_0^4 \Big[4x^3 - x^4 + \frac{1}{3}$

$\times (64x^3 - 48x^4 + 12x^5 - x^6) \Big] dx$

$= \rho \int_0^4 \left[\frac{76}{3} x^3 - 17x^4 + 4x^5 - \frac{1}{3} x^6 \right] dx$

$= \rho \left[\frac{19}{3} x^4 - \frac{17}{5} x^5 + \frac{2}{3} x^6 - \frac{1}{21} x^7 \right]_0^4$

$= \frac{9472}{105} \rho \approx 90.21\rho$

(b) As in Exercise 17, $m = \frac{32}{3}\rho$, so $r_0 = \sqrt{\frac{I_0}{m}} =$

$\sqrt{\frac{9472}{105} \cdot \frac{3}{32}} = \sqrt{\frac{296}{35}} \approx 2.91$

24. **(a)** $I_0 = \rho \int_1^2 \int_{1/x}^{x^2} (x^2 + y^2)\ dy\ dx$

$= \rho \int_1^2 \left[x^2 y + \frac{1}{3} y^3 \right]_{1/x}^{x^2} dx$

$= \rho \int_1^2 \left[x^4 + \frac{1}{3} x^6 - x - \frac{1}{3} x^{-3} \right] dx$

$= \rho \left[\frac{1}{5} x^5 + \frac{1}{21} x^7 - \frac{1}{2} x^2 + \frac{1}{6} x^{-2} \right]_1^2$

$= \rho \Big[\left(\frac{32}{5} + \frac{128}{21} - 2 + \frac{1}{24} \right)$

$- \left(\frac{1}{5} + \frac{1}{21} - \frac{1}{2} + \frac{1}{6} \right) \Big]$

$= \frac{8923}{840} \rho \approx 10.6226\rho$

(b) As in Exercise 19, $m = 1.6402\rho$, so $r_0 =$

$\sqrt{\frac{I_0}{m}} = \sqrt{\frac{10.6226}{1.6402}} \approx 2.54.$

CHAPTER 30 REVIEW

1.   A plane whose intercepts are (6, 0, 0), (0, 2, 0), and (0, 0, 3)

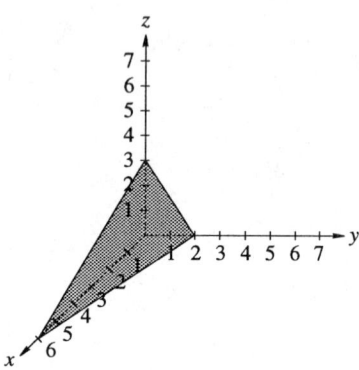

2.   A plane whose intercepts are (3, 0, 0), (0, −4, 0) and (0, 0, −6)

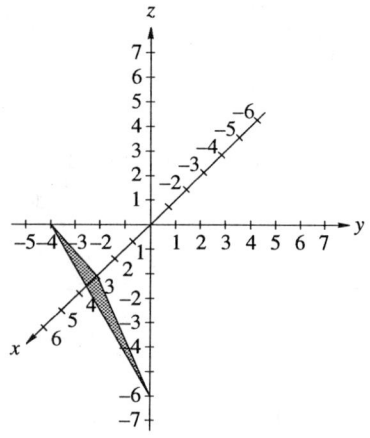

3.   An elliptical cylinder whose axis is the x-axis, intercepts are (0, ±1, 0), (0, 0, ±2)

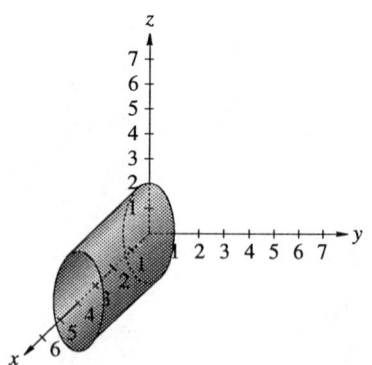

4.   Sphere of radius 4.

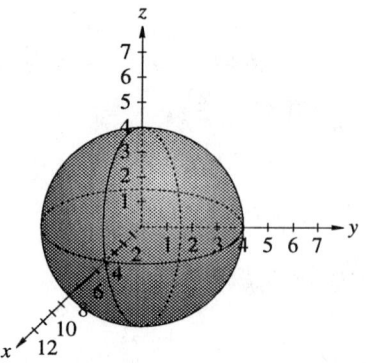

5.   A parabolic cylinder whose axis is the y-axis and has the trace in the xy-plane of $y = 4x^2$

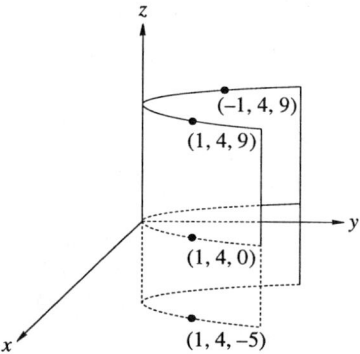

6.   A hyperbolic cylinder whose axis is the x-axis and has the trace in the xy-plane of $9x^2 - 4y^2 = 1$.

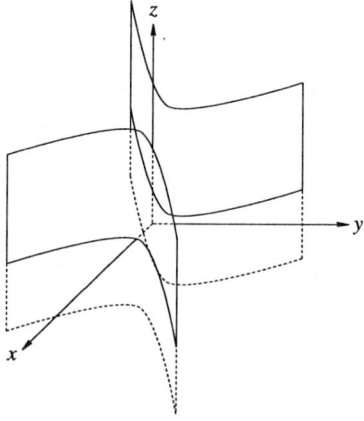

7. Hyperboloid of one-sheet. The trace in the yz-plane is the ellipse $\frac{x^2}{16} + \frac{y^2}{4} = 1$. The trace in the yz-plane is the hyperbola $\frac{y^2}{4} - \frac{z^2}{16} = 1$ and the trace in the xz-plane is the hyperbola $\frac{x^2}{16} - \frac{z^2}{16} = 1$.

8. An elliptic cone

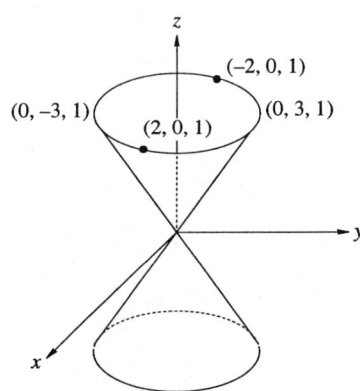

9. $z = 3x^2 + 6xy^3$; $\frac{\partial z}{\partial x} = 6x + 6y^3$; $\frac{\partial z}{\partial y} = 18xy^2$; $\frac{\partial^2 z}{\partial x^2} = 6$; $\frac{\partial^2 z}{\partial y^2} = 36xy$; $\frac{\partial^2 z}{\partial x \partial y} = 18y^2$

10. $z = \frac{x^2+y^2}{y} = x^2y^{-1}+y$; $\frac{\partial z}{\partial x} = \frac{2x}{y}$; $\frac{\partial z}{\partial y} = \frac{-x^2}{y^2}+1 = \frac{y^2-x^2}{y^2}$; $\frac{\partial^2 z}{\partial x^2} = \frac{2}{y}$; $\frac{\partial^2 z}{\partial y^2} = \frac{2x^2}{y^3}$; $\frac{\partial^2 z}{\partial x \partial y} = \frac{-2x}{y^2}$

11. $z = e^x \cos y - e^y \sin x$; $\frac{\partial z}{\partial x} = e^x \cos y - e^y \cos x$; $\frac{\partial z}{\partial y} = -e^x \sin y - e^y \sin x$; $\frac{\partial^2 z}{\partial x^2} = e^x \cos y + e^y \sin x$; $\frac{\partial^2 z}{\partial y^2} = -e^x \cos y - e^y \sin x$; $\frac{\partial^2 z}{\partial x \partial y} = -e^x \sin y - e^y \cos x$

12. $z = \ln\sqrt{x^2 + y^3} + \sin^2(3xy)$;
$$\frac{\partial z}{\partial x} = \frac{2x}{2\left(x^2 + y^3\right)} + 2\sin(3xy)\cdot \cos 3xy \cdot 3y$$
$$= \frac{x}{x^2 + y^3} + 6y\sin(3xy)\cos(3xy)$$

$$\frac{\partial z}{\partial y} = \frac{3y^2}{2\left(x^2 + y^3\right)} + 6x\sin 3xy \cos 3xy$$
$$\frac{\partial^2 z}{\partial x^2} = \frac{x^2 + y^2 - 2x^2}{(x^2 + y^3)^2} + 18y^2 \cos^2 3xy$$
$$\quad - 18y^2 \sin^2 3xy$$
$$= \frac{y^3 - x^2}{(x^2 + y^3)^2} + 18y^2(\cos^2 3xy - \sin^2 3xy)$$
$$= \frac{y^3 - x^2}{(x^2 + y^3)^2} + 18y^2 \cos 6xy$$
$$\frac{\partial^2 z}{\partial y^2} = \frac{2(x^2 + y^3)6y - 3y^2 \cdot 2 \cdot 3y^2}{4(x^2 + y^3)^2}$$
$$\quad + 18x^2\left(\cos^2 3xy - \sin^2 3xy\right)$$
$$= \frac{6x^2y - 3y^4}{2(x^2 + y^3)^2} + 18x\cos 6xy$$
$$\frac{\partial^2 z}{\partial x \partial y} = \frac{12xy^2}{4(x^2 + y^3)^2} + 6\sin 3xy \cos 3xy$$
$$\quad + 18xy \cos^2 3xy - 18xy \sin^2 3xy$$
$$= \frac{3xy^2}{(x^2 + y^3)^2} + 6\sin 3xy \cos 3xy$$
$$\quad + 18xy \cos 6xy$$

13. $\int_0^2 \int_0^{x^3} xy\ dy\ dx = \int_0^2 \frac{xy^2}{2}\Big|_0^{x^3} dx = \int_0^2 \frac{x^7}{2}dx = \frac{x^8}{18}\Big|_0^2 = 16$

14. $\int_0^{\pi/2} \int_0^{\pi} \sin x \cos y\ dy\ dx$
$$= \int_0^{\pi/2} \sin x[\cos \pi - \cos 0]dx$$
$$= -2\int_0^{\pi/2} \sin x\ dy = -2[\cos x]_0^{\pi/2}$$
$$= -2\left[\cos\left(\frac{\pi}{2}\right) - \cos 0\right] = 2$$

15. $\int_0^1 \int_0^4 (x + y)^{1/2}dy\ dx = \int_0^1 \frac{2}{3}(x + y)^{3/2}\big|_0^4 dx = \frac{2}{3}\int_0^1 \left[(x + 4)^{3/2} - x^{3/2}\right]dx = \frac{2}{3}\left[\frac{2}{5}(x + 4)^{5/2} - \frac{2}{5}x^{5/2}\right]_0^1 = \frac{4}{15}\left(5^{5/2} - 1 - 4^{5/2}\right) = \frac{4}{15}\left(5^{5/2} - 33\right) \approx 6.1071.$

16. $\int_0^4 \int_0^9 xy\sqrt{x^2 + y^2}\ dy\ dx$: Let $u = x^2 + y^2$, and then $du = 2y\ dy$. When $y = 0 \Rightarrow u = x^2$, and when $y = 9 \Rightarrow u = x^2 + 9^2$. Substitution yields $\frac{1}{2}\int_0^4 \int_{x^2}^{x^2+9^2} x(u)^{1/2}du = \frac{1}{2}\int_0^4 \frac{2}{3}xu^{3/2}\big|_{x^2}^{x^2+9^2} dx = \frac{1}{2}\int_0^4 (x(x^2+9^2)^{3/2} - x^4)dx = \frac{1}{3}\left[\frac{1}{2}\cdot\frac{2}{5}(x^2+9^2)^{5/2} - \frac{x^5}{5}\right]_0^4 = \frac{1}{15}\left[(4^2 + 9^2)^{5/2} - 4^5 - 9^5\right] = 2172.9935.$

17. $\int_0^{\ln 4} \int_0^{\ln 10} e^{x+2y} dy \, dx = \int_0^{\ln 4} \frac{1}{2} e^{x+2y} \Big|_0^{\ln 10} dx =$

$\int_0^{\ln 4} \left(\frac{1}{2} e^{x+2\ln 10} - \frac{1}{2} e^x \right) dx = \int_0^{\ln 4} 49\frac{1}{2} e^x \, dx =$

$49.5 e^x \big|_0^{\ln 4} = 49.5(4-1) = 148.5.$

18. $\int_0^1 \int_0^{x^2} x \, dy \, dx = \int_0^1 x^3 \, dx = \frac{x^4}{4} \Big|_0^1 = \frac{1}{4}$

19. $\left(4, \frac{\pi}{6}, 2 \right), x = 4\cos\frac{\pi}{6} = 2\sqrt{3} \approx 3.4641, y =$

$4\sin\frac{\pi}{6} = 2$, and $z = 2$; rectangular: $(2\sqrt{3}, 2, 2)$; ρ

$= \sqrt{(2\sqrt{3})^2 + 2^2 + 2^2} = \sqrt{12+4+4} = \sqrt{20}$

$= 2\sqrt{5} \approx 4.4721.$ $\theta = \tan^{-1} \frac{2}{2\sqrt{3}} = \frac{\pi}{6}, \phi =$

$\cos^{-1} \frac{2}{2\sqrt{5}} = 1.1071;$ spherical $\left(2\sqrt{5}, \frac{\pi}{6}, 1.1071 \right)$

20. $x = 9\cos\frac{\pi}{6} = -\frac{9\sqrt{3}}{2} \approx -7.7942, y = 9\sin\frac{5\pi}{6} =$

$\frac{9}{2} = 4.5; z = -3;$ rectangular: $(-7.7942, 4.5, -3).$

$\rho = \sqrt{\left(\frac{9\sqrt{3}}{2} \right)^2 + \left(\frac{9}{2} \right)^2 + 3^2} = 9.4868; \theta = \tan^{-1}$

$\left(-\frac{1}{\sqrt{3}} \right) = \frac{5\pi}{6}, \phi = \cos^{-1} \frac{-3}{9.4868} = 1.8925;$ spher-

ical: $\left(9.4868, \frac{5\pi}{6}, 1.8925 \right).$

21. $x = 4\sin\frac{\pi}{6}\cos\frac{3\pi}{4} = 2 \cdot \cos\frac{3\pi}{4} = -\sqrt{2}, y =$

$4\sin\frac{\pi}{6}\sin\frac{3\pi}{4} = \sqrt{2}, z = 4\cos\frac{\pi}{6} = 2\sqrt{3};$ rectan-

gular: $(-\sqrt{2}, \sqrt{2}, 2\sqrt{3}).$ $r = \sqrt{(-\sqrt{2})^2 + \sqrt{2}^2} =$

$2, \theta = \tan^{-1} \frac{\sqrt{2}}{-\sqrt{2}} = \frac{3\pi}{4};$ cylindrical: $\left(2, \frac{3\pi}{4}, 2\sqrt{3} \right).$

22. $x = 5\sin\frac{2\pi}{3}\cos\frac{7\pi}{6} = 5 \cdot \frac{\sqrt{3}}{2} \cdot \frac{-\sqrt{3}}{2} = \frac{-15}{4} =$

$-3.75; y = 5\sin\frac{2\pi}{3}\sin\frac{7\pi}{6} = 5 \cdot \frac{\sqrt{3}}{2} \cdot \left(\frac{-1}{2} \right) =$

$\frac{-5\sqrt{3}}{4} \approx -2.1651; z = 5\cos\frac{2\pi}{3} = 5\left(\frac{-1}{2} \right) =$

$-2.5;$ rectangular: $(-3.75, -2.1651, -2.5).$ $r =$

$\sqrt{(3.75)^2 + (-2.1651)^2} = 4.3301, \theta = \theta = \frac{7\pi}{6}, z$

$= z = -2.5;$ cylindrical: $\left(4.3301, \frac{7\pi}{6}, -2.5 \right).$

23. $\bar{f} = \frac{1}{4 \cdot 4} \int_0^4 \int_{-2}^2 (x^2 + y^4) \, dy \, dx$

$= \frac{1}{16} \int_0^4 \left[x^2 y + \frac{1}{5} y^5 \right]_{-2}^2 dx$

$= \frac{1}{16} \int_0^4 \left[4x^2 + \frac{64}{5} \right] dx$

$= \frac{1}{16} \left[\frac{4}{3} x^3 + \frac{64}{5} x \right]_0^4$

$= \frac{16}{3} + \frac{16}{5} = \frac{128}{15} \approx 8.53$

24. $\bar{f} = \frac{1}{4 \cdot 4} \int_1^5 \int_0^4 (x^3 + 2y) \, dy \, dx$

$= \frac{1}{16} \int_1^5 \left[x^3 y + y^2 \right]_0^4 dx$

$= \frac{1}{16} \int_1^5 \left[4x^3 + 16 \right] dx$

$= \frac{1}{16} \left[x^4 + 16x \right]_1^5$

$= \frac{1}{16} [(625 + 80) - (1 + 16)] = \frac{688}{16} = 43$

25. $V = \ell \cdot w \cdot h = 300$ so $h = \frac{300}{\ell \cdot w}$. $A = \ell \cdot w +$

$2\ell h + 2wh = \ell \cdot w + \frac{600}{w} + \frac{600}{\ell}; A_\ell = w - \frac{600}{\ell^2} = 0;$

$A_w = \ell - \frac{600}{w^2} = 0.$ Solving the first equation for

w we have $w = \frac{600}{\ell^2}$. Substituting this into the

second equation, we get $\ell - \dfrac{600}{\left(\frac{600}{\ell^2} \right)^2} = \ell - \frac{\ell^4}{600} = 0$

or $\ell = \frac{\ell^4}{600} \Rightarrow \ell^3 = 600$ or $\ell = \sqrt[3]{600} \approx 8.434.$

in. Now $w = \frac{600}{\ell^2} = \frac{600}{\sqrt[3]{600^2}} = \sqrt[3]{600} \approx 8.434$ in.

$h = \frac{300}{\sqrt[3]{600}} = \frac{300\sqrt[3]{600}}{600} = \frac{\sqrt[3]{600}}{2} \approx 4.217$ in.

26. $R = \frac{R_1 R_2}{R_1 + R_1}; \frac{\partial R}{\partial R_1} = \frac{(R_1 + R_2)R_2 - R_1 R_2}{(R_1 + R_2)^2} = \frac{R_2^2}{(R_1 + R_2)^2}.$

$\frac{\partial R}{\partial_1 R_2} = \frac{R_1^2}{(R_1 + R_2)^2}, dR_1 = 300 \times 0.01 = 3, dR_2 =$

$600 \times 0.01 = 6.$ $dR = \frac{\partial R}{\partial R_1} dR_1 + \frac{\partial R}{\partial R_2} dR_2 =$

$\frac{R_2^2}{(R_1 + R_2)^2} dR_1 + \frac{R_1^2}{(R_1 + R_2)^2} dR_2 = \frac{600^2}{(900)^2} \times 3 + \frac{300^2}{900^2} \times 6 =$

$2\ \Omega.$

27. $\frac{\partial R}{\partial V} = \frac{1}{I}, \frac{\partial R}{\partial I} = -VI^{-2}, dR = \frac{\partial R}{\partial V} dV + \frac{\partial R}{\partial I} dI =$

$\frac{1}{I} dV - \frac{V}{I^2} dI = \frac{1}{2} \times 0.2 - \frac{116 \times 0.01}{2} = -0.19\ \Omega.$

28. $V = \ell \cdot w \cdot h = 25$, so $h = \frac{25}{\ell w}$. $A = \ell w +$

$2\ell h + 2wh = \ell w + \frac{50}{w} + \frac{50}{\ell}.$ $A_\ell = w - \frac{50}{\ell^2}; A_w =$

$\ell - \frac{50}{w^2}; A_\ell = A_w = 0; \ell - \frac{50}{w^2} = 0 \Rightarrow \ell =$

$\frac{50}{w^2}; w - \dfrac{50}{\left(\frac{50}{w^2} \right)^2} = w - \frac{w^4}{50} = 0; w = \frac{w^4}{50}; 50 =$

w^3 or $w = \sqrt[3]{50} \approx 3.684$ m. In like manner,

$\ell = \sqrt[3]{50} = 3.684$ m. Since $h = \frac{25}{\ell \cdot w}$, we get

$h = \frac{25}{\sqrt[3]{50^2}} = \frac{25\sqrt[3]{50}}{50} = \frac{\sqrt[3]{50}}{2} \approx 1.842$ m.

29. $PV = nRT \Rightarrow V = \frac{nRT}{P}; T = \frac{PV}{nR} = \frac{4 \times 1000}{8R} =$

$\frac{500}{R}; dV = \frac{\partial V}{\partial T} dT + \frac{\partial V}{\partial P} dP = \frac{nR}{P} dT - \frac{nRT}{P^2} dP =$

$\frac{8R}{4}(0.5) - \dfrac{8R \cdot \frac{500}{R}}{4^2}(0.4) = R - 100$ cm^2/min.

30. $V = \frac{1}{3}\pi r^2 h$; $dV = \frac{\partial V}{\partial r}dr = \frac{1}{3}\pi r^2 \cdot dh + \frac{2}{3}\pi rh\, dr =$
$\frac{1}{3}\pi(180)^2 15 + \frac{2}{3}\pi(180)(270)10 = 486{,}000\pi$ cm^2/s.

31. $m = \int_1^5 \int_0^{6x} dy\ dx = \int_1^5 6x\ dx = 3x^2\big|_1^5 = 75 -$
$3 = 72$, $M_x = \int_1^5 \int_0^{6x} y\ dy\ dx = \int_1^5 \frac{y^2}{2}\big|_0^{6x} dx =$
$\int_1^5 18x^2 dx = 6x^3\big|_1^5 = 744$. $M_y = \int_1^5 \int_0^{6x} x\ dy\ dx$
$= \int_1^5 6x^2\ dx = 2x^2\big|_1^5 = 248$, $\bar{x} = \frac{M_y}{m} = \frac{248}{72} =$
3.4444; $\bar{y} = \frac{M_x}{m} = \frac{744}{72} \approx 10.3333$.

32. $m = \int_1^2 \int_1^{x^4} dy\ dx = \int_1^2 x^4 dx = \frac{x^5}{5}\big|_1^2 = \frac{31}{5}$, $M_x =$
$\int_1^2 \int_0^{x^4} y\ dy\ dx = \int_1^2 x^4 dx = \frac{y^2}{2}\big|_0^{x^4} dx = \int_1^2 \frac{x^8}{2}dx =$
$\frac{x^9}{18}\big|_{1_y}^2 = \frac{511}{18}$, $M_y = \int_1^2 \int_0^{x^4} x\ dy\ dx = \int_1^2 x^5 dx =$
$\frac{x^6}{6}\big|_1^2 = \frac{63}{6}$, $\bar{x} = \frac{63/6}{31/5} = \frac{105}{62} \approx 1.6935$, $\bar{y} =$
$\frac{511/18}{31/5} = \frac{2555}{558} = 4.579$

33. $m = \int_0^1 \int_{x^3}^{\sqrt[3]{x}} dy\ dx = \int_0^1 \left(x^{1/3} - x^3\right) dx = \frac{3}{4}x^{4/3} -$
$\frac{1}{4}x^4\big|_0^1 = \frac{1}{2}$; $M_x = \int_0^1 \int_{x^3}^{\sqrt[3]{x}} y\ dy\ dx = \int_0^1 \frac{y^2}{2}\big|_{x^3}^{\sqrt[3]{x}}dx$
$= \frac{1}{2}\int_0^1 \left(x^{2/3} - x^6\right) dx = \frac{1}{2}\left[\frac{3}{5}x^{5/3} - \frac{1}{7}x^7\right]_0^1 = \frac{1}{2}\times$
$\left(\frac{3}{5} - \frac{1}{7}\right) = \frac{8}{35}$; $M_y = \int_0^1 \int_{x^3}^{\sqrt[3]{x}} x\ dy\ dx = \int_0^1 \left(x^{4/3}\right.$
$\left. -x^4\right) dx = \left[\frac{4}{7}x^{7/4} - \frac{1}{5}x^5\right]_0^2 = \frac{8}{35}$; $\bar{x} = \bar{y} =$
$\frac{8/35}{1/2} = \frac{16}{35} \approx 0.4571$

34. These two graphs intersect when $x^2 = 9 - x^2$ or
$2x^2 = 9 \Rightarrow x = \pm\frac{3/\sqrt{2}}{2} = \pm\frac{3}{\sqrt{2}} \approx \pm 2.1213$. $m =$
$\int_{-3/\sqrt{2}}^{3/\sqrt{2}} \int_{x^2}^{9-x^2} dy\ dx = \int_{-3/\sqrt{2}}^{3/\sqrt{2}} \left(9 - 2x^2\right) dx = 9x -$
$\frac{2}{3}x^3\big|_{-3\sqrt{2}}^{3/\sqrt{2}} = \frac{54}{\sqrt{2}} - \frac{9}{2\sqrt{2}} = 27\sqrt{2} - 9\sqrt{2} = 18\sqrt{2}$; M_x
$= \int_{-3/\sqrt{2}}^{3/\sqrt{2}} \int_{x^2}^{9-x^2} y\ dy\ dx = \int_{-3/\sqrt{2}}^{3/\sqrt{2}} \frac{y^2}{2}\big|_{x^2}^{9-x^2} dx =$
$\frac{1}{2}\int_{-3/\sqrt{2}}^{3/\sqrt{2}} \left(81 - 18x^2\right) dx = \frac{1}{2}\left[81x - 6x^3\right]_{-3/\sqrt{2}}^{3/\sqrt{2}}$
$= \frac{243}{\sqrt{2}} - \frac{81}{\sqrt{2}} = 81\sqrt{2}$. $M_y = \int_{-3/\sqrt{2}}^{3/\sqrt{2}} \int_{x^2}^{9-x^2} x\ dy\ dx$
$= \int_{-3/\sqrt{2}}^{3/\sqrt{2}} \left(9x - 2x^3\right) dx = 0$. This integral is 0
since $9x - 2x^3$ is an odd function and we are in-
tegrating from $-\frac{3}{\sqrt{2}}$ to $\frac{3}{\sqrt{2}}$. $\bar{x} = 0$, $\bar{y} = \frac{81\sqrt{2}}{18\sqrt{2}} =$
$\frac{9}{2} = 4.5$. The answer is exactly what you would
expect to get due to the symmetry of the region.

35. $I_x = \rho\int_0^5 \int_0^{6x} y^2 dy\ dx = \rho\int_0^5 \frac{y^3}{3}\big|_0^{6x} dx =$
$\rho\int_0^5 72x^3\ dx = \rho 18x^4\big|_0^5 = 11{,}250\rho$. $m =$
$\rho\int_0^5 dy\ dx = \rho\int_0^5 6x\ dx = \rho 3x^2\big|_0^5 = 75\rho$.
$r_x = \sqrt{\frac{I_x}{m}} = \sqrt{\frac{11{,}250\rho}{75\rho}} = \sqrt{150} \approx 12.2474$

36. $I_y = \rho\int_0^4 \int_0^{4-x} x^2 dy\ dx = \rho\int_0^4 \left(4x^2 - x^3\right) dx =$
$\rho\left[\frac{4}{3}x^3 - \frac{x^4}{4}\right]_0^4 = 21.3333\rho$ or $\frac{64}{3}\rho$. $m =$
$\rho\int_0^4 \int_0^{4-x} dy\ dx = \rho\int_0^4 (4 - x)dx = \rho\left[4x - \frac{x^2}{2}\right]_0^4$
$= 8\rho$. $r_y = \sqrt{\frac{I_y}{m}} = \sqrt{\frac{\frac{64}{3}\rho}{8}} = \sqrt{\frac{8}{3}\rho} \approx 1.6330$.

37. $I_x = \rho\int_0^1 \int_0^{x^{2/3}} y^2 dy\ dx = \rho\int_0^1 \frac{y^3}{3}\big|_0^{x^{2/3}} dx =$
$\rho\int_0^1 \frac{x^2}{3}dx = \rho\frac{x^3}{8}\big|_0^1 = \frac{1}{9}\rho$. $m = \rho\int_0^1 \int_0^{x^{2/3}} dy\ dx =$
$\rho\int_0^1 x^{2/3}dx = \rho\frac{3}{5}x^{5/3}\big|_0^1 = \frac{3}{5}\rho$. $r_x = \sqrt{\frac{1/9\rho}{3/5\rho}} =$
$\sqrt{\frac{5}{27}} \approx 0.4303$

38. $I_y = \rho\int_0^1 \int_{x^3}^{\sqrt[3]{x}} y^2 dy\ dx = \rho\int_0^1 \frac{y^3}{3}\big|_{x^3}^{\sqrt[3]{x}} dx =$
$\rho\int_0^1 \left(\frac{x}{3} - \frac{x^9}{3}\right) dx = \rho\left[\frac{x^2}{6} - \frac{x^{10}}{30}\right]_0^1 = \left(\frac{1}{6} - \frac{1}{30}\right)\rho$
$= \frac{4}{30}\rho = \frac{2}{15}\rho$. $m = \rho\int_0^1 \int_{x^3}^{\sqrt[3]{x}} dy\ dx = \rho\int_0^1 \left(x^{1/3} - \right.$
$\left. x^3\right) dx = \rho\left(\frac{3}{4}x^{4/3} - \frac{x^4}{4}\right)\big|_0^1 = \frac{1}{2}\rho$. $r_y = \sqrt{\frac{\frac{2}{15}\rho}{\frac{1}{2}\rho}} =$
$\sqrt{\frac{4}{15}} \approx 0.5164$

39. **(a)** $I_0 = \rho\int_0^8 \int_0^{10x} \left(x^2 + y^2\right) dy\ dx$
$= \rho\int_0^8 \left[x^2 y + \frac{1}{3}y^3\right]_0^{10x} dx$
$= \rho\int_0^8 \left[10x^3 + \frac{1000}{3}x^3\right] dx$
$= \rho\int_0^8 \left[\frac{1030}{3}x^3\right] dx$
$= \rho\left[\frac{1030}{12}x^4\right]_0^8 = \frac{1{,}054{,}720}{3}\rho$
(b) $m = \rho\int_0^8 \int_0^{10x} dy\ dx = \rho\int_0^8 10x\ dx = \rho\left[5x^2\right]_0^8$
$= 320\rho$, so $r_0 = \sqrt{\frac{I_0}{m}} = \sqrt{\frac{1{,}054{,}720}{3\cdot 320}} = \sqrt{\frac{3{,}296}{3}}$
≈ 33.15.

40. **(a)** $I_0 = \rho\int_0^6 \int_{x^2}^{6x} \left(x^2 + y^2\right) dy\ dx$
$= \rho\int_0^6 \left[x^2 y + \frac{1}{3}y^3\right]_{x^2}^{6x} dx$
$= \rho\int_0^6 \left[6x^3 + 72x^3 - x^4 - \frac{1}{3}x^6\right] dx$
$= \rho\left[\frac{3}{2}x^4 + 18x^4 - \frac{1}{5}x^5 - \frac{1}{21}x^7\right]_0^6$
$= \frac{363528}{35}\rho$

(b) $m = \rho \int_0^6 \int_{x^2}^{6x} dy\ dx = \rho \int_0^6 \left(6x - x^2\right) dx =$

$\rho \left[3x^2 - \frac{1}{3}x^3\right]_0^6 = 36\rho$, so $r_0 = \sqrt{\frac{I_0}{m}} = \sqrt{\frac{363528}{35\cdot36}} =$

$\sqrt{\frac{10,098}{35}} \approx 16.99$.

41. $\int_0^2 \int_0^x \left(x^2 - y\right) dy\ dx + \int_2^4 \int_0^{4-x} \left(x^2 - y\right) dy\ dx =$

$\int_0^2 \left(x^2 y - \frac{y^2}{2}\right) \Big|_0^x dx + \int_2^4 \left(x^2 y - y\frac{2}{2}\right) \Big|_0^{4-x} dx =$

$\int_0^2 \left(x^3 - \frac{x^2}{2}\right) dx + \int_2^4 \left(4x^2 - x^3 - 8 + 4x - \frac{x^2}{2}\right)$

$\times dx = \left[\frac{x^4}{4} - \frac{x^3}{6}\right]_0^2 + \left[\frac{4}{3}x^3 - \frac{x^4}{4} - 8x + 2x^2 - \frac{x^3}{6}\right]_2^4$

$= 4 - \frac{8}{6} + \frac{256}{3} - 64 - 32 + 32 - \frac{64}{4} - \frac{32}{3} + 4 + 16 -$

$8 + \frac{8}{6} = 16$.

42. $\int_{-\sqrt{5}}^{\sqrt{5}} \int_{x^2+x-5}^{x} (x - 4y) dy\ dx = \int_{-\sqrt{5}}^{\sqrt{5}} (xy -$

$2y^2) \big|_{x^2+x-5}^{x}\ dx = \int_{-\sqrt{5}}^{\sqrt{5}} \left[x^2 - 2x^2 - \left(x^3 + x^2 -\right.\right.$

$\left.\left.5x\right) + 2\left(x^2 + x - 5\right)^2\right] dx = \int_{-\sqrt{5}}^{\sqrt{5}} \left[-x^3 - 2x^2 +\right.$

$\left.5x + 2\left(x^4 + 2x^3 - 9x^2 - 10x + 25\right)\right] dx = \int_{-\sqrt{5}}^{\sqrt{5}}$

$\times \left(2x^4 + 3x^3 - 20x^2 - 15x + 50\right) dx = \left[\frac{2}{5}x^5 +\right.$

$\left.\frac{3}{4}x^4 - \frac{20}{3}x^3 - \frac{15}{2}x^2 + 50x\right]_{-\sqrt{5}}^{\sqrt{5}} = 2\left(\frac{2}{5} \cdot 25\sqrt{5} -\right.$

$\left.\frac{20}{3} \cdot 5\sqrt{5} + 50\sqrt{5}\right) = 2\left(10\sqrt{5} - \frac{100}{3}\sqrt{5} + 50\sqrt{5}\right)$

$= \frac{160}{3}\sqrt{5} \approx 119.2570$ (Note: to evaluate from

$-\sqrt{5}$ to $\sqrt{5}$, we can double the odd terms and

cancel the even.)

CHAPTER 30 TEST

1. A plane whose intercepts are $(4, 0, 0)$, $(0, -8, 0)$, and $(0, 0, 2)$

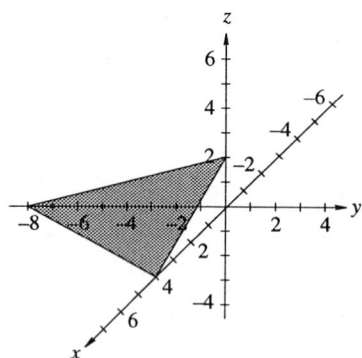

2. Sphere of radius 3

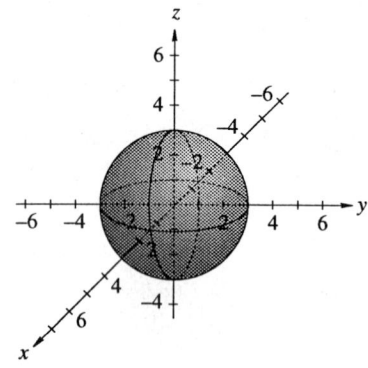

3. Hyperboloid of one-sheet. The trace in the xy-plane is the ellipse $\frac{x^2}{4} + \frac{y^2}{9} = 1$. The trace in the yz-plane is the hyperbola $\frac{y^2}{9} - \frac{z^2}{4} = 1$ and the trace in the xz-plane is the hyperbola $\frac{x^2}{4} - \frac{z^2}{4} = 1$.

4. Elliptic paraboloid with intercepts $(\pm 2, 0, 0)$, $(0, \pm 3, 0)$, and $(0, 0, -36)$

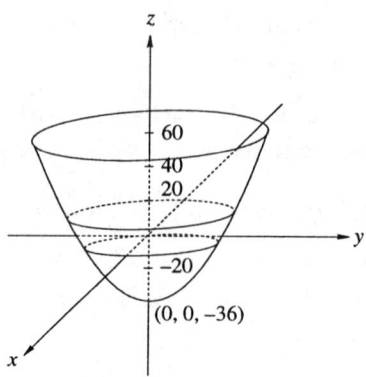

5. Treating y as a constant and differentiating z with respect to x, produces $\frac{\partial z}{\partial x} = 12x^2 - 10xy^4$. Treating x as a constant and differentiating z with respect to y, produces $\frac{\partial z}{\partial y} = -20x^2y^3$

6. Rewriting z as $z = \left(x^2 + 4y^3\right)^{1/2} + \ln\left(x^2 y\right)$, treating y as a constant and differentiating z with respect to x, produces $\frac{\partial z}{\partial x} = \frac{1}{2}\left(x^2 + 4y^3\right)^{-1/2}(2x) + \frac{2xy}{x^2 y}$ which simplifies to $\frac{\partial z}{\partial x} = x\left(x^2 + 4y^3\right)^{-1/2} + \frac{2}{x}$. Treating y as a constant and differentiating $\frac{\partial z}{\partial x}$ with respect to x produces $\frac{\partial^2 z}{\partial x^2} = x\left[-\frac{1}{2}\left(x^2 + 4y^3\right)^{-3/2}(2x)\right] + \left(x^2 + 4y^3\right)^{-1/2} - \frac{2}{x^2} = -x^2\left(x^2 + 4y^3\right)^{-3/2} + \left(x^2 + 4y^3\right)^{-1/2} - \frac{2}{x^2}$

7.
$$\int_0^4 \int_0^{\pi/4} x\sin y\; dy\; dx = \int_0^4 \left. (-x\cos y)\right|_{y=0}^{y=\pi/4} dx$$
$$= \int_0^4 -x\left(\frac{\sqrt{2}}{2} - 1\right) dx$$
$$= \left(1 - \frac{\sqrt{2}}{2}\right)\int_0^4 x\; dx$$
$$= \left(1 - \frac{\sqrt{2}}{2}\right)\left.\frac{1}{2}x^2\right|_0^4$$
$$= 8 - 4\sqrt{2} \approx 2.3431$$

8. $\int_0^2 \int_0^{x^2} 3xy^2\, dy\; dx = \int_0^2 \left. xy^3\right|_0^{x^2} dx = \int_0^2 x^7 dx = \left.\frac{1}{8}x^8\right|_0^2 = \frac{256}{8} = 32$

9. First we will convert these coordinates from cylindrical coordinates to rectangular coordinates. A point with the cylindrical coordinates $(r,\ \theta,\ z)$, has the rectangular coordinates $(x,\ y,\ z)$, where $x = r\cos\theta$, $y = r\sin\theta$, $z = z$. We begin with $x = r\cos\theta = 3\cos\frac{5\pi}{6} = 3\left(-\frac{\sqrt{3}}{2}\right) = -\frac{3\sqrt{3}}{2}$. Next, $y = r\sin\theta = 3\sin\frac{5\pi}{6} = 3\left(\frac{1}{2}\right) = \frac{3}{2}$. Finally, $z = z = 8$. So, the cylindrical coordinates $\left(3,\ \frac{5\pi}{6},\ 8\right)$ are equivalent to the rectangular coordinates $\left(-\frac{3\sqrt{3}}{2},\ \frac{3}{2},\ 8\right)$. Next, we convert these rectangular coordinates to spherical coordinates. A point with the rectangular coordinates $(x,\ y,\ z)$ has the spherical coordinates $(\rho,\ \theta,\ \phi)$, where $\rho = \sqrt{x^2 + y^2 + z^2}$, $\tan\theta = \frac{y}{x}$, $\cos\phi = \frac{z}{\rho}$. First, we obtain $\rho = \sqrt{x^2 + y^2 + z^2} = \sqrt{\left(-\frac{3\sqrt{3}}{2}\right)^2 + \left(\frac{3}{2}\right)^2 + 8^2} = \sqrt{\frac{27}{4} + \frac{9}{4} + 64} =$

$\sqrt{73}$. Next we determine $\tan\theta = \frac{y}{x} = \frac{\frac{3}{2}}{\frac{-3\sqrt{3}}{2}} = -\frac{1}{\sqrt{3}}$ and so $\theta = \frac{5\pi}{6}$. Finally, we determine $\cos\phi = \frac{z}{\rho} = \frac{8}{\sqrt{73}}$ and so $\phi = \cos^{-1}\frac{8}{\sqrt{73}} \approx 0.3588$. [The value of θ is the same in both the cylindrical and spherical coordinate systems, so we could have used the given value $\theta = \frac{5\pi}{6}$.]

10. First we will convert these coordinates from spherical coordinates to rectangular coordinates. We have $\rho = 4$, $\theta = \frac{7\pi}{6}$ and $\phi = \frac{3\pi}{4}$. Using the formulas in (3), we have $x = \rho\sin\phi\cos\theta = 4\sin\frac{3\pi}{4} \times \cos\frac{7\pi}{6} = 4\left(\frac{\sqrt{2}}{2}\right)\left(-\frac{\sqrt{3}}{2}\right) = -\sqrt{6}$; $y = \rho\sin\phi\sin\theta = 4\left(\frac{\sqrt{2}}{2}\right)\left(-\frac{1}{2}\right) = -\sqrt{2}$; and $z = \rho\cos\phi = 4\left(-\frac{\sqrt{2}}{2}\right) = -2\sqrt{2}$. So, the spherical coordinates of $\left(4,\ \frac{7\pi}{6},\ \frac{3\pi}{4}\right)$ are, in rectangular coordinates, $(-\sqrt{6},\ -\sqrt{2},\ -2\sqrt{2})$.
Next, we convert these rectangular coordinates to cylindrical coordinates. A point with the rectangular coordinates $(x,\ y,\ z)$, has the cylindrical coordinates $(r,\ \theta,\ z)$, where $r = \sqrt{x^2 + y^2}$, $\tan\theta = \frac{y}{x}$, $z = z$. First, $r = \sqrt{x^2 + y^2} = \sqrt{(-\sqrt{6})^2 + (-\sqrt{2})^2} = \sqrt{6 + 2} = \sqrt{8} = 2\sqrt{2}$. Next, $\tan\theta = \frac{y}{x} = \frac{-\sqrt{2}}{-\sqrt{6}} = \frac{1}{\sqrt{3}}$. Since the coordinates of x and y are in the third quadrant of the xy-plane, we have $\theta = \pi + \tan^{-1}\frac{1}{\sqrt{3}} = \frac{7\pi}{6}$. [Notice that since θ is the same in both the spherical and cylindrical coordinate systems, we could have used the given value of *theta*]. Finally, $z = z = -2\sqrt{2}$. So, the spherical coordinates of $\left(4,\ \frac{7\pi}{6},\ \frac{3\pi}{4}\right)$ are, in cylindrical coordinates, $\left(2\sqrt{2},\ \frac{7\pi}{6},\ -2\sqrt{2}\right)$.

11. Let $z = 0$. Then the base of the region in the xy-plane is the ellipse $x^2 + 4y^2 = 4$ or $\frac{x^2}{4} + y^2 = 1$. Solving this for y, we see that $y = \pm\sqrt{\frac{4 - x^2}{4}}$ and so the bounds for y are $-\sqrt{(4 - x^2)/4} \le y \le \sqrt{(4 - x^2)/4}$ and $-2 \le x \le 2$. Thus, the volume is $V = \int_{-2}^2 \int_{-\sqrt{(4-x^2)/4}}^{\sqrt{(4-x^2)/4}} \left(4 - x^2 - 4y^2\right) dy\; dx = \int_{-2}^2 \left[4y - x^2 y - \frac{4}{3}y^3\right]_{-\sqrt{(4-x^2)/4}}^{\sqrt{(4-x^2)/4}} dx = \frac{2}{3}\int_{-2}^2 \left(4 - x^2\right)^{3/2} dx$. To integrate this, we use Formula

#28, Appendix C, and obtain $\frac{2}{3}\left[\frac{x}{4}\left(x^2-10\right)\right.$

$\left.\sqrt{4-x^2}+\frac{3}{2}\sin^{-1}\frac{x}{2}\right]_{-2}^{2} \approx 12.5664 = 4\pi$

12. The two functions intersect at the points $(-1, 4)$ and $(3, 4)$ as shown in Figure 30.0. So, we have $f(x) = 4$, $g(x) = x^2 - 2x + 1$, $a = -1$ and $b = 3$. The area of this region is $A = \int_{-1}^{3}\int_{x^2-2x+1}^{4} dy\ dx$
$= \int_{-1}^{3} y|_{x^2-2x+1}^{4}\ dx = \int_{-1}^{3}\left(3+2x-x^2\right) dx =$
$3x + x^2 - \frac{1}{3}x^3\Big]_{-1}^{3} = 9 - \left(-\frac{5}{3}\right) = \frac{32}{3}$. So, the

mass is $\frac{32}{3}\rho$. Next, we find the first moments of this region. $M_y = \rho\int_{-1}^{3}\int_{x^2-2x+1}^{4} y\ dy\ dx =$
$\rho\int_{-1}^{3}\left(3x+2x^2-x^3\right) dx = \rho\left[\frac{3}{2}x^2+\frac{2}{3}x^3-\right.$

$\left.\frac{1}{4}x^4\right]_{-1}^{3} = \frac{32}{3}\rho$ and $M_x = \rho\int_{-1}^{3}\int_{x^2-2x+1}^{4} y\ dy\ dx$
$= \rho\int_{-1}^{3}\int_{(x-1)^2}^{4} y\ dy\ dx = \rho\int_{-1}^{3}\frac{y^2}{2}\Big|_{(x-1)^2}^{4}\ dx =$
$\rho\int_{-1}^{3}\frac{1}{2}\left[16-(x-1)^4\right] dx = \frac{\rho}{2}\left[16x - \frac{1}{5}\times\right.$

$\left.(x-1)^5\right]_{-1}^{3} = 25.6$. So, $\overline{x} = \frac{M_y}{m} = \frac{32}{3}/\frac{32}{3} = 1$
and $\overline{y} = \frac{M_x}{m} = 25.6/\frac{32}{3} = 2.4$. The centroid is $(\overline{x}.\ \overline{y}) = (1,\ 2.4)$.

13. $\overline{f} = \frac{1}{3\cdot 4}\int_{-1}^{2}\int_{0}^{4} x^2 y^3 dy\ dx$

$= \frac{1}{12}\int_{-1}^{2}\left[\frac{1}{4}x^2 y^4\right]_{0}^{4} dx = \frac{1}{12}\int_{-1}^{2} 64x^2 dx$

$= \frac{1}{12}\left[\frac{64}{3}x^3\right]_{-1}^{2} = 16$

14. **(a)** $I_0 = \rho\int_{0}^{2}\int_{x}^{6-x^2}\left(x^2+y^2\right) dy\ dx$

$= \rho\int_{0}^{2}\left[x^2 y + \frac{1}{3}y^3\right]_{x}^{6-x^2} dx$

$= \rho\int_{0}^{2}\left[6x^2 - x^4 + \frac{1}{3}\left(216 - 108x^2\right.\right.$

$\left.\left. +18x^4 - x^6\right) - x^3 - \frac{1}{3}x^3\right] dx$

$= \rho\int_{0}^{2}\left[72 - 30x^2 - \frac{4}{3}x^3 + 5x^4 - \frac{1}{3}x^6\right] dx$

$= \rho\left[72x - 10x^3 - \frac{1}{3}x^4 + x^5 - \frac{1}{21}x^7\right]_{0}^{2}$

$= \frac{592}{7}\rho$

(b) $m = \rho\int_{0}^{2}\int_{x}^{6-x^2} dy\ dx = \rho\int_{0}^{2}(6-x^2-x)dx =$

$\rho\left[6x - \frac{1}{3}x^3 - \frac{1}{2}x^2\right]_{0}^{2} = \frac{22}{3}\rho$, so $r_0 = \sqrt{\frac{I_0}{m}} =$

$\sqrt{\frac{592}{7}\cdot\frac{3}{22}} = \sqrt{\frac{888}{77}} \approx 3.40$.

CHAPTER

31

Infinite Series

≡ 31.1 MACLAURIN SERIES

1. We first find the derivatives as $f(x) = \sin x$, $f'(x) = \cos x$, $f''(x) = -\sin x$, $f'''(x) = -\cos x$, $f^{(4)}(x) = \sin x$, $f^{(5)}(x) = \cos x$, $f^{(6)}(x) = -\sin$ and $f^{(7)}(x) = -\cos x$. Then, we evaluate f and its derivatives at $x = 0$, with the following results: $f(0) = 0$, $f'(0) = 1$, $f''(0) = 0$, $f'''(0) = -1$, $f^{(4)}(0) = 0$, $f^{(5)}(0) = 1$, $f^{(6)}(0) = 0$, and $f^{(7)} = -1$. Hence, $a_0 = 0$, $a_1 = 1$, $a_2 = 0$, $a_3 = -1$, $a_4 = 0$, $a_4 = 0$, $a_5 = 1$, $a_6 = 0$, and $a_7 = -1$. Using these results we get $\sin x = x - \frac{x^3}{3!} + \frac{x^5}{5!} - \frac{x^7}{7!}$.

2. First we find the derivatives of $f(x) = \sinh x$, $f'(x) = \cosh x$, $f''(x) = \sinh(x)$, etc. Then, we evaluate f and its derivatives at $x = 0$, with the following results: $f(0) = 0$, $f'(0) = 1$, $f''(0) = 0$, $f'''(0) = 1$, etc. Using these results, we get $\sinh(x) = x = +\frac{x^3}{3!} + \frac{x^5}{5!} - \frac{x^7}{7!}$.

3. Finding the derivatives of $\cosh(x) = f(x)$, we obtain $f'(x) = \sinh x$, $f''(x) = \cosh x$, etc. Evaluating f and its derivatives at $x = 0$ produces $f(0) = 1$, $f'(0) = 0$, $f''(0) = 1$, $f'''(0) = 0$, $f^{(4)}(0) = 1$, etc. Thus we obtain the desired result $\cosh(x) = 1 + \frac{x^2}{2!} + \frac{x^4}{4!} + \frac{x^6}{6!}$.

4. Finding the derivatives of $f(x) = \ln(1 + x)$, we get $f'(x) = \frac{1}{1+x}$, $f''(x) = -1(1+x)^{-2}$, $f'''(x) = 2(1 + x)^{-3}$, $f^{(4)}(x) = -6(1 + x)^{-4}$. Evaluating f and its derivatives at $x = 0$ produces $f(0) = 0$, $f'(0) = 1$, $f''(0) = -1$, $f'''(0) = 2$, $f^{(4)}(0) = -6$, etc. Hence, we obtain $\ln(1 + x) = x - \frac{x^2}{2!} + \frac{2x^3}{3!} - \frac{6x^4}{4!} = x - \frac{x^2}{2} + \frac{x^3}{3} - \frac{x^4}{4}$.

5. Here $f(x) = e^{3x}$, $f'(x) = 3e^{3x}$, $f''(x) = 9e^{3x}$, and $f'''(x) = 27e^{3x}$. Thus, $f(0) = 1$, $f'(0) = 3$, $f''(0) = 9$, and $f'''(0) = 27$. Using these results, we obtain the Maclaurin series $e^{3x} = 1 + 3x + \frac{9x^2}{2!} + \frac{27x^3}{3!} = 1 + 3x + \frac{9}{2}x^2 + \frac{9}{2}x^3$.

6. Here we have

$$f(x) = \sin^3 x, \qquad\qquad f(0) = 0$$
$$f'(x) = 3\sin^2 x \cos x, \qquad\qquad f'(0) = 0$$
$$f''(x) = 6\sin x \cos^2 x - 3\sin^3 x, \qquad\qquad f''(0) = 0$$
$$f'''(x) = 6\cos^3 x - 12\sin^2 x \cos x - 9\sin^2 x \cos x$$
$$= 6\cos^3 x - 21\sin^2 x \cos x, \qquad\qquad f'''(0) = 6$$
$$f^{(4)}(x) = -18\cos^3 x \sin x - 42\sin x \cos^2 x + 21\sin^3 x$$
$$= -60\cos^2 x \sin x + 21\sin^3 x, \qquad\qquad f^{(4)}(0) = 0$$
$$f^{(5)}(x) = 120\cos x \sin^2 x - 60\cos^3 x + 63\sin^2 x \cos x$$
$$= 183\cos x \sin^2 x - 60\cos^3 x, \qquad\qquad f^{(5)}(0) = -60$$
$$f^{(6)}(x) = -183\sin^2 x + 366\cos^2 x \sin x + 180\cos^2 x \sin x$$
$$= -183\sin^3 x + 546\cos^2 x \sin x, \qquad\qquad f^{(6)}(0) = 0$$
$$f^{(7)}(x) = -549\sin^2 x \cos x - 1092\cos x \sin^2 x + 546\cos^3 x$$
$$= -1641\cos x \sin^2 x + 546\cos^3 x, \qquad\qquad f^{(7)}(0) = 546$$
$$f^{(8)}(x) = -4920\sin x \cos^2 x + 1641\sin^3 x, \qquad\qquad f^{(8)}(0) = 0$$
$$f^{(9)}(x) = -4920\cos^3 x + 14763\sin^2 x \cos x, \qquad\qquad f^{(9)}(0) = -4920$$

Hence, we have $\sin^3 x = \frac{6x^3}{3!} - \frac{60x^5}{5!} + \frac{546x^7}{7!} + \frac{-4920x^9}{9!} = x^3 - \frac{1}{2}x^5 + \frac{13}{120}x^7 - \frac{4920}{9!}x^9 = x^3 - \frac{1}{2}x^5 + \frac{13}{120}x^7 - \frac{41}{3024}x^9$.

7. Here we have

$$f(x) = \ln\left(1 + x^2\right), \qquad\qquad f(0) = 0$$

$$f'(x) = \frac{2x}{\left(x^2 + 1\right)}, \qquad\qquad f'(0) = 0$$

$$f''(x) = \frac{-2\left(x^2 - 1\right)}{\left(x^2 + 1\right)^2}, \qquad\qquad f''(0) = 2$$

$$f'''(x) = 4x\left(x^2 - 3\right)/\left(x^2 + 1\right)^3, \qquad\qquad f'''(0) = 0$$

$$f^{(4)}(x) = \frac{-12\left(x^4 - 6x^2 + 1\right)}{\left(x^2 + 1\right)^4}, \qquad\qquad f^{(4)}(0) = -12$$

$$f^{(5)}(x) = \frac{48x\left(x^4 - 10x^2 + 5\right)}{\left(x^2 + 1\right)^3}, \qquad\qquad f^{(5)}(0) = 0$$

$$f^{(6)}(x) = \frac{-240\left(x^6 - 15x^4 + 15x^2 - 1\right)}{\left(x^2 + 1\right)^6}, \qquad\qquad f^{(6)}(0) = 240$$

$$f^{(7)}(x) = \frac{1440\left(x^7 - 21x^5 + 35x^3 - 7x\right)}{\left(x^2 + 1\right)^7}, \qquad\qquad f^{(7)}(0) = 0$$

$$f^{(8)}(x) = \frac{-10{,}080\left(x^8 - 28x^6 + 70x^4 - 28x^2 + 1\right)}{\left(x^2 + 1\right)^8}, \qquad f^{(8)}(0) = -10{,}080$$

Hence, we have $\ln\left(1 + x^2\right) = \frac{2x^2}{2!} + \frac{12x^4}{4!} + \frac{240x^6}{6!} + \frac{-10{,}080x^8}{8!} = x^2 - \frac{x^4}{2} + \frac{x^6}{3} - \frac{x^8}{4} = x^2 - \frac{1}{2}x^4 + \frac{1}{3}x^6 - \frac{1}{4}x^8$.

8. Here we have

$$f(x) = e^{-x}, \qquad f(0) = 1$$
$$f'(x) = -e^{-x}, \qquad f'(0) = -1$$
$$f''(x) = e^{-x}, \qquad f''(0) = 1$$
$$f'''(x) = -e^{-x}, \qquad f'''(0) = -1$$

Thus, $e^{-x} = 1 - x + \frac{x^2}{2!} - \frac{x^3}{3!} + = 1 - x + \frac{1}{2}x^2 - \frac{1}{6}x^3$.

9. This time we have

$$f(x) = \cos x^2, \qquad\qquad f(0) = 1$$
$$f'(x) = -2x \sin x^2, \qquad\qquad f'(0) = 0$$
$$f''(x) = -4x^2 \cos x^2 - 2 \sin x^2, \qquad\qquad f''(0) = 0$$
$$f'''(x) = 8x^3 \sin x^2 - 12x \cos x^2, \qquad\qquad f'''(0) = 0$$
$$f^{(4)}(x) = 16x^4 \cos x^2 - 12 \cos x^2 + 48x^2 \sin x^2, \qquad\qquad f^{(4)}(0) = -12$$
$$f^{(5)}(x) = 160x^3 \cos x^2 - 32x^5 \sin x^2 + 120x \sin x^2, \qquad\qquad f^{(5)}(0) = 0$$
$$f^{(6)}(x) = -64x^6 \cos x^2 + 720x^2 \cos x^2 - 480x^4 \sin x^2,$$
$$+ 120 \sin x^2, \qquad\qquad f^{(6)}(0) = 0$$
$$f^{(7)}(x) = -1{,}344x^5 \cos x^2 + 1{,}680x \cos x^2 + 128x^7 \sin x^2$$
$$- 3{,}360x^3 \sin x^2, \qquad\qquad f^{(7)}(0) = 0$$
$$f^{(8)}(x) = 256x^8 \cos x^2 - 13{,}440x^4 \cos x^2 + 1{,}680 \cos x^2$$
$$+ 3{,}584x^2 \sin x^2 - 13{,}440x^2 \sin x^2, \qquad\qquad f^{8)}(0) = 1680$$
$$f^{(9)}(x) = -512x^9 \sin x^2 + 9{,}216x^7 \cos x^2 + 48{,}384x^5 \sin x^2$$
$$- 80{,}640x^3 \cos x^2 - 30240x \sin x^2 \qquad\qquad f^{(9)}(0) = 0$$
$$f^{(10)}(x) = -1{,}024x^{10} \cos x^2 - 23{,}040x^8 \sin x^2 + 161{,}280x^6 \cos x^2$$
$$+ 403{,}200x^4 \sin x^2 - 302{,}400x^2 \cos x^2 - 30{,}240 \sin x^2, \qquad\qquad f^{(10)}(0) = 0$$
$$f^{(11)}(x) = 2{,}048x^{11} \sin x^2 - 56{,}320x^9 \cos x^2 - 506{,}880x^7 \sin x^2$$
$$+ 1774080x^5 \cos x^2 + 2{,}217{,}600x^3 \sin x^2 - 665{,}280x \cos x^2, \qquad f^{(11)}(0) = 0$$

$$f^{(12)}(x) = 4{,}096x^{12}\cos x^2 + 135{,}168x^{10}\sin x^2 - 1{,}520{,}640x^8\cos x^2$$
$$- 7{,}096{,}320x^6\sin x^2 + 13{,}305{,}600x^4\cos x^2$$
$$+ 7{,}983{,}360x^2\sin x^2 - 665{,}280\cos x^2, \qquad f^{(12)}(0) = -665{,}280$$

Hence, $\cos x^2 = 1 - \frac{12x^4}{4!} + \frac{1680x^8}{8!} - \frac{665{,}280x^{12}}{12!} = 1 - \frac{1}{2}x^4 + \frac{1}{24}x^8 - \frac{1}{720}x^{12} = 1 - \frac{1}{2!}x^4 + \frac{1}{4!}x^8 - \frac{1}{6!}x^{12}$.

10. Here we have

$$\begin{aligned}
f(x) &= e^{-x^2}, & f(0) &= 1 \\
f'(x) &= -2xe^{-x^2}, & f'(0) &= 0 \\
f''(x) &= 4x^2 e^{-x^2} - 2e^{-x^2}, & f''(0) &= -2 \\
f'''(x) &= 12xe^{-x^2} - 8x^3 e^{-x^2}, & f'''(0) &= 0 \\
f^{(4)}(x) &= 16x^4 e^{-x^2} - 48x^2 e^{-x^2} + 12e^{-x^2}, & f^{(4)}(0) &= 12 \\
f^{(5)}(x) &= -32x^5 e^{-x^2} + 160x^3 e^{-x^2} - 96xe^{-x^2} - 24e^{-x^2} & f^{(5)}(0) &= 0 \\
f^{(6)}(x) &= 64x^6 e^{-x^2} - 480x^4 e^{-x^2} + 672x^2 e^{-x^2} + 48xe^{-x^2} - 120e^{-x^2} & f^{(6)}(0) &= -120
\end{aligned}$$

Hence, we have $e^{-x^2} = 1 - \frac{2x^2}{2!} + \frac{12x^4}{4!} - \frac{120x^6}{6!} = 1 - x^2 + \frac{1}{2}x^4 - \frac{1}{6}x^6$.

11. Here we have

$$\begin{aligned}
f(x) &= e^x \sin x, & f(0) &= 0 \\
f'(x) &= e^x \sin x + e^x \cos x, & f'(0) &= 1 \\
f''(x) &= e^x \sin x + e^x \cos x + e^x \cos x - e^x \sin x \\
&= 2e^2 \cos x, & f''(0) &= 2 \\
f'''(x) &= 2e^x \cos x - 2e^x \sin x, & f'''(0) &= 2 \\
f^{(4)}(x) &= 2e^x \cos x - 2e^x \sin x - 2e^x \sin x - 2e^x \cos x \\
&= -4e^x \sin x, & f^{(4)}(0) &= 0 \\
f^{(5)}(x) &= -4e^x \sin x - 4e^x \cos x, & f^{(5)}(0) &= -4
\end{aligned}$$

Thus, $e^x \sin x = x + \frac{2x^2}{2!} + \frac{2x^3}{3!} - \frac{4x^5}{5!} = x + x^2 + \frac{1}{3}x^3 - \frac{x^5}{30}$.

12. Here we have

$$\begin{aligned}
f(x) &= xe^x, & f(0) &= 0 \\
f'(x) &= xe^x + e^x, & f'(0) &= 1 \\
f''(x) &= xe^x + 2e^x, & f''(0) &= 2 \\
f'''(x) &= xe^x + 3e^x, & f'''(0) &= 3 \\
f^{(4)}(x) &= xe^x + 4e^x, & f^{(4)}(0) &= 4
\end{aligned}$$

Thus, $xe^x = x + \frac{2x^2}{2!} + \frac{3x^2}{3!} + \frac{4x^2}{4!} = x + x^2 + \frac{x^3}{2} + \frac{x^4}{6}$.

13. This time we have

$$\begin{aligned}
f(x) &= x^2 e^{-x^2}, & f(0) &= 0 \\
f'(x) &= (2x - 2x^3)e^{-x^2}, & f'(0) &= 0 \\
f''(x) &= (4x^4 - 10x^2 + 2)e^{-x^2}, & f''(0) &= 2 \\
f'''(x) &= (-8x^5 + 36x^3 - 24x)e^{-x^2}, & f'''(0) &= 0 \\
f^{(4)}(x) &= (16x^6 - 112x^4 + 156x^2 - 24)e^{-x^2}, & f^{(4)}(0) &= -24 \\
f^{(5)}(x) &= (-32x^7 + 320x^5 - 760x^3 + 360x)e^{-x^2}, & f^{(5)}(x) &= 0 \\
f^{(6)}(x) &= (64x^8 - 864x^6 + 3120x^4 - 3000x^2 + 360)e^{-x^2}, & f^{(6)}(0) &= 360 \\
f^{(7)}(x) &= (-128x^9 + 2240x^7 - 11424x^5 + 18{,}480x^3 - 6720x)e^{-x^2}, & f^{(7)}(0) &= 0 \\
f^{(8)}(x) &= (256x^{10} - 5632x^8 + 38{,}528x^6 - 94{,}080x^4 + 68{,}880x^2 - 6720)e^{-x^2}, & f^{(8)}(0) &= -6720
\end{aligned}$$

Hence, $x^2 e^{-x^2} = \frac{2x^2}{2!} - \frac{24x^2}{4!} + \frac{360x^6}{6!} - \frac{6720x^8}{8!} = x^2 - x^4 + \frac{1}{2}x^6 - \frac{1}{6}x^8 = x^2\left(1 - x^2 + \frac{1}{2}x^4 - \frac{1}{6}x^6\right)$.

14. For this problem, we have

$$f(x) = e^{\sin x}, \qquad\qquad\qquad\qquad\qquad f(0) = 1$$
$$f'(x) = \cos x e^{\sin x}, \qquad\qquad\qquad\qquad f'(0) = 1$$
$$f''(x) = \cos^2 x \cdot e^{\sin x} - \sin x e^{\sin x}, \qquad\qquad f(0) = 1$$
$$f'''(x) = -2\cos x \sin x e^{\sin x} + \cos^3 x e^{\sin x} - \cos x e^{\sin x} - \sin x \cos x e^{\sin x}, \qquad f'''(0) = 0$$
$$f^{(4)}(x) = 3\sin^2 x e^{\sin x} - 3\cos^2 x e^{\sin x} - 3\cos^2 x \sin x e^{\sin x}$$
$$\qquad = 3\cos^2 x \sin x e^{\sin x} + \cos^4 x e^{\sin x} + \sin x e^{\sin x} - \cos^2 x e^{\sin x}$$
$$\qquad = 3\sin^2 x e^{\sin x} - 4\cos^2 x e^{\sin x} - 6\cos^2 x \sin e^{\sin x} + \cos^4 x e^{\sin x}$$
$$\qquad\quad + \sin x e^{\sin x}, \qquad\qquad\qquad\qquad f^{(4)}(0) = -3$$

Thus, $e^{\sin x} = 1 + x + \frac{x^2}{2!} - \frac{3x^4}{4!}$

15. Here we have

$$f(x) = x\sin 3x, \qquad\qquad\qquad f(0) = 0$$
$$f'(x) = 3x\cos 3x + \sin 3x, \qquad\qquad f'(0) = 0$$
$$f''(x) = 6\cos 3x - 9x\sin 3x, \qquad\qquad f''(0) = 6$$
$$f'''(x) = -27x\cos 3x - 27\sin 3x, \qquad f'''(0) = 0$$
$$f^{(4)}(x) = 81x\sin 3x - = 108\cos 3x, \qquad f^{(4)}(0) = -108$$
$$f^{(5)}(x) = 243x\cos 3x + 405\sin 3x, \qquad f^{(5)}(0) = 0$$
$$f^{(6)}(x) = 1458\cos 3x - 729x\sin 3x, \qquad f^{(6)}(0) = 1458$$
$$f^{(7)}(x) - 5103\sin 3x - 2187x\sin 3x \qquad f^{(7)}(0) = 0$$
$$f^{(8)}(x) = -17{,}496\cos 3x - 6561x\cos 3x \qquad f^{(8)}(0) = -17{,}496$$

Hence, we have $x\sin 3x = \frac{6x^2}{2!} - \frac{108x^4}{4!} + \frac{1458x^6}{6!} - \frac{17{,}496x^8}{8!} = 3x^2 - \frac{9}{2}x^4 + \frac{81}{40}x^6 - \frac{243}{560}x^8$.

16. Here we have

$$f(x) = e^{-x}\cos x, \qquad\qquad\qquad\qquad f(0) = 1$$
$$f'(x) = -e^{-x}\cos x - e^{-x}\sin x, \qquad\qquad f'(0) = -1$$
$$f''(x) = e^{-x}\cos x + e^{-x}\sin x + e^{-x}\sin x - e^{-x}\cos x$$
$$\qquad = 2e^{-x}\sin x, \qquad\qquad\qquad\qquad f''(0) = 0$$
$$f'''(x) = -2e^{-x}\sin x + 2e^{-x}\cos x, \qquad\qquad f'''(0) = 2$$
$$f^{(4)}(x) = 2e^{-x}\sin x - 2e^{-x}\cos x - 2e^{-x}\cos x - 2e^{-x}\sin x$$
$$\qquad = -4e^{-x}\cos x, \qquad\qquad\qquad\qquad f^{(4)}(0) = -4$$

Combining these results, we see that $e^{-x}\cos x = 1 - x + \frac{2}{3!}x^3 - \frac{4}{4!}x^4 = 1 - x + \frac{1}{3}x^3 - \frac{1}{6}x^4$.

17. As in example 31.5, let $f(t) = A'(t)$. Hence $f(t) = \frac{-60}{t^2+30}$, then $f(0) = -2$; $f'(t) = \frac{120t}{(t^2+30)^2}$, then $f'(0) = 0$, and $f''(t) =$

$$\frac{120(t^2+30)^2 - 120t(2)(t^2+30)2t}{(t^2+30)^4} = \frac{120t^2+120\cdot30 - 480t^2}{(t^2+30)^3},$$ then

$f''(0) = \frac{4}{30} = \frac{2}{15}$. This gives the first two terms of the Maclauren series for $A'(t)$ as $A'(t) = -2 + \frac{2}{15}\frac{t^2}{2} = -2 + \frac{t^2}{15}$. Thus, $A(t) = \int A'(t)dt = \int\left(-2 + \frac{t^2}{15}\right)dt = -2t + \frac{t^3}{45} + C$. Since the initial wound was 12cm^2, $C = 12$. Hence, $A(t) = 12 - 2t + \frac{t^3}{45}$, and so $A(2) = 12 - 2\cdot2 + \frac{2^3}{45} = 12 - 4 + \frac{8}{45} = 8\frac{8}{45} \approx 8.178$ cm^2.

18.

$$i(0) = 2 + \cos 0 + 2\sin 0 = 2 + 1 + 0 = 3$$
$$i'(\alpha) = -\sin\alpha + 2\cos\alpha$$
$$i'(0) = -\sin 0 + 2\cos 0 = 0 + 2 = 2$$
$$i''(\alpha) = -\cos\alpha - 2\sin\alpha$$
$$i''(0) = -\cos 0 - 2\sin 0 = -1 - 0 = -1$$
$$i'''(\alpha) = \sin\alpha - 2\cos\alpha$$
$$i'''(0) = \sin 0 - 2\cos 0 = 0 - 2 = -2$$

$$i(\alpha) = i(0) + i'(0)\alpha + \frac{i''(0)}{2!}\alpha^2 + \frac{i'''(0)}{3!}\alpha^3$$

$$= 3 + 2\alpha + \frac{-1}{2}\alpha^2 + \frac{-2}{6}\alpha^3$$

$$= 3 + 2\alpha - \frac{1}{2}\alpha^2 - \frac{1}{3}\alpha^3$$

19. $E(x) = \sec x;$ $E(0) = 1$

$E'(x) = \sec x \tan x;$ $E'(0) = 0$

$E''(x) = \sec^3 x + \sec x \tan^2 x;$ $E''(0) = 1$

$E'''(x) = 3\sec^3 x \tan x + \sec x \tan^3 x$

$\qquad + 2\sec^3 x \tan x$

$\qquad = 5\sec^3 x \tan x + \sec x \tan^3 x$ $E'''(0) = 0$

$E^{(4)}(x) = 15\sec^3 x \tan^2 x + 5\sec^5 x$

$\qquad + \sec x \tan^4 x + 3\sec^3 x \tan^2 x;$ $E^{(4)}(0) = 5$

$E(x) = 1 + \frac{1}{2!}x^2 + \frac{5}{4!}x^4 = 1 + \frac{1}{2}x^2 + \frac{5}{24}x^4.$

20.   **(a)** Since the initial wound had an area of 16 cm^2, the constant term of the Maclaurin series is 16.

$A'(t) = \dfrac{-75}{t^2 + 25}$ $A'(0) = \dfrac{-75}{25} = -3$

$A''(t) = -75(-1)(t^2 + 25)^{-2}(2t)$

$\qquad = 150t(t^2 + 25)^{-2}$ $A''(0) = 0$

$A'''(t) = 150(t^2 + 25)^{-2} - 2(150t)(t^2 + 25)^{-3}(2t)$

$\qquad = 150(t^2 + 25)^{-2} - 600t^2(t^2 + 25)^{-3}$ $A'''(0) = \dfrac{6}{25}$

$A^{(4)}(t) = -300(t^2 + 25)^{-3}(2t) - 1200t(t^2 + 25)^{-3}$

$\qquad + 1800t^2(t^2 + 2t)^{-4}(2t)$

$\qquad = -1800t(t^2 + 25)^{-3} + 3600t^3(t^2 + 25)^{-4}$ $A^{(4)}(0) = 0$

$A^{(5)}(t) = -1800(t^2 + 25)^{-3} + 5400t(t^2 + 25)^{-4}(2t)$

$\qquad + 10800t^2(t^2 + 25)^{-4} - 14400t^3(t^2 + 25)^{-5}(2t)$

$\qquad = -1800(t^2 + 25)^{-3} + 21600t^2(t^2 + 25)^{-4}$

$\qquad - 28800t^4(t^2 + 25)^{-5}$ $A^{(5)}(0) = \dfrac{-1800}{25^3} = \dfrac{-72}{625}$

$A(x) = 16 - 3t\dfrac{6}{25}\dfrac{t^3}{3!} - \dfrac{72}{625}\dfrac{t^5}{5!}$

$\qquad = 16 - 3t + \dfrac{1}{25}t^3 - \dfrac{3}{3125}t^5$

(b) $A(3) = 16 - 3\cdot 3 + \frac{27}{25} - \frac{3\cdot 3^5}{3125} \approx 7.85$ cm

≡ 31.2 OPERATIONS WITH SERIES

1.   Since $e^x = 1 + x + \frac{x^2}{2!} + \frac{x^3}{3!}$, we see that $e^{3x} =$

$1 + 3x + \frac{(3x)^2}{2} + \frac{(3x)^3}{3!} = 1 + 3x + \frac{9x^2}{2} + \frac{9x^3}{2}.$

2.   Since $e^x = 1 + x + \frac{x^2}{2!} + \frac{x^3}{3!}$, then $e^{-4x} = 1 +$

$(-4x) + \frac{(-4x)^2}{2!} + \frac{(-4x)^3}{3!} = 1 - 4x + 8x^2 - \frac{32}{3}x^3.$

3.   Since $\cos x = 1 - \frac{x^2}{2!} + \frac{x^4}{4!} - \frac{x^6}{6!}$, we can find

$\cos \frac{x}{2} = 1 - \frac{x^2}{8} + \frac{x^4}{16\cdot 4!} - \frac{x^6}{2^6 6!} = 1 - \frac{x^2}{8} + \frac{x^4}{384} - \frac{x^6}{46{,}080}$

4.   Since $\sin x = x - \frac{x^3}{3!} + \frac{x^5}{5!} - \frac{x^7}{7!}$, then, we determine

that $\sin x^3 = x^3 - \frac{(x^3)^3}{3!} + \frac{(x^3)^5}{5!} - \frac{(x^3)^7}{7!} = x^3 - \frac{x^9}{3!} +$

$\frac{x^{15}}{5!} - \frac{x^{21}}{7!}.$

5.   Since $\cos x = 1 - \frac{x^2}{2!} + \frac{x^4}{4!} - \frac{x^6}{6!}$, and so $\cos x^3 =$

$1 - \frac{x^6}{2!} + \frac{x^{12}}{4!} - \frac{x^{18}}{6!}.$

6. $\sin 3x = 3x - \frac{(3x)^3}{3!} + \frac{(3x)^5}{5!} - \frac{(3x)^7}{7!} = 3x - \frac{9}{2}x^3 +$

$\frac{3^5 x^5}{5!} - \frac{3^7 x^7}{7!}$

7. $\sin 2x^2 = 2x^2 - \frac{(2x^2)^3}{3!} + \frac{(2x^2)^5}{5!} - \frac{(2x^2)^7}{7!} = 2x^2 - \frac{4}{3}x^6 + \frac{2^5 x^{10}}{5!} - \frac{2^7 x^{14}}{7!} = 2x^2 - \frac{8x^6}{3!} + \frac{32x^{10}}{5!} - \frac{128x^{14}}{7!}$

8. $\ln(1-x) = (-x) - \frac{(-x)^2}{2} + \frac{(-x)^3}{3} - \frac{(-x)^4}{4} = -x - \frac{x^2}{2} - \frac{x^3}{3} - \frac{x^4}{4}$

9. $\sin x = x - \frac{x^3}{3!} + \frac{x^5}{5!} - \frac{x^7}{7!} + \frac{x^9}{9!} + \cdots$, and hence $\frac{d}{dx}\sin x = 1 - \frac{x^2}{2!} + \frac{x^4}{4!} - \frac{x^6}{6!} + \frac{x^8}{8!} + \cdots$, which is $\cos x$.

10. $\cos x^2 = 1 - \frac{x^4}{2!} + \frac{x^8}{4!} - \frac{x^{10}}{5!} + \frac{x^{12}}{6!}$; $\frac{d}{dx}\cos x^2 = -\frac{4x^3}{2!} + \frac{8x^7}{4!} - \frac{10x^9}{5!} + \frac{12x^{11}}{6!} - \cdots = -2x^3 + \frac{2x^7}{3!} - \frac{2x^9}{4!} + \frac{2x^{11}}{5!} - \cdots = -2x\left(x^2 - \frac{x^6}{3!} + \frac{x^8}{4!} - \frac{x^{10}}{5!}\right) = -2x\sin x^2$

11. $e^{2x} = 1 + 2x + \frac{(2x)^2}{2!} + \frac{(2x)^3}{3!} + \frac{(2y)^4}{4!} = 1 + 2x + \frac{4x^2}{2!} + \frac{8x^3}{3!} + \frac{16x^4}{4!} + \cdots$. Hence, $\frac{d}{dx}e^{2x} = 2 + 4x + \frac{8x^2}{2!} + \frac{16x^3}{3!} + \cdots = 2\left(1 + 2x + \frac{4x^2}{2!} + \frac{8x^3}{3!} + \cdots\right) = 2\left(1 + 2x + \frac{(2x)^2}{2!} + \frac{(2x)^3}{3!} + \cdots\right) = 2e^{2x}$.

12. By problem #6, $\sin 3x = 3x - \frac{3^3 x^3}{3!} + \frac{3^5 x^5}{5!} - \frac{3^7 x^7}{7!} + \cdots$. Hence, $\frac{d}{dx}\sin 3x = 3 - \frac{3^3 x^2}{2!} + \frac{3^5 x^4}{4!} - \frac{3^7 x^6}{6!} = 3\left(1 + \frac{3^2 x^2}{2!} + \frac{3^4 x^4}{4!} - \frac{3^6 x^6}{6!}\right) = 3\cos 3x$.

13. $\int_0^1 \sin x^2 dx \approx \int_0^1 \left(x^2 - \frac{x^6}{3!} + \frac{x^{10}}{5!}\right)dx = \left[\frac{x^3}{3} - \frac{x^3}{7 \cdot 3!} + \frac{x^{11}}{11 \cdot 5!}\right]_0^1 = \frac{1}{3} - \frac{1}{42} + \frac{1}{1320} \approx 0.3103$.

14. Using the expansion for e^x, we have $e^x - 1 = x + \frac{x^2}{2!} + \frac{x^3}{3!} + \frac{x^4}{4!}$ and dividing by x produces $\frac{e^x - 1}{x} = 1 + \frac{x}{2!} + \frac{x^2}{3!} + \frac{x^3}{4!}$. Hence, we have $\int_0^{0.1} \frac{e^x - 1}{x} dx \approx \int_0^1 \left(1 + \frac{x}{2!} + \frac{x^2}{3!}\right)dx \approx \left[x + \frac{x^2}{4} + \frac{x^3}{18}\right]_0^{0.1} = 0.1 + \frac{0.1^2}{4} + \frac{(0.1)^3}{18} \approx 0.1026$.

15. $\int_0^{0.2} \sin\sqrt{x}\, dx = \int_0^{0.2}\left(x^{1/2} - \frac{x^{3/2}}{3!} + \frac{x^{5/2}}{5!}\right)dx = \left[\frac{2}{3}x^{3/2} - \frac{2x^{5/2}}{5 \cdot 31} + \frac{2x^{7/2}}{7 \cdot 5!}\right]_0^{0.2} = \frac{2}{3}(0.2)^{3/2} - \frac{1}{15}(0.2)^{5/2} + \frac{2}{7 \cdot 5!}(0.2)^{7/2} \approx 0.05844$.

16. $\int_0^1 \frac{\sin x}{x}dx \approx \int_0^1 \left(1 - \frac{x^2}{3!} + \frac{x^4}{5!}\right)dx = \left[x - \frac{x^3}{3 \cdot 3!} + \frac{x^5}{5 \cdot 5!}\right]_0^1 = 1 - \frac{1}{18} + \frac{1}{600} \approx 0.9461$

17. We know that $e^{-x} = 1 - x + \frac{x^2}{2!} - \frac{x^3}{3!} + \frac{x^4}{4!}$ and that $\cos x = 1 - \frac{x^2}{2!} + \frac{x^4}{4!} - \frac{x^6}{6!}$. Multiplying produces $e^{-x}\cos x = \left(1 - x + \frac{x^2}{2!} - \frac{x^3}{6} + \frac{x^4}{4!}\right)\left(1 - \frac{x^2}{2} + \frac{x^4}{4!} - \cdots\right) - 1 - x - \frac{x^2}{2} + \frac{x^2}{2} + \frac{x^3}{2} - \frac{x^3}{6} + \frac{x^4}{24} - \frac{x^4}{4} + \frac{x^4}{4!} = 1 - x + \frac{2x^3}{3!} - \frac{4x^4}{4!} + \cdots$.

18. Here we subtract the expansion for e^{-x} from that of e^x and multiply this answer by $\frac{1}{2}$. The result is

$$\frac{1}{2}\left(e^x - e^{-x}\right)$$
$$= \frac{1}{2}\left[\left(1 + x + \frac{x^2}{2!} + \frac{x^3}{3!} + \frac{x^4}{4!}\right) - \left(1 - x = \frac{x^2}{2!} - \frac{x^3}{3!} + \frac{x^4}{4!} + \cdots\right)\right]$$
$$= \frac{1}{2}\left[2x + 2\frac{x^3}{3!} + 2\frac{x^5}{5!} + 2\frac{x^7}{7!} + \cdots\right]$$
$$= x + \frac{x^3}{3!} + \frac{x^5}{5!} + \frac{x^7}{7!} + \cdots$$

This is the same as Exercise #2 in Exercise Set 30.1.

19. $e^{-x^2} = 1 - x^2 + \frac{x^4}{2!} + \frac{x^6}{3!} + \frac{x^8}{4!} + \cdots$. Hence $x^2 e^{-x^2} = x^2 - x^4 + \frac{x^6}{2!} - \frac{x^8}{3!} + \frac{x^{10}}{4!} - \cdots$

20. $\tan x = \frac{\sin x}{\cos x} = \dfrac{x - \frac{x^3}{3!} + \frac{x^5}{5!} - \frac{x^7}{7!} \cdots}{1 - \frac{x^2}{2!} + \frac{x^4}{4!} - \frac{x^6}{6!} \cdots}$. To compute this we must use a procedure like long division. We get the following:

$$1 - \tfrac{1}{2}x^2 + \tfrac{1}{24}x^4 - \cdots \overline{\smash{\big)}\, x - \tfrac{1}{6}x^3 + \tfrac{1}{120}x^5 - \cdots}$$

$$\begin{array}{r} x + \tfrac{1}{3}x^3 + \tfrac{2}{15}x^5 + \cdots \\[4pt] x - \tfrac{1}{2}x^3 + \tfrac{1}{24}x^5 - \cdots \\ \hline \tfrac{1}{3}x^3 - \tfrac{1}{30}x^5 + \cdots \\ \tfrac{1}{3}x^3 - \tfrac{1}{6}x^5 + \cdots \\ \hline \tfrac{2}{15}x^5 + \cdots \end{array}$$

Hence, $\tan x = x + \frac{x^3}{3} + \frac{2}{15}x^5 + \cdots$.

21. $5e^{0.8j} = 5(\cos 0.8 + j\sin 0.8) = 3.4835 + 3.5868j$

22. $5 - 12j$; $r = \sqrt{5^2 + 12^2} = 13$, and $\theta = \tan^{-1}\frac{-12}{5} = -1.176$. Thus, $5 - 12j = 13\left[\cos(-1.176) + j\sin(-1.176)\right]$

23. $6\operatorname{cis}\frac{4\pi}{3} = 6e^{\frac{4\pi}{3}j} = 6e^{4\pi j/3}$

24. $(-3 + 4j)$; $r = 5$, and $\theta = \tan^{-1}\frac{4}{-3} = 2.2143$. Thus, $(-3 + 4j) = 5e^{2.2143j}$

25. $y = e^{x^2} = 1 + x^2 + \frac{x^4}{2} + \cdots$. Hence, we obtain $A = \int_0^1 \left(1 + x^2 + \frac{x^4}{2}\right)dx = \left[x + \frac{x^3}{3} + \frac{x^5}{10}\right]_0^1 = 1 + \frac{1}{3} + \frac{1}{10} = \frac{43}{30} = 1.4333$.

26. $y = \cos x^2 = 1 - \frac{x^4}{2!} + \frac{x^8}{4!} - \cdots$. As a result, we obtain $\int_0^{\pi/2} \left(1 - \frac{x^4}{2!} + \frac{x^8}{4!}\right) dx = \left[x - \frac{x^5}{10} + \frac{x^9}{9 \cdot 4}\right]_0^{\pi/2} = \frac{\pi}{2} - \frac{\left(\frac{\pi}{2}\right)^5}{10} + \frac{\left(\frac{\pi}{2}\right)^9}{9 \cdot 4!} \approx 0.8840$. (The actual value is 0.8491.)

27. $x^2 e^2 = x^2 + x^3 + \frac{x^4}{2}$. Integrating, we obtain $\int_0^{0.2} \left(x^2 + x^3 + \frac{x^4}{2}\right) dx = \left[\frac{x^3}{3} + \frac{x^4}{4} + \frac{x^5}{5}\right]_0^{0.2} = \frac{(0.2)^3}{3} + \frac{(0.2)^4}{4} + \frac{(0.2)^5}{10} = 0.0031$.

28. $e^{-x^2} = 1 - x^2 + \frac{x^4}{2} + \cdots$. Hence, we get $\int_0^1 \left(1 - x^2 + \frac{x^4}{2}\right) dx = \left[x - \frac{x^3}{3} + \frac{x^5}{10}\right]_0^1 = 1 - \frac{1}{3} + \frac{1}{10} = \frac{23}{30} = 0.7667$.

29. $\cos 0.1 = \frac{(0.1)^2}{2!} + \frac{(0.1)^4}{4!} - \frac{(0.1)^6}{6!} + \frac{(0.1)^8}{8!} \approx 0.995004$. This produces $15\cos(0.1) = 15 \times 0.9950 = 14.9251$.

30. $\sinh x + \cosh x = \left[\left(x + \frac{x^3}{3!} + \frac{x^5}{5!} + \cdots\right) + \left(1 + \frac{x^2}{2!} + \frac{x^4}{4!} + \cdots\right)\right] = 1 + x + \frac{x^2}{2!} + \frac{x^3}{3!} + \frac{x^4}{4!} + \frac{x^5}{5!} + \cdots = e^x$.

31. **(a)** $i = \sin t^2 = t^2 - \frac{t^6}{3!} + \frac{t^{10}}{5!} - \frac{t^{14}}{7!} + \cdots$
(b) Charge $= \int_0^{0.02} \left(t^2 - \frac{t^6}{3!} + \frac{t^{10}}{5!} - \frac{t^{14}}{7!}\right) dx =$

$\left[\frac{t^3}{3} - \frac{t^7}{7 \cdot 3!} + \frac{t^{11}}{11 \cdot 5!} - \frac{t^{15}}{15 \cdot 7!}\right]_0^{0.02} = \frac{(0.02)^3}{3} - \frac{(0.02)^7}{7 \cdot 3!} + \frac{(0.02)^{11}}{11 \cdot 5!} = 2.7 \times 10^{-6} = 0.0000027$.

32. To find this volume, we can use the shell method. $V = 2\pi \int_0^{0.1} \left(xe^{-x}\right) dx$. Since $xe^{-x} = x - x^2 + \frac{x^3}{2} - \cdots$, we get $V = 2\pi \int_0^{0.1} \left(x - x^2 + \frac{x^3}{2}\right) dx = 2\pi \left[\frac{x^2}{2} - \frac{x^3}{3} + \frac{x^4}{8}\right]_0^{0.1} = 2\pi \left[\frac{0.01}{2} - \frac{0.001}{3} + \frac{0.0001}{8}\right] = 0.009358\pi \approx 0.02940$.

33. **(a)** Since we know that $e^x = 1 + x + \frac{x^2}{2!} + \frac{x^3}{3!}$, then

$e^{-2t} = 1 + (-2t) + \frac{(-2t)^2}{2!} + \frac{(-2t)^3}{3!}$

$= 1 - 2t + 2t^2 - \frac{4}{3}t^3$

and $e^{-t} = 1 - t + \frac{t^2}{2} - \frac{t^3}{6}$

Hence,

$P = P_0\left(e^{-2t} + e^{-t}\right)$

$= P_0\left(1 - 2t + 2t^2 - \frac{4}{3}t^3 + 1 - t + \frac{t^2}{2} - \frac{t^3}{6}\right)$

$= P_0\left(2 - 3t + \frac{5}{2}t^2 - \frac{3}{2}t^3\right)$

(b) $\frac{dP}{dt} = P_0\left(-3 + 5t - \frac{9}{2}t^2\right)$.

34. **(a)** Since we know that $e^x = 1 + x + \frac{x^2}{2!} + \frac{x^3}{3!}$, then

$e^{-0.05t} = 1 + (-0.05t) + \frac{(-0.05t)^2}{2!} + \frac{(-0.05t)^3}{3!}$
$= 1 - 0.05t + 0.00125t^2 - 0.000\,020\,83t^3$
$p(t) = 0.05e^{-0.05t} = 0.05 - 0.0025t + 0.0000625t^2 - 1.04167 \times 10^{-6}t^3$

(b) $\int_0^2 p(t)dt = \int_0^2 \left(0.05 - 0.0025t + 0.0000625t^2 - 1.04167 \times 10^{-6}t^3\right)dt$

$= 0.05t - 0.00125t^2 + 2.083 \times 10^{-5}t^3 - 2.604175 \times 10^{-7}t^4\Big]_0^2$
$= 0.05(2) - 0.00125(4) + 2.083 \times 10^{-5}(8) - 2.604175 \times 10^{-7}(16)$
$= 0.09516$

≡ 31.3 NUMERICAL TECHNIQUES USING SERIES

1. $e^{-0.3} = 1 - 0.3 + \frac{0.3^2}{2!} - \frac{(0.3)^3}{3!} + \frac{(0.3)^4}{4!} = 1 - 0.3 + 0.045 - 0.0045 + 0.0003375$. Using the first four terms and rounding we get $e^{-0.3} = 0.7408$.

2. $\sin(0.1) = 0.1 - \frac{(0.1)^3}{3!} + \frac{(0.1)^5}{5!} = 0.1 - 0.000167 + 0.000000083$. Using only the first two terms and rounding we get $\sin(0.1) = 0.0998$.

3. $\cos(0.1) = 1 - \frac{(0.1)^2}{2!} + \frac{(0.1)^4}{4!} - \frac{(0.1)^6}{6!} = 1 - 0.005 + 0.00000417$. Using the first two terms we get 0.9950

4. $e^{0.2} = 1 + 0.2 + \frac{(0.2)^2}{2!} + \frac{(0.2)^3}{3!} + \frac{(0.2)^4}{4!} = 1 + 0.2 + 0.02 + 0.00133 + 0.000067$. Using the first four terms and rounding we obtain 1.2213. If we use the first five terms we get 1.2214.

5. $5° = \frac{\pi}{36}$. Hence, $\sin 5° = \sin\frac{\pi}{36} = \frac{\pi}{36} - \frac{\left(\frac{\pi}{36}\right)^3}{3!} -$

$\frac{\left(\frac{\pi}{36}\right)^5}{5!} - 0.087266 - 0.00011 = 0.0872$

6. $\ln(1.5) = \ln(1+0.5) = 0.5 - \frac{(0.5)^2}{2} + \frac{(0.5)^3}{3} - \frac{(0.5)^4}{4} +$

$\frac{(0.5)^5}{5} + \cdots = 0.5 - 0.125 + 0.041667 - 0.015626 +$

$0.00625 - 0.002605 + 0.001116 - 0.000488 +$

$0.000217 - 0.000097 = 0.4054$

7. $\ln(0.97) = \ln(1-0.03) = -0.03 - \frac{(0.03)^2}{2} - \frac{(0.03)^3}{3}$

$- \frac{(0.03)^4}{4} = -0.03 - 0.00045 = -0.0305$

8. $\sqrt{e} = e^{1/2} = 1 + \frac{1}{2} + \frac{(0.5)^2}{2!} + \frac{(0.5)^3}{3!} + \frac{(0.5)^4}{4!} =$

$1 + 0.5 + 0.125 + 0.020833 + 0.002604 + 0.00026 +$

$0.0000217 = 1.6487$

9. $\ln(0.5) = \ln(1-0.5) = -0.5 - \frac{(0.5)^2}{2} - \frac{(0.5)^3}{3} -$

$\frac{(0.5)^4}{4} - \cdots = -0.5 - 0.125 - 0.041667 - 0.015625 -$

$0.00625 - 0.002604 - 0.001116 - 0.000488 -$

$0.000217 - 0.000097 = -0.6931$

10. $\cos 5° = \cos\frac{\pi}{36} = 1 - \frac{\left(\frac{\pi}{36}\right)^2}{2!} + \frac{\left(\frac{\pi}{36}\right)^4}{4} = 1 -$

$0.003808 + 0.0000024 = 0.9962$

11. First we must find a series for $\frac{1-\cos x}{x}$. Since

$\cos x = 1 - \frac{x^2}{2!} + \frac{x^4}{4!} \cdots$, we have then $1 - \cos x =$

$\frac{x^2}{2!} - \frac{x^4}{4!} + \frac{x^6}{6!}$, and $\frac{1-\cos x}{x} = \frac{x}{2!} - \frac{x^3}{4!} + \frac{x^5}{6!}$. As a result, we get

$\int_0^{0.5} \frac{1-\cos x}{x} dx$

$= \int_0^{0.5} \left(\frac{x}{2!} - \frac{x^3}{4!} + \frac{x^5}{6!} + \cdots \right) dx$

$= \frac{x^2}{4} - \frac{x^4}{4\cdot 4!} + \frac{x^6}{6\cdot 6!} - \frac{x^8}{8\cdot 8!} \Big|_0^{0.5}$

$= \frac{(0.5)^2}{4} - \frac{(0.5)^4}{4\cdot 4!} + \frac{(0.5)^6}{6\cdot 6!} - \frac{(0.5)^8}{8\cdot 8!}$

$= 0.0526 - 0.000651 + 0.0000036$

$= 0.0619.$

12. Since $\frac{x-\sin x}{x} = \int_0^1 \frac{x - \left(x - \frac{x^3}{3!} + \frac{x^5}{5!}\right)}{x} dx =$

$\frac{\frac{x^3}{3!} - \frac{x^5}{5!} + \frac{x^7}{7!}}{x} = \frac{x^2}{3!} - \frac{x^4}{5!} + \frac{x^6}{7!}$, we have $\int_0^{0.5} \frac{x-\sin x}{x}$

$dx = \int_0^{0.5} \left(\frac{x^2}{3!} - \frac{x^4}{5!} + \frac{x^6}{7!}\right) dx = \left[\frac{x^3}{3\cdot 3!} - \frac{x^5}{5\cdot 5!} + \frac{x^7}{7\cdot 7!}\right]_0^{0.5}$

$= 0.006944 - 0.000052 = 0.00689$ or 0.0069. (Note: we only needed the first term for the desired accuracy.)

13. $\int_0^1 \frac{\sin x - x}{x^2} dx = \int_0^1 \frac{-x}{3!} + \frac{x^3}{5!} - \frac{x^5}{7!} dx = \left[\frac{-x^2}{12} + \frac{x^4}{4\cdot 5!}\right.$

$\left. + \frac{x^6}{6\cdot 7!}\right]_0^1 = -\frac{1}{12} + \frac{1}{480} - 0.000033 = -0.0813.$

14. $\int_0^{0.5} \sqrt{x}\cos\sqrt{x}\, dx$

$= \int_0^{0.5} \sqrt{x}\left(1 - \frac{x}{2!} + \frac{x^2}{4!} - \frac{x^3}{6!} + \cdots\right) dx$

$= \int_0^{0.5} \left(x^{1/2} - \frac{x^{3/2}}{2!} + \frac{x^{5/2}}{4!} - \frac{x^{7/2}}{6!}\right) dx$

$= \left[\frac{2}{3}x^{3/2} - \frac{1}{5}x^{5/2} + \frac{2x^{7/2}}{7\cdot 4!} - \frac{2x^{9/2}}{9\cdot 6!}\right]_0^{0.5}$

$= 0.235702 - 0.035355 + 0.001052$

$- 0.0000316 \approx 0.2014$

15. $\int_0^{0.2} \frac{e^x - 1}{x}$

$= \int_0^{0.2} \left(1 + \frac{x}{2!} + \frac{x^2}{3!} + \frac{x^3}{4!} + \cdots\right) dx$

$= \left[x + \frac{x^2}{4} + \frac{x^3}{3\cdot 3!} + \frac{x^4}{4\cdot 4!}\right]_0^{0.2}$

$= 0.2 + 0.01 + 0.000444 + 0.0000167$

$= 0.2105$

16. $\int_{0.5}^1 \frac{\ln(1+x)}{x} dx$

$= \int_{0.5}^1 \left(1 - \frac{x}{2} + \frac{x^2}{3} - \frac{x^3}{4} + \frac{x^4}{5}\right) dx$

$= \left[x - \frac{x^2}{4} + \frac{x^3}{9} + \frac{x^4}{16} + \frac{x^5}{25}\right]_{0.5}^1$

$= \left(1 - \frac{1}{4} + \frac{1}{9} - \frac{1}{16} + \frac{1}{25} - \frac{1}{36} + \frac{1}{49}\right)$

$- \left(0.5 - \frac{(0.5)^2}{4} + \frac{(0.5)^3}{9}\right) = 0.3828$

17. **(a)** $\sin\theta = \theta - \frac{\theta^3}{3!} + \frac{\theta^4}{5!} - \cdots$, so $\sin\theta = \theta$ when $\frac{\theta^3}{3!}$ is negligible.

(b) If we want accuracy to 0.0001, then set $\frac{\theta^3}{3!} < 0.0001$ or $\theta^3 < 0.0006$ or $\theta < \sqrt[3]{0.0006} \approx 0.0843$.

18. To solve this, we need to find $\frac{1}{C}\int_0^{0.2} i\, dt$. Since

$1\,\mu\text{F}\ 10^{-6}\text{F}$, this integral is $\frac{1}{10^{-6}}\int_0^{0.2} \frac{0.1(t-\sin t)}{t} dt =$

$10^5 \int \left(\frac{t^2}{3!} - \frac{t^4}{5!} + \frac{t^6}{7!}\right) dt = 10^5 \left[\frac{t^3}{18} - \frac{t^5}{5\cdot 5!} + \frac{t^7}{7\cdot 7!}\right]_0^{0.2} =$

$10^5\left[4.\overline{4}\times 10^{-4} - 5.33\times 10^{-7}\right) = 10^5 \times 4.43911\times$

$10^{-4} = 44.3911$ V.

19. The Maclaurin series for e^x is $1 + x + \frac{x^2}{2!} + \frac{x^3}{3!} + \cdots$,

so $e^{-0.05t^2} = 1 - 0.05t^2 + \frac{0.0025t^4}{2!} - \frac{0.000125t^6}{3!} + \cdots$.

Thus,

$$q = \int_0^{0.5} \left(1 - e^{-0.05t^2}\right) dt$$

$$= \int_0^{0.5} \left(0.05t^2 - \frac{0.0025t^4}{2!} + \frac{0.000125t^6}{3!} + \cdots\right) dt$$

$$= \left[\frac{0.05}{3}t^3 - \frac{0.0025t^5}{5 \cdot 2!} + \frac{0.000125t^7}{7 \cdot 3!} + \cdots\right]_0^{0.5}$$

$$\approx 0.002083 - .000008 + \cdots \approx 0.0021$$

This charge is about $0.0021\ \mu C$.

20. The Maclaurin series for e^x is $1 + x + \frac{x^2}{2!} + \frac{x^3}{3!} + \cdots$, so $e^{-0.25t^2} = 1 - 0.25t^2 + \frac{0.0625t^4}{2!} - \frac{0.015625t^6}{3!} + \cdots$. Thus,

$$q = \int_0^{1.5} \left(1 - e^{-0.25t^2}\right) dt$$

$$= \int_0^{1.5} \left(0.25t^2 - \frac{0.0625t^4}{2!} + \frac{0.015625t^6}{3!} + \cdots\right) dt$$

$$= \left[\frac{0.25}{3}t^3 - \frac{0.0625t^5}{5 \cdot 2!} + \frac{0.015625t^7}{7 \cdot 3!} + \cdots\right]_0^{1.5}$$

$$\approx 0.28125 - 0.0474609 + 0.006356 - 0.000695$$

$$\approx 0.23945$$

The charge is about $0.23945\ \mu F$.

21. **(a)** The Maclaurin series for e^x is $1 + x + \frac{x^2}{2!} + \frac{x^3}{3!} + \cdots$, so the desired Maclaurin series is $V = 0.75e^{0.75t} = 0.75\left(1 + 0.75t + \frac{(0.75)^2}{2!}t^2 + \frac{(0.75)^3}{3!}t^3 + \cdots\right)$;

(b) When $t = 0.45$, this becomes $0.75(1 + 0.3375 + 0.05695 + 0.00641 + 0.00054\cdots) \approx 0.75(1.4014) \approx 1.05105$.

22. **(a)** $\cos x = 1 - \frac{x^2}{2!} + \frac{x^4}{4!} - \frac{x^6}{6!}$

$$\cos t^2 = 1 - \frac{(t^2)^2}{2!} + \frac{(t^2)^4}{4!} - \frac{(t^2)^6}{6!}$$

$$= 1 - \frac{t^4}{2} + \frac{t^8}{24} - \frac{t^{12}}{720}$$

(b) $c(x) = \int_0^x \cos t^2\, dt$

$$= \int_0^x \left(1 - \frac{t^4}{2} + \frac{t^8}{24} - \frac{t^{12}}{720}\right) dt$$

$$= \left[t - \frac{t^5}{10} + \frac{t^9}{9 \cdot 24} - \frac{t^{13}}{13 \cdot 720}\right]_0^x$$

$$= x - \frac{1}{10}x^5 + \frac{1}{216}x^9 - \frac{1}{9360}x^{13}$$

(c) $c(0.3) = (0.3) - \frac{1}{10}(0.3)^5 + \frac{1}{216}(0.3)^9 - \frac{1}{9360}(0.3)^{13}$ ≈ 0.2998

23. **(a)** $e^x = 1 + x + \frac{x^2}{2!} + \frac{x^3}{3!}$

$$e^{-t^2} = 1 + \left(-t^2\right) + \frac{\left(-t^2\right)^2}{2!} + \frac{\left(-t^2\right)^3}{3!}$$

$$= 1 - t^2 + \frac{1}{2}t^4 - \frac{1}{6}t^6$$

(b) $V(x) = V_0 \int_0^x e^{-t^2}\, dt$

$$= V_0 \int_0^x \left(1 - t^2 + \frac{1}{2}t^4 - \frac{1}{6}t^6\right) dt$$

$$= V_0 \left[t - \frac{1}{3}t^3 + \frac{1}{10}t^5 - \frac{1}{42}t^7\right]_0^x$$

$$= V_0 \left[x - \frac{1}{3}x^3 + \frac{1}{10}x^5 - \frac{1}{42}x^7\right]$$

(c) $V(1.4) = V_0 \left[(1.4) - \frac{1}{3}(1.4)^3 + \frac{1}{10}(1.4)^5 - \frac{1}{42}(1.4)^7\right] \approx 0.7722V_0$.

(d) The next nonzero term in the series of e^x is $\frac{x^4}{4!}$ and so the next nonzero term in the series of e^{-t^2} is $\frac{(-t^2)^4}{4!} = \frac{t^8}{24}$ and $\int \frac{t^8}{24}\, dt = \frac{t^9}{9 \cdot 24}$. When $t = 1.4$, then $\frac{t^9}{9 \cdot 24}V_0 \approx 0.0957V_0$. The error is less than 0.0957 since this is an oscillating series.

24.

$$e^x = 1 + x + \frac{x^2}{2!} + \frac{x^3}{3!}$$

$$e^{0.1t^2} = 1 + \left(0.1t^2\right) + \frac{\left(0.1t^2\right)^2}{2!} + \frac{\left(0.1t^2\right)^3}{3!}$$

$$= 1 + 0.1t^2 + 0.005t^4 + 0.000167t^6$$

$$1 - e^{0.1t^2} = -0.1t^2 - 0.005t^4 - 0.000167t^6$$

$$q = \int_0^{1.0} \left(-0.1t^2 - 0.005t^4 - 0.000167t^6\right) dt$$

$$= \left[-.03333t^3 - 0.001t^5 - 0.0000238t^7\right]_0^{1.0}$$

$$= -.03333 - 0.001 - 0.0000238$$

$$= -0.03436$$

Thus, the charge is about -0.0344 mC.

≡ 31.4 TAYLOR SERIES

1. $f(x) = e^x$, $f(2) = e^2$, $f'(x) = e^x$, $f'(2) = e^2$, etc. $e^x = e^2 + e^2(x - 2) + e^2 \frac{(x-2)^2}{2!} + e^2 \frac{(x-2)^3}{3!} + \cdots = e^2 \left[x - 1 + \frac{(x-2)^2}{2!} + \frac{(x-2)^3}{3!} + \cdots \right]$

2. $g(x) = e^{2x}$, $g(-1) = e^{-2}$, $g'(x) = 2e^{2x}$, $g'(-1) = 2e^{-2}$, $g''(x) = 2^2 e^{2x}$, $g''(-1) = 2^2 e^{-2}$, ... Hence, $e^{2x} = e^{-2} + 2e^{-2}(x + 1) + 2e^{-2} \frac{(x+1)^2}{2!} + \cdots = e^{-2} \left[1 + 2(x + 1) + \frac{2^2(x+1)^2}{2!} + \frac{2^3(x+1)^3}{3!} + \cdots \right]$.

3. $h(x) = \sin x$, $h\left(\frac{\pi}{4}\right) = \frac{\sqrt{2}}{2}$, $h'(x) = \cos x$, $h'\left(\frac{\pi}{4}\right) = \frac{\sqrt{2}}{2}$, $h''(x) = \sin x$, $h''\left(\frac{\pi}{4}\right) = -\frac{\sqrt{2}}{2}$, $h'''(x) = -\cos x$, $h'''\left(\frac{\pi}{2}\right) = -\frac{\sqrt{2}}{2}$. As a result we get

$\sin x = \frac{\sqrt{2}}{2} = \frac{\sqrt{2}}{2}\left(x - \frac{\pi x}{4}\right) - \frac{\sqrt{2}}{2} \frac{\left(x - \frac{\pi}{2}\right)^2}{2!}$
$- \frac{\sqrt{2}}{2} \frac{\left(x - \frac{\pi}{4}\right)3}{3!} + \cdots$
$= \frac{\sqrt{2}}{2} \left[1 + \left(x - \frac{\pi}{4}\right) - \frac{\left(x - \frac{\pi}{4}\right)^2}{2!} \right.$
$\left. - \frac{\left(x - \frac{\pi}{4}\right)^3}{3!} + \frac{\left(x - \frac{\pi}{4}\right)^4}{4!} + \cdots \right]$

4. $j(x) = \cos x$, $j\left(\frac{\pi}{4}\right) = \frac{\sqrt{2}}{2}$, $j'(x) = -\sin x$, $j'\left(\frac{\pi}{4}\right) = -\frac{\sqrt{2}}{2}$, $j''(x) = -\cos x$, $j''\left(\frac{\pi}{4}\right) = -\frac{\sqrt{2}}{2}$, $j'''(x) = \sin x$, $j'''\left(\frac{\pi}{4}\right) = \frac{\sqrt{2}}{2}$. Hence,

$\cos x = \frac{\sqrt{2}}{2} - \frac{\sqrt{2}}{2}\left(x - \frac{\pi}{2}\right) - \frac{\sqrt{2}}{2} \cdot \frac{\left(x - \frac{\pi}{4}\right)^2}{2!}$
$+ \frac{\sqrt{2}}{2} \cdot \frac{\left(x - \frac{\pi}{4}\right)^3}{3!} + \cdots$
$= \frac{\sqrt{2}}{2} \left[1 - \left(x - \frac{\pi}{4}\right) = \frac{1}{2!}\left(x - \frac{\pi}{4}\right)^2 \right.$
$\left. + \frac{1}{3!}\left(x - \frac{\pi}{4}\right)^3 + \frac{1}{4!}\left(x - \frac{\pi}{4}\right)^4 - \cdots \right]$

5. $k(x) = \frac{1}{x}$, so $k(2) = \frac{1}{2}$, $k'(x) = -x^{-2}$, so $k'(2) = -\frac{1}{4} = \frac{-1}{2^2}$, $k''(x) = 2x^{-3}$, so $k''(2) = \frac{2}{2^3}$, and $k'''(x) = -6x^{-4}$, so $k'''(2) = \frac{-6}{2^4}$. Hence, we see that $\frac{1}{x} = \frac{1}{2} - \frac{(x-2)}{2^2} + \frac{2(x-2)^2}{2^3 2!} - \frac{6(x-2)^3}{2^4 3!} + \frac{4!(x-2)^4}{2^5 4!} -$

$\cdots = \frac{1}{2}\left[1 - \frac{x-2}{2} + \frac{(x-2)^2}{2^2} - \frac{(x-2)^3}{2^3} + \frac{(x-2)^4}{2^4} - \cdots \right]$.

6. $m(x) = \sqrt{x}$, $m(4) = 2$, $m'(x) = \frac{1}{2}x^{-\frac{1}{2}}$, $m'(4) = \frac{1}{4} = \frac{1}{2^2}$, $m''(x) = -\frac{1}{4}x^{-\frac{3}{2}}$, $m''(4) = \frac{-1}{32}$, $m'''(x) = \frac{3}{8}x^{-\frac{5}{2}}$, $m'''(4) = \frac{3}{2^8}$. Hence, $\sqrt{x} = 2 + \frac{1}{4}(x - 4) - \frac{1}{32}\frac{(x-4)^2}{2!} + \frac{3}{2^8}\frac{(x-4)^3}{3!} - \cdots = 2 + \frac{1}{2^2}(x-4) - \frac{1}{2^5}\frac{(x-4)^2}{2!} + \frac{3}{2^8}\frac{(x-4)^3}{3!} - \frac{15(x-4)^4}{2^{11}4!} + \cdots$.

7. $f(x) = \frac{1}{(x+1)^2} = (x + 1)^{-2}$, $f(0) = 1$, $f'(x) = -2(x + 1)^{-3}$, $f'(0) = -2$ $f''(x) = 6(x + 1)^{-4}$, $f''(0) = 6$, $f'''(x) = -24(x + 1)^{-5}$, $f'''(0) = -24$. As a result, we obtain $\frac{1}{(x+1)^2} = 1 - 2(x) + \frac{6(x)^2}{2!} - \frac{24(x)^3}{3!} + \cdots = 1 - 2(x) + 3(x)^2 - 4(x)^3 + 5(x)^4 - \cdots$.

8. $g(x) = e^{1+x}$, $g(1) = e^2$, $g'(x) = e^{1+x}$, $g'(1) = e^2$, ... Hence, we get $e^{1+x} = e^2 + e^2(x - 1) + e^2 \frac{(x-1)^2}{2!} + e^2 \frac{(x-1)^2}{3!} + \cdots = e^2 [1 + (x - 1) + \frac{(x-1)^2}{2!} + \frac{(x-1)^3}{3!} + \cdots]$.

9. $h(x) = e^{-x}$, so $h(1) = e^{-1}$, and $h'(x) = -e^{-x}$, so $h'(1) = -e^{-1}$, and $h''(x) = e^{-x}$, so $h''(1) = e^{-1} \cdots$. Combining these results, we see that $e^{-x} = e^{-1}e^{-1}(x - 1) + e^{-1}\frac{(x-1)^2}{2!} - e^{-1}\frac{1}{3!}(x - 1)^3 \cdots = e^{-1}\left[1 - (x - 1) + \frac{1}{2!}(x - 1)^2 - \frac{1}{3!}(x - 1)^3 + \cdots \right]$.

10. $j(x) = \ln(x)$, $j(4) = \ln 4$, $j'(x) = \frac{1}{x}$, $j'(4) = \frac{1}{4}$, $j''(x) = -x^{-2}$, $j''(4) = \frac{-1}{4^2}$, $j'''(x) = 2x^{-3}$, $j'''(4) = \frac{2}{4^3}$. Hence, $\ln(x) = \ln 4 + \frac{1}{4}(x - 4) - \frac{1}{4^2}\frac{(x-4)^2}{2!} + \frac{2}{4^3}\frac{(x-4)^3}{3!} - \cdots = \ln 4 + \frac{1}{4}(x - 4) - \frac{1}{4^2}\frac{(x-4)^2}{2} + \frac{1}{4^3}\frac{(x-4)^3}{3} - \frac{1}{4^4}\frac{(x-4)^4}{4} + \cdots$

11. $k(x) = \sqrt[3]{x}$, $k(8) = 2$, $k'(x) = \frac{1}{3}x^{-2/3}$, $k'(8) = \frac{1}{3} \cdot \frac{1}{2^2}$, $k''(x) = \frac{-2}{9}x^{-5/3}$, $k''(8) = \frac{-2}{9} \cdot \frac{1}{2^5} = \frac{-1}{9 \cdot 2^4}$, $k'''(x) = \frac{10}{27}x^{-8/3}$. Hence, we obtain $\sqrt[3]{x} = 2 + \frac{1}{3} \cdot \frac{1}{2^2}(x - 8) - \frac{1}{9 \cdot 2^4} \cdot \frac{(x-8)^2}{2!} + \frac{10}{27} \cdot \frac{1}{2^8} \cdot \frac{(x-8)^3}{3!} + \cdots$.

12. $m(x) = \tan x$, $m\left(\frac{\pi}{4}\right) = \tan\left(\frac{\pi}{4}\right) = 1$, $m'(x) =$ $\sec^2 x$, $m'\left(\frac{\pi}{4}\right) = 2$, $m''(x) = 2\sec^2 x \tan x$, $m''\left(\frac{\pi}{4}\right) = 4$, $m'''(x) = 4\sec x \tan x + 2\sec^4 x$, $m'''\left(\frac{\pi}{4}\right) = 8 + 8 = 16$. As a result, we obtain

$$\tan x = 1 + 2\left(x - \frac{\pi}{4}\right) + 4\frac{\left(x - \frac{\pi}{4}\right)^2}{2!} + 16\frac{\left(x - \frac{\pi}{4}\right)^3}{3!} + \cdots.$$

13. Using Exercise #1, with $x = 2.1$, and $c = 2$, we see that $x - c = 0.2$. As a result, we obtain $e^{2.1} = e^2\left[1 + (0.1) + \frac{(0.1)^2}{2!} + \frac{(0.1)^3}{3!} + \cdots\right] = e^2[1.1052] = 8.1662.$

14. Here, we use Exercise #2, with $x = -0.4$, $c = -1$, and $x - c = x + 1 = -0.4 + 1 = 0.6$. As a result, we get $e^{-0.8} \approx e^{-2}\left[1 + 2(0.6) + \frac{2^2(0.6)^2}{2!} + \frac{2^3(0.6)^3}{3!} + \frac{2^4(0.6)^4}{4!}\right] = e^{-2}(3.2944) = 0.4458.$

15. In this exercise we use the result from Exercise #3. We convert $43°$ to $\frac{43\pi}{180}$ radians. We have $x = \frac{\pi}{4}$ and $c = \frac{43\pi}{180}$, with the result that $x - c = \frac{43\pi}{180} - \frac{\pi}{4} = -\frac{\pi}{90}$. Thus, $\sin 43° = \sin\frac{43\pi}{180}\sin\left(-\frac{\pi}{90}\right) =$

$$\frac{\sqrt{2}}{2}\left[1 + \left(-\frac{\pi}{90}\right) - \frac{\left(-\frac{\pi}{90}\right)^2}{2!} - \frac{\left(-\frac{\pi}{90}\right)^3}{3!} + \frac{\left(-\frac{\pi}{90}\right)^4}{4!}\right] = \frac{\sqrt{2}}{2}[0.9645] = 0.6820.$$

16. $\cos 43° = \cos\left(-\frac{\pi}{90}\right) = \frac{\sqrt{2}}{2}\left[1 - \left(-\frac{\pi}{90}\right) - \frac{1}{2!}\times\left(-\frac{\pi}{90}\right)^2 + \frac{1}{3!}\left(-\frac{\pi}{90}\right)^3 + \frac{1}{4!}\left(-\frac{\pi}{90}\right)^4\right] = \frac{\sqrt{2}}{2}\times[1.0343] = 0.7314.$

17. Using Exercise #5, with $x = 1.8$, $c = 2$, and $x - c = -0.2$, we get $\frac{1}{1.8} \approx \frac{1}{2}\left[1 - \frac{(-0.2)}{2} + \frac{(-0.2)^2}{2^2} - \frac{(-0.2)^3}{2^3} + \frac{(0.2)^4}{2^4}\right] = \frac{1}{2}[1 + 0.1 + 0.01 + 0.001 + 0.0001] = \frac{1}{2}[1.1111] = 0.5555$

18. $\sqrt{3.9} = \sqrt{4 - 0.1} \approx 2 + \frac{1}{2^2}(-0.1) - \frac{1}{2^6}(-0.1)^2 = 2 - 0.025 - 0.000156 = 1.9748$

19. $\frac{1}{1.12^2} = \frac{1}{(0.12+1)^2} = 1 - 2(0.12) + 3(0.12)^2 - 4(0.12)^3 = 1 - 0.24 + 0.0432 - 0.006912 + 0.0010368 - 0.000149 = 0.7972$

20. $e^{1.9} = e^{1+.9} \approx e^2\left[1 + (-0.1) + \frac{(-0.1)^2}{2!} + \frac{(-0.1)^3}{3!} + \frac{(-0.1)^4}{4!}\right] = e^2[0.9048] = 6.6859.$

21. $e^{-0.8} \approx e^{-1}\left[1 - (-0.2) + \frac{1}{2!}(-0.2)^2 - \frac{1}{3!}(-0.2)^3\right] = e^{-1}[1.2214] = 0.4493.$

22. $\ln(4.1) = \ln(4 + 0.1) = \ln 4 + \frac{1}{4}(0.1) - \frac{1}{16}\frac{(0.1)^2}{2} = \ln 4 + 0.025 - 0.0003125 = 1.4110$

23. $\sqrt[3]{7.8} = \sqrt[3]{8 - 0.2} = 2 + \frac{1}{3}\cdot\frac{1}{2^2}(-0.2) - \frac{1}{9}\cdot\frac{1}{2^4}\frac{(-0.2)^2}{2} = 2 - 0.01667 - 0.000139 = 1.9832$

24. $\tan 43° = \tan\left(-\frac{\pi}{90}\right) \approx 1 + 2\left(-\frac{\pi}{90}\right) + 2 \times \left(-\frac{\pi}{90}\right)^2 + \frac{16}{6}\left(-\frac{\pi}{90}\right)^3 = 1 - 0.069813 + 0.002437 - 0.0001134 = 0.9325$

25. From Example 31.0, we know that $\frac{\sin \pi t}{t} = \pi\left(\frac{t-2}{t}\right) - \frac{\pi^3}{3!}\frac{(t-2)^3}{t} + \cdots$. When we let $t = 2.1$, we get $\frac{\sin \pi t}{t} = \pi\left(\frac{0.1}{2.1}\right) - \frac{\pi^3}{3!}\frac{(0.1)^3}{2.1} + \cdots \approx 0.14960 - (0.00246) + \cdots \approx 0.14714$. A calculator gives $\frac{\sin(2.1\pi)}{2.1} \approx 0.14715$.

26. Using a procedure similar to the one used in Example 31.0, we determine the following fundamental information needed for the Taylor's expansion of $\sin x$ around $t = 1$. $f(x) = \sin 2\pi x$, so $f(1) = \sin 2\pi = 0$, $f'(x) = 2\pi\cos 2\pi x$, so $f'(1) = 2\pi\cos 2\pi = 2\pi$, $f''(x) = -4\pi^2\sin 2\pi x$, so $f''(1) = -4\pi^2\sin 2\pi = 0$, $f'''(x) = -8\pi^3 \times \cos 2\pi x$, so $f'''(1) = -8\pi^3\cos 2\pi = -8\pi^3$. Using these values we obtain

$$\sin 2\pi t = 0 + 2\pi(t - 1) - 0\cdot\frac{1}{2!}(t - 1)^2$$
$$- 8\pi^3\frac{1}{3!}(t - 1)^3 + \cdots$$
$$= 2\pi(t - 1) - \frac{8\pi^3}{3!}(t - 1)^3 + \cdots$$

Dividing by t produces $\frac{\sin 2\pi t}{t} = 2\pi\left(\frac{t-1}{t}\right) - 8\frac{\pi^3}{3!}\frac{(t-1)^3}{t} + \cdots$ and when we let $t = 1.05$, we get $\frac{\sin 2\pi t}{t} = 2\pi\left(\frac{0.05}{1.05}\right) - 8\pi\frac{\pi^3}{3!}\frac{(0.05)^3}{1.05} + \cdots \approx 0.2991993 - (0.0049216) + \cdots \approx 0.294278$. A calculator gives $\frac{\sin 2(1.05\pi)}{1.05} \approx 0.294302$.

☰ 31.5 FOURIER SERIES

1. $f(x) = \begin{cases} 2, & -\pi < x < 0 \\ -2, & 0 < x < \pi \end{cases}$, period 2π. $2L = 2\pi$, so $L = \pi$, First, $a_0 = \frac{1}{2\pi} \int_{-\pi}^{\pi} f(x)dx = \frac{1}{2\pi} \int_{-\pi}^{0} (-2)dx + \frac{1}{2\pi} \int_{0}^{\pi} 2\,dx = \frac{1}{2\pi}[-2x]_{-\pi}^{0} + \frac{1}{2\pi}[2x]_{0}^{-\pi} = \frac{-2\pi}{2\pi} + \frac{2\pi}{2\pi} = 0$. We have $a_k = \frac{1}{\pi} \int_{-\pi}^{\pi} f(x)\cos\frac{k\pi x}{\pi}dx = \frac{1}{\pi} \int_{-\pi}^{0} 2\cos kx\,dx + \int_{0}^{\pi} -2\cos kx\,dx = \frac{2}{\pi}\left[\frac{\sin kx}{d}\right]_{-\pi}^{0} - \frac{2}{\pi}\left[\frac{\sin kx}{k}\right]_{0}^{\pi} = 0 - \frac{2}{k\pi}\sin(-k\pi) - \frac{2}{k\pi}\sin k\pi = 0$. We also have $b_k = \frac{1}{\pi} \int_{-\pi}^{\pi} f(x)\sin\frac{k\pi x}{\pi}dx = \frac{1}{\pi} \int_{-\pi}^{0} 2\sin kx\,dx + \frac{1}{\pi} \int_{0}^{\pi} = 2\sin kx\,dx = \frac{-2}{\pi}\left[\frac{\cos kx}{k}\right]_{-\pi}^{0} + \frac{2}{\pi}\left[\frac{\cos kx}{k}\right]_{0}^{\pi} = \frac{-2}{k\pi}[1 - \cos k\pi] + \frac{2}{k\pi}[\cos k\pi - 1] = \frac{-4}{k\pi}(1 - \cos k\pi) = \begin{cases} 0 & \text{if } k \text{ is even} \\ \frac{-8}{k\pi} & \text{if } k \text{ is odd} \end{cases}$. Hence $f(x) = \frac{-8}{\pi}\sin x + \frac{-8}{3\pi}\sin 3x + \frac{-8}{5\pi}\sin 5x + \cdots = \frac{-8}{\pi}\left(\sin x + \frac{1}{3}\sin 3x + \frac{1}{5}\sin 5x + \cdots\right) = \frac{-8}{\pi}\sum_{k=1}^{\infty} \frac{1}{2k-1}\sin(2k-1)x$.

2. $g(x) = 1, -2 < x < 2$, period 4, $L = 2$. First, $a_0 = \frac{1}{4} \int_{-2}^{2} 1dx = \frac{1}{4}x\big|_{-2}^{2} = \frac{1}{4}(2+2) = 1$ and $a_k = \frac{1}{2} \int_{-2}^{2} \cos\frac{k\pi x}{2}dx = \frac{1}{2}\frac{2}{k\pi}\sin\frac{k\pi x}{2}\big|_{-2}^{2} = \frac{1}{k\pi}[\sin k\pi - \sin(-k\pi)] = 0$ for all k. We also have $b_k = \frac{1}{2} \int_{-2}^{2} \sin\frac{k\pi x}{2}dx = 0$ since sine is an odd function. Hence $g(x) = 1$.

3. $h(x) = \begin{cases} 0, & -3 < x < 0 \\ 2, & 0 < x < 3 \end{cases}$, period 6 so $L = 3$. First, $a_0 = \frac{1}{6} \int_{-3}^{3} h(x)dx = \frac{1}{6} \int_{0}^{3} 2\,dx = 1$. We have $a_k = \frac{1}{3} \int_{-3}^{3} h(x)\cos\frac{k\pi x}{3}dx = \frac{1}{3} \int_{0}^{3} 2\cos\frac{k\pi x}{3} \times dx = \frac{2}{3}\frac{3}{k\pi}\left[\sin\frac{k\pi x}{3}\right]_{0}^{3} = \frac{2}{k\pi}\sin k\pi = 0$ for all k. We also have $b_k = \frac{1}{3} \int_{-3}^{3} h(x)\sin\frac{k\pi x}{3}dx = \frac{1}{3} \int_{0}^{3} 2\sin\frac{k\pi x}{3}dx = \frac{2}{3}\cdot\frac{3}{k\pi}\left[-\cos\frac{k\pi x}{3}\right]_{0}^{3} = \frac{2}{k\pi}[-\cos$

$k\pi + 1] = \begin{cases} 0 & \text{if } k \text{ is even} \\ \frac{4}{k\pi} & \text{if } k \text{ is odd} \end{cases}$. Hence, $h(x) = 1 + \frac{4}{\pi}\sin\frac{\pi x}{3} + \frac{4}{3\pi}\sin\frac{3\pi x}{3} + \frac{4}{5\pi}\sin\frac{5\pi x}{3} + \cdots$.

4. $j(x) = \begin{cases} 0, & -1 < x < 0 \\ x & 0 < x < 1 \end{cases}$, period $= 2$ so $L = 1$. First, $a_0 = \frac{1}{2} \int_{-1}^{1} j(x)dx = \frac{1}{2} \int_{0}^{1} x\,dx = \frac{1}{2}\frac{x^2}{2}\big|_{0}^{1} = \frac{1}{4}$. We have $a_k = \frac{1}{1} \int_{-1}^{1} j(x)\cos k\pi x\,dx = \int_{0}^{1} x\cos k\pi x\,dx = \left(\frac{1}{k\pi}\right)^2 \int_{0}^{1} (k\pi x)\cos k\pi x\,dx = \left(\frac{1}{k\pi}\right)^2 \times \left[\cos k\pi x + k\pi x\sin k\pi x\right]_{0}^{1} = \left[\left(\frac{1}{k\pi}\right)^2 \cos k\pi x + \frac{x}{k\pi}\sin k\pi x\right]_{0}^{1} = \left(\left(\frac{1}{k\pi}\right)^2 \cos k\pi + 0\right) - \left(\frac{1}{k\pi}\cos 0 + 0\right) = \begin{cases} 0 & \text{if } k \text{ is even} \\ \frac{-2}{(k\pi)^2} & \text{if } k \text{ is odd} \end{cases}$. We also have $b_k = \int_{-1}^{1} j(x)\sin k\pi x\,dx = \int_{0}^{1} x\sin k\pi x\,dx = \left[\frac{1}{(k\pi)^2}\sin k\pi x - \frac{x}{k\pi}\cos k\pi x\right]_{0}^{1} = -\frac{1}{k\pi}\cos k\pi = \begin{cases} -\frac{1}{k\pi}, & \text{if } k \text{ is even} \\ \frac{1}{k\pi}, & \text{if } k \text{ is odd} \end{cases}$. So,

$$j(x) = \frac{1}{4} - \frac{2}{\pi^2}\cos \pi x - \frac{2}{9\pi^2}\cos 3\pi x$$
$$- \frac{2}{25\pi^2}\cos 5\pi x - \cdots$$
$$+ \frac{1}{\pi}\sin \pi x - \frac{1}{2\pi}\sin 2\pi x + \frac{1}{3\pi}\sin 3\pi x$$
$$- \frac{1}{4\pi}\sin 4\pi x + \cdots$$
$$= \frac{1}{4} + \sum_{k=1}^{\infty}\left[\frac{-2}{(2k-1)^2\pi^2}\cos(2k-1)\pi x\right.$$
$$\left. + (-1)^{k+1}\frac{1}{k\pi}\sin k\pi x\right].$$

5. $f(x) = \begin{cases} -x & -\pi < x < 0 \\ x, & 0 < x < \pi \end{cases}$, period $= 2\pi$ so $L = \pi$. First, $a_0 = \frac{1}{2\pi} \int_{-\pi}^{\pi} f(x)dx = \frac{1}{2\pi} \int_{-\pi}^{0} -x \, dx + \frac{1}{2\pi} \int_{0}^{\pi} x \, dx = \frac{1}{2\pi} \frac{-x^2}{2} \Big|_{-\pi}^{0} + \frac{1}{2\pi} \frac{x^2}{2} \Big|_{0}^{\pi} = \frac{\pi^2}{4\pi} + \frac{\pi^2}{4\pi} = \frac{\pi}{2}$.
Next, $a_k = \frac{1}{\pi} \int_{-\pi}^{\pi} f(x) \cos kx \, dx = \frac{1}{\pi} \int_{-\pi}^{0} -x \cos kx \, dx + \frac{1}{\pi} \int_{0}^{\pi} x \cos kx \, dx = \frac{2}{\pi} \int_{0}^{\pi} x \cos kx \, dx = \frac{2}{\pi} \left[\frac{\cos kx}{k^2} + \frac{x \sin kx}{k} \right]_{0}^{\pi} = \frac{2}{\pi k^2} [\cos k\pi - 1] = \begin{cases} 0 & \text{if } k \text{ is even} \\ \frac{-4}{k^2 \pi} & \text{if } k \text{ is odd} \end{cases}$. Also, we see that $b_k = \frac{1}{2\pi} \times \int_{-\pi}^{\pi} f(x) \sin kx = 0$ since $f(x) \sin k\pi$ is an odd function. Hence, $f(x) = \frac{\pi}{2} - \frac{4}{\pi} \cos x - \frac{4}{9\pi} \cos 3x - \frac{4}{25\pi} \cos 5\pi - \cdots = \frac{\pi}{2} - \frac{4}{\pi} \sum_{k=1}^{\infty} \frac{1}{(2k-1)^2} \cos(2k - 1)x$.

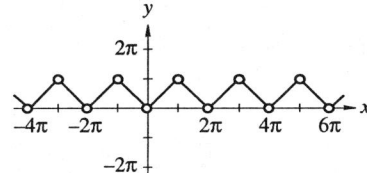

6. $g(x) = \begin{cases} 3, & 0 < x < 4 \\ -3, & 4 < x < 8 \end{cases}$, period $= 8$ so $L = 4$.
First, $a_0 = \frac{1}{8} \int_{0}^{8} g(x)dx = \frac{1}{8} \int_{0}^{4} 3 \, dx + \frac{1}{8} \int_{4}^{8} (-3)dx = 0$. Next, $a_k = \frac{1}{4} \int_{0}^{8} g(x) \cos \frac{k\pi x}{4} dx = \frac{1}{4} \int_{0}^{4} 3 \cos \frac{k\pi x}{4} dx - \frac{1}{4} \int_{4}^{8} 3 \cos \frac{k\pi x}{4} dx = \frac{3}{4} \cdot \frac{4}{k\pi} \sin \frac{k\pi x}{4} \Big|_{0}^{4} - \frac{3}{4} \cdot \frac{4}{k\pi} \sin \frac{k\pi x}{4} \Big|_{4}^{8} = 0$. We also have $b_k = \frac{1}{4} \int_{0}^{4} 3 \sin \frac{k\pi x}{4} dx - \frac{1}{4} \int_{4}^{8} 3 \sin \frac{k\pi x}{4} dx = -\frac{3}{4} \cdot \frac{4}{k\pi} \cos \frac{k\pi x}{4} \Big|_{0}^{4} + \frac{3}{4} \cdot \frac{4}{k\pi} \cos \frac{k\pi x}{4} \Big|_{4}^{8} = -\frac{3}{k\pi} [\cos k\pi - 1] + \frac{3}{k\pi} [1 - \cos k\pi] = \frac{6}{k\pi} [1 - \cos k\pi] = \begin{cases} 0 & \text{if } k \text{ is even} \\ \frac{12}{k\pi} & \text{if } k \text{ is odd} \end{cases}$. Thus, $g(x) = \frac{12}{\pi} \sin \frac{\pi x}{4} + \frac{12}{3\pi} \sin \frac{3\pi x}{4} + \frac{12}{5\pi} \sin \frac{5\pi x}{4} \cdots = \frac{12}{\pi} \sum_{k=1}^{\infty} \frac{1}{2k-1} \sin \frac{(2k-1)\pi x}{4}$

7. $h(x) = x$, $-\pi < x < \pi$, period 2π so $L = \pi$. First, $a_0 = \frac{1}{2\pi} \int_{-\pi}^{\pi} x \, dx = 0$. Next, $a_k = \frac{1}{\pi} \int_{-\pi}^{\pi} x \cos kx \, dx = 0$, since $[x \cos kx]$ is an odd function. Also, $b_k = \frac{1}{\pi} \int_{-\pi}^{\pi} x \sin kx \, dx =$

$\frac{1}{\pi} \left[\frac{\sin kx}{k^2} - \frac{x \cos kx}{k} \right]_{-\pi}^{\pi} = \frac{1}{\pi} \left[\left(-0 - \frac{\pi \cos k\pi}{k} \right) - \left(0 + \frac{\pi \cos k\pi}{k} \right) \right] = \begin{cases} \frac{-2}{k} & \text{if } k \text{ even} \\ \frac{2}{k} & \text{if } k \text{ is odd} \end{cases}$. Hence, $h(x) = 2 \sin \pi - \frac{2}{2} \sin 2\pi + \frac{2}{3} \sin 3\pi - \cdots = 2 \sum_{k=1}^{\infty} \frac{(-1)^{k+1}}{k} \sin kx$.

8. $j(x) = \begin{cases} 0, & -\frac{\pi}{2} < x < 0 \\ 2 & 0 < x < \frac{\pi}{2} \end{cases}$, period $= \pi$, so $L = \frac{\pi}{2}$. First, $a_0 = \frac{1}{\pi} \int_{-\pi/2}^{\pi/2} j(x)dx = \frac{1}{\pi} \int_{0}^{\pi/2} 2 \, dx = \frac{1}{\pi} 2 \Big|_{0}^{\pi/2} = 1$. Next, we have $a_k = \frac{2}{\pi} \int_{-\pi/2}^{\pi/2} j(x) \cos \frac{k\pi x}{\pi/2} dx = \frac{2}{\pi} \int_{0}^{\pi/2} 2 \cos 2kx \, dx = \frac{4}{\pi} \cdot \frac{1}{2k} \left[\sin 2kx \right]_{0}^{\pi/2} = \frac{2}{k\pi} (\sin \pi - \sin 0) = 0$. Also, $b_k = \frac{2}{\pi} \int_{0}^{\pi/2} 2 \sin 2k, x \, dx = \frac{4}{\pi} \cdot \frac{1}{2k} \left[-\cos 2kx \right]_{0}^{\pi/2} = \frac{2}{k\pi} [-\cos k\pi + \cos 0] = \frac{2}{k\pi} [1 - \cos k\pi] = \begin{cases} 0 & \text{if } k \text{ is even} \\ \frac{4}{k\pi} & \text{if } k \text{ is odd} \end{cases}$. Thus, $j(x) = 1 + \frac{4}{\pi} \sin 2x + \frac{4}{3\pi} \sin 6x + \frac{4}{5\pi} \sin 10x + \cdots = 1 + \frac{4}{\pi} \sum_{k=1}^{\infty} \frac{1}{2k-1} \sin(4k - 2)x$.

9. $f(x) = \begin{cases} 1, & 0 < x < \frac{2\pi}{3} \\ 0, & \frac{2\pi}{3} < x < \frac{4\pi}{3} \\ -1 & \frac{4\pi}{3} < x < 2\pi \end{cases}$, period $= 2\pi$, so $L = \pi$. First, $a_0 = \frac{1}{2\pi} \int_{0}^{2\pi} f(x)dx = \frac{1}{2\pi} \times \left[\int_{0}^{2\pi/3} 1 \, dx + \int_{4\pi/3}^{2\pi} -1 \, dx \right] = 0$. Then, we have $a_k = \frac{1}{\pi} \int_{0}^{2\pi} f(x) \cos kx \, dx = \frac{1}{2} \left[\int_{0}^{2\pi/3} \cos kx \, dx - \int_{4\pi/3}^{2\pi} \cos kx \, dx \right] = \frac{1}{k\pi} \left[\sin kx \Big|_{0}^{2\pi/3} - \sin kx \Big|_{4\pi/3}^{2\pi} \right] = \frac{1}{\pi k} \left[\sin \frac{2k\pi}{3} - 0 - 0 + \sin \frac{4\pi k}{3} \right]$. This is always 0. Now, we have $b_k = \frac{1}{\pi} \int_{0}^{2\pi} f(x) \sin kx \, dx = \frac{1}{\pi} \int_{0}^{2\pi/3} \sin kx - \frac{1}{\pi} \int_{4\pi/3}^{2\pi} \sin kx \, dx = \frac{-1}{k\pi} \cos kx \Big|_{0}^{2\pi/3}$

$+ \frac{1}{k\pi}\cos kx \Big|_{4\pi/3}^{2\pi} = \frac{-1}{k\pi}\left[1 - \cos\frac{2k\pi}{3} + 1 - \cos\frac{4k\pi}{3}\right]$.

Thus, $b_1 = \frac{3}{\pi}$, $b_2 = \frac{3}{2\pi}$, $b_3 = 0$, $b_4 = \frac{3}{4\pi}$, $b_5 = \frac{3}{5\pi}$, $b_6 = 0$, and we obtain $f(x) = \frac{3}{\pi}\sin x + \frac{3}{2\pi}\sin 2x + 0\cdot\sin 3x + \frac{3}{4\pi}\sin 4x + \cdots \frac{3}{\pi}\sum_{k=1}^\infty \times \left(\frac{\sin(3k-2)x}{3k-2} + \frac{\sin(3k-1)x}{3k-1}\right)$.

10. $g(x) = \begin{cases} x, & 0 < x \le \pi \\ \pi & \pi < x < 2\pi \end{cases}$, period $= 2\pi$ so $L = \pi$. Thus, $a_0 = \frac{1}{2\pi}\int_0^{2\pi} g(x)dx = \frac{1}{2\pi}\int_0^\pi x\,dx + \frac{1}{2\pi}\int_\pi^{2\pi}\pi\,dx = \frac{1}{2\pi}\cdot\frac{x^2}{2}\Big|_0^\pi + \frac{1}{2\pi}(\pi x)\Big|_\pi^{2\pi} = \frac{\pi}{4} + \pi - \frac{\pi}{2} = \frac{3\pi}{4}$. Next we have $a_k = \frac{1}{\pi}\int_0^{2\pi} g(x)\cos kx\,dx = \frac{1}{\pi}\int_0^\pi x\cos kx\,dx + \frac{1}{\pi}\int_\pi^{2\pi}\pi\cos kx\,dx = \frac{1}{\pi}\left[\frac{\cos kx}{k^2} + \frac{x\sin kx}{k}\right]_0^\pi + \left[\frac{1}{k}\sin kx\right]_\pi^{2\pi} = \frac{1}{\pi k^2}[\cos\pi - 1] = \begin{cases} 0 & \text{if } k \text{ is even} \\ \frac{-2}{\pi k} & \text{if } k \text{ is odd} \end{cases}$

We also have $b_k = \frac{1}{\pi}\int_0^\pi x\sin kx\,dx + \frac{1}{\pi}\int_\pi^{2\pi}\pi\sin kx\,dx = \frac{1}{\pi}\left[\frac{\sin kx}{k^2} - \frac{x\cos kx}{k}\right]_0^\pi + \frac{1}{k}(-\cos kx)\Big|_\pi^{2\pi} = \frac{1}{\pi}\left(-\frac{\pi\cos k}{k}\right)\frac{1}{k}(\cos k\pi - \cos 2\pi k) = -\frac{\cos k\pi}{k} + \frac{\cos k\pi}{k} - \frac{\cos 2\pi k}{k} = \frac{1}{k}$. Hence, $g(x) = \frac{3\pi}{4} - \frac{2}{\pi}\left(\cos x + \frac{1}{9}\cos 3x + \frac{1}{25}\cos 5x + \cdots\right) - \left(\sin x + \frac{1}{2}\sin 2x + \frac{1}{3}\sin 3x + \cdots\right) = \frac{3\pi}{4} - \frac{2}{\pi}\sum_{k=1}^\infty \frac{\cos(2k-1)x}{(2k-1)^2} - \sum_{k=1}^\infty \frac{\sin k\pi}{k}$

11. $h(x) = \begin{cases} 0, & \text{if } -\pi < x < 0 \\ \sin x, & \text{if } 0 < x < \pi \end{cases}$, period $= 2\pi$, so $L = \pi$. First, $a_0 = \frac{1}{2\pi}\int_{-\pi}^\pi h(x)dx = \frac{1}{2\pi}\int_0^\pi \sin x\,dx = \frac{1}{2\pi}[-\cos x]_0^\pi = \frac{1}{2\pi}[1+1] = \frac{1}{\pi}$. For a_1 we have $a_1 = \frac{1}{\pi}\int_0^\pi \sin x\cos x\,dx = \frac{1}{2\pi}\int_0^\pi 2\sin x\cos x\,dx = \frac{1}{2\pi}\int_0^\pi \sin 2x\,dx = \left[-\frac{1}{4\pi}\cos 2x\right]_0^\pi = 0$. For a_k, $k > 1$, we have $a_k = \frac{1}{\pi}\int_{-\pi}^\pi h(x)\cos kx\,dx = \frac{1}{\pi}\int_0^\pi \sin x\cos kx\,dx = \frac{1}{\pi}\left[-\frac{\cos(1-k)x}{2(1-k)} - \frac{\cos(1+k)x}{2(1+k)}\right]_0^\pi$

$= -\frac{1}{\pi}\left[\frac{\cos(1-k)\pi}{2(k-1)} + \frac{\cos(1+k)x}{2(1+k)} - \frac{1}{2(1-k)} - \frac{1}{2(1+k)}\right]$. If k is odd, $\cos(1-k)\pi = \cos(1+k)\pi = 1$, so $a_k = 0$. If k is even, $\cos(1-k)\pi = \cos(1+k)\pi = -1$, so $a_k = \frac{1}{\pi}\left[\frac{1}{1-k} + \frac{1}{1+k}\right] = \frac{1}{\pi}\frac{2}{1-k^2} = -\frac{2}{(k^2-1)\pi}$. In particular, $a_2 = -\frac{2}{(2^2-1)\pi} = -\frac{2}{3\pi}$, $a_4 = -\frac{2}{(4^2-1)\pi} = -\frac{2}{15\pi}$, $a_6 = -\frac{2}{35\pi}$. For b_1 we have $b_1 = \frac{1}{\pi}\int_0^\pi \sin^2 x\,dx = \frac{1}{\pi}\left[\frac{x}{2} - \frac{1}{4}\sin 2x\right]_0^\pi = \frac{1}{2}$. For $k > 1$, we have $b_k = \frac{1}{\pi}\int_0^\pi \sin x\sin kx\,dx = \frac{1}{\pi}\left[\frac{\sin(1-k)x}{2(1-k)} - \frac{\sin(1+k)x}{2(1+k)}\right]_0^\pi = 0$. Hence, $h(x) = \frac{1}{\pi} + \frac{1}{2}\sin x - \frac{2}{\pi}\left[\frac{1}{3}\cos 2x + \frac{1}{15}\cos 4x + \cdots\right] = \frac{1}{\pi} + \frac{1}{2}\sin x - \frac{2}{\pi}\sum_{k=1}^\infty \frac{\cos 2kx}{4k^2-1}$.

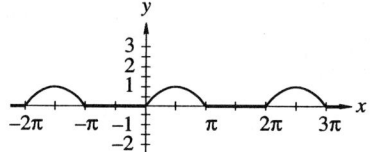

12. $f(x) = \begin{cases} \sin x, & 0 < x < \pi \\ -\sin x, & \pi < x < 2\pi \end{cases}$, period 2π, so $L = \pi$. First $a_0 = \frac{1}{2\pi}\int_0^{2\pi} f(x)dx = \frac{1}{2\pi}\int_0^\pi \sin x\,dx + \frac{1}{2\pi}\int_\pi^{2\pi}(-\sin x)dx = \frac{1}{2\pi}\left[-\cos x\Big|_0^\pi + \frac{1}{2\pi}\cos x\Big|_\pi^{2\pi}\right] = \frac{1}{2\pi}(1 + 1 + 1 + 1) = \frac{2}{\pi}$. For a_1 we have $\frac{1}{\pi}\int_0^\pi \sin x\cos x\,dx - \frac{1}{\pi}\int_\pi^{2\pi}\sin x\cos x\,dx = \frac{1}{2\pi}\times\int_0^\pi 2\sin x\cos x\,dx - \frac{1}{2\pi}\int_\pi^{2\pi} 2\sin x\cos x\,dx = \frac{1}{2\pi}\times\int_0^\pi \sin 2x\,dx - \frac{1}{2\pi}\int_\pi^{2\pi}\sin 2x\,dx = \frac{1}{4\pi}[-\cos 2x]_0^\pi - \frac{1}{4\pi}[-\cos 2x]_\pi^{2\pi} = 0$. For $k > 1$, $a_k = \frac{1}{\pi}\int_0^\pi \sin x\cos kx\,dx - \frac{1}{\pi}\int_\pi^{2\pi}\sin x\cos kx\,dx = -\frac{1}{\pi}\left[\frac{\cos(1-k)x}{2(1-k)} + \frac{\cos(1+k)x}{2(1+k)}\right]_0^\pi + \frac{1}{\pi}\left[\frac{\cos(1-k)x}{2(1-k)} + \frac{\cos(1+k)x}{2(1+k)}\right]_\pi^{2\pi} = -\frac{1}{\pi}\times\left[\frac{\cos(1-k)\pi}{2(1-k)} + \frac{\cos(1+k)\pi}{2(1+k)} - \frac{1}{2(1-k)} - \frac{1}{2(1+k)}\right] + \frac{1}{\pi}\left[\frac{1}{2(1-k)} + \frac{1}{2(1+k)} - \frac{\cos(1-k)\pi}{2(1-k)} - \frac{\cos(1-k)\pi}{2(1-k)}\right]$. If k is odd, $\cos(1-k)\pi = \cos(1+k)\pi1$, so $a_k = 0$. If k is even, $\cos(1-k)\pi = \cos(1+k)\pi = -1$, so $a_k = \frac{1}{\pi}\left[\frac{2}{1-k} + \frac{2}{1+k}\right] = \frac{4}{\pi(1-k^2)}$. In particular, $a_2 = \frac{4}{\pi(1-2^2)} = -\frac{4}{3\pi}$, $a_4 = \frac{4}{\pi(1-4^2)} = -\frac{4}{15\pi}$, $a_6 = -\frac{4}{35\pi}$. In a similar manner we find that $b_1 = \frac{1}{\pi}\int_0^\pi \sin^2 x\,dx - \frac{1}{\pi}\int_\pi^{2\pi}\sin^2 x\,dx = \frac{1}{\pi}\int_0^\pi \sin^2 x\,dx - \frac{1}{\pi}\int_{-\pi}^0 \sin^2 x\,dx$. Since $\sin^2 x$ is an even function, these two integrals are equal, so their difference is 0. The same argument holds for b_k when $k > 1$. Hence,

$f(x) = \frac{2}{\pi} - \frac{4}{3\pi}\cos 2x - \frac{4}{15\pi}\cos 4x - \cdots = \frac{2}{\pi}\left(1 - \sum_{k=1}^{\infty}\frac{2\cos 2kx}{4k^2-1}\right).$

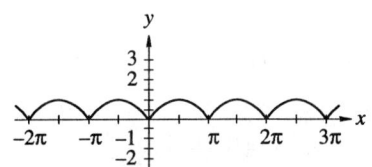

CHAPTER 31 REVIEW

1. Since $\cos x = 1 - \frac{x^2}{2!} + \frac{x^4}{4!} - \frac{x^6}{6!} + \cdots$ we have $1 - \cos x = \frac{x^6}{2!} - \frac{x^4}{4!} + \frac{x^6}{6!} - \cdots$ and $\frac{1-\cos x}{x}$ equals $\frac{x}{2!} - \frac{x^3}{4!} + \frac{x^5}{6!} - \frac{x^7}{8!} + \frac{x^9}{10!} - \frac{x^{11}}{12!} + \cdots$.

2. Since $e^x = 1 + x + \frac{x^2}{2!} + \frac{x^3}{3!} + \cdots$ we get $e^{2x} = 1 - 2x + \frac{4x^2}{2!} - \frac{8x^3}{3!} + \cdots$ and so $xe^{-2x} = x - 2x^2 + \frac{4x^3}{2!} - \frac{8x^4}{3!} + \frac{16x^5}{4!} - \frac{32x^6}{5} + \cdots$

3. $\sin x \cos x = \left(x - \frac{x^3}{3!} + \frac{x^5}{5!} - \frac{x^7}{7!} + \cdots\right)\left(1 - \frac{x^2}{2!} + \frac{x^4}{4!} - \frac{x^6}{6!} + \cdots\right) = x - \frac{x^3}{2} - \frac{x^3}{3} + \frac{x^5}{4!} + \frac{x^5}{6} + \frac{x^5}{6} - \frac{x^7}{3!4!} - \frac{x^7}{2\cdot5!} - \frac{x^7}{6!} - \frac{x^7}{7!} + \cdots = x - \frac{5x^3}{3!} + \frac{16x^5}{5!} - \frac{64x^7}{7!} + \cdots$

4. $\ln(2 + x) = \ln(x + 1 + 1) = x + 1 - \frac{(x+1)^2}{2} + \frac{(x+1)^3}{3} - \frac{(x+1)^4}{4} + \cdots$

5. $f(x) = (1+x)^{3/4}$, $f(0) = 1$; $f'(x) = \frac{3}{4}(1+x)^{-1/4}$; $f'(0) = \frac{3}{4}$; $f''(x) = -\frac{3}{16}(1+x)^{-5/4}$; $f''(0) = -\frac{3}{16}$; $f'''(x) = \frac{15}{16}(1+x)^{-9/4}$; $f'''(0) = \frac{15}{64}$. Hence $(1+x)^{3/4} = 1 + \frac{3}{4}(1+x) - \frac{3}{16}\frac{(1+x)^2}{2!} + \frac{15}{64}\frac{(1+x)^3}{3!} - \frac{135}{256}\frac{(1-x)^4}{4!} + \cdots$.

6. Since $e^x = 1 + x + \frac{x^2}{2!} + \frac{x^3}{3!} + \cdots$ we get $e^{-x^4} = 1 - x^4 + \frac{x^8}{2!} - \frac{x^{12}}{3!} + \frac{x^{16}}{4!} - \frac{x^{20}}{5!} + \cdots$

7. Since $\sin x = x - \frac{x^8}{3!} + \frac{x^5}{5!} - \frac{x^7}{7!} + \cdots$ we obtain $\sin x^2 = x^2 - \frac{x^6}{3!} + \frac{x^{10}}{5!} - \frac{x^{14}}{7!} + \frac{x^{18}}{9!} - \cdots$

8. $1 - e^x = -x - \frac{x^2}{2!} - \frac{x^3}{3!} - \frac{x^4}{4!} - \cdots$ and we get $\frac{1-e^x}{x} = -1 - \frac{x}{2!} - \frac{x^2}{3!} - \frac{x^3}{4!} - \frac{x^4}{5!} - \cdots$

9. First we obtain the Maclaurin series for $\frac{\sin x}{\sqrt{x}}$ which is $x^{1/2} - \frac{x^{5/2}}{3!} + \frac{x^{9/2}}{5!} - \cdots$. Now we integrate $\int_0^1\left(x^{1/2} - \frac{x^{5/2}}{3!} + \frac{x^{9/2}}{5!} - \frac{x^{13/2}}{7!}\right) = \left[\frac{2}{3}x^{3/2} - \frac{2x^{3/2}}{7\cdot3!} + \right.$

$\left.\frac{2}{11}\frac{x^{11/2}}{5!} - \frac{2x^{15/2}}{15\cdot7!}\right]_0^1 = \frac{2}{3} - \frac{2}{42} + \frac{2}{1320} - \frac{2}{75600} = 0.66667 - 0.04762 + 0.001515 - 0.000026 = 0.6205$

10. $x^3 e^{-x^3} = x^3 - x^6 + \frac{x^9}{2!} - \frac{x^{12}}{3!} + \frac{x^{15}}{4!} \cdots$. Hence, we obtain $\int_0^1 x^3 e^{-x^3} dx = \int_0^1\left(x^3 - x^6 + \frac{x^9}{2!} - \frac{x^{12}}{3!} + \cdots\right)dx = \left[\frac{x^4}{4} - \frac{x^7}{7} + \frac{x^{10}}{20} - \frac{x^{13}}{13\cdot13!} + \frac{x^{16}}{16\cdot4!} - \frac{x^{19}}{19\cdot5!} + \cdots\right]_0^1 = \frac{1}{4} - \frac{1}{7} + \frac{1}{20} - \frac{1}{78} + \frac{1}{384} - \frac{1}{2280} + \frac{1}{15840} = 0.1466.$

11. $e^x = 1 + x + \frac{x^2}{2!} + \frac{x^3}{3!} + \frac{x^4}{4!} + \cdots$. Hence, $e^{0.25} = 1 - 0.25 + \frac{0.25^2}{2!} - \frac{0.25^3}{3!} + \cdots = 1 - 0.25 + 0.03125 - 0.002604 + 0.000163 - 0.000008 = 0.7788.$

12. $\cos x^4 = 1 - \frac{x^8}{2!} + \frac{x^{16}}{4!} - \cdots$. Hence, $\int_0^1\cos x^4 dx = \int_0^1\left(1 - \frac{x^8}{2!} + \frac{x^{16}}{4!} - \frac{x^{24}}{6!} + \cdots\right)dx = \left(x - \frac{x^9}{18} + \frac{x^{17}}{17\cdot4!} - \frac{x^{25}}{25\cdot6!} + \cdots\right)\Big|_0^1 = 1 - \frac{1}{18} + \frac{1}{408} - \frac{1}{18000} = 0.9468.$

13. $f(x) = \cos x$, $\cos\frac{\pi}{6} = \frac{\sqrt{3}}{2}$, $f'(x) = -\sin x$, $f'\left(\frac{\pi}{6}\right) = -\frac{1}{2}$, $f''(x) = -\cos x$, $f''\left(\frac{\pi}{6}\right) = -\frac{\sqrt{3}}{2}$, $f'''(x) = \sin x$, $f'''\left(\frac{\pi}{6}\right) = \frac{1}{2}$. Hence, we obtain $\cos x = \frac{\sqrt{3}}{2} - \frac{1}{2}\left(x - \frac{\pi}{6}\right) - \frac{\sqrt{3}}{2}\cdot\frac{\left(x-\frac{\pi}{6}\right)^2}{2!} + \frac{1}{2}\frac{\left(x-\frac{\pi}{6}\right)}{3!} + \cdots$

14. $f(x) = \sinh x$, $f(\ln 2) = \frac{3}{4}$, $f'(x) = \cosh x$, $f'(\ln(2)) = \frac{5}{4}$; $f''(x) = f(x)$, $f'''(x) = f'(x)$, etc. Hence, $\sinh x = \frac{3}{4} + \frac{5}{4}(x - \ln 2) + \frac{3}{4}\cdot\frac{(x-\ln 2)^2}{2!} + \frac{5}{4}\frac{(x-\ln 2)^3}{3!} + \cdots$

15. $f(x) = x^{1/2}$, $f(9) = 3$; $f'(x) = \frac{1}{2}x^{-1/2}$; $f'(9) = \frac{1}{6}$, $f''(x) = -\frac{1}{4}x^{-3/2}$; $f''(9) = \frac{-1}{4\cdot27} = \frac{-1}{108}$; $f'''(x) = \frac{3}{8}x^{-5/2}$, $f'''(9) = \frac{3}{8\cdot3^5}$. Thus, we obtain $\sqrt{x} = 3 + \frac{1}{6}(x - 9) - \frac{1}{4\cdot3^3}\frac{(x-9)^2}{2!} + \frac{3}{8\cdot3^5}\frac{(x-9)^3}{3!} - \cdots$.

16. $f(x) = \ln x$, $f(1) = 0$, $f'(x) = \frac{1}{x}$, $f'(1) = 1$, $f''(x) = -x^{-2}$, $f''(1) = -1$, $f'''(x) = 2x^{-3}$;

$f'''(1) = 2$, $f^{(4)}(x) = -6x^{-4}$; $f^{(4)} = -6$. Thus, we get $\ln x = (x-1) - \frac{(x-1)^2}{2} + \frac{(x-1)^3}{3} - \frac{(x-1)^4}{4} + \cdots$.

17. $\cos 29° = \cos(30-1)^2 = \cos\left(\frac{\pi}{6} - \frac{\pi}{180}\right) = \frac{\sqrt{3}}{2} - \frac{1}{2}\left(-\frac{\pi}{180}\right) - \frac{\sqrt{3}}{4}\left(\frac{-\pi}{180}\right)^2 + \frac{1}{2}\left(\frac{-\pi}{180}\right)^3 + \cdots = 0.866025 + 0.008727 - 0.00013 = 0.8746$

18. $\sinh(\ln 1.9) = \sinh(\ln(2-0.1)) = \sinh\left(\frac{\ln 2}{\ln 0.1}\right)$; $\sinh(\ln 1.9) = \frac{3}{4} + \frac{5}{4}(\ln 1.9 - \ln 1) + \frac{3}{4}\frac{(\ln 1.9 - \ln 2)^2}{2} + \cdots = 0.75 - 0.06412 + 0.00099 - 0.00003 = 0.6868$.

19. $\sqrt{9.1} = 3 + \frac{1}{6}(.1) - \frac{1(0.1)^2}{4 \cdot 3^2 2} = 3 + 0.016667 - 0.000046 = 3.0166$

20. $\ln(1.1) = 0.1 - \frac{0.1^2}{2} + \frac{0.1^3}{3} - \frac{0.1^4}{4} = 0.1 - 0.005 + 0.0003333 - 0.000025 = 0.09531$

21. $f(x) = \begin{cases} 1 & -\pi < x < 0 \\ 0 & 0 < x < \pi \end{cases}$, period $= 2\pi$ so $L = \pi$. First, $a_0 = \frac{1}{2\pi}\int_{-\pi}^{\pi} f(x)dx = \frac{1}{2\pi}\int_{-\pi}^{0} 1\ dx = \frac{1}{2\pi}x\Big|_{-\pi}^{0} = \frac{1}{2}$. Next, we have $a_k = \frac{1}{\pi}\int_{-\pi}^{\pi} f(x)\cos kx\ dx = \frac{1}{\pi}\int_{-\pi}^{0}\cos kx\ dx = \frac{1}{k\pi}\sin kx\Big|_{-\pi}^{0} = 0$. Also, $b_k = \frac{1}{\pi}\int_{-\pi}^{0}\sin kx\ dx = \frac{1}{k\pi}(-\cos kx)\Big|_{-\pi}^{0} = \frac{1}{k\pi}(-1+\cos(-k\pi)) = \begin{cases} 0, & \text{if } k \text{ is even} \\ \frac{-2}{k\pi} & \text{if } k \text{ is odd} \end{cases}$. Thus, $f(x) = \frac{1}{2} - \frac{2}{\pi}\sin x + \frac{2}{3\pi}\sin 3x - \frac{2}{5\pi}\sin 5x - \cdots = \frac{1}{2} - \frac{2}{\pi}\sum_{k=1}^{\infty}\frac{\sin(2k-1)x}{2k-1}$.

22. $g(x) = \begin{cases} 1 & -\pi < x < 0 \\ 2 & 0 < x < \pi \end{cases}$, period 2π so $L = \pi$. First, $a_0 = \frac{1}{2\pi}\int_{-\pi}^{\pi} g(x)dx = \frac{1}{2\pi}\int_{-\pi}^{0} 1\ dx + \frac{1}{2\pi}\int_{0}^{\pi} 2\ dx = \frac{1}{2\pi}(0+\pi) + \frac{1}{2\pi}(2\pi) = \frac{1}{2} + 1 = \frac{3}{2}$. Next, $a_k = \frac{1}{\pi}\int_{-\pi}^{\pi} g(x)\cos kx\ dx = \frac{1}{\pi}\int_{-\pi}^{0}\cosh dx + \frac{1}{\pi}\int_{0}^{\pi} 2\cos kx\ dx = \frac{1}{k\pi}\sin kx\Big|_{-\pi}^{0} + \frac{2}{k\pi}\sin kx\Big|_{0}^{\pi} = 0$. We also have $b_k = \frac{1}{\pi}\int_{-\pi}^{0}\sin kx\ dx + \frac{1}{\pi}\int_{0}^{\pi} 2\sin kx\ dx = \frac{-1}{k\pi}(\cos kx)\Big|_{-\pi}^{0} + \frac{-2}{k\pi}\cos kx\Big|_{0}^{\pi} = \frac{-1}{k\pi}(1 - \cos(-k\pi)) - \frac{2}{k\pi}(\cos k\pi - 1) = \begin{cases} 0 & \text{if } k \text{ is even} \\ \frac{2}{k\pi} & \text{if } k \text{ is odd} \end{cases}$. This gives $g(x) = \frac{3}{2} + \frac{2}{\pi}\sin x + \frac{2}{3\pi}\sin 3x + \cdots = \frac{3}{2} + \frac{2}{\pi}\sum_{k=1}^{\infty}\frac{\sin(2k-1)x}{2k-1}$.

23. $h(x) = x^2$, $-\pi < x < \pi$, period 2π. First, $a_0 = \frac{1}{2\pi}\int_{-\pi}^{\pi} x^2 dx = \frac{x^3}{6\pi}\Big|_{-\pi}^{\pi} = \frac{\pi^2}{3}$. Next, $a_k = \frac{1}{\pi}\int_{-\pi}^{\pi} x^2\cos kx\ dx = \frac{x^2}{\pi k}\sin kx - 2\int x\sin kx = \frac{x^2}{k\pi}\sin kx + \frac{2x}{k^2\pi}\cos kx - \frac{2}{\pi}\int\cos kx = \left[\frac{x^2}{k\pi}\sin kx + \frac{2x}{k^2\pi}\cos kx - \frac{1}{k^3\pi}\sin kx\right]_{-\pi}^{\pi} = \begin{cases} -\frac{4}{k^2}, & k \text{ odd} \\ \frac{4}{k^2}, & k \text{ even} \end{cases}$. Also, $b_k = 0$ since $x^2\sin kx$ is an odd function. Hence, we obtain $h(x) = \frac{\pi^2}{3} - 4\cos x + \frac{4}{2^2}\cos\cos 2x - \frac{4}{3^2}\cos 3x + \frac{4}{4^2}\cos 4x - \cdots = \frac{\pi^2}{3} + 4\sum_{k=1}^{\infty}\frac{(-1)^k}{k^2}\cos kx$.

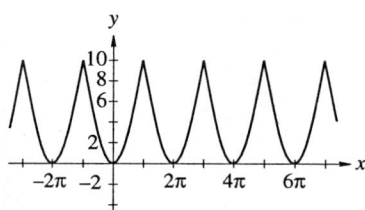

24. $j(x) = 4 \cdot x^2$, $-\pi < x < \pi$, period $= 2\pi$. First, $a_0 = \frac{1}{2\pi}\int_{-\pi}^{\pi}\left(4 - x^2\right)dx = \frac{1}{2\pi}\left(4x - \frac{x^3}{3}\right)\Big|_{-\pi}^{\pi} = \frac{1}{\pi}\left(4\pi - \frac{\pi^3}{3}\right) = 4 - \frac{\pi^2}{3}$. (Note: the function is even).

$a_k = \frac{1}{\pi}\int_{-\pi}^{\pi}\left(4 - x^2\right)kx\ dx$

$= \frac{1}{\pi}\int_{-\pi}^{\pi} 4\cos kx - \frac{1}{\pi}\int_{-\pi}^{\pi} x^2\cos kx\ dx$

$= \left[\frac{4}{\pi k}\sin kx - \frac{x^2}{k\pi}\sin kx + \frac{x}{k^2\pi}\cos kx - \frac{1}{k^3\pi}\cos kx\right]_{-\pi}^{\pi}$

$= \begin{cases} \frac{-4}{k^2} & k \text{ even} \\ \frac{4}{k^2} & k \text{ odd} \end{cases}$

$b_k = 0$ since $\left(4 - x^2\right)\sin kx$ is an odd function. Hence $j(x) = 4 - \frac{\pi^2}{3} + 4\sum_{k=1}^{\infty}\frac{(-1)^{k+1}}{k^2}\cos kx$.

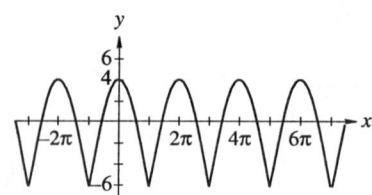

CHAPTER 31 TEST

1. Since $\sin x = x - \frac{x^3}{3!} + \frac{x^5}{5!} - \frac{x^7}{7!} + \cdots$, we get
$\sin x^2 = x^2 - \frac{x^6}{3!} + \frac{x^{10}}{5!} - \frac{x^{14}}{7!} + \cdots$

2. Since $\cos x = 1 - \frac{x^2}{2!} + \frac{x^4}{4!} - \frac{x^6}{6!} + \cdots$, we obtain
$\cos \sqrt{x} = 1 - \frac{x}{2!} + \frac{x^2}{4!} - \frac{x^3}{6!} + \cdots$

3. $h(x) = \frac{1}{x}$; $h(2) = \frac{1}{2}$; $h'(x) = -x^2$; $h'(2) = -\frac{1}{4}$; $h''(x) = 2x^{-3}$; $h''(2) = \frac{1}{4}$; $h'''(x) = -6x^{-4}$; $h'''(2) = \frac{-6}{2^4}$. Thus, we see that $h(x) = \frac{1}{x} = \frac{1}{2} - \frac{1}{4}(x-2) + \frac{1}{4} \cdot \frac{(x-2)^2}{2!} - \frac{6}{2^4} \cdot \frac{(x-2)^3}{3!} + \cdots = \frac{1}{2} - \frac{(x-2)}{4} + \frac{(x-2)^2}{8} - \frac{(x-2)^3}{16} + \cdots = \sum_{k=0}^{\infty} = \frac{(-1)^k (x-2)^k}{2^{k+1}}$.

4. $\frac{1}{2.1} = \frac{1}{2} - \frac{0.1}{4} + \frac{(0.1)^2}{8} - \frac{(0.1)^3}{16} = 0.5 - 0.025 + 0.00125 - 0.0000625 = 0.47619$

5. $j(x) = x^{1/3}$; $j(-1) = -1$; $j'(x) = \frac{1}{3}x^{-2/3}$; $y'(-1) = \frac{1}{3}$; $y''(x) = -\frac{2}{9}(x)^{-5/3}$; $y''(-1) = \frac{2}{9}$; $y'''(x) = \frac{10}{27}x^{-8/3}$; $y'''(-1) = \frac{10}{27}$. Hence, $j(x) = -1 + \frac{1}{3}(x+1) + \frac{2}{9} \cdot \frac{(x+1)^2}{2!} + \frac{10}{27}\frac{(x+1)^3}{3!} = -1 + \frac{(x+1)}{3} + \frac{(x+1)^2}{9} + \frac{5(x+1)^3}{81}$.

6. $\sqrt[3]{0.9} = -1 + \frac{(0.1)}{3} + \frac{(0.1)^2}{9} + \frac{5(0.1)^3}{81} = -1 + 0.03333 + 0.001111 + 0.000062 = -0.96549$

7. $f(x) = x$, $0 < x < 2\pi$, period is 2π so $L = \pi$. First, $a_0 = \frac{1}{2\pi}\int_0^{2\pi} x\, dx = \frac{1}{2\pi}\frac{x^2}{2}\Big|_0^{2\pi} = \pi$. Next, we see that $a_k = \frac{1}{\pi}\int_0^{2\pi} x\cos kx\, dx = \frac{1}{k^2\pi}\int_0^{2\pi} kx\cos kx\, k\, dx = \frac{1}{k^2\pi}[\cos kx + kx\sin kx]_0^{2\pi} = 0$. We also determine that $b_k = \frac{1}{\pi}\int_0^{2\pi} x\sin kx\, dx = \frac{1}{k^2\pi}\int_0^{2\pi} kx\sin kx\, k\, dx = \frac{1}{k^2\pi}[\sin kx - kx\cos kx]_0^{2\pi} = \frac{1}{k^2\pi}[-2k\pi] = \frac{-2}{k}$. Hence, we obtain $f(x) = \pi - 2\left(\sin x + \frac{\sin 2x}{2} + \frac{\sin 3x}{3} + \cdots\right) = \pi - 2\sum_{k=1}^{\infty}\frac{\sin kx}{k}$.

8. $g(x) = \begin{cases} -\cos x & -\pi < x < 0 \\ \cos x & 0 < x < \pi \end{cases}$, period is 2π so $L = \pi$. First, $a_0 = \frac{1}{2\pi}\int_{-\pi}^{\pi} g(x)dx = \frac{1}{2\pi}\int_{-\pi}^{0}$

$-\cos x\, dx + \frac{1}{2\pi}\int_0^{\pi}\cos x\, dx = \frac{1}{2\pi}(-\sin x)\Big|_{-\pi}^0 + \frac{1}{2\pi}(\sin x)\Big|_0^{\pi} = 0$. Then, we have $a_k = \frac{1}{\pi} \times \int_{-\pi}^{\pi} g(x)\cos kx\, dx = \frac{1}{\pi}\int_{-\pi}^{0} -\cos x\cos kx\, dx + \frac{1}{\pi}\int_0^{\pi}\cos x\cos kx\, dx$. These integrals add up to 0 since the cosine function is even and the product of the cosine functions is even. Hence, $\int_{-\pi}^{0} = \int_0^{\pi}$ and, since one of the integrals is negative, their sum is 0. We also have, for $k > 1$, $b_k = \frac{1}{\pi}\int_{-\pi}^{0} -\cos x\sin kx\, dx + \frac{1}{\pi}\int_0^{\pi}\cos x\sin kx\, dx$. By form 72 this is $\frac{1}{\pi}\left[\frac{\cos(k-1)x}{2(k-1)} + \frac{\cos(k+1)x}{2(k+1)}\right]_{-\pi}^{0} + \frac{1}{\pi} \times \left[-\frac{\cos(k-1)x}{2(k-1)} - \frac{\cos(k+1)x}{k+1}\right]_0^{\pi} = \frac{1}{\pi}\left[\frac{-\cos(k-1)(-\pi)}{2(k-1)} - \frac{\cos(k+1)(-\pi)}{2(k+1)} + \frac{1}{2(k-1)} + \frac{1}{2(k+1)} - \frac{\cos(k-1)\pi}{2(k-1)} - \frac{\cos(k+1)\pi}{2(k+1)} + \frac{1}{2(k-1)} + \frac{1}{2(k+1)}\right]$. When k is odd, $\cos(k-1)(-\pi) = \cos(k+1)(-\pi) = \cos(k-1)\pi = \cos(k+1)\pi = 1$, and so $b_k = 0$. When k is even, $\cos(k-1)(-\pi) = \cos(k+1)(-\pi) = \cos(k-1)\pi = \cos(k+1)\pi = -1$, and so we get $b_k = \frac{1}{\pi}\left[\frac{4}{2(k-1)} + \frac{4}{2(k+1)}\right] = \frac{4k}{\pi(k^2-1)}$. In particular, $b_2 = \frac{4\cdot2}{\pi(2^2-1)} = \frac{8}{3\pi}$, $b_4 = \frac{4\cdot4}{(4^2-1)\pi} = \frac{16}{15\pi}$, $b_6 = \frac{24}{35\pi}$. Now, $k = 1$ yields $b_1 = \frac{1}{\pi}\int_{-\pi}^{0}(-\cos x\sin x\, dx) + \frac{1}{\pi}\int_0^{\pi}\cos x\sin x\, dx = 0$. Putting it all together, we get $g(x) = \frac{8}{\pi}\left(\frac{\sin 2x}{3} + \frac{2\sin 2x}{15} + \frac{3\sin 2x}{35} + \cdots\right) = \frac{8}{\pi}\sum_{k=1}^{\infty}\frac{k\sin 2x}{4k^2-1}$.

9. The Maclaurin series for $\cos x$ is $1 - \frac{x^2}{2!} + \frac{x^4}{4!} + \cdots$, so $\cos x^3 = 1 - \frac{(x^3)^2}{2!} + \frac{(x^3)^4}{4!} + \cdots = 1 - \frac{x^6}{2!} + \frac{x^{12}}{4!} + \cdots$. Thus, $\int_1^2 \cos x^3 dx = \int_1^2\left(1 - \frac{x^6}{2!} + \frac{x^{12}}{4!} + \cdots\right)dx = \left[x - \frac{x^7}{7\cdot2!} + \frac{x^{13}}{13\cdot4!} + \cdots\right]_1^2 \approx 19.1136 - 0.9318 = 18.1818$.

10. The Maclaurin series for e^x is $1 + x + \frac{x^2}{2!} + \frac{x^3}{3!} + \cdots$, so $e^{-0.30t^2} = 1 - 0.30t^2 + \frac{0.09t^4}{2!} - \frac{0.027t^6}{3!} + \cdots$. Thus, $q = \int_0^{0.15} e^{-0.30t^2} dt = \int_0^{0.15}\left(1 - 0.30t^2 + \frac{0.09t^4}{2!} - \frac{0.027t^6}{3!} + \cdots\right)dt = \left[t - 0.10t^3 + 0.009t^5 - \frac{0.027t^7}{7\cdot3!} + \cdots\right]_0^{0.15} \approx 0.149663\ \mu C$.

32

First-Order Differential Equations

≡ 32.1 SOLUTIONS OF DIFFERENTIAL EQUATIONS

1. $y = e^x$; $y' = e^x$, $y'' = e^x$. Hence $y'' - y = e^x - e^x = 0$

2. $y = e^{2x} + e^x$; $y' = 2e^{2x} = e^x$; $y'' = 4e^{2x} + e^x$. Hence $y'' - 4y' + 4y = (4e^{2x} + e^x) - 4(2e^{2x} + e^x) + 4(e^{2x} + e^x) = 4e^{2x} + e^x - 8e^{2x} - 4e^x + 4e^{2x} + 4e^x = (4 - 8 + 4)e^{2x} + (1 - 4 + 4)e^x = e^x$.

3. $y = 2x^3$, $y' = 6x^2$. Hence $xy' = x6x^2 = 6x^3 = 3 \cdot 2x^3 = 3y$

4. $y = x^3 + 3$, $y' = 3x^2$. Hence $3y - xy' = 3(x^3 + 3) - x(3x^2) = 3x^3 + 9 - 3x^3 = 9$

5. $y = 5\cos x$; $y' = -5\sin x$. Hence $y' + y\tan x = -5\sin x + 5\cos x\tan x = -5\sin x + 5\sin x = 0$

6. $y = e - 1$, $y' = e^x$. Hence $y' - y = e^x - (e^x - 1) = 1$

7. $y = x^2 + 3x$; $y' = 2x + 3$; $y'' = 2$. Hence $xy' - 2y + 3xy'' - 3x = x(2x + 3) - 2(x^3 + 3x) + 3x(2) - 3x = 2x^2 + 3x - 2x^2 - 6x + 6x - 3x = 0$

8. $y = 4x^2$, $y' = 8x$. Hence $xy' = x(8x) = 8x^2 = 2 \cdot 4x^2 = 2y$

9. $y = 2e^x + 3e^{-x} - 4x$, $y' = 2e^x - 3e^{-x} - 4$, $y'' = 2e^x + 3e^{-x}$. Hence $y'' - y = (2e^x + 3e^{-x}) - (2e^x + 3e^{-x} - 4x) = 4x$

10. $x^2 - y^2 + 2xy = 9$. Using implicit differentiation we get $2x - 2yy' + 2xy' + 2y = 0$ or $y'(2x - 2y) = -2x - 2y$ or $y' = \frac{2x + 2y}{2y - 2x} = \frac{y + x}{y - x}$

11. $y = 9x^2 + 6x - 54$; $y' = 18x + 6$; $y'' = 18$; $y''' = 0$

12. $xy^2 = 7x - 3$ so $y^2 = 7 - 3x^{-1}$, Using implicit differentiation we get $2yy' = 3x^{-2}$ or $y' = \frac{3}{2yx^2} = \frac{3}{2}y^{-1}x^{-2}$, $y'' = -\frac{3}{2}y^{-2}y'x^{-1} - 3y^{-1}x^{-3} = -\frac{3}{2}y^{-2}(\frac{3}{2}y^{-1}x^{-2})x^{-2} - 3y^{-1}x^{-3} = -\frac{9}{4}y^{-3}x^{-4} - 3y^{-1}x^{-3}$. Hence $x(y')^2 + 2yy' + xyy'' = x(\frac{3}{2}y^{-1}x^{-2})^2 + 2y(\frac{3}{2}y^{-1}x^{-2}) + $

$xy(-\frac{9}{4}y^{-3}x^{-4} - 3y^{-1}x^{-3}) = \frac{9}{4}y^{-2}x^{-3} + 3x^{-2} + \frac{9}{4}y^{-2}x^{-3} - 3x^{-2} = 0$

13. $y = x^2 + 4$; $y' = 2x$

14. $2y = \sin x$, $y' = \frac{1}{2}\cos x$, $y'' = -\frac{1}{2}\sin x$. Hence $y'' + 5y = -\frac{1}{2}\sin x + 5 \cdot \frac{1}{2}\sin x = 2\sin x$

15. $y = 5e^{-3x} + 2e^{2x}$, $y' = -15e^{-3x} + 4e^{2x}$, $y'' = 45e^{-3x} + 8e^{2x}$. Hence $y'' + y' - 6y = (45e^{-3x} + 8e^{2x}) + (-15e^{-2x} + 4e^{2x}) - 6(5e^{-3x} + 2e^{2x}) = (45 - 15 - 30)e^{-3x} + (8 + 4 - 12)e^{2x} = 0$

16. $y = 5\cos x$; $y' = -5\sin x$, $y'' = -5\cos x$. Hence $y'' + y = -5\cos x + 5\cos x = 0$.

17. $y' = \int y'' dx = \int 6x^2 dx = 2x^3 + C_1$. Since $y'(1) = -2, 2(1)^3 + C_1 = 2 + C_1 = 2 + C_1 = -2$ or $C_1 = -4$ so $y' = 2x^3 - 4$. Next, we see that $y = \int y' dx = \int (2x^3 - 4)dx = \frac{x^4}{2} - 4x + C_2$. Since $y(1) = 3$, $\frac{1}{2} - 4 + C_2 = 3$ or $C_2 = 6\frac{1}{2}$. Hence $y = \frac{1}{2}x^4 - 4x + 6\frac{1}{2}$

18. $y' = \int y'' dx = \int \sin x \, dx = -\cos x + C_1$, since $y'(0) = 2$, $-\cos 0 + C_1 = 2 \Rightarrow C_1 = 3$. Thus, $y = \int y' dx = \int(-\cos x + 3)dx = -\sin x + 3x + C_2$. Since $y(\pi) = \pi$, $-\sin \pi + 3\pi + C_2 = \pi \Rightarrow C_2 = -2\pi$. $y = -\sin x + 3x - 2\pi$.

19. $y = \int \frac{dy}{dx} dx = \int \sec^2 x \, dx = \tan x + C$. Since $y(\frac{\pi}{4}) = -6$, $\tan \frac{\pi}{4} + C = -6 \Rightarrow C = -7$ so $y = \tan x - 7$.

20. $y' = \int y'' dx = \int e^{-x} dx = -e^{-x} + C_1$, since $y'(0) = 0$, $-e^0 + C_1 = 0 \Rightarrow -1 + C_1 = 0$ or $C_1 = 1$. So, $y = \int y' dx = \int (-e^{-x} + 1)dx = e^{-x} + x + C_2$. Since $y(1) = -1$, $e^{-1} + 1 + C_2 = -1 \Rightarrow C_2 = -2 - e^{-1}$ so $y = e^{-x} + x - 2 - e^{-1}$.

21. As in Example 32.0, we have $a(t) = -30$, so $v(t) = -30t + C_1$. We were given $v(0) = 65$ mph $= 95\frac{1}{3}$ ft/s and so $v(t) = -30t + 95\frac{1}{3}$ ft/s. By integrating, we obtain $s(t) = -15t^2 + 95\frac{1}{3}t + C_2$. Since $s(0) = 0$, we see that $C_2 = 0$ and thus $s(t) = -15t^2 + 95\frac{1}{3}t$.

(a) When the car stops $v(t) = -30t + 95\frac{1}{3} = 0$. Solving for t, we see that the car will stop in about $t = 3.178$ s.

(b) Substituting $t = 3.178$ s in the equation for $s(t)$, we get $s(3.178) \approx 151.474$. The car takes about 151.474 ft to stop.

22. Here $a(t) = -28$, so $v(t) = -28t + C_1$. Since $v(0) = 70$ mph $= 102\frac{2}{3}$ ft/s, we have $v(t) = -28t + 102\frac{2}{3}$. Integrating, we get $s(t) = -14t^2 + 102\frac{2}{3}$.

(a) Setting $v(t) = 0$ and solving for t, we see that it takes $t = 3\frac{2}{3}$ s to stop the car.

(b) Substituting $t = 3\frac{2}{3}$ s in the equation for $s(t)$, we get $s\left(3\frac{2}{3}\right) = 188\frac{2}{9}$ ft. It takes the car $188\frac{2}{9}$ ft to stop.

23. **(a)** The cars will strike when $s(t) = 151$. Solving $s(t) = -15t^2 + 95\frac{1}{3}t = 151$ or $15t^2 - 95\frac{1}{3}t + 151 = 0$ for t we get $t = 3$ s.

(b) When they strike $v(3) = -30(3) + 95\frac{1}{3} = 5\frac{1}{3}$. So, the first car is going $5\frac{1}{3}$ mph.

24. **(a)** The cars hit when $s(t) = 184.5$. Solving $s(t) = -14t^2 = 102\frac{2}{3} = 184.5$, we get $t \approx 3.15$ s.

(b) At $t = 3.15$, $v(3.15) = -28t + 102\frac{2}{3} = -28(3.15) + 102\frac{2}{3} \approx 14.47$. The car is moving at around 14.47 mph when the cars hit.

25. **(a)** Acceleration is gravity, $g = -32$ ft/s^2. So, $v(t) = -32t + C_1$. Since $v(0) = 64$, we have $C_1 = 64$ and $v(t) = -32t + 64$. Thus, the ball's position is $s(t) = -16t^2 + 64t + C_2$. At $t = 0$, we are given $s(0) = 192$, so $C_2 = 192$ and $s(t) = -16t^2 + 64t + 192 = -16\left(t^2 - 4t - 12\right) = -16(t - 6)(t + 2)$. The ball will strike the ground at $t = 6$ s.

(b) At $t = 6$, $v(6) = -32(6) + 64 = -128$, so it is going 128 ft/s downward when it strikes the ground.

26. **(a)** Since $a(t) = -9.8 \text{ m/s}^2$ we have $v(t) = -9.8t + C_1$. Since $v(0) = 29.4$, we have $v(t) = -9.8t +$

29.4. Integrating, we obtain $s(t) = -4.9t^2 + 29.4t + C_2$. We are given $s(0) = 34.3$ m, so $s(t) = -4.9t^2 + 34.3 = -4.9\left(t^2 - 6t - 7\right) = -16(t - 7)(t + 1)$. The ball will strike the ground at $t = 7$ s.

(b) When it hits the ground, its velocity is $v(7) = -9.8(7) + 29.4 = -39.2$, so it is going 39.2 m/s downward when it strikes the ground.

27. **(a)** Let $p(t)$ be the number of mg of pain reliever still in her system t hours after taking the tablet. We are given $\frac{dp(t)}{dt} = -75$ as the rate of metabolism. Integrating, we get $p(t) = -75t + C$. At $t = 0$, there are 500 mg in the system, so $C = 500$ and $p(t) = -75t + 500$.

(b) We want to find t so that $p(t) = -115t + 500 = 0$. Solving for t, we get $t = 6\frac{2}{3}$ h.

28. We begin with the result of Exercise 0 with $p(t) = -75t + 500$. At $t = 3$, she has 275 mg from the first tablet plus 500 mg from the second tablet. Now, solving $p(t) - 75 + C$ when $t = 3$ and $p(t) = 775$ we get $C = 1000$. Putting these results with those from Exercise 0 we find that

$$p(t) = \begin{cases} -75t + 500 & \text{if } 0 \le 3 \\ -75t + 1000 & \text{if } t \ge 3 \end{cases}$$

where t represents the number of hours since the first pill was taken. Solving $p(t) = -75t + 1000$ for t we get $t = 13\frac{1}{3}$ h after the first pill was taken or $10\frac{1}{3}$ h after the second pill was taken before all the pain reliever is metabolized.

29. Let $c(t)$ be the number of mg of cortisone still in the patient's system t hours after taking the 5 mg dose. We are given $\frac{dc(t)}{dt} = -\frac{1}{2\sqrt{t}} = -\frac{1}{2}t^{-1/2}$ as the rate of metabolism. Integrating, produces $c(t) = -t^{1/2} + C$. At $t = 0$, there are 5 mg in the system, so $C = 5$ and $c(t) = -t^{1/2} + 5$. We want to find t so that $c(t) = -t^{1/2} + 5 = 0$. Solving for t, we get $\sqrt{t} = 5$ and so $t = 25$ h.

30. **(a)** Differentiating the given value of N, we see that the left-hand side of the equation is $\frac{dN}{dt} = \frac{4{,}000Ce^{-4t}}{\left(1 + Ce^{-4t}\right)^2}$. Substituting the given value of N in the right-hand side, we obtain $\frac{1}{250}\left(\frac{1{,}000}{1 + Ce^{-4t}}\right)\left(1{,}000 - \frac{1{,}000}{1 + Ce^{-4t}}\right)$ which simplifies to $\frac{4{,}000Ce^{-4t}}{\left(1 + Ce^{-4t}\right)^2}$. Since this is the same as the left-hand side, we have shown that $N(t) = \frac{1,000}{1+Ce^{-4t}}$ is a general solution of the general differential equation $\frac{dN}{dt} = \frac{1}{250}N(1,000 - N)$.

(b) Substituting 0 for t and setting $N(0) = 200$, we obtain $\frac{1,000}{1+Ce^{-4\cdot 0}} = \frac{1,000}{1+C} = 200$ or $1000 = 200(1 + C) = 200 + 200C$ or $200C = 800$ and $C = 4$.

31. If $i = \frac{V}{R} + ke^{-Rt/L}$, then $\frac{di}{dt} = -\frac{kR}{L}e^{-Rt/L}$. Substituting this into the original differential equation, we obtain $L\left(-\frac{kR}{L}e^{-Rt/L}\right) + \left(\frac{V}{R} + ke^{-Rt/L}\right) =$

$-kRe^{-Rt/L} + V + kRe^{-Rt/L} = V$, which was to be shown.

32. If $N(t) = 200 - Ce^{-0.5t}$, then $\frac{dN}{dt} = 0.5Ce^{-0.5t}$. Substituting this into the differential equation, produces

$$0.5Ce^{-0.5t} = 100 - 0.5\left(200 - Ce^{-0.5t}\right)$$
$$= 100 - 100 + 0.5Ce^{-0.5t}$$
$$= 0.5Ce^{-0.5t}$$

Thus, $0.5Ce^{-0.5t} = 0.5Ce^{-0.5t}$, which was to be shown.

≡ 32.2 SEPARATION OF VARIABLES

1. $y' = \frac{dy}{dx} = \frac{2x}{y} \Rightarrow 2x\,dx - y\,dy = 0$. Integrating we get $x^2 - \frac{1}{2}y^2 = C_1$ or $\frac{1}{2}y^2 - x^2 = C_2$ or $y^2 - 2x^2 = C_3$.

2. $\frac{dy}{dx} = \frac{2y}{x} \Rightarrow 2y\,dx - x\,dy = 0$. The integrating factor $I(x, y)$ is $\frac{1}{xy}$. Multiplying and cancelling we get $\frac{2}{x}\,dx - \frac{1}{y}\,dy = 0$. Integration yields $2\ln|x| - \ln|y| = \ln C \Rightarrow \ln x^2 - \ln y = \ln C \Rightarrow \ln\left(\frac{x^2}{y}\right) = \ln C \Rightarrow \frac{x^2}{y} = C$. Hence $x^2 = Cy$ or $y = Cx^2$.

3. $5x\,dx + 3y^2\,dy = 0$. Integrating we get $\frac{5}{2}x^2 + y^3 = C$

4. $5y\,dx + 3y^2\,dy = 0$. Multiply first by $I(x, y) = \frac{1}{y}$. This yields $5\,dx + 3y\,dy = 0$. Now by integration $5x + \frac{3}{2}y^2 = C$.

5. $5y\,dx + 3\,dy = 0$. Multiply by $I(x, y) = \frac{1}{y}$ to get $5\,dx + \frac{3}{y} = 0$. Now by integrating we get $5x + 3\ln|y| = C$

6. $4y^5\,dx - 3x^2\,dy = 0$. Multiplying by $\frac{1}{x^2y^5}$ we get $\frac{4}{x^2}\,dx - \frac{3}{y^5}\,dy = 0$. Integration yields $-4x^{-1} + \frac{3}{4}y^{-4} = C$.

7. $4y^5\,dx - 3x^2y\,dy = 0$. Multiplying by $I(x, y) = \frac{1}{y^5x^2}$ we get $\frac{4}{x^2}\,dx - \frac{3}{y^4}\,dy = 0$. Now integration yields $-4x^{-1} + y^{-3} = C$.

8. $xyy' + \sqrt{1 + y^2} = 0$ is the same as $\frac{dy}{dx} + \frac{\sqrt{1+y^2}}{xy} = 0$. Hence we get $\sqrt{1 + y^2}\,dx + xy\,dx = 0$. $I(x, y)$

$= \frac{1}{\sqrt{1+y^2}x}$ and multiplying by this integrating factor gives $\frac{1}{x}\,dx + \frac{y}{\sqrt{1+y^2}}\,dy = 0$. Integration now yields $\ln|x| + \frac{1}{2}\frac{2}{1}\left(1+y^2\right)^{1/2} + C$ or $\ln|x| + \sqrt{1 + y^2} = C$.

9. $x^3\left(y^2 + 4\right) + yy' = 0$ is the same as $x^3\left(y^2 + 4\right)dx + y\,dy = 0$. Multiplying by $\frac{1}{y^2+4}$ yields $x^3\,dx + \frac{y}{y^2+4} = 0$. Now integration gives us $\frac{1}{4}x^4 + \frac{1}{2}\ln\left(y^2 + 4\right) = C_1$ or $\frac{1}{4}x^4 + \ln\sqrt{y^2 + 4} = C_1$ or $\frac{1}{2}x^4 + \ln\left(y^2 + 4\right) = C_2$.

10. $x^4\left(y^3 - 3\right)dx + y^2\left(x^5 - 2\right)dy = 0$ becomes $\frac{x^4}{x^5-2}\,dx + \frac{y^2}{y^3-3}\,dy = 0$. Integration yields $\frac{1}{5}\ln\left|x^5 - 2\right| + \frac{1}{3}\ln\left|y^3 - 3\right| = \ln|C|$. This is the same as $\left|x^5 - 2\right|^{1/5} \cdot \left|y^3 - 3\right|^{1/3} = C$

11. $ye^{x^2}\,dy = 2x\left(y^2 + 4\right)dx$. The integrating factor is $I(x, y) = \frac{1}{\left(y^2+4\right)3^{x^2}}$. Multiplying by $I(x, y)$ yields $\frac{y}{y^2+4}\,dy = \frac{2x}{e^{x^2}}\,dx$. Now integration yields $\frac{1}{2}\ln\left(y^2 + 4\right) = -e^{-x^2} + C$ or $\ln\sqrt{y^2 + 4} = -e^{-x^2} + C$ or $\ln\left(y^2 + 4\right) = -2e^{-x^2} + C_2$.

12. $e^{x^2}\,dy + x\sqrt{4 - y^2}\,dx = 0$ is the same as $\frac{1}{\sqrt{4-y^2}}\,dy + \frac{x}{e^{x^2}}\,dx = 0$. Integration yields $\arcsin\frac{y}{2} - \frac{1}{2}e^{-x^2} + C = 0$.

13. $2y + e^{-3x}y' = 0 \Rightarrow 2y\,dx + e^{-3x}\,dy = 0 \Rightarrow 2e^{3x}\,dx + \frac{1}{y}\,dy = 0$. Integration yields $\frac{2}{3}e^{3x} + \ln|y| = C$

14. $\sin^2 y\,dx + \cos^2 x\,dx = 0$. First multiply by $\frac{1}{\sin^2 y \cos^2 x}$ to obtain $\frac{1}{\cos^2 x}\,dx + \frac{1}{\sin^2 y}\,dy = 0$ or $\sec^2 x\,dx + \csc^2 y\,dy = 0$. This integrates to $\tan x - \cot y = C$

15. $(y^2 - 4)\cos x\,dx + 2y\sin x\,dy = 0$. Multiplying by the integrating factor yields $\frac{\cos x}{\sin x}\,dx + \frac{2y}{y^2 - 4}\,dy = 0$. Integration yields $\ln|\sin x| + \ln|y^2 - 4| = \ln C$. This simplifies to $|y^2 - 4||\sin x| = C$ or $\left|(y^2 - 4)\sin x\right| = C$.

16. $(1 + x)^2 y' = 1 \Rightarrow (1 + x)^2 \frac{dy}{dx} = 1$ or $dy = (1 + x)^{-2}\,dx$. Integration yields $y = -(1 + x)^{-1} + C$, or $y = \frac{1}{1+x} + C$.

17. $y' = \frac{x^2 y - y}{y^2 + 3} = \frac{y(x^2 - 1)}{y^2 + 3} = \frac{dy}{dx}$. Hence $(x^2 - 1)\,dx = \frac{y^2 + 3}{y}\,dy$. Integrating we get $\frac{1}{3}x^3 - x = \frac{y^2}{2} + 3\ln|y| + C$ or $\frac{1}{2}y^2 + 3\ln|y| = \frac{1}{3}x^3 - x + C$.

18. $\frac{dy}{dx} = \frac{y}{x^2 + 6x + 9}$ is equivalent to $\frac{1}{y}\,dy = \frac{1}{(x+3)^2}\,dx$. Integration yields $\ln|y| = -(x + 3)^{-1} + C$ or $\ln|y| + \frac{1}{x+3} = C$.

19. $\sec 3x\,dy + y^2 e^{\sin 3x}\,dx = 0$ is equivalent to $\frac{1}{y^2}\,dy + \cos 3x e^{\sin 3x}\,dx = 0$. Integration yields $-\frac{1}{y} + \frac{1}{3}e^{\sin 3x} = C$

20. $2xy + (1 + x^2)y' = 0$. The integrating factor is $\frac{1}{y(1+x^2)}$. Multiplying yields $\frac{2x}{1+x^2}\,dx = \frac{1}{y}\,dy = 0$. Integrating yields $\ln(1 + x^2) + \ln|y| = \ln C$. This simplifies to $(1 + x^2)|y| = C$

21. $y' + y^2 x^3 = 0$ becomes $\frac{1}{y^2}\,dy + x^3\,dx = 0$. Integration yields the general solution $-\frac{1}{y} + \frac{x^4}{4} + C$. Substituting -1 for y and 2 for x we get $1 + 4 = C$ so $C = 5$. The particular solution is $-\frac{1}{y} + \frac{1}{4}x^4 = 5$.

22. $x\,dy + y\,dx = 0$ becomes $\frac{1}{y}\,dy + \frac{1}{x}\,dx = 0$. Integration yields the general solution $\ln|y| + \ln|x| = \ln C$ or $xy = C$. Substituting 3 for y and 2 for x we get $C = 6$. The particular solution is $xy = 6$.

23. $\frac{dy}{dx} = \frac{3x + xy^2}{y + x^2 y} = \frac{x(3 + y^2)}{y(1 + x^2)}$. This is equivalent to $\frac{y}{3 + y^2}\,dy = \frac{x}{1 + x^2}\,dx$. Integration yields the general solution $\frac{1}{2}\ln(3 + y^2) = \frac{1}{2}\ln(1 + x^2) + \frac{1}{2}\ln C$ or $(3 + y^2)/(1 + x^2) = C$. Substituting 3 for y and 1

for x we get $\frac{12}{2} = 6 = C$. The particular solution is $\frac{3 + y^2}{1 + x^2} = 6$.

24. $\sin^2 y\,dx + \cos x\,dy = 0$ is the same as $\sec^2 x\,dx + \csc^2 y\,dy = 0$. Integration yields the general solution $\tan x - \cot y = C$. Substituting $\frac{3\pi}{4}$ for y and $\frac{\pi}{4}$ for x we get $\tan \frac{\pi}{4} - \cot \frac{3\pi}{4} = 1 + 1 = 2 = C$. The particular solution is $\tan x - \cot y = 2$.

25. $y' = \frac{dy}{dx} = 3x^2 y - 2y = y(3x^2 - 2)$. Hence we get $\frac{1}{y}\,dy = (3x^2 - 2)\,dx$. Integration yields the general solution $\ln|y| = x^3 - 2x + C$. Substituting $y = x = 1$ we get $\ln 1 = 1 - 2 + C$ or $C = 1$. The particular solution is $\ln|y| = x^3 - 2x + 1$

26. $\sqrt{1 + 9x^2}\,dy = y^4 x\,dx$. Multiplying by the integrating factor $\frac{1}{y^4 \sqrt{1 + 9x^2}}$ we get $\frac{1}{y^4}\,dy = \frac{x}{\sqrt{1 + 9x^2}}\,dx$. Integrating to get the general solution produces $\frac{-1}{3y^3} = \frac{1}{9}(1 + 9x^2)^{1/2} + C$. This is equivalent to $\sqrt{1 + 9x^2} + \frac{3}{y^3} = C$. Substituting $y = -1$ and $x = 0$ we get $\sqrt{1} + (-3) = -2 = C$. Hence the particular solution is $\sqrt{1 + 9x^2} + \frac{3}{y^3} = -2$ or $\sqrt{1 + 9x^2} = \frac{3}{y^3} + 2 = 0$.

27. **(a)** As in Example 32.13, the general solution is $T = T_m + (T_0 - T_m)e^{-kt}$, where $T_m = 325°F$ and $T_0 = 10°F$. Hence, $T = 325 + (-315)e^{-kt}$. After $t = 1$ hour, $T = 75°F$. Substituting, we solve for k

$$75 = 325 - 315e^{-k}$$
$$e^{-k} = \frac{75 - 325}{-315} \approx 0.793651$$
$$-k = \ln 0.79365$$
$$k \approx 0.231112$$

The particular solution is $T = 325 - 315e^{-0.231112t}$.
(b) Setting $T = 145$, we solve for t.

$$145 = 325 - 315e^{-0.231112t}$$
$$e^{-0.231112t} = \frac{145 - 325}{-315} \approx 0.5714286$$
$$-0.231112t = \ln 0.5714286 = -0.559616$$
$$t \approx 2.42$$

It will take about 2.42 hr $\approx$ 2 hr 25 min.

28. **(a)** $T = 120 + (40 - 120)e^{-kt} = 120 - 80e^{-kt}$. After 45 min, $T = 90$. Substituting this value of T into the formula, and solving for k produces

$$90 = 120 - 80e^{-45k}$$

$$e^{-45k} = \frac{90 - 120}{-80} = 0.375$$

$$-45k = \ln 0.375 \approx -0.980829$$

$$k \approx 0.021796$$

Hence, $T(t) = T_m + (40 - T_m)e^{-0.021796t}$, where t is in minutes

(b)

$$90 = 150 + (40 - 150)e^{-0.021796t}$$

$$e^{-0.021796t} = \frac{90 - 150}{-110} \approx 0.545454$$

$$-0.021796t = \ln 0.545454$$

$$t = 27.8$$

It will take about 27.8 min.

(c) It will never reach 90°F.

29.

$$\frac{dx}{dt} = \frac{5}{x} \text{ or } dx = 5\,dt$$

$$\int x\,dx = \int 5\,dt$$

$$\frac{1}{2}x^2 = 5t + C_1$$

$$x^2 = 10t + C$$

$$x = \sqrt{10t + C}$$

When $t = 0$ (6:00 A.M.), $x = 2$ and so, $C = 4$.
We want to solve for t when $x = 6$.

$$6 = \sqrt{10t + 4}$$

$$36 = 10t + 4$$

$$10t = 32$$

$$t = 3.2$$

The ice will be 6 in. thick 3.2 hours or about 3 hr 24 min after 6:00 A.M., or at about 9:24 A.M.

30. (a) $\frac{dA}{dt} = -kA$

(b) $\frac{dA}{A} = -k\,dt$

$$\ln A = -kt + C_1$$

$$A = e^{-kt + C_1} = e^{kt}e^{C_1} = Ce^{kt}$$

(c) At $t = 0$ and $A = 20$, we see that $C = 20$ and since the amount of the drug decreases at a rate of 2% per hour, then $\frac{dA}{dt} = -0.02A$ and $k = -0.02$.
Hence, this particular solution is $A = 20^{-0.02t}$.

(d) Six hours after the injection, there is $A = 20^{-0.02(6)} = 20^{-0.12} \approx 17.74$ cc left in the body.

31. (a) If i represents the number of people who are infected, then $\frac{di}{dt} = ki(500{,}000 - i)$

(b)

$$\frac{di}{dt} = ki(500{,}000 - i)$$

$$\frac{1}{i(500{,}000 - i)}di = k\,dt$$

By form #48 in Appendix C, $\int \frac{1}{i(500{,}000 - i)}di = \frac{1}{500{,}000}\ln\left|\frac{i}{500{,}000 - i}\right|$, so we get

$$\frac{1}{500{,}000}\ln\left|\frac{i}{500{,}000 - i}\right| = kt + C_1$$

$$\ln\left|\frac{i}{500{,}0000 - i}\right| = 500{,}000kt + C_2$$

$$\frac{i}{500{,}000 - i} = e^{(500{,}000kt + C_2)} = C_3 e^{500{,}000kt}$$

$$i = (500{,}000 - i)C_3 e^{500{,}000kt}$$

$$= 500{,}000C e^{500{,}000kt} - iC_3 e^{500{,}000kt}$$

$$i\left(1 + C_3 e^{500{,}000kt}\right) = 500{,}000C_3 e^{500{,}000kt}$$

$$i = \frac{500{,}000C_3 e^{500{,}000kt}}{1 + C_3 e^{500{,}000kt}}$$

$$= \frac{500{,}000C_3}{e^{-500{,}000kt} + C_3}$$

$$= \frac{500{,}000}{Ce^{-500{,}000kt} + 1}$$

$$= \frac{500{,}000}{1 + Ce^{-500{,}000kt}} = \frac{500{,}000}{1 + Ce^{-Bt}}$$

(c) When $t = 0$, we know that $i = 100$, and so $100 = \frac{500,000}{1+Ce^{-B(0)}} = \frac{500,000}{1+C}$. Solving this equation for C, produces $C = 4,999$. Thus, $i = \frac{500,000}{1+4,999e^{-Bt}}$. When $t = 10$, we are given $i = 750$, and so $750 = \frac{500,000}{1+4,999e^{-B(10)}}$ or $750(1 + 4,999e^{-B(10)}) = 500,000$ or $(750)4,999e^{-10B} = 3,749,250e^{-10B} = 499,250$ and $e^{-10B} = \frac{499,250}{3,749,250}$ and $-10B = \ln\left(\frac{499,250}{3,749,250}\right)$ 2 which results in $B \approx 0.201620$. Thus, a particular solution is $i = \frac{500,000}{1+4,999e^{-0.201620t}}$.

(d) Evaluating $i = \frac{500,000}{1+4,999e^{-0.201620t}}$ when $t = 20$ produces $i \approx 5,578$. Thus, around 5,578 people will be infected after 20 days.

32. (a) Let N be the number of people who have learned about the new product. Then $\frac{dN}{dt} = k(250,000 - N)$ or $\frac{dN}{250,000-N} = k\,dt$. Integrating produces $-\ln(250,000 - N) = kt + C_1$, so $250,000 - N = e^{-kt+C_1} = Ce^{-kt}$ and $N = 250,000 - Ce^{-kt}$. At $t = 0$, $N = 0$, so $C = 250,000$ and the formula is $N = 250,000(1 - e^{-kt})$. At $t = 7$, we are given $N = 40,000$, so $40,000 = 250,000(1 - e^{-k(7)})$ or $0.16 = 1 - e^{-7k}$ and $e^{-7k} = 0.84$. Taking the natural logarithm of both sides produces $-7k = \ln 0.84$ and thus, $k \approx 0.0249$. Hence, $N = 250,000(1 - e^{-0.0249t})$. When $t = 14$, $N = 250,000(1 - e^{-0.0249(14)}) \approx 73,581$ or about 73,600 people will know about the product after 14 days.

(b) Here $N = 125,000 = 250,000(1 - e^{-0.0249t})$ and we want to solve for t. Dividing by 250,000 produces $0.5 = 1 - e^{-0.0249t}$ or $e^{-0.0249t} = 0.5$. Thus, $\ln e^{-0.0249t} = -0.0249t = \ln 0.5$ and $t = \frac{\ln 0.5}{-0.0249} \approx 27.84$. It will take almost 28 days for 125,000 people to become aware of the product.

≡ 32.3 INTEGRATING FACTORS

1. No, $\frac{\partial(x^2-y)}{\partial y} = -1$, $\frac{\partial x}{\partial x} = 1$

2. No, $\frac{\partial(x^2+y^2)}{\partial y} = 2y$, $\frac{\partial xy}{\partial x} = y$

3. Yes, $\frac{\partial(x^2+y^2)}{\partial y} = 2y = \frac{\partial 2xy}{\partial x}$

4. No, $\frac{\partial(x+\sin y)}{\partial y} = \cos y$, $\frac{\partial(-\sin x)}{\partial x} = -\cos x$

5. Yes, $\frac{\partial(x+y\cos x)}{\partial y} = \cos x = \frac{\partial \sin x}{\partial x}$

6. Yes, $\frac{\partial(2ye^{2x})}{\partial y} = 2e^{2x} = \frac{\partial(e^{2x}-1)}{\partial x}$

7. Yes, $\frac{\partial(2xy+x)}{\partial y} = 2x = \frac{\partial(y+x^2)}{\partial x}$

8. No, $\frac{\partial(4x^3y^2)}{\partial y} = 8x^3y$, $\frac{\partial(3x^4y+5x^4)}{\partial x} = 12x^3y + 20x^3$

9. No, $\frac{\partial(x-y)}{\partial y} = -1$, $\frac{\partial(y+x)}{\partial x} = 1$

10. No, $\frac{\partial(y\sin x+x\cos y)}{\partial y} = \sin x - s\sin y$,

$\frac{\partial\left(\frac{1}{2}y^2\cos x+\sin y\right)}{\partial x} = -\frac{1}{2}y^2\sin x$.

11. $y\,dx - x\,dy = 0$. By Table 32.1 we can use one of four Integrating factors. Choosing $-\frac{1}{xy}$ we get $-\frac{y\,dx-x\,dy}{xy} = 0$. Integration yields $\ln\frac{y}{x} = C_1$ or letting $C_2 = e^{C_1}$ we get $\frac{y}{x} = C_2$.

12. $y\,dx - x\,dy - x^3\,dx = 0$. This time choose $-\frac{1}{x^2}$ to be the integration factor. Multiplication yields $-\frac{y\,dx-x\,dy}{x^2} + x\,dx = 0$. Integrating we get $\frac{y}{x} + \frac{1}{2}x^2 = C$.

13. $y\,dx - x\,dy - 5y^4\,dy = 0$. Choose $I(x,y) = \frac{1}{y^2}$. Multiplying by $\frac{1}{y^2}$ we get $\frac{y\,dx-x\,dy}{y^2} - 5y^2\,dy = 0$. Integration yields $\frac{x}{y} - \frac{5}{3}y^3 = C$.

14. $y\,dx - x\,dy - x^2y\,dx = 0$. Choose $I(x,y) = -\frac{1}{xy}$ giving $-\frac{y\,dx-x\,dy}{xy} + x\,dx = 0$. Integrating we get $\ln\frac{y}{x} + \frac{1}{2}x^2 = C$.

15. $y\,dx + y^2\,dx = x\,dy - x^2\,dx$ is equivalent to $y\,dx - x\,dy + x^2\,dx + y^2\,dx = 0$. This time we choose $I(x,y) = \frac{-1}{x^2+y^2}$ and multiplying we get $-\frac{y\,dx-x\,dy}{x^2+y^2} - \frac{x^2+y^2}{x^2+y^2}\,dx = 0$. Now integration yields $\arctan\frac{y}{x} - x = C$.

16. $y\,dx - x\,dy - 3\,dx = 0$. Multiplying by $I(x,y) = -\frac{1}{x^2}$ we get $-\frac{y\,dx-x\,dy}{x^2} + \frac{3}{x^2}\,dx = 0$. Integration yields $\frac{y}{x} - \frac{3}{x} = C$ or $\frac{y-3}{x} = C$.

17. $y\,dx + x\,dy - 3x^3y\,dx = 0$. The only integrating factor is $\frac{1}{xy}$. Multiplying by this we get $\frac{y\,dx+x\,dy}{xy} - 3x^2dx = 0$. Integration yields $\ln(xy) - x^3 = C$.

18. $x\,dx+y\,dy-3x^2\,dy-3y^2\,dy = 0$. Using $I(x,y) = \frac{1}{x^2+y^2}$ we get $\frac{x\,dx+y\,dy}{x^2+y^2} - 3\,dy = 0$. Integrating yields $\frac{1}{2}\ln\left(x^2 + y^2\right) - 3y = C$.

19. $xy^2\,dx + x^2y\,dy = 0$. This does not fit a set form but using $I(x,y) = \frac{1}{x^2y^2}$ we get $\frac{1}{x}dx + \frac{1}{y}dy = 0$. Integrating we get $\ln x + \ln y = C_1$ or $\ln(xy) = C_1$ or $xy = C_2$ where $C_2 = e^{C_1}$.

20. $4y\,dx - 3x\,dy - x^{-3}y^2dy = 0$. The first two terms fit form 8 of Table 32.1. Using $I(x,y) = x^3y^{-4}$ we get $(4y\,dx - 3x\,dy)x^3y^{-4} - x^{-3}y^2\left(x^3y^{-4}\right)dy = 0$ or $x^3y^{-3}dx - 3x^4y^{-4}dy - y^{-2}dy = 0$. Integration yields $x^4y^{-3} + y^{-1} = C$.

21. $2y\,dx + x\,dy - 3x\,dx = 0$. Using $I(x,y) = x$ we get $2xy\,dx + x^2dy - 3x^2dx = 0$. Integrating we have $x^2y - x^3 = C$.

22. $x^2\,dx + 2xy\,dy - y^2\,dx = 0$. The second two terms are somewhat similar to the terms of form 4 in Table 32.1. Multiplying by $\frac{1}{x^2}$ we get $dx + \frac{2xy\,dy-y^2\,dx}{x^2} = 0$. Integrating we get $x + \frac{y^2}{x} = C$.

23. $3x^2\left(x^2+y^2\right)dx+y\,dx-x\,dy = 0$. Using $I(x,y) = \frac{-1}{x^2+y^2}$ we get $-3x^2dx - \frac{y\,dx-x\,dy}{x^2+y^2} = 0$. Integration yields $-x^3 + \arctan\frac{y}{x} = C$.

24. $\left(x^2 + y^2\right)dx = 4x\,dx + 4y\,dy = 4(x\,dx + y\,dy)$. Using $I(x,y) = \frac{1}{x^2+y^2}$ we get $dx = 4\frac{x\,dx+y\,dy}{x^2+y^2}$. Integration yields $x = 4 \cdot \frac{1}{2}\ln\left(x^2 + y^2\right) + C$ or $x - 2\ln\left(x^2 + y^2\right) = C$.

25. $\tan\left(x^2 + y^2\right)dy = x\,dx + y\,dy$. First we can multiply both sides by $\cot\left(x^2 + y^2\right)$ giving us $dy = $ $\cot\left(x^2 + y^2\right)(x\,dx+y\,dy)$. Upon examination you will discover that $(x\,dx + y\,dy)$ is the derivative of $x^2 + y^2$. Hence we can take the integral of both sides giving us $y = \ln\left|\sin\left(x^2 + y^2\right)\right| + C$.

26. $8x\,dy+4y\,dy+9x^{-3}y\,dy = 0$. Let $I(x,y) = x^3y^7$ and multiplying yields $\left(8x^4y^7dy + 4x^3y^8dx\right) + 9y^8dy = 0$. Integration yields $x^4y^8 + y^9 = C$.

27. $2x\,dy + 2y\,dx = 3x^3y\,dx$. Multiplying by $\frac{1}{xy}$ we get $2\frac{x\,dy+y\,dx}{xy} = 3x^2\,dx$ so the general solution is $2\ln(xy) = x^3 + C$ or $\ln x^2y^2 = x^3 + C$ or $x^2y^2 = e^{x^3} + C$. Now letting $x = 1$ and $y = 3$ and solving for C we get $9 = e^1+C$ or $C = 9-e$. Hence the particular solution is $x^2y^2 = e^{x^3} +9-e$.

28. $y\,dx - x\,dy = x^2\,dx$. Multiplying by $-\frac{1}{x^2}$ we get $-\frac{y\,dx-x\,dy}{x^2} = -dx$. Integration yields the general solution $\frac{y}{x} = -x + C$. Now substituting $y = 4$ and $x = 2$ we solve for C. $\frac{4}{2} = -2+C$ or $C = 4$. Hence the particular solution is $\frac{y}{x} = -x + 4$ or $\frac{y}{x} + x = 4$.

29. $\left(x^2 + y + y^2\right)dx = x\,dy$ is equivalent to $\left(x^2 + y^2\right)dx = x\,dy-y\,dx$. Now let $I(x,y) = \frac{-1}{x^2+y^2}$ and multiplying yield $-dx = \frac{dyd-x\,dy}{x^2+y^2}$. Integration yields $-x = -\arctan\frac{y}{x} + C$ or $\arctan\left(\frac{y}{x}\right) = x + C$. Now setting $y = \frac{\pi}{3}$ and $x = \frac{\pi}{3}$ we solve for C $\arctan 1 = \frac{\pi}{3} + C$ or $C = -\frac{\pi}{12}$. Hence the particular solution is $\arctan\frac{y}{x} = x - \frac{\pi}{12}$.

30. $2y\,dx+3x\,dy = 0$. Let $I(x,y) = xy^2$ and multiply to get $2xy^3\,dx + 3x^2y^2\,dy = 0$. Integration yields $x^2y^3 = C$. Now setting $y = -2$ and $x = 2$ we get $C = 2^2(-2)^3 = -32$. The particular solution is $x^2y^3 = -32$.

$\equiv$ 32.4 LINEAR FIRST-ORDER DIFFERENTIAL EQUATIONS

1. $dy + 2xy\,dx = 6x\,dx$ is equivalent to $\frac{dy}{dx} + 2xy = 6x$. The integrating factor is $I(x,y) = e^{\int 2x\,dx} = e^{x^2}$. Multiplying we get $\frac{dy}{dx}e^{x^2} + 2xye^{x^2} = 6xe^{x^2}$. Now taking the integral of both sides we get $ye^{x^2} = 3e^{x^2} + C$.

2. $y' +y = 2$ or $\frac{dy}{dx} +y = 2$, $I(x,y) = e^x$. Multiplying we get $\frac{dy}{dx}e^x + ye^x = 2e^x$. Integrating yields $ye^x = 2e^x + C$.

3. $y' - 2y = 4$ or $\frac{dy}{dx} - 2y = 4$. $I(x,y) = e^{-2x}$ and multiplying we get $\frac{dy}{dx}e^{-2x} - 2ye^{-2x} = 4e^{-2x}$. Integrating we get $ye^{-2x} = -2e^{-2x} + C$.

4. $\frac{dy}{dx} - 3y = e^{4x}$, so $I(x,y) = e^{-3x}$ and multiplying we get $\frac{dy}{dx}e^{-3x} = 3ye^{-3x} = e^x$. Now integration yields $ye^{-3x} = e^x + C$.

5. $\frac{dy}{dx} + \frac{y}{x} = x^2$. $I(x,y) = e^{\int x^{-1}dx} = e^{\ln x} = x$. Multiplying we get $\frac{dy}{dx}x + y = x^3$. Integration yields $xy = \frac{1}{4}x^4 + C$.

6. $y' - 6y = e^x$, so $I(x,y) = e^{-6x}$. Multiplying we get $\frac{dy}{dx}e^{-6x} - 6ye^{-6x} = e^{-5x}$. Now integration yields $ye^{-6x} = -\frac{1}{5}e^{-5x} + C$.

7. $y' - 6y = 12x$. $I(x,y) = e^{-6x}$ and multiplying we get $\frac{dy}{dx}e^{-6x} - 6ye^{-6x} = 12xe^{-6x}$. The left-hand side integrates to ye^{-6x}. The right-hand side we can integrate of parts with $u = 12x$, $du = 12\,dx$, $dv = e^{-6x}dx$, $v = \frac{-1}{6}e^{-6x}$. Hence $\int 12xe^{-6x}dx = 12x\left(\frac{-1}{6}e^{-6x}\right) - \int \frac{-1}{6}e^{-6x}\cdot 12\,dx = -2xe^{-6x} - \frac{1}{3}e^{-6x}$. Putting it all together we get $ye^{-6x} = -\frac{1}{3}e^{-6x}(-6x - 1) + C$.

8. $\frac{dy}{dx} + x^2y = x^2$. $I(x,y) = e^{x^3/3}$ and multiplying we get $\frac{dy}{dx}e^{x^3/3} + x^2ye^{x^3/3} = x^2e^{x^3/3}$. Integration yields $ye^{x^3/3} = e^{x^3/3} + C$.

9. $\frac{dy}{dx} = e^x - \frac{y}{x}$ is equivalent to $\frac{dy}{dx} + \frac{y}{x} = e^x$. $I(x,y) = e^{\int \frac{1}{x}dx} = e^{\ln x} = x$. Multiplying we get $\frac{dy}{dx}x + y = xe^x$. Integration yields $xy = e^x(x - 1) + C$. (Use form 84 from Appendix C).

10. $\frac{dy}{dx} - 6xy = 0$, $I(x,y) = e^{-3x^2}$, hence we get $\frac{dy}{dx}e^{-3x^2} - 6xye^{-3x^2} = 0$. Integration yields $ye^{-3x^2} = C$.

11. $y' - \frac{1}{x^2}y = \frac{1}{x^2}$. $I(x,y) = e^{\int \frac{-1}{x^2}dx} = e^{1/x}$. Now multiplying by $e^{1/x}$ we get $\frac{dy}{dx}e^{1/x} - \frac{1}{x^2}ye^{1/x} = \frac{1}{x^2}e^{1/x}$. Integration yields $ye^{1/x} = -e^{1/x} + C$.

12. $y' - 4xy = xe^{x^2}$. $I(x,y) = e^{-2x^2}$ and multiplication yields $\frac{dy}{dx}e^{-2x^2} - 4xye^{-2x^2} = xe^{-x^2}$. Now integration yields $ye^{-2x^2} = -\frac{1}{2}e^{-x^2} + C$.

13. $y\,dy - 2y^2\,dx = y\sin 2x\,dx$. Dividing by $y\,dx$ we get $\frac{dy}{dx} - 2y = \sin 2x$. $I(x,y) = e^{-2x}$ and multiplying we get $\frac{dy}{dx}e^{-2x} - 2ye^{-2x} = e^{-2x}\sin 2x$. Integrating using form 87 we get $ye^{-2x} = \frac{e^{-2x}}{8}(-2\sin 2x + 2\cos 2x) + C$ or $ye^{-2x} = -\frac{1}{4}e^{-2x}(\sin 2x + \cos 2x) + C$.

14. $x\,dy - y\,dx = 4x\,dx$. Dividing by $x\,dx$ we get $\frac{dy}{dx} - \frac{y}{x} = 4$. $I(x,y) = e^{\int \frac{-1}{x}dx} = e^{-\ln x} = e^{\ln \frac{1}{x}} = \frac{1}{x}$. Multiplying by this we get $\frac{dy}{dx}\cdot\frac{1}{x} - \frac{y}{x^2} = \frac{4}{x}$. Integration now yields $\frac{y}{x} = 4\ln x + C$.

15. $4x\,dy - y\,dx = 8x\,dx$. Dividing by $4x\,dx$ we get $\frac{dy}{dx} - \frac{y}{4x} = 2$. $I(x,y) = e^{-\int \frac{1}{4x}dx} = e^{-\frac{1}{4}\ln x} = e^{\ln x^{-\frac{1}{4}}} = x^{-\frac{1}{4}}$. Multiplying by this we get $\frac{dy}{dx}x^{-1/4} - \frac{y}{4x}x^{-1/4} = 2x^{-1/4}$. Now integrating we get $yx^{-1/4} = \frac{8}{3}x^{3/4} + C$.

16. $\sec x\,dy = (y - 1)dx$. Dividing by $\sec x\,dx$ we get $\frac{dy}{dx} = (y - 1)\cos x$ or $\frac{dy}{dx} - y\cos x = -\cos x$. The integrating factor is $I(x,y) = e^{-\sin x}$ and multiplying by this we get $\frac{dy}{dx}e^{-\sin x} - ye^{-\sin x} = -\cos xe^{-\sin x}$. Integration yields $ye^{-\sin x} = e^{-\sin x} + C$.

17. $\frac{dy}{dx} - y\sec^2 x = \sec^2 x$. $I(x,y) = e^{\int -\sec^2 x\,dx} = e^{-\tan x}$. Multiplying we get $\frac{dy}{dx}e^{-\tan x} - y\sec^2 xe^{-\tan x} = \sec^2 xe^{-\tan x}$. Now integration yields $ye^{-\tan x} = -e^{-\tan x} + C$.

18. $x\frac{dy}{dx} = y + \left(x^2 - 1\right)^2$ is equivalent to $\frac{dy}{dx} - \frac{y}{x} = \frac{(x^2+1)^2}{x}$. $I(x,y) = e^{-\int \frac{1}{x}} = e^{-\ln x} = \frac{1}{x}$. Multiplying we get $\frac{dy}{dx}\cdot\frac{1}{x} - \frac{y}{x^2} = \frac{(x^2-1)^2}{x^2} = \frac{x^4-2x^2+1}{x^2}$. Integrating we get $\frac{y}{x} = \frac{1}{3}x^3 - 2x - \frac{1}{x} + C$.

19. $x\,dy + (1 - 4x)y\,dx = 4x^2e^{4x}dx$. Dividing by $x\,dx$ we obtain $\frac{dy}{dx} + \frac{1-4x}{x}y = 4xe^{4x}$. To get the integrating factor we must find $\int \frac{1-4x}{x}dx = \left(\frac{1}{x} - 4\right)dx = \ln x - 4x$. Hence $I(x,y) = e^{\ln x - 4x}$. Multiplying by this we obtain the equation $\frac{dy}{dx}e^{\ln x - 4x} + \frac{1-4x}{x}ye^{\ln x - 4x} = 4xe^{4x}e^{\ln x - 4x} = 4xe^{\ln x} = 4x^2$. Now integration yields $ye^{\ln x - 4x} = \frac{4}{3}x^3 + C$. The left side of this equation is $ye^{\ln x} \cdot e^{-4x} = yx \cdot e^{-4x} = \frac{yx}{e^{4x}}$. Hence the answer is $\frac{yx}{e^{4x}} = \frac{4}{3}x^3 + C$.

20. $x\,dy - y\,dx = x^3 \sin x^2 dx$. Dividing by $x\,dx$ we get $\frac{dx}{dx} - \frac{y}{x} = x^2 \sin x^2$. $I(x,y) = \frac{1}{x}$ and multiplying by this we get $\frac{dy}{dx} \cdot \frac{1}{x} - \frac{y}{x^2} = x \sin x^2$. Integration yields $\frac{y}{x} = -\frac{1}{2} \cos x^2 + C$.

21. $\frac{dy}{dx} + \frac{2y}{x} = x$. $I(x,y) = e^{\int \frac{2}{x}} = e^{2 \ln x} = e^{\ln x^2} = x^2$. Multiplying we obtain $\frac{dy}{dx}x^2 + 2yx = x^3$. Integration yields the general solution $yx^2 = \frac{1}{4}x^4 + C$. Substituting $y = 3$ and $x = 2$ we solve for C. $3 \cdot 4 = \frac{1}{4} \cdot 2^4 + C$ or $12 = 4 + C$ or $C = 8$. The particular solution is $yx^2 = \frac{1}{4}x^4 + 8$.

22. $\frac{dy}{dx} + 3y = e^{-2x}$ has the integrating factor e^{3x}. Multiplying we get $\frac{dy}{dx}e^{3x} + 3ye^{3x} = e^x$. Integration yields the general solution $ye^{3x} = e^x + C$. Substituting $y = 2$ and $x = 0$ we solve for C. $2e^0 = e^0 + C$ or $C = 1$. The particular solution is $ye^{3x} = e^x + 1$.

23. $y' - \frac{2y}{x} = x^2 \sin 3x$. The integrating factor is $e^{\int \frac{-2}{x}} = e^{-2 \ln x} = \frac{1}{x^2}$. Multiplying by $\frac{1}{x^2}$ yields $\frac{dy}{dx} \cdot \frac{1}{x^2} - \frac{2y}{x^3} = \sin 3x$. Integrating we get the general solution $\frac{y}{x^2} = \frac{-1}{3} \cos 3x + C$. Substituting $y = \pi^2$ and $x = \pi$ we solve for C. $1 = \frac{-1}{3} \cos 3\pi + C$. $1 = \frac{1}{3} + C$ or $C = \frac{2}{3}$. The particular solution is $yx^{-2} = -\frac{1}{3} \cos 3x + \frac{2}{3}$.

24. $\sin x\,dy + (y \cos x - 1)dx = 0$ is equivalent to $\frac{dy}{dx} + y \cot x = \csc x$. The integrating factor is $e^{\int \cot x\,dx} = e^{\ln |\sin x|} = \sin x$. Multiplying by $\sin x$

we obtain the equation $\frac{dy}{dx} \sin x + y \cos x = 1$. Integration yields the general solution $y \sin x = x + C$. Now substituting $y = \pi$ and $x = \frac{\pi}{6}$ we solve for C and obtain $\pi \sin \frac{\pi}{6} = \frac{\pi}{6} + C$ and so $\frac{1}{2}\pi - \frac{\pi}{6} = C$ or $C = \frac{\pi}{3}$. The particular solution is $y \sin x = x + \frac{\pi}{3}$.

25. **(a)** The general solution is $w(t) = \frac{2}{35}C + ke^{-0.005t}$. Substituting 160 for $w(0)$ and 2,100 for C, we can solve for k. $160 = \frac{2}{35}(2100) + ke^{-0.005(0)} = 120 + k$, hence $k = 40$. The particular solution is $w(t) = 120 + 40e^{-0.005t}$.
(b) $w(30) = 120 + 40e^{-0.005(30)} \approx 154.4$ lb
(c) $w(t) = 150 = 120 + 40e^{-0.005t}$
$$30 = 40e^{-0.005t}$$
$$0.75 = e^{-0.005t}$$
$$\ln 0.75 = -0.005t$$
$$t = \frac{\ln 0.75}{-0.005} \approx 57.5$$
It will take about 58 days for this person to lose 10 lb.
(d) $\lim_{t \to \infty} w(t) = \lim_{t \to \infty} \left(120 + 40e^{-0.005t}\right) = 120 + 40(0) = 120$.
(e) This person's weight will approach 120 lb if this diet is maintained for a long time.

26. **(a)** Here $w(0) = 200$. Substituting 200 for $w(0)$, 0 for t, and 2,100 for C, we can solve for k. $200 = \frac{2}{35}(2100) + ke^{-0.005(0)} = 120 + k$, hence $k = 80$. The particular solution is $w(t) = 120 + 80e^{-0.005t}$.
(b) $w(30) = 120 + 80e^{-0.005(30)} \approx 188.9$ lb
(c) $w(t) = 190 = 120 + 80e^{-0.005t}$
$$70 = 80e^{-0.005t}$$
$$0.875 = e^{-0.005t}$$
$$\ln 0.875 = -0.005t$$
$$t = \frac{\ln 0.875}{-0.005} \approx 26.7$$
It will take about 27 days for this person to lose 10 lb.

≡ 32.5 APPLICATIONS

1. As in example 32.22, $\frac{dN}{dt} - kN = 0$ has general solution $N = Ce^{kt}$. At $t = 0$, $N = 100$ which gives us that $C = 100$. Now when $t = 15$ days, $N = 75$ g and we can solve for k. $75 = 100e^{15k}$, so $\frac{75}{100} = e^{15k}$ and taking natural logarithms we get $15k = \ln \left(\frac{3}{4}\right)$ or $k = -0.0192$. The equation is

$N = 100e^{-0.0192t}$ where N is in grams and t in days.

2. The half-life is when $N = 50$, hence we get $50 = 100e^{-0.192t} \Rightarrow \frac{1}{2} = e^{-0.0192t} \Rightarrow -0.0192t = \ln \frac{1}{2} \Rightarrow t = 36.1413$ days $= 36$ days 3 hr 23 min 29.41 s.

3. If the half-life is 1000 years, we get $\frac{1}{2} = e^{k1000}$

or $k = \dfrac{\ln \frac{1}{2}}{1000} = -6.93147 \times 10^{-4}$. When 10% has decayed we have $0.9 = e^{-6.93147 \times 10^{-4}t}$ or $-6.93147 \times 10^{-4}t = \ln(0.9) \Rightarrow t = 152.0031$ yr.

4. As in example 32.22 $\frac{dN}{dt} = kN$ or $N = Ce^{kt}$. Now substituting the particular values we get $2000 = Ce^{k}$ and $6000 = Ce^{4k}$. Dividing the second equation by 2000 or Ce^{k} we get $3 = e^{3k}$ so $k = 0.3662$. Now back substituting into the first equation we get $2000 = Ce^{0.3662}$ or $C = \frac{2000}{e^{0.3662}} = 1386.7$ or 1387. The equation is $N = 1387e^{0.3662t}$, t in hours.

5. When $t = 0$, $N = 1387$.

6. $e^{0.3662t} = 2 \Rightarrow 0.3662t = \ln 2$ or $t = \frac{\ln 2}{0.3662} = 1.8928$ h.

7. First we must find k in the equation $N = Ce^{kt}$. Since the half-life is 5600 years, $\frac{1}{2} = e^{5600k}$ or $k = -1.23776 \times 10^{-4}$. Now we solve for t in the equation $\frac{1}{500} = e^{-1.23776 \times 10^{-4}t}$, with the result $t = \dfrac{\ln\left(\frac{1}{500}\right)}{-1.23776 \times 10^{-4}} = 50208.4$ years ≈ 50210 years.

8. Since $\frac{dI}{dt} = kI$ we get $I = Ce^{kt}$ where I is the light intensity and t is the thickness. At 3 ft, $I = 25\%$ of the original intensity. Hence $0.25 = e^{3k}$ or $k = \frac{\ln(0.25)}{3} = -0.4621$. At 18 ft, $\frac{I}{C} = e^{-0.4521 \times 18} = 2.4414 \times 10^{-4} = 0.0244\%$ of the original intensity.

9. **(a)** As in example 32.23 we have $\frac{dT}{dt} + kT = 30k$. Solving this differential equation we get $Te^{kt} = 30e^{kt} + C$ or $T = 30 + Ce^{kt}$. When $t = 0$, $T = 10°$, hence $C = -20$ or $T = 30 - 20e^{kt}$. Now when $t = 10$, $T = 15°$ or $15 = 30 - 20e^{-k10}$ or $\frac{-15}{-20} = e^{10k}$. Hence $k = \frac{\ln(0.75)}{-10} = 0.02877$. The equation is $T = 30 - 20e^{-0.02877t}$

(b) $22 = 30 - 20e^{-0.02877t}$, or $\frac{-8}{-20} = e^{-0.02877t}$, and so $t = \frac{\ln(0.4)}{-0.02877} = 31.85$ min.

(c) $t = 30 - 20e^{-0.02877 \times 60} = 26.44°$C

10. As in example 32.23, $T = 40 + Ce^{-kt}$. Substituting the particular values we get $0 = 40 + Ce^{10k}$ and $10 = 40 + Ce^{-20k}$. These are equivalent to $\frac{-40}{C} = e^{-10k}$ and $\frac{-30}{C} = e^{-20k}$, respectively. Squaring both sides of $\frac{-40}{C} = e^{-10k}$ we

get $\frac{1600}{C^2} = e^{-20k} = \frac{-30}{C}$. Hence $1600C = -30C^2$ or $C = \frac{1600}{-30} = -53.33$. This yields the equation $T = 40 - 53.33e^{-kt}$. When $t = 0$, $T = 40 - 53.33 = -13.33°$F.

11. As in example 32.23, $T = 40 + Ce^{-kt}$. At $t = 0$, $T = 375°$F so $C = 375 - 40 = 335°$F. Now, when $t = 15$ min, $T = 280°$F so $280 = 40 + 335e^{-15k}$ or $\frac{240}{335} = e^{-15k}$, and so $k = 0.02223$. Thus, we get $T = 40 + 335e^{-0.02223t}$. Hence $75 = 40 + 335e^{-0.02223t}$, or $\frac{35}{335} = e^{-0.02223t}$, and as a result,

$$t = \frac{\ln\left(\frac{35}{335}\right)}{-0.02223} = 101.6 \text{ min or } 1 \text{ hr } 41.6 \text{ min.}$$

12. As in Example 32.23, $T = 20 + Ce^{-kt}$. When $t = 0$, $T = -20°$ so $-20 = 20 + C$ or $C = -40°$, and hence $T = 20 - 40e^{kt}$. At $t = 1$, $T = -16°$ or $-16 = 20 - 40e^{-k}$, or $\frac{36}{40} = e^{-k}$, or $k = 0.1054$, so $T = 20 - 40e^{-0.1054t}$. Now substituting 15 for T we get $15 = 20 - 40e^{-0.1054t}$ or $\frac{1}{8} = e^{-0.1054t}$ or $t = 19.74$ min $= 19$ min 44 s.

13. This is an RL circuit and so $\frac{dI}{dt} + \frac{R}{L}I = \frac{E}{L}$ or $\frac{dI}{dt} + \frac{10}{2}I = \frac{100}{2}$ or $\frac{dI}{dt} + 5I = 50$. The general solution is $I = Ce^{-5t} + 10$. At $t = 0$, $I = 0$ so $C = -10$. The particular solution is $I = -10e^{-5t} + 10$.

14. In this problem, $\frac{dI}{dt} + 5I = 10\cos 5t$. The integrating factor is e^{5t} and so we get $\left(\frac{dI}{dt} + 5I\right)e^{5t} = 10e^{5t}\cos 5t$. Integrating we get $Ie^{5t} = \frac{10e^{5t}}{50}(5\cos 5t + 5\sin 5t) + C$ or $I(\cos 5t + \sin 5t) + Ce^{-5t}$. When $t = 0$, $I = 0$, so $0 = 1 + C$ or $C = -1$. The solution is $I = (\cos 5t + \sin 5t) - e^{-5t}$.

15. $\frac{dI}{dt} + 50I = 5$. The integrating factor is e^{50t} and we get $Ie^{50t} = \frac{1}{10}e^{50t} + C$. Hence $I = 0.1 + Ce^{50t}$. When $t = 0$, $I = 0$, so $C = -0.1$ and the particular solution is $I = 0.1 - 0.1e^{-50t}$

16. $\frac{dI}{dt} + 20I = 6\sin 2t$. The integrating factor is e^{20t} so we get $\frac{dI}{dt}e^{20t} + 20Ie^{20t} = 6e^{20t}\sin 2t$. Integrating we obtain $Ie^{20t} = \frac{6e^{20t}}{404}(20\sin 2t - 2\cos 2t) + C$ or $I = \frac{3}{101}(10\sin 2t - \cos 2t) + Ce^{-20t}$. When $t = 0$, $I = 6$ so $6 = \frac{3}{101}(-1) + C$, and so $C = \frac{606}{101} + \frac{3}{101} = \frac{609}{101}$. The particular solution is $I = \frac{609}{101}e^{-20t} + \frac{3}{101}(10\sin 2t - \cos 2t)$

≡ 32.6 MORE APPLICATIONS

1. $x^2 + y^2 = C$ is a family of circles with center at the origin. Taking the derivative implicitly we get $2x + 2yy' = 0$ or $\frac{dy}{dx} = -\frac{x}{y} = f(x,y)$. The orthogonal trajectories satisfy the equation $\frac{dy}{dx} = \frac{-1}{f(x,y)}$ or $\frac{dy}{dx} = \frac{y}{x}$. Separating variables we get $\frac{dy}{y} = \frac{dx}{x}$. Integration gives $\ln y = \ln x + C_1$ or $y = e^{\ln x + C_1} = e^{C_1} x$. Letting $e^{C_1} = k$ we get the family of curves $y = kx$ or the set of lines through the origin.

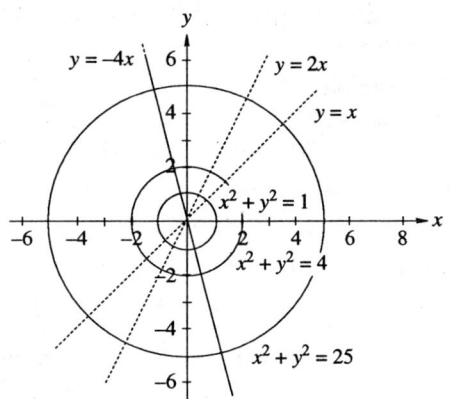

2. $xy = C$ is a family of hyperbolas with the x and y axis as asymptotes. Taking the derivative implicitly we obtain $xy' + y = 0$ or $y' = -\frac{y}{x} = f(x,y)$. The family we need satisfies $\frac{dy}{dx} = \frac{-1}{f(x,y)}$ or $\frac{dy}{dx} = \frac{x}{y}$. Separating variables we have $y\,dy = x\,dx$. Integration yields $\frac{1}{2}y^2 = \frac{1}{2}x^2 + C_1$ or equivalently $y^2 - x^2 = k$.

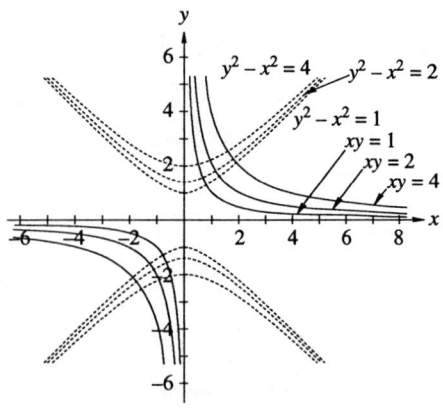

3. $y = Ce^x$ had derivative $\frac{dy}{dx} = Ce^x$. Substituting ye^{-x} for C we obtain $\frac{dy}{dx} = y$. The family of

curves we seek satisfies the equation $\frac{dy}{dx} = -\frac{1}{y}$ or $y\,dy = -dx$. Integrating we get $\frac{1}{2}y^2 = -x + C$ or $y^2 = -2x + k$ or $y^2 + 2x = k$.

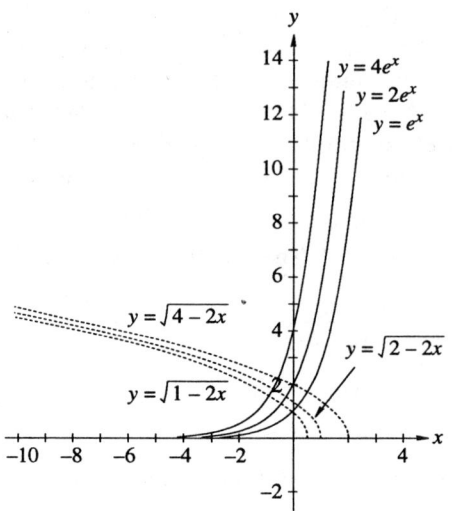

4. $x^2 - y^2 = C$ has implicit derivative $2x - 2yy' - 0$ or $y' = \frac{x}{y}$. The equation we need to solve is $\frac{dy}{dx} = -\frac{y}{x}$ or $\frac{dy}{y} = -\frac{dx}{x}$. This has solution $\ln y = -\ln x + C$ or $y = k \cdot \frac{1}{x}$ or $xy = k$.

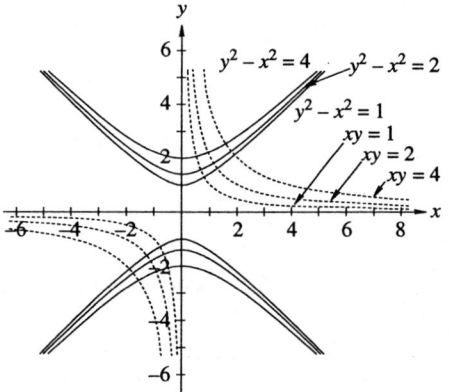

5. $x^2 + \frac{1}{2}y^2 = C^2$ has implicit derivative $2x + yy' = 0$ or $y' = \frac{-2x}{y}$. The orthogonal trajectories satisfy the equation $\frac{dy}{dx} = \frac{y}{2x}$. Separating variables we get $\frac{dy}{y} = \frac{dx}{2x}$. Integration yields $\ln y = \frac{1}{2}\ln x + C_1$ or $\ln y^2 = \ln x + C_2$ or $y^2 = kx$.

6. $x^2 + 2y^2 = C$ has the implicit derivative $2x + 4yy' = 0$ or $y' = \frac{-x}{2y}$. We seek the solution to $\frac{dy}{dx} = \frac{2y}{x}$. Separating variables we obtain $\frac{dy}{y} =$

$2\frac{dx}{x}$. Integration yields $\ln y = 2\ln x + C_1$ or $\ln y = \ln x^2 + C_1$ or $y = kx^2$, where $k = \ln C_1$.

7. As in example 32.26 we start with the equation $\frac{dv}{dt} + \frac{kv}{m} = g$ or $\frac{dv}{dt} + \frac{0.2v}{10} = 9.8$. This is of linear form and the integrating factor is $e^{0.02t}$. Hence we obtain the equation $\frac{dv}{dt}\left(ve^{0.02t}\right) = 9.8e^{0.02t}$. Integration yields $ve^{0.02t} = 490e^{0.02t} + C_1$ or $v = 490 + C_1e^{-0.02t}$. Since the initial velocity is 0, $C_1 = -490$ and the equation becomes $v(t) = 490\left(1 - e^{-0.02t}\right)$ m/s or $v(t) = 490\left(1 - e^{-\frac{t}{50}}\right)$ m/s.

8. After 10 seconds $v = 490\left(1 - e^{-\frac{1}{5}}\right) = 88.82$ m/s

9. **(a)** Using the general solution from problem 7 and substituting 1 for v and 0 for t we find $C_1 = 489$ so the particular solution is $v(t) = 490 - 489e^{-\frac{t}{50}}$,

(b) $v(10) = 490 - 489e^{-\frac{1}{5}} = 89.64$ m/s,

(c) $v_L = \frac{10 \times 9.8}{0.2} = 490$ m/s. (Note: $\lim_{t\to\infty} v(t) = 490$).

10. **(a)** Using $v_L = \frac{mg}{k}$ we can solve for k. $k = \frac{mg}{v_L} = \frac{10 \cdot 32}{320} = 1$. This yields the differential equation $\frac{dv}{dt} + \frac{v}{10} = 32$. The integrating factor is $e^{0.1t}$ and we get $\frac{dv}{dt}\left(ve^{0.1t}\right) = 32e^{0.1t}$ which has general solution $ve^{0.1t} = 320e^{0.1t} + C_1$. Since $v = 0$ when $t = 0$, $C_1 = -320$ and the particular solution is $v(t) = 320\left(1 - e^{-0.1t}\right)$ ft/s.
(b) The position with respect to where it was dropped is $\int v(t)\,dt = \int 320\left(1 - e^{-0.1t}\right)dt = 320\left(t + 10e^{-0.1t} + C_1\right)$. Since the position is 0 when $t = 0$, then $C_1 = -10$, or $d(t) = 320\left(t + 10e^{-0.1t}\right)$. From the ground, $h(t) = 1000 - d(t) = 1000 - 320\left(t + 10e^{-0.1t} - 10\right) = 1000 + 3200 - 320\left(t - 10e^{-0.1t}\right) = 4200 - 320\left(t + 10e^{-0.1t}\right) = 320\left(13.125 - t - 10e^{-0.1t}\right)$.

11. As in Example 32.27, $\frac{dQ}{dt} + \frac{2}{50 + (2-2)t}Q = 10$ or $\frac{dQ}{Q} = -\frac{dt}{25}$. Separating variables we get $\frac{dQ}{Q} + \frac{dt}{25} = 0$. Integration yields $\ln Q = -\frac{1}{25}t + C$ or $Q = ke^{-t/25}$. When $t = 0$, $Q = 25$ so $k = 25$ and $Q = 25e^{-t/25}$. After 15 minutes $Q = 25e^{-15/25} = 13.72$ lb.

12. First we must find a formula for Q. As in Example 32.27 we get $\frac{dQ}{dt} + \frac{5}{100}Q = 0$ or $\frac{dQ}{dt} = -\frac{1}{20}\,dt$. Integrating we get $\ln Q = -\frac{1}{20}t + C$ or $Q = ke^{-t/20}$.

Since $Q = 40$ when $t = 0$ the particular solution is $Q(t) = 40e^{-t/20}$
(a) $Q(10) = 40e^{-10/20} = 24.26$ lb,
(b) $Q(30) = 40e^{-30/20} = 8.93$ lb,
(c) $Q(1 \text{ hr}) = Q(60) = 1.99$ lb.

13. **(a)** $\frac{dQ}{dt} + \frac{4}{8}Q = 12$. This equation is linear and the integrating factor is $e^{t/2}$. Multiplying yields $\frac{dQ}{dt}\left(Qe^{t/2}\right) = 12e^{t/2}$. Integrating we get $Qe^{t/2} = 24e^{t/2} + C$ or $Q = 24 + Ce^{-t/2}$. Since $Q = 2$ when $t = 0$, $C = -22$ so the particular solution is $Q(t) = 24 - 22e^{-t/2}$
(b) $Q(8) = 24 - 22e^{-4} = 23.597$ lb.

14. $\frac{dQ}{dt} + \frac{2}{60}Q = 6$; $\frac{dQ}{dt}\left(Qe^{t/30}\right) = 6e^{t/30}$; $Qe^{t/30} = 180e^{t/30} + C_1$ or $Q = 180 + Ce^{-t/30}$. When $t = 0$, $Q = 60 \times 2 = 120$, so $C = -60$, and we have $Q = 180 - 60e^{-t/30}$. Now setting $Q = 150$ we solve for t. $150 = 180 - 60e^{-t/30}$, or $\frac{1}{2} = e^{-t/30}$; or $\frac{t}{30} = \ln\frac{1}{2}$, and $t = -30\ln\frac{1}{2} = 20.79$ min.

15. **(a)** This is a mixture problem with $r_1 = r_2 = 10\text{m}^3/\text{hr}$, $V_0 = 200\text{m}^3$, $a = 10$ kg, and $b = 1 \text{ kg}/5 \text{ m}^3 = 0.2 \text{ kg/m}^3$. Thus,

$$\frac{dQ}{dt} + \frac{10}{200 + (10 - 10)t}Q = (0.2)10$$
$$\frac{dQ}{dt} + \frac{1}{20}Q = 2$$

The integrating factor is $e^{t/200}$, so $\frac{d}{dt}\left(e^{t/200}Q\right) = 2e^{t/200}$, which leads to

$$e^{t/200}Q = \int 2e^{t/200}dt$$
$$e^{t/200}Q = 400e^{t/200} + C$$
$$Q = 400 + Ce^{-t/200}$$

Since $Q(0) = 10$, $C = -390$. Thus, the quantity of waste in the tank at any time t, in hr, is given by $Q(t) = 400 - 390e^{-t/200}$.
(b) When $t = 3$ hr, we have $Q(3) = 400 - 390e^{-3/200} \approx 15.806$ kg of waste in the tank.

16. Using separation of variables on $\frac{dV}{dt} = \lambda e^{-\alpha t}V$ produces $\frac{dV}{V} = \lambda e^{-\alpha t}dt$. Integrating, we obtain $\ln V = -\frac{\lambda}{\alpha}e^{-\alpha t} + k$ and so $V = e^{-(\lambda/\alpha)e^{-\alpha t} + k} = e^{-(\lambda/\alpha)e^{-\alpha t}}e^k$. When $t = 0$, we have $V(0) = V_0$, or $V_0 = e^{-(\lambda/\alpha)e^{-\alpha(0)}}e^k = e^{-\lambda/\alpha}e^k$, so $e^k = V_0e^{\lambda/\alpha}$. Substituting, we get

$$V(t) = e^{-\lambda e^{-\alpha t}/\alpha}V_0e^{\lambda/\alpha}$$
$$= V_0e^{\lambda\left(1 - e^{-\alpha t}\right)/\alpha}$$

CHAPTER 32 REVIEW

1. We have $y' = x + y$ or $\frac{dy}{dx} - y = x$. This is linear with integrating factor e^{-x}. Hence we get $\frac{dy}{dx}(ye^{-x}) = xe^{-x}$. Integrating with form 84 yields $ye^{-x} = e^{-x}(-x-1) + C$ or $y = Ce^x - (x+1)$.

2. $x\,dx - y^2\,dy = 0$ or $\frac{1}{2}x^2 - \frac{1}{3}y^3 = C$

3. $y' = y^2x^3$ or $\frac{dy}{y^2} = x^3\,dx$. Integrating produces $-y^{-1} = \frac{1}{4}x^2 + C_1$ or $x^4 + 4y^{-1} = C$.

4. $y' = 8y$ or $\frac{dy}{y} = 8\,dx$ and integrating gives us $\ln y = 8x + C$, or $y = ke^{8x}$.

5. $e^x\,dx - 2y\,dy = 0$ and integrating produces $e^x - y^2 = C$.

6. $(y^2 - y)dx + x\,dy = 0$ or $\frac{dx}{x} + \frac{dy}{y^2 - y} = 0$. Integrating by Form 48 we get $\ln x - \ln\left|\frac{y}{y-1}\right| = C$ or $\ln\frac{x(y-1)}{y} = C_1$ or $-x + \frac{x}{y} = C$.

7. $(y - xy^3)dx + (x - 2x^2y^2)\,dy = 0$ is equivalent to $y\,dx + x\,dy - xy^3\,dx - 2x^2y^2\,dy = 0$. Now we multiply by $\frac{1}{xy}$ to get $\frac{y\,dx + x\,dy}{xy} - (y^2\,dx + 2xy\,dy) = 0$. The first part has integral $\ln|xy|$. The second part has integral $-xy^2$. Putting it together we get $\ln|xy| - xy^2 = C$.

8. $(y+1)dx - x\,dy = 0$ is equivalent to $\frac{dx}{x} - \frac{dy}{y+1} = 0$. This has solution $\ln x - \ln(y+1) = \ln C$ or $\frac{x}{y+1} = C$ or $\frac{y+1}{x} = C$.

9. $y\,dx + (2-x)dy = 0$ is equivalent to $\frac{dx}{2-x} + \frac{dy}{y} = 0$. This has solution $-\ln|2 - x| + \ln y = \ln C$ or $\ln\frac{y}{x-2} = \ln C$ or $\frac{y}{x-2} = C$ or $\frac{x-2}{y} = C$.

10. $(y + x^3y)dx + x\,dy = 0$ or $y\,dx + x\,dy = -x^3y\,dx$. Multiplying by the integrating factor $\frac{1}{xy}$, we get $\frac{y\,dx + x\,dy}{xy} = \frac{-x^3y}{xy}\,dx = -x^2\,dx$. Integrating, we get $\ln xy = -\frac{1}{3}x^3 + C$

11. $y^2\,dx + xy\,dy = 0$. Multiplying by $\frac{1}{xy^2}$ to separate variables we get $\frac{dx}{x} + \frac{dy}{y} = 0$. Integration yields $\ln|x| + \ln|y| = \ln C$ or $\ln|xy| = C$.

12. $dy - 3y\,dx = 6\,dx$ is linear first-order and has integrating factor e^{3x}. Multiplying we get $e^{-3x}dy = $

$3ye^{-3x}dx = 6e^{-3x}dx$ and integrating we have $ye^{-3x} = -2e^{-3x} + C$ or $y = -2 + Ce^{3x}$

13. $y' - 4xy = x$ in linear first-order and has integrating factor e^{-2x^2}. Its solution is $ye^{-2x^2} = \int xe^{-2x^2}dx$ or $ye^{-2x^2} = \frac{-1}{4}e^{-2x^2} + C$ or $y = \frac{-1}{4} + Ce^{2x^2}$.

14. $y' + \frac{12y}{x} = x^{12}$ is linear first-order with integrating factor $e^{12\ln x} = x^{12}$. Multiplying and integrating we get $yx^{12} = \int x^{24}dx$ or $yx^{12} = \frac{x^{25}}{25} + C$.

15. $y' + y = \sin x$ is linear first-order with integrating factor e^x. Multiplying and integrating we get $ye^x = \int e^x \sin x\,dx$. Using form 87 this becomes $ye^x = \frac{e^x}{2}(\sin x - \cos x) + C$ or $y = \frac{1}{2}(\sin x - \cos x) + Ce^{-x}$.

16. $y' - 7y = e^x$ is linear first-order with integrating factor e^{-7x}. Multiplying and integrating we have $ye^{-7x} = \int e^{-6x}dx$ or $ye^{-7x} = \frac{1}{6}e^{-6x} + C$ which is equivalent to $y = -\frac{1}{6}e + Ce^{7x}$.

17. $\cos x\,dx + y\,dy = 0$ has general solution $\sin x + \frac{1}{2}y^2 = C$. Substituting $y = 2$ and $x = 0$ we find $C = \sin 0 + \frac{1}{2}\cdot 2^2 = 2$. The particular solution is $\sin x + \frac{1}{2}y^2 = 2$.

18. $(x^2 + y + y^2)dx - x\,dy = 0$ is equivalent to $y\,dx - x\,dy + (x^2 + y^2)dx = 0$. Using the integrating factor $\frac{-1}{x^2+y^2}$ we get the general solution $x + \arctan\frac{y}{x} = C$. Substituting $y = \frac{\pi}{3}$ and $x = \frac{\pi}{3}$ we solve for C. $C = -\frac{\pi}{3} + \arctan 1 = -\frac{\pi}{3} + \frac{\pi}{4} = -\frac{\pi}{12}$. The particular solution is $x + \arctan\frac{y}{x} = -\frac{\pi}{12}$.

19. $y' + \frac{2}{x}y = x$ is of linear first-order form with integrating factor x^2. Multiplying and integrating we get the general solution $yx^2 = \int x^3\,dx$ or $yx^2 = \frac{1}{4}x^4 + C$. Substituting $x = 1$ and $y = 0$ we find that $C = -\frac{1}{4}$. The particular solution is $yx^2 = \frac{1}{4}x^4 - \frac{1}{4}$ or $4yx^2 = x^4 - 1$.

20. $dy + 6xy\,dx = 0$ This is linear first-order with integrating factor e^{3x^2}. Multiplying and integrating we get $ye^{3x^2} = C$. Substituting $y = 5$ and $x = \pi$

we get $C = 5e^{3\pi^2}$. Hence the particular solution is $ye^{3x^2} = 5e^{3\pi^2} \approx 3.6134 \times 10^{13}$.

21. **(a)** As in Example 32.22 we get $N = Ce^{kt}$. After 2 years $\frac{N}{C} = 0.9$ so $e^{2k} = 0.9$, and so $k = -0.05268$. $N(t) = Ce^{-0.0527t}$ where $C = N(0)$;
(b) $e^{-0.0527t} = \frac{1}{2} \Rightarrow -0.0527t = \ln\frac{1}{2}$ or $t = 13.15$ years or 13 years 57 days 12 hours.

22. $y = cx^2$ had derivative $\frac{dy}{dx} = 3cx^2$. Substituting $\frac{y}{x^3}$ for c we get $\frac{dy}{dx} = 3 \cdot \frac{y}{x^3} \cdot x^2 = \frac{3y}{x}$. The family we seek satisfies $\frac{dy}{dx} = -\frac{x}{3y}$ or $3y\,dy = -x\,dx$ or $\frac{3}{2}y^2 = -\frac{1}{2}x^2 + k$ or $\frac{3}{2}y^2 + \frac{1}{2}x^2 = k$ or $3y^2 + x^2 = C$.

23. As in example 32.33 we get $T = 72 + Ce^{-kt}$. When $t = 0$, $T = 0$ so $C = -72$ and hence $T = 72 - 72e^{-kt}$. When $t = 15$ min, $T = 20°F$, so we get $20 = 72 - 72e^{-15t}$ or $e^{-15t} = \frac{52}{72}$ or $t = \dfrac{\ln\frac{52}{72}}{-15} = 0.021695$. This gives $T = 72\left(1 - e^{-0.021695t}\right)$.
(a) Substituting 50 for T and solving we get $1 - \frac{50}{72}e^{-0.021695t}$ or $t = \dfrac{\ln\left(\frac{20}{72}\right)}{0.021695} = 59.04$ min.
(b) $T(60) = 72\left(1 - e^{-0.021695 \times 60}\right) = 52.41°F$.

24. Using the formula $\frac{dI}{dt} + \frac{R}{L}I = \frac{E}{L}$ we get $\frac{dI}{dt} + 30I = \frac{25}{2}$. This equation is linear first-order with integrating factor e^{30t}. Multiplying and integrating we get $Ie^{30t} = \frac{25}{60}e^{30t} + C$ or $I = \frac{5}{12} + Ce^{-30t}$. The initial current is 0, so $C = -\frac{5}{12}$ and the particular solution is $I = \frac{5}{12}\left(1 - e^{-30t}\right)$.

25. Using the equation $\frac{dq}{dt} + \frac{1}{RC}q = \frac{E}{R}$ we get $\frac{dq}{dt} + 1.25q = 5\cos 2t$. This is linear first-order with integrating factor $e^{1.25t}$. Multiplying and integrating we get $qe^{1.25t} = 5\int e^{1.25t}\cos 2t$. Using form 88 we get $qe^{1.25t} = \frac{5e^{1.25t}}{17.5625}(1.25\cos 2t + 2\sin 2t) + C$ or $q = \frac{1}{3.5125}\left(1.25\cos 2t + 2\sin 2t + ke^{-1.25t}\right)$. When $t = 0$, $q = 0$ so $k = -1.25$. Finally, multiplying by $\frac{80}{80}$ to clear up fractions and decimals we get $q = \frac{1}{281}\left(100\cos 2t + 160\sin 2t - 100e^{-1.25t}\right)$.

26. First we convert 192 lb to slugs: $\frac{192}{32} = 6$ slugs. Since the limiting velocity is 16 ft/s, we get $k = \frac{192}{16} = 12$. Now using the equation $\frac{dv}{dt} + \frac{kv}{m} = g$, we have $\frac{dv}{dt} + \frac{12}{6}v = 32$. This is linear first-order with integrating factor e^{2t}. Multiplying and integrating we get $ve^{2t} = 16e^{2t} + C$ or $v = 16 + Ce^{-2t}$. Since $v_0 = 0$, $C = -16$ and $v(t) = 16\left(1 - e^{-2t}\right)$
(a) $v(1) = 13.83$ ft/s.
(b) $15 = 16\left(1 - e^{-2t}\right) \Rightarrow e^{-2t} = \frac{1}{16}$ or $t = \dfrac{\ln\left(\frac{1}{16}\right)}{-2} = 1.386$ s.

27. **(a)** As in example 32.28, $\frac{dQ}{dt} + \frac{100}{2000}Q = 0.035$. This is linear first-order with integrating factor $e^{0.05t}$. Multiplying and integrating we get $Qe^{0.05t} = 0.7e^{0.05t} + C$ or $Q = 0.7 + Ce^{-0.05t}$. The initial quantity of CO_2 is 2000×0.001 so $C = 1.3$. The particular solution is $Q(t) = 0.7 + 1.3e^{-t/20}$ ft^3,
(b) $Q(60) = 0.7547$ ft^3,
(c) $Q(180) = 0.7002$ ft^3.

28. **(a)** As in example 32.28, $\frac{dQ}{dt} + \frac{1}{250}Q = 0.0005$. The general solution is $Qe^{t/250} = 0.125e^{t/250} + C$ or $Q = 0.125 + Ce^{-t/250}$. Since $Q_0 = 2.5$, $C = 2.375$ and the solution is $Q(t) = 0.125 + 2.375e^{-t/250}$ ft^3,
(b) $Q(60) = 1.8807$ ft^3.

29. $\frac{dQ}{dt} + \frac{25}{100000}Q = 0.0125$ has solution $Qe^{t/4000} = 50e^{-t/4000} + C$ or $Q = 50 + ce^{-t/4000}$. Since $Q_0 = 100{,}000 \times 0.0001 = 10$, $C = -40$ and $Q(t) = 50 - 40e^{-t/4000}m^3$, **(b)** Solving $20 = 50 - 40e^{-t/4000}$ fot t we get $\frac{3}{4} = e^{-t/4000}$ or $t = -4000\ln\left(\frac{3}{4}\right) = 1151$ days or 3 years 56 days.

30. **(a)** $\frac{dQ}{dt} + \frac{500}{10^6}Q = 0.00074 \times 500$ or $\frac{dQ}{dt} + \frac{Q}{20000} = 0.37$ has solution $Qe^{t/20\,000} = \int 0.37e^{t/20\,000} = 7400e^{t/20\,000} + C$ or $Q = 7400 + Ce^{-\frac{t}{20\,000}}$. Since $Q_0 = 10^3$, $C = -6400$. Hence $Q(t) = 7400 - 6400e^{-t/20\,000}$,
(b) Solving $2000 = 7400 - 6400e^{-t/20\,000}$ for t we get $\frac{54}{64} = e^{-t20\,000}$ or $t = -20000\ln\frac{54}{64} = 3398$ days or 9 years 3.6 months,
(c) As t approaches infinity, Q approaches 7400 m^3.

CHAPTER 32 TEST

1. Differentiating $y = 4e^{2x} + 2e^{-3x}$ we get $\frac{dy}{dx} = 8e^{2x} - 6e^{-3x}$ and $\frac{d^2y}{dx^2} = 16e^{2x} + 18e^{-3x}$.

Substituting these in the given differential equation, we get $\left(16e^{2x}+18e^{-3x}\right)+\left(8e^{2x}-6e^{-3x}\right)-6\left(4e^{2x}+2e^{-3x}\right) = (16+8-24)e^{2x}+(18-6-12)e^{-3x} = 0.$

2. Here $M = y^2$ and $N = 2xy$. Both M and N are continuous. Since $\frac{\partial M}{\partial y} = 2y = \frac{\partial N}{\partial x}$, the equation is exact.

3. Here $y' = \frac{dy}{dx} = \frac{5x^4}{y^3}$ can be rewritten as $y^3\,dy = 5x^4\,dx$. Integrating, we obtain $\frac{1}{4}y^4 = x^5 + C$.

4. Dividing this equation by x^2, we obtain $\frac{x^2+1}{x^2}dx + y^2\,dy = 0$ or $\left(1+\frac{1}{x^2}\right)dx + y^2\,dy = 0$. Integrating produces the solution $x - \frac{1}{x} + \frac{1}{3}y^3 = C$.

5. Rewriting the given equation as $\left(1 + y^2\right)dx + \left(x^2+1\right)dy = 0$ we see that the equation is separable. If the variables are separated by dividing by $\left(1+y^2\right)\left(x^2+1\right)$, we obtain $\frac{dx}{x^2+1}+\frac{y}{1+y^2}dy = 0$. Integrating we obtain the solution $\arctan x + \frac{1}{2}\ln\left(1+y^2\right) = C$.

6. Dividing by $\sin x$ we obtain $\frac{dy}{dx} = \frac{1}{\sin x} - 2y\frac{\cos x}{\sin x} = \csc x - 2y\cot x$, which can be written in the standard form for a linear first-order differential equation as $\frac{dy}{dx} + (2\cot x)y = \csc x$. This has an integrating factor of $e^{\int 2\cot x\,dx} = e^{\ln|\sin x|} = \sin x$ and the solution is $y\sin x = \int(\sin x)(\csc x)dx = \int dx = x + C$, and so the solution is $y\sin x = x + C$.

7. The equation contains the combination $x\,dy - y\,dx = -(y\,dx - x\,dy)$. Using integrating factor (4) from Table 32.1, we multiply the given equation by $-\frac{1}{x^2}$ to obtain $\frac{y\,dx - x\,dy}{x^2} = \frac{x^6}{x^2}\,dx = 0$ or $\frac{y\,dx - x\,dy}{x^2} + x^4\,dx = 0$. The left-hand term is the derivative of $-\frac{y}{x}$ so integrating produces the solution $-\frac{y}{x} + \frac{1}{5}x^5 = C$.

8. The given equation contains the combination $x\,dy + y\,dx$. Using integrating factor (1) from Table 32.1, we multiply the given equation by $I(x,y) = \frac{1}{xy}$ obtaining $4\left(\frac{x\,dy+y\,dx}{xy}\right) + \frac{6x^2y}{xy}\,dx = 4\left(\frac{x\,dy+y\,dx}{xy}\right) + 6x\,dx = 0$. Integrating, we get $4\ln xy + 3x^2 = C$. We are given $x = 1$ when $y = e$. Substituting, we obtain $4\ln e + 3 = 7 = C$, so the particular solution is $4\ln xy + 3x^2 = 7$.

9. We have $F(x,\,y,\,c) = y - ce^{-2x}$. Differentiating the given equation, we get $y' = -2ce^{-2x}$. Solving the given equation for c, we obtain $c = ye^{2x}$ and so $y' = -2\left(ye^{2x}\right)e^{-2x} = -2y$. The orthogonal trajectories are the solutions of $\frac{dy}{dx} = \frac{-1}{-2y} = \frac{1}{2y}$ or $2y\,dy - dx = 0$. The solution of this differential equation is $y^2 - x = k$.

10. Let N represent the amount of the radioactive substance present at time t. Because the rate of decay is proportional to the amount of substance remaining we have $\frac{dN}{dt} = kN$ or $\frac{dN}{dt} - kN = 0$. Using the integrating factor e^{-kt} and integrating, we obtain $Ne^{kt} = C$ or $N = Ce^{-kt}$. If N_0 represents the amount when $t = 0$, the equation becomes $N = N_0e^{-kt}$. We are told that $N = 0.70$ when $t = 40$ years, and so we get $0.7N_0 = N_0e^{-40k}$ or $0.7 = e^{-40k}$ and so $k = \frac{\ln 0.7}{-40} \approx 0.008\,917$ and the desired equation is $N = N_0e^{-0.008\,917t}$. To find the half-life, we solve $e^{-0.008\,917t} = \frac{1}{2}$ and obtain $t \approx 77.73$ years.

11. The body satisfies the equation $\frac{dv}{dt} + \frac{0.2v}{5} = 32$ or $\frac{dv}{dt} = 32 - 0.04v$. Separating variables, we obtain $\frac{dv}{32-0.04v} = dt$ and integrating produces $-25\ln(32-0.04v) = t + C_1$ which can be written as $\ln(32 - 0.04v) = -\frac{t}{25} + C$ where $-25\ln C = C_1$. This is equivalent to $32 - 0.04v = Ce^{-t/25}$. At $t = 0$, we are given $v = 0$ so $C = 32$ which makes the equation $32 - 0.04v = 32e^{-t/25}$ or $v = \frac{3s}{0.04}\left(1 - e^{-t/25}\right) = 800\left(1 - e^{-t/25}\right)$. When $t = 6$ s, $v = 800\left(1 - e^{-6/25}\right) \approx 170.70$ ft/s.

12. If T represents the temperature of the object at some time t minutes and T_m the temperature of the surrounding medium, then according to Newton's law of cooling, we have $\frac{dT}{dt} = -k\left(T - T_m\right)$. Here $T_m = 30°C$ and so $\frac{dT}{dt} + kT = 30k$. The solution of this differential equation is $Te^{kt} = 30e^{kt} + C$ or $T = 30 + Ce^{-kt}$. Since $T = 50°C$ when $t = 0$, we get $C = 20$ and the equation becomes $T = 30 + 20e^{-kt}$. When $t = 20$, we are given $T = 45$ which gives $45 = 30 + 20e^{-20k}$ or $0.75 = e^{-20k}$. Taking the natural logarithm we get $\ln = 0.75 = -20k$ or $k = \frac{0.75}{-20} \approx 0.0144$. The desired equation is $T = 30 + 20e^{-0.0144t}$.

CHAPTER

33

Higher-Order Differential Equations

≡ 33.1 HIGHER-ORDER HOMOGENEOUS EQUATIONS WITH CONSTANT COEFFICIENTS

1. $\left(D^2 - 7D + 10\right)y = 0$ has auxiliary equation $m^2 - 7m + 10 = 0$ which factors into $(m-2)(m-5) = 0$ and so has roots $m_1 = 2$ and $m_2 = 5$. The general solution is thus $y = c_1 e^{2x} + c_2 e^{5x}$.

2. $\left(D^2 + 7D = 12\right)y = 0$ has auxiliary equation $m^2 + 7m + 12 = 0$ which has roots -3 and -4. The solution is $y = c_1 e^{-3x} + c_2 e^{-4x}$.

3. $D^2 y - 8Dy + 12y = 0$ has auxiliary equation $m^2 - 8m + 12 = 0$ which has roots 2 and 6. The solution is $y = c_1 e^{2x} + c_2 e^{6x}$.

4. $6D^2 y + Dy = y$ has auxiliary equation $6m^2 + m - 1 = 0$ which factors into $(3m - 1)(2m + 1) = 0$ and has roots $\frac{1}{3}$ and $-\frac{1}{2}$. Hence the solution is $y = c_1 e^{\frac{1}{3}x} + C_2 e^{-\frac{1}{2}x}$.

5. $2\frac{d^2 y}{dx^2} + \frac{dy}{dx} - 5y = 0$ has auxiliary equation $2m^2 + 9m - 5 = 0$ which factors into $(2m-1)(m+5) = 0$. The roots are $\frac{1}{2}$ and -5 so the solution is $y = c_1 e^{\frac{1}{2}x} + c_2 e^{-5x}$.

6. $\frac{d^2 y}{dx^2} = 9$ has auxiliary equation $m^2 - 9 = 0$. This has roots -3 and 3 so the solution is $y = c_1 e^{-3x} + c_2 e^{3x}$.

7. $4D^2 y + 7Dy - 2y = 0$ has auxiliary equation $4m^2 - 7m - 2 = 0$. This factors into $(4m+1)(m-2) = 0$ and has roots $-\frac{1}{4}$ and 2. The solution is $y = c_1 e^{-\frac{1}{4}x} + c_2 e^{2x}$.

8. $4y'' + 11y' - 3y = 0$ has auxiliary equation $4m^2 + 11m - 3 = 0$ which factors into $(4m - 1)(m + 3) = 0$. The roots are $\frac{1}{4}$ and -3 so the solution is $y = c_1 e^{\frac{1}{4}x} + c_2 e^{-3x}$.

9. $y'' - 5y' + 4y = 0$ has auxiliary equation $m^2 - 5m + 4 = 0$. This has roots 1 and 4 so the solution is $y = c_1 e^x + c_2 e^{4x}$.

10. $\frac{d^2 y}{dx^2} + \frac{dy}{dx} + 3y = 0$ has auxiliary equation $m^2 + 4m + 3 = 0$ with roots -1 and -3. Hence the solution is $y = c_1 e^{-x} + c_2 e^{-3x}$.

11. $y'' - y = 0$ has auxiliary equation $m^2 - 1 = 0$. The roots are 1 and -1 so the solution is $y = c_1 e^x + c_2 e^{-x}$.

12. $D^2 y - Dy + 30y = 0$ has auxiliary equation $m^2 - m - 30 = 0$ with roots 6 and -5. The solution is $y = c_1 e^{6x} + c_2 e^{-5x}$.

13. $y'' = 6y$ has auxiliary equation $m^2 - 7 = 0$. This has roots $\sqrt{7}$ and $-\sqrt{7}$ so the solution is $y = c_1 e^{\sqrt{7}x} + c_2 e^{-\sqrt{7}x}$.

14. $D^2 y + Dy - y = 0$ has auxiliary equation $m^2 + m - 1 = 0$. Using the quadratic formula we get the roots $\frac{-1+\sqrt{5}}{2}$. Hence the solution is $y = c_1 e^{\frac{-1+\sqrt{5}}{2}x} + c_2 e^{\frac{-1-=\sqrt{5}}{2}x}$.

15. $2\frac{d^2 y}{dx^2} + \frac{dy}{dx} - y = 0$ has auxiliary equation $2m^2 + m - 1 = 0$. This factors into $(2m - 1)(m + 1) = 0$ and has roots $\frac{1}{2}$ and -1. Hence the solution is $y = c_1 e^{x/2} + c_2 e^{-x}$.

16. $D^2 y + 2Dy - y = 0$ has auxiliary equation $m^2 + 2m - 1 = 0$. Using the quadratic formula we get roots $\frac{-2+\sqrt{8}}{2} = -1 \pm \sqrt{2}$. Hence the solution is $y = c_1 e^{(-1+\sqrt{2})x} + c_2 e^{(-1-\sqrt{2})x}$.

17. $D^3 y - 6D^2 y + 11Dy - 6y = 0$ has auxiliary equation $m^3 - 6m^2 + 11m - 6 = 0$. This factors into $(m-1)(m-2)(m-3) = 0$ and has roots 1, 2 and 3. The solution is $y = c_1 e^x + c_2 e^{2x} + c_3 e^{3x}$.

18. $D^3 y - D^2 y - 4Dy + 4y = 0$ has auxiliary equation $m^3 - m^2 - 4m + 4 = 0$. Factoring by grouping we get $m^2(m - 1) - 4(m - 1)$ or $(m^2 - 4)(m - 1)$ or $(m + 2)(m - 2)(m - 1) = 0$. The solution is $y = c_1 e^x + c_2 e^{2x} + c_3 e^{-2x}$.

19. $y''' + 6y'' + 11y' + 6y = 0$ has auxiliary equation $m^3 + 6m^2 + 11m + 6 = 0$. This has only negative roots. With synthetic division we can verify that the roots are -1, -2, and -3. Hence the solution is $y = c_1 e^{-x} + c_2 e^{-2x} + c_3 e^{-3x}$.

20. $D^3 y - D^2 y - 17Dy = 15y$ has auxiliary equation $m^3 - m^2 - 17m - 15 = 0$. Using synthetic division we see 5 is a root. The other two roots are -3 and -1. Hence the solution is $y = c_1 e^{-x} + c_2 e^{-3x} + c_3 e^{5x}$.

21. $D^2 y + 2Dy - 15y = 0$ has auxiliary equation $m^2 + 2m - 15 = 0$. This has roots 3 and -5 and so the general solution is $y = c_1 e^{3x} + c_2 e^{-5x}$. When $x = 0$ we have $y = 2$ so $2 = c_1 + c_2$ or $c_1 = 2 - c_2$ finding $y' = 3c_1, e^{3x} - 5c_2 e^{-5x}$ we also have that $y' = 6$ when $x = 0$. Hence $6 = 3c_1 - 5c_2$. Substituting $2 - c_2$ for c_1 we get $6 = 3(2 - c_2) - 5c_2$ or $6 = 6 - 3c_2 - 5c_2$. Hence $c_2 = 0$ and $c_1 = 2$. The particular solution is $y = 2e^{3x}$.

22. $3D^2 y - 14Dy + 8y = 0$ has auxiliary equation $3m^2 - 14m + 8 = 0$. This factors into $(3m - 2)(m - 4) = 0$. Hence the general solution is $y = c_1 e^{4x} + c_2 e^{2x/3}$. When $x = 0$, $y = 3$ so $3 = c_1 + c_2$ or $c_1 = 3 - c_2$. Also, $y' = 4c_1 e^{4x} + \frac{2}{3}c_2 e^{2x/3}$ and $y' = 12$ when $x = 0$. This gives $12 = 4c_1 + \frac{2}{3}c_2$. Substituting this for c_1 we have $12 = 4(3 - c_2) + \frac{2}{3}c_2$ of $12 = 12 - 3\frac{2}{3}c_2$. Hence, $c_2 = 0$ and $c_1 = 3$. The particular solution is $y = 3e^{4x}$.

23. $D^2 y + 2Dy - 8y = 0$ has auxiliary equation $m^2 + 2m - 8 = 0$. Its roots are -4 and 2. Hence the general solution is $y = c_1 e^{2x} + c_2 e^{-4x}$. When $x = 0$ we are given that $y = 0$ which produces

$$0 = c_1 + c_2 \tag{1}$$

Differentiating that general solution gives $y' = 2c_1 e^{2x} - 4c_2 e^{-4x}$. When $x = 0$, we know that $y' = 6$, and so

$$\begin{aligned} 6 &= 2c_1 - 4c_2 \\ \text{or} \quad 3 &= c_1 - 2c_2 \end{aligned} \tag{2}$$

Subtracting equation (2) from equation (1) yields $3c_2 = -3$, and so $c_2 = -1$. Back-substitution into equation (1) gives $c_1 = 1$. The particular solution is $y = e^{2x} - e^{-4x}$.

24. $D^2 y + 2Dy - 8y = 0$ has auxiliary equation $m^2 + 2m - 8 = 0$. Its roots are -4 and 2. Hence, the general solution is $y = c_1 e^{2x} + c_2 e^{-4x}$. When $x = 0$ we are given that $y = 3$ which produces

$$3 = c_1 + c_2 \tag{1}$$

Differentiating the general solution gives $y' = 2c_1 e^{2x} - 4c_2 e^{-4x}$. When $x = 0$, we know that $y' = -12$, and so

$$\begin{aligned} -12 &= 2c_1 - 4c_2 \\ \text{or} \quad -6 &= c_1 - 2c_2 \end{aligned} \tag{2}$$

Subtracting equation (2) from equation (1) yields $3c_2 = 9$, and so $c_2 = 3$. Back-substitution into equation (1) gives $c_1 = 0$. The particular solution is $y = 3e^{-4x}$.

25. The differential equation $y'' + 2y' - 15y = 0$ has auxiliary equation $m^2 + 2m - 15 = 0$. Its roots are 3 and -5. Hence, the general solution is $y = c_1 e^{3x} + c_2 e^{-5x}$. When $x = 0$ we are given that $y = 5$ which produces

$$c_1 + c_2 = 5 \tag{1}$$

Differentiating the general solution gives $y' = 3c_1 e^{3x} - 5c_2 e^{-5x}$. When $x = 0$, we know that $y' = -1$, and so

$$3c_1 - 5_2 = -1 \tag{2}$$

Multiplying equation (1) by 5 and adding it to equation (2) produces $8c_1 = 24$, $c_1 = 3$. Back-substitution gives $c_2 = 2$. The particular solution is $y = 3e^{3x} + 2e^{-5x}$.

26. The differential equation $y'' - 4y' - 32y = 0$ has auxiliary equation $m^2 - 4m - 32 = 0$ with roots 8 and -4. The general solution is $y = c_1 e^{8x} + c_2 e^{-4x}$ and $y' = 8c_1 e^{8x} - 4c_2 e^{-4x}$. When $x = 0$, then $y = 15$ and $y' = 0$, giving the equations

$$\begin{aligned} c_1 + c_2 &= 15 \tag{1} \\ 8c_1 - 4c_2 &= 0 \tag{2} \end{aligned}$$

Multiplying equation (1) by 4 and adding it to equation (2) produces $12c_1 = 60$ or $c_1 = 5$. Back-substitution gives $c_2 = 10$. The particular solution is $y = 5e^{8x} + 10e^{-4x}$.

27. The differential equation $y''' + y'' - 4y' - 4y = 0$ has auxiliary equation $m^3 + m^2 - 4m - 4 = 0$. Removing common factors gives $m^2(m+1) - 4(m+1) = (m^2 - 4)(m + 1)$. So, the roots are -1, 2, and -2. The general solution is $y = c_1 e^{-x} + c_2 e^{2x} + c_3 e^{-2}$, its derivative is $y' = -c_1 e^{-x} + 2c_2 e^{2x} - 2c_3 e^{-2}$, and its second derivative is $y'' = c_1 e^{-x} + 4c_2 e^{2x} + 4c_3 e^{-2}$. When $x = 0$, we are given $y = 3$, $y' = 7$, and $y'' = -3$. Substituting these in the appropriate equations generate the system of equations

$$c_1 + c_2 + c_3 = 3$$
$$-c_1 + 2c_2 - 2c_3 = 7$$
$$c_1 + 4c_2 + 4c_3 = -3$$

The solutions of this system are $c_1 = 5$, $c_2 = 2$, and $c_3 = -4$. The particular solution is $y = 5e^{-x} + 2e^{2x} - 4e^{-2x}$.

28. The differential equation $y''' + y'' - 17y' + 15y = 0$ has auxiliary equation $m^3 + m^2 - 17m + 15 = 0$. Inspection shows that 1 is a root. Synthetic division produces

```
1 | 1   1   -17   15
  |     1    2    -15
  -------------------
    1   2   -15    0
```

Thus, $m^2 + m^2 - 17m + 15 = (m-1))(m^2 + 2m - 15) = (m+1)(m+5)(m-3)$. The roots are 1, 3, and -5. Hence, the general solution is $y = c_1 e^x + c_2 e^{3x} + c_3 e^{-5x}$, its derivative is $y' = c_1 e^x + 3c_2 e^{3x} - 5c_3 e^{-5x}$, and its second derivative is $y'' = c_1 e^x + 9c_2 e^{3x} + 25c_3 e^{-5x}$. When $x = 0$, we are given $y = 0$, $y' = -12$, and $y'' = 0$ which leads to the system of equations

$$c_1 + c_2 + c_3 = 0$$
$$c_1 + 3c_2 - 5c_3 = -12$$
$$c_1 + 9c_2 + 25c_3 = 0$$

The solutions of this system are $c_1 = 2$, $c_2 = -3$, and $c_3 = 1$. The particular solution is $y = 2e^x - 3e^{3x} + 3^{-5x}$.

33.2 AUXILIARY EQUATIONS WITH REPEATED OR COMPLEX ROOTS

1. $D^2 y + 2Dy + y = 0$ has auxiliary equation $m^2 + 2m + 1 = 0$ This has -1 as a double root. Hence the general solution is $y = c_1 e^{-x} + c_2 x e^{-x}$ or $y = (c_1 + c_2 x) e^{-x}$.

2. $(D^2 y - 2D + 1) y = 0$ has auxiliary equation $m^2 - 2m + 1 = 0$. This has 1 as a double root. Hence the general solution is $y = c_1 e^x + c_2 x e^x$ or $y = (c_1 + c_2 x) e^x$.

3. $D^2 y - 4Dy + 4y = 0$ has auxiliary equation $m^2 - 4m + 4 = 0$. This has 2 as a double root and hence the solution is $y = (c_1 + c_2 x) e^{2x}$.

4. $y'' - 10y' + 25y = 0$ has auxiliary equation $m^2 - 10m + 25 = 0$. 5 is a double root and the solution is $y = (c_1 + c_2 x) e^{5x}$.

5. $9y'' - 6y' + y = 0$ has auxiliary equation $9m^2 - 6m + 1 = 0$. $\frac{1}{3}$ is a double roots and the solution is $y = (c_1 + c_2 x) e^{x/e}$.

6. $16'' + 8y'' + y = 0$ has auxiliary equation $16m^2 + 8m + 1 = 0$. $-\frac{1}{4}$ is a double root and so the solution is $y = (c_1 + xc_2) e^{-x/4}$.

7. $4y'' + 12y' + 9y = 0$ has auxiliary equation $4m^3 + 12m + 9 = 0$. This has $-\frac{3}{2}$ as a double root and hence the solution is $y = (c_1 + c_2 x) e^{-3x/2}$.

8. $9y'' - 12y' + 4y = 0$ has $9m^2 - 12m + 4 = 0$ as its auxiliary equation. This has $\frac{2}{3}$ as a double root and hence the solution is $y = (c_1 + xc + 2) e^{2x/3}$.

9. $y'' + 4y' + 5y = 0$ has $m^2 + 4m + 5 = 0$ as its auxiliary equation. By the quadratic formula, this has $\frac{-4 \pm \sqrt{16-20}}{2} = -2 \pm j$ as its complex roots. Hence the solution is $y = e^{-2x} (c_1 \cos x + c_2 \sin x)$.

10. $y'' + 4y = 0$ has auxiliary equation $m^2 + 4 = 0$. This has complex roots $2j$ and $-2j$ so $a = 0$ and $b = 2$. The solution is $y = c_1 \cos 2x + c_2 \sin 2x$.

11. $D^2 y + 9Dy = 0$ has auxiliary equation $m^2 + 9m = 0$. This has 0 and -9 for its roots and hence the solution is $y = c_1 + c_2 e^{-9x}$.

12. $D^2 y = 0$ has auxiliary equation $m^3 = 0$. This has 0 as a triple roots so the solution is $y = c_1 + c_2 x + c_3 x^2$.

13. $D^3 y = y$ has auxiliary equation $m^3 - 1 = 0$. 1 is real solution and synthetic division yields $m^2 + m + 1$ as the depressed quotient. The other two roots are $-\frac{1}{2} \pm \frac{\sqrt{3}}{2} j$. Hence the solution is $y = c_1 e^x + e^{-x/2} \left(c_2 \cos \frac{\sqrt{3}}{2} x + c_3 \sin \frac{\sqrt{3}}{2} x \right)$.

14. $\frac{d^2 y}{dx^3} + y = 0$ has auxiliary equation $m^3 + 1 = 0$. We can determine that -1 is a real root and, using synthetic division, we get $m^2 - m + 1$ as the depressed quotient. This has complex roots $\frac{1}{2} \pm \frac{\sqrt{3}}{2} j$. Hence the solutions are $y = c_1 e^{-x} + e^{x/2} \left(c_2 \cos \frac{\sqrt{3}}{2} x + c_3 \sin \frac{\sqrt{3}}{2} x \right)$.

15. $\frac{d^3 y}{dx^3} + 8y = 0$ has auxiliary equation $m^3 + 8 = 0$. We can see that -2 is one root. Using synthetic division we get the depressed quotient m^2-2m+4 which has roots $1\pm\sqrt{3}j$. Hence the solution is $y=c_1e^{-2x}+e^x\left(c_2\cos\sqrt{3}+c_3\sin\sqrt{3}x\right)$.

16. $\frac{d^3y}{dx^3}-27y=0$ has auxiliary equation $m^3-27=0$. This equation factors into $(m-3)\left(m^2+3m+9\right)=0$ so the roots are -3, $\frac{-3}{2}\pm\frac{3\sqrt{3}}{2}j$. The solution is $y=c_1e^{-3x}+e^{-3x/2}\left(c_2\cos\frac{3\sqrt{3}}{2}x+c_3\sin\frac{3\sqrt{3}}{2}x\right)$.

17. $y'''-6y''+11y'-6y=0$ has auxiliary equation $m^3-6m^2+11m-6=0$. Using the rational root theorem and synthetic division we find the roots are 1, 2 and 3. Hence the solution is $y=c_1e^x+c_2e^{2x}+c_3e^{3x}$.

18. $D^4y-9D^2y+20y=0$ has auxiliary equation $m^4-9m^2+20=0$. This factors as $\left(m^2-4\right)\left(m^2-5\right)=0$ and the roots are 2, -2, $\sqrt{5}$ and $-\sqrt{5}$. Hence the solution is $y=c_1e^{2x}+c_2e^{-2x}+c_3e^{\sqrt{5}x}+c_4e^{-\sqrt{5}x}$.

19. $D^4y+8D^3y+24D^2y+32Dy+16y=0$ has auxiliary equation $m^4+8m^3+24m^2+32m+16=0$. This has -2 as a root four times. Hence the solution is $y=\left(c_1+c_2x+c_3x^2+c_4x^3\right)e^{-2x}$.

20. $y'''-y''+y'-y=0$ has auxiliary equation $m^3-m^2+m-1=0$. The real root is 1 and using synthetic division we get the depressed quotient m^2+1 which has complex roots $\pm j$. Hence the solution is $y=c_1e^x+c_2\cos x+c_3\sin x$.

21. $(D-1)^2(D+2)^3y=0$ has auxiliary equations with 1 as a double root and -2 as a triple root. Hence the solution is $y=(c_1+c_2x)e^x+\left(c_3+c_4x+c_5x^2\right)e^{-2x}$.

22. $(D-2)^4(D+5)y=0$ has auxiliary equation with roots 2 of multiplicity four and -5. The solution is $y=\left(c_1+c_2x+c_3x^2+c_4x^3\right)e^{2x}+c_5e^{-5x}$.

23. $(D-3)^2\left(D^2-6D-y\right)y=0$ has auxiliary equation with 3 as a double root and irrational roots $\frac{6\pm\sqrt{36+36}}{2}=3\pm3\sqrt{2}$. Hence the solution is $y=(c_1+c_2x)e^{3x}+c_3e^{(3+3\sqrt{2})x}+c_4e^{(3-3\sqrt{2})x}$.

24. $(D+1)^2\left(D^2+4D+9\right)y=0$ has auxiliary equation with -1 as a double root and complex roots $\frac{-4\pm\sqrt{16-36}}{2}=-2\sqrt{5}j$. Hence the solution is given by $y=(c_1+c_2x)e^{-x}+e^{-2x}\times\left(c_3\cos\sqrt{5}x+c_4\sin\sqrt{5}x\right)$.

25. $\left(3D^2-2D+1\right)y=0$ has $3m^2-2m+1=0$ as its auxiliary equation. This equation has complex roots $\frac{2\pm\sqrt{4-12}}{6}=\frac{1}{3}\pm\frac{\sqrt{2}}{3}j$. Hence the solution is $y=e^{x/3}\left(c_1\cos\frac{\sqrt{2}}{3}x+c_2\sin\frac{\sqrt{2}}{3}x\right)$.

26. $\frac{d^2y}{dx^2}-2\frac{dy}{dx}+3y=0$ has auxiliary equation m^2-2m+3. This has complex roots $\frac{2\pm\sqrt{4-12}}{2}=1\pm\sqrt{2}j$. Hence the solution is $y=e^x\times\left(c_1\cos\sqrt{2}+c_2\sin\sqrt{2}x\right)$.

27. $y''-2y'+5y=0$ has auxiliary equation $m^2-2m+5=0$. This has complex roots $1\pm2j$. Hence the general solution is $y=e^x\left(c_1\cos2x+c_2\sin2x\right)$. When $x=0$, y is 4 and so $c_1=4$. When $x=0$, $y'=7$ so $y'=e^x\left(c_1\cos2x+c_2\sin2x\right)+e^x\left(-2c_1\sin2x+2c_2\cos2x\right)=e^x\left[(c_1+2c_2)\times\cos2x+(c_2-2c_1)\sin2x\right]$. Hence $c_1+2c_2=7$ or $4+2c_2=7$ or $c_2=\frac{3}{2}$. The particular solution is $y=e^x\left(4\cos2x+\frac{3}{2}\sin2x\right)$.

28. $y''+8y'+16y=0$ has auxiliary equation $m^2+8m+16=0$. We see that -4 is a double root and the general solution is $y=(c_1+c_2x)e^{-4x}$. When $x=0$, $y=4$ and so $c_1=4$. $y'=c_2e^{-4x}-4(c_1+c_2x)e^{-4x}$. When $x=0$, $y'=c_2-4c_1=-6$ so $-6+16=10$. The particular solution is $y=(4+10x)e^{-4}$.

29. $\left(4D^2-12D+9\right)y=0$ has auxiliary equation with double root $\frac{3}{2}$. Hence the general solution is $y=(c_1+c_2x)e^{3x/2}$. When $x=0$, $y=1$ so $c_1=1$ or $y=(1+c_2x)e^{3x/2}$, which means that $c_2=-1$. Hence the particular solution is $y=(1-x)e^{\frac{3}{2}x}$.

30. $\left(D^2-4D+5\right)y=0$ has auxiliary equation with complex roots $\frac{4\pm\sqrt{-4}}{2}=2\pm j$. Hence the general solution is $y=e^{2x}\left(c_1\cos x+c_2\sin x\right)$. When $x=0$, $y=1$ yields $c_1=1$, so $y=e^{2x}(\cos x+c_2\sin x)$. Now when $x=\frac{\pi}{2}$, $y=3e^\pi$ so $3e^\pi=3^\pi\left(\cos\frac{\pi}{2}+c_2\sin\frac{\pi}{2}\right)$ and so $c_2=3$. Hence the particular solution is $y=e^{2x}(\cos x+3\sin x)$.

31. The differential equation $y''+2y'+y=0$ has auxiliary equation $m^2+2m+1=0$. This equation has -1 as a double root. Hence, the general solution is $y=c_1e^{-x}+c_2xe^{-x}$ and its derivative is $y'=-c_1e^{-x}-c_2e^{-x}-c_2xe$. When $x=0$, we are given $y=1$ and $y'=-3$ which leads equations $c_1=1$ and $-c_1+c_2=-3$. Hence, $c_2=-2$ and the particular solution is $y=e^{-x}-2xe^{-x}$.

32. The differential equation $y'' - 10y' + 25y = 0$ has the auxiliary equation $m^2 - 10m + 25 = 0$. This equation has -5 as a double route. Hence, the general solution is $y = c_1 e^{-5x} + c_2 x e^{-5x}$ and its derivative is $y' = -5c_1 e^{-5x} + c_2 e^{-5x} - 5c_2 x 3^{-5x}$. When $x = 0$, we are given $y = -4$ and $y' = 4$. These lead to $c_1 = -4$ and $-5c_1 + c_2 = 20 + c_2 = 4$, and so $c_2 = -16$ and the particular solution is $y = -4e^{-5x} - 16xe^{-5x}$.

33. $(D^2 - 6D + 34)y = 0$ has the auxiliary equation $m^2 - 6m + 34 = 0$. This has the complex roots $3 \pm 5j$, so the general solution is $y = e^{3x}(c_1 \cos 5x + c_2 \cos 5x)$ and $y' = 3e^{3x}(c_1 \cos 5x + c_2 \sin 5x) + e^{3x}(-5c_1 \sin 5x + 5c_2 \cos 5x)$.

When $x = 0$, we are given $y = 0.5$ and $y' = 1.5$ and so, $c_1 = 4$ and $3c_1 + 5c_2 = -3$. Hence, $c_2 = -3$ and the particular solution is $y = e^{3x}(4\cos 5x - 3\sin 5x)$.

34. $(D^2 + 10D + 29)y = 0$ has the auxiliary equation $m^2 + 10m + 29 = 0$. This has the complex roots $-5 \pm 2j$, so the general solution is $y = e^{-5x}(c_1 \cos 2x + c_2 \sin 2x)$ and $y' = -5e^{3x} \times (c_1 \cos 2x + c_2 \sin 2x) + e^{-5x}(2c_1 \sin 2x + 2c_2 \cos 2x)$. When $x = 0$, we are given $y = 0.5$ and $y' = 1.5$ and so, $c_1 = 0.5$ and $-5c_1 + 2c_2 = 1.5$. Hence, $c_2 = 2$ and the particular solution is $y = e^{-5x}(0.5 \cos 2x + 2 \sin 2x)$.

≡ 33.3 SOLUTIONS OF NONHOMOGENEOUS EQUATIONS

1. $(D^2 - 10D + 25)y = 4$. To find y_c we solve the auxiliary equation $m^2 - 10m + 25 = 0$. This has double root 5. Hence $y_c = (c_1 + xc_2)e^{5x}$. For y_p, we use Case 1 and $y_p = A_0$, $y_p' = 0$. Substituting into the original equation we get $25A_0 = 4$ or $A_0 = \frac{4}{25}$. Hence the solution is $y = (c_1 + c_2 x)^{5x} + \frac{4}{25}$.

2. $(D^2 - 10D + 25y)y = 4x^2$. Using the auxiliary equation $m^2 - 10m + 25 = 0$, we get a double root 5. Hence $y_c = (c_1 + c_2 x)x^{5x}$. To find y_p we use Case 1 and set $y_p = A_2 x^2 + A_1 x + A_0$, $y' = 2A_2 x + A_1$, and $y'' = 2A_2$. Substituting into the original equation we get $2A_2 - 10(2A_2 + A_1) + 25(A_2 x^2 + A_1 x + A_0) = 4x^2$ or $25A_2 x^2 + (25A_1 - 20A_2)x + 25A_0 - 10A_1 + 2A_2 = 4x^2$. Hence $A_2 = \frac{4}{25}$, $A_1 = \frac{20A_2}{25} = \frac{80}{625} = \frac{16}{125}$, and $A_0 = \frac{10A_1 - 2A_2}{25} = \frac{\frac{160}{125} - \frac{8}{25}}{25} = \frac{24}{625}$. The solution is $= (c_1 + c_2 x)e^{5x} + \frac{4}{25}x^2 + \frac{16}{125} + \frac{24}{625}$

3. $(D^2 - 10D + 25)y = e^{3x}$ has the same complentary solution as problems 1 and 2, namely $y_c = (c_1 + c_2 x)e^{5x}$. To find the particular solution y_p, we use Case 2 with $y_p = A_0 e^{3x}$, $y' = 3A_0 e^{3x}$, and $y'' = 9A_0 e^{3x}$. Substituting into the original equation we get $9A_0 e^{3x} - 10 \cdot 3A_0 e^{3x} + 25A_0 e^{3x} = 3^{3x}$ or $4A_0 e^{3x} = e^{3x}$. Hence $A_0 = \frac{1}{4}$. The solution is $y = (c_1 + c_2 x)e^{5x} + \frac{1}{4}e^{3x}$.

4. $(D^2 - 10D + 25)y = 2xe^{3x}$. Again $y_c = (c_1 + c_2 x)e^{5x}$. To find y_p we use Case 2 with $y_p = e^{3x}(A_1 x + A_0)$, $y_p' = 3e^{3x}(A_1 x + A_0) + A_1 e^{3x} =$

$3A_1 xe^{3x} + (A_1 + 3A_0)e^{3x}$, and $y_p'' = 0A_1 xe^{3x} + 3A_1 e^{3x} + 3(A_1 + 3A_0)e^{3x}$. Substituting into the original equation we get $9A_1 xe^{3x} + (6A_1 + 3A_0)e^{3x} - 30A_1 xe^{3x} - (10A_1 + 30A_0)e^{3x} + 25(A_1 x + A_0) \times e^{3x} = 2xe^{3x}$. Hence $9A_1 - 30A_1 + 25A_1 = 2$ and $6A_1 + 9A_0 - 10A_1 - 30A_0 + 25A_0 = 0$. So $A_1 = \frac{1}{2}$ and $4A_0 = 4A_1 = 2$, so $A_0 = \frac{1}{2}$. The solution is $y = (c_1 + c_2 x)e^{5x} + (\frac{1}{2}x + \frac{1}{2})e^{3x}$.

5. $(D^2 - 10D + 25)y = 10 + e^x$. Again $y_c = (c_1 + c_2 x)e^{5x}$. The y_p has two parts. Case 1 $y_{p1} = A_0$ and we get $25A_0 = 10$ or $A_0 = \frac{2}{5}$. Case 2, $y_{p2} = B_0 e^x$, $y_{p2}' = B_0 e^x$, $y_{p2}'' = B_0 e^x$. Substituting we get $16B_0 e^x = e^x$ or $B_0 = \frac{1}{16}$. The solution is $y = (c_1 + c_2 x)e^{5x} + \frac{2}{5} + \frac{1}{16}e^x$.

6. $(D^2 - 10D + 25)y = 3\sin 2x$. This is Case 3 so we set $y_p = A\sin 2x + B\cos 2x$, $y_p' = 2A\cos 2x - 2B\sin 2x$, $y_p'' = -4A\sin 2x - 4B\cos 2x$. Substituting into the original equation and collecting like terms we get $(-4A + 20B + 25A)\sin 2x + (-4B - 20A + 25B)\cos 2x = 3\sin 2x$. Hence $21A + 20B = 3$ and $-20A + 21B = 0$. Eliminating the B term we get $841A = 63$ or $A = \frac{63}{841}$. Also, $B = \frac{20A}{21} = \frac{60}{841}$. The solution is $y = (C_1 + C_2 x)e^{5x} + \frac{63}{841}\sin 2x + \frac{60}{841}\cos 2x$.

7. $(D^2 - 10D + 25)y = 29\sin 2x$. Again $y_c = (C_1 + C_2 x)e^{5x}$. Using Case 3 we set $y_p = A\sin 2x + B\cos 2x$, and so $y_p' = 2A\cos 2x - 2B\sin 2x$, and $y_p'' = -4A\sin 2x - 4B\cos 2x$. Substituting into the original equation we get $(-4A + 20B + $

$25A)\sin 2x + (-4B - 20A + 25B)\cos 2x = 29\sin 2x$. Hence $21A + 20B = 29$ and $-20A + 21B = 0$. Eliminating B we get $(441 + 440)A = 29 \cdot 21$ or $A = \frac{609}{841} = \frac{21}{29}$. Also, $B = \frac{20A}{21} = \frac{20}{21} \cdot \frac{21}{29} = \frac{20}{29}$. The solution is $y = (C_1 + C_2 x)e^{5x} + \frac{21}{29}\sin 2x + \frac{20}{29}\cos 2x$.

8. $(D^2 - 9)y = 3\cos x$ has complementary solution $y_c = c_1 e^{3x} + c_2 e^{-3x}$. To find y_p we use Case 4 with $y_p = A\sin x + B\cos x$, $y_p' = A\cos x - B\sin x$ and $y_p'' = -A\sin x - B\cos x$. Substituting into the original equation we get $-10A\sin x - 10B\cos x = 3\cos x$. Solving we get $A = 0$ and $B = -\frac{3}{10}$. The solution is $y = c_1 e^{3x} + c_2 e^{-3x} - \frac{3}{10}\cos x$.

9. $(D^2 - 4) = x^2 e^x - 3x$ has complementary auxiliary equation with roots ± 2. Hence $y_c = c_1 e^{3x} + c_2 e^{-2x}$. y_p will have 2 parts. $x^2 e^x$ is Case 2 so we set $y_{p1} = e^x(A_2 x^2 + A_1 x + A_0)$, with $y_{p1}' = e^x(A_2 x^2 + A_1 x + A_0) + e^x(2A_2 x + A_1) = A_2 e^x x^2 + (A_1 + 2A_2)e^x x + (A_0 + A_1)e^x$ and $y_{p1}'' = A_2 e^x x^2 + 2x A_2 e^x + (A_1 + 2A_2)e^x x + (A_1 + 2A_2)e^x + (A_0 + A_1)e^x = A_2 e^x x^2 + (2A_2 + A_1 + 2A_2)e^x x + (2A_2 + 2A_1 + A_0)e^x$. Substititing into the original equation yields $(A_2 - 4A_2)e^x x^2 + (4A_2 + A_1 - 4A_1)e^x x + (2A_2 + 2A_1 + A_0 - 4A_0)e^x = x^2 e^x$. Hence $-3A_2 = 1$ or $A_2 = -\frac{1}{3}$. Next, $-\frac{4}{3} - 3A_1 = 0$ or $A_1 = -\frac{4}{9}$ and finally $3A_0 = 2A_2 + 2A_1$ and so $A_1 = \dfrac{-\frac{2}{3} - \frac{8}{9}}{3} = \frac{-14}{27}$. For the second part of y_p, we have $y_{p2} = B_1 x + B_0$, $y_{p2}' = B$ and $y_{p2}'' = 0$. Hence $-4B_1 x = -3x$ and $B_1 = \frac{3}{4}$ $B_0 = 0$. The solution is $y = c_1 e^{2x} + c_2 e^{-2x} - \left(\frac{1}{3}x^2 + \frac{4}{9} + \frac{14}{29}\right)e^x + \frac{3}{4}x$.

10. $(D^2 + r)y = 6x + 3$. The auxiliary equation for y_c has complex roots $\pm 2j$. Hence $y_c = c_1\cos 2x + c_2\sin 2x$. For y_p we use Case 1 with $y_p = A_1 x + A_0$, $y_p' = A_1$, and $y_p'' = 0$. Substituting into the original equation we get $4(A_1 x + A_0) = 6x + 3$. So $4A_1 = 6$ gives $A_1 = \frac{3}{2}$ and $4A_0 = 3$ gives $A_0 = \frac{3}{4}$. The solution is $y = c_1\cos 2x + c_2\sin 2x + \frac{3}{2}x + \frac{3}{4}$.

11. $(D^2 + 9)y = 4\cos x + 2\sin x$. y_c has auxiliary equation with roots $\pm 3j$. Hence $y_c = c_1\cos 3x - c_2\sin 3x$. y_p fits Cases 3 and 4 so we set $y_p = A\cos x + B\sin x$, $y_p' = -A\sin x + B\cos x$ and $y_p'' = -A\cos x - B\sin x$. Substituting into the

original equation we get $(-A\cos x - B\sin x) + 9(A\cos x + B\sin x) = 4\cos x + 2\sin x$ so $8A = 4$ and $8B = 2$ giving $A = \frac{1}{2}$ and $B = \frac{1}{4}$. The solution is $y = c_1\cos 3x + c_2\sin 3x + \frac{1}{2}\cos x + \frac{1}{4}\sin x$.

12. $y'' - 5y' + 6y = 9x + 2e^x$. y_c has auxiliary equation with roots 2 and 3. Hence $y_c = c_1 e^{2x} + c_2 e^{3x}$. y_p fits Case 1 and 2. We set $y_{p1} = A_1 x + A_0$ with $y_{p1}' = A_1$, and $y_{p1}'' = 0$. Substituting we get $-5(A_1) + 6(A_1 x + A_0) = 9x$. Hence $6A_1 = 9$, and so $A_1 = \frac{3}{2}$ and $6A_0 - 5A_1 = 0$, which means that $A_0 = \frac{5A_1}{6} = \frac{5}{4}$. Now we set $y_{p2} = Be^x$ and $y_{p2}' = y_{p2}'' = Be^x$. Substituting into the original equation and collecting like terms, we obtain $2Be^x = 2e^x$ so $B = 1$. Putting these all together we get $y = c_1 e^{2x} + c_2 e^{3x} + \frac{3}{2}x + \frac{5}{4} + e^2$.

13. $y'' + y = 2e^{3x}$. y_c has auxiliary equation with roots $\pm j$. Hence $y_c = c_1\cos x + c_2\sin x$, y_p fits Case 2 so we set $y_p = Ae^{3x}$, $y_p' = 3Ae^{3x}$, and $y_p'' = 9Ae^{3x}$. Substituting we get $9Ae^{3x} + Ae^{3x} = 2e^{3x}$ or $10A = 2 \Rightarrow A = \frac{1}{5}$. The solution is $y = c_1\cos x + c_2\sin x + \frac{1}{5}e^{3x}$.

14. $(D^2 + 2D + 1)y = 4\sin 2x$. The auxiliary equation $m^2 + 2m + 1 = 0$ has -1 for a double roots. Hence $y_c = (c_1 + c_2 x)e^{-x}$. Now, y_p fits Case 3, so we set $y_p = A\sin 2x + B\cos 2x$, with $y_p' = 2A\cos 2x - 2B\sin 2x$, and $y_p'' = -4A\sin 2x - 4B\cos 2x$. Substituting we get $(-4A\sin 2x - 4B\cos 2x) + 2 \times (2A\cos 2x - 2B\sin 2x) + (A\sin 2x + B\cos 2x) = 4\sin 2x$. Hence $-4A - 4B + A = 4$ and $-4B + 4A + B = 0$. These simplify to $-3A - 4B = 4$ and $A = -3B = 0$ with $A = -\frac{12}{25}$ and $B = -\frac{16}{25}$. The solution is therefore $y = (c_1 - c_2 x)e^{-x} - \frac{12}{25}\sin 2x - \frac{16}{25}\cos 2x$.

15. $y'' - 4y = 8x^2$. First, $y_c = c_1 e^{2x} + c_2 e^{-2x}$. Now set $y_p = A_2 x^2 + A_1 x + A_0$, $y_p' = 2A_2 x + A_1$, and $y_p'' = 2A_2$. Substituting we get $2A_2 - 4(A_2 x^2 + A_1 + A_0) = 8x^2$. Hence $-4A_2 = 8$ and $A_2 = -2$. Also, $A_1 = 0$. Finally, $2A_2 - 4A_0 = 0$, so $A_0 = \frac{1}{2}A_2 = -1$. The solution is $y = c_1 e^{2x} + c_2 e^{-2x} - 2x^2 - 1$.

16. $y'' + 4y' + 5y = e^{-x} + 10x$. The auxiliary equation for y_c is $m^2 + 4m + 5$ and has complex roots $-2 \pm j$. Hence $y_c = e^{-2x}(c_1\cos x + c_2\sin x)$. y_p fits Cases 1 and 2. Set $y_{p1} = -Ae^{-x}$, $y_{p1}' = -Ae^{-x}$, and

$y''_{p1} = Ae^{-x}$. Substituting we get $Ae^{-x} - 4Ae^{-x} + 5Ae^{-x} = e^{-x}$ or $2A = 1$ and so $A = \frac{1}{2}$. Thus, $y_{p1} = \frac{1}{2}e^{-x}$. Now set $y_{p2} = B_1 x + B_0$, with $y'_{p2} = B_1$ and $y''_{p2} = 0$. Substituting we get $4B_1 + 5(B_1 x + B_0) = 10x$. Hence $5B_1 = 10 \Rightarrow B_1 = 2$ and $4B_1 + 5B_0 = 0$ and so $B_0 = \frac{4}{5}B_1 = -\frac{8}{5}$. Hence $y_{p2} = 2x - \frac{8}{5}$. Putting y_c, y_{p1}, and y_{p2} together we get $y = e^{-2x}(c_1 \cos x + c_2 \sin x) + \frac{1}{2}e^{-x} + 2x - \frac{8}{5}$.

17. $D^2 y + y = 4 + \cos 2x$. $y_c = c_1 \cos x + c_2 \sin x$. Set $y_{p1} = A$ and we get $A = 4$ so $y_{p1} = 4$. Now set $y_{p2} = A \sin 2x + B \cos 2x$, $y'_{p2} = 2A \cos 2x - 2B \sin 2x$, and $y''_{p2} = -4A \sin x - 4B \cos 2x$. Hence, $-4A + A = 0$ and $-4B + B = 1$. Thus, we see that $A = 0$ and $B = -\frac{1}{3}$. The solution is $y = c_1 \cos x + c_2 \sin x - \frac{1}{3} \cos 2x + 4$.

18. $3D^2 y + 2Dy - y = 4 + 2x + 6x^2$. The auxiliary equation for y_c is $3m^2 + 2m - 1 = 0$ which factors into $(3m - 1)(m + 1) = 0$ so it has roots $\frac{1}{3}$ and -1. Hence $y_c = c_1 e^{-x} + c_2 e^{x/3}$. Now we set $y_p = Ax^2 + Bx + C$, with $y'_p = 2Ax + B$, and $y''_p = 2A$. Substituting we get $3 \cdot 2A + 2(2Ax + B) - (Ax^2 + Bx + C) = 4 + 2x + 6x^2$. Collecting like terms we get $-Ax^2 + (4A - B)x + (6A + 2B - C) = 4 + 2x + 6x^2$. Hence $-A = 6$, or $A = -6$. We also have $4A - B = 2$ or $B = 4A - 2 = -26$ and $6A + 2B - C = 4$ or $C = 6A + 2B - 4 = -36 - 52 - 4 = -92$. So $y_p = -6x^2 - 26x - 92$. The solution is $y = y_c + y_p = c_1 e^{-x} + c - 2e^{x/3} - 6x^2 - 26x - 92$.

19. $3D^2 y + 2Dy + y = 4 + 2x + 6x^2$. Here the auxiliary equation is $3m^2 + 2m + 1 = 0$ has complex roots $\frac{-1}{3} \pm \frac{\sqrt{2}}{3} j$ so $y_c = e^{-x/3}\left(c_1 \cos \frac{\sqrt{2}}{3}x + c_2 \sin \frac{\sqrt{2}}{3}x\right)$. As in problem 18 set $y_p = Ax^2 + Bx + C$ with $y'_p = 2Ax + B$ and $y''_p = 2A$. Now substituting, we get $3 \cdot 2A = 2(2A + B) + (Ax^2 + Bx + C) = 4 + 2x + 6x^2$. Collecting like terms we obtain $Ax^2 + (4A + B)x + (6A + 2B + C) = 4 + 2x + 6x^2$. So $A = 6$, $4A + B = 2$, and so $B = 2 - 4A = 2 - 24 = -22$. We also obtain $6A + 2B + C = 4$ or $C = 4 - 6A - 2B = 4 - 36 + 44 = 12$. The solution is $y = e^{-x/3}\left(c_1 \cos \frac{\sqrt{2}}{3}x + c_2 \sin \frac{\sqrt{2}}{3}x\right) + 6x^2 - 22x + 12$.

20. $y'' + 9y' - y = x^2 + 6e^{2x}$. The auxiliary equation is $m^2 + 9m - 1 = 0$ and the quadratic formula yields the roots $\frac{-9 \pm \sqrt{85}}{2}$ which are real. Hence

$y_c = c_1 e^{\frac{-9+\sqrt{85}}{2}x} + c_2 e^{\frac{-9-\sqrt{85}}{2}x}$. Now set $y_{p1} = Ax^2 + Bx + C$, $y'_{p1} = 2Ax + B$ and $y''_{p2} = 2A$. Substituting we get $2A + 9(2Ax + B) - (Ax^2 + Bx + C) = x^2$. Collecting like terms $-Ax^2 + (18A - B)x + (2A + 9B - C) = x^2$. Hence $-A = 1$ or $A = -1$, $18A - B = 0$, and so $B = -18$ and $2A + 9B - C = 0$, which means that $C = -2 - 162 = -164$. So $y_{p1} = -x^2 - 18x - 164$. Now set $y_{p2} = Ae^{2x}$, $y'_{p2} = 2Ae^{2x}$, $y''_{p2} = 4Ae^{2x}$. Substituting we have $(4A + 9 \cdot 2A - A)e^{2x} = 6x^{2x}$ or $21A = 6$, or $A = \frac{6}{21} = \frac{2}{7}$. So $y_{p2} = \frac{2}{7}e^{2x}$. The answer is $y' = c_1 e^{\frac{-9+\sqrt{85}}{2}x} + c_2 e^{\frac{-9-\sqrt{85}}{2}x} + \frac{2}{7}e^{2x} - x^2 - 18x - 164$.

21. $y'' - 2y' + x = x^2 - 1$. The auxiliary equation $m^2 - 2m + 1 = 0$ has 1 as a double root. Hence $y_c = (c_1 + c_2 x)e^x$. Now set $y_p = Ax^2 + Bx + C$, $y'_p = 2Ax + B$, and $y''_p = 2A$. Substituting we get $2A - 2(2Ax + B) + Ax^2 + Bx + C = x^2 - 1$. Collecting like terms we have $Ax^2 + (-4A + B)x + (2A - 2B + C) = x^2 - 1$. Hence, $A = 1$, $-4A + B = 0$, and $2A - 2B + C = -1$. $B = 4A = 4$, and $C = -1 + 2B - 2A = -1 + 8 - 2 = 5$. Thus, $y_p = x^2 + 4x + 5$. Now to find c_1 and c_2 of y_c when $x = 0$, $y = (c_1 + 0)1 + 5 = 7$ and so $c_1 = 2$. $y' = (2 + c_2 x)e^x + c_2 e^x + 2x + 4$ and when $x = 0$, $y' = 2 + c_2 + 4 = 15$, or $c_2 = 9$. Hence the particular solution is $y = (2 + 9x)e^x + x^2 + 4x + 5$.

22. $y'' - 2y' + y = 10$. Here $y_c = (c_1 + c_2 x)e^x$. Also, $y_p = A$, $y'_p = 0$, so $A = 10$. Thus, $y = (c_1 + c_2 x)e^x + 10$. When $x = 0$, $y = c_1 + 10 = 20$, so $c_1 = 10$. Also $y' = (10 + c_2 x)e^x + c_2 e^x$. When $x = 0$, $y' = 10 + c_2 = 5$, and so $c_2 = -5$. The particular solution is $y = (10 - 5x)e^x + 10$.

23. $y'' - y' - 2y = \sin 2x$. $y_c = c_1 e^{2x} + c_2 e^{-x}$. Now set $y_p = A \sin 2x + B \cos 2x$, with $y'_p = 2A \cos 2x - 2B \sin 2x$ and $y''_p = -4A \sin 2x - 4B \cos 2x$. Substituting we obtain the equation $(-4A \sin 2x - 4B \cos 2x) - (2A \cos 2x - 2B \sin 2x) - 2(A \sin 2x + B \cos 2x) = \sin 2x$. This yields $-4A + 2B - 2A = 1$ or $-6A + 2B = 1$ and $-4B - 2A - 2B = 0$ or $-2A - 6B = 0$. Solving we get $A = -\frac{3}{20}$, $B = \frac{1}{20}$. Hence $y = c_1 e^{2x} + c_2 e^{-x} - \frac{3}{20} \sin 2x + \frac{1}{20} \cos 2x$. When $x = 0$, $y = c_1 + c_2 + \frac{1}{20} = 1$, or $c_1 + c_2 = \frac{19}{20}$. Differentiating, we obtain $y' = 2c_1 e^{2x} - c_2 e^{-x} - \frac{6}{20} \cos 2x - \frac{1}{10} \sin 2x$, and when $x = 0$, we know that $y' = 2c_1 - c_2 - \frac{6}{20} = \frac{7}{4}$ or $2c_1 - c_2 = \frac{41}{20}$. So, $3c_1 = \frac{60}{20}$ and $c_1 = 1$; also

$c_2 = \frac{19}{20} - 1 = -\frac{1}{20}$. The particular solution is

$y = e^{2x} + \frac{1}{20}e^{-x} - \frac{3}{20}\sin 2x + \frac{1}{20}\cos 2x$.

24. $y'' - y' - 2y = e^{3x}$, $y_c = c_1 e^{2x} + c_2 e^{-x}$. Now set $y_p = Ae^{3x}$, with $y_p' = 3Ae^{3x}$, and $y_p'' = 9Ae^{3x}$. Substituting we get $9Ae^{3x} - 3Ae^{3x} - 2 \cdot AE^{3x} - e^{3x}$ so $9A - 3A - 2A = 1$ and $A = \frac{1}{4}$. Hence the general solution is $y = c_1 e^{2x} + c_2 e^{-x} + \frac{1}{4}e^{3x}$.

When $x = 0$, $y = c_1 + c_2 + \frac{1}{4} = 2$ and so, $c_1 + c_2 = \frac{7}{4}$. Differentiating, we obtain $y' = 2c_1 e^{2x} - c_2 e^{-x} + \frac{3}{4}e^{3x}$. When $x = 0$, $y' = 2c_1 - c_2 + \frac{3}{4} = 11$, so $2c_1 - c_2 = \frac{41}{4}$. This gives $3c_1 = \frac{48}{4} = 12$ or $c_1 = 4$. Thus, $c_2 = \frac{7}{4} - 4 = \frac{-9}{4}$. The particular solution is $y = 4e^{2x} - \frac{9}{4}e^{-x} + \frac{1}{4}e^{3x}$.

25. The differential equation $y'' + 2y' + y = 3x^2 + 2x - 1$ has the auxiliary equation $m^2 + 2m + 1 = 0$. This has the double root -1, so $y_c = c_1 e^{-x} + c_2 x e^{-x} = (c_1 + c_2 x)e^{-x}$. This is Case 1, so $y_p = A_2 x^2 + A_1 x + A_0$, with $y_p' = 2A_2 x + A_1$, and $y_p'' = 2A_2$. Substituting, we get
$$2A_2 + 2(2A_2 x + A_1) + A_2 x^2 + A_1 x + A_0 = 3x^2 + 2x - 1$$
Eliminating parentheses and collecting like terms produces
$$A_0 x^2 + (4A_1 + A_1)x + 2A_2 + 2A_1 + A_0 = 3x^2 + 2x - 1$$
Hence, $A_2 = 3$, $4A_1 + A_1 = 2$ or $4(3) + A_1 = 2$ and $A_1 = -10$, and $2A_2 + 2A_1 + A_0 = -1$ or

$2(3) + 2(-10) + A_0 = -1$ and $A_0 = 13$. This gives the general solution
$$y = (c_1 + c_2 x)e^{-x} + 3x^2 - 10x + 13$$
and its derivative is
$$y' = -(c_1 + c_2 x)e^{-x} + c_2 e^{-x} + 6x - 10$$
We are told that when $x = 0$, then $y = 3$ and $y' = 5$. Substituting these in the general solution and its derivative produce $c_1 + 13 = 3$ or $c_1 = -10$. Next, $10 + c_2 - 10 = 5$ or $c_2 = 5$. The particular solution is $y = (-10 + 5x)e^{-x} + 3x^2 - 10x + 13$ or $y = -10e^{-x} + 5xe^{-x} + 3x^2 - 10x + 13$.

26. The differential equation $y'' - 5y' - 6y = e^{3x}$ has the auxiliary equation $m^2 - 5m - 6 = 0$. This has the roots 6 and -1, so $y_c = c_1 e^{6x} + c_2 e^{-x}$. This is Case 2, so $y_p = A_0 e^{3x}$, with $y_p' = 3A_0 e^{3x}$, and $y_p'' = 9A_0 e^{3x}$. Substituting, we obtain the equation
$$9A_0 e^{3x} - 15A_0 e^{3x} - 6A_0 e^{3x} = e^{3x}$$
or $9A_0 - 15A_0 - 6A_0 = 1$ and $A_0 = -\frac{1}{12}$. Hence, we obtain $y = c_1 e^{6x} + c_2 e^{-x} - \frac{1}{12}e^{3x}$ and its derivative $y' = 6c_1 e^{6x} - c_2 e^{-x} - \frac{1}{4}e^{3x}$. When $x = 0$, we are given $y = 1$ and $y' = 1$. Substituting these in the general solution and its derivative produce the equations $c_1 + c_2 - \frac{1}{12} = 2$ and $6c_1 - c_2 - \frac{1}{4} = 1$. Adding these equations leads to $7c_1 - \frac{1}{3} = 3$ or $7c_1 = 3\frac{1}{3} = \frac{10}{3}$ and so $c_1 = \frac{10}{21}$. Back-substituting into the first equation yields $\frac{10}{21} + c_2 - \frac{1}{12} = 2$ or $c_2 = \frac{45}{28}$. Hence, the particular solution is
$$y = \frac{10}{21}e^{6x} + \frac{45}{28}e^{-x} - \frac{1}{12}e^{3x}.$$

27. The differential equation $y'' + 9y = 8\cos x$ has the auxiliary equation $m^2 + 9 = 0$ with roots $\pm 3j$. Hence, $y_c = c_1 \cos 3x + c_2 \sin 3x$. This is a Case 4 equation, so $y_p = A_0 \sin x + B_0 \cos x$, with $y_p' = A_0 \cos x - B_0 \sin x$ and $y_p'' = -A_0 \sin x - B_0 \cos x$. Substituting, we obtain the equation

$$-A_0 \sin x - B_0 \cos x + (A_0 \sin x + B_0 \cos x) = 8\cos x$$

Hence, $8A_0 = 0$ and $A_0 = 0$ as well as $8B_0 = 8$ and $B_0 = 1$. This leads to

$$y = y_c + y_p = c_1 \cos 3x + c_2 \sin 3x + \cos x$$
$$y' = -3c_1 \sin 3x + 3c_2 \cos 3x - \sin x$$

When $x = \frac{\pi}{2}$, we are given $y = -1$, so $-c_2 = -1$ and $c_2 = 1$. We are also told that when $x = \frac{\pi}{2}$, then $y' = 1$ and so $-3c_1 - 1 = 1$ and $c_1 = \frac{2}{3}$. Putting it all together, we see that the particular solution is $y = \cos x + \frac{2}{3}\cos 3x + \sin 3x$.

28. $y'' - 5y' + 6y = e^x(2x - 3)$ has the auxiliary equation $m^2 - 5m + 6 = 0$ with roots 2 and 3. Hence $y_c = c_1 e^{2x} + c_2 e^{3x}$. This is a Case 2 problem, so $y_p = e^x(A_1 x + A_0)$, with $y_p' = e^x(A_1 x + A_0) + A_1 e^x = e^x(A_1 x + A_1 + A_0)$ and $y_p'' = e^x(A_1 x + A_1 + A_0) + A_1 e^x = e^x(A_1 x + 2A_1 + A_0)$. Substituting, we obtain

$$e^x\left(A_1x + 2A_1 + A_0\right) - 5e^x\left(A_1x + A_1 + A_0\right) + 6e^x\left(A_1x + A_0\right) = e^x(2x - 3)$$

Collecting like terms and dividing by e^x leads to

$$\left(A_1 - 5A_1 + 6A_1\right)x + \left[(2A_1 + A_0) + -5(A_1 + A_0) + 6A_0\right] = (2x - 3)$$

Hence, $2A_1 = 2$ and $A_1 = 1$ and $-3A_1 + 2A_0 = -3$ and $A_2 = 0$ giving $y_p = xe^x$.

$$y = y_c + y + p = c_1 e^{2x} + c_2 e^{3x} + xe^x$$

≡ 33.4 APPLICATIONS

1. Since $F = kx$, we have $k = \frac{F}{x} = \frac{20\text{lb}}{5\text{in.}} = 4\text{lb}/\text{in.} = 48$ lb/ft.

2. Since $F = kx$, we have $k = \frac{F}{x} = \frac{15}{3}\frac{\text{lb}}{\text{in.}} = 5\text{lb}/\text{in.} = 60$ lb/ft.

3. Here $F = mg$ and $F = kx$ or $k = \frac{F}{x} = \frac{mg}{x} = \frac{5 \times 98}{0.2}\frac{\text{N}}{\text{m}} = 245$ N/m.

4. $k = \frac{mg}{x} = \frac{8 \times 98}{0.4} = 176$ N/m

5. We know that $m = \frac{20}{32} = 0.625$ Slugs, $\ell = 2$ and $k = 48$, $2b = \frac{\ell}{m} = 3.2$, and $\omega^2 = \frac{k}{m} = \frac{48}{0.625} = 76.8$. So the auxiliary equation is $m^2 + 3.2m + 76.8 = 0$. This has solutions $-1.6 \pm \sqrt{1.6^2 - 76.8} = -1.6 \pm \sqrt{75.2}j$. Hence the general solution is $x = e^{-1.6t}(c_1 \cos \sqrt{75.2}t + c_2 \times \sin \sqrt{75.2}t)$. Since $x(0) = 0$, $c_1 = 0$, and $x'(t) = -1.6c_2e^{-1.6t}\sin \sqrt{75.2}t + c_2\sqrt{75.2}e^{-1.6t}\cos \sqrt{75.2}t$ and since $x'(0) = c_2\sqrt{75.2} = 4$, then $c_2 = \frac{4}{\sqrt{75.2}}$. So, $x = \frac{4}{\sqrt{75.2}}e^{-1.6t}\sin \sqrt{75.2}t \approx 0.4613e^{-1.6t}\sin 8.6178t$.

6. $m = \frac{F}{g} = \frac{15}{32} = 0.46875$ slugs, $\ell = 1$, $k = 60$, so $2b = \frac{\ell}{m} = \frac{32}{15}$ and $\omega^2 = \frac{k}{m} = 128$. The auxiliary equation is $m^2 + \frac{32}{15}m + 128 = 0$ has roots $-\frac{16}{15} \pm \sqrt{\frac{16^2}{15} - 128} = \frac{-16}{15} \pm \sqrt{\frac{1094}{15}}j$. Hence, the general solution is $x = e^{-16t/15}(c_1 \cos \sqrt{\frac{1904}{15}}t + c_2 \sin \sqrt{\frac{1904}{15}}t)$. Since $x(0) = 0$, we find that $c_1 = 0$, and since $x'(t) = -\frac{16}{15}e^{-16t/15}c_2\sin \sqrt{\frac{1904}{15}}t + c_2\sqrt{\frac{1904}{15}}e^{-16/15t}\cos \sqrt{\frac{1904}{15}}t$ and $x'(0) = c_2\sqrt{\frac{1904}{15}} = 2$, we determine that $c_2\frac{2\sqrt{15}}{\sqrt{1904}}$. Thus, we have

found that the particular solution is $x = \frac{2\sqrt{15}}{\sqrt{1904}} \times e^{-16t/15}\sin\sqrt{\frac{1904}{15}}t \approx 0.1775e^{-1.0667t}\sin 11.2665t$.

7. $m = 5$, $\ell = 40$, and $k = 245$, so $2b = 8$ and $\omega^2 = 49$. The auxiliary equation is $m^2 + 8m + 49 = 0$ has solutions $-4 \pm \sqrt{16 - 49} = -4 \pm \sqrt{33}j$ and the general solution is $x = e^{-4t}(c_1 \cos \sqrt{33}t + c_2 \sin \sqrt{35}t)$. Since $x(0) = 0$, $c_1 = 0$, and we have $x'(t) = -4e^{-4t}c_2 \sin \sqrt{33}t + c_2\sqrt{33}e^{-4t} \times \cos \sqrt{33}t$. Also, $x'(0) = c_2\sqrt{33} = 2$, so $c_2 = \frac{2}{\sqrt{33}}$. The particular solution is $x = \frac{2}{\sqrt{33}}e^{-4t} \times \sin \sqrt{33}t \approx 0.3482e^{-4t}\sin 5.7446t$.

8. $m = 8$, $\ell = 80$, and $k = 196$, so $2b = 10$ and $\omega^2 = 24.5$. The auxiliary equation is $m^2 + 10m + 24.5 = 0$ has solutions $-5 \pm \sqrt{25 - 24.5} = -5 \pm \sqrt{0.5}$. The general solution is $x = c_1e^{(-5+\sqrt{0.5}t)} + c_2e^{(-5-\sqrt{0.5}t)}$. Since $x(0) = 0$, then $c_1 + c_2 = 0$. Also, $x'(t) = (-5 + \sqrt{0.5})c_1e^{(-5+\sqrt{0.5})t} - (5 + \sqrt{0.5})c_2e^{(-5-\sqrt{0.5})t}$. It is given that $x'(0) = 3$, and so $(-5 + \sqrt{0.5})c_1 - (5 + \sqrt{0.5})c_2 = 3$, or, after substituting $-c_1$ for c_2, we have $(-5 + \sqrt{0.5})c_1 + (5 + \sqrt{0.5})c_1 = 3$. Hence, $2\sqrt{0.5}c_1 = 3$, and so $c_1 = \frac{3}{2\sqrt{0.5}}$, and $c_2 = -\frac{3}{2\sqrt{0.5}}$. The particular solution is $x = \frac{3}{2\sqrt{0.5}}e^{(-5+\sqrt{0.5})t} - \frac{3}{2\sqrt{0.5}}e^{(-5-\sqrt{0.5})t}$ or $x = 2.1213e^{-4.2929t} - 2.1213e^{-5.7071t}$.

9. $E(t) = L\frac{d^2q}{dt^2} + R\frac{dq}{dt} + \frac{q}{c}$ so $50 \sin 100t = 0.1\frac{d^2q}{dt^2} + 6\frac{dq}{dt} + 100q$. q_c has auxiliary equation $0.1m^2 + 6m + 100 = 0$ with roots $\frac{-6 \pm \sqrt{36 - 40}}{0.2} = -30 \pm 10j$. The general complementary solution is $q_c = e^{-30t}(c_1 \cos 10t + c_2 \sin 10t)$. For q_p we get $q_p = A\sin 100t + B\cos 100t$, $q_p' = 100A\cos 100t - 100B \sin 100t$, and $q_p'' = -10^4A\sin 100t - 10^4 \cos 100t$. Substituting we get $\left(-10^3 A \sin 100t - 10^3 \cos 100t\right) + 6(100A\cos 100t - 100B\sin 100t) + 100(A\sin 100t + B\cos 100t) = 50\sin 100t$ so $-1000A - 600B + 100A = 50$ and $-1000B + 600A + 100B = 0$; $-900A - 600B = 50$, so

$-18A - 12B = 1$. Also, $600A - 900B = 0$ or $2A - 3B = 0$. Hence, $A = -\frac{1}{26}$, $B = -\frac{1}{39}$.

So $q_p = -\frac{1}{26}\sin 100t - \frac{1}{39}\cos 100t$. The general solution is $q = e^{-30t}(c_1 \cos 10t + c_2 \sin 10t) - \frac{1}{26}\sin 100t - \frac{1}{39}\cos 100t$. Since $q(0) = c_1 - \frac{1}{39} = 0$, then $c_1 = \frac{1}{39}$. Also, $i(t) = q'(t) = -30e^{-30t} \times (c_1 \cos 10t + c_2 \sin 10t) + e^{-30t}(-10c_1 \sin 10t + 10c_2 \cos 10t) - \frac{100}{26}\cos 100t + \frac{100}{39}\sin 100t$. Since $q'(0) = -30c_1 + 10c_2 - \frac{100}{26} = 0$, we have $c_2 = \frac{\frac{30}{39} - \frac{100}{26}}{10} = \frac{3}{39} - \frac{10}{26} = -\frac{24}{78} = -\frac{4}{13}$. The particular solution is $q(t) = e^{-30t}(\frac{1}{39}\cos 10t - \frac{4}{13}\sin 10t) - \frac{1}{26}\sin 100t - \frac{1}{39}\cos 100t$.

10. $E(t) = L\frac{d^2q}{dt^2} + R\frac{dq}{dt} + \frac{q}{c}$ so we get the equation $\frac{1}{8}\frac{d^2q}{dt^2} + 5\frac{dq}{dt} + 100q = \sin t$. The auxiliary equation is $\frac{1}{8}m^2 + 5m + 100 = 0$ is equivalent to $m^2 + 40m + 800 = 0$ and it has solutions $-20 \pm 20j$ and so $q_c = e^{-20t}(c_1 \cos 20t + c_2 \sin 20t)$. Now we set $q_p = A\sin t + B\cos t$, $q'_p = A\cos t - B\sin t$, and $q''_p = -A\sin t - B\cos t$. Substituting, we have $\frac{1}{8}(-A\sin t - B\cos t) + 5(A\cos t - B\sin t) + 100(A\sin t + B\cos t) = \sin t$. Hence $-\frac{A}{8} - 5B + 100A = 1$ and $-\frac{B}{8} + 5A + 100B = 0$ or $\frac{799}{8}A - 5B = 1$ and $5A + \frac{799}{8}B = 0$ or $799A - 40B = 8$ and $40A = 799B = 0$. Using Cramer's rule $A = \frac{6392}{640001}$ and $B = \frac{-320}{640001}$. The general solution is $q(t) = e^{-20t}(c_1 \cos 20t + c_2 \sin 20t) + \frac{6392}{640001}\sin t - \frac{320}{640001}\cos t$. Since $q(0) = 0$, $c_1 = \frac{320}{640001}$, $q'(t) = -20e^{-20t}(c_1 \cos 20t + c_2 \sin 20t) + e^{-20t}(-20c_1 \sin 20t + 20c_2 \cos 20t) + \frac{6392}{640001}\cos t + \frac{320}{640001}\sin t$. $q'(0) = -20c_1 + 20c_2 + \frac{6392}{640001} = 0$ means that $c_2 = \frac{320}{640001} - \frac{319.6}{640001} = \frac{0.4}{640001}$. The particular solution is $q(t) = e^{-20t}(\frac{320}{640001}\cos 20t + \frac{0.4}{640001}\sin 20t) + \frac{6392}{640001}\sin t - \frac{320}{640001}\cos t$.

11. **(a)** $E(t) = L\frac{d^2q}{dt^2} + R\frac{dq}{dt} + \frac{q}{c}$ so we get $\frac{1}{10}\frac{d^2q}{dt^2} + \frac{100}{4}q = 180\cos 60t$. The auxiliary equation is $0.1m^2 + \frac{1000}{4} = 0$, so $m = \pm 50j$ and as a result, $q_c = c_1 \cos 50t + c_2 \sin 50t$. Now we get $q_p = A\sin 60t + B\cos 60t$, $q'_p = 60A\cos 60t - 60B\sin 60t$, and $q''_p = -3600A\sin 60t - 3600 \times$

$\cos 60t$. Substituting we get $-360A\sin 60t - 360B\cos 60t + 250A\sin 60t + 250B\cos 60t = 180\cos 60t$. Hence we have the equations $-360A + 250A = 0$ or $A = 0$ and $-360B + 259B = 180$ or $B = \frac{180}{-110} = -\frac{18}{11}$. The general equation is $q(t) = c_1 \cos 50t + c_2 \sin 50t - \frac{18}{11}\cos 60t$. Since $q(0) = 0$, $c_1 = \frac{18}{11}$. Also, since $q'(t) = -50c_1 \sin 50t + 50c_2 \cos 50t + \frac{18 \cdot 60}{11}\sin 6t$, and $q'(0) = 50c_1 = 0$, we find that $c_2 = 0$. The particular solution is $q(t) = \frac{18}{11}\cos 50t - \frac{18}{11}\cos 60t$.

(b) The steady-state current is $i = \frac{dq_p}{dt} = \frac{d}{dt} \times (\frac{-18}{11}\cos 60t) = \frac{1080}{11}\sin 60t$.

12. **(a)** $1\frac{d^2q}{dt} + 10\frac{dq}{dt} + 100q = 50\cos 10t$. The auxiliary equation is $m^2 + 10m + 100 = 0$ which has roots $-5 \pm 5\sqrt{3}j$. Hence $q_c = e^{-5t}(c_1 \cos 5\sqrt{3}t + c_2 \times \sin 5\sqrt{3}t)$. To find q_p we set $q_p = A\sin 10t + B\cos 10t$, $g'_p = 10A\cos 10t - 10B\sin 10t$, and $q''_p = -100A\sin 10t - 100B\cos 10t$. Substituting, we get $(-100A\sin 10t - 100B\cos 10t) + 10(10A \times \cos 10t - 10B\sin 10t) + 100(A\sin 10t + B\cos 10t) = 50\cos 10t$. This yields the linear equations $-100A - 100B + 100A = 0$ or $B = 0$ and $-100B + 100A + 100B = 50$ or $A = \frac{1}{2}$. The general solution is $q(t) = e^{-5t}(c_1 \cos 5\sqrt{3}t + c_2 \sin 5\sqrt{3}t) + \frac{1}{2}\sin 10t$.

(b) The steady-current is $\frac{dq_p}{dt} = 5\cos 10t$.

13. The differential equation $\frac{d^2g}{dt^2} + 2\alpha\frac{dg}{dt} + \omega_0^2 g = 0$ has auxiliary equation $m^2 + 2\alpha m + \omega_0^2 = 0$. Completing the square, we get $m^2 + 2\alpha m + \alpha^2 = \alpha^2 - \omega_0^2$ and so $(m + \alpha)^2 = \alpha^2 - \omega_0^2$ and $m + \alpha = \pm\sqrt{\alpha^2 - \omega_0^2}$, which means that $m = -\alpha \pm \sqrt{\alpha^2 - \omega_0^2}$. Since $\alpha^2 - \omega_0^2 < 0$, these are complex roots with $m = -\alpha \pm \sqrt{\omega_0^2 - \alpha^2}j$. Thus, the general solution is

$$g(t) = e^{-\alpha t}(c_1 \cos\sqrt{\omega_0^2 - \alpha^2}t + c_2 \sin\sqrt{\omega_0^2 - \alpha^2}t)$$

14. The differential equation $\frac{d^2z}{d\theta^2} + z = k$ has auxiliary equation $m^2 + 1 = 0$ with roots $\pm j$. Hence, $z_c = c_1 \cos\theta + c_2 \sin\theta$. Here, $z_p = A_0$ and substitution yields $A_0 = k$. Hence, the general solution is $z = c_1 \cos\theta + c_2 \sin\theta + k$.

15. Substituting L, R, and C, produces the differential equation $1.2\frac{d^2q}{dt^2} + 5\frac{dq}{dt} + \frac{1}{0.025}q = 15$. This has auxiliary equation $1.2m^2 + 5m + 40 = 0$ with roots $-2.083 \pm 5.3845j$ or about $-2.1 \pm 5.4j$. Hence, $q_c = e^{-2.1t}(c_1 \cos 5.4t + c_2 \sin 5.4t)$. Here, $q_p = A_0$ and $40A_0 = 15$ so $A_0 = \frac{15}{40} = 0.375$. Hence, the general solution is $q(t) = e^{-2.1t}(c_1 \cos 5.4t + c_2 \sin 5.4t) + 0.375$.

16. Substituting the given data produces $\frac{1}{20}\frac{d^2q}{dt^2} + 1\frac{dq}{dt} + 505q = 0$ or $\frac{d^2q}{dt^2} + 20\frac{dq}{dt} + 10,000q = 0$. Its auxiliary equation is $m^2 + 20m + 10,100 = 0$ which has roots $-10 \pm 100j$. Hence, $q = e^{-10t}(c_1 \cos 100t +$

$c_2 \sin 100t)$. Thus, $i = \frac{dq}{dt} = -10e^{-10t}(c_1 \cos 100t + c_2 \sin 100t) + e^{-10t}(-100c_1 \sin 100t + 100c_2 \cos 100t)$. When $t = 0$, we were given $q = 1$ and $i = 0$. Substituting these values into the equations for q and i produces $c_1 = 1$ and $-10c_1 + 100c_2 = 0$ which leads to $c_2 = 0.1$. So, the particular solutions

$$q = e^{-10t}(\cos 100t + 0.1 \sin 100t)$$
$$i = -10e^{-10t}(\cos 100t + 0.1 \sin 100t)$$
$$+ e^{-10t}(-100 \sin 100t + 10 \cos 100t)$$

When $t = 0.01$, we get $q(0.01) = 0.565$ or about 0.57 C and $i(0.01) = -76.9$. Hence, at $t = 0.01$ s, the charge is 0.57 C and the current is -76.9 A.

CHAPTER 33 REVIEW

1. $D^2y = 0$ has auxiliary equation $m^2 = 0$ with 0 as a double root. Hence $y = (c_1 + c_2x)e^{0x}$ or more simply $y = c_1 + c_2x$.

2. $(D^2 - 9)y = 0$. The auxiliary equation $m^2 - 9 = 0$ has roots 3 and -3. Hence $y = c_1e^{3x} + c_2e^{-3x}$.

3. $(D^2 - 5D + 6)y = 0$. The auxiliary equation $m^2 - 5m + 6 = 0$ has roots 2 and 3. Hence $y = c_1e^{2x} + c_2e^{3x}$.

4. $(D^2 - 5D - 14)y = 0$. The auxiliary equation $m^2 - 5m - 14 = 0$ has roots 7 and -2. Hence the solution is $y = c_1e^{7x} + c_2e^{-2x}$.

5. $(D^2 - 6D + 25)y = 0$. The auxiliary equation $m^2 - 6m + 25 = 0$ has complex roots $3 + 4j$. Hence the solution is $y = e^{3x}(c_1 \cos 4x + c_2 \sin 4x)$

6. $(D^2 + 2D - 2)y = 0$. The auxiliary equation $m^2 + 2m - 2 = 0$ has roots $-1 \pm \sqrt{3}$. Hence $y = c_1e^{(-1+\sqrt{3})x} + c_2^{(-1-\sqrt{3})x}$

7. $(D^3 - 8)y = 0$; $m^3 - 8$ factors as $(m - 2)(m^2 + 2m + 4)$ and has roots 2, $-1 \pm \sqrt{3}j$. Hence $y = c_1e^{2x} + e^{-x}(c_2 \cos \sqrt{3}x + c_3 \sin \sqrt{3}x)$

8. $y''' + y'' - y' - y = 0$ has auxiliary equation $m^3 + m^2 - m - 1 = 0$. This factors by grouping into $(m^2 - 1)(m + 1) = 0$ and has roots -1 twice and 1. Hence, the solution is $y = c_1e^x + c_2e^{-x} + c_3xe^{-x}$

9. $(D^2 + 3D - 4)y = 9x^2$. The auxiliary equation is $m^2 + 3m - 4 = 0$ and it has roots -4 and 1. Hence $y_c = c_1e^x + c_2e^{-4x}$. Now we set $y_p = Ax^2 + Bx + C$, $y_p' = 2Ax + B$, and $y_p'' = 2A$. Substituting we get $2A + 3(2A + B) - 4(Ax^2 + Bx + C) = 9x^2$ so

$-4A = 9$ or $A = -\frac{9}{4}$, and also $6A - 4B = 0$ so $B = \frac{6A}{4} = \frac{-27}{8}$. Now, since $2A + 3B - 4C = 0$, we see that $C = \frac{2A+3B}{4} = \frac{-\frac{9}{2} - \frac{81}{8}}{4} = \frac{-117}{32}$. So $y_p = -\frac{9}{4}x^2 - \frac{27}{8}x - \frac{117}{32}$ and $y = y_c + y_p = c_1e^x + c_2e^{-4x} - \frac{9}{4}x^2 - \frac{27}{8}x - \frac{117}{32}$.

10. $(D^2 - 4D - 5)y = 3e^{2x}$. The auxiliary equation has roots 5 and -1 so $y_c = c_1e^{5x} + c_2e^{-x}$. Now we set $y_p = Ae^{2x}$, $y_p' = 2Ae^{2x}$ and $y_p'' = 4Ae^{2x}$. Substituting we get $4Ae^{2x} - 4(2Ae^{2x}) - 5(Ae^{2x}) = 3e^{2x}$. Hence $4A - 8A - 5A = 3$ or $-9A = 3$ or $A = -\frac{1}{3}$. The solution is $y = c_1e^{5x} + c - 2e^x - \frac{1}{3}e^{2x}$

11. $(D^2 + 7D + 12)y = \cos 5x$. The auxiliary equation $m^2 + 7m + 12 = 0$, has roots -3 and -4. So $y_c = c_1e^{-3x} + c_2e^{-4x}$. Now we set $y_p = A \sin 5x + B \cos 5x$, $y_p' = 5A \cos 5x - 5B \sin 5x$, and $y_p'' = -25A \sin 5x - 25B \cos 5x$. Substituting we get $(-25A \sin 5x - 25B \cos 5x) + 7(5A \cos 5x - 5B \sin 5x) + 12(A \sin 5x + B \cos 5x) = \cos 5x$. This yields the linear equations $-25A - 35B + 12A = 0$ and $-25B + 35A + 12B = 1$. These simplify to $-13A - 35B = 0$ and $35A - 13B = 1$. Using Cramer's Rule we get $A = \frac{35}{1394}$ and $B = \frac{-13}{1394}$. So $y_p = \frac{35}{1394} \sin 5x - \frac{13}{1394} \cos 5x$. The solution is

$y = y_c + y_p = c_1e^{-3x} + c_2e^{-4x} + \frac{35}{1394} \sin 5x - \frac{13}{1394} \cos 5x$.

12. $(D^2-6D+8)y = 6x^2-2$. The auxiliary equation has roots 2 and 4 so $y_c = c_1e^{2x} + c_2e^{4x}$. Now we set $y_p = Ax^2+Bx+C$, $y_p' = 2Ax+B$, $y_p'' = 2A$. Substituting we get $2A - 6(2Ax + B) + 8(Ax^2 + Bx + C) = 6x^2 - 2$. Hence $8A = 6$ or $A = \frac{3}{4}$ and since $-12A+8B = 0$, we obtain $B = \frac{12A}{8} = \frac{3}{2}\cdot\frac{3}{4} = \frac{9}{8}$, and since $2A - 6B + 8C = -2$, we find $C = \frac{-2-2A+6B}{8} = \frac{-2-\frac{3}{2}+\frac{27}{4}}{8} = \frac{13}{32}$. So we get $y_p = \frac{3}{4}x^2 + \frac{9}{8}x + \frac{13}{32}$ and $y = y_c + y_p = c_1e^{2x} + c_2e^{4x} + \frac{3}{4}x^2 + \frac{9}{8}x + \frac{13}{32}$.

13. $(D^2+4D+4)y = 4x+e^{2x}$. The auxiliary equation has double root at -2, so $y_c = (c_1 + c_2x)e^{-2x}$. Now y_p has two parts. Set $y_{p1} = Ax + B$, with $y_{p1}' = A$, and $y_{p1}'' = 0$. Substituting we get $4A + 4Ax+4B = 4x$. so $A = 1$ and $B = -1$. Next we set $y_{p2} = Ae^{2x}$ so $y_{p2}' = 2Ae^{2x}$ and $y_{p2}'' = 4Ae^{2x}$. Substituting we have $(4A + 8A + 4A)e^{2x} = e^{2x}$ or $A = \frac{1}{16}$. $y = y_c + y_{p1} + y_{p2} = (c_1 + c_2x)e^{-2x} + x - 1 + \frac{1}{16}e^{2x}$.

14. $(D^2 + 4D - 9)y = x\sin 2x$. The auxiliary equation $m^2 + 4m - 9 = 0$ has solutions $-2 \pm \sqrt{3}$ and so $y_c = c_1e^{(-2+\sqrt{13})x} + c_2e^{(-2-\sqrt{13})x}$. Now we set $y_p = A_1 \sin 2x + A_0 \sin 2x + B_1 x \cos 2x + B_0 \cos 2x$, $y_p' = 2A_1x \cos 2x + A_1 \sin 2x + 2A_0 \times \cos 2x - 2B_1x\sin 2x + B_1 \cos 2x - 2B_0 \sin 2x$, and $y_p'' = 2A_1x \cos 2x - 4A_1x \sin 2x + 2A_1 \cos 2x - 4A_0x \sin 2x - 2B_1 \sin 2x - 4B_1x \cos 2x - 2B_1 \sin 2x - 4B_0 \cos 2x$. Substituting we get $y_p'' + 4y_p' - 9y = x\sin 2x$. Collecting like terms we have the following four linear equations.

$$-4A_1 + 4(-2B_1) - 9A_1 = 1 \quad x\sin 2x$$
$$-4B_1 - 4A_0 + 4$$
$$\times(-2B_0 + A_1) - 9A_0 = 0 \quad \sin 2x$$
$$-4B_1 + 4\cdot 2A_1 - 9B_1 = 0 \quad x\cos 2x$$
$$-4B_0 + 4A_1 + 4(B_1 + 2A_0) - 9B_0 = 1 \quad \cos 2x$$

These simplify to

$$-13A_1 - 8B_1 = 1$$
$$4A_1 - 13A_0 - 4B_1 - 8B_0 = 0$$
$$8A_1 - 13B_1 = 0$$
$$4A_1 + 8A_0 + 4B_1 - 13B_0 = 0$$

Solving the first and third equations with Cramer's Rule yields $A_1 = \frac{-13}{233}$ and $B_1 = \frac{-8}{223}$. Substituting these results into the 2nd and 4th equations we get

$$13A_0 + 8B_0 = \frac{20}{233}$$
$$8A_0 - 13B_0 = \frac{84}{233}$$

Again using Cramer's Rule we have $A_0 = \frac{932}{54289}$ and $B_0 = \frac{932}{54289}$. Putting this all together we have

$$y = c_1e^{(-2+\sqrt{13})x} + c_2e^{(-2-\sqrt{13})x} - \frac{x}{233}(13\sin 2x + 8\cos 2x) + \frac{932}{54289}(\sin 2x + \cos 2x).$$

15. First we find the spring constant $k = \frac{mg}{x} = \frac{4\cdot 9.8}{0.2} = 196$ N/m. So $m = 4$ kg, $k = 196$ N/m and $\ell = 10$. Hence we get auxiliary equation $4m^2 + 10m + 196 = 0$. This auxiliary equation has roots $\frac{-5}{4} \pm \frac{\sqrt{759}j}{4}$. Hence $x = e^{-5t/4}\left(c_1 \cos \frac{\sqrt{759}}{4}t + c_2 \times \sin \frac{\sqrt{759}}{r}t\right)$. Assuming $x(0) = 0$, we obtain $c_1 = 0$. Now, differentiating, we see that $x'(t) = -\frac{5}{4}e^{-5t/4}\left(c_1 \sin \frac{\sqrt{759}}{4}t\right) + c_2\frac{\sqrt{759}}{4}e^{-5t/4} \cos \frac{\sqrt{759}}{4}t$. Since $v(0) = x'(0) = c_2\frac{\sqrt{759}}{4} = 1$, we have $c_2 = \frac{4}{\sqrt{759}}$. Hence the solution is $x = \frac{4}{\sqrt{759}}e^{-5t/4} \sin \frac{\sqrt{759}}{4}t$.

16. **(a)** The given information yields the equation $\frac{d^2q}{dt^2} + 6\frac{dq}{dt} + 2500q = 16\cos 10t$. The auxiliary equation $m^2 + 6m + 2500$ has solutions $-3 \pm \sqrt{2491}j$. So, we see that $q_c = e^{-3t}(c_1 \cos \sqrt{2491}t + c_2 \times \sin \sqrt{2491}t)$. To find q_p we set $q_p = A\sin 10t + B\cos 10t$, $q_p' = 10A \cos 10t - 10B \sin 10t$, and $q_p'' = 100A \cos 10t - 100B \sin 10t$. Substituting we get $(-100A \sin 10t - 100B \cos 10t) + 6(10A \times \cos 10t - 10B \sin 10t) + 2500(A\sin 10t + B\cos 10t) = 16\cos 10t$. As a result, $-100A - 60B + 2500A = 0$ and $-100B + 60A + 2500B = 16$. These simplify to $2400A - 60B = 0$ and $60A + 2400B = 16$. Using Cramer's Rule we get $A = \frac{960}{5763600}$ and $B = \frac{38400}{5763600}$ which reduce to $A = \frac{4}{24015}$ and $B = \frac{160}{24015}$. Thus, $q_p = \frac{1}{24015}(4\sin 10t + 160\cos 10t)$. Hence the general solution is $q(t) = e^{-3t}(c_1 \cos \sqrt{2491}t + c_2 \sin \sqrt{2491}t) + \frac{1}{24015}(4\sin 10t + 160\cos 10t)$. Since $q(0) = c_1 + \frac{160}{24015} = 0$, $c_1 = \frac{-160}{24015}$. Also $i(t) = q'(t) = -3e^{-3t}(c_1 \cos \sqrt{2491}t + c_2 \sin \sqrt{2491}t + e^{-3t}(-\sqrt{2491}c_1 \sin \sqrt{2491}t + \sqrt{2491}c_2 \cos \sqrt{(2491t)}) + \frac{40}{24105} \cos 10t - \frac{1600}{24015} \sin 10t$.

So $q'(0) = -3c_1 + \sqrt{2491}c_2 + \frac{40}{24015} = 0$ and so $c_2 = \frac{-520}{24015\sqrt{2491}}$. Putting it all together we get

$q(t) = \frac{1}{24015}\left[e^{-3t}\left(-160\cos\sqrt{2491}t - \frac{520}{\sqrt{2491}} \times \sin\sqrt{2491}t\right) + 4\sin 10t + 160\cos 10t\right]$.

(b) The steady state current is $\frac{dq_p}{dt} = \frac{1}{24015} \times [40\cos 10t - 1600\sin 10t]$.

CHAPTER 33 TEST

The auxiliary equation is formed by replacing the operators, D, with place-holders for the roots, usually m and replacing the variable, y, with 1. In this problem the resulting auxiliary equation is $5m^2 + 2m - 3 = 0$.

1. The auxiliary equation is $m^2 - 25 = 0$. This factors as $m^2 - 25 = (m-5)(m+5) = 0$ and has roots $m_1 = 5$ and $m_2 = -5$. So, the general solution of this differential equation is $y = c_1 e^{5x} + c_2 e^{-5x}$.

2. The auxiliary equation is $m^2 + 8m - 20 = 0$ which factors as $(m+10)(m-2) = 0$. The roots of the auxiliary equation are $m_1 = -10$ and $m = 2$. The solution of the differential equation is $y = c_1 e^{-10x} + c_2 e^{2x}$.

3. The auxiliary equation is $(m-5)^3(m^2+4) = 0$. This has the real root 5 with multiplicity 3 and the complex roots $\pm 2j$. So, the general solution is $y = (c_1 + c_2 x + c_3 x^2)e^{5x} + c_4 \cos 2x + c_5 \sin 2x$.

4. The auxiliary equation is $m^2 + 6m + 13 = 0$. Using the quadratic formula we get the roots $m = \frac{-6\pm\sqrt{6^2-4(13)}}{2} = \frac{-6\pm\sqrt{-16}}{2} = -3\pm 2j$. So the general solution to this differential equation is $y = e^{-3x}(c_1\cos 2x + c_2\sin 2x)$.

5. The auxiliary equation is $m^2 + 6m - 7 = 0$ which factors as $(m+7)(m-1) = 0$. The roots of this auxiliary equation are $m = -7$ and $m = 1$. The general solution is $y = c_1 e^{-7x} + c_2 e^x$.

6. From Exercise 0 we have the complementary solution $y_c = c_1 e^{-7x} + c_2 e^x$. The given equation is a nonhomogeneous equation with $f(x) = 6e^{2x}$ and so this is a Case 2 problem using the Method of Undetermined Coefficients with $a = 2$ and $p_n = 6$ so $y_p = A_0 e^{2x}$. Differentiating, we obtain $y_p' = 2A_0 e^{2x}$ and $y_p'' = 4A_0 e^{2x}$. Substituting this in the given differential equation produces $4A_0 e^{2x} + 12A_0 e^{2x} - 7A_0 e^{2x} = 6e^{2x}$. Solving for A_0, we determine that $A_0 = \frac{2}{3}$ and so the general solution is $y = c_1 e^{-7x} + c_2 e^x + \frac{2}{3}e^{2x}$.

7. Again the complementary solution is $y_c = c_1 e^{-7x} + c_2 e^x$. This is a nonhomogeneous equation with $f(x) = 3\cos x$, a Case 4 problem using the Method of Undetermined Coefficients with $a = 0$, $p_n = 3$, and $b = 4$, so $y_p = A_0\sin 4x + B_0\cos 4x$. Differentiating, produces $y_p' = 4A_0\cos 4x - 4B_0\sin 4x$ and $y_p'' = -16A_0\sin 4x - 16B_0\cos 4x$. Substituting in the given differential equation we get $(-16A_0\sin 4x - 16B_0\cos 4x) + 6(4A_0\cos 4x - 4B_0\sin 4x) - 7(A_0\sin 4x + B_0\cos 4x) = 3\cos 4x$. Multiplying and collecting terms produces $-23\times(A_0 + B_0)\sin 4x + 24(A_0 - B_0)\cos 4x = 3\cos 4x$. Thus, $A_0 + B_0 = 0$ and $A_0 - B_0 = 3$. Solving these two equations we determine that $A_0 = \frac{3}{2}$ and $B_0 = -\frac{3}{2}$ and so the general solution is $y = c_1 e^{-7x} + c_2 e^x + \frac{3}{2}\sin 4x - \frac{3}{2}\cos 4x$.

8. Again the complementary solution is $y_c = c_1 e^{-7x} + c_2 e^x$. This is a nonhomogeneous equation with $f(x) = 8x$, a Case 1 problem using the Method of Undetermined Coefficients, so $y_p = A_1 x + A_0$. Differentiating, we obtain $y_p' = A_1$ and $y_p'' = 0$. Substituting these in the given differential equation produces $0 + 6(A_1) - 7(A_1 x + A_0) = 8x$. Multiplying and collecting terms produces $-A_1 x + A_0 = 8x$. Thus, $A_1 = -8$ and $A_0 = 0$ and so the general solution is $y = c_1 e^{-7x} + c_2 e^x - 8x$.

9. Once again the complementary solution is $y_c = c_1 e^{-7x} + c_2 e^x$. This is a nonhomogeneous equation with $f(x) = e^{4x}$, a Case 2 problem using the Method of Undetermined Coefficients with $a = 4$, and $p_n = 1$, so $y_p = A_0 e^{4x}$. Differentiating, we get $y_p' = 4A_0 e^{4x}$ and $y_p'' = 16A_0 e^{4x}$. Substituting these in the given differential equation produces $16A_0 e^{4x} + 6(4A_0 e^{4x}) - 7(A_0 e^{4x}) = e^{4x}$. Collecting terms produces $33A_0 e^{4x} = e^{4x}$. Thus, $A_0 = \frac{1}{33}$ and the general solution is $y = c_1 e^{-7x} + c_2 e^x + \frac{1}{33}e^{4x}$. To find the particular solution, we find $y' = -7c_1 e^{-7x} + c_2 e^x + \frac{4}{33}e^{4x}$. Substituting $x = 0$ and $y = \frac{4}{11}$ in the general solution produces

$\frac{4}{11} = c_1 + c_2 + \frac{1}{33}$ and substituting $x = 0$ and $y' = \frac{5}{11}$ in the derivative gives $\frac{5}{11} - 7c_1 + c_2 + \frac{4}{33}$. Solving these two equations we find $c_1 = 0$ and $c_2 + \frac{1}{3}$. Thus, the particular solution of the given differential equation is $y = \frac{1}{3}e^x + \frac{1}{33}e^{4x}$.

10. Here $E(t) = 5\sin 10t$ and the desired differential equation is $5\sin 10t = 0.5\frac{d^2q}{dt^2} + 30\frac{dq}{dt} + \frac{q}{2\times 10^{-3}} = 0.5\frac{d^2q}{dt^2} + 30\frac{dq}{dt} + 500q$. The auxiliary equation is $0.5m^2 + 30m + 500 = 0$ or $m^2 + 60m + 1000 = 0$. Using the quadratic equation we get $m = -30 \pm 20j$. These are complex roots and so the complementary solution is $q_c = e^{-30t}(c_1\cos 20t + c_2 \times \sin 20t)$. For the particular solution we have a case 3 situation with $a = 0$, $b = 10$, and $p_n = 5$,

and so $q_p = A_0\sin 10t + B_0\cos 10t$. Differentiating, we get $q_p' = 10A_0\cos 10t - 10B_0\sin 10t$ and $q_p'' = -100A_0\sin 10t - 100B_0\cos 10t$. Substituting these in the differential equation, we have $0.5(100A_0\sin 10t - 100B_0\cos 10t) + 30(10A_0\times \cos 10t - 10B_0\sin 10t) + 500(A_0\sin 10t + B_0\times \cos 10t) = 5\sin 10t$. Multiplying and collecting terms, we get $(450A_0 - 300B_0)\sin 10t + (450B_0 + 300A_0)\cos 10t = 5\sin 10t$. Hence, $450A_0 - 300B_0 = 5$ and $450B_0 + 300A_0 = 0$. Solving these two equations simultaneously gives $A_0 = 0.02$ and $B_0 = \frac{1}{75}$. Thus, the general solution is $q(t) = e^{-30t}(c_1\cos 20t + c_2\sin 20t) + 0.02\sin 10t + \frac{1}{75}\times \cos 10t$. The steady-state solution is $q_p = 0.02\times \sin 10t + \frac{1}{75}\cos 10t$. Consequently, the steady-state current is $i = \frac{dq_p}{dt} = 0.2\cos 10t - \frac{10}{75}\sin 10t$.

34

Numerical Methods and LaPlace Transforms

≡ 34.1 EULER'S OR THE INCREMENT METHOD

1. See *Computer Programs* in the main text.

2.

	y		Correct solution
x	$h = 0.1$	$h = 0.05$	$y = \dfrac{1}{2}x^2 + 2x + 1.5$
1.00	4	4	4
1.05	–	4.15	4.15125
1.10	4.3	4.3025	4.305
1.15	–	4.575	4.46125
1.20	4.61	4.615	4.62
1.25	–	4.775	4.78125
1.30	4.93	4.9375	4.95
1.35	–	5.1025	5.11125
1.40	5.26	5.27	5.28
1.45	–	5.44	5.45125
1.50	5.6	5.6125	5.625
1.55	–	5.7875	5.780125
1.60	5.95	5.965	5.98
1.65	–	6.145	6.16125
1.70	6.31	6.3275	6.345
1.75	–	6.5125	6.53125
1.80	6.68	6.7	6.72
1.85	–	6.89	6.91125
1.90	7.06	6.0825	7.105
1.95	–	7.2775	7.30125
2.00	7.45	7.475	7.5

3.

	y		Correct solution
x	$h = 0,1$	$h = 0.05$	$y = \dfrac{1}{2}e^x + \dfrac{1}{2}e^{3x}$
0	0	0	0
0.5	–	0.5	0.05528
0.10	0.1	0.1101	0.12234
0.15	–	0.1818	0.2032
0.20	0.2405	0.2672	0.3004
0.25	–	0.3683	0.4165
0.30	0.4348	0.4878	0.5549
0.35	–	0.6285	0.7193
0.40	0.7002	0.7937	0.9141
0.45	–	0.9873	1.1446
0.50	1.0595	1.2138	1.4165
0.55	–	1.4784	1.7369
0.60	1.5422	1.7868	2.1138
0.65	–	2.1459	2.5566
0.70	2.1871	2.5636	3.0762
0.75	–	3.0488	3.6854
0.80	3.0046	3.6120	4.3988
0.85	–	4.2650	5.2337
0.90	4.1805	5.0218	6.2101
0.95	–	5.8980	7.3510
1.00	5.6807	6.9120	8.6836

4.

	y	Correct solution
x	$h = 0.1$	$y = -1 + 2e^{x^2}$
0	1	1
0.1	1	1.0201
0.2	1.04	1.0816
0.3	1.1216	1.1883
0.4	1.2489	1.3470
0.5	1.4288	1.5681
0.6	1.6717	1.8667
0.7	1.9923	2.2646
0.8	2.4112	2.7930
0.9	2.9570	3.4958
1.0	3.6693	4.4366

5.

x	y $h = 0.05$	Correct solution $y = \dfrac{1}{53}\left(55e^{7x} - 2\cos 2x - 7\sin 2x\right)$
0	1	1
0.05	1.35	1.4219
0.10	1.8275	2.0265
0.15	2.4770	2.8904
0.20	3.3588	4.1220
0.25	4.5538	5.8753
0.30	6.1717	8.3686
0.35	8.3600	11.9117
0.40	11.3182	16.9442
0.45	15.3154	24.0898
0.50	20.7149	34.2336

6.

x	y
0	0
0.1	0.5
0.2	1.005
0.3	1.5251
0.4	2.0709
0.5	2.6537
0.6	3.2864
0.7	3.9836
0.8	4.7624
0.9	5.6434
1.0	6.6513

7.

x	y
0	1
0.1	1.1
0.2	1.222
0.3	1.3753
0.4	1.5735
0.5	1.8371
0.6	2.1995
0.7	2.7193
0.8	3.5078
0.9	4.8023
1.0	7.1895

8.

x	y
0	1
0.1	1.4
0.2	2.185
0.3	4.0987
0.4	10.8274
0.5	57.7364
0.6	1391.1571
0.7	175,518.399
0.8	2.406×10^{19}
0.9	2.316×10^{22}
1.0	

9.

x	y
0	0
0.1	0.1
0.2	0.2010
0.3	0.3051
0.4	0.4147
0.5	0.5327
0.6	0.6633
0.7	0.8121
0.8	0.9887
0.9	1.2092
1.0	1.5062

10.

x	y
0	1
0.05	1.1
0.1	1.2007
0.15	1.3022
0.20	1.4046
0.25	1.5080
0.30	1.6126
0.35	1.7185
0.40	1.8258
0.45	1.9345
0.50	2.0449

11.

x	y
$\pi/2 \approx 1.57$	1
1.62	1.0270
1.67	1.0507
1.72	1.0711
1.77	1.0882
1.82	1.1022
1.87	1.1130
1.92	1.1209
1.97	1.1259
2.02	1.1281
2.07	1.1277
2.12	1.1249
2.17	1.1197
2.22	1.1123
2.27	1.1028
2.32	1.0913
2.37	1.0779

12.

x	y
$\pi \approx 3.14$	2
3.19	2.05
3.24	2.1026
3.29	2.1579
3.34	2.2164
3.39	2.2782
3.44	2.3437
3.49	2.4131
3.54	2.4868
3.59	2.5651
3.64	2.6483
3.69	2.7368
3.74	2.8311
3.79	2.9313
3.84	3.0381
3.89	3.1518
3.94	3.2728

13. See *Computer Programs* in the main text.

14.

(Problem #3)		
		y
x	$h = 0.1$	$h = 0.05$
0	0	0
0.05	–	0.0513
0.1	0.1053	0.1131
0.15	–	0.1871
0.2	0.2547	0.2755
0.25	–	0.3805
0.3	0.4636	0.5048
0.35	–	0.6517
0.4	0.7517	0.8246
0.45	–	1.0279
0.5	1.1455	1.2664
0.55	–	1.5457
0.6	1.6798	1.8722
0.65	–	2.2535
0.7	2.4007	2.6982
0.75	–	3.2163
0.8	3.3689	3.8194
0.85	–	4.5207
0.9	4.6644	5.3358
0.95	–	6.2823
1.00	6.3926	7.3808

(Problem #4)	
	y
x	$h = 0.1$
0	1
0.1	1.021
0.2	1.0837
0.3	1.1911
0.4	1.3492
0.5	1.5674
0.6	1.8592
0.7	2.2440
0.8	2.7485
0.9	3.4105
1.0	4.2826

(Problem #5)	
	y
x	$h = 0.5$
0	1
0.05	1.3612
0.10	1.8571
0.15	2.5356
0.20	3.4624
0.25	4.7263
0.30	6.4479
0.35	8.7913
0.40	11.9793
0.45	16.3143
0.50	22.2077

15. Substituting $T_m(t) = 70°$ in the given formula, produces $\frac{dT}{dt} = k\left(T_m{}^4 - T^4\right) = 40^{-4}\left(70^4 - T^4\right)$. By Euler's method, we have $T_{n+1} = 40^{-4}\left(70^4 - T_n{}^4\right)h + T_n$

n	t	$T(t)$	
0	0	100	
1	0.1	97.0316	
2	0.2	94.5068	
$\vdots$			
10	1	82.6936	$T(1) \approx 82.694°$
11	1.1	81.8049	
$\vdots$			
20	2	76.4459	$T(2) \approx 76.446°$

16. Here, $\frac{dT}{dt} = -k\left(T - T_m\right)$, or $dT = -k\left(T - T_m\right)dt$. By Euler's method, we have $T_{n+1} = T_n - \left(T_n - 70\right)0.1$, which leads to the following table.

n	t	$T(t)$	
0	0	100	
1	0.1	97	
2	0.2	94.3	
$\vdots$			
10	1.0	80.4604	$T(1) \approx 80.460°$
$\vdots$			
20	2.0	73.6473	$T(2) \approx 73.647°$

≡ 34.2 SUCCESSIVE APPROXIMATIONS

1. $y' = x + 2$, $(1, 4)$, $x = 2$. Integrating we get $y = \int (x+2)dx = \frac{1}{2}x^2 + 2x + C$. Now substituting 1 for x and 4 for y we obtain $C = 4 - 2\frac{1}{2} = \frac{3}{2}$. Hence $y = \frac{1}{2}x^2 + 2x + \frac{3}{2}$, $y(2) = 7.5$ the exact value.

2. $y' = -3y + e^x$. Substituting 0 for y and integrating we have $y = \int e^x dx = e^x + C_1$. Substituting $(0, 0)$ we obtain $C_1 = -1$ so the first approximation is

$y = e^x - 1$. Substituting this for y in the original equation we have $y' = 3\left(e^x - 1\right) + e^x$ or $y' = 4e^x - 3$. Integrating we have $y = \int \left(4e^x - 3\right)dx = 4e^x - 3x + C_2$. Again substituting $(0, 0)$ we obtain $C_2 = -4$ so $y = 4e^x - 3x - 4$ is the second approximation. Substituting this expression for y in the original equation yields $y' = 3\left(4e^x - 3x - 4\right) + e^x = 13e^x - 9x - 12$. Integrating once more $y = 13e^x - \frac{9}{2}x^2 - 12x + C_3$. Substituting $(0, 0)$ yields $C_3 = -13$. The third approximation is $y =$

$13e^x - \frac{9}{2}x^2 - 12x = 13.$ $y(1) = 13e - 29.5 \approx$ 5.8377. This answer is not very close to the correct solution, 8.6836.

3. $y' = 2xy + 2x$. Substituting 1 for y we have $y' = 4x$ so $y = 2x^2 + C_1$. Substituting $(0, 1)$ we obtain $C_1 = 1$. The first approximation is $y = 2x^2 + 1$. Substituting this for y in the original equation we have $y' = 2x(2x^2 + 1) + 2x = 4x^3 + 4x$. Integration yields $y = x^4 + 2x^2 + C_2$. Substituting $(0, 1)$ again we get $C_2 = 1$ so the second approximation is $y = x^4 + 2x^2 + 1$. Substituting for y once more in the original equation we have $y' = 2x(x^4 + 2x^2 + 1) + 2x = 2x^5 + 4x^3 + 4x$. Integration yields $y = \frac{1}{3}x^6 + x^4 + 2x^2 + C_3$. Again $C_3 = 1$ so the third approximation is $y = \frac{1}{3}x^6 + x^4 + 2x^2 + 1$. $y(1) = 4\frac{1}{3}$ compares favorably with 4.4366, the correct solution.

4. $y' = 7y + \sin 2x$. Substituting 1 for y we have $y' = 7 + \sin 2x$. Integration yields $y = 7x - \frac{1}{2}\cos 2x + C_1$. Substituting $(0, 1)$ we obtain $C_1 = \frac{3}{2}$ so the first approximation is $y = 7x - \frac{1}{2}\cos 2x + \frac{3}{2}$. Substituting this into the original equation we have $y' = 49x - \frac{7}{2}\cos 2x + \frac{21}{2} + \sin 2x$. Integration yields $y = \frac{49}{2}x^2 - \frac{7}{4}\sin 2x + \frac{21}{2}x - \frac{1}{2}\cos 2x + C_2$. Substituting $(0, 1)$ we obtain $C_2 = \frac{3}{2}$ and the second approximation is $y = \frac{49}{2}x^2 - \frac{7}{4}\sin 2x + \frac{21}{2}x - \frac{1}{2}\cos 2x + \frac{3}{2}$. Again substituting into the original equation we get $y' = \frac{343}{2}x^2 - \frac{49}{4}\sin 2x + \frac{147}{2}x - \frac{7}{2}\cos 2x + \frac{21}{2} + \sin 2x$. Integration yields $y = \frac{343}{6}x^3 + \frac{49}{8}\cos 2x + \frac{147}{4}x^2 - \frac{7}{4}\sin 2x + \frac{21}{2}x - \frac{1}{2}\cos 2x + C_3 = \frac{343}{6}x^3 + \frac{45}{8}\cos 2x + \frac{147}{4}x^2 - \frac{7}{4}\sin 2x + \frac{21}{2}x + C_3$. Substituting $(0, 1)$ we obtain $C_3 = -\frac{37}{8}$ and the third approximation is $y = \frac{343}{6}x^3 + \frac{45}{8}\cos 2x + \frac{147}{4}x^2 - \frac{7}{4}\sin 2x + \frac{21}{2}x - \frac{37}{8}$. Thus, $y(0.5) = 18.5250$. This compares poorly with the correct answer 34.2750.

5. $y' = x^2 + y^2$. Substituting 1 for y we have $y' = x^2 + 1$. By integration $y = \frac{1}{3}x^3 + x + C_1$. Substituting $(0, 1)$ we get $C_1 = 1$. The first approximation is $y = \frac{1}{3}x^3 + x + 1$. Substituting this for y in the original equation yields $y' = x^2 + \frac{1}{9}x^6 + \frac{2}{3}x^4 + \frac{2}{3}x^3 + x^2 + 2x + 1$. Integration

yields $y = \frac{1}{63}x^7 + \frac{2}{15}x^5 + \frac{1}{6}x^4 + \frac{2}{3}x^3 + x^2 + x + C_2$. Again, $C_2 = 1$ so the second approximation is $y = \frac{1}{63}x^7 + \frac{2}{15}x^5 + \frac{1}{6}x^4 + \frac{2}{3}x^3 + x^2 + x + 1$.

6. $y' = ye^x$. Substituting 1 for y we have $y' = e^x$. Integration yields $y = e^x + C_1$. Substituting $(0, 1)$ we get $C_1 = 0$, so $y = e^x$. Substituting this into the original equation we have $y' = e^{2x}$. Integrating we get $y = \frac{1}{2}e^{2x} + C_2$. This time $C_2 = \frac{1}{2}$. The second approximation is $y = \frac{1}{2}e^{2x} + \frac{1}{2}$ or $\frac{1}{2}(e^{2x} + 1)$.

7. $y' = 4 + xy$. Substituting 1 for y and integrating we have $y = 4x + \frac{1}{2}x^2 + C_1$. With the point $(0, 1)$ we get $C_1 = 1$ and $y = 4x + \frac{1}{2}x^2 + 1$. Substituting this for y in the original equation yields $y' = 4 + 4x^2 + \frac{1}{2}x^3 + x$. Integrating we get $y = \frac{1}{8}x^4 + \frac{4}{3}x^3 + \frac{1}{2}x^2 + 4x + C_2$. Again $C_2 = 1$ so the second approximation is $y = \frac{1}{8}x^4 + \frac{4}{3}x^3 + \frac{1}{2}x^2 + 4x + 1$.

8. $y' = x^2 + 4y^2$, $(0, 1)$. Substituting 1 for y we have $y' = x^2 + 4$. Integration yields $y = \frac{1}{3}x^3 + 4x + C_1$ and $C_1 = 1$. Substituting into the original equation yields $y' = x^2 + 4\left(\frac{1}{9}x^6 + \frac{8}{3}x^4 + \frac{2}{3}x^3 + 16x^2 + 8x + 1\right)$ or $y' = \frac{4}{9}x^6 + \frac{32}{3}x^4 + \frac{8}{3}x^3 + 65x^2 + 32x + 4$. Integration yields $y = \frac{4}{63}x^7 + \frac{32}{15}x^5 + \frac{8}{12}x^4 + \frac{65}{3}x^3 + 16x^2 + 4x + C_2$. Again $C_2 = 1$ so the second approximation is $y = \frac{4}{63}x^7 + \frac{32}{15}x^5 + \frac{2}{3}x^4 + \frac{65}{3}x^3 + 16x^2 + 4x + 1$.

9. $y' = \sin 2x + y$; $\left(\frac{\pi}{2}, 1\right)$. Substituting 1 for y and integrating we get $y = -\frac{1}{2}\cos 2x + x + C_1$. Using $\left(\frac{\pi}{2}, 1\right)$ we get $C_1 = 1 - \frac{1}{2} - \frac{\pi}{2} = \frac{1-\pi}{2}$. The first approximation is $y = -\frac{1}{2}\cos 2x + x + \frac{1-\pi}{2}$. Substituting this for y in the original equation we get $y' = \sin 2x - \frac{1}{2}\cos 2x + x + \frac{1-\pi}{2}$. Integration yields $y = -\frac{1}{2}\cos 2x - \frac{1}{4}\sin 2x + \frac{1}{2}x^2 + \frac{1-\pi}{2}x + C_2$. Using $\left(\frac{\pi}{2}, 1\right)$ again we get $C_2 = 1 - \frac{1}{2} - \frac{\pi^2}{8} - \frac{1-\pi}{2} \cdot \frac{\pi}{2} = \frac{4-2\pi-\pi^2}{8}$. The second approximation is $y = -\frac{1}{2}\cos 2x - \frac{1}{4}\sin 2x + \frac{1}{2}x^2 + \frac{1-\pi}{2}x + \frac{4-2\pi-\pi^2}{8}$.

10. $y' = y + \cos x$, $(\pi, 2)$. $y' = 2 + \cos x$ so $y = 2x + \sin x + C_1$. Using $(\pi, 2)$ we have $C_1 = 2 - 2\pi$. The first approximation is $y = 2x + \sin x + 2 - 2\pi$. Substituting this into the first equation we get $y' =$

$2x + \sin x + 2 - 2\pi + \cos x$. Integration yields $y = x^2 - \cos x + (2 - 2\pi)x + \sin x + C_2$. Using $(\pi, 2)$ again we obtain $C_2 = 2 - \pi^2 - 1 - 2\pi + 2\pi^2$ or $C_2 = 1 - 2\pi + \pi^2$. The second approximation is $y = x^2 + (2 - 2\pi)x - \cos x + \sin x + 1 - 2\pi + \pi^2$.

11. The Maclaurin series for e^{-x} is $1 - x + \frac{x^2}{2!} - \frac{x^3}{3!} + \frac{x^4}{4!} - \frac{x^5}{5!} + \cdots$. Substituting into $3e^{-x} + x - 1$ we obtain $3\left(1 - x + \frac{x^2}{2} - \frac{x^3}{6} + \frac{x^4}{24} - \frac{x^5}{120}\right) + x - 1 = 2 - 2x + \frac{3x^2}{2} - \frac{x^3}{2} + \frac{x^4}{8} - \frac{x^5}{40}$. The first four terms are the same.

12. The Maclaurin series for e^{x^2} is $1 + x^2 + \frac{x^4}{2!} + \frac{x^6}{3!} + \frac{x^8}{4!} + \cdots$. Substituting into $2e^{x^2} - 1$ we get $1 + 2x^2 + \frac{x^6}{3} + \frac{x^8}{12}$. The first four terms are the same.

13. $I'(t) = -2I(t) + \sin t$ is like $I' = -2I + \sin t$. Substitute 0 for I and integrate we get $I = -\cos t + C_1$. Using $(0, 0)$ we get $C_1 = 1$. The first approximation is $I = -\cos t + 1$. Substituting into the original equation to get $I' = -2(-\cos t + 1) + \sin t = 2\cos t + \sin t - 2$. Integrating again yields $I = 2\sin t - \cos t - 2t + C_2$. Using $(0, 0)$ again we obtain $C_2 = 1$. The second approximation is $I = 2\sin t - \cos t - 2t + 1$. Substituting once more we have $I' = -3\sin t + 2\cos t + 4t - 2$. Integrating we get $I = 3\cos t + 2\sin t + 2t^2 - 2t + C_3$. Using $(0, 0)$ we obtain $C_3 = -3$. The third approximation is $I(t) = 3\cos t + 2\sin t + 2t^2 - 2t - 3$. $I(0.5) \approx 0.0916$.

14. Since it is dropped from rest, when $t = 0$, $v = 0$. Substituting we have $v' = 9.8$ so $v = 9.8t + C_1$. $C_1 = 0$ and the first approximation is $v = 9.8t$. Substituting this into the original we get $v' = 9.8 - 0.4(9.8t)^2 = 9.8 - 38.416t^2$. Integrating we have $v = 9.8t - 12.8053t^3 + C_2$. Again $C_2 = 0$. The second approximation is $v(t) = 9.8t - 12.8053t^3$. Substituting again we have $v' = 9.8 - 0.4(9.8t - 12.8053t^3)^2 = 9.8 - 38.416t^2 + 100.39355t^4 - 65.59028t^6$. Integrating we obtain $v = 9.8t - 12.8053t^3 + 20.0787t^5 - 9.37t^7 + C_3$. Again $C_3 = 0$, $v(0.5) = 3.85359$ m/s.

≡ 34.3 LAPLACE TRANSFORMS

1. $f(t) = n$, $L(f) = L(n) = \int_0^\infty e^{-st} n\, dt$

$$= \lim_{b \to \infty} \int_0^b e^{-st} n\, dt$$

$$= \lim_{b \to \infty} \left[\frac{-n}{s} e^{-st}\right]_0^n$$

$$= \lim_{n \to \infty} \frac{-n}{b} e^{-sb} + \frac{n}{s} = \frac{n}{s}$$

2. $f(t) = e^{-at}$, $L(f) = L(e^{-at}) \int_0^\infty e^{-st} e^{-at} dt$

$$= \lim_{b \to \infty} \int_0^b e^{(-s-a)t} dt$$

$$= \lim_{b \to \infty} \left[\frac{-1}{s+a} e^{-(s+a)t}\right]_0^b$$

$$= \lim_{b \to \infty} \frac{-1}{s+a} e^{-(s+a)b} + \frac{1}{s+a} = \frac{1}{s+a}$$

3. $f(t) = \cos at$, $L(f) = L(\cos at) = \int_0^\infty e^{-st} \times \cos at\, dt = \lim_{b \to \infty} \int_0^b e^{-st} \cos at\, dt$. By form 88, we see that this is $\lim_{b \to \infty} \left[\frac{e^{-st}}{s^2 + a^2}(-s\cos at + a\sin at)\right]_0^b$

$$= \lim_{b \to \infty} \frac{e^{-sb}}{s^2 + a^2}(-s\cos ab + a\sin ab) - \frac{1}{s^2 + a^2}(-s)$$

$$= \frac{s}{s^2 + a^2}.$$

4. $f(t) = te^{at}$, $L(f) = L(te^{at}) = \int_0^\infty e^{-st} te^{at} dt = \lim_{b \to \infty} \int_0^t te^{(a-s)t} dt$. By form 84, we find that this is $\lim_{b \to \infty} \left[\frac{1}{(a-s)^2}((a-s)t - 1)e^{(a-s)t}\right]_0^b = \lim_{b \to \infty} \left[\frac{1}{(a-s)^2} \times ((a-s)b - 1)e^{(a-s)b}\right] - \left[\frac{1}{(a-s)^2}(-1)\right] = \frac{1}{(a-s)^2}$ or $\frac{1}{(s-a)^2}$.

5. $f(t) = t^3 = 3! \cdot \frac{t^3}{3!}$. $l(f) = 3!L\left(\frac{t^3}{3!}\right) = 3!\frac{1}{s^4} = \frac{3!}{s^4}$. (form 3)

6. $f(t) = e^{2t}$, $L(f) = \frac{1}{s-2}$. (form 5)

7. $f(t) = \sin 6t$, $L(f) = \frac{6}{s^2 + 36}$. (form 8)

8. $f(t) = e^{2t} \sin 6t$, $L(e^{2t} \sin 6t) = \frac{6}{(s-2)+36}$. (form 19)

9. $f(t) = t^3 e^{5t}$, $L(f) = \frac{3!}{(s-5)^4}$. (form 7)

10. $f(t) = e^{-5t} \sin 10t$, $L(f) = \frac{10}{(s+5)^4 + 100}$. (form 19)

11. $f(t) = \sin 4t + 4t\cos 4t$, $L(f) = \frac{2 \cdot 4s^2}{\left(s^2 + 4^2\right)^2} = \frac{8s^2}{\left(s^2 + 16\right)^2}$. (form 14)

12. $f(t) = 5\cos 75$, $L(f) = \frac{s^2-7^2}{\left(s^2+7^2\right)^2} = \frac{s^2-49}{\left(s^2+49\right)^2}$.
(form 15)

13. $L(\sin 5t+\cos 3t) = L(\sin 5t)+L(\cos 3t) = \frac{5}{s^2+25} + \frac{s}{s^2+9}$. (forms 8 and 9)

14. $L\left(e^{3t}\sin 5t + 4t^8\right) = L\left(e^{3t}\sin 5t\right) + 4L\left(t^8\right) = \frac{5}{(s-3)^2+25} + \frac{4\cdot 8!}{s^9}$. (forms 3 and 19)

15. $L(3t\sin 5t - \cos 4t + 1) = 3L(t\sin 5t) - L(1 - \cos 4t) = \frac{30s}{\left(s^2+25\right)^2} - \frac{16}{s\left(s^2+16\right)}$. (forms 10 and 13)

16. $L\left(t^3 e^{4t} + 4e^{3t}\right) = L\left(t^3 e^{4t}\right) + 4L\left(e^{3t}\right) = \frac{3!}{(s-a)^4} + \frac{4}{s-3}$. (forms 7 and 5)

17. $L\left(e^{-2t}\cos 3t - 2e^{5t}\sin t\right) = L\left(e^{-2t}\cos 3t\right) - 2L\times\left(e^{5t}\sin t\right) = \frac{s+2}{(s+2)^2+9} - \frac{2}{(s-5)^2+1}$.
(forms 19 and 20)

18. $L(t\cos 3t - 2t\sin 2t) = L(t\cos 3t) - 2L(t\sin 2t) = \frac{s^2-9}{\left(s^2+9\right)^2} = \frac{8s}{\left(s^2+4\right)^2}$

19. $L\left(4 + 3t + 2e^t\right) = 4L(1) + 3L(t) + 2L\left(e^t\right) = \frac{4}{s} + \frac{3}{s^2} + \frac{2}{s-1}$. (forms 1, 2 and 5)

20. $L\left(6 - 4t + 2\sin 10t + e^{5t}\right) = \frac{6}{s} - \frac{4}{s^2} + \frac{20}{s^2+100} + \frac{1}{s-5}$

21. We need to use the trigonometric identity $\cos^2\theta = \frac{1}{2}(1 + \cos 2\theta)$ with $\theta = 3t$. This leads to $\cos^2 3t = \frac{1}{2} + \frac{1}{2}\cos 6t$, and so we have $L\left(\cos^2 3t\right) = L\left(\frac{1}{2} + \frac{1}{2}\cos 6t\right)$. By Table 34.1, $L\left(\frac{1}{2}\right) = \frac{1}{2}L(1) = \frac{1}{2}\left(\frac{1}{s}\right)$ and $L\left(\frac{1}{2}\cos 6t\right) = \frac{1}{2}L(\cos 6t) = \frac{1}{2}\left(\frac{s}{s^2+6^2}\right) = \frac{1}{2}\left(\frac{s}{s^2+36}\right)$. Thus, $L\left(\cos^2 3t\right) = L\left(\frac{1}{2} + \frac{1}{2}\cos 6t\right) = \frac{1}{2}\left(\frac{1}{s}\right) + \frac{1}{2}\left(\frac{s}{s^2+36}\right) = \frac{s^2+18}{s\left(s^2+36\right)}$.

22. Using the linearity property, we have $L\left(5e^{2t} - t^3\right) = L\left(5e^{2t}\right) - L\left(t^3\right) = 5L\left(e^{2t}\right) - L\left(t^3\right)$. By transform 5 of Table 34.1, $L\left(e^{2t}\right) = \frac{1}{s-2}$ and by transform 3, $L\left(t^3\right) = \frac{3!}{s^{3+1}} = \frac{6}{s^4}$. Thus, $L\left(5e^{2t} - t^3\right) = 5L\left(e^{2t}\right) - L\left(t^3\right) = 5\left(\frac{1}{s-2}\right) - \left(\frac{6}{s^4}\right) = \frac{5}{s-2} - \frac{6}{s^4}$.

≡ 34.4 INVERSE LAPLACE TRANSFORMS AND TRANSFORMS OF DERIVATIVES

1. $\frac{4}{s-5}$. $s - a$ appears in the denominator of form 5. $L^{-1}(F) = 4e^{5t}$.

2. $\frac{6}{s^2+4} = 3\frac{2}{s^2+2^2}$. $L^{-1}(F) = 3\sin 2t$ (form 8)

3. $\frac{16}{s^3+16s} = \frac{16}{s\left(s^2+16\right)} = \frac{4^2}{s\left(s^2+4^2\right)}$ fits form 10.
$L^{-1}(F) = 1 - \cos 4t$

4. $\frac{4}{(s+9)^2} = 4\cdot\frac{1}{(s-(-9))^2}$ fits form 6. $L^1(F) = 4te^{-9t}$.

5. $\frac{3s}{s^2+9} = 3\cdot\frac{s}{s^2+3^2}$. $L^{-1}(F) = 3\cos 3t$. (form 9)

6. $\frac{s+6}{(s+16)^2} = \frac{2+16-10}{(s+16)^2} = \frac{1}{s+16} - \frac{10}{(s+16)^2}$. By forms 5 and 6 we get $L^{-1}(F) = e^{-16t} - 10te^{-16t} = e^{-16t}(1 - 10t)$

7. $\frac{5s}{s^2+9s+14} = \frac{5s}{(s+7)(s+2)}$. Using form 18 with $a = -2$ and $b = -7$ we get $-2e^{-2t} + 7e^{-7t}$

8. $\frac{s^2-16}{\left(s^2+4\right)^2} = \frac{s^2-4-12}{\left(s^2+4\right)^2} = \frac{s^2-4}{\left(s^2+4\right)^2} - \frac{12}{\left(s^2+4\right)^2} = \frac{s^2-2^2}{\left(s^2+2^2\right)^2} - \frac{3}{4}\frac{2\cdot 2^3}{\left(s^2+2^2\right)^2}$. By forms 15 and 12, $L^{-1}(F) = t\cos 2t - \frac{3}{4}(\sin 2t - 2t\cos 2t)$.

9. $\frac{2s+6s^2}{\left(s^2+25\right)^2} = \frac{1}{5}\cdot\frac{2\cdot 5s}{\left(s^2+5^2\right)^2} = \frac{6}{10}\frac{2\cdot 5s^2}{\left(s^2+5^2\right)^2}$. By forms 13 and 14, $L^{-1}(F) = \frac{1}{5}t\sin 5t + \frac{3}{5}(\sin 5t + 5t\cos 5t)$.

10. $\frac{s+5}{s^2+10s+26} = \frac{s+5}{s^2+10s+25+1} = \frac{s+5}{(s+5)^2+1}$. By form 20, $L^{-1}(F) = e^{-5t}\cos t$

11. $L(f'') + 4L(f') = \left[s^2L(f) - sf(0) - f'(0)\right] + 4[sL(f) - f(0)] = \left(s^2 + 4s\right)L(f) - s - 4 = \left(s^2 + 4s\right)L(f) - (s + 4)$

12. $L(f'') - 6L(f) = s^2L(f) - sf(0) - f'(0) - 6L(f) = \left(s^2 - 6\right)L(f) - 3s - 2 = \left(s^2 - 6\right)L(f) - (3s + 2)$

13. $2L(f'') - 3L(f') + L(f) = 2\left[s^2L(f) - sf(0) - f'(0)\right] - 3[sL(f) - f(0)] + L(f) = \left(2s^2 - 3s + 1\right)L(f) - (2s - 3)$

14. $L(f'') - 4L(f') = s^2L(f) - sf(0) - f'(0) - 4[sL(f) - f(0)] = \left(s^2 - 4s\right)L(f) - 2s - 1 + 8 = \left(s^2 - 4s\right)L(f) - (2s - 7)$

15. $L(f'') + 6L(f') - L(f) = \left[s^2L(f) - sf(0) - f'(0)\right] + 6[sL(f) - f(0)] - L(f) = \left(s^2 + 6s - 1\right)L(f) - s + 1 - 6 = \left(s^2 + 6s - 1\right)L(f) - (s + 5)$

16. $L(f'') - 3L(f') + 4L(f) = s^2 L(f) - sf(0) - f'(0) - 3[sL9f) - f(0)] + 4L(f) = (s^2 - 3s + 4)L(f) +$

$2s - 1 - 6 = (s^2 - 3s + 4)L(f) + (2s - 7)$

34.5 PARTIAL FRACTIONS

1. $\frac{1}{x(x+1)} = \frac{A}{x} + \frac{B}{x+1}$. Multiplying by $x(x+1)$ we get $1 = A(x+1) + Bx$. If $x = 0$ we get $A = 1$. If $x = -1$, we get $B = -1$. The partial fraction decomposition is $\frac{1}{x} + \frac{-1}{x+1}$.

2. $\frac{4}{x(x-3)} + \frac{A}{x} + \frac{B}{x-3}$. Multiplying by $x(x-3)$ we get $4 = A(x-3) + Bx$. Setting $x = 0$ we get $A = -\frac{4}{3}$. Setting $x = 3$ we get $B = \frac{4}{3}$. The partial fraction decomposition is $\frac{-4/3}{x} + \frac{4/3}{x-3}$.

3. $\frac{4}{x(x-1)(x+1)} = \frac{A}{x} + \frac{B}{x-1} + \frac{C}{x+1}$. Multiplying by the LCD we get $4 = A(x^2 - 1) + B(x^2 + x) + C(x^2 - x) = (A + B + C)x^2 + (B - C)x - A$. Hence $A = -4$. Substituting this we have $B + C - 4 = 0$ and $B - C = 0$. Solving we get $B = 2$ and $C = 2$. The partial fraction decomposition is $\frac{-4}{x} + \frac{2}{x-1} + \frac{2}{x+1}$.

4. $\frac{5x}{(x-2)(x+3)} = \frac{A}{(x-2)} + \frac{B}{(x+3)}$. Multiplying by the LCD we get $5x = A(x+3) + B(x-2)$. Setting $x = 2$ we find $A = \frac{10}{5} = 2$. Setting $x = -3$ we obtain $B = \frac{-15}{-15} = 3$. The answer is $\frac{2}{x-2} + \frac{3}{x+3}$.

5. $\frac{3x}{x^2 - 8 + 15} = \frac{3x}{(x-3)(x-5)} = \frac{A}{x-3} + \frac{B}{x-5}$. Multiplying by the LCD we get $3x = A(x-5) + B(x-3)$. Setting $x = 3$ yields $A = \frac{9}{-2} = -\frac{9}{2}$. Setting $x = 5$ yields $B = 15/2$. Hence the answer is $\frac{-9/2}{x-3} + \frac{15/2}{x-5}$.

6. $\frac{4x}{(x+1)(x+2)(x+3)} = \frac{A}{x+1} + \frac{B}{x+2} + \frac{C}{x+3}$. This is equivalent to $4x = A(x^2 + 5x + 6) + B(x^2 + 4x + 3) + C(x^2 + 3x + 2)$. Collecing like terms we have $(A+B+C)x^2 + (5A+4B+3C)x + 6A+3B+2C = 4x$. This gives the linear systems $A + B + C = 0$; $5A + 4B + 3C = 4$; and $6A + 3B + 2C = 0$. The solution is $A = -2$, $B = 8$ and $C = -6$. The partial fraction decomposition is $\frac{-2}{x+1} + \frac{8}{x+2} + \frac{-6}{x+3}$.

7. $\frac{x-1}{x^2 + 2x + 1} = \frac{x-1}{(x+1)^2} + \frac{A}{x+1} + \frac{B}{(x+1)^2}$. This is equivalent to $x - 2 = A(x+1) + B$. Setting $x = 1$ yields $B = -2$. Substituting this we have $x - 1 = Ax + A - 2$. Using the x-term we see $A = 1$. The answer is $\frac{1}{x+1} + \frac{-2}{(x+1)^2}$.

8. $\frac{x}{(x+1)^2(x-2)} = \frac{A}{x+1} + \frac{B}{(x+1)^2} + \frac{C}{x-2}$. This is equivalent to $x = A(x+1)(x-2) + B(x-2) + C(x+1)^2$. Setting $x = -1$ we obtain $B = \frac{-1}{-3} = \frac{1}{3}$. Setting $x = 2$ we obtain $2 = 9C$ or $C = \frac{2}{9}$. The expansion of the equation (1) yields an x^2-term of $(A + C)x^2 = 0$ so $A = -\frac{2}{9}$. The partial fraction decomposition is $\frac{-2/9}{x+1} + \frac{1/3}{(x+1)^2} + \frac{2/9}{x-2}$.

9. $\frac{x^3 + x^2 + x + 2}{(x^2+1)(x^2+2)} = \frac{Ax+B}{x^2+1} + \frac{Cx+D}{x^2+2}$ is equivalent to $x^3 + x^2 + x + 2 = (Ax + B)(x^2 + 2) + (Cx + D)(x^2 + 1) = (A + C)x^3 + (B + D)x^2 + (2A + C)x + (2B + D)$. So $A + C = 1$ and $2A + C = 1$ yield $A = 0$ and $C = 1$. Also, $B + D = 1$ and $2B + D = 2$ yield $B = 1$ and $D = 0$. The partial fraction decomposition is $\frac{1}{x^2+1} + \frac{x}{x^2+2}$.

10. $\frac{3x^2 + 5}{(x^2+1)^2} = \frac{Ax+B}{x^2+1} + \frac{Cx+D}{(x^2+1)^2}$ is equivalent to $3x^2 + 5 = (Ax + B)(x^2 + 1) + Cx + D = Ax^3 + Bx^2 + Ax + B + Cx + D = Ax^3 + Bx^2 + (A+C)x + (B+D)$. Hence $A = 0$, $A + C = 0 \Rightarrow C = 0$, $B = 3$, and $B + D = 5$ so $D = 5 - 3 = 2$. The partial function decomposition is $\frac{3}{x^2+1} + \frac{2}{(x^2+1)^2}$.

11. $L^{-1}\left[\frac{1}{s(s+1)}\right] = L^{-1}\left[\frac{1}{s}\right] + L^{-1}\left[\frac{-1}{s+1}\right] = 1 - e^{-t}$.

12. $L^{-1}\left[\frac{4}{s(s-3)}\right] = L^{-1}\left[\frac{-4/3}{s}\right] + L^{-1}\left[\frac{4/3}{s-3}\right] = -\frac{4}{3} + \frac{4}{3}e^{3t}$.

13. $L^{-1}\left[\frac{4}{s(s-1)(s+1)}\right] = L^{-1}\left[\frac{-4}{s}\right] + L^{-1}\left[\frac{2}{s-1}\right] + L^{-1}\left[\frac{2}{s+1}\right] = -4 + 2e^t + 2e^{-t}$.

14. $L^{-1}\left[\frac{5s}{(s-2)(s+3)}\right] = L^{-1}\left[\frac{2}{s-2}\right] + L^{-1}\left[\frac{3}{s+3}\right] = 2e^{2t} + 3e^{-3t}$.

15. $L^{-1}\left[\frac{3s}{s^2 - 8s + 15}\right] = L^{-1}\left[\frac{-9/2}{s-3}\right] + L^{-1}\left[\frac{15/2}{s-5}\right] = -\frac{9}{2}e^{3t} + \frac{15}{2}e^{5t}$.

16. $L^{-1}\left[\frac{4s}{(s+1)(s+2)(s+3)}\right] = L^{-1}\left[\frac{-2}{s+1}\right] + L^{-1}\left[\frac{8}{s+2}\right] + L^{-1}\left[\frac{-6}{s+3}\right] = -2e^{-t} + 8e^{-2t} - 6e^{-3t}$.

17. $L^{-1}\left[\frac{s-1}{s^2+s+1}\right] = L^{-1}\left[\frac{1}{s+1}\right] + L^{-1}\left[\frac{-2}{(s+1)^2}\right] = e^{-t} - 2te^{-t}$

18. $L^{-1}\left[\frac{s}{(s+1)^2(s-2)}\right] = L^{-1}\left[\frac{-2/9}{s+1}\right] + L^{-1}\left[\frac{1/3}{(s+1)^2}\right] + L^{-1}\left[\frac{2/9}{s-2}\right] = -\frac{2}{9}e^{-t} + \frac{1}{3}te^{-t} + \frac{2}{9}e^{2t}$.

19. $L^{-1}\left[\frac{s^3+s^2+s+2}{(s^2+1)(s^2+2)}\right] = L^{-1}\left[\frac{1}{s^2+1}\right] + L^{-1}\left[\frac{s}{s^2+2}\right] = \sin t + \cos\sqrt{2}t$.

20. $L^{-1}\left[\frac{3s+5}{(s^2+1)^2}\right] = L^{-1}\left[\frac{3}{s^2+1}\right] + L^{-1}\left[\frac{2}{(s^2+1)^2}\right] = 3\sin t + \sin t - t\cos t = 4\sin t - t\cos t$.

21. First we must find the partial fraction decomposition of $\frac{2}{s^2(s-1)}$. We set it equal to $\frac{A}{s} + \frac{B}{s^2} + \frac{C}{s-1}$. Multiplying by the LCD we obtain $2 = As(s-1) + B(s-1) + Cs^2$. Setting $s = 0$ yields $B = -2$. Setting $s = 1$ yields $c = 2$. Expanding the s^2-term is $(A+C)s^2$ so $A = -C = -2$. $L^{-1}\left[\frac{2}{s^2(s-1)}\right] = L^{-1}\left[\frac{-2}{s}\right] + L^{-1}\left[\frac{-2}{s^2}\right] + L^{-1}\left[\frac{2}{s-1}\right] = -2 - 2t + 2e^t$.

22. $\frac{s}{(s+1)(s^2+1)} = \frac{A}{s+1} + \frac{Bs+C}{s^2+1}$ is equivalent to $s = A(s^2+1) + (Bs+C)(s+1) = (A+B)s^2 + (B+C)s + (A+C)$. So $A + B = 0$, $B + C = 1$, and $A + C = 0$. Solving we get $A = -\frac{1}{2}$, $B = \frac{1}{2}$ and $C = \frac{1}{2}$. Hence $L^{-1}\left[\frac{s}{(s+1)(s^2+1)}\right] = L^{-1}\left[\frac{-1/2}{s+1}\right] + L^{-1}\left[\frac{s/2+1/2}{s^2+1}\right] = -\frac{1}{2}e^{-t} + \frac{1}{2}\cos t + \frac{1}{2}\sin t$.

23. $\frac{1}{(s+4)(s^2+9)} = \frac{A}{s+4} + \frac{Bs+C}{s^2+9}$ is equivalent to $1 = A(s^2+9) + (Bs+C)(s+4) = (A+B)s^2 + (4B+C)s + 9A + 4C$. Hence $A+B = 0$, $4B+C = 0$ and $9A+4C = 1$. Solving we find $A = \frac{1}{25}$, $B = -\frac{1}{25}$,

and $C = \frac{4}{25}$. $L^{-1}\left[\frac{1}{(s+4)(s^2+9)}\right] = L^{-1}\left[\frac{1/25}{s+4}\right] + L^{-1}\left[\frac{-s/25+4/25}{s^2+9}\right] = \frac{1}{25}e^{-4t} - \frac{1}{25}\cos 3t + \frac{4}{25}\sin 3t$.

24. $\frac{1}{(s^2+1)(s^2-1)} = \frac{1}{(s^2+1)(s+1)(s-1)} = \frac{A}{s+1} + \frac{B}{s-1} + \frac{Cs+D}{s^2+1}$. Multiplying by the LCD we get $1 = A(s^2+1)(s-1) + B(s+1)(s^2+1) + (Cs+D)(s^2-1)$. Setting $s = 1$ we obtain $1 = B(2\cdot 2)$ or $B = \frac{1}{4}$. Setting $s = -1$ we obtain $1 = A(2)(-2)$ or $A = -\frac{1}{4}$. The s^3-term of the expansion is $(A+B+C)s^3$ and this is 0 so $C = -A - B = 0$. The constant term of the expansion is $-A + B - D = 1$ so $D = B - A - 1 = -\frac{1}{2}$. $L^{-1}\left[\frac{1}{(s^2+1)(s^2-1)}\right] = L^{-1}\left[\frac{-1/4}{s+1}\right] + L^{-1}\left[\frac{1/4}{s-2}\right] + L^{-1}\left[\frac{-1/2}{s^2+1}\right] = -\frac{1}{4}e^{-t} + \frac{1}{4}e^t - \frac{1}{2}\sin t$.

25. $\frac{1}{(s^2+1)(s-1)^2} = \frac{A}{s-1} + \frac{B}{(s-1)^2} + \frac{Cs+D}{s^2+1}$. Multiplying by the LCD we obtain $1 = A(s-1)(s^2+1) + B(s^2+1) + (Cs+D)(s-1)^2$. Setting $s = 1$ we get $B + \frac{1}{2}$. Expanding the right-hand side of the equation we get $As^3 - As^2 + As - A + Bs^2 + B + Cs^3 - 2Cs^2 + Cs + Ds^2 - 2Ds + D$. Collecting like terms we have $(A+C)s^3 + (-A+B-2C+D)s^2 + (A+C-2D)s + (A+B+D)$. So $A + C = 0$, $(-A+B-2C+D) = 0$, $(A+C-2D) = 0$, and $(-A+B+D) = 1$. Solving we get $A = -\frac{1}{2}$, $B = \frac{1}{2}$, $C = \frac{1}{2}$, and $D = 0$. Hence $L^{-1}\left[\frac{1}{(s^2+1)(s-1)^2}\right] = L^{-1}\left[\frac{-1/2}{s-1}\right] + L^{-1}\left[\frac{1/2}{(s-1)^2}\right] + L^{-1}\left[\frac{s/2}{s^2+1}\right] = -\frac{1}{2}e^t + \frac{1}{2}te^t + \frac{1}{2}\cos t$.

26. $\frac{1}{(s^2+4s+7)(s+1)} = \frac{A}{s+1} + \frac{Bs+C}{s^2+4S+7}$. Multiplying by the LCD we get $1 = A(s^2+4s+7) + (Bs+C)(s+1)$. Expanding we have $1 = As^2 + 4As + 7A + Bs^2 + Bs + Cs + c$. Collecting like terms we have $(A+B)s^2 + (4A+B+C)s + (7A+C) = 1$. Hence $A + B = 0$, $4A + B + C = 0$, and $7A + C = 1$. Solving this system we get $A = \frac{1}{4}$, $B = -\frac{1}{4}$, and $C = -\frac{3}{4}$. Thus, $L^{-1}\left[\frac{1}{(s^2+4s+7)(s+1)}\right] = L^{-1}\left[\frac{\frac{1}{4}}{s+1}\right] + L^{-1}\left[\frac{-\frac{s}{4}-\frac{3}{4}}{s^2+4s+7}\right]$. The first inverse Laplace

transform is $\frac{1}{4}e^{-t}$ but the second needs work. So, $\frac{-\frac{1}{4}s-\frac{3}{4}}{s^2+4s+7} = \frac{-\frac{1}{4}s-\frac{3}{4}}{s^2+4s+4+3} = \frac{-\frac{1}{4}s-\frac{3}{4}}{(s+2)^2+\sqrt{3}^2} =$

$-\frac{1}{4}\left(\frac{s+3}{(s+2)^2+\sqrt{3}^2}\right) = -\frac{1}{4}\left(\frac{s+2+1}{(s+2)^2+\sqrt{3}^2}\right) = -\frac{1}{4}\left(\frac{s+2}{(s+2)^2+(\sqrt{3})^2}\right) - \frac{1}{4\sqrt{3}}\left(\frac{\sqrt{3}}{(s+2)^2+(\sqrt{3})^2}\right)$. Hence, $L^{-1}\left[\frac{-\frac{1}{4}s-\frac{3}{4}}{s^2+4s+7}\right] =$

$-\frac{1}{4}L^{-1}\left[\frac{s+2}{(s+2)^2+\sqrt{3}^2}\right]-\frac{1}{4\sqrt{3}}L^{-1}\left[\frac{\sqrt{3}}{(s+2)^2+\sqrt{3}^2}\right]=-\frac{1}{4}e^{-2t}\cos\sqrt{3}t-\frac{1}{4\sqrt{3}}e^{-2t}\sin\sqrt{3}t.$ Putting these answers together we get $\frac{1}{4}e^{-t}-\frac{1}{4}e^{-2t}\cos\sqrt{3}t-\frac{1}{4\sqrt{3}}e^{-2t}\sin\sqrt{3}t.$

≡ 34.6 USING LAPLACE TRANSFORMS TO SOLVE DIFFERENTIAL EQUATIONS

1. $y'-y=1$, $y(0)=0$, $L(y')=sL(y)-y(0)=sL(y)$. Hence the Laplace Transform is $sY-Y=\frac{1}{s}$ or $Y(s-1)=\frac{1}{s}$ or $Y=\frac{1}{s(s-1)}$. The partial fraction decomposition for $\frac{1}{s(s-1)}=\frac{A}{s}+\frac{B}{s-1}$ or $1=A(s-1)+Bs$. Setting $s=0$ yields $A=-1$. Setting $s=1$ yields $B=1$ so $Y=\frac{-1}{s}+\frac{1}{s-1}$ and $L^{-1}(Y)=L^{-1}\left(\frac{-1}{s}\right)+L^{-1}\left[\frac{1}{s-1}\right]$ or $y=-1+e^t$.

2. $y'+2y=t$, $y(0)=-1$. The Laplace Transform is $L(y')+2L(y)=L(t)$ or $sL(y)-y(0)+2L(y)=\frac{1}{s^2}$ or $(S+2)Y+1=\frac{1}{s^2}$. This yields $(s+2)Y=\frac{1}{s^2}-1=\frac{1-s^2}{s^2}$ or $Y=\frac{1-s^2}{s^2(s+2)}$. The partial fraction decomposition is as follows: $\frac{1-s^2}{s^2(s+2)}=\frac{A}{s+2}+\frac{B}{s}+\frac{C}{s^2}$ or $1-s^2=As^2+B(s+2)s+C(s+2)$. Setting $s=-2$ yields $A=-\frac{3}{4}$. Setting $s=0$ yields $C=\frac{1}{2}$. The s^2-term of the expansion yields $-1=A+B$ so $B=-1-A=-\frac{1}{4}$. Thus, $Y=\frac{-\frac{3}{4}}{s+2}+\frac{-\frac{1}{4}}{s}+\frac{\frac{1}{2}}{s^2}$ and $y=-\frac{3}{4}e^{-2t}-\frac{1}{4}+\frac{1}{2}t=-\frac{1}{4}+\frac{1}{2}t-\frac{3}{4}e^{-2t}$.

3. $y'+6y=e^{-6t}$, $y(0)=2$. The Laplace Transform is $sY-y(0)+6Y=\frac{1}{s+6}$ or $(s+6)Y-2=\frac{1}{s+6}$. So $(s+6)Y=\frac{1}{s+6}+2=\frac{1+2s+12}{s+6}=\frac{2s+13}{s+6}$. Hence $Y=\frac{2s+13}{(s+6)^2}=\frac{A}{s+6}+\frac{B}{(s+6)^2}$. Multiplying by the LCD we have $2s+13=A(s+6)+B=As+6A+B$. Hence $A=2$ and $6A+B=13$ or $B=1$, $y=L^{-1}\left[\frac{2}{s+6}\right]+L^{-1}\left[\frac{1}{(s+6)^2}\right]=2e^{-6t}+te^{-6t}$.

4. $y'-y=\sin 10t$, $y(0)=0$. The Laplace Transform is $sY-y(0)-Y=\frac{10}{s^2+100}$ or $(s-1)Y=\frac{10}{s^2+100}$. Hence $Y=\frac{10}{(s-1)(s^2+100)}=\frac{A}{s-1}+\frac{Bs+C}{s^2+100}$. This yields $10=A(s^2+100)+(Bs+c)(s-1)$. Letting $s=1$ we get $A=\frac{10}{101}$. Expanding the s^2-term is $(A+B)s^2$ which is zero so $B=-A=-\frac{10}{101}$. The s-term is $(-B+C)s$ which is also 0 so $C=B=\frac{-10}{101}$. Hence $y=\frac{10}{101}\left(\frac{1}{s-1}+\frac{-1s-1}{s^2+100}\right)$ and

$y=\frac{10}{101}L^{-1}\left[\frac{1}{s-1}-\frac{s+1}{s^2+100}\right]=\frac{10}{101}\left[e^t-\cos 10t-\frac{1}{10}\sin t\right].$

5. $y'+5y=te^{-5t}$, $y(0)=3$. The Laplace Transform is $sY-y(0)+5Y=\frac{1}{(s+5)^2}$. Hence $(s+5)Y=\frac{1}{(s+5)^2}+3=\frac{1+3s^2+30s+75}{(s+5)^2}=\frac{3s^2+30s+76}{(s+5)^2}$. This yields $Y=\frac{3s^2+30s+76}{(s+5)^3}=\frac{A}{s+5}+\frac{B}{(s+5)^2}+\frac{C}{(s+5)^3}$ or $3s^2+30s+76=A(s+5)^2+B(s+5)+C$. Letting $s=-5$ we get $C=1$. The s^2-term gives $A=3$. The s-term yields $10A+B=30$, so $B=30-10A=30-30=0$. Hence $Y=\frac{3}{s+5}+\frac{1}{(s+5)^3}$ and $y=L^{-1}\left[\frac{3}{s+5}+\frac{1}{(s+5)^3}\right]=3e^{-5t}+\frac{1}{2}t^2e^{-5t}$.

6. $y'+y=\sin t$, $y(0)=-1$. The Laplace Transform is $sY+1+Y=\frac{1}{s^2+1}$ so $(s+1)Y=\frac{1}{s^2+1}-1=\frac{-s^2}{s^2+1}$ and $Y=\frac{-s^2}{(s+1)(s^2+1)}=\frac{A}{s+1}+\frac{Bs+C}{s^2+1}$. Multiplying by the LCD yields $-s^2=A(s^2+1)+(Bs+C)(s+1)$. Setting $s=-1$ we get $A=-\frac{1}{2}$. Expanding, the s^2-term yields $A+B=-1$, so $B=-\frac{1}{2}$. The constant term is $A+C=0$ so $C=\frac{1}{2}$. Hence $Y=\frac{-\frac{1}{2}}{s+1}+\frac{-\frac{1}{2}s}{s^2+1}+\frac{\frac{1}{2}}{s^2+1}$ and $y=-\frac{1}{2}L^{-1}\left[\frac{1}{s+1}+\frac{s}{s^2+1}-\frac{1}{s^2+1}\right]=-\frac{1}{2}\left[e^{-t}+\cos t-\sin t\right]$

7. $y''+4y=0$, $y'(0)=3$, $y(0)=2$. The Laplace Transform is $s^2Y-sy(0)-y(0)+4Y=0$ or $(s^2+4)Y-2s-3=0$. Hence $Y=\frac{2s+3}{s^2+4}=2\frac{s}{s^2+2^2}+\frac{3}{s^2+2^2}$ and $y=2\cos 2t+\frac{3}{2}\sin 2t$.

8. $y''-5y'+4y=0$ has Laplace Transform $s^2Y-sy(0)-y'(0)-5sY+5y(0)+4y=0$ or $(s^2-5s+4)Y+2s-4-10=0$. Hence $Y=\frac{-2s+14}{s^2-5s+4}=\frac{-2s+14}{(s-4)(s-1)}=\frac{A}{s-4}+\frac{B}{s-1}$. Multiplying by the LCD yields $-2s+14=A(s-1)+B(s-4)$. Setting $s=4$ we get $3A=6$ or $A=2$. Setting $s=1$ we get $-3B=12$ or $B=-4$. Hence $Y=\frac{2}{s-4}+\frac{-4}{s-1}$. The inverse Laplace transform yields $y=2e^{4t}-4e^t$.

9. $y'' + 4y = e^t$, $y(0) = y'(0) = 0$. The Laplace transform is $s^2Y + 4y + \frac{1}{s-1}$ or $Y = \frac{1}{(s-1)(s^2+4)} =$

$\frac{A}{s-1} + \frac{Bs+C}{s^2+4}$. Multiplying by the LCD we get $A(s^2+4) + (Bs+C)(s-1) = 1$. Setting $s = 1$ we get $A = \frac{1}{5}$. Expanding we get $(A+B)s^2 + (-B+C)s + (4A-C) = 1$. Hence $B = -A = -\frac{1}{5}$, and $C = B = -\frac{1}{5}$. So the Laplace becomes $Y = \frac{1}{5}\left(\frac{1}{s-1} + \frac{-s-1}{s^2+4}\right) = \frac{1}{5}\left(\frac{1}{s-1} - \frac{s}{s^2+2^2} - \frac{1}{s^2+2^2}\right)$ and the inverse Laplace transform yields $y = \frac{1}{5}\left(e^t - \cos 2t - \frac{1}{2}\sin 2t\right)$.

10. $y'' + 2y' - 3y = 5e^{2t}$, $y(0) = 2$, $y'(0) = 3$. The Laplace transform is $s^2Y - sy(0) - y'(0) + 2sY - 2y(0) - 3Y = \frac{5}{s-2}$ of $(s^2+2s-3)Y - 2s - 3 - 4 = \frac{5}{s-2}$ or $(s^2+2s-s)Y = \frac{5}{s-2} + (2s+7) = \frac{5+2s^2+3s-14}{s-2}$ so $Y = \frac{2s^2+3s-9}{(s-2)(s+3)(s-1)} = \frac{A}{s-2} + \frac{B}{s+3} + \frac{C}{s-1}$. Multiplying by the LCM we get $2s^2 + 3s - 9 = A(s+3)(s-1) + B(s-2)(s-1) + C(s-2)(s+3)$. Setting $s = 2$ we get $A = \frac{5}{5} = 1$. Setting $S = -3$, $B = \frac{0}{20} = 0$ and setting $s = 1$, $C = \frac{-4}{-4} = +1$. Hence $Y = \frac{1}{s-2} + \frac{0}{s+3} + \frac{1}{s-1}$ and the inverse Laplace transform yields $y = e^{2t} + e^t$.

11. $y'' - 6y' + 9y = t$, $y(0) = 0$, $y'(0) = 1$. The Laplace Transform is $s^2Y - sy(0) - y'(0) - 6(sY - y(0)) + 9 = \frac{1}{s^2}$ or $(s^2-6s+9)Y - 1 = \frac{1}{s^2}$ or $(s^2-6s+9)Y = \frac{1}{s^2} + 1 = \frac{1+s^2}{s^2}$. Hence $Y = \frac{s^2+1}{s^2(s-3)^2} = \frac{A}{s} + \frac{B}{s^2} + \frac{C}{s-3} + \frac{IQ}{(s-3)^2}$. Multiplying by the LCD we have $s^2 + 1 = As(s-3)^2 + B(s-3)^2 + C(s-3)s^2 + Ds^2$. Setting $s = 0$, we find $B = \frac{1}{9}$, and setting $s = 3$, we get $D = \frac{10}{9}$. The s-term yields $9A - 6B = 0$ so $A = \frac{6B}{9} = \frac{6}{81} = \frac{2}{27}$. The s^3-term yields $A + C = 0$ so $C = -A = -\frac{2}{27}$. $Y = \frac{2}{27}\frac{1}{s} + \frac{1}{9}\frac{1}{s^2} - \frac{2}{27}\frac{1}{s-3} + \frac{10}{9}\cdot\frac{1}{(s-3)^2}$. The inverse Laplace Transform yields $y = \frac{2}{27} + \frac{1}{9}t - \frac{2}{27}e^{3t} + \frac{10}{9}te^{3t}$.

12. $y'' + y = 1$, $y(0) = y'(0) = 1$. The Laplace Transform is $s^2Y - s\cdot 1 - 1 + Y = \frac{1}{s}$ so $(s^2+1)Y = \frac{1}{s} + s + 1 = \frac{s^2+s+1}{s}$ and $Y = \frac{s^2+s+1}{s(s^2+1)} = \frac{A}{s} + \frac{Bs+C}{s^2+1}$. Multiplying we get $s^2 + s + 1 = A(s^2+1) + (Bs+C)(s) = (A+B)s^2 + Cs + A$. So $A = 1$, $C = 1$,

and $A + B = 1 \Rightarrow B = 0$. $Y = \frac{1}{s} + \frac{1}{s^2+1}$. The inverse Laplace Transform is $y = 1 + \sin t$.

13. $y'' + 6y' + 13y = 0$, $y(0) = 1$, $y''(0) = -4$. The Laplace Transform is $s^2Y - s\cdot 1 + 4 + 6(sY - 1) + 13Y = 0$ which simplifies to $(s^2+6s+13)Y = s + 2$, so $Y = \frac{s+2}{s^2+6s+13} = \frac{s+2}{s^2+6s+9+4} = \frac{s+3-1}{(s+3)^2+2^2} = \frac{s+3}{(s+3)^2+2^2} - \frac{1}{2}\frac{2}{(s+3)^2+2^2}$. The inverse Laplace Transform yields $y = e^{-3t}\left(\cos 2t - \frac{1}{2}\sin 2t\right)$.

14. $y'' + 2y' + 5y = 8e^t$, $y(0) = y'(0) = 0$. The Laplace Transform is $s^2Y + 2sY + 5Y = \frac{8}{s-1}$, so $Y = \frac{8}{(s-1)(s^2+2s+s)} = \frac{A}{s-1} + \frac{Bs+C}{s^2+2s+5}$, and $8 = A(s^2+2s+5) + (Bs+C)(s-1) = (A+B)s^2 + (2A-B+C)s + 5A - C$. So $A + B = 0$, $2A - B + C = 0$, and $5A - C = 8$. This system yields $A = 1$, $B = -1$, $C = -3$ so $Y = \frac{1}{s-1} - \frac{s+3}{s^2+2s+5} = \frac{1}{s-1} - \frac{s+2+1}{s^2+2s+1+4} = \frac{1}{s-1} - \frac{s+2}{(s+1)^2+2^2} - \frac{1}{2}\cdot\frac{2}{(s+1)^2+2^2}$. The inverse Laplace Transform yields $y = e^t - e^{-t}\cos 2t - \frac{1}{2}e^{-t}\sin 2t$.

15. $y'' - 4y = 3\cos t$, $y(0) = y'(0) = 0$. The Laplace Transform is $s^2Y - 4Y = \frac{3s}{s^2+1}$ so $Y = \frac{3s}{(s^2+1)(s^2-1)} = \frac{3s}{(s^2+1)(s+2)(s-2)} = \frac{A}{s+2} + \frac{B}{s-2} + \frac{Cs+D}{s^2+1}$. Multiplying by the LCD we get $3s = A(s-2)(s^2+1) + B((s+2)(s^2+1) + (Cs+D)(s^2-4)$. Setting $s = -2$, we get $A = \frac{-6}{(-4)(5)} = \frac{3}{10}$. Setting $s = 2$, we get $B = \frac{6}{4\cdot 5} = \frac{3}{10}$. The s^3-term of the expansion is $A + B + C = 0$ so $C = -\frac{6}{10} = -\frac{3}{5}$. The constant term of the expansion is $-2A + 2B - 4D = 0$ so $D = \frac{-2A+2B}{4} = 0$. Hence $Y = \frac{3}{10}\cdot\frac{1}{s+2} + \frac{3}{10}\cdot\frac{1}{s-2} - \frac{3}{5}\cdot\frac{s}{s^2+1}$ and the inverse Laplace Transform is $y = \frac{3}{10}e^{-2t} + \frac{3}{10}e^{2t} - \frac{3}{5}\cos t$.

16. $y'' - y = e^t$, $y(0) = 1$, $y'(0) = 0$. The Laplace Transform is $s^2Y - s - Y = \frac{1}{s-1}$. This is equivalent to $(s^2-1)Y = \frac{1}{s-1} + s\frac{s^2-s+1}{s-1}$ or $Y = \frac{s^2-s+1}{(s^2-1)(s-1)} = \frac{s^2-s+1}{(s-1)(s-1)^2}$. Finding the partial fraction decomposition we have $\frac{s^2-s+1}{(s-1)(s-1)^2} = \frac{A}{s+1} + \frac{B}{s-1} + \frac{C}{(s-1)^2}$. This becomes $s^2 - s + 1 = A(s-1)^2 + B(s^2-1) + C(s+1)$. Setting $s = -1$ we get $A = \frac{3}{4}$. Setting $s = 1$ we get $C = \frac{1}{2}$. The s^2-term

yields $A + B = 1$ or $B = 1 - A = 1 - \frac{3}{4} = \frac{1}{4}$.
Hence $Y = \frac{3}{4} \cdot \frac{1}{s+1} + \frac{1}{4} \cdot \frac{1}{s-1} + \frac{1}{2} \cdot \frac{1}{(s-1)^2}$. The inverse Laplace transform yields the answer $y = \frac{3}{4}e^{-t} + \frac{1}{4}e^t + \frac{1}{2}te^t$.

17. $y'' + 2y' + 5y = 3e^{-2t}$, $y(0) = y'(0) = 1$. The Laplace transform is $\left(s^2Y - s - 1\right) + 2sY - 2 + 5Y = \frac{3}{s+2}$. This is equivalent to $\left(s^2 + 2s + 5\right)Y - (s + 3) = \frac{3}{s+2}$. So $\left(s^2 + 2s + 5\right)Y = \frac{3}{s+2} + (s + 3) = \frac{3 + \left(s^2+5s+6\right)}{s+2} = \frac{s^2+5s+9}{s+2}$ Hence $Y = \frac{\left(s^2+5s+9\right)}{(s+2)\left(s^2+2s+5\right)} = \frac{A}{s+2} + \frac{Bs+C}{s^2+2s+5}$. This gives $\left(s^2+5s+9\right) = A\left(s^2+2s+5\right) + (Bs+C)(s+2) = (A+B)s^2 + (2A+2B+C)s + 5A + 2C$. And we have linear equations $A + B = 1$, $2A + 2B + C = 5$ and $5A + 2C = 9$. Solving we obtain $A = \frac{3}{5}$, $B = \frac{2}{5}$, and $C = 3$. So $Y = \frac{3}{5} \cdot \frac{1}{s+2} + \frac{\frac{2}{5}s + 3}{s^2 + 2s + 5}$. The second fraction is
$\frac{2}{5}\frac{s + \frac{15}{2}}{s^2 + 2s + 1 + 4} = \frac{2}{5}\frac{s + 1 + \frac{13}{2}}{(s + 1)^2 + 2^2} = \frac{2}{5}\frac{s+1}{(s+1)^2+2^2} = \frac{13}{10}\frac{2}{(s+1)^2+2^2}$. The inverse Laplace transform is $y = \frac{3}{5}e^{-2t} + e^{-t}\left(\frac{2}{5}\cos 2t + \frac{13}{10}\sin 2t\right)$.

18. $y'' - 6y' + 9y = 12t^2e^{3t}$, $y(0) = y'(0) = 0$. The Laplace transform is $s^2Y - 6sY + 9Y = \frac{24}{(s-3)^3}$ or $\left(s^2 - 6s + 9\right)Y = \frac{24}{(s-3)^3}$. Hence $Y = \frac{24}{(s-3)^5} = \frac{4!}{(s-3)^5}$. The inverse Laplace Transform yields the answer $y = t^4e^{3t}$.

19. $y'' + 2y' - 3y = te^{2t}$, $y(0) = 2$, $y'(0) = 3$. The Laplace transform is $\left(s^2Y - 2s - 3\right) + 2sY - 4 - 3Y = \frac{1}{(s-2)^2}$. Regrouping we get $\left(s^2+2s-3\right)Y - (2s+7) = \frac{1}{(s-2)^2}$. Hence $\left(s^2+2s-3\right)Y = \frac{1}{(s-2)^2} + (2s+7) = \frac{2s^3-s^2-20s+29}{(s-2)^2}$ or $Y = \frac{2s^3-s^2-20s+29}{(s-2)^2(s+3)(s-1)} = \frac{A}{s+3} + \frac{B}{s-1} + \frac{C}{s-2} + \frac{Q}{(s-2)^2}$. Multiplying by the LCD we get $2s^3 - s^2 - 20s + 29 = A(s-1)(s-2)^2 + B(s+3)(s-2)^2 + C(s+3)(s-1)(s-2) + D(s+3)(s-1)$. Setting $s = -3$ we get $A = \frac{26}{-100} = \frac{-13}{50}$. Setting $s = 1$ we get $B = \frac{10}{4} = \frac{5}{2}$. Setting $s = 2$, we get $D = \frac{1}{5}$. the s^3-term yields $A + B + C = 2$ or $C = 2 - A - B = \frac{6}{25}$. Hence we have $Y = -\frac{13}{50} \cdot \frac{1}{s+3} + \frac{5}{2} \cdot \frac{1}{s-1} - \frac{6}{25} \cdot \frac{1}{s-2} + \frac{1}{5} \cdot \frac{1}{(s-2)^2}$.

The inverse Laplace Transform gives the answer
$y = -\frac{15}{50}e^{-3t} + \frac{5}{2}e^t - \frac{6}{25}e^{2t} + \frac{1}{5}te^{2t}$.

20. $y'' + 2y' - 3y = \sin 2t$, $y(0) = y'(0) = 0$. The Laplace Transform is $s^2Y + 2sY - 3Y = \frac{2}{s^2+4}$. This is equivalent to $\left(s^2 + 2s - 3\right)Y = \frac{2}{s^2+4}$ or $Y = \frac{2}{\left(s^2+4\right)(s+3)(s-1)} = \frac{A}{s+3} + \frac{B}{s-1} + \frac{Cs+D}{s^2+4}$. Multiplying by the LCD we have $2 = A(s - 1)\left(s^2 + 4\right) + B(s + 3)\left(s^2 + 4\right) + (Cs + D)(s + 3)(s - 1)$. Setting $s = -3$, we get $A = \frac{2}{-52} = \frac{-1}{26}$. Setting $s = 1$, we get $B = \frac{2}{20} = \frac{1}{10}$. The s^3-term yields the equation $A + B + C = 0$ or $C = \frac{1}{26} - \frac{1}{10} = \frac{-16}{260} = \frac{-4}{65}$. The constant term yields the equation $-4A + 12B - 3D = 2$ or $D = \frac{-4A-12B+2}{-3} = \frac{-\frac{2}{13} - \frac{6}{5} + 2}{-3} = \frac{-14}{65}$. Hence $Y = -\frac{1}{26} \cdot \frac{1}{s+3} + \frac{1}{10} \cdot \frac{1}{s-1} - \frac{4}{65} \cdot \frac{s}{s^2+4} - \frac{7}{65} \cdot \frac{2}{s^2+4}$. The inverse Laplace Transform yields the answer $y = -\frac{1}{26}e^{-3t} + \frac{1}{10}e^6 - \frac{4}{65}\cos 2t - \frac{7}{65}\sin 2t$.

21. $y'' + 2y + 5y = 10\cos t$, $y(0) = 2$, $y'(0) = 1$. The Laplace transform is $(s^2Y - 2s - 1) + (2sY - 4) + 5Y = \frac{10s}{s^2+1}$. This is equivalent to $\left(s^2 + 2s + 5\right)Y - (2s + 5) = \frac{10s}{s^2+1}$ or $\left(s^2 + 2s + 5\right)Y = \frac{10s}{s^2+1} + (2s + 5) = \frac{2s^3+5s^2+12s+5}{s^2+1}$ or $Y = \frac{2s^3+5s^2+12s+5}{\left(s^2+1\right)\left(s^2+2s+5\right)} = \frac{As+B}{s^2+1} + \frac{Cs+D}{s^2+2s+5}$. Multiplying by the LCD we get $2s^3 + 5s^2 + 12s + 5 = (As + B)\left(s^2 + 2s + 5\right) + (Cs + D)\left(s^2 + 1\right) = As^3 + (2A + B)s^2 + (5A + 2B)s + 5B + Cs^3 + ds^2 + Cs + D$. The yields the linear equations $A + C = 2$, $2A + B + D = 5$, $5A + 2B + C = 12$, and $5B + D = 5$. Solving this system we get $A = 2$, $B = 1$, $C = 0$, and $D = 0$. Hence the partial fraction decomposition is $Y = \frac{2s+1}{s^2+1} = 2\frac{s}{s^2+1} + \frac{1}{s^2+1}$. The inverse Laplace transform yields the answer $y = 2\cos t + \sin t$. (Note: $2s^3 + 5s^2 + 12s + 5$ factors into $(2s + 1)\left(s^2 + 2s + 5\right)$ so we could have reduced $\frac{2s^3+5s^2+12s+5}{\left(s^2+1\right)\left(s^2+2s+5\right)}$ to $\frac{2s+1}{s^2+1}$ and not used partial fractions.)

22. $y'' + 2y' + 5y = 10\cos t$, $y(0) = 0$, $y' = (0) = 3$. The Laplace transform is $\left(s^2Y - 3\right) + 2sY + 5Y = \frac{10s}{s^2+1}$ of $\left(s^2 + 2s + 5\right)Y - 3 = \frac{10s}{s^2+1}$. Thus $\left(s^2 + 2s + 5\right)Y = \frac{10s}{s^2+1} + 3 = \frac{3s^2+10s+3}{s^2+1}$. This gives $Y = \frac{3s^2+10s+3}{\left(s^2+1\right)\left(s^2+2s+5\right)} = \frac{As+B}{s^2+1} + \frac{Cs+D}{s^2+2s+5}$.

Multiplying we get $3s^2 + 10s + 3 = (As+B)(s^2 + 2s + 5) + (Cs + D)(s^2 + 1) = As^3 + (2A + B)s^2 + (5A+2B)s + 5B + Cs^3 + Ds^2 + Cs + D$. The four like terms yield the four linear equations $A + C = 0$, $2A + B + D = 3$, $A + 2B + C = 10$, and $5B + D = 3$. Solving this system we get $A = 2$, $B = 1$, $C = -2$, and $D = -2$. So the partial fraction decomposition of $Y = \frac{2s+1}{s^2+1} - 2$.

$\frac{s+1}{s^2+2s+5} = 2\frac{s}{s^2+1} + \frac{1}{s^2+1} - 2\frac{s+1}{(s+1)^2+2^2}$. The inverse Laplace transform yields the answer $y = 2\cos t + \sin t - 2e^{-t}\cos 2t$.

23. We start with the equation $L\frac{d^2q}{dt^2} + R\frac{dq}{dt} + \frac{1}{C}q = E(t)$ which becomes $q'' + 20q' + 200q = 150$. The Laplace transform is $s^2Q + 20sQ + 200Q = \frac{150}{s}$. Hence $(s^2 + 20s + 200)Q = \frac{150}{s}$ or $Q = \frac{150}{s(s^2+20s+200)} = \frac{A}{s} + \frac{Bs+C}{s^2+20s+200}$. Multiplying by the LCD we get $10 = A(s^2 + 20s + 200) + (Bs + C) = (A + B)s^2 + (20A + C)s + 200A$. The constant term yields $A = \frac{150}{200} = \frac{3}{4}$. The s^2-term yields $A + B = 0$ so $B = -\frac{3}{4}$. The s-term yields $20A + C = 0$ or $C = -15$. Hence $Q = \frac{3}{4} \cdot \frac{1}{s} + \frac{-\frac{3}{4}s - 15}{s^2 + 20s + 200} = \frac{3}{4} \cdot \frac{1}{s} + \frac{-\frac{3}{4}(s + 20)}{s^2 + 20s + 100 + 100} = \frac{3}{4} \cdot \frac{1}{s} - \frac{3}{4} \cdot \frac{s+10+10}{(s+10)^2+10^2} = \frac{3}{4}\frac{1}{s} - \frac{3}{4}\frac{s+10}{(s+10)^2+10^2} - \frac{3}{4}\frac{10}{(s+10)^2+10^2}$. The inverse Laplace transform yields the answer $q(t) = \frac{3}{4} - \frac{3}{4}e^{-10t}(\cos 10t + \sin 10t)$.

$i(t) = q'(t) = \frac{15}{2}e^{-10t}(\cos 10t + \sin 10t) - \frac{15}{2}e^{-10t} \times (-\sin 10t + \cos 10t) = 15e^{-10t}\sin 10t$. The steady state current is 0.

24. We begin with the equation $i' + 10i = \sin t$. The Laplace transform is $SI + 10I = \frac{1}{s^2+1}$. Hence $I = \frac{1}{(s^2+1)(s+10)} = \frac{A}{s+10} + \frac{Bs+C}{s^2+1}$. Multiplying by the LCD we get $1 = A(s^2+1) + (Bs+C)(s+10) = (A+B)s^2 + (10B+C)s + (A+10C)$. This yields the linear equation $A + B = 0$, $10B + C = 0$, and $A + 10C = 1$. Solving we get $A = \frac{1}{101}$, $B = \frac{-1}{101}$, and $C = \frac{10}{101}$. So $I = \frac{1}{101} \cdot \frac{1}{s+10} + \frac{1}{101} \cdot \frac{-s+10}{s^2+1} = \frac{1}{101}\left(\frac{1}{s+10} - \frac{s}{s^2+1} + \frac{10}{s^2+1}\right)$. The inverse Laplace transform yields the answer $i(t) = \frac{1}{101}\left(e^{-10t} - \cos t + 10\sin t\right)$.

25. We begin with the equation $0.9q'' + 6q' + 50q = 6$. The Laplace transform is $0.1s^2Q + 6sQ + 50 = \frac{6}{s}$

so $Q = \frac{6}{s(0.1s^2+6s+50)}$ or $Q = \frac{60}{s(s^2+60s+500)}$ or $Q = \frac{60}{s(s+10)(s+50)} = \frac{A}{s} + \frac{B}{s+10} + \frac{C}{s+50}$. Multiplying by the LCD we get $60 = A(s + 10)(s + 50) + Bs(s + 50) + Cs(s + 10)$. Setting $s = 0$, we get $A = \frac{60}{500} = \frac{3}{25}$. Setting $s = -10$, we get $B = \frac{60}{-400} = -\frac{3}{20}$. Setting $s = -50$, yields $C = \frac{60}{2000} = \frac{3}{100}$. Hence we get $Q = \frac{3}{25} \cdot \frac{1}{s} - \frac{3}{20} \cdot \frac{1}{s+10} + \frac{3}{100} \cdot \frac{1}{s+50}$. The inverse Laplace transform yields the answer $q(t) = \frac{3}{25} - \frac{3}{20}e^{-10t} + \frac{3}{100}e^{-50t}$. Thus, $i(t) = q'(t) = \frac{3}{2}e^{-10t} - \frac{3}{2}e^{-50t} = \frac{3}{2}\left(e^{-50t} - e^{-50t}\right)$.

26. We begin with the equation $0.05q'' + 20q' + 10{,}000q = 100\cos 200t$. Multiplying by 20 to eliminate decimals yields $q'' + 400q' + 200{,}000q = 200 \times \cos 200t$. The Laplace transform is $s^2Q + 400sQ + 200{,}000Q = \frac{2{,}000s}{s^2+(200)^2}$. Solving for Q, we get $Q = \frac{2{,}000s}{(s^2+40000)(s^2+400s+200{,}000)} = \frac{As+B}{s^2+40{,}000} + \frac{Cs+D}{s^2+400s+200{,}000}$. Multiplying by the LCD we get $2000s = (As+B)(s^2+400s+200{,}000) + (Cs+D)(s^2+40{,}000) = (A+C)s^3 + (400A+B+D)s^2 + (200{,}000A + 400B + 40{,}000C) + (200{,}000B + 40{,}000D)$. This yields the linear equations $A + C = 0$, $400A + B + D = 0$, $200000A + 400B + 40000C = 2000$, and $200000B + 4000D = 0$. Solving we get $A = \frac{1}{100}$, $B = 1$, $C = \frac{-1}{100}$, and $D = -5$. Hence $Q = \frac{1}{100}\left(\frac{s+100}{s^2+200^2}\right) - \frac{1}{100} \times \left(\frac{s+500}{s^2+400s+200{,}00}\right) = \frac{1}{100}\frac{s}{s^2+200s^2} + \frac{1}{200}\frac{200}{s^2+200^2} - \frac{1}{100} \times \left(\frac{s+200+300}{(s+200)^2+160{,}000}\right) = 0.01\frac{s}{s^2+200^2} + 0.005\frac{200}{s^2+200^2} - 0.01 \cdot \frac{s+200}{(s+200)^2+400^2} - 0.0075\frac{400}{(s+200)^2+400^2}$. The inverse Laplace transform yields $q(t) = 0.01\cos 200t + 0.005\sin 200t - 0.01e^{-200t}\cos 400t - 0.0075e^{-200t}\sin 400t = 0.01\cos 200t + 0.005\sin 200t - \frac{1}{400}e^{-200t}(4\cos 400t + 3\sin 400t)$. Thus, $i(t) = q'(t) = -2\sin 200t + \cos 200t + \frac{1}{2}e^{-200t}(4\cos 400t + 3\sin 400t) - e^{-200t}(-4\sin 400t + 3\cos 400t) = -2\sin 200t + \cos 200t + e^{-200t} \times \left(\frac{11}{2}\sin 400t - \cos 400t\right)$. The steady-state current is $-2\sin 200t + \cos 200t$ or $\cos 200t - 2\sin 200t$.

27. First, a 16 pound weight has a mass of $\frac{1}{4}$ slug. The differential equation is $\frac{1}{2}\frac{d^2x}{dt^2} = -4x - 64\frac{dx}{dt}$, we need not bother changing to slugs. The differential equation is $16\frac{d^2x}{dt^2} = -4x - 64\frac{dx}{dt}$ which is equivalent to $x'' + 128x' + 8x = 0$. This has auxiliary

equation $m^2 + 128m + 18 = 0$, which has roots $\frac{-128 \pm \sqrt{128^2 - 32}}{2} = -64 \pm \sqrt{4088} = -64 \pm 2\sqrt{1022}$. Hence the general solution is $x = c_1 e^{(-64 + 2\sqrt{1022})t} + c_2 e^{(-64 - 2\sqrt{1022})t}$. When $t = 0$, then $x = 2$, so $c_1 + c_2 = 2$. When $t = 0$, $x' = 0$, so $(-64 + 2\sqrt{1022})c_1 + (-64 - 2\sqrt{1022})c_2 = 0$. Substituting $2 - c_2$ for c_1, we get $(-64 + 2\sqrt{1022})(2 - c_2) + (-64 - 2\sqrt{1022})c_2 = 0$ or $-128 + 4\sqrt{1022} + 64c_2 - 2\sqrt{1022}c_2 - 64c_2 - 2\sqrt{1022} = 0$ or $-4\sqrt{1022}c_2 = 128 - 4\sqrt{1022}$, and so $c_2 = 1 - \frac{32}{\sqrt{1022}}$. Thus, $c_1 = 2 - c_2 = 2 - \left(1 - \frac{32}{\sqrt{1022}}\right) = 1 + \frac{32}{\sqrt{1022}}$. Thus, the final solution is $x = \left(1 + \frac{32}{\sqrt{1022}}\right)e^{(-64 + 2\sqrt{1022})t} + \left(1 - \frac{32}{\sqrt{1022}}\right)e^{(-64 - 2\sqrt{1022})t}$.

28. As in Exercise #27, we have the differential equation $\frac{1}{2}x'' + 64x' + 4x = \cos 4t$. This has the same complementary solution as in Exercise #27. To complete the particular solution, we assume $x = 4\cos 4t + B\sin 4t$. Then, $x' = -4A\sin 4t + 4B\cos 4t$ and $x'' = -16A\cos 4t - 16B\sin 4t$. Substituting, we obtain $-8A\cos 4t - 8B\sin 4t - 256A\sin 4t + 256B\cos 4t + 4A\cos 4t + 4B\sin 4t = \cos 4t$. Collecting like functions, we obtain the equations

$$-8A + 256B + 4A = 1$$

and

$$-8B - 256A + 4B = 0$$

or

$$-4A + 256B = 1$$

and

$$-256A - 4B = 0$$

Solving the last equation for B, we have $B = -64A$. Substituting, we get $-4A + 256(-64A) = 1$ or $A = \frac{-1}{16,388}$. Back substituting, we find $B = \frac{64}{16,388}$. Hence, the final solution is $x = \left(1 + \frac{32}{\sqrt{1022}}\right)e^{(-64 + 2\sqrt{1022})t} + \left(1 - \frac{32}{\sqrt{1022}}\right)e^{(-64 - 2\sqrt{1022})t} - \frac{1}{16,388}\cos 4t + \frac{64}{16,388}\sin 4t$.

29. The Laplace transform of the left-hand side of the differential equation $y'' + 2y' + 17y = e^{-t}\cos 4t$ is $s^2 Y - sy(0) - y'(0) + 2[sY - y(0)] + 17Y = s^2 Y + 2sY + 17Y = Y(s^2 + 2s + 17)$. The Laplace transform of the right-hand side is $\frac{s+1}{(s+1)^2 + 4^2} = \frac{s+1}{s^2 + 2s + 17}$. Hence, $Y(s^2 + 2s + 17) = \frac{s+1}{s^2 + 2s + 17} = \frac{s+1}{(s+1)^2 + 4^2}$ or

$$Y = \frac{s+1}{\left(s^2 + 2s + 17\right)^2} = \frac{s+1}{\left[(s+1)^2 + 4^2\right]^2} = \frac{s}{\left[(s+1)^2 + 4^2\right]^2} + \frac{1}{\left[(s+1)^2 + 4^2\right]^2}.$$

Using formulas 21 and 22 in Table 34.1, we obtain

$$y = L^{-1}\left(\frac{s+1}{\left[(s+1)^2 + 4^2\right]^2}\right)$$

$$= L^{-1}\left(\frac{s+1}{\left[(s+1)^2 + 4^2\right]^2}\right)$$

$$+ L^{-1}\left(\frac{s+1}{\left[(s+1)^2 + 4^2\right]^2}\right)$$

$$= \frac{1}{128}e^{-t}\left[4t\cos(4t) + 16t\sin(4t) - \sin(4t)\right]$$

$$+ \frac{1}{128}e^{-t}\left[\sin(4t) - 4t\cos(4t)\right]$$

$$= \frac{1}{8}te^{-t}\sin 4t$$

CHAPTER 34 REVIEW

1.

	y	
x	$h = 0.1$	$h = 0.05$
0	2	2
0.05	–	1.8
0.10	1.6	1.6299
0.15	–	1.4838
0.20	1.3184	1.3572
0.25	–	1.2467
0.30	1.1098	1.1495
0.35	–	1.0636
0.40	0.9497	0.9873
0.45	–	0.9190
0.50	0.8234	0.8578
0.55	–	0.8026
0.60	0.7217	0.7527
0.65	–	0.7074
0.70	0.6384	0.6661
0.75	–	0.6284
0.80	0.5691	0.5938
0.85	–	0.5621
0.90	0.5108	0.5329
0.95	–	0.5059
1.00	0.4612	0.4809

2.

	y	
x	$h = 0.1$	$h = 0.05$
0	1	1
0.05	–	1.1
0.10	1.2	1.2052
0.15	–	1.3157
0.20	1.4218	1.4313
0.25	–	1.5520
0.30	1.6649	1.6778
0.35	–	1.8088
0.40	1.9288	1.9448
0.45	–	2.0859
0.50	2.2133	2.2321
0.55	–	2.3833
0.60	2.5183	2.5395
0.65	–	2.7001
0.70	2.8436	2.8671
0.75	–	3.0385
0.80	3.1892	3.2149
0.85	–	3.3963
0.90	3.5549	3.5827
0.95	–	3.7742
1.00	3.9408	3.9706

3.

x	y
	$h = 0.05$
0	60
0.05	59.9
0.10	59.8
0.15	59.6990
0.20	59.5961
0.25	59.4905
0.30	59.3812
0.35	59.2678
0.40	59.1496
0.45	59.0263
0.50	58.8975
0.55	58.7630
0.60	58.6228
0.65	58.4767
0.70	58.3249
0.75	58.1675
0.80	58.0047
0.85	57.8366
0.90	57.6635
0.95	57.4857
1.00	57.3034

4.

x	y
	$h = 0.01$
0	0
0.01	0.06
0.02	0.1200
0.03	0.1800
0.04	0.2400
0.05	0.2999
0.06	0.3598
0.07	0.4197
0.08	0.4796
0.09	0.5394
0.10	0.5991

5. $y' = xy + 4$, $(0, 1)$. Substituting 1 for y we get $y' = x + 4$. Integrating we have $y = \frac{1}{2}x^2 + 4x + C_1$. Solving for C_1 we get $C_1 = 1$, so the first approximation is $y = \frac{1}{2}x^2 + 4x + 1$. Now we substitute this for y and get $y' = \frac{1}{2}x^3 + 4x^2 + x + 4$. Integrating this we get $y = \frac{1}{8}x^4 + \frac{4}{3}x^3 + \frac{1}{2}x^2 + 4x + C_2$. Again $C_2 = 1$ so the second approximation is $y = \frac{1}{8}x^4 + \frac{4}{3}x^3 + \frac{1}{2}x^2 + 4x + 1$.

6. $y' = 2x + y$. Substituting 0 for y we have $y' = 2x$. Integration yields $y = x^2 + C_1$. $C_1 = 0$ so the first approximation is $y = x^2$. Substituting into the original equation we get $y' = 2x + x^2$. Integration we have $y = x^2 + \frac{1}{3}x^3 + C_2$. Again $C_2 = 0$ so the second approximation is $y = x^2 + \frac{1}{3}x^3$.

7. $y' = y\sqrt{x - 9}$, $(10, 1)$. Substituting 1 for y we get $y' = \sqrt{x - 9} = (x - 9)^{1/2}$. Integration yields $y = \frac{2}{3}(x - 9)^{\frac{3}{2}} + C_1$. Substituting $(10, 1)$ we get $1 = \frac{2}{3}(1) + C_1$ or $C_1 = \frac{1}{3}$. The first approximation is $y = \frac{2}{3}(x - 9)^{3/2} + \frac{1}{3}$. Now substituting this for y in the original equation we obtain $y' = \left(\frac{2}{3}(x - 9)^{3/2} + \frac{1}{3}\right)(x - 9)^{1/2} = \frac{2}{3}(x - 9)^2 + \frac{1}{3}(x - 9)^{1/2}$.

Integrating again we get $y = \frac{2}{9}(x - 9)^3 + \frac{2}{9}(x - 9)^{\frac{3}{2}}$. Solving for C_2 we have $1 - \frac{2}{9} - \frac{2}{9} = \frac{5}{9}$. The second approximation is $y = \frac{2}{9}(x - 9)^3 + \frac{2}{9}(x - 9)^{\frac{3}{2}} + \frac{5}{9}$.

8. $y' = y + \cos x$, $\left(\frac{\pi}{2}, 0\right)$. Substituting 0 for y we have $y' = \cos x$ so $y = \sin x + C_1$. Using $\left(\frac{\pi}{2}, 0\right)$ we obtain $C_1 = -1$ so $y = \sin x - 1$. Substituting this in the original equation we get $y' = \sin x + \cos x - 1$. Integration yields $y = -\cos x + \sin x - x + C_2$. $C_2 = \frac{\pi}{2} - 1$ and the second approximation is $y = -\cos x + \sin x - x + \frac{\pi}{2} - 1$.

9. $2y' - y = 4$, $y(0) = 1$ has Laplace transform $2sY - 2 - Y = \frac{4}{s}$ or $(2s - 1)Y = \frac{4}{s} + 2 = \frac{4 + 2s}{s}$. So we get $Y = \frac{4 + 2s}{s(2s - 1)} = \frac{A}{s} + \frac{B}{2s - 1}$. Multiplying by the LCD we have $4 + 2s = A(2s - 1) + Bs$. Setting $s = 0$, we obtain $A = \frac{4}{-1} = -4$. Setting $s = \frac{1}{2}$, we obtain $B = \frac{5}{1/2} = 10$. So we have $Y = -4\frac{1}{s} + 10 \cdot \frac{1}{2s - 1} = -4\frac{1}{s} + 5\frac{1}{s - \frac{1}{2}}$. The

inverse Laplace transformation yields the solution $y = -4 + 5e^{t/2}$.

10. $3y' + y = t$, $y(0) = 2$. The Laplace transform is $3sY - 6 + Y = \frac{1}{s^2}$ so $(3s + 1)Y = \frac{1}{s^2} + 6 = \frac{1 + 6s^2}{s^2}$ and $Y = \frac{1 + 6s^2}{s^2(3s + 1)} = \frac{A}{3s + 1} + \frac{B}{s} + \frac{C}{s^2}$. Multiplying by the LCD we have $1 + 6s^2 = As^2 + Bs(3s + 1) + C(3s + 1)$. Setting $s = -\frac{1}{3}$, we obtain $A = \frac{5/3}{1/9} = 15$. Setting $s = 0$, we obtain $C = 1$. The s^2-term yields $6 = A + 3B$ so $B = \frac{6 - A}{3} = \frac{-9}{3} = -3$. So the partial fraction decomposition is $Y = \frac{15}{3s + 1} - \frac{3}{s} + \frac{1}{s^2} = \frac{5}{s + \frac{1}{3}} - 3 \cdot \frac{1}{s} + \frac{1}{s^2}$. The inverse Laplace transform yields the answer $y = 5e^{-t/3} - 3 + t$.

11. $y' + 2y = e^t$, $y(0) = 1$. The Laplace transform is $sY - 1 + 2Y = \frac{1}{s - 1}$ or $(s + 2)Y = \frac{1}{s - 1} + 1 = \frac{s}{s - 1}$. So $Y = \frac{s}{(s - 1)(s + 2)} = \frac{A}{s - 1} + \frac{B}{s + 2}$. This gives $s = A(s + 2) + B(s - 1)$. Setting $s = 1$, we get $A = \frac{1}{3}$. Setting $s = -2$, we get $B = \frac{2}{3}$. Hence $Y = \frac{1}{3} \cdot \frac{1}{s - 1} + \frac{2}{3} \cdot \frac{1}{s + 2}$. The inverse Laplace transform yields $y = \frac{1}{3}e^t + \frac{2}{3}e^{-2t}$.

12. $y' + 5y = 0$, $y(0) = 1$. The Laplace transform is $sY - 1 + sY = 0$ or $(s + 5)Y = 1$ or $Y = \frac{1}{s + 5}$. The inverse Laplace transform yields $y = e^{-5t}$.

13. $y'' - y = 0$, $y(0) = y'(0) = 1$. The Laplace transform is $s^2Y - s - 1 - Y = 0$ or $(s^2 - 1)Y = s + 1$. Hence $Y = \frac{s + 1}{s^2 - 1} = \frac{1}{s - 1}$. The inverse Laplace transform yields $y = e^t$.

14. $y'' - y = e^t$, $y(0) = y'(0) = 0$. The Laplace transform is $s^2Y - Y = \frac{1}{s - 1}$ or $Y = \frac{1}{(s - 1)(s^2 - 1)} = \frac{1}{(s - 1)^2(s + 1)}$. Getting partial fractions we have $\frac{1}{(s - 1)^2(s + 1)} = \frac{A}{s - 1} + \frac{B}{(s - 1)^2} + \frac{C}{s + 1}$. Multiplying we get $1 = A(s^2 - 1) + B(s + 10) + C(s - 1)^2$. Setting $s = 1$, we obtain $B = \frac{1}{2}$. The s^2-term yields $A + C = 0$. The constant term yields $-A + B + C = 1$ or $-A + C = \frac{1}{2}$. Solving, we get $C = \frac{1}{4}$ nd $A = -\frac{1}{4}$. Hence $Y = -\frac{1}{4}\frac{1}{s - 1} + \frac{1}{2} \cdot \frac{1}{(s - 1)^2} + \frac{1}{4} \cdot \frac{1}{s + 1}$. The inverse Laplace transform yields $y = -\frac{1}{4}e^t + \frac{1}{2}te^t + \frac{1}{4}e^{-t}$.

15. $y'' + 2y' + 5y = 0$, $y(0) = 1$, $y'(0) = 0$. The Laplace transform is $s^2Y - s + 2sY - 2 + 5y = 0$ or $(s^2 + 2s + 5)Y = s + 2$. So $Y = \frac{s+2}{s^2+2s+5} = \frac{s+2}{(s+1)^2+2^2} = \frac{s+1}{(s+1)^2+2^2} + \frac{1}{2}\frac{2}{(s+1)^2+2^2}$. The inverse Laplace transform yields the answer $y = e^{-t}\cos 2t + \frac{1}{2}e^{-t}\sin 2t$.

16. $y'' + y' + y = 0$, $y(0) = 4$, $y'(0) = -2$. The Laplace transform is $(s^2Y - 4s + 2) + (sY - 4) + Y = 0$ or $(s^2 + s + 1)Y = 4s + 2$. Hence $Y = $

$$\frac{4s+2}{s^2+s+1} = \frac{4s+2}{s^2+s+\frac{1}{4}+\frac{3}{4}} = 4\frac{s+\frac{1}{2}}{\left(s+\frac{1}{2}\right)^2+\frac{3}{4}}.$$ The inverse Laplace transform gives $y = 4e^{-t/2}\cos\frac{\sqrt{3}}{2}t$.

17. $y'' + 2y' + 5y = 3e^{-2t}$, $y(0) = y'(0) = 1$. The Laplace transform is $(s^2Y - s - 1) + 2sY - 2 + 5 = \frac{3}{s+2}$ or $(s^2 + 2s + 5)Y - (s+3) = \frac{3}{s+2}$. Hence $(s^2 + 2s + 5)Y = \frac{3}{s+2} + s + 3 = \frac{s^2+5s+9}{s+2}$ and $Y = \frac{s^2+5s+9}{(s+9)(s^2+2s+5)} = \frac{A}{s+2} + \frac{Bs+C}{s^2+2s+5}$. Multiplying we obtain $s^2 + 5s + 9 = A(s^2 + 2s + 5) + (Bs + C)(s + 2) = (A + B)s^2 + (2A + 2B + C)s + (5A + 2C)$. Hence we get the linear equation $A + B = 1$, $2A + 2B + C = 5$, and $5A + 2C = 9$. Solving we obtain $A = \frac{3}{5}$, $B = \frac{2}{5}$, and $C = 3$. Thus we have $Y = \frac{3}{5}\cdot\frac{1}{s+2} + \frac{2}{5}\cdot\frac{s}{s^2+2s+5} + \frac{3}{s^2+2s+5} = \frac{3}{5}\cdot\frac{1}{s+2} + \frac{2}{5}\frac{s+1}{(s+1)^2+2^2} + \frac{13}{10}\frac{2}{(s+1)^2+2^2}$. The inverse Laplace transform yields the answer $y = \frac{3}{5}e^{-2t} + \frac{2}{5}e^{-t}\cos 2t + \frac{13}{10}e^{-t}\sin 2t$.

18. $y'' + 2y' - 3y = 5e^{2t}$, $u(0) = 2$, $y'(0) = 3$. The Laplace transform is $(s^2Y - 2s - 3) + (2sY - 4) - 3Y = \frac{5}{s-2}$ or $(s^2 + 2s - 3)Y - (2s + 7) = \frac{5}{s-2}$ so $(s^2 + 2s - 3)Y = \frac{5}{s-2} + (2s + 7) = \frac{2s^2+3s-9}{s-2}$. This yields $Y = \frac{2s^2+3s-9}{(s-2)(s^2+2s-3)} = \frac{(2s-3)(s+3)}{(s-2)(s+3)(s-1)} = \frac{2s-3}{(s-2)(s+1)} = \frac{A}{s-2} + \frac{B}{s-1}$. Multiplying we have $2s - 3 = A(s - 1) + B(s - 2)$. Setting $s = 2$ we get $A = 1$. Setting $s = 1$, we get $B = 1$. The partial fraction decomposition of Y is $\frac{s}{s-2} + \frac{1}{s-1}$. The inverse Laplace transform yields the answer $y = e^{2t} + e^t$.

19. $y'' + 4 = \sin t$, $y(0) = y'(0) = 0$. The Laplace transform is $s^2Y + \frac{4}{s} = \frac{1}{s^2+1}$. So $s^2Y = \frac{1}{s^2+1} - \frac{4}{s}$ and $Y = \frac{1}{(s^2+1)s^2} - \frac{4}{s^3}$. The inverse Laplace transform (by forms 11 and 3) is $y = t - \sin t - 2t^2$.

20. $y'' + 4 = t$, $y(0) = -1$, $y'(0) = 0$. The Laplace transform is $s^2Y + s + \frac{4}{s} = \frac{1}{s^2}$. So $s^2Y = \frac{1}{s^2} - \frac{4}{s} - s$ and $Y = \frac{1}{s^4} - \frac{4}{s^3} - \frac{1}{s}$. The inverse Laplace transform is $y = \frac{t^3}{6} - 2t^2 - 1$.

21. We begin with the equation $\frac{1}{2}q'' + 10q' + 1001 = 10\sin t$. The Laplace transform is $\frac{1}{2}(s^2Q - 10) + 10(sQ) + 100Q = \frac{10}{s^2+1}$ or $(\frac{1}{2}s^2 + 10s + 100)Q = \frac{10}{s^2+1} + 5 = \frac{5s^2+15}{s^2+1}$. Multiplying by 2 to clear fractions we have $(s^2 + 20s + 200)Q = \frac{10s^2+30}{s^2+1}$. Hence $Q = \frac{10s^2+30}{(s^2+1)(s^2+20s+100)} = \frac{As+B}{s^2+1} + \frac{Cs+D}{s^2+20s+200}$. Multiplying by the LCD we obtain the equation $10s^2 + 30 = (As + b)(s^2 + 20s + 200) + (Cs + D)(s^2 + 1) = (A + C)s^3 + (20A + B + D)s^2 + (200A + 20B + C)s + 200B + D$. Hence we have linear equations $A + C = 0$, $20A + B + D = 10$, $200A + 20B + C = 0$, and $200B + D = 30$. Solving this system, we obtain $A = \frac{-400}{40,001}$, $B = \frac{3980}{40,001}$, $C = \frac{400}{40,001}$ and $D = \frac{404,030}{40,001}$. Substituting, we obtain $Q = \frac{10}{40,001}\left[\frac{-40s+398}{s^2+1} + \frac{40s+40,403}{s^2+20s+200}\right]$. Working each part separately, we have: $\frac{-40s+398}{s^2+1} = -40\frac{s}{s^2+1} + 398\frac{1}{s^2+1}$. The inverse Laplace transform yields $q = -40\cos t + 398\sin t$. For $\frac{40s+40,403}{s^2+20s+200}$, we complete the square in the denominator, getting $\frac{40s+40,403}{s^2+20s+200} = \frac{40s+40,403}{s^2+20s+100+100} = \frac{40s+40,403}{(s+10)^2+10^2} = 40\frac{s+10}{(s+10)^2+10^2} + \frac{40,003}{(s+10)^2+10^2} = 40\frac{s+10}{(s+10)^2+10} + \frac{40,403}{10}\cdot\frac{10}{(s+10)^2+10^2}$. The inverse Laplace transform yields $40e^{-10t}\cos 10t + \frac{40,003}{10}e^{-10t}\sin 10t$. Hence, we get

$$q(t) = \frac{10}{40,001}\Big(-40\cos 5 + 398\sin 5 + 40e^{-10t}\cos 10t + \frac{40,003}{10}e^{-10t}\sin 10t\Big)$$

Thus, we have

$i(t) = q'(t)$

$$= \frac{10}{40,001}\Big(40\sin t + 398\cos t$$
$$- 400e^{-10t}\cos 10t$$
$$- 400e^{-10t}\sin 10t - 40{,}003e^{-10t}\sin 10t$$
$$+ 40{,}003e^{-10t}\cos 10t\Big)$$

$$= \frac{10}{40,001}\Big(40\sin t + 398\cos t$$
$$+ 39{,}603e^{-10t}\cos 10t$$
$$- 40{,}403e^{-10t}\sin 10t\Big)$$

The steady-state current is $\frac{10}{40,001}(40\sin t + 398\cos t)$.

22. The initial differential equation is $20x'' + 90x' + 700x = 5\sin t$. This is equivalent to $2x'' + 9x' + 70x = \frac{1}{2}\sin t$ with auxiliary equation $2m^2 + 9m + 70 = 0$. This has solutions $\frac{-9\pm\sqrt{81-560}}{4} = -\frac{9}{4} \pm \frac{\sqrt{479}}{4}j$. Hence, the general complementary solution is $x_c = e^{-9t/4}\big(C_1\cos\frac{\sqrt{479}}{4}t + C_2\sin\frac{\sqrt{479}}{4}t\big)$. For x_p we set $x_p = A\sin t + B\cos t$. Thus, $x_p' = A\cos t - B\sin t$. Substituting these expressions into the equivalent equation above, we have $2(-A\sin t - B\cos t) + 9(A\cos t - B\sin t) + 70(A\sin t + B\cos t) = \frac{1}{2}\sin t$. Collecting like

terms produces $-2A - 9B + 70A = \frac{1}{2}$ and $-2B + 9A + 70B = 0$. These equations are equivalent to $68A - 9B = \frac{1}{2}$ and $9A + 68B = 0$. Solving, we get $A = \frac{68}{9410} = \frac{34}{4705}$ and $B = -\frac{9}{9410}$. This produces $x_p = \frac{1}{9410}(68\sin t - 9\cos t)$. Hence, the general solution is $x = x_c + x_p = e^{-9t/4}\big(c_1\cos\frac{\sqrt{479}}{4}t + c_2\sin\frac{\sqrt{479}}{4}t\big) + \frac{1}{9410}(68\sin t - 9\cos t)$. Now, $x(0) = 0$ yields $c_1 = \frac{9}{9410}$. We find that $x'(t) = -\frac{9}{4}e^{9t/4}\times\big(\frac{9}{9410}\cos\frac{\sqrt{479}}{4}t + c_2\sin\frac{\sqrt{479}}{4}t\big) + e^{-9t/4}\big(-\frac{9\sqrt{479}}{9410\cdot4}\times\sin\frac{\sqrt{479}}{4}t + \frac{\sqrt{479}}{4}c_2\cos\frac{\sqrt{479}}{4}t\big) + \frac{1}{9410}(68\cos t + 9\sin t)$. Now, using the fact that $v(0) = -1$ or $x'(0) = -1$, (recall, down is positive), we obtain $-1 = -\frac{9}{4}\cdot\frac{9}{9410} + \frac{\sqrt{479}}{4}c_2 + \frac{68}{9410}$ and we obtain $c_2 \approx -0.183692$. The particular solution is

$$x(t) = e^{-9t/4}\Big(\frac{9}{9410}\cos\frac{\sqrt{479}}{4}t$$
$$- 0.1837\sin\frac{\sqrt{479}}{4}t\Big)$$
$$+ \frac{1}{9410}(68\sin t - 9\cos t)$$
$$\approx e^{-2.25t}(0.0009564\cos 5.4715t$$
$$- 0.1837\sin 5.4715t) + 0.007226\sin t$$
$$- 0.0009564\cos t$$

▤ CHAPTER 34 TEST

---

1. From Entry 5 in Table 34.1, we see that the Laplace transform of e^{at} is $\frac{1}{s-a}$. Here $a = -3$, so the Laplace transform is $5\big(\frac{1}{s+3}\big) = \frac{5}{s+3}$.

2. From Entry 3, we have the Laplace transform of $2t^3 = 6\big(\frac{t^3}{3}\big)$ as $6\big(\frac{1}{s^4}\big)$ and from Entry 8, the Laplace transform of $\sin 8t$ is $\frac{8}{s^2+8^2}$. Adding these together, we get the Laplace transform of $2t^3 + \sin 8t$ as $\frac{6}{s^4} + \frac{8}{s^2+64}$.

3. From Entry 20, we have $a = -5$ and $b = 4$, so the Laplace transform is $\frac{s-5}{(s-5)^2+4^2}$.

4. From Entry 7, we have $n - 1 = 9$ and $a = 5$, so the Laplace transform is $8\big(\frac{9}{(s-5)^{10}}\big) = \frac{72}{(s-5)^{10}}$.

5. We see that this is of the form in Entry 19 of Table 34.1 with $a = 6$ and $B = \sqrt{7}$. So, $L^{-1}(F) = e^{-6t}\sin\sqrt{7}t$.

6. Completing the square of the denominator we get $F(s) = \frac{s}{(s-6s+9)+4} = \frac{s}{(s-3)^2+2^2}$. This does not satisfy any of the entries in Table 34.1, but it is close to Entry 20. If we rewrite $F(s)$ by adding $0 = -3 + 3$ to the numerator and regrouping, we obtain $F(s) = \frac{s-3+3}{(s-3)^2+2^2} = \frac{s-3}{(s-3)^2+2^2} + \frac{3}{(s-3)^2+2^2} = \frac{s-3}{(s-3)^2+2^2} + \frac{3}{2}\big(\frac{2}{(s-3)^2+2^2}\big)$. Applying Entries 20 and 19 of Table 34.1, respectively, we ggt $L^{-1}(F) = e^{-3t}\cos 2t + \frac{3}{2}e^{-3t}\sin 2t$.

7. Using partial fraction decomposition we have $\frac{5}{(s+3)(s^2+1)} = \frac{A}{s+3} + \frac{Bs+C}{s^2+1}$. Multiplying by $(s+3)(s^2+1)$ produces $5 = A(s^2+1) + (Bs+C)(s+3) = As^2 + A + Bs^2 + Cs + 3Bs + 3C = (A+B)s^2 + (3B+C)s + (A+3C)$. Solving the simultaneous equations $A + B = 0$, $3B + C = 0$, and $A + 3C = 5$ we obtain $A = 0.5$, $B = -0.5$, and $C = 1.5$. So we can rewrite $F(s)$ as

$\frac{0.5}{s+3} + \frac{-0.5s+1.5}{s^2+1} = \frac{0.5}{s+3} - \frac{0.5s}{s^2+1} + \frac{1.5}{s^2+1}$. From Entries 5, 9, and 8 of Table 34.1, respectively, we obtain $L^{-1}(F) = 0.5e^{-3t} - 0.5\cos t + 1.5\sin t$.

8. Taking Laplace transforms of both sides produces $L(y') + 5L(y) = L(0)$. Since $L(y') = sY - y(0) = sY - 2$, this equation becomes $sY - 2 + 5Y = 0$ or $Y(s+5) = 2$ and so $Y = \frac{2}{s+5}$. Taking the inverse Laplace transform of both sides we obtain $y = L^{-1}\left(\frac{2}{s+5}\right) = e^{-5t}$.

9. Taking the Laplace transform of the given equation produces $L(y'') + 2L(y) = L\left(e^{-2t}\right)$. Since $L(y'') = s^2Y - sy(0) - y'(0)$, $y(0) = 0$ and $y'(0) = 0$ and $y'(0) = 0$, we can rewrite this as $s^2Y - s + 2Y = \frac{1}{s+2}$ or $\left(s^2 + 2\right)Y = s + \frac{1}{s+2}$ and so $Y = \frac{s}{s^2+2} + \frac{1}{(s^2+2)(s+2)}$. Using partial fractions, we determine that $\frac{1}{(s^2+2)(s+2)} = \frac{1/6}{s+2} + \frac{-s/6+1/3}{s^2+2} = \frac{1}{6}\left(\frac{1}{s+2}\right) - \frac{1}{6}\left(\frac{s}{s^2+2}\right) + \frac{1}{3}\left(\frac{1}{s^2+2}\right)$. Combining this partial fraction with the first $\frac{s}{s^2+2}$, we obtain $Y = \frac{1}{2}\left(\frac{1}{s+2}\right) + \frac{5}{6}\left(\frac{s}{s^2+2}\right) + \frac{1}{3}\left(\frac{1}{s^2+2}\right)$. So, $y = \frac{1}{6}L^{-1}\left(\frac{1}{s+2}\right) + \frac{5}{6}L^{-1}\left(\frac{s}{s^2+2}\right) + \frac{1}{3}L^{-1}\left(\frac{1}{s^2+2}\right) = \frac{1}{6}L^{-1}\left(\frac{1}{s+2}\right) + \frac{5}{6}L^{-1}\left(\frac{s}{s^2+2}\right) + \frac{1}{3\sqrt{2}}L^{-1}\left(\frac{\sqrt{2}}{s^2+2}\right)$. Using Entries 5, 9, and 8 in Table 34.1 we determine that $y = \frac{1}{6}e^{-2t} + \frac{5}{6}\cos\sqrt{2}t + \frac{1}{3\sqrt{2}}\sin\sqrt{2}t$.

10. Taking the Laplace transform of the given equation produces $L(y'') + 2L(y') + L(y) = 0$. Since $L(y') = sY - y(0) = sY - 1$ and $L(y'') = s^2Y - sy(0) - y'(0) = s^2Y - s + 1$, we can rewrite this as $\left(s^2Y - s + 1\right) + 2sY + Y = 0$ or $\left(s^2 + 2s + 1\right)Y = s + 1$ and so $Y = \frac{s+1}{s^2+2s+1} = \frac{1}{(s+1)}$ and $y = L^{-1}(Y) = L^{-1}\left(\frac{1}{(s+1)}\right)$. By form 5, we get $y = e^{-t}$.

11. Using the formula $L\frac{d^2q}{dt^2} + R\frac{dq}{dt} + \frac{q}{C} = E(t)$ we have $0.2q'' + 8q' + \frac{q}{0.0025} = 0.2q'' + 8q' + 400q = 12\sin 20t$. Taking the Laplace transform of this equation, we get $0.2L(q'') + 8L(q') + 400L(q) = L(12\sin 20t)$. Since $L(q') = sQ - q(0) = sQ$, $L(q'') = s^2Q - sq(0) + q'(0) = s^2Q$, and $L \times (12\sin 20t) = 12\left(\frac{20}{s^2+20^2}\right) = \frac{240}{s^2+400}$, we can rewrite our equation as $0.2s^2Q + 8sQ + 400 = \frac{240}{s^2+400}$ or $\left(0.2s^2 + 8s\right)Q = -400 + \frac{240}{s^2+400}$ and so $Q = \frac{-400}{0.2s^2+8s} + \frac{240}{\left(0.2s^2+8s\right)\left(s^2+20^2\right)} = \frac{-2000}{s^2+40s} + \frac{1200}{\left(s^2+40s\right)\left(s^2+20^2\right)}$. Using partial fraction decomposition, we can rewrite this as $Q = \left(\frac{-50}{s} + \frac{50}{s+40}\right) + \left(\frac{0.075}{s} + \frac{-0.015}{s+40} - \frac{0.06s+0.6}{s^2+400}\right) = \frac{-49.925}{s} + \frac{49.985}{s+40} - \frac{0.06s}{s^2+400} - \frac{0.6}{s^2+400}$. Taking the inverse Laplace transform, we obtain $q = L^{-1}(Q) = -49.925 + 49.985e^{-40t} - 0.06\cos 20t - 0.03\sin 20t$.

12. Using Euler's method where $y_{n+1} = y_n + f(x_n, y_n)h$ we have $f(x, y) = 2x + y$. Thus we get the formula $y_{n+1} = y_n + 0.2(2x_n + y_n)$. We are given $x_0 = 0$, $y_0 = 2$, and $h = 0.2$. Using this data and applying the results successively, we obtain the following:

x	y $h = 0.2$
0	2
0.2	2.4
0.4	2.96
0.6	3.712
0.8	4.6944
1.0	5.95328

Solutions for Additional Computer Exercises

Technical Mathematics contains 32 exercises which ask students to write a computer program and *Technical Mathematics with Calculus* contains an additional 6 such exercises. This appendix consists of programs written in GWBASIC that serve as answers to each of those exercises. Just as the accompanying textbook increases in sophistication and complexity as you progress from beginning to end, the computer programs in this Program Book also increase in sophistication and complexity.

The name of each program corresponds to its location in the text. Each name consists of the three letters PRG followed immediately by five digits. The first two digits represent the chapter in the text. Thus, Chapter 9 is 09 and chapter 25 is 25. The next (third) digit denotes the section in the chapter; and the last two digits, the exercise in that section where the program was assigned.

Using the above description, the program for Chapter 9, Section 5, Exercise 43, has the name PRG09543. Similarly, the computer program for Chapter 17, Section 3, Exercise 22, is named PRG17322.

* * * * *

You may wish to write your own versions of these programs. If you do, then use the following programs as guidelines. In some cases, we have added additional steps to the programs to make them easier to use and more "user friendly." You may have some better ideas on how this might be done.

However, it is important that anyone programming a computer remember that a computer must be told what to do and how it is done. If the directions are wrong, the computer will give a wrong answer.

There are several special commands, statements, and functions that are used in BASIC. In fact, most of these, with some variation such as SQRT rather than SQR, are used in other computer programming languages. This is not intended as a programming course, and so we assume that you are familiar with BASIC.

Commands, such as LOAD, RUN, LIST, and SAVE are not part of a program, but allow you to tell the computer to do something with a program.

Program *statements* used in these programs include PRINT, PRINT USING, INPUT, REM, GOTO (or GO TO), DEF FN, DIM, FOR...NEXT, IF...THEN, GOSUB...RETURN, CLS, CLEAR, and END.

Functions used in these programs include ABS, ATN, COS, EXP, INT, LEFT$, LOG, SGN, SIN, GQR, TAB, and TAN. You should remember that a computer evaluates trigonometric functions expressed in radians and not degrees. You will need to include conversion factors in any programs where you want to enter, or have the computer print, angles given in degrees.

You may also have noticed that BASIC includes only the sine, cosine, tangent, and arctangent functions. If your programs include any other trigonometric formulas, you will have to define them in that particular program. Naturally, since these are programs involving mathematics, you will need to use the symbols for the mathematical *operations*. The computer uses $+$, $-$, $*$, and $/$ for addition, subtraction, multiplication, and division, respectively. For exponentiation, these BASIC programs use $\wedge$. (Some computers and some languages use either $**$ or $\uparrow$.)

* * * * *

Not everyone wishes to take the time to write a computer program. Thus, each of these programs is available on an IBM formatted disk in either $5\frac{1}{4}''$ or $3\frac{1}{2}''$ sizes.

At the end of these program listings is an exercise set of 30 exercises. Some of the exercises ask you to modify a program so that it will do a better job of performing the desired task or so that it can be used in some additional situations. The remaining exercises ask you to use one of the programs with a given set of numbers or equations.

A solution to each of these additional exercises is given in the *Instructor's Solutions Guide*; solutions to the odd numbered exercises are in the *Student's Solutions Guide*. Whenever possible, each solution contains a copy of the exact wording and format used when the computer printed each solution.

1. If you select $x = -1.9$, you get $x = -1.90625$, $y = 1.160156$. If you select $y = 1.16$, you get $x = -1.90625$, $y = 1.1625$.

2. (a) Using 1.414 for $\sqrt{2}$ and 0.3780 for $\frac{1}{\sqrt{7}}$ produces an answer of 1.463,
   (b) If you let $\pi = 3.1416$ and $-\sqrt{19} = -4.3590$, you get the result -13.68424.

3. (a) $= -12.000000$,
   (b) If you let $\pi = 3.1416$, $\frac{4}{3} = 1.3333$, $-\frac{17}{6} = -2.8333$, $1 + \sqrt{2} = 2.4142$, and $\sqrt{11} = 3.3166$, you get the result is 13.879020.

4. In order to avoid any places in PRG05517 that directs the computer to go to lines 440 or 450, change lines 440 and 450 to

```
440 '
450 '
```

Add the following lines to program PRG05517:

```
481 IF ABS (DET)>= .00001
        THEN GOTO 490
482 IF ABS (DX) < .00001 AND ABS (DY)
    < .00001 AND ABS (DZ) < .00001
    THEN PRINT "THIS SYSTEM IS
    DEPENDENT.": GOTO 520
483 PRINT "THIS SYSTEM IS
    INCONSISTENT.": GOTO 520
```

5. (a) THIS SYSTEM IS DEPENDENT.
   (b) THE SOLUTIONS ARE: X = 0, Y = 0, Z = 0.
   (c) THIS SYSTEM IS INCONSISTENT.

6. Replace line 70 and add new lines 72–78, as shown below

```
70 Y = SQR (ABS(D))
72 X1 = -B/(2*A)
74 X2 = Y/(2*A)
76 PRINT:PRINT"THE ROOTS ARE NOT
```

```
REAL NUMBERS. THE ROOTS ARE:
";X1;" + ";X2"J AND "X1;

    " - ";X2"J"
78 GOTO 140
```

7. (a) THE ROOTS ARE: 1.75 AND .5
   (b) THE ROOTS ARE: 3.651635 AND - 6.480089
   (c) THE ROOTS ARE BOTH THE SAME: 2.3 AND 2.3
   (d) THE ROOTS ARE NOT REAL NUMBERS. THE ROOTS ARE: 1.109375 + .6523321J AND 1.109375 - .6523321J

8. In PRG09236, change line 70 to FOR X = 0 TO 3.2 STEP .1. (You may also want to change lines 30 and 50 to read "from 0 to 3.2 …".)
   In PRG09237, change line 70 to FOR I = 1 TO 180. (You may also want to change line 30 and 50 to read "from 0 to 180 …".)

9. (a) THE HORIZONTAL VECTOR HAS MAGNITUDE: 5.17638 AND DIRECTION: 0 RADIANS OR 0 DEGREES THE VERTICAL VECTOR HAS MAGNITUDE: 19.31852 AND DIRECTION 1.570796 RADIANS OR 90 DEGREES.
   (b) THE HORIZONTAL VECTOR HAS MAGNITUDE: 13.56877 AND DIRECTION: 3.141593 RADIANS OR 180 DEGREES THE VERTICAL VECTOR HAS MAGNITUDE: 8.478708 AND DIRECTION 4.712389 RADIANS OR 270 DEGREES.
   (c) THE HORIZONTAL VECTOR HAS MAGNITUDE: 4.687785 AND DIRECTION: 0 RADIANS OR 0 DEGREES THE VERTICAL VECTOR HAS MAGNITUDE: 17.79283 AND DIRECTION 4.712389 RADIANS OR 270 DEGREES.
   (d) THE HORIZONTAL VECTOR HAS MAGNITUDE: 14.32789 AND DIRECTION: 3.141593 RADIANS OR 180 DEGREES THE VERTICAL VECTOR HAS MAGNITUDE: 18.8786 AND DIRECTION 1.570796 RADIANS OR 90 DEGREES.

10. **(a)** THE RESULTANT VECTOR HAS
MAGNITUDE = 222.6022
AND DIRECTION = .9193124 RADIANS
OR 52.67272 DEGREES
(b) THE RESULTANT VECTOR HAS
MAGNITUDE = 41.06048
AND DIRECTION = 1.790467 RADIANS
OR 102.5862 DEGREES
(c) THE RESULTANT VECTOR HAS
MAGNITUDE = 58.91126
AND DIRECTION = 1.234034 RADIANS
OR 70.70492 DEGREES

11. **(a)** -4.96 + 2.25 J,
(b) -5.709999 + .8699999 J

12. **(a)** THE PRODUCT IS:
-198.1568 + 165.6024 J
(b) THE PRODUCT IS:
.0744 + 2.2719 J

13. **(a)** THE QUOTIENT IS:
-.0450621 + -.7665063 J
(b) THE QUOTIENT IS:
.1509434 + .5283019 J

14. **(a)** THE NUMBER IS
-12.4 + 3.7 J
OR
12.94025 CIS 2.851617 RADIANS
12.94025 CIS 163.3856 DEGREES
(b) THE NUMBER IS
-1.641071 + -3.86612 J
OR
4.2 CIS 4.310963 RADIANS
4.2 CIS 247 DEGREES
(c) THE NUMBER IS
4.562753 + -10.28318 J
OR
11.25 CIS 5.13 RADIANS
11.25 CIS 293.9274 DEGREES

15. **(a)** THE TOTAL IS -198.1568 + 165.6024
OR
258.2446 CIS 2.445451 RADIANS
258.2446 CIS 140.114 DEGREES
(b) THE TOTAL IS
18.89179 + -10.47188 J
OR
21.6 CIS
5.77704 RADIANS
21.6 CIS 331 DEGREES
(c) THE TOTAL IS
-17.64838 + -20.43367 J
OR
27 CIS 4 RADIANS
27 CIS 229.1831 DEGREES

16. **(a)** THE TOTAL IS
-.0450621 $+ -$.7665063 J
OR
.7678297 CIS 1.512075 RADIANS
.7678297 CIS 86.63551 DEGREES
(b) THE TOTAL IS
.9270507 $+ -$2.853169 J
OR
3 CIS 5.026549 RADIANS
3 CIS 288 DEGREES
(c) THE TOTAL IS
-3.413423 + -2.085316 J
OR
4 CIS 3.69 RADIANS
4 CIS 211.4214 DEGREES

17. **(a)** THE TOTAL IS
-28.31983 + -153.408 J
OR
156.0001 CIS 4.52984 RADIANS
156.0001 CIS 259.5407 DEGREES
(b) THE FIRST ROOT IS
1.271921 $+ -$.2530007 J
OR
1.29684 CIS 6.086836 RADIANS
1.29684 CIS 348.75 DEGREES
THE 2ND ROOT IS
.2530007 + 1.271921 J
OR
1.29684 CIS 1.374447 RADIANS
1.29684 CIS 78.75 DEGREES
THE 3RD ROOT IS
-1.271921 + .2530008 J
OR
1.29684 CIS 2.945243 RADIANS
1.29684 CIS 168.75 DEGREES
THE 4TH ROOT IS
-.253001 + -1.271921 J
OR
1.29684 CIS 4.51604 RADIANS
1.29684 CIS 258.75 DEGREES

18. **(a)** THE TOTAL IMPEDANCE FOR A
SERIES CIRCUIT IS
3 + -2 J,
3.605551 CIS -.5880026 RADI-
ANS OR
3.605551 CIS -33.69007 DEGREES.
AND FOR A
PARALLEL CIRCUIT IS
5 + .9999999 J,
5.09902 CIS .1973955 RADIANS OR
5.09902 CIS 11.30993 DEGREES.
(b) THE TOTAL IMPEDANCE FOR A
SERIES CIRCUIT IS
10 + -5 J,

11.18034 CIS -.4636476 RADI-
ANS OR
11.18034 CIS -26.56505 DEGREES.
AND FOR A
PARALLEL CIRCUIT IS
-5.692307 + .5384617 J,
5.717718 CIS -9.431399E-02 RADI-
ANS OR
5.717718 CIS -5.403794 DEGREES.
(c) THE TOTAL IMPEDANCE FOR A
SERIES CIRCUIT IS
4.069313 + 5.451568 J,
6.80286 CIS .9295722 RADI-
ANS OR
6.80286 CIS 53.26057 DEGREES.
AND FOR A
PARALLEL CIRCUIT IS
1.401921 + .8910406 J,
1.661125 CIS .5661751 RADIANS OR
1.661125 CIS 32.43944 DEGREES.
(d) THE TOTAL IMPEDANCE FOR A
SERIES CIRCUIT IS
2.799219 + 2.23625 J,
3.582798 CIS .6740596 RADI-
ANS OR
3.582798 CIS 38.62077 DEGREES.
AND FOR A
PARALLEL CIRCUIT IS
.6744572 + .9550899 J,
1.169226 CIS .9559405 RADIANS OR
1.169226 CIS 54.77136 DEGREES.

19. (a) THE ANSWER IS
     7.7000 -2.0000
    -2.9000 10.0000
    11.1000   6.0000
(b) THE ANSWER IS
    2.4000 -1.3000   3.5000
    5.7000   8.2000 -2.0000

20. THE ANSWER IS
    8.8500   9.0000
  -14.7000   7.5000
   13.3000 13.0000

21. THE ANSWER IS
    42.8600   94.1100 -206.0900
   108.8600 -23.3600   -64.0500

22. (a) THE INVERSE MATRIX IS
    0.1818 -0.6182
    0.9091 -2.4242
(b) THIS MATRIX IS SINGULAR. IT DOES
NOT HAVE AN INVERSE.

(c) THE INVERSE MATRIX IS
   0.4484   0.3139 -0.3049   0.1928
  -1.9058   1.1659 -0.7040 -0.0695
  -4.4843   1.8610   0.0493   0.0717
   2.9821 -1.4126   0.3722   0.1323

23. (a) THE MATRIX IS SINGULAR. THIS
SYSTEM HAS NO SOLUTION.
(b) THE INVERTED MATRIX IS
   -.067   -.333   .267
   -.067   .667 -.233
    .333   -.333   .167
THE ANSWER IS
    x(1) = 2
    x(2) = -3
    x(3) = 3

24. LEFT ENDPOINT = ? 0
RIGHT ENDPOINT = ? 1
THE ROOT IS APPROXIMATELY .6850733
THE ACTUAL ROOT IS BETWEEN .6850733
AND 1
THE ROOT IS APPROXIMATELY .7827885
THE ACTUAL ROOT IS BETWEEN .7827885
AND 1
THE ROOT IS APPROXIMATELY .7982691
THE ACTUAL ROOT IS BETWEEN .7982691
AND 1
THE ROOT IS APPROXIMATELY .8006438
THE ACTUAL ROOT IS BETWEEN .8006438
AND 1
INTERPOLATION COMPLETED.
THE ROOT IS APPROXIMATELY .8006438
AND F(C) = 8.363128E-04

25. THE TOTAL NUMBER IN THE SAMPLE IS:
150
THE MEAN IS: 2.833333
THE VARIANCE IS:
1.938889
AND THE STANDARD DEVIATION IS:
1.39244
THE NUMBER OF SCORES WITHIN ONE
STANDARD DEVIATION OF THE MEAN IS:
109 OR 72.66666 PERCENT OF THE
TOTAL VALUE.

AND THE NUMBER WITHIN TWO S.D. OF
THE MEAN IS: 144 OR 96 PERCENT.

26. THE FOLLOWING ARE THE TOTALS
M = 1.886307
B = 37.4829
R = .8217226

27. ENTER THE LEFT END POINT, A=?   0

NEXT ENTER THE RIGHT POINT,
B =?   3.1416

AND FINALLY, THE NUMBER OF SEGMENTS,
N= ?   50
```
A    B         N
0    3.1416    50
```
THE AREA IS 3.915187

28.
```
I    X(I)        F(X(I))         F'(X(I))
1    .75         9.592449E-02    -2.268887
2    .7922783    1.709938E-02    -2.282505
3    .7997697    2.546013E-03    -2.283336
4    .8008848    3.644228E-04    -2.283416
5    .8010444    5.185604E-05    -2.283427
6    .8010671    7.331372E-06    -2.283429
7    .8010703    1.013279E-06    -2.283429
8    .8010707    2.384186E-07    -2.283429
```
THE ROOT IS APPROXIMATELY .8010707

29. THE AREA UNDER F(X) OVER THE INTER-
VAL FROM 0 TO 3.1416 IS
3.91427 BY SIMPSON'S RULE AND
3.912377 BY THE TRAPEZOIDAL RULE

30. **(a)** CYLINDRICAL COORDINATES OF THE
RECTANGULAR POINT (-2, 5, -4.5) ARE
R = 5.385165
THETA = 1.951303 = .621119 *PI
Z = -4.5

THE SPHERICAL COORDINATES OF THE
RECTANGULAR POINT (-2, 5, -4.5) ARE
RHO = 7.017834
THETA = 1.951303 = .621119 *PI
PHI = 2.266888 = .7215729 *PI

(b) THE RECTANGULAR COORDINATES OF
THE CYLINDRICAL COORDINATE POINT
(3, 1 * PI/6, -2) ARE
X = 2.598076
Y = 1.5
Z = -2

THE SPHERICAL COORDINATES OF THE
CYLINDRICAL COORDINATE POINT
(3, 1 * PI/6, -2) ARE
RHO = 3.605551
THETA = .5235988 = 1.666667 * PI
PHI = 2.158799 = .6871671 * PI

(c) THE RECTANGULAR COORDINATES OF
THE SPHERICAL COORDINATE POINT
(5, 1 * PI/6, 1 * PI/4) ARE
X= 3.061863
Y= 1.767767
Z= 3.535534

THE CYLINDRICAL COORDINATES OF THE
SPHERICAL COORDINATE POINT
(5, 1 * PI/6, 1 * PI/4) ARE
R = 3.535534
THETA = .5235988 = .1666667 * PI
Z = 3.535534